W9-CZY-065

WELDING:
PRINCIPLES AND PRACTICES

Revised

Bethlehem Steel Co.

Structures on the St. Louis riverfront that have been constructed by welding. High voltage cable towers, all-welded steel barges, a bridge across the Mississippi, the all-welded excursion steamer The Admiral, *and the all-welded stainless steel Gateway Arch are visible.*

New York, New York Columbus, Ohio Woodland Hills, California Peoria, Illinois

WELDING: PRINCIPLES AND PRACTICES

Revised

RAYMOND J. SACKS

Welder and Machinist, Dist. Lodge 837, Int. Assn. of Machinists

Formerly *Director* of St. Louis, MO, Vocational, Technical & Adult Education, C.E.T.A. Training Program, and Manpower Training Program; *Department Head* and *Welding Instructor* in the St. Louis Public Schools; *Consultant* to Labor Management Training Programs in St. Louis

Industrial Experience: Plant Superintendent, Welding Foreman, Journeyman Combination Welder—experienced in aircraft, pipe, tank, structural, and general maintenance welding

Glencoe/McGraw-Hill
A Division of The McGraw-Hill Companies

Printed in the United States of America

Send all inquires to:
Glencoe/McGraw-Hill
3008 W. Willow Knolls Drive
Peoria, IL 61614-1083

ISBN 0-02-666140-3

11 12 13 14 15 16 17 027 02 01

Preface

WELDING: PRINCIPLES AND PRACTICES is the third revision and an expansion of THEORY AND PRACTICE OF ARC WELDING, which was first published in 1943. The previous editions have enjoyed success during these years as major texts used in the training of welders by industry and the schools.

This book is designed specifically to be used as the principal text for welding training in secondary vocational schools, in area trade schools, and for technical education programs in junior colleges and engineering schools. It is also suitable for on-the-job training and apprenticeship training programs. It can serve as a supplementary text for classes in building construction, metalworking, and industrial arts.

WELDING: PRINCIPLES AND PRACTICES provides a course of instruction in welding, other joining processes, and cutting that will enable students to begin with the most elementary work and progressively study and practice each process until they are skilled. Both principles and practice are presented so that the student can combine the "why" and the "how" for complete understanding.

The text provides a two-fold approach. In Part I, Welding Principles, students are introduced to fundamentals that will enable them to understand what is taking place in the application of the various processes. In Part II, Welding Practices, they learn the necessary manual skills.

WELDING: PRINCIPLES AND PRACTICES presents the fundamental theory of and practice in gas, arc and semi-automatic welding, brazing, soldering, and plastic welding processes in accordance with current industrial practices. The content is based upon the studies and recommendations of the American Welding Society and other leading welding authorities.

Included is a comprehensive treatment of equipment, electrodes, joints and welds, testing and inspection, metals and their welding characteristics, safety practices, and the fundamentals of print reading. Questions are presented at the end of each chapter.

There are many photographs and drawings showing the latest advances in techniques and equipment. Course outlines are provided for each of the major processes that make it possible for the student to be trained in a sequence of exercises that have proven their worth in the training of thousands of welders. The emphasis is on learning by doing. The jobs and exercises presented are the most useful ones for helping students to acquire basic skills and understanding.

The job descriptions are concise and emphasize actual performance. They provide an adequate basis for demonstration and coaching by the instructor. Evaluation of the work of each student can be carried out by observation and by testing the work. Students are urged to experiment, to practice a variety of manipulative procedures, and to compare the results. Thus, they learn through firsthand experiences how changes in welding conditions produce changes in results.

The course outlines have great flexibility. If instructors need to train a person to do a specific operation for a production job in industry, they can readily choose those jobs which will provide the necessary experiences. If, on the other hand, a welding trainee must qualify for code welding in plate or pipe for pressure vessels or for pressure piping (Navy Test, ASME or API) in all positions, the complete course should be assigned. If it is not necessary to qualify in all positions, the course may include only those jobs involving the particular position of welding needed.

Welding is both an art and a trade requiring the skillful manipulation of the hands and a thorough knowledge of welding processes and the characteristics of metals. Students can be assured of success if they are willing to spend the time required in actual practice work and in the study of the principles presented in this text until they thoroughly understand their significance. Faithful adherence to this course of study will enable them to master these modern industrial joining processes thoroughly.

Throughout this book, the student may be referred to by the masculine pronoun. The use of *he* (*him, his*) is not meant to imply that all welders are male. Today there are thousands of women in welding careers, and their number will no doubt increase as equal opportunity employment becomes a reality. In this text, the use of masculine pronouns in reference to students has been kept to a minimum. When such references do appear, they are to be considered as generic terms encompassing both sexes, rather than as references to male students exclusively.

Acknowledgments

The author gratefully acknowledges the assistance and counsel of the many people and organizations that have contributed to the preparation of this text.

Mr. Harry Brown has prepared the drawings for publication.

Credit is due Harvey Johnson, John D. Orr and Harry E. Raidt, instructors of welding, O'Fallon Technical Center, for serving as consultants and for their assistance in the development of the several course outlines.

Since the author has drawn freely from the works of the American Welding Society, he is especially indebted to that organization.

ACF Industries
Airco Welding Products Co.
American Iron and Steel Institute
American Petroleum Institute
American Pullmax Co.
American Society for Metals
American Society of Mechanical Engineers
American Standards Association
The Arcair Co.
Arcos Corp.
Atlas Welding Accessories Co.
The Auto Arc-Weld Mfg. Co.
Baldor Electric Co.
Bernard Welding Equipment Co.
Black and Decker Manufacturing Co.
Branson Instruments Co.
Central Fire Truck Co.
Chemetron Corp.
Cincinnati Milacron
Clausing Corp.
Combustion Engineering Co.
Contour Sales Corp.
Crane Co.
CRC-Crose International
Dept. of Health, Education and Welfare
De-Sta Co.
Do-ALL Co.
Dreis & Krump Mfg. Co.
Duggan Corp.
Empire Abrasive Equipment Corp.
Enerpac
Eriez Magnetics
ESAB
Eutectic Corp.
Fibre-Metal Products Co.
Stanley G. Flagg & Co.
General Dynamics
General Electric Co.
General Welding & Equipment Co.
Glenn Pacific Co.
Handy & Harman
Harnischfeger Corp.
Harsfeld Manufacturing Co.
Heating, Piping & Air Conditioning Contractors National Association
Hobart Brothers Co.
Industrial Plastic Fabricators
International Acetylene
International Nickel Co. and Bructon Industries
Iron Age Magazine
Jackson Products Co.
Jewel Manufacturing Co.
Kaiser Aluminum & Chemical Corp.
Kamweld Products Co.
King Tester Corp.
Laramy Products Co.
Lenco, Inc.
The Lincoln Electric Co.
Linde Division, Union Carbide Corp.
Magnaflux Corp.
Manpower Magazine, U.S. Department of Labor
McDonnell Douglas Corp.
Mechanical Contractors of America
Merrill, McEnroe & Associates
Metal Fabricating Institute
Metals Handbook
Miller Electric Mfg. Co.
Mobile Homes Manufacturers Association
Mobil Oil Corp.
Modern Engineering Co.
B. C. Motl and A. O. Smith Corp.
National Certified Pipewelding Bureau
National Cylinder Gas Division, Chemetron Corp.
National Welding Equipment Co.
Niagara Machine and Tool Works
Nooter Corp.
Ohio State University
Pandjiris Weldment Co.
Penn-Chem Corp.
Phoenix Products Co.
Picker Industrial
Pipefitters and Plumbers Local No. 553—Alton, Ill.
Ransome Co.
Republic Steel Co.
Revere Copper and Brass, Inc.
Rexarc

Robvan Backing Ring Co.
Safety Appliances Co.
St. Louis Car Co.
Seelye Plastics
The Shore Instrument & Mfg. Co.
South Bend Lathe Co.
L. S. Starrett Co.
Stahl Equipment Co.
Steelways
Thermacote Co.
Towers Fire Apparatus Co.
Tube Turns
United States Steel Corp.
Unitron Instrument Co.
Victor Equipment Co.
WABCO
E. H. Wachs Co.
Welding Data Book
The Welding Journal
Weld Tooling Corp.
Wells Manufacturing Co.
Westinghouse Electric Corp.
Widder Corp.
Willson Products
Wilson Instrument Division, ACCO
Worthington Corp.
Wyatt Industries, Inc.
Zephyr Manufacturing Co.

And to my beloved wife, Lillian, whose calm assurance and encouragement spurred me on, I feel the deepest gratitude.

Raymond J. Sacks
St. Louis, Missouri

UNION RECOGNITION

Recognition is due the Pipefitter Unions in St. Louis, Missouri and in Alton, Illinois for their assistance in providing information and materials for those chapters concerned with the welding of pipe. This assistance insures that the practices presented in the text are current.

ABOUT THE AUTHOR

The author has had wide experience as a journeyman welder and a teacher of welding. He has been head of the welding department at the O'Fallon Technical Center, consultant for seventeen Labor Management apprenticeship committees, director of the Pipefitter School in St. Louis and the Pipefitter and Plumber School in Alton, Illinois, and a vocational administrator in the public schools of St. Louis. He was director of the War Training Program in St. Louis during World War II and has served as supervisor of adult education and as principal of the O'Fallon Technical Center. He is presently Director of Vocational, Technical, and Adult Education in St. Louis. He has also had experience as one of the owners of a metal products business, a fabricator of a diversified line of welded components. He has been a member of the American Welding Society for a number of years.

AUTHOR'S NOTE

Careful consideration has been given to the safety regulations as outlined by OSHA—Occupational Safety Health Act. Use is also made of the American Welding Society's newly approved welding process names. However, references such as MIG for gas metal-arc welding and TIG for gas tungsten-arc welding are also used because they have a conversational informality that seems to appeal to welding personnel. These terms are convenient, easily spoken, and enjoy widespread understanding in a particular area of operation.

TABLE OF CONTENTS

TABLE OF CONTENTS

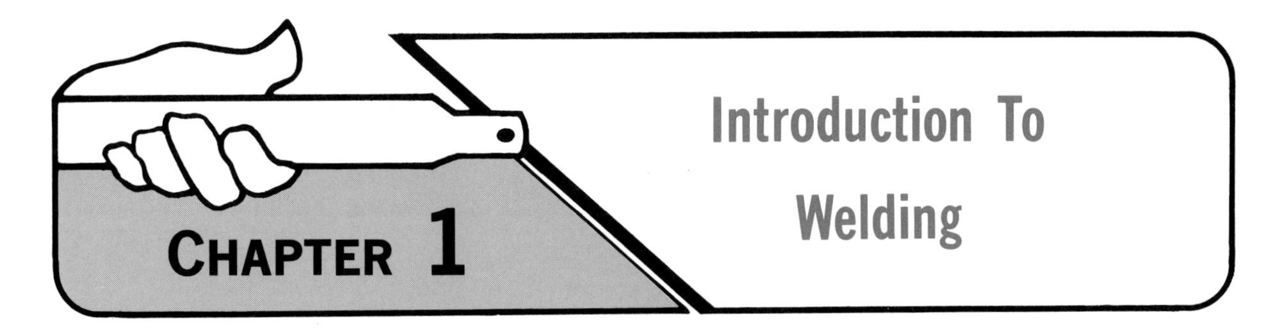

Chapter 1 Introduction To Welding

You are about to begin the learning process of preparing yourself for a position in one of the fastest growing industries in the world of work—the welding industry. *Welding* is the joining together of two pieces of metal by heating to a temperature high enough to cause softening or melting, with or without the application of pressure, and with or without the use of filler metal. Any filler metal used either has a melting point approximately the same as the metals being joined, or a melting point which is below these metals but above 800 degrees F.

New methods, new applications, and new systems have been developed in the last several years. Continuing research makes welding a dynamic leader in industrial processes. The industry has made tremendous progress in a short period of time. Furthermore, it has made a major contribution toward raising the standard of living of the American people. By simplifying and speeding up industrial processes and making it possible to develop new industries, such as the nuclear power and space industries, it has increased the world's supply of goods.

THE HISTORY OF METALWORKING

Metalworking began when primitive people found that they could shape rocks by chipping them with other rocks. The first metal to be worked was probably pure copper, since it is a soft, ductile metal which was widely available. Excavations in Egypt indicate the use of copper as early as 4000 B.C., and in what is now the United States, before 2000 B.C. More than 4000 years ago copper mines on the peninsula of Sinai and the island of Cyprus were worked. Welding began more than 3000 years ago when hot or cold metals were hammered to obtain a forge weld.

Archaeologists have determined that bronze was developed some time between 3000 and 2000 B.C. Iron became known to Europe about 1000 B.C., several thousand years after the use of copper. About 1300 B.C. the Philistines had four iron furnaces and a factory for producing swords, chisels, daggers, and spearheads. The Egyptians began to make iron tools and weapons during the period of 900–850 B.C. After 800 B.C. iron replaced bronze as the metal used in the manufacture of utensils, armor, and other practical applications.

The famous Damascus swords and daggers were made in Syria about 1300 B.C. These were sought after because of their strength and toughness. Their keen edge was said to be capable of severing heavy iron spears or cutting the most delicate fabric floating in the air. The swords were made by forge-welding iron bars of different degrees of hardness, drawing them down, and repeating the process many times.

Thus the working of metals —copper, bronze, silver, gold, and iron—closely followed one another in the great ancient civilizations. By the time of the Roman Empire, the use of iron was common in Europe, the Near East, and the Far East. The Chinese developed the ability to make steel from wrought iron in 589 A.D. The Belgians were responsible for most of the progress made in Europe, due to the high degree of craftsmanship developed by their workers. By the eighth century the Japanese manufactured steel by repeated welding and forging and controlled the amount of carbon in steel by the use of fluxes. They produced the famous Samurai sword with a blade of excellent quality and superior workmanship.

The blast furnace was developed for melting iron about the years 1000–1200 A.D. One such furnace was in the Province of Catalonia, in Spain. The fourteenth and fifteenth centuries saw great improvements in the design of blast furnaces. The first cast iron cannon was produced in the early 1600s.

About the middle of the eighteenth century, a series of

inventions in England revolutionized the methods of industry and brought on what later came to be known as the Industrial Revolution. Our present factory system of mass production was introduced. An American, Eli Whitney, developed the idea of interchanging parts in the manufacture of arms. By the beginning of the nineteenth century, the working of iron with the use of dies and molds became commonplace.

***1-1.** The ability to make multipass welds such as this one, on plate and pipe, led to the growth of the industry. Welds are sound and have uniform appearance.*

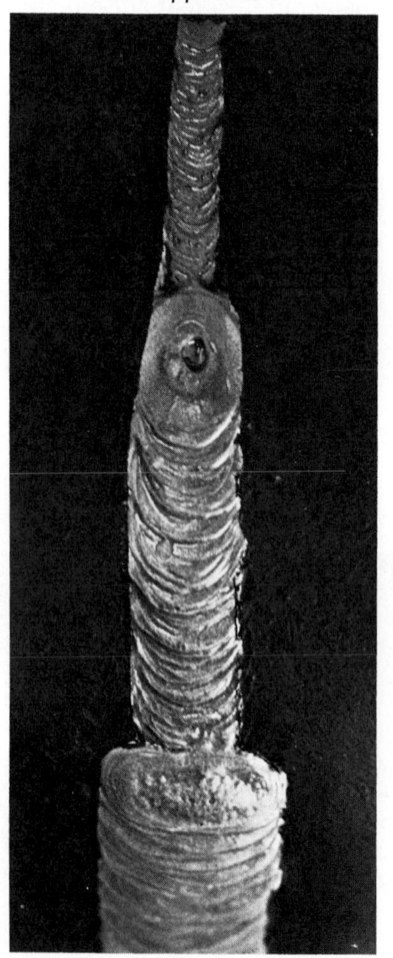

Early Developments in Welding

At the beginning of the nineteenth century, Edmund Davy discovered acetylene, a gas which was later used in oxyacetylene welding. The electric arc was first discovered by Sir Humphry Davy in 1801 while conducting experiments in electricity. He was concerned primarily with the possibilities of the use of the arc for illumination. By 1809 he had demonstrated that it was possible to maintain a high voltage arc for varying periods of time. By the middle of the nineteenth century, workable electrical-generating devices were invented and developed on a practical basis. These inventions were the forerunner of the present arc welding process.

The first documented instance of fusion welding was done by Auguste de Meritens in 1881. He welded lead battery plates together with a carbon electrode. Two of his pupils, N. Benardos and S. Olszewski, saw the possibilities of this discovery and experimented with the arc powered by batteries which were charged from high voltage dynamos. After four years of work, they were issued a British patent for a welding process using carbon electrodes and an electric power source. Applications of the process included the fusion welding of metals, the cutting of metals, and the punching of holes in metal. Although they experimented with solid and hollow carbon rods filled with powdered metals, the solid electrodes proved more successful. Repair welding was the primary goal of the inventors.

Bare metal electrode welding was introduced in 1888 by N. G. Slavianoff, a Russian. His discovery was first recognized in western Europe in 1892. C. L. Coffin was one of the pioneers of the welding industry in the United States. In 1889 he received a patent on the equipment and process for flash-butt welding. In 1890 he received additional patents for spot-welding equipment. In 1892, working without knowledge of Slavianoff's work, he received a patent for the bare metal electrode *arc welding* process. By the turn of the century welding was a common method of repair. At this time welding was given added impetus by the development of the first commercial *oxyacetylene welding* torch by two Frenchmen, Foresche and Picard. Bare electrode welding became the prevailing electric arc welding method used in the United States until about 1920.

Bare metal electrode welding was handicapped because the welds produced by these electrodes were not as strong as the metal being welded and the welding arc was very unstable. In 1907 Kjellberg, a Swedish engineer, received a patent covering the electrode-coating process. The coating was thin and acted only as a stabilizer of the arc rather than as a purifier of the weld metal. It produced welds that were little better than those made with bare electrodes. In 1912 Kjellberg received another patent for an electrode with a heavier coating made of asbestos with a binder of sodium silicate. See Fig. 1-1. Benardos patented a process in 1908 that has come into popular use in the past few years. This is the *electroslag* process of welding thick plates in one pass.

Welding technology and its industrial application progressed rather slowly until World War I. Prior to that time it was used chiefly as a means of maintenance and repair. The demands of the war for an increased flow of goods called for improved methods of fabrication. After 1919 welding was widely accepted. Research on coated electrodes through the 1920s resulted in electrode coatings and improved core wire. The development of X-ray examination of weld metal made it possible to examine the internal soundness of welded joints.

The Development of Modern Welding

During the postwar period the design of welding machines was changed very little. Since welding was first done with d.c. current from battery banks, it was only natural that as welding machines were developed, they would be d.c. machines. In the late 1920s and during the 1930s, considerable research was carried on with a.c. current for welding. The use of a.c. welding machines increased through the early 1930s. One of the first high frequency, stabilized a.c. industrial welding machines was introduced in 1936 by the Miller Electric Manufacturing Company. The a.c. welding machines have since become popular because of the high rate of metal deposition and the absence of arc blow.

World War II spurred the development of inert gas welding, thus making it possible to produce welds of high purity and critical application. A patent was issued in 1930 to Hobart and Devers for the use of the electric arc within an inert gas atmosphere. The process was not well received by industry because of the high cost of argon and helium and the lack of suitable torch equipment.

Russell Merideth, an engineer for the Northrop Aircraft Company, was faced with the task of finding an improved means of welding aluminum and magnesium in the inert atmosphere. Because of a high burnoff rate, the magnesium procedure was replaced by a tungsten electrode, and a patent was issued in 1942. Later in 1942 the Linde Company obtained a license to develop the *gas tungsten-arc (GTAW or TIG) process* used today, Fig. 1-2. The company perfected a water-cooled torch capable of high amperage.

TIG welding was first done with rotating d.c. welding machines. Later a.c. units, with built-in high frequency, were developed. About 1950 selenium rectifier type d.c. welding machines came into use, and a.c.-d.c. rectifier welding machines with built-in frequency for TIG welding became available in 1950. Since that time the Miller Electric Manufacturing Company has developed the Miller controlled-wave a.c. welder for critical welds on aircraft and missiles.

The use of aluminum and magnesium increased at a rapid rate as a result of (1) the development of TIG welding, and (2) the desirable characteristics of reduced weight and resistance to corrosion. As the size of weldments increased, thicker materials were employed in their construction. It was found that for aluminum thicknesses above $\frac{1}{4}$ inch, TIG welding required preheating. Since this was costly and highly impractical for large weldments, a number of welding equipment manufacturers engaged in the search for another welding process. This concentration of effort produced the *gas metal-arc (GMAW or MIG) process*, and a patent was issued in 1948. The MIG process, Fig. 1-3, was a new concept in welding. The process concentrates

1-2. *An aluminum weld done with the TIG process. The welding of aluminum is no longer a problem and can be done with the same ease as steel.*

1-3. *A fillet weld made with the MIG process and flux-cored wire.*
St. Louis Car. Co.

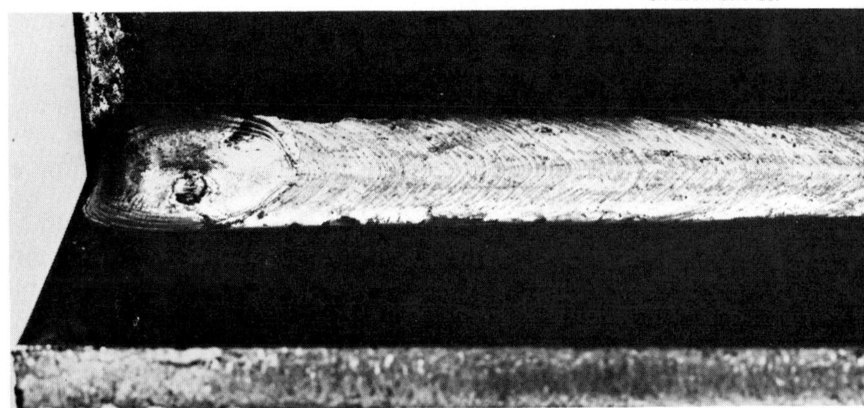

high heat at a focal point, producing deep penetration, narrow bead width, a small heat-affected zone, faster welding speeds, and minimum postweld cleaning. Less warpage and distortion of the joint results. The use of MIG welding has increased very rapidly; it is now used in many industries.

Recent Advances

A number of special welding processes have been developed in the past few years. These methods have limited application and have been designed to fill a special need.

The following processes involve the use of the electric arc:

- Arc spot welding
- Atomic-hydrogen welding
- Electroslag and electrogas
- Plasma arc welding
- Stud welding
- Submerged-arc welding
- Underwater arc welding

Other specialized processes include:

- Cold welding
- Electron beam welding
- Explosive welding
- Forge welding
- Friction welding
- Laser welding
- Self-generating oxyhydrogen welding
- Thermit welding
- Ultrasonic welding
- Welding of plastics

Today there are over forty welding processes in use. The demands of industry in the future will force new and improved developments in machines, torches, electrodes, procedures, and technology. The space and nuclear industries conduct constant research for new metals, which in turn spurs research in welding. For example, the ability to join metals with nonmetallic materials is the subject of much effort at the present time. As industry expands and improves its technology, processes of welding will play an indispensable part in progress.

1-4. *A Gemini spacecraft in a 30-foot vacuum chamber. Welding is used extensively in the construction of both units.*

McDonnell Douglas Corp.

WELDING AS AN OCCUPATION

The job you get when you leave school will depend upon the skills and knowledge you have to offer to industry. The variety of jobs available today was beyond the wildest dreams of the high school graduates of the 20s and 30s, and even those entering the workforce as recently as ten years ago. The new fields of space exploration (Fig. 1-4), air conditioning, jet aircraft, and nuclear energy represent many job opportunities to be filled by trained workers. Welding is one of the major occupations involved in all of these industries. The most recent population census established that over 900,000 men and women in the United States are employed as welding and/or flame-cutting operators.

Because welders hold key positions in the major industries; they are important to the economic welfare of our country. Without welding, the metal industry would be seriously restricted; many of the scientific feats of the past and the future would be impossible. As long as there are metal products, welders will be needed to fabricate and repair them. It is an occupation that many women find highly remunerative and satisfying, Fig. 1-5. Thousands of women are so employed throughout the country.

Welding is done in every civilized country in the world. You may wish to work in the oil fields of the Near East or in our own country. You may wish to work in some jungle area of South America or Africa, constructing buildings, pipelines, or bridges. Our many military installations throughout the world offer jobs for civilian workers. Employment opportunities for welders are plentiful in all parts of the United States.

Keep in mind that the field of welding can offer you prestige and security. It can offer you a future of continuous employment with steady advancement at wages that are equal to other skilled trades and are better than average. It can offer you employment in practically any industry you choose and travel to all parts of the world. It is an expanding industry, and your chances for advancement are excellent. The welder has the opportunity to participate in many phases of industrial processes, thus giving him the broad knowledge of the field necessary for advancement to a supervisory or technical position.

INDUSTRIAL WELDING APPLICATIONS

Welding is not a simple operation. The more than forty different welding processes are divided into three major types: arc, gas, and resistance welding. A number of other types such as induction, forge, thermit, flow welding, and brazing are used to a somewhat lesser extent.

Resistance welding includes spot welding, seam welding, flash welding, and other similar processes that are performed on machines. These welding areas are not the subject of this text. Because of the specialized nature of the machines, operators are usually taught on the job. They are semiskilled workers who do not need specific welding skills or knowledge.

Arc and gas welders, on the other hand, have almost complete control of the process. Much of their work demands manipulative skill and independent judgment that can be gained only through training and a wide variety of job experience. In a sense welders are scientists. They must know the properties of the metals they weld, which welding process to use, and how to plan, measure, and assemble their work. They must be precise, logical, and able to use their heads as well as their hands. Most welders are expected to be able to weld in the vertical and overhead positions, Figs, 1-6 and 1-9, as well as in the flat and horizontal positions.

Gas welders may specialize in oxyacetylene or oxyhydrogen welding, but a qualified worker is able to do both. *Arc welders* may use open carbon or metallic arcs in room atmosphere or arcs shielded by an atmosphere of argon, helium, carbon dioxide, hydrogen, and other inert gases. Some welders are skilled in all the processes. You should acquire competence in both gas and arc welding.

1-5. *This woman is a shielded metal-arc welder.*

1-6. *Welding a butt joint in the vertical position inside a tank.*
General Electric Co.

Qualifications and Personal Characteristics

The standards are high in welding. In doing work in which lives may depend on the quality of the welding—high-rise buildings, bridges, tanks and pressure vessels of all kinds, aircraft, spacecraft, and pipelines—a welder must be certified for his ability to do the work, and his work is inspected, Figs. 1-7 and 1-8. Welders are required to pass periodic qualification tests established by various code authorities, insurance companies, the military, and other governmental inspection agencies. Certifications are issued according to the kind and gauge of metal and the specific welding process, technique, or procedure used. Some welders hold several different certifications simultaneously.

The welder must perform certain basic tasks and possess certain technical information in order to perform the welding operation. In making a gas weld, the welder attaches the proper tip to the torch and adjusts the welding regulators for the proper volume and pressure of the gases. He must also regulate the flame according to the needs of the job.

For electric welding, the operator must be able to regulate the welding machine for the proper welding current and select the proper electrode as to size, type, and shielding gas.

Welding requires a steady hand. The welder must hold the torch or electrode at the proper angle, a uniform distance from the work, and move it along the line of weld at a uniform speed. Although much of the work is single pass, welds made on heavy material often require a number of passes side-by-side and in layers according to the specified weld procedure.

Welders must also be able to cut metals with the oxyacetylene cutting torch and with the various cutting procedures involving the electric arc. Flame cutting is often the only practical method for cutting parts out of steel plate or in making repairs.

Safety and Working Conditions

Welders work on many kinds of jobs in almost any environment. They may do light or heavy welding, indoors or out-of-doors, in spacious surroundings or cramped quarters. Often they work in awkward positions,

1-7. *Using a method of weld inspection known as magnetic-particle testing in building construction. Weld testing and inspection give proof to the soundness of welds.*

Magnaflux Corp.

1-8. *The welds in these tanks must meet X-ray requirements and pass a dye penetrant test. These tanks are often lined with a $\frac{1}{16}$-inch thickness of pure silver. Welders who do this type of work are true craftsmen who deal with a wide range of materials.*

Nooter Corp.

sometimes lying down as they make overhead welds, Fig. 1-9. In boiler shops, shipyards, and on tank construction the work may be extremely noisy, and welders may have to work on scaffolds high off the ground, Fig. 1-10. On some jobs there may be considerable lifting, tugging, and pushing as equipment and materials are placed in position.

A large number of unsafe situations must be of concern to the welder who is conscious of the need to work in a safe environment. Very often accidents are caused as a result of some small, relatively unimportant condition. Extremely dangerous hazards usually get the attention of the welder and are, therefore, rarely a cause of accidents.

Job hazards may include burns, "sunburn" from electric arcs, noxious fumes from materials vaporized at high temperatures, eyestrain, and electric shock. These hazards can be minimized or eliminated by the use of the proper protective clothing, face shields, goggles, and adequate ventilation. In doing a job, the welder always takes precautionary measures for his own safety and the safety of others in the area.

You are encouraged to study the various safety practices and regulations presented in this text. Safety precautions related to specific processes are presented in the chapters in Part I in which the equipment and processes are explained. Safe welding technique and the safe use of equipment are given in the practice chapters in Part II. Before you begin your welding practice, you should read Chapter 15, Safety, which summarizes the safety measures described

The Lincoln Electric Co.

1-10. *Welders in the construction industry are called upon to weld in many unusual positions.*

1-9. *This welder is working in the overhead position on the bottom of a barge.*

General Electric Co.

1-11. *A large amount of art metalwork is done with welding processes.*

The Lincoln Electric Co.

elsewhere and presents the precautions to be followed both in the school shop and in industry.

The master welder is a master craftsman, Fig. 1-11. Such a person is able to weld all of the steels and their alloys, as well as nickel, aluminum, tantalum, titanium, zirconium, and their alloys and claddings. From heavy pressure vessels requiring 4-inch plate to the delicate welding of silver and gold, the welds are of the highest quality and can be depended upon to meet the requirements of the job.

The following welding occupations require a high school education:

- Welding operator
- Welder (may be combination gas, arc, MIG, TIG)
- Welder-fitter
- Specialist welder
- Welding supervisor
- Welding analyst
- Welding technician
- Inspector
- Welding foreman
- Welding superintendent
- Job or fabrication shop owner
- Equipment sales
- Sales demonstrator
- Sales troubleshooter
- Welding instructor

Certain welding occupations also require some college.

- Welding engineer (metallurgical)
- Welding development engineer
- Welding research engineer
- Welding engineer
- Technical editor
- Welding teacher (college-level)
- Corporation executive
- Owner of welding business
- Sales engineer

Many people in the positions listed above entered the industry as welders and were able to improve their positions by attending evening college.

You have selected an excellent field of work. Prepare yourself well and apply yourself diligently to your tasks. Keep up-to-date on the improvements in the field by reading trade journals, service manuals, and trade catalogues. Set your sights high. It is better to be equipped for the top position and reach second place than to be frustrated by lack of knowledge and skills.

REVIEW QUESTIONS

1. When was bronze developed?
2. Is welding a recent industrial process? Explain your answer.
3. Name four metals that were used by early metal workers. Which metal was the first to be worked?
4. When did the manufacture of steel begin?
5. When was fusion welding first developed as we know it?
6. Before WWI, welding was used chiefly for maintenance and repair. True. False.
7. Name six important welding processes.
8. Name ten occupational classifications in the welding industry.
9. When were electric arc welding and oxyacetylene welding developed?
10. In what country was the patent for electric arc welding first issued?
11. What invention gave electric arc welding its greatest boost?
12. When were MIG and TIG welding developed?
13. Which welding process contributed most to aluminum welding?
14. Name three welding occupations that require a college degree.
15. Is it true that industry uses MIG welding only for special applications because of its instability? Explain your answer.

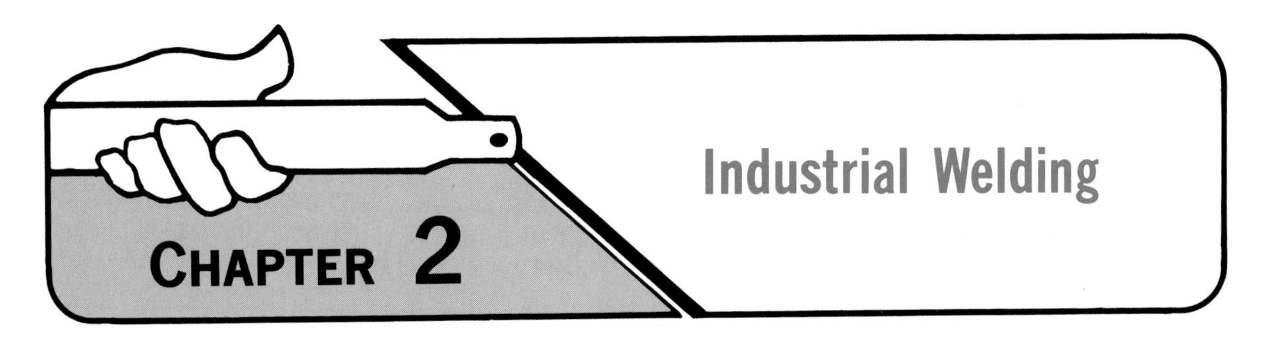

Chapter 2 Industrial Welding

It may be said that welding has two major functions in industry: (1) as a means of fabrication and (2) for maintenance and repair. It would be difficult to find a single industry that does not use welding in either of these classifications. It is common for a great many industries to use the process in both capacities.

The following industries have found welding to be an advantage:

- Aircraft
- Automotive
- Bridge building
- Construction equipment
- Farm equipment and appliances
- Furnaces and heating equipment
- Guided missiles and spacecraft
- Jigs and fixtures
- Machine tools
- Military equipment
- Mining equipment
- Oil drilling and refining equipment
- Ornamental iron work
- Piping
- Quantity food service equipment
- Railroad equipment
- Residential, commercial, and industrial construction
- Sheet metal
- Steel mill equipment
- Tanks and boilers
- Tools and dies
- Watercraft

FABRICATION

Welding fabrication has grown rapidly because of its speed and economy. Welding plays an important part in manufacturing processes for the following reasons:

- *Greater design flexibility and lower design costs.* Welded design affords an easy, quick means of meeting the functional requirements. This freedom results in lower costs and improvements in the service life of the product.
- *An elimination of patterns.* Welded designs are built directly from standard steel shapes. Since patterns are not required, this saves the cost of pattern drawings, pattern making, storage, and repairing.
- *Lower cost of material.* Rolled steel is a stronger, stiffer, more uniform material than castings. Therefore, fewer pounds are required to do an equivalent job. And since rolled steel costs one-quarter to one-half as much as a casting, the cost of material is cut by as much as three-quarters by using welded steel fabrication. By replacing riveting, welded fabrication saves more than 35 percent. Welded steel construction eliminates connecting members such as gusset plates, simplifies drafting, and cuts material costs and weight from 15 to 25 percent. Welding permits the use of simple jigs and fixtures for speeding layouts and fabrication 10 to 35 percent.
- *Fewer man-hours of production.* Roundabout casting procedures are eliminated, machining operations are minimized, and man-hours are not wasted because of defective castings which must be rejected. Special machines can be produced with only slight modifications of the standard design, saving time and money. With welding and today's highly efficient fabrication methods, production is straightforward and faster.
- *Absorption of fixed charges.* Instead of purchasing parts from an outside concern, the company may make them in its own shop. The additional work applied to existing production facilities and supervision absorbs overhead costs, thereby resulting in more efficient production. This extra work means more jobs for the welding operator.
- *Minimized inventory and obsolescence charges.* Inventory for welding is approximately 10 percent of that required for casting. The standard steel parts used for welding may be purchased on short notice from any steel mill or jobber. New design developments do not make stocked material obsolete.

MAINTENANCE AND REPAIR

Hundreds of companies save thousands of dollars by using

welding for maintenance and repair as follows:

- *The addition of new metal.* Hardfacing worn parts usually produces a part that is more serviceable than the original at a substantial saving over replacement cost.
- *Repair and replacement of broken parts.* Immediate repair by welding forestalls costly interruptions in production and saves expensive replacements.
- *Special needs.* Production equipment, shop fixtures, and structures of many kinds can be adapted to meet particular production requirements.

INDUSTRIES

Aircraft

Wood, fabric, and wire were the principal materials to be found in the first airplanes. With the change to metal construction, the industry became one of the foremost users of the welding process in all its forms. Aircraft welding was first tried in 1911 and used in warcraft production by the Germans in World War I. Some few years after the war, the tubular steel fuselage was developed, and production lines were set up.

Today welding is universally accepted by the aircraft, missile, and rocket industries. The development of supersonic aircraft and missiles, involving increased stresses, higher temperatures, and high speeds, have presented fabrication problems that only welding can meet.

Welding processes now employed in the aerospace industries, Figs. 2-1, 2-2, and 2-3, include all types of fusion and solid state welding, resistance welding, brazing, and soldering. Aircraft welders performing manual welding operations are required to have qualification certification.

Welding is used in the fabrication of the following aircraft units:

- Pulley brackets
- Structural fitting, mufflers, and exhaust manifolds
- Axle and landing gear parts and assemblies
- Struts, fuselages
- Engine bearers, mounts, parts
- Brackets
- Control columns, quadrants, levers
- Armor plate assemblies
- Gun mounts
- Gas and liquid tanks
- Fuselage wing and tail assemblies
- Engine cowling
- Wheel boots
- Ammunition boxes and chutes
- Mounting brackets for radio and other equipment
- Oil coolers and heaters
- Nose cones and rocket shells
- Space capsules

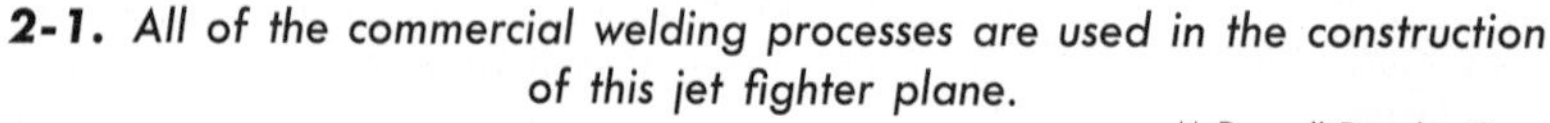

2-1. *All of the commercial welding processes are used in the construction of this jet fighter plane.*

McDonnell Douglas Corp.

Automotive

Welding processes for the manufacture of passenger cars were first introduced during World War II. Since that time, the automobile industry has employed welding on a large scale.

Welding is the method of fabrication for the whole automobile, Fig. 2-4. It is the joining process used to build the body, frame, structural brackets, much of the running gear, and parts of the engine. Welding also is a necessary process in the service and repair of automotive equipment.

A comparatively recent welding application is in the construction of fire-fighting equipment, Figs. 2-5 and 2-6. Modern fire trucks have welded tanks, bodies, frames, aerial supports, and a number of brackets.

Welding is used extensively in military automotive construction and in the construction of all types of trucks, Fig. 2-7. Many people are not familiar with the many types of military automotive equipment since much of this equipment is of a secret nature. However, this is not the case with freight carriers. Most people recognize the many types of trucks that may be seen on the highways. You may have wondered about the great variety of designs, sizes, and shapes. These would not be possible but for the flexibility of the welding process and the use of standard sheets and bars. Individual designs and needs may be handled on a mass production basis.

Many dollars have been saved by passenger car owners and truck operators by the application of welding as a method of repair. Alert mechanics who were quick to realize the utility of the process have applied it to automotive repairs of all types: engine heads, engine blocks, oil pans, cracked and broken frames, engine and body brackets, and body and fender repairs.

Construction Machinery

The highway to Alaska was built in nine months. This 1,600-mile highway, through what had been considered impassable terrain, would have been impossible without imagination, daring, and proper equipment. Construction crews encountered many serious problems. Huge tonnage had to be hauled over frozen tundra, mud, and gumbo. Replacement and repair centers were located hundreds of miles away from some construction sites. Both the climate and the job required equipment that could take heavy punishment.

Equipment manufacturers met the challenge with welded equipment: pullers, pushers, scrapers, diggers, rollers, haulers, and graders. This equipment had to meet several requirements. Irregularly shaped parts and movable members had

2-2A. *Welded jigs used in spot welding of a unit for a space capsule.* McDonnell Douglas Corp.

2-2B. *Welded jigs used to assemble the total space capsule.* McDonnell Douglas Corp.

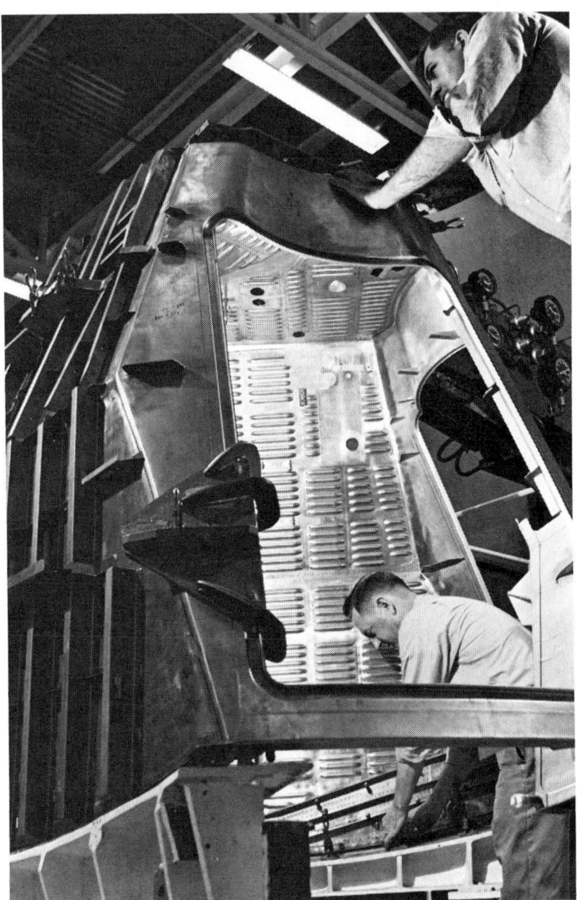

McDonnell Douglas Corp.

2-3. *Fusion-welded strip of 0.010-inch thick titanium for a spacecraft.*

to be very strong, yet light so that economical motive power could be employed. Relatively low first cost was important, but more important was the need for strength, rigidity, and light weight.

The design of an earth-moving unit, fabricated entirely by the welded method from mill-run steel plates and shapes reduced the weight of the total earth-moving machine from 15 to 20 percent over the conventional method of manufacturing. The welded joint, which fuses the edges of the parts, was substituted for heavy reinforcing sections involved in the other common methods of joining parts. The greater strength and rigidity of rolled steel, combined with the fact that a welded joint unifies the parts to produce a rigid and permanent unit, the joints of which are at least as strong as

Central Fire Truck Co.

2-5. *An all-welded pumper.*

2-4. *Carbon-dioxide-gas-shielded, metal-arc welding of trunk of automobile body.*

Hobart Brothers Co.

2-6. *An all-welded pumper with an aerial extension for reaching fires on the upper floors of buildings.*

Towers Fire Apparatus Co.

the section joined, paved the way for more powerful tractors and for larger earth-moving units. A further reduction of weight is now affected by the use of high tensile, low alloy steels. The additional savings of 15 to 25 percent may be added to the load-carrying capacity of the machines. Welded, rolled-steel construction also lends itself to repair and maintenance. Breakdowns can be repaired in the field by the welding process.

America is engaged in a vast road-building program over the entire country and is also clearing vast land areas for many types of construction projects. These jobs would be impossible to accomplish but for the tremendous advances that have been made in earth-moving equipment. Today thousands of yards of earth can be moved in a fraction of the time and for a fraction of the cost formerly required with nonwelded equipment.

Some idea of the tremendous job that is being done can be gained from the following comparisons. One of the giant truck dirt haulers is powered by 930-horsepower diesel engines. The body and box-beam frame are constructed of high-tensile strength steel, completely fabricated by welding. When fully loaded, it holds 119 cubic yards of earth, weighing 160 tons. The loaded gross weight of the truck is over half a million pounds, Fig. 2-8. Assuming that one wheelbarrow will hold 150 pounds of dirt, it would take 2,135 loads to remove the same quantity of earth. Of course, there is no way to compare the hauling distance these trucks can cover in a unit of time compared with the wheelbarrow. Figure 2-9 shows a scraper that can move 31 cubic yards of dirt when fully loaded. If the same amount of earth were moved in the wheelbarrow cited above, it would take 950 loads to finish the job.

2-7. *Truck body beds of aluminum being welded with the gas tungsten-arc (TIG) process.*
The Lincoln Electric Co.

WABCO

2-8. *An off-highway truck used for moving earth. Note its size in comparison with the workers.*

2-9. *A diesel scraper which can move up to 31 cubic yards of earth.*
WABCO

Household Equipment

The application of welding to household equipment is a fairly recent development. Welded fabrication is popular in those industries engaged in the manufacture of tubular metal furniture, kitchen and sink cabinets, sinks, furnaces, ranges, refrigerators, and various kinds of ornamental ironwork. Much of this work is done with the inert-gas welding processes and brazing.

Welded fabrication permits the use of stainless steels, aluminum, and magnesium —materials which can provide light weight, strength and rigidity, and long life. Welding is adaptable to mass-production methods and saves material. In addition, it permits flexibility of design and contributes to the pleasing appearance of a product, Fig. 2-10.

Jigs and Fixtures

Welding is a valuable aid to tooling in meeting requirements for mass production and, it affords outstanding economies. For these reasons welded jigs are used universally in the aerospace industry, Fig. 2-2.

Among the advantages of welded steel jigs and fixtures are maximum strength and accuracy for close tolerances. Cost saving as high as 75 percent, time saving as much as 85 percent, and weight saving as much as 50 percent can be achieved. Welding permits a wide range of application, simplifies design, and minimizes machining. Modifications to meet design changes can be easily made.

Machine Tools

For the manufacture of machine tools, steel has certain advantages over cast iron:

- Steel is two to three times stiffer.
- Steel has about four times the resistance to fatigue. Cast iron has a fatigue resistance of about 7,500 pounds per square inch (p.s.i.) whereas that of steel is 28,000–32,000 p.s.i.
- Steel costs one-quarter to one-half as much. Rolled steel costs approximately 40 percent as much per pound as cast iron and 25 percent as much as cast steel.
- Steel is three to six times stronger in tension. A test of two equal-sized bars showed that the cast iron bar broke at 26,420 p.s.i., but the mild steel bar withstood 61,800 p.s.i.
- Steel can withstand heavy impacts. In a test, one blow of a 9-pound sledge shattered a cast iron part. Twenty blows of the sledge merely bent the duplicate part, which was built of steel.
- Steel is uniform and dependable. Its homogeneous structure, devoid of blowholes and uneven strains, makes possible more economical and more structurally sound designs.
- Steel can be welded without losing desirable physical properties. A weld in steel, made by the shielded metal-arc process, has physical properties equal to or better than those of mild rolled steel, Table 2-1.

Today some manufacturers make and feature full lines of broaching, drilling, boring, and grinding machines with welded members, Fig. 2-11. Presses, brakes, and numerous types of handling machinery are now weldments. Other manufacturers are using welding for many of the accessories, pipelines, chip pans, and subassemblies.

The flexibility of welding in this work is exemplified by a milling machine bed made in one piece and constructed without any manufacturing difficulty in lengths varying from 10 feet to 25 feet. Broaching machines, which are used for a great variety of accurate machining of plane and curved surfaces, depend on welded construction. The manu-

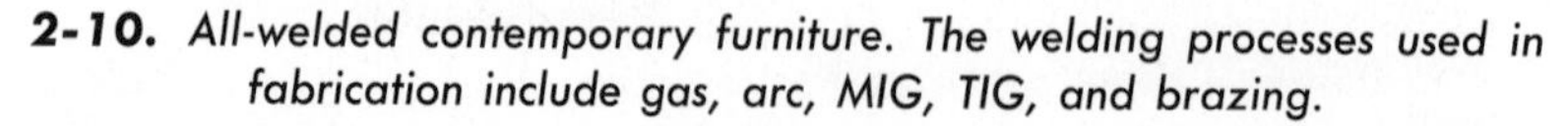

2-10. *All-welded contemporary furniture. The welding processes used in fabrication include gas, arc, MIG, TIG, and brazing.*

International Nickel Co.

facture of machine tools requires the use of bars, shapers, and heavy plate up to 6 inches in thickness. Bed sections for large planers and profilers may be as long as 98 feet, Figs. 2-12 and 2-13. A horizontal cylinder block broach, among the largest (almost 35 feet long) equipment of its kind made, has construction features that would be almost impossible without welding. For example, a rough bed for a certain broach would have necessitated a casting of over 40,000 pounds. Heat-treated alloy parts, mild carbon plate, sheet steel, iron and steel castings, and forgings were used where suitable. Because of welded fabrication, the bed weighed only 26,000 pounds. Each member was made of material of the correct mechanical type, and the whole was a rigidly dependable unit.

Nuclear Power

Nuclear power depends upon the generation of large, concentrated quantities of heat energy, rapid heat removal, a highly radioactive environment, and changes in the properties of radioactive materials. The heart of the process is a reactor pressure vessel to contain the nuclear reaction, Fig. 2-14.

In addition to the reactor vessel, the system includes a heat exchanger or steam generator and associated equipment such as piping, valves, and pumps, as well as purification, storage, and waste disposal equipment. Welding is necessary for the fabrication of all of these units. The production of nuclear energy would not be possible but for the highly developed processes of today.

Piping

High pressure pipeline work, with its headers and other fittings, is a vast field in which welding has proved itself. The number of ferrous and nonferrous alloys used as piping materials is increasing. Industry re-

Table 2-1
Physical properties of a steel weldment

Tensile strength	65,000–85,000 p.s.i.
Ductility	20%–30% elongation/2 in.
Fatigue resistance	25,000–32,000 p.s.i.
Impact resistance	50–80 ft-lbs. (Izod)

2-11. *The welding department of a large manufacturer of machine tools. Note the variety of weldments being fabricated.*

Cincinnati Milacron

quires better materials to meet the high heat and high pressure operating conditions of power plants, nuclear plants, oil refineries, chemical and petrochemical plants, and many other manufacturing plants where steam, air, gas, or liquids are used. Pressures of over 1,000 p.s.i. and temperatures ranging from −200 to +1,200 degrees F. are not uncommon in high pressure pipelines, Figs. 2-15 and 2-16. Marine lines and generator stations have installations operating at 1,250 p.s.i. with 950 degrees F. at turbine throttles. Demands for equipment in the steel mills, oil refineries, and other industries in which such lines operate emphasize reductions in size and weight and streamlining the appearance of piping as well as the flow. The lines are becoming increasingly complex: recirculation units, boosters, headers, and miscellaneous accessories and fittings are introduced into the lines, making them take on the appearance of complex electrical lines. Small pipe is connected with large pipe; *Ts*, bends, return valves, and other fittings are introduced into the lines, Fig. 2-17.

The design of fittings for welded pipe is flexible and simple. Many fittings required by mechanically connected systems can be eliminated. The absence of projections inside the pipe

Cincinnati Milacron

2-12. *A large three-spindle, vertical-bridge, numerically controlled profiler.*

2-13. *A mechanic grinding welds on the bed of the profiler shown in Fig. 2-12.*

Cincinnati Milacron

2-14. *This 80-ton nuclear superheat reactor vessel is designed to withstand a pressure of 1,250 p.s.i. and a temperature of 650 degrees F. Walls are 3½ inches thick with stainless steel liner. Flame cutting is used extensively.*

Nooter Corp.

produce less resistance to flow, Fig. 2-18. Because welded piping systems have permanently tight connections of greater strength and rigidity, maintenance costs are reduced. Other advantages of welded fabrication include a more pleasing appearance and easier, cheaper application of insulation.

With the development of welded fittings, the pipe fabricator realized the possibility of easily making any conceivable combination of sizes and shapes. Practically all overland pipeline is welded, Fig. 2-19. Piping is used for the transportation of crude petroleum and its derivatives, gas and gasoline, in all parts of the country. Overland welded-pipe installations are both efficient and economical. Successive lengths of pipe are put together so cheaply in the field that total construction costs are materially reduced. These lines can be welded to the older lines without difficulty.

No better example of the extreme reliability and speedy construction available in the welding of pipeline can be cited than the oil line running from Texas to Illinois. This is almost 2,400 miles of solid pipe joined by welding.

Railroad Equipment

Welding is the principal method of joining materials used by the railroad industry. The railroads first made use of the process as a maintenance tool. In recent years it has been extended to the building of all rolling stock. It is also used extensively in the construction and repair of equipment on the right-of-way.

Railroad units fabricated by welding include streamlined diesel and electric locomotives (Figs. 2-20 and 2-21), passenger cars, subways, freight cars, tank cars, refrigerator cars, and many other special types. Most new track is of the continuous welded type, and battered rail ends of old track are built up with the process.

Weight is an important consideration in the design of freight

2-15. *Y-branches fabricated for the Illinois Power Company for use in hot reheat piping.* Crane Co.

2-16. *High pressure welded ell being fabricated. The pipe wall is 5 inches thick, and the inside diameter is 26 inches. Approximately 100 passes are required to weld the ell.* Crane Co.

cars. In an entire trainload, a freight engine may be called upon to haul a string of up to 100 freight cars whose deadweight alone amounts to 2,400–3,400 tons. Welding and alloy steel construction have reduced the weight of 50-ton boxcars from a light weight (empty) of 48,000 pounds to 36,000 pounds, a reduction of approximately 25 percent. These cars are also able to carry a 25 percent greater load than their former 50-ton capacity.

Large cars have been designed recently which are 50 feet in length and have a capacity of up to 100 tons and an empty weight of approximately 61,500 pounds. This is an increase of 100 percent in load-carrying capacity and only 35 percent in empty weight. Cars constructed for automobile shipping service are 89 feet in length.

The most recent development in freight car construction, is the super-size aluminum gondola. The car is a third longer and almost twice as high as the ordinary gondola. The car is 90 feet long and weighs 96,400 pounds: some 60 pounds per foot less than ordinary steel gondolas of similar design. The underframe is steel, and the extensive use of aluminum saves 11,000 pounds in weight. The principle welding processes are shielded metal-arc, gas metal-arc, and gas tungsten-arc. A considerable amount of gas and arc cutting is also used in the fabrication of the plate.

In addition to the weight-saving and the greater payloads, the tight joints which welding makes possible have virtually eliminated infiltration of dust and cinders into the cars, thus reducing damage claims against railroads. This can be effected only through welded fabrication, for no matter how carefully

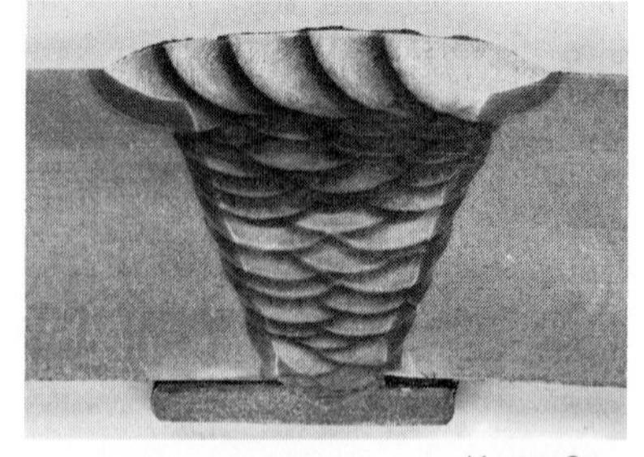

Nooter Corp.

***2-18.** An etched cross section of a multipass weld in chromemoly pipe. Note the sequence of weld beads.*

***2-17.** A pipe header with a number of branches coming in at all angles. This unit is called a mixer.* Crane Co.

***2-19.** A crew of welders are making a butt joint on an overland pipeline. Note that a welder works from each side of the joint.*

Hobart Brothers Co.

riveted joints are made, the constant jolting to which freight cars are subjected loosens the joints and opens seams that permit dust infiltration. Welded freight cars have a life of 20 to 40 years with very little maintenance required.

By employing low alloy steel and welded fabrication, car builders have been able to reduce the weight of a refrigerator car by as much as $6\frac{1}{2}$ tons. This leakproof construction, made possible by welding and the corrosion-resistant properties of the low-alloy steel which make up the car body, make it possible to keep the car in active service because the damage done to riveted refrigerator cars by leaking brine solutions is prevented.

Tank cars, too, have benefited through the use of corrosion-resistant steels and the permanently tight joints. The old type of construction caused many leakage problems. Often liquids are carried which require preheating to make them flow easily when the cars are being emptied. The alternate heating and cooling to which the riveted joints were subjected pulled them loose. Many of the tank cars being built today are 61 feet long and have a capacity of 30,300 gallons.

A fast-growing means of freight transportation in the U.S. is the highway trailer carried by railroad flat cars, commonly called piggyback service. Automobile carriers are also a recent development. As many as twelve standard cars can be hauled on one 89-foot flat car equipped with triple decks. These flat cars weigh 56,300 pounds and can transport 130,000 pounds.

Hopper cars have been redesigned with a saving in weight and the development of interior smoothness which permits free discharge of the load. With riveted construction it was difficult to clean these cars when unloading, because coal, carbon black, sulfur, sand, and cement adhered to the rivet heads and laps.

The increase in the demand for covered hopper cars to transport many powdered and granu-

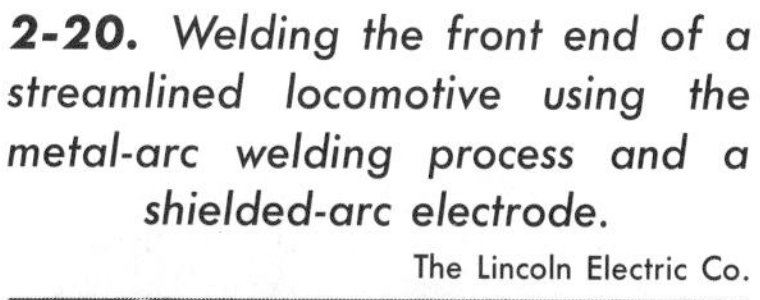

2-20. *Welding the front end of a streamlined locomotive using the metal-arc welding process and a shielded-arc electrode.*

The Lincoln Electric Co.

2-21. *Carbon-dioxide-gas-shielded metal-arc welding of diesel engine cabs. Note wire feeders overhead for MIG welding.*

Hobart Brothers Co.

lar food products, as well as chemicals, has resulted in the Center Flow car of modern welded design, Fig. 2-22. These cars are 50 or more feet in length and have capacities of from 2,900 to 5,250 cubic feet and loading capacities up to 125 tons.

Many hopper cars are now made of aluminum. A steel hopper car weighs 72,500 pounds, but one made of aluminum weighs only 56,400 pounds. This makes it possible to carry heavier loads and thus reduce the cost of moving freight. Steel is still needed, however, for highly stressed parts like the underframe.

Hopper cars have two, three, or four compartments. Each compartment has a loading hatch and hopper outlet, providing for up to four different kinds of materials in the load. A 4-compartment hopper car requires 4,000 feet of weld. Welding is the largest and most important production operation, Fig. 2-23. One-piece side plates are 112 inches by 49 feet by $^{5}/_{16}$ inch thick. Steel members are welded with the submerged-arc process, and the aluminum is welded with the gas metal-arc process. A considerable amount of gas-tungsten and shielded metal-arc welding is also used.

2-22. *A crew of 23 arc welders supply their own light for the photographer as they apply 100 feet of weld in a matter of minutes to the underframe of a Center Flow dry bulk commodities railroad car, with a capacity of 5,250 cubic feet.*

ACF Industries

Passenger coaches have come in for their share of improvement. A standard steel coach has a weight of about 160,000 pounds. Experimentation with welded construction for this type of coach has produced a coach weight of 98,000 pounds. In addition, the modern welded coaches are stronger, safer, and considerably more comfortable than former types.

Not only has the rolling stock of the railroads been improved through the use of welding, but the rails they roll on are also undergoing great change. The old 90- to 100-pound rail sections of 39-foot lengths are giving way on several roads to continuous rail of $^{1}/_{4}$ mile lengths. This so-called continuous rail is not actually continuous but consists of a number of lengths welded together at the joints.

Track maintenance men are continually faced with the problem of batter in rail ends which occurs at the joints. Battered rail ends cause a jolting when the train passes over them, which in turn means discomfort for passengers, shifting freight load, and wear on rolling stock. Battered rail ends can be eliminated by one of two methods: either replace the rail or build up the battered end. By employing welding and building up battered ends to a common level somewhere near the rail body level, both rails may remain in service. However, as old rails are replaced by continuous welded rail, service life is longer, and joint maintenance is decreased.

Inside the railroad shop, welding is likewise doing yeoman service. In locomotive maintenance it is used on pipe and tubing repair, frames, cylinders, hub liners, floors, housings, tender tanks, and for streamlining shrouds. On freight cars, posts and braces, bolster and center plate connections, and other underframe and superstructure parts are repaired, strengthened, or straightened. Passenger car bolsters and cracked truck sides are repaired without difficulty. Vestibule and baggage-car side doors and inside trim are weld-repaired as standard practice.

Shipbuilding

The Naval Limitation Treaties of the twenties and thirties were the impetus behind the research program which led to a new conception of welding in ship construction. Under these treaties the various nations agreed to limit not only the number of capital vessels built, but also their weight. The Navy's reaction, therefore, was to build the most highly effective ships possible by any method within the limitations of the treaties. A capital ship must be light in weight and highly maneuverable, but it must have adequate defensive armor plate, gun power, and strength. It must be built to take as well as to give punishment.

That welded ships can take it is borne out by the story of the U.S.S. *Kearney* which limped into port on October 18, 1941. This fighting ship, blasted amid-ships by a torpedo, came home under its own power, putting the stamp of approval on a type of construction in which our Navy had been a leader for years. It is highly improbable that any other than a welded type of ship could have reached home, and it was impossible that any other could have rejoined its command, as did the *Kearney,* a few months later. Since World War II a large number of similar occurrences involving military and nonmilitary ships have been recorded.

Military watercraft fabricated by welding include aircraft carriers, battleships, destroyers, cruisers, atomic-powered submarines, and many others.

The standard specifications for Navy welding work, which cover all welding done for the Bureau of Ships are concerned with a variety of structures, such as watertight and oiltight longitudinals, bulkheads, tanks, turret assemblies, rudder crossheads, pressure vessels, and pipelines. Air, steam, oil, and water lines in various systems are all of homogeneous welded construction, Fig. 2-24.

Some idea of the immensity of these units may be gathered from the fact that gun turrets of a 35,000-ton battleship are built from welded materials ranging from one-half to several inches thick. The units weigh 250 tons

2-23. *The end view of a hopper car, showing butt, lap, and T-joints and the use of groove and fillet welds.*

The Lincoln Electric Co.

each. The sternposts weigh 70 tons; and the rudders, 40 tons. The welded rudder of the carrier U.S.S. *Lexington* weighed 129 tons and was a $12\frac{1}{2}$-foot thick (not $12\frac{1}{2}$ inches) fabrication.

It is now possible to construct submarine hulls with a seam efficiency of 100 percent, as against the 70 percent efficiency of riveted hulls. Caulking is unnecessary because the hull is permanently leakproof. Hull production time is reduced by approximately 25 percent, and the total weight of the hull is reduced by about 15 percent because of the use of butt-welded plate. The smooth lines of the welded plate make hulls more streamlined and, therefore, faster and more maneuverable. They foul less quickly because of their smooth lines and can stay away from bases longer.

The hulls and power plants of nuclear submarines are also constructed of all-welded alloy steel plate instead of castings. Reductions in weight and size are accomplished along with improved structural strength. Between these savings and the weight reductions possible with welded piping and accessories, the modern submarine is made into a fabrication of far greater potential use. Hull strength for longer underwater runs, resistance to depth bombs, and deeper dives; increased power plant efficiency; and an over-all decrease in weight per horsepower make the submarine an outstanding example of a unit welded for its purpose.

2-24. *The main steam system piping for the engine room of a nuclear submarine.* Crane Co.

Ships differ widely in type and conditions of service. They range from river barges to large cargo and passenger vessels, Fig. 2-25. The adoption of the construction methods used in building ocean-going "Liberty ships" during World War II has reduced construction time from keel laying to launching by more than 20 percent. Prefabrication, preassembly, and welding are the reasons for the dramatic reduction in building time, Fig. 2-26. Parts and substructures are shaped in advance. Accessories, pipelines, and necessary preassemblies are constructed in many cases far away from the scene of the actual building of the ship's hull. After completion they are transported to the site and then installed as units into the vessel.

A completely riveted freighter would require in its construction thousands of rivets, averaging about 1 pound each. From a labor and timesaving standpoint, there is a reduction of 20 to 25 percent in deadweight which can be used largely for cargo carrying. In many ships today, there are only 200 rivets. A welded ship uses approximately 18 percent less steel than one that is riveted. In other words, in every six 10,000-ton vessels built, enough steel is saved to build another ship. Today's cargo ships weigh from 10 to 15 percent less than their 1918 counterparts, despite the fact that their deadweight capacity is 2,000 tons greater. Smoothness of hull construction has materially increased the speed of the vessels and reduced hull maintenance costs by 25 percent.

Oil tankers are of such size that only the welding process with its great saving in weight and strength makes construction possible. A recently constructed tanker is more than 100 feet longer than one of the world's largest passenger liners, the *Queen Elizabeth II,* which is 1,031 feet long and 119 feet wide. The deck would dwarf a football field. The tanker is 105 feet high. It is powered by an 18,720-horsepower diesel engine, with a second one in reserve. It is designed to carry 276,000 tons of cargo and cost $20 million. Tankers now on the drawing boards will have a capacity of 600,000 tons.

Structural Steel Construction

The welding process has been fairly recently applied to the construction of hydroelectric units, power generation units, bridges, commercial buildings, and private dwellings. The construction of such super-power projects as the Bonneville, Grand Coulee, Hoover Dam, and the Tennessee Valley Authority projects called for entirely new methods of construction for water turbine parts and water power machinery.

Welded construction was first used extensively for the 74,000-horsepower hydraulic turbines in the Bonneville project on the Columbia River near Portland, Oregon. The units alone of this power project involved 80,000 feet of gas cutting, 118,000 linear feet of welding, 286,000 pounds of electrodes, and 6,450,000 pounds of rolled plate steel. The Bonneville project demonstrated that the following advantages were realized by the use of welded members:

- A large number of patterns could be eliminated.
- Parts were ready more quickly for machining.
- Because of the use of steel plate, there was the practical assurance that machine work would not expose defects with resultant replacement and delay. This was important because of the necessity for quick delivery.
- Weights could be figured accurately, allowing close estimates for material costs.

2-25. *An excursion steamer docked on the Mississippi River at St. Louis, Missouri. This is an all-welded steel steamer.*

2-26. *Carbon-dioxide-gas-shielded metal-arc welding in the shipyard. Note the overhead position.*

Hobart Brothers Co.

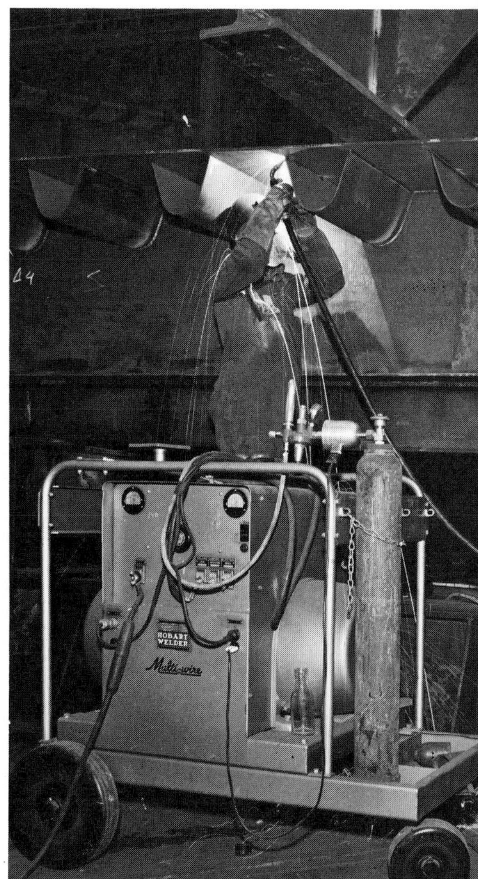

- The amount of metal allowed for machining was reduced, at the same time saving the time necessary for machining.
- Exact scale models could be made and tested under the same conditions as large units.
- Composite construction could be used. This type of construction involves the welding together of plate steel and castings or forgings, a combination of mild steel and alloy steel, or a combination of two alloy steels.
- Welding was also responsible for the usual saving in weight, together with greater strength, improved quality, increased efficiency, and flexibility of design.

BRIDGES. Bridges are now being constructed wholly or in part by the welding process, Fig. 2-27. For over 30 years, steel bridges, both highway and railroad, have been constructed by this means, and the number of welded-steel bridges is increasing.

Typical of the weight reduction possible in bridge construction is a saving of 42½ tons in a bascule span of a highway bridge built in Florida. One hundred tons were eliminated in the counterweights. Fixed and expansion bridge shoes had welded rolled-steel slabs for strength, reliability, and economy.

Savings in typical steel bridges, resulting from welded construction, range up to 20 percent. If these savings were extended to the long-range road-building program that has been initiated by the federal government, enough steel could be saved to build a highway girder bridge approximately 800 miles long. Cost comparisons of actual rivet construction and welded construction have demonstrated that there is a 5.5 to 1 advantage in cost for welding construction.

Although cost and weight are important considerations, the strength of welded steel tips the scale in its favor. A welded-butt joint is the best type of joint. It has the greatest strength and the most uniform stress distribution. The flow of stress in a riveted joint, however, is not uniform; it has a number of stress concentrations at various points. Just the punching of a hole in a plate for the rivet causes high stress concentrations when the plate is loaded.

Most rivets are driven hot. A hot rivet always shrinks upon cooling after being driven. This means that all rivets tend to shrink lengthwise, thus producing locked-up tensile stress in the rivet body, even without an external load. It also means that the rivet shrinks transversely so that it never quite fills the hole. The holes must be reamed so that the rivet is not deformed by holes that do not line up. This operation adds extra cost to the job.

The foundation pilings of many bridges have cutting edges made of welded steel plate. Tower caissons are made in sections and, because they are watertight, are floated to the site and filled with concrete. All-welded bridge floors are fairly common. Reinforcing girders (Fig. 2-28), crossbeams, and other members have been constructed with a saving of as much as 50 percent in both weight and time.

2-27. *The New York State Thruway bridge at Suffern, New York. The center span is 234 feet long with all-welded construction.*

The Lincoln Electric Co.

INDUSTRIAL AND COMMERCIAL BUILDINGS. All types of buildings are welded during construction. Welding has become a major method of making joints in structures. The fact that there are no holes needed for rivets is an advantage in the design of trusses and plate girders. Flange angles are not needed in plate girders, and single plates can be used for stiffeners instead of angles. Rigid frame structures are possible, permitting the bent-rib type of roof construction which gives maximum headroom, no diagonal cross-bracing members, and no shadow lines from truss members, Fig. 2-29. In multiple-story buildings, the rigid frame permits shallow beam depths which allow lower story heights.

Welding reduces construction and maintenance costs due to smooth lines of construction, decreased weight of moving elements such as cranes, and ease of making alterations and new additions. First cost is materially less because of a saving in weight of materials which may be as much as 10 to 30 percent. Many building units can be fabricated in the shop under controlled conditions, thus reducing expensive on-the-site work. Interiors are open and unrestricted; there are no columns in the way.

Excavation is speeded up by the use of digging equipment with abrasion-resisting teeth, made economically possible by welding. Piling sections and reinforcing steel are flame-cut and welded. Welding replaces riveting in the shop fabrication and field erection of columns, beams, and girder sections. Flame-cutting is used to prepare gussets and perform field trimming operations. Incidentally, most of the construction equipment used on the job such as cranes, bulldozers, and concrete mixers, is welded.

After the structural steel framework of the building is complete, continued use of welding also speeds up the mechanical installations. Pipelines and electrical conduits are welded into continuous lengths. Air ducts and smoke risers are fabricated to the required shapes by welding and cutting. Welded electrical junction and panel boxes are secured to the columns and beams by welding. Transformers, switchboards, furnaces, ventilating equipment, tanks, grating, railing, and window sash are partially or completely prefabricated. Once located, their installation and connections are made with the aid of welding. Changes or additions to the building or its equipment are greatly aided by this method.

The construction industry has long felt the need to solve the problems of creating housing for a mass market. Some architects have turned to a steel-fabricated welded structure as a solution. Such prefabricated housing has the following advantages:

2-28. *Joining a welded girder frame to the column of a bridge span. Another example of the unusual positions a welder may be called upon to weld in. The joint is in the horizontal position.*

The Lincoln Electric Co.

2-29. *Industrial building interiors take on an entirely new appearance. Arc-welded rigid frames replace conventional truss sawtooth framing. Note the absence of columns and the improved headroom.*

The Lincoln Electric Co.

- The construction method uses factory-produced materials of many kinds that are standard, readily available, and accurate.
- A large part of the construction can be shop fabricated under controlled conditions and mass produced, thus requiring less site labor.
- Site erection is fast, thus providing for an overall reduction in cost. Mobile homes, Fig. 2-30, are also enjoying increased popularity.
- Construction materials weigh less, are stronger, and lend themselves to accoustical treatment more easily than standard materials.
- Prefabricated modules provide flexibility of design and floor plan arrangements.
- A higher factor of earthquake, flood, and wind resistance is possible.

Mobile Home Manufacturers Association

2-30. *A welded mobile home. Welding is a principal factor in the growth of the mobile home industry.*

Tanks and Pressure Vessel Construction

The growth of cities and towns has increased both the number and the size of tanks needed for the storage of water, oil, natural gas, and propane. The increase in the number of automobiles, trucks, and aircraft has increased the need for storage facilities for petroleum products. In addition, our space and missile programs have created the need for the storage of oxygen, nitrogen, and hydrogen in large quantities. The fertilizer industry requires volume storage facilities for ammonia. The basic materials for many industries, supplying such diverse products as tires, fabrics, soap, and food products, are stored in pressure vessels. Tanks and vessels of all types have become one of the principal applications of welding.

2-31. *Riveted construction formerly used in constructing pressure vessels. Each rivet was a point of breakdown. Compare with today's all-welded vessel shown in Fig. 2-32.*

Combustion Engineering Co.

2-32. *This steam generator plant has a capacity of 127,000 pounds and contains more than 9 miles of tubing. The plant produces steam from controlled nuclear fission.*

Nooter Corp.

Welding replaced riveting in the fabrication of pressure vessels approximately forty years ago, Figs. 2-31 and 2-32. This improved the service performance of a pressure vessel through the elimination of two common areas of service failure in riveted vessels: leakage and corrosion around rivets.

The construction and maintenance costs of both welded tanks and pressure vessels are also reduced. Less material is used in the construction of a welded vessel. A riveted joint develops a strength equal to only 80 percent of the tank plate, whereas a welded joint develops a strength of 20 to 30 percent greater than the plate. It is, therefore, possible to reduce the plate thickness and still obtain the same design strength by welding. Some of the heavier pressure vessels, 3 to 5 inches in thickness, cannot be fabricated in any other way because it is impossible to rivet plates of this thickness with any degree of success. In addition, there is further saving because it is unnecessary to punch the plates and caulk the seams of a welded joint. Maintenance costs of welded tanks are practically negligible, and the joints are permanently tight, Figs. 2-33 and 2-34.

One of the leading pressure vessel manufacturers points to the following in support of welded construction:

- Elimination of thickness limit of about $2\frac{3}{4}$ inches for successful riveting, and elimination of leakage at high pressure.
- Elimination of thickness limit for forge and hammer welding, which was about 2 inches.
- Elimination of caustic embrittlement in riveted boiler drums.
- Economy in weight through higher joint efficiency and elimination of butt-straps and rivets.
- A reduction in size to meet the same service requirements.
- Greater flexibility of design, permitting uniform, or at least gradual, stress distribution.
- Elimination of all fabricating stresses in the completed vessel by heat treatment.

To these achievements of welding in the fabrication of pressure vessels, might be added increased speed of fabrication (Fig. 2-35), reduction of corrosion for longer life, and smooth interiors of chemical and food vessels for sanitation, Fig. 2-36. Finally, by eliminating the size limit on pressure vessels, welding has made a direct contribution to our expanding productive capacity and technology.

Miscellaneous Applications

A few miscellaneous applications are illustrated in Figs. 2-37 through 2-40 so that the student may appreciate the flexibility of the welding process.

Nooter Corp.

2-33. *A water tank constructed of plate $1\frac{1}{2}$ inches thick, which is 240 feet in diameter and has a capacity of 11 million gallons of water.*

2-34. *World's largest titanium tower—10 feet in diameter. A considerable amount of gas metal-arc welding is used on this type of work.*

Wyatt Industries

Nooter Corp.

2-35. *An oil refinery sphere being constructed in the field indicates the mobility and flexibility of the welding process.*

2-36. *Automatic gas-shielded, metal-arc welding of brewery vessels.*

Hobart Brothers Co.

2-37. *All-welded fabricated gear. The parts of the gear were flame cut.*

The Lincoln Electric Co.

The Lincoln Electric Co.

2-38. *Constructing a minute-man missile base. Welding and cutting are used extensively.*

2-39. *Gear-reducing unit. All parts were flame cut, and unit includes all types of joints and welds.*

General Electric Co.

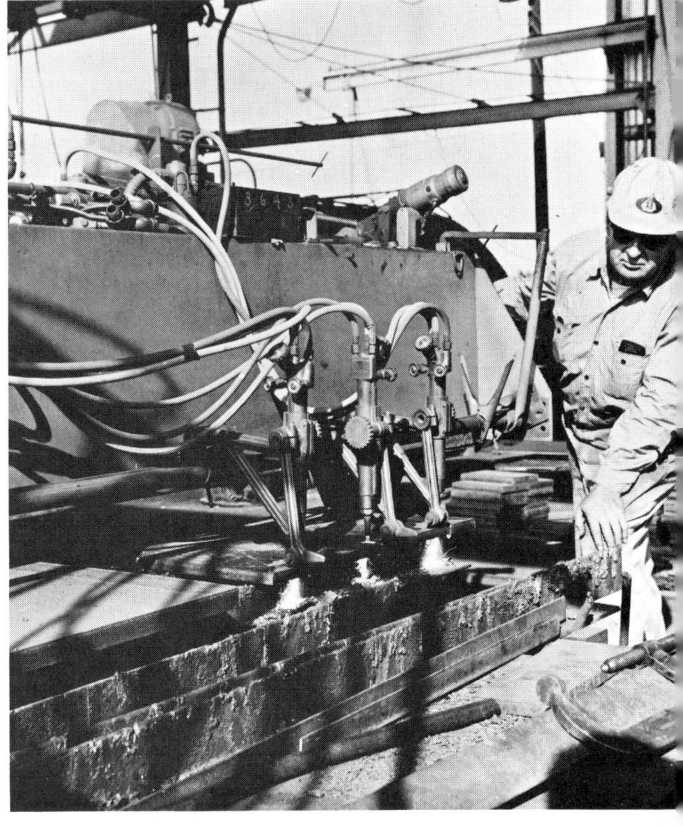

Linde Division, Union Carbide Corp.

2-40. *All of the fabrications shown in this chapter used gas and arc cutting as a fabricating tool. Shown here is a multiple-torch application, burning natural gas and oxygen, which is cutting out parts that will later become part of a weldment.*

REVIEW QUESTIONS

1. List the advantages of welding as a means of fabrication.
2. List the advantages of welding when used for maintenance and repair.
3. Steel has several advantages for construction. Name them.
4. How is welding used by the railroads?
5. What are some of the advantages of welding in pressure and overland piping?
6. Is welding limited in its application to piping and pressure vessels because the process is not dependable at high pressures and temperatures? Explain your answer.
7. List some of the types of watercraft that are fabricated by welding. How is it possible to finish a cargo ship in three weeks?
8. List some of the advantages of welded construction when applied to pressure vessels.
9. List at least five advantages which can be gained in the application of welding to building construction.
10. How has welding in bridge construction progressed in recent years?
11. Identify three weaknesses of rivet construction.
12. List at least ten products used by the military that are manufactured wholly or in part by welding.
13. Can a tank with a wall thickness of over 3 inches be riveted? Can it be welded?
14. Do bridges commonly have all-welded construction?
15. What features of welded construction make it resistant to earthquakes, floods, and high winds?

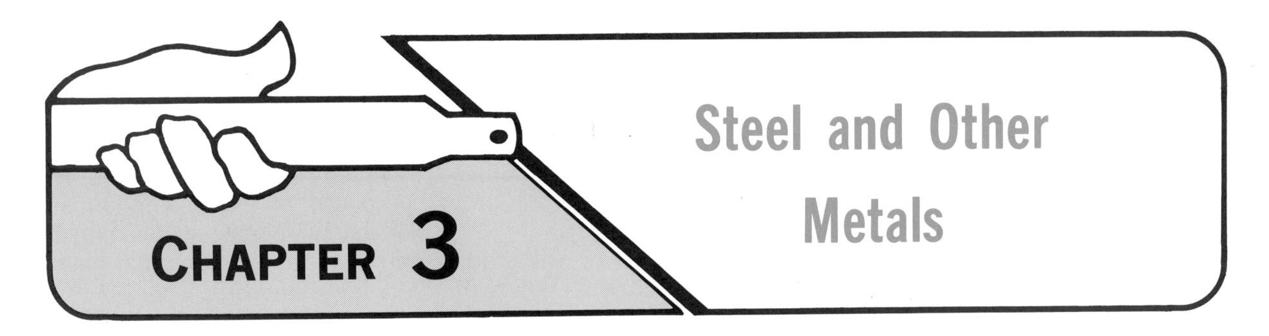

Chapter 3 Steel and Other Metals

The welding process joins metals, plastics, and glass without the use of mechanical fastening devices. In this study we are primarily concerned with its application to metals. Metals are separated into two major groups: ferrous metals and nonferrous metals.

Ferrous metals are those metals which have a high iron content. They include the many types of steel and its alloys, cast iron, and wrought iron.

A *nonferrous* metal is one that is almost free of iron. The nonferrous group includes such common metals as copper, lead, zinc, aluminum, nickel, tungsten, manganese, brass, and bronze. The precious metals (gold, platinum, and silver) and radioactive metals such as uranium and radium are also nonferrous.

Steel is a combination of iron and carbon. Iron is a pure chemical element. Oxides of iron are found in nature, and iron ore is abundant throughout the world. Because iron is not strong enough and hard enough to be used in structural members, it must be combined with carbon to produce the characteristics necessary for steel forms. Up to a certain point, the more carbon steel contains, the stronger and harder the steel will be. Although it is possible to weld nearly all of the ferrous and nonferrous metals and alloys, this chapter will be concerned principally with steels in the low and medium carbon ranges. These are the steels that the student will be primarily concerned with in the practice of welding. It has been estimated that nearly 80 percent of all weldments are fabricated from steel and that 85 percent of the total amount of steel welded is in the mild (low carbon) steel classification.

HISTORY OF STEEL

The ancient Assyrians are credited with the first recorded use of iron about 3700 B.C. Since use of iron in making weapons gave them an advantage over other nations, they became the most powerful nation of their time. From about 1350 B.C. to 1300 A.D. all of the iron tools and weapons were produced directly from iron ore.

Low carbon iron was first produced in relatively flat hearth furnaces. Gradually the furnaces were increased in height, and the charge was introduced through the top. These shaft furnaces produced molten high carbon iron. Shaft furnaces were used in Europe after 1350 A.D. The modern blast furnace is a shaft furnace.

Accurate information about the first process for making steel is not available. Tools with hardened points and edges have been found that date back to 1000 to 500 B.C. Early writers mention steel razors, surgical instruments, files, chisels, and stone-cutting tools several hundred centuries before the Christian era.

Prior to the Bessemer process of making steel, only two methods were used. The cementation process increased the carbon content of wrought iron by heating it in contact with hot carbon in the absence of air. The crucible process consisted of melting wrought iron in crucibles to which carbon had been added. Both of these processes were known and used by the ancients.

During the Middle Ages both the cementation and crucible processes were lost to civilization. The cementation process was revived in Belgium about the year 1600 A.D. while the crucible process was rediscovered in England in 1742. The crucible process eventually came to be used chiefly for making special steels. The cementation process was highly developed and was also used extensively in England during the eighteenth and nineteenth centuries. This process is still used to a limited extent. The crucible process has been replaced by the various electric furnace processes for making special alloy steels and carbon tool steels.

Steelmaking in the United States

The history of the iron and steel industry in North America

extends back over 300 years, beginning with a successful ironworks in Saugus, Massachusetts, about 20 miles northeast of Boston. It operated from 1646 to 1670. Through the support of the American Iron and Steel Institute, this site has been restored and is open to the public.

Very little steel was manufactured in America during the early days. The first patent was issued in Connecticut in 1728. A succession of events spurred the growth of the steel industry:

- New uses for iron
- The discovery of large iron ore deposits in northern Michigan
- The development of the Bessemer and open hearth processes
- The Civil War and America's explosive industrial growth following the war
- The expansion of the railroads
- World Wars I and II

More than half of the steel made in this century has been produced since the early 1940s. Today the steel industry in the United States is the largest in the world. Other large producers are France, Japan, Russia, United Kingdom, West Germany, Belgium-Luxembourg, and Italy.

Annual steel production in this country has exceeded 100 million tons. Steelmaking facilities include about 250 blast furnaces for iron production and more than 1,220 furnaces for the making of steel. Nearly 40 percent of all industrial jobs in this country involve the making of steel or the use of steel.

The perfection of the welding process as a means of joining metals has speeded up and expanded the use of steel. The adaptability of steel to manufacturing processes and its ability to join with many other metals to give a wide variety of alloys have also contributed to its widespread use.

With the rapid development of gas tungsten-arc and gas metal-arc welding, the welding of aluminum, stainless steel, and other alloys have become routine production applications.

In this chapter you will study the important characteristics of iron, steel, and other metals so that you will have a basic understanding of the nature of metals and the various results of the welding process.

RAW MATERIALS FOR THE MAKING OF STEEL

Huge quantities of raw materials are needed to produce the vast amount of steel needed by mankind. The United States is well-supplied with the basic resources such as iron ore, limestone, and coal. The principal supplies of manganese, tin, nickel, and chromium necessary to the making of steel and its alloys are found in other countries.

As stated earlier, iron occurs in nature in the form of compounds with oxygen. In order to obtain the iron, the oxygen is removed in a blast furnace by contact with carbon. Coke, a coal product, is the usual source of carbon. In this process the iron becomes contaminated with some of the carbon. This extra carbon is in turn removed in the steelmaking processes by using controlled amounts of oxygen. The resulting product is formed into various shapes by rolling or other processes. Steel may also be heat treated.

Iron Ore

Iron is a metallic element which is the most abundant and most useful of all metals. It occurs in the free state only in limited quantities in basalts and in meteorites. Combined with oxygen and other elements in the form of an ore mixed with rocks, clay, and sand, iron is found in many parts of the world. About 5 percent of the earth's crust is composed of iron compounds. For economic reasons it is mined only in those locations that have very large deposits.

In the United States heavy deposits are mined around the Lake Superior region, in Alabama, and in various western states. Mines are now being developed in Missouri. Extensive deposits are also located in Brazil, India, Venezuela, Liberia, and Labrador. The purest iron ore comes from Sweden and is responsible for the high quality of Swedish steel.

The following iron ores are listed in the order of their iron content:

- *Magnetite* (Fe_3O_4) is a brownish ore which contains about 65 to 70 percent iron. This is the richest iron ore and the least common.
- *Hematite* (Fe_2O_3), known as red iron, contains about 70 percent iron. It is widely mined in the United States.
- *Limonite* ($2Fe_2O_3 \cdot H_2O$) contains from 52 to 66 percent iron.
- *Siderite* ($FeCO_3$) contains about 48 percent iron.
- *Taconite* is a low grade ore which contains from 22 to 40 percent iron and a large amount of silica. It is green in color.
- *Jasper* is iron-bearing rock. The ore is predominantly magnetite or hematite.

Iron ore is mined by the underground and open pit methods. The method chosen primarily depends on the depth of the ore body below the surface and the character of the rock surrounding the ore body.

In underground mining, a vertical shaft is sunk in the rock

next to the ore body. Tunnels are drilled from this shaft and blasted horizontally into the ore body at a number of levels. In open pit mining the mineral is lying relatively near the surface. The earth and rock covering the ore body are first removed. Blast holes are then drilled, and explosives shatter the ore to permit easy digging. The loosened ore is hauled out of the pit by truck, train, or conveyor belt.

For many years in the past, the great bulk of iron ore had only to be mined and shipped directly to the blast furnace. This is no longer the case. With the great drain on the ore bodies due to the rapidly expanding steel industry, the supply of high grade ore suitable for direct shipment was seriously depleted by the end of World War II. To solve this problem, steel producers and mining companies started upgrading ore quality by crushing, screening, and washing the ore in order to obtain a more suitable feed for the blast furnace. Figures 3-1 and 3-2 detail the complex treatment that low grade ore receives before it is ready for the blast furnace.

Oxygen

Oxygen is the most abundant element on earth. Almost half the weight of the land, 21 percent by weight of the air, and nearly 90 percent by weight of the sea consist of oxygen. Most of the oxygen for commercial purposes is made through the electrolysis of water and the liquefaction and subsequent distillation of air.

The steel industry is a major consumer of oxygen. The gas goes into most of the standard steel mill processes from the blast furnace to the finished product. Oxygen is used in the steelmaking process to purify the material. When directed onto the surface of molten iron, it oxidizes the carbon, silicon, manganese, and other undesirable elements. It also speeds up the steelmaking process by supporting the combustion of other fuels. These oxy-fuel flames provide much higher temperatures than fuels burned in air. Figure 3-3 (Pages 50 and 51) illustrates the processing of oxygen for industrial use.

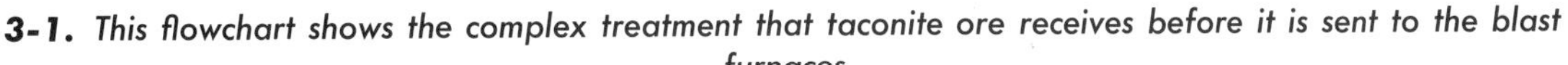

3-1. *This flowchart shows the complex treatment that taconite ore receives before it is sent to the blast furnaces*

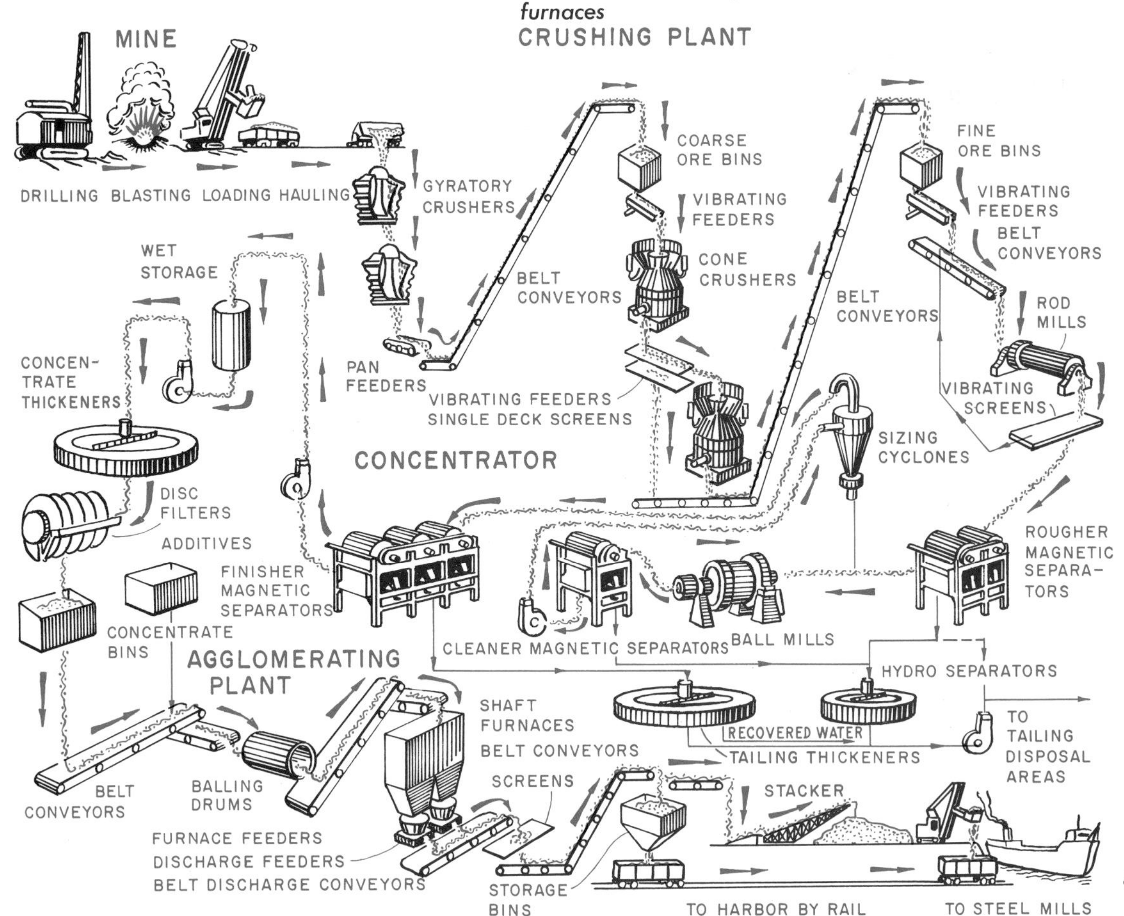

American Iron and Steel Institute

3-2. *Ore processing methods.*

IMPROVING IRON ORE

Crude ore, varying in size from dust to boulders and mixed with earth and sand, is processed before use in ironmaking.

Big ore chunks are crushed. Smaller pieces are sorted by size and, if high grade, may go through screens (right) to blast furnace.

SCREENS

Screened, properly sized high-grade ore needs only washing.

This flow chart shows three major ore processing methods, omitting possible variations. The flow lines above trace ore that requires relatively little treatment before use in ironmaking. Flow lines to the left and below trace ore that is fine, either naturally or because of crushing. Those fine particles usually are made into either sinter or pellets. Sintering is described at far right, and pelletizing at lower right.

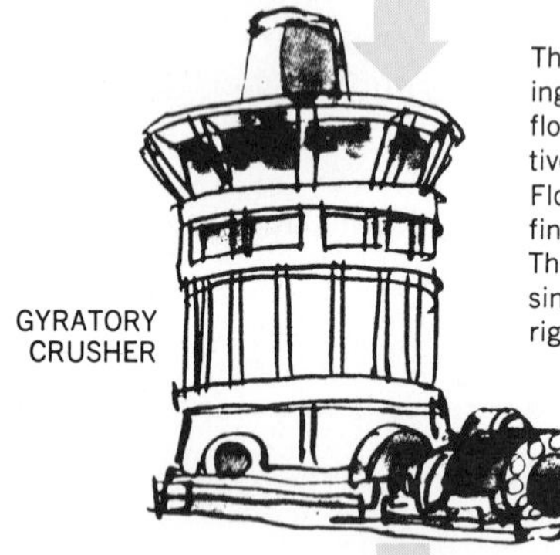

GYRATORY CRUSHER

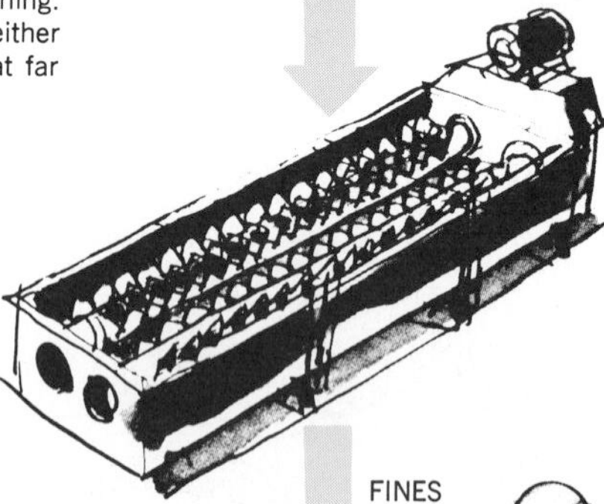

LOG WASHER

Machine commonly used for washing iron ores.

FINES

Rod mill, using whirling steel rods, typifies equipment used to crush ore finer.

ROD MILL

HIGH GRADE & LEAN, NON-MAGNETIC ORES

FINES FOR SINTER

CLASSIFIER

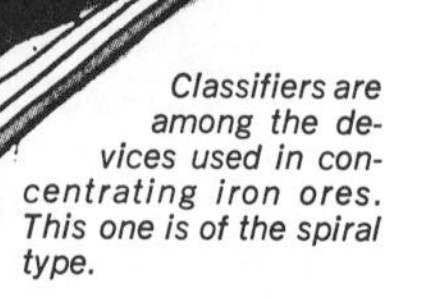

Classifiers are among the devices used in concentrating iron ores. This one is of the spiral type.

FINES FOR PELLETS

MAGNETIC TACONITE

Ore containing magnetic iron goes to the magnetic separator, where iron-bearing particles are recovered. Nonmagnetic ore goes direct from rod mill to classifier or ball mill.

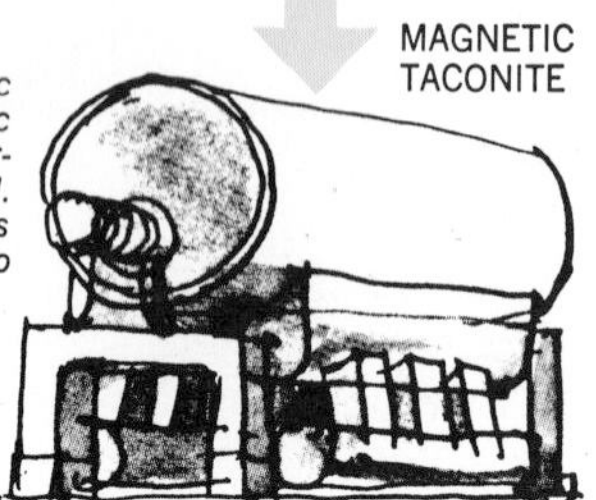

MAGNETIC SEPARATOR

TACONITE

BALL MILL

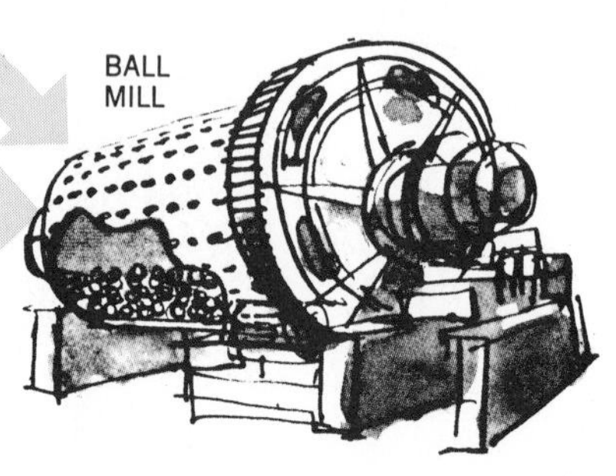

Ball mills like this grind ore to powder. Cyclone classifiers may thereafter be used to separate fines.

In the United States, the element iron is abundant in the earth's crust. However, locked in its ore, it comes in widely assorted sizes and mixed with large quantities of dirt and sand. To break it down or build it up to sizes that the industry can use efficiently, and to concentrate the iron content of low grade ores is the job of the ore processing plants. Some ore is simply washed, screened and sorted; some is crushed to fine powder, has dirt removed and is made into pellets or other solid form; some is mixed in powder form with other materials and caked to make sinter. The major processes are indicated here.

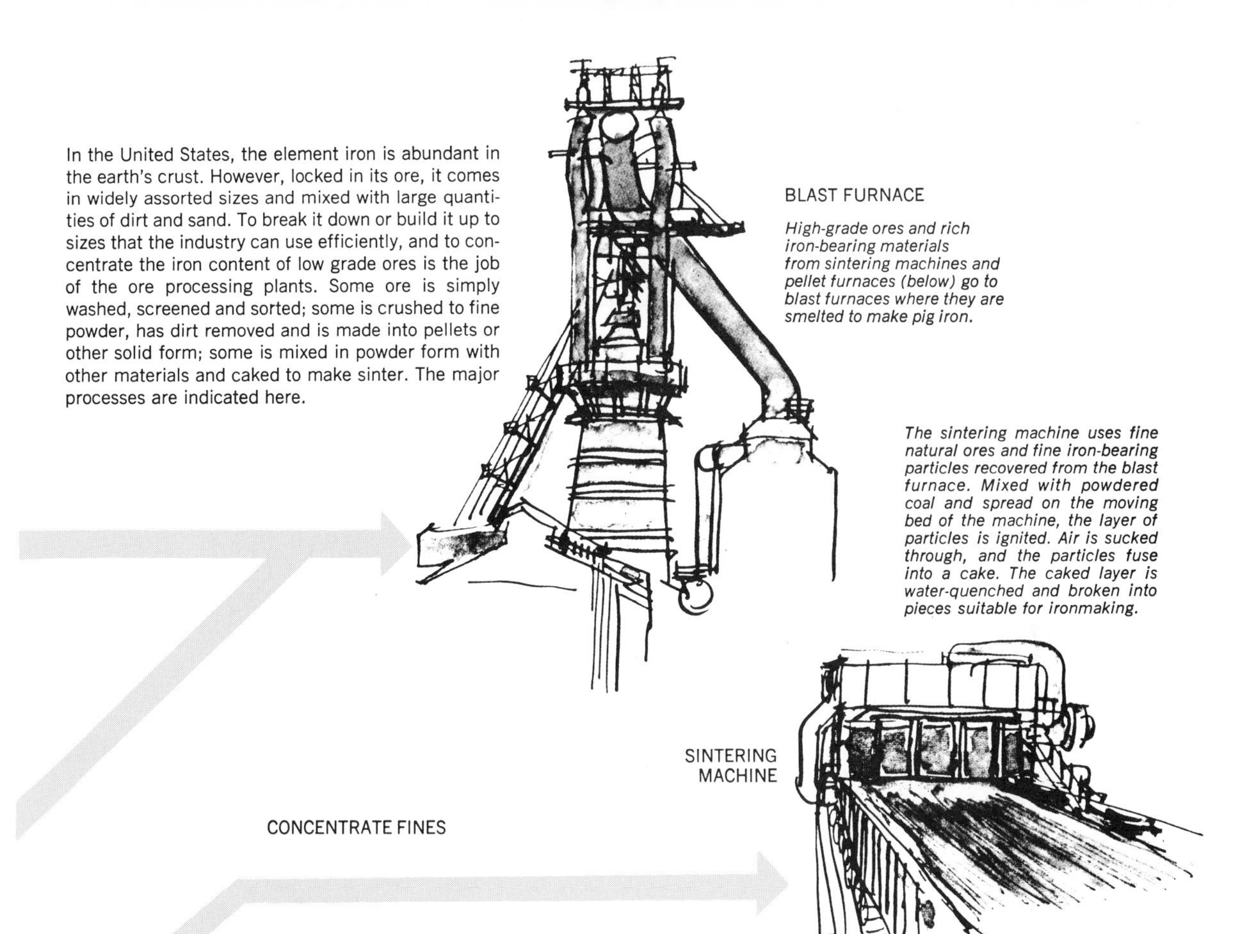

BLAST FURNACE
High-grade ores and rich iron-bearing materials from sintering machines and pellet furnaces (below) go to blast furnaces where they are smelted to make pig iron.
The sintering machine uses fine natural ores and fine iron-bearing particles recovered from the blast furnace. Mixed with powdered coal and spread on the moving bed of the machine, the layer of particles is ignited. Air is sucked through, and the particles fuse into a cake. The caked layer is water-quenched and broken into pieces suitable for ironmaking.
SINTERING MACHINE
CONCENTRATE FINES

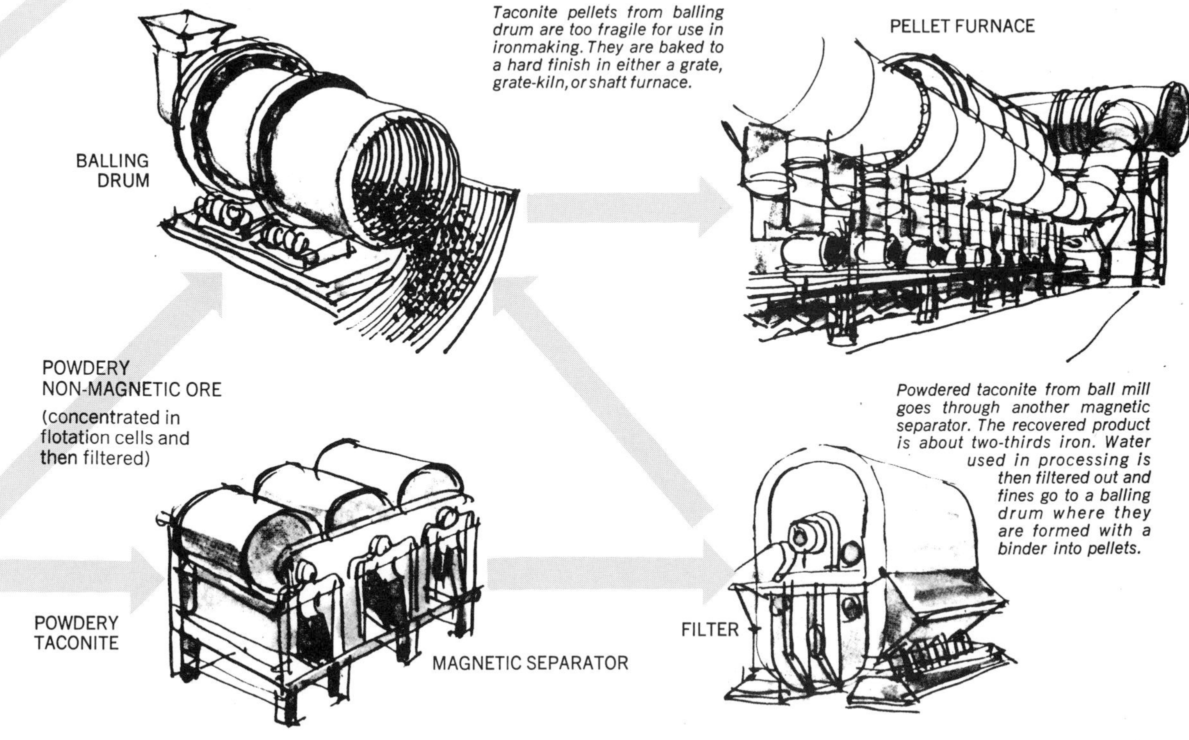

Taconite pellets from balling drum are too fragile for use in ironmaking. They are baked to a hard finish in either a grate, grate-kiln, or shaft furnace.
PELLET FURNACE
BALLING DRUM
POWDERY NON-MAGNETIC ORE
(concentrated in flotation cells and then filtered)
Powdered taconite from ball mill goes through another magnetic separator. The recovered product is about two-thirds iron. Water used in processing is then filtered out and fines go to a balling drum where they are formed with a binder into pellets.
POWDERY TACONITE
MAGNETIC SEPARATOR
FILTER

Dry air is nearly 21 percent oxygen, about 78 percent nitrogen and the remainder includes argon and numerous other rare gases. An early step in producing oxygen is filtering dust and other particles out of the air.

CENTRIFUGAL COMPRESSOR

Filtered air must be compressed. The compressors used to accomplish this usually are centrifugal rather than the piston types used earlier.

The heat exchangers are usually enclosed in a "cold box" which is the dominant structure in an oxygen plant. Warm compressed air goes in one end of the exchanger and is cooled down to about minus 270° F by cold gases from the process. The very cold temperature readies the compressed air for liquifaction. After the air has become liquid, adsorption processes remove other impurities.

WATER WASH TOWER

The water wash tower removes some of the remaining impurities in the compressed air.

HEAT EXCHANGER

LIQUID AIR

Nitrogen gas at very low temperature goes back into the air-chilling process in the heat exchangers.

CRUDE ARGON GAS

CRUDE NEON GAS

GASEOUS OXYGEN

MAIN LIQUID OXYGEN STREAM

LIQUID NITROGEN

Gaseous oxygen to be warmed in heat exchangers.

Gaseous oxygen warmed and returned to pipeline.

AIR SEPARATION COLUMN

The separation of air into its component parts occurs in an air separation column (by distillation) only after extreme cold has liquified the air in the exchangers. Gases separate out of liquid at varying degrees of cold—oxygen at about minus 297° F; nitrogen at about minus 320° F; argon at about minus 303° F.

American Iron and Steel Institute

3-3. *Oxygen for steelmaking.*

STEEL'S APPETITE FOR OXYGEN

The modern iron and steel industry uses more oxygen than any other. Oxygen can be considered as much a raw material as the specially prepared iron ore, coke and lime that the industry consumes. Like them it is a product of complicated and costly processing. The plants which "make" oxygen—and, increasingly, nitrogen and argon—for the steel industry distill ordinary air into its components, which they sell to steelmakers and others. Despite the fast growing demand for oxygen, there is no danger of depletion in the air we breathe. Industry uses a very small fraction of one percent of the 400 billion tons of oxygen that nature produces annually.

Liquid oxygen may be shipped by special rail car or trailer for steel, or other industry, consumption.

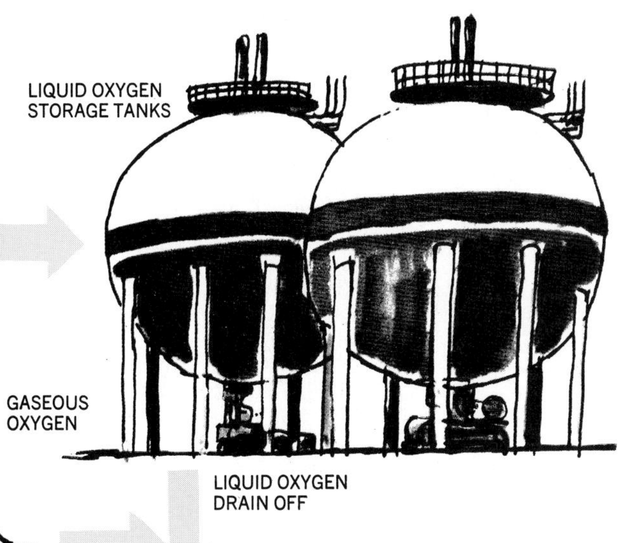

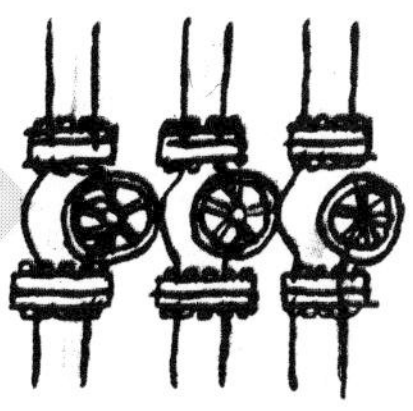

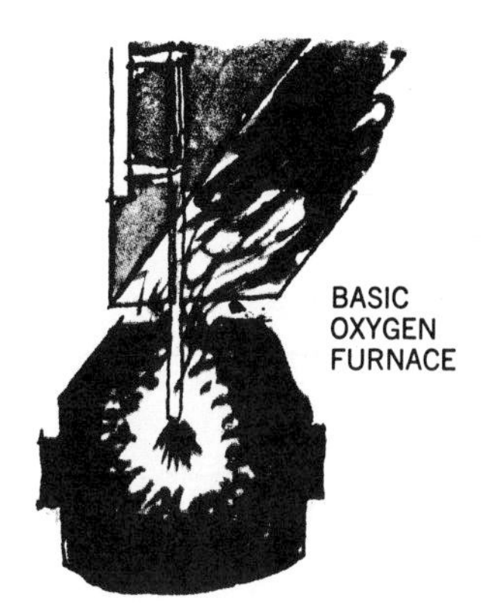

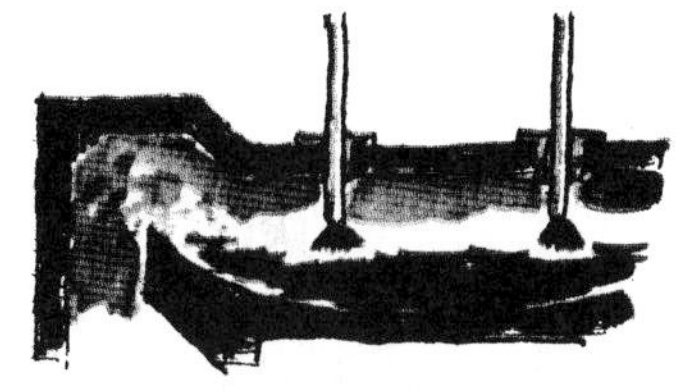

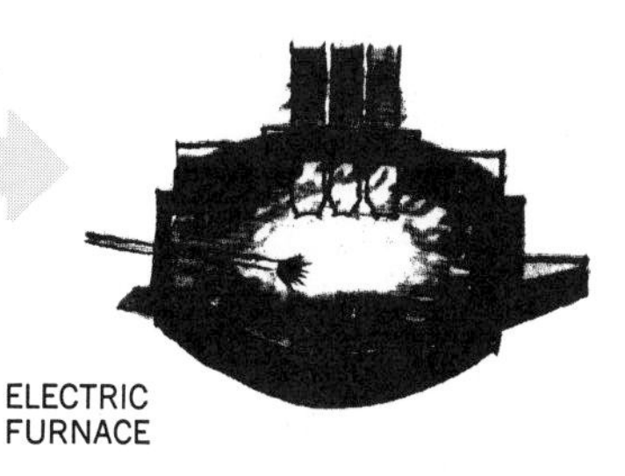

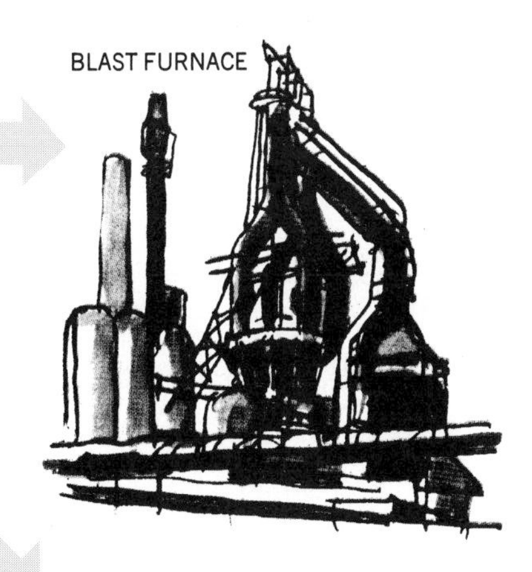

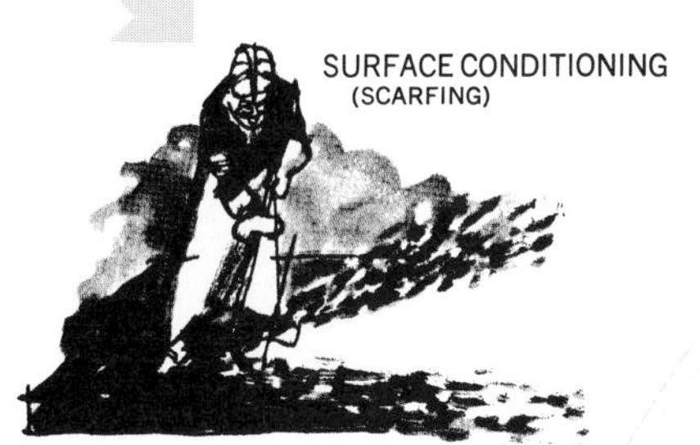

Bulk gas trailers carry tube banks of compressed oxygen gas to steel and other industries, for a large variety of uses.

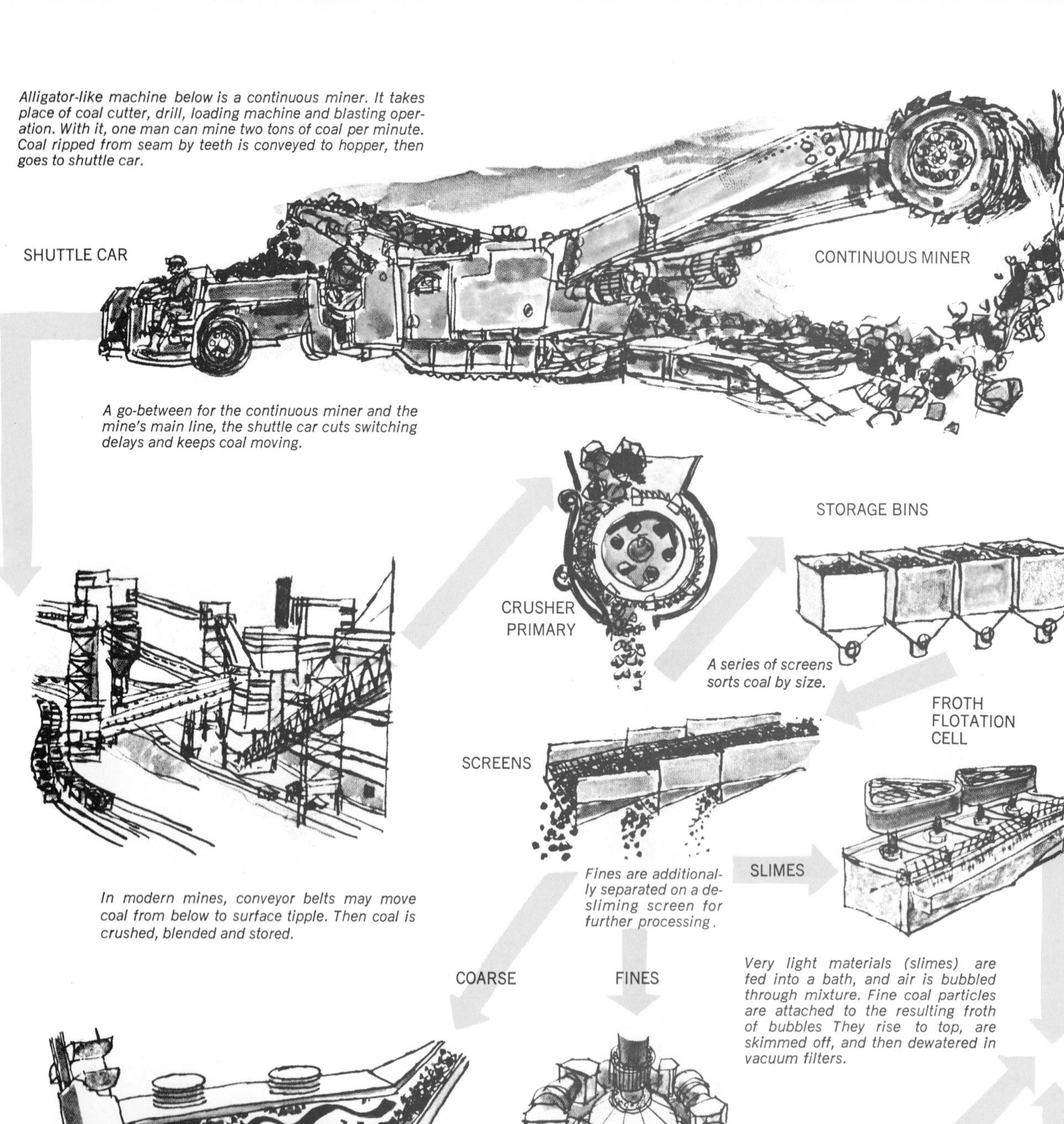

Alligator-like machine below is a continuous miner. It takes place of coal cutter, drill, loading machine and blasting operation. With it, one man can mine two tons of coal per minute. Coal ripped from seam by teeth is conveyed to hopper, then goes to shuttle car.

A go-between for the continuous miner and the mine's main line, the shuttle car cuts switching delays and keeps coal moving.

In modern mines, conveyor belts may move coal from below to surface tipple. Then coal is crushed, blended and stored.

A series of screens sorts coal by size.

Fines are additionally separated on a desliming screen for further processing.

Very light materials (slimes) are fed into a bath, and air is bubbled through mixture. Fine coal particles are attached to the resulting froth of bubbles They rise to top, are skimmed off, and then dewatered in vacuum filters.

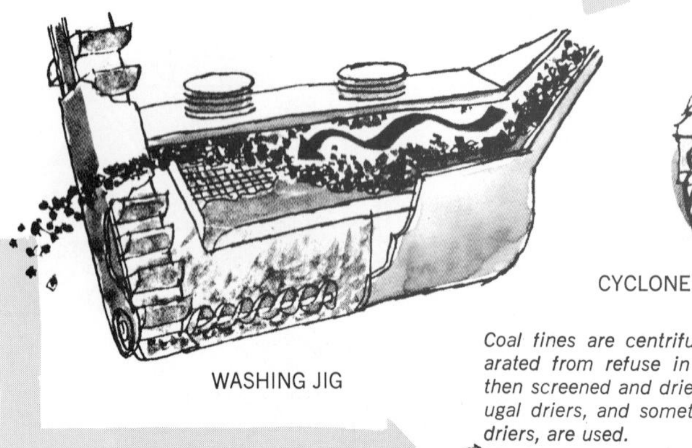

Slate and other refuse are washed from coarse coal in equipment typified by a washing jig. Jig stratifies feed into layers—light coal on top, refuse on bottom. Coal passes off end of jig, while riffles guide refuse to side.

Washed coal is dewatered on screens, then discharged to a clean-coal belt for delivery to the bins.

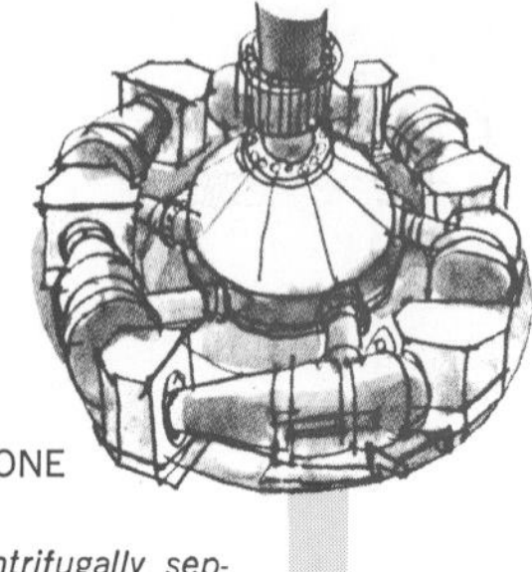

Coal fines are centrifugally separated from refuse in cyclones, then screened and dried. Centrifugal driers, and sometimes heat driers, are used.

American Iron and Steel Institute

3-4. *Producing coke.*

Twelve to 18 hours after the coal has gone into the oven the doors are removed and a ram shoves the coke into a quenching car for cooling.

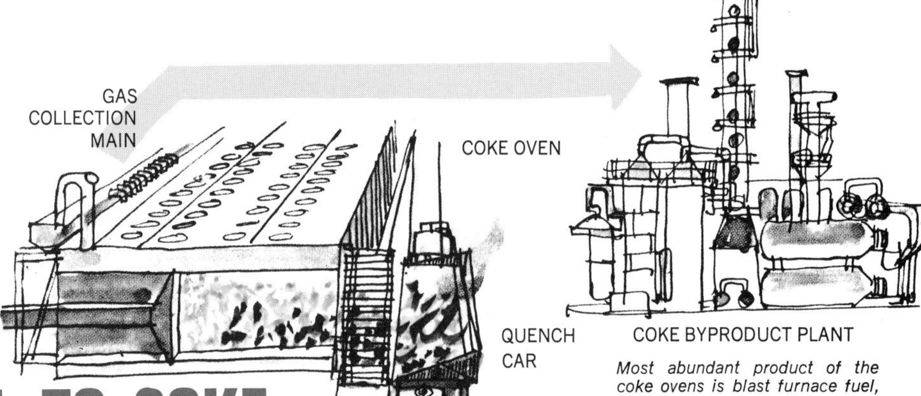

Most abundant product of the coke ovens is blast furnace fuel, but there are many byproducts, from ammonia to xylol.

FROM COAL TO COKE

HOW FUEL IS BAKED FOR THE BLAST FURNACE.

Byproduct coke plants bake solid bituminous coal until it is porous. This fuel, called coke, is just right for use in the blast furnaces which make iron. Coke, unlike coal, burns inside as well as outside. It does not fuse in a sticky mass. It retains strength under the weight of iron ore and limestone charged with it into blast furnaces.

The coke oven is delicate. Lined with silica brick, it must be warmed gradually at start-up to avoid damage. Averaging 40 feet in length and up to 20 feet in height each oven is very narrow, 12 to 22 inches in width. In a battery of such ovens, gas burning in flues in the walls heats the coal to temperatures as high as 2,000 degrees Fahrenheit. The heat drives off gas and tar. Regenerator chambers beneath the ovens use some exhaust gases to preheat air. Coal is loaded into the ovens from the top and the finished coke is pushed out from one side of the oven out the other.

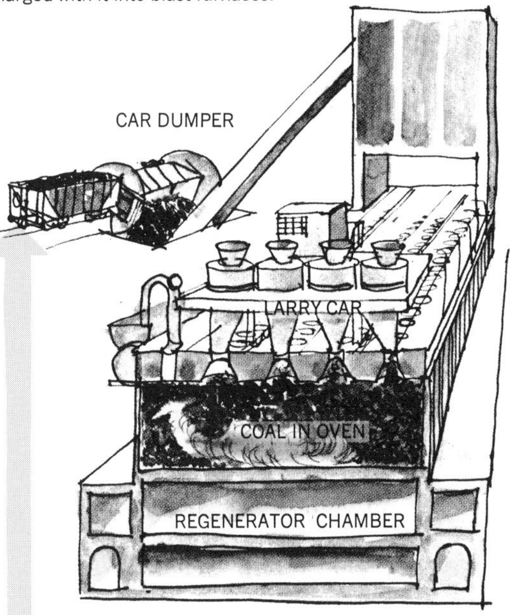

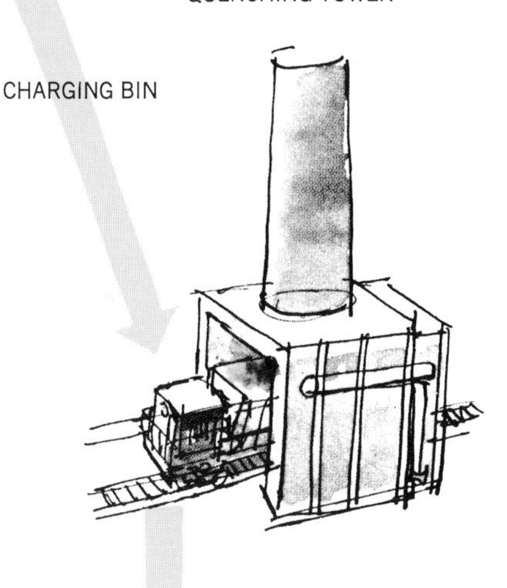

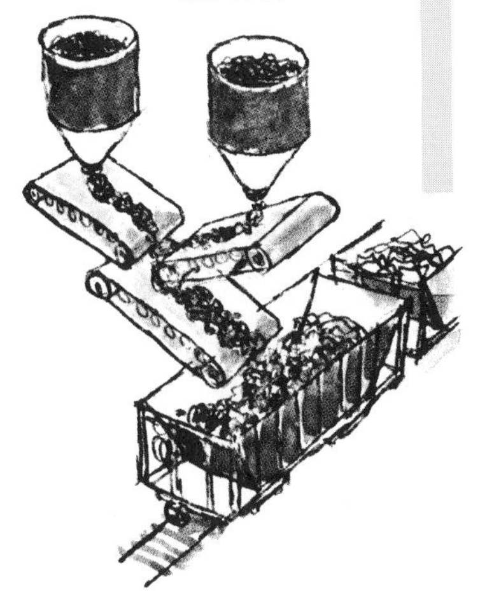

Fuels

Heat is indispensable in the manufacture of iron and steel. It is also essential in making steel mill products. To supply the heat required, the steel industry depends on three major natural fuels—coal, oil, and natural gas. Coal is the most important of these fuels.

COAL. Coal supplies more than 80 percent of the iron and steel industry's total heat and energy requirements. More than 100 million tons have been consumed by this industry in one year. This is enough to supply more than 15 million homes with their average yearly supply of fuel for heating. A large part of the coal is used in making coke for use in the blast furnace. About 1,300 pounds of coke are used for each ton of pig iron produced.

Not all types of bituminous coal can be used to make coke. Coke must be free from dust, the right size to permit rapid combustion, strong enough to carry the charge in the blasting furnace, and as free as possible from sulfur. Coal of coking quality is mined in 24 states; however, West Virginia, Pennsylvania, Kentucky, and Alabama supply nearly 90 percent of the coal used in the steel industry.

OIL. Oil is used extensively by the industry both as a fuel and as a lubricant for machinery and products. The heaviest grades of oil are most commonly consumed in steel plants. About 70 percent of the fuel oil used by the steel industry is consumed in melting iron, mostly in open hearth furnaces. More than 20 percent of the industry's fuel oil is burned in heating and annealing furnaces where steel products are given special heat treatments. The remainder is used in a wide variety of applications.

NATURAL GAS. Natural gas is burned in open hearth furnaces and reheating furnaces and in other places where clean heat is necessary. It contains almost no objectionable constituents, leaves no wastes or residues, and has a flame temperature as high as 3,700 degrees F.

Natural gas contains more heating value than all other gases employed: it delivers 1,000 British thermal units (Btu.) per cubic foot as compared with about 500 Btu. for coke oven gas, 300 Btu. for blue water gas, and 85 Btu. for blast furnace gas. At peak capacity the industry consumes over 400 billion cubic feet of natural gas per year. About 50 percent of this amount is consumed in heat-treating and annealing furnaces.

COKE. The heat required for smelting iron in blast furnaces is obtained from the burning of coke. *Coke* may be defined as the solid residue obtained when coal is heated to a high temperature in the absence of air. This causes the gases and other impurities to be released. Coke is a hard, brittle substance consisting chiefly of carbon, together with small amounts of hydrogen, oxygen, nitrogen, sulfur, and phosphorus. In recent years it has found some use as a smokeless domestic fuel.

The Manufacture of Coke. Prior to 1840 charcoal was the only fuel used in the United States for iron smelting. In 1855 anthracite coal became the leading blast furnace fuel because it was readily available, and charcoal was becoming more difficult to obtain. Another natural fuel, raw bituminous coal, was first burned in 1780 following the opening of the Pittsburgh coal seam. It was discovered in 1835 that by coking this coal, a product more suited to the needs of the blast furnace could be produced. In 1875, coke succeeded anthracite as the major blast furnace fuel.

By 1919 the coal chemical process of producing coke was developed, and coke became the leading fuel of the steel industry. The process, in addition to recovering the chemicals in coal, makes possible the production of stronger coke from a greater variety of coals. Figure 3-4 illustrates the mining of coal and its manufacture into coke. A coke oven is shown in Fig. 3-5. The volatile products which pass out of the ovens are piped to the chemical plant where they are treated to yield gas, tar, ammonia liquor, ammonium sulfate, and light oil. Further refinement of the light oil produces benzene, toluene, and other chemicals.

From these basic chemicals are produced such varied products as aviation gasoline, nylon, printing inks, pharmaceuticals, perfumes, dyes, TNT, sulfa drugs, vitamins, soaps, and synthetic flavors.

Coke production in the United States exceeds 64 million tons per year of which more than 92 percent is consumed as blast furnace fuel.

Steel Scrap

The earliest methods of making steel could not make use of scrap. The open hearth furnace made the melting of scrap possible. Today nearly 85 percent of all steel ingots produced are made in open hearth furnaces, and the amount of scrap consumed as a raw material has grown to such an extent that nearly as much scrap as pig iron is needed. About 70 percent of the scrap goes into open hearth furnaces, electric furnaces take 10 to 15 percent, and blast furnaces utilize the remainder.

Steel mills are their own best suppliers of steel scrap. This scrap comes from croppings of ingots, billets, and other rolled products; clippings from plates and sheets; borings; and defective products that have been rejected. It is estimated that 30 to 60 percent of the scrap required by the steel industry is generated in its own steel plants and that the remainder is purchased from outside sources. In mills utilizing only scrap, as much as 75 percent may need to be purchased.

Limestone

Limestone is used as a flux in the blast furnace. It is a sedimentary rock commonly found all over the world. There are large deposits in many parts of the United States, especially in the Appalachian Mountains, the Rocky Mountains, and the Mississippi Valley.

Limestone consists largely of *calcium carbonate* in varying degrees of purity. Common chalk is a form of pure limestone. Color changes with the presence of different types of impurities. It is white when pure and may also be found as gray, yellow, or black due to such impurities as iron oxide and organic matter. The properties of the rock change if certain compounds are present: silica makes it harder, clay softer, and magnesium carbonate turns it to dolomite, which is pinkish in color. Limestone may contain many fossils and loosely cemented fragments of shells.

Limestone is one of the chief fluxes used in steelmaking to separate the impurities from the iron ore. Many of the impurities associated with iron ores are of a highly *refractory* nature, that is, they are difficult to melt. If they remained unfused, they would retard the smelting operation and interfere with the separation of metal and the impurities. The primary function of limestone is to make these substances more easily fusible. Figure 3-6 shows the steps taken to process limestone.

Refractory Materials

Refractory materials may be defined as nonmetallic materials which can tolerate severe or destructive service conditions at high temperatures. They must withstand chemical attack, molten metal and slag erosion, thermal shock, physical impact, catalytic actions, pressure under load in soaking heat, and other rigorous abuse. Melting or softening temperatures of most refractory materials range from 2,600 degrees F. for light duty fireclay to 5,000 degrees F. for brick made from magnesia in its purest commercial form.

Refractory materials have an almost unlimited number of applications in the steel industry. Among the most important are linings for blast furnaces, steel-

3-5. *A schematic diagram of a coal-chemical coke oven. Coal falls from bins into a hopper car, which runs on top of many narrow ovens, dropping in coal. Heat, in the absence of air, drives gases from the coal to make coke. The collected gases are valuable byproducts for chemicals.*

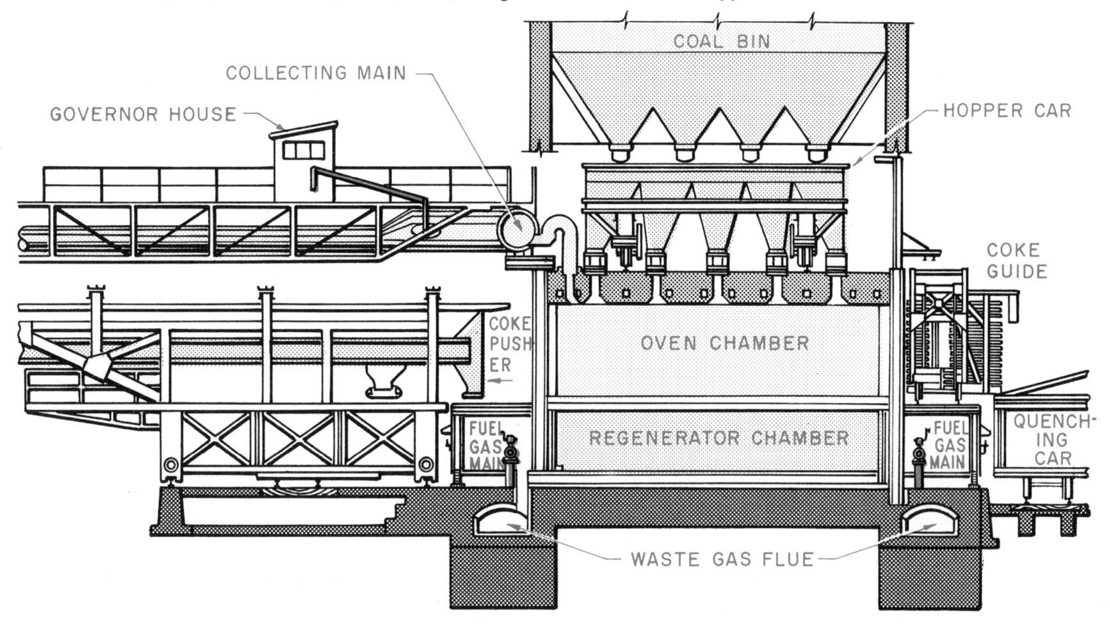

American Iron and Steel Institute

3-6. *Limestone.*

Quarrying limestone for preparation and use in iron and steelmaking furnaces is a large-scale operation. Most states have limestone deposits, but much of the more than 30 million tons consumed annually as fluxing material by the steel industry comes from Michigan, Pennsylvania and Ohio. The stone is blasted from its formation, loaded into trucks and taken to skip hoists which carry it to a processing plant near the quarry.

Blasting creates pieces of limestone of random size, many of which are too big for use as flux in furnaces. The stone goes to primary crushers—enormously strong steel equipment capable of fragmenting boulders. A jaw-type crusher is shown here, and its product is then screened and sorted to matching sizes.

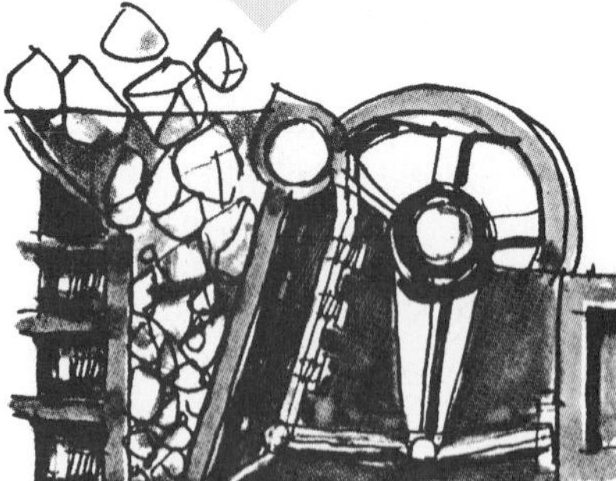

JAW CRUSHER
(Primary)

VERTICAL LIME KILN

The coarser stone from the screening operation may go to vertical kilns to be processed into lime. Some smaller material is further broken-up in secondary crushers, many of which are of the gyratory type.

The material resulting from secondary crushing is again screened to various sizes. Some of the limestone pieces may be processed through rotary kilns to make lime. Other small pieces of stone are either used directly in blast furnace ironmaking or are dried and pulverized for use largely in the sinter plants which beneficiate iron ore for blast furnaces. Open hearth steelmaking furnaces are also important users of crushed limestone as a flux.

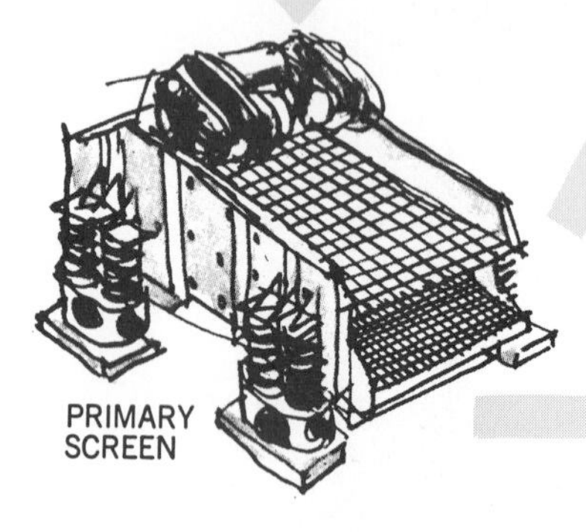

PRIMARY
SCREEN

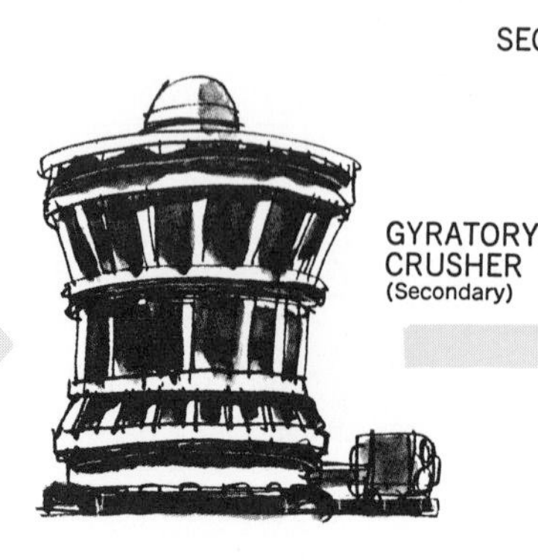

GYRATORY
CRUSHER
(Secondary)

SECONDARY SCREEN

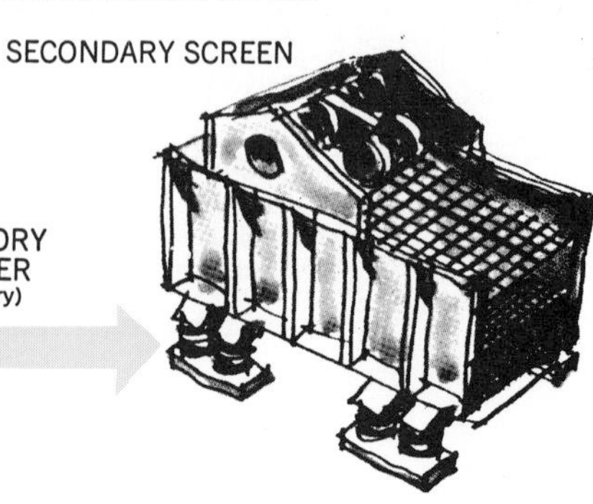

THE PURIFYING STONE

Marine animals and shellfish once lived and died at the bottom of seas which are now dry land. From their calcareous remains comes limestone that, in the steel industry, is used primarily to remove impurities from iron ore in blast furnaces. Limestone is also processed in kilns to make quicklime, a flux used to help remove impurities from the molten metal in steelmaking furnaces. The temperatures at which the industry's furnaces ordinarily operate would not melt the impurities, but limestone and lime make them fusible, combine with them and carry them off as slag.

Limestone is also used for purposes other than fluxing in the steel industry. For example, hydrated lime is used for wire drawing, water treatment, waste pickle liquor treatment, etc. But by far the largest use of limestone is in the industry's furnaces as described in this chart.

To derive lime from limestone, carbon dioxide is driven off by high temperatures in either vertical kilns (left) or horizontal rotary kilns (below).

Limestone is used in much greater quantities than lime in the iron and steel industry. It is chemically effective and physically strong. However, lime works quicker than the stone as a flux and is necessary in the fast-producing basic oxygen process where it is consumed at the rate of about 150 pounds per ton of raw steel produced.

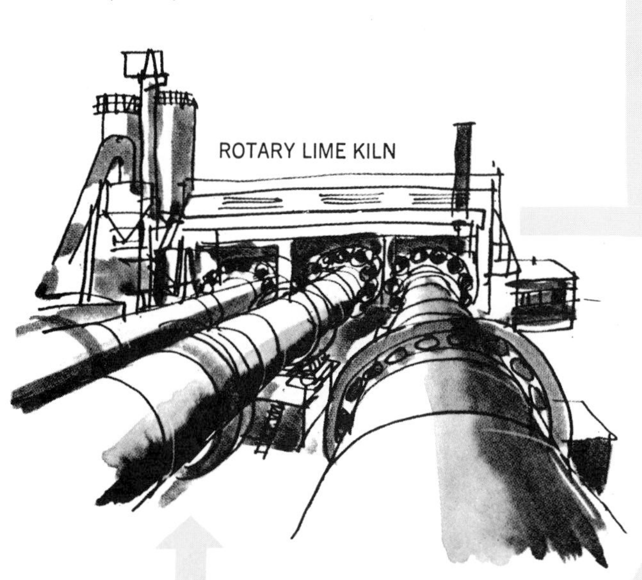

In making lime, horizontal rotary kilns process small limestone pieces that would pass through vertical kilns too rapidly to be thoroughly calcined.

ELECTRIC FURNACE

OPEN HEARTH FURNACE

Although blast furnaces, sinter plants and open hearths, in that order, are the primary users of limestone, small amounts are also used in basic oxygen and electric furnaces.

making furnaces, soaking pits, reheating furnaces, heat-treating furnaces, ladles, and submarine cars.

Refractory materials are produced from quartzite, fireclay, alumina (aluminum oxide), magnesia (magnesium oxide), iron oxide, natural and artificial graphites, and various types of coal, coke, and tar. The raw materials are crushed, ground, and screened to proper sizes for use in making bricks and other forms of linings. They are combined with certain binders, and the prepared batches are fed to the forming machines. The most common methods for forming refractory bricks are power pressing, extrusion, and hand molding. Most refractory bricks are fired in kilns at high temperature to give them permanent strength.

Iron Blast Furnace Slag

Slag is the residue produced from the interaction of the molten limestone and the impurities of the iron. It contains the oxides of calcium, silicon, aluminum, and magnesium, small amounts of iron oxide, and sulfur. Slag may be processed for use in the manufacture of cement and concrete blocks, road materials, insulating roofing material, and soil conditioner.

Carbon

Carbon is a nonmetallic element which can form a great variety of compounds with other elements. Compounds containing carbon are called organic compounds.

In union with oxygen, carbon forms carbon monoxide and carbon dioxide. When carbon combines with a metal, it may form compounds such as calcium carbide and iron carbide.

Three forms of pure carbon exist. The diamond is the hard crystalline form, and graphite is the soft form. Carbon black is the amorphous form.

In addition to being important as an ingredient of steel, carbon is used for industrial diamonds and abrasives and arc carbons of all kinds. As graphite, it forms a base for lubricants and lining for blast furnaces.

THE SMELTING OF IRON

The Blast Furnace

The first step in the conversion of iron ore into steel takes place in the blast furnace, Figs. 3-7 and 3-8. In this towering cylindrical structure, iron is freed from most of the impurities associated with it in the ore.

3-7. This new blast furnace will produce over 1,800 tons of pig iron daily. The furnace stack and other accessories, fabricated by welding, contain over 2,400 tons of steel plate and structurals.

Nooter Corp.

The furnace is charged with iron ore, limestone, and coke. A blast of preheated air burns the coke, producing heat to melt the iron, which falls to the bottom of the furnace. The molten limestone combines with most of the impurities in the ore to form a slag which separates from the liquid iron because it is lighter and floats. The liquid iron and the liquid slag are removed periodically from the bottom of the furnace. This is a continuous process: as a new charge is introduced at the top, the liquid iron and slag are removed at the bottom. The progress of the charge through the furnace from the time it enters the top until it becomes iron is gradual; five to eight hours are required. The process is illustrated in Figs. 3-9 and 3-10 (Page 62).

The liquid iron is poured into molds to form what is known as *pigs* of iron. *Pig iron* is hard and brittle. It contains considerable amounts of dissolved carbon, manganese, silicon, phosphorus, and sulfur. Steelmaking is the process of removing impurities from pig iron and then adding certain elements in predetermined amounts to arrive at the properties desired in the finished metal. While several of the elements added are the same as those removed, the proportions differ.

Nearly all of the pig iron produced in blast furnaces remains in the molten state and is loaded directly into steelmaking furnaces. A small amount is solidified and transported to iron foundries which remelt it. Then the iron is cast into a wide variety of products ranging from toys to cylinder heads for automobile engines.

A modern blast furnace may be as much as 250 feet in height and 28 feet in diameter. The furnace shaft is brick lined, and this lining is water cooled to withstand high temperatures. Flame temperatures as high as 3,500 degrees F. and gas temperatures of 700 degrees F. are generated. As much as 10 to 12 million gallons of water per day may be used to cool a furnace. A furnace may operate for several years before relining is necessary.

The number of blast furnaces in the United States has declined over the past 30 years, but the total annual pig iron production has increased greatly. Enlarged furnaces, refined and controlled raw materials, and much higher blast temperatures are responsible for increased production. The number of furnaces probably will continue to decrease as the production rate for leading furnaces exceeds 3,000 net tons per day.

STEELMAKING PROCESSES

We have read that steel was used in a primitive form for several thousand years. However, this early steel was not strong nor did it have the variety of properties necessary for extensive use. It was produced by the cementation and the crucible processes. In recent times two major developments have made it possible to produce large quantities of steel with a variety of properties at a competitive cost.

The first of these developments was the Bessemer furnace invented in 1856 in both Europe and the United States. The second was the open hearth furnace which was invented 12 years later in the United States. Figure 3-11 (Pages 64, 65) shows the modern steelmaking process from raw materials to finished product.

Cementation Process

Cementation is the oldest method of steelmaking. It consists of heating wrought iron with carbon in a vacuum. This increases the carbon content of surfaces and edges which can then be hardened by heating and quenching. The metal is not molten during steelmaking. Hence impurities are not removed from the iron, and only the surface of the metal is affected. It is probable that most of the steel of ancient times was produced in this way.

A later improvement of this process was the stacking of alternate layers of soft, carbon-free iron with iron containing carbon. The layers were then heated so that the pieces could be worked. The layers of soft and hard metal strengthened the internal structure of the steel. Much of this steelmaking was centered in Syria during the Middle Ages, and the steels became known as the famous Damascus steels, used widely for swords and spears of the highest quality.

The steel made by this process was further improved by the crucible process which came into use in the eighteenth century.

Crucible Process

The crucible process was revived in England during the early 1740s. Steel produced by the cementation process was melted in a clay crucible to remove the impurities. While fluid, the slag was skimmed off the top. Then the metal was poured into a mold where it solidified into a mass which could be worked into the desired shape. In the United States graphite crucibles, with a capacity of about 100 pounds of metal, were used in a gas-fired furnace. This process produced

3-8. *U.S. Steel's "Dorothy" blast furnace and related units. A considerable amount of welding is employed in the construction of these units.*

United States Steel Corp.

BLAST FURNACE IRONMAKING

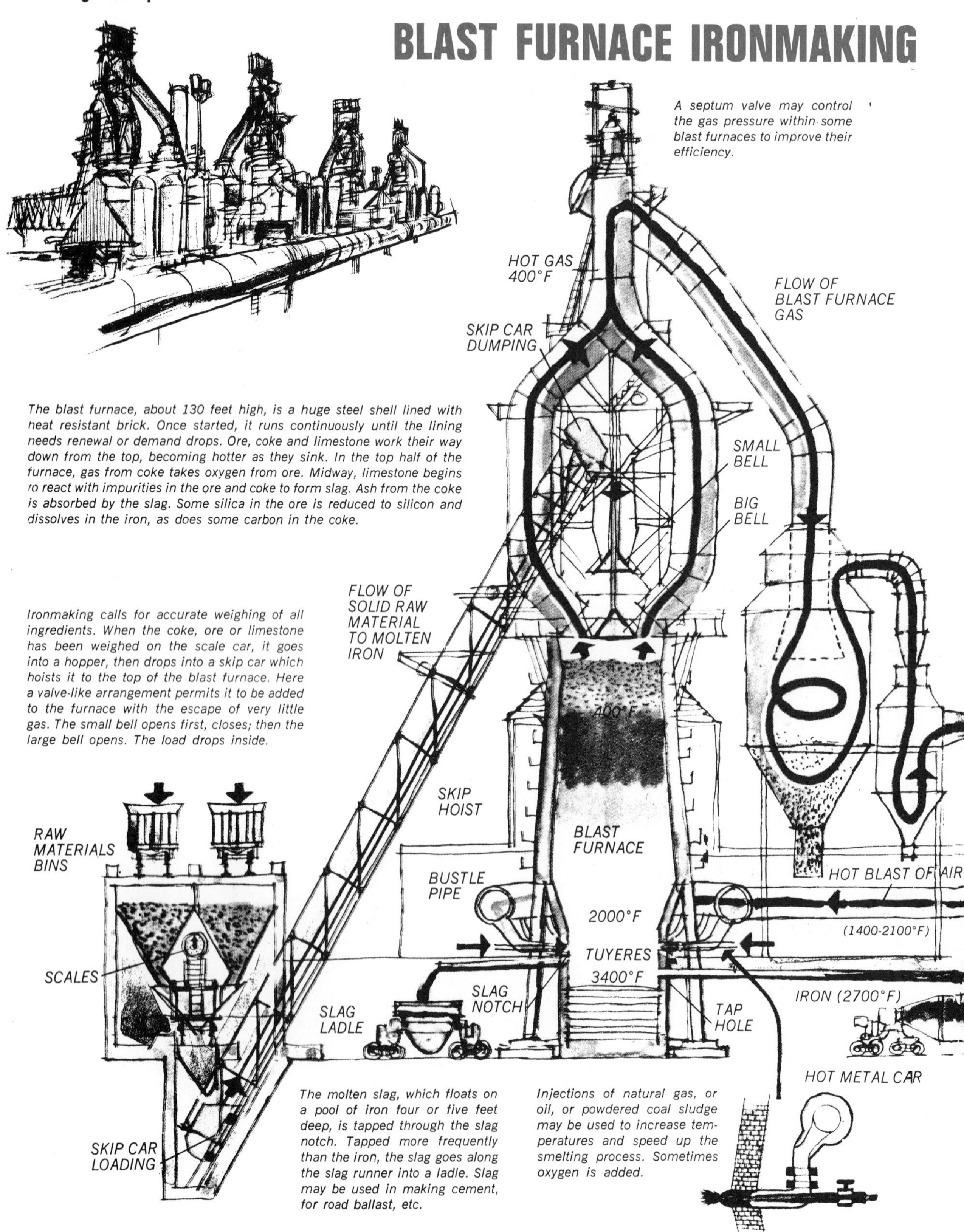

A septum valve may control the gas pressure within some blast furnaces to improve their efficiency.

The blast furnace, about 130 feet high, is a huge steel shell lined with heat resistant brick. Once started, it runs continuously until the lining needs renewal or demand drops. Ore, coke and limestone work their way down from the top, becoming hotter as they sink. In the top half of the furnace, gas from coke takes oxygen from ore. Midway, limestone begins to react with impurities in the ore and coke to form slag. Ash from the coke is absorbed by the slag. Some silica in the ore is reduced to silicon and dissolves in the iron, as does some carbon in the coke.

Ironmaking calls for accurate weighing of all ingredients. When the coke, ore or limestone has been weighed on the scale car, it goes into a hopper, then drops into a skip car which hoists it to the top of the blast furnace. Here a valve-like arrangement permits it to be added to the furnace with the escape of very little gas. The small bell opens first, closes; then the large bell opens. The load drops inside.

The molten slag, which floats on a pool of iron four or five feet deep, is tapped through the slag notch. Tapped more frequently than the iron, the slag goes along the slag runner into a ladle. Slag may be used in making cement, for road ballast, etc.

Injections of natural gas, or oil, or powdered coal sludge may be used to increase temperatures and speed up the smelting process. Sometimes oxygen is added.

American Iron and Steel Institute
3-9. The blast furnace process.

Hot air is indispensable in a blast furnace. As much as four and one-half tons of it may be needed to make one ton of pig iron. It pours in at the bottom of the furnace and roars up through the charge of iron ore, coke, and limestone that has been dumped in from the top.

Fanned by the air, the coke burns. Its gases reduce the ore to metallic iron by removing oxygen from it while the limestone causes the earthy matter of the ore to flow. Freed, the heavy metal settles to the bottom. From there, 300 to 600 tons of pig iron are drawn off every three to five hours.

Air for the blast furnace is heated in huge stoves. At least two stoves are needed for each blast furnace. One stove heats while the other blows hot air into the bustle pipe and through tuyers to the bottom of the furnace. In a combustion chamber in the stove being heated, cleaned exhaust gases from the blast furnace are mixed with air and burned to raise the temperature of refractory brick.

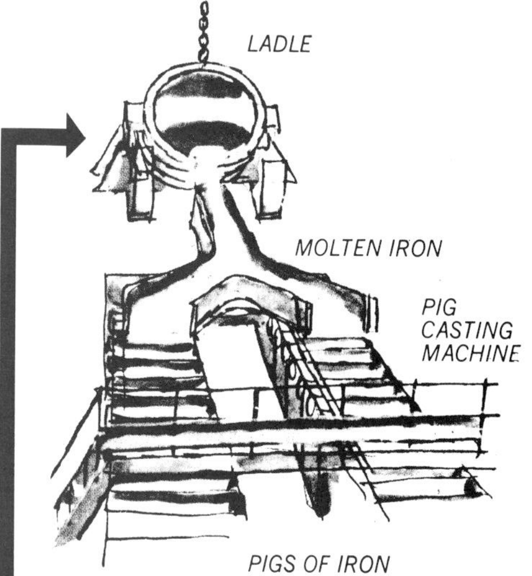

For convenience in shipping, liquid iron is ladled off into continuously moving molds, is then quenched and turned out in pig form. Each year, a small percentage of the pig iron output is shipped in solid pigs to thousands of foundries where it is made into a variety of castings. Solid pigs are also used by steel mills that do not have blast furnaces.

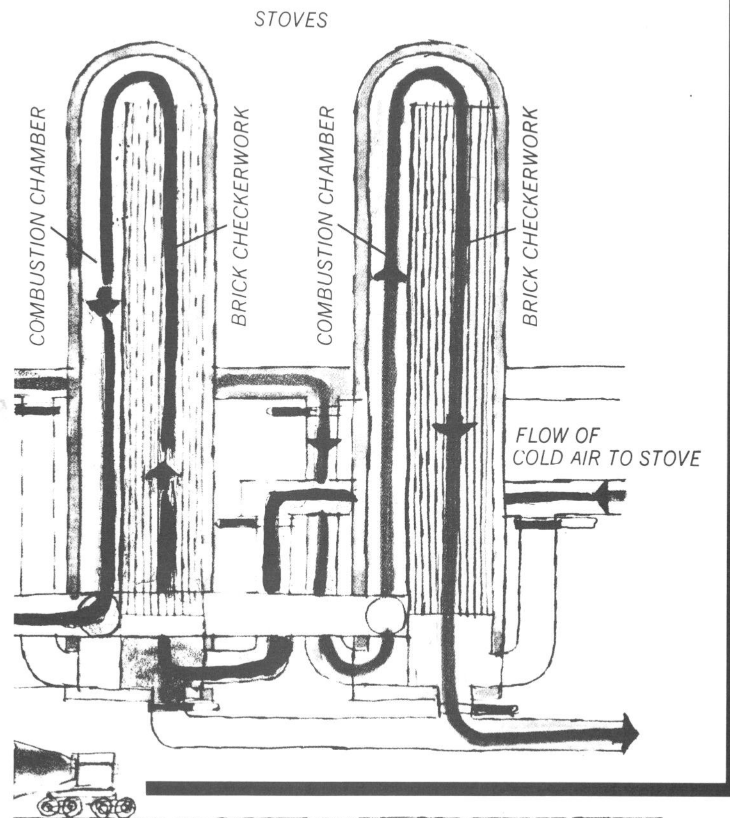

When the blast furnace is tapped for its store of iron, the molten metal is channeled into a hot metal car, a gigantic drum lined with refractory brick. A hot metal car holds about 160 tons of liquid iron, insulating it like a gigantic vacuum bottle. Most molten iron goes to open hearth or basic oxygen steelmaking facilities, but some goes to a casting machine where it is made into solid "pigs."

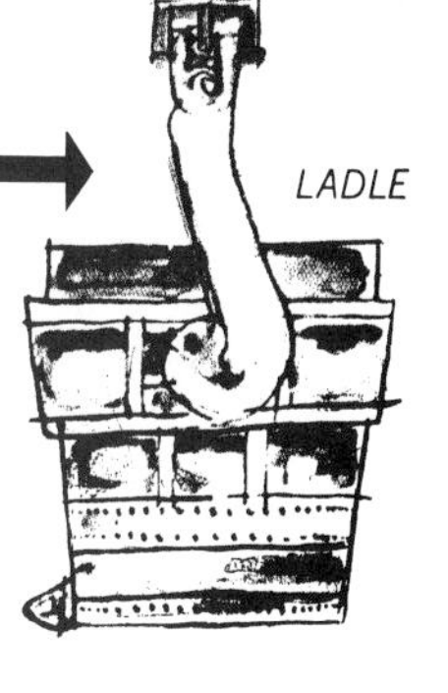

A ladle full of molten iron joins limestone, scrap steel and alloying materials in a basic oxygen furnace or in an open hearth to form a special heat of steel meeting rigid specifications.

a steel of uniform quality which was free of slag and dust.

Bessemer Process

The Bessemer process was invented at about the same time by two men working independently in two countries. One was an American, William Kelly, in Eddyville, Kentucky, and the other was an Englishman, Henry Bessemer. Bessemer applied for his patent in 1856: and Kelly, a year later. The first plant in the United States, a $2\frac{1}{2}$ ton converter, was erected in 1864 and produced steel for the first rails made in the United States. By 1880, a number of plants were producing a total of 852,000 tons of steel and exceeding the production of iron for the first time. Some steel is still made this way.

The Bessemer process consists of blowing compressed air through a refractory lined vessel containing molten pig iron, Fig. 3-12 (Page 66). Nearly all of the silicon and manganese, most of the carbon, and some of the iron are oxidized by the oxygen in the air that is blown into the furnace. The oxidation reaction furnishes the heat for the process, and the oxidized iron, silicon, and manganese forms the slag. As the impurities are burned out, the color of the flame changes. An experienced steelmaker can tell when the molten steel is ready to be poured into large ingots. These ingots are either worked immediately into the desired shapes or stored in a soaking pit where they are kept at a temperature suitable for later working.

Bessemer converters usually range from 25 to 40 tons in ca-

3-10. *Schematic diagram of a blast furnace, hot blast stove, and skiploader. Ore, limestone, and coke are fed in at the top of the furnace. Preheated air, delivered at the bottom, burns the coke and generates gases and heat required to separate iron from the ore.*

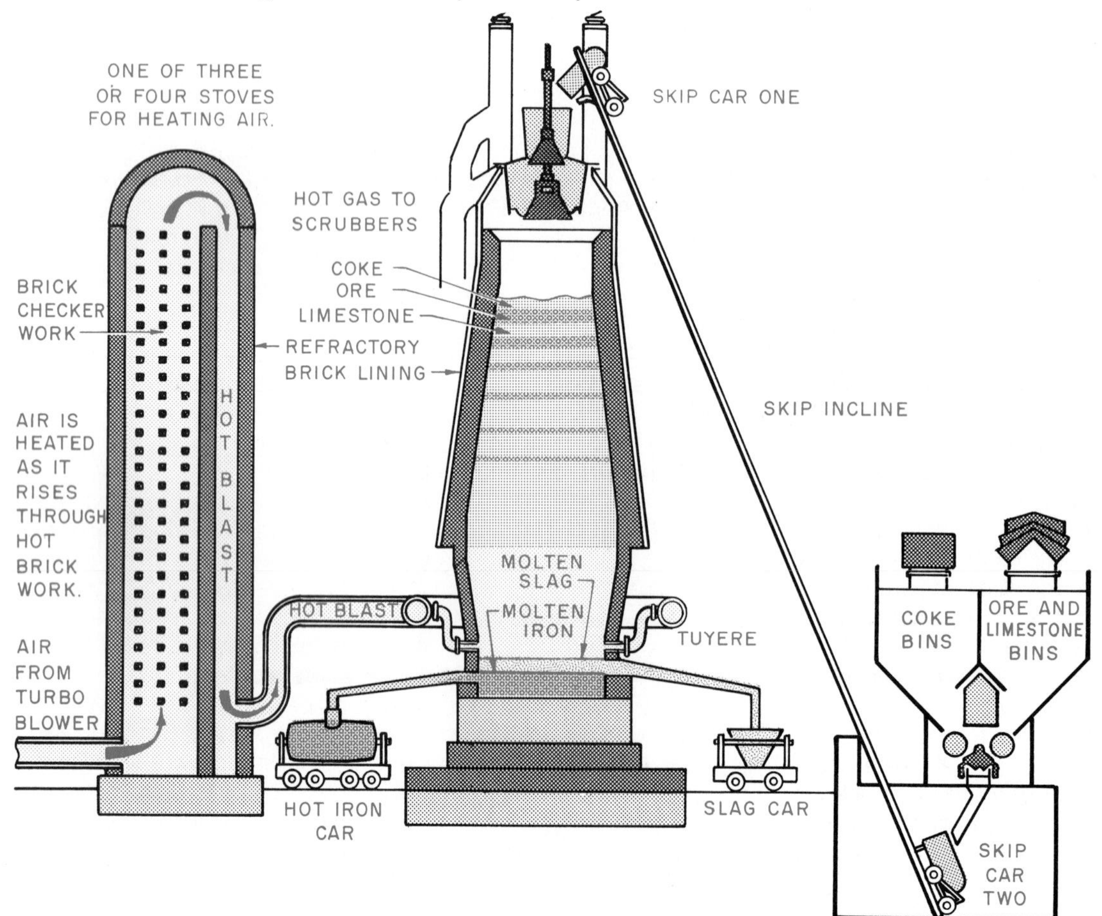

pacity. Until 1907 this was the principal method of making steel in the United States. Today it accounts for less than 3 percent of the total steel production. Today Bessemer steel is used mainly for butt-welded and seamless pipe, easily machinable steel, wire, and castings.

Open Hearth Process

The open hearth process was invented by Karl Siemens in 1868 and eventually outstripped the Bessemer process in steel production. This process was given the name *open hearth* because the steel, although melted on a hearth under a roof (Fig. 3-13), was accessible through furnace doors for visible inspection and sampling.

In this process a rectangular carved hearth holds the charge of pig iron and scrap. Figure 3-14 shows the charging floor of the furnace. Much of the heat needed to purify the melt is provided by passing burning fuel gas over the top of the charge. A system of brick checkerwork absorbs the heat of the exhaust gases leaving the hearth and in turn preheats the incoming stream of fuel gases and air, Fig. 3-15. Preheating the incoming gases increases the temperature in the combustion chamber. The open hearth process is an oxidation process. Impurities in the ore—carbon, silicon, and manganese—are oxidized by the oxygen in the furnace atmosphere and by the oxygen contained in the resin which is added to the bath. Figure 3-16 (Pages 68, 69) summarizes the process.

In the most commonly used open hearth process, the hearth is lined with magnesite brick with a layer of magnesite or burned dolomite on top. The magnesite permits the charging of limestone which combines chemically with the phosphorus and sulfur in the iron so that these impurities can be removed as slag. This chemical reaction is important since it permits the utilization of the large iron ore deposits in the United States which have a high phosphorus content.

The open hearth produces about 85 percent of the steel used in the United States. The process has the following advantages:

- Scrap steel and iron may be used.
- Pig iron may have high phosphorus content.
- The proportions of pig iron and scrap steel may be adjusted as necessary.
- High and low carbon steels and low alloy steels can be produced.
- Furnaces provide high production capacity.

Open hearth furnaces range in capacity from 100 to 350 tons although in some cases the capacity may be as small as 40 tons or as large as 600 tons.

Electric Furnace Processes

Electric furnaces are of two types: (1) the electric arc type and (2) the induction furnace. The first *electric arc furnace* had a capacity of 4 tons. It was put into operation in France by the French metallurgist Paul Heroult in 1899 and introduced into the United States in 1904. The modern furnace, Fig. 3-17 (Pages 70, 71), has a charge of 80 to 100 tons. A few furnaces hold a charge of 200 tons and produce more than 800 tons of steel in 24 hours. These large furnaces are made possible by the increase of electric power capacity, the production of large graphite (carbon) electrodes, the development of improved refractory materials for linings, and better furnace design.

Electricity is used solely for the production of heat and does not impart any properties to the steel. These furnaces have three electrodes ranging in size from 4 to 24 inches in diameter. They produce a direct arc with three-phase power and are supplied with electric current through a transformer. The electrodes enter the furnace through the roof. The roof is removable and can be swung aside to charge the furnace. The charge consists almost entirely of scrap with small amounts of burned lime and mill scale. The furnaces are circular and can be tilted to tip the molten steel into a ladle, Fig. 3-18 (Page 72). They may be lined with either basic (magnesite, dolomite) or acid (silica brick) refractory materials, Fig. 3-19.

Before World War II practically all alloy, stainless, and tool steels were produced in electric furnaces. Today, however, ordinary steels may also be produced in those areas where there are large supplies of scrap and favorable electric power rates.

The *electric induction furnace*, Fig. 3-20, is essentially a transformer with the molten metal acting as the core. It consists of a crucible, usually made of magnesia, surrounded by a layer of tamped-in magnesia refractory. Around this is a coil made of copper tubing, forming the primary winding which is connected to the current source. The whole is encased in a heavy box with a silica brick bottom lining. A lip is built into the box to allow the metal contents to run out as the furnace is tilted forward.

The charge consists of scrap of the approximate composition desired plus necessary ferroalloys to give final chemical composition within specifications. Scrap may be any size which will fit into the furnace. A 1,000

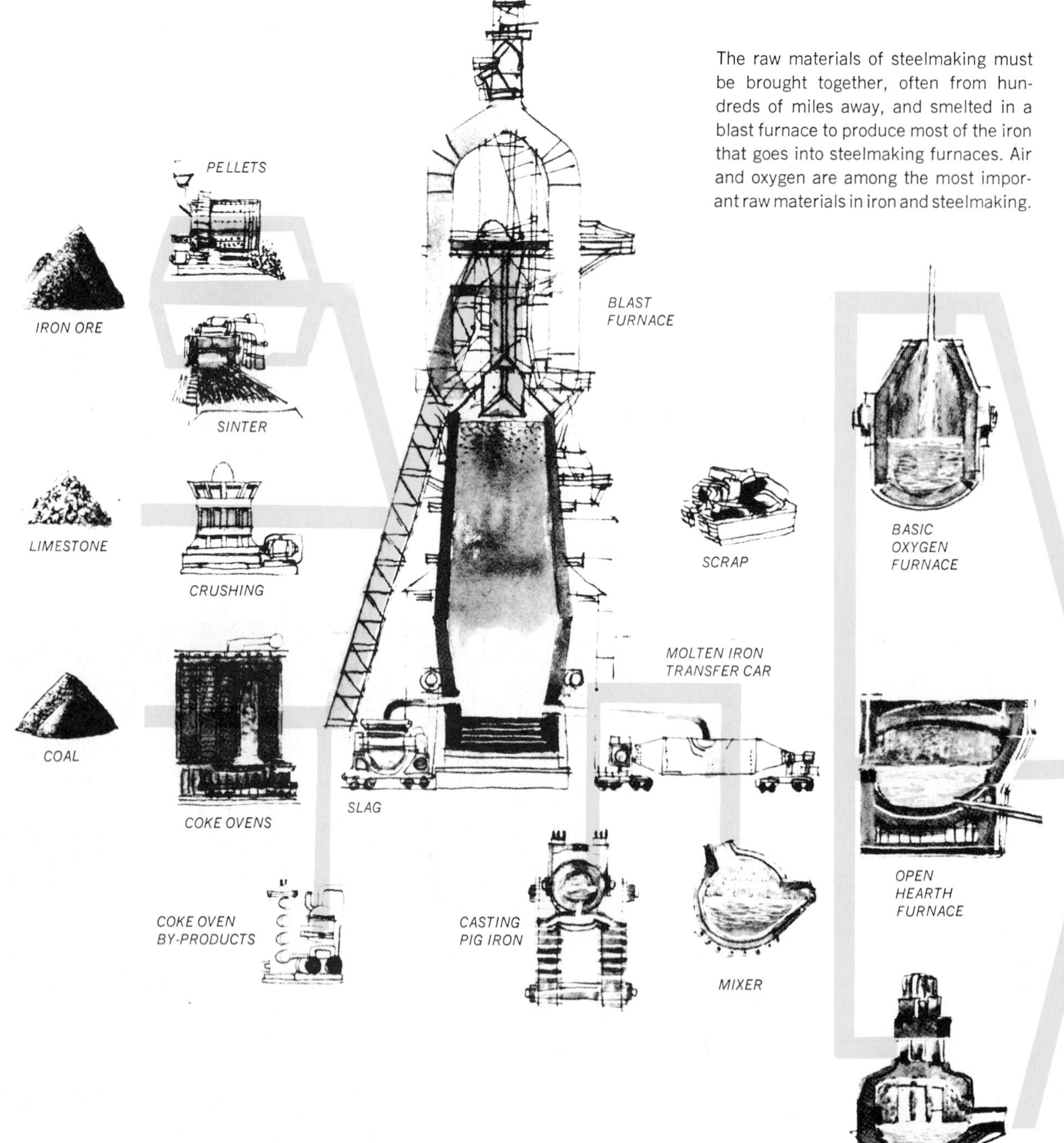

A FLOWLINE ON STEELMAKING

This is a simplified road map through the complex world of steelmaking. Each stop along the routes from raw materials to mill products contained in this chart can itself be charted. From this overall view, one major point emerges: Many operations—involving much equipment and large numbers of men—are required to produce civilization's principal and least expensive metal.

American Iron and Steel Institute

3-11. *A "roadmap" from raw materials to mill products.*

Molten steel must solidify before it can be made into finished products by the industry's rolling mills and forging presses. The metal is usually formed first at high temperature, after which it may be cold-formed into additional products.

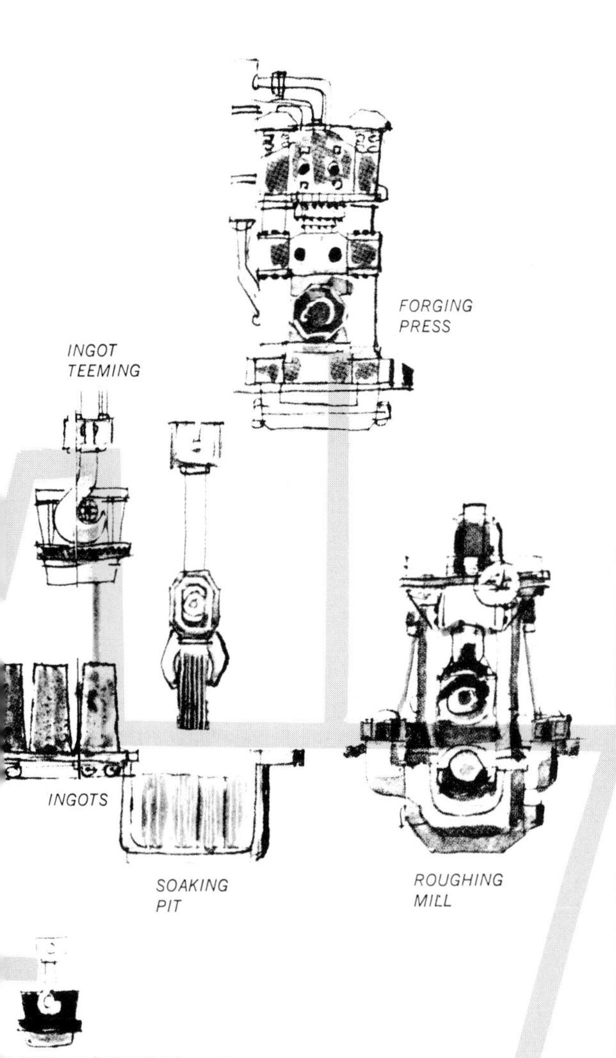

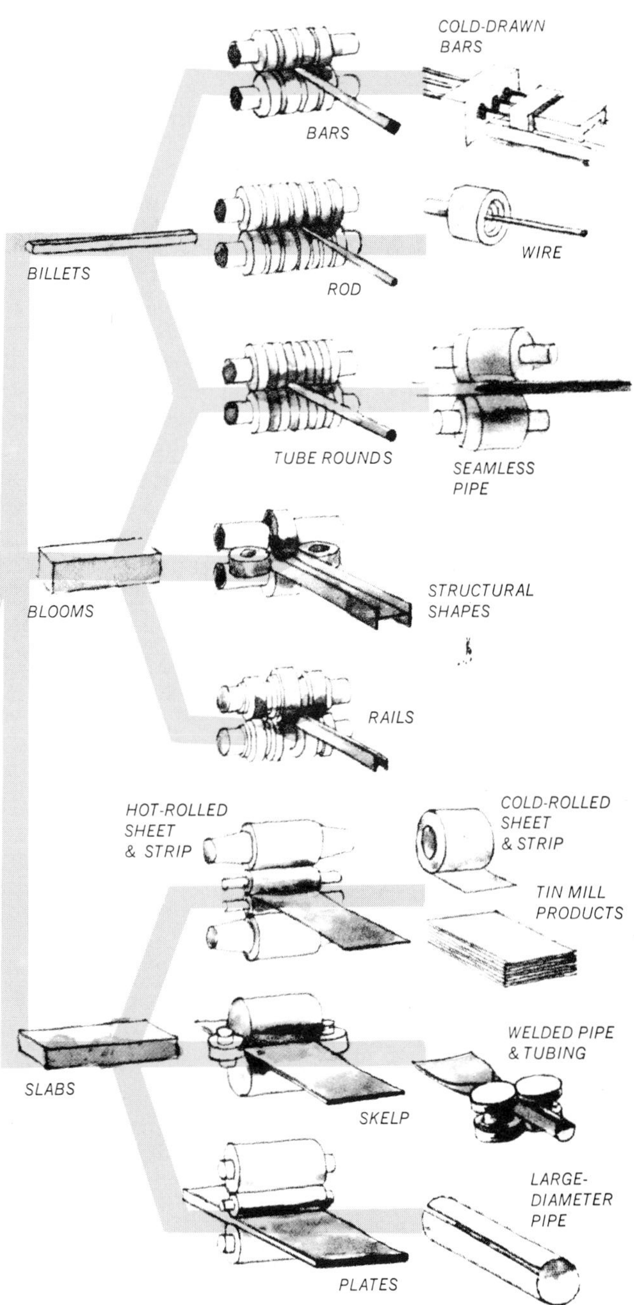

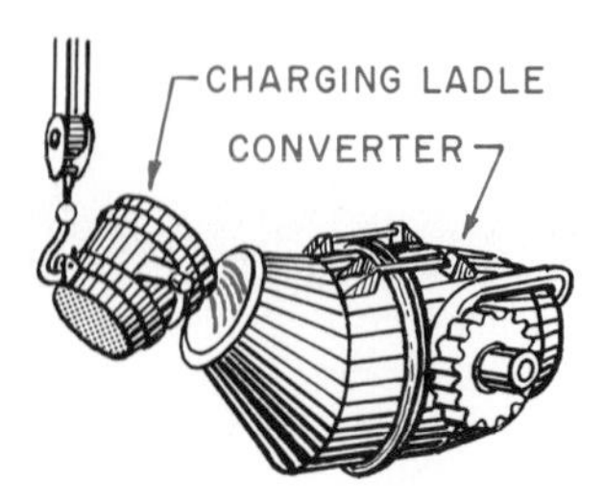

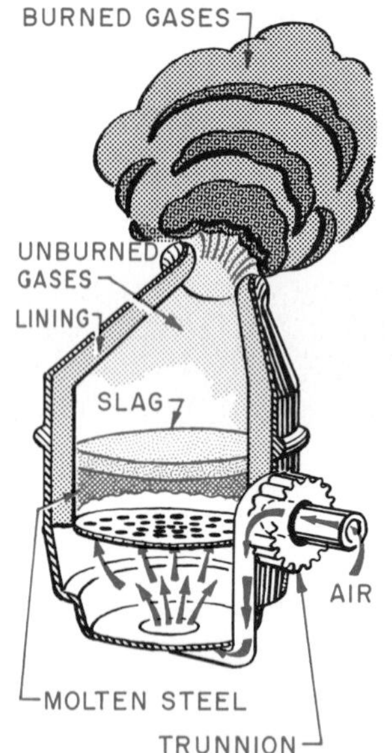

3-12. *Bessemer process: oxygen from a blast of air burns out unwanted elements from steel.*

pound charge can be melted down in 45 minutes. After melting is complete, the metal is further heated to the tapping temperature in about 15 minutes. During this time small additions of alloys or deoxidizers are added. When the proper temperature is obtained, the furnace is tilted and liquid metal runs out over the lip into a ladle or directly into a mold.

Oxygen Process

The oxygen process, also known as the *Linz-Donawitz* process, was first established in Linz, Austria in 1952 and in Donawitz, Austria a short time later. The process was first used in the United States in 1954.

The Linz-Donawitz process is a method of pig iron and scrap conversion whereby oxygen is injected downwards over a bath of metal. A fairly large amount of hot metal is necessary to start the oxidizing reaction so that the scrap content is limited to about 30 percent of the charge instead of the usual 50 percent of scrap in the open hearth charge. A pear-shaped vessel is charged with molten pig iron and scrap while the vessel is in a tilted position, Fig. 3-21 (Page 73). Then the vessel is turned upright. Fluxes are added, and high purity oxygen is directed over the surface of the molten metal bath by the insertion of a water-cooled lance into the vessel's mouth, Fig. 3-22.

The chemical reaction of the oxygen and fluxes refines the pig iron and scrap into steel. The temperature reaches 3,000 degrees F., and the refining continues for 20 to 25 minutes.

When the refining is complete, the lance is withdrawn. The furnace is tilted, and the steel is tapped through a hole in the side near the top. The slag is also removed, and the furnace is ready for another charge. The complete process is shown in Figs. 3-23 (Page 74) and 3-24 (Pages 76, 77).

The main advantage claimed for the process is that it takes only 45 minutes to complete.

3-13. *Cross section of an open hearth shop. The action begins at the left where scrap and other solid materials enter. The open hearth furnace and pouring platform are in the center, and the mold yard is at the right.*

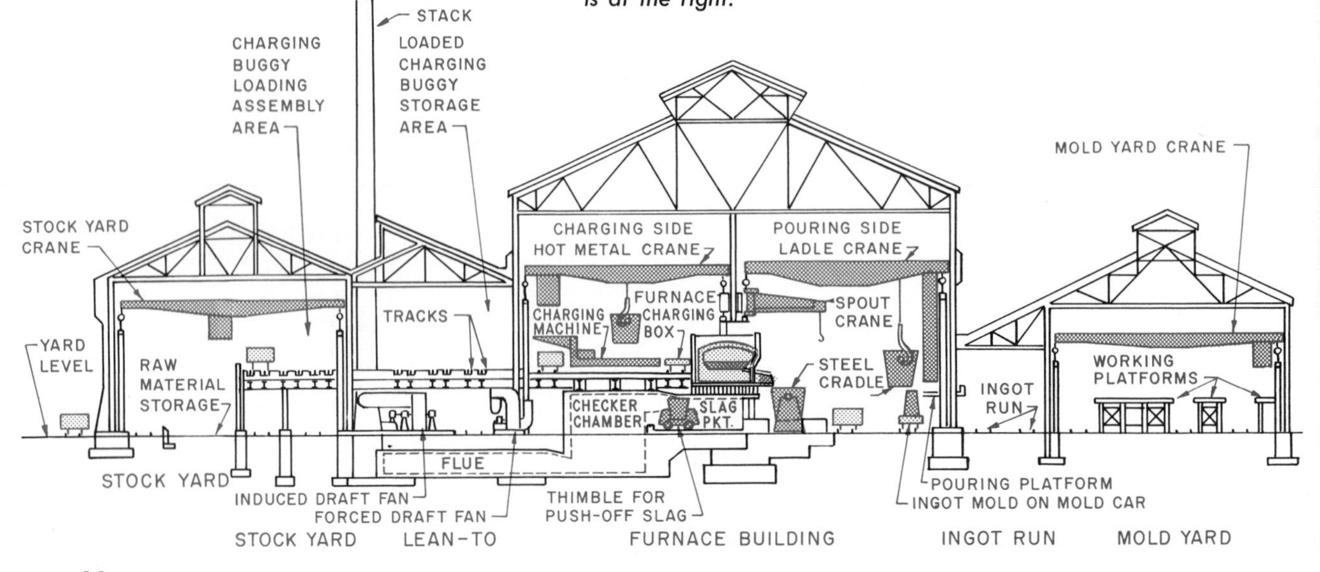

Heats as large as 300 tons are made, and the product is equivalent to open hearth steel in every respect. Steels of any carbon content can be produced. Somewhat lower amounts of phosphorus, sulfur, and nitrogen in the finished product are reported than are normally secured with the open hearth process. While alloy and stainless steels have been made by the oxygen process, the holding time in the vessel to obtain the desired chemical composition largely eliminates the short time-cycle advantage and, in general, only carbon steels are produced.

The Stora-Kaldo Process

The *Stora-Kaldo* process is a development of Sweden's Stora company, one of the oldest continuously operating steelmaking companies in the world. Its history goes back in an unbroken line for more than 600 years, tracing its recorded existence from the year 1288, some 200 years before Columbus discovered America.

The Stora-Kaldo process is also an oxygen process. The main difference between it and the basic oxygen process previously discussed is that the entire furnace rotates as well as tilts. Thus it is possible to utilize a higher percentage of scrap charge (up to 50 percent). Other advantages claimed for the process are greater flexibility in grades of steel produced and cleaner steels.

United States Steel Corp.

3-14. *Charging floor of an open hearth furnace.*

3-15. *Open hearth furnace cutaway shows "bath" at left where the metal is melted. Air for combustion enters at lower right. Gases of combustion leave via the stack at the upper right.*

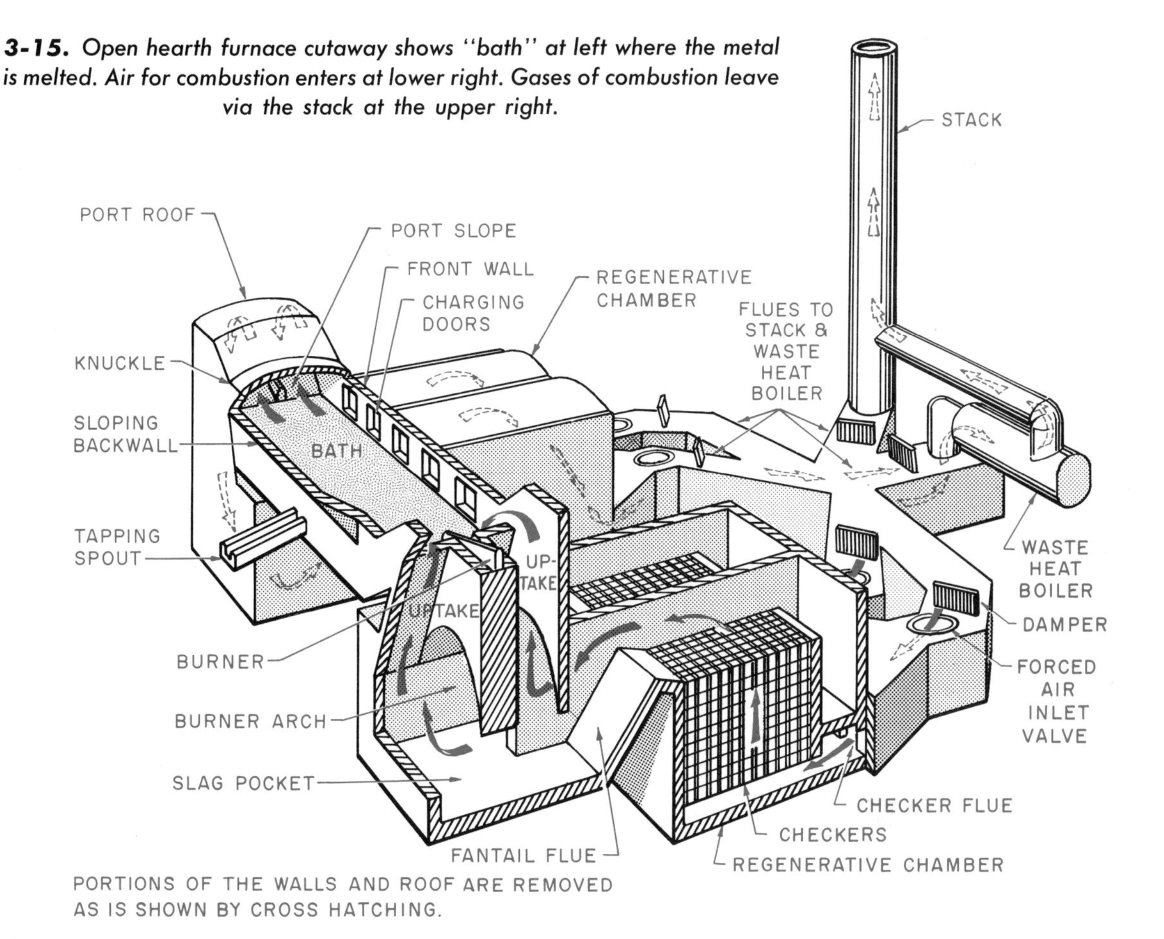

OPEN HEARTH STEELMAKING

Open hearth furnaces are so named because the limestone, scrap steel and molten iron charged into the shallow steelmaking area (the hearth) are exposed (open) to the sweep of flames. A furnace that will produce a fairly typical 350 tons of steel in five to eight hours may be about 90 feet long and 30 feet wide.

The cutaway drawing below shows several steps simultaneously that would normally occur in sequence. First the long-armed charging machine picks up boxes of limestone and steel scrap, thrusts them through the furnace doors and dumps the contents. The flame of burning fuel oil, tar or gases partially melts the solid charge, after which molten iron (lower right) is poured into the furnace. High-temperature reactions cause several unwanted elements to combine with the limestone to form a slag.

When tests of samples show the steel to be of specified chemistry, the tap hole is opened by an explosive charge and the steel runs into a ladle. The slag, which is lighter than steel, floats on the metal and overflows into a slag thimble during pouring. Alloy additions are made to the steel in the ladle.

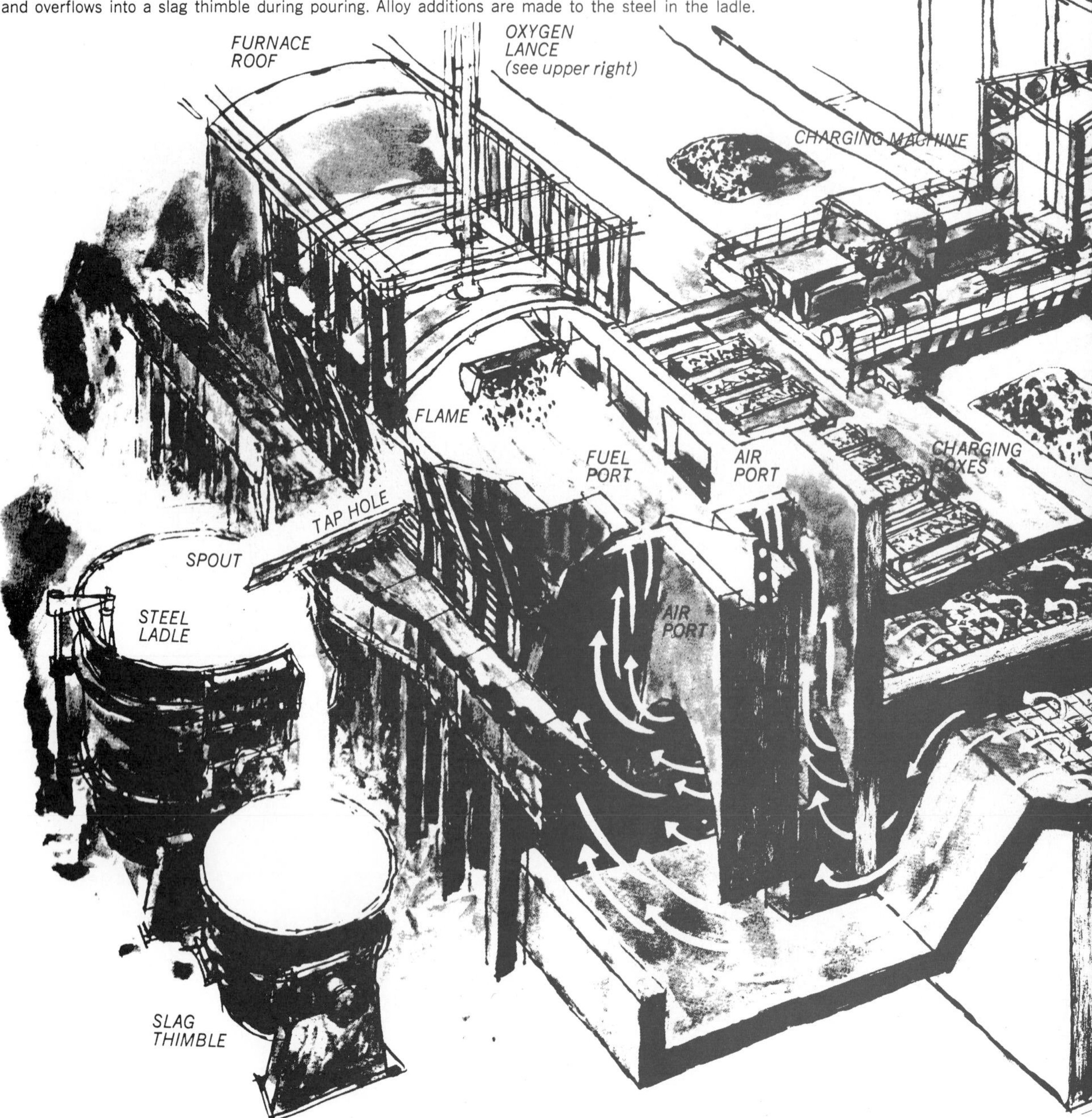

American Iron and Steel Institute
3-16. ***The open hearth process.***

In recent years practically all open hearth furnaces have been converted to the use of oxygen. The gas is fed into the open hearth through the roof by means of retractable lances. The use of gaseous oxygen in the open hearth increases flame temperature, and thereby speeds the melting process.

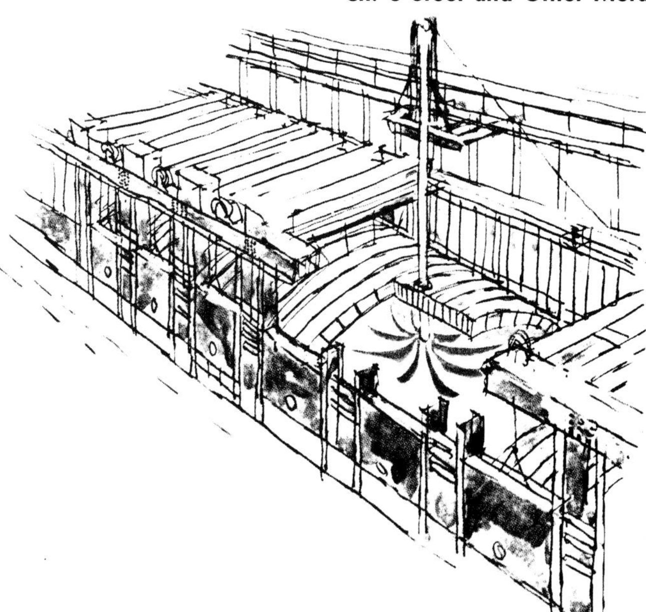

Molten iron from a blast furnace is a major raw material for the open hearth furnaces. A massive "funnel" is wheeled to an open hearth door and the contents of a ladle of iron are poured through it into the furnace hearth. The principal addition of molten metal is made after the original scrap charge has begun to melt.

Brick checker chambers are located on both ends of the furnace. The bricks are arranged to leave a great number of passages through which the hot waste gases from the furnace pass and heat the brickwork prior to going through the cleaner and stack. Later on, the flow is reversed and the air for combustion passes through the heated bricks and is itself heated on its way to the hearth.

BRICK
CHECKER
CHAMBERS

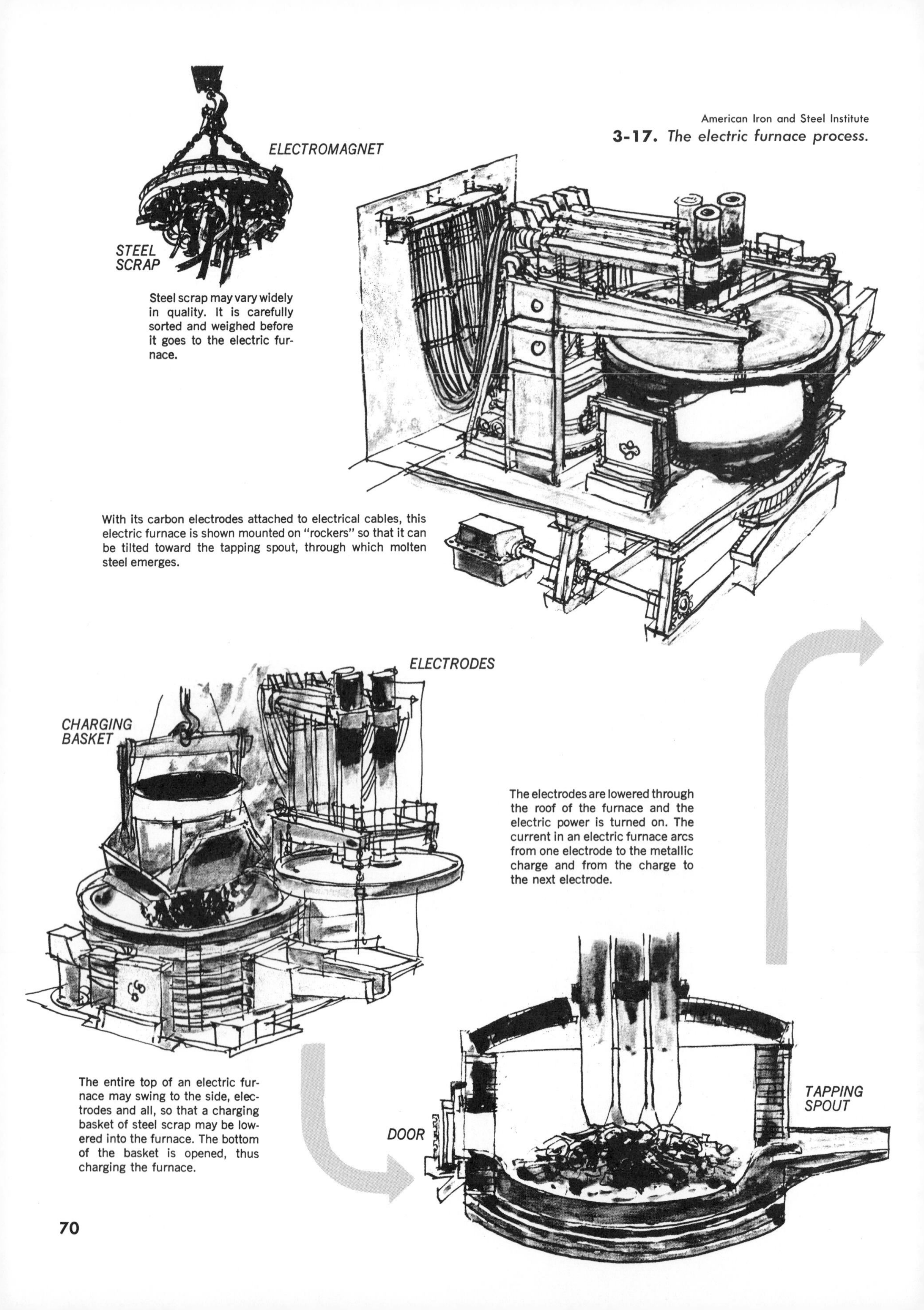

American Iron and Steel Institute
ELECTROMAGNET
STEEL SCRAP
Steel scrap may vary widely in quality. It is carefully sorted and weighed before it goes to the electric furnace.
With its carbon electrodes attached to electrical cables, this electric furnace is shown mounted on "rockers" so that it can be tilted toward the tapping spout, through which molten steel emerges.
ELECTRODES
CHARGING BASKET
The electrodes are lowered through the roof of the furnace and the electric power is turned on. The current in an electric furnace arcs from one electrode to the metallic charge and from the charge to the next electrode.
The entire top of an electric furnace may swing to the side, electrodes and all, so that a charging basket of steel scrap may be lowered into the furnace. The bottom of the basket is opened, thus charging the furnace.
DOOR
TAPPING SPOUT

3-17. *The electric furnace process.*

ELECTRIC FURNACE STEELMAKING

A long-deserved reputation for producing alloy, stainless, tool, and other specialty steels belongs to America's electric furnaces. Operators have also learned to make larger heats of carbon steels in these furnaces; this development helps account for the record tonnage outputs of recent years.

The heat within the electric furnace is intense and rigidly controlled. Modern electric furnaces have top sections that can be moved away so that special containers can charge scrap (and sometimes pig iron) from above.

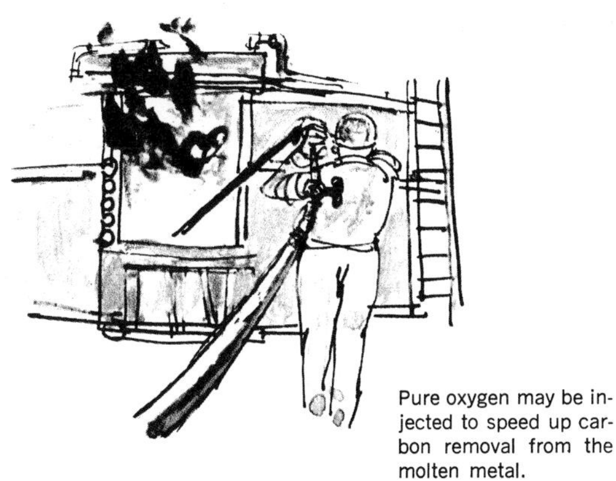

Pure oxygen may be injected to speed up carbon removal from the molten metal.

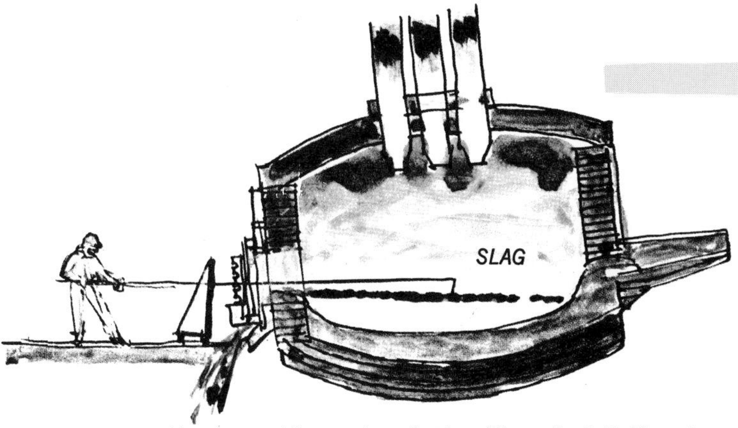

Limestone and flux are charged on top of the molten bath. Through a chemical interaction, impurities in the steel rise into the molten slag, which floats on top of the metal. The furnace is tilted slightly and the slag is raked off. Electric furnace steel can be made either with a single-slag or a double-slag practice. In the double-slag method, an oxidizing slag is first formed, raked off, and a reducing slag formed.

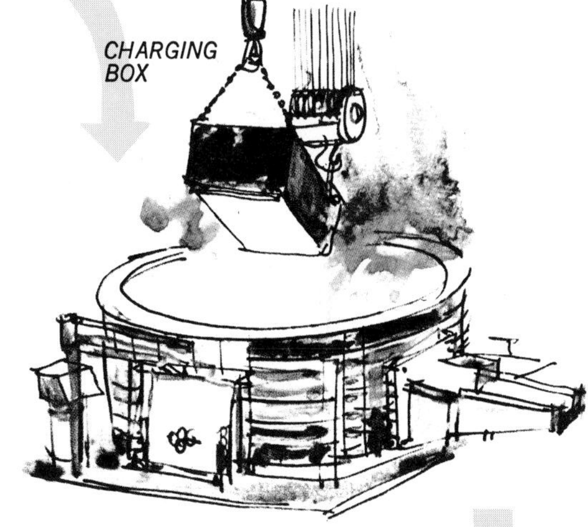

Alloying elements, which come from many parts of the world, are usually added to the molten steel in the form of ferroalloys. Typical elements include chromium from the Philippines, tungsten from Brazil, nickel from Canada, and cobalt from Africa.

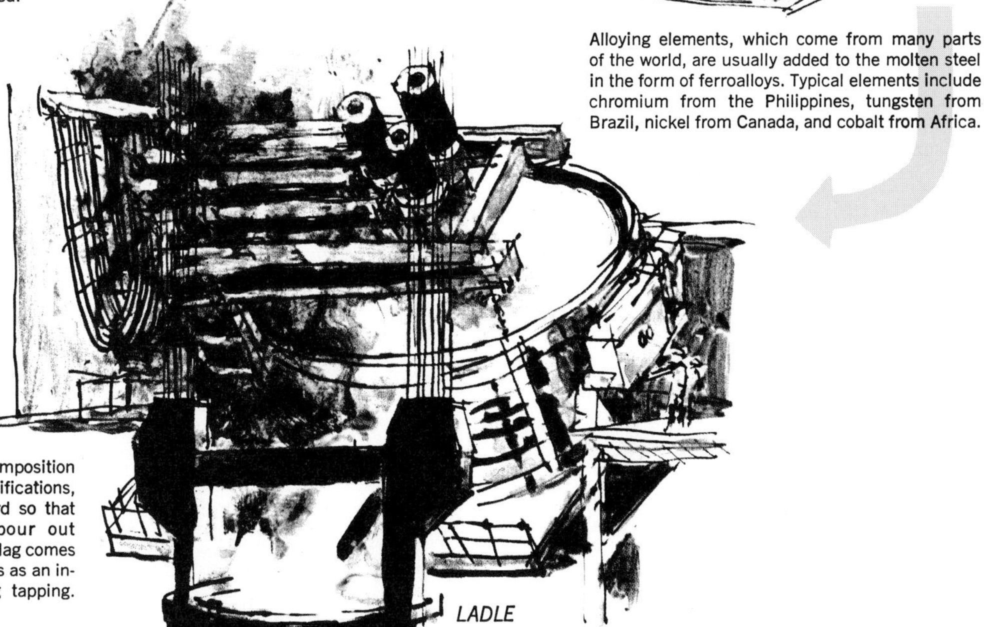

When the chemical composition of the steel meets specifications, the furnace tilts forward so that molten metal may pour out through the spout. The slag comes after the steel and serves as an insulating blanket during tapping.

Practically all grades of steel can be made by the Stora-Kaldo process although carbon steels usually are produced due to the longer length of time required to analyze alloy compositions while the charge is still in the furnace.

The greatest single difficulty with the process is the short life of the refractory furnace lining. Refractory linings have been improved, and deterioration is currently not considered a serious problem.

A steelmaking facility for the Stora-Kaldo process began producing steel in the United States in 1962. The plant has two fur-

United States Steel Co.

***3-18.** Making a "pour" from an electric furnace. Note the large electrodes through which the electric current flows to provide the arcing that produces the heat to melt the metal.*

***3-19.** The electric arc furnace produces heat through arcing action from electrodes to metal. Electrodes move down as metal melts.*

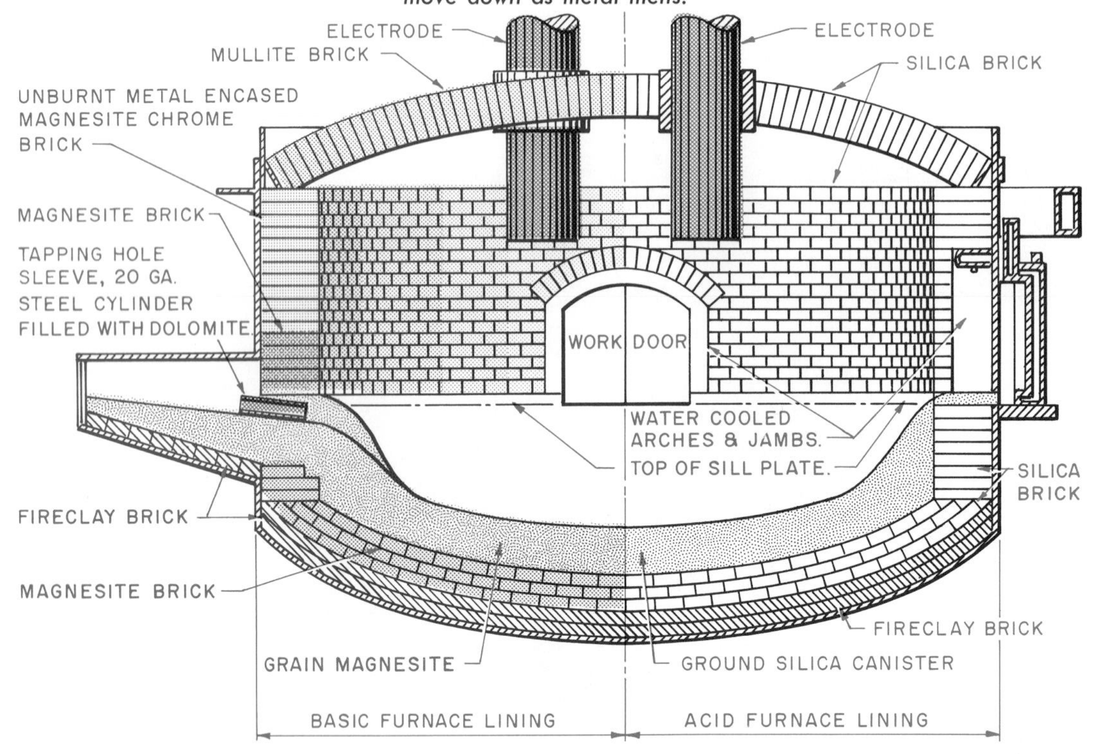

naces with an annual capacity of approximately one million ingot tons. A third furnace now under construction will raise the capacity proportionally.

Vacuum Furnaces and Degassing Equipment

The melting of steel and other alloys in a vacuum reduces the gases in the metal and produces metal with a minimum of impurities. The gases formed in a vacuum furnace are pulled out of the metal by vacuum pumps. Figure 3-25 (Pages 78, 79) illustrates the various vacuum melters and degassers. There are two general types of furnaces used for vacuum melting. The two processes are called vacuum induction melting and consumable electrode vacuum arc melting.

VACUUM INDUCTION MELTING. Vacuum induction melting was first used in the 1940s. The charge is melted in a conventional induction furnace contained within an airtight, water-cooled steel chamber, Fig. 3-26 (Page 80). The furnace resembles induction furnaces used for air-melt processes. Advantages of the vacuum induction process include freedom from air contamination, close control of heat, and fewer air inclusions.

3-21. *Charging hot metal into a 150-ton basic oxygen furnace.*

3-20. *Cross section of an electric induction furnace. Heat is generated by means of transformer action, where the bath of molted metal (B) acts as the core of the secondary winding; water-cooled copper coil (C) carries the primary electric current.*

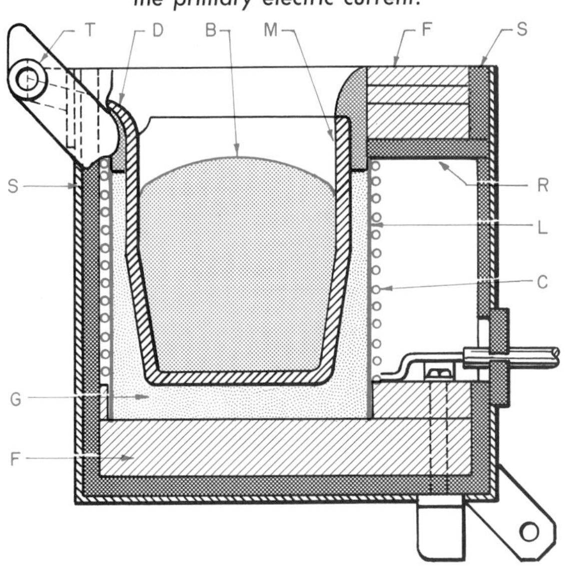

3-22. *Basic oxygen steelmaking furnace. After scrap and hot metal are charged into the furnace, the dust cap is put on, and oxygen is blown through the lance to the surface of the molten metal in order to burn out impurities.*

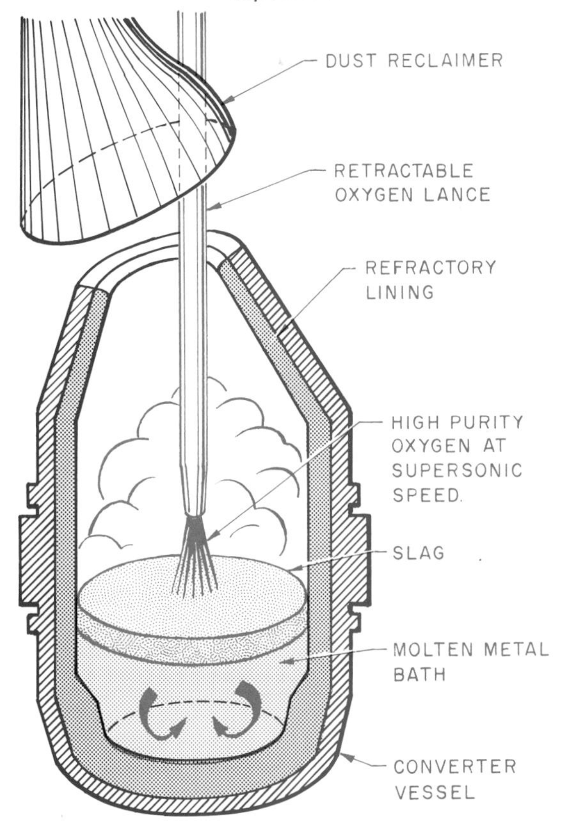

CONSUMABLE ELECTRODE VACUUM ARC MELTING. Consumable electrode melting is a refining process for steel prepared by other methods. Steel electrodes of a predetermined composition are remelted by an electric arc in an airtight, water-cooled crucible. The principle of operation is similar to arc welding. (Refer to Chapter 7, pages 195–226.)

The furnace consists of a water-cooled copper crucible, a vacuum system for removing air from the crucible during melting, and a d.c. power source for producing the arc, Fig. 3-27 (Page 80). The electrode is attached to an electrode holder which feeds the electrode during the remelting operation to maintain the arc. The copper crucible is enclosed by a water jacket, which provides the means of controlling ingot solidification.

In general, both of these processes produce high quality steel and steel alloys. The equipment has the following advantages:

- Production of alloys too expensive to manufacture by air-melt processes
- Use of reactive elements
- Decreased amounts of hydrogen, oxygen, and nitrogen in the finished product
- Improved mechanical properties
- Close heat control
- Better hot and cold workability

VACUUM DEGASSING. The vacuum degassing of molten steel is a refining operation. Its purpose is to reduce the amounts of hydrogen, oxygen, and nitrogen in steel. The process is carried out after the molten metal is removed (tapped) from the furnace and before it is poured into ingots and castings. It is based on the principle that the solubility of a gas in liquid steel decreases as pressure decreases. There are three processes used today.

Stream Degassing. Steel is poured into a tank from which the air has been already removed. After degassing, it is collected in an ingot mold or ladle, Fig. 3-28.

Ladle Degassing. A ladle of molten steel is placed in a tank and then air is removed from the tank, thus exposing the metal to the vacuum, Fig. 3-29 (Page 82). This method has the advantage of being able to process smaller amounts of steel than stream degassing.

Vacuum Lifter Degassing. A vacuum is created in a chamber suspended above a ladle of steel. The metal is forced upward into the vacuum chamber through nozzles by means of atmospheric pressure, Fig. 3-30.

The following benefits are generally derived from the degassing operation:

- The reduction of hydrogen eliminates flaking of the steel.
- The reduction of oxygen pro-

3-23. Cross section of a basic oxygen steel plant. The furnace (converter vessel) nearly 18 feet in diameter and 27 feet high, is just left of center. The charging box at the right of the converter is 25 feet above floor level. The entire steelmaking cycle takes about 45 minutes from top to top.

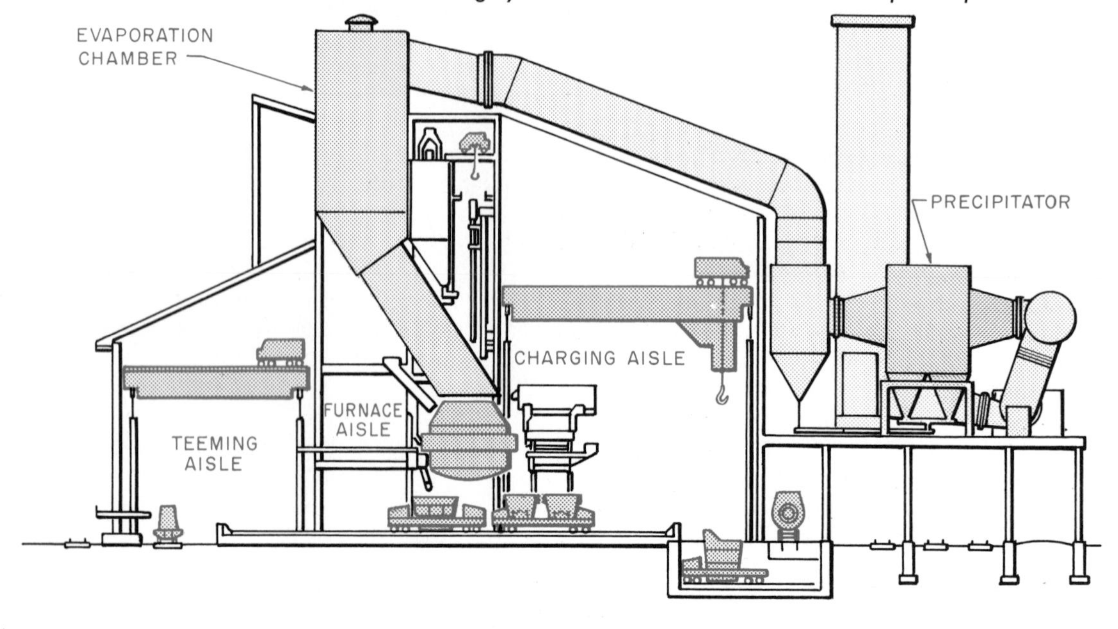

motes internal cleanliness. Oxygen reduction, however, is not as low as that achieved in vacuum-melted steels.

- Nitrogen content is reduced slightly.
- The transverse ductility (flexibility across the grain of the metal) of most degassed forged products is nearly double that of air-cast steel.

Solidification

CASTING AND SOAKING INGOTS. If molten steel from any of the steel furnaces mentioned above were to be cast into molds having the shape of the desired product, we would always be dealing with cast steel in our structures. Since *cast steel* is generally inferior to *wrought steel* (metal which is to be worked mechanically), the molten steel is poured into ingot molds and allowed to cool until solidified. To give the inside a chance to become solid and still keep the outside from cooling off too much, the ingot is lowered into a furnace called a *soaking pit,* which heats the steel for rolling. Figure 3-31 (Pages 84, 85) illustrates the processes used in solidifying steel.

DEOXIDATION. In most steelmaking processes the primary reaction involved is the combination of carbon and oxygen to form a gas. Proper control of the amount of gas evolved during solidification of the ingot determines the type of steel. If no gas is evolved, the steel is termed *killed* because it lies quietly in the molds. Increasing degrees of gas evolution result in killed, semikilled, capped, and rimmed steel.

Killed Steels. Because killed steels are strongly deoxidized, they are characterized by a relatively high degree of uniformity in composition and properties. This uniformity of killed steel renders it most suitable for applications involving such operations as forging, piercing, carburizing, and heat treatment.

Semikilled Steels. Semikilled steels are intermediate in deoxidation between killed and rimmed grades. Consequently, there is a greater possibility that the carbon will be unevenly distributed than in killed steels, but the composition is more uniform than in rimmed steels. Semikilled steels are used where neither the cold-forming and surface characteristics of rimmed steel nor the greater uniformity of killed steels are essential requirements.

Capped Steels. The duration of the deoxidation process is curtailed for capped steels so that they have a thin low-carbon rim. The remainder of the cross section, however, approaches the degree of uniformity typical of semikilled steels. This combination of properties has resulted in a great increase in the use of capped steels over rimmed steels in recent years.

Rimmed Steels. Rimmed steels have the surface and cold-forming characteristics of capped steels. They are only slightly deoxidized so that a brisk evolution of gas occurs as the metal begins to solidify. The low-carbon surface layer of rimmed steel is very ductile. Rolling rimmed steel produces a very sound surface. Consequently, rimmed grades are adaptable to applications involving cold forming and when the surface is of prime importance.

METALWORKING PROCESSES

After steel has been cast into ingot molds and solidified, it may be put through one or more of several metalworking processes to shape it and to further improve its characteristics. *Forging* and *rolling* serve two fundamental purposes. They serve the purely mechanical purpose of getting the steel into the desired shape, and they improve the mechanical properties by destroying the cast structure. This breaking up of the cast structure, also called "refining the grain," is important chiefly in that it makes the steel stronger, more ductile, and gives it a greater shock resistance.

Forging

The method of reducing metal to the desired shape is known as forging. It is usually done with a steam hammer. The piece is turned and worked in a manner similar to the process used by the blacksmith when hand forging.

Considerable forging is done today with hydraulic presses instead of with hammers. The press can take cooler ingots and can work to closer dimensions.

Another forging process is that known as *drop forging,* in which a piece of roughly shaped metal is placed between die-shaped faces of the exact form of the finished piece. The metal is forced to take this form by drawing the dies together. Many automobile parts are made in this way.

Rolling

Steel is nearly always rolled hot except for finishing passes on sheet. After rolling, ingots are known as blooms, billets, or slabs, depending on their size and shape.

- A *bloom* is square or oblong with a minimum cross-sectional area of 36 square inches.
- A *billet* is also square or oblong, but it is considerably smaller than a bloom. A bloom

This schematic drawing of a BOF facility shows the emphasis the steel industry places on air quality control. A hood over the furnace catches the dirty waste gases from the steelmaking process. The gases are conducted to air treatment facilities which occupy most of the space to the left of the crane-held ladle in the diagram.

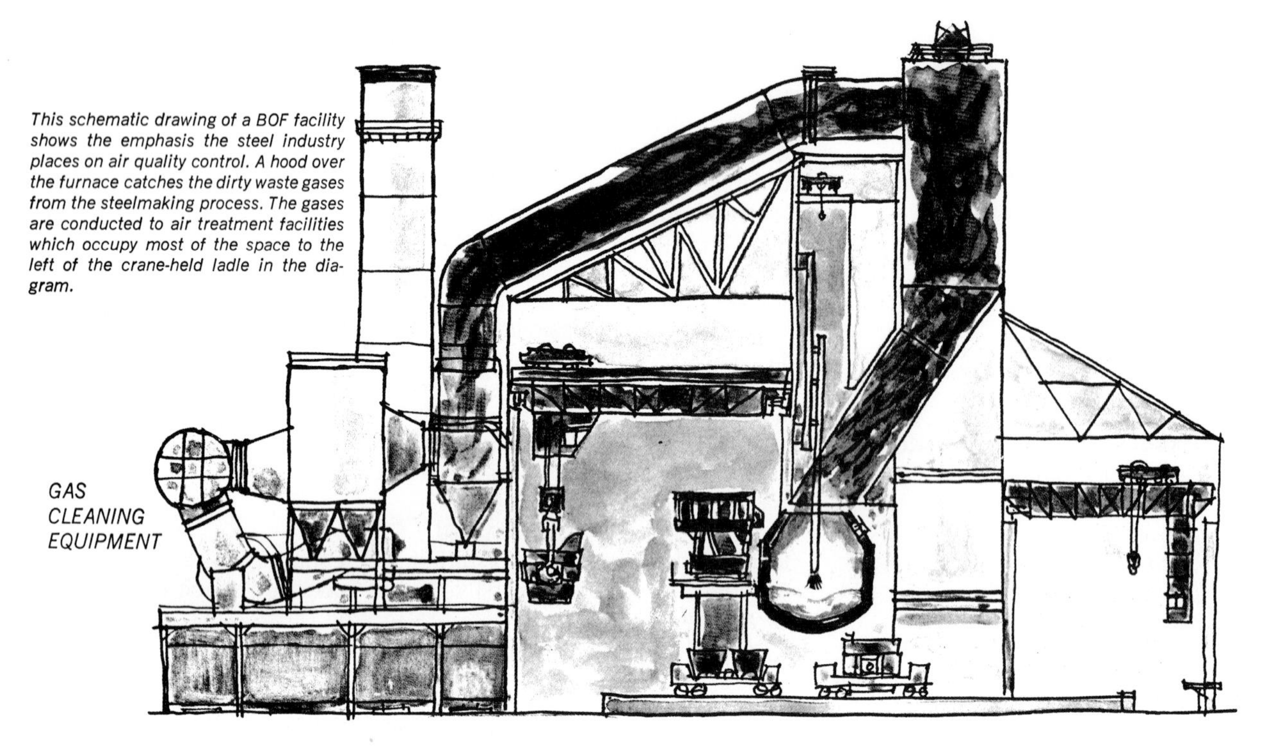

The principal material used in manufacturing steel by the basic oxygen process is molten iron. Therefore, most BOF facilities are built near blast furnaces. Some scrap steel is used in the process. Oxygen producing facilities are usually built in the same plant.

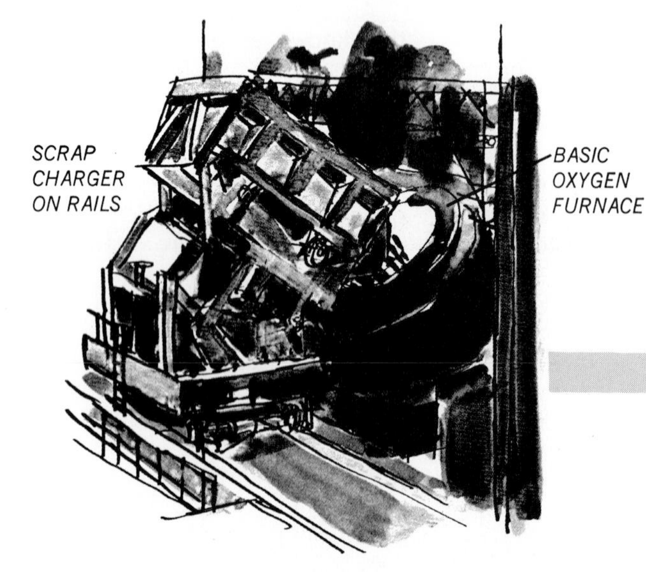

The first step for making a heat of steel in a BOF is to tilt the furnace and charge it with scrap. The furnaces are mounted on trunnions and can be swung through a wide arc.

Molten pig iron accounts for between 65% and 80% of the charge and is poured from a ladle into the top of the tilted furnace.

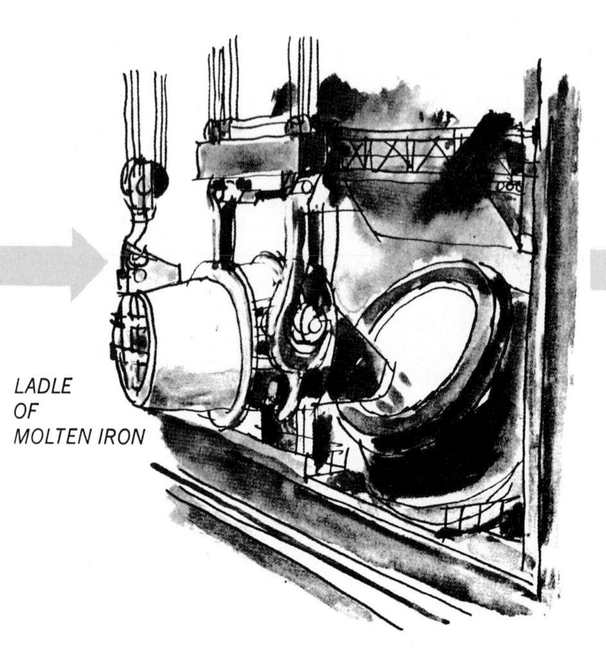

American Iron and Steel Institute

3-24. *The basic oxygen process.*

BASIC OXYGEN STEELMAKING

America's capability to produce steel by the basic oxygen process has grown enormously from small beginnings during the middle 1950's. The high tonnage of steel now made in basic oxygen furnaces—commonly called BOF's—requires the consumption of large amounts of oxygen to provide operational heat and to promote the necessary chemical changes. No other gases or fuels are used.

The basic oxygen process produces steel very quickly compared with the other major methods now in use. For example, a BOF may produce up to 300-ton batches in 45 minutes as against 5 to 8 hours for the older open hearth process. Most grades of steel can be produced in the refractory-lined, pear-shaped furnaces.

FLUX CHARGE

During the oxygen blow, lime is added as a flux to help carry off the oxidized impurities as a floating layer of slag. Lime is consumed at a rate of about 150 pounds per ton of raw steel produced.

Oxygen combines with carbon and other unwanted elements, eliminating these impurities from the molten charge and converting it to steel.

The furnace is returned to upright position. A water cooled oxygen lance is lowered into the furnace and high purity oxygen is blown onto the top of the metal at supersonic speed.

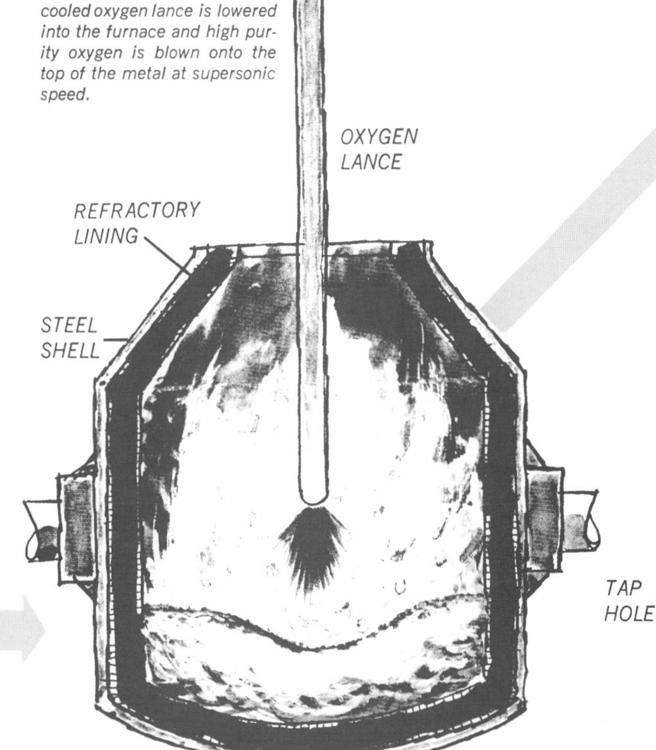

After steel has been refined, the furnace is tilted and molten steel pours into a ladle. Alloy additions are made into the ladle.

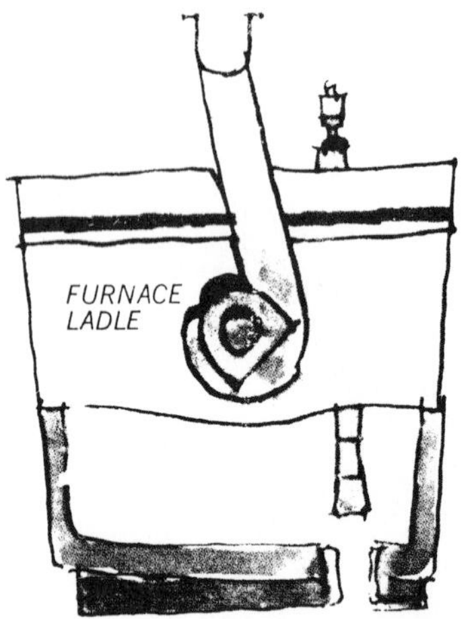

VACUUM PROCESSING OF STEEL

Steels for special applications are often processed in a vacuum to give them properties not otherwise obtainable. The primary purpose of vacuum processing is to remove such gases as oxygen, nitrogen, and hydrogen from molten metal to make higher-purity steel.

Many grades of steel are degassed by processes similar to those shown on this page. Even greater purity and uniformity of steel chemistry than is available by degassing is obtained by subjecting the metal to vacuum melting processes like those shown on the facing page.

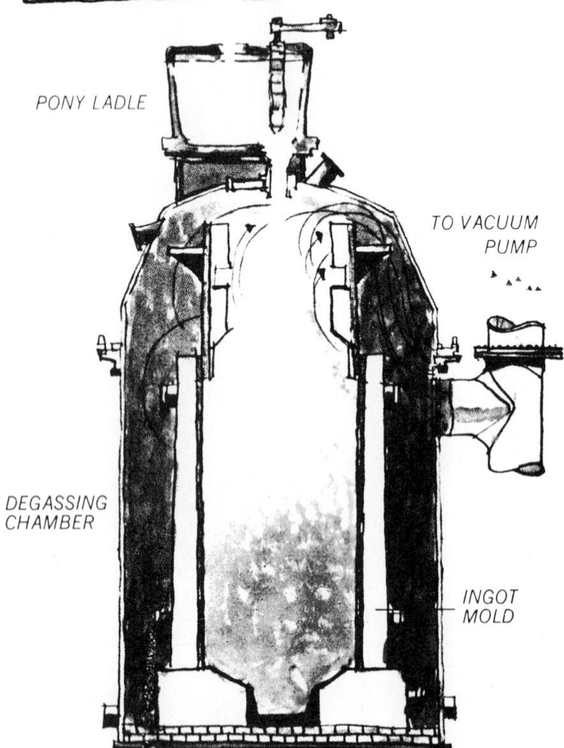

The Vacuum Degassers

In vacuum stream degassing (left), a ladle of molten steel from a conventional furnace is taken to a vacuum chamber. An ingot mold is shown within the chamber. Larger chambers designed to contain ladles are also used. The conventionally melted steel goes into a pony ladle and from there into the chamber. The stream of steel is broken up into droplets when it is exposed to vacuum within the chamber. During the droplet phase, undesirable gases escape from the steel and are drawn off before the metal solidifies in the mold.

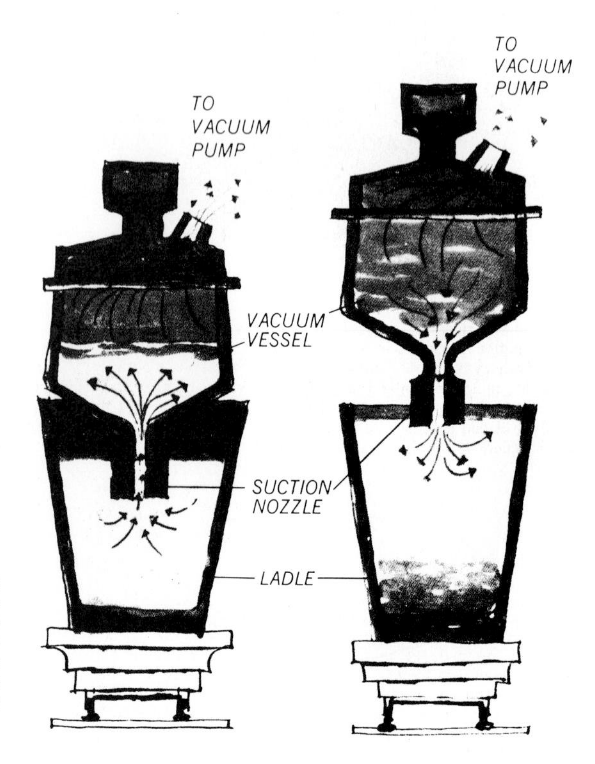

Ladle degassing facilities (right) of several kinds are in current use. In the left-hand facility, molten steel is forced by atmospheric pressure into the heated vacuum chamber. Gases are removed in this pressure chamber, which is then raised so that the molten steel returns by gravity into the ladle. Since not all of the steel enters the vacuum chamber at one time, this process is repeated until essentially all the steel in the ladle has been processed.

American Iron and Steel Institute

3-25. *Vacuum degassing and melting.*

The Vacuum Melters

Vacuum melting by either of the two processes shown on this page has helped make possible steels for many advances in space flight, nuclear science, electronics, and industry. A third process called electroslag remelting is coming into increasing favor; it is an extension, in some ways, of the consumable-electrode method described below.

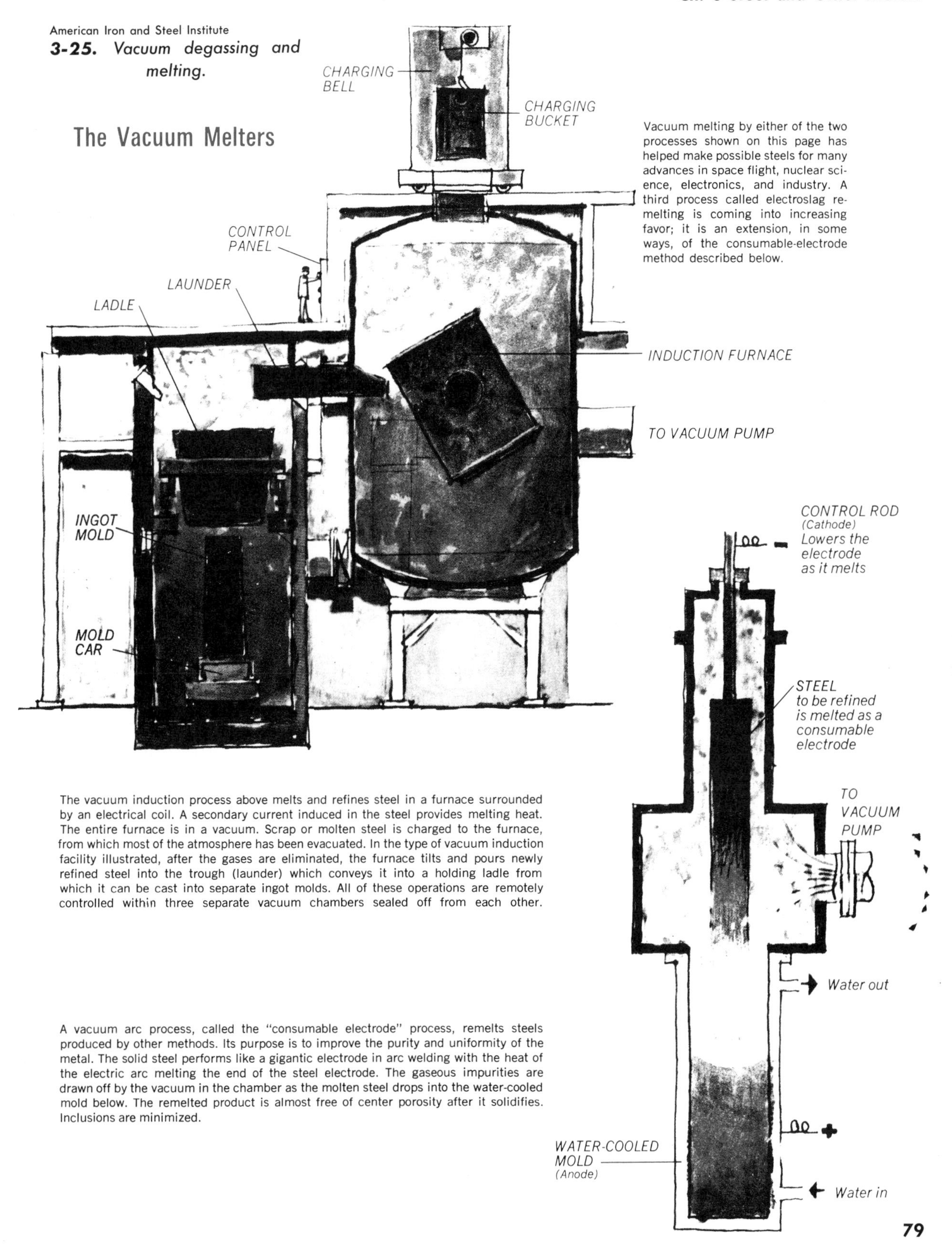

The vacuum induction process above melts and refines steel in a furnace surrounded by an electrical coil. A secondary current induced in the steel provides melting heat. The entire furnace is in a vacuum. Scrap or molten steel is charged to the furnace, from which most of the atmosphere has been evacuated. In the type of vacuum induction facility illustrated, after the gases are eliminated, the furnace tilts and pours newly refined steel into the trough (launder) which conveys it into a holding ladle from which it can be cast into separate ingot molds. All of these operations are remotely controlled within three separate vacuum chambers sealed off from each other.

A vacuum arc process, called the "consumable electrode" process, remelts steels produced by other methods. Its purpose is to improve the purity and uniformity of the metal. The solid steel performs like a gigantic electrode in arc welding with the heat of the electric arc melting the end of the steel electrode. The gaseous impurities are drawn off by the vacuum in the chamber as the molten steel drops into the water-cooled mold below. The remelted product is almost free of center porosity after it solidifies. Inclusions are minimized.

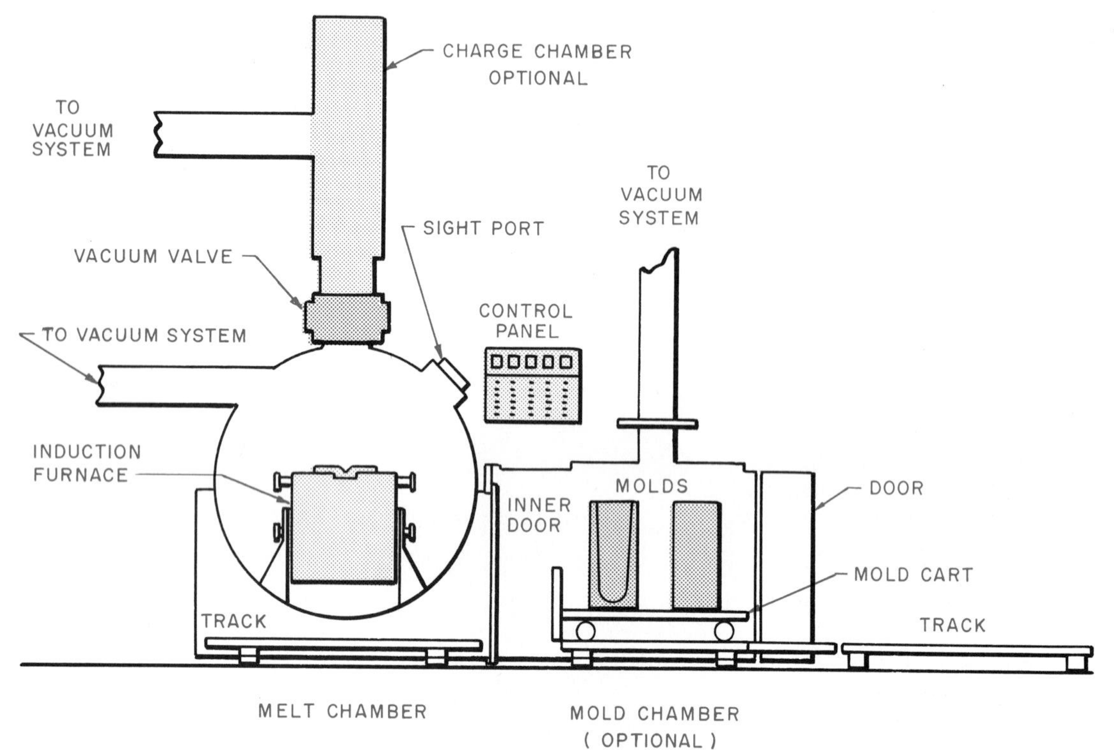

3-26. *Cross section of a typical vacuum induction furnace inside a vacuum chamber.*

3-27. *Schematic drawing of a consumable electrode remelting furnace. Direct current produces an arc which melts a single electrode. Circulating water cools the ingot mold.*

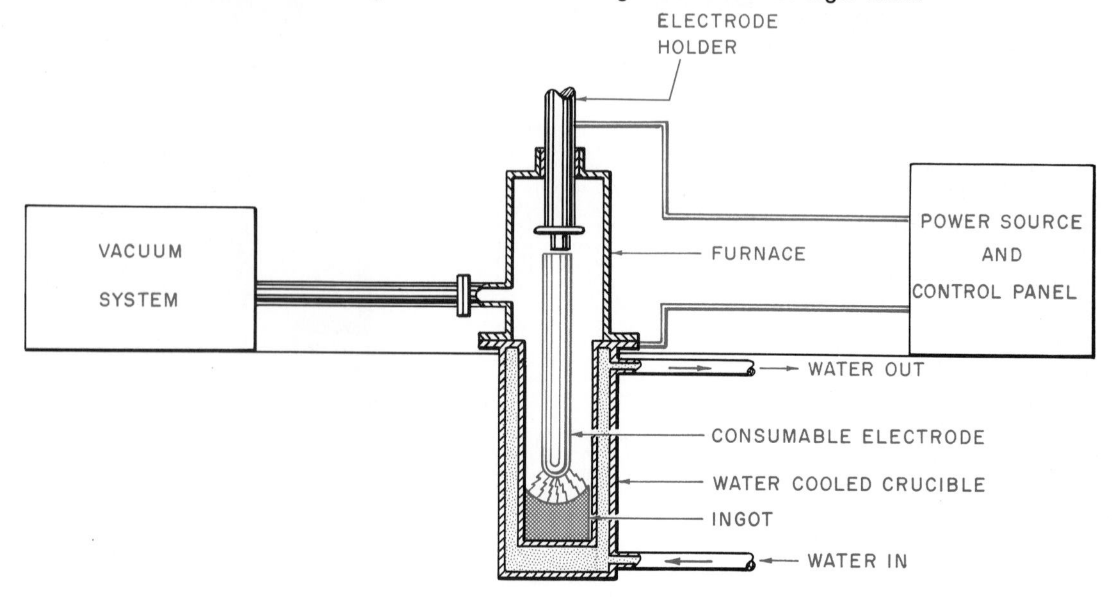

may sometimes be preheated and rolled into billets.

- A *slab* is oblong. It varies in thickness from 2 to 6 inches and in width from 5 to 6 feet.

Steel may be also rolled into bars of a wide variety of shapes such as angles, rounds, squares, round-cornered squares, hexagons, and flats as well as pipe and tubing. Figure 3-32 (Page 86) illustrates the various shapes produced by hot rolling.

About half the rolled steel products made in the United States are flat rolled. These include such items as plates, sheet, and strip. Plates are usually thicker and heavier than strip and sheet. Figure 3-33 summarizes the processes for rolling steel.

Flat-rolled steel is divided into two major categories: hot rolled and cold rolled. Hot-rolled steel is usually finished at temperatures between 900 and 2,400 degrees F. Untreated flat steel which is hot rolled is known as *black iron.* Cold-rolled products are reduced to their final thickness by rolling at room temperature. The surface finish is smooth and bright. If the sheets are coated with zinc, they are known as *galvanized* sheets; if they are coated with tin, they are known as *tin plate. Terne plate* is sheet coated with an alloy of lead and tin.

Tubular steel products are classified according to two principal methods of manufacture: the welded and seamless methods. Welded tubing and pipe are made by *flash welding* steel strip. In this process, the metal pieces are heated until the contacting surfaces are in a plastic (semisolid) state and then forced together quickly under pressure. Seamless tubing or pipe is made from billets by two processes known as piercing and cupping. In *piercing,* a heated steel bar is pierced by a mandrel and rolled to the desired diameter and wall thickness, Fig. 3-34 (Page 88). In the *cupping process,* heated plate is formed around cup-shaped dies.

Steel may also be shaped into wire, bars, forgings, extrusions, rails, and structured shapes, Fig. 3-35. These are the basic steel shapes with which the welder fabricator works.

3-28. *Cross section of a vacuum degassing unit shows principal components. Molten steel at the top pours into a pony ladle which measures steel into the vacuum unit, permitting the escape of hydrogen and other gases.*

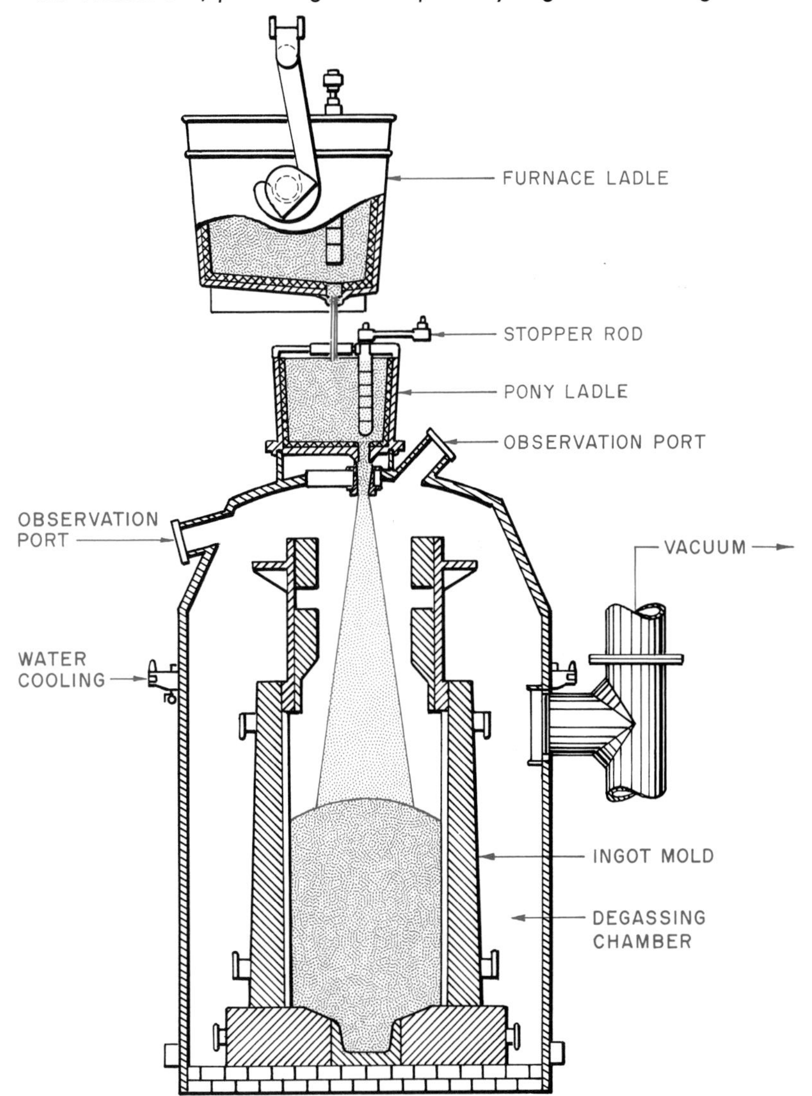

Drawing

Drawing is the operation of reducing the cross section and increasing the length of a metal bar or wire by drawing it through a series of conical, tapering holes in a die plate. Each hole is a little smaller than the preceding one. Shapes varying in size from the finest wire to sections having a cross-sectional area of several square inches are drawn.

Extrusion

Some metals lend themselves to forming by pressing through an opening, rather than by drawing or rolling. Brass rod is usually formed in this way. Perfectly round rods are obtainable. The metal to be extruded is placed in a closed chamber fitted with an opening at one end and a piston at the other end and is forced out through the opening by hydraulic pressure.

Cold Working

Cold working is the shaping of metals by working at ordinary temperatures. They may be hammered, rolled, or drawn.

Heat Treatment

Heat treatment, Fig. 3-36 (Page 90), is a process of heating and cooling a metal for the purpose of improving its structural or physical properties. Very often this is done to remove stresses caused by welding, casting, or heavy machining. Through various processes of heat treatment we can make a metal easier to machine, draw, or form by making it softer, or we can increase the hardness so that it will have wear resistance.

The important variables in any heat treatment process are (1) carbon content, (2) temperature of heating, (3) time allowed for cooling, and (4) the cooling medium, (water, oil, or air).

HARDENING. Hardening is a process in which steel is heated above its critical point and then cooled rapidly. The critical point is the point at which the carbon, which is the chief hardening agent, changes its structure. This produces a hardness that is superior to that of the steel before heating and cooling. Only medium, high, and very high carbon steel can be treated in this way. The 24-ton vessel shown in Fig. 3-37 is ready for quenching. It has just been heated to 1,950 degrees F. and will be dunked for immersion cooling. The furnace uses low sulfur gas for heating and can reach a temperature of 2,300 degrees F.

CASE HARDENING. Case hardening is a process which gives to steel a hard, wear-resistant surface while leaving the interior soft and tough. The process chosen may be cyaniding, carburizing, nitriding, flame hardening, hard surfacing by welding, or metal spraying. Plain carbon steels and alloy steels are often case hardened.

Cyaniding. Cyaniding is a method of surface-hardening low carbon steels. Carbon and nitrogen are absorbed in the outer layer of the steel to a depth of 0.003–0.020 inch.

The process can be liquid or gas. Liquid cyaniding involves heating the parts in a bath of cyanide salts at a temperature of 1,550–1,600 degrees F. The steel is held at this temperature for up to two hours, depending upon

3-29. *The ladle degassing process substitutes a ladle for the ingot mold used in the stream degassing process.*

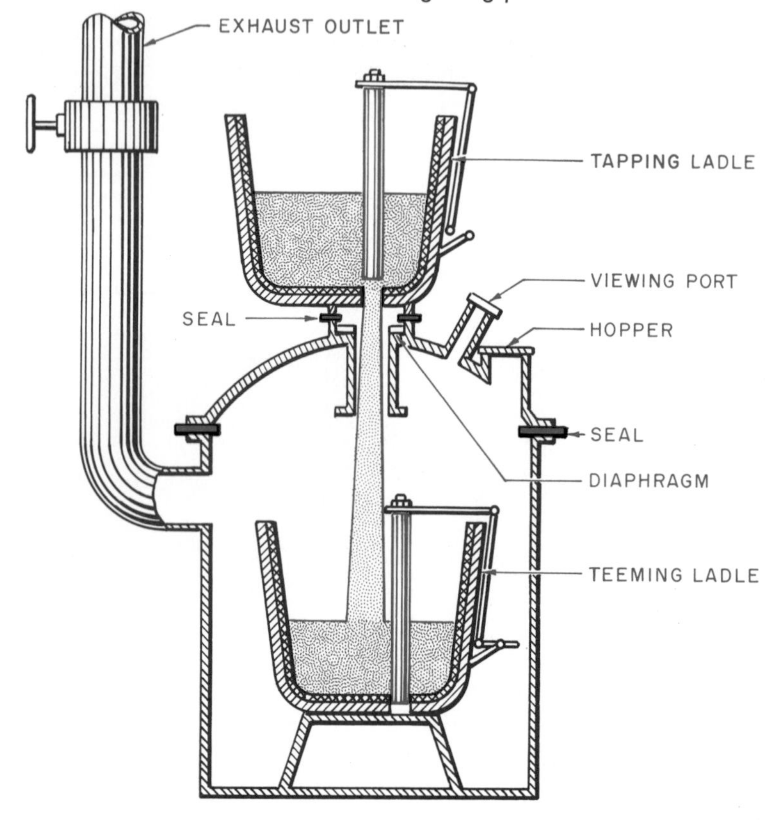

the depth of hardening desired. Then it is quenched in brine, water, or oil. Gas cyaniding involves case-hardening low carbon steels in a gas carburizing atmosphere that contains ammonia. The steel is heated to a temperature of 1,700 degrees F. and quenched in oil. These processes form a hard, but very thin, surface over the steel, beneath which the rest of the metal is still in a relatively soft condition.

Carburizing. Carburizing is a process whereby low carbon steel is made to absorb carbon in its outer surface so that it can be hardened. The depth to which the carbon will penetrate depends upon the time a heat is held, the temperature which is reached, and the carburizing compound used.

Carburizing may be done with carbonaceous solids (solid substances containing carbon), cyanidizing liquids, or hydrocarbon gases. The carburizing process selected depends on the nature of the job, the depth of hardening desired, and the type of heat-treatment equipment available. In gas welding, the surface of a weld may be hardened by the use of a carbonizing flame while welding or heating.

Nitriding. Nitriding is a case-hardening process that is used only with a group of low alloy steels. These steels contain elements such as vanadium, chromium, or aluminum which will combine with nitrogen to form nitrides. The nitrides act as a super hard skin on the surface of the steel.

The parts are heated in a nitrogenous atmosphere, usually ammonia gas, to a temperature of 900–1,000 degrees F. Nitrogen is slowly absorbed from the ammonia gas. Because of the low temperature and the fact that quenching is unnecessary, there is little distortion or warpage.

Flame Hardening. Flame hardening is the most recent of the hardening processes. It permits localized treatment with complete control.

The steel must contain enough carbon for hardening to take place. The article is heat treated and drawn. Then the surface to be hardened is exposed to a multiple-tipped oxyacetylene flame which heats it quickly to a high temperature. It is cooled quickly by water. The depth of hardness can be controlled by the temperature of the water.

Hardening the surfaces of gear teeth is an example of flame hardening. Flame hardening can also be used on certain types of cast iron. Flame hardening has the advantage in that it can be used on parts that are too bulky to put into a furnace.

ANNEALING. Annealing includes several different treatments. The purpose of annealing may be:

- To remove stresses
- To induce softness for better machining properties
- To alter ductility, toughness, or electrical, magnetic, or other physical properties
- To refine the crystalline structure
- To produce a definite microstructure

The changes that take place in a metal depend on the annealing temperature, the rate of cooling, and the carbon content.

When it is desired to produce maximum softness and grain refinement in previously hardened steel, the steel is heated

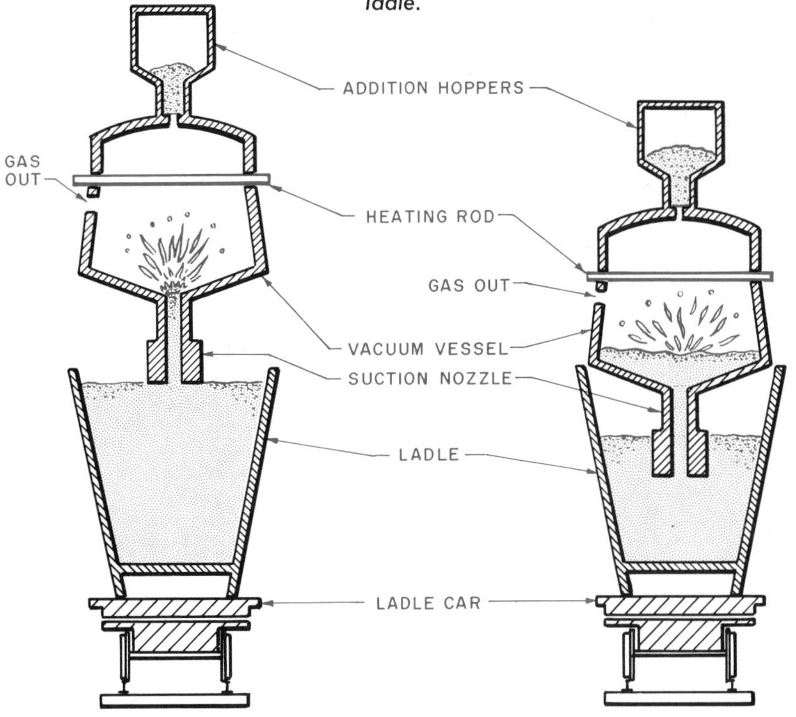

3-30. *Vacuum lifter degassing works on the principle of atmospheric pressure pushing steel upwards into a newly created vacuum. After the steel is exposed to the vacuum for the proper time, it is returned to the lower ladle.*

THE FIRST SOLID FORMS OF STEEL

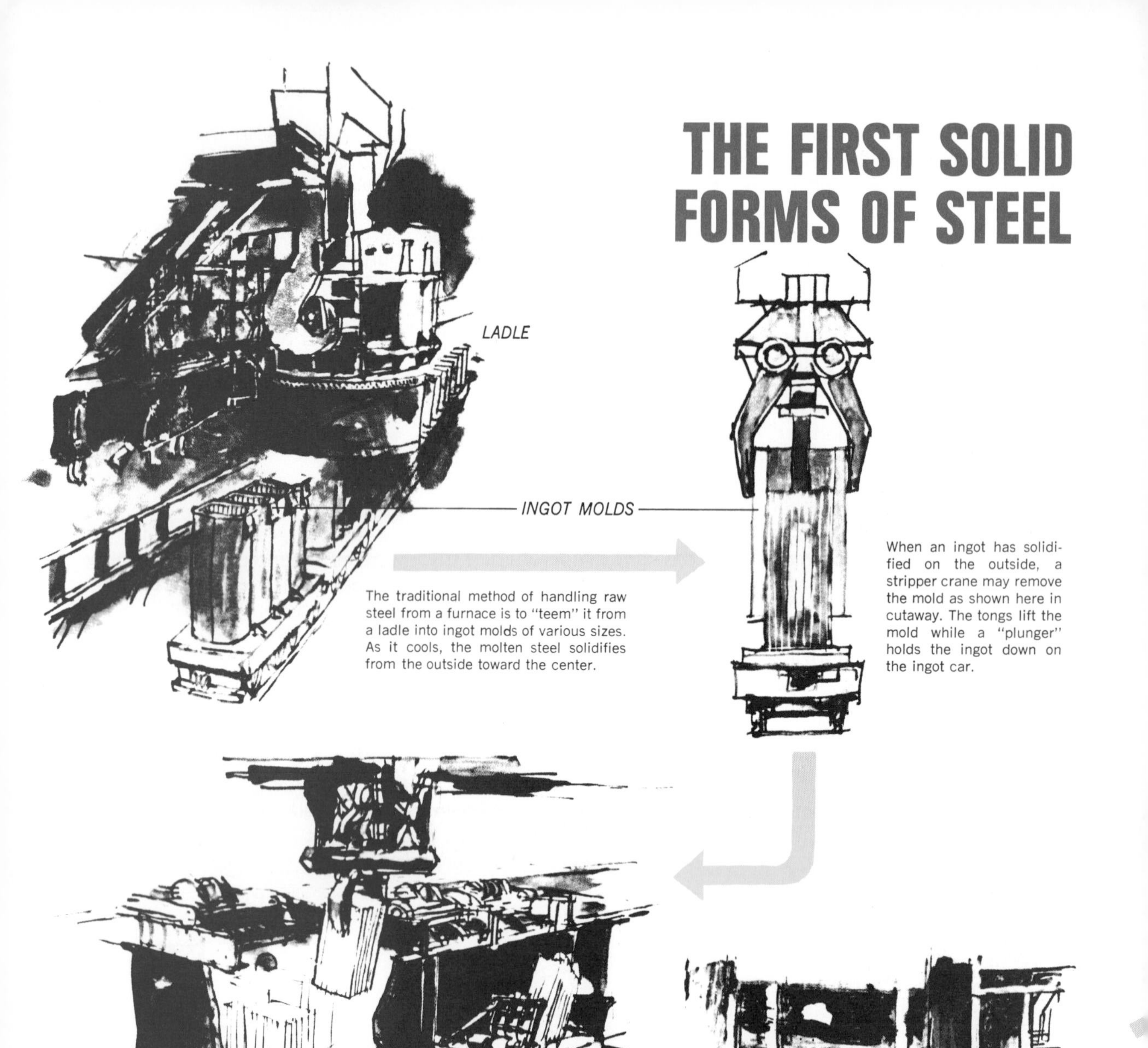

The traditional method of handling raw steel from a furnace is to "teem" it from a ladle into ingot molds of various sizes. As it cools, the molten steel solidifies from the outside toward the center.

When an ingot has solidified on the outside, a stripper crane may remove the mold as shown here in cutaway. The tongs lift the mold while a "plunger" holds the ingot down on the ingot car.

Ingots are taken to soaking pits (above) where they are "soaked" until they are of uniform temperature throughout. As each ingot is required at the roughing mill (right) it is lifted from the soaking pit and carried towards the huge facility. The almost-square end section of the steel emerging from the rolls at the right identifies it as a bloom. Another kind of semifinished steel is wider than it is high and is called a slab.

American Iron and Steel Institute

3-31. *Producing solid steel.*

Molten steel from the nation's basic oxygen, open hearth, and electric furnaces flows into ladles and then follows either of two major routes towards the rolling mills that make most of the industry's finished products. Both processes shown on these pages provide solid, semifinished steel products to the finishing mills. The first step in the traditional method is shown at the left. A much newer method—strand casting—is diagrammed here.

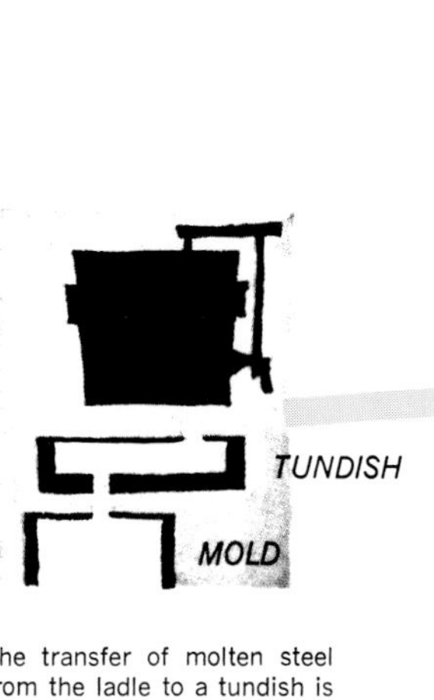

The transfer of molten steel from the ladle to a tundish is important. The tundish provides an even pool of molten metal to be fed into the casting machine. The tundish also allows an empty ladle to be removed and a full ladle to be positioned and to start pouring without interrupting the flow of metal to the casting machine.

Strand casting is a newer method by which many steps in traditional operations are bypassed. No ingot teeming, stripping, soaking, or rolling is required. Molten steel is lifted in a ladle to the top of a strand caster, which may produce either slabs, blooms, or billets, depending upon its design. (The one shown at the left produces slabs.) The molten steel from the ladle drops into a tundish and from there into a strand caster. Cooling water quickly forms a solid "skin" on the outside of the metal; this skin becomes thicker as the column of steel descends through the cooling system. The steel cools toward the center and eventually becomes solid throughout. In the machine at the left the descending slab is turned to a horizontal position. A traveling torch cuts off sections of desired length. In other types of strand casters the steel is cut to length while still in a vertical position.

Blooms (left) may go directly to finishing mills. Some are further reduced in cross section in special mills to make billets. These billets are a form of semifinished steel from which smaller finished products are made.

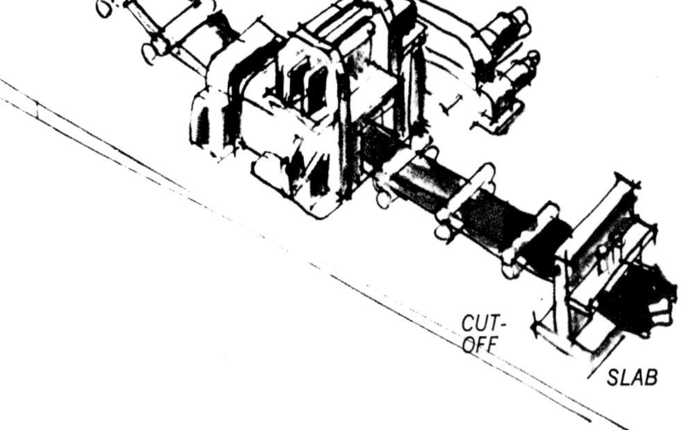

slightly above the critical range and cooled slowly. The metal is allowed to cool either in the furnace, which is gradually cooled, or it is buried in lime or some other insulating material.

Another form of annealing is stress-relief annealing. It is usually applied only to low carbon steels. The purpose here is to relieve the stress caused by working of the steel, such as in welding. The material is heated to a point just below the critical range and allowed to cool normally.

It is important to note here that the difference between hardening and softening of steels is due to the rate of cooling. Fast cooling hardens, and slow cooling softens. Both tempering and annealing reduce the hardness of a material.

TEMPERING. Tempering is a process wherein the hardness of a steel is reduced after heat treatment. It is also used to relieve the stresses and strains caused by quenching. This is usually done by heating the hardened steel to some predetermined temperature between room temperature and the critical temperature, holding it at that temperature for a length of time, and cooling it in air or water. The reduction of hardness depends upon the following three factors: (1) the tempering temperature, (2) the amount of time the steel is held at this temperature, and (3) the carbon content of the steel. Generally, as the temperature and time are increased, the hardness will be decreased. The higher the carbon content at a given temperature and time, the higher the resulting hardness.

NORMALIZING. The purpose of normalizing is to improve the grain structure of a metal and return it to normal by removing stresses after welding, casting, or forging. These stresses are caused by the uneven cooling following these operations.

Normalizing is done by heating the steel to a temperature higher than that used for annealing and then cooling it in still air. Normalizing requires a faster rate of cooling than that employed for annealing, and it results in harder, stronger metal than that which is obtained by annealing.

PHYSICAL PROPERTIES OF METALS

It is very important for the welder to be familiar with the physical properties of metals and the terms and measurements used to describe them. For convenience, the definitions of common properties have been divided into three general classifications: those related to the absorption and transmission

3-32. *Bar mill roll passes. Each vertical line of roll passes indicates the steps in rolling the bar or section shown at the bottom.*

ROUND SECTIONS SHAPED BY 3 SYSTEMS OF ROLL PASSING SHOW COMPARATIVE REDUCING ABILITIES.
OVAL & SQUARE
DIAMOND & SQUARE
FLAT & EDGE
12 STAND BAR MILL
ROLL PASSES 1 TO 12
MOST FREQUENTLY USED SYSTEM. HEAVY BUT GOOD REDUCTION.
MODERATELY SEVERE REDUCTION. USED MOSTLY FOR MED. BARS.
GENERALLY USED TO ROLL LARGE DIAMETER BARS.
SHAPING & FINISHING PASSES FOR VARIOUS SECTIONS.
SQUARE
HEXAGON
CHANNEL
ANGLE
SMALL STRUCTURAL SHAPES MAY BE FORMED BY A WIDE VARIETY OF PASSING PROCEDURES.

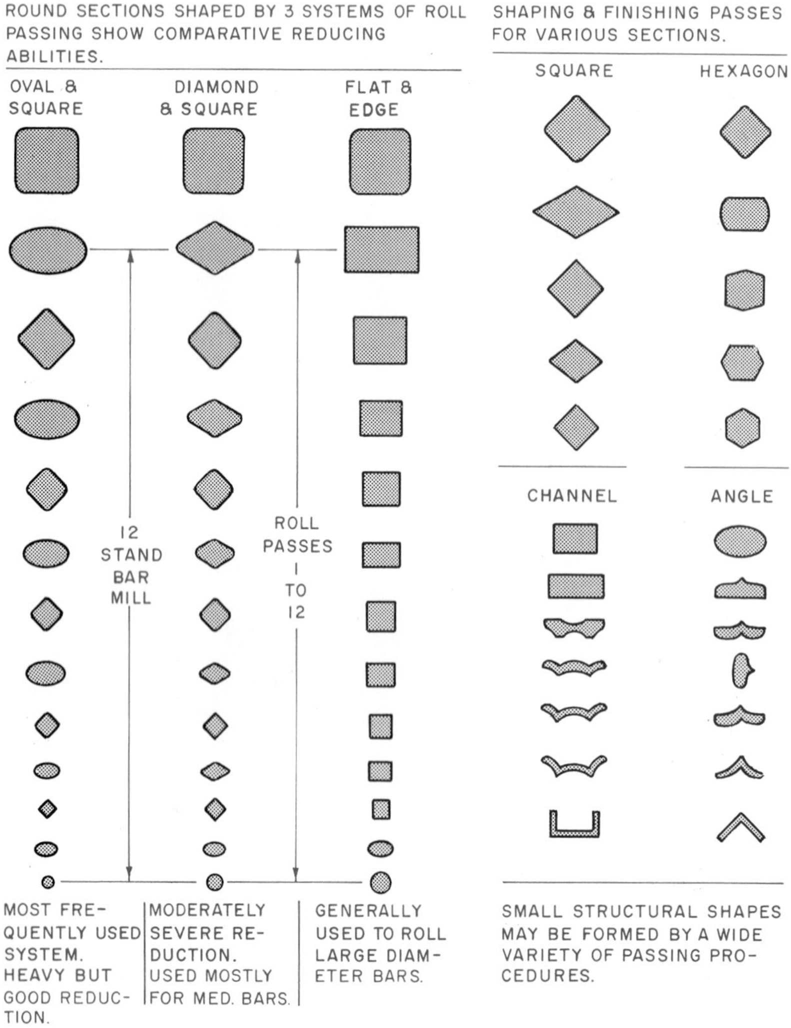

of energy, the internal structure of the metal, and resistance to stress.

Properties Related to Energy

MELTING POINT. The melting point is the temperature at which a substance passes from a solid to a liquid condition. For water this is 32 degrees F. Steel has a melting point around 2,700 degrees F., depending upon the carbon range. The higher the carbon content is, the lower will be the melting point. The higher the melting point, the greater the amount of heat needed to melt a given volume of metal. The temperature of the heat source in welding must be above the melting point of the material being welded. For example, the temperature of a flame produced by the burning of acetylene with air is not as high as the temperature of the flame produced by the burning of acetylene with oxygen. Thus it does not have the ability to melt the same materials that the oxyacetylene flame has.

WELDABILITY. Weldability is the capacity of a metal substance to form a strong bond of adherence while under pressure or during solidification from a liquid state.

FUSIBILITY. Fusibility is the ease with which a metal may be melted. In general, soft metals are easily fusible, whereas harder metals melt at higher temperatures. For example, tin, lead, and zinc are more easily fused than iron, chromium, and molybdenum.

VOLATILITY. Volatility is the ease with which a substance may be vaporized. A metal which has a low melting point is more volatile than a metal with a high melting point. Volatility is measured by the degree of temperature at which a metal boils under atmospheric pressure.

ELECTRICAL CONDUCTIVITY. The electrical conductivity of a substance is the ability of the substance to conduct electrical current.

ELECTRICAL RESISTANCE. The opposition to electric current as it flows through a wire is termed the resistance of the wire. Resistance is measured by a unit called the *ohm.* Lead has ten times the resistance of copper. This means that lead wire would have to be ten times as large as the cooper wire to carry the same amount of current without loss. A poor conductor heats up to a greater extent than a good conductor when the same amount of current is passed through each.

THERMAL CONDUCTIVITY. The thermal conductivity of a sub-

3-33. *Schematic diagram of a series of steel mill processes for the production of hot- and cold-rolled sheet and strip steel.*

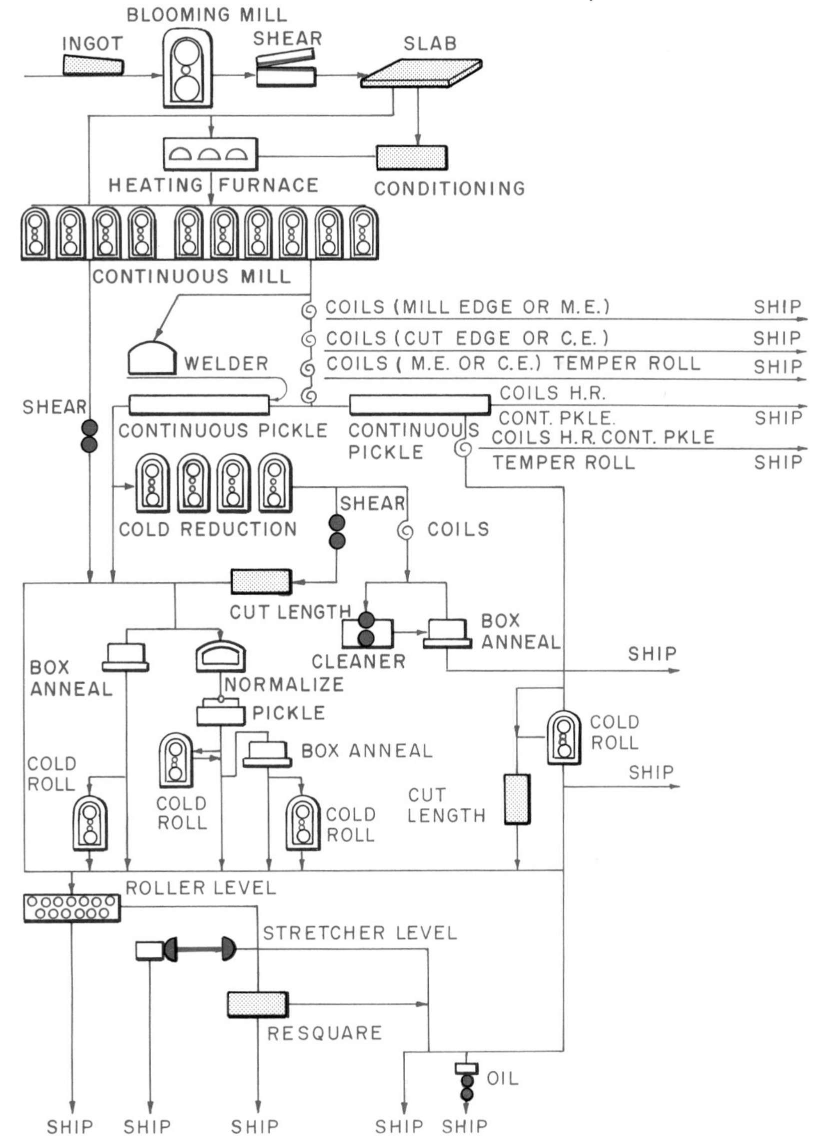

stance is the ability of a substance to carry heat. The heat that travels to both sides of the scarves during the welding of a bevel butt joint is proof that metals conduct heat. The heat is rapidly conducted away from the scarves in a good thermal conductor, but slowly in a poor one. Copper is a good conductor, and iron is a poor conductor. This accounts for the fact that copper requires more heat for welding than iron although its melting point (1,981 degrees F.) is lower than the melting point of iron (2,750 degrees F.).

COEFFICIENT OF THERMAL EXPANSION. The coefficient of thermal expansion is the amount of expansion a metal undergoes when it is heated and the amount of contraction that occurs when it is cooled. The increase in the length of a bar 1 inch long when its temperature is raised 1 degree Celsius is called the *linear* coefficient of thermal expansion. The higher the coefficient, the greater the amount of expansion; and, therefore, the greater the contraction upon cooling. Expansion and contraction will be discussed in more detail under Effects of Welding on Metal, pages 102–114.

HOT SHORTNESS. Hot shortness is brittleness in metal when hot. This characteristic should be kept in mind in the handling of hot metals and in jig construction and clamping.

OVERHEATING. A metal is said to be overheated when the temperature exceeds its critical range, that is, it is heated to such a degree that its properties are impaired. In some instances it is possible to destroy the original properties of the metal through heat treatment. If the metal does not respond to further heat treatment, it is considered to be burned and cannot meet the requirements of a heavy load. In arc welding, excess welding cur-

3-34. *Path of a solid steel round on its route toward becoming a tube. Included are heating, piercing, rolling, and sizing operations.*

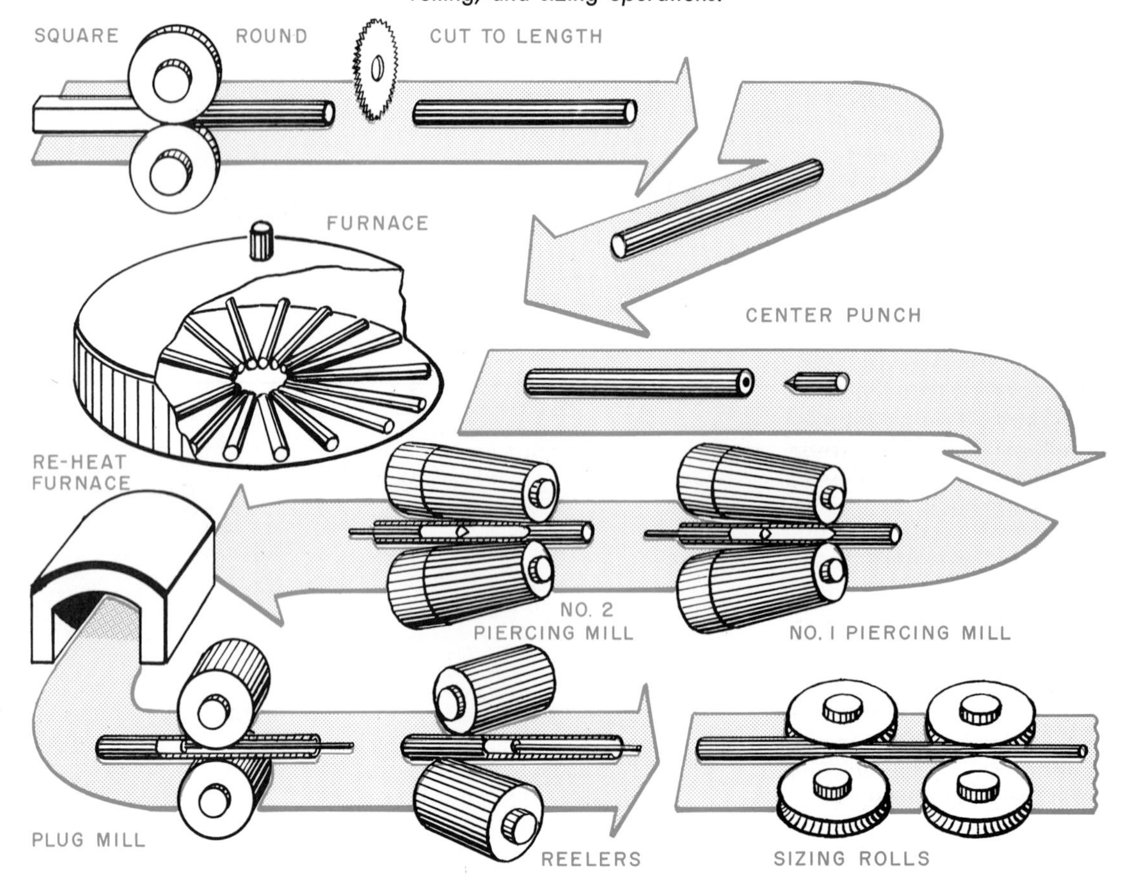

rent may cause overheating in the weld deposit.

Properties Related to Internal Structure

SPECIFIC GRAVITY. Specific gravity is a unit of measurement based on the weight of a volume of material compared with an equal volume of water. Aluminum has a specific gravity of 2.70: thus it is almost $2\frac{3}{4}$ times heavier than water. When two molten metals are mixed together, the metal with the lower specific gravity will be forced to the top, and the metal with the higher specific gravity will sink to the bottom.

DENSITY. A metal is said to be dense when it is compact and does not contain such defects as slag, inclusions, and gas pockets. Density is expressed as the quantity per unit volume. The density of low carbon steel, for example, is 0.283 pounds per cubic inch. The density of aluminum, a much lighter metal, is only 0.096 pounds per cubic inch.

POROSITY. Porosity is the opposite of density. Some materials are porous by their very nature and allow liquids under pressure to leak through them. Materials that are porous have an internal structure that lacks compactness or has other defects that leave voids in the metal.

Properties Related to Stress Resistance

An important physical property of a metal is the ability of that material to perform under certain types of stress. Stresses to which metal fabrications are subjected during both welding and service include the following:

- *Compression:* squeezing
- *Shear:* strain on a lap joint pulled in opposite directions
- *Bending:* deflection as a result of a compressive force
- *Tension:* pulling in opposite directions
- *Fatigue:* result of a combination of forces working in all directions
- *Torsion:* twisting force in opposite directions

These typical stresses are illustrated in Fig. 3-38.

PLASTICITY. The ability of a material to deform without breaking is its plasticity. Strength combined with plasticity is the most important combination of properties a metal can have. Metals having these properties can be used in structural fabrications. For example, if a member of a bridge structure becomes overloaded, the property of plasticity allows the overloaded member to flow so that the load becomes redistributed to other parts of the bridge structure.

3-35. *Commercial structural steel shapes.*

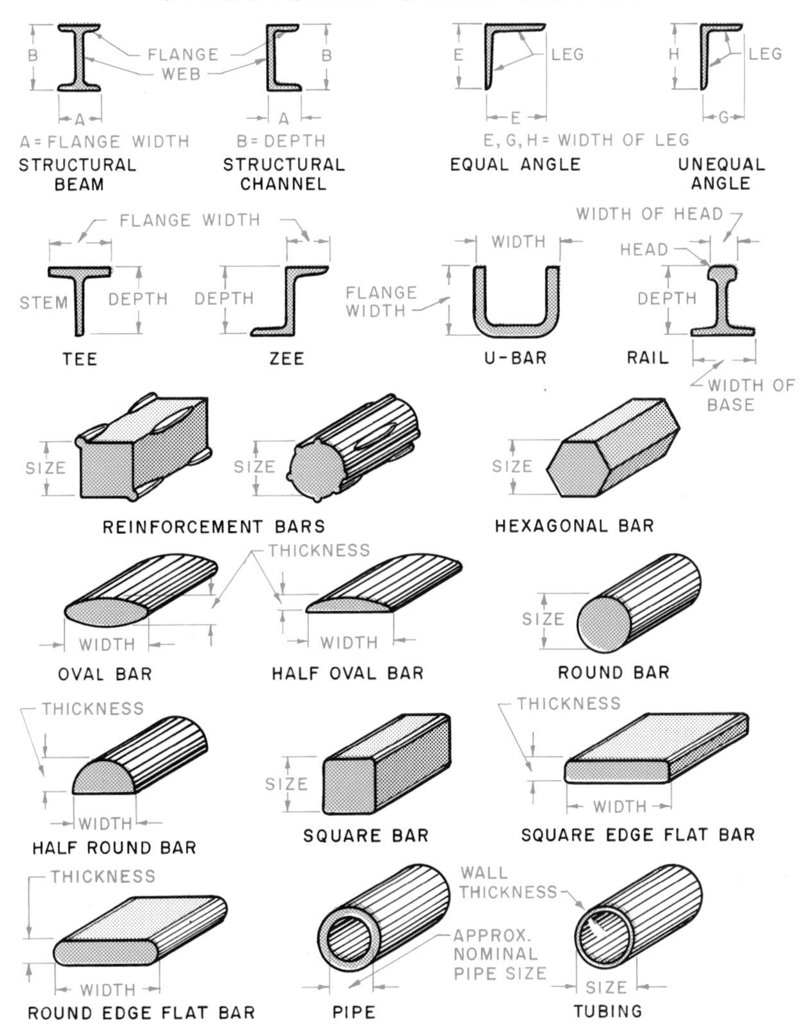

STRENGTH. Strength is the ability of a material to resist deformation. It is usually expressed as the *ultimate tensile strength* in pounds per square inch. The ultimate tensile strength of a material is its resistance to breaking. Cast iron has an approximate tensile strength of 15,000 p.s.i. One type of stainless steel, on the other hand, has reached a strength of 400,000 p.s.i.

TOUGHNESS. Although there is no direct method of measuring the toughness of materials accurately, a material may be assumed to be tough if it has high tensile strength and the ability to deform permanently without breaking. Toughness may be thought of as the opposite of brittleness since a tough metal gives warning of failure through deformation whereas a brittle material breaks without any warning. Copper and iron are tough materials.

SHOCK RESISTANCE. Shock resistance may be defined as the ability of a material to withstand a maximum load applied suddenly. The shock resistance of a material is often taken as an indication of its toughness.

BRITTLENESS. Brittle materials fail without any warning such as deformation, elongation, or a change of shape. It may be said that a brittle material lacks plasticity and toughness. A piece of chalk is very brittle.

HARDNESS. The ability of one material to penetrate another material without fracture of either is known as hardness. The greater the hardness, the greater the resistance to marking or deformation. Hardness is usually measured by pressing a hardened steel ball into the material. In the Brinell hardness test the diameter of the impression is measured, and in the Rockwell hardness test the depth of the impression is measured. A hard material is also a strong material, but it is not very ductile. The opposite of hardness is softness.

MALLEABILITY. The ability a material possesses to deform permanently under compression without breaking or fracturing is known as the malleability of the metal. Metals that possess this characteristic can be rolled or hammered into thinner forms. Metals must have malleability in order to be forged.

ELASTIC LIMIT. Loading a material will cause it to change its shape. The ability of the material to return to its original shape after the load has been removed is known as *elasticity.* The *elastic limit* is the greatest load that may be applied after which the material will return to its original condition. For practical purposes the elastic limit is required in designing because it is usually more important to know what load will deform a structure than what load will cause a fracture or break.

MODULUS OF ELASTICITY. Some materials require higher stresses to stretch than others do. In other words some materials are stiffer than others. To compare the stiffness of one metal with that of another, we must determine what is known as the *modulus of elasticity* for each of them. The modulus of elasticity is the ratio of the stress to the strain. It is a measure of relative stiffness. If the modulus is high, the material is more likely to resist movement or distortion. A material that stretches easily has a low modulus.

YIELD POINT. When a sample of low or medium carbon steel is subjected to a tension test, a curious thing happens. As the

3-36. *All-welded pressure vessel being removed from a heat-treating furnace.* Nooter Corp.

load on the test specimen is increased slowly, a point is found at which a definite increase in the length of the specimen occurs with no increase in the load. The load at this point, expressed as pounds per square inch, is called the yield point of the material. Nonferrous metals and types of steel other than low and medium carbon steels do not have a yield point.

RESILIENCE. Resilience (springiness) is the energy stored in a material under strain within its elastic limit that causes it to resume its original shape when the load is removed. Resilience is a property of all spring steels.

DUCTILITY. Ductility is the ability of a material to be permanently deformed (stretched) by loading and yet resist fracture. When this happens, both elongation and reduction in area take place in the material. The amount of stretching is expressed as *percent of elongation.* Metals with high ductility may be stretched, formed, or drawn without tearing or cracking. Gold, silver, copper, and iron are metals with good ductility. A ductile metal is not necessarily a soft metal.

FATIGUE FAILURE. Failure of metals under repeated or alternating stresses is known as fatigue failure. When a metal is broken in a tensile machine, it is found that a certain load is required to break it. The same material, however, will fail when a much smaller load has been applied and removed many times. A spring, for example, may fail after it has been in service for months even though its loading has not been changed or increased. In designing parts subjected to varying stresses, the fatigue limit of a material is more important than its tensile strength or elastic limit. The *fatigue limit* is that load, usually expressed in pounds per square inch, which may be applied for an indefinite number of cycles without causing failure. If a load can be applied 100 million times without causing failure, it is assumed that can be applied indefinitely without failure.

RESISTANCE TO CORROSION. The ability of metals to resist atmospheric corrosion and corrosion by liquids or gases is often very important. Corrosion is the gradual wearing away or disintegration of a material by a chemical process. The action of oxygen on steel to form rust is a form of slow corrosion. Corrosion may

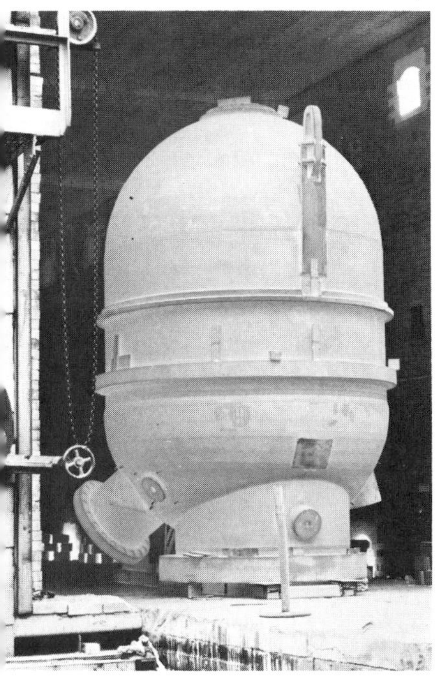

3-37. Vessel ready for quenching, or quick cooling. The metal changes from white-hot to black in less than three minutes.

Nooter Corp.

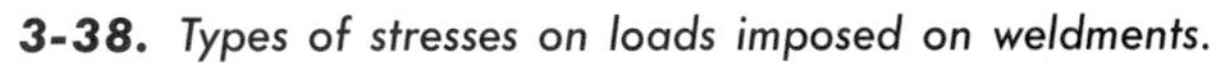

3-38. Types of stresses on loads imposed on weldments.

COMPRESSION – THE APPLICATION OF PRESSURE. ________

TENSION – A PULLING OR STRETCHING ACTION. ________

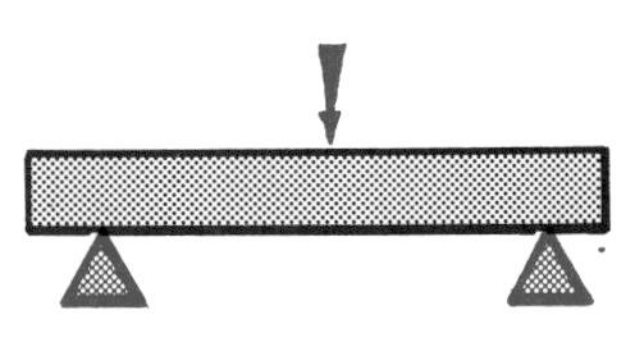

BENDING – PRESSURE APPLIED TO FORCE AWAY FROM A STRAIGHT LINE. ________

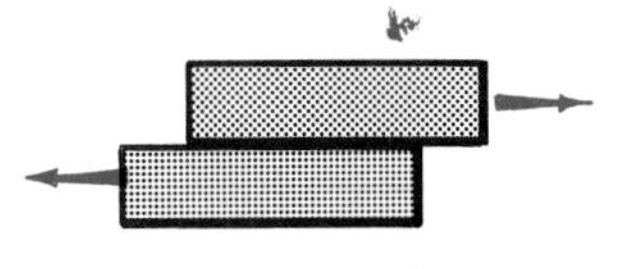

SHEAR – A PULLING ACTION CAUSING TWO BODIES TO SLIDE ON EACH OTHER, PARALLEL TO THEIR PLANE OF CONTACT. ________

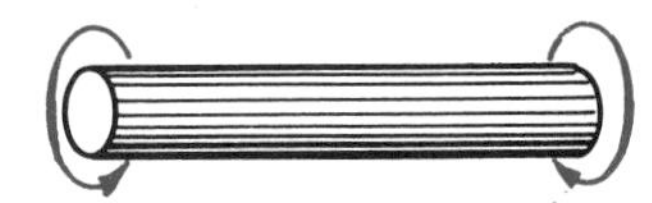

TORSION – A TURNING OR TWISTING ACTION.

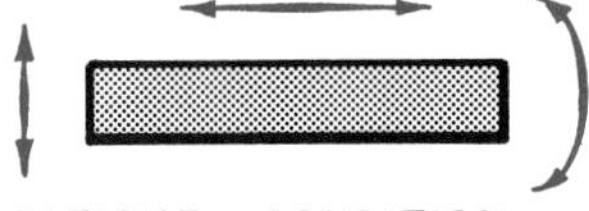

FATIGUE – CONDITION CAUSED BY REPEATED STRETCHING, TWISTING, COMPRESSION, WHILE IN SERVICE.

Table 3-1 *Metals and their properties.*

Metal	Melting point—degrees fahrenheit	Linear expansion per 10 ft. length per 100 degrees fahrenheit rise in temperature in inches	Heat conductivity British thermal units per hour per square foot per inch of thickness per degree fahrenheit	Density—pounds per cubic inch	Shrinkage allowance in castings inches per foot	Brinell hardness		Approximate tensile strength pounds per square inch	Approximate analysis of chemical composition (percent)	Remarks
						Hard	Soft			
Allegheny metal	2640	0.115		0.283			140	90,000[1]–120,000		[1] Annealed
Aluminum—cast—8% copper	1175				0.1875			20,000	Aluminum 92; copper 8	
Aluminum—pure	1218	0.148	1393	0.096		40	23	12,000–28,000	Aluminum	
Aluminum—5% silicon	1117	0.146		0.093				18,000	Aluminum 95; silicon 5	
Ambrac—A	2100	0.109		0.310				50,000–130,000	Copper 75; nickel 20; zinc 5	
Antimony	1166	0.075		0.245				1,000	Antimony	
Bismuth	520								Bismuth	
Boron	3992			0.094					Boron	
Brass—commercial high	1660	0.115	756	0.306	0.1875			46,000	Copper 66; zinc 34	
Bronze—tobin	1625	0.119		0.304				54,000	Copper 60; zinc 39; tin 1	
Bronze—muntz metal	1625								Copper 60; zinc 40	
Bronze—manganese	1598	0.119		0.302			95	60,000		
Bronze—phosphor	1922	0.119		0.321		160	100	45,000	Copper 96; tin 3.75; phosphorus 0.25	
Cadmium	610								Cadmium	
Carbon	6332[2]								Carbon	[2] Greater than
Chromium	2740			0.235					Chromium	
Cobalt	2700	0.082		0.312				34,400	Cobalt	
Copper—deoxidized	1981	0.118		0.307	0.25			32,000–55,000	Copper 99.99	
Copper—electrolyte	1981	0.106	2640	0.322		107	42	20,000–70,000	Copper 99.93; oxygen 0.07	
Duriron	2310	0.104		0.253	0.1875				Silicon 14.5; carbon 0.85; manganese 0.35; iron (base)	Acid resisting iron
Everdur	1866	0.113	871	0.306	0.1875	180	80	52,000–100,000	Copper 94.8–96; silicon 3–4; manganese, 1–1.2	Wt. cast 0.294 lbs. cu. in.
Gold	1945	0.094	2046	0.697				14,000	Gold	
Iron—cast	2300	0.067	338	0.260	0.125		193	15,000		
Iron—malleable	2300			0.268	0.125			53,000		
Iron—pure	2786	0.078	467	0.283			84	38,500	Iron	
Iron—wrought	2900	0.078	419	0.278			90	48,000	Iron; slag	
Lead—pure	620	0.181	240	0.411	0.312		6	1,780	Lead	
Lead—chemical	620	0.193		0.410	0.312			1,780	Lead 99.92; copper 0.08	
Manganese	2246			0.268					Manganese	
Molybdenum	4532		100	0.309				42,500–154,000	Molybdenum	
Monel metal	2480	0.093	174	0.318		190	015	79,000–109,000[3]	Nickel 67; copper 28; iron; manganese; silicon; carbon; sulfur	[3] Sheets from soft to full hard
Nichrome	2460	0.091		0.295				100,000	Nickel 60; iron 24; chromium 16; carbon 0.1	[4] Sheets from soft to full hard
Nickel	2646	0.083	413	0.319	0.25		112	61,000–109,000[4]	Nickel	
Nickel silver—18%	1955	0.122		0.309		158	77	58,000–95,000	Copper 65; nickel 18; zinc 17	
Platinum	3218						65		Platinum	
Slicon	2588								Silicon	
Silphos	1300							33,000[5]	Silver 15; copper 80; phosphorus 5	Welding rods
Silver—pure	1762	0.126	2919	0.380			65	36,000	Silver	[5] Joint
Steel—hard (0.40–0.70% carbon)	2500	0.076	312[6]	0.283		590	240	75,000		[6] 1% carbon
Steel—low carbon (less than 0.15%)	2700	0.076		0.283			138	50,000		
Steel—medium (0.15–0.40 carbon)	2600	0.076		0.283			180	60,000		
Steel—manganese	2450					255		70,000–100,000	Carbon 1.0–1.45; manganese 12–15; silicon 0.10–0.20; iron	
Steel—amsco nickel manganese	2450								Carbon 1.3; manganese 14; nickel 5; silicon 0.35; iron	Welding rods
Steel—nickel—3½%	2600			0.284				80,000	Nickel 3.25–3.75; carbon 0.15–0.25; manganese 0.50–0.80	
Steel—cast	2600			0.282	0.25		144	58,000		
Stainless steel—18–8	2550	0.073		0.279			140	89,000–100,000	Chromium 18; nickel 8; carbon 0.16; iron (base)	
Stainless steel—18–8 low carbon	2640			0.280			140	80,000[7]	Chromium 18; nickel 8; carbon 0.07; iron (base)	[7] Annealed
Tin	450	0.139	450	0.263	0.083	9[10]	15	5,000	Tin	
Tungsten	6152		1381	0.678			442		Tungsten	
Tungsten carbide	[9]									[9] Cannot be melted with torch
Vanadium	3182			0.199					Vanadium	
Zinc—cast	786	0.169		0.248	0.312		45	9,000	Zinc	[10] Moh's scale
Zinc—rolled	786	0.169	770	0.258				31,000[11]	Zinc	[11] Av. hard rolled

be measured by (1) determining the loss in strength of tensile samples, (2) determining loss in weight of materials that dissolve in the corroding medium, or (3) determining gain in weight when a heavy coating of rust is formed.

Table 3-1 lists types of metals and their physical and chemical properties. Study this table carefully in order to acquire a basic understanding of the differences in metals and the part these differences play in choosing a metal for a particular job.

EFFECTS OF COMMON ELEMENTS ON STEEL

Nonmetals

CARBON. Pure carbon is found in its native state both as diamond, a very hard material, and as graphite, a soft material. Carbon is a part of coal, petroleum, asphalt, and limestone. Commercially it can be obtained as lampblack, charcoal, and coke.

As indicated previously, the amount of carbon present in steel determines its hardness and has serious implications for welding. Increased carbon content increases the tensile strength of steel but reduces its ductility and weldability. If the carbon content is above 0.25 percent, sudden cooling from the welding temperature may produce a brittle area next to the weld. The weld itself may be hard and brittle if an excess of carbon is picked up from the steel being welded. If other alloying elements are added to promote high tensile strength, good weld qualities can be retained. In general, an effort is made to use steels of low or medium carbon content.

BORON. Boron is a nonmetallic element which is plentiful and occurs in nature in combination with other elements, as in borax. Pure boron is a gray, extremely hard solid with a melting point in excess of 400 degrees F. It increases the hardenability of steel, that is, the depth to which the steel will harden when quenched. Boron's effectiveness is limited to sections whose size and shape permit liquid quenching.

Boron also intensifies the hardenability characteristics of other elements present in the steel. It is very effective when used with low carbon steels. Its effect, however, is reduced as the carbon content of the steel increases.

SILICON. Silicon is the main substance in sand and sandstone. It forms about one-fourth of the earth's crust. This element is added mainly as a deoxidizing agent to produce soundness during the steelmaking process. A large amount of silicon may increase the tensile strength. If the carbon content is also high, however, the addition of silicon increases the tendency to cracking. Large amounts are alloyed with steel to produce certain magnetic qualities for electrical and magnetic applications.

PHOSPHORUS. Phosphorus is usually present in iron ore. Small amounts improve the machinability of both low and high carbon steel. It is, however, an impurity as far as welding is concerned, and the content in steel should be kept as low as possible. Over 0.04 percent phosphorus makes welds brittle and increases the tendency to cracking.

SULFUR. Sulfur is considered to be a harmful impurity in steel because it makes steel brittle and causes cracking at high temperatures. Steel picks up some sulfur from coke used in the blast furnace. Most of the sulfur present in the blast furnace is fluxed out by the lime in the furnace.

The sulfur content in steel should be kept below 0.05 percent. Sulfur increases the tendency of the weld deposit to crack when cooling and may also cause extreme porosity if the weld penetration is deep. Sulfur does, however, improve the machinability of steel. Steels containing sulfur may be readily welded with low hydrogen electrodes.

SELENIUM. This element is used interchangeably with sulfur in some stainless steels to promote machinability.

Metals

MANGANESE. Manganese is a very hard, grayish-white metal with a reddish luster. In its pure state it is so hard that it can scratch glass. It was first used to color glass during glassmaking. Today it is one of the most useful metals for alloying steel. The addition of manganese increases both tensile strength and hardness. The alloy is a steel that can be readily heat treated. Special care must be exercised in welding since manganese steels have a tendency to porosity and cracking.

High manganese steels are very resistant to abrasion. In amounts up to 15 percent, manganese produces very hard, wear-resistant steels which cannot be cut or drilled in the ordinary way. They must be machined with carbide-tipped tools. Because of their high resistance to abrasion, manganese steels are used in such equipment as rock crushers, grinding mills, and power shovel scoops.

MOLYBDENUM. Molybdenum is a silvery white metal that increases the toughness of steel. Since it also promotes tensile strength in steels that are subject to high temperatures, it is an alloying element in pipe where high pressure and high temperature are common. Molybdenum also increases the corrosion resistance of stainless steels.

Molybdenum steels may be readily welded if the carbon content is low. Preheating is required for welding if the carbon content is above 0.15 percent. It can be hardened by quenching in oil or air rather than water.

CHROMIUM. Chromium is a hard, brittle, grayish-white metal which is highly resistant to corrosion. It is the principal element in the straight chromium and nickel-chromium stainless steels. The addition of chromium to low alloy steels increases the tensile strength, hardness, and resistance to corrosion and oxidation. Ductility is decreased. This is true at both high and low temperatures. Chromium is used as an alloying element in chrome steel and as plating metal for steel parts such as auto bumpers and door handles. The depth of hardness is increased by quenching chromium in oil.

Steels that contain chromium are easily welded if the carbon content is low. However, the presence of a high percentage of carbon increases hardness. Thus preheating and sometimes postheating are required to prevent brittle weld deposits and fusion zones.

NICKEL. Nickel is a hard, silvery white element. It is used extensively for plating purposes and as an alloying element in steel. In combination with chromium, nickel is an important alloy in stainless steels. Nickel increases the strength, toughness, and corrosion resistance of steel. Nickel-chromium steels are readily welded. Heavy sections should be preheated.

COLUMBIUM. The use of columbium in steel has been largely confined to stainless steels in which it combines with carbon and improves the corrosion resistance. More recently it has been added to carbon steel as a means of developing higher tensile strength.

COBALT. Cobalt is a tough, lustrous, silvery white metal. It is usually found in nature with iron and nickel. Cobalt is used as an alloying metal in high speed steel and special alloys when high strength and hardness must be maintained at high temperatures. It is also added to some permanent magnet steels in amounts of 17 to 36 percent. Cobalt is being used increasingly in the aerospace industry.

COPPER. Copper is a soft, ductile, malleable metal that melts at 1,984 degrees F. It has an expansion rate $1\frac{1}{2}$ times greater than that of steel, and its thermal conductivity is 10 times greater. Only silver is a better conductor of heat and electricity. Copper is also highly corrosion resistant.

Copper is added to steel to improve its resistance to atmospheric corrosion. In the small amounts used (0.10 to 0.40 percent) copper has no significant effect on physical properties. Its other effects are undesirable, particularly the tendency to promote hot shortness (brittleness when hot), thereby lowering surface quality.

Copper is used for roofing, plumbing, electrical work, and in the manufacture of such alloys as brass, bronze, and German silver. When used in silver and gold jewelry, it increases hardness.

Brass is the most common class of copper alloys. Zinc is the alloying element in brass. Bronzes are produced when other alloying elements such as zinc, tin, silicon, aluminum, phosphorus, and beryllium are added to copper.

ALUMINUM. Aluminum is never found in nature in its pure state. It is derived chiefly from bauxite, an aluminum hydroxide. It is one of the lightest metals: its weight is about one-third that of iron. It is a good conductor of heat and electricity and is highly resistant to atmospheric corrosion. Aluminum is ductile and malleable. It can be easily cast, extruded, forged, rolled, drawn, and machined. Aluminum can be joined by welding, brazing, soldering, adhesive bonding, and mechanical fastening. Aluminum melts at about 1,200 degrees F.

Aluminum is used in both carbon and alloy steels. When used in the making of alloy steels, it has several important functions. Because it combines easily with oxygen, it is a reliable deoxidizer and purifier. It also produces fine austenitic grain size. When aluminum is present in amounts of approximately 1 percent, it promotes high hardness in steel. Inert gas welding of aluminum is very successful because the gas protects the weld from oxide formation.

Aluminum finds many commercial uses, especially in aircraft, trucks, trains, and in the construction industry. Types of aluminum and aluminum alloys are discussed on page 101, 102.

TITANIUM AND ZIRCONIUM. These metals are sometimes added in small amounts to certain high-strength, low-alloy steels to deoxidize the metal, control fine grain size, and improve physical properties.

LEAD. Lead sulfide is the most important lead ore. Lead is a soft, malleable, heavy metal. It has a very low melting point: approximately 620 degrees F. Lead has little tensile strength. It is highly resistant to corrosion.

Additions of lead to carbon and alloy steels improve machinability without significantly affecting physical properties. Increases of 20 to 40 percent in machinability ratings are normal, and in many instances even greater improvement is realized. The economic factors of each job must be studied. For leaded steels to be economical, the machining must involve considerable removal of metal, and the machine tools must be able to take advantage of the increased cutting speeds.

To date, leaded carbon steels have been used mainly for stock which is to be free machined. Lead is added to a base composition with high phosphorus, carbon, sulfur, and nitrogen content to obtain the optimum in machinability.

Lead is used extensively in the plumbing industry, on cable coverings, and in batteries. It is also used in making such alloys as solder, bearing metals, type metal for printing, and terne plate. Oxides of lead are used in paints.

TUNGSTEN. Tungsten is a steel-gray metal which is more than twice as heavy as iron. It has a melting point above 6,000 degrees F.

Tungsten improves the hardness, wear resistance, and tensile strength of steel. In amounts from 17 to 20 percent and in combination with chromium and molybdenum, it produces a steel that retains its hardness at high temperature. Tungsten is a common element in high speed and hot-worked steels and in hard-surfacing welding rods which are used for building up surfaces that are subject to wear.

VANADIUM. Vanadium increases the toughness of steel and gives it the ability to take heavy shocks without breaking. Vanadium also has a high resistance to metal fatigue and high impact resistance, thus making steel containing vanadium excellent for springs, gears, and shafts. When heat treated, a vanadium steel has fine grain structure. Vanadium steel may require preheat for welding.

TYPES OF STEEL

Carbon Steels

Steel may be defined as refined pig iron or an alloy of iron and carbon. Besides iron, steel is made up of carbon, silicon, sulfur, phosphorus, and manganese. Carbon is the most important alloying ingredient in steel. An increase of as little as 0.1 percent carbon can materially change all of the properties of steel.

The carbon content of the steel has a direct effect on the physical properties of steel. Increases in the carbon content reduce the melting point of steel. It becomes harder, has a higher tensile strength, and is more resistant to wear. Harder steel has the tendency to crack if welded, and it is more difficult to machine. As carbon increases, steel also loses some of its ductility and grows more brittle. The addition of carbon makes it possible to heat treat steel. High and medium carbon steels usually cost more than low carbon steel.

The physical properties of steel are so dependent upon its carbon content and its final heat treatment that types of steels range from the very soft steels, such as those used in the manufacture of wire and nails, to the tool steels, which can be hardened and made into cutting tools to cut the softer steels and other metals. Carbon steels are usually divided into low carbon steels, medium carbon steels, high carbon steels, and tool steels.

LOW CARBON STEELS. Those steels whose carbon content does not exceed 0.30 percent and may be as low as 0.03 percent are low carbon steels. They are also referred to as *mild steels* and *plain steels*. These steels may be quenched very rapidly in water or brine and do not harden to any great extent. General purpose steels of 0.08 to 0.25 percent carbon content are found in this classification, and they are available in standard shapes from steel warehouses. Machine steel (0.08–0.30 percent carbon) and cold-rolled steel (0.08–0.30 percent carbon) are the most common low carbon steels. Low carbon steels are produced in greater quantities than all other steels combined, and they make up the largest part of welded fabrication. Thus they are the type of steel which will be stressed in this text.

The weldability of low carbon steels is excellent. They go into most of the structures fabricated by welding such as bridges, ships, tanks, pipes, buildings, railroad cars, and automobiles.

MEDIUM CARBON STEELS. Medium carbon steels are those steels which have a carbon content ranging from 0.30 to 0.60 percent. They are considerably stronger than low carbon steels and have higher heat-treat qualities. Some hardening can take place when the steel is heated and quenched. They can be welded with shielded metal-arc electrodes and other processes. More care must be taken in the welding operation, and best results are obtained if the steel is preheated before welding and normalized after welding. This insures maximum tensile strength and ductility.

Medium carbon steels are used in many of the same structures indicated above for mild steel, except that these structures are subject to greater stress and higher load demands.

HIGH CARBON STEELS. Steels whose carbon content ranges from 0.60 to 1.7 percent are known as high carbon steels. They are more difficult to weld than low or medium carbon steels. They can be heat treated for maximum hardness and wear resistance. Preheating and heat treatment after welding eliminate hardness and brittleness at the fusion zone. These steels are used in springs, punches, dies, tools, military tanks, and structural steel.

Alloy Steels

Steel is classified as alloy steel when the content of alloying elements exceeds certain limits. The amounts of alloying elements lie within a specified range for commercial alloy steels. These elements are added to obtain a desired effect in the finished product as described on pages 96–98. Alloy steels are readily welded by welding processes such as MIG and TIG.

HIGH-STRENGTH, LOW-ALLOY STEELS. The high-strength, low-alloy steels make up a group of steels with chemical compositions specially developed to give higher physical property values and materially greater corrosion resistance than are obtainable from the carbon steel group. These steels contain, in addition to carbon and manganese, other alloying elements which are added to obtain greater strength, toughness, and hardening qualities.

High-strength, low-alloy steel is generally used when savings in weight are important. Its greater strength and corrosion resistance require less reinforcement and, therefore, fewer structural members than fabrications made with carbon steel. Its better durability is also an advantage in these applications. Among the steels in this classification are oil-hardening steel, air-hardening steel, and high speed steel.

High-strength, low-alloy steel is readily adaptable to fabrication by shearing, gas cutting, punching, forming, riveting, and welding without quenching and tempering heat treatment by the fabricator.

STAINLESS AND HEAT-RESISTING STEELS. As the name implies, stainless and heat-resisting steels possess unusual resistance to corrosion at both normal and elevated temperatures. This superior corrosion resistance is accomplished by the addition of chromium to iron. The corrosion resistance of the stainless steels generally increases with increasing chromium content. It appears that when chromium is present, a thin layer of chromium oxide is bonded to the surface, and this oxide prevents any further *oxidation* (ordinary rusting, which is the most common kind of corrosion). Four percent chromium is generally accepted as the dividing line between low alloy steel and stainless steel. Although other elements such as copper, aluminum, silicon, nickel, and molybdenum also increase the corrosion resistance of steel, they are limited in their usefulness.

Some stainless steels have practically an indefinite life even without cleaning. Stainless steels are also resistant to corrosion at elevated temperatures which are the result of oxidation, carburization, and sulfidation, (deterioration of the surface caused by the action of oxygen, carbon, and sulfur respectively). Users of stainless steel have experienced some difficulty with pitting. This usually occurs when the material

Table 3-2
Typical compositions of martensitic stainless steels.

AISI type	Composition[1](%)		
	Carbon	Chromium	Other[2]
403	0.15	11.5–13.0	0.5 silicon
410	0.15	11.5–13.5	—
414	0.15	11.5–13.5	1.25–2.5 nickel
416	0.15	12.0–14.0	1.25 manganese, 0.15 sulfur (min.), 0.060 phosphorus, 0.60 molybdenum (opt.)
416Se	0.15	12.0–14.0	1.25 manganese, 0.060 phosphorus, 0.15 selenium (min.)
420	0.15 (min)	12.0–14.0	—
431	0.20	15.0–17.0	1.25–2.5 nickel
440A	0.60–0.75	16.0–18.0	0.75 molybdenum
440B	0.75–0.95	16.0–18.0	0.75 molybdenum
440C	0.95–1.20	16.0–18.0	0.75 molybdenum

The Lincoln Electric Co.

[1]Single values denote maximum percentage unless otherwise noted.
[2]Unless otherwise noted, other elements of all alloys listed include maximum contents of 1.0% manganese, 1.0% silicon, 0.040% phosphorus, and 0.030% sulfur. Balance is iron.

is exposed to chlorides, or at points where the steel is in contact with other materials, such as leather, glass, and grease. Pitting can be materially reduced by treating the area with strong oxidizing agents such as some chromates or phosphates. The addition of molybdenum to austenitic nickel-chromium steels also helps to control pitting.

The uses for stainless steels are many, and there are many varieties to choose from. Stainless steels have the following advantages.

- They resist corrosion and the effects of high temperatures.
- They maintain the purity of materials in contact with them.
- They permit greater cleanliness than other types of steel.
- Stainless steel fabrications usually cost little to maintain.
- Low strength-to-weight ratios are possible both at room and elevated temperatures.
- They are tough at low temperatures.
- They have high weldability.
- They are highly pleasing in appearance and require a minimum of finishing.

In general, stainless steels are produced in either the electric arc or the induction furnace. The largest tonnages by far are melted in electric arc furnaces.

Stainless and heat-resisting steels are commonly produced in finished forms such as plates, sheets, strip, bars, structural shapes, wire, tubing, semi-finished castings, and forgings. These steels fall into five general classifications according to their characteristics and alloy content.

1. Five percent chromium, hardenable (martensitic) 500 series
2. Twelve percent chromium, hardenable (martensitic) 400 series
3. Seventeen percent chromium, nonhardenable (ferritic) 400 series
4. Chromium-nickel (austenitic) 300 series
5. Chromium-nickel-manganese (austenitic) 200 series

Series 400 and 500 (Martensitic). Steels in these two groups are primarily heat resisting and retain a large part of their properties at temperatures up to 1,100 degrees F. They are somewhat more resistant to corrosion than alloy steels, but they are not considered true stainless steels.

These steels contain carbon, chromium, and sometimes nickel in such proportions that they will undergo hardening and annealing. Chromium content in this group ranges from 11.5 to 18 percent; and carbon, from 0.15 to 1.20 percent, Table 3-2.

Because of their lower chromium content, steels in the martensitic groups do not offer quite as much corrosion resistance as types in the ferritic and austenitic groups. They are satisfactory for mildly corrosive conditions. They are suitable for applications requiring high strength, hardness, and resistance to abrasion and wet or dry erosion. Thus they are suitable for coal-handling equipment, steam and gas turbine parts, bearings, and cutlery.

These steels are satisfactory for both hot and cold working. They are air hardening and must be cooled slowly or annealed after forging or welding to prevent cracking.

Series 400 (Ferritic). The chromium content of this group ranges from 11.5 to 27 percent, and the carbon content is low, generally under 0.20 percent, Table 3-3. There is no nickel. Ferritic stainless steels cannot be hardened by heat treatment although hardness may be increased by cold working. Suitable hot or cold working, followed by annealing, is the only means of refining the grain and improving ductility.

Stainless steels in the ferritic group have a low coefficient of thermal expansion and good resistance to corrosion. They are adaptable to high temperatures. Since their ductility is fair, they can be fabricated by the usual methods such as forming, bending, spinning, and light drawing. Welding is possible, but the

Table 3-3
Typical compositions of ferritic stainless steels.

AISI type	Composition[1] (%)			Other[2]
	Carbon	Chromium	Manganese	
405	0.08	11.5–14.5	1.0	0.1–0.3 aluminum
430	0.12	14.0–18.0	1.0	—
430F	0.12	14.0–18.0	1.25	0.060 phosphorus, 0.15 sulfur (min.), 0.60 molybdenum (opt.)
430FSe	0.12	14.0–18.0	1.25	0.060 phosphorus, 0.060 sulfur, 0.15 selenium (min.)
442	0.20	18.0–23.0	1.0	—
446	0.20	23.0–27.0	1.5	0.25 nitrogen

The Lincoln Electric Co.

[1] Single values denote maximum percentage unless otherwise noted.
[2] Unless otherwise noted, other elements of all alloys listed include maximum contents of 1.0% silicon, 0.40% phosphorus, and 0.030% sulfur. Balance is iron.

welds have low toughness and ductility, which can be improved somewhat by heat treatment. These steels may be buffed to a high finish resembling chromium plate.

For the most part, ferritic stainless steels are used for automotive trim, applications involving nitric acid, high temperature service requiring resistance to scaling, and uses which call for low thermal expansion.

Series 200 and 300 (Austenitic). The chromium content of the austenitic groups ranges from 16 to 26 percent, the nickel from 3.5 to 22 percent, and the carbon from 0.15 to 0.08 percent, Table 3-4. These steels are more numerous and more often used than steels of the 400 series. They differ widely from the chromium alloys due principally to their stable structure at low temperatures. They offer a low yield point with high ultimate tensile strength at room temperatures, a combination which makes for ductility. They are not hardenable by heat treatment, but they harden when cold worked to a degree varying with each type.

Austenitic stainless steels provide the maximum resistance to corrosion, and they are well-suited to standard fabrication. They have the ductility required for severe deep drawing and forming. They are easily welded. By controlling the chromium-nickel ratio and degree of cold reduction, a material with high tensile strength is produced which is especially suitable for lightweight welded structures.

At high temperatures, the chromium-nickel types have good oxidation resistance and high rupture and creep-strength values. They are very satisfactory for high temperature equipment because of their relatively high coefficient of thermal expansion.

Table 3-4
Typical compositions of austenitic stainless steels.

AISI type	Composition[1] (%)			
	Carbon	Chromium	Nickel	Other[2]
201	0.15	16.0–18.0	3.5–5.5	0.25 nitrogen, 5.5–7.5 manganese, 0.060 phosphorus
202	0.15	17.0–19.0	4.0–6.0	0.25 nitrogen, 7.5–10.0 manganese, 0.060 phosphorus
301	0.15	16.0–18.0	6.0–8.0	—
302	0.15	17.0–19.0	8.0–10.0	—
302B	0.15	17.0–19.0	8.0–10.0	2.0–3.0 silicon
303	0.15	17.0–19.0	8.0–10.0	0.20 phosphorus, 0.15 sulfur (min.), 0.60 molybdenum (opt.)
303Se	0.15	17.0–19.0	8.0–10.0	0.20 phosphorus, 0.06 sulfur, 0.15 selenium (min.)
304	0.08	18.0–20.0	8.0–12.0	—
304L	0.03	18.0–20.0	8.0–12.0	—
305	0.12	17.0–19.0	10.0–13.0	—
308	0.08	19.0–21.0	10.0–12.0	—
309	0.20	22.0–24.0	12.0–15.0	—
309S	0.08	22.0–24.0	12.0–15.0	—
310	0.25	24.0–26.0	19.0–22.0	1.5 silicon
310S	0.08	24.0–26.0	19.0–22.0	1.5 silicon
314	0.25	23.0–26.0	19.0–22.0	1.5–3.0 silicon
316	0.08	16.0–18.0	10.0–14.0	2.0–3.0 molybdenum
316L	0.03	16.0–18.0	10.0–14.0	2.0–3.0 molybdenum
317	0.08	18.0–20.0	11.0–15.0	3.0–4.0 molybdenum
321	0.08	17.0–19.0	9.0–12.0	titanium (5 × % carbon min.)
347	0.08	17.0–19.0	9.0–13.0	Columbium + tantalum (10 × % carbon min.)
348	0.08	17.0–19.0	9.0–13.0	Columbium + tantalum (10 × % carbon min., but 0.10 tantalum max.), 0.20 cobalt

The Lincoln Electric Co.

[1]Single values denote maximum percentage unless otherwise noted.
[2]Unless otherwise noted, other elements of all alloys listed include maximum contents of 2.0% manganese, 1.0% silicon, 0.045% phosphorus, and 0.030% sulfur. Balance is iron.

TOOL STEELS. Tool steels are either carbon or alloy steels capable of being hardened and tempered. They are produced primarily for machine tools which cut and shape articles used in all types of manufacturing operations. Tool steels vary in chemical composition depending upon the end use. They range from plain carbon types with no appreciable alloying elements to high-speed cutting types containing as much as 45 percent of alloying elements.

There are many different types of tool steel including high speed, hot work, cold work, shock-resisting, mold, special purpose, and water-hardening tool steels. They have a carbon range from 0.80 to 1.50 percent carbon and may also contain molybdenum, tungsten, and chromium.

Tool steels are usually melted in electric furnaces, in comparatively small batches, to meet special requirements. They are produced in the form of hot and cold finished bars, special shapes, forgings, hollow bar, wire, drill rod, plate, sheets, strip, tool bits, and castings.

Tool steels may be used for certain hand tools or mechanical fixtures for cutting, shaping, forming and blanking materials at normal or elevated temperatures. They are also used for other applications when wear resistance is important.

Tool steels are rarely welded and must be preheated to do so. After-treatment is also necessary. Tool steel is most often welded to resurface cutting tools and dies. Special hard-surfacing electrodes are required for this work, depending upon the type of deposite required. (See Chapter 7, pages 219–222).

SAE/AISI STEEL NUMBERING SYSTEM

The various types of steels are identified by a numbering system developed by the Society of Automotive Engineers (SAE) and the American Iron and Steel Institute (AISI). It is based on a chemical analysis of the steel. This numbering system makes it possible to use numerals on shop drawings and blueprints which indicate the type of steel to be used in fabrication.

In the case of the simple alloy steels, the second digit generally indicates the approximate percentage of the predominant alloying element in the steel. Usually the last two or three digits indicate the average carbon content in *points,* or hundredths of 1 percent. Thus the digit *2* in *2340* identifies a nickel steel. The digit *3* denotes approximately 3 percent nickel (3.25 to 3.75), and *40* indicates 0.40 percent carbon (0.35 to 0.45). The digit *7* in *71360* indicates a tungsten steel of about 13 percent tungsten (12 to 15) and 0.60 percent carbon (0.50 to 0.70).

In order to avoid confusion in some instances, it has been necessary to depart from this system of identifying the approximate alloy composition of a steel by varying the second and third digits of the number. An instance of such departure is the steel numbers selected for several of the corrosion- and heat-resisting alloys.

The prefix *X* is used to denote variations in the range of manganese, sulfur, and chromium.

The prefix *T* is used with the high manganese steels (1300 series) to avoid confusion with steels of a somewhat different manganese range that have been identified by the same numerals without the prefix. For example, 1340 is a free-cutting steel with 1.35 to 1.65 percent manganese. T1340 is a high manganese steel with 1.60 to 1.90 percent manganese.

In general, the AISI code system is like that of the SAE code system except that it has letter prefixes added to the number designations. These letter designations include:

A. Open hearth alloy steel
B. Acid Bessemer carbon steel
C. Basic open hearth carbon steel
D. Acid open hearth carbon steel

Table 3-5
SAE/AISI classification system. Note that the prefixes and numbers are in the first and second places. The xs in the table represent the numbers indicating alloy percentages. The range of these percentages for each classification is given in tables 3-6 through 3-16.

Type of steel	Numerals
Carbon steels	1xxx
Plain carbon	10xx
Free cutting, (screw stock)	11xx
Free cutting, manganese	X13xx
High manganese	T13xx
Nickel steels	2xxx
0.50 percent nickel	20xx
1.50 percent nickel	21xx
3.50 percent nickel	23xx
5.00 percent nickel	25xx
Nickel chromium steels	3xxx
1.25 percent nickel, 0.60 percent chromium	31xx
1.75 percent nickel, 1.00 percent chromium	32xx
3.50 percent nickel, 1.50 percent chromium	33xx
3.00 percent nickel, 0.80 percent chromium	34xx
Corrosion- and heat-resisting steels	30xxx
Molybdenum steels	4xxx
Chromium	41xx
Chromium nickel	43xx
Nickel	46xx and 48xx
Chromium steels	5xxx
Low chromium	51xx
Medium chromium	52xxx
Corrosion- and heat-resisting	51xxx
Chromium vanadium steels	6xxx
Tungsten steels	7xxx and 7xxxx
Silicon manganese steels	9xxx

E. Electric furnace steel

The number designations for the various types of SAE/AISI steels are given in Table 3-5. The specific classification numbers and the alloy amounts they denote are given in Tables 3-6 through 3-16.

Consult Table 3-17 (Page 104) which gives the mechanical properties of various ferrous metals. Note that in the case of steel, the tensile strength and hardness increases, and the ductility decreases as the carbon content increases. (Carbon is indicated by last two digits in the specification number.) Note too that in the case of cast irons, the alloying element is the important factor that increases tensile strength in the metal.

Refer to Table 3-18 to become familiar with the various uses of the different grades of steel.

TYPES OF CAST IRON

Cast iron is an iron-based material containing 91 to 94 percent iron and such other elements as carbon (2.0 to 4.0 percent), silicon (0.4 to 2.8 percent), manganese (0.25 to 1.25 percent), sulfur (0.2 percent maximum), and phosphorus (0.6 percent maximum), Table 3-19. Alloyed cast irons are produced by adding chromium, copper, nickel, and molybdenum. One of the differences between cast iron and steel is in the amount of carbon present. Most cast irons contain from 2.5 to 3.5 percent carbon.

Cast iron cannot be formed by forging, rolling, drawing, bending, or spinning because of its low ductility and lack of malleability. Gray cast iron produces castings that have low ductility and low tensile strength. The material fractures readily when subjected to bending or pulling stresses, successive shocks, or sudden temperature changes.

While the name "cast iron" refers to a wide variety of materials, the four major classes of cast iron generally used today are gray iron, white iron, nodular iron, and malleable iron. Gray cast iron and malleable iron are the most common types used commercially.

Gray Iron

Gray cast iron may be fusion welded or braze welded without difficulty if preheating before welding and cooling after welding are controlled. It is low in ductility and has moderate tensile strength and high compression strength. Corrosion resistance and tensile strength can be improved by adding nickel, copper, and chromium as alloying materials. Gray cast iron has high machinability.

White Iron

White cast iron is produced through a process of rapid cooling which causes the carbon to combine with the iron. There is no free carbon as in gray cast iron. This causes white cast iron to be hard, brittle, and very difficult to machine except with special cutting tools. It is so difficult to weld that it is considered unweldable. White cast iron is not generally used for castings. It is the first step in the making of malleable iron. White iron has a fine grain structure and a silvery white appearance when fractured.

Malleable Iron

Malleable iron forms when white cast iron has been heat treated by a long annealing process that changes the combined iron and carbon structure

Table 3-6
Carbon steels.

SAE no.	Carbon range	Manganese range	Phosphorus, max.	Sulfur, max.
1010	0.05–0.15	0.30–0.60	0.045	0.055
1015	0.10–0.20	0.30–0.60	0.045	0.055
X1015	0.10–0.20	0.70–1.00	0.045	0.055
1020	0.15–0.25	0.30–0.60	0.045	0.055
X1020	0.15–0.25	0.70–1.00	0.045	0.055
1025	0.20–0.30	0.30–0.60	0.045	0.055
X1025	0.20–0.30	0.70–1.00	0.045	0.055
1030	0.25–0.35	0.60–0.90	0.045	0.055
1035	0.30–0.40	0.60–0.90	0.045	0.055
1040	0.35–0.45	0.60–0.90	0.045	0.055
1045	0.40–0.50	0.60–0.90	0.045	0.055
1050	0.45–0.55	0.60–0.90	0.045	0.055
1055	0.50–0.60	0.60–0.90	0.040	0.055
1060	0.55–0.70	0.60–0.90	0.040	0.055
1065	0.60–0.75	0.60–0.90	0.040	0.055
X1065	0.60–0.75	0.90–1.20	0.040	0.055
1070	0.65–0.80	0.60–0.90	0.040	0.055
1075	0.70–0.85	0.60–0.90	0.040	0.055
1080	0.75–0.90	0.60–0.90	0.040	0.055
1085	0.80–0.95	0.60–0.90	0.040	0.055
1090	0.85–1.00	0.60–0.90	0.040	0.055
1095	0.90–1.05	0.25–0.50	0.040	0.055

into iron and free carbon. The tensile strength, impact strength, ductility, and toughness are higher than that of gray or white cast iron. The material may be bent and formed to a certain degree. In many respects the mechanical properties approach those of low-carbon steel. Fusion welding destroys the properties of malleable iron in the weld area. Because of fast cooling, the area reverts back to chilled cast iron and must be heat treated. Braze welding is recommended because of the relatively lower temperature (1,500 degrees F.) used. If a piece of malleable iron is broken, the fracture will show a white rim and dark center.

Nodular Iron

This material is also referred to as *ductile iron.* Amounts of magnesium and/or cerium are added to the iron when it is produced. Without these alloys, the graphite (free carbon) produces a notch effect which lowers the tensile strength, toughness, and ductility of the iron. The alloys change the shape of the graphite particles from flakes to spheroids and so reduce the notch effect. The silicon content of nodular iron is higher than that in other irons. Nodular iron approaches the tensile strength and ductility of steel. It has excellent machinability, shock resistance, thermal shock resistance, wear resistance, and rigidity.

Nodular iron is readily fusion welded with a filler rod containing nickel. Both preheating and postheating are necessary, and the weldment must be cooled slowly.

TYPES OF ALUMINUM

A new 4-digit numbering system is used to identify pure aluminum and wrought aluminum alloys, Table 3-20. The first digit indicates the major alloying group. For example IXXX identifies an aluminum that is at least 99.00 percent pure; 2XXX is an aluminum with copper as the major alloying element.

The three categories of aluminum which find the most welded applications are commercially pure aluminum, wrought aluminum alloys, and aluminum casting alloys.

- *Commercially pure wrought aluminum* (old 2S—new 1100) is

Table 3-7
Free-cutting steels.

SAE no.	Carbon range	Manganese range	Phosphorus range	Sulfur range
1112	0.08–0.16	0.60–0.90	0.09–0.13	0.10–0.20
X1112	0.08–0.16	0.60–0.90	0.09–0.13	0.20–0.30
1115	0.10–0.20	0.70–1.00	0.045 max.	0.075–0.15
X1314	0.10–0.20	1.00–1.30	0.045 max.	0.075–0.15
X1315	0.10–0.20	1.30–1.60	0.045 max.	0.075–0.15
X1330	0.25–0.35	1.35–1.65	0.045 max.	0.075–0.15
X1335	0.30–0.40	1.35–1.65	0.045 max.	0.075–0.15
X1340	0.35–0.45	1.35–1.65	0.045 max.	0.075–0.15

Table 3-8
Manganese steels.

SAE[1] no.	Carbon range	Manganese range	Phosphorus, max.	Sulfur, max.
T1330	0.25–0.35	1.60–1.90	0.040	0.050
T1335	0.30–0.40	1.60–1.90	0.040	0.050
T1340	0.35–0.45	1.60–1.90	0.040	0.050
T1350	0.45–0.55	1.60–1.90	0.040	0.050

[1]Silicon range of all SAE basic open hearth alloy steels shall be 0.15–0.30. For electric and acid open hearth alloy steels, the silicon content shall be 0.15 minimum.

Table 3-9
Nickel steels.

SAE[1] no.	Carbon range	Manganese range	Phosphorus, max.	Sulfur, max.	Nickel range
2315	0.10–0.20	0.30–0.60	0.040	0.050	3.25–3.75
2330	0.25–0.35	0.50–0.80	0.040	0.050	3.25–3.75
2340	0.35–0.45	0.60–0.90	0.040	0.050	3.25–3.75
2345	0.40–0.50	0.60–0.90	0.040	0.050	3.25–3.75
2515	0.10–0.20	0.30–0.60	0.040	0.050	4.75–5.25

[1]Silicon range of all SAE basic open hearth alloy steels shall be 0.15–0.30. For electric and acid open hearth alloy steels, the silicon content shall be 0.15 minimum.

Table 3-10
Nickel-chromium steels.

SAE[1] no.	Carbon range	Manganese range	Phosphorus, max.	Sulfur, max.	Nickel range	Chromium range
3115	0.10–0.20	0.30–0.60	0.040	0.050	1.00–1.50	0.45–0.75
3120	0.15–0.25	0.30–0.60	0.040	0.050	1.00–1.50	0.45–0.75
3130	0.25–0.35	0.50–0.80	0.040	0.050	1.00–1.50	0.45–0.75
3135	0.30–0.40	0.50–0.80	0.040	0.050	1.00–1.50	0.45–0.75
3140	0.35–0.45	0.60–0.90	0.040	0.050	1.00–1.50	0.45–0.75
X3140	0.35–0.45	0.60–0.90	0.040	0.050	1.00–1.50	0.60–0.90
3145	0.40–0.50	0.60–0.90	0.040	0.050	1.00–1.50	0.45–0.75
3150	0.45–0.55	0.60–0.90	0.040	0.050	1.00–1.50	0.45–0.75
3215	0.10–0.20	0.30–0.60	0.040	0.050	1.50–2.00	0.90–1.25
3220	0.15–0.25	0.30–0.60	0.040	0.050	1.50–2.00	0.90–1.25
3240	0.35–0.45	0.30–0.60	0.040	0.050	1.50–2.00	0.90–1.25
3245	0.40–0.50	0.30–0.60	0.040	0.050	1.50–2.00	0.90–1.25
3250	0.45–0.55	0.30–0.60	0.040	0.050	1.50–2.00	0.90–1.25
3312	max. 0.17	0.30–0.60	0.040	0.050	3.25–3.75	1.25–1.75
3415	0.10–0.20	0.30–0.60	0.040	0.050	2.75–3.25	0.60–0.95

[1]Silicon range of all SAE basic open hearth alloy steels shall be 0.15–0.30. For electric and acid open hearth alloy steels, the silicon content shall be 0.15 minimum.

Table 3-11
Molybdenum steels.

SAE[1] no.	Carbon range	Manganese range	Phosphorus, max.	Sulfur, max.	Chromium range	Nickel range	Molybdenum range
X4130	0.25–0.35	0.40–0.60	0.040	0.050	0.80–1.10	. . .	0.15–0.25
4140	0.35–0.45	0.60–0.90	0.040	0.050	0.80–1.10	. . .	0.15–0.25
4150	0.45–0.55	0.60–0.90	0.040	0.050	0.80–1.10	. . .	0.15–0.25
4320	0.15–0.25	0.40–0.70	0.040	0.050	0.30–0.60	1.65–2.00	0.20–0.30
X4340	0.35–0.45	0.50–0.80	0.040	0.050	0.50–0.80	1.65–2.00	0.20–0.30
4615	0.10–0.20	0.40–0.70	0.040	0.050	. . .	1.65–2.00	0.20–0.30
4620	0.15–0.25	0.40–0.70	0.040	0.050	. . .	1.65–2.00	0.20–0.30
4640	0.35–0.45	0.50–0.80	0.040	0.050	. . .	1.65–2.00	0.20–0.30
4815	0.10–0.20	0.40–0.60	0.040	0.050	. . .	3.25–3.75	0.20–0.30
4820	0.15–0.25	0.40–0.60	0.040	0.050	. . .	3.25–3.75	0.20–0.30

[1]Silicon range of all SAE basic open hearth alloy steels shall be 0.15–0.30. For electric and acid open hearth alloy steels, the silicon content shall be 0.15 minimum.

Table 3-12
Chromium steels.

SAE[1] no.	Carbon range	Manganese range	Phosphorus, max.	Sulfur, max.	Chromium range
5120	0.15–0.25	0.30–0.60	0.040	0.050	0.60–0.90
5140	0.35–0.45	0.60–0.90	0.040	0.050	0.80–1.10
5150	0.45–0.55	0.60–0.90	0.040	0.050	0.80–1.10
52100	0.95–1.10	0.20–0.50	0.030	0.035	1.20–1.50

[1]Silicon range of all SAE basic open hearth alloy steels shall be 0.15–0.30. For electric and acid open hearth alloy steels, the silicon content shall be 0.15 minimum.

99 percent pure aluminum with just a little iron and silicon added. It is easily welded, and the welds have strengths equal to the material being welded.

- *Wrought aluminum-manganese alloy* (old 3S—new 3003) contains about 1.2 percent manganese and a minimum of 97 percent aluminum. It is stronger than the 1100 type and is less ductile. This reduces its workability. It can be welded without difficulty, and the welds are strong.
- *Aluminum-silicon-magnesium-chromium alloy* (old A51S—new 6151) has silicon and magnesium as its main alloys. The welds are not as strong as the material being welded, but weld strength can be improved by heat treatment.
- *Aluminum-magnesium-chromium alloy* (old 52S—new 5052) is strong and highly resistant to corrosion. Good ductility permits the material to be worked. Cold working will produce hardness.
- *Aluminum-magnesium-silicon alloy* (old 53S—new 6053) is readily welded and can be heat treated.

EFFECTS OF WELDING ON METAL

Expansion and Contraction

All materials when loaded or stressed will deform, shrink, or stretch. Stress may also be caused by the welding process. During welding, the heat of the gas flame or electric arc causes the metal to expand. This causes a strain on the article being welded. The metal expands, but it is not free to move because of other welds, tacking, jigs, and the design of the weldment. The metal contracts when it cools. This combination of expansion due to heat, contraction due to cooling, and the conditions of restraint present in every weldment causes stresses to build up in the weldment during fabrication.

Distortion and Stress in Welding

There are two major aspects of contraction in all types of welding, namely, *distortion* and *stress*. *Distortion,* also called *shrinkage,* usually means the overall motion of parts being welded from the position occupied before welding to that occupied after welding. *Stress,* on the other hand, is a force that will cause distortion later unless it is relieved. For example, the *distortion* of a weld made between two parts, held everywhere with absolute rigidity during welding, is probably zero, but the *stresses* due to shrinkage may be great, Fig. 3-39 (Page 106).

Temporary distortion and stress occur while the welding is in progress. Stresses that remain after the welded members have cooled to normal temperature are those that affect the weldment. These are referred to as *residual stresses.* These stresses must be relieved or they will cause cracking or fracture in the weld and/or the plate. The shrinkage of a completely stress-relieved weld between two parts free to move during welding may be large, whereas the residual stresses may be small. The joint in Fig. 3-39 may be high in stresses, and yet there is no distortion, whereas that in Fig. 3-40 shows marked distortion and may contain very little stress.

Stress and distortion are affected by the physical properties of the base metal, the welding process, and the operator's welding technique. The remainder of this chapter will be devoted to the causes and control

Table 3-13

Corrosion- and heat-resisting alloys.

SAE[1] no.	Carbon, max.	Manganese, max.	Silicon, max.	Phosphorus, max.	Sulfur, max.	Chromium range	Nickel range
30905	0.08	0.20–0.70	0.75	0.030	0.030	17.00–20.00	8.00–10.00
30915	0.09–0.20	0.20–0.70	0.75	0.030	0.030	17.00–20.00	8.00–10.00
51210	0.12	0.60	0.50	0.030	0.030	11.50–13.00	...
X51410	0.12	0.60	0.50	0.030	0.15–0.50	13.00–15.00	...
51335	0.25–0.40	0.60	0.50	0.030	0.030	12.00–14.00	...
51510	0.12	0.60	0.50	0.030	0.030	14.00–16.00	...
51710	0.12	0.60	0.50	0.030	0.030	16.00–18.00	...

[1]Silicon range of all SAE basic open hearth alloy steels shall be 0.15–0.30. For electric and acid open hearth alloy steels, the silicon content shall be 0.15 minimum.

Table 3-14

Chromium-vanadium steels.

SAE[1] no.	Carbon range	Manganese range	Phosphorus, max.	Sulfur, max.	Chromium range	Vanadium	
						Min.	Desired
6135	0.30–0.40	0.60–0.90	0.040	0.050	0.80–1.10	0.15	0.18
6150	0.45–0.55	0.60–0.90	0.040	0.050	0.80–1.10	1.15	0.18
6195	0.90–1.05	0.20–0.45	0.030	0.035	0.80–1.10	0.15	0.18

[1]Silicon range of all SAE basic open hearth alloy steels shall be 0.15–0.30. For electric and acid open hearth alloy steels, the silicon content shall be 0.15 minimum.

Table 3-15

Tungsten steels.

SAE[1] no.	Carbon range	Manganese, max.	Phosphorus, max.	Sulfur, max.	Chromium range	Tungsten range
71360	0.50–0.70	0.30	0.035	0.040	3.00–4.00	12.00–15.00
71660	0.50–0.70	0.30	0.035	0.040	3.00–4.00	15.00–18.00
7260	0.50–0.70	0.30	0.035	0.040	0.50–1.00	1.50–2.00

[1]Silicon range of all SAE basic open hearth alloy steels shall be 0.15–0.30. For electric and acid open hearth alloy steels, the silicon content shall be 0.15 minimum.

Table 3-16

Silicon-manganese steels.

SAE[1] no.	Carbon range	Manganese range	Phosphorus, max.	Sulfur, max.	Silicon range
9255	0.50–0.60	0.60–0.90	0.040	0.050	1.80–2.20
9260	0.55–0.65	0.60–0.90	0.040	0.050	1.80–2.20

[1]Silicon range of all SAE basic open hearth alloy steels shall be 0.15–0.30. For electric and acid open hearth alloy steels, the silicon content shall be 0.15 minimum.

Table 3-17

Mechanical properties and chemical composition of various ferrous metals.

Material	Spec.	Chemical analysis			Mechanical properties			
		Carbon	Others		Yield strength	Tensile strength	Elongation % in 2 in.	Brinell hardness
Cast iron, gray, grade 20	ASTM A48-56	3.00–4.00			. . .	20,000	. . .	163
gray grade 30	ASTM A48-56	3.00–4.00			. . .	30,000	. . .	180
nickel		2.00–3.50	Nickel 0.25–0.50		. . .	40,000	. . .	310
chrome–nickel		2.00–3.50	Nickel 1.00–3.00	Chromium 0.50–1.00	. . .	53,000	. . .	510
white		2.00–4.00	Silicon 0.80–1.50		. . .	46,000	. . .	420
malleable	ASTM A47-52	1.75–2.30	Silicon 0.85–1.20		35,000	53,000	18	140
Iron, wrought, plates	ASTM A42-55	0.08	Silicon 0.15	Slag 1.20	26,000	46,000	35	105
Iron, wrought, forgings	ASTM A75-55	0.01–0.05	Iron 99.45–99.80		25,000	44,000	30	100
Steel, cast, low carbon		0.11	Manganese 0.60	Silicon 0.40	35,000	60,000	22	120
Steel, cast, medium carbon		0.25	Manganese 0.68	Silicon 0.32	44,000	72,000	18	140
Steel, cast, high carbon		0.50			40,000	80,000	17	182
Steel, rolled, carbon	SAE 1010	0.05–0.15			28,000	56,000	35	110
Steel, rolled, carbon	SAE 1015	0.15–0.25			30,000	60,000	26	120
Steel, rolled, carbon	SAE 1025	0.20–0.30			33,000	67,000	25	135
Steel, rolled, carbon	SAE 1035	0.30–0.40			52,000	87,000	24	175
Steel, rolled, carbon	SAE 1045	0.40–0.50			58,000	97,000	22	200
Steel, rolled, carbon	SAE 1050	0.45–0.55			60,000	102,000	20	207
Steel, rolled, carbon	SAE 1095	0.90–1.05			100,000	150,000	15	300
Steel, rolled, nickel	SAE 2315	0.10–0.20	Nickel 3.25–3.75		90,000	125,000	21	230
Steel, rolled, nickel–chromium	SAE 3240	0.35–0.45	Nickel 1.50–2.00	Chromium 0.90–1.25	113,000	136,000	21	280
Steel, rolled, molybdenum	SAE 4130	0.25–0.35	Chromium 0.50–0.80	Molybdenum 0.15–0.25	115,000	139,000	18	280
Steel, rolled, chromium	SAE 5140	0.35–0.45	Chromium 0.80–1.10		128,000	150,000	19	300
Steel, rolled, chromium–vanadium	SAE 6130	0.25–0.35	Chromium 0.80–1.10	Vanadium 0.15–0.18	125,000	150,000	18	310
Steel, rolled, silicon–manganese	SAE 9260	0.55–0.65	Manganese 0.60–0.90	Silicon 1.80–2.20	180,000	200,000	12	390

The James F. Lincoln Arc Welding Foundation

of stress and distortion. The student should refer to Chapter 4, Basic Joints and Welds, as necessary.

PHYSICAL PROPERTIES OF METAL AND DISTORTION. Distortion is the result of heating and cooling and involves stiffness and yielding. Heat changes the physical properties of metals, and these changes have a direct effect on distortion. When the temperature of a metal increases during welding, the physical properties change as follows.

- The yield point lowers.
- The modulus of elasticity decreases.
- The coefficient of thermal expansion increases.
- The thermal conductivity decreases.
- Specific heat increases.

The differing physical properties of the various metals also affect the amount of distortion and stress which can be expected.

Yield Point. The yield point of steel is the point at which it will stretch and elongate under load even though the load is not increased. The higher the yield point of the weld and the base metal next to the weld, the greater the amount of residual stress that can act to distort the assembly. The lower the yield point, the less likely or severe the residual stress is.

Coefficient of Thermal Expansion. The coefficient of thermal expansion is the amount of expansion a metal undergoes when heated and the amount of contraction that occurs when it is cooled. If the metals we weld did not change in length when they were welded, there would be no distortion of the part being welded and no shrinkage. A high coefficient tends to increase the shrinkage of the weld metal and the base metal next to the weld, thus increasing the possibility of distortion in the weldment. If a metal also has a high thermal conductivity, thus allowing the spread of heat over a larger area, the problems of distortion are greater.

Thermal Conductivity. Thermal conductivity is a measure of the flow of heat through a metal. A metal with low thermal conductivity retards the flow of heat from the weld, thus causing a

Table 3-18
Uses for steel by carbon content.

Carbon class	Carbon range %	Typical uses
Low	0.05–0.15	Chain, nails, pipe, rivets, screws, sheets for pressing and stamping, wire
	0.15–0.30	Bars, plates, structural shapes
Medium	0.30–0.45	Axles, connecting rods, shafting
High	0.45–0.60	Crankshafts, scraper blades
	0.60–0.75	Automobile springs, anvils, bandsaws, drop hammer dies
Very high	0.75–0.90	Chisels, punches, sand tools
	0.90–1.00	Knives, shear blades, springs
	1.00–1.10	Milling cutters, dies, taps
	1.10–1.20	Lathe tools, woodworking tools
	1.20–1.30	Files, reamers
	1.30–1.40	Dies for wire drawing
	1.40–1.50	Metal cutting saws

The James F. Lincoln Arc Welding Foundation

Table 3-19
Compositions of cast irons (percentage of constituents).

	Iron	Total carbon	Silicon	Sulfur	Phosphorus	Manganese
Gray iron	Balance	2.0–4.0	1.0 min	0.2	0.6	1.0 max
Malleable iron	Balance	2.0–3.0	0.9–1.8	0.2 max	0.2 max	0.25–1.25
Nodular iron	Balance	3.2–4.1	1.8–2.8	0.03 max	0.1 max	0.80 max
White iron	Balance	2.5–4.0	0.4–1.6	0.15	0.4	0.3–0.8

The Lincoln Electric Co.

Table 3-20
Designations for aluminum alloy groups.

Major alloying element	Designation[1]
99.0% minimum aluminum and over	1xxx
Copper	2xxx
Manganese	3xxx
Silicon	4xxx
Magnesium	5xxx
Magnesium and silicon	6xxx
Zinc	7xxx
Other element	8xxx
Unused series	9xxx

American Welding Society

[1]Aluminum Association designations.

concentration of heat at the weld area. This increases the shrinkage of the weld and the plate next to it.

Modulus of Elasticity. The modulus of elasticity is a measure of the relative stiffness of a metal. If the modulus is high, the material is more likely to resist movement and distortion.

CAUSES AND CONTROL OF DISTORTION. Distortion is one of the serious problems that the welder must contend with. Very often the ability to control distortion in a weldment is the difference between a satisfactory and an unsatisfactory job. A great deal of welding engineering and welding experience has been devoted to this subject. The great advance in the welding industry is a tribute to the success of these studies.

The student welder is urged to learn as much as possible about the control of distortion. During your welding practice you are urged to experiment with the various welding methods and techniques that will be presented in this text. On the job you should be capable of making welds free of defects which have good physical properties. No operator should attempt welds that he feels he may not be able to do well enough to meet requirements.

THE TYPES OF DISTORTION.

Lengthwise Shrinkage (Longitudinal Contraction). If a weld is deposited lengthwise along a strip of steel which is not clamped or held in any way, the strip will bow upward at both ends as the weld cools, Fig. 3-41. This is due to the contraction of the weld reinforcement above the plate surface. Weld beads that are small, or that have deep penetration and are flat, do not cause as much deformation as those that are convex.

If a welding procedure can be followed that will keep the heat on both sides of the plate nearly the same, very little distortion will occur. By depositing weld beads on the opposite side of the strip, it can be brought back to its original form. Excess weld deposit is to be avoided since it adds nothing to the strength of the joint and increases the cost of welding.

Crosswise Shrinkage (Transverse Contraction). If two plates are being butt-welded and are free to move during welding, they will be drawn together at the opposite end due to the contraction of the weld metal upon cooling, Fig. 3-42. This is known as *transverse contraction.* Transverse contraction can be controlled. If the seam to be welded is short, it may be tack-welded at the opposite end, Fig. 3-43. If the seam is long, it may have to be tacked in several places, Fig. 3-44. The frequency and size of the tack welds depend upon the thickness of plate and the type of edge preparation. They are usually twice as long as the thickness of plate and spaced at intervals of 8 to 12 inches. The tacks also prevent the plates from buckling out of plane to each other.

Long seams can also be controlled by the use of clamping devices and wedges. The wedge is advanced along the seam ahead of the weld during the weld operation. Clamping devices help to keep the plates on the same plane.

Prespacing is another means of controlling expansion and contraction. The amount of spacing depends upon a number of variables such as length of seam, thickness of plate, and speed of welding. The operator will learn from experience the proper spacing for the job at hand.

Warping (Contraction of Weld Deposit). In welding beveled edges of thick plates such as single V-butt joints and U-groove joints, the plates will be pulled out of line with each other, Fig. 3-45. This is so because the opening at the top of the groove is greater than at the bottom, resulting in more weld deposit at the top and hence more contraction. The greater the number of passes, the greater will be this warping. Warping can be counteracted by setting up plates before welding so that they bow in the opposite direction, Fig. 3-46.

3-39. *Butt-welded joint that shows no distortion but may be high in stresses.*

3-40. *Butt-welded joint shows evidence of extreme distortion, but it may contain no shrinkage stresses.*

This is not always possible, however, and the use of various clamping devices may be necessary.

Angular Distortion. Fillet welds contain both longitudinal and transverse stresses, Fig. 3-47. When a fillet weld is used in a T-joint, it will pull the vertical member of the joint toward the side that is welded. It will also bow somewhat because of the longitudinal stresses. Here again the greater the size of the weld, the greater the number of passes, and the slower the speed of welding, the greater the amount of distortion. Figure 3-48 illustrates this angular distortion.

EFFECT ON BUTT WELDS. The following factors affect the shrinkage *perpendicular* to the weld.

- Cross-sectional area of weld for a given thickness of plate: the larger the cross section, the greater the shrinkage.
- The angle of the scarf is not nearly as important as the *free distance* (spacing) between roots in causing distortion perpendicular to the weld.
- Total heat input: the greater the total heat input, the greater the amount of distortion.
- *Rate of heating:* other factors being equal, a greater rate of heat input results in less distortion.
- *Wandering* and *step-back* procedures in welding lessen the amount of distortion. These procedures will be explained in the following pages.
- *Peening,* properly used, is effective in reducing the amount of distortion, but it is not recommended except under very careful control and supervision. Peening will be discussed under Control of Residual Stress, page 112.

EFFECTS OF ANGULAR DISTORTION. The following factors affect angular distortion.

- The angular distortion of V-joints free to move increases with the number of layers.
- Angular distortion is greatest in V-welds, next in U-welds, and least in double-V and double-U welds.
- Angular distortion may be controlled by peening every layer to a suitable extent.
- Angular distortion may be practically eliminated in double-V and double-U welds by welding

3-41. *Dotted lines indicate position before welding. Solid lines indicate position after welding.*

3-42. *Position of plates before welding is indicated by dotted lines. Solid lines show position after welding.*

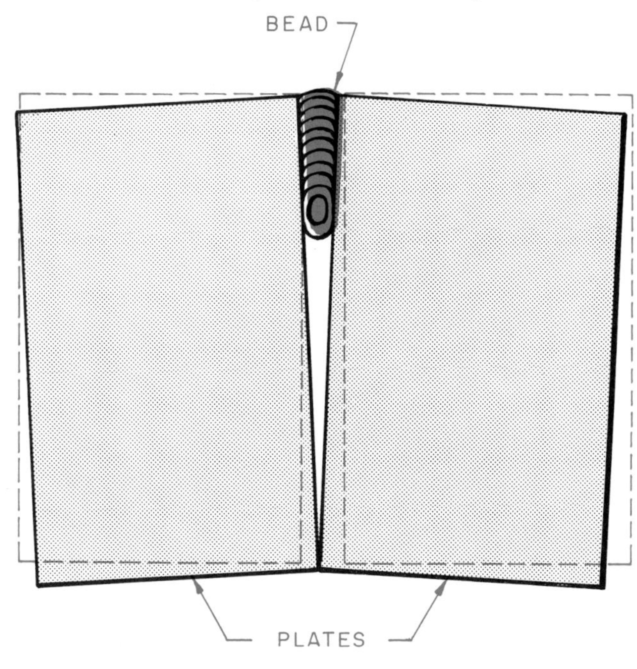

3-43. *Short seam, tack welded.*

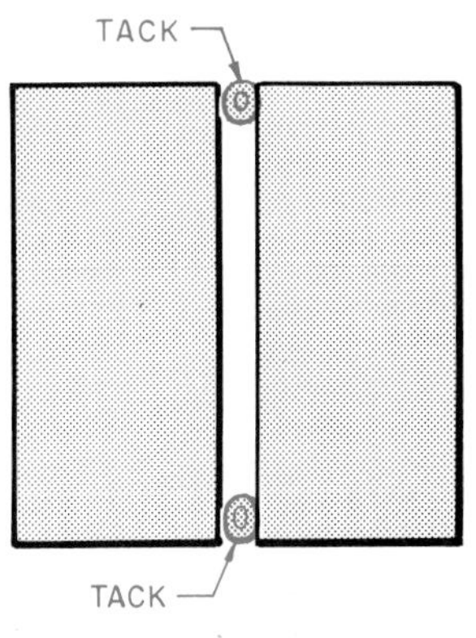

alternately on both sides in multilayer welding.

- The time of welding and size of electrode have an important bearing on angular distortion.
- Rate of heating: other factors being equal, a greater rate of heat input results in less angular distortion.

EFFECT ON FILLET WELDS. Distortion affects fillet welds in the following ways.

- Shrinkage increases with the size of the weld and decreases as the rate of heat input increases.
- If the weld is intermittent, shrinkage is proportional to the length of the weld.
- Shrinkage may be decreased materially by choosing suitable sequences and procedures of welding and peening.
- Transverse shrinkage is less for a lap joint than for a V-butt joint. Angular distortion is reduced by preheating and by suitably arranging the sequence of welding and staggering.

Prevention of Distortion Before Welding

DESIGN. Joints should be such that they require a minimum amount of filler metal, and they should be arranged so that they

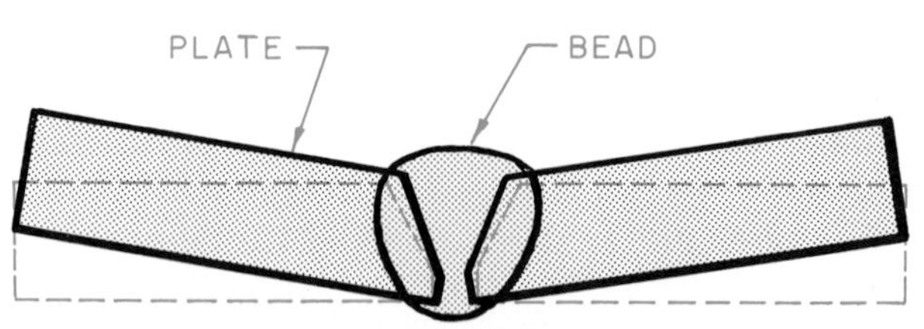

3-45. *Warpage after welding. Dotted lines indicate original position. Position after welding is indicated by solid lines.*

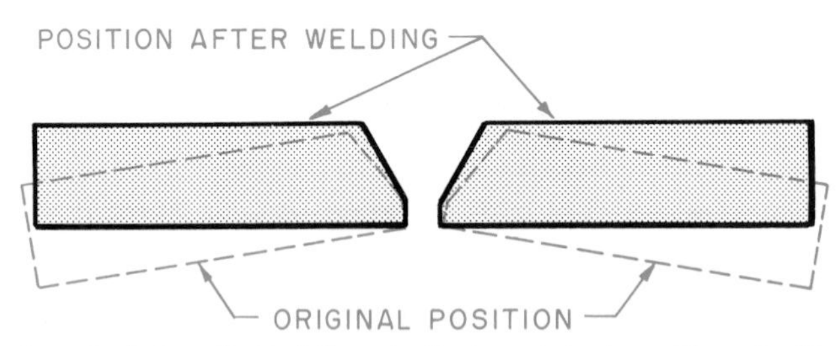

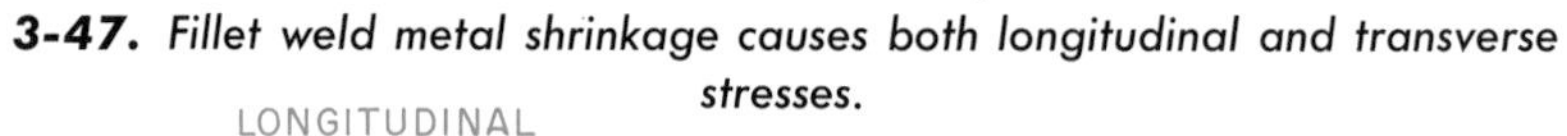

3-46. *Preset plates. Dotted lines indicate original position. Position after welding is indicated by solid lines.*

3-44. *Long seam, tack welded.*

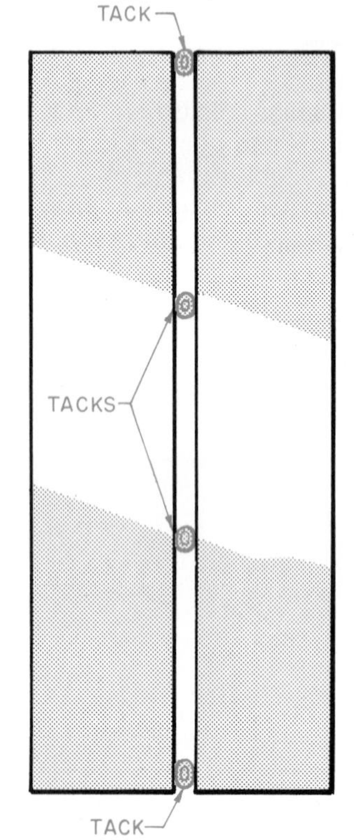

3-47. *Fillet weld metal shrinkage causes both longitudinal and transverse stresses.*

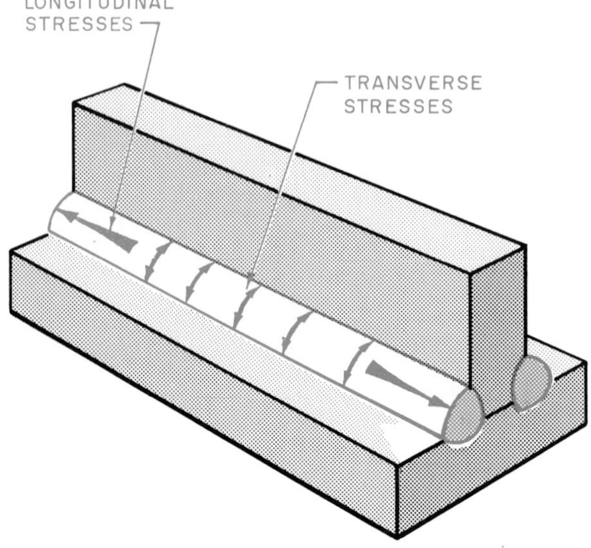

balance each other, avoiding localized areas of extensive shrinkage. For example, when welding heavy materials the double-V butt joint should be used instead of the single-V joint whenever possible.

SELECTION OF PROCESS AND EQUIPMENT. Achieving higher welding speeds through the use of powdered-iron manual electrodes or semi-automatic and full automatic submerged-arc welding processes reduce the amount of base metal that is affected by the heat of the arc, thereby decreasing the amount of distortion.

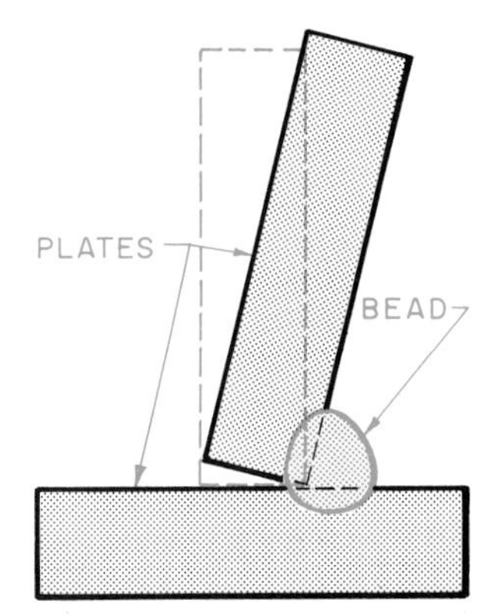

3-48. *How T-joints may be affected by welding. Dotted lines indicate original position. Position after welding is indicated by solid lines.*

PREBENDING. Shrinkage forces can be put to work by prebending the parts to be welded, Fig. 3-49. The plates are bent in a direction opposite to the side being welded. The shrinkage of the weld metal is restrained during welding by the clamps. Clamping reduces warping and is more effective when the welded members are allowed to cool in the clamps. However, clamping does not entirely eliminate warping. When the clamps are removed after cooling, the plates spring back so that they are pulled into alignment.

SPACING OF PARTS. Another method is to space parts out of position before welding, Fig. 3-50. Experience is necessary to determine the exact amount to allow for a given job. The arms will be pulled back to the proper spacing by the shrinkage forces of the welding. Figure 3-51 shows the vertical plate of a T-joint out of alignment. This is done before welding. When the completed weld shrinks, it will pull the vertical plate into the correct position. Before welding, the operator should make sure that all joints are fitted properly and do not have too great a gap. If, however, bad fitup does occur, *dutchmen* or *metal fillers* should be used with caution for they may produce serious consequences.

JIGS AND FIXTURES. Jigs and fixtures prevent warping by holding the weldment in a fixed position to reduce movement. They are widely used in production welding since, in addition to reducing warping, they permit positioning of the weldment and its parts, thus materially increasing the speed of welding.

Strong Backs. Strong backs are temporary stiffeners for the purpose of increasing the resistance to distortion. They are re-

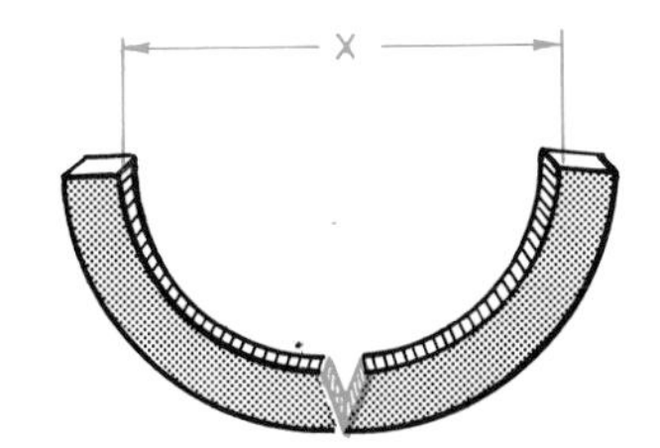

3-50. *Parts are spread before welding to reduce distortion.*

3-49. *Parts are prebent and restrained.*

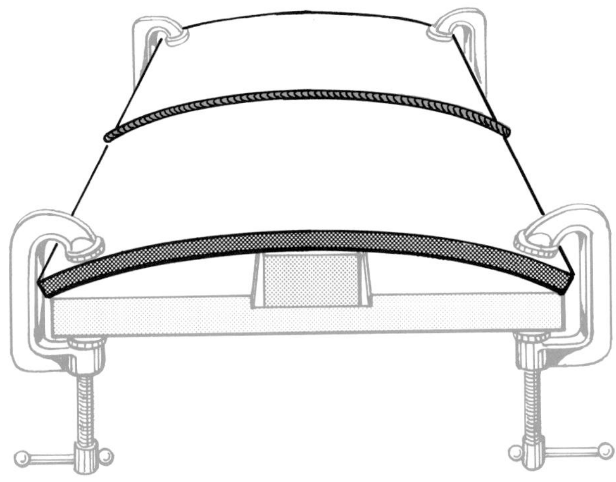

3-51. *The plate is set up out-of-square away from the weld. The weld shrinkage will bring it back to square.*

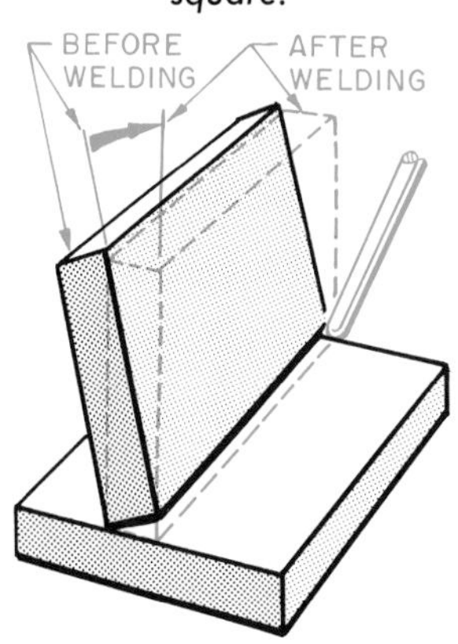

moved after the welding is completed and cooled. This method is used in shipbuilding and for other large structures.

Distortion Control During Welding

Distortion may be reduced by using a sequence of welding known as *wandering* that provides for making welds at different points of the weldment. The shrinkage set up by one weld is counteracted by the shrinkage set up by another. This is accomplished by employing either the method shown in Fig. 3-52, called *chain intermittent fillet welds*, or that shown in Fig. 3-53, called *staggered intermittent fillet welds*. If the job requires a continuous weld, the unwelded spaces may be welded in the same order.

If it is found advisable to distribute the stresses or to help prevent their accumulation, the *step-back* method of welding may be employed. This method consists of breaking up the welds in short sections and depends upon welding in the proper direction. The general progression of welding is from left to right, but each bead is deposited from right to left, Fig. 3-54. Step-back welding reduces locked-up stresses and warping.

Distortion may also be held at a minimum through the use of the *skip-stop step-back* method, a combination of skip and step-back welding. This is done in the sequence shown in Fig. 3-55. The direction of welding is the same as that employed in the step-back method except that the short welds are not made in a continuous sequence. One is made at the beginning of the joint, a section is skipped, and the second weld applied near the center. Then a third weld is applied after further spacing. After the end of the seam has been reached, a return is made to the beginning, and the unwelded sections are completed.

3-52. *Chain intermittent fillet welds.*

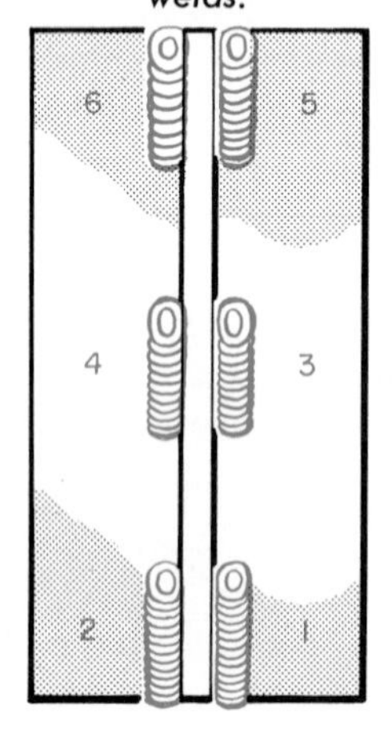

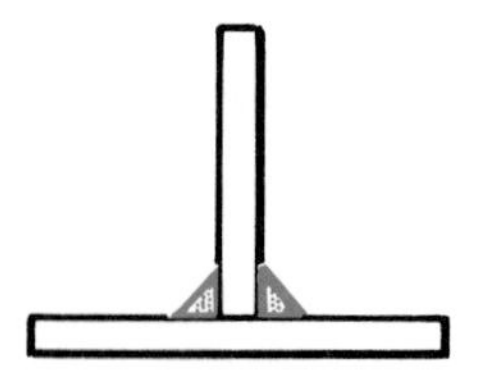

3-53. *Staggered intermittent fillet welds.*

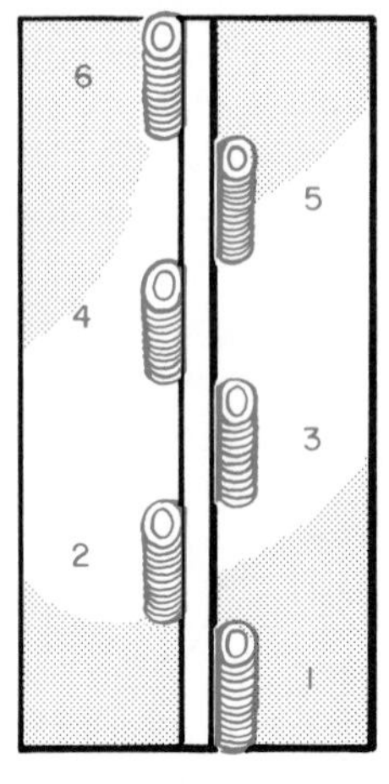

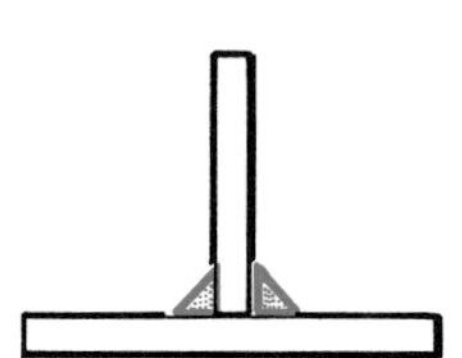

In a *balanced welding sequence* an equal number of operators weld on opposite sides of a structure at the same time, thus introducing balanced stresses. The wandering techniques given above and balanced welding all contribute to the simultaneous completion of welded connections in large fabrications. Thus distortion caused by restraint and reduction in local heating is reduced.

Correction of Distortion After Welding

If warping has occurred in a structure, the following corrective measures may be used.

- *Shrinkage.* This consists of alternate heating and cooling, frequently accompanied by hammering or mechanical working, thus shrinking excess material in a wrinkle or buckle.
- *Shrink Welding.* This is a variation of shrinkage in which the heat is applied by running beads of weld metal on the convex side of a buckled area. On cooling, the combined shrinkage of the heated base metal and the added weld metal remove the distortion. The beads of weld metal may then be ground off if a smooth surface is desired.
- *Added Stiffening.* This technique can be used only on plate. It consists of pulling the plate into line with strong backs and welding additional stiffeners to the plate to make it retain its plane. Benefit is also derived from the shrinkage in the connecting welds.

Summary

Following is a summary of the basic means that can be applied in the control of distortion.

● Some metals expand more than others. A metal with a high coefficient of expansion distorts more than one with a lower coefficient. For example, aluminum has a high coefficient, and care must be taken to keep distortion at a minimum.

● The kind of welding process has an influence on distortion. Gas welding produces more distortion than shielded metal-arc welding and the forms of automatic and semi-automatic welding. Use higher deposition rate processes. Use higher speed welding methods—iron-powder coated electrodes or mechanized welding. Use welding methods that give deeper penetration and thus reduce the amount of weld metal needed for the same strength and amount of heat input.

● Use welding positioners to achieve maximum amount of downhand welding, thus allowing the use of larger diameter electrodes or welding procedures with higher deposition rates and faster welding speeds.

● One shrinkage force can be balanced with another by prebending and presetting in a direction opposite to the movement caused by weld shrinkage. Shrinkage will pull the material back into alignment.

● Effective control can be achieved by restraining the parts forcibly through the use of clamps, fixtures, and tack welds. Use strong backs and tack welds to maintain fitup and alignment. Control fitup at every point. The welder must be careful not to overrestrain the parts. This increases the stresses during welding and the tendency for cracking. Weld toward the unrestrained part of the member.

● During fabrication the parts can be clamped or welded to a heavy fixture which can be stress relieved with the weldment, thus insuring dimensional tolerance and stability.

● Distribute the welding heat as evenly as possible through a planned welding sequence and planned weld positions. Weld the more flexible sections first so that they can be straightened, if necessary, before final assembly. Sequence subassemblies and final assemblies so that the welds being made continually balance each other around the neutral axis of the section.

● As the heat input is increased, the welding speed should be increased.

● All other things being equal, a decrease in speed and an increase in the number of passes increases warping.

● Welding from both sides reduces distortion, Fig. 3-56. Welding from both sides at the same time all but eliminates distortion.

● The direction of welding should be away from the point of restraint and toward the point of maximum freedom.

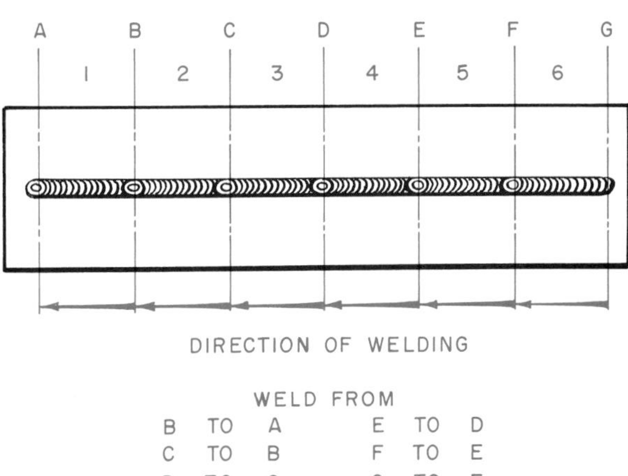

3-54. *Example of procedure and sequence of welding by the step-back method.*

3-55. *Example of procedure and sequence of welding by the skip-stop, step-back method.*

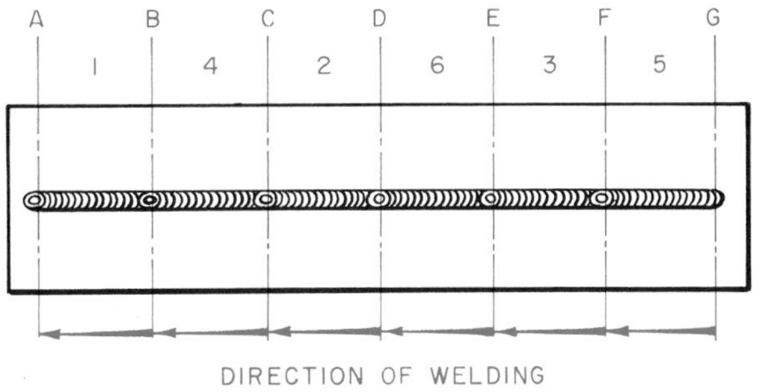

- The employment of wandering sequences of welding, such as skip welding and step-back welding, prevents a local buildup of heat and so reduces shrinkage.
- To make sure that welds will not fail at the end and carry on into the joint, the welds on the ends of members should be fixed or welded around the end.
- Don't overweld. Too much welding increases distortion. Excessive weld size and too many weld passes cause additional heat input, Fig. 3-57. A stringer bead produces less distortion than a weave bead, and a single pass is better than several passes. Use the smallest leg size permissible when fillet welding. Use fewer weld passes.
- Excessive widths of groove welds add nothing to strength, but they increase weld shrinkage as well as welding costs. The root opening, included angle, and reinforcement should be minimum, Fig. 3-58. Select joints that require little weld metal. For example, choose a double-V joint instead of a single-V joint. Weld those joints that cause the most contraction first.
- Weak welds or cracked tack welds should be chipped or melted out before proceeding with the weld.
- Peening the weld is effective. Too much peening, however, causes a loss of ductility and impact properties.

3-56. *Distortion is reduced by welding from both sides.*

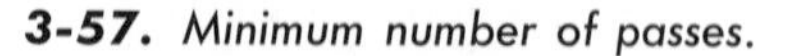

3-57. *Minimum number of passes.*

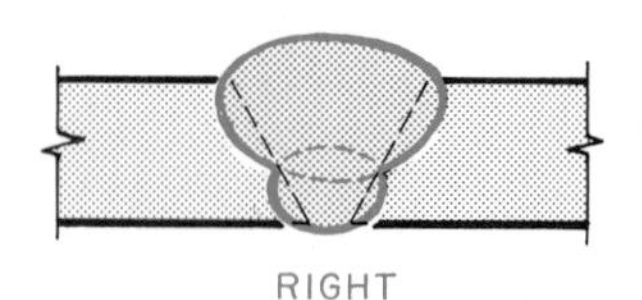

Control of Residual Stress

A welded structure develops many internal stresses which may lead to cracking or fracture. Under normal conditions, these stresses are not a threat to the structure. However, for certain kinds of code welding requirements, and on those structures where there is chance of cracking, stress relieving is necessary, Fig. 3-59. Both hot and cold processes are used.

3-58. *Correct edge preparation and good fitup.*

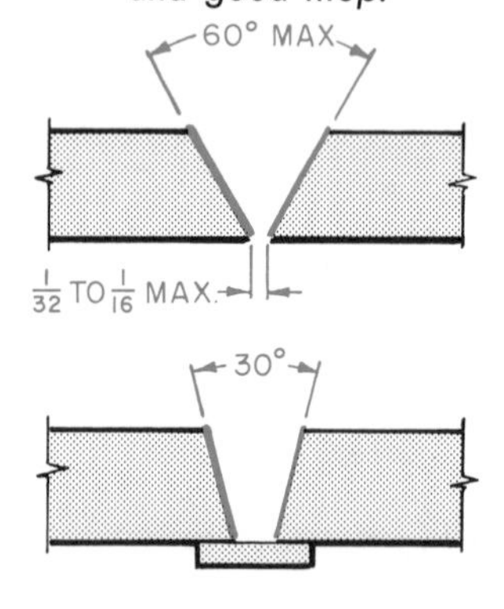

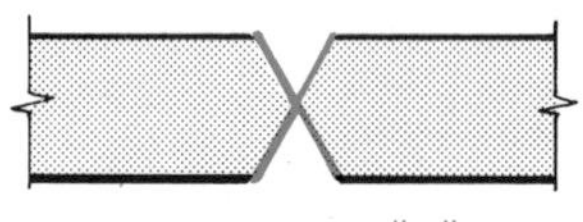

PREHEATING. It is often necessary to control or reduce the rate of expansion and contraction in a structure during the welding operation. This is done by preheating the entire structure before welding and maintaining the heat during welding. Considerable care must be taken to make sure that preheat is uniform throughout the structure. If one part of the structure is heated to a higher temperature than another, internal stresses will be set up, thus offsetting the advantages sought through preheating. After the weld is completed, the structure must be allowed to cool slowly. Preheating may be done in a furnace with an oxyacetylene or carbon flame.

POSTHEATING. The most common method of stress relieving is postheating. This kind of heat treatment must be done in a furnace capable of uniform heating under temperature control. The method of heating must not be injurious to the metal being treated. The work must be supported so that distortion of any part of the weldment is prevented. The rise in temperature must be gradual, and all parts of the weldment must be heated at a uniform rate. Mild steel is usually heated to about 1,100 to 1,200 degrees F. Other steels may require a higher temperature, depending upon the yield characteristics of the metal. Some alloy steels are brought up to a temperature of 1,600 degrees F. or higher.

When the weldment reaches the maximum temperature, it is permitted to soak. The length of time depends upon the thickness of plate and the plasticizing rate of the steel of which it is made. The rate is usually 1 hour per inch of thickness. The weldment should be permitted to soak long enough to insure relief to the thickest part.

The reduction of temperature must be gradual and at a rate that will insure approximately uniform temperatures throughout all parts. Structures of different thicknesses sometimes may require as long as 48 hours to cool.

The temperature at which the work may be withdrawn from the

3-59. *A large press brake frame during welding. Internal stresses set up by welding must be relieved before the bed is ready for machining to close tolerances.*

Niagara Machine and Tool Works

furnace depends upon the varying thicknesses and rigidity of the members. This may be as low as 200 degrees F. or as high as 600 degrees F. It is important that the air surrounding the furnace be quiet enough to insure uniform temperature.

It is not always possible to heat the entire structure, as for example, pipelines. In such cases it may be possible to relieve stress by heating only one portion of the structure at a time. It is important that the member be able to expand and contract at will. Otherwise, additional stresses will be introduced into the structure which may be greater than the original stresses being treated.

FULL ANNEALING. Annealing is superior to all other methods, but it is very difficult to handle. Work that is fully annealed must be heated up to 1,600–1,650 degrees F. This causes the formation of a very heavy scale and there is danger of collapse on some types of weldments.

COLD PEENING. In cold peening, the bead is hammered to stretch it and counteract shrinkage due to cooling. Cold peening of the weld metal causes plastic flow, thereby relieving the restraint which causes the residual stress. Effective peening requires considerable judgment. Peening is identical to cold working the steel and, if overdone, will cause it to lose ductility and become work hardened. Overpeening may result in cracks or the introduction of new residual stresses.

The weld should be peened at a temperature low enough to permit the hand to be placed upon it. The root and face layers of the weld should not be peened. A pneumatic chisel with a blunt, rounded edge is used. Hand peening cannot be controlled properly.

MECHANICAL LOADING. In mechanical loading, the base metal is stressed just at the point of yielding by applying internal pressure to pressure vessel: This procedure works well with a simple weldment. On vessels where the plate is not of uniform thickness or where there are reinforcements, however, all of the metal cannot be stretched to the same extent. It is important that only a very small yielding take place. Otherwise, there is danger of strain-hardening or embrittling the steel. Hydraulic pressure rather than air pressure is used because of the dangers in connection with air pressure if the vessel should rupture.

REDUCING STRESS THROUGH WELDING TECHNIQUE. While it is impossible to completely control the residual stresses due to the welding operation without preheating or postheating, the amount of stress can be minimized if the following welding procedures are employed.

- The product should be designed to incorporate the types of joints having the lowest residual stress.
- The degree of residual stress caused by the welding process should be considered when choosing a process.
- Plan assembly welding sequences which permit the movement of component parts during welding, which do not increase joint fixity, and in which joints of maximum fixity are welded first.
- Avoid highly localized and intersecting weld areas.
- Use electrodes that have an elongation of at least 20 percent in 2 inches.
- Peening is an effective method of reducing stresses and partly correcting distortion and warping. Root and face layers and layers more than 1/8 inch should not be peened.

REVIEW QUESTIONS

A. Steelmaking and Metalworking Processes

1. Metals are separated into two major groups. Name these groups and explain the difference between them.

2. What is iron?

3. For how long has iron been used by man for weapons and other articles?

4. What is steel?

5. Iron was first produced in America about 150 years ago. True. False.

6. Name at least five events that happened in this country to spur the growth of steelmaking.

7. Name the basic raw materials necessary for making steel.

8. Name at least four steelmaking processes.

9. Most of the steel made in the United States is made using the Bessemer process. True. False.

10. Outline briefly the Bessemer process for the making of steel.

11. Name the three types of carbon steels and give the range of carbon content for each.

12. Name at least six common elements that are added to steel to change its properties.
13. Name six metalworking processes.
14. What are ingots, billets, and blooms?

B. Properties of Metals

Supply the missing word or term in the following definitions:

1. ________________ is the capacity of a metallic substance to form a strong bond of adherence under pressure or when solidifying from a liquid state.
2. ________________ is the ease with which a metal may be melted.
3. ________________ is the ease with which a substance may be vaporized.
4. The ________________ is the temperature at which a substance passes from a solid to a liquid state.
5. The ________________ of a substance is the ability of the substance to conduct electrical current.
6. The opposition to electric current as it flows through a wire is termed the ________________ of the wire.
7. The ________________ of a substance is the ability of a substance to conduct heat.
8. The expansion of metals is called the ________________.
9. ________________ is the capacity of a metal to be compact.
10. ________________ is the weight of a volume of material compared with an equal volume of water.
11. ________________ is the opposite of density.
12. ________________ is the ability of a material to resist deformation.
13. The ability of a material to deform without breaking is its ________________.
14. Loading of a material will cause it to lose its form. The ability of the material to return to its original shape after the load has been removed is known as its ________________.
15. ________________ is the ability of a material to be permanently deformed by loading and resist fracture.
16. ________________ is the ability of a material to deform permanently without breaking and have high strength.
17. It may be said that a ________________ material lacks plasticity and toughness.
18. The ability a material possesses to deform permanently under compression without breaking or fracturing is known as the ________________ of the metal.
19. ________________ or springiness is the energy stored in a material.
20. The ability of one material to penetrate another material without fracture of either is known as ________________.
21. The resistance of a material to a maximum load applied suddenly may be defined as ________________.
22. Failure of materials under repeated or alternating stresses is known as ________________.
23. ________________ is the gradual wearing away or disintegration of a material by chemical process.
24. Brittleness in metal when hot is called ________________.

C. Expansion and Contraction

1. What happens when metal is heated? When it is cooled?
2. Two pieces to be welded should never be spaced since this increases expansion and contraction. True. False.
3. What are the two major aspects of contraction in all types of welding? Explain.
4. If a weld is deposited lengthwise along a strip of steel, what effect will it have upon the piece of steel when it cools?
5. List eight methods that may be employed to control expansion and contraction.
6. List eight methods that may be employed to control residual stress.
7. What methods may be employed to control shrinkage?
8. What methods may be employed to control warping?
9. How may warping be corrected if found in a welded structure?
10. Peening a weld is effective. True. False.
11. Name three ways to avoid overwelding.
12. Name three factors that are important in the choice of weld joints.
13. In welding from both sides, one side should contain more welding in order to counteract distortion. True. False.
14. Never weld on both sides of a joint at the same time. True. False.

15. Because of the even application of heat over the entire weld area, gas welding produces less distortion than other forms of welding. True. False.

16. All metals expand at about the same rate. True. False.

17. Preheating the part increases residual stress after welding. True. False.

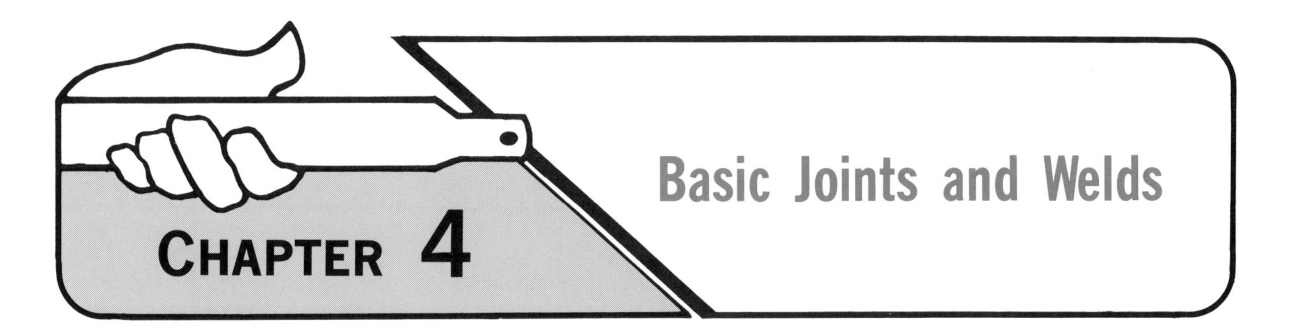

Chapter 4 Basic Joints and Welds

The fabrication of welded structures is becoming a highly competitive business. In order to survive, the shop doing this type of work must take advantage of every economy possible. Thus it adopts the latest improvements and designs, makes full use of materials, and eliminates unnecessary operations. The selection of the wrong type of joint may result not only in a great loss of time and money but may also contribute to the breakdown of the weldment in use, thus specifically damaging the reputation of the manufacturer and, in general, contributing to a distrust in welding. Only men who have practical and technical training along these lines can possibly hope for satisfactory results.

Although joint design and joint selection are the responsibility of the engineering department, this does not mean that the welding operator should not be concerned about joint design and welding procedures. Recognition of the requirements for a particular type of joint or weld will lead to work of higher quality and accuracy. It is the welder's responsibility to understand fundamental joint construction and welding procedures. A welder who has a practical understanding of the values of joint design and the characteristics of different types of welds can be of great aid and assistance to the supervisory personnel and the engineering department. Best results are obtained where this kind of cooperation is found.

TYPES OF JOINTS

There are only five basic types of joints: the edge joint, the butt joint, the corner joint, the lap joint, and the T-joint (Fig. 4-1). Some of the variations that are formed from these basic joints are illustrated in Table 4-1. The most common joints illustrated in the table will be described in terms of their use, advantages and disadvantages, joint preparation, and economy.

Open and Closed Roots

Open roots are spaces between the edges of the members to be welded. They are used to secure complete root penetration in butt joints and to secure attachment to a backing member, Fig. 4-2 (Page 120).

The term *penetration* refers to the depth to which the base metal is melted and fused with the metal of the filler rod or electrode. In those cases in which there is no root opening, some of the weld metal from the first pass is partly removed by chipping or machining. Whether a certain type of joint should be set up as an open or closed root depends upon the following factors:

- The thickness of the base metal

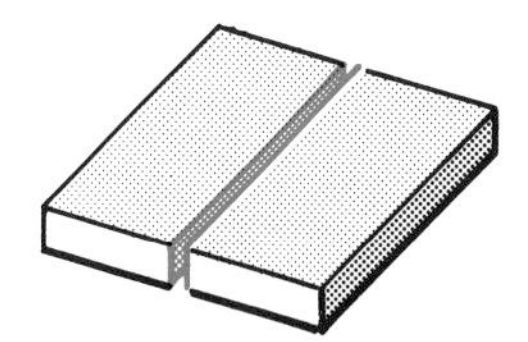

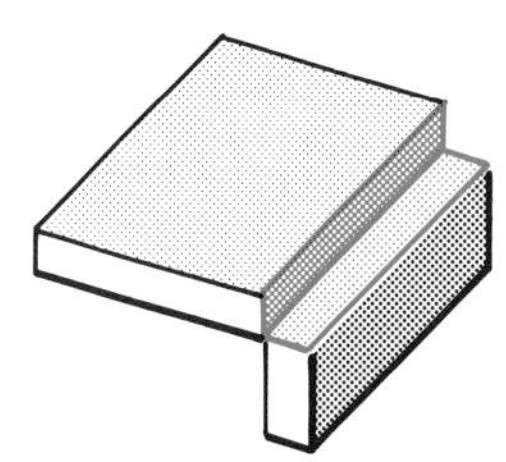

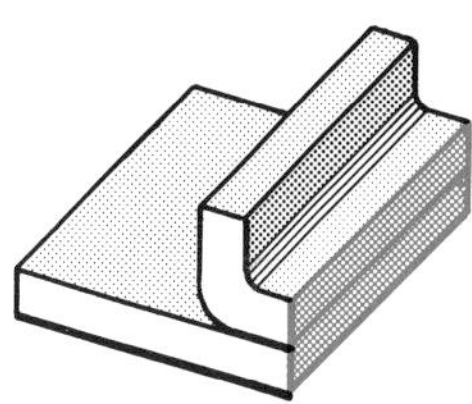

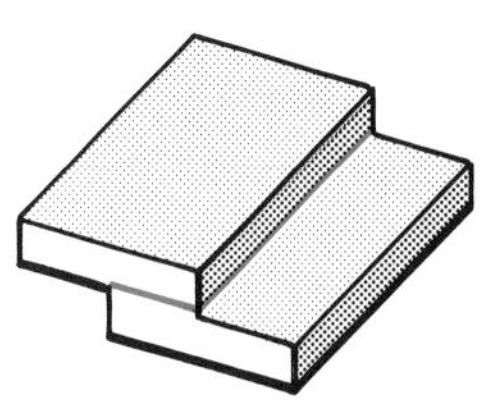

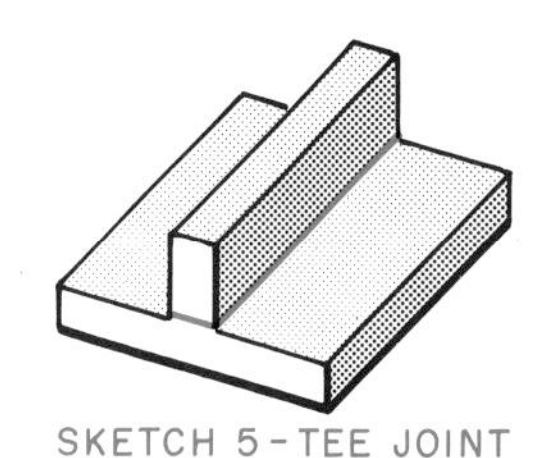

4-1. *Basic types of joints.*

- The kind of joint
- The nature of the job
- The position of welding
- The type and size of electrode
- The structural importance of the joint in the fabrication (whether it is a prime load-carrying joint)
- The physical properties required of the weld

Edge Joints

The edge joint, Fig. 4-3, is economical for noncode work since the cost of preparation is low. It is not suitable, however, for severe load conditions. This joint should not be used if either member is subject to direct tension or bending at the root. Very deep penetration is impossible. The edge joint should be used only on $\frac{1}{4}$-inch metal or thinner.

Butt Joints

CLOSED SQUARE BUTT JOINT. The square butt joint can be welded in several different ways. It is satisfactory for all usual load conditions. Preparation of the joint is simple and inexpensive since it requires only the butting together of the plate edges.

Complete penetration of the base metal is necessary if the closed butt joint is to be used for code work. Welding from one side, Fig. 4-4A, cannot secure complete penetration in most stock. Because of the unwelded root, the joint is weak at this point. Welding from both sides, Fig. 4-4B, materially increases its strength. Constant and severe loading, however, causes failure of the joint because of the unwelded areas at the root.

On metal $\frac{1}{8}$ inch or thinner, complete penetration can be obtained by welding from one side, Fig. 4-4C. On metal $\frac{3}{16}$ inch or thinner, complete penetration is possible by welding from both sides, Fig. 4-4D. Metal electrode

Table 4-1
Forms of weld joints.

Edge joints	Fig. no.
Leaf edge joint (An edge joint in which the parts are flanged into the same plane on completion of welding.)	
Flanged edge joint	1
Butt joints	**Fig. no.**
Closed single-flanged butt joint	2
Open single-flanged butt joint	3

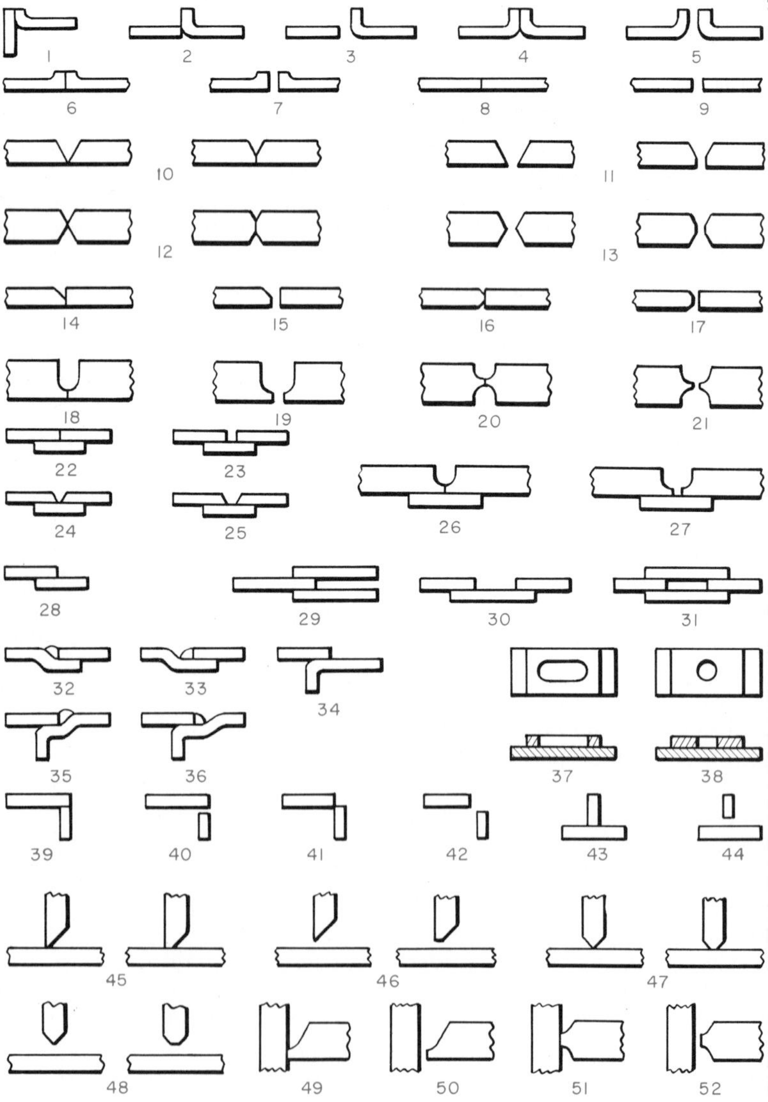

FORMS OF WELD JOINTS, ADAPTED FROM REPORT OF COMMITTEE ON NOMENCLATURE, ETC., AMERICAN WELDING SOCIETY.

Table 4-1 (Continued)

Butt joints	Fig. no.
Closed double-flanged butt joint	4
Open double-flanged butt joint	5
Closed upset butt joint	6
Open upset butt joint	7
Closed square butt joint	8
Open square butt joint	9
Closed single-V butt joint	10
Open single-V butt joint	11
Closed double-V butt joint	12
Open double-V butt joint	13
Closed single-bevel butt joint	14
Open single-bevel butt joint	15
Closed double-bevel butt joint	16
Open double-bevel butt joint	17
Closed single-U butt joint	18
Open single-U butt joint	19
Closed double-U butt joint	20
Open double-U butt joint	21
Strapped closed square butt joint	22
Strapped open square butt joint	23
Strapped closed single-V butt joint	24
Strapped open single-V butt joint	25
Strapped closed single-U butt joint	26
Strapped open single-U butt joint	27
Lap joints	**Fig. no.**
Single lap joint	28
Double lap joint	29
Single-strap lap joint	30
Double-strap lap joint	31
Closed joggled single lap joint	32
Open joggled single lap joint	33
Flanged single lap joint	34
Flanged closed joggled single lap joint	35
Flanged open joggled single lap joint	36
Linear slotted lap joint	37
Circular slotted lap joint	38
Corner joints	**Fig. no.**
Closed lapped corner joint	39
Open lapped corner joint	40
Closed corner joint	41
Open corner joint	42
T-joints	**Fig. no.**
Closed square T-joint	43
Open square T-joint	44
Closed single-bevel T-joint	45
Open single-bevel T-joint	46
Closed double-bevel T-joint	47
Open double-bevel T-joint	48
Closed single J/T joint	49
Open single J/T joint	50
Closed double J/T joint	51
Open double J/T joint	52

Adapted from *Report of Committee on Nomenclature,* American Welding Society

welding may be used on metal ¼ inch thick and the submerged-arc welding process on metal ⅝ inch thick. For welds in which complete penetration is necessary on metal more than 3/16 inch thick, it is recommended that the root of the first pass be chipped out to sound metal before depositing the second weld.

OPEN SQUARE BUTT JOINTS. Securing penetration on open square butt joints, Fig. 4-5A, is easier than on closed square butt joints. Because of this fact, heavier sections can be welded. It is possible to weld 3/16-inch material or less from one side, Fig. 4-5B, and up to ¼ inch from both sides, Fig. 4-5C, with complete penetration. If complete penetration is not achieved, however, (Figs. 4-5D and E), the open square butt joint will not be any stronger than the closed type, and it will have the same possibility of failure at the root of the weld under load.

Metal ⅜ inch thick may be welded with the shielded metal-arc process, and metal ¾ inch thick, with the submerged-arc welding process if the root of the first pass is chipped out to sound metal before depositing the second weld. The cost of joint preparation is the same as for closed square butt joints. Oftentimes it is a little more difficult to line up the work with the proper gapping all along the entire length of the joint.

SINGLE-V BUTT JOINT. Single-V butt joints, Fig. 4-6, are superior to square butt joints and are used a great deal for important work. They provide for 100 percent penetration and offer a better plate edge preparation for welding than square butt joints. Metal preparation, however, is more costly than for the square butt joint, and a greater amount of electrode deposit is used in welding. The single-V type is ordinarily used on plate thicknesses ranging from $\frac{1}{4}$ inch to $\frac{5}{8}$ inch. If welding is to be from one side only, Fig. 4-7A, full penetration to the root of the weld must be obtained. Failure to do so will cause a fracture if the joint is subjected to severe loading.

Joints welded from both sides with complete joint penetration provide full strength and meet the requirements of code welding. Welding from both sides, Fig. 4-7B, can be accomplished only where the work will permit the operator to weld from both sides of the plate. It is easier to obtain complete penetration through the entire thickness in this way. If a backing strip is used, Fig. 4-7C, it is possible to weld faster and use larger electrodes, especially on the first or root pass. A removable backing is also used when welding from one side with the submerged-arc process. Metal thickness up to $1\frac{1}{2}$ inch can be welded in this manner.

DOUBLE-V BUTT JOINT. The double-V butt joint, Fig. 4-8, is suitable for most severe load conditions. It is used on heavier plate than single-V butt joints, usually $\frac{3}{4}$ inch to $1\frac{1}{2}$ inch thick. For metal thicknesses greater than $1\frac{1}{2}$ inch, the double-U butt joint is recommended because less electrode metal is needed. The cost of joint preparation is greater than for the single-V butt joint, but the amount of filler metal needed in welding is less.

It is essential that complete root penetration be achieved. The work must permit welding from both sides, and the back side of the first pass must be chipped before applying the second pass from the other side. Welding from both sides permits an even distribution of heat through the joint, thus reducing the concentration of stress at the joint and the amount of warpage and distortion.

BEVELED-BUTT JOINTS. Single-bevel butt joints (Figs. 4-9 and 4-10) and double-bevel butt joints (Fig. 4-11) are used in some areas. They are suggested for work where load demands are greater than can be met by square butt joints and less than values requiring V-butt joints. They join metal up to $\frac{3}{4}$ inch thick, and less filler metal is required than for a V-butt joint, thus reducing the number of electrodes needed. The cost of preparation is less than for V-butt joints since it is necessary to bevel only one plate edge. For full strength the root of the first pass should be chipped to sound metal before depositing the second pass. The operator will find it difficult to obtain good fusion to the perpendicular wall of the square plate and to secure good root penetration.

SINGLE-U BUTT JOINT. Single-U butt joints, Fig. 4-12, are used for very important work, such as fired and nonfired pressure vessels. The cost of preparation is greater than for bevel and V-butt joints, but fewer electrodes are required in welding. The single-U butt joint is used on plate thick-

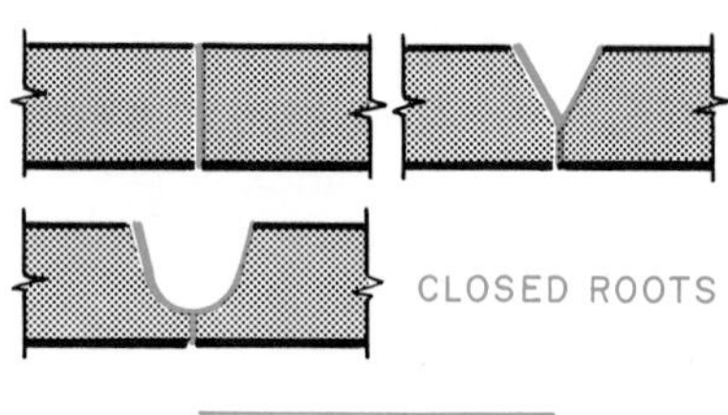

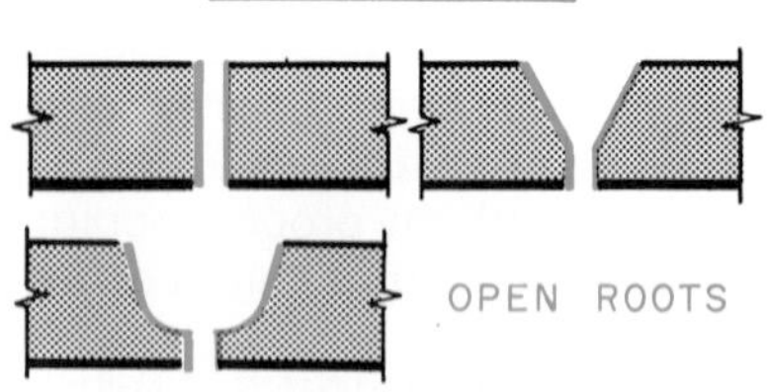

4-2. *Closed and open roots.*

4-3. *Edge joints.*

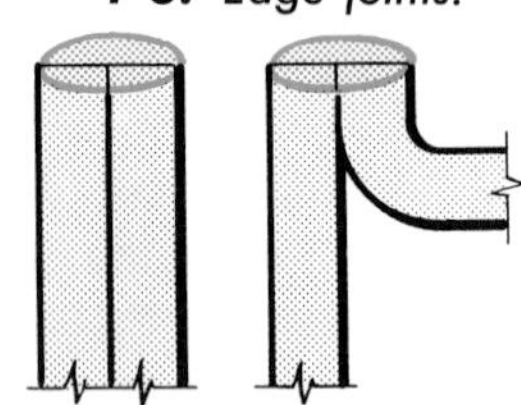

4-4. *Closed square butt joints.*

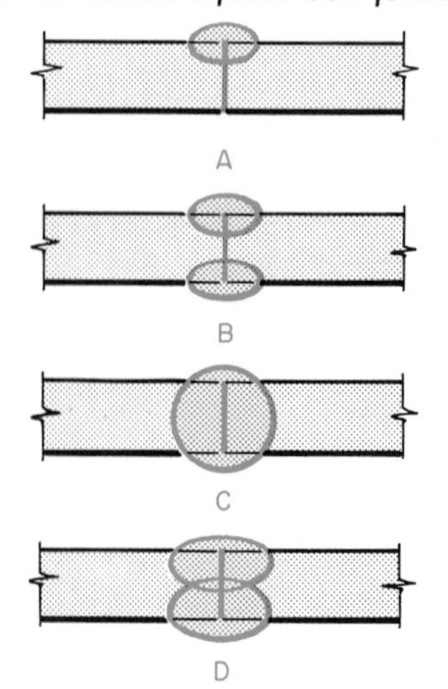

4-5. *Open square butt joints.*

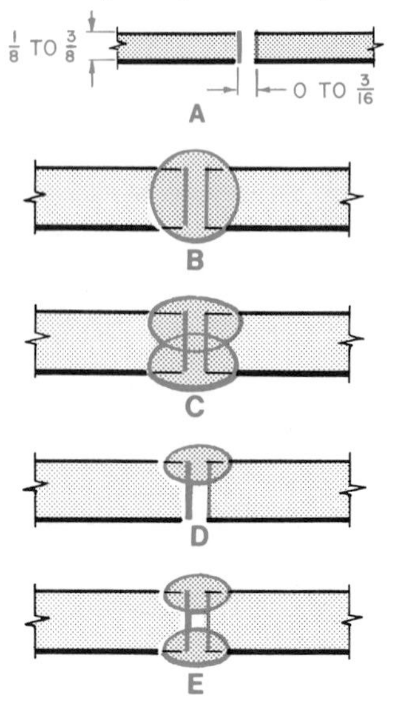

nesses ranging from $\frac{1}{2}$ inch to $\frac{3}{4}$ inch. Heavier metal may be welded with the submerged-arc process. It is often the practice to make the first pass with the shielded metal-arc or the MIG process. Complete penetration is necessary for the single-U butt joint to give satisfactory service. It is easier to obtain complete penetration on single-U butt joints welded from both sides (Fig. 4-12C), and on joints with a backup strip (Fig. 4-12D), than on joints welded from one side only (Fig. 4-12B). The joint is usually welded with free-flowing electrodes.

DOUBLE-U BUTT JOINT. Double-U butt joints, Fig. 4-13, are used on work of the same nature as single-U butt joints but when plate thicknesses are greater and welding can be done from both sides. Plate thicknesses range up to $\frac{3}{4}$ inch. Although the cost of preparation is greater than for single-U butt joints, double joints may be welded with fewer electrodes. Welding from both sides permits a more even distribution of stress and reduces distortion.

The choice between double-U and double-V butt joints should be made on the basis of the relative costs of metal preparation and welding.

J-BUTT JOINTS. Single J-butt joints, Fig. 4-14, and double J-butt joints, Fig. 4-15, are used on work similar to that requiring U-butt joints, but when load conditions are not as demanding. The cost of preparation is less since only one plate edge must be prepared. Less filler metal is required to fill the groove. It is difficult to secure good fusion and thorough penetration because of the perpendicular wall of the square member.

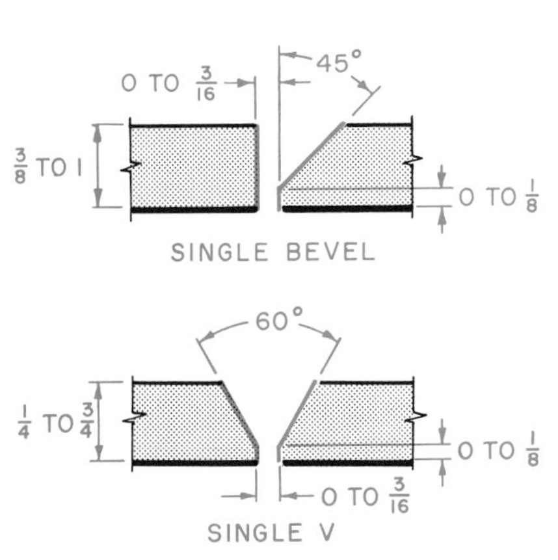

4-6. *Proportions for single-V butt joints.*

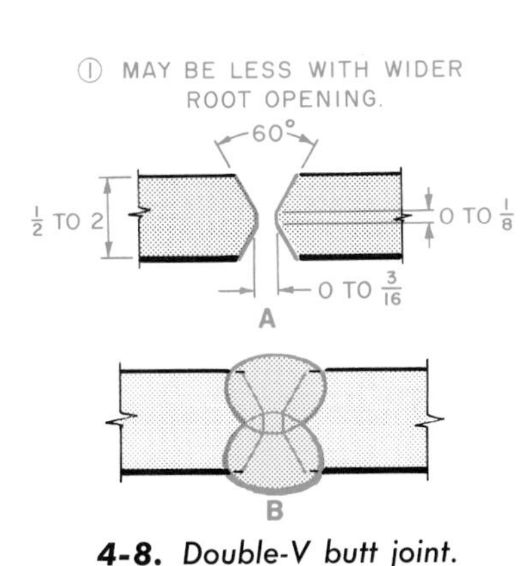

4-8. *Double-V butt joint.*

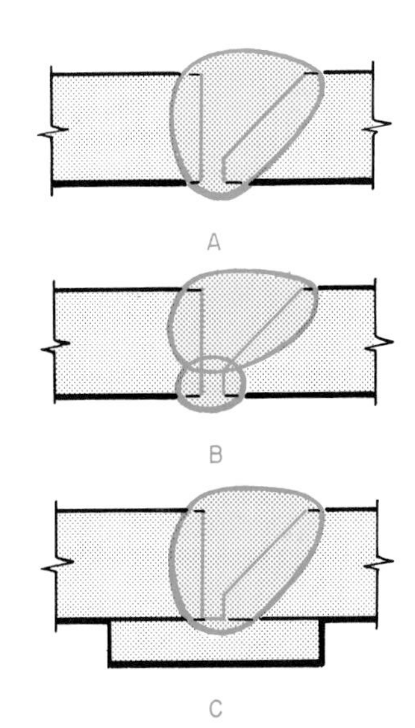

4-10. *Single-bevel butt joints.*

4-7. *Single-V butt joints.*

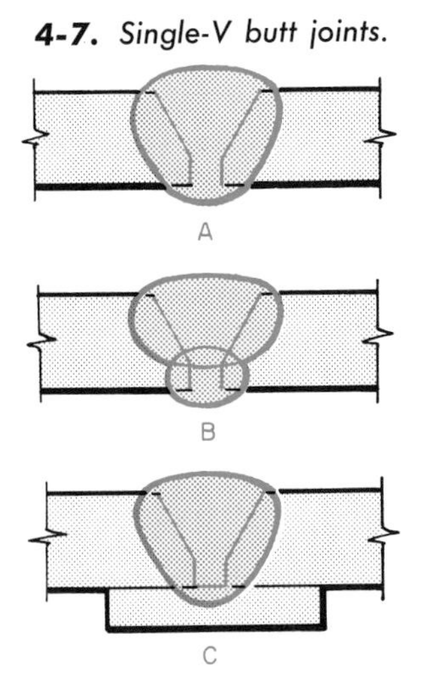

4-9. *Proportions for single-bevel butt joints.*

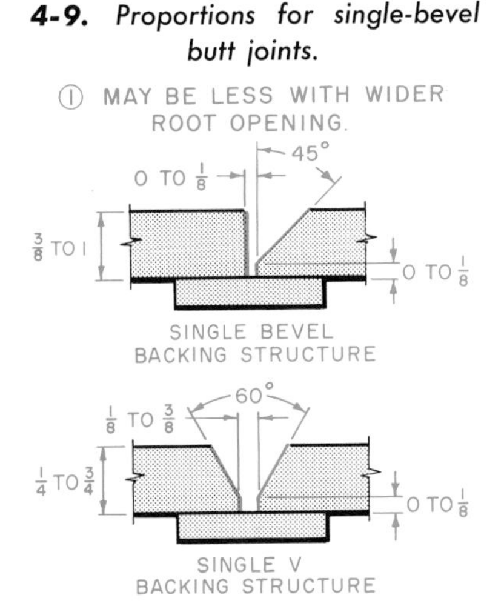

4-11. *Double-bevel butt joint.*

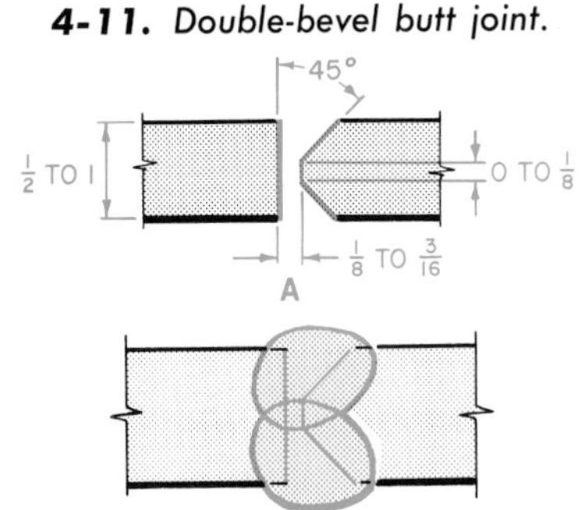

Lap Joints

Lap joints are used frequently on all kinds of work. There is no plate preparation involved. The single-fillet lap joint (Fig. 4-16A), while not as strong as the double-fillet lap joint (Fig. 4-16B), is more often used on noncode work. Single-fillet lap joints should not be used if the root of joint is to be subjected to bending. In both cases penetration to the root of the joint is necessary. The welder must make sure that the edge of the upper plate is not burned away. A lap joint should never replace the butt joint on work under severe load.

LINEAR SLOTTED AND CIRCULAR SLOTTED LAP JOINT. The linear slotted lap joint and the circular slotted lap joint, Fig. 4-17, are used infrequently. They join one plate or bracket to another when it is desirable to conceal the weld or when there is a lack of an edge to weld on. In order to withstand a heavy load, the unit to be welded requires a series of these welds, and the cost of preparation is high. If the slots are small, it is difficult for the operator to make welds that are free of porosity and slag inclusions.

Corner Joints

Flush corner joints, Fig. 4-18A, can be used on light gauge sheet metal, usually under 12 gauge. Heavier plates can be welded if load is not severe and if there is no bending action at the root of the weld. No edge preparation is needed, and fitup is usually simple.

Half-open corner joints, Fig. 4-18B, may be used on 12-gauge to 7-gauge plate. This type of joint forms a groove and permits weld penetration to the root and good appearance. No edge preparation is required, and fitup is usually simple.

Full-open corner joints, Fig. 4-18C, can be used on any plate thickness. If welding is to be from one side, penetration must be secured through the root of

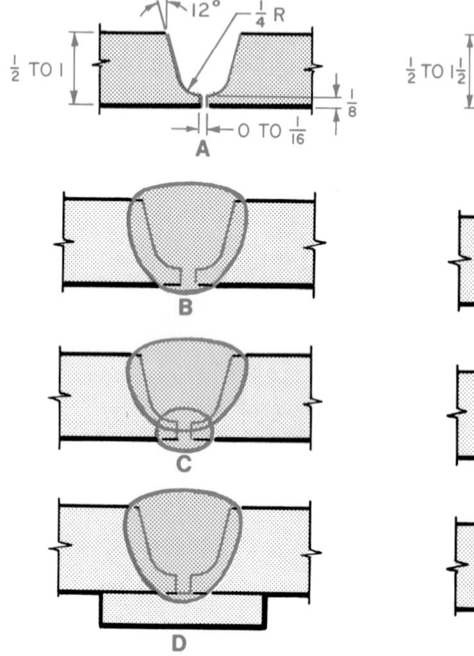

4-12. *Single-U butt joints.*

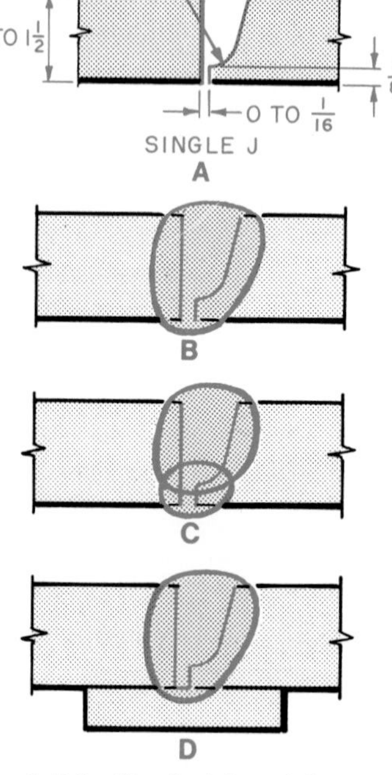

4-14. *Single-J butt joints.*

4-13. *Double-U butt joint.*

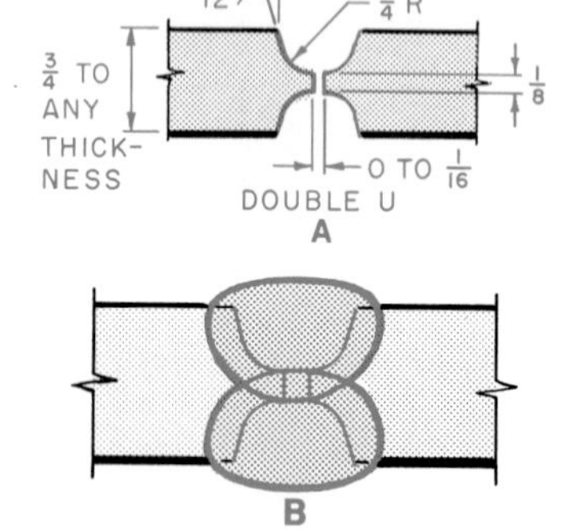

4-15. *Double-J butt joint.*

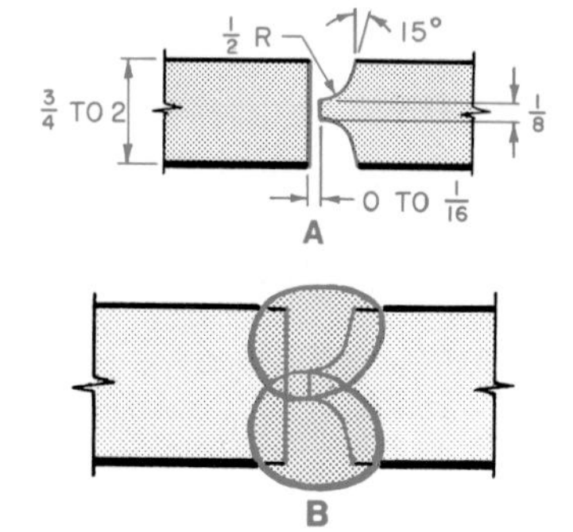

4-16. *Lap joints.*

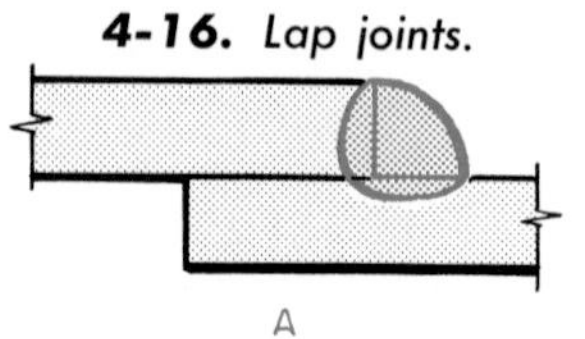

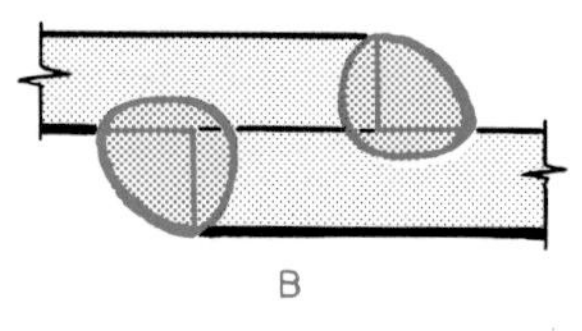

the weld. If welded from both sides, the joint is suitable for severe loads. It has a good stress distribution, and no edge preparation is required.

More filler metal is required than for the half-open joint, and fitup is likely to be difficult. Plates must be cut absolutely square, and suitable clamping and holding devices are often needed to facilitate fitup. This type of joint is used in production welding.

T-Joints

SQUARE T-JOINT. The square T-joint, Fig. 4-19, may be used on ordinary plate thicknesses up to $\frac{1}{2}$ inch. Preparation of the plate is not necessary, and fitup can be fast and economical. Electrode costs are high.

The single-fillet T-joint will not withstand bending action at the root of the weld and should be used with caution. If it is possible to weld from both sides, Fig. 4-19B, the joint will withstand high load conditions.

SINGLE-BEVEL T-JOINT. The single-bevel T-joint, Fig. 4-20, is able to withstand more severe loads than the square T-joint. It can be used on plate thicknesses ranging from $\frac{3}{8}$ to $\frac{5}{8}$ inch. Plate of greater thickness can be welded with the submerged-arc process. Cost of preparation is greater than for the square T-joint, and fitup is likely to take longer. Electrode costs are less.

If it is possible to weld from one side only, Fig. 4-20A, full penetration to the root of the weld must be obtained so that bending does not cause failure. If welding can be done from both sides, Fig. 4-20B, the load resistance of the joint is materially increased.

4-17. *Linear and circular slotted lap joints.*

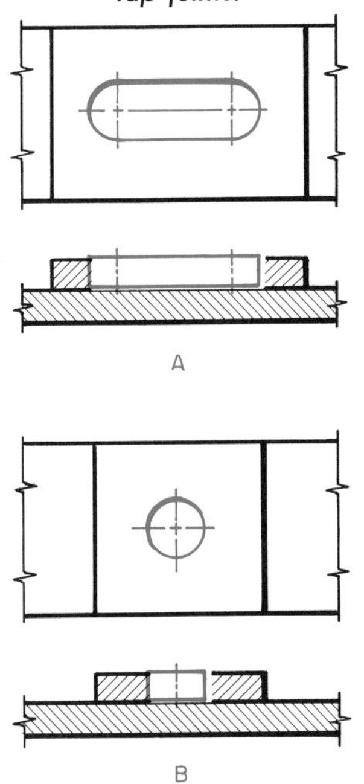

4-18. *Corner joints.*

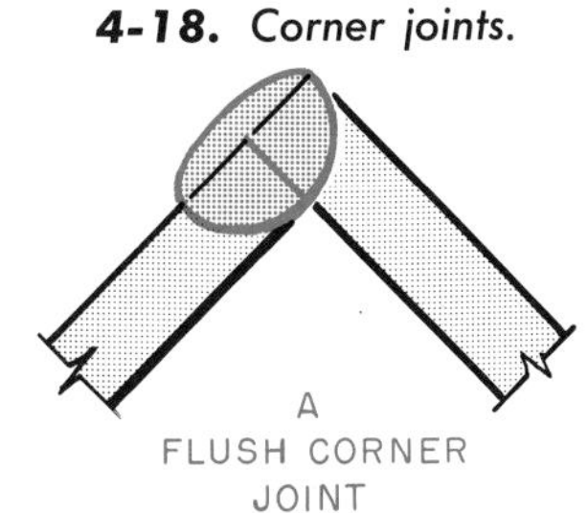

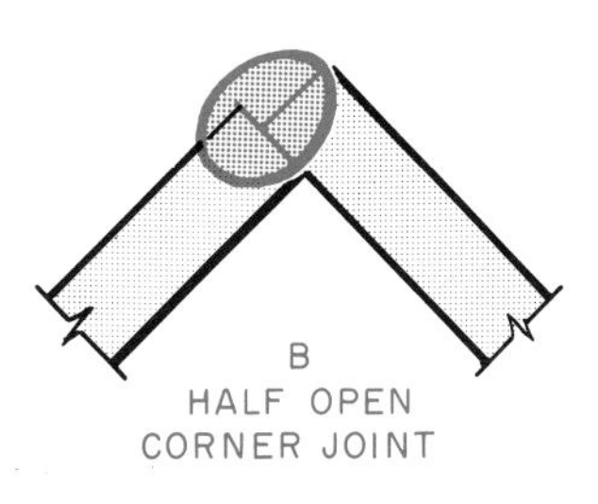

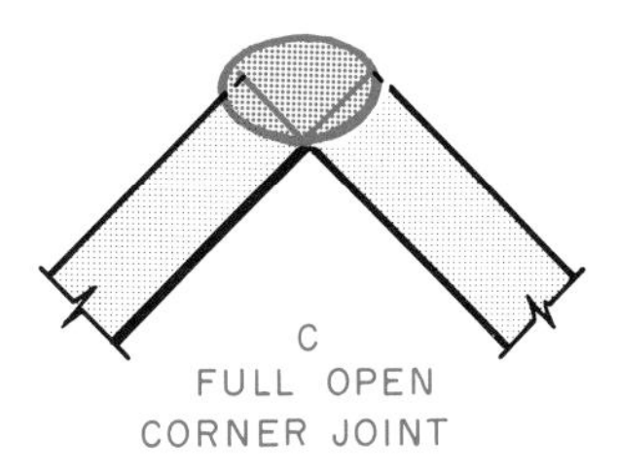

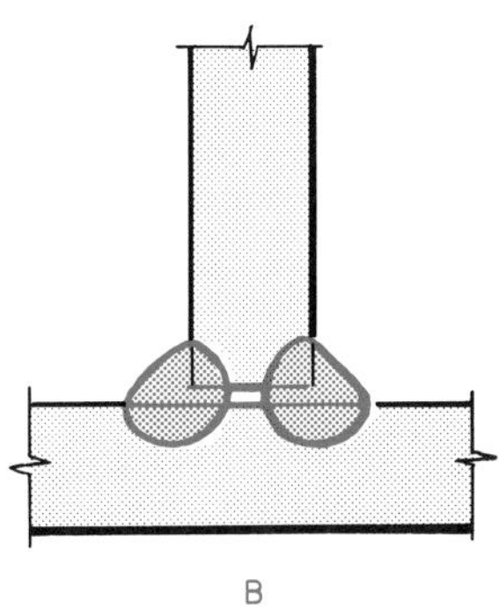

4-19. *Square T-joints.*

4-20. *Single-bevel T-joints.*

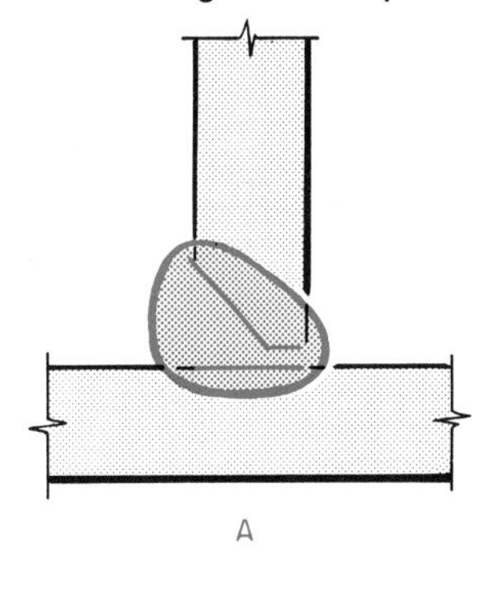

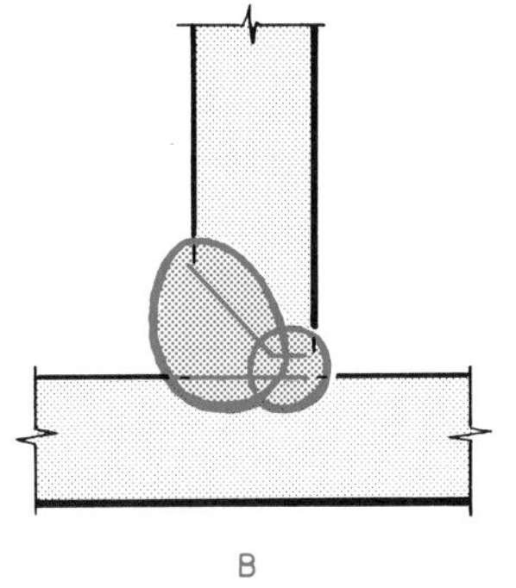

DOUBLE-BEVEL T-JOINT. The double-bevel T-joint, Fig. 4-21, is used for heavy plate thicknesses up to 1 inch. Welding is done from both sides of the plate. This joint may be used for severe loads. The welder must make sure that fusion is obtained with both the flat and vertical plates. Good root penetration is necessary. Joint preparation is more expensive than for the square T- or single-bevel joint, but electrode costs are less.

SINGLE J/T JOINT. Single J/T joints, Fig. 4-22, can be used for most severe load conditions. They are generally used on plates 1 inch or heavier. If welding is to be done from one side, Fig. 4-22A, great care should be taken to secure good root penetration. If welding from both sides is possible, efficiency of the joint can be increased materially by putting the bead on the side opposite *J*, Fig. 4-22B. This reduces the tendency of failure at the root as a result of load at this point. The cost of plate edge preparation is higher than for the bevel T-joint, but there is a saving in electrode costs.

DOUBLE J/T JOINT. The double J/T joint, Fig. 4-23, will withstand the most severe load conditions. It is used on plates $1\frac{1}{4}$ inch or heavier. The operator must be able to weld from both sides of the plate. Good root penetration and surface fusion are essential to prevent failure under severe load. Plate edge preparation is higher than for V/T joints and single J/T joints; however, electrode costs are lower. Frequently both *J*s are not of equal dimensions, Fig. 4-23B.

Summary

The material you have just studied indicates the importance of joint design and the weld requirements. In general, the soundness and service life of the product will depend upon the proper joint design and flawless welding. We have seen that proper plate preparation and setup depends upon a number of variables such as the thickness of plate, the plate edge preparation, and the root preparation. (See Figs. 4-5A through 4-15A, pages 120–122.) These variables are determined by the nature of the components to be welded and the service that the weldment will be expected to give. Correct joint design and proper welding procedure keep distortions to a minimum, reduce cracking due to shrinkage, and make it easier for the welder to produce sound welds with good appearance at the lowest possible cost.

Figures 4-24 to 4-29 (Pages 125–130) show the proper preparation of plate edges as recommended by the American Welding Society. Study these elements of joint design carefully because they form the basis for understanding welded construction.

TYPES OF WELDS

We have seen that there are many combinations of joints that can be formed from the five basic joints. The types of welds

4-22. *Single J/T joints.*

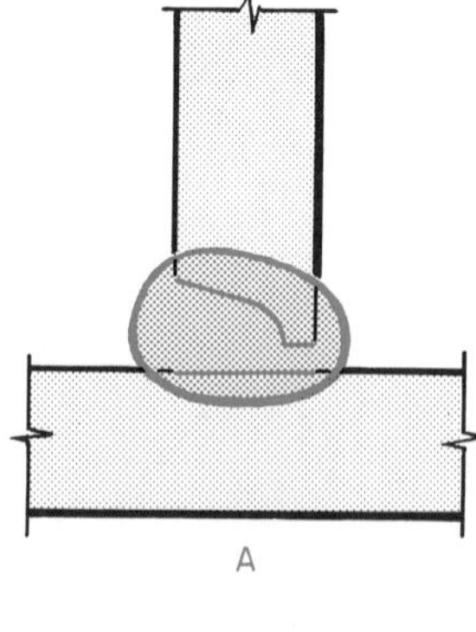

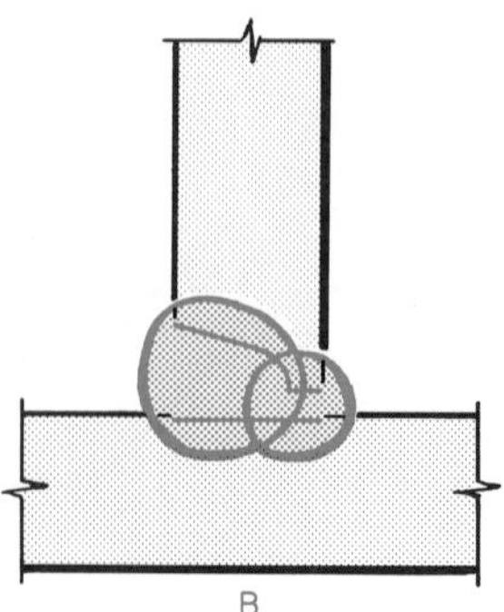

4-23. *Double J/T joints.*

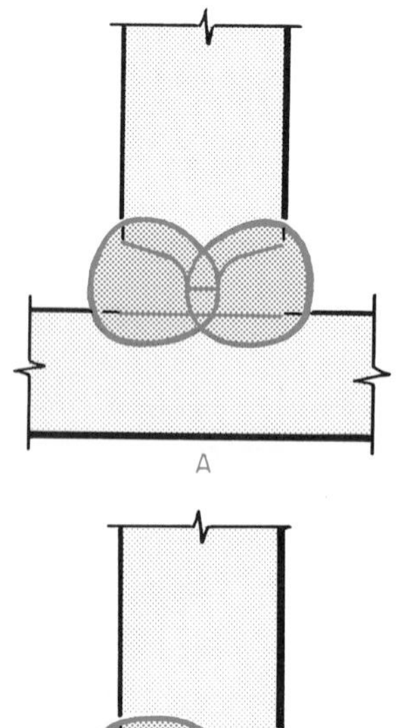

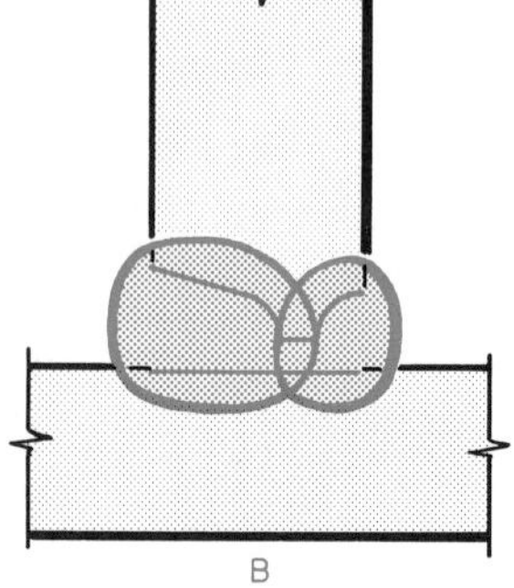

4-21. *Double-bevel T-joint.*

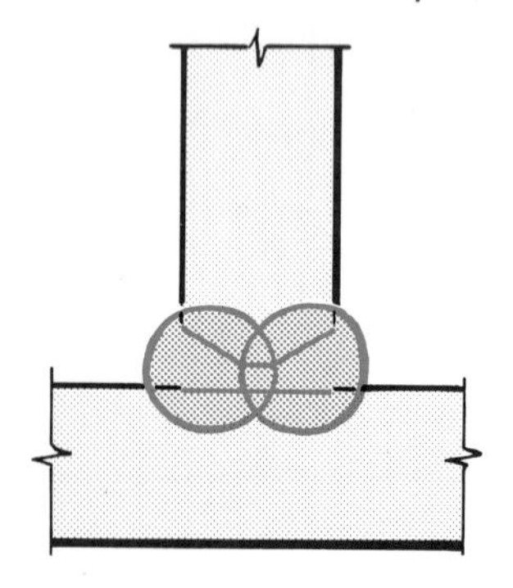

needed to join the many forms of welded joints, however, are not numerous. There are only four basic types of welds: the bead weld, the groove weld, the fillet weld, and the plug weld.

- *Bead welds* are single-pass deposits of weld metal, Fig. 4-30A. They are used to build up a pad of metal and to replace metal on worn surfaces. Bead welds are also used for square butt joints.
- *Fillet welds* consist of one or more beads deposited in the right angle formed by two plates, Fig. 4-30B. They are used for lap joints and T-joints.
- *Groove welds* consist of one or more beads deposited in a groove, Fig. 4-30C. Groove welds are used for bevel butt joints and V-, J-, and U-butt joints. Open-corner joints are also welded with groove welds.
- *Plug welds* are used for slotted and circular lap joints, Fig. 4-30D. They are like fillet welds except that the entire area of the slot or hole is filled. Plug welds are not common.
- *Tack welds* join separate parts to hold the assembly before welding. They are a series of short welds spaced at intervals.

Structural Welds

STRENGTH WELDS. Strength welds, Fig. 4-31, are the main welds of a structure, that is, those welds that must be depended upon to carry the load. They are usually designed to possess the maximum physical characteristics of the base metal. Strength welds that must meet code requirements, such as the butt joint in piping, must have

4-24. *Recommended dimensions of grooves for shielded metal-arc welding, gas metal-arc welding, and gas welding (except pressure gas welding). Note: Dimensions marked * are exceptions which apply specifically to designs for gas metal-arc welding.*

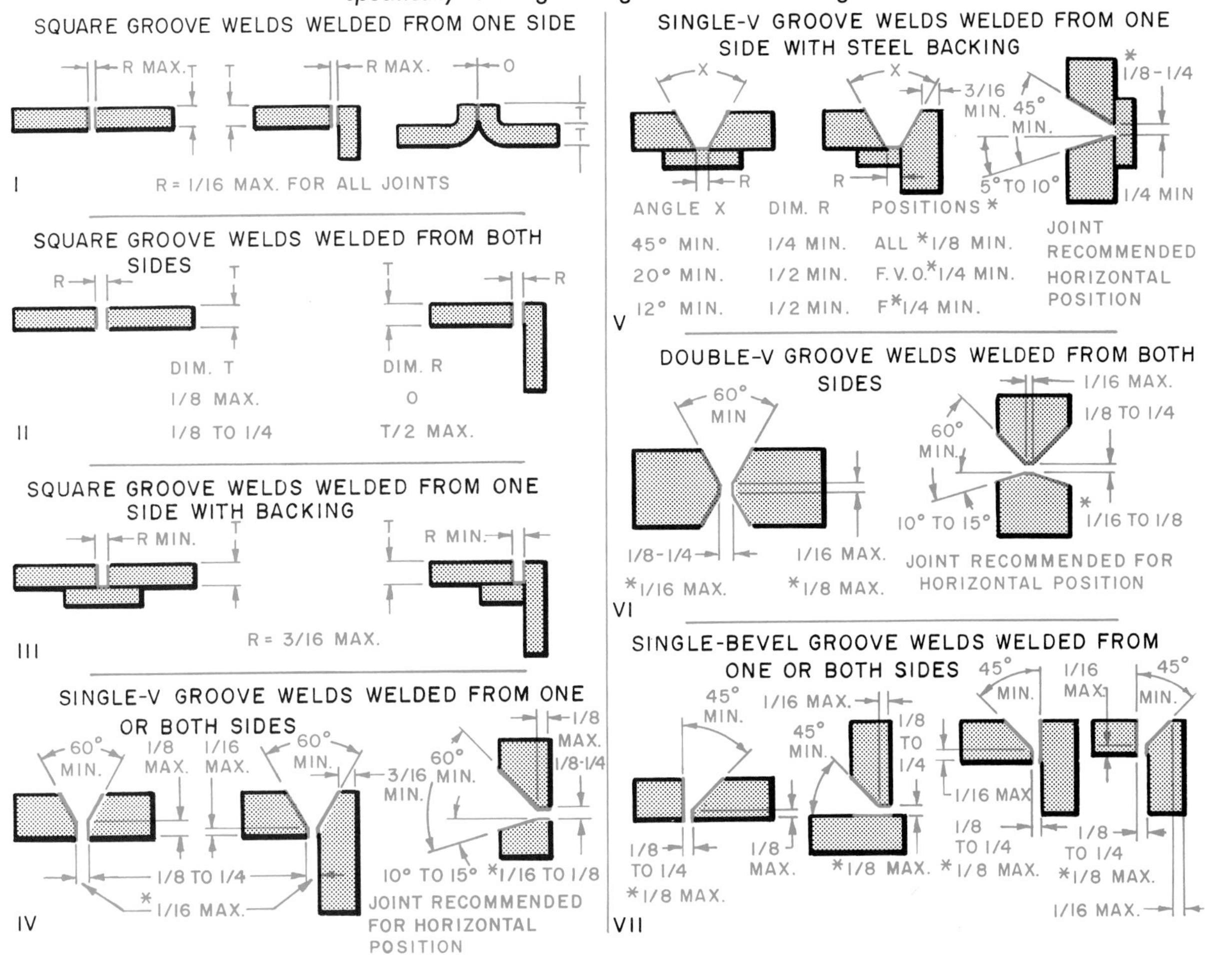

better physical properties than those used in the fabrication of noncode production components.

Strength welds may be bead, groove, or fillet welds. At least the minimum size welds called for by the designer must be made for this type of weld.

CAULKING WELDS. On many structures, such as certain types of tanks, the strength of the joint may be derived from its riveted construction. However, to make sure that the joints will not leak when the tank is full, continuous welds are run the entire length of the joints. These welds are called caulking welds. They are usually single-pass welds deposited along the root of the joint, Fig. 4-32 (Page 131). They must be sound, but they are not expected to carry a heavy load.

COMPOSITE WELDS. Welds that must be strong enough to meet load requirements and at the same time must not leak are known as composite welds. They are a combination of strength and caulking welds. A butt weld in pipe is this kind of weld, Fig. 4-31. It must be strong enough to withstand the pressure and vibration under which the line operates and yet it must also hold liquids or gas without leaking.

CONTINUOUS WELDS. Continu-

4-25. *Recommended dimensions of grooves for shielded metal-arc welding, gas metal-arc welding, and gas welding (except pressure gas welding). Note: Dimensions marked * are exceptions which apply specifically to designs for MIG welding.*

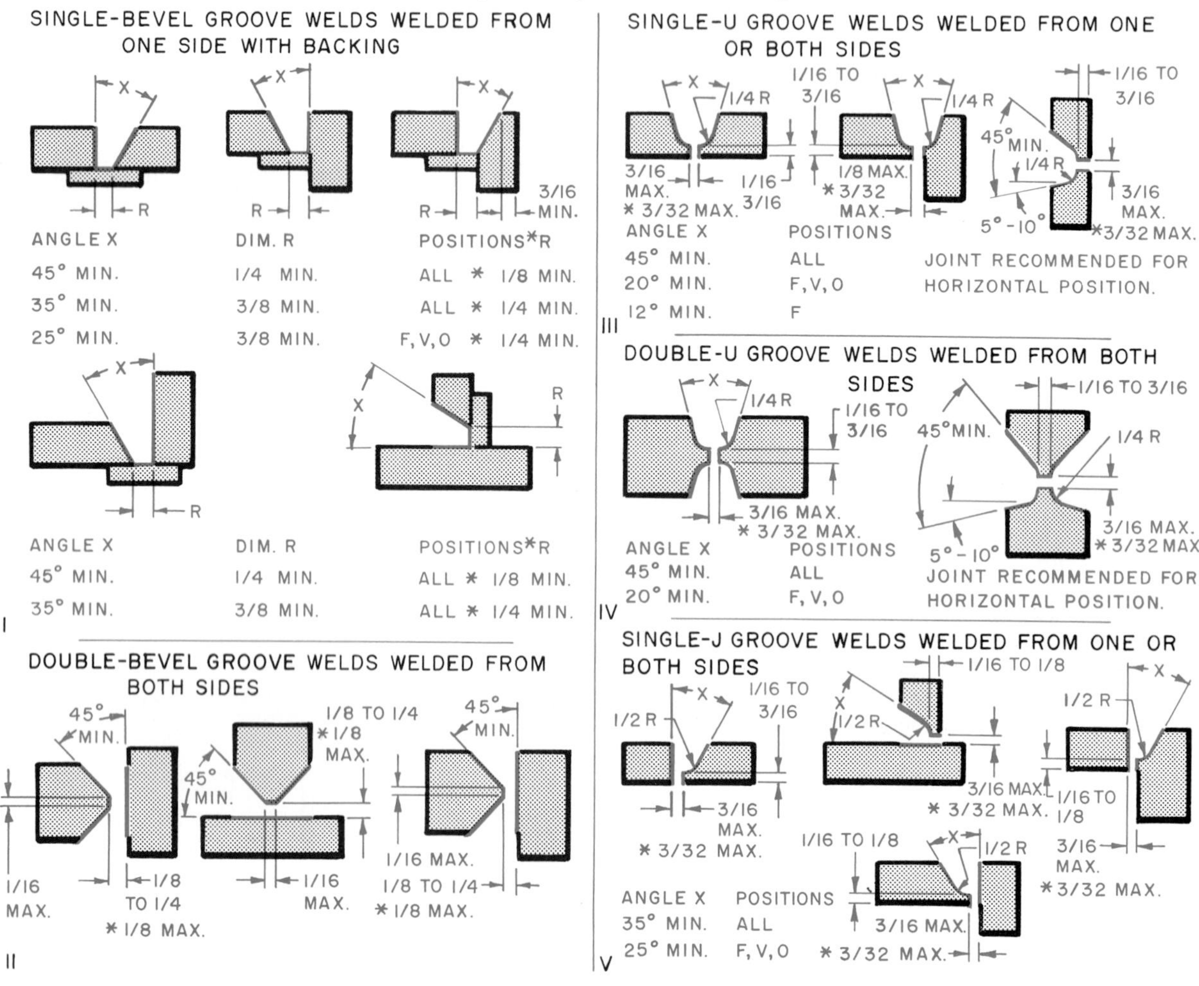

ous welds, Fig. 4-33, extend across the entire length of the joint from one end to the other. On structures that are to develop maximum strength and tightness, it is necessary to weld all of the seams completely.

INTERMITTENT WELDS. Intermittent welds, illustrated in Fig. 4-34, are a series of short welds spaced at intervals. They cannot be used where maximum strength is required or where it is necessary that the work be water- or air-tight. However, on work that is not critical, the cost of welding can be considerably reduced by the use of intermittent welds.

The frequency, length, and size of the welds depend upon the thickness of the plates, the type of joint, the method of welding, and the service requirements of the job. Intermittent welds are usually employed in lap and T-joints. They are sometimes used for square butt joints, but rarely, if ever, for groove joints.

TACK WELDS. Welded fabrications are often made up of many parts. In the process of assembly before welding, some means is necessary to join the parts to the whole. This is done by a series of short welds spaced at intervals called tack welds, Fig. 4-35.

Some operators do not attach enough importance to the tack welding procedure and the remelting of tack welds in the major welding operation. There are many instances when an operator has failed an important test because he was careless in tack welding. Tack welds must be strong. Not only must they be able to hold the part in the position in which it is to be welded, but they must also be able to resist the stress exerted on them when expansion and contraction occur during welding. Cold working, which is often necessary, imposes a severe load on the tack welds.

The number and size of the tack welds depend upon the thickness of the plate, the length of the seam, the amount of cold working to be done, and the nature of the welding operation. Tack welds must have good fusion and good root penetration. They should be flat and smooth—not convex and lumpy.

***4-26.** Recommended dimensions of grooves for shielded metal-arc welding, gas metal-arc welding, and gas welding (except pressure gas welding). Note: Dimensions marked * are exceptions which apply specifically to gas metal-arc welding.*

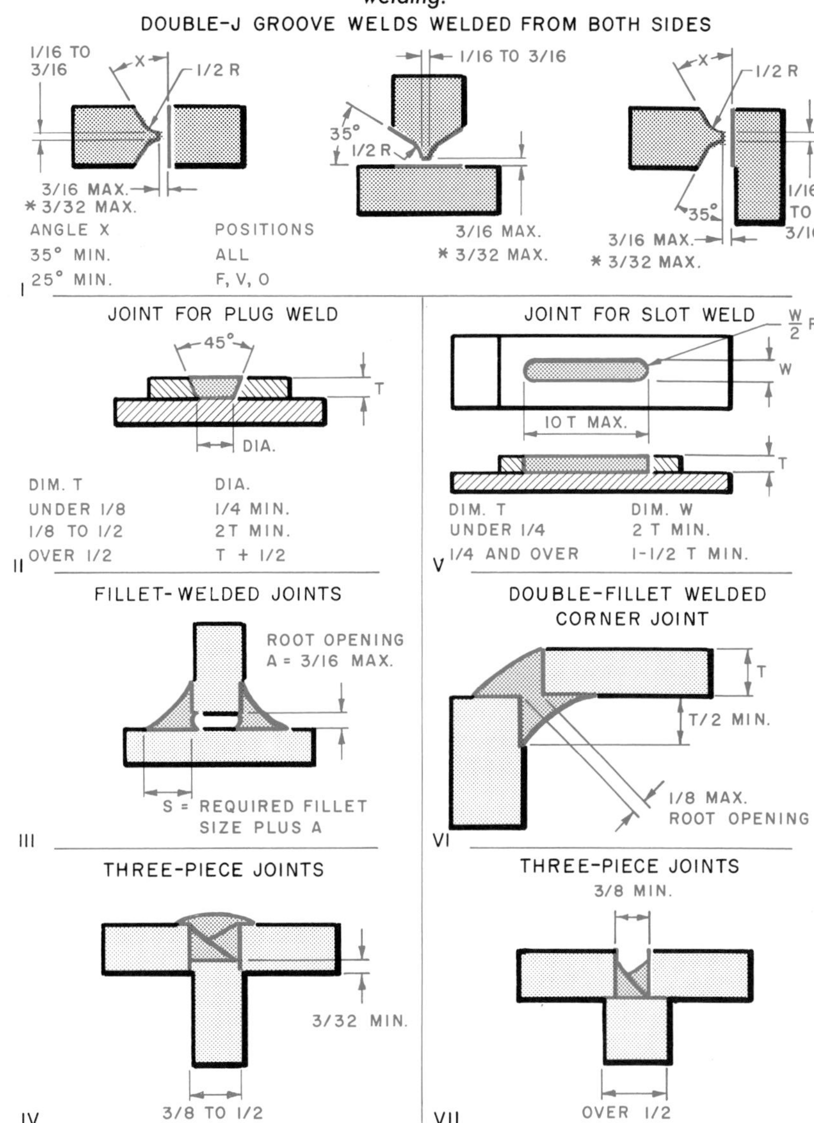

It is advisable to use more heat for tack welding than for the major welding operation.

POSITIONS OF WELDING

There are only four basic positions of welding. Examine the positions in Fig. 4-36 (Page 132) and identify them in Fig. 4-37 as you study the following definitions:

- *Flat position:* a position of welding in which the filler metal is deposited from the upper side of the joint with the face of the weld horizontal, Fig. 4-36A. For this position, the welding end of the electrode or torch is pointed downward.
- *Horizontal position:* a position of welding in which the weld is deposited on the upper side of a horizontal surface and against a vertical surface, Fig. 4-36B. The welding end of the electrode or torch is pointed back toward the weld. Thus it is at an angle opposed to the direction of travel.
- *Vertical position:* a position of welding in which the axis of the weld is in a vertical plane and deposited on a vertical surface, Fig. 4-36C. Welding travel may be up or down. When travel is up, the welding end of the electrode or torch is pointed upward at an angle, ahead of the weld. When the travel is down, the end of the electrode or torch is pointed up and at an angle to the weld pool.
- *Overhead position:* a position of welding in which the weld is deposited from the underside of the joint, and the face of the weld is horizontal. The welding end of the electrode or torch is pointed upward, Fig. 4-36D. The overhead position is the reverse of the flat position.

The trend in most shops is toward welding in the flat and horizontal positions wherever possible. Welding in these positions increases the speed of welding, allows more flexibility in the choice of electrodes, and insures work of better appearance and quality. Vertical and overhead welding find their widest application in those industries where the fabrications are large and permanent. Such conditions exist in shipyards, on construction projects, and in piping installations. Vertical welding is done more often than overhead welding, and most welders find it a less difficult position to weld in. However, operators must be able to weld in all positions. Inability to do so limits their possibilities of advancement to a higher job classification and prevents them from taking advantage of all the job opportunities they may encounter.

THE MEASUREMENT OF WELDS

A groove weld is measured by the depth of the throat, Fig. 4-38. Where the plates are of unequal thickness, the thickness of the lighter plate determines the size of the groove weld. Metal extend-

4-27. *Recommended dimensions of mixed grooves for arc welding.*

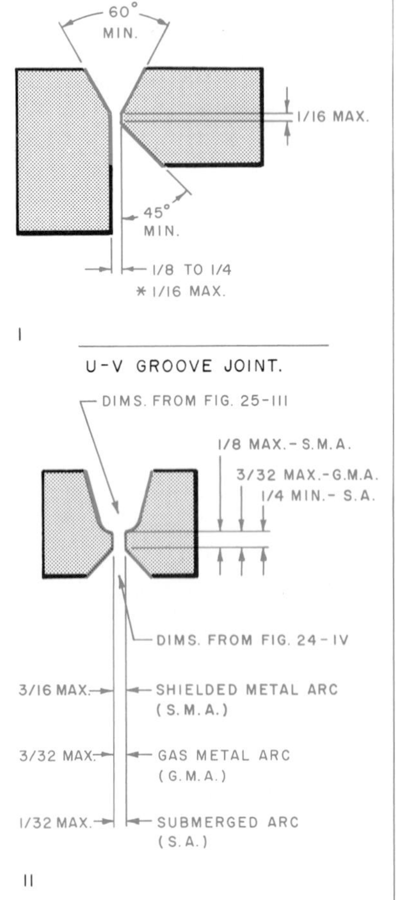

MODIFIED SQUARE GROOVE.
1/8 MAX.
1.0 MAX.
THERMAL GOUGE BEFORE DEPOSITING SECOND WELD.
III

MODIFIED SINGLE-V GROOVE.
45° MIN.
1/8 MAX.
THERMAL GOUGE BEFORE DEPOSITING SECOND WELD.
IV

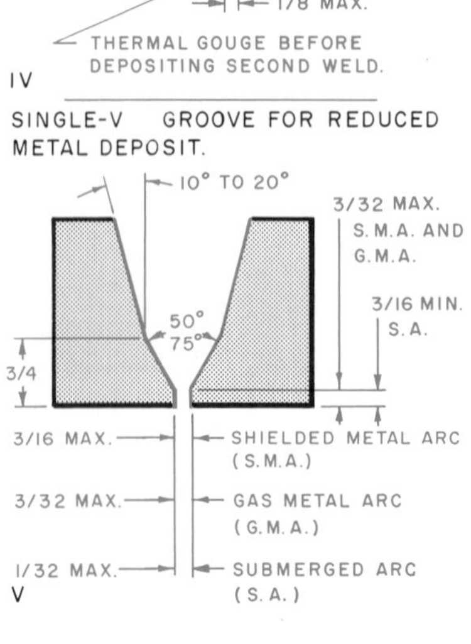

ing above the surfaces of the plate is called *reinforcement.*

Many welders believe that high reinforcement increases the strength of the welded joint. This is not true. Not only is excessive reinforcement above the allowable limits a waste of hard-to-get material, but it also decreases the working strength of the joint because of a concentration of stresses at the toe of the weld. It is obvious, on the other hand, that a lack of reinforcement or insufficient penetration through the root of the weld decreases the size of the weld. Proper reinforcement for grooved welds is shown in Fig. 4-39 (Page 134).

The width of a groove weld should not be more than $\frac{1}{4}$ inch greater than the distance across the joint from edge to edge. This allows for a deposition of metal $\frac{1}{8}$ inch past the shoulder on each side of the joint. Metal deposited beyond these limits is a waste of time and electrodes. The excess deposit increases cost and adds nothing to the strength of the welded joint.

The size of a fillet weld is often designated as its shortest leg length. (The *leg* of a fillet weld is the distance from the root to the toe of the weld.) While this is true for flat and convex fillet welds (Fig. 4-40A and B) and is satisfactory for all practical purposes, the true size of a fillet weld is the leg length of the largest inscribed isosceles right triangle (Fig. 4-40C). The *throat thickness* is the distance from the root to the face of the weld. By referring to Fig. 4-40C, it can readily be seen that for the same size weld, the leg length of the concave fillet is greater, yet this additional length adds nothing to the throat thickness. The weld shown demonstrates that it is foolhardy to attempt to gain strength through excessive convexity of the weld deposit. As in the case of the groove weld, excessive reinforcement wastes

4-28. *Recommended dimensions of grooves for metal-arc welding processes to obtain controlled and complete penetration. Note: for steel except as noted.*

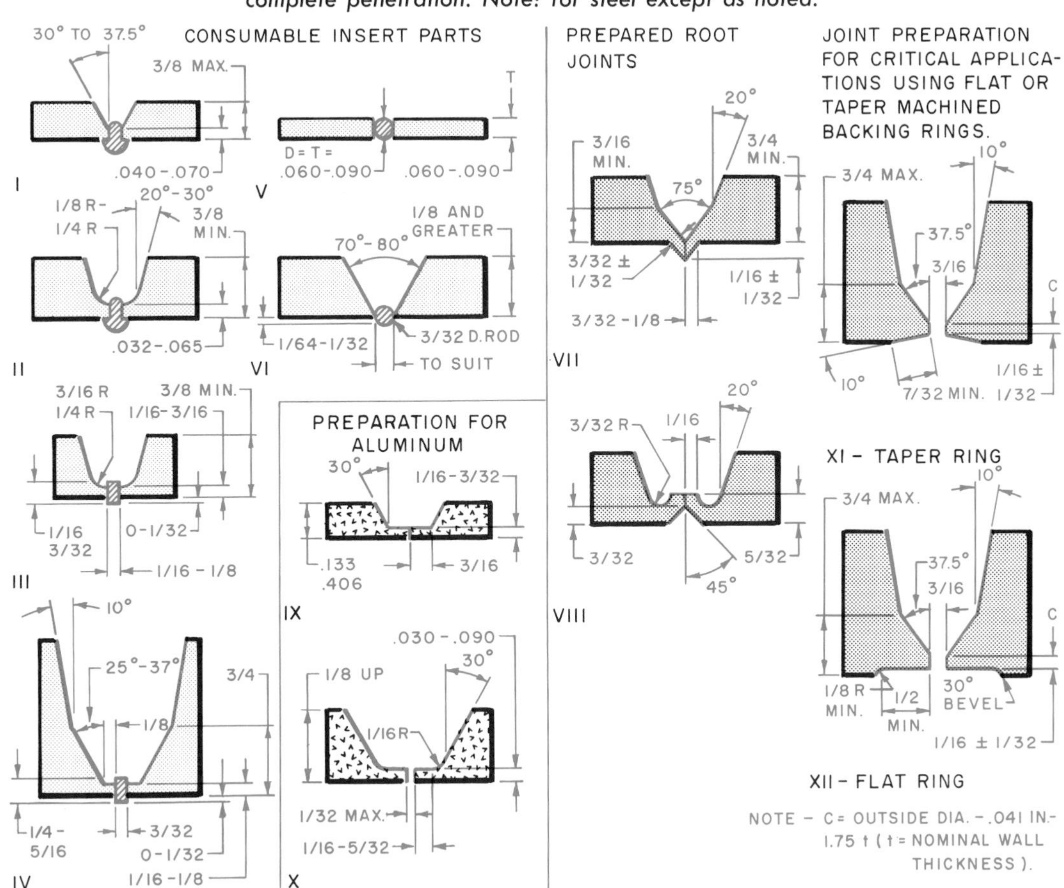

materials, increases cost, and concentrates stress to cause a reduction in strength. The ideal fillet weld, Fig. 4-41 (Page 134), has a flat or slightly convex face and equal leg length.

One advantage of the concave fillet as compared with the flat fillet is its gradual change in contour at the toe. There is some tendency, however, toward shrinkage cracking. Because the weld is concave, throat size is the critical dimension. Although stress distribution is improved, the concave fillet weld requires more material than the flat fillet weld for the same size welds and increases cost.

The convex fillet, in contrast to the concave and flat fillets, has less tendency to crack as a result of shrinkage set up by cooling. The welder will also find it easier to keep from undercutting at the weld edges.

All three types—flat, concave, and convex—are widely used. The position of welding, type of electrode, type of joint, and job requirements are some of the factors which determine the type of fillet to be used.

STRENGTH OF WELDS

In general, welded joints are as strong as, or stronger than, the base metal being welded. It is not always necessary that this be so. Good welding design specifies welds that require the minimum amount of weld metal which is adequate for the job at hand. Weld metal costs a good deal more than base metal and requires labor costs for its application.

The strength of a welded joint depends upon the following factors:

- Strength of the weld metal
- Type of joint preparation
- Type of weld
- Location of the joint in relation to the parts joined
- Load conditions under which weld will be put

4-29. *Recommended dimensions of grooves for the welding of tubes to tube sheets.*

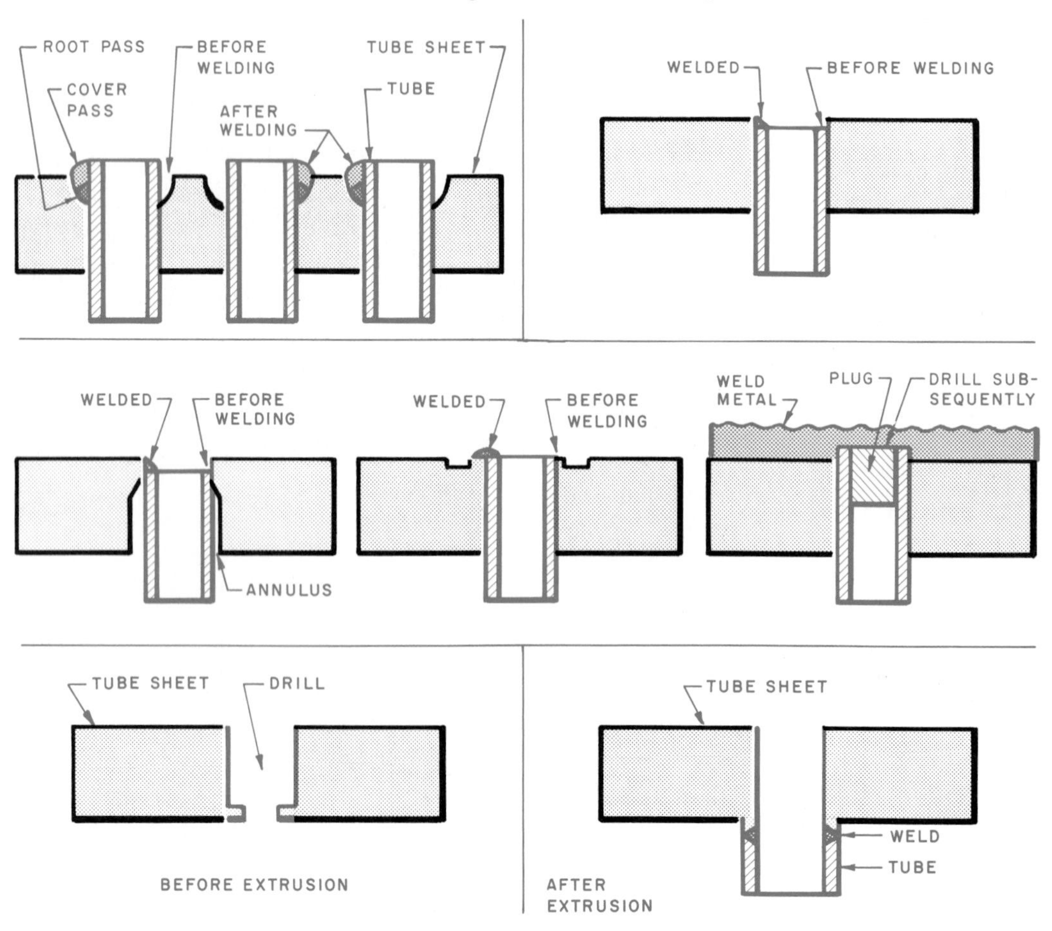

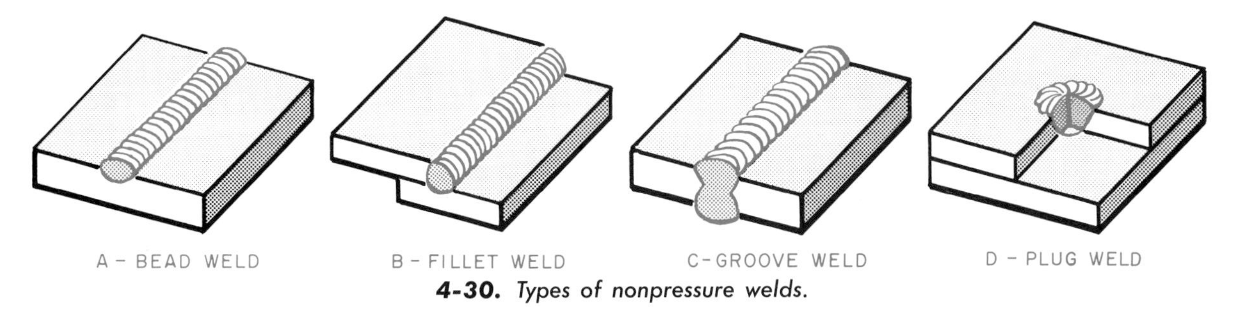

4-30. *Types of nonpressure welds.*

Crane Co.

4-31. *This code-welded ell is a strength and composite groove weld in a butt joint.*

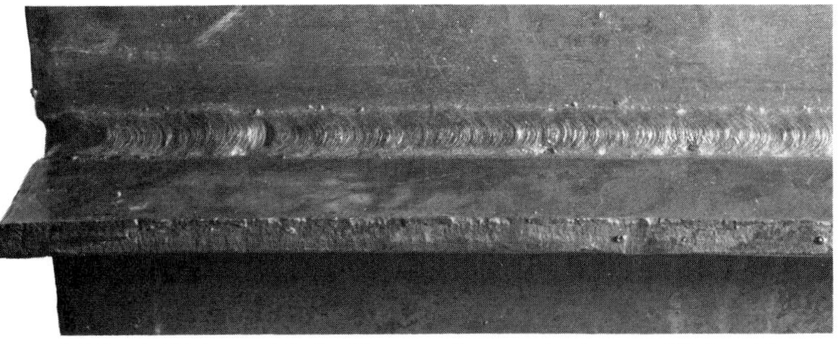

4-33. *A T-joint welded with a continuous single-pass fillet weld.*

4-34. *A T-joint with intermittent welds of equal length and equally spaced.*

4-32. *Riveted and caulked lap joint.*

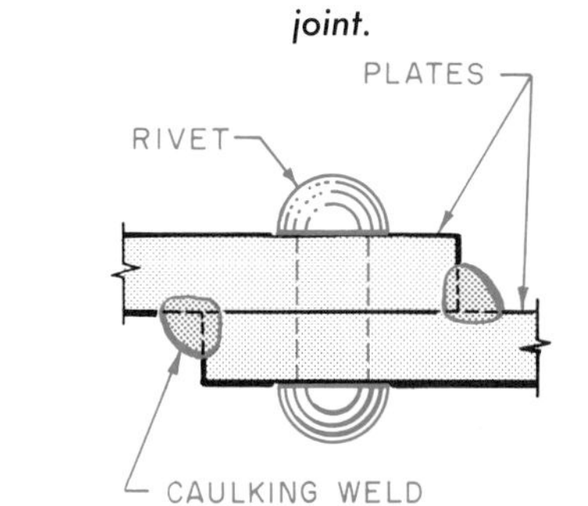

4-35. *Tack welds.*

- Welding process and procedure
- Heat treatment
- Skill of the operator

Approximately $\frac{1}{4}$ inch should be added to the designed length of fillet welds for starting and stopping the arc. The craters in the welds should be filled.

The location of the welds in relation to the parts joined, in many cases, has an effect on the strength of the welded joint. Repeated tests reveal that, when other factors are equal, welds having their linear dimension transverse (at right angles) to the lines of stress are approximately 10 to 15 percent stronger per average unit length than welds which have their linear dimension parallel to the lines of stress, Fig. 4-42.

Resistance to a turning effect of one member at a joint is best obtained by welds which are well-separated, rather than by a single weld or welds close together. In Fig. 4-43, a single weld at *A* is not as effective as welds at both *A* and *B* in resistance to turning effect. Two small welds at *A* and *B* are much more effective than a large single weld at *A* or *B* only. If possible, welded joints should be designed so that bending or prying action is minimized.

In some designs it may be desirable to take into account the stress distribution through the welds in a joint. Any abrupt change in surface (for example, a notch or saw cut in a square bar under tension) causes stress concentration and increases the possibility of fracture. As an illustration of this principle, the weld shown in Fig. 4-44 would have considerably more concentration of stress than that in Fig. 4-45. The weld shown in Fig. 4-46 allows a minimum of stress concentration and improved service. Under many load conditions, the stress is greater at the ends of the weld than in the mid-

4-36. *Positions of welding.*

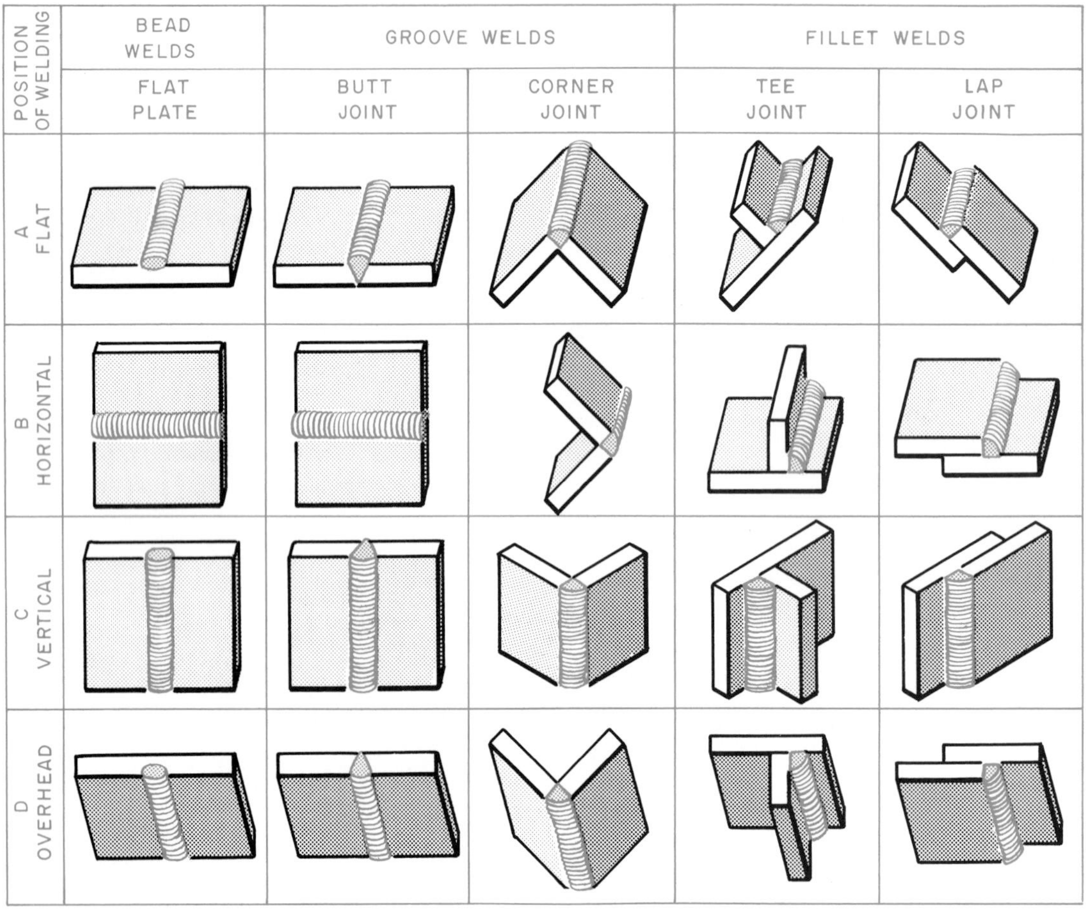

POSITION OF WELDING	BEAD WELDS	GROOVE WELDS		FILLET WELDS	
	FLAT PLATE	BUTT JOINT	CORNER JOINT	TEE JOINT	LAP JOINT
A FLAT					
B HORIZONTAL					
C VERTICAL					
D OVERHEAD					

dle. Therefore, it is advisable in such cases to hook the bead around the joint as indicated in Fig. 4-47. When this is done, far greater resistance to a tearing action on the weld is obtained.

COMMON WELD DEFECTS

Fillet Welds

Figure 4-48 shows flat and concave fillet weld profiles that are considered desirable. Figure 4-49 illustrates a slightly convex profile that is also acceptable. Thus we are again reminded that the welder should try to avoid excess convexity. Convex fillet welds are acceptable, providing the convexity is within the limits indicated by Fig. 4-49.

Figure 4-50 (Page 136) shows profiles of weld defects which result from poor welding technique.

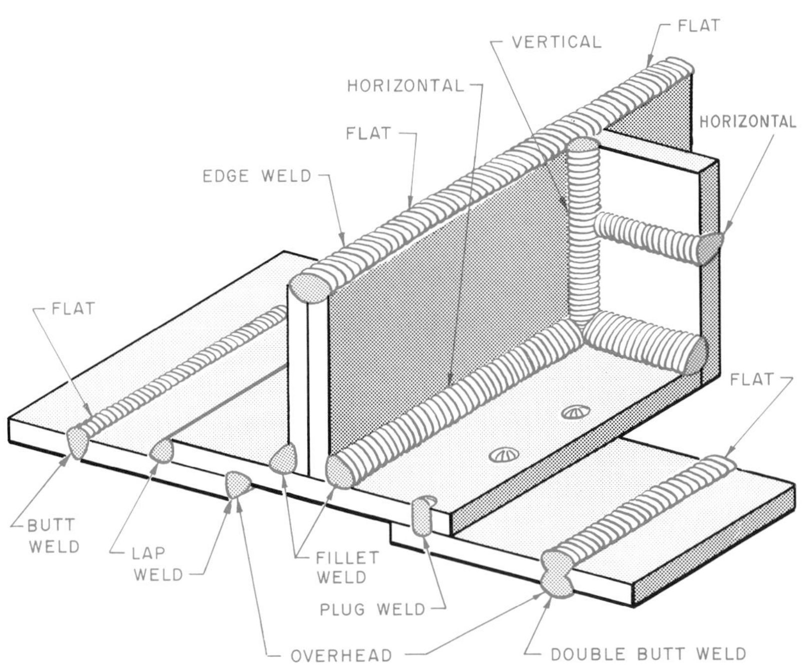

4-37. *Examples of welds and positions of welding.*

FILLET WELD WITH INSUFFICIENT THROAT. Reduction of the effective throat, Fig. 4-50A, materially reduces the size of the weld. The abrupt change in the face concentrates stress at the center. The smaller size of the weld and stress concentration weaken the weld and invite joint failure. The defect is usually caused by too fast travel and excessive welding current.

FILLET WELD WITH EXCESSIVE CONVEXITY. The weld metal in this type of defect, Fig. 4-50B, may contain a great deal of porosity due to slag and gas entrapment. There may also be poor penetration at the root of the weld and poor fusion of the weld metal to plate surfaces, Figs. 4-51 and 4-52. Stress concentrates at the toe of the weld. This weld defect is usually caused by low welding current and a slow rate of travel.

FILLET WELD WITH UNDERCUT. This is a cutting away of one of the plate surfaces at the edge of the weld, Fig. 4-50C. It decreases the thickness of the plate at that point. Any material reduction in plate thickness also reduces plate strength, thus inviting joint failure since the designed load of the joint is based on the original plate thickness. The possibilities of failure at this point are increased when undercut-

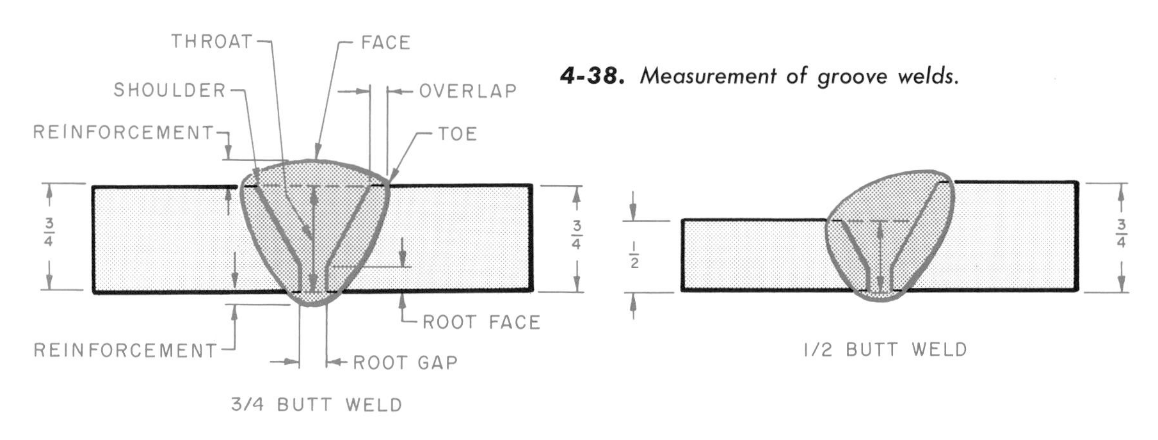

4-38. *Measurement of groove welds.*

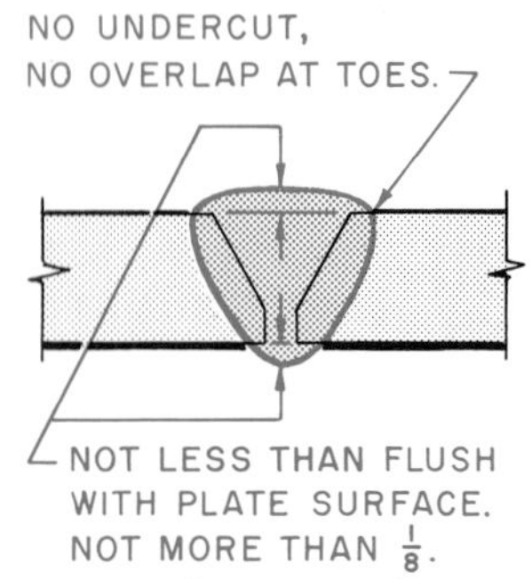

4-39. *Reinforcement for groove welds.*

4-40. *Measurement of fillet welds.*

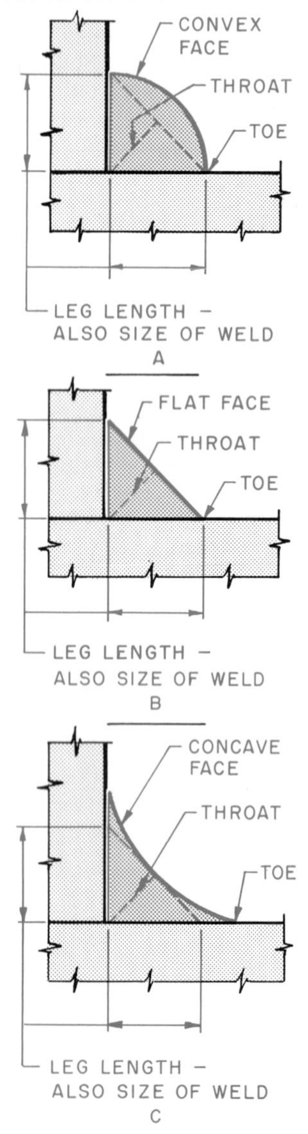

ting occurs at the toe of the weld, a point where there is high stress concentration. This weld defect is usually caused by improper arc manipulation, fast travel, and excessive welding current.

FILLET WELD WITH OVERLAP. Overlap, Fig. 4-50D and 4-51, is a sign of poor fusion (poor bond) between weld metal and base metal. It can be likened to applying a wad of chewing gum to a surface. When load is applied to the gum, it will peel from the surface. The welded joint will act in the same way under load, and the result will be weld failure. It is obvious that good fusion must be obtained if we are to prevent a peeling off of the weld metal when load is applied. Failure of the joint is certain when the overlap is located at the toe of the weld. This is a serious defect and should be avoided. It may be caused by low welding current, fast travel, or improper electrode manipulation.

FILLET WELD WITH INSUFFICIENT LEG. A reduction in leg length, Fig. 4-50E, means a reduction in the size of the fillet weld. If the demands of a joint require a fillet of a certain size, any reduction of that size results in a weld that does not possess the physical properties needed for safe operation. Failure is sure to result. This defect is usually caused by improper electrode angle and faulty electrode manipulation. In addition, these faults in welding technique may be accompanied by too fast travel.

FILLET WELD WITH POOR PENETRATION AND FUSION. This defect is usually found at the root of the weld and at the plate surfaces, Figs. 4-51 and 4-52. Stress is concentrated at the toe of the weld. Poor penetration and fusion are usually caused by welding with the current too low, the speed of travel too fast, and/or by improper electrode manipulation. When these conditions exist during welding, the deposited weld metal may become porous due to slag and gas entrapment.

4-41. *Ideal fillet weld shape.*

DEPTH OF FUSION OR BOND
LEG
THROAT
FACE
TOE
LEG
ROOT
ROOT PENETRATION

Butt Weld Profiles

Figure 4-53 shows acceptable butt weld profiles. It should be noted that the recommended reinforcement does not extend more than $\frac{1}{8}$ inch above the surface of the plate. Figure 4-54 shows defective butt weld profiles.

BUTT WELD WITH INSUFFICIENT THROAT. A decrease of the throat size, Fig. 4-54A, reduces the size of the butt weld. The thickness at the throat of the weld is less than the thickness of the plate, and the weld will not be as strong as the plate. Failure of the weld under maximum load is certain. This defect is usually caused by a combination of high welding current and travel that is too fast. Although penetration at the root of the weld may be complete and fusion to the plate surfaces may be excellent, these desirable characteristics cannot overcome insufficient weld size.

BUTT WELD WITH EXCESSIVE CONVEXITY. This is the opposite of a concave profile, Fig. 4-54B.

THIS WELD APPROXIMATELY 30% STRONGER THAN
THIS WELD PER UNIT LENGTH.

4-42. *Transverse welds are stronger than welds parallel to lines of stress.*

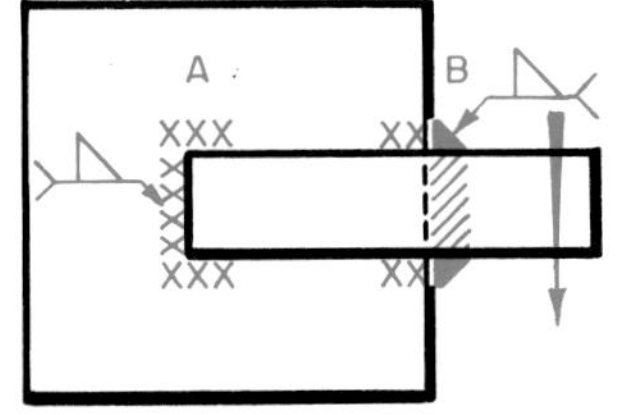

4-43. *Example of proper placement of welds to resist turning effect of one member at the joint.*

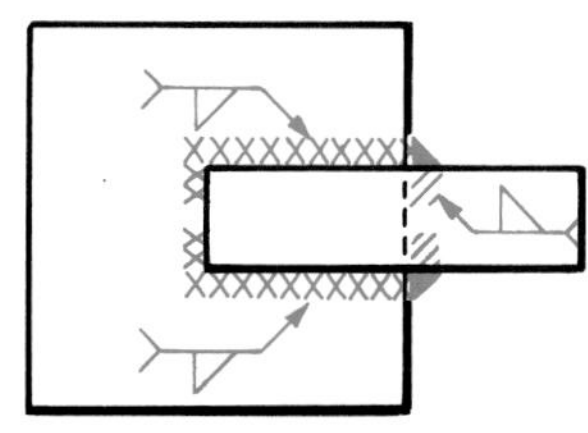

4-47. *Example of welds hooked around the corners to obtain resistance to tearing action on welds when subjected to eccentric loads.*

4-44. *A lap weld having poor distribution of stress through the weld. Excessive reinforcement.*

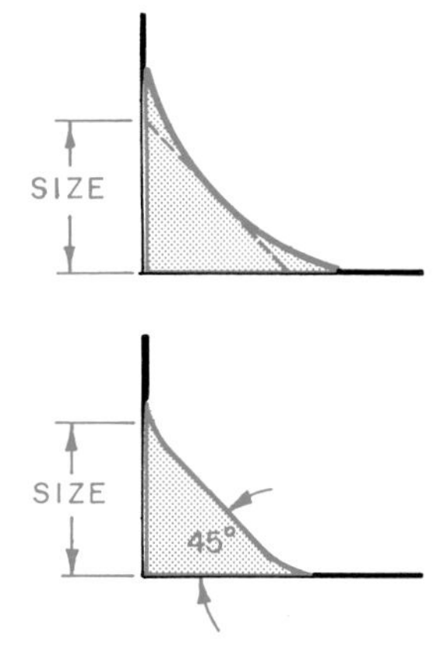

4-48. *Desirable fillet weld profiles.*

4-45. *A lap weld having a more even distribution of stress through the weld than that shown in Fig. 4-44.*

4-49. *Acceptable fillet weld profile.*

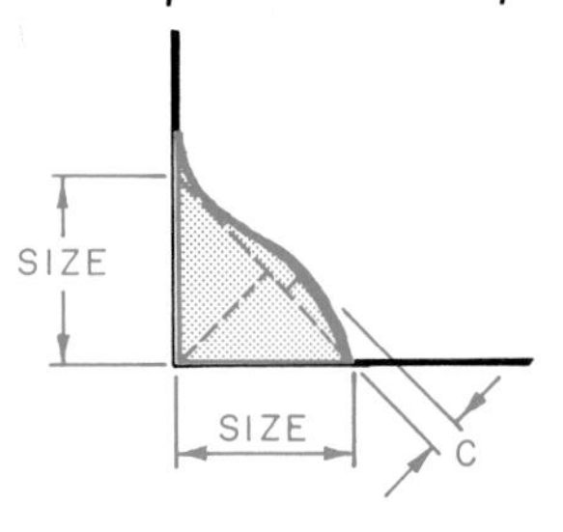

CONVEXITY, C, SHALL NOT EXCEED 0.15 ± 0.03 INCHES.

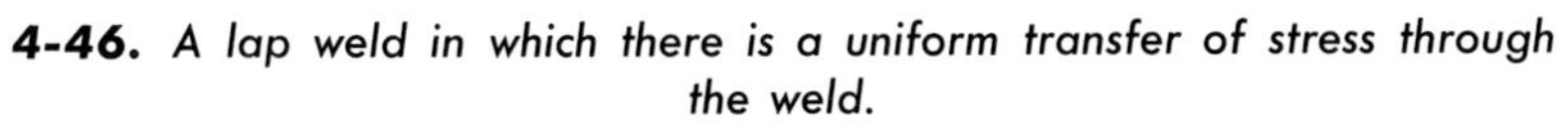

4-46. *A lap weld in which there is a uniform transfer of stress through the weld.*

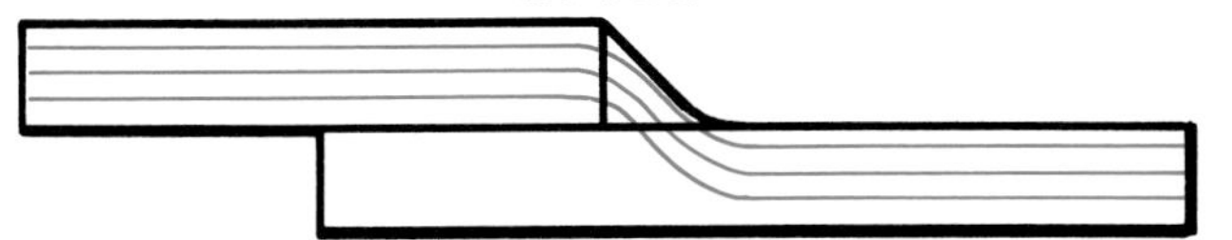

It may be less strong than the weld with insufficient throat, due to concentration of stress in the weld. Comparative strength, of course, depends upon the degree of convexity of one weld and the throat insufficiency of the other.

4-50. *Defective fillet weld profiles.*

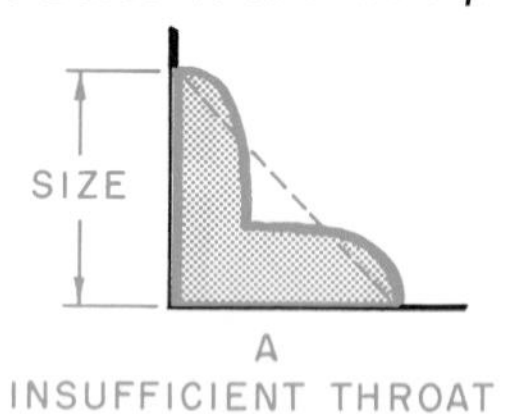

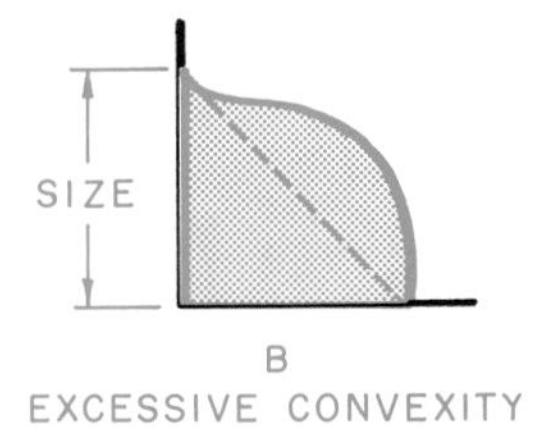

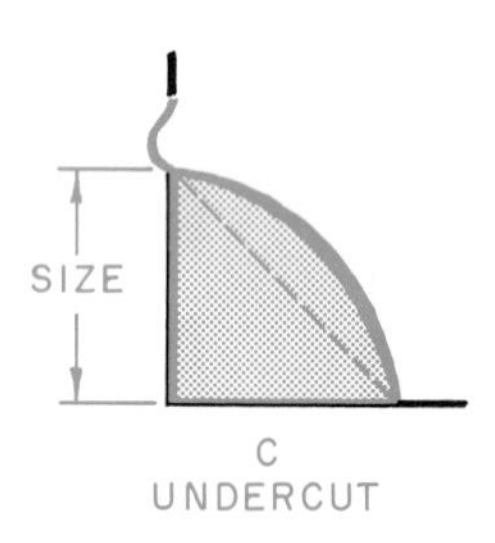

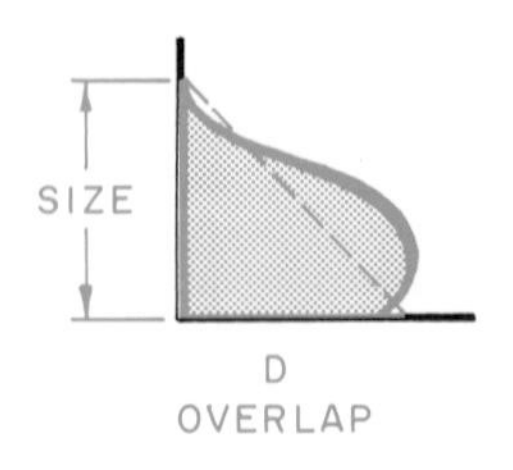

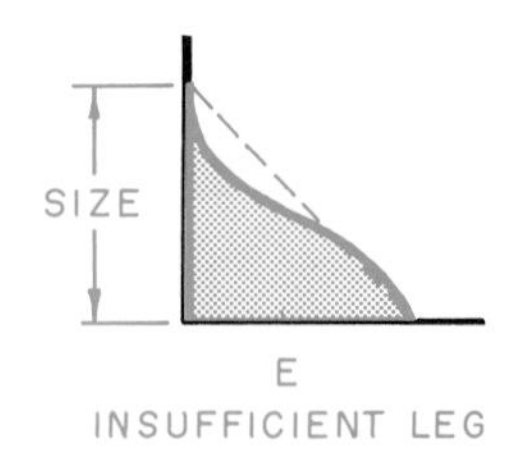

Excessive convexity may be caused by travel that is too slow or low welding current. Even though complete penetration and good fusion may exist, these desirable characteristics cannot overcome the loss of strength due to extreme convexity. There is also the possibility of porosity in the weld metal due to slag inclusion and gas pockets. The defect wastes material and time, thus increasing costs. Very poor appearance will also result.

BUTT WELD WITH UNDERCUT. As with the fillet weld, a cutting away of the plate surface at the toe of the weld results in a reduction of actual plate thickness, Fig. 4-54C. The reduction in plate surface, together with the concentration of stress at the toe due to the sharp corner, may cause failure of the welded joint at this point.

Undercutting may be considered the most serious defect of all. It is usually caused by high welding current, travel that is too fast, or improper electrode manipulation.

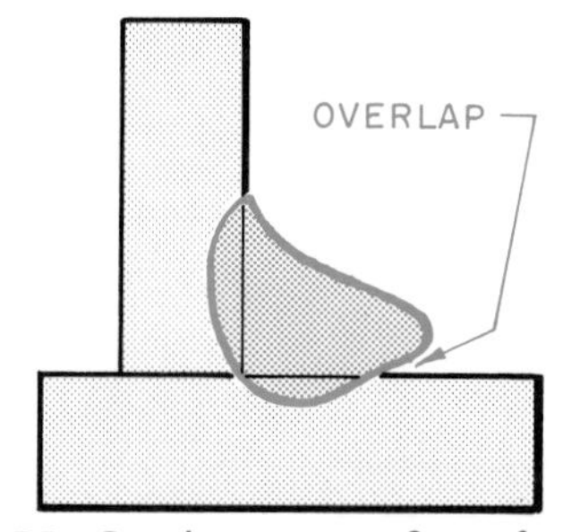

4-51. *Overlap, an overflow of weld metal beyond the end of fusion.*

4-52. *Incomplete fusion in fillet welds. B is often termed "bridging."*
American Welding Society

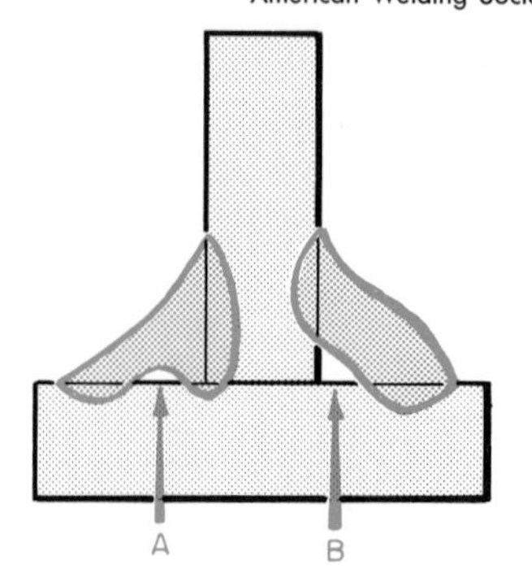

BUTT WELD WITH OVERLAP. Overlap, Fig. 4-54D, is the result of poor fusion. If this defect is present, there is every reason to believe that poor penetration and porosity in the weld metal will also be present. Overlap is usually caused by low welding current, fast rate of travel, or improper electrode manipulation. A weld with excessive convexity and overlap usually contains a certain amount of porosity and poor fusion. Figures 4-55 through 4-58 illustrate the defects that may be found alone or in combination.

Defects and Inspection

Welding students often ask the following questions about weld defects. Which defect contributes to joint failure? Which ones may pass inspection? Are there jobs on which weld defects are permitted? Knowing the construction and use of the weldment helps the operator to answer these questions. In general, his goal should be to avoid all defects: they all contribute to

4-53. *Acceptable butt weld profile.*

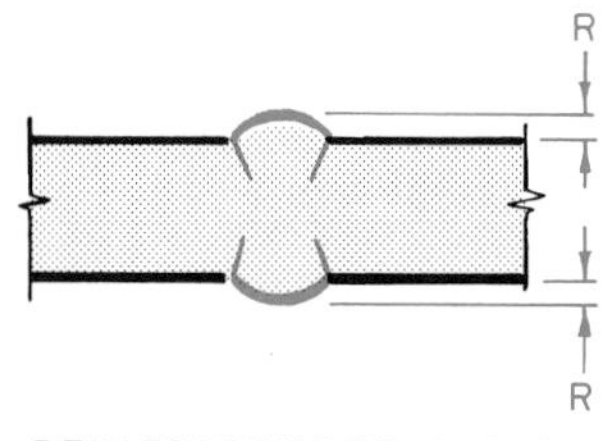

weld and joint failure. Minor allowances may be permitted if the work is not critical. No tolerance is permitted in critical or code work because high strength is necessary due to load conditions of heat, pressure, or stress.

If the welder expects to be recognized as a craftsman, he will always strive to do work that is sound and of good appearance. Inability to do so with the high quality equipment available suggests carelessness on the part of the welder. It will surely lessen the regard that the shop foreman and other workmen have for him or her.

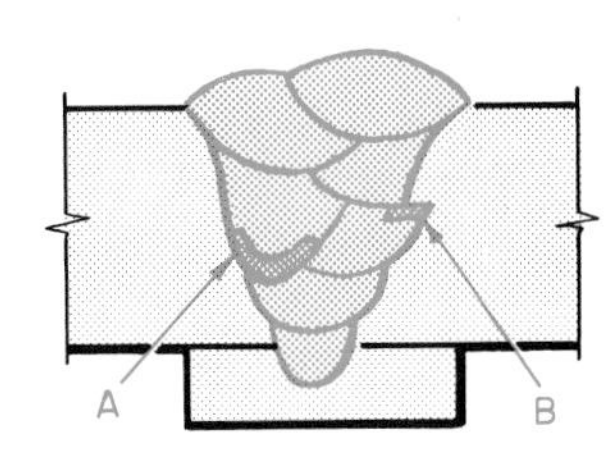

4-56. *Slag inclusions, between passes at A, and at undercut at B.*

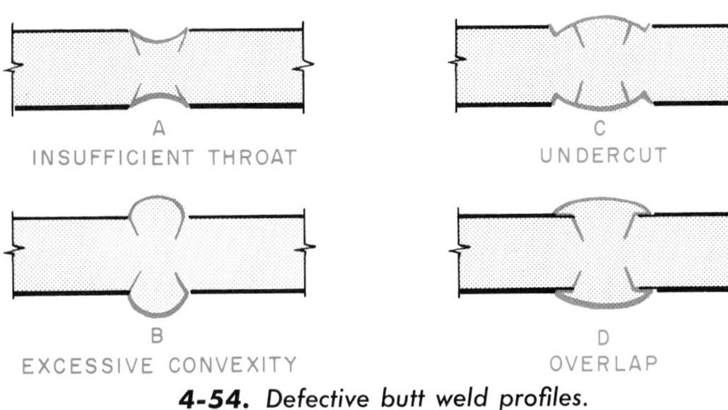

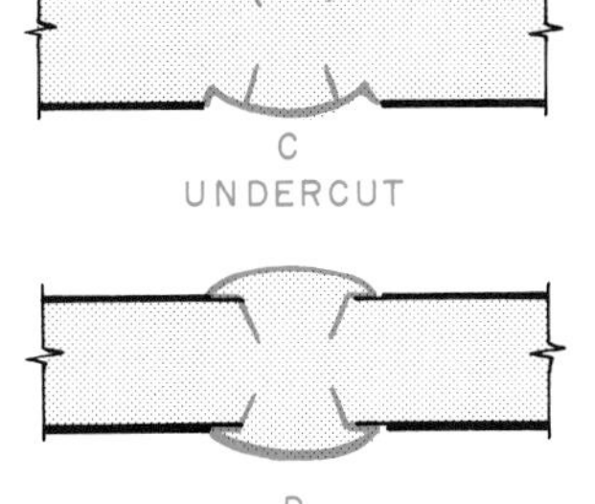

4-54. *Defective butt weld profiles.*

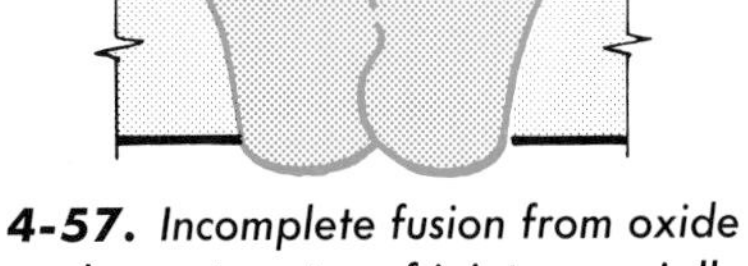

4-57. *Incomplete fusion from oxide or dross at center of joint, especially in aluminum.*

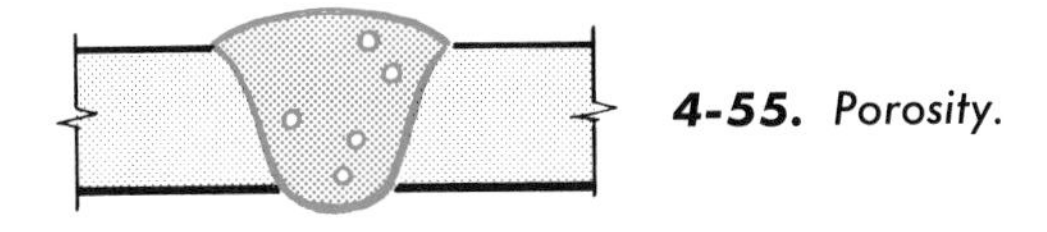

4-55. *Porosity.*

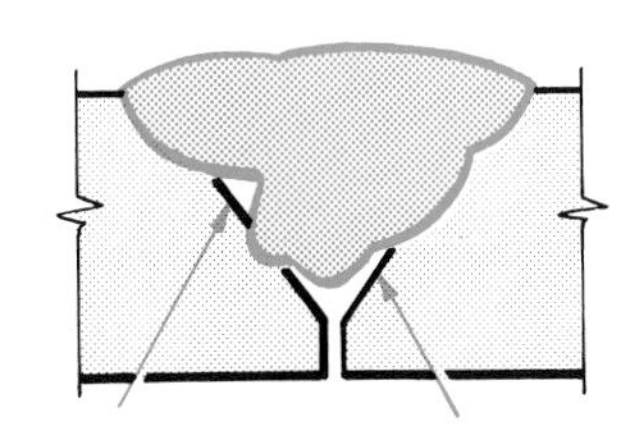

4-58. *Incomplete fusion in a groove weld.*

REVIEW QUESTIONS

1. Name five fundamental types of joints.
2. Describe the essential properties of a good butt joint.
3. The purpose of all bevels is preparation for necessary weld penetration. True. False.
4. The face of a weld is the top of the finished bead. True. False.
5. A triple-V joint is in common use and has very high strength characteristics. True. False.
6. Lap joints welded from one side are as strong as butt joints. True. False.
7. Name the four fundamental types of welds.
8. What are the factors that determine whether a joint is to have open or closed roots?
9. What is the purpose of tack welding?
10. Give a short explanation of the following: (a) continuous welds, (b) intermittent welds, (c) strength welds, (d) caulking welds, and (e) composite welds.
11. Name the four fundamental welding positions and explain each.
12. How are groove welds measured? Fillet welds?
13. A high reinforcement on a groove weld insures maximum strength. True. False.

14. A convex fillet weld is the strongest of all possible fillet welds. True. False.

15. Make a simple sketch of a groove weld in a single-bevel butt joint and name all of its parts.

16. Make a simple sketch of a fillet weld in a T-joint and list the names of all its parts.

17. Name three defects that may be found in fillet welds.

18. Name three defects that may be found in butt welds.

19. The face of a fillet weld may take on three face characteristics. Name them and describe each type.

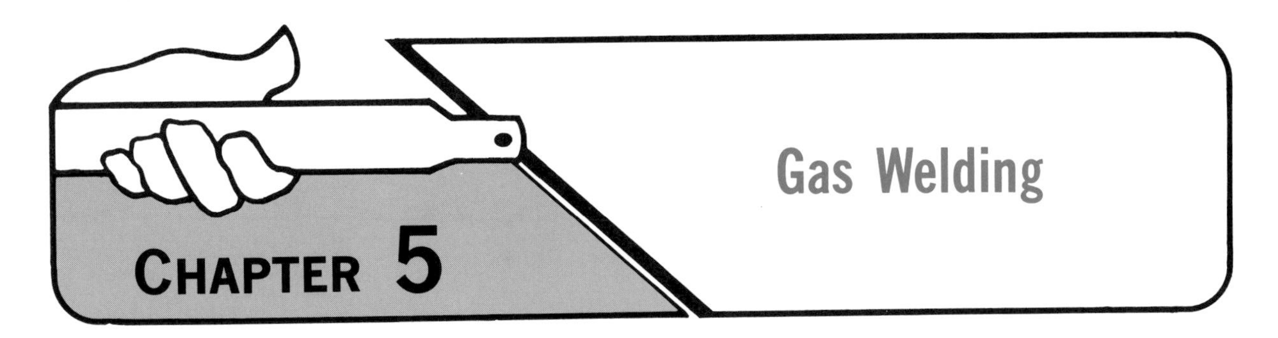

Gas Welding

The cooperation of the Linde Division, Union Carbide Corporation, and the Modern Engineering Company in providing information and photographs for this chapter is gratefully acknowledged.

OXYACETYLENE WELDING

Oxyacetylene welding (OAW), Fig. 5-1, is a way to join metal by heating the surfaces to be joined to the melting point with a gas flame, fusing the molten metal into a homogeneous mass, and then letting it solidify into a single unit. The flame at the cone reaches temperatures as high as 5,800 to 6,300 degrees F. It is produced by burning oxygen and acetylene gases mixed in the proper proportions in a welding torch. A filler rod may or may not be used to intermix with the molten pool of the metal being welded.

During the first part of this century, oxyacetylene welding became the major welding process both for fabrication and construction and for maintenance and repair. It had wide application since it can be used to weld practically all of the major metals. Today, however, we find that its use is limited for industrial production purposes. It is slower than the other welding processes, and many of the prime metals such as aluminum and stainless steel can be welded more easily with other processes. The oxyacetylene process is still used for performing such operations as brazing, soldering, and metalizing; welding metals with low melting points; and general maintenance and repair work, Fig. 5-2. Some production welding on light gauge materials and welding on pipe with small diameters is still being done with the oxyacetylene process, Fig. 5-3.

While the oxyacetylene process is not used as much as it once was, it has a wide enough application to make it a necessary skill. It is an excellent means through which the student welder can observe the effect of heat and the flow of molten metal. It also develops the coordination of both hands that is the basis of good technique for all welding processes. In industry many jobs require the ability to weld with both the oxyacetylene process and the electric arc processes.

The History of Oxyacetylene Welding

The oxyacetylene process had its beginning many centuries ago. The early Egyptians, Greeks, and Romans used an alcohol or oil flame to fuse metals.

In the nineteenth century various gases were tested in experimental welding. They were used in the laboratory and in working with precious metals. In 1847

5-1. *Welds made with the oxyacetylene process on steel plate. They were welded in the vertical-up position. Gas welding produces sound welds of good appearance. Welds can be applied in all positions.*

Robert Hare of Philadelphia fused platinum with an oxyhydrogen flame. In 1880 the production of oxygen and hydrogen through the electrolysis of water made possible the distribution of these gases of cylinders under pressure. Experiments were also done with oxygen-coal gas and air-hydrogen flames in the late 1800s.

A number of discoveries lead to the development of the oxyacetylene process.

- In 1836 Edmund Davey discovered acetylene gas.
- In 1862 acetylene gas was produced from calcium carbide.
- In 1895 Thomas L. Wilson began to produce calcium carbide commercially. It was first used for residential lighting.
- In 1895 LeChatelier, a French chemist, announced his discovery that the combustion of acetylene with oxygen produced a flame hotter than any other gas flame.
- In 1900, Edmond Fouche, a Frenchman, invented a high pressure acetylene torch. He later designed a low pressure torch that worked on the injector principle.
- In 1906 Eugene Bourbonville brought the first welding torch to this country. The process was first used for maintenance and repair.
- During World War I oxyacetylene welding came into its own as a production tool.

GASES

The oxyacetylene welding process makes use of two principal gases: oxygen and acetylene. However, a number of other fuel gases are also used. These include propane, natural gas, and a comparatively new gas called by the trade name of Mapp® gas. At one time hydrogen was also used extensively. Study the relative gas temperatures given in Table 5-1. Figure 5-4 shows the type of cylinders for the different gases, their sizes, and capacities.

5-2. *Student preparing to gas weld damaged automobile fender.*

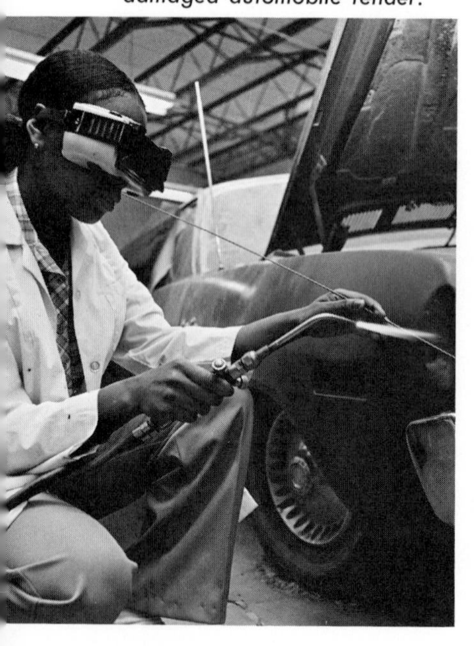

5-3. *Oxyacetylene welding a small diameter pipe on a construction site. This is a radiant heating installation in the floor.* Tube Turns

Oxygen

Oxygen is the gaseous chemical element in the air which is necessary for life. It is the most abundant chemical element in the crust of the earth. It has no color, odor, or taste. It does not burn, but by combining with other elements, it supports their combustion. Substances that burn in air do so much more vigorously in pure oxygen. Other substances, such as iron, that do not burn in air will do so in oxygen. It is this property that makes it useful in cutting iron and steel.

Oxygen Production and Distribution

There are two commercial processes used in the production of oxygen. One of these is the separation of air into oxygen and nitrogen by liquefying the air. The other method is the separation of water into oxygen and hydrogen by the *electrolysis* of water, or in other words, by passing an electric current through it. By far the greatest part of the oxygen used commercially today is manufactured by the liquefaction process.

Oxygen is distributed in steel cylinders, Fig. 5-5, which are made from a single plate of high grade steel which has been heat treated to develop maximum strength and hardness. Internal construction is shown in Fig. 5-6. Because of the high pressure involved, these cylinders undergo rigid testing and inspection. Oxygen is also distributed in bulk tanks which are made according to the same specifications.

VALVE MECHANISMS. Each cylinder has a valve which must be opened to release the oxygen. Some manufacturers use a double-seated valve which is perfectly tight when completely open or closed, Fig. 5-7. Valves that are not fully opened or closed cause oxygen to leak through the stem and result in a serious waste of materials. Other manufacturers use a type of valve that requires only a turn or two to open. The valve is protected from damage by an iron cap that screws on the neck ring of the cylinder. This cap should always be in place except when the cylinder is in use.

Cylinders are charged with oxygen at a pressure of about 2,200 p.s.i. at 70 degrees F. An increase in the temperature of the gas causes it to expand and increase the pressure within the tank. A decrease in temperature causes the gas to contract and reduce the pressure within the tank. To prevent excess pressure, every cylinder valve has a safety device to blow off the oxygen long before there is any danger of explosion. Nevertheless, cylinders should not be stored where they might become overheated. If this occurs, the safety disk will burst, and the oxygen will be lost.

CAPACITY OF CYLINDER. There are three cylinder sizes generally used for welding and cutting. Gas suppliers fill their tanks with varying amounts of gas. The large size contains 220 to 244 cu. ft. of oxygen. It weighs about 148 to 152 lbs. when full and 130 to 133 lbs. when empty. The middle size contains 110 to 122 cu. ft. of oxygen. It weighs 89 to 101 lbs. when full and 79 to 93 lbs. when empty. A small size has 55 to 80 cu. ft. of gas and weighs about 67 lbs. when full and 60 lbs. when empty. All of the oxygen in the cylinder may be used.

SAFETY IN HANDLING. It is important to remember that pure oxygen under pressure is an active substance. It will cause oily and greasy materials to burst into flame with almost explosive violence. The following safety precautions should be strictly observed.

- Take special care to keep oil and grease away from oxygen. Never store oxygen cylinders near oil, grease, or other combustibles.
- When using the cylinders, do not place them where oil might drop on them from overhead bearings or machines.
- Never use oxygen in pneumatic tools or to start internal combustion engines.
- Never use oxygen to blow out pipe or hose lines, dust clothes, or to create head pressure in a tank of any kind.
- Do not store oxygen cylinders near an acetylene generator, carbide, acetylene, or other fuel-gas cylinders.
- Do not drop or strike cylinders.
- Do not use a hammer or wrench to open cylinder valves.
- Do not use the cylinder as a roller or lift it by the cap.
- Keep cylinders away from the welding operation and close the cylinder valve when work is completed.
- Keep cylinders away from any electrical contact.

Acetylene

Acetylene is the most widely used of all the fuel gases, both for welding and cutting. It is generated as the result of the chemical reaction that takes place when calcium carbide comes in contact with water. Laboratory tests have shown oxyacetylene flame temperatures up to approximately 6,300 degrees F. Thus it has a very rapid rate of preheating. Acetylene has a peculiar odor. If lighted, it burns with a smoky flame, and gives off a great deal of carbon as soot.

Acetylene Production and Distribution

Commercial acetylene is made from calcium carbide, which is

Table 5-1
Various fuel gas efficiencies.

Fuel gas	Btus. cu. ft.	Combined intensity (usable heat) Btu. (sec./sq. ft. of flame cone area	Flame temp.[1]	Oxygen per cu. ft. of fuel[1]	Approx. normal velocity ft./sec.
Acetylene	1433	12,700	5,420°F.	1.04	17.7
Mapp®[2]	2381	5,540	5,301°F.	2.4	7.9
Propane	2309	5,500	5,190°F.	4.00	11.9
Natural gas[3] (Mpls./St. Paul)	918	5,600	5,000°F.	1.50	15.2
Hydrogen	275	7,500	4,600°F.	0.25	36

Linde Division, Union Carbide Corp.

[1] All the figures above are based on neutral flame conditions. Slightly higher flame temperatures can be attained for most of these gases by burning them in an oxidizing flame. Oxidizing flames require higher oxygen consumption. If the often quoted 6,300° temperature for acetylene were used, then figures for the other four gases would be respectively higher.

[2] Mapp® gas data supplied by Dow Chemical Co.

[3] Natural gas data varies with exact composition in different geographic areas. City gas is as low as 4400°F.

commonly referred to as *carbide.* Carbide is a gray, stonelike substance, which is the product of smelting coke and lime in an electric furnace. It is distributed in standard steel drums containing 100 pounds of carbide for use in acetylene generators. Several sizes are available: lump ($3\frac{1}{2} \times 2$ inches); egg, nut, quarter, or pea ($\frac{1}{4} \times \frac{1}{2}$ inches); rice and 14 ND. (0.055×0.0173 inches).

Carbide is also available in a special form known as *carbic-processed carbide,* which has been compressed in briquet form. The cakes are used in acetylene generators and floodlights. They are 4×3 inches and weigh about $2\frac{1}{2}$ pounds. They are packed in drums containing 40 cakes each.

Like oxygen, acetylene is distributed in cylinders. (See Fig. 5-5.) These cylinders are constructed differently from oxygen cylinders because of the fact that free acetylene should not be stored at a pressure above 15 p.s.i. After much study, the problem of combining safety with capacity was solved by packing the cylinders with a porous material filled with *acetone,* a liquid chemical having the property of dissolving or absorbing many times its own volume of acetylene. The cylinder itself is a strong steel container. It is packed completely full. In such cylinders, the acetylene is perfectly safe. Figure 5-8 shows three basic types of acetylene cylinders and their internal construction.

VALVE MECHANISMS. Acetylene is drawn off through a valve which in some cylinders is located in a recessed top; and in others, on a convex top. This valve does not have to stand the high pressure that the oxygen valve is subjected to and is,

5-4. *Gas cylinder sizes and capacities.*

Modern Engineering Co.

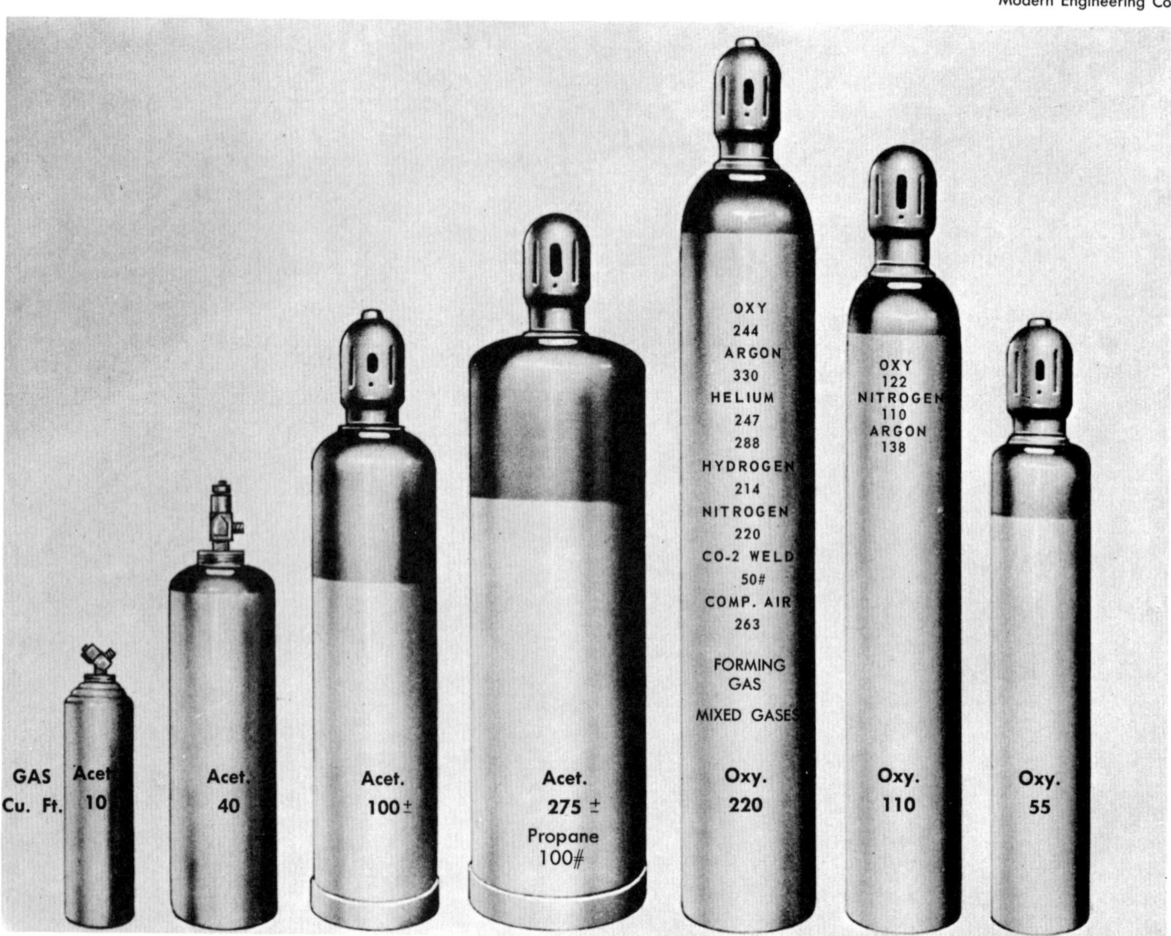

therefore, much simpler in construction. It needs to be opened only about $1\frac{1}{2}$ turns. This is done so that the cylinder can be turned off quickly if a fire starts in any part of the welding apparatus. Safety fuse plugs are also provided.

CAPACITY OF CYLINDER A full cylinder of acetylene has a pressure of about 225 p.s.i. Two sizes are generally used for welding and cutting. The large size contains about 300 cubic feet of acetylene and weighs about 232 pounds when full and 214 pounds when empty. The small size contains about 100 cubic feet of acetylene and weighs about 91 pounds when full and 85 pounds when empty. Two special sizes containing 10 cubic feet and 40 cubic feet are also available.

Not all the acetylene in the cylinder can be used. The maximum practical use of the gas is reached when the oxyacetylene flame begins to lengthen and loses much of its heat. The acetylene pressure regulator reading will be about 35 p.s.i. This varies with temperature.

SAFETY IN HANDLING. Remember that acetylene will burn, and like any other combustible gas, it will form an explosive mixture with air. The following precautions should be observed:

- Do not leave acetylene cylinders on their sides. Store and use them with the valve end up.
- Store the cylinders in a well-protected, ventilated, dry location. They should not be near highly combustible material such as oil or excelsior, or stoves, radiators, furnaces, and other sources of heat. Keep the valve cap on when the cylinder is not in use.
- If the outlet valve becomes clogged with ice, thaw it with warm (not boiling) water applied only to the valve. The fusible safety plugs with which all cylinders are provided melt at the boiling point of water. Never use a flame for thawing the valve.
- Handle acetylene cylinders carefully. Rough handling, knocks, and falls may damage the cylinder, valve, or fuse plugs and cause leakage.
- If acetylene leaks around the valve spindle when the valve on the cylinder is opened, close the spindle and tighten the gland nut around it. This compresses the packing around the spindle. If the leakage continues, stop using the cylinder and advise the supplier.

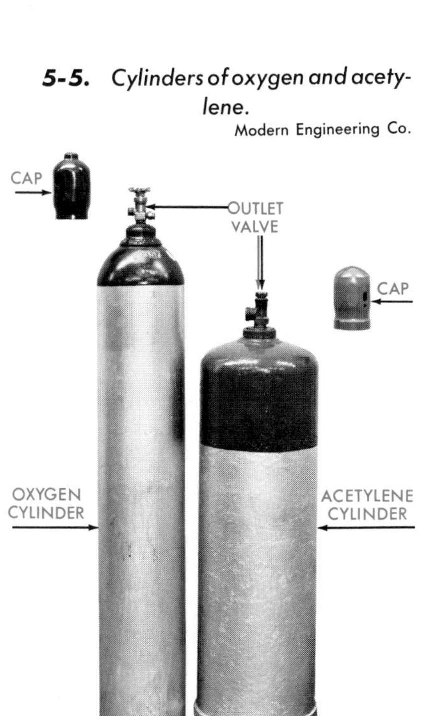

5-5. *Cylinders of oxygen and acetylene.*
Modern Engineering Co.

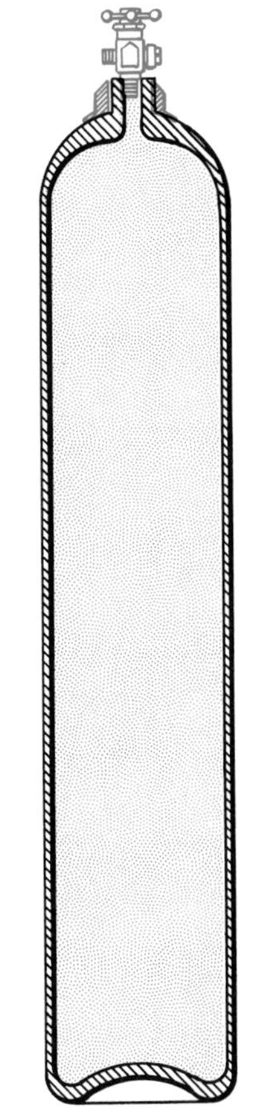

5-6. *Sectional view of an oxygen cylinder. Oxygen cylinders are seamless, drawn-steel vessels, having a malleable iron neck ring shrunk on at the top and a cylinder valve screwed into the neck.*

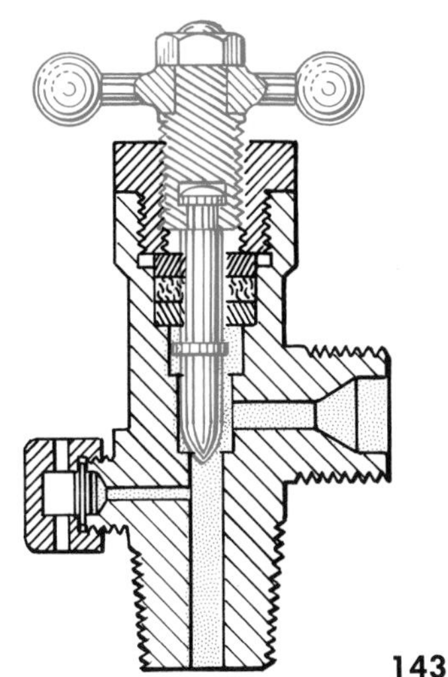

5-7. *This type of oxygen cylinder valve is double seated to prevent leakage when open. It should always be opened all the way. The valve is constructed to operate efficiently under high pressure.*

- Never tamper with the fuse plugs.

Propane Gas

Propane is a *hydrocarbon* present in petroleum and natural gas. Its use in gas welding and cutting is limited. It is sold and transported in steel cylinders containing from 20 to 100 pounds of the liquified gas. It can also be supplied by tank car and bulk delivery. The oxypropane flame temperature is approximately 5,190 degrees F. Because this temperature is less than that of oxyacetylene, it takes longer to bring the steel to the melting point. Propane is used extensively for soldering and alloy brazing.

Mapp® Gas

This is a liquified acetylene compound which is a comparatively new fuel gas for oxygas welding and cutting. It has a strong smell which is an aid in discovering leaks. When mixed with oxygen, the flame has a temperature of 5,301 degrees F. This temperature is higher than that of the oxypropane flame but not as high as that for the oxyacetylene flame. Although the welding and cutting may be somewhat slower due to the lower temperature, users indicate that overall expenses are lower due to reduced handling costs and lower gas costs. The use of this gas as a fuel gas for welding and cutting is growing.

MAPP® GAS DISTRIBUTION. Mapp® gas is distributed in bulk or in steel cylinders similar in appearance to acetylene cylinders. They have a shutoff valve similar to that on acetylene cylinders.

CAPACITY OF CYLINDER. Although Mapp® gas is an acetylene product, it is liquified and stabilized so that it can be used at pressures as high as 375 p.s.i. at 170 degrees as compared with 15 p.s.i. for acetylene. The explosive limits of Mapp® gas are lower than those of acetylene. Since the gas can be stored in its free state and at high pressures, a 120-pound cylinder contains as much gas as five 240-pound cylinders of acetylene. This reduces handling cost for the user.

SAFETY IN HANDLING. Mapp® gas, like all other gases, forms an explosive mixture with the air. The same general precautions that have been recommended in the handling of acetylene cylinders should be observed in the handling of these cylinders. It is the safest of industrial fuels. The explosive limits of Mapp® gas vapor in air and oxygen are much narrower than acetylene and about the same as propane and natural gas. You can smell it at concentrations as low as 0.01 percent, and find leaks that cannot be detected with other gases.

Manifold Distribution

Where it is necessary to supply a number of work stations and conserve space, both oxygen and the fuel gas may be supplied by a manifold system or an acetylene generator, Figs. 5-9 and 5-10. The acetylene manifolds must be equipped with a water seal flash arrester, Figs. 5-11 and 5-12, to prevent flashback through the manifold into the cylinders. In addition, each cylinder is connected to the manifold by means of an individual pigtail flash arrester and backcheck valves. There is the full cylinder pressure of 2,000 p.s.i. in the oxygen manifold pipes. The manifold regulator reduces the pressure to 50 or 75 pounds in the line which goes to the various station outlets in the shop. Figures 5-13 and 5-14 present an oxygen manifold installation.

5-8. Three basic types of acetylene cylinders and their internal construction.

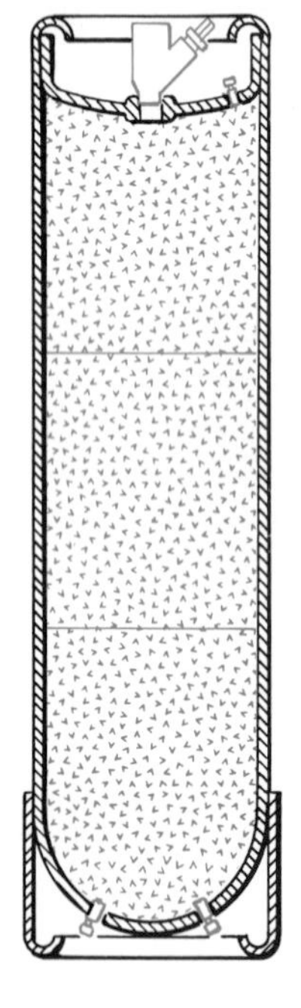
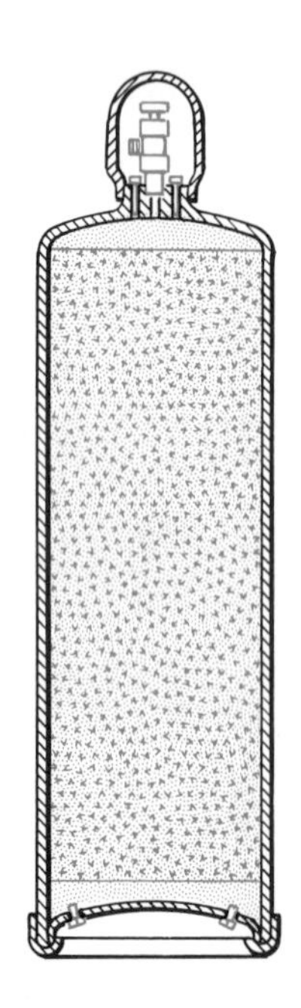
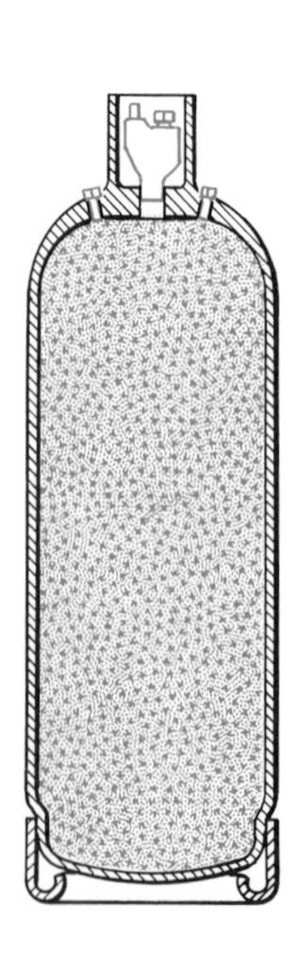

Both types of manifolds have a pressure regulator for the purpose of reducing and controlling the pressure of the gas to the work station. The work station is also equipped with an acetylene and oxygen regulator for further pressure control at the welding or cutting torch, as shown in Fig. 5-10.

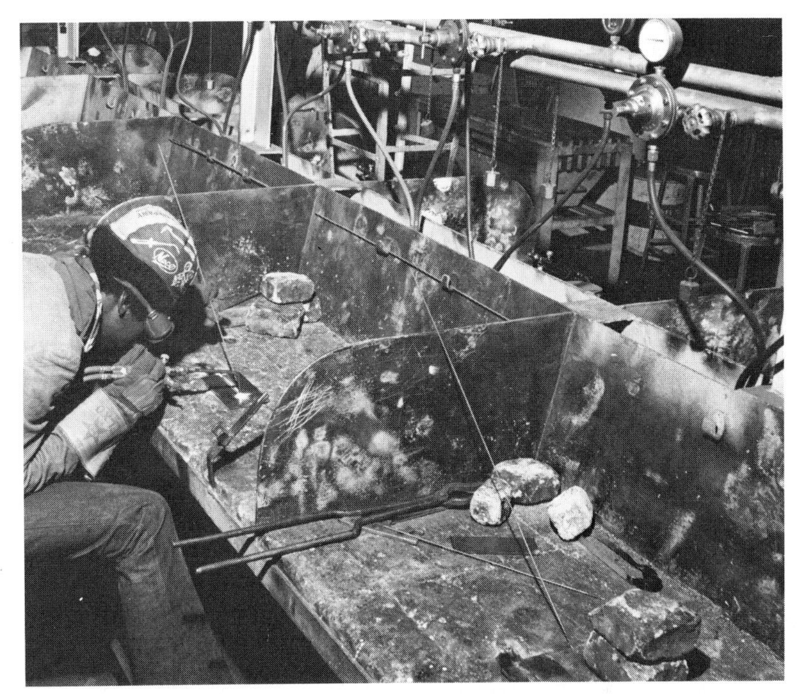

5-9. *School installation of manifold distribution to the work stations.*

Acetylene Generators

Large users of acetylene generate their own gas in an acetylene generator, Fig. 5-15 (Page 148). There are two general types. In one type the carbide is dropped into the water, and in the other the water is allowed to drip on the carbide. Miner's lamps are an example of the water-to-carbide principle.

The carbide-to-water generators are used to produce acetylene gas for welding and cutting. Small amounts of calcium carbide are fed into relatively large amounts of water. The water absorbs the heat given off by the chemical reaction, and the gas is cooled and purified by bubbling through the water, Figs. 5-16 and 5-17 (Pages 149, 150).

There are two classes of generators: low-pressure, in which the acetylene pressure is less than 1 p.s.i., and medium pressure, in which the acetylene pressure is from 1 to 15 p.s.i. You will recall that free acetylene cannot be used at a pressure higher than 15 p.s.i.

Generators can be further classified as stationary or portable. Generating capacities range from 30 cubic feet of acetylene per hour for a small portable, medium-pressure generator to 6,000 cubic feet of acetylene per hour for stationary types used in large industrial installations. Compare the large generator in Fig. 5-15 with the portable one shown in Fig. 5-18 (Page 151).

5-10. *A typical station outlet for both oxygen and acetylene pipelines from the generator house. Stations for the attachment of oxygen and acetylene regulators are placed at convenient points around the shop or plant where welding or cutting is to be done.*

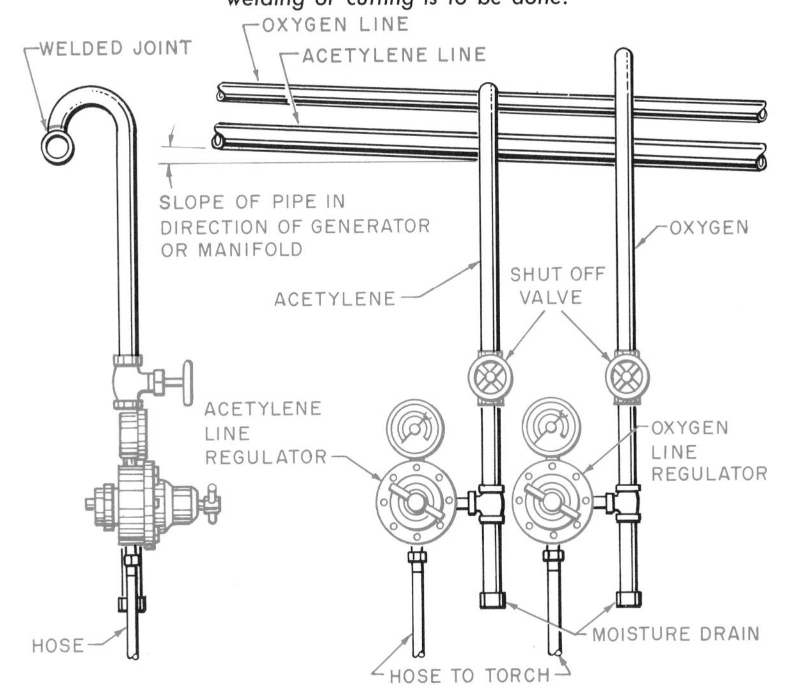

Modern Engineering Co.

5-11. *An acetylene gas manifold system.*

5-12. *Acetylene cylinder manifolds are constructed of extra heavy pipe and fittings to conform to the regulations of the National Board of Fire Underwriters.*

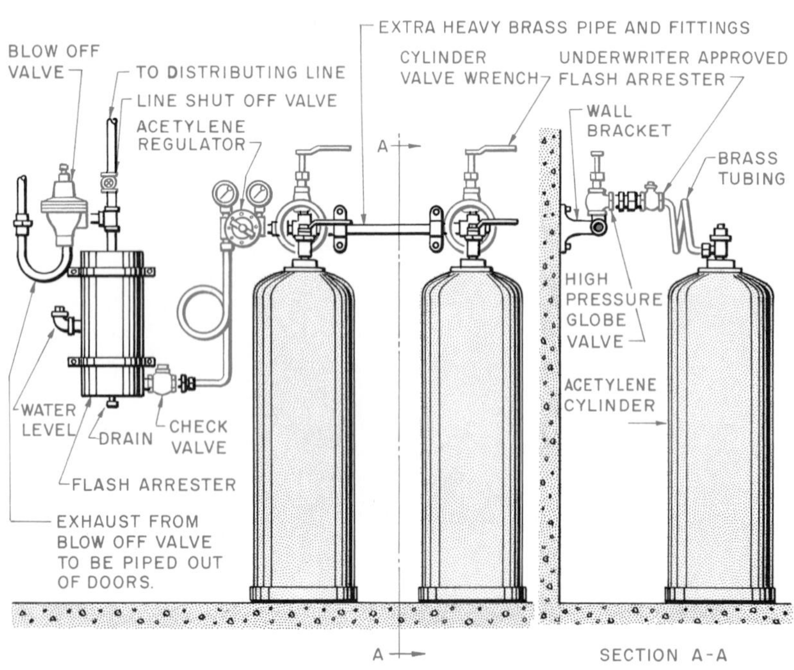

WELDING EQUIPMENT

Oxyacetylene welding requires the following equipment:

- Oxygen regulator
- Acetylene regulator
- Oxygen welding hose
- Acetylene welding hose
- Hose couplings
- Single-purpose cutting torch or welding torch

Much of this equipment is illustrated in Fig. 5-19 (Page 151). Note that the welding torch comes with a wide assortment of tips which provide a choice in the volume of heat desired. When welding or flame cutting, the operator must wear a pair of protective goggles to prevent harm to the eyes from hot particles of metal, sparks, and glare. A flint lighter is also required to light the torch. The equipment is set up for welding in Fig. 5-20 (Page 152). Portable tank outfits are mounted on a truck similar to that shown in Fig. 5-21 (Page 153).

Pressure Regulators

The pressure at which oxygen and acetylene is compressed into cylinders is much too high for direct use in welding and cutting. Some means must be provided to reduce the high internal cylinder pressure—about 2,200 p.s.i. in the case of oxygen—to the relatively low pressure of 0 to 45 p.s.i. required for welding and cutting. The flame must also be steady and uniform. This can be accomplished only if the gas pressures do not fluctuate. Pressure regulators carry out both of these all-important functions. They reduce the high cylinder pressure to that used for welding, and they can maintain that pressure without variation during the welding operation.

DESIGN. Figure 5-22 illustrates regulator design. They have a union nipple (A) for attachment to the cylinders and an outlet connection (B) for the hose leading to the torch. Two pressure gages are mounted on the body of the regulator: one (C) shows the pressure in the cylinder, and the other (D) shows the pressure of the gas being supplied to the torch. The pressure gauge that shows the pressure in the cylinder is also useful in indicating to the operator the amount of gas remaining in the cylinder.

The working pressure is adjusted by means of a hand screw (E). When this screw is turned to the left, or counterclockwise, the valve mechanism inside the regulator is shut off, and gas cannot pass through the regulator to the torch. Turning the pressure-adjusting screw to the right, or clockwise, presses it against the regulator mechanism. The valve opens, and gas passes through the regulator to the torch at the pressure shown on the working pressure gauge. Any pressure can be set up by turning the handle until the desired pressure is indicated. Figure 5-23 (Page 154) illustrates the internal mechanism of a typical single stage gas regulator.

SAFETY PRECAUTIONS. The following precautions regarding the use of regulators should be observed.

- Before opening the valve of a cylinder to which a regulator has been attached, be sure that the pressure-adjusting screw has been completely released by turning it to the left. Failure to do this allows the full cylinder pressure to hit the engaged mechanism of the regulator. Not only will the regulator be damaged, but it may also burst and injure the operator.
- Never attempt to connect an acetylene regulator to a cylinder containing oxygen or vice versa. Do not force connections that do

Modern Engineering Co.

5-13. *Oxygen manifold system. Note master regulator and flexible copper tubing leading from each tank.*

5-14. *An oxygen manifold. It is often convenient to place all of the oxygen cylinders at one point where they may be readily handled and to connect them together to an oxygen manifold as illustrated.*

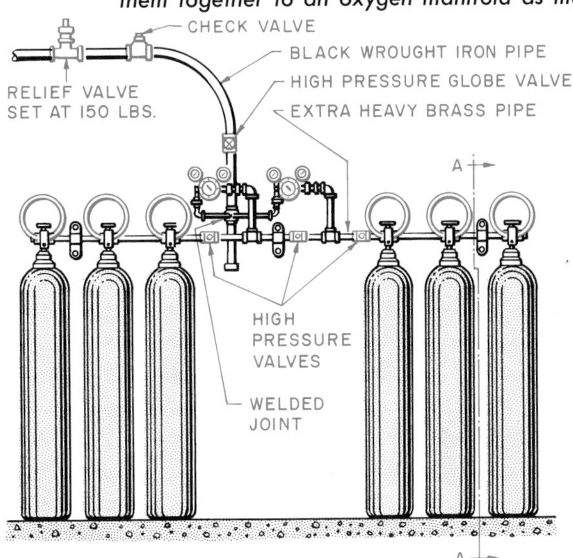

not fit, and be sure that all connections are tight.

- Use regulators only for the gas and pressures for which they are intended.
- Have regulators repaired only by skilled mechanics who have been properly trained.
- Inspect all nuts and connections before use to detect faulty seats which may cause leakage of gas.

Regulator Construction. To better understand the workings of the internal mechanism of a gas regulator, study Fig. 5-24 (Page 155). Let us consider an oxygen regulator as an example. The oxygen—under high pressure and coming directly from the oxygen cylinder—enters the regulator at the left at 2,200 p.s.i., and it must leave the regulator at the desired pressure for welding, perhaps 10 p.s.i.

Before the oxygen is allowed to enter at the left, tension on the spring (S) puts pressure on the flexible diaphragm, deflecting it to the left, which in turn causes it to contact the valve stem, opening the valve (V). If the oxygen tank is opened and the gas is permitted to flow into the chamber (C), the full force of 2,200 p.s.i. pressure is exerted against the diaphragm, causing it to move to the right. Thus the pressure against the valve stem is removed so that the valve (V) closes.

When, because of the requirements of the welding or cutting flame, oxygen has been withdrawn from the chamber (C), the pressure falls below a certain point. Therefore, the tension spring (S) becomes the greater force and deflects the diaphragm to the left, reopening the valve to permit more high pressure oxygen to enter the chamber. Bear in mind that the force which opens the valve is provided by the tension of the spring, while the force which closes the valve is provided by the high gas pressure from the tank. When these two forces are balanced, a constant flow of oxygen to the torch results. When the tension of the spring is properly adjusted by means of the regulator-adjusting screw (Figs. 5-22 and 5-23), the constant pressure desired is maintained in the chamber (C). Thus a constant pressure can be withdrawn from the chamber, providing an even flow of oxygen to the torch.

Regulators are designated as single-stage and two-stage regulators. The operation of the single-stage regulator has just been explained. In the two-stage regulator the pressure reduction is accomplished in two stages. In the first stage the spring tension has been set so that the pressure in the high pressure chamber is a fixed amount. For example, it may be set at 150 p.s.i. The gas then passes into a second reducing chamber that has a screw adjustment similar to that in the single-stage regulator. This adjustment makes it possible to obtain any desired pressure. The principle of the two-stage regulator is shown in Fig. 5-24, in which the upper sketch represents the first stage; and the lower, the second. Figure 5-25 (Page 156) illustrates three typical regulator designs.

For the most part single-stage regulators are used with manifold systems. In these systems the 2,200 p.s.i. pressure is reduced at the manifold before it enters the piping system through a heavy duty regulator on the manifold. The relatively low manifold pressure is further low-

5-15. *A large acetylene generator for use with multi-welding station installations. It can use carbide sizes from lump to 14 ND.* Rexarc

ered at the work station to the required pressure by the single-stage regulator.

Regulator springs are made of a good grade of spring steel. The diaphragm may be made of brass, sheet spring steel, stainless steel, or rubber.

When a number of stations are serviced by a line gas system, individual oxygen and acetylene regulators are required at each station. These regulators are smaller than cylinder regulators because they are not subject to high cylinder pressure. Figure 5-26 (Page 157) shows the difference in the internal construction of a tank and a line regulator.

Although the mechanical details of regulator construction vary among different manufacturers, the fundamental operating principles are the same for all oxyacetylene regulators used for welding and cutting.

Welding Torches

The welding torch is an apparatus for mixing oxygen and acetylene in the proportions necessary to carry on the welding operation. It also provides a handle so that the welder can hold and direct the flame while welding. The handle has two inlet gas connections: one for oxygen and the other for acetylene. Each inlet has a valve that controls the volume of oxygen or acetylene passing through. By means of these valves, the desired proportions of oxygen and acetylene are allowed to flow through the torch where they are thoroughly mixed before issuing from the torch tip. Each torch can be supplied with a wide range of welding tip sizes so that a large number of flame types and sizes can be set up for the various thicknesses of metal to be welded.

Two types of oxyacetylene welding torches are in common use: the injector and the equal (balanced pressure) types. Figure 5-27 (Page 158) shows an in-

5-16. Diagrammatic sectional view of a large carbide-to-water acetylene generator.

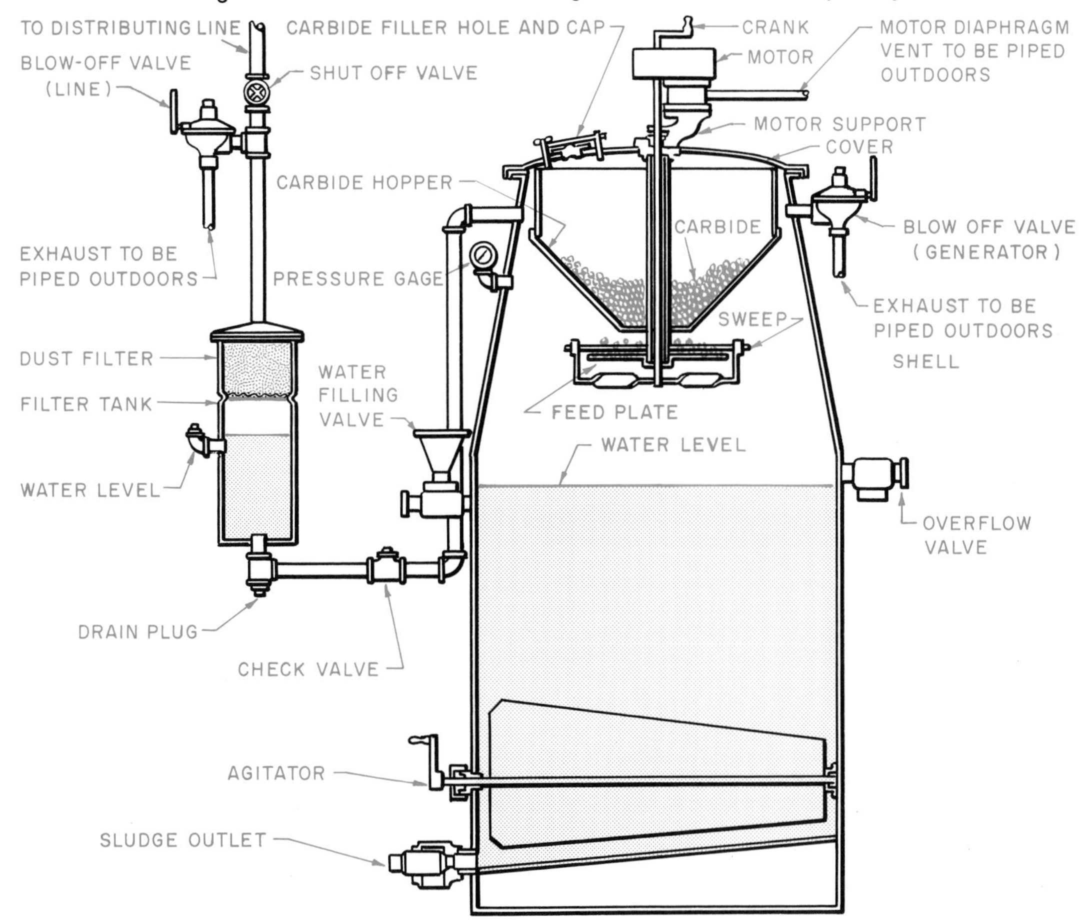

ternal view of the injector torch. The acetylene is carried through the torch and tip at low pressure by the suction force of the higher oxygen pressure passing through the small orifice of the injector nozzle, Fig. 5-28 (Page 158). The mixing head and injector are usually made as an integral part of the tip, and they are designed to correspond to the various tip sizes. These torches can be used with low and medium pressure oxyacetylene generators. The oxygen pressure used with this type of torch is considerably higher than that used with the equal pressure torch.

In the equal (balanced pressure) torch, Fig. 5-29, both gases are delivered through the torch to the tip at essentially equal pressures. The mixer or mixing head is usually a separate replaceable unit in the body of the torch into which a variety of tips may be fitted. See Fig. 5-30 (Page 160) for a comparison of the equal pressure and injector types of mixing chambers. A standard mixer can provide for the variety of tip sizes. These torches can be used with medium pressure acetylene generators.

Light gauge sheet metal welding and aircraft welding are usually done with a torch equipped with a smaller handle than the standard torch. This type of welding requires a delicate touch, and the heavier standard torch would be clumsy. An internal view of the balanced-pressure torch and the cutting attachment for light welding is shown in Fig. 5-31 (Page 161). Note the high pressure oxygen valve (poppet).

TORCH TIPS. Oxyacetylene welding requires a variety of flame sizes. For this reason a series of interchangeable heads or tips of different sizes and types are available. The size of the tip is governed by the diameter of the opening at the end of it. Manufacturers vary in their system of numbering. The tip size is marked on the side of the tip. The most common system consists of numbers which may range from 000 to 15. In this system the larger the number, the larger the hole in the tip and the greater the volume of heat that is provided. An increased volume of gas is also required for the flame. Table 5-2 gives information concerning one manufacturer's balanced pressure torch and tip sizes.

You will recall that tips for injector torches are provided with individual mixers so that the mixer and tip are one unit. Tips for medium pressure torches do not have the mixer as part of the tip. Mixers are part of the torch and serve a number of tip sizes. Also available are universal mixers designed to give proper gas mixtures for a full range of tip sizes.

Most welding tips are made of pure drawn copper because of the ability of this material to dissipate heat rapidly. Since copper is soft and tips are subject to considerable wear, both at the tip and threaded ends, tips must

5-17. *Carbide-to-water sight-feed acetylene generator.* Rexarc

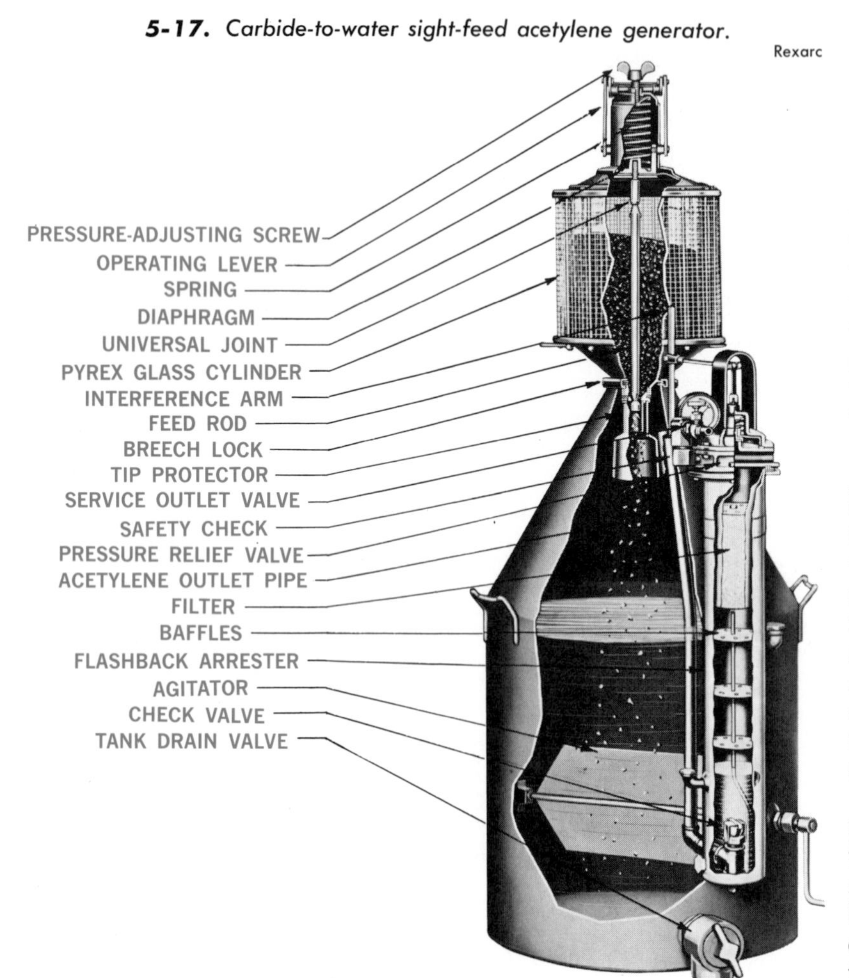

be handled carefully. The following precautions should be observed:

- Do not remove a tip with pliers. Pliers make heavy gouge marks on the tip. Manufacturers provide a wrench for their tip design which should be used at all times.
- Never insert or remove a tip while the tip tube is hot. Allow the tip and the tip tube to cool first.
- Keep the hole at the end of the tip clean at all times. During welding, weld spatter, scale, and molten metal may partially close the hole and cause the welding flame to be very uneven. The tip will also erode unevenly from the heat of the flame. The hole should be cleaned often with tip cleaners. Fig. 5-32. (Page 162). Do not scratch the tip end on the firebrick or the metal you

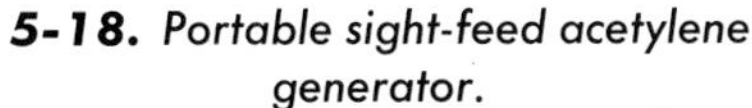

5-18. *Portable sight-feed acetylene generator.*

Rexarc

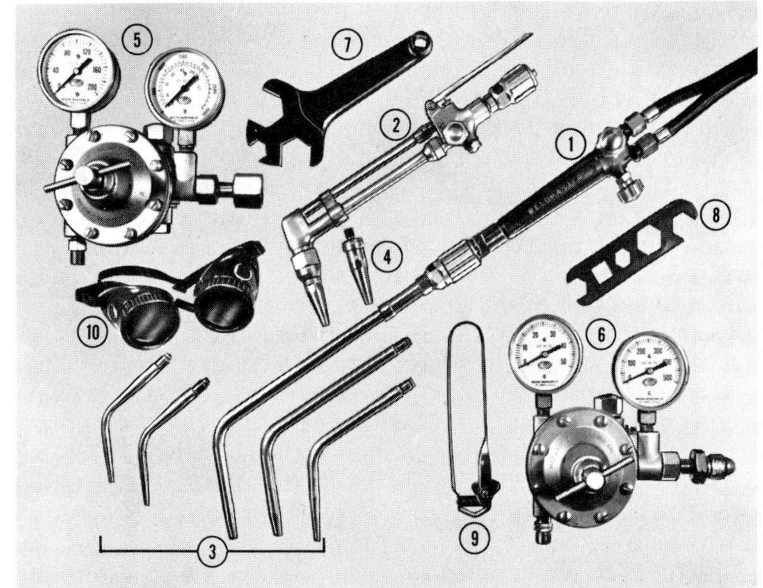

Modern Engineering Co.

5-19A. *Standard balanced-pressure welding torch and other equipment required for gas welding. Used for general gas welding and pipe welding. Note the cutting attachment: 1. Welding torch. 2. Cutting attachment. 3. Welding tips. 4. Cutting tip. 5. Oxygen regulator. 6. Acetylene regulator. 7. Tank and regulator wrench. 8. Torch wrench. 9. Lighter. 10. Goggles.*

5-19B. *Small balanced-pressure welding torch and supporting equipment. (Compare with Fig. 5-19A to identify items.) This size torch is used for light gauge sheet metal and aircraft welding. Note the cutting attachment.*

Modern Engineering Co.

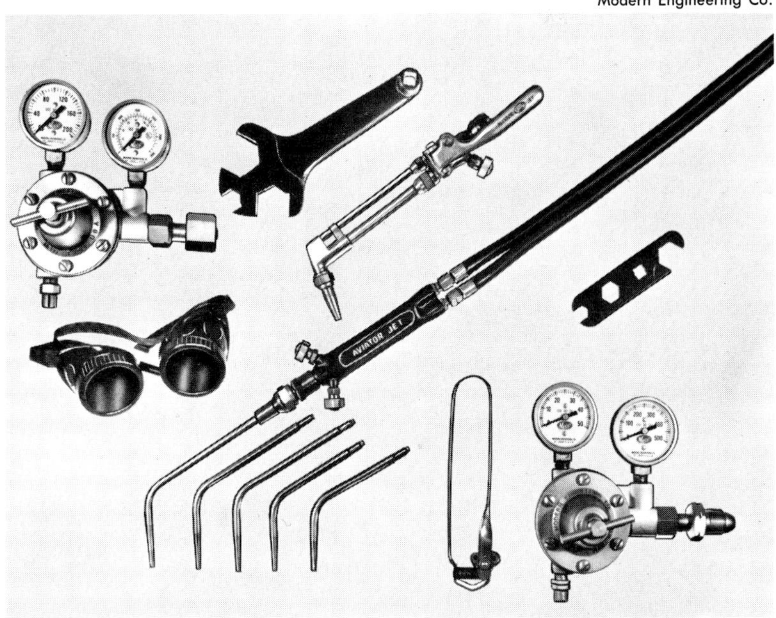

are welding. Some welders like to use a wood block for cleaning. The block removes contamination on the outside of the tip end, but it does not remove the particles on the inside of the hole.

- Do not use the tip as a hammer. This is the quickest way to destroy a tip.
- Protect the seat of the tip. If the tip is nicked through dropping or other rough treatment, it will leak at the joint and be dangerous or impossible to use.

Gas Economizer

The gas economizer, Fig. 5-33 (Page 162), is a widely used device which not only saves gas, but also provides a place to hang the torch and eliminates the need of constantly relighting the torch and readjusting the flame. Without the economizer the welder tends to lay the lighted torch down on the table or work, thus creating a serious hazard to himself and those around him.

The economizer is equipped with a pilot light and has passages through which the oxygen and acetylene pass before entering the torch. When you hang the lighted torch on the hook, the gases shut off automatically. When you take the torch off the hook and pass it over the pilot light, the gases turn on and the torch relights to its previous flame setting.

Oxygen and Acetylene Hose

The hose used for welding and cutting, Fig. 5-34 (Page 162), is especially manufactured for the purpose. It must be strong enough to resist internal wear, flexible enough so that it does not interfere with the welder's movement, and able to withstand a great deal of abrasive wear on the job.

Hose is usually made of three layers of construction. The inner lining is composed of a very good grade of gum rubber. This is surrounded by layers of rubber-impregnated fabric. The outside cover is made of a colored vulcanized rubber which is plain or ribbed to provide maximum life. The oxygen hose is green, and the word *oxygen* is sometimes molded on the hose. The acetylene hose is red, and the word *acetylene* is sometimes molded on the hose.

A black hose is used for inert gas and air. Different-sized hoses may be required for different types of welding and cutting operations, depending upon the amount of gas that is required, the length of hose used, and the pressures that are needed. Hoses can be obtained in sizes of $\frac{3}{16}$, $\frac{1}{4}$, $\frac{3}{8}$, and $\frac{1}{2}$ inch. The $\frac{3}{16}$ I.D. (inside diameter) hose is very flexible and light; it is used for light welding such as aircraft welding and as a whip for pipe welding. The $\frac{1}{2}$-inch I.D. is usually used for heavy cutting. Hose may be single or double. The

5-20. *Diagrammatic sketch showing cylinders, regulators, hose, and welding torch properly connected and ready to operate.*

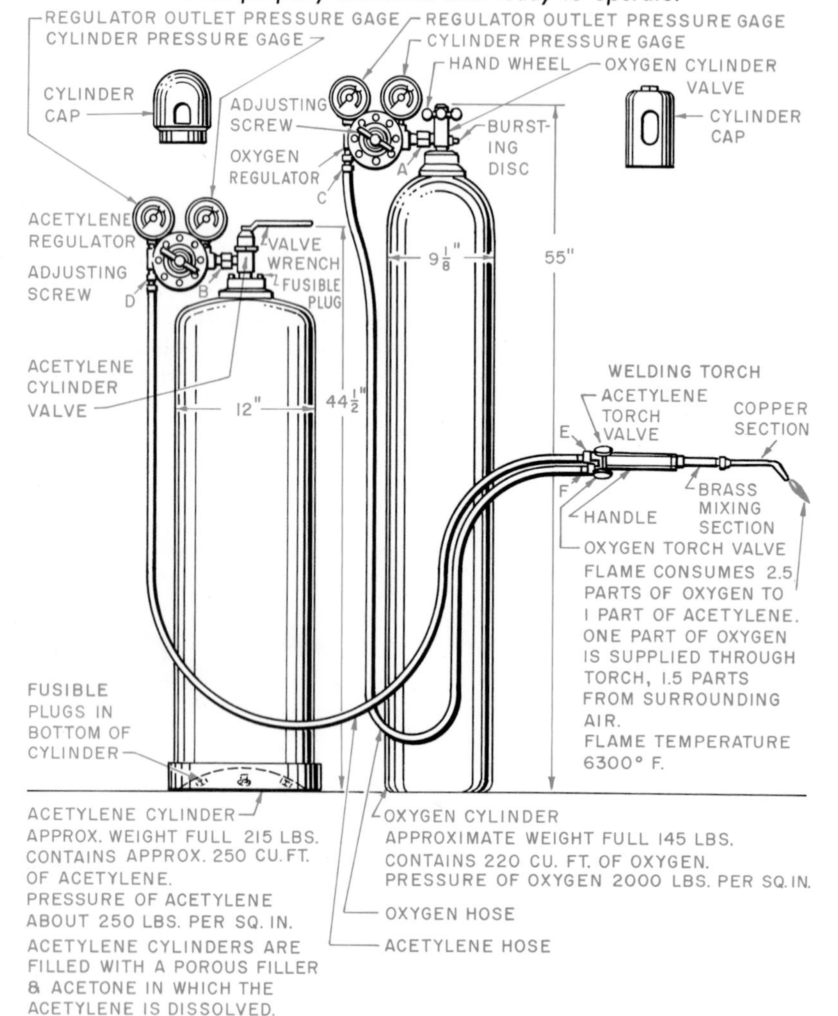

double hose is actually two pieces joined by a web which prevents tangling the hose.

HOSE CONNECTIONS. Only hose connections made for the purpose should be used for connecting the hose to the regulator and the welding and cutting torch. A standard hose connection, Fig. 5-35 (Page 162), consists of a nipple, which is inserted in the end of the hose, and a nut which attaches the nipple to the torch or regulator. So that there is no danger of attaching the wrong hose to the wrong regulator or torch connection, the oxygen coupling has a right-hand thread, and the acetylene coupling has a left-hand thread. The hose connection nuts are marked *STD. OXY* for the oxygen and *STD. ACET* for the acetylene. In addition, the acetylene nuts have a groove cut around their center to indicate a left-hand thread. Clamps or ferrules connect the hose tightly to the nipple to insure a leakproof connection.

CARE OF HOSE. The welder must use and care for hose correctly in the interest of both economy and safety. Hose is subject to a great deal of wear even under normal conditions, and if it is abused, the wear is considerable. The welder is urged to take the following precautions.

- Always use a hose to carry only one kind of gas. A combustible mixture may result if it is used for first one gas and then another.
- Test the hose for leaks frequently by immersing the hose at normal working pressure under water. A leaking hose is a serious hazard and a waste of gas. If there is a leak in the connection, cut off the hose a few inches back and remake the connection. Leaks in other locations should be repaired by cutting off the bad section and inserting a hose coupling as a splice.
- Clamp all of the hose connections or fasten them securely so that they will withstand a pressure of at least 300 p.s.i. without leakage.
- Hose showing leaks, burns, worn places, or other defects are unfit for service and must be repaired or replaced.
- Do not attempt to repair a hose with tape. Tape has a tendency to break down the hose material, and it is not a permanent repair.
- Handle the hose carefully when welding. Avoid dragging it on a greasy floor. Hose should not be allowed to come in contact with flame or hot metal. The hose should be protected from falling articles, from vehicles running over it, and from being stepped on. It should not be kinked sharply.
- At the end of the day or at the end of the job, roll up the hose and hang it where it will be out of the way. Spring-loaded hose reels are available.

Lighters

The welding torch should be lighted with a friction spark lighter, Fig. 5-36. The flints for friction lighters can be easily replaced at small cost when worn out. Matches should never be used because the operator's hand has to be too close to the torch tip and may be burned when the gases ignite. If the operator carries matches in his

5-21. *Steel handtruck used to carry oxygen and acetylene tank and welding equipment. A portable outfit offers wide flexibility in the use of gas welding in both shop and field.*
Modern Engineering Co.

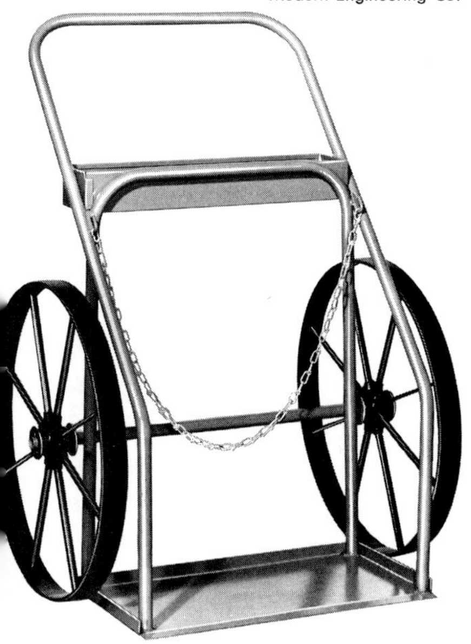

5-22. *Oxygen and acetylene welding regulators for use with individual tanks.*
Modern Engineering Co.

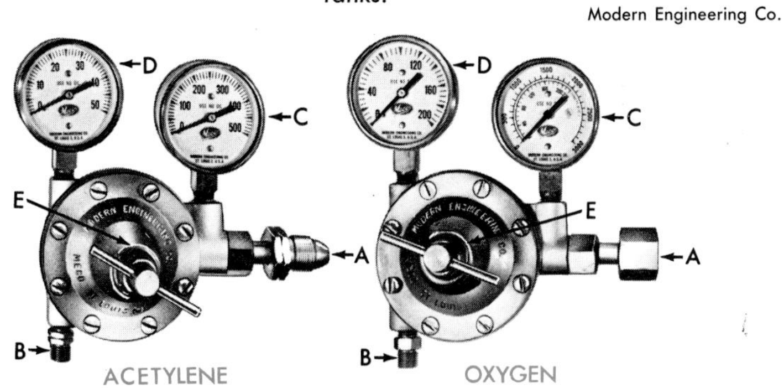

pockets, there is also the danger that they will ignite while he is welding and cause severe burns.

Filler Rod

While a great deal of oxyacetylene welding is done by merely fusing the metal edges together, most gas welding is done with the addition of a filler rod. The filler rod provides the additional metal necessary to form a weld bead. Filler rods are available for the welding of mild steel, cast iron, stainless steel, various brazing alloys, and aluminum. The usual rod length is 36 inches, and the diameters are $\frac{1}{16}$, $\frac{3}{32}$, $\frac{1}{8}$, $\frac{5}{32}$, $\frac{3}{16}$, $\frac{1}{4}$, $\frac{5}{16}$, and $\frac{3}{8}$ inch. Welding rods are available in bundles of 50 or 100 pounds net weight and in boxes of 10, 50, 100, and 300 pounds net weight.

Steel rods are copper coated to keep them from rusting. Some types of aluminum rods are flux coated to improve their working characteristics. Both steel and aluminum rods are 28 inches in length.

The American Welding Society has set up the following AWS classification numbers for steel gas welding rods: RG-65, RG-60, and RG-45. The letter *R* indicates a welding rod and the *G* indicates that it is used with gas welding. The numbers designate the approximate tensile strength of the weld metal produced in thousand pounds per square inch. For example, 45 designates a rod with a tensile strength of approximately 45,000 p.s.i.

Gas welding rods are often classified as GB-45, GA-50, GA-60, GB-60, GA-65, and GB-65. In this system the letter *G* stands for gas welding; the letter *A*, for welding rod of higher ductility; and the letter *B*, for welding rod of lower ductility. The numbers designate the approximate tensile strength of the weld metal produced.

ROD CHARACTERISTICS. Gas welding rods produce welds of varying tensile strengths depending upon the nature of the base

5-23. *Internal construction of single-stage cylinder regulator.*

Modern Engineering Co.

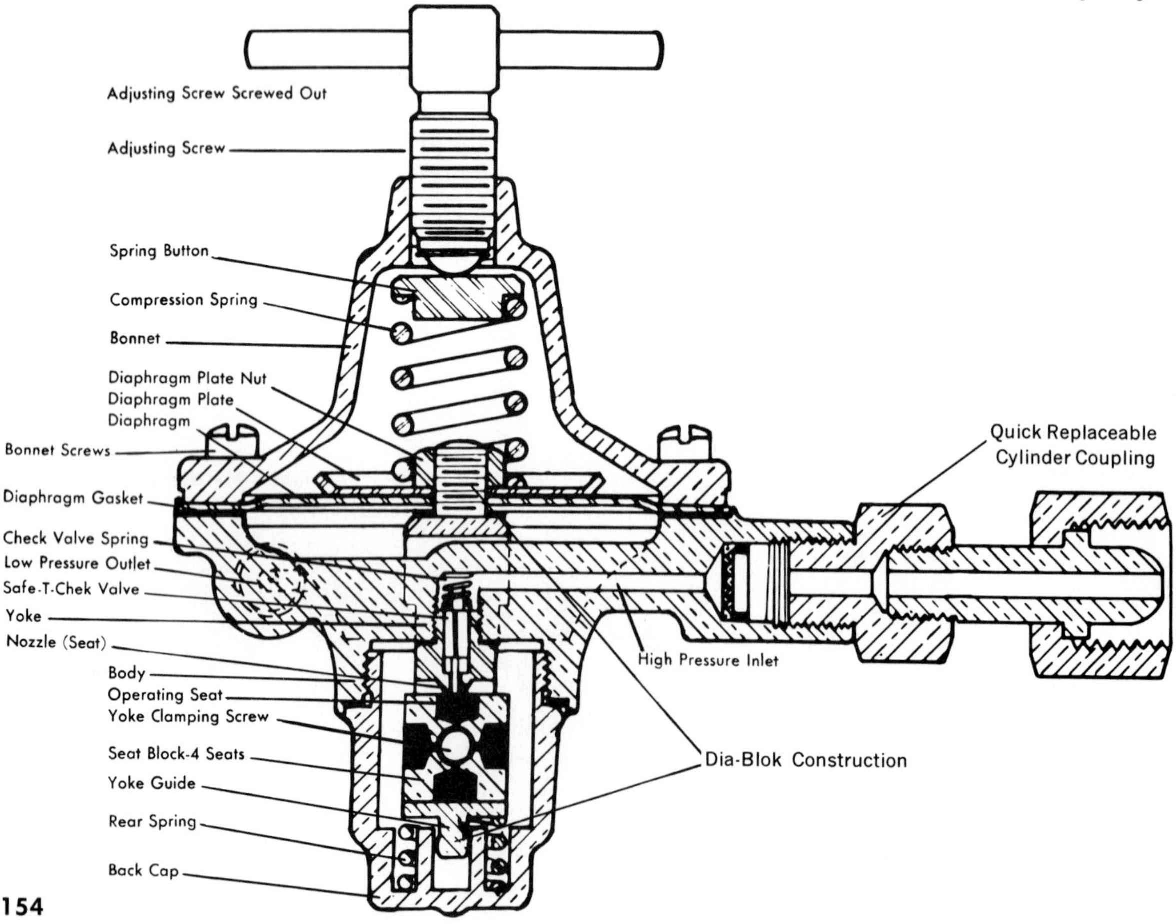

metal. Welds made on alloy steels will produce weld composition between that of the base metal and the filler metal. Gas welding rods can be used in all positions, limited only by the skill of the welder.

Class RG-65. RG-65 gas welding rods are of low-alloy steel composition and may be used to weld sheet, plate, tubes, and pipes of carbon and low-alloy steels. They produce welds in the range of 65,000 to 75,000 p.s.i.

Class RG-60. RG-60 gas welding rods are of low-alloy composition and may be used to weld carbon steel pipes for power plants, process piping, and other severe service conditions. They produce welds in the range of 50,000 to 65,000 p.s.i. This type of welding rod is used extensively and is considered a general-purpose welding rod. It produces highly satisfactory welds in such materials as carbon steels, low-alloy steels, and wrought iron.

Class RG-45. RG-45 gas welding rods are general-purpose welding rods of low-carbon steel composition. They may be used to weld mild steels and wrought iron, and they produce welds in the range of 40,000 to 50,000 p.s.i.

Fluxes

A flux is a cleaning agent used to dissolve oxides, release trapped gases and slag, and cleanse metal surfaces for welding, soldering, and brazing.

The oxides of all commerical metals, except steel, have higher melting points than the metals themselves and do not flow away readily. The function of the flux is to combine with oxides to form a fusible slag having a melting point lower than the metal so that it flows away from the weld area. Since there is a wide variation in the chemical characteristics and the melting points of the different oxides, there is no one flux that is satisfactory for all metals.

The melting point of a flux must be lower than that of either the metal or the oxides formed so that it will be liquid. The ideal flux has exactly the right fluidity at the welding temperature so that it can blanket the molten metal to protect it from atmospheric oxidation. It must remain close to the weld area instead of flowing all over the job.

Fluxes are available as dry powders, pastes, thick solutions, and coatings on filler rod. Powdered fluxes are packed in tin cans or glass jars. When not in use, they should be stored in a closed container because they lose their welding properties if exposed too long to the atmosphere.

Fluxes differ in their composition and the way in which they work according to the metals with which they are to be used. In cast iron welding, slag forms on the surface of the puddle, and the flux breaks the slag up. In welding aluminum the flux combats the tendency for the heavy slag to mix with the melted aluminum and weaken the weld. Flux will be explained in more detail in connection with the welding of those metals that require the use of a flux.

Other Gas Welding Processes

OXYHYDROGEN WELDING. Oxyhydrogen welding (OHW) is a

5-24. *Connecting the outlet shown in the top sketch to the inlet of the lower demonstrates the principle of two-stage regulators.*

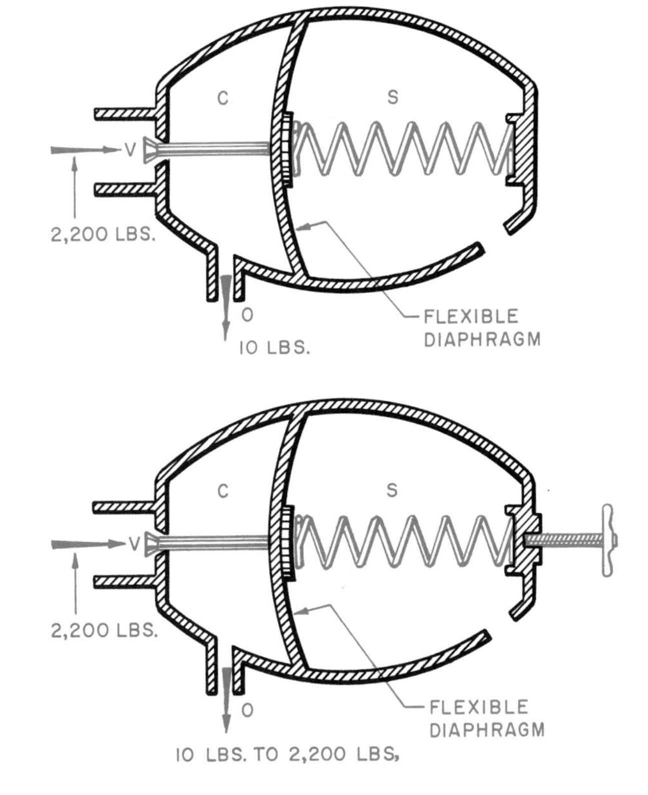

form of gas welding that was once used extensively. Today it has only a limited use.

The oxyhydrogen flame is produced by burning two volumes of hydrogen (H_2) with one volume of oxygen (O_2). The result is a flame with a temperature of approximately 4,100 degrees F. The flame is almost invisible, making it somewhat difficult to adjust the welding torch.

The equipment necessary is very similar to that used for oxyacetylene welding. The same torches, mixers, tips, and hose are used for both. There is some difference in the regulators. A standard oxygen regulator is used on the oxygen cylinder. A regulator specifically designed for use with hydrogen must be used with the hydrogen cylinder.

Because of the relatively low flame temperature, the process is used principally in the welding of metals having low melting points, such as aluminum, magnesium, and lead. It is also used to a limited extent in the welding of very light gauge steel and brazing operations.

The oxyhydrogen flame is still used extensively in the welding of lead. The lower flame temperature is ideal for lead because it has a low melting point. Oxyhydrogen is used for welding thicknesses of lead up to $\frac{1}{4}$ or $\frac{3}{8}$ inch. For greater thicknesses the oxyacetylene flame is generally used because of the greater heat input required. Another advantage of oxyhydrogen welding is that there is no deposit of carbon, which accelerates corrosion in welded assemblies.

FUEL GASES. Standard oxyacetylene welding equipment can also be used with propane, butane, city gas, and natural gas. It is necessary to have a special fuel gas regulator for use with these gases. Suitable heating and cutting tips are available in a variety of sizes. City gas and natural gas are supplied by pipelines, while propane and butane are stored in cylinders or delivered in liquid form to storage tanks on the user's property.

Because of the oxidizing nature of the flame and the relatively low flame temperature, these gases are not suitable for welding ferrous materials. They are used extensively for both manual and mechanized brazing and soldering operations. The plumbing, refrigeration, and electrical trades use propane in small cylinders for many heating

5-25. *Typical gas regulators (A) single-stage stem type, (B) single-stage nozzle type, and (C) two stage.*

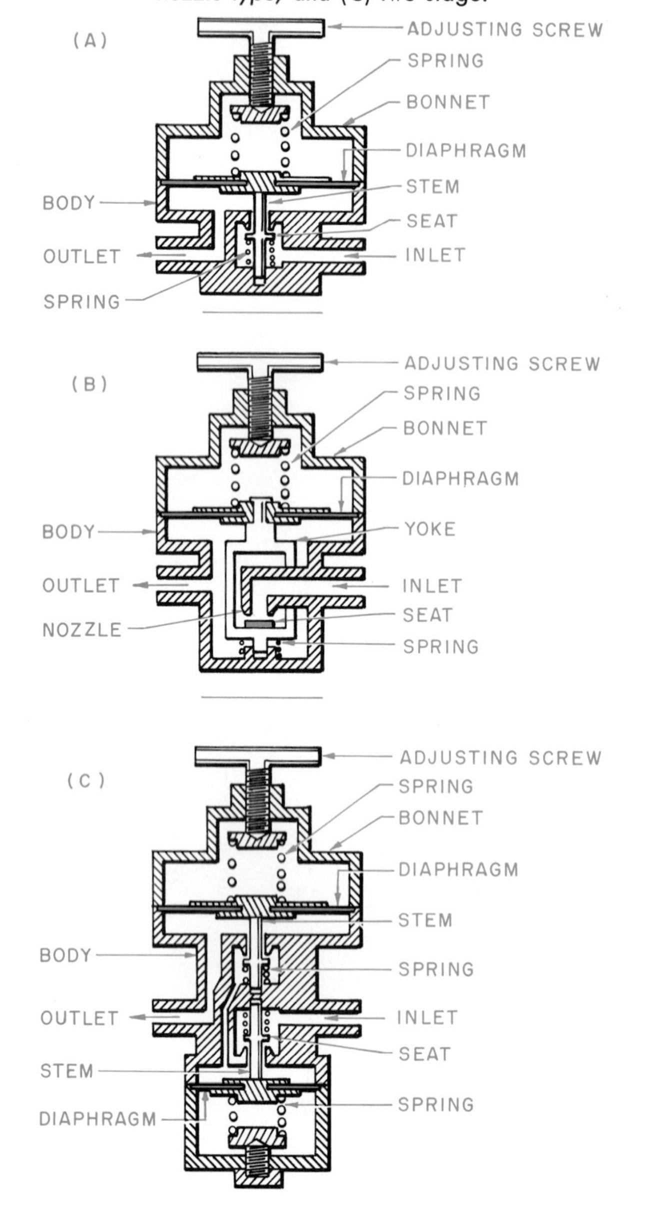

REGULATORS

Below are illustrated the two major types of regulators; namely A) The Single Stage Cylinder (or Master) Regulator and B) the Single Stage Station Regulator.

A. Single Stage Cylinder Regulator—most commonly used, however, a manifold design of this same regulator, the Master Regulator is used when gases are manifolded.

B. Single Stage Station Regulator—used to reduce the outlet pressure of the Master Regulator to the pressures commonly used for welding and cutting at the individual stations.

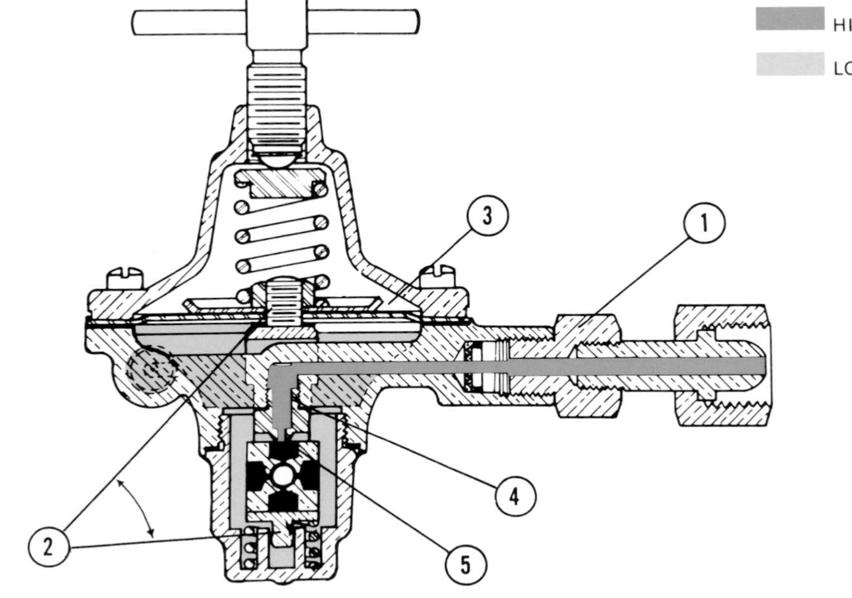

Type "O" SINGLE STAGE CYLINDER (OR MASTER) REGULATOR

1. Quick Replacement Cylinder Coupling—used on cylinder type regulators only, for added strength.
2. Dia-Blok Construction—diaphragm and seat are positively connected by means of a yoke, so that both the seat and diaphragm move at the same time—providing long seat life and minimum pressure fluctuation.
3. Diaphragm—Stainless Steel Diaphragms are used in Single Stage Cylinder Regulators and Master Regulators, whereas Reinforced Rubber Diaphragms are used in the lower pressure Station Regulators.

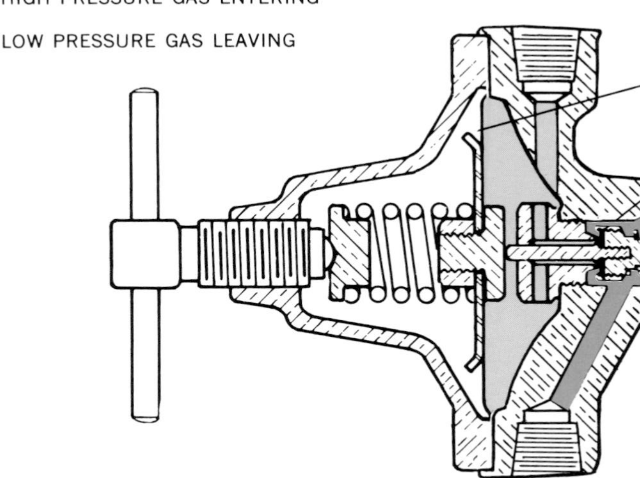

Type "O" SINGLE STAGE STATION REGULATOR

4. Safe-T-Chek Valve—located in the nozzle of both Single Stage Cylinder and Master Regulators, this valve will automatically close—should the seat of the regulator be off of the nozzle, when full cylinder pressure enters—protects against seat failure.
5. Regulator Seat—a Multi-Seat Block is used on Cylinder Type Single Stage Regulators, which can be rotated for seat change. A single seat is used on the lower pressure Single Stage Station Regulator.

Modern Engineering Co.

5-26. *Internal construction of tank and line regulator.*

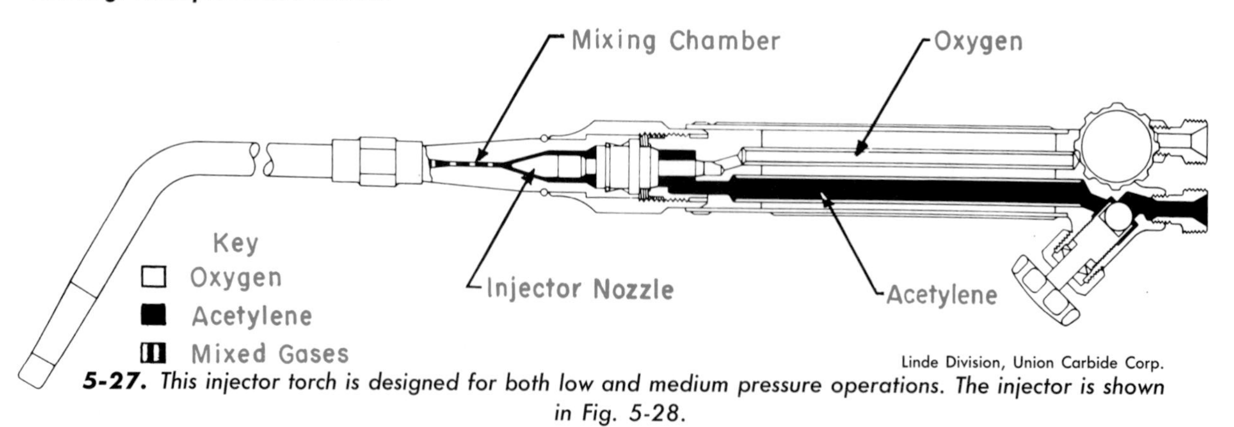

Linde Division, Union Carbide Corp.

5-27. *This injector torch is designed for both low and medium pressure operations. The injector is shown in Fig. 5-28.*

and soldering applications. The torches are designed to be used with air as the combustion-supporting gas.

AIR-ACETYLENE WELDING (AAW). Fuel gas burned with air has a lower flame temperature than that obtained when the same gas is burned with oxygen. You will recall that oxygen mixed with a fuel gas produces the hottest flame temperature. Air contains approximately $^4/_5$ nitrogen by volume, which is neither a fuel gas nor a supporter of combustion. Thus acetylene burned with air produces lower flame temperatures than the other gas combinations. The total heat content is also lowered.

Torches for use with air-acetylene are generally designed to draw in the proper quantity of air from the atmosphere to provide combustion. The acetylene flows through the torch at a supply pressure of 2 to 15 p.s.i. and serves to suck in the air. For light work the acetylene is usually supplied from a small cylinder that is easily transportable.

The air-acetylene flame is used for welding lead up to approximately $^1/_4$ inch in thickness. The greatest application is in the plumbing and electrical industry, where it is used extensively for soldering and brazing copper tubing with sweat-type joints.

5-28. *As the oxygen issues at relatively high velocity from the tip of the injector, it draws the proper amount of acetylene into the stream. The oxygen and acetylene are thoroughly mixed before issuing from the torch tip.*

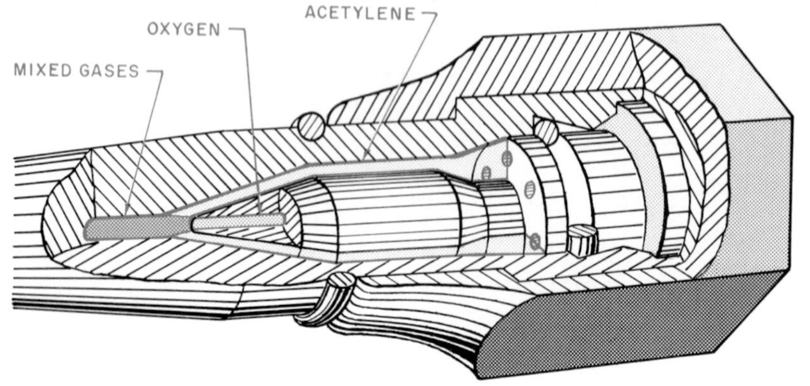

SUPPORTING EQUIPMENT

The welding shop should be equipped with a great deal of equipment that is needed for the preparation of the work and is necessary to the welding process.

- A welding table with either a cast iron top, slotted to permit the use of hold-down clamps, or a firebrick top, Fig. 5-37 (Page 163).
- C-clamps, carpenter clamps, various other types of clamps, straightedges, metal blocks, V-blocks, and a steel square for holding and lining up parts.
- Grinders, air-chisels, files, and hand chisels for beveling plate.
- A cutting torch, gas or electric, for beveling or for repair work.
- Carbon, in the form of rods, plates, or paste. It is highly fire-resistant and is useful to protect surfaces and holes, to back-up welds, to control and shape the flow of metal, and to support and align broken parts.
- Preheating equipment, Fig. 5-38, and a number of materials

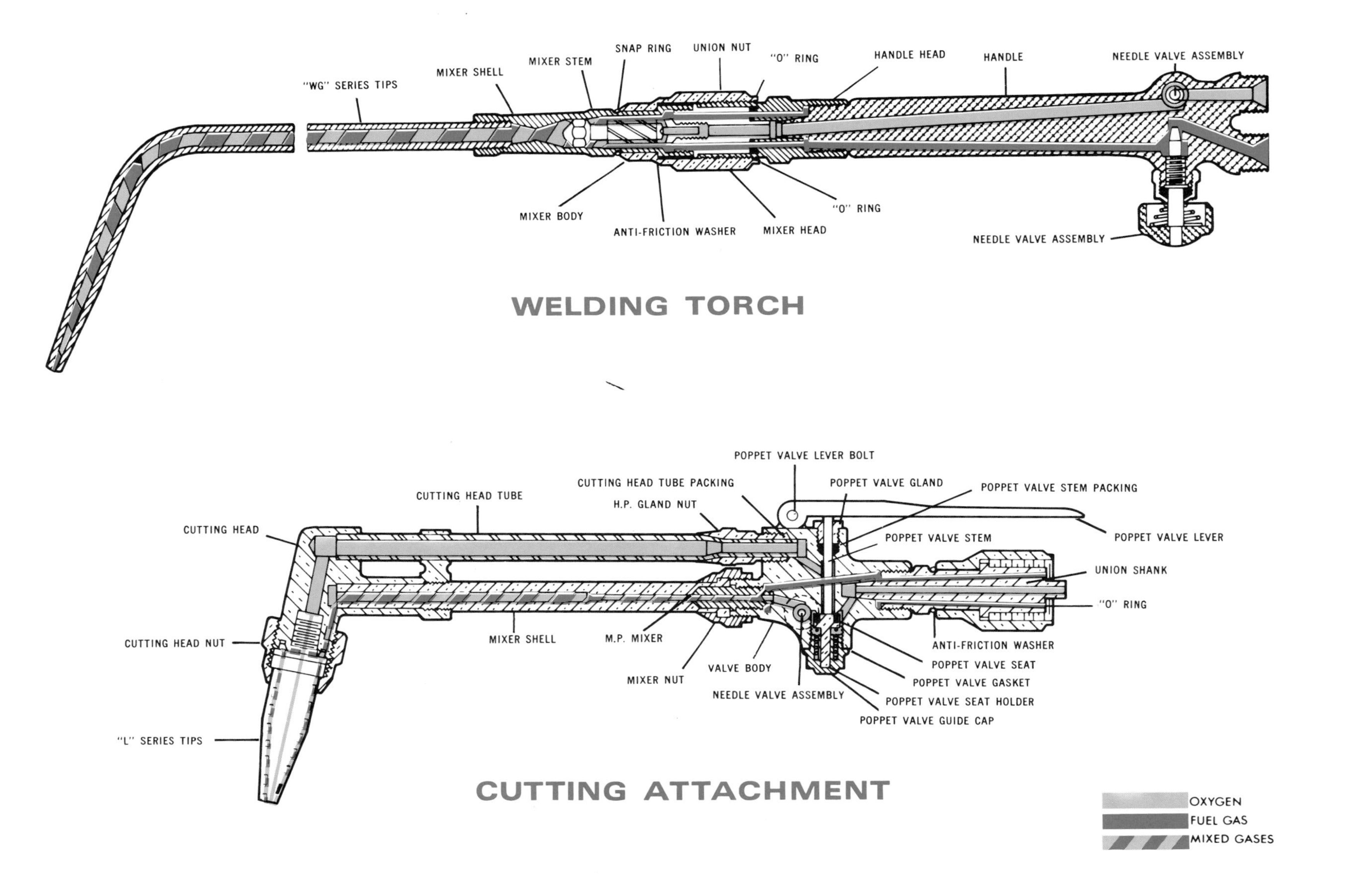

Modern Engineering Co.

5-29. *Internal view of balanced-pressure welding torch and cutting attachment.*

to provide for slow cooling, such as dry asbestos, fiber cement, hydrated lime, and asbestos paper.

- A wire power brush for cleaning scale and slag.
- Some jobs may need no finishing of any kind, while others may require filing, grinding, drilling, or even considerable machining. The welder should know the use of the part and the type of finishing that is going to be done before the job is welded. With this information a good welder can keep the amount of excess metal deposited close to the finish requirements, thus keeping the amount of postwelding work to a minimum.

SAFETY EQUIPMENT

Welding and Cutting Goggles

The worker must wear specially designed goggles to protect his eyes from infrared and ultraviolet rays, flying sparks, and particles of hot metal, Fig. 5-39. The eyes must also be protected from the heat coming from the job. Just any kind of colored lens does not give the necessary protection.

Filter lenses are made of special optical glass of various diameters and tinted green or brown. These lenses filter out the harmful rays, but they also minimize the effect of glare to permit the operator to see his work clearly.

The shade of the lens must be selected carefully. Lenses can be obtained in light, medium, and dark shades, Table 5-3. Lenses that are too dark cause eye strain since the operator is not able to see his work clearly. Lenses that are too light cause the eyes to suffer the effects of light and heat.

The outer lens is a clear glass or plastic of optical quality and $^{3}/_{64}$ to $^{1}/_{16}$ inch thick. This cover lens protects the filter lens from spatter. When the outer lens becomes pitted and reduces vision, it should be replaced. Treated lenses that have longer life than the untreated type can be purchased.

Goggle frames are usually made of a tough, heat-resistant material similar to Bakelite®. They should be light in weight and fit the face so that they are comfortable. Although they should provide adequate ventilation, it is important that they do not leak light.

It is foolhardy to attempt to save a few cents by purchasing an inferior grade of goggles. Purchase only the best. Never use oxyacetylene welding and cutting goggles for welding with the electric arc.

Some welders prefer to use eye shields, Fig. 5-40. These shields also have the clear cover lens and the colored filter lens. The lenses are the same size as those used in the arc welding helmets. These shields provide a wide range of vision and can be used over eyeglasses. They may be fitted with a headband and work on a swivel.

It is also possible to obtain eyeglass-type frames that have the correct lenses. Their use is

5-30. *Basic elements of welding torch mixers: (A) positive (balanced-pressure) type, (B) injector type.*

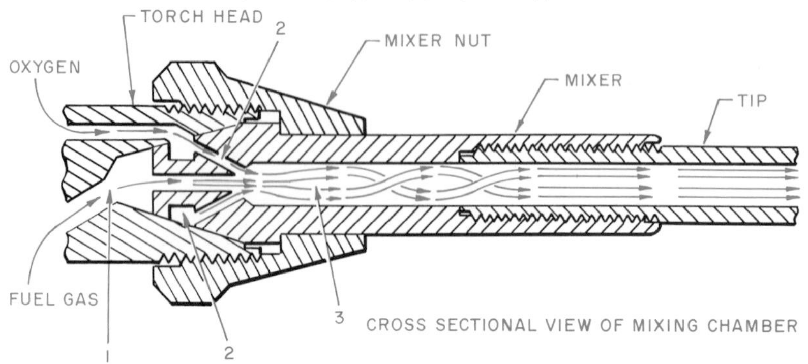
TORCH HEAD
2
MIXER NUT
OXYGEN
MIXER
TIP
FUEL GAS
3
1
2
CROSS SECTIONAL VIEW OF MIXING CHAMBER
THE TWO GASES ARE FED IN AT POINTS (1) FUEL GAS AND (2) OXYGEN, THERE BEING A MULTIPLICITY OF HOLES FOR THE LATTER. THE GASES ARE MIXED, BEGINNING AT POINT (3), THROUGHOUT THE MIXING CHAMBER, AS SHOWN BY THE VARIOUS ARROWS.

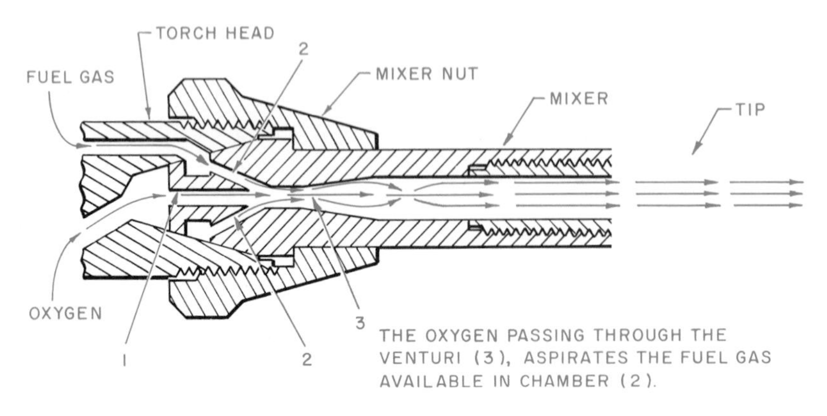
TORCH HEAD
2
FUEL GAS
MIXER NUT
MIXER
TIP
OXYGEN
3
1
2
THE OXYGEN PASSING THROUGH THE VENTURI (3), ASPIRATES THE FUEL GAS AVAILABLE IN CHAMBER (2).

not recommended for welding or cutting because they do not give any protection from the sides. Thus the eyes are exposed to injurious rays, glare, sparks, and hot particles of metal. They may be used by inspectors or onlookers since they are usually a safe distance away from the work.

Protective Clothing and Gloves

Welders must always remember that they are working with fire and that they should avoid any clothing that is highly flammable. Sparks, molten bits of metal, and hot scale are hazards. Under no circumstances should a sweater be worn. The body should be protected by an apron, a shop coat, or coveralls that resist fire. The head and hair should be protected by a leather welder's cap. For overhead welding and cutting, ear protection is desirable. Avoid wearing low-cut shoes and clothing with cuffs and open pockets.

Women and men wear the same kinds of clothing on the job: jeans (without cuffs), heavy shirts, high-top shoes, and the necessary protective clothing.

The heat coming from the welding job may be very intense. There will also be a shower of sparks, hot material to handle, and a hot welding torch to hold. This makes it necessary for welders to protect their hands with gloves. Gloves should be made of asbestos or other nonflammable material. However, for light welding jobs it is the common practice to wear an ordinary canvas glove with a cuff, which can be purchased at very small cost. Gloves must be kept free from grease and oil because of the danger involved in contact with oxygen.

When there is danger of sharp or heavy falling objects and when

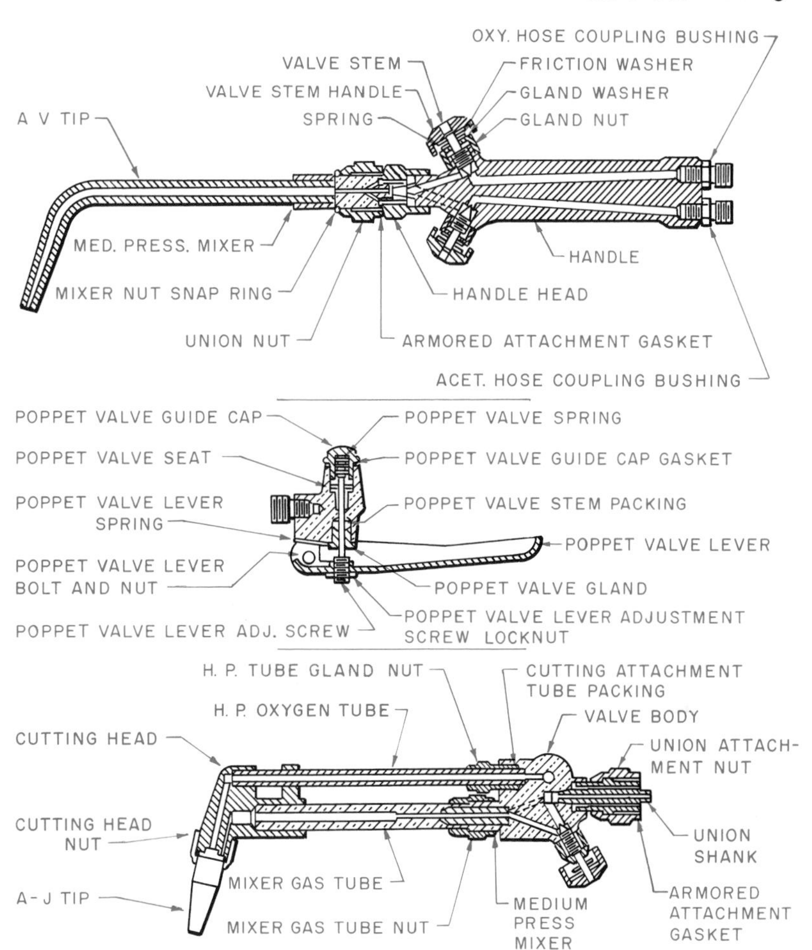

5-31. *Internal view of the balanced-pressure torch used for light welding. A detachable cutting head is also shown.*

Table 5-2

Gas pressures for different size welding tips.

Tip no.	Thickness of metal, inches	Acetylene pressure, lbs.	Oxygen pressure, lbs.	Oxygen consumption per hour	Lineal ft. welded per hour
1	1/32	1/2	1/2	7.80	30
2	1/16	1	1	7.90	25
3	3/32	1	1 1/2	8.10	20
4	1/8	1	2	9.75	15
5	3/16	1 1/2	2 1/2	16.80	9
6	5/16	2	2 1/2	26.40	6
7	3/8	3	5	39.35	5
8	1/2	5	8	51.15	4
9	5/8	8	14	69.10	3
10	3/4 & up	10	18	80.00	2

Modern Engineering Co.

Note:—Consumption given above, taken with Hydrex Flow Indicator, with maximum size flame. They are intended for estimating purposes only, and should amply cover adverse conditions.

the working space is confined, hard hats or head protectors should be worn.

Flash Guard Check Valves

These prevent the reverse flow of mixed gases in torch hoses or regulators. One valve is required for the acetylene hose and one for the oxygen hose. They may be attached to the torch or to the regulators.

More information about equipment can be found in Chapter 11.

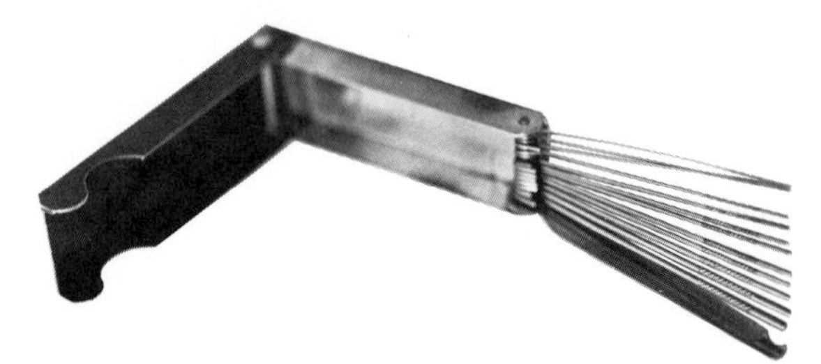

5-32. *Tip cleaners used to clean out the holes of welding and cutting tips.*

Modern Engineering Co.

5-33. *A gas economizer permits the welder to interrupt and restart the welding operation without readjusting the torch flame each time.*

Modern Engineering Co.

5-34. *Gas welding hose with fittings attached.*

5-35. *Standard oxygen and acetylene hose connections. Oxygen: right-hand thread; acetylene: left-hand thread.*

Modern Engineering Co.

Table 5-3

The proper shade of welding lens to use for different types of welding and cutting and the percent of rays transmitted.

Shade	Recommended uses	Percent of rays transmitted		
		Noninjurious visible rays	Injurious infrared	Injurious ultraviolet
Shade 2	Reflected glare and low temperature furnace work	28.0%	0.87%	1.075%
Shade 3	Light brazing and lead burning	16.0%	0.43%	1.035%
Shade 4	Acetylene burning and brazing	6.5%	None	0.097%
Shade 5	Light acetylene welding and cutting.	2.0%	None	0.046%
Shade 6	Standard shade for acetylene gas welding	0.8%	None	None
Shade 8	Heavy acetylene welding, electric arc cutting and welding up to 75 amperes	0.25%	None	None
Shade 10	Electric arc cutting and welding between 75 and 250 amperes	0.014%	None	None
Shade 12	Electric arc cutting and welding above 250 amperes	0.002%	None	None
Shade 14	Carbon arc cutting and welding	0.0003%	None	None

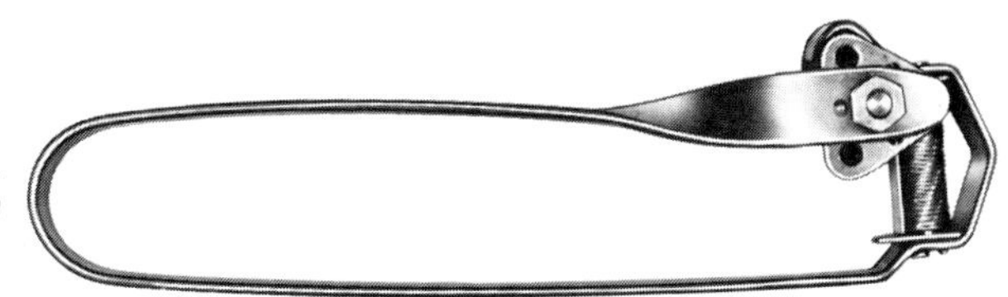

Modern Engineering Co.

5-36. *Triple-flint torch friction lighter.*

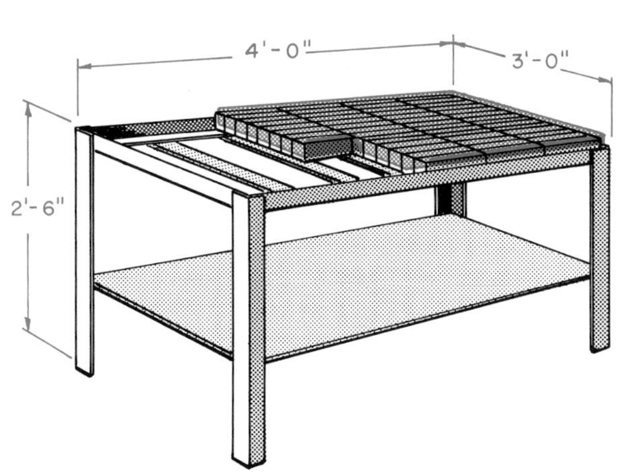

5-37. *Welding table: angle iron construction with welded joints and firebrick top.*

Modern Engineering Co.

5-39. *Deep cup oxyacetylene welding goggles. This type can be worn over eyeglasses.*

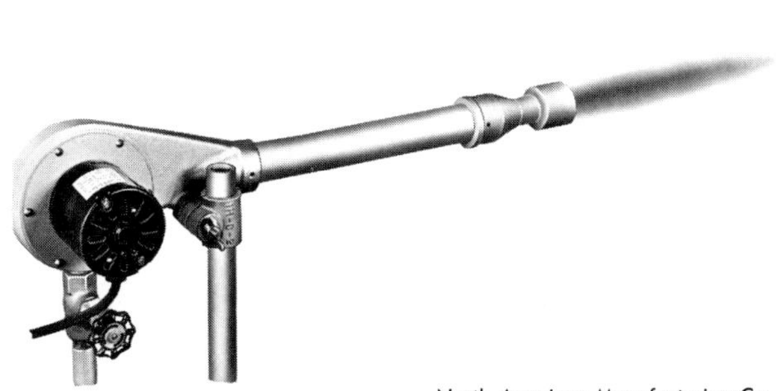

North American Manufacturing Co.

5-38. *A typical gas-electric blowtorch used for preheating.*

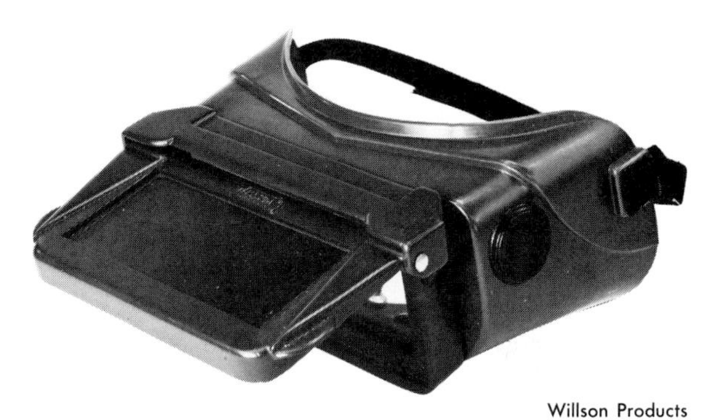

Willson Products

5-40. *Multiweld, flip-front window eye shields. This type can be worn over eyeglasses.*

REVIEW QUESTIONS

1. Oxyacetylene welding can only be used to weld the basic metals such as steel and cast iron. True. False.
2. Name four gases used in various forms of gas welding.
3. How is acetylene gas produced?
4. How is oxygen distributed and in what quantities? What safety precautions are necessary for safe handling?
5. How is acetylene distributed and in what quantities? What safety precautions are necessary for safe handling?
6. Explain the special precautions that must be taken to protect the body during welding operations.
7. Explain the function of the welding regulator.
8. What is the function of the welding torch?

9. Explain the difference in the design of the balanced-pressure torch and the injector torch.

10. List four precautions in the use and care of the welding tips.

11. Is it absolutely necessary that the eyes of the welder be protected while performing the welding operation? Explain your answer.

12. Gas welding filler rods are designated as RG-65, RG-60, and RG-45. What do the numbers 65, 60, and 45 indicate?

13. What is the approximate temperature of the oxyacetylene flame?

14. The oxyhydrogen flame is hotter than the oxyacetylene flame. True. False.

15. Explain the function of the two gauges on the welding regulator.

16. Explain the function of the gas economizer.

17. What kinds of protective clothing are worn for oxyacetylene welding?

Chapter 6 Shielded Metal-Arc Welding Principles

Shielded metal-arc welding—SMAW—is manual arc welding in which the heat for welding is generated by an electric arc established between a flux-covered consumable metal rod called the *electrode* and the work. For this reason, the process is also called *stick electrode welding.* The combustion and decomposition of the electrode creates a gaseous shield which protects the electrode tip, weld puddle, arc, and the highly heated work from atmospheric contamination, Fig. 6-1. Additional shielding is provided for the molten metal in the weld puddle by a covering of molten slag (flux).

The weld metal is supplied by the metal core of the consumable electrode, and the physical properties vary with the electrode classification. Certain electrodes also have metal powder mixed with the electrode coverings. The shielding and filler metal largely control the mechanical, chemical, metallurgical, and electrical characteristics of the weld. Characteristics of electrodes for shielded metal-arc welding are presented in Chapter 7.

PROCESS CAPABILITY

Shielded metal-arc welding is by far the most widely used of the various electric arc welding processes. Rather than replacing shielded metal-arc welding, the popularity of the other processes is causing the total amount of welding performed to increase. Equipment for shielded metal-arc welding is less complex, more portable, and less costly than that for other arc welding processes. This type of welding can be done indoors and outdoors, in any location, and in any position.

Shielded metal-arc electrodes are available to match the properties and strength of most base metals. Metals most easily welded by the shielded metal-arc process are carbon and low alloy steels, stainless steels, and heat-resisting steels. Cast iron and the high strength and the hardenable types of steel can also be welded providing the proper preheating and postheating procedures are employed. Copper and nickel alloys can be welded by this process although gas metal-arc (GMAW or MIG) welding and gas tungsten-arc (GTAW or TIG) welding are preferred.

The shielded process is not used for welding the softer metals such as zinc, lead, and tin, which have low melting and boiling temperatures.

Processes that employ continuously fed electrode wires (gas metal-arc) or no filler metal (gas tungsten-arc) have some advantage in deposition efficiency. After each pass, the slag formed by the shielded metal-arc electrode must be removed. Deslagging causes some loss of deposition efficiency.

OPERATING PRINCIPLES

Shielded metal-arc welding equipment sets up an electric circuit which includes the welding machine, the work, electric cables, the electrode holder and

6-1. Schematic representation of the shielded metal arc.
American Welding Society

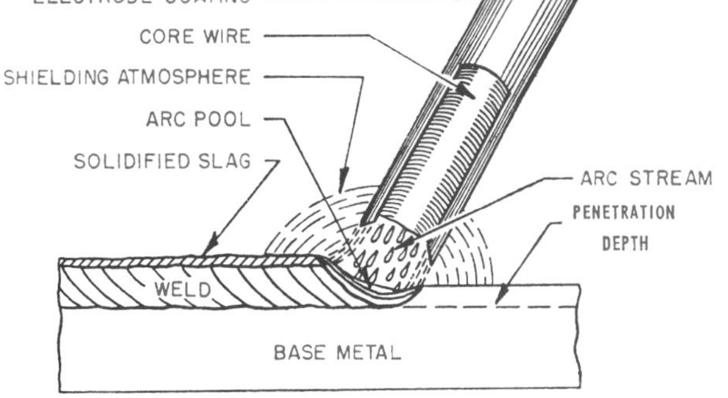

electrodes, and a ground clamp, Fig. 6-2. The heat of the electric arc brings the work to be welded and the consumable electrode to a molten state. The heat of the arc is intense. Temperatures as high as 9,000 degrees F. (5,000 degrees C.) have been measured at its center.

Welding begins when an electric arc is started by striking the work with the metal electrode. The heat of the arc melts the electrode and the surface of the base metal next to the arc. Tiny globules of molten metal form on the tip of the electrode and are transferred by the arc into the molten pool on the work surface. In flat welding the transfer is induced by the force of gravity, molecular attraction, and surface tension. When the welding is in the vertical and overhead positions, molecular attraction and surface tension are the forces that cause the metal transfer. Because of the high temperature of the arc, melting takes place almost instantaneously as the arc is applied to the weld. After the weld is started, the arc is moved along the work, melting and fusing the metal as it progresses.

WELDING POWER SOURCES

The successful application of the welding process requires proper tools and equipment. A welder will find it impossible to do good work if he has inferior equipment, if he lacks equipment, or if he chooses the wrong type of equipment for a particular job. The welding industry has spent many years in experimentation and practical application, as well as millions of dollars in developing the equipment necessary to carry on welding processes.

Each welding process reaches its maximum efficiency when used with the power source designed for it. Each type of power source has certain fundamental electrical differences that make it best suited for a particular process.

A welding machine must be able to meet changing arc load and environmental conditions instantly. It must deliver the exact amount of electrical current at precisely the right time to the welding arc in order to control arc behavior and make welds of consistent quality.

The arc welding process requires enough electrical current (measured in *amperes*) to produce melting of the work and the metal electrode and the proper voltage (measured in *volts*) to maintain the arc. Depending on their size and type, electrodes require 17 to 45 volts and approximately 100 to 500 amperes. The current can be alternating or direct, but it must be provided through a source that can be controlled to meet the many conditions encountered on the job.

Welding machines are available in a wide variety of types and sizes to suit the demands of different welding processes, operations, and types of work. There are a number of different makes, each having its own particular design. The control panel of each model is different in many respects. The selection of a particular machine may also depend upon cost, portability, and personal preference.

Welding power sources are also known as *power supplies, welders,* and *welding machines.* All machines may be classed by (1) output slope, whether constant current or constant voltage, and (2) power source type, such as transformer, transformer/rectifier, or generator.

Type of Output Slope

Welding machines have two basic types of output slope—constant current and constant voltage, which is also referred to as "constant potential."

Output slope, sometimes called the *volt-ampere characteristic* or *curve,* is the relationship between the output voltage and output current (amperage) of the machine as the current or welding workload is increased or decreased, Fig. 6-3.

Output slope largely determines how much the welding current will change for a given

6-2. *Elements of a typical electric circuit for shielded metal-arc welding.* American Welding Society

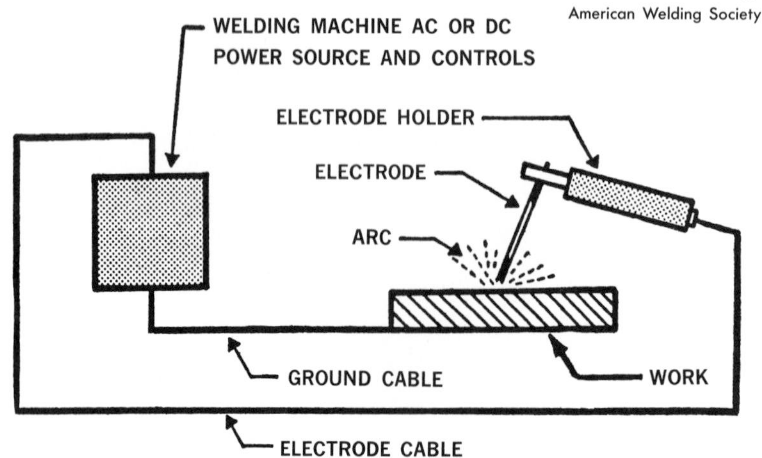

change in arc load voltage. Thus it permits the welding machine to control the welding heat and maintain a stable arc.

The output slope of the machine indicates the type and amount of electrical current it is designed to produce.

Each arc welding process has characteristic output slope:

- *TIG and stick electrode welding* require a steep output slope from a constant current welding machine. Constant current is necessary to control the stability of the arc properly.
- *MIG welding* requires a relatively flat output slope from a constant voltage power source to produce a stable arc.
- *Submerged-arc welding* is adaptable to either slope, depending on the application and extra control equipment.

Some d.c. welding machines may combine the two basic types of output slope in a single unit. By turning a selector switch on the machine, a steep slope (constant current) or a flat slope (constant voltage) can be produced.

Type of Power Source

Constant current welding machines are also classified by the manner in which they produce welding current. There are three general types:

- *Motor-generators* are driven by an a.c. or d.c. electric motor or an internal combustion engine, and they may produce d.c., a.c., or a.c./d.c. welding current.
- *Transformer-rectifiers* use a basic electrical transformer to step down the a.c. line power voltage to a.c. welding voltage. The welding voltage is in turn passed through a rectifier to convert the a.c. output to d.c. welding current. Transformer-rectifiers may be either d.c. or a.c./d.c. machines.
- An *a.c. transformer* uses an electrical transformer to step down the a.c. line power voltage to a.c. welding voltage.

It is important to select the right power source for each job, Table 6-1. A study of the job will usually indicate whether alternating or direct current should be used. This will determine the selection of either an a.c. or d.c. power source or a combination a.c./d.c. power source. For shielded metal-arc welding and gas tungsten-arc welding, the performance characteristics of any of these three power sources must be the constant current rather than the constant voltage type. The constant voltage machine is preferred for semiautomatic and automatic arc welding processes such as gas metal-arc.

Constant Current Characteristics

Constant current welding machines are primarily used for shielded metal-arc welding and gas tungsten-arc welding because current remains fairly constant regardless of changes in arc length. They are also called *drooping voltage* or *droopers* because the load voltage decreases, or droops, as the welding current increases.

These machines can also be adapted to spray-arc MIG welding (d.c.) or submerged-arc welding (a.c. or d.c.) with a special arc-voltage sensing control and variable-speed wire feeder. This control senses the changes in arc voltage and corresponding arc length and varies the wire feed speed to maintain a constant arc length voltage. Constant current welding machines,

6-3. *Typical output slopes for constant current and constant voltage power sources.*

American Welding Society

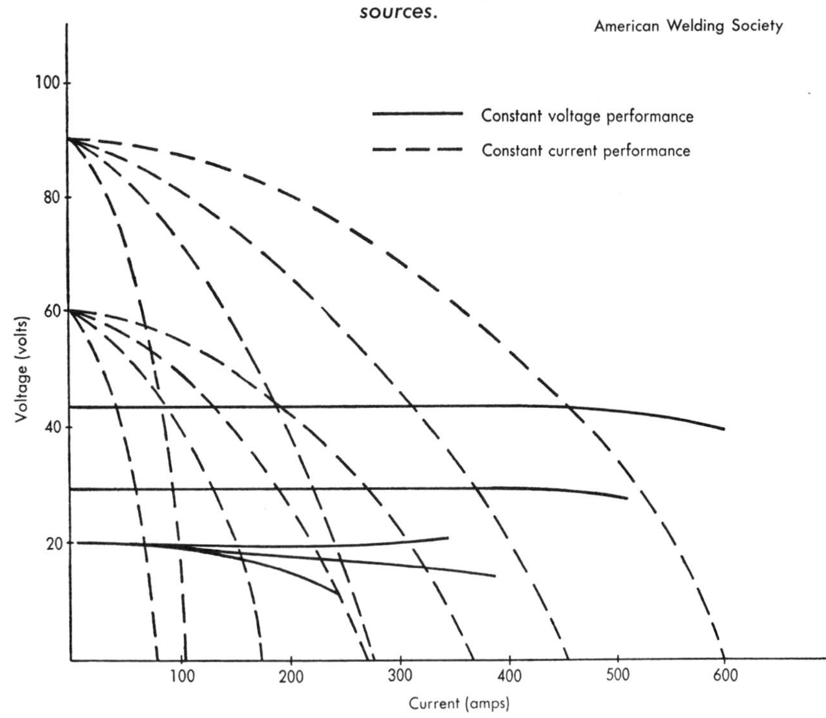

Table 6-1
Common types and uses of arc welding machines.[1]

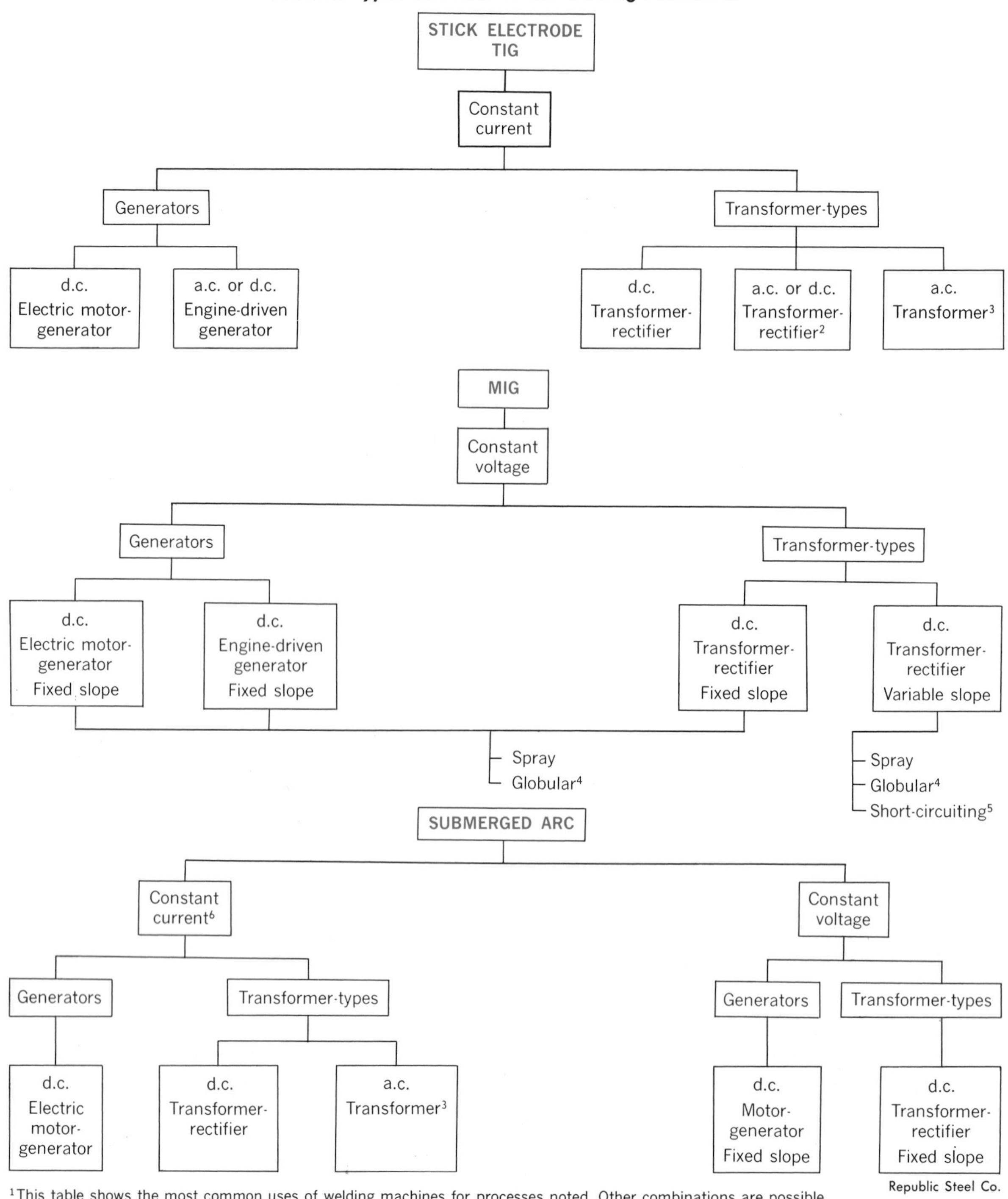

Republic Steel Co.

[1]This table shows the most common uses of welding machines for processes noted. Other combinations are possible.
[2]The a.c. can be used only on single-phase line power. The d.c. is most commonly used on 3-phase power, except for most a.c.-d.c. units.
[3]Single phase only.
[4]Mild steel carbon dioxide welding—not normally used for stainless.
[5]Inductance usually added. Some shorting-arc machines have a fixed slope set at a steeper angle with adjustment made by varying the inductance.
[6]Used with voltage-sensing control and variable speed wire-fed unit.

however, are not as satisfactory for spray-arc MIG welding as constant voltage machines because of the poorer arc starting and more complicated equipment.

OUTPUT SLOPE. Constant current welding machines have a steep output slope and are available in both d.c. and a.c. welding current. The steeper the slope of the volt-ampere curve within the welding range, the smaller the current change for a given change in arc voltage, Fig. 6-4. Some jobs require a steep volt-ampere curve (*A*) while other jobs require the use of a less steep volt-ampere curve (*B*). The constant current power source enables the operator to control welding current in a specific range by simply changing the length of the arc as welding progresses.

Open Circuit and Arc Voltage. *Open circuit voltage* is the voltage generated by the welding machine when no welding is being done. The machine is running idle. *Arc voltage* is the voltage generated between the electrode and the work during welding. Open circuit voltage generally runs between 50 and 100 volts. Arc voltage runs between 18 and 36 volts. The two open circuit voltages expressed in Fig. 6-4 are approximately 90 volts (*A*) and slightly below 50 volts (*B*). The arc voltages range from 32 volts (long arc) to 22 volts (short arc).

Open circuit voltage drops to arc voltage when the arc is struck, and the welding load is on the machine. Arc voltage is determined by the arc length held by the operator and the type of electrode used. When the arc is lengthened, the arc voltage increases and the current decreases. If the arc is shortened, the arc voltage decreases and the current increases. Note that in Fig. 6-4 there is a difference of 40 amps between the long arc and the short arc. Just how much change takes place depends on the open circuit voltage setting. Although the total current range between the long arc and the short arc in Fig. 6-4 is 40 amps, the range which the two open circuit voltages have in common is only 15 amps. The minimum open circuit voltage (*B*) produces a much wider current range than the maximum circuit voltage (*A*).

Open circuit voltage on constant current machines is higher (about 80 volts) than on most constant voltage machines (about 50 volts). *Arc voltage* depends on the physical arc length at the point of welding and is controlled by the operator in stick electrode and gas tungsten-arc welding. Arc voltage is much lower than open circuit voltage: normally about 20 to 30

6-4. *Two possible output slopes for a constant current welding power source. The steep slope (A) gives minimum current variation. The flatter slope (B) indicates the variation in current by changing the length of the arc. Note that changing the open circuit voltage (OCV) changes the slope of the curve.*

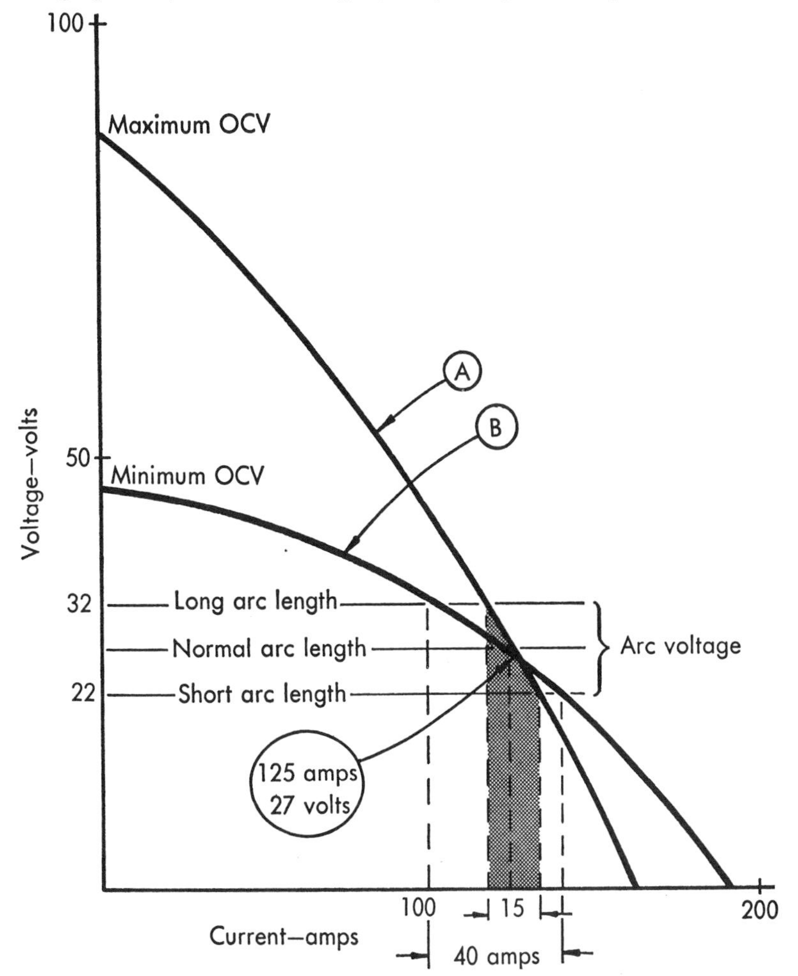

volts for stick electrode, 10 to 15 volts for gas tungsten-arc, and 25 to 35 volts for submerged arc.

There is no voltage adjustment dial on constant current welding machines. However, open circuit voltage can be adjusted with the fine current adjustment dial on welding machines having dual control.

In manual stick electrode and gas tungsten-arc welding, the operator cannot hold the arc length truly constant. Rather wide changes in arc length and, therefore, voltage sometimes occur. With a constant voltage machine, relatively small changes in current (amperage) result from changes in arc length. Thus the welding heat and burnoff rate of the electrode are influenced very little, and the operator is able to maintain good control of the weld puddle and the welding operation.

6-5. A motor-generator d.c. welding machine. The machine features multi-range dual control—the large wheel controls the welding current and has ten settings. The small knob in the center enables the operator to make fine adjustments. 1. Sturdy steel lifting eye. 2. Large voltmeter. 3. Calibrated dial for easy, fine adjustment of heat and relation between voltage and current. 4. Outer wheel and dial for selecting desired welding range. 5. Stop button. 6. Ground cable connector. 7. Optional steel guard. 8. Pressed steel bearing cap. 9. Heavy-duty ball bearings. 10. Arc-welded copper squirrel cage rotor. 11. Motor stator. 12. Heavy steel fan. 13. Steel frame. 14. Fourpole "Multi-Range" generator. 15. Heavy-duty metallic graphite brushes. 16. Large commutator. 17. Heavy-duty ball bearings. 18. Pressed steel bearing cap. 19. Arc-welded steel frame. 20. Removable steel covers. 21. Heavy-duty single unit steel shaft. 22. Welding cable connector. 23. Polarity switch. 24. Large ammeter. 25. Steel turret top with removable cover.

Hobart Brothers Co.

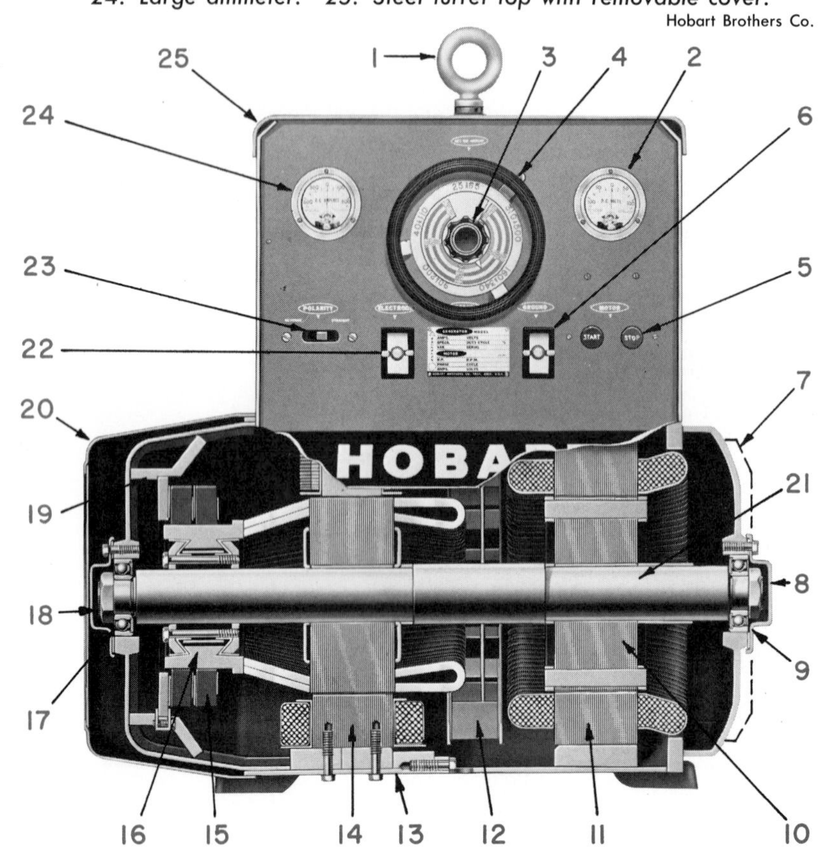

MACHINES FOR SHIELDED METAL-ARC WELDING

D.C. Motor-Generator Welding Machines

Motor-generators usually supply only direct current although they can be made to supply alternating current. Most d.c. motor-generators are the constant current type. They are used chiefly for stick electrode and gas-tungsten-arc welding. These models are shown in Figs. 6-5, 6-6, and 6-8. Figure 6-7 shows a d.c. motor-generator in use.

Generators are classified by the type of motor which drives the generator. The common motor-generator welding machine consists of an a.c. motor, a d.c. generator, and an exciter built on a single shaft. The a.c. current from the power line drives the electric motor which in turn drives the generator that provides the direct current for welding. These parts are shown in Fig. 6-8.

Other d.c. generators, which are used in the field where electric power supply is not readily available, are driven by a gasoline or diesel internal combustion engine like the one shown in Fig. 6-9 (Page 174). Generators used in the field are illustrated in Figs. 6-10 and 6-11.

The d.c. motor-generators have a very wide use due chiefly to their desirable characteristics for many arc welding operations.

- They have a forceful penetrating arc.

- They are versatile; they can be used to weld all metals that are weldable by the arc process.
- They are flexible; with the proper electrode, the d.c. welding can be used in all positions.
- They are durable; d.c. welders have a long machine life.

SIZES. Machine sizes are determined on the basis of amperage (current output). They range from 100-ampere rated machines for home use to more than 1,500-ampere rated machines for use with automatic submerged-arc welding equipment.

In the manual welding operation, the machine is not required to provide current continuously. The machine is idle part of the time while the operator changes electrodes, adjusts the heat setting, sets up the work, cleans the welds between passes, and shifts welding positions. For machines rated at 200 amperes or more, the output rating is based on a 60-percent duty cycle. This means that the machine can deliver its rated load output for six out of every ten minutes. Machines rated at 100 amperes or less are usually rated on the basis of a 50-percent duty cycle. In many cases these are portable machines, which are not used on a continuous basis. Fully automatic power supply units are usually rated at a 100-percent duty cycle.

A welding machine will deliver a higher current value than its rating indicates. Thus, a welder rated at 200 amperes will deliver from 200 to 250 amperes for welding. The actual output is always somewhat higher than the rated output. However, a machine should not be used at or beyond its maximum capacity over an extended period. A larger machine is indicated. These machines provide a steady supply of current over a wide range of welding voltage.

MAINTENANCE. The d.c. motor-generator machines are very durable and give long service if

6-6. A d.c. motor-generator welding machine that can be used for stick electrode and semi-automatic processes. It has continuous control. The Lincoln Electric Co.

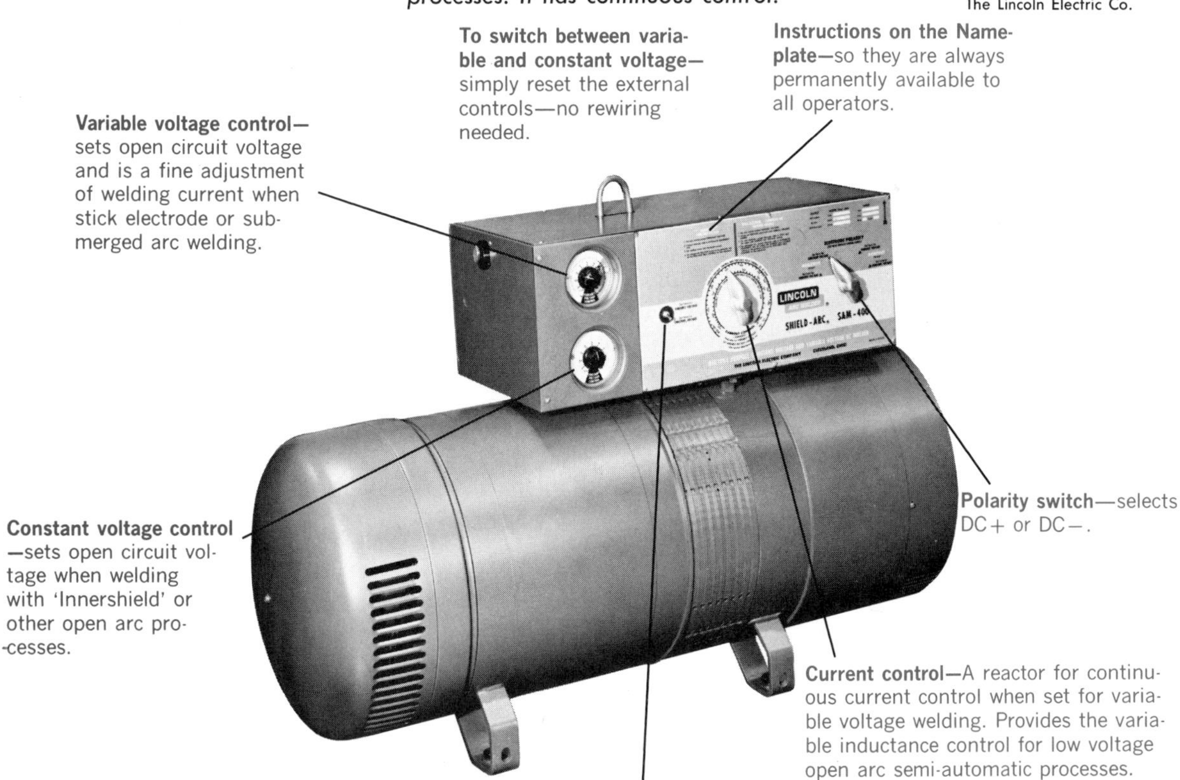
To switch between variable and constant voltage—simply reset the external controls—no rewiring needed.
Instructions on the Name-plate—so they are always permanently available to all operators.
Variable voltage control—sets open circuit voltage and is a fine adjustment of welding current when stick electrode or submerged arc welding.
LINCOLN
SHIELD-ARC, SAM-400
Constant voltage control—sets open circuit voltage when welding with 'Innershield' or other open arc processes.
Polarity switch—selects DC+ or DC−.
Current control—A reactor for continuous current control when set for variable voltage welding. Provides the variable inductance control for low voltage open arc semi-automatic processes.
Toggle switch—selects constant or variable voltage output.

properly maintained. Repair costs may be somewhat higher than for other types of welding machines because many moving parts provide opportunity for wear and malfunction. The contacts of the starter switch and the control rheostat should be inspected, cleaned frequently, and replaced when necessary. Brushes need frequent inspection for wear and must be replaced when worn down. Brushes should ride free in the brush holders. When inspecting the brushes, check the commutator for wear or burning. This condition can be corrected by cleaning with fine sandpaper. If the commutator is badly worn, it must be rewound and turned on a lathe.

The main bearings on the shaft should be inspected and greased at each six-month period. Clean old grease out of bearing housings.

6-7A. *A d.c. motor-generator welding machine: vertical type, continuous voltage control, continuous current control. It can be operated by remote control. The operator is resurfacing a worn part.*
The Lincoln Electric Co.

Controls of the D.C. Motor-Generator

START AND STOP BUTTONS. These controls are for the purpose of starting and stopping the motor that drives the generator of the welding machine. The start button is black, and the stop button is red. It is important that the buttons be engaged firmly when starting. Snap-switching should be avoided: it damages the contact points and may burn out fuses. Identify these controls in Fig. 6-5, page 170.

POLARITY SWITCH. Straight and reverse polarity are used in d.c. welding. When welding with *straight polarity* (DCSP), the electrode is connected to the negative terminal of the power source, and the work is connected to the positive terminal. When welding with *reverse polarity* (DCRP), the electrode is connected to the positive terminal of the power source, and the work is connected to the negative terminal.

The polarity switch changes polarity to either reverse or straight. An alternate name for reverse polarity is *positive polarity*. Other terms for straight polarity include *negative, standard,* and *normal* polarity. Any of these terms may be used on the polarity switch. Identify the polarity switch in Fig. 6-6, page 171.

6-7B. *Control panel of the d.c. machine shown in Fig. 6-7A.*
The Lincoln Electric Co.

VOLT-AMPERE METERS. Meters sometimes serve a dual purpose. On some machines they indicate polarity as well as current values in volts and amperes. Some machines, such as that shown in Fig. 65, page 170, have individual meters for volts and amperes. Others have single meters that indicate both the volt and ampere readings. A switch or button must be engaged to obtain individual readings on a combination volt-ampere meter. In order to obtain a true indication of the volt and ampere values being used, the operator must have some other person check the meters while he is welding.

Because the increased demand for additional devices for other purposes has raised the prices of welding machines, meters have been discontinued by some manufacturers.

CURRENT CONTROLS. The student will better understand the function of the current controls if he keeps in mind that the ampere rating can be compared with the amount of water flowing through a pipe. Amperage is the quantity of current. Amperage determines the amount of heat produced at a weld. Voltage is like pressure behind the water in the pipe. It is a measure of the force of the current. Voltage determines the ability to strike an arc and maintain its consistency. If voltage is too high, the arc is too harsh and may produce arc blow. On the other hand, if it is too low, it is very difficult to maintain the arc.

Dual control welding generators provide a wide range of

6-8A. *Instructor showing students how to set the controls of the welding machine.*

The Lincoln Electric Co.

6-8B. *External and internal construction of a d.c. motor-generator welding machine with continuous control of both open circuit voltage and amperage. The machine has a capacity of 400 amperes.*

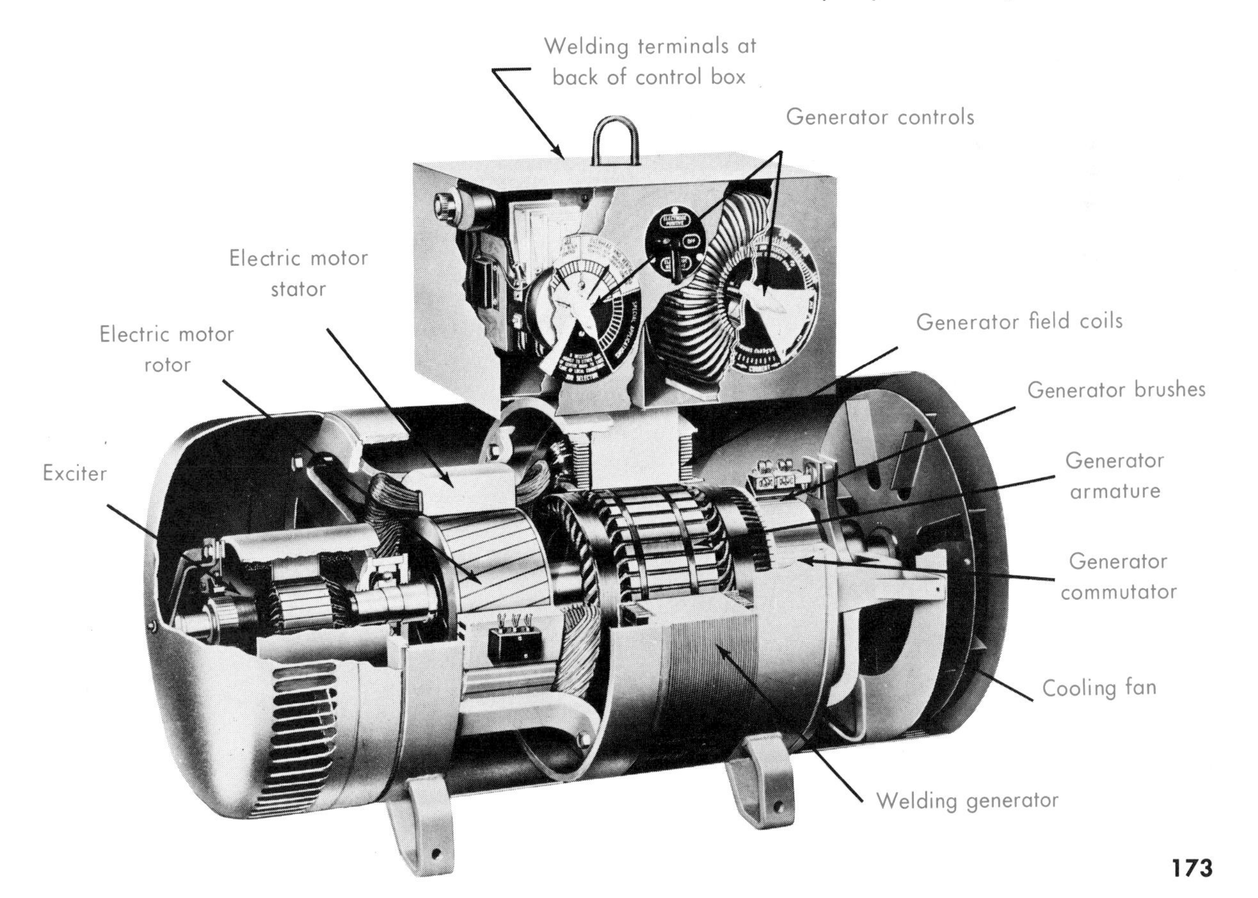

welding current selection and make it possible to provide precise heat values. Good arc control is possible because the output slope of the welding machine can be readily adjusted. A wheel or lever permits the operator to set the current at a definite rate.

There are two types of dual control generators: tapped-step current control and continously variable current control.

Dual Tapped-Current Control. On generators with fixed current steps, the coarse adjustment dial selects the current range, called *steps* or *taps.* It is impossible to secure a current value between two steps by setting the dial between them. The dial must be set squarely on the desired step to prevent burning out the control device and taps.

The fine adjustment dial trims the current between the steps. Whether the fine adjustment dial is to be set high or low depends on the type and size of the electrode, the thickness of the metal, and the position of welding.

The welding machine shown in Fig. 6-5, page 170, has dual tapped-current control.

Dual Continuous Control. In constant current generators with dual continuous control, the coarse adjustment dial continuously adjusts current. (Some manufacturers refer to the coarse adjustment dial as the *job selector* or *electrode selector.*)

The fine dial adjusts both the current (amperage) and the open circuit voltage. The operator adjusts the output slope for a given current setting by manipulating both the coarse and fine adjustment setting by manipulating both the coarse and fine adjustment dials together. The machines shown in Figs. 6-6 through 6-13 have dual continuous control.

A wheel or knob on both the amperage and the voltage setting devices gives the operator continuous control of both the amperage and voltage ratings. This separate, complete control of both voltage and amperage makes it possible to set up any volt-ampere combination within the output range of the particular machine being used from minimum voltage at minimum current to maximum voltage at maximum current. Thus, a welding arc having various characteristics may be secured.

REMOTE CONTROL. Welding machines may be obtained that are equipped with remote-control, current-control units. The welding machine may be

6-10. Heavy-duty d.c. engine-driven welding machine with a 400-ampere capacity. It has continuous fine control between major current settings and may be used for stick and MIG welding. It may be fitted with a trailer undercarriage. Note the eye bracket for crane lifting.
Airco Welding Products Division, Air Reduction Co.

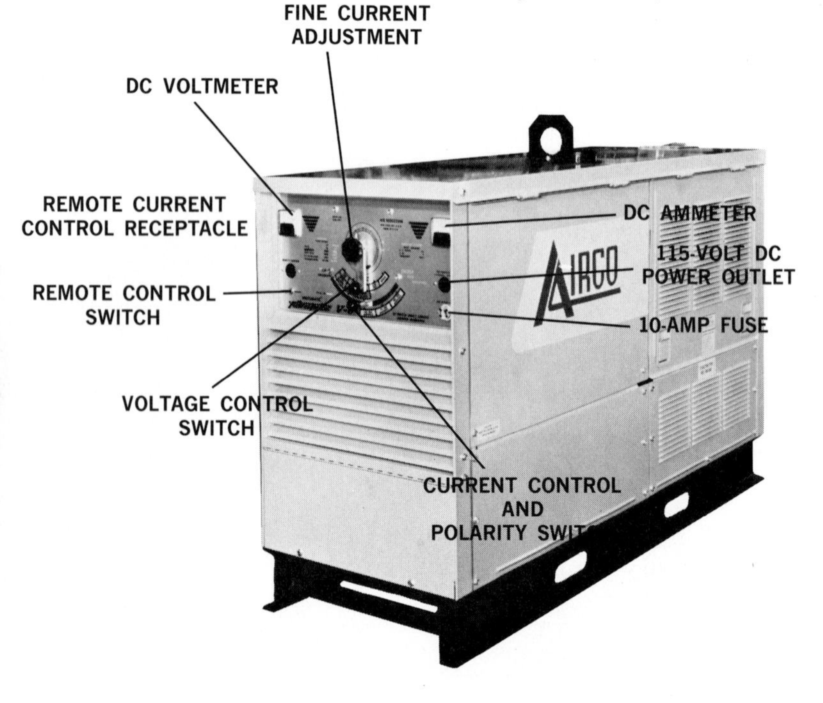

6-9. A water-cooled Jeep® engine which powers the engine-driven welding machine shown in Fig. 6-10.
Airco Welding Products Division, Air Reduction Co.

installed in any remote part of the plant, and the operator may adjust the current without leaving the welding station or welding job. This is timesaving on work where it is necessary for the welder to leave the fabrication to readjust current.

AIR FILTERS. Wear in arc welding machines is costly since it necessitates not only the cost of replacement parts and labor, but also the loss of production due to nonuse of the machine. Bearing wear is critical and may be reduced through the use of an air filter fitted on the suction end of a motor-generator machine. The filter is cleaned regularly with high pressure air, a commercial solvent, or steam.

D.C. Transformer-Rectifier Welding Machines

Transformer-rectifier welding machines, Figs. 6-14 and 6-15, have many designs and purposes. Flexibility is one reason for their wide acceptance by industry. These machines can deliver either straight or reverse polarity direct current. They may be used for stick electrode welding, gas tungsten-arc welding, submerged-arc welding, multiple-operator systems, and stud welding.

All transformer-rectifier machines have two basic parts: (1) a transformer for producing and regulating the alternating current which enters the machine and (2) a rectifier which converts the alternating current to direct current. A third important part is a ventilating fan. The fan keeps the rectifier from overheating and thus shortening its life.

The simple design of the transformer-rectifier improves arc stability and makes it easy to hold a short arc. The arc itself is soft and steady. As with d.c. motor-generators, continuous current control is available over a wide range. Transformer-rectifiers have no major rotating parts so that they consume little power while idle and operate quietly.

A.C.-D.C. Transformer-Rectifier Welding Machines

An a.c.-d.c. welding machine is used for stick electrode welding and gas tungsten-arc welding in which alternating current is required for most nonferrous metals and direct current is required for stainless steel. The a.c.-d.c. welders have the versatility of the d.c. welder and the special advantages of the a.c. welder in a single machine.

The a.c.-d.c. machines permit the operator to select either alternating or direct current and either straight or reverse polarity. They are essentially a.c. transformer-rectifiers, Fig. 6-16. A switch permits the operator to use only the transformer part of the machine for a.c. welding. By flipping a switch or turning a dial, output current is directed through the rectifier. This converts it to direct current welding. The rectifier circuit is similar to all other transformer-rectifier machines.

High frequency arc-starting devices, water and gas flow controls, d.c. component filters for a.c. operation, and other welding controls such as remote control are often built into the machine. Controls for two models are identified in Figs. 6-17 and 6-18 (Page 178). Figure 6-19 shows

6-11. *Heavy-duty d.c. engine-driven welding machine. Like the machine shown in Fig. 6-10, it may be fitted with a trailer undercarriage. It has dual continuous control of both current and voltage.*

The Lincoln Electric Co.

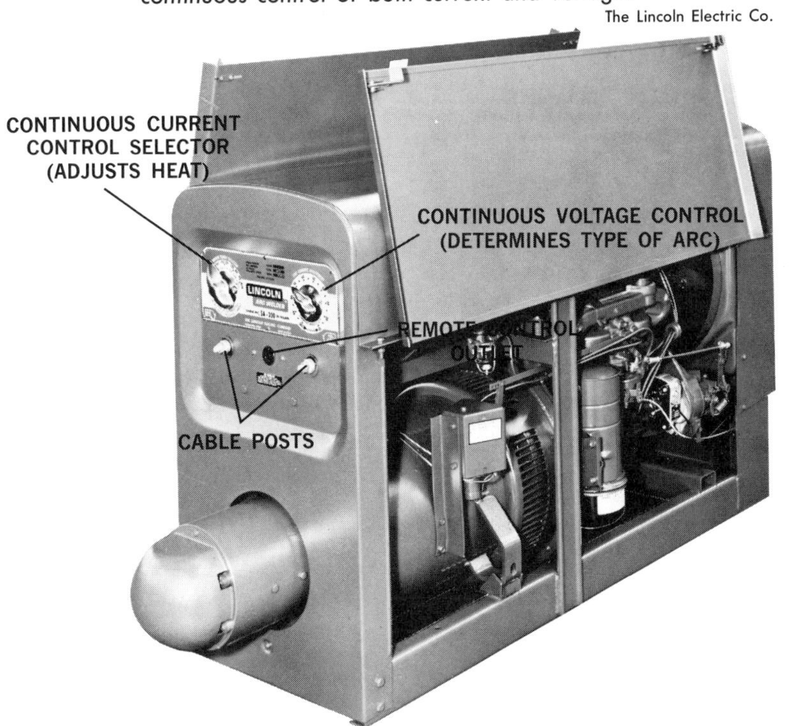

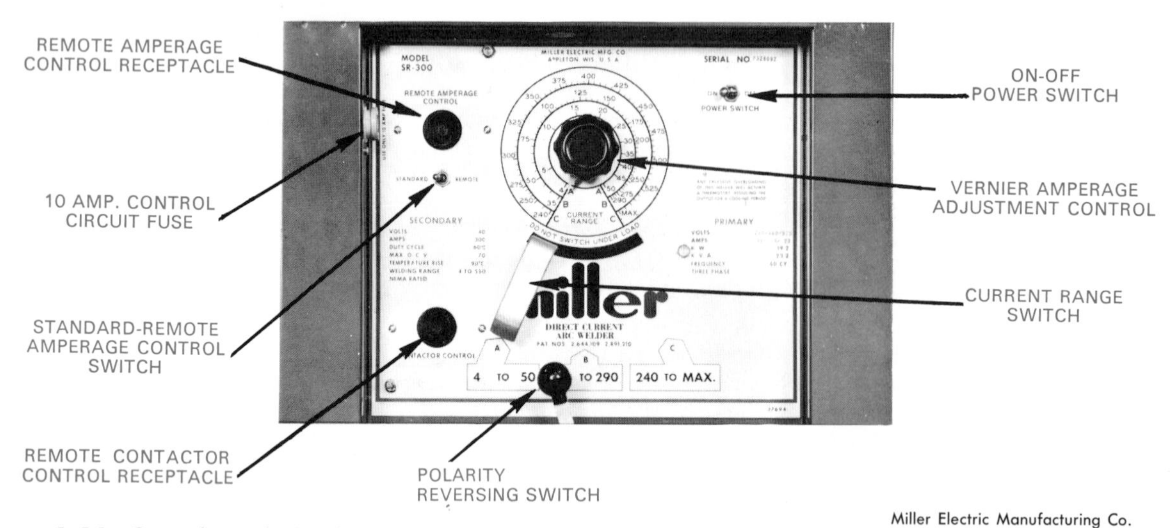

Miller Electric Manufacturing Co.

6-12. *Control panel of a d.c. rectifier arc welder for shielded metal-arc welding. The machine has three current ranges with infinite current control in each range.*

6-13. *Control panel of an engine-driven a.c. welder and power plant. The welder can provide a.c. welding current and at the same time put out 115-volt auxiliary d.c. current for powering drills, other power tools, and lights while welding. There is continuous current control within each of the five current ranges.*

Miller Electric Manufacturing Co.

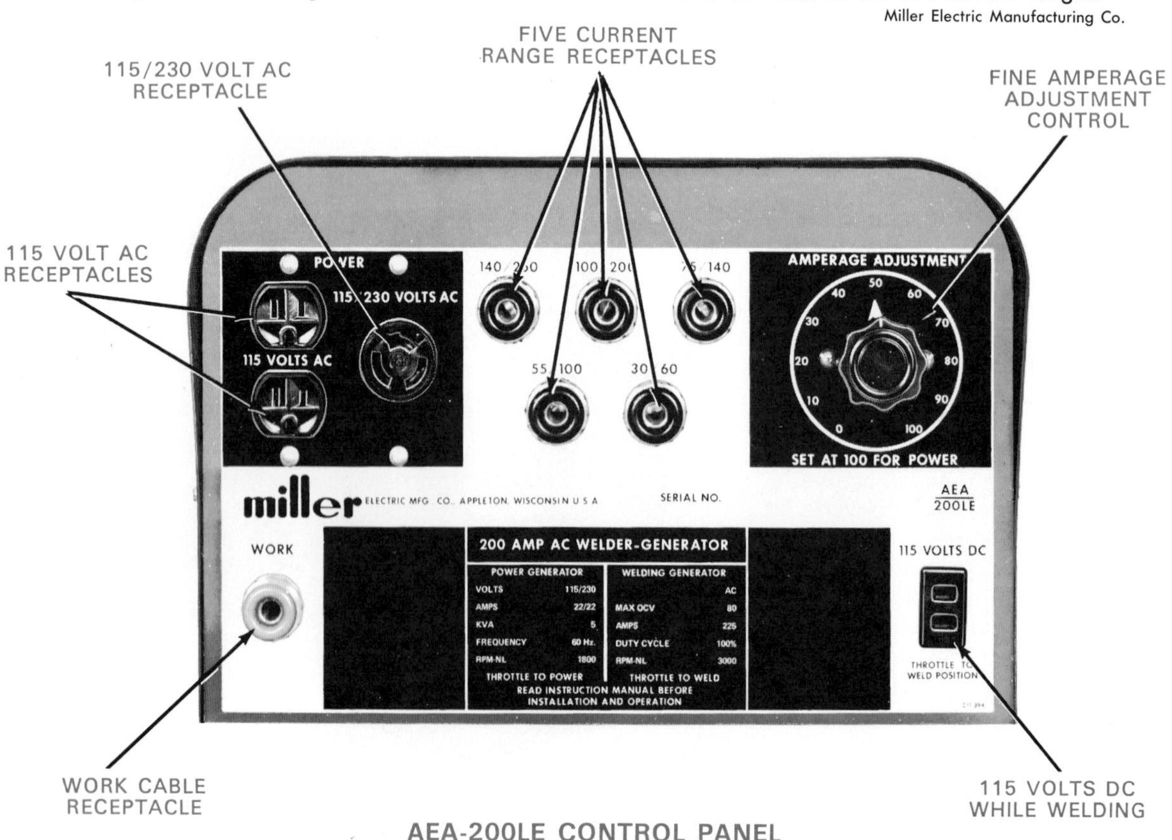

AEA-200LE CONTROL PANEL

Westinghouse Electric Corp.

6-14. *Silicon rectifier d.c. welding machine. Major current settings are fixed and are set by closing a switch. The rheostat provides vernier adjustments between breaker settings.*

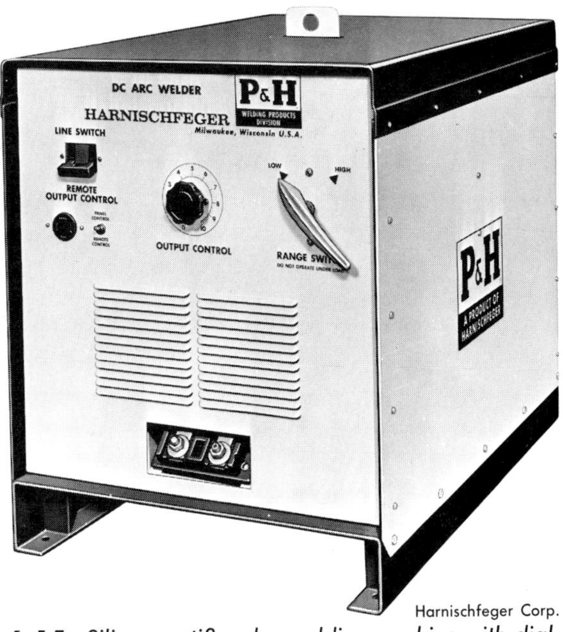

Harnischfeger Corp.

6-15. *Silicon rectifier d.c. welding machine with dial-lectric continuous amperage control.*

6-16. *Internal construction of an a.c.-d.c. transformer-rectifier welding machine.*

Hobart Brothers Co.

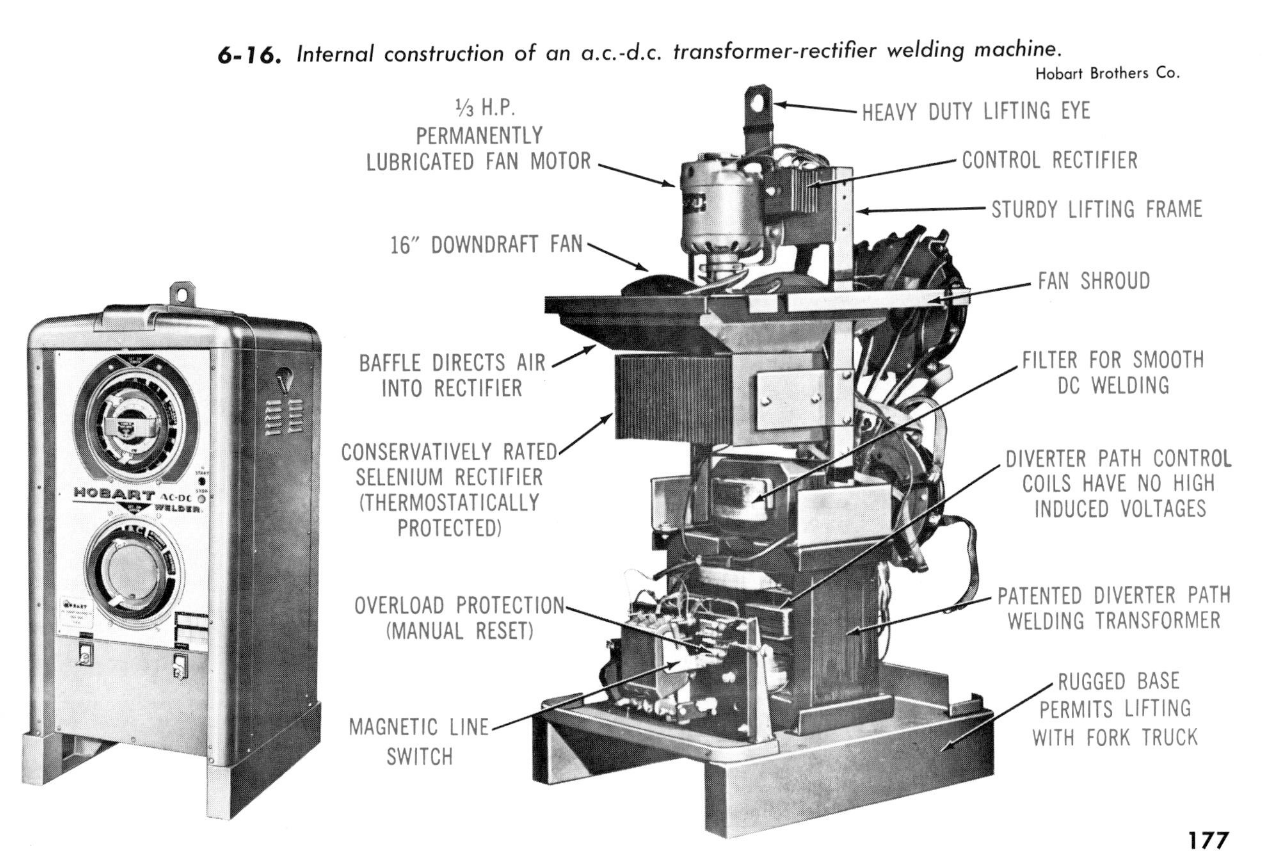

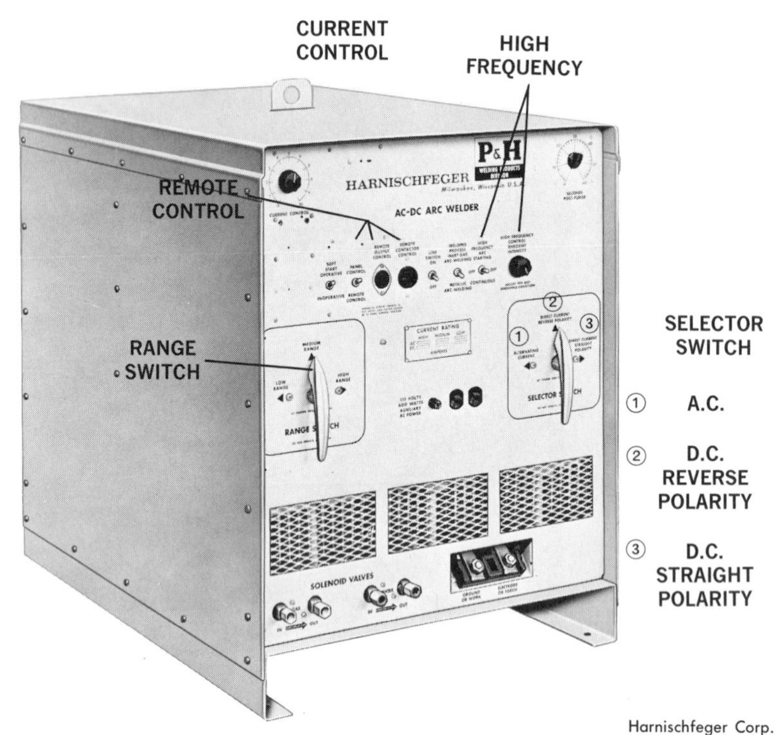

Harnischfeger Corp.

6-17. An a.c.-d.c. high frequency transformer-rectifier welding machine. This model has dial-lectric continuous current control between current ranges. It may be operated by remote control.

one type of a.c.-d.c. welder in use.

A.C. Welding Machines

The a.c. welding machines are gaining rapidly in popularity because of the advantage of a.c. current for welding, the great improvement in the performance of machines, and the development of new electrodes for use with these machines.

The characteristics of a.c. current are such that there is a reversal of current each $\frac{1}{120}$ of a second. This constant reversal of current keeps the effect of the magnetic field at a minimum, thus reducing the tendency to arc blow. Arc blow results in excessive spatter and interferes with metal transfer.

While the arc may be somewhat more difficult to start than one produced by direct current,

6-18. An a.c.-d.c. transformer-rectifier welding machine. The handwheel permits continuous current control with fine adjustment in each range. This model is available in 300-, 400-, and 500-ampere capacities.

Airco Welding Products Division, Air Reduction Co.

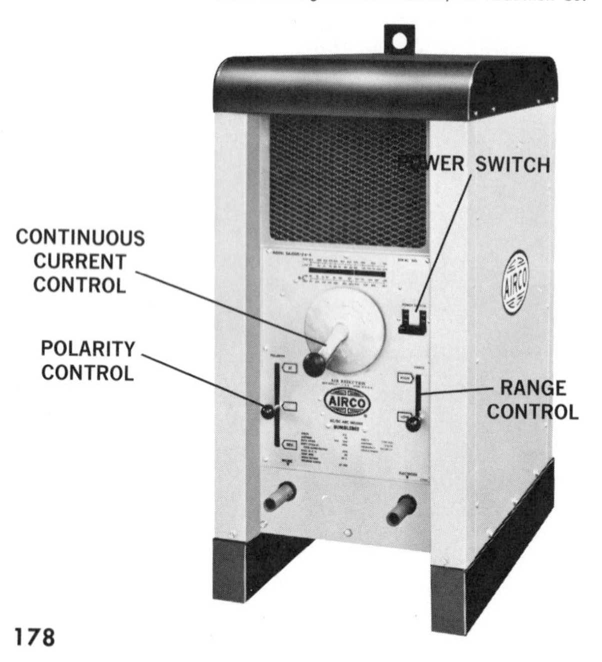

6-19. An a.c.-d.c. transformer-rectifier welding machine. In this model, which has a 300-ampere capacity, a handwheel permits continuous control of current. The a.c.-d.c. current selector is at the top right. A direct-reading current indicator is at the bottom of the machine.

The Lincoln Electric Co.

the absence of arc blow and higher voltage makes the arc easy to hold once it is obtained. This condition also permits the use of larger electrodes, resulting in faster speeds on heavy materials.

Other advantages of a.c. welders include lower cost, decreased power consumption, high overall electrical efficiency, noiseless operation, and reduced maintenance.

Best welding performance can be obtained in the downhand and horizontal positions on groove and fillet welds with certain types of heavily coated shielded arc electrodes. An a.c. welder is especially suited for heavy work.

The two basic types of a.c. welding machines are the transformer type and the motor-generator type.

A.C. Transformer Welding Machines

The transformer type is the most popular a.c. welding machine, Figs. 6-20 and 6-21. The function of the transformer is to step down the high voltage of the input current (for example, 120, 220, or 440 volts) to the high amperage, low voltage current required for welding.

The inner construction of the transformer, Fig. 6-22, consists of (1) a doughnut-shaped or square block-shaped core, (2) a primary coil which carries the input current, and (3) a secondary coil which is connected to the electrode holder. The core, which is made of laminated-steel plates with magnetic properties, steps down the voltage from the primary to the secondary coil. The secondary coil carries the welding current. Many types are also equipped with capacitors to give high current surges to improve striking the arc and a motor-fan to aid in cooling.

CURRENT CONTROLS: A.C. TRANSFORMER WELDING MACHINES. The controls which adjust current values vary. Some machines have plugs or switches. Others have continuous control adjustment over the entire current range. Tapped-current control is often used on small utility machines. Continuously variable current controls, which may be mechanical or electrical, are on most industrial welders. Current controls of the various a.c. transformers and a.c.-d.c. transformers fall into five general classifications:

- *Plug-in control:* this type of welder does not provide continuous current control over the en-

6-20. *An a.c. transformer welding machine with continuous current control.*
The Lincoln Electric Co.

6-21. *An a.c. transformer welding machine for the farm or small shop. This machine has a 12-position current control.*
The Lincoln Electric Co.

CURRENT CONTROL

tire range. Welding current can be chosen only in steps, Fig. 6-23. Each current step is tapped into different turns on the reactor coil. Current steps larger than 30 amperes do not provide for flexibility in adjustment.

- *Movable cores:* the welding current is controlled by changing the distance between the primary and secondary coils of the transformer, Fig. 6-24. The closer together the coils, the greater the current output. As the primary coil is moved away from the secondary coil, the current output is decreased.
- *Magnetic shunt control:* a movable, magnetic, laminated-iron shunt moves between the fixed primary and secondary coils and thereby varies the current output, Fig. 6-25.
- *Reactor control:* a variable reactor is connected between the secondary transformer winding and the welding load, Fig. 6-26. By changing the magnetic characteristics of the reactor, the welding output is regulated. Normally, the reactor characteristic is changed by mechanically adjusting an air gap in the reactor core.
- *Dial-lectic control:* this form of control is similar to reactor control except that current variation is achieved by a rheostat, Fig. 6-27. This can be located at some distance from the machine so that the operator has remote control of the current output. Adjustment can also be made by means of a foot control.

Electrical, rather than mechanical, current controls are used for critical gas tungsten-arc welding and other a.c. or d.c. uses because of their adaptability to remote control and control of arc behavior.

A.C. Motor-Generator Welding Machines

There are two general types of a.c. motor-generators: one is driven by an electric motor; and the other, by a gasoline engine.

The electric motor-driven machine makes use of a single-phase or three-phase a.c. motor to drive an alternating current generator. The generator is similar to those which generate electrical power for industry and the home. This type of welder is not widely used.

The engine-driven machines are powered by gas engines of various sizes. They are quite popular because of their portability, and they have the same flexibility that the motor-driven d.c. welder has. They are especially useful on those types of field jobs in which arc blow is a particular problem. The genera-

6-22. *An a.c. transformer welding machine. It provides continuous control between major current settings.*

Hobart Brothers Co.

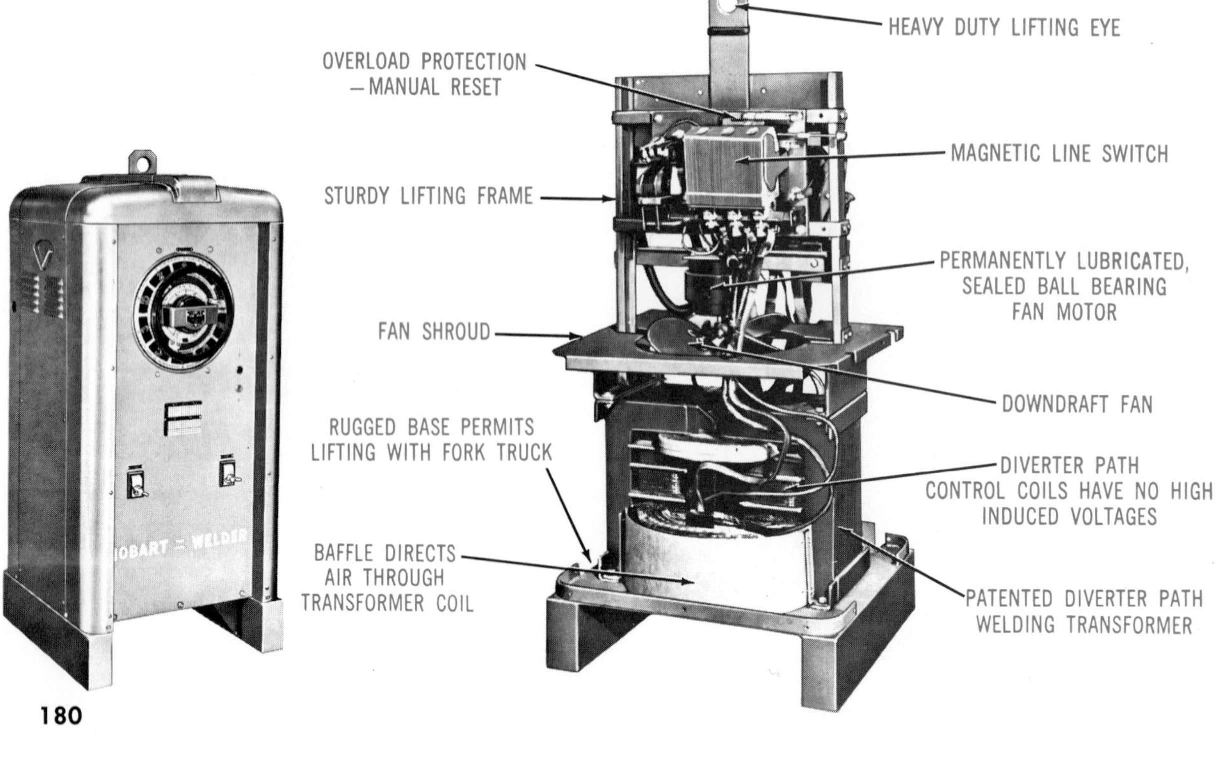

HEAVY DUTY LIFTING EYE
OVERLOAD PROTECTION —MANUAL RESET
MAGNETIC LINE SWITCH
STURDY LIFTING FRAME
PERMANENTLY LUBRICATED, SEALED BALL BEARING FAN MOTOR
FAN SHROUD
DOWNDRAFT FAN
RUGGED BASE PERMITS LIFTING WITH FORK TRUCK
DIVERTER PATH CONTROL COILS HAVE NO HIGH INDUCED VOLTAGES
BAFFLE DIRECTS AIR THROUGH TRANSFORMER COIL
PATENTED DIVERTER PATH WELDING TRANSFORMER

tors of these welders are similar in construction to d.c. welding generators.

On some construction and repair jobs it is not economical nor practical to secure electric power from supply lines. Very often electric power is not available. It is, therefore, necessary to use a gasoline or diesel engine as the means of power to drive the welding generator. The welding controls of these welding machines are similar to those on the d.c. motor-generator machines. The units are usually built on a chassis with automobile-type wheels and tires so that they can be readily moved to various locations on the job, Fig. 6-28.

The engine fuel is usually gasoline, diesel oil, or propane gas.

Cost Comparisons: Arc Welders

There are three main areas of cost in considering the type of

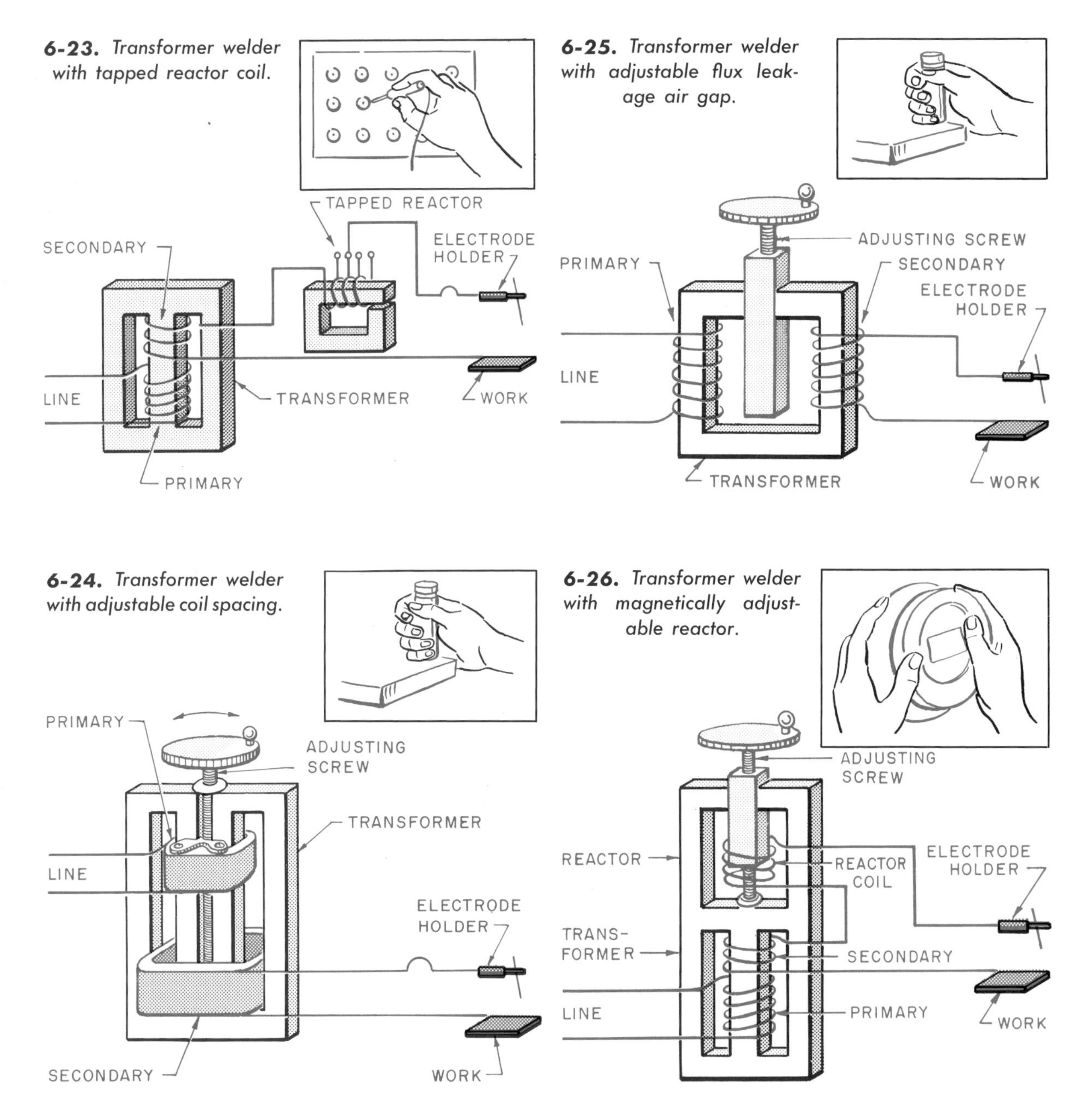

6-23. *Transformer welder with tapped reactor coil.*

6-24. *Transformer welder with adjustable coil spacing.*

6-25. *Transformer welder with adjustable flux leakage air gap.*

6-26. *Transformer welder with magnetically adjustable reactor.*

machine to purchase: the cost of purchasing the equipment, operating efficiency, and maintenance. A recent check indicated all types cost nearly the same.

It is in the second area—operating efficiency—that real cost advantages appear. Various authorities indicate that the motor-generator machine operates at 52 to 65 percent efficiency. In a comparative study, transformer-rectifier welders operated at 64 to 72 percent efficiency. Thus, the cost of operating a motor-generator machine is higher. An added reason for the difference lies in the lower idle-time power consumption of the transformer-rectifier.

The third area for cost comparison is maintenance. Maintenance on motor-generator welders requires replacing brushes, cutting the commutator, and lubricating bearings. In a transformer-rectifier type, on the other hand, there are no moving parts except for the cooling fan. For practical purposes, the transformer-rectifier machine has no maintenance problem because it has no rotating parts to wear. Formerly, however, rectifier life was a problem. Improvement in design and construction, the use of selenium and silicon, and general refinements in cooling design have solved the problem of rectifier burnouts.

Consult Table 6-2 for a comparison of the overall characteristics of the three basic types of welding machines.

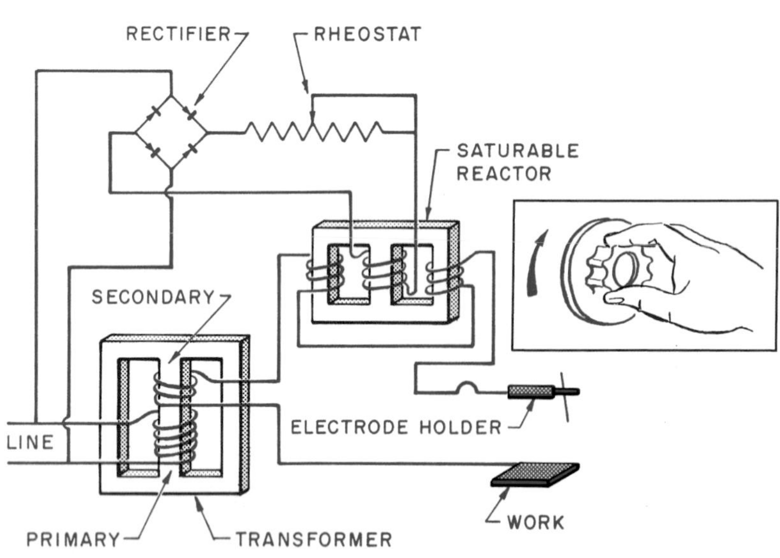

6-27. *Transformer welder with electrically adjustable reactor.*

6-28. *This a.c. engine-driven welding machine also provides a.c. power for emergency lighting and power tools. It is most suitable for farms and small shops.*

The Lincoln Electric Co.

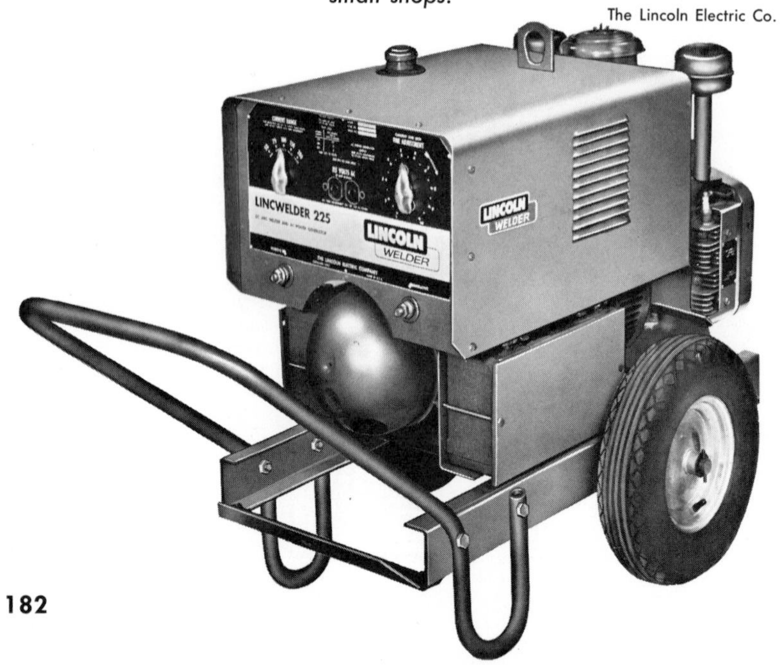

MULTIPLE-OPERATOR SYSTEMS

On construction jobs, in steel mills, and in shipyards, it is often necessary for a large number of welders to work within a limited area. If single-operator machines are used for each welder, the space becomes too crowded for efficient work and may be highly dangerous. A multiple-operator power unit, Fig. 6-29, can be installed away from the work site and be connected to control panels located close to the welding operator, Fig. 6-30. A large number of welding stations may be supplied from each unit. When using direct current, all welders must weld with the same polarity.

The welding current may be supplied by rotating motor-generators, static-rectifiers, or transformers. Most installations are direct current although many shipyards are using alternating current. Power sources may range in size from 600 to 2,500 amperes. A multiple-operator installation costs less, saves space and cable, lowers operating costs, and requires less maintenance.

POWER SUPPLY RATINGS

Welding power sources are rated by standards which have been set by The National Electrical Manufacturers Association (NEMA). These standards provide guidelines for the manufacture and performance of power sources.

Welding power sources are rated by current output, open circuit voltage, duty cycle, efficiency of output, and power factor.

Current Output. Welding machines are rated on the basis of current output in amperes. Amperage may range from 200 amperes or less for light or medium work to over 2,000 amperes for submerged-arc welding.

Open Circuit Voltage. The maximum allowable open circuit voltage of machines used for manual welding is set at 80 volts for safety reasons. Some constant current machines used for machine (automatic) welding are rated up to 125 open circuit voltage. Constant voltage types are normally rated from 15 to 50 open circuit voltage.

Duty Cycle. The duty cycle is the percentage of any given 10-minute period that a machine can operate at its rated current without overheating or breaking down.

A machine that can be used at its rated amperage on a continuous basis is rated as 100 percent duty cycle. Continuous, automatic machine welding requires this type of machine.

A machine that can be used at its capacity 6 out of every 10 minutes without damage is rated as 60 percent duty cycle. This type of machine is satisfactory for heavy stick electrode welding and TIG welding.

Machines used for welding in small shops or on farms for light repair usually have a 20 percent duty cycle.

Efficiency. Efficiency is the relationship of secondary power output to primary power input and is indicated in percent. It is determined by losses through the machine when actually welding at rated current and voltage.

Most d.c. generator welding machines have an efficiency of about 50 percent while a d.c. transformer-rectifier averages about 70 percent. The lower welding efficiency of the d.c. generator welding machine is due to power losses in transfer from mechanical to electrical energy and other losses.

Power Factor. The power factor is a measure of how effectively the welding machine makes use of a.c. primary line power. The primary power used is divided by the amount total drawn and is usually expressed in percent.

Three-phase d.c. transformer-rectifiers have a power factor of about 75 percent as compared with single-phase a.c. power units with a power factor of about 55 percent.

CABLES AND FASTENERS

Power Cable

For welding machines connected to an electric power line, conductors of ample capacity and adequately insulated for the voltage transmit the power. It is necessary to ground the frame of the welding machine so that a break in the insulation of the cable or the machine does not cause an electrical shock. The machine is often grounded by a portable cable with an extra conductor which is fastened to the machine frame on one end and a solid ground on the other. Since the voltage on the power side of the machine is much

Table 6-2

Characteristics of three basic types of welding machines. (300-ampere capacity).

Characteristics	d.c. Motor-generator	d.c. Transformer-rectifier	a.c. Transformer
Weight (lb.)	765	427	573
Floor area (sq. in.)	960	440	576
Volume (cu. ft.)	12.8	12.8	16.7
Efficiency (rated load)	60.70	84.70	64.70
No-load input (watts)	2850	576	1060
Safety	Best	Fair	Good
Electrode choice	Good	Least	Good
Electrode cost	Standard	Premium	Standard
Current change with input voltage	Least	Large	Most
Current change with warm up	Most	Least	Low
Arc blow	High	Low	Moderate
Noise	High	Low	Low
Maximum output (%)	153	140	138
Welder life	Good	Best	Good
Weld quality	Good	Good	Good
Maintenance cost	Most	Least	Medium

greater than on the output side of the machine, it is important that portable power cable be adequately insulated with tough abrasion-resisting insulation so that it will stand up under the rough usage ordinarily encountered in dragging it around welding shops.

Electrode and Ground Cable

To complete the electric circuit between the welding machine and the work, two cables of adequate current-carrying capacity are required. An *electrode cable,* also called a welding cable, is attached to the electrode holder. A *ground cable* is attached to the work. Rubber-covered multistrand copper cable, Fig. 6-31, especially designed for welding, is generally used. Multistrand aluminum cable is also becoming popular.

Cable, especially electrode cable, must have high flexibility so that it will not interfere with and tire the welder when working. Vertical and overhead welding, by the very nature of the position that the operator must maintain, imposes a severe strain. An electrode cable with poor flexibility increases this difficulty and encourages fatigue which, in turn, makes it difficult or impossible for the operator to maintain speed and good quality of workmanship.

High flexibility is the result of the core construction. The cable is woven of thousands of very fine, almost hairlike strands of copper wire. The greater the number of strands in a given-size cable, the more flexible it is. Figure 6-32 shows the inner construction of rubber-covered copper cable. Note the following components:

A. Thousands of hairlike copper wires are stranded in a special manner for extra flexibility.

B. Paper wrapping around stranded wires allows the conductor to slip within the rubber covering when the cable is bent sharply.

C. Extra strength is obtained by the open-braided reinforcement of extra cotton cords imbedded in the covering.

D. The special composition and curing of the heavy rubber covering makes it tough, durable, elastic, and waterproof.

Aluminum electrode cable is also multistrand; however, each strand in a cable is larger than a strand in a copper cable of the same size. Thus the next size larger aluminum cable should be selected for a given condition. Despite the extra size, the aluminum cable and covering weigh about half as much as the copper cable.

It is not necessary for ground cable to have the flexibility of the electrode cable since it is not handled by the operator in the act of welding. Nevertheless, the same type of cable is often used for the ground cable.

The amperage of the welding machine and its distance from the work are important considerations when selecting the size of welding cable, Table 6-3. The greater the amperage and the greater the distance between the work and the welding machine, the larger the cable size must be. Resistance increases as the diameter of the cable decreases. Therefore, if the cable is too

6-29. *Multiple-operator power source. A.c.-d.c. rectifier can be used for shielded metal-arc, gas tungsten-arc, and gas metal-arc welding, arc-air gouging, carbon arc burning, and steel welding.* (1) Simplified controls, (2) *high efficiency silicon rectifiers,* (3) *fan and motor,* (4) *heavy-duty transformer.*

Harnischfeger Corp.

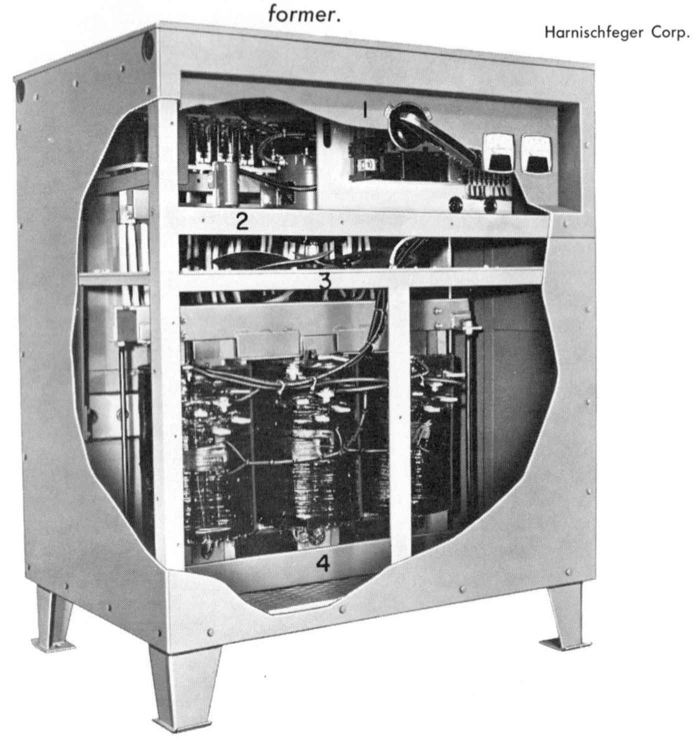

small, it will overheat and affect the welding operation. Resistance also increases as the length of the cable increases. The welding machine should be set up as close to the work as possible. Welding cables that are too long may cause a voltage drop so great that it has serious effects on the welding current and arc.

Cable Lugs and Ground Clamps

Suitable lugs, Fig. 6-33, are required on both the electrode cable and the ground cable to connect them to the welding machine. These lugs should be soldered or fastened mechanically to the cables. The connections that attach the cable lugs to the terminal posts on the welding machine must be tight. Loose connections cause the lugs to become hot, melting the solder which holds the cables to the lugs. Then the connector posts and cable lugs will burn and interfere with the welding current.

The other end of the electrode cable is connected to the electrode holder. The other end of the ground cable should have a means of connecting it to the work. If this connection is not secure, arcing across the connection will burn up the connection and may cause unsatisfactory welding conditions.

If the work is of such a nature that it can be done on a welding bench or in a permanent fixture, the ground cable is usually bolted to the bench or fixture. When the operator must work on a variety of structures in different parts of the shop, many types of ground devices may be used, including a copper hook, a heavy metal weight, a C-clamp, and specialized ground clamps, Figs. 6-34 and 6-35.

Figure 6-36 (Page 188) shows a type of rotating ground clamp. This clamp stops the twisting and turning of welding cable attached to welding fixtures and positioners where the work rotates. The C-clamp permits the ground to be attached in seconds to almost any shape. This clamp is generally used in fabricating tanks and pressure vessels, and on weld positioners.

When the work is of such nature that different lengths of cable leads are necessary, cables may be made up in specific lengths and equipped with cable connectors so that they can be quickly and easily attached to make up any desired length. The connector shown in Fig. 6-37 has

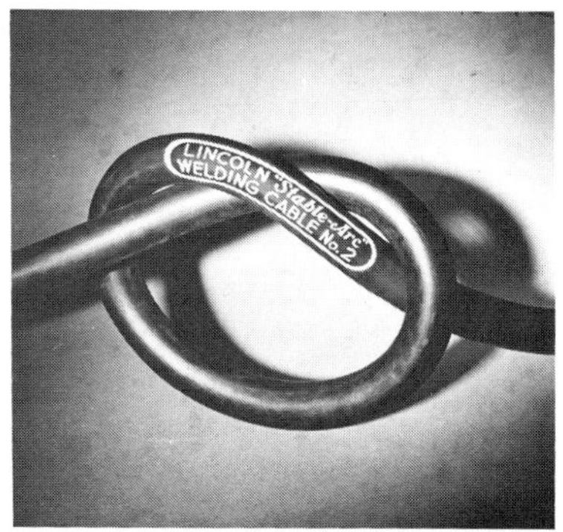

The Lincoln Electric Co.

6-31. *Welding cable with thousands of fine wires has greater flexibility. A specially finished, durable paper wrapper allows the conductor to slip easily when the cable is bent.*

6-30. *Multiple-operator control at the work site. Units can be stacked in a frame rack.* Harnischfeger Corp.

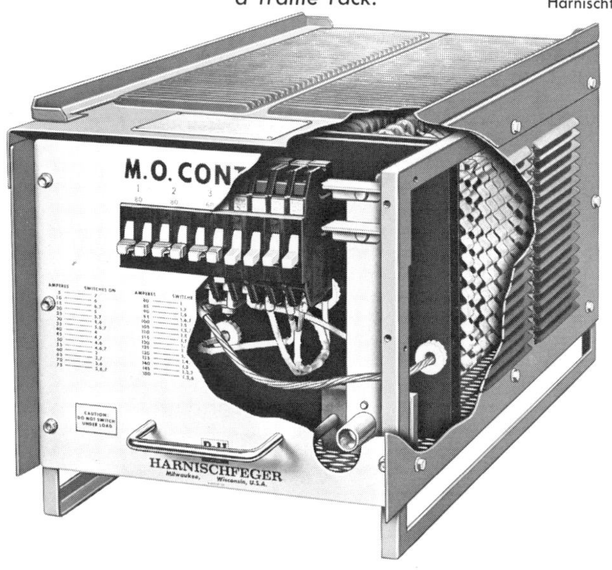

6-32. *Internal construction of welding cable.*

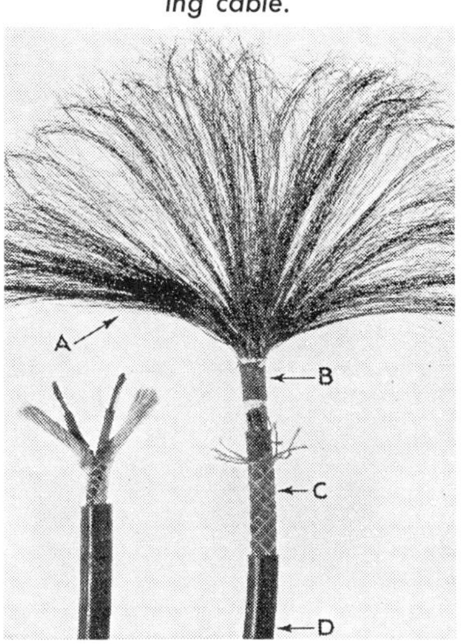

a cam-type action that insures a positive stop and lock and cannot come loose or accidentally fall apart.

ELECTRODE HOLDERS

Metal Electrode Holders

A metal *electrode holder,* is the device used for holding the electrode mechanically. It conveys electric current from the welding cable to the electrode, and it has an insulated handle to protect the operator's hand from heat.

The jaws of the holder should be designed so that they may grip the electrode firmly at any desired angle. They should be made of a metal that has high electrical conductivity and possesses the ability to withstand high temperatures, Figs. 6-38 and 6-39. On many holders the jaws can be replaced with new ones when they become so badly burned that they do not grip securely. The holder should be light in weight, well-balanced, and have a comfortable grip. Although it should be easy to change the electrode, the holder must be sturdy enough to withstand rough usage. The current-carrying parts must be large enough to prevent overheating, which causes the handle to become too hot for the operator to hold. For the same reason, the size of the holder must be in line with the size of the welding machine; that is to say, a larger electrode holder is required for a 400-ampere welding machine than for a 200-ampere machine. Most holders are fully insulated and may be laid anywhere on the work without danger of a short circuit. This is especially convenient for work in close quarters.

Figure 6-40 (Page 190) shows an electrode holder of relatively new design. It holds electrodes worn to a very short stub. Its twist-type locking device permits electrode-gripping power in excess of 2,000 pounds. There is never danger of dislodging the electrode when attempting to break through slag to start the arc. The electrode position is always known. There is positive contact between the holder and electrode, thus reducing both heating up of the holder and electrode waste. The handle is fully insulated so that it stays cool even with high duty cycles.

One of the newest holders is the angle-head, screw-clamp, fully insulated holder, Fig. 6-41. It is available in sizes of 400 and 600 amps and takes electrodes from $\frac{1}{16}$ through $\frac{5}{16}$ inch in diameter. Such holders are efficient and reduce costs since electrode stub loss is at a minimum, and holder maintenance is low. The screw-on head can be replaced to make a "new" holder.

The holder is usually attached to the welding cable by means of solderless connections. It is important that good contact be maintained to prevent overheating.

Table 6-3

Choosing the right size welding cable.

Using the right size welding cable on a welder is vitally important in obtaining sound welds. For example a 100-ft. lead of 4/0 cable at 500 amperes will have a voltage drop of 4 volts. At an output voltage of 40 volts, the power loss would amount to 10%. The current setting should be raised 10% to compensate for this loss. The following table gives the recommended sizes of cable to be used as approved by the American Welding Society.

Maximum length of cable to be used

Welder cap. in amps.	50	75	100	125	150	175	200
100	2	2	2	2	1	1/0	1/0
150	2	2	1	1/0	2/0	2/0	3/0
200	2	1	1/0	2/0	3/0	4/0	4/0
250	2	1/0	2/0	3/0	4/0		
300	1	2/0	3/0	4/0			
350	1/0	3/0	4/0				
400	1/0	3/0	4/0				
450	2/0	3/0					
500	3/0	4/0					

American Welding Society

OTHER ELECTRIC ARC PROCESSES

The electric arc generates the heat necessary for several major welding processes. Chapter 9 will present the principles of gas tungsten-arc (GTAW) welding, and

Chapter 10 will present the principles of gas metal-arc (GMAW) welding. Arc cutting, which utilizes much of the same equipment as arc welding, will be presented in Chapter 8.

Carbon-Arc Welding

In the carbon-arc process, welding heat comes from an arc formed between the base metal and a carbon electrode or an arc formed between two carbon electrodes. Depending on the job, the welding may be done with or without the addition of filler rod.

The carbon-arc electrode is a flexible tool which can be used for various welding applications. Examples include the welding of aluminum and copper alloys, brazing and soldering, heating, bending, and straightening.

Only copper-coated and cored carbon electrodes are used in the process. These electrodes keep the holder from overheating and allow the carbon to burn evenly. Carbon electrodes are available in sizes of $\frac{1}{4}$, $\frac{5}{16}$, and $\frac{3}{8}$ inch. Alternating current is used in a range of 30 to 125 amperes.

Holders designed for metal electrodes are not suitable for carbon-arc welding. The carbon-arc electrode becomes hotter than the metal electrode does and so requires a different kind of clamping mechanism. The carbon electrode holder, Fig. 6-42, is larger than the metal electrode holder. It has a metal shield to protect the operator's hand from the intense heat. Heavy-duty, continuous carbon-arc operations require a holder that is water cooled.

Atomic-Hydrogen Arc Welding

Atomic-hydrogen arc welding is a process in which the electric arc is surrounded by an atmosphere of hydrogen. The gas shields the molten metal from oxidation and contamination from the air. It also transfers heat from the electrode to the work. The arc is formed between two electrodes. The temperature produced by the arc is about 7,500 degrees F. Welding current is supplied by an a.c. welding transformer, and the hydrogen is supplied in cylinders. The electrode holder for the process is shown in Fig. 6-43.

Metal of practically the same analysis as that being welded can be deposited. Welds may be

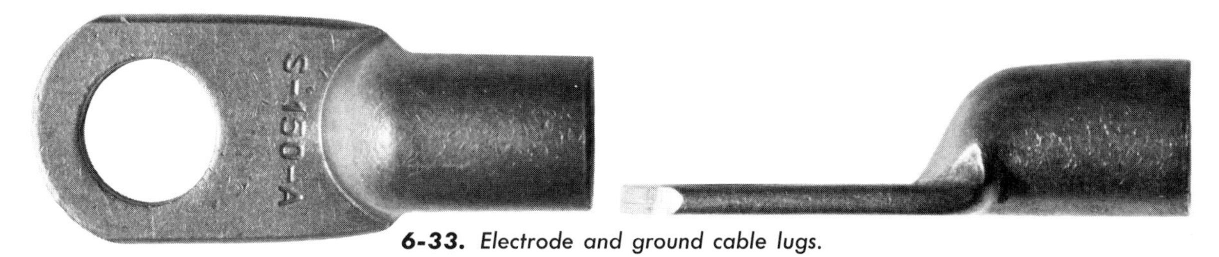

6-33. *Electrode and ground cable lugs.*

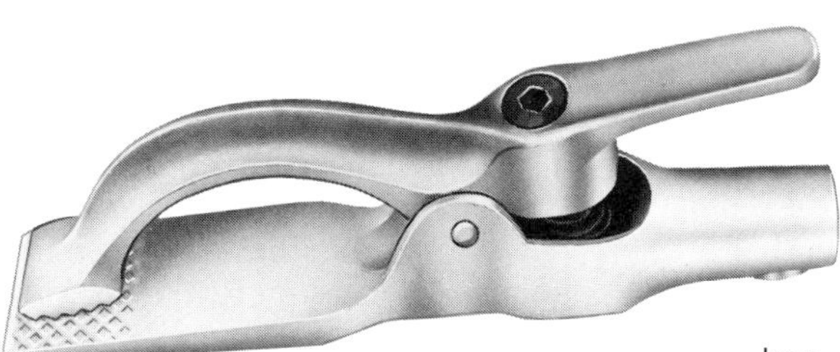

Lenco

6-34. *High amperage ground clamp. A good ground helps to eliminate arc blow and the flexibility of this clamp makes it easy to change.*

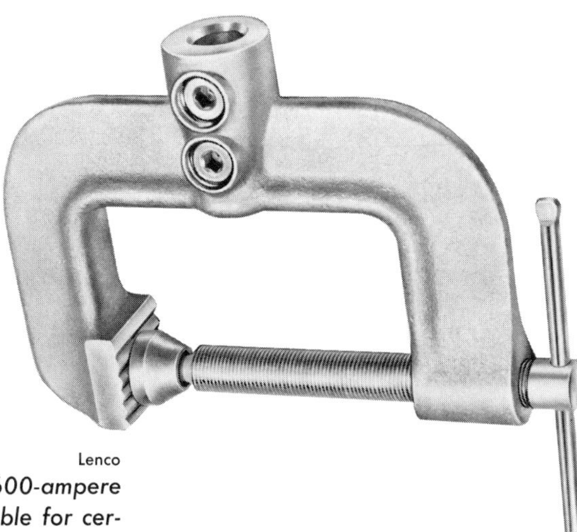

Lenco

6-35. *C-clamp type of ground clamp (600-ampere capacity.) The deep throat makes it desirable for certain work.*

heat treated. They are unusually smooth, ductile, nonporous, and free from impurities. The surface of the weld is free from scale, and the work is not pitted by weld spatter. Undercutting is avoided easily. Small sections of cutting edges can be built up without reworking the entire edge, thus reducing the amount of machining necessary.

Such hard-to-weld metals as chrome, nickel, molybdenum steels, Inconel,® Monel,® and thin sheets of stainless steel may be welded by the atomic-hydrogen process.

Advantages of atomic-hydrogen welding include increased production, low operating cost, and low maintenance cost. Although the process may be used in the fabrication of new products, it is particularly useful for repairing dies and molds, circular saws, and high grade machine tools.

OPERATOR EQUIPMENT

Hand and Head Shields

The brilliant light caused by an electric arc contains two kinds of invisible rays which injure the eyes and skin unless protection is provided. One of these invisible rays is known as *ultraviolet* and the other as *infrared.* Repeated exposure, either directly or indirectly, results in painful but not permanent injury to the eyes. Welders speak of this as "hot sand in the eyes." The rays may also produce a severe case of sunburn and sometimes infection. The rays affect the eyes at any distance within 50 feet; and the skin, at any distance within 20 feet.

Shields protect the operator not only against harmful light rays, but also against the hot globules of metal that the welding operation gives off, especially in the vertical and overhead positions. The hand shield, Fig. 6-44, has a handle so that the person using it may hold the shield in front of his face. Welding inspectors and supervisors use this type of shield. It is not suitable for the welding operator since he can work with only one hand while using it. It is impossible to manipulate the electrode and perform other necessary operations at the same time with the other hand.

The head shield, Fig. 6-45, also called a hood or helmet, is worn like a helmet. It is attached to an adjustable headband, Fig. 6-46, which allows it to be moved up or down as the wearer desires. Both hands are free to grasp the electrode holder and carry out accompanying operations. Partial protection is provided for the top of the head, but the operator must also wear a leather cap for adequate protection. This cap should be smooth and have no pockets or turned-up edges that will hold hot globules of metal.

Both the hand shield and the head shield are constructed of a heat-resisting, pressed-fiber insulating material. The shields are

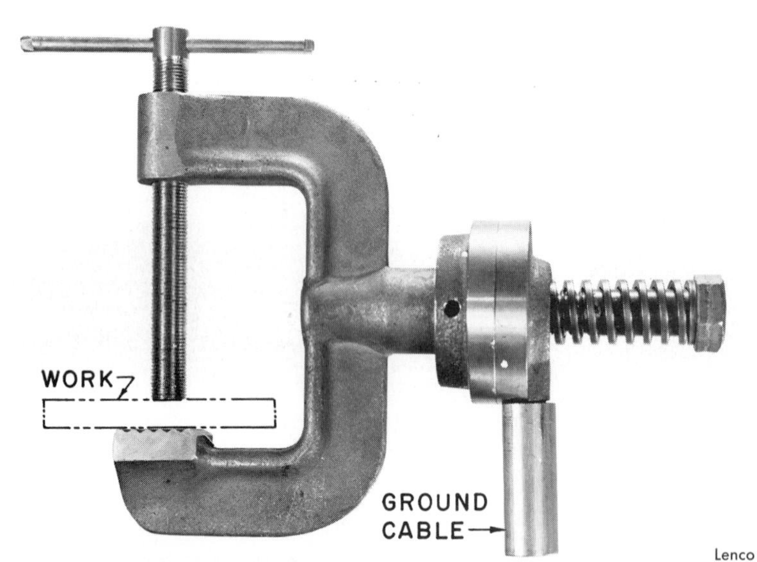

Lenco

6-36. *This rotary clamp is used when work must be turned. The clamp turns with the work, but the cable does not turn and twist.*

6-37. *Cable connectors with positive cam-type locking action cannot come loose. Covers are made of special plastic to resist weld spatter and corrosion. Inserts are brass.*

Lenco

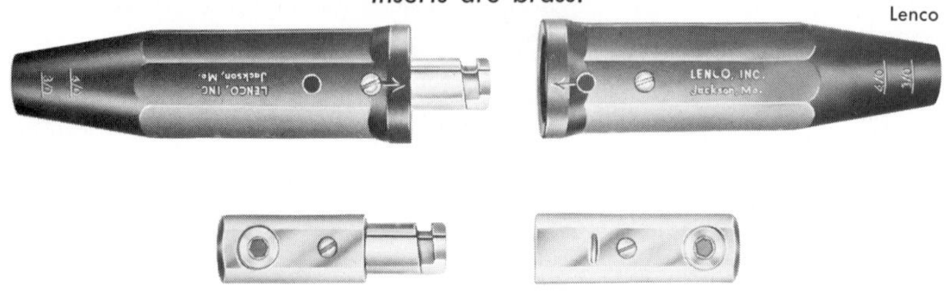

fully molded at the top and bottom to protect the head and neck from metal particles, fumes, and dangerous light rays. They are black to reduce reflection and have a window frame for holding the protective lens which permits the operator to look at the arc safely.

The size of the protective lens is $2 \times 4\frac{1}{2}$ inches. It is colored so that it can screen out ultraviolet rays, infrared rays, and most of the visible rays from the electric arc. A variety of shades of color may be obtained. The density of the color chosen depends on the brilliance of the arc, which varies with the size of the electrode and the volume of current. The shade for welding with metallic electrodes with current values up to 300 amperes is shade no. 10. Shade no. 12 is recommended for current values beyond 300 amperes and for gas-arc welding.

Good quality lenses are guaranteed to absorb 99.5 percent or more of the infrared rays and 99.75 percent or more of the ultraviolet rays present. The lenses have absorbed 100 percent of these rays as reported in tests conducted by the U.S. Bureau of Standards. The purchase of cheap filter lenses is to be discouraged.

The side of the protective glass exposed to the weld pool is protected by a clear glass or plastic cover lens. This is to protect the more costly filter lens from molten metal spatter and breakage. Cover glasses are comparatively inexpensive and may be obtained with or without a transparent coating. Chemically treated cover glasses last longer than the untreated type. When the cover glass becomes dirty from smoke and fumes and pitted from weld spatter, it should be replaced. Welding with clouded glass impairs vision and causes eye strain. Many operators prefer to protect both sides of the colored lens with clear glass.

Another type of head shield is shown in Fig. 6-47. This differs from the others only in that it is equipped with a flip front. The holder, which contains the protective and clear cover lenses, flips up at a touch of the finger. A clear cover glass in the stationary section protects the eyes from hot scale when inspecting hot welds and from flying particles of slag and steel during cleaning. This type is especially useful when working in close quarters where it is difficult to raise the helmet.

SPECIAL HELMETS. The following helmets are designed to fit specific conditions.

- *Wide vision lenses* for better viewing may be featured in standard helmets, Fig. 6-48 (Page 192). The size of the lenses are $4\frac{1}{2} \times 5\frac{1}{4}$ inches.
- *Chrome leather helmets,* Fig. 6-49, are used in hard-to-get-into areas where there is not enough room for a standard hard helmet. The helmet has small, round ventilating screens at the top.
- *A fiber-metal helmet and safety cap combination,* Fig. 6-50, offers practical, dependable protection when there is danger of falling or flying objects.
- *The air-conditioned helmet* shown in Fig. 6-51 has a supply

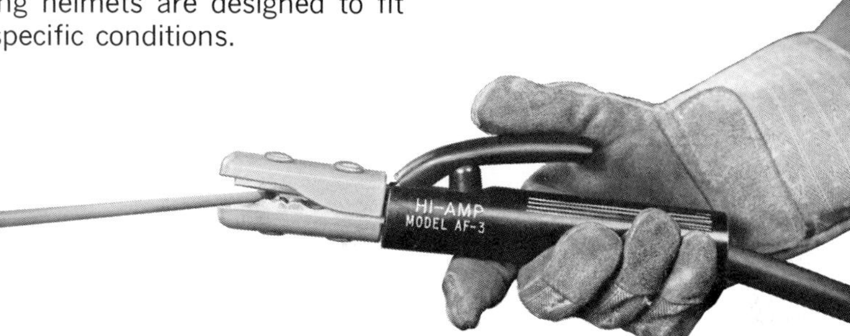

Lenco

6-38. *Fully insulated electrode holder. It may be laid on the work without fear of short circuit.*

Lenco

6-39. *The construction of an insulated electrode holder. There is a solderless connection to the cable. An Allen set screw applies pressure to an internal copper plate.*

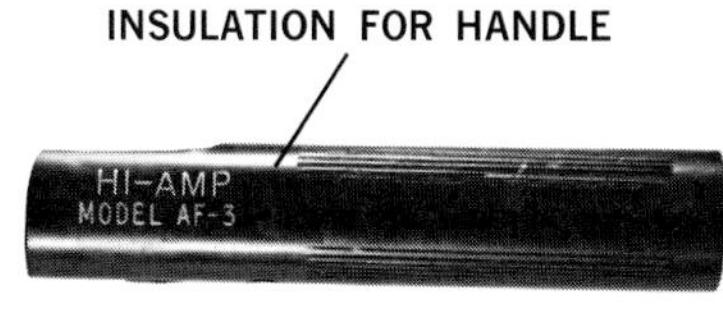

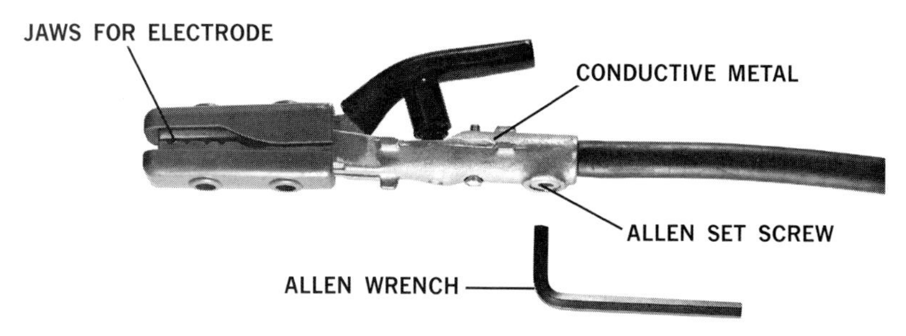

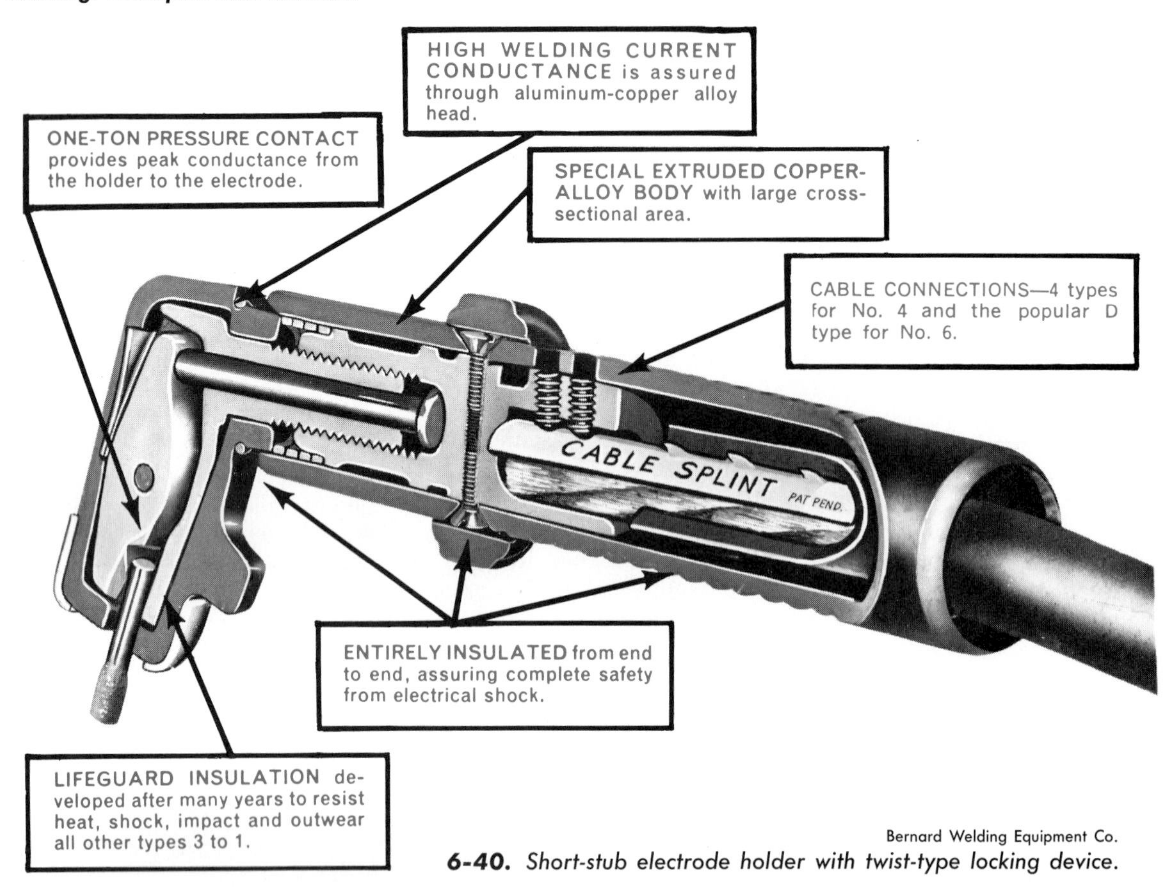
HIGH WELDING CURRENT CONDUCTANCE is assured through aluminum-copper alloy head.
ONE-TON PRESSURE CONTACT provides peak conductance from the holder to the electrode.
SPECIAL EXTRUDED COPPER-ALLOY BODY with large cross-sectional area.
CABLE CONNECTIONS—4 types for No. 4 and the popular D type for No. 6.
CABLE SPLINT
PAT PEND.
ENTIRELY INSULATED from end to end, assuring complete safety from electrical shock.
LIFEGUARD INSULATION developed after many years to resist heat, shock, impact and outwear all other types 3 to 1.

Bernard Welding Equipment Co.

6-40. *Short-stub electrode holder with twist-type locking device.*

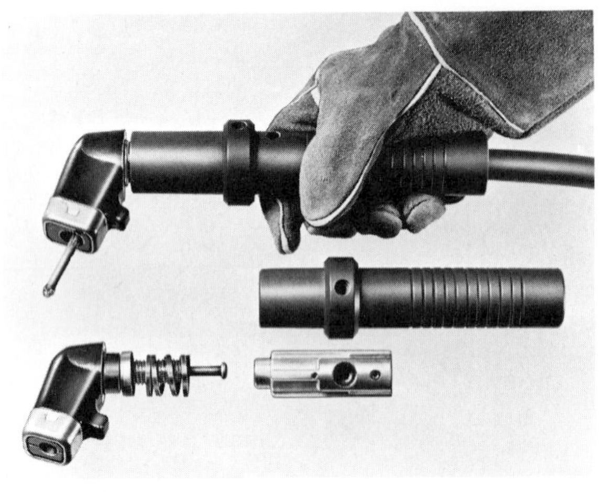

Jackson Products Co.

6-41. *This type of electrode holder has an angled head, and it is shorter and lighter than other holders of comparable capacity.*

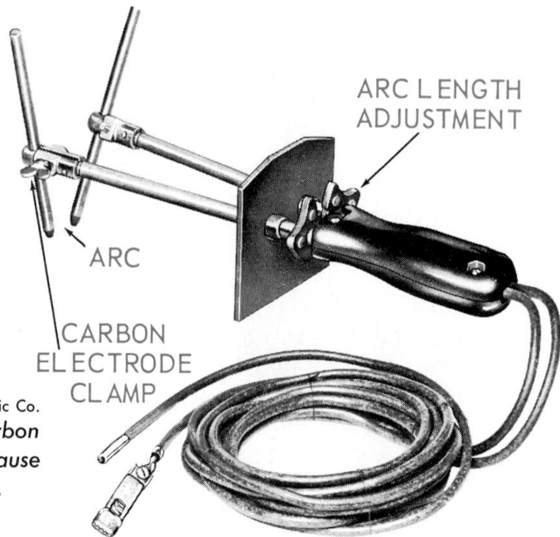
ARC LENGTH ADJUSTMENT
ARC
CARBON ELECTRODE CLAMP

The Lincoln Electric Co.

6-42. *A twin carbon electrode holder with carbon electrodes in place. Two leads are required because the arc is created between the two electrodes.*

of fresh air hose-fed directly to the immediate breathing zone. It reduces noise, and no annoying air jets pass the eyes. It is used in maintenance and repair when smoke, heat, and fumes may be a problem.

● *Special features* may be built inside of the helmet, such as volt-ampere meters to keep a close watch on current values while welding. Lights inside the helmet are often used to check the condition of the arc. Arc welders in training frequently wear these helmets.

● *Air line respirators,* Fig. 6-52, provide two-way safety for the welder. First, they provide cool, clean air to the operator, guarding against toxic fumes. Second, they have special exhalation valves that actually flush fumes from the helmet to provide unobstructed vision. An aluminum floating-yoke suspension adjusts easily to individual facial features for a comfortable fit. The mask's turned-in lip construction and deep chin cup also aid in obtaining a good face seal. The total weight of the unit is 20 ounces. A flow-control valve provides the welder with optimum flow control; the valve has a simple, quick disconnect.

FLASH GOGGLES. The practice of wearing flash goggles, Fig. 6-53, behind the hood is becoming increasingly popular. This is especially true in shops where there are a number of welders working close to each other. It is almost impossible under these conditions to prevent severe arc flash without flash goggles. Flash

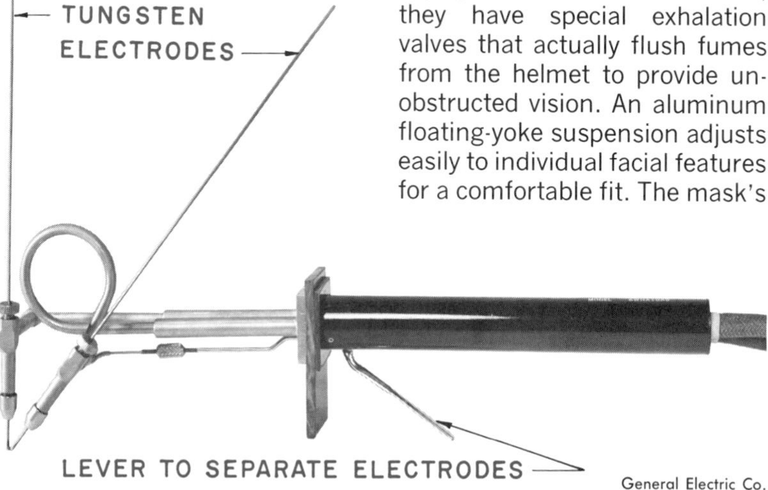

General Electric Co.

6-43. *The atomic-hydrogen electrode holder is lightweight, efficient, and easy to handle.*

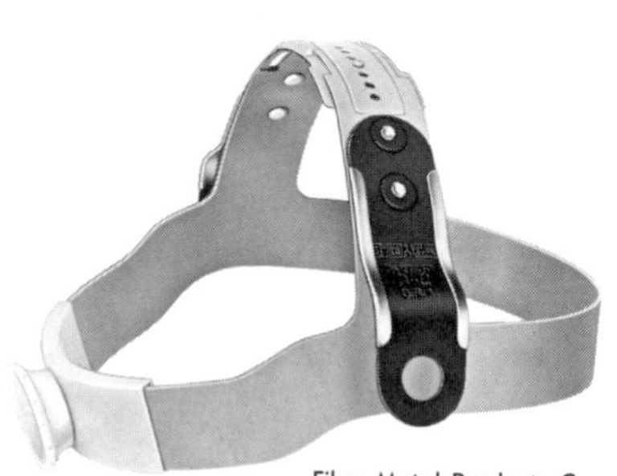

Fibre-Metal Products Co.

6-46. *An adjustable free-floating headband with adjustable crown.*

6-44. *The hand shield usually used by foremen and inspectors.*

Fibre-Metal Products Co.

6-45. *The standard head shield or hood.*

Fibre-Metal Products Co.

6-47. *The flip-front welding helmet permits the welder to inspect and brush the weld without lifting the hood.*

Willson Products

goggles also protect the welder's eyes while inspecting the recently completed weld, and when cleaning slag, grinding, and chipping. They should also be worn by helpers, foremen, inspectors, and others who work with welders. The goggles should be light in weight, well ventilated, and comfortable. In order to protect the eyes from side glare, they should be provided with blinders, and the lenses should have a tint.

Fibre-Metal Products Co.

6-48. *A welding helmet fitted with a wide vision lens. It is suitable for the welder who wears eyeglasses.*

Fibre-Metal Products Co.

6-50. *Fiber-metal helmet and safety cap combinations offer dependable protection on construction jobs.*

6-49. *Chrome leather helmets are ideal for those hard-to-get-into areas.*

Fibre-Metal Products Co.

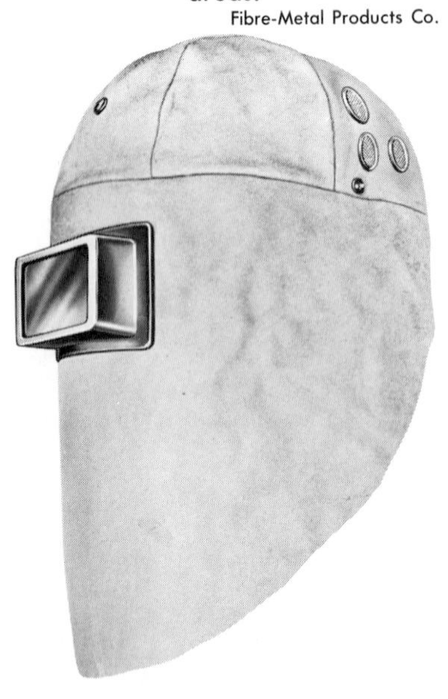

6-51. *Air-conditioned helmet to be used when smoke, heat, or fumes are a special problem.*

Fibre-Metal Products Co.

Protective Clothing

A continuous shower of sparks and hot globules of molten metal are thrown out by the arc during the welding operation. These will cause severe burns if they are allowed to contact unprotected skin. Ordinary clothing is not heavy enough, and the material is not fire resistant. The operator should wear work clothing made of a heavy material to protect the body from the rays emitted from the arc. Figure 6-54 shows a student welder who is well-protected from sparks, rays, and heat. It is not always necessary to wear all the protective clothing described for all kinds of welding and all positions. The welder can be the best judge of the protective clothing needed for the particular job. Gloves are necessary to protect the hands. Gloves are made of leather, asbestos, or some other type of fire-resistant material, Fig. 6-55. Leather sleeves, shoulder garments, and aprons protect the clothing and body of the welder from harm, Fig. 6-56. They are really necessary when the welder is called upon to do vertical and overhead work. If much of the work is to be done in a sitting position, the welder should wear overalls or a spit-type apron, because full aprons form a lap for hot particles. Chrome leather overalls may be worn with leather jackets, Fig. 6-57. High-top shoes must always be worn when welding. The feet and legs can be further protected by the use of leggings and spats. Burns on the feet are quite painful, become infected easily, and are slow to heal. Rolled sleeves and turned up trouser cuffs provide lodging places for hot metal and should be avoided.

Safety Appliances Co.

6-52. *Welder's air line respirator provides cool, clean air to the operator working in confined quarters.*

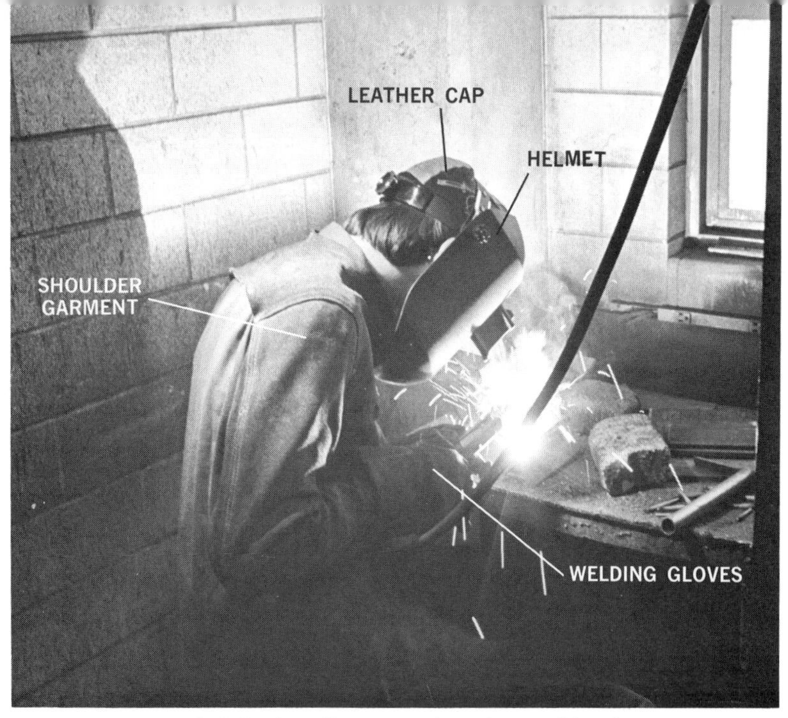

6-54. *A well-protected student welder.*

6-53. *This welder is wearing flash goggles behind her hood. The goggles protect the eyes from ultra-violet rays and the glare of the molten pool.*

Department of Labor

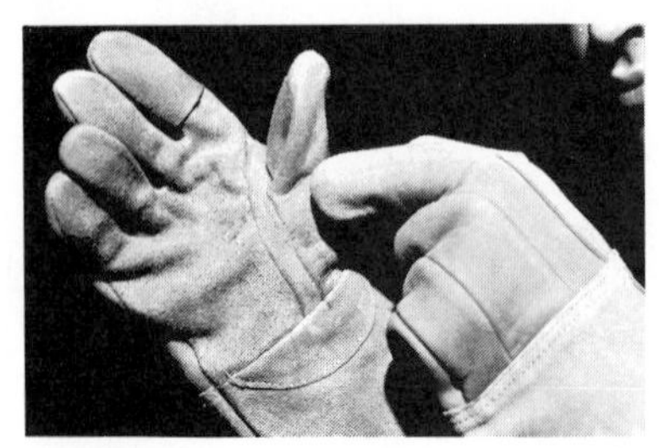

The Lincoln Electric Co.
6-55. *Leather welding gloves.*

Department of Labor
6-56. *An apron helps protect the welder from sparks and heat.*

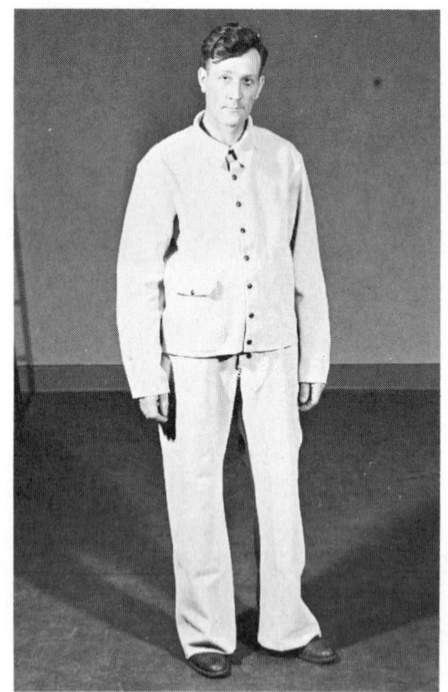

General Electric Co.
6-57. *A full leather coat and leather overalls give total body protection.*

REVIEW QUESTIONS

1. Name the three basic types of arc welding machines. Which type is used most?

2. What is the difference between a motor-generator welding machine and an engine-generator welding machine? Give the general use of each.

3. Name three articles of protective clothing worn by welders.

4. How is the size of an arc welding machine indicated?

5. The maximum welding current that can be used with a single arc is 200 amperes. True. False.

6. List the usual controls that are likely to be found on a d.c. motor-generator welding machine.

7. What is meant by single control? By dual control with fixed current steps? By dual continuous control?

8. What is the purpose of the polarity switch?

9. What is the function of meters on an arc welding machine?

10. List three advantages of a.c. arc welding.

11. Describe the construction of transformer-rectifier arc welding machines.

12. Compare the characteristics of motor-generator, transformer, and transformer-rectifier welding machines.

13. Explain the use of multiple-operator systems.

14. Do a.c. welding machines have more wearing parts than d.c. machines? Explain.

15. Are d.c. welding machines capable of welding faster than a.c. machines? Explain.

16. How is the current delivered to the electrode and the work? Explain.

17. What device is used to hold the electrode? Give a general description of this device.

CHAPTER 7 Shielded Metal-Arc Welding Electrodes

INTRODUCTION

Shielded metal-arc welding (SMAW), often referred to in this text as *arc welding* and by welders in the field as *stick electrode welding,* has become one of the most widely used welding processes in the fabrication and repair of metals. This popularity has resulted from the development of flux-covered electrodes capable of making welds having physical properties that are equal to or superior to the material being welded.

Leading manufacturers of welding equipment and supplies are constantly developing new welding electrodes and improving existing ones. Two other groups have had a major part in research and development: the *American Welding Society* and the *American Society for Testing and Materials.* These societies are continually improving the specifications and methods of classifying electrodes and filler rods so that the operator can readily select the best electrode for a particular job. Electrodes are designed to meet the needs of each welding application.

It is not the purpose of this chapter to discuss in detail all of the different types of electrodes that are used in the welding process. At this time we are concerned chiefly with those steel electrodes that will be used for the practice jobs in Part II of this text. Such electrodes comprise about 80 percent of the total number used by industry. Many welders have spent years as journeymen and have never had occasion to use anything but a low carbon steel electrode.

It is, of course, your objective to be able to deposit welds that have the most desirable physical and chemical properties, soundness, and appearance that the electrodes are capable of giving. Study Fig. 7-1 carefully. Learn to recognize good and bad welds and to understand the factors involved in good and bad welding. After you are thoroughly familiar with the characteristics of the electrodes discussed in this chapter, you should consult the catalogues of electrode suppliers and the materials available from the American Welding Society for information on other types of electrodes, such as those for high tensile steel, alloy steels, nonferrous materials, and surfacing materials.

TYPES OF ARC WELDING ELECTRODES

The general definition of a metal arc welding electrode, as given in the American Welding Society's "Definitions of Welding Terms and Master Chart of Welding Processes," is as follows:

- *Metal arc welding electrode:* filler metal in the form of a wire or rod, either bare or covered, through which current is conducted between the electrode holder and the arc.

It is important to note that the definition mentions bare and covered electrodes. These are two distinct classifications:

- *Bare (lightly coated) electrode:* a solid metal electrode with no coating other than that incidental to the manufacture of the electrode or with a light coating.
- *Covered (shielded arc) electrode:* a metal electrode which has a relatively thick covering material serving the dual purpose of stabilizing the arc and improving the properties of the weld metal.

Comparison of Bare and Covered Electrodes

Not very many years ago, bare electrodes were used exclusively in arc welding. The welds were very poor in appearance. Although they had satisfactory tensile strength, they had low ductility and low resistance to fatigue and impact. The defects were caused by the vaporization of certain important elements of the weld metal and by the presence of oxides and nitrides resulting from atmospheric contamination of the weld metal.

With the commercial introduction of the shielded metal-arc electrode, electric arc welding soon became the important metalworking process it is today. The use of the shielded-arc electrode overcomes the deficiencies of the bare electrode since the covering on the core wire shields both the arc and the weld metal from the atmosphere during the entire range of liquid action and solidification. Thus it made possible the deposition of weld metal with physical properties comparable to those of the base metal.

The type of covering influences the degree of penetration of the arc and the crater depth. The proper electrode selection, therefore, makes it possible to obtain sound welds in close-fitting joints and to avoid burning through poorly fitted joints. Since the covering influences the extent of heat penetration, it affects the extent of recrystallization and annealing of previously deposited layers. This characteristic improves the internal (X-ray) quality of the weld. The low electrical conductivity of the covering permits the use of electrodes in narrow grooves. Bare electrodes may cause short circuiting in these situations if the sides of the electrode arc against the base metal. The covering also reduces weld spatter.

FUNCTIONS OF ELECTRODE COVERINGS

Protective Gaseous Atmosphere and Slag Covering

The covering materials on the electrode provide an automatic cleansing and deoxidizing action in the molten weld crater. By supplying a protective gaseous atmosphere and blanket of molten slag for the weld metal, the covering excludes harmful oxygen and nitrogen, Fig. 7-2. The extent of gaseous and slag protection depends upon the type of covering. In addition to protection, slag performs the following functions:

- Acts as a scavenger in removing oxides and impurities
- Slows down the freezing rate of molten metal
- Slows down the cooling rate of solidified weld metal
- Controls the shape and appearance of the deposit

Alteration or Restoration of Base Metal

To a large extent, the covering controls the composition of the weld metal, either by maintaining the original composition of

7-1. *Plan and elevation views of welds made with shielded-arc electrodes under various conditions.*

Hobart Brothers Co.

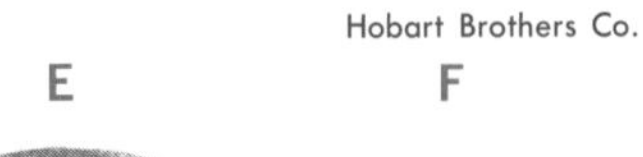

A	B	C	D	E	F
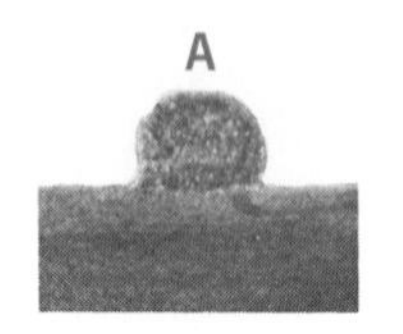	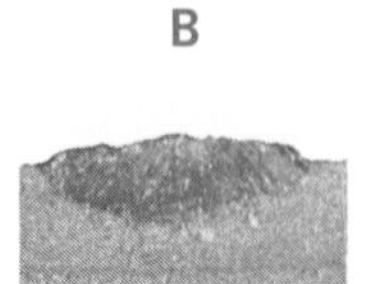	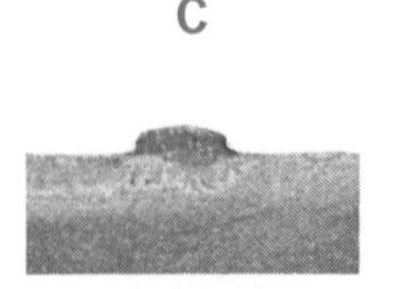	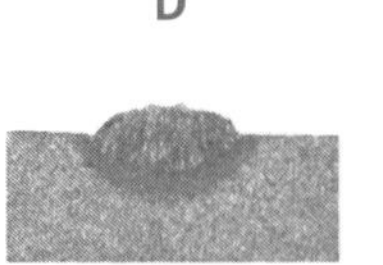	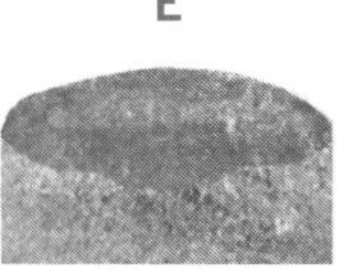	
Welding Current Too Low	*Welding Current Too High*	*Arc Too Long (Voltage Too High)*	*Welding Speed Too Fast*	*Welding Speed Too Slow*	*Proper Current Voltage & Speed*
Excessive piling up of weld metal.	**Excessive spatter to be cleaned off.**	**Bead very irregular with poor penetration.**	**Bead too small, with contour irregular.**	**Excessive piling up of weld metal.**	**A smooth, regular, well-formed bead.**
Overlapping bead has poor penetration.	**Undercutting along edges weakens joint.**	**Weld metal not properly shielded.**	**Not enough weld metal in the cross section.**	**Overlapped without penetration at edges.**	**No undercutting, overlapping, or piling up.**
Slows up progress.	**Irregular deposit.**	**An inefficient weld.**	**Weld not strong enough.**	**Too much time consumed.**	**Uniform in cross section.**
Wasted electrodes and productive time.	**Wasted electrodes and productive time.**	**Wasted electrodes and productive time.**	**Wasted electrodes and productive time.**	**Wasted electrodes and productive time.**	**Excellent weld at minimum material and labor cost.**
	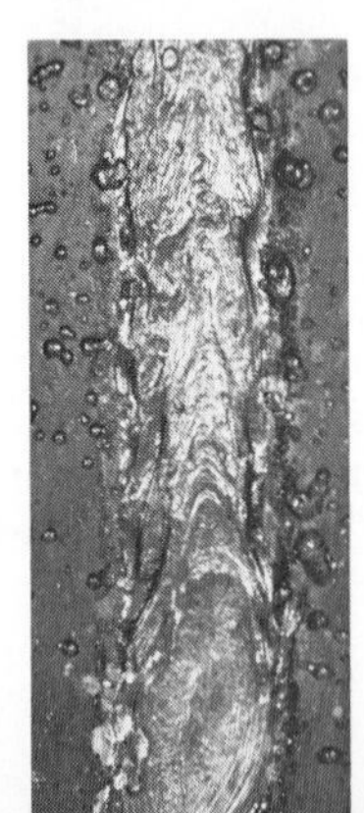	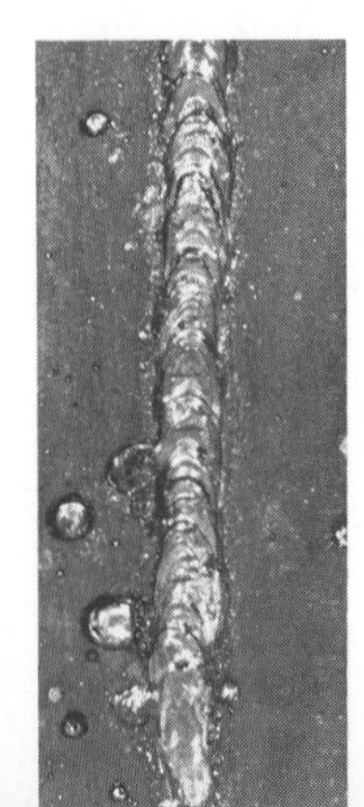	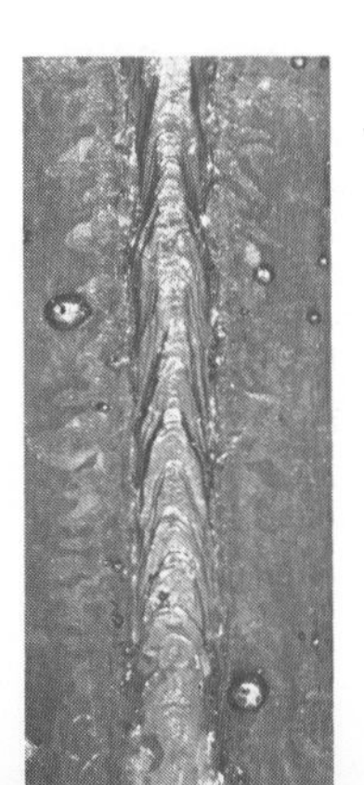	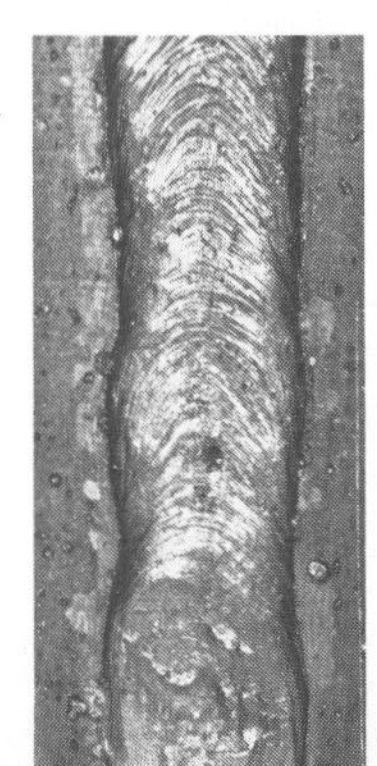	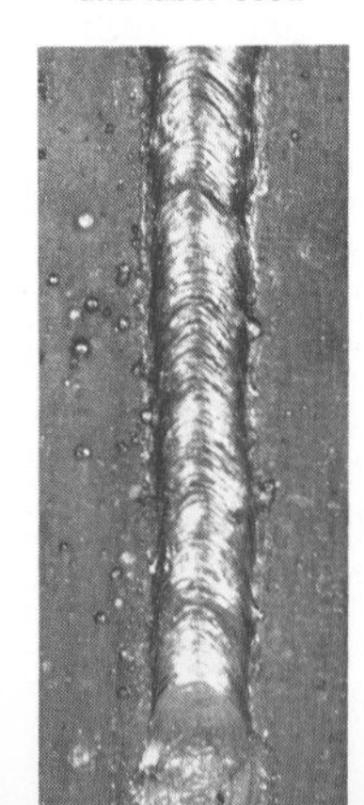

the core wire of the electrode or through the introduction of additional elements. In this way alloying elements are added to the weld metal or lost elements are restored. For example, molybdenum and vanadium may be added to the covering because these alloys produced better physical properties in the weld metal than those possessed by the core wire. On the other hand, manganese is usually included in the covering to maintain the same manganese content in the weld metal that was present in the core wire. The coating may also be balanced to adjust the carbon and silicon content of the weld deposit.

A comparatively recent development is the addition of powdered iron to the covering. In the intense heat of the arc, the iron powder is converted to steel and contributes metal to the weld deposit. When added in large amounts, the speed of welding is increased, and the weld appearance is improved.

The introduction of low hydrogen electrodes has greatly improved the welding of high carbon and alloy steels, high sulfur steels, and phosphorus-bearing steels. Such steels tend to be porous and crack under the bead. The reduction of the hydrogen content in the weld eliminates these harmful characteristics.

Control of Arc Characteristics

The covering makes it easier to start the arc when beginning to weld and to maintain a steady arc during the entire operation. It also serves as an insulator for the core wire of the electrode. Shielded-arc electrodes have less tendency to freeze to the base metal than bare electrodes do, and they allow greater variation in arc length. Better arc control permits the use of higher currents and larger electrodes.

The covering concentrates the heat of the arc on the work and thus causes an increased melting rate. The flux contains elements that ionize at the temperature of the arc and form a conducting path between the poles during the periods when the metal arc is extinguished by the transfer of globules of metal. The decreased welding time usually more than offsets the time necessary to remove the slag from the shielded-arc electrode welds.

COMPOSITION OF ELECTRODE COVERINGS

The type of covering affects the arc length and the welding voltage as well as the position of welding in which the electrode can be used. Its composition is very important. The covering should have a melting point lower than that of both the core wire and the base metal. The slag must have a lower density than the solidifying weld metal in order to be expelled quickly and thoroughly. The slag must also be able to solidify quickly when the electrode is used for overhead and vertical welding. These are all essential functions that insure the formation of sound welds having the physical and chemical properties required for the job. The coating is responsible for the differences between electrodes; the core wires are usually made of the same type of steel.

Materials for Electrode Coverings

Materials used in the coverings of shielded-arc electrodes may be classified according to their purpose: (1) fluxes, (2) deoxidizers, (3) slagging ingredients, (4) alloying ingredients, (5) gas reducers, (6) binders, (7) arc stabilizers, and (8) shielding gas.

7-2. Cross section of arc action, gaseous shielding, and metal flow when welding with covered electrode.

The Lincoln Electric Co.

Table 7-1
Comparison of physical properties of welds: bare and covered electrodes.

	Bare electrode weld metal	Covered electrode weld metal	Base metal
Ultimate strength, lbs. per sq. in.	50–60,000	60–75,000	55–70,000
Yield point, lbs. per sq. in.	38–45,000	45–60,000	30–32,000
Elongation, % in 2 in.	5.0–10.0	20.0–40.0	30.0–40.0
Elongation-free bend, %	10.0–20.0	35.0–60.0	. . .
Reduction of area, %	8.0–20.0	35.0–65.0	60.0–70.0
Density, grams/cm. ¾	7.5–7.7	7.80–7.85	7.85
Endurance limit, lbs. per sq. in. .	12–18,000	26–30,000	26–30,000
Impact (Izod), ft./lb.	5–15	40–70	50–80

American Welding Society

Table taken from "Welding Handbook" of American Welding Society.

Table 7-2
Comparison of core wire composition: filler metal, bare, and covered electrodes.

Element (Percentage)	Filler metal core wire	Bare electrode weld metal	Shielded-arc weld metal
Carbon	0.10–0.15	0.02–0.07	0.08–0.15
Manganese .	0.40–0.60	0.05–0.35	0.30–0.50
Silicon	0.025 max.	0.025 max.	0.05–0.30
Sulfur	0.04 max.	0.035 max.	0.035 max.
Phosphorus .	0.04 max.	0.035 max.	0.035 max.
Oxygen	0.06 max.	0.15–0.30	0.04–0.10
Nitrogen . . .	0.006 max.	0.10–0.15	0.01–0.03

American Welding Society

Table taken from "Welding Handbook" of American Welding Society.

Table 7-3
Weld metal properties afforded by types of covered electrodes.

	Minimum weld metal properties		
Property no.	Tensile strength p.s.i.	Yield point p.s.i.	Percentage of elongation
E60XX	62,000	50,000	17–25
E70XX	70,000	57,000	22
E80XX	80,000	67,000	19
E90XX	90,000	77,000	14–17
E100XX	100,000	87,000	13–16
E120XX	120,000	107,000	14

American Welding Society

Sodium and potassium silicates are universally used as binders; some organic gums also have a limited use for this purpose. Ferro-alloys and pure metals serve as deoxidizers and alloying ingredients. The alkaline earth metals are the best arc stabilizers. Wood flour, wood pulp, refined cellulose, cotton linters, starch, sugar, and other organic materials provide a shield of reducing gases. Fluxes and slagging ingredients include silica, alumina, clay, iron ore, rutile, limestone, magnesite, asbestos, mica, and many other minerals, as well as some man-made materials, such as potassium titanate and titanium dioxide.

Polarity Interchangeability

The composition of the covering determines the best polarity of the electrode used for d.c. applications. Some coverings operate more efficiently with straight polarity (electrode negative). Other coverings make the electrode more efficient with reverse polarity (electrode positive). Both types have advantages which make them preferable for certain applications. Coverings have also been developed which operate equally well on either polarity. These types are also efficient in a.c. welding.

Tables 7-1 through 7-3 demonstrate the important function of the covering and its direct effect upon the chemical and physical properties of the weld. Study carefully the properties of the various types of electrodes described on the following pages and in Table 7-4. It will also be beneficial at this time to become familiar with the current values given in Table 7-5.

IDENTIFYING ELECTRODES

Two basic systems for identifying steel electrodes are used in

Table 7-4

Welding characteristics of mild steel electrodes.

	Type of coating	Position of welding	Type of current[1]	Penetration	Rate of deposition	Appearance of bead	Spatter	Slag removal	Minimum tensile strength	Yield point	Minimum elongation in 2 in.
E6010	High cellulose sodium	All positions	DCRP	Deep	Average rate	Rippled and flat	Moderate	Moderately easy	62,000 p.s.i.	50,000 p.s.i.	22%
E6011	High cellulose potassium	All positions	DCRP, a.c.	Deep	Average rate	Rippled and flat	Moderate	Moderately easy	62,000 p.s.i.	50,000 p.s.i.	22%
E6012	High titania sodium	All positions	DCSP, a.c.	Medium	Good rate	Smooth and convex	Slight	Easy	67,000 p.s.i.	55,000 p.s.i.	17%
E6013	High titania potassium	All positions	DCRP, DCSP, a.c.	Mild	Good rate	Smooth and flat to convex	Slight	Easy	67,000 p.s.i.	55,000 p.s.i.	17%
E7014	Iron powder titania	All positions	DCRP, DCSP, a.c.	Medium	High rate	Smooth and flat to convex	Slight	Easy	70,000 p.s.i.	60,000 p.s.i.	17%
E7015	Low hydrogen sodium	All positions	DCRP	Mild to medium	Good rate	Smooth and convex	Slight	Moderately easy	70,000 p.s.i.	60,000 p.s.i.	22%
E7016	Low hydrogen potassium	All positions	DCRP, a.c.	Mild to medium	Good rate	Smooth and convex	Slight	Very easy	70,000 p.s.i.	60,000 p.s.i.	22%
E6020	High iron oxide	Flat hor. fillets	Flat: d.c., a.c. hor. fillets: DCSP, a.c.	Deep	High rate	Smooth and flat to concave	Slight	Very easy	62,000 p.s.i.	50,000 p.s.i.	25%
E7024	Iron powder titania	Flat hor. fillets	DCSP, DCRP, a.c.	Mild	Very high rate	Smooth and slightly convex	Slight	Easy	72,000 p.s.i.	60,000 p.s.i.	17%
E6027	Iron powder iron oxide	Flat hor. fillets	Flat: d.c., a.c. hor. fillets: DCSP, a.c.	Medium	Very high rate	Flat to concave	Slight	Easy	62,000 p.s.i.	50,000 p.s.i.	25%
E7018	Iron powder low hydrogen	All positions	DCRP, a.c.	Mild	High rate	Smooth and flat to convex	Slight	Very easy	72,000 p.s.i.	60,000 p.s.i.	22%
E7028	Iron powder low hydrogen	Flat hor. fillets	DCRP, a.c.	Mild	Very high rate	Smooth and slightly convex	Slight	Very easy	72,000 p.s.i.	60,000 p.s.i.	22%

[1] DCRP means direct current, reverse polarity (electrode positive).
DCSP means direct current, straight polarity (electrode negative).

National Cylinder Gas Division, Chemetron Corp.

the welding industry. The first is the color identification system for both bare and covered electrodes. This means of identification was developed by the National Electrical Manufacturers Association (NEMA) and the American Welding Society (AWS). Figures 7-3 and 7-4 indicate the locations of the color markings on covered electrodes, and Fig. 7-5 indicates the location on bare electrodes. Tables 7-6 and 7-7 list the colors and their locations according to electrode classifications. This is a complicated system and is little used by the welder. Color code charts are available through the National Electrical Manufacturers Association in New York.

The second system of identifying electrodes is used on covered arc-welding electrodes only. It requires that the electrode classification number be imprinted or stamped on the electrode covering, Fig. 7-6.

AWS Classification of Carbon Steel Electrodes

The many types of electrodes that are available to weld carbon and alloy steels are classified in the AWS-ASTM *Tentative Specifications for Iron and Steel Welding Electrodes.* A copy of this booklet can be purchased from the American Welding Society.

Other agencies concerned with electrode approvals, specifications, and classifications include:

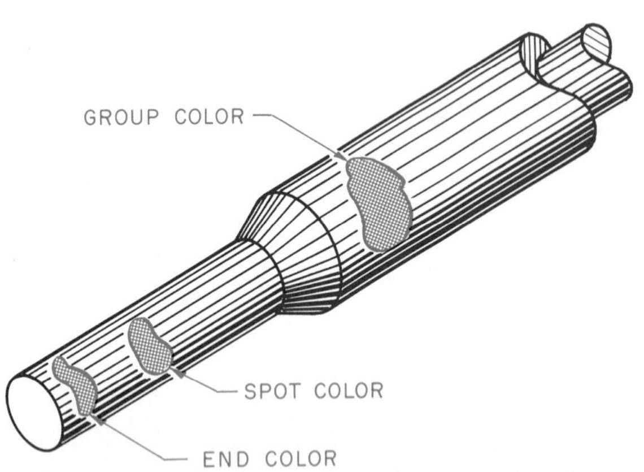

7-3. *Location of color markings for covered end-grip welding electrodes.*

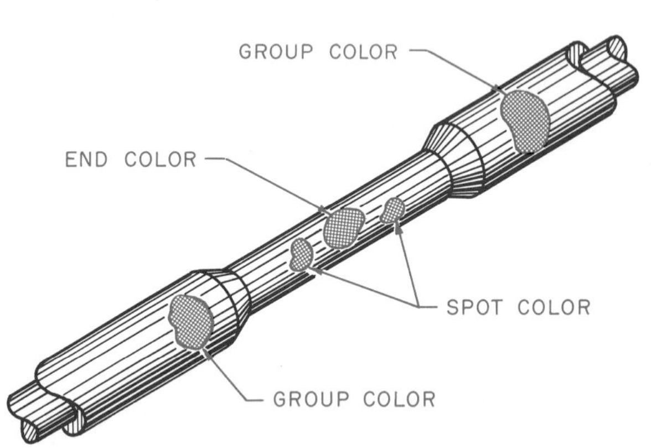

7-4. *Location of color markings for covered center-grip welding electrodes.*

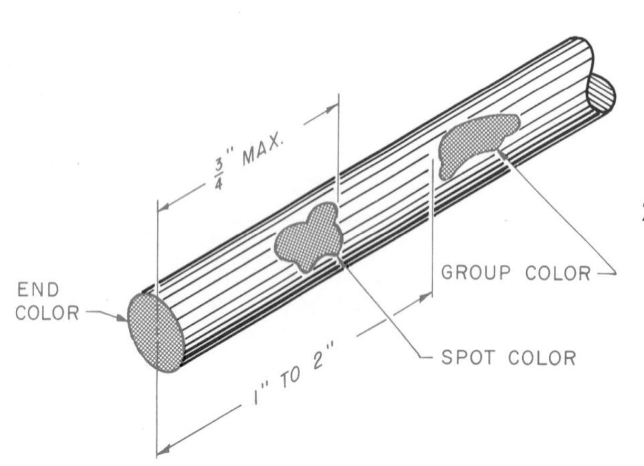

7-5. *Location of color markings for bare welding electrodes.*

7-6. *Location of electrode classification number for covered end-grip welding electrodes.*

Table 7-5

Current range for mild and low alloy steel electrodes.

Electrode diameter (in.)	Current range (amp)								
	Electrode type								
	E6010, E6011 DC+	E6012	E6013	E6020	E6027	E7014	E7015, E7016	E7018	E7024, E7028
1/16 5/64 3/32	— — 40–80	20–40 25–60 35–85	20–40 25–60 45–90	— — —	— — —	— — 80–125	— — 65–110	— — 70–100	— — 100–145*
1/8 5/32 3/16	75–125 110–170 140–215	80–140 110–190 140–240	80–130 105–180 150–230	100–150 130–190 175–250	125–185 160–240 210–300	110–160 150–210 200–275	100–150 140–200 180–255	115–165 150–220 200–275	140–190 180–250 230–305
7/32 1/4 5/16	170–250 210–320 275–425	200–320 250–400 300–500	210–300 250–350 320–430	225–310 275–375 340–450	250–350 300–420 375–475	260–340 330–415 390–500	240–320 300–390 375–475	260–340 315–400 375–470	275–365 335–430 400–525*

*These values do not apply to the E7028 classification.

American Welding Society

- American Society of Mechanical Engineers (ASME)
- American Bureau of Shipping (Bureau of Ships)
- U.S. Department of Defense (for the Army, Navy, and Air Force)
- U.S. Coast Guard
- Canadian Welding Bureau, Division of Canadian Standards Association
- A large variety of municipal, county, and state organizations responsible for welding standards

The various electrodes are classified as a system of numbers and organized on the basis of the mechanical properties of the deposited weld metal, type of covering, recommended welding position of the electrode, and type of current required for best results. Compare the various characteristics of carbon steel electrodes in Tables 7-3 and 7-4, pages 198, 199.

The numbering system provides a series of four- or five-digit numbers prefixed with the letter *E*. The *E* indicates electrode for electric welding. The first two digits (three digits in a five-digit

Table 7-6

Color identification for covered mild steel and low alloy steel electrodes.

Group color—no color				
XX10, XX11, XX14, XX24, XX27, XX28 and all 60 XX				
Spot color \ End color	**No color**	**Blue**	**Black**	**Orange**
No color White	E6010 E6012	E7010G E7010-A1		EST ECI
Brown	E6013		E7014	
Green	E6020			
Blue	E6011	E7011G		
Yellow		E7011-A1	E7024	
Black			E7028	
Silver	E6027			
Group color—silver				
All XX13 and XX20 except E6013 and E6020				
Brown				
White				
Green		E7020G		
Yellow		E7020-A1		

The Lincoln Electric Co.

number) multiplied by one thousand expresses the minimum tensile strength in thousands of pounds per square inch. For example, *60* in E6010 electrodes means 60,000 p.s.i.; *70* in E7010 electrodes means 70,000 p.s.i.; and *100* in E10010 electrodes means 100,000 p.s.i. The next-to-last digit indicates the position of welding. Thus the *1* in E6010 means that welding can be done in all positions—flat, horizontal, vertical, and overhead. The *2* in E7020 means that the electrode should be used for flat and horizontal fillet welding. The last digit indicates the type of current and the type of covering on the electrode. Further interpretation of the classification numbers is presented in Tables 7-8 and 7-9.

ELECTRODE SELECTION

The selection of the proper electrode for a given welding job is one of the most important decisions that the welding operator is faced with. The nature of the deposited weld metal and its suitability as a joining material for the pieces being welded depend, of course, on the proper selection of the type of electrode. The tremendous progress of the shielded metal-arc welding process is due in great measure to the high quality of the electrodes available at this time.

Electrodes may be generally grouped according to (1) operating characteristics, (2) type of covering, and (3) characteristics of deposited metal. Size is also an important consideration.

It is important to understand that many electrode classifications contain the same basic core wire. The differences in operating characteristics and in the physical and chemical nature of the deposited weld metal are mostly determined by the materials in the covering.

You are urged to study carefully the characteristics of the basic welding electrodes described in this section. A thorough knowledge will assist you

Table 7-7

Color identification for covered low hydrogen, low-alloy steel electrodes.

Group color—green										
XX15, XX16 and XX18 except E6015 and E6016										
Spot color \ End color	No color	Blue	Black	White	Gray	Brown	Violet	Green	Red	Orange
Red	E7015G	E7015			E8015G	E9015G		E10015G		E12015G
White		E7015-A1	E9015-B3L			E9015-D1				
Brown										
Green			E8015-B2L			E9015-B3				
Bronze			E8015-B4L			E8015-B4				
Orange	E7016G	E7016	E7018	E8016-C3		E9016G		E10016G		E12016G
Yellow		E7016-A1	E7018-A1	E8016G		E9016-D1		E10015-D2	E11016G	
Black			E8018-C3	E8016-B1	E8018-B1		E9018-B3			
Blue	E7018G		E8018G	E8016-C1	E8018-C1	E9016-B3	E9018G	E10018G	E11018G	E12018G
Violet				E8016-C2	E8018-C2	E8016-B4	E9018-D1	E10018-D2		
Gray			E8018-B4	E8016-B2	E8018-B2			E10016-D2		
Silver			Mil-12108							

The Lincoln Electric Co.

not only in acquiring the technical knowledge necessary to choose the proper electrode for a given job, but also in mastering the manipulative techniques of the welding operation.

Size of Electrodes

Selecting the proper size of electrode to use on a given job is as important as selecting the right classification of electrode. The following points should be taken into consideration:

- *Joint design:* A fillet weld can be welded with a larger electrode than an open butt weld.
- *Material thickness:* It is obvious that a larger size electrode can be used as the thickness of material increases.
- *Thickness of weld layers:* The thickness of the material to be welded and the position of welding are factors. More weld metal can be deposited in the flat and horizontal positions than in the vertical and overhead positions.
- *Welding position:* A larger size electrode can be used in the flat and horizontal positions than in the vertical and overhead positions.
- *Amount of current:* The higher the current value used for welding, the larger the electrode size.
- *Skill of the operator:* Some welders become so expert that they can handle large size electrodes in the vertical and overhead positions.

All classes of covered electrodes are designed for multiple-pass welding. The size varies with the types of joints and welding position.

- The first pass for pipe welding and other bevel butt joints should be welded with 1/8-inch or 5/32-inch electrodes. This is necessary in order to obtain good fusion at the root and to avoid excessive burn-through. The remaining passes may be welded with 5/32-inch or 3/16-inch electrodes in all positions and 3/16 inch or larger in the flat position.
- For flat position welding of beveled-butt joints that have a backup strip, a 3/16-inch electrode can be used for the first pass and an electrode 7/32 inch or larger for the remainder.

Table 7-8

AWS electrode classification system.

Digit	Significance	Example
1st two or 1st three	Min. tensile strength (stress relieved)	E-60xx = 60,000 p.s.i. (min) E-110xx = 110,000 p.s.i. (min)
2nd last	Welding position	E-xx1x = all positions E-xx2x = horizontal and flat E-xx3x = flat
Last	Power supply, type of slag, type of arc, amount of penetration, presence of iron powder in coating	See Table 7-9
Note: Prefix "E" (to left of a 4 or 5-digit number) signifies arc welding electrode		

American Welding Society

Table 7-9

Interpretation of the last digit in the AWS electrode classification system.

Designation	Current	Covering type
E6010	DCRP only	Organic
E6011	DCRP or a.c.	Organic
E6012	DCSP or a.c.	Rutile
E6013	DCSP, DCRP, or a.c.	Rutile
E7014	DCSP, DCRP, or a.c.	Rutile, iron-powder (approx. 30%)
E7015	DCRP only	Low hydrogen
E7016	DCRP or a.c.	Low hydrogen
E7018	DCRP or a.c.	Low hydrogen, iron powder (approx. 25%)
E6020	DCSP, DCRP, or a.c.	High iron oxide
E7024	DCSP, DCRP, or a.c.	Rutile, iron powder (approx. 50%)
E6027	DCSP, DCRP, or a.c.	Mineral, iron powder (approx. 50%)
E7028	DCRP or a.c.	Low hydrogen, iron powder (approx. 50%)

American Welding Society

- For fillet welds in the flat position $^{3}/_{16}$-inch, $^{7}/_{32}$-inch, or $^{1}/_{4}$-inch electrodes may be used. Extra heavy plates can be welded with larger electrodes.
- Out-of-position fillet and butt welding is usually done with $^{5}/_{32}$-inch electrodes. For certain jobs $^{3}/_{16}$-inch electrodes may be used.
- The sizes of low hydrogen electrodes generally used for vertical and overhead welding are $^{1}/_{8}$ inch and $^{5}/_{32}$ inch. Electrodes for flat and horizontal welding may be $^{3}/_{16}$ inch or larger.

Job Requirements

The requirements of the job are the basis for the proper selection of electrodes. Look the job over carefully to determine just what the electrode must do. Study the variable factors given in Table 7-10. Following are a few conditions to check:

- Skill of the welder
- Code requirements (if any)
- Properties of base metal
- Position of the joint
- Type and preparation of joint
- Heat-treating requirements
- Environmental job conditions
- Expansion and contraction problems
- Amount of weld required
- Tightness of fitup
- Type of welding current available
- Thickness and shape of base metal
- Specifications and service conditions
- Demands of production and cost considerations

Operating Characteristics of Electrodes

The nature of the materials that go into the covering of an electrode usually determines not only the physical and chemical content of the weld deposit, but also the operating characteristics of the electrode. Different electrodes require different welding techniques. Thus electrodes may be grouped according to their operating characteristics and the requirements of the joints to be welded as *fast fill, fast follow,* and *fast freeze.*

FAST FILL. The fast-fill electrode deposits weld metal rapidly. It is the opposite of the fast-freeze electrode. The fast-fill group includes the heavy-coated, iron powder electrodes which are widely used for fillet and deep groove deposition. The fast-fill electrode is especially designed for fast downhand welding. It has high metal deposition and permits easy slag removal. There is little undercutting. It burns with a soft arc and has shallow penetration which causes little mixing of the base metal and

Table 7-10
Relative ratings of factors affecting the preliminary selection of electrodes.

Variable factors (a)	Electrode class										
	E6010	E6011	E6012	E6012X	E6013	E7014	E7016	E7018	E6020	E7024	E6027
1 Groove butt welds, flat (>¼ in.)	4	5	3	2	8	8	7	9	10	9	10
2 Groove butt welds, all positions (>¼ in.)	10	9	5	4	8	8	7	6	(b)	(b)	(b)
3 Fillet welds, flat or horizontal	2	3	8	7	7	7	5	7	10	10	7
4 Fillet welds, all positions	10	9	6	4	7	7	8	6	(b)	(b)	(b)
5 Current	DCRP	a.c. DCRP	a.c. DCSP	a.c. DCSP	a.c. d.c	d.c. d.c.	a.c. DCRP	a.c. d.c.	a.c. d.c.	a.c. d.c.	a.c. d.c.
6 Thin material (<¼ in.)	5	, 7	8	10	9	8	2	2	(b)	7	(b)
7 Heavy plate or highly restrained joint	8	8	6	(b)	8	8	10	9	8	7	8
8 High-sulfur or off-analysis steel	(b)	(b)	5	4	3	3	10	9	(b)	5	(b)
9 Deposition rate	5	5	7	7	7	9	5	8	9	10	10
10 Depth of penetration	10	9	6	5	5	5	7	7	8	4	8
11 Appearance, undercutting	6	6	8	7	9	9	7	10	9	10	10
12 Soundness	6	6	3	3	5	5	10	8	9	8	9
13 Ductility	6	7	4	3	5	5	10	10	10	5	10
14 Low-temperature impact strength	8	8	4	4	5	5	10	10	8	9	9
15 Low spatter loss	1	2	6	6	7	7	6	8	9	10	10
16 Poor fit-up	6	7	10	10	8	8	4	4	(b)	8	(b)
17 Welder appeal	7	6	8	8	9	10	6	8	9	10	10
18 Slag removal	10	8	6	6	8	8	4	7	8	8	8

Hobart Brothers Co.

(a) Rating (for same size electrodes) is on a comparative basis for electrodes listed in this table: 10 is highest value. Ratings may change with change in size. (b) Not recommended.

weld metal. Bead appearance is very smooth. It has a flat to slightly convex face and there is little spatter, Fig. 7-7, welds H and I.

Certain electrodes have been developed for out-of-position welding that have faster freezing characteristics. An example of this type of electrode is the EXX14. The EXX24 and EXX27 electrodes are generally used for flat fillets and groove welding, Fig. 7-7.

FAST FOLLOW. This group is also known as fill-freeze electrodes. They have characteristics which in some degree combine both fast-freeze and fast-fill requirements. In making lap welds or light-gauge sheet metal welds, little additional weld metal is needed to form the weld. The most economical way to make the joint is to move rapidly. Because it is necessary to make the crater follow the arc as rapidly as possible, the electrode is called fast-follow. It burns with a moderately forceful arc and has medium penetration. This, together with the lower current and lower heat input, reduces the problem of burn-through.

Fast-follow electrodes are commonly referred to as straight polarity electrodes even though they may also be used with alternating current. These electrodes have complete slag coverage, and beads are formed with distinct, even ripples, Fig. 7-7, welds D and E. Many production shops use them as a general purpose electrode, and they are also used widely for repair work. While they may be used for all-position work, the fast-freeze electrodes are preferable. Many shops that engage in light-gauge sheet steel fabrication use the fast-follow electrode for vertical position, travel-down, welding. Examples of these electrodes are the EXX12 for d.c. welding and the EXX13 for a.c. welding.

FAST FREEZE. Fast-freeze electrodes have the ability to deposit weld metal which solidifies, or freezes, rapidly. This is important when there is some chance of slag or weld metal spilling out of the joint and when welding in the vertical or overhead positions.

These electrodes have a snappy, deep-penetrating arc. They are commonly referred to as *reverse polarity electrodes* even though some can be used with alternating current. They have little slag and produce flat beads, Fig. 7-7, welds B and C. With few exceptions they produce X-ray quality weld deposits and are used for pipe and pressure vessel code work. They are widely used for all-position welding, in both fabrication and repair.

COMBINATION TYPES. Some joints require the characteristics of both fast-fill and fast-freeze electrodes. When fast-freeze is required, the best electrodes are the EXX10 and EXX11 types. An electrode with a share of both characteristics is the all-position, iron powder electrode EXX14. These electrodes do not have as much fast-fill as an EXX24, nor do they have the degree of fast-freeze of the EXX10, but they are a compromise between the two types.

LOW HYDROGEN. Low hydrogen electrodes are those which have coverings containing practically no hydrogen. They produce welds that are free from underbead and microcracking and have exceptional ductility, Fig.

7-7. Comparative appearance of weld beads made with different types of electrodes.

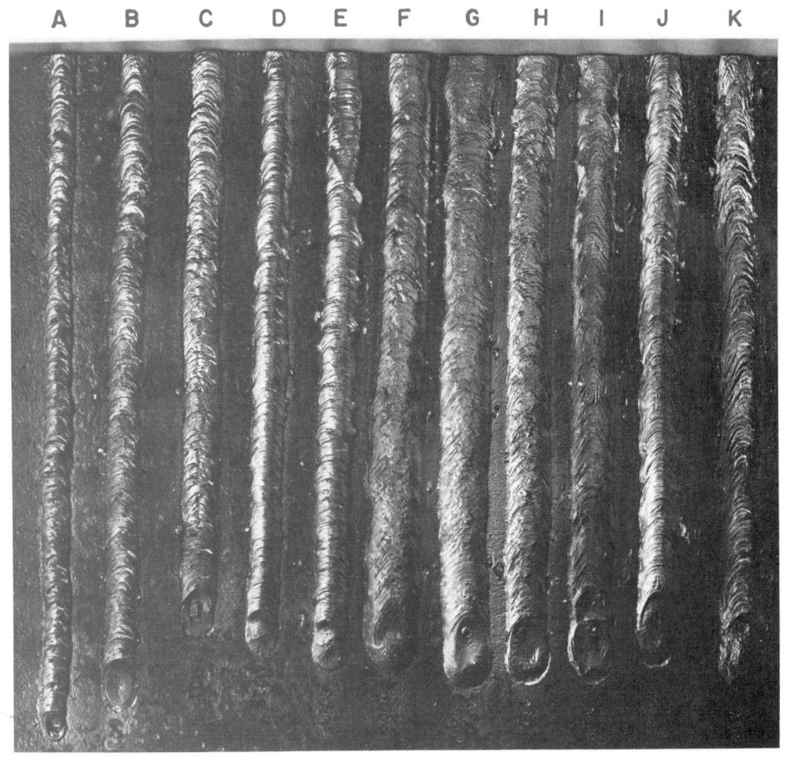

7-7, welds F and K. The electrodes eliminate porosity in sulfur-bearing steels and insure X-ray quality deposits. Because they reduce preheat requirements, their chief use is in the welding of hard-to-weld steels and high-tensile alloy steels. Examples of these types of electrodes are the EXX18 and EXX28 classifications.

IRON POWDER. Iron powders are added to the covering of many types of electrodes. In the intense arc heat the iron powder is converted to steel, thus contributing additional metal to the weld deposit. When iron powder has been added to the electrode covering in relatively large amounts (30 percent or more), the speed of welding is appreciably increased, the arc is stabilized, spatter is reduced, and slag removal is improved, Fig. 7-8. The weld appearance is very smooth, Fig. 7-7, welds H and I. Examples of these types of electrodes are the EXX24 and EXX27.

7-8. A thicker coating of iron powder on an electrode creates a crucible effect at the tip of the electrode, making more efficient use of arc energy.

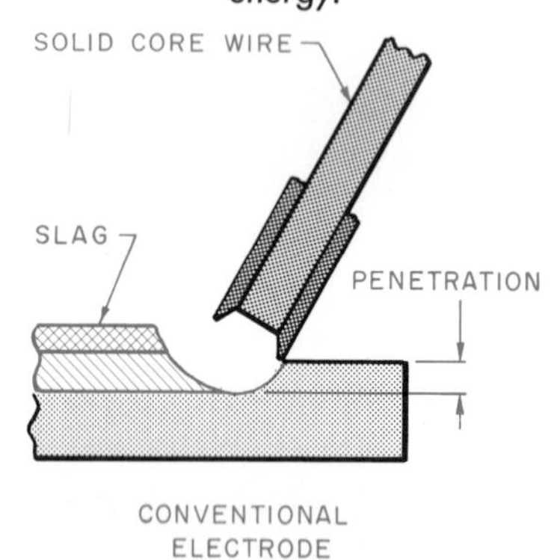

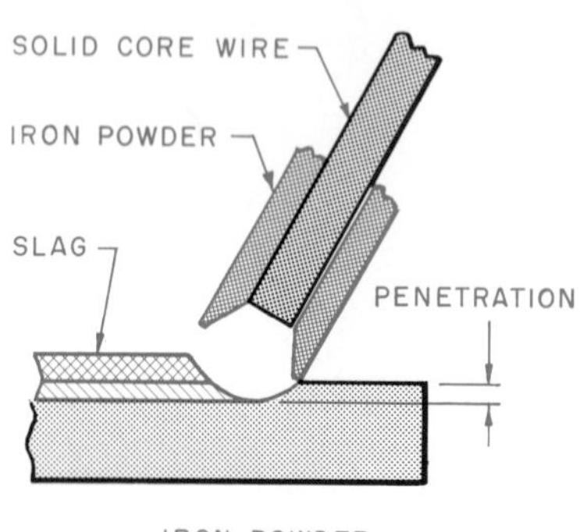

Type of Base Metal

The nature of the material to be welded is of prime importance. Satisfactory welds cannot be made if the weld metal deposited does not have the same physical and chemical qualities as the material being welded.

If the material specifications are not known, simple tests such as the spark test, torch test, chip test, magnetic test, color test, fracture test, and sound test may be performed. The information in Tables 7-11 through 7-14 will help you to recognize the various metals according to their characteristics. Such information can, however, identify only broad categories of material. The operator may be able to tell, for example, that the material is steel rather than cast iron, but there are so many types of alloy and stainless steels that it is necessary to know the correct analysis of the steel as designated by the manufacturer. For example, if the steel is high in sulfur or carbon or if it contains certain alloys, the E7015, E7016, E7018, or E7028 electrode should be chosen. These electrodes reduce the tendency towards underbead cracking which is characteristic of such steels. They produce welds of high tensile strength and ductility without stress-relieving, and they reduce or eliminate the need for preheat.

Nature of Welding Current

Welding machines produce two types of welding current, alternating current and direct current. Direct current has two kinds of polarity: straight and reverse. The nature of the current available also influences the selection of the electrode. For example, if only a.c. equipment is available, E6010 and E7015 electrodes are eliminated from consideration since they are designed to operate on direct current. If only d.c. equipment is available, current is not a limiting factor. While primarily designed for use with alternating current, E6011, E6013, E7016, E7014, E7018, and E7028 electrodes perform adequately with direct current.

Thickness and Shape of Material to be Welded

Whether the material is of heavy gauge or light gauge partially determines electrode size. As a general rule, never use an electrode having a diameter that is larger than the thickness of the material being welded.

For light gauge sheet metal work ($^{3}/_{32}$ inch and thinner) the E6013 electrode is usually the choice although E6012 may be used. The E6013 electrode was designed for this type of work, and it has the least penetration of any electrode in the E60XX series.

Joint Design and Fitup

There are many kinds of welded joints, and each type has particular requirements for welding. Welding fillets differs a great deal from welding butt joints. Butt joints may be either square or deep grooved. The fitup may have large gaps, or it may be too tight.

In cases in which poor fitup is unavoidable, the E6012 electrode should be chosen for use with the d.c. equipment; and E6013, with a.c. equipment.

Both of these electrodes bridge gaps very well due to the globular transfer of metal through the arc stream.

Welding Position

The position of welding is a very important consideration in the choice of an electrode. Certain types of electrodes can be used only in the flat position; others perform equally well in all positions. The type of position also has an influence on costs. Welding is most economical in the flat position, then horizontal, and then vertical; the overhead position is the least economical. As you develop skill in welding, you will understand the limitations that welding in the vertical and overhead positions places on the choice of an electrode.

The size of the electrode to be used is strongly influenced by the position of welding. V-butt joints in the vertical and overhead positions are usually welded with a small diameter electrode in order to obtain complete penetration at the root of the weld. In multilayer welding, the other passes can be made with large electrodes. Welding in the vertical and overhead positions should never be attempted with an electrode larger than $\frac{3}{16}$ inch in diameter.

For production work the largest electrode size that can be handled should be used. This permits higher welding current values, thus increasing the speed of welding. Higher deposition rates are also achieved. The larger the diameter of the electrode, the greater the quantity of weld deposited in a unit of time. The cost of labor is also reduced because fewer stops are necessary to change electrodes.

If welding must be done in an overhead, vertical, or horizontal position, electrodes of the EXX20, EXX24, EXX27, and EXX28 classifications cannot be used, leaving the choice to be made among the remaining electrodes in the series. The EXX15, EXX16, and EXX18 electrodes, though classified for all positions, are more difficult to

Table 7-11
Temperature data for metals and alloys.

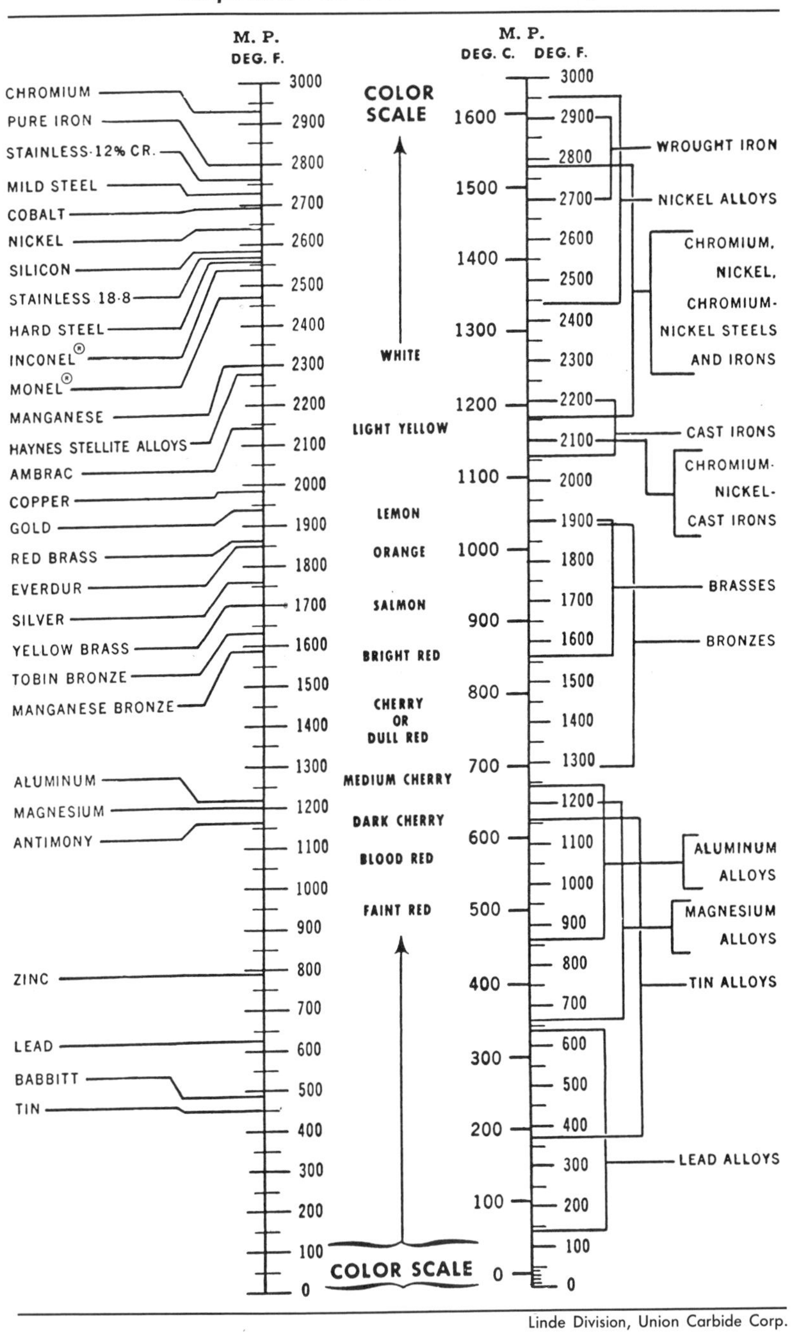

Linde Division, Union Carbide Corp.

handle in the vertical and overhead positions.

In general, the welder will find that electrodes in the EXX12, EXX13, EXX20, EXX24, EXX27, and EXX28 classifications are easiest to handle in the horizontal and flat positions. Vertical and overhead welding is easiest with the EXX10 and EXX11, except in vertical-down welds for which the EXX12 and EXX13 are easier.

Conditions of Use

The service requirements are of utmost importance. The type of structure and the stress that it will encounter in use must be considered. Tensile strength, ductility, and fatigue resistance are important weld characteristics that help to determine choice of electrode. Note the variation in weld metal properties among the electrodes compared in Table 7-3, page 198.

Engineering Specifications

All code requirements and engineering specifications must be noted carefully in any determination of the correct electrode to be used. The type of electrode to be used is specified in the code requirements.

Production Efficiency

Several electrode classifications have high deposition char-

Table 7-12
Identification of metals by appearance.

	Alloy steel	Copper	Brass and bronze	Aluminum and alloys	Monel®	Nickel	Lead
Fracture	Medium gray	Red color	Red to yellow	White	Light gray	Almost white	White; crystalline
Unfinished surface	Dark gray; relatively rough; rolling or forging lines may be noticeable	Various degrees of reddish brown to green due to oxides; smooth	Various shades of green, brown, or yellow due to oxides; smooth	Evidence of mold or rolls; very light gray	Smooth; dark gray	Smooth; dark gray	Smooth; velvety; white to gray
Newly machined	Very smooth; bright gray	Bright copper red color dulls with time	Red through to whitish yellow; very smooth	Smooth; very white	Very smooth; light gray	Very smooth; white	Very smooth; white

	White cast iron	Gray cast iron	Malleable iron	Wrought iron	Low carbon steel and cast steel	High carbon steel
Fracture	Very fine silvery white silky crystalline formation	Dark gray	Dark gray	Bright gray	Bright gray	Very light gray
Unfinished surface	Evidence of sand mold; dull gray	Evidence of sand mold; very dull gray	Evidence of sand mold; dull gray	Light gray; smooth	Dark gray; forging marks may be noticeable; cast—evidences of mold	Dark gray; rolling or forging lines may be noticeable
Newly machined	Rarely machined	Fairly smooth; light gray	Smooth surface; light gray	Very smooth surface; light gray	Very smooth; bright gray	Very smooth; bright gray

American Welding Society

Table 7-13 ***Identification of metals by flame test. (Footnotes are listed below Table 7-14.)***

	Alloy[1] steel	Copper	Brass and bronze	Aluminum and alloys[2]	Monel®	Nickel	Lead[3]
Speed of melting (from cold state		Slow	Moderate to fast	Faster than steel	Slower than steel	Slower than steel	Very fast
Color change while heating		May turn black and then red; copper color may become more intense	Becomes noticeably red before melting	No apparent change in color	Becomes red before melting	Becomes red before melting	No apparent change
Appearance of slag		So little slag that it is hardly noticeable	Various quantities of white fumes, though bronze may not have any	Stiff black scum	Gray scum; considerable amounts	Gray scum; less slag than Monel®	Dull gray coating
Action of slag		Quiet	Appears as fumes	Quiet	Quiet; hard to break	Quiet; hard to break	Quiet
Appearance of molten puddle		Has mirror-like surface directly under flame	Liquid	Same color as unheated metal; very fluid under slag	Fluid under slag	Fluid under slag film	White and fluid under slag
Action of molten puddle under blowpipe flame		Tendency to bubble; puddle solidifies slowly and may sink slightly	Like drops of water; with oxidizing flame will bubble	Quiet	Quiet	Quiet	Quiet; may boil if too hot

	White cast iron[4]	Gray cast iron	Malleable iron[5]	Wrought iron	Low-carbon steel and cast steel	High-carbon steel
Speed of melting (from cold state)	Moderate	Moderate	Moderate	Fast	Fast	Fast
Color change while heating	Becomes dull red before melting	Becomes dull red before melting	Becomes red before melting	Becomes bright red before melting	Becomes bright red before melting	Becomes bright red before melting
Appearance of slag	A medium film develops	A thick film develops	A medium film develops	Oily or greasy appearance with white lines	Similar to molten metal	Similar to molten metal
Action of slag	Quiet; tough, but can be broken up	Quiet; tough, but possible to break it up	Quiet; tough, but can be broken	Quiet; easily broken up	Quiet	Quiet
Appearance of molten puddle	Fluid and watery; reddish white	Fluid and watery; reddish white	Fluid and watery; straw color	Liquid; straw color	Liquid; straw color	Lighter than low-carbon steel; has a cellular appearance

(Continued on page 210)

(Continued from page 209)

	White cast iron[4]	Gray cast iron	Malleable iron[5]	Wrought iron	Low-carbon steel and cast steel	High-carbon steel
Action of molten puddle under blowpipe flame	Quiet; no sparks; depression under flame disappears when flame is removed	Quiet; no sparks; depression under flame disappears when flame is removed	Boils and leaves blowholes; surface metal sparks; interior does not	Does not get viscous; generally quiet; may be slight tendency to spark	Molten metal sparks	Sparks more freely than low-carbon steel

Table 7-14 *Identification of metals by chips.*

	Alloy[1] steel	Copper	Brass and bronze	Aluminum and alloys[2]	Monel®	Nickel	Lead[3]
Appearance of chip		Smooth chips; saw edges where cut	Smooth chips; saw edges where cut	Smooth chips; saw edges where cut	Smooth edges	Smooth edges	Any shaped chip can be secured because of softness
Size of chip		Can be continuous if desired	Can be continuous if desired	Can be continuous if desired	Can be continuous if desired	Can be continuous if desired	Can be continuous if desired
Facility of chipping		Very easily cut	Easily cut; more brittle than copper	Very easily cut	Chips easily	Chips easily	Chips so easily it can be cut with penknife

	White cast iron[4]	Gray cast iron	Malleable iron[5]	Wrought iron	Low-carbon steel and cast steel	High-carbon steel
Appearance of chip	Small broken fragments	Small partially broken chips but possible to chip a fairly smooth groove	Chips do not break short as in cast iron	Smooth edges where cut	Smooth edges where cut	Fine grain fracture; edges lighter in color than low-carbon steel
Size of chip		$\frac{1}{8}$ in.	$\frac{1}{4}$–$\frac{3}{8}$ in.	Can be continuous if desired	Can be continuous if desired	Can be continuous if desired
Facility of chipping	Brittleness prevents chipping a path with smooth sides	Not easy to chip because chips break off from base metal	Very tough, therefore harder to chip than cast iron	Soft and easily cut or chipped	Easily cut or chipped	Metal is usually very hard, but can be chipped

American Welding Society

[1] Alloy steels vary so much in composition and consequently in results of tests that experience is the best solution to identification problems.
[2] Due to white or light color and extremely light weight, aluminum is usually easily distinguishable from all other metals; aluminum alloys are usually harder and slightly darker in color than pure aluminum.
[3] Weight, softness, and great ductility are distinguishing characteristics of lead.
[4] Very seldom used commercially.
[5] Malleable iron should always be braze welded.

acteristics. Compare the rates of deposition given in Table 7-4, page 199. Full advantage cannot always be made of these characteristics because of the nature of the material, the type of joint, and the position of the work. For example, the electrodes classified as having a very high rate of deposition in Table 7-4 can be used only for flat and horizontal fillet welds.

The principal factor in the cost of a welding job is the speed with which the welding can be done. Electrode cost is small by comparison. E7024 and E6027 electrodes permit the highest rate of deposition. E6020, E6012, E6013, E6011, and E6010 follow in the order given. Type of steel, not speed, should govern the choice of the E7015 or E7018. The E7028 is similar to the E7018, but it has a much thicker coating which contains a higher percentage of iron powder; thus, its deposition rate is much higher.

Welding speed is increased by using electrodes with large diameters, particularly in flat and horizontal position welding. The E7024 and E6020 classifications, followed by the E6027, give the highest increases in welding speeds as the diameters of the electrodes are increased.

Job Conditions

Is the material clean, rusty, painted, or greasy? What is the type of surface treatment required for the finished job? Is the completed job to be stress relieved or heat treated? Are the welds in a prominent location so that weld appearance is important? Only an operator with a thorough knowledge of electrode characteristics can answer these questions by choosing the best electrode for the job. It is important to study manufacturer's specifications.

Summary of Factors Affecting The Selection of Electrodes

The foregoing are just a few of the considerations that make it necessary for you to become highly familiar with the nature of the different electrode classifications.

Selection of the proper electrode size and type for a given welding job requires a thorough knowledge of electrodes coupled with common sense. Careful study of the physical, chemical, and working characteristics of electrodes and practice in their use will enable you to make the proper selection without difficulty.

In summary, the following factors are the most important to be considered:

- Type of joint and position of welding
- Type of welding current
- Properties of the base metal
- Thickness of the base metal
- Depth of penetration desired
- Weld appearance desired
- Whether or not the work is required to meet code specifications
- Tensile strength, ductility, and impact strength required of the weld
- Design and fitup of the joint to be welded
- Nature of slag removal

SPECIFIC ELECTRODE CLASSIFICATIONS

Carbon Steel Electrodes

The carbon steel electrodes for welding low and medium carbon steels carry AWS classification numbers E4510 and E4511, E4520, E4521, and E6010, -11, -12, -13, -14, -15, -16, -18, -20, -24, -27, and -28. The E4510, -11, -20, and -21 are bare electrodes which are used very little today; the others are all shielded-arc covered electrodes.

LIGHTLY COATED, MILD STEEL ELECTRODES. E4510 Electrodes with Coating Applied before Drawing (All-Position). This type of electrode is commonly known as a *sul-coated* or *sul-finish* electrode. It is often referred to as a bare electrode although the sul-coating is definitely effective in promoting arc stability. During the manufacture of the wire, the steel rod is pickled in acid and held in a spray of water until a desired coating of rust is formed. The rod is dipped in a lime bath, after which it is dried and drawn to wire. The coating, therefore, consists of a mixture of rust and lime, and the quantity is small. Lime is the more effective ingredient in promoting arc stability. The core is usually a plain low-carbon steel.

These electrodes are adapted for welding in all positions and are suitable for a wide variety of types of repair and structural welding and for beginning student training. The thin coating contains arc-stabilizing materials, but it does not provide any important gaseous or slag protection of the weld metal from the air. Therefore, the tensile strength, ductility, impact values, and other physical properties of the deposit are low in comparison to those of shielded-arc electrodes.

Electrodes of this type are satisfactory only when service conditions are not severe. The electrodes are usable only with direct current, straight polarity (electrode negative). The arc is wild, not particularly strong, and penetration is not deep. Since the coating contains practically no slag-forming materials, only a thin scale of oxide covers the finished weld, Fig. 7-7A, page 205.

E4511 Electrodes with Light Coating Applied after Drawing (All-Position). E4511 electrodes are used for the same types of welding as E4510. The arc stability is better, the electrodes are somewhat easier to use, and they produce slightly smoother and more uniform welds. The physical properties of the weld are only slightly better than those of E4510 because the coating is still designed mainly for improved arc characteristics. It does not afford adequate protection to the molten metal from the air. Like E4510 electrodes, this type can be used in all-position welding and only with direct current, on straight polarity. This type of electrode is preferable to the E4510 for student training if it can be obtained. It is recommended that practice with bare electrodes not exceed 20 hours.)

The core wire is often the E4510 sul-coated wire which already contains lime on the surface although bright finish wire may also be used. The additional light coating is applied by tumbling, dipping, or other means. Of the many coating materials used, the most common are iron oxide and calcium carbonate. Calcium carbonate is the principal arc-stabilizing material.

E4520 Electrodes with Coating Applied before Drawing (Horizontal Fillet and Flat Position). This electrode is adapted only to welding in the flat position and to horizontal fillets. In comparison with E4510 and E4511 electrodes, the E4520 operates with a higher arc voltage and has a very high melting rate. Thus it is impractical for vertical and overhead welding. The amount and fluidity of molten metal area are too great to permit its staying in place in these positions. Physical properties are similar to those of the other bare electrode classifications.

E4521 Electrodes with Light Coating Applied after Drawing (Horizontal Fillet and Flat Position). This electrode overlaps with the E4520 electrode in usage and in operation although it is generally limited to automatic welding. The coating materials, core wire, and welding characteristics may be considered the same as those of E4520.

HEAVILY COVERED MILD STEEL, SHIELDED-ARC ELECTRODES. **E6010: All-Position, Direct Current, Reverse Polarity (Fast-Freeze Type).** This electrode is the best adapted of the shielded-arc types for vertical and overhead welding. It is, therefore, the most widely used electrode for the welding of steel structures which cannot be readily positioned and which require considerable multiple-pass welding in the vertical and overhead positions. While the majority of applications are on mild steel, E6010 electrodes may be used to advantage on galvanized plate and on some low alloy steels. In welding steel, the forceful arc bites through the galvanizing and the light slag to reduce bubbling and prevent porosity. Typical applications include shipbuilding, structures such as buildings and bridges, storage tanks, pipe welding, tanks, and pressure vessels.

The quality of the weld metal is of a high order, Fig. 7-7B, page 205, and the specifications for this classification are correspondingly rigid. The essential operating characteristics of the electrode are:

- Strong and penetrating arc, enabling penetration beyond the root of the butt or fillet joint.
- Quickly solidifying weld metal, enabling the deposition of welds without excessive convexity and undercutting.
- Low quantity of slag with low melting and low density characteristics so that it does not become entrapped nor interfere with oscillating and weaving techniques.
- Adequate gaseous atmosphere to protect molten metal during welding. Electrodes of this type are usable only with direct current, reverse polarity (electrode positive).

The E6010 electrode is commonly classified as the *cellulosic* type. The electrode coating contains considerable quantities of cellulose, either in a treated form or as wood flour or other natural forms. During welding the cellulose is changed to carbon dioxide and water vapor, forming the gaseous envelope which excludes the harmful oxygen and nitrogen in the air. The water vapor from minerals containing water of hydration and the vapor retained by the binder are also liberated during welding.

The slag-forming materials of the E6010 covering include titanium dioxide and either magnesium or aluminum silicates. Ferromanganese is used as a *deoxidizer,* or *degasifier,* as it is often called. Since there is usually no increase in manganese in the weld deposit over that of the core wire, the manganese enters the slag as an oxide. The common binder for the coating materials is sodium silicate solution, which also is a slag-forming material. The core wire is low-carbon rimmed steel. It usually contains 0.10–0.15 percent carbon, 0.40–0.60 percent manganese, a maximum of 0.40 percent sulfur and phosphorus, and a maximum of 0.025 percent silicon.

E6011: All-Position, Alternating Current (Fast Freeze Type). The operating characteristics, mechanical properties, and welding applications of the E6011 resemble those of the E6010, but the E6011 requires alternating current. Although it may also be used with direct current, reverse polarity, it loses many of its beneficial characteristics with this polarity.

The penetration, arc action, slag, and fillet-weld appearance are similar to those obtained with the E6010 type, Fig. 7-7C, page 205. The weld deposit is free from porosity, holes, and pits. The slag can be removed readily. Fillet and bead contours are flat rather than convex. E6011 electrodes may be used in all-position welding.

The E6011 coverings are classified as the high-cellulose potassium type. Small quantities of calcium and potassium are present in addition to the other ingredients usually found in the E6010-type coverings. The core wire is identical to that used for E6010 electrodes.

E6012: All-Position, Alternating and Direct Current, Straight Polarity (Fill-Freeze Type). The E6012 may also be used with direct current, reverse polarity, and alternating current. This electrode is often referred to as a poor fitup electrode because of its ability to bridge wide gaps in joints. It is particularly well-adapted for the single-layer welding of horizontal fillets. The E6012 is used extensively in most kinds of steel fabrication because it offers economy due to ease of application and high welding speeds. The weld metal deposited by the E6012 is lower in ductility and higher in yield strength than weld metal from either the E6010 or E6011.

The E6012 is suitable for use with direct current, with either straight (electrode negative) or reverse (electrode positive) polarity, and with alternating current. Straight polarity is preferred because of a more direct and stable arc. With the larger electrodes alternating current is generally preferred because of the relative freedom from arc blow.

The E6012 provides adequate penetration in order to reach the root of the fillet and other joints, but it does not penetrate as deeply as the E6010. The small diameters, such as $^3/_{32}$ inch and $^1/_8$ inch, are especially suited for sheet metal welding since they do not burn through the sheets.

The slag is more abundant and covers more of the molten pool than the E6010 does, but it is neither as abundant nor as fluid as the slag of the E6020 and E6030 types. The slag soldifies very rapidly just below the freezing point of the metal. It is generally dense and adheres to the deposit.

The molten metal is slightly more fluid than that of the E6010 electrode, but not to the extent that the E6012 cannot be used in all-position welding. The molten metal and slag characteristics control the shape of the weld deposit, Fig. 7-7D, page 205. Thus the E6012 is especially suitable for horizontal fillet welding, in which it produces flat or slightly convex beads without undercutting. Many of the E6012 electrodes are suitable for vertical welding in the downward direction although for some purposes, the penetration and throat thicknesses are insufficient. In vertical-up welding, small welds are more convex and have wider spaced ripples than those made with E6010 electrodes. When welds are large enough to permit the deposition of a substantial shelf upon which to build the deposit, the shape and appearance are more satisfactory.

The covering of the E6012 is high in titania and sodium. Titania and rutile are other terms for titanium dioxide. Thus the E6012 is often known as the *rutile* type because it contains titanium dioxide as the primary slag-forming constituent. In addition to rutile, various siliceous materials such as feldspar and clay are used. Since the electrode depends mainly on slag for protection of the weld metal, only a small amount of combustible material, usually cellulose, is present. Ferromanganese is the common dioxidizer. Sodium silicate solutions are used as the binder. Calcium compounds are often incorporated to produce satisfactory arc characteristics on straight polarity.

E6013: All-Position, Alternating and Direct Current, Straight Polarity (Fill-Freeze Type). The E6013 electrode, although similar to the E6012, differs in some important respects. Slag removal is better, and the arc can be established and maintained more readily. This is especially true of electrodes with small diameters ($^1/_{16}$, $^5/_{64}$, and $^3/_{32}$ inch). Consequently, it permits satisfactory operation with lower open-circuit voltage. Originally, this electrode was designed specifically for light-gauge sheet metal work and for vertical welding from the top down. The larger sizes, however, are being used on many kinds of work previously welded with E6012 electrodes.

The covering of the E6013, like that of the E6012, contains rutile, siliceous materials, cellulose, ferromanganese, potassium, and liquid silicate binders. An important difference is that easily ionized materials are incorporated in the covering. This feature permits the establishment and maintenance of an arc with alternating current at low welding currents and low open-circuit voltages.

The E6013 is similar to the E6012 classification in operation and in the appearance of the deposit, Fig. 7-7E, page 205. The arc action tends to be quieter, and the bead surface is smoother with a finer ripple. These electrodes are suitable for making fillet welds and butt welds with a flat or slightly convex appearance.

E7014: All-Position, Alternating and Direct Current, Straight Polarity (Fast-Fill Type). The covering of this electrode is similar to that of the E6012 and E6013, but the addition of iron powder makes it much thicker. The deposition rate is somewhat higher.

The E7014 is suitable for welding mild steel and low alloy steels in all positions. The weld beads have a smooth surface with fine ripples, and the slag is easily removed. The fillet welds made with the E7014 are flat to slightly convex. It is a good electrode for production welding on plate of medium thickness.

LOW HYDROGEN ELECTRODES. Low hydrogen electrodes are a

result of research during World War II. The object of this research was to find an electrode for welding armor plate which would require the use of less strategic alloys. Today these electrodes are no longer considered emergency tools. They are used because they have superior mechanical properties and because many are custom-made to match the heat-treating properties of alloy steels.

The name stems from the fact that the coatings are free of minerals that contain hydrogen. The lack of hydrogen is an important characteristic because hydrogen causes underbead cracking in high carbon and alloy steels. By eliminating hydrogen, underbead cracking is prevented and difficult steels can be welded with little or no preheat. These electrodes also produce porosity-free welds in high sulfur steels and eliminate hot-shortness in phosphorus-bearing steels. The addition of iron powder in the coating increases the deposition (melting) rate.

These electrodes are low in hydrogen-bearing compounds so that only traces of hydrogen and moisture are present in the arc atmosphere. The core contains from 0.08 to 0.13 percent carbon, 0.40 to 0.60 percent manganese, and a maximum of 0.04 percent sulfur and phosphorus. A typical analysis of the deposit from a low hydrogen electrode is 0.08 percent carbon, 0.56 percent manganese, and 0.25 percent silicon.

A low hydrogen electrode has a core of mild steel or low alloy steel. The mineral covering consists of alkaline earth carbonates, fluorides, silicate binders, and ferro-alloys. This covering produces the desired weld metal analysis and mechanical properties. E6018 and E6028 have iron powder in their covering. During welding the covering forms a carbon dioxide shield around the arc. On the job these electrodes must not be exposed to humid air because of their tendency to absorb a considerable amount of moisture.

A wide range of weld properties is possible by adding a number of alloying elements to the covering. Such additions may include carbon, manganese, chromium, nickel, molybdenum, and vanadium.

The arc is not harsh and is moderately penetrating. The slag is heavy, friable (easily crumbled), and easily removed. A short arc must be used in welding. A long arc permits *hydrogen pickup* (an increase in moisture) which causes porosity and slag inclusions in the weld deposit. High quality, radiographically sound welds can be made with proper welding techniques. See Fig. 7-7. Welds F, J, and K.

Although low hydrogen electrode sizes up to and including $\frac{5}{32}$ inch may be used in all positions, they are not truly all-position welding electrodes such as those in the E6010 classification. The larger diameters are useful for horizontal fillet welds in the horizontal and flat positions.

The mechanical properties produced by these electrodes are far superior to those furnished by conventional electrodes, such as the E6010 and E6012. Tensile strengths of 120,000 p.s.i. and better with high ductility are obtainable in the as-welded condition. With heat treatment tensile strength can be increased to as much as 300,000 p.s.i.

Following is a description of each electrode in this class.

E7015: All-Position, Direct Current, Reverse Polarity (Low Hydrogen). The coating of this electrode is high in calcium compounds and low in hydrogen, carbon, manganese, sulfur, and phosphorus. It contains a trace of silicon. It is referred to as the low hydrogen, sodium type because sodium silicate is used as a binder for the covering.

The arc is moderately penetrating. The slag is heavy, friable, and easily removed. The deposited metal is flat and may be somewhat convex. Welding with a short arc is essential for low hydrogen, high quality deposits. Welding in all positions is possible with sizes up to $\frac{5}{32}$ inch. Larger electrodes can be used in the horizontal and flat positions.

These electrodes are recommended for the welding of alloy steels, high carbon steels, high sulfur steels, malleable iron, sulfur-bearing steels, steels that are to be enameled, spring steels, and for the welding of the mild steel side of clad plates. Very often pre- and postheating may be eliminated by the use of this electrode.

E7016: All-Position, Alternating and Direct Current, Reverse Polarity (Low Hydrogen). The E7016 electrode has all the characteristics of the E7015 type. An added advantage is that it may be used with either alternating or direct welding current. The core wire and covering are similar to the E7015 except for the addition of potassium silicate or other potassium salts. The potassium compounds make these electrodes suitable for alternating current. A typical weld produced by this electrode is shown in Fig. 7-7J, page 205.

E7018: All-Position, Alternating and Direct Current, Reverse Polarity (Low Hydrogen, Iron Powder). The coating of this electrode contains a high percentage of iron powder, from 25 to 40 percent, in combination with low hydrogen ingredients. The covering of the E7018 is similar to, but thicker than, the covering of the E7015 and E7016 electrodes. The slag is heavy, friable, and easily removed. The deposited metal is flat, Fig. 7-7K, page 205, and its appearance is better than welds made with the E7015. The deposit may be slightly convex in a fillet or groove weld.

Welding may be done in all positions with electrode sizes up to $^{5}/_{32}$ inch. Larger diameters are used for fillet and groove welds in the horizontal and flat positions. A short arc must be held at all times, and special care taken to keep the covering in contact with the molten pool when welding in the vertical-up position. A long arc will cause porosity in the weld deposit. The deposition rate of E7018 is somewhat higher than the E7015.

The strength of the deposited weld metal can be improved through the addition of certain alloys to the coverings rather than by changing the composition of the core wire. Adding alloying elements to the coating is more economical and can be better controlled. These electrodes are available in the E8018 through E12018 classifications (tensile strength 80,000 p.s.i. through 120,000 p.s.i.).

Usually the applications of the E7018 electrodes require specific mechanical and chemical properties so that the weld metal will meet the requirements of the base metal, adjust to stress-relieving and heat treatment, withstand extreme loading and fatigue, and resist cracking.

E7028: Horizontal and Flat Positions, Alternating and Direct Current, Reverse Polarity (Low Hydrogen, Iron Powder). These electrodes are similar to the E7018 electrodes with some differences. The E7028 is suitable only for horizontal and flat position welding, whereas the E7018 can be used in all positions. The coating of the E7028 contains a higher percentage of iron powder (50 percent) than the E7018 and, as a result, is thicker and heavier. The deposition rate of the E7028 is higher than the rate of the E7018.

The penetration is not deep, and the weld appearance is flat to concave, with a smooth fine ripple, Fig. 7-7F, page 205. The slag coating is heavy and easily removed. The E7028 has the characteristics of the fast-fill type of electrode.

OTHER COVERED ELECTRODES.

E6020: Horizontal Fillet and Flat Position, Alternating and Direct Current, Either Polarity, High Iron Oxide. The E6020 is designed to be used with alternating current or direct current (straight polarity) for horizontal fillet welding and with alternating current or direct current (either polarity) for flat position welding. It is used mainly with alternating current. Penetration is moderate with normal welding currents and deep when used with the high currents required for deep fillet welding.

The E6020 is essentially a mineral-covered electrode. The slag covers the weld so completely and the slag-metal reactions are of such nature that the electrode does not ordinarily depend on gaseous atmosphere protection. The covering contains a high percentage of iron oxide, manganese compound, and silicates. It produces a slag coating containing iron oxide, manganese oxide, and silica. Other constituents containing the oxides of aluminum, magnesium, or sodium modify the slag as desired. Ferromanganese is used as the main deoxidizer. The quantities of basic oxides, acid silica, silicates, and deoxidizers must be carefully controlled to produce satisfactory operation and good weld metal. Sodium silicate is used as the binder. The core wire is the usual low carbon steel wire used with other electrodes. A sample bead is shown in Fig. 7-7G, page 205.

E6020 electrodes may be used with high heat, resulting in deep penetration. Thus they are used for numerous applications in which very high quality weld metal is required and when work is positioned for downhand welding. Such applications include pressure vessels, tanks, machine bases, heavy equipment units, and structural sections.

The essential operating characteristics are as follows:

- The E6020 electrode is usable with either alternating or direct current. It is used mainly with alternating current, but when used with direct current, straight polarity is preferred, especially for the welding of horizontal fillets.
- The main requirement of the electrode is to produce horizontal fillets of flat or concave surface without undercutting. Therefore, the molten metal and slag must be comparatively fluid, the metal must be quick-freezing, and the slag must cover the back portion of the pool continuously and actually wet the molten metal.
- Slags of this type are not quick-freezing: they remain as a plastic glass for some time after the molten metal has solidified. The slag and metal are both too fluid to permit general welding in the vertical and overhead positions. After the slag has cooled, it can be readily removed.
- While specifically designed to meet horizontal fillet welding requirements, the electrode is also adapted to the welding of butt

and other deep groove, flat position joints.

- The physical properties of welds can meet exacting specifications, especially in reducing elongation. Radiographs show that properly made welds are practically perfect.
- The E6020 electrode cannot be used to weld light-gauge metals.

E7024: Horizontal Fillet and Flat Position, Alternating and Direct Current, Either Polarity (Iron Powder, Titania). The E7024 classification indicates a covering with a high percentage of iron powder, usually 50 percent of the weight of the covered electrode, in combination with fluxing ingredients similar to those commonly found in the E6012 and E6013 electrodes. The E7024 electrodes may be used for those applications that usually require E6012 or E6013.

E7024 electrodes are also referred to as *contact electrodes* since the electrode coating may rest on the surface of the joint to be welded. During actual welding, the electrode "drags," resulting in an effective shielding of the weld pool from the atmosphere. Many welders prefer to hold a short, free arc. In addition to melting the core wire and the base metal, the arc melts iron powder in the electrode coating to provide greater deposition of metal per ampere, Fig. 7-8, page 206. Hence, it makes possible greater welding speed. It has been determined that one-third of the weld metal deposited comes from the covering.

In comparing the E7024 electrodes with conventional electrodes, it is claimed that contact welding has the following advantages:

- Less weld spatter and, therefore, higher deposition efficiencies even at high welding current
- Lower nitrogen content within the weld metal
- Sounder metal with less tendency for such defects as cracking, porosity, and slag inclusions.
- Welds that are practically self-cleaning and have greater freedom from spatter
- Smoother weld appearance, approaching the bead quality obtained by automatic welding

E7024 electrodes are well-suited for fillet welds in mild steel. The welds are slightly convex in profile. They have a very

7-9. A multipass groove weld, bevel butt joint in ¾-inch plate welded with an iron powder electrode (E7024).

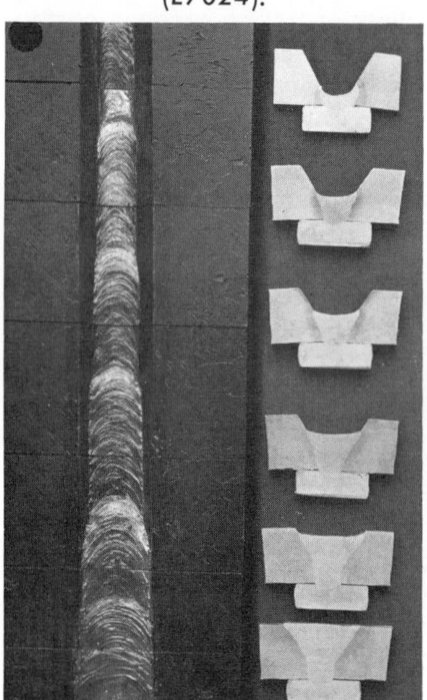

7-10. Iron powder electrodes provide high welding speeds with good appearance. The top joint was made with a single iron powder electrode, and the bottom was welded with a single E6012 electrode. The iron powder electrode was 36 percent faster and gave the weld a better appearance.

The Lincoln Electric Co.

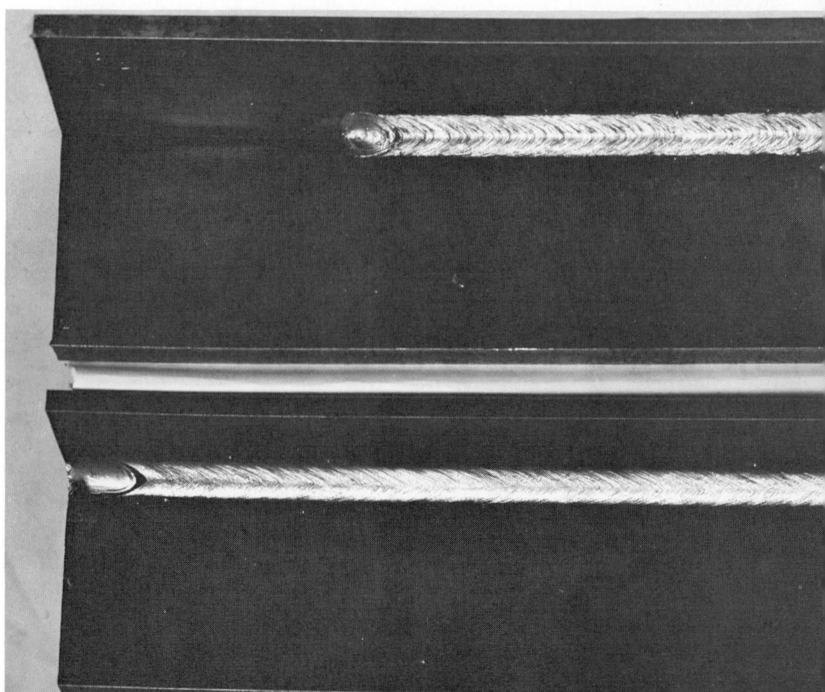

smooth surface and an extremely fine ripple that approaches the appearance of machine-made welds, Figs. 7-9 and 7-10. The electrode is characterized by a smooth quiet arc, very low spatter, and low penetration. It can be used at high lineal speed. Moreover, it may also be used to weld low alloy, medium, and high carbon steels. It operates with alternating or direct current (either polarity) although alternating current is preferable.

E6027: Horizontal Fillet Welds and Flat Position, Alternating and Direct Current, Straight Polarity (Iron Powder, Iron Oxide). The E6027 electrode has a covering that contains a high percentage of iron powder in combination with fluxing ingredients similar to those commonly found in the E6020 electrode. The welding action, however, resembles that of the E7024 electrode.

The covering of the E6027 is very heavy, due to the high percentage of iron powder, which usually amounts to about 50 percent of the total weight of the electrode. The iron powder in the covering increases the metal deposition rate. The iron oxide is a slag former and produces a heavy slag. The core wire is identical to that used in E6010 and E6020 electrodes.

The E6027 electrodes, like the E7024 electrodes, are almost always used with a "drag" technique. The E6027 electrode is designed to produce high quality fillet and groove welds with alternating current or direct current (straight polarity) in the horizontal position, and with alternating current or direct current (either polarity) in the flat position.

The E6027 classification has a spray-type metal transfer and deposits metal at high lineal speeds. Penetration is medium, and spatter loss is very low. Welds vary in profile from flat to slightly concave with a smooth, fine, even ripple and with metal wash up the sides of the joint, Fig. 7-11. The covering produces a heavy slag which is honeycombed on the underside and covers the weld deposit. It is quite friable and easy to remove. The E6027 is well-suited for the welding of heavy structures. While welding speed is slower than that possible with the E7024 type, the weld deposit properties are superior and possess excellent radiographic quality.

Alloy Steel Electrodes

Due to the high strength requirements of many industrial fabrications, the use of high strength alloy steels has increased a great deal. The ability to fabricate by welding is a major factor in this increase. A shielded-arc electrode capable of producing weld deposits with a tensile strength exceeding 100,000 p.s.i. has been developed. The core wire of this electrode is composed of alloy steel instead of low carbon steel. The electrode covering has the lime-ferritic nature typical of the low hydrogen types, and it may contain powdered iron.

Operating characteristics of electrodes for welding alloy steel resemble those of the low hydrogen electrodes in the E6015, -16, and -18 classifications. They are available, however, in tensile-strength classifications of 70XX, 80XX, 90XX, 100XX, 110XX, and 120XX.

You will recall that the first two or three digits of the electrode identification number indicate tensile strength. Thus the E11018 electrode is rated at a tensile strength of 110,000 p.s.i. while the E6018 electrode has a tensile strength of only 60,000 p.s.i. Both electrodes weld in all positions, use alternating or direct current, reverse polarity, and have 30 percent iron powder in their covering. They have medium arc force and penetration, and their slag is heavy, friable, and easily removed.

THE STAINLESS STEEL ELECTRODES. *Stainless steel* is the popular term for the chromium and chromium-nickel steels. It is a tough, strong material which is highly resistant to corrosion, high temperatures, oxidation, and scaling. There are a large variety of stainless steels and electrodes to weld them. Both the base metals and the electrodes are expensive and should be handled with care.

Metallurgically, the stainless steels are classified as martensitic, ferritic, and austenitic. In order to understand the various uses for these steels, it is important that you have an understanding of these terms.

- The *martensitic* type is an air-hardening steel containing chromium as its principal alloying element in amounts ranging

7-11. *Deep groove welds have an excellent appearance and wash-in, easy slag removal. E6027 iron powder electrodes produce a smooth, clean, flat cover pass.*

from 4 to 12 percent. It is normally hard and brittle. Martensitic stainless steel requires both preheating and postheating for welding.

- The *ferritic* type is a magnetic, straight-chromium steel, containing 14 to 26 percent chromium. It is normally soft and ductile, but it may become brittle when welded. Preheating and postheating are necessary for successful welding.
- The *austenitic* type contains both chromium and nickel. The nickel content usually ranges from 3.5 to 22 percent; and the chromium content, from 16 to 26 percent. It is strong, ductile, and resistant to impact. Austenitic stainless steel is nonmagnetic when annealed but slightly magnetic when cold worked. Heat treatment is not necessary during welding.

The selection of the proper type of electrode for stainless steel welding is a critical choice. A different welding method is required for each type. The resultant weld must have the required tensile strength, ductility, and corrosion resistance equal to the base metal. There are also the problems of matching the color of the base metal and producing a smooth bead which will require a minimum amount of grinding.

The identification numbers for stainless steel electrode classifications are somewhat different from those used in the AWS system for carbon steel electrodes. They are based on the AISI (American Iron and Steel Institute) classifications of metal alloys, Table 7-15. The first digits in the identification number refer to the AISI metal classification number instead of the tensile strength. Thus the electrodes in the E308XX–E309XX series are suitable for austenitic stainless steels. (See Chapter 3, pages 99, 100, for more information on the AISI classification system.) The last two digits of the electrode identification number refer to the position of welding and operating characteristics as given in the AWS classification system. E309-15, for example, may be used in all positions with

Table 7-15

Identification of stainless steel electrodes.

AISI designation	Popular trade names	AWS electrode recommended
Austenitic or chromium–nickel types AISI 300 series		
302, [2]303, 304, 308	18/8 or 19/9	E308-15, E308-16
309	25/12	E309-15, E309-16, [1]E309Cb-15
310	25/20	E310-15, E310-16, [1]E310Cb-15, [1]E310Cb-16, [1]E310Mo-15, [1]E310Mo-16
312	29/9	E312-15, E312-16
316	18/12Mo	E316-15, E316-16, [1]E316Cb-15, E316Cb-16
317	18/12Mo	E317-15, E317-16
318	18/12Mo Cb	
330	15/35	E330-15, E330-16
347	18/8Cb	E347-15, E347-16
Straight-chromium types AISI 400 series		
410	12 chromium	E410-15
430	16 chromium	E430-15
442	18 chromium	E442-15
446	28 chromium	E446-15
	AISI 500 Series	
502	5 chromium	E502-15, 502-18, 505-18
505	9 chromium	E502-16, 505-18

American Welding Society

[1]Columbium and/or molybdenum added to these coatings to aid in prevention of carbide precipitation or to improve strength at elevated temperatures (decrease creep rate) respectively.

[2]AISI 303 has additional sulfur added to improve machinability. Use lime type E308-15 when welding.

direct current, reverse polarity. It is a low hydrogen electrode with a medium arc and penetration.

Stainless steel electrodes are designated as having *lime, lime-titania* or *titania* coverings. In general, a lime-type covering is one whose chief mineral ingredients include limestone and fluorspar. It contains minor amounts of titanium dioxide (up to 8 percent). Coverings containing more than 20 percent titanium dioxide are usually considered as the titania type; those containing between 8 and 20 percent are considered to be the lime-titania type.

The lime-type covering is designed for direct current, reverse polarity only. These electrodes produce welds that have a convex face and are desirable for root passes in which the full throat section prevents cracking. They may be used in the vertical and overhead positions. The slag completely covers the weld, provides a rapid wetting action, and produces welds with a minimum amount of spatter. The coating also produces a flux which drives the impurities from the weld, thereby insuring a weld that is free of porosity and has the mechanical and corrosion properties expected. These electrodes are in the E3XX-15 classification.

The titania-type covering is designed for either alternating or direct current, reverse polarity. Electrodes with such coverings are preferred to the lime-type electrodes because of the smooth arc action, fine bead appearance, and very easy slag removal. They produce slightly concave welds which require a minimum of cleaning, grinding, and polishing time. These electrodes are in the E3XX-16 classification. They are also available for direct current, reverse polarity only. Electrodes with the lime-titania covering may be used with direct current only or both alternating and direct current. They are all-position electrodes. They weld the straight chromium (ferritic) and chromium-molybdenum stainless steels and, to some extent, the chromium-nickel (austenitic) stainless steels.

Lime is used for the electrode covering since it tends to eliminate hydrogen, which causes underbead cracking. Since chromium has an affinity for carbon and materials high in carbon, carbon compounds such as alkaline earth carbonatis, which is used to eliminate hydrogen in low hydrogen electrodes, cannot be put in the covering for stainless steel electrodes. Manganese and silicon are included to reduce oxidation. Titanium promotes arc stability, produces an easily removable slag, and prevents carbon precipitation. Columbium also prevents carbon precipitation.

Hard-facing Electrodes

Hard-facing is the deposition of an alloy material on a metal part by one of several welding processes to form a protective surface. This can also be done by metal spraying. Depending upon the alloy added, the surface resists abrasion, impact, heat, corrosion, or a combination of these factors.

- *Abrasion* is associated with surfaces that are subjected to continuous grinding, rubbing, or gouging action. The forces move parallel to the surface of the component.
- *Impact* causes metal to be lost or deformed as a result of chipping, upsetting, cracking, or crushing forces. The force is a striking action that is perpendicular to the absorbing members.
- *Corrosion* is the destruction of a surface from atmospheric chemical contamination and from oxidation or scaling at elevated temperatures.

Hard-facing may be applied to new parts to improve their resistance to wear during service or to worn parts to restore them to serviceable condition. The wear-resistant material is applied only to those surfaces of a component where maximum wear takes place.

Alloys for hard surfacing are usually available as bare cast or tubular rod, covered solid or tubular electrodes, or solid wire and powder. Electrodes for metal arc welding have been classified on the basis of the following service conditions:

1. Resistance to severe impact
2. Resistance to severe abrasion
3. Resistance to corrosion and abrasion at high temperature
4. Resistance to severe abrasion with moderate impact
5. Resistance to abrasion with moderate-to-heavy impact

In most cases only a single type of electrode is needed. However, there are some conditions that require two types of electrodes. For example, when severe abrasion is encountered in combination with severe impact, a Class 1 electrode should be used for buildup metal. A second type of electrode is used to deposit a material that has high abrasion characteristics. The first material cushions impact loads and supports the hard deposit which resists abrasion.

There are many different kinds of hard-facing materials. Generally they have a base of iron, nickel, copper, or cobalt. Alloying elements may be carbon, chromium, molybdenum, tungsten, silicon, manganese, nitrogen, vanadium and titanium. The alloying elements form hard carbides which contribute to the properties of the hard-facing metals. A high percentage of tungsten or chromium with a high carbon content forms a carbide that is harder than quartz. Materials with a high chromium content have high resistance to oxidation and scaling. Nickel, cobalt, and chromium have high corrosion resistance.

Hard-facing electrodes are classified on the basis of the type of service they perform and are divided between ferrous and nonferrous base alloys, Table 7-16.

FERROUS BASE ALLOYS. Austenitic Steel Electrodes. There are two major types of austenitic steel electrodes: those containing a high percentage of manganese and those containing chromium, nickel, and iron. The latter are known as stainless steels. Electrodes made of austenitic manganese steel provide metal-to-metal wear resistance coupled with impact, and surface protection or replacement of worn areas where abrasion is associated with severe impact. They have been widely used to resurface battered-down railway trackwork. Surfacing of power shovel dippers is another common application.

The stainless steel types of hard-facing electrodes are used for corrosion-resistance overlays and for joining or buildup purposes. Certain types are good heat-resistant alloys and serve for surface protection against oxidation up to 2,000 degrees F.

Martensitic Steel Electrodes. The carbon in martensitic steels is the major determinant of their characteristics. These steels are inexpensive and tough. The alloys with little carbon are tougher and more crack-resistant than the types with greater amounts of carbon. When used as electrodes, they can be built up to form thick, crackfree deposits of high strength and some ductility. Abrasion resistance is moderate and increases with carbon content and hardness. The deposits with little hardness may be machinable with tools while grinding is advisable for those with higher hardnesses. These electrodes are used for building up the surfaces of shafts, rockers, and other machined surfaces.

Iron Electrodes. Iron-base alloys with high carbon content are called *irons* because they have the characteristics of cast iron and are used for facing heavy cast iron machinery parts. They have a moderate-to-high alloy content of chromium, molybdenum, or nickel. Irons resist abrasion better than the austenitic and martensitic steels up to the point at which they lack toughness to withstand the associated impact. Hard-facing with iron-base electrodes is usually limited to one or two layer overlays. Cracking often results from the deformation of a soft base under the harder overlay. Proper support of the tough iron-base alloys is important.

NONFERROUS BASE ALLOYS. Cobalt-base Surfacing Metals (Rods and Electrodes). Cobalt-base alloys usually contain 26 to 33 percent chromium, 3 to 14 percent tungsten, and 0.7 to 3.0 percent carbon. Three grades are available in which hardness, abrasion resistance, and crack sensitivity increase as the carbon and tungsten content increases. They have high resistance to oxidation, corrosion, and heat. Metals surfaced with cobalt-base alloys stand up under some types of service at temperatures as high as 1,800 degrees F. Thus they are often used in the manufacture of exhaust valves for internal combustion engines.

Composite Tungsten Carbide Materials. Tungsten carbide hard-facing material is supplied in the form of mild steel tubes filled with crushed and sized granules of cast tungsten carbide, usually in proportions of 60 percent carbide and 40 percent tungsten, by weight. The carbide is very hard, tough, and abrasion resistant. Deposits containing large, undissolved amounts of tungsten carbide have more resistance to all types of abrasion than any other welded overlay.

As the heat of welding melts the steel tube, the molten metal dissolves some of the tungsten carbide to form a matrix of high tungsten steel or iron. This serves as an anchor and support for the undissolved granules of carbide. The highest abrasion resistance is achieved with oxyacetylene welding. Some of the carbide is dissolved by shielded metal-arc welding, thus decreasing abrasion resistance.

Copper-base Surfacing Metals (Rods and Electrodes). Various alloys of copper with aluminum, silicon, tin, and zinc are used for corrosion resistance as well as wear applications. Alloys with aluminum contain 9 to 15 percent aluminum and up to 5 percent iron. They are the hardest of the copper-surfacing alloys, achieving a reading of 380 on the Brinell scale. For this reason, copper-base surfacing alloys are used extensively to minimize metal-to-metal wear. Alloys with tin and zinc are also used for bearing surfaces.

Nickel-base Surfacing Metals (Rods and Electrodes). Nickel-base surfacing metals contain a relatively high degree of chromium and lesser amounts of

Table 7-16
Properties of hard-surfacing electrodes.

Abrasion	Friction	Impact	Hardness	Ductility	Machinability	Corrosion	Hard-surfacing guide
1	2	5	1	6	6	1	Carbide type—powder form
1	1	5	1	5	6	1	Carbide type—high abrasion resistance
3	2	5	1	5	6	2	Carbide type—good abrasion, moderate toughness
4	3	4	1	5	5 *4	5	Semi-austenitic type—high carbon, chromium alloy
5	4	1	6	2	4	6	Austenitic type—11 to 14% manganese
5	5	1	6	1	4	1	Austenitic type—18-8 and 25-20 stainless
5	2	5	1	6	6 *4	4	Martensitic type—high speed tool steel
2	2	5	1	6	6 *4	4	Martensitic type—5% chrome tool steel
5	4	3	2	4	5 *3	5	Martensitic type—low carbon chrome, Manganese
5	5	3	3	4	4 *3	5	Martensitic type—medium carbon, chrome, manganese
6	6	2	6	3	3 *1	6	Ferritic type—low carbon with 5% molybdenum
6	6	1	6	1	1	6	Ferritic type—conventional low carbon electrode

Scale 1 to 6
1. Excellent
2. Very good
3. Good
4. Fair
5. Poor
6. Very poor
*Annealed condition

carbon, boron, silicon, and iron. Their hardness and abrasion resistance increase with the carbon, boron, silicon, and iron content. In general nickel-base alloys are hard and have satisfactory resistance to abrasion, oxidation, corrosion, and heat. Hot strength and resistance to high-stress abrasion, however, is somewhat lower than for the cobalt-base group.

Aluminum Electrodes

Aluminum is the most widely fabricated metal after steel. This popularity is due primarily to such factors as its wide availability, strength, light weight, good workability, and pleasing appearance. The increase in many applications is due principally, however, to the development of dependable, high speed welding processes for joining the majority of aluminum alloys. More than two dozen major welding processes are used: gas metal-arc and gas tungsten-arc are used more than the others. A limited amount of shielded metal-arc welding is used. Aluminum can be welded throughout a thickness range of 0.00015 inch in foil to 6 inches in plate.

The high strength liquid and airtight joints now produced at high speeds in welding aluminum have extended the use of this lightweight material for freight cars of all types, tank cars, hopper cars, ships, barges, trucks, piping, and many other industrial applications.

The major alloying elements used in aluminum electrodes are magnesium (with zinc and with or without manganese) and silicon (with or without copper). See Table 7-17.

The two types of covered electrodes generally available and used a great deal are the 1100 and 4043 alloys. Alloy 1100 is commercially pure aluminum (99 percent aluminum) giving a weld deposit with a minimum tensile strength of 12,000 p.s.i. Alloy 4043 contains approximately 95 percent aluminum and 5 percent silicon. It has an ultimate tensile strength of 30,000 p.s.i. The 4043 (AL-2) electrode is a general purpose wire. It gives better fluidity during welding than the 1100 (AL-43) electrode. A large number of other aluminum electrodes are available to meet specific job requirements.

In applications where corrosive factors are important, an electrode should be selected which has a composition as close to that of the base metal as practical.

The presence of moisture in the electrode coating is a major cause of a porous weld structure. Since it is essential that the covering be completely dry, it is advisable to bake all "doubtful" electrodes and those taken from previously opened packages at 350 to 400 degrees F. for an hour before welding. After baking, they should be stored in a heated cabinet until used.

Ductility and cracking, known as *hot-shortness*, are problems in the welding of aluminum. These conditions are influenced by the degree of dilution between

Table 7-17

Composition of weld metal: aluminum and aluminum alloy welding rods and bare electrodes.

AWS Classification	Silicon, percent	Iron, percent	Copper, percent	Manganese, percent	Magnesium, percent	Chromium, percent	Nickel, percent	Zinc, percent	Titanium, percent	Other elements, percent		Aluminum, percent
										Each	Total	
ER1100 . . .	. . .	. . .	0.05–0.20	0.05	. . .	. . .	. . .	0.10	. . .	0.05	0.15	99.00 min.
ER1260 . . .	. . .	. . .	0.04	0.01	. . .	. . .	. . .	. . .	. . .	0.03	. . .	99.60 min.
ER2319 . . .	0.20	0.30	5.8–6.8	0.20–0.40	0.02	. . .	. . .	0.10	0.10–0.20	0.05	0.15	remainder
ER4145 . . .	9.3–10.7	0.8	3.3–4.7	0.15	0.15	0.15	. . .	0.20	. . .	0.05	0.15	remainder
ER4043 . . .	4.5–6.0	0.8	0.30	0.05	0.05	. . .	. . .	0.10	0.20	0.05	0.15	remainder
ER4047 . . .	11.0–13.0	0.8	0.30	0.15	0.10	. . .	. . .	0.20	. . .	0.05	0.15	remainder
ER5039 . . .	0.10	0.40	0.03	0.30–0.50	3.3–4.3	0.10–0.20	. . .	2.4–3.2	0.10	0.05	0.10	remainder
ER5554 . . .	. . .	. . .	0.10	0.50–1.0	2.4–3.0	0.05–0.20	. . .	0.25	0.05–0.20	0.05	0.15	remainder
ER5654 . . .	. . .	. . .	0.05	0.01	3.1–3.9	0.15–0.35	. . .	0.20	0.05–0.15	0.05	0.15	remainder
ER5356 . . .	. . .	. . .	0.10	0.05–0.20	4.5–5.5	0.05–0.20	. . .	0.10	0.06–0.20	0.05	0.15	remainder
ER5556 . . .	. . .	. . .	0.10	0.50–1.0	4.7–5.5	0.05–0.20	. . .	0.25	0.05–0.20	0.05	0.15	remainder
ER5183 . . .	0.40	0.40	0.10	0.50–1.0	4.3–5.2	0.05–0.25	. . .	0.25	0.15	0.05	0.15	remainder
R-C4A . . .	1.5	1.0	4.0–5.0	0.35	0.03	. . .	. . .	0.35	0.25	0.05	0.15	remainder
R-CN42A . .	0.7	1.0	3.5–4.5	0.35	1.2–1.8	0.25	1.7–2.3	0.35	0.25	0.05	0.15	remainder
R-SC51A . .	4.5–5.5	0.8	1.0–1.5	0.50	0.40–0.60	0.25	. . .	0.35	0.25	0.05	0.15	remainder
R-SG70A . .	6.5–7.5	0.6	0.25	0.35	0.20–0.40	. . .	. . .	0.35	0.25	0.05	0.15	remainder

American Welding Society

base metal and filler metal. Joints that require a great amount of filler metal should be avoided. Cracking is generally due to the low strength of some weld metal compositions at elevated temperatures. This can be overcome through the use of one of the higher magnesium-content aluminum alloys, since they give the weld higher strength during solidification. The type of joint and the welding technique also influence cracking.

Specialized Electrodes

There are a large variety of specialized electrodes to meet the conditions presented by such metals as nickel and high nickel alloys, copper and copper alloys, magnesium and magnesium alloys, and titanium and titanium alloys.

High nickel electrodes have been developed for the welding of gray iron castings, ductile iron, malleable iron, and other iron-base metals. Special high-nickel alloy, filler-metal compositions are capable of welding dissimilar metal combinations. A number of electrodes, which contain 50 percent or more nickel, are used in the welding of nickel and its alloys. These electrodes include the following combinations of alloys:

- Nickel-copper alloys
- Monel®-nickel-copper alloys
- Age-hardenable Monel®-nickel-copper alloys
- Age-hardenable Inconel®
- Inconel® - nickel - chromium-iron alloys
- High-nickel alloy filler metal

Copper electrodes contain tin and silicon in addition to copper. These copper alloys are classified with the prefix *E* plus the chemical abbreviations of the metals they contain. For example, a copper-aluminum electrode would be designated ECUAL. (E = electrode; CU = copper; AL = aluminum.)

Copper-silicon alloys (ECUSI) are often referred to as *silicon bronzes.* The core wire contains about 3 percent silicon and may contain small percentages of manganese and tin. Copper-silicon electrodes weld copper-silicon metal and copper-zinc (brass).

Copper-tin alloys (ECUSN-A) are usually referred to as *phosphor bronzes.* Phosphor bronze *A* contains about 8 percent tin. Both copper and tin are deoxidized with phosphorus. Phosphor bronze electrodes weld copper, bronze, brass, and cast iron. They are also used for overlaying steel. These electrodes require preheat, especially on heavy sections. They are used with direct current, reverse polarity.

The copper-nickel (ECUNI) electrodes contain 70 percent copper and 30 percent nickel. They are used for the standard copper-nickel alloys and must not be preheated or allowed to overheat. They are used with direct current, reverse polarity.

Copper-aluminum (ECUAL) electrodes are of two types: the copper-aluminum type used with gas welding, and the copper-aluminum-iron type for shielded metal-arc welding. They weld aluminum bronzes, manganese bronzes, some nickel alloys, many ferrous metals and alloys, and combinations of dissimilar metals. They are used with direct current, reverse polarity.

Also available is a copper-alloy electrode containing 11 to 12 percent aluminum and 3 to 4.25 percent iron. This electrode produces a deposit with higher tensile strength, yield strength, and hardness and lower ductility than a pure copper-aluminum electrode. It is used for joining aluminum-bronze plate and sheet, repairing castings, and for joining nonferrous metals such as silicon bronze to steel. It is used with direct current, reverse polarity.

Two comparatively new aluminum-bronze electrodes are composed of nickel-aluminum-bronze and the manganese-nickel-aluminum-iron alloys. They are used to weld similar base metals for ship propellers and ship fittings. They are used with direct current, reverse polarity.

The selection of the proper type of electrode for a particular welding job is vital to the successful completion of the job. A welder will never be in the position of knowing all there is to know about the various types of electrodes. You are urged to consult the various welding suppliers, catalogues and the excellent material available through the American Welding Society in order to develop a comprehensive understanding of the differences and uses of electrodes.

PACKING AND PROTECTION OF ELECTRODES

Standard Sizes and Lengths

The standard sizes and lengths of end-gripping electrodes are shown in Table 7-18. In all cases, standard size refers to the diameter of the core wire exclusive of the coating.

In 18-inch and 36-inch lengths, center gripping of the electrode is standard. End gripping is standard for all other lengths.

Packing

Electrodes are always suitably packed, wrapped, boxed, or crated to insure against injury during shipment or storage as follows:

- Bundles of 50 pounds net weight
- Boxes of 25 or 50 pounds net weight
- Coils, reels, or spools of 200 pounds or less

Marking

All bundles, boxes, coils or reels usually contain the following information:

- Classification
- Manufacturer's name and trade designation
- Standard size and length (weight instead of length in case of reels and coils)
- Guarantee

The welder should always give careful attention to the manufacturer's recommendations concerning heat settings, type and preparation of joint, base metal, welding technique, welding position, and nature of the welding current. Any references to moisture control are also important.

Note: You should turn to Table 7-15 on page 218 and study the classification of electrodes as they apply to the different manufacturers. It is important that you be familiar with the various trade names of electrodes in the same classification.

Moisture Control

A perfectly dry electrode is the first requirement for a perfect welding job, especially when the job requires low hydrogen electrodes or other moisture-prone electrodes. If the electrode is not dry, the operator cannot be sure of a sound weld. Welding with moist electrodes leads to increased arc voltage, spatter loss, undercutting, and poor slag removal. The weld deposit may suffer from porosity, underbead cracking, and rough appearance. When present, these conditions cause rejects and expensive reworking and, if undetected, contribute to produce failure.

All mineral-covered electrodes are "thirsty." The minute they are unpacked, they start absorbing moisture—too much moisture for a sound weld. Outside of a laboratory it is impossible to tell when an electrode has absorbed enough moisture from the air to be unsafe. Electrodes, therefore, require antimoisture protection. This is especially true when field welding is being done. A number of building and construction codes contain the specific provision that some kind of electrode holding and conditioning equipment is provided on the job site.

ELECTRODE OVENS. Electrode manufacturers recommend oven storage at specified holding temperatures to preserve and maintain factory baked-in quality. Oven protection is not only recommended, but mandatory for the storage of low-hydrogen and hard-surfacing electrodes and others made from special alloys such as the following:

- Iron powder
- Stainless steel
- Aluminum
- Inconel®
- Monel®
- Brass
- Bronze

The covering of low hydrogen electrodes, for example, is reduced to less than 0.2 percent at manufacture, and the electrodes are packaged in moisture-proof containers. Within 2 hours at 80 percent humidity, the electrodes may contain up to 13 times the allowable moisture content for U.S. Government specifications. Within 24 hours they may contain up to 26 times the 0.2 percent allowed.

Electrode drying ovens have capacities varying from 12 to 1,000 pounds and temperature controls to 1,000 degrees F. The smaller ovens, Fig. 7-12, are portable, making them convenient for shop or field welding. The larger ovens, Figs. 7-13 and 7-14, provide for central storage and baking for the entire shop.

The increased use of the submerged-arc process has caused a problem in regard to flux moisture. Flux that is unprotected will pick up moisture that results in welds with hydrogen inclusions. Figure 7-15 shows a portable type of holding oven

Table 7-18
Sizes and lengths of end-gripping electrodes.

Standard sizes diameter of core wire (inches)	Standard lengths length of core wire (inches)
1/16	9
5/64	9, 12
3/32	12
1/8	14
5/32	14
3/16	14
7/32	14, 18
1/4	18
5/16	18
3/8	18

Phoenix Products Co.

7-12. *The electric dry-rod electrode oven on the job.*

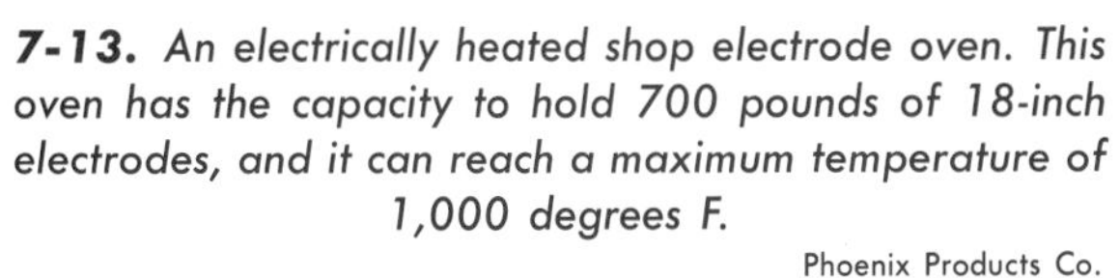

7-13. *An electrically heated shop electrode oven. This oven has the capacity to hold 700 pounds of 18-inch electrodes, and it can reach a maximum temperature of 1,000 degrees F.*

Phoenix Products Co.

Phoenix Products Co.

7-14. *An electrically heated shop electrode oven. This oven has the capacity to hold 350 pounds of 18-inch electrodes.*

7-15. *This portable flux holding oven may also be used for flux-covered wire.*

Phoenix Products Co.

that can be used to protect flux or flux-covered wire.

Many welders protect their electrodes with a small leather electrode carrier that is strapped to the operator's waist. These carriers offer limited protection for moisture, and they are not satisfactory for the electrodes listed previously. The oven shown in Fig. 7-12 is portable and can be plugged into auxiliary current in the field. Such ovens also offer good protection after the current has been turned off.

REVIEW QUESTIONS

1. Give a definition for the metal arc welding electrode.
2. Name the two major classifications of arc welding electrodes and explain each.
3. What is the function of the coating on a shielded metal-arc electrode?
4. What are the advantages of a shielded metal-arc electrode?
5. Shielded metal-arc electrodes have higher tensile strength than bare electrodes because the core wire has a higher tensile strength. True. False.
6. Give the AWS classification numbers for the following electrode characteristics:
   a. Bare
   b. Reverse polarity
   c. Straight polarity
   d. Iron powder
   e. Low hydrogen
7. Explain the functions of fast-fill, fast-follow, and fast-freeze electrodes.
8. In general terms, compare tensile strength and elongation in welds made with bare and shielded metal-arc electrodes. How do these properties compare with those of the base metal?
9. Explain the difference between a reverse and a straight polarity mild steel electrode.
10. What is the purpose of heavy slag-type, free-flowing electrodes? Name some of the other types of electrodes used for shielded metal-arc welding.
11. List the standard sizes of electrodes.
12. An E6010 electrode has a digging arc. True. False.
13. An E6012 electrode is primarily an a.c. electrode. True. False.
14. The same type of core wire is used in E6010 and E6012 electrodes. True. False.
15. Because of the heavy coating on iron powder electrodes, it is almost impossible to have too high a welding current. True. False.
16. List the important considerations in selecting the type of electrode to use.
17. Which type of electrode is best for welding light-gauge sheet steel?
18. Some iron powder electrodes can be used in all positions. True. False.
19. An E6013 electrode is the same as an E6012 except that it is hotter. True. False.
20. An E7028 electrode has high spatter loss, and the weld is rough. True. False.
21. Stainless steel electrodes can be used only with direct current, reverse polarity. True. False.
22. Stainless steel electrodes can be used in all positions. True. False.
23. Weld cracking is a problem in welding aluminum. True. False.
24. List the service conditions for hard-facing electrodes.
25. A slight amount of moisture is good for electrodes since it introduces extra oxygen into the weld pool. True. False.
26. How are electrodes best protected from moisture in the field?

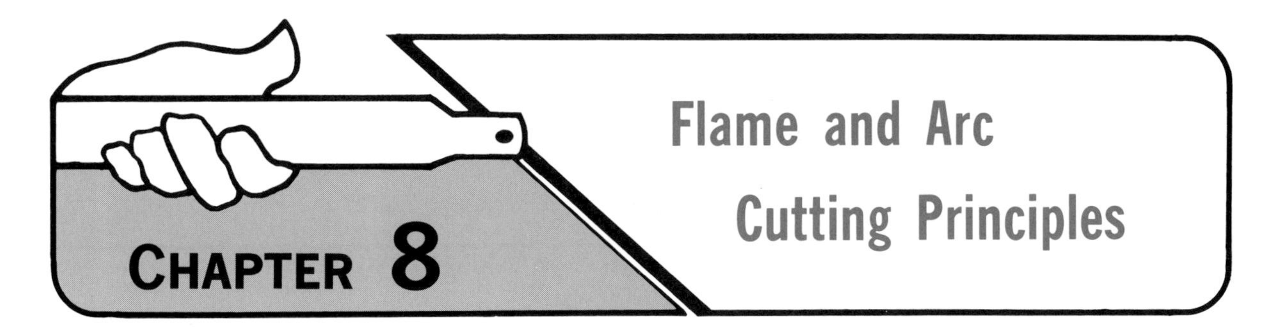

CHAPTER 8 Flame and Arc Cutting Principles

In the majority of fabricating shops and in the field, the welder must be able to do manual flame and arc cutting, Fig. 8-1. The oxy-fuel gas cutting torch and the various arc cutting devices have become universal tools. They are widely used in foundries for the removal of risers from castings, for cleaning castings, and as a means of scrapping obsolete metal structures. Cutting devices are also used in the fabrication of metal structures, Fig. 8-2.

Cutting processes have made it possible to fabricate structures requiring heavy thicknesses of metal from rolled steel. Formerly, these structures had to be cast. The combination of cutting and welding processes created an industry devoted to the fabrication of heavy machinery and equipment from rolled steel. Oxy-fuel gas and arc cutting increase the speed of fabrication and eliminate many costly joining, shaping, and finishing operations.

OXY-ACETYLENE AND OTHER FUEL GAS CUTTING

Oxyacetylene cutting (OFC-A) is limited to the cutting of ferrous materials (materials containing iron). Stainless steel, manganese steels, and nonferrous materials are not readily cut because they do not oxidize rapidly. Many of these materials may be cut by the arc process. Most ferrous materials have an affinity for oxygen. Even under normal conditions the oxygen in the air attacks these materials to form an iron oxide which we recognize as rust. Thus, the rusting process is a slow form of oxygen cutting. At elevated temperatures the oxidation process is increased. Oxygen cutting requires that the part to be cut be raised to a temperature of 1,500 to 1,600 degrees F. (usually a cherry red). A stream of pure oxygen is directed onto the hot metal causing it to burn rapidly.

Steel burns in pure oxygen after having reached its kindling temperature just as paper burns in air. The main difference is that burning paper gives off carbon dioxide and water vapor. These products of combustion are gaseous and pass off into the air. When steel burns, however, it gives off iron oxide which is solid at room temperature. Its melting point is below the melting point of steel. The heat generated by the burning iron and the oxyacetylene flame is high enough to melt the iron oxide so that it runs off as molten slag, exposing more iron to the oxygen jet. Thus the jet can be moved along to produce a clean cut. Cutting is really a process of very rapid rusting.

The oxy-fuel gas cutting process makes use of several other fuel gases in addition to acetylene. These fuel gases include propane, natural gas, and a

8-1. *Welding students being instructed in the proper use of the oxy-fuel gas cutting torch.*
Department of Health, Education, and Welfare

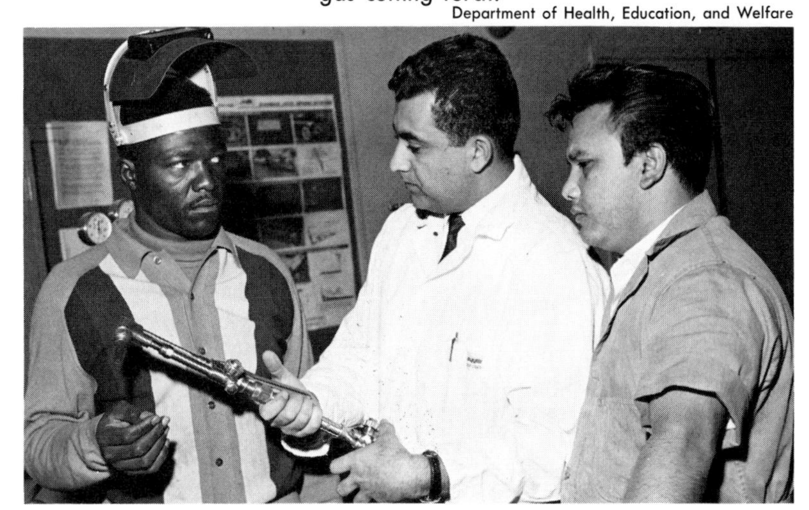

comparative new gas called by the trade name of Mapp® gas. Refer to Chapter 5 for a review of the characteristics of these gases and their distribution.

Equipment

Oxy-fuel gas cutting requires the following equipment:

- Single-purpose cutting torch, Figs. 8-3 and 8-4, or a welding torch to which an adaptable cutting head has been attached
- Flint lighter
- Oxygen regulator
- Acetylene regulator
- Oxygen welding hose with couplings attached to each end
- Acetylene welding hose with couplings attached to each end

Portable tank outfits are mounted on a truck similar to that shown in Fig. 5-21, page 153. Review pages 146–155 in Chapter 5 concerning the regulators, hose, and other equipment.

CUTTING TORCH. The cutting torch, Fig. 8-4, mixes oxygen and acetylene or other fuel gases in the proportions necessary for cutting. It consists of a handle (A), connecting tubes (B), and the cutting head (C). The handle is equipped at the rear with hose connections for oxygen (D) and acetylene (E). A needle valve in the acetylene inlet connection controls the supply. The oxygen furnished to the preheating flame is regulated by a preheat valve on the side of the handle. A high pressure oxygen valve operated by a lever (F) controls the oxygen.

On some torches the preheating oxygen and acetylene mix in the cutting tip. In others mixing takes place in a mixing chamber in the torch body. The cutting head may be at a 45- or 90-degree angle to the body of the torch. Torches with straight heads may also be obtained.

Figure 8-5 illustrates the internal construction of a standard cutting torch.

On some jobs, such as pipe welding and repair work, cutting is a small part of the welding operation. For such jobs, the adaptable cutting attachment is used for cutting. In a few seconds it can be attached to the welding torch without disconnecting the hose from the torch. The construction and operation of the cutting attachment is similar to the type of cutting torch in which the preheating gases mix in the handle. Figure 8-6 illustrates the internal construction of a standard adaptable cutting attachment.

CUTTING TIPS. The cutting tip, Fig. 8-7, has a central hole through which the high pressure oxygen flows. Around this center hole are a number of preheating flame holes. Cutting tips may be obtained in various shapes and sizes. The thicker the metal that is to be cut, the larger the size

8-2. *Cutting a large I-beam which is part of the structural steel on a building construction job.*

Linde Division, Union Carbide Corp.

8-3. *Equipment for oxyacetylene and other fuel gas cutting. 1. Cutting torch. 2. Cutting tip. 3. Oxygen regulator. 4. Acetylene regulator. 5. Torch wrench. 6. Tank and regulator wrench. 7. Lighter. 8. Goggles.*

Modern Engineering Co.

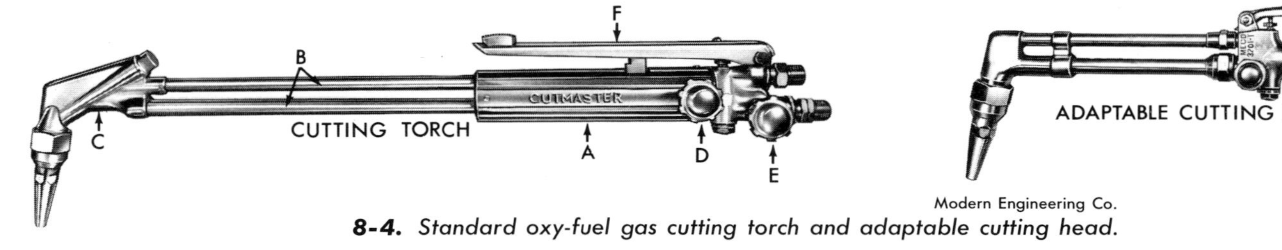

Modern Engineering Co.

8-4. *Standard oxy-fuel gas cutting torch and adaptable cutting head.*

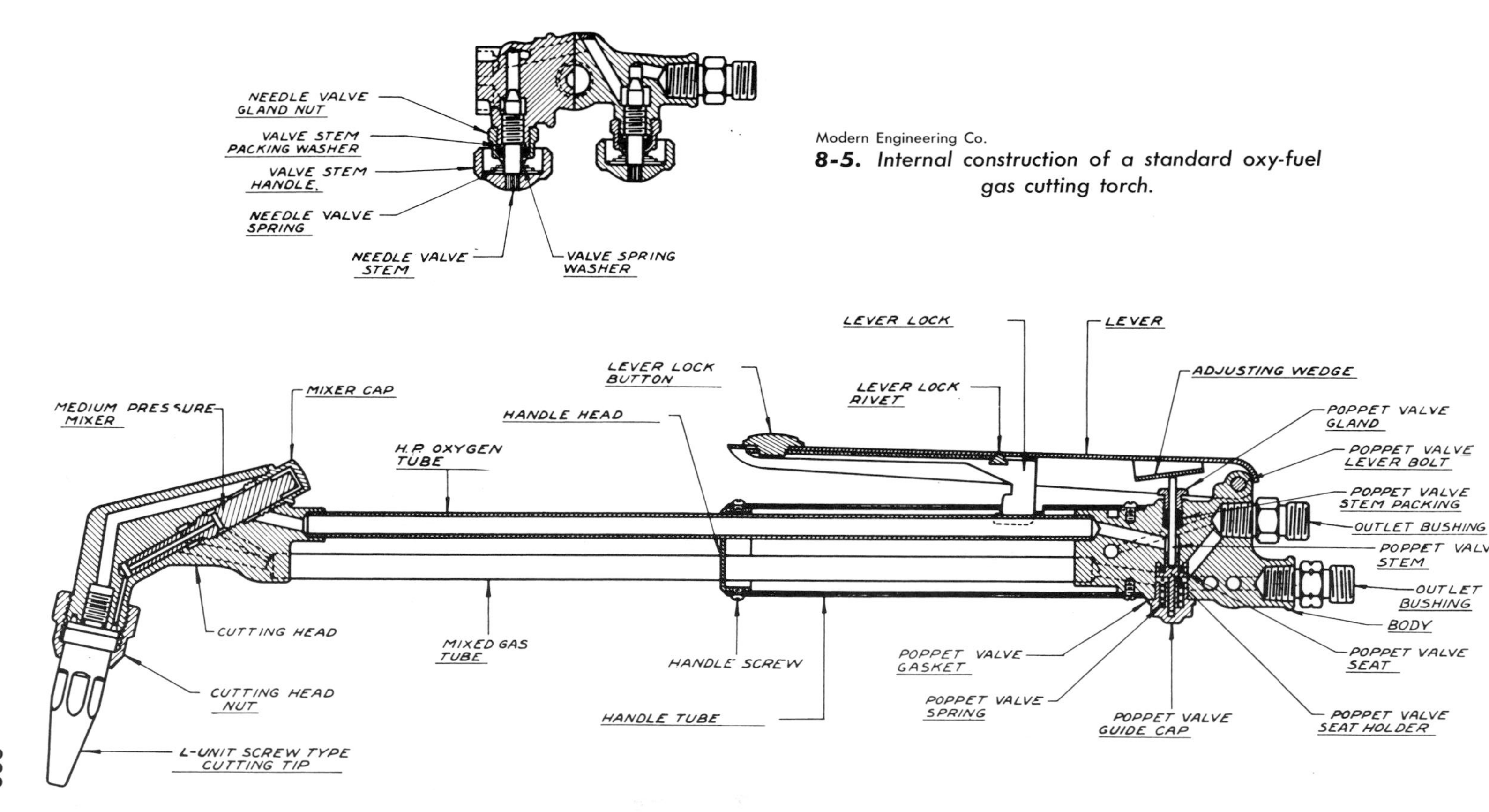

Modern Engineering Co.

8-5. *Internal construction of a standard oxy-fuel gas cutting torch.*

of the center hole must be. Table 8-1 lists different size cutting tips and the gas pressures appropriate for them. Tips may be obtained for flame machining, gouging, scarfing, and rivet cutting, Fig. 8-8.

LIGHTERS. The cutting torch should be lighted with a friction lighter, Fig. 8-3, page 228. The flints of friction lighters can be easily replaced at small cost when worn out. Matches should never be used because the operator's hand has to be too close to the torch tip and may be burned when the gases ignite. There is also the danger that the supply of matches that the operator may be carrying in his pockets will ignite and cause severe burns.

GOGGLES. The operator must wear protective goggles to prevent harm to his eyes from sparks, hot particles of metal, and glare. Suitable goggles and lenses are discussed in Chapter 5, pages 160, 161.

GLOVES. The heat coming from a cutting job may be very intense. There may also be a shower of sparks and hot material which makes it necessary for the operator to protect his hands with gloves. For the best protection gloves should be of asbestos or other nonburnable material. It is common practice, however, to wear an ordinary canvas glove with a cuff, which can be purchased at very small cost.

Gloves should be kept free from grease and oil because of the danger involved in contact with oxygen.

MAGNETIC BURNING SQUARE. The magnetic burning square, Fig. 8-9 makes it possible to cut straight lines with a high degree of accuracy. It can be used for structural burning and layout work in general. The base of the tool is a perfect square, with two strong, cast-in magnets for holding the tool in the overhead, vertical, angular, and horizontal positions. The tool is inscribed with a 90-degree protractor in 1-degree increments, allowing a swing of 180 degrees with the blade or tool. The top part of the tool is cast with an 18-inch blade in $\frac{1}{8}$-inch increments, and it has a large knob for setting and holding the blade in the desired degree position. With the exception of the magnets, the tool is made of aluminum.

Oxy-Fuel Gas Cutting Machines

A very large part of the cutting done today is performed by oxy-fuel gas cutting machines, Fig. 8-10. These machines have a device to hold the cutting torch and guide it along the work at a uniform rate of speed. It is possible to produce work of higher quality and at a greater speed than with the hand cutting torch. Machines may be used for cutting straight lines, bevels, circles, and other cuts of varied shape.

Small portable cutting machines are used with only one torch. Large permanent installations can make use of several cutting torches to make a number of similar shapes at the same time. A multiple-torch cutting machine and its automatic controls are shown in Figs. 8-11 and 8-12.

8-6. *Internal construction of a standard oxy-fuel gas adaptable cutting attachment.* Modern Engineering Co.

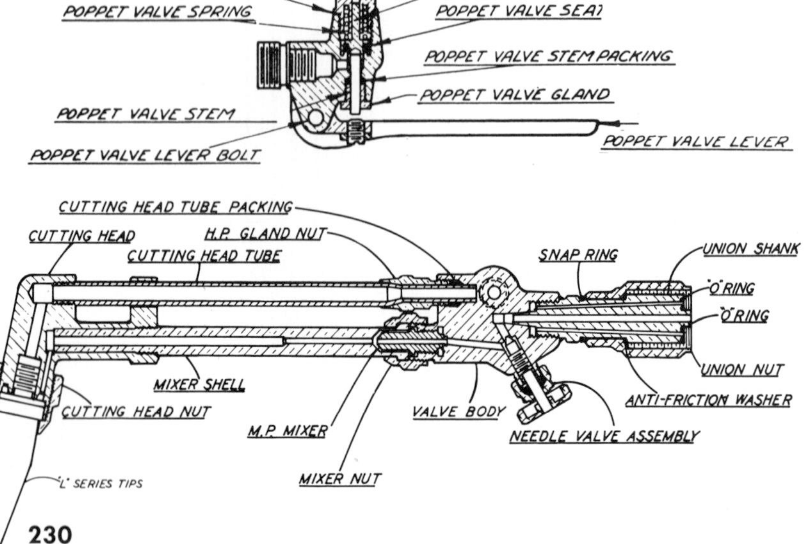

8-7. *Standard cutting tip attached to cutting torch head.* Modern Engineering Co.

Maximum productive capacity is achieved through the use of stationary cutting machines developed for production cutting of regular and irregular shapes of practically any design. These machines can be particularly adapted to operations in which the same pattern or design is to be cut repeatedly, Fig. 8-13 (Page 234). A number of cutting torches are mounted on the machine so that a number of parts of the same shape can be cut simultaneously. These machines can be used for straight-line or circle cutting. They can be guided by hand or a template.

Cutting machines may be guided by various types of tracing devices. One type follows a

Table 8-1
Gas pressures for different size cutting tips.

Tip no.	Thickness of metal, inches	Acetylene pressure, lbs.	Oxygen pressure, lbs.	Acetylene consumption cubic feet per hour	Oxygen consumption cubic feet per hour
L-00	1/8	4	10	9.5	35
L-0	1/4	4	15	9.5	40
L-1	3/8	4	20	9.5	45
L-1	1/2	4	25	9.5	50
L-2	3/4	5	30	15.3	60
L-2	1	5	40	15.3	100
L-2	1 1/2	5	50	15.3	150
L-2	2	5	60	15.3	200
L-3	3	6	70	25.2	275
L-3	4	6	80	25.2	350
L-3	5	6	90	25.2	425
L-4	6	7	100	27.3	550
L-5	8	7	130	27.3	825
L-6	10	8	150	28.2	1000

Modern Engineering Co.

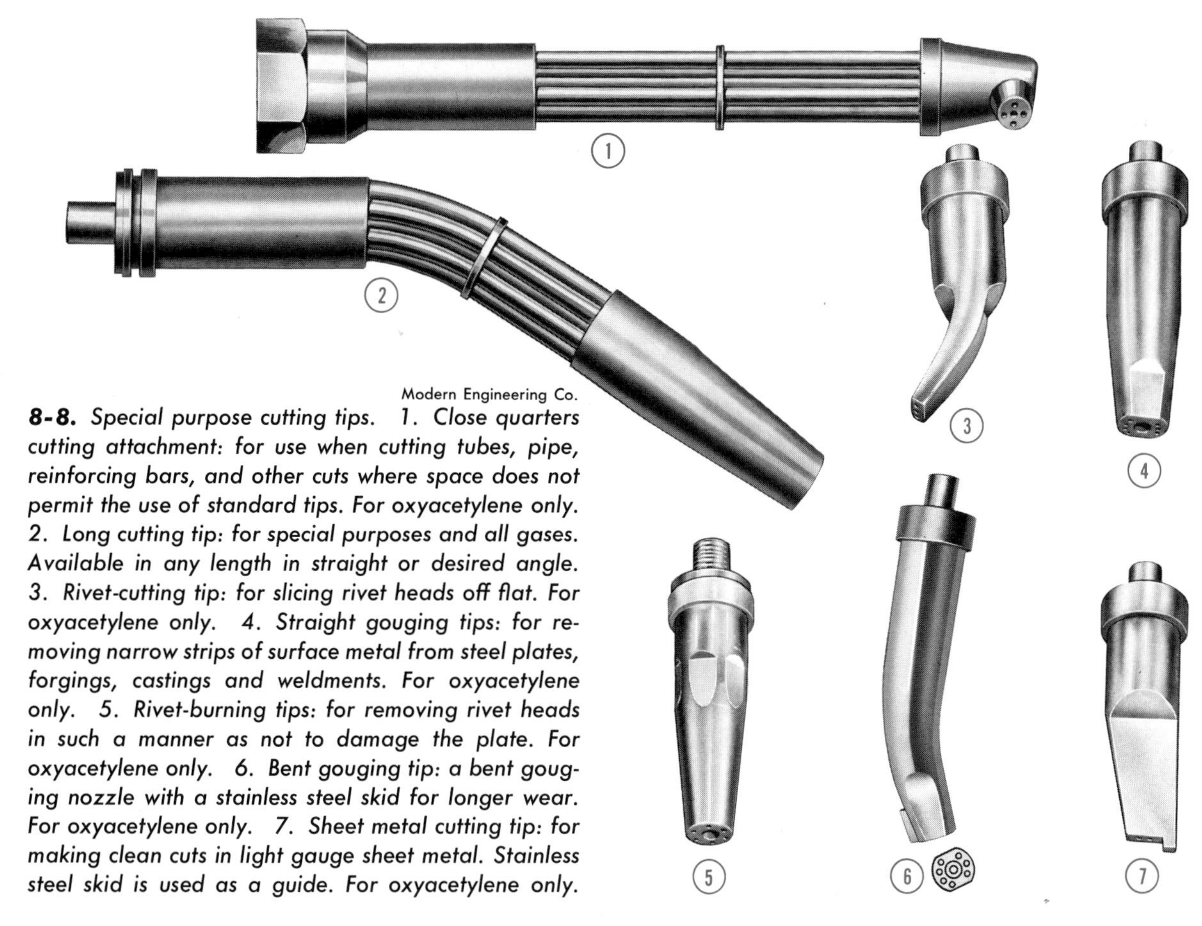

Modern Engineering Co.

8-8. *Special purpose cutting tips. 1. Close quarters cutting attachment: for use when cutting tubes, pipe, reinforcing bars, and other cuts where space does not permit the use of standard tips. For oxyacetylene only. 2. Long cutting tip: for special purposes and all gases. Available in any length in straight or desired angle. 3. Rivet-cutting tip: for slicing rivet heads off flat. For oxyacetylene only. 4. Straight gouging tips: for removing narrow strips of surface metal from steel plates, forgings, castings and weldments. For oxyacetylene only. 5. Rivet-burning tips: for removing rivet heads in such a manner as not to damage the plate. For oxyacetylene only. 6. Bent gouging tip: a bent gouging nozzle with a stainless steel skid for longer wear. For oxyacetylene only. 7. Sheet metal cutting tip: for making clean cuts in light gauge sheet metal. Stainless steel skid is used as a guide. For oxyacetylene only.*

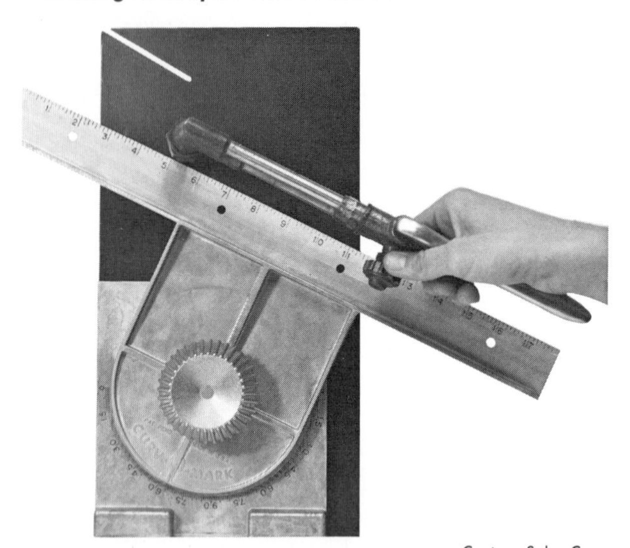

Contour Sales Corp.

8-9. *Cutting a straight line with the aid of the magnetic burning square.*

8-10. *This oxy-fuel gas cutting machine is used for straight line and bevel cutting. Protective goggles should be worn when working with this machine.*

patterns line of brown tracer ink and electrically controls the movement of the torch by means of a servomechanism.

Some units make use of a tracer roller, which is magnetized and kept in contact with a steel pattern, Fig. 8-14. The tracer follows the exact outer contour of the pattern and causes the cutting tools to produce a cut in exactly the same shape. Figure 8-15 shows a pattern tracer in use.

Stack Cutting

In addition to cuts made through a single thickness of material, oxygen cuts are made through several thicknesses at the same time. This is a machine cutting process known as stack cutting. The plates in the stack must be clean and flat and have their edges in alignment where the cut is started. The plates must be in tight contact so that there is a minimum of air space between them. It is usually necessary to clamp them together.

Stack cutting is particularly suitable for cutting thin sheets. A sheet $\frac{1}{8}$ inch thick or less warps, and the edge is rough with slag if it is cut singly. If stack cut, the edges are straight and smooth and free from slag. Stack cutting may be used on plates up to $\frac{1}{2}$ inch thick.

Beam Cutter

A recent development in gas cutting machines is the beam

8-11. *Multiple-torch shape-cutting machine cuts squares, rectangles, circles, and other geometric shapes without templates.*

Linde Division, Union Carbide Corp.

cutter, Fig. 8-16. The beam cutter is a portable structural fabricating tool. From one rail setting the operator can trim, bevel and cope beams, channels and angles. The beam rail is positioned across the flanges. Two permanent magnets lock and square the rail in position. Variable speed power units are used on both the horizontal and vertical drives. A squaring gauge enables the operator to adjust the tip quickly from bevel to straight trim cuts.

This all-position flame-cutting machine weighs only 60 pounds. It is easily moved and set in po-

8-12. *The automatic digital control unit for the machine shown in Fig. 8-11.*

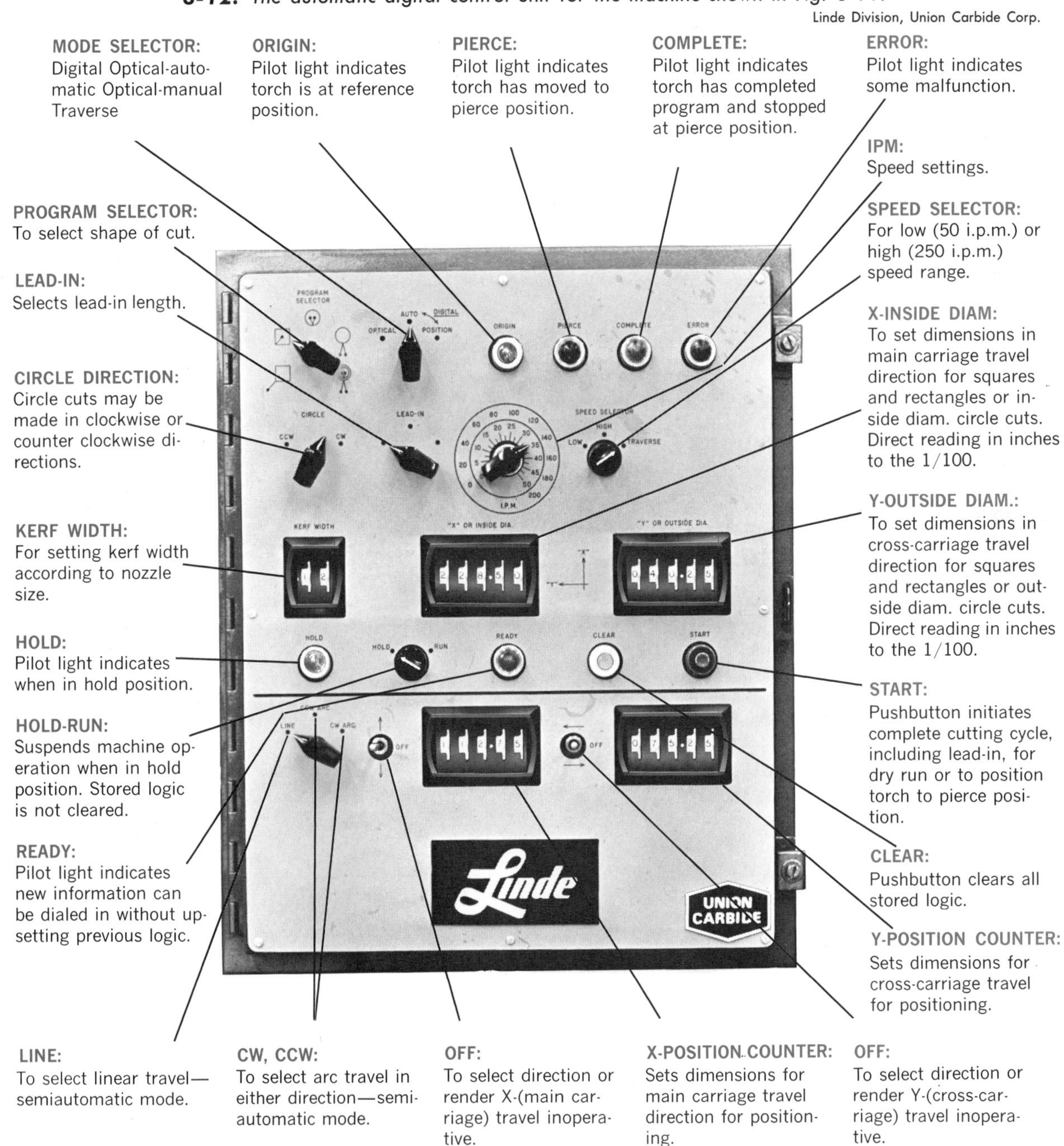
Linde Division, Union Carbide Corp.
MODE SELECTOR: Digital Optical-automatic Optical-manual Traverse
ORIGIN: Pilot light indicates torch is at reference position.
PIERCE: Pilot light indicates torch has moved to pierce position.
COMPLETE: Pilot light indicates torch has completed program and stopped at pierce position.
ERROR: Pilot light indicates some malfunction.
IPM: Speed settings.
SPEED SELECTOR: For low (50 i.p.m.) or high (250 i.p.m.) speed range.
PROGRAM SELECTOR: To select shape of cut.
LEAD-IN: Selects lead-in length.
CIRCLE DIRECTION: Circle cuts may be made in clockwise or counter clockwise directions.
X-INSIDE DIAM: To set dimensions in main carriage travel direction for squares and rectangles or inside diam. circle cuts. Direct reading in inches to the 1/100.
KERF WIDTH: For setting kerf width according to nozzle size.
Y-OUTSIDE DIAM.: To set dimensions in cross-carriage travel direction for squares and rectangles or outside diam. circle cuts. Direct reading in inches to the 1/100.
HOLD: Pilot light indicates when in hold position.
HOLD-RUN: Suspends machine operation when in hold position. Stored logic is not cleared.
START: Pushbutton initiates complete cutting cycle, including lead-in, for dry run or to position torch to pierce position.
READY: Pilot light indicates new information can be dialed in without upsetting previous logic.
CLEAR: Pushbutton clears all stored logic.
Y-POSITION COUNTER: Sets dimensions for cross-carriage travel for positioning.
LINE: To select linear travel—semiautomatic mode.
CW, CCW: To select arc travel in either direction—semiautomatic mode.
OFF: To select direction or render X-(main carriage) travel inoperative.
X-POSITION COUNTER: Sets dimensions for main carriage travel direction for positioning.
OFF: To select direction or render Y-(cross-carriage) travel inoperative.
Linde
UNION CARBIDE

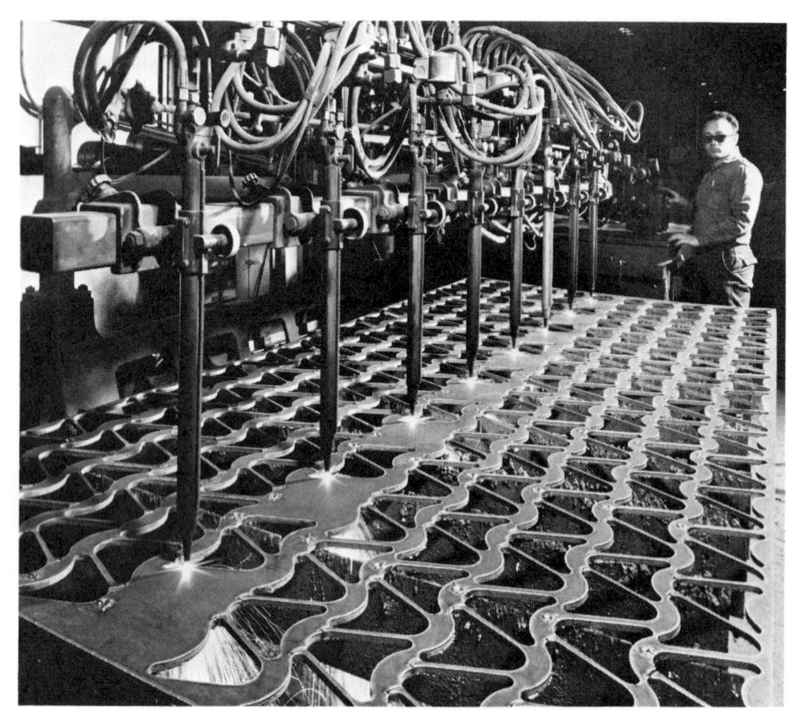

Metal Fabricating Institute

8-13. *Multiple-gas cutting by machine control.*

8-14. *A tracer machine with a template follower in the torch arm of the machine. (A) provides tip control; (B) controls the speed of the cut; (C) is the cutting torch; (D) is the magnet for precise contact with the template.*

Victor Equipment Co.

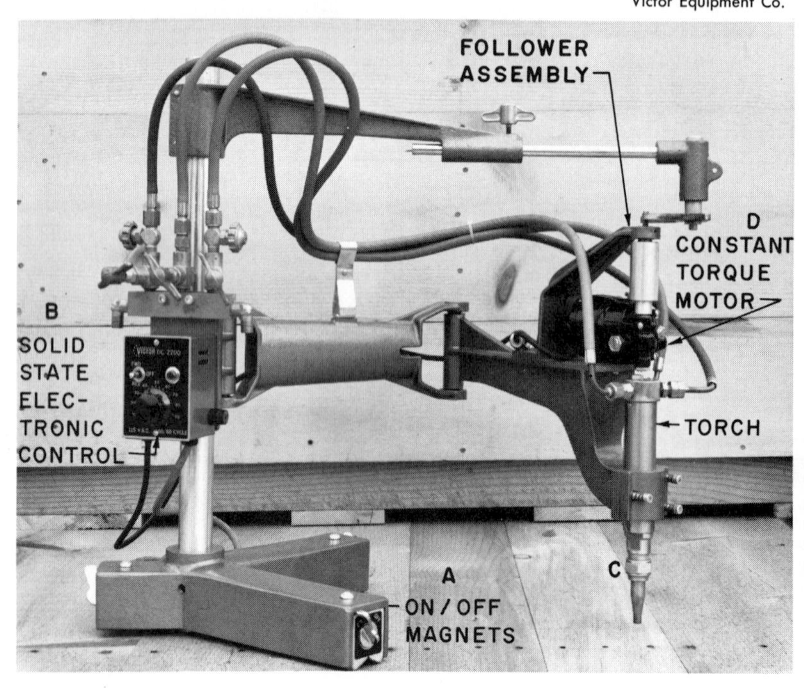

sition by one operator, so, setup time is minimal. The beam cutter provides clean, accurate cuts in a minimum amount of time. It has become a very important tool for use in the construction industry.

OXYGEN LANCE CUTTING

Oxygen lance cutting (LOC) is a method of cutting heavy sections of steel that would be very difficult by any other means. The lance is merely a length of black iron pipe fitted with a valve at one end to which an oxygen hose is connected.

Oxygen pressure of 75 to 100 p.s.i. is used. The pipe size may vary from $\frac{1}{4}$ to $\frac{3}{8}$ inch.

In order to start the cut, it is necessary to preheat the area at which the cut is to begin. This is

8-15. *A welder is using a machine-mounted cutting torch on a pattern tracer to cut scale components from ¼-inch mild steel plate.*

Linde Division, Union Carbide Corp.

done with an oxyacetylene welding torch. When the starting point has been brought to a bright red heat, the end of the lance is brought against it, and the oxygen valve on the lance opened. Once the operation is started, sufficient heat is liberated to continue the cut. The lance is slowly consumed during the operation and must be replaced from time to time.

The oxygen lance is useful for piercing holes in heavy thicknesses of steel, cutting off large risers in the foundry, and opening holes in steelmaking equipment which have become plugged with frozen metal.

Weld Tooling Corp.

8-16. *Beam cutter used in heavy steel construction.*

ARC CUTTING

Arc cutting processes melt metal along a desired line of cut with the heat generated by an electric arc. A number of the processes also use oxygen, compressed air, or one of the inert gases in addition to the arc. Several of the newer arc processes compare favorably with oxy-fuel gas cuts in quality of cut, but generally it is not possible to do work of the same quality and smoothness. The primary advantage of arc cutting is that it can be used on all types of metals. Some of its applications include cast iron, scrap, nonferrous metals, and stainless steels. Arc cutting equipment is also used for hole piercing, rivet cutting, gouging, and other special uses.

Arc cutting is done with both the carbon arc and the metal arc.

Carbon Arc Cutting (CAC)

Metals that can be welded without postheat treatment may be cut by the carbon arc process. Any alloy with high hardenability will be in a nonmachinable condition after the cut. Since the temperature of the arc is high, the method is used for the cutting of cast iron. It is also used in scrapyards and for the tearing down of steel structures. Carbon arc cutting is not used if a smooth and accurate cut is desired.

In carbon arc cutting, an electric arc is drawn between a carbon or graphite electrode and the material to be cut. The heat of the arc melts away the material. The material is removed from the cut by the force of the arc and by gravity. This is done with a direct current welding machine on straight polarity. Tables 8-2 and 8-3 give various heat settings for different size carbon electrodes.

Because of the heavy current values used in arc cutting, it is best to use graphite or hard carbon electrodes. Soft carbon electrodes do not give very long service life. Graphite and hard carbon electrodes also retain a

Table 8-2

Current values for graphite electrodes for carbon arc cutting.

Electrode diameter, in.	Current, amp.
1/4	Up to 200
3/8	200–400
1/2	300–600
5/8	400–700
3/4	600–800
7/8	700–1000
1	800–1200

American Welding Society

Table 8-3

Recommended current ranges for carbon arc cutting electrodes.

Electrode size, in. →	5/32	3/16	1/4	5/16	3/8	1/2	5/8	3/4
Minimum amp.	80	110	150	200	300	400	600	800
Maximum amp.	150	200	350	450	550	800	1000	1600

American Welding Society

point longer than soft carbon electrodes. This is very important if clean, smooth cuts are expected.

If a great deal of cutting is to be done, a special electrode holder must be used. An ordinary electrode holder overheats because of the high currents used in the operation. The special holders usually have a hand shield for protecting the hand against the intense heat. They may also be water cooled.

CARBON ARC CUTTING TECHNIQUE. The manipulation of the cutting arc is simple. The arc is advanced along the section to be cut at a rate equal to that at which the metal is melted away. The best cuts are made with the plates in the flat position, but cuts may be made in the vertical position. Cuts in the overhead position are a hazard to the operator because of the molten slag.

The speed of cutting depends upon the size of electrode being used, the thickness of the material, and the amount of current being used. The width of the cut increases with the increased current and larger electrode diameters. The skill of the operator also influences cutting speed.

8-17A. Operator completing the cut of a 3/16-inch, 18/8 stainless steel center sill of a railroad car with the oxygen arc process. Arcos Corp.

Shielded Metal-Arc Cutting

Although a great deal of arc cutting is done by the carbon arc process, it is not always possible to have carbon arc equipment on the job. Equipment for shielded metal-arc cutting (SMAC) is more versatile. Surfaces cut by the shielded metal-arc process are less ragged than those produced by the carbon arc.

The shielded metal-arc process is used for cutting risers and nonferrous gates in foundries and nonferrous materials in scrapyards. This process may also be used for underwater cutting. A specially constructed, fully insulated holder is necessary.

The shielded metal-arc process consists of drawing an arc between a covered electrode and the material being cut. The covering forms a gaseous shield which protects the metal being cut from the atmosphere. Material is removed from the cut by gravity and the force of the arc.

The electrode covering serves as electrical insulation, permitting deep penetration of the electrode into the cut on heavy sections without short circuiting the sides of the electrode. On those jobs where it is necessary to pierce holes through the material, the coated electrodes are a distinct advantage over carbon electrodes because of the insulation afforded by the coating. The covering also helps to stabilize the arc so that the heat can be concentrated at the front of the cut.

It is customary to select electrodes ranging from 5/32 to 1/4 inch in diameter since the smaller electrodes heat up too quickly and melt away, Table 8-4. Electrodes with both gaseous and slag-type coatings may be used, but the latter type gives better results because it can handle higher current values. Either direct or alternating current may be used although direct current, straight polarity is preferred.

It is necessary to use considerably more current for cutting than is used for welding with an electrode of the same size. Heavy duty holders should be used with electrodes having a diameter of 3/16 inch and larger.

As with carbon electrode cutting, the speed of cut and width of cut depend upon the size of electrode being used, the current values employed, the thickness of the material, the kind of material, and the skill of the operator.

Oxygen Arc Cutting

Oxygen arc cutting (AOC) is a method of cutting, piercing, and gouging metals with an electric arc and a stream of oxygen.

In the oxygen arc process, Fig. 8-17, a stream of oxygen is directed into a pool of molten metal. The pool is formed and kept molten by an arc established between the base metal

Table 8-4

Electrode sizes for the shielded metal-arc cutting of steel plate.

Electrode diameter, in.	Plate thickness, in.	Current, amps
1/4	1/4 1/2 3/4	400 400 400
3/16	1/4 1/2	300 300
3/16	1/4 1/2 3/4	400 400 400

American Welding Society

and a tubular coated electrode, which is consumed during the cutting operation, Fig. 8-18. In addition to providing the source of the arc, the coating on the electrode provides insulation, acts as an arc stabilizer, and aids the flow of molten metal from the cut.

The cutting action is fast, and preheating is not required. The cut material is not contaminated in any way. The finished cuts require very little grinding or machining. The cut surfaces, however, are usually rougher than those produced by oxyacetylene cutting. Experienced welders are able to use the process for the first time with very little practice.

The oxygen arc process can be used to cut those metals that have always been considered nearly impossible by standard methods. Electrodes were first developed primarily for use in underwater cutting. It is possible to cut ferrous and nonferrous metals in any thickness and position. Such metals as stainless steels, nickel-clad steel, bronze, copper, brass, aluminum, and cast iron are cut without difficulty.

8-17B. *Cutting a circle in a tank head of 3/8-inch Monel® with the oxygen arc process.*

Arcos Corp.

Equipment includes an oxygen arc electrode holder, oxygen arc coated tubular electrodes, an a.c. or d.c. welding machine, a tank of oxygen, and oxygen-regulating gauges.

The oxygen arc cutting electrode is a ferrous metal tube covered with a nonconductive coating. The function of the tube is to conduct current for the establishment and maintenance of the arc. The bore of the tube directs the oxygen to the metal being cut. These electrodes are available in 3/16 and 5/16 inch diameters with bores of 1/16 and 1/10 inch respectively. They are 14 and 18 inches long.

The electrode is held by a special holder, Fig. 8-19. This holder is similar in appearance to a welding electrode holder. When used for cutting under water, a fully insulated holder equipped with a suitable flashback arrester is required. Both current and oxygen are fed to the electrode through the holder. It is easy to insert the electrode into the holder and remove the stub end after cutting. The flow of oxygen is controlled by a valve built into the holder and triggered by the operator with the hand that grasps the holder.

Both the cutting and piercing operations are begun by tapping the tip of the electrode on the work to establish an arc and release the oxygen. For piercing, the electrode is pushed into and through the plate. For cutting, the electrode is dragged along the plate surface at the same rate as the cut. The coating is kept in contact with the material being cut at all times. The coating burns off slower than the electrode, thus maintaining the arc. Figure 8-20 illustrates the process at the point of cutting.

The speed of cut varies with the thickness and composition of the plate, the oxygen pressure, the current, and the size of the electrode, Fig. 8-21. Table 8-5 gives the ampere settings and oxygen pressures for various types of metal that can be cut by the oxygen arc process.

Gouging is performed by inclining the electrode until it is almost parallel to the plate surface and pointed away from the operator along the line of the gouge.

8-18. *Schematic of the oxygen arc electrode in operation*

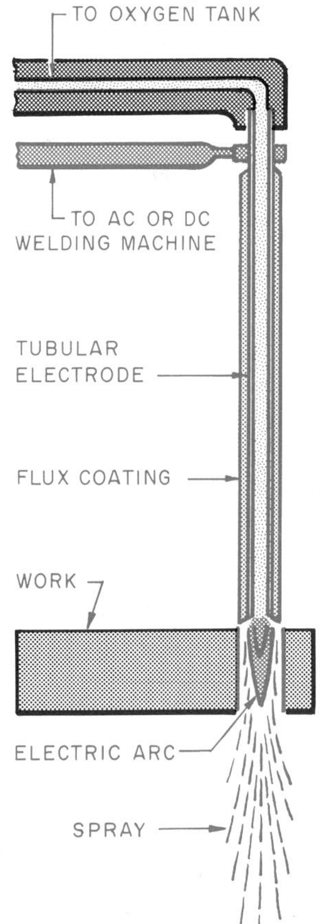

Air Carbon Arc Cutting

The air carbon-arc process, Fig. 8-22, is a method of cutting and gouging by melting the work with an electric arc and blowing away the molten metal with a strong jet of compressed air. Because the process does not depend on oxidation, it can be used on virtually all metals. The metals are not harmed by the process.

An arc is struck between the carbon electrode and the metal to be cut. The metal melts instantly, and high velocity jets of air blast the molten metal away. The air blast is continuous and directed behind the point of arcing. The electrode is pushed forward at a rapid rate. The travel speed depends on the size of electrode, type of material, amperage, and air pressure. The depth and contour of the groove is controlled by the electrode angle, travel speed, and current. The width is determined by the size of the electrode. Two plates may be grooved simultaneously, Fig. 8-23.

The air carbon-arc process is commonly used in steel foundries to remove defects from castings. It is also used a good deal in metal fabrication. The process can remove weld defects, clean out the roots of welds, widen grooves that have pulled together during welding, and make U-grooves in plates. As a maintenance process, it is used by the railroads to gouge out cracks before damaged railroad cars and tracks are welded. In refineries it cuts stainless steel pipe, removes stainless steel liners, and patches tank bottoms. Air carbon-arc cutting also prepares metal parts for hard facing and repair welding.

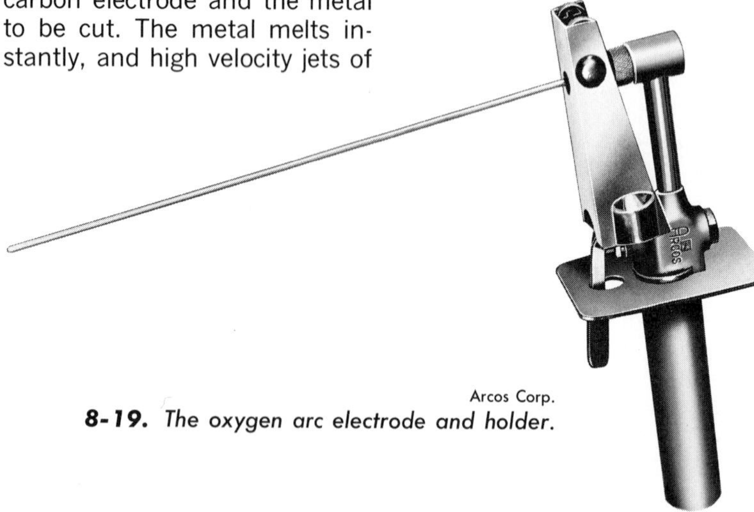

Arcos Corp.

8-19. *The oxygen arc electrode and holder.*

8-20. *The oxygen arc process at the point of cutting.*

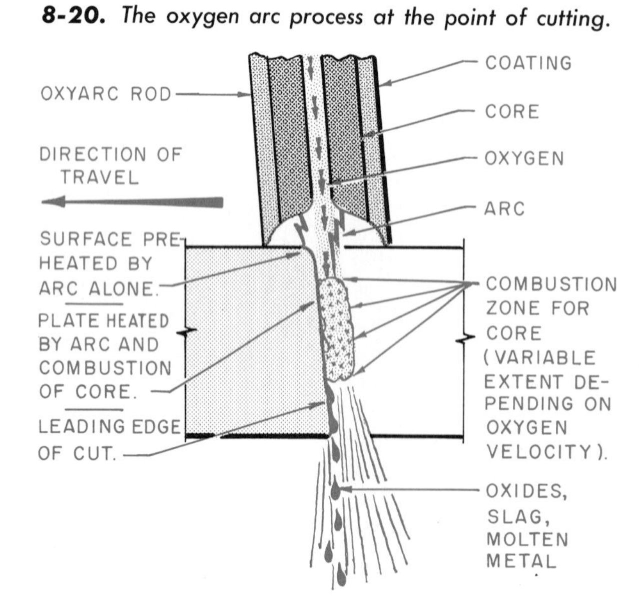

8-21. *The oxygen arc cutting of cast iron risers in the foundry. Risers are 5 inches square. The operation required 5 minutes cutting time.*

Arcos Corp.

The process can be used on all carbon, manganese, and stainless steels, copper and nickel alloys, cast iron, and other hard-to-cut metals.

The air carbon-arc electrode is made of carbon and graphite, and it is coated with copper. Sizes range from 5/32 to 1 inch. A flat electrode, Fig. 8-24, allows greater flexibility. Table 8-6 lists types of current and polarities for different metals.

The compressed air passes through holes in the electrode holder which directs it parallel to the electrode, Fig. 8-25. The electrode holder contains an air control valve and a cable which carries both the current and air. The holder may be air or water cooled. The cable is connected to a welding machine and a source of compressed air.

Table 8-5

Ampere settings (a.c. or d.c.) and oxygen pressures.

Chrome nickel, Straight chrome, Monel® nickel			Brass, bronze, copper			Cast iron			Aluminum			Low alloy (alloy content less than 12%)			Carbon steel and low alloy high tensile		
Thickness	Amps	Oxygen pounds per square inch	Thickness	Amps	Oxygen pounds per square inch	Thickness	Amps	Oxygen pounds per square inch	Thickness	Amps	Oxygen pounds per square inch	Thickness	Amps	Oxygen pounds per square inch	Thickness	Amps	Oxygen pounds per square inch
1/4	175	3-5	1/4	180	10-15	1/4	180	10	1/4	200	30	1/4	175	30-35	1/4	175	75
1/2	185	5-10	1/2	185	10-15	1/2	185	10-15	1/2	200	30	1/2	180	35-40	1/2	175	75
3/4	195	10-15	3/4	190	15-20	3/4	190	15-20	3/4	200	30	3/4	190	40-45	3/4	175	75
1	200	15-20	1	200	20-25	1	200	20-25	1	200	30	1	200	45-50	1	175	75
1 1/4	210	20-25	1 1/4	210	25-30	1 1/4	210	25-30	1 1/4	200	35	1 1/4	205	50-55	1 1/4	200	75
1 1/2	215	25-30	1 1/2	215	30-35	1 1/2	215	30-35	1 1/2	200	35	1 1/2	210	55-60	1 1/2	200	75
1 3/4	220	35-40	1 3/4	220	35-40	1 3/4	220	35-40	1 3/4	200	40	1 3/4	215	60-65	1 3/4	200	75
2	220	40-45	2	225	40-45	2	225	40-45	2	200	40	2	220	65-70	2	200	75
2 1/4	225	45-50	2 1/4	225	45-50	2 1/4	225	45-50	2 1/4	175	45	2 1/4	225	70-75	2 1/4	225	75
2 1/2	225	50-55	2 1/2	230	50-55	2 1/2	230	50-55	2 1/2	175	45	2 1/2	225	75	2 1/2	225	75
2 3/4	230	55-60	2 3/4	230	55	2 3/4	230	55-60	2 3/4	175	45	2 3/4	230	75	2 3/4	225	75
3	230	60	3	235	55	3	235	60	3	175	45	3	230	75	3	225	75

Arcos Corp.

8-22. *Cutting a groove in the outside of a tank with the air carbon-arc process.* Arcos Corp.

8-23. *Nature of the cut that is possible with the air carbon-arc torch. Both plates were set up as a square butt joint, and a U-groove was cut into the joint. The groove is ready for welding.* The Arcair Co.

The regular welding machine may be used. Direct current, reverse polarity is used for most applications, but an electrode for alternating current is available. There is equipment for manual, semi-automatic, and automatic operation. Figure 8-26 shows a water-cooled, manual unit. Figure 8-27 illustrates automatic equipment in operation.

Ordinary compressed air is supplied by a compressor. Most applications require from 40 to 125 p.s.i., depending upon the thickness of the material. The air pressure must be of such volume that a clean, slag-free surface is assured.

Another form of air carbon-arc cutting involves the use of the standard shielded-arc electrode holder and copper-coated carbon electrodes. A low cost attachment, Fig. 8-28 and 8-29, converts a.c. and d.c. arc welders into an easy-to-use tool for cutting, piercing, gouging, and beveling.

The air attachment is quickly attached to any standard jaw-type metal electrode holder. An air-curtain action provided by specially designed orifices gives pinpoint control to blast away the molten metal. The air-curtain surrounds the electrode to keep it cool.

A regular welding machine with enough power to produce the current required can be used as the power source. The minimum electrode size is $^{5}/_{32}$ inch, and requires a minimum of 80 amperes of current. The maximum electrode size is $^{5}/_{16}$ inch and requires a maximum of 400 amperes of current. For maximum efficiency, air pressure must be at least 80 p.s.i. For field or emergency use, nitrogen or other inert gas can be substituted for air.

Bare Metal-Arc Cutting (MIG Cutting) (BMAC)

Bare metal-arc cutting was developed from gas metal-arc (MIG) welding. In the cutting process heat is obtained from an electric arc formed between the electrode, which is a continuous, flat metal wire, and the plate. The wire becomes white hot and vaporizes, establishing an arc between itself and the workpiece. *Inert gas* (gas which is chemically inactive) shields the

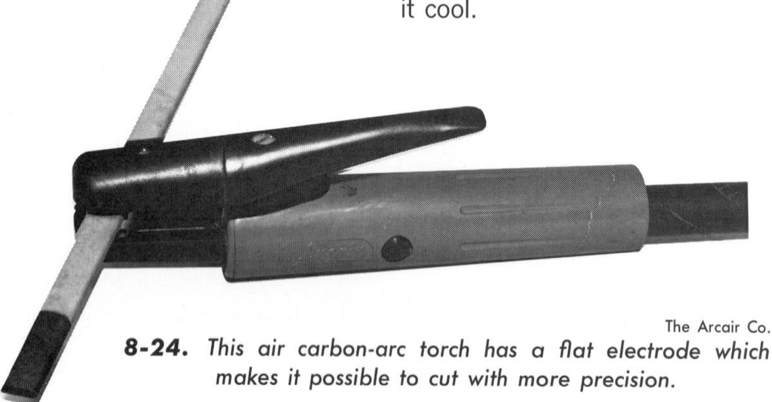

The Arcair Co.

8-24. *This air carbon-arc torch has a flat electrode which makes it possible to cut with more precision.*

8-25. *Air carbon-arc torch for manual operation (air cooled).*

The Arcair Co.

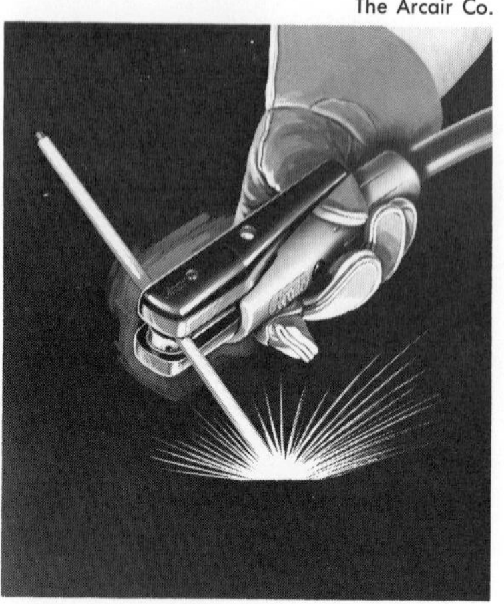

8-26. *Air carbon-arc torch for manual operation. The torch is water cooled. The water-cooling unit is also shown.*

The Arcair Co.

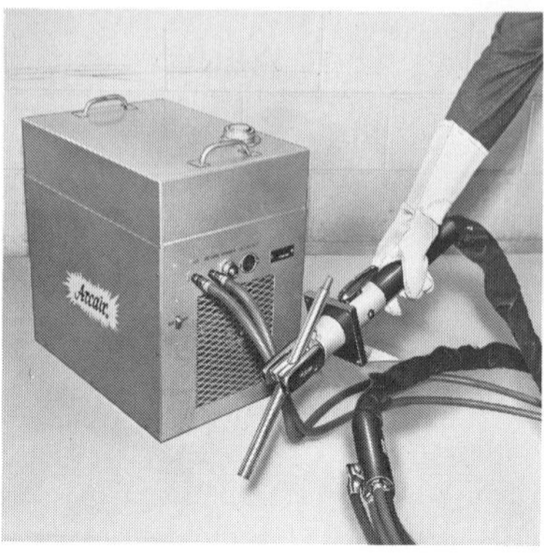

8-27. *Air carbon-arc torch and carriage for automatic machine operation. The current is d.c., reverse polarity, variable voltage. Electrode sizes are $^{5}/_{16}$ through $^{5}/_{8}$ inch.*

The Arcair Co.

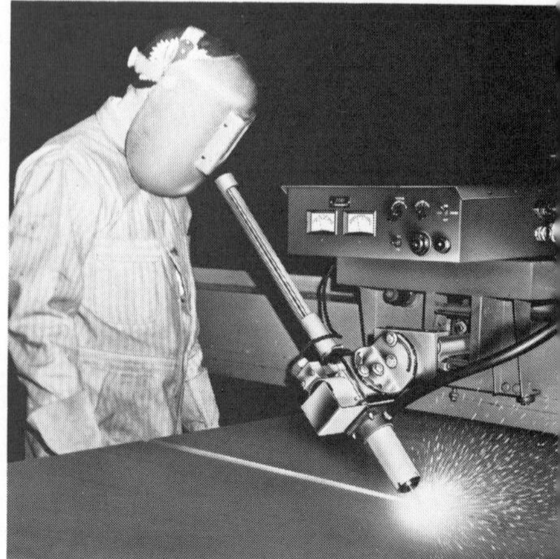

operation and the work from contamination.

In bare metal-arc cutting, the metal runs from the kerf as a result of the force produced by the pressure of the shielding gas, the jumping action of the arc, and the metal vapor from the electrode. The arc extends between the leading edge of the wire and the edge of the plate along the entire depth of the cut.

The nature of the cut is influenced by the voltage and wire-feed speed. The width of the kerf increases with the voltage. The cutting speed determines the amount of current necessary. As the cutting speed is increased, the current must be increased although the job may require that the voltage remain the same.

The conventional MIG welding equipment described in Chapter 10 can be used with some modifications for application to cutting. Alternating current instead of direct current is used for low speed cutting.

Because of its lower cost, steel wire is usually used. Stainless steel wire and aluminum wire are also available. A variety of gases, including air, may be used for shielding the arc, but argon, an inert gas, is the most common.

The process is not used extensively. Bare metal-arc cutting consumes a good deal of wire and cannot penetrate thick stock. Smooth surfaces require operation at speeds of 30 to 50 inches per minute. The maximum thickness that can be cut with standard wire feed is about $1\frac{1}{2}$-inch thick stainless steel and 3-inch thick aluminum.

Two recent developments in bare metal-arc cutting have increased the range to low travel speeds and have eliminated the high wire consumption. It has been found that satisfactory cuts at low speeds can be obtained with alternating current. This also cuts down the cost of the power. The other development, which reduces the rate of wire consumption, uses a band saw principle. A round or rectangular cutting electrode is rotated at high speed. The cutting arc is established between the workpiece and the high speed rotating band.

Gas Tungsten-Arc Cutting (TIG Cutting) (GTAC)

This cutting process was developed from the gas tungsten-arc (TIG) welding process. In both processes an electric arc is struck between a tungsten electrode, which does not melt, and the metal workpiece to provide intense heat. In TIG cutting, a

Table 8-6

Copper-clad electrode type and current for air carbon-arc cutting.

Material	Electrode	Current	Polarity
Steel	d.c.	d.c.	Reverse
	a.c.	a.c.	
Stainless	d.c.	d.c.	Reverse
	a.c.	a.c.	
Iron (cast, ductile & malleable)	d.c.	d.c. (high amp)	Reverse
	a.c.	a.c. or d.c.	Straight
Copper alloys	a.c.	a.c. or d.c.	Straight
	d.c.	d.c.	Reverse
Nickel alloys	a.c.	a.c. or d.c.	Straight

American Welding Society

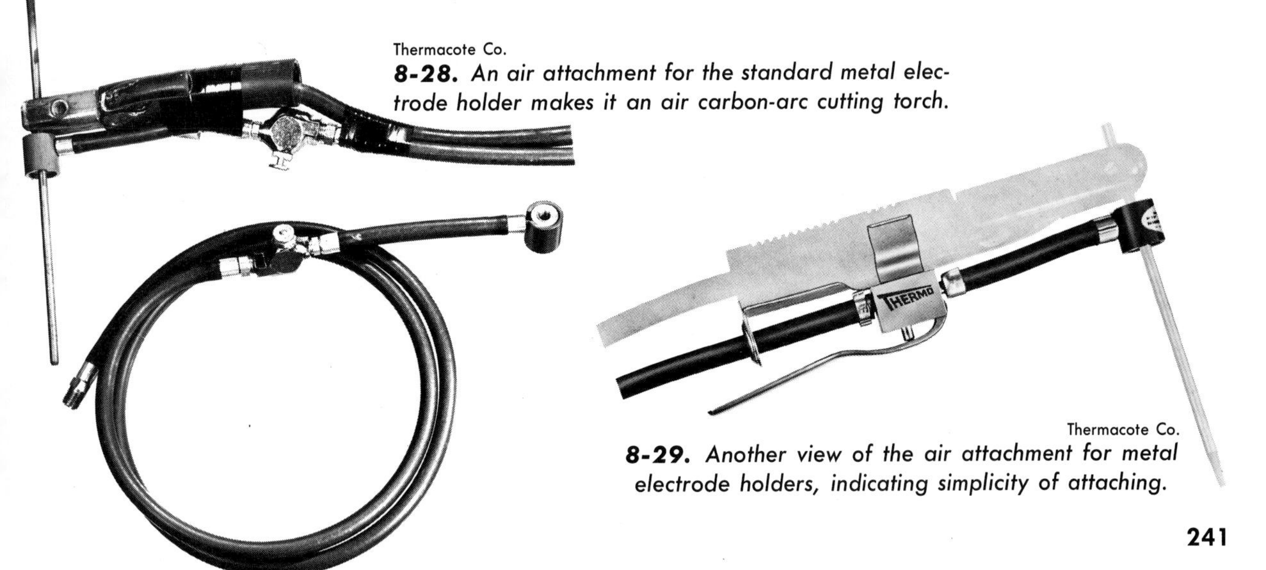

Thermacote Co.

***8-28.** An air attachment for the standard metal electrode holder makes it an air carbon-arc cutting torch.*

Thermacote Co.

***8-29.** Another view of the air attachment for metal electrode holders, indicating simplicity of attaching.*

high velocity jet of gas comes through the nozzle of the torch and ejects the melted metal, thus producing the cut.

The arc is started either by a high frequency spark or by making contact with the plate. As the torch is moved over the plate, the heat of the arc melts the plate. The molten metal is blown away by the stream of gas to form the kerf. Higher currents are used for gas tungsten-arc cutting than for gas tungsten-arc welding.

The quality of the cut is good. The process brings all the advantages of oxygen cutting to the cutting of nonferrous metals and stainless steel. Metals which can be cut are aluminum, magnesium, copper, silicon, bronze, nickel, copper-nickel and stainless steels. The gas tungsten-arc process can be used for manual or mechanized cutting. Materials as thick as ½-inch aluminum may be cut, Fig. 8-30.

The same general equipment is used for both TIG welding and cutting. A d.c. motor-generator or transformer-rectifier welding machine with straight polarity is recommended. Alternating current causes the loss of tungsten from the electrode at the high currents required for cutting. Alternating current with high frequency, however, may be used for cutting metal thicknesses up to ¼ inch.

Gas tungsten-arc welding torches can be used for cutting. Although the current should be held within the current rating of the torch, currents in excess of the rating may be used for a short period. A cap size of ⅜ inch is recommended. Caps that are too large reduce the force of the gas flow. Nitrogen can be used as the shielding gas. Cutting quality is increased by the use of an argon-hydrogen gas mixture containing 65 to 80 percent argon. Current requirements and gas mixtures for cutting metal plate are given in Table 8-7.

Plasma Arc Cutting (PAC)

Plasma arc cutting, Fig. 8-31, is similar in many respects to the gas tungsten-arc (TIG). This process was introduced by the Linde Division of Union Carbide in 1955. Both automated and manual equipment produce economical, high quality, ready-to-weld cuts.

An oxyacetylene cutting torch cannot cut aluminum, magnesium, or stainless steels because these metals form oxides when exposed to oxygen. The oxides resist further oxidation—the basis of oxyacetylene cutting. The plasma arc process, Fig. 8-32, can cut these metals because the arc stream is much hotter than the melting temperatures of both the metals and their oxides. It is also a high speed process, Fig. 8-33.

The only requirement for plasma cutting is that the metal being cut must be able to conduct electricity. By blowing out molten metal, the forceful plasma cutting jet forms the kerf.

Plasma is the fourth state of matter; the others are gas, liquid, and solid. Unlike gas, plasma is ionized so that it can conduct electric current. (An *ion* is an atom or group of atoms which has lost or gained electrons so that it can carry a positive or negative electrical charge.) Thus the purpose of the gas used in the plasma arc process is different from that used in the oxyacetylene process and the inert gas processes (TIG and MIG).

8-30. *Cutting aluminum plate by the gas tungsten-arc process. TIG cutting produces high quality cuts in nonferrous metals.*

Linde Division, Union Carbide Corp.

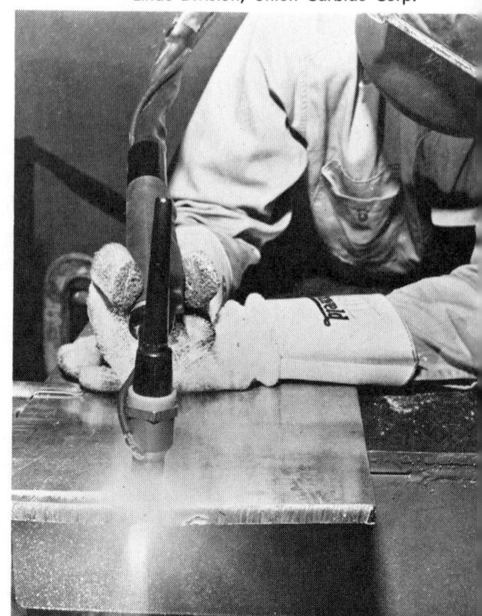

Table 8-7

Typical conditions for gas tungsten-arc cutting.

Material	Thickness, in.	Travel speed, i.p.m.	Current, d.c.s.p. amps	Type of gas
Stainless steel	⅛	20	350	80% argon + 20% hydrogen
Stainless steel	¼	20	500	65% argon + 35% hydrogen
Stainless steel	½	15	600	65% argon + 35% hydrogen
Aluminum	⅛	30	200	80% argon + 20% hydrogen
Aluminum	¼	20	300	65% argon + 35% hydrogen
Aluminum	½	20	450	65% argon + 35% hydrogen

American Welding Society

Instead of producing a flame for cutting the work or shielding the operation from the atmosphere, the gas is superheated so that it can actually maintain an electric arc.

In order for matter to pass from one state to another, *latent heat* is necessary. Just as water requires latent heat to become steam, so gas requires latent heat to become plasma. The plasma torch supplies this energy. When the plasma changes back into a gas, heat is released. The heat for cutting in the plasma arc process, therefore, comes from two sources: the electric arc maintained by the plasma and the heat discharged by the plasma as it changes to a gas.

Because maximum transfer of heat to the work is essential, plasma arc torches use a *transferred arc* for cutting in which both the material being cut and the torch act as electrodes in the electrical circuit, Fig. 8-34. The work is thus subjected to both plasma heat and arc heat.

In the plasma arc torch, the tip of the electrode is located within the nozzle. The nozzle has a relatively small opening which constricts the arc. The high pressure gas must flow through the electric arc where it is heated to the plasma temperature range of approximately 15,000 degrees F., a temperature far higher than that of any flame. Instead of expanding as a result of the heat, the gas is forced through the small opening in the nozzle where it emerges in the form of a sonic jet at high velocity. The hot jet of gas can melt any known metal. The high velocity gas blasts the molten metal through the kerf to produce a high quality cut, which is free of oxide slags. Argon, hydrogen, and occasionally helium are used in the plasma arc process.

Good ventilation should be provided when doing plasma arc cutting to remove all visible fumes from the area of the cutting action.

MECHANIZED PLASMA ARC CUTTING. The equipment which is needed for the mechanized plasma arc cutting system includes a torch, a cutting control box, gas regulators, power supply, a carriage unit, a supply of cutting gases, and water.

Torch. The typical torch, Fig. 8-35, can be used for cutting all metals. It is water cooled and may be equipped with a variety of nozzles to permit the use of different gases. The amount of current also affects nozzle selection. Single-port nozzles, Fig. 8-34, are generaly used with argon-hydrogen and nitrogen-hydrogen mixtures, but multiport and dual flow nozzles, Fig. 8-36, produce better results when compressed air or oxygen is the cutting gas.

Controls. The control unit, Fig. 8-37, provides the sequence of operations and control of all functions such as arc starting, gas flow, carriage travel and flow of water.

Electrodes. Standard thoriated tungsten electrodes, held in place by a collet, are used in the cutting torch. The knob on the back of the torch permits rapid, positive electrode centering.

Power supply. Direct current, straight polarity is used. A power source with an open circuit volt-

8-31. *Heavy cut in carbon steel with the plasma arc cutting process.*
Linde Division, Union Carbide Corp.

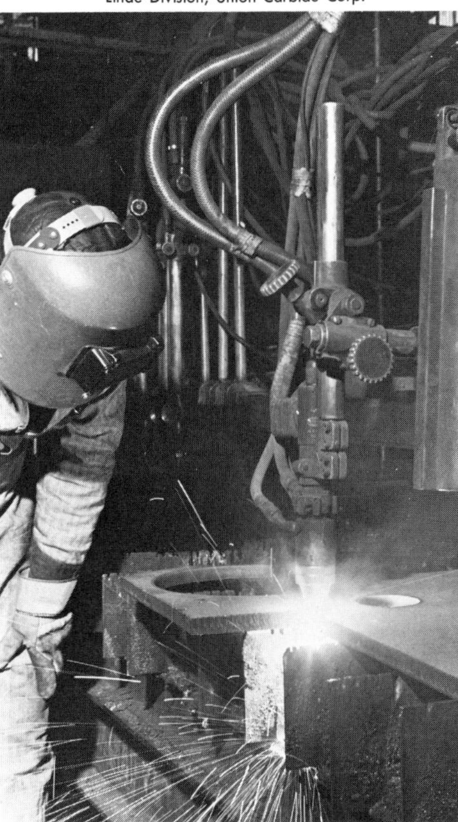

8-32. *Plasma arc cutting makes shape and line cuts on hard-to-work metals such as this 1¾-inch stainless steel plate.*
Linde Division, Union Carbide Corp.

age of at least 100 volts is required, Fig. 8-38. For heavy cutting a machine capable of producing 400 open-circuit volts and 200 kilowatts or more may be necessary. Both motor-generator and transformer-rectifier power units are available. These units may also be connected in series to meet higher voltage requirements.

Regulators. Gas pressure regulators and flowmeters are required for controlling the flow of the cutting gas. See Chapter 5, pages 146–149. All gas goes through the main port of the cutting torch at 60 to 350 cubic feet per hour. A shielding gas is not needed in addition to the gas used for the plasma.

Carriage Unit. The carriage unit is similar to that used for other machine cutting processes. The unit runs on a track. The torch may also be mounted on a standard multitorch shape-cutting machine, or it may be guided by any regular tracing equipment. The minimum contour radius that can be cut is determined by the limitations of the shape-cutting equipment.

Cutting Gases. Aluminum, stainless steels, and other nonferrous metals require a nonoxidizing gas for cutting, such as a mixture of argon-hydrogen or nitrogen-hydrogen, Fig. 8-39A. Carbon steel, cast iron, and certain alloy steels require an oxidizing gas which provides additional heat from the iron-oxygen reaction at the cutting point. Separately supplied nitrogen, oxygen, or compressed air may be used for these metals. The life of electrodes operating in oxygen is short. Oxygen is usually supplied as an outer sheath surrounding a nitrogen plasma, Fig. 8-39B. Typical cutting conditions are given in Table 8-8.

Water Supply. Water at high pressure and a high rate of flow is necessary to dissipate the heat generated in the torch. Usually it is necessary to install a circulatory system in order to provide an adequate flow.

MANUAL PLASMA ARC CUTTING. Manual plasma arc cutting, Fig. 8-40, is used for workpieces which cannot be adapted to a mechanized setup, for remote locations, and for specialized work on odd-shaped pieces. Typical manual cutting capabilities include 2-inch thick stainless steel and 3-inch thick aluminum. The equipment needed includes a manual torch, supply of electrodes, power source, and a control unit. The electrodes and power source for the mechanized process may also be used for manual cutting.

Torch. A water-cooled torch with a 400-ampere capacity is a popular model. It has a right-angle head. The pilot arc is established by pressing the switch on the torch handle. The cutting arc is established when the torch is brought to within ½ inch of the workpiece. The arc is immediately extinguished when the operator releases the torch switch or when the torch is raised from the workpiece.

Control Unit. The control unit for manual cutting regulates the electrical, gas, and water supplies. Separate flowmeters are provided for argon and hydro-

***8-33.** Cutting access holes in a Polaris launch tube jacket with the plasma arc process. The cutting speed is 108 inches per minute. The diameters of the holes vary from 3 to 26 inches. The jackets are made of ⅛-inch thick, high-strength, low-alloy steel.*

Linde Division, Union Carbide Corp.

***8-34.** Schematic diagram of plasma arc cutting equipment.*

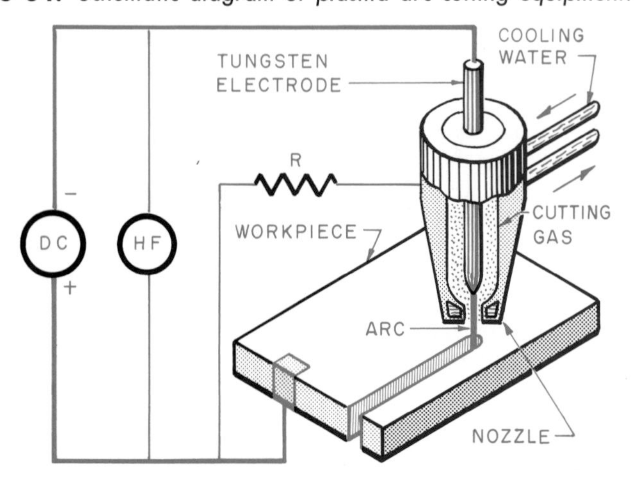

gen. Solenoid valves provide for flow of pure argon to the torch to establish the pilot arc at the start of the cutting cycle. The manual torch and control unit are shown in Fig. 8-41.

Cutting Gases. The manual torch is designed to use nonoxidizing cutting gases and, therefore, it is not recommended for carbon steel cutting in which an oxidizing gas is needed for optimum cut quality. A mixture of 80 percent argon and 20 percent hydrogen is recommended. The two gases are fed separately to the control unit. Nitrogen is not recommended for manual cutting.

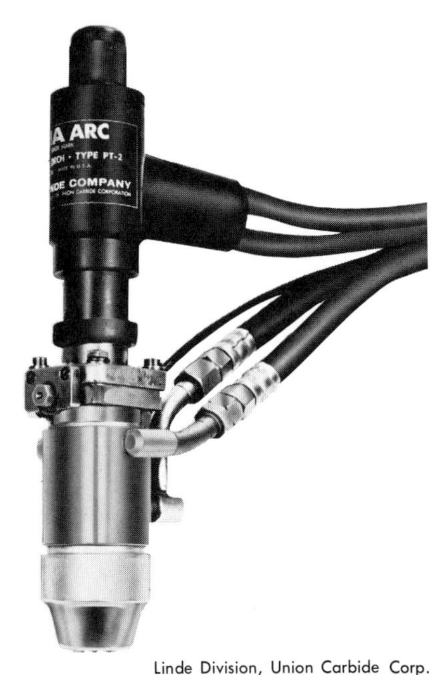

Linde Division, Union Carbide Corp.

8-35. Plasma arc cutting head used with machine operation.

ADVANTAGE OF PLASMA CUTTING. Both the mechanized and manual plasma cutting processes produce economical, high speed, ready-to-weld cuts. They are intended to replace less efficient, slower methods such as sawing, powder-cutting, and oxyacetylene cutting on some applications. The plasma arc process have the following advantages:

- Slag-free cuts on carbon and stainless steels, nickel, Monel®, Inconel®, cast iron, clad steels, aluminum, copper, and magnesium
- Clean cuts on most metals up to 5 inches thick
- Precision cuts with a narrow kerf
- Minimum heat-affected zone
- Cutting speeds up to 300 inches per minute (many times faster than oxyacetylene cutting)
- Cuts of such quality that machining or finishing is not needed in many cases
- There is almost no distortion of metals when using the plasma arc cutting processes. There is no bowing or cambering. Except for a microscopically thin layer at the cut surface, magnetic permeability and hardness are not affected.

Plasma arc cutting is finding increased use as a fabricating tool in the transportation, nuclear power, and chemical industries. It is also used in the forging and casting industry for removing risers and gates. The stack cutting of several sheets of $\frac{1}{16}$- to $\frac{1}{4}$-inch thickness is possible. In addition to cutting, the process can be used for the piercing of holes and gouging, including pad washing and scarfing.

Cutting Carbon Steel. The plasma arc processes produce slag-free cuts in carbon steel with smooth surfaces and sharp edges. No preheating is required. Stack cutting of sheets produces cuts comparable to those obtained when cutting one sheet of equal thickness. The edges are not fused when cutting carbon steel. Net cost per foot compares favorably with other processes in thicknesses up to 2 inches.

8-36. (A) Multiport plasma arc cutting nozzle. (B) Dual flow plasma arc cutting nozzle.

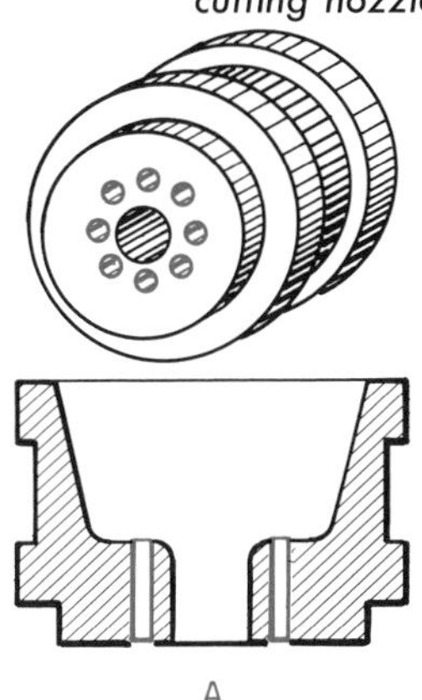

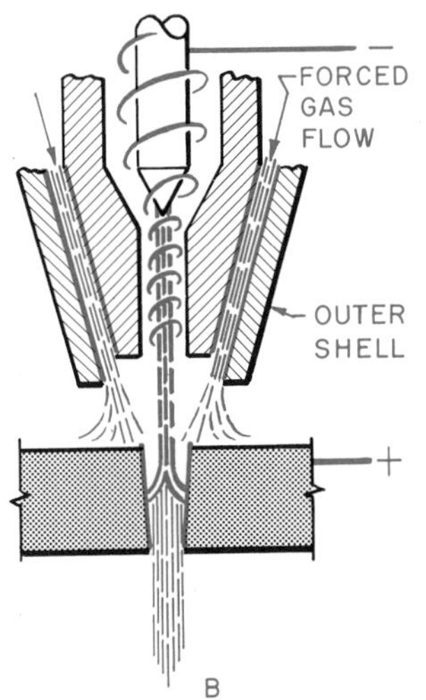

8-37. Control unit for the mechanized plasma arc cutting head.

Linde Division, Union Carbide Corp.

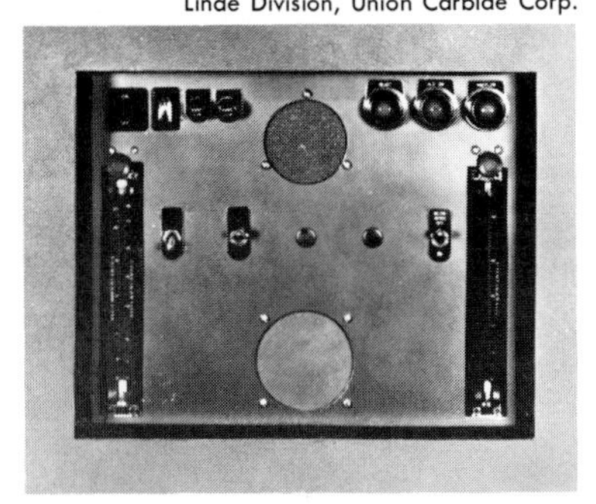

While it is possible to obtain fairly good results if inert gases are used to cut carbon steel, superior results are obtained when nitrogen and oxygen are used. The cutting speed is higher than for oxyacetylene cuts.

Cutting Stainless Steel. Completely slag-free plasma arc cuts in stainless steel up to 2 inches thick have eliminated the need for further finishing. X-ray quality welds can be produced without further cleaning of cut surfaces. Cut quality of high-strength alloys, including those with a high nickel or cobalt content, is similar to those of stainless steel.

Stainless steel and nonferrous metals are generally cut with mixtures of argon and hydrogen or with nitrogen mixtures.

Cutting Aluminum. Plasma arc cutting methods provides equal or better quality at much faster speeds than other cutting methods. Prior to its develop-

Table 8-8

Typical conditions for plasma arc cutting.

	Thickness, in.	Speed, i.m.p.	Orifice diameter, in.	Power, kw.	Gas flow, c.f.h.
ALUMINUM	1/4	300	1/8	60	80 argon–40 hydrogen or 130 nitrogen–30 hydrogen
	1/2	200	1/8	50	65 argon–35 nitrogen or 140 nitrogen–60 hydrogen
	1	90	5/32	80	65 argon–35 hydrogen or 140 nitrogen–60 hydrogen
	2	20	5/32	80	65 argon–35 hydrogen or 140 nitrogen–60 hydrogen
	3	15	3/16	90	130 argon–70 hydrogen or 140 nitrogen–60 hydrogen
	4	12	3/16	90	130 argon–70 hydrogen or 140 nitrogen–60 hydrogen

Linde Division, Union Carbide Corp.

8-38. *Transformer-rectifier power supply (400 volts, open circuit).*

Linde Division, Union Carbide Corp.

8-39. *(A) Plasma-arc inert-gas cutting. (B) Plasma-arc nitrogen-oxygen cutting.*

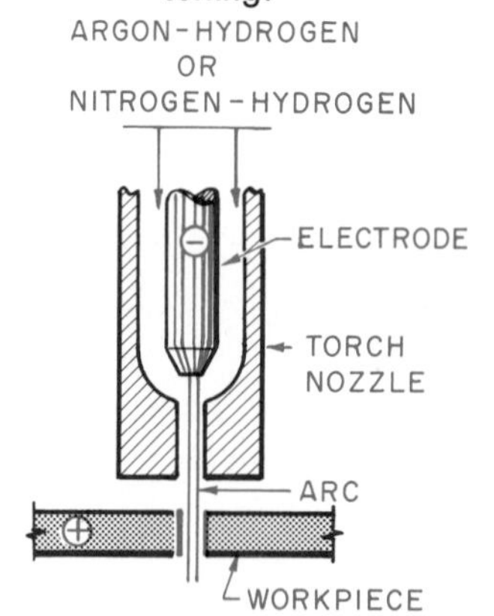

A PLASMA ARC INERT GAS CUTTING.

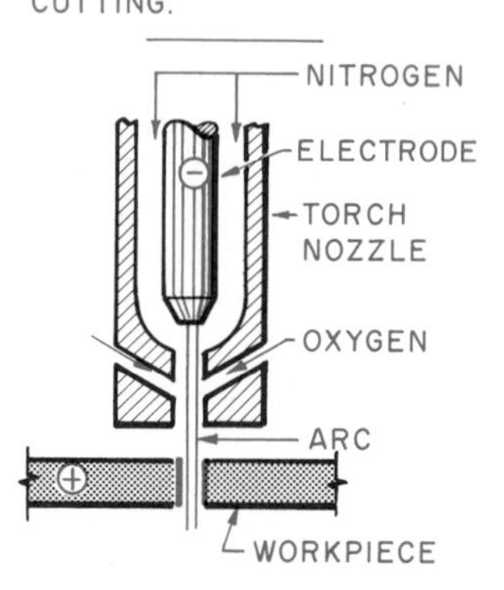

B PLASMA ARC NITROGEN-OXYGEN CUTTING.

ment other flame cutting attempts were unsatisfactory on a production basis. Plasma arc cuts in aluminum are slag-free for thicknesses up to 5 inches. Excellent cuts are also obtained on magnesium at higher cutting speeds than aluminum.

Metal Powder Cutting (POC)

Powder cutting is used as a means of cutting stainless steel, alloy steel, cast iron, nickel, aluminum, and copper. These materials are difficult to cut with oxygen because of their resistance to oxidation. Their resistance is overcome by feeding iron powder into the standard oxyacetylene cutting torch flame.

Powder cutting adapters can be attached to a conventional cutting torch, and special powder cutting torches are available. A powder dispenser is necessary as a source for the powder. The powder is fed to the torch under constant pressure. The pressure is supplied through the use of air or nitrogen. The heat of the oxyacetylene flame is increased as a result of the combustion of the iron powder.

Other powders such as aluminum may also be mixed with the iron powder. In cutting stainless steels, the powder prevents insulating chromium oxides from forming. Powder cutting of stainless steels is as fast as oxygen cutting of carbon steels of the same thickness. An unusual application of this process is in the cutting of concrete. Thicknesses up to 18 inches thick can be cut with the torch, and thicknesses up to 12 feet thick can be cut with powder cutting.

The Burning Bar

One of the more recent developments for cutting, burning, and melting, is a device called

Linde Division, Union Carbide Corp.

8-40. *Cutting a carbon steel tank head manually with the plasma arc process. The process has the same flexibility as oxyacetylene cutting.*

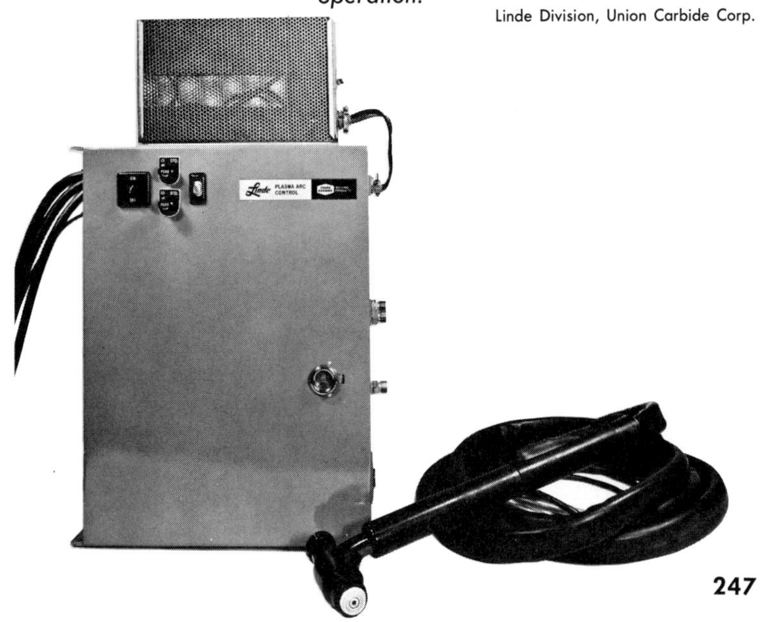

8-41. *Water-cooled plasma arc cutting torch and control unit for manual operation.*

Linde Division, Union Carbide Corp.

Modern Engineering Co.

8-42. *Cutting a steel column with the burning bar.*

the thermal burning bar, Fig. 8-42. The investment in equipment is small, and it opens new fields for burning, drilling, cutting, and other applications.

The principle of the burning bar is that very high temperatures, when concentrated on one spot, can melt or burn through almost any material. The burning bar makes it possible to perform many jobs which, until its invention, had been impossible to accomplish.

The burning bar is a combination of 11 special alloys in an iron bar, which reaches a temperature of more than 4,500 degrees F. when combined with oxygen. The burning bar is approximately 10 feet long with a diameter of $\frac{3}{8}$ inch. The bar is placed in a holder which is connected to a cylinder of oxygen equipped with a two-stage regulator set at approximately 85 p.s.i. The end of the burning bar is heated by an oxyacetylene flame until the metal begins to burn. Each bar burns down in approximately five minutes. The only equipment needed for this process is oxygen, a two-stage regulator, an oxygen welding hose and fittings, the bar holder, and the burning bars. The operator must wear a nonflammable suit and a face protector.

Cutting materials is a simple procedure since the burning-through process is concentrated on one spot. For example, only one small area is affected when burning a hole in reinforced concrete even though it is well seasoned. Some smoke is developed during the process. The presence of moisture speeds up the operation.

OPERATION. The procedure for cutting with the burning bar is as follows:

1. Connect the burning bar holder to the oxygen hose.

2. Insert a bar into the holder and tighten the lock by hand.

3. Set the oxygen regulator at about 85 pounds pressure.

4. Open the valve, causing the oxygen to flow, and bring the bar into contact with the oxy-acetylene torch flame. The bar will ignite and reach a temperature of 4,500 degrees F.

5. Contact the material to be cut with the bar. The cut proceeds at a very rapid rate.

ADVANTAGES. The burning bar is an important tool in the building trades because it can cut an object without vibration. Vibration weakens structural strength. Another important advantage is that the burning bar can be used in buildings in which the use of pneumatic tools, drills, and saws would be less than satisfactory because of dust, noise, or use of water. Thus it is especially useful in the repairing and remodeling of hospitals, banks, and any other building in which the regular routine must continue normally. It is also widely used in the demolition of old buildings, Fig. 8-43.

The burning bar is also advantageous for concrete construction where many holes or trenches may be required for piping, plumbing, wiring, heating, and air conditioning. It may be used for the boring of deep holes, the cutting of doorways, and almost any other kind of work in concrete.

Underwater demolition and construction have profited greatly by the invention of the burning bar. There is a great saving in time, and the mobility of the equipment lends itself to wide flexibility.

In rescue work done by the police and fire departments, the mobility and time-saving value of the burning bar make it invaluable.

Modern Engineering Co.

8-43. *Cutting concrete in the demolition of an old building.*

The burning bar also has some industrial applications. Although it is not a precision cutting tool, it is useful in the rough cutting of a wide variety of materials such as reinforced concrete, steel, mineral slag, cast iron, marble, and basalt. In the steel industry it is used for cutting openings in blast furnaces and burning off slag and other waste products.

REVIEW QUESTIONS

1. Explain the chemical reaction that takes place in the cutting of steel with the oxyacetylene cutting torch.
2. What type of metals can be cut with the oxyacetylene cutting torch?
3. Make a list of the equipment necessary for oxyacetylene cutting.
4. How does the oxyacetylene cutting torch work?
5. Explain the design of the oxyacetylene cutting tip and the purpose of the multiholes.
6. Describe the equipment necessary for oxyacetylene machine cutting.
7. Name other fuel gases that can be used with oxygen for cutting purposes.
8. It is impossible to cut more than one layer of steel at a time. True. False.
9. Name five cutting processes in addition to the oxyacetylene method. Briefly describe each.
10. Name two processes that may be used for the cutting of nonferrous metals. Briefly describe each.
11. Name and describe the cutting process used for the flame cutting of concrete.
12. Air carbon-arc cutting can only be used with ferrous metals. True. False.

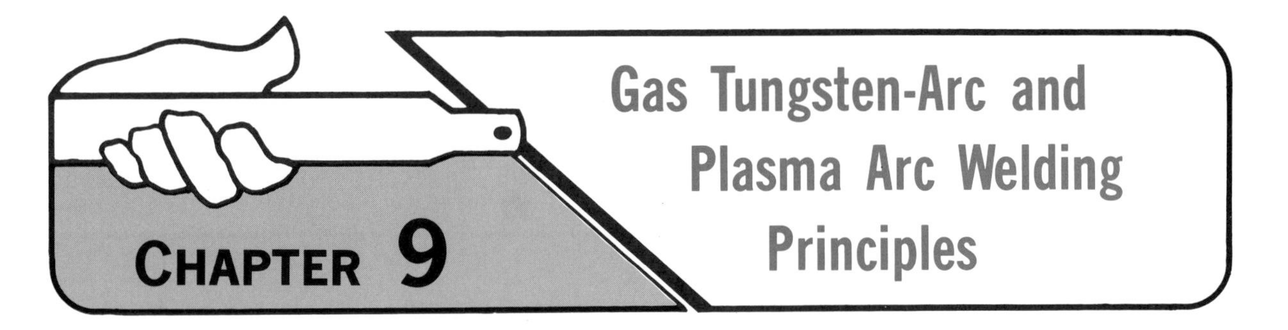

Gas Tungsten-Arc and Plasma Arc Welding Principles

GAS-SHIELDED ARC WELDING PROCESSES

In the gas-shielded arc welding processes the weld is produced by the arc flame maintained between the end of a metal electrode and the part to be welded. The electrode may be consumable or nonconsumable. A shield of gas ejected from the torch surrounds the arc and weld region. In the manual process the shielding gas may or may not be *inert* (chemically inactive), and pressure is not used. Filler metal may or may not be added.

History

You have learned that a vital characteristic of sound welding depends upon the shielding of the electric arc and the weld pool from the contamination of the surrounding air. The first experiments with gas shielding were carried on in the early 1920s. The development of coated electrodes in the 1930s eliminated interest in the process at that time. Welds produced by coated electrodes are actually gas-shielded welds since the electric arc produces a gas as it burns off the coating.

Covered electrode welding served industry well through World War II. But, when the war ended and industry began retooling for consumer products, competition spurred intensive research and experimentation to develop new welding methods. This effort led to the development of the inert-gas-shielded arc welding processes.

In 1930 Henry M. Hobart and Phillip K. Devers of the General Electric Company were granted patents that covered the basic principle of gas-shielded arc welding. In 1944 Russell Meredith of the Northrop Aircraft Company was issued a patent on the welding of magnesium and magnesium alloys by an electric arc in helium- and argon-shielded atmospheres with a tungsten electrode. This is the gas tungsten-arc (GTAW, or TIG) process. The process was originally developed for welding corrosion-resistant and other difficult-to-weld metals such as magnesium, stainless steel, and aluminum. Today it is a standard process, widely used in both factory and field to weld practically all commercial metals, Figs. 9-1 and 9-2.

Continued experimentation to make the process faster led to the development of the gas metal-arc (GMAW, or MIG) welding process in 1948. It was first applied to the welding of aluminum and was found to be several times faster than the TIG method. By 1951 it was learned that by adding a small amount of oxygen to the argon, the arc action was much improved and could be applied to the welding of carbon and stainless steels. Carbon dioxide was also found to be effective as a shielding gas. Recently, fluxes have been added to the core wire or applied to the outside of the wire.

Overview of the Processes

There are two gas-shielded arc welding processes in which the welding student should develop skill. They have several of the characteristics of both oxyacetylene and shielded metal-arc welding.

- The *nonconsumable* electrode process, Fig. 9-3, is known as inert-gas-shielded, tungsten-arc welding. It is more often referred to as gas tungsten-arc welding (GTAW) or by the shop term TIG.
- The *consumable* electrode process, shown in Fig. 9-4, is known as inert-gas-shielded, metal-arc welding. It is more often referred to as gas metal-arc welding (GMAW) or by the shop term MIG.

There is a third process known as *inert-gas-shielded, carbon-arc welding* (GCAW), often referred to as carbon inert-gas welding or its abbreviation CIG. A fourth process, *gas-shielded-arc spot-welding,* is a semi-automatic process that does not require a lot of practice.

GAS TUNGSTEN-ARC. In the gas tungsten-arc process, the heat necessary to melt the metal is provided by an intense electric arc which is struck between a nonconsumable tungsten electrode and the metal workpiece, Fig. 9-5 (Page 254). The electrode does not melt and become a part of the weld.

An edge or a corner joint may be fused together without the addition of filler metal, using a technique similar to that used with the oxyacetylene flame. On joints where filler metal is required, a welding rod is fed into the weld zone and melted with the base metal in the same manner as that used with oxyacetylene welding. The weld pool is protected by shielding it with an inert gas such as helium or argon, or a mixture of the two. The gas is fed through the welding torch.

Gas tungsten-arc welding will be our major concern in this chapter. It is known by such trade names as *Heliarc®* and *Heliwelding.®* This process can be used to weld such difficult-to-weld metals as aluminum, magnesium, and titanium, and it can weld dissimilar metals. Because there is no flux, there is no contaminating residue and no cleaning problems. This in itself results in considerable cost saving, especially when multipass welding is necessary. Since the shielding gas is transparent, the welder can observe the weld pool clearly as it is formed and carried along during welding. The almost total absence of smoke, fumes, and sparks contributes to neater and sounder welds and to the comfort of the welder.

Welding can be performed in all positions. Because heat concentration and amperage can be more closely controlled in TIG welding than in stick electrode welding, there is less distortion of the base metal near the weld, less weld cracking, and fewer locked-up stresses.

A TIG weld is sound, smooth, strong, ductile, uniform and bright. These characteristics made it ideal for the food, chemical, hospital equipment, and aerospace industries.

GAS METAL-ARC PROCESS. Gas metal-arc welding, also known by such trade names as *Aircomatic,®* *Sigma,®* *Millermatic®* and *Micro Wire®* welding, is a consumable electrode process. It uses direct current and a shield of argon or helium, a mixture of both gases, or a mixture with other gases. Carbon dioxide is used extensively.

A small diameter wire serves both as electrode and filler metal. It is fed into the welding gun automatically, and then into the weld pool at high speed. MIG

9-1. *Welding an aluminum cylinder with the gas tungsten-arc process. Note the inert gas cylinders behind the welding machine.*

Miller Electric Manufacturing Co.

9-2. *A welder putting the finishing touches on a chemical reactor. He is welding a lining of 1/16 inch thick pure silver with the gas tungsten-arc process.*

Nooter Corp.

welding will be studied in the next chapter. Compare Figs. 9-3 and 9-4 for a basic understanding of TIG and MIG Welding.

THE CIG WELDING PROCESS. The gas carbon-arc (CIG) process employs a carbon electrode, and the arc is surrounded by an inert gas such as helium or argon. It is used for welding thin gauge sheet metal and for welding those types of joints in which it is necessary to fuse only the edges of the base metal. Although CIG welding is not used very widely, it is excellent for student practice since the arc is easy to start and the action of the flame on the weld pool may be studied.

THE GAS-SHIELDED ARC SPOT-WELDING PROCESS. The gas-shielded arc spot-welding process is produced with an arc spot pistol-grip gun, Fig. 9-6. This gun may be provided with a tungsten electrode or a consumable electrode. The gun makes a weld in the manner shown in Fig. 9-7 (Page 256). An arc is drawn to the surface of the metal under the protection of the shielding gas. The time of the current flow is controlled so that the arc melts through the surface plate to the plate beneath it and forms a plug weld.

Gas-shielded arc spot-welding is used mainly in the welding of light gauge sheet metal up to $1/8$ inch thick. It can be used to weld a wide range of metals such as carbon, alloy, and stainless steels; and most nonferrous metals. It is used in assembling auto bodies, refrigerators, and other home appliances.

Gas-shielded arc spot-welding has the following advantages over resistance spot-welding.

- It is necessary to have access only to the front side of the joint.
- The equipment is portable and commonly available since it is also used for MIG and TIG welding.
- There is a minimum of spatter, smoke, and sparks.
- There is little distortion of the workpiece.

Comparison of the Gas-Shielded and Shielded Metal-Arc Welding Processes

In any type of welding, the best weld is that which has about the same chemical, metallurgical, and physical properties as the base metal. To obtain such conditions, the molten weld puddle must be protected from the atmosphere during the welding operation. Otherwise, atmospheric oxygen and nitrogen combine with the molten weld metal and cause a weak, porous weld.

In gas-shielded arc welding, the weld zone is protected from

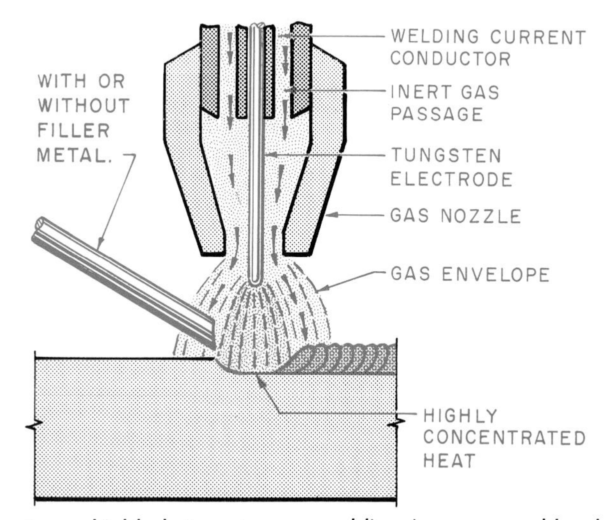

9-3. *Inert-gas-shielded, tungsten-arc welding (nonconsumable electrode).*

9-4. *Inert-gas-shielded, metal-arc welding (consumable electrode).*

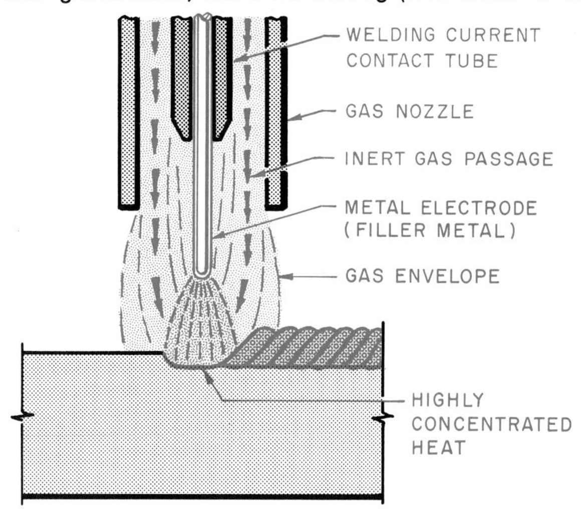

the atmosphere by an inert gas which is fed through the welding torch. Because of this complete protection from the atmosphere, welds made by a gas-shielded arc process are stronger, more ductile, and more resistant to corrosion than those made by the shielded metal-arc (stick electrode) process. The gas-shielded arc processes have the following additional advantages:

- The welding heat is confined to a small area. Thus there is a narrow heat-affected zone, and faster welding speeds are possible. Distortion of the welded joint is reduced.
- Welding takes place without spatter, sparks, and fumes. Therefore, weld finishing is kept to a minimum. Welds usually require no finishing at all so that production costs are kept low.
- There is no need for flux. The absence of slag not only reduces the amount of weld finishing, but also produces smoother, nonporous welds. The welder can see (and thus control) the weld pool better.
- The addition of filler metal is by hand and is independent of maintaining the arc. Thus it may be added only when necessary. Filler metal can often be eliminated when welding thin and medium stock in which good fitup is secured. If filler metal is not added, the operator can see the weld pool clearly, thereby improving the appearance of the weld.
- Fusion welds can be made in nearly all common metals. Metals which can be welded by a gas-shielded arc process include plain carbon and low alloy steels, cast iron, aluminum and aluminum alloys, stainless steels, brass, bronze, and even silver. Combinations of dissimilar metals can also be welded. Hardfacing and surfacing materials can be applied to steel.

GAS TUNGSTEN-ARC WELDING

The material outlined in this chapter concerns gas tungsten-arc (TIG) welding. Chapter 10 provides information concerning the consumable electrode process, gas metal-arc (MIG) welding. If you have mastered oxyacetylene and shielded metal-arc

9-5. Essentials of the gas tungsten-arc welding process.

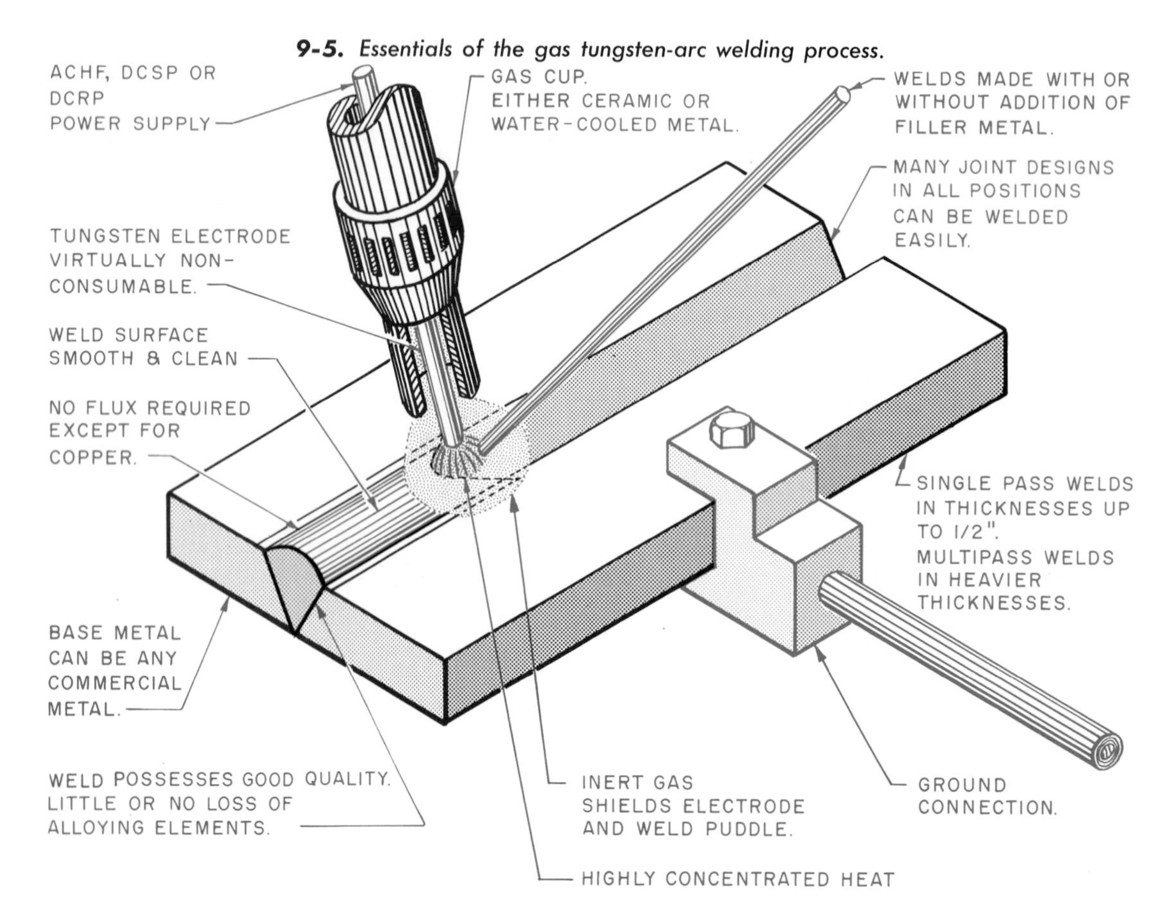
ACHF, DCSP OR DCRP POWER SUPPLY
GAS CUP. EITHER CERAMIC OR WATER-COOLED METAL.
WELDS MADE WITH OR WITHOUT ADDITION OF FILLER METAL.
MANY JOINT DESIGNS IN ALL POSITIONS CAN BE WELDED EASILY.
TUNGSTEN ELECTRODE VIRTUALLY NON-CONSUMABLE.
WELD SURFACE SMOOTH & CLEAN
NO FLUX REQUIRED EXCEPT FOR COPPER.
SINGLE PASS WELDS IN THICKNESSES UP TO 1/2". MULTIPASS WELDS IN HEAVIER THICKNESSES.
BASE METAL CAN BE ANY COMMERCIAL METAL.
WELD POSSESSES GOOD QUALITY. LITTLE OR NO LOSS OF ALLOYING ELEMENTS.
INERT GAS SHIELDS ELECTRODE AND WELD PUDDLE.
GROUND CONNECTION.
HIGHLY CONCENTRATED HEAT

welding, you will have little difficulty learning to weld with both of the common gas-shielded arc processes.

Figure 9-8 illustrates the essential equipment needed for manual TIG welding:

- TIG torch (A)
- Supply of inert gas (B)
- Gas regulator and flowmeter (C)
- Welding generator or transformer (D)
- Water supply (E) and return

A supply of tungsten electrodes, hose for the gas and water, electrical cable, and a fuse assembly or shutoff valve to protect the torch from overheating are also necessary.

Shielding Gases

From your previous studies, you are familiar with the function of the shielding gas. It permits welding to take place in a controlled atmosphere. In shielded metal-arc welding, this is accomplished by placing a coating on the electrode which produces a protective atmosphere as it burns in the welding arc. TIG and MIG welding are similar in that they surround the arc with gases. In the gas-shielded arc welding processes, however, the gas is supplied by the torch instead of the electrode. The air in the arc area is displaced by the gas before the arc is struck.

WELD CONTAMINATION. By volume, air is made up of 21 percent oxygen, 78 percent nitrogen, 0.94 percent argon, and 0.06 percent other gases (primarily carbon dioxide). The atmosphere also contains a certain amount of water vapor, depending upon its humidity at any given time. There is hydrogen in the water vapor. Of all of the elements that are in the air, the three which cause the most contamination of weld deposits are oxygen, nitrogen, and hydrogen.

Oxygen. Oxygen combines readily with other elements in the weld pool to form unwanted oxides and gases. Such deoxidizers as manganese and silicon combine with the oxygen to form a light slag which floats to the top of the weld pool. In the absence of a deoxidizer, the oxygen combines with iron to form compounds which are trapped in the weld metal and reduce its mechanical properties. Free oxygen also can combine with the carbon in the metal to form carbon monoxide. If carbon monoxide is trapped in the weld metal as it cools, it collects in pockets which causes porosity in the weld.

Nitrogen. Nitrogen is one of the most serious problems in the welding of steel materials because there is so much of it in the air. Nitrogen forms compounds called *nitrides* during the welding of steel which cause hardness, a decrease in ductility, and lower impact resistance. These conditions often lead to cracking in and next to the weld metal. In large amounts nitrogen also causes serious porosity in the weld metal.

Hydrogen. The presence of hydrogen when welding carbon steels produces an erratic arc and affects the soundness of the weld metal. As the weld pool solidifies, the hydrogen is rejected. Part of it, however, is trapped

9-6. *Gun for gas-shielded arc spot-welding.* Hobart Brothers, Co.

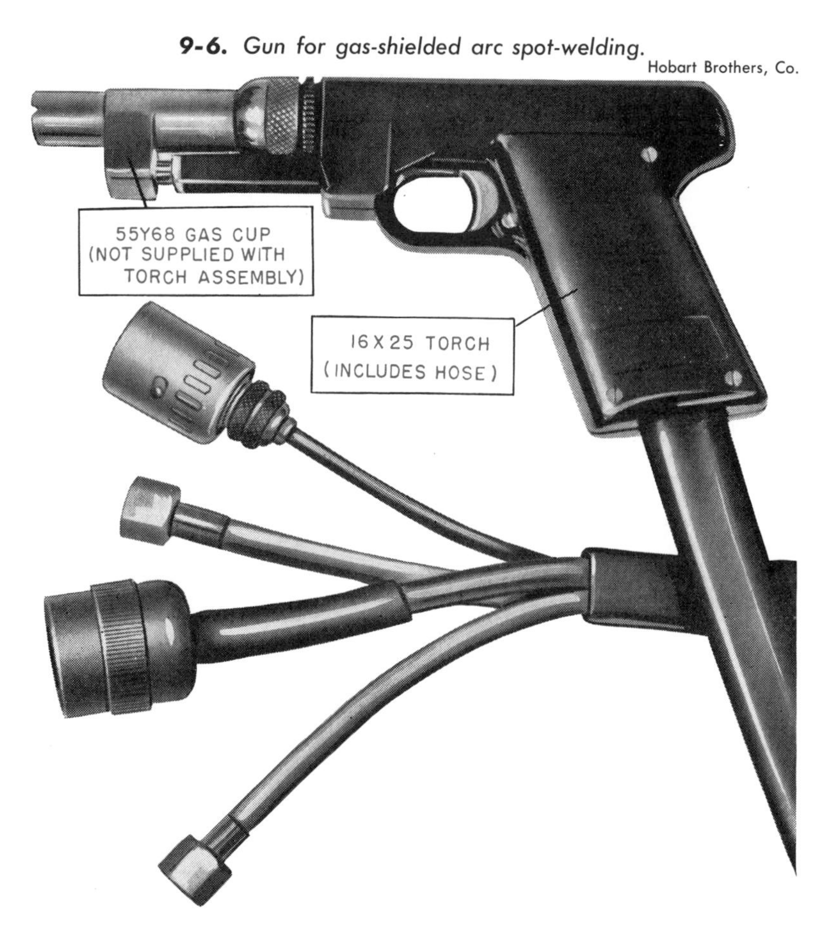

and collects at certain points where it causes stresses. These stresses lead to small cracks in the weld metal that may later enlarge. Other defects such as *fish eyes* and *underbead cracking* may also result from hydrogen contamination.

INERT GASES AND GAS MIXTURES. Generally, argon or helium serves well for TIG welding. Mixtures containing oxygen may be used on some MIG jobs. Oxygen is not used for TIG welding because tungsten electrodes have a low tolerance for oxygen.

Argon and helium are both chemically inert, that is, they do not form compounds with other materials. Thus they do not affect tungsten electrodes or the work material.

The welding arc in an atmosphere of either argon or helium is remarkably smooth and quiet. Many other gases and gas mixtures have been experimented with, but they all have some deficiencies such as causing rapid erosion of the tungsten electrode, porosity in the weld metal, arc instability, or high cost.

Some noninert gases react chemically with some metals, but not with others. For example, nitrogen can form an effective shield for welding copper. However, it cannot be used for welding mild steel because at arc temperatures nitrogen causes

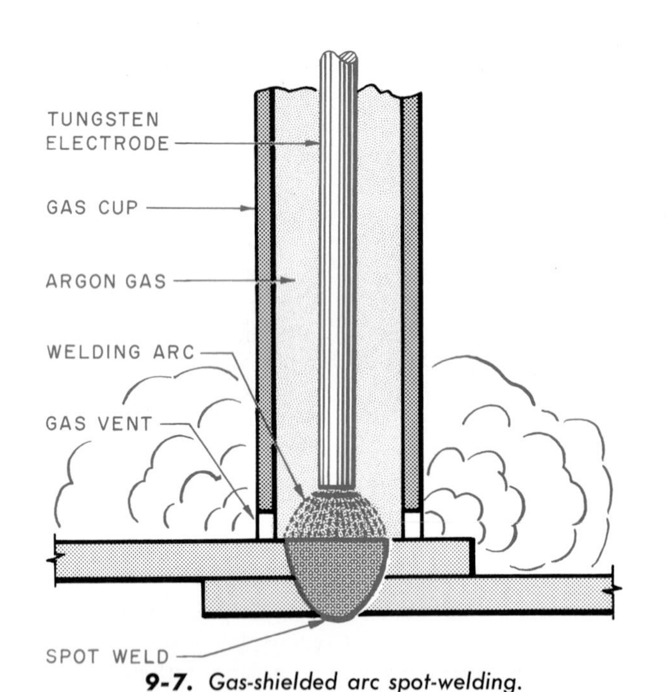

9-7. *Gas-shielded arc spot-welding.*

9-8. *Typical gas tungsten-arc installation.*

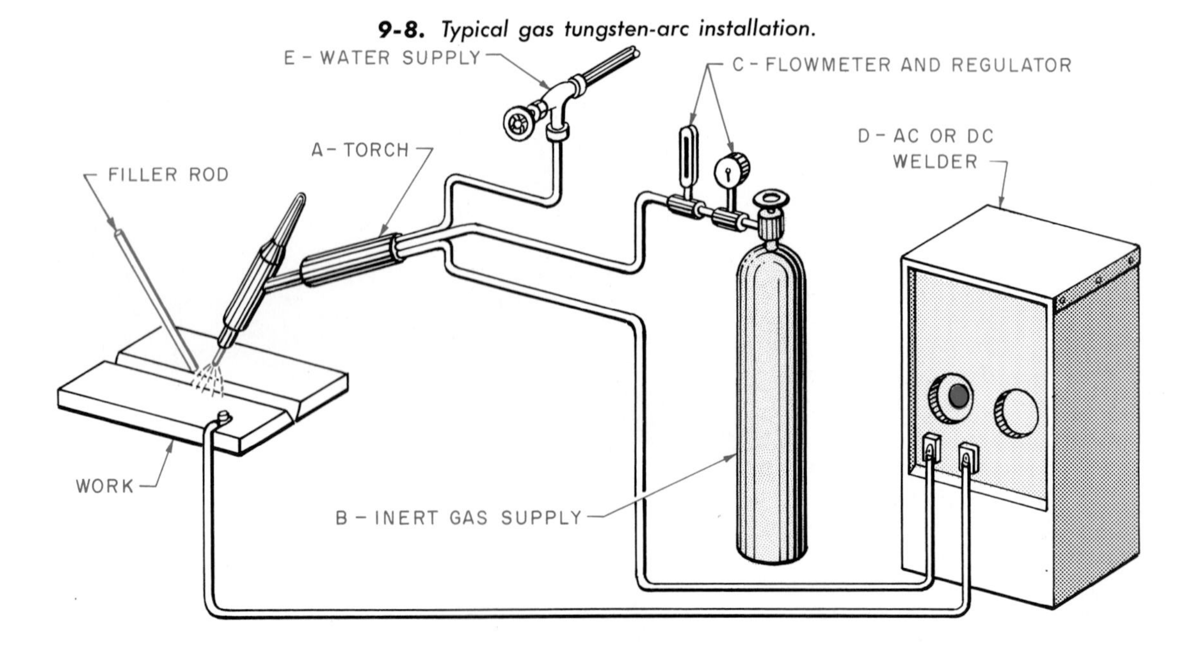

porosity in the steel and also forms iron nitrides which embrittle the weld.

If one gas will not provide all the traits desired, a mixture will often improve the action of one gas, or combine the best features of two or more gases. For example, mixing argon with helium gives a balance between penetration and arc stability. Adding 25 percent argon to helium makes penetration deeper than that obtained with argon alone, and the arc stability approaches that of pure argon. Tables 9-1 and 9-2 list the performance characteristics of helium and argon.

Helium and Argon. Helium is a product of the natural gas industry. It is found in large quantities in oil and natural gas fields. Helium is distributed in steel cylinders similar to oxygen cylinders. At one time it was used extensively for inflating balloons because of its lightness (next to hydrogen, helium is the lightest of all gases) and because it is not inflammable.

Argon forms a little less than 1 percent of the earth's atmosphere. It is a byproduct of the liquefaction of air during the production of oxygen. Argon is distributed in cylinders like helium. The cylinders are usually yellow or brown.

Although helium and argon are alike in that they do not react chemically with other materials, they have differences which are important in welding.

Argon is used more extensively for TIG welding than helium. It has a smoother, quieter arc and greater cleaning action in the welding of such metals as aluminum and magnesium with alternating current.

Helium is more expensive than argon. Its lightness is also a

Table 9-1

Suitability of argon and helium for use as shielding gases in gas tungsten-arc welding of various metals.

(Welding with direct current, straight polarity, unless noted otherwise)

Aluminum alloys. Argon (with alternating current) preferred; offers arc stability and good cleaning action. Argon plus helium (with alternating current) gives less stable arc than argon, but good cleaning action, higher speed, and greater penetration. Helium (with DCSP) gives a stable arc and high welding speed on chemically clean material.

Aluminum bronze. Argon reduces penetration of base metal in surfacing (for which aluminum bronze is used).

Brass. Argon provides stable arc, little fuming.

Cobalt-base alloys. Argon provides good arc stability, is easy to control.

Copper-nickels. Argon provides good arc stability, is easy to control. Used also in welding copper-nickels to steel.

Deoxidized copper. Helium preferred; gives high heat input to counteract thermal conductivity. A mixture of 75% helium and 25% argon provides a stable arc, gives lower heat input than helium alone, and is preferred for thin work metal ($^{1}/_{16}$ in. or less).

Inconel.® Argon provides good arc stability, is easy to control. Helium is preferable for high speed automatic welding.

Low-carbon steel. Argon preferred for manual welding; success depends on welder skill. Helium preferred for high speed automatic welding; gives more penetration than argon.

Magnesium alloys. Argon (with alternating current) preferred; offers arc stability and good cleaning action.

Maraging steels. Argon provides good arc stability, is easy to control.

Molybdenum-0.5 titanium alloy. Purified argon or helium equally suitable; welding in chamber preferred, but not necessary if shielding is adequate. For good ductility of the weld, the nitrogen content of the welding atmosphere must be kept below 0.1%, and the oxygen content below 0.005%.

Monel.® Argon provides good arc stability, is easy to control.

Nickel alloys. Argon provides good arc stability, is easy to control. Helium is preferred for high speed automatic welding.

PH stainless steels. Helium preferred; provides more uniform root penetration than argon. Argon and argon-helium mixtures also have been used successfully.

Silicon bronze. Argon minimizes hot shortness in the base metal and weld deposit.

Silicon steels. Argon provides good arc stability, is easy to control.

Stainless steel. Helium preferred. Provides greater penetration than argon, with fair arc stability.

Titanium alloys. Argon provides good arc stability, is easy to control. Helium is preferred for high speed automatic welding.

American Society for Metals

disadvantage, since it has a tendency to rise from the weld very rapidly. Two or three volumes of helium are required to give the equivalent protection of one volume of argon. Thus helium is consumed in larger quantities, increasing the cost disadvantage. Furthermore, since argon is denser than helium, the use of argon materially decreases cylinder handling and reduces the amount of space required for cylinder storage.

More heat is liberated at the arc with helium than argon. Welds of deeper penetration are produced, and welding is faster. Helium is preferred, therefore, for the welding of thick sections of steel and for metals with high thermal conductivity such as aluminum and copper. Argon, on the other hand, is used extensively for welding thin metal and dissimilar metals. Its low thermal conductivity results in a weld deposit with a relatively wide top bead.

In TIG welding helium produces greater arc voltage (40 percent) per unit of arc length. This results in a hotter arc, deeper penetration, and greater welding speeds. It also reduces the effects of heat on the work. For these reasons helium is preferred for automatic high production welding.

The lower arc voltage of an argon-shielded arc permits changes in the arc length during welding without breaking the arc. With alternating current, helium gives inconsistant starting of the TIG arc. When mixed in the correct proportions with argon, however, helium produces increased speeds, greater current density, and deeper penetration than pure argon. The mixture must be used with balanced alternating current or alternating current with continuous high frequency. A mixture of the two gases combines the arc stability of argon shielding and the higher heat input of helium. Refer to Tables 9-1, page 257, and 9-2 for additional information.

Hydrogen. Hydrogen used with TIG welding has resulted in increased arc potential. The effect is more pronounced when hydrogen is mixed with helium than with argon. However, an argon-hydrogen mixture also improves weld shielding at low gas flow rates.

As much as 35 percent hydrogen has been added to argon. A 20-percent addition produces an arc voltage equal to that obtained with helium alone for the same rate of gas flow. The argon-hydrogen mixture is limited to welding stainless steels.

Nitrogen. Nitrogen has been used to TIG weld deoxidized copper. But since this gas dissolves in steel at welding temperatures and causes aging effects, it seems unlikely that it will ever be used to weld mild steel. Additions of 5 percent hydrogen to nitrogen have been used as a backup shield for stainless steel, to produce brighter weld root beads.

Refer to Chapter 10, Gas Metal-Arc Welding Principles, for more detailed information on shielding gases.

Gas Control Equipment

Gas control equipment should provide a uniform flow of the desired quantity of gas around the arc and weld pool during welding. The gas is usually supplied in high pressure cylinders or by a piped system supplied by a cylinder manifold.

FLOWMETER. Flowmeters indicate the rate of flow of inert gas to the torch. They are calibrated to indicate flows in cubic feet per hour.

Table 9-2

Characteristics and comparative performance of argon and helium as shielding gases.

Argon
Low arc voltage. Results in less heat; thus, argon is used almost exclusively for manual welding of metals less than $1/16$ in. thick.
Good cleaning action. Preferred for metals with refractory oxide skins, such as aluminum alloys, or ferrous alloys containing a high percentage of aluminum.
Easy arc starting. Particularly important in welding of thin metal.
Arc stability is greater than with helium.
Low gas volume. Being heavier than air, argon provides good coverage with low gas flows, and it is less affected by air drafts than helium.
Vertical and overhead welding. Sometimes preferred because of better weld-puddle control, but gives less coverage than helium.
Automatic welding. May cause porosity and undercutting with welding speeds of more than 25 in. per min. Problem varies with different metals and thicknesses, and can be corrected by changing to helium or a mixture of argon and helium.
Thick work metal. For welding metal thicker than $3/16$ in., a mixture of argon and helium may be beneficial.
Welding dissimilar metals. Argon is normally superior to helium.

Table 9-2 (Continued)

Helium
High arc voltage. Results in a hotter arc, which is more favorable for welding thick metal (over $^3/_{16}$ in.) and metals with high heat conductivity.
Small heat-affected zone. With high heat input and greater speeds, the heat-affected zone can be kept narrow. This results in less distortion and often in higher mechanical properties.
High gas volume. Helium being lighter than air, gas flow is normally $1\frac{1}{2}$ to 3 times greater than with argon. Being lighter, helium is more sensitive to small air drafts, but it gives better coverage for overhead welding, and often for vertical-position welding.
Automatic welding. With welding speeds of more than 25 in. per min, welds with less porosity and undercutting may be attained (depending on work metal and thickness).

American Society for Metals

Conventional welding regulators reduce the gas pressure, but they do not provide any measure of gas flow. As a result, a gas flowmeter has been incorporated with a regulator, Fig. 9-9. A separate flowmeter may also be used with a standard gas regulator, Fig. 9-10.

Gas Flow Rates. It is difficult to predict the exact gas flow rate for a given welding application. A general rule to follow is that the gas flow rate should be high enough to protect the weld pool and the end of the tungsten electrode.

Too much gas is wasteful, and it can cause porosity in the weld deposit because the gas cannot flow from the weld metal fast enough during cooling. Not enough gas flow permits contamination of the weld pool by the surrounding air and causes oxidation of the weld deposit.

Careful inspection of the end of the tungsten electrode will indicate if the proper quantity of gas is being provided. Proper gas flow leaves the tungsten electrode smooth and rounded on the tip. Improper gas flow causes burning, scale, and deformation of the electrode tip.

The correct gas flow for TIG welding depends on the following variables:

- Type of shielding gas
- Type and design of the weld joint
- Distance of the nozzle from the work
- Size and shape of the nozzle
- Size of the weld puddle
- Amount and type of welding current
- Position of welding
- Position of the torch in relation to the work
- Design and type of jigs
- Speed of welding
- Air currents in the welding area

This last condition is very important since a strong draft across the weld area can dispel most of the shielding gas.

SHUTOFF VALVES. In most welding operations, the arc is maintained for only short periods, and consequently there are many starts and stops during a day's operation. In order to avoid wasting gas, it should be very convenient for the welder to turn the gas off and on at the proper time. Some electrode holders are equipped with gas shutoff valves in the handle. For the most part, an electric solenoid gas valve or quick-acting, mechanical shutoff valve is installed in the gas line. It is placed so that it can be controlled at or near the point of welding. For manual work, the simplest and most practical device is usually the modified "gas-saver" shown in Fig. 9-11. This device controls the flow of gas and cooling water.

Power Sources

Alternating current and direct current with either straight or reverse polarity are used in gas tungsten-arc welding. Straight polarity is used more than reverse polarity.

Welding power sources are often classified as being either *rotating* or *static.* The rotating types are *generators,* driven by either an electric motor or an internal combustion engine. They put out direct current or both alternating and direct current. The static types are either *transformers* with a.c. output or *transformer-rectifiers* with d.c. or a.c.-d.c. output.

The machines used for TIG welding are classified as constant current (variable voltage) machines. A constant current arc-welding power supply is one which has a drooping output slope. It produces a relatively constant current with a limited change in load voltage. This type of supply can also be used for stick electrode welding. (For a detailed explanation of the output slope, refer to Chapter 6, pages 165–194.)

In a constant current power source, the welding current remains substantially constant even though the arc length varies and the arc voltage changes. Thus it is highly suited to most manual operations, including TIG welding, where variations in arc length are most apt to occur. This type of power source is also called *drooping voltage* because output voltage goes down, or droops, as the current (amperage) goes up. You will note in Fig. 9-12 that when the voltage is 80, amperage is 100. When voltage drops to 55, amperage increases to 400.

Power sources for precision TIG welding on thin metals may be rated as low as 50 amperes and provide current selections separated by ½-ampere steps, Fig. 9-13. These machines can make perfect welds on metals as thin as 0.001 inch. For the general run of work, machines of 150 to 300 amperes are generally used. The open circuit (no load) voltage on most constant current machines is usually between 55 and 125 volts. During welding the closed circuit voltage (operating voltage) usually drops to between 15 to 40 volts on constant current welders.

One of the most important ratings for a welding power source, the *duty cycle,* expresses as a percentage the portion of time that the power supply must deliver its rated output in each 10-minute interval. For manual TIG welding, industrial units rated at a 60-percent duty cycle are generally satisfactory. For automatic processes, a 100-percent duty cycle is required. Limited service and limited input power supplies rated at a 20-percent duty cycle are available for light work, such as in garages, repair shops, and on farms.

Special TIG power sources are available that incorporate gas and water-flow controls, stop-and-start switches, and in the case of a.c. machines, a high frequency stabilizer and the choice of high, medium, low, and special low current ranges. Remote amperage controls, either hand- or foot-operated, are available as optional extras. Refer to pages 170–175 for a description of a typical TIG power source.

WELDING CURRENT CHARACTERISTICS. The different types

9-9. Combination regulator and flowmeter unit.

Linde Division, Union Carbide Corp.

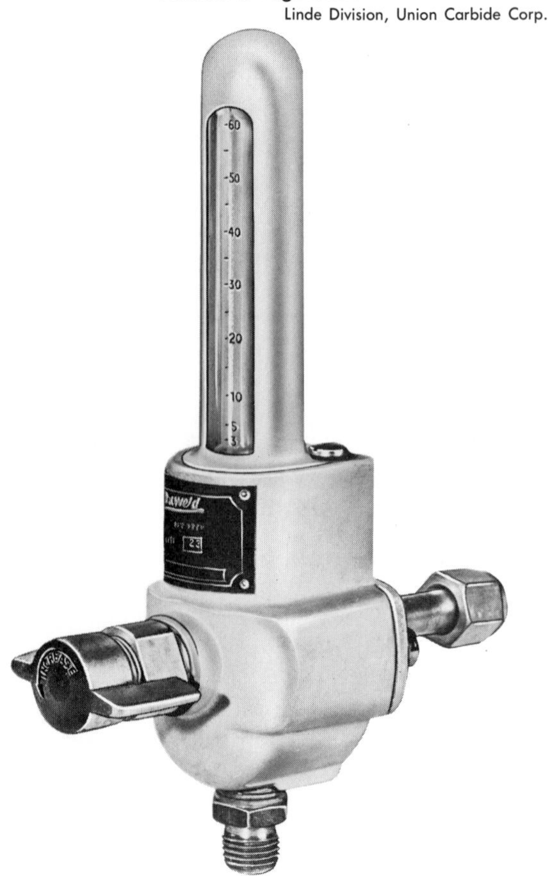

9-10. An individual flowmeter can be used with a standard regulator.

Linde Division, Union Carbide Corp.

of operating current directly affect weld penetration, contour, and metal transfer.

In d.c. welding, the welding current circuit may be hooked up as either *straight polarity* or *reverse polarity.* The machine connection for direct current, straight polarity (DCSP) welding is electrode negative and work positive. In other words, the electrons flow from the electrode to the work, concentrating heat at the joint, as shown in Fig. 9-14. For direct current, reverse polarity (DCRP) welding, the connections are just the opposite. Electrons flow from the work to the electrode, as shown in Fig. 9-15.

Direct Current, Straight Polarity. In straight polarity (DCSP) welding, the electrons hitting the plate at high velocity exert a considerable heating effect on the plate. The work tends to melt rapidly since it receives the greater part of the heat. The positive gas ions are directed toward the negative electrode at low velocity so that it is comparatively cool. DCSP provides no cleaning action for the removal of oxides from the metal.

With straight polarity, higher currents may be used on tungsten electrodes of the same size. Increased amperage yields deeper penetration, permits higher speeds, and forms a narrow, deep bead, Fig. 9-16A. The narrower bead is due to the smaller electrode used, while the increased penetration is due to the concentration of the electrons. The narrower bead also results in a narrower heat-affected zone surrounding the weld. Because there are fewer contraction stresses, less trouble with hot cracking of some metals is encountered. The DCSP type of operating current is preferred for heavier materials.

Direct Current, Reverse Polarity. In reverse-polarity (DCRP) welding, the tungsten electrode acquires extra heat instead of the work. The heat tends to melt the tip of the electrode. Because of this excess heating of the electrode, a larger size electrode is required for DCRP than for DCSP. Electrode size limits welding to relatively low current values. Large diameter electrodes are undesirable because they tend to cut down the operator's visibility and to increase the instability of the arc.

With reverse polarity the work stays comparatively cool, and shallow penetration results, Fig. 9-16B. This makes DCRP welding very desirable for thin sections.

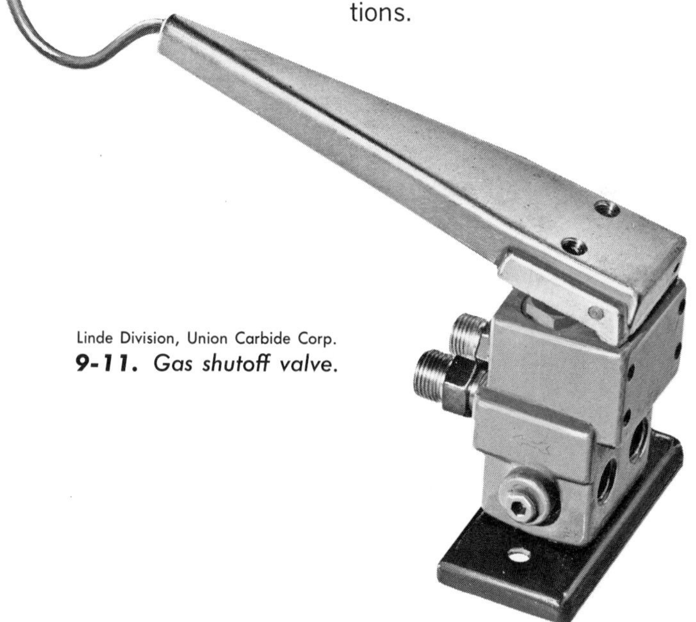

Linde Division, Union Carbide Corp.
9-11. *Gas shutoff valve.*

9-12. *The volt-ampere curve of a constant current welder.*

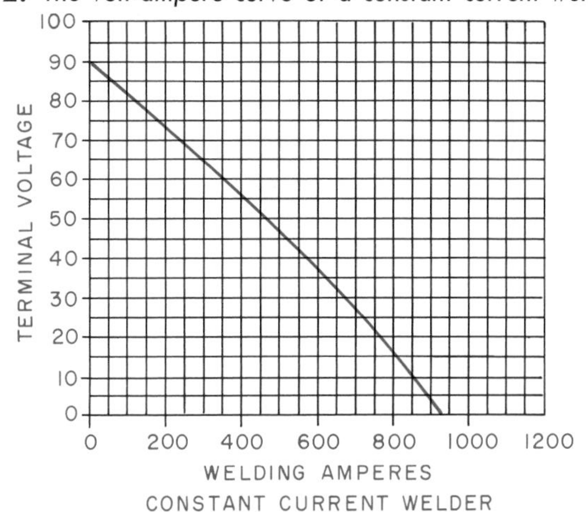

An important characteristic of DCRP welding is its plate-cleaning ability, expecially with such metals as aluminum, beryllium-copper, and magnesium. It makes possible the welding of these metals without the use of flux. This cleaning effect also takes place with the reverse-polarity cycle of high-frequency alternating current.

Although the exact reason for the surface-cleaning action is not known, it seems probable that either the electrons leaving the plate or the gas ions striking the plate tend to break up the surface oxides, scale, and dirt.

Alternating-Current Welding. Alternating current is widely used for TIG welding. It is recommended principally for light metals such as aluminum, magnesium, and beryllium-copper. Theoretically, a.c. welding is a combination of DCSP and DCRP welding. This can be best explained by showing the three current waves visually. As shown in Fig. 9-17, half of each complete a.c. cycle is DCSP; the other half is DCRP.

Actually, however, moisture, oxides, and scale on the surface of the plate tend to prevent the flow of current in the reverse polarity direction. This is called *rectification.* For example, if no current at all flowed in the reverse polarity direction, the current wave would look like one-half of the a.c. wave shown in Fig. 9-17. The arc would be unstable and sometimes even go out. To prevent this from happening, it is a common practice to introduce into the welding current a high voltage, high frequency, low power additional current, Fig. 9-18. This high-frequency current jumps the gap between the electrode and the workpiece. It pierces the oxide film, thereby forming a path for the welding current to follow.

Since the heat is evenly distributed at the two ends of the arc, the depth of weld penetration is less than that obtained with DCSP but more than that which would be obtained with DCRP. The contour of the weld lies between that of the deep, narrow type produced by straight polarity and the wide, shallow type produced by reverse polarity, Fig. 9-19.

High frequency, stabilized alternating current has the following advantages over straight alternating current:

- The arc may be started without touching the electrode to the work. Thus less contamination is picked up on the tungsten electrode so that the weld deposit is cleaner.
- Better arc stability is obtained.
- A long arc is possible, leading to easier and less tiring welding. The long arc is also useful in surfacing and hard-surfacing operations.
- There is an increase in electrode life of almost 100 percent.
- The electrode does not absorb as much heat.
- The use of wider current ranges for a specific diameter electrode is possible. Also, less current is required for a given weld.
- It is easier to make all position welds.

CURRENT SELECTION FOR TIG WELDING. Table 9-3 is a handy guide as to the type of current that should be used for a given job. It also indicates the different kinds of metals that can be welded with TIG process. The

9-13. *A 50-ampere, precision TIG welder. Due to its coarse and fine adjustments, current can be set to within a fraction of an ampere. The machine is used to weld capillary tubes, surgical instruments, and pressure-sensitive devices.*

Auto Arc-Weld Mfg. Co.

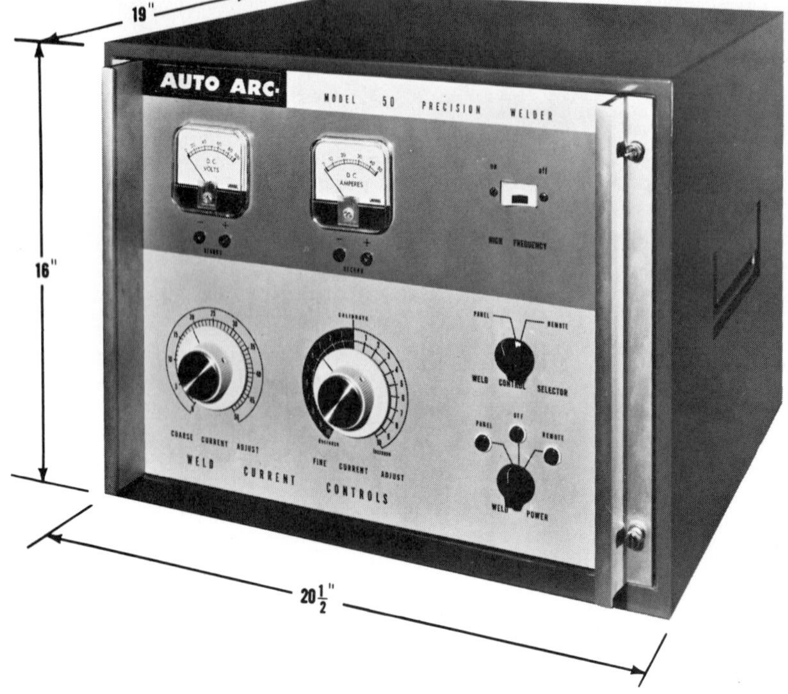

training course, as outlined on page 800, is limited to welding on mild steel, stainless steel, and aluminum. After these have been mastered, other metals may be welded easily.

D.C. POWER SOURCES FOR TIG WELDING. For d.c. welding power, any standard motor-driven welding generator or transformer-rectifier with an adequate amperage capacity for the job may be used. It is important that the motor-generator or transformer-rectifier have good current control at the lower end of its current range. Arc stability is essential, particularly for welding thin-gauge materials. Refer to Chapter 6, pages 165–194 for a detailed discussion of d.c. welding machines used for both TIG and stick electrode welding.

A.C. POWER SOURCES FOR TIG WELDING. The common requirement of all a.c. TIG welding machines is that they produce a constant current output and have high frequency stabilization. Standard a.c. welding transformers cannot be used successfully for TIG welding without certain changes because the open circuit voltage required to start current flowing is far above that supplied by the ordinary a.c. transformer.

A.c. transformer welders are designed especially for the TIG welding process. They are equipped with high frequency controls and gas and water valves as well as solenoids and other current controls.

Usually the power source has a switch with three selections for

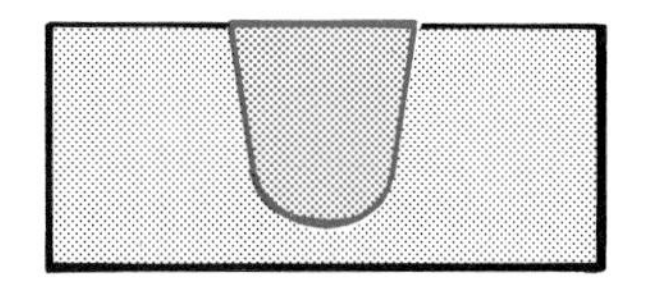

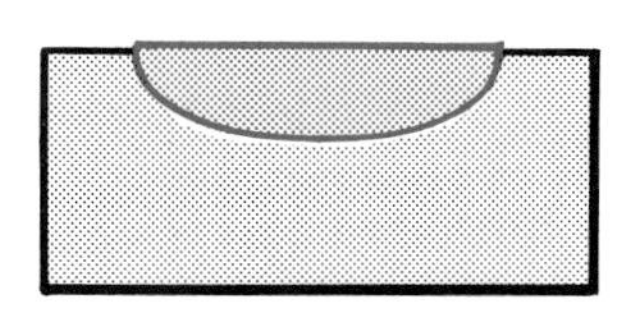

***9-16.** Effect of polarity on weld shape. Comparative weld contours.*

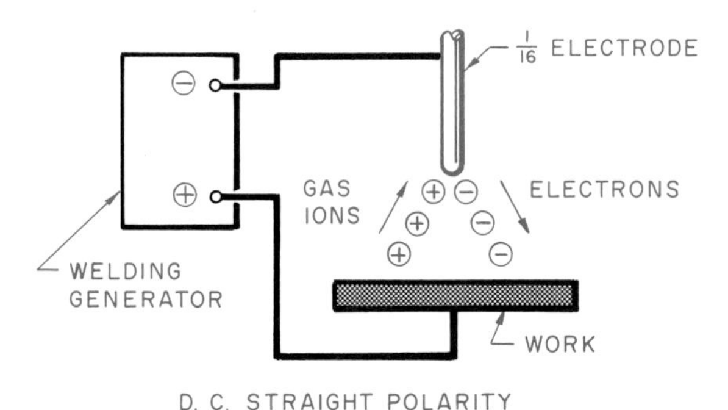

***9-14.** DCSP current characteristics and flow.*

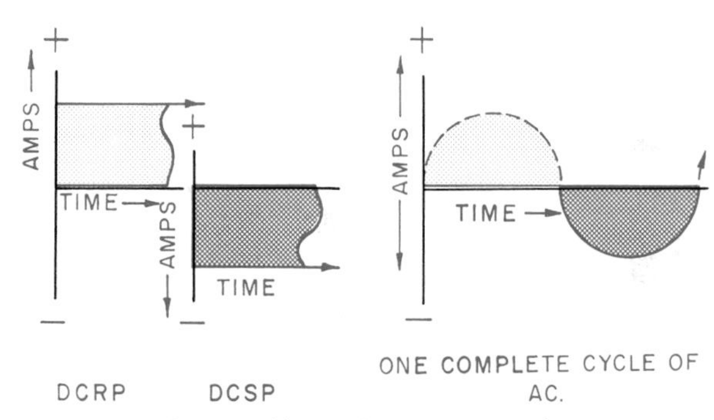

***9-17.** Alternating current cycle.*

***9-15.** DCRP current characteristics and flow.*

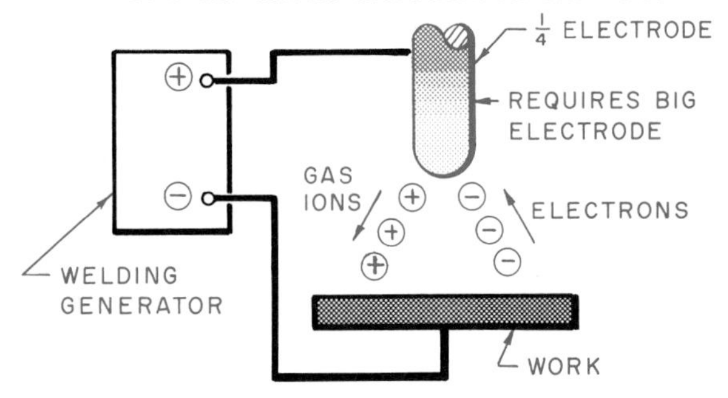

***9-18.** Rectified or interrupted a.c. current cycle.*

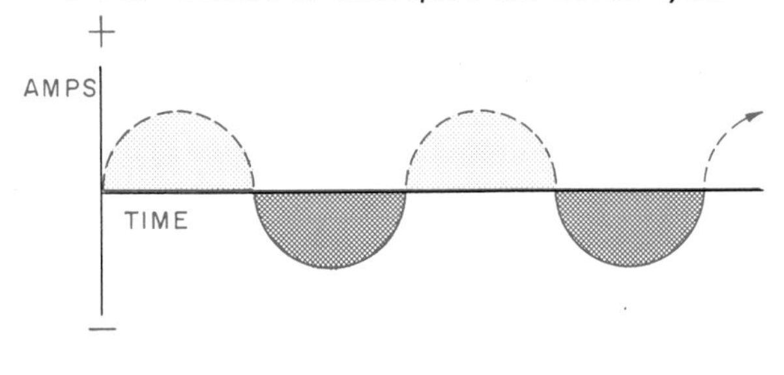

continuous high frequency, high frequency at the *start only* (this is sometimes called automatic), and *arc high frequency.* When used at the start only, the high frequency is on when the foot control is depressed. It helps to establish the arc and then shuts off automatically. When continuous, the high frequency is on as long as the foot pedal is depressed. Most a.c. welders also have a control which varies the high frequency intensity. Experienced TIG welders usually set this control wide open.

Many of these machines can be used for stick electrode, TIG, TIG spot, and automatic welding. Figs. 9-20 through 9-28 on the next eight pages present various types of constant current welding machines used for TIG welding. These machines can be controlled by remote control, Figs. 9-29 and 9-30.

9-19. *Comparison of weld contours.*

WELD RESULT SUMMARY

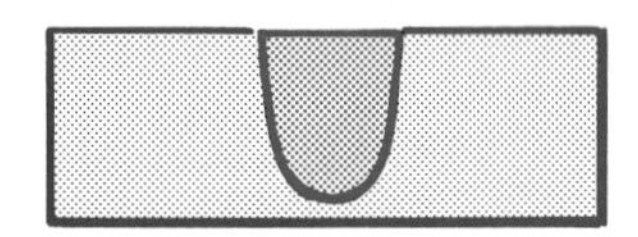

D. C. STRAIGHT POLARITY

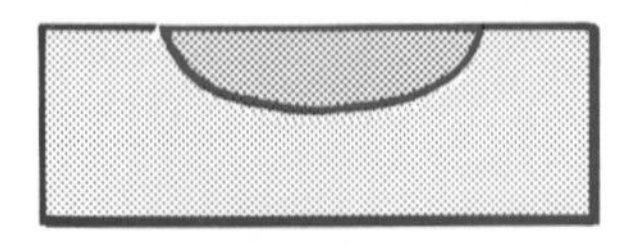

D. C. REVERSE POLARITY

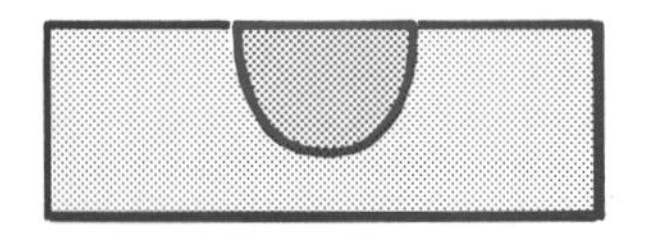

A. C. WELDING

EXAMPLE OF MACHINE CONTROLS. The following outline is presented so that the student can better understand the purpose and function of each control on a typical combination gas tungsten-arc and shielded metal-arc welding machine. Proper setting of the controls is necessary to produce sound welds of good appearance. Figure 9-26, page 271, illustrates the controls listed below.

Transformer-Rectifier A.C.-D.C. Arc Welding with Solid State Controls (300-Ampere Capacity).

A. Capabilities
   1. For both TIG and stick electrode welding
   2. Joins all ferrous and nonferrous alloys that can be arc welded
   3. Current range of 2-375 amps, a.c. and d.c., handles stock ranging in thickness from gauge metal through heavy plate.
   4. Almost no tungsten spitting (melting of electrode tip)
   5. Optional remote control

B. Advantages of solid state electronic controls
   1. Constant arc heat produced during welding because internal circuits

Table 9-3

Current selection for TIG welding.

Material	Alternating current*	Direct current	
	With high frequency stabilization	Straight polarity	Reverse polarity
Magnesium up to $\frac{1}{8}$ in. thick	1	N.R.	2
Magnesium above $\frac{3}{16}$ in. thick	1	N.R.	N.R.
Magnesium castings	1	N.R.	2
Aluminum up to $\frac{3}{32}$ in. thick	1	N.R.	2
Aluminum over $\frac{3}{32}$ in. thick	1	N.R.	N.R.
Aluminum castings	1	N.R.	N.R.
Stainless steel	2	1	N.R.
Brass alloys	2	1	N.R.
Silicon copper	N.R.	1	N.R.
Silver	2	1	N.R.
Hastelloy alloys	2	1	N.R.
Silver cladding	1	N.R.	N.R.
Hard-facing	1	1	N.R.
Cast iron	2	1	N.R.
Low carbon steel, 0.015 to 0.030 in.	2**	1	N.R.
Low carbon steel, 0.030 to 0.125 in.	N.R.	1	N.R.
High carbon steel, 0.015 to 0.030 in.	2	1	N.R.
High carbon steel, 0.030 in. and up	2	1	N.R.
Deoxidized copper***	N.R.	1	N.R.

Linde Division, Union Carbide Corp.

Key: 1. Excellent operation.
2. Good operation.
N.R. Not recommended.

*Where a.c. is recommended as a second choice, use about 25 percent higher current than is recommended for DCSP.

**Do not use a.c. on tightly jigged part.

***Use brazing flux or silicon bronze flux for $\frac{1}{4}$ inch stock and thicker.

compensate for the following:

a. Changes of input line voltage
b. Changes in operating temperature to maintain output current as set by the controls

2. Provides instantaneous output response with every control adjustment
3. Current setting is easy and accurate because the effect of fine current control adjustments are essentially linear over the entire range.

C. Complete welding controls

9-20. *A 300-ampere constant current power supply offers a choice of a.c., DCRP, or DCSP current. The machine can be used for TIG, MIG, and stick electrode welding.*

Linde Division, Union Carbide Corp.

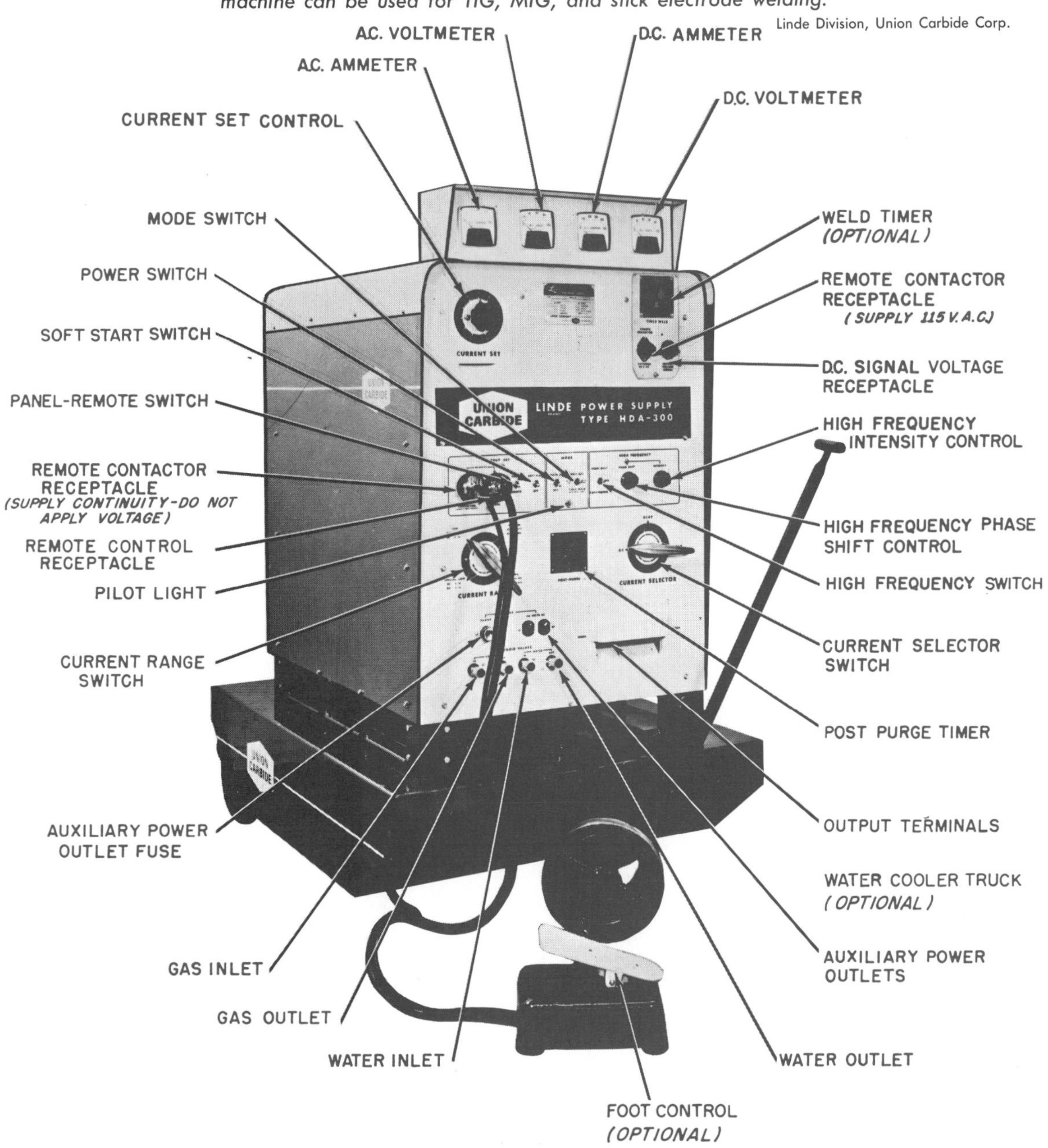

1. Current range selector with five overlapping ranges as shown in Table 9-4 (Page 273)
2. Fine adjustment current control
3. A.c., DCSP, DCRP selector switch
4. Remote control switch and receptacle for connecting current control, arc start switch, or arc fade control
5. Soft-start switch reduces current during starting
6. Spark intensity rheostat adjusts strength of high frequency spark.
7. The high frequency and gas and water control switches provide any one of the following combinations:
   a. Continuous high frequency while welding with gas and water flow control

9-21A. *Right side view of the interior construction of the welder shown in Fig. 9-20.*

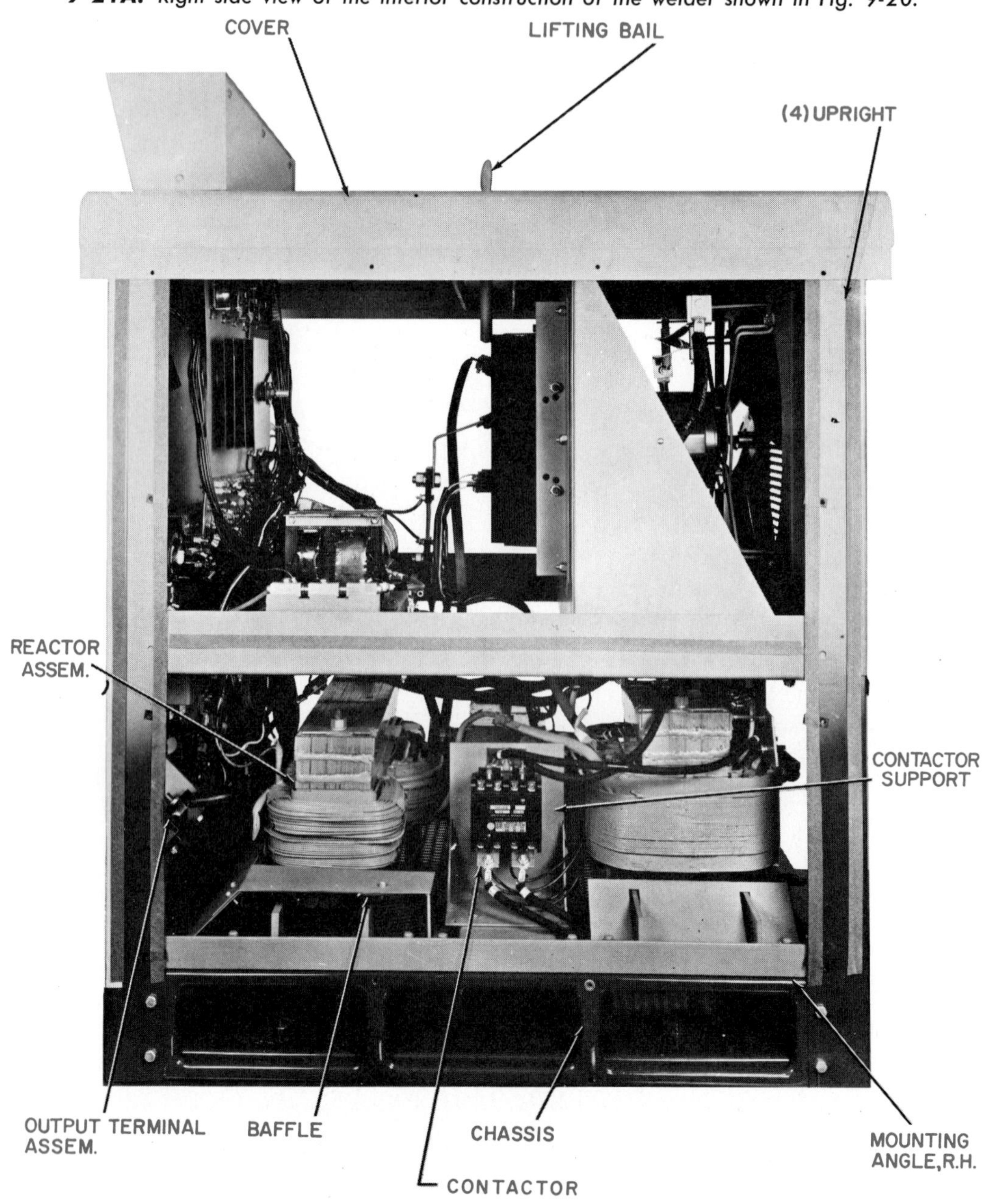

b. High frequency for start only with gas and water flow control
c. No high frequency but with gas and water flow control
d. No high frequency and no gas and water flow
8. Afterflow timer controls gas and water flow after welding.
9. Gas and water inlets and outlets
10. 115-volt a.c. power (1,700 watts) available from receptacles in the nameplate
11. 115-volt on-off pushbutton operates line contactor.
12. Pilot light indicates when machine is on.

D. Current Control
   1. Varies current while welding for
      a. Making critical TIG or stick electrode welds
      b. Filling craters
   2. Built-in arc start switch automatically starts and stops both the high frequency welding power

9-21B. *Left side view of the interior construction of the welder shown in Fig. 9-20.*

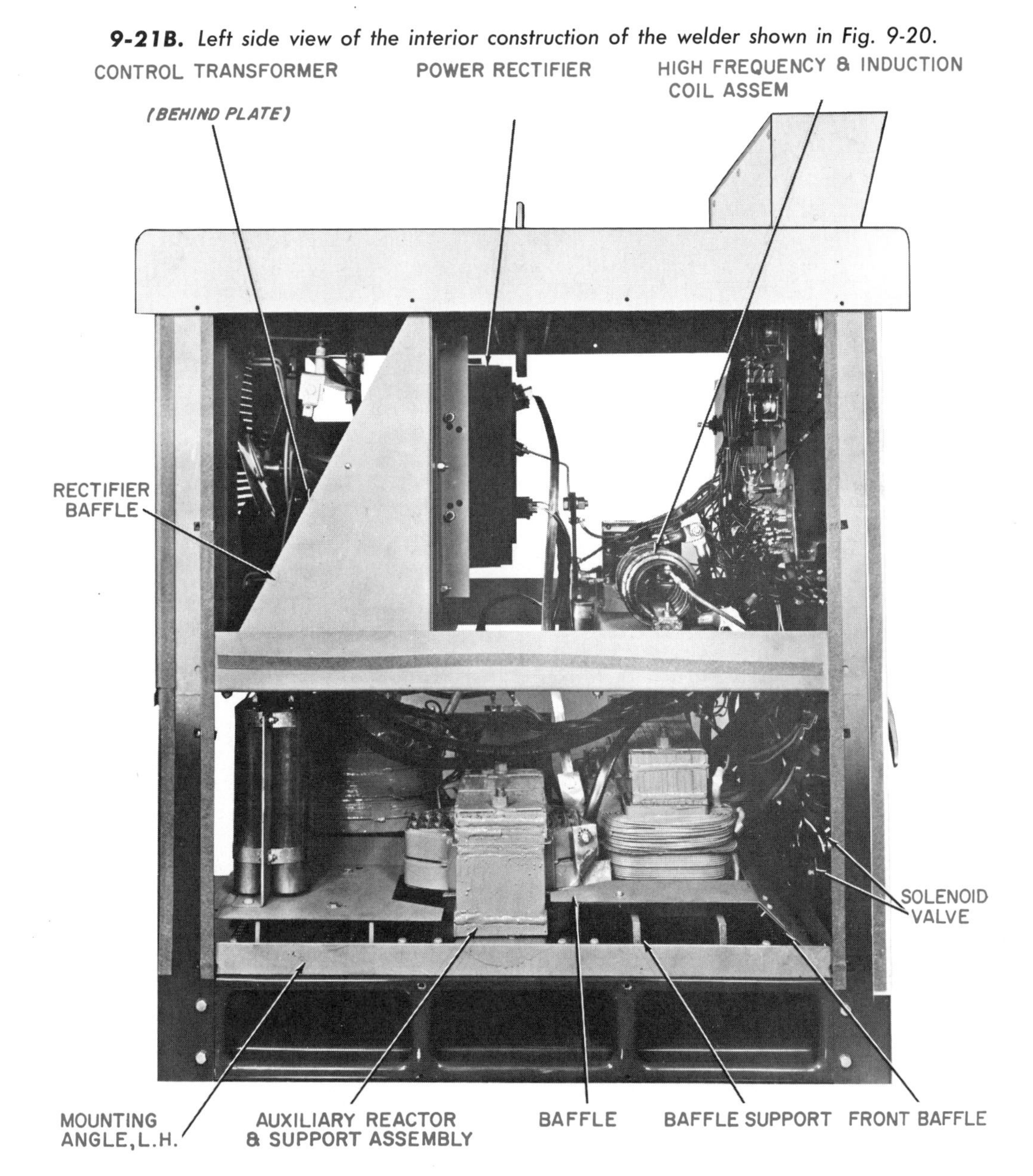

and the gas and water flow.

3. Stacking adapters permit mounting one welding machine on top of another.
4. Hand control fastens to the torch for fingertip control.
5. Fast control provider; maximum freedom for the hands
6. Twenty-five foot lead lets the operator take the control to the work.

E. Optional features

1. Arc start switch required for TIG welding when the current control is not used.
2. Automatic arc fade control is a separate attachment that automatically controls crater filling. It adjusts the rate of current fade-out from 2 to 60 seconds, sets the current level cut-off, and provides remote current control.
3. Arc polarizer improves arc characteristics on two types of aluminum a.c. TIG welding applications:
   a. The feather-edge butt welds, such as the root pass of a pipe weld, when a consistent, smooth back bead is required
   b. Low current (less than 40 amps) applications which require exceptional cleaning action, such as edge and corner welds
4. The arc polarizer contains three 6-volt batteries and a charging circuit.
5. The condenser for power factor correction reduces operating costs by improving the supply line power factor.
6. The undercarriage consists of a frame and three 9-inch solid rubber tires or bare steel wheels.

The TIG Torch

The TIG torch (electrode holder) is one of the most important items of apparatus. Figure 9-31 (Page 274) is a sectional drawing of a holder which identifies the essential working parts.

9-21C. *Rear view, lower deck of the interior construction of the welder shown in Fig. 9-20.*

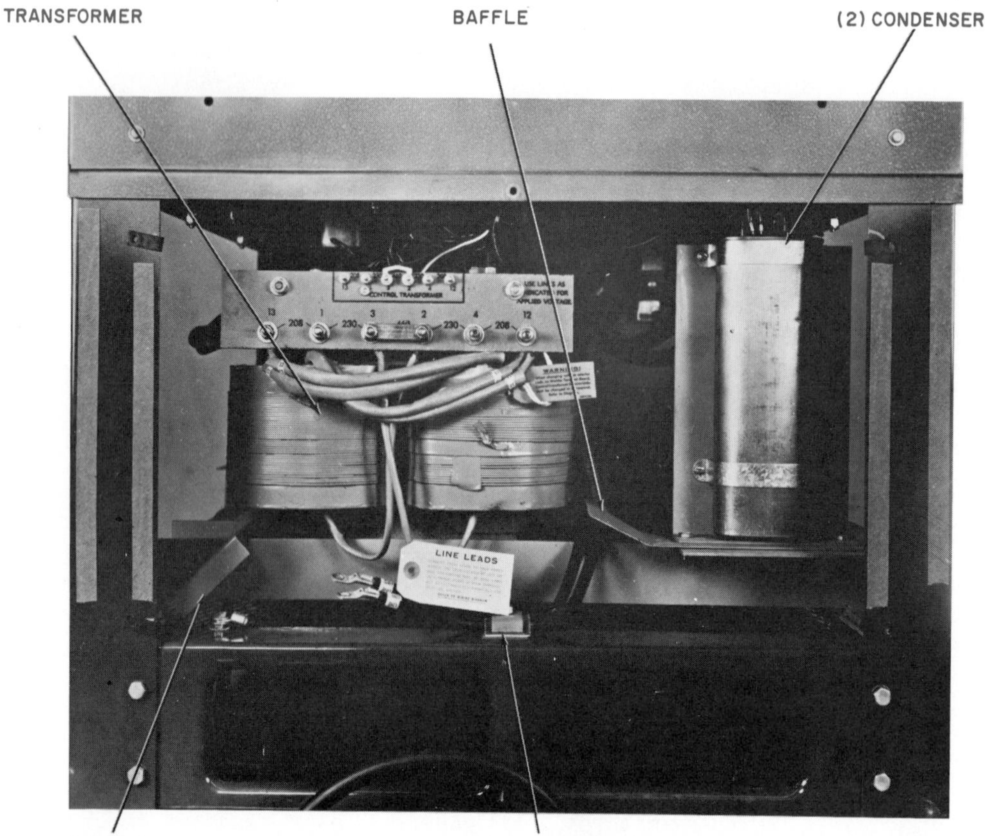

The torch feeds both the current and the inert gas to the weld zone. The torch is made up of a torch body, a collet, a ceramic collet cup, a nozzle and the electrode, Fig. 9-32. A hose and cable assembly is attached to the holder that connects the torch to the power source, the shielding gas supply, and the source of water it requires for cooling. Some torches have a manually operated switch on the handle to control the gas. All torches are properly insulated to handle safely the maximum current for which they are designed.

The current is fed to the welding zone through the tungsten electrode which is held firmly in place by a steel collet in the electrode holder. The electric arc is the intense source of heat for welding. Most of the heat of the arc is absorbed by the base metal, but a portion of the heat of the arc goes into the electrode. The electrode material should be practically nonconsumable, and only two materials, tungsten and carbon, have been found to be usable. Tungsten electrodes are more generally used, but carbon electrodes have found a limited application for low current welding. Various sizes of electrodes may be inserted into the holders for a range of welding heats.

The shielding gas is fed to the weld zone through a gas cup or nozzle at the head of the torch. A large variety of torch heads are available. Some have straight heads, others have heads at different angles, and a few have movable heads that can be adjusted to any angle. The material of the nozzle must be highly heat-resistant and capable of conducting the heat away from the lower edge so that the nozzle does not overheat by radiation from the electrode. Gas nozzles are usually made of a refractory ceramic composition, glass or metal, for welding up to 250 amperes. Metal nozzles are used for high welding currents over 250 amperes, and the torches are usually water cooled. A glass cup eliminates blind spots by permitting the operator to watch the weld pool.

Ceramic cups that become charred and burned from overheating, or that have a buildup of contaminated material on the inside, should be discarded. If used, charred cups will spread contamination to the weld deposit. Metal cups that become contaminated can be wire-brushed clean and reused.

A number of manufacturers are indicating the size of nozzles by numbers such as 4, 5, 6, 7 and 8. These numbers indicate the nozzle size in sixteenths of an inch. For example, a size No. 4 nozzle indicates $\frac{4}{16}$ ($\frac{1}{4}$) inch diameter. In general, select a nozzle with a diameter that is from four to six times the diameter of the electrode. Nozzles with

9-22. *This 300-ampere constant current power supply can be used for a.c. and d.c. stick electrode welding, manual and automatic TIG welding, and d.c. TIG spot welding.*
Miller Electric Manufacturing Co.

9-23. *Closeup of the operating panel of the welder shown in Fig. 9-22.*

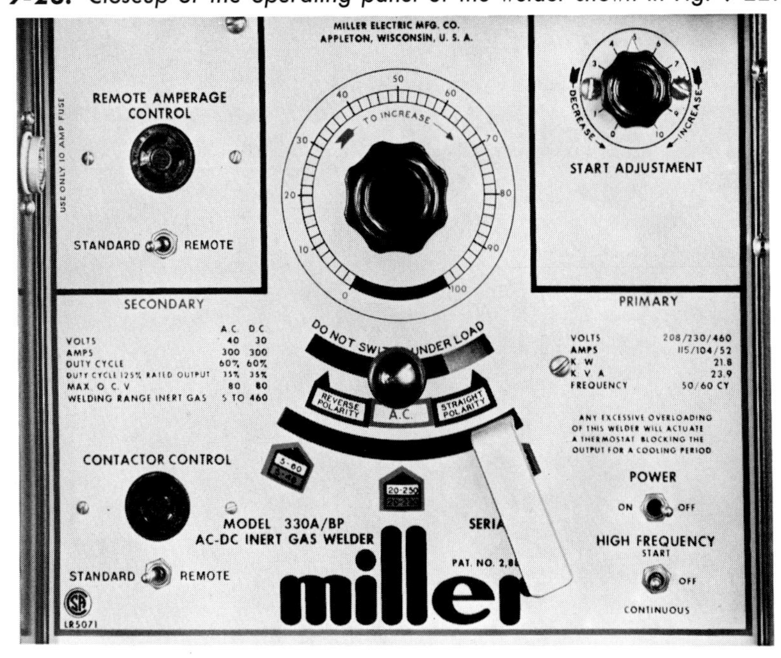

too small a hole size tend to overheat and break. A hole that is too large provides poor gas shielding and wastes gas.

Electrode holders are water cooled, air cooled, or gas cooled. Air- and gas-cooled holders are generally confined to very light and intermittent duty. They are particularly desirable, however, for work requiring currents up to 100 amperes because of their light weight, small size, and simplicity of operation. Holders above 100 amperes in size are water cooled. (Electrode holders with capacities up to 500 ampers are available.)

A recent development for TIG welding torches is the *gas lens*, Fig. 9-33 (Page 274). This is a permeable barrier of concentric fine-mesh stainless steel screens which produce an unusually stable stream of shielding gas. This device prevents turbulence of the gas stream which tends to pull in air, causing weld contamination. The source of gas turbulence is the torch nozzle where the velocity of the entering gas is much greater than that of the exit gas. The gas lens reduces the difference in velocities and directs the gas into a coherent stream.

When a torch is equipped with a gas lens, the shielding gas can be projected to a greater distance. Welding is possible at nozzle elevations up to 1 inch. The tungsten electrode can be extended well beyond the cup. This improves visibility by eliminating the blind spot at the weld puddle, Fig. 9-34. Inside corners and other hard-to-reach places can be readily welded.

Electrodes

It is important that the correct type and size of electrode be used for each welding application. The proper electrode is essential to good welding.

CARBON ELECTRODES. Carbon electrodes are generally limited to use when welding with low currents. Easy arc starting and good arc stability favor their use. However, among the disadvantages of carbon electrodes are the high rate of consumption and ease of breaking.

TUNGSTEN ELECTRODES. Tungsten is a metal with a melting point higher than any other metal; of all the elements it is second only to carbon. Carbon has a melting point of 6,740 degrees F., and tungsten, 6,170 degrees F. Tungsten electrodes are available as pure tungsten, with 1 and 2 percent thoria added, or with 1 percent zirconium added. They have been found to be practically nonconsumable.

Pure tungsten is generally used for a.c. welding. The zirconium type is also excellent for a.c. welding. It has a high resistance to contamination, produces a more stable arc than pure tungsten, and has a higher current-carrying capacity.

THORIATED TUNGSTEN ELECTRODES. Thoriated tungsten electrodes are available for direct current, straight polarity welding.

9-24. *This machine is recommended for pipe welding. It can be used for manual or automatic welding and for gas tungsten-arc or shielded metal-arc welding. It has two current ranges: 3–30 amperes and 3–300 amperes.*

Hobart Brothers Co.

9-25. *View of the programmer control panel of the machine shown in Fig. 9-24.*

Hobart Brothers Co.

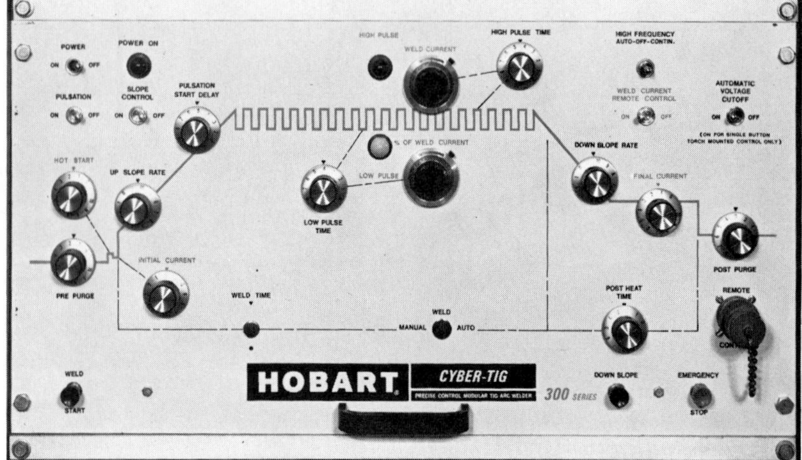

Although they are more costly than pure tungsten electrodes, their lower rate of consumption makes them more economical to use. Thoriated tungsten electrodes run cooler, and their tips do not become molten so that the arching end is kept square and intact. When it is accidentally touched by the electrode, the work does not become contaminated. Touch starting is facilitated and the arc stability is improved, especially at low currents.

STRIPED ELECTRODES. A new type of tungsten electrode has been developed called a *striped* electrode. The striped electrode differs physically from a standard thoriated electrode in that a solid stripe of 2 percent thoria is inserted in a wedge the full length of the electrode, Fig. 9-35. The combination of tungsten and thoria as separated physical elements results in entirely dif-

9-26. *A 300-ampere a.c.-d.c., constant current transformer with solid state controls. This machine can be used for a.c.-d.c. TIG and a.c.-d.c. stick electrode welding.*

The Lincoln Electric Co.

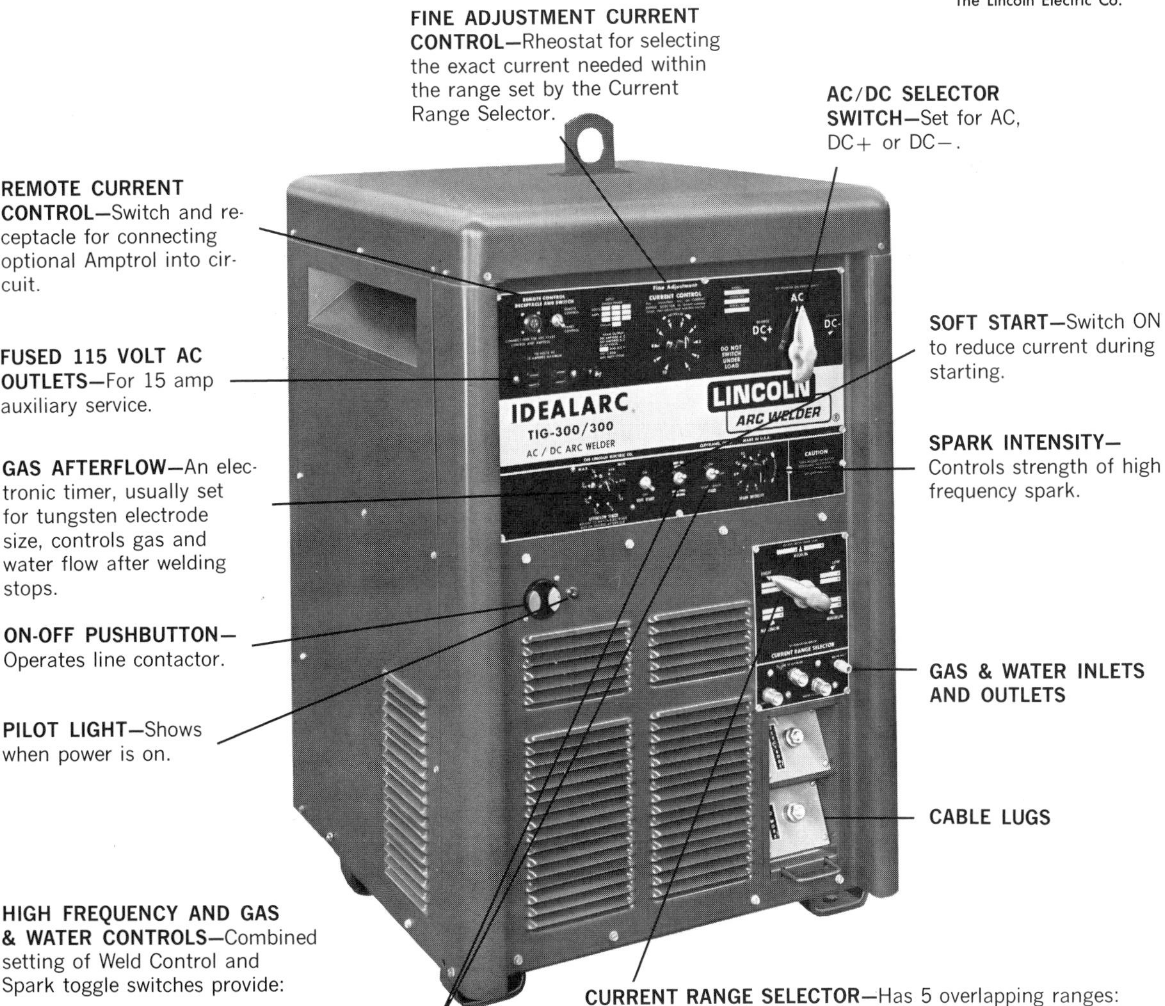
FINE ADJUSTMENT CURRENT CONTROL—Rheostat for selecting the exact current needed within the range set by the Current Range Selector.
AC/DC SELECTOR SWITCH—Set for AC, DC+ or DC−.
REMOTE CURRENT CONTROL—Switch and receptacle for connecting optional Amptrol into circuit.
SOFT START—Switch ON to reduce current during starting.
FUSED 115 VOLT AC OUTLETS—For 15 amp auxiliary service.
IDEALARC
TIG-300/300
AC / DC ARC WELDER
LINCOLN
ARC WELDER
SPARK INTENSITY—Controls strength of high frequency spark.
GAS AFTERFLOW—An electronic timer, usually set for tungsten electrode size, controls gas and water flow after welding stops.
ON-OFF PUSHBUTTON—Operates line contactor.
GAS & WATER INLETS AND OUTLETS
PILOT LIGHT—Shows when power is on.
CABLE LUGS
HIGH FREQUENCY AND GAS & WATER CONTROLS—Combined setting of Weld Control and Spark toggle switches provide:
1. Continuous high frequency while welding with gas and water flow control.
2. High frequency for start only with gas and water flow control.
3. No high frequency but gas and water flow control.
4. No high frequency or gas and water flow.
CURRENT RANGE SELECTOR—Has 5 overlapping ranges:

Range	Min.	Low	Medium	High	Max.
AC Amps.	2-27	8-93	13-150	25-250	60-375
DC Amps.	2-27	8-93	15-165	35-270	100-375

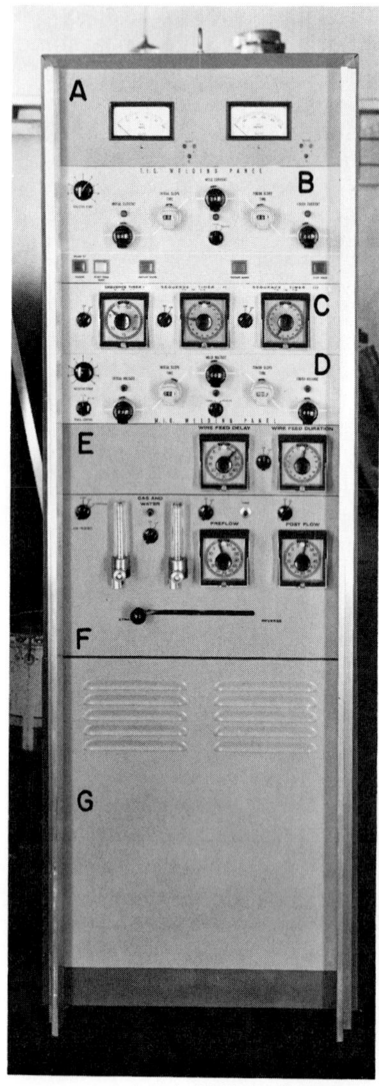

Auto Arc-Weld Manufacturing Co.

9-27. *TIG weld programmer has digital readout controls for preselection of current schedules. Starting currents, weld currents, and finish currents can be preselected by multi-turn dials. Wire-feed speeds and gas and water flow can also be programmed. All controls operate independently. A. Meter panel. B. TIG welding panel. C. Sequence timer panel. D. Accessory panel. E. Wire-feed timer panel. F. Gas, water, and high frequency panel. G. Welding power supply may be 300, 400, or 800 amperes.*

ferent operating characteristics such as the following:

- Use of smaller diameters with higher currents
- Positive arc stability providing uniform beads
- Low electrical resistance with superior heat resistance
- Superior balling and anti-spitting
- Uniform voltage and easy arc-starting
- Compatibility with all metals
- Better performance than pure tungsten
- Maximum electrode life
- Little contamination of the weld by pure tungsten
- No aluminum contamination
- High current capacity with better a.c. arc stability

ELECTRODE SIZES AND CAPACITIES. Tungsten electrodes are available in diameters of 0.010, 0.020, 0.040, $\frac{1}{16}$, $\frac{3}{32}$, $\frac{1}{8}$, $\frac{5}{32}$, $\frac{3}{16}$, and $\frac{1}{4}$ inch. They are manufactured in lengths of 3, 6, 7, 18 and, in some instances, 24 inches. Six inches is the convenient size for most manual operations. They have either a chemically cleaned finish or a ground finish that holds the diameter to a close tolerance. The ground finish costs somewhat more than the cleaned finish.

The current-carrying capacity of tungsten electrodes depends on a number of variables such as the following:

- Type of shielding gas
- Length of the electrode extending beyond the collet
- Effectiveness of the cooling afforded by the electrode holder
- Position of welding
- Polarity of the current

Note the difference in the current capacity of electrodes with DCSP, DCRP, and high frequency, alternating current (ACHF) indicated in Table 9-5 (Page 276). Slightly higher current values are used with thoriated tungsten electrodes than with regular tungsten electrodes.

These heats are only approximations. A good working rule to follow is to select an electrode with the smallest diameter that can be employed without dripping molten tungsten into the weld pool. Too high a current density with helium may result in a black deposit of tungsten on the work. Too low a current density results in an arc that is hard to direct and may be hard to start.

ELECTRODE PREPARATION AND MAINTENANCE. There is some disagreement as to the practice of grinding an electrode to a pencil point. Experience has indicated that grinding should be

9-28. *This power source can be used for stick electrode welding, manual TIG welding, and semi-automatic and automatic TIG welding. The machine gives complete control of all functions.*

Auto Arc-Weld Manufacturing Co.

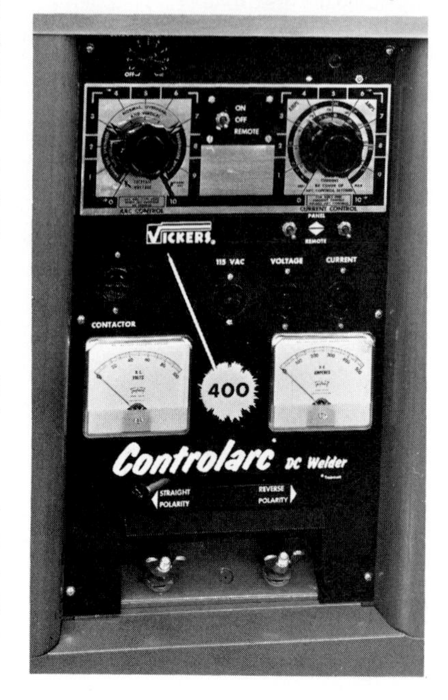

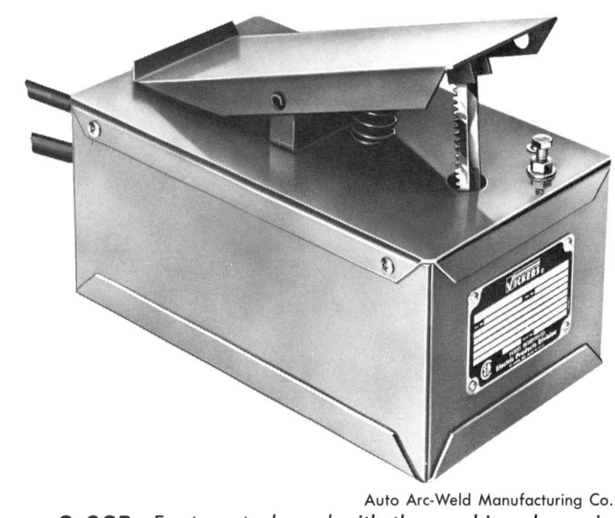

Auto Arc-Weld Manufacturing Co.

9-29B. *Foot control used with the machine shown in Fig. 9-28.*

Auto Arc-Weld Manufacturing Co.

9-29A. *Remote control unit used with the machine shown in Fig. 9-28.*

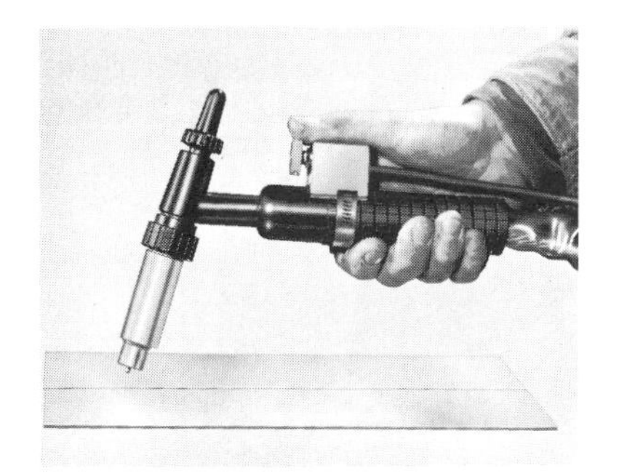

Linde Division, Union Carbide Corp.

9-30A. *Remote control in a torch handle.*

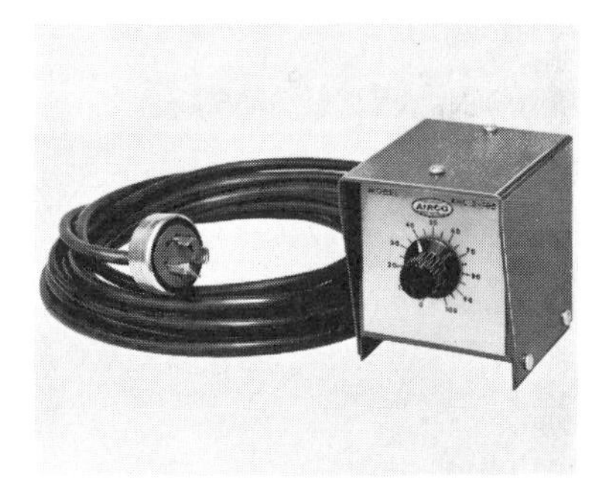

Linde Division, Union Carbide Corp.

9-30B. *Hand remote control.*

Table 9-4

Current ranges for an a.c.-d.c. transformer-rectifier welder.

Range	Min.	Low	Medium	High	Max.
A.c. amps.	2–27	8–93	13–150	25–250	60–375
D.c. amps.	2–27	8–93	15–165	35–275	100–375

The Lincoln Electric Co.

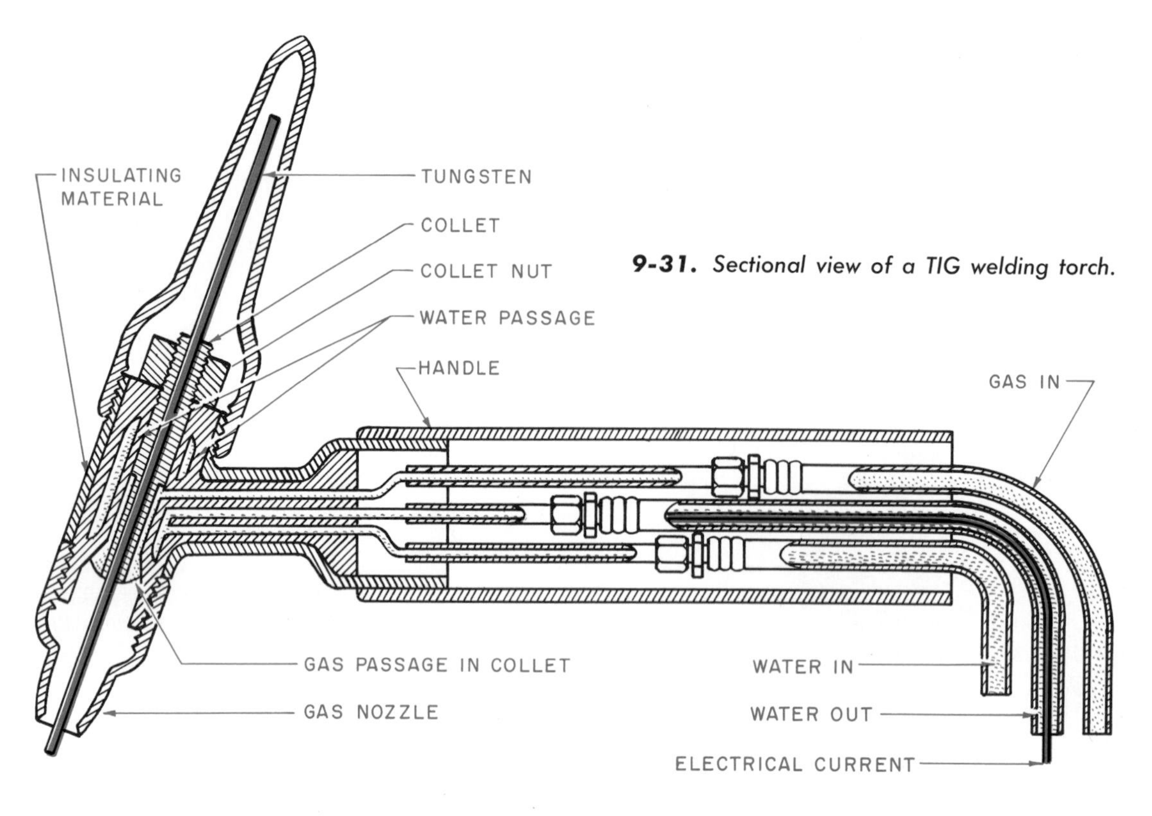

9-31. *Sectional view of a TIG welding torch.*

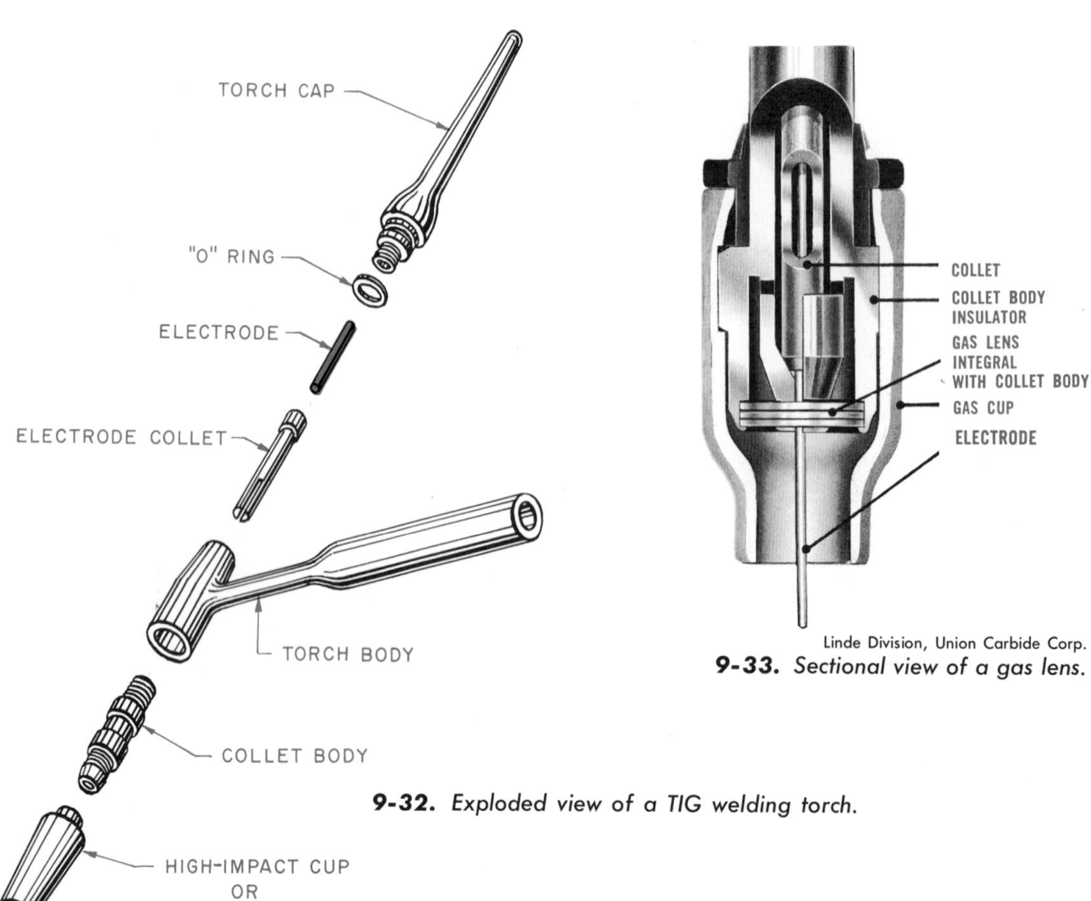

9-32. *Exploded view of a TIG welding torch.*

Linde Division, Union Carbide Corp.

9-33. *Sectional view of a gas lens.*

done only when welding thin materials. A sharpened point is also used with DCSP for carbon and stainless steels. Grinding should be done on a fine-grit, hard abrasive wheel. This wheel should not be used to grind any other material to prevent contaminating the tungsten electrode.

Another method of electrode preparation is to ball the electrode so that a rounded end forms on the electrode. The ball end is formed by using DCRP and striking an arc on a piece of copper. This ball must not be over $1\frac{1}{2}$ times the diameter of the electrode. (A larger ball indicates that the current is too high.) The balled point is used with alternating current for aluminum and aluminum-magnesium alloys.

Maximum electrode life can be obtained if proper operating conditions are employed in the welding process. The following factors shorten tungsten electrode life.

- *Erosion* due to low temperature is caused by operation at too low a current density.
- *Melting and dripping* are caused by operation at too high a current density.
- *Oxidation* is usually caused by improper shielding during the cooling period when the gas flow has been shut off too soon. When cool, tungsten electrodes should show a mirror-bright arcing end.
- *Contamination* by other metal is caused by the hot electrode touching filler rod or base metal. This is a particular problem with aluminum. The aluminum apparently vaporizes and disrupts the arc, causing sputtering and resulting in a black splotch on the work. This continues until the aluminum is removed. During this process the weld may become contaminated with the tungsten that is thrown off.

Water Supply

The tremendous heat of the arc and high current usually necessitate water cooling of the torch and power cable. If the water flow is stopped or reduced, the torch and cable will overheat and burn. Although adequate protection from heat is important, the equipment must also be lightweight and flexible for easy handling.

There are two methods of preventing overheating:

- A safety switch which opens and shuts off the electrical power source until the flow of water resumes
- A water-cooled, fusible link in the welding current circuit, Fig. 9-36

When the temperature begins to rise, the switch closes or the fuse blows, cutting off all power. When welding in close quarters, however, care must be taken so that the torch does not become overheated due to the heat generated by welding.

The cooling water must be clean. Otherwise, blocked passages may cause overheating and damage the equipment. It is

9-34. *Electrode extension and improved visibility are possible with use of the gas lens.*

Linde Division, Union Carbide Corp.

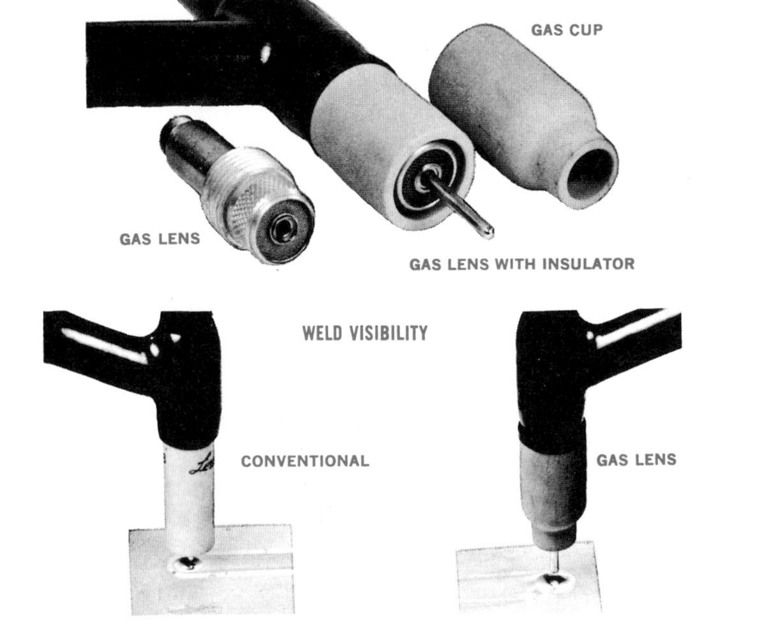

9-35. *Construction of the striped tungsten electrode.*

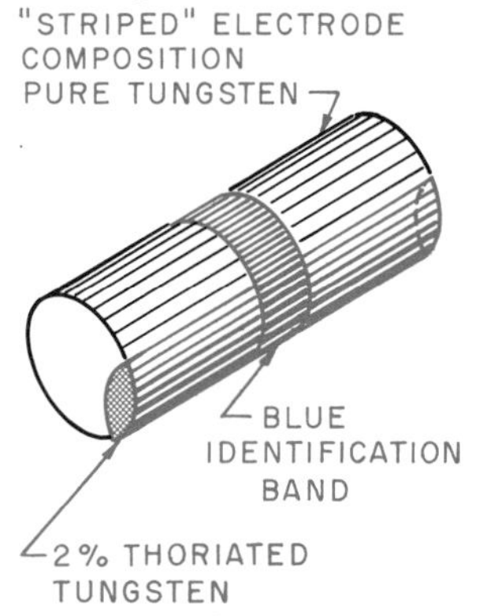

advisable to use a water strainer or filter at the water supply source. This prevents scale, rust, and dirt from entering the hose assembly. If the fuse should become clogged, it should be disassembled and cleaned.

The rate of water flow through the torch is important. Rates that are too low may decrease cooling efficiency. Rates that are too high damage the torch and service line.

Cables and Hoses

Tungsten electrode torches must be supplied with electric current and shielding gas, and, if water cooled, they must be provided with a supply of cooling water. Electricity, gas, and water are conducted to the torch by means of a copper cable and flexible hose. When the torch is a manual one, it is usually desirable to cool the current-conductor cable by enclosing it in one of the water supply hoses, which permits a very good flexible conductor, Fig. 9-31, page 274.

Plastic hoses are used for inert gases since helium will diffuse through ordinary rubber and fabric welding hose. Thus air can enter from the outside, causing contamination.

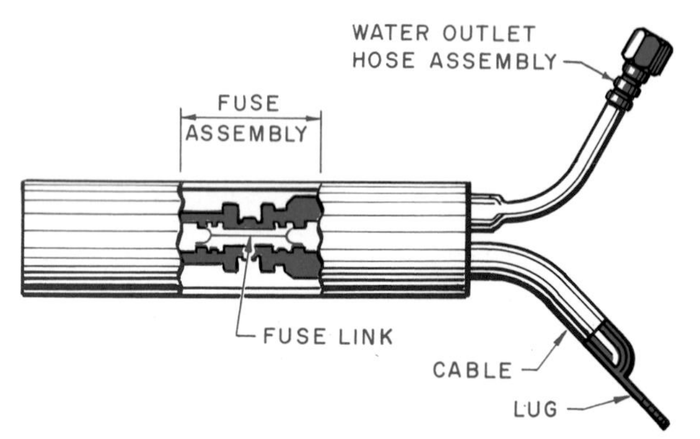

9-36. *Fuse and hose assembly.*

TIG HOT WIRE WELDING

TIG hot wire welding is a mechanized gas tungsten-arc welding process. The filler wire is preheated so that it is in a molten state as it enters the weld puddle. The molten filler metal is added behind the arc to form the weld, Fig. 9-37. The heat of the

Table 9-5
Current ranges for tungsten electrodes.

Electrode diameter (in.)	ACHF current[1]		DCSP current[2]				DCRP[3] current
	Pure tungsten	Thoriated	Pure tungsten		Thoriated		Either gas
	Argon	Argon	Argon	Helium	Argon	Helium	
0.010	8–15	8–20	8–15	8–20	10–25	10–30	—
0.020	10–20	10–25	10–30	15–35	15–35	15–45	—
0.040	15–60	15–80	20–70	25–80	25–80	30–90	—
1/16	40–140	50–150	50–140	60–150	50–150	60–160	10–20
3/32	90–160	130–250	120–230	130–240	135–240	140–250	15–30
1/8	140–210	220–340	220–360	230–390	225–380	250–400	25–40
5/32	180–275	280–440	360–450	380–500	370–480	390–500	30–50
3/16	240–340	380–540	440–740	480–780	450–780	490–800	40–80
1/4	280–490	480–790	740–950	750–1000	760–1000	780–1100	80–125

[1]Recommended for welding aluminum, magnesium, and their alloys
[2]Recommended for welding steels, stainless steels, and other metals
[3]Recommended only when minimum penetration and maximum surface cleaning are desired. Seldom used.

arc, therefore, is concentrated on the weld—not the wire.

TIG hot wire welding produces TIG quality welds at the higher speed of MIG welding. The hot wire process is up to four times faster than TIG cold wire welding. Welding progresses faster since the heat of the arc is concentrated entirely on the workpiece. Arc power is not used for melting the filler wire. Production time is further reduced by the fewer number of passes required to complete the weld, Fig. 9-38. Refer to Table 9-6 for a comparison of the TIG hot wire, TIG cold wire, and MIG processes.

When too much cold filler wire is fed under the arc, taking up most of the arc heat, cold lapping may result. In hot wire welding, the molten filler wire is introduced at the trailing edge of the weld. Cold lapping does not occur for two reasons: (1) the filler metal never passes under the arc; and (2) the arc heat concentrates directly on fusing the base metal.

9-37. *The weld pool and the addition of hot wire at the trailing edge. Note the highly fluid weld pool.*

Linde Division, Union Carbide Corp.

The view of the weld puddle is not blocked by the cold wire-feed assembly. Fusion of the workpiece under the arc is clearly visible since the hot wire torch follows the TIG torch. The welder can see the weld of the base metals without filler metal from one side and with filler metal from the other side. He also has a clear front view.

The welds are smooth, strong and ductile. Postweld cleaning is not necessary because oxidation and contamination are eliminated. This type of welding offers two important benefits: (1) porosity is eliminated since preheating tends to drive off contaminants; and (2) transfer efficiency approaches 100 percent because alloys are not burned up by passing under the arc.

The following equipment is needed for the process:

- Torch which transfers the welding current to the wire, guides the filler metal into the puddle, and provides inert-gas shielding as the wire heats, Fig. 9-39A and Fig. 9-40.
- Wire-feeding unit which provides instant change of wire feed speed from 50 to 825 inches per minute, Fig. 9-39B.
- Constant voltage a.c. power source, Fig. 9-39C.

TIG hot wire welding can be used to weld a variety of steels, titanium, and nickel alloys. It is not recommended for aluminum or copper due to the low resistance of filler wires. It is ideal for the fabrication of heavy wall missiles, rocket motor casings, pressure vessels, hydrospace equipment, and corrosion-resistant piping.

PLASMA ARC WELDING

This process (PAW), Fig. 9-41 (Page 280), is a fairly recent development in welding. Plasma arc cutting was introduced first in 1955, making it possible to cut aluminum and nonferrous metals with speed and precision. Plasma arc surfacing became available in 1960. It permits the deposition of a wide range of wear resistant alloys faster and

9-38. *A cross-sectional etched view of hot wire and cold wire TIG welds. Note that the hot wire process required fewer passes and appears to have a finer grain structure than the cold wire process.*

Linde Division, Union Carbide Corp.

WELD COMPARISON
5/8-in. thick 18% maraging steel

	Cold wire	Hot wire
deposition rate (lb./hr.)	3	11½
travel speed (in./min.)	8	12
number of passes	9	5
welding time (min./ft.)	14.4	5.3

TIG COLD WIRE

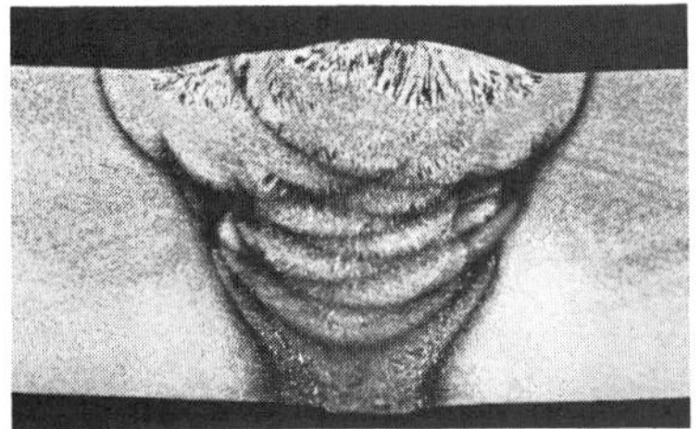

TIG HOT WIRE

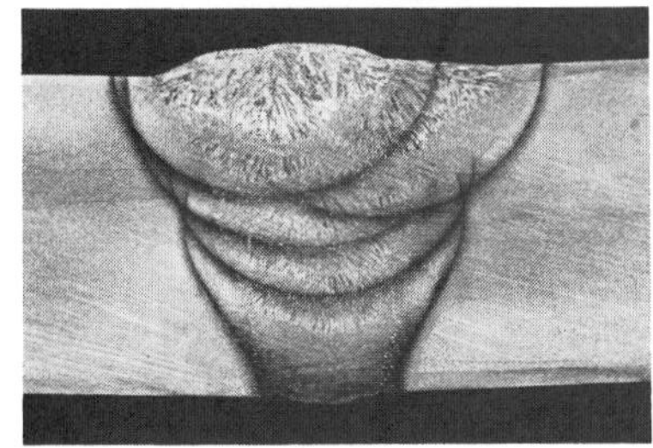

with better control than practically any other surfacing method. Then in 1963, welding on a commercial basis was developed. Today the plasma arc process is being used for cutting, plating or coating, weld surfacing, and welding. Plasma arc welding is a highly suitable process for the tough-to-weld metals such as stainless steels, titanium, and zirconium.

Plasma arc and gas tungsten-arc are much alike electrically. Both protect electrode and puddle with inert gas. Both use tungsten electrodes. There is a big difference, however, in the constriction (tightness) of the arc and in the use of the tungsten electrode. The TIG electrode sticks out of the gas cup on the torch. Its arc is conical, and its heat pattern on the work is wide with shallow penetration. In TIG welding a small change in torch standoff changes the heat pattern at the work a great deal, Fig. 9-42A. By comparison, the concentrated heat and the forceful jet of plasma arc produce a deep, but narrow, weld, Fig. 9-42B. Variations in torch standoff change little in the area of the arc spot on the work.

Plasma Arc Characteristics

What occurs in a plasma torch, Fig. 9-43, is that a plasma-forming gas is passed through a d.c. arc maintained between the *cathode* (negative pole), which is the tungsten electrode, and an *anode* (positive pole). The nozzle of the torch is small enough so that the gas is forced into intimate contact with the arc. As explained in Chapter 8, the in-

Table 9-6
Typical welding conditions for low and high alloy steels and heat- and corrosion-resistant materials.

	MIG	TIG hot wire	TIG cold wire
Arc current, amps.	350	450	350
Voltage, volts	26–28	12–14	11–13
Welding speed, in./min.	12–15	12–15	5–8
Deposition rate, lb./hr.	10–12	11–16	2–4
No. of passes, 1-in. plate	15	12	35
Weld soundness	Usually good. Wire quality is factor.	Excellent. Wire preheat eliminates major cause of porosity.	Very good.
Filler metals	Pass through arc, may be harmful to transfer efficiency.	Commercial alloys satisfactory.	High transfer efficiency. Commercial alloys usually satisfactory.
Heat input	High (spray) or low (short arc).	Controllable.	High.
Shielding gas	Argon-oxygen, helium, carbon dioxide singly or in combination.	Inert: argon-helium	Inert: argon-helium.
Equipment	Manual or mechanized.	Mechanized.	Manual or mechanized.
Process application	Limited by transfer efficiency and reactive gases.	Same as TIG cold wire plus exceptional weld soundness.	Very broad; transfer efficiency and inert gas retain desired properties.

Linde Division, Union Carbide Corp.

tense heat changes the gas into plasma. Its molecules are broken down into ionized atoms with a high energy content. Present working temperatures for the process fall in the 10,000 to 50,000 degree F. range, depending upon the application. Laboratory process have created temperatures up to 100,000 degrees F. Compare these flame temperatures with the 5,600 to 6,300 degrees F. of the oxyacetylene and arc flames.

In the plasma torch, the arc never touches the nozzle wall, and a layer of cool nonionized gas insulates the arc. Total power consumption is not so important as power concentration. (If 50 kilowatts are delivered through a nozzle $\frac{1}{8}$ inch in diameter, a power concentration of 3 megawatts p.s.i. results.)

TRANSFERRED AND NONTRANSFERRED ARCS. There are two types of arcs employed by the plasma equipment: transferred and nontransferred arcs. The *transferred* arc, Fig. 9-44, travels between the electrode and the work, which acts as an anode. Thus the arc heats the work with electrical energy and hot gas. This arc is used for plasma welding, weld surfacing and cutting.

The nontransferred arc is struck over a short distance and is entirely contained within the torch housing. The arc is struck between a tungsten cathode and usually a water-cooled copper anode. The transferred arc puts more heat on the work and is used mostly in plasma welding.

Fusion welding is performed with a transferred arc. The workpiece is connected into the electrical circuit by means of a

9-39. *A typical TIG hot wire installation: (A) torch, (B) wire feeder, and (C) power supply.*

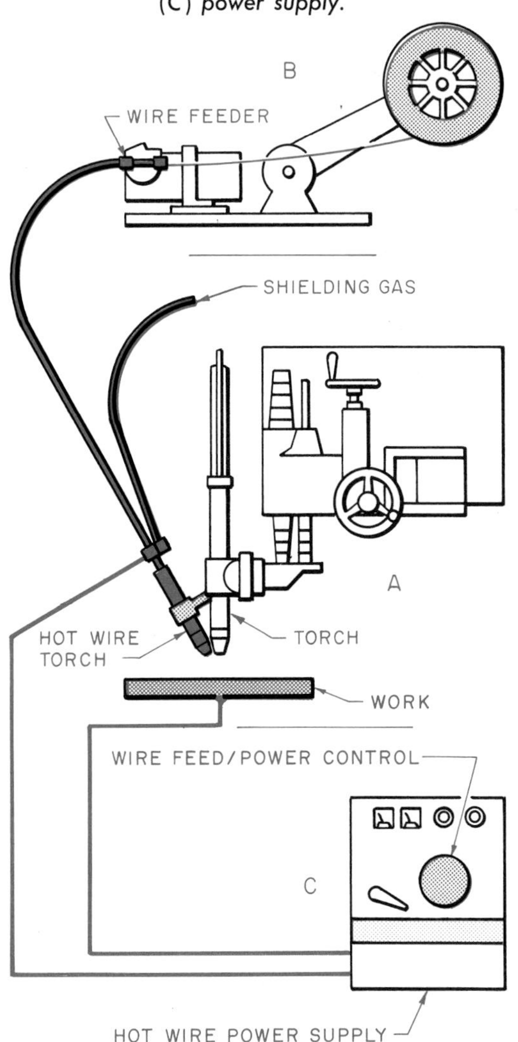

9-40. *The TIG hot wire welding system employs a special wirefeed torch mounted behind the TIG welding torch to supply molten filler metal to the weld puddle.*

Linde Division, Union Carbide Corp.

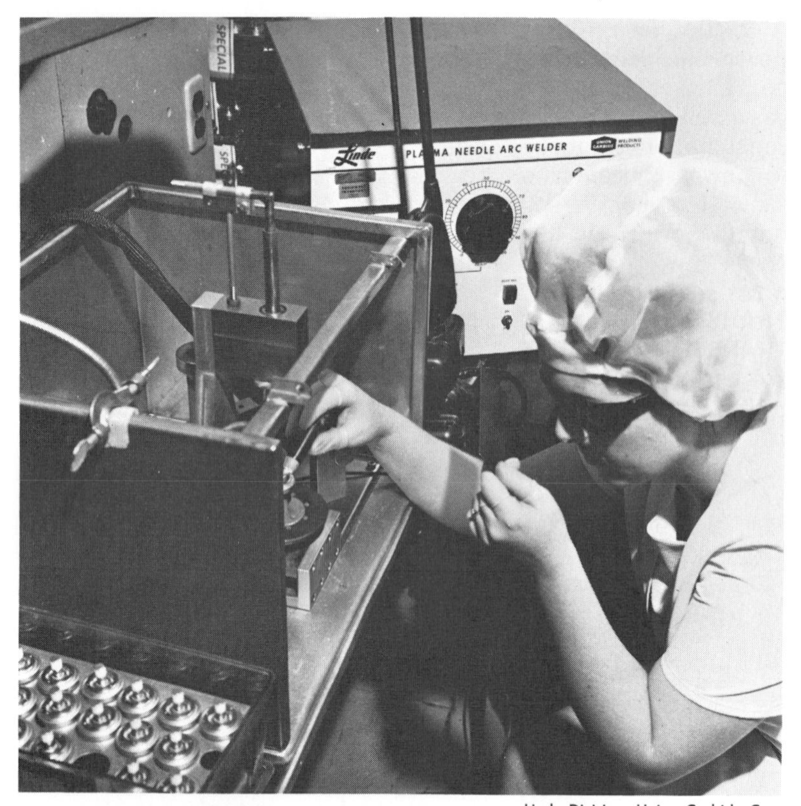

Linde Division, Union Carbide Corp.

9-41. *Welding with the manual plasma needle-arc process. Note how clean and cool the young woman appears to be. This is one of the advantages of the process.*

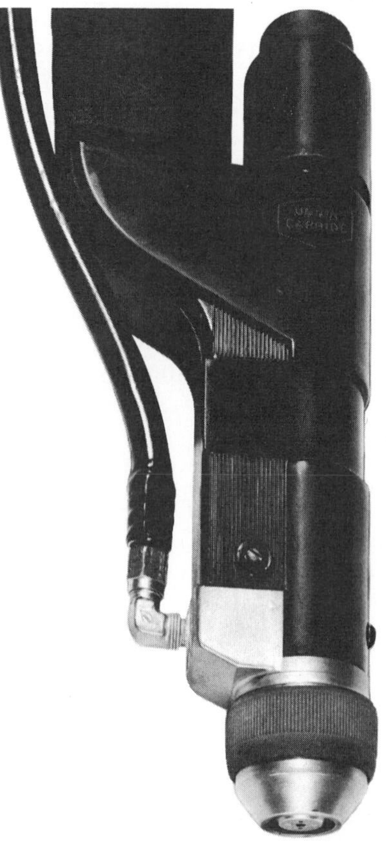

Linde Division, Union Carbide Corp.

9-43. *Plasma arc welding torch.*

A. TIG WELDING

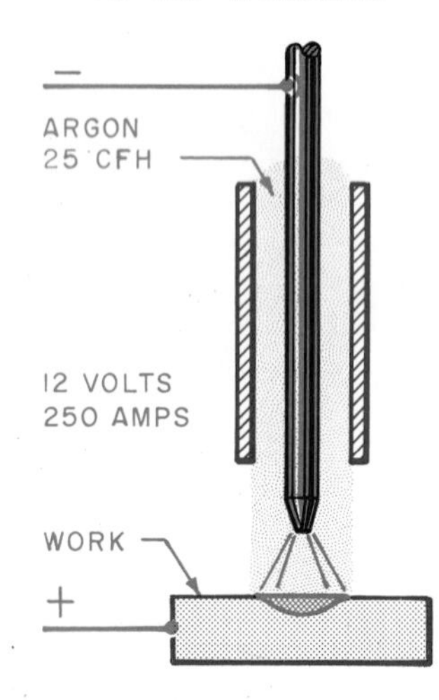

9-42. *Except for the pilot arc, gas tungsten-arc and plasma arc welding torches are nearly alike electrically. The difference between them is the constricted arc of plasma which forces the arc into a column and concentrates the heat at the work.*

B. PLASMA ARC WELDING

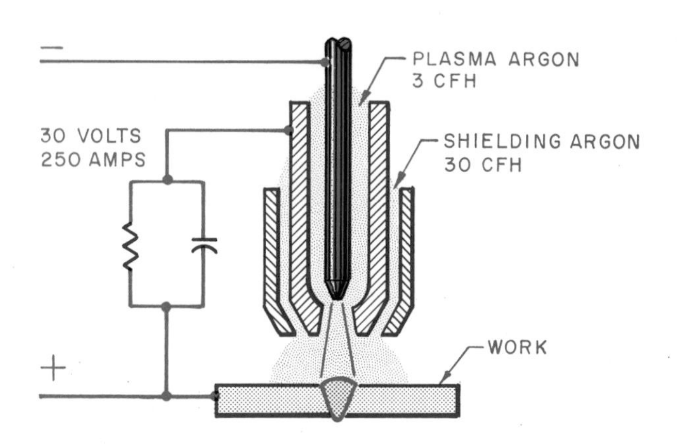

ground clamp. If the workpiece is not grounded, however, the arc is not transferred to the work but remains within the torch nozzle. Only the superheated jet of gas emerges from the torch nozzle and provides an effective heating tool for both metals and nonconductive materials. The nontransferred arc may be used in special welding applications when a lower heat concentration is desirable. Its main use is in joining or cutting nonconductive materials and for plating or coating.

Unique to plasma welding is the *keyhole effect* in which the plasma jet penetrates the workpiece completely. Plasma keyholes the front of the puddle because the jet blows aside the molten metal and lets the arc pass through the seam. As the torch moves, molten metal, supported by surface tension, flows in behind and fills the hole. The keyhole assures the operator of full-depth welds of uniform quality.

Equipment

In addition to the torch, the plasma flame system includes the power supply, control console (Fig. 9-45) and the welder's remote control station. The controls provide simplicity of operation. Pushbutton starting and stopping of the torch is remotely controlled. A water-cooling pump is usually needed to assure a controlled flow of water to the torch at a steady pressure. This is necessary to cool the torch nozzle and electrode. Study Fig. 9-46 which shows the relationship of the various equipment units needed to complete the plasma system.

The power supply shown in Fig. 9-47 is a heavy duty d.c. rectifier welder that can be used for high current plasma welding, TIG welding, and stick electrode welding. (Most plasma welding is performed using direct current, straight polarity.) It can be used as a plasma welder for welding steel or stainless steel in thicknesses from 0.001 inch to 0.125 inch in one pass. The welder includes water and gas solenoids, a postweld shielding gas timer, and a high frequency generator for TIG welding. Remote and foot controls are available. The controls contain cir-

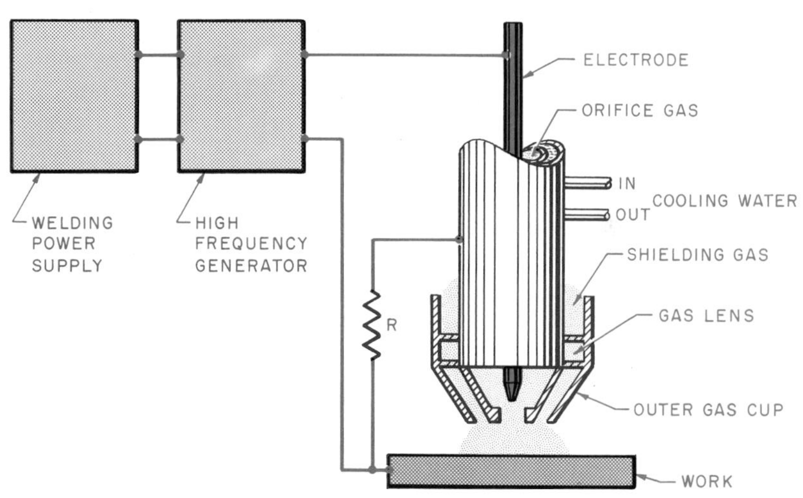

9-44. *A cutaway view of the plasma torch and plasma arc welding system. Note that the work is part of the electric circuit so that the arc is transferred.*

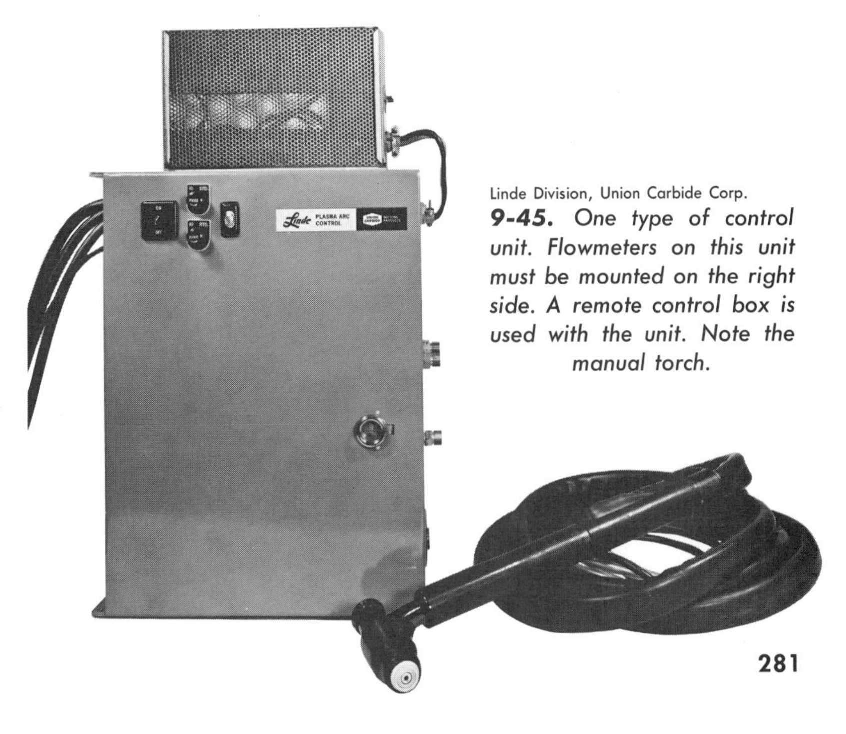

Linde Division, Union Carbide Corp.

9-45. *One type of control unit. Flowmeters on this unit must be mounted on the right side. A remote control box is used with the unit. Note the manual torch.*

cuitry for plasma needle-arc welding (low-current) and are mounted on top of the power supply. The water-cooled torch operates at 75 amps on a 100-percent duty cycle or 100 amps on a 60 percent duty cycle.

The plasma arc must be mechanized to utilize its advantages of speed and penetration. The torch is mounted on a carriage and is operated with voltage control units or wire feeders. The size of the main port in the nozzle depends on the amount of welding current it must carry: the higher the current, the bigger the port.

A multiport nozzle, Fig. 9-48, changes the shape of the arc jet and improves certain applications. The multiport nozzle converts the cylindrical plasma arc jet into an oval or elongated shape. The resulting change in heat pattern allows an additional increase in welding speed of 50 to 100 percent with a narrower heat-affected zone for many applications.

Welding Gases

In plasma welding, the shielding gas is usually the same as the plasma gas. Argon can be used with all metals, but it is not always the best gas for all jobs. Hydrogen mixed with argon gives a hotter arc and better transfer of heat, and it is faster. Too much hydrogen, however, causes porosity. Argon-hydrogen is used for plasma and shielding gases for keyhole welding stainless steel, Inconel,® nickel, and copper-nickel alloys. The amount of hydrogen in the mixture varies from 5 to 15 percent. In general, the thinner the work, the more hydrogen that can be used.

Argon should be used when welding the reactive alloys such as titanium and zirconium and for carbon and high strength steels.

Helium has limited use as a plasma gas. It overheats the nozzle, shortens its life, and cuts its current-carrying capacity. Helium is used only for melt-in welds. In a 50-percent mixture with argon, it gives a hotter flame for a given current. Argon-helium mixtures are generally used for filler and cover passes.

Plasma Needle-Arc Welding

The plasma needle-arc welding process is a manual operation. It uses a small diameter, constricted arc in a water-cooled nozzle. The arc is exceptionally stable, and in the low current range it can be maintained from 15 amps down to less than $\frac{1}{10}$ amp. Since the needle arc is cylindrical in shape, variations in torch-to-work distance do not substantially change the area of arc action on the work. This makes it excellent for welding contoured work.

Plasma needle-arc welding in the higher current range (15 to 100 amps) is a manual version of plasma arc welding. The system functions in both the melt-in (conventional fusion welding) and keyhole modes.

Keyholing provides 100-percent penetration on $\frac{1}{16}$- to $\frac{1}{8}$-inch metal. Arc focusing, brought about by arc constriction, makes more efficient use of arc heat. This results in narrower welds, and destortion is reduced because less total heat passes into the work.

With the exception of aluminum and magnesium, all metals that can be welded with the gas tungsten arc can be welded with the plasma needle arc. But even with a groove, the thickest plate that can be welded with plasma

9-46. *A typical plasma arc welding installation, showing the relationship of each unit.*

Linde Division, Union Carbide Corp.

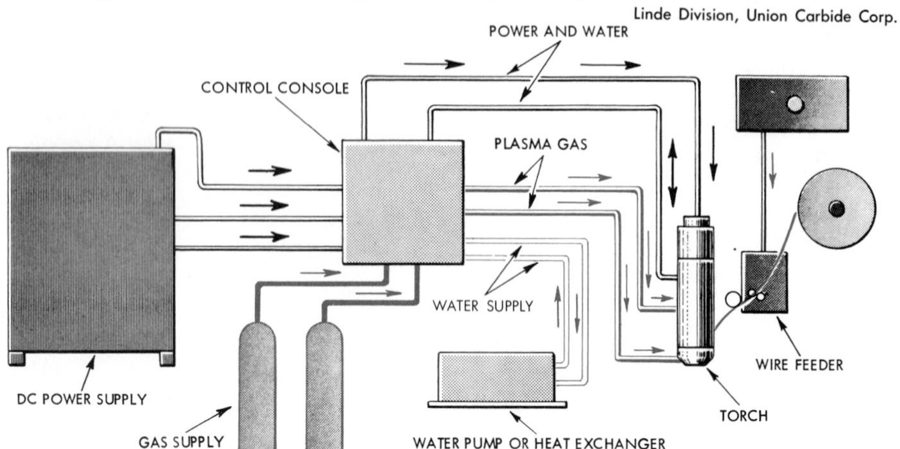

is 1 inch. Pipe of all kinds can be welded in all positions. Wall thicknesses of 1/4 inch and over must be beveled.

Table 9-7 gives the welding conditions for butt joints welded with the plasma needle-arc process. Figure 9-49 shows the torch position when welding a butt joint in the flat position with the plasma needle-arc process. Figure 9-50 shows the proper position for an edge joint.

Since the electrode is recessed in the torch nozzle, Fig. 9-51, there is no danger of electrode contamination or tungsten inclusions in the weld puddle. These factors, combined with exceptional arc stability and noncritical torch-to-work distance, make the system exceptional for many applications.

Figure 9-52 shows a low current needle-arc welder and accessories. The control-power supply unit includes the power supply and controls in a compact cabinet designed for bench mounting. Connections for water and gas supplies are at the rear

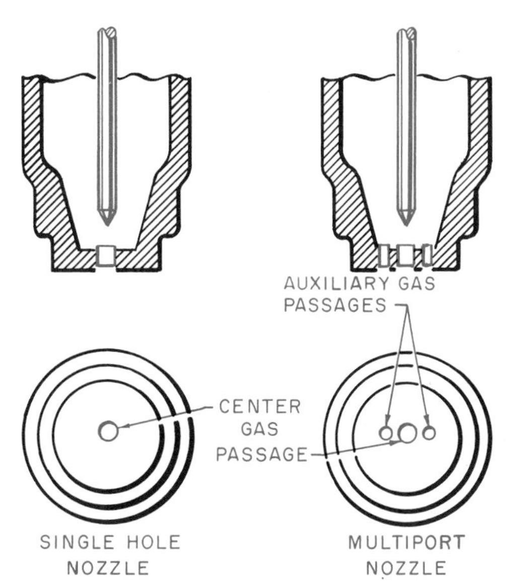

9-48. *Multiport and single-hole nozzles for plasma arc welding.*

9-47. *A plasma welding unit: d.c. rectifier welder, power control for needle arc circuitry, and foot control.*
Linde Division, Union Carbide Corp.

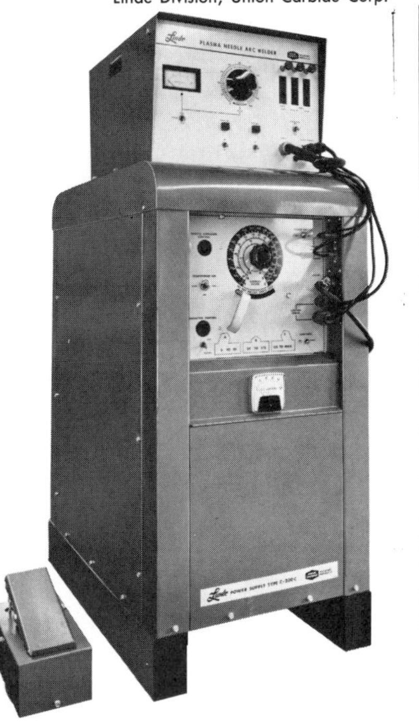

Table 9-7

Typical conditions for welding butt joints with the plasma needle-arc process.

Material	Thickness inches	Arc current amp. DCSP	Shielding gas mix	Welding speed i.p.m.
Stainless steel	0.031	10	Argon	3
	0.030	10	1% hydrogen-argon	5
	0.010	6	1% hydrogen-argon	8
	0.010	5.6	3% hydrogen-argon	15
	0.005	2	1% hydrogen-argon	5
	0.005	1.6	50% helium-argon	13
	0.003	1.6	1% hydrogen-argon	6
	0.001	0.3	1% hydrogen-argon	5
Titanium	0.022	10	75% helium-argon	7
	0.015	5.8	argon	5
	0.008	5	argon	5
	0.003	3	50% helium-argon	6
Inconel®—718	0.016	3.5	1% hydrogen-argon	6
	0.012	6	75% helium-argon	15
Hastelloy X	0.020	10	Argon	10
	0.010	5.8	Argon	18
	0.005	4.8	Argon	10
Copper	0.003	10	75% helium-argon	6

Linde Division, Union Carbide Corp.

of the cabinet. A built-in water pressure switch shuts the unit off automatically if the water supply fails. All control elements are located on the front panel. Three separate flow meters regulate nozzle and shielding gases and permit precise mixing of the shielding gases for best welding performance. A foot switch is plugged into the front panel to provide on-off control of the arc current. Figure 9-53 is a plasma needle-arc interconnection diagram.

The torch is lightweight and water cooled. The torch head is angled so that the welder can inspect the arc and workpiece continuously. Electrode replacement is easy, requiring only the removal of the knurled cap on top of the torch head. The torch can be adapted to mechanized welding operations.

ADVANTAGES. In general, plasma welding is designed for direct current, straight polarity welding of metals other than aluminum and for direct current, reverse polarity when welding aluminum.

Plasma needle-arc welding has many advantages.

1. A characteristic of plasma welding is high heat concentration, which provides the following advantages.
   a. Narrower weld beads restrict the heat-affected zone. This reduces the expansion and contraction of adjacent areas, resulting in less distortion and stress.
   b. The constricted arc provides greater penetration.
   c. Thicknesses from 1.8 inch down to 0.001 inch

Linde Division, Union Carbide Corp.

9-49. Welding with the manual plasma needle-arc welding process.

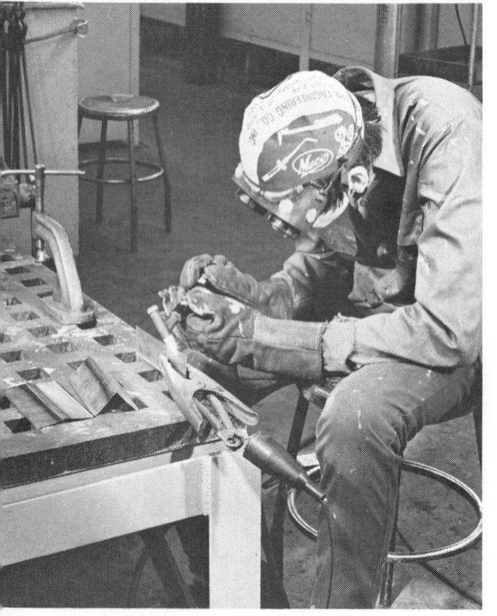

9-50. A student making an edge joint with the plasma needle arc.

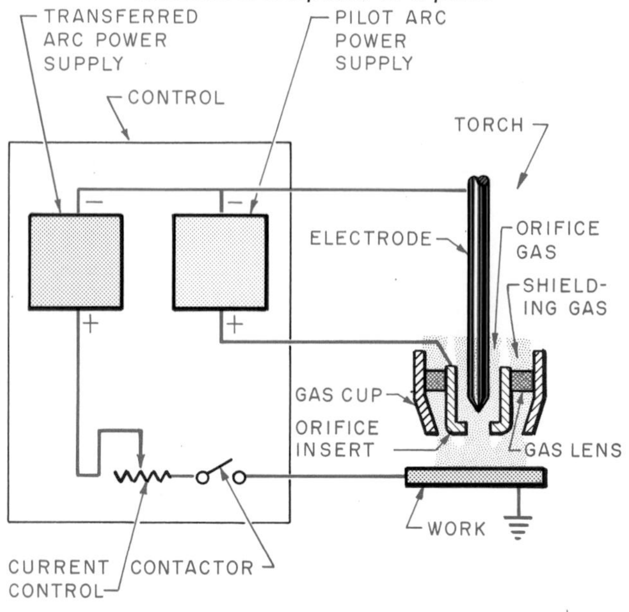

9-51. Plasma needle-arc operation schematic. Note that the tungsten electrode is sharpened to a point.

can be welded. Dissimilar metals of varying thicknesses can be welded.

d. There are no spatters, sparks, fumes, or flux.

e. A minimum of current is required.

2. A second characteristic of plasma welding is that greater standoff distance is permissible between torch and work. Standoff distance provides the following advantages.

a. Pilot arc starting is possible.

b. Less operator skill is required, and it is easy for a person with welding experience to learn the process.

c. Torch-to-work distance and torch angle are not critical.

d. The operator can see what he is welding. The long arc permits visual inspection, and the pilot arc spotlights welds.

e. Filler wire or overlay material can be added more easily.

f. The process can be either manual or mechanized.

3. A third characteristic of plasma welding is the protection of the tungsten electrode by the plasma torch. Electrode protection provides the following advantages.

a. Tungsten contamination of the weld is eliminated. Noncontamination is especially important when welding reactive metals.

b. Erosion of the electrode tip is retarded, and electrode life is prolonged.

c. Less downtime is required to change and adjust electrodes.

d. The lightweight torch is manuverable and minimizes welder fatigue.

e. Electrode protection is a factor in reliable arc starting.

4. A fourth characteristic of plasma welding is the soft and stiff arcs that may be obtained. Variable arc characteristics provide the following advantages:

a. A stiff arc is easy to direct. There is no wandering.

b. A stiff arc is usable in areas which have minor air drafts and magnetic fields.

c. The stiff arc is used to obtain maximum penetration.

d. The soft arc is used for fusion welding.

Plasma Arc Surfacing

Metal surfacing is done to give the surface of a part greater resistance to corrosion, abrasion, or impact. The extent to which the operation is successful depends upon keeping dilution with the base metal at a minimum. Study the process comparison in Table 9-8.

The plasma process is really a weld surfacing process. It permits precision-controlled, thin overlays of metals resistant to heat, wear, and corrosion. The deposit is fully bonded. Plasma arc surfacing provides close thickness tolerances that require less grinding and machining time to finish than deposits applied by other surfacing processes.

9-52. *A plasma needle-arc welding outfit with a power control unit, torch, ground clamp, and an on-and-off foot switch. A remote control device for arc current may be added.*

Linde Division, Union Carbide Corp.

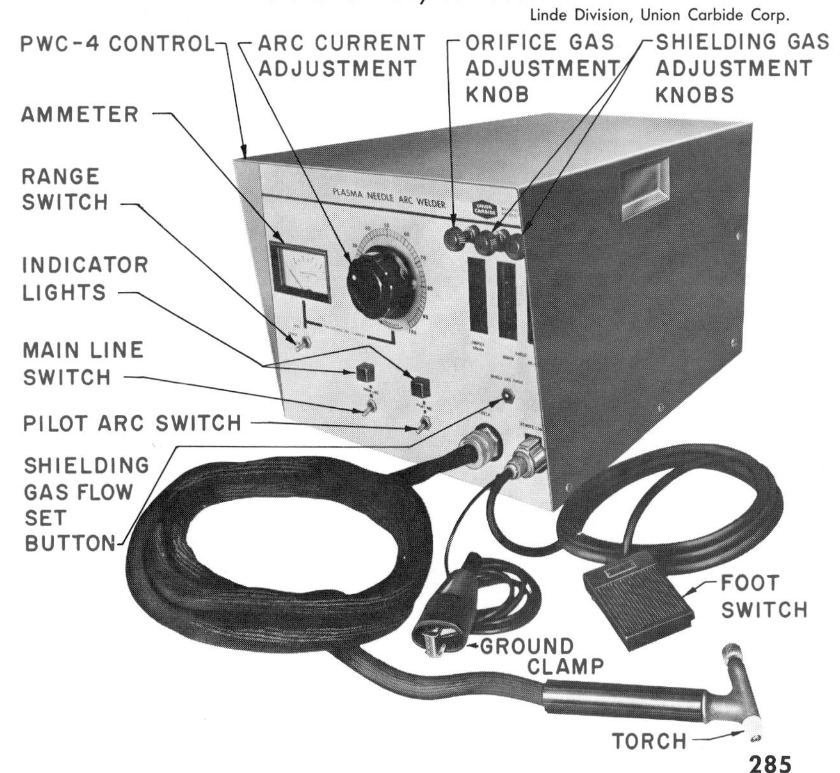

Because plasma arc surfacing uses powdered metals and alloys, it is not limited by wire availability. All metals can be powdered, but not all of them can be made into rod or wire. The plasma arc can be tailored to meet specific requirements and can be used to apply the following materials to carbon steel: cobalt, nickel- and iron-base hard-facing alloys, stainless steels, copper, and tin. Tungsten-carbide particles in a cobalt or nickel-base alloy matrix can also be applied.

The source of heat is the constricted tungsten arc. The powder is carried to the torch from a hopper. The arc melts the powder and fuses it to the work. Dilution is low, and deposition is high. Plasma arc surfacing is a true welding process, not a metal spray process. A direct-current, straight polarity power source, produces the transferred arc and controls the amount of heat delivered to the work. Argon forms the plasma, transports the metal powder, and shields both the overlay and base metal from oxidation and harmful gas pickup, Fig. 9-54.

Plasma arc surfacing works best on high production jobs. Typical surfacing applications include wear rings and plates, valve parts, wedges, slides, tillage and mower blades, and oil drill tool joints. Plasma arc surfacing has the following advantages:

- The depth of penetration into the base metal is precisely controlled to within 0.005 inch.
- The dilution of the surfacing material by the base metal is precisely controlled. The range of dilution is 5 to 50 percent.
- Layers as thin as 0.010 inch or thick as 1/4 inch can be deposited in each pass.
- Strips as narrow as 1/8 inch or wider than 1/4 inch can be deposited.
- Speeds over 20 i.p.m. can be attained with 95 percent deposition efficiency.

9-53. *An interconnection diagram for the plasma needle-arc welder shown in Fig. 9-52.*

Linde Division, Union Carbide Corp.

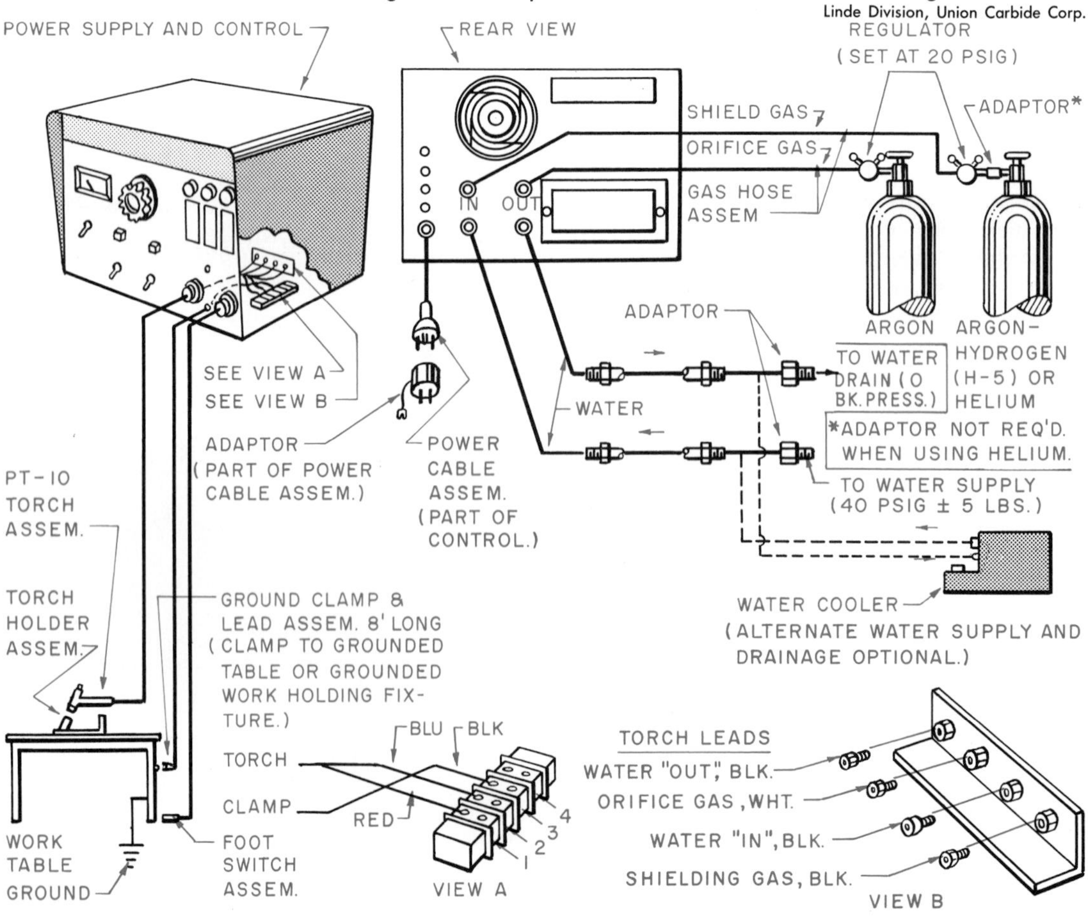

- Flat, smooth deposits require little finishing.
- There is a wide choice of alloy powders for surfacing.

LASER WELDING (LBW)

The development of the laser beam as a welding process is in its initial stages. The word *laser* means light amplification by stimulated electromagnetic radiation. We can all recall using a magnifying glass to concentrate the rays from the sun on a single spot in order to char or burn paper. The laser energy source is a refinement of this process.

The laser beam and carbon dioxide gas are used for welding, cutting, slitting, and drilling materials. The process joins microcircuit components, hairsprings, terminal connections, intriçate medical components, dissimilar metals, and heat-sensitive materials. The articulating arm shown in Fig. 9-55 can move in several directions, somewhat like a dentist's drill to manipulate the beam accurately by hand. The laser can drill minute holes in brittle, hard-to-machine materials and in dimensionally unstable nonmetals, as well as the more common metals. Figure 9-56 shows controlled cutting with the laser.

A number of limitations are encountered in the use of the laser beam for welding. Highly reflective metals such as aluminum and stainless steel are not easily welded because the bright surfaces of these metals reflect the greater part of the radiant energy of the laser. Another problem is associated with precise, accurate control of the pulse duration. Fluctuating values cause the laser to punch a hole through the metal instead of making a weld. Another very serious problem concerns the power source. More heat is generated in the power source than in the work. The laser is so inefficient that only about 2 percent of the input power can be transferred to the workpiece.

This process will undergo a considerable amount of improvement as more users have experience with it. Its ability to weld such reactive metals as tantalum and tungsten without an inert gas shield is of considerable value.

PRACTICE JOBS

The practice of gas tungsten-arc welding is covered in Chapter

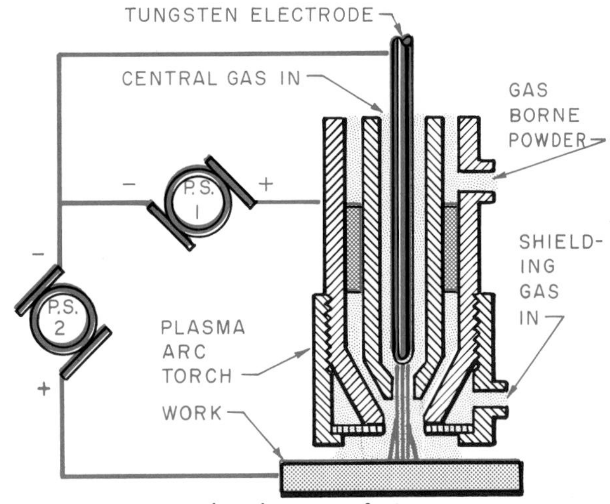

9-54. *The plasma surfacing system.*

Table 9-8

Surfacing process comparisons.

Process	Average deposition rate (lb./hr.)	Minimum weld dilution (%)	Minimum deposit thickness (in.)	Surfacing material form	Type of operation
Plasma arc weld surfacing	7	5	0.010	Powder	Mechanized
Oxy-acetylene	4	1	$\frac{1}{32}$	Rod	Mechanized & manual
TIG	5	10	$\frac{3}{32}$	Rod, wire	Mechanized & manual
Submerged-arc single wire	15	20	$\frac{1}{8}$	Wire	Mechanized & semi-automatic
Submerged-arc series circuit	30	15	$\frac{3}{16}$	Wire	Mechanized
MIG single wire	12	30	$\frac{1}{8}$	Wire	Mechanized & semi-automatic
MIG with aux. wire	25	20	$\frac{3}{16}$	Wire	Mechanized

Linde Division, Union Carbide Corp.

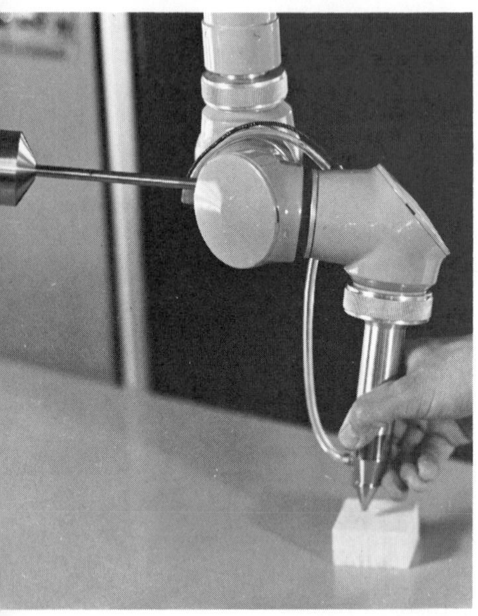

American Optical Corp.

9-55. Hand control of the carbon dioxide laser. Note the articulating arm. The laser head can be used for both welding and cutting.

24 for the welding of plate and Chapter 25 for the welding of pipe.

Practice in this process should not begin until you have mastered the gas welding process and the shielded metal-arc welding process and have successfully passed at least the welder qualification tests in the flat, horizontal, and vertical positions.

You will find that welding with the gas tungsten-arc process will be relatively easy because of the thorough practice you have had in the course of oxyacetylene welding and shielded metal-arc welding. Once you have mastered the handling of the flame and the arc, understand the equipment, and are able to tell when fusion and penetration are taking place, you will have no further difficulty with any other welding process.

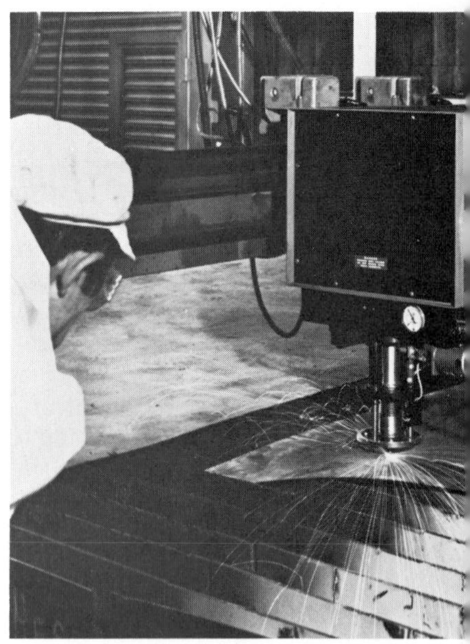

American Optical Corp.

9-56. Carbon dioxide laser machine with controlled cutting action. The unit is being guided by an optical tracing system.

REVIEW QUESTIONS

1. Give a brief account of the development of the gas tungsten-arc welding process.
2. Describe the gas tungsten-arc process in one or two sentences.
3. What are the advantages of gas tungsten-arc welding?
4. Name the metals that may be welded by the gas tungsten-arc welding process.
5. List the basic equipment needed for gas tungsten-arc welding.
6. List four important factors which shorten tungsten electrode life.
7. How are nozzle or cup sizes indicated? Explain.
8. Welds made with the gas tungsten-arc process are somewhat less ductile than those made with the shielded metal-arc process. True. False.
9. Explain the difference between the consumable electrode and the nonconsumable electrode processes.
10. In gas tungsten-arc welding a long electrode extension is highly desirable. True. False.
11. What is the function of the electrode holder? Describe its construction.
12. What are the functions of the regulator, flowmeter, and shutoff valve?
13. What are the principal gases that are used with gas tungsten-arc welding? Describe their function and characteristics.
14. When argon is the shielding gas, more gas is required than with other gases. True. False.
15. What is plasma arc welding?
16. What equipment is required for plasma arc welding?
17. List four major characteristics of plasma arc welding.
18. One of the disadvantages of plasma arc process is that it can be used only for welding. True. False.
19. The plasma arc process cannot be used for the welding of light gauge sheet metal. True. False.
20. Constant current welding machines are used for gas tungsten-arc welding. True. False.
21. What three types of nonconsumable electrodes are used in gas tungsten-arc welding? Compare them.
22. List the advantages of high-frequency, stabilized alternating current.

Chapter 10

Gas Metal-Arc Welding Principles

OVERVIEW

The gas metal-arc process, GMAW commonly referred to by welders as MIG, is becoming increasingly popular today.

Gas metal-arc welding, a continuous wire process, was introduced to industry in the early 1950s. At that time the process was limited by high cost, consumable wire electrodes; relatively crude wire-feeding systems; and welding machines that were not suited for MIG welding. The process was used principally for the welding of stainless steel and aluminum. The wires used were within the $\frac{1}{16}$ inch to $\frac{3}{32}$ inch range. Practical minimum stock thickness was $\frac{1}{4}$ inch.

Today MIG welding has become one of the more flexible all-round welding tools in the fabricating industry. A great number of types and sizes of low-cost wires have been developed, and simplified wire-feeding systems and torches are available. A significant step in power source improvement was the invention of the constant voltage welding machine. Thus the door was opened to the short-circuiting arc with its greater flexibility. For the first time, all-position MIG welding was possible, and sheet metal as thin as 22 gauge could be welded. Constant voltage machines have been further improved by the addition of variable voltage control, variable slope control, and variable inductance control. Control of these three factors permits the welder to fine tune the welding current for any desired condition. The introduction of carbon dioxide as a shielding gas extended the application of the process to a wide variety of mild steels on an economical basis. Figures 10-1 and 10-2 show shop applications of the MIG process.

The complete name for MIG welding is *inert-gas-shielded, metal-arc welding.* It also carries the trade names of various manufacturers of welding equipment, such as *Sigma Welding®* (Linde), *Millermatic Welding®* (Miller), *In-*

10-1. *Welding with the gas metal-arc process. Note the constant voltage welding machine, wire feeder, shielding gas cylinders, flow regulator, and torch.*

Miller Electric Manufacturing Co.

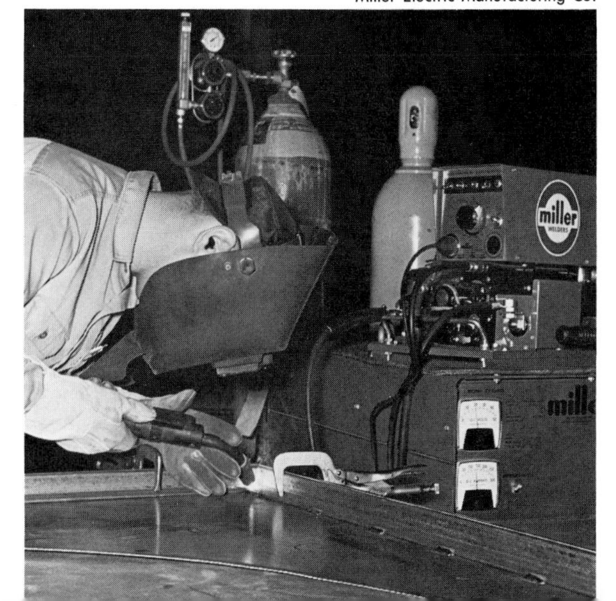

10-2. *Carbon-dioxide-shielded, metal-arc welding. The operator is making a splice for the top flange of bridge girders.*

Hobart Brothers Co.

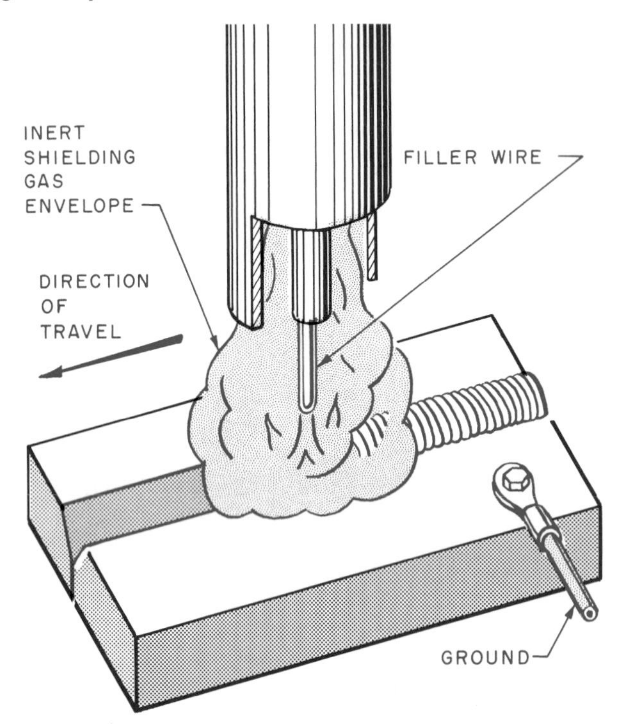

10-3. *Gas metal-arc shielding.*

nershield® (Lincoln), *Micro-wire Welding*® (Hobart) and *Aircomatic Welding*® (Airco). The short-circuiting method of metal transfer is known by the trade name of Dipmatic,® and submerged-arc CO_2 welding is derived from the use of carbon dioxide (CO_2) as a shielding gas.

Gas metal-arc welding is similar to gas tungsten-arc in some respects. For example, inert gas shields the weld area in both processes. In gas metal-arc welding, however, an electrode filler wire or a consumable bare electrode wire is used instead of a nonconsumable tungsten electrode. The wire is fed continuously into the weld by a wire-feed mechanism through a torch or gun. The weld area is surrounded by an inert gas blanket to protect it from atmospheric contamination, Fig. 10-3. The

10-4. *Schematic diagram of a gas metal-arc installation.*

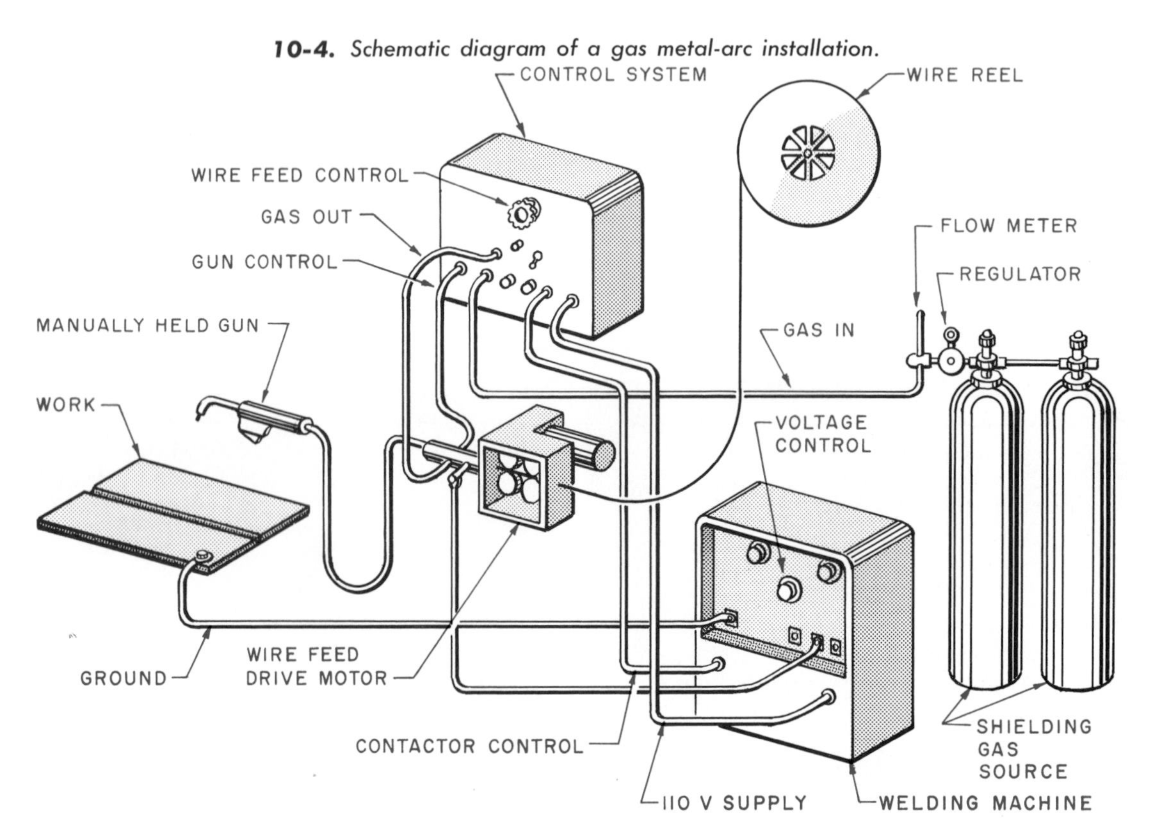

basic system is shown in Fig. 10-4.

Flux plays a part in some forms of MIG welding. It is commonly used as an improving agent in soldering, brazing, welding, and cutting processes. There are three ways in which flux may be used in MIG welding:

- It can be mixed as a fine powder in the inert gas, Fig. 10-5A.
- It can be used as a flux powder coating on the solid metal electrode as it leaves the torch nozzle. The electrode is magnetized so that iron particles in the flux cling to it, Fig. 10-5B.
- It constitutes the core of flux-cored tubular wire, Fig. 10-5C.

Welding can be an automatic process or a semi-automatic, manually controlled process. When completely automatic, the wire feed, power setting, gas flow, and travel of the welding head over the work are preset and proceed without manual direction. When semi-automatic, the wire feed, power setting, and gas flow are preset, but the torch is manually controlled over the work by the operator.

Advantages of Gas Metal-Arc Welding

Gas metal-arc welding produces high quality welds at high speeds without the use of flux or the necessity for postweld-cleaning. It is very desirable for both small jobs and high production metal joining. It frequently replaces another joining process such as riveting, brazing, silver-soldering, or resistance welding. It may be used instead of the following fusion welding processes: oxyacetylene welding, shielded metal-arc welding, submerged-arc welding, composite electrode automatic welding, and gas tungsten-arc welding.

When selecting a welding process for a given job, the welding method chosen is that which will do the best job at the lowest net cost. The decision is based upon a consideration of the costs of labor, equipment, electrodes and gas, material preparation, actual arc time, and postweld cleaning as well as the importance of weld soundness and appearance. The following are a few of the advantages of MIG welding:

- Welding operators who are proficient in the use of other welding processes can be readily trained in the MIG process. The equipment is simple to set up, and control of the process is incorporated into the welding equipment. The welder must watch the angle of the torch relative to the workpiece, the speed of travel, and the gas-shielding pattern.
- The welder can weld as fast as he is able. One of the principal advantages is the elimination of weld starting and stopping due to the changing of electrodes. This prevents weld failures due to slag inclusions, cold lapping, poor penetration, crater cracking, and poor fusion, which may result from starting and stopping to change electrodes.
- There is no welding slag or flux to remove when solid wire is used, and weld spatter is at a minimum. The process gives a smooth and good-appearing weld surface. These characteristics provide substantial cost savings since metal finishing is frequently a costly production item. Many manufacturers are able to paint or plate over MIG welds without additional surface preparation.
- Electrode costs are somewhat less due to the very small amount of electrode loss. It is estimated that when using shielded metal-arc welding electrodes, there is an average stub end loss of up to 17 percent and spatter and flux coating losses up to 27 percent. In figuring their costs, many fabricators provide for an electrode use of about 60 percent. In MIG welding, approximately 95 percent of the consumable electrode wire is deposited in the welded joint. The stub loss is totally eliminated.

10-5. *Three basic types of gas metal-arc welding processes.*

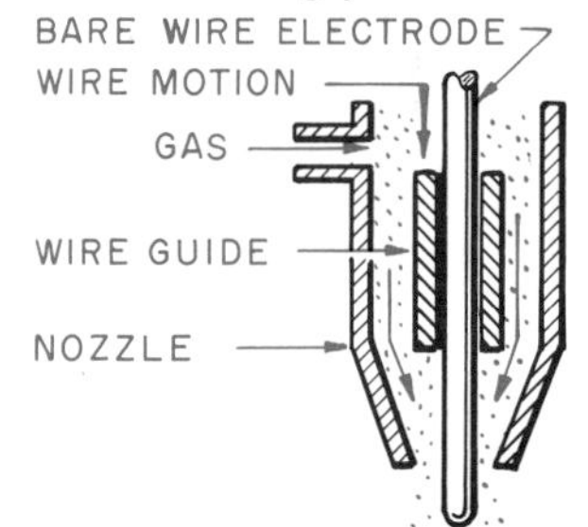

A GAS METAL-ARC WELDING

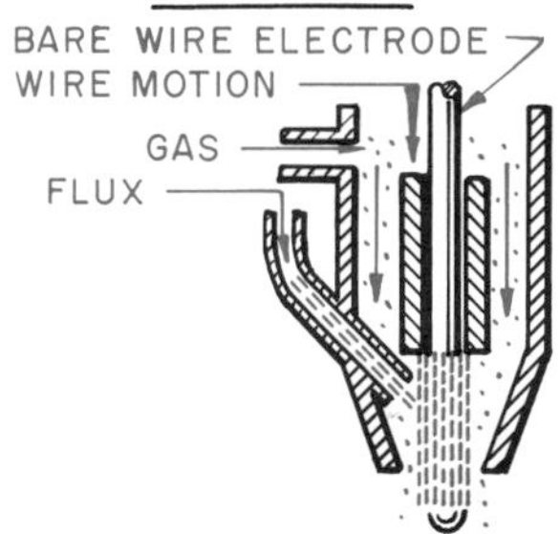

B GAS METAL-ARC WELDING WITH MAGNETIZED FLUX

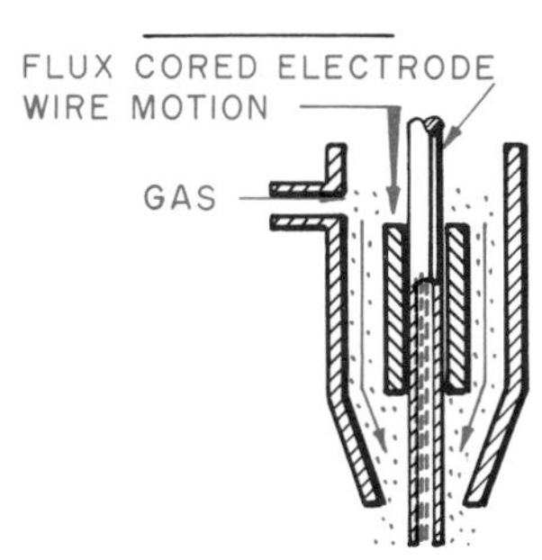

C GAS METAL-ARC WELDING WITH FLUX CORED WELDING WIRE.

- Welding may be done in all positions and on both light and heavy gauge materials. Weldable metal thicknesses range from 24 gauge to $\frac{1}{4}$ inch without edge preparation and with single-pass welding. Thicknesses above $\frac{1}{4}$ inch may be welded with edge preparation and multiple passes. For heavy materials a narrower beveled joint with a thicker root face can be used. A narrow groove for the weld joint reduces preparation time and the amount of filler metal deposit needed. A reduction of welding time and heat input keeps distortion at a minimum.
- The concentration of current (current density) is higher for MIG than for stick electrode welding, Fig. 10-6. *Current density* may be defined as the amperage per square inch of the cross-sectional area of the electrode. An arc with a high current density concentrates more energy at one point than one with low density. The MIG arc has a high current density, but the metal arc produced by the stick electrode does not. The stream of the MIG arc is sharp and incisive, but the stream of the metal arc is relatively soft and widespread.

The weld deposit made with a stick electrode is wider and more bowl-shaped than a deposit made with the MIG process, Fig. 10-7. The bead width-to-depth ratio is greater in shielded metal-arc welding and the heat input is less. Because of this characteristic, the speed of shielded metal-arc welding is slower, and more heat is applied per linear inch of weld. Since there is a greater volume of electrode deposited per linear inch of weld with the shielded metal-arc process than the gas metal-arc process, the heat input is greater for stick electrodes. With gas metal-arc welding there is greater penetration into the workpiece because there is higher current density at the electrode tip. The width of the normal shielded metal-arc weld deposit on $\frac{1}{4}$-inch plate is about $2\frac{1}{2}$ to 3 times the thickness of the plate, whereas the width of the gas metal-arc bead is about $1\frac{1}{4}$ the thickness of the plate when the weld is shielded with carbon dioxide.

Forms of Metal Transfer

In gas metal-arc welding an electric arc is struck between the metal being welded and a consumable wire electrode that is fed continuously through the torch at a controlled speed. At the same time a shielding gas is fed through the torch into the weld zone to protect the molten weld pool. MIG is a semi-automatic process. Welding current and wire-feed speed are electrically interlocked so that the welding arc is self-correcting.

The operator first sets the controls to give him the correct length of welding arc for the job at hand. During welding, if he holds the torch too close to the work, the current automatically increases, and the wire is burned off faster than it is fed until the correct arc length is re-established. If he raises the torch too high over the work, the current automatically decreases and the wire feeds faster than it is burned off, shortening the arc to the correct length.

In MIG welding, filler metal is transferred directly through the welding arc. There are two basic types of metal transfer. In the *open arc method* the molten metal is separated from the welding electrode, moved across the arc gap, and deposited as weld metal in the joint. In the *short circuit method* the weld metal is deposited by direct contact of the welding electrode with the base metal.

OPEN ARC TRANSFER. There are several types of open arc transfer methods: spray transfer, globular transfer, pulsed-

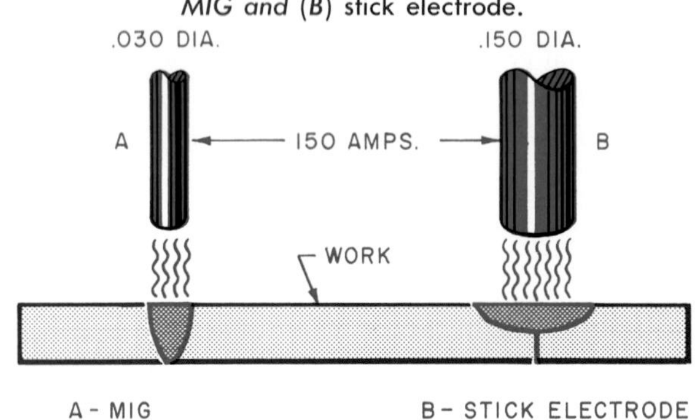

10-6. *Arc concentration and deposit comparison with the same current: (A) MIG and (B) stick electrode.*

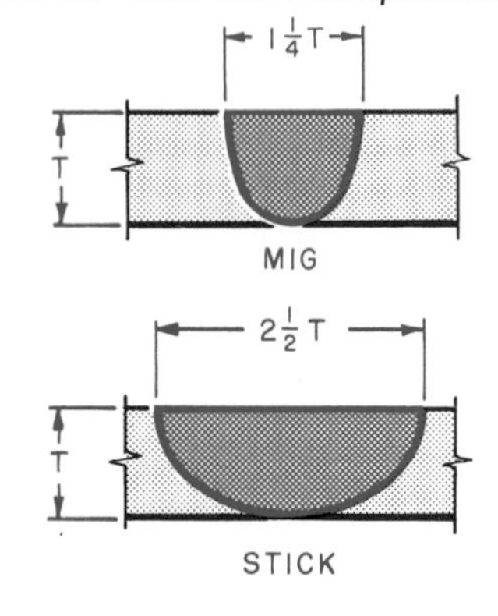

10-7. *Weld bead comparisons.*

current transfer, and submerged-arc transfer. The particular type of metal transfer depends upon the electrode wire size, the shielding gas, the welding current, and the arc voltage.

Spray transfer. Spray transfer is a high heat method with rapid deposition of weld metal. *Spray* describes the way in which the molten metal is transferred to the work. Very fine droplets of electrode metal are transferred rapidly through the arc from the electrode to the workpiece. The droplets are equal to or smaller than the diameter of the filler wire. There is almost a constant spray of metal, Fig. 10-8.

Spray transfer is used with inert gas shields and, when bare electrodes are used, only with direct current, reverse polarity. However, if the filler wire has been treated, direct current, straight polarity and alternating current have certain limited applications. The spray arc is almost spatter-free, provides deep weld penetration, and has self-regulating characteristics. Filler wire diameters for spray arc transfer are generally between 0.030 and $^3/_{32}$ inch in diameter. For each wire diameter, there is a certain minimum welding current that must be exceeded to achieve spray transfer, Table 10-1.

Because of its high deposition rate, spray transfer is recommended for materials which are $^1/_8$ inch or thicker, stock requiring heavy, single or multipass welds, and for any filler pass application where speed is advantageous. Due to high arc stability and the high rate of metal transfer, the spray transfer can be directed by the operator. Thus it is suitable for welding in the vertical and overhead positions.

Globular Transfer. Current range has a considerable influence on the type of metal transfer. At high currents filler metal is transferred as a spray, and at very low currents the filler metal is transferred in large *globules* (drops). In globular transfer the molten ball at the end of the electrode can grow in size until its diameter may reach two or three times the diameter of the filler wire before it separates from the electrode and is transferred across the arc, Fig. 10-9. Because it is a low voltage, low amperage method, the force of the arc is materially reduced. The physical forces react on the globule of metal and cause it to be highly unstable. There is a great deal of spatter with this method. The globular method of transfer is used with low currents and carbon dioxide shielding gas for the welding of thin gauge mild steel.

Pulsed-Arc Transfer. Pulsed-arc transfer is the latest MIG welding process to be developed. It is achieved by pulsing the current back and forth between the globular and spray transfer current ranges. It permits the use of spray transfer at much lower current levels than usual. It provides for the spray transfer to take place in evenly spaced pulses rather than continuously. The metal transfer from the electrode tip to the base metal occurs at regular intervals. The pulsing current has its peak in the spray transfer current range and its minimum value in the globular transfer current range.

Two levels of current are employed to achieve the pulsed level of transfer, Fig. 10-10. Both lev-

Table 10-1
Current and wire sizes for spray arc transfer.

Wire Dia, in.	Transition welding current amperes, DCRP
0.030	150
0.035	175
0.045	200
$^1/_{16}$	275
$^3/_{32}$	350

American Welding Society

Argon containing 5 percent oxygen

10-8. *Spray transfer.*

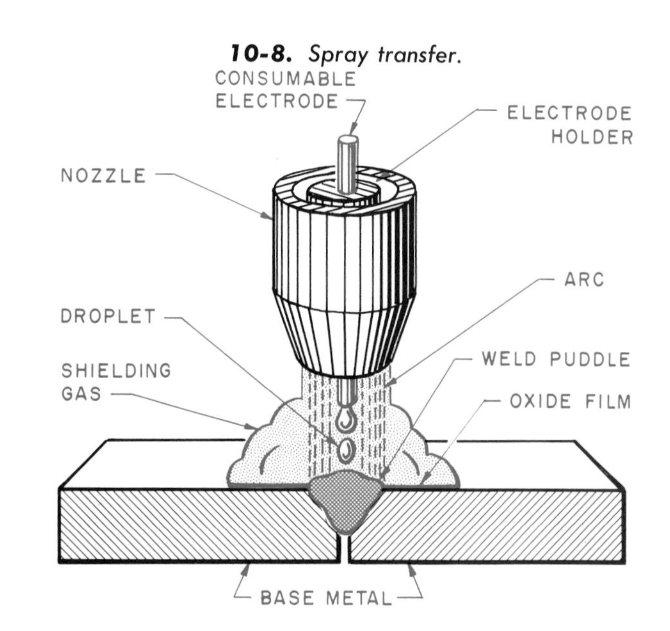

els of current come from a single power source, which, in fact, combines two power units which work together as a single unit. A standard three-phase power unit (See Fig. 10-28, page 305) supplies a steady background current which is just enough to maintain the arc between pulses while heating the filler wire and weld joint. The current level is supplied by a single-phase power unit which provides a high-density pulsing current at the rate of 60 pulses per second. This pulsed-current level should be set just above the minimum required for spray transfer.

During the pulsed-current cycle, the weld metal is sprayed smoothly across the arc and deposited in the weld joint. This type of transfer differs from the normal spray transfer in that the molten drops of filler metal are separated from the tip of the electrode at a regular frequency, corresponding to the frequency of current pulses supplied from the power source.

Pulsed-arc transfer permits use of the spray transfer of filler metal at low current levels than usual. This extends the advantages of the MIG process to all-position welding and to lighter gauges of metal than previously possible. It is effective in the welding of thin materials where burn-through is a problem due to high heat. It is also used for out-of-position work where it is difficult to maintain the molten pool of weld metal, Fig. 10-11. All of the advantages of open arc transfer are possible at average current levels, from the minimum possible for continuous spray transfer down to current values low in the globular transfer range. The method produces essentially spatter-free metal transfer.

The Submerged—Arc Transfer. The submerged-arc method of transfer is a process in which the metal transfer occurs below the surface of the base metal. Relatively high current and voltage are necessary. It is not unusual for a filler wire 0.045 inch in diameter to be run at 400 to 425 amperes at 35 to 37 volts. As a result of the high current values, the force of the arc digs a crater into the material being welded. The crater acts as a crucible for

10-9. Globular transfer.

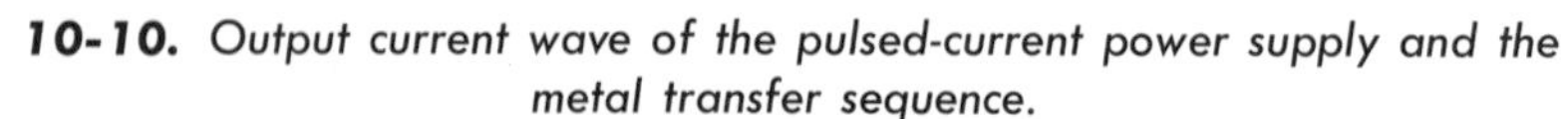

***10-10.** Output current wave of the pulsed-current power supply and the metal transfer sequence.*

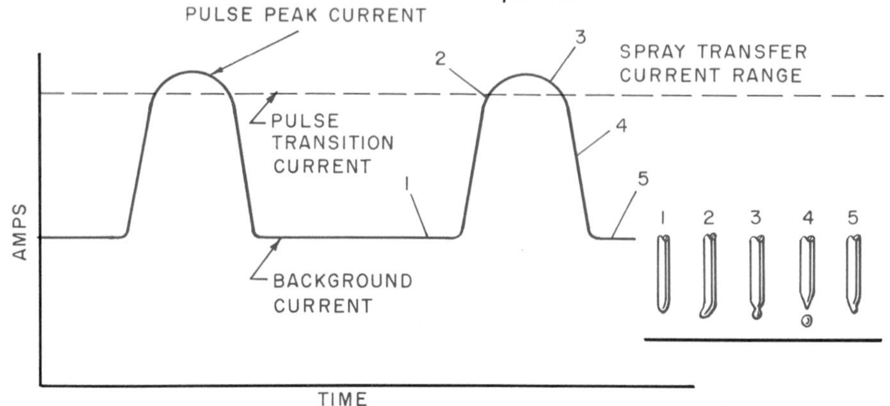

the weld metal and reduces spatter. This method of transfer is used with carbon dioxide as the shielding gas to weld mild steel.

SHORT CIRCUIT TRANSFER. In the early development of the gas metal-arc process, spray transfer was used almost exclusively. Welding with a short-circuiting arc is a relatively new application. No metal is transferred across the welding arc. Metal transfer takes place only when the filler wire makes contact with the material being welded. Metal deposited in this manner is less fluid and less penetrating than that formed with the spray arc.

At the start of the short arc cycle, the high temperature electric arc melts the filler wire into a globule of liquid metal. The electrode wire is fed at such a high rate that the molten tip of the filler wire contacts the workpiece before the globule can separate from the electrode. The contact with the workpiece causes a short circuit, the arc is extinguished, and metal transfer begins due to gravity and surface tension, Fig. 10-12. The process is actually a series of periodic short circuits that occur as the molten globule of metal contacts the workpiece and momentarily extinguishes the arc. *Pinch force,* a squeezing action common to all current carriers, breaks the molten metal bridge at the tip of the electrode, and a small drop of molten metal transfers to the weld puddle. Electrical contact is broken, causing the arc to reignite, and the short arc cycle begins again.

Short circuit transfer employs low currents, low voltages, and small-diameter filler wires. Currents range from 50 to 225 amperes; and voltages, from 12 to 22 volts. Filler wires with diameters of 0.030, 0.035, and 0.045 inch are used. Shorting occurs at a steady rate of 20 to over 200

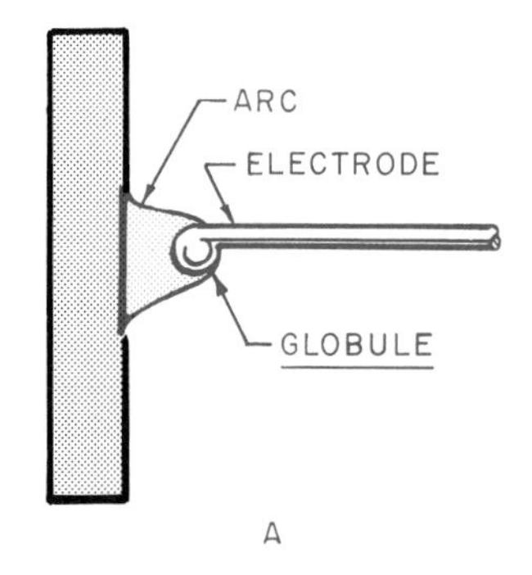

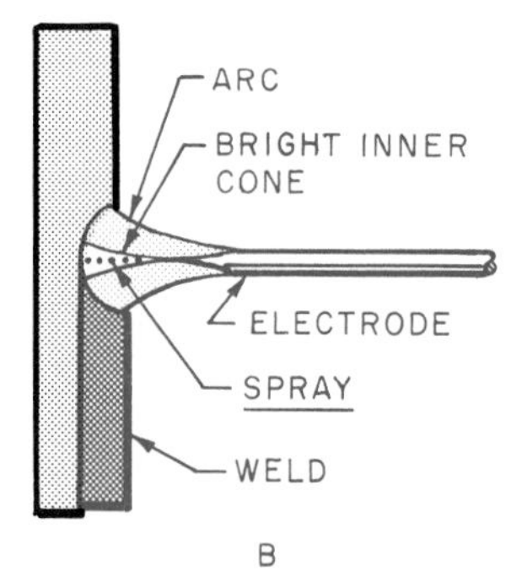

10-11. *An arc transferring metal by the (A) globular and (B) continuous and pulsed-spray modes is applied in the vertical welding position.*

10-12. *Complete short-arc cycle.*

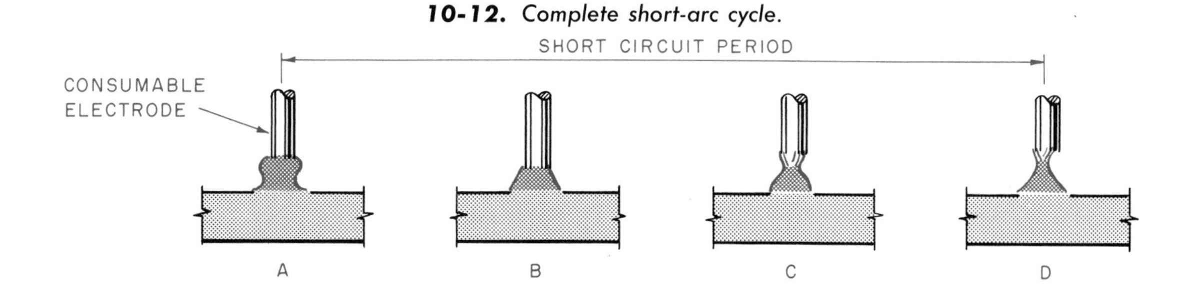

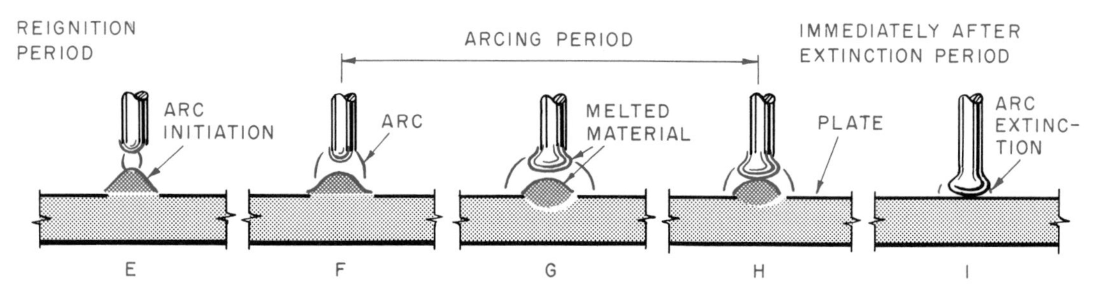

times a second according to preset conditions. The faster the wirefeed speed, the more short circuits per second.

The low heat input of this technique minimizes distortion, burn-through, and spatter. It is particularly useful for welding thin gauge materials in all positions. Short circuit transfer is finding increased use in the welding of heavy thicknesses of quenched and tempered steels. It is used for the welding of carbon steels, low alloy steels, stainless steels, and light gauge aluminum. When there is relatively poor fitup on a job, the short circuit transfer method permits the welder to bridge the wide gaps.

A full range of short circuit transfer cannot be obtained with constant current power supply units. Constant voltage machines are available with adjustable slope and appropriate voltage and inductance controls that produce the specific current surges which are needed for short circuit metal transfer over a full range.

All shielding gases may be used with the short circuit transfer method. Pure argon and helium and their mixtures are used with thin aluminum. Carbon dioxide or a mixture of 25 percent carbon dioxide and 75 percent argon may be used in the welding of carbon steel. There is a growing use of helium-argon-carbon dioxide mixtures for the welding of stainless steels.

MIG WELDING EQUIPMENT

The equipment necessary for MIG welding is in many respects similar to that needed for the TIG process. The success of the process depends upon the proper design and the matching of the various components. (Refer to Fig. 10-4, page 290.) The basic equipment includes the following:

- Welding machine or power source. This may be a motor-generator or transformer-rectifier with constant voltage output. (Constant voltage is also known as *constant potential.*)
- Consumable electrode wire-feed unit, including the controls
- Welding torch, also called a *gun.* A source of cooling water and flow controls are necessary for most guns.
- Interconnecting hose and cable assemblies leading to the gun. They provide for the gas, water, wire, and cable control wire needed for the process.
- Cylinder of shielding gas, cylinder regulator, and flowmeter
- Ground lead and ground clamp
- A reel of consumable electrode wire of the proper type and size for the particular welding job being performed.

Constant Voltage Characteristics

It is important that the student understand how the constant voltage system functions.

In MIG welding as in all arc welding, the heat required is generated by the arc that is produced between the workpiece and the end of the electrode. To sustain this arc for MIG welding, control of five items is necessary:

- The proper arc voltage on the wire
- Current going through the wire
- Appropriate wire-feed speed to replace continuously the wire melted by the heat of the arc to form the weld deposit
- Burnoff rate proportional to the current
- Arc action proportional to the arc voltage

Keeping these five factors in mind, let us see what happens when the welding operation is started. The welding power source is turned on, and when the electrode wire touches the grounded workpiece, current flows in the closed circuit. For an instant the arc voltage is zero, and because the power source is designed to maintain a preset voltage, it sends a tremendous surge of current (short circuit current) to the wire. Instantaneously, the wire gets white hot. It turns to molten metal and vaporizes to establish the arc. The nature of the arc is determined by the voltage setting set by the operator on the power source. These flawless, instantaneous arc starts are a characteristic of a constant voltage (potential) machine.

Once the arc is started, it is self-regulating over a wide range. The current is capable of wide variations while the arc voltage remains constant. Thus the wire can be fed at a constant speed. The welding current automatically adjusts itself to maintain the constant physical arc length and arc voltage. The power source supplies enough current to burn the wire off as fast as it is fed and to maintain the proper arc dictated by the voltage adjustment.

The constant voltage machine is flexible in meeting amperage and voltage demands over a wide area. Two students are urged to try this experiment: using any size welding electrode wire, set the proper wire feed speed and voltage for welding. One student should weld while the other observes the machine's amperage and voltage meters. When the student who is welding begins, the other will observe that amperage and voltage readings hold steady as long as the stickout remains about the same. The student welder should then lengthen the stickout without halting the welding operation and breaking the arc. The observer will note that as the stickout is lengthened, the amperage reading is reduced but the voltage reading remains constant. It should also be noted that the wire feed speed does not change. Thus it can be seen that voltage stays constant as set and that the machine makes internal changes in amperage.

Thus constant voltage welders insure uniform welds due to three inherent characteristics:

- The flow of welding current to the work is automatically adjusted to the rate at which the electrode wire is fed to the work.
- The wire-feed speed is constant.
- Instant arc starting eliminates defective spots at the start of the weld since the wire is practically vaporized by the instant current surge. Instant arc starting prevents the wire from sticking to the work, while the quick recovery to normal current prevents burn-back of the wire.

Output Slope

As stated previously, the power source is referred to as *constant voltage* or *constant potential* because the voltage at the output terminals varies relatively little over wide ranges of current.

The volt-ampere curve of the constant current power source, generally used for stick electrode welding, has a drooping output slope, Fig. 10-13A. When the arc length is constant, the welding current that will be obtained is shown by the point at which the voltage reading intersects the output slope. In shielded metal-arc welding, the operator is not able to maintain a constant arc length and, therefore, the voltage may deviate from the normal value. The deviation may be high or low, depending on arc manipulation. Figure 10-14 shows that a rather large increases or decrease in welding voltage, due to change in arc length when welding, does not result in a large change in current output. The operator can control the arc length by varying the speed at which he feeds the electrode into the welding arc. The constant current power source changes its voltage output in order to maintain a constant welding current level.

The volt-ampere curve of the constant voltage power source used for MIG welding has a relatively flat output slope, Fig. 10-13B. Small changes in arc voltage result in relatively large changes in welding current. Figure 10-14 shows that when arc length shortens slightly, a large increase in welding current occurs. This increases the burnoff rate, which brings the arc length back to the desired level.

The constant voltage power source puts out enough current so that the burnoff rate equals the wire-feed rate. If the wire-feed rate is increased, the power source puts out more welding

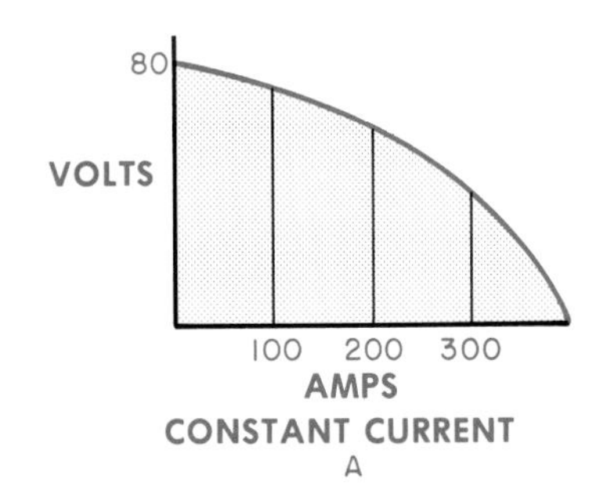

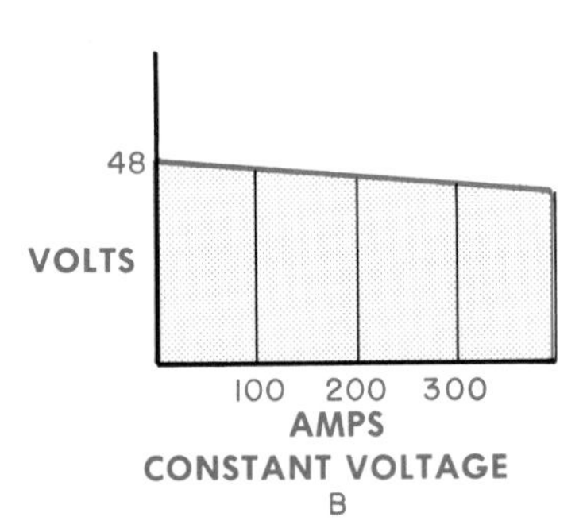

10-13. *Comparable volt-ampere curves.*

10-14. *Comparison of constant current and constant voltage characteristic volt-ampere curves.*

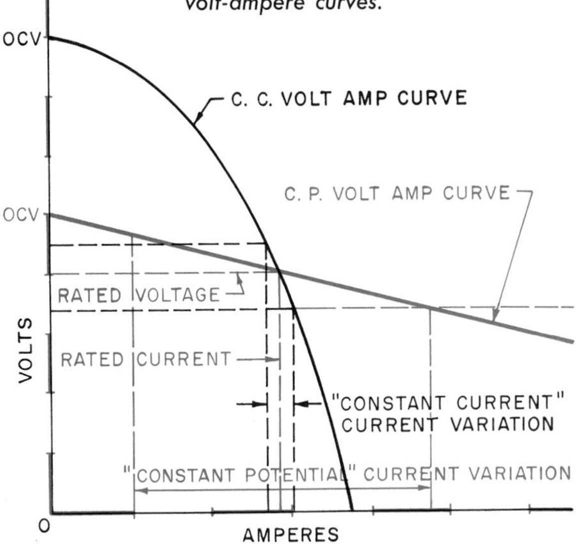

current so that the burnoff rate and the wire-feed rate are again balanced. The arc length is controlled by setting the voltage level on the welding power source, and the welding current is controlled by adjusting the wire-feed speed. Voltage during welding (arc voltage) does not vary much from the voltage produced while the machine is running idle (open circuit voltage). Arc voltage is set by the open circuit voltage control.

Short circuit current can be as high as several thousand amperes, usually 6 to 8 times the rated current capacity.

Types of Controls

All constant voltage machines have voltage control. Open circuit voltage on these machines is lower than on constant current machines. The maximum is usually about 50 volts.

Once the voltage has been set, the machine automatically maintains an arc length while the electrode wire is fed continuously at a set speed. Changes in filler metal, shielding gas, or granular flux may require a voltage adjustment.

Self-regulation of arc length is possible because small changes in voltage cause the machine to produce large changes in welding current. Thus any change in arc length (voltage) results in a large change in current and a corresponding, almost instantaneous change, in the burnoff rate of the electrode wire. This regulates the current at the arc.

Constant voltage welders do not have a heat adjustment like the constant current machines. They do have an arc voltage control and slope and inductance adjustments for arc stabilization. A small transformer is included to supply 115-volt, single-phase power for operating auxiliary

10-15. *A 300-ampere, d.c. constant voltage rectifier with continuous slope and voltage controls.*
Glenn Pacific Co.

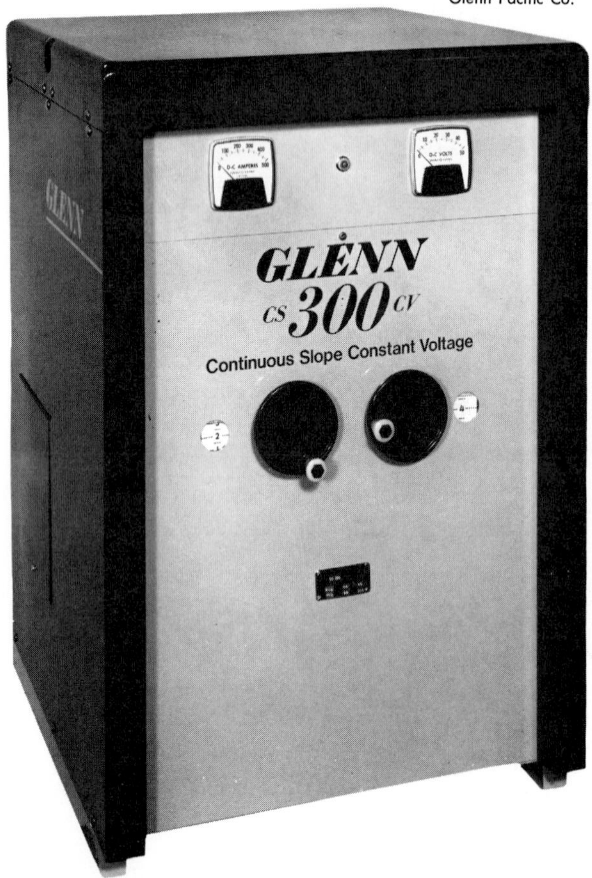

10-16. *Internal construction of the rectifier shown in Fig. 10-15. (A) Variable transformer, (B) variable reactor, (C) silicon rectifier assembly, (D) cooling fan at the bottom right. The small transformer (E) is for voltage control.*
Glenn Pacific Co.

equipment such as the wire drive motor, coolant pump, and gas and water solenoid valves, Figs. 10-15 and 10-16. A built-in primary contactor controls the flow of the welding current and voltage to the electrode when the trigger on the gun is pulled.

Constant voltage power supplies provide extreme flexibility. With variable slope, voltage, and inductance, each unit can be adjusted, within its current and voltage range, to provide ideal arc characteristics for any specific application. A single machine can be used for short-circuiting applications, spray arc, TIG, submerged arc, stick electrode, and any other welding process within its range of operation. These machines combine the two basic types of power supplies in one single unit. By turning a selector switch on the machine, a steep slope (constant current) or a flat slope (constant voltage) can be produced.

VOLTAGE CONTROLS. Voltage controls may be either tapped (stepped) or continuously variable. *Tapped-voltage control* is sometimes used for more common types of spray-arc MIG welding. A voltage range tap switch is employed. The control panel shown in Fig. 10-17 has the following controls.

- *Voltmeter and ammeter set:* These meters permit accurate, repeatable selection of weld current and voltage.
- *Voltage range switch:* This switch offers the operator a choice of three open circuit voltage ranges: high (39–51), medium (30–39), and low (24–31). Wire size and work will determine which voltage range is required.
- *Fine voltage adjustment:* The hand control on the panel of the welder provides infinite fine voltage adjustment within the range selected. Fine voltage changes can be made *while welding.* This permits the operator to obtain exact welding settings under welding load conditions.
- *Voltage indicator:* This indicator shows open circuit voltage settings on the welding machine for the particular range selected.
- *115-volt a.c. power receptacle:* This receptacle provides 115-volt power for operation of the wire drive equipment. Power for this circuit is supplied from a built-in control transformer.
- *Contactor control receptacle and switch:* Provisions are made for a 2-conductor lead from the wire drive control to the contactor control receptacle. This completes the circuitry necessary for remote control of the welding power. The actuating mechanism for closing the circuit is the gun trigger switch. When it is closed, the primary contactor in the power source is energized, and welding power is available.
- *115-volt duplex receptacles:* these receptacles provide power for water coolant systems, fixtures, etc. The power for them comes from the same control transformer that supplies power for the wire drive equipment.
- *On-off power switch:* this switch controls the electrical power to the fan motor and the 115-volt power outlets.

10-17. *Control panel of a typical constant voltage MIG welding power source.*
Miller Electric Manufacturing Co.

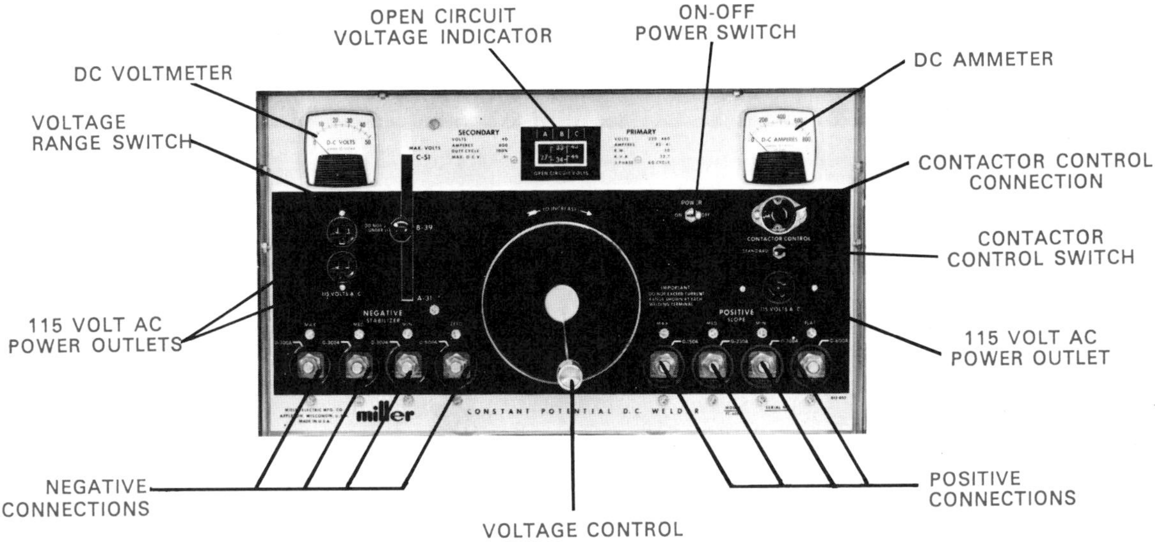

OPEN CIRCUIT VOLTAGE INDICATOR
ON-OFF POWER SWITCH
DC VOLTMETER
DC AMMETER
VOLTAGE RANGE SWITCH
CONTACTOR CONTROL CONNECTION
CONTACTOR CONTROL SWITCH
115 VOLT AC POWER OUTLETS
115 VOLT AC POWER OUTLET
NEGATIVE CONNECTIONS
POSITIVE CONNECTIONS
VOLTAGE CONTROL

- *Cooling fan:* the fan is driven by a prelubricated, completely sealed ball bearing motor and provides adequate cooling to all components.
- *Current surge protector:* this device guards the transformer and rectifier in the event of abnormal operation of welder.

Continuously variable voltage control is commonly used for more critical spray-arc MIG welding, and short-circuit MIG weldings, Figs. 10-18 and 10-19. On machines of 300 amperes or more, a tap switch usually sets the voltage range for the variable voltage control.

CURRENT CONTROLS. There is no current adjustment dial on a constant voltage unit. Current is adjusted by the wire-feed speed setting on the wire drive unit.

SLOPE CONTROLS. Slope controls the amount of current change. It is caused by an impedance to the current flowing through the welding power circuit. (Impedance means the slowing down of a moving object.) Since voltage is the force that causes current to flow but does not flow itself, the impedance is directed toward limiting the flow of current. It is not intended that the current be stopped, only limited. As more impedance is added to the welding circuit, there is a steeper slope to the volt-ampere curve. The steeper slope limits the available short circuit current and slows the machine's rate of response to changing arc conditions. Constant voltage d.c. transformer-rectifiers are available with fixed slope or variable slope.

Fixed-slope machines are generally used for ordinary types of spray-arc MIG welding and the welding of mild steel with a carbon dioxide shield. A certain amount of fixed slope is designed into the machine, and no adjustment is needed for most jobs, Fig. 10-20.

In the past, the term slope was used only in connection with the gas tungsten-arc and the shielded metal-arc processes. An increase of welding current was referred to as *upslope;* and a decrease of current, as *downslope.* The operating characteristics and versatility of the constant voltage power sources used for gas metal-arc welding have been accomplished through

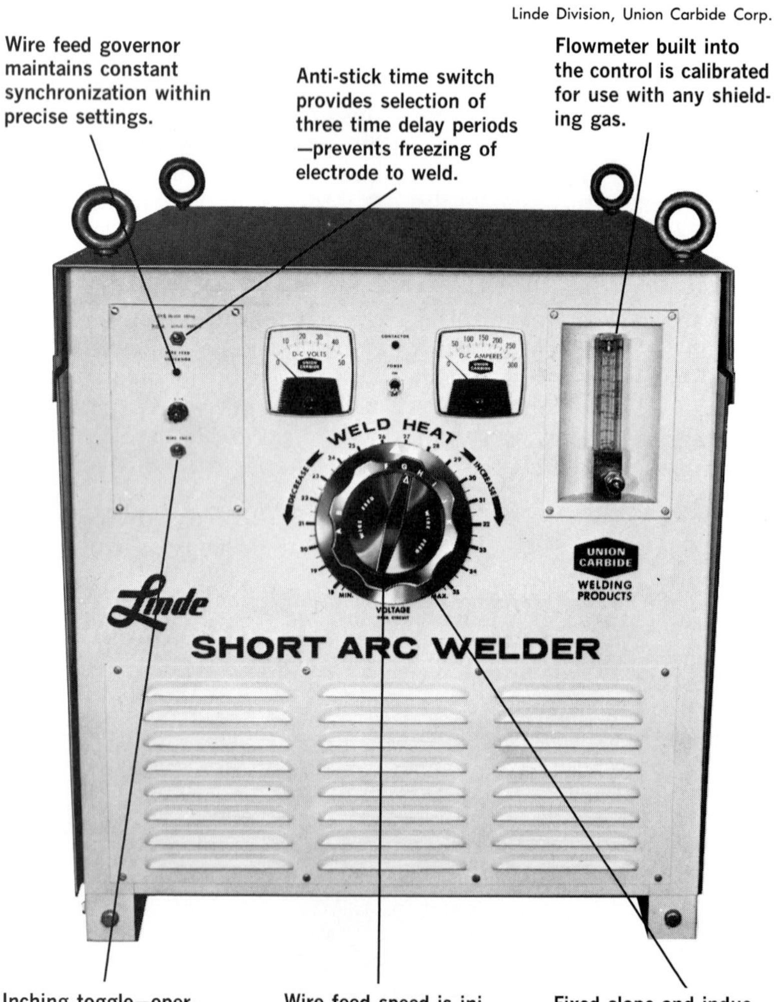
Wire feed governor maintains constant synchronization within precise settings.
Anti-stick time switch provides selection of three time delay periods —prevents freezing of electrode to weld.
Flowmeter built into the control is calibrated for use with any shielding gas.
D-C VOLTS
D-C AMPERES
WELD HEAT
DECREASE
INCREASE
UNION CARBIDE
WELDING PRODUCTS
Linde
SHORT ARC WELDER
Inching toggle—operator can inch wire without energizing power supply.
Wire feed speed is initially set on inner dial. Thereafter only outer dial is used to adjust weld heat—automatically maintains best wire feed current ratios.
Fixed slope and inductance are preset for ideal short arc welding.

10-18. *A 200-ampere, d.c. constant voltage transformer-rectifier for short arc MIG welding. The single-knob control changes both voltage and wire-feed speed automatically in perfect synchronization over the carbon steel arc range. The machine also has preset inductance and slope.*

Linde Division, Union Carbide Corp.

the introduction of variable slope control.

Variable *slope machines,* Fig. 10-21, permit the use of a wider range of wire types and sizes under a variety of MIG welding conditions. By changing the slope of the flat volt-ampere curve, the short-circuiting transfer method of welding has been improved. The sudden surge that takes place when the electrode makes contact in starting the weld is decreased, and the weld puddle can be kept more fluid. Continuous adjustment of the slope from flat to steep permits precise control of short circuit current. Slope adjustments also reduces weld spatter. Variable slope machines are preferred for the welding of stainless steel.

Variable slope controls may be tapped or continuously variable, and they are similar to those controlling voltage. Tapped-slope machines are more versatile, but continously variable slope controls are more precise. Figure 10-22 indicates the current curves possible with a machine having seven tapped-slope settings. Figure 10-23 illustrates the current curves possible with a machine equipped with continously variable control.

The reactor used for variable slope control consists of one or more current-carrying coils placed around an iron core. In tapped-current control, the operator adjusts the slope by means of the mechanical connections through taps that are connected with the coil. In continuously variable control, contact brushes move over the face of one side of the coil that has been machined.

INDUCTANCE CONTROLS. Inductance controls the rate of current change. Adjustment of inductance is most common with short-arc MIG welding to control weld spatter. *Fixed inductance* is common on many constant voltage machines, and no further adjustment is normally required. Figure 10-16, page 298 is a machine of this type. *Variable inductance* may be tapped or continuously variable. Controls are similar to those used for voltage and slope.

Variable voltage and inductance controls give precise control of the weld pool, regulate the frequency of short circuiting in short circuit arc transfer, and reduce spatter.

OTHER CONTROLS. A weld timer, most often offered as an optional plug-in unit, permits MIG spot welding and other timed welding.

Remote on-off control of welding current, Figs. 10-24 and 10-25, is an important feature to have on a machine. The dual schedule control shown in Fig. 10-25 (Page 304) has the following applications.

- Hot starts give extra penetration.
- The control provides a low range for root passes and a high range for hot and cover passes.
- It is ideal for tack welding and can bridge extra-wide gaps.
- The operator can select one condition for overhead or vertical welding and switch to another

10-19. *Close-up of single-knob control shown in Fig. 10-16.*

Linde Division, Union Carbide Corp.

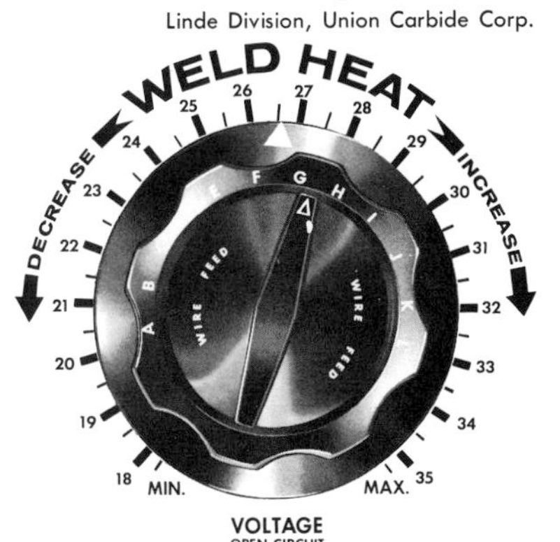

10-20. *A 200-ampere, d.c. constant voltage transformer-rectifier with built-in predetermined slope control.*

Westinghouse Electric Co.

AMMETER
VOLTMETER
ON-OFF SWITCH
HIGH-LOW VOLTAGE CONTROL
VOLTAGE CONTROL
CONTACTOR CONTROL
GROUND TAP
WIRE FEEDER TAP
115 AC OUTLET

preselected condition for flat welding.

- A change in the welding condition from one size of material to another can be made without readjustment.

The operator can change from Schedule A to Schedule B or vice versa with the gun trigger or foot switch at any time during the course of the weld without interrupting the arc. A pilot light on the panel indicates which setting is being used.

High frequency adjustment, gas and water control, and post-purge timing are other important features to have on a machine.

Types of Constant Voltage Welding Machines

Constant voltage machines are classified by the manner in which they produce welding current. There are two general types:

- *Transformer-rectifiers* which have an electrical transformer to step down the a.c. line power voltage to a.c. welding voltage and a rectifier to convert the alternating current to direct current
- *Motor-generators* which have an electric motor or internal combustion engine to drive a generator that produces direct current

D. C. TRANSFORMER - RECTIFIERS. Constant voltage transformer-rectifiers may be used for MIG, submerged arc, and multi-

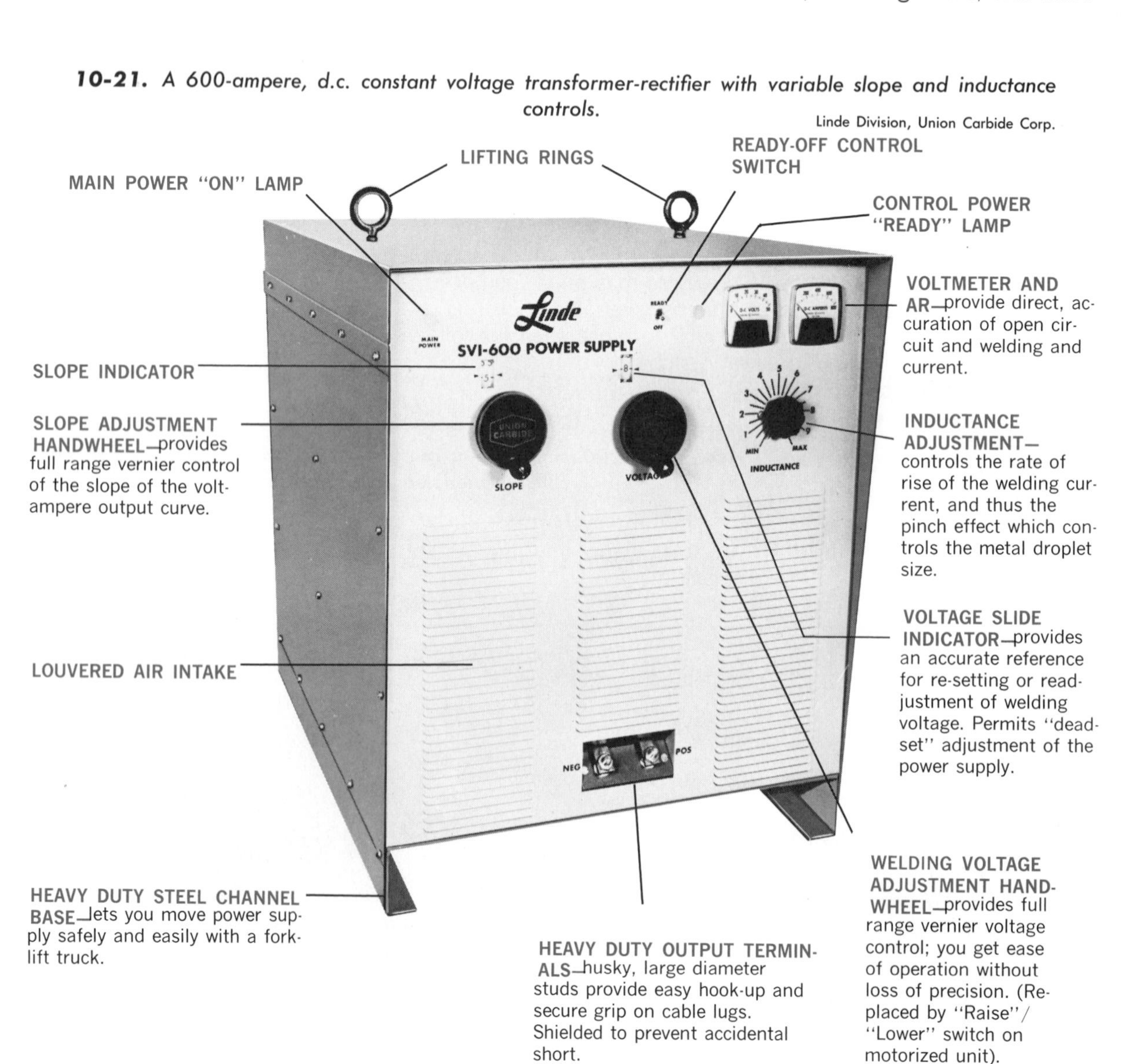
LIFTING RINGS
READY-OFF CONTROL SWITCH
MAIN POWER "ON" LAMP
CONTROL POWER "READY" LAMP
VOLTMETER AND AR—provide direct, accuration of open circuit and welding and current.
SLOPE INDICATOR
SLOPE ADJUSTMENT HANDWHEEL—provides full range vernier control of the slope of the volt-ampere output curve.
INDUCTANCE ADJUSTMENT—controls the rate of rise of the welding current, and thus the pinch effect which controls the metal droplet size.
VOLTAGE SLIDE INDICATOR—provides an accurate reference for re-setting or readjustment of welding voltage. Permits "dead-set" adjustment of the power supply.
LOUVERED AIR INTAKE
HEAVY DUTY STEEL CHANNEL BASE—lets you move power supply safely and easily with a fork-lift truck.
HEAVY DUTY OUTPUT TERMINALS—husky, large diameter studs provide easy hook-up and secure grip on cable lugs. Shielded to prevent accidental short.
WELDING VOLTAGE ADJUSTMENT HANDWHEEL—provides full range vernier voltage control; you get ease of operation without loss of precision. (Replaced by "Raise"/"Lower" switch on motorized unit).
Linde
SVI-600 POWER SUPPLY
SLOPE
INDUCTANCE
MIN
MAX
NEG
POS

***10-21.** A 600-ampere, d.c. constant voltage transformer-rectifier with variable slope and inductance controls.*

Linde Division, Union Carbide Corp.

ple-operator systems. They are three-phase power supplies with voltage control, slope, and/or inductance control. Selection depends on the type of welding process, the type and size of filler wire used, the end use of the weldment, cost, and operator preference. Figures 10-26 to 10-29 show the transformer-rectifiers for different MIG welding applications. The welder shown in Fig. 10-29 (Page 306) can be used for stick, MIG, and TIG welding. This is possible because of a controlled-slope characteristic. Controlled slope permits 15 separate output settings plus fine voltage adjustment control of all positions.

MOTOR-GENERATORS. Constant voltage welding generators usually supply d.c. welding current. They are used for MIG, some submerged-arc welding, and as multiple-operator power sources.

Generally, the transformer-rectifier is preferred to the generator because of its faster response time. Old motor-generator sets have a response time two to three times slower than the transformer-rectifier units. However, new motor-generators are being introduced that have highly improved performance in this respect. One of the benefits of faster response time is quicker arc starts so that there is less possibility of cold-lapping at the beginning of the weld. Motor-generators are classified by the type of motor or engine driving the generator.

The common motor-generator, Fig. 10-30, is driven by an a.c. electric motor connected to primary line power.

Engine-driven welding generators are used in the field and for emergency work when electric line power is not available. The engine is usually a gasoline or diesel type. Fuel is gasoline, diesel oil, or propane gas.

These generators differ from constant current generators in their controls. As previously described, constant voltage generators have controls for voltage instead of current. Controls for slope and inductance are also available. Controls may be continuous or tapped.

Machine Selection

Constant voltage machines are used with the MIG and submerged-arc welding processes. They are better suited for welding processes in which the consumable electrode wire is fed continuously into the arc at a fixed rate of speed because they permit self-regulation of arc length.

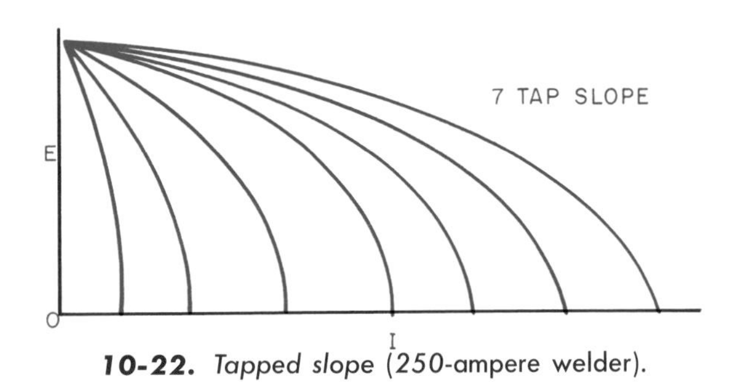

10-22. *Tapped slope (250-ampere welder).*

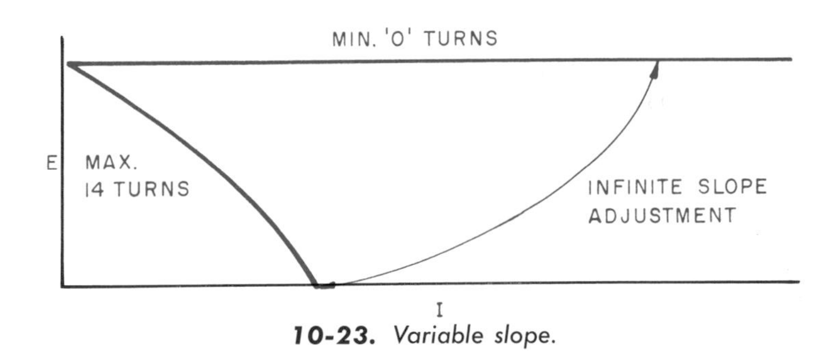

10-23. *Variable slope.*

Westinghouse Electric Co.

10-24. *Remote on-off foot control for semiautomatic welding.*

MIG power supply units for both manual and automatic operation are almost always d.c. constant voltage machines. Alternating current has been found to be generally unsuitable because the burnoff rate is erratic. A few a.c. applications are possible with a coated filler wire.

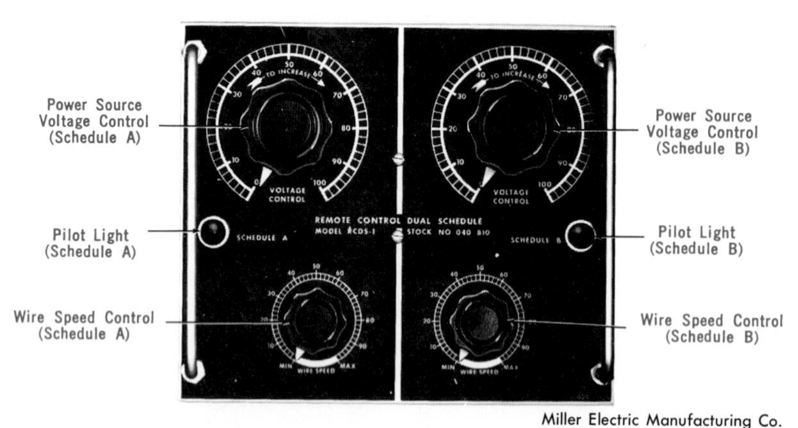

Miller Electric Manufacturing Co.

10-25. *Dual schedule control for use with the Miller constant voltage d.c. welder. It provides two individual preset voltage and amperage (wire speed) welding settings with one machine and one wire feeder.*

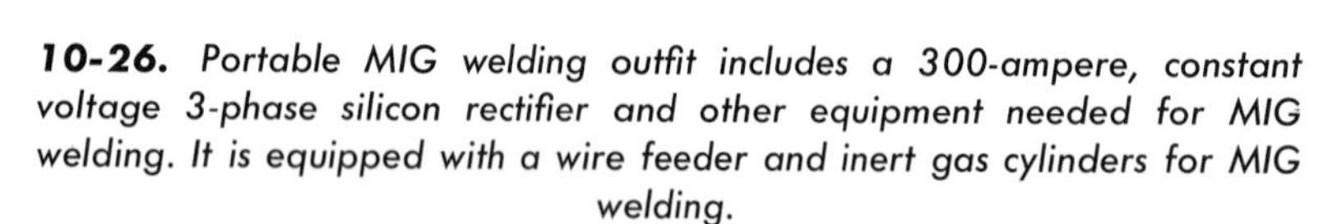

10-26. *Portable MIG welding outfit includes a 300-ampere, constant voltage 3-phase silicon rectifier and other equipment needed for MIG welding. It is equipped with a wire feeder and inert gas cylinders for MIG welding.*

Hobart Brothers Co.

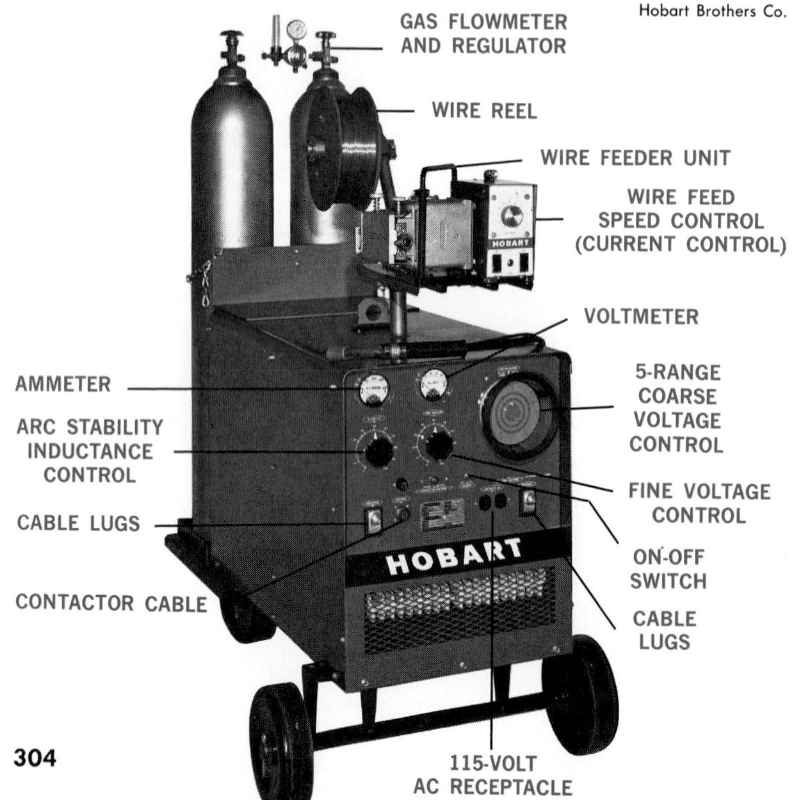

Most MIG welding is done with reverse polarity. DCRP gives better melting control, deeper penetration, and good cleaning action. This is especially important in welding such oxide-forming metals as aluminum and magnesium. Straight polarity is seldom used. The wire burnoff rate is greater with DCSP, and the arc instability and spatter are problems for most applications.

The selection of the proper machine size for a particular job is always a problem. It should be pointed out, however, that each machine has a wide range of application.

Spray-arc welding generally requires from 100 to 400 amperes for semi-automatic applications. Automatic installations may range from 50 to 600 amperes.

Flux-cored wire utilizes from less than 200 up to 500 amperes in applications without shielding gas. With carbon dioxide shielding gas in automatic setups, the current ranges from 350 to 750 amperes.

Automatic submerged-arc welding with bare wire and carbon dioxide shielding requires from 75 to 1,000 amperes. The *semi-automatic, short-circuiting* process with the same shielding gas requires only 25 to 250 amperes.

Wire-Feeding Unit and Control

The wire-feeding unit, also called a *wire feeder* or *wire drive,* drives the electrode wire from the coil automatically, Figs. 10-31 and 10-32. It feeds the wire through the cable assembly to the gun, arc, and weld pool. Since a constant rate of wire feed is required, the wire feeder is

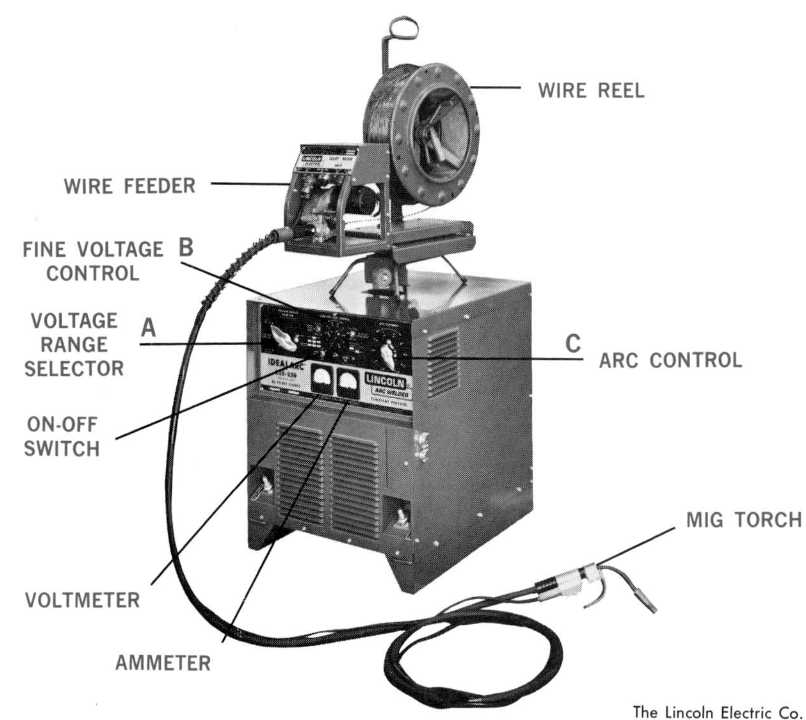

The Lincoln Electric Co.

10-27. *This 250-ampere, constant voltage transformer-rectifier with solid state controls can be used for MIG and stick electrode welding. A VOLTAGE RANGE SELECTOR has two overlapping voltage ranges: 12 to 25 volts and 20 to 35 volts. B FINE VOLTAGE CONTROL adjusts the welding voltages within the two ranges for precise settings. C ARC CONTROL permits fine tuning the arc characteristics as desired for each job.*

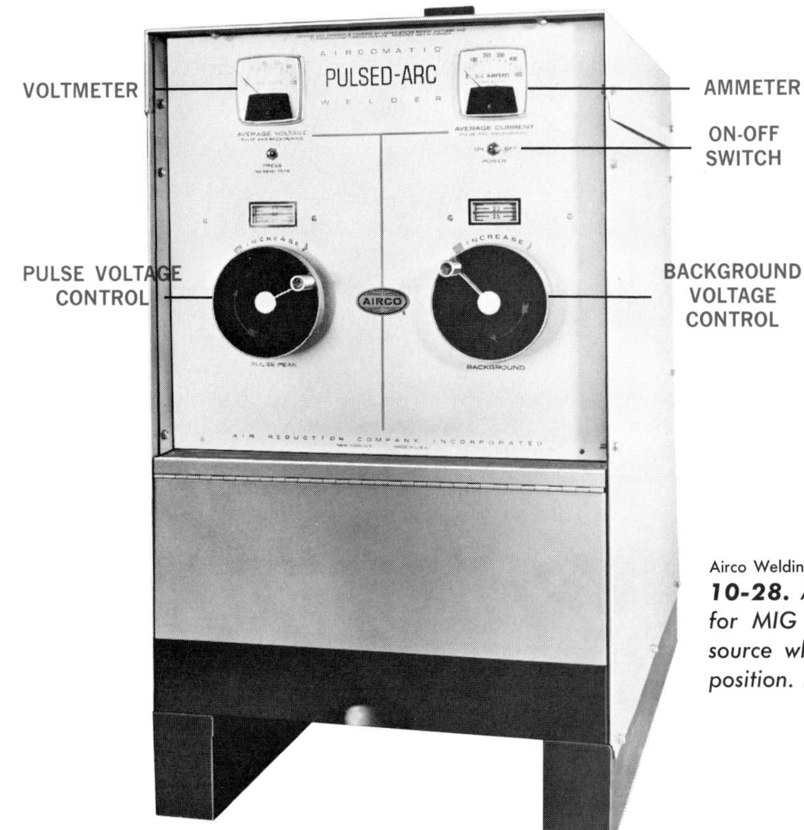

Airco Welding Products, Air Reduction Co.

10-28. *A 300-ampere d.c. transformer-rectifier for MIG welding. This is a pulsed-arc power source which makes it possible to weld out-of-position. It is particularly suitable for thin gauge materials.*

adjustable to provide for different welding currents.

Usually, the machine is a one-piece unit. A control cabinet wire feeder and wire mount are attached to an all-welded frame. The wire feeder is mounted with the power source in a portable unit, Fig. 10-33 (Page 308). In some cases the unit is separate from the power source and is mounted on an overhead crane to allow the operator to cover a greater area. For field or maintenance work, the unit is small and portable and may be a great distance from the power source, Fig. 10-34.

The welding operator has trigger control over gas flow, wire feed, and the welding-current contactor right at the work. When he presses the torch trigger, gas and water flow and wire feed start automatically. The arc strikes as the wire touches the workpiece. When the trigger is released, gas, water, and wire feed stop. The wire-feed rate and electrical variables are set on the wirefeeder before welding and automatically controlled during welding.

Wire drives are manufactured so that different sizes of drive rolls can be installed quickly. All sizes of wire can be used, thus making it possible to weld a wide range of metal thicknesses. Different rolls are required for hard wires than for soft wires.

Most wire-feeding units push the wire through the cable to the torch. The push type of drive is generally used with hard wires

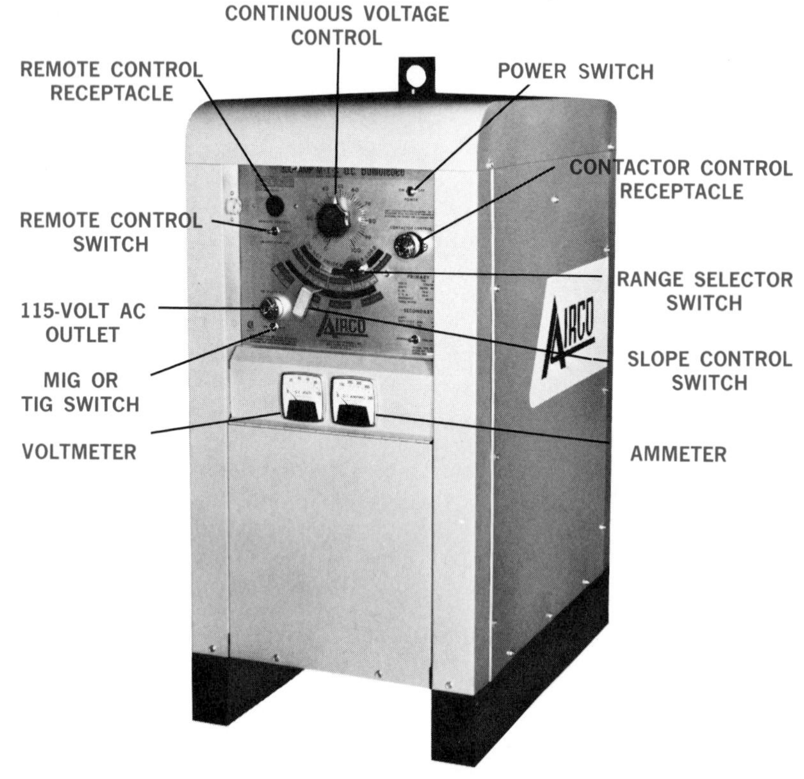

Airco Welding Products, Air Reduction Co.

10-29. *This 300-ampere d.c. transformer-rectifier has varied slope control that can be used for stick, MIG, or TIG welding.*

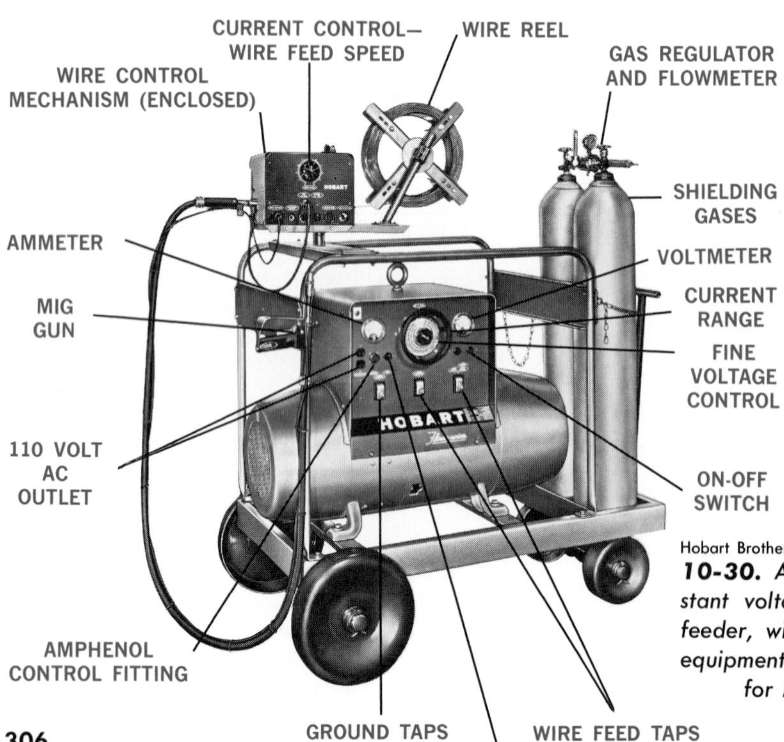

Hobart Brothers Co.

10-30. *A 600-ampere, dual purpose d.c. constant voltage motor-generator, including wire feeder, wire coil, gun, and shielding gas. The equipment is on a portable unit. It can be used for MIG or stick electrode welding.*

such as carbon steel and stainless steel. A number of torches have an internal mechanism that pulls the wire from a wire reel unit through the torch to the arc. Figures 10-35 and 10-36 show two types of wire-drive, pull-type guns. The pull type of drive is preferable for use with soft wires such as aluminum and magnesium. Other torches have a small spool of wire and a drive motor in the torch which feeds the wire directly through the torch, Figs. 10-37 and 10-38 (Page 310). Still another type of torch acts as a feeder gun and operates in connection with a canister, Fig. 10-39.

Welding Torches and Accessories

The function of the MIG welding torch, Fig. 10-40, is to deliver the electrode wire and the shielding gas from the wire feeder and the welding current from the power source to the welding area. The MIG torch is comparable to the electrode holder used for stick electrode welding.

Torches are designed for different types of service. An air-cooled torch may be used for light duty service and with carbon dioxide as the shielding gas, Figs. 10-41 and 10-42. Carbon dioxide actually assists in the cooling of the torch. For heavy duty service and inert gas, the torch may be water cooled, Figs. 10-43 and 10-44 (Page 312 and 313). Water cooling is usually necessary when currents of over 250 amperes are used. Welding

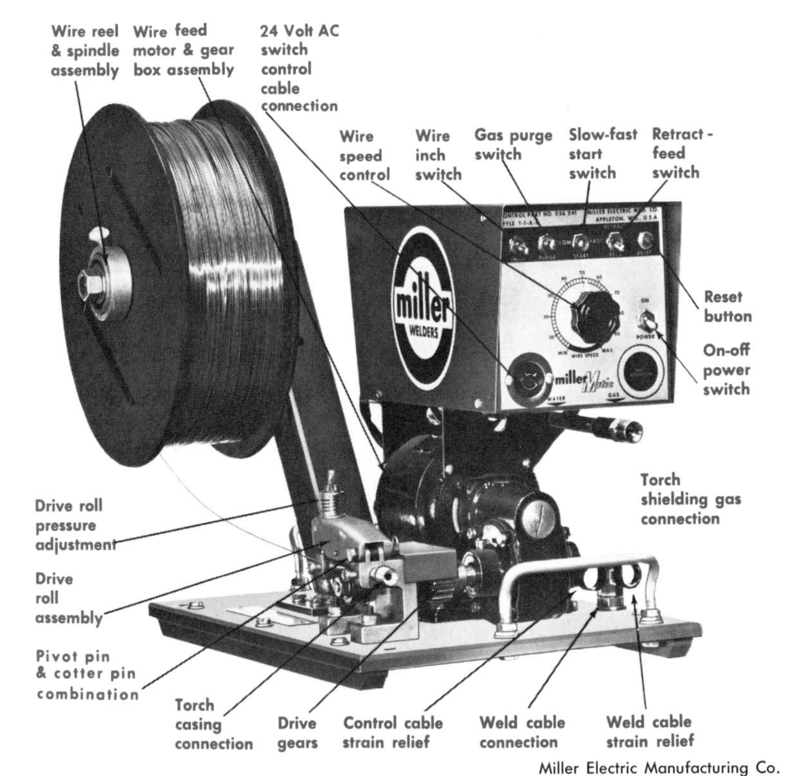

Miller Electric Manufacturing Co.

10-31. *A typical wire feeder is designed for the popular short-circuiting and spray methods of metal transfer. It accommodates hard wires 0.030 and 0.035 inch in diameter and operates from a d.c. constant voltage welding machine.*

Miller Electric Manufacturing Co.

10-32. *The internal construction of the wire feeder in Fig. 10-31. The relationship of the wire guides to the wire drive rolls is shown. These guides must be changed to suit the diameter of wire being used for welding.*

continuously with high current generates a great amount of heat in the torch. Portable water-cooling equipment, Fig. 10-45, provides a positive system of cooling the torch. It is also used when the welding location is not near a fresh water supply line. These units have an internal cooling system and force water through the torch.

MIG torches for manual operation may be shaped like a pistol, or have a straight handle or a handle with a gooseneck nozzle end. All types have a trigger switch that turns on the controls for wire feed, shielding gas, and water flow.

NOZZLES. Many different types of materials are used for nozzles: copper, wood, asbestos, ceramics, glass, and glass-coated metals. The glass nozzles give the operator a better view of the welding operation. Many men feel, however, that clean, smooth, shiny copper does the best job of all.

CABLE ASSEMBLIES. The cable assemblies leading from the gun to the wire feeder are available in several types and lengths. Table 10-2 lists the recommended cable sizes and lengths for various current settings. One type of assembly is made up of separate hoses or cables. Another type has a single molded conductor that encloses all of the necessary components. In this type, the electrode wire is guided through a flexible liner contained in the cable assembly. The liner is made of spring steel for strength and protection. Plastic liners are sometimes used. The liner serves as a guide for the electrode wire and provides uninterrupted travel of the wire without buckling or kinking from the drive rolls to the contact tip of the gun. The cable also contains tube passages for shielding gas flow.

Dirt and foreign material gradually build up inside the conductor cable and interfere with the free flow of the electrode wire through the cable. Cables should be blown out each time the electrode wire spool is changed.

10-33. *A typical wire feeder mounted on a d.c. constant voltage transformer-rectifier power source. It can be made into a portable unit and can be operated by remote control.*

National Cylinder Gas and Harnischfeger Corp.

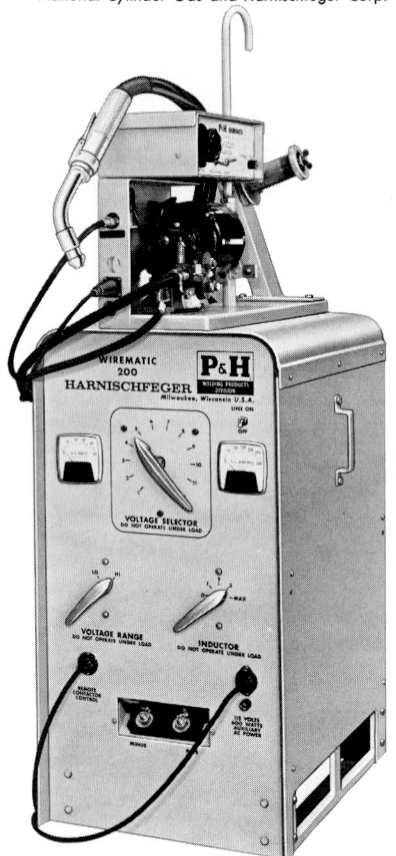

The Shielding Gas System

The shielding gas system supplies and controls the flow of gas which shields the arc area from contamination by the surrounding atmosphere. The system consists of the following equipment:

- One or more gas storage cylinders
- Hose
- Pressure-reducing regulator
- Flowmeter
- Solenoid control valves

The gas from the cylinder is controlled by means of a pressure-reducing regulator, flowmeter, and the solenoid control valves. It is fed to the torch through an innerconnecting hose system. Some power supplies have certain of these units incorporated into the machine. It is important that the correct regulators and flowmeters be used with the particular shielding gas being employed. The gas cylinders can be replaced with a central gas supply system when there is a sufficient volume of MIG welding being done.

GAS FLOW RATES. In order to provide a good gas shield, the gas must be fed through the nozzle at such a rate as to ensure *laminar flow* (a straight line of flow), (Fig. 10-46, Page 313). If the flow rate is too high, it causes *turbulence* (swirling). The turbulent gas shield mixes with the surrounding air under the nozzle and causes contamination of the weld deposit. If the gas shield fails to cover the arc or is contaminated, porosity results. Almost all porosity in MIG welds is caused by poor gas coverage. Too much is just as bad as too little.

The gas flow rate is a function of the shape and cross-sectional area of the gas passages in the nozzle. The recommended gas flow rate for a particular nozzle and a specified gas is suitable for all wire-feed speeds. Always refer to the manufacturer's instructions for recommendations concerning a particular nozzle.

Specific Gravity. The chief factor influencing the effectiveness of a gas for arc shielding is the specific gravity of the gas. This in turn determines the flow rate selected for the welding operation. The *specific gravity* of a gas is a measure of its density relative to the density of air. The density of air is fixed as one. The specific gravity of helium is 0.137, or approximately $\frac{1}{10}$ as heavy as air. Thus helium tends to rise quickly from the weld area, and a high flow rate is needed to provide adequate shielding. A gas of low density such as helium is also greatly affected by crossdrafts of air across the arc and weld pool.

Argon and carbon dioxide are much heavier than helium, and lower flow rates are required for equivalent shielding. The specific gravity of argon is 1.38, and the specific gravity of carbon dioxide is 1.52. Both gases, therefore, are heavier than air. They blanket the weld area. Carbon dioxide has the advantage of being heavier than argon, thus insuring good protection. It is also highly resistant to crossdrafts.

Other Factors Affecting Flow Rates. Flow rates may vary from 15 to 50 cubic feet per hour depending upon the type and thickness of the material to be welded, the position of the weld, the type of joint, the shielding gas, and the diameter of the electrode wire. The operator can tell if the amount of shielding gas is correct by the rapid cracking and sizzling arc sound. Inade-

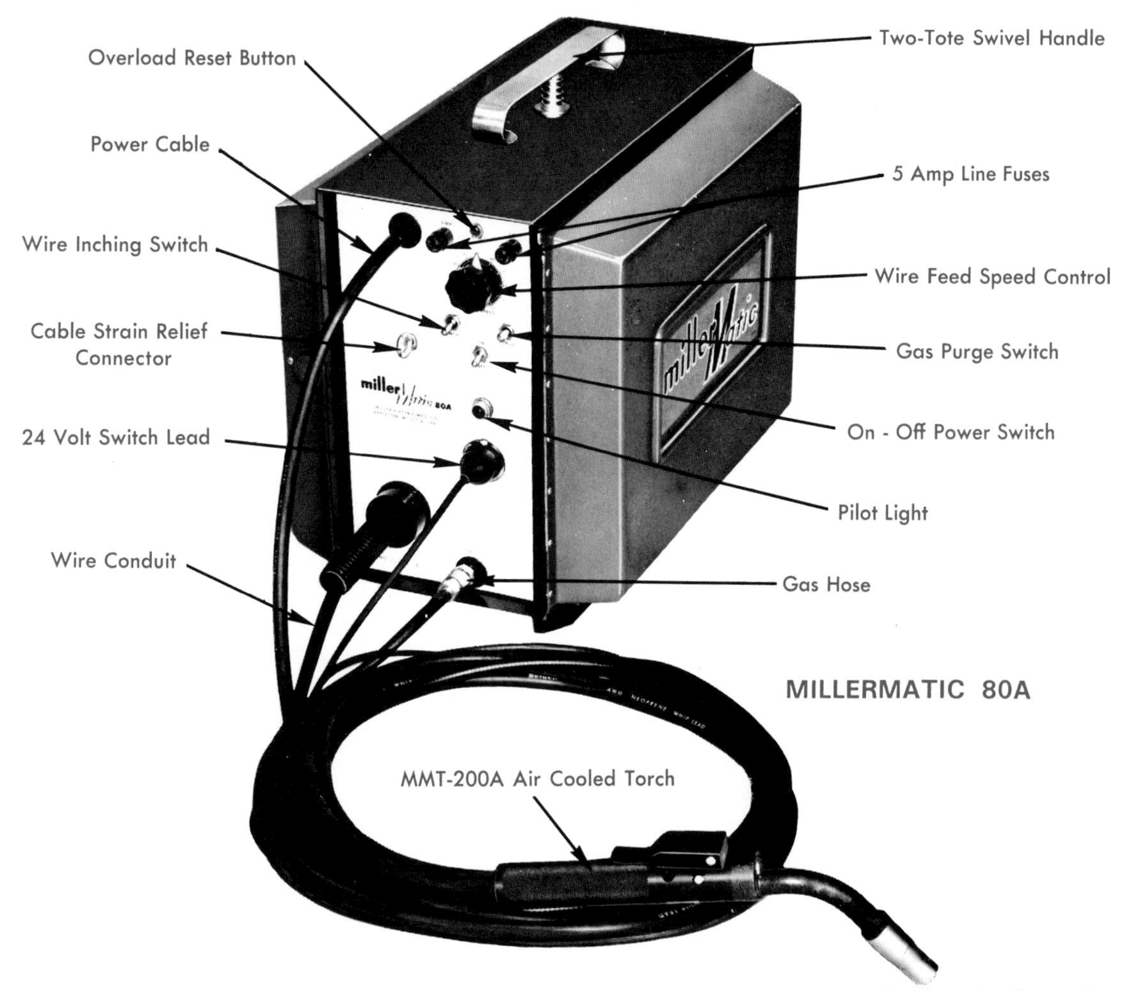

Miller Electric Manufacturing Co.

10-34. *Portable control-feeder for MIG solid wire welding. It is used with 0.030- and 0.045-inch wire sizes. It operates with a constant voltage power source and is designed for short-circuiting and spray-type metal transfer. The unit can be taken to the work site by the operator for convenient control.*

National Cylinder Gas and Harnischfeger Corp.

10-35. *The wire feeder is in the handle. A motor in the handle pulls the wire through the feed conduit from standard size spools, which are mounted on the control unit. The control unit also has a threading motor which threads wire automatically through the conduit to the torch. This gun can be operated 25 feet from the welding unit. This unit is known as a "push-pull" unit.*

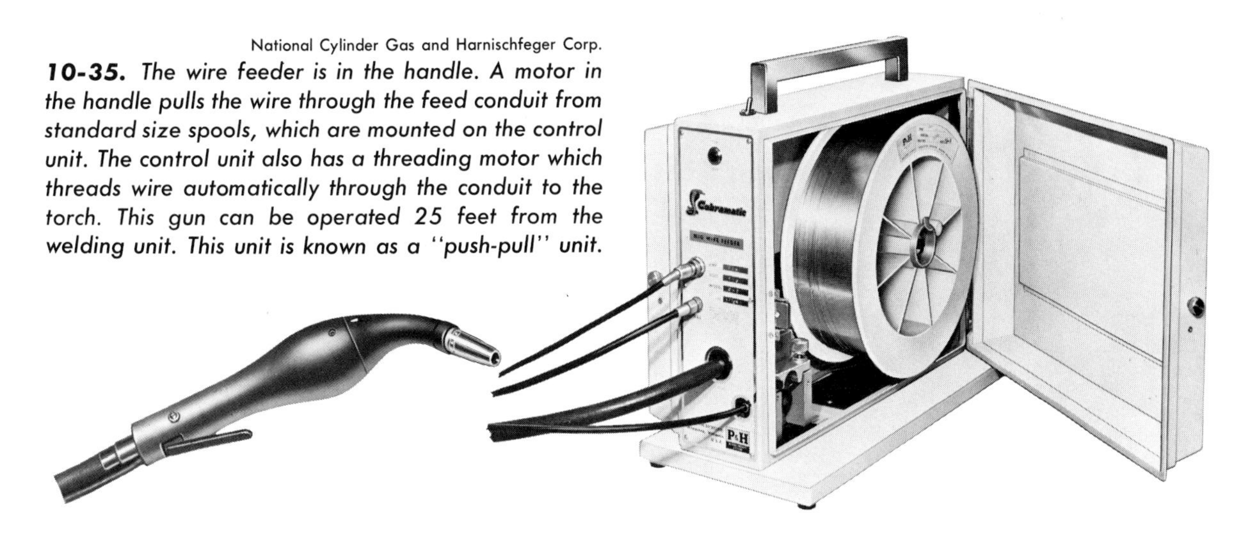

Airco Welding Products Division, Air Reduction Co.

10-36. *This pull-type gun handles soft wire and small diameter hard wire. Its drive rolls, located in the gun, are driven by a flexible shaft coupled to the wire feeder. The gun is water cooled.*

10-37. *A gun which contains the wire and the drive motor. Small coils of wire are used, usually 1-pound spools. Since the wire is pushed only 6 inches from the drive coils to the arc, the possibility of wire snarls is reduced. The gun also permits prepurge and post-purge gas coverage.*

Westinghouse Electric Co.

10-38. *The wire-drive gun shown in Fig. 10-37 attached to a weld-pacing unit. The unit gives precise control of travel speed when welding light gauge sheet metal. The speed control box is also shown.*

Westinghouse Electric Co.

quate shielding produces a popping, erratic sound and a considerable amount of spatter. The weld is also discolored and has a high degree of porosity.

The following additional factors also have an influence on effective gas shielding:

- Size and shape of the nozzle
- Nozzle-work distance
- Size of the weld pool
- Amount of welding current
- Air currents
- Design of welding fixture
- Welding speed

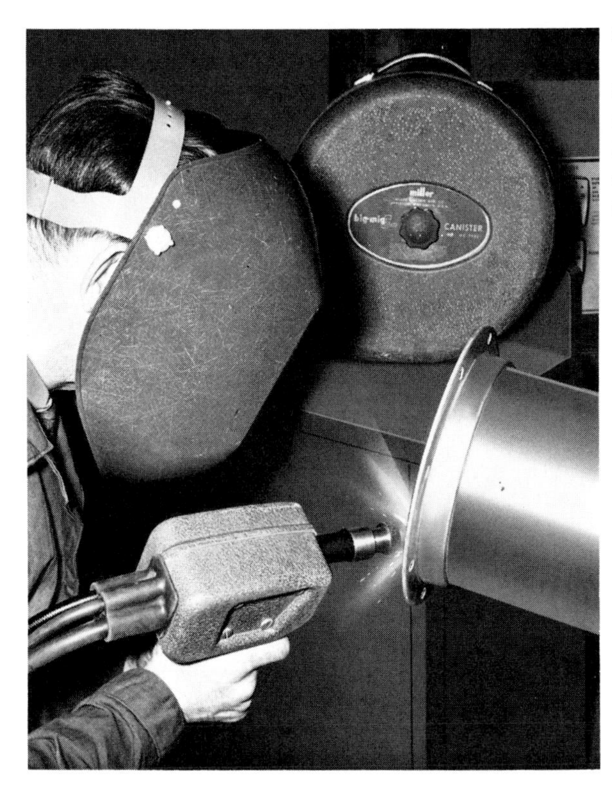

Miller Electric Manufacturing Co.

10-39. *A wire-drive gun and canister used for MIG spray and short-circuit welding with either aluminum or hard wire. The small portable gun is fed by a 25-pound spool in the portable canister. The gun is attached to the canister by a 10-foot cable. The total unit can operate a distance of 50 feet from the power source to allow extreme flexibility.*

Bernard Welding Equipment Co.

10-40. *Internal construction of a MIG welding gun.*

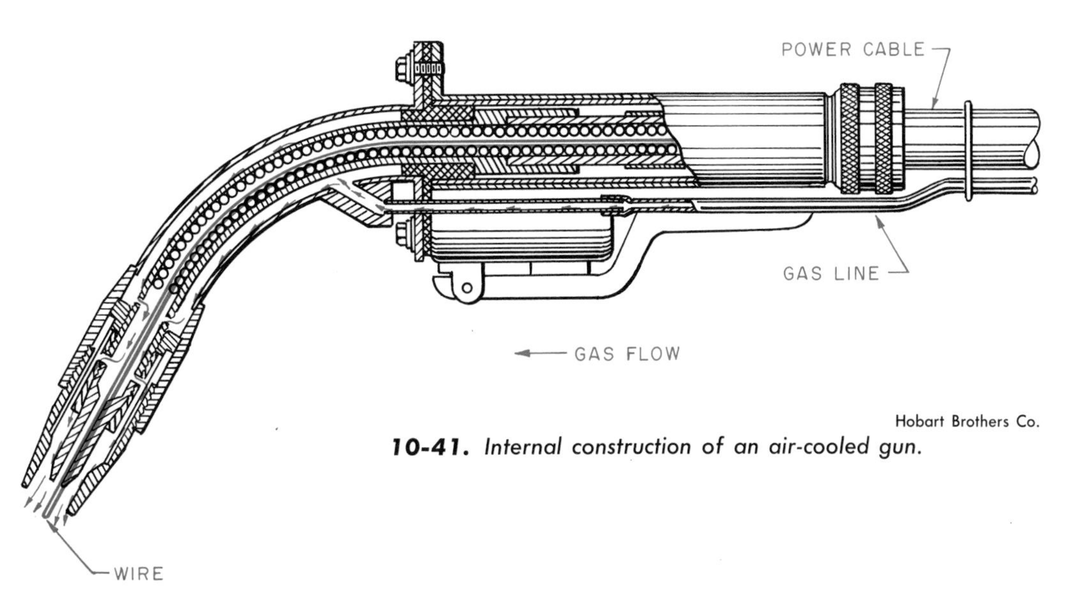

Hobart Brothers Co.

10-41. *Internal construction of an air-cooled gun.*

- Inclination of the torch
- Excessive shielding gas pressure
- Leaking gas fittings

Shielding Gases

Modern metals are meeting the stringent requirements of the space age with their high corrosion resistance, excellent strength-to-weight ratios, and good physical properties. Add the advantages of welded fabrication, and the metals become an engineer's dream come true.

However, in the molten state, many of the metals absorb oxygen, nitrogen, and hydrogen from the air. These gases cause porosity and brittleness. The prime consideration when welding them is protection of the weld zone from atmospheric elements. In the preceding chapter, we have seen that this is accomplished by a shielding gas.

Selecting a shielding gas for MIG welding can be a problem. Arc atmosphere governs, to a large extent, arc stability, bead shape, depth of penetration, freedom from porosity, and al-

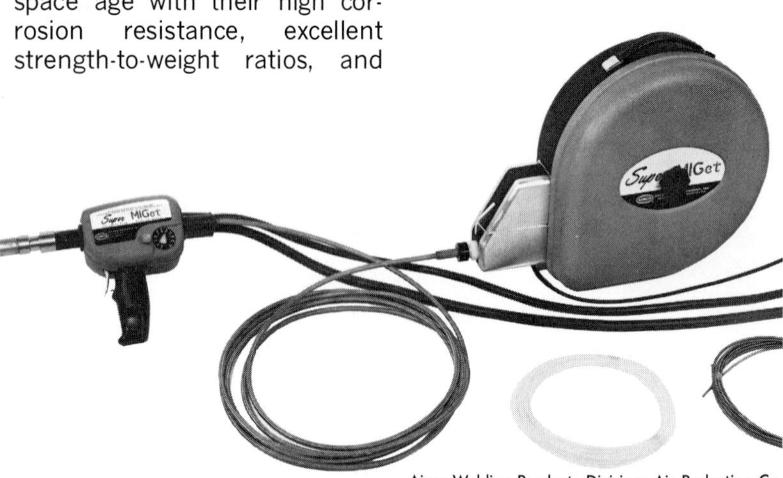

Airco Welding Products Division, Air Reduction Co.

10-42. *Air-cooled, push-pull gun in this package can be operated up to 20 feet away from its wire canister. A positive, high torque, feed-drive system—including a pull motor in the gun and a push motor at the canister—eliminates problems of feeding fine-diameter soft and hard wires caused by coiling or kinking of the casing.*

Table 10-2
Recommended cable size for welding cable from welding machine to workpiece for MIG welding.

Amperes	Max. cable length	Cable size
200	15 feet	No. 1
300	15 feet	No. 2/0
400	15 feet	No. 3/0
500	15 feet	No. 4/0
600	15 feet	No. 2–2/0

Miller Electric Manufacturing Co.

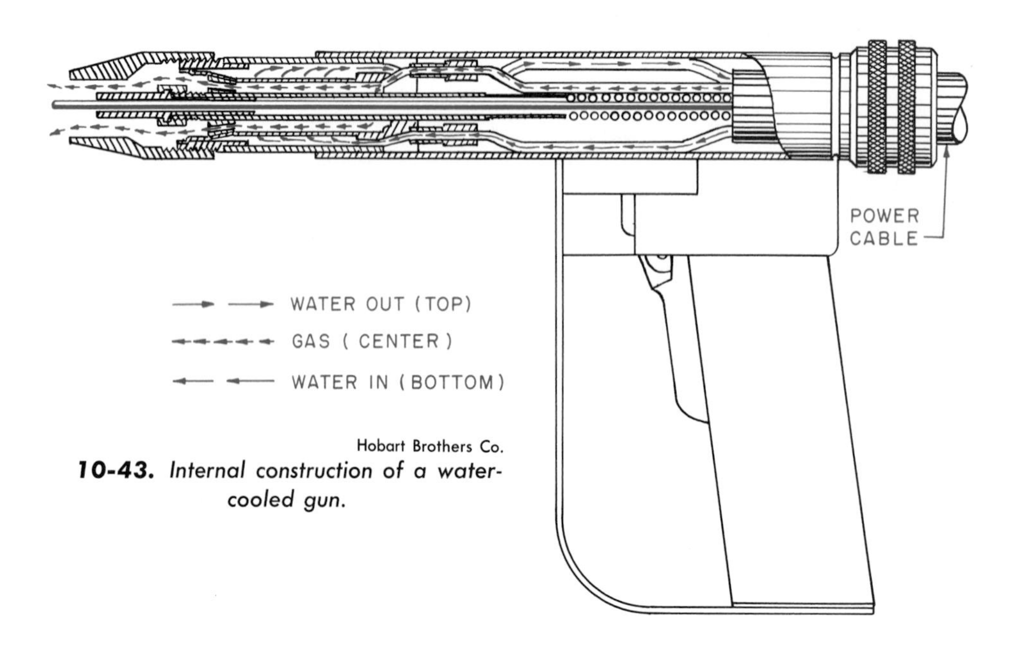

Hobart Brothers Co.

10-43. *Internal construction of a water-cooled gun.*

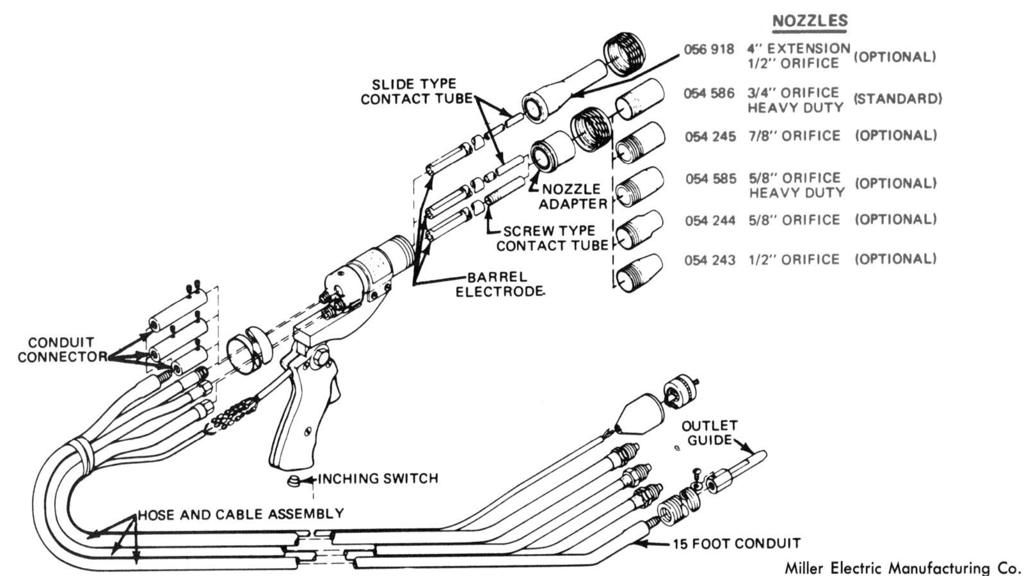

Miller Electric Manufacturing Co.

10-44. *Exploded view of a 500-ampere, water-cooled MIG welding gun. It handles solid core or flux-cored wire.*

lowable welding speed. Considering these points along with the cost of various gases will lead to the best choice.

The three principal gases for TIG and MIG welding are argon, helium, and carbon dioxide. These three gases coupled with the two welding processes are producing the porosity-free welds specified for the missiles, rockets, jets, and nuclear reactors that are fabricated from space-age metals.

For MIG welds, gases used singly or in mixtures can be completely inert. If such an atmosphere is desired, argon or helium with a purity greater than 99.9 percent may be best. Other inert gases (xenon, krypton, radon and neon) are too rare and expensive.

Study Tables 10-3 through 10-6 carefully, and you will have a good understanding of the proper shielding gas to use for a given situation.

ARGON. Argon is in abundant supply. The amount of air covering one square mile of the earth's surface contains approximately 800,000 pounds of argon. The gas is a byproduct of the manufacture of oxygen.

In the argon shield, the welding arc tends to be more stable than in other gas shields. For this reason, argon is often used as a mixture with other gases for arc shielding. The argon gives a quiet arc and thereby reduces spatter.

10-45. *Water-circulating unit to cool heavy duty MIG torches.*

Bernard Welding Equipment Co.

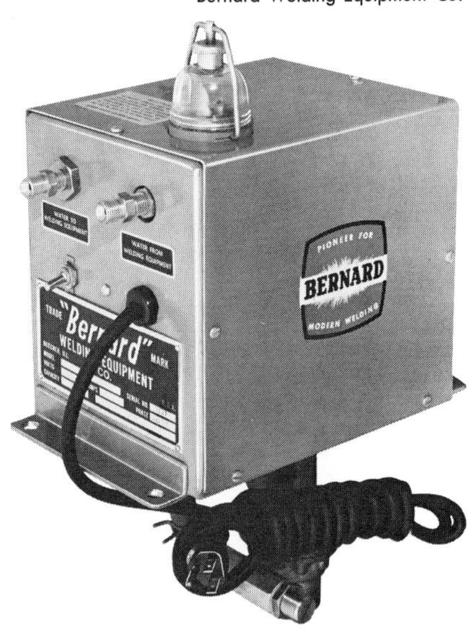

10-46. *For an effective gas shield, the gas should be fed through the torch nozzle in a laminar flow (left). A swirling gas shield results in contamination of the weld deposit (right).*

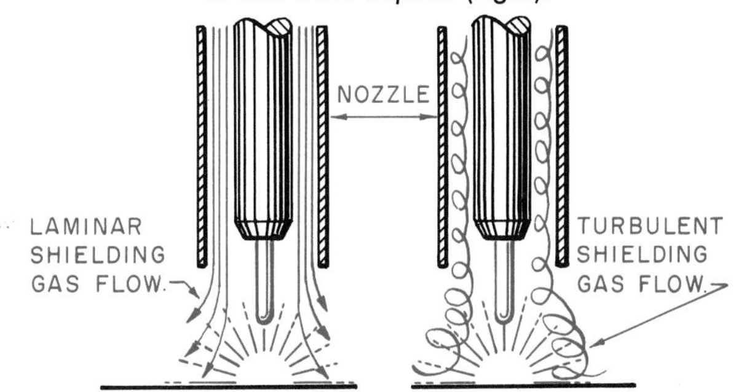

Argon has low *thermal* (heat) *conductivity.* The arc column is constricted so that high arc densities are produced. A high arc density permits more of the available arc energy to go into the work as heat. The result is a relatively narrow weld bead with deep penetration at the center of the deposit, Fig. 10-47. In some welding applications, however, argon does not provide the penetration characteristics needed for thicker metals.

Argon ionizes more readily than helium, and thus can transmit some electrical energy. Therefore, argon requires a lower arc voltage for a given arc length than helium does. Since the result is less heat, it is believed that the argon arc is not hot as the helium arc. The cooler arc produced with argon makes it preferable for use in welding light gauge metal and materials of low thermal conductivity.

Pure argon is seldom used in the MIG welding of such materials as carbon steel since it causes poor penetration, undercutting, and poor bead contour. Its primary use is in the welding of nonferrous metals such as copper and alloys containing aluminum and magnesium.

The combination of constricted penetration and reduced spatter makes argon desirable when welding in other than the flat position. Argon is recommended for manual operations because changes in arc length do not produce as great a change in arc voltage and heat input to the work as when helium is used.

HELIUM. Helium is produced from natural gas. Natural gas is cooled and compressed. The hydrocarbons are drawn off first, then nitrogen, and finally helium. The gases are liquified until helium is produced at −345 degrees F. In the past, helium has been in short supply due to government restrictions. Recent discoveries of natural gas deposits with high helium content have made helium more readily available for use in the inert gas-shielded welding processes.

Helium is a light gas which tends to rise in a turbulent fashion and disperse from the weld region. Therefore, higher flow rates are generally required with helium than with argon shielding. Helium, because of greater arc heat, is usually preferable for welding heavy sections and materials with high thermal conductivity. It produces a wider weld bead with slightly less penetration than argon, Fig. 10-48.

When pure helium is used with MIG and reverse polarity, there is no cleaning action. Metal transfers from the electrode in large drops and tends to spatter. The beads are broad and flat, and penetration is uniform. Helium lacks the deep central penetration of argon. For a given current and arc length, arc voltage is higher and wire burnoff is greater than with argon.

Helium is used more with automatic and mechanized processes than argon because control of arc length is not a

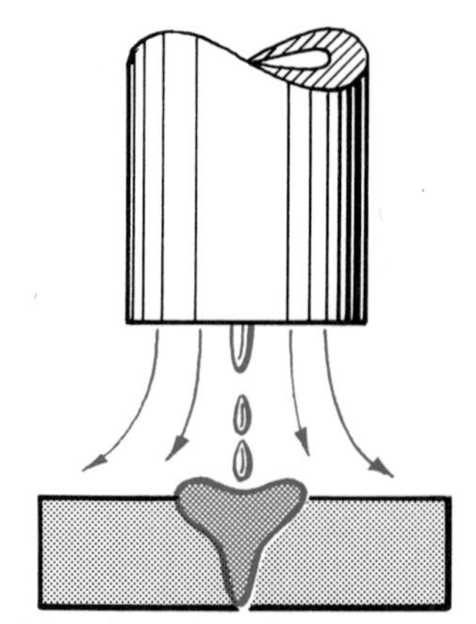

***10-47.** Argon-shielded weld.*

Table 10-3

Proper wire and gas combinations.

	Mild steel solid wire	Mild steel & alloy cored wire	Aluminum	Copper	Stainless	Bronze
Carbon dioxide	X	X			X	X
75% argon + 25% carbon dioxide	X				X	X
100% argon			X	X		
95–98% argon + 5–2% oxygen	X				X	X
100% helium			X	X		
75% helium + 25% argon			X	X		

Table lists wire and gas combinations for semi-automatic welding.

American Welding Society

***10-48.** Helium-shielded weld.*

problem. It is used primarily for the nonferrous metals such as aluminum, magnesium, and copper and their alloys.

CARBON DIOXIDE. Carbon dioxide (CO_2) is not an inert gas such as argon or helium. It is a chemical compound composed of one part carbon (C) and two parts oxygen (O_2). It is manufactured from flue gases which are given off by the burning of natural gas, fuel oil, and coke. Carbon dioxide is also a byproduct of the manufacture of ammonia and the fermentation of alcohol. The carbon dioxide from these processes is almost 100 percent pure.

The use of carbon dioxide as a shielding gas is increasing. It is cheaper than the inert gases and has several desirable characteristics. It produces a wide weld pattern and deep penetration, thus making it easier to avoid lack of fusion. The bead contour is good, and there is no tendency towards undercutting.

One disadvantage of carbon dioxide is the tendency for the arc to be unstable and cause weld spatter. This is particularly serious on thin materials when appearance is important. The amount of weld spatter may be reduced by maintaining a very close arc. On heavier materials, the arc may be buried in the workpiece. Spatter can also be reduced by using flux-cored wire.

The oxidation effect of the carbon dioxide shield is about equal to a mixture of 91 percent argon and 9 percent oxygen. Under a carbon dioxide shield, about 50 percent of the manganese and 60 percent of the silicon in the welding wire are converted to oxides as it passes through the arc. It is for this reason that the electrode wire contains a balance of deoxidizers and that a short arc is main-

Table 10-4

Shielding gases and gas mixtures used for MIG welding.

No.	Shielding gas	Chemical behavior	Uses and usage notes
1	Argon	Inert	Welds virtually all metals except steel
2	Helium	Inert	Aluminum and copper alloys; for greater heat and to minimize porosity
3	Argon and helium (20–80 to 50–50%)	Inert	Aluminum and copper alloys; for greater heat and to minimize porosity but with quieter, more readily controlled arc action
4	Argon and chlorine (trace chlorine)	Essentially inert	Aluminum alloys; to minimize porosity
5	Nitrogen	Reducing	On copper; very powerful arc
6	Argon and 25–30% nitrogen	Reducing	On copper; powerful but smoother operating, more readily controlled arc than with pure nitrogen
7	Argon and 1–2% oxygen	Oxidizing	Stainless and alloy steels; also for some deoxidized copper alloys
8	Argon and 3–5% oxygen	Oxidizing	Plain carbon, alloy and stainless steels (generally requires highly deoxidized wire)
9	Argon and 5–10% oxygen	Oxidizing	Various steels, deoxidized wire
10	Argon and 20–30% carbon dioxide	Oxidizing	Various steels; used chiefly with short-circuiting arc
11	Argon and 5% oxygen and 15% carbon dioxide	Oxidizing	Various steels; deoxidized wire; used chiefly in Europe.
12	Carbon dioxide	Oxidizing	Plain carbon and low alloy steels; deoxidized wire essential
13	Carbon dioxide and 3–10% oxygen	Oxidizing	Various steels, deoxidized wire, used chiefly in Europe
14	Carbon dioxide and 20% oxygen	Oxidizing	Steels, favored and chiefly used in Japan

American Welding Society

tained. With the proper care, X-ray quality welds can be consistently produced in carbon steels and some low alloy steels.

In the heat of the arc, carbon dioxide tends to break down into its component parts. At normal arc lengths, about 7 percent of the total volume of the gas shield converts to carbon monoxide (a compound made up of one atom carbon and one atom oxygen). At excessive arc lengths the quantity reaches 12 percent. Because carbon monoxide is poisonous, welding should never be done in unventilated areas. The operator should be able to see the welding operation clearly without getting so close to the shielding gas that he inhales the carbon monoxide.

Carbon dioxide is available in liquid or solid form for welding purposes. Because of the small gas requirement of a single welding arc, liquid carbon dioxide in cylinders is usually the most economical. Bulk systems for distributing carbon dioxide at low pressure to various welding areas are also in use.

Cylinders are filled on a basis of weight. The liquid level on a full cylinder is about two-thirds

10-49. *Standard carbon dioxide cylinder.*

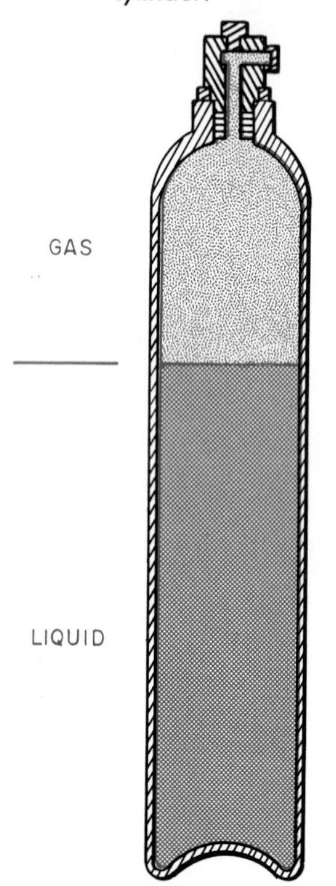

Table 10-5

Selection of gases for MIG welding with drop transfer.

Metal	Shielding gas	Advantages
Aluminum	Argon	0–1 in. thick: Best metal transfer and arc stability; least spatter.
	35% argon—65% helium	1–3 in. thick: Higher heat input than straight argon.
	25% argon—75% helium	3 in. + in thickness: Highest heat input; minimizes porosity.
Magnesium	Argon	Excellent cleaning action.
Carbon steel	Argon-oxygen (3–5%)	Improves arc stability; produces a more fluid and controllable weld puddle; good coalescence and bead contour; minimizes undercutting; permits higher speeds, compared with argon.
	Carbon dioxide	High-speed mechanized welding; low-cost manual welding.
Low-alloy steel	Argon-oxygen (2%)	Eliminates undercutting; provides good toughness.
Stainless steel	Argon-oxygen (1%)	Improves arc stability; produces a more fluid and controllable weld puddle, good coalescence and bead contour; minimizes undercutting on heavier stainless steels.
	Argon-oxygen (2%)	Provides better arc stability, coalescence and welding speed than 1% oxygen mixture for thinner stainless steel materials.
Copper, nickel and their alloys	Argon	Provides good wetting; decreases fluidity of weld metal (for thicknesses up to $\frac{1}{8}$ in.).
	Argon-helium	Higher heat inputs of 50 and 75% helium mixtures offset high heat conductivity of heavier gages.
Titanium	Argon	Good arc stability; minimum weld contamination. Inert gas backing is required to prevent air contamination on back of weld area.

American Welding Society

up from the bottom, Fig. 10-49. Space for gas over the liquid is necessary. The liquid in the cylinder absorbs heat from the atmosphere and boils. Gas is formed until the pressure in the cylinder raises the boiling temperature above the temperature of the liquid. Boiling, and consequently gas production, ceases until either the atmospheric temperature increases or some gas is drawn off, lowering the pressure in the cylinder. The pressure in the cylinder is a function of the atmospheric temperature. Typical pressures for various temperatures are given in Table 10-7.

At a constant temperature, the pressure in the cylinder remains constant as long as there is liquid remaining. Continued use of the gas after the liquid has completely evaporated noticeably reduces cylinder pressure. Cylinders should be changed when the pressure goes down to 200 p.s.i.

Since even small amounts of moisture in carbon dioxide cause porosity in the weld deposit, it is important that only welding grade carbon dioxide be used. Welding grade gas is supplied in specially treated cylinders that prevent moisture. (See Figs. 10-50 and 10-51.) Cylinders should be marked *welding grade.*

There is a considerable refrigeration effect in the high pressure section of the regulator. Sometimes, on very heavy duty cycles, frost forms on the outside of the regulator. This indicates only that the regulator is getting cold. When frost appears, the small amount of moisture in the gas can form ice in the regulator and either stop the gas flow entirely or break off and be blown into the weld. In either case porosity can, and usually does, occur in the weld metal. If regulator freezing is a problem, use a gas heater or a common heat lamp on the regulator. Bulk systems are not prone to freezing because of the small expansion ratio and the large heat absorption area of piping.

Table 10-6

Selection of gases for MIG welding with short-circuiting transfer.

Metal	Shielding Gas	Advantages
Carbon steel	75% argon–25% carbon dioxide	Less than 1/8 in. thick: high welding speeds without burn-through; minimum distortion and spatter.
	75% argon—25% carbon dioxide	More than 1/8 in. thick: minimum spatter; clean weld appearances; good puddle control in vertical and overhead positions.
	Carbon dioxide	Deeper penetration; faster welding speeds.
Stainless steel	90% helium—7.5% argon 2.5% carbon dioxide	No effect on corrosion resistance; small heat-affected zone; no undercutting; minimum distortion.
Low-alloy steel	60–70% helium—25–35% argon—4–5% carbon dioxide	Minimum reactivity; excellent toughness.
	75% argon—25% carbon dioxide	Excellent arc stability, wetting characteristics and bead contour; little spatter. Fair toughness; excellent arc stability, wetting characteristics and bead contour; little spatter.
Aluminum, copper, magnesium, nickel and their alloys	Argon and argon-helium	Argon satisfactory on sheet metal; argon-helium preferred on thicker sheet material.

American Welding Society

Table 10-7

Cylinder pressure of carbon dioxide at various temperatures.

Degrees fahrenheit	Pounds per square inch
100	1,450
70	835
30	476
0	290

The normal gas requirement for MIG welding is 25 to 30 feet per hour. A single cylinder may not be large enough to support gas production at this rate. It is generally necessary to connect two or more cylinders to a manifold.

Experience has shown the above gas flow rates to be proper for most applications. Higher than normal flow rates are usually necessary on very high speed applications or procedures utilizing high contact tube heights. Too low a flow rate results in a partial or complete lack of gas coverage, which causes a high degree of porosity in the weld. Too much gas flow produces turbulence about the arc so that the carbon dioxide mixes with air. This also causes porosity of the weld deposit.

GAS MIXTURES. As has been shown in Tables 10-3 through 10-6, gases are mixed in specified percentages to obtain desired characteristics.

Figures 10-52 and 10-53 show combination gas mixers and flow controllers. The mixer in Fig. 10-52 will mix three gases or any two of the three gases the model was designed to mix. For example, if a model is designed to mix argon, carbon dioxide, and helium, three mixtures of two gases (argon-carbon dioxide, argon-helium, and helium-carbon dioxide) can also be produced.

The mixer has three flow control knobs, one for each of the three gases to be mixed. By turning a control knob while he watches the flowmeter above it, the operator can adjust the flow of that gas independently. This gives him the ability to change mixtures when he changes jobs or experiment with the mixture on any job until he obtains the best weld.

It is necessary to use a gas regulator with the mixer in order to reduce the high tank pressure to a steady 50 p.s.i. working pressure, Fig. 10-54.

Argon-Helium. For many MIG welding applications the advantages of both argon and helium may be achieved through the use of mixtures of the two gases. The mixture is especially desirable whenever a completely inert gas is essential or desirable for MIG welding. Mixing argon and helium gives a balance between penetration and arc stability. Adding 25 percent argon to helium makes penetration deeper than that obtained with argon alone, and the arc stability approaches that of pure argon.

The percentage of helium to argon may be 20—80 percent to 50—50 percent. The gases are mixed by the user. Separate cylinders of each gas are connected to a control unit and the amount of gas used is regulated through a flowmeter.

When helium is too hot or argon too cool for the desired arc characteristics, they may be mixed to obtain any combination of properties. When argon is added in amounts above 20 percent, the arc becomes more stable. Adding 75 percent helium to the mix practically eliminates any spatter tendency in MIG welding. Bead contour shows a

10-50. *This weld was made using carbon dioxide which contained moisture. Moisture poisoned the gas shield, causing a weak, porous structure.*

National Cylinder Gas Division, Chemetron Corp.

10-51. *This high quality weld was made by the same operator who made the weld in Fig. 10-50. He used the same equipment and flux-cored electrodes, but the carbon dioxide was free of moisture.*

National Cylinder Gas Division, Chemetron Corp.

penetration pattern characteristic of both gases.

Argon-helium mixtures are used to weld aluminum and copper-nickel alloys. The mixtures provide greater heat and produce less porosity than argon, but they have a quieter, more readily controlled arc action than helium. These mixtures are especially useful in the welding of heavy sections of nonferrous metals. The heavier the material, the greater the percentage of helium in the mixture.

Argon-Carbon Dioxide. Shielding gas mixtures are not confined to the inert gases themselves. Argon and helium can be mixed with other gases to improve welding characteristics.

The addition of carbon dioxide to argon stabilizes the arc, promotes metal transfer, and reduces spatter. The penetration pattern changes when welding carbon and low alloy steels. The molten weld pool is highly fluid along the fusion edges so that undercutting is prevented. For this reason, argon-carbon dioxide mixtures are used principally in the welding of mild steel, low alloy steel, and some types of stainless steel.

Some difference of opinion exists in regard to the amount of carbon dioxide that should be used in the mixture. Some welding engineers believe that the mixture should not exceed 20 to 30 percent carbon dioxide. Others feel that mixtures containing up to 70 percent produce good results. Electrode filler wires used with an oxygen-bearing shielding gas such as carbon dioxide must contain deoxidizers to offset the effects of the oxygen. Up to about 25 percent carbon dioxide permits spray transfer with a solid steel wire.

One of the important reasons for using as much carbon dioxide as possible in the mixture is to reduce the cost of welding. Carbon dioxide costs about 85 percent less than argon. Premixed cylinders of argon and carbon dioxide sell for the same price as pure argon. It is a good practice to purchase the shielding gases in separate cylinders and mix the gases on the job.

Helium-Argon-Carbon Dioxide. This mixture of inert gases and a compound gas is used mostly for the welding of austenitic stainless steels with the short circuit method of metal transfer. It is usually purchased premixed from the supplier in cylinders that contain 90 percent helium, $7\frac{1}{2}$ percent argon, and $2\frac{1}{2}$ percent carbon dioxide. You will note that the percentage of helium is high.

This combination of gases produces a weld with very little buildup of the top bead profile.

10-52. Three-module gas mixer and flow controller.
National Welding Equipment Co.

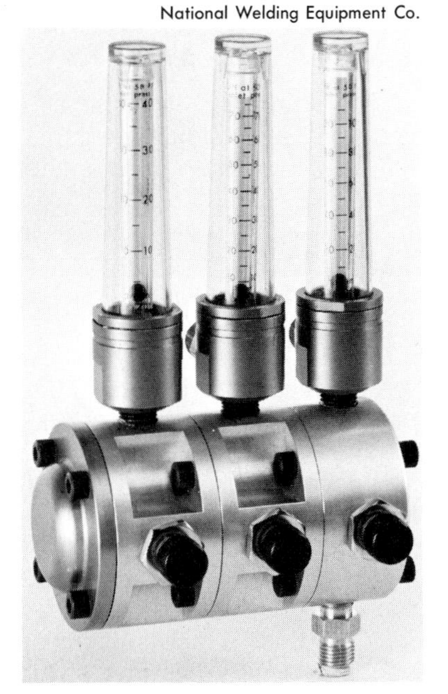

10-53. *The controls and working parts of a two-module gas mixer similar to that shown in Fig. 10-52.*

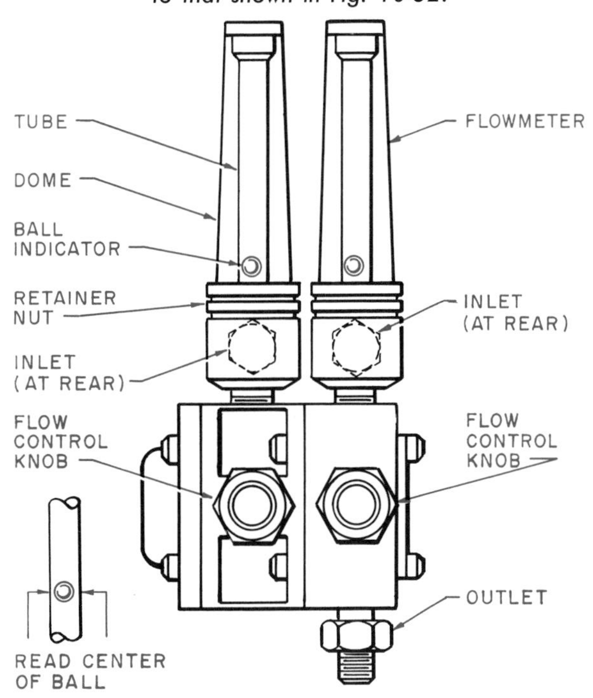

Such a weld is highly desirable when high bead buildup is detrimental to the weldment. A low bead profile also reduces or eliminates postweld grinding. This gas mixture has found considerable use in the welding of stainless steel pipe with the MIG process.

Argon-Oxygen. The addition of small amounts of oxygen, usually 1- to 5-percent mixtures, improves and expands the use of argon. Oxygen provides wider penetration, improves bead contour, and eliminates the undercut at the edge of the weld that is obtained with pure argon when welding steel, Fig. 10-55. Argon-oxygen mixtures are used principally in the welding of aluminum alloys, low alloy steel, carbon steels, and stainless steels.

Best results with shielding gas mixtures, other than argon-helium mixtures, have been obtained with oxygen. The M-1 gas is a mixture of 1 percent oxygen with argon, while the M-5 is a 5-percent oxygen mixture. Both are used only in MIG welding since oxygen would cause rapid loss of a tungsten electrode. The oxygen oxidizes the base plate slightly but evenly and provides a cathode surface with uniform emission properties. It also eliminates undercutting tendencies. A maximum of 2 percent oxygen delivers these advantages on low alloy steel. Above this percentage, there are no additional benefits, and a more expensive wire with higher deoxidation characteristics is required for welding. A mixture of 5 percent oxygen improves arc stability when welding stainless steels with DCSP.

Research has shown that the droplet rate (metal transfer) can be materially increased with these percentages of oxygen. The increased rate, with no change in current density, permits welding at higher speeds without undercutting. At the increased speed, coalescence of the weld metal is improved so that lower current densities are possible, and larger diameter wires can be used for a given welding current.

A comparison of penetration characteristics for the three main shielding gases and the two oxygen mixtures are shown in Fig. 10-56.

Argon-oxygen mixtures can be used on direct current, straight polarity jobs, but transfer is not as good as with reversed polarity. Straight polarity applications of the gas mixture are generally limited to overlay work. In such cases, the reduced penetration and dilution of the base metal are advantages.

Other Gases. In some applications of aluminum welding, a small percentage of *chlorine* has been added to argon or helium to reduce the tendency toward porosity. This practice is not widespread because chlorine gas is both poisonous and extremely corrosive.

Nitrogen has been used as a shielding gas, either in pure form or in combination with argon, for the welding of copper and copper alloys. When used as a mixture, it is usually in the percentage of 70 percent argon and 30 percent nitrogen. The addition of argon smoothes out the arc and reduces the agitation in the welding pool.

10-54. *Regulator used with gas mixer to reduce the tank pressure to a steady 50 p.s.i. for welding.*

National Welding Equipment Co.

10-55. *Typical MIG welds in stainless steel.*

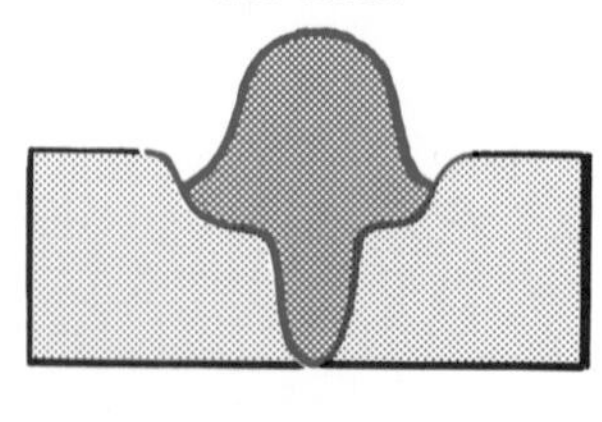

ARGON GAS

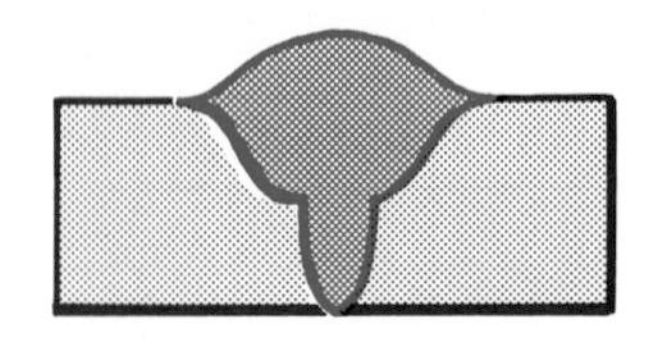

ARGON–OXYGEN GAS

Electrode Filler Wire

The solid filler wires used for gas metal-arc welding are generally quite smaller in diameter than those used for stick electrode welding. Wires range in size from 0.020 to $\frac{1}{8}$ inch in diameter. Thin materials are usually welded with 0.020, 0.030, or 0.035 inch. Medium thickness materials call for wire sizes 0.045 or $\frac{1}{16}$ inch, and thick materials require wire sizes $\frac{3}{32}$ or $\frac{1}{8}$ inch. The position of welding is important in selecting the proper wire size. Generally, smaller wire sizes are used for out-of-position work such as vertical or overhead welding. Cost is another factor. The smaller the diameter of the electrode wire, the more its cost per pound. Metal deposition rates, however, must also be considered.

Due to the small wire size and the high currents used in MIG welding, the filler metal melts at a rapid rate. The rate ranges from 40 to over 900 inches per minute. Because of this high burnoff rate and the need to be able to weld without interruption, the wires are provided as spools or in coils. Standard spools usually carry 1, 2, 3, 10, or 25 pounds of wire. To take advantage of lower prices, very large users purchase their filler wire in drums which contain 300, 700, or 1000 pounds of coiled wire.

The specific type of filler metal is selected for the following reasons:

- To match the composition of the base metal
- To control various weld properties
- To deoxidize the weld deposit
- To promote arc stability and desirable metal transfer characteristics

Wire composition usually matches that of the base metal being welded. In many cases, however, it is necessary to use a filler wire of completely different composition. This is because some alloys lose some of their characteristics as weld deposits. Filler metal alloys which are both favorable for welding and produce the required weld metal properties must be chosen. Among the materials that require filler wires of different composition than the base metal are copper and zinc alloys, high strength aluminum, and high strength steel alloys.

In addition to the alloying elements used in filler wires, deoxidizers and other scavenging agents are nearly always added. This is to prevent porosity or damage to the mechanical properties of the weld metal. When welding steel, some shielding gases such as carbon dioxide cause severe oxidation losses of alloying elements across the arc. The deoxidizers most frequently used in steel filler wires are manganese, silicon, and aluminum. Titanium and silicon are the principal ones used in nickel alloys. Copper alloys may be deoxidized with titanium, silicon, or phosphorus. Deoxidants are not used in titanium, zirconium, aluminum, and magnesium filler wires since these metals are highly reactive. They must be welded with oxygen-free inert gas and with complete shielding, or in closed chambers filled with inert gas.

FLUX-CORED FILLER WIRES. Flux protects the weld pool during the process of solidification. This is done in several ways. In submerged-arc welding and in most brazing operations, the flux is supplied as a bulk powder and added during welding. Stick electrodes carry their flux on the outside of the wire filler metal. Flux-cored filler wires used in MIG welding carry their flux as the core of the filler wire, Fig. 10-57. Experimentation with this type of wire began in the 1930s, but it was not until 1954 that serious development took place.

Flux-cored filler wires are available in diameters of .045, .052, .068, 1/16, 5/64, 3/32, 7/64, 1/8 and 5/32 inch, and are provided in continuous coils. They usually require high current values and CO_2 or $ArCO_2$ shielding gas. Shielding may be omitted, but there is some loss of ductility, penetration, and weld toughness.

The flux is primarily a method of carrying the alloying and deoxidizing elements to the weld. This is less costly than providing these elements through the filler wire. Flux-cored wire provides the following advantages over stick electrodes:

- Lower cost to manufacture
- Less chance for handling and shipping damage

10-57. Solid and fabricated flux-cored electrode wire.

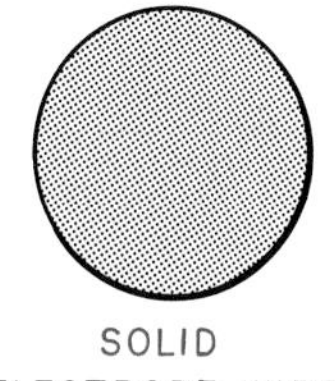

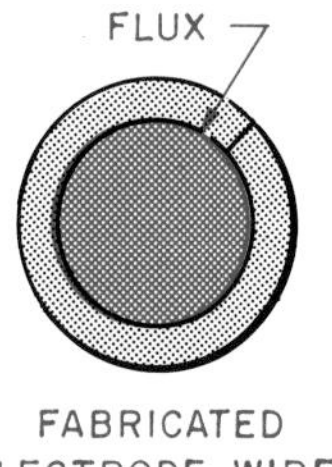

10-56. Cross section of welds made with the three main shielding gases and two mixtures show penetration differences.

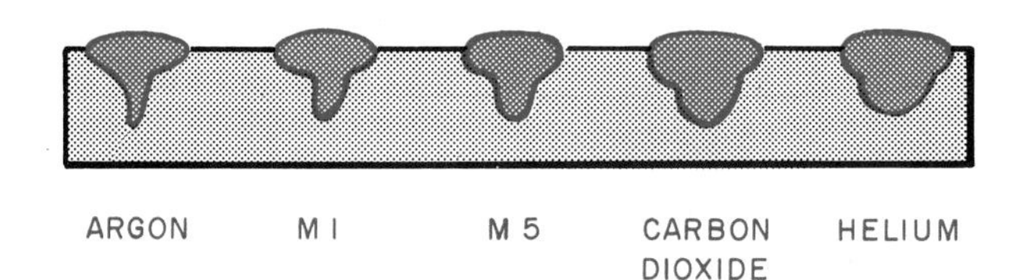

- Continuous welding when used with MIG and other continuous wire processes
- Deeper penetration. Fewer passes are necessary when welding heavy stock.
- Greater arc time and no stub loss
- V-joints with a 30-degree angle are possible instead of the 60-degree angle required with manual methods. Thus there is a saving in welding time and filler metal costs.

Flux-cored filler wires have not been developed for the welding of all metals. At the present time these filler wires are available for the welding of mild and low alloy steel and a few stainless steels. They have limited hard-surfacing applications. They are particularly useful in the welding of heavy plate sections and when high deposition rates are desired. They are most suitable for the downhand and horizontal positions. Flux-cored wire welding is a high-heat process that is fast and economical. It can deposit metal at a higher rate than any other manual or semi-automatic process except manual submerged arc.

FILLER WIRE CLASSIFICATIONS. The American Welding Society worked many years in its attempts to classify continuous wires on a basis similar to those established for stick electrodes. For stick electrodes the fabricator specifies an AWS classification, such as E7010 or E6024. For continuous wires, he can now specify a wire classification such as E70S-3 or E70S-4.

The specifications cover mild steel electrodes for MIG welding. They include gas-shielded solid wires, gas-shielded flux-cored wires and gasless cored wires. The classification of solid wires carry the letter S, as in E70S-3. All flux-cored wires have the letter *T* for tubular, as in E70T-1. The lone exception to the rule is the electrode classified as E70U-1. This is the emissive coated solid wire. Electrode specifications booklets can be ordered from the American Welding Society. The booklet numbers are AWS A5.18-69 for solid wires and AWS A5.20-69 for flux-cored wires. The classification code is demonstrated in Fig. 10-58. In many respects it is somewhat similar to that used for shielded-arc electrodes.

Tentative specifications covers bare mild steel electrodes and fluxes for submerged-arc welding. Nine welding wires, divided into three groups according to manganese content, and a number of companion powder fluxes are listed in the tentative specifications. These specifications are also available from the American Welding Society. The booklet number is AWS A5.17-69.

In the welding of mild steel, there are enough continuous wires to replace any stick electrode used in shielded metal-arc welding. In the downhand position, the E70T-1 electrode, for instance, is a good replacement for the E7010 stick electrode. The gasless cord wire, E70T-4, can replace the E6024. Fine wire, such as E70S-3, is used for MIG welding in all positions.

The first classification in each of the three types (E70S-1, E70T-1 and E70U-1) contains minimum amounts of deoxidizers (manganese and silicon). Study Table 10-8 and the following material carefully for an understanding of the nature of each filler wire.

E70S-1 (formerly E60S-1). These wires contain the lowest silicon content of the solid electrode classification. They work best on killed or semikilled steels. An argon-oxygen shield-

10-58. *New gas metal-arc electrode classifications.*

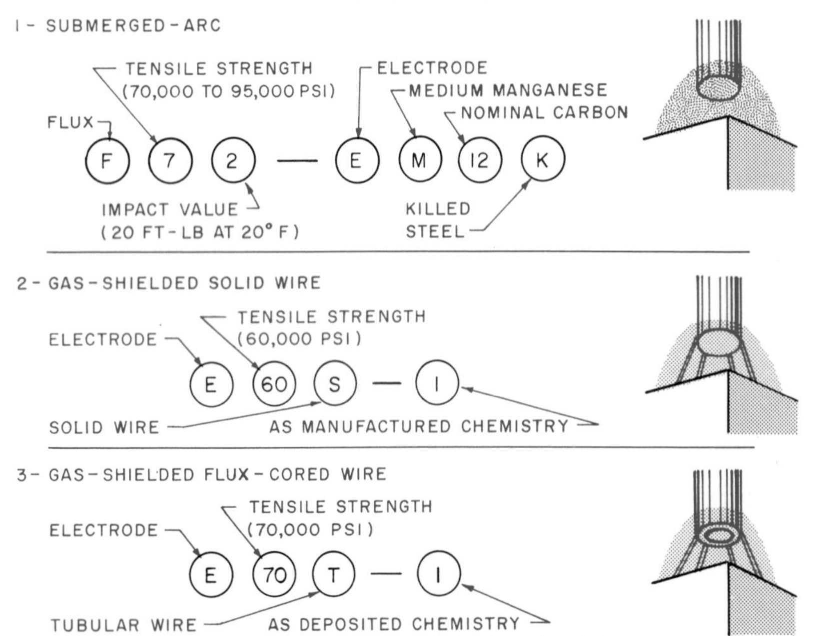

ing gas mixture is generally necessary to meet the best physical requirements. However, carbon dioxide may be used when weld quality is not paramount and cost is a factor.

E70S-2 (formerly E60S-2). The E70S-2 classification covers a grade of solid wire with considerably greater amounts of deoxidizers than E70S-1. Consequently, it can be used for welding steels with a rusty or dirty surface. There is some slight sacrifice of weld quality, depending on the degree of contamination. E70S-2 electrodes can be used to weld semikilled, rimmed, and killed steels. They can be used with argon-oxygen mixtures, carbon dioxide, and argon-carbon dioxide mixtures. They are preferred for out-of-position welding by the short-circuiting type of metal transfer because of ease of operation.

E70S-3 (formerly E60S-3). The wires in the E70S-3 classification are very similar to those in the E70S-1 classification except that greater amounts of deoxidizers (manganese and silicon) are included. They may be used with either carbon dioxide or argon-oxygen as a shielding gas. These wires are widely used in production welding on single-pass weldments although they can be used on multipass welds in killed or semikilled steels. They can be used for out-of-position welding with small diameter electrodes, the short-circuiting type of metal transfer, and pure carbon dioxide or argon-carbon dioxide mixtures as gas shielding.

E70S-4. These wires contain slightly greater quantities of silicon and produce a weld deposit of higher tensile strength than the three classifications above. They are used with carbon dioxide shielding and a somewhat longer arc.

Table 10-8
Continuous wire data.

Electrodes	Comments	Shielding gas	Minimum tensile strength, psi	Minimum yield strength at 0.2% offset, psi
Solid wires				
E70S-1	Solid wire with the lowest silicon content. Meets AWS requirements with argon-oxygen gas only.	Argon-oxygen	70,000	60,000
E70S-2	Triple-deoxidized wire (aluminum, zirconium, titanium). Premium quality for welding over dirty surfaces.	Carbon dioxide, argon-oxygen	70,000	60,000
E70S-3	Most popular solid wire. Good substitute for E6012. Single and multipass welds. Similar to E60S-1, but contains more silicon. Meets AWS requirements with argon-oxygen or carbon dioxide.	Carbon dioxide	70,000	60,000
E70S-4	Higher silicon than E60S-3. Also, higher tensile. Good where longer arcs or more deoxidants are needed.	Carbon dioxide	70,000	60,000
E70S-5	Another triple-deoxidized wire (aluminum, manganese, silicon). Welds rusty steels; not available in fine wire.	Carbon dioxide	70,000	60,000
E70S-6	Premium wire. Used in Europe like E70S-3 is used here. Demand is building up for this single and multipass wire.	Carbon dioxide	70,000	60,000
E70S-G	A catch-all classification, but it includes this ½% moly wire, the second most popular solid wire. Used on pipelines. Used mostly in fine wire. Makes multipass welds.	Carbon dioxide	70,000	60,000
E70S-1B	Has deoxidizers. Produces radiographic quality welds in low carbon and low alloy steels. Out-of-position welding. Single and multipass welds.	Carbon dioxide	70,000	60,000

(Continued on page 324)

Table 10-8 (Continued)

Electrodes	Comments	Shielding gas	Minimum tensile strength, p.s.i.	Minimum yield strength at 0.2% offset, p.s.i.
Solid wires (Continued)				
E70S-GB	Have deoxidizers and alloy additions for those applications not covered by E70S-1B.	Carbon dioxide	70,000	60,000
Flux-cored wires				
E60T-7	Used with internally applied gas shielding. Single and multipass welds.	None	67,000	55,000
E60T-8	Same as for E60T-7. Less tensile strength.	None	62,000	50,000
E70T-1	One of the most popular flux-cored wires. It is fast and produces little spatter. Requires clean surfaces. Single and multipass welds.	Carbon dioxide	70,000	60,000
E70T-2	Another very popular flux-cored wire. For single pass welds. Some overlap between this wire and E70T-1.	Carbon dioxide	70,000	not req'd
E70T-3	Open arc welding; no gas shielding is used. For single-pass welds on light gage steel.	None	70,000	not req'd
E70T-4	The most popular gasless cord wire. Single and multipass welds. Crack insensitive.	None	70,000	60,000
E70T-5	Can be used with or without gas shielding.	None, carbon dioxide	70,000	60,000
E70T-6	Similar to E70T-5 without externally applied shielding gas.	None	72,000	60,000
E70T-G	Composite electrodes not included in preceding classes. With or without gas shielding. Single or multipass work.	With or without	72,000	60,000
Emissive-coated wire				
E70U-1	Premium wire. Operates on DCSP. (All of the above wires operate on dc reverse polarity.) No spatter.	Argon-oxygen, argon	70,000	60,000

Iron Age Magazine

E70S-5. The E70S-5 classification contains aluminum in addition to manganese and silicon as deoxidizers. It can be used when welding rimmed, killed, or semikilled steels with carbon dioxide shielding gas and high welding currents. E70S-5 wires can be used on dirty steels. They are not used with the short-circuiting type of metal transfer.

E70S-6. These wires contain additional manganese and still higher amounts of silicon. Welds have a high resistance to impact when shielded with carbon dioxide. E70S-6 wires may be used with high currents for welding rimmed steels and on sheet metal where smooth weld beads are desired. They can be used for out-of-position welding with the short-circuiting type of metal transfer. They are suitable for moderately dirty steels. These wires exhibit outstanding mechanical properties and increased hot-short crack resistance, especially on high carbon steels such as rail steel.

E70S-G (general). This classification includes those solid electrodes which are not included in the above mild steel classification. They include certain types of alloy electrodes.

E70S-1B. These electrodes contain a properly balanced chemical content with adequate deoxidizers to control porosity during welding with carbon dioxide as the shielding gas. They give X-ray quality welds to both standard and difficult-to-weld low carbon and low alloy steels. They can be used in out-of-position welding and for single-and multiple-pass welding. Like the E70S-6 they may be used on dirty steels.

E70S-GB. This classification includes those electrodes to which alloys that are not covered by the E70S-1B classification have been added for deoxidation and usability improvement.

E70U-1. This classification includes dust-coated solid wires that can be used with direct current, straight polarity (DCSP) and argon shielding. These wires are of minor consideration today. The original intention was to reduce the penetrating characteristic of direct current, reverse polarity (DCRP) by going to DCSP. The wires are designed to give a spatter-free, spray-type weld metal transfer. E70U-1 wires may be used in the flat and horizontal positions for fillet or multipass welds on rimmed steels.

The following electrode descriptions cover those mild steel composite electrodes used for the flux-cored MIG welding of mild and low alloy steels.

E60T-7. These wires are used with externally applied gas shielding and may be used for single- and multiple-pass applications. The weld deposits have a low sensitivity to cracking.

E60T-8. These wires are used without externally applied gas shielding and may be used for single- and multiple-pass applications in the flat and horizontal positions. Weld deposits made without shielding gas have a tensile strength ranging from 50,000 to 62,000 p.s.i. and little sensitivity to cracking.

E70T-1. These wires are used with carbon dioxide shielding gas for single- and multiple-pass welding in the flat position and for horizontal fillets. In order to secure X-ray quality, joints must be clean and free from oil, excessive oxide, and scale. The wires produce a quiet arc with little spatter. The deposition rate is high. A flat-to-slightly convex bead and easily removable slag are characteristic of the weld deposits. The deposits have good impact properties.

E70T-2. These wires are used with carbon dioxide shielding gas. They are suitable for single-

Table 10-9
General recommendations for the gas metal-arc welding of various metals.

Base metal	Shielding gas	Specific alloy to be welded	Filler metal	Elect. diameter	Current ranges
Aluminum and its alloys	Pure argon or helium-argon mixture	1100 3003, 3004 5050 5052 5154, 5254 5083, 5084, 5456 6061	1100, 4043 4043 4043, 5554 5554, 5154 5554, 5154 5556, 5356 4043, 5556	0.030 0.045 1/16 3/32 1/8	50–175 90–250 160–350 225–400 350–475
Magnesium and its alloys	Pure argon or helium-argon mixture	AZ31B, 61A, 81A, ZE10XA ZK20XA AZ31B, 61A, 63A, 80A, 81A, 91C, 92A, 100A, AM80A, ZE10XA XK20XA AZ63A	AZ61A AZ92A AZ63A	0.045 1/16 3/32	220–280 240–390 330–420
Copper	Pure argon or helium-argon mixtures	Deoxidized copper	Deoxidized copper silicon .25% tin .75% manganese .15%	1/16	300–470
Copper-nickel alloy	Pure argon	Copper-nickel alloys 70–30 90–10	Deoxidized titanium 70–30 90–10	1/16	250–300
Plain low carbon steel	Carbon dioxide argon-carbon dioxide argon-oxygen	Hot or cold rolled sheet ASTM A7, A36, A285, A373 or equivalent	Deoxidized carbon steel	0.030 0.035 0.045 1/16 5/64	50–150 75–230 100–300 300–450 300–500
Low alloy steel	Argon-oxygen Argon-carbon dioxide	Hot or cold rolled sheet	Deoxidized low alloy	0.030 0.035 0.045 1/16	50–150 75–230 100–350 300–450
Stainless steel	Argon-oxygen	302, 304 321, 347 309, 310 316, etc.	Elect. to match base metal	0.030 0.035 0.045 1/16	75–150 100–160 140–310 280–350
Nickel and nickel alloys	Argon or helium-argon mixtures	Nickel Monel® Inconel®	Deoxidized titanium to match base metal	0.035 0.045 1/16	100–150 150–260 200–400
Bronzes	Argon helium-argon argon-oxygen	Manganese-bronze aluminum-bronze nickel-aluminum-bronze tin-bronze	Aluminum-bronze aluminum-bronze aluminum-bronze phosphorus-bronze	1/16 5/64	225–300 275–350

American Welding Society

Table 10-10
MIG process selection.

	Submerged arc	Inert gas shielded	Carbon dioxide shielded: Large wire	Carbon dioxide shielded: Small wire	Tubular wire: Open arc	Tubular wire: Shielded arc
Metals to be welded	Low carbon and medium carbon steels low alloy high strength steels quenched and tempered steels many stainless steels	Aluminum and aluminum alloys stainless steels and phosphorus steels nickel and nickel alloys copper alloys, titanium, etc. as well as carbon steels	Low carbon and medium carbon steel Low alloy high strength steels Some stainless steels		Rebuilding and hardsurfacing most steels	Low carbon and medium carbon steels, low alloy high strength steels
Metal thickness	16 gauge (0.062″) to 1/2-inch (No preparation); single pass to 1 1/2″ and thicker; maximum thickness practically unlimited	12 gauge (0.109″) to 3/8-inch (No preparation); single pass welding up to 1″; maximum thickness practically unlimited	10 gauge (0.140″) up to 1/2-inch (No preparation); practical max. 1 inch	20 gauge (0.038″) up to 1/4-inch; above 1/4-inch not economical	Primarily for hardsurfacing	1/4-inch to 1/2-inch (No preparation); maximum thickness practically unlimited
Welding positions	Flat and horizontal	All positions	Flat and horizontal	All positions	Flat	Flat and horizontal
Major advantages	Highest deposition rate; wide range of materials; wide range of thickness; no visible arc; (No arc shielding necessary)	Welds most nonferrous metals; no flux required; all positions (with small wire); visible arc; negligible clean-up	Low cost; high speed; deep penetration; visible arc	All positions; thin material; will bridge gaps; minimum clean-up; visible arc	Fast deposition; simple; rugged	Smooth surface; Deep penetration; sound welds; visible arc
Limitations	Flux removal and handling; visibility obstructed; possible flux entrapment	Minimum thickness limited; cost of gases; spatter removal sometimes required	Uneconomical in heavy thicknesses; gas coverage essential		Minimum thickness limited	Lower efficiency; Slag removal
Quality	Good properties, X-ray	Good properties, X-ray	Good properties, X-ray		High hardness	X-ray
Appearance of weld	Very smooth surface; no spatter; flat contour	Fairly smooth convex surface	Relatively smooth; some spatter	Smooth surface; relatively minor spatter	Relatively smooth	Smooth surface; some spatter
Travel speeds	Up to 200 i.p.m.	Up to 100 i.p.m.	Up to 300 i.p.m.	Max. 30 i.p.m. (semi-automatic)	Up to 20 lbs. per hr.	Up to 150 i.p.m.
Range of wire sizes—inches	Diameter—1/16, 5/64, 3/32, 1/8, 5/32, 3/16, 7/32, 1/4	Diameter—0.035, 0.045, 1/16, 3/32, 1/8	Diameter 0.045, 1/16, 5/64, 3/32, 1/8	Diameter 0.030, 0.035, 0.045	Diameter 3/32, 7/64, 1/8	
Range of welding current	50 min.—1200 max. amps.	50 min.—600 max. amps.	75 min.—900 max. amps.	25 min.—200 max. amps.	200 min.—500 max. amps.	300 min.—700 max. amps.

Hobart Brothers Co.

Table 10-10 (Continued)

		Submerged arc	Inert gas shielded	Carbon dioxide shielded: Large wire	Carbon dioxide shielded: Small wire	Tubular wire: Open arc	Tubular wire: Shielded arc
	Electrode and shielding costs	Least expensive electrode wire; flux relatively inexpensive	Expensive electrode wires; relatively expensive gas	Reasonably priced electrode wires; least expensive gas		Fairly expensive electrode wire	Relatively expensive wire; least expensive gas
	Overall welding costs	Least expensive for heavy metal	Least expensive for nonferrous metals	Least expensive for medium thickness	Least expensive for thin material	Less expensive than manual	Least expensive on low alloy steels
Equipment required	Welding power	Constant voltage motor generator or rectifier conventional for semi-automatic 500—900 amps.	Constant voltage motor generator or rectifier 200, 300, 500, 900 amps.	Constant voltage motor generator or rectifier 500, 900 amps.	200, 300 amps. (and 350A gas engine)	Conventional or constant voltage Motor generator or rectifier 500, 900 amp.	
Equipment required	Wire feeder	Constant speed or arc voltage control (semi-automatic)	Constant speed head with gas valve	Constant speed head with gas valve		Constant speed or arc voltage control head	Constant speed, head with gas valve
Equipment required	Nozzle or gun required — Automatic	Flux nozzle (waterless)	Concentric gas nozzle (water-cooled)	Side-delivery gas nozzle (waterless)		Waterless nozzle	
Equipment required	Nozzle or gun required — semi-automatic	Flux flow or flux gun and hopper	Pistol-grip gun (water-cooled)	Water-cooled or waterless guns		Open arc gun (waterless)	Pistol-grip gun (water-cooled or waterless)
Equipment required	Shielding and regulation	Flux hopper or flux flow flux recovery unit	Gas regulator and flowmeter	Gas regulator and flowmeter		None	Gas regulator and flowmeter
Consumables	Wire	Match base metal	Match base metal	Match base metal		Specified usage	Match base metal
Consumables	Shielding medium	Pellitized flux	Argon or helium (and inert mixtures)	Carbon dioxide and argon-carbon dioxide mixtures		None	Carbon dioxide

pass welding in the flat position and for horizontal fillets. Electrodes with high manganese content may require multiple passes. The E70T-2 wires may be used on dirty steels and still produce welds of X-ray quality.

E70T-3. These wires are used without externally applied gas shielding. They are intended primarily for depositing single-pass, high-speed welds in the flat and horizontal positions on light plate and sheet metal. They should not be used on heavy sections or for multiple-pass applications.

E70T-4. These wires are used without externally applied gas shielding. They may be used for single- and multiple-pass applications in the flat and horizontal positions. Due to low penetration and other properties, and the weld deposits have a low sensitivity to cracking.

E70T-5. These wires are designed for flat fillet or groove welds with or without externally applied shielding gas. Horizontal fillet welds can be made at lower deposition rates. They may be used with or without externally applied gas shielding. Welds made with carbon dioxide shielding gas have better quality than those made without shielding gas. E70T-5 wires may be used for single-pass welding with little surface preparation and for multipass welding. They have a globular metal transfer with shallow penetration. The deposit has a convex bead and a thin, easily removable slag. It can meet the highest impact requirements.

E70T-6. These wires are similar to those of the E70T-5 classification, but they are designed for use without an externally applied shielding gas.

E70T-G. This classification includes those composite electrodes which are not included in the preceding classes. They may be used with or without gas shielding and may be used for single-or multiple-pass work.

The specifications for mild steel electrodes and fluxes used for submerged-arc welding are described in Chapter 27.

SUMMARY

You are now ready to begin practice with the MIG process. Because of the thorough training you have had up to this point with the other welding processes, you will find that MIG welding will be relatively easy to learn. The learning situation can be improved by mastering the material covered in this chapter. Any skill is more readily mastered if the technical and related information is understood so that it can be applied while practicing the skill.

The information presented in Tables 10-9 and 10-10, on pages 325–327, summarizes the basic information concerning the various MIG processes and the application of the various gases and filler rods. Become so familiar with the information contained in these tables that you will know how to meet each welding problem as it develops in shop and field.

REVIEW QUESTIONS

1. Give the complete name for the MIG process.
2. What type of welding machine is used for MIG welding?
3. How do MIG and TIG welding differ?
4. List at least six of the advantages of the MIG welding process.
5. List the equipment necessary for MIG welding.
6. The constant potential welding machine differs from the constant voltage welding machine in that the volt-ampere curve is different. True. False.
7. The main control on the MIG welding machine is the amperage adjustment. True. False.
8. Explain the characteristics of the pulsed-arc process.
9. Changing the wire-feed speed has no effect on the current value. True. False.
10. Of what materials are welding torch nozzles made?
11. Name the various shielding gases that can be used with the MIG process.
12. Give the sizes of MIG filler wires.
13. Name the two basic types of filler wire and explain each.
14. Name the various forms of metal transfer and explain each.
15. The MIG welding process is never used for vertical or overhead welding. True. False.
16. Pulsed-arc welding should never be used for thin gauge materials. True. False.
17. Pulsed-arc welding should never be used for vertical or overhead welding. True. False.
18. Slope, voltage, and inductance controls improve the characteristics of which form of transfer?
19. Shielding gases should never be mixed. True. False.
20. The proper shielding gas flow for MIG welding is from 5 to 10 c.f.h. True. False.

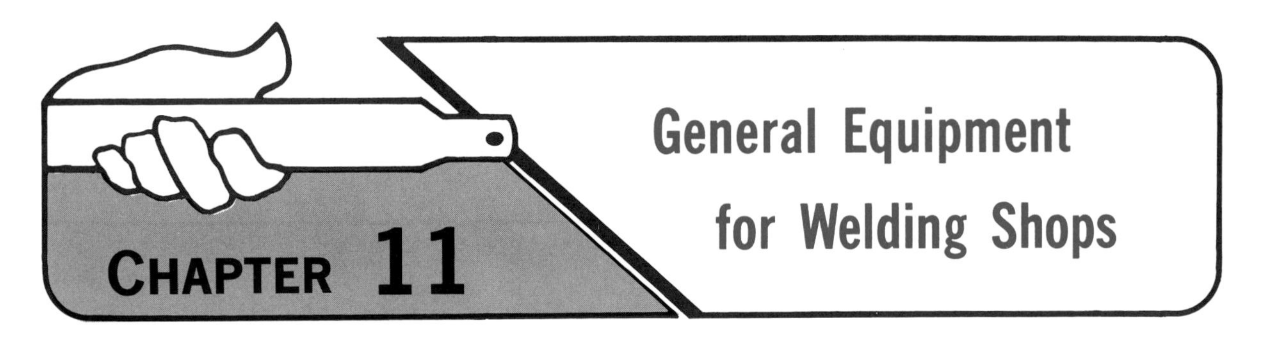

Chapter 11

General Equipment for Welding Shops

Companies that do welded fabrication make use of a variety of equipment. Some of this equipment is used in carrying out the welding operation. Other types are used for the related processes such as metal forming, cutting, and finishing that are necessary in fabricating welded structures. It is important for you, as a welding student, to learn how to use the tools and equipment presented in this chapter so that you will be able to perform these processes on the job.

SCREENS AND BOOTHS

Whenever welding is done in a shop where others are doing other jobs, these workers must be protected from the effects of arc rays, the spatter of molten metal, and sparks. In areas given over to the welding of small parts, permanent booths are erected. They are made of sheet metal or heavy canvas, and they are painted with a special protective paint. The booths are often equipped with exhaust fans for removing fumes and ducts for introducing fresh air.

A portable screen is used to shield large work when the welding equipment must be taken to the job. The screen is made of noninflammable material such as fireproofed jute or canvas. The screen is painted to absorb ultraviolet rays. Special ventilating systems are not required for large work because the screen blocks out the air circulation from one side only. Thus the fumes are permitted to rise to the ceiling and pass on. If the welding is to be done within the confines of an enclosure such as a tank or compartment, however, suction blowers and fans are installed to provide adequate ventilation.

WORK-HOLDING DEVICES

Welding Positioners

Welding positioners, Fig. 11-1, permit the placing of a weldment in a position suitable for downhand welding. They have plane table areas which can be tilted and rotated in any direction. On the smaller positioners the table is moved by handwheels or gears; and on the larger positioners, by electric gear drives. Parts to be welded, and in some cases, production jigging are secured to the plane table of the positioner.

The use of positioners to enable welding in the downhand position has increased production, reduced costs, and promoted

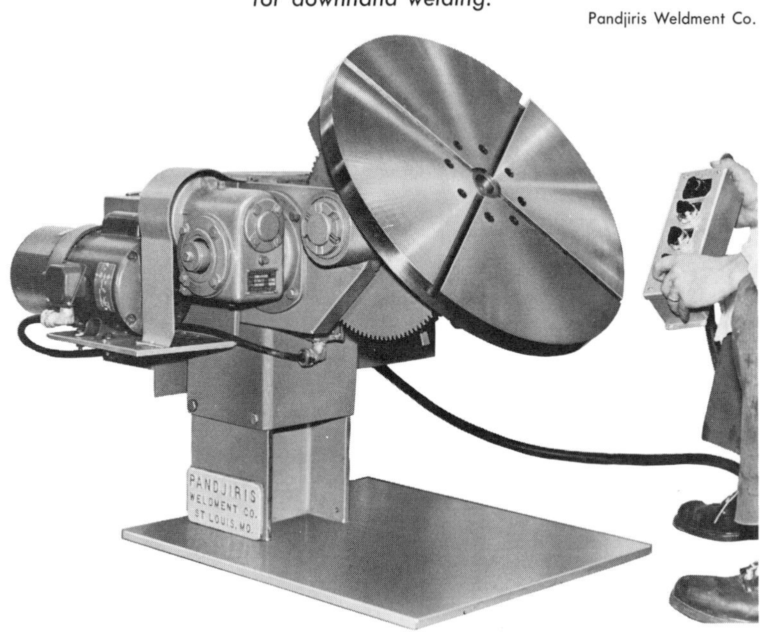

11-1. *A weld positioner can be tilted or rotated in any direction to provide for downhand welding.*

Pandjiris Weldment Co.

safety in both production and repair welding. Downhand welding permits the use of larger electrodes, higher currents, fewer passes, and better control of the weld pool. A vertical-up weld takes almost three times as long to do as the same weld done in the downhand position.

Weldments vary in size and weight. For this reason welding positioners are available in a wide range of sizes. Standard capacities run from 100 to 60,000 pounds. Some positioners have a capacity of 200,000 pounds. The positioning table may be rotated or tilted as desired, Fig. 11-2. It permits welding on the sides, top, and bottom of a job. The table may be tilted through 135 degrees (45 degrees beyond the vertical position). Regardless of the angle of tilt, the whole table may be rotated through a complete circle. Rotation allows all joints to be welded in the downhand position without resetting or handling the weldment.

Since welding positioners enable average welders to do skilled work, their use today is timely when so many new welders are entering the industry.

11-2. *The positioner can be set at any angle and rotated. A. Pieces may be positioned at any angle. B. The positioner is set at 45 degrees so that fillet welding may be done flat. The arrow indicates the point of welding. C. Sometimes it is necessary to position at 135 degrees (45 degrees beyond the vertical). D. Set up with a roller stand, the positioner rotates pipe or long cylindrical objects.*

The Welding Encyclopedia

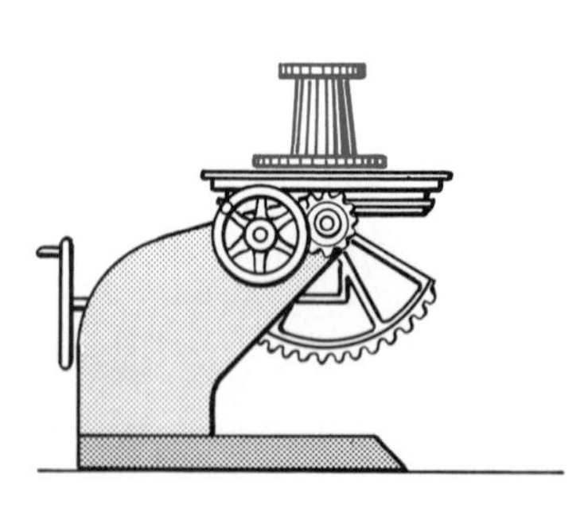

A
PIECES MAY BE POSITIONED AT ANY ANGLE.

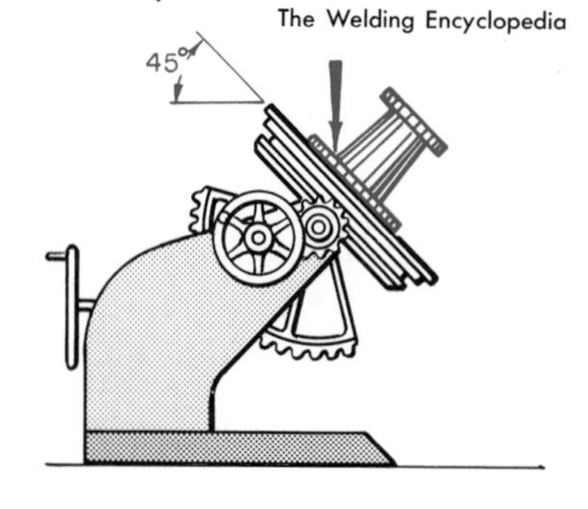

C
SOMETIMES IT IS NECESSARY TO POSITION AT 135° (45° BEYOND VERTICAL) TO DO THE JOB.

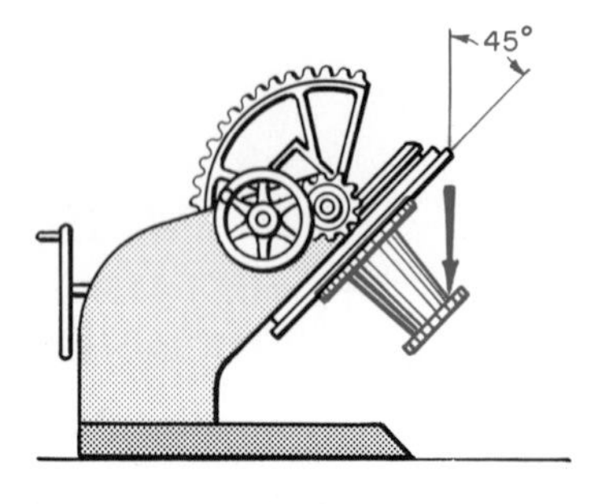

B
THE POSITIONER IS SET AT 45° SO THAT FILLET WELDING MAY BE DONE FLAT. ARROW INDICATES POINT OF WELDING.

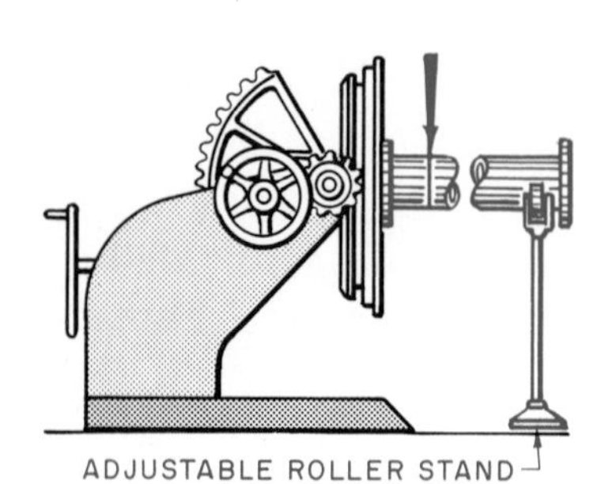

D
IN CONJUNCTION WITH A ROLLER STAND, THE POSITIONER ROTATES PIPE OR LONG CYLINDRICAL OBJECTS.

Turning Rolls

Welding fabricators who do a great deal of tank fabrication depend on turning rolls to make *circumferential seams* in the downhand position without interruption in the travel. Turning rolls, Fig. 11-3, support the workpiece on its outside diameter and rotate it.

There are two main types of turning rolls: those with steel or rubber-tired rolls and those with roller chain slings.

The wheel-type turning rolls are more common. They come in two varieties: the separate driver-idler type and the unit frame. In the *driver-idler* type, the driver turns the workpiece. The idler is a matching unit which is not powered. Several idlers can be used with a single driver. They range in capacity from 600 to over 1 million pounds. *Unit-frame* turning rolls have one fixed-location driving axle. The other axle is adjustable for various center distances which handle diameters from 3 inches to 6 feet or more. Weight capacities range up to 30,000 pounds; and frame lengths, up to 12 feet or more. Models that rotate and tilt are available.

Sling-type turning rolls have a roller chain sling with rubber or brass feet. The chain cradles the workpiece so that large areas of thin-walled cylinders can be supported. The load capacity is as high as 27 tons.

Figure 11-4 shows a larger tank being rotated on a pair of

powered rolls. The submerged-arc welding process is being used to weld a *circumferential seam.* The welding wire size is $^{3}/_{32}$ inch, and the welding current is 24 volts at 300 amperes. The welding speed is 50 inches per minute on both longitudinal and circumferential seams.

Weld Grippers

A weld gripper is a welding work-holding device that has 3 movable jaws like the jaws of a lathe chuck. The workpiece—a tank or pipe section—can be gripped on the inside diameter or the outside diameter. The action is fast. The gripper is mounted on a turntable or weld positioner for universal movement, Fig. 11-5.

Headstock-Tailstock Positioners

Like weld grippers and rollers, headstock-tailstock positioners rotate weldments about a horizontal axis and permit welding in the downhand position. They support the workpiece at each end. These positioners can be used for angular weldments as well as cylindrical ones.

The headstock is powered and may have a constant or variable speed. The tailstock is not powered. When used in pairs, the headstock and tailstock may have a capacity as high as 160,000 pounds. When used singly, they can support 80,000 pounds. The positioners shown in Fig. 11-6 are available with capacities ranging from 100 to 120,000 pounds.

Manipulators

Manipulators are applicable to both the positioned fillet and the simultaneous double fillet welds. They are also widely used for *circumferential* butt joints and other types. Automatic and submerged-arc welding units are mounted on the horizontal arm, Fig. 11-7. The manipulator provides vertical and horizontal travel and may be rotated through a full 360 degrees. Large units may be mounted on a double-rail track for maximum usage in the shop. On large installations, seating is provided for the operator at the welding head. Operation may be remotely controlled.

Turntable

A turntable provides powerized table rotation in either direction at adjustable speeds. Ro-

Pandjiris Weldment Co.

11-3. *Car-mounted tandem power rolls. Each roll has a 90-ton turning capacity and a 30-ton weight capacity. Both rolls are driven by synchronous motors to give a combined rotating capacity of 180 tons for high eccentric loading with a weight capacity of 60 tons. The fixed center distance handles workpieces 72 through 120 inches in diameter. Manual cars facilitate setup changes for vessels of various lengths.*

Pandjiris Weldment Co.

11-4. *A manipulator teamed with turning rolls to produce 100-gallon oil tanks with lap joint construction from 10-gauge mild steel. The submerged-arc welding process is used. Welding wire required is $^{3}/_{32}$-inch, and the power supplied is 24 volts at 300 amperes.*

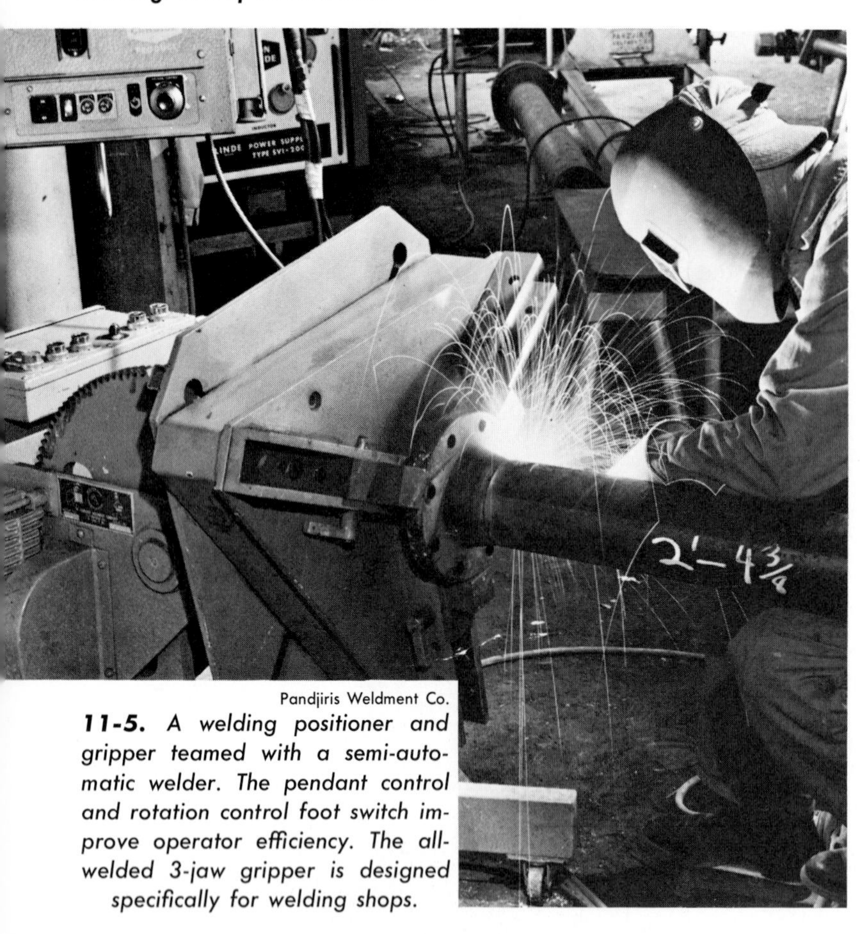

Pandjiris Weldment Co.

11-5. A welding positioner and gripper teamed with a semi-automatic welder. The pendant control and rotation control foot switch improve operator efficiency. The all-welded 3-jaw gripper is designed specifically for welding shops.

11-6. Headstock and tailstock positioners have tables 42 inches square with T-slots. The headstock has variable speed rotation, and the tailstock has a brake. All drive components are housed within the column structure.

Pandjiris Weldment Co.

tation is controlled by the operator, Fig. 11-8, and may be set for variable speeds. The work is positioned on the turntable and rotates under the welding gun.

Weld Seamers

Seamers, Fig. 11-9, provide accurate clamping and backup for welding longitudinal seams in tanks and box forms. This equipment is well-suited to semi-automatic and automatic welding processes. There are two basic types of seamers: one permits welding on the inside so that the weld reinforcement is on the inside of the workpiece, Fig. 11-10; the other permits the welding on the outside so that the weld reinforcement occurs on the outside, Fig. 11-11.

Weld Elevator

This mechanical unit permits the welder to be raised or lowered along a vertical surface so that he is always in the best possible position in the shortest possible time. It is a platform type controlled by the welder, Fig. 11-12 (Page 336). The unit may be mounted on rollers for movement in the shop, and it may also be attached to the walls of a tank being constructed in the field. When mounted on the workpiece, Fig. 11-13, it moves up the side walls of the tank as it is being erected. Weld elevators provide maximum safety for the welder. Figure 11-14 is a view of a pipe-fabricating shop using several of the work-holding devices discussed above.

Magnetic Grip Fixtures

There are a number of clamps and holding devices that take advantage of magnetic attraction to speed the work of the welder. One type is shown in Fig. 11-15.

Pandjiris Weldment Co.

11-7. *A manipulator with an automatic welding head assembly mounted on vertical and horizontal ball bearing precision slides. It has an 11-foot arc height and a 10-foot horizontal arc travel. Weld speed may vary from 3 to 113 inches per minute.*

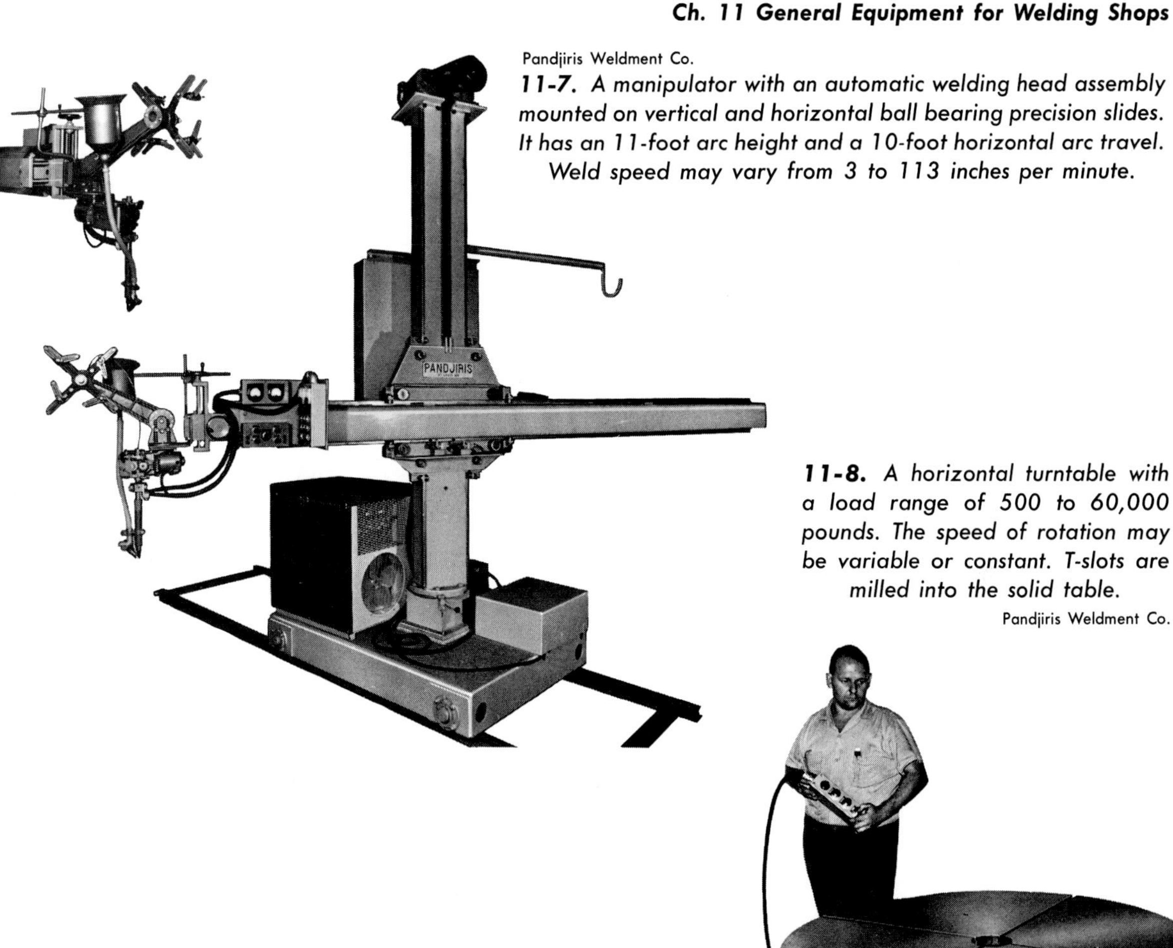

11-8. *A horizontal turntable with a load range of 500 to 60,000 pounds. The speed of rotation may be variable or constant. T-slots are milled into the solid table.*

Pandjiris Weldment Co.

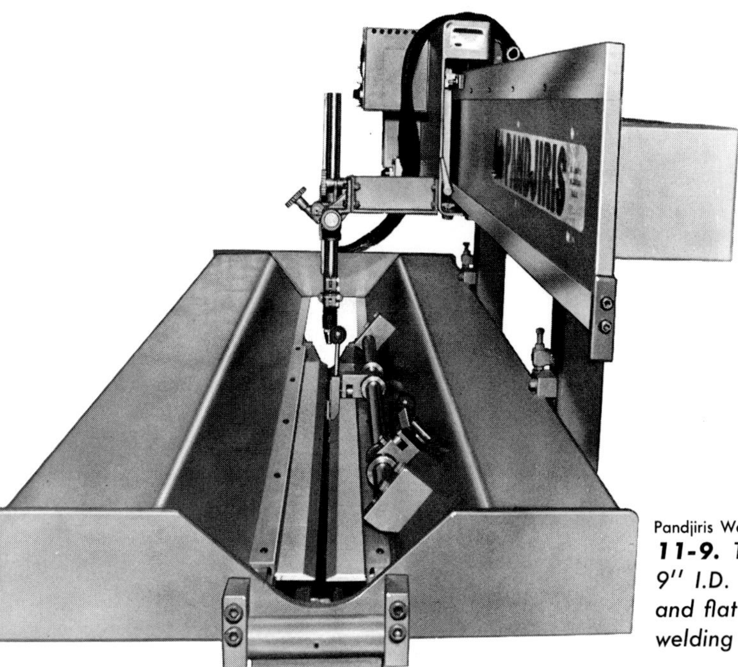

Pandjiris Weldment Co.

11-9. *This seamer has 10′6″ clamping length for 9″ I.D. through 36″ I.D. cylinders, other shapes, and flat sheets. It may be used with any standard welding process manual, semi-automatic, or automatic.*

This fixture is a permanent, nonelectric magnet which can be turned on and off, and yet it requires no outside power sources nor internal batteries. Such fixtures are portable and are available in three sizes of 2¾ pounds, 5¼ pounds, and 31 pounds. They provide safe holding strengths from 6 to more than 250 pounds. Whenever holding and positioning are required, this type of fixture can speed the welding operation and increase production.

Magnetic grip devices are equipped with 90-degree and 180-degree protractors and V-shaped shoes. They act as fixtures, jigs, and hold-down tools. They provide safe, positive holding and positioning of sheets, parts, rods, and tubes during welding and other fabricating operations. Magnetic fixtures eliminate tacking, clamps, and makeshift holding setups. The position of the work can be changed quickly since it can be held or released instantly.

The grip fixture has magnetic work-holding surfaces. Two are on opposite sides and parallel to each other, and the third is square with the two parallel surfaces.

Bug-O-Verlay

Figure 11-16 (Page 338), the Bug-O-Verlay,® is an automatic device that can be used with various welding processes for metal surfacing. It provides accurate and consistent travel speed and flame position. Precision metal coverage is possible, and a minimum of base metal dilution takes place. This is an important benefit because dilution of the base metal seriously affects such characteristics of the surfacing material as resistance to corrosion, abrasion, and impact.

Miscellaneous Equipment

In addition to the positioning equipment previously described, cranes, chain hoists, jacks, clamps, and tongs are required for handling and positioning of the work. A generous supply of C-clamps of all types and sizes, hold-down clamps, wedges, bars and blocks is necessary for the proper spacing and lining up of

11-10. Weld seamer procedure for internal welding.

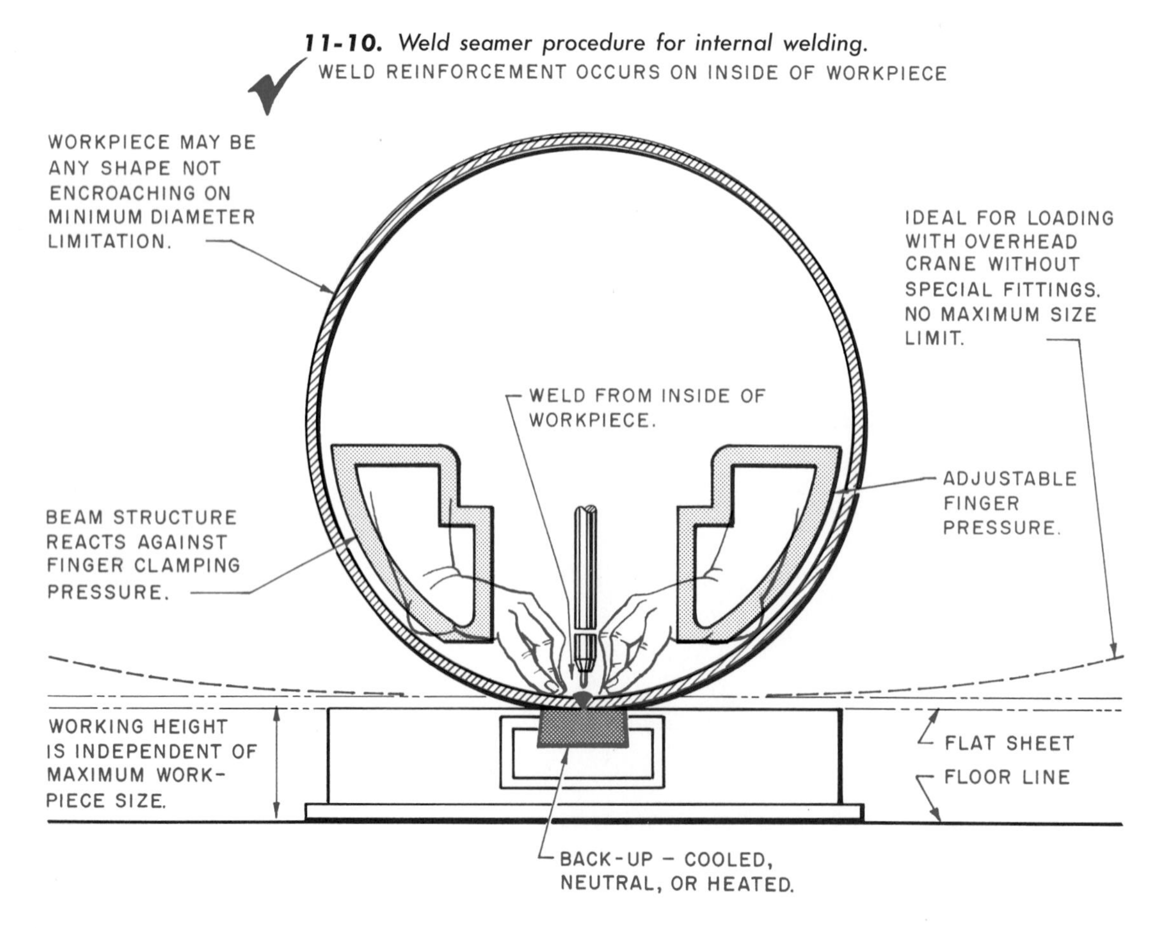
WELD REINFORCEMENT OCCURS ON INSIDE OF WORKPIECE
WORKPIECE MAY BE ANY SHAPE NOT ENCROACHING ON MINIMUM DIAMETER LIMITATION.
IDEAL FOR LOADING WITH OVERHEAD CRANE WITHOUT SPECIAL FITTINGS. NO MAXIMUM SIZE LIMIT.
WELD FROM INSIDE OF WORKPIECE.
ADJUSTABLE FINGER PRESSURE.
BEAM STRUCTURE REACTS AGAINST FINGER CLAMPING PRESSURE.
WORKING HEIGHT IS INDEPENDENT OF MAXIMUM WORK-PIECE SIZE.
FLAT SHEET
FLOOR LINE
BACK-UP – COOLED, NEUTRAL, OR HEATED.

parts. One type of hold-down clamp is shown in Fig. 11-17 (Page 338).

Backing Materials

The quantity, quality, and appearance of welds are improved by the use of backing materials of all kinds. Copper in bars, strips, and blocks supports the molten metal when welding some types of joints in certain metals. Copper strips laid along the parallel edges of a seam help to keep the welds of uniform width. By carrying away some of the heat, they also reduce distortion. Copper is chosen because of its high heat conductivity and because it resists fusion with the base metal being welded.

Shops that do a great amount of maintenance work have shapes of all kinds in carbon, plastic, asbestos, and fireclay. These materials form dams or molds on operations that require the building up of pads and shoulders to certain limits. Round carbon sticks are used to retain the shape of holes in hot metal and to protect threads in tapped holes from the heat of welding.

PREHEATING AND ANNEALING EQUIPMENT

As discussed in Chapter 3, heat treating, annealing, and normalizing are specific metal-working processes required for some types of work. Often the equipment is permanent, and it is found in most big shops. It may be large enough to handle only small parts, or it may be designed to take very large pressure vessels. Sometimes the preheating is done in temporary ovens built of firebrick. Heat-treating ovens can be fired by electricity, gas or oil burners, coal, or coke fires.

When heat treating metal, it is necessary to control the rate of cooling. Many of the shops are equipped with sand and lime pits to be used in delayed cooling, and they have containers of water, oil, or pickling solutions for various hardening processes. A generous supply of asbestos

11-11. *Weld seamer procedure for external welding.*

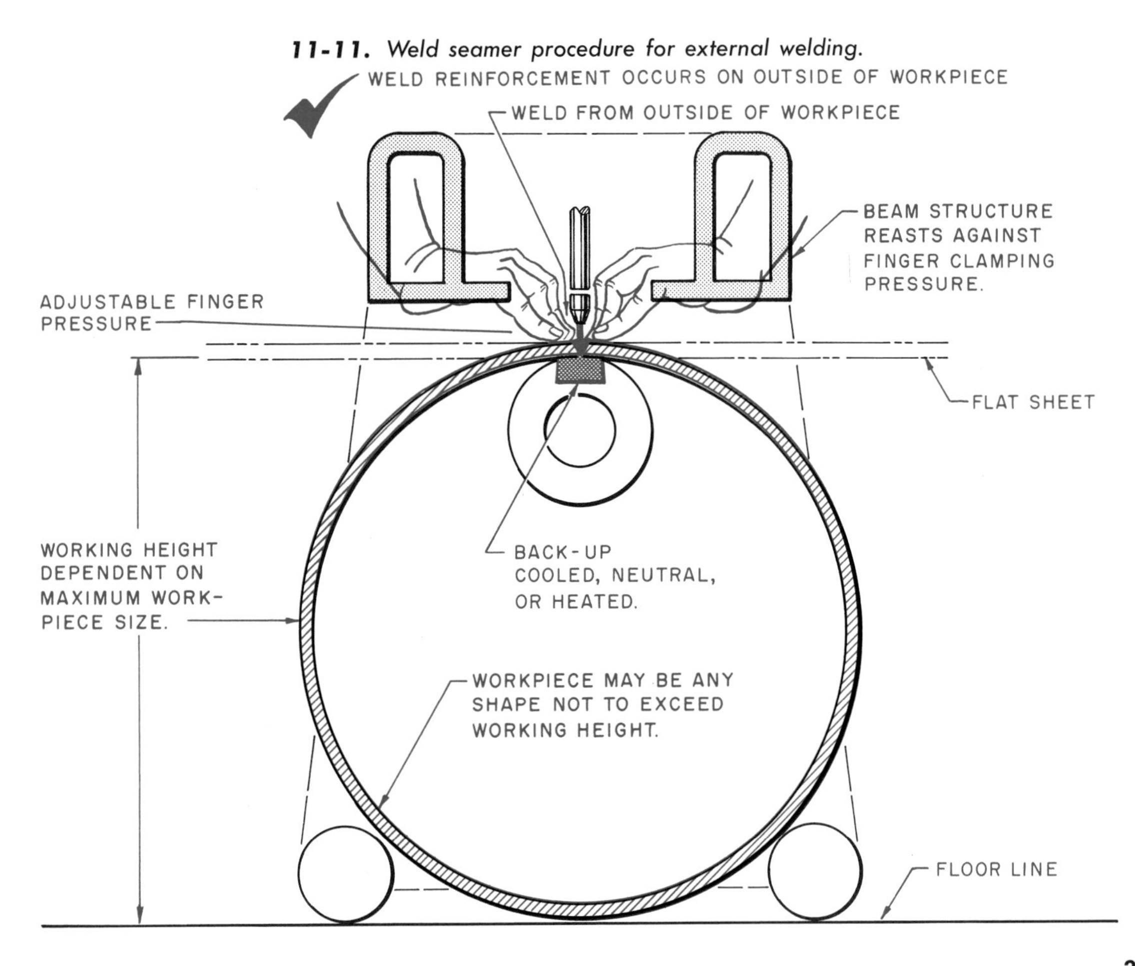
WELD REINFORCEMENT OCCURS ON OUTSIDE OF WORKPIECE
WELD FROM OUTSIDE OF WORKPIECE
BEAM STRUCTURE REASTS AGAINST FINGER CLAMPING PRESSURE.
ADJUSTABLE FINGER PRESSURE
FLAT SHEET
WORKING HEIGHT DEPENDENT ON MAXIMUM WORK-PIECE SIZE.
BACK-UP COOLED, NEUTRAL, OR HEATED.
WORKPIECE MAY BE ANY SHAPE NOT TO EXCEED WORKING HEIGHT.
FLOOR LINE

powder and asbestos sheet may also be required.

SANDBLASTING EQUIPMENT

Shops that do work that requires a great deal of preweld and postweld treatment are equipped with sandblasting equipment to clean the surface for welding and remove scale, slag, and rust after welding.

Cleaning the material to be welded is one of the important operations that must be performed in preparation for welding. Clean base metal makes a considerable difference in the physical and chemical qualitites of the weld.

The ease and simplicity of abrasive cleaning reduces the physical effort of the worker to a minimum. The operator simply directs the blasting stream on the surface to be cleaned and moves as fast as the foreign material is removed. When combined with the tremendous force of compressed air, the small abrasive particles remove the most stubborn foreign material from areas too confining for other cleaning tools. Abrasive cleaning makes cleaning around screw and bolt heads, in narrow corners, and in deep indentations as easy and simple as doing any flat surface.

Cabinet blasters, Fig. 11-18, eliminate such cleaning processes as scraping, sanding, wire brushing, chipping, and etching. The cleaning particles may be aluminum oxide, metal grit, or sand which produce an etched or frosted finish on the pieces being blasted. The finish may be fine, medium, or coarse depending on the size of the abrasive. Abrasives such as walnut shells, corn cobs, and glass beads remove foreign material without affecting the surface being blasted.

11-12. *This elevator has a 12-foot maximum elevation. It is available with a carrying capacity of 1,000 or 2,000 pounds.*

Pandjiris Weldment Co.

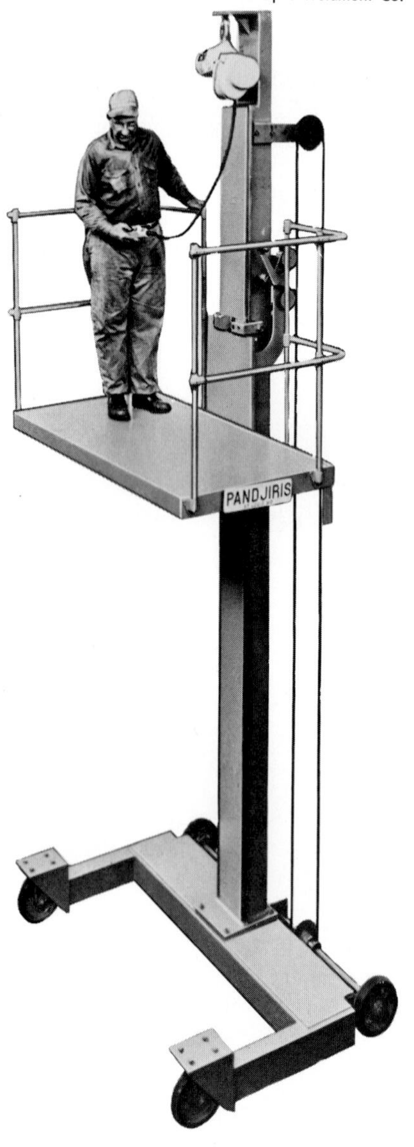

11-13. *For outside and/or inside girth welds of tanks with a minimum diameter of 20 feet. This elevator is used mostly for storage tanks, blast furnaces, and other field-welded tanks. The unit is equipped with submerged-arc welding heads. The variable speed carriage is also adaptable to gas metal-arc welding.*

Pandjiris Weldment Co.

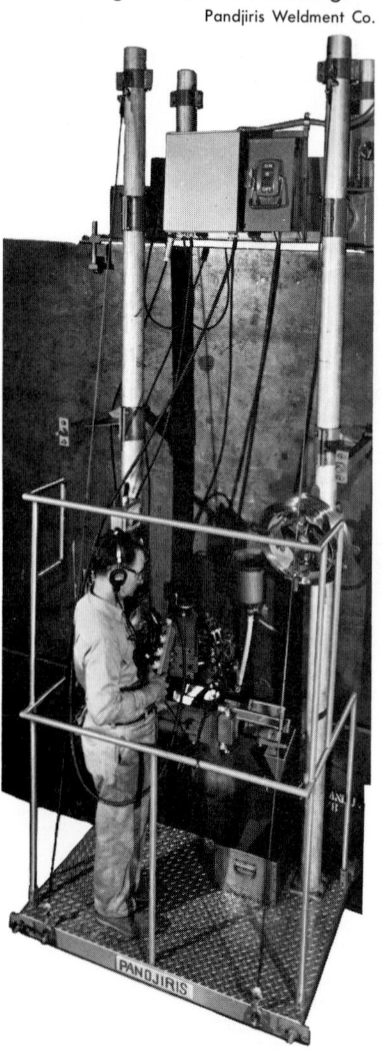

SPOT WELDER

Most welding shops have a spot-welding and/or seam welding resistance welder in the shop. The spot welder is the most common of the resistance welding machines. It is generally used for the welding of light gauge sheet metal and offers great flexibility in the fabrications of metal parts.

Spot welding is a process where two lapped pieces are

welded together by heat and pressure. The two pieces to be joined are pressed together between two metal electrodes. These electrodes carry the low-voltage, high-density electric current needed and also provide the pressure needed. The pressure may be applied by compression spring forces, hydraulic forces, pneumatic forces, or magnetic forces. The pressure varies from a few ounces to hundreds of pounds for different size spot welders.

The machine may be controlled by a foot pedal, Fig. 11-19. Spot welders are available in a great variety of sizes from small bench units to very large machines. The capacity of a spot welder, that is, the thicknesses of metal that can be spot welded together, depends on the KVA (kilovolt-ampere) rating of the machine.

Pandjiris Weldment Co.

11-14. *A pipe fabrication shop. Work-holding devices include a positioner with an adjustable L-plate, a boom-type manipulator, turning rolls, leveling and elevating tables, a headstock rotator, and a powerized weld elevator.*

HYDRAULIC TOOLS

To a great extent, hydraulic tools have replaced hand tools in today's welding shop. The demands of production, fabrication, testing, maintenance, and setup operations are such that special tools must be used. Hydraulic tools can do anything that the hands can do, but faster, with tons of controlled force. High pressure hydraulic units can package 5 tons of linear force in less than 2 cubic inches of space. One person can carry a hydraulic unit capable of generating over 50 tons with precision control. These tools are used in the shop and on the construction site for the following applications:

- Pressing
- Bending
- Forming
- Clamping
- Pulling
- Straightening
- Lifting
- Materials handling
- Holding
- Spreading
- Pushing
- Positioning
- Testing

Principles of Hydraulics

The basic function of hydraulic tools is simple. Hydraulic tools multiply force and put it to work. The required hydraulic pressure is easily generated by piston-type hand or power pumps which transmit oil from the pump reservoir into a closed system. High pressure flexible hoses with plug-in couplers form the union in the line. The line transmits the oil under pressure from the pump

11-15. *A permanent magnetic grip welding fixture.*

Eriez Magnetics

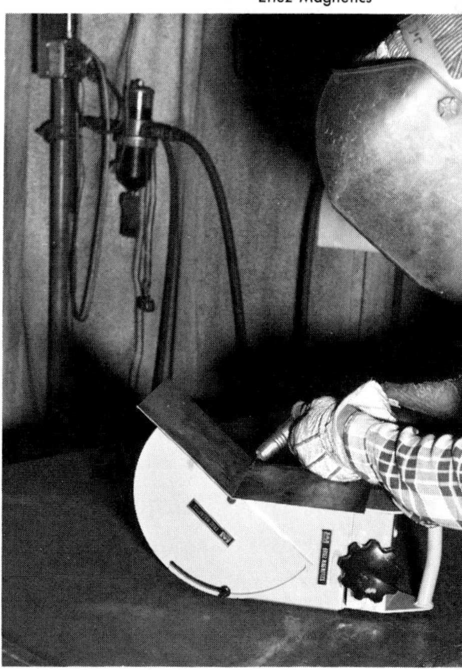

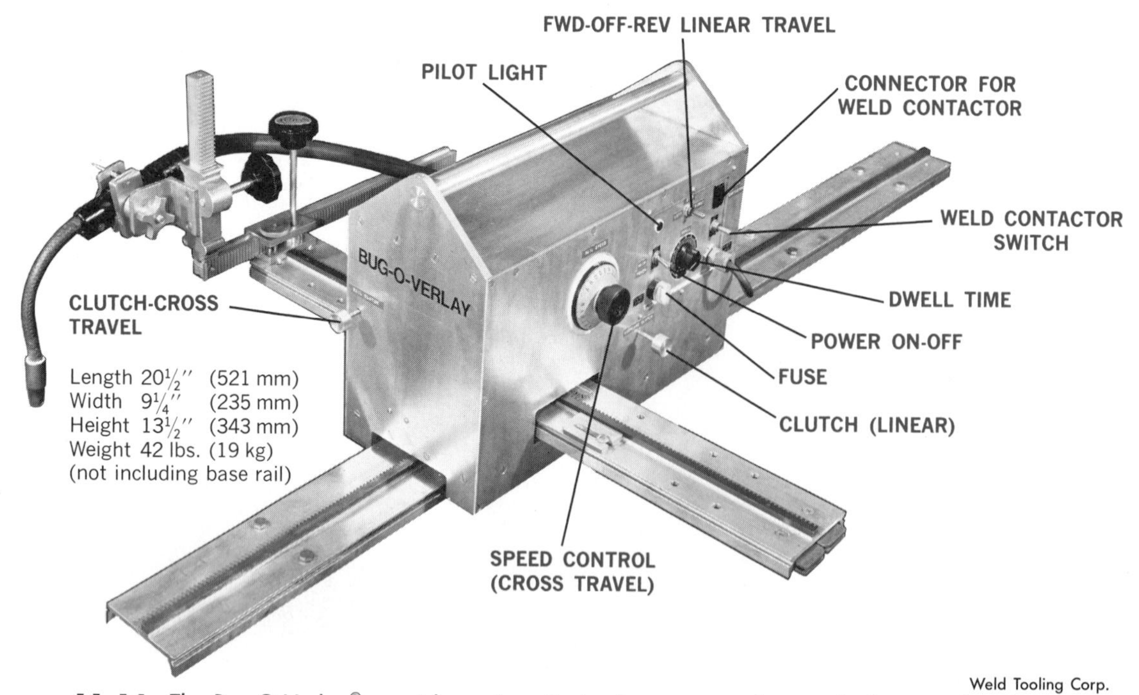

Weld Tooling Corp.

11-16. *The Bug-O-Verlay® provides automatic torch movement for overlaying metal surfaces.*

11-17. *A hold-down clamp.*

De-Sta Co.

11-18. *Sandblasting equipment can be very useful in a welding shop.*

Empire Abrasive Equipment Corp.

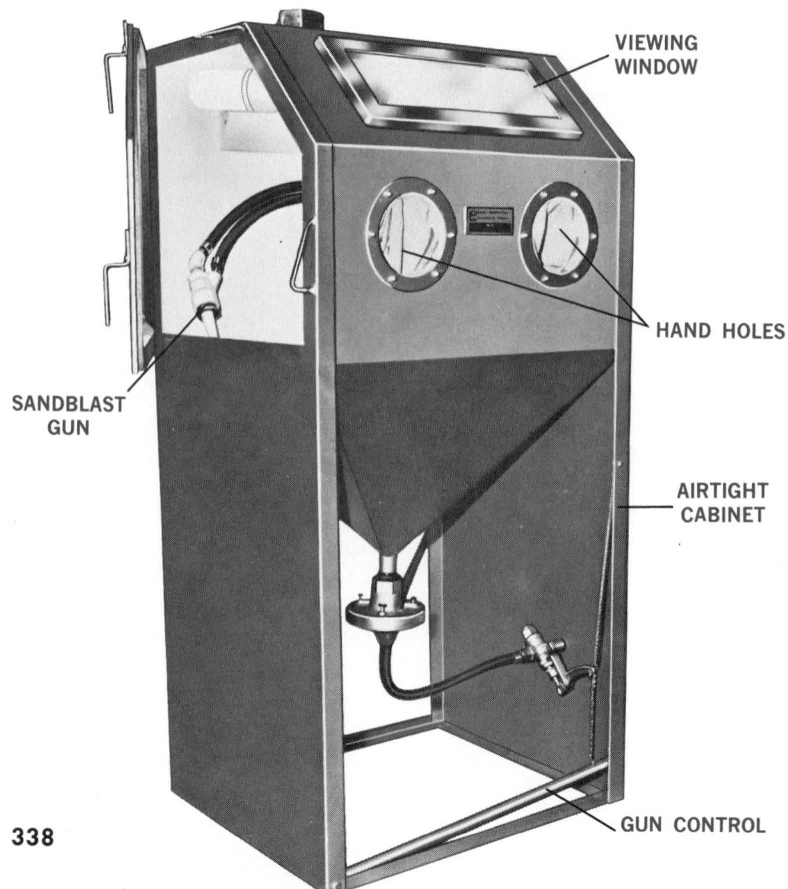

to the main cylinder doing the work. Gauges permit accurate reading of the forces that are generated. Pressure generated by the pump is converted by a hydraulic cylinder into an applied force which is hundreds of times greater than the input force at the pump.

Figure 11-20 shows the variety of hydraulic tools that are available. In Fig. 11-21 a hydraulic-powered C-clamp is being used in a trailer body manufacturing plant for pulling and squeezing I-beam rings around trailer tanks before welding. The rings give added strength to the tank shells for handling heavy loads of fluids and chemicals. An air-hydraulic system clamps work instantly in tack-welding operation so that it reduces setup time and labor costs, Fig. 11-22. As shown in Fig. 11-22A, clamping cylinders are directly mounted on a shop-built welding fixture, which swings down and locks into the closed position. The welder then activates a hand or foot valve to provide instant clamping of the work to be welded, Fig. 11-22B. Deactivation of the valve, in turn, provides instant release of work after welding is completed.

Hydraulic Bending Machines

The manufacture of tanks and cylinders comprises a large amount of the welding fabrication being done in this country. In order to produce tanks and

11-19. *A welding student using a spot-welding machine.*

11-20. *Assortment of hydraulic tools available to the welder.*

Enerpac Inc.

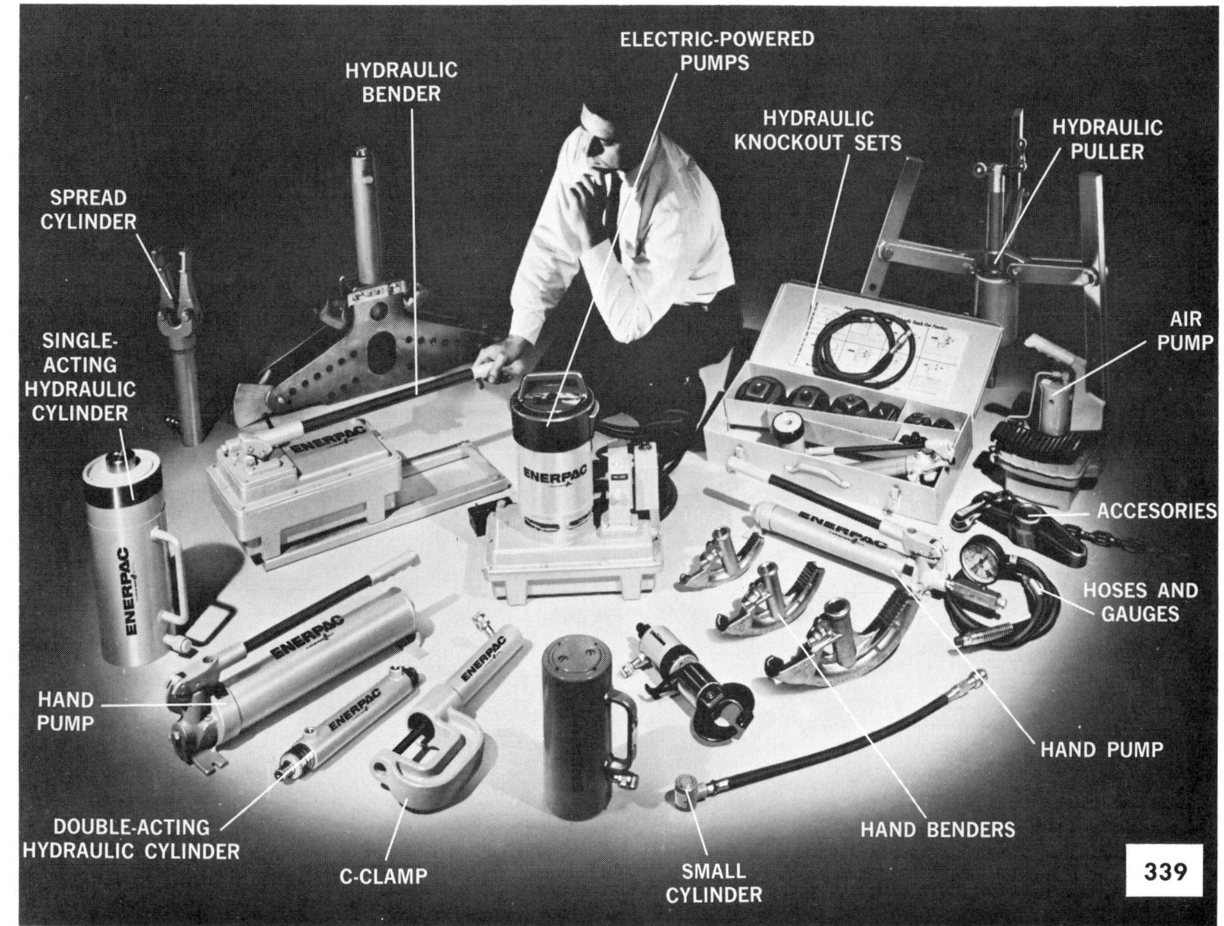

Enerpac Inc.

11-21. *A welder with a hydraulic C-clamp. Safety glasses should always be worn when operating such a C-clamp.*

11-22. *The use of hydraulics in connection with jigs and fixtures makes setting up and tacking easy.*
Enerpac Inc.

A

B

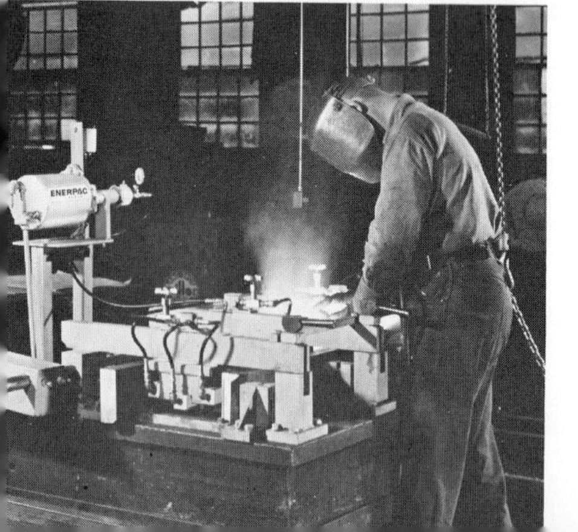

cylinders, the shop must have the capacity to bend and roll plate. The ability to do quality plate-rolling efficiently and economically often spells the difference between profit and loss in a job.

Figure 11-23 shows a hydraulic bending machine, also called a roll, rolling ½-inch stainless steel plate into a cylinder. The steps necessary to carry out the complete process from flat plate to finished cylinder are shown in Fig. 11-24.

Bending Brakes

Bending and forming machines for sheet metal and plate are natural descendants of the wooden cornice brakes used in fabricating architectural metals

11-23. *A hydraulic bending machine with a capacity of 12′ × ¾″ thick mild steel. The machine is bending 8′ × ½″ stainless steel plate. Note the use of the hydraulic loading support.*
Widder Corp.

as early as the 1830s. Massive carved stone and terra-cotta building blocks used for topping out buildings in those days were both expensive and unwieldy. The introduction of the cornice brake made it possible to bend lightweight, easy-to-handle sheet metals for cornices and face trim. Modern hand and powered bending machines are the response to the demand of welding fabricators for a line of machines to form both light and heavy gauge metals.

Power Press Brakes

Power press brakes like those shown in Figs. 11-25 and 11-26 have a workpower capacity of 30 to 45 tons and a metal-bending capacity of 10 gauge sheet 48 inches wide. Press brakes with hydraulic controls have a capacity of from 200 to 2,000 tons. In addition to their capacity to bend materials, power press brakes can punch, blank, form, and notch sheet metal. These brakes are all-welded in their construction. This is another example of the use of tools for welding fabrication in which the tools themselves are also of welded construction.

Hand Bending Brake

A hand brake, Fig. 11-27, makes bending and forming easier for the sheet metal fabricator. A great deal of welding fabrication is concerned with

11-25. *Welding student in front of a 45-ton power press brake, which forms metals for welding fabrication. Safety glasses should always be worn when operating this machine.*

R
F
LOWER ROLLER SET AT PLATE THICKNESS. FRONT ROLLER SET FOR PREBENDING OPERATION.
A

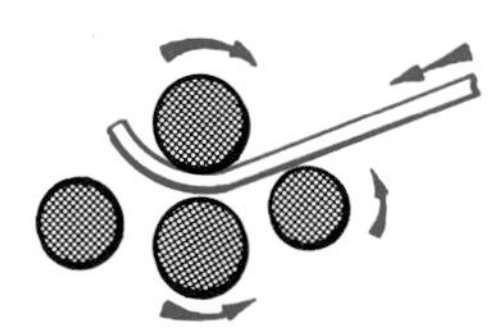

PRE-BENDING OPERATION COMPLETED.
B

C

R

F

REAR SIDE ROLLER SET TO GIVE REQUIRED DIAMETER. FRONT ROLLER LOWERED.

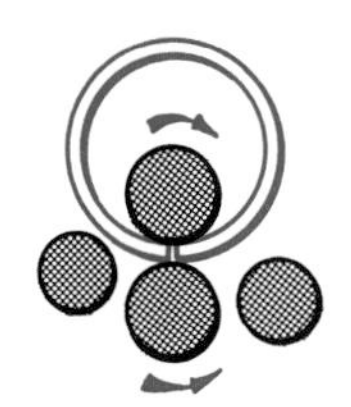

BENDING OPERATION COMPLETED IN ONE PASS.
D

ALL LOWER ROLLERS DROP TO ALLOW CYLINDER TO BE EJECTED.
E

11-24. *Steps in rolling steel sheet. This is the process performed by the bending machine shown in Fig. 11-23.*

11-26. *Welding student forming a metal part on a power press brake prior to welding.*

WARNING

boxes and pans of all types that can be formed on these machines. Figure 11-28 shows examples of metalwork made with the hand bending brake.

Hand Box and Pan Brakes

A box and pan brake, Fig. 11-29, incorporates all of the features of a standard hand brake. Because it has removable, sectioned fingers, nose bars, and bending bars, as well as greater clearances than the bending brake, it is more versatile. For example, the brake can form a box or pan having four sides and a bottom from one sheet of metal. Radius bends, such as those in modern metal furniture and cabinets, can be made with the round nose bars. Figure 11-30 shows examples of the various shapes that can be fabricated. Corners are usually welded.

***11-27.** A welding student bending a part in hand bending brake.*

Universal Bender

The universal bender is an indispensable piece of equipment for a welding fabrication shop. It is a bending and forming tool that can bend radii and angles on a wide variety of shapes ranging from small rod through pipe and tubing, to flat stock and angle iron. Its versatility, capacity, and fast, easy setup make it ideal for short run production, custom fabrication, and maintenance work.

No tools are necessary for the assembly of dies which bend pipe, rounds, flats, squares, tubing, and angle iron to specifications. Finger adjustment of mounting pins is all that is required. Figure 11-31 shows tubing being bent to a radius. Figure 11-32 shows the bender being used to bend flat stock edgewise to a radius. Figure 11-33 shows how a round rod can be bent into a U-shape. The machines shown are hand operated. Hard-to-pull bends can be accomplished by adding a hydraulic power attachment. Simple, positive foot control of the hydraulic power makes it possible to use both hands for controlling stock.

The universal bender is found in most job shops, maintenance departments, ornamental iron shops and on construction jobs that require a large amount of steel fabrication.

POWER SQUARING SHEARS

The power squaring shear, Fig. 11-34, is an indispensable piece of equipment in the welding shop or school. The fabrication of metals requires that they be cut with accuracy, speed, economy, and safety. In the school shop, square edges on the material to be welded are necessary for quality welding and the efficient use of training time.

The power squaring shears shown in Fig. 11-35 have a capacity that ranges from 16 gauge

***11-28.** Forms that can be produced in sheet metal on a hand bending brake. These may become a part of a total weldment.*

Dreis and Krump Mfg. Co.

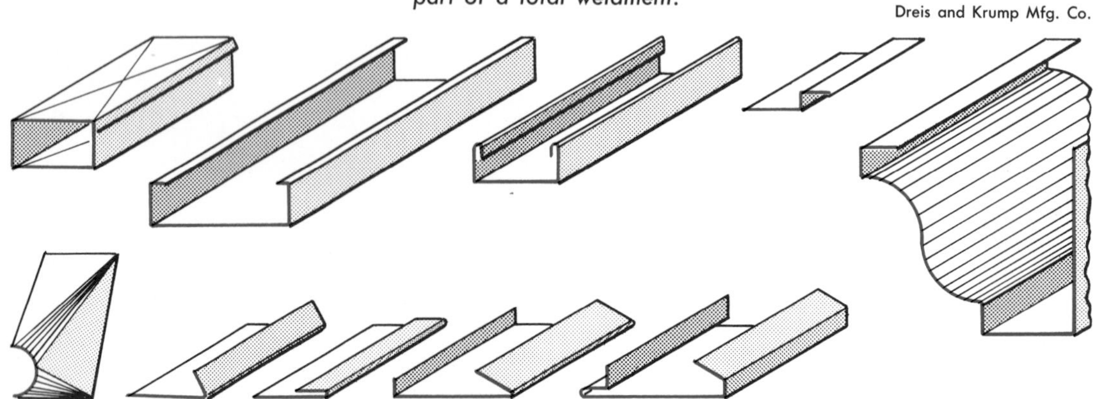

sheet metal to $\frac{1}{2}$-inch thick plate. Shears with a capacity of 1 inch are available. The machine is of welded construction. Every effort has been made in the interest of safety of operation. Drives are fully enclosed, and the rear of the machine is free from exposed, rotating shafts and eccentrics that pose potential hazards. Flange-mounted motors, direct-connected through splines, eliminate belts, guards, and overhanging flywheels.

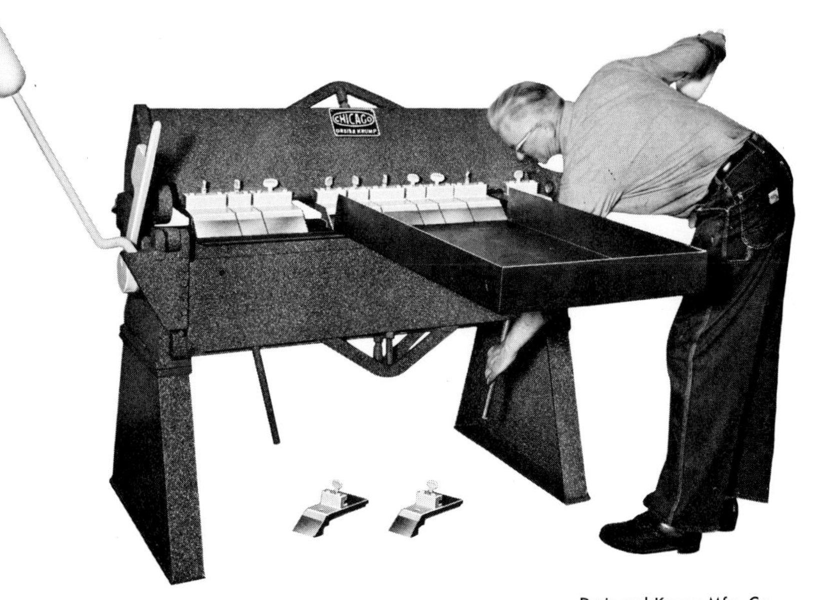

Dreis and Krump Mfg. Co.

11-29. *A hand box and pan brake with a 12-gauge bending capacity.*

SMALL HAND TOOLS

The welder must frequently use one of several methods to clean the work. The surfaces of the joint to be welded must be free of rust, oil, grease, and paint. These materials cause brittleness and gas pockets in the welded joint. Multipass welding requires cleaning between each pass. Cleaning after the joint has been welded is also important for the sake of appearance and in preparation for painting. Steel wire scratch brushes (Fig. 11-36), chisels, and cleaning hammers (Fig. 11-37) remove slag, rust, and other dry material from metal surfaces.

The cleaning hammer has one end shaped like a chisel for general cleaning. The other end is pointed in order to reach the slag in corners, along the edges, and in deep ripples. Some hammers are mounted with a wire brush, Fig. 11-38. Electric or pneumatic tools may also be used for preweld and postweld cleaning. These include grinders, wire wheels, and powered chipping hammers. Sandblasting is required for a flawless surface. Nonporous metals are chemically cleaned, especially when a high grade paint or enamel finish is desired.

A welding operator will also have use for the hand tools

11-30. *Bending profiles produced in sheet metal on a box and pan brake.*

Dreis and Krump Mfg. Co.

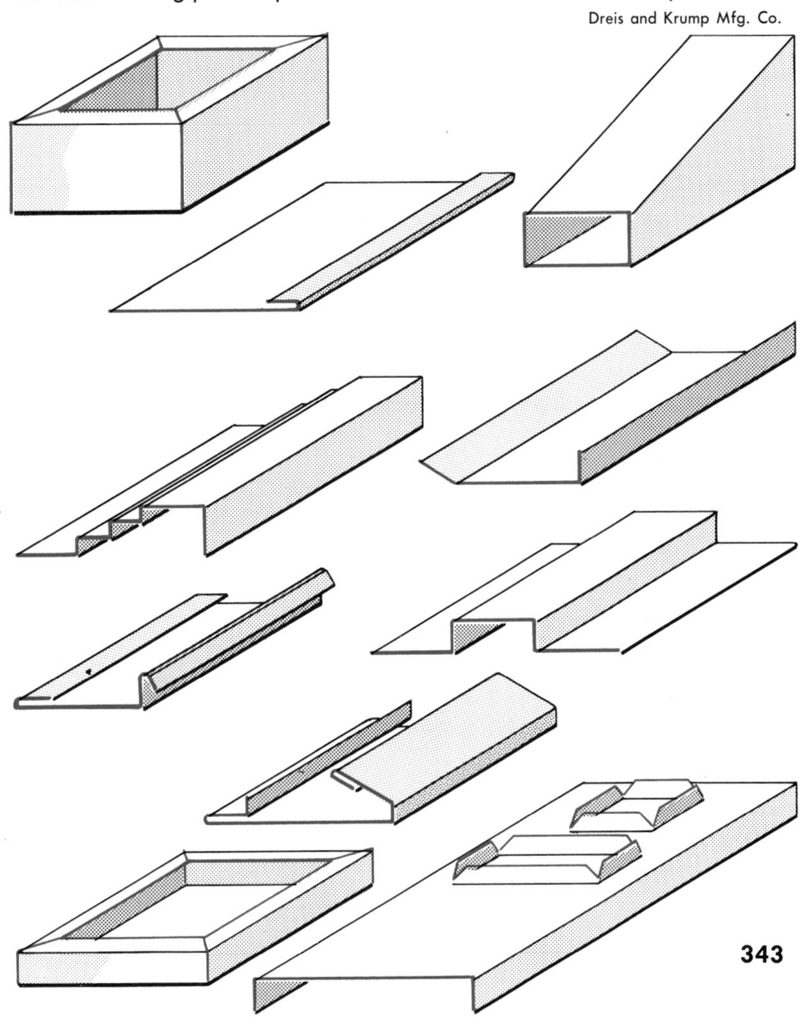

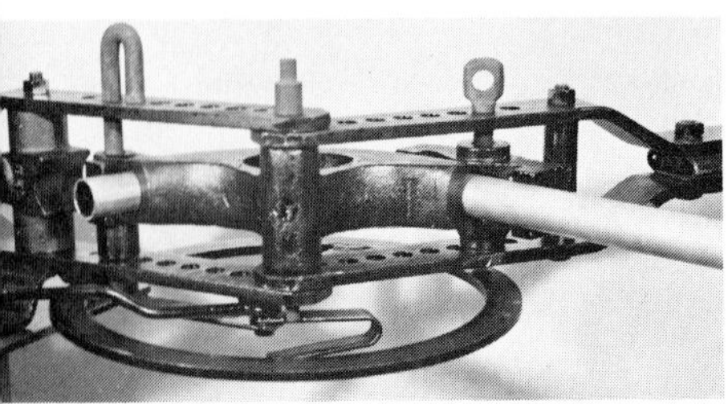

Hossfeld Manufacturing Co.

11-31. *Bending tubing.*

Hossfeld Manufacturing Co.

11-32. *Bending flat strip.*

Hossfeld Manufacturing Co.

11-33. *Bending round rod.*

shown in Fig. 11-39, which are identified as follows:

A. Identification stamps are necessary in shops where critical welding is done. They are for the purpose of identifying the welder with the job.

B. Weld gauges make it possible for the welder to measure the size of welds and determine their correctness according to specifications.

C. The combination square and scale may be useful in squaring stock, squaring angles, marking off and measuring dimensions, and doing general layout work.

D. The bevel protractor is essential in work requiring the tacking up and welding of fittings at an angle.

E. Thickness gauges and micrometers determine the thickness of material. On some jobs only a thickness gauge is necessary. For jobs that involve a great deal of light gauge material and when accuracy is necessary as in aircraft construction, the welder will use the micrometer for checking thicknesses.

F. The flexible scale is useful for all types of measuring, especially around radii and other hard-to-secure measurements.

G. The scribe and center punch mark lines and other layout information accurately on steel. The scribe, also called a *scratch awl,* has a hardened steel point in order to make an impression in the metal. After the line has been made, it is often found that the finely scribed line must be further marked to be seen clearly. This is especially true of lines that serve as a guide for cutting torch operations. The additional marking is made with the center punch. A center punch is also useful in spotting a point that is to be drilled. The impression made by the punch prevents the drill from wandering all over the surface to be drilled.

H. The cold chisel is handy for preparing the joint for welding and for removing slag, burrs, spatter, high places in the weld, and evidence of poor fusion after welding.

I. The hand hammer is an indispensable tool. It provides the force for using the center punch in layout work and the cold chisel in cleaning.

J. Pliers are used for carrying and handling hot metal and for holding parts that are to be tack welded to the main structure.

11-34. *A welding instructor demonstrating the use of the power squaring shear to high school students.*

11-35. *Welding student at a power squaring shear capable of cutting ½-inch thick steel plate. Safety glasses always should be worn when operating such a machine.*

K. The center head is used to find the center of round bars and pipe.

A hand metal-cutting hacksaw (not illustrated) is useful for cutting a small piece of round or flat steel.

PORTABLE POWER TOOLS

The fabrication of metal and the preparation of plate for welding require a variety of hand power tools such as chipping hammers, peening hammers, hand and bench drills, hand and floor grinders, edge preparation tools, and weld-shaving devices.

Efficient and safe use of portable power tools requires a basic understanding of their construction and their power source as well as an appreciation of the advantages, disadvantages, and limitations of each type.

Power tools have universal electric, pneumatic (compressed air), hydraulic, or high-frequency electric power. All four types do the same work. The productivity of each tool, however, varies with the application. The operator has constant control over speed and load. Tool size has an effect on operator fatigue and hence, efficiency.

Types of Power Source

UNIVERSAL ELECTRIC. Universal electric tools are the most common portable tools. They operate on standard 110- or 220-volt a.c. or d.c. single-phase cur-

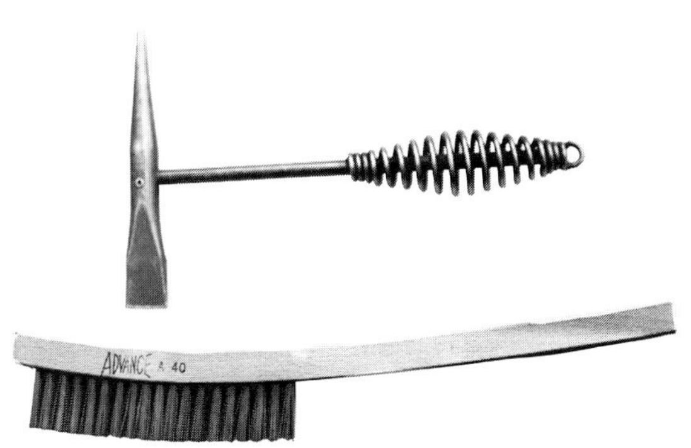

11-36. *Wire scratch brush for cleaning welds.*

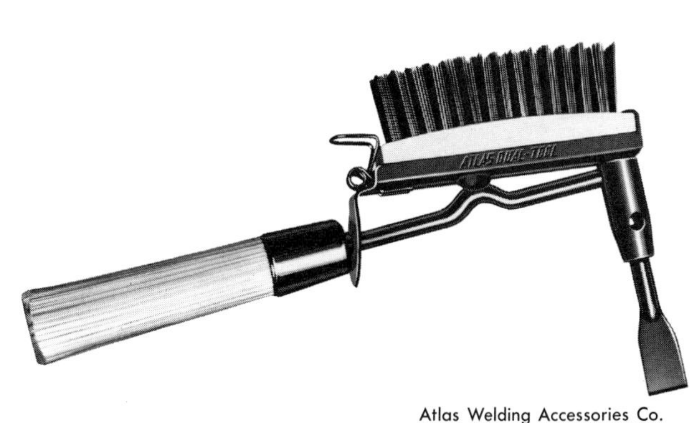

Atlas Welding Accessories Co.

11-38. *Combination wire brush and cleaning hammer.*

11-37. *Weld cleaning hammer.*

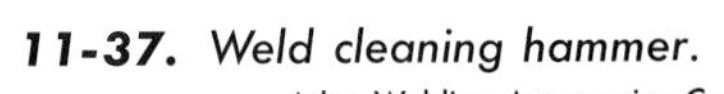

Atlas Welding Accessories Co.

11-39. *Set of hand tools needed by a journeyman welder for layout, fitting up, material checking, and weld checking.*

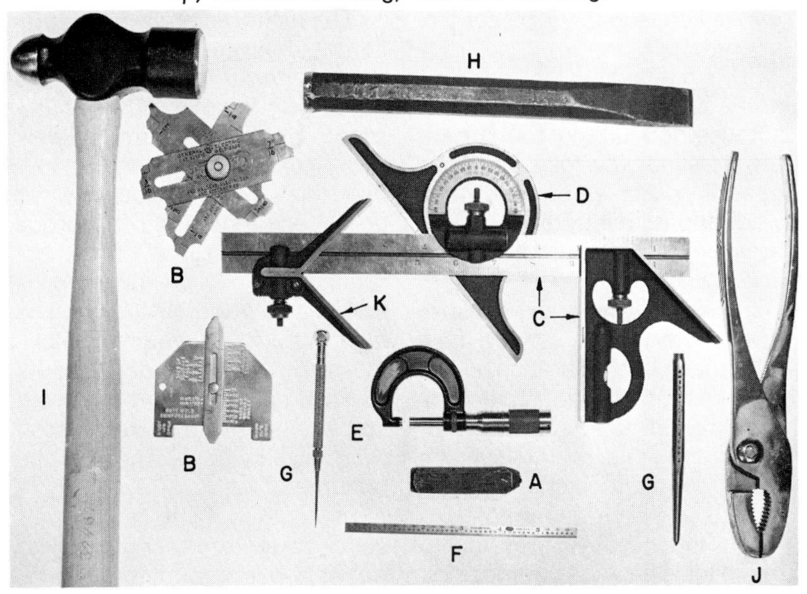

rent, the cheapest power available. These tools are best suited for intermittent operation, typified by maintenance, installation, and field work.

Universal electric tools have some drawbacks. One of the most significant is that the units have commutators and carbon brushes as well as armature windings, all of which require attention. The maintenance of universal electric tools is costly and they should, under no circumstances, be used continuously as in assembly line operations.

PNEUMATIC. Pneumatic power tools require compressed air to operate. The most common air compressors are electrically driven by a three-phase, 220-volt or 440-volt motor. On construction sites, these units are often powered by either diesel or gasoline engines.

There are two types of pneumatic motors: the turbine and the piston. Piston motors are rarely used except on certain types of reciprocating tools. The turbine type is better suited for short stroke applications of 3 inches or less. The turbine motor is used in about 95 percent of pneumatic tools.

The pneumatic tool has several advantages over the universal electric tool. Most important are its suitability for continuous operation such as assembly line work and its relatively low maintenance cost. Pneumatic power is ideal for impact tools and chipping hammers because pneumatic motors have a high tolerance for vibration. The motor is small, light, and cool-running. Speed is infinitely variable, and the motor can be stalled without damage. Most models are explosion-proof.

On the negative side, pneumatic tools have a low efficiency. Torque under constant load is low, and motor power decreases with time and use. The larger motors produce a loud exhaust noise and have such a high air consumption that operating costs become excessive. Moreover, the average pneumatic installation has a 20-percent power loss between the compressor and the tool, and an additional 5- to 15-percent loss due to deterioration from aging. Finally, pneumatic tools are not suited for operation in cold weather, particularly at the higher altitudes.

HIGH FREQUENCY ELECTRIC. High frequency electric tools employ a frequency converter and have a generator and a rotor on a common shaft. They operate on three-phase current, which has important advantages. Instead of an armature, they use a rotor which cannot be worn out or broken. This also eliminates commutators and brushes. The only similarity between high frequency and universal electric tools is that they both have field coils: the high frequency has six fields, and the universal electric has two.

The high frequency tool represents a high degree of technical development. Tools of this type are widely used throughout Europe. The power source is well-suited for production and assembly line work, especially in operations requiring high torque under constant load. The most notable use of high frequency is in offhand grinding for which a high output is required. Maximum slowdown under load is about 6 percent, compared with 25 percent for the universal electric tool and 36 percent for the pneumatic tool.

In addition to its more constant power and torque, fewer working parts, and less noise, the high frequency tool has the advantage of a completely sealed motor. Dust and dirt from work operations are prevented from entering the tool so that service life is extended. Finally, a high frequency installation requires a lower initial cost and less maintenance than does comparable pneumatic installation.

The high frequency tool also has its limitations. Speed can be varied only through the gear changes. Tool weight is about 25 percent higher than for comparable pneumatic units. Stalling of the motor may damage the fields unless they are protected by circuit breakers.

HYDRAULIC. The hydraulic power tool is employed most in applications requiring extremely high torque at very low speeds, such as tapping over $\frac{3}{4}$ inch, drilling over $1\frac{1}{4}$ inch, reaming, and tube expanding. Operation is somewhat similar to that of the pneumatic tool except that a hydraulic fluid is used instead of air. A pump takes the place of the compressor. For the most part, hydraulic tools are used for continuous use and special applications.

SUMMARY. When operation is intermittent and the tool is used for different jobs, the universal electric tool is the best choice. High frequency and pneumatic tools are more desirable for production line work when operation is sustained. Of the latter two, the high frequency tool is the more versatile and efficient. The pneumatic tool excells as a small tool for continuous, light duty applications.

Portable Electric Hand Drills

The electric hand drill is a tool that the welder uses frequently in fabrication and maintenance work. As indicated previously, drill motors may be electric or pneumatic. They can take drill sizes from the $\frac{1}{32}$ inch to $1\frac{1}{4}$ inch in diameter. Figure 11-40 shows the internal construction of a $\frac{3}{4}$ inch, heavy duty electric drill. Study the photograph carefully to understand the electrical and mechanical functions involved.

Electric Hammers

Figure 11-41 shows the internal construction of a heavy duty portable electric hammer. This type of hammer is used by the welder for chipping, peening, channeling, and masonry drilling. It is used with punches, chisels, seaming tools, scaling tools, and many sizes and types of masonry drills.

A small riveting hammer is one of the hand tools with which the welding student should have some experience. Many fabricated units are joined with a combination of welding and riveting. In large building construction, the mixing of welding and riveting is a common practice. This is also true for those items in which light and medium gauge sheet metal is used. Figure 11-42 shows two students riveting two aluminum plates together with a pneumatic riveting hammer.

Sanders and Grinders

Sanders and grinders are abrasive tools which perform almost every kind of surfacing job for the welder. Equipped with abrasive disks, they handle all sanding, from fast material removal to satin-smooth finishing. Equipped with flaring cup wheels and depressed center wheels, they smooth welds and casting ridges and cut off studs, bolts, and rivets. Equipped with wire cup brushes, they remove paint, rust, and scale from welds and clean castings, tanks, sheet metal, and soldered joints.

Angle sanders and grinders are usually rated to take abrasives with a diameter of 6 to 9 inches. They have a speed of 4,000 to 8,000 r.p.m. Straight grinders may be used with wheels of $\frac{1}{2}$ to 6 inches in diameter. Wheels from 6 inches to $2\frac{1}{2}$ inches are run at speeds of from 3,750 to 14,500 r.p.m. The larger the wheel, the slower the running speed. Wheels from 2 inches to $\frac{1}{2}$ inch are run at speeds of from 14,500 to 30,000 r.p.m.

The internal construction of a portable electric angle sander and grinder is shown in Fig. 11-43. Figure 11-44 shows a high frequency electric angle grinder grinding a weld. The pneumatic angle grinder in Fig. 11-45 (Page 350) is grinding a weld bead at a speed of 8,500 r.p.m. The flame-cut edge in Fig. 11-46 is being finished with a pneumatic straight grinder.

11-40. *Internal construction of a heavy duty portable electric drill. 1. Half-inch hardened-steel key-operated drill chuck. 2. Spindle mounted on ball bearings for long life. 3. Heavy duty ball bearings. 4. Heat-treated gears mounted with ball and roller bearings. 5. Powerful ventilating fan maintains cool operating temperatures. 6. Universal motor mounted with ball bearings. 7. Handle for one-handed operation of tool. 8. Light weight, extra rugged aluminum housing. 9. Heavy duty trigger switch. 10. Heavy duty three-wire cable.*

Black and Decker Manufacturing Co.

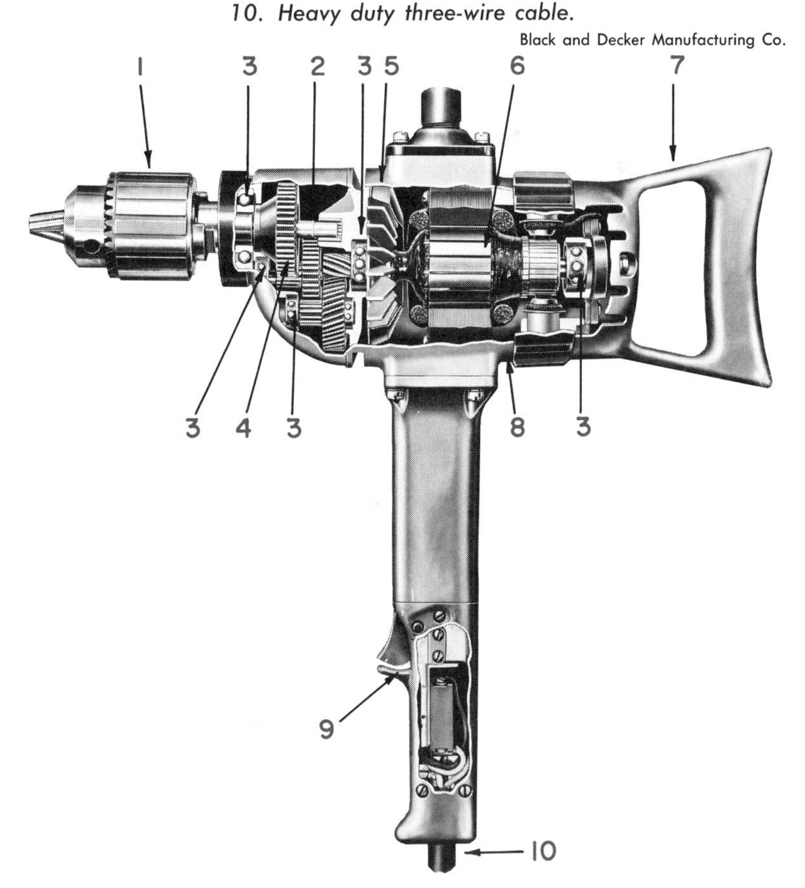

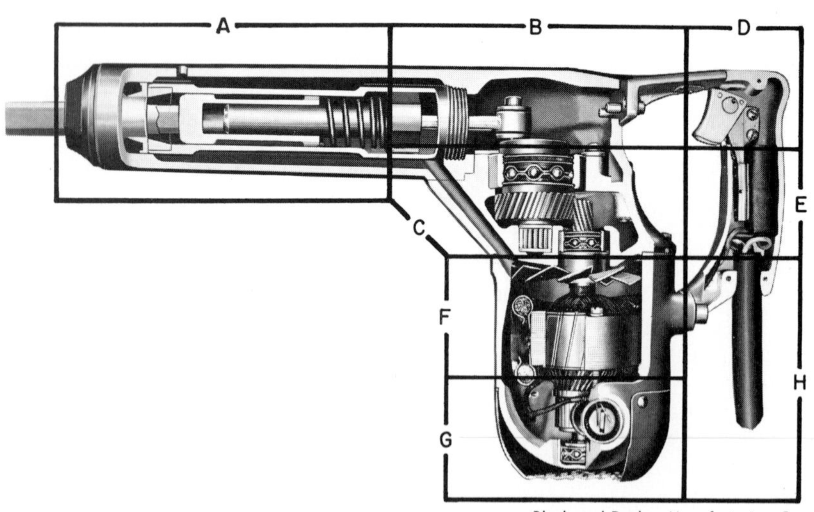

Black and Decker Manufacturing Co.

11-41. *Internal construction of a heavy duty portable electric hammer. A. Heat-treated piston and ram. B. Heat-treated crank and connecting rod. C. Heat-treated helical gears, ball-and-roller-bearing mounted. D. Heavy-duty trigger switch. E. Pistol-grip handle. F. Universal motor, ball-bearing mounted. G. Aluminum housings. H. Three-conductor cable.*

11-42. *Welding students using a pneumatic riveting hammer.*

Shears and Nibblers

The welder who works in fabrication or repair frequently cuts metal into various shapes and sizes. Sometimes flame and arc cutting methods cannot be used because they may cause color change or warpage in the work or they may create a fire hazard. In such cases, metal must be cut by a shearing method. Most portable electric shears and nibblers can cut steel as thick as 8 gauge and aluminum up to $\frac{1}{4}$ inch. These machines are fast cutting and produce a smooth edge without distorting the body of the metal. The nibbler uses a straight up-and-down punching action, and the shears cut with a powerful scissor-like action.

Figure 11-47 shows the internal construction of a portable electric nibbler. The nibbler in Fig. 11-48 is cutting $\frac{1}{8}$-inch steel plate at a speed of 4 feet per minute. Figure 11-49 shows portable electric shears cutting a $\frac{3}{8}$-inch I-beam at a speed of 6 feet per minute. It has a capacity of $\frac{3}{8}$ inch in mild steel and $\frac{15}{32}$ inch in aluminum.

Magnetic-Base Drill Press

On many construction and maintenance jobs, it is necessary to drill a hole in a weldment which is too big to be taken to a drill press for drilling. Because of the size of the hole or the need for accuracy, hand drilling with a portable hand drill may be too slow or inaccurate. The magnetic-base drill press, Fig. 11-50 (Page 352), is a powerful heavy duty drill mounted on a base with great magnetic holding force. This unit attaches to steel surfaces in the horizontal, vertical, and overhead positions. Thus the machine provides both the speed and the accuracy of a regular bench- or floor-type drill press. The unit is available with either electric or pneumatic power.

Beveling Machine

The edge preparation of plate, tubing, and pipe before welding has always been a problem because quality cuts are usually costly. Industry has used the traditional methods of sawing, shearing, grinding, planing, and flame and arc cutting with various degrees of success. Each method works well within its limitations. Beveling by machine has the advantage of low initial purchase and operating cost. Beveling machines are available as electric or pneumatic tools. They are capable of beveling mild steel, alloy steel, and stainless steel as well as aluminum and soft metals. Beveled angles ranging from 15 degrees to 55 degrees are possible. These machines have the following features:

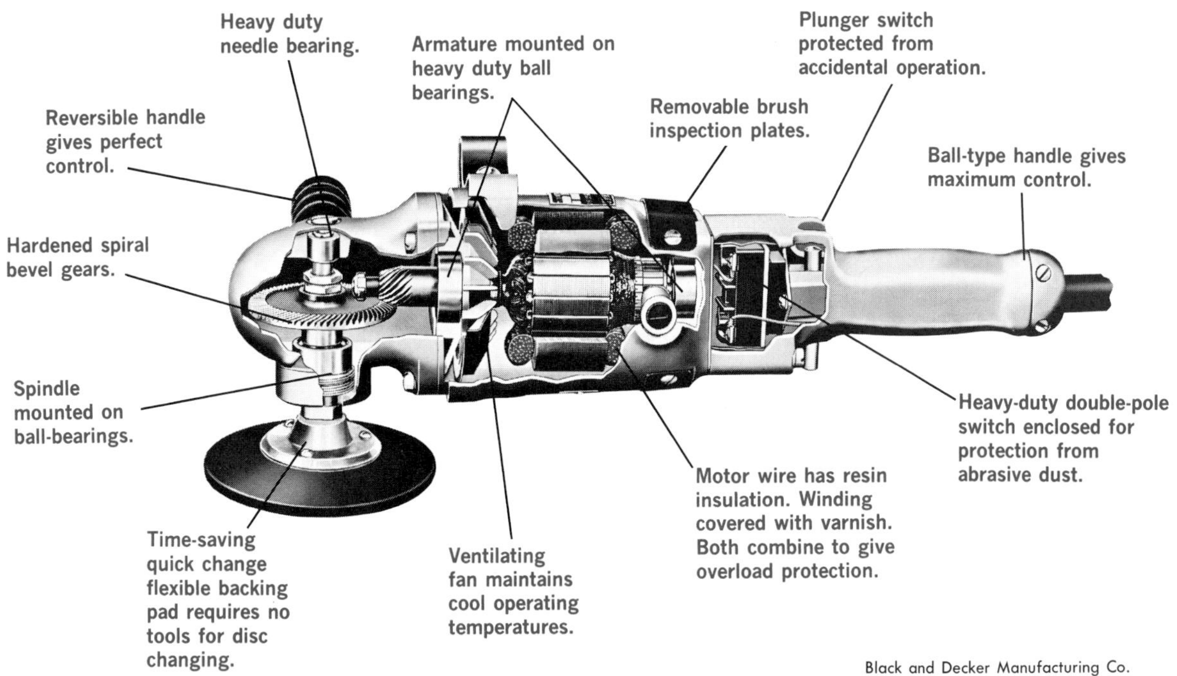
Heavy duty needle bearing.
Armature mounted on heavy duty ball bearings.
Plunger switch protected from accidental operation.
Reversible handle gives perfect control.
Removable brush inspection plates.
Ball-type handle gives maximum control.
Hardened spiral bevel gears.
Spindle mounted on ball-bearings.
Heavy-duty double-pole switch enclosed for protection from abrasive dust.
Motor wire has resin insulation. Winding covered with varnish. Both combine to give overload protection.
Time-saving quick change flexible backing pad requires no tools for disc changing.
Ventilating fan maintains cool operating temperatures.

Black and Decker Manufacturing Co.

11-43. *Internal construction of a heavy duty portable electric angle sander and grinder.*

- They are portable and can be hand held or mounted in a stationary stand or vise.
- Curves can be beveled. Any angle can be followed on convex curves. On concave curves the machines can handle diameters as small as $1\frac{1}{2}$ inch.
- All types of pipes can be beveled.
- The angle of cut can be varied from 10 to 55 degrees.
- Any thickness of metal can be worked on.
- It is possible to maintain a constant and smooth feed rate and to work forward and backward.
- The machine can start and stop at any desired point.
- The machine can operate upside down so that large pieces do not have to be turned when preparing X- or K-weld edges.
- The cutting tool is easily sharpened and inexpensive to replace. Figure 11-51 indicates the extreme flexibility of the tool.

Floor-type, plate-beveling machines, Fig. 11-52, can produce bevels at a rate of 12 feet per minute. They can form 1-inch bevel widths on plate thicknesses up to $2\frac{1}{2}$ inches with angles of 25 to 55 degrees on either straight or curved edges. Cleaning time is reduced because the bevels have little or no burr. The machine is highly flexible in its use and can be used upright, on its back, upside down, or hung from a crane.

Weld Shavers

The weld shaver is a comparatively new tool that is meeting with wide acceptance. It was designed specifically for aircraft, missile, and industrial operations which require the precision shaving, milling, and grooving of high strength materials. The tool is suitable for stainless steel, titanium, aluminum, Monel,® magnesium, and copper. Shavers replace other grinding,

11-44. *Using a high frequency electric angle grinder on a welded joint.*

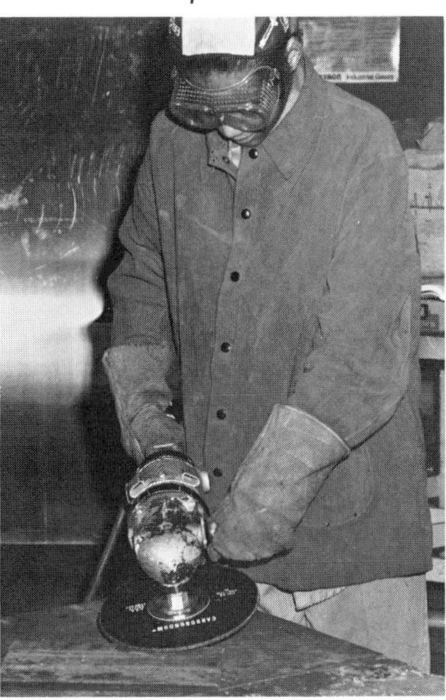

buffing, and finishing tools. They cut medium size weldments up to $^3/_4$ inch wide down to surface flushness or to a preselected height controlled to within 0.0005 inch. The depth of cut per pass depends on the hardness of the material being cut. Also available are grooving cutters capable of grooving to a maximum of 0.290 inch. Shavers and grooving cutters are pneumatic tools.

The shaver is fitted with carbide and high speed steel cutters which are precision ground and fluted for hard usage and long life. Cutters are easily removed and replaced, and they may be reground many times. Figure 11-53 shows the tool removing a weld bead reinforcement. Figure 11-54 is an example of the weld finish after removal of the reinforcement.

MACHINE TOOLS

Many large fabricating plants have completely equipped machine shops where metal machining of all kinds can be done. Those welders who wish to work in the fields of fabrication, maintenance, and repair will find it an advantage to be able to operate such standard machine tools as the engine lathe, the shaper, and the milling machine.

Basically, a *machine tool* is a machine which cuts metal. It holds both the material being worked on and the cutting tool. Its purpose is to produce metal parts by changing the shape, size, and finish of metal pieces.

In addition to maintenance work, these machines can be used to cut and bevel pipe, bevel plate, and machine-test weld specimens.

Engine Lathe

The lathe, Fig. 11-55, is a turning tool which shapes metal by revolving the workpiece against the cutting edge of the tool. The lathe performs many kinds of external and internal machining operations. Turning, Fig. 11-56, can produce straight, curved, and irregular cylindrical shapes. Other operations include knurling, thread cutting, drilling, boring, and reaming. The lathe can be used to cut and bevel pipe for welding.

Shaper

Shapers are used primarily for machining flat surfaces. The workpiece is held securely in a machine vise, and the cutting tool, which is mounted in a toolholder, is moved back and forth in a straight line by a ram, Fig. 11-57. The cutting tool peels off a chip each time the ram moves forward on a cutting stroke. The shaper is an excellent tool for shaping weld test specimens and removing the excess weld on the face and root of the specimen.

Milling Machine

A milling machine is a machine tool which cuts metal with a multiple-tooth cutting tool called a *milling cutter*. The workpiece is mounted on the milling machine table and is fed against the revolving milling cutter.

There are two basic types: (1) the horizontal milling machine, (Fig. 11-58, Page 354) and (2) the vertical milling machine (Figs. 11-59 and 11-60). These machines can be used to machine flat surfaces, shoulders, T-slots, dovetails, and keyways. Irregular or curved surfaces can also be formed with specially designed cutters. One of the important uses is in the cutting of gear teeth. The vertical miller can also be used for drilling, reaming, countersinking, boring, and counterboring.

A milling machine is an excellent machine to prepare weld test specimens and remove excess weld metal where the work is portable.

Pedestal Grinder

The pedestal grinder, Fig. 11-61, is one of the most important pieces of equipment in the welding shop. It polishes or cuts metal with an abrasive wheel. It prepares metal parts for welding, grinds weld metal, removes rust

11-45. *Using a pneumatic angle grinder to grind a weld. Wheel speed is 8,500 r.p.m.*

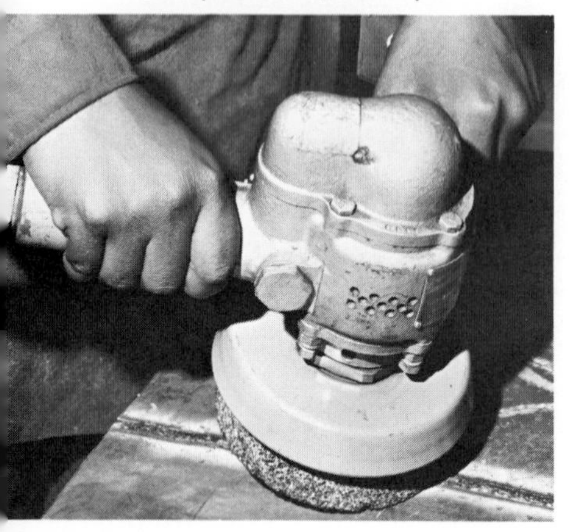

11-46. *Using a portable pneumatic straight grinder on a flame-cut edge. The speed of the grinder is 7,000 r.p.m.*

Widder Corp.

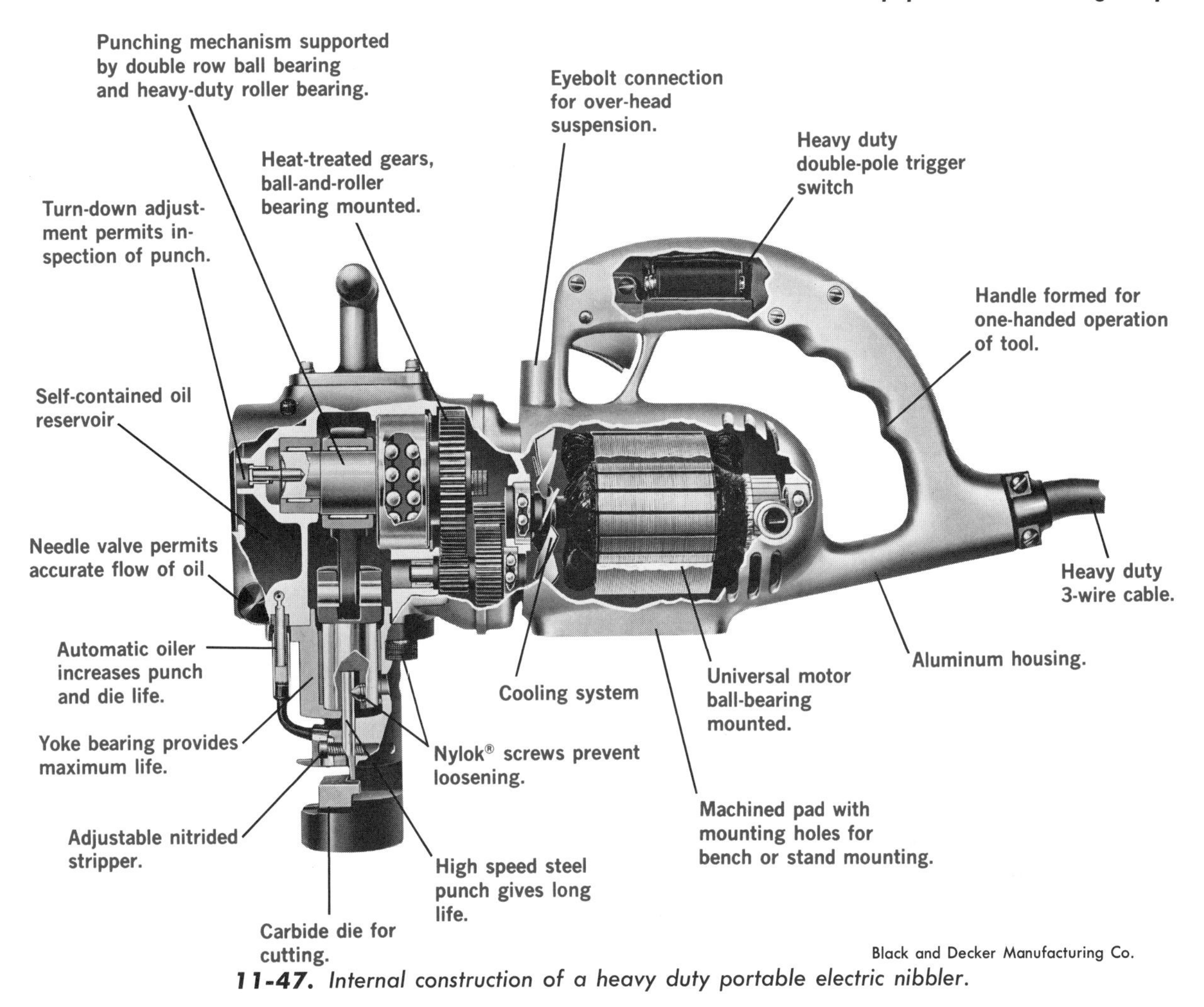
Punching mechanism supported by double row ball bearing and heavy-duty roller bearing.
Eyebolt connection for over-head suspension.
Heavy duty double-pole trigger switch
Heat-treated gears, ball-and-roller bearing mounted.
Turn-down adjust-ment permits in-spection of punch.
Handle formed for one-handed operation of tool.
Self-contained oil reservoir
Needle valve permits accurate flow of oil
Heavy duty 3-wire cable.
Automatic oiler increases punch and die life.
Cooling system
Universal motor ball-bearing mounted.
Aluminum housing.
Yoke bearing provides maximum life.
Nylok® screws prevent loosening.
Machined pad with mounting holes for bench or stand mounting.
Adjustable nitrided stripper.
High speed steel punch gives long life.
Carbide die for cutting.

Black and Decker Manufacturing Co.

11-47. *Internal construction of a heavy duty portable electric nibbler.*

11-48. *Using a portable electric nibbler. Its cutting capacity is ⅛-inch mild steel plate at a cutting speed of 4 feet per minute.*

Widder Corp.

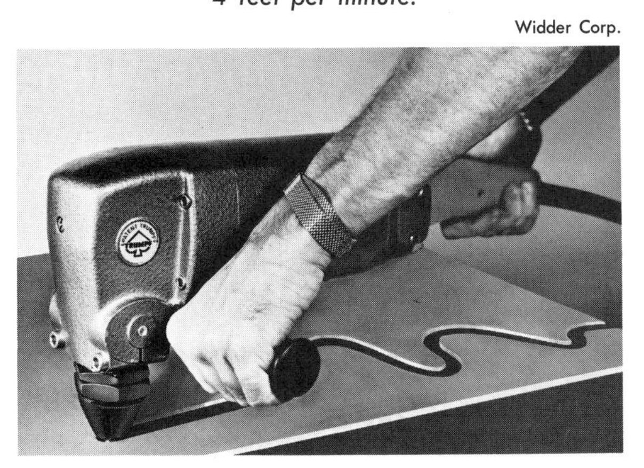

11-49. *Using a portable electric shear on a ⅜-inch thick I-beam. The cutting speed is 6 feet per minute.*

Widder Corp.

and scale, and sharpens tools. The floor-type grinder usually has a grinding wheel on each end of a shaft which extends through an electric motor.

A grinding wheel is made of abrasive grains cemented or bonded together to form a wheel. There are three types of wheels which are classified according to the means of bonding:

- *Vitrified wheels* in which the bonding is a kind of earth or clay. The wheels are baked at about 3,000 degrees F. These wheels are as large as 36 inches in diameter. They are usually used for rough grinding.
- *Silicate wheels* in which the bond is silicate or water glass. They are as large as 60 inches in diameter. They may be used for fine grinding.
- *Elastic wheels* in which the bond is rubber, shellac, or bakelite. Very thin, strong wheels are made with this bond so that they can be run at very high speed.

Wheels are made in many different shapes such as cylindrical, straight, tapered, recessed, cup, and dish. Generally, the straight, tapered, and cup wheels are used in the shop. They may range from 1 to 3 inches in thickness and up to 14 inches in diameter.

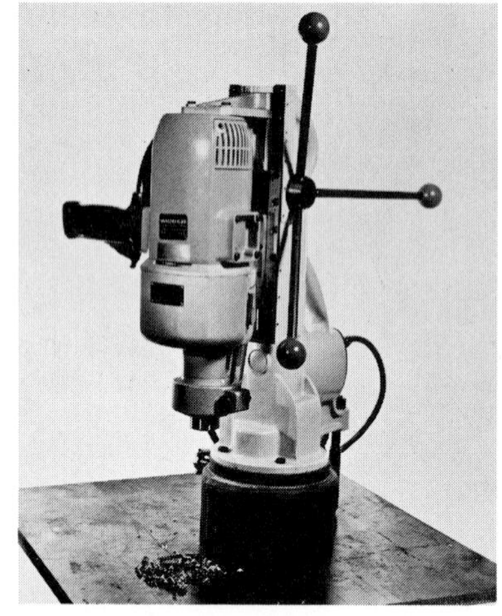

Widder Corp.

11-50. *A magnetic base drill press with a 4-speed drill.*

11-51. *A portable electric beveling tool. It can cut along straight edges, convex and concave curves, and circles. The machine's capacity is 5/8 inch for single-V bevels and 1 1/4 inch for K- or double-K bevels.*

Widder Corp.

Surface Grinder

The preparation of such weld test coupons as the tensile, face, and root bend coupons requires that the surface of the coupons be ground parallel to their length. A smooth surface is necessary. Coarse grinding or surface cuts may cause a weld failure when the coupon is tested. After the coupon is cut from the original test specimen, the surface grinder is the ideal tool to produce the smooth surface, Figs. 11-62 and 11-63. These grinders can handle tough materials to close tolerances.

Drill Press

In a shop that does a great deal of fabrication and repair, a drill press is an indispensable machine tool. Many of the holes

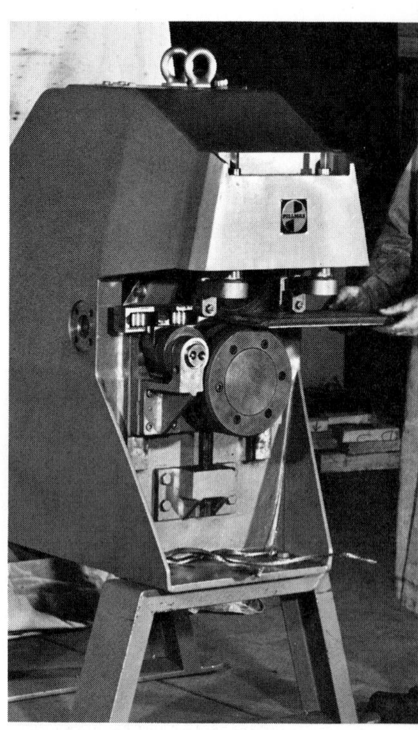

American Pullmax Co.

11-52. *Adjustable-angle joint beveler.*

11-53. *Removing weld reinforcement with a pneumatic portable weld shaver.*

Zephyr Manufacturing Co.

that need to be drilled can be done with a portable power drill. In heavy plate thicknesses and when accuracy is required, however, a bench or pedestal drill press is necessary. The pedestal press shown in Fig. 11-64 has a 20-inch throat, and can drill at speeds from 150 to 2,000 r.p.m. Figure 11-65 shows a welding student using a bench-type drill press.

Power Punch

Another method of making holes in metal is with a power punch. It is important for the shop to provide a machine that can punch holes in different shapes and sizes with a minimum of tool change time. The power turret punch shown in Fig. 11-66 (Page 356) makes it possible for the welder to punch a wide variety of holes rapidly. Just a "twirl of the turrets" changes punches and dies to different sizes and shapes. Sizes range from $\frac{1}{32}$ inch to 2 inches in diameter. Burr-free holes are possible in the thinnest material. Micro-twin front-operated back

11-56. *Welding student beveling a 10" diameter section of steel pipe on the lathe in preparation for welding.*

Zephyr Manufacturing Co.

11-54. *Weld reinforcement has been removed flush with the plate surface. Note how smooth the finish is.*

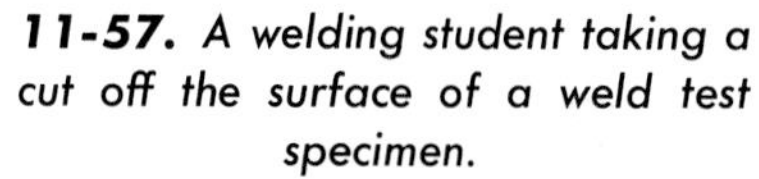

11-57. *A welding student taking a cut off the surface of a weld test specimen.*

11-55. *The parts of a standard engine lathe.*

South Bend Lathe Co.

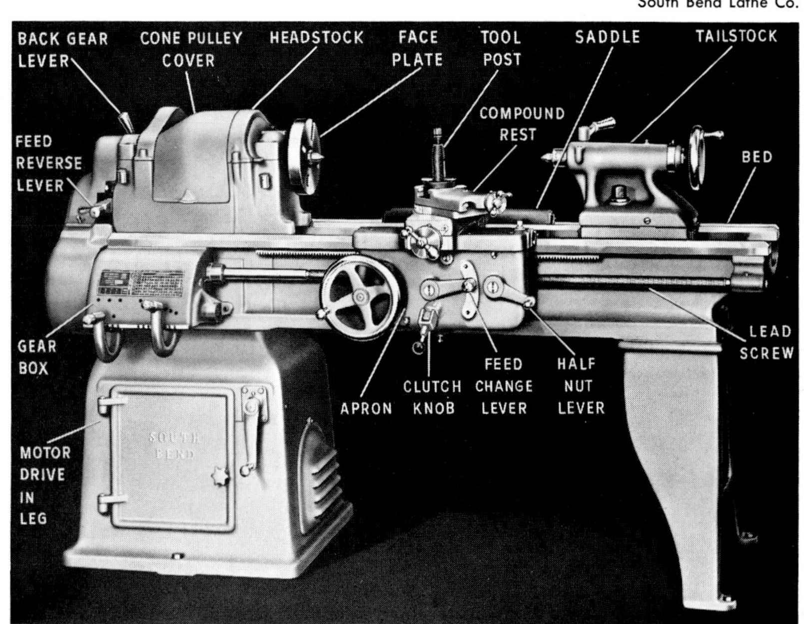

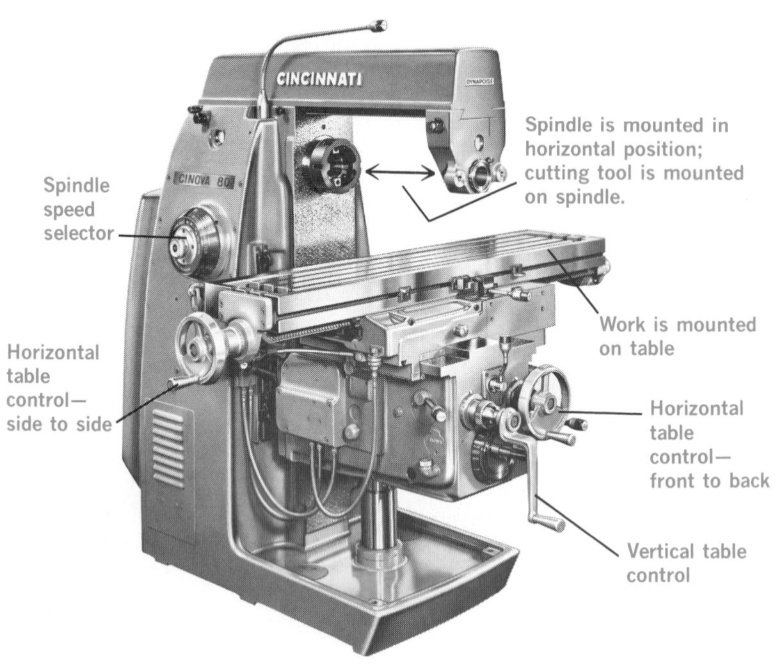

Cincinnati Milacron

11-58. *Horizontal milling machine. The cutting tool is mounted on the horizontal head.*

11-60. *Student using a vertical milling machine.*

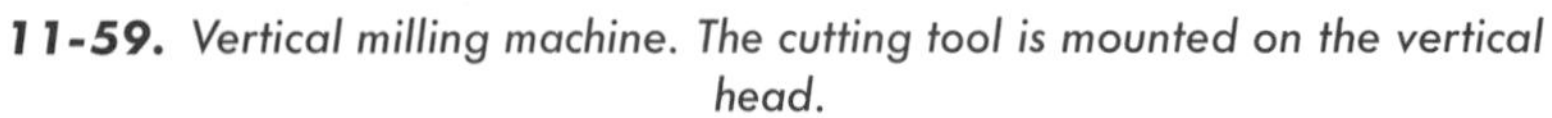

11-59. *Vertical milling machine. The cutting tool is mounted on the vertical head.*

Cincinnati Milacron

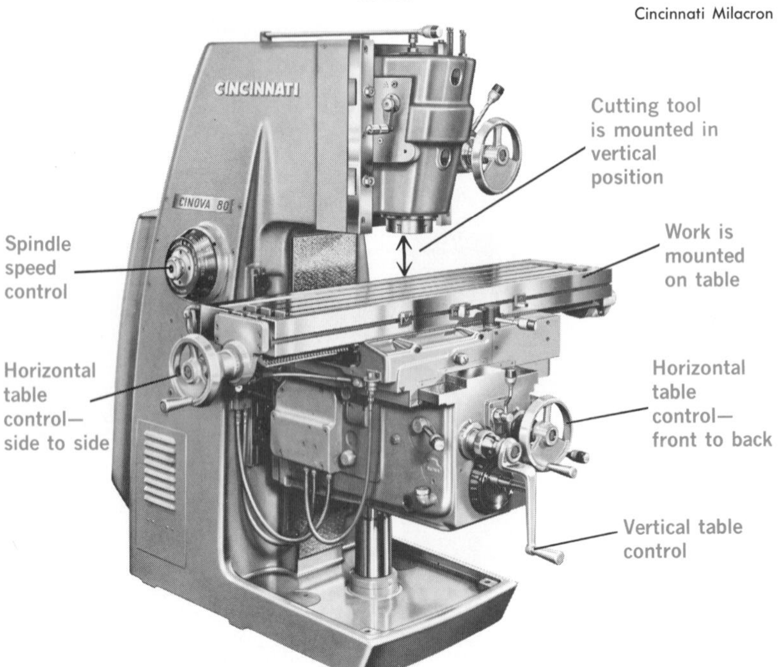

11-61. *A pedestal grinder (2 h.p., 1,725 r.p.m.) takes a wheel 12 inches in diameter with a 2¼-inch face. It should be equipped with eyeshields and lights.*

Baldor Electric Co.

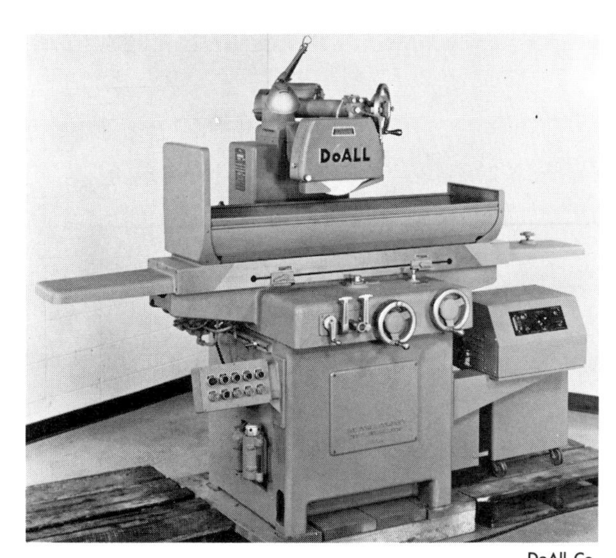

DoAll Co.

11-62. *A precision surface grinder can be used to machine weld test specimens.*

11-63. *Welding students using a surface grinder to prepare weld test specimens.*

and side gauges assure fast gauge setup and accuracy.

Metal-Cutting Band Saws

A power saw is used more than any other tool in the welding shop. These saws cut rectangular and round bars and angle iron. The metal-cutting band saw shown in Fig. 11-67 is a portable machine that is actually two machines in one. As a horizontal cut-off saw, it features a quick-action vise, which swivels to 45 degrees; an adjustable blade guide; and an automatic shutoff switch. Only seconds are required to convert the tool for vertical use. The head is swung into an upright position, a worktable is attached, and the saw is ready for cutting angles, slots, notches, and bevels, Fig. 11-68. Figure 11-69 shows prewelding students using a vertical bandsaw and a scroll saw. Others are doing layout work at the bench.

A more sophisticated model of the vertical band saw is shown in Fig. 11-70. This type of ma-

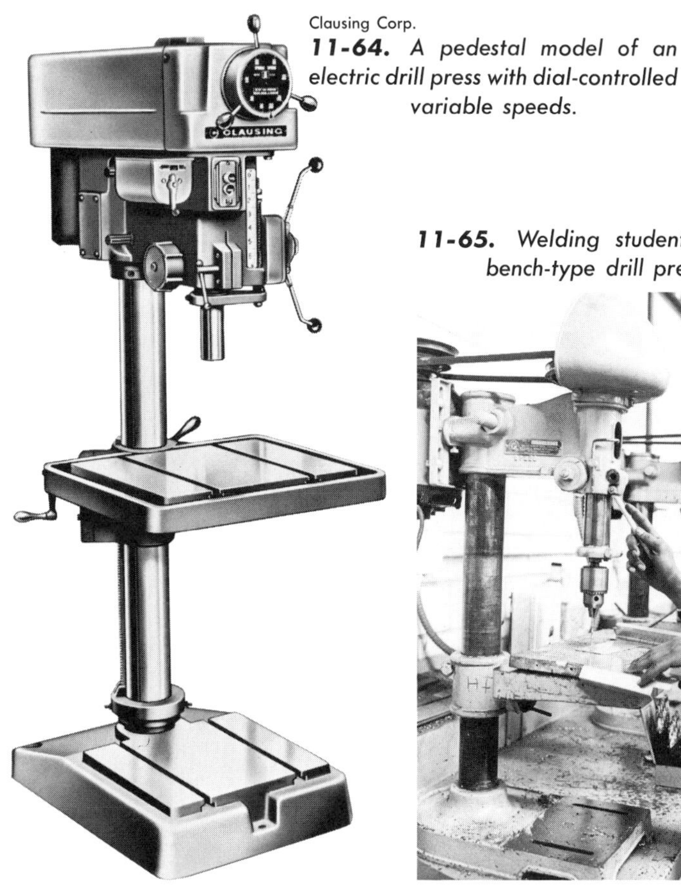

Clausing Corp.

11-64. *A pedestal model of an electric drill press with dial-controlled variable speeds.*

11-65. *Welding student using a bench-type drill press.*

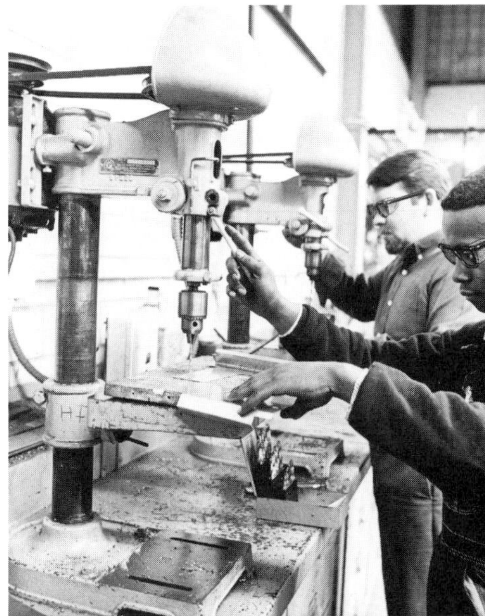

chine is often called a *band machine* because of its continuous sawblade. Many people in industry feel that it offers many advantages over other types of cutting tools. Unlike other machine tools, it cuts directly to a layout line and removes material in sections instead of chips. Chipless cutting saves both time and material.

The cutting tool on a band saw is a continuous band in which each single-point tooth is a precision cutting tool. It cuts continuously and fast. Wear spreads evenly over all the teeth

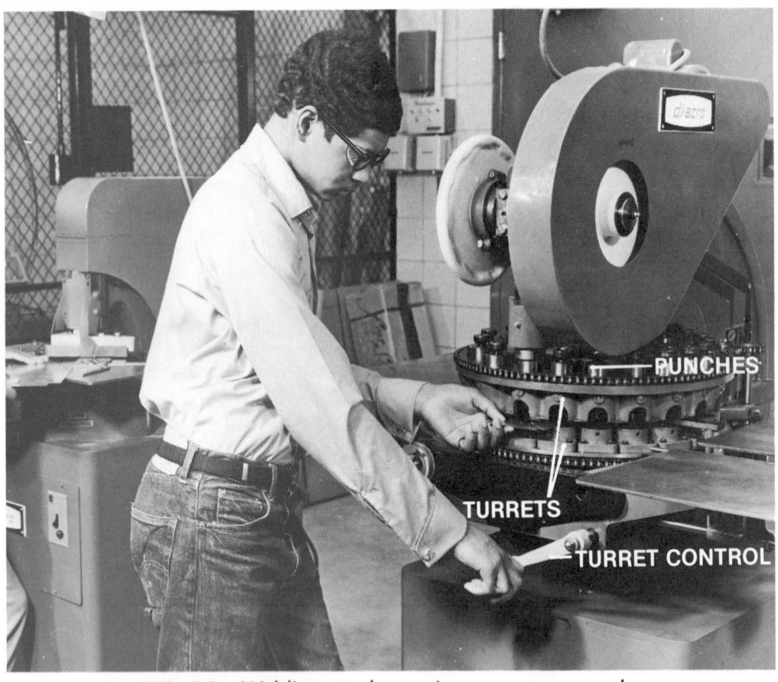

11-66. *Welding student using a power punch.*

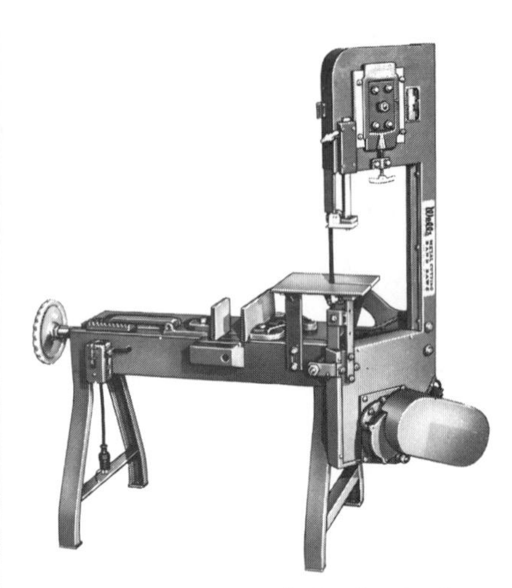

Wells Manufacturing Co.

11-68. *A metal-cutting band saw in the vertical position.*

11-67. *A horizontal metal-cutting band saw for cutting bars, angles and pipe.*

Wells Manufacturing Co.

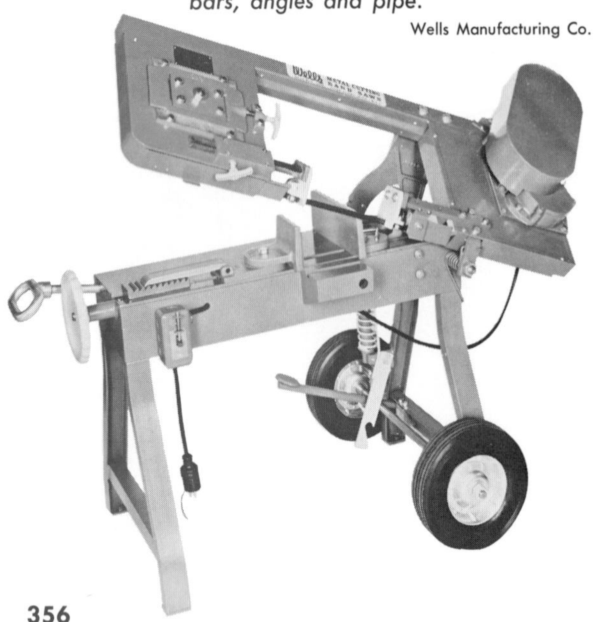

11-69. *Prewelding students working in the shop.*

to extend tool life. Because reciprocating machines waste motion on the return stroke, the band machine's continuous action allows it to do more work in the same amount of time.

The thin band tool, a fraction of an inch thick, saves material. Because of its size, it takes less horsepower to overcome material resistance than do the bigger, wider cutters on other machine tools.

A significant advantage of band machining is its unrestricted versatility. There is no limit to the length, angle, contour, or radius that can be cut. The constant downward force of the band holds the work to the table so that it is easy to hold or fixture work for production runs. Making a setup or changeover is fast, operation is easy, and cost per cut is low. Figure 11-71 summarizes the advantages of band machining.

Since much of the work that can be done on band saws and band machines can also be done with arc or flame cutting, the welder or fabricator will have to determine which method is better for a particular job to be done.

DoALL

DoAll Co.

11-70. *A metal-cutting band saw and the many forms that can be cut with this machine.*

11-71. *Advantages of the band machine.*

LESS HORSEPOWER
NARROW TOOTH KERF
MINIMUM OF MATERIAL REDUCED TO CHIPS.

LEAST MATERIAL WASTE
REMOVES WHOLE SECTIONS
CUTS DIRECTLY TO FINISH LINE

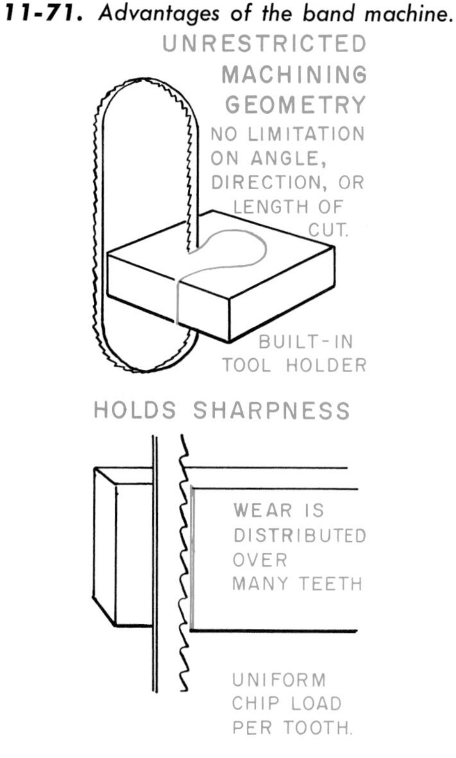
UNRESTRICTED MACHINING GEOMETRY
NO LIMITATION ON ANGLE, DIRECTION, OR LENGTH OF CUT.
BUILT-IN TOOL HOLDER
HOLDS SHARPNESS
WEAR IS DISTRIBUTED OVER MANY TEETH
UNIFORM CHIP LOAD PER TOOTH.

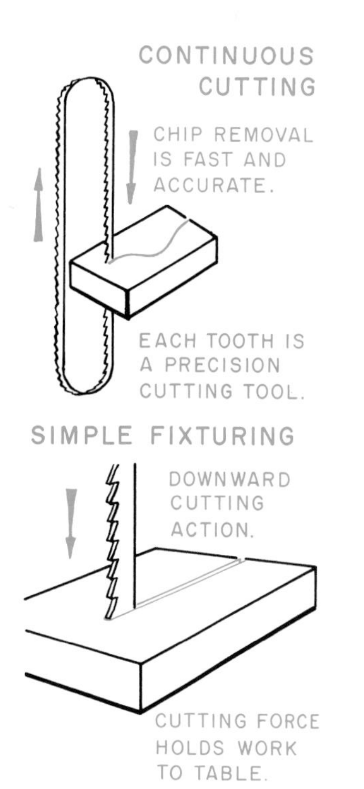
CONTINUOUS CUTTING
CHIP REMOVAL IS FAST AND ACCURATE.
EACH TOOTH IS A PRECISION CUTTING TOOL.
SIMPLE FIXTURING
DOWNWARD CUTTING ACTION.
CUTTING FORCE HOLDS WORK TO TABLE.

REVIEW QUESTIONS

1. List four types of hydraulic tools and explain their use.
2. List at least five different types of portable hand tools used in a welding fabricating shop.
3. List the differences among pneumatic, electric, and hydraulic power tools.
4. Name five small hand tools required by the operator, and describe the use of each.
5. What is a weld positioner and how is it used?
6. Of what use are turning rolls?
7. How is a welding manipulator used?
8. List at least six backing materials used in the welding shop, and explain the purpose of each.
9. What is a spot welder? How does it work?
10. Explain the use of the universal bender.
11. List five machine tools used in the welding shop.
12. List at least ten different types of major equipment used in a welding fabricating shop. (Do not include those listed above.)
13. Name the two basic types of weld seamers and explain their use.
14. The Bug-O-Verlay® is the name given to the speed control of the automatic gas cutting machine. True. False.
15. A beveling machine can only be used on steel plate. True. False.
16. A beveling machine cannot be used on thicknesses greater than $\frac{1}{4}$ inch. True. False.
17. Pneumatic hand tools are operated with oxygen. True. False.
18. The electric hand tool is best suited to production line applications. True. False.
19. A nibbler is used for removing the reinforcement from the weld bead of a butt joint. True. False.

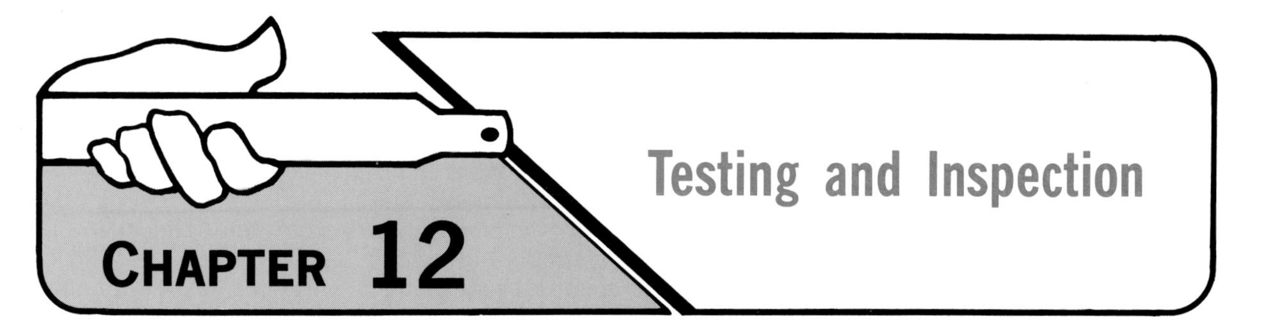

Testing and Inspection

Welding was first used as a means of patching and repairing. It was rarely employed as a means of fabricating pressure vessels, pipelines, and other structures such as buildings and bridges which would be a hazard to life and property if they failed. As welding began to be used as a fabrication process, it became essential for welded joints to be strong enough to meet the service requirements for which they were designed. Methods for testing both the quality of the weld and the ability of the welder had to be devised.

For a long time it had been considered necessary only to look at a completed weld in order to judge its quality and the welder's ability. If carried out by a competent inspector or welder, visual inspection may be satisfactory for welds which are designed primarily to hold parts together and which are not subject to high stress in service. This kind of inspection is limited since there is no way of knowing if the weld metal has internal defects. The outer appearance of the weld may be entirely satisfactory, and yet it may be porous, contain cracks, and lack both complete fusion and penetration. The weld metal may have been damaged seriously by poor welding technique.

Critical welding demands that the weld metal and joint be tested for strength, soundness, and other physical qualities required in the design. The reliability of the welded joint can be determined by the degree to which the weld metal is kept free of foreign materials such as slag and by the degree to which it is fused to the base material.

CODE WELDING

It is usually the practice of a particular industry to develop a code. A *code* is a set of regulations governing all of the elements of welded construction in a certain industry. Codes provide for human safety and protect property against failure of the weldment. A universal testing procedure for all types of welding in all locations does not exist. Thus in the field of piping, we find that welding must conform to a number of standards, depending on the nature of the installation. The welding of pressure piping conforms to the *Code for Pressure Piping* of the American Standards Association. The welding of piping connected to boilers and pressure vessels conforms to the *Code for Boilers and Pressure Vessels,* Section IX, developed by the American Society of Mechanical Engineers (ASME). The welding of pipelines conforms to the "Standard for Welding Pipelines and Related Facilities" (API Standard 1104) developed by The American Petroleum Institute. The welding of buildings, bridges, and aircraft must conform to the codes and standards set for these structures. Generally, these standards are set by the federal, state, and local governments, insurance companies, and various professional organizations. Many state and local bridge and building codes accept the procedure and operator qualification provisions of the "AWS Structural Welding Code D1.1-62." A code of growing national importance in the piping and tubing field is the "AWS Standard For Qualification of Welding Procedures and Welders for Piping and Tubing D10.9-69." Both the structural and piping codes can be obtained from the American Welding Society.

The welder does not have to be thoroughly informed about the details of all of the existing codes. The employer, through the engineering and production departments, makes sure that the work meets the standards required for it. The welder should, however, have a good understanding of the different weld tests and know what to look for in any visual inspection.

There are two broad categories of welding tests. A *procedure qualification test* is a test conducted for the purpose of determining the correctness of the method of welding for a specific welding project. The American Welding Society and various code authorities have established standard procedures for welding. The welding procedure meets specifications for base metal filler metal, joint preparation, position of welding, the welding process, and welding techniques. Such requirements as current setting, electrode size, electrode manipulation and preheat, interpass, and postheat temperatures are also specified. Following a particular procedure assures uniform results. Tests which certify welders for codework may be known as *operator qualification tests* or *performance qualification tests.* They are given to find out whether or not the welder has the knowledge and the skill to follow and apply a procedure of welding as developed for the class of work at hand. Testing may be with hand-held welding equipment (the stick-electrode holder, the gas welding torch, or the semi-automatic welding gun) or with fully automatic welding equipment. Every effort has been made to simplify these tests so that they can be administered at low cost and still establish the soundness of the weld.

Reliability of welding is based on the use of appropriate inspection controls. The methods of testing which determine the quality of a weld are divided into three very broad classifications: (1) nondestructive testing, (2) destructive testing, and (3) visual inspection.

A *nondestructive* test, as the name implies, is any test that does not damage the weld or the finished product. Modern inspection equipment and techniques have made nondestructive testing an effective inspection tool. Nondestructive testing is usually done by specialists who have been trained in the use of the equipment and the interpretation of test results. Although the original cost of the equipment may be high, nondestructive testing can be the fastest and least expensive of the inspection methods.

Destructive testing, also called *mechanical testing,* usually requires that a test specimen be taken from the fabrication or that sample plates be welded from which the test specimens are cut. The weld is damaged beyond use. Destructive testing is normally used to determine the mechanical and metallurgical qualities of a proposed welding procedure and to determine the ability of a welding operator before the actual production work is started.

In *visual inspection* the surface of the weld and the base metal are observed for visual imperfections. Certain inspection tools and gauges may be used with the observation procedure. Visual inspection is the testing method most commonly

Magnaflux Corp.

12-1. *A portable magnetic particle testing unit. This method speeds inspection of welds and stressed areas during fabrication and repair. The unit can be used both in shop and field.*

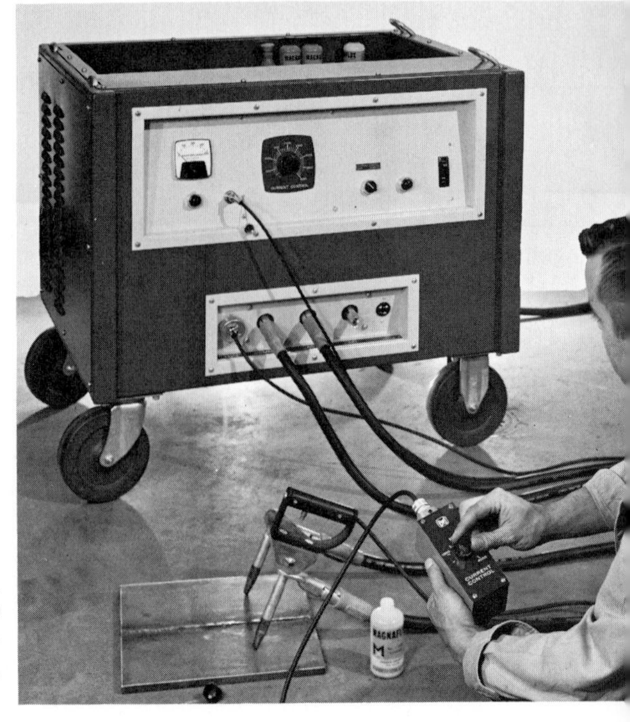

Magnaflux Corp.

12-2. *Remote control used with a magnetic particle test unit. Welds can be inspected for lack of penetration and fusion, slag inclusions, crater cracks, porosity, and gas pockets.*

used by the welder himself as well as the welding inspector and welding supervisor. This method is highly satisfactory for noncritical (noncode) work.

NONDESTRUCTIVE TESTING

Magnetic Particle Testing

Magnetic particle testing is one of the most easily used nondestructive tests. It is used to inspect plate edges before welding for surface imperfections. It tests welds for such defects as surface cracks, lack of fusion, porosity, undercut, poor root penetration, and slag inclusions. This method is limited to magnetic materials such as steels and cast iron. It cannot be used with such nonmagnetic materials as the stainless steels, aluminum, and copper. Very often the method is referred to as the *Magnaflux*® method. Magnaflux® is the name of a particular brand of testing equipment, Fig. 12-1.

Magnetic particle testing, Fig. 12-2, detects the presence of internal and surface cracks too fine to be seen by the naked eye. Defects can be detected to a depth of $\frac{3}{4}$ to $\frac{3}{8}$ inch below the surface of the weld. Defects much deeper than this are not likely to be found.

The part to be examined must be smooth, clean, dry, and free from oil, water, and excess slag from the welding operation. Wire brushing or sandblasting is usually satisfactory preparation for most welds. The part is magnetized by using an electric current to set up a magnetic field within the material or by putting the piece in an electric coil. The magnetized surface is covered with a thin layer of magnetic powder such as blast resin or red iron oxide, Fig. 12-3. A new product known as *Magnaglo,*® is a fluorescent powder which glows in a black light. These powders can be applied dry or they may flow over the surface if they are in a suspension of oil, water, or any other low viscosity liquid. The layer of powder can be blown off the surface when there are no defects. If there is a defect, the powder is held to the surface at the defect because the powerful magnetic field in the workpiece sets up a north magnetic pole at one end of the defect and a south magnetic pole at the other. These poles have a stronger attraction for the magnetic particles than the surrounding surface of the material.

Figure 12-4 gives a simple explanation of the principles of magnetic particle testing. As shown in Part 1, an open magnet has two poles: north and south. The magnetic field between the two poles attracts and holds a nail. If the magnet is bent until the poles almost touch, the magnetic field between the two poles attracts iron powder (Part 2). If the magnet is completely closed as in Part 3, it will not attract or hold iron powder because the magnetic field is in a circle inside the ring and there can be no polarity. This is the reason that the powder can be blown off the surface of a weld without defects. If there is a crack in the weld, it will cause polarity and a magnetic field is built up at the crack which will hold the powder (Part 4).

Cracks must be at an angle to the magnetic lines of force in order to show. Two general procedures are used. In the method illustrated by Fig. 12-4, probes are placed on each side of the area to be inspected, and a high amperage current is passed through the workpiece. A magnetic field is produced at right angles to the flow of current, which may be represented by circular lines of force within the workpiece (Part 3). When these lines of force encounter a longitudinal (lengthwise) crack, they leak through and become points of surface attraction for the magnetic powder dusted on the surface (Part 4). A transverse (crosswise) crack would not show because the lines of force would be parallel with the crack, Fig. 12-5. The other method is to magnetize a workpiece by putting it inside a solenoid. In this method, the magnetic lines of force are longitudinal and parallel with the workpiece so that transverse cracks show up, Fig. 12-6.

***12-3.** A portable magnetic particle test unit checking critical welds during the construction of a Detroit bank building. Note the use of magnetic powder as the unit is applied.*

Magnaflux Corp.

The volume and variety of products tested by the magnetic particle method is enormous. In addition to weldments of all types, finished steel products, castings, pipe and tubing, racing cars, aircraft, and missiles can be checked for structural defects.

Radiographic Inspection

Radiography is a nondestructive test method that shows the presence and type of microscopic defects in the interior of welds. The method utilizes either the X-ray or gamma ray. The source of X-rays is the X-ray tube. Gamma rays are similar to X-rays except for their shorter wavelengths. They are produced by the atomic disintegration of radium or one of the several commercially available radioisotopes. While gamma rays, because of their short wavelengths, can penetrate a considerable thickness of material, exposure time is much longer than for X-rays. The films obtained by the use of X-rays are called *exographs* and those by use of gamma rays, *gammagraphs.* Both are generally referred to as *radiographs.*

The size of X-ray equipment is rated on the basis of its electrical energy which in turn determines the intensity of the X-ray produced. The voltage across the tube controls the X-ray wavelength and penetrating power. Voltage is measured in kilovolts: 1 kilovolt equals 1,000 volts. Generally, industrial applications run from 50 kilovolts (50,000 volts), which is used for microradiography, to 2,000 kilovolts (2,000,000 volts) which can penetrate 9-inch thick steel. These thickness limits increase with softer metals. One-hundred fifty kilovolts (150,000 volts) have a thickness limit of 1 inch in steel and $4\frac{1}{2}$ inches in aluminum. Special units have a capacity of 24,000 kilovolts (24,000,000 volts).

In radiographic testing, a photograph is taken of the internal condition of the weld metal. The photographic film is placed on the side opposite the source of radiation, Fig. 12-7. The distance

12-4. *Circular magnetization produces the lines of force shown in parts 3 and 4. Compare these patterns with those evident in the common horseshoe and closed magnets.*

CIRCULAR MAGNETIZATION

OPEN MAGNET

PARTIALLY CLOSED MAGNET

COMPLETELY CLOSED MAGNET

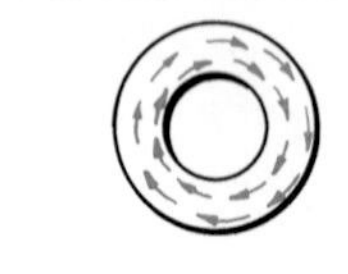

CRACKED MAGNET

12-5. *When a current is passed through the workpiece, the magnetic lines of force are at right angles to the current, and discontinuities that are angled against the lines of force will create the diversion needed to produce magnetic poles on the surface. Under the illustrated arrangement, a transverse crack would not give an indication, but by changing the position of the probes 90 degrees, it would be at right angles to the lines of force and would show.*

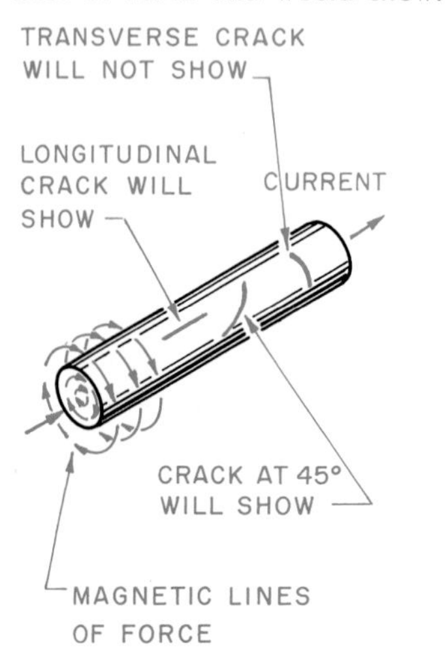

12-6. *When the magnetic field is produced with a solenoid coil, the lines of force are parallel and longitudinal. A longitudinal crack will not show, but a crack angled against the lines of force is indicated.*

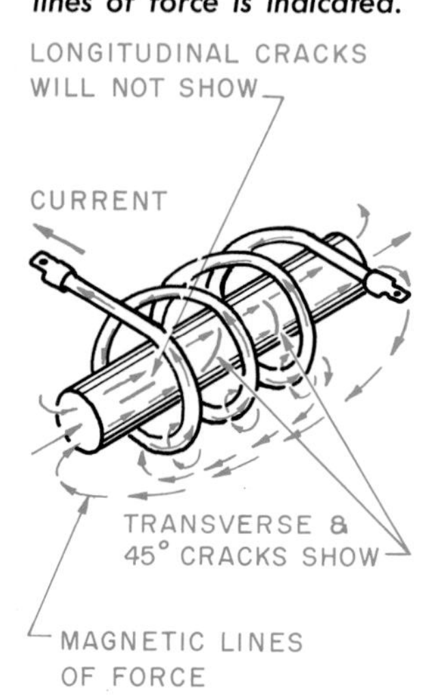

between the film and the surface of the workpiece must not be greater than 1 inch. The radiation rays penetrate the metal and produce an image on the film. Different materials absorb radiation at different rates. Since slag absorbs less radiation than steel, the presence of slag permits more radiation to reach the film. Thus the area of the slag inclusion shows up darker than the steel on the film, and this indicates a flaw in the weld metal. A radiograph can establish the presence of a variety of defects and record their size, shape, and location.

Figure 12-8 shows common defects as they appear in radiographs. Porosity, which is caused by gas pockets in the weld deposit, shows up as small dark spots (Fig. 12-8A). A nonmetallic inclusion, such as slag, usually shows up as an irregular shape (Fig. 12-8B). These images may be fine or coarse, and they may be widely scattered or closely grouped. Cracks show up as a line darker than the film background (Fig. 12-8C). Both longitudinal and transverse cracks are visible. Incomplete fusion gives dark shadows (Fig. 12-8D). Incomplete root penetration is indicated as a straight dark line (Fig. 12-8E). Undercutting shows as a dark, linear shadow at the edge of the weld (Fig. 12-8F). Icicles and burn-through show as light circular areas or as darkened areas of rounded contour surrounded by light rings (Fig. 12-8G).

Penetrant Inspection

Penetrant inspection is a nondestructive method for locating defects open to the surface. Like radiographs, it can be used on nonmagnetic materials such as stainless steel, aluminum, magnesium, tungsten, and plastics. The penetrant method cannot detect interior defects.

RED DYE PENETRANT. The surface to be inspected must be clean and free of grease, oil, and other foreign materials. It is

12-7. *Typical arrangements of the X-ray source and film in weld radiography. The angle of exposure and the geometry of the weld influence the interpretation of the negative. Note that multiple exposure may be necessary for pipe welds.*

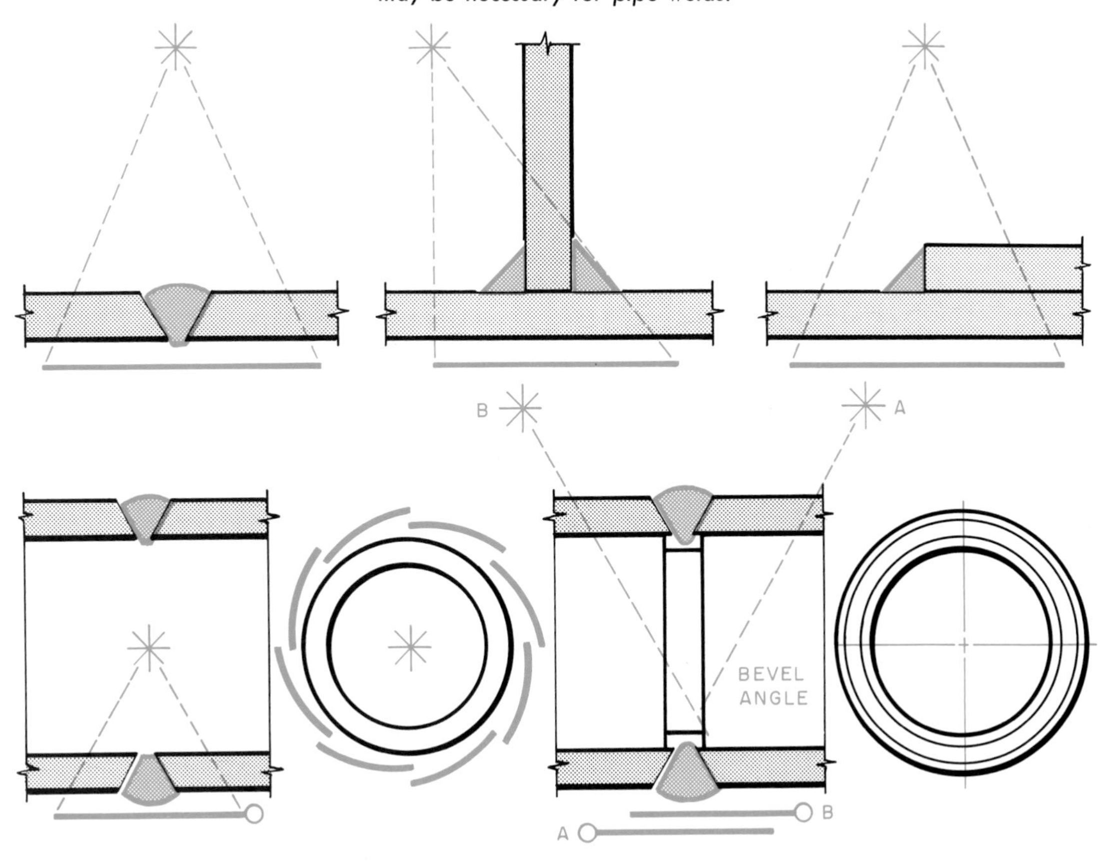

sprayed with the dye penetrant which penetrates into cracks and other irregularities. The excess penetrant is wiped clean. Then the part is sprayed with a highly volatile liquid which contains a fine white powder. This is known as a *developer*.

Evaporation of the liquid will leave the dry white powder which has a blotting paper action on the red dye left in the cracks, drawing it out by capillary action so that the defects are marked clearly in red.

Red dye penetrant is used on pressure and storage vessels and in critical piping applications to check the initial weld pass made by the TIG welding process for hairline cracks. It is often applied to the inspection of aircraft jet engines and weldments made of aluminum, magnesium, and stainless steels.

SPOTCHECK.® *Spotcheck,®* Fig. 12-9, is a dye penetrant test for defects open to the surface. Like other dye penetrants it relies on penetration of the defect by a dye, removal of the excess dye, and development of the indication.

Spotcheck® dye penetrant is a highly sensitive process. Small cracks show up against the white developer background. It locates cracks, pores, leaks, and seams invisible to the unaided eye. It marks them clearly and distinctly in red, right on the part of material.

Spotcheck® is used on almost all materials. Examples include steel, aluminum, brass, carbides, glass and plastics. Figures 12-10A and 12-10B shows inspection with Spotcheck® at Disneyland. Spotcheck® readily locates such defects as cracks from shrinkage, fatigue, grinding, and heat treating, porosity, and cold shuts. It also shows up seam defects and forging laps and bursts as well as lack of bond between joined metals and through-leaks in welds. Spotcheck® is particularly recommended for the testing of moderate numbers of medium-to-small parts.

Spotcheck® offers the following advantages:

- Complete portability for critical inspection at remote shop or field locations

12-8. *The manner in which weld defects show on radiographs.*

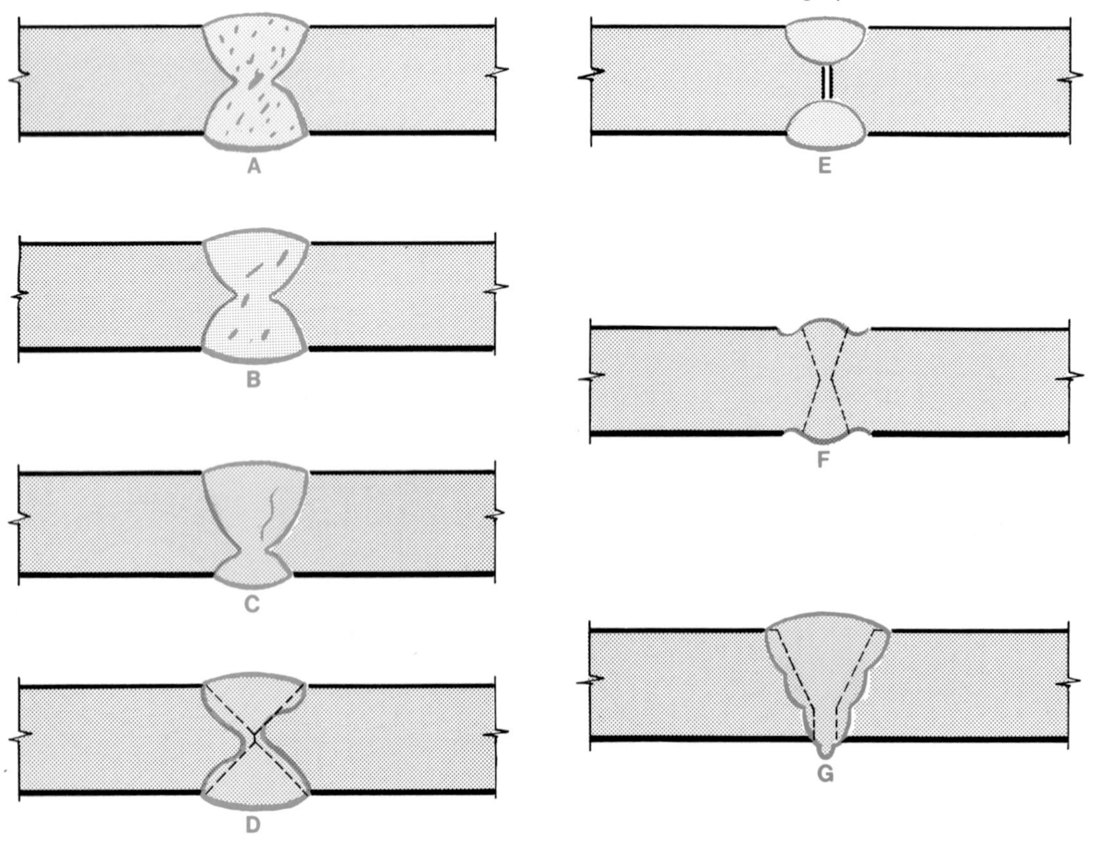

- Fast inspection of small, critical sections suspected of being defective
- Ease of application and dependable interpretation of results
- Low initial investment and low per part cost in moderate volume uses

FLUORESCENT PENETRANT. Fluorescent penetrants may be used instead of the red dye penetrant, and the technique is similar to that used in the dye method. The treated metal surface is examined under ultraviolet or black light in semidarkness. This method is referred to as *fluorescent penetrant inspection.* The sharp contrast between the fluorescent material and the base or weld metal background clearly indicates cracks or other defects in the metal. Fluorescent penetrant is most useful for leak detection in lined or clad vessels. It is also used for brazed joints. The joint in Fig. 12-11 is being cleaned in preparation for fluorescent inspection.

Ultrasonic Inspection

Utrasonic inspection is one of the most recent developments in the field of nondestructive testing. Ultrasonic inspection is rapid and has the ability to probe deeply without damaging the weldment. Because it can be closely controlled, it is able to supply precise information without elaborate test setups. It can detect, locate, and measure both surface and subsurface defects in the weld or base metal.

Ultrasonic inspection requires considerable skill and experience both in its application and in the interpretation of the echoes that appear on the screen. For example, a properly welded backing ring may produce the same pattern as a weld with unacceptable lack of penetration or root cracking. Only a highly experienced inspector may be able to tell the difference between serious defects and normal weld conditions. This disadvantage limits its application. Many companies

Magnaflux Corp.

12-10A. *Applying Spotcheck® to a track fitting for the inspection and maintenance of the Matterhorn at Disneyland.*

12-9. *Spotcheck® test kit.*

Magnaflux Corp.

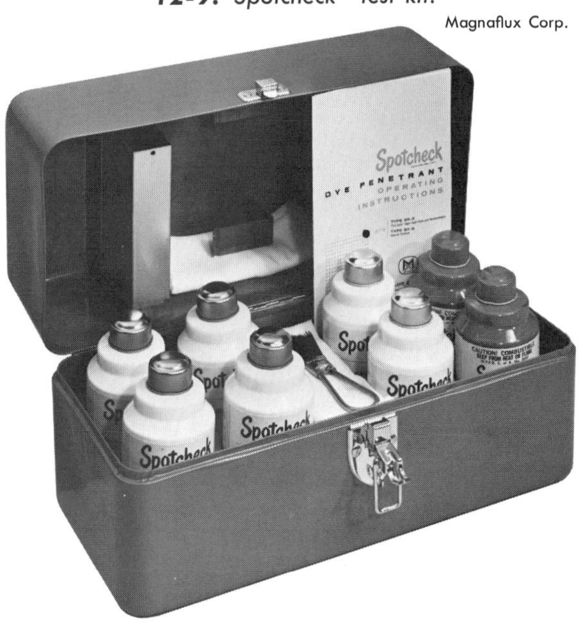

12-10B. *Spotcheck® inspection of a welded joint in the Tahitian tree at Disneyland. The tree trunk is a welded fabrication.*

Magnaflux Corp.

use this method to locate defects and then apply the radiographic method to those areas which are doubtful.

Ultrasonic testing is done by means of an electrically timed wave which is similar to a sound wave, but of higher pitch and frequency. The term *ultrasonic* comes from the fact that these frequencies are above those heard by the human ear. The ultrasonic waves are passed through the material being tested and are reflected back by any density change. Three basic types of waves are used: shear (angle) beams, longitudinal (straight) beams for surface and subsurface flaws, and surface waves for surface breaks and cracks. The waves are generated by a unit similar to a high fidelity amplifier, to which a search unit is attached. The reflected signals appear on a screen as vertical reflections of the horizontal baseline, Fig. 12-12.

When the search unit is applied to the material, two reference pips appear on the screen. The first pip is the echo from the surface contacted, and the second pip is the echo from the bottom or opposite surface of the material. The distance between these pips is carefully calibrated, and this pattern indicates that the material is in satisfactory condition. When a defect is picked up by the search unit, it produces a third pip, which registers on the screen between the first and second pip (Fig. 12-13) since the flaw must be located between the top and bottom surfaces of the material. The distance between the pips and the relative height indicate the location and the severity of the defect.

This method of testing is finding increased use by the piping industry not only for detecting defects, but also for checking on the progress of corrosion and wear by taking periodic measurements of metal thickness. As the pipe material is reduced in thickness by corrosion and wear, the two pips move closer together, and the rate of deterioration can be measured. Applications of ultrasonic testing in both field and shop are shown in Figs. 12-14 through 12-16.

Process Comparison

The welding student is urged to study Table 12-1 carefully. It is a condensed reference guide to the use of the nondestructive methods just described on steel weldments. It should be noted that proof of the existence of a flaw does not indicate that the flaw will seriously affect the service of the weldment. Standards and correlations as to the

12-11. *Preparing the nose wheel fork of an aircraft for fluorescent penetrant. Note the black light in the inspection kit.*

Magnaflux Corp.

12-12. *Portable ultrasonic testing instrument. This unit detects internal and surface flaws in fillet welds.*

Magnaflux Corp.

effect of the flaw on service must be set up, or the nondestructive test may well give no more information than someone's opinion as to what is good or bad.

Eddy Current Testing

Like magnetic particle testing, the eddy current method makes use of electromagnetic energy to detect defects in the material. Testing equipment is shown in Fig. 12-17.

When a coil that has been energized with alternating current at high frequency is brought close to a conductive material, it will produce eddy currents in the material. *Eddy currents* are secondary currents induced in a conductor (in this case the metal being tested). Eddy currents are caused by a variation in the magnetic field. A search coil is used in addition to the energizing coil. The search coil may be connected to meters, recorders, or oscilloscopes, which pick up the signals from the weldment. A level of individual current or voltage is noted for the material. Any defect in the material distorts the magnetic field and is indicated on the recording instrument. The size of the defect is shown by the amount of this change.

For example, in the inspection of butt-welded tubing, the tubing is passed through the energizing coil to induce an a.c. eddy current in the tube. The search coil also circles the tube. This coil is connected to a sensitive instrument. The eddy currents induced in the tubing induce eddy currents in the search coil, and the meter records the current or voltage in the coil. If the tube is sound throughout, a steady level of current or voltage is induced in the search coil and the meter reading holds steady. A flaw in the tubing distorts the magnetic

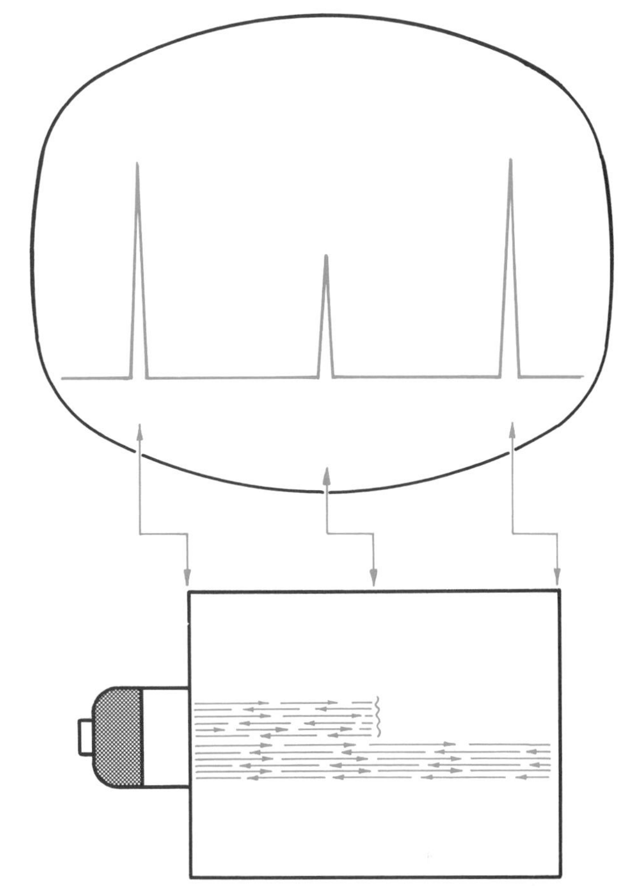

12-13. *Short pulses appear as pips and register on the ultrasonic testing screen.*

12-14. *Ultrasonic testing the bond of the copper liner to the base metal of a copper-clad reactor. The welds were X-rayed with gamma rays, and a chemical analysis was made of the weld deposit. Testing was done in the shop.*

Nooter Corp.

Table 12-1

Reference guide to major methods for the nondestructive testing of welds.

Inspection method	Equipment required	Enables detection of	Advantages	Limitations	Remarks
Visual	Magnifying glass Weld-size gauge Pocket rule Straightedge Workmanship standards	Surface flaws—cracks, porosity, unfilled craters, slag inclusions Warpage, under-welding, over-welding, poorly formed beads, misalignments, improper fitup	Low cost Can be applied while work is in process, permitting correction of faults Gives indication of incorrect procedures	Applicable to surface defects only Provides no permanent record	Should always be the primary method of inspection, no matter what other techniques are required Is the only "productive" type of inspection Is the necessary function of everyone who in any way contributes to the making of the weld
Radiographic	Commercial X-ray or gamma units, made especially for inspecting welds, castings, and forgings Film and processing facilities Fluoroscopic viewing equipment	Interior macroscopic flaws—cracks, porosity, blow holes, nonmetallic inclusions, incomplete root penetration, undercutting, icicles, and burn-through	When the indications are recorded on film, gives a permanent record When viewed on a fluoroscopic screen, a low cost method of internal inspection	Requires skill in choosing angles of exposure, operating equipment, and interpreting indications Requires safety precautions. Not generally suitable for fillet weld inspection	X-ray inspection is required by many codes and specifications. Useful in qualification of welders and welding processes Because of cost, its use should be limited to those areas where other methods will not provide the assurance required.
Magnetic particle	Special commercial equipment Magnetic powders in dry or wet form; may be fluorescent for viewing under ultraviolet light	Excellent for detecting surface discontinuities, especially surface cracks	Simpler to use than radiographic inspection Permits controlled sensitivity Relatively low cost method	Applicable to ferromagnetic materials only Requires skill in interpretation of indications and recognition of irrelevant patterns Difficult to use on rough surfaces	Elongated defects parallel to the magnetic field may not give pattern; for this reason the field should be applied from two directions at or near right angles to each other
Liquid penetrant	Commercial kits, containing fluorescent or dye penetrants and developers Application equipment for the developer A source of ultraviolet light if fluorescent method is used	Surface cracks not readily visible to the unaided eye Excellent for locating leaks in weldments	Applicable to magnetic & nonmagnetic materials Easy to use Low cost	Only surface defects are detectable Cannot be used effectively on hot assemblies	In thin-walled vessels, will reveal leaks not ordinarily located by usual air tests Irrelevant surface conditions (smoke, slag) may give misleading indications.
Ultrasonic	Special commercial equipment, either of the pulse-echo or transmission type Standard reference patterns for interpretation of RF or video patterns	Surface and subsurface flaws, including those too small to be detected by other methods. Especially for detecting subsurface lamination-like defects.	Very sensitive Permits probing of joints inaccessible to radiography	Requires high degree of skill in interpreting pulse-echo patterns Permanent record is not readily obtained	Pulse-echo equipment is highly developed for weld inspection purposes. The transmission-type-equipment simplifies pattern interpretation where it is applicable.

The Lincoln Electric Co.

field induced in the tubing by the energizing coil. When the flow passes within range of the search coil, the needle on the search meter moves.

The eddy current method is suitable for both ferrous and nonferrous materials, and it is being used extensively in testing welded tubing and pipe, Figs. 12-18 and 12-19. It can determine the physical characteristics of a material, the wall thickness in tubing, and the thickness of various coatings. It can check for porosity, slag inclusions, cracks, and lack of fusion. Testing of tubing on a factory production line may be accomplished at a rate of approximately 3,000 feet per minute.

Leak Tests

Leak tests are made by means of pneumatic or hydraulic pressure. A load is applied that is equal to or greater than that expected in service. If the test is to determine leakage, it is not necessary to test beyond the working load of the weldment. If, however, failure of the weldment might cause great property loss or personal injury, the test pressure applied usually exceeds the working pressure. This method of testing is usually used to test pressure vessels and pipelines. If used as a destructive method, pressure is applied until the unit bursts, Fig. 12-20.

Water is usually used to test for leaks, but very small leaks are not always detected. When air or oil of low viscosity is used, it indicates leaks that water cannot. Hydrogen will leak where oil will not, but hydrogen is a fire hazard and must be used with great care.

When the expansive force of air or gas is high, adequate safety measures must be taken to guard against explosion. When air or gas is used, the weld seam may be painted with liquid soap or another chemical solution to cause the formation of bubbles at the point of leakage. Small tanks may be immersed in a water or liquid bath. On work that is of a critical nature, the pressure should be held for a period of time and the pressure noted on a gauge that is connected to the vessel.

Hardness Tests

It is often important to test the hardness of the weld deposit or the base metal in the area of the weld. It is important to know the hardness of the weld deposit if the weld is going to be machined or if it will be subject to surface wear of one kind or another.

There are a number of nondestructive hardness tests, and the choice depends upon the material being tested. The most common tests are the Brinell, Rockwell, Vickers, and Shore Scleroscope.

The preparation of the specimen is very important. The surface should be flat and free of scratches.

BRINELL. The Brinell hardness test consists of impressing a hardened steel ball into the metal to be tested at a given pressure for a predetermined time, Fig. 12-21. The diameter of the impression is measured, and this indicates a Brinell number on a chart.

The ball is 10 ± 0.0025 millimetres. It is forced into the spec-

***12-15.** Ultrasonic testing of a weld in a large steel beam in the field.*

Branson Instruments Co.

***12-16.** Using the portable ultrasonic instrument shown in Fig. 12-12 to check a structural weld on the seventy-sixth floor of the John Hancock Building in Chicago.*

Magnaflux Corp.

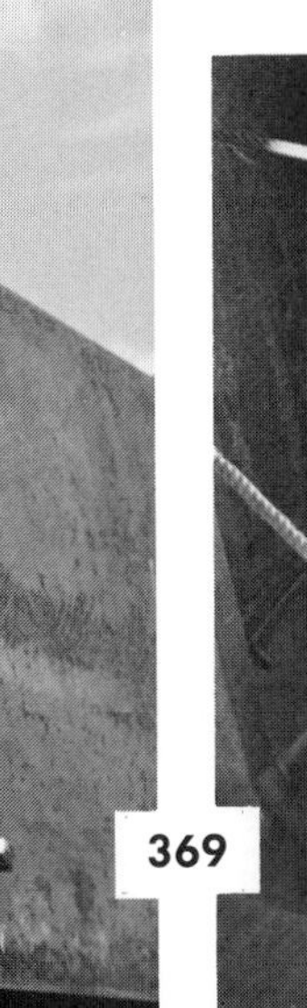

imen by hydraulic pressure of about 3,000 kilograms for a period of 15 seconds. The load for soft metals such as brasses and bronzes is 500 kilograms applied for the same amount of time.

Because the impression the Brinell ball makes is large, it can be used only for obtaining values over a large area and when the impression on the surface is not objectionable.

ROCKWELL. The Rockwell system is similar to the Brinell system, but it differs in that the readings can be obtained from the dial. The Rockwell tester, Fig. 12-22, measures the depth of residual penetration made by a small hardened steel ball or diamond cone. A minor load of 10 kilograms is applied. This seats the penetrator (ball or cone) in the surface of the specimen and holds it in position. The full load

12-17. *The eddy current testing unit provides inspection for nonmagnetic rod, tube, and wire. It is used to test for seams, cracks, and splits in welded copper and aluminum tubing.*

Magnaflux Corp.

Magnaflux Corp.

12-18. *Eddy current testing. The tubing passes from the wheel unit (right) through the eddy current coil, which is connected with the unit shown in Fig. 12-17, to the marking system which sprays paint on defective areas.*

12-19. *Testing pipe welds with a movable-yoke eddy current unit. The sensing probe can be rotated around the pipe to scan the passing weld.*

Magnaflux Corp.

of 150 kilograms is then applied. After the major load is removed, the hardness number is indicated on the dial gauge. The hardness numbers are based on the difference of penetration between the major and minor loads.

The two Rockwell scales are known as the B-scale and the C-scale. The C-scale is used for the harder metals. A cone-shaped diamond penetrator is used instead of a ball, and it is applied at a load of 150 kg. The B-scale is used for the softer metals, and the penetrator is a hardened steel ball $\frac{1}{8}$ or $\frac{1}{16}$ inch in diameter applied at a lesser load of 100 kg., Fig. 12-23. A lesser load is applied to the ball than to the cone.

VICKERS. The Vickers hardness test consists of impressing a diamond penetrator into the surface of a specimen under a predetermined load. The Vickers number is the ratio of the impressed load to the area of the resulting indentation. Because the Vickers penetrator is a diamond, it can test the hardest materials. The load varies with the hardness of the material to be tested. The impression is a square. The reading is obtained across the diagonals of the square, and the hardness figure is then obtained from the chart. Both the Rockwell and the Vickers tests make less of a surface indentation than the Brinell test does.

SHORE SCLEROSCOPE. The scleroscope, Fig. 12-24, is a portable machine which consists of a vertical glass tube in which a small cylinder with a diamond point slides freely. This cylinder is allowed to fall on the sample being tested and to rebound from the test specimen. The distance which it rebounds, measured on a scale on the glass tube, is the hardness reading. This testing method is highly satisfactory for hard materials, but it is not applicable to soft materials. Table 12-2 compares the readings of the three major methods. Only a few readings are given, going from low to high.

DESTRUCTIVE TESTING

Destructive testing is the mechanical testing of weld samples to determine their strength and

Crane Co.

12-20. A piping unit which has been subjected to hydraulic testing. The weld failed at 6,200 p.s.i.

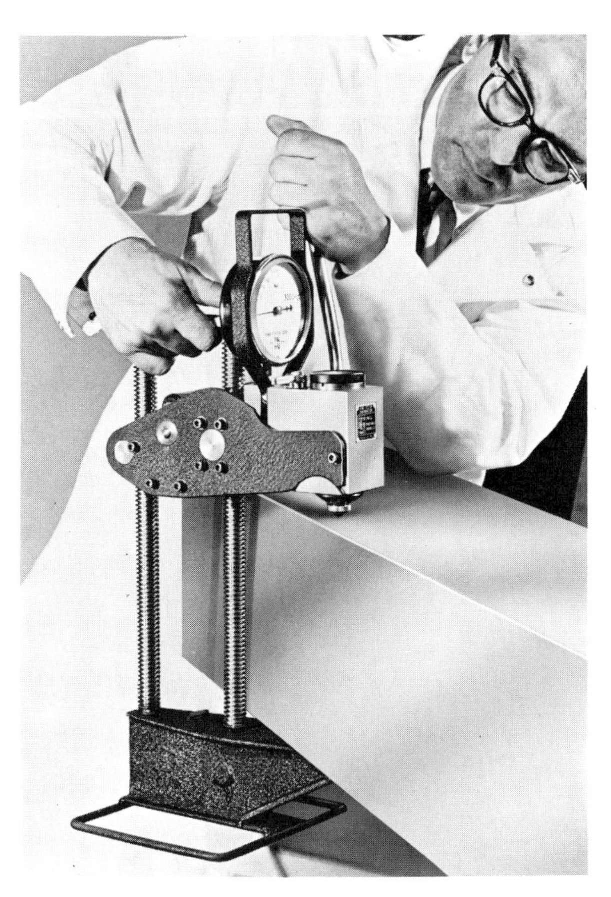

King Tester Corp.

12-21. A Brinell hardness tester being used on the job. The equipment provides accuracy to within ½ of 1 percent. It is an easy test to do, and the surface being tested does not have to be polished.

other properties. Testing methods are relatively inexpensive and highly reliable. Therefore, they are widely used for operator and procedure qualification. Destructive testing is usually performed on test specimens that are taken from a welded plate that duplicates the material and the weld procedures used on the job. Certain general test procedures have been developed by the American Welding Society which are the standard for the industry and various code-making bodies. While tests may differ somewhat in detail and application from industry to industry, they all make use of the basic procedures and types of test specimens recommended by the American Welding Society. The following list summarizes the more common mechanical tests which will be presented in this chapter.

Groove Welds

- Reduced-section tension test: determines tensile strength; used for procedure qualification.
- Free-bend test: determines ductility; used for procedure qualification.
- Root-bend test: determines soundness; used widely for operator qualification; also used for procedure qualification.
- Face-bend test: determines soundness; used widely for operator qualification; also used for procedure qualification.
- Side-bend test: determines soundness; used widely for operator qualification; also used for procedure qualification.
- Nick-break test: determines soundness; at one time used widely for operator qualification; used infrequently today.

Fillet Welds

- Longitudinal and transverse shear tests: determines shear strength; used for procedure qualification.
- Free-bend test: determines ductility; used for procedure qualification.
- Fillet weld soundness test: determines soundness; used extensively for operator qualification.
- Fillet weld break test: determines soundness; used infrequently today.
- Fillet weld fracture and macro test: determines soundness;

12-22. *Rockwell hardness tester.*
Wilson Instrument Division, ACCO

Table 12-2
Typical readings for standard hardness tests.

Material	Rockwell C	Rockwell B	Brinell	Scleroscope
Hard	69		755	98
	60		631	84
	51		510	71
	39		370	52
	24	100	245	34
	16	95	207	29
↓	10	89	179	25
Soft	0	79	143	21

12-23. *Spheroconical diamond penetrator.*
Wilson Instrument Division, ACCO

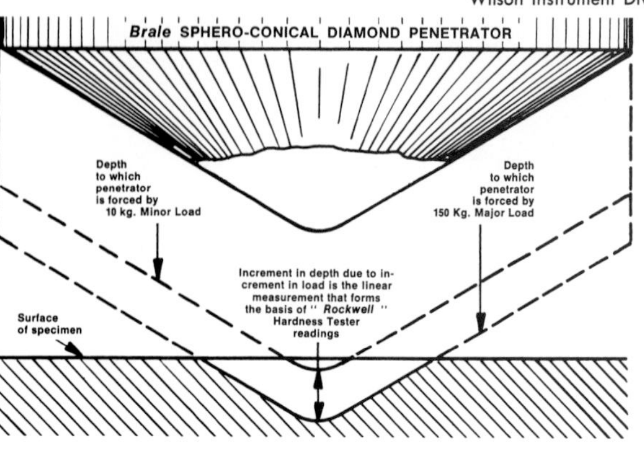

Sphero-conical
Diamond Penetrator

ASME test for procedure qualification and operator qualification.

Table 12-3 lists the minimum physical properties of welds recommended by three code-making organizations. These values are obtained by destructive testing.

General Requirements

The material that follows will give the student the information

12-24. *The Shore scleroscope can test the hardness of all metals, polished or unpolished, and flat, curved, or tubular surfaces. The device is portable, fast, and simple to operate.*

The Shore Instrument and Mfg. Co.

Table 12-3

Minimum physical properties of welds.

	Class of welding	1	2	3
	Material	Low carbon steel, B4A12C or equiv.	Low carbon steel, B4A8 or equiv.	Low carbon steel, B4A8 or equiv.
Joint	Tensile strength p.s.i.	55,000	52,000	45,000
Joint	Yield point p.s.i.	40,000	40,000	35,000
Joint	Free bend % elongation	30%	20%	10%
Joint	Corrosion resistance relative to plate stock	100%	100%	50%
All Weld Metal	Tensile strength p.s.i.	60,000	60,000	50,000
All Weld Metal	Yield point p.s.i.	45,000	45,000	40,000
All Weld Metal	Elongation in 2 inches	20%	20%	10%
All Weld Metal	Density	7.80	7.78	7.75
All Weld Metal	Charpy impact Ft-Lbs.	30	20	2
All Weld Metal	Endurance limit p.s.i.	28,000	25,000	12,000
Fillet weld ultimate shear strength in pounds per linear inch	Size			
	$\frac{1}{4}$''		11,000	10,000
	$\frac{3}{8}$''	14,000	14,000	13,000
	$\frac{1}{2}$''	18,000	18,000	16,000
	$\frac{5}{8}$''	21,000	21,000	19,000
	$\frac{3}{4}$''	25,000	25,000	21,000
References		(IPP-102) *ASME Power Boiler Code*	(IP518) *ASA Code For Pressure Piping*	(AWS) *Fusion Welding & Gas Cutting In Building Construction*
		(IPU-68) *ASME Code For Unfired Pressure Vessels*	(IPU-69) *ASME Code For Unfired Pressure Vessels*	(IPU-70) *ASME Code For Unfired Pressure Vessels*

American Welding Society

NOTE: Above Values are Ultimate. Use a factor of safety. Values are for the "As Welded" or "Stress-Relieved" Condition.

needed to understand the specifications for the various test specimens, their preparation, and method of testing.

In general, all codes require essentially the same qualifying procedures for plate or pipe. Each position of welding, type of joint, and weld has a designated number of identification. Study Table 12-4 for an understanding of test requirements for the various positions of welding for plate and pipe. Generally face-, root-, and side-bend test specimens are required for groove welds in plate or pipe, and T-joint break or macro-etch test specimens are required for fillet welds in plate. The required number and type of test specimens vary with the thickness of material.

Under most welding codes, the tests remain in effect indefinitely unless

- The welder does not work with the welding process for which he is qualified for a period of more than 6 months. Requalification is required only on $\frac{3}{8}$ inch thick plate.
- There is reason to be dissatisfied with the work of the welder. If a welder fails his test, he may retest as follows.
  - An immediate retest consists of two test welds of each type he failed. All test specimens must pass.
  - A complete retest may be made if the welder has had further training or practice since his last test.

The AWS code for piping provides for three levels of qualifications termed Acceptance Requirements 1, 2, and 3 (abbreviated AR-1, AR-2, and AR-3 respectively). Typical applications are as follows.

- *AR-1* is for the highest weld quality applications such as weldments used for nuclear energy and space exploration and in the fabrication of high-pressure, high-temperature chemical and gas systems.
- *AR-2* is for high weld quality applications such as those required for pipelines and most pressure piping systems.
- *AR-3* is for a nominal degree of weld quality as required for domestic water and heating pipelines.

PROCEDURE QUALIFICATION TESTS. These tests are conducted for the purpose of determining the correctness of the welding method. The American Welding Society and various code authorities have established standard procedures for welding various types of structures. The welding procedure test should cover base metal specifications, filler metals, joint preparation, position of welding, and welding process, techniques and characteristics, including current setting, electrode size and electrode manipulations, preheat, interpass and postheat temperatures. The close adherence to the procedures established will insure uniform results.

Following are the procedure qualification requirements for different types of welds.

Groove Welds. The specimens required for groove welds in $\frac{3}{8}$-inch plate or less and up to, but not including, $\frac{3}{4}$-inch plate are 2 tension, 2 face-bend, and 2 root-bend test coupons. For plate thicker than $\frac{3}{4}$ inch, 2 tension specimens and 4 side-bend specimens are required, Fig. 12-25. In most cases groove weld tests qualify the welding procedure for use with both groove and fillet welds within the range tests.

Fillet Welds. Longitudinal shear test specimens are welded as shown in Fig. 12-26 and prepared for testing as shown in Fig. 12-27. Transverse shear test

12-25. *Test specimens for ASME procedure qualification: 2 tensile, 2 face-, and 2 root-bend specimens. Note that the tensile specimens are pulled until broken.*

Nooter Corp.

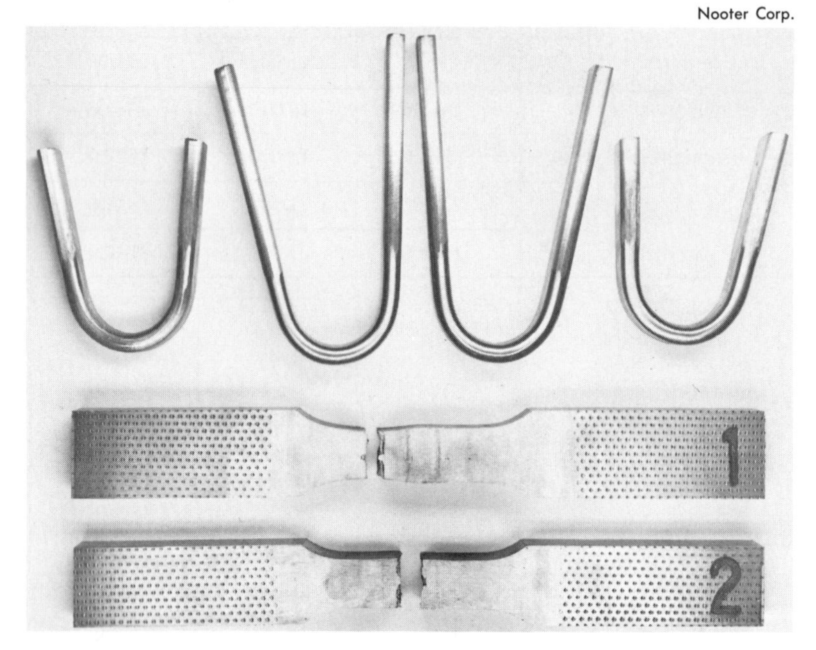

specimens are made as shown in Fig. 12-28.

The test weld for the free-bend and soundness tests is made as shown in Fig. 12-29. Two free-bend test specimens are taken from the test weld which are prepared for testing as shown in Fig. 12-30 for plate and Fig. 12-31 for pipe. Two specimens for the fillet weld soundness test are also taken from the test weld and prepared as shown in Fig. 12-32. Another form of fillet weld test is that required by the ASME code for procedure qualification. This is shown in Fig. 12-33 (Page 378). The specimen is cut as indicated and etched.

OPERATOR QUALIFICATION TESTS (PERFORMANCE). This is also referred to as a Performance Qualification Test. The test is conducted for the purpose of determining whether the operator has the knowledge and skill to make sound welds and to follow and apply a procedure of welding as determined for the class of work at hand. Most of the test specimens are similar to those used for the procedure qualification test. However, not as many are required, and the preparation of the test specimen is simplified. The cost of testing is reduced and the determination of weld soundness is achieved.

It is the general practice for code welding to qualify operators on the groove weld tests. Passing of these tests also permits the operator to weld fillet welds. Following are the operator qualification requirements for different types of welds.

Groove Welds. One face-bend specimen and one root-bend specimen are required for material $\frac{3}{8}$ inch or less and up to, but not including, $\frac{3}{4}$ inch. For material $\frac{3}{4}$ inch and over, 2 side-bend specimens are required. (Face- and root-bend specimens may be substituted for the side-bend specimens.)

Fillet Welds. There are three different kinds of fillet weld test specimens to serve a variety of purposes.

Many companies use the fillet weld test procedure as repre-

Table 12-4

General qualifying requirements for plate and pipe welding.

Material	Type of weld	Position test specimen	Position number designation[1]	Qualifying position[2]	
				Grooves	Fillets
Plate	Groove	Flat	1 G	F	F, H
Plate	Groove	Horizontal	2 G	F, H	F, H
Plate	Groove	Vertical	3 G	F, H, V	F, H, V
Plate	Groove	Overhead	4 G	F, OH	F, H, OV
Plate	Groove	Vertical & overhead	3 G & 4 G	—	All
Plate	Fillet	Flat	1 F	—	F
Plate	Fillet	Horizontal	2 F	—	F, H
Plate	Fillet	Vertical	3 F	—	F, H, V
Plate	Fillet	Overhead	4 F	—	F, H, OH
Plate	Fillet	Vertical & overhead	3 F & 4 F	—	All
Pipe	Groove	Horizontal rolled	1 G	F, rolled	F, H[3]
Pipe	Groove	Vertical fixed	2 G	F, H	F, H
Pipe	Groove	Horizontal fixed	5 G	F, V, OH	F, V, OH
Pipe	Groove	Inclined 45° fixed	6 G	All	All
Pipe	Groove	Vertical & Horizontal fixed	2 G & 5 G	All	All

American Welding Society

[1] Indicates position number identification and position of test unit for specific qualifications
[2] Indicates position of welding for which the operator is qualified
[3] Plate welding only

12-26. *Longitudinal fillet weld shearing specimen after welding.*

DIMENSIONS				
SIZE OF WELD F, IN.	$\frac{1}{8}$	$\frac{1}{4}$	$\frac{3}{8}$	$\frac{1}{2}$
THICKNESS t, IN., MIN.	$\frac{3}{8}$	$\frac{1}{2}$	$\frac{3}{4}$	1
THICKNESS T, IN., MIN.	$\frac{3}{8}$	$\frac{3}{4}$	1	$1\frac{1}{4}$
WIDTH W, IN.	3	3	3	$3\frac{1}{2}$

sented by Fig. 12-34. Welds are made in each position for which the operator is to be qualified. Two root-bend tests are made.

Another form of fillet weld test that may be used for welder operators who wish to qualify for the ASME code is shown in Figs. 12-35 and 12-36 (Page 380). Test specimens may be subjected to a fracture test and etched.

When some knowledge of the welder's ability for general welding and noncritical work is wanted, the fillet weld test shown in Fig. 12-37 is used. The weld is fractured as shown in Fig. 12-38 (Page 381).

PREPARATION OF TEST SPECIMENS. The material that follows will give the student the information needed to understand the nature of the various test specimens, their preparation, use, method of testing, and results required for passing. A word of caution to the student regarding the welding of a test specimen: While these tests are designed to determine the capability of welders, many welders have failed them for reasons not related to their welding ability. This is due principally to carelessness in the application of the weld and in the preparation of the test unit and the test specimen.

Selecting and Preparing Plates. It is important that both

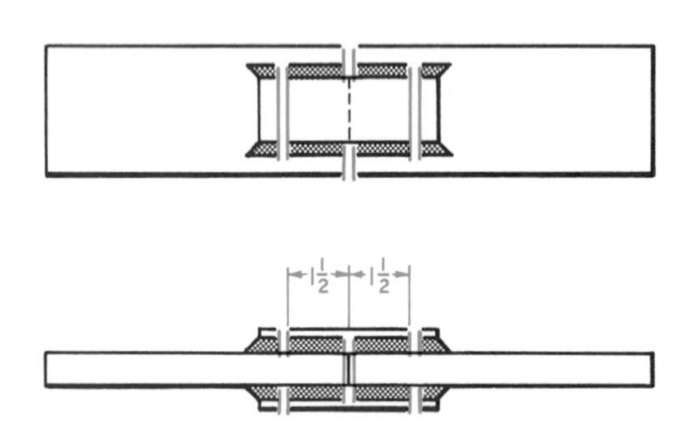

FOR OTHER DIMENSIONS SEE PRECEDING FIG.

12-27. *Longitudinal fillet weld shearing specimen after machining.*

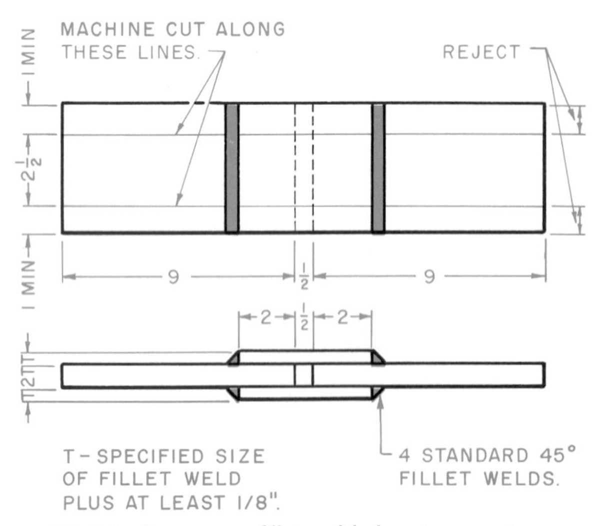

12-28. *Transverse fillet weld shearing specimen.*

FREE FACE BEND SPECIMEN
GUIDED ROOT BEND SPECIMEN (SOUNDNESS TEST)
FREE FACE BEND SPECIMEN
GUIDED ROOT BEND SPECIMEN (SOUNDNESS TEST)
DISCARD
1/4" ALLOWED FOR ALL CUTS
7 3/4
3/16
9/16
1 1/2
9/16
1 1/2
3/16
DISCARD
3/8
5
15/16
5
3/8
2
MAX. SIZE OF SINGLE PASS FILLET WELD (NOT OVER 3/8")
BACKING STRIP TO BE IN INTIMATE CONTACT WITH BASE MATERIAL.

12-29. *Test welds for free-bend and soundness tests.*

the test plate and backup strip (where it is used) are weldable, ductile low carbon steel. The test is designed so that both the plate and the weld bend and stretch during the test. If the plate has tensile strength much above that of the weld metal, it will not stretch at all and will force the weld metal to stretch far beyond its yield point and fail.

Welding The Plates. Proper electrode selection is the first step in producing a sound weld. Since plates are generally welded in all positions, they should be welded with an all-position electrode which has good ductility. Electrodes should be in the E6010, E6011, E6018 or E7018 classifications.

The most important part of the entire job is the first pass in the grooved butt weld and the root pass in the fillet test. Every effort should be made to get especially good penetration, fusion, and sound weld metal on these root beads. The welder should not let his desire to have a bead of good appearance interfere with the objective of securing good penetration and fusion. These weld characteristics are the difference between a satisfactory weld and failure.

No preheat or postheat treatment is permissible to pass the test. It is helpful to keep the plates fairly hot and allow them to cool slowly after the welding is completed. The welder should

DIMENSIONS										
t, IN.	1/4	3/8	1/2	5/8	3/4	1	1 1/4	1 1/2	2	2 1/2
W, IN.	3/8	9/16	3/4	15/16	1 1/8	1 1/2	1 7/8	2 1/4	3	3 3/4
L MIN, IN.	6	8	9	10	11	12	13 1/2	15	18	21
B*MIN, IN.	1 1/4	1 1/4	1 1/4	2	2	2	2	2	2	3

* SEE FIG. 47.
NOTE: THE LENGTH IS SUGGESTIVE ONLY, NOT MANDATORY.

IF FLAME CUT, NOT LESS THAN 1/8" SHALL BE MACHINED FROM EDGES.
WELD REINFORCEMENT SHALL BE MACHINED FLUSH WITH BASE METAL.
EDGE OF WIDEST FACE OF WELD.
R=0.1t MAX.
GAGE LINES
1/16 MIN.
1/8 MIN.
W=1 1/2 t
L/2
L
t

12-30. *Free-bend specimen (plate).*

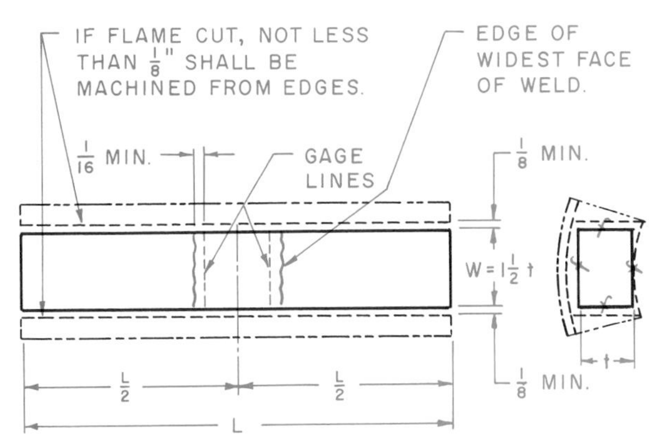
IF FLAME CUT, NOT LESS THAN 1/8" SHALL BE MACHINED FROM EDGES.
EDGE OF WIDEST FACE OF WELD.
1/16 MIN.
GAGE LINES
1/8 MIN.
W=1 1/2 t
t
L/2
L/2
L
1/8 MIN.
NOTE: FOR DIMENSIONS SEE PRECEDING FIGURE.

12-31. *Free-bend specimen (pipe).*

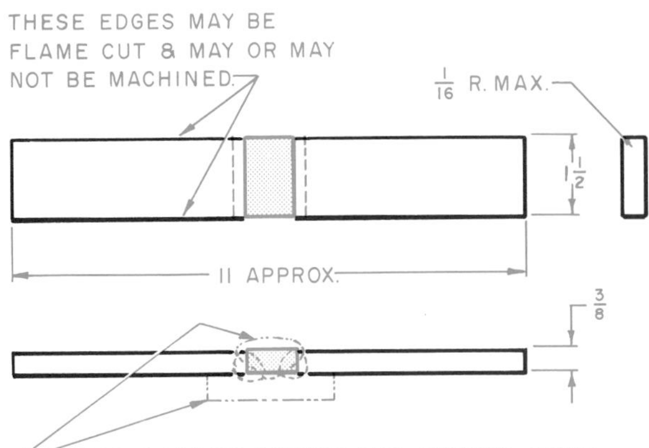
THESE EDGES MAY BE FLAME CUT & MAY OR MAY NOT BE MACHINED.
1/16 R. MAX.
1 1/2
11 APPROX.
3/8
WELD REINFORCEMENT AND BACKING STRIP SHALL BE REMOVED FLUSH WITH BASE METAL. FLAME CUTTING MAY BE USED FOR THE REMOVAL OF THE MAJOR PART OF THE BACKING STRIP, PROVIDED AT LEAST 1/8" OF ITS THICKNESS IS LEFT TO BE REMOVED BY MACHINING OR GRINDING.

12-32. *Fillet weld soundness test specimen.*

not, under any conditions, quench the plates in cold water or in any other way accelerate the cooling rate.

Finishing the Specimen. It is important that care be taken in finishing the specimen because poor finishing can cause a sound weld to fail the test. As shown in Fig. 12-39 (Page 382), all grinding and machining marks must be lengthwise on the sample. Otherwise they produce a notch effect which may cause failure. The smoother the finish, the better the chance of passing the test. Even a slight nick across the sample may open up under the severe stress of the test. Unless there is a distinct lack of penetration, the surface should be ground or machined until the entire bend area cleans up, leaving no low or irregular spots.

Any bead reinforcement either on the face or on the root side must be removed, Fig. 12-40. The edges of the specimen should have a smooth $1/16$ inch radius, which can be done with a file. When grinding specimens, do not quench them in water when hot. This may create small surface cracks that become larger during the bend test.

After the test specimens have been bent, the convex surface should be examined for surface cracks, breaks, and other open defects. Any specimen in which a crack or other open defect is present after bending, exceeding $1/8$ inch measured in any direction, has failed. Cracks occurring in the corners of the specimen during testing are not considered.

Groove Weld Soundness Tests

REDUCED-SECTION TENSION TEST. Purpose. The reduced-section tension test determines the tensile strength of the weld metal.

It is used only for procedure qualification tests, but the test procedure gives spectacular results and can be used to test welders in school. The pulling and breaking of the specimen is of great interest to students.

The test is suitable for a groove butt joint in plate or a rolling butt joint in pipe having a nominal diameter exceeding 2 inches. It is recommended for wall thicknesses $3/8$ inch or more.

Usual Size and Shape of Specimens. Refer to Figs. 12-41 through 12-43 (Pages 382, 383).

Method of Testing. The test is made by subjecting the specimen to a longitudinal load great enough to break it or pull it apart, Fig. 12-25, page 374. This is usually accomplished by means of a tensile-testing machine designed for this purpose, Fig. 12-44 (Page 383). Before testing, the least width and corresponding thickness of the reduced section is measured in inches. The specimen is ruptured under tensile load and the maximum load in pounds is determined. The cross-sectional area is obtained as follows:

12-33. *Fillet weld soundness test for procedure qualification.*

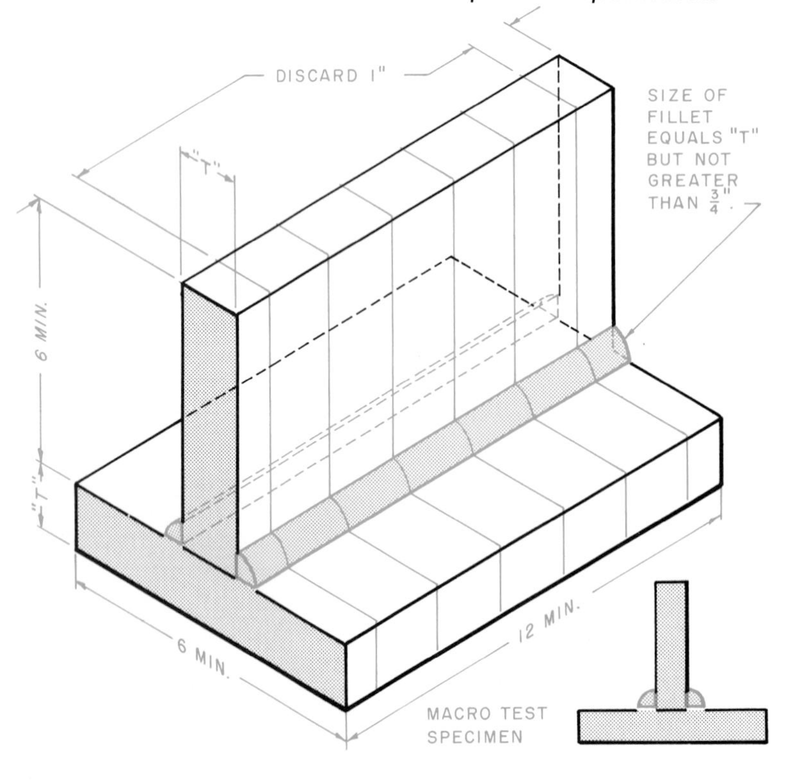

"T" MAXIMUM THICKNESS OF BASE MATERIAL IN THE VESSEL AT POINT OF WELDING OR 1", WHICHEVER IS SMALLER.

MACRO TEST: THE FILLET SHALL SHOW FUSION AT THE ROOT OF THE WELD BUT NOT NECESSARILY BEYOND THE ROOT. THE WELD METAL AND HEAT AFFECTED ZONE SHALL BE FREE OF CRACKS. BOTH LEGS OF THE FILLET SHALL BE EQUAL TO WITHIN $\frac{1}{8}$".

Cross-sectional area
= width × thickness

The tensile strength in pounds per square inch is obtained by dividing the maximum load by the cross-sectional area.

On jobs for which a quick test is desired and for which the testing agent is interested only in determining whether the joint is stronger than the plate, the weld reinforcement is not removed and the specimen is pulled until failure occurs. No calculations are made when this type of test is done.

Usual Test Results Required. The specimen shall have a tensile strength equal to or greater than:

- The minimum specified tensile strength of the base material.
- The lower of minimum specified tensile strengths of dissimilar materials.
- The specified tensile strength of the weld metal if the weld metal is of a lower strength than the base metal.
- Five percent below the specified minimum tensile strength of the base metal if the specimen breaks in the base metal outside the weld.

FREE-BEND TESTS. Purpose. Free-bend tests are for the purpose of determining the ductility of the weld metal. They are used only for procedure qualification tests. The free-bend test is applied to groove welds of butt joints in plate and pipe.

Usual Size and Shape of Specimens. Refer to Figs. 12-30 and 12-31, page 377.

Method of Testing. The gauge lines indicated in Figs. 12-30 and 12-31 are scribed lightly on the face of the weld. The gauge length (the distance between the gauge lines) is approximately $\frac{1}{8}$ inch less than the width of the face of the weld. It is measured in inches to the nearest 0.01 inch.

For single-groove welds, the gauge lines are on the face of the weld. For double-groove welds, the gauge lines on one-half of the specimen are on one face of the weld; and on the other half of the specimen, on the other face.

Each specimen may be bent initially in a fixture complying with the requirements of Fig. 12-45 (Page 383). The surface of the specimen containing the gauge lines is directed toward the supports with the face of the weld down. After the initial bend has been made, the specimen is placed in the vise, press, or testing machine and the bending is continued until a crack or open defect exceeding $\frac{1}{16}$ inch in any direction appears on the convex surface of the specimen. If no crack appears, the specimen is bent double. Cracks less than $\frac{1}{16}$ inch occurring on the corners of the specimen during testing are not considered. A device like that shown in Fig. 12-46 (Page 384) is recommended. It prevents the ends of the specimen from slipping as it is bent. Personal injury or property damage may occur if the specimen slips. The minimum distance between the gauge lines is measured along the convex surface of the weld to the nearest 0.01 inch. The elongation is determined by subtracting the initial gauge length. The percentage of elongation is obtained by dividing the elongation by the initial gauge and multiplying by 100. The measurements can be

12-34. *Fillet weld soundness test for operator qualification only.*

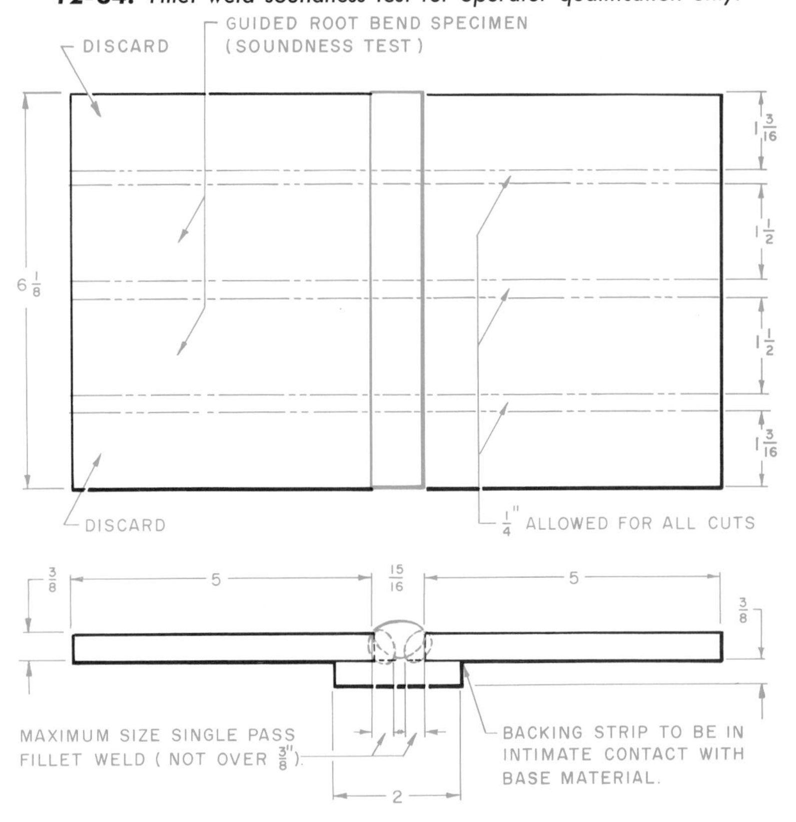

taken with a flexible scale graduated in hundredths of an inch.

Usual Test Results Required. The elongation cannot be less than 30 percent for stress-relieved welds nor less than 25 percent for nonstress-relieved welds. On some classes of work the elongation may be less.

ROOT-, FACE-, AND SIDE-BEND SOUNDNESS TESTS. Purpose. These tests are for the purpose of revealing lack of soundness, penetration, and fusion in the weld metal. They are procedure and operator qualification tests which are applied to groove welds in both plate and pipe. The face-bend test checks the quality of fusion to the side walls and the face of the weld joint, porosity, slag inclusion, gas pockets, and other defects. It also measures the ductility of the weld. The root-bend test checks the penetration and fusion throughout the root of the joint. The side-bend test also checks for soundness and fusion.

Usual Size and Shape of Specimens. Refer to Figs. 12-47 through 12-49 (Pages 384, 385).

Method of Testing. The tests are rather severe, and the ability to make welds that can meet these test requirements is a mark of good craftsmanship. These tests lend themselves to school and shop applications. The testing equipment needed is inexpensive and can be built in the school shop. See Figs. 12-50 through 12-52 (Pages 386, 387) for three designs that students find easy to work with in testing their weld specimens.

Each specimen is bent in a jig having the contour and other features shown in Fig. 12-50. You will note that there is a size for plate and pipe. Any convenient means—manual, mechanical, electrical, or hydraulic—may be used for moving the male member in relation to the female member.

The specimen is placed on the female member of the jig with the weld at midspan. Face-bend specimens are placed with the face of the weld directed toward the gap. Root-bend specimens are placed with the root of the weld directed toward the gap. Side-bend specimens are placed with the side showing the greater defects, if any, directed toward the gap. The two members of the jig are forced together until the curvature of the specimen is such that a $^{1}/_{32}$ inch diameter wire cannot be passed between the curved portion of the male member and the specimen. Then the specimen is removed from the jig.

Usual Test Results Required. The convex surface of the specimen is examined for the appearance of cracks or other defects. Any specimen in which a crack or other defect exceeding $^{1}/_{16}$ to $^{1}/_{8}$ inch, measured in any direction, is present after bending fails the test. Cracks occurring on the corners of the specimen are not considered as failing unless they exceed $^{1}/_{16}$ to $^{1}/_{8}$ inch and show of evidence of slag inclusion or internal defects.

Consult Figs. 12-53 and 12-54 (Page 387) for a comparison of

12-35. *Fillet weld specimen for operator qualification test.*

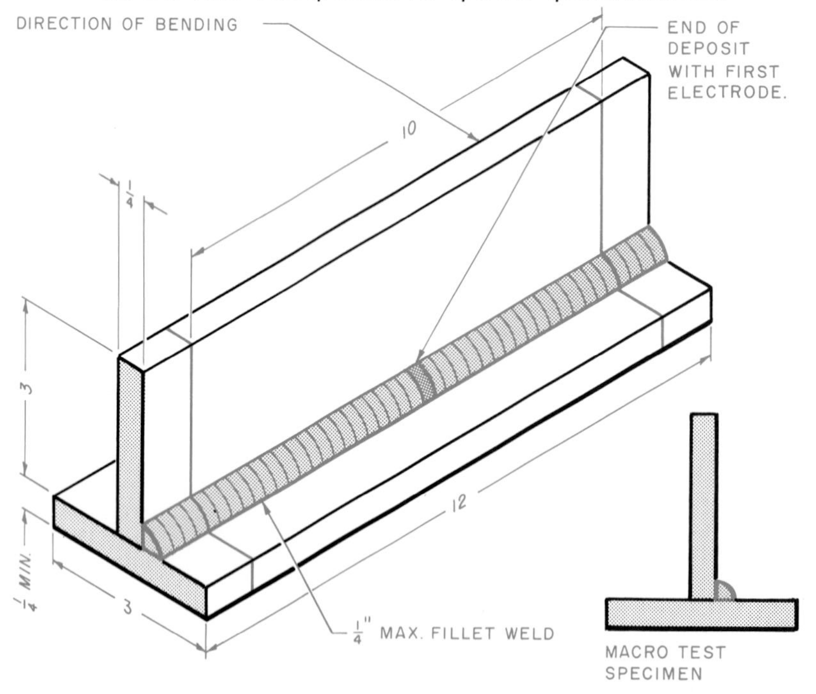

FILLET WELD SOUNDNESS TEST FOR PERFORMANCE QUALIFICATION OF WELDERS.

FRACTURE TEST: MAXIMUM PERMISSIBLE DEFECTS SUCH AS SLAG, NON-FUSION, ETC. – 20 PER CENT OR 2 INCHES. EVIDENCE OF CRACKING OR FILLET SHALL CONSTITUTE GROUNDS FOR REJECTION.

MACRO TEST: THE FILLET SHALL SHOW FUSION TO THE ROOT OF THE WELD BUT NOT NECESSARILY BEYOND THE ROOT. CONVEXITY AND/OR CONCAVITY OF THE FILLET SHALL NOT EXCEED $\frac{1}{16}$ INCH. BOTH LEGS OF THE FILLET SHALL BE EQUAL TO WITHIN $\frac{1}{16}$ INCH.

satisfactory and defective specimens.

NICK-BREAK TEST. Purpose. The nick-break test is for the purpose of determining the soundness of the weld. At one time it was used rather widely. Today, however, very few people are qualified to evaluate the grain structure of the cross section of the fractured weld metal. Thus this test is not as reliable as the others.

Usual Size and Shape of Specimens. Refer to Figs. 12-55 and 12-56 (Page 388).

Method of Testing. The weld reinforcement is not removed from the specimen to be tested. The specimen is notched in the sides by cutting it with a saw. Then it is supported and struck with quick, sharp blows by a hammer or heavy weight, Fig. 12-57 (Page 388). This causes a failure to occur between the saw

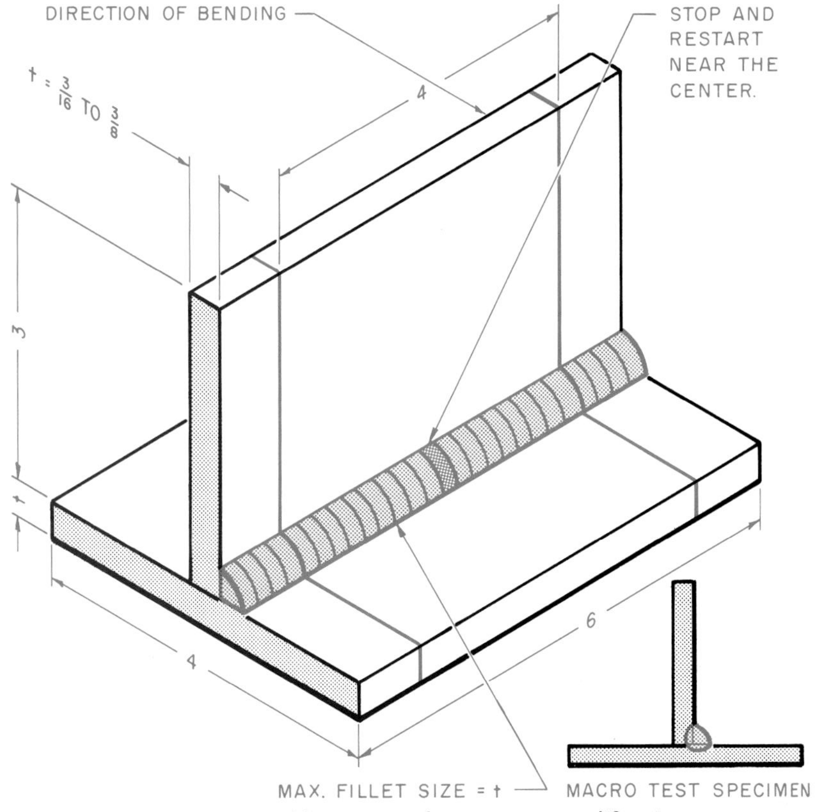

12-36. Fillet weld specimen for operator qualification test.

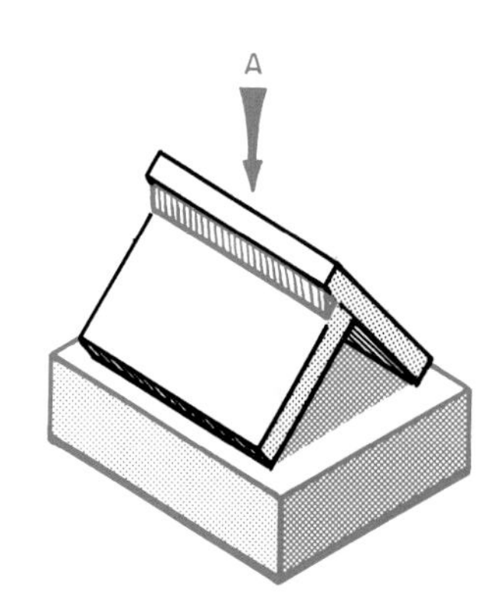

12-38. *Method of rupturing fillet weld break specimen.*

12-37. *Fillet weld break specimen.*

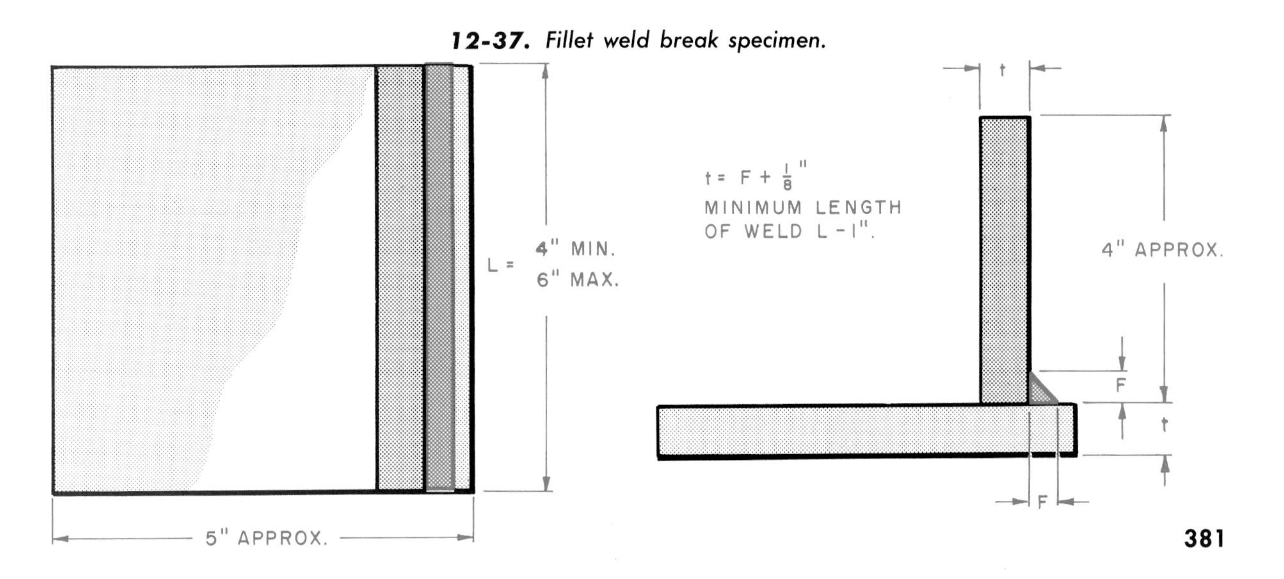

cuts. The weld metal is examined for defects such as slag and oxide inclusions, gas pockets, and lack of fusion. Fig. 12-58 (Page 388).

Usual Test Results Required. The requirements for passing this test are that the fracture shall disclose a degree of porosity not to exceed six gas pockets per square inch of weld area. The maximum dimension of any such pockets shall not be larger than $\frac{1}{16}$ inch in any direction. Further, the examination must show complete fusion for the entire weld, which must be free of slag inclusions.

Fillet Weld Soundness Tests

LONGITUDINAL AND TRANSVERSE SHEAR TESTS. Purpose. These tests determine the shearing strength of fillet welds. They are ordinarily used for procedure qualifications.

Usual Size and Shape of Specimens. Refer to Figs. 12-26, 12-27, and 12-28, pages 375, 376, for standard specifications. Figures 12-59 and 12-60 (Page 388) show the prepared longitudinal weld specimens.

Method of Testing. The specimen is ruptured by pulling in a tensile testing machine, and the

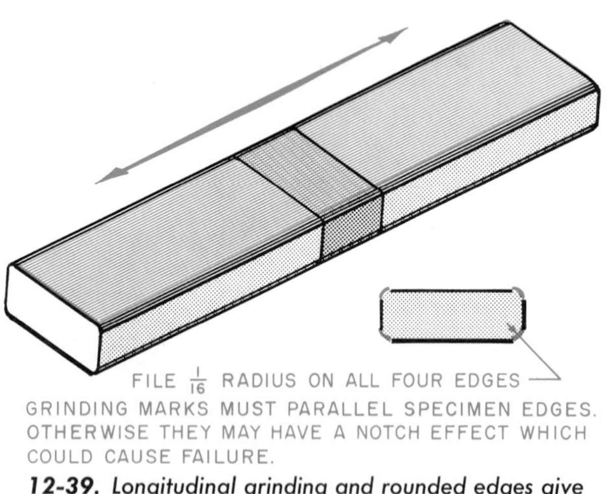

12-39. Longitudinal grinding and rounded edges give the welder a fair chance and reduce failures due to causes beyond his control.

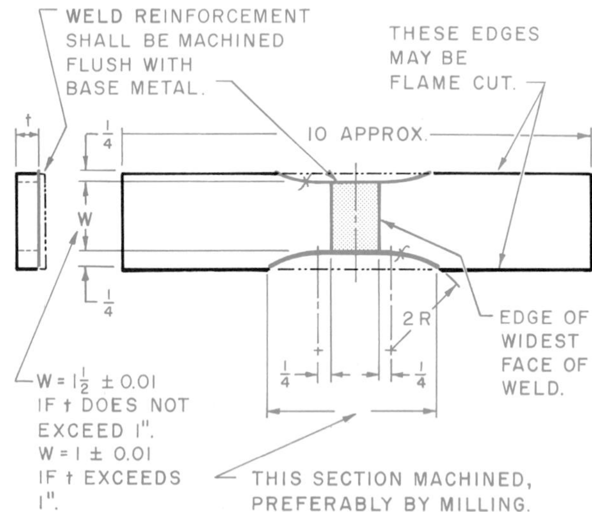

12-41. Tension specimen (plate).

12-40. AWS specifies that bead reinforcement be removed by grinding or machining. The drawing shows why.

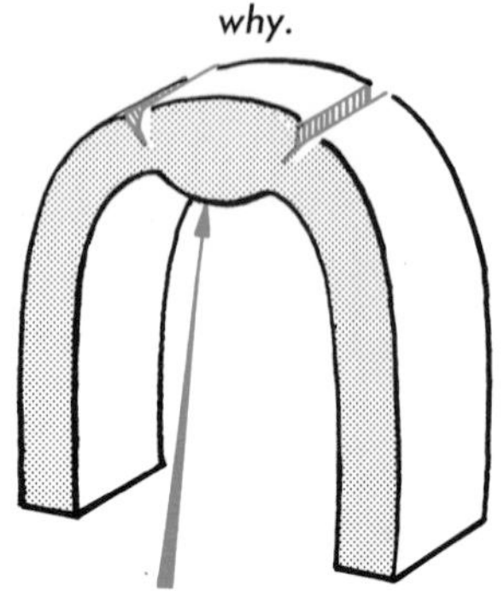

IF WELD REINFORCEMENT IS NOT REMOVED BEFORE ROOT BEND TEST, STRETCHING IS CONCENTRATED IN TWO PLACES AND FAILURE RESULTS.

12-42. Tension specimen (pipe) exceeding 2 inches in diameter.

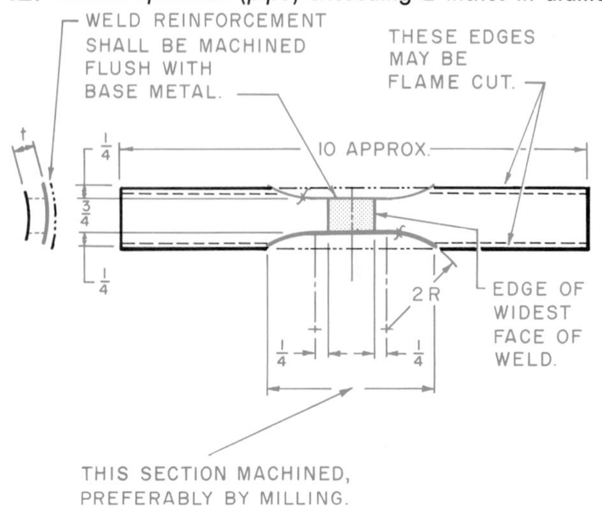

maximum load in pounds is determined, Fig. 12-61 (Page 389).

a. *Transverse Welds.* The shearing strength of transverse fillet welds is determined in the following manner. Before pulling the specimen, the width of the specimen is measured in inches, and the size of the fillet welds is recorded. The shearing strength of the welds in pounds per *linear* inch is obtained by dividing the maximum force by twice the width of the specimen.

Shearing strength (lb./in.)

$$= \frac{\text{maximum load}}{2 \times \text{width of specimen}}$$

The shearing strength of the welds in pounds per *square* inch is obtained by dividing the shearing strength in pounds per linear inch by the average theoretical throat dimension of the welds in inches. The theoretical throat dimension may be obtained by multiplying the size of the fillet weld by 0.707.

Shearing strength (p.s.i.)

$$= \frac{\text{shearing strength (lb./in.)}}{\text{throat dimension of weld}}$$

12-43. *Tension specimen (pipe) not exceeding 2 inches in diameter.*

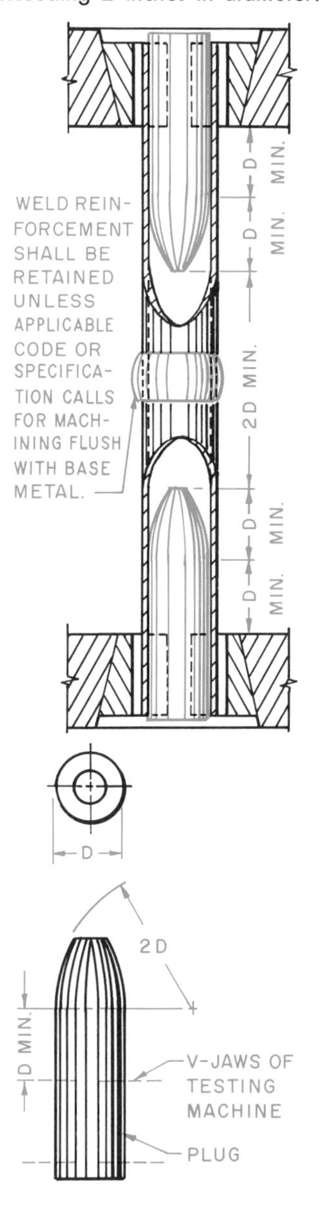
WELD REINFORCEMENT SHALL BE RETAINED UNLESS APPLICABLE CODE OR SPECIFICATION CALLS FOR MACHINING FLUSH WITH BASE METAL.
D
MIN.
2D MIN.
2D
D MIN.
V-JAWS OF TESTING MACHINE
PLUG

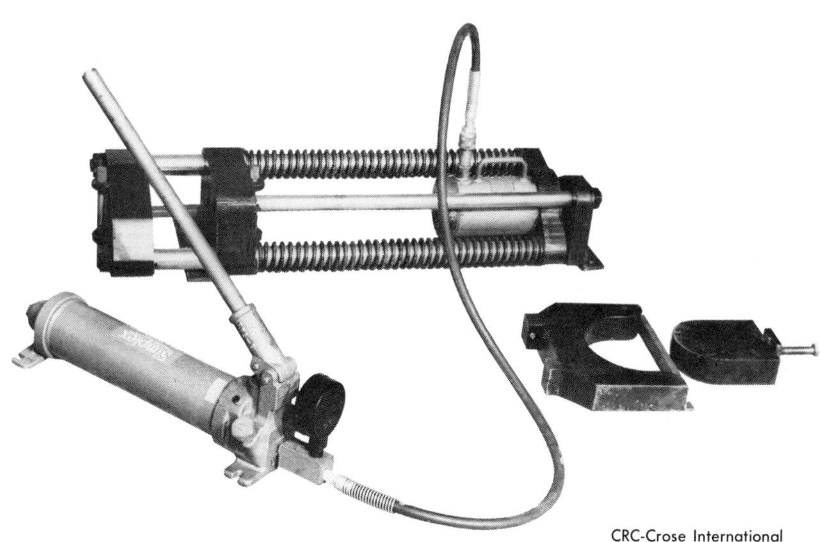

CRC-Crose International

12-44. *Tensile and guided-bend testing machine. It is portable and has a range of 60,000 to 225,000 p.s.i.. A large, direct-reading dial gauge shows the amount of pressure applied.*

12-45. *Initial bend for free-bend specimens.*

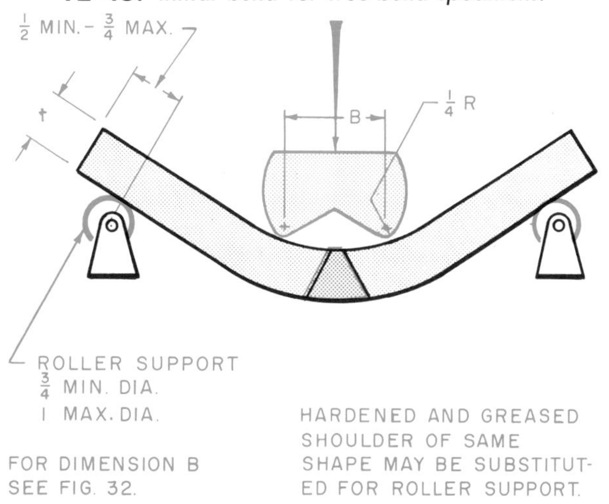
$\frac{1}{2}$ MIN. – $\frac{3}{4}$ MAX.
t
B
$\frac{1}{4}$ R
ROLLER SUPPORT
$\frac{3}{4}$ MIN. DIA.
1 MAX. DIA.
FOR DIMENSION B SEE FIG. 32.
HARDENED AND GREASED SHOULDER OF SAME SHAPE MAY BE SUBSTITUTED FOR ROLLER SUPPORT.

b. *Longitudinal Welds.* The shearing strength of longitudinal fillet welds, Figs. 12-59 and 12-60, is determined in the following manner. Before pulling the specimens, the length of each weld is measured in inches. The shearing strength of the welds in pounds per *linear* inch is obtained by dividing the maximum force by the sum of the lengths of the welds which ruptured.

$$\text{Shearing strength (lb./in.)} = \frac{\text{maximum load}}{\text{sum of lengths}}$$

Usual Test Results Required. For the longitudinal shear test specimen, the shearing strength of the welds in pounds per square inch cannot be less than two-thirds of the minimum specified tensile range of the base material. For the transverse shear test specimens, the shearing strength of the welds in pounds per square inch cannot be less than seven-eighths of the minimum specified tensile range of the base material.

FILLET WELD GUIDED-BEND TEST. Purpose. This test reveals lack of soundness in the weld metal of fillet welds.

Usual Size and Shape of Specimens. Refer to Fig. 12-32, page 377.

Method of Testing. Each specimen is bent in a jig having the contour and other features shown in Fig. 12-50. Any convenient means may be used for moving the male member with relation to the female member. The specimen is placed on the female member of the jig with the weld at midspan. The specimen is placed with the root of the

12-46. *Recommended device for the final bending for free-bend specimens.*

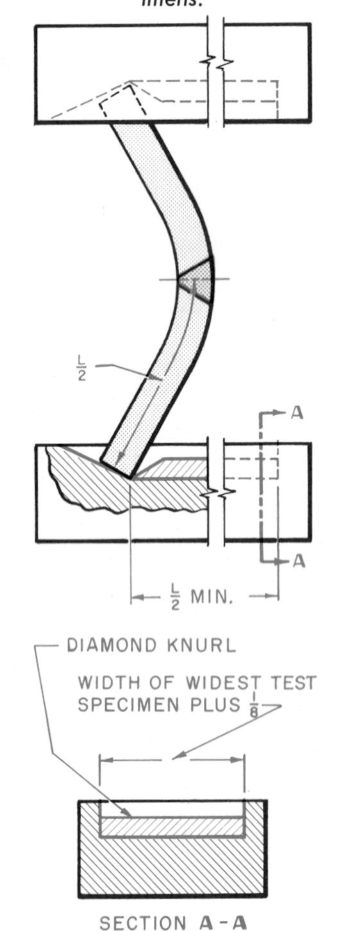

12-47. *Face- and root-bend specimens (plate).*

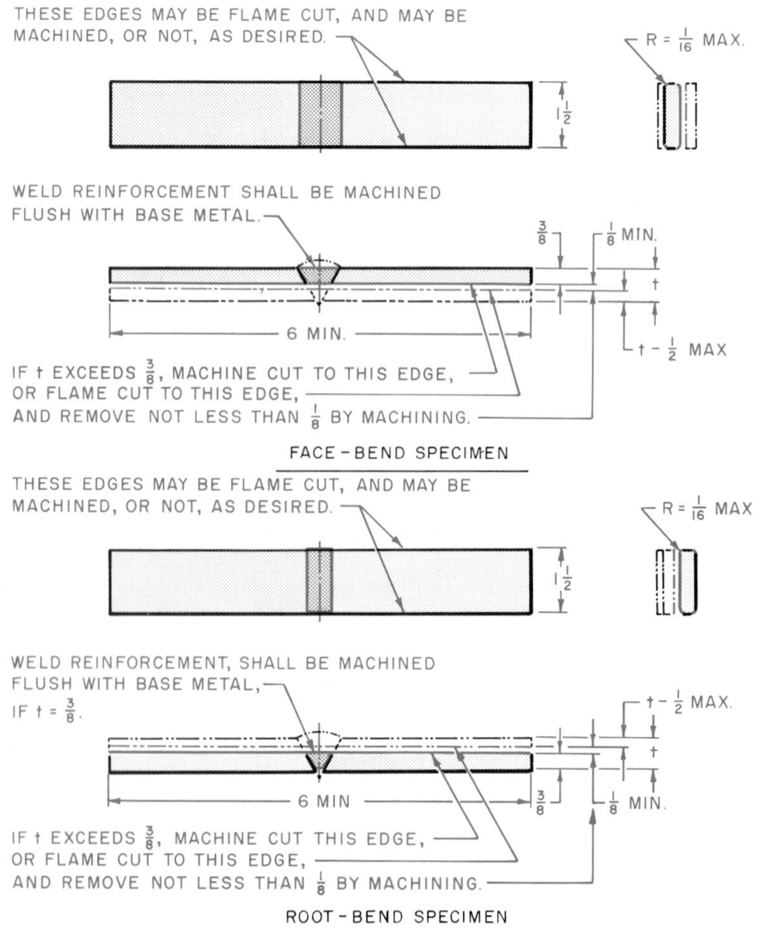

weld towards the gap. The two members of the jig are forced together until the curvature of the specimen is such that a $\frac{1}{32}$-inch diameter wire cannot be passed between the curved portion of the male member and the specimen. Then the specimen is removed from the jig.

Usual Test Results Required. The convex surface of the specimen is examined for the appearance of cracks or other defects. Any specimen in which a crack or other defect exceeding $\frac{1}{16}$ to $\frac{1}{8}$ inch measured in any direction is present after the bending fails the test. Cracks less than $\frac{1}{16}$ inch which occur on the corners of the specimen during testing are not considered.

FILLET-WELD BREAK TEST. Purpose. This test is for the purpose of determining the soundness of a fillet weld.

There are several different types of fillet weld tests. One is the form recommended by the American Welding Society. It is a quick, easy method of testing for operator qualification and is widely used, especially by schools and industries that are not doing code work.

The other forms are ASME tests for fillet welds, for both performance (operator) and procedure qualification. These are the tests which qualify an operator for code welding.

Usual Size and Shape of Specimen. Refer to Fig. 12-37, page 381, for the AWS test for performance qualification. The ASME test forms are shown in Fig. 12-33, page 378 for procedure qualification on ferrous metals, in Fig. 12-35, page 380, for performance (operator) qualification for ferrous metals, and in Fig. 12-36, page 381, for performance (operator) qualification for nonferrous metals.

Method of Testing (AWS form, Fig. 12-37). The specimen is formed by fillet welding one flat plate or bar at right angles to another. The specimen is then fractured by application of pressure from a press, testing machine, or hammer. The fillet weld is fractured at the root, Fig. 12-38, page 381. The fractured weld metal is then examined for defects such as slag and oxide inclusions, gas pockets, lack of fusion, lack of root penetration, and uneven distribution of weld metal.

Usual Test Results Required (AWS). The weld must fracture through the throat of the weld, Fig. 12-62. It should have complete penetration to the root of the weld and no inclusions. The number of gas pockets cannot exceed six per square inch of weld area. The maximum dimension of any such pocket cannot be larger than $\frac{1}{16}$ inch.

Method of Testing (ASME forms, Figs. 12-33, 12-35, and 12-36).

a. *Fracture Test.* Load the center plate with the root of the

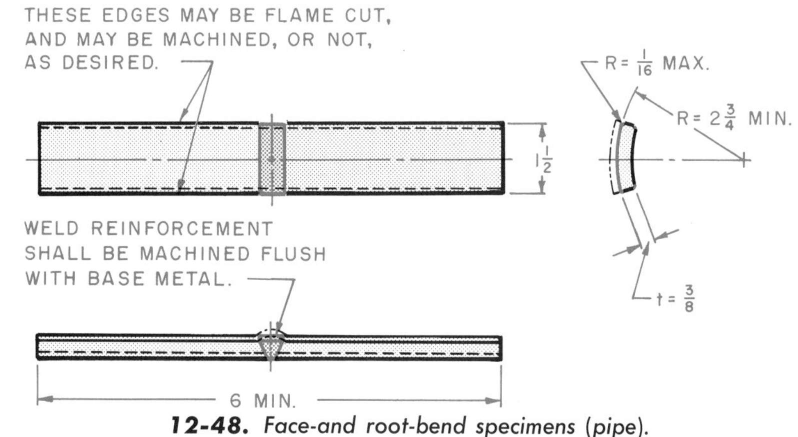

12-48. *Face-and root-bend specimens (pipe).*

12-49. *Side-bend specimen.*

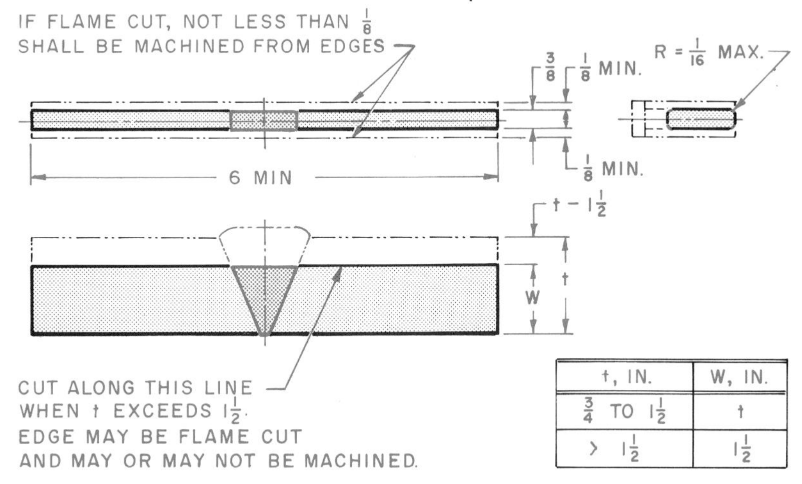

t, IN.	W, IN.
3/4 TO 1 1/2	t
> 1 1/2	1 1/2

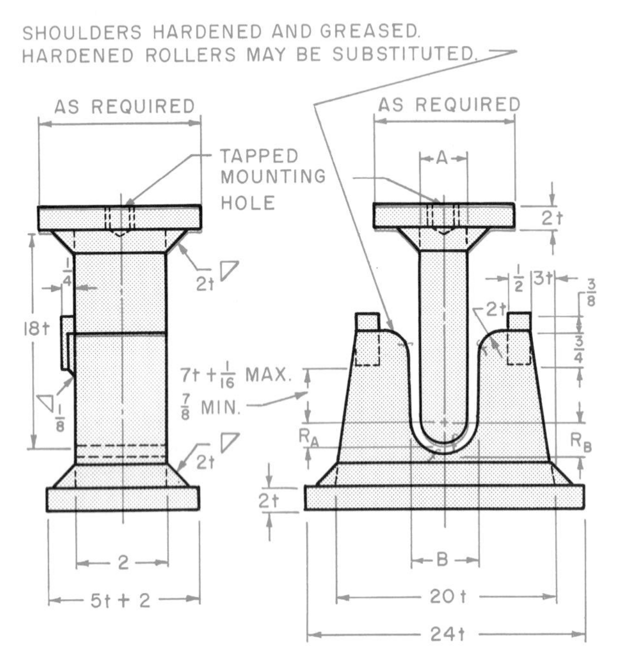

12-50. *Jig for guided-bend test used to qualify welders for work done under AWS and API specifications. It can be made in the school shop.*

weld in tension. Bend over flat or until it fractures.

b. *Macro Test* (*Etching*). Section the test weld so that the cross section of weld metal can be etched and examined.

Usual Test Results Required.

a. *Fracture Test.* Fracture specimens must be free from cracks and incomplete fusion at the root: slag inclusions and gas pockets must not measure more than a total of 20 percent or 2 inches. For the nonferrous specimen, slag inclusions must not measure more than $^{3}/_{4}$ inch.

b. *Macro Test.* Visually inspect etched cross section and reject any specimen that shows cracks and incomplete fusion at the root. The concave or convex face of the fillet weld must not exceed $^{1}/_{16}$ inch from a flat face

12-51. *A school-built guided-bend tester.*

The Lincoln Electric Co.

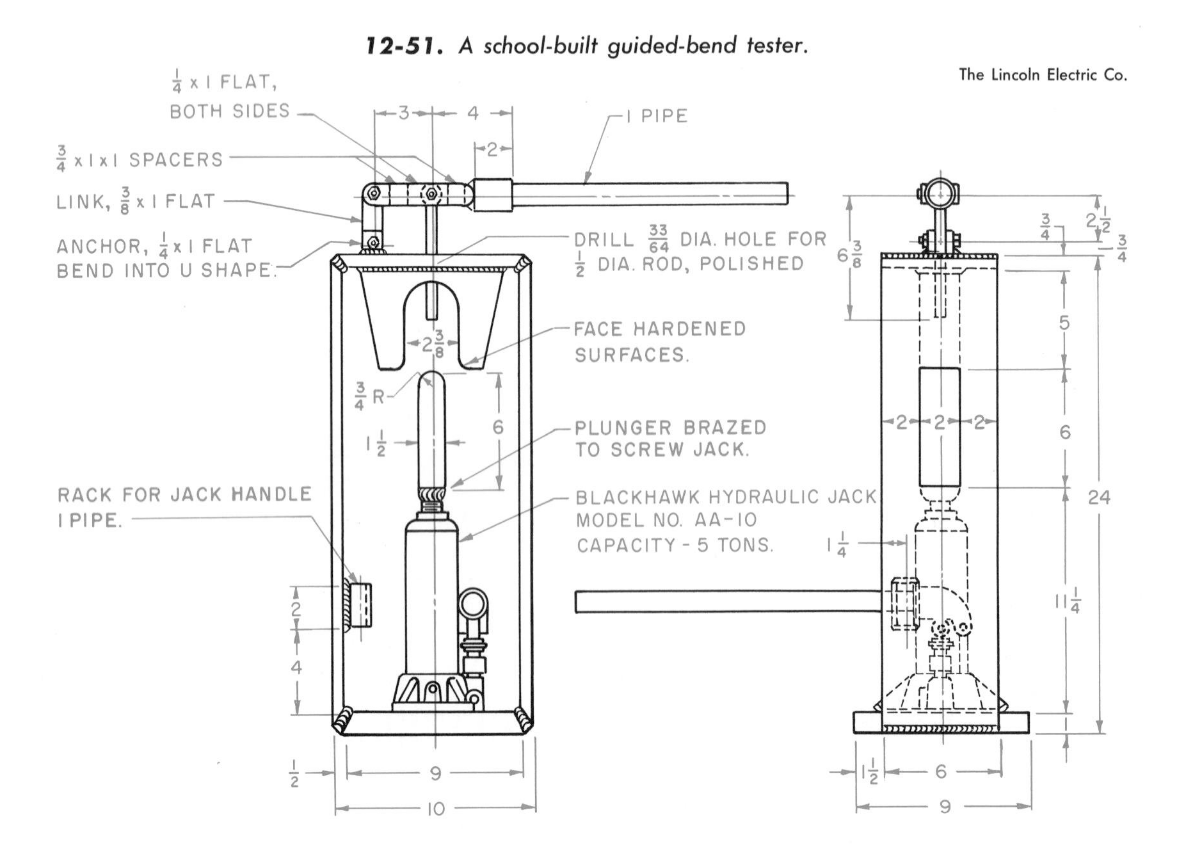

plane. Leg length of the fillet weld must be within $^1/_{16}$ inch of being equal for the performance specimen and $^1/_8$ inch if being equal for the procedure test specimen.

Other Tests

In addition to the destructive test outlined in the foregoing section, the welding industry makes use of such other methods of testing as etching, impact, fatigue, corrosion and specific gravity testing.

ETCHING. Often a weld must be sectioned and etched to be examined for defects. Etching reveals the penetration and soundness of a weld cross section, Fig. 12-63. The process has the following objectives:

- To determine the soundness of a weld
- To make visible the boundaries between the weld metal and the base metal and between the layers of weld metal
- To determine the location and the depth of penetration of the weld.
- To determine the location and number of weld passes
- To examine the metallurgical structure of the heat-affected zone

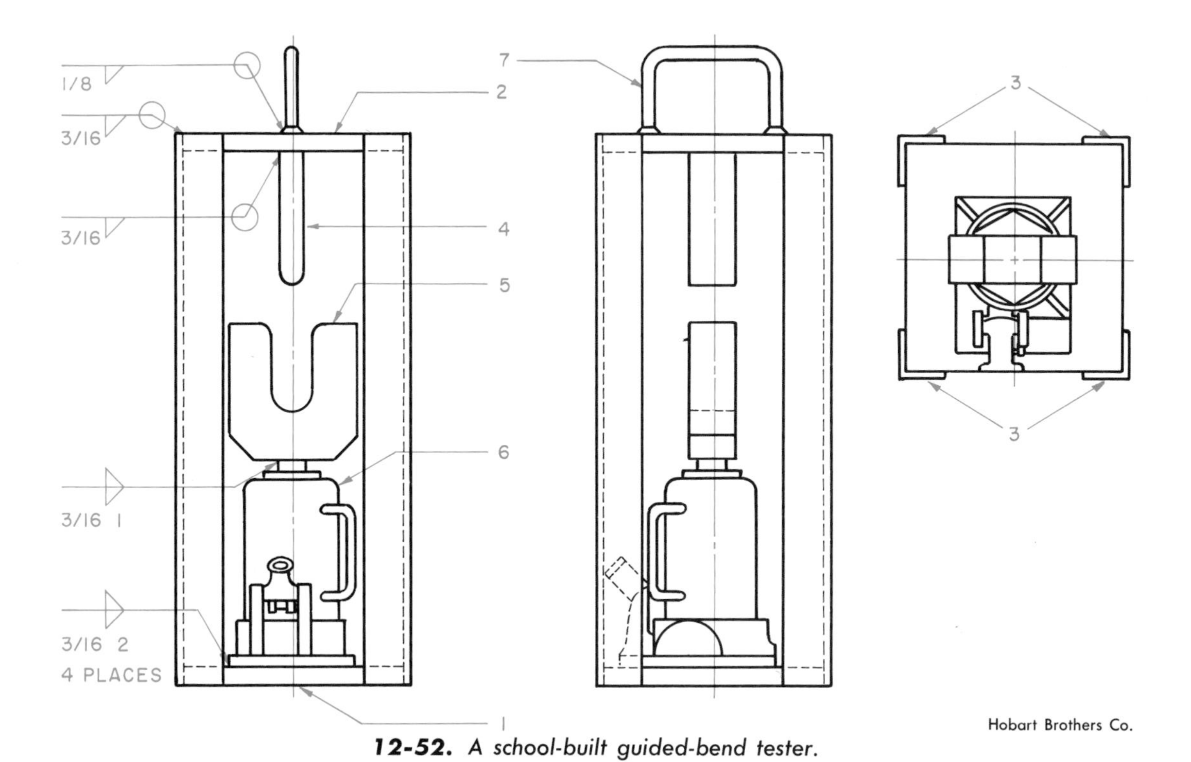

Hobart Brothers Co.

12-52. *A school-built guided-bend tester.*

12-53. *Face-bend specimens. The specimen on the right withstood the test satisfactorily whereas the specimen on the left cracked and failed before the bending could be completed.*

12-54. *Root-bend specimens. The specimen on the right withstood the test satisfactorily whereas the specimen on the left cracked and failed before the bending could be completed. Note the full penetration in the specimen that passed.*

A deep etch greatly exaggerates defects such as normally harmless small cracks or porosity. Therefore, the surface examination must be made as soon as the weld is clearly defined so that over-etching will not destroy the value of the sample. The surface may be preserved with a thin, clear lacquer. It may be inspected with a polarizing microscope, Fig. 12-64 or photo-

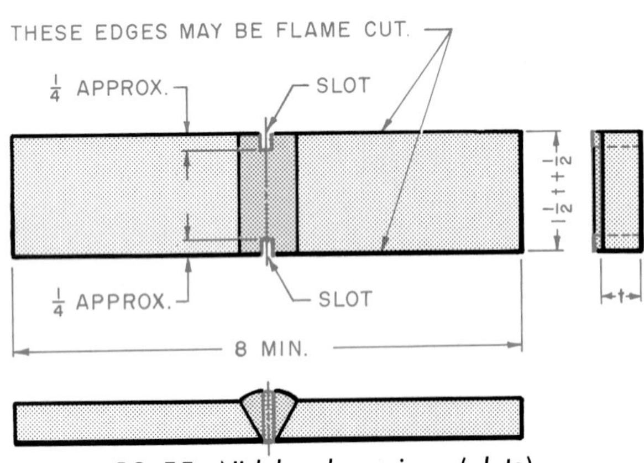

12-55. *Nick-break specimen (plate).*

12-58. *A nick-break specimen after it has been broken for inspection of the weld deposit. A sound weld is free of slag and oxide inclusions and gas porosity and shows complete fusion.*

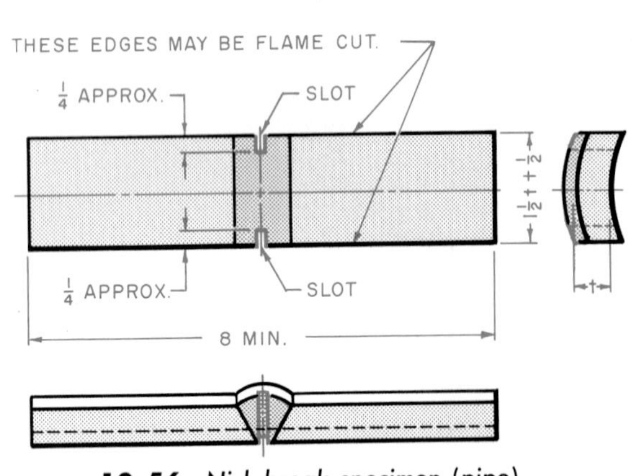

12-56. *Nick-break specimen (pipe).*

12-59. *Top side of a longitudinal fillet weld specimen.*

12-57. *Method of rupturing nick-break specimens.*

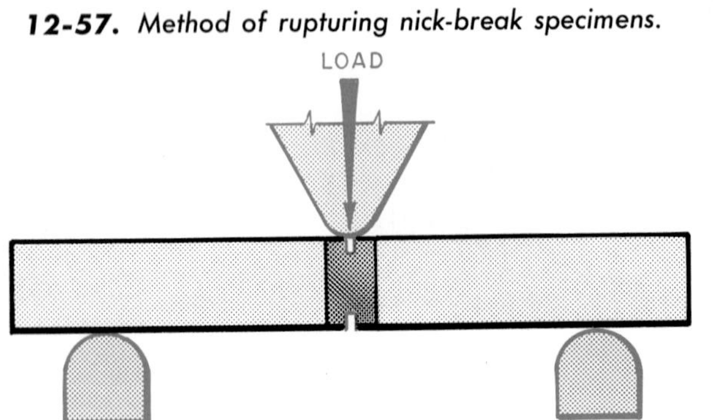

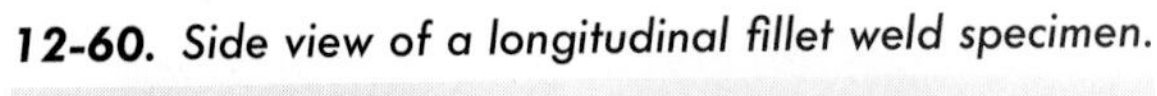

12-60. *Side view of a longitudinal fillet weld specimen.*

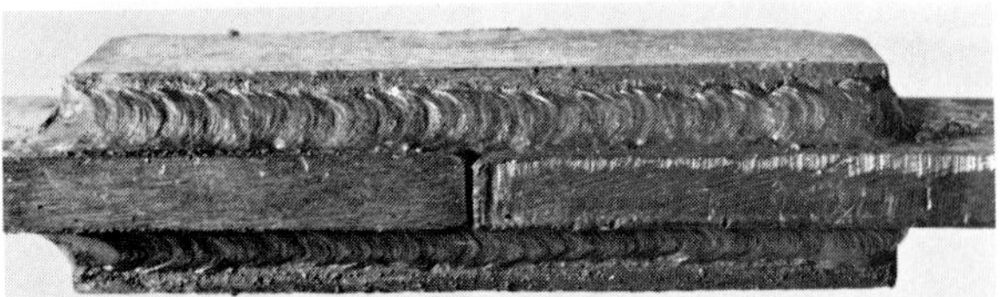

graphed with a metallograph, Fig. 12-65.

A transverse section is cut from the welded joint to be etched. Cutting with a fine tooth saw is preferred. The face of the weld and the base material should be filed to a smooth surface and then polished with fine emery cloth. The surface is exposed to one of the following etching solutions:

Iodine and Potassium Iodide. This solution is made by mixing 1 part of powdered iodine (solid) with 12 parts of a solution of potassium iodide by weight. The resulting solution should consist of 1 part of potassium iodide to 5 parts of water by weight.

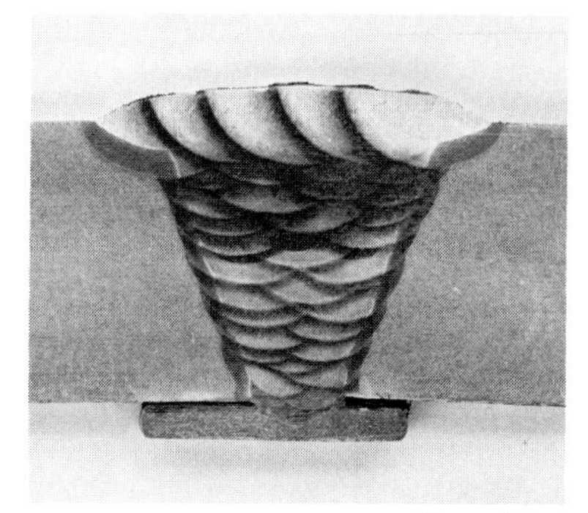

Nooter Corp.

12-63. *Etched chromium steel specimen showing multiple-pass groove welds in a single-V butt joint.*

12-61. *The bottom photograph shows the construction of a transverse fillet weld specimen. The top photograph shows a specimen after it has been ruptured by pulling. Note that the specimen withstood a tensile pull of 81,660 pounds.*

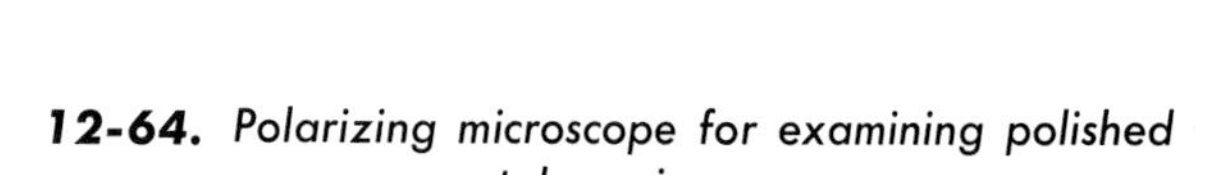

12-64. *Polarizing microscope for examining polished metal specimens.*

Unitron Instrument Co.

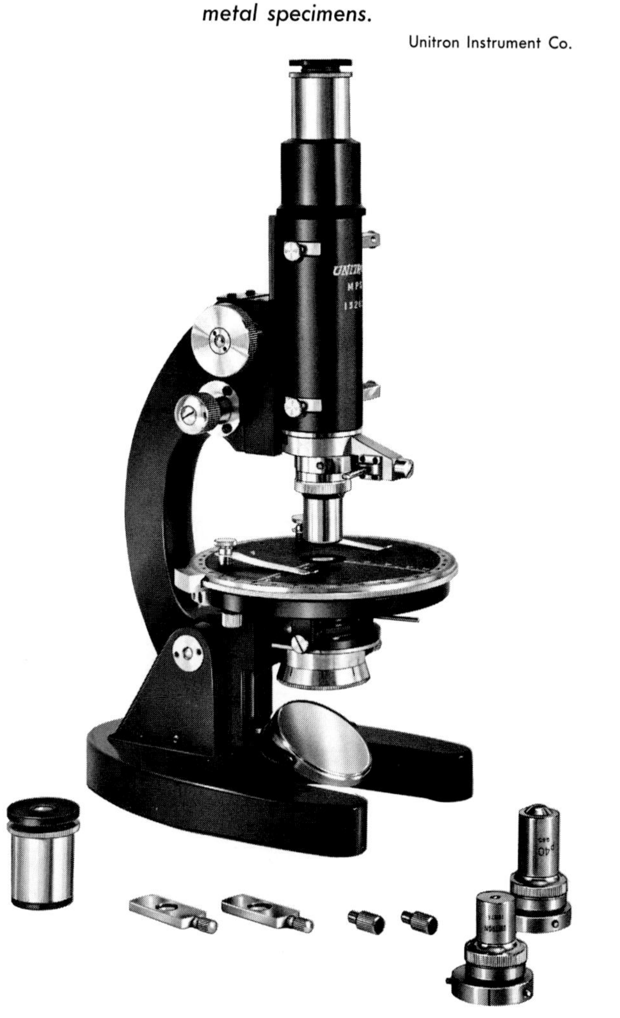

12-62. *A ruptured fillet weld break-test specimen. Note that the weld broke through the center. This indicates a weld of even proportion and penetration.*

Brush the surface of the polished weld with this reagent at room temperature.

Nitric Acid. This solution is made by mixing 1 part of concentrated nitric acid to 3 parts of water by volume. You must be cautious in the use of nitric acid because it causes bad stains and severe burns. *Always pour the acid into the water when mixing.*

Nitric acid solution may be applied to the surface of the polished weld with a glass stirring rod at room temperature, or the specimen may be immersed in the boiling solution. Make sure that the room is well-ventilated. Nitric acid etches rapidly.

After etching, the specimen should be immediately rinsed in clear hot water. Remove the excess water and immerse the etched surface in ethyl alcohol. Then remove and dry the specimen in a warm air blast. The appearance may be preserved by coating with a thin clear lacquer.

Hydrochloric Acid. This etching solution consists of equal parts by volume of concentrated hydrochloric acid, also called *muriatic acid,* and water. The weld specimen is immersed in the solution at a temperature near boiling. Both polished and unpolished surfaces will react to the acid treatment. The solution usually dissolves slag inclusions, thus causing cavities, and enlarges gas pockets.

You must be cautious in the use of any acid because it causes bad stain and severe burns. *Always pour the acid into the water when mixing.*

Ammonium Persulphate. A solution is made by mixing 1 part of solid ammonium persulphate with 9 parts of water by weight. The surface of the polished weld specimen is rubbed vigorously with cotton that has been saturated with this reagent at room temperature.

IMPACT TESTING. The purpose of impact testing is to determine the impact strength of welds and base metal in welded products. The impact strength is the ability of a metal to withstand a sharp, high velocity blow. The test provides the information for comparing the toughness of the weld metal with the base metal. A weld or metal may show high tensile strength and high ductility in a tension stress and yet break if subjected to a sharp, high velocity blow.

There are two standard methods of testing: the *Izod* and the *Charpy* tests. The specimen is broken by a single blow. The impact strength is measured in foot-pounds which is a unit of work. (*Work* in physics is defined as the product of a force and the distance through which it moves.) The two types of specimens used for these tests and the method of applying the load are shown in Fig. 12-66. Both tests are made in an impact-testing machine.

The Izod specimen is notched about 1 inch from the end. It is supported between two vise jaws in a vertical position, and the free end above the notch is subject to a swinging blow.

The Charpy specimen is notched in the center. The speci-

12-65. *Metallograph for photographing polished metal specimens.*

Unitron Instrument Co.

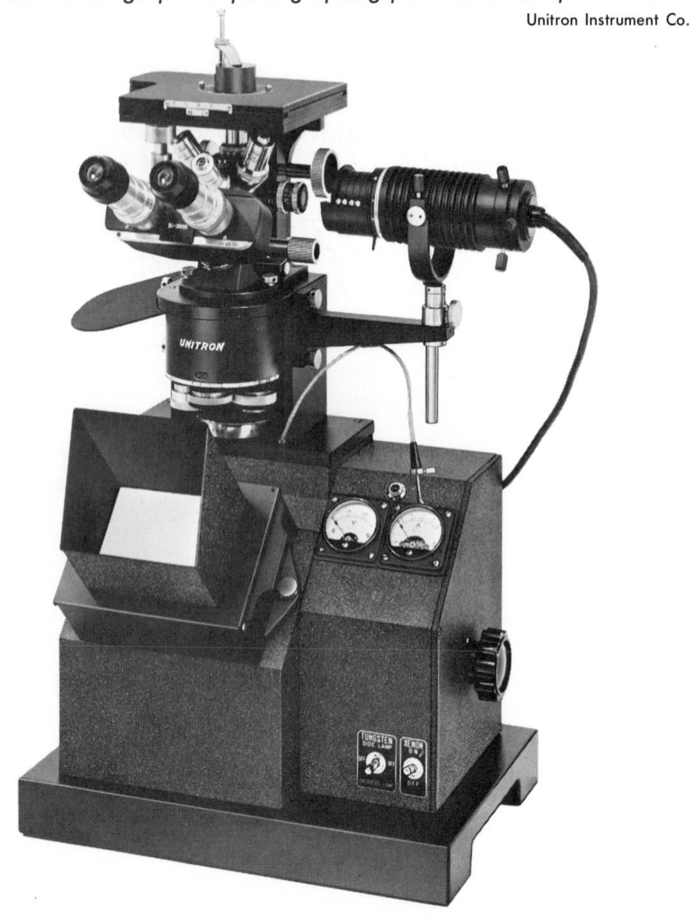

men is supported horizontally between two anvils, and a weight (pendulum) swings against the specimen opposite the notch. The amount of energy in the falling pendulum is known. The distance through which the pendulum swings after breaking the specimen indicates how much of the total energy was used in breaking it. The shorter the distance traveled by the pendulum, the higher the reading on the indicating scale and the tougher the metal. The tougher the metal specimen, the shorter the distance traveled by the pendulum and the higher the reading on the indicating scale.

A welder or inspector may test steel welds for toughness and roundness in the shop without expensive equipment. This is called a *nick-break* test. A notch is sawed in the weld or steel specimen and placed in a vise, Fig. 12-67. The specimen is hit with a hammer on the notch side. The impact strength is indicated by the force required to break the sample, the amount of bending at the break, and the appearance of the fractured surfaces.

FATIGUE TESTING. Fatigue testing is for the purpose of finding out how well a weld can resist repetitive stress as compared with the base metal. Improperly made welds that contain such defects as porosity, slag inclusions, lack of penetration, and cracks are very low in fatigue strength as compared with the base metal. The improper use of heat in welding also reduces the fatigue resistance of the base metal next to the weld.

Two principal methods of testing are used. In one, the specimen is bent back and forth in a regular fatigue-testing machine. In the second method, the specimen is rotated under load in a testing machine. In each case the specimen is subject to tension and compression stresses. The number of cycles imposed before failure is recorded. In both methods careful attention is given to duplicating service conditions.

CORROSION TESTING. The environmental conditions under which many weldments are subject require the use of corrosion-resisting materials such as brass, copper, stainless steels, and other alloy steels. In welding a corrosion-resistant material, great care must be taken to insure that the weld metal is equal to or better than the base material.

In testing for corrosion resistance, the weld metal and base metal are subjected to similar corrosive conditions, and the materials are compared for their resistance to corrosion. Any defect in the weld material shows as a difference in the rate of corrosion when compared to the base metal.

SPECIFIC GRAVITY. The specific gravity test is carried out to make sure that no fine porosity exists. It is performed in a laboratory. The test specimen is a cylinder of all-weld metal $\frac{5}{8}$ inch in diameter and 2 inches long taken from the weld bead. The weight and volume of the specimen are determined in metric units. The specific gravity is obtained by dividing the weight in grams, by

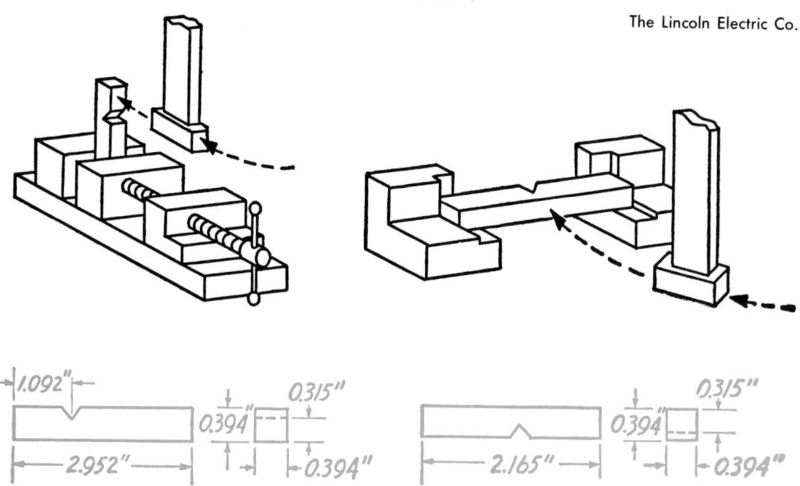

***12-66.** Typical Izod (left) and Charpy (right) impact test specimens and methods of holding and applying the test load. The V-notch specimens (shown) have an included angle of 45 degrees and a bottom radius of 0.010 inch in the notch.*

The Lincoln Electric Co.

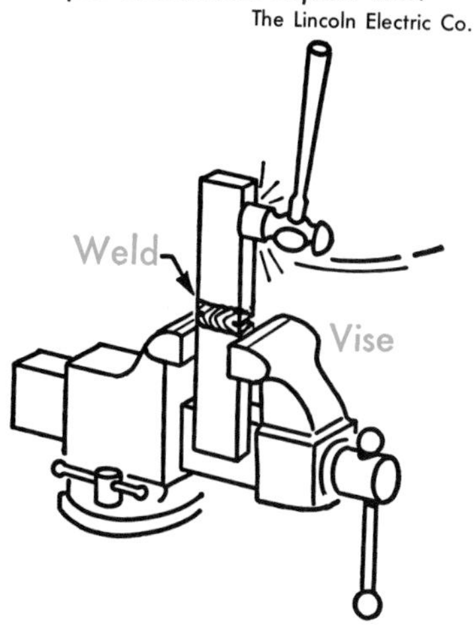

***12-67.** Procedure for making a simple nick-break impact test.*

The Lincoln Electric Co.

the volume in cubic centimetres. A weld specimen of high quality will have a specific gravity of 7.80 g per cm^3. The smallest defects materially affect the results.

VISUAL INSPECTION

Visual inspection is probably the most widely used of all inspection methods. It is quick and does not require expensive equipment. A good magnifying glass (×10 or less) is recommended. A great deal can be learned from the surface condition of a weld, and a careful evaluation of the appearance can determine the suitability for a given job. The service conditions to which a weldment may be subject must also be considered.

Visual inspection of the weldment is usually satisfactory on noncode work that is not critical and when it is not important that the internal condition of the weld be known. This method of inspection should be employed by the welder, the welding inspector, and supervisor from the beginning to the end of the welding job.

12-68. *Lack of penetration and incomplete fusion through the back side of a bevel butt joint. This will cause failure when the joint is subjected to load.*

Principal Defects

Every welder should be familiar with the principal defects that contribute to the failure of welded joints. This will aid him in his effort to produce work of the highest quality. It takes little additional time to inspect a weld, and it often prevents embarrassment when the work is finally inspected by the shop foreman or inspector. When taking important tests, the knowledge of these defects and an awareness of the disastrous results that usually follow when defects are present will increase the chances of the welder to meet the required specifications.

The material to be welded should be inspected carefully for surface defects and the presence of contaminating materials. The faces and edges of the material should be free of laminations, blisters, nicks, and seams. Heavy scale, oxide films, grease, paint, and oil should be removed. Pieces of material that are warped or bent should be corrected or rejected. The material should be checked for size, edge preparation, and angle of bevel. Make sure that the material is a type suitable for welding. In setting up or assembling the job, all parts should fit and be in alignment. It is important to understand the purpose of the design, the use of the weldment, and the welding procedure to be followed.

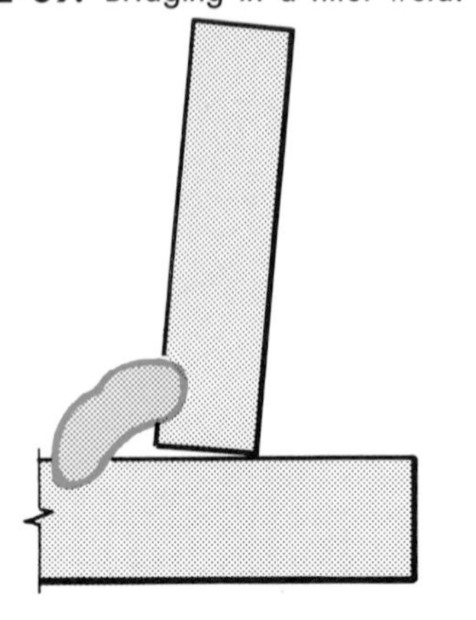

12-69. *Bridging in a fillet weld.*

It will be helpful for the student to review the weld characteristics described in Chapter 4, pages 117–138, at this time.

The following defects are commonly found in welded steel joints:

- Incomplete penetration
- Lack of fusion
- Undercutting
- Slag inclusions
- Porosity
- Cracking
- Dimensional defects
- Brittle welds

INCOMPLETE PENETRATION. This term is used to describe the failure of the filler and base metal to fuse together at the root of the joint, Fig. 12-68. The root-face sections of a welding groove may fail to reach the melting temperature for their entire depth, or the weld metal may not reach the root of the fillet joint, leaving a void caused by the bridging of the weld metal from one plate to the other, Fig. 12-69. Bridging occurs in groove welds when the deposited metal and base metal are not fused at the root of the joint because the root face has not reached fusion temperature along its entire depth.

Although in a few cases incomplete penetration may be due to unclean surfaces, the heat transfer conditions at the root of the joint are a more frequent cause. If the metal being joined first reaches the melting point at the surfaces above the root of the joint, molten metal may bridge the gap between these surfaces and screen off the heat source before the metal at the root melts. In metal-arc welding, the arc establishes itself between the electrode and the closest part of the base metal. All other areas of the base metal receive heat mainly by conduction. If the portion of the base metal closest to the electrode is far from the root of the joint, the conduction of heat may be insufficient to raise the temperature of the metal at the root to the melting point.

Incomplete penetration will cause weld failure if the weld is subjected to tension or bending stresses such as those produced in tensile and bend testing. Even though the service stresses in the completed structure may not require tension or bending at the point of incomplete penetration, distortion and shrinkage stresses in the parts during welding frequently cause a crack at the unfused section. Such cracks may grow as successive beads are deposited until they extend through almost the entire thickness of the weld.

The most frequent cause of incomplete penetration is a joint design which is not suitable for the welding process or the conditions of construction. When the groove is welded from one side only, incomplete penetration is likely to result under the following conditions:

- The root face dimension is too big even though the root opening is adequate, Fig. 12-70. For example, the proper root face dimension for V-groove joints in $3/8$-inch plate is about $1/16$ to $3/32$ inch. Root face dimensions of $3/16$ or $1/4$ inch are too big and keep the root surfaces from melting, thus preventing fusion and penetration from taking place.
- The root opening is too small, Fig. 12-71. For example, the opening at the root of a V-groove joint in $3/8$-inch plate should not be less than $3/32$ to $1/8$ inch. A root opening of $1/16$ inch or less would make it difficult to melt through to the other side, causing incomplete penetration.
- The included angle of a V-groove is too small, Fig. 12-72. For example, the included angle of V-groove joints in $3/8$-inch plate should be about 60 degrees. An angle of 40 degrees does not allow for enough freedom of movement for the electrode to be manipulated at the root of the joint, causing the root face to be burned away. Fusion also takes place on the surfaces of the side walls, bridging the root gap as shown in Fig. 12-69, page 392.

Even if the joint design is adequate, incomplete fusion will result from the following errors in technique:

- The electrode is too large. For example, running the root pass in a V-groove joint in $3/8$-inch plate in the overhead position should be done with a $1/8$- or $5/32$-inch electrode; a $3/16$- or $1/4$-inch electrode is too large.
- The rate of travel is too high. Traveling too fast causes the metal to be deposited only on the surface above the root.
- The welding current is too low. If there is not enough current or if the current setting is incorrect, the weld metal cannot be forced from the electrode to the root of the joint and the arc is not strong enough to melt the metal at the root.

LACK OF FUSION. Many welders confuse lack of fusion with lack of penetration. Lack of fusion is the failure of a welding process to fuse together layers of weld metal or weld metal and base metal. (See Figs. 12-73 through 12-75.) In Fig. 12-73 note that the weld metal just rolls over the plate surfaces. This is generally referred to as *overlap.* Failure to obtain fusion may occur at any point in the welding groove. Overlap at the toe of the weld,

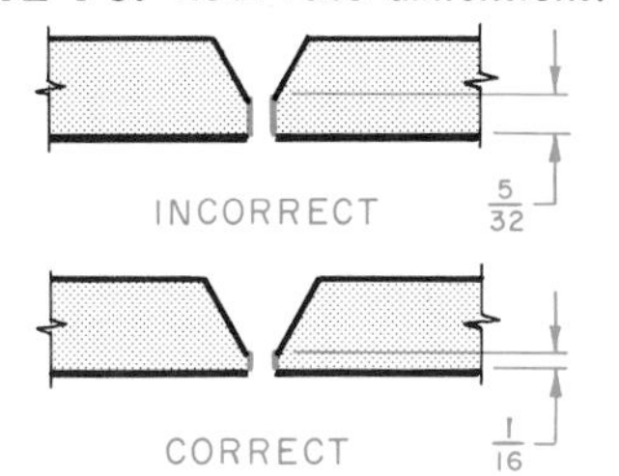

12-70. *Root face dimensions.*

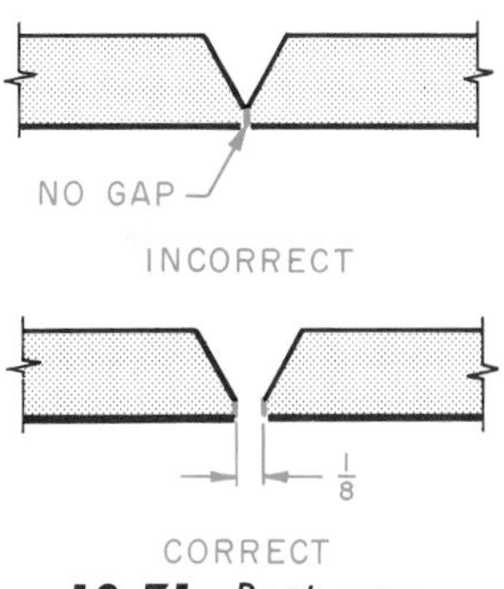

12-71. *Root gaps.*

12-72. *Bevel angle.*

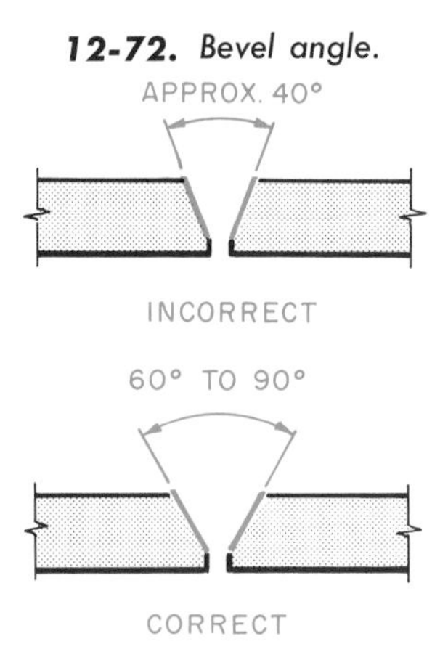

Fig. 12-74, is a form of lack of fusion. Very often the weld has good fusion at the root of the weld and the plate surface but because of poor welding technique and heat conduction, the toe of the weld does not fuse.

Lack of fusion is caused by the following conditions:

- Failure to raise the temperature of the base metal or the previously deposited weld metal to the melting point. Reasons for this failure include (1) an electrode that is too small, (2) a rate of travel that is too fast, (3) an arc gap that is too close, and (4) welding current that is too low.
- Improper fluxing, which fails to dissolve the oxide and other foreign material from the surfaces to which the deposited metal must fuse. Lack of fusion is not common in metal-arc welding unless the surfaces being welded are covered with a material which prevents the molten weld metal from fusing to them.

Lack of fusion is avoided by

- Making sure that the surfaces to be welded are free of foreign material
- Selecting the proper type and size of electrodes
- Selecting the correct current adjustment
- Using good welding technique

The welder does not have to melt away large portions of the side walls of the groove in order to be sure of obtaining fusion. It is only necessary to bring the surface of the base metal to the melting temperature to obtain a bond between the base metal and the weld metal.

UNDERCUTTING. Undercutting is the burning away of the base metal at the toe of the weld, Fig. 12-76 (top). On multilayer welds, it may also occur at the juncture of a layer with the wall of a groove, Fig. 12-76 (bottom). Figure 12-77 shows severe undercutting on the vertical plate of a T-joint.

Undercutting of both types is usually due to improper electrode manipulation. Different types of electrodes have varying tendencies to undercut. For example, reverse polarity electrodes have a greater tendency to undercut than straight polarity electrodes, and a different technique of welding must be employed if undercut is to be avoided. With some electrodes even the most skilled welder may be unable to avoid undercutting under certain conditions. In addition to poor welding technique and the type of electrode required, undercutting may be caused by:

- Current adjustment that is too high
- Arc gap that is too long
- Failure to fill up the crater completely with weld metal. This permits the arc to range over surfaces that are not to be covered with weld metal.

Undercutting is a very serious defect. To prevent any serious

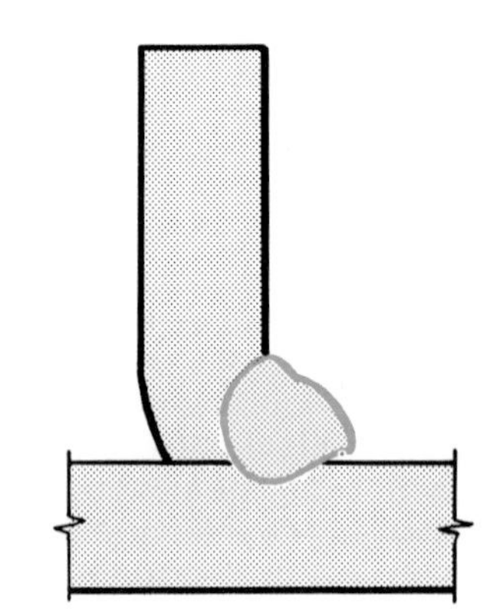

12-74. *Overlap. Overlap may be attributed generally to improper welding technique or improper welding heat.*

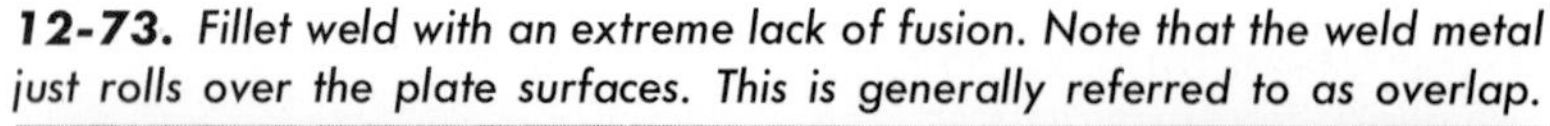

12-73. *Fillet weld with an extreme lack of fusion. Note that the weld metal just rolls over the plate surfaces. This is generally referred to as overlap.*

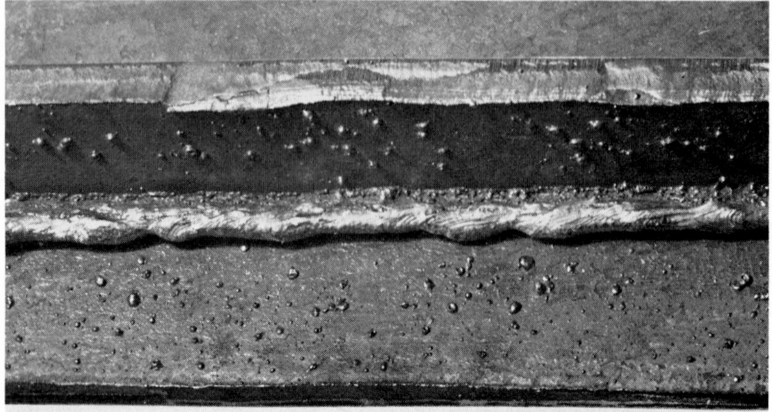

12-75. *Incomplete fusion.*

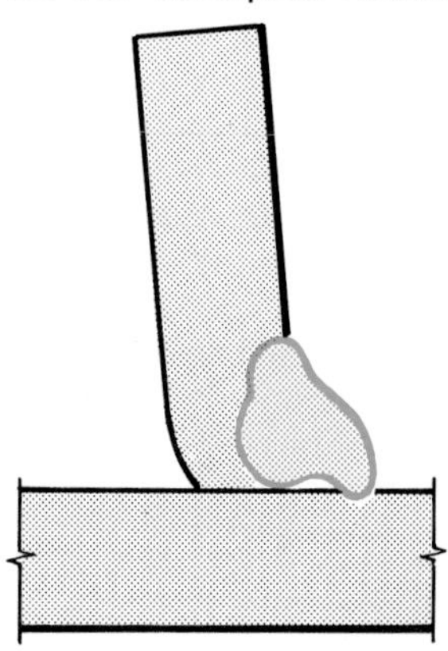

effect upon the completed joint, it must be corrected before depositing the next bead. A well-rounded chipping tool is used to remove the sharp recess which might otherwise trap slag. If the undercutting is slight and the welder is careful in applying the next bead, it may not be necessary to chip.

Undercutting at the surface of a joint should not be permitted since it materially reduces the strength of the joint, Fig. 12-78. For example, if an undercut is $1/16$ inch deep in $1/4$-inch plate, the real strength of the plate has been reduced to less than that of a $3/16$-inch plate. Fortunately, the welder can always see this type of undercutting when he examines the weld, and he can usually correct it by the deposition of additional metal. A good craftsman develops such skill in welding that he rarely causes undercutting of the base metal.

SLAG INCLUSIONS. Slag inclusions are elongated or globular pockets of metallic oxides and other solid compounds. They produce porosity in the weld metal, Figs. 12-79 and 12-80. Slag inclusions may be caused by contamination of the weld metal by foreign bodies. In arc welding, slag inclusions are generally made up of electrode coating materials or fluxes. In multilayer welding operations, failure to remove the slag between the layers causes slag inclusions.

During the deposition and solidification of the weld metal, chemical reactions occur between the metal, the air, and the electrode coating materials or the gases produced by the arc flame. Some of the products of these reactions are metallic compounds which are only slightly soluble in the molten weld metal. The slag may be forced below the surface by the stirring action of the arc, or it may flow ahead of the arc, causing the metal to be deposited over the slag.

When slag is once present in the molten metal from any cause, it tends to rise to the surface because of its lower density. Any factors such as high viscosity of the weld metal, rapid chilling, or too low a temperature may prevent its release. Slag is frequently trapped in the weld when weld metal is deposited by the metal-arc process over a sharp V-shaped recess, such as that produced by a diamond point tool. Under such conditions, the arc may fail to raise the temperature of the bottom of the recess high enough for the metal to fill it and allow the slag, which is ahead of the arc, to float out. The same problem arises if a sharp recess is present due to undercutting by the previous bead. Slag inclusions of this type are usually elongated, and if individual inclusions are of considerable size or are closely spaced, they may reduce the strength of the joint. It is usually not desirable to remove small or isolated inclusions.

Slag forced into the metal by the arc or formed there by chemical reactions usually appears as finely divided or globular inclusions. Inclusions of this type are likely to be a particular problem in overhead welding.

Most slag inclusion can be prevented by:

- Preparing the groove and weld properly before each bead is deposited
- Taking care to avoid leaving any contours which will be difficult to penetrate fully with the arc
- Making sure that all slag has been cleaned from the surface of the previous bead.

The release of slag from the molten weld metal is aided by all factors which make the metal

12-76. *Undercutting.*

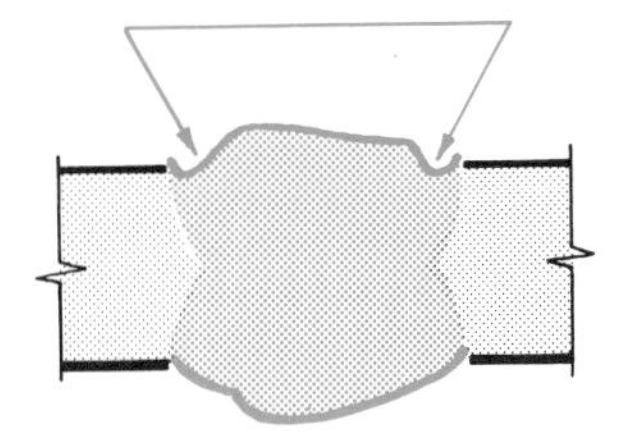

12-77. *Fillet weld that is undercut severely along the upper edge of the weld.*

less viscous (thick) and which delay its solidification, such as preheating and high heat input per inch per unit of time.

POROSITY. Porosity, Fig. 12-81, is the presence of pockets which do not contain any solid material. These pockets are another kind of inclusion resulting from the chemical reactions which take place during welding. They differ from slag inclusions in that the pockets contain gas rather than a solid. The gases forming the voids are derived from:

- Gas released by the cooling weld metal because of its reduced solubility as the temperature drops
- Gases formed by chemical reactions in the weld

Excessive porosity in metal-arc welds has a serious effect on the mechanical properties of the joint. Certain codes permit a specified maximum amount of porosity. Pockets may be found scattered uniformly throughout the entire weld, isolated in small groups, or concentrated at the root.

Porosity is best prevented by avoiding:

- Overheating and underheating of the weld metal
- Too high a current setting
- Too long an arc

A metal temperature which is too high increases unnecessarily the amount of gas dissolved in the molten metal. This excess gas is available for release from solution upon cooling. If the welding current and/or the arc length is excessive, the deoxidizing elements of the electrode coating are used up during welding so that there are not enough of them left to combine with the gases in the molten metal during cooling. Underheating does not permit the weld pool to be molten long enough to allow the trapped gases to escape.

CRACKING. Cracks are linear ruptures of metal under stress. When they are large, they can be seen easily, but they are often very narrow separations. Cracks may occur in the weld metal, in the plate next to the weld, or in the weld deposit itself. Common weld metal and base metal cracks are shown in Fig. 12-82.

Cracking results from localized stress which exceeds the ultimate strength of the material. Little deformation is apparent because the cracks relieve stress when they occur during or as a result of welding.

There are three major classes of cracking: hot cracking, cold cracking, and microfissuring.

Hot cracking occurs at elevated temperatures during cooling shortly after the weld has been deposited and has started to solidify. Stress must be present to induce cracking. Slight stress causes very small cracks that can be detected only with some of the nondestructive testing techniques such as radiographic and liquid penetrant inspection. Most welding cracks are hot cracks.

Cold cracking refers to cracking at or near room temperature. These cracks may occur hours or

12-78. *Undercutting in a fillet weld reduces the cross section of the members and acts as a stress raiser.*

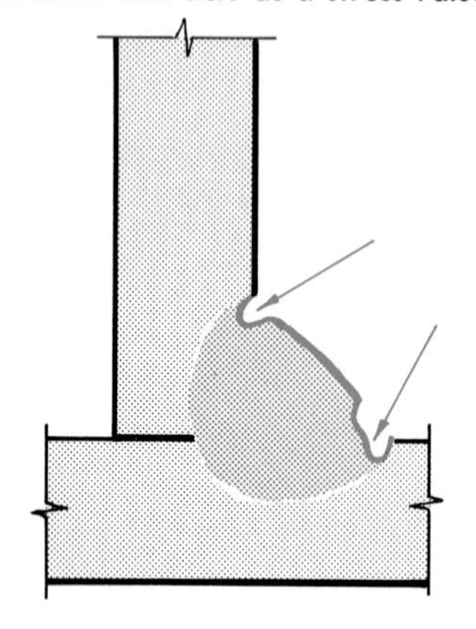

12-79. *Cross section of a fillet weld, exposing internal defects caused by slag inclusions in the weld metal deposit. These voids reduce the mechanical properties of the weld metal.*

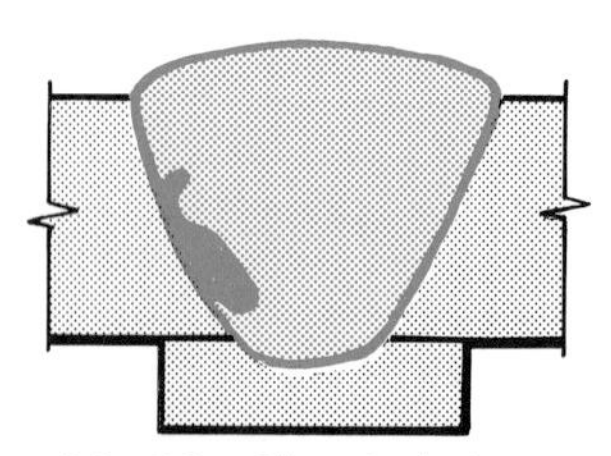

12-80. *Slag inclusions.*

12-81. *Porosity.*

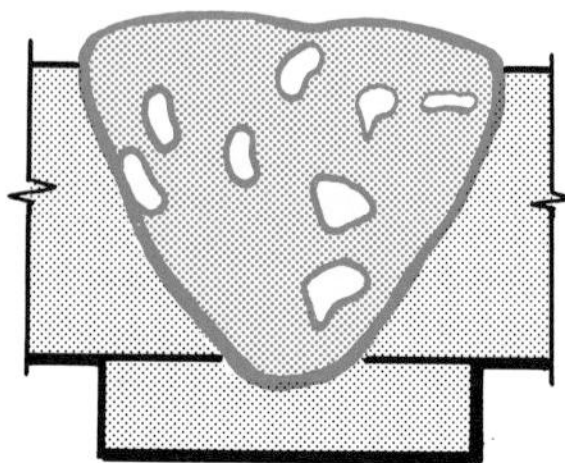

days after cooling. These cracks usually start in the base metal in the heat-affected zone. They may appear as underbead cracks parallel to the weld or as toe cracks at the edge of the weld. Cold cracking occurs more often in steels than in other metals.

Microfissures (microcracks) may be either hot or cold cracks. They are too small to be seen with the naked eye and are not detectable at magnifications below 10 diameters. These cracks usually do not reduce the service life of the fabrication.

Weld Metal Cracking. Three different types of cracks occur in weld metal: transverse cracks, longitudinal cracks, and crater cracks.

Transverse cracks run across the face of the weld and may extend into the base metal. They are usually caused by excessive restraint during welding.

Longitudinal cracks are usually confined to the center of the weld deposit. They may be the continuation of crater cracks or cracks in the first layer of welding. Cracking of the first pass is likely to occur if the bead is thin. If this cracking is not eliminated before the following layers are deposited, the crack will progress through the entire weld deposit. Longitudinal cracking can be corrected by:

- Increasing the thickness of the root pass deposit
- Controlling the heat input
- Decreasing the speed of travel to allow more weld metal to build up
- Correcting electrode manipulation
- Preheating and postheating

Whenever the welding operation is interrupted, there is a tendency for the formation of a crack in the crater. *Crater cracks* usually proceed to the edge of the crater and may be the starting point for longitudinal weld cracks.

Base Metal Cracking. Base metal cracking usually occurs within the heat-affected zone of the metal being welded. The possibility of cracking increases when working with hardenable materials. These cracks usually occur along the edges of the weld and through the heat-affected zone into the base metal. The *underbead crack,* limited mainly to steel, is a base metal crack usually associated with hydrogen. Toe cracks in steel can be of similar origin. In other metals, including stainless steel, cracks at the toe are often termed *edge-of-weld cracks.* They are caused by not cracking in or near the fusion line.

Incorrect welding procedure causes cracking in the base

12-82. *Cracks in welded joints.*

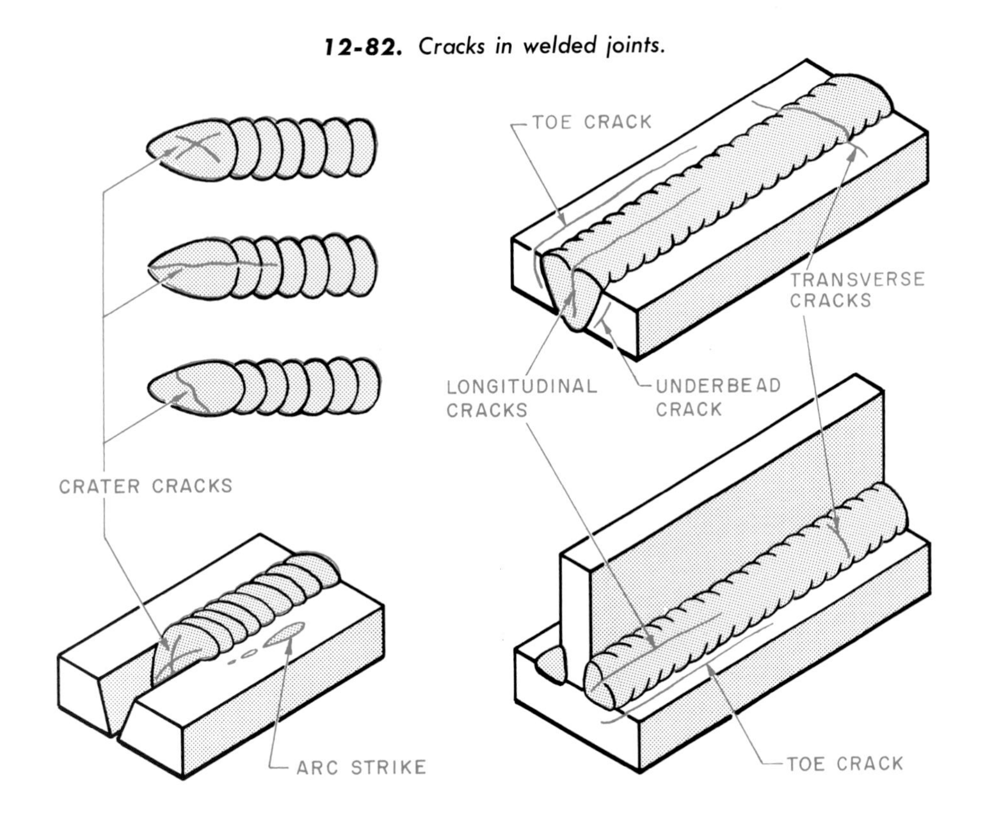

metal. Arc strikes—accidental touching of the electrode to the work—may cause small cracks. If the weld is started at edge of the plate and proceeds into the plate, a crack will occur along the edge of the weld at the toe. Cracks can also be started as a result of undercutting. Root cracks, which start from the root of the weld and progress through the weld metal, often produce cracking in the plate on the side opposite the weld.

A nonfused area at the root of a weld may also crack without apparent deformation if this area is subject to tensile stress. In welding two plates together, the root of the weld is subject to such tensile stress as successive layers are deposited that the nonfused root will permit a crack to start which may progress through practically the entire thickness of the weld.

Care must be taken in the type of steel selected for a particular job and in the electrode chosen for welding. Improper design with little regard for expansion and contraction also contributes to cracking.

BRITTLE WELDS. A brittle weld has poor elongation, a very low yield point, very poor ductility, and poor resistance to stress and strain. Brittle welds are highly subject to failure and may fail without warning any time during the life of the weldment.

One of the principal causes of brittle welds is the use of excessive heat which burns the metal. Multilayer welds are recommended to avoid brittleness. Such welds have a tendency to anneal the previously deposited weld beads and the base metal weld area. Careful selection of the material and the electrode also insure against brittle weld deposits.

DIMENSIONAL DEFECTS. Dimensional defects include longitudinal contraction, transverse contraction, warping, and angular distortion. They are caused by improper welding procedure and/or technique. The use of such controls as welding jigs, proper welding sequences, correct welding procedure, suitable joint design and preheat and postheat processes prevent distortion or keep it at an acceptable minimum. Refer to Chapter 3, pages 45–116, for a detailed discussion of the causes and control of distortion.

Weld Gauges

Incorrect weld size and contour are defects that can be detected by visual inspection with the aid of weld gauges. Weld gauges are tools that the welder can use to make sure that the completed weld is within the limits specified by the engineering design and weld procedure.

Figure 12-83 shows three types of weld gauges: (A) combination butt and fillet weld gauge, (B) fillet weld gauge, and (C) a second type of fillet weld gauge made in the shop.

Figure 12-84 illustrates the method of using the combination butt and fillet weld gauge shown in Fig. 12-83A. Figures 12-85 and 12-86 illustrate the method of using the fillet weld gauge shown in Fig. 12-83B. Figure 12-87 illustrates the method of using the shop-made fillet weld gauge shown in Fig. 12-83C. Note that there are two designs: one for measuring flat fillets and another for measuring convex fillets. The gauge for convex fillets must have a concave face. A shop-made gauge cannot measure concave fillet welds accurately.

SUMMARY

Welding demands constant visual examination during the entire operation. In order to make corrections during the welding operation, all welding requires careful attention to such considerations as the welding process, cleaning, preheating, joint preparation, the proper electrode, chipping, peening and postheating, proper weld size and surface defects. Special attention should be given to the root pass because it tends to solidify quickly and may trap slag or gas. The root pass is also susceptible to cracking, which may continue through all layers.

12-83. *Three types of weld gauges commonly used by welders and inspectors for determining weld size.*

A B C

The completed weld should be examined carefully for proper size and contour, surface defects, and warping and distortion of the weldment. Special attention should be given to unfilled craters, poor starts and stops, crater cracks, and cracking at the edge of the welds. Table 12-5 is a summary of the various weld and base metal defects that may be encountered, and the various methods of testing that will establish the presence of these defects. In Table 12-6 are the recommended inspection methods to be used for evaluating fillet and butt joints. An understanding of these tables, and Table 7-1, page 198, will give you adequate knowledge of testing and its application to weld and base metal. The following points should be remembered:

- Visual inspection is the most convenient method. Everyone can participate in it. Fig. 12-88, p. 402.
- Radiographic inspection permits looking into the weld for defects which fall within the sensitivity range of the process. It provides a permanent record of the results.
- Magnetic particle inspection is outstanding for detecting surface cracks and is used to advantage on heavy weldments and assemblies.
- Dye penetrant is easy to use for detecting surface cracks. Its indications are readily interpreted.
- Ultrasonic inspection is excellent for detecting subsurface discontinuities, but it requires expert interpretation.
- Hydrostatic testing determines the tightness of welds in fabricated vessels. Various liquids and gases may be used which, when escaping in minute amounts, are easily detected. The method is simple to apply.

12-84. *Method of using butt and fillet weld gauge. (A) To determine the size of a convex fillet weld, place the gauge against the toe of the shortest leg of the fillet. Slide the pointer out until it touches the structure. Read "size of convex fillet" on the face of the gauge. (B) To determine the convexity of a convex fillet weld after its size has been determined, place the gauge against the structure, and slide the pointer out until it touches the face of the fillet weld. The maximum convexity should not be greater than indicated by "maximum convexity" for the size of fillet being checked. (C) To determine the size of a concave fillet weld, place the gauge against the structure, and slide the pointer out until it touches the face of the fillet weld. Read "size of concave fillet" on the face of the gauge. (D) To determine the reinforcement of a butt weld, place the gauge so that the reinforcement will come between the legs of the gauge and slide the pointer out until it touches the face of the weld. The permissible reinforcement is that indicated on the face of the gauge. (E) To determine the root opening (1/8 or 3/16 inch) of a butt joint, place one leg of the gauge in the space separating the parts. If the gauge fits snugly, the root opening is that indicated on the face of the gauge.*

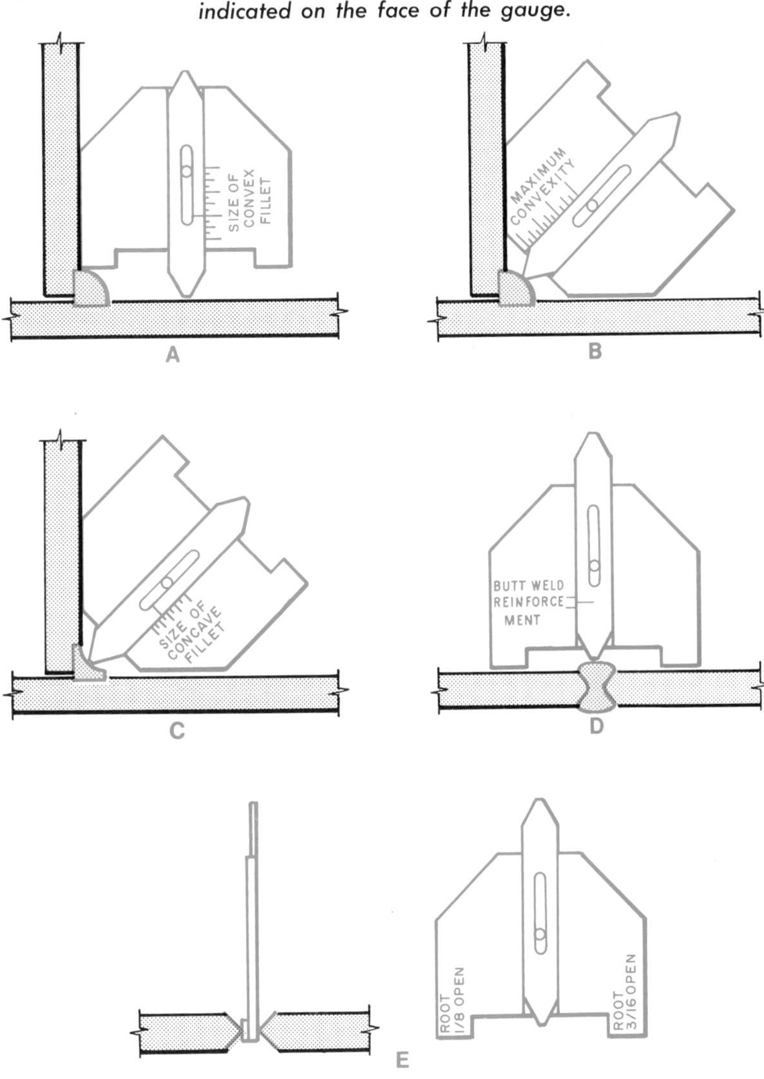

- Hardness tests indicate the approximate tensile strengths of metals and show whether or not the base metal and weld metal strengths are matched. The method of testing is fairly easy to apply.
- Destructive testing gives an absolute measure of the strength of the sample tested. Samples may not always be representative of all of the welding on a component. They do, however, have the potential of predicting the serviceability of a weldment. Destructive testing is a good means of determining a welder's ability. Results are permanent. The method is fairly easy to apply in shop and field.

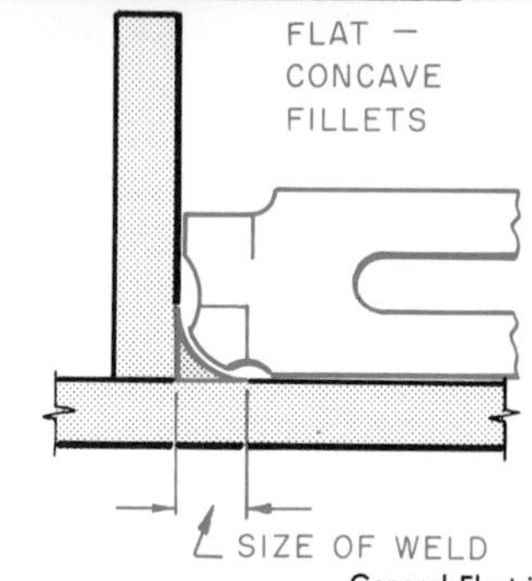

General Electric Co.

12-85. *Method of using fillet weld gauge to determine the size of flat and concave fillet welds.*

12-86. *Method of using fillet weld gauge to determine the size of a convex fillet weld.*

General Electric Co.

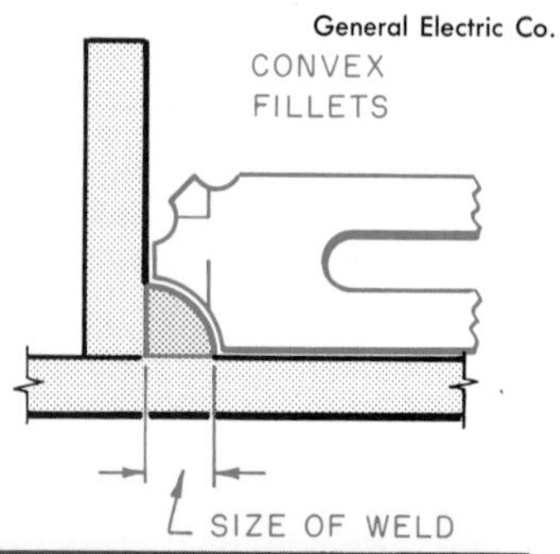

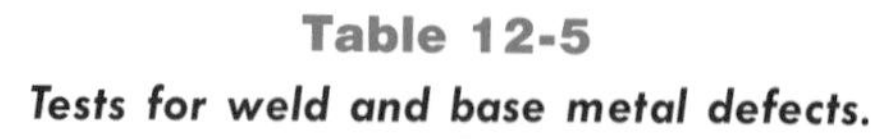

Table 12-5

Tests for weld and base metal defects.

Defects	Methods of testing
Dimensional defects:	
Warpage	Visual inspection with proper mechanical gauges and fixtures
Incorrect joint separation	Visual inspection with proper mechanical gauges and fixtures
Incorrect weld size	Visual inspection with approved weld gauge
Incorrect weld profile	Visual inspection with approved weld gauge
Structural discontinuities:	
Porosity	Radiographic, fracture, microscopic, macroscopic, ultrasonic
Nonmetallic inclusions	Radiographic, fracture, microscopic, macroscopic, ultrasonic
Incomplete fusion	Radiographic, fracture, microscopic, macroscopic, ultrasonic
Cracking	Visual inspection, bend tests, radiographic, microscopic, macroscopic, magnetic particle, penetrating oil and ultrasonic
Undercutting	Visual inspection, bend tests, radiographic and ultrasonic
Surface defects	Visual inspection
Inadequate joint penetration	Radiographic, fracture, microscopic, macroscopic, ultrasonic
Defective properties:	
Low tensile strength	All-weld-metal tension test, transverse tension test, fillet-weld shear test, base metal tension test
Low yield strength	All-weld-metal tension test, transverse tension test, base metal tension test
Low ductility	All-weld-metal tension test, free bend test, guided-bend test, base metal tension test
Improper hardness	Hardness tests
Impact failure	Impact tests
Incorrect composition	Chemical analysis
Improper corrosion resistance	Corrosion tests

American Welding Society

Table 12-6

Inspection methods for fillet and butt joints.

	Recommended inspection for	
Type of defect	**Fillet joints**	**Butt joints**
Undersize weld	Visual[1]	Visual
Surface porosity	Visual	Visual
Internal porosity	Destructive	Radiographic
Undercut	Visual	Visual
Cracks	Magnetic particle	Magnetic particle
	Dye penetrant	Dye penetrant
	Visual	Visual
	Destructive[2]	Ultrasonic
		Radiographic[3]
Lack of penetration	Destructive	Radiographic
	Ultrasonic	Ultrasonic
Slag inclusions	Destructive	Radiographic
	Ultrasonic	Ultrasonic

The Lincoln Electric Co.

[1] Use fillet gauges.
[2] Destructive test will reveal subsurface cracks.
[3] Radiographic inspection has its limitations in revealing crack-type defects.

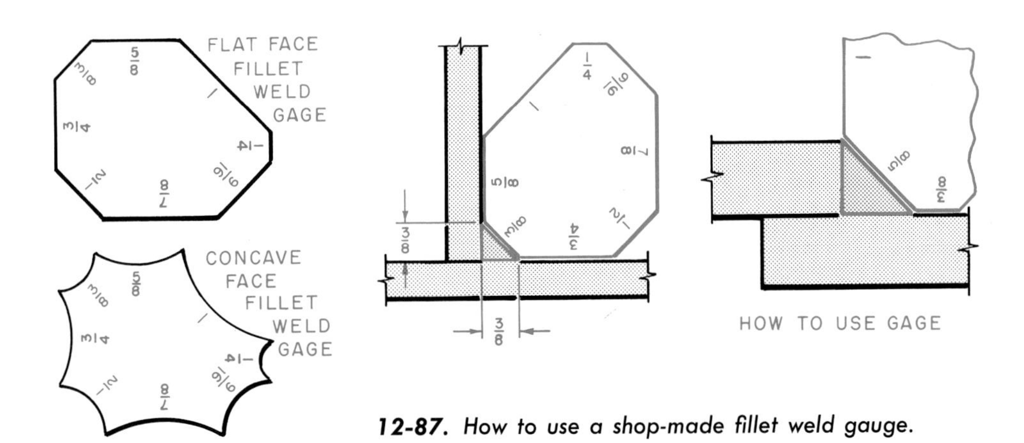

***12-87.** How to use a shop-made fillet weld gauge.*

REVIEW QUESTIONS

1. Name three basic types of nondestructive tests and explain each.
2. List six destructive tests employed in testing groove welds and the purpose of each.
3. List four destructive tests employed in testing fillet welds and the purpose of each.
4. Explain the difference between procedure qualification and operator qualification tests.
5. What is the essential difference between the face and root bend tests?
6. List three principal defects that may be found in welded joints. Describe each.
7. What are the contributing causes of the weld defects given above?
8. How many welds should be measured for size by the operator?
9. List four types of hardness tests and give a brief explanation of each.
10. List the weld defects that may be detected by visual inspection.

11. Destructive test specimens should be ground across the weld. True. False.

12. Liquid penetrant inspection can detect surface and interior defects. True. False.

13. Describe a nondestructive test that is fast, low in cost, and simple to apply.

14. Magnetic particle inspection is not suitable for pipe welding inspections. True. False.

15. The basic purpose of inspection is quality control. True. False.

16. Only a red dye penetrant can be used with the liquid penetrant method. True. False.

17. Name four acid solutions that can be used for weld etching.

18. Once a welder has qualified he does not have to be tested again. True. False.

19. Once a welder is qualified he can work on any job. True. False.

20. Draw a rough sketch showing the general outlines of each type of weld defect.

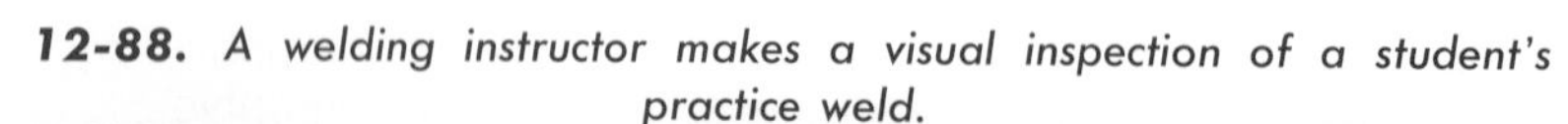

12-88. *A welding instructor makes a visual inspection of a student's practice weld.*

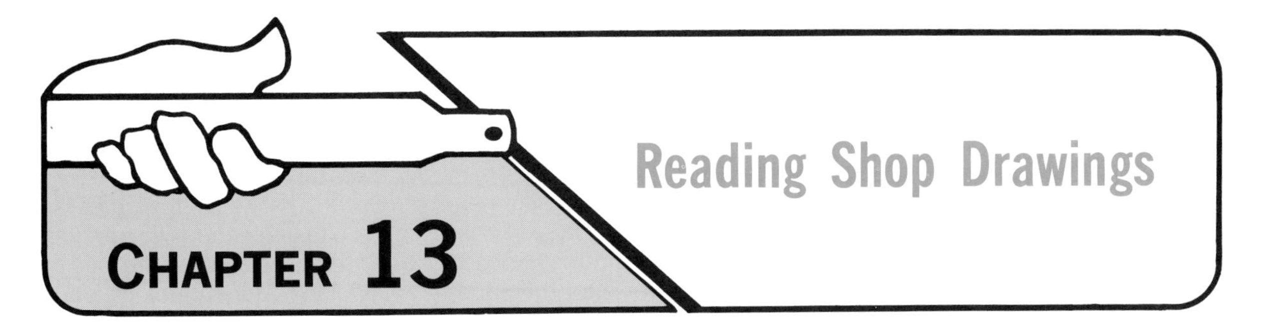

Chapter 13 Reading Shop Drawings

INTRODUCTION

Blueprints and shop sketches are the sign language of industry. They transmit ideas from design to finished product.

Drawings are the universal language of crafts people the world over. It is possible for workers who can read blueprints to produce identical parts even though they work in different lands. Tanks, pipelines, ships, buildings, airplanes, and machines are made from the information given on blueprints.

The Purpose of Shop Drawings

Pictures and notes on drawings are only a method of passing information from one person to another in a condensed form. Information is passed from the engineer who designs the job to the worker who fabricates it. The engineer under whose direction the drawings are prepared must have a thorough knowledge of design, mathematics, mechanics, physics, chemistry, and drafting. Workers who interpret the finished drawings do not need such theoretical knowledge, but they must understand the drawing well enough to carry out all instructions shown.

Skilled welders are not only competent in welding, but they are also able to read the drawings which describe the work that they must do. Welders are much more valuable to the employer if they can read drawings. With this knowledge, they can learn from a print just what is to be built, and they can picture the object in their minds as it will be after it is completed. They must also be able to interpret the notations on the print. The print serves as an instruction sheet to which they can refer from time to time without having to ask the foreman question after question.

The ability to read prints is a common reason for promotion to a position of more responsibility. When welding supervisors are chosen, the one who can interpret prints has a better chance than other qualified welders. This person understands the design requirements because he or she understands the designer's language, and can translate that language into welding which meets the requirements.

Welders should learn to read and interpret plans for the following reasons:

- A journeyman welder who can interpret plans is always in demand on any job, and is likely to be given responsibility.
- The welder has an accurate idea of what the weldment should look like when completed. An understanding of the total job makes assembly easier and guides the choice of welding procedure.
- The welder is able to fabricate a weldment that duplicates the engineering design as closely as possible.
- The welder can check on the equipment and supplies that are needed for the job.
- In field work welders can avoid interference with other trades by knowing the extent of the work even before they start. They can begin preparations for some of their work while carpenters are building their forms or masons are getting ready to pour concrete.

Since sketches and prints are used repeatedly in this text, you must be able to read prints properly and understand them, Fig. 13.1. As part of your welding training, you will also be called upon to do production jobs in order to experience real job situations. Very often these jobs come to the school welding shop as a print. Such prints are usually very simple, and you will have to learn only the fundamentals of blueprint reading in order to interpret them. After you have completed this chapter, you are urged to study a book on advanced blueprint reading so that you will be able to read the more complicated drawings with which you will work in industry.

Methods of Printing Drawings

Blueprints are copies of drawings made by drafters with drawing instruments. The original tracings are seldom used by the workers. They are given prints which are reproductions of the original. The following are conventional methods for preparing plans.

PENCIL TRACINGS. The drafter prepares an original drawing on transparent paper or cloth in pencil to permit revisions and corrections. Originals must not be taken out of the drafting room since hours of hard work are required for their preparation and they soil and tear easily. Prints for field use are made from pencil tracings.

INKED TRACINGS. Inked tracings are pencil tracings on which no more changes are anticipated. They are inked for making better prints and to make them more permanent. Inked tracings never go to the field.

13-1. This young welder, who has just completed his training course, is reading a print of the job he is working on.

U.S. Department of Labor

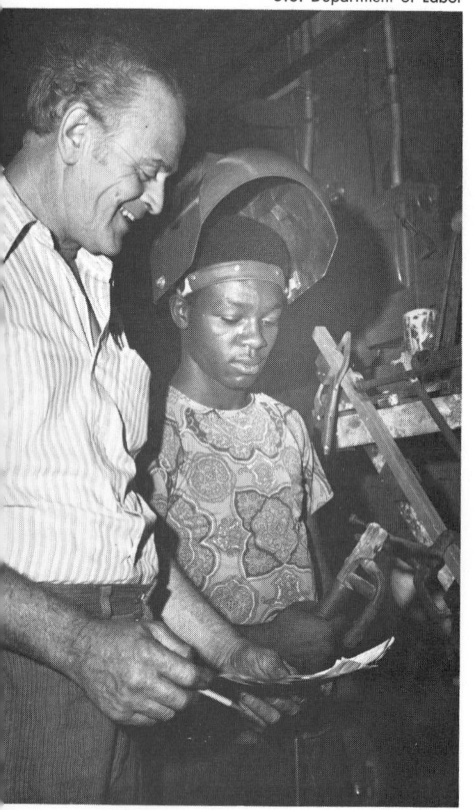

BLUEPRINTS. Blueprinting is a photographic process in which an original tracing acts as a negative. The tracing is placed over sensitized paper and is exposed to artificial light. Then the paper is washed in water, potassium dichromate, or other solutions and dried.

Blueprints have white lines on a blue background. The background is easy on the eyes and does not soil easily. These prints can be produced in large quantities at low cost. Blueprints, however, fade easily in sunlight. The wet process used in making them shrinks the paper so that scaling to close tolerances cannot be done.

BLUE-LINE PRINTS. In blue-line prints the background is white, and the lines are dark blue. The process resembles that for making blueprints except that a transparent brown-and-white print known as a *Van Dyke* is made from the original drawings. The Van Dyke is the negative from which the final prints are made.

Blue-line prints are neat-looking, and because they have a white background, penciled notes are legible on them. The intermediate print is a disadvantage, and the process is more expensive than that for blueprints. The double-wet process increases shrinkage, and blue-line prints are hard to keep clean in the field.

OZALID PRINTS. Ozalid prints are made by placing the original tracing over sensitized paper and passing both the tracing and the paper through a reproducing machine called a separation white printer where they are exposed to a mercury vapor light. The tracing image is reproduced on the sensitized paper, which is then exposed to an ammonia vapor wash. The finished product is the ozalid print.

Because it is a dry process, this is a fast, inexpensive method with no shrinkage. Ozalid prints have black, dark blue, or dark purple lines on a white background. They do not fade easily in the sun.

PHOTOGRAPHIC PROCESSES.
Photostats. Photostats are made by photographing a drawing and then using the negative to make prints. The neat, permanent prints can be reduced or enlarged to any desired size, but they are too costly and impractical for field use.

Microfilm and Microfiche. The majority of industries use microfilm and microfiche for their drawings. The great size reduction common with these processes and the need for clear, sharp lines have caused a return to the use of ink in making the original drawings. Ink drawings are made on plastic film for the most part because of its stability in varying weather conditions, its durability, and permanence. Film may also be washed, cleaned, and reused.

PRINTED PLANS. Prints can be made by photographing the tracing and making metal plates for printing on a press. Like photostats, printed plans are neat, permanent, and expensive. They are practical only when they are made in large quantities.

Care of Prints

Any kind of drawing or print should be handled carefully in the field. Avoid thumbprints by handling by the "thumb room" space left along the edge. Do not put any marks on the drawing.

They may lead to costly mistakes or delays. Any notations added to a drawing on the job should be made only by a person in authority, and only if the information will aid in doing better work.

STANDARD DRAWING TECHNIQUES

Types of lines

To read blueprints, it is first necessary to understand what the different lines on the drawings represent. The line is the basis for all industrial drawings. Lines show an object in such detail that the worker has no difficulty in visualizing the shape of the object when fabricated. Figure 13-2 illustrates the several types of commonly used lines. Note the different weights (thicknesses) of the lines.

13-2. *Standard lines.*

BORDER LINE

OBJECT LINE

EXTENSION LINE

6"

DIMENSION LINE

CENTER LINE

INVISIBLE EDGE LINE

PHANTOM LINE

A A

CUTTING PLANE LINE

LEADER

OBJECT LINES. The shape of an object is shown on a drawing by unbroken heavy lines known as object lines or *visible edge lines.* The lines may vary in thickness depending on the size of the object. A heavy line is also used as a border line around the edge of drawings.

EXTENSION LINES. Extension lines extend from the object lines of an object to a dimension. Thus they express the size of the object. They extend away from the object and are not a part of the object. An extension line is medium in thickness and unbroken.

DIMENSION LINES. Dimension lines are medium, solid lines, which are unbroken except where the dimension is placed. They have arrowheads at their ends. The distance determined by the dimension line is the distance from the end of one arrowhead to the end of the other or the distance between the extension lines to which the dimension refers.

CENTER LINES. Center lines are often referred to as dot-and-dash lines, but they are actually light broken lines made up of long and short dashes spaced alternately. Center lines are not a part of the drawing of the object. They indicate the center of a whole circle, a part of a circle, or certain dimensions.

HIDDEN LINES. A drawing must include lines that represent all the edges and intersections of the object. In some views it is not possible to see all of the surfaces. These surfaces are represented by lines of medium thickness made up of a series of short dashes. These lines are referred to as hidden or *invisible object lines* since they represent edges or surfaces which cannot be seen.

CUTTING PLANE LINES. A cutting plane, also called a *sectional line,* is a line of medium thickness which indicates where an imaginary cut is made through the object. The line is a series of one long and two short dashes alternately spaced. Arrowheads are placed at the ends of the line pointing towards the direction in which the section will be shown. The cutting plane line also has letters to identify it with its respective section on the same or on a different drawing.

LEADERS. Leaders usually point to a particular surface to show a dimension or a note. The leader is usually drawn at an angle, and it is a light line with an arrowhead at the pointing end.

BREAK LINES. Break lines are used to show a break in an object or to indicate that only a part of an object is shown. A short break is shown as a heavy irregular line drawn freehand, Fig. 13-3A. A

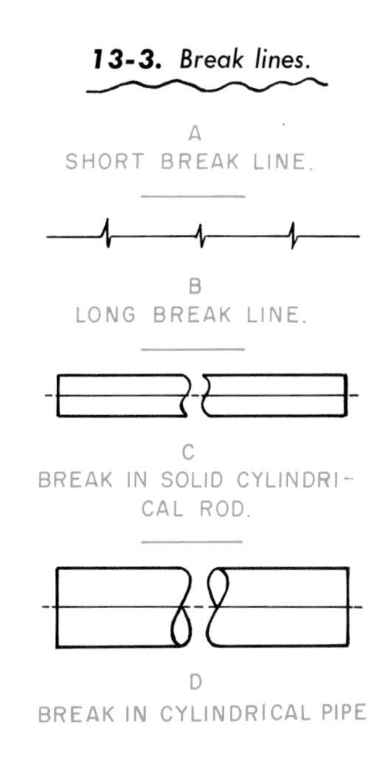

13-3. *Break lines.*

long break is shown as a ruled light line with zigzags in it, Fig. 13-3B. A break in a solid bar or pipe is shown as a curved line, Fig. 13-3C and D.

PROJECTION LINES. Projection lines show the relationship of surfaces in one view with the same surfaces in other views. Projection lines on shop drawings are helpful to welders when laying out pipe joints. They do not appear on drawings unless a part is complicated so that it is necessary to show how certain detail on the drawing are obtained. Projection lines are fine unbroken lines running from a point in one view to the same point in another view, Fig. 13-4.

Dimensioning

THE ENGLISH AND SI METRIC MEASUREMENT SYSTEMS. The two most commonly used systems of measurement throughout the world are the English, or customary, system (sometimes called the inch-pound system) and the metric system which is based on the metre, litre, and gram. The metric system is a decimal system expressed in multiples of ten. The base unit is ten, and all other units are ten times the basic unit.

The English system is currently being used in the United States and a few other countries. The International System of units, abbreviated SI, is being used by 92 percent of the world's population, and 65 percent of all world production and trade is based on this system. It is a modernized version of the metric system that has been established by international agreement. Many attempts have been made to introduce the SI metric system into the United States.

The United States Congress is expected to pass legislation that will direct the government to convert our present English system of measurement to the SI metric system of measurement. This will have special significance for welders since dimensions on prints, rod sizes, and metal sizes will all have to be identified by SI metric measurements. The teaching of this system will be required in our schools.

Understanding the SI metric system of measurement involves chiefly the mastery of a new vocabulary. The most commonly used units for length, for example, are metres, centimetres, millimetres, and kilometres; for mass they are grams, milligrams, and kilograms; for volume, litres and millilitres. Converting from the traditional English units to SI metric units, or vice versa, involves complicated computations. However, as the SI metric system comes into full use, converting back and forth will not be necessary. Metric concepts will be introduced in the schools at an early grade level, and students will learn them as they traditionally have learned inches and ounces.

The only quantities that must be learned as equivalents in metric calculation are 10, 100 and 1 000. On the other hand the student is forced to contend with 29 combinations in the English system. Many of them are multidigit: for example, 5,280 ft. = 1 mile. Table 13-1 lists the common SI metric units of measure, and Tables 13-2 and 13-3 supply approximate conversion factors.

The common metric ruler is divided into small increments using the millimetre (one-thousandth of a metre) as a fine unit of measurement. Major divisions or bold figures are marked in units of 10 mn.

Multiples and Submultiples. SI metric measurements are worked from a base of 10 and increase by powers of 10. The powers of 10 are referred to as *multiples* and *submultiples*. They are expressed as shown in Table

13-4. *Projection lines.*

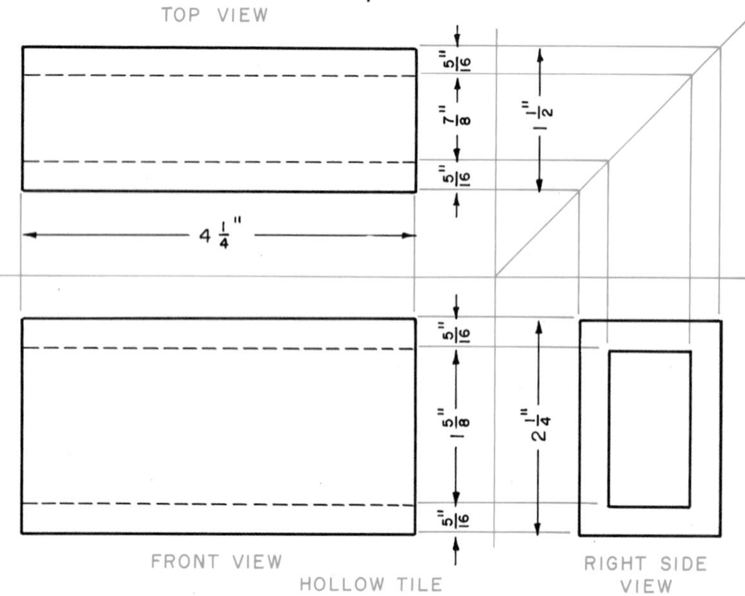

13-4. Table 13-5 demonstrates the multiples and submultiples of the metre.

Application to Drawings. Figure 13-5 shows a simple drawing in the standard English linear measurement system. Figure 13-6 (Page 410) shows the same drawing in the SI metric linear measurement system. You will note that both the drawings are very similar. The only difference is in the use of millimetres instead of inches for the dimensions. It is anticipated that the system of showing all drawings in metric, that is, with only SI metric measurements shown on the drawing itself will become standard practice. Some companies also place a small chart in the upper left-hand corner of the drawing which shows their English equivalents. This helps the worker to begin to "think metric."

At the present time many companies that do business with foreign countries have a dual dimensioning system on their drawings. This permits the part to be manufactured with either the English or SI measurements. Drawing practice is to put the English measurement first, and the SI metric measurement in parenthesis directly behind or slightly below it. The order may be reversed with the SI measurement first. Because of the extra cost involved and the complex nature of the drawing, it is not likely that this system will continue to be used once the changeover to SI metric begins in earnest.

METHODS OF DIMENSIONING. There are several ways in which dimensions are shown on drawings. The size and shape of the object being drawn determine which method is chosen for each dimension. The location of a dimension is indicated by an arrowhead. See Fig. 13-7 for correct and incorrect forms of arrowheads. The placement of the dimension itself depends on the amount of space there is.

- Parts A, B, and C of Figure 13-8 show the method used when there is enough space at the location for the dimension line, arrowhead, and number.
- Figure 13-8D shows the method used when there is space for the number but not the dimension line and arrowhead.

Table 13-1

Common SI metric measurements.

Quantity	Unit of measurement	Symbol
Length	millimetre (1/1000th of a metre)	mm
	metre	m
	kilometre (one thousand metres)	km
Area	square metre	m^2
	hectare (ten thousand square metres)	ha
Volume	cubic centimetre	cm^3
	cubic metre	m^3
	millilitre (1/1000th of a litre)	ml
	litre (1/1000th of a cubic metre)	l
Mass	gram (1/1000th of a kilogram)	g
	kilogram	kg
	ton (one thousand kilograms)	t

Table 13-2

Approximate conversions from English (customary) measures to SI metric measures.

When you know	Multiply by	To find	Symbol
Length			
in. inches	25.4	millimetres	mm
in. inches	2.54	centimetres	cm
ft. feet	30	centimetres	cm
ft. feet	0.3048	metres	m
yd. yards	0.9	metres	m
mi. miles	1.6	kilometres	km
Area			
$in.^2$ square inches	6.5	sq. centimetres	cm^2
$ft.^2$ square feet	0.09	sq. metres	m^2
$yd.^2$ square yards	0.8	sq. metres	m^2
$yd.^3$ cubic yards	0.764	cubic metres	m^3
acres	0.4	hectares	ha
$mi.^2$ square miles	2.6	square kilometres	km^2
Mass (weight)			
oz. ounces (avdp.)	28	grams	g
lb. pounds (avdp.)	0.45	kilograms	kg
t. short tons	0.9	tonnes	t

Table 13-3
Approximate conversions from SI metric measures to English (customary) measures

Symbol	When you know	Multiply by	To find	Symbol
Length				
mm	millimetres	0.04	inches	in.
cm	centimetres	0.4	inches	in.
m	metres	3.3	feet	ft.
m	metres	1.1	yards	yd.
km	kilometres	0.6	miles	mi.
Area				
cm^2	sq. centimetres	0.16	sq. inches	in.^{2}
m^2	sq. metres	1.2	sq. yards	yd.^{2}
km^2	sq. kilometres	0.4	sq. miles	mi.^{2}
m^3	cubic metres	1.3	cubic yards	yd.^{3}
ha	hectares (10 000 m^2)	2.5	acres	

Table 13-4
Multiples and submultiples of base 10.

Multiplication factors	Prefix	Symbol
1 000 000 000 000 = 10^{12}	tera	T
1 000 000 000 = 10^9	giga	G
1 000 000 = 10^6	mega	M
1 000 = 10^3	kilo	k
100 = 10^2	hecto	h
10 = 10^1	deka	da
1 = 10^0		
0.1 = 10^{-1}	deci	d
0.01 = 10^{-2}	centi	c
0.001 = 10^{-3}	milli	m
0.000 001 = 10^{-6}	micro	μ
0.000 000 001 = 10^{-9}	nano	n
0.000 000 000 001 = 10^{-12}	pico	p
0.000 000 000 000 001 = 10^{-15}	femto	f
0.000 000 000 000 000 001 = 10^{-18}	atto	a

Table 13-5
Multiples and submultiples of the metre.

Multi-factor		Quantity		Symbol
10^3 metres	=	kilometre	=	km
10^2 metres	=	hectometre	=	hm
10^1 metres	=	dekametre	=	dam
10^0 metre	=	metre	=	m
10^{-1} metre	=	decimetre	=	dm
10^{-2} metre	=	centimetre	=	cm
10^{-3} metre	=	millimetre	=	mm
10^{-6} metre	=	micrometre	=	μm

- Parts E and F of Figure 13-8 show two methods used when there is no space for the dimension line, arrowhead, and number.
- Parts A and B of Figure 13-9 show two methods used for dimensioning large circles.
- Parts C, D, and E of Figure 13-9 show two methods used for small circles when there is space for the dimension neither within the circle nor projected from its diameter. Note that diameter is abbreviated as *DIA*.
- Parts A and B of Figure 13-10 show two methods used for radii (rounded corners) when there is space for the dimension line, arrowhead, and number.
- Parts C and D of Figure 13-10 show two methods used for radii when there is space for the arrowhead but not the dimension line and number. Note that radius is abbreviated as *R*.
- Figure 13-11 shows the methods for dimensioning a bevel (A); a chamfer (B); an included angle (C); a 45-degree bevel in a butt-joint (D); and a 45-degree bevel in a T-joint (E).

ANGULAR DIMENSIONS. Angles are measured in degrees, not in feet and inches. Each degree is one-three hundred sixtieth ($\frac{1}{360}$) of a circle. The degree may be divided into smaller units called *minutes*. There are 60 minutes in 1 degree. The minute may be divided into smaller units called *seconds*. There are 60 seconds in each minute.

The symbol for degrees is a small circle (°), for minutes a single apostrophe (') mark, and for seconds a double apostrophe (''). For example, 10 degrees, 15 minutes, 8 seconds would be written thus: 10° 15' 8''.

Figure 13-12 (Page 412) illustrates the measurement of angles in degrees. As shown in Fig. 13-12A, if a wheel makes one complete turn or revolution, a point on the wheel would travel 360 degrees. Measurement in degrees can also be thought of in terms of the movement of the hands on a clock, Fig. 13-12A. If the minute hand moves from 12 around to 12, it travels 360 degrees. The hour hand moves 90 degrees from 12 to 3; 45 degrees from 3 to half past 4, 30 degrees from 6 to 7; 60 degrees from 9 to 11; and 15 degrees from 11 to half past 11.

DIMENSIONS OF HOLES. Drilled Holes. The size of drilled holes is often given in notes such as the one found in Fig. A-2, page 419. The note "Drill $\frac{1}{4}$ hole" means that the holes are made with a $\frac{1}{4}$-inch drill and will be $\frac{1}{4}$ inch in diameter. The note also refers to the hole near the bottom of the lock keeper.

Counterbored Holes. Counterbored holes are usually designated by giving (1) the diameter of the drilled hole, (2) the diameter of the counterbore, (3) the depth of the counterbore, and (4) the number of holes to be drilled. For example, the note in Fig. 13-13 reads "$\frac{1}{2}$ drill, $\frac{3}{4}$ c'bore, $\frac{3}{8}$ deep, 3 holes."

Countersunk Holes. Countersunk holes are designated by giving (1) the diameter of the drilled hole, (2) the angle at which the hole is to be countersunk, (3) the diameter at the large end of the hole, and (4) the number of holes to be countersunk. See Fig. 13-14.

Screw Threads and Threaded Holes. The two main types of screw threads are called the *National Coarse Thread Series* and the *National Fine Thread Series.* Figure 13-15 shows two ways to represent screw threads on a bolt. The note under *A,* "$1\frac{1}{2}$"—NC" means that the diameter is $1\frac{1}{2}$ inches, there are 6 threads for each inch of the threaded rod, and the thread is

13-5. *Three-view drawing of a dovetailed slide with the dimensions given in the standard English linear measurement system.*

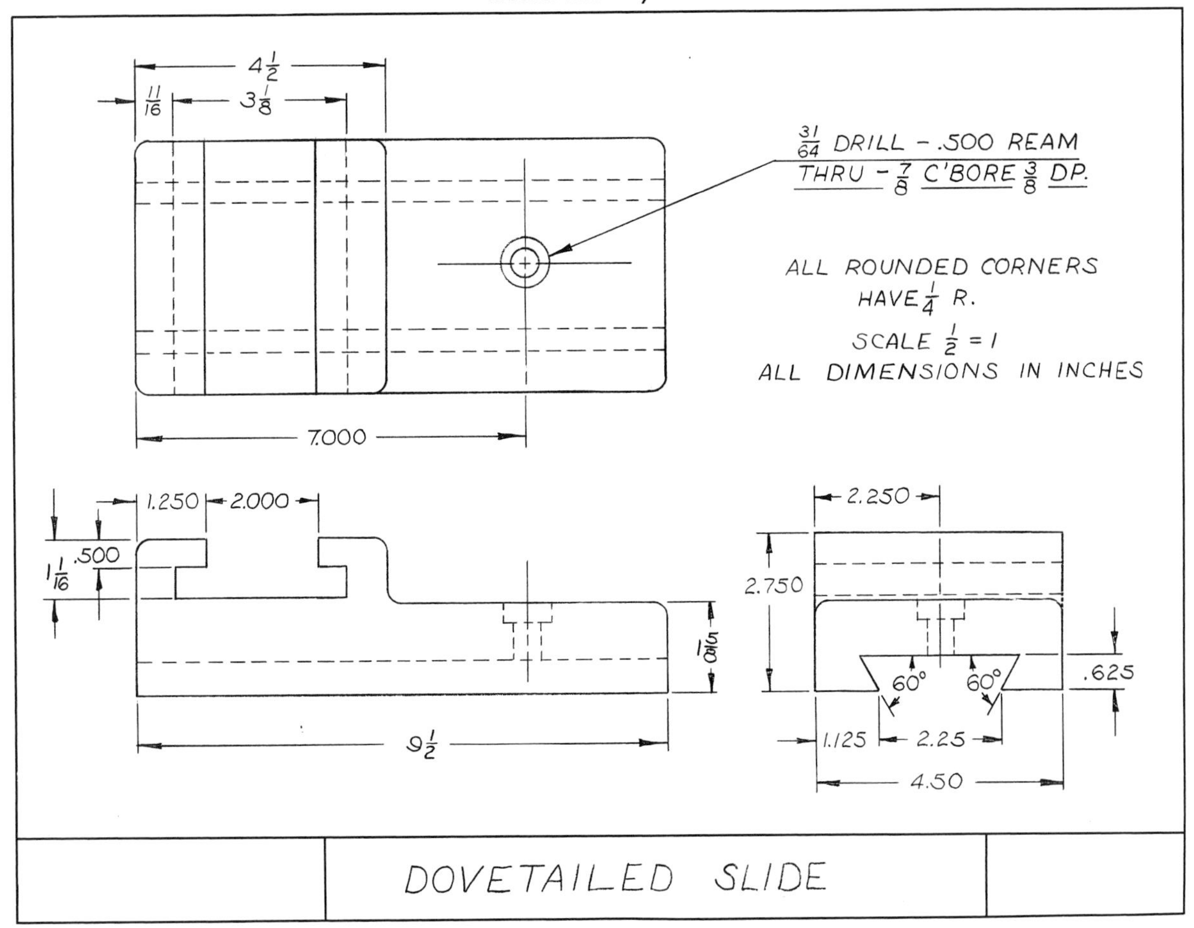

in the National Coarse Thread Series. The note under *B*, "⅝″—NF" means that the diameter is ⅝ inch, there are 18 threads per inch of threaded length of rod, and the thread is in the National Fine Thread Series.

Figure 13-16 shows the cross section of blocks of metal in which the threaded holes do not go through the blocks. Part A shows the depth of the drilled hole (dimension *X*) and the depth of the threads (dimension *Y*). In this case the hole has not been threaded the full length of the drilled hole. In Part B, the threads have been tapped the full length of the drilled hole, dimension *Z*. The thread is ⅝″—11 NC (National Coarse).

TOLERANCES. Most drawings indicate the *limits of tolerance*, which is the permissible range of variation in the dimensions of the completed job, Fig. 13-17. A drawing may specify limits of tolerance as $\pm \frac{1}{32}$ inch. This indicates that the dimensions of the object must be held to any size between one thirty-second of an inch larger or one thirty-second of an inch smaller than the size specified on the print. Tolerances may be expressed in fractional, decimal, or SI metric dimensions.

Scale

When practical, drawings are made full size, that is, the drawing is made the same size as the object. The scale of a drawing is the ratio between the actual size of the drawing and the actual size of the object which the drawing represents. The scale of a drawing made full size is indicated as follows: *full size* or *scale 12″ = 1′*. In other words, the ratio of a full-size drawing to the object is 1 to 1. The scale notation is often omitted for full-size drawings.

Many objects are so large that full-size drawings of them are not practical. In such cases, the

13-6. *Three-view drawing of the same dovetailed slide with the dimensions given in the SI metric linear measurement system.*

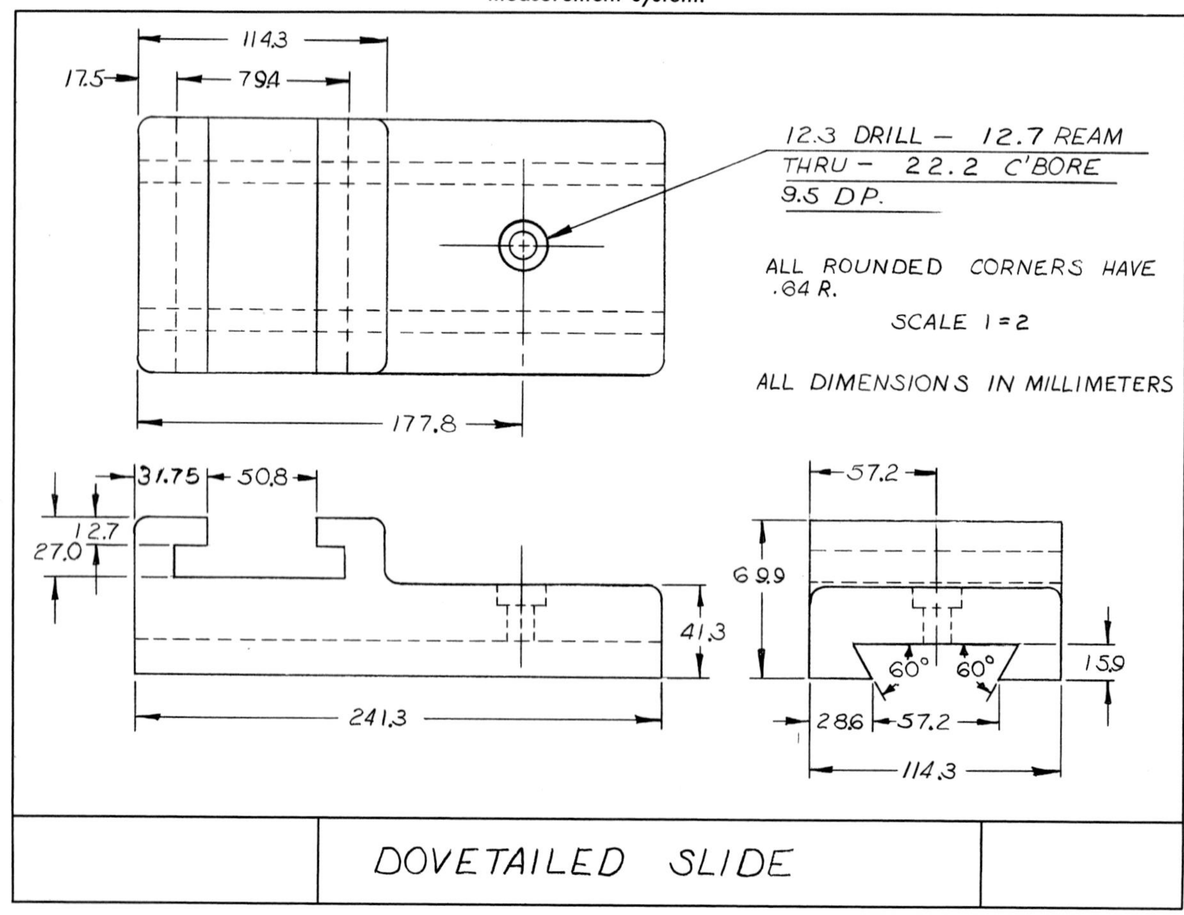

drawings are made smaller than the object, that is, to scale. For example, if the scale of a drawing is $\frac{1}{2}'' = 1''$, the drawing is one-half the actual size of the object. If the object is 12 inches long, the drawing is only 6 inches long. The notation scale $\frac{1}{4}'' = 1''$ on a drawing means that the drawing is made one-fourth the size of the object.

Your attention is called to the fact that the dimensions given on a drawing must be the actual dimensions of the object and are to be read as such. Never measure a drawing to determine the dimension to use. Paper shrinkage and dimension revisions may have occurred. Always use the dimension called for in the drawing.

Symbols and Abbreviations

Plans are specifically designed for the purpose of passing on information in the clearest and briefest manner possible. Sometimes the drawing must be supplemented with additional information. To avoid lengthy notes, symbols and abbreviations have been devised.

A *symbol* is a simple picture which represents an object or an idea. Its purpose is to condense information. The addition sign (+) is a common symbol. Except in a few instances, a drawing symbol usually shows the outline of an object, and it can be easily recognized as the object it represents. A drawing symbol also represents the same object to all trades. The weld symbols given in Chapter 14 are an example of these two features of drawing symbols.

Abbreviations are letters which are used for brevity. Like symbols, abbreviations are standardized so that they have the same meaning for all trades. The drawing symbol for inch is ″. The abbreviation is *in.* Except for a few unusual cases, the letters in an abbreviation are contained in the words they stand for.

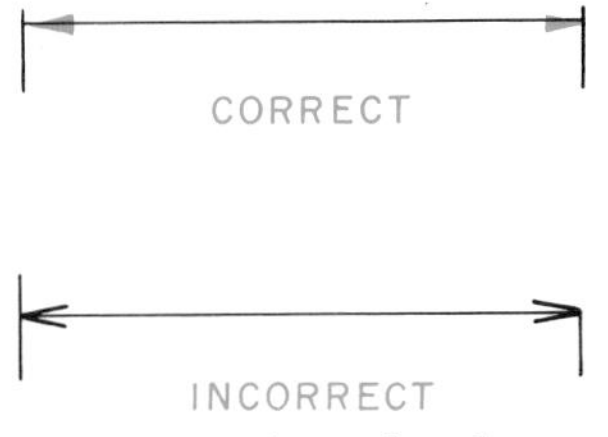

13-7. *Arrowheads.*

13-8. *Common methods of dimensioning.*

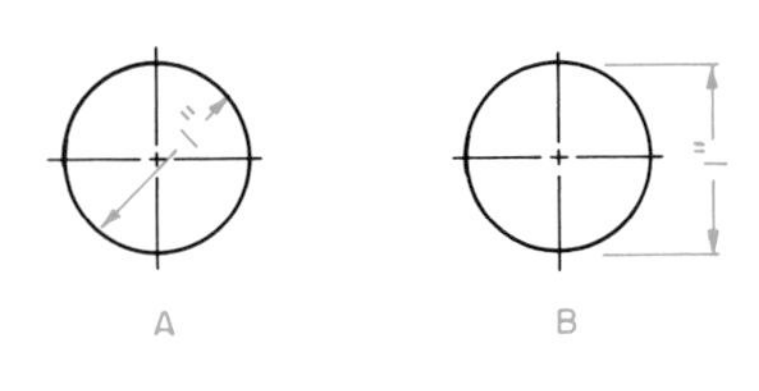

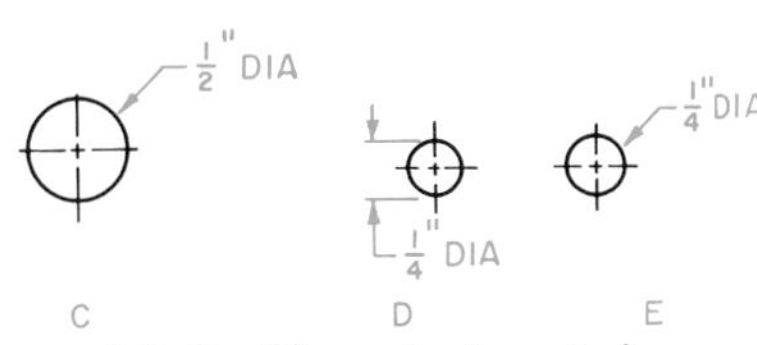

13-9. *Dimensioning circles.*

13-10. *Dimensioning radii.*

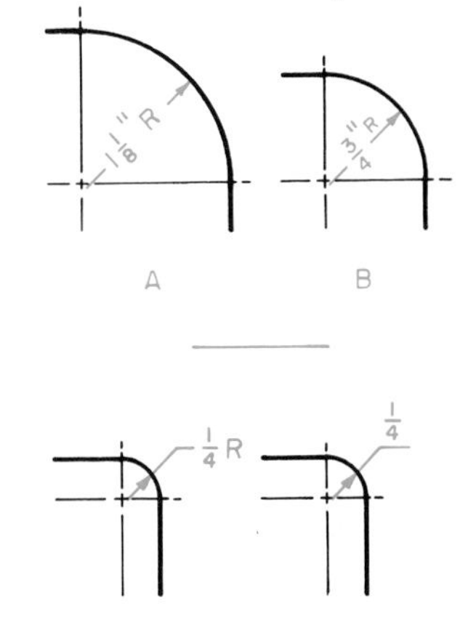

13-11. *Dimensioning angles.*

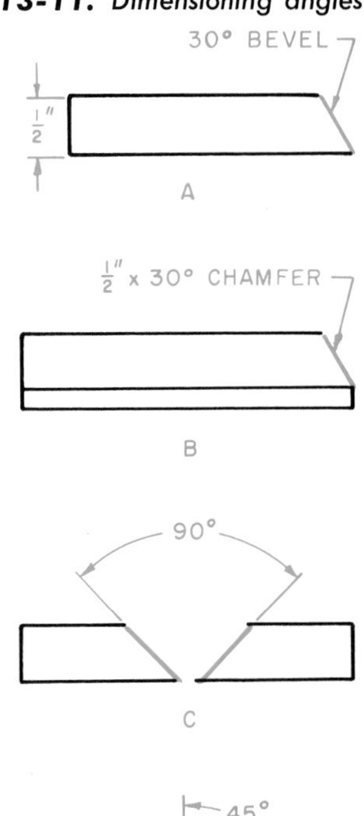

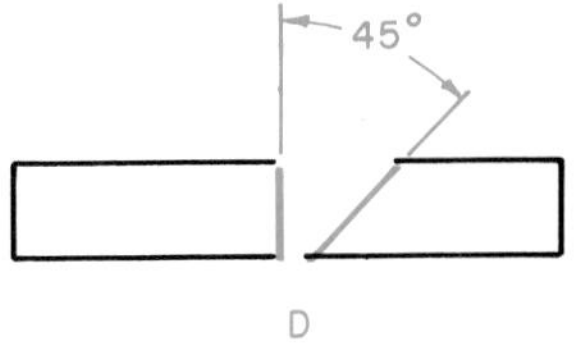

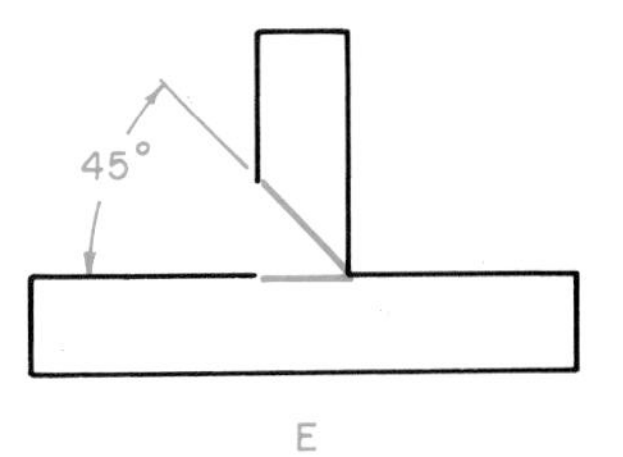

No matter how many standards are developed, there are always some engineers and architects who, through years of service, have developed symbols of their own. They usually issue an index sheet with a set of drawings, and this sheet defines the meaning of their own symbols and abbreviations. It is always good practice to check these index sheets on a job before interpreting the drawing.

Table 13-6 lists symbols and abbreviations generally used in industry. Some occur quite frequently, while others are so seldom seen that they can be easily forgotten.

Geometric Shapes

The following geometric shapes are combined in different ways to indicate the shape of the object in a drawing.

- *Circle:* Parts of a circle are drawn as shown in Fig. 13-18. The lines *B* running through the center of the circle are center

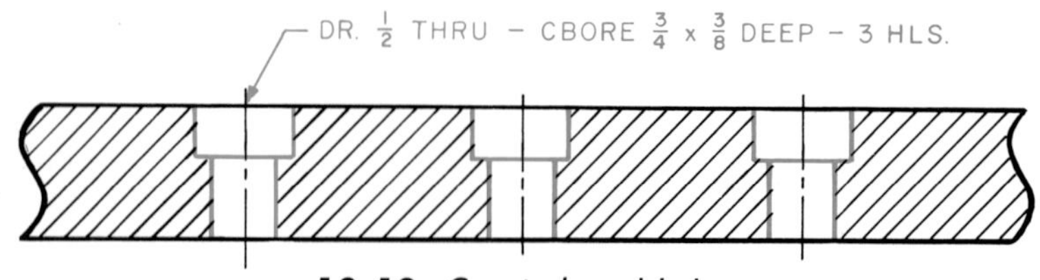

13-13. *Counterbored holes.*

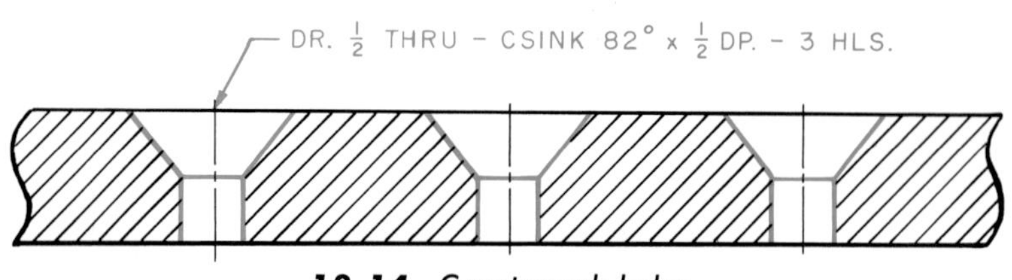

13-14. *Countersunk holes.*

13-12. *Angular dimensions.*

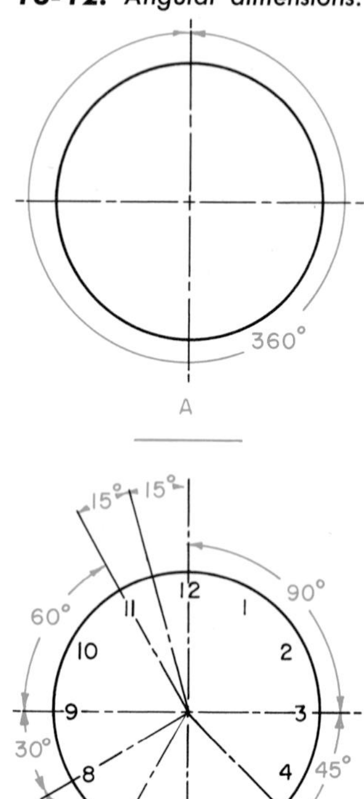

13-15. *Coarse and fine screw threads.*

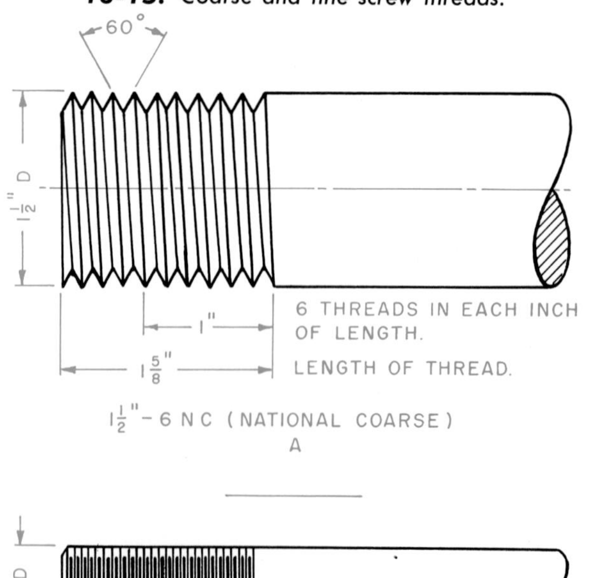

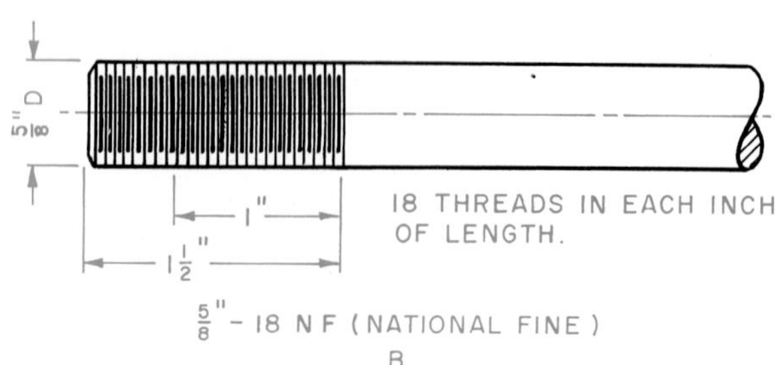

lines. The line *C* is the *circumference* of the circle (the distance around the circle). The line *D* is the *diameter* of the circle (the distance from one edge of the circle to the other). The diameter may be thought of as the width of the circle. The line *R* is the *radius* of the circle (the distance from the center of the circle to the circumference). The radius is one-half the length of the diameter. A *right angle* is one-fourth (90 degrees) of a circle. Lines *B* form a right angle.

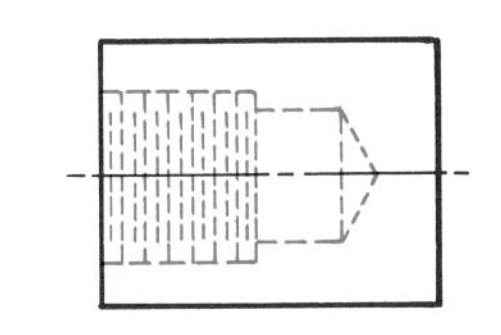

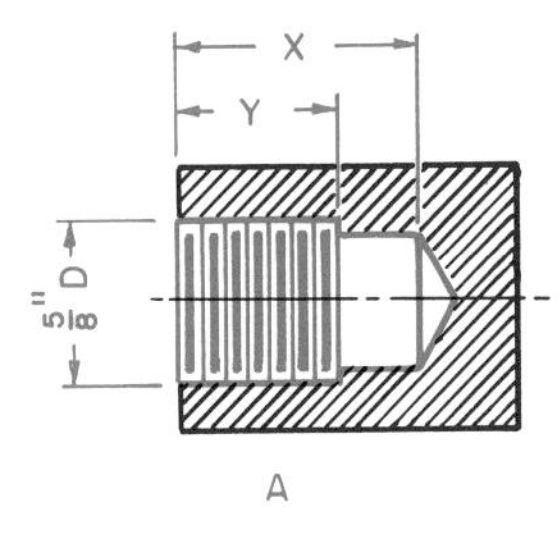

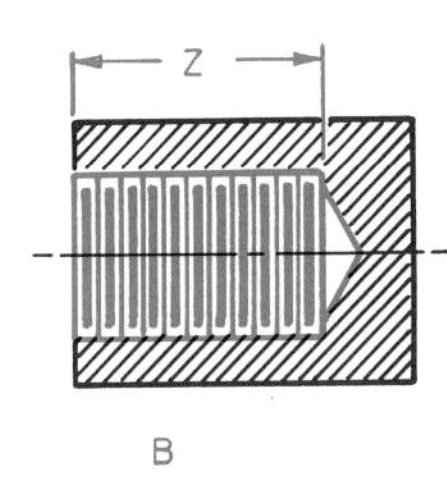

5/8" - 11 NC (NATIONAL COARSE)
13-16. *Threaded holes.*

- *Rectangle:* A rectangle, Fig. 13-19A, is a figure with four right angle (square) corners. The opposite sides are parallel. Two of the sides are of unequal length.
- *Square:* A square, Fig. 13-19B, is a rectangle with four sides of equal length. The corners of a square are right angles.
- *Diagonal:* A straight line connecting the opposite corners of a rectangle or square is called a diagonal, Fig. 13-19C.
- *Triangle:* A triangle, Fig. 13-19D and E, is a flat figure having three straight sides.
- *Equilateral triangle:* An equilateral triangle, Fig. 13-19D, is a triangle whose three sides are equal and whose three angles are equal.
- *Right Triangle:* A right triangle, Fig. 13-19E, is a triangle with one right (square) angle.

GAUGES

There are two gauges in common use for measuring the thickness of metal. The *U.S.S.* (United States Standard) gauge designates the thickness of sheet steel or sheet iron. The *B. & S.* (Brown and Sharpe) gauge designates the thickness of sheet brass and copper wire. In both cases, the larger the gauge number, the thinner the material. For example, 16 gauge is thinner than 9 gauge. Gauge sizes are given in the Appendix.

TYPES OF VIEWS

Drawings are made so that they describe the object in enough detail to be fabricated. A worker must learn to visualize what the object looks like from studying a drawing. It is important for him to determine immediately what his relative position is in regards to the picture. Is he looking down at the object, at its side, or at the inside? He can obtain a mistaken impression by failing to orient himself.

Three-View Drawings

In the United States the general practice has been to show first the front view, then the top view directly above it, and a side view; see the core box Fig. 13-20. These three basic views are enough to show all the details of an object. The inside or hidden surfaces which cannot be seen in the front and side views are

NO	PART NO	NAME	SIZE	MATL	SPEC
TOLERANCE–FRACTIONAL LIMITS 1/8"			DECIMAL LIMITS 0.125		
PART NAME			TYPE OF WELD		
SCALE	POSITION OF WELDING		PART NUMBER		

13-17. *Title block. Note 1/8-inch tolerance.*

Table 13-6

Drawing symbols and abbreviations.

Term	Abbreviation
Alternating current	AC
American Society for Testing Materials	ASTM
American Standard Association	ASA
American Welding Society	AWS
American Wire Gauge	AWG
Ampere	AMP
And	&
Angle	∠
Approximate	APPROX
Asbestos	ASB
Average	AVG
Barrel	BBL
Basement	BASMT
Bevel	BEV
Birmingham Wire Gauge	BWG
Brass	BRS
Brass, SAE #	BRS #
British Thermal Unit	BTU
Bronze, S.A.E.	BRZ #
Brown & Sharpe	B&S
Building	BLDG
Butt welded	BW
Cast iron	CI
Center to center	C to C
Centerline	℄
Chamfer	CHAM
Channel	L or CHAN
Circular	CIR
Clean out	CO
Cold-rolled steel	CRS
Column	COL
Composition	COMP
Concrete	CONC
Construction	CONST
Counterbore	CBORE
Countersink	CSK
Copper	COP
Cubic feet	CU FT
Cubic feet per minute	CFM
Cylinder, cylindrical	CYL
Degree	DEG or °
(Example)	30 DEG. or 30°
Detail	DET
Diagonal	DIAG
Diameter	DIA
Dimension	DIM
Direct current	DC
Ditto	DO
Down	DN
Drawing	DWG
Drill	DR
Electric	ELEC
Elevation	EL or ELEV
Extension	EXTEN
Exterior	EXT
Extra heavy	XHVY
Fabricate	FAB
Feet	FT or ′
(Example)	10 FT or 10′
Feet and inches (Example)	5 FT-11 IN or 5′-11″
Field weld	FW
Figure	FIG
Finish symbol	√ or ƒ
Flange	FLG
Floor	FL
Floor drain	FD or FLDR
Forging dies	FOD
Front view	FV
Gallon	GAL
Galvanized iron	GI
Gauge	GA
Glass	GL
Hexagonal	HEX
Holes	HLS
Horsepower	HP
Inches	IN. or ″
(Example)	10 IN. or 10″
Inside diameter	ID
Insulation	INS
Interior	INT
I-beam	I
I-beam column	I COL
(Example) 8″ I-beam, 30 pounds per ft.	8″ 30 # 1 COL
Kilowatt	KW
Lap-Weld	LW
Left hand	LH
Left side view	LSV
Linear feet	LIN FT
Malleable iron	MI
Maximum	MAX
Mechanical	MECH
Mild steel	MSM STL
Minimum	MIN
National coarse (screw threads)	NC
National fine (screw threads)	NF
National form	N
Number	NO. or #
Object	OBJ
Opening	OPNG
Ounce	OZ
Outside diameter	OD
Overall	OA
Oxidize	OXD
Perpendicular	PERP or ⊥
Piece(s)	PC, PCS
Pitch	P
Plate	⅊
Point	PT
Pound	LB or #
Pounds per square inch	PSI
Projection	PROJ
Radius	R or RAD
Reference	REF
Required	REQD
Revision	REV
Revolutions per minute	RPM
Right hand	RH
Right side view	RSV
Room	RM
Round	RD or ∅
Screwed	SCRD
Seamless	SMLS
Section	SECT
Sheet Steel	SHT STL
Side View	SV
Society of Automotive Engineers	SAE
Specification	SPEC
Square	SQ
Square feet	SQ FT or ⧄ FT
Square inch(es)	SQ IN
Steel	STL
Steel casting	STL CSTG
Standard	STD
Technical	TECH
Thread(s)	THD, THDS
Threaded	THRD
Through	THRU
Tongue and groove	T&G
Top view	TV
Typical	TYP
United States Standard	USS
Volt	V
Volume	VOL
Watts	W
Weight	WT
Wrought iron	WI
Yard	YD

shown by the use of invisible edge lines (L).

It is also possible to show the inside construction of an object by resorting to what are called sectional views. Figs. 13-20 & 24.

The pictorial drawing is rather easy for the untrained worker to understand. It is useful in showing the general appearance of simple objects. Since, however, most manufactured articles are rather complicated, it is necessary to use a system of three-view drawings. In addition to the three views, a pictorial drawing is used in this text to help make clear to you the approximate appearance of the object when seen from a corner. In general, isometric pictorial drawings are used only as an aid in learning to read working drawings. Pictorial drawings are seldom found on industrial blueprints.

You should try to visualize the completed object as you study each view shown on the three-view drawing along with the pictorial drawing. Thus you will understand better the relationship between the object as you are used to seeing it and the way this object is represented on a working drawing. The ability to look at a three-view drawing and visualize the object as it actually appears when looked at from any angle is the basis for reading blueprints.

The combination of front, top, and right side views shown in Fig. 13-21 represents the three views most commonly used by draftsmen to describe simple fabrications. In the pictorial view we can see the top surface, the front surface, and the right side surface of the camera. If you looked directly down on the camera with the front of it toward you, you would see the top as

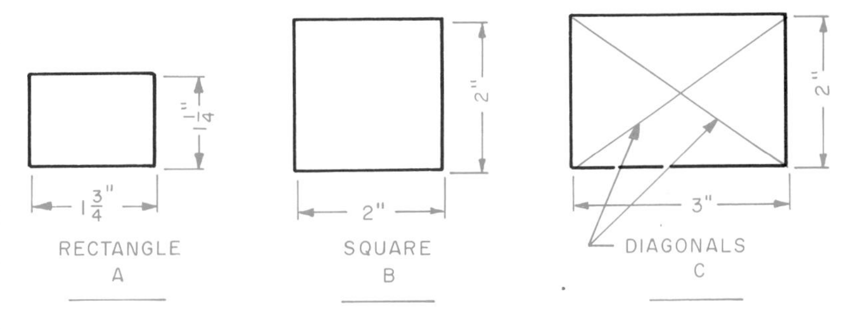

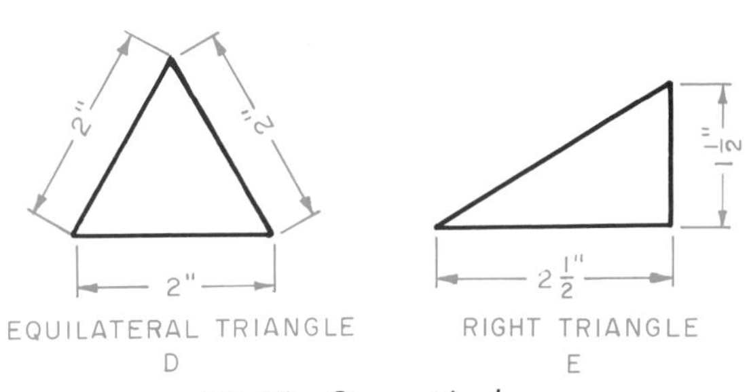

13-19. *Geometric shapes.*

13-18. *Parts of a circle.*

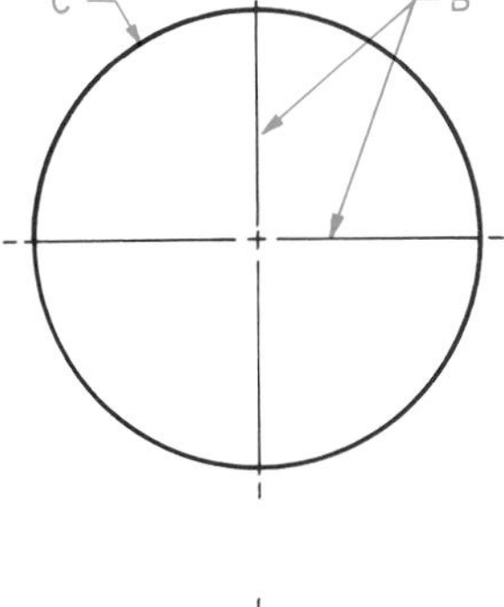

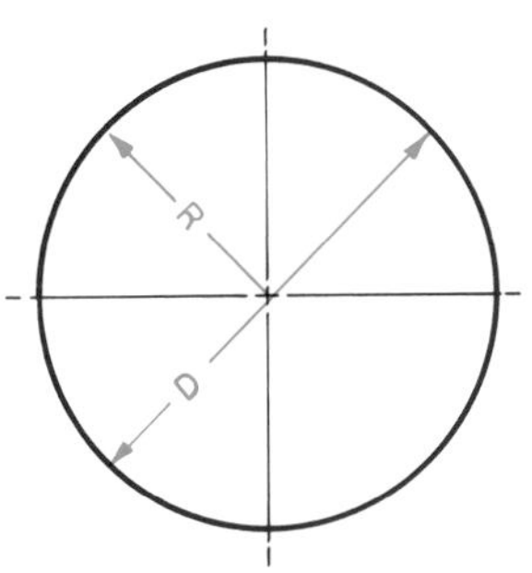

13-20. *Pictorial view of a core box and three-view drawing.*

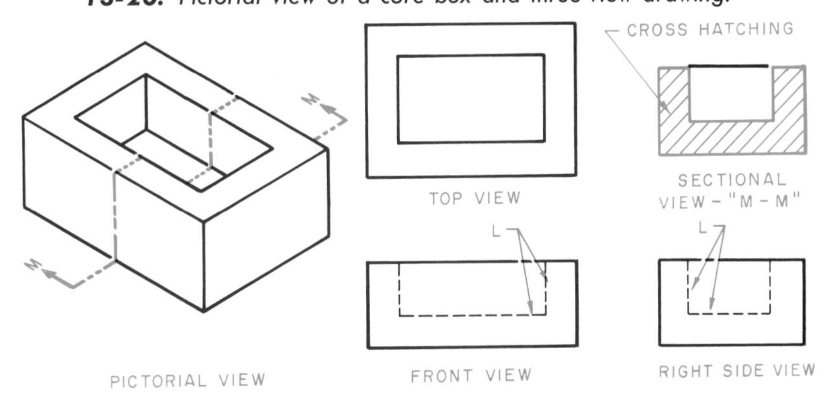

shown in the *top view* of Fig. 13-21. If you looked directly at the front of the camera, you would see the front as shown in the *front view.* If you looked directly at the right side, you would see it as shown in the *right side view.*

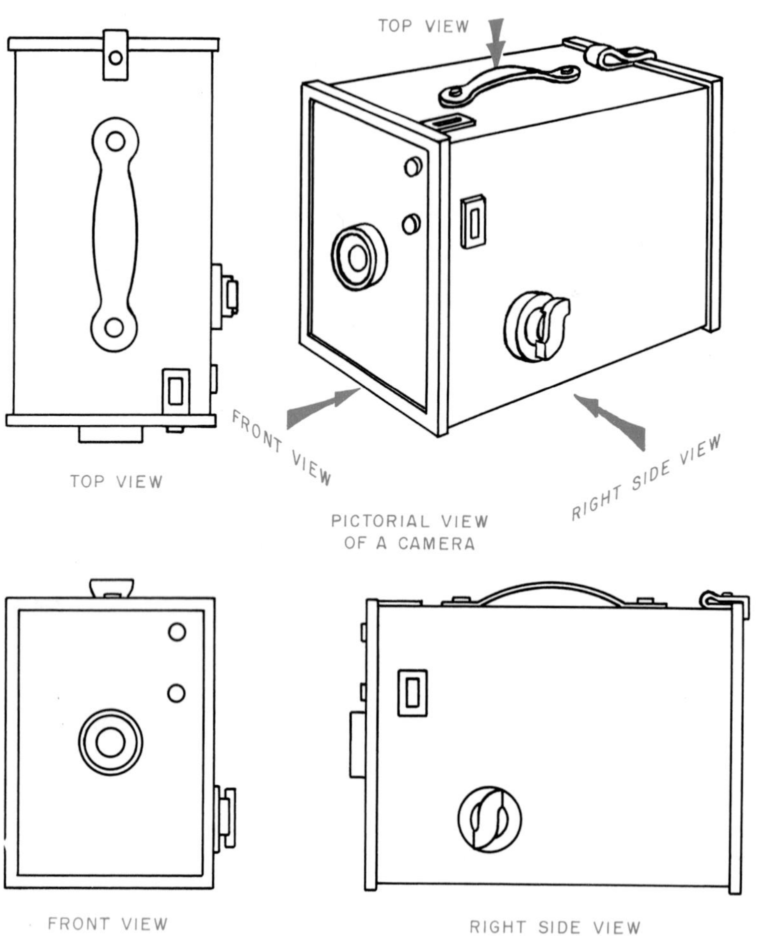

13-21. *Three-view drawing of a box camera.*

13-22. *Pictorial view of a steel plate and typical three-view drawing.*

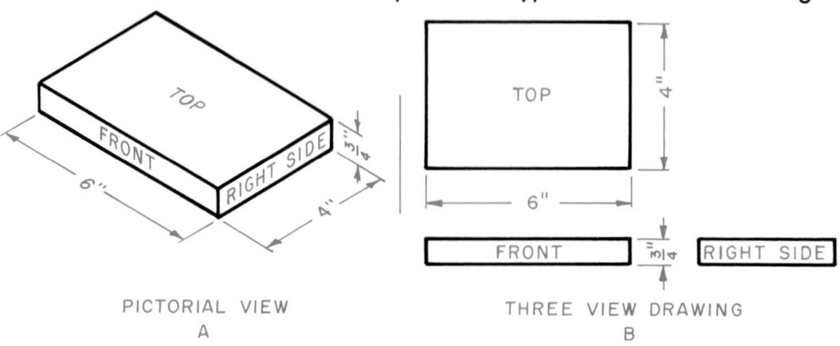

In Fig. 13-22 a steel plate is shown in a pictorial view and the normal three views. The top surface, the front surface, and the right side or end surface are shown in the pictorial view. If you were above the plate looking down on it, you would see the top surfaces as shown in the top view. If you were to see the plate from the front, you would see the front surface as it is shown in the front view. If you were to the right of the plate looking at the right side, you would see the right side surface as it is shown in right side view.

Sometimes the left side view is shown instead of or in addition to the right side view. Fig. 13-23A is a pictorial view of the same steel plate showing the top, front, and left surfaces. These three faces of the plate are shown in the three-view drawing, Fig. 13-23B.

In the American system of industrial drawings, the top, front, and right side views, Fig. 13-22B, are generally preferred to the top, front and left side views, Fig. 13-23B. However, since this text is a manual of basic welding operations, emphasis is placed on instructing you with the aid of complete and clear illustrations. Assuming a right-handed operator, the photographic and pictorial illustrations show each series of welding operations clearly in the top, front and left side views. The working drawings employ the top, front, and left side views to help you associate the working drawings with the accompanying sketches and photographs.

If the object is very complicated, it may be necessary to use more than three views. Sometimes the top, front, right and left side views, as well as the back and bottom views and special detail views are needed to present the shape, dimensions, and shop operations involved in making the object.

When all the required details can be shown in two views, the third view can be omitted. There are times when only one view shows all the necessary details. In this case the best view is shown on a drawing. Although it is good practice to show an object in the most convenient manner, an attempt is usually made to show it in its natural position.

Sectional and Detail Views

Sometimes in order to give the worker the necessary information to make the article, other views such as details and sections which show the inside of the object are given.

It is possible to show the inside construction of an object with *sectional views.* If we assume an imaginary cut on sectional line *M,* Fig. 13-24, we would have sectional view M-M instead of the usual side view. Line *M,* with arrows at its ends, shows not only the location of the imaginary cut in the top view, but also the direction in which the observer would look in order to see the sectional view M-M. Line *M* is called a *section line.* The diagonal lines indicate that part of the solid section of the box which would be cut through for the purpose of showing the sectional view. The diagonal lines are called "cross-hatching."

Detail views must be used when the scale of a drawing is so small as to make it impossible to show a clear picture of an object. When this happens, the object is drawn in a simplified form on the main drawing, and then the reader is referred to a drawing of the object in detail by a notation such as *SEE DETAIL A.* By using a much larger scale, the detail drawing can show the object more clearly. This method is also used when large numbers of an object are shown and when a certain assembly is used in several places. The main drawing then gives the location of the assembly and a typical detail is drawn somewhere else to show it more clearly.

The welder should study each view of a drawing carefully because each of them shows structural features necessary in the completed weldment.

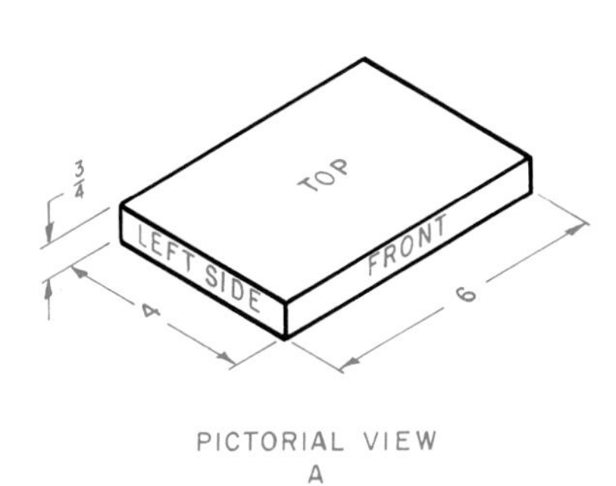

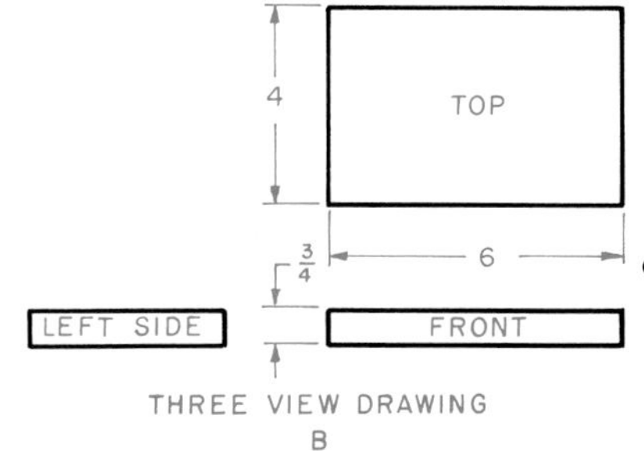

13-23. *Pictorial view of steel plate and unorthodox three-view drawing.*

13-24. *Pictorial view of a core box and method of sectioning.*

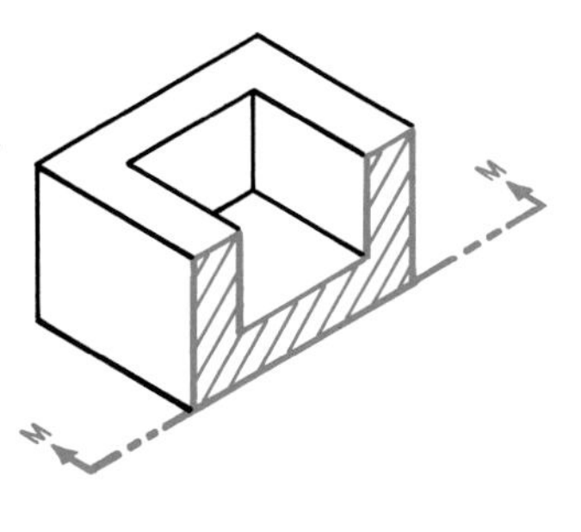

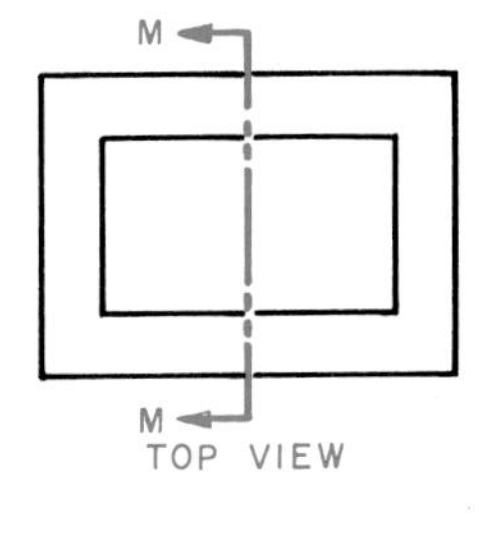

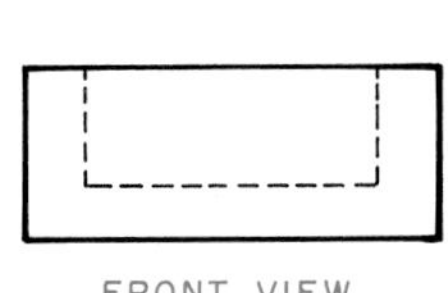

CROSS HATCHING

SECTIONAL VIEW - "M - M"

REVIEW QUESTIONS

Part I. Reading Assignment

A.
1. Is it important that a welder be able to read blueprints? Explain.
2. What is the purpose of a print?
3. Name and explain at least four methods of printing drawings.
4. What is the meaning of "scale" as applied to drawings?
5. Name the two gauges used to measure the thickness of metal.
6. What views of an object are usually drawn?
7. A single-view drawing is never used in industry. True. False.
8. What is a "detail view"? Explain.
9. Abbreviations are never used on a drawing. True. False.
10. By what type of line are hidden surfaces indicated?
11. Draw a simple three-view sketch of a rectangular block 4 inches long, 3 inches wide and 2 inches thick. Mark dimensions on the views.
12. How may the inside construction of an object be shown?
13. A square block is cut through the center. How may this be indicated on the drawing?
14. A circle contains (a) 180 degrees, (b) 210 degrees, (c) 360 degrees, (d) 280 degrees.
15. Blueprints and shop sketches are the sign language of industry. True. False.
16. The metric system is difficult and will never be used in the United States. True. False.
17. Only a very small part of the world's population uses the metric system. True. False.
18. Name the two main types of screw threads.
19. What sign is used on drawings to indicate the allowable tolerance?
20. Welding symbols should never be shown on a drawing. True. False.
21. In order to be sure of the dimensions on a drawing, it is good practice to scale a drawing. True. False.

B. *Draw the appropriate line for the following terms:*
1. Border line
2. Visible edge line
3. Invisible edge line
4. Extension line
5. Center line
6. Section line
7. Pointer
8. Dimension line
9. Break line

C. *Draw the appropriate figure for the following:*
1. Rectangle
2. Square
3. Diagonal
4. Triangle
5. Right triangle

Part II. Interpreting Drawings

Following are a number of illustrations that present the views common to shop drawings. Study them carefully and answer the questions which follow each drawing.

A. ONE-VIEW DRAWINGS. Figures A-1 through A-4 show four different objects that are presented as single-view drawings. Flat objects of this kind can be shown with all dimensions given, and the thickness and type of material are given by a note.

A.1 Washer.

- *Center lines:* Your attention is called to lines *B* and *C* which are drawn through the center of the circles. These are center lines and are always used on views of drawings representing circles and generally on drawings representing parts of circles. They are called dot-and-dash lines, but their construction actually consists of short and long dashes. Center lines are not a part of the drawing of the object.

- *Diameter and radius:* Line G is a dimension line indicating the outside diameter of the washer. Diameter is often abbreviated *DIA*. One-half of the diameter is the radius, abbreviated *R*.

1. Of what material is the washer made?
2. What is the abbreviation for wrought iron?
3. What is the thickness of the washer in decimals? This is approximately $\frac{5}{32}$ inch.
4. What is the outside diameter of the washer? Do not include the word diameter or its abbreviation in your answer.
5. What is the diameter of the hole?
6. What is the radius of the outside of the

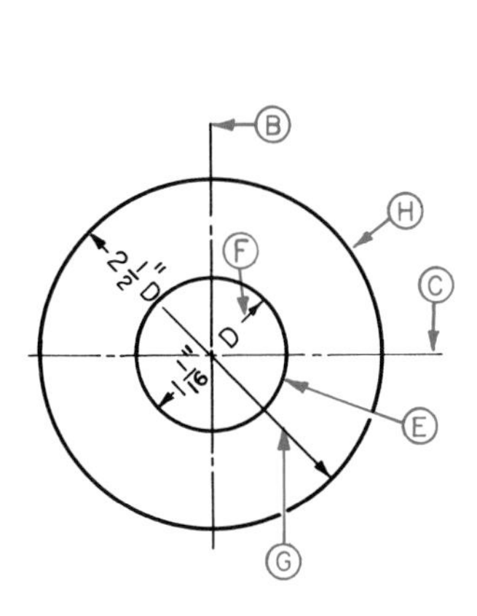

WASHER
WROUGHT IRON (W. I.)
0.1563" THICK

A-1. *Washer.*

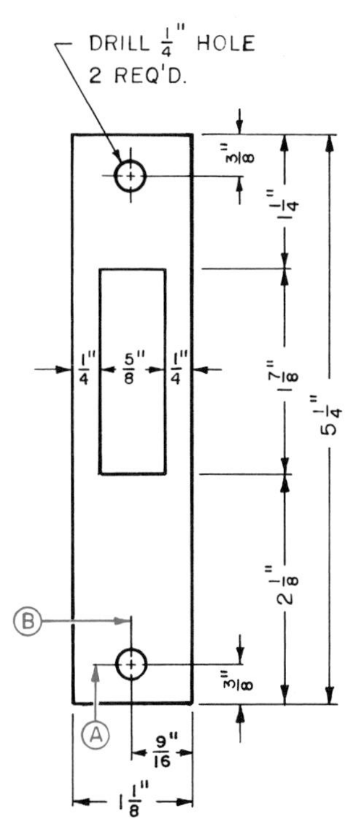

A-2. *Lock keeper.*

LOCK KEEPER
SHEET BRASS
No. 14 (0.064") B. & S. GA.

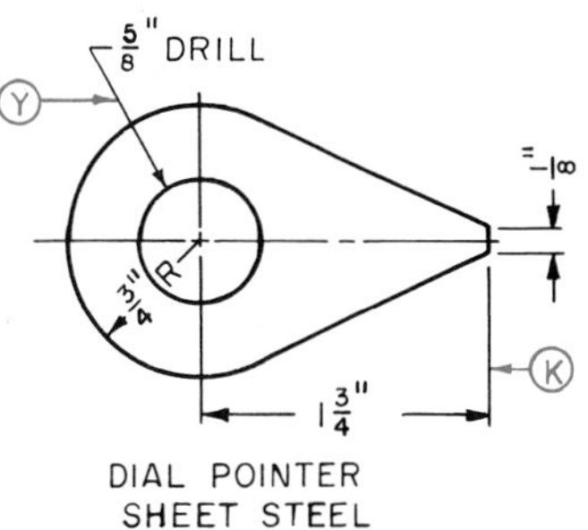

DIAL POINTER
SHEET STEEL
No. 16 (0.0625") U. S. S. GA.

A-3. *Dial pointer.*

A-4. *Spacing washer.*

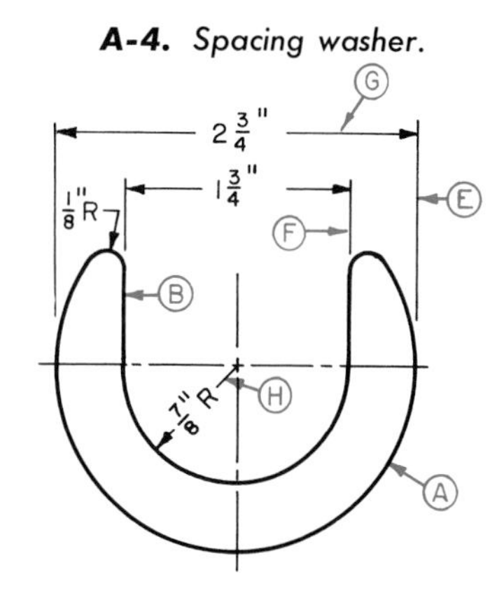

SPACING WASHER
SHEET STEEL
No. 28 (0.0156") U.S.S. GA.

washer? Do not include the word radius or its abbreviation in your answer.

7. What is the distance from the edge of the hole to the outside of the washer?

8. What kind of lines are lines *F* and *G*?

9. What kind of lines are lines *H* and *E*?

10. What kind of lines are lines *B* and *C*?

11. What do we call lines which refer to or point to some part of a drawing?

12. For what purpose is a dot-and-dash line used?

A.2 Lock Keeper. Figure A-2 is a drawing of a lock keeper which is fastened to a doorjamb or doorframe to receive the bolt from the lock.

● *Brown and Sharpe Gauge:* The notation "No. 14 (0.064") B. & S. Ga." means under 14 gauge, Brown and Sharpe. The brass in this size is approximately $\frac{1}{16}$ inch thick. This is another one of the several gauges, such as U.S.S., for designating thickness of metals. The Brown and Sharpe gauge is the most common one for sheet brass and copper wire.

● *Dimensioning:* Where a space is too small for the dimension line and the dimension, the dimension lines are placed outside the extension lines. Note the $\frac{3}{8}$-inch dimensions at the bottom and the top of the drawing. In this case, the distance dimensioned is from the center line to the extension line, or from arrowhead to arrowhead. A similar example is that of the $\frac{1}{4}$-inch dimension which locates the rectangular hole from the sides of the object.

● *Drilled Holes:* The size of drilled holes is often given in notes such as the one found on this drawing. The note "DRILL $\frac{1}{4}$" HOLE 2 REQ'D." means that the holes are made with a $\frac{1}{4}$-inch drill and will be $\frac{1}{4}$ inch in diameter. This note refers not only to the hole to which it points but also to the hole near the bottom of the lock keeper.

Note that lines *A* and *B* are center lines which are also used as extension lines for dimensions.

1. Of what material is the lock keeper made?
2. What is the overall length?
3. What is the overall width?
4. What is the thickness in decimals of an inch?
5. What number gauge metal is used?
6. What kind of lines are lines *A* and *B*?
7. What is the diameter of the drilled holes?
8. How many drilled holes are there?
9. What is the distance from the center of the drilled hole, near the bottom, to the bottom of the keeper?
10. What is the distance from the center of the drilled holes to the left side of the keeper?
11. What is the length of the rectangular hole?
12. What is the width of the rectangular hole?
13. What is the distance from the right side of the keeper to the right side of the rectangular hole?
14. What is the distance from the bottom of the rectangular hole to the bottom of the keeper?

A.3 Dial Pointer. Figure A-3 is a dial pointer used on an oil-burning engine. The left end of the pointer is partly circular in shape with a circular hole. The other end tapers to a blunt point. The pointer is made of 16-gauge metal which is 0.0625 inch thick.

1. What kind of line is line *K*?
2. What kind of line is line *Y*? (This is not a dimension line.)
3. What is the distance from the center line of the drilled hole to the right end of the pointer?
4. What is the radius of the circular end of the pointer?
5. What is the overall length of the pointer?
6. What is the overall width of the pointer?
7. How wide is the pointer at the right end?
8. What is the diameter of the drilled hole?

A.4 Spacing Washer. The spacing washer shown in Fig. A-4 is a part of an electric motor. The washer is made of 28-gauge steel which is 0.0156 inch thick. Its general shape is somewhat like the letter *U*.

1. What is the outside diameter for the washer?
2. What kind of lines are lines *E* and *F*?
3. What kind of lines are lines *A* and *B*?
4. What kind of lines are lines *G* and *H*?
5. What is the radius of the small ends at the back of the washer (at the top of the *U*)?
6. What is the distance between the straight edges forming the back part of the opening?

B. TWO-VIEW DRAWINGS. Only two views, such as the top and front views or the front and side views may be adequate for some simple objects. The drawing of the spring cap, Fig. B-1, consists of only two views—the front view and the side view. When two views of an object are the same, or if the third view does not give additional information which an experienced shopman could not get from the other two views, only two views are given. Since the drawing of the top view of the cap would be exactly the same as the drawing of the front view, the top view is omitted.

B.1 Spring Cap. Figure B-1 is a two-view drawing of a spring cap used on an engine.

1. Which lines in the front view represent the cylindrical hole in the body?
2. Which circle in the side view represents the hole in the body?
3. Which circle in the side view represents the short cylinder between the flange and the body?
4. Which circle in the side view represents the outside cylindrical surface of the flange?
5. Which circle in the side view represents the outside surface of the body?
6. Which lines in the front view represent the cylindrical surface of the larger hole at the right end of the spring cap?
7. Which line in the side view represents the larger cylindrical hole?
8. Which line in the front view represents the shoulder (offset) that is formed where the larger hole and the smaller hole meet?
9. Between which two circles in the side view does the shoulder lie?
10. Which line in the front view represents the shoulder formed where the flange and the short cylinder meet (the left side of the flange)?
11. Between which two lines in the side view should the shoulder lie?
12. How many surfaces are visible in the side view?
13. If you were to look at the left end of the spring cap, how many flat surfaces would be visible?
14. What is the diameter of the hole in the body?
15. What is the diameter of the larger hole in the right end?
16. What is the length of the body?
17. What is the radius of the profile (the outline of the widened part at the right end of the body)?

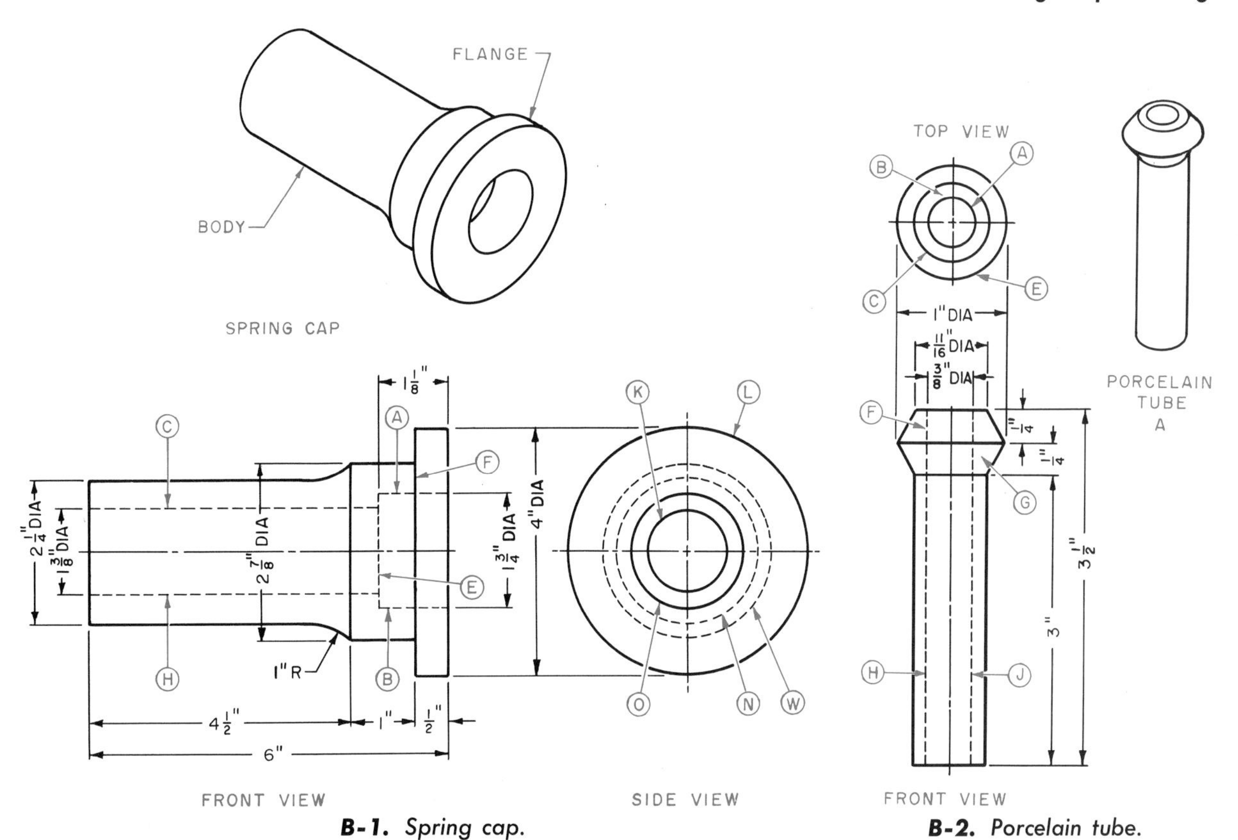

B-1. *Spring cap.*

B-2. *Porcelain tube.*

18. What is the length of the cylinder between the flange and the body?

19. What is the thickness of the flange (the distance from the left to right)?

20. How long is the large cylindrical hole at the right end?

21. What is the diameter of the body?

22. What is the thickness of the wall at the left end of the body?

23. What is the width of the shoulder (offset) between the two cylindrical holes? Note that the diameter of the large hole includes the diameter of the small hole plus two shoulders.

24. What is the diameter of the short cylinder next to the flange?

25. If the left side view of the spring cap were drawn, how many cylindrical surfaces would be represented by dotted lines?

B.2 Porcelain Tube. The porcelain tube, Fig. B-2, is made of clay and baked in a kiln. Tubes were formerly used in electrical work when wires were run through joists, studs, partitions, or any wooden member of a building. The building laws of almost all large cities prohibit this type of wiring and require metal pipes called conduits. Knob and tube wiring is still permitted in some districts. The porcelain tube, however, is an excellent object to be studied as a two-view drawing. Center lines are used on drawings of cylindrical or cone-shaped objects.

The outside diameter of the top face of the tube and the outside diameter of the cylindrical part are the same, $^{11}/_{16}$ inch. If these diameters were different, the surface of the cylinder would have to be shown in the top view by a dotted circle. Since they are the same, the visible (solid) circle covers the dotted circle. When looking at the top view, you must understand that circle *C* represents both the outside edge of the top face of the tube and the outside surface of the cylindrical part.

1. What is the largest diameter?

2. What is the length of the cylindrical part?

3. What is the length of each of the conical parts? Length in this case is taken along the axis or center of the tube.

4. What is the diameter of the base of the conical parts?
5. What is the outside diameter of the cylindrical part?
6. What is the diameter of the hole?
7. What is the thickness of the wall of the cylinder? (Note that the outside diameter of this cylinder includes the diameter of the hole and the thickness of the two walls.)
8. Which circle in the top view represents the hole in the tube?
9. Between which two circles in the top view does the conical surface *F* in the front view lie?
10. Which lines in the front view represent the hole?
11. Which circle in the top view represents the outside edge of the top surface of the tube?
12. Which circle in the top view represents the outside surface of the cylinder?
13. Between which two circles in the top view does the conical surface *G* in the front view lie?
14. Which surface in the top view is the top of the tube?
15. How many circles in the top view represent cylindrical surfaces?

C. THREE-VIEW DRAWINGS. It has been previously shown that simple objects can be presented as one-view or two-view drawings. Complicated objects must be presented by three-view drawings for the welder to be able to fabricate the final component. The views are arranged in a group to show the shape and construction of the object.

C.1 Channel. The channel shown in Fig. C-1 is a steel structural member used in building construction, shipbuilding, railroad work, and the machine metal trades. Channels of various shapes are used.

The front and back legs of the channel are called *flanges,* and the other member is called the *web.*

In the top view, surface *B* is the top of the web. This surface is represented by line *V* in the side view. In the front view, the top surface of the web is not visible. Therefore, line *N,* representing this surface, is a dotted line to indicate a hidden surface. Line *G* in the top view and *U* in the side view represent the inside flat surface of the front flange. Lines *E* in the top view and *W* in the side view represent the inside flat surface of the back flange. Neither of these surfaces is visible in the front view.

1. Which line in the top view represents the front of the channel?
2. Which line in the side view represents the front of the channel?
3. Which surface in the front view is the front?
4. Which line in the top view represents the back of the channel?
5. Which line in the side view represents the back of the channel?
6. Which surface in the side view is the right end?
7. Which line in the side view represents the inside flat surface of the back flange?
8. Which line in the top view represents the inside flat surface of the back flange?
9. Which line in the side view represents the curved surface at the top of the back flange?
10. Which surface in the top view is the curved surface at the top of the back flange?
11. Which line in the side view represents the upper flat surface of the web?
12. Which surface in the top view is the upper flat surface of the web?
13. Which line in the front view represents the upper flat surface of the web?
14. Which line in the side view represents the rounded corner where the back flange and the web meet?
15. Which surface in the top view is the rounded corner where the back flange and the web meet?
16. What is the distance from the top of the flat surface of the web to the top of the channel?
17. What is the thickness of the web?
18. What is the thickness of the flanges?
19. What is the distance between the inside flat surfaces of the flanges?
20. What is the radius of the curved surfaces at the top of the flanges?
21. What is the radius of the curved surfaces where the web and the flanges meet?
22. What is the height (width) of the inside flat surface of the front flange?
23. What is the width of the upper flat surface of the web?
24. Line *E* in the top view represents a surface. Which line in the side view also represents this surface?
25. Is the surface referred to in question 24 visible in the front view?
26. Which line in the top view represents surface *T* in the side view?
27. Which surface does line *N* in the front view represent in the top view?

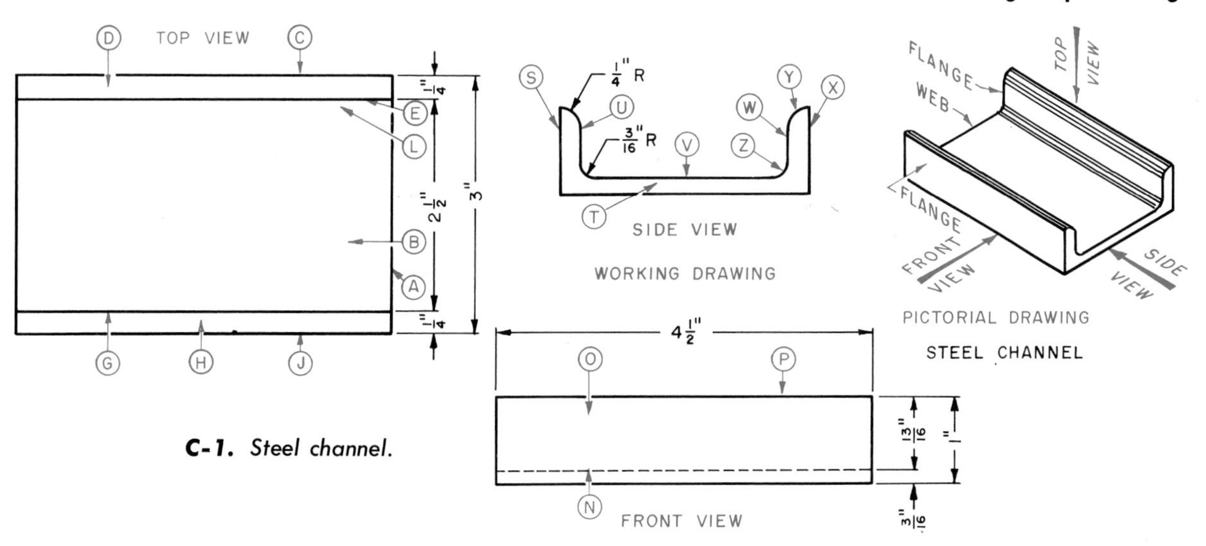

C-1. *Steel channel.*

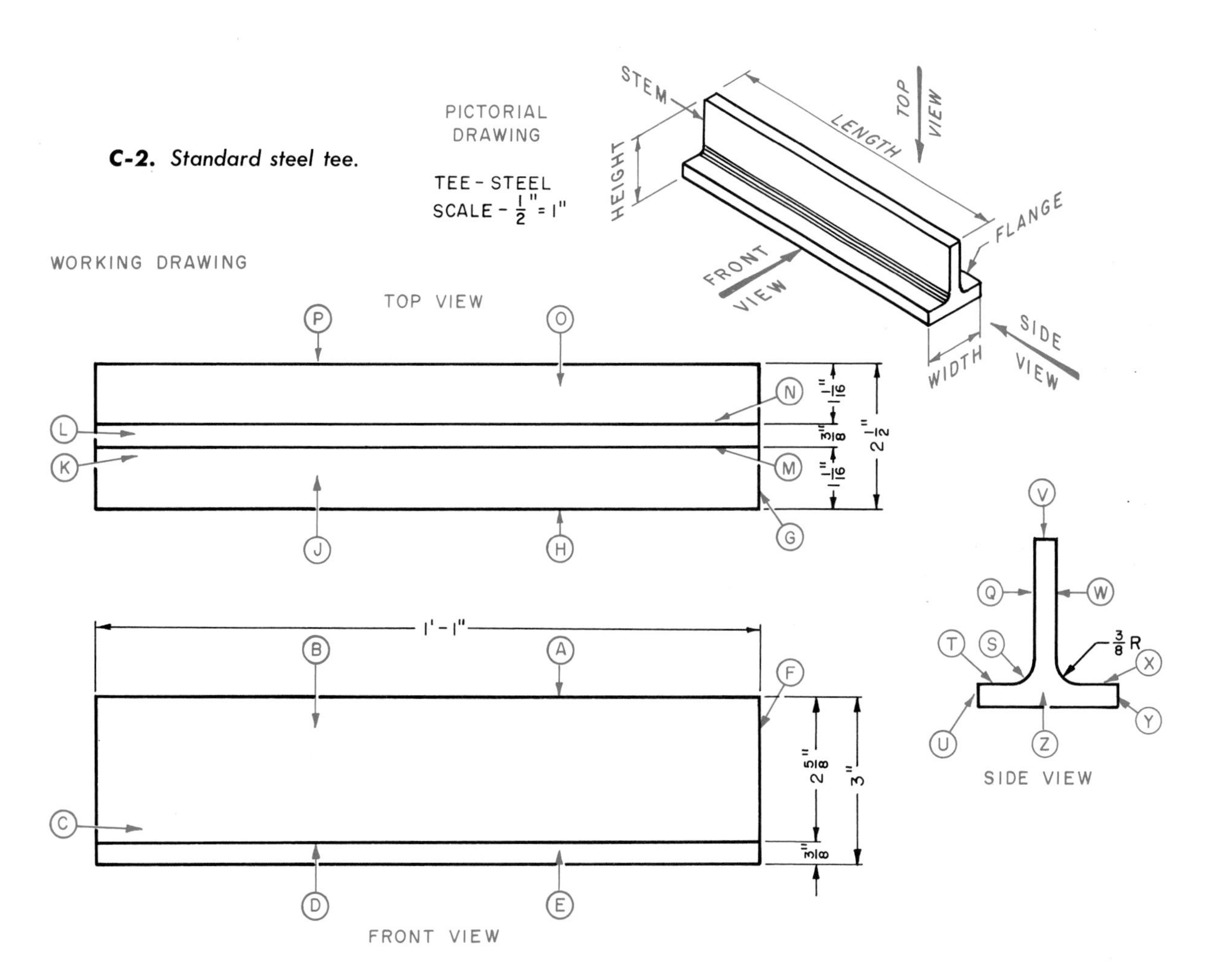

C-2. *Standard steel tee.*

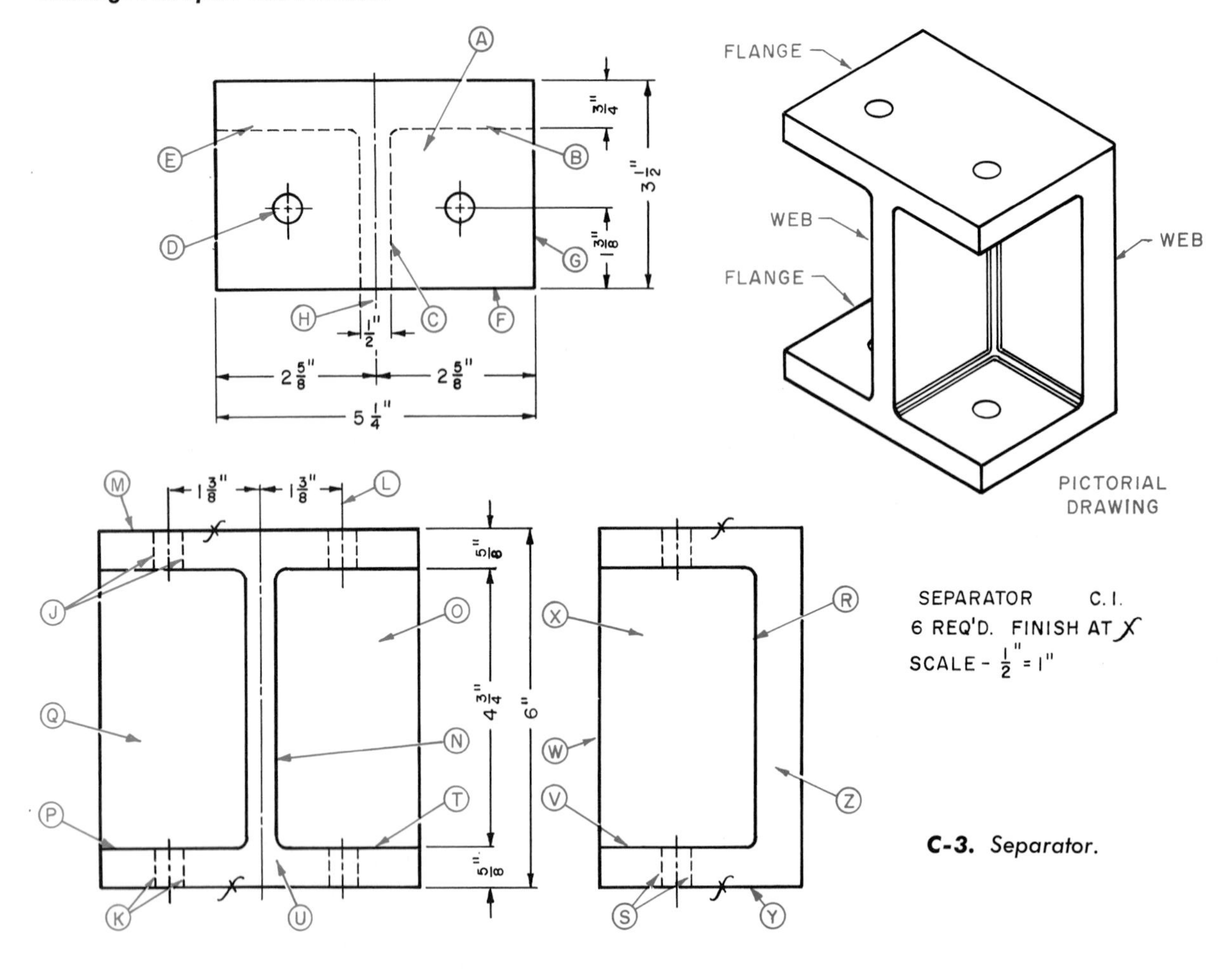

C-3. *Separator.*

28. Which line in the side view represents surface *B* in the top view?

29. Which line in the front view represents line *V* in the side view?

30. Which surface in the top view represents line *Z* in the side view?

31. Which line in the top view represents line *U* in the side view?

32. Is the surface referred to in question 31 visible in the front view?

33. Which line in the top view represents surface *O* in the front view?

34. Which line in the side view represents surface *D* in the top view?

35. Which line in the front view represents the tops of the front and back flanges?

36. How many flat surfaces are visible in the top view?

37. How many curved surfaces are visible in top view?

C.2 Standard Steel Tee. Figure C-2 is a drawing of a standard steel tee, so called because its end view resembles the letter *T*. Tees are used in structural members of buildings and steel bridges. T-shaped parts are often used in machinery.

The upright member of the tee is known as the *stem*. The bottom member is known as the *flange*.

1. What is the scale of the drawing?
2. Of what material is the tee made?
3. What is the overall length?
4. What is the overall width?
5. What is the overall height?
6. Which line in the top view represents the back of the flange?
7. Which line in the top view represents the back of the stem?
8. Which line in the front view represents the top of the front part of the flange?

9. Which line in the side view represents the front of the stem?

10. Which flat surfaces (note there are two) in the top view are the top of the flange?

11. Which line in the front view represents the right side of the tee?

12. Which surface in the top view is the top of the stem?

13. Which surface in the front view is the front of the stem?

14. Which surface in the front view is the front of the flange?

15. What is the thickness of the flange?

16. What is the thickness of the stem?

17. What is the distance from the front of the flange to the front flat surface of the stem?

18. What is the distance from the top of the stem to the top flat surface of the flange?

19. Note the rounded or curved corner where the flange and the stem meet. This makes the connection between the flange and the stem much stronger. What is the radius of this curve?

20. The distance from the front of the flange to the front surface of the stem is $1\frac{1}{16}$ inch. Since $\frac{3}{8}$ inch of this distance is taken up by the curved surface, the flat surface of the top of the flange at the front is $\frac{11}{16}$ inch wide. This is also true of the flat surface at the back of the flange. What is the width (height) of the flat surface of the front of the stem?

21. Which surface in the front view does line *Q* in the side view represent?

22. Which line in the front view does surface *J* in the top view represent?

23. Which line in the top view does surface *B* in the front view represent?

24. Which line in the side view represents line *P* in the top view?

25. Which surface in the top view does line *X* in the side view represent?

26. Which line in the side view represents surface *L* in the top view?

27. Which surface in the top view does line *T* in the side view represent?

28. Which surface in the front view does line *H* in the top view represent?

29. Which surface in the top view does line *A* in the front view represent?

30. Which line in the top view represents line *W* in the side view?

31. Which line in the top view represents surface *Z* in the side view?

32. Which line in the side view represents surface *E* in the front view?

33. The rounded corner (curved surface) where the front of the stem meets the top of the flange is represented in the side view by line S. Which letter indicates this surface in the top view?

34. Which letter in the front view indicates the rounded corner where the front of the stem meets the top of the flange?

35. If the scale of a drawing were $1'' = 1''$, and the length of the drawing of the object were $5''$, what would be the actual length of the object?

36. If the length of an object were 20 inches, and the scale of the drawing were $\frac{1}{2}'' = 1''$, the length of the drawing would be one-half the length of the object, or 10 inches. If the scale of a drawing were $\frac{1}{12}'' = 1''$, and the length of the drawing were 7 inches, what would be the actual length of the object?

37. If the actual length of this tee were 13 inches, and the scale of the drawing were $\frac{1}{2}'' = 1''$, what would be the length of the drawing of the tee?

38. What is the actual length of the tee represented on this drawing?

39. What is the actual height of the tee represented on this drawing?

40. What is the width of the drawing of the tee represented on this drawing? Do not use a rule to measure the drawing, but get your information from the dimensions on the drawing and the scale of the drawing.

C.3 Separator. Figure C-3 is a drawing of a machine part called a separator. It is often used as a shelf for bearings.

In previous problems, center lines were used only for circles or parts of circles. In the top and front views of this drawing, the center line of the separator itself is used for the purpose of dimensioning.

The note "6 *Req'd.*" (required) means that 6 pieces are required of the kind represented on the drawing.

1. What is the scale of the drawing?
2. What is the overall height?
3. What is the overall depth, front to back?
4. What is the overall width?
5. What kind of lines are lines *W* and *R* in the side view?
6. What kind of lines are lines *B* and *C* in the top view: extension lines, projection lines, object lines, or invisible lines?
7. What kind of object lines are lines *J* in the front view?
8. What kind of line is line *H* in the top view?

9. What kind of line is line *L* in the front view?
10. Which two lines in the top view represent the front surfaces of the back web?
11. Which two lines in the front view represent the top surfaces of the bottom flange?
12. Which surface in the side view is the right surface of the center web?
13. How many holes are represented by line *D* in the top view?
14. How many holes are represented by lines *S* in the side view?
15. What is the thickness of the back web?
16. What is the thickness of the center web?
17. What is the thickness of the top flange?
18. What is the distance, top to bottom, of the inside surfaces of the flanges?
19. What is the distance from the centers of the holes in the bottom flange to the back of the separator?
20. What is the distance from the center of the right hole in the upper flange to the right side of the separator?
21. What is the center-to-center distance of the holes in the lower flange?
22. What is the depth of the drilled holes?
23. How many drilled holes are there?
24. Which surface in front view is represented by line *E* in top view?
25. Line *V* in the side view represents two surfaces. Which lines in the front view represent these surfaces?
26. Which line in the top view represents line *W* in the side view?
27. Which surface in the side view represents line *G* in the top view?
28. Which surface in the front view represents line *R* in the side view?
29. The left hole in the bottom flange is represented by line *D* in the top view. Which lines in the side view represent this hole?
30. Which surface in the front view represents line *F* in the top view?
31. Which line in the top view represents surface *X* in the side view?
32. Which line in the side view represents the surfaces indicated by lines *E* and *B* in the top view?
33. Which surface in the side view does line *N* in the front view represent?
34. Which surface in the top view does line *M* in the front view represent?
35. Which line in the top view represents surface *O* in the front view?
36. Which line in the side view represents surface *U* in the front view?
37. Which line in the front view represents the surface indicated by line *C* in the top view?
38. Lines *K* in the front view represent a hole in the bottom flange. Which line in the top view represents this hole?
39. Which of the following surfaces are to be finished: *O* and *U* in the front view, *A* in the top view, or *X* in the side view?
40. Which of the following lines in the side view represent surfaces to be finished: *R*, *V*, or *Y*?

Chapter 14 Welding Symbols

The following material has been developed by the American Welding Society and is being reprinted here through the courtesy of that organization. The symbols presented here include the latest revisions. For those students desiring a more detailed reference, the booklet *Symbols for Welding and Nondestructive Testing* (AWS A2.4-79) published by the American Welding Society is recommended.

Welding cannot take its proper place as a production tool unless designers can put across their ideas to the workers. Noncode work may not require precise or detailed information on drawings. When the failure of welded structures would endanger life and property, however, simple and specific instructions must be given to the shop. Such practices as writing "To be welded throughout" or "To be completely welded" on the bottom of a drawing, in effect, transfer the design of all attachments and connections from the designer to the welder who cannot be expected to know what strength is necessary. In addition to being highly dangerous, this practice is also costly because certain shops, in their desire to be safe, use much more welding than is necessary.

Welding symbols give complete welding information. They indicate quickly to the designer, draftsman, production supervisor, and the welder, which welding technique is needed for each joint to satisfy the requirements for material strength and service conditions.

Many companies develop their own system of symbols, references, and legends with the AWS symbols as a base. Very often they will need only a few of the symbols presented in this chapter.

In the past the terms "far side" and "near side" led to a great deal of confusion. They were not clear because when joints are shown in section, all welds are equally distant from the reader and the words "near" and "far" are meaningless. In the revised symbol system the joint is the basis of reference. Any welded joint indicated by a symbol will always have an *arrow side* and *other side*. Accordingly, the terms arrow side, other side, and both sides are used to locate the weld with respect to the joint.

Another feature is the use of the tail of the reference line for giving the specifications for welding. The size and type of the weld is only part of the information needed to make the weld. The welding process; type and brand of filler metal; whether or not peening, chipping, or preheating are required; and other information must also be known. The specifications placed in the tail of the reference line usually conform to the practices of each company. If notations are not used, the tail of the symbol may be omitted.

The AWS standard makes a distinction between the terms weld symbol and welding symbol. The *weld symbol,* Figs. 14-1 (Page 430) and 14-2, is the graphic symbol which indicates the weld required for the job. The assembled *welding symbol,* Fig. 14-3, consists of the following eight elements or as many of these elements as are necessary:

- Reference line
- Arrow
- Basic weld symbols
- Dimensions and other data
- Supplementary symbols
- Finish symbols
- Tail
- Other specifications

The standard welding symbols are given in Table 14-1. Although it may appear at first that there are a very large number of different symbols, the system of symbols is broken down into a few basic elements. These elements can be combined to describe any set of conditions governing a welded joint.

The symbols, together with the specification references, provide for a shorthand system whereby a large volume of information may be communicated

Table 14-1

AMERICAN WELDING SOCIETY

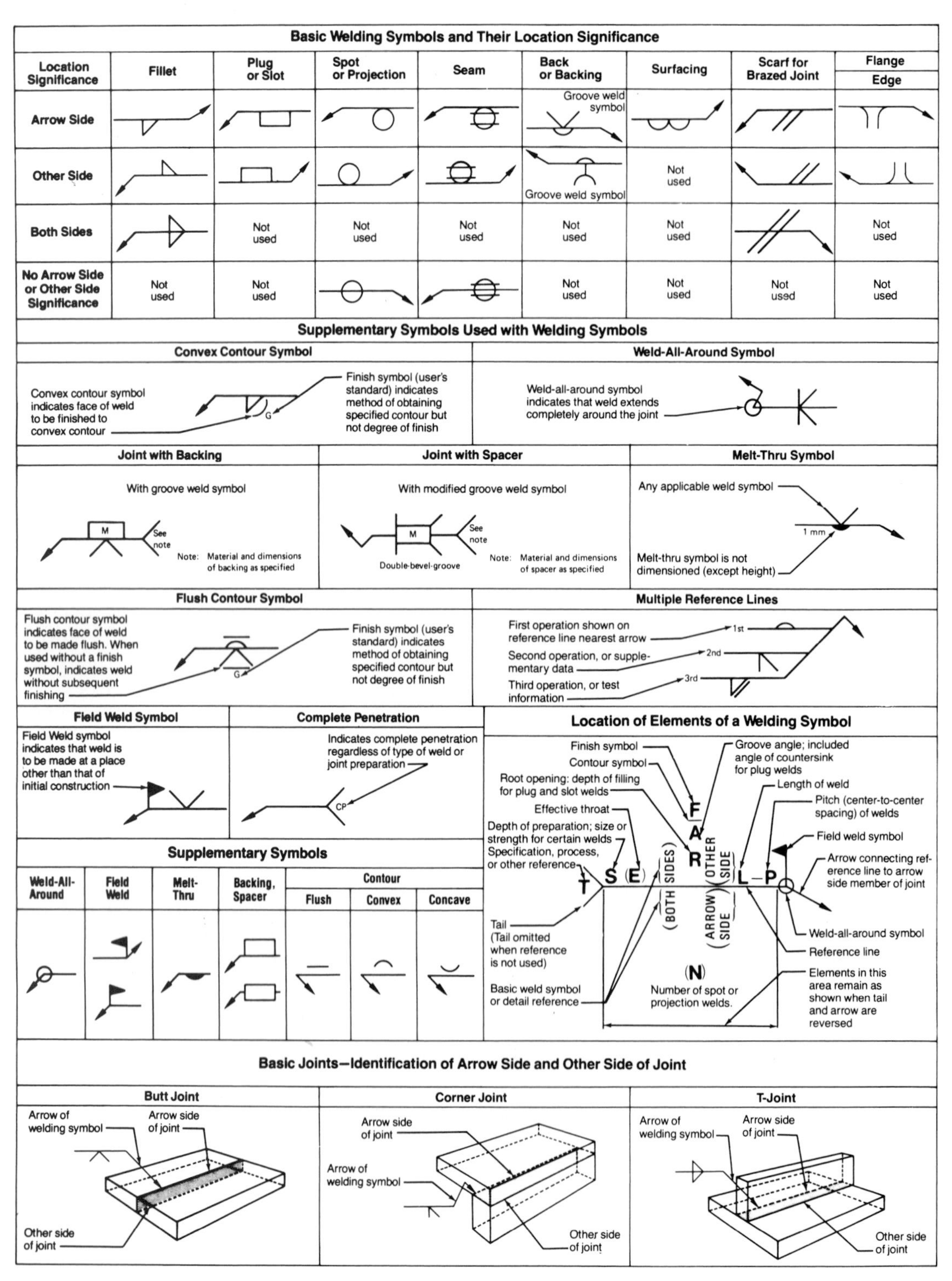

STANDARD WELDING SYMBOLS

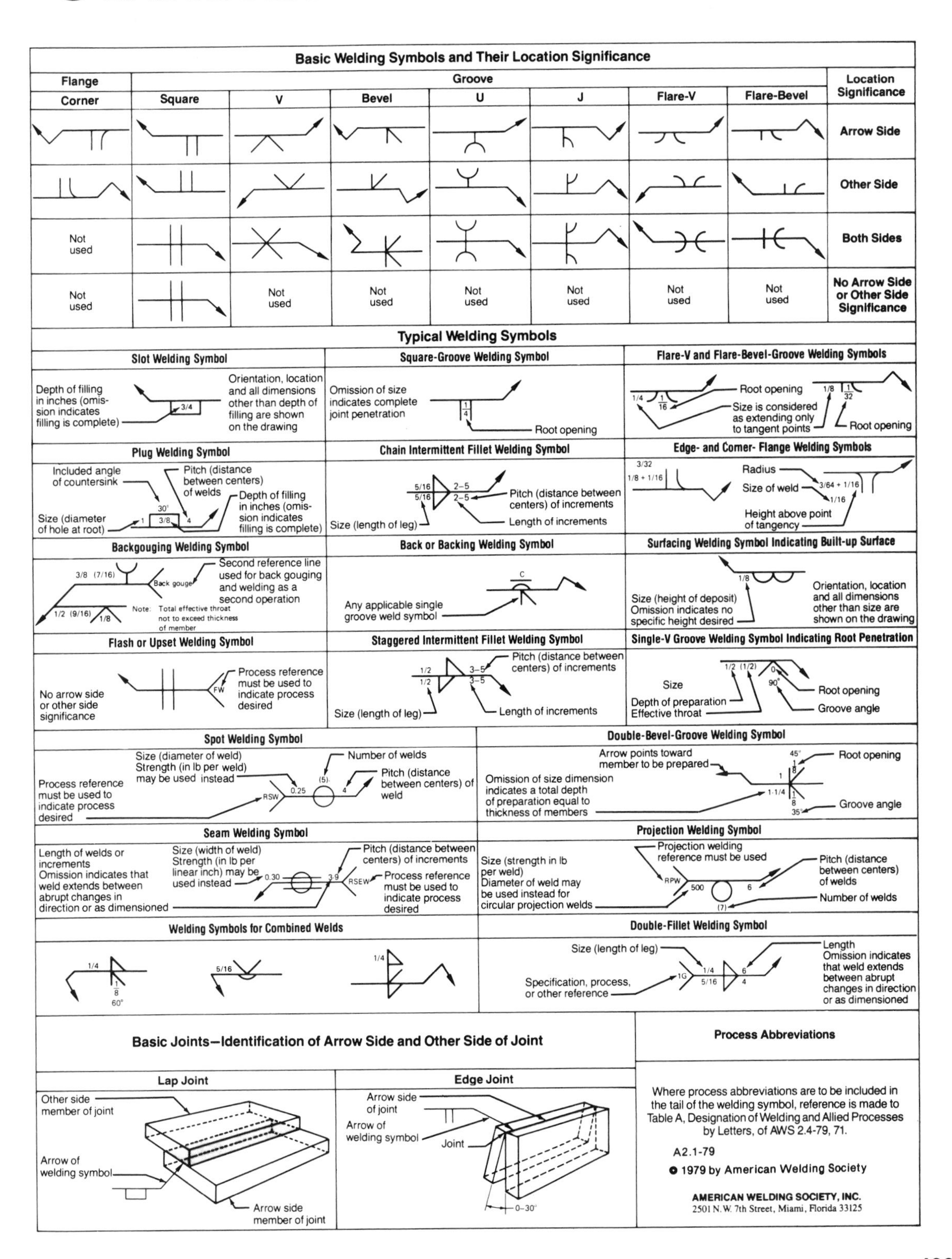

Groove							
Square	Scarf*	V	Bevel	U	J	Flare-V	Flare-bevel

American Welding Society

14-1. *Basic weld symbols.*

Fillet	Plug or slot	Spot or projection	Seam	Back or backing	Surfacing	Flange	
						Edge	Corner

*Used for brazed joints only

NOTE: The field weld symbol has been changed from a dot to a flag since the first editions of this text and its student guide were published. If you are using the *1977 edition of the student guide,* disregard questions 5 and 17 of Chapter 14 in that guide.

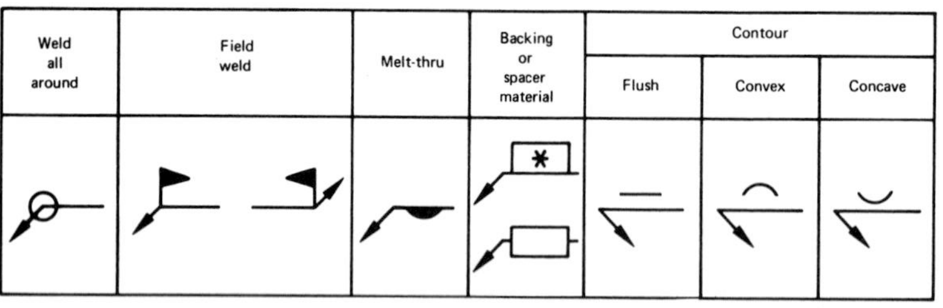

Weld all around	Field weld	Melt-thru	Backing or spacer material	Contour		
				Flush	Convex	Concave

American Welding Society

14-2. *Supplementary symbols.*

14-3. *Standard location of elements of a welding symbol.*

American Welding Society

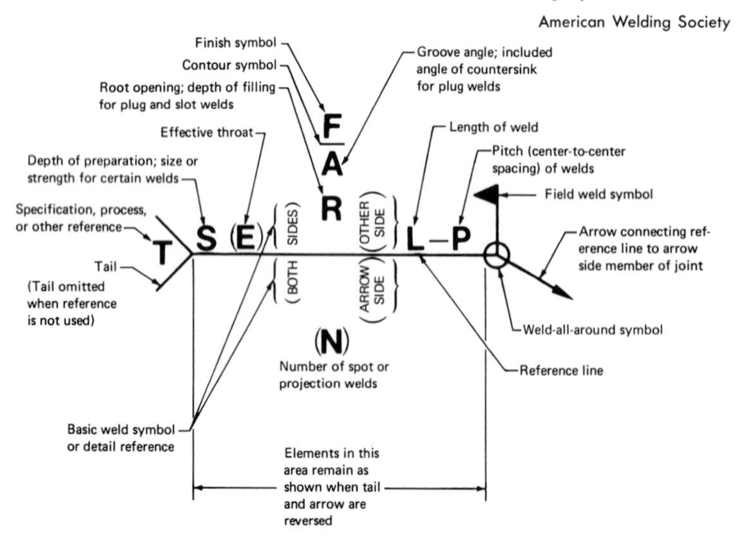

14-4. *Symbolic method of expressing welding information.*

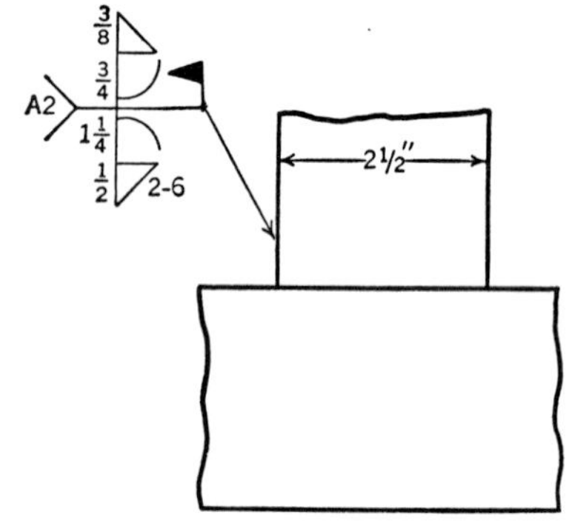

Sketch 1

Interpretation of Symbol:

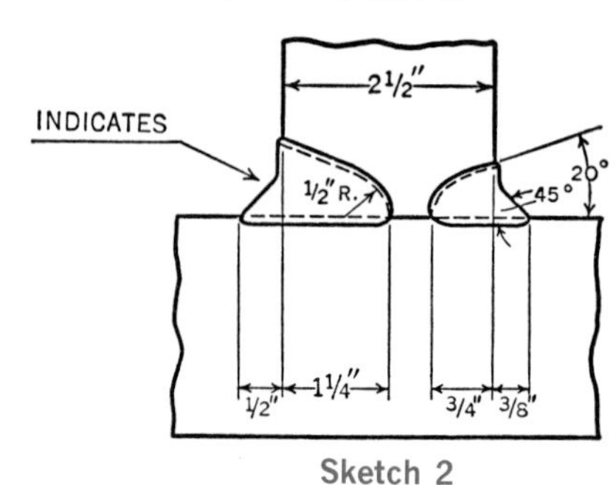

Sketch 2

accurately with only a few lines, abbreviations, and numbers. The amount of information which can be expressed by a welding symbol is illustrated in Fig. 14-4. Written specifications for the same weld would read as follows:

> Double fillet welded, partially grooved double-J T-joint with incomplete penetration. Grooves of standard proportions $\frac{1}{2}$ in. R 20° included angle, edges in contact before welding $\frac{3}{4}$ in. deep for other-side weld and $1\frac{1}{4}$ in. deep for arrow-side weld. $\frac{3}{8}$ in. continuous other-side fillet weld and $\frac{1}{2}$ in. intermittent arrow-side fillet weld with increments 2 in. long, spaced 6 in. center-to-center. All fillets standard 45° fillets. All welding done in field in accordance with welding specification number A2, which requires that the weld be made by manual d.c. shielded metal-arc process using high grade, covered, mild steel electrode; that the root be unchipped; and that the welds be unpeened but that the joint be preheated before welding.

Table 14-1 and the illustrations on the following pages are taken from the American Welding Society standards. Only those symbols concerned with the manual and semi-automatic welding processes are presented. Many students find that they can learn to read and understand these symbols more readily through the use of the illustrations than through the use of written instructions. You are urged to study the table and the illustrations carefully in order to be able to recognize the symbols on sight. The welder who can interpret welding symbols accurately is valuable to the employer, and is more likely to be selected for promotion than those unable to read shop drawings.

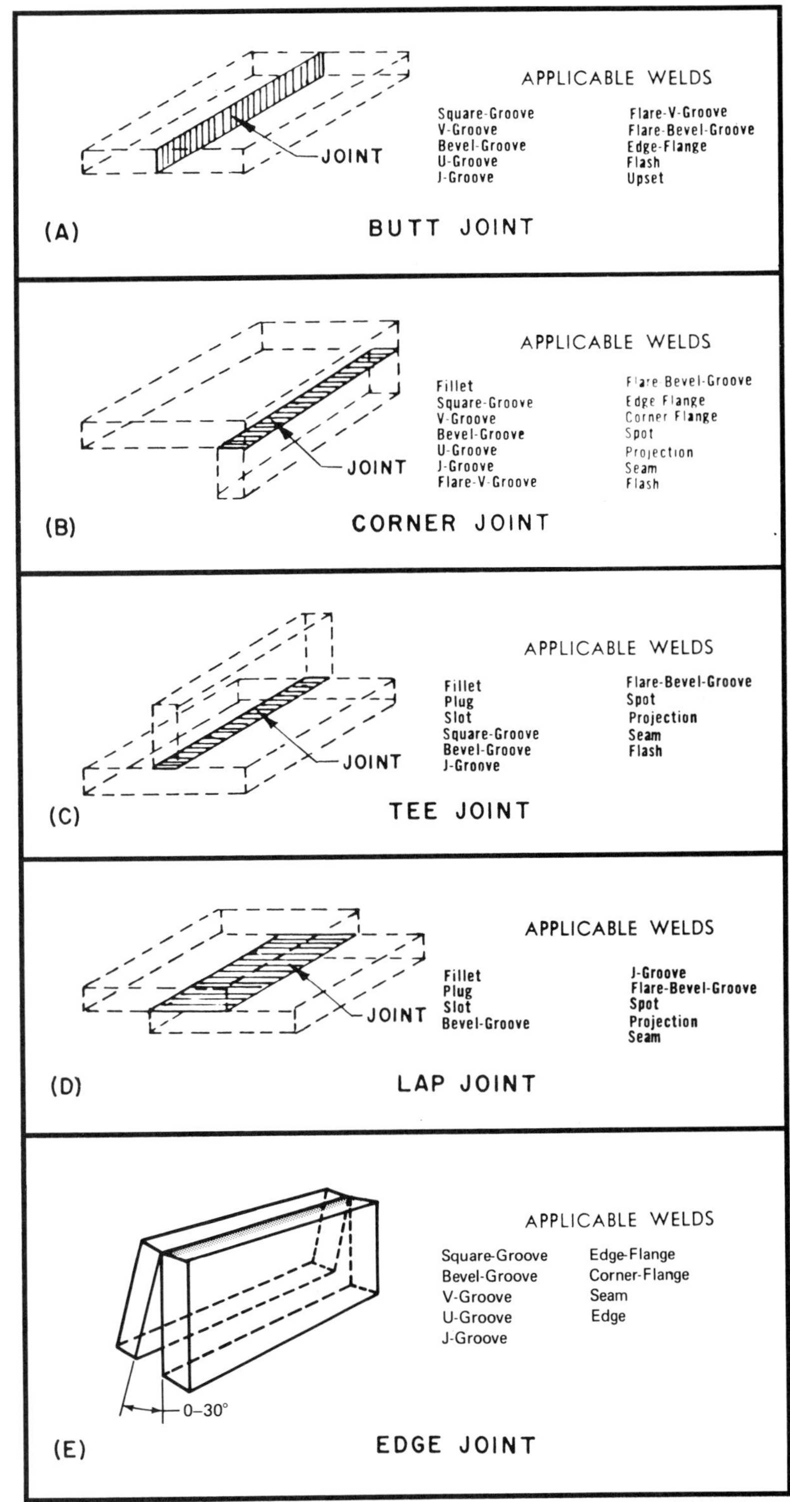

American Welding Society

***14-5.** Basic types of joints.*

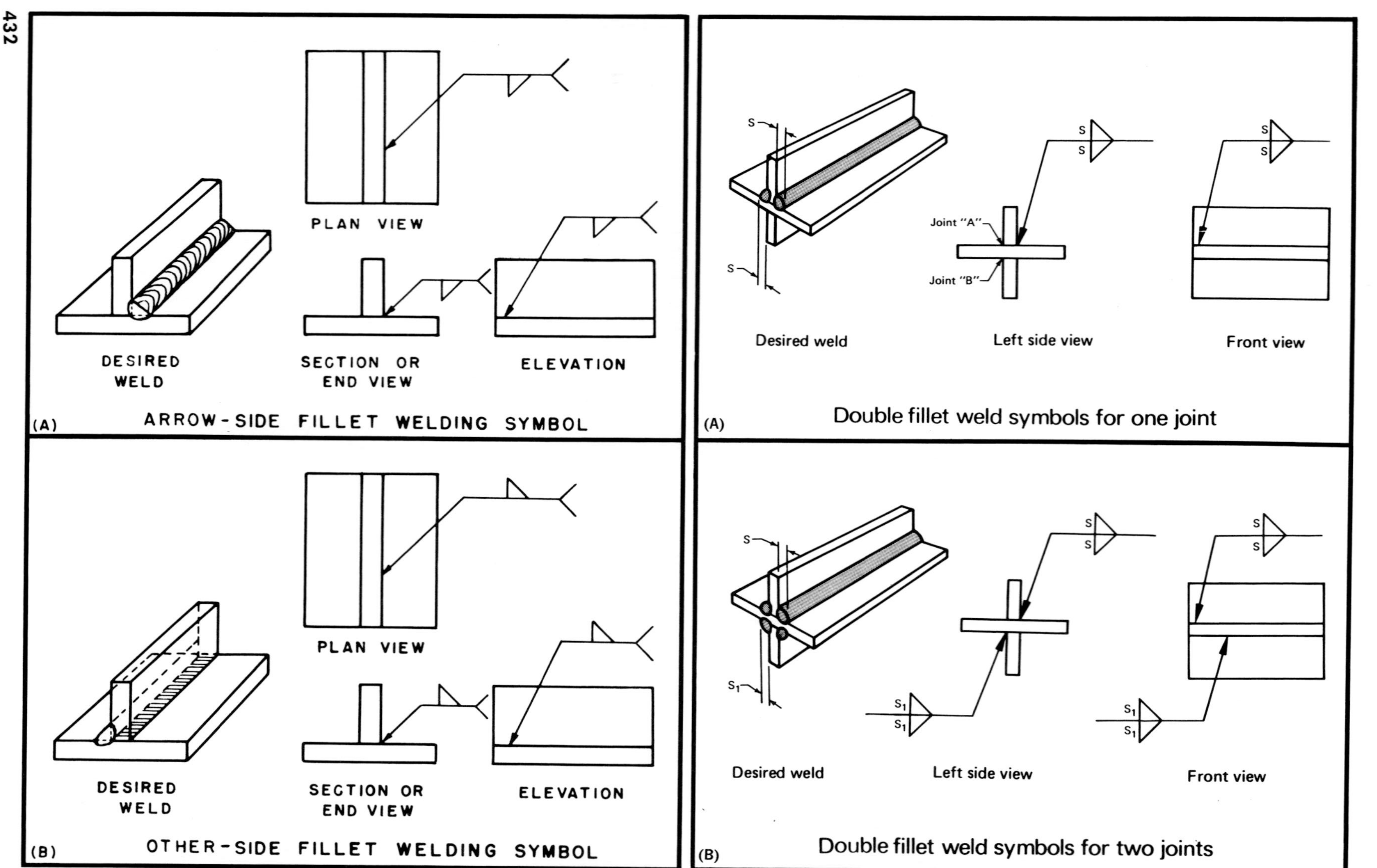

American Welding Society

14-6. *Application of fillet welding symbol.*

American Welding Society

14-7. *Application of fillet welding symbol.*

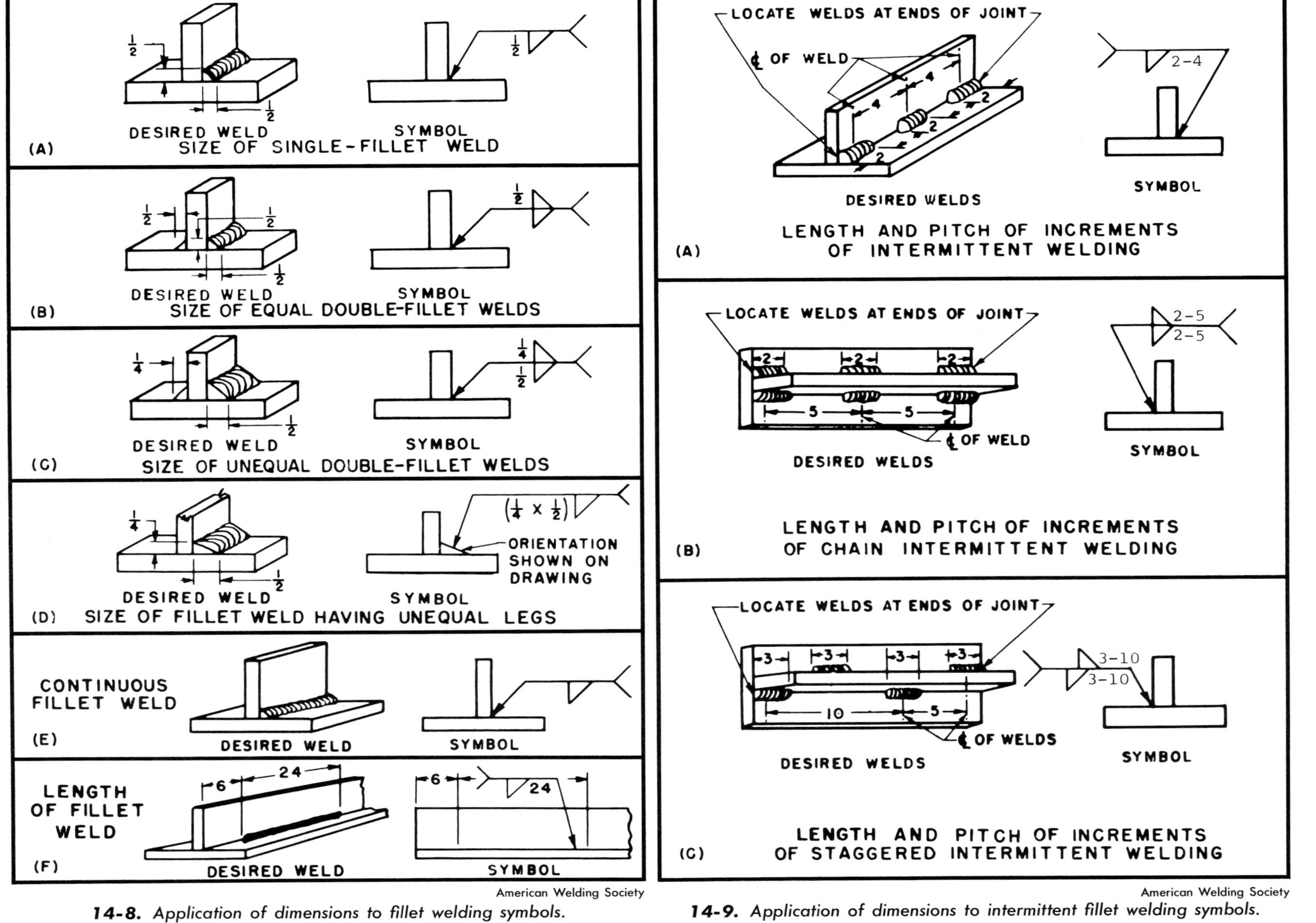

14-8. *Application of dimensions to fillet welding symbols.*

14-9. *Application of dimensions to intermittent fillet welding symbols.*

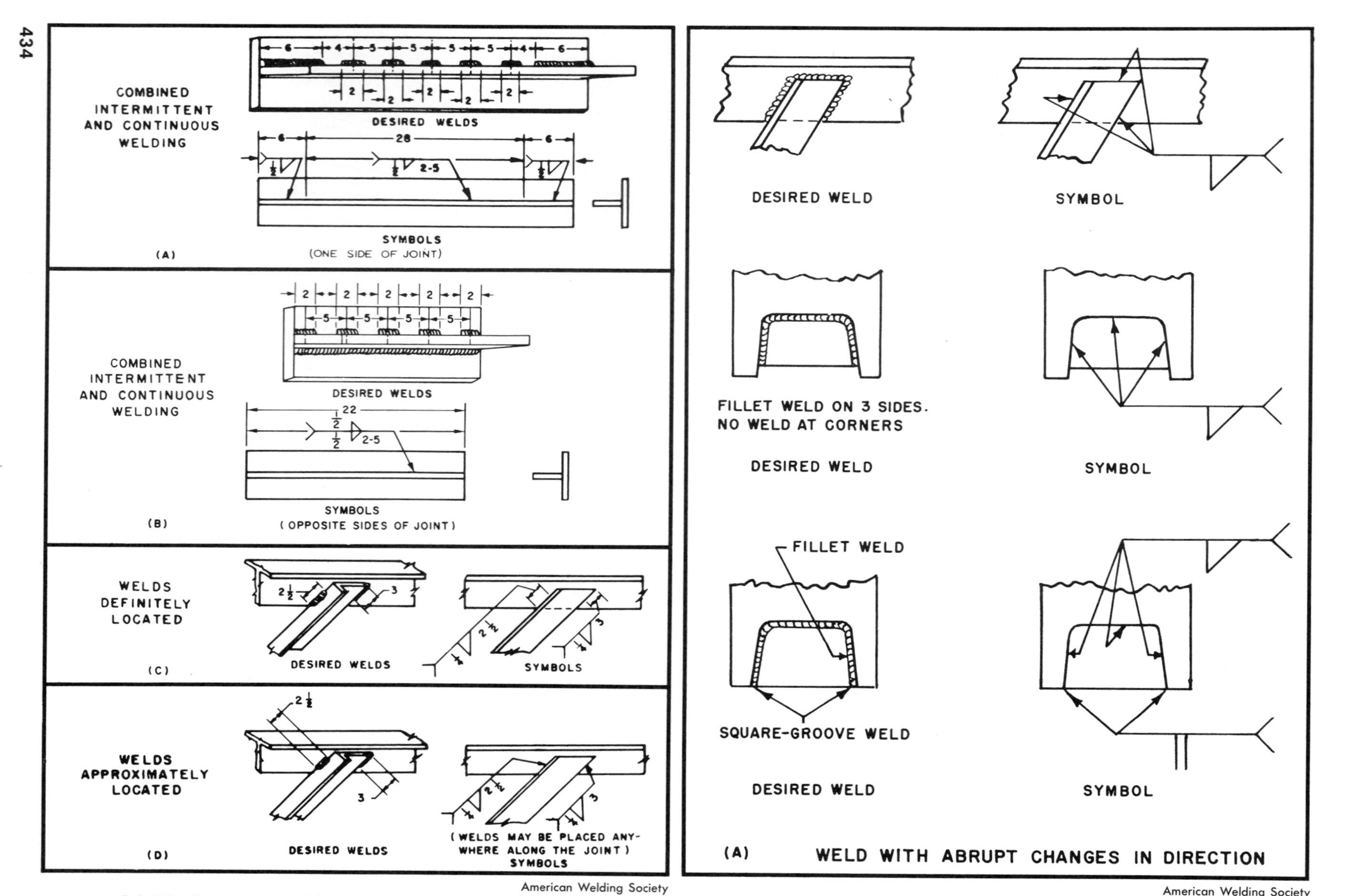

American Welding Society

14-10. *Designation of location and extent of fillet welds.*

American Welding Society

14-11. *Designation of extent of welding.*

DESIRED WELD

SYMBOL

DESIRED WELD

SYMBOL

DESIRED WELD

SYMBOL

(B)

WELD ALL-AROUND SYMBOL

American Welding Society

14-12. *Designation of extent of welding.*

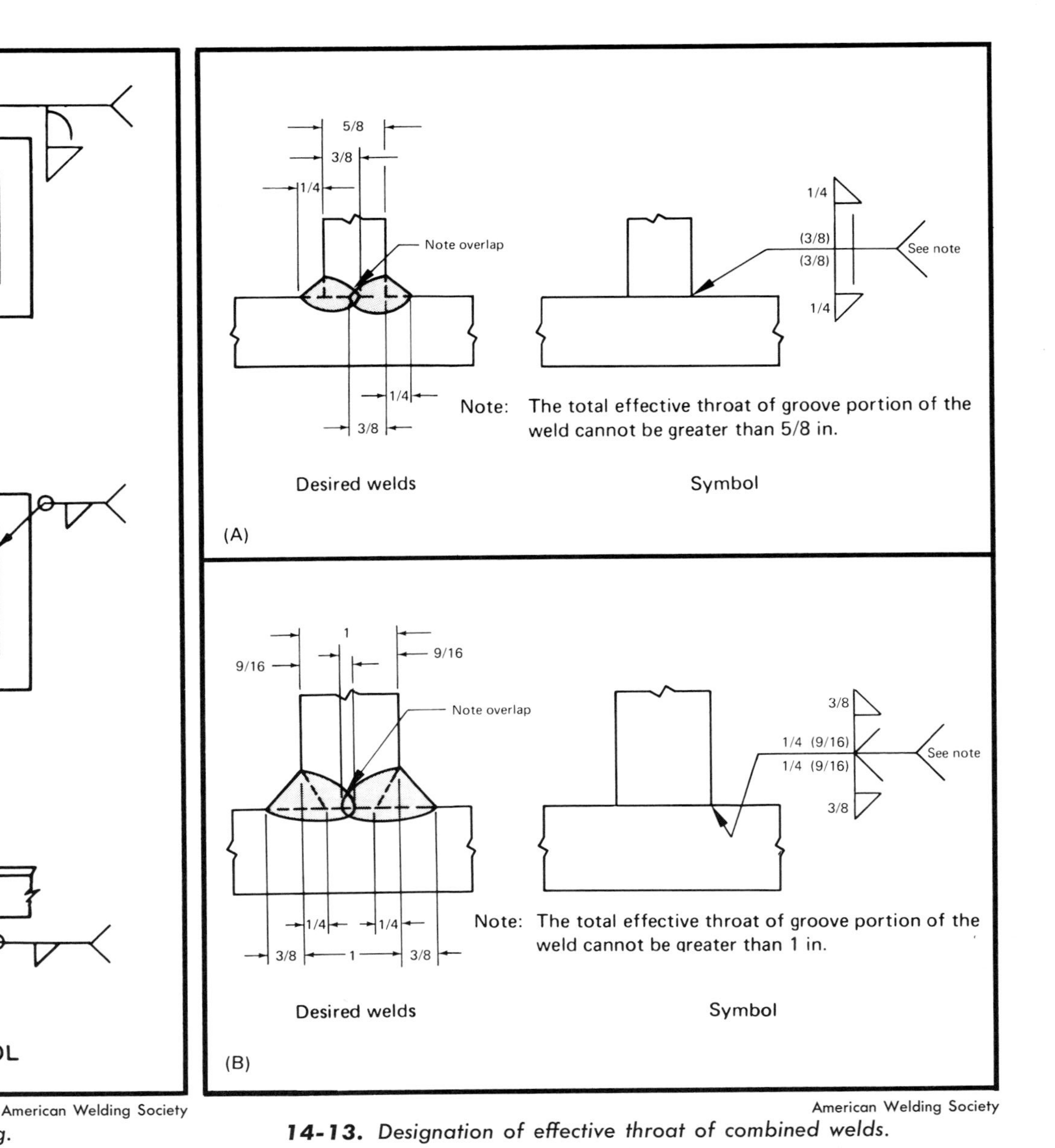

American Welding Society

14-13. *Designation of effective throat of combined welds.*

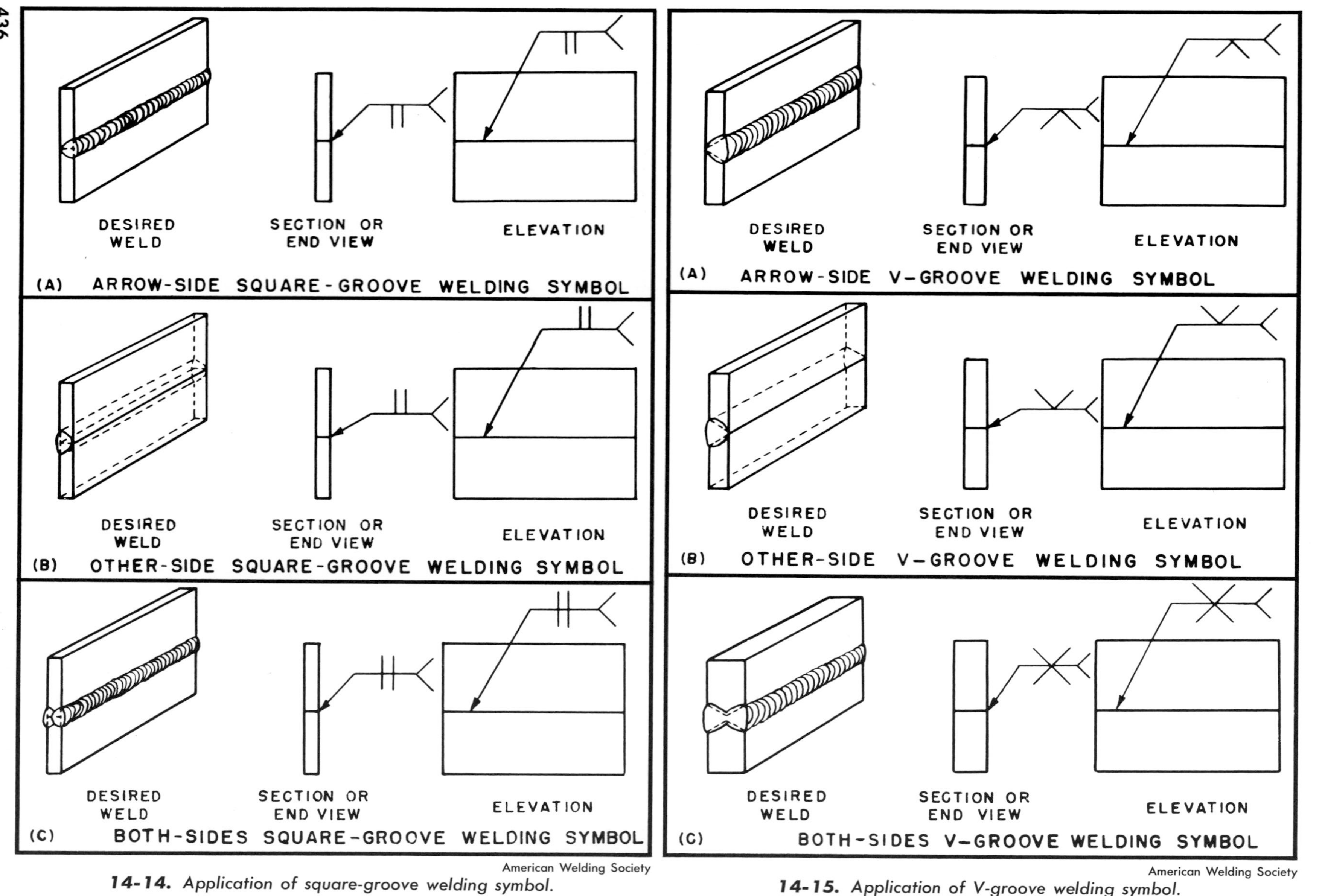

American Welding Society

14-14. *Application of square-groove welding symbol.*

American Welding Society

14-15. *Application of V-groove welding symbol.*

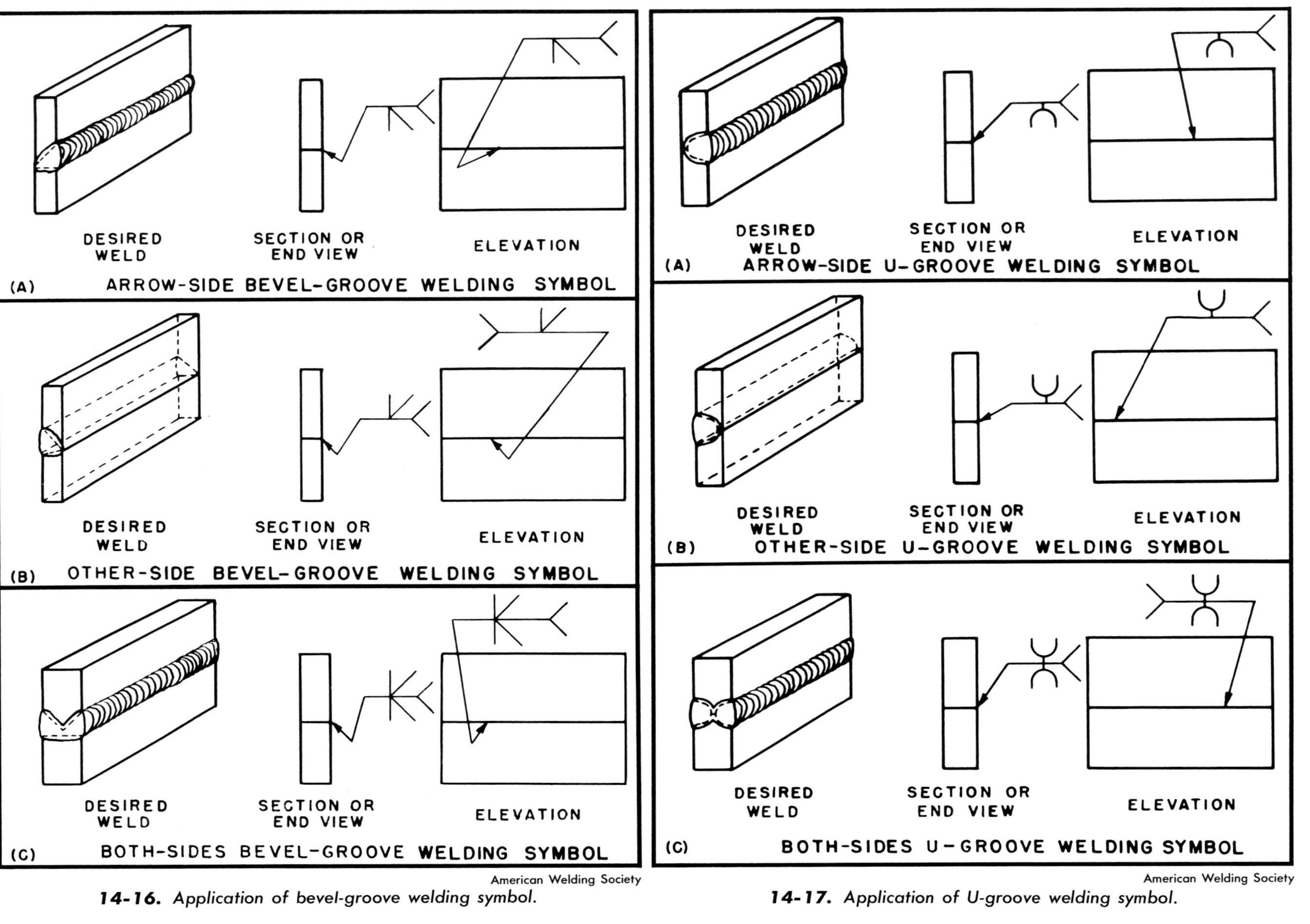

14-16. *Application of bevel-groove welding symbol.*

14-17. *Application of U-groove welding symbol.*

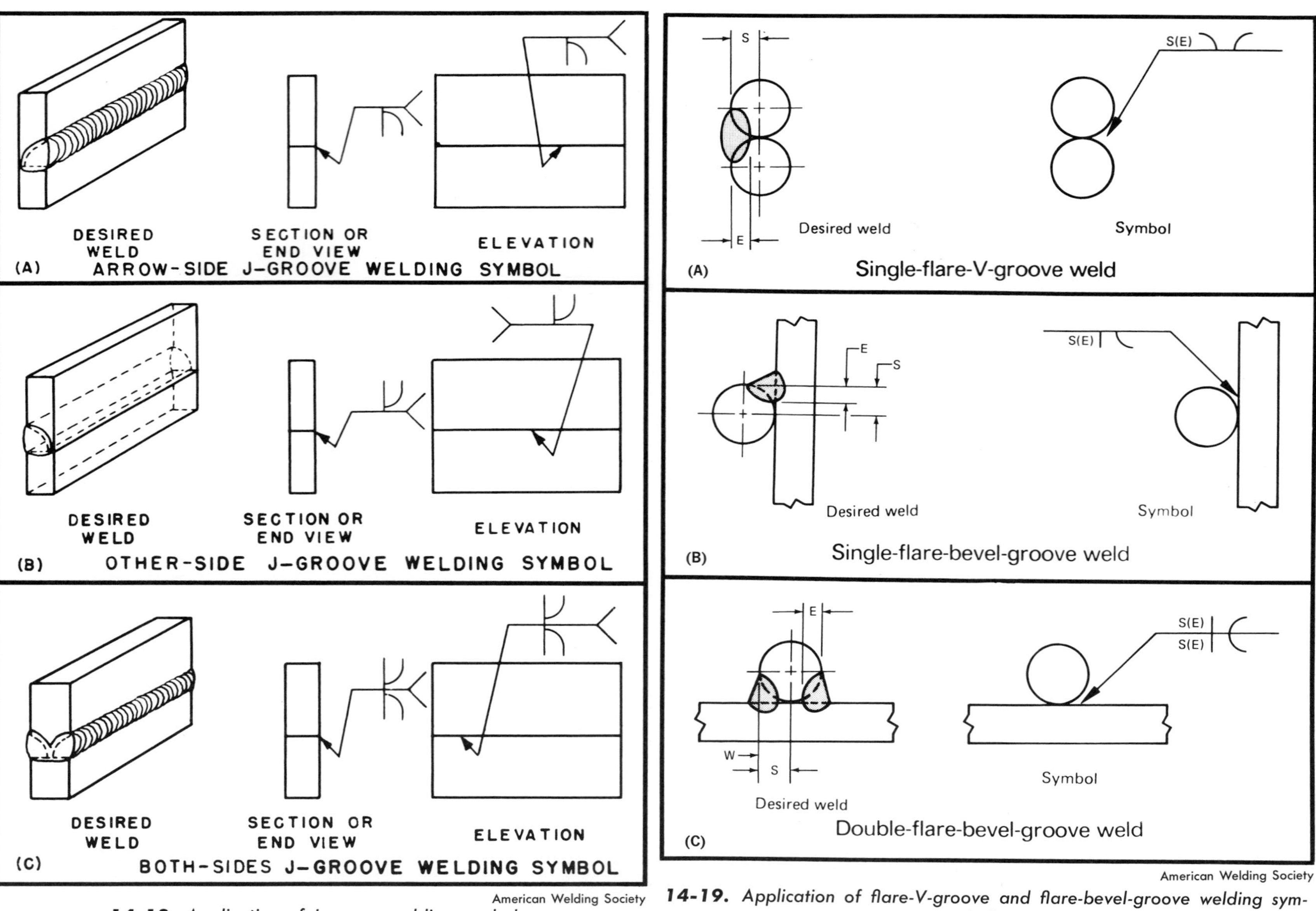

American Welding Society

14-18. *Application of J-groove welding symbol.*

American Welding Society

14-19. *Application of flare-V-groove and flare-bevel-groove welding symbols.*

Desired weld

E

S

W

S(E)

Symbol

E – Effective throat

S – Radius of bar

(A) Two single-flare-bevel-groove welds

Effective throat

E

S

S = Radius from point of tangency to top of member

Desired weld

S(E)

Symbol

For complete penetration weld
$S \leq E$

(B) Flare-V-groove weld with complete penetration

American Welding Society

14-20. *Application of flare-V-groove and flare-bevel-groove welding symbols.*

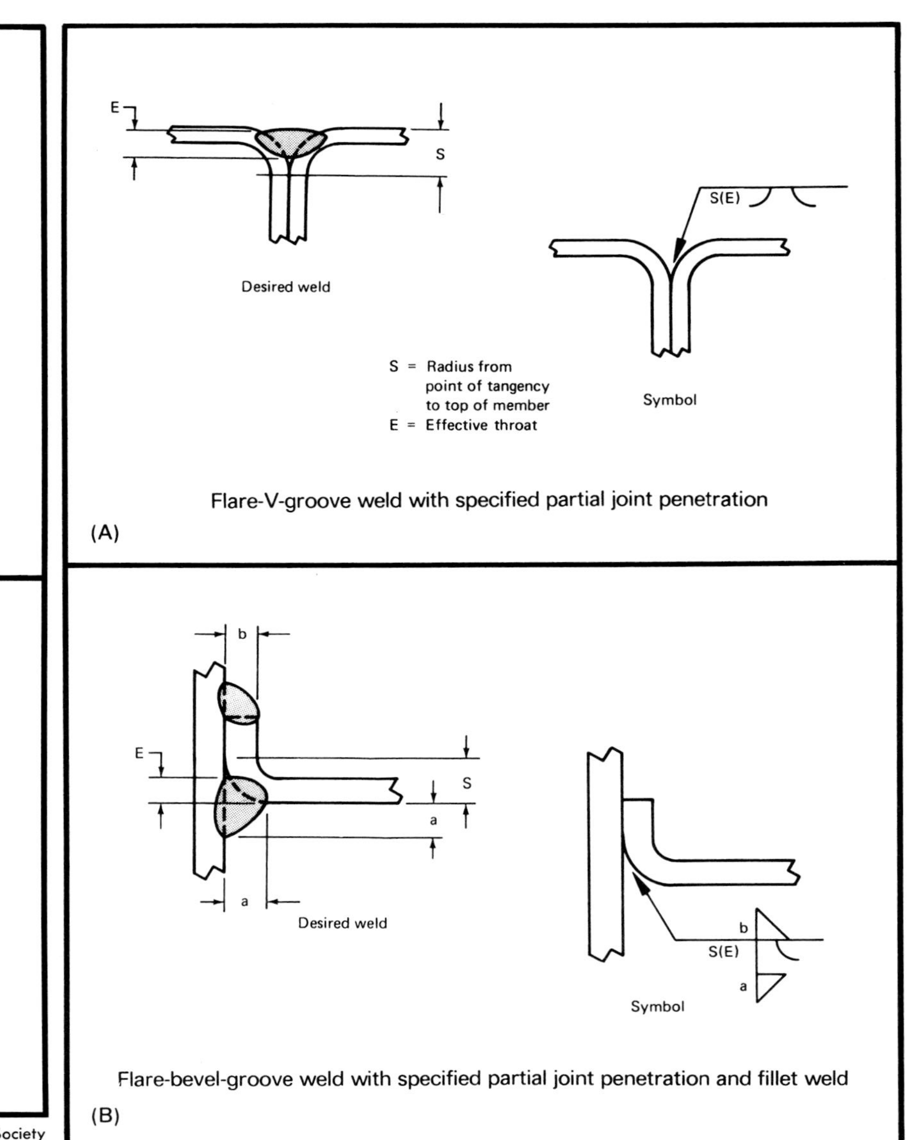

American Welding Society

14-21. *Application of flare-bevel- and flare-V-groove welding symbols.*

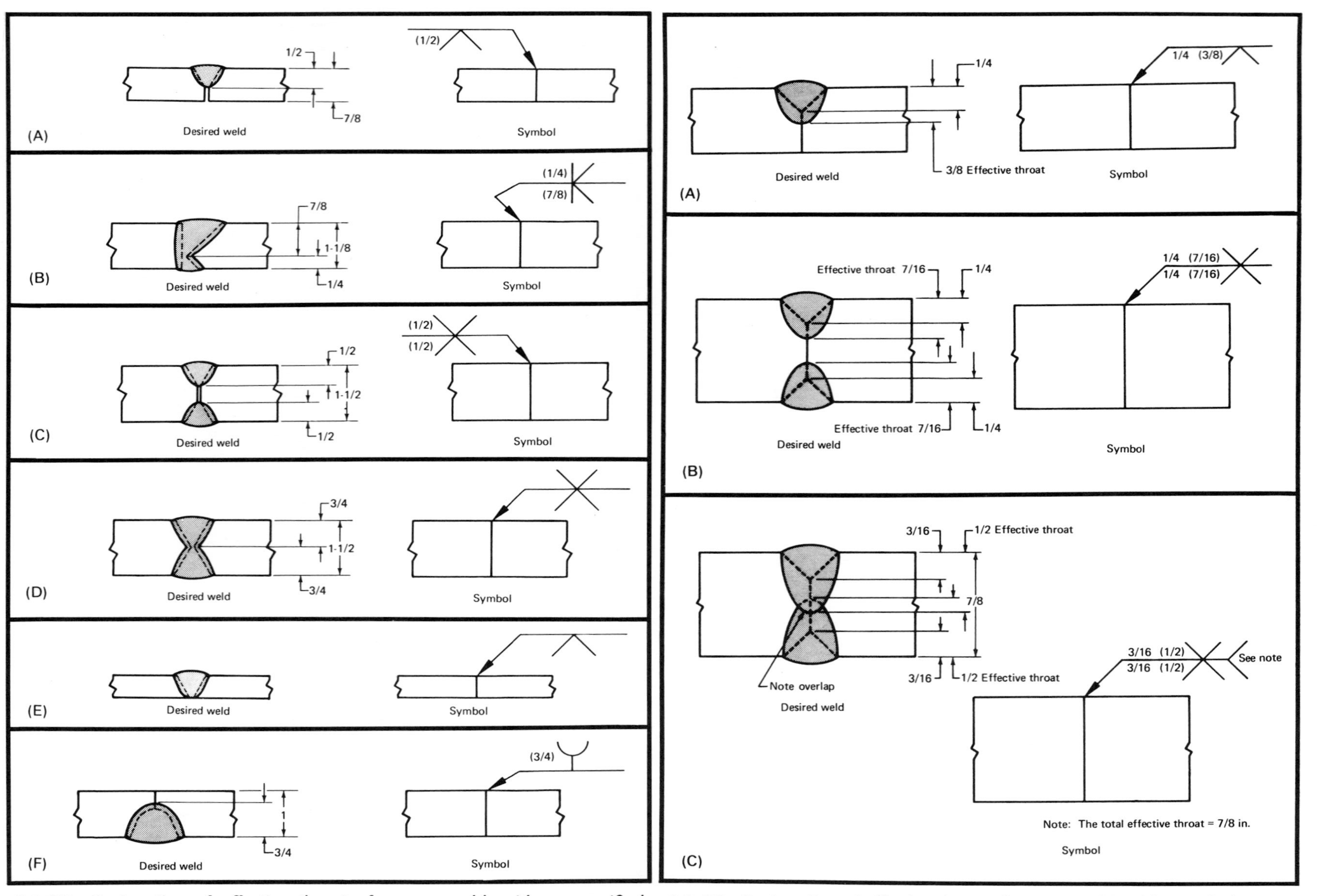

14-22. *Designation of effective throat of groove welds with no specified groove preparation.*

14-23. *Designation of effective throat of groove welds with specified groove preparation.*

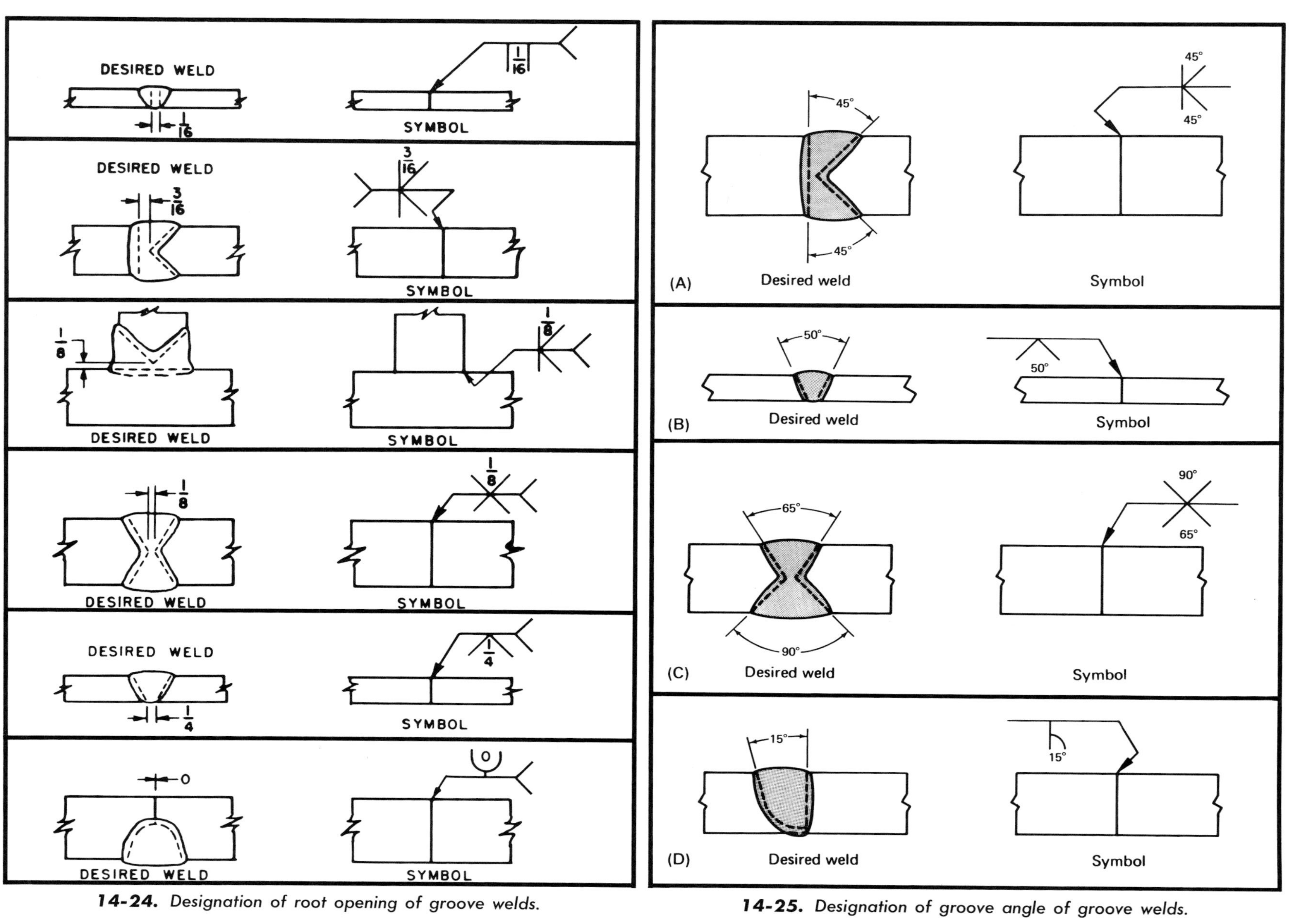

14-24. *Designation of root opening of groove welds.*

14-25. *Designation of groove angle of groove welds.*

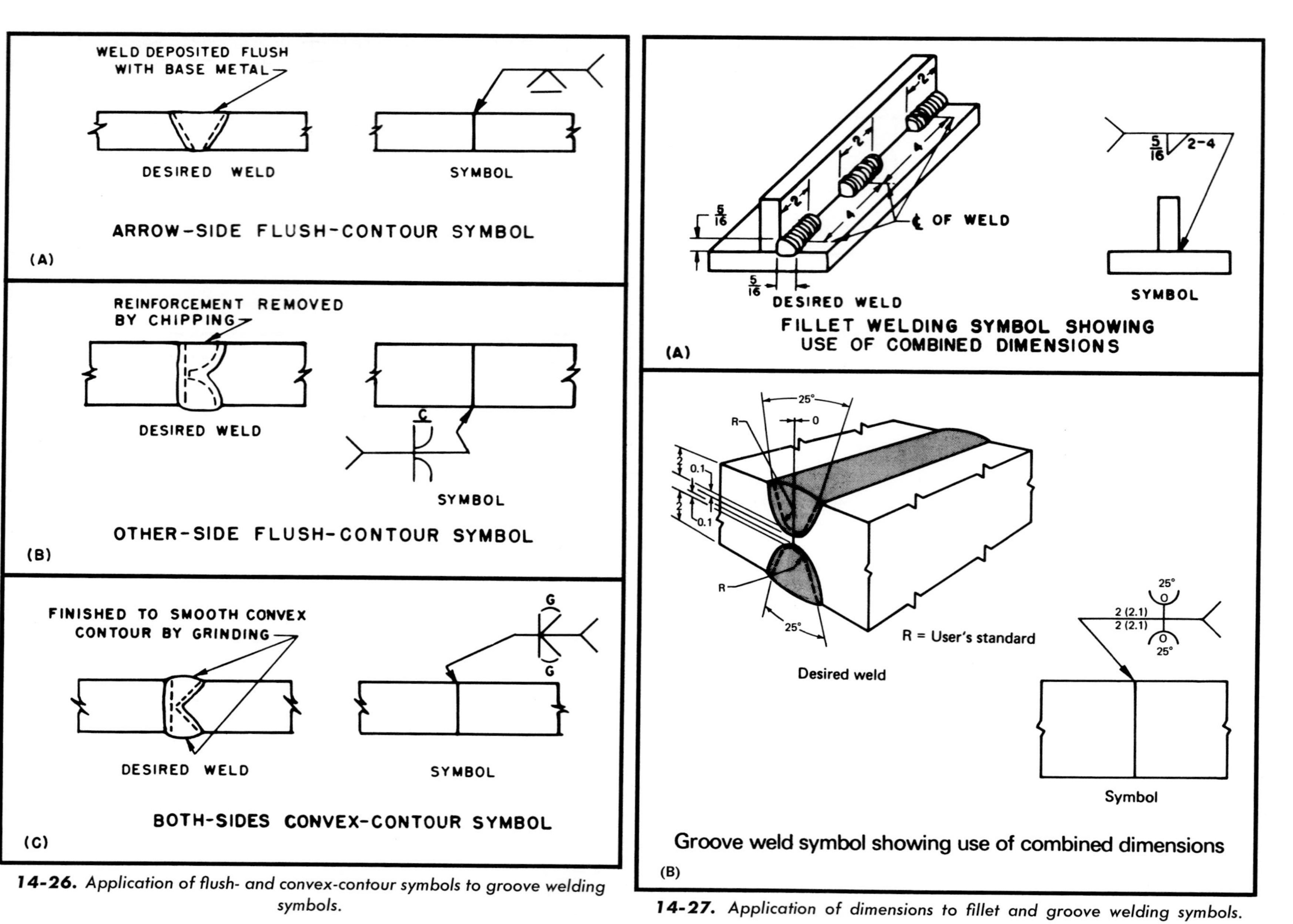

14-26. *Application of flush- and convex-contour symbols to groove welding symbols.*

14-27. *Application of dimensions to fillet and groove welding symbols.*

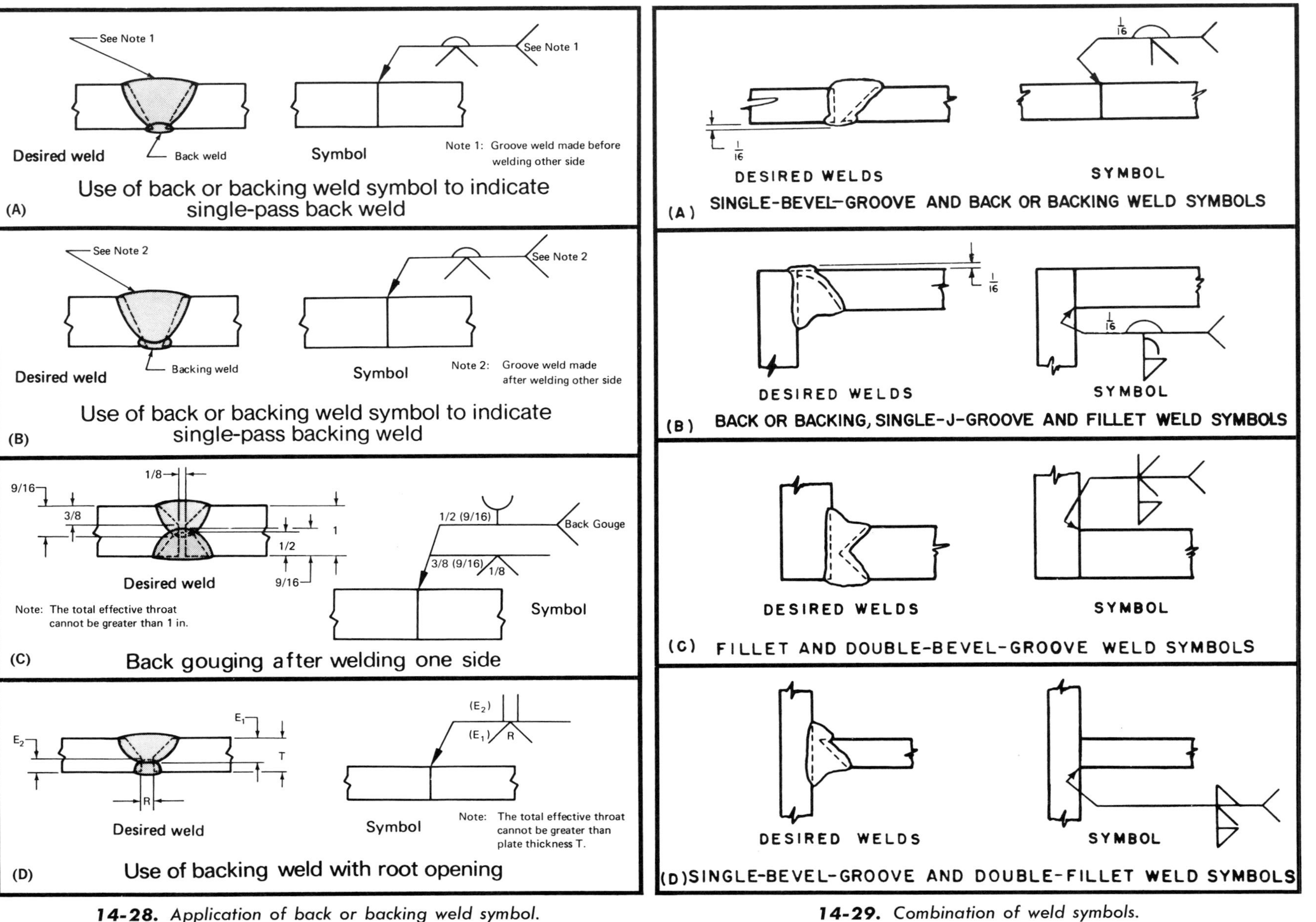

14-28. *Application of back or backing weld symbol.*

14-29. *Combination of weld symbols.*

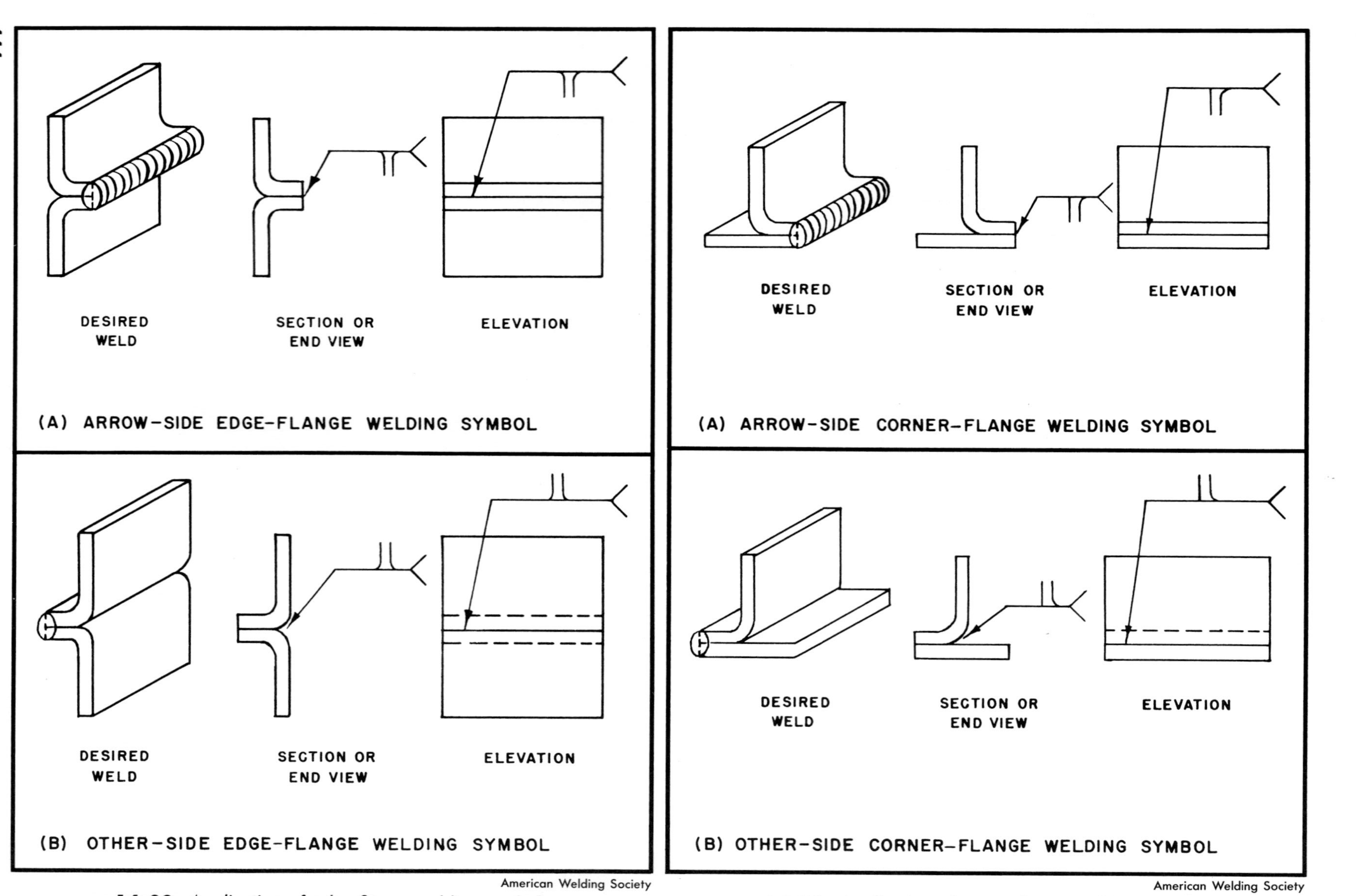

American Welding Society

14-30. *Application of edge-flange welding symbol.*

American Welding Society

14-31. *Application of corner-flange welding symbol.*

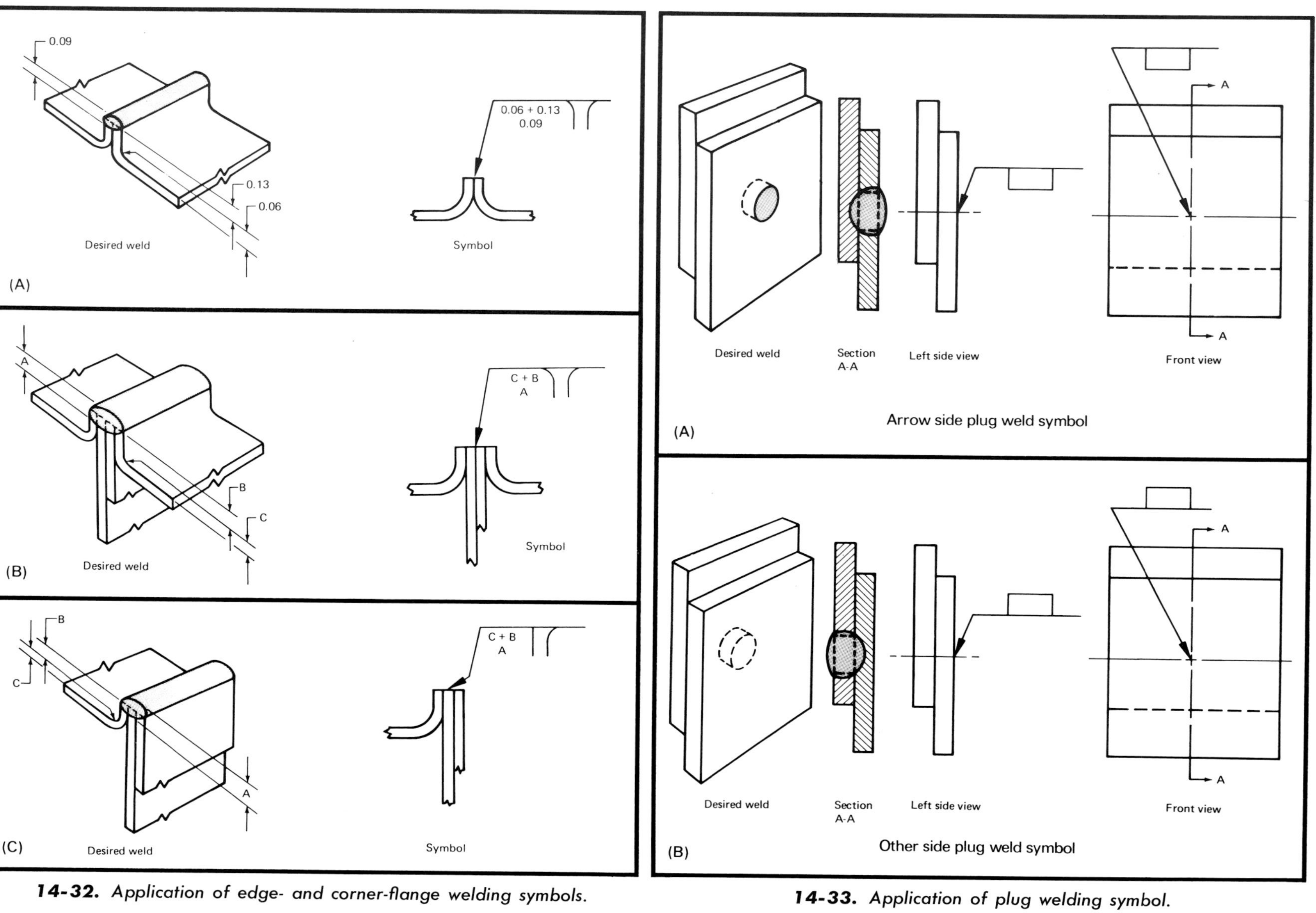

14-32. *Application of edge- and corner-flange welding symbols.*

14-33. *Application of plug welding symbol.*

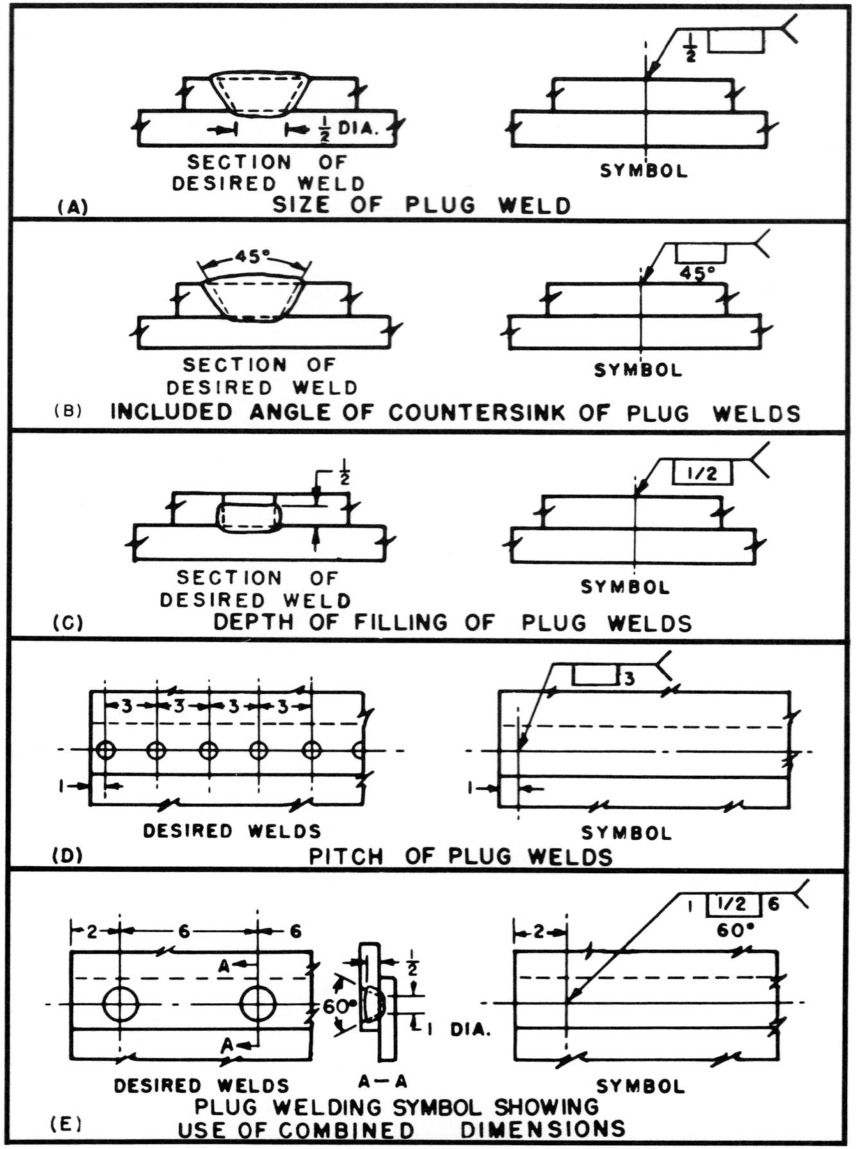
SECTION OF
DESIRED WELD
1/2 DIA.
SYMBOL
1/2
(A)
SIZE OF PLUG WELD
45°
SECTION OF
DESIRED WELD
SYMBOL
45°
(B)
INCLUDED ANGLE OF COUNTERSINK OF PLUG WELDS
1/2
SECTION OF
DESIRED WELD
SYMBOL
1/2
(C)
DEPTH OF FILLING OF PLUG WELDS
3
3
3
3
1
DESIRED WELDS
3
1
SYMBOL
(D)
PITCH OF PLUG WELDS
2
6
6
A
A
60°
1/2
1 DIA.
DESIRED WELDS
A—A
2
1
1/2
6
60°
SYMBOL
(E)
PLUG WELDING SYMBOL SHOWING
USE OF COMBINED DIMENSIONS

American Welding Society

14-34. *Application of dimensions to plug welding symbols.*

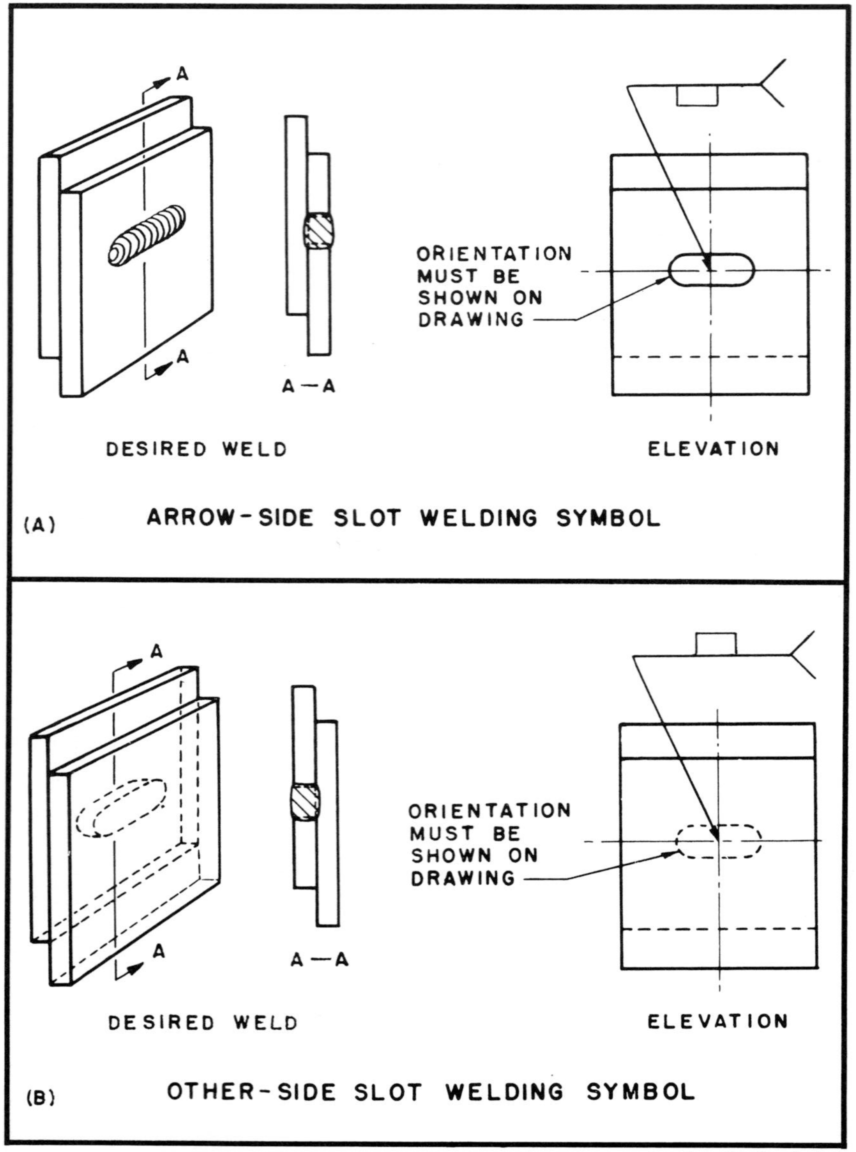
A
A
A—A
ORIENTATION
MUST BE
SHOWN ON
DRAWING
DESIRED WELD
ELEVATION
(A)
ARROW-SIDE SLOT WELDING SYMBOL
A
A
A—A
ORIENTATION
MUST BE
SHOWN ON
DRAWING
DESIRED WELD
ELEVATION
(B)
OTHER-SIDE SLOT WELDING SYMBOL

American Welding Society

14-35. *Application of slot welding symbol.*

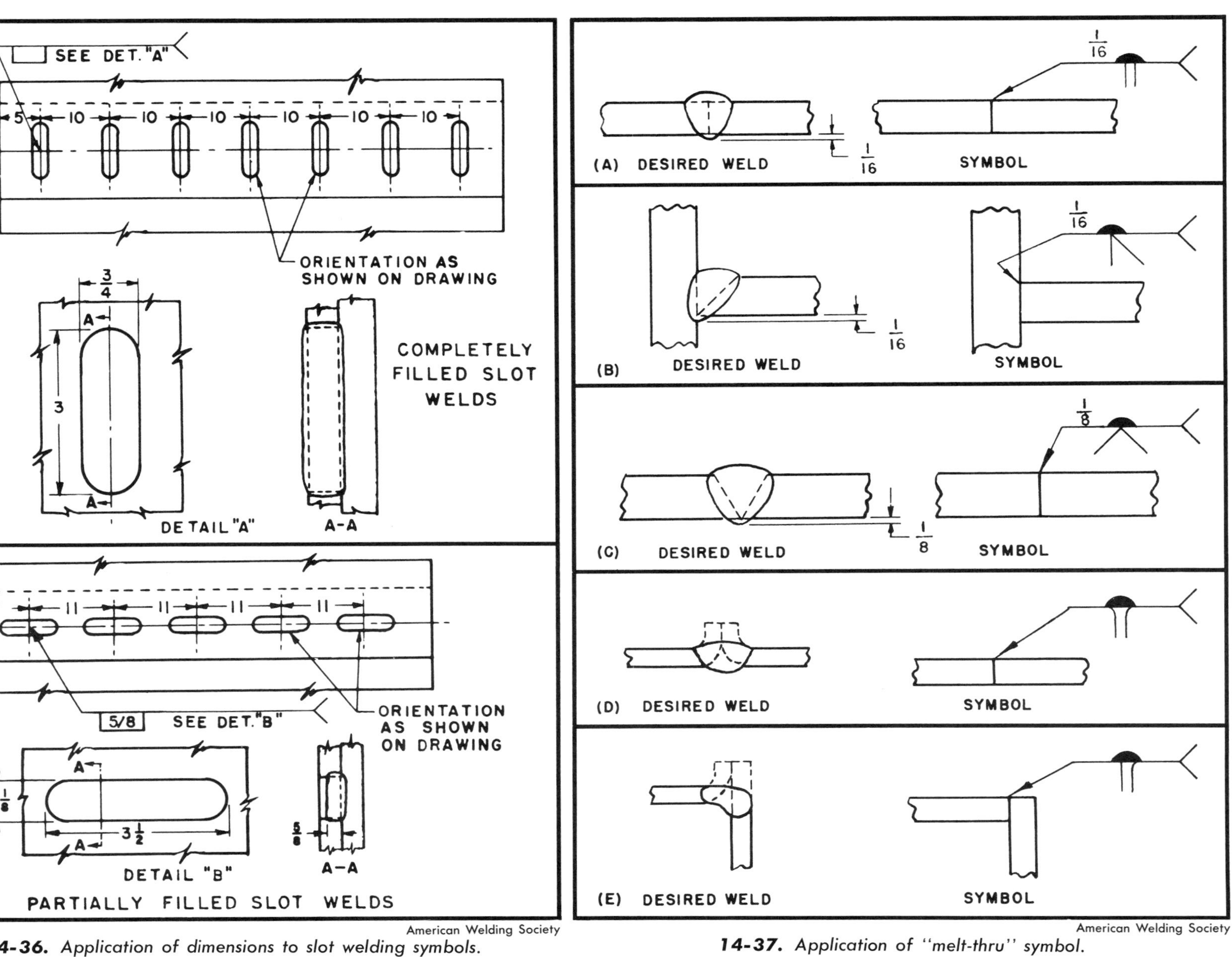

American Welding Society

14-36. *Application of dimensions to slot welding symbols.*

American Welding Society

14-37. *Application of "melt-thru" symbol.*

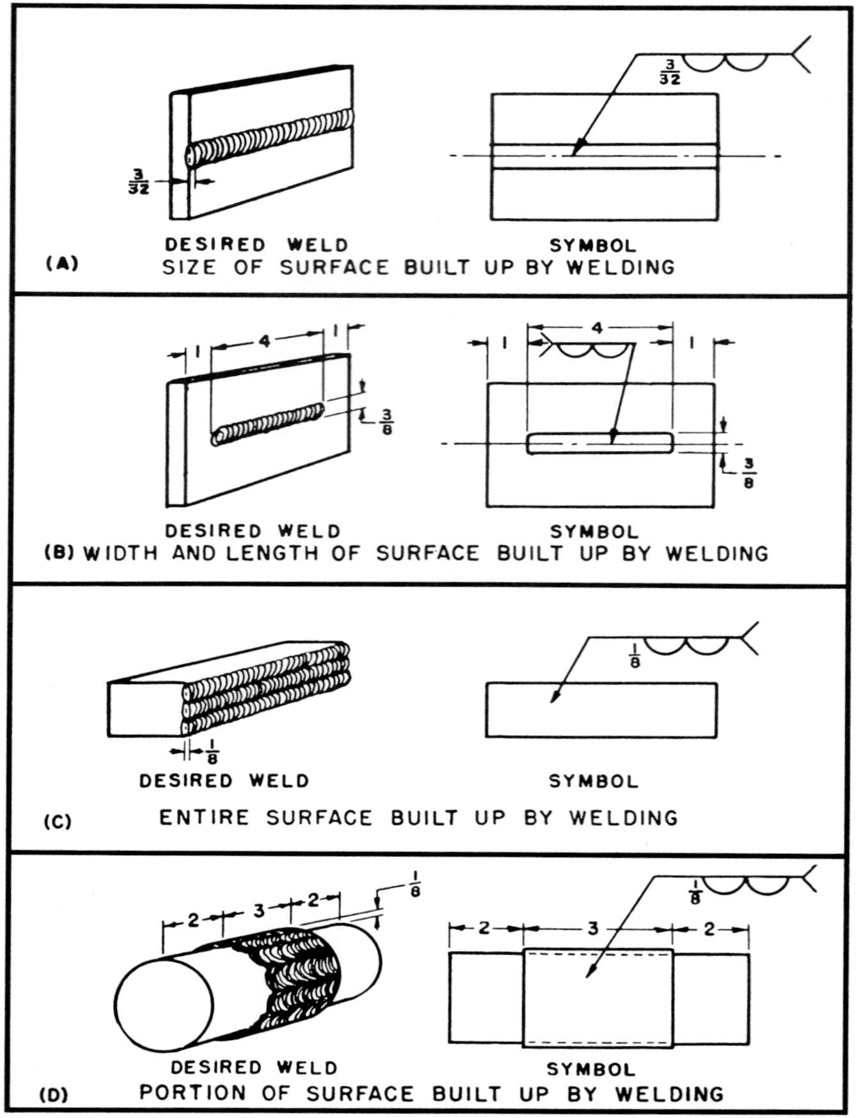

3/32
3/32
DESIRED WELD
SYMBOL
(A) SIZE OF SURFACE BUILT UP BY WELDING
1
4
1
3/8
4
1
1
3/8
DESIRED WELD
SYMBOL
(B) WIDTH AND LENGTH OF SURFACE BUILT UP BY WELDING
1/8
1/8
DESIRED WELD
SYMBOL
(C) ENTIRE SURFACE BUILT UP BY WELDING
2
3
2
1/8
1/8
2
3
2
DESIRED WELD
SYMBOL
(D) PORTION OF SURFACE BUILT UP BY WELDING

American Welding Society

14-38. *Application of surfacing weld symbol to indicate surfaces built up by welding.*

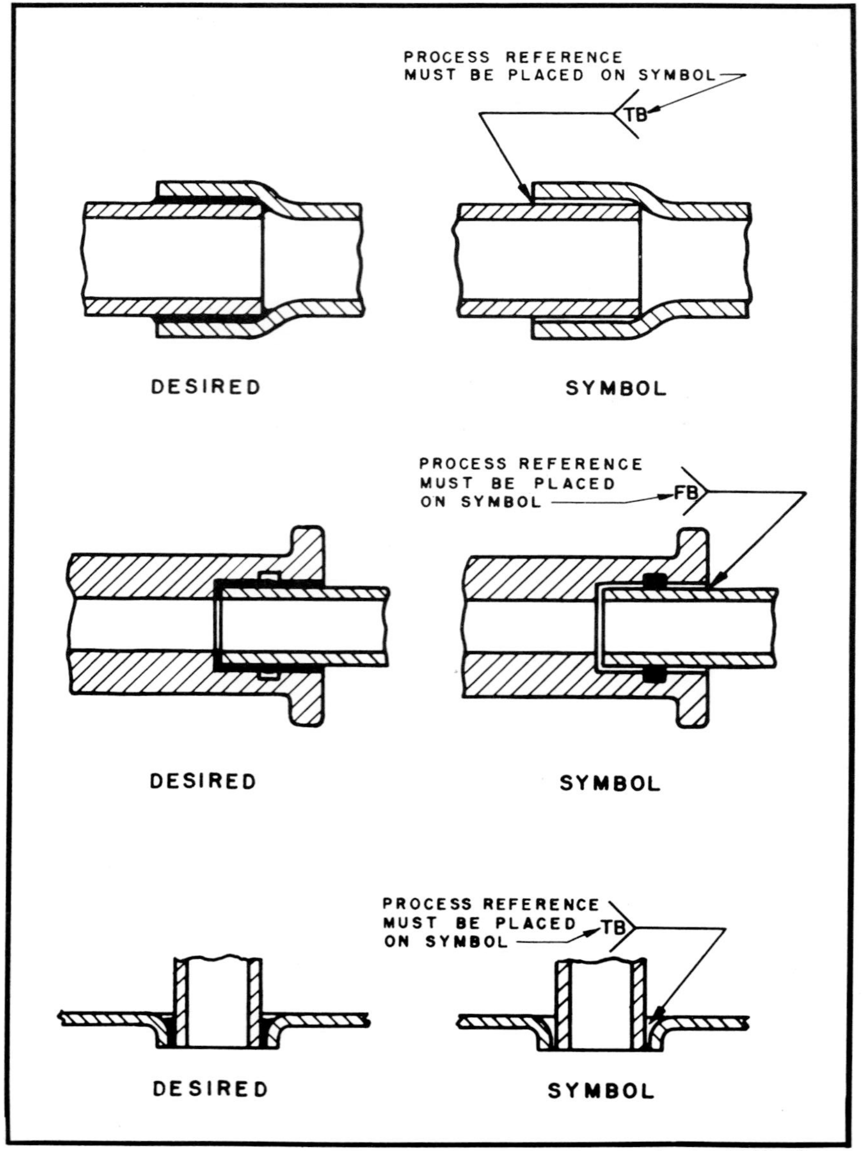

PROCESS REFERENCE MUST BE PLACED ON SYMBOL
TB
DESIRED
SYMBOL
PROCESS REFERENCE MUST BE PLACED ON SYMBOL
FB
DESIRED
SYMBOL
PROCESS REFERENCE MUST BE PLACED ON SYMBOL
TB
DESIRED
SYMBOL

American Welding Society

14-39. *Application of brazing symbols.*

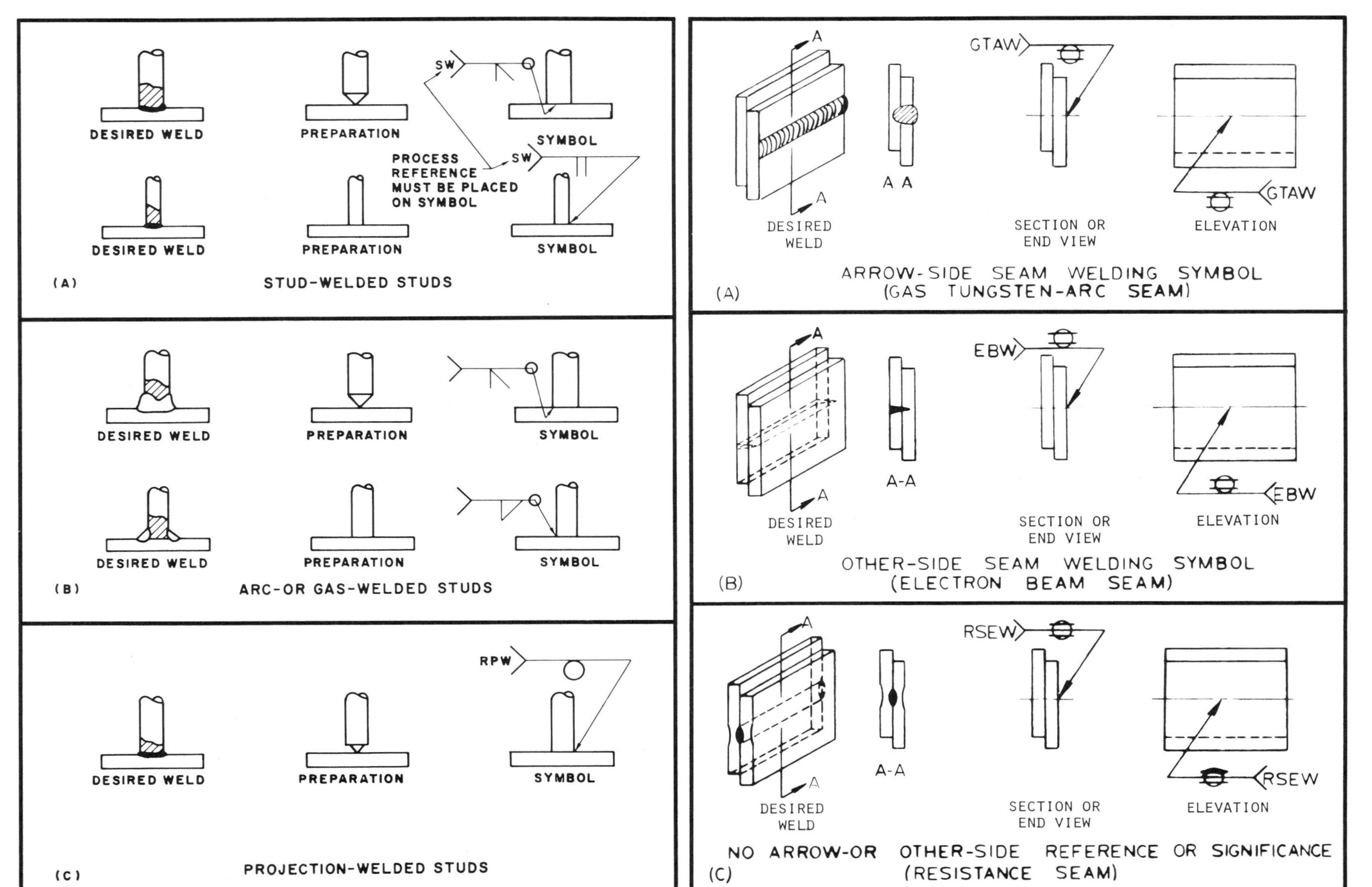

American Welding Society

14-40. *Use of welding symbols to indicate the welding of studs.*

American Welding Society

14-41. *Application of seam welding symbol.*

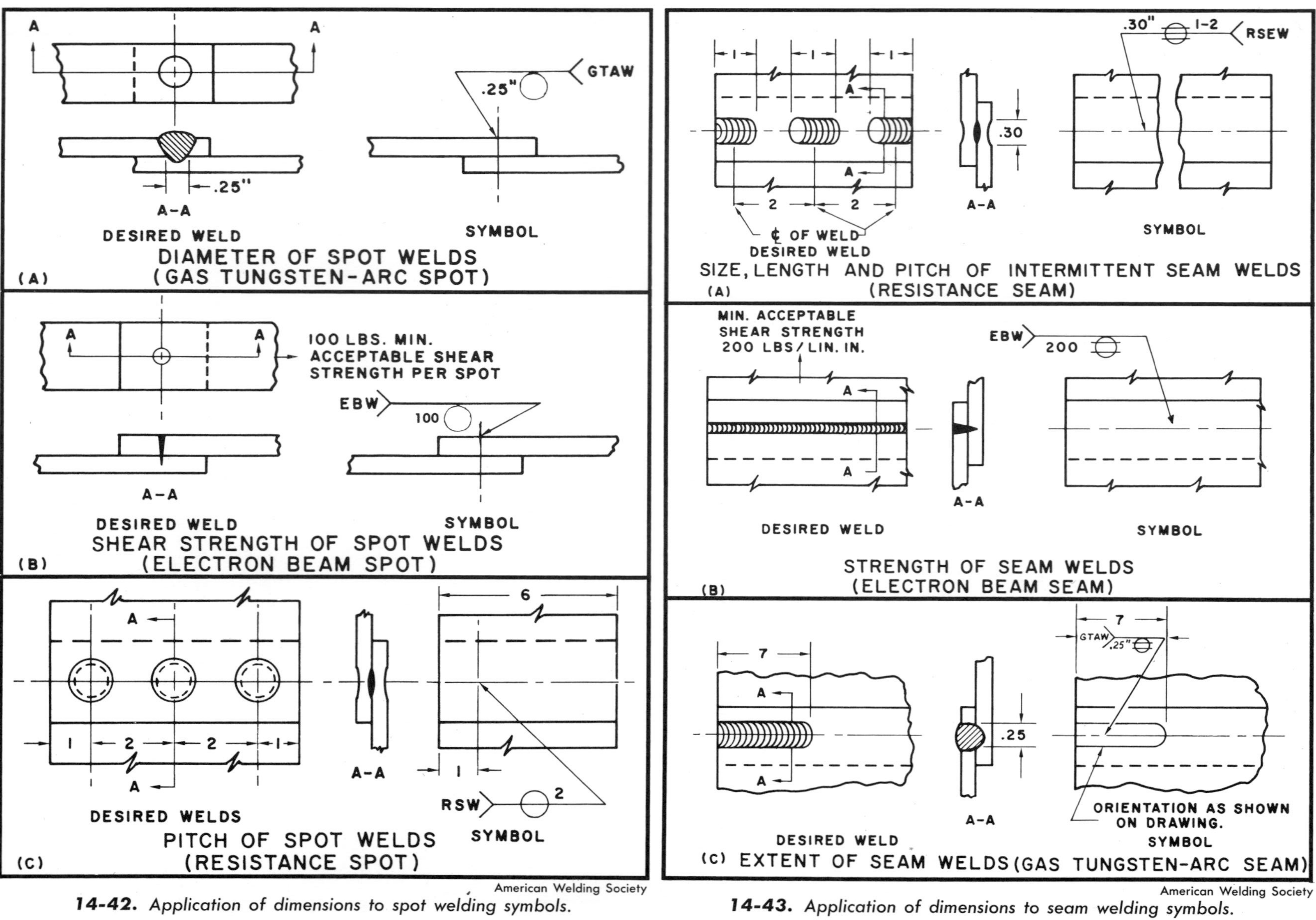

American Welding Society

14-42. *Application of dimensions to spot welding symbols.*

American Welding Society

14-43. *Application of dimensions to seam welding symbols.*

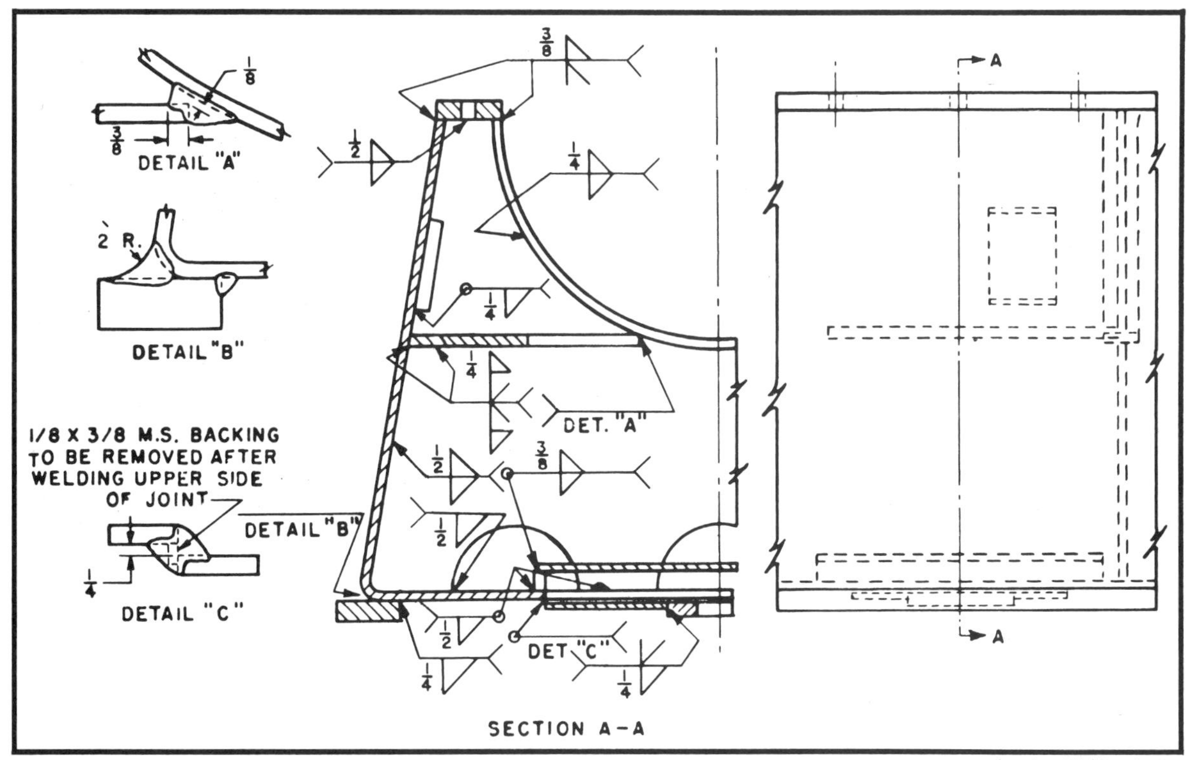

American Welding Society

14-44. Use of welding symbols on machinery drawing.

REVIEW QUESTIONS

1. Why are symbols used on drawings of welded fabrications?

2. Formerly locations for welds were referred to as "far side" and "near side." By what terms are these locations referred to now?

3. List eight elements that may be contained in a welding symbol.

4. The tail of the reference arrow should never be used to make a reference on the drawing. True. False.

5. Companies are never permitted to deviate in any manner from the official AWS symbol system. True. False.

6. Symbols have not been developed for use with the gas tungsten-arc and gas metal-arc processes. True. False.

7. Draw the symbol for the following types of welds:
   a. bead
   b. fillet
   c. plug and slot

8. Draw the symbol for the following types of grooves:
   a. square
   b. V
   c. bevel
   d. J

9. Draw the symbol for the following weld specifications:
   a. field weld
   b. weld all-around
   c. flush weld

10. How are specification references shown on the symbol?

11. Draw an arrow and show the location of each of the elements of a welding symbol.

12. Draw the proper symbol for the following specifications:
   a. $^3/_8$-inch fillet weld on T-joint, welding from one side only (near side), single pass

b. $\frac{1}{2}$-inch fillet weld on T-joint, welding on both sides, 3 passes on tank

c. V-groove weld, 60-degree included angle, $\frac{3}{8}$-inch plate, $\frac{1}{8}$-inch root opening, 3 passes, to be welded all-around from one side of plate

d. $\frac{1}{4}$-inch intermittent fillet welds on T-joint, length of welds 2 inches, welds spaced 4 inches center-to-center, welding on both sides of joint

e. A pad is built up on both sides with stringer beads to a height of $\frac{1}{2}$ inch

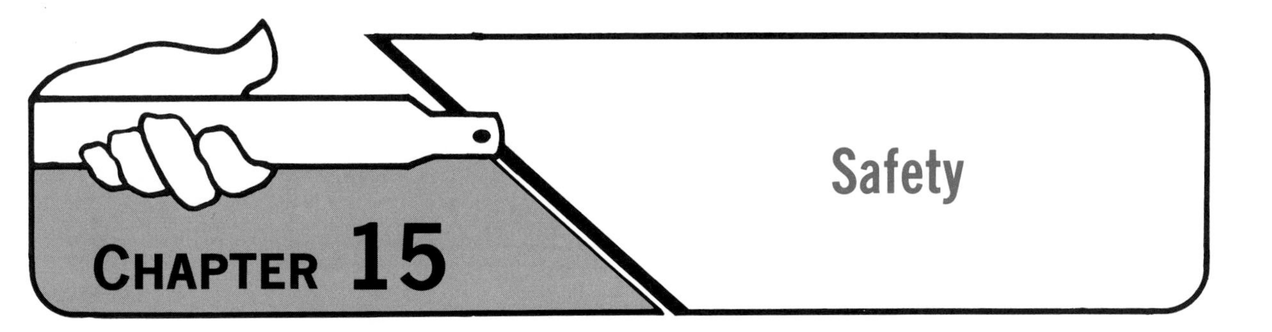

Chapter 15 Safety

The first attempts at establishing standards for the operation of electric and gas welding and cutting equipment was organized in March 1946 under the sponsorship of the American Welding Society. The committee which carried out this assignment was made up of users and suppliers of welding equipment and welded products, insurance companies, government agencies, and other organizations interested in welding and cutting.

Although the hazards associated with the welding processes are not greater than the hazards connected with any other form of industrial work, it is important that the welder recognize them. He should know and practice the safety procedures that prevent injury to himself and to others. The safety recommendations that follow are for the protection of persons from injury and illness and the protection of property and equipment from damage by fire and other hazards caused by the installation, operation, and maintenance of welding and cutting equipment. All of the welding processes included in this book are given consideration.

The welding student is urged to secure a copy of the complete welding safety standards published by the American Welding Society, entitled *Safety in Welding and Cutting* (Standard 249.1-1967).

SAFETY PRACTICES: ELECTRIC WELDING PROCESSES

Machines

Electric welding machines are subjected to all kinds of abuse. They receive little expert attention as compared with electric equipment for the usual power purposes. The operator should make sure that such machines are well-protected against accidental contact by himself or other workmen. The following precautions should be routine.

- The welder should never attempt to install or repair welding equipment. A qualified electrician should be in charge.
- In many cases the welding machines are moved from one operation to another. Consequently, convenient primary power receptacles should be provided about the shop or construction operation. Welding transformers should not be attached to lighting circuits.
- Welding machines should be furnished with a grounding connection for grounding the frame and case. This should not be confused with the work lead. The ground protects against electrical shock if there is a short circuit in the machine.
- Welding machines should not be operated above the current ratings specified by the manufacturer.
- When several operators are working on one structure, they should avoid touching two electrode holders at the same time. If this cannot be done, all d.c. machines should be connected with the same polarity and all a.c. machines should be set at the same phase of the supply circuit and with the same instantaneous polarity.
- Before starting operations, all connections to the machine shall be checked to make certain they are secure.
- Equipment should be checked for leaks of water, shielding gas, and engine fuels.
- When the welder stops work for an extended period or when the machine is to be moved, the power supply switch should be disconnected from the source of power.
- If gasoline-powered generators are used inside buildings and confined areas, arrangements must be made for the engine exhaust to be vented outside. Otherwise, enough carbon monoxide and other toxic gases may accumulate to harm the workers.
- A.c. transformers are usually air cooled. If transformers are cooled with a liquid, the liquid should be nonflammable. The transformers should be covered on the top and sides with enclosures of reasonably heavy sheet steel or other suitable material. This prevents injury to transformer windings and other parts and also prevents accidental

contact with uninsulated live parts.

Cables and Electrical Connections

Welding cable is subjected to severe abuse since it may be dragged over the work and across sharp corners and even run over by shop trucks. Special welding cables of high quality insulation should be used and kept in good condition. The fact that welding circuit voltages are low may lead to carelessness in keeping the welding cable in good repair. In large shops where the cable is tended by an electrical department, operators should observe and report the condition of their cable. In smaller shops, operators should learn how to make their own repairs. In all shops, workers should observe the following precautions.

- When several lengths of cable are coupled together, the cable connectors should be insulated on both the ground line and the electrode holder line.
- On large jobs there is apt to be a considerable amount of loose cable lying around. Operators must keep this cable arranged neatly and not put it where it could cause a stumbling hazard or become damaged. When possible, it should be strung overhead high enough to permit free passage of persons and vehicles.
- Special care should be taken to see that welding supply cables are not near or across the power supply cables and other high tension wires.
- Welding cables should be the extra flexible type designed for welding service and be the correct size for current and duty, Table 15-1.
- Steel conduits containing electric wiring, chains, wire ropes, cranes, hoists, and elevators should not carry welding current.
- If the metal structure of a building carries welding current, make sure that proper electrical contact exists at all joints.
- All ground connections should be checked to make sure that they are mechanically strong and electrically adequate for the required current.
- Welding cables must be kept dry and free from grease and oil. They should be arranged so that they do not lie in water or oil or in ditches or on tank bottoms. Rooms in which electric welding is done regularly should be wired with enough outlets so that it is not necessary to have extension cables strewn about the work place, Fig. 15-1.

Table 15-1
Cable sizes for arc welding machines based on safe operating temperatures.

Welding current (amperes)	Copper cable size no.	Aluminum cable size no.
100	4	4
150	3	2
200	2	0
300	1/0	3/0
400	2/0	4/0
500	3/0	—
600	4/0	—

Electrode Holders

The two hazards presented by electrode holders are overheating and electric shocks. Electrode holders frequently become hot during welding operations. This is usually caused by using holders for heavy welding that were designed for lighter work. Loose connections between the cable and the holder and poor ground connections may also overheat the holder.

- Care should be taken to use holders of the proper design for the work being performed.
- If the holder becomes hot, it should be allowed to cool. An extra holder should be provided for each welder to permit one to cool while the other is in use. The practice of dipping hot electrode holders in water must be prohibited.
- Both holder and cable must be well-insulated. If they are not, the operator may get an electric shock if he removes his gloves or if his clothing is damp.
- Metal and carbon electrodes should be removed from holders when not in use to prevent injuring workers, igniting tanks of fuel gas, or defacing the work. Tungsten electrodes should be removed or retracted within holders. Wire electrodes in semi-automatic holders should be retracted or cut off.
- When the welder is working in a sitting or prone position, he should make sure that his body is protected by a mat made of dry insulating material.
- Water-cooled holders for TIG and MIG welding must not be used if they leak.
- The power supply must be turned off when changing electrodes in TIG electrode holders and when threading coiled electrodes into MIG electrode equipment.

Maintenance of Equipment

All arc welding equipment should be maintained in safe working order at all times. Defective equipment should not be used. The following routine care is essential to keeping a welding shop safe.

- Commutators should be kept clean to prevent excessive flashing. Fine sandpaper should be used instead of flammable liquids for cleaning commutators.

- After welding, the entire welding machine should be blown out with clean, dry compressed air.
- Fuel systems on engine-driven machines should be checked regularly for possible leaks.
- Rotating and moving components should be kept properly lubricated.
- Air filters are not recommended. The reduction of air flow caused by even a clean filter may cause overheating, and a dirty filter reduces air flow dangerously.
- Outdoor welding equipment should be protected from bad weather. The protective cover, however, must not obstruct the ventilation system which prevents overheating of the machine. When not in use, the equipment should be stored in a clean, dry place. Machines which have become wet must be thoroughly dried and tested before being used.

Operator Protection

Hazards in electric welding processes include burns from sparks and molten metal, glare, electrical shocks, and harmful fumes. The operator protects himself by wearing articles such as face shields and aprons, Fig. 15-2. He is also careful to keep his clothing and the working area dry. Good ventilation must be provided.

EYE PROTECTION. The electric arc produces a high intensity of ultraviolet and infrared rays which have a harmful effect on the eyes and skin under continued and repeated exposure. It is necessary, therefore, to have full protection at all times both while engaged in actual welding and while observing welding operations. Ultraviolet rays do not usually cause permanent injury to the eyes unless by continued and repeated exposure, but temporary effects may be quite painful. Even short exposures have caused painful results and disability. Infrared rays are the heat rays of the spectrum. They do not cause permanent injury to the eyes except from excessive exposure. The distance at which persons can observe an electric arc without receiving eye injury is not known. The intensity of ultraviolet light varies inversely with the square of the distance between the source of light and the eyes. While the effects are lessened as the distance is increased, it is not known when the hazard becomes zero. Workers have been affected when arcs have been reflected from walls, ceilings, and even screens having highly reflective surfaces. The direct rays of the electric arc affect the skin like sunburn. The skin may become uncomfortable and even painful, but the rays do not cause permanent injury.

Because of the intensity of ultraviolet and infrared rays from the electric arc, the operator and his assistant must use a handshield or helmet. The device protects the skin of the face and neck and is also equipped with a filter glass to provide adequate eye protection. The selection of filter glass depends upon laboratory tests since the transmission of ultraviolet and infrared radiation cannot be determined by visual inspection. Depth of color does not necessarily indicate removal of the invisible radiation

15-1. *A welding shop with individual stations. Note that the welding cables are suspended so that they are free of the floor.*

Nooter Corp.

which may injure the eyes. Reliable dealers are able to supply filter glasses which have been shown by tests to conform to requirements. Table 15-2 is a guide for the selection of the proper shade numbers. These recommendations may be varied to suit the individual's needs.

The following precautions regarding helmets and handshields should be observed.

- All handshields and helmets should have a clear cover glass to protect the filter lenses from spatter. The cover glass should be free from defects which cause eye strain. Ordinary window glass is not usually suitable. The cover glass should be discarded when enough spatter has accumulated to interfere with vision.
- Helmets, handshields, and goggles must be made of a material which is not readily flammable and which is an insulator for heat and electricity. It must not corrode readily nor discolor the skin and be capable of sterilization.
- Helmets and goggles should not be transferred from one person to another without being sterilized.
- Although some shops employ helpers to chip and clean welds, the welding operator frequently does his own chipping and cleaning. Because of the colored filter glass in the face shields and helmets, it is necessary to raise them in order to see the welds properly during cleaning. Additional protection is needed to protect the eyes from flying particles. In many cases arc welding operators wear *chippers,* which are goggles with clear lenses under the shield or helmet. In other cases a clear lens is a part of the helmet, Fig. 15-3. It is mounted directly under the hinged filter glass holder. Both methods give good protection.
- Welds are frequently cleaned by using a power-driven wire brush. Such a brush should be guarded with a hood guard, and the operator should wear eye protection approved for such operations.
- When more than one welder is working on the job at the same time, all operators should wear

***15-2.** A welder properly protected for the MIG process. He is wearing a helmet, a leather sleeve on his left arm, and leather gloves. He has rolled down the cuffs of his pants so that they cannot catch particles of hot metal.*

Linde Division, Union Carbide Corp.

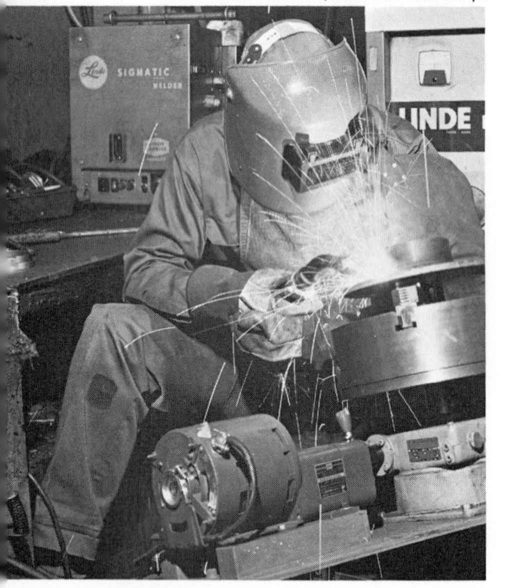

Table 15-2

Filter lens shades for different electric welding processes.

Welding operation	Shade number
Shielded metal-arc welding—1/16, 3/32, 1/8, 5/32 inch electrodes	10
[1] Gas metal-arc welding (nonferrous)—1/16, 3/32, 1/8, 5/32 inch electrodes	11
Gas metal-arc welding (ferrous)—1/16, 3/32, 1/8, 5/32 inch electrodes	12
Shielded metal-arc welding—3/16, 7/32, 1/4 inch electrodes	12
5/16, 3/8 inch electrodes	14
Atomic hydrogen welding	10–14
Carbon arc welding	14
Soldering	2
Torch brazing	3 or 4
[2] Light cutting, up to 1 inch	3 or 4
Medium cutting, 1 inch to 6 inches	4 or 5
Heavy cutting, 6 inches and over	5 or 6
Gas welding (light) up to 1/8 inch	4 or 5
Gas welding (medium) 1/8 inch to 1/2 inch	5 or 6
Gas welding (heavy) 1/2 inch and over	6 or 8

[1] The intensity of ultraviolet radiation emitted during MIG welding ranges from 5 to 30 times higher than with covered electrodes. The increased brightness makes it necessary to use filter plates one or two shades darker than usual to prevent eye strain, burn or fatigue. The use of flash protection goggles with side shields under the helmet is strongly recommended.

[2] In gas welding or oxygen cutting where the torch produces a high yellow light, it is desirable to use a filter or lens that absorbs the yellow or sodium line in the visible light of the operation.

goggles under their helmets to give added protection. These goggles should have shaded lenses with the same shade of sideshield attached to them.

- Helpers working with arc welding operators should be equipped with goggles and gloves. If they are exposed to the arc, they should wear welders' helmets or handshields.
- Special precautions must be taken to protect other workers from the harmful rays given off from arc welding operations and from flying chips during cleaning operations. It is preferable to locate welding jobs in special rooms or booths. If this is impractical, the operations should be screened or enclosed, not only to prevent workers from looking directly at the arc, but also to protect them from reflected rays as much as possible. Further protection against reflected rays is frequently provided by applying paint of low reflective qualities to the screen or enclosure and to other nearby surfaces. Zinc oxide combined with lamp black to produce a bluish gray color is satisfactory for this purpose.

PROTECTION AGAINST BURNS. The following precautions should be taken to avoid burns from sparks and molten metal.

- On account of the intense heat of the welding arc and flying particles of hot metal, aprons made of asbestos or other nonflammable material are desirable. Aprons should not be worn while climbing ladders. Long sleeves and gloves with long gauntlets made of leather, asbestos, or other nonflammable material should be worn to protect the arms and hands. Rubber gloves with leather protectors are desirable when working in damp locations and when perspiration causes leather or cotton gloves to become wet. Shirt sleeves should not be rolled up. When welding overhead, capes or shoulder covers made of leather or other suitable material should be worn. Leather skull caps are sometimes worn under helmets to prevent head burns.
- Welding operators and their helpers should wear leather shoes high enough to protect their ankles and fitted with bellows tongues to prevent hot metal and scale from getting into the shoes and causing burns. Low-cut shoes should not be worn. In some cases, spats are necessary to prevent hot metal from entering shoe tops. Trouser cuffs should not be turned up on the outside. Woolen clothing is preferable to cotton because it is not so easily ignited. Fireproof outer clothing may be needed in some instances. Welding operators should not wear oil-soaked clothing because such clothing ignites easily.
- In production work a sheet metal screen in front of the worker's legs provides further protection against sparks and molten metal in cutting operations.
- For overhead welding or welding in extremely confined spaces, the ears should be shielded by inserting wool or rubber plugs and covering them with wire screen protectors.

PROTECTION AGAINST SHOCK. Voltages required for arc welding are low and normally do not cause injury or severe shock. Nevertheless, under some circumstances, voltages may be dangerous to life. The severity of the shock is determined largely by the amount of current flowing through the body, and the current, in turn, is determined by both the voltage and the contact resistance of the skin. Clothing damp from perspiration or wet working conditions reduces contact resistance. Low contact resistance increases a low current to a value high enough to cause such violent muscular contraction that the welder cannot let go of the live part. The following safety precautions reduce the possibility of severe electrical shocks.

- The welder must make sure his machine is grounded. While welding, he must protect himself from electrical contact with the work and the ground.
- The welder should never permit the live metal parts of an electrode holder to touch his bare skin or wet clothing.
- Electrode holders must not be cooled by immersion in water.
- Water-cooled holders for TIG and MIG welding must not be used if there is a water leak.
- The welding machine supplying power to the arc must always be turned off when changing electrodes in TIG electrode hold-

15-3. *A clear glass viewing shield is part of the helmet. Also note the leather jacket and gloves.*

Modern Engineering Co.

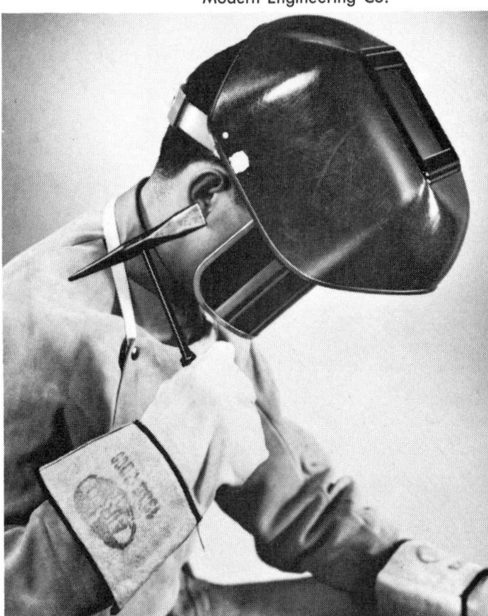

ers and when threading coiled electrodes into MIG equipment.

- Special precautions should be taken to prevent shock-induced falls when the operator is working above ground level. When working above ground, the welder should not coil or loop welding electrode cable around parts of his body. He should not use cables with splices or repaired insulation within 10 feet of the holder.
- Although insulated holders are used and the electrode coatings provide insulation, the operator is nevertheless exposed to the open circuit voltage when changing electrodes and at any other time when the arc is extinguished. He should avoid standing on wet floors or coming into contact with a grounded surface.
- The danger of electric shock is increased during periods of high temperature and high humidity. Under these conditions the physical reserve of the individual worker is lowered. The electrical resistance of the skin and clothing is reduced by perspiration. Thus, an environment is set up in which electric shock is much more of a potential hazard.

SAFETY-PRACTICES: OXYACETYLENE WELDING AND CUTTING

Care of Cylinders

Oxyacetylene welding and cutting require the use of a mixture of flammable gases and air which is highly explosive. Generally, compressed gas cylinders are safe for the purpose for which they are intended. Few accidents have occurred, but those that have occurred have been the result of abuse and mishandling in the use and/or storing of flammable gases. The following precautions should be taken routinely.

Handling Cylinders

- Cylinders may be lifted by a crane or derrick only if a cradle or platform is used. Never use a sling or attach a hook to the valve protection caps. Never drop cylinders nor permit them to strike each other violently.
- Cylinders may be moved by tilting and rolling them on their bottom edges. Never drag or slide them. If transferred in a carrier, they should be fastened securely.
- Cylinders should be fastened securely while in use.
- Never move unattached cylinders with the regulators on. Cylinder valves must be closed and capped before being moved. The valves of empty cylinders must be closed, the valve protection caps secured in place, and the cylinder marked MT(empty).
- Cylinder valves must be closed when work is finished.
- Keep cylinders away from welding and cutting operations so that sparks, hot slag, and flame cannot reach them.
- Do not place cylinders near an electrical circuit. They must not be used as a ground in arc welding nor to strike an arc.
- Tampering with the numbers and markings stamped on the cylinders is illegal. Never tamper with the valves or safety devices on valves or cylinders.
- Never attempt to refill a cylinder or to mix gases in a cylinder.

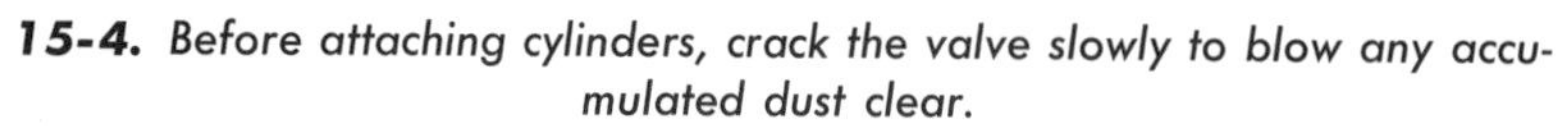

15-4. *Before attaching cylinders, crack the valve slowly to blow any accumulated dust clear.*

Linde Division, Union Carbide Corp.

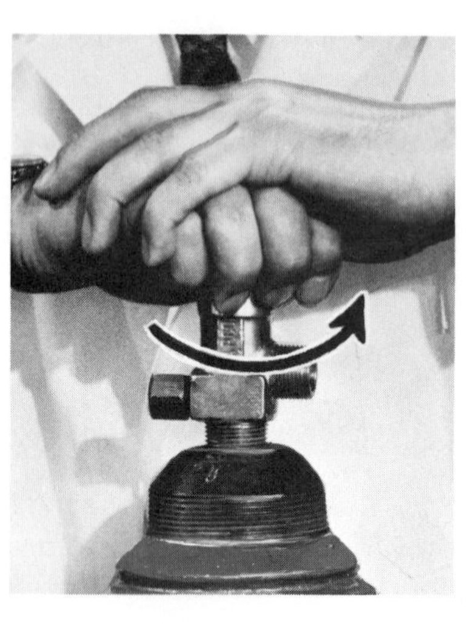

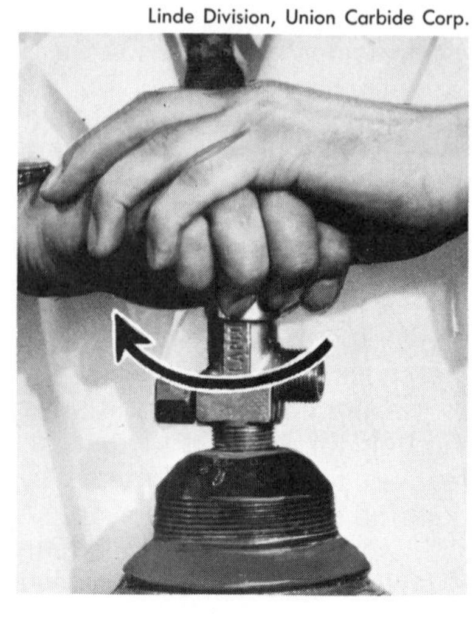

Storage of Cylinders

- Cylinders are stored in a dry, well-ventilated room away from open flame.
- Cylinders should be protected against the weather. Ice and snow should not accumulate on them. They should be screened against the continuous direct rays of the sun to prevent excessive rises in temperature.
- Cylinders must not be stored near highly flammable substances such as oil and gasoline.
- Cylinders shall be stored away from locations where heavy moving objects may strike or fall on them.

Acetylene Cylinders

- Acetylene is a fuel gas and should be referred to by its

proper name and not by the word *gas.*

- Always store and use acetylene cylinders valve end up.
- Keep sparks and open flame away from cylinders.
- Handle carefully. A damaged cylinder may leak and cause a fire.
- Do not use acetylene directly from the cylinder. Install a regulating device between the cylinder and the torch. Before connecting a regulator to a cylinder valve, open the valve slightly and close it immediately, Fig. 15-4. This is called *cracking.* Cracking removes dust and dirt from the valve. Never crack a cylinder near other welding work or near sparks, flame, and other sources of combustion.
- Before removing a regulator from a cylinder valve, close the valve and release the gas from the regulator.
- Never connect two or more cylinders together with a manifold or other connecting device that has not been approved for this purpose.
- The cylinder valve should be opened about $1\frac{1}{2}$ turns. The wrench used for opening the cylinder should always be kept on the valve when the cylinder is in use.
- Never test for acetylene leaks with an open flame. Use soapy water.

Oxygen Cylinders

- Always refer to oxygen by its proper name, *oxygen,* and not as *air.*
- Although oxygen itself does not burn, it supports and speeds up combustion. Never permit oil and grease to come in contact with oxygen cylinders or any of the equipment being used in the welding or cutting operation. A jet of oxygen should never be permitted to strike a substance that has oil or grease on it.
- Never use oxygen as a substitute for compressed air.
- Do not store oxygen cylinders near combustible materials or fuel gases.
- Do not use oxygen directly from the cylinder; the pressure must be reduced through regulation.
- Never use a hammer or wrench on the oxygen cylinder valve. If the valve cannot be opened with the hands, return the tank to the supplier.
- Before connecting the regulator, open the cylinder valve for an instant to remove dirt and dust. After attaching the regulator and before cylinder valve is opened, be sure that the adjusting screw of the regulator is released. Do not open the cylinder valve suddenly. Open the valve completely when the cylinder is in use.
- When connecting two or more cylinders, always use a manifold designed and approved for the purpose, Fig. 15-5.
- Never try to mix gases in an oxygen cylinder.
- Never interchange equipment made for use with oxygen with equipment intended for use with other gases.

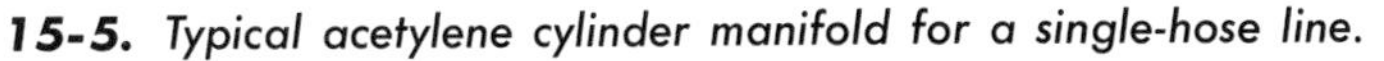

15-5. *Typical acetylene cylinder manifold for a single-hose line.*

Linde Division, Union Carbide Corp.

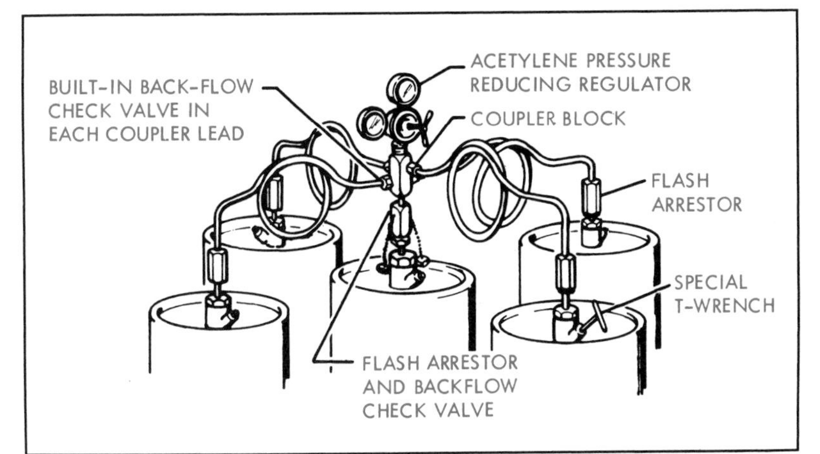

Operator Protection

Fire is the greatest hazard to the welder when using the oxyacetylene process. Accordingly, he should wear protective clothing and be very careful not to introduce fire hazards in the welding area. The following precautions should be observed.

- Never perform a cutting operation without goggles fitted with lenses of the proper shade.
- The head and hair should be protected by a cap.
- The hands should be protected by gloves.
- The arms should be protected by long sleeves.
- The feet should be protected by high-top shoes, and pants legs should be cuffless.
- Keep clothing free from oil and grease.

Torch

The torch and hose assembly must be put together carefully and correctly to insure safe operation. Figure 15-6 shows the complete assembly. The following precautions are routine.

- Connect the oxygen hose from the oxygen regulator to the hose connection on the torch marked *oxygen.*
- Connect the acetylene hose from the acetylene regulator to the hose connection on the torch marked *acetylene.*
- Select the proper welding tip or cutting nozzle and screw it carefully, but not too tightly, on the torch.
- When changing torches, shut off the gases at the pressure regulators and not by crimping the hose.
- Use a friction lighter instead of a match for lighting the torch. Make sure that the torch is not directed toward another person while you are lighting it.
- Never attempt to light or relight a torch from hot metal in a small hole where gas may accumulate. Light the torch with a friction lighter and adjust the flame before inserting it into the hole.
- When the welding or cutting is finished, extinguish the flame by closing the acetylene valve first and the oxygen valve next.
- When stopping the operation for a few minutes, it is permissible to close only the torch valves. Always extinguish a torch when it is not in your hands.
- When stopping for a longer period (during the lunch hour or overnight), close the cylinder valves. Then release all gas pressure from the regulators by opening the torch valves momentarily. Close the torch valves and release the pressure-adjusting screws. If the equipment is to be taken down, make sure that all pressure-adjusting screws are turned to the left until free.
- Do not leave unlighted torches connected for use in boiler tubes, tanks, and other confined spaces during the lunch hour or when leaving the job for other reasons.
- If leakage develops around the torch valve stems, tighten the packing nuts and repack them if necessary. Use only packing supplied by or recommended by the manufacturer of the torch. Never use oil on the packing or elsewhere.
- If a torch valve does not shut off completely, shut off the gas supply and remove the valve assembly. Wipe the seating portion of the valve stem and the body or replaceable seat with a clean cloth. If the valve still leaks, new parts should be used or the valve body should be reseated.
- If the holes in the torch tip or nozzle become clogged, clean them with the proper size drill or a copper or brass wire. A sharp hard tool, which would enlarge or bellmouth the holes, must not be used. Clean the holes from the inner end wherever possible.

Hose

It is very important to have the correct hose for the kind of gas used. The generally recognized colors are red for acetylene and other fuel-gases, green for oxygen, and black for inert gas and air. Hose must be kept in good repair to prevent fires and explosions. The following precautions should be taken for proper use and maintenance.

- Protect the hose from flying sparks, hot slag, and hot objects.
- Do not allow hose to come in contact with oil and grease. These materials cause the rubber to deteriorate and increase the danger of fire when in contact with oxygen.
- Hose should be stored in cool locations. Be careful not to put it on greasy floors or shelves because the rubber absorbs grease and oil.
- New hose is dusted on the inside with fine talc. Blow this dust out with compressed air before using.
- All hose should be examined carefully at frequent intervals for

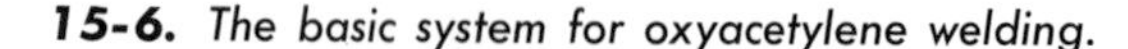

15-6. *The basic system for oxyacetylene welding.*

Linde Division, Union Carbide Corp.

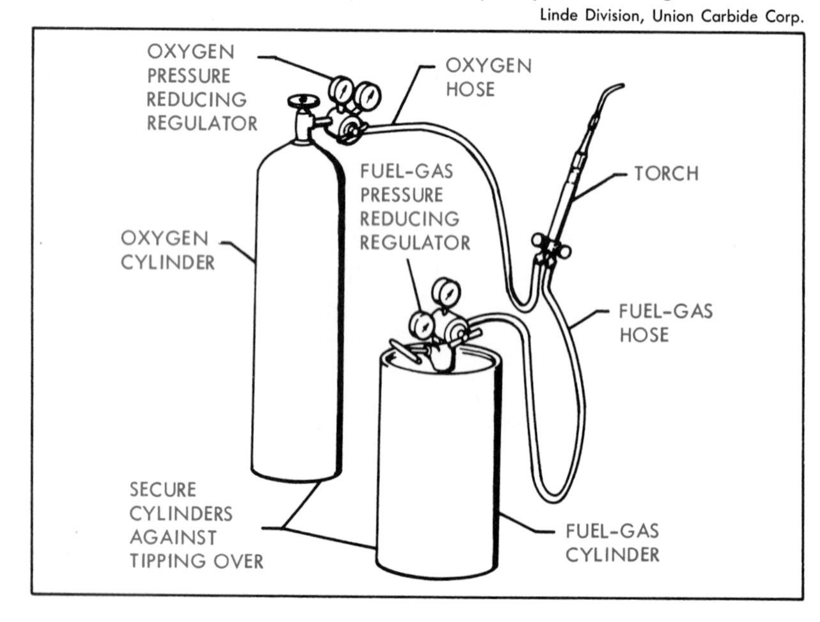

leaks, worn places, and loose connections. This can be done by immersing the hose in water under normal working pressure. Hose and hose connections may also be tested with oil-free air or an oil-free inert gas at twice the normal pressure to which it is subjected in service. In no case should this be less than 200 p.s.i. for hose and 300 p.s.i. for hose connections.

- Leaks must be repaired at once by cutting the hose and inserting a splice. Acetylene escaping from defective hose may ignite and start a serious fire. It may also set fire to the operator's clothing and cause severe burns.
- When hose shows wear at a connection, cut off the worn portion and reinsert connections securely.
- Do not repair hose with tape.
- If a flashback occurs and burns inside the hose, discard that length of hose. A flashback of this sort makes a piece of hose unsafe because it burns the inner walls. Sooner or later this part of the hose will disintegrate and cause trouble by clogging or otherwise interfering with the hose and operation of the torch.
- Hose connections should be of the regulation type conforming to the standards of the International Acetylene Association, Fig. 15-7.

Pressure-Reducing Regulators

Pressure-reducing regulators, both adjustable and nonadjustable, must be used only for the gas and at pressures for which they are intended. The following safety precautions are observed in their use.

- Always use the right regulator for the gas being used.
- Clear all passages before applying pressure.
- Never force a regulator onto or into a cylinder valve.
- Never use a creeping regulator.
- Never use oil on the regulator for any purpose.
- Pressure-adjusting screws on regulators should always be fully released and the regulator drained of gas before the regulator is attached to a cylinder and before the cylinder valve is opened.
- Only skilled mechanics properly instructed in the work should repair regulators and parts of regulators such as gauges.
- The working or low-pressure gauges attached to regulators should be periodically tested by the supplier to insure accuracy.
- Union nuts and connections on regulators should be inspected before use to detect faulty seats which may cause leakage of gas when the regulators are attached to the cylinder valves. They should be removed from service if damaged.

Welding and Cutting Operations

The welder must take the following precautions when welding or cutting with the oxyacetylene process.

- Do not use the welding or cutting flame where an open flame of any kind would be dangerous such as in or near rooms containing flammable vapors or liquids, dust, and loose combustible stock.
- Do not use welding or cutting equipment near dipping or spraying rooms.
- Be careful when welding or cutting around a sprinkler system.
- If the work can be moved, it is better to take it to a safe place rather than perform the work in a hazardous location.
- When the work requires that torches be used near wooden construction and in locations where the combustible material cannot be removed or protected, station extra men with small hoses, chemical extinguishers, or fire pails nearby. It is advisable to carry a fire extinguisher as regular equipment.
- Never do any hot work such as welding or cutting used drums, barrels, tanks or other containers until they have been cleaned so thoroughly that it is

15-7. *Two types of hose nipples are commonly used for oxyacetylene welding: the push type has a serrated nipple and requires a hose clamp, and the screw type has a ferrule placed over the hose end and a coarsely threaded nipple turned into the hose end.*

Linde Division, Union Carbide Corp.

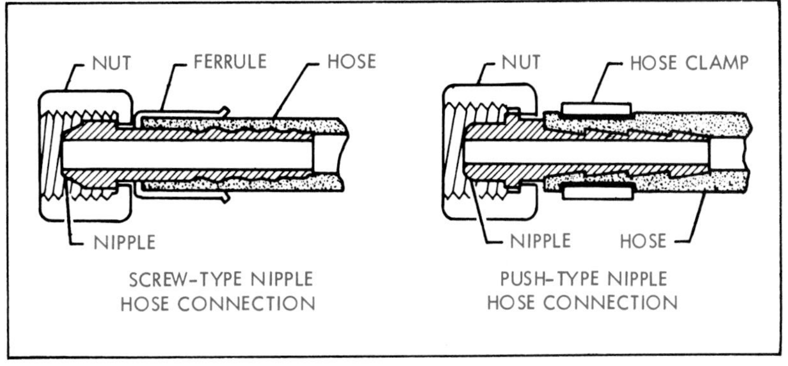

absolutely certain that no flammable materials are present.

- Never put down a torch unless the oxygen and acetylene have been completely shut off. Never hang torches from regulators or other equipment so that the flame can come in contact with the oxygen or acetylene cylinders even though you think the valves have been shut off.
- Mount cylinders on portable equipment securely.
- Never support work on compressed gas cylinders.
- Do not cut materials in such a position as to permit the severed section to fall on your legs or feet. Protect your legs and feet from sparks and hot slag.
- Do not allow showers of sparks from a welding or cutting operation to fall upon persons who may be working below. Make sure that sparks do not fall on flammable material or into the unprotected head of an acetylene cylinder. The sparks may melt the fusible plugs of the cylinder head and ignite the escaping contents. Such carelessness is both dangerous and wasteful.
- When welding or cutting is being performed in a confined space, such as the interior of a boiler, always leave the cylinders on the outside with an attendant and lead the gas in through the hose to the point where the work is being done.
- Be careful when beginning a cut on a closed container. The air pressure inside will cause it to blow out as soon as a hole is made. Keep your face to one side.
- Never work directly on a concrete floor because when it is heated, the concrete may spall (chip) and fly about, possibly injuring you and other workers.
- Make sure that the room is well-ventilated.
- Do not set the oxygen pressure higher than that required to do the job, Fig. 15-8.

BACKFIRE AND PREIGNITION. Sometimes the welding or cutting operation will be interrupted by a series of popping sounds at the torch. This is caused by a momentary retrogression (backfire) of the flame into the torch tip. Backfire results from the preignition of the gases. Preignition is caused by a number of conditions that can be avoided.

- It may be caused by touching the torch tip to the work. If this happens, the torch can be relighted instantly if the metal being welded or cut is hot enough to ignite the gases. Otherwise, a lighter should be used. Never relight the torch from hot metal in a small hole.
- The pressure at the regulators may be too low. Adjust the regulator to a higher pressure before relighting.
- The flame may be too small for the tip size. Relight the torch and increase the size of the flame.
- The tip may have carbon deposits or metal particles inside the hole. They become overheated and act as ignitors of the gas before it passes through the hole.
- If welding or cutting in a confined area such as a corner, the tip may become overheated and preignition will take place inside the tip. Correct this condition by cooling the tip.

FLASHBACK (SUSTAINED BACKFIRE). Sustained backfire, known as *flashback,* occurs when the retrogression of the flame back into the mixing chamber is ac-

15-8. *Oxygen pressure that is too high can blow sparks twice as far as need be. They can easily start fires, and oxygen is wasted.*

Modern Engineering Co.

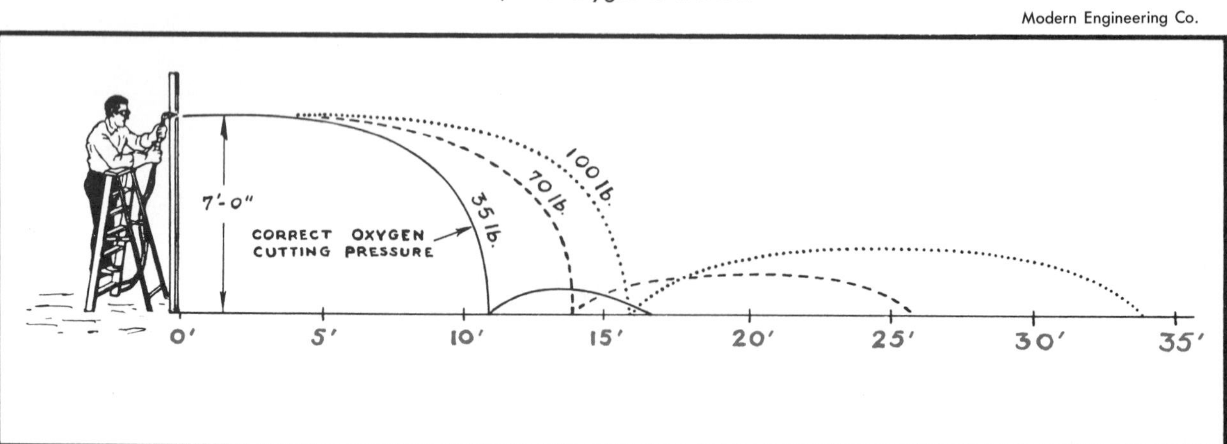

companied by a hissing or squealing sound and a characteristic smoky, sharp-pointed flame of small volume. The gas supply should be cut off immediately to prevent overheating and the possible destruction of the torch head. Generally, the torch head must be cooled before it can be reignited.

If the flame flashes back (burns back inside the torch), immediately shut off the torch oxygen valve which controls the flame. Then close the acetylene valve. After a moment, relight the torch in the usual manner. Even with improper handling of a torch, flashback rarely occurs. When one happens, it indicates that something is seriously wrong with the torch or the manner of operating it. In order to prevent flashbacks, proper delivery pressure of both gases should be maintained.

GENERAL PRECAUTIONS

Ventilation

When welding is carried on outdoors or in large, well-ventilated shops and nontoxic materials are involved, welding operators suffer no harmful effects. When toxic concentrations of gases, fumes, and dust are generated, however, protection must be provided.

The amount of contamination to which welders may be exposed depends on the following factors:

- The dimensions of the space in which welding is to be done. The height of the ceiling is of major importance.
- The number of welders working in a unit of space
- The nature of the hazardous fumes, gases, or dust resulting from the metals being welded
- Atmospheric conditions such as humidity
- Heat generated by the welding operation
- The presence of volatile solvents. (A volatile substance evaporates rapidly, thus polluting the air.)

The following precautions are taken to assure a supply of fresh air in the work area when natural ventilation is not effective.

- Local exhaust systems, Fig. 15-9, remove toxic substances. When welding operations are permanently located, the entire booth may be ventilated with a system like those provided for spray-coating booths. Portable exhaust systems are also available for this purpose. When large quantities of air are removed by exhaust systems, it is replaced by fresh air.
- Supplied-air respirators, such as air line respirators, hose masks with or without blowers, and self-contained oxygen-breathing apparatus, are recommended for confined areas and other locations where there are high concentrations of toxic substances, Fig. 15-10.
- If screens are used, they should be arranged so that ventilation is not seriously restricted. Screens should have a space at the bottom of about 2 feet above the floor. The operator should make sure, however, that the screen protects from the arc glare.
- Oxygen from a cylinder or torch is never used instead of air for ventilation.
- Extreme care must be taken when welding or cutting the following materials:
    - Fluxes, electrode coverings, and fusible granular materials containing fluorine compounds.
    - Zinc-bearing base and filler metals and metals coated with zinc-bearing materials

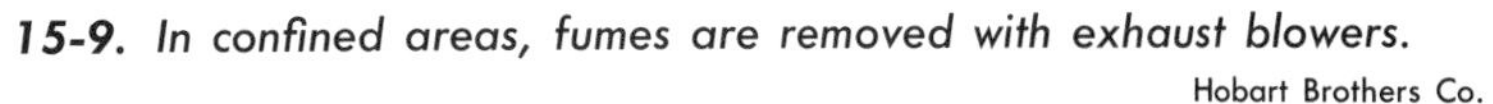

***15-9.** In confined areas, fumes are removed with exhaust blowers.*

Hobart Brothers Co.

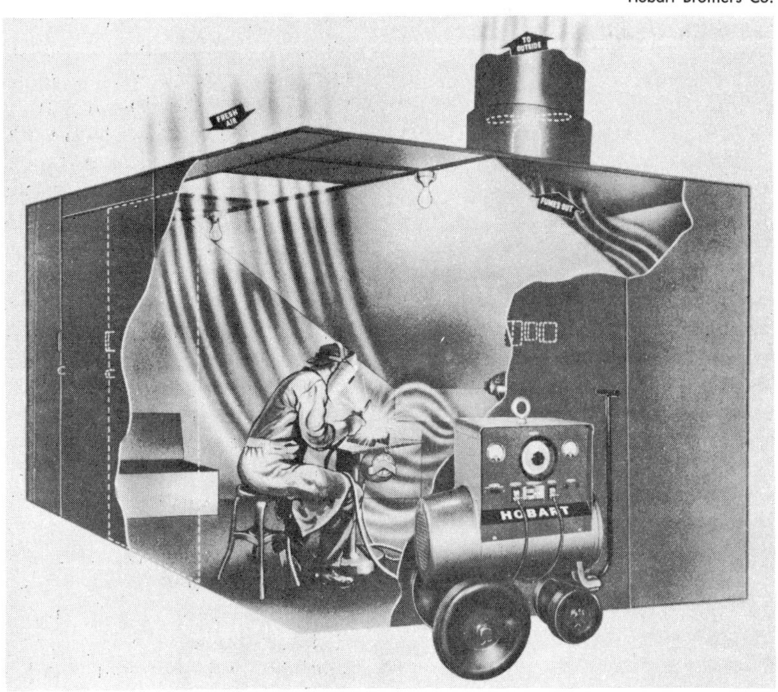

- Lead base metals and metals coated with lead-bearing materials such as paint
- Base and filler metals containing beryllium
- Cadmium-coated base metals and filler metals
- Metals coated with mercury-bearing materials, including paint
- Fluxes, coverings, and other materials containing antimony, arsenic, bismuth, chromium, cobalt, copper, nickel, manganese, magnesium, molybdenum, thorium, vanadium and their compounds

MIG SMOKE EXHAUST SYSTEM. The MIG smoke extractor system, Fig. 15-11, increases the operator's visibility, thereby improving efficiency and weld quality. Integral tubing permits him to weld wherever he can reach without repositioning a separate exhaust duct.

Smoke swirls up and away from the arc and then is directed through the gun to the filter. Visible smoke particles are captured within the filter and kept from re-entering the work area, thus leading to a cleaner work area and less maintenance to other equipment, Fig. 15-12. The MIG welding gun also runs cooler due to air flow through internal ducting, Fig. 15-13.

Work in Confined Spaces

A confined space is intended to mean a relatively small space such as a tank, boiler, small compartment of a weldment, or a small room, Fig. 15-14. The following protection should be provided for welders.

- Ventilation is necessary when working in confined spaces.
- Where it is impossible to provide local ventilation, air-supplied respirators or hose masks are provided. A workman is stationed outside of the confined space. He services the power and ventilation lines to insure the safety of those who are working within.
- Gas cylinders and welding machines are left outside of confined spaces. Portable equipment mounted on wheels must be securely blocked to prevent accidental movement.
- Where a welder must enter a confined space through a movable or other small opening, means must be provided for removing him quickly in case of emergency. When safety belts and lifelines are used for this purpose, they are attached to the welder's body in such a way that his body cannot be jammed in the small exit opening.
- In order to prevent the accumulation of gas because of leaks or improperly closed valves, the gas supply to the torch should be positively shut off outside the confined area whenever the torch is not to be used for a period of time such as during the lunch hour or overnight. If possi-

15-10. Air line respirator.
Mine Safety Appliances Co.

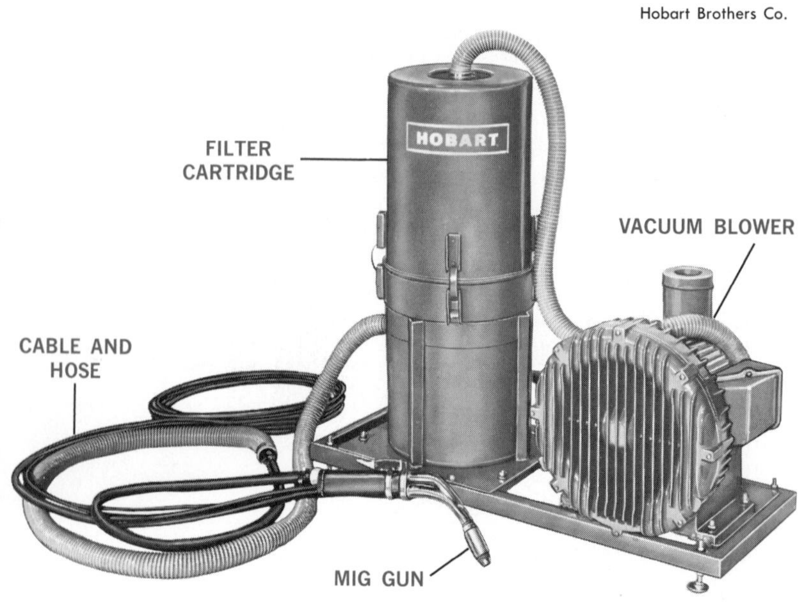

15-11. The filter cartridge system used in connection with the MIG smoke exhaust system.
Hobart Brothers Co.

ble, remove the torch and hose from the confined space.

● After welding operations are completed the welder marks the hot metal or provides some other means of warning other workers.

Fire Protection

Many safety practices for electric and oxyacetylene welding and cutting are required for fire prevention. In addition to these special practices, the following general precautions should be taken wherever any welding or cutting is being done.

● Welding operations should be done in permanent locations free from fire hazards. Booths should be constructed of nonflammable materials such as asbestos and sheet metal. If it is possible to move the job, it is preferable to take it to a safe location for cutting and welding rather than to perform the work in a hazardous location.

● Fire is a particular hazard when portable welding equipment is used. Before welding operations in which portable equipment is used are started, the location should be thoroughly inspected by a competent employee to determine what fire protection equipment is necessary. It is advisable to require written permits issued by the welding supervisor, a member of the plant fire department, or some other qualified person before welding operations are started. This is particularly important in a hazardous location. In small plants this responsibility is usually given to a competent welding operator.

● Welding operations should not be permitted in or near rooms containing flammable vapors, liquids, or dust until all fire and explosion hazards have been eliminated. If welding is necessary in such locations, the area should be thoroughly ventilated. Sufficient draft should be maintained during welding and cutting operations to prevent the accumulation of explosive concentrations of such substances.

● Tanks, drums, and pipelines which have contained flammable liquids should be cleansed of all solid or liquid flammable material, purged of all flammable gases and vapors, and tested for the presence of flammable gases before welding operations are started.

● Welding operations should not be performed on spray booths or ducts that may contain combustible deposits without first making sure that such places are free of flammable materials.

● Where welding has to be done in the vicinity of combustible material, special precautions should be taken to make certain that sparks or hot slag do not reach such material and thus start a fire. If the work cannot be moved, exposed combustible materials should, if possible, be moved a safe distance away. Otherwise, they should be covered with asbestos curtains or sheet metal during welding operations.

● Wood floors should be swept clean before welding operations

15-12. *(Left) Volume of smoke with the smoke exhaust holder and system; (right) volume of smoke without the smoke exhaust holder and system.*

Hobart Brothers Co.

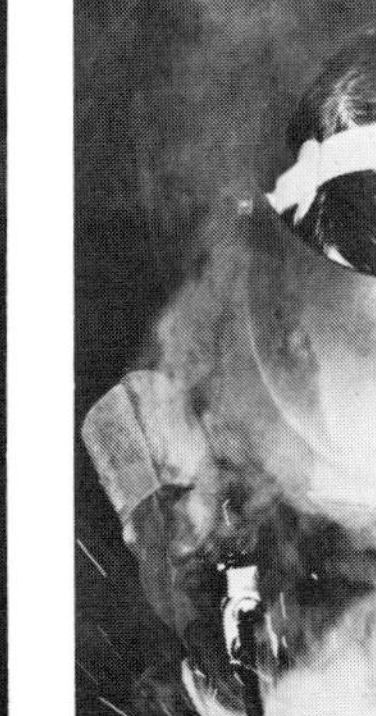

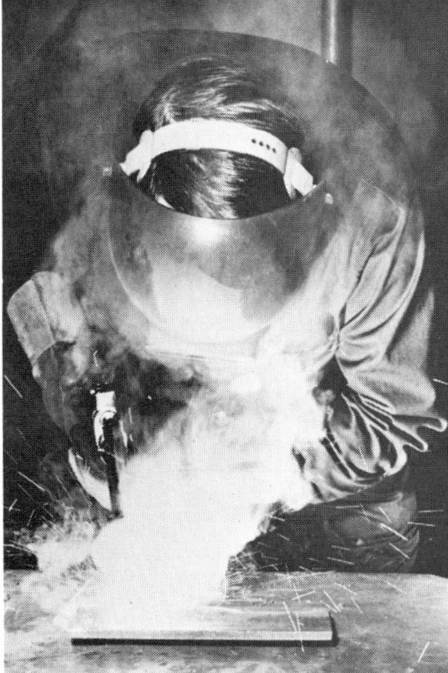

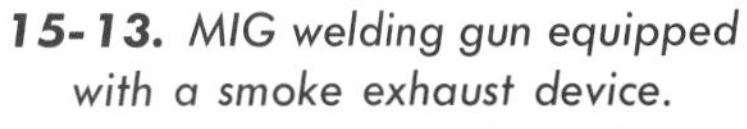

15-13. *MIG welding gun equipped with a smoke exhaust device.*

Hobart Brothers Co.

are started. Wood floors should be covered with metal or other noncombustible material where sparks of hot metal are likely to fall. In some cases, it is advisable to wet down the floors. Hot metal or slag should not be allowed to fall through cracks in the floor or other openings on combustible materials on the floor below. Particular attention should be taken to see that hot slag or sparks do not fall into machine tool pits.

- Similarly, sheet metal guards or asbestos curtains should be used to guard cracks and holes in walls, open doorways, and open or broken windows. Make certain that there is no opening where the curtain meets the floor.
- A workman equipped with fire-extinguishing equipment should be stationed at or near welding operations in hazardous locations. He makes sure that sparks do not lodge in floor cracks or pass through floor or wall openings. The workman should be kept at the job site for as long as 30 minutes after the job is completed to make sure that smoldering fires have not been started.

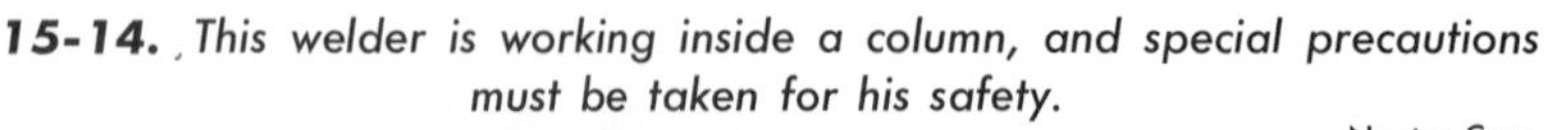

15-14. *This welder is working inside a column, and special precautions must be taken for his safety.*

Nooter Corp.

- The usual precautions in handling electric power should be observed. The insulation of the welding cable may be burned off if leads are too small to carry the necessary current. The insulation may be cut through by dragging the leads across sharp objects. Electrode holders may be carelessly dropped. Welding sets may be shortened due to mishandling. These hazards can largely be eliminated by good maintenance of equipment.

Other Precautions for Safe Working Conditions

A large number of potentially hazardous miscellaneous situations must also be of concern to the welder who is conscious of the need to work in a safe environment. Very often accidents are caused by a relatively unimportant condition that has not been taken care of. Extremely dangerous hazards usually draw the attention of the welder and are therefore rarely a cause of accidents.

The welding operator should be thoroughly instructed in the performance of his work regarding the protection of himself and others working nearby. He should also realize that good workmanship in making sound welds is essential so that others may not be injured because of the failure of the welded part. The American Welding Society has prepared standard codes for welding procedures and operator qualifications which are generally accepted by industry.

The following suggestions reduce unexpected hazards.

- If it is necessary for a welder to work at an elevation of more than 5 feet, adequate provision should be made to prevent falling in case of electric shock or other injury. This can be accomplished by the use of railings, safety belts and lifelines, or some other equally effective safeguard. Lifebelts and similar devices should be of a type that will permit quick escape.
- If welding is performed in confined spaces, such as in tanks and the hulls of ships, some means should be provided for quickly removing the welder in case of emergency. An attendant should always be stationed where he can readily give assistance to welders working in such places.
- When the operator has occasion to leave his work or to stop work for any length of time, he should always open the main switch in the equipment.
- Before operations are started, heavy portable equipment mounted on wheels should be securely blocked to prevent accidental movement.
- Welding equipment should be maintained in good mechanical and electrical condition to avoid unnecessary electrical hazards. Commutators should be kept clean to prevent excessive flashing. Gasoline or other flammable liquids should not be used for cleaning commutators.
- Welding equipment used in the open should be protected from inclement weather conditions. When not in use, the equipment should be stored in a clean, dry place. Cables should be neatly coiled and stored where they will not be damaged or create stumbling hazards to employees.
- After welding operations are completed, the operator should mark the hot metal or provide some warning sign to prevent other employees from touching them and getting burned.
- Report all injuries at once.
- Good housekeeping should be maintained on all welding jobs. Operators should not discard electrode stubs on the floor or leave tools or other objects where they will constitute hazards. Each operator should have a receptacle in which to keep his daily supply of electrodes. The same receptacle can be used for disposing of electrode stubs. Structural operators should be provided with a receptacle with a strap for attachment to a belt to facilitate carrying electrodes.
- Material-handling equipment, such as cranes and hoists, should be maintained in a safe operating condition. Cables, chains, and slings used in moving heavy parts to be welded should receive special attention.
- All tools such as hammers, chisels, brushes, bars, and other hand tools should be maintained in safe condition. A tool box should be used by each welder in which all tools should be kept when not in use.

REVIEW QUESTIONS

A. Safety Precautions: Electric Welding Processes

1. Name four safety precautions that have a bearing on the safe handling of the electric arc welding machine.
2. How may the cable be handled?
3. An electrode holder becomes hot during welding. List the contributing causes.
4. List six precautions that may be taken to guard against electric shock.
5. The amount of contamination to which welders may be exposed depends on a number of factors. Name at least five of these.
6. Name the light rays that are involved with the electric arc welding process. Explain the effect of these rays on the operator.
7. List five precautions that should be observed when welding in confined areas.
8. Arc welding current is never strong enough to injure the welder. True. False.

B. Safety Precautions: Oxyacetylene Welding and Flame Cutting

1. How may the operator protect himself from harm while carrying out gas welding and cutting operations?
2. List six precautions that should be observed when handling gas cylinders.
3. Both the oxygen and the acetylene cylinder can be laid on the ground during use. True. False.
4. List at least six precautions that should be observed in the use of the acetylene cylinder.
5. List at least six precautions that should be observed in the use of the oxygen cylinder.

6. How may the welding and cutting hose be protected from damage?

7. List some of the safety precautions that should be observed by the operator when hooking up the cutting torch.

8. When the valves on the cutting torch freeze or become hard to operate, should oil be applied to free them? Explain.

9. List five conditions that may cause backfire or preignition in the gas welding torch.

10. Make a list of ten safety precautions that may be employed in connection with gas welding and cutting operations.

C. General Precautions

1. Name six precautions that may be employed as a protection against fire during welding.

2. If it is necessary to weld at an elevation of more than 5 feet, what precautions may be taken to protect the operator from falling?

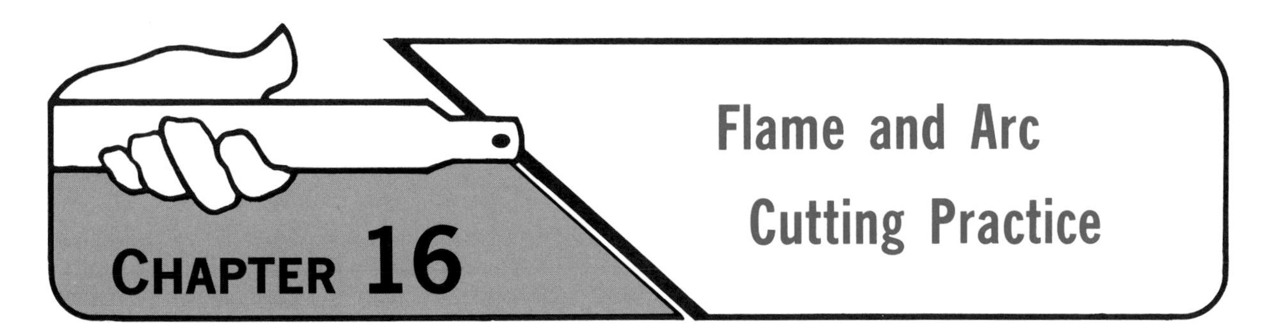

Flame and Arc Cutting Practice

The ability to do flame (oxy-acetylene) manual cutting, Fig. 16-1, is essential if the welder expects to enjoy success as a skilled craftsman. A number of flame-cutting procedures are commonly used in the preparation and treatment of metal in a wide variety of industries. The process is used in such different industries as multiple-piece production in manufacturing, shipbuilding, steel production, construction, scrapping and salvage, and for rescue work by police and fire departments.

It is important that the welding student learn the art of flame cutting before he is introduced to the various electric welding processes. Much of the material used for electric welding is heavier than that used for gas welding practice, and edge preparation becomes a problem. In those schools that do not have stocks of prepared material available or machine-cutting equipment to prepare plates, the student must be able to cut so that he can prepare his own material in the shortest possible time with the least waste.

In many cases, however, the training program may be such that the student need have no knowledge of cutting. The purpose of the training may be to train a welder for certain production jobs requiring limited skill. If this is true, the instructor may ask you to begin with the arc welding course.

Since most oxyacetylene cutting that the welder will be called upon to do in industry will be with straight carbon steel, the emphasis in this chapter will be on steel. It is important, however, for the skilled welder to know about the wider applications of oxyacetylene cutting and to understand basic principles and techniques.

REVIEW OF FLAME CUTTING PRINCIPLES

The cutting of metals by the oxyacetylene, oxy-hydrogen, and oxy-fuel gas processes is based on the fact that all metals oxidize to a greater or lesser degree, depending upon the physical conditions around them. Wrought iron and steel oxidize quite rapidly, even under ordinary atmospheric conditions. When oxidation occurs in the air around us, it is called rusting and can readily be recognized by the oxide that forms on the metal. Rusting is, of course, a slow process, but it illustrates the tendency of ferrous metals to combine with oxygen.

After iron or steel is heated to a red heat and cooled, it is covered with a thick scale. These metals oxidize more rapidly when hot than when cold. If the temperature of the steel is raised even higher to a white heat and a stream of pure oxygen is directed against the white hot spot, it burns rapidly. This can be demonstrated in the shop by taking a thin piece of steel wire and heating it to a red heat and then submerging it in a bottle containing oxygen. The heated end will immediately burst into

16-1. These two young welders have just completed their training course and are now on the job. They are being asked to cut a structural steel I-beam.

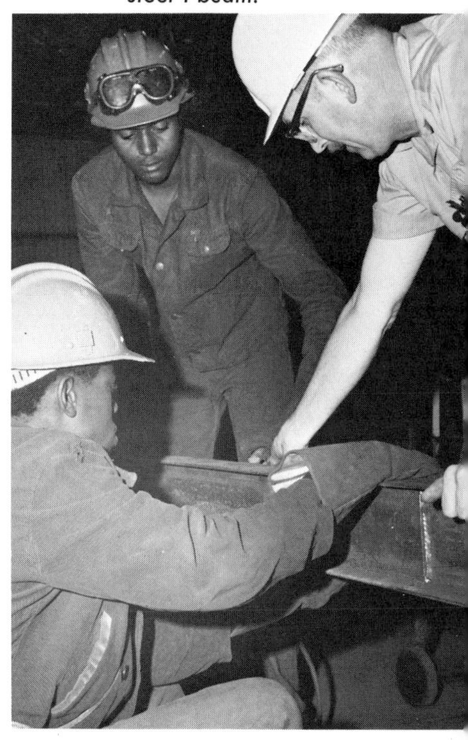

flame and burn vigorously until the wire is burned away or the oxygen is consumed.

The reaction between oxygen and iron causes a considerable amount of heat. It forms a molten oxide which flows or is blown away, exposing more metal to the action of the oxygen. In order to keep this action going, it is necessary to supply heat to the point of cutting from an external source.

The function of the cutting torch, Fig. 16-2, is to provide a flame to heat the metal to a red heat, to maintain that heat, and direct a stream of oxygen on the heated point of cutting. Always bear in mind that the high pressure oxygen jet does the cutting. Practically all trouble in cutting is caused by the tip nozzle becoming burred or obstructed by small particles of molten slag adhering to it. An unobstructed, cylindrical jet of oxygen always produces a smooth cut. Any obstruction either in the bore or at the end of the tip retards the speed of the cut and produces a rough cut.

Flame-Cutting Effects On Steel

For many years after the development of the oxyacetylene cutting process, steel fabricators were reluctant to use the process in the preparation of metals for welding. They felt that harmful metallurgical changes took place on the surface of the cut, that surface cracking occurred, and that locked-up stresses were present. Engineering investigation has proved that the oxyacetylene-cut edge is superior to that of some mechanical procedures which, because of violent contact, damage the plate edge.

Care must be exercised in flame-cutting steels with high carbon content or alloy content that have air-hardening properties. When cut, these steels have a tendency for hardened zones and cracking along the edge of the cut. They can be successfully flame cut by preheating, postheating, or both. The cut should be allowed to cool slowly.

When applied to large areas, the cutting procedure produces a certain amount of expansion and contraction which may cause distortion if it is not controlled. This problem is not as serious as in welding, and the usual techniques used to control distortion when welding may be applied to the cutting procedure. See Chapter 3, pages 103–114, for ways to control distortion.

Comparison of Fuel Gases

The function of the fuel gas is to feed the preheat flames that maintain the temperature of the materials being cut so that the cutting process can continue without interruption. For many years acetylene gas was the only gas used for flame cutting. Today a number of different gases are being used, such as acetylene, propane, Mapp® gas, natural gas, hydrogen, and acetogen. While acetylene is still the workhorse of the industry, each of the other gases has its particular advantage, depending upon the application. For example, hydrogen was once used exclusively for underwater cutting since it can be safely compressed to high pressures and regulated to overcome the pressure exerted by the water at salvage operation depths. Natural gas, which can be similarly compressed, is replacing hydrogen for underwater cutting.

The choice of the type of fuel gas to use depends upon the following factors:

- Cost of the gas
- Availability of the gas
- Flame temperature needed

16-2. *An oxyacetylene cutting torch is designed to supply mixed acetylene for the heating flames and a stream of high purity oxygen to do the actual cutting.*

Linde Division, Union Carbide Corp.

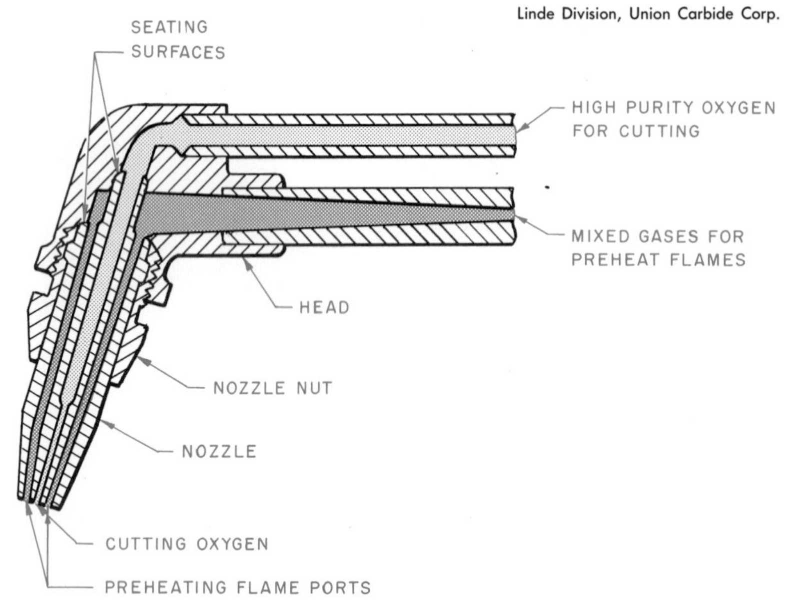

- Length of the cutting procedure
- Length of cut required
- Speed of cutting

When used as a fuel gas, the flame temperature of acetylene is the highest, followed by acetogen, Mapp® gas, propane, natural gas, and hydrogen in that order. All of the gases give satisfactory cuts when used under the proper conditions. For best results, it is essential that the torches and tips employed be designed for the particular fuel gas used.

FLAME TEMPERATURE. Of the four principal gases used today, acetylene gives the hottest flame—350 degrees F. hotter than its nearest rival. Acetylene has a flame temperature of 5,650 degrees F.; Mapp® gas, 5,300 degrees F.; propane, 4,800 degrees F.; and natural gas, 4,800 degrees F. Acetylene burns much faster than the other gases and much closer to the tip of the torch. Therefore, it can be adjusted to a very intense, concentrated flame, which is exactly what is needed for cutting. The oxyacetylene flame is ready to cut in less than half the time of its nearest competitor and five times as fast as the slowest.

HEAT CONCENTRATION. In a comparison of the heat produced by the various gases, measured in British thermal units (Btus), Mapp® gas has 2,400; propane, 2,400; acetylene, 1,450; and natural gas, 1,000. Although acetylene is third in Btus, it concentrates the most Btus in the smallest area of any gas because of the way it burns. In the oxyacetylene flame Btus are created mainly in the inner cone. With the other gases, Btus are largely dispersed through the outer flame envelope, Fig. 16-3. Thus, acetylene concentrates heat at the point of cutting.

COST. Acetylene is among the higher priced fuel gases. Its cost is offset, however, by the reduction in the use of oxygen in the cutting process. Acetylene requires less oxygen per cubic foot of fuel than any of the other three fuel gases. The following amounts of oxygen are needed for 1 unit of fuel gas: acetylene 1.5 to 1; natural gas 2 to 1; Mapp® gas 3.4 to 1; and propane 4.7 to 1.

SPECIAL APPLICATIONS. Although acetylene has many advantages, no one gas can do everything best. There are a number of applications in which propane, Mapp® gas, and natural gas are better choices than acetylene.

For cutting plate 6 inches or heavier, propane or Mapp® gas is the best choice. Because a deep area must be heated, the dispersed Btus, the "bushy" flame, and the high total Btu content of these gases do a good job in heavy plate cutting, Fig. 16-4.

Natural gas is widely used in steel mills for removing surface defects because it is the most economical fuel gas. Most mills use fuel gas with air.

Acetylene is ideal for cutting steel plate. In the $\frac{1}{4}$-inch to 1-inch range, acetylene preheats up to 50 percent faster and allows cutting 25 percent faster than its closest competitor. In repetitive cutting, acetylene is far superior because this kind of cutting requires frequent preheats and starts.

Bevel cutting and cutting on rounded surfaces are natural applications for acetylene. The flame is held at an angle to the surface to that the heat tends to bounce off the work at the opposite angle, Fig. 16-5. In order to get started, a concentrated flame is needed. On surfaces that are covered with rust, scale, grease, or paint, the concentrated acetylene flame cuts through the barriers. Figure 16-6 presents the applications for which the acetylene flame works best.

CUTTING DIFFERENT METALS

Oxyacetylene cutting finds its widest use in the cutting of carbon and low alloy steels. The normal oxygen cutting methods must be varied for metals with high alloy content such as cast iron and stainless steel. You may think that this is peculiar since all metals oxidize. Some metals, however, do not form an oxide that melts at a lower temperature than the base metal. This high temperature oxide protects the metal surface and prevents further oxidation. Generally, as the amounts of alloying materials, including carbon, increase, the oxidation rate decreases. This is one reason why cast iron, which has a high carbon content, is more difficult to cut than steel.

16-3. *The Btu. concentration of the oxyacetylene flame.*

ACETYLENE

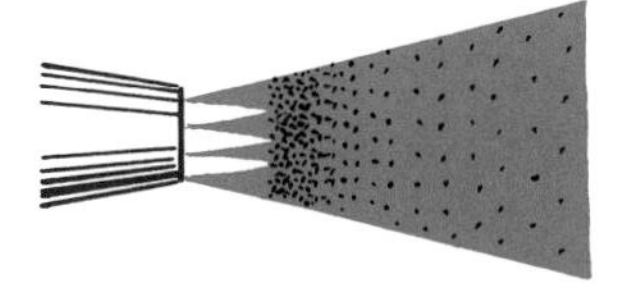

BTU'S CONCENTRATED

OTHER GASES

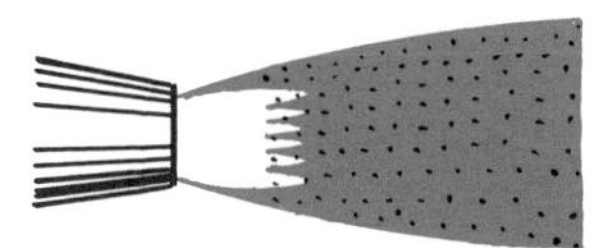

BTU'S DISPERSED

There are a number of methods that overcome to some degree the effects of carbon and alloying materials that interfere with the cutting process.

PREHEATING. It has been pointed out that high heat increases the rate of oxidation. As a consequence, many materials which are difficult to cut can be cut more easily if they are heated to a temperature approaching their melting point.

WASTER PLATE. A low carbon steel plate may be clamped along the line of cut. The melting of this plate causes a great amount of iron oxide to be formed, making it possible for the cut to continue without interruption. A welding bead of low carbon steel may also be deposited along the line of cut to serve the same purpose.

WIRE FEED. Low carbon steel wire may be fed into the preheat flames to produce rapid oxidation. The heat that is generated by the burning of the wire rapidly brings the surface of the plate to ignition temperature. This method can be used to obtain rapid starts in heavy plate. It can also be used for cutting cast iron. Short lengths of wire can be fed manually, or special equipment can be attached to the cutting torch for continuous iron wire feed.

OSCILLATORY MOTION. When the torch is moved from side to side, Fig. 16-7, more material is heated to the ignition temperature so that additional material is oxidized and blown out of the kerf. This is one method used in the cutting of cast iron, and produces a rough cut.

Difficult-to-cut materials may also be cut using other methods such as oxygen-arc cutting, flux cutting, and powder cutting. These methods are explained in Chapter 8.

CUTTING TECHNIQUE

The welder must manipulate the torch when cutting so that he creates the proper kerf and drag for the material being cut.

Kerf

Kerf is the gap created as the material is removed by cutting. Control of the kerf is important to the accuracy of the cut and the squareness of the face of the cut. The width of the kerf is determined by the size of the cutting tip, speed of cutting, oxygen pressure, and torch movement. Since oxygen pressure is directly affected by the thickness of the material being cut, the width of the kerf increases as the thickness of the material increases. A rough cut also increases the width of the kerf.

Drag

When steel is cut, lines form on the face of the work which are caused by the flow of the high pressure oxygen. When the torch

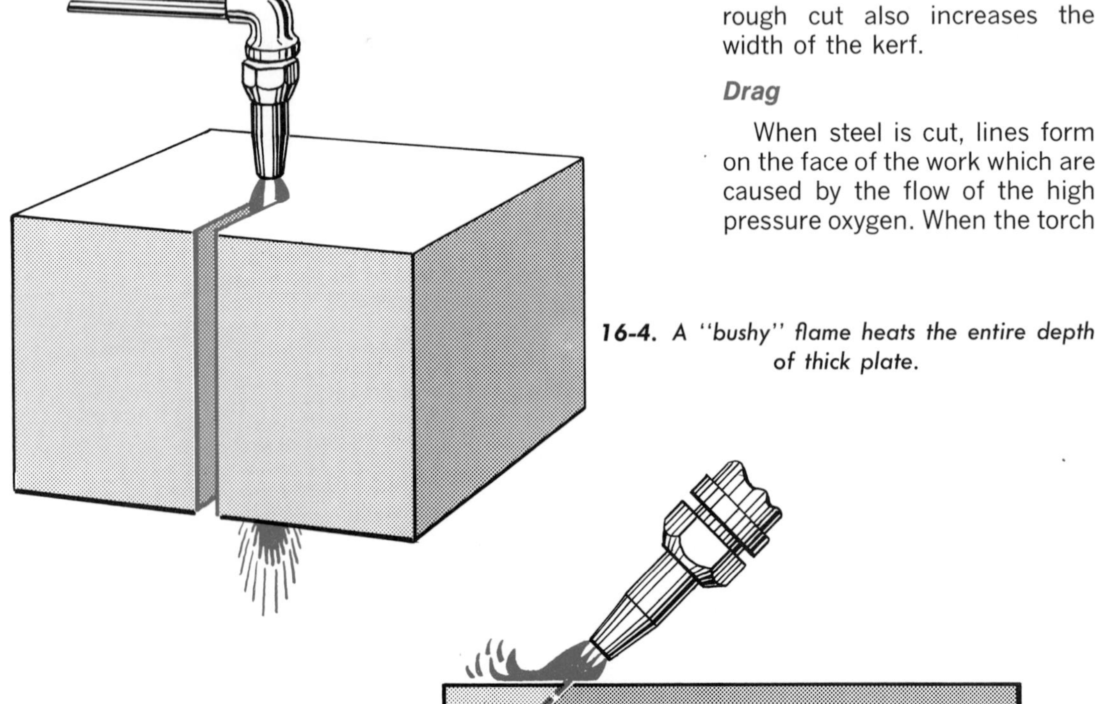

16-4. *A "bushy" flame heats the entire depth of thick plate.*

16-5. *A concentrated flame is needed for bevel cutting.*

is held in a vertical position and the cutting conditions are correct, this line is vertical from top to bottom, Fig. 16-18 (1), page 478. This condition is referred to as *zero drag.* If the speed of cutting is increased or if the oxygen pressure is not set high enough for the thickness of the material being cut, the drag lines at the bottom of the kerf lag behind the top of the kerf. The amount of lag is usually expressed as a percentage of the plate thickness, Fig. 16-8. In reverse drag, the drag lines at the top of the kerf lag behind those at the bottom. Reverse drag may result from too much oxygen or a speed of travel that is too slow. Both drag and reverse drag must be avoided since they affect the quality of the cut and may even cause the cut to be lost.

STRAIGHT LINE CUTTING. Cutting along a straight line that has been laid out on the material to be cut may be done with a great degree of accuracy, depending upon the skill of the operator. Very often a straightedge or angle iron is clamped along the line of cut to act as a guide. See Table 16-2, page 483, for hand cutting data. The torch is held perpendicular to the plate with the holes positioned as shown in Fig. 16-9. Machine cutting is somewhat more accurate than manual cutting.

BEVEL CUTTING. Bevel cutting is one of the common operations used in beveling the edges of plate and pipe for welding. The torch tip is held sideways at the angle desired with the holes positioned as shown in Fig. 16-9. Cuts may be made either by hand or machine in straight or irregular lines.

In order to do the jobs that are outlined for advanced gas welding, the shielded metal-arc, gas tungsten-arc, and gas metal-arc welding of plate and pipe, it will be necessary to prepare bevel joints.

For plate beveling the student will use the flame-cutting machine provided in the welding shop, Fig. 16-10. A commercial machine, Fig. 16-11, may be available for the beveling of pipe.

A pipe-beveling machine may also be constructed in the school shop. The machine shown in Fig. 16-12 was built by students. It is manually powered. The driving power is an old worm gear arrangement. The rolls are cold-rolled steel with bearing ends that are old motor-generator

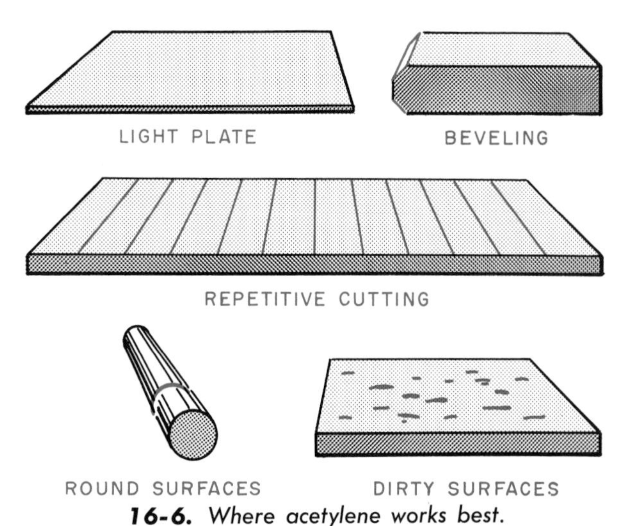

16-6. *Where acetylene works best.*

16-7. *Typical cutting torch manipulation for cutting thin cast iron (top) and heavy cast iron (bottom).*

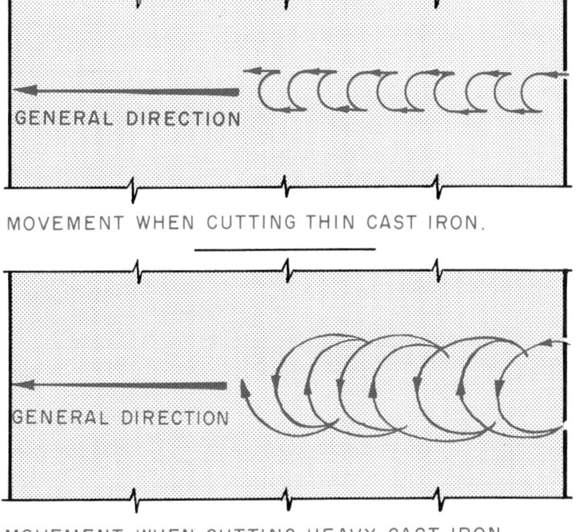

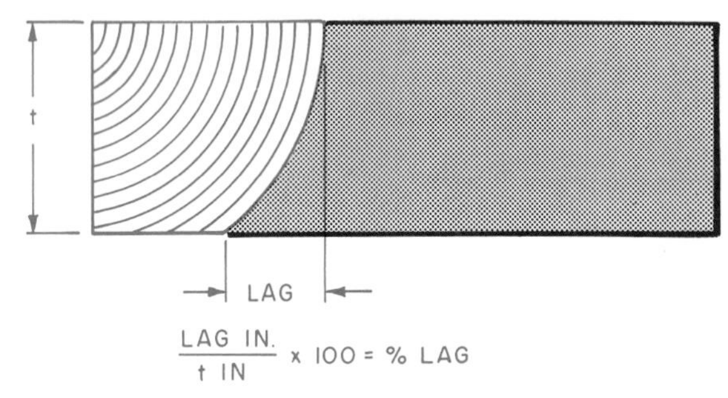

16-8. Lag is the amount by which the bottom of the cut lags behind the top. Lag is usually expressed as a percent of the plate thickness. If this were a 1 inch thick plate and the amount of lag were ½ inch, there would be a 50-percent lag.

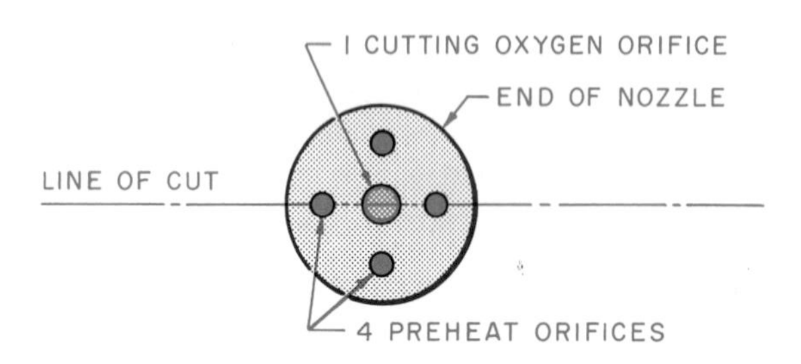

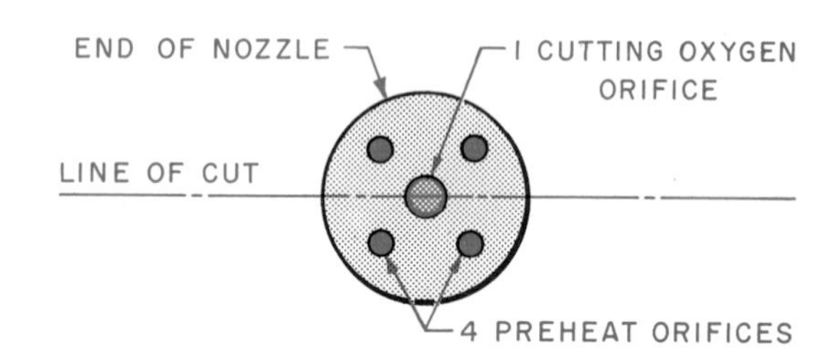

16-9. For square and bevel cutting, the nozzle should be set so that the preheat orifices are positioned as shown. Cutting tips with more than four preheat orifices should be positioned in the same way.

16-10. A standard flame-cutting machine for plate shape-cutting and beveling. A machine of this type has extreme flexibility.

General Welding & Equipment Co.

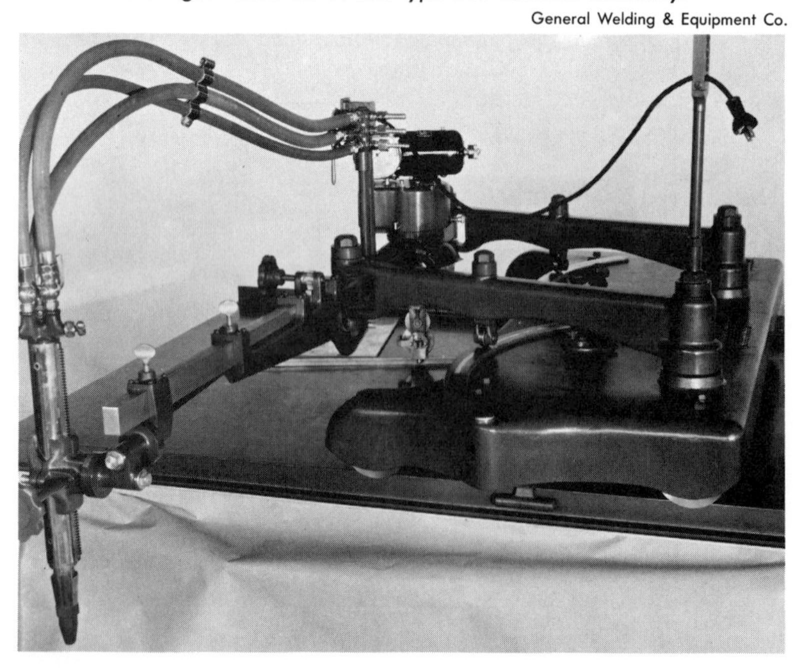

16-11. Commercial pipe beveling machine.

CRC-Chose International Co.

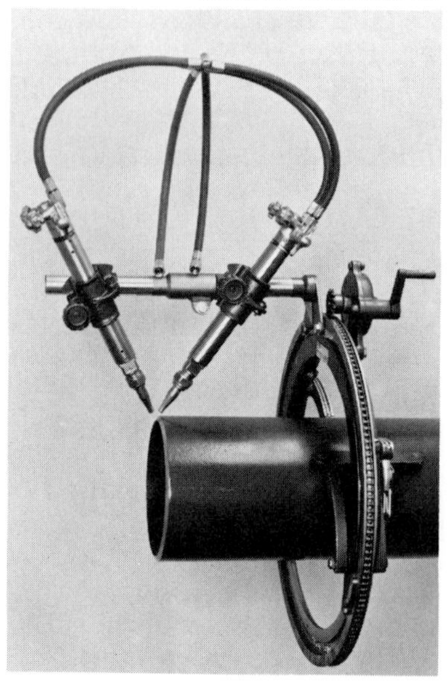

bearings or old automobile bearings. The torch may be purchased from any welding equipment distributor. Study Table 16-1 for machine cutting data.

FLAME PIERCING. Piercing is the process of forming a hole in a metal part. This may be for the purpose of producing a bolt hole or starting a cut in from the edge of a plate.

Piercing takes more time than edge starting to bring the metal to the temperature required for cutting. A spot is heated to a bright red, the torch is raised slowly to about 1/2 inch above the plate, and a limited amount of oxygen is turned on. As soon as the plate is pierced, the cut can continue in the regular manner. Care must be taken so that the slag which is formed on the surface of the plate is not blown back into the torch orifice. Figure 16-13 shows the following sequence of operations:

16-12. *A welding student is beveling pipe with a shop-built beveling machine.*

Table 16-1

Machine flame cutting data for clean mild steel (not preheated).

Thickness of steel, in.	Diameter of cutting orifice, in.	Oxygen pressure, p.s.i.	Cutting speed, in./min.	Gas consumption	
				Oxygen, cu. ft./hr.	Acetylene, cu. ft./hr.
1/8	0.0200–0.0400	15–30	22–32	17–55	5– 9
1/4	0.0310–0.0595	11–35	20–28	36–93	6–11
3/8	0.0310–0.0595	17–40	19–26	46–115	6–12
1/2	0.0310–0.0595	20–55	17–24	63–125	8–13
3/4	0.0380–0.0595	24–50	15–22	117–159	12–15
1	0.0465–0.0595	28–55	14–19	130–174	13–16
1 1/2	0.0670–0.0810	22–55	12–15	185–240	14–18
2	0.0670–0.0810	22–60	10–14	185–260	16–20
3	0.0810–0.0860	30–50	8–11	207–332	16–23
4	0.0810–0.0860	40–60	6.5–9	293–384	21–26
5	0.0810–0.0860	50–65	5.5–7.5	347–411	23–29
6	0.0980–0.0995	45–65	4.5–6.5	400–490	26–32
8	0.0980–0.0995	60–90	3.7–4.9	505–625	31–39
10	0.0995–0.1100	70–90	2.9–4.0	610–750	37–45
12	0.1100–0.1200	69–105	2.4–3.5	720–880	42–52

American Welding Society

In this table acetylene pressures have been omitted, since these pressures are mainly a function of equipment design and are not directly related to the thickness of the material to be cut. It is suggested that for acetylene pressure settings, the charts of the equipment manufacturer be consulted.

1. The cutting torch is held still until a spot on the surface just begins to melt.

2. The cutting torch is raised until the end of the nozzle is about $\frac{1}{2}$ inch above the plate.

3. Meanwhile the cutting oxygen lever is slowly depressed, and the cutting torch moved slightly to one side at the same time to start a small spiral motion. As the cutting action starts, the slag will be blown out on the opposite side of the puddle.

4. When the cut has pierced all the way through the plate, the cutting torch is lowered and the cut is continued with a spiral motion until the desired diameter of hole has been obtained.

FLAME SCARFING. Flame scarfing is a process used mainly in the steel mills to remove cracks, surface seams, scabs, breaks, decarburized surfaces, and other defects on the surfaces of unfinished steel shapes. The operation is done before the steel reaches the final finishing departments.

FLAME WASHING. Washing is a procedure similar to that of scarfing. It is used in the steel mills to remove unwanted metal such as fins on castings and to blend in riser pads and sand washouts in castings. It is also used to remove rivets without damage to the riveted parts. A skilled operator can remove a rivet from a hole, a threaded bolt from a threaded hole, or a piece of pipe from a threaded flange without destroying the hole or the threads, Figs. 16-14, 16-15 and 16-16.

FLAME GOUGING. Flame gouging provides a means for quickly and accurately removing a narrow strip of surface metal from steel plate, forgings, and castings without penetrating through the entire thickness of metal. The process is generally used for the removal of weld defects, tack welds, and metal from the back side of weld seams. When used for removing defects, the procedure is called *spot gouging.* Care must be taken so that only a small area of the weld is affected. Edge preparation of plate for welding is also frequently done by flame gouging.

Figure 16-17 shows the following sequence of operations:

1. To start a gouge at the edge of a piece of plate, the plate edge is preheated to a bright red.

2. As the nozzle angle is reduced to the operating angle, the cutting oxygen is turned on to make a gradual starting cut.

***16-13.** The sequence of operations to be followed in piercing a hole in steel plate.*

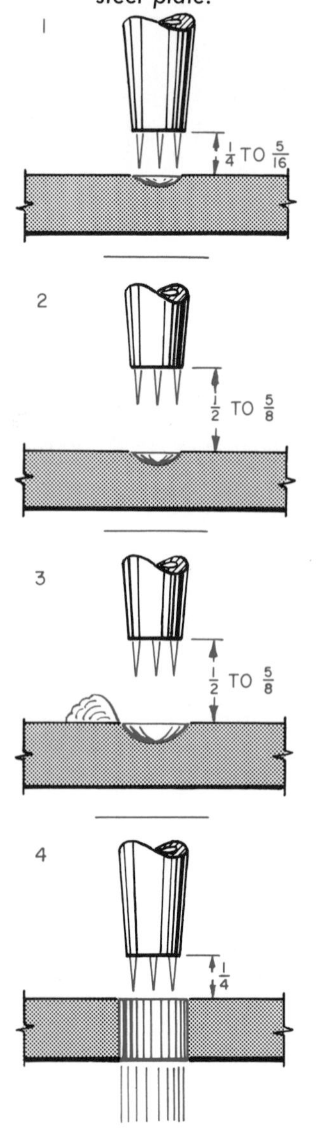

***16-14.** The step-by-step procedure for removing countersunk rivet heads with a special rivet-cutting nozzle.*

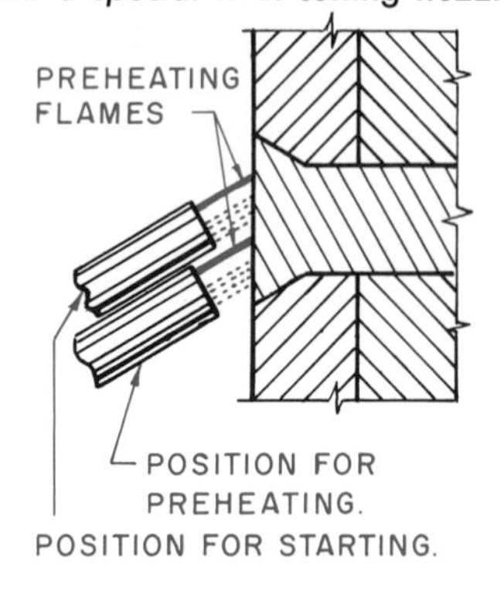

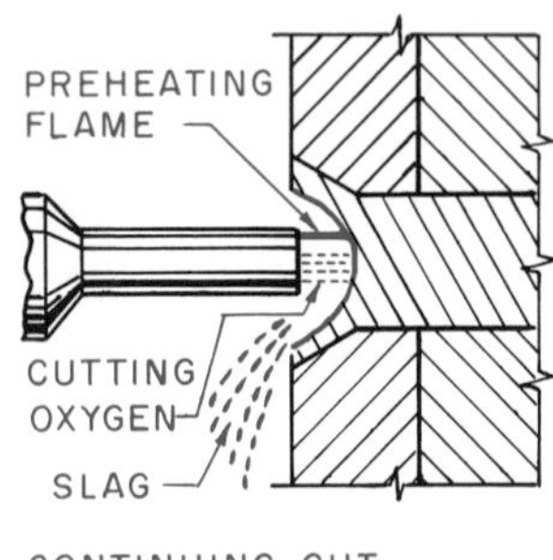

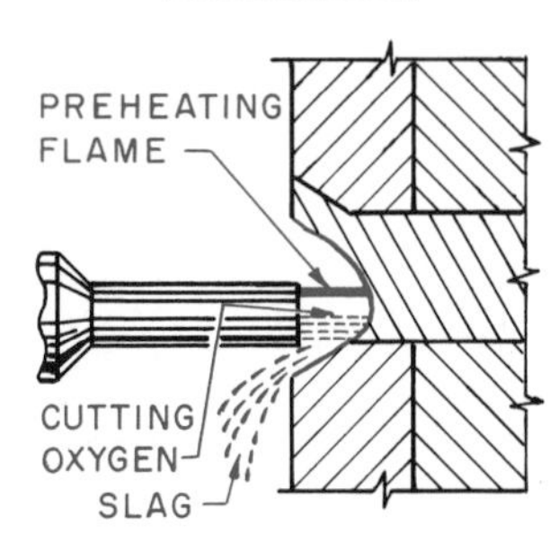

3. and 4. When the desired depth of groove is reached, the gouging nozzle is moved along the line of cut.

There are two methods of preparing a U-type butt joint for welding by flame gouging. In the first method the plates are tacked together in the form of a square butt joint. Then they are welded on one side, and the opposite side is gouged out to the root of the weld to form a U-groove. In the second method the plates are tacked together in the form of a square butt joint. The U-groove is cut to a depth within $\frac{1}{8}$ inch of the thickness of the plate, and the plate is welded with as much penetration as possible through to the back side. The reverse side is then welded with an electrode having deep penetration. U-grooves may also be formed with the arc cutting processes.

16-15. *The step-by-step procedure for removing buttonhead rivets with a special rivet-cutting nozzle.*

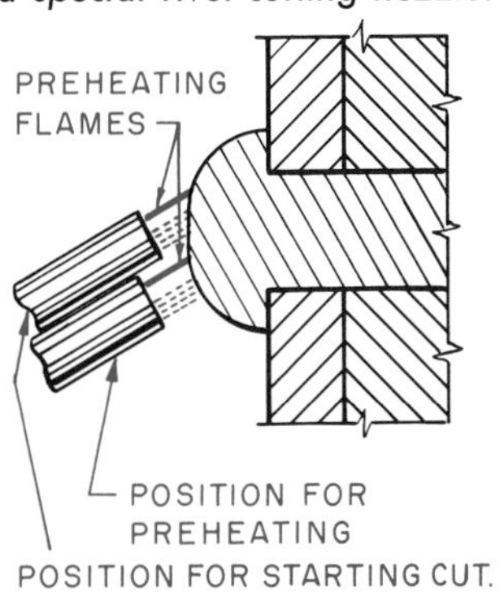

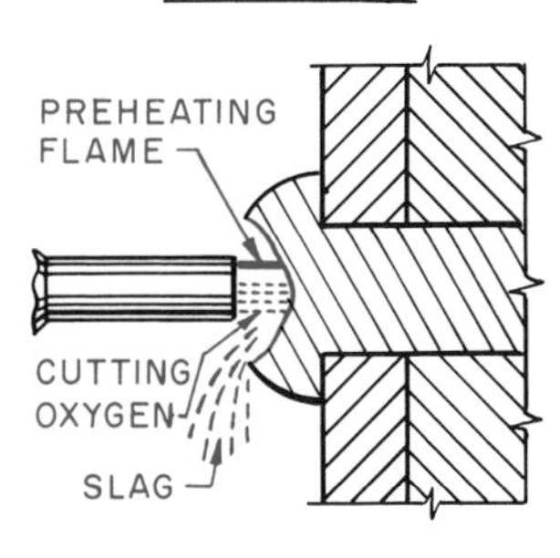

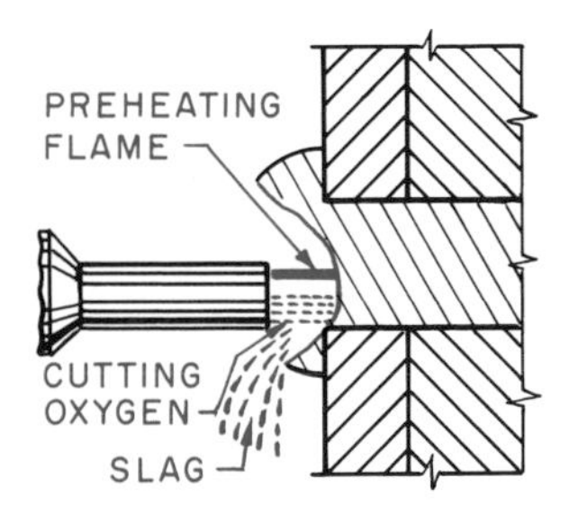

16-16. *Buttonhead rivet heads can be removed with a standard cutting nozzle by following this step-by-step procedure.*

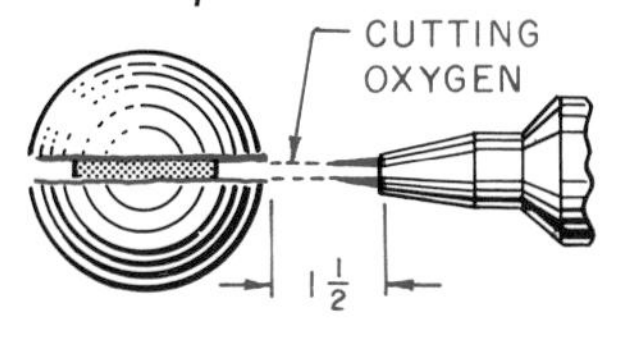

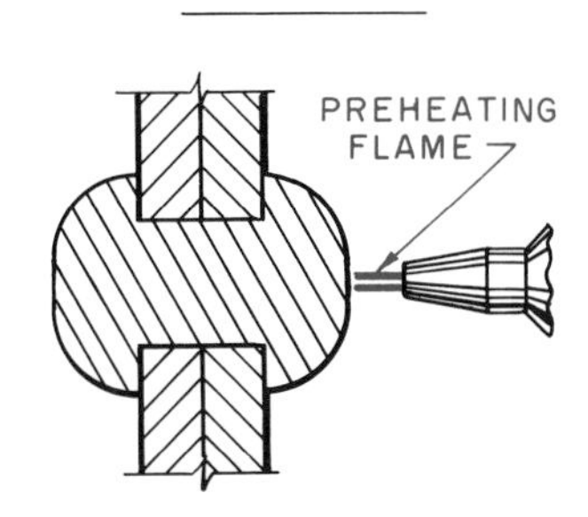

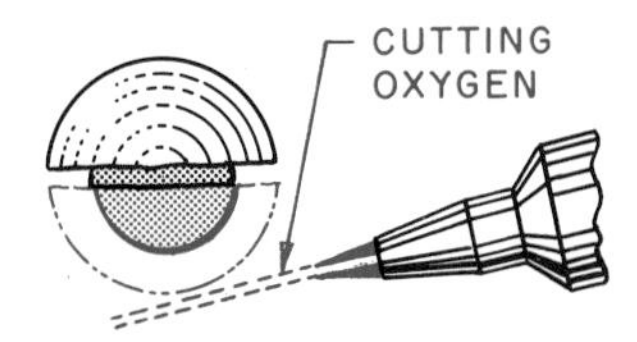

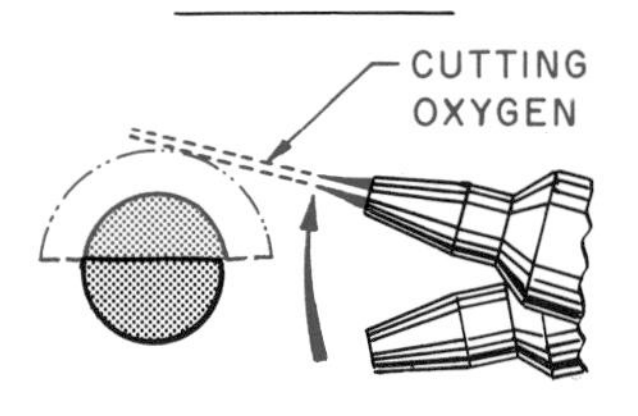

SURFACE APPEARANCE OF HIGH QUALITY FLAME CUTS

Since one of the principal fields for oxyacetylene flame cutting is in the preparation of plate and pipe for welded fabrications, it is important that the cuts be accurate, smooth, and free of slag that will contaminate the

16-17. *The sequence of operations to be followed in flame-gouging a piece of plate.*

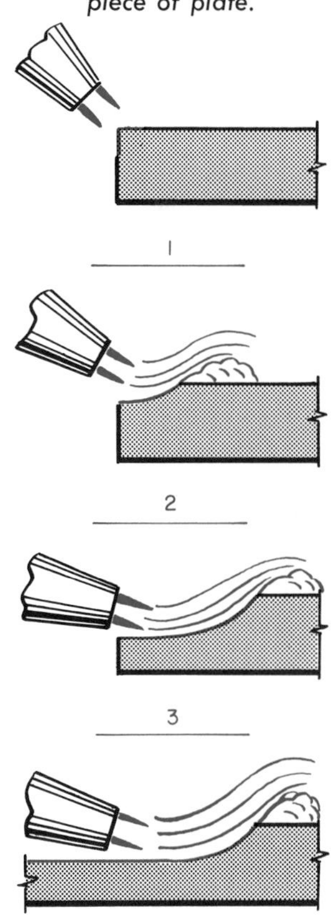

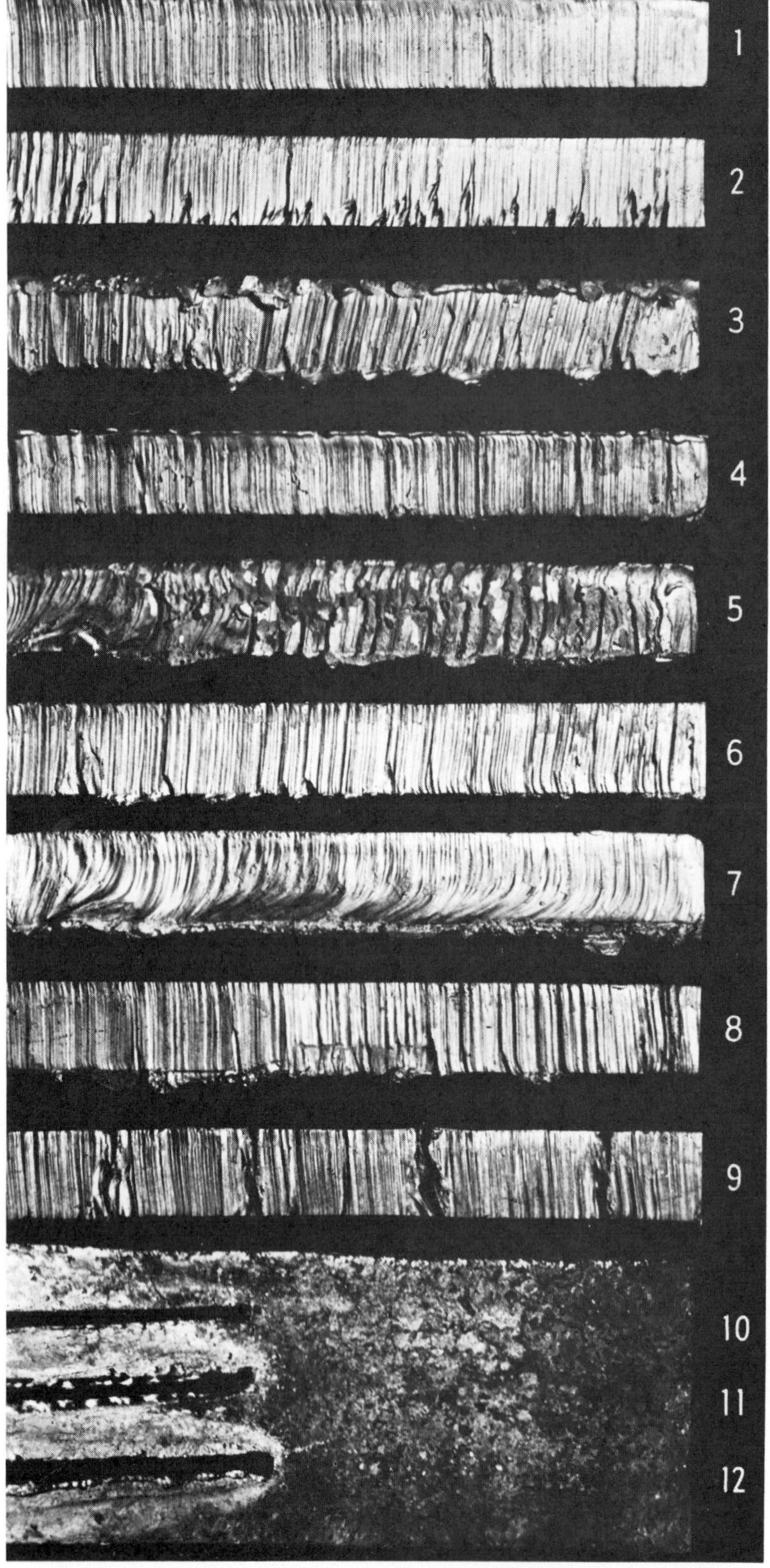

Linde Division, Union Carbide Corp.

16-18. *These photographs show good cuts and the common faults that may occur in hand cutting.*

weld or cause the welder to spend a great deal of time in grinding or chipping.

The shoulder at the top edge of the plate should be sharp and square. The face of the cut should be smooth and not deeply grooved, fluted, or ragged. If it is square cut, the face should be square with the plate surface. Bevel cuts should be approximately the degree specified. The slag on the back side should be at a minimum and should be easily removed.

Figure 16-18 shows high quality cuts and a variety of defects in the following order:

1. This is *a correctly made cut* in 1-inch plate. The edge is square, and the draglines are vertical and not too pronounced. There is little slag along the bottom edge.
2. *Preheat flames were too small* for this cut so that the cutting speed was too slow, causing bad gouging at the bottom.
3. *Preheat flames were too long.* The top surface has melted over, the cut edge is irregular, and there is too much slag.
4. *Oxygen pressure was too low.* The top edge has melted over because of too slow cutting speed.
5. *Oxygen pressure was too high and the tip size too small* so that the entire control of the cut has been lost.
6. *Cutting speed was too slow* so that irregularities of the draglines are pronounced.
7. *Cutting speed was too high* so that there is a pronounced break and lag to the dragline, the cut is irregular, and there is too much slag on the bottom side.
8. *Torch travel was unsteady* so that the cut is wavy and irregular.
9. *Cut was lost* and not carefully restarted. Bad gouges

were caused at the restarting point.

10. *Correct prodecure* was used in making this cut.

11. *Too much preheat was used and the tip was held too close to the plate* so that bad melting at the top edge occured.

12. *Too little preheat was used and the flames were held too far from the plate* so that the heat spread opened up the kerf at the top. The kerf is too wide at the top and tapers in.

Conditions Affecting Quality of Cut

The following is a list of important conditions which should be met in order to make accurate and smooth cuts. Study these recommendations thoroughly so that you will be able to follow them during your flame-cutting practice.

- *Uniformity of oxygen pressure regulation,* especially when cutting heavy plate thicknesses. Make sure that the oxygen and acetylene regulators are working properly and do not creep or fluctuate.
- *High oxygen purity.* If the oxygen contains impurities, the chemical reaction that takes place in the cutting operation will produce poor quality cuts. This is not a serious problem today since the manufacturers of the gas take great care in its production.
- *Low oxygen operating pressure.* High oxygen operating pressures usually do not improve the quality of the cut, and they are a serious waste of oxygen.
- *Proper tip size.* It the wrong tip size is being used, cutting is sure to be unsatisfactory. Too large a tip wastes oxygen and acetylene. Too small a tip results in poor cutting and wastes time.
- *Smoothness of bore and proper cutting orifice.* The bore may become out of round or nicked by improper cleaning methods. Always use a tip drill of the proper size. Do not use makeshift reamers for this purpose, and do not scrape the tip on the metal being cut. Remember that the tip is made of soft copper, and it will be damaged by the much harder steel.
- *Cleanliness of cutting holes.* If the cutting holes become dirty or clogged with oxide, the preheat flames will be distorted and the jet of high pressure oxygen will not be directed squarely along the line of cut. This produces ragged and uneven cuts.
- *Uniformity of steel being cut.* Any variation in the thickness of the material, its shape, or the plate surface condition alters the cutting procedure.
- *Low enough speeds for light and moderate thicknesses.* A common error is to proceed at excessive speeds which results in a cut of poor quality.
- *A gradual increase in speed* when starting the cut in very heavy thickness.
- *High enough speeds for heavier thicknesses.* If the speed is not high enough, the lower portion of the cut will be ragged. To increase the speed, it is necessary to increase the tip size.
- *Uniformity of torch movement.* The operator should make every effort to develop a smooth, even torch movement. Uneven movements make rough, ragged cuts.

ARC CUTTING

In the large welding fabricating shops there is a growing use of the various arc cutting processes in the fabrication of heavy weldments of various kinds. Consult Chapter 8 for a review of these processes and practice the processes available in your school welding shop. You are strongly urged to become as skilled as possible in their use.

PRACTICE JOBS

Setting up Equipment

Secure the following equipment:

- Tank of acetylene
- Tank of oxygen
- Oxygen regulator
- Acetylene regulator
- Oxygen hose
- Acetylene hose
- Cutting torch with tip, Fig. 16-19
- Pair of cutting goggles
- Acetylene tank wrench
- Torch wrench
- Friction lighter

Before setting up the equipment, review page 250 in Chapter 8, Flame and Arc Cutting Principles, and pages 458–463 in Chapter 15, Safety.

The equipment should be set up in the following manner.

1. During storage some dirt and dust will collect in the cylinder valve outlet. To make sure that this dirt and dust will not be carried into the regulator, clear the cylinder valves by blowing them out. Crack the valves only slightly, Fig. 16-20. Open the acetylene cylinder outdoors and away from open flame.

2. Attach the oxygen regulator, Fig. 16-21, and the acetylene regulator, Fig. 16-22, to the respective tanks.

Note that the oxygen regulator has a right-hand thread and that the acetylene regulator has a left-hand thread. Make sure that regulator nipples are in line with the valves of the tank so that they may seat properly.

3. Attach the welding hose to the regulators, Fig. 16-23 (Page 482). The oxygen hose is green and has a right-hand thread. The acetylene hose is red and has a left-hand thread. Make sure that the nipples of the hose are in

line with the connections of the regulator to insure correct seating. Tighten nuts securely and make sure that the connections do not leak.

4. It is important that no grit or dirt be allowed to remain in the regulators or the hose. This can be prevented by blowing out regulator and hose. After having made sure that the regulator-adjusting screw is fully released, open the cylinder valves slowly, so as not to damage the regulator. Adjust the regulators by turning the adjusting screws to the right, Fig. 16-24, to permit the gases to flow through both hoses. Turn the adjusting screws to the left again after enough gas has been allowed to flow to clear regulators and hose.

5. When connecting the hose to the torch, Fig. 16-25, connect the green oxygen hose to the torch outlet marked *oxygen* and the red acetylene hose to the torch outlet marked *acetylene*. Make sure that nipples of the hose connections are in line with the torch outlets for proper seating. Be certain that connections do not leak.

6. To adjust the working pressure, Table 16-2, open the torch valves and set the pressure on the regulator by turning the adjusting screw to the right, Fig. 16-24. Close the torch valves when pressure has been set.

7. To light the torch, open the acetylene preheat valve and light the gas with a friction lighter, Fig. 16-26. The acetylene will burn with a smoky yellow flame (Fig. 1, page 480A) and give off fine black soot.

8. Open the oxygen preheat valve slowly. The flame will gradually change in color from yellow to blue, Fig. 2, page 480A. It will show the characteristics of the excess acetylene flame, also called the *carburizing* or *reducing* flame. There will be three distinct parts to the flame: a brilliant but feathery edged inner cone surrounded by a secondary cone, and a bluish outer envelope forming a third zone. As you adjust the oxygen valve, you will notice that the secondary cone gets smaller and smaller until it finally disappears. Just at this

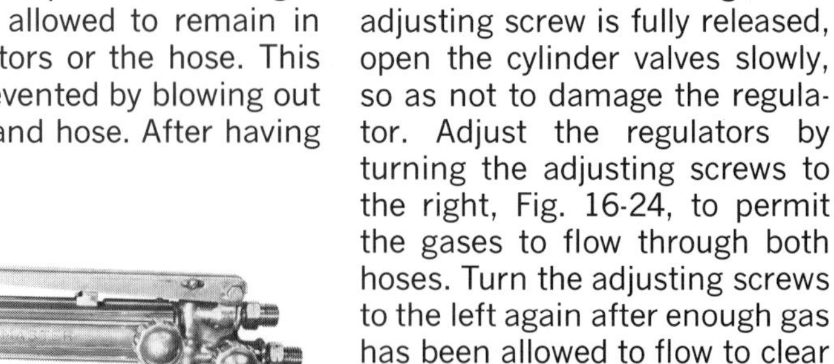

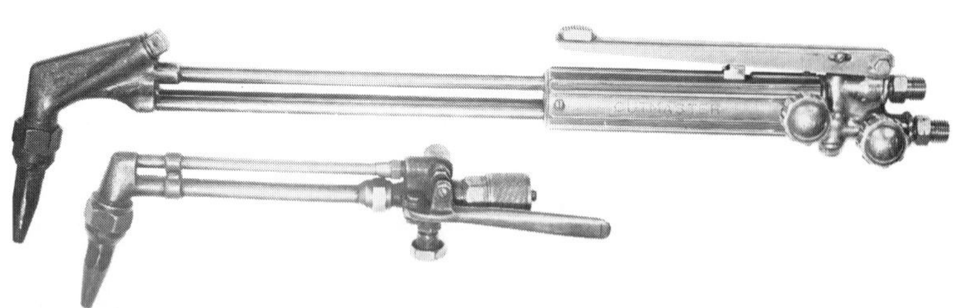

Modern Engineering Co.

16-19. *Standard oxyacetylene cutting torch and adaptable cutting head.*

16-20. *Oxygen and acetylene cylinders. Note the difference in the valves on each cylinder. The oxygen valve has a male connection, and the acetylene valve has a female connection.*

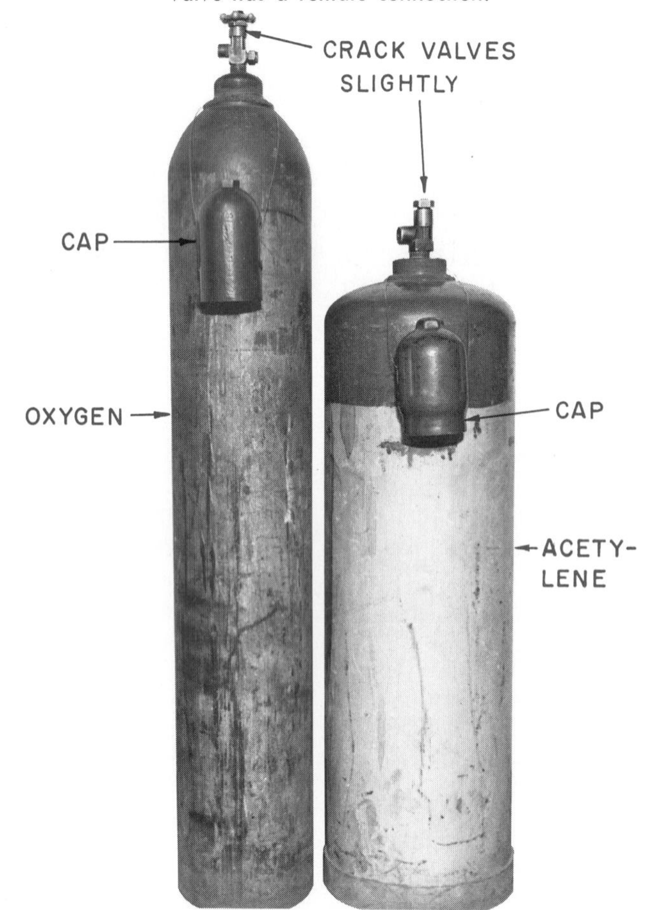

Flames: Oxy-Acetylene Cutting

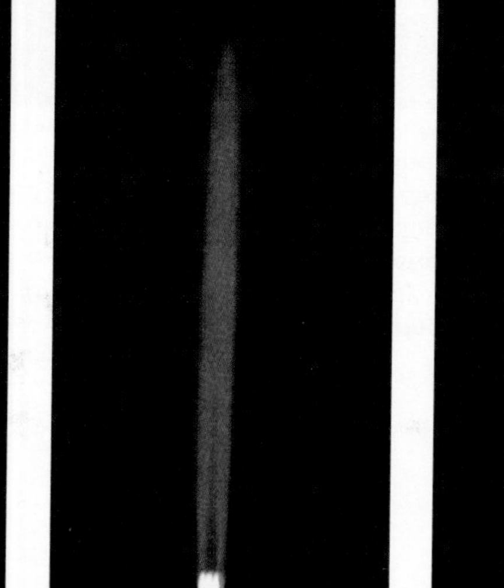

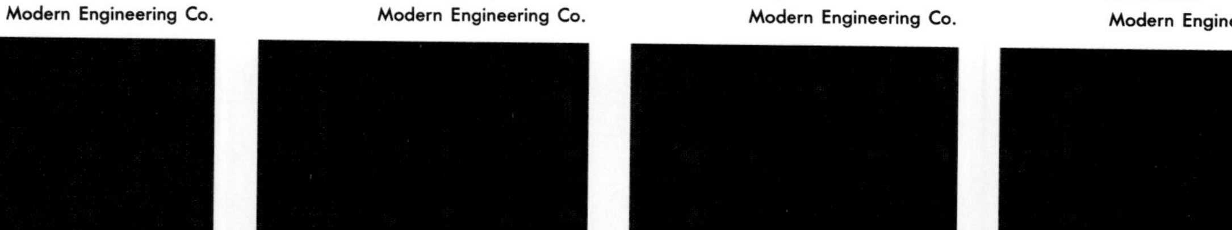

Fig. 1
Atmospheric burning of acetylene. Open acetylene valve until smoking of the flame disappears.
Modern Engineering Co.

Fig. 2
Carburizing cutting flame. It is an excess acetylene flame. More oxygen is required for preheat flames.
Modern Engineering Co.

Fig. 3
Neutral cutting flame without excess of oxygen or acetylene. At proper adjustment, the temperature of the flame is approximately 6,300 degrees F.
Modern Engineering Co.

Fig. 4
Oxidizing cutting flame. It is an excess oxygen flame used only for rough work. It reduces preheat time.
Modern Engineering Co.

Fig. 5
Neutral flame with the cutting lever depressed. The cutting jet is straight and smooth.
Modern Engineering Co.

Flames: Oxy-Acetylene Welding

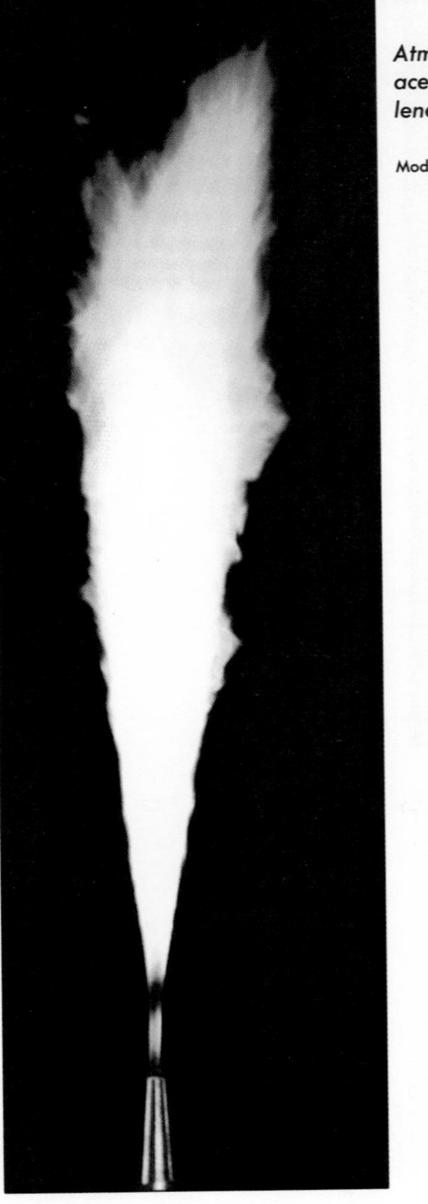

Fig. 1
Atmospheric burning of acetylene. Open the acetylene valve until the smoking disappears.
Modern Engineering Co.

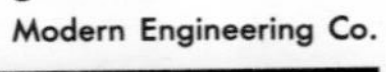
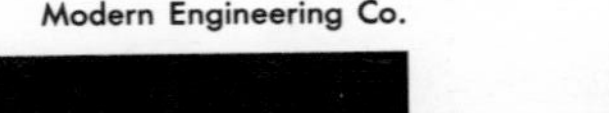

Fig. 2
Neutral welding flame does not have an excess of oxygen or acetylene. It is used for welding steel and cast iron. Its temperature is approximately 6,300 degrees F.
Modern Engineering Co.

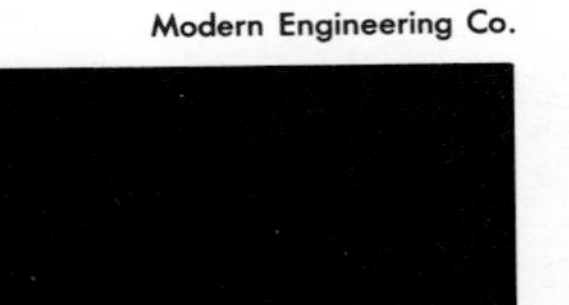

Fig. 3
Carburizing welding flame commonly used for welding white metal, stainless steel, hard facing, some forms of pipe welding, and soldering and brazing.
Modern Engineering Co.

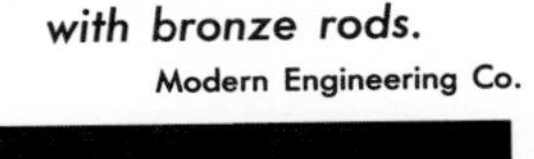

Fig. 4
Oxidizing welding flame commonly used for brazing with bronze rods.
Modern Engineering Co.

point of complete disappearance the neutral flame is formed, Fig. 3, page 480A. If you continue to adjust the oxygen valve, an *oxidizing* (excess oxygen) flame will be formed. The entire flame will decrease in size, and the inner cone will become much less sharply defined, Fig. 4, page 480A.

9. Practice adjusting the flame several times until you are completely familiar with the neutral flame. Fig. 16-27 (Page 484) summarizes the different flame characteristics.

The fastest preheat time is achieved with an oxidizing flame because it is hotter and more concentrated than the carburizing flame and because it speeds up the oxidizing processes, Fig. 16-28. If the cut surface is to be used for welding, the neutral flame is recommended. The cut surface will also have a cleaner appearance with the use of the neutral flame.

10. Check the preheat flame by pressing the high pressure lever shown in Fig. 16-25 and observe any change in the flame, Fig. 5, page 480A. Excess acetylene feather should not appear.

11. Hold the torch as indicated in Fig. 16-29 for both preheating and cutting. Figures 16-30 and 16-31 illustrate two of the several positions the operator may have to use when cutting.

ATTACHING THE ADAPTABLE CUTTING HEAD TO THE OXY-ACETYLENE TORCH. The adaptable cutting head is attached in the following manner:

1. Detach the welding tip and neck from the welding torch handle by unscrewing the shell nut, Fig. 16-32.

2. Attach the adaptable cutting head to the welding torch handle by screwing the shell nut to the handle, Fig. 16-33.

16-21. *Attaching an oxygen regulator to an oxygen cylinder. Note that a gauge with high pressure calibrations is necessary because of the high pressure in the oxygen cylinder.*

16-22. *Attaching an acetylene regulator to an acetylene cylinder. Note the relatively low pressure gauge, which is necessary because of the low pressure in acetylene cylinders.*

3. Light the torch by opening oxygen valve *A*, Fig. 16-33, on the welding torch handle wide to permit the passage of oxygen to oxygen preheat valve *C* on the adaptable cutting head. Open acetylene valve *B*, Fig. 16-33, on the welding torch handle and light the gas with a friction lighter.

4. Open the oxygen preheat valve *C* on the adaptable cutting head until the proper flame is obtained.

5. Check the preheat flame by pressing the high pressure lever *D*, Fig. 16-33, on the adaptable cutting head and observe the effect upon the flame. You are now ready to begin cutting.

16-23. *Attaching the oxygen hose to the oxygen regulator. The acetylene hose is already attached to the acetylene regulator.*

Closing Down Equipment

When the cutting operation is finished, the equipment may be taken down in the following manner:

1. Close the acetylene valve on the torch. Next close the oxygen valve on the torch.

2. Close the acetylene and oxygen cylinder valves.

3. Open the acetylene and the oxygen valves on the torch to permit the trapped gas to pass out of the regulators and the hose. Then release the regulator-adjusting screws by turning to the left.

4. Disconnect the hose from the torch.

16-24. *Regulator-adjusting screw is turned to the right to cause the flow of gas through the regulator and hose to the torch.*

16-25. *Connecting the oxygen hose to the oxygen outlet on the cutting torch. The acetylene hose is already connected.*

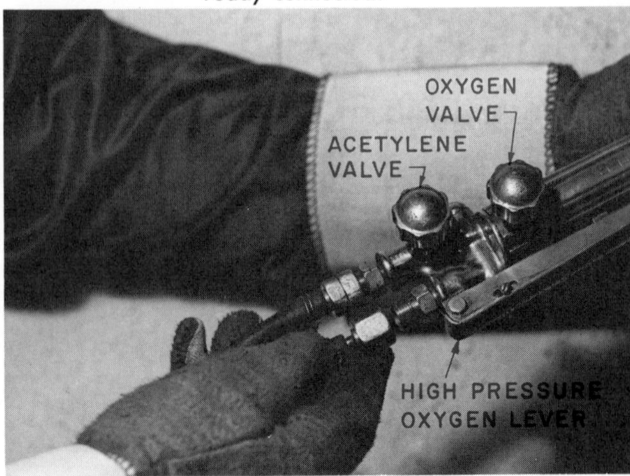

5. Disconnect the hose from the regulator.

6. Remove the regulators from the cylinders and put protective caps on the cylinders.

Instructions for Completing Practice Jobs

Jobs 16-J1* (Page 486) and 16-J2 (Page 488) are necessary for you to practice in order to be able to cut low carbon steel. Before you begin each job, study the print which accompanies it. The title block gives the size and type of stock. A pictorial view of the job is shown to help you interpret the side and top views. These jobs also provide experience in layout. Before making the cuts, check your final layout with the print for accuracy.

*To distinguish job numbers from illustration numbers the letter *J* is used. Job 16-J1 is thus Job #1, Chapter 16.

Job 16-J3 is an optional activity which may be assigned by your instructor. A print has not been given for this job. Any casting of a suitable grade of cast iron may be used for practice.

JOB 16-J1 STRAIGHT LINE AND BEVEL CUTTING

Objective

To make straight line and bevel cuts on flat plate with the oxyacetylene hand cutting torch.

General Job Information

Almost all cutting that the welder does on the job, whether it be a straight line or shape cut, is made with either a square edge or a beveled edge. On many of the jobs in manufacturing plants, it is not necessary for the operator to be able to cut. It is essential, however, that he be able to cut in fabricating plants and on outside construction jobs. Cutting is a quick and economical method of shaping a

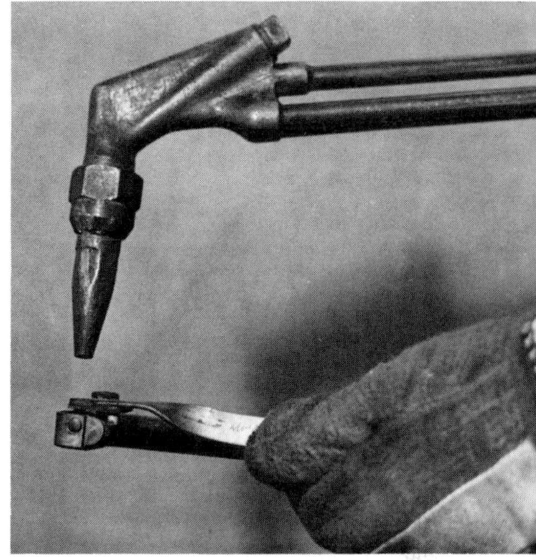

16-26. *Lighting the cutting torch with a friction lighter.*

Table 16-2

Hand flame cutting data for mild steel (not preheated).

Thickness of steel, in.	Diameter of cutting orifice, in.	Oxygen pressure, p.s.i.	Cutting speed, in./min.	Gas consumption	
				Oxygen, cu. ft./hr.	Acetylene, cu. ft./hr.
1/8	0.0200–0.0400	15–30	20–30	18–55	6– 9
1/4	0.0310–0.0595	11–20	16–26	37–93	7–11
3/8	0.0310–0.0595	17–30	15–24	47–115	7–12
1/2	0.0400–0.0595	20–31	12–22	66–125	10–13
3/4	0.0465–0.0595	24–35	12–20	117–143	12–15
1	0.0465–0.0595	28–40	9–18	130–160	13–16
1 1/2	0.0595–0.0810	30–45	6–12	150–225	15–20
2	0.0670–0.0810	22–50	6–13	185–231	16–20
3	0.0670–0.0810	33–55	4–10	207–290	16–23
4	0.0810–0.0860	42–60	4–8	235–388	20–26
5	0.0810–0.0860	49–70	3.5–6.4	281–437	20–29
6	0.0980–0.0995	36–80	3.0–5.4	400–567	25–32
8	0.0995–0.1100	57–77	2.6–4.2	505–625	30–39
10	0.0995–0.1100	66–96	1.9–3.2	610–750	36–46
12	0.1100–0.1200	58–86	1.4–2.6	720–905	42–55

American Welding Society

Acetylene pressures have been omitted from the table above, since they depend mainly on torch and tip design rather than on external factors such as thickness of the material. It is therefore recommended that for specific acetylene pressure settings particularly, and also for oxygen pressure values, reference be made to equipment manufacturer's charts for the apparatus being used.

plate for fabrication. Refer to Fig. 16-44, page 491, for the many shapes that can be flame cut by hand or machine.

Cutting Technique

It is assumed that you have your equipment set up, have selected the proper tip size, and have practiced flame adjustment. Study Table 16-2, page 483, which lists gas pressures. In the actual cutting, the left hand is used to steady the torch, the elbow or forearm is rested on a convenient support, and the right hand is used to move the torch along the line of cut, Fig. 16-34. The thumb on the right hand is used to operate the oxygen cutting lever. The tips of the preheating flame cones are held about $\frac{1}{16}$ inch above the surface of the material.

Hold the torch steady. When a spot of metal at the edge of the plate has been heated to a cherry red, press the lever controlling the oxygen cutting jet and start the cutting action. Move the torch slowly but steadily along the line of cut. If the torch head wavers, a wide cut will result. Besides producing a poor cut, this is a waste of time and material.

For square cuts the cutting head must be held perpendicular, Fig. 16-35A. For bevel cuts the cutting head should be tilted at an angle corresponding to the angle of bevel desired, Fig. 16-35B. Fig. 16-36 shows the torch positioned for bevel cuts as instructed in Fig. 16-35B.

If the cut has been started properly, a shower of sparks will fall from the underside of the plate. This indicates that the

16-27. *Three basic types of flames: neutral, carburizing, and oxidizing.*

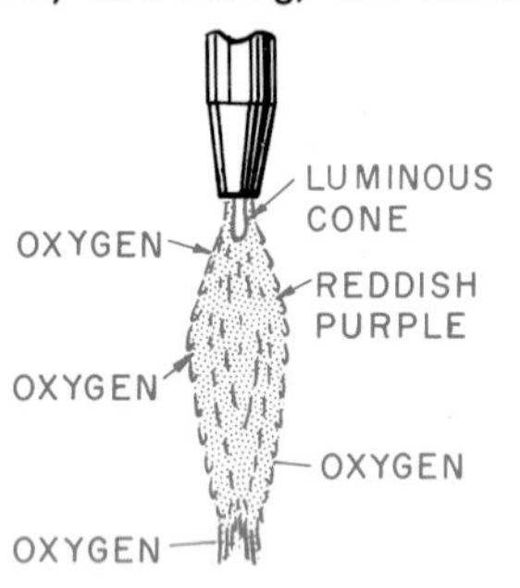

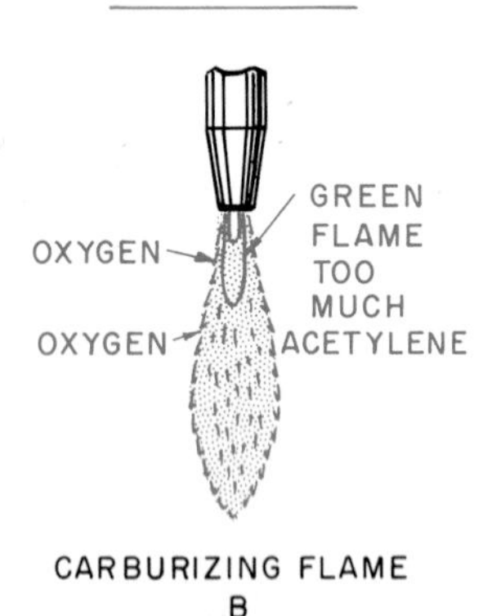

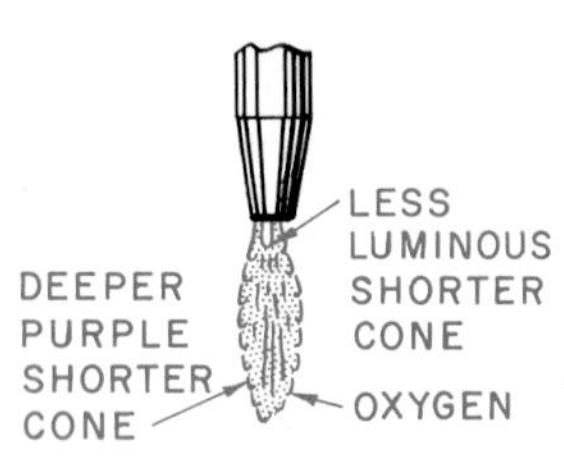

16-28. *Comparative preheat time for two flame types.*

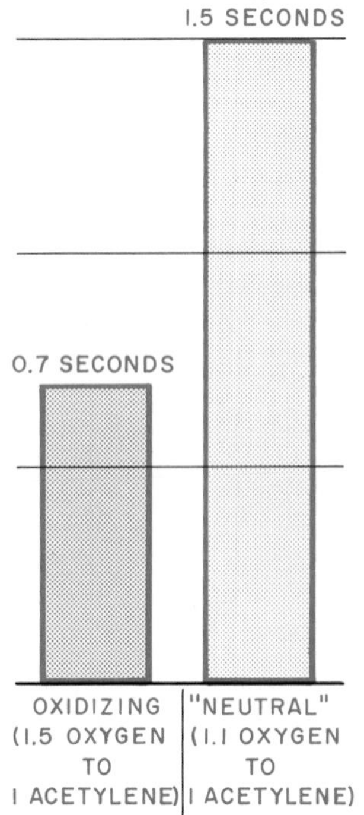

16-29. *Position while cutting on the bench.*

16-30. *Plates can be cut at the point of fabrication and need not be carried to another machine for cutting and back again for fabrication.*

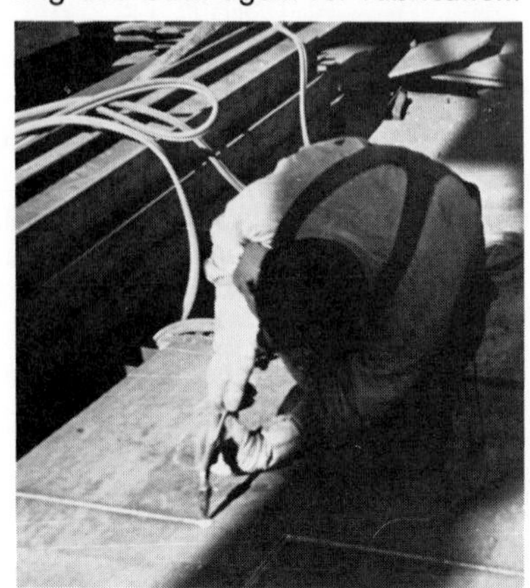

cutting action is penetrating through the work. The forward movement should be just fast enough so that the cut continues to penetrate the full plate thickness.

If the torch is moved forward too rapidly, the cutting jet will fail to go clear through the plate, the molten pool will be lost, and the cutting will stop. If this happens, close the cutting valve immediately and preheat the joint where the cut stopped until it is a bright red. Open the cutting valve to restart the cut.

If, on the other hand, the torch is not moved forward rapidly enough, the preheating flames will tend to melt the edges of the cut. This may produce a ragged edge or at times fuse the metal together again.

Watch the cutting action closely. Make sure that the slag flows freely from the kerf and does not gather at the bottom of the groove, thus hindering the cut and producing rough edges.

The speed of travel is determined by the cutting action and varies with the thickness of metal and the tip size being used.

16-31. *Making a cut in large diameter pipe. It would be impossible to find a shape that could not readily be cut with the hand cutting torch.*

Operations

1. Obtain plate ($\frac{1}{4} \times 7 \times 8$ inches).
2. Obtain a square head and scale, center punch, and scribe from the toolroom.
3. Lay out parallel lines as shown on the job print.
4. Mark the lines with a center punch.
5. Set up the oxyacetylene cutting equipment. (Refer to page 458.)
6. Adjust the regulators for 5 pounds acetylene pressure and 20 pounds oxygen pressure.
7. Light the torch, adjust it to the proper flame, and proceed to cut as shown on the job print.
8. Chip the slag from the cut and inspect it.
9. Practice until you can produce even, clean cuts consistently.
10. Take down the oxyacetylene cutting equipment.

Inspection

The top edges of the cut should be sharp and in a straight line—not ragged. The bottom edge will have some oxide adhering to it, but it should not be

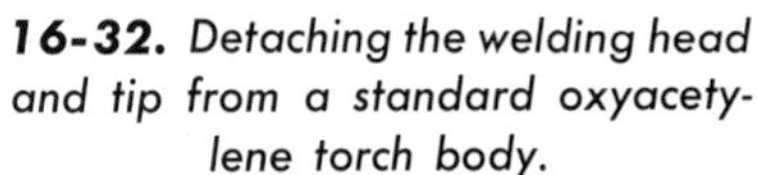

16-32. *Detaching the welding head and tip from a standard oxyacetylene torch body.*

16-33. *Attaching an adaptable cutting head to the welding torch body. The welding torch may now be used as a cutting tool.*

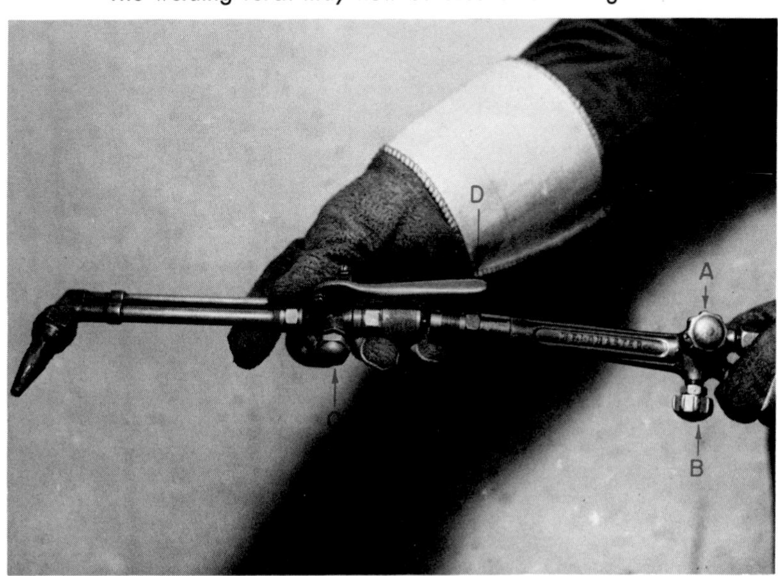

excessive. The face of the cuts should be square and smooth. Every effort should be made to remove gouging on the surface of the cut. If a clean cut has been made, the piece will fall when the cut is finished. There will be no tendency for the plate to weld together at the bottom of the cut. Compare your cut with Fig. 16-18 (1), page 478.

Disposal

Cut strips should be discarded in the waste bin.

JOB 16-J2 LAYING OUT AND CUTTING ODD SHAPES

Objectives

1. To lay out simple shapes on flat plate.

2. To cut out shapes on flat plate, using the oxyacetylene hand cutting torch.

General Job Information

You will do many jobs for which you must be able to lay off squares, parallel lines, and circles. You must also be prepared to cut in other than a straight line. Forms will take on many odd shapes and if you develop in school the flexibility to follow shaped lines, you will have little difficulty on the job. Study the examples in Fig. 16-44, page 491, carefully.

Cutting Technique

Hold the torch head in the position shown in Fig. 16-37. The cutting technique is basically the

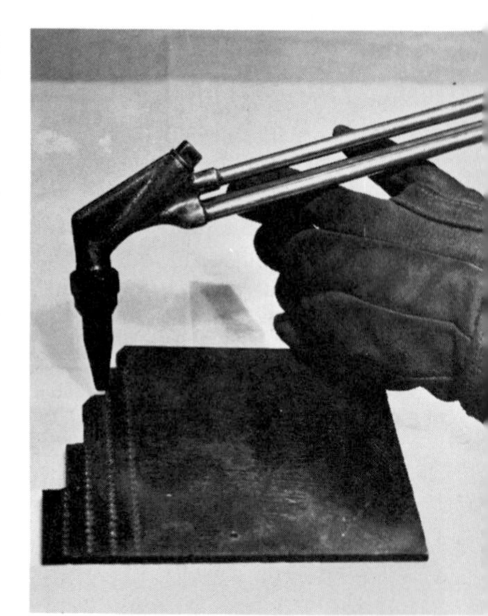

***16-34.** Position of the cutting torch when making square cuts in the flat position.*

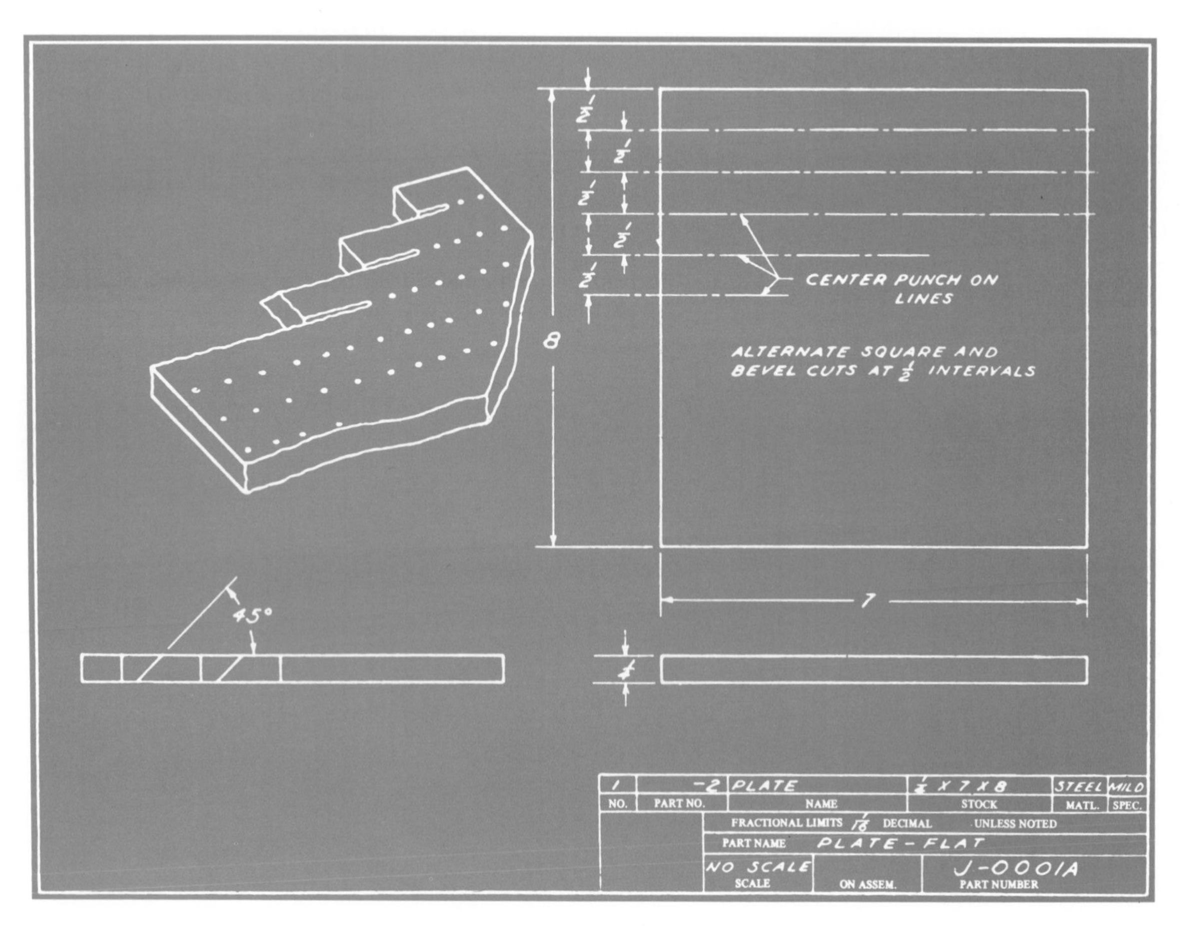

same as in the previous job except that starts will have to be made in the plate away from the edge. This requires more time to bring the plate up to the kindling temperature.

After the spot has been brought up to a cherry red, raise the torch about ½ inch above the plate and turn on the cutting oxygen slowly. When the plate has been pierced through, lower the torch to the normal cutting height and complete the cut. Here again cut edges may be square or beveled, depending upon the nature of the cut. When cuts must be precise, drill a hole along the edge of the line and start the cut in the hole.

Operations

1. Obtain a flat plate ¼ × 8½ × 11 inches).
2. Obtain a square head and scale, dividers, scribe, and center punch from the toolroom.
3. Lay out the shapes as shown on the job print.
4. Mark the lines with a center punch.
5. Set up the oxyacetylene cutting equipment.
6. Adjust the regulators for 5 pounds acetylene pressure and 20 pounds oxygen pressure.
7. Light the torch, adjust it to the proper flame, and proceed to cut as shown in the job print.
8. Chip the slag from the cut and inspect it.
9. Practice these cuts until you can produce even, clean cuts consistently. Compare them with the cuts in Figs. 16-18 (1), page 478, and Fig. 16-38.
10. Take down the oxyacetylene cutting equipment.

Inspection

The cut edge of both the pieces cut from the plate and the plate itself should be as clean as possible. The top edges should be sharp. They should show no evidence of raggedness. Inspect the bottom of the cuts for excessive slag. The faces of cut should be square, smooth, and free of gouging. If the cuts have been made cleanly, the pieces will fall from the plate when entirely cut through.

Disposal

Return the plate to the scrap bin. It may be used for Job 20-J1, Striking the Arc, and Short Stringer Beading, in Chapter 20.

JOB 16-J3 CUTTING CAST IRON

Objective

To make cuts in cast iron using a carburizing flame and an oscillatory motion of the torch.

General Job Information

Most industrial welders do not have occasion to cut cast iron with oxyacetylene flame. If skill in this technique is necessary in your area, you should complete the job after you have had some experience with the hand and machine cutting of low carbon steel.

Cutting cast iron with the oxyacetylene flame is somewhat more difficult than the cutting of steel. You will recall that the oxidation process slows down as the amount of carbon and other alloy materials is increased. Cast iron, of course, is relatively high in carbon compared with low carbon steel.

Cutting Technique

The technique for cutting cast iron is quite different from that for cutting steel. The preheat flame must be larger for cast iron and should be adjusted for an excess of acetylene. The oxygen pressure required is from 25

16-35. Position of the cutting head when making (A) straight and (B) bevel cuts in the flat position.

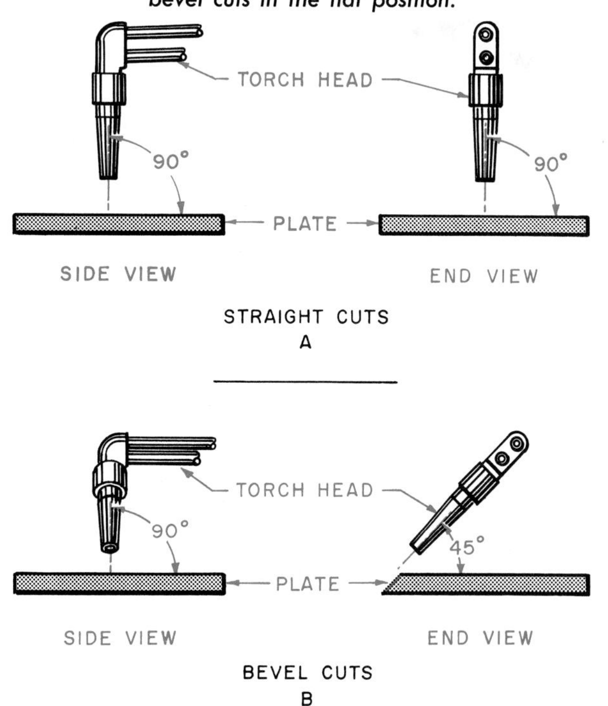

16-36. *Position of the cutting torch when making bevel cuts in the flat position.*

to 200 percent greater than that for cutting steel of the same thickness. The torch tip is moved back and forth constantly across the line of cut.

The quality of the cast iron also determines the ease with which it can be cut. Good quality gray iron, such as that usually used for machinery parts, is less difficult to cut than the furnace iron which is for cheap castings.

When cutting a poor grade of cast iron, the slag does not flow

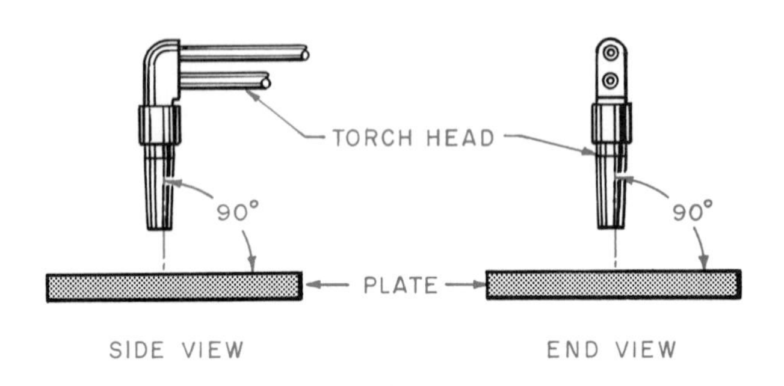

ODD SHAPES

16-37. *Position of the cutting torch head when making square cuts.*

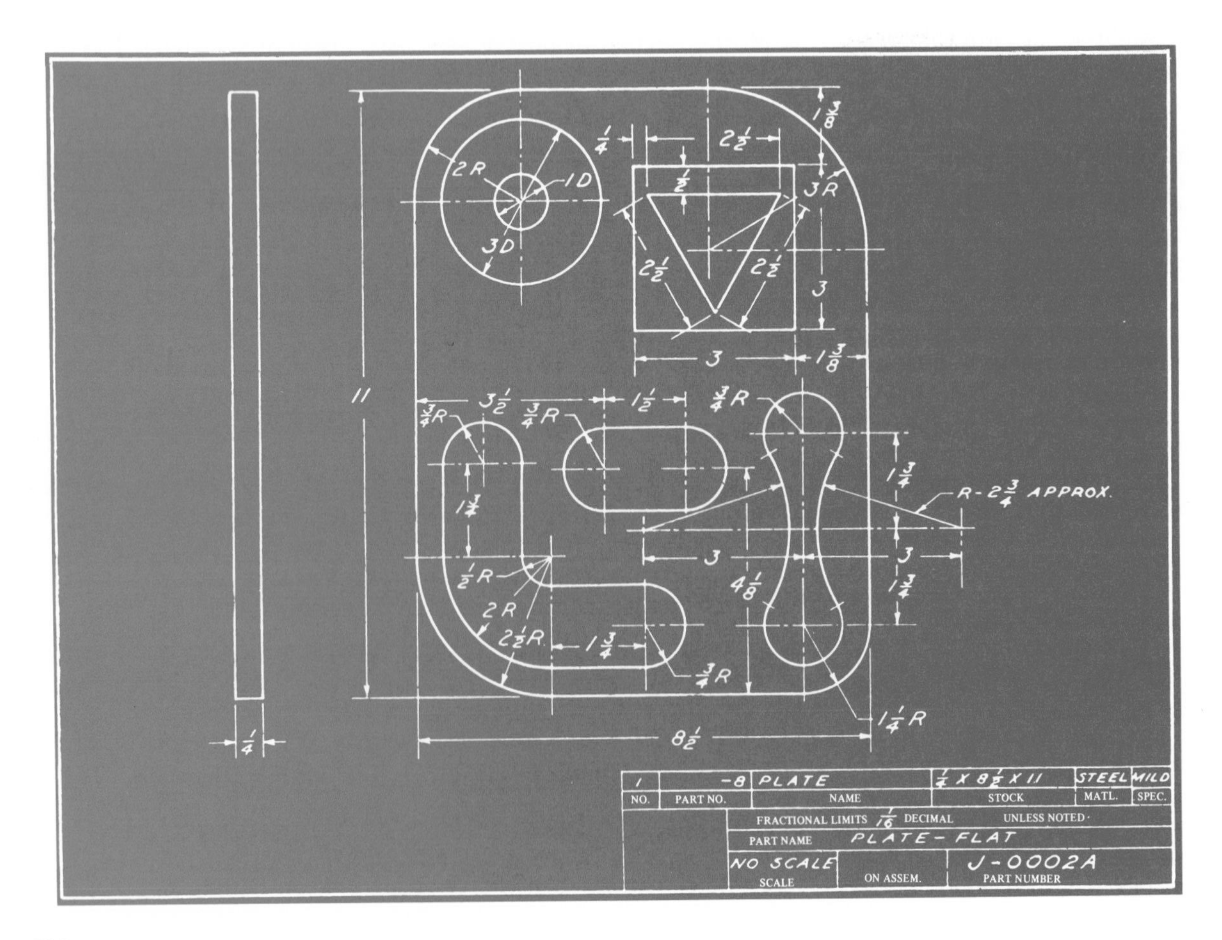

freely. The fluidity of the slag can be increased by mixing it with a flux. Feeding a steel rod into the kerf is a form of flux cutting. This may be done at the start of the cut and may have to be continued for the length of the cut, Fig. 16-39.

Operations

1. Select a piece of high quality gray cast iron of whatever thickness is available.
2. Adjust the excess acetylene preheat flames with the cutting oxygen valve open.
3. Close the cutting oxygen valve and begin preheating along the front edge of the casting from top to bottom, Fig. 16-40. This makes starting and continuing the cut easier.
4. When the line of cut is heated, position the torch tip at about a 45-degree angle with the

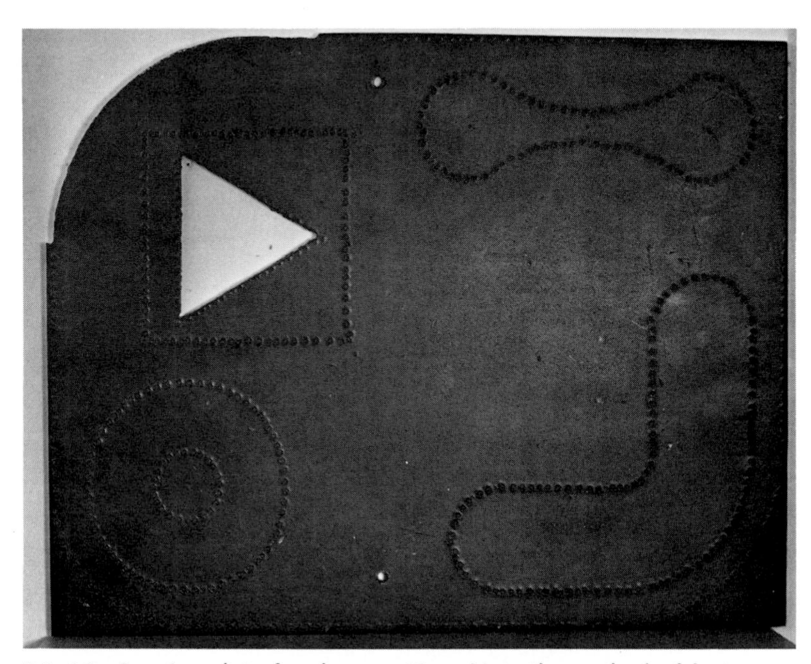

16-38. *Practice plate for shape cutting. Note the method of laying out guidelines and center punching.*

16-39. *The use of a steel welding rod for starting a cut on cast iron is an example of flux-cutting. This method is used frequently on low-grade castings, as well as on those that are somewhat oxidation-resistant.*

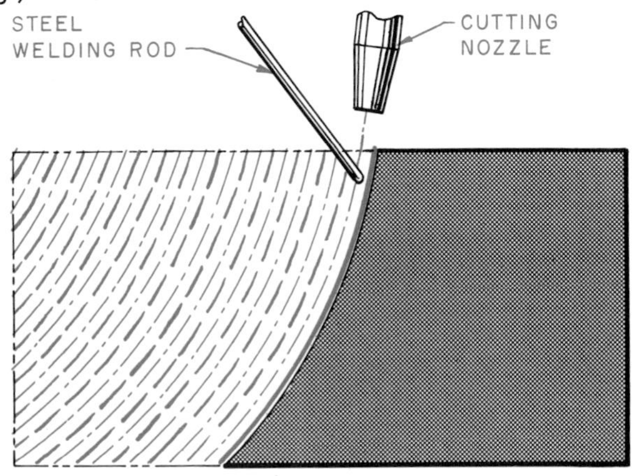

16-40. *The first step in cast iron cutting is to preheat the entire length of the edge to be cut. The nozzle is then directed at the starting point at the top and the side-to-side motion begun.*

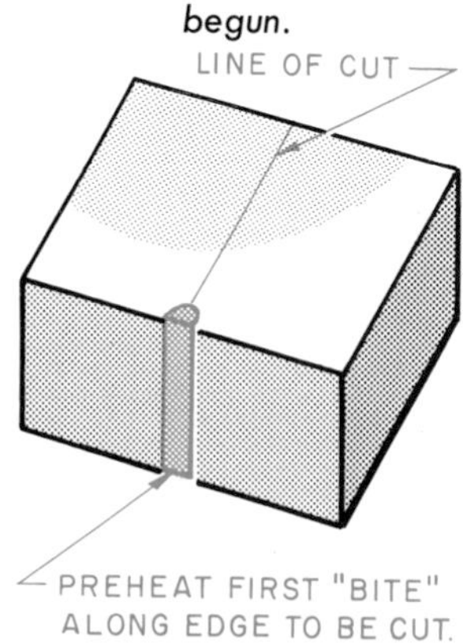

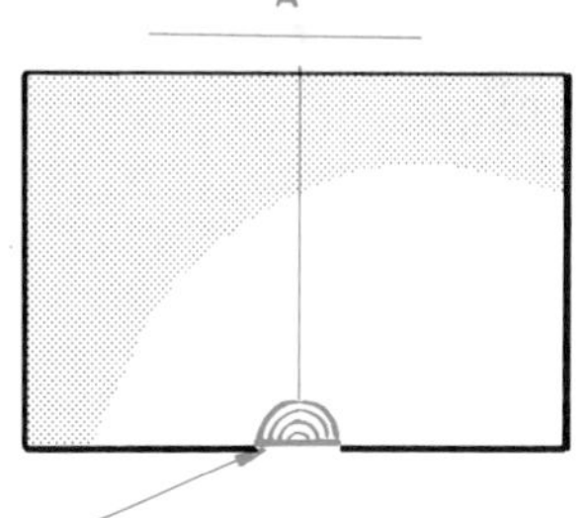

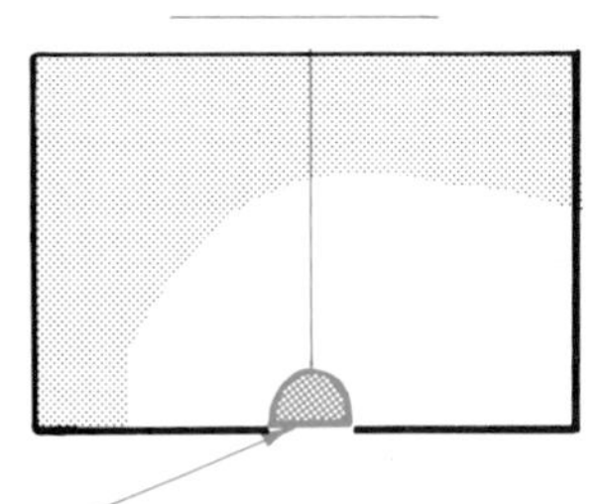

inner cones of the flame about $\frac{1}{8}$ to $\frac{1}{4}$ inch above the surface of the casting, Fig. 16-41.

5. Heat a semicircular area about $\frac{3}{4}$ inch in diameter until it is molten, Fig. 16-40. Open the cutting oxygen valve to start the cut. Use an oscillating (back and forth) motion throughout the cut, Fig. 16-41C. When the cut extends through the bottom of the casting, change the tip angle to approximately 75 degrees, Fig. 16-41B. Be sure to keep the metal hot so that the cut is not lost. If the cut is lost, heat up one side and part of the end of the cut again just as if starting a new cut. When cutting low-grade castings, the entire cut may have to be done as a series of definite steps in this way, Fig. 16-42.

6. There is always considerable lag when cutting cast iron even if the nozzle is brought up to the vertical position. When the cutting has reached the far edge of the plate, carry it down the far side at about the same angle as the angle of lag until the cut is completed, Fig. 16-43.

7. Practice these cuts until you can produce even, clean cuts consistently.

8. Take down the oxyacetylene cutting equipment.

Inspection

Inspect the surface of the cut and compare it with Fig. 16-18 (1), page 478. The face of the cut will not be as smooth as that possible with steel plate. It should, however, be in a fairly straight line and square from top to bottom.

Disposal

Cut strips should be discarded in the waste bin.

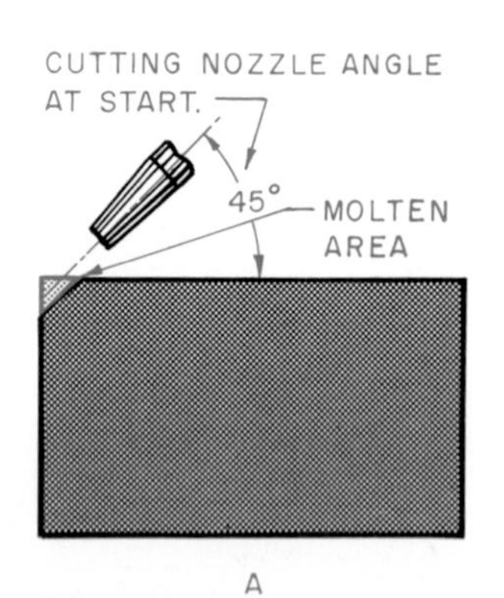

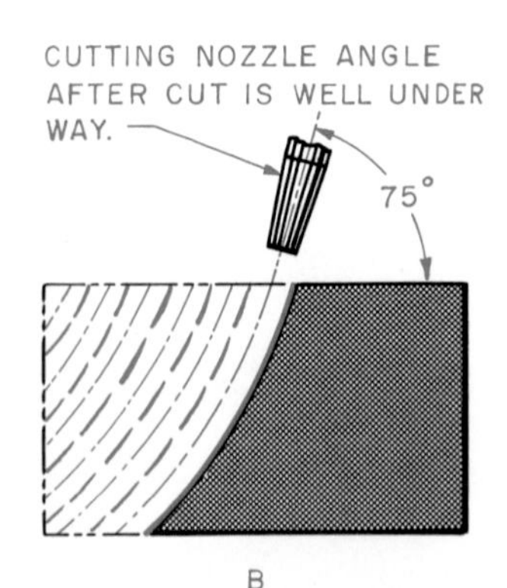

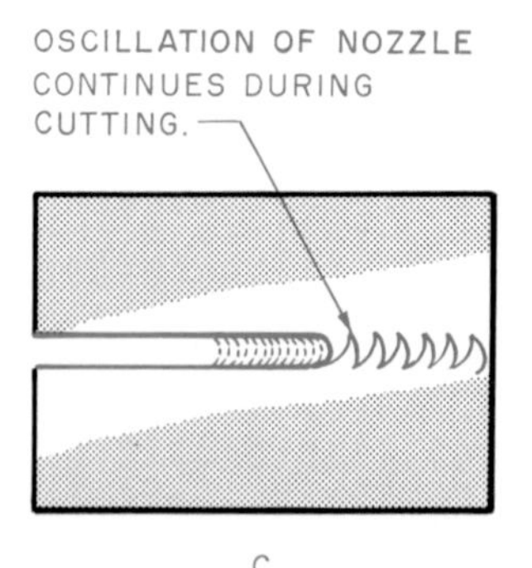

16-41. *At the start of cutting, the nozzle makes an angle of about 45 degrees with the top surface. As the cutting oxygen lever is gradually depressed so that the cutting will start, the side-to-side motion of the nozzle is continued.*

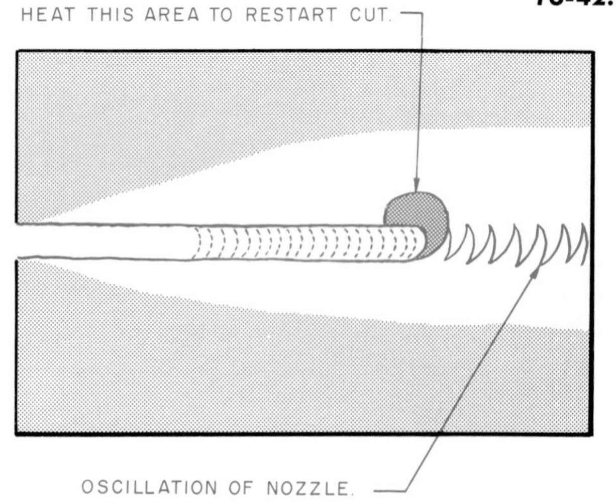

16-42. *Restarting the cut.*

16-43. *Finishing the cut.*

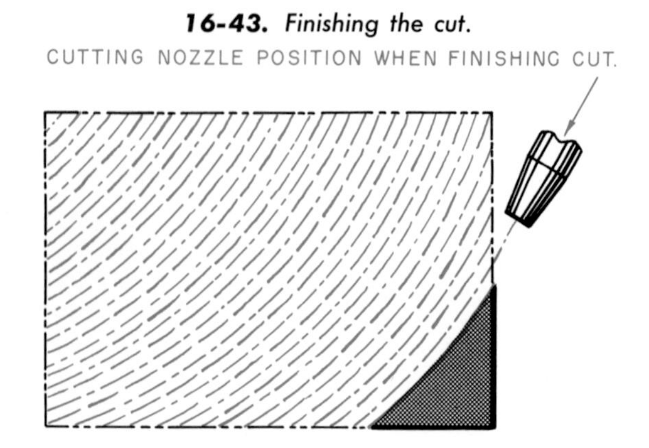

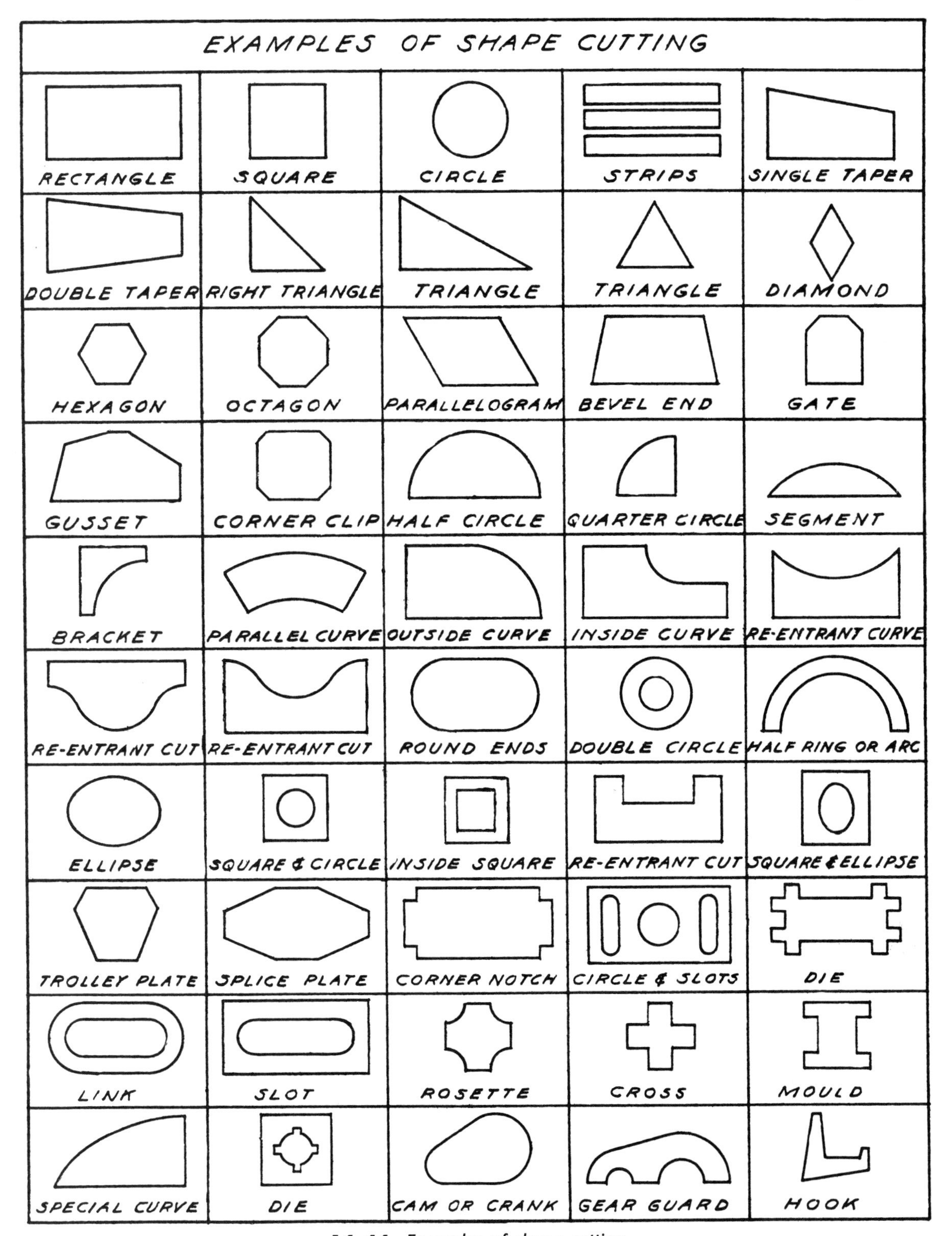

16-44. Examples of shape cutting.

REVIEW QUESTIONS

1. Explain the reaction that takes place in the cutting of ferrous metals with the oxy-fuel gas processes.
2. Name four fuel gases that can be used in the oxy-fuel gas process.
3. Oxy-fuel gas and arc cutting are used chiefly as a means of ________________ and as a ________________ tool.
4. List five factors that should be considered in the choice of a fuel gas.
5. List the procedures that may be used to improve the cutting of cast iron.
6. What is meant by *kerf?*
7. Mapp® gas gives the highest flame temperature. True. False.
8. Acetylene gas has the highest Btu. rating. True. False.
9. Explain the term *drag* when referring to a cut surface.
10. In the flame-cutting process, a number of procedures and techniques are commonly used in the preparation and treatment of metal and metal objects. Name five of these and briefly explain them.
11. Name at least nine conditions that affect the quality of the cut.
12. Explain the function of the high pressure lever on the gas cutting torch.
13. The neutral flame is the hottest preheat flame. True. False.
14. A special welding torch is necessary in order to use the adaptable cutting torch. True. False.
15. When turning off the cutting torch, the oxygen valve is closed first. True. False.
16. A special tip is necessary for bevel cutting. True. False.
17. When cutting cast iron, the oxygen pressure is less than for steel of the same thickness. True. False.
18. The excess oxygen flame is best suited for the cutting of cast iron. True. False.
19. It is not important for the modern welder to learn how to cut. True. False.
20. The preheat flames of the cutting torch should be in contact with the metal surface while the cut is in progress. True. False.
21. What is the position of the cutting head when making a square cut?
22. What is the position of the cutting head when making a bevel cut?
23. Why is it necessary to prevent the torch from wavering along the line of cut?
24. Explain the appearance of a good cut.
25. Is it possible to weld on an edge that has been cut with a cutting torch?

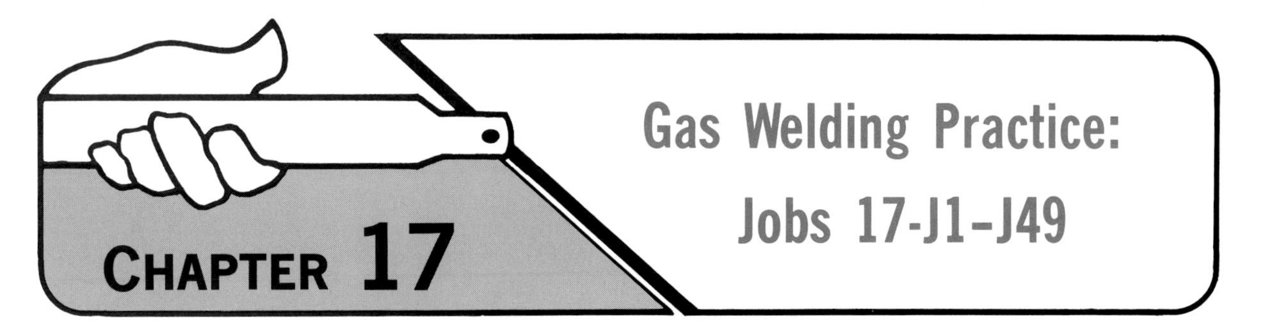

Chapter 17 Gas Welding Practice: Jobs 17-J1–J49

The cooperation of the Linde Division, Union Carbide Corporation, and the Modern Engineering Company in providing information and photographs for this chapter is gratefully acknowledged.

As a new student of welding, these practice jobs in oxyacetylene welding are your first experience in actually performing a welding operation. The practice jobs in this text have been carefully selected, according to the recommendations of a number of skilled welders, supervisors of welding, and teachers of welding. Thus the sequence represents many years of experience in the welding industry. In the Course Outline, the jobs are organized on the basis of learning difficulty in an orderly sequence. If you complete the assigned reading and welding practice, you will acquire enough skill and knowledge to qualify for a job in industry in any one of the many branches of welding.

It is desirable for the student to begin practice with the oxyacetylene welding process because it offers many learning advantages. The student is able to develop that coordination of mind, eye, and hands that is the basis for all welding practice. The action of a molten pool of metal can be clearly observed. The significance of fusion and penetration are better understood than when the beginner practices with the electric arc processes. The ability to control the size, direction, and appearance of the molten pool is quickly acquired.

The undivided attention of the welder to his work is required at all times. He must not only see that he is fusing the weld metal properly to the base metal, but he must also observe the welding flame itself to see that it is not burning an excess of oxygen or acetylene.

Review Chapter 5 before beginning practice so that you have a thorough knowledge of the equipment used in the process. Before you begin welding, you must also learn the following information in this chapter:

- How to set up the equipment in a safe and practical manner
- How to adjust the equipment for the best operation
- How to light and adjust the torch for the proper flame
- How to manipulate the torch for sound welding

It is also important that you recognize the basic characteristics of a sound weld of good appearance.

SOUND WELD CHARACTERISTICS

Fusion

Fusion is the complete blending of the two edges of the base metal being joined or the blending of the base metal and the filler metal being added during welding. When there is a lack of fusion, we have the condition referred to as *adhesion,* Fig. 17-1.

Adhesion may be caused by adding molten metal to solid metal or by lack of fluidity in the molten weld pool. Adhesion may also be caused by improperly chamfering the pieces to be welded, by improper inclination of the torch, by improper use of the filler rod, or by faulty adjustment and manipulation of the welding flame.

Beginning welding students frequently do not prepare the plates properly for welding. The *chamfering* (grooving) is not deep enough to extend entirely through the section to be welded, or it is not wide enough.

A common fault of the beginning welder is that he adds filler rod to the surfaces of the material being joined before they are in the proper condition for fusion to take place. Sometimes one surface is in fusion, but the other is not. When a lack of fusion takes place on only one side, the overall strength of the weld is

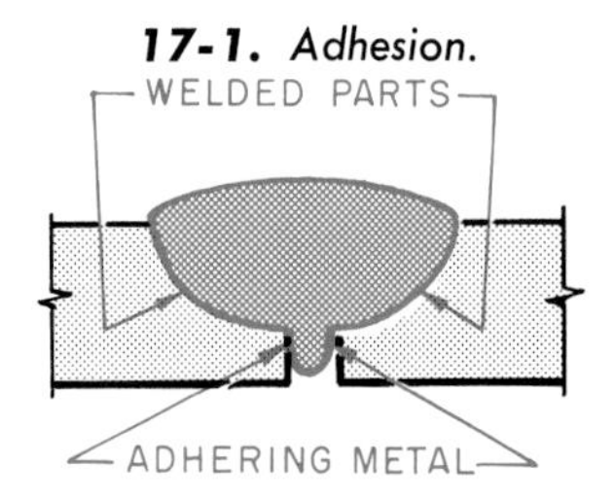

17-1. *Adhesion.*

lessened just as if a lack of fusion were present on both sides.

In some cases the surfaces of the metal are brought to a state of fusion too soon so that oxide has a chance to form on the edges of the weld. When the filler rod is added, adhesion occurs because a film of oxide separates the surface and the added filler material.

Often through faulty torch manipulation, some of the molten weld metal is forced ahead. The surface of this metal is not in the proper state of fusion, and adhesion will occur.

Penetration

The term *penetration* refers to the depth to which the base metal is melted and fused. *Fusion* is the essential characteristic of good weld. It is possible to have penetration through the joint, and at the same time have a condition of poor fusion, Fig. 17-2.

A sound weld must penetrate through to the root of the joint. The cross section of weld metal should be as thick as the material being welded.

In his desire to complete the weld as soon as possible, the beginning welder has the tendency to hasten over the most important part of the work, which is to penetrate to the root of the weld. Incomplete penetration reduces the thickness of the metal at the weld and provides a line of weakness that will break down when the joint is under stress and strain.

17-2. *Poor penetration.*

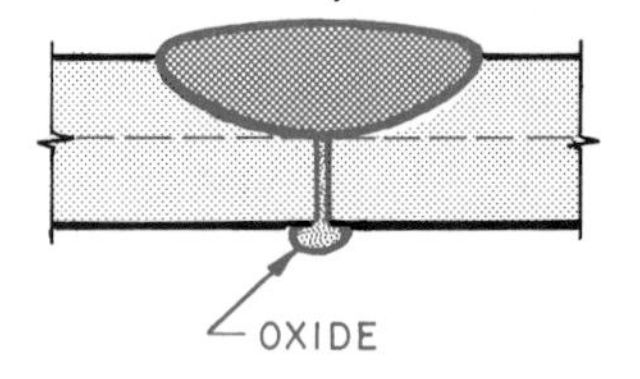

The principal cause of incomplete penetration is improper joint preparation and alignment. The material may not be beveled in the proper manner, the root face may be too thick, or the spacing at the root may be too much or too little. Improper heat, poor welding technique, or using a filler rod that is too large may also cause incomplete penetration.

Weld Reinforcement

A weld is said to be *reinforced* when the weld metal is built up above the surface of the metal being welded. The beginning welding student may build up weld metal above the surface of the base metal because he feels that the more weld metal that is added to the bead, the stronger the weld. Metal above the surface of the plate adds little strength to the joint. In fact, too much reinforcement reduces the strength of the welded joint by introducing stresses at the center and edges of the weld bead. If the weld is sound from the root of the joint to the surface of the joint, there is no need to add a great amount of bead reinforcement.

Some reinforcement is desirable to offset possible pockets and other defects on the surface of the weld. It should not be more than $\frac{1}{16}$ inch above the surface of the plate. There should be an absence of undercutting where the weld bead joins the plate. The edge of the weld bead should flow into the plate surface with an appearance of perfect fusion.

Satisfactory Weld Appearance

Careful evaluation of each bead that the student welder completes is necessary before going on to the next bead. He must be familiar with the characteristics of a good weld in order to set the standard of performance that he will try to attain. At this time we will limit our concern with visual inspection. It is assumed that the student is familiar with the chapter on inspection and testing.

- The weld should be of consistent width throughout. The two edges should form straight parallel lines.
- The face of the weld should be slightly convex with a reinforcement of not more than $\frac{1}{16}$ inch above the plate surface. The convexity should be even along the entire length of the weld. It should not be high in one place and low in another.
- The face of the weld should have fine, evenly spaced ripples. It should be free of excessive spatter, scale, and pitting.
- The edges of the weld should be free of undercut or overlap. The edges should be fused into the plate surface so that they do not have a distinct line of demarcation. The weld should appear to have been molded into the surface.
- Starts and stops should blend together so that it is difficult to determine where they have taken place.
- The crater at the end of the weld should be filled and show no porosity or holes. There is always a tendency to undercut at the end of the weld because of the high heat that has built up. Proper torch and filler rod manipulation prevents undercutting.

If the joint is a butt joint, check the back side for complete penetration through the root of the joint. A slight bead should form on the back side.

- The root penetration and fusion of lap and T-joints can be checked by putting pressure on the upper plate until it is bent double. If the weld has not penetrated through the root, the plate will crack open at the joint as it is being bent. If it breaks, observe the extent of the penetration and fusion at the root. It will probably be lacking in fusion and penetration.

THE OXYACETYLENE WELDING FLAME

Before setting up the welding equipment and adjusting the flame, the student should have a good understanding of the various types of oxyacetylene flames and their welding applications. A knowledge of the chemistry of the flame is also important.

The tool of the oxyacetylene welding process is not the welding torch, but the flame it produces. The sole purpose of the various items of equipment is to enable the welder to produce an oxyacetylene flame suited for the work at hand. The flame must be the proper size, shape, and chemical type to operate with maximum efficiency.

There are three chemical types of oxyacetylene flames, depending upon the ratio of the amounts of oxygen and acetylene supplied through the welding torch. These flames are the neutral flame, the excess acetylene flame, and the excess oxygen flame. Their applications are given in Tables 17-1 and 17-2.

The Neutral Flame

The name of this flame is derived from the fact that there is no chemical effect of the flame on the molten weld metal during welding. The metal is clean and clear and flows easily.

The neutral flame, Fig. 17-3, is obtained by burning an approximately one-to-one mixture of acetylene and oxygen. The pale blue core of the flame, Fig.

Table 17-1

The various types of flame which are recommended for different metals and alloys. This chart was modified from one of the publications of the Oxy-Acetylene Committee of the International Acetylene Association.

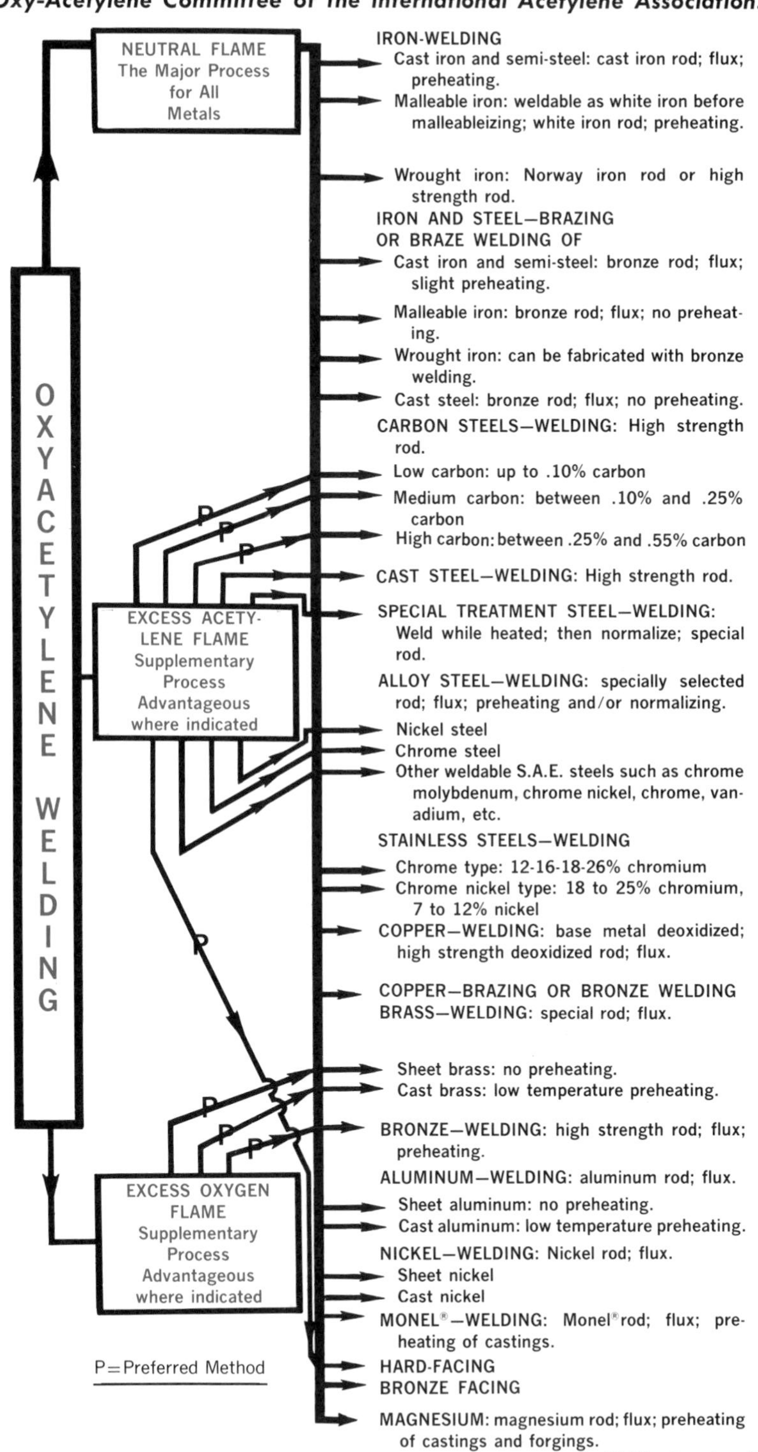

The Welding Encyclopedia

2, page 480B, is known as the inner cone. The oxygen required for the combustion of the carbon monoxide and hydrogen in the outer envelope of the flame is supplied from the surrounding air, Fig. 17-4. The temperature at the inner cone ranges from 5,800 to 6,300 degrees F.

The neutral flame is used for the majority of oxacetylene welding and cutting operations. It serves as a basic point of reference for making other flame adjustments.

The Excess Acetylene Flame

The excess acetylene flame has an excess of acetylene gas in its mixture. The flame has three zones: the inner cone, the excess acetylene feather, and the outer envelope, Fig. 17-3, page 480B. This acetylene feather contains white hot carbon particles, some of which are introduced into the weld pool during welding. For this reason, the flame adjustment is also referred to as a *carburizing flame.* It is also known as a *reducing flame* since it tends to remove the oxygen from iron oxides when welding steel.

The excess acetylene flame is used for many gas welding applications, high test pipe welding, and for certain surfacing applications. Its effect on steel is to cause the weld pool to boil and be very cloudy. Welds are hard and brittle. The temperature of this flame ranges between 5,500 and 5,700 degrees F., depending upon the oxygen content. The temperature increases as the amount of oxygen increases.

Table 17-2
Gas welding data for the welding of ferrous metals.

Base metal	Flame adjustment	Flux	Welding rod
Steel, cast	Neutral	No	Steel
Steel pipe	Neutral	No	Steel
Steel plate	Neutral	No	Steel
Steel sheet	Neutral Slightly oxidizing	No Yes	Steel Bronze
High carbon steel	Reducing	No	Steel
Manganese steel	Slightly oxidizing	No	Base metal composition
Cromansil steel	Neutral	No	Steel
Wrought iron	Neutral	No	Steel
Galvanized iron	Neutral Slightly oxidizing	No Yes	Steel Bronze
Cast iron, gray	Neutral Slightly oxidizing	Yes Yes	Cast iron Bronze
Cast iron, malleable	Slightly oxidizing	Yes	Bronze
Chromium nickel	Slightly oxidizing	Yes	Bronze
Chromium nickel steel castings	Neutral	Yes	Base metal composition 25-12 chromium nickel
Chromium nickel (18-8) and (25-12)	Neutral	Yes	Columbium stainless steel or base metal composition
Chromium steel	Neutral	Yes	Columbium stainless steel or base metal composition
Chromium iron	Neutral	Yes	Columbium stainless steel or base metal composition

The Excess Oxygen Flame

This type of flame adjustment has an excess of oxygen in the mixture. The flame has only two zones like the neutral flame, but the inner cone is shorter, and may be sharper. It is nicked on the sides and has a purplish tinge. See Fig. 17-3 page 497, and Fig. 4, p. 480B. At high temperatures the excess oxygen in this flame mixture combines readily with many metals to form oxides which are hard, brittle, and of low strength, thus seriously reducing weld quality.

A slightly oxidizing flame is used in braze welding and bronze surfacing. A more strongly oxidizing flame is used in fusion welding certain brasses and bronzes. On steel it causes excessive foaming and sparking of the molten weld pool. Welds have very poor strength and ductility. The temperature of this flame ranges between 6,000 and 6,300 degrees F., depending upon the oxygen content. Beyond a certain point, an increase in the oxygen content reduces the flame temperature.

SETTING UP THE EQUIPMENT

If you are a student in a welding school, the oxyacetylene

welding shop is probably equipped with a line system of welding stations. They have the regulators and welding torch already set up and attached, ready for welding. It is essential, however, that you know how to set up a portable welding outfit for welding. This type of equipment is found on construction jobs, in repair shops, and wherever maintenance welding is performed. Figure 17-5 shows a dolly that can be used to move oxygen cylinders safely.

Review pages 481–482 in Chapter 16 before setting up the welding equipment. The equipment necessary to perform oxyacetylene welding includes the following:

- Oxygen cylinder
- Acetylene cylinder
- Oxygen regulator
- Acetylene regulator
- Welding torch (also called a *blowpipe*)
- Oxygen hose
- Acetylene hose
- Wrenches
- Friction lighter
- Filler rod
- Gloves and goggles

Use the following procedure to set up the equipment.

1. Set up the cylinders. Secure them on a portable hand truck or fasten them to a wall so that they cannot fall over. Remove the protector caps from the cylinders.

2. Dirt and dust collect in the cylinder valve outlets during storage. To make sure that dirt and dust will not be carried into the regulators, clear the cylinder valves by blowing them out. Crack the valves only slightly. Open the acetylene cylinder outdoors and away from open flame.

3. Attach the oxygen regulator (Fig. 16-21, page 481) and the acetylene regulator (Fig. 16-22, page 481) to the tanks. Observe that the oxygen regulator has a right-hand thread and that the acetylene regulator has a left-hand thread. Make sure that the regulator nipples are in line with the valves of the tanks so that they may seat properly. Be careful not to cross the threads in the connecting nuts.

4. Attach the welding hoses to the regulators, Fig. 16-23, page 482. The oxygen hose is green or black and has a right-hand thread. The acetylene hose is red and has a left-hand thread. Make

17-3. *Typical oxyacetylene flames.*

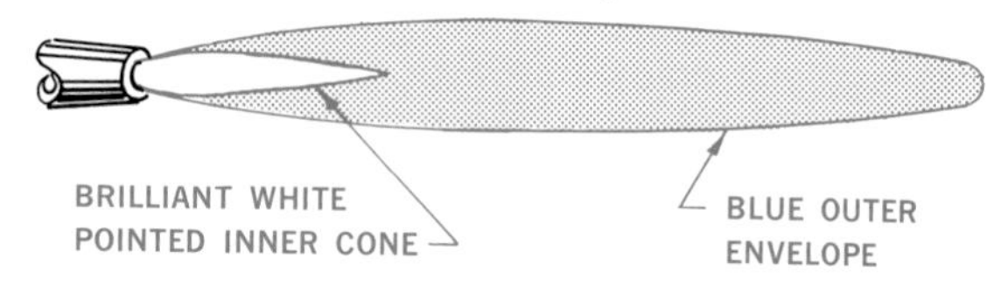

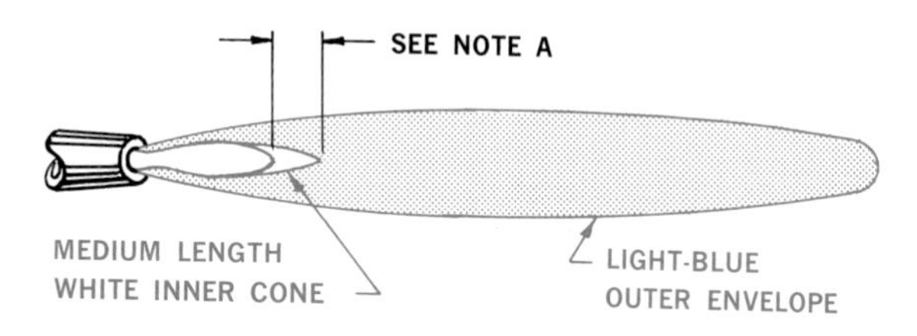

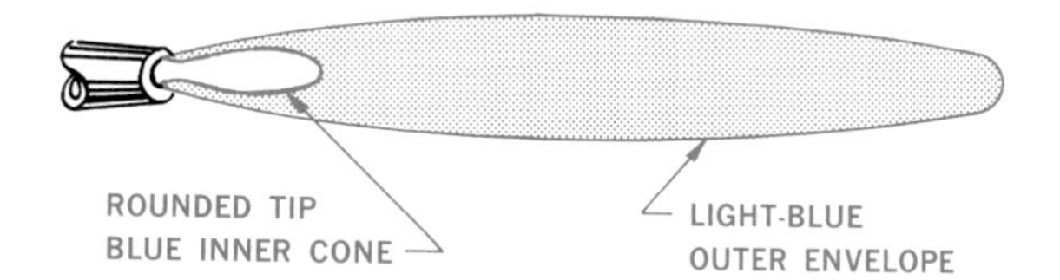

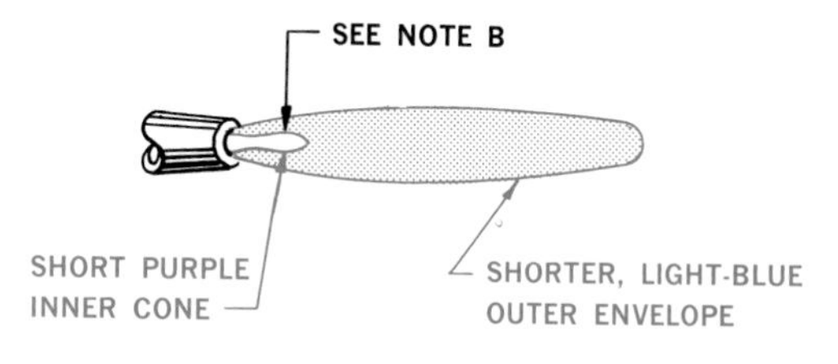

NOTE A – FEATHER OF C_2H_2 FOR WELDING ALUMINUM, FEATHER LENGTH SHOULD RANGE FROM $\frac{1}{4}$ TO $\frac{1}{2}$ THE LENGTH OF INNER CONE.

NOTE B – NECKING OF INNER CONE IS CHARACTERISTIC OF OXIDIZING FLAME.

sure that the nipples of the hose are in line with the connections on the regulator to insure correct seating. Make sure that the threads in the nuts are not crossed. Tighten the nuts securely with a wrench and make sure the connections do not leak.

5. It is important that no grit or dirt be allowed to remain in the regulators or the hoses. This can be prevented by blowing them out. After having made sure that the regulator-adjusting screw is fully released so that the regulator will not be damaged, turn on the tanks. Be sure that you use the proper acetylene tank wrench. Adjust the regulator by turning the adjusting screw to the right (Fig. 16-24, page 482) to permit the gases to flow through. (After turning off the tanks, turn the regulator-adjusting screw to the left again after enough gas has flowed to clear the regulator and hose.)

6. Connect the green or black oxygen hose to the torch outlet marked *oxygen* and the red acetylene hose to the torch outlet marked *acetylene,* Fig. 16-25, page 482. Make sure that the nipples of the hose connections are in line with the torch outlets for proper seating. Make sure that the threads in the nuts are not crossed. Tighten the nuts securely with a wrench and make sure the connections do not leak.

7. Open the valve of the oxygen and acetylene tanks. Be sure you have the proper acetylene tank wrench.

8. Select the proper welding tip for the job at hand and adjust the oxygen and acetylene regulators for the proper working pressure, Table 17-3. To adjust the working pressure, open the torch valves and set the correct pressure on the regulators by turning the adjusting screw to the right, Fig. 16-24, page 482. Close the torch valves when the pressure has been set. You are now ready to light the torch and adjust the flame.

FLAME ADJUSTMENT

Proper flame adjustment for the job at hand is essential. If the flame is not correct, the properties of the weld metal will be affected, and a sound weld of maximum strength and durability will be impossible. The oxidizing flame causes oxides to form in the weld metal, and an excess acetylene introduces carbon particles into the weld. Both of these conditions are detrimental to good welding under certain conditions. The most common fault is the presence of too much oxygen in the welding flame. It is not easily detected because of its similarity to the neutral flame. Too much acetylene is readily seen by the presence of the slight feather.

A flame may also be too harsh. This is caused by setting the regulator at too high gas pressure. The gas flows in greater quantity and with greater force than it should. A harsh flame disturbs the weld pool, causing the metal to spatter around it. High pressure is often the major cause of poor fusion. It is impossible to make a weld of good appearance with a high pressure flame. The flame is noisy, and the inner cone is sharp.

If the gas pressure is correct for the tip size, the flame is quiet and soft. The molten weld pool is not disturbed. Spatter and

17-4. *The products of combustion of the oxyacetylene flame.*

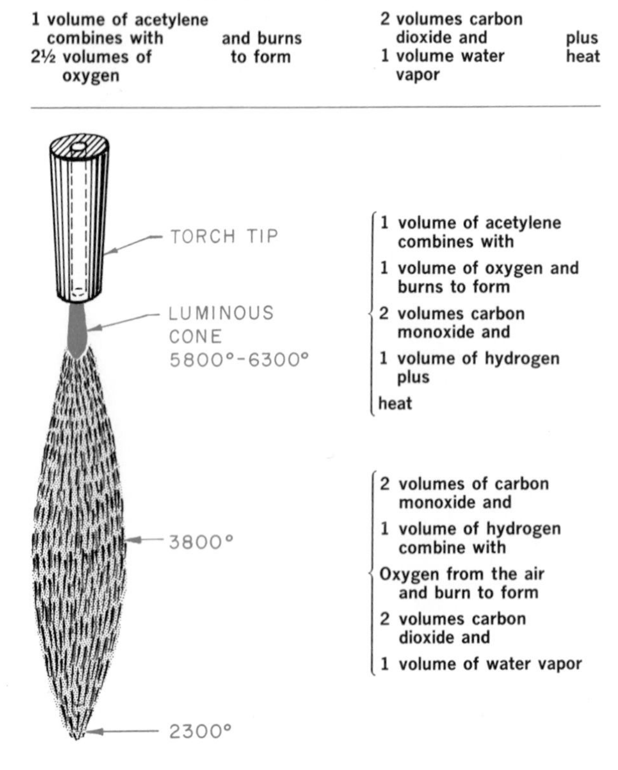

Torch Flame—Complete Reaction
1 volume of acetylene combines with 2½ volumes of oxygen
and burns to form
2 volumes carbon dioxide and 1 volume water vapor
plus heat
TORCH TIP
LUMINOUS CONE 5800°-6300°
3800°
2300°
1 volume of acetylene combines with
1 volume of oxygen and burns to form
2 volumes carbon monoxide and
1 volume of hydrogen plus
heat
2 volumes of carbon monoxide and
1 volume of hydrogen combine with
Oxygen from the air and burn to form
2 volumes carbon dioxide and
1 volume of water vapor

sparks are at a minimum. Good fusion results, and the weld pool may be advanced along the weld joint with little risk of adhesions.

The condition of the welding tip also helps in securing a quiet, even flame. The hole in the tip end must be absolutely round, and the internal passage of the tip must be free of dirt and metal particles. An out-of-round hole or foreign matter in the tip changes the direction of gas flow, causing a harsh flame with an imperfect inner cone.

Table 17-3

Tip size for steel plates of different thicknesses.

Tip size	[1] Diameter of hole inches	Thickness of plate	Average length of flame, inches	Approximate pressure of regulators		Approximate no. of cubic feet of gas used per hour		Diameter of rod, inches
				Acetylene	Oxygen	Oxygen	Acetylene	
1	0.037	22 ga. to 16 ga.	3/16	1	1	4.0	4.0	1/16
2	0.042	1/16″ to 1/8″	1/4	2	2	5.0	5.0	1/16 to 1/8
3	0.055	1/8″ to 3/16″	5/16	3	3	8.0	8.0	1/8
4	0.063	3/16″ to 5/16″	3/8	4	4	12.0	12.0	3/16
5	0.076	5/16″ to 7/16″	7/16	5	5	19.0	19.0	3/16
6	0.086	7/16″ to 5/8″	1/2	6	6	23.0	23.0	1/4
7	0.098	1/2″ to 3/4″	1/2	7	7	35.0	35.0	1/4
8	0.1065	5/8″ to 1″	9/16	8	8	48.0	48.0	1/4
9	0.116	1″ or over	5/8	9	9	57.0	57.0	1/4
10	0.140	Heavy duty	3/4	10	10	95.0	95.0	1/4
11	0.147	Heavy duty	7/8	10	10	100.0	100.0	1/4
12	0.149	Heavy duty	7/8	10	10	110.0	110.0	1/4

Modern Engineering Co.

[1] There is no standardization in tip sizes so that this table gives approximations only. The pressures are correct for the hole size indicated.

17-5. *This cylinder support dolly makes it easy to move cylinders about the shop. It can also be used at the welding station. The cylinder cannot fall over. It can be tipped to approximately 35 degrees and will return to vertical position.*

Penn-Chem Corp.

Procedure for Lighting the Torch

1. Keeping the tip facing downward, open the acetylene valve and light the gas with the friction lighter. The acetylene will burn with a smoky yellow flame and will give off quantities of fine black soot, Fig. 1, page 480B.

2. Open the oxygen valve slowly. The flame will gradually change in color from yellow to blue and will show the characteristics of the excess acetylene flame. There will be three distinct parts to the flame: (1) a brilliant but feathery edged inner cone, (2) a secondary cone, and (3) a bluish outer envelope of flame, Fig. 3, page 480B.

3. If you continue turning the oxygen valve, you will notice that the secondary cone gets smaller and smaller until it finally disappears completely. Just at this point of complete disappearance, the neutral flame is formed, Fig. 2, page 480B. If you continue to turn the oxygen valve, the flame will go past the neutral stage, and an excess oxygen (oxidizing) flame will be formed. The entire flame will decrease in size. The inner cone will become shorter and much bluer in color. It is usually more pointed than the neutral flame, Fig. 4, page 480B.

4. Practice adjusting the flame several times until you are familiar with all three flames and can adjust the neutral flame with precision.

Backfire

Improper operation of the welding torch or defective equipment may cause the flame to go out with a loud snap or pop. This is called *backfire*. The torch should be turned off at the valves. Check the equipment and the job to see which of the following causes may be at fault:

- *Operating the torch at lower pressure than that required for the tip size used.* The gases are flowing too slowly through the tip, and they are burning faster than the speed at which they flow out of the tip. Increasing the gas pressure will correct this.
- *Touching the tip to the work.* This smothers the flame and does not allow for combustion to take place. The welder should always keep the inner cone of the flame from touching the work.
- *The tip may become overheated.* This can result from overuse, from welding in a corner, or from being too close to the weld. The tip must be cooled before going ahead with the work. The job conditions should also be corrected.
- *A loose tip or head.* A loose tip may be the result of overheating, which causes expansion of the threaded joint. The obvious correction is to tighten the tip. The screw threads on the tip may also be crossed.
- *The inside of the tip may have carbon deposits or small metal particles inside the hole.* These particles become hot and cause preignition of the gases. Clean the tip out with standard tip cleaners.
- *The seat of the tip may have dirt on it, or it may have become nicked through careless handling.* Dirt and nicks prevent a positive seat at the joint where the tip is screwed into the stem. Gas escapes at this point, causing preignition. The dirt should be cleaned out. If the seat is nicked, it may be possible to reseat it with a seating tool. A new tip may have to be used.

Flashback

A *flashback* occurs when the flame burns back inside the torch and causes a shrill hissing or squealing. The oxygen torch valve should be closed at once and then the acetylene valve. The torch should be allowed to cool. Before relighting, blow oxygen through the torch tip for a few seconds to clear out any soot that may have formed in the passage.

If a flashback travels back into the hoses or the regulators, it may cause a fire inside the equipment. The parts affected must be repaired before they are used again.

Repeated flashbacks indicate that there is a serious defect in the equipment or it is being used incorrectly. The orifice, barrel, or mixing chamber may be clogged. Oxides may have formed, or gas pressures may be incorrect. Never use gas pressures higher than those recommended by the manufacturer of the equipment you are using.

CLOSING DOWN THE EQUIPMENT

When the welding practice is finished for the day, take down the equipment. If you are welding with a portable outfit on the job, use the following method.

1. Turn off the flame by first closing the acetylene valve on the torch. This will extinguish the flame. Then close the oxygen valve on the torch.
2. Close the oxygen and acetylene cylinder valves.
3. Open the acetylene and the oxygen valves on the torch to permit the trapped gases to pass out of the regulators and the hoses.
4. Release the regulator-adjusting screws by turning them to the left.
5. Disconnect the hoses from the torch.
6. Disconnect the hoses from the regulators.
7. Remove the regulators from the cylinders and replace the protective caps on the cylinders.

If you are working on a line system in the school shop, use the following procedure:

1. Turn off the flame as stated in step 1 above.
2. Close the oxygen and acetylene valves on the line system.
3. Open the acetylene and the oxygen valves on the torch to permit the trapped gases to pass out of the regulators and the hoses.
4. Release the regulator-adjusting screws by turning them to the left.
5. Close the torch valves and hang the torch on the gas economizer.

SAFETY

At this point it would be well to review a few important safety tips for gas welding and cutting.

- Wear goggles with the right filter lenses.
- Wear gauntlet gloves of heat-resistant leather or asbestos; keep them away from oil and grease.
- Don't wear oily or greasy clothes. Oil or grease, plus oxygen, will burn. Don't blow off your clothes or the work with oxygen. It's wasteful and dangerous.
- Woolen clothes are better than cotton. They burn less easily and protect better against heat.
- Wear fire-resistant apron, sleeves, and leggings when doing heavy cutting. Keep cuffs rolled down and pockets closed. Don't wear low oxfords.
- Oxygen should never be substituted for compressed air.

- Do not use matches for lighting the torch.
- When welding or cutting material containing, or coated with, lead, zinc, aluminum, mercury, cadmium, or beryllium, wear an air-supply respirator. Fluoride-type fluxes give off poisonous fumes. Avoid heavy smoke.
- Wear a respirator when welding galvanized iron, brass, or bronze in confined areas. Even if ventilation is good, the respirator is a good idea.
- Don't let your clothing become saturated with oxygen or air-rich in oxygen; you're chancing burns if you do.
- Don't work with equipment that you suspect is defective. Have leaking or damaged equipment repaired by a qualified person.
- Never use oil on, or around, oxygen regulators, cylinder connections, or torches. Keep your hands clean.
- Blow water or dust out of new welding hose with compressed air or oxygen—never acetylene.
- Never use oily compounds on hose connections.
- Never interchange hose connections. Standard oxygen hose is green; fittings have right-hand threads. Standard hose for acetylene is red; fittings have left-hand threads and a grooved nut. Don't force connections.
- Unnecessarily long hoses tend to kink and leak.
- Store hose on reels. Inspect hose every week for cracks, wear, and burns.
- Test hose for leaks at normal working pressure by submerging it in water. Test connections for leaks with soapy water.
- Do not hang the torch and hose on the regulators and cylinders.
- Cut off parts of hose burned by flashback and remake connections. Flashback weakens the inner wall.
- Don't repair hose with friction tape.

PRACTICE JOBS

Instructions for Completing Practice Jobs

Your instructor will assign appropriate practice in gas welding from the jobs listed in the Course Outline, pages 522, 523. Before you begin a job, study the specifications given in the Course Outline. Then turn to the pages indicated in the Text Reference column for that job and study the welding technique described. For example, the specifications for Job 14 are given in the Course Outline, page 522. According to the outline, you are to practice beading on a 1/8-inch mild steel plate. You are to use a type RG45 mild steel filler rod with 3/32-inch diameter. The weld is to be done in the overhead position. Information on the welding technique proper for this job is given on pages 508, 509.

LOW CARBON STEEL PLATE

It is common practice to limit the application of oxyacetylene welding of low carbon steel to steel plate with a maximum thickness of about 11 gauge (1/8 inch). The electric arc welding processes such as shielded metal-arc, gas metal-arc, and gas tungsten-arc are used for heavier plate. The oxyacetylene process is still used a good deal in maintenance and repair, for welding cast iron, braze-welding, brazing, and soft soldering. It is also used for the welding of small-diameter steel pipe.

It was pointed out earlier that the skills learned in the practice of oxyacetylene welding are the basis for learning techniques in the other welding processes. These gas welding practice jobs provide practice in the coordinated use of both hands. They also provide the means for observing the flow of molten metal and the appearance of fusion and penetration taking place. You are urged to develop the highest possible skill in torch manipulation and visual inspection.

Carrying a Puddle Without Filler Rod

The first steps in learning to weld are to get the "feel" of the torch and to observe the action of the flame on steel and the flow of molten metal. Probably the best way to accomplish these important steps is to practice carrying a puddle without melting-through.

RIPPLE WELDING, FLAT POSITION: JOB 17-J1. An important factor in carrying a pool of molten metal across the plate is the speed of travel of the flame. The weld should move forward only when the pool of molten metal is formed. If travel is too fast, the bead will be narrow, and you will lose the molten pool. Poor fusion may also result. If travel is too slow, the bead will become too wide, and it will melt through the plate. A puddle which has been moved at the proper rate will produce a weld with closely formed ripples and be of uniform width.

1. Check the Course Outline for plate thickness. Follow the manufacturer's recommendations for tip size and study, Table 17-3.
2. Adjust the torch for a neutral flame (Review pages 499, 500 as necessary.)
3. Grasp the torch like a pencil, Fig. 17-6. If it is more comfortable for you, you may

grip it like a fishing pole or hammer. The hose should be slack so that the torch is in balance. Hold the torch so that the tip is at an angle of 45 to 60 degrees with the plate.

4. Hold the torch in one spot so that the inner cone of the flame is about 1/8 inch above the surface of the plate until a molten pool of metal is formed which is about 3/16 to 3/8 inch in diameter.

5. Carry the molten pool across the plate, forming a bead of even width and even ripples. Be sure the weld pool is fluid so that adhesion does not result. Develop the technique illustrated in Figs. 17-7 and 17-8.

6. Examine the other side of the plate for even penetration. There must be no evidence of burn-through. Also note the effect of expansion and contraction on the plate.

7. Practice the job until you can run beads that are passed by the instructor. Acceptable beads are shown in Fig. 17-9A and D.

8. Experiment with carburizing and oxidizing flame adjustments. Notice the effect of each flame on the weld pool. These flames are shown in Figs. 2 and 4, page 480B.

Linde Division, Union Carbide Corp.

17-6. *Ripple welding. A light welding torch can be held like a pencil. A large torch is held like a hammer.*

RIPPLE WELDING, VERTICAL POSITION: JOB 17-J2. This job should be practiced with flame adjustment of different types and sizes. It is only through first-hand experience and observation that you will learn the amount of heat required to obtain a uniform, smooth weld.

1. Set the plate up in a vertical position. The direction of welding is from the bottom to the top.

2. Adjust the torch for a neutral flame.

3. Grip the torch like a hammer and point it in the direction of travel. As in welding in the flat position, the tip should be at an angle of 45 to 60 degrees with the plate.

4. Form a molten pool about 3/16 to 3/8 inch in diameter as in welding in the flat position.

5. After forming the pool, begin to travel up the plate. Use a short, side-to-side motion across the weld area and move up the plate with a semicircular motion. This technique provides equal heat distribution and uniform control of the molten metal.

6. When you find that the weld is getting too hot, pull the torch away so that the flame does not play directly on the weld pool. Begin welding again after the weld has cooled somewhat. If the puddle gets too hot, it will burn through or the weld pool will become so fluid that it runs down the face of the plate.

7. Practice this job until you can consistently produce beads that are uniform and have smooth ripples.

17-7. *Proper tip angle when welding in the flat position.*

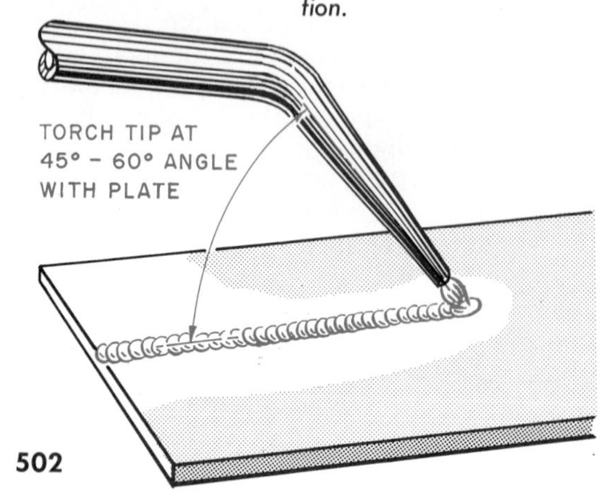

17-8. *The recommended torch motion when carrying a puddle on flat plate.*

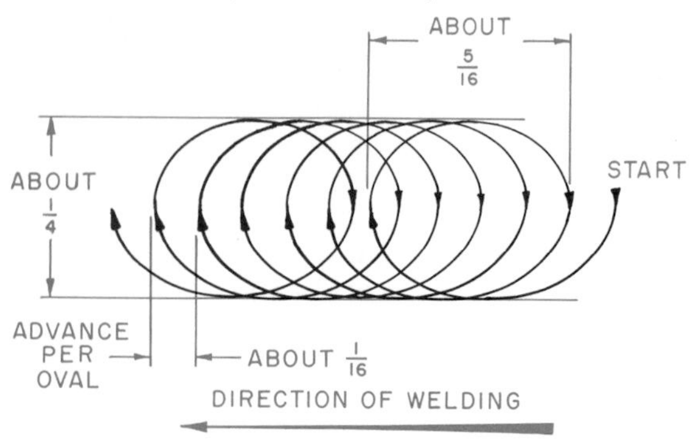

Making a Bead Weld With Filler Rod

Although edge and corner joints may be welded without filler rod, filler rod is necessary for most joints to build up a weld bead to the required strength. The independent but coordinated motion of both hands is necessary to weld with a filler rod. The welding torch is held in the right hand, and the filler rod is held in the left hand. Each hand has a specific function in the welding operation.

BEADING, FLAT POSITION: JOB 17-J3.

1. Check the Course Outline for the thickness of the plate, and size and type of filler rod. The rod size to use for a particular job usually depends upon the thickness of the plate. A general rule to follow is to use a rod diameter equal to the thickness of the metal being welded. Follow the manufacturer's recommendations for tip size.
2. Adjust the torch for a neutral flame.
3. Hold the torch so that the tip is at an angle of 45 to 60 degrees with the plate. Hold the filler rod in your left hand. Slant it away from the torch tip at an angle of about 45 degrees, Figs. 17-10 and 17-11.
4. Heat an area about $\frac{1}{2}$ inch in diameter to the melting temperature. Hold the end of the filler rod near the flame so that it will be heated while the plate is being heated.
5. When the molten pool has been formed, touch it with the tip of the filler rod. Always dip the rod in the center of the molten weld pool. Do not permit filler metal to drip from above the pool through the air. Oxygen in the air will attack the molten drop of filler metal and oxidize it.
6. Move the torch flame straight ahead. Rotate the torch

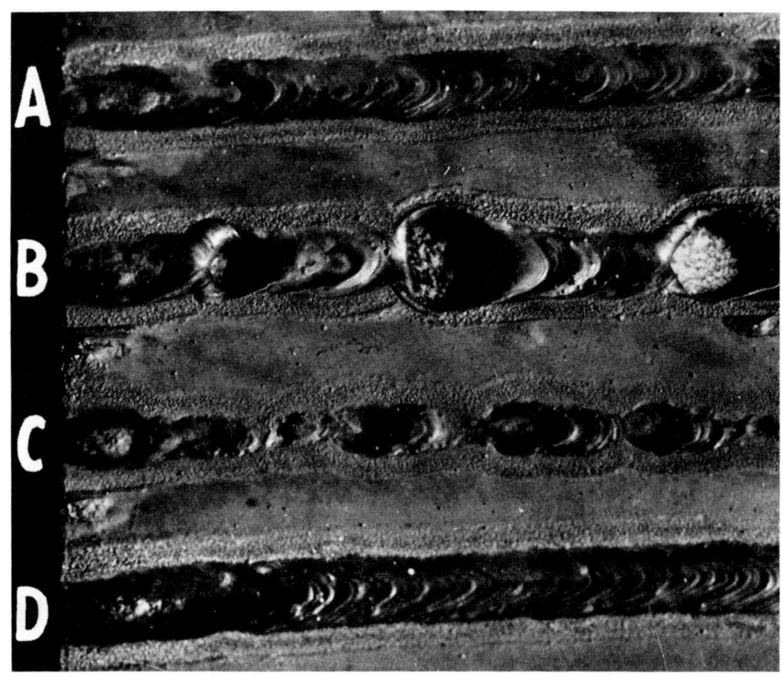

Linde Division, Union Carbide Corp.

17-9. *Correct welding technique will result in strips which look like A and D. Excess heat will cause holes to be burned into the sheet as in B. Insufficient heat and uneven torch movement will result in a strip like C.*

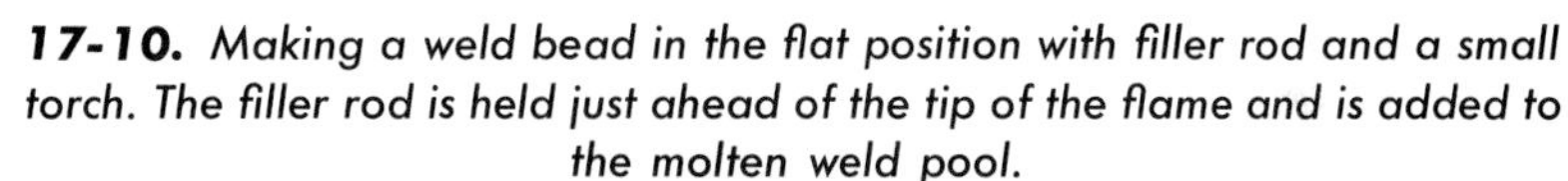

17-10. *Making a weld bead in the flat position with filler rod and a small torch. The filler rod is held just ahead of the tip of the flame and is added to the molten weld pool.*

to form overlapping ovals, Fig. 17-12. Add filler rod as needed as the molten pool moves forward. Raise and lower the rod, and add just enough filler metal to supply the metal necessary to build up the weld to the desired width and height. When the welding rod is not in the molten pool, hold the end in the outer envelope of the flame. Remember that the cone of the flame should not touch the molten weld pool. It should be about $\frac{1}{8}$ inch above the surface of the weld pool. In the beginning you will have trouble with the filler rod sticking. Either the weld pool is not hot enough, or the rod has been applied to a cold area outside of the molten pool. Release the rod by placing the flame directly on it.

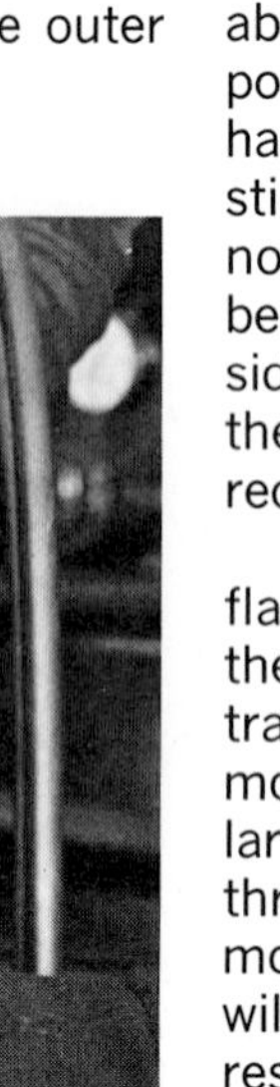

Department of Health, Education, and Welfare

17-11. *Torch and filler rod position when welding in the flat position with a large torch. The torch may also be held like a hammer.*

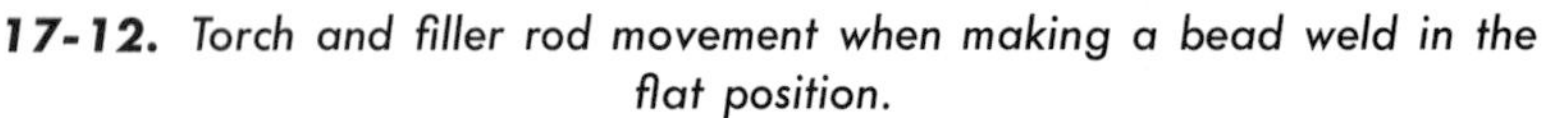

17-12. *Torch and filler rod movement when making a bead weld in the flat position.*

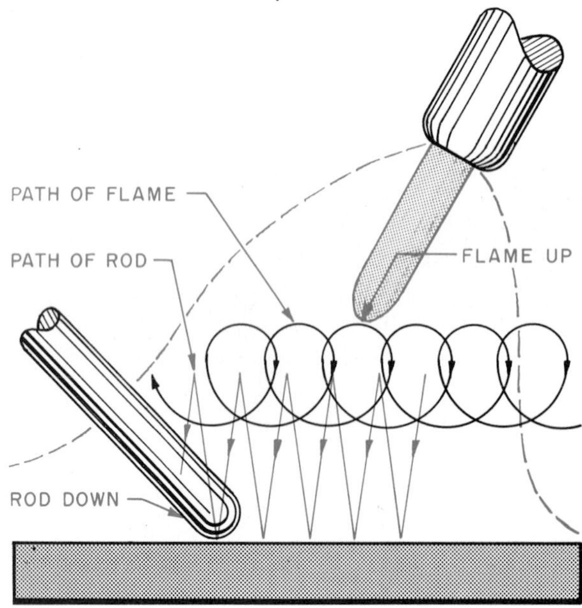

7. Practice with different flame adjustments. Also observe the effect of various speeds of travel. If the molten pool is moved too slowly, it will grow so large and fluid that it burns through, Fig. 17-13B. If it is moved too rapidly, the filler rod will be deposited on cold metal, resulting in a lack of fusion, Fig. 17-13C. The bead will also be very narrow. Make sure that the flame protects the molten weld pool so that it will not be contaminated by the surrounding air.

8. Continue practicing until the beads conform to the desirable characteristics shown in Figs. 17-13A and D and 17-14 and are acceptable to your instructor.

BEADING, VERTICAL POSITION: JOB 17-J4. Applying a bead weld in the vertical position is somewhat more difficult than welding in the flat position. After you have mastered the vertical position, however, you may find that it is your favorite position of welding. In this position you will learn how to control molten metal that has a tendency to run with the torch flame and filler rod. You must time the application of heat, the movement of the filler rod, and the movement upward to obtain a uniform weld bead.

1. Check the Course Outline for plate thickness and size and

type of filler rod. Follow the manufacturer's recommendations for tip size.

2. Adjust the torch for a neutral flame.

3. Place the plate in the vertical position. Grip the torch like a hammer and point it upward in the direction of travel at about a 45-degree angle.

4. Heat an area about $\frac{1}{2}$ inch in diameter at the bottom of the plate to the melting point. Place the tip of the filler rod near the torch flame so that it will become preheated while the plate is being heated.

5. When a molten pool has been formed, apply the tip of the filler rod to the forward edge of the molten pool of metal. Move the torch flame and filler rod away from and toward each other so that they cross at the center of the weld deposit, Figs. 17-15 and 17-16. The distance of cross movement should not exceed three times the diameter of the filler rod. A slight circular motion of the torch flame may be easier for you than a straight side-to-side movement. The rod can be used straight or bent as desired. A bend of about 45 to 90 degrees in the filler rod a few inches from the tip will enable you to keep your hand away from the heat and also move the rod freely. In the beginning, you may find the bent rod difficult to manipulate and may prefer to practice with the straight rod.

6. Weld evenly up the plate, manipulating the torch as instructed in step 5. Do not allow the pool to become too hot. If the molten metal becomes highly fluid, or if there is an excess of metal in the weld pool, it will spill out of the weld pool. If the weld pool seems to be getting too hot, raise the flame slightly so that it will not play directly upon the puddle and give it a chance to chill slightly. The flame can also be directed momentarily on the filler rod, but be careful not to burn it. You must not bring the rod to a white heat or molten condition unless you are ready to add rod to the weld pool. Add only enough rod to form the weld size desired.

7. Practice this job with flames of different types and sizes in order to observe their

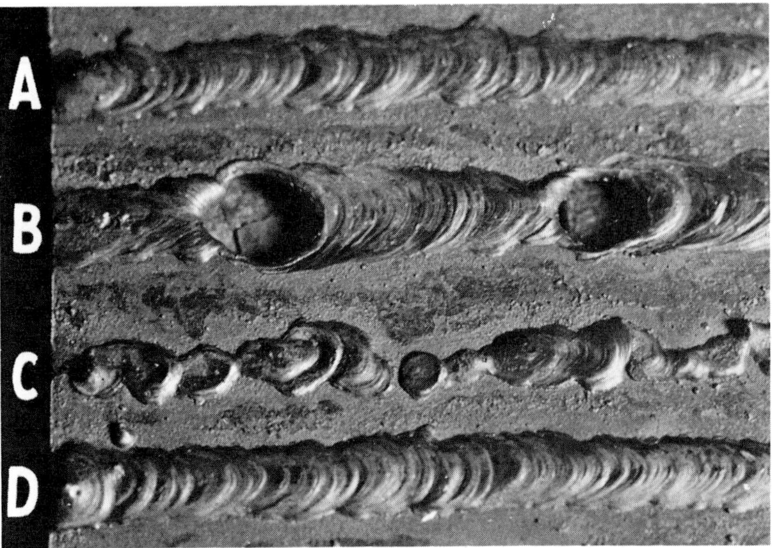

Linde Division, Union Carbide Corp.

17-13. *Welds A and D are satisfactory for beginners. Weld B shows overheating, and weld C indicates insufficient heating, little fusion, and improper melting of the filler rod.*

17-14. *Weld beads in ⅛-inch mild steel plate welded in the flat position. A no. 3 tip and a $\frac{3}{32}$-inch filler rod were used. The flame was neutral.*

effects on the molten pool of weld metal. Practice until the beads pass inspection. Satisfactory beads are shown in Fig. 17-17.

RIPPLE WELDING, HORIZONTAL POSITION: JOB 17-J5. Some trainees find welding in the horizontal position the most difficult of all positions to master. Effort to apply any material (such as plaster, paint, adhesive, weld metal, etc.) to a vertical surface requires careful manipulation of the material to keep it of uniform thickness. The tendency of the material is to sag toward the bottom. Close observation of the molten weld pool and careful torch movement will overcome this difficulty and make it possible to produce welds that have somewhat the same appearance as welds made in the flat, vertical, or overhead position.

1. Check the Course Outline for plate thickness. Select the proper tip size.
2. Adjust the torch for a neutral flame.
3. Place the plate in a vertical position. Welding should be from right to left. Hold the torch as shown in Fig. 17-18.
4. Heat the plate until a molten pool is formed. Carry the pool forward with a semicircular torch motion. There will be a tendency for the pool to sag on the lower edge. This can be prevented by directing the flame upward and forward. It is important that the weld pool does not become too liquid, since this will cause the pool to sag at the lower edge.
5. Experiment with different flame types and sizes.
6. Practice this job until you can produce weld beads that are of uniform width and have smooth ripples.

BEADING, HORIZONTAL POSITION: JOB 17-J6.

1. Check the Course Outline for plate thickness and type and size of filler rod. Follow the manufacturer's recommendations for tip size.
2. Adjust the torch for a neutral flame.
3. Place the plate in a vertical position. Welding progresses from right to left. Hold the torch and rod as shown in Fig. 17-18.
4. After the usual molten pool has been formed, carry the pool forward with a semicircular torch motion. As the weld bead moves along, the bead will sag and build up on the lower edge. To correct this, direct the flame upward and in the direction of welding while adding filler rod to the upper edge of the weld pool.
5. Experiment as before with different flame types and sizes. Practice until you can do satisfactory work. Good quality beads are shown in Fig. 17-19.

RIPPLE WELDING, OVERHEAD POSITION: JOB 17-J7. Most welding students will find welding in the overhead position no more difficult than welding in the vertical and horizontal positions. It will take some time to get used to the position in which your arms and body are placed. You may find it very tiresome to hold

17-15. Manipulation of the torch and rod for beading in the vertical position.

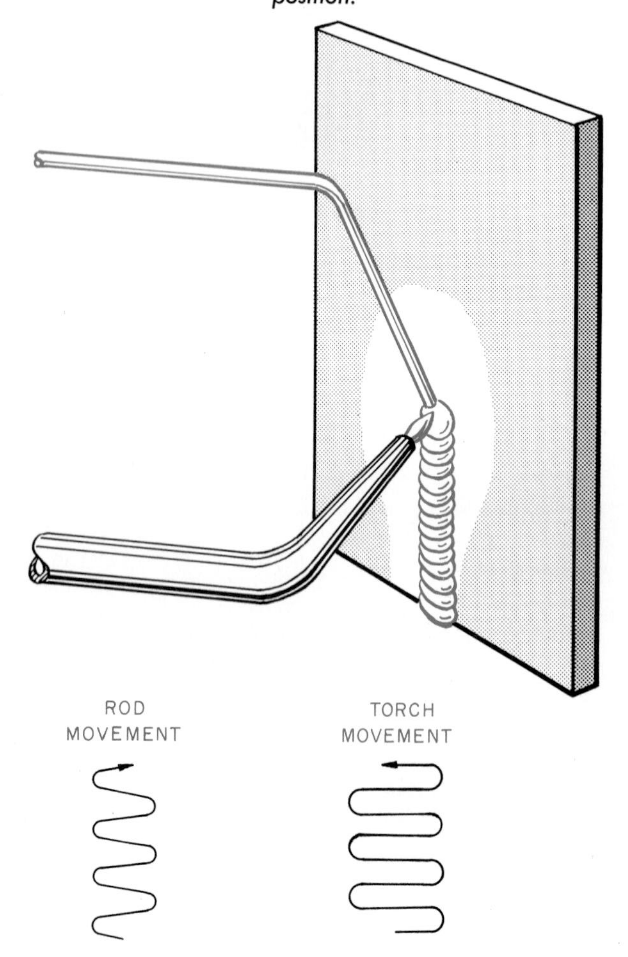

your torch arm above your head. Grip your torch lightly and be as relaxed as possible. If the torch is held too tightly, you will not have free movement of the torch and you will tire rapidly.

1. Check the Course Outline for plate thickness. Select the proper tip size.

2. Adjust the torch for a neutral flame.

3. Place the plate in the overhead position. Welding progresses from right to left. Hold the torch as shown in Fig. 17-20.

4. Form the weld pool as before. The side-to-side torch movement will determine the width of the ripples. The weld pool will not fall as long as it is not permitted to become too fluid and too large. The torch should also be directed in the direction of welding at an angle of about 45°.

5. Experiment with various flame types and sizes. Observe the effect that different flames have on the molten pool of metal.

6. Check the ripple appearance carefully. It should be smooth and of even width the entire length of the ripple weld.

BEADING, OVERHEAD POSITION: JOB 17-J8. In doing this job you will be required to hold both arms above your head. Both hands will need to be coordinated in their movement overhead and this will probably be a little difficult for a short time. You will also experience some difficulty getting used to the position in which your arms and body are placed. Again you are cautioned to grip the torch and welding rod lightly and to be as relaxed as possible. If you are tense, you will not be able to respond readily to the demands of the weld bead.

1. Check the Course Outline for plate thickness and type and

17-16. *Welding in the vertical position.*

17-17. *Weld beads in ⅛-inch mild steel plate welded in the vertical position. A no. 3 tip and a 3/32-inch high-test filler rod were used. The flame was neutral.*

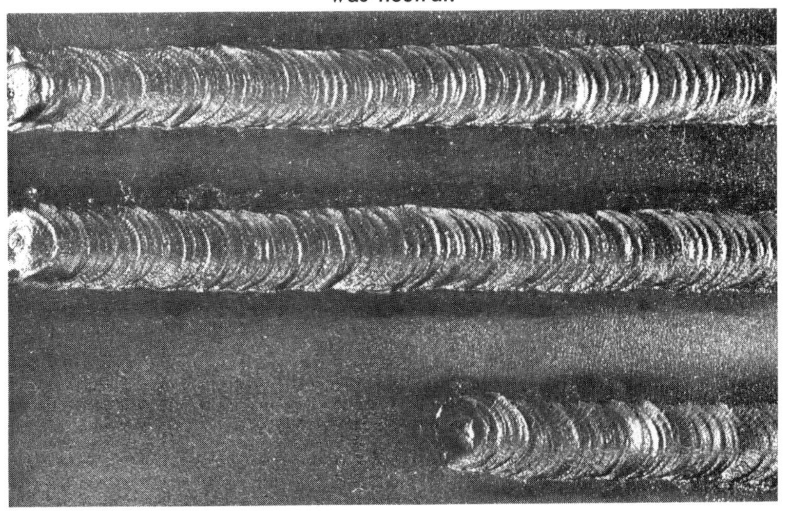

size of filler rod. Follow the manufacturer's recommendations for tip size.

2. Adjust the torch for a neutral flame.

3. Place the plates in the overhead position. The torch and the filler rod have the same relation to each other as in other positions except that they are overhead, Fig. 17-20.

17-18. *Positions of the torch and rod for horizontal welding. The direction of travel is from right to left.*

4. Start the weld pool as before. Control the filler rod and the size of the pool. The weld pool will not fall as long as it is not allowed to get too large. The weld pool must be kept shallow and under control. Do not permit it to become highly fluid. If the pool is kept fairly small and not allowed to form a drop, you will have little difficulty. Add filler rod sparingly. Too much filler rod will cause the weld metal to drip. Make sure that the hot filler rod and the weld pool are protected by the torch flame at all times to prevent oxidation.

5. Continue practice until the beads have the characteristics shown in Fig. 17-21 and are acceptable to your instructor.

Welding Edge and Corner Joints: Jobs 17-J9 through J16

Edge joints are sometimes referred to as *flange joints*. The flange is formed by turning over the edge of the sheet to a height equal to the thickness of the plate. Flange, edge, and corner joints are used a great deal on sheet metal less than 16 gauge in thickness. These joints may be welded without filler rod by fusing the two plates together. Filler rod should be used, however, when maximum strength is desired.

1. Check the Course Outline for plate thickness and type and size of filler rod. Follow the manufacturer's recommendations for tip size.

2. Set up and tack the plates about every 2 inches.

3. Adjust the torch for a neutral flame.

4. Welding will be in the flat position. The positions of the torch and filler rod are similar to those used for welding flat plate in the flat position. Figure 17-22 shows the torch without the filler rod. Note that the flame should be directed at the inside edges of the plate.

5. Form the molten puddle and move it along the seam. Make sure that the edge surfaces are molten and that you are melting through to the root of the weld. The corner joint should show penetration through to the back side. Do not add too much filler rod to the weld. These joints do not require much reinforcement on the bead. Move the torch with a slight crosswise or circular motion over the surface of the joint. If the movement is

17-20. *Welding in the overhead position on steel plate. A small torch and steel filler rod are used.*

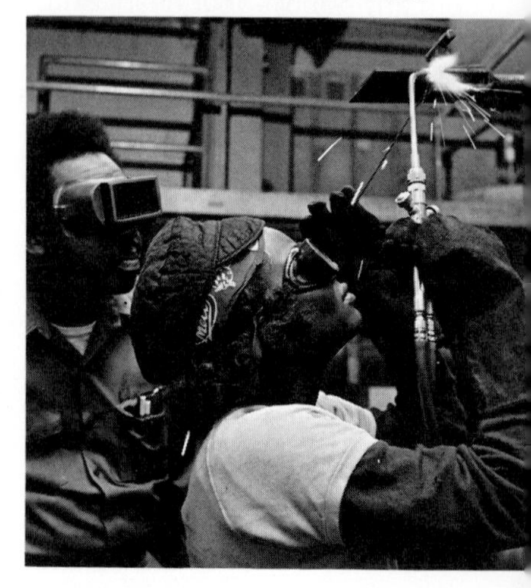

17-19. *Weld beads in 1/8-inch mild steel plate welded in the horizontal position. A no. 3 tip and a 3/32-inch mild steel filler rod were used. The flame was neutral.*

too wide and too much filler rod is added, the weld metal will spill over the sides of the joint. Add the filler rod to the center or forward edge of the molten weld pool. Make sure that the pool is protected at all times with the torch flame to prevent the formation of oxides and nitrides.

6. Continue practice until the welds have the characteristics shown in Figs. 17-23 and 17-24 and are acceptable to your instructor.

7. Test the welds by bending the pieces flat so that the back side of the weld is under tension. After the sheet has been bent over as far as you can hammer, place the joint in a vise and squeeze the sheets on each side of the weld down on top of each other. Inspect the underside of the weld bead carefully. The weld should not crack or break.

8. After you are able to make welds of good appearance and high strength in the flat position, practice in the other positions.

9. Weld a test joint in each of the positions being practiced. Weld one edge only and pry the plates apart to check penetration and fusion at the inside corners of both plates and along the edge surface.

Welding a Square Butt Joint: Jobs 17-J17 and J18

Square butt joints require no edge preparation. For this reason the joint is seldom used for material more than 12 gauge in thickness. In order to secure 100-percent penetration in thicker stock, welding must be done from both sides.

1. Check the Course Outline for plate thickness and type and size of filler rod. Follow the manufacturer's recommendations for tip size.

2. Adjust the torch for a neutral flame.

3. Space the plates so that penetration can be secured through the root of the joint and so that expansion can take place during welding. The plates will draw together if not spaced. They should be spaced about 1/16 inch apart at the starting end and about 1/8 inch apart at the other end, Fig. 17-25.

4. Tack the weld at each end and in the center. Be sure that the edge of each plate is molten before you add filler rod to bridge the gap.

5. Play the flame on the first tack weld and remelt a small puddle on its surface. This puddle should be increased in size by adding a limited amount of filler rod. Be sure that the pool is molten before adding the filler rod.

6. Move the molten pool along the open seam from right to left with a small circular motion. As the flame moves from side to side, it should play directly first on the edge of one sheet and then on the edge of the other so that they are brought to the state of fusion at the same time. Preheat the rod as you move along. The weld pool must not be too fluid, or you will burn a hole in the plate, and

17-21. Weld beads in 1/8-inch mild steel plate welded in the overhead position. A no. 3 tip and a 3/32-inch, high-test filler were used. The flame was neutral.

17-22. Position of the torch when making an edge weld in the flat position.

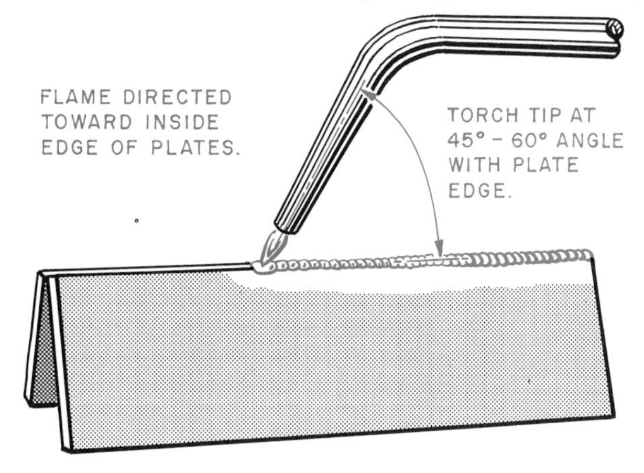

excessive weld metal will hang underneath. Keep the following important points in mind as you weld.

- The joint crack should be the center of the molten weld bead.
- The plate edges must be completely fused, and penetration must extend through to the root.
- Weld reinforcement should not be higher than $\frac{1}{16}$ inch. It should be of uniform height and width and have smooth, even ripples.

7. Continue practice until the welds conform to the characteristics described in step 6 and are acceptable to your instructor. A good weld will show a slight bead on the back side which indicates that complete penetration and fusion have taken place, Fig. 17-26. You should not be able to see the edges of the sheet metal on the back side.

8. To test your welds, clamp the sheet in a vise so that the weld is parallel to the jaws of the vise and about $\frac{1}{8}$ inch above the jaws. The back side of the weld should be towards you. Then hammer the top sheet away from you so that the root of the weld is under stress. After the sheet has been bent over as far as you can hammer it, place the joint in the vise and squeeze the sheets on each side of the weld until they are back to back, Fig. 17-27. The weld should not crack or break. If it cracks or breaks, you will probably notice that the plate

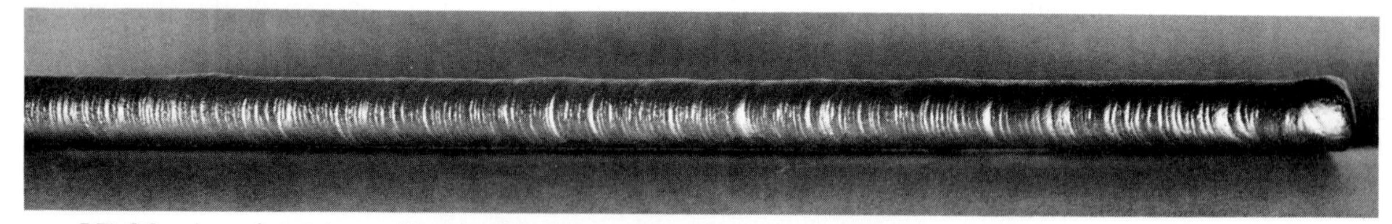

17-23. *An edge joint welded in the flat position. Note the even, smooth weld. It is made in ⅛-inch steel plate. A no. 3 tip and a 3/32-inch filler rod were used. The flame was neutral.*

17-24. *An edge joint welded in the vertical-up position. It is made in ⅛-inch steel plate. A no. 3 tip and a 3/32-inch filler rod were used. The flame was neutral.*

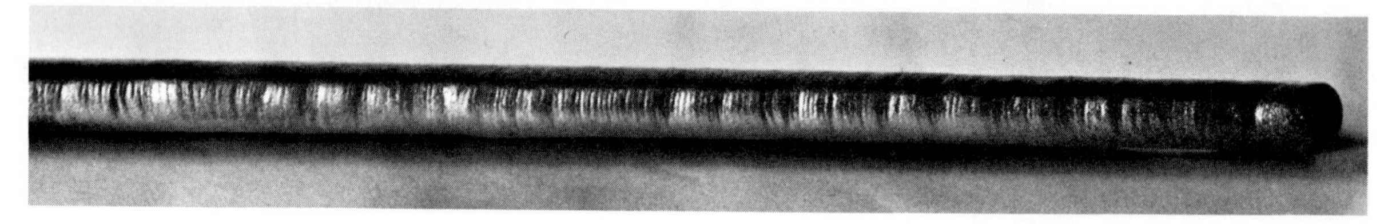

17-25. *Proper spacing of the plates for a butt weld.*

Linde Division, Union Carbide Corp.

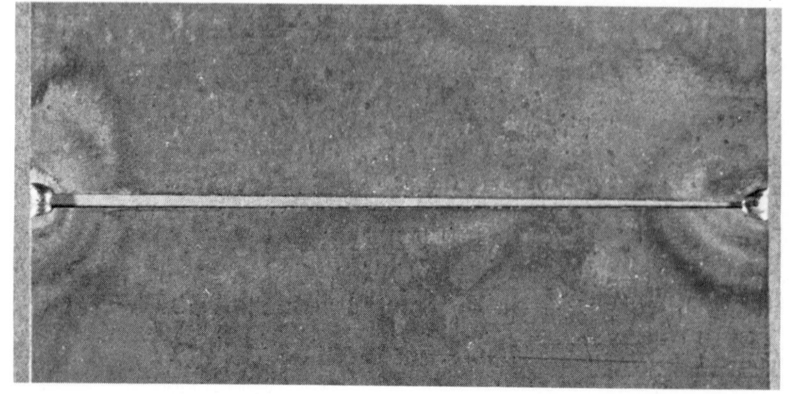

17-26. *Root penetration through to the back side of a butt joint is essential for welds of maximum strength.*

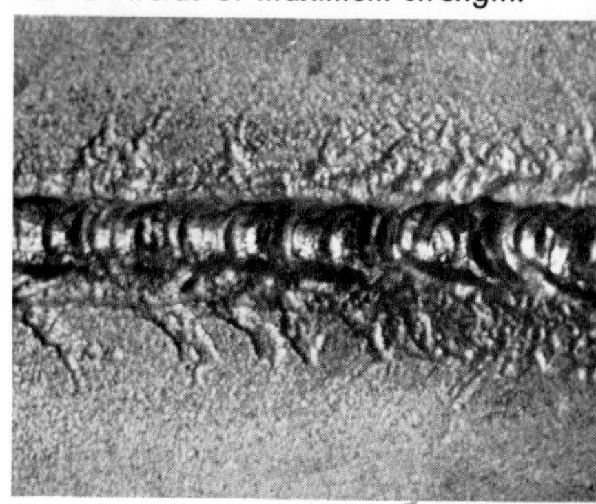

edge was not fused and that it is in its original square-cut condition. This indicates that the plate edges were not brought to a molten condition and that the filler rod was shoved through a cold seam.

9. After you are able to make welds of good appearance and good quality in the flat position, practice in the other positions. Note the vertical butt weld shown in Fig. 17-28.

Welding a Lap Joint: Jobs 17-J19 through J22

The lap joint is made with a fillet weld. It is one of the most difficult to weld with the oxy-acetylene process because it requires careful attention to the distribution of heat on all surfaces to be welded. The top sheet may be so hot that it burns away while the bottom sheet may not be hot enough. This adds to the difficulty of obtaining fusion at the root of the joint. If the joint is welded from one side only, the weld is under shearing stress, and the joint cannot reach maximum strength. Welding from both sides strengthens the joint, but this is not always possible on the job.

1. Check the Course Outline for plate thickness and type and size of filler rod. Follow the manufacturer's recommendations for tip size.

2. Adjust the torch for a neutral flame.

3. Set up the plates in the flat position with one plate overlapping the other.

4. Tack weld all four corners of the two plates.

5. Weld from right to left. Position the torch and filler rod as shown in Fig. 17-29. You will find that you will have to vary the position of both the torch and the filler rod depending upon the condition of the upper plate edge and the molten pool. Move the torch with a slight semicircular motion. The edge of the upper plate will have a tendency to melt before the root of the joint and the surface of the lower plate have had a chance to melt. This can be overcome by directing the flame to the lower plate and also

17-27. A butt weld in sheet metal is tested by bending the piece completely over in a vise. A good weld will bend from 90 to 180 degrees without failure. In this view both the base metal and the weld have withstood maximum bending.

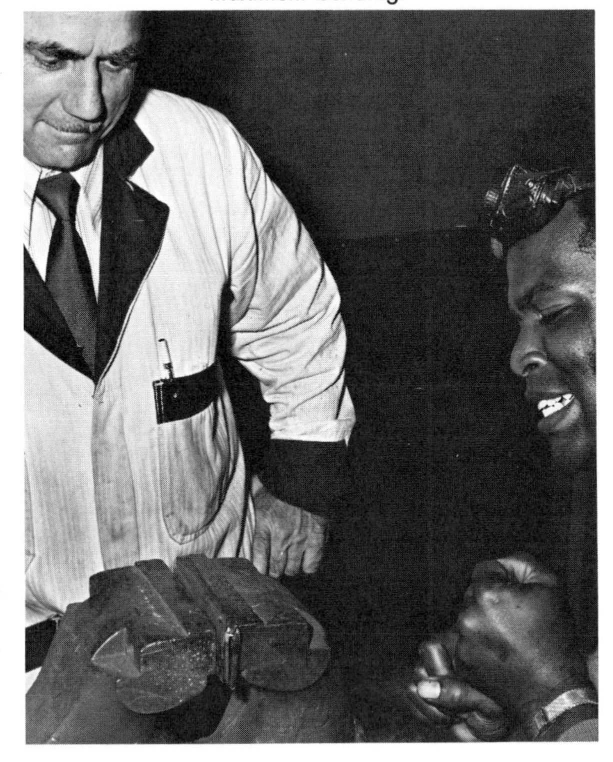

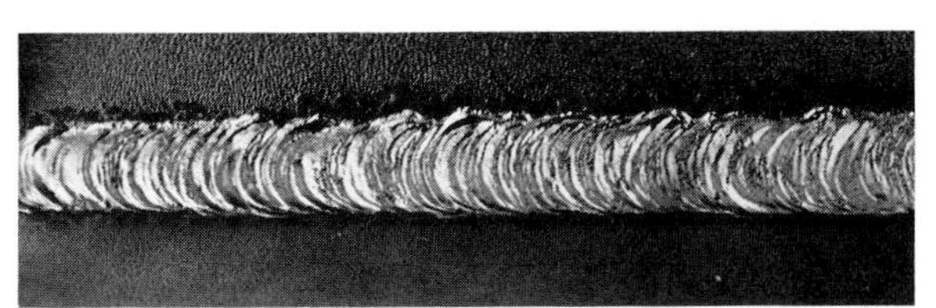

17-28. A vertical butt weld in 1/8-inch plate. A no. 3 tip and a 3/32-inch, high-test filler rod were used. The flame was neutral.

17-29. Positions of the torch and filler rod when making a lap weld in flat position. The upper plate edge is protected by the filler rod.

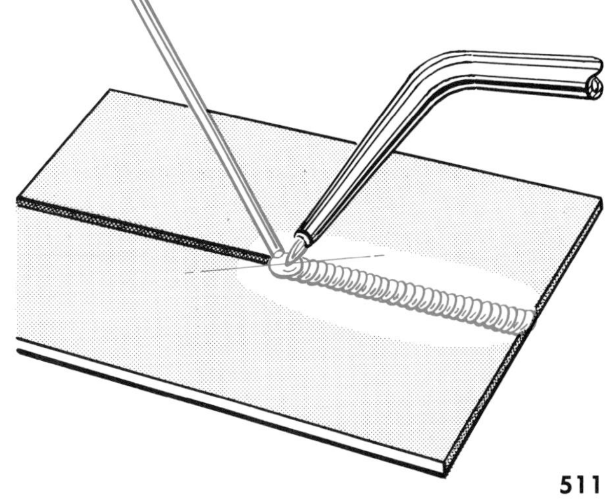

by protecting the upper plate with the welding rod as shown in Fig. 17-29. The rod absorbs some of the heat to prevent excessive melting of the top plate. Torch movement must be uniform and not too wide, or the completed weld will be too wide, and the appearance will be rough. The weld face should be slightly convex.

6. Continue practice until the welds have the characteristics shown in Fig. 17-30 and are acceptable to your instructor.

7. To test your weld, place it in a vise, Fig. 17-31, and pry the two sheets apart. Bend the upper sheet backward until it forms a *T* with the bottom plate. Inspect the root of the weld for fusion and complete penetration. The weld should be thoroughly fused to the bottom plate and not pull away.

8. After you are able to make welds of good appearance and of good quality in the flat position, practice in the other positions.

Welding a T-Joint: Jobs 17-J23 through J25

A T-joint also requires fillet welds. In welding a T-joint from both sides, you will have to apply more heat than for the other joints and manipulate the filler rod more. It is a good joint to practice in preparation for the welding of small diameter pipe with the oxyacetylene process.

1. Check the Course Outline for plate thickness and type and size of filler rod. Follow the manufacturer's recommendations for tip size.

2. Adjust the torch for a neutral flame.

3. Set up the plates and tack weld them at each end of the joint.

4. Place the joint on the table so that welding will be in the horizontal position.

5. The torch and filler rod should be held in about the same position as that used when welding a lap joint, Fig. 17-32.

6. Weld from right to left. Pay careful attention to flame control. There is a tendency for the vertical plate to melt before the flat plate. This causes undercutting along the edge of the weld on the vertical plate. Direct more heat toward the bottom plate and toward the corner of the joint in order to secure fusion at the root. Be careful that you

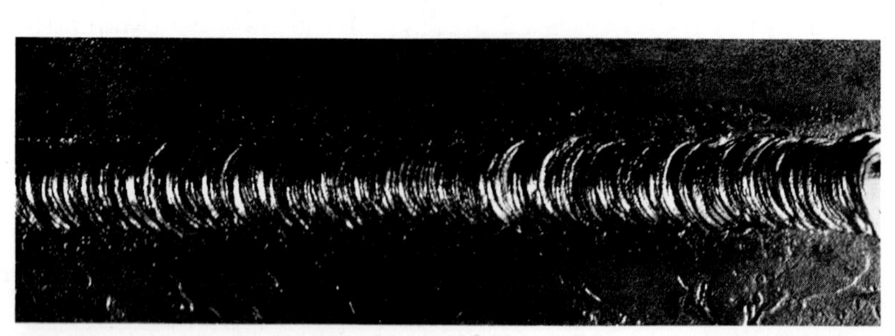

17-30. *A lap joint welded in the vertical-up position in ⅛-inch steel plate. A no. 3 tip and 3/32-inch mild steel filler rod were used. The flame was neutral.*

17-31. *To test a fillet weld on a lap joint, pry the sheet apart and inspect the fusion and penetration.*

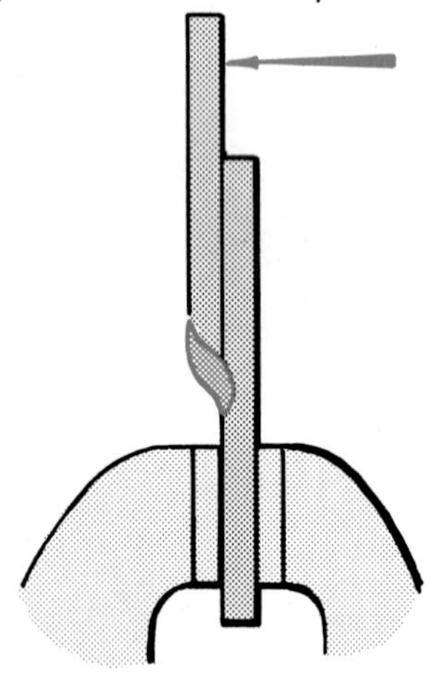

17-32. *Positions of the torch and filler rod when welding a T-joint in the horizontal position.*

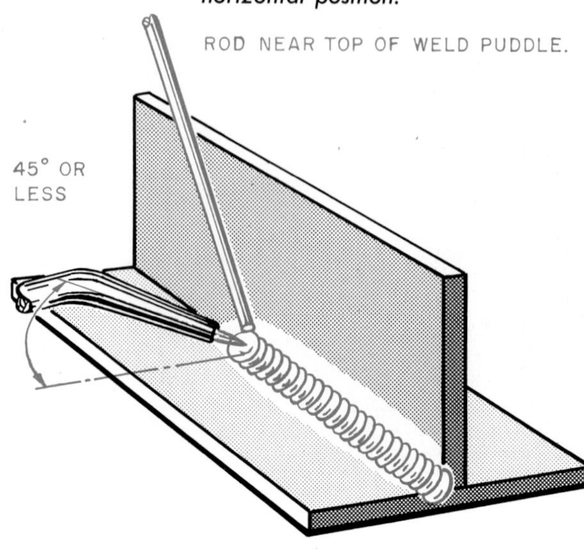

do not form a pocket in the corner and thus overheat the tip. Move the torch back and forth slightly or in a semicircular movement as required for proper heat distribution. Feed the filler rod just above center at the forward edge of the molten pool so that it comes between the flame and the vertical plate. Thus the filler rod absorbs some heat and prevents undercut in the vertical plate. The weld bead should be equally divided between the two plates and have a slightly convex face.

7. Clean the scale produced by the first weld from the back side of the joint. This can be done by moving the flame rapidly back and forth along the joint so that the oxide scale expands and pops from the surface.

8. Weld the joint from both sides. You will find the second side somewhat harder than the first side because of the weld on the other side. There is more material to bring to the welding temperature. The welding technique is different because the bottom plate has a tendency to melt before the root of the weld and the vertical plate. You must direct more heat toward the root of the weld and the vertical plate than on the flat plate. Move the weld pool slowly along the line of weld and pay careful attention to the state of fusion at the root of the joint. Bridging from plate to plate is a common fault because root penetration is difficult.

9. Continue practice until the welds have the characteristics shown in Fig. 17-33 and are acceptable to your instructor.

10. Make up a test joint welded only from one side. Bend the vertical plate toward the face of the weld. Examine the root of the weld for penetration and fusion. The weld should be thoroughly fused to the flat plate and should not peel away. Inspect the edge of the vertical plate. It should show that fusion has taken place. The raw, square-cut edges should not be intact.

11. After you are able to make welds of good appearance and of good quality in the flat position, practice in the other positions. A T-joint welded in the vertical-up position is shown in Fig. 17-34.

LIGHT GAUGE SHEET STEEL

Whether a steel sheet is called "sheet steel" or "steel plate" depends upon the thickness of the steel. There is not total agreement in the industry regarding the dividing line. For all practical purposes, all rolled materials less than $\frac{1}{8}$ inch in thickness are known as *sheet*, while materials $\frac{1}{8}$ inch or heavier are referred to as *plate*. These designations do not only apply to steel, but also include aluminum, brass, copper, and other metals.

A great deal of gas welding is done on light gauge sheet steel. Large industries that specialize in sheet metal fabrications, the aircraft and space industries, and the automotive industry do a considerable amount of gas welding. Light gauge gas welding has wide application in auto body repair.

17-34. *A fillet weld in a T-joint welded in the vertical position travel up. Stock is ⅛-inch plate. A no. 4 tip and ⅛-inch mild steel filler rod were used. The flame was neutral.*

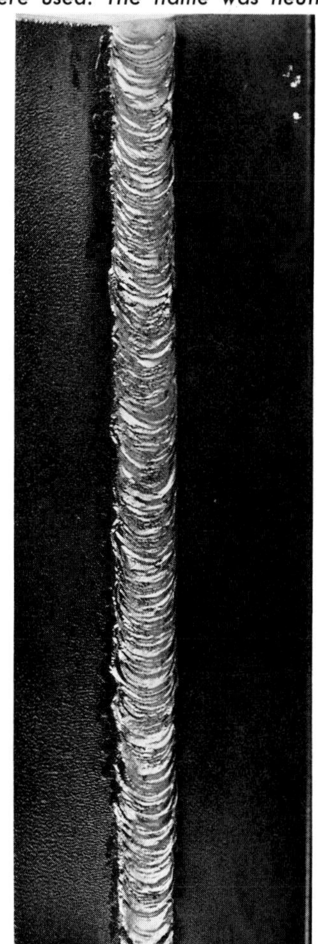

17-33. *A fillet weld in a T-joint welded in the horizontal position. Stock is ⅛-inch plate. A no. 4 tip and ⅛-inch mild steel filler rod were used. The flame was neutral.*

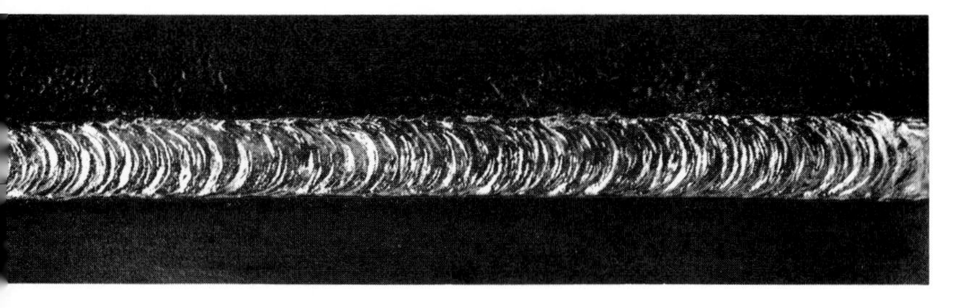

Light Gauge Sheet Steel Welding Practice: Jobs 17-J26 through J36

The jobs in this series should be welded with an aircraft-type welding torch. The torch should be held like a pencil if you are sitting down or like a hammer if you are standing up. A relaxed position while welding is less tiring for you and produces the best quality work.

The major concern in making these welds is to use the proper tip size so that the flame is not too hot and the speed of travel is not too slow. Correct flame temperature and speed of travel help to prevent burn-through, a defect that is always possible when welding light gauge materials.

When making fillet welds in lap and T-joints in light gauge sheet metal, proper feed of the filler rod is most important in order to avoid plate burn-away and undercut. The top plate of a lap joint and the vertical plate of a T-joint must be protected by the filler rod.

A neutral flame adjustment is critical. An oxidizing flame burns the metal and causes excessive penetration and oxidation. A large shower of sparks indicates that the flame is oxidizing.

Check the Course Outline regarding the material to be welded and the type and size of filler rod. Practice on all types of joints and in all positions. Use the same welding technique as that used in the welding of ⅛-inch plate. Since burning through is a constant problem in welding light gauge sheet metal, the correct tip size and flame adjustment is critical.

Inspect the completed bead for all the characteristics of a good weld. The weld bead should be small and narrow. The weld reinforcement should be of uniform width and height and have the smooth, close ripple that was obtained in welding ⅛-inch plate.

Make a test weld of a butt, lap, and T-joint. Use the same testing procedure outlined for plate welding. Compare root penetration with Fig. 17-26, page 510 and face appearance with the photo of the particular joint in ⅛-inch plate in the previous section.

HEAVY STEEL PLATE AND PIPE

Heavy Plate Welding Practice: Jobs 17-J37 through J44

These jobs will give you the opportunity to practice oxyacetylene welding on heavier steel. You will gain experience in handling a greater amount of heat and larger pools of molten metal.

Practice Jobs 17-J37–J44 on heavy plate with high test welding rod before actual practice on pipe. Practice in heavy plate welding will reduce the number of hours in pipe practice necessary.

All of the welding techniques learned while completing the gas welding jobs on ⅛-inch plate apply to heavier plate. Your principal concern in welding heavy plate will be that the plate surface is melted and that the weld pool is fluid so that fusion, not adhesion, can take place. In making a V-butt weld, make sure that the plate edges at the bottom of the V are in a molten state when the filler rod is applied. There must be fusion through to the back side. Make sure that the tip is the proper size so that the flame is large enough to provide enough heat to melt the plate. A flame that is too large for the tip size will be harsh and cause the completed weld to be rough. Proper regulator pressure is also important to insure a soft flame.

Backhand Welding of Plate

Up to this point all of your welds were made with the forehand method of welding. The direction of travel was from right to left, and the flame and filler rod were in front of the completed weld bead. The backhand method is another technique which has certain advantages over the forehand method, especially when welding heavy plate and pipe.

In the welding of plate by the backhand method, the welding progresses from left to right. The flame is directed backward at the partially completed weld. The welding rod is held between the partially completed weld and the flame, Figs. 17-35 and 17-36. This permits the flame to be directed at all times on the surfaces of the material to be welded and ahead of the weld pool with little motion of the torch and filler rod. In welding beveled pipe or plate, a narrower angle may be used than for forehand welding. This speeds up the welding operation. The surface appearance of a backhand weld is not as smooth as that of a forehand weld. The ripples are coarse, and there are fewer of them per unit of weld length. The penetration bead on the back side at the root of the weld should look about the same as in forehand welding. Welders usually find it easier to get complete penetration through the back side of a V-butt joint with the backhand method.

Pipe Welding Practice: Jobs 17-J45 through J49

Pipe welding was first done with the oxyacetylene process. This was the only process used for a number of years both in construction and overland pipeline. For the most part, pipe now is welded with the shielded metal-arc, gas metal-arc, or gas tungsten-arc processes. You will have an opportunity to practice with these other processes in the later chapters of this text. Oxyacetylene welding is now generally confined to small diameter piping and field installations where there is no power source for electric welding. Oxyacetylene welding of small pipe up to 3 inches in diameter is faster than with the electric arc. Welding may be done in all positions with a single or multipass procedure and either the forehand or backhand technique. The choice of procedure and welding technique varies with the type of joint, the diameter and wall thickness of the pipe, the position of welding, and the size of the filler rod. The number of weld passes necessary for a particular pipe joint should be determined by the rule of one weld pass for each $\frac{1}{8}$-inch thickness of pipe wall over $\frac{3}{16}$-inch.

Study Fig. 17-37, which presents standard pipe joints.

On each pipe fabrication job, the welding procedure to be followed in order to comply with the ASA Code for Pressure Piping and the ASME Boiler and Pressure Vessel Code is prescribed. The procedure includes the following:

- Welding process to be used
- The nature of the pipe material
- The type of filler metal to be used in welding
- The position of welding
- The preparation of the material, whether by machining, flame cutting, chipping or grinding, or a combination of these
- Use or nonuse of backup rings. The welder must qualify for each.
- Size of the welding tip and oxygen and acetylene pressure
- The nature of the flame
- Method of welding, whether forehand or backhand
- The number of layers of welding for the various pipe thicknesses
- The cleaning process between each layer
- Treatment of weld defects
- Heat treatment or stress relieving if required

Prepare the pipe for welding according to the following procedure.

1. Check the Course Outline for size and type of pipe and filler rod.
2. Before making a butt weld on pipe that is over $\frac{1}{8}$ inch thick, bevel the ends of each pipe at an angle of 30 to $37\frac{1}{2}$ degrees to form a 60- to 75-degree included angle. Space the pipe at the root a distance of $\frac{3}{32}$ to $\frac{1}{8}$ inch, Fig. 17-38. Pipe which has a wall thickness less than $\frac{1}{8}$ inch may be successfully welded without beveling by spacing the pipe ends apart a distance equal to the wall thickness and fusing the pipe ends ahead of the weld deposit. This is a similiar technique to that practiced in welding square butt welds.
3. Adjust the torch for a neutral flame. Some high test filler rods may be run with a slight carburizing flame.
4. Place two sections of beveled pipe on a supporting jig like

17-35. The backhand method of welding from left to right. The flame precedes the welding rod, but the weld advances from left to right instead of from right to left. Note rod manipulation.

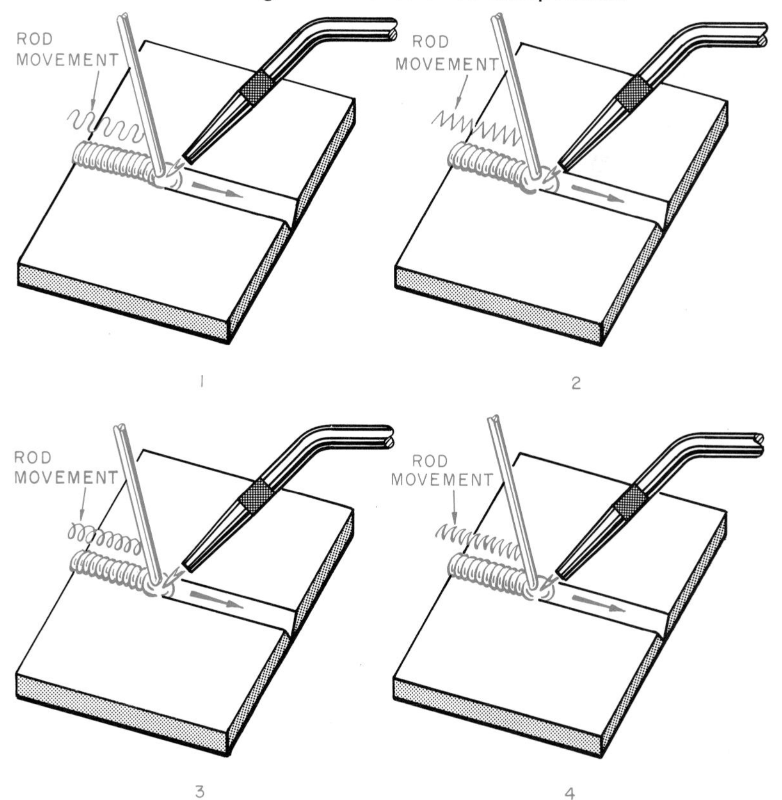

that shown in Fig. 17-39. Space the beveled edges at least $\frac{1}{8}$ inch all the way around.

5. Make a tack weld on the top of the pipe. Make sure that the tack is fused to the surface of each pipe bevel and deep into the root of the joint.

6. When the tack weld has cooled to a black heat, roll the pipe around and make another tack weld on the opposite side. Before making this tack, make sure that the spacing is equal all the way around the pipe.

7. Make tack welds at the quarter points as instructed in steps 5 and 6 so that the joint is held by four equally spaced tack welds.

ROLL POSITIONS: JOB 17-J45. Even though you are able to gas weld in all positions as a result of the practice you have had up to this point, it is recommended that your first pipe practice be a roll weld. Welding around the circumference of a section of stationary pipe requires torch and rod angles that are changing constantly, and this is not like any of the other positions of welding.

1. Place the tacked pipe sections in the channel iron jig used for tack welding.

2. Start the weld between the 2 and 3 o'clock positions. Hold the torch so that it is pointed toward the top of the pipe and along the line of the joint at about a 45-degree angle. This is somewhat similar to the position used in the vertical position of welding on plate.

3. Begin the weld by heating an area slightly wider than the width of the bevel. Add the filler rod when the bevel surfaces of both pipe ends have been brought to the molten state. The weld should be built up to the size desired before moving on. Move the torch in a short, semicircular movement, and the rod with a short, crosswise motion. The torch motion and the rod motion should cross at the center of the joint. A tack weld should be thoroughly fused with the advancing weld pool. Make sure that fusion is obtained at the root of the joint and on the bevel surfaces. Penetration must be through to the back side, and

17-37. *Pipe joints.*

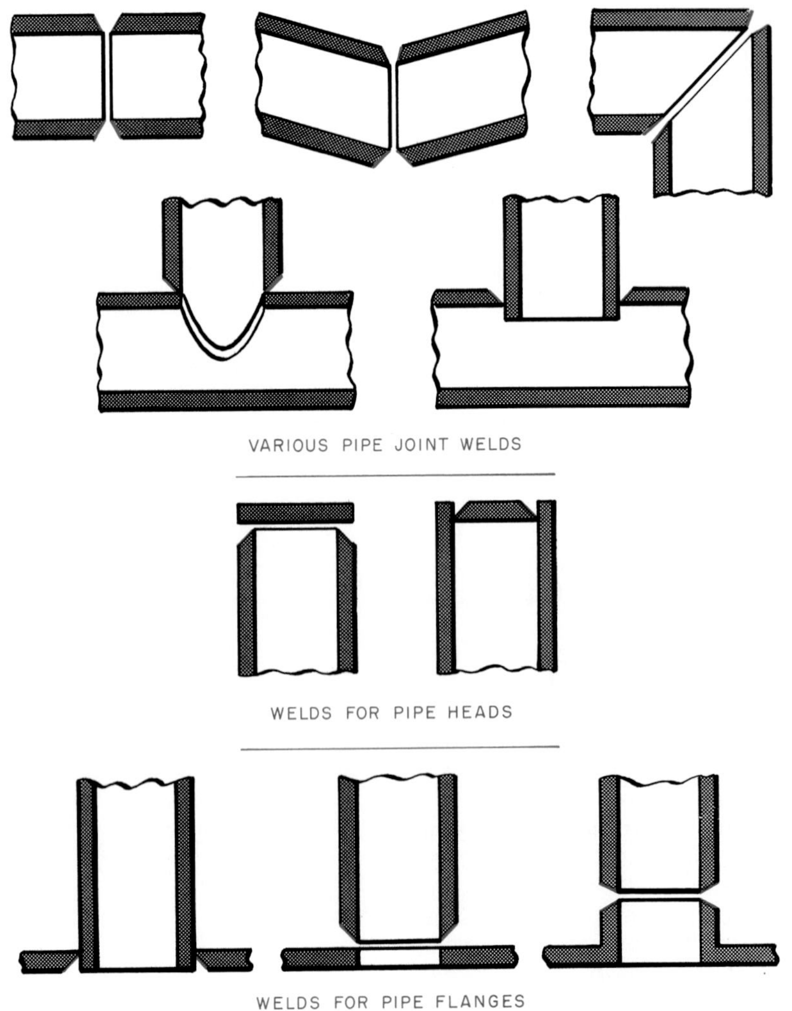

17-36. *Welding a butt joint in the overhead position, using the backhand welding technique.*

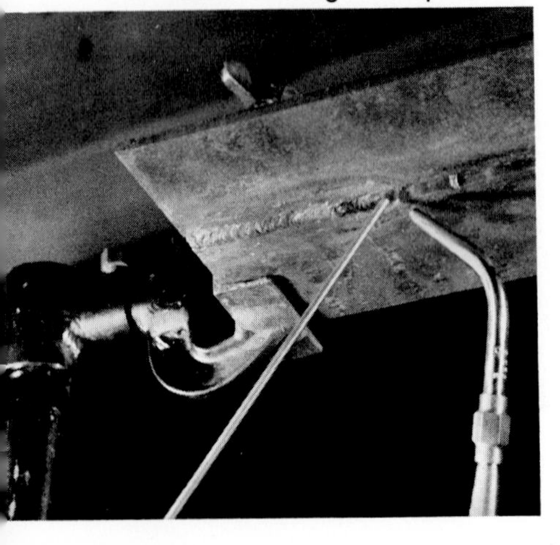

a small bead should be formed on the inside of the pipe.

4. Carry the weld up on the pipe to a point slightly below the 12 o'clock position at top center. Give the pipe a one-quarter turn toward you and begin a new weld at the end of the previous weld.

5. Repeat the procedure until the entire pipe is welded. The pipe is welded in quarter sections. Rotate each quarter section toward you as it is completed, Fig. 17-40. Make sure that fusion is obtained at the root of the weld and on the bevel surfaces. The weld must penetrate the back side, and a small bead should be formed on the inside of the pipe.

6. Continue practice until the welds have the appearance shown in Fig. 17-41 and the characteristics described in step 8, page 518.

7. Make up a test weld and remove coupons as shown in Figs. 23-39 and 23-30, page 742. Subject them to face- and root-bend tests as described under Root-, Face-, and Side-Bend Soundness Tests, pages 380, 381. Figure 17-42 shows the typical appearance of larger pipe specimens after testing.

17-38. *Students beveling the edge of a pipe with a standard beveling machine.*

HORIZONTAL FIXED POSITION (FOREHAND TECHNIQUE); JOB 17-J46.

1. Tack the pipe joint in the same manner as directed for roll welding. Place it in the horizontal fixed position.

2. Start welding at the bottom of the pipe between two tack welds at the 5 or 7 o'clock position, depending upon the side of the pipe you are welding from. Use the welding technique described in step 3 for the roll positions, page 516. Welding is in the overhead position. The pool must not become too fluid, and just enough filler rod must be added to form the bead size desired.

3. As the weld progresses upward, change the technique to that learned in making vertical welds. The weld pool must not be too fluid, and control of the pool must be maintained through proper torch and rod manipulation. The force and pressure of the torch flame is used to keep the weld pool in place. When you find that the weld is getting too hot, pull the torch away so that the flame does not play directly on the weld pool. Begin welding again after the weld has cooled somewhat. Careful use of the filler rod is necessary.

4. As you approach the top of the joint, the welding technique will merge into that used for welding in the flat position. Travel will be a little faster, and you must be careful not to burn through.

5. After you have reached a point a little past center on the top, return to the bottom of the pipe. Concentrate the flame on the end of the previously completed weld and melt the front surface so that good fusion and penetration is secured at this point. The new weld bead must be thoroughly fused with the previously laid bead.

6. Carry the welding puddle up to the top of the pipe. Use the same welding technique employed on the first side. Make sure that all tack welds are melted and become a part of the weld pool.

7. When you are about $\frac{1}{4}$ inch away from the end of the completed weld at the top of the

17-39. *Tack welding a beveled-pipe joint.*

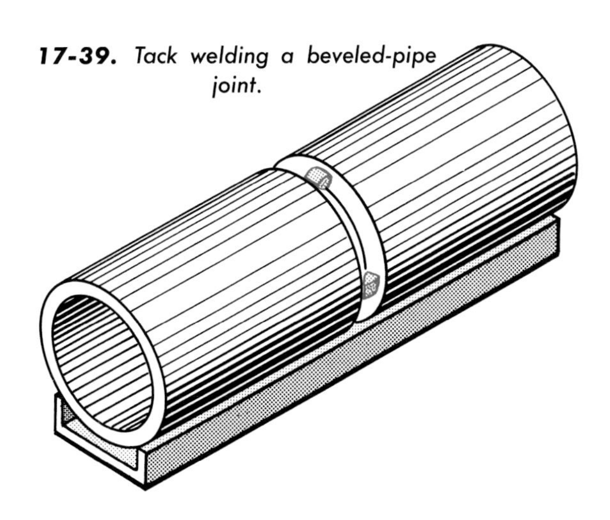

pipe, concentrate the flame on the face of the completed weld and at the bottom of the *V* in order to establish a weld pool that will join the two welds.

8. Inspect the weld. The face of the weld should be only about $\frac{1}{8}$ inch wider on each side than the width of the bevel shoulders. Weld reinforcement should be about $\frac{3}{32}$ inch high. It should be of uniform width and height without high or low areas. Ripples should be smooth and uniform. It should be very difficult to find the spot at the top and bottom where the welding started and ended. The edges of the weld should be of even width and parallel all around the pipe with no undercut. The edges should flow into the pipe surfaces as if molded.

9. Continue practice until the welds have the appearance shown in Fig. 17-41, the characteristics described in step 8 above, and are acceptable to your instructor.

10. Make a test weld and cut it up into face and root bevel test coupons. Use the testing procedure described in step 7 in the procedure for welding pipe in the roll position, page 517.

Backhand Welding of Pipe

It has already been pointed out that the backhand method of welding for heavy plate provides the advantage of increased speed of welding along with a decrease in the amount of filler rod and gases consumed. This technique was improved when Linde developed its backhand welding method for pipe. Practice with this process increases the student's skill in the use of the gas welding torch. Backhand pipe welding is based on the principle that hot steel will absorb carbon, which lowers the melting point. Thus less heat is required for fusion so that welding speed increases.

The flame is adjusted for excess acetylene or carburizing. The acetylene feather is about $1\frac{1}{2}$ times the length of the inner cone. When the steel is heated, it picks up carbon from the excess acetylene feather, thus lowering the melting point of the base metal at the surface. This action takes place to a slight depth only since carbon cannot penetrate very deeply into the steel within the time required to complete the weld.

The lowering of the melting point means that a thin layer of metal at the surface is ready for fusion sooner than it would be if a neutral flame were being used. The result is that thorough fusion between the base metal and weld metal is obtained without deep melting. During the actual welding, the carbon that was absorbed by the surface of the base metal becomes distributed through the weld deposit. Larger flames and welding rods can be used. The V-angle is smaller, and

17-41. *A butt weld in pipe. Note the complete fusion and penetration at the leading point of the weld. Note also the size in relation to the width of the joint.*

Linde Division, Union Carbide Corp.

17-40. *Follow this sequence for roll pipe welding with a gas torch.*

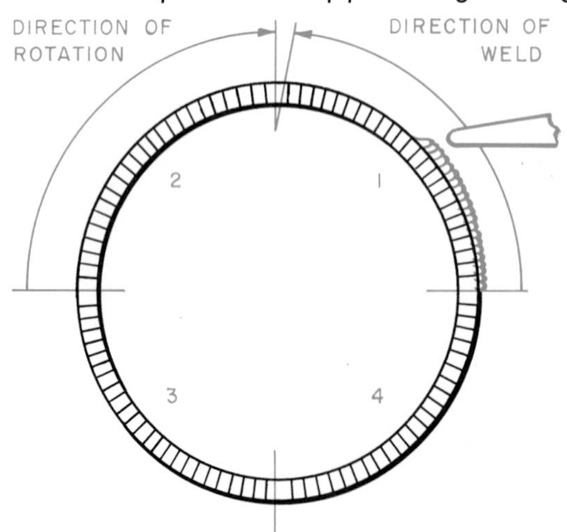

the root of the weld can be fused readily without serious burn-through.

VERTICAL WELDS IN PIPE IN THE HORIZONTAL FIXED POSITION: JOB 17-J48 (BACKHAND TECHNIQUE).

1. Check the Course Outline for pipe size and type and size of filler rod. Follow the manufacturer's recommendations for tip size.

2. Adjust the torch for a neutral or carburizing flame, depending on the method of welding you wish to practice and the type of filler rod used.

3. Tack weld two pieces of beveled pipe and set them up in the horizontal fixed position.

4. Start the weld at the top of the joint at about the 12 o'clock position and carry it down one side to the bottom. Do not start the weld at a tack weld. Hold the torch and filler rod as shown in Fig. 17-43. Note the backhand position of the torch. The method of depositing the bead is shown in Fig. 17-44. Advance the filler rod with a circular motion, and at the same time move the flame back and forth in the direction of travel. Time these movements so that the rod and flame part and then come together like an accordion. When you reach the 3 o'clock position, you will be welding in the vertical position. The pressure of the burning gases, the flame, and careful rod manipulation will keep the weld pool from running out, Fig. 17-45. As you approach the 6 o'clock position, you will be welding in the overhead position. The most important factor in puddle control in this position is to prevent the formation of a drop of metal. The metal will not fall until a drop is formed. The cohesiveness of the weld pool will

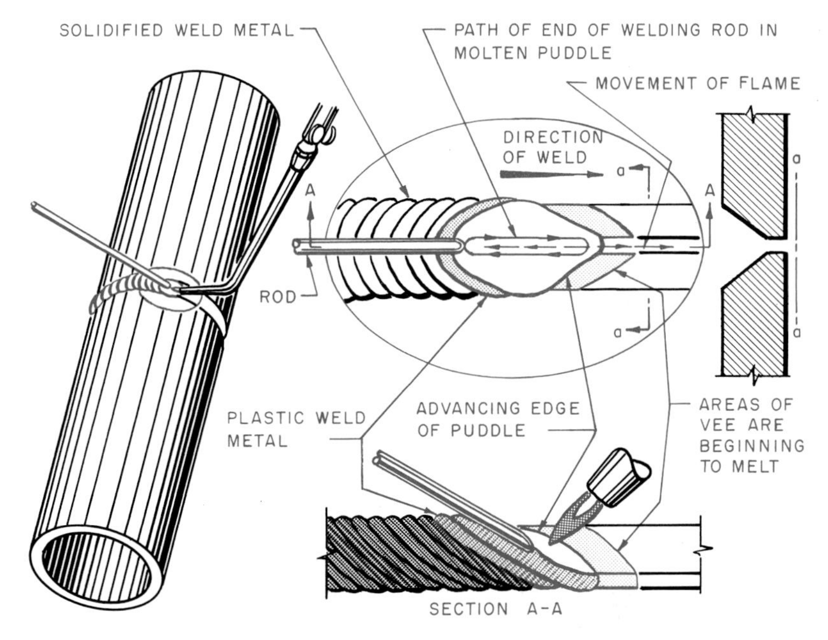

17-43. *The essential elements in the backhand method for pipe welding.*

17-42. *A good weld in low-carbon steel will bend at least 90 degrees without cracking or fracturing. These specimens have been bent double with no sign of failure.*

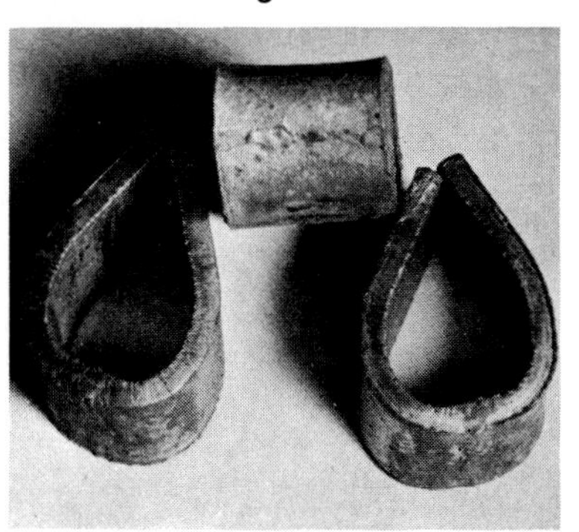

17-44. *This detail of Fig. 17-43 demonstrates the motion of flame and rod in backhand welding.*

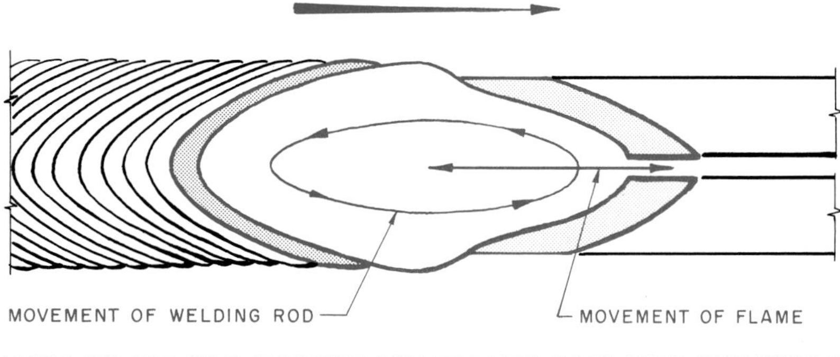

also assist in keeping the molten metal from sagging and dripping.

5. Restart the weld at the top and carry it down the opposite side to meet the first weld at the bottom of the joint. Repeat the welding technique practiced on the second side. The changes in torch position for control of the pool are summarized in Fig. 17-46.

6. Continue practice until the welds have the desirable weld appearance shown in Fig. 17-42, the characteristics described in step 8, page 518, and are acceptable to your instructor.

7. Weld a test joint in the same position and following the same welding procedure used for this practice job. Take standard face- and root-bend coupons from the test joint and subject them to bending tests.

HORIZONTAL WELDS IN PLATE AND PIPE IN THE VERTICAL FIXED POSITION: (BACKHAND TECHNIQUE) JOBS 17-J47 and 17-J49. Horizontal welding is the most difficult of all gas welding positions and becoming proficient will take a great deal of practice. In order to save time and cut the cost of materials, you are asked to practice J-47 on plate. After you have mastered welding on plate, you will find that welding pipe in this position is relatively easy.

1. Check the Course Outline for plate or pipe size and type and size of filler rod. Follow the manufacturer's recommendations for tip size.

2. Adjust the torch for a neutral or carburizing flame, depending on the method of welding you wish to practice.

3. Tack weld two pieces of beveled plate or pipe and set them up in the vertical fixed position.

4. In making this weld, travel from left to right. The flame is ahead of the filler rod and points back at the completed weld. Apply heat to the lower side of the *V*. The weld has a tendency to build up on the lower panel of the joint, and a shelf must be provided on the lower bevel to support the puddle. The backhand motion of the torch and filler rod causes the molten weld metal to assume a diagonal shape. The completed weld has coarse diagonal ripples that are not close together, Fig. 17-47. The support that is supplied by the built-up metal on the lower bevel makes it possible to keep the puddle somewhat larger than in a vertical weld and fairly molten. This permits rapid welding. Manipulate the filler rod more than in forehand welding and use a circular motion to keep the welding puddle spread out rather thin and to prevent too much metal sagging at the lower side of the weld. In multilayer welding, starting and stopping points of beads are staggered alternately with successive beads.

5. Continue practice until the welds conform to the desirable weld appearance shown in Fig. 17-42, the characteristics described in step 8, page 518, and are acceptable to your instructor. Make sure that you have good penetration and fusion on the inside of the pipe.

6. Weld a test joint in the same position and following the same welding procedure which you used for this practice job. Take standard face- and root-bend coupons from the test joint and subject the coupons to bending tests.

***17-45.** In making a position weld in pipe, part of the weld is done in the vertical position. In the backhand technique, welding in the vertical position can be done from top to bottom. The path of the end of the welding rod in the molten puddle is changed to keep the puddle spread out. The puddle itself is kept small.*

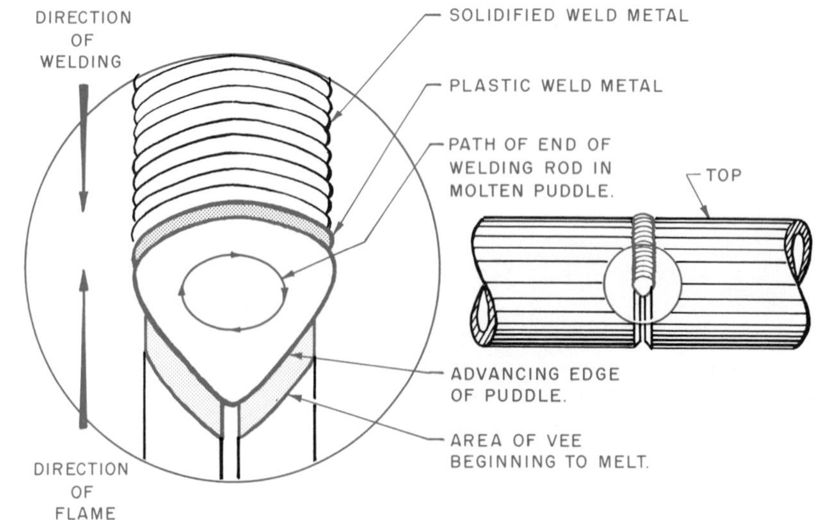

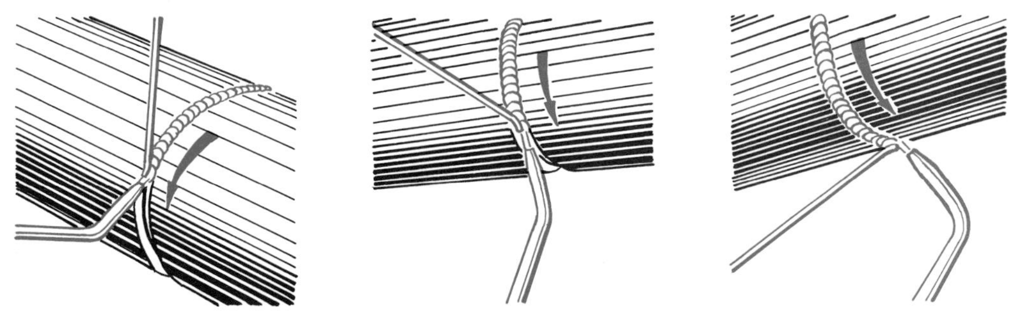

17-46. *The angle of the torch flame is changed as a position pipe weld is made so that as much advantage as possible can be taken of the pressure of the burning gases in controlling the puddle.*

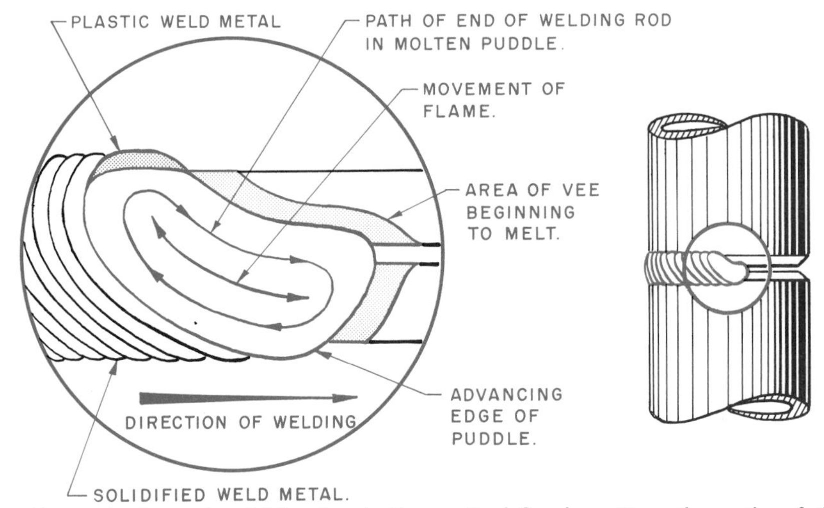

17-47. *In making a horizontal weld in pipe in the vertical fixed position, the paths of the torch and the welding rod tip are altered to assist in the control of the puddle.*

REVIEW QUESTIONS

1. Most of the welding being done in industry at the present time is gas welding. True. False.
2. Give a brief explanation of the following:
   a. fusion
   b. penetration
   c. weld reinforcement
   d. undercutting
3. Name the three stages of the oxyacetylene flame.
4. What is the excess acetylene flame called?
5. Braze welding and bronze surfacing are usually performed with what type of flame?
6. What tool is used to light the gas welding torch?
7. Name at least four causes of torch back-fire.
8. What is a flashback?

(Continued on page 523)

Table 17-4
Course Outline: Gas Welding Practice: Jobs 17-J1–49

Recommended job order[1]	Number in text	Joint	Operation	Material		Filler rod		Work position	Text reference
				Type	Thickness	Type	Size		
1st	17-J1	Flat plate	Ripple welding	Mild steel	1/8	None		Flat	501
2nd	17-J2	Flat plate	Ripple welding	Mild steel	1/8	None		Vertical	502
3rd	17-J3	Flat plate	Beading	Mild steel	1/8	Mild steel RG45	3/32	Flat	503
4th	17-J9	Edge joint	Flange welding	Mild steel	1/8	Mild steel RG45	3/32	Flat	508
5th	17-J10	Corner joint	Groove welding	Mild steel	1/8	Mild steel RG45	3/32	Flat	508
6th	17-J4	Flat plate	Beading	Mild steel	1/8	Mild steel RG45	3/32	Vertical	504
7th	17-J11	Edge joint	Flange welding	Mild steel	1/8	Mild steel RG45	3/32	Vertical	508
8th	17-J12	Corner joint	Groove welding	Mild steel	1/8	Mild steel RG45	3/32	Vertical	508
9th	17-J5	Flat plate	Ripple welding	Mild steel	1/8	None		Horizontal	506
10th	17-J6	Flat plate	Beading	Mild steel	1/8	Mild steel RG45	3/32	Horizontal	506
11th	17-J13	Edge joint	Flange welding	Mild steel	1/8	Mild steel RG45	3/32	Horizontal	508
12th	17-J14	Corner joint	Groove welding	Mild steel	1/8	Mild steel RG45	3/32	Horizontal	508
13th	17-J7	Flat plate	Ripple welding	Mild steel	1/8	None	3/32	Overhead	506
14th	17-J8	Flat plate	Beading	Mild steel	1/8	Mild steel RG45	3/32	Overhead	507
15th	17-J15	Edge joint	Flange welding	Mild steel	1/8	Mild steel RG45	3/32	Overhead	508
16th	17-J16	Corner joint	Groove welding	Mild steel	1/8	Mild steel RG45	3/32	Overhead	508
17th	17-J19	Lap joint	Fillet welding	Mild steel	1/8	Mild steel RG45	3/32	Flat	511
18th	17-J20	Lap joint	Fillet welding	Mild steel	1/8	Mild steel RG45	3/32	Vertical	511
19th	17-J17	Square butt joint	Groove welding	Mild steel	1/8	Mild steel RG45	3/32	Flat	509
20th	17-J18	Square butt joint	Groove welding	Mild steel	1/8	Mild steel RG45	3/32	Vertical	509
21st	17-J21	Lap joint	Fillet welding	Mild steel	1/8	Mild steel RG45	3/32	Horizontal	511
22nd	17-J23	T-joint	Fillet welding	Mild steel	1/8	Mild steel RG45	3/32	Horizontal	512
23rd	17-J22	Lap joint	Fillet welding	Mild steel	1/8	Mild steel RG45	3/32	Overhead	511
24th	17-J24	T-joint	Fillet welding	Mild steel	1/8	Mild steel RG45	3/32	Vertical	512
25th	17-J25	T-joint	Fillet welding	Mild steel	1/8	Mild steel RG45	3/32	Overhead	512
26th	17-J26	Flat plate	Beading	Mild steel	1/16	Mild steel RG45	1/16	Flat	514
27th	17-J27	Flat plate	Beading	Mild steel	1/16	Mild steel RG45	1/16	Vertical	514
28th	17-J28	Square butt joint	Groove welding	Mild steel	1/16	Mild steel RG45	1/16	Flat	514
29th	17-J29	Square butt joint	Groove welding	Mild steel	1/16	Mild steel RG45	1/16	Vertical	514
30th	17-J30	Lap joint	Fillet welding	Mild steel	1/16	Mild steel RG45	1/16	Flat	514
31st	17-J31	Lap Joint	Fillet welding	Mild steel	1/16	Mild steel RG45	1/16	Vertical	514
32nd	17-J32	Lap joint	Fillet welding	Mild steel	1/16	Mild steel RG45	1/16	Horizontal	514
33rd	17-J33	T-joint	Fillet welding	Mild steel	1/16	Mild steel RG45	1/16	Horizontal	514
34th	17-J34	T-joint	Fillet welding	Mild steel	1/16	Mild steel RG45	1/16	Vertical	514
35th	17-J35	Lap joint	Fillet welding	Mild steel	1/16	Mild steel RG45	1/16	Overhead	514

Table 17-4 (Continued)

Recommended job order[1]	Number in text	Joint	Operation	Material		Filler rod		Work position	Text reference
				Type	Thickness	Type	Size		
36th	17-J36	T-joint	Fillet welding	Mild steel	$\frac{1}{16}$	Mild steel RG45	$\frac{1}{16}$	Overhead	514
37th	17-J37	Flat plate	Beading-backhand	Mild steel	$\frac{3}{16}$	Steel, high test RG-60	$\frac{1}{8}$	Flat	514
38th	17-J38	Flat plate	Beading	Mild steel	$\frac{3}{16}$	Steel, high test RG-60	$\frac{1}{8}$	Vertical	514
39th	17-J39	Lap joint	Fillet welding—backhand	Mild steel	$\frac{3}{16}$	Steel, high test RG-60	$\frac{1}{8}$	Flat	514
40th	17-J40	Flat plate	Beading—backhand	Mild steel	$\frac{3}{16}$	Steel, high test RG-60	$\frac{1}{8}$	Horizontal	514
41st	17-J41	Lap joint	Fillet welding—backhand	Mild steel	$\frac{3}{16}$	Steel, high test RG-60	$\frac{1}{8}$	Horizontal	514
42nd	17-J42	Lap joint	Fillet welding	Mild steel	$\frac{3}{16}$	Steel, high test RG-60	$\frac{1}{8}$	Vertical	514
43rd	17-J43	Beveled-butt joint	Groove welding—backhand	Mild steel	$\frac{3}{16}$	Steel, high test RG-60	$\frac{1}{8}$	Flat	514
44th	17-J44	Beveled-butt joint	Groove welding	Mild steel	$\frac{3}{16}$	Steel, high test RG-60	$\frac{1}{8}$	Vertical	514
45th	17-J45	Beveled-butt joint	Groove welding	2- to 4-inch pipe	Standard	Steel, high test RG-60	$\frac{1}{8}$ or $\frac{5}{32}$	Roll	516
46th	17-J46	Beveled-butt joint	Groove welding—forehand	2- to 4-inch pipe	Standard	Steel, high test RG-60	$\frac{1}{8}$ or $\frac{5}{32}$	Horizontal Fixed	517
47th	17-J48	Beveled-butt joint	Groove welding—backhand	2- to 4-inch pipe	Standard	Steel, high test RG-60	$\frac{5}{32}$ or $\frac{3}{16}$	Horizontal Fixed	519
48th	17-J47	Beveled-butt joint	Groove welding—backhand	Mild steel plate	$\frac{3}{16}$	Steel, high test RG-60	$\frac{1}{8}$ or $\frac{5}{32}$	Vertical	520
49th	17-J49	Beveled-butt joint	Groove welding—backhand	2- to 4-inch pipe	Standard	Steel, high test RG-60	$\frac{1}{8}$ or $\frac{5}{32}$	Vertical	520

[1] It is recommended that the student do the jobs in this order. In the text, the jobs are grouped according to the type of operation to avoid repetition.

(Continued from page 521)

9. When shutting off the gas welding torch, the oxygen valve should be closed down first. True. False.
10. If the regulating adjusting screw becomes hard to operate, which lubricant should be used, oil or grease?
11. In making a gas weld, the filler rod should not touch the weld pool. True. False.
12. What is the temperature of the oxyacetylene flame?
13. Name two AWS classification numbers for gas welding filler rods.
14. Backhand welding is especially suited for light gauge sheet metal. True. False.
15. It is the general practice to weld pipe with the oxyacetylene process. True. False.
16. Do tack welds have to be melted out, or can they become a part of the completed weld?
17. What are the important factors in control of the puddle in the vertical position?
18. In tacking up a pipe joint, make sure that the pipe ends are touching. True. False.
19. Backhand welding is never used to weld pipe joints. True. False.
20. What is meant by a roll position weld?

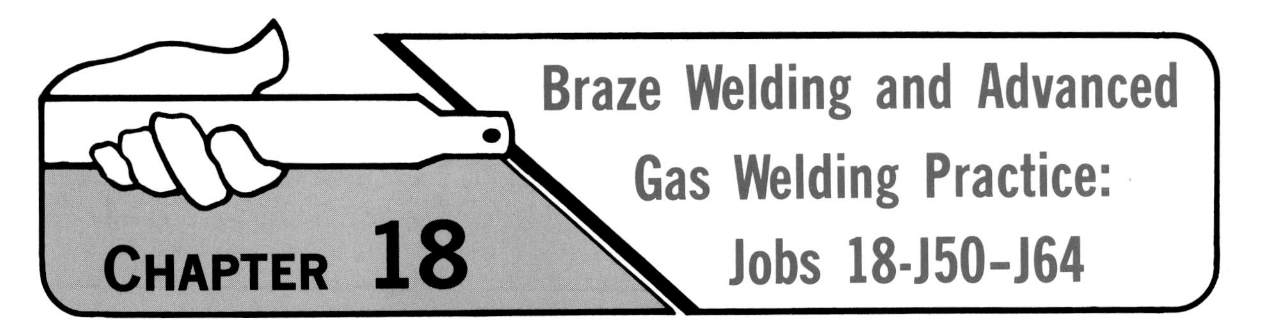

Chapter 18 Braze Welding and Advanced Gas Welding Practice: Jobs 18-J50–J64

The cooperation of the Linde Division, Union Carbide Corporation, and the Modern Engineering Company in providing information and photographs for this chapter is gratefully acknowledged.

Much of the gas welding done in industry today is on the ferrous and nonferrous metals and alloys covered in this chapter. The practice jobs have been carefully selected on the basis of the practices being followed in industry. As a beginning welder, you may not have an opportunity on the job to weld all of these materials until you gain more experience. Mastery of the practice jobs listed in the Course Outline, however, will provide you with the variety of gas welding skills that is the mark of the skilled combination welder. Because this unit continues the discussion of gas welding practice begun in Unit 17, the practice jobs are numbered consecutively. Thus the first job in this unit will be 18-J50.

BRAZE WELDING

Braze welding is a form of torch brazing (TB). It is often referred to as *bronze welding,* "a process of welding whereby a groove, fillet, plug, or slot weld is made using a nonferrous filler metal, having a melting point below that of the base metals but above 800 degrees F." (*American Welding Society, Terms and Definitions:* AWS A3.069). Fusion of the base metal is not required. Unlike soldering and brazing, the filler metal is not distributed in the joint by capillary attraction. In torch brazing a bronze filler rod supplies the weld metal, and the oxyacetylene flame furnishes the heat.

The joint design for braze welding is the same as for fusion welds. The welding technique is the same except that the base metal is not melted. It is, in fact, raised only to the tinning temperature. A *bond* (molecular union) is formed between the bronze welding rod and the prepared surface of the work. The welds compare favorably with fusion welds, Fig. 18-1.

Heavy materials have to be beveled, and the weld must be made with several passes to insure a weld throughout the total thickness of the metal. Each pass should be cleaned before applying the next layer of weld metal.

Industrial Uses

Braze welding is particularly adaptable to the joining and repairing of such metals as cast iron, malleable iron, copper, brass, and dissimilar metals. It is also used for the building up of worn surfaces, Fig. 18-2, a process which is referred to as *bronze surfacing.* Parts that are going to be subjected to high stress or high temperature should not be braze welded. Bronze loses its strength when heated to a temperature above 500 degrees F. The process also cannot be used when a color match is desired.

Braze welding has the following commercial advantages.

- The speed of welding is increased because less heat is required than for fusion welding.
- The welding of cast iron is greatly simplified because the base metal is not melted. Preheating is reduced or eliminated.
- Since the work does not have to be brought up to a high tem-

18-1. *A braze-welded butt joint in ¼-inch steel plate. The weld was made with a no. 4 tip and a 5⁄32-inch bronze rod in the flat position.*

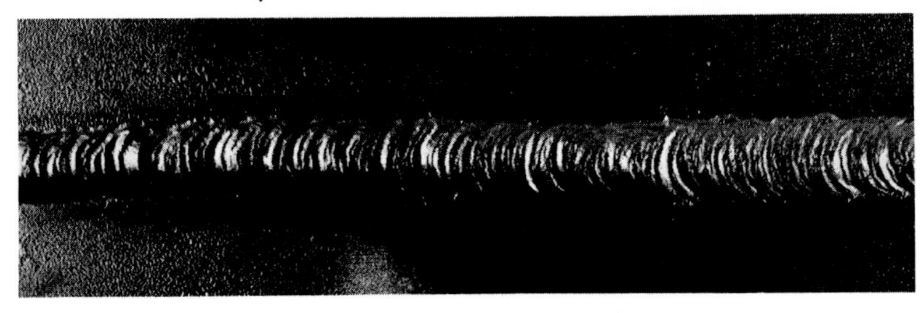

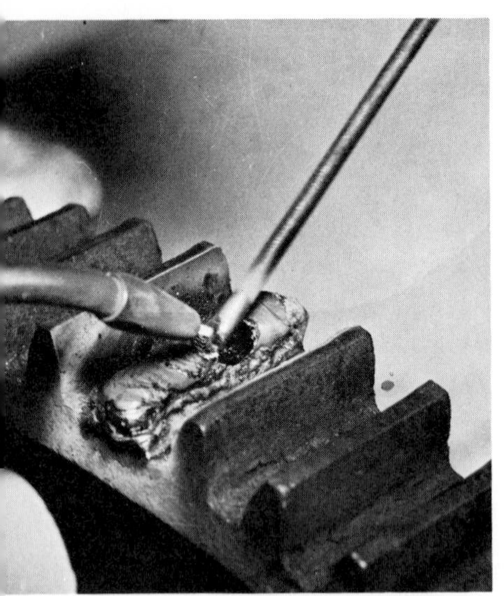

Linde Division, Union Carbide Corp.

18-2. *Building up a gear tooth with bronze in a series of layers, each of which is about 1/16-inch thick.*

18-3. *The brazing action. After a spot at the edge of the plate has been heated to a visible red, some bronze from the fluxed end of the rod is melted on the hot area. The bronze filler metal spreads and runs evenly over the hot surface. This is known as tinning, and appears in this photograph as the small, dark area under the flame.*

Linde Division, Union Carbide Corp.

perature, expansion and contraction are reduced.

- The bronze weld yields as the work cools until the temperature reaches about 500 degrees F. This reduces locked-up stresses.
- Malleable cast iron can be welded only by braze welding since the higher heat of fusion welding destroys the malleability.
- The coating on galvanized iron is less affected by the low temperatures of braze welding. This in turn minimizes the distortion in sheet metal fabrication.
- Braze welding is a good method for joining dissimilar metals such as cast iron and steel.

Bronze Filler Rod

The melting temperature of bronze filler rod is approximately 1,600 degrees F. This is considerably below the melting point of such metals as steel, cast iron, and aluminum. The filler rods are copper alloys. They usually contain a little less than 60 percent copper and 40 percent zinc. These metals provide high tensile strength and high ductility. Small quantities of tin, iron, manganese, and silicon are also added. These metals produce a rod that is free flowing and has a deoxidizing effect on the weld. They decrease the tendency to fume and increase the hardness of the weld metal for greater wear resistance. Special filler rods are also available for use as a bearing bronze.

Flux

Braze welding depends on the fact that molten metal with low surface tension flows easily and evenly over the surface of a heated and chemically clean base metal, Fig. 18-3. The surface of the base metal contains oxides that can be removed only through the use of a flux. Since brazing is an adhesion process, the surface of the base material must be free of contamination of any kind.

Two methods are in general use for applying this flux. In the first method, the end of the filler rod is heated and dipped into a supply of powdered flux. This causes the flux to adhere to the rod. It flows over the work surface as the rod is melted. Note the flux on the rod in Fig. 18-3. In the second method, liquid flux is added to the stream of acetylene flowing to the torch through a special unit designed for that purpose, Figs. 18-4 and 18-5. The flux becomes an integral part of the welding flame. This second method applies the flux evenly and increases the speed of welding. This process is used by those manufacturing industries that engage in mass production.

Braze Welding Practice: Jobs 18-J50 through 18-J55

Practice in braze welding includes running beads on plate in the flat and vertical positions, lap and T-joints, and single-bevel butt joints. The metals should include steel, gray cast iron, and copper or brass if available.

1. No edge preparation is necessary if beads are being run on flat plate. Generally, joint design for braze welds is the same as for fusion welds in base metals of like thickness. Bevel the edges of metal thicker than 1/8 inch to provide a groove for welding.
2. Clean the surface of the metal with a wire brush to remove any foreign substances.
3. Position the work so that it runs slightly uphill. This prevents the molten bronze from flowing ahead to plate surfaces that are not hot enough.

4. Adjust the torch for a slightly oxidizing flame.

5. Heat the filler rod and dip it into the flux. Heat the base metal to just the right temperature. The proper temperature is indicated when the base metal begins to glow. At this point melt a small amount of the rod and let it spread over the joint. This is referred to as *tinning.* If filler rod is applied before it is hot enough, the molten bronze will not flow over the surface. It will form drops that will not adhere to the material being welded. If the base metal is too hot, the molten bronze will boil and form little balls on the surface of the material being welded. Bronze weld metal that is too hot will also burn and give off a white smoke. This is the zinc which is burning with the oxygen in the air. It forms zinc oxide. The bronze will flow readily over the surface when the work has reached the correct temperature.

6. Carry the weld forward with a slightly circular movement of the torch like that used in fusion welding while adding filler rod to the weld pool. As the weld progresses, keep the end of the filler rod well-fluxed. Keep in mind that the proper heating of the base metal is critical.

7. If more than one layer of beads is necessary, as in the welding of a second pass in a groove, make sure that you obtain through-fusion between the new bronze weld and the previously deposited bronze bead, and a good bond on the bevel and plate surfaces. Be careful that there are no inclusions of slag or oxide and other contaminations.

8. Continue to practice until you can make braze welds that are of uniform width and height. They should be smooth with fine ripples and free of pits and other porosity. They should be brightly colored. The edges should flow into the plate without overlap or other signs of lack of adhesion. A deposit of white residue on the weld indicates overheating and burning.

9. Test the welds by making butt, lap, or T-joints, and testing them as you did your fusion welds in steel. Examine them for evenness and depth of bond. Find out whether the bronze has flowed to the root of the joint. Good braze welds often tear out cast iron when tested.

Powder Brazing

Powder brazing is a comparatively new form of brazing that is finding increased use, especially in mass production industries. A specially designed torch, Fig. 18-6, or a small hopper attachment unit that can be attached to a standard welding torch provides the powder. The mixer matches the powder flow range to the gas flow range of the tip to provide maximum performance with each tip size.

The powder brazing process can be used as a means of hard surfacing, brazing, and buildup welding. It produces high quality work at high speeds. A wide selection of overlay powders is available with a large variety of hardness ratings, machineability, and resistance to heat and corrosion. Brazing alloys are also available for use in many applications.

WELDING CAST IRON

Characteristics of Cast Iron

You will recall that cast irons are classified as gray iron, white iron, nodular iron, and malleable iron. Both gray cast iron and malleable iron are used commercially. We will be concerned here with the fusion welding of gray cast iron. Malleable iron can not be fusion welded with the oxyacetylene process and must be braze welded.

Gray cast iron is an alloy of iron, carbon, and silicon. The carbon in gray cast iron may be present in two forms: in a carbon and iron solution and as a free carbon in the form of graphite.

18-4. *Gas fluxer introduces flux into the flame for bronze and silver brazing. It uses a highly volatile liquid flux and can be used with any fuel gas.*

The Gasflux Co.

When it is broken, the fractured surface has a gray look due to the presence of the graphite particles. Gray cast iron is easy to machine because graphite is a fine lubricant. The presence of graphite also causes gray cast iron to have low ductility and tensile strength.

Welding Applications

Gray cast iron may be braze welded or fusion welded. For the most part, cast iron is welded in maintenance and repair work. Welding is seldom used as a fabricating process. Braze welding is preferred since it can be applied at a low temperature, and the bronze weld is highly ductile. Fusion welding is used when the color of the base metal must be retained and when the welded part is to be subjected to service temperatures over 500 degrees F. Table 18-1 summarizes the various cast iron welding procedures. Before practice welding, review Chapter 3, pages 100, 101, for additional information concerning the nature of cast iron.

Preheating

Control of expansion and contraction is very important in cast iron welding. The bulk and shape of the casting and whether or not light sections join heavy sections affect preheating and welding technique. When fusion welding cast iron, all parts of the casting must be able to expand equally to prevent cracking and locked-in stress in the job. If the torch is applied directly to the cold casting and the joint to be welded is raised to the melting point, the expansion of the heated metal will cause a break or crack in the relatively cold casting surrounding the weld. If it does not break, severe internal stresses are locked in that may later cause a failure under service.

Small castings can be preheated with the oxyacetylene

18-5. *How the liquid fluxer is connected to the torch and gas supply.*

The Gasflux Co.

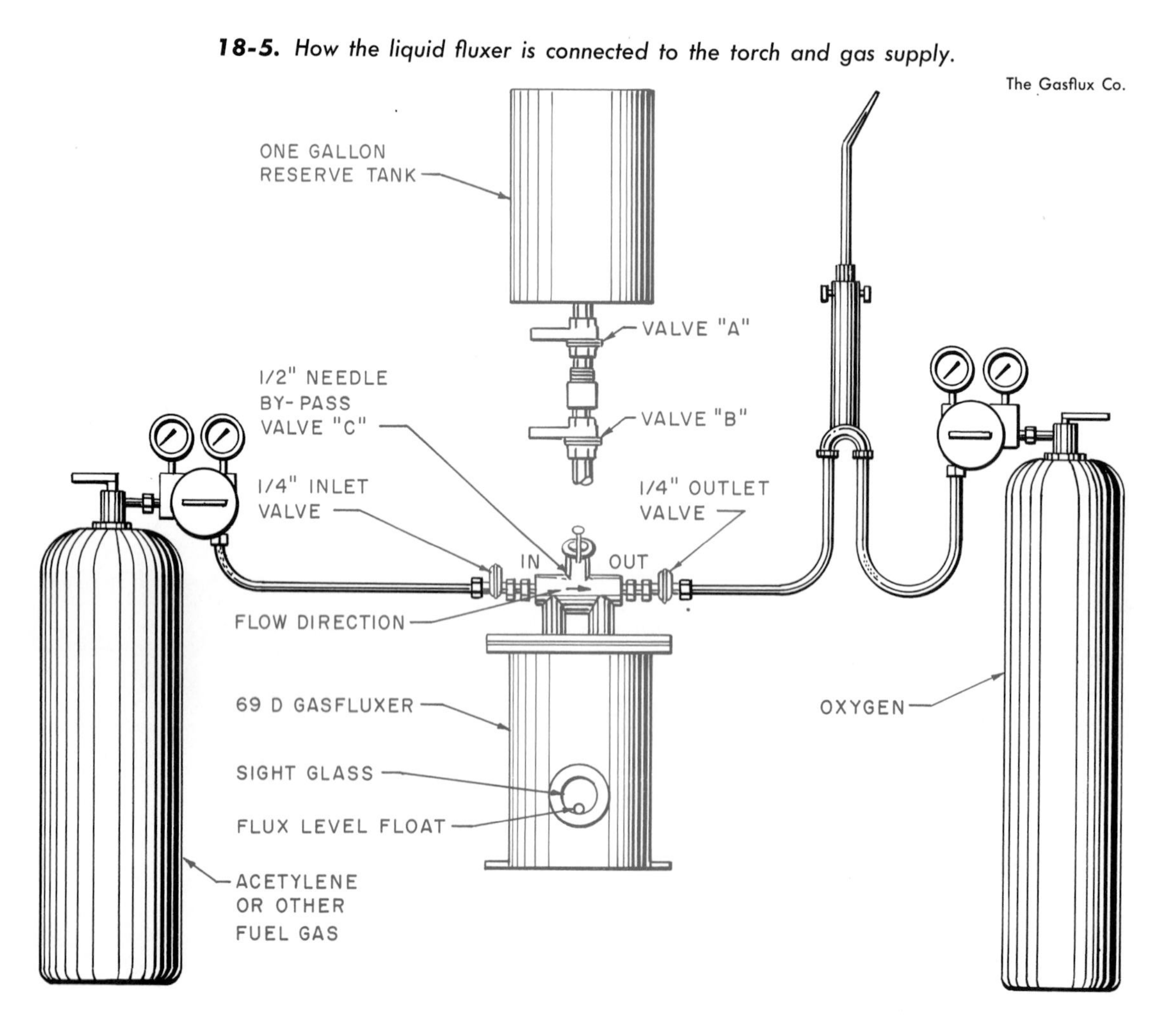

flame during the welding operation if the entire casting is heated evenly. Large castings may have to be preheated in a firebrick furnace built around the casting. Heating is usually done with gas- or oil-fired burners. The furnace is covered with asbestos paper to retain the heat and keep out the cold drafts. The casting is welded through a large hole torn in the asbestos paper. When the weld is completed, the casting is again raised to an even heat all over. Then it is buried in asbestos and allowed to cool very slowly.

Filler Rod

Fusion welding requires the use of a good grade of cast iron filler rod that matches the material being welded. The rod must contain enough silicon to replace the silicon which tends to burn out during welding. Silicon assists in the flow of the molten metal during welding and retards oxidation that may lead to the formation of slag inclusions and blowholes. Good cast iron welding rods contain 3 to 4 percent silicon.

Cast iron filler rods are supplied in diameters of $\frac{3}{16}$, $\frac{1}{4}$, $\frac{3}{8}$, and $\frac{1}{2}$ inch and in lengths of 12 to 18 inches. They carry the AWS-ASTM classification of RCI, RCI-A, and RCI-B. They may be round, square, or hexagonal in shape. If the welder needs a longer or heavier rod, he can weld two or more rods together.

Flux

The problem in welding cast iron, as in welding other metals, is to prevent the formation of oxide, and when it is formed, to remove it from the weld. The flux dissolves the oxide, floats off other impurities such as sand, scale, and dirt, and increases the fluidity of the molten metal.

The student welder must learn to apply flux properly. Too much flux can cause as much trouble as too little. Excessive flux becomes entrapped in the molten metal and causes blowholes and porosity. Also, the molten iron will combine with certain elements in the flux if it is applied in excess. You will learn by experience the right amount to use. The amount that adheres to the hot end of the welding rod when it is dipped in the flux is usually enough. Do not throw additional quantities into the weld as you are welding.

Cast Iron Welding Practice: Jobs 18-J56 and 18-J57

The same general welding procedure may be used for beading and groove welding. The following steps are for groove welding.

1. Prepare the metal as in other welding. If the material is more than $\frac{1}{8}$ inch thick, bevel the edges, leaving about $\frac{1}{8}$ inch thickness at the root face. Clean

18-6. *Powder brazing torch.*

Smith Welding Equipment Co.

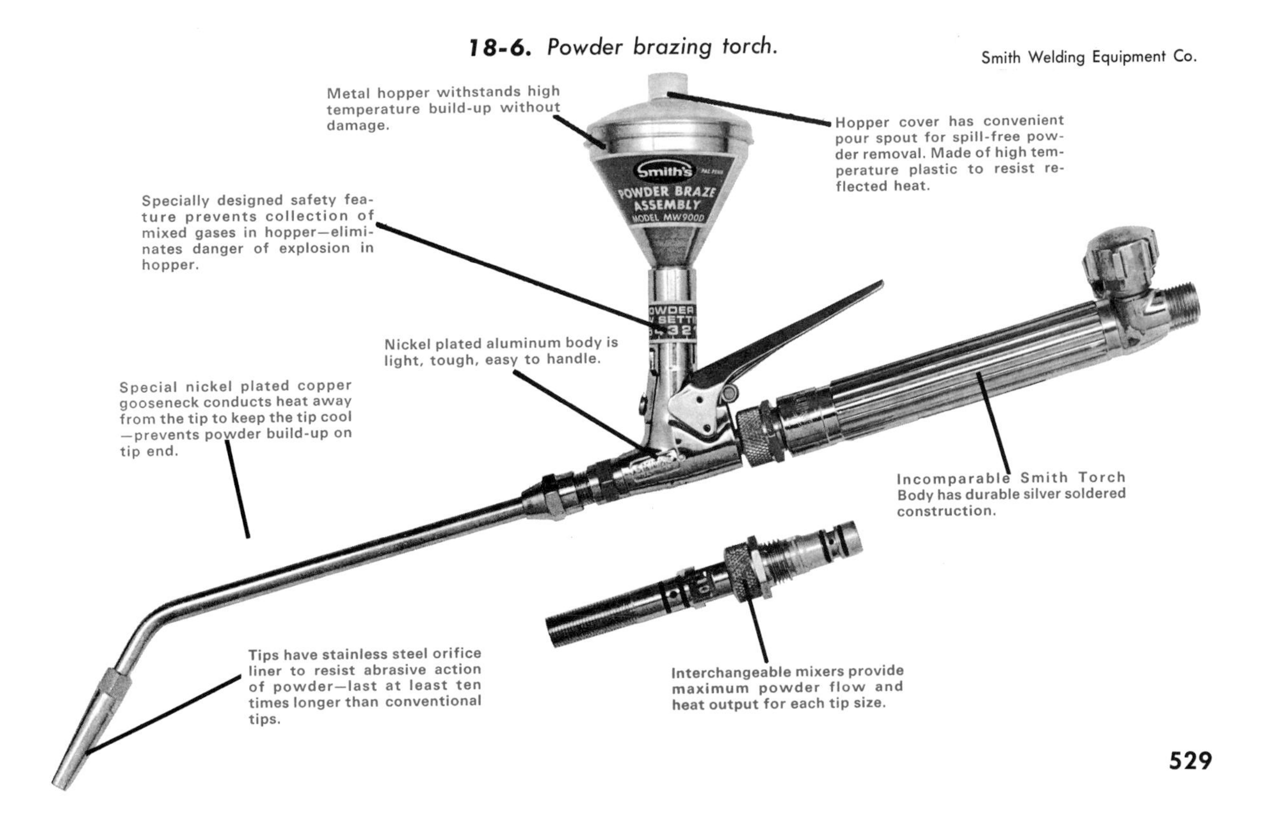
Metal hopper withstands high temperature build-up without damage.
Hopper cover has convenient pour spout for spill-free powder removal. Made of high temperature plastic to resist reflected heat.
Smith's
POWDER BRAZE ASSEMBLY
MODEL MW900D
Specially designed safety feature prevents collection of mixed gases in hopper—eliminates danger of explosion in hopper.
Nickel plated aluminum body is light, tough, easy to handle.
Special nickel plated copper gooseneck conducts heat away from the tip to keep the tip cool —prevents powder build-up on tip end.
Incomparable Smith Torch Body has durable silver soldered construction.
Tips have stainless steel orifice liner to resist abrasive action of powder—last at least ten times longer than conventional tips.
Interchangeable mixers provide maximum powder flow and heat output for each tip size.

all dirt, rust and scale from the surface. It is assumed that you are practicing with small pieces of casting that have been beveled.

2. Select the proper size tip and adjust the torch for a neutral flame.

3. Play the torch flame over the entire work until the entire joint has been preheated. Then starting at the right edge, direct the flame at the bottom of the groove until the metal there has been melted.

4. Heat the bottom of the groove and the side walls until they are molten, and the metal flows to the bottom of the groove. If the metal gets too hot and the molten puddle tends to run away, raise the flame slightly. Hold the torch at 90 degrees to the work and using a motion like that for welding steel plate. Keep the sides and the bottom in a molten condition. If the torch is held at an angle, the flame will blow the molten metal ahead of the weld, and adhesion will result. Make sure that the surfaces of the groove are fused ahead of the weld pool so that the molten metal is not forced ahead to colder plate surfaces to form an adhesion.

5. Filler rod that has been fluxed should be added to the weld pool to fill up the groove. Never hold the filler rod above the weld and melt it drop by drop into the molten puddle.

6. Gas bubbles or impurities can be floated to the surface of the weld by the addition of flux and the use of the flame. Skim these impurities off the surface of the weld with the filler rod. Impurities left in the weld are defects which weaken the joint.

7. After the weld has been completed, heat the entire casting to the same temperature throughout and allow it to cool very slowly.

8. Continue to practice until you can make good quality welds that are satisfactory to the instructor. Be particularly concerned about fusion along the

Table 18-1

Summary of cast iron welding procedures.

Cast-iron type	Procedure	Treatment	Properties
Gray iron	Weld with cast iron	Preheat and cool slowly	Same as original
Gray iron	Braze weld	Preheat and cool slowly	Weld better; heat-affected zone as good as original
Gray iron	Braze weld	No preheat	Weld better; base metal hardened
Gray iron	Weld with steel	Preheat if at all possible	Weld better; base metal may be too hard to machine; if not preheated, needs to be welded intermittently to avoid cracking
Gray iron	Weld with steel around studs in joint	No preheat	Joint as strong as original
Gray iron	Weld with nickel	Preheat preferred	Joint as strong as original; thin hardened zone; machinable
Malleable iron	Weld with cast iron	Preheat, and postheat to repeat malleableizing treatment	Good weld, but slow and costly
Malleable iron	Weld with bronze	Preheat	As strong, but heat-affected zone not as ductile as original
White cast iron	Welding not recommended		
Nodular iron	Weld with nickel	Preheat preferred; postheat preferred	Joint strong and ductile, but some loss of original properties; machinable; all qualities lower in absence of preheat and/or postheat

The James F. Lincoln Arc Welding Foundation

edges of the weld, a smooth surface without holes and depressions, and good penetration and fusion on the back side.

9. Test one of your completed welds. Use the same procedure used in testing butt welds in sheet metal. Place the specimen in a vise with the weld above the vise jaws. Break it in pieces with a hammer. A sound weld will cause the break to take place in the casting. You may also wish to break or saw through the weld to examine the weld metal. The weld metal should be sound and have no slag inclusions or blowholes.

WELDING OF ALUMINUM

While aluminum is one of the newest metals developed for commercial use, going back only about 50 years, it has become important to industry. It is lightweight (about one-third as heavy as steel) and yet it has a high strength-to-weight ratio. It is a good conductor of electricity and has high resistance to corrosion.

Aluminum can be welded by gas welding and many other welding processes. It is comparatively easy to fabricate. Much of the aluminum used in industry is welded with the shielded metal-arc and the gas metal-arc and gas tungsten-arc processes. It is highly desirable, however, for the welding student to practice welding aluminum with the oxyacetylene process so that he can better judge the flow of metal and heat ranges.

Characteristics of Aluminum

The three categories of aluminum which have the most welded applications are commercially pure aluminum, wrought aluminum alloys, and aluminum casting alloys.

Pure aluminum melts at 1,220 degrees F., while weldable commercial aluminum alloys have a melting range of 900 degrees to 1,220 degrees F. Compare these temperatures with steel, which melts at about 2,800 degrees F., and copper, which melts at about 1,980 degrees F. Aluminum does not change color during heating, and when the melting temperature is reached, it collapses suddenly. This characteristic is called *hot-shortness.* Because aluminum is hot-short, it requires support when hot.

Aluminum is a good *thermal conductor.* It conducts heat about three times faster than iron and requires higher heat input than that used in welding steel.

Aluminum expands during heating. For this reason aluminum welds have a tendency to crack because of the shrinkage that takes place in the weld metal when cooling. Too much restraint of the parts during cooling may also result in weld cracking. The speed of welding is also important. Welding at a slow rate of speed causes more heat input into the part being welded. This increases the rate of expansion and contraction.

The aluminum weld pool oxidizes very rapidly. It forms an oxide with a melting point of 5,400 degrees F., which must be removed either chemically with a flux or mechanically with a paddle.

Before practice welding, review Chapter 3, pages 101, 102, for additional information concerning the nature of aluminum.

Filler Rod

The selection of the proper filler rod for aluminum welding depends on the composition of the base metal. The AWS-ASTM classification R1100 is recommended for welding commercially pure aluminum sheet (1100); and R1260 and R4043, for other types of aluminum. Castings are welded with R-C4A, R-CN42A, R-SC51A, R-SG70A, and R-2G61A.

Aluminum filler rods are obtainable in sizes of 1/16, 3/32, 1/8, 5/32, 3/16, and 1/4 inch in diameter. Standard lengths are 36 inches.

As a rule, the diameter of the filler rod should equal the thickness of the metal being welded. A rod that is too large melts too slowly. Thus it retards the fluid action of the weld pool and may cause a lack of fusion. On the other hand, a rod that is too small melts too fast and may be burned. There is also not enough filler metal for the weld pool. This overheats the weld and may cause burn-through.

Flux

The welding of aluminum requires the use of flux to remove the aluminum oxides that are on the surface and those that are formed during welding. The flux-oxide compound forms a slag which is expelled from the weld pool by the action of the flame and the molten metal.

A number of commercial fluxes are available. Those in powdered form, which are mixed with alcohol or water to make a paste, are the most practical. Mixing should be done in a glass or ceramic container or an aluminum or stainless steel container. Containers of steel, copper, or brass may contaminate the flux and should not be used.

When welding sheet aluminum, a brush is normally used to spread the flux on the welding area and to coat the welding rod. When welding cast aluminum, the flux is applied to the end of the filler rod by dipping its heated tip into the flux. During welding the flux melts and forms a protective coating over the molten metal and the heated end of the filler rod. The flux is melted in advance of the weld puddle and thus cleans the base metal before it is welded. The flux that is applied to the base metal should be heated slowly before welding to insure that excess moisture is driven off. This is done to prevent spattering and porosity.

Some welders prefer filler rods that are prefluxed or flux-covered electrodes of the type used for the shielded metal-arc welding of aluminum. When such electrodes are used, it is not always necessary to apply flux to the base metal. Better results are generally obtained, however, if flux is applied to the work. The period between fluxing and welding should not exceed 45 minutes.

Flame Adjustment

Acetylene is the most popular gas to use with oxygen for welding aluminum because of its high heat and wide availability. Hydrogen is also used with oxygen for the welding of aluminum. Because of its lower heat, it is used chiefly in the welding of the thinnest gauges of aluminum sheet.

In order to produce clean, sound welds and maximum welding speed, a neutral flame is ideal and a slightly reducing flame (carburizing) is highly satisfactory. You will recall that aluminum oxidizes readily. The reason for the excess acetylene flame is to make sure that the gas mixture does not stray to the excess oxygen side. An oxidizing flame causes a bead that balls up. The weld has poor fusion, poor penetration, and much porosity.

Aluminum may also be welded with the oxyhydrogen flame. The same torch and type of tip are used. Because of the lower temperature of the oxyhydrogen flame, a larger tip size is required. You may have some difficulty adjusting the flame. It is very similar, however, to the adjustment of the oxyacetylene flame, Fig. 18-7.

The neutral oxyhydrogen flame is a white to pale violet color, fairly large, and rather indistinct. The small, well-defined cone in the center has a bluish tinge.

The reducing flame is larger and ragged. Its color varies from pale blue to reddish-violet, depending upon the amount of excess hydrogen. There is no well-defined cone at the center. This flame is usually used for welding.

The oxidizing flame has a very short, blue inner cone. The flame is thinner than the reducing flame and ranges in color from white to transparent. Because these differences are very hard to see, a flowmeter may be the best means for assuring the desired hydrogen-oxygen ratio.

A soft flame is essential in all oxy-gas welding of aluminum. A strong or noisy flame causes turbulence and lowers weld quality. The proper pressure adjustment at the regulators and precise torch adjustment are a must.

Aluminum Welding Technique

EDGE PREPARATION. As with all other metals, the type of edge preparation depends upon the thickness of the metal being welded. On aluminum butt joints up to a thickness of $\frac{1}{16}$ inch, no edge preparation is necessary. A flange-type joint may also be used. Material from $\frac{1}{16}$ inch to $\frac{5}{32}$ inch can be welded in the form of a square butt joint, but the plate edges should be notched. For aluminum plate $\frac{3}{16}$ to $\frac{7}{16}$ inch thick, the edge preparation is that for a single-bevel butt joint with a 90-degree *V*. For thicknesses of $\frac{1}{2}$ inch and over, the double-bevel butt joint with a 90-degree *V* on each side is used if both sides are accessible. Table 18-2 lists the rod and tip sizes and gas pressures for welding various thicknesses of aluminum.

CLEANING. Since grease, oil, and dirt cause weld porosity and interfere with welding, they should be removed from the welding surfaces. Commercial degreasing agents are available for this purpose. Wire brushing may also be used to remove heavy oxide films from the plate surfaces. Removal of oxides permits more effective fluxing action.

PREHEATING. Like cast iron, cast aluminum requires careful preheating before welding and slow cooling after the weld is completed. Preheating is always recommended for gas welding aluminum when the mass of the base metal is such that heat is conducted away from the joint area faster than it can be supplied to produce fusion. A large difference in temperature between the surrounding metal and the weld area increases welding difficulties. The preheat temperature must stay below 700 degrees F. Preheat temperature-indicating compounds can be used to advantage. These are available in both crayon and liquid form. They are accurate and are made in almost any temperature range.

JIGS AND FIXTURES. Aluminum and many of its alloys are weak when hot. Aluminum parts, especially thin stock, should be supported adequately. Jigs and fixures maintain alignment and reduce buckling and distortion. It is especially important for the edges of the joint to be spaced and aligned correctly. Tack welding before final welding also aids in maintaining alignment and minimizing distortion. Tack welds should be placed carefully. They must be small enough not to interfere with welding, and they must have good penetration and fusion.

WELDING POSITIONS. Aluminum welding can be done in all positions. Overhead welding, however, is difficult and should be attempted on the job only by highly experienced welders. The flat position is preferable and should be used whenever possible.

Horizontal welding technique is similar to that for flat welding. Special manipulation of the rod and torch is necessary, however, to offset tendency of the molten metal to sag and build up the lower edge of the weld bead. Butt joints are not practical in the horizontal position.

Vertical welding is usually performed on materials $\frac{1}{4}$ inch or thicker. It is almost impossible to keep thinner materials from burning through, and the weld pool becomes too fluid and spills over. Torch and welding rod angles are about the same as for all other vertical welding.

POSTWELD CLEANING. Thorough cleaning after welding aluminum is very important. Flux residues on gas-welded sections corrode aluminum if moisture is present. Small sections may be cleaned by a 10- to 15-minute immersion in a cold 10 percent sulfuric acid bath, or a 5- to 10-minute immersion in a 5 percent sulfuric acid bath held at 150 degrees F. Acid cleaning should always be followed by a hot or cold water rinse. Steam cleaning also may be used to remove flux residue, particularly on parts that are too large to be immersed. Brushing may also be necessary to remove adhering flux particles.

Table 18-2

Rod sizes, tip sizes, and gas pressures for the welding of aluminum.

Aluminum thickness (inch)	Filler rod diameter (inch)	Oxyacetylene welding			Oxyhydrogen welding		
		Orifice diameter in tip of torch (inch)	Oxygen pressure (p.s.i.)	Acetylene pressure (p.s.i.)	Orifice diameter in tip of torch (inch)	Oxygen pressure (p.s.i.)	Hydrogen pressure (p.s.i.)
0.020	$\frac{3}{32}$	0.025	1	1	0.035	1	1
0.032	$\frac{3}{32}$	0.035	1	1	0.045	1	1
0.050	$\frac{3}{32}$	0.045	2	2	0.065	2	1
0.064	$\frac{3}{32}$	0.045	2	2	0.065	2	1
0.080	$\frac{1}{8}$	0.055	3	3	0.075	2	1
$\frac{1}{8}$	$\frac{1}{8}$	0.065	4	4	0.095	3	2
$\frac{3}{16}$	$\frac{5}{32}$	0.065	5	5	0.095	3	2
$\frac{1}{4}$	$\frac{3}{16}$	0.075	5	4	0.105	4	2
$\frac{5}{16}$	$\frac{3}{16}$	0.085	5	5	0.115	4	2
$\frac{3}{8}$	$\frac{3}{16}$	0.095	6	6	0.125	5	3
$\frac{1}{2}$	$\frac{1}{4}$	0.100	7	7	0.140	8	6
$\frac{5}{8}$	$\frac{1}{4}$	0.105	7	7	0.150	8	6

Kaiser Aluminum & Chemical Corp.

18-7. *A neutral flame (center) tending toward a slightly reducing flame (top) is recommended for the oxyhydrogen welding of aluminum.*

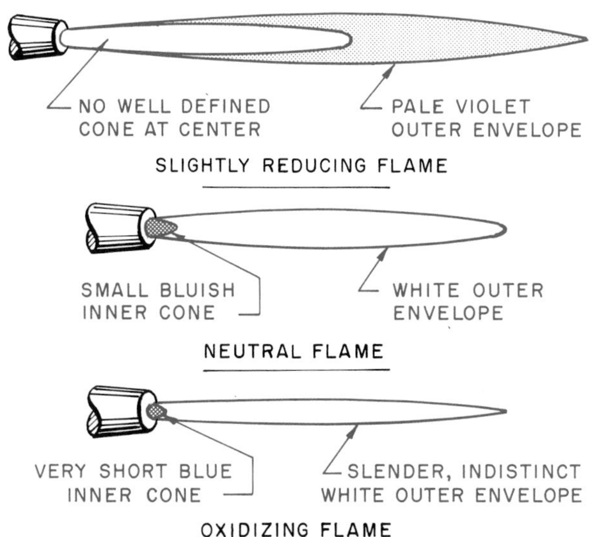

Sheet Aluminum Welding Practice: Jobs 18-J58, J59, and J60

1. Prepare, clean, and tack the joint. It is recommended that you practice on aluminum sheet ranging from 14 gauge to about $\frac{3}{16}$ inch thick. Select the proper tip size for the thickness of the metal being welded. Select the

filler rod according to the type of aluminum being welded.

2. Apply flux to the filler rod and the work.

3. The welding technique used to weld aluminum is not much different than that used in the welding of steel. Remember that aluminum does not change color when heated, it burns through readily, it forms oxides rapidly, and it has a high rate of expansion and contraction. Start welding about 1½ inches in from the edge of the sheet and travel to the opposite edge. Heat the entire plate or joint evenly.

4. After a small pool has been formed, point the flame in the direction of welding at an angle of about 30 degrees, Fig. 18-8. This position heats the plate ahead of you, reduces the problem of burning through, and increases the welding speed. Do not let the cone of the flame touch the weld pool. A distance of ⅛ to ¼ inch is advised. Make sure the weld pool is in a molten state before adding filler rod. Stir the filler rod in the weld pool. This action causes the oxides to rise to the surface of the weld pool and reduces porosity. Side-to-side movement depends upon the thickness of the plate and the size of the weld bead desired. Side-to-side movement should not be used for thin sheets. On heavier sheets the amount of movement depends upon the size of the weld bead desired.

As the end of the joint is approached, flatten out the angle of the torch until the flame is reaching mainly the welding rod. This is particularly recommended for the thinner gauges of aluminum.

5. Return to the beginning of the short weld and restart the weld. Weld towards the other edge of the plate.

6. Clean the weld and base metal by brushing with hot water.

7. Continue to practice these welds on various thicknesses of aluminum until you can make welds of satisfactory appearance that are acceptable to your instructor. Note the weld characteristics shown in Fig. 18-9. Be particularly concerned about fusion along the edges of the weld, a smooth surface without pits, holes and depressions, and good penetration on the back side. The beads should not be too wide, and their edges should be parallel.

8. Test a few of these welds as you did for steel by bending the plates back. The weld should not fracture or peel off the surface of the metal.

Cast Aluminum Welding Practice: Job 18-J61

1. Prepare the weld and filler rod as instructed in steps 1 and 2 of the preceding procedure.

2. Preheat the work carefully. Weld cracking is a greater problem in welding cast aluminum than sheet aluminum. This is especially true if there are varying thicknesses of cross section in the casting. You will recall that small castings may be heated with the torch but that large units must be heated in a furnace. Welding should begin at the center of the casting and continue to the outer edges.

3. In general, the welding techniques for welding cast aluminum are like those for sheet and plate aluminum. The formation of oxides in the weld presents an even greater problem in the welding of cast aluminum than

18-8. *Typical angles for torch and filler rod range from 30 to 45 degrees in the welding of aluminum.*

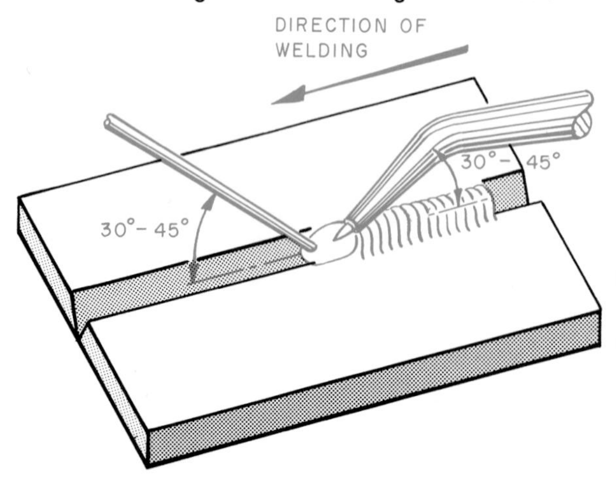

18-9. *A corner weld in aluminum sheet. The ripples should be smooth, and the weld should show no overhang.*

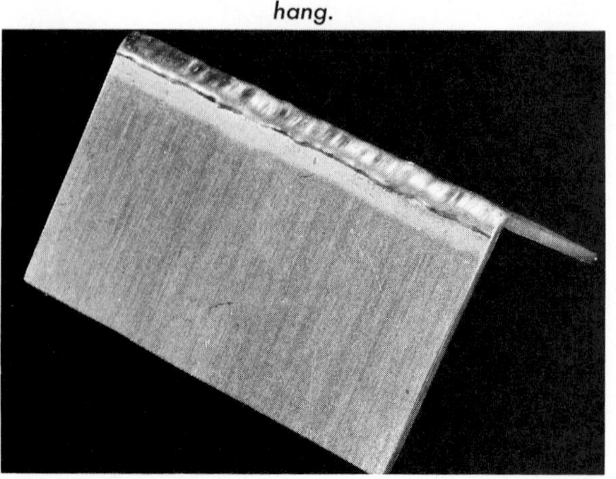

it does in the welding of other forms of aluminum. The oxide has a higher melting point than the aluminum being welded, and this temperature difference allows the aluminum to melt before the oxide film. The oxide film prevents fusion and must be removed by a chemical cleaner such as flux or mechanical means such as the steel paddle, Fig. 18-10. The paddle is used to stir up the molten weld pool, to work out the oxides and other impurities from the weld, and to smooth out the surface of the weld.

4. Clean the weld and base metal by brushing with hot water.

5. Continue to practice these welds with various thicknesses of cast aluminum until you can make welds of satisfactory appearance that are acceptable to your instructor.

6. Test a few of these welds as you did for cast iron. The welds should not break.

WELDING OTHER METALS WITH THE OXYACETYLENE PROCESS

Stainless Steel: Jobs 18-J62, J63, and J64

Because of its strength, durability, high temperature service, and resistance to corrosion, stainless steel is used in ever-increasing quantities. The common terms for the popular types are *18-8, 25-12,* and *25-20.* In each of these the first figure refers to the percentage of chromium and the second figure, to the percentage of nickel. Almost without exception the commercial types of stainless can be welded with the oxyacetylene process, but most of this welding is being done with the shielded metal-arc and the gas metal-arc and gas tungsten-arc processes. Oxyacetylene welding of stainless steel is still used for light gauge sheets.

Before practice welding, review Chapter 3, pages 45–116, for additional information concerning the nature of stainless steel.

The following factors must be considered in the welding of stainless steel with the oxyacetylene process.

- Joint design and preparation are like those used for steel. Thicknesses above $\frac{1}{8}$ inch must be beveled.
- For welding the high-chromium steels, the welding tip should be one or two sizes smaller than that used for the same thickness of carbon steel because as the content of chromium increases, the heat conductivity decreases.
- A flux is necessary because the chromium in stainless steel oxidizes rapidly and acts as a barrier between the flame and the surface of the material being welded. Most fluxes are powder which must be mixed to a paste in a nonmetallic container. The flux is painted on the filler rod and both sides of the joint being welded.
- Some chromium is lost as a result of oxidation during welding. This loss is restored to the weld by the use of a filler rod which contains at least 1 to $1\frac{1}{2}$ percent more chromium than the base metal.
- *Group 1, martensitic steels* (400 and 500 series, straight chromium steels, 12- to 18-percent chromium, basic type 410), which have the characteristic of air hardening, should be preheated locally or completely to 300 to 500 degrees F. This prevents excessive air hardening and possible cracking. Parts should also be annealed. R308, R309, or R310 filler wire is used for welding.
- *Group 2, ferritic steels* (400 series, straight high-chromium steels 15- to 27-percent chromium, no nickel, basic type 430), which are not hardenable, should be preheated to 300 to 500 degrees F., especially in cold weather. Parts should also be annealed. Filler wire R308, R309, or R310 is used for welding.
- *Group 3 austenitic chromium-nickel steels* (200 and 300 series, which includes the basic type 18-8 [types 302 and 304], 18-percent chromium and 8-percent nickel), should be heat treated if the part is to be put into severe chemical service. For many less severe general uses, heat treatment is not necessary. Filler wire R308 is used for welding.
- A precise neutral flame must be used. Since chromium oxidizes so readily, the oxidizing flame would burn out increased amounts of chromium and reduce the resistance to corrosion. An excess acetylene flame introduces an amount of carbon in the weld metal which reduces its quality. Check the flame frequently to make sure that it does not get into the oxidizing range. Some welders prefer a very slight acetylene feather to make sure that the flame does not become oxidizing.
- The welding technique is like that used for steel. The flame should not touch the weld metal, and it should be held at a steeper angle than for steel. When adding filler rod, hold it close to the flame. When rod is not needed, it should be withdrawn from the flame. Be careful not to overheat the filler rod or the weld pool. Avoid going back over a weld since this increases the amount of heat used and thus reduces corrosion resistance. Use 18-8 columbium-bearing filler rods.

- Depending upon the final use, it may be necessary to grind the weld smooth and polish it to match the base metal. If a polish is not needed, you may have to remove scale by sandblasting or by pickling with acid.

Copper and Copper Alloys

Copper and its alloys in the form of wire, sheets, rods, tubes, and castings are of great commercial importance. There are over a hundred alloys of copper, such as yellow brass, red brass, Tobin bronze, Everdur bronze, aluminum bronze, and nickel silver.

The fusion welding of copper with the oxyacetylene process is not widely used. The process must be restricted to deoxidized copper base metals. These metals have had the oxygen removed from them.

The following factors must be taken into consideration in the welding of deoxidized copper with the oxyacetylene process.

- Joint design is essentially the same as that necessary in the welding of steel.
- Because of the high heat conductivity of copper, a tip one or two sizes larger than is used for steel of the same thickness is necessary.
- Flux is not generally required for fusion welding since the absence of oxygen in the base metal reduces oxide formation. Occasionally, a brazing flux assists the welding action.
- The filler rod should be completely deoxidized and have a melting point slightly below that of the base metal.
- The quality of the completed weld can be improved by peening and then heating it to a red heat and allowing it to cool slowly. This increases the tensile strength and the ductility of the weld.
- Copper is a high conductor of heat and has a high expansion rate. This high expansion rate when heated and the high contraction rate during cooling causes locked-up stresses and distortion. These can be materially reduced by preheating, postheating, and slow cooling.
- Copper cannot be heat treated, but it will harden when cold worked.
- The flame adjustment should be neutral. The flame must not touch the molten pool. The position of torch and filler rod and the welding technique is like that for welding steel. The metal is solid until it actually reaches the melting point of 1,980 degrees F. Then it goes immediately into a highly fluid liquid state. Complete puddle control can be maintained by skillful manipulation of the torch flame and the filler rod.
- Copper can also be braze welded.

A number of copper alloys can be gas welded. These alloys include the copper-zinc alloys (brasses), the copper-tin alloys (bronzes), and the copper-silicon alloys. These alloys are welded with an oxidizing flame. With the exception of the copper-silicon alloys, they require the use of a flux. The silicon content in the copper-silicon alloys provides the fluxing action when welding this alloy. In all cases joint preparation and welding technique is similar to that for steel.

Nickel and Nickel Alloys

NICKEL. Nickel has high strength and high corrosion resistance. It is often used in pipe that will be subjected to high temperature and high pressure service. This includes piping installations in nuclear power plants, refineries, and chemical plants. The metal that is now being used commercially is 99.4 percent nickel. The remaining 0.6 percent is made up of copper, iron, manganese, silicon, and carbon. It can be heat treated.

The following factors must be taken into consideration in the welding of nickel with the oxyacetylene flame.

- Joint design is essentially the same as that for a corresponding thickness of steel. A *V* with an angle of at least 75 degrees is required.
- The filler rod should have the same composition as the base metal.
- A slightly excess acetylene flame is used for welding. This is largely to prevent the flame from becoming oxidizing. The flame should be soft.
- No flux is required.

Nickel is highly susceptible to embrittlement by materials that contain lead, sulfur, or phosphorus. It should be cleaned by brushing with hot water on both sides of the joint being welded.

- Welding technique is similar to that used for steel. The molten puddle is somewhat sluggish, but it should not be stirred with the flame or filler rod. Welds should be made in a single pass.

MONEL.® Monel® is an alloy of nickel. It is composed of nickel and copper with small percentages of other elements. The metal can be heat treated. The following factors must be taken into consideration in the welding of Monel® with the oxyacetylene flame:

- Joint design is essentially the same as that for a corresponding thickness of steel. A vise with an angle of at least 75 degrees is required.
- The filler rod should have the same composition as the base metal.
- A Monel® flux that is mixed with water to the consistency of

a paste is used. The rod and both sides of the base metal should be painted.
- The welding flame should be slightly carburizing.
- The tip size is one size larger than that used for the same thickness of steel.
- Welding technique is like that for welding of steel. The weld pool is fluid.

INCONEL.® Inconel® is an alloy of nickel, chromium, iron, and small percentages of other elements. The tensile strength of annealed Inconel® is higher than that of steel. It can be cold-worked to give a tensile strength as high as 185,000 p.s.i. in wire form. Molten Inconel® is more sluggish than Monel,® but not as sluggish as nickel. This characteristic causes the ripples to be somewhat coarser. Conditions of welding are the same as those listed for Monel.®

K-MONEL.® This material is a wrought alloy of nickel, copper, aluminum, and titanium. It possesses the corrosion resistance that is characteristic of Monel,® but it has greater strength and hardness. It can be cold-worked and heat-treated. Conditions of welding are the same as those listed for Monel.®

Nickel alloys may also be joined by braze welding, silver brazing, and silver soldering.

Magnesium Alloys

Magnesium is used extensively by the aircraft industry. It is two-thirds as heavy as aluminum and less than one-quarter as heavy as steel. It has a melting point of 1,202 degrees F. It burns with a brilliant light and is used for flares and photographic flashes. When magnesium is used in a fabrication that is to be welded, it is usually alloyed with aluminum. Magnesium may be welded by the gas and shielded metal-arc processes, but gas metal-arc and gas tungsten-arc are widely used today.

Magnesium has a high coefficient of expansion that increases as the temperature increases. Distortion and internal stresses are conditions that must be provided for. Sheet metal should be welded with one pass.

Pure magnesium has little strength. Its strength is increased when alloyed with aluminum, manganese, or zinc. It has a high resistance to corrosion.

The following factors must be taken into consideration in the welding of magnesium alloys with the oxyacetylene flame:
- Joint design is similar to that for a corresponding thickness of aluminum. Thicknesses up to ⅛ inch do not need to be beveled. Thicknesses above ⅛ inch should have a 45-degree bevel.
- Lap and fillet welds are not recommended due to the possibility of flux entrapment.
- Magnesium alloys have low strength at a temperature just below the melting point. The joints must be well-supported so that they will not collapse.
- The filler rod must be of the same composition as the base metal.
- A flux is used to prevent oxidation. It is a powder and must be mixed with water to form a paste. The filler rod and both sides of the base metal should be coated with flux. Use flux sparingly since too much may cause flux inclusions in the weld and makes final cleaning much more difficult.
- A neutral flame is recommended for welding. The flame must never enter the oxidizing stage.
- Oil, grease, and dirt near the weld should be removed by a degreasing solvent that does not contain fluorides. The oxide film on the surface of the base metal should be removed by wire brushing or by filing. Preheating is not recommended.
- Welding technique is similar to that used in the welding of aluminum. The inner cone of the flame should be just above the weld puddle. The tip should be held at an angle of 30 to 45 degrees, depending on the thickness of the material being welded. Thin materials require a smaller angle than heavier materials do. The heat should be concentrated in order to prevent buckling and cracking.
- When the weld is completed, the flux should be removed by brushing with hot water. Welded parts should be treated with a solution of 0.5 percent sodium dichromate. Small parts are boiled in the solution for 2 hours. Large parts are heated to about 150 degrees F. and brushed with the hot solution. The parts should be rinsed with clear water and dried with a hot air blast. This treatment gives the material a yellowish appearance.

Lead

Pure lead is a heavy, soft metal with a dull gray surface appearance. When cut, the cross section has a bright metallic luster. Lead is highly malleable and

18-10. A puddling rod, made by forging a steel welding rod into the approximate shape shown in the sketch, is helpful in welding aluminum castings.

Linde Division, Union Carbide Corp.

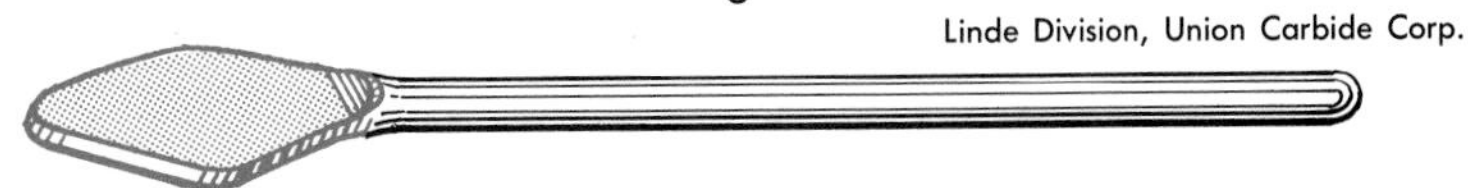

highly ductile. It has very little tensile or compression strength. Antimony may be added to increase the strength of pure lead. Lead is a poor conductor of electricity.

Lead is used in making pipe and containers for corrosive liquids. It is an element in many useful alloys including solder, type metal for printing, and antifriction metals. Lead is also widely used in oil refineries, chemical plants, paper mills, and the storage battery industry.

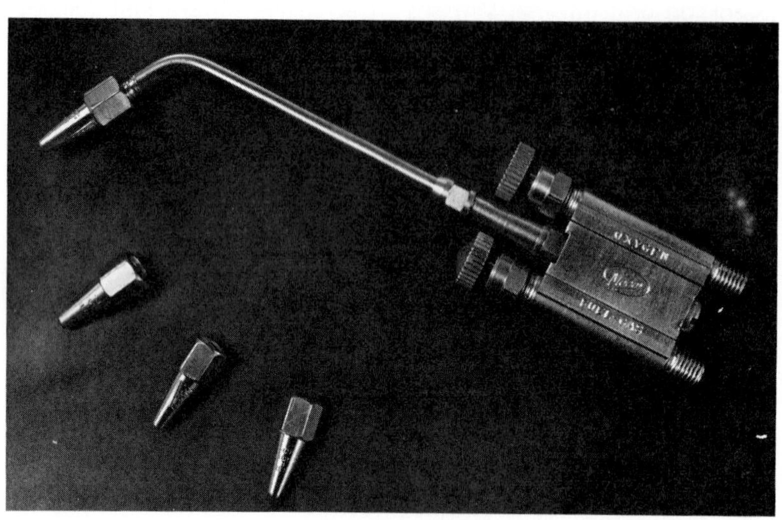

Modern Engineering Co.

18-11. *Midget welding and brazing torch for lead welding.*

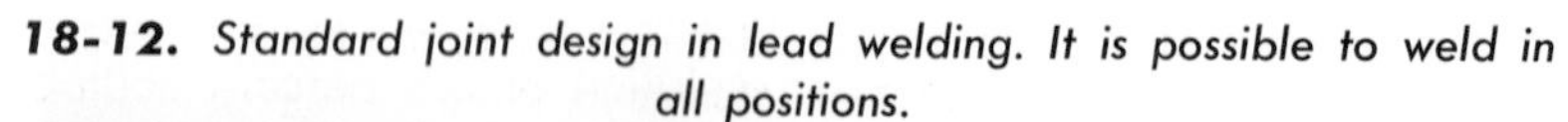

18-12. *Standard joint design in lead welding. It is possible to weld in all positions.*

Lead welding is commonly referred to as "lead burning." This term is incorrect because the lead is not burned. The welding process produces true fusion of the base metal. Lead has a low melting point, about 600 degrees F. It can be welded with oxyacetylene, oxy-propane, oxy-natural gas, or oxyhydrogen. A special small torch, Fig. 18-11, with small tip sizes is used for welding lead with oxyacetylene.

The following factors must be taken into consideration in the welding of lead with the oxyacetylene flame.

- Lead is suitable for all of the standard welded joints, Fig. 18-12. Edge preparation is easy since the metal is soft. The main problem is to make sure that the surface to be welded is clean. This is done with a special scraping tool. Welding may be done in all positions.
- Because of the soft nature of the material, the work must be well-supported to prevent collapse and severe distortion.
- Edge and flange joints may be welded without the addition of filler rod. When filler rod is necessary, it can be made by cutting lead sheet into strips. Special molds are also available for casting rod. The *V* of an angle iron can be used as a mold.
- No flux is required.
- Flame adjustment should be slightly excess acetylene.
- The torch technique is somewhat different from that used with any other metal. Lead has a low melting point and will quickly fall through or develop holes. When filler rod is added to

the weld pool, it is essential that both reach the state of fusion at the same time. Care must also be taken that the weld pool does not get too large and too fluid. Some welders withdraw the torch when the pool becomes too fluid and let it solidify. Then they form a new pool that overlaps the first and again withdraw the torch at the proper time. This bead may be likened to a series of overlapping spot welds. Other lead welders have developed such skill that they have complete control of the molten metal, and they carry a continuous bead with a torch technique somewhat like that used for welding aluminum.

- Lack of fusion is not a problem in lead welding. The chief difficulties are excess fusion and burn-through.

HARD FACING (SURFACING)

Hard facing is the process of welding on wearing metal surfaces a coating, edge, or point of metal which is highly capable of resisting abrasion, corrosion, erosion, high temperature, or impact. The process is also called *metal surfacing.* A surface that is worn away can be restored to its original state, or additional qualities can be given to the material. The part to be surfaced may be prepared as shown in Fig. 18-13.

Hard facing has the following advantages.

- Hard-faced surfaces will last two to forty times as long as low carbon steel, depending on the type of surfacing alloy and the service required.
- The service life of a part is greatly increased. Production schedules are maintained due to fewer replacements and more continuous operation of equipment.
- Surfacing materials need be applied only to those surfaces that will be subjected to severe conditions. Cheaper material may be used to make the part being surfaced.
- Parts can be rebuilt again and again without being replaced.
- Special cutting edges can be formed for special situations.

Welding Processes

A great deal of hard facing is done with the oxyacetylene process, Fig. 18-14. This is the oldest process used. Hard facing is also being done today with the gas tungsten-arc, gas metal-arc, shielded metal-arc, atomic hydrogen, metal spraying, and plasma arc welding processes.

Oxyacetylene may be used for most applications. When a flawless surface is required, the gas tungsten-arc process is highly desirable. The shielded metal-arc process is used when a heavy deposit of metal is necessary. When heavy sections are to be covered with a light coating of surfacing material, the atomic hydrogen process is preferred. Metal spraying can be used for a thin deposit on almost all materials. The plasma arc process will bond almost any material to any other. Ceramics may even be bonded to metals.

Surfacing Materials

There are many types of hard-surfacing rods to be used with the oxyacetylene process. Nearly all of the surfacing alloys have a base of iron, nickel, cobalt, or copper.

- The *cobalt-base rods* are composed of cobalt-chromium-tungsten alloys. This material has high resistance to oxidation, corrosion, abrasion, and heat. The rods have a high degree of hardness at service temperatures of 1,500 degrees F. They recover full hardness at room temperature. There are a number of rod types in this category to meet specific conditions.
- *Nickel-base rods* are recommended for applications involving metal-to-metal wear, temperatures to 1,000 degrees F., and corrosion. They recover full hardness at room temperature after being heated. Several types are available.
- *Iron-base rods* are designed to resist abrasion and impact in varying degrees depending upon the type used. Many of these rods have a high chromium content. They are available as cast rods and tubing.
- *Copper-base alloys* are employed chiefly to resist corrosion and metal-to-metal wear. They are excellent materials for bearing surfaces. They have poor abrasion resistance.
- *Tungsten carbide rods* are made in two forms: a tube type and a composite type. The tube type is a steel tube containing crushed tungsten carbide particles in a choice of grain sizes. When deposited, the tungsten carbide grains are held in a matrix of tough steel. The tube rod provides excellent resistance to severe abrasion and good resistance to impact. The composite type is a cast rod containing controlled grain sizes of crushed tungsten carbide. One is a nickel-base alloy with excellent corrosion resistance and severe abrasion resistence. Another tungsten carbide composite rod is a cobalt-base alloy with excellent high temperature hardness and good corrosion resistance. A third rod has large, hard, sharp tungsten particles in a matrix of bronze. This rod has good corrosion and wear resistance.
- A wide variety of spray hard-surfacing powders are available for all types of service. Some are nickel-base alloys that resist

abrasion and corrosion. Others have a cobalt base and resist abrasion, high temperature, erosion, and corrosion. Still others have a tungsten carbide and nickel base and resist severe abrasion. These powders are applied with standard oxyacetylene spraying equipment.

Application of Hard-Facing Materials to Steel

Hard-surfacing materials may be applied to ordinary steels, high carbon and alloy steels, cast iron, and malleable iron.

1. Thoroughly clean the surface of the base metal by grinding, machining, filing, or wire brushing. Round all sharp corners.
2. Preheat the parts to be hard faced. Small parts can be preheated with the welding torch. Large parts will have to be preheated in a furnace. The parts should be heated to a very dull red, about 800 degrees F. Be careful not to heat the parts too much, or oxidation in the form of scale will form on the surface of the part to be surfaced.
3. Select a tip size that allows slow, careful deposition and good heat control without overheating the base metal.
4. Adjust the torch to an excess acetylene flame. This flame adjustment causes the deposited metal to spread freely. Too little acetylene will cause the material to foam or bubble. On the other hand, too much acetylene will leave a heavy black carbon deposit. The inner cone of the flame should not be more than $1/8$ inch from the surface of the base metal. The excess acetylene flame prepares the steel surface by melting an extremely thin surface layer. This gives the steel the watery, glazed appearance known as "sweating" which is necessary to the hard-surfacing application.
5. Direct the flame at an angle of 30 to 60 degrees to the base metal. Heat a small area until sweating appears, Fig. 18-15. This indicates that the base metal is ready for the application of the hard-surfacing material.
6. Withdraw the flame and bring the end of the hard-surfacing rod between the inner cone of the flame and the hot base metal. The tip of the inner cone should just about touch the rod, and the rod should lightly touch the sweating area, Fig. 18-16. The end of the rod will melt. The backhand welding technique may also be used, Fig. 18-17. If the base metal is not hot enough, the surfacing material will not spread uniformly. If the conditions are correct, the deposited metal will spread and flow as bronze does in braze welding.
7. The pressure of the flame should be used to move the puddle along. Do not stir the weld pool with the rod. Avoid melting too much base metal because it dilutes the hard-surfacing material, thus reducing its service characteristics. Pinholes may be present in the deposited metal. These may be caused by particles of rust or scale that have not floated to the surface with the flame. Pinholes are also caused when there is not enough acetylene in the flame, when the hard-surfacing rod is applied be-

18-14. *Welder uses oxyacetylene torch and Hayne-Stellite no. 6, a hard-surfacing rod, to hard-surface edges of a screw conveyor for handling crushed limestone.*

Linde Division, Union Carbide Corp.

18-13. *A sectional view showing the method of machining steel edges and corners before hard-facing with Haynes-Stellite alloy rod.*

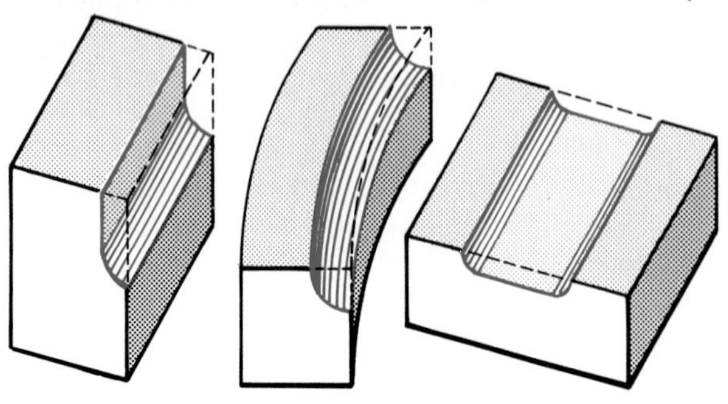

fore the base metal has reached a sweating surface heat, and when the welding flame is removed suddenly from the pool of molten metal.

8. Cool the part slowly. This may be done by furnace cooling or covering it with powdered lime, powdered asbestos, warm dry sand, wood ashes, or an asbestos blanket.

9. After the part has cooled, it may be necessary to remove the high spots or form the part by grinding. Do not use a speed that is slower than 2,800 nor more than 4,200 surface feet per minute.

10. Practice hard facing until you can produce deposits that have the hardness claimed for the rod. Use a hardness-testing machine.

Special Hard-Surfacing Techniques

CAST IRON. Cast iron does not sweat like steel, and less acetylene should be used in the flame. A cast iron flux is used. The hard-surfacing material is applied in a thin layer, and then additional layers are deposited to build up the desired thickness. Do not melt the base metal too deeply.

ALLOY STEELS. High manganese steels, silicon steels, and some forms of stainless steels are difficult to hard face because of their tendency to crack and reluctance to sweat. These materials require special care in preheating and postcooling.

Because of the high rate of expansion of stainless steel, it must be heated evenly and cooled slowly to prevent uneven internal stresses.

COPPER. Copper is relatively difficult to surface. Brass and other alloys with low melting points cannot be hard-faced satisfactorily.

HIGH SPEED STEEL. Hard-surfacing high speed steel is not satisfactory because many times cracks will be formed in the steel below the coating. If it is attempted, the steel must be fully annealed. Heat must be kept even over the entire part. When the application is completed, the entire part must be brought to an even red heat and permitted to cool very slowly.

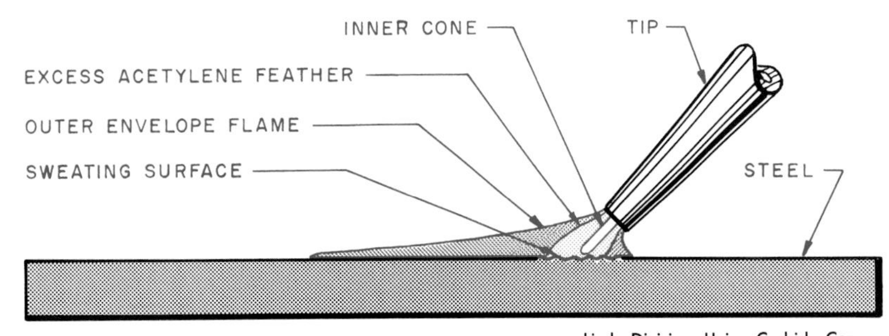

Linde Division, Union Carbide Corp.

18-15. *The proper angle at which to hold the torch and the position of the flame for producing sweating.*

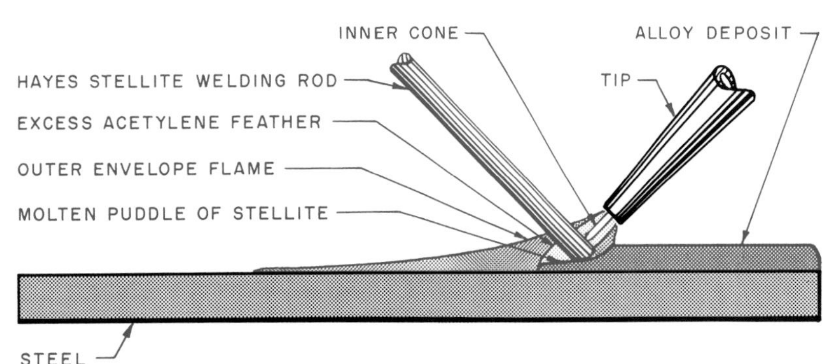

18-16. *The approximate relationship of torch tip, hard-facing rod, puddle, and base metal during hard-facing with the forehand technique.*

18-17. *The backhand technique for hard facing, as in welding, has the hard-facing rod between the flame and the completed deposit. The approximate relationship of torch tip, hard-facing rod, deposit, puddle, and base metal are shown.*

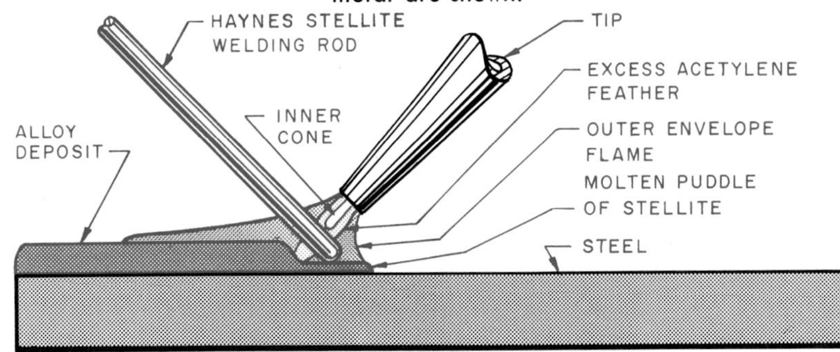

REVIEW QUESTIONS

1. What is braze welding?
2. Malleable cast iron can only be welded with a high test steel filler rod. True. False. Explain your answer.
3. Because the weld pool is highly fluid, braze welding cannot be done in the vertical position. True. False.
4. What is the function of the flux used in braze welding?
5. Braze welding is carried out with an excess acetylene flame. True. False.
6. What is powder brazing?
7. Gray cast iron can only be braze welded. True. False.
8. What is the reason for preheating a casting prior to welding?
9. What is meant by the term *hot-shortness* as applied to aluminum?
10. Aluminum melts at approximately 2,800 degrees F. True. False.
11. It is not necessary to use a flux in the welding of aluminum. True. False.
12. Acetylene is the only fuel gas that can be used in the welding of aluminum. True. False.
13. Cast aluminum cannot be welded with the gas welding process. True. False.
14. Stainless steel is welded almost exclusively with the gas welding process. True. False.
15. Name the three basic types of stainless steel.
16. The fusion welding of all types of copper with the oxyacetylene process is growing rapidly. True. False.
17. Copper welds often contain locked-up stresses and distortion. Why is this so?
18. What is Monel®?
19. Magnesium is not weldable. True. False.
20. What is lead burning?
21. What is the approximate melting point of lead?

Table 18-3

Course Outline: Advanced Gas Welding and Braze Welding Practice: Jobs 18-J 50-64

Job No.	Joint	Operation	Material		Filler rod		Work position	Text reference
			Type	Thickness	Type	Size		
18-J50	Flat plate	Beading (braze welding)	Mild steel	1/8	Bronze RCuZn-C	1/8	Flat	526
18-J51	Lap joint	Fillet (braze welding)	Mild steel	1/8	Bronze RCuZn-C	1/8	Flat	526
18-J52	T-joint	Fillet (braze welding)	Mild steel	1/8	Bronze RCuZn-C	1/8	Horizontal	526
18-J53	Beveled-butt joint	Groove (braze welding)	Mild steel	3/16	Bronze RCuZn-C	1/8	Flat	526
18-J54	Flat casting	Beading (braze welding)	Cast iron	3/16	Bronze RCuZn-C	1/8	Flat	526
18-J55	Beveled-butt joint	Groove (braze welding)	Cast iron	3/16 to 1/4	Bronze RCuZn-C	1/8	Flat	526
18-J56	Casting	Beading (fusion)	Cast iron	3/16 to 1/4	Cast iron RCI	3/16 or 1/4	Flat	529
18-J57	Beveled butt joint	Groove welding (fusion)	Cast iron	1/4 to 1/2	Cast iron RCI	3/16 or 1/4	Flat	529
18-J58	Flat plate	Beading	Sheet aluminum	1/8	Aluminum R1100 or R4043	3/32 or 1/8	Flat	533
18-J59	Outside corner joint	Groove welding	Sheet aluminum	1/8	Aluminum R1100 or R4043	1/8	Flat	533
18-J60	Square butt joint	Groove welding	Sheet aluminum	1/8	Aluminum R1100 or R4043	1/8	Flat	533
18-J61	Beveled-butt joint	Groove welding	Cast aluminum	3/8 to 1/2	Aluminum R4043	3/16	Flat	534
18-J62	Flat plate	Beading	Stainless steel	1/16	Available	1/16	Flat	535
18-J63	Lap joint	Fillet welding	Stainless steel	1/16	Available	1/16	Flat	535
18-J64	Square butt joint	Groove welding	Stainless steel	1/16	Available	1/16	Flat	535

At this point continue welding practice with those other metals described in the chapter that are available in the school shop. Hardfacing should also be practiced.

It is very important that you become proficient in gas cutting. (See Chapter 16.) Practice should include straight line, shape, and bevel cutting with both the hand cutting torch and the machine cutting torch. This is a skill that is also necessary for the electric arc welder.

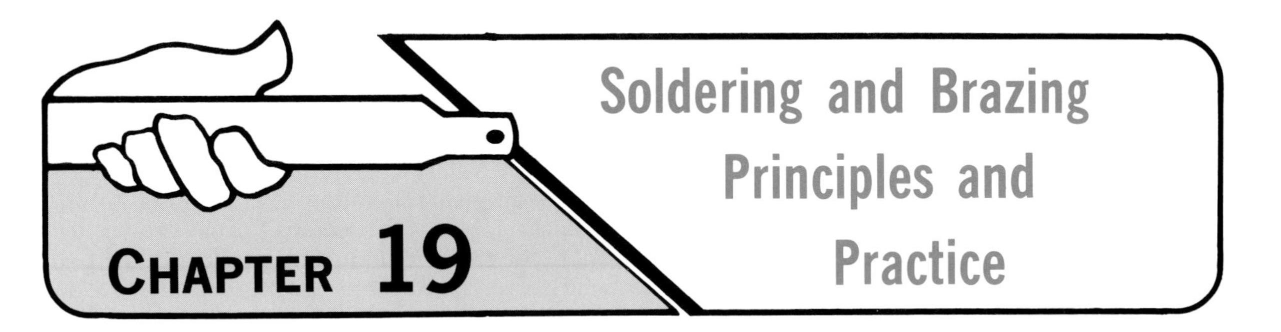

Chapter 19 Soldering and Brazing Principles and Practice

SOLDERING AND BRAZING COPPER TUBING

This chapter deals mainly with the soldering and brazing of copper tubing.

Copper was one of the first metals used. People made tools for handicraft and agriculture, weapons for hunting and war, and decorative and household articles from copper. Pieces of copper pipe, buried for centuries, have been found in excellent condition. This is a testimonial to copper's durability and resistance to corrosion.

Today copper in the forms of pipe and tubing is used in the installation of plumbing and heating, Fig. 19-1. It is light, strong, and corrosion resistant and is available in hard and soft tempers. Tubing comes in a wide variety of diameters and wall thicknesses with clean, efficient fittings to serve every purpose. Joints are made simply and effectively by soldering and brazing.

Copper tubing and pipe are widely used in shipbuilding, oil refineries, chemical plants and, in general, by those industries in which corrosion and scaling are problems. Applications include saltwater lines, oil lines, refrigeration systems, vacuum lines, chemical lines, air lines and low pressure steam lines.

19-1. *Student soldering copper pipe in a plumbing installation. Note the use of the air-acetylene torch.*

Definitions of Soldering and Brazing

The following definitions for soldering and brazing are adapted from those established by the American Welding Society as given in the *Welding Society Handbook.*

Soldering: A joining process in which coalescence is produced by heating, generally to a temperature *below* 800 degrees F., and by using a nonferrous filler metal (solder) that has a melting point below that of the base metal. The molten filler metal is distributed between the properly fitting joints by capillary action. Capillary action is the flow of a liquid when it is drawn into a small space between wet surfaces.

Brazing: A group of welding processes in which coalescence is produced by heating to a suitable temperature *above* 800 degrees F. and by using a nonferrous filler metal having a melting point below that of the base metals. Unlike *braze welding,* described in Chapter 18, the filler metal is distributed between the closely fitted surfaces of the joint by capillary attraction, Fig. 19-2.

Other terms describing soldering and brazing are defined in the Glossary, pages 941–955. You are urged to look up any words you do not know.

SOLDERING

Filler Metals

Selection of the proper materials is an important preliminary step in the soldering of copper tube joints. Selection depends on the metals to be joined, the expected service, operational temperatures, and the expansion, contraction, and vibration that will be experienced during service.

Under ordinary circumstances, a properly made joint will be stronger than the tube itself for stresses of short duration. Because solder is somewhat plastic, however, it may give when stress is maintained at high temperatures over long periods of time. The stress which causes failure under these conditions is less than that which would produce a break with short-time loads. This condition is known as *creep,* and the creep strength of various types of solder varies widely.

Tables 19-1 and 19-2 list the method of classifying solders, their composition, and their melting and solid temperatures. The *pasty range* given in Table 19-2 is the difference between the melting and solid temperatures. Solder is semisolid in the pasty range.

TIN-LEAD SOLDERS. Tin-lead and tin-lead-other alloy solders are the most widely used solders, and they are suitable for joining most metals. A good grade of 50–50 tin-lead solder is generally used for all service at room temperatures and for low pressure steam (up to 15 p.s.i.) as well as for moderate pressures with temperatures up to 250 degrees F. This solder is classified as 50A. It is molten at 418 degrees F. and solid at 362 degrees F.

When the tin content is increased, the flow of the solder and its wetting characteristics are increased. Because the pasty range is wide, however, care must be taken to keep the joint from being moved when it is cooling.

TIN-ANTIMONY-LEAD SOLDERS. Antimony is added for higher strength. The gain in higher tensile strength and creep strength is offset because these solders are more difficult to work with than tin-lead solders. They have poorer flow and capillarity characteristics. The tin-antimony-lead solders may be used for joints in operating temperatures around 300 degrees F. They are not recommended for aluminum, zinc-coated steels, or other alloys having a zinc base.

TIN-ANTIMONY SOLDERS. If higher strengths than those provided by tin-lead solders and tin-lead-antimony solders are required, a 95–5 tin-antimony solder should be used for temperatures up to 250 degrees F. It has a melting point of 464 degrees F. and is completely solid at 452 degrees F. Its very narrow pasty range (12 degrees) makes it difficult to use in the vertical position. The absence of lead, a toxic substance, makes it highly desirable for food handling equipment.

TIN-ZINC SOLDER. These solders are used for joining aluminum. Their melting point ranges from 390 to 708 degrees F., and they are solid at 390 degrees F. As the zinc content increases, the melting temperature increases. The type containing 91 percent tin and 9 percent zinc both melts and solidifies at 390 degrees F. It wets aluminum readily, flows easily, and possesses a high resistance to corrosion with aluminum.

CADMIUM-SILVER SOLDER. The most common solder in this classification is 95 percent cadmium and 5 percent silver. It has a melting temperature of 740 degrees F. and is solid at 640 degrees F. When it is used for butt joints in copper tube, the joint has a tensile strength of 2,600 p.s.i. at temperatures up to 425 degrees F.

CADMIUM-ZINC SOLDERS. The cadmium-zinc solders are used

19-2. *Heat sucks the brazing alloy into the gap where it wets both metal surfaces.*

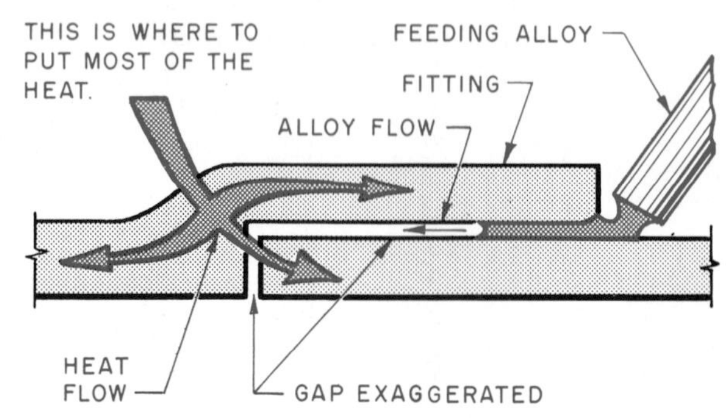

to join aluminum with joints of wide clearance and provide a strong, corrosive-resistant joint. They have a melting range of 509 to 750 degrees F. and solidify at 509 degrees F. The solder containing 90 percent zinc has a wide pasty range of 241 degrees.

ZINC-ALUMINUM SOLDER. The 95 percent zinc, 5 percent aluminum solder is in common use. It is a high temperature solder which melts at 720 degrees. Because of the high zinc content, it also has high resistance to corrosion.

PASTE SOLDERS. These solders are composed of finely granulated solder, generally 50–50 lead-tin, which is in suspension in a paste flux. The flux paste makes cleaning the copper unnecessary. Care must be taken in making a vertical joint because the solder and flux have a tendency to run down the tube.

FORMS OF SOLDERS. The solder in general use for the soldering of copper tubing is commercially available in solid wire form. The wire comes in diameters of 0.010 to 0.30 inch on spools weighing 1, 5, 10, 20, 25, and 50 lbs. Flux may be incorporated with the solder in single or multiple hollows or in external parallel grooves.

Other forms available for special applications include pig, slabs, cakes, bars, paste, tape, powder, foil, and sheet. An unlimited range of sizes and shapes may be preformed to meet special requirements.

Fluxes

A soldering flux is a liquid, solid, or gaseous material which, when heated, improves the wetting of metals with solder. The flux does not clean the base metal. If the base metal has been cleaned, however, the flux removes the tarnish films and oxides from both the metal and solder. When applied to a properly cleaned surface, flux performs the following functions:

- Protects the surface from oxidation during heating
- Permits easy displacement by the filler metal so that it flows into the joint
- Floats out the remaining oxides ahead of the molten solder
- Increases the wetting action of the molten solder by lowering its surface tension

Fluxes are classified into three general groups: highly corrosive, intermediate, and noncorrosive, Table 19-3. The type of flux to use depends upon the metal being soldered, the oxidation rate of the metal, and the resistance of the oxide to removal. Such metals as aluminum, stainless steels and high alloy steels, and aluminum bronzes form a hard oxide film when exposed to air. They require a highly active and corrosive flux. A milder flux can be used with copper because of its slow rate of oxidation and the ease of removal of the oxide. Good soldering practice requires the selection of the mildest flux that will perform satisfactorily in a specific application.

HIGHLY CORROSIVE FLUXES. The highly corrosive fluxes consist of such inorganic acids and salts as zinc chloride, ammonium chloride, sodium chloride, potassium chloride, hydrochloric acid and hydrofluoric acid. They are available as liquids, pastes, and dry salts. Corrosive fluxes

Table 19-1

Melting range of tin-lead solders.

Solder classif.	Percentage of tin %	Percentage of lead %	Completely molten °F	Completely solid °F	Pasty range °F
	0	100	621	621	Pure lead
5A	5	95	596	572	24°
10A	10	90	573	514	59°
15A	15	85	553	437	116°
	19.5	**80.5**	**535**	**362**	173°
20A	20	80	533	362	171°
25A	25	75	514	362	152°
30A	30	70	496	362	134°
35A	35	65	478	362	116°
40A	40	60	460	362	98°
45A	45	55	440	362	78°
50A	50	50	418	362	56°
	55	45	397	362	35°
60A	60	40	374	362	12°
	61.9	**38.1**	**362**	**362**	Eutectic
	65	35	367	362	5°
70A	70	30	376	362	14°
	75	25	385	362	23°
	80	20	396	362	34°
	85	15	408	362	46°
	90	10	421	362	59°
	95	5	434	362	72°
	98.4	**1.6**	**444**	**362**	**82°**
	99	1	446	414	32°
	100	0	450	450	Pure tin

American Welding Society

Table 19-2
Melting ranges of solders containing other metals.

Tin-antimony-lead solders						
Solder classif.	Percent tin	Percent antimony	Percent lead	Completely molten °F	Completely solid °F	Pasty range °F
20C	20	1.0	79.0	517	363	154
25C	25	1.3	73.7	504	364	140
30C	30	1.6	68.4	482	364	118
35C	35	1.8	63.2	470	365	105
40C	40	2.0	58.0	448	365	83

Tin-antimony solder				
Percent tin	Percent antimony	Completely molten °F	Completely solid °F	Pasty range °F
95	5.0	464	452	12

Tin-zinc solder				
Percent tin	Percent zinc	Completely molten °F	Completely solid °F	Pasty range °F
91	9	390	390	0
80	20	518	390	128
70	30	592	390	202
60	40	645	390	255
30	70	708	390	318

Cadmium-silver solder				
Cadmium	Silver	Completely molten °F	Completely solid °F	Pasty range °F
95	5	740	640	100

Cadmium-zinc solders				
Cadmium	Zinc	Completely molten °F	Completely solid °F	Pasty range °F
82.5	17.5	509	509	0
40	60	635	509	126
10	90	750	509	241

Zinc-aluminum solder				
Zinc	Aluminum	Completely molten °F	Completely solid °F	Pasty range °F
95	5	720	720	0

American Welding Society

are recommended for those metals requiring a rapid and highly active fluxing action. These fluxes leave a residue that is chemically active after soldering and will cause severe corrosion at the joint if not removed.

INTERMEDIATE FLUXES. Intermediate fluxes are weaker than the inorganic salt types. They consist mainly of such mild organic acids and bases as citric acid, lactic acid, and benzoic acid. They are very active at soldering temperatures, but this activity is very short since they are also highly volatile at soldering temperatures. These fluxes are useful for quick soldering operations. The residue does not remain active after the joint has been soldered, and it can readily be removed with materials requiring a mild flux.

NONCORROSIVE FLUXES. The electrical industry is a large user of a noncorrosive flux composed of water and white resin dissolved in an organic solvent such as abietic acid or benzoic acid. The residue from these fluxes does not cause corrosion. Noncorrosive fluxes are effective on copper, brass, bronze, nickel, and silver.

PASTE FLUXES. Many types of paste flux ranging from noncorrosive to corrosive are available. A paste flux can be localized at the joint and will not spread to other parts of the work where it would be harmful. The body of the flux is composed of petroleum jelly, tallow, lanolin and glycerin, or other moisture-retaining substances.

Joint Design

Although our emphasis in this chapter concerns the soldering of pipe joints, it is well to consider briefly the other joints that are soldered in industry. Figure 19-3 presents typical soldered joint designs and should be studied carefully. Generally, the joint design depends on the service requirements of the assembly. Other factors include the heating method, assembly requirements before soldering, the number of items to be soldered, and the method of applying the solder. If the service conditions are severe, the design should be such that the strength of the joint is equal to or greater than the load-carrying capacity of the weakest member of the assembly. The joint must be accessible because the solder is normally face-fed into the joint. For high production parts solder in the

Table 19-3
Suggested flux types for soldering metals.

Metal	Suggested flux
Copper	Noncorrosive
Copper-tin alloys	Activated rosin or intermediate
Copper-zinc alloys	Activated rosin or intermediate
Copper-nickel alloys	Intermediate and corrosive
Copper chromium and beryllium-copper	Intermediate and corrosive
Copper-silicon alloys	Corrosive (special purpose)
Steel	Corrosive
Stainless steel	Corrosive (special purpose)
Nickel	Corrosive (special purpose)
Monel®	Corrosive (special purpose)
Durinickel	Corrosive (special purpose)
Aluminum	Corrosive (special purpose)

American Welding Society

19-3. *Typical solder joint designs.*

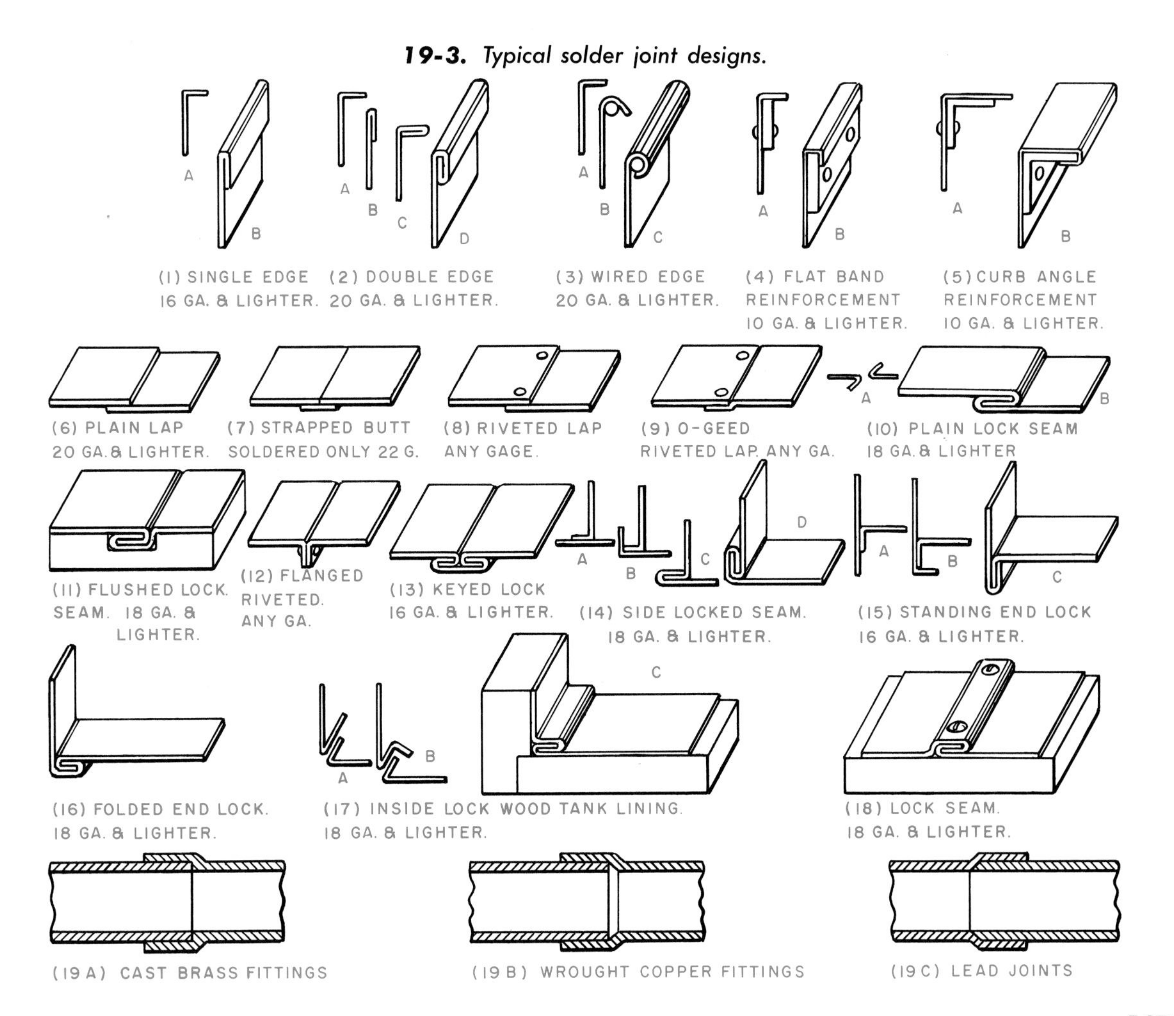

form of wire, shims, strip, powder, a precoat, or solder flux-paste may be preplaced.

Clearance between the parts being joined should be such that the solder can be drawn into the space between them by capillary action, but not so large that the solder cannot fill the gap. Capillary attraction cannot function well if the clearance is greater than 0.010 inch. A clearance range of 0.003 to 0.005 inch is recommended. A joint's tensile strength is reduced as the clearance increases beyond the recommended amount.

Heating Methods

Obviously, heat is necessary to carry out the soldering application. The solder must melt while the surface is heated to permit the molten solder to flow over the surface. Heat may be applied in one of several ways, depending upon the application. Methods include soldering irons, dip soldering, induction heating, resistance heating, oven heating, spray gun heating, and flame heating. In our soldering practice, we will be concerned with flame heating.

The type of the torch to be used depends upon the size, mass, and design of the assembly. Time is also an important factor. Fast soldering requires a high temperature flame and large tip size. Slower techniques require a low temperature flame and small tip size. Fuel gas that burns with oxygen will also burn with air. The highest flame temperatures are reached with acetylene; and lower temperatures, with propane, butane, natural gas, and manufactured gas, in the order named. Care must be taken to avoid a sooty flame since the carbon deposited on the base metal prevents the solder from flowing. In general, the oxyacetylene welding torch, the air-acetylene torch (Fig. 19-4) or the propane torch is used, depending upon the amount of heat necessary. A small hand-held pressure tank with an attached stem is available for small jobs.

19-4. *The air-acetylene torch and accessories used for soldering and light brazing. Note the various tip sizes.*

Modern Engineering Co.

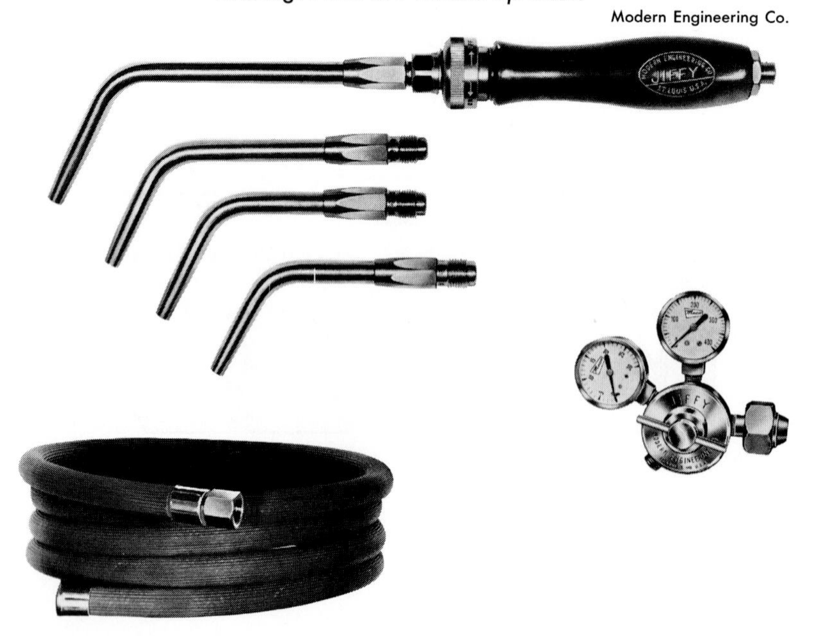

Preparation for Soldering

This practice course will deal with the soldering of copper tubing since this is the type of work a journeyman welder will probably do in industry. Other forms of soldering are usually special applications which are not done by welders.

JOINT PREPARATION. The material covered here applies to the preparation of copper tubing for both soldering and for brazing.

In order to make the assembly, the copper tubing is cut to various lengths and soldered to copper fittings. The end of the tube to be soldered should be square and free from burrs. The outer surface of the end of the tube should be round and within 0.001–0.002 inch of the specified diameter for a distance of 1 inch.

Use a hacksaw with a straightedge ring jig or a square-end sawing vise to cut ends off, Fig. 19-5. A bandsaw equipped for making perfectly square cuts will also do a good job. A pipe cutter, Fig. 19-6, may be used, but you must be careful not to put so much pressure on the cutter that it deforms the tube, Fig. 19-7. A pipe cutter also makes reaming necessary. Regardless of the method of cutting, all burrs on the outside and inside of the tubing should be removed. A hand file can be used for the purpose. If the end of the tube is out of round, a plug sizing can be used to round it off, Fig. 19-8.

PRECLEANING AND SURFACE PREPARATION. Care in cleaning the surface of the material to be soldered is essential. A dirty surface impairs the wetting and alloying action because it prevents the solder from flowing as a thin film. All foreign materials such as oil, paint, pencil markings, lubricants, general atmospheric dirt, and oxide films must be removed before soldering. The strength and adherence of the solder is a function of the surface contact area of the solder to the base metal. Contact may be improved by roughening the surface of the base metal.

Two methods of surface cleaning are employed: mechanical and chemical. The mechanical method is more widely used in soldering and brazing of tube or pipe.

Mechanical Cleaning. Several methods of cleaning may be used, depending upon the nature of the job and the availability of the equipment. These methods include grit or shotblasting, mechanical sanding or grinding, filing or hand sanding (Fig. 19-9), cleaning with steel wool, wire brushing (Fig. 19-10), and scraping with a knife or shave hook. Sandcloth is the most widely used method of mechanical cleaning for copper, brass, and the softer metals.

The end of the tube should be cleaned for a distance only slightly more than that required for the full insertion of the tube into the cup of the fitting. The cup of the fitting should also be cleaned. Cleaning beyond these areas may result in a waste of filler metal and permit the solder to flow beyond the desired areas. Excessive cleaning may reduce the outer diameter of the tubing, and thus cause clearance problems. If the tubing is very dirty, it may require both mechanical and chemical cleaning.

Chemical Cleaning. Chemical cleaning is usually done for production operations. Either solvent or alkaline degreasing is recommended. The vapor condensations of trichlorethylene solvents probably leave the least residual film on the surface. Because of the equipment required for this process, a less expensive method of dipping the tube ends into a liquid solvent such as carbon tetrachloride, trichlorethylene, or methyl chloroform is widely used.

Acid cleaning, also called *pickling,* removes rust, scale, oxides, and sulfides. The inorganic acids—hydrochloric, sulfuric, phosphoric, nitric, and hydrofluoric—are used singly or mixed. Hydrochloric and sulfuric acid are the most frequently used. The tubing should be thoroughly washed in hot water after pickling and dried as quickly as possible.

PRACTICE JOBS: SOLDERING

Instructions for Completing Practice Jobs

The following procedure will serve as a guide in soldering practice. Use copper tubing with diameters of $\frac{1}{2}$ to 1 inch and copper fittings to match the tube diameters. Select a good grade of 50–50 and/or 95–5 solder. Heat may be applied by a gas-air torch or a blowtorch. The gas-air torch

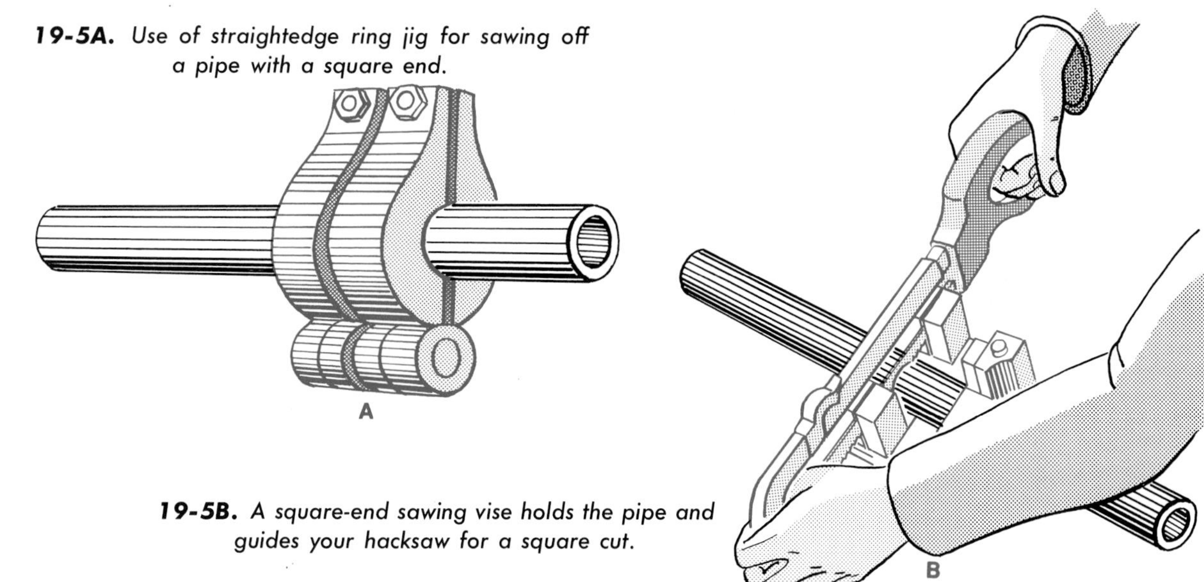

19-5A. *Use of straightedge ring jig for sawing off a pipe with a square end.*

19-5B. *A square-end sawing vise holds the pipe and guides your hacksaw for a square cut.*

is preferred. Practice in all positions.

1. Cut the tubing into 12-inch lengths that can be recut for each new joint. (Check the information on cutting.)

2. Remove the burrs and straighten up the ends of the tubing. Clean the surface thoroughly. (Check the information previously given for these operations.) Figure 19-11 illustrates properly prepared tubing.

3. Flux the tube and fitting surfaces as soon as possible after cleaning. The preferred flux is one that is mildly corrosive and contains zinc and ammonium chlorides in a petrolatum base. Because the chemicals have a tendency to settle from long standing, stir the paste thoroughly when you open a new can. Use solder flux brushes to apply the flux evenly, Fig. 19-12.

4. Assemble the joint by inserting the tube into the fitting. Make sure that the tube is hard up against the stop of the socket. A small twist helps to spread the flux over the two surfaces. The joint is now ready for soldering. The joint may be held in position in a vise. Frequently, a large number of joints are cleaned, fluxed, and assembled before soldering. Do not let the assembling get more than two or three hours ahead of the soldering, and never leave prepared joints unfinished overnight.

5. Adjust the flame as shown in Fig. 19-13. Play the flame on the fitting. Keep it moving so as to heat as large an area as possible, Fig. 19-14. Do not point the flame into the socket. When the metal is hot enough, move the flame away.

6. When the joint is the correct temperature, touch the end of the solder wire to the joint. If the joint has been made properly, a ring of solder will be observed almost instantly all the way around the joint. Opinions differ as to whether or not a fillet is desirable. Never apply the flame directly on the solder. It

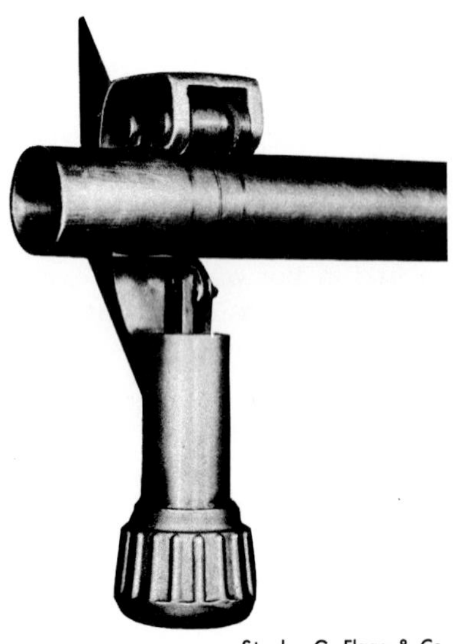

Stanley G. Flagg & Co.

19-6. *Making a square cut in copper tubing with a pipe cutter.*

19-7. *The top sketch shows, with some exaggeration, why using a pipe cutter makes extra work in preparing ends for brazing.*

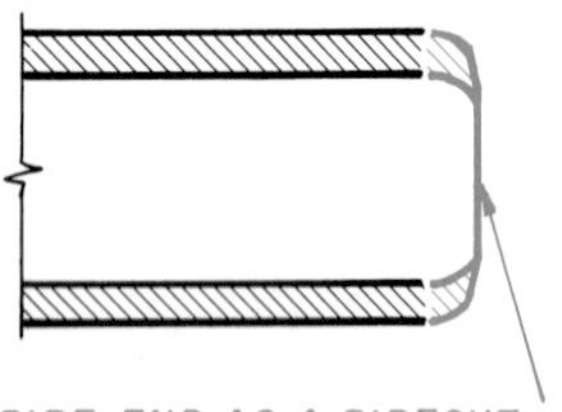

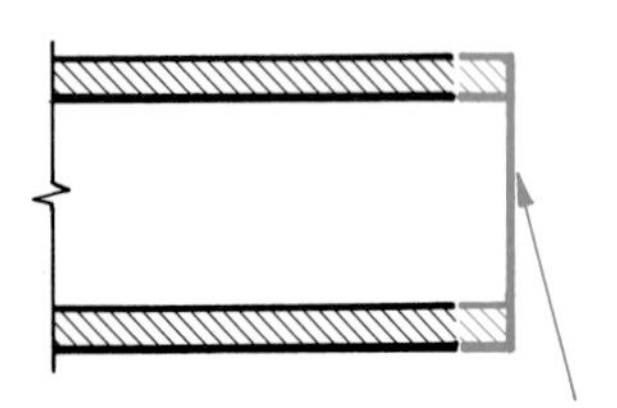

19-8. *For tubing such as copper type B tubing that has walls made to close tolerances, a plug-type tool is very effective in getting the outside diameter to size, true, and round.*

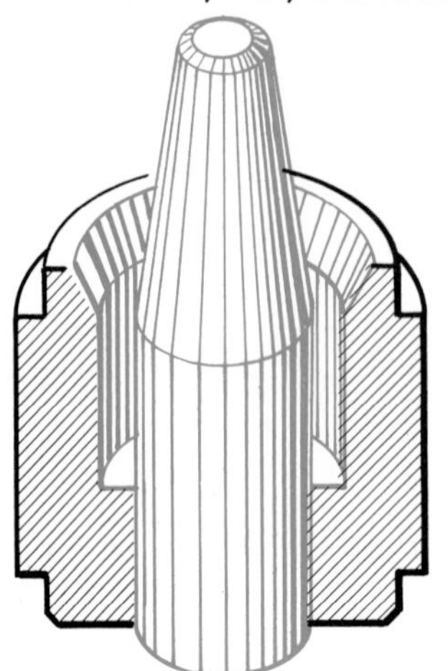

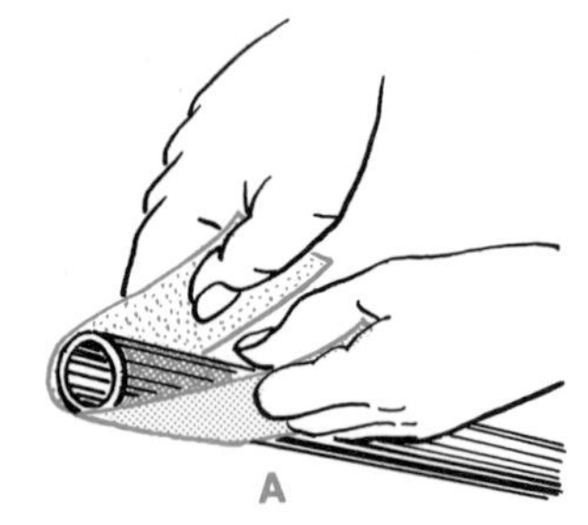

19-9A. *Use abrasives to get scale and dirt off the end of the pipe.*

19-9B. *Clean out the cup and chamfer of each fitting outlet with abrasives.*

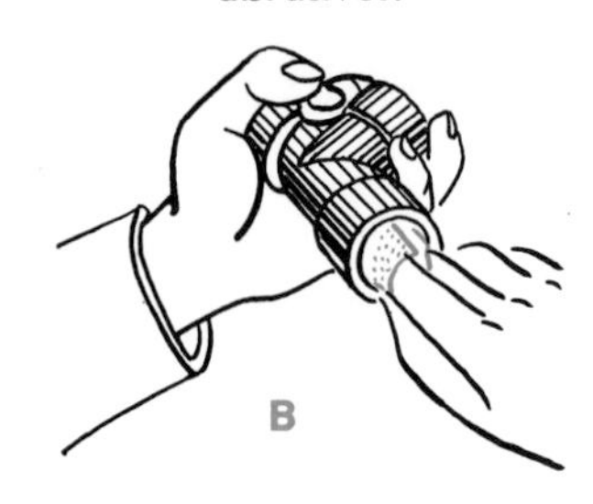

should melt on contact with the surface of the base metal and be drawn into the joint by the natural force of capillary attraction, regardless of whether the solder is being fed upward, downward, or sideways. If the solder does not melt, remove the solder and apply more heat. Then apply the solder again. Avoid overheating, which may burn the flux and destroy its effectiveness. If the flux has been burned, the solder will not enter the joint, and the joint must be opened, recleaned, and refluxed. Overheating of cast fittings may also cause the fittings to crack.

7. While the joint is still hot, remove surplus solder and flux with a rag or brush, Fig. 19-15. This improves the appearance of the assembly and removes any chance of continued corrosive action by the surplus flux.

8. Allow the joint to cool naturally for some time before applying water, particularly if cast fittings are used. Too rapid cooling has been known to crack cast fittings.

9. Practice these joints in all positions until you can make joints with relative ease and have mastered the control of heat and the flow of solder. You are now ready for a check test.

Check Tests

You will perform two kinds of tests to check the soundness of your soldered joints: surface inspection and a water pressure test.

SURFACE INSPECTION.

1. Select a 1-inch T-connection and solder a length of tubing about 12 inches long into each opening of the connection. Following the procedure outlined above, complete the solder joints. Solder the three cups of the fitting in different positions: horizontal, vertical, and overhead.

2. After the joints have cooled and the excess flux has been cleaned off, cut the tubing at each of the ends. Leave about 1

Stanley G. Flagg & Co.

19-10. *Cleaning the inside of the pipe with a wire brush.*

19-11. *This sketch is a checklist of the requirements for a pipe that has been cleaned and sized for soldering.*

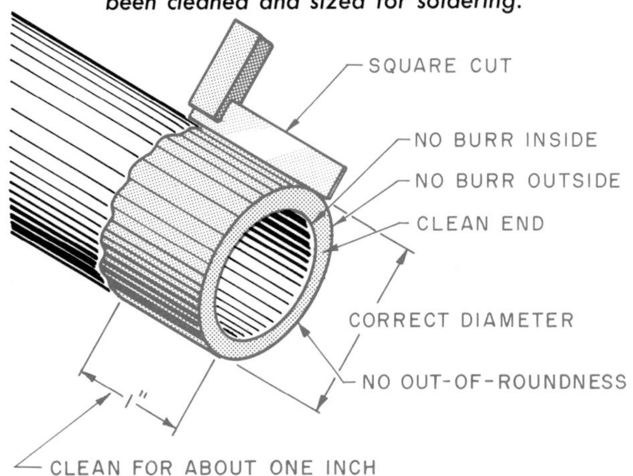

19-12. *Brush flux on the end of the pipe as soon as you have cleaned it. Flux all the cups on a fitting as soon as they are clean. Brush more flux around the joint after you have fitted the two parts together.*

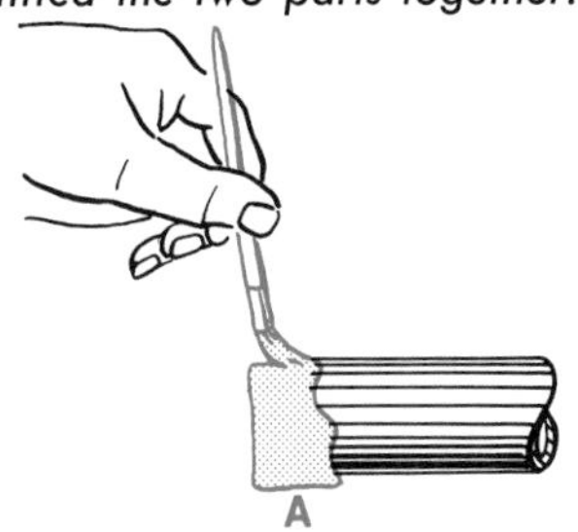

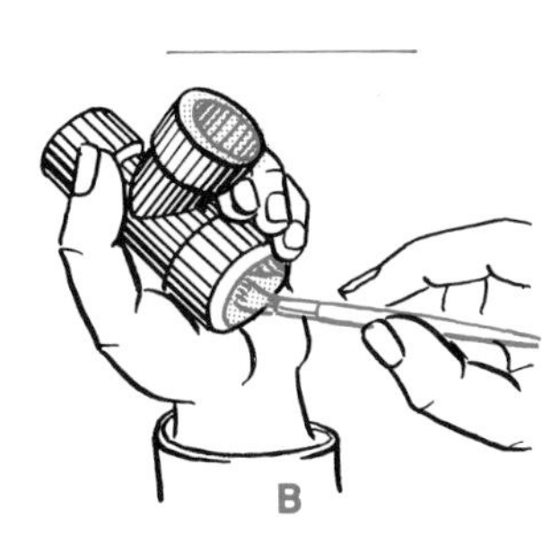

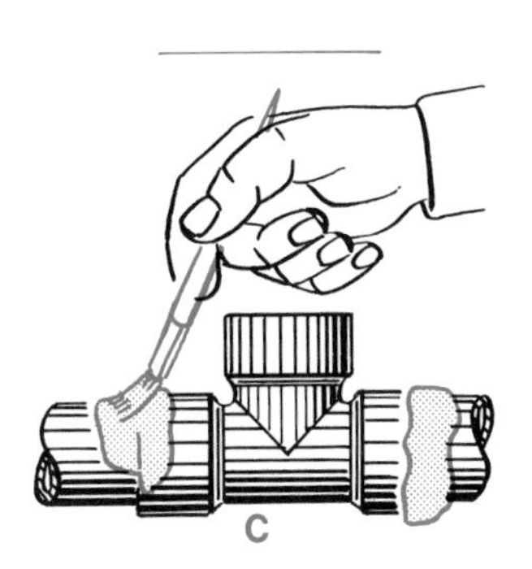

inch of tubing sticking out. Cut the three joints lengthwise along the center line with a hacksaw, Fig. 19-16.

3. Place the half tube and fitting in a vise, with the tube end down and the face of the fitting cup flush against the jaws of the vise. Tighten the vise until the tube end is flattened. Pull the pipe away from the fitting, Fig. 19-17.

4. Inspect the soldered surfaces of the tubing and the fitting. A perfect joint will have the entire cup area of the fitting completely covered with solder, which will have a grayish appearance. Defects include flux inclusions and unsoldered areas.

- *Unsoldered areas:* Fittings are designed to allow an ample safety factor in the joint and permit a bare spot or two if they do not cause leaks. Bare spots running circumferentially may not leak, but they weaken the joint. A bare spot of the same size running lengthwise may leak. Unsoldered areas may be the result of improper fluxing and heating or improper cleaning if the surface does not have the glassy appearance of dried flux. If the unsoldered area is covered with flux, it is caused by a flux inclusion.
- *Flux inclusions:* A flux inclusion indicates that the flux had no chance to flow ahead of the solder. It may be caused by feeding the solder into the joint improperly. On small tubing, the solder should be fed at one point. Shiny areas indicate that although the metal is tinned on both surfaces, there is a flux inclusion between them. This is as serious as though there were no solder at all on these areas. The cause may be that the fitting was too loose on the tube.

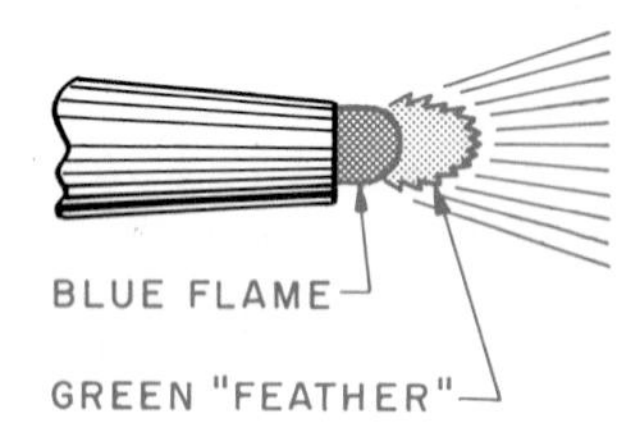

19-13. *Add enough acetylene to the mixture to produce a slight green "feather" on the blue cone. This is a reducing flame.*

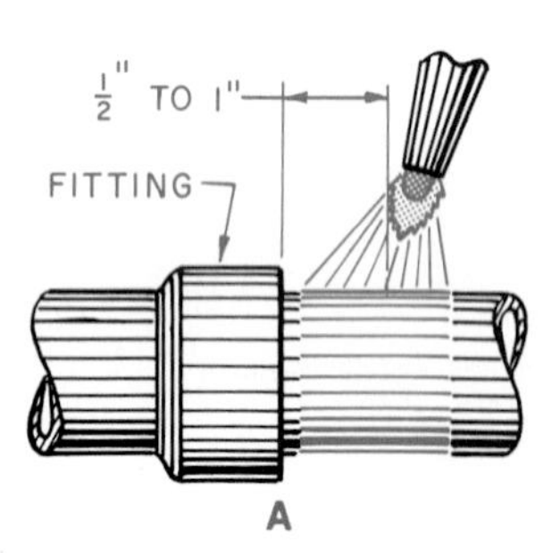

19-14A. *Heat the pipe near the fitting all around its circumference.*

19-14B. *After the pipe is warmed up, shift the heat to the fitting. Keep the flame pointed toward the pipe.*

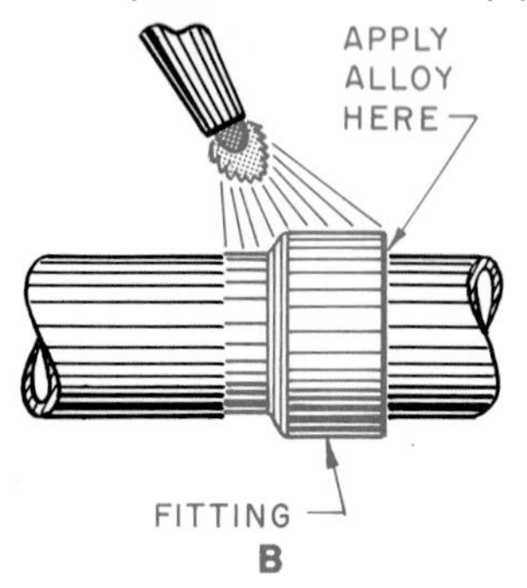

19-15. *Use water and a stiff brush to remove excess flux after completing a joint.*

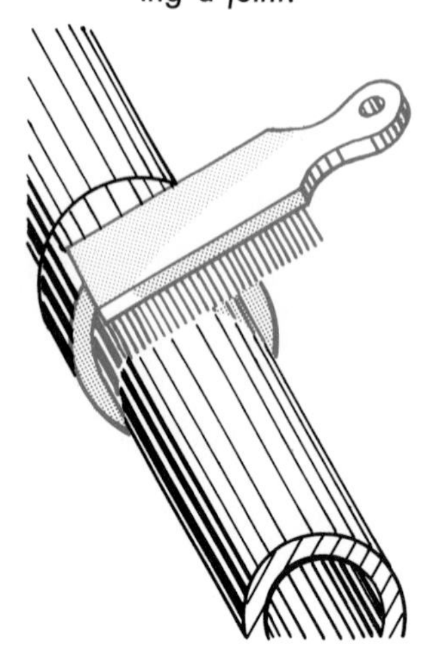

WATER PRESSURE TEST.

1. Make up a closed-line assembly composed of joints made in all positions. Solder the connections with both the silver alloys and the copper-phosphorus alloys.

2. Braze a male-to-female fitting into the line so that you can introduce water under pressure into the tubing. Although the line can be tested with air pressure, the water-pressure test detects more leaks. If flux has sealed any pinholes on the inside of the assembly, water dissolves it and leaks through.

3. Pump water into the assembly at line pressure and observe for leaks. This may require from several hours to a day under test pressure.

TORCH BRAZING (TB)

Brazing is one of the oldest joining processes. It was first used to join ornamental gold fabrications with gold-silver and gold-copper-silver alloys as filler metals. At the beginning of the iron age, copper-zinc alloys, called *spelter*, were developed for joining iron and steel. These alloys are strong and easily melted. They have a vigorous wetting action on clean, fluxed ferrous metals. The early silver solders survive today as silver-base brazing alloys and are of great importance. Silver alloys are used extensively in brazing joints in copper tubing.

The essential differences between brazing and soldering are the much higher melting temperatures of the brazing filler metals and the special fluxes used for brazing. Brazing makes a joint stronger than soft solder, but the higher temperature necessary to melt the brazing filler metal anneals the copper tube in the heat-affected zone.

Strong, leaktight brazed connections for copper water tube may be made with brazing alloys melting at temperatures between 1,100 and 1,500 degrees F. These are sometimes referred to as *hard solders,* a term not universally accepted.

The highest temperature at which a brazing material is completely solid is known as the *solidus temperature.* At the *liquidus temperature* the brazing material is completely melted. This is the minimum temperature at which brazing will take place. The difference between the solidus and liquidus temperatures is known as the *melting range.* The melting range may be important in the selection of the brazing material, particularly as an indication of the rapidity with which the alloy will "freeze" after brazing.

Industrial Applications

The brazing process is used in the joining of copper and other metals. The process has the following advantages.

- Brazed joints are stronger than threaded joints because the pipe or tube is not notched or mutilated. The joints are as strong as the fittings themselves, Fig. 19-18.
- Vibration does not loosen brazed joints. If a system is damaged, the brazed joint will hold together longer than the threaded joint.
- Brazed joints do not leak. A sound joint will stay pressure- or vacuum-tight throughout its service life.
- Corrosion resistance is one of the main requirements of the kinds of piping and fittings commonly assembled by brazing. When copper, brass, or copper-nickel alloy is used to combat rust and deterioration, the joining material must resist corrosion too. The silver alloy filler metal used for brazing is gener-

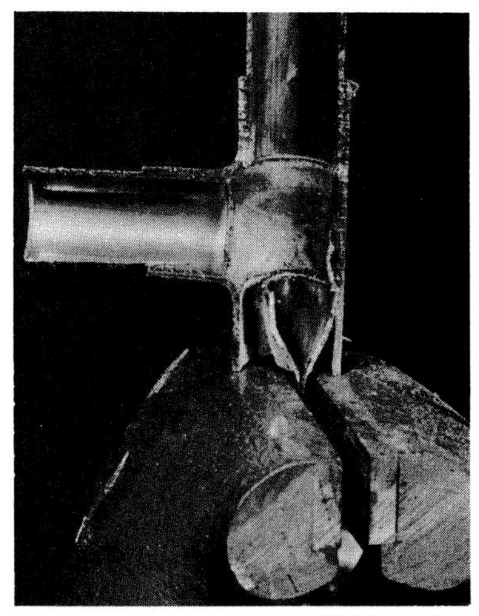

American Welding Society

19-17. *Separate the tube from the fitting.*

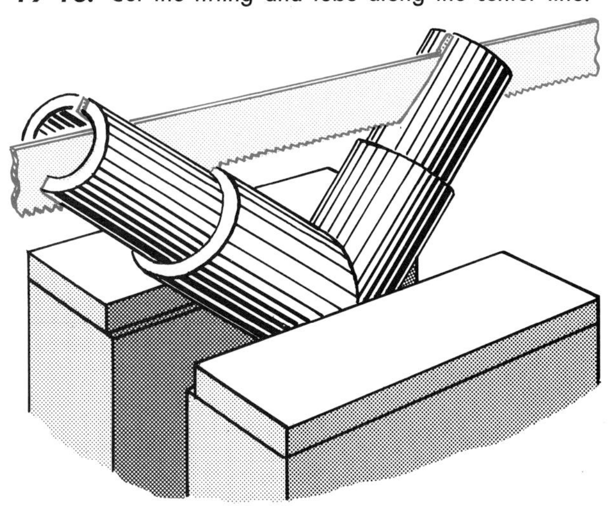

19-16. *Cut the fitting and tube along the center line.*

19-18. *Brazed joints have all the strength of pipe because they have no notches to cause weak spots.*

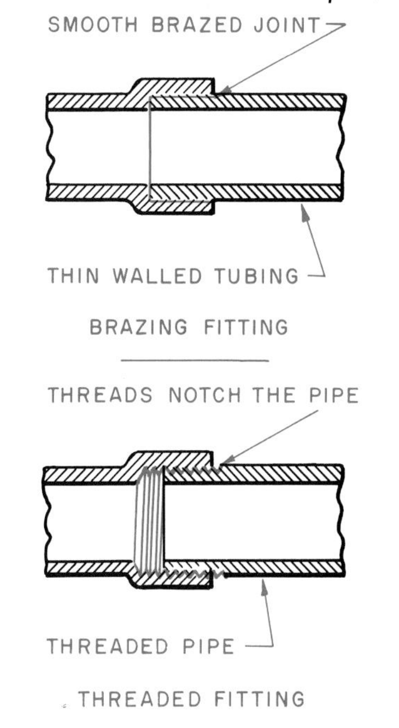

ally as resistant to attack as are these metals themselves.

- Streamlined design, which brazing makes possible, means that there will be less pressure drop, reduction of dead weight, less clogging, and reduced tendency to pit or erode piping near the fittings.
- Accurate assemblies can be made by brazing. Pipe or tubes can be cut to exact dimensions because no guesswork allowance for threading is necessary, Fig. 19-19. The angle of a fitting on a pipe can be preset. There is never a need for overtightening or slacking off in order to line parts up.
- Temporary or emergency piping can be assembled rapidly by brazing.
- Brazed piping can be taken apart, and all the pieces can be reused.

Filler Metals

The American Welding Society lists filler metals under seven classifications: (1) aluminum-silicon, (2) copper-phosphorus, (3) silver, (4) gold, (5) copper and copper-zinc, (6) magnesium, and (7) nickel. The composition and melting ranges of filler metals in these classifications are given in Table 19-4.

ALUMINUM-SILICON FILLER METALS (BAlSi). These are used exclusively for brazing aluminum. They require flux. Type 2 is used as a cladding and applied with dip and furnace brazing. Type 3 is a general-purpose metal for dip and furnace application. Type 4 is used for torch brazing and dip and furnace brazing. It is highly corrosion resistant. Type 5 is used for dip and furnace brazing at temperatures lower than type 2.

COPPER-PHOSPHORUS FILLER METALS (BCuP). We will be concerned with these types of filler metals in the practice course. They are used primarily for joining copper and copper alloys, but they may also be used for joining other nonferrous metals. With copper, these types of filler metals are self-fluxing, but fluxes are recommended for other metals and copper alloys. Type 1 is used for preplacing in joints, and it is suited for resistance and furnace brazing. Types 2, 3, and 4 are all highly fluid filler metals, which are suited for close clearance. Type 5 is used for joints where the clearance is less.

19-19. You can work to accurate measurements by brazing because the cut ends of every piece of pipe are seated against precision-machined shoulders in the fitting.

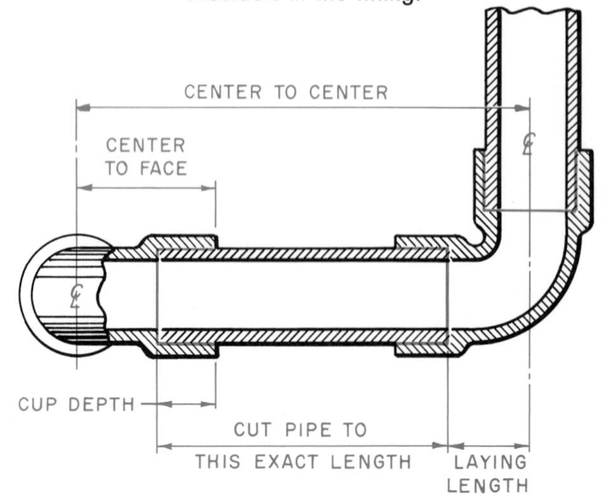

GOLD FILLER METALS (BAu). Gold alloys are used to join parts in electron tube assemblies and for missile components. They are suitable for induction, furnace, and resistance brazing. They require a flux.

COPPER (BCu) AND COPPER-ZINC (RBCuZn) FILLER METALS. These filler metals are used for joining both ferrous and nonferrous metals with borax-boric acid flux. Since copper and copper-zinc alloys are extremely fluid, they require close fits. Overheating will cause volatilization of the zinc. These filler metals should not be used to join copper alloys or stainless steels because of interior corrosion resistance. This group is used for joining ferrous metals, nickel, and copper-nickel alloys.

MAGNESIUM FILLER METALS (BMg). Magnesium alloys are used for joining magnesium with the torch, dip, and furnace brazing processes.

NICKEL FILLER METALS (BNi). These materials are used when extreme heat and corrosion resistance are required. Typical applications include jet and rocket engines, food and chemical processing equipment, automobiles, cryogenic and vacuum equipment, and nuclear reactor components. Nickel alloys are very strong and may have high or low ductility depending upon the brazing method. The filler metal is supplied as a powder, paste, or sheet, or it is formed with binder materials into wire and strip.

Type 1 is highly corrosive and cannot be used with thin sheets. Type 2 has the lowest melting point and is the least corrosive of the group. Type 3 is a chromium-free alloy with a narrow melting range and free-flowing characteristics. Type 4 is similar to type 3. It is used for joints containing large gaps and for forming large ductile fillets. Type 5 resists oxidation up to 2000 degrees F. It is used for high strength joints needed in elevated temperature service, as in nuclear reactor components. Type 6 is a free-flowing filler metal. It produces a minimum amount of corrosion with most nickel and iron-base metals. Type 7 makes a strong leak-proof joint at relatively low temperatures. It is used for thin wall tube assemblies. Ductility increases with time.

SILVER FILLER METALS (BAg). Silver and copper-phosphorus filler metals are the materials we are primarily concerned with in this practice course. Silver alloys are used for joining virtually all ferrous and nonferrous metals with the exception of aluminum, magnesium, and several other metals with low melting points. They are generally free-flowing. Best results are obtained when the clearance between the tube or pipe and the bore of the fitting is held between 0.002 and 0.005 inch. A flux is required.

Types 1 and 1a are general purpose metals, which are free

Table 19-4
Summary of brazing filler metals.

Filler metal	Nominal composition, %[1]											Temperature, °F	
Classification	Copper	Silver	Phosphorus	Zinc	Cadmium	Gold	Nickel	Aluminum	Chromium	Silicon	Boron	Other	Solidus	Liquidus
Aluminum-silicon (BAlSi)-2	. . .	. . .	. . .	. . .	. . .	. . .	. . .	92.5	. . .	7.5	. . .	. . .	1070	1135
-3	4	. . .	. . .	. . .	. . .	. . .	. . .	86	. . .	10	. . .	. . .	970	1085
-4	. . .	. . .	. . .	. . .	. . .	. . .	. . .	88	. . .	12	. . .	. . .	1070	1080
-5	. . .	. . .	. . .	. . .	. . .	. . .	. . .	90	. . .	10	. . .	. . .	1070	1095
Copper-phosphorus (BCuP)-1	95	. . .	5	. . .	. . .	. . .	. . .	. . .	. . .	. . .	. . .	. . .	1310	1650
-2	92.75	. . .	7.25	. . .	. . .	. . .	. . .	. . .	. . .	. . .	. . .	. . .	1310	1460
-3	89	5	6	. . .	. . .	. . .	. . .	. . .	. . .	. . .	. . .	. . .	1190	1485
-4	86.75	6	7.25	. . .	. . .	. . .	. . .	. . .	. . .	. . .	. . .	. . .	1190	1335
-5	80	15	5	. . .	. . .	. . .	. . .	. . .	. . .	. . .	. . .	. . .	1190	1475
Silver (BAg)-1	15	45	. . .	16	24	. . .	. . .	. . .	. . .	. . .	. . .	. . .	1125	1145
-1a	15.5	50	. . .	16.5	18	. . .	. . .	. . .	. . .	. . .	. . .	. . .	1160	1175
-2	26	35	. . .	21	18	. . .	. . .	. . .	. . .	. . .	. . .	. . .	1125	1295
-3	15.5	50	. . .	15.5	16	. . .	3	. . .	. . .	. . .	. . .	. . .	1170	1270
-4	30	40	. . .	28	. . .	. . .	2	. . .	. . .	. . .	. . .	. . .	1240	1435
-5	30	45	. . .	25	. . .	. . .	. . .	. . .	. . .	. . .	. . .	. . .	1250	1370
-6	34	50	. . .	16	. . .	. . .	. . .	. . .	. . .	. . .	. . .	. . .	1270	1425
-7	22	56	. . .	17	. . .	. . .	. . .	. . .	. . .	. . .	. . .	Tin–5	1145	1205
-8	28	72	. . .	. . .	. . .	. . .	. . .	. . .	. . .	. . .	. . .	. . .	1435	1435
-8a	27.8	72	. . .	. . .	. . .	. . .	. . .	. . .	. . .	. . .	. . .	Lithium–0.2	1410	1410
-13	40	54	. . .	5	. . .	. . .	1	. . .	. . .	. . .	. . .	. . .	1325	1575
-18	30	60	. . .	. . .	. . .	. . .	. . .	. . .	. . .	. . .	. . .	Tin–10	1115	1325
-19	7.3	92.5	. . .	. . .	. . .	. . .	. . .	. . .	. . .	. . .	. . .	Lithium–0.2	1435	1635
Gold (BAu)-1	62.5	. . .	. . .	. . .	. . .	37.5	. . .	. . .	. . .	. . .	. . .	. . .	1815	1860
-2	20	. . .	. . .	. . .	. . .	80	. . .	. . .	. . .	. . .	. . .	. . .	1635	1635
-3	62	. . .	. . .	. . .	. . .	35	3	. . .	. . .	. . .	. . .	. . .	1785	1885
-4	. . .	. . .	. . .	. . .	. . .	82	18	. . .	. . .	. . .	. . .	. . .	1740	1740
Copper (BCu)-1	99.90 min.	. . .	. . .	. . .	. . .	. . .	. . .	. . .	. . .	. . .	. . .	. . .	1980	1980
-1a	99 min.	. . .	. . .	. . .	. . .	. . .	. . .	. . .	. . .	. . .	. . .	. . .	1980	1980
-2	86.5 min.	. . .	. . .	. . .	. . .	. . .	. . .	. . .	. . .	. . .	. . .	. . .	1980	1980
Copper-zinc (RBCuZn)-A	59.25	. . .	. . .	40	. . .	. . .	. . .	. . .	. . .	. . .	. . .	Tin–0.75	1630	1650
-D	48	. . .	. . .	42	. . .	. . .	10	. . .	. . .	. . .	. . .	. . .	1690	1715
Magnesium[2] (BMg)-1	. . .	. . .	. . .	2	. . .	. . .	. . .	9	. . .	. . .	. . .	Magnesium–89 Magnesium–0.1	830	1110
-2	. . .	. . .	. . .	5	. . .	. . .	. . .	12	. . .	. . .	. . .	Magnesium–83	770	1050
-2a[2]	. . .	. . .	. . .	5	. . .	. . .	. . .	12	. . .	. . .	. . .	Magnesium–83	770	1050
Nickel (BNi)-1	. . .	. . .	. . .	. . .	. . .	. . .	73.25	. . .	14	4	3.5	Iron–4.5, carbon–0.75	1790	1900
-2	. . .	. . .	. . .	. . .	. . .	. . .	82.4	. . .	7	4.5	3.1	Iron–3	1780	1830
-3	. . .	. . .	. . .	. . .	. . .	. . .	90.9	. . .	. . .	4.5	3.1	Iron–1.5 Max.	1800	1900
-4	. . .	. . .	. . .	. . .	. . .	. . .	93.4	. . .	. . .	3.5	1.6	Iron–1.5 Max.	1800	1950
-5	. . .	. . .	. . .	. . .	. . .	. . .	70.9	. . .	19	10.1	. . .	. . .	1975	2075
-6	. . .	. . .	11	. . .	. . .	. . .	89	. . .	. . .	. . .	. . .	. . .	1610	1610
-7	. . .	. . .	10	. . .	. . .	. . .	77	. . .	13	. . .	. . .	. . .	1630	1630

American Welding Society

[1] Values shown in this table should not be used for purposes of specification. Reference should be made to AWS A5.8.
[2] Contains 0.0005% beryllium for use in furnace brazing.

flowing and low melting. Type 2 is suitable for general purposes at higher temperatures. Type 3 is used for brazing carbide tool tips to shanks and for corrosion-resistant joints in stainless steel. Type 4 is also for brazing carbide tip brazing, but at higher temperatures. Types 5 and 6 are general purpose metals for higher brazing temperatures. Type 7 is a cadmium-free filler metal with a low melting point used for furnace brazing. Type 8 is a silver-copper eutectic used for vacuum tube parts. It is free flowing but does not wet well on ferrous metals. Type 8a resembles type 8, but the addition of lithium makes it self-fluxing on ferrous metals and alloys in a dry, protective atmosphere. Type 13 is a filler metal with a high melting point which is used in aircraft and aircraft engine construction. Type 18 does not contain cadmium and zinc. It is made of sterling silver, and lithium is added to promote self-fluxing. It has a low melting point. This alloy is used for brazing stainless steel for ultra-high-speed aircraft.

UNCLASSIFIED FILLER METALS. There are also available a number of special purpose alloy filler metals based on such uncommon metals as gold, platinum, and palladium. They are used for the most part in brazing vacuum tube components. Such new industrial fields as jet and rocket propulsion and nuclear energy will require new filler metals to be developed.

FORMS OF FILLER METALS. Most of the filler metals are ductile so that they can be rolled or drawn to wire or strip in various standard forms. They may also be supplied as powders. Nickel filler metals are available only as powders although the powders can be bonded into wire or strip with plastic binder materials.

Fluxes

You will recall the importance of fluxing in the soldering process. It carries the same or greater responsibility in the brazing process. The importance of using the right flux becomes clear when we consider all the things the flux is called upon to do. It has to stay on the tube without blowing or washing away while being heated. The flux prevents oxidation from spoiling the clean metal surfaces when they are being heated for brazing. As the brazing alloy flows in, the flux has to flow out of the joint without leaving impurities or inclusions. Finally, it should be possible to clean all flux off the parts easily after the joint has been made. Recommended types can be washed off with water. Fluxes are available in the following forms: paste or liquid, powder, solid coating preapplied on the brazing filler metal, and vapor.

Entirely apart from the protection job it performs on the joint, flux also acts as a temperature indicator. Were it not for this feature, it would be difficult to get the base metal hot enough for brazing without overheating it. When fluxed parts are warmed up, the water in the paste boils off at 212 degrees F. Further heating, to about 600–700 degrees F., makes the flux begin to work (bubble). At 800 degrees F., the flux begins to melt. At 1,100 degrees F., the flux is a clear, water-like fluid, and the bright metal brazing surface can be seen beneath it. At 1,150 to 1,300 degrees F., the brazing temperature has been reached. If the parts reach 1,600 degrees F., the flux will lose its protective qualities. See Fig. 19-20.

19-20. *Here is how flux behaves as the temperature rises.*

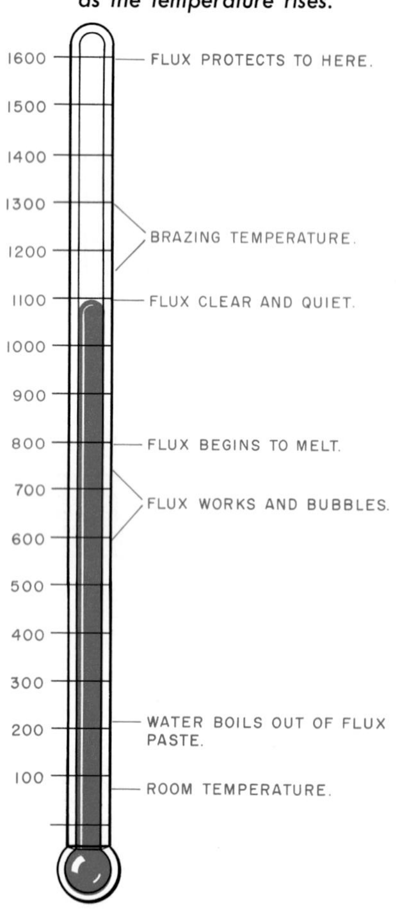

Brazing fluxes include the following materials.

- *Borates.* Sodium, potassium, and lithium borate compounds are used in high temperature fluxes whose melting range is 1,400 degrees F. Their oxide-dissolving characteristics are good.
- *Fused borax.* This material is used as an active flux at high temperatures.
- *Fluoborates.* These are compounds of fluorine, boron, and active metals such as sodium and potassium. They have better flow properties and oxide removal properties than the borates. Protection against oxidation is of short duration.

- *Fluorides.* Fluorides containing sodium, potassium, lithium, and other elements form active fluxes and react readily with even the most stable oxides at elevated temperatures. They work well in dissolving the refractory metal oxides, and they assist in brazing with silver filler metals. Fluorides increase the capillary flow of brazing filler metals.
- *Chlorides* are similar in their fluxing characteristics to fluorides although they are less effective.
- *Boric acid* is a commonly used base for brazing fluxes and is used principally as a cleaning agent. It assists in removing the flux residue from the metal surface after brazing.
- *Alkalies.* These are hydroxides of sodium and potassium. They elevate the temperature at which fluxes are effective. Alkalies have the ability to absorb moisture from the air which limits their usefulness.
- *Wetting agents.* Chemical wetting agents are commonly used in paste and liquid fluxes to improve contact between the flux and the metal interfaces.
- *Water.* Water is present in all fluxes. Hard water cannot be used effectively and if no other water is available, alcohol should be used. Study Table 19-5 for the correct use of commercially available brazing fluxes.

Joint Design

As in our study of the soldering process, our main concern in brazing practice will be joints in copper tubing and piping. It is well, however, to consider briefly the other types of joints that are brazed in industry.

The design of a joint to be brazed depends upon a number of factors, the most important of which are

- *Composition of base and filler metals.* Members to be brazed may be of similar or dissimilar materials.
- *Types of joints.* In selecting the type of joint for brazing, the following conditions must be given consideration: the brazing process to be used, the fabrication or manufacturing techniques required before brazing, the quantity of production, the method of applying filler metal, and service requirements such as pressure, temperature, corrosion, and tightness. The two basic types of brazed joints are the butt joint and the lap joint. Lap joints have the highest efficiency, but they have the disadvantage of increasing the thickness of the joint. The strength of butt joints is usually less than that of lap joints. Figure 19-21 illustrates the various types of brazed joints.
- *Service Requirements.* Your attention is again called to the importance of the service a part is expected to provide. The use to which an assembly is subject is the determining factor. The characteristics of the material to

19-21. *Typical joints for brazing.*

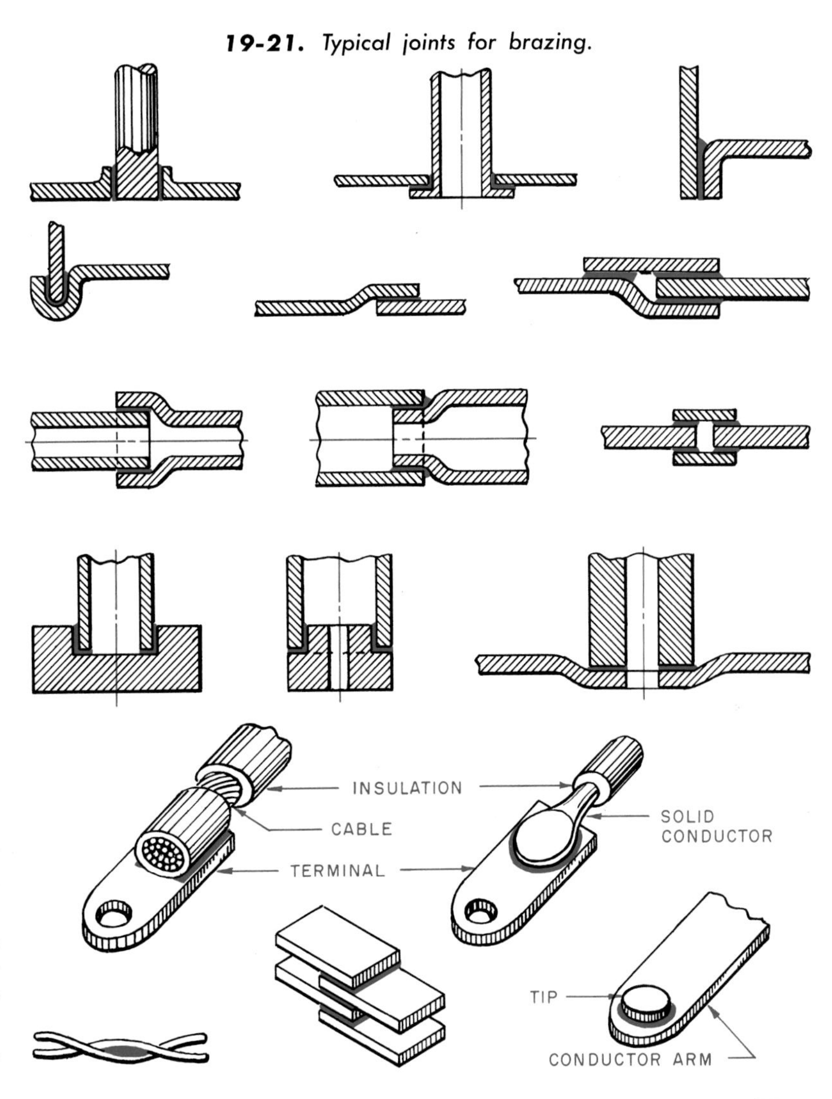

be carried in the system affects joint design. Is it liquid, solid, or gas? What are its temperature and pressure ranges? Is it corrosive? The joint must maintain the properties in the base metal and filler metal which the material demands. These properties include tensile strength and resistance to impact, fatigue, and extremes of temperature and pressure.

- *Stress Distribution.* Joints should be designed to avoid stress concentrations at the brazed area that may cause tearing.
- *Placement of brazing filler metal.* Before designing a brazed joint, it is necessary to select the brazing process and the manner in which the filler metal will be applied to the joint. In most torch-brazed joints the filler metal is simply face-fed. Mass factory production, however, may require the use of automatic equipment for the preplacement of brazing filler metal. See Figs. 19-22, 19-23, and 19-24.
- *Electric conductivity.* In brazing an electrical joint, consideration must be given to the resistance set up by the brazed joint. In general, brazing filler materials have a lower electrical

Table 19-5

Information on applications for brazing fluxes.[1]

AWS brazing flux type no.	Metal combinations for which various fluxes are suitable		Effective temperature range of flux, °F	Major constituents of flux	Physical form	Methods of application[2]
	Base metals	Filler metals				
1	Aluminum and aluminum alloys	Aluminum-silicon (BAlSi)	700–1190	Fluorides; chlorides	Powder	1,2,3,4,
2	Magnesium alloys	Magnesium (BMg)	900–1200	Fluorides; chlorides	Powder	3,4,
3A	Copper and copper-base alloys (except those with aluminum) iron base alloys; cast iron; carbon and alloy steel; nickel and nickel base alloys; stainless steels; precious metals (gold, silver, palladium, etc.)[3]	Copper-phosphorus (BCuP) Silver (BAg)	1050–1600	Boric acid, borates, fluorides, fluoborate wetting agent	Powder Paste Liquid	1,2,3,
3B	Copper and copper-base alloys (except those with aluminum); iron base alloys; cast iron; carbon and alloy steel; nickel and nickel base alloys; stainless steels; precious metals (gold, silver, palladium, etc.)	Copper (BCu) Copper-phosphorus (BCuP) Silver (BAg) Gold (BAu) Copper-zinc (RBCuZn) Nickel (BNi)	1350–2100	Boric acid Borates Fluorides Fluoborate Wetting agent	Powder Paste Liquid	1,2,3,
4	Aluminum-bronze; aluminum-brass[4]	Silver (BAg); copper-zinc (RBCuZn); copper-phosphorus (BCuP)	1050–1600	Borates Fluorides Chlorides	Powder Paste	1,2,3,
5	Copper and copper-base alloys (except those with aluminum) nickel and nickel-base alloys; stainless steels; carbon and alloy steels; cast iron and miscellaneous iron-base alloys; precious metals (except gold and silver)	Copper (BCu); copper-phosphorus (BCuP) Silver (BAg 8–19) Gold (BAu); copper-zinc (RBCuZn) Nickel (BNi)	1400–2200	Borax Boric acid Borates	Powder Paste Liquid	1,2,3,

American Welding Society

[1] This table provides a guide for classification of most of the proprietary fluxes available commercially. For additional data consult AWS specification for brazing filler metal A5.8 ASTM B260; consult also AWS Brazing Manual, 1963 Ed.
[2] 1–Sprinkle dry powder on joint; 2–dip heated filler metal rod in powder or paste; 3–mix to paste consistency with water, alcohol, monochlorobenzene, etc.; 4–molten flux bath.
[3] Some Type 3A fluxes are specifically recommended for base metals listed under Type 4.
[4] In some cases Type 1 flux may be used on base metals listed under Type 4.

conductivity than copper. One approach is to use a shorter lap, thus reducing the bulk of the joint.

- *Pressure tightness.* Wherever possible, the lap joint should be used in the fabrication of pressure-tight assemblies. In making these joints, the entire surface that is to be joined must have uniform coverage. There must be no channels or bare spots through which leakage can occur. It is also important in brazing a closed assembly that it be vented in some way to provide an outlet for the air or gases enclosed. If gases are not allowed to escape during brazing, they will create a pressure on the filler metal flowing through the joint and retard capillary attraction.

Heating Methods

Several methods of heating are used to produce brazed joints. In selecting the method to be used, consideration should be given to the heat requirements of the joint and the materials being brazed, the accessibility to the joint, production quotas, compactness and lightness of design, and the mass of the component.

TORCH BRAZING (TB). Four different kinds of torches are used for the brazing process, depending upon the fuel-gas mixtures. Mixtures include air-gas, air-acetylene, oxyacetylene, oxyhydrogen, and other oxy-fuel gases such as city gas, natural gas, propane, and butane.

Air-gas torches provide the lowest flame temperatures and the least heat. Both air-gas and air-acetylene torches can be used to advantage on small parts and thin sections.

Torches which employ oxygen with city gas, natural gas, propane, or butane provide a higher flame temperature. Like air-gas torches, they are suitable for small components, lower heating speeds, and certain brazing alloys.

Oxyhydrogen torches are often used for brazing aluminum and other nonferrous alloys. The temperature produced is higher than those of the torches previously considered and lower than the oxyacetylene torch. The danger of overheating is reduced. Excess hydrogen provides the joint with additional cleaning and protection during brazing.

Oxyacetylene torches provide the widest range of heat control and the highest temperatures of all of the torches considered. They may be used in a variety of situations and with most filler materials. Because of the high heat possible, extreme skill must be exercised to avoid local overheating. It is an advantage to have the torch constantly moving over the work.

Torch heating is limited to those brazing filler metals which may be used with flux or are

19-22. Preplacement of brazing filler metal in shim form. The bad shim is locked in so that there is no room for expansion nor for gases to escape.

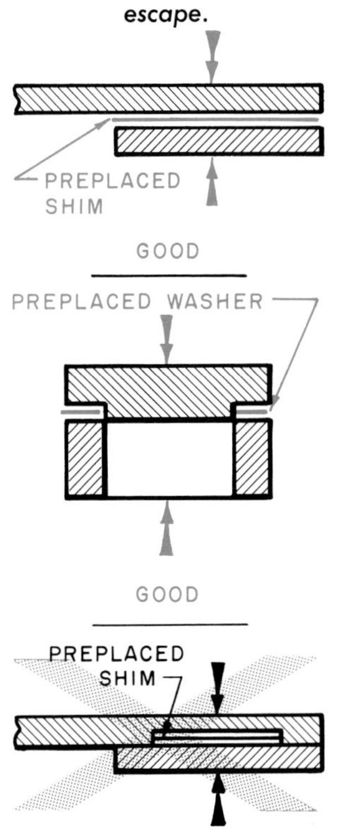

19-23. Methods of preplacing brazing filler metal in wire form.

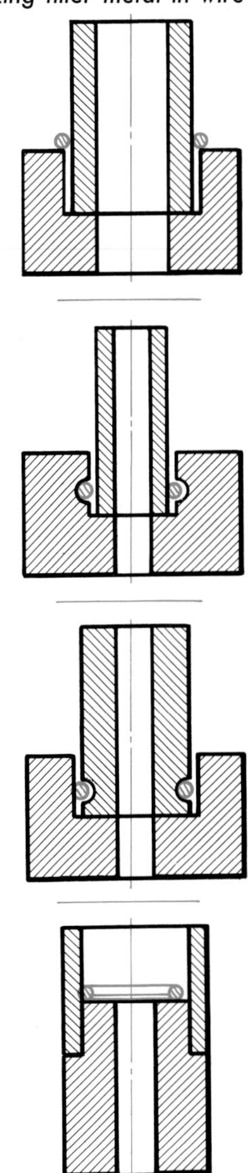

self-fluxing. This includes the aluminum-silicon, silver, copper-phosphorus, and copper-zinc classifications.

FURNACE BRAZING (FB). Furnace brazing is used widely when (1) the parts to be brazed can be preassembled or jigged to hold them in position, (2) the brazing filler metals can be preplaced, and (3) exacting atmosphere control is necessary. The method of heating varies according to the application and relative cost of the fuel. Some furnaces are heated by gas or oil, but the majority are heated electrically. The parts to be brazed are usually assembled and placed in trays for loading into the furnace. The flame does not make direct contact with the parts being brazed. Loading and unloading may be either manual or automatic.

INDUCTION BRAZING (IB). Induction heating is used on parts that are self-jigging or that can be fixtured in such a manner that effective heating will not be reduced by the fixture. Parts to be brazed act as a short-circuited resistance unit in the electrical circuit and are heated as a result. Most of the heat generated by this method is relatively near the surface. The interior is heated by thermal conduction from the hot surface.

19-24. *Preinserted silver-brazing alloy in a groove flows both ways through the cup to make a joint.*

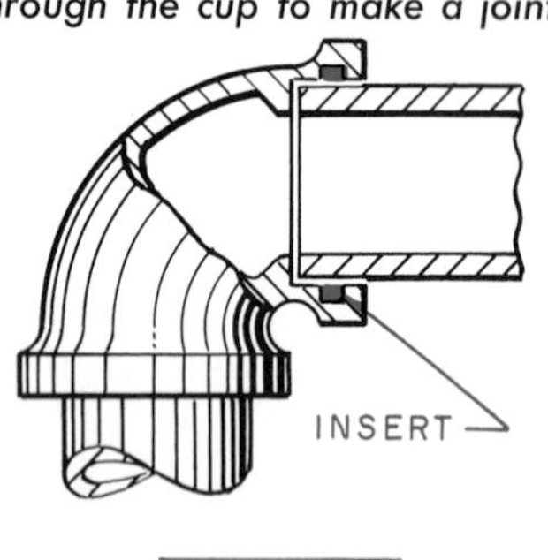

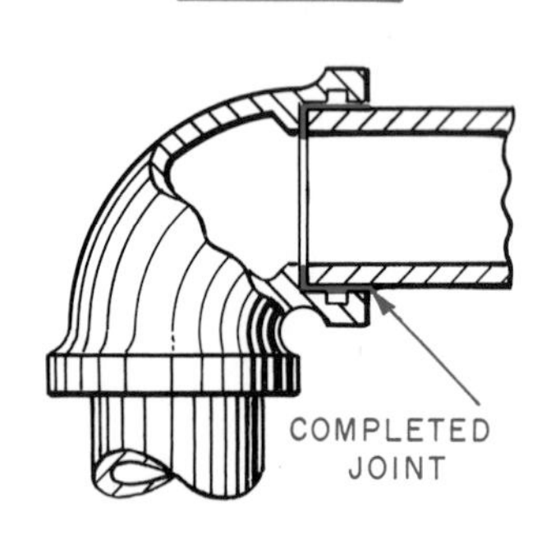

RESISTANCE BRAZING (RB). This process is used when small areas are to be brazed and the material is high in electrical conductivity. The heat is provided by the resistance of the parts to the flow of low-voltage, high-current power supplied by a transformer which is brought to the brazing area through conductors made of carbon, molybdenum, tungsten, or steel. A typical application is the brazing of conductors into the commutator slots in large electrical motors or generators.

DIP BRAZING (DB). Two methods are in common use: (1) molten metal dip brazing and (2) molten chemical (flux) dip brazing.

The molten metal dip process is limited to the brazing of small assemblies such as wire connections and metal strips. The filler metal is melted in a graphite crucible, which is externally heated. A cover of flux is maintained over the molten filler metal. Clean parts are immersed into the molten metal. Care must be taken to make sure that the mass to be brazed does not lower the temperature of the molten metal below that necessary for brazing.

The molten chemical dip process is also limited to small assemblies that can be dipped. The flux is heated in a metal or ceramic container to a fluid condition. Heating may be applied externally or by means of resistance heating of the flux itself. Parts should be cleaned, assembled, and preferably held in jigs before dipping. Brazing filler metal is preplaced as rings, washers, slugs or cladding on the base metal. After dipping the flux is drained off while the parts are hot. Any flux remaining must be removed by water or chemical means after cooling.

INFRARED BRAZING (IRB). This process is used mostly for brazing of small parts. The heat source is a high intensity quartz lamp which produces a radiant heat. These lamps can produce up to 5,000 watts of radiant energy.

DIFFUSION BRAZING (DFB). The demands of the nuclear and aerospace industries have forced the development of many special processes. These processes have been given the general term *diffusion brazing.* They employ a filler metal which diffuses into the base metal under a specific set of conditions of time, temperature, and pressure. The joint brazed by diffusion bonding has a higher melting point than those which have been joined by the normal brazing process. Thus diffusion bonding permits the high service temperatures so necessary to these types of assemblies.

Brazeable Metals

It has been pointed out earlier in the chapter that the purpose here is to provide practice in the soldering and brazing of copper tubing and pipe. These are the kinds of jobs that you will be likely to encounter as a welder. The other forms of brazing discussed in this chapter are production processes performed by semiskilled factory workers. It will add to your confidence and worth as a welder, however, to have some knowledge of the reaction of other metals when brazed.

ALUMINUM AND ALUMINUM ALLOYS. Aluminum and most of its alloys can be brazed. Aluminum alloys with high magnesium content are difficult to braze because of poor fluxing and wetting. Aluminum filler metals usually contain silicon and belong to the BAlSi group. A flux is usually required.

MAGNESIUM AND MAGNESIUM ALLOYS. Most forms can be brazed with filler metals of the BMg group. Torch and furnace brazing have limited applications, but dip brazing can be used for all magnesium alloys. Corrosion resistance depends upon the thoroughness of flux removal.

COPPER. The two types of copper for industrial use are oxygen-bearing copper and oxygen-free copper. The oxygen-bearing coppers contain a small percentage of oxygen in the form of cuprous oxide. At temperatures above 1,680 degrees F., the oxide is active and reduces the ductility of the brazing materials. The tensile strength is not affected by heating.

The oxygen-free coppers do not contain copper oxide and are not subject to oxygen migration or hydrogen embrittlement during brazing. Filler metals of the BAg and BCuP groupings are used.

LOW CARBON AND LOW ALLOY STEELS. These steels can be brazed without difficulty. All of the processes may be used. The BAg, BCu, and BCuZn groups are the best suited filler metals. A flux is necessary.

STAINLESS STEELS. This category covers a wide range of base metals. Most of the BAg, BCu, and BCuZn filler metals may be used. The BAg grades that contain nickel are generally best for corrosion resistance. Filler metals containing phosphorus should be avoided on highly stressed parts because brittle nickel and iron phosphides may be formed at the joint interface. The BNi filler materials should be used for all applications above 800 degrees F. to obtain maximum corrosion resistance. A flux is necessary.

HIGH CARBON AND HIGH SPEED TOOL STEELS. The brazing of high carbon steel is best accomplished prior to or at the same time as the hardening operation. The hardening temperatures for carbon steels range from 1,400 to 1,500 degrees F. Therefore filler metals having brazing temperatures above 1,500 degrees should be used after hardening. When brazing and hardening are done at the same time, filler metals having a solidus at or below the hardening temperature should be used. A flux is necessary.

NICKEL AND HIGH NICKEL ALLOYS. These metals may be brazed by the standard processes. They are subject to embrittlement when mixed with sulfur and metals with low melting points such as lead, bismuth, and antimony. Nickel and its alloys are subject to stress corrosion cracking when brazing and should therefore be annealed before brazing. When high corrosion resistance is desired, the BAg groups are used. The BNi filler metals offer the greatest corrosion and oxidation resistance and elevated-temperature strength. The BCu materials may also be used.

CAST IRON. The brazing of cast iron is somewhat difficult. It requires thorough cleaning with electrochemical flame, grit-blasting, or chemical methods. When the percentage of silicon and graphitic carbon in cast iron is relatively low, the brazing alloys wet without difficulty. Where the percentage is high, wetting is difficult. The high temperature filler metals such as BCu may be used. When silver brazing alloys with low melting points are used, they reduce the oxidation effect and make wetting easier. Cast iron with high carbon content has a low melting point, and care must be taken to use as low a temperature as possible to avoid melting the surface area.

OTHER METALS. Aircraft and missile development have brought into use a number of metals and materials that lend themselves to brazing. These include such metals as tungsten, tantalum, molybdenum, columbium, beryllium, titanium, and zirconium. Most of these can be brazed with the silver-based filler metals as well as other filler metals to meet special conditions.

In addition to the reactive and refractory metals listed above, brazing may also be applied to certain types of ceramics and join graphite to graphite and graphite to metals. Another expanding application is the brazing of dissimilar metals. There appears to be no limit to the uses of the various brazing processes.

PRACTICE JOBS: BRAZING

Brazing materials suitable for joining copper tubing may be divided into two classes: the silver alloys and the copper-phosphorous alloys. The two classes have fairly wide differences in their melting and flowing characteristics. The welder should consider these characteristics as well as the time required to make a joint when he selects filler metal. For joining copper tube with copper or bronze capillary fittings, any brazing alloy in the BAg and BCuP classifications provides needed strength and tightness.

Instructions for Completing Practice Jobs

The following general procedure should be followed for brazing copper tubing and pipe. Brazing should be done in each position until joints of acceptable quality are produced.

1. Select copper tubing with diameters ranging from $\frac{1}{2}$ to $1\frac{1}{2}$ inches and copper fittings to match the tube diameters. Cut the tubing into 12-inch lengths that can be recut for each new joint. Review the information concerning cutting methods on page 548.

2. Refer to pages 549–554 for the proper way of measuring, cutting, and cleaning the tube ends. Tube ends and sockets must be thoroughly cleaned and free of burrs.

3. Choose and apply a flux in accordance with the recommendations of the manufacturer of the brazing alloy. Apply the flux with a brush to the cleaned area of the tube end and the fitting socket. Do not use too much flux and do not apply flux to areas outside of the brazing area. Be especially careful not to get flux into the inside of the tube. Apply flux immediately after cleaning even though the parts are not going to be brazed immediately. The flux prevents oxidation which gradually affects clean metal surfaces even at ordinary room temperatures. If the fluxed surfaces dry out, add some fresh flux when you are ready to braze. Use flux which has the consistency of mucilage or honey. Add water if it gets stiff. A great deal of brazing trouble can be prevented by keeping the flux just right—not too thick and not too watery. Never use a dirty brush for applying flux.

4. Assemble the joint by inserting the tube into the socket of the fitting hard against the stop. Support the two parts so that they are lined up true and square. The joint may be positioned with the aid of a vise, V-blocks, clamps, or other devices. Avoid massive metal supports or clamps with large areas in contact near the joint because they will tend to suck the heat out of the pipe and the fitting.

The strength of a brazed copper tube joint depends to a large extent upon the maintenance of the proper clearance between the outside of the tube and the socket of the fitting. Check the clearance gap carefully. Figure 19-25 indicates that as the clearance increases, the tensile strength of the brazed joint is reduced. Because copper tubing and braze-type fittings are accurately made for each other, the tolerances permitted for each assure that the capillary space will be kept within the limits so necessary for a joint of satisfactory strength.

19-25. *Tensile tests with brazing alloy having a strength of 42,000 p.s.i. show that a small clearance gap makes a joint with as much as three times the strength of the alloy itself.*

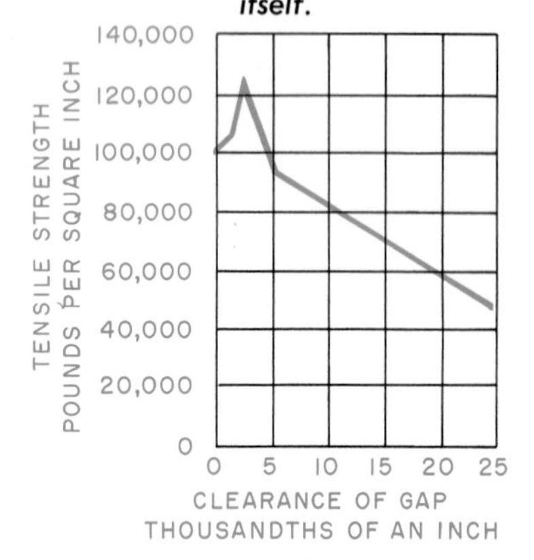

5. Brush additional flux at the joint around the chamfer of the fitting. A small twist of the tube and fitting helps to spread the flux over the two surfaces. The joint is now ready for brazing.

6. Use an oxyacetylene torch for brazing and adjust the flame to slightly excess acetylene. Propane and other gases are sometimes used on small assemblies.

7. Heat the tube, beginning at about 1 inch from the edge of the fitting. Sweep the flame around the tube in short strokes up and down at right angles to the run of the tube. It is very important that the flame be in constant motion to avoid burning through the tube. If the flame is permitted to blow on the tube, it may also wash the flux away. Heating the tube first makes it expand. This causes the tube to press against the cup, and some heat gets carried through to warm up the fitting. Generally, the flux may be used as a guide as to how long to heat the tube. Continue heating after the flux starts to bubble and until the flux becomes quiet and transparent like clear water. This indicates that the tube has reached the brazing temperature.

8. Switch the flame to the fitting at the base of the cup. Heat uniformly by sweeping the flame from the fitting to the tube until the flux on the fitting stops bubbling. Avoid overheating the fittings.

9. When the flux appears liquid and transparent on both the tube and the fitting, start sweeping the flame back and forth along the axis of the joint to maintain heat on the parts to be joined, especially toward the base of the cup of the fitting. The flame must be kept moving to avoid burning the tube or fitting.

10. It is helpful to brush some flux on the brazing rod and to play the flame on the rod briefly to warm it up. Apply the brazing

rod at a point where the tube enters the socket of the fitting. Because the temperature of the joint is hot enough to melt the brazing alloy, keep the flame away from the rod as it is fed into the joint. Keep both the fitting and the tube heated by moving the flame back and forth from one to the other as the filler alloy is drawn into the joint, Fig. 19-26. When the proper temperature is reached, the alloy will flow readily into the space between the tube outer wall and the fitting socket, drawn in by the natural force of capillary attraction, Fig. 19-27. When the joint is filled, a continuous fillet of brazing alloy will be visible completely around the joint. Stop feeding as soon as the joint is filled. If the alloy fails to flow or has a tendency to ball up, it indicates either oxidation of the metal surfaces, or insufficient heat on the parts to be joined. If the work starts to oxidize during heating, there is not enough flux or the flux is too thin. If the brazing alloy flows over the outside of either member of the joint, one member is overheated, the other member is underheated, or both members are at the wrong temperature.

11. After the brazing alloy has set, clean off the remaining flux with a wet brush or cloth. Wrought fittings may be chilled quickly. It is advisable, however, to allow cast fittings to cool naturally to some extent before cleaning. All fluxes must be removed before inspection or testing.

12. Practice these joints in all positions until you have mastered the control of heat and the flow of the brazing material. You are now ready for a check test.

13. Use the same tests outlined for testing a soldering joint, pages 551, 552. All of the conditions and results are similar. Carry out both the surface inspection and the water pressure test.

Brazing Pipe

For tube or pipe 1 inch or larger in diameter, it is difficult to bring the whole joint up to the proper temperature at the same time. A double-tip torch maintains the proper temperature over the larger areas. A mild preheating of the whole assembly, both pipe and fitting, is recommended for the larger sizes. The heating then can proceed as outlined in the steps above.

If difficulty is encountered in getting the entire joint up to the desired temperature at the same time, a portion of the joint is heated to the brazing temperature and the alloy applied, Fig. 19-28. The joint is divided into sectors, and each is given individual treatment. The size of the pipe and fitting determine how much of the fitting cup circumference can be heated and brazed successfully. At the proper brazing temperature, the alloy is fed into the joint and then the torch is moved to the next sector. The process is repeated, overlapping the previous sector. This procedure is continued until the joint is complete all around.

Horizontal and Vertical Joints

Joints in the horizontal and vertical positions can be brazed with the same ease as those in the flat position. This is possible

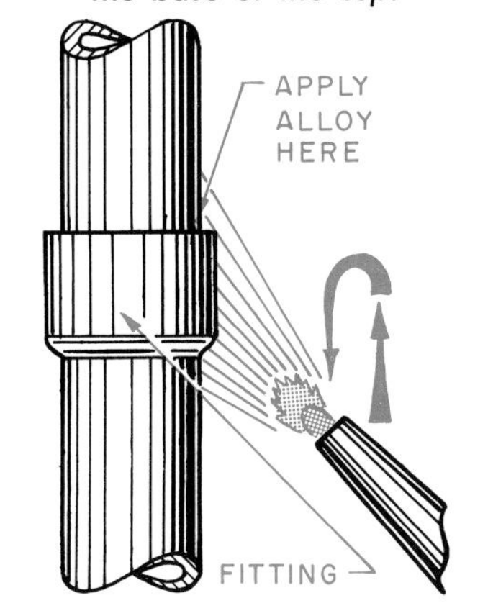

19-26. *Pull the alloy in with a brushing motion of the torch. Concentrate a good deal of the heat on the base of the cup.*

19-27. *Wet metal surfaces causes brazing alloy to flow into the gap.*

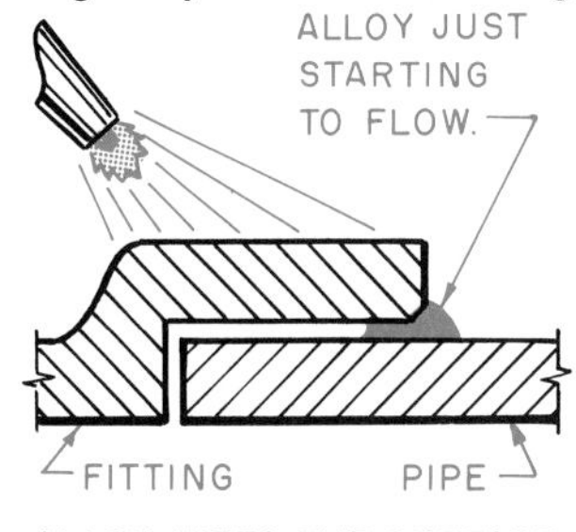

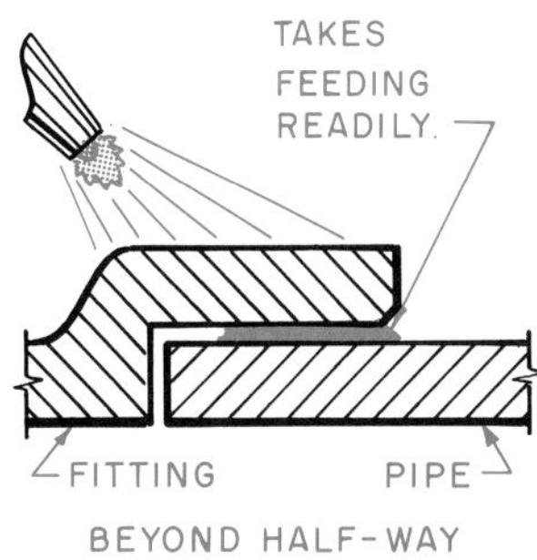

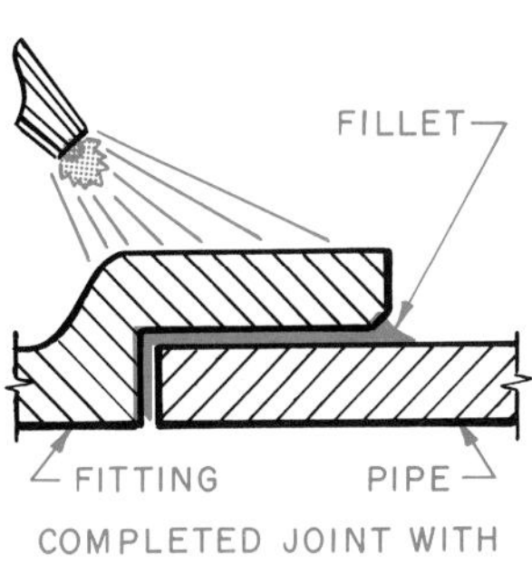

since the filler material is drawn into the joint by capillary attraction and not the action of gravity. The major problem is overheating. If the surface of the bare metal is too hot, the brazing alloy will run out of the joint.

When making horizontal joints, it is preferable to start applying the filler metal at the top, then the two sides, and finally the bottom. Make sure that the filler alloy overlaps. On vertical joints it does not matter where the start is made. If the opening of the socket of the fitting is pointed down, care should be taken to avoid overheating the tube. This may cause the alloy to run down the tube. If the alloy runs, take the heat away and allow the alloy to set. Then reheat the band of the fitting to draw up the alloy. Filler metal is added in the manner indicated in Fig. 19-29.

Brazed joints can be made in close quarters where screwed piping or flanges would be difficult or even impossible to handle. The torch can be formed so that you can reach hard-to-braze joints. Bend the extension on the torch or the tip, if it is a gooseneck type, to direct the flame where you want it, Fig. 19-30. Find the most comfortable place to stand or sit where you can keep the torch in a generally horizontal position. It may be helpful to rig a polished metal mirror that will enable you to observe the flux on the far side of the joint. The torch flame will supply illumination. It may also be necessary to protect all of the surfaces around the joint. Wet rags, sheet metal, or asbestos sheeting may be used as a protective shield.

19-28. *Work in overlapping sectors around the pipe until you have completed the fillet.*

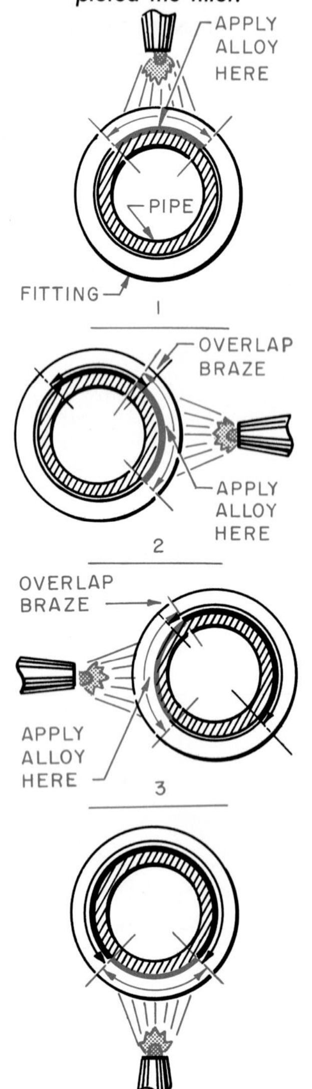

Fittings That Can Be Brazed

Pipe fittings and valves of all types are commercially available for joining copper tubing and pipe by brazing. The following fittings require special techniques.

- *Couplings, Ls, Ts, and crosses* are available in a variety of sizes and types. The cups must be cleaned thoroughly and well-fluxed.
- *Fittings with both brazed and threaded ends:* Do not braze next to a screwed joint. Heat damages the compound which seals the threads, and the fitting may leak. Make up the brazed end first and then the threaded end, Fig. 19-31.
- *Unions:* Protect the ground sealing surfaces of unions with a generous supply of flux, Fig. 19-32. This will keep them from tarnishing and the surface from being damaged when heat is ap-

19-29. *Concentrate heat at the base of the cup to draw the alloy upward.*

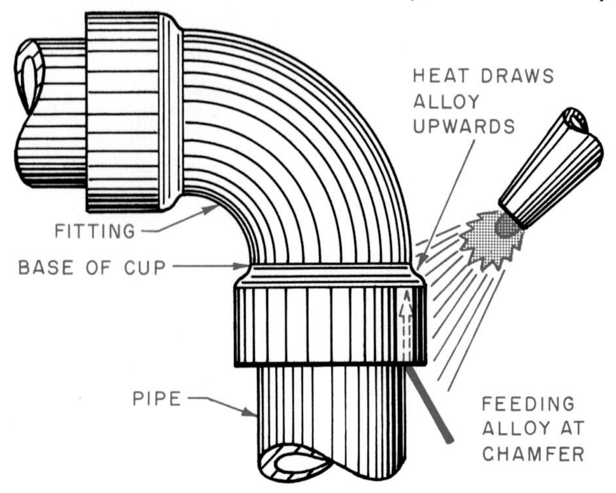

plied to make the joints. Do not play the torch flame directly on the ground surfaces. If the union is assembled during brazing, run the nut up only by hand.

- *Flanges:* Heat should be applied to the hub of the flange, Fig. 19-33. Heat warps a seating surface. Large flanges may need preheating from another source in order to make sure the flange is evenly heated all over. The entire assembly must be cooled slowly.
- *Return bends* must be free at one end as the brazing operation is carried on in order to provide for expansion.
- *Valves:* Do not remove the valve bonnet. It helps to stiffen the valve during brazing. The valve should be opened wide and then backed off just a little so that it will not be jammed in the open position. Wrap a wet cloth around the bonnet to protect it.

Each cup should be cleaned, fluxed, and brazed like any other fitting. A light coat of flux on the valve seat and on the plug will protect these surfaces, Fig. 19-34. Do not direct the flame toward the body of the valve. Keep it directed near the ends and toward the pipe. Most of the heat should be at the base of the cup.

Aids to Good Brazing

Many of the points set out here have been mentioned elsewhere in this chapter. They are of such importance, however, that they are condensed and repeated for you to review.

FLUXING. Fluxing is a very important operation. Oxidation forms surface deposits. Flux reduces this action and soaks up the few oxides that may form.

Do not expect flux to clean a dirty metal surface. The surface must be cleaned by chemical or mechanical means. The flux will protect after the surface is cleaned.

High pressure joints that take longer to braze than standard connections and metals like copper, nickel, and steel, which have a greater tendency to oxidize than other metals, require a thicker coating of flux. One way to apply more flux is to keep the mixture thick.

Watch the flux closely during brazing. The clear, still, watery look it gets at about 1,100 degrees F. tells you that the metal is almost hot enough to melt a silver brazing alloy. Overheating destroys the effectiveness of the flux.

SUPPORT. Keep a piping assembly well-supported so that there are no strains on it while you work. Allow for the expansion that results from the heat of the joint.

POOR CAPILLARY ACTION. If the alloy runs down the pipe, the joint may be too hot, the surfaces may not be clean or sufficiently fluxed, or the joint clearance may be too great, Fig. 19-35. Correction is obvious.

TIMING. Timing—knowing when to flux, when to heat, and when

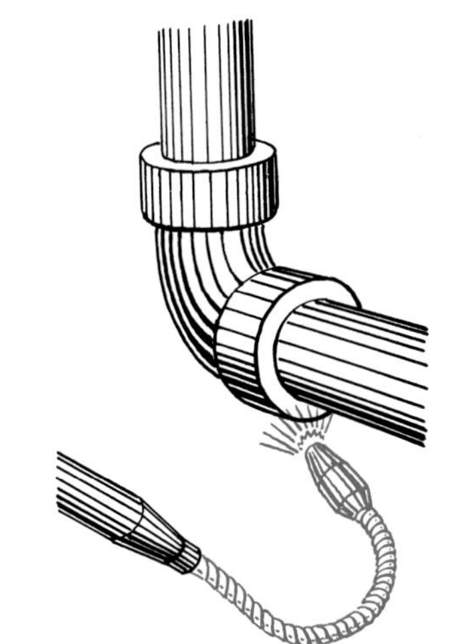

19-30. *Bend the flexible extension on your torch tip to put the flame where you want it.*

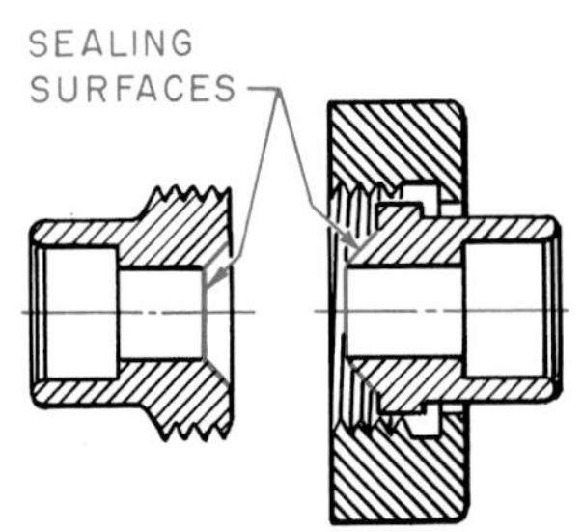

19-32. *Flux the sealing surfaces of a union to prevent them from tarnishing in the heat required during brazing.*

19-31. *Always make up the silver-brazed joint first because heat will damage pipe-thread compound. See that necessary unions are provided for disassembly and servicing.*

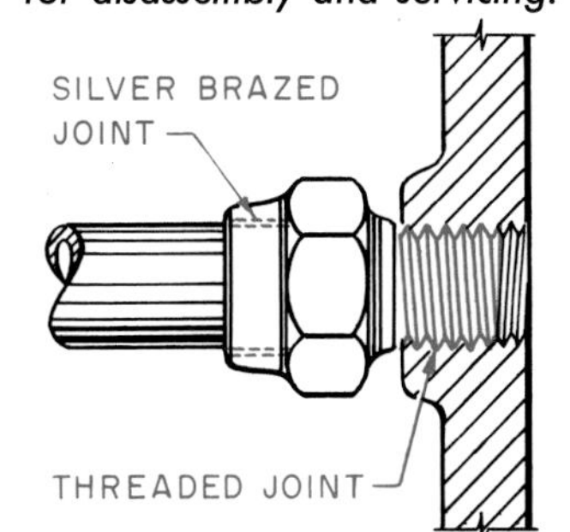

19-33. *When brazing a flange, heat the base of the cup—not the outer part of the fitting.*

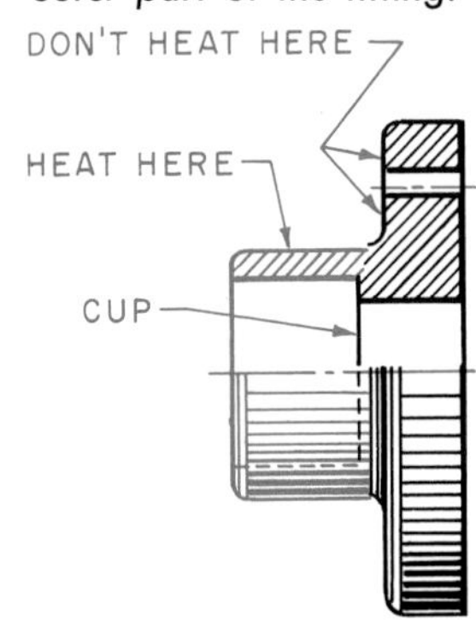

to braze—is extemely important. Brazing is a speedy operation. In fact, one of the worst things you can do is to take too long.

The time to put on flux is right after a surface has been cleaned. If the flux has dried out, put on more moist flux.

When you begin the heating operation, be careful not to heat the workpiece beyond the brazing temperature. It is important to know when to stop heating. When the flux shows that the metal is hot enough, start feeding the filler alloy. Work as rapidly as the alloy can be flowed in. Take the torch away as soon as the fillet is formed. Nothing is gained by more heating.

Two points, then, are most important: apply flux immediately after cleaning and do not heat the assembly any longer than necessary.

AFTERCARE. Be sure to remove the flux after you have finished brazing. This may be done while the joint is hot or you can wait until it cools. Scrub the joints with water. The inside of piping systems should be flushed with water to take away the flux inside.

DISASSEMBLY. It is often necessary to take brazed piping apart. First, brush flux around the area of the fillet at the edge of the cup. Put the pipe in a vise and heat the joint as you would to braze it. It takes about as much time and heat to take a joint apart as it did to braze it. When it is up to brazing heat, pull the tubing out of the fitting. If the alloy is melted, an easy pull does the job.

Pipe and fittings can be used again after they are taken apart. Wipe the molten alloy off the pipe and out of the cup before it sets. After the parts have been cleaned, they can be used again like new material.

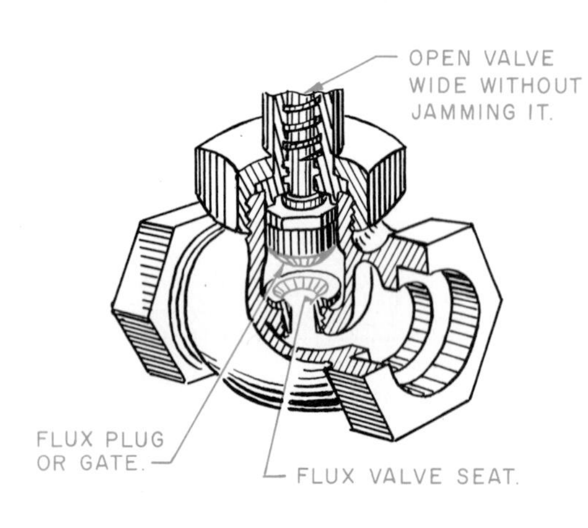

***19-34.** Open a valve wide and then close it part of a turn. Flux the sealing surfaces to prevent damage while you are heating the valve.*

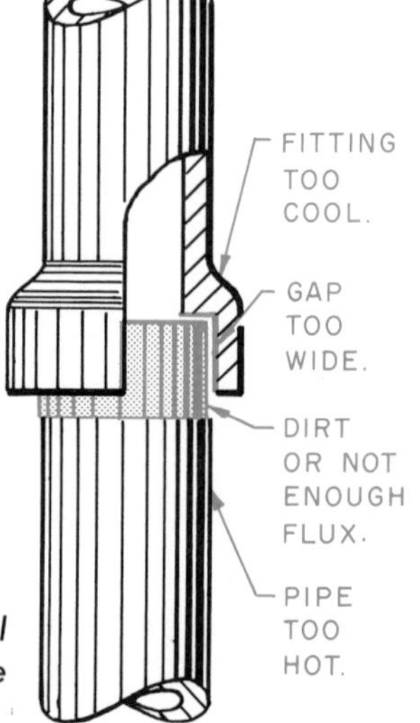

***19-35.** Some of the things that will make silver brazing alloy flow the wrong way.*

REVIEW QUESTIONS

1. List at least five uses of copper in industry.
2. What is the major difference between soldering and brazing?
3. What is capillary action?
4. One of the advantages of soldering is that the joint is strong at high temperatures. True. False.
5. Name at least six types of solders.
6. What is the purpose of a flux in the soldering and brazing operation?
7. What types of joints may be soldered?
8. List the steps necessary to prepare a joint for soldering.
9. What type of torch is used for soldering?

10. List the defects that may be present in a soldered joint and explain the cause of each.

11. Brazing is one of the most recent of joining processes. True. False.

12. Give the melting temperature range of brazing alloys.

13. List at least six advantages of brazing copper tubing.

14. List the seven classifications of brazing filler metals and their abbreviations.

15. Gold is an element in some types of brazing filler metals. True. False.

16. List at least five types of fluxes used for brazing.

17. List the types of fuel gas used for brazing.

18. List at least four types of brazing. Give a brief explanation of each.

19. List at least six types of metals that may be joined by the brazing process.

20. Name the two types of brazing alloys that are used to braze copper tubing.

21. One of the disadvantages of the brazing process is that it cannot be used for vertical and overhead brazing. True. False.

22. Name the types of copper fittings that can be brazed.

23. Name at least four important points to remember in carrying out the brazing operation.

24. Solidus is the lowest temperature at which the metal is completely liquid. True. False.

Success in industry as a welder depends upon (1) the mastery of the manual skills involved in the welding operation; (2) an understanding (gained through careful study of the textbook) of the technical and related information on the equipment and materials used in the welding process; and (3) the ability to get along with others on the job.

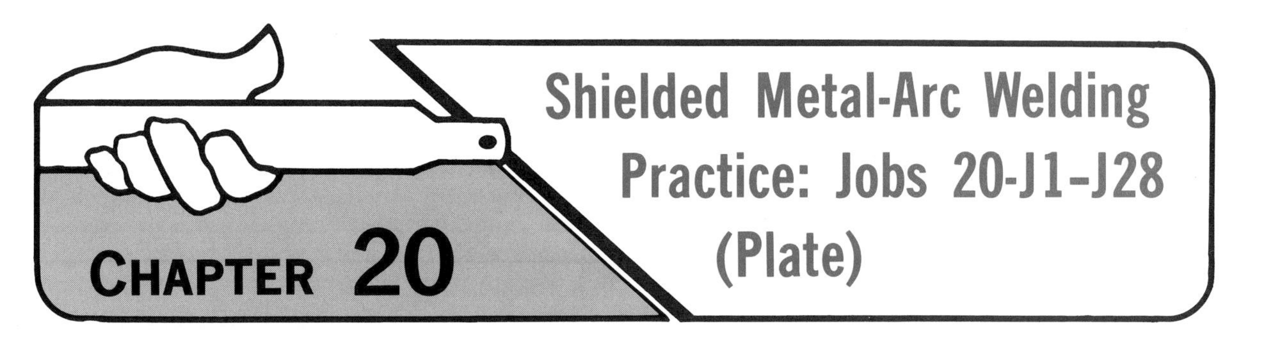

Chapter 20 Shielded Metal-Arc Welding Practice: Jobs 20-J1–J28 (Plate)

INTRODUCTION

Practice in shielded-metal arc welding (SMAW), also called stick electrode welding, is the basis for learning the other types of welding, Fig. 20-1. Approximately 70 percent of the total amount of welding is done with stick electrodes. MIG welding and submerged-arc welding follow stick electrode welding in popularity.

Stick welding has great versatility. It is a dependable process which welds any steel in any position. It is readily available at modest cost. It will probably be used as long as any type of welding is being done.

20-1. *A young welder who has completed training is welding a steel fabrication on the job with the shielded metal-arc welding process.*

The practice jobs in this and the next two chapters are the basis for the mastery of welding. The jobs are numbered consecutively and include all of the basic joints and welds in all positions and provide experience with all of the commonly used electrodes. As you practice you will also learn how to set the welding machine for all kinds of welding situations. A welder must be able to set the welding machine to meet the varying conditions of the job in order to produce satisfactory welding.

The jobs in this chapter provide practice on a number of fundamental joints and welds. Welded joints include the edge, lap, and T-joint. You will learn how to run stringer and weaved beads and make fillet welds in the flat, horizontal, and vertical positions. You will experience the different characteristics and heat requirements of various types of electrodes such as bare, shielded arc, straight and reverse polarity, and alternating current. You will also begin to read shop drawings and become familiar with welding symbols. The satisfactory completion of these jobs will provide a sound basis for the more advanced jobs that require more skill to master.

APPROACH TO THE JOB

Be relaxed. This is necessary if you are to avoid fatigue. When fatigue sets in, it is almost impossible to do good work. Grip the electrode holder lightly. If the electrode holder is held too tightly, your arm will tire, making it difficult to produce a sound weld. The electrode cable should be draped around the shoulder or across the lap to take the weight of it off the holder. When sitting or standing, assume the most comfortable position possible, Fig. 20-2. No part of the body should be under the slightest

20-2. *Student welder is making a fillet weld on a T-joint. He is seated in order to be as relaxed and comfortable as possible. Note that he is wearing all of the usual protective clothing.*
Department of Health, Education, and Welfare

strain. It is equally important that you are at all times protected by use of the proper protective equipment.

Review of Operating Characteristics

Below are described those basic arc welding techniques that the welding student must be able to apply in order to improve with practice. Discuss these points with your instructor to make sure that you understand them.

POLARITY. In a d.c. circuit the current always flows in one direction. One line of the circuit is the positive side, and the other line is the negative side. Polarity determines which line is positive and which line is negative. When welding is being performed with *reverse polarity* (DCRP), the electrode lead becomes the positive (+) side of the circuit and the ground lead becomes the negative (−) side of the circuit, Fig. 20-3A. When welding is being performed with straight polarity (DCSP), the electrode lead becomes the negative (−) side of the circuit and the ground lead becomes the positive (+) side of the circuit, Fig. 20-3B. The positive side of the circuit is the hottest side. The type of electrode determines the kind of polarity necessary for operation.

Reverse polarity is also called positive polarity. Straight polarity is also called *negative, normal,* and *standard* polarity.

In a.c. welding there is no choice of polarity since the circuit is alternately positive: first on one side and then on the other.

ARC LENGTH AND ARC VOLTAGE. A fairly short arc is necessary for metal-arc welding. A short arc permits the heat to be concentrated on the work, and it is more stable since it reduces the effect of arc blow. The flame, consisting of vapors from the arc, surrounds the electrode metal and the arc pool and prevents air from destroying the weld metal.

With a long arc a great deal of the heat is lost into the surrounding area, preventing good penetration and fusion. The flame is very unstable, the effect of arc blow is increased, and the flame has a tendency to blow out. The greater arc length permits air to reach the molten globule of metal as it passes from the electrode to the weld and the weld pool itself. Thus the weld metal absorbs oxygen and nitrogen, both of which are harmful. Weld metal deposited with a long arc has low strength, poor ductility, and high porosity. The weld shows poor fusion and excessive spatter.

The length of arc gap that the operator should maintain depends on the type and size of electrode, the position of welding, and the amount of current.

A shorter arc is maintained for vertical, horizontal, and overhead welding than for flat welding. The normal arc voltage in the flat position for $\frac{1}{8}$-inch to $\frac{1}{4}$-inch bare electrodes is 14 to 25 volts. For covered electrodes it is 22 to 40 volts. Voltages for the other positions are 2 to 5 volts less. Since the arc voltage increases with the arc length, a shorter arc gap is necessary when welding in the vertical, horizontal, and overhead positions. A 14-volt arc with bare or lightly coated electrodes is approximately $\frac{1}{8}$ inch long, and a 40-volt covered electrode arc is approximately $\frac{1}{4}$ inch long. The type of coating on the electrode also has some effect on the arc length.

A good welder can tell if he is welding with the proper arc gap by the appearance of the metal transfer and the weld pool and

20-3A. *Direction of current when welding is being done with reverse polarity.*

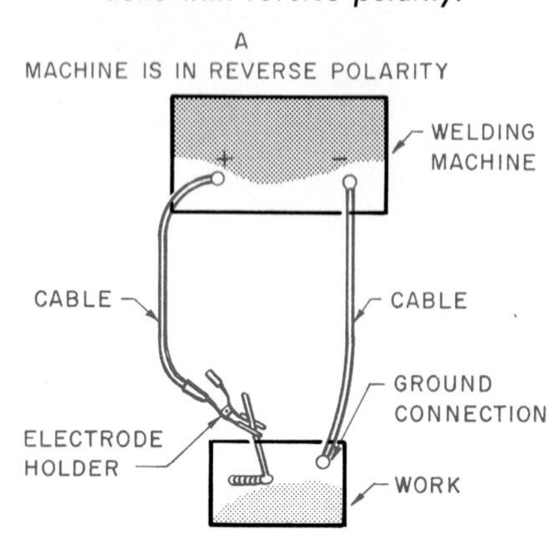

20-3B. *Direction of current when welding is being done with straight polarity.*

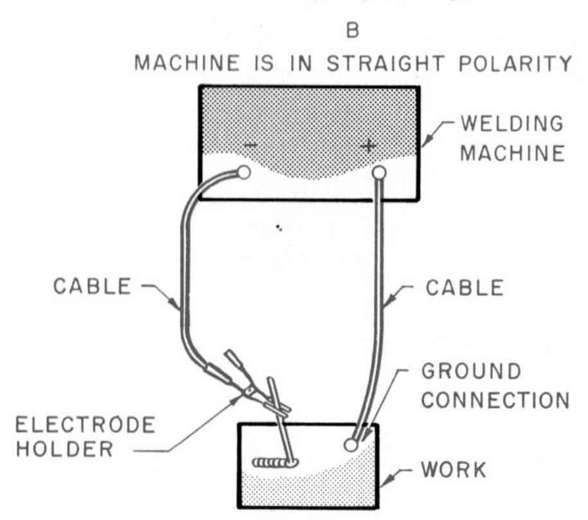

by listening to the sound of the arc. With a long arc he can see the globule of molten metal flutter from side to side, throwing off small particles of metal in all directions. The proper arc gap transfers the metal in a steady stream that is not easily apparent. A short arc makes a sharp crackling sound like grease frying in a pan. A long arc is identified by a hissing sound and explosions at steady intervals.

ARC BLOW. The welding student will find arc blow a disturbing feature of arc welding. Arc blow is the swerving of an electric arc from its normal path because of magnetic forces. It may be likened to a number of magnets all pulling at a ball of steel. The ball will waver between the different magnets and will finally be attracted towards the magnet with the strongest pull.

Arc blow is especially noticeable when welding with bare electrodes. The shielding on a covered electrode has a protective effect on the molten globule of weld metal, thus reducing the force of the pull. Excessive arc blow may cause lack of fusion, porosity, and excessive spatter.

Arc blow may be controlled somewhat by moving the ground to another location and by changing the position of the electrode.

CURRENT VALUES. It is impossible to give a table that will indicate the exact amount of current needed to burn a certain size of electrode for every welding condition. The best that can be done is to offer a basic heat table, the values of which can be increased or decreased for the different conditions. The following have a bearing on the amount of current:

- *Type and size of electrode:* A 3/16-inch electrode requires higher heat than a 1/8-inch electrode. A straight polarity electrode requires more heat than a reverse polarity electrode.
- *Material to be welded:* Materials that are good conductors require more heat than materials that are poor conductors.
- *Size and shape of the work:* A piece of material 20 inches square and 1 inch thick requires more heat for welding than one which is 10 inches square and 1/8 inch thick.
- *Type of joint:* The first pass in a T-joint can be welded with a higher current than the first pass of an open, single-V butt joint.
- *Position of the weld:* Welding in the flat and horizontal positions can be done with a higher current than welding in the vertical and overhead positions.
- *Obstructions around the weld site:* A weld that must be made behind obstructions requires a change in heat.
- *Kind and condition of equipment:* No two machines can be set in exactly the same manner. Even two machines of the same size and make differ somewhat.
- *Speed required:* Speed has a direct bearing on the current setting.
- *Skill of the operator:* Some operators are able to use higher heats than others on the same type of work.

Table 20-1 is a simplified guide to approximate current settings for bare and covered electrodes. Note that the decimal equivalent of the fractional sizes also gives some indication of the heat setting. Thus 5/32 inch is 0.156, and the corresponding current range is ±150. Table 20-2 lists both voltage and current settings according to covered electrode classification and size.

ELECTRODE ANGLE. The angle of the electrode to the work affects the quality of the weld. The proper angle prevents undercut and slag inclusions and makes it easier to deposit the filler metal in the weld. Because it influences surface tension and the force of gravity on the molten metal, the correct angle promotes unformity of fusion and weld contour in vertical and overhead welding.

LEARNING WELDING SKILLS

Study Table 20-3, the Welder's Troubleshooter, thoroughly before you begin welding practice. Look up your weld defects in the table during practice and correct them as instructed before beginning the next job.

PRACTICE JOBS

Welding With Bare Electrodes

Bare electrodes are seldom used in industry. You may wonder why time is spent learning how to weld with this type of electrode. The bare electrode provides a useful introduction to the electric arc welding processes because you are able to see what takes place in the arc

Table 20-1

How heat ranges coincide with the decimal equivalent of the fractional electrode size.

Diameter of electrode, in.	Decimal equivalent	Current range, amperes
1/16	0.062	± 60
3/32	0.093	± 75
1/8	0.125	±100
5/32	0.156	±150
3/16	0.187	±200
1/4	0.250	±250
5/16	0.312	±300
3/8	0.375	±400

and arc pool. Arc blow is especially noticeable. You will also observe fusion of the weld metal and base metal so that you will gain a good understanding of its action and importance. The unstable arc requires that the operator hold an almost perfect arc gap for a weld of good appearance and quality. Practice with bare electrodes will help you develop a steady hand and make it easier for you to do good welding with covered electrodes.

Study Fig. 20-4 (Page 576) and Table 20-4 (Page 578) to learn what operating and weld characteristics you should strive for and what you should avoid when welding with bare electrodes.

Only four practice jobs with bare electrodes have been included. As soon as you have gained minimal mastery of electrode manipulation, you will begin actual practice with shielded metal-arc welding electrodes.

Instructions for Completing Practice Jobs

Your instructor will assign appropriate practice in arc welding from the jobs listed in the Course Outline, pages 632–633. Before you begin a job, study the specifications given in the Course Outline. The Text Reference column lists the pages in the chapter on which the job print and job description appear. Pay particular attention to the print which accompanies the job description. The title block gives the size and type of stock, the number, type and size of electrodes, and the tolerances required. A pictorial view of the job is shown to help you interpret the top, front, and left side views. Refer to Chapter 14, Welding Symbols, as necessary to interpret the welding symbols shown on the prints.

As an example of how to find the information necessary to complete a job, turn to the Course Outline and find Job 20-J14, Stringer Beading. The Course Outline states that the weld is to be made on flat plate, the direction of travel is down, and the position of welding is vertical. The welding current may be either direct current, reverse polarity or alternating current. Use the E6010 electrode with direct current, reverse polarity and the E6011 electrode with

Table 20-2

Current variations with covered electrode classifications and sizes.

Electrode dia.	Electrode classification									
	E 6010 E 6011	E 6012	E 6013	E 6020	E 6024	E 6027	E 7016	E 7018	E 7028	E 7024
1/16		20–40 A[1] 17–20 V[2]	20–40 A 17–20 V	No Size	No Size	No Size	No Size	No Size	No Size	No Size
3/32	30–80 A 22–24 V	30–80 V 17–21 V	30–80A 17–21 V	No Size	80–130 A 20–24 V	No Size	60–100 A 17–22 V	70–120 A 17–21 V	No Size	100–145 A 20–24 V
1/8	80–120 A 24–26 V	80–130 V 18–22 V	70–120 A 18–22 V	100–140 A 24–28 V	120–200 A 21–25 V	125–185 A 24–28 V	80–120 A 18–22 V	100–150 A 18–22 V	140–190 A 21–25 V	140–190 A 21–25 V
5/32	120–160 A 24–26 V	120–180 A 18–22 V	120–170 A 18–22 V	120–180 A 26–30 V	180–240 A 22–26 V	160–240 A 26–30 V	140–190 A 20–24 V	120–200 A 20–24 V	180–250 A 22–26 V	180–250 A 22–26 V
3/16	140–220 A 26–30 V	140–250 A 20–24 V	140–240 A 20–24 V	175–250 A 30–36 V	240–300 A 23–27 V	210–300 A 27–32 V	170–250 A 21–25 V	200–275 A 21–25 V	230–305 A 23–27 V	230–305 A 23–27 V
7/32	170–250 A 26–30 V	200–320 A 20–24 V	210–300 A 21–25 V	225–310 A 30–36 V	300–310 A 23–28 V	250–350 A 30–36 V	240–320 A 23–27 V	260–340 A 22–26 V	275–365 A 23–28 V	275–365 A 27–28 V
1/4	200–300 A 28–32 V	200–400 A 20–24 V	200–350 A 22–26 V	250–400 A 30–36 V	300–400 A 24–29 V	300–420 A 30–36 V	300–400 A 24–28 V	300–400 A 23–27 V	335–430 A 24–29 V	335–430 A 24–29 V
5/16	250–450 A 28–32 V	250–500 A 22–26 V	250–450 A 23–27 V	350–450 A 32–38 V	400–525 A 24–30 V	375–475 A 32–38 V	375–475 A 24–28 V	375–470 A 23–28 V	400–525 A 24–30 V	400–525 A 24–30 V

[1]A indicates current in amperes.
[2]V indicates voltage, arc volts.

Table 20-3

The welder's troubleshooter.

Trouble	Cause	Cure
Distortion	A. Shrinkage of deposited metal pulls parts together and changes relative positions. B. Nonuniform heating of parts during welding causes them to distort before welding is finished. Final welding of parts in distorted position prevents the maintenance of proper dimensions. C. Improper welding sequence.	A. Clamp or tack parts properly to resist shrinkage. B. Separate or preform parts to compensate for shrinkage of welds. C. Distribute welding to prevent excessive local heating. Preheating is desirable in some heavy structures. D. Removal of rolling or forming strains before welding is sometimes helpful. E. Study structure and develop a definite sequence of welding.
Warping (Thin plates)	A. Shrinkage of deposited weld metal. B. Excessive local heating at the joint. C. Improper preparation of joint. D. Improper welding procedure. E. Improper clamping of parts.	A. Select electrode with high welding speed and moderate penetrating properties. B. Weld rapidly to prevent overheating the plates adjacent to the weld. C. Do not have wide spaces between the parts to be welded. D. Clamp parts next to the joint properly. Use backup strip to cool parts rapidly. E. Use step-back or skip welding sequence. F. Hammer joint edges thinner than rest of plate before welding. This elongates edges, and the weld shrinkage causes them to pull back to the original shape.
Welding stresses	A. Joints too rigid. B. Improper welding procedure. C. Stress occurs in all welds, especially in heavy parts.	A. Slight movement of parts during welding will reduce welding stresses. B. Make weld in as few passes as practical. C. Peen each deposit of weld metal. D. Heat finished product at 1,100–1,200° F. for one hour per inch of thickness. E. Develop welding procedure that permits all parts to be free to move as long as possible.
Spatter	A. Characteristic of some electrodes. B. Excessive welding current for the type or diameter of electrode used. C. Coated electrodes produce larger spalls than bare type electrodes.	A. Select proper type of electrode. B. Do not use too much welding current. C. Paint parts next to weld with whitewash. This prevents spalls from welding to parts, and they can be removed easily.
Cracked welds	A. Joint too rigid. B. Welds too small for size of parts joined. C. Improper welding procedure. D. Poor welds. E. Improper preparation of joints.	A. Design the structure and develop a welding procedure to eliminate rigid joints. B. Do not use too small a weld between heavy plates. Increase the size of welds. C. Do not make welds in string beads. Make weld full size in short section 8 to 10 inches long. D. Plan welding sequence to leave ends free to move as long as possible. E. Insure welds are sound and the fusion is good. F. Preheating parts to be welded is sometimes helpful. G. Prepare joints with a uniform free space of the proper width. In some cases a free space is essential. In other cases a shrink or press fit may be required.

(Continued on pages 574, 575)

Table 20-3 (Continued)

Trouble	Cause	Cure
Poor weld appearance	A. Poor welding technique, improper current setting or electrode manipulation. B. Characteristic of electrode used. C. Welding in improper position for which electrode is designed. D. Improper joint preparation.	A. Use the proper welding technique for the electrode. B. Use an electrode designed for the type of weld and the position in which the weld is to be made. C. Do not make fillet welds with downhand-electrodes unless the parts are positioned. D. Do not use too high welding currents. E. Use a uniform weave or rate of travel at all times. F. Prepare all joints properly.
Undercut	A. Too high welding current. B. Improper manipulation of electrode. C. Attempting to weld in a position for which the electrode is not designed.	A. Use a moderate welding current and do not try to travel too rapidly. B. Do not use too large an electrode. If the puddle of molten metal becomes too large, undercut may result. C. Avoid too much weaving. D. A uniform weave will aid greatly in preventing undercut in butt welds. E. If an electrode is held too near the vertical plane when making a horizontal fillet weld, undercut will occur on the vertical plate.
Poor fusion	A. Improper diameter of electrode. B. Improper welding current. C. Improper welding technique. D. Improper preparation of joint.	A. When welding in narrow Vs, use an electrode small enough to reach the bottom. B. Use sufficient welding current to deposit the metal and penetrate into the plates. Heavier plates require more current for a given electrode than light plates. C. Be sure the weave is wide enough to thoroughly melt the sides of a joint. D. The deposited metal should sweat on the plates and not curl away from it.
Incomplete penetration	A. Improper preparation of joint. B. Too large an electrode. C. Too low welding current. D. Too fast a welding speed.	A. Be sure to allow the proper free space at the bottom of a weld. B. Do not expect too much penetration from an electrode. C. Use small diameter electrodes in a narrow welding groove. D. Use enough welding current to obtain proper penetration. Do not weld too rapidly. E. Use a backup bar if possible. F. Chip or cut out the back of the joint and deposit a bead of weld metal at this point.
Porous welds	A. Characteristic of some electrodes. B. Improper welding procedure. C. Not sufficient puddling time to allow entrapped gas to escape. D. Poor quality base metal.	A. Some electrodes produce sounder welds than others. Be sure the proper electrodes are used. B. Puddling keeps the weld metal molten longer and often insures sounder welds. C. A weld made of a series of strung beads is apt to contain minute pinholes. Weaving will often eliminate this trouble. D. Do not use welding currents that are too high. E. In some cases the base metal may be at fault. Check it for segregations and impurities.

Table 20-3 (Continued)

Trouble	Cause	Cure
Brittle welds	A. Unsatisfactory electrode. B. Excessive welding current causing coarse-grained and burnt metal. C. High carbon or alloy base metal which has not been taken into consideration.	A. Bare electrodes produce brittle welds. Covered electrodes must be used if ductile welds are required. B. Do not use welding current that is too high. It may cause coarse grain structure and oxidized deposits. C. A single-pass weld may be more brittle than a multiple-layer weld because it has not been refined by successive layers of weld metal. D. Welds may absorb alloy elements from the base metal and become hard. Do not weld a steel unless the analysis and characteristics are known.
Brittle joints	A. Air hardening base metal. B. Improper welding procedure. C. Unsatisfactory electrode.	A. When welding medium carbon steel or certain alloy steels, the fusion zone may be hard as a result of rapid cooling. Preheat at 300–500° F. before welding. B. Multiple-layer welds will tend to anneal hard zones. C. Heating at 1,100–1,200° F. after welding will generally reduce hard areas formed during welding. D. Austenitic electrodes will often work on special steels, but the fusion zone will generally contain an alloy which is hard.
Corrosion	A. Type of electrode used. B. Improper weld deposit for corrosive media. C. Metallurgical effect of welding. D. Improper cleaning of weld.	A. Bare electrodes produce welds that are less resistant to corrosion than the base metal. B. Covered electrodes produce welds that are more resistant to corrosion than the base metal. C. Do not expect more from the weld than you do from the base metal. On stainless steels use electrodes that are equal to or better than the base metal. D. When welding 18–8 austenitic stainless steel, be sure the analysis of the steel and welding procedure is correct so that the butt welding does not cause carbide precipitations. Condition can be corrected. Anneal at 1,900–2,100° F. and quench. E. Certain metals such as aluminum require careful cleaning of all slag to prevent corrosion.
Arc blow	A. Magnetic field causing a d.c. arc to blow away from the point at which it is directed. Arc blow is particularly noticeable at ends of joints and in corners.	A. Proper location of the ground on the work. Placing the ground in the direction the arc blows from the point of welding is often helpful. B. Separating the ground in two or more parts is helpful. C. Weld toward the direction the arc blows. D. Hold a short arc. E. Arc blow is not present with a.c. welding.

Westinghouse Electric Co.

alternating current. The job print and job description may be found on pages 603, 604. Turn to page 604 and study the job print. The title block states that the stock is $\frac{3}{16} \times 7 \times 8$ mild steel plate, the fractional limits are $\frac{1}{8}$-inch, and ten $\frac{5}{32}$-inch electrodes are required. The front view indicates $\frac{1}{16}$-inch reinforcement, $\frac{1}{16}$-inch penetration, and beads which are $\frac{5}{16}$ inch wide. The top view shows the direction of travel and instructs you to make beads on both sides of the plate. The welding symbol shown in the pictorial view tells you that these are surface beads and directs you to read note *A* below the view for additional information.

Complete the jobs in the order assigned by your instructor. At certain points in your practice, you will test a sample weld for soundness. Refer to Job 20-J17, page 610, for an example of such a check test. As you progress, you will be asked to pass performance tests for operator qualification. After completing Job 21-J32, for example, you will take Test 1 in the flat position. Depending on local requirements, your instructor may also ask you to pass Test 3 in the flat position. At the conclusion of the course in shielded metal-arc welding (plate) you will be required to take the complete series of operator qualification tests. These tests are described on pages 714, 720.

JOB 20-J1 STRIKING THE ARC AND SHORT STRINGER BEADING

Objective

To strike the arc and deposit short stringer beads on flat steel plate in the flat position with bare electrodes (AWS-ASTM E4510).

General Job Information

Striking the arc is a basic action throughout the entire welding operation. It will occur each time the welding operation is started.

Welding Technique

The arc is established by lightly touching the plate with the electrode and then withdrawing it; Fig. 20-5. Current flows immediately, and the arc flame will be maintained if the gap is not too great. The heat of the arc flame causes the plate and the electrode to melt, making it possible to weld. If the electrode is not withdrawn fast enough, it will stick to the plate and may be freed by a quick twist of the wrist.

Striking the arc may be accomplished by a straight up-and-down motion (Fig. 20-6A), or by a scratching motion (Fig. 20-6B). Use the method that causes you the least trouble.

After you have mastered striking the arc, practice depositing short stringer beads. Hold the electrode at the starting point for a short time to allow fusion to

20-4. *Plan and elevation views of welds made with bare electrodes under various operating conditions. (A) Current, voltage, and speed normal. (B) Current too low. (C) Current too high. (D) Voltage too low, short arc. (E) Voltage too high, long arc. (F) Speed of travel too low. (G) Speed of travel too high. See Table 20-4 for additional information.*

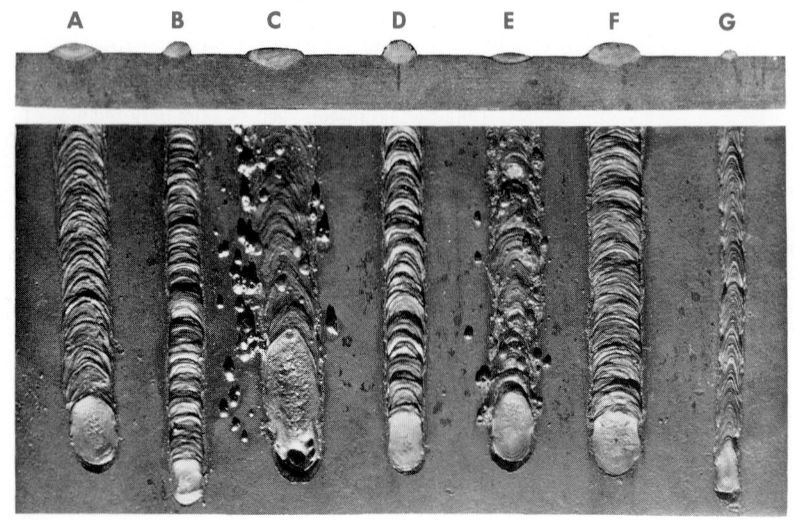

20-5. *Position of the electrode and electrode holder when striking the arc.*

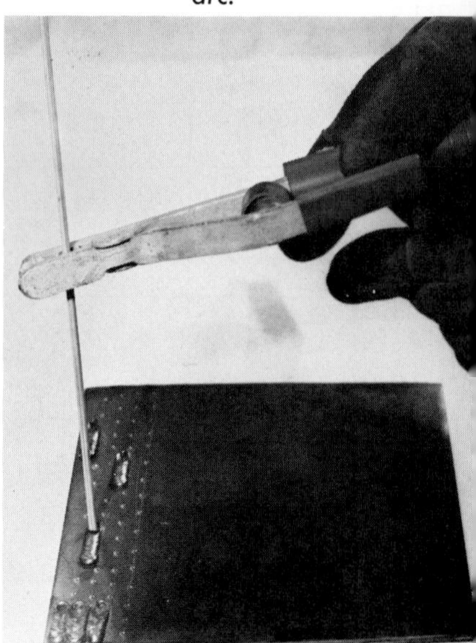

occur and the bead to form. Advance the electrode at a uniform rate of travel, feeding steadily downward as the electrode is melted into the puddle.

It is important that the correct arc gap be maintained. A sharp crackling sound, an even transfer of the molten ball of metal across the arc gap, and lack of spatter are indications of correct arc gap. An arc that is held too short is very erratic. It will stick to the plate and go out. A long arc is recognized by a steady hiss like escaping steam. Penetration is poor, and overlap occurs along the sides of the bead. There is a considerable amount of spatter.

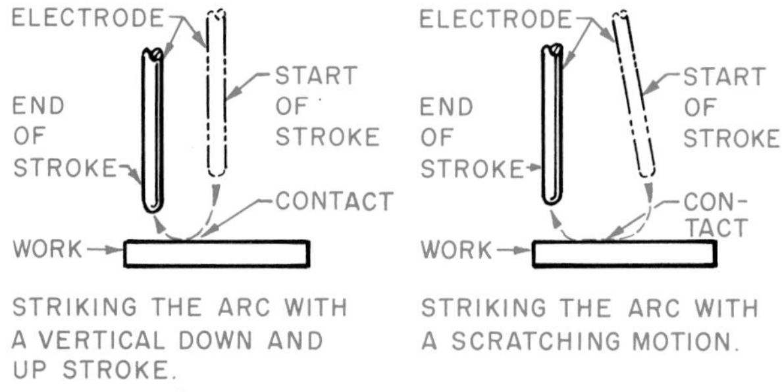

20-6. *Two methods of striking the arc.*

Operations

1. Obtain plate; check the job print for size.
2. Obtain a square head and scale, dividers, scribe, and center punch from the toolroom.
3. Lay out parallel lines as shown in the job print.

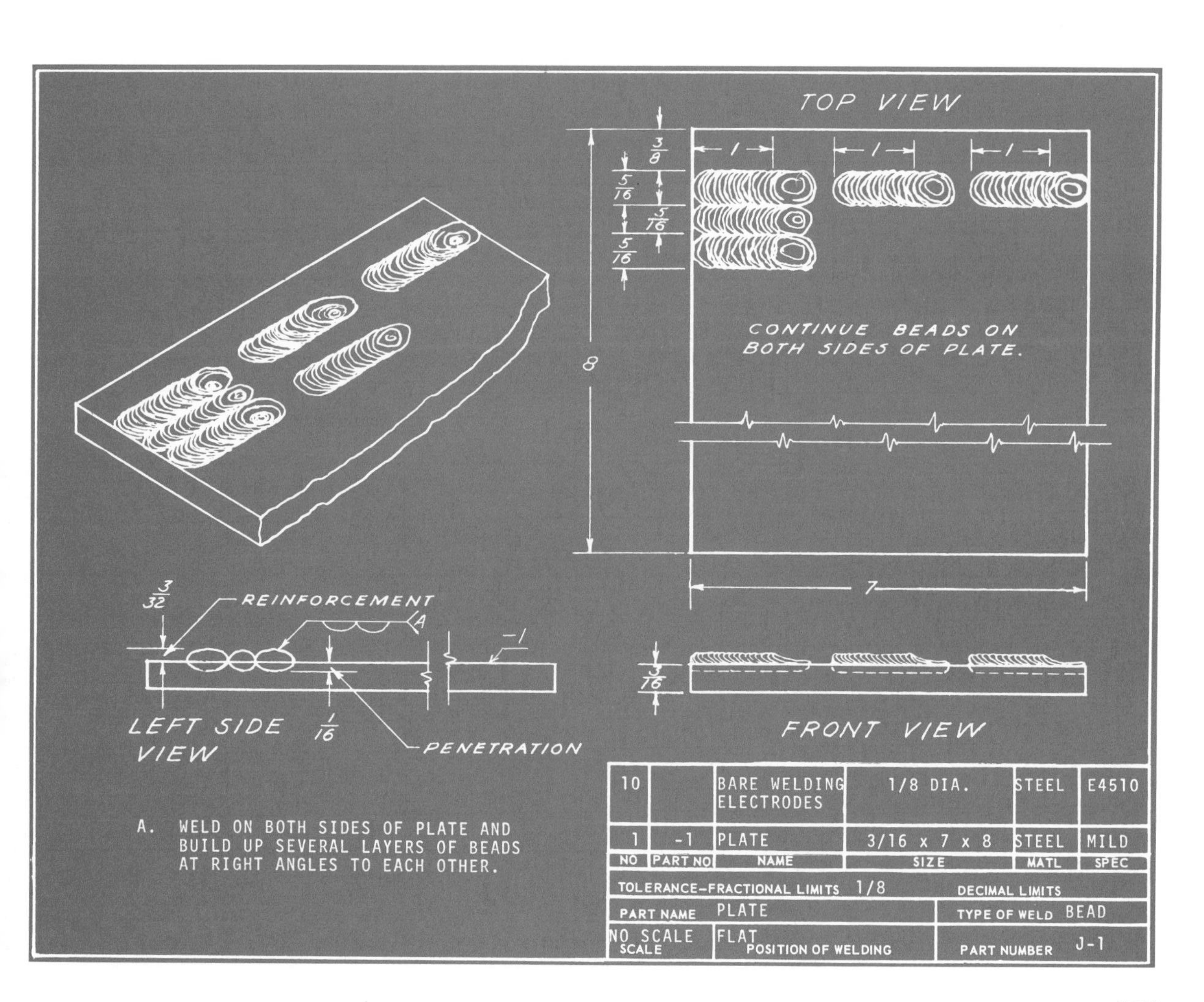

NO	PART NO	NAME	SIZE	MATL	SPEC
10		BARE WELDING ELECTRODES	1/8 DIA.	STEEL	E4510
1	-1	PLATE	3/16 x 7 x 8	STEEL	MILD

TOLERANCE–FRACTIONAL LIMITS 1/8	DECIMAL LIMITS
PART NAME PLATE	TYPE OF WELD BEAD
SCALE NO SCALE / POSITION OF WELDING FLAT	PART NUMBER J-1

4. Mark the lines with a center punch. They should look like those in Fig. 20-7.

5. Obtain electrodes of each quantity, type, and size specified in the job print.

6. Set power source for 90 to 120 amperes. Set a d.c. power source for straight polarity.

7. Lay the plate in the flat position on the welding table. Make sure it is well-grounded.

8. Practice striking the arc and make beads as instructed in the job print. Hold the electrode in the position shown in Fig. 20-5. Strike the arc by touching the plate with either of the motions shown in Fig. 20-6.

9. Brush the beads and inspect. Refer to Inspection, below.

10. Practice this job until you can strike an arc freely, without sticking, and where desired.

20-7. *Method of marking parallel lines and typical appearance of short stringer beads welded in the flat position with bare electrodes.*

Inspection

Compare the beads with Fig. 20-7 and check them for the following weld characteristics:

Width and height: Uniform

Table 20-4

Weld characteristics obtained when proper welding procedure is used except as indicated. This tabulation applies only to welding of mild rolled steel in flat position with bare or washed electrodes. See Fig. 20-4 for welds A through G.

	Resulting weld characteristics				
Operating variables	**Arc sound**	**Penetration—fusion**	**Burn off of electrode**	**Appearance of bead**	**Comments**
Normal polarity, correct current, voltage and speed of travel (A)	Steady, sharp crackling sound	Good—crater averages $\frac{1}{16}$″ deep for $\frac{1}{8}$″ rod	Stable	Smooth, no overlaps	
Low current (B)	Pulsating "low energy"	Poor—very shallow crater		Bead piles up—overlap	
High current (C)	Explosions crackling	Poor weld characteristics, porosity		Flat	Electrode becomes red hot—splatter—weld porous
Low voltage (across arc) (D)	Steady sputter	Poor—crater shallow	Rod apt to freeze to work	High	
High voltage (across arc) (E)	Whistling or hissing and crackling	Very little fusion—crater shallow	Bubble on end of rod—arc wanders	Flat	Splatter—pockets in weld—weld oxidizes
Low speed of travel (F)		Crater excessively deep		Piles up and rolls over—bluish color	Electrode becomes red at tip—splatter
High speed of travel (G)		Almost no crater—no fusion		Irregular in width—flat	
Reverse polarity	Sputter	No crater—porosity		Irregular	Splatter

Appearance: Somewhat rough with coarse ripples; free of voids and high spots
Size: Refer to the job print.
Face of beads: Slightly convex
Edges of beads: Good fusion, no overlap, no undercut
Beginnings and endings: Full size, craters filled
Penetration and fusion: To plate surfaces
Surrounding plate surfaces: Free of spatter

Disposal

Discard completed plates in the waste bin. Plates must be filled with beads on both sides before disposal.

JOB 20-J2 CONTINUOUS STRINGER BEADING

Objectives

1. To deposit continuous stringer beads on flat steel plate in the flat position with bare electrodes (AWS-ASTM E4510).
2. To develop the technique of restarting the arc.

General Job Information

These beads are identical to those formed in Job 1 except that they are longer. You will probably find it difficult to maintain a steady arc gap for the entire length of the weld, and the beads will be uneven and crooked. Practice will overcome these problems.

Welding Technique

Hold the electrode at a 90-degree angle to the plate, Fig. 20-8A. Then tilt it 10 to 20 degrees in the direction of travel, Fig. 20-8B. Current setting must not be so high as to produce arc blow. Using the marked line as a guide, deposit continuous beads across the plate. Bear in mind that the arc gap must be constant and that the rate of travel must be such as to produce welds of uniform height and width. If travel is too fast, beads will be narrow with little

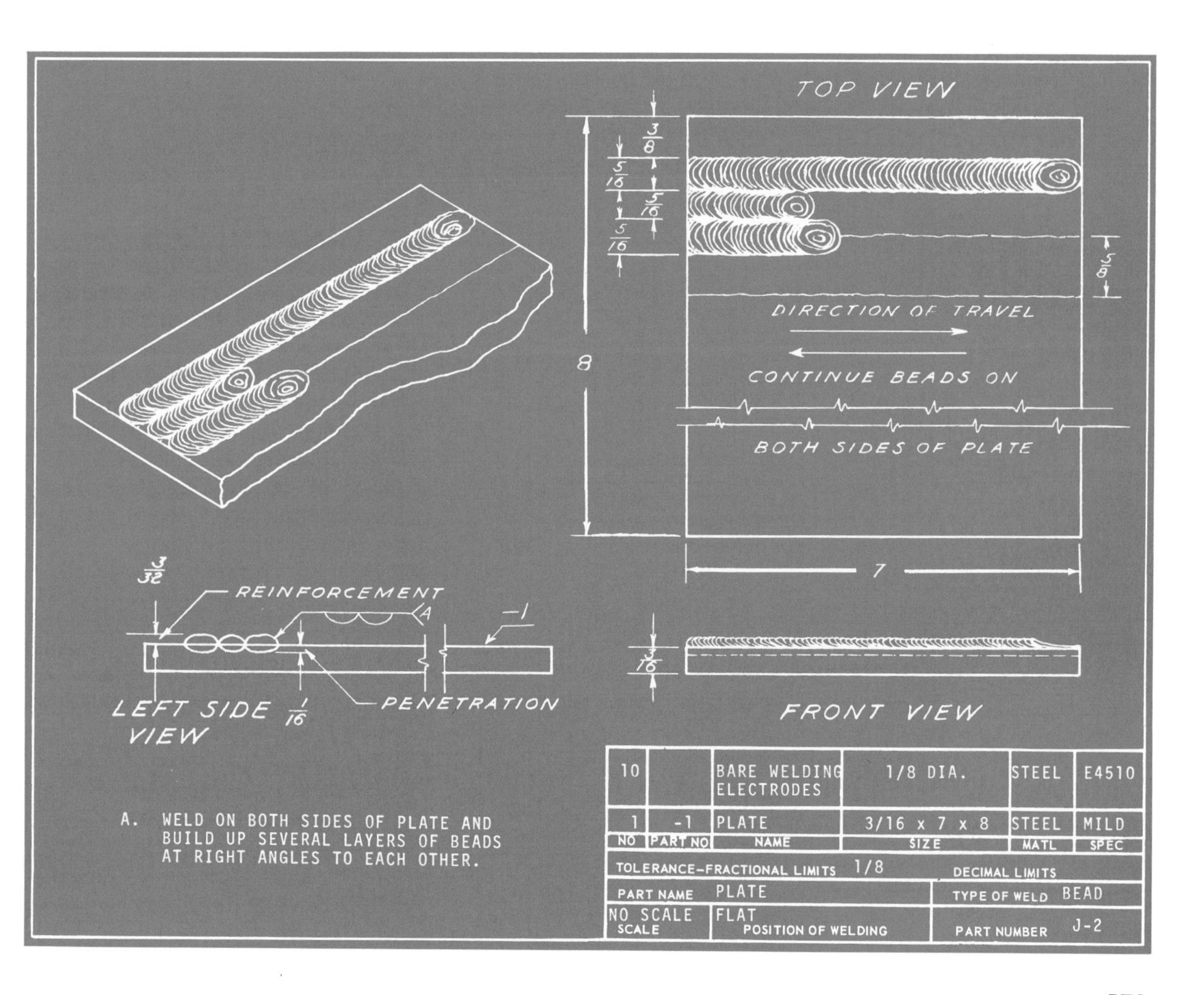

reinforcement, and ripples will be coarse. If travel is too slow, beads will be large, rough, and show overlap along the edges.

If the arc is broken for any reason, a special procedure must be used to insure a good start with good fusion and appearance. Restart the arc at the forward edge of the crater. Bring it back across the crater to the edge of already deposited metal, and then forward again in the direction of welding, Fig. 20-9. If you do not bring the electrode back far enough, there will be a depression between the starting and stopping points. If you bring it back too far, the weld metal will pile up in a large lump.

You will probably notice that deep craters will be formed at the end of each weld bead and at each stopping point. These craters are a helpful indication of penetration. They should be filled at the finish of the weld bead. This is done by *choking the arc,* that is, holding a very short arc gap within the crater until it is completely filled.

Weld on both sides of the plate. When the plate is filled, deposit additional beads at right angles to the beads already on the plate. Practice traveling in all directions: left to right, right to left, and to and from the operator.

Operations

1. Obtain plate; check the job print for size.
2. Obtain a square head and scale, dividers, scribe, and center punch from the toolroom.
3. Lay out parallel lines as shown on the job print.
4. Mark the lines with a center punch. They should look like those in Fig. 20-10.
5. Obtain electrodes of each quantity, type, and size specified in the job print.
6. Set the power source for 90 to 120 amperes. Set a d.c. power source for straight polarity.
7. Lay the plate in the flat position on the welding table. Make sure that it is well-grounded.
8. Make stringer beads on 5/8-inch center lines as shown on the job print. Hold the electrode as shown in Fig. 20-8 and manipulate it as shown in Fig. 20-9.
9. Brush the beads and inspect. Refer to Inspection, below.
10. Make additional beads between the already deposited beads as shown on the job print.
11. Brush the beads and inspect. Refer to Inspection, below.
12. Practice these beads until you can produce uniform beads consistently.

Inspection

Compare the beads with Fig. 20-10 and check them for the following weld characteristics:

Width and height: Uniform
Appearance: Somewhat rough with coarse ripples; free of voids and high spots. Restarts should be difficult to locate.
Size: Refer to the job print.
Face of beads: Slightly convex
Starts and stops: Free of depressions and high spots

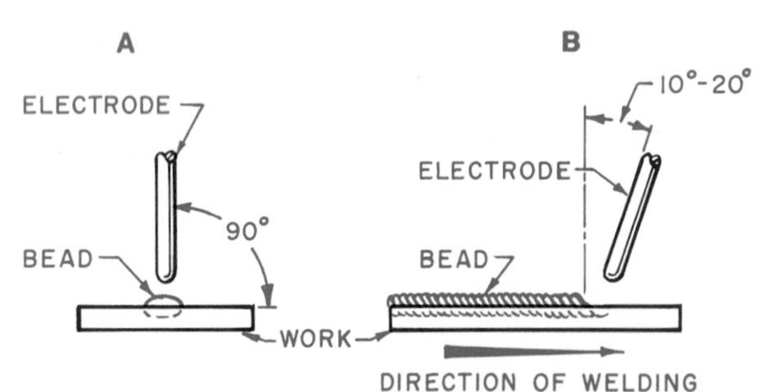

20-8. *Electrode position for stringer beads in the flat position.*

20-9. *Method of restarting after stopping.*

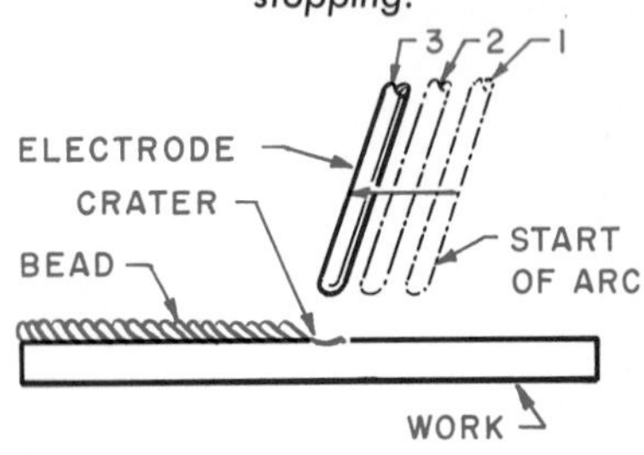

20-10. *Method of marking parallel lines and typical appearance of continuous stringer beads welded in the flat position with bare electrodes.*

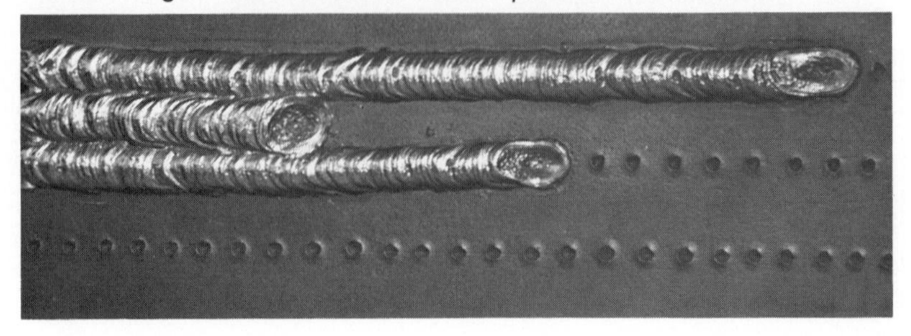

Beginnings and endings: Full size, craters filled
Penetration and fusion: To plate surfaces and adjacent beads
Surrounding plate surfaces: Free of spatter

Disposal

Discard completed plates in the waste bin. Plates must be filled with beads on both sides.

JOB 20-J3 WEAVED BEADING

Objective

To deposit weaved beads on flat steel plate in the flat position with bare electrodes (AWS-ASTM E4510).

General Job Information

It is often necessary to deposit weld beads wider than the widths that can be obtained with stringer beads. Deep groove joints and multipass fillet welds in flat position usually require such beads. They are called *weaved* beads.

Welding Technique

Figure 20-11 illustrates the correct electrode position and technique for weaved beading. Weave the electrode from side to side for the full width of the bead as shown in Fig. 20-11A. The electrode is held at a 90-degree angle to the plate and tilted 10 to 20 degrees in the direction of travel, Fig. 20-11B. The zigzag pattern of electrode movements is shown in Fig. 20-11C. Note the positions of the electrode and electrode holder in Fig. 20-12. Watch the weld pool closely for height, width, and fusion. Do not travel too rapidly. Hesitate an instant at the sides on each weave. The width of the weave should be slightly less than the desired width of the final weld bead.

You will have to stop and restart many times during the practice of this job. Use the same technique for restarting that you practiced in Job 20-J2.

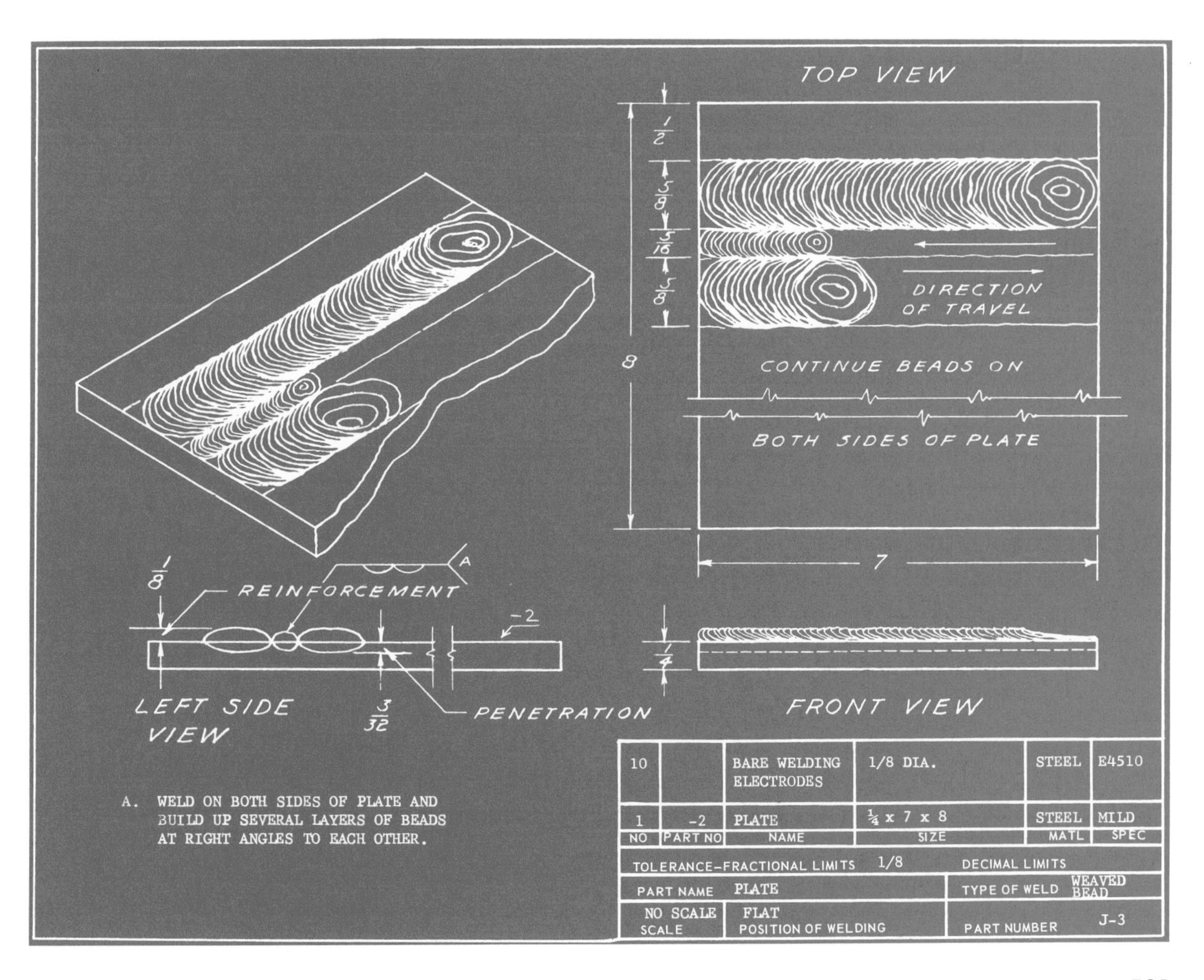

NO	PART NO	NAME	SIZE	MATL	SPEC
10		BARE WELDING ELECTRODES	1/8 DIA.	STEEL	E4510
1	-2	PLATE	¼ x 7 x 8	STEEL	MILD

TOLERANCE–FRACTIONAL LIMITS 1/8	DECIMAL LIMITS	
PART NAME PLATE	TYPE OF WELD WEAVED BEAD	
SCALE NO SCALE	POSITION OF WELDING FLAT	PART NUMBER J-3

Weld on both sides of the plate. When the plate is completely filled, deposit additional beads at a right angle to the beads already on the plate. Practice traveling in all directions.

Operations

1. Obtain plate; check the job print for size.
2. Obtain a square head and scale, scribe, and center punch from the toolroom.
3. Lay out parallel lines as shown on the job print.
4. Mark the lines with a center punch. They should look like those in Fig. 20-13.
5. Obtain electrodes of each quantity, type, and size specified in the job print.
6. Set the power source for 90 to 120 amperes. Set a d.c. power source for straight polarity.
7. Lay the plate in the flat position on the welding table. Make sure that the plates are well-grounded.
8. Make weaved beads between parallel lines as shown on the job print. Manipulate the electrode as instructed in Fig. 20-11.
9. Brush the beads and inspect. Refer to Inspection.
10. Make additional stringer beads between the weaved beads already deposited.
11. Practice these beads until you can produce uniform beads consistently.

Inspection

Compare the beads with Fig. 20-13 and check them for the following weld characteristics:

Width and height: Uniform
Appearance: Somewhat rough with coarse ripples; free of voids and high spots. Restarts should be difficult to locate.
Size: Refer to the job print. Check with a butt weld gauge.
Face of beads: Slightly convex
Edges of beads: Good fusion, no overlap, no undercut
Starts and stops: Free of depressions and high spots
Beginnings and endings: Full size, craters filled
Penetration and fusion: To plate surfaces and adjacent beads
Surrounding plate surfaces: Free of spatter

Disposal

Discard completed plates in the waste bin. Plates must be filled with beads on both sides.

JOB 20-J4 WELDING A LAP JOINT

Objective

To weld a lap joint with a single-pass fillet weld in the horizontal position with bare electrodes (AWS-ASTM E4510).

General Job Information

This type of joint is often used on welded structures. It is usually welded with shielded metal-arc electrodes. This is the last bare electrode application, and it is an excellent opportunity for you to observe the action of the weld pool between two surfaces.

Welding Technique

In this practice job, plates are tacked at each corner to form a series of lap joints, Fig. 20-15, page 585. The lap joint is made with a fillet weld. The electrode is tilted at a 45-degree angle to the lower (flat) plate, Fig. 20-14A, and at a 60-degree angle to the direction of travel, Fig. 20-14B. Figure 20-15 shows this position for an actual weld.

Although it is not necessary to weave the electrode, a slight back-and-forth motion will be helpful. Pay close attention to the penetration at the root of the weld and on the vertical and flat plate surfaces. (Consult the job print.) The resulting weld should be full. The strength of a fillet weld depends upon the throat thickness.

20-11. *Electrode position for weaved beads.*

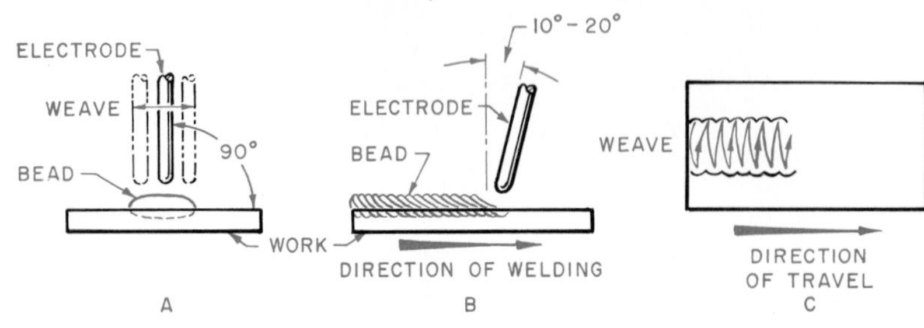

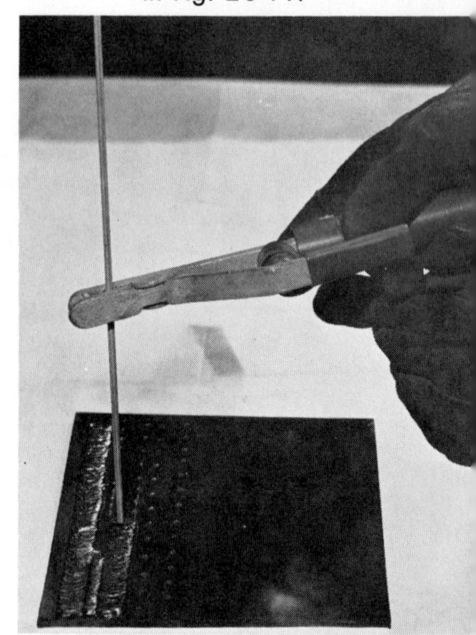

20-12. *Weaved beading in the flat position. The electrode and electrode holder are positioned as instructed in Fig. 20-11.*

Filling craters and starts and stops may be practiced using the methods described for Job 20-J2 under Welding Technique. Practice traveling in all directions.

Operations

1. Obtain plates; check the job print for quantity and size. See Fig. 20-16 for tacking up.
2. Obtain electrodes of each quantity, type, and size specified in the job print.
3. Set the power source for 100 to 130 amperes. Set a d.c. power source for straight polarity.
4. Set up the plates and tack weld them at each corner.
5. Place the joint in the flat position on the welding table. Make sure that the plates are well-grounded.)
6. Make $^{3}/_{16}$-inch fillet welds as shown on the job print. Manipulate the electrode as instructed in Fig. 20-14.

20-13. *Method of marking guide lines and typical appearance of weaved beads welded in the flat position with bare electrodes.*

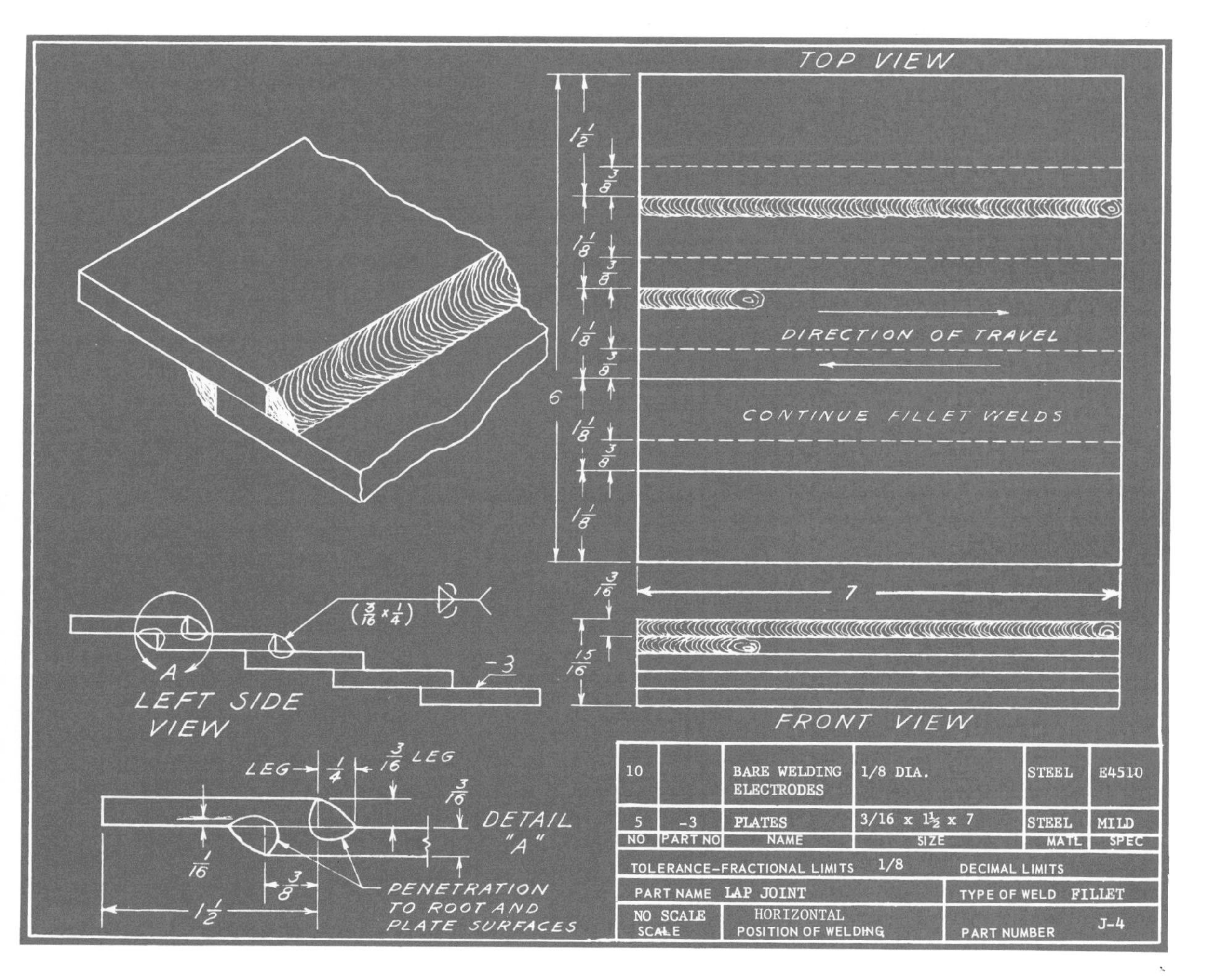

7. Brush the welds and inspect. Refer to Inspection, below.
8. Practice these welds until you can produce uniform welds consistently.

Inspection

Compare the welds with Fig. 20-15 and check them for the following characteristics:

Width and height: Uniform
Appearance: Somewhat rough with coarse ripples; free of voids and high spots. Restarts should be difficult to locate.
Size: Refer to the job print. Check with a fillet weld gauge.
Face of welds: Slightly convex
Edges of welds: Good fusion, no overlap, no undercut
Starts and stops: Free of depressions and high spots
Beginnings and endings: Full size, craters filled at end
Penetration and fusion: To the root of the joint and plate surfaces
Surrounding plate surfaces: Free of spatter

Check Test

After you are able to make welds that are satisfactory in appearance, make up a specimen similar to that shown in Fig. 20-17. Use the plate thickness, welding technique, and electrode type and size specified for this job. Weld on one side only.

Put the joint in a vise and separate the plates by driving a wedge between them as shown in Fig. 20-18. Examine the surface of the fracture for soundness. The weld must not be porous and must show good fusion and penetration at the root and to the plate surfaces. The joint should break through the throat of the weld, and the weld should not peel away from the plate.

Disposal

Put completed joints in the scrap bin so that they will be available for further use. Unwelded plate surfaces can be used for beading.

Welding with Shielded Metal-Arc Electrodes

Molten metal combines with the oxygen and nitrogen in the surrounding air to form oxides and nitrides. Bare electrode welding does not afford any protection from these formations. Thus it is almost impossible to make welds in mild steel plate as strong or ductile as the base metal with bare electrodes, Table 20-5.

All industries use shielded-arc (covered) electrodes for metal

Table 20-5
Comparison of properties of weld metal deposited with bare and covered electrodes.

Shielded arc-properties of weld metal	
Tensile strength	65,000 to 77,000 p.s.i.
Ductility or elongation in 2″	20 to 26%
Impact	30 to 70 ft.-lbs. (Izod)
Fatigue	28,000 to 32,000 p.s.i
Density	7.84 to 7.86

Resistance to corrosion is better than mild steel.
It should be noted that these characteristics permit the weld metal to be flanged or bent cold with success.
As deposited (not stress relieved).

Unshielded-arc properties of weld metal (Bare or lightly coated mild steel electrodes)	
Tensile strength	45,000 to 55,000 p.s.i.
Ductility or elongation in 2″	5 to 10%
Impact	8 to 15 ft.-lbs. (Izod)
Fatigue	12,000 to 15,000 p.s.i.
Density	7.5 to 7.7

Poor resistance to corrosion.
Impossible to be flanged cold or bent cold to any degree.

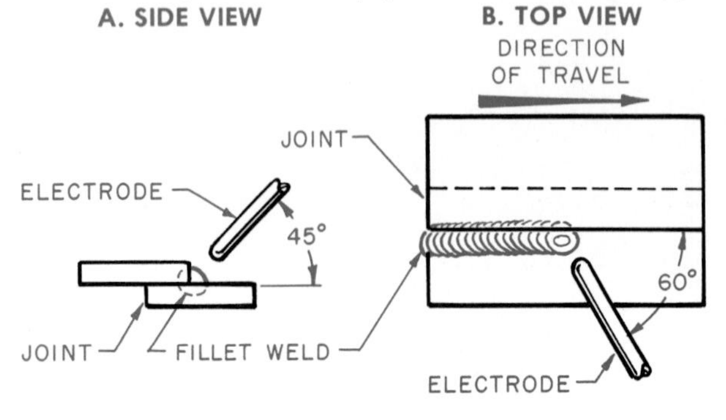

20-14. *Electrode position for lap joints in the horizontal position.*

arc welding because they reduce the formation of oxides and nitrides by protecting the molten weld pool. Protection is accomplished in the following manner.

- The covering is consumed at a lower rate than the metallic core wire. Thus it projects over the core wire so that it affords mechanical protection to the wire and the arc and controls the direction of the arc, Fig. 20-19A.
- A protective shield of gases is formed as the coating burns, and the atmosphere is thus excluded, Fig. 20-19B.
- Slag is formed over the deposited weld metal, Fig. 20-19C. It protects the deposit from the atmosphere while it is cooling.

Shielded-arc electrodes produce weld metal characteristics that are superior to the base metal.

In this course we will be concerned with those types of shielded-arc electrodes that are widely used in industry. Figure 20-20 shows welds made with a bare electrode and covered electrodes in the following AWS-ASTM classifications:

A. *E4510:* bare electrode; DCSP or a.c.; all position
B. *E6010:* covered electrode; DCRP only; all position
C. *E6011:* covered electrode; DCRP or a.c.; all position
D. *E6012:* covered electrode; DCSP only; all position
E. *E6013:* covered electrode; DCSP or a.c.; all position
F. *E7028:* iron powder, low hydrogen; DCRP or a.c.; flat and horizontal position welding
G. *E6020:* covered electrode; DCSP or a.c. for horizontal fillet welds; DCSP, DCRP, or a.c. for flat position welding
H. *E6024:* iron powder; DCSP, DCRP, or a.c. for flat and horizontal position welding

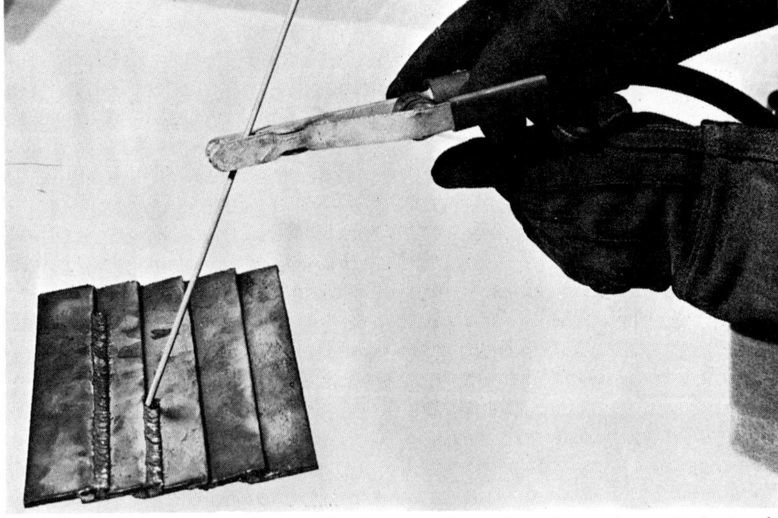

20-15. *Welding lap joints in the flat position. The electrode and electrode holder are positioned as instructed in Fig. 20-14.*

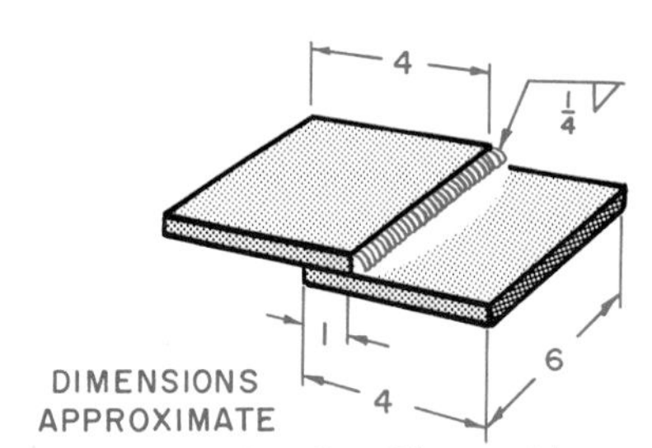

20-17. *Joint for fillet weld test specimen.*

20-16. *Typical fillet weld appearance of a lap joint welded in the flat position with bare electrodes.*

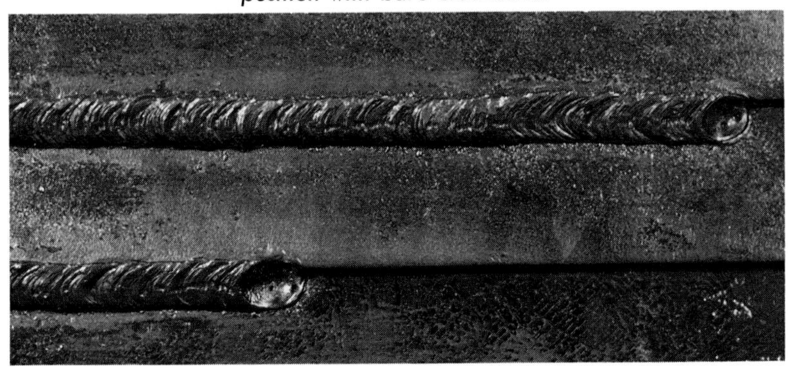

20-18. *Wedge test for fillet welds on lap joints.*

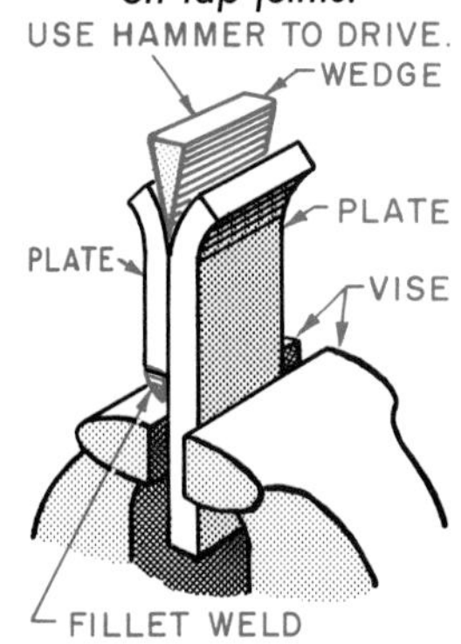

I. *E6027:* iron powder; DCSP or a.c. for horizontal fillet welding; DCSP, DCRP, or a.c. for flat position welding

J. *E7016:* low hydrogen; DCRP or a.c.; all position

K. *E7018:* iron powder, low hydrogen; DCRP or a.c.; all position

All electrodes in other tensile strength classifications produce welds resembling those in the 60XX and 70XX classifications shown in Fig. 20-20. Study Table 20-6 carefully. It lists operating variables and weld characteristics that will be helpful as you begin practice with covered electrodes.

JOB 20-J5 STRINGER BEADING

Objective

To deposit stringer beads on flat plate in the flat position with DCSP and/or a.c. shielded-arc electrodes (AWS-ASTM E6012-E6013).

General Job Information

This is the first job in which shielded-arc (covered) electrodes are used. You will find them much easier to weld with than bare electrodes. Arc blow will almost completely vanish. Welds will have a more uniform and pleasing appearance. It has been found beneficial to start practice with medium-sized electrodes ($\frac{1}{8}$-and $\frac{5}{32}$-inch diameters) although the tendency in industry today is towards the use of larger sized electrodes ($\frac{3}{16}$- and $\frac{1}{4}$-inch diameters). You should practice with $\frac{3}{16}$-inch diameter electrodes after you have mastered the use of the $\frac{5}{32}$-inch size.

This job requires both E6012 and E6013 electrodes. The power source may be either DCSP or alternating current. The electrodes are very similar in their application, but there are some important differences. Study the section in Chapter 7 concerning each of these electrodes. Although both types are classified as all-position electrodes, they are used for the most part in the flat and horizontal positions. Vertical-down welding is also a major use.

E6012 electrodes have a quiet arc with medium penetration and no spatter. The welds have good appearance. They are slightly convex and relatively free from undercut. E6013 electrodes used with a.c. current require lower welding currents and lower open circuit voltages than the E6012. They are particularly suited to welding thin metals. Their arc is soft, and penetration is very light. The bead is smoother, and the ripple is finer than that produced by the E6012. Travel is somewhat slower. E6013 electrodes produce a flat fillet weld. They are suitable for groove welds because of the flat bead shape and easily removed slag. The radiographic quality of welds made with the E6013 is better than those welded with the E6012.

The principal differences between alternating current and direct current welding are (1) the establishing of the arc and (2) the length of the arc maintained when welding. The a.c. arc is established by scratching or dragging the electrode over the work. The constant reversal of alternating current causes starting to be somewhat difficult and makes it necessary to hold a longer arc than when welding with a direct current. The longer arc makes it more difficult to weld in the vertical and overhead positions.

One major advantage of welding with alternating current is the absence of arc blow. This permits the use of larger electrodes and more current so that heavy steels can be welded faster. The same basic techniques used in welding with direct current may also be used in welding with alternating current. A.c. welds have good penetration, are of high quality, and are similar in appearance to those produced with direct current.

20-19. *Action of the shielded-arc electrode.*

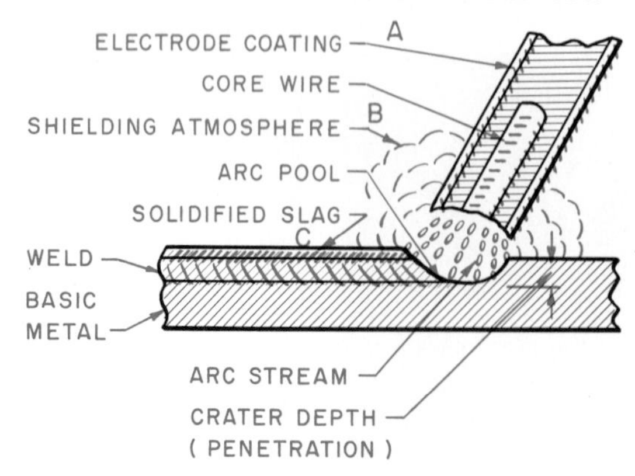

Welding Technique

Pay close attention to your current setting. Electrode position should be approximately that shown in Fig. 20-8, page 580. The arc gap may be somewhat longer than that required for bare electrodes. Too short an arc gap allows the coating to contact the molten weld pool and causes slag inclusions. If the arc is too long, the metal is transferred in large irregular drops and is exposed to the surrounding air. This produces rough welds, excessive spatter, and oxidation. As with bare electrode welding, the sound of the arc and the action of the weld pool indicate the correct arc gap.

Starting the arc may be done in much the same manner as with bare electrodes. When starting the bead, strike a long arc and hold it on one spot long enough to heat the base metal. Gradually close the arc gap until a little pool of molten metal of the proper size is formed. The pool should be fused well into the base metal before moving forward with the bead. A slight back-and-forth motion along the line of weld may be used. Most operators, however, prefer to advance along the line of weld without this motion. They pace their travel by the formation of the bead and feed the electrode downward with a slow steady movement to maintain a constant arc gap.

In restriking the arc to continue a bead, use the same method employed with bare electrodes, but remove the slag from the crater before starting. To fill the crater, pause a short time over the crater. Then draw the electrode slowly away from it and lengthen the arc until it breaks.

Weld on both sides of the plate. When the plate is completely filled, deposit additional beads at right angles to the beads already on the plate. Be sure to practice traveling in all directions.

Operations

1. Obtain plate; check the job print for size.
2. Obtain a square head and scale, dividers, scribe, and center punch from the toolroom.

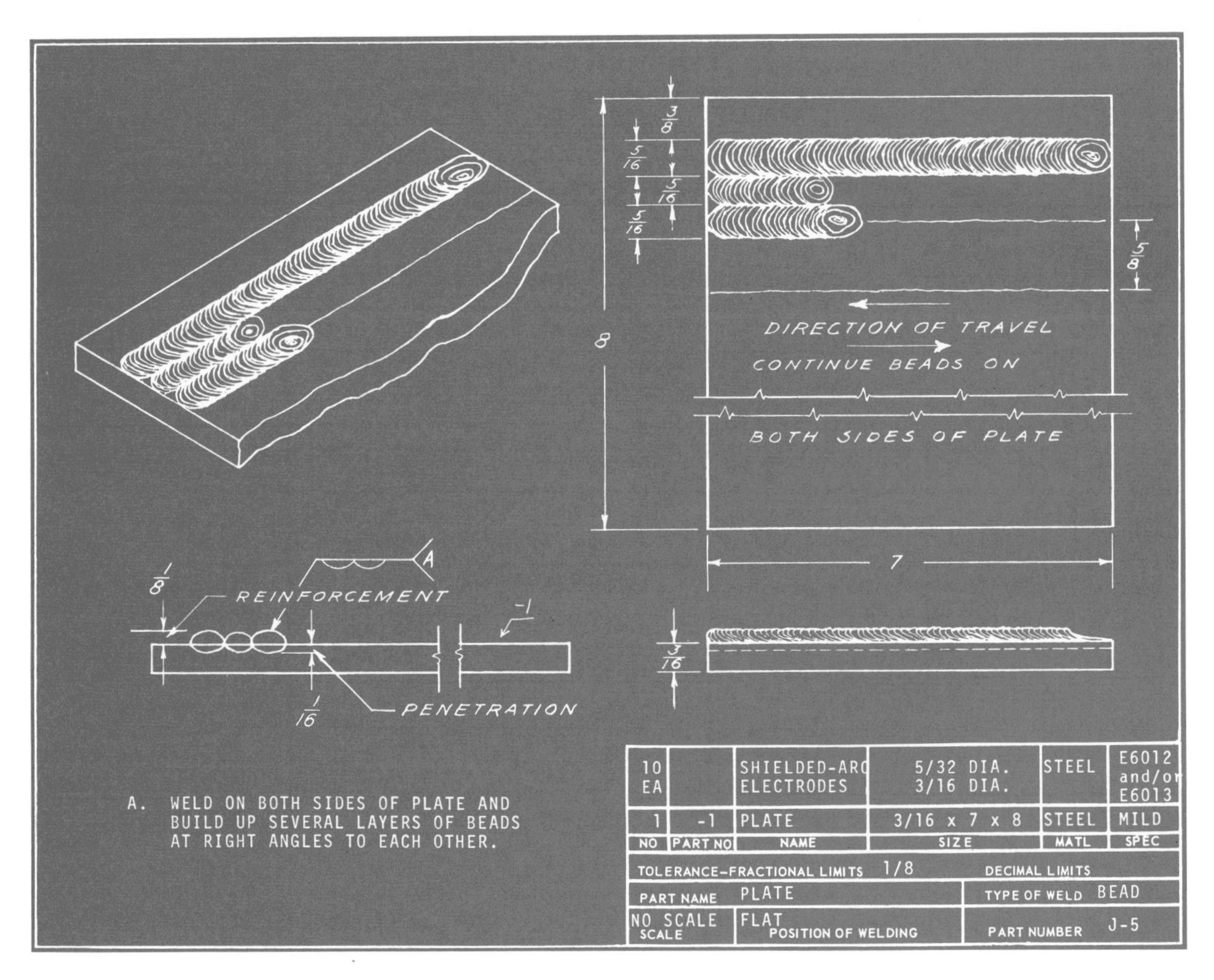

3. Lay out parallel lines as shown on the job print.

4. Mark the lines with a center punch. They should look like those in Fig. 20-21.

5. Obtain electrodes of each quantity, type, and size specified in the job print.

6. Set a d.c. power source for straight polarity at 110 to 190 amperes, or an a.c. power source at 120 to 210 amperes.

7. Lay the plate in the flat position on the welding table. Make sure that the plates are well-grounded.

8. Make beads on $\frac{5}{8}$-inch center lines as shown on the job print. Hold the electrode at a 90-degree angle to the plate and then tilt it 10 to 20 degrees in the direction of welding, Fig. 20-8, page 580.

9. Chip the slag from the beads, brush, and inspect. Refer to Inspection.

10. Make additional beads between the already deposited beads as shown on the job print.

11. Chip the slag from the beads, brush, and inspect. Refer to Inspection.

12. Practice these beads until you can produce uniform beads consistently with both types of electrodes.

Table 20-6

Weld characteristics obtained in mild rolled steel with heavily coated, shielded-arc electrodes.

Operating variables	Resulting weld characteristics			
	Arc sound	Penetration and fusion	Electrode burnoff	Appearance of bead
Normal amps, normal volts, normal speed	Sputtering hiss plus irregular energetic crackling sounds	Fairly deep and well-defined	Normal appearance Coating burns evenly	Smooth, well-defined bead and no undercutting. Excellent fusion—no overlap
Low amps, normal volts, normal speed	Very irregular Sputtering Few crackling sounds	Poor: not very deep nor defined	Not greatly different from above	Bead lies on top of plate. Note there is not the overlap produced by bare rod; excessive pilling
High amps, normal volts, normal speed	Rather regular explosive sounds	Deep, long crater	Coating is consumed at irregular high rate—watch carefully	Broad, rather thin bead; good fusion; excessive spatter and undercutting; irregular deposit
Low volts, normal speed, normal amps	Hiss plus steady sputter	Small	Coating too close to crater. Touches molten metal and results in porosity	Bead lies on top of plate but not so pronounced as for low amps. Somewhat broader
High volts, normal speed, normal amps	Very soft sound plus hiss and few crackles	Wide and rather deep	Note drops at end of electrode. Flutter and then drop into crater	Wide, spattered, irregular bead; pilling
Low speed, normal amps, normal volts	Normal	Crater normal	Normal	Wide bead—large overlap base metal and bead heated for considerable area; pilling
High speed, normal amps, normal volts	Normal	Small, rather well-defined crater	Normal	Small bead—undercut. The reduction in bead size and amount of undercutting depend on ratio of speed and amps.

Inspection

Compare the beads with Fig. 20-21 and check them for the following characteristics:

Width and height: Uniform
Appearance: Smooth with close ripples; free of voids and high spots. Restarts should be difficult to locate.
Size: Refer to the job print.
Face of beads: Slightly convex
Edges of beads: Good fusion, no overlap, no undercut
Starts and stops: Free of depressions and high spots
Beginnings and endings: Full size, craters filled
Penetration and fusion: To plate surfaces and adjacent beads
Surrounding plate surfaces: Free of spatter
Slag formation: Full coverage, easily removable

Disposal

Discard completed plates in the waste bin. Plates must be filled with beads on both sides.

JOB 20-J6 WEAVED BEADING

Objective

To deposit weaved beads on flat steel plate in the flat position with DCSP and/or a.c. shielded-arc electrodes (AWS-ASTM E6012-E6013).

General Job Information

It is often necessary to deposit beads wider than stringer beads. In order to do this, the electrode must be weaved. Weaved beads are usually necessary in the welding of deep-groove joints and multipass fillet welds.

Welding Technique

Pay close attention to your current setting. The electrode position should be approximately

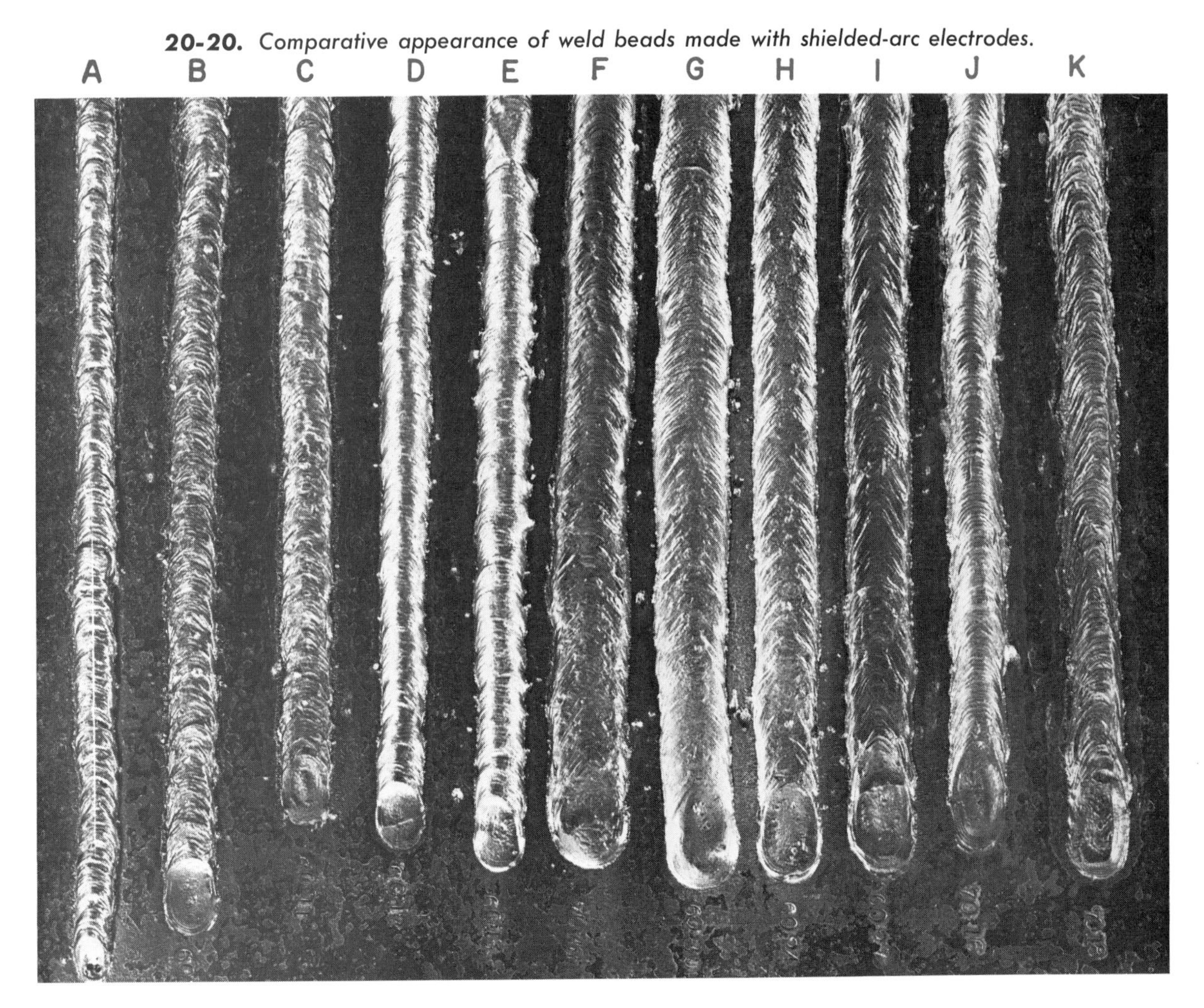

20-20. *Comparative appearance of weld beads made with shielded-arc electrodes.*

that shown in Fig. 20-22. Hold the electrode at a 90-degree angle to the plate, Fig. 20-22A. Then tilt it 10 to 20 degrees in the direction of welding, Fig. 20-22B. Move the electrode back and forth, advancing it no more than $\frac{1}{8}$ inch with each weave, Fig. 20-22C. Hesitate at the sides of the weld to prevent undercutting along the edges of the weld with the arc.

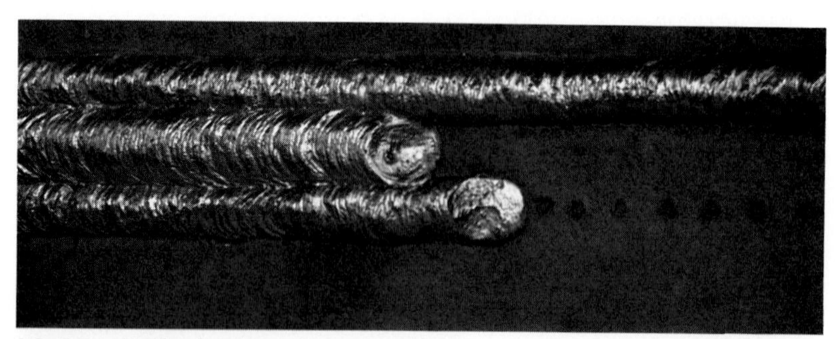

20-21. Method of marking parallel lines and typical appearance of stringer beads welded in the flat position with DCSP or a.c. shielded-arc electrodes.

It is not good practice to weave beads wider than three times the diameter of the electrode. The weld deposit must be molten until the desired shape is formed. Advancing the arc too far ahead and then delaying the return to the molten pool allows the pool to cool. This causes trapped slag and poor appearance. You will not have difficulty with undercutting with these electrodes, but it is well to hesi-

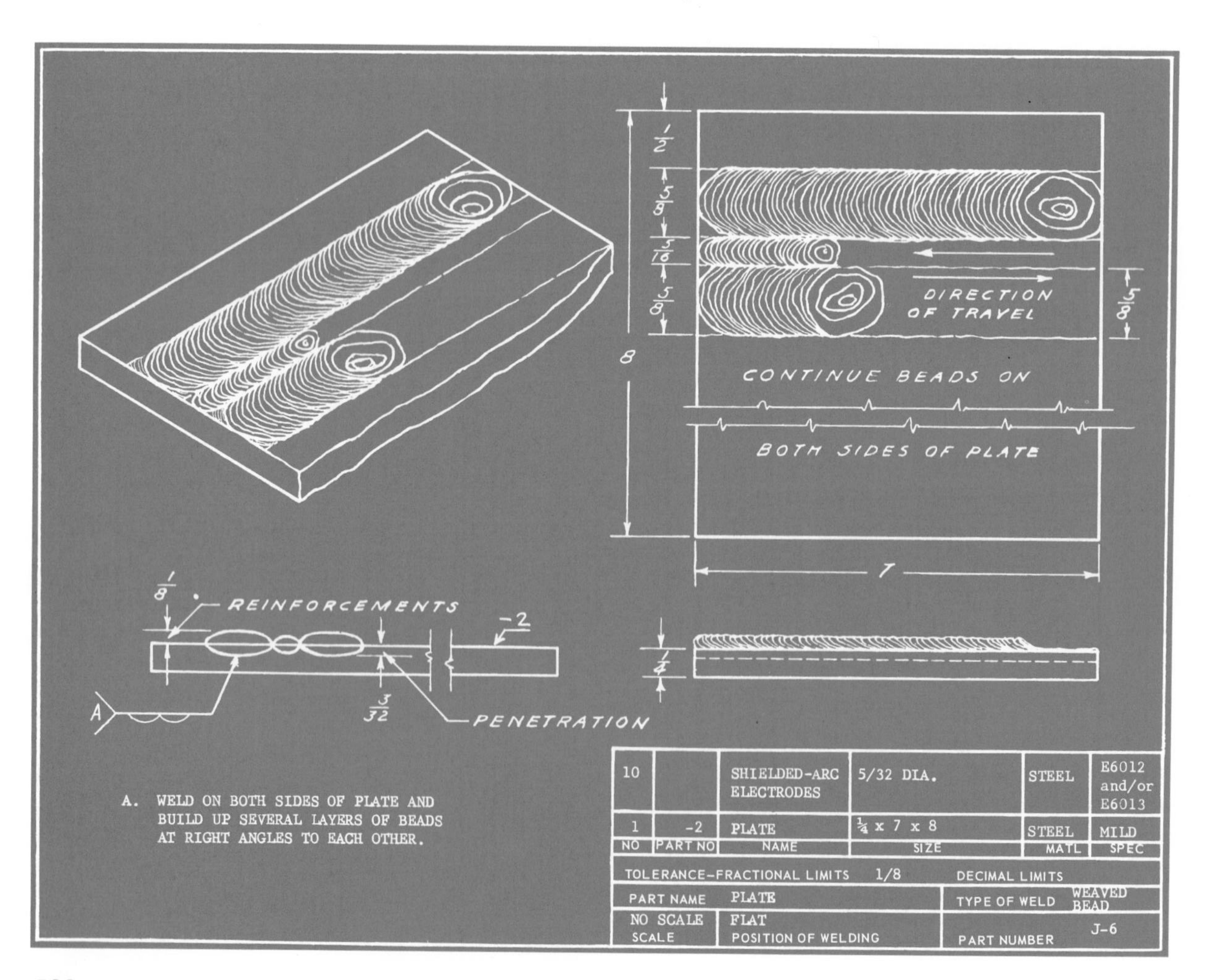

tate an instant at the sides. Electrode motion should be smooth and rhythmic along the entire weld.

Starts and stops are made and craters are filled as practiced in previous jobs.

Weld on both sides of the plate. When the plate is completely filled, deposit additional beads at right angles to the beads already on the plate. Practice traveling in all directions.

Operations

1. Obtain plate; check the job print for size.
2. Obtain a square head and scale, dividers, scribe, and center punch from the toolroom.
3. Lay out parallel lines as shown on the job print.
4. Mark the lines with a center punch. They should look like those in Fig. 20-23.
5. Obtain electrodes of each quantity, type, and size specified.
6. Set a d.c. power source for straight polarity at 110 to 190 amperes, or an a.c. power source at 120 to 210 amperes.
7. Lay the plate in the flat position on the welding table. Make sure that the plates are well-grounded.
8. Make weaved beads between parallel lines as shown on the job print. Hold the electrode as shown in Fig. 20-22.
9. Chip the slag from the beads, brush, and inspect. Refer to Inspection.
10. Make stringer beads between the weaved beads already deposited.
11. Chip the slag from the beads, brush, and inspect. Refer to Inspection.
12. Practice weaved beads and stringer beads until you can produce uniform beads consistently with both types of electrodes.

Inspection

Compare the beads with Figs. 20-23 and 20-24 and check them for the following weld characteristics:

Width and height: Uniform
Appearance: Smooth with close ripples; free of voids and high spots. Restarts should be difficult to locate.
Size: Refer to the job print.
Face of beads: Slightly convex
Edges of beads: Good fusion, no overlap, no undercut
Starts and stops: Free of depressions and high spots
Beginnings and endings: Full size, craters filled
Penetration and fusion: To plate surfaces and adjacent beads
Surrounding plate surfaces: Free of spatter

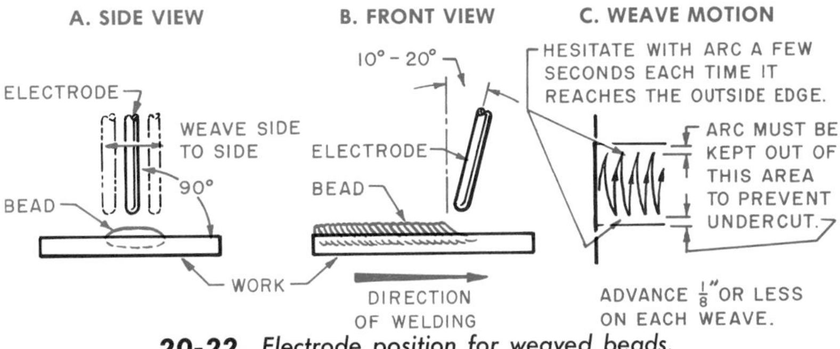
A. SIDE VIEW
B. FRONT VIEW
C. WEAVE MOTION
ELECTRODE
WEAVE SIDE TO SIDE
90°
BEAD
WORK
10° - 20°
ELECTRODE
BEAD
DIRECTION OF WELDING
HESITATE WITH ARC A FEW SECONDS EACH TIME IT REACHES THE OUTSIDE EDGE.
ARC MUST BE KEPT OUT OF THIS AREA TO PREVENT UNDERCUT.
ADVANCE 1/8" OR LESS ON EACH WEAVE.

20-22. *Electrode position for weaved beads.*

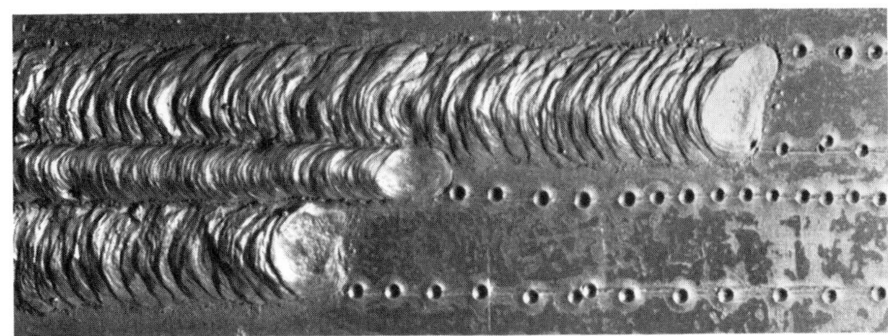

20-23. *Method of marking parallel lines and typical appearance of weaved beads welded in the flat position with DCSP shielded-arc electrodes.*

20-24. *Typical appearance of weaved beads welded in the flat position with the E6013 a.c. type of shielded-arc electrode.*

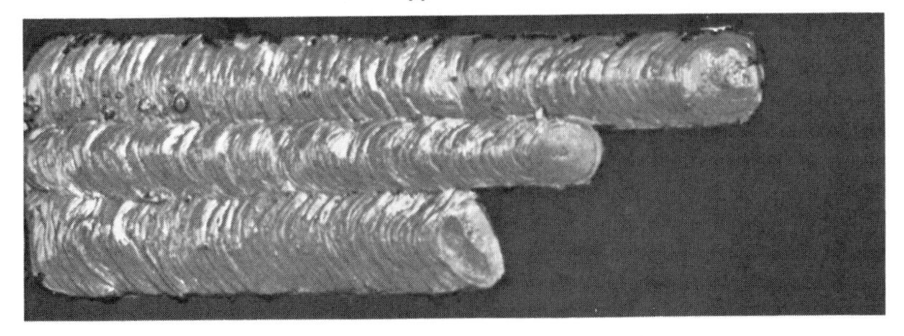

Slag formation: Full coverage, easily removable

Disposal

Discard completed plate in the waste bin. Plates must be filled with beads on both sides.

JOB 20-J7 STRINGER BEADING

Objective

To deposit stringer beads on flat steel plate in the flat position with DCRP and/or a.c. shielded-arc electrodes (AWS-ASTM E6010-E6011).

General Job Information

This job is basically the same as Job 20-J5, but different types of electrodes are used, resulting in different arc characteristics and deposition.

This job requires either a DCRP or a.c. power source and both E6010 and E6011 electrodes. Study the section in Chapter 7 concerning these electrodes.

The electrodes produce welds of the highest quality in all positions. Because their spray-type arc produces deep penetration, precise electrode manipulation is necessary in order to control undercut and weld spatter. The thickness of the covering is held to a minimum. The weld pool wets and spreads well, but it sets up very quickly. Thus these electrodes have fine out-of-position welding characteristics. Fillet and groove welds are flat in profile and have a rather coarse, unevenly spaced ripple. The current and voltage ranges are similar for both electrodes. The coating is slightly heavier on the E6011. The ductility, tensile strength, and yield strength of the deposited weld metal are higher than can be obtained with an E6010 electrode. Both of these electrodes are used whenever welding of high quality is required, such as in shipyards, field construction, pressure vessel fabrication, piping, and other code welding applications.

Welding Technique

Pay close attention to your heat setting. The electrode position should be approximately that shown in Fig. 20-8, page 580. The arc gap for the E6010 electrode (reverse polarity) may be less than for the E6012 (straight polarity) electrode. Do not permit the coating to contact the molten pool. Observe the weld metal transfer closely and listen to the sound of the arc for correct arc gap.

Start the arc and proceed with the bead in much the same manner as practiced in Job 20-J7. A slight forward and backward motion along the line of weld may be used. These electrodes will have a tendency to undercut the plate along the edges of the weld. To prevent undercutting, do not move the arc past the forward end of the crater until enough metal has been deposited to advance a weld of the required size and all undercut has been filled.

Starts and stops and filling craters are performed as practiced in previous jobs.

Weld on both sides of the plate. When the plate is completely filled, deposit additional beads at right angles to the beads already on the plates. Practice traveling in all directions. You should also practice with the $\frac{3}{16}$-inch diameter electrodes after you have mastered the $\frac{5}{32}$-inch size.

20-25. *Method of marking parallel lines and typical appearance of stringer beads welded in the flat position with DCRP or a.c. shielded-arc electrodes.*

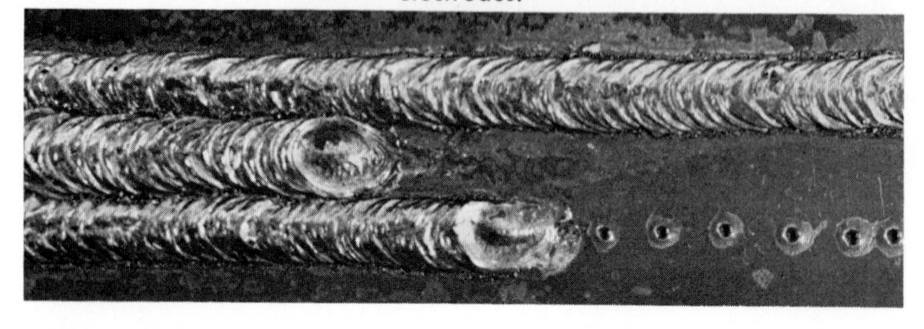

Operations

1. Obtain plate; check the job print for size.
2. Obtain a square head and scale, dividers, scribe, and center punch from the toolroom.
3. Lay out parallel lines as shown on the job print.
4. Mark the lines with a center punch. They should look like those in Fig. 20-25.
5. Obtain electrodes of each quantity, type, and size specified in the job print.
6. Set the power source for 110 to 170 amperes. Set a d.c. power source for reverse polarity.
7. Lay the plate in the flat position on the welding table. Make sure it is well-grounded.
8. Make stringer beads on $\frac{3}{4}$-inch center lines as shown on the job print. Hold the electrode as shown in Fig. 20-8, page 580.
9. Chip the slag from the beads, brush, and inspect. Refer to Inspection.
10. Make additional beads between the beads already de-

posited as shown on the job print.

11. Chip the slag from the beads, brush, and inspect. Refer to Inspection below.

12. Practice these beads until you can produce uniform beads consistently with both the E6010 and E6011 electrodes.

Inspection

Compare the beads with Fig. 20-25 and check for the following weld characteristics:

Width and height: Uniform
Appearance: Smooth with close ripples; free of voids and high spots. Restarts should be difficult to locate.
Size: Refer to the job print.
Face of beads: Slightly convex
Edges of beads: Good fusion, no overlap, no undercut
Starts and stops: Free of depressions and high spots
Beginnings and endings: Full size, craters filled
Penetration and fusion: To plate surfaces and adjacent beads
Surrounding plate surfaces: Free of spatter
Slag formation: Full coverage, easily removable

Disposal

Discard completed plates in the waste bin. Plates must be filled with beads on both sides.

JOB 20-J8 WEAVED BEADING

Objective

To deposit weaved beads on flat steel plate in the flat position with DCRP and/or a.c. shielded-arc electrodes (AWS-ASTM E6010-E6011).

General Job Information

This job is basically the same as Job 20-J6, except that different types of electrodes are used resulting in different arc and deposition characteristics. In industry you will do more weaved beading with reverse polarity electrodes than with straight polarity electrodes.

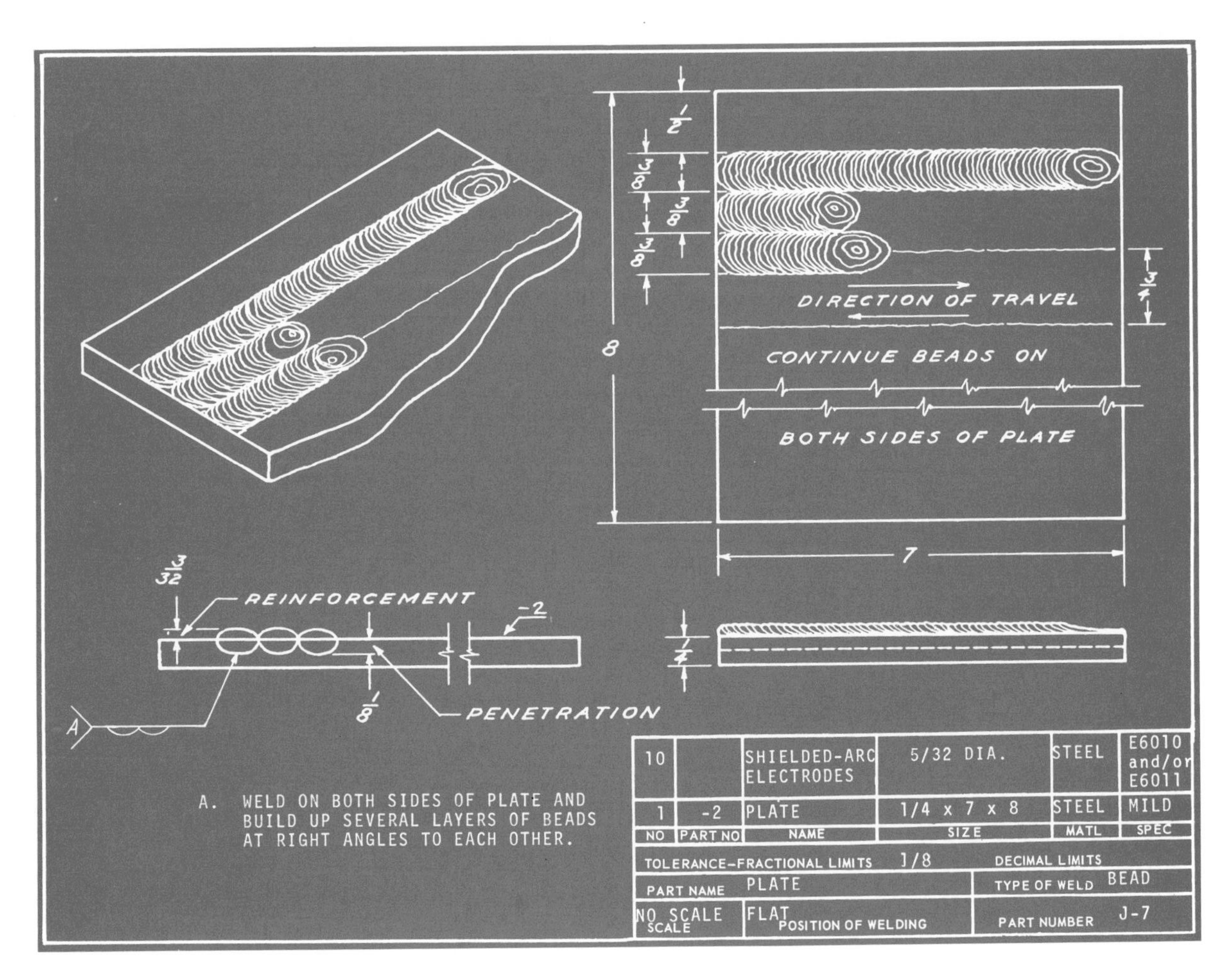

NO	PART NO	NAME	SIZE	MATL	SPEC
10		SHIELDED-ARC ELECTRODES	5/32 DIA.	STEEL	E6010 and/or E6011
1	-2	PLATE	1/4 x 7 x 8	STEEL	MILD

TOLERANCE–FRACTIONAL LIMITS 1/8	DECIMAL LIMITS
PART NAME PLATE	TYPE OF WELD BEAD
NO SCALE SCALE / FLAT POSITION OF WELDING	PART NUMBER J-7

Welding Technique

Pay close attention to the current setting. Study the electrode position and weaving technique shown in Fig. 20-22, page 591. The method of depositing beads is the same as for straight polarity electrodes except that undercutting is a problem. This can be overcome only by longer hesitation at the sides of the weld and by keeping the movement of the electrode within the confines of the bead width. Bead reinforcement will not be as high as with straight polarity electrodes, and penetration will be deeper. There will be less slag present. It will be less fluid and will solidify more quickly.

Starts and stops are made and craters are filled as practiced in previous jobs.

Weld on both sides of the plate. When the plate is completely filled, deposit additional beads at right angles to the beads already on the plate. Practice traveling in all directions.

Operations

1. Obtain plate; check the job print for size.
2. Obtain a square head and scale, dividers, scribe, and center punch from the toolroom.
3. Lay out parallel lines as shown on the job print.
4. Mark the lines with a center punch. They should look like those shown in Fig. 20-26.
5. Obtain electrodes of each quantity, type, and size specified in the job print.
6. Set a d.c. power source for reverse polarity at 110 to 170 amperes or an a.c. power source at 110 to 190 amperes.
7. Lay the plate in the flat position on the welding table. Make sure it is well-grounded.
8. Make weaved beads between parallel lines as shown on the job print. Hold the electrode

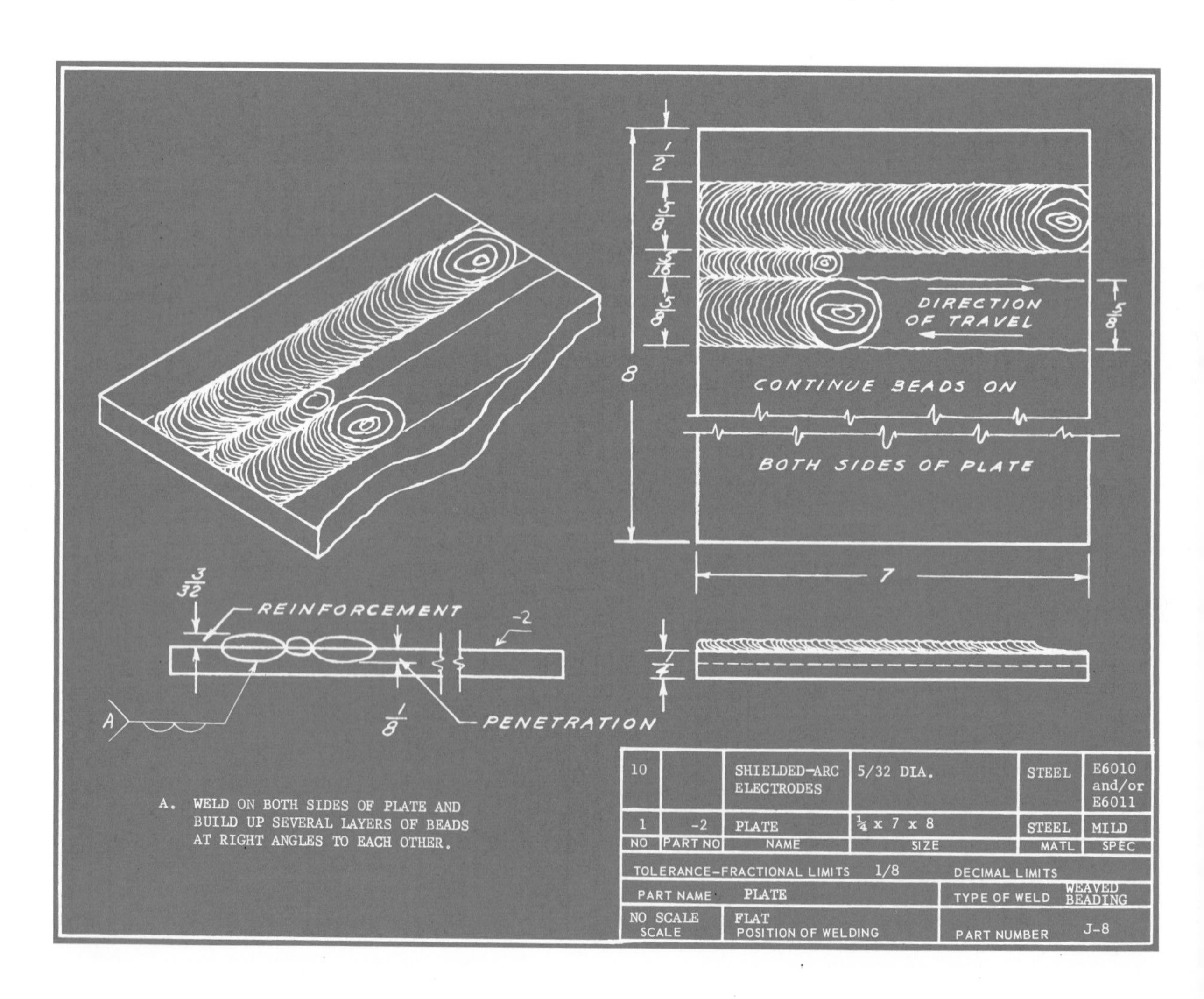

NO	PART NO	NAME	SIZE	MATL	SPEC
10		SHIELDED-ARC ELECTRODES	5/32 DIA.	STEEL	E6010 and/or E6011
1	-2	PLATE	¼ x 7 x 8	STEEL	MILD

TOLERANCE-FRACTIONAL LIMITS 1/8		DECIMAL LIMITS
PART NAME PLATE		TYPE OF WELD WEAVED BEADING
NO SCALE SCALE	FLAT POSITION OF WELDING	PART NUMBER J-8

as shown in Fig. 20-22, page 591.

9. Chip the slag from the beads, brush, and inspect. Refer to Inspection, below.

10. Make stringer beads between the weaved beads already deposited.

11. Chip the slag from the beads, brush, and inspect. Refer to Inspection, below.

12. Practice these beads until you can produce uniform beads consistently with both types of electrodes.

Inspection

Compare the beads with Figs. 20-26 and 27 and check them for the following characteristics:

Width and height: Uniform
Appearance: Smooth with close ripples; free of voids and high spots. Restarts should be difficult to locate.
Size: Refer to the job print.
Face of beads: Slightly convex
Edges of beads: Good fusion, no overlap, no undercut
Starts and stops: Free of depressions and high spots
Beginnings and endings: Full size, craters filled
Penetration and fusion: To plate surfaces and adjacent beads
Surrounding plate surfaces: Free of spatter
Slag formation: Full coverage, easily removable

Disposal

Discard completed plates in the waste bin. Plates must be filled with beads on both sides before disposal.

JOB 20-J9 WELDING AN EDGE JOINT

Objective

To weld an edge joint in the flat position with DCSP and/or a.c. shielded-arc electrodes (AWS-ASTM E6012-E6013).

General Job Information

The edge joint is used extensively on vessels and tanks that are not subjected to high pressures. The joint will not stand very much load when subjected to tension and bending.

It is an economical joint to set up. The amount of electrode needed for welding is small since some of the base metal is melted off and combines with the electrode deposit to form the weld. It is an easy joint to weld.

The edge joint is often welded with the air carbon-arc process.

Welding Technique

The edge joint can be welded over a wide range of current settings. Hold the electrode at a 90-degree angle to the plates to be welded, Fig. 20-28A. Then tilt it 10 to 20 degrees in the direction of travel, Fig. 20-28B. Note the positions of the electrode and electrode holder in Fig. 20-29. It is good practice to weave slightly so that there is complete coverage of the joint area. It is important, however, not to weave so wide that the metal hangs over the sides of the plate. Be very careful to fuse both edges and to secure good penetration along the center of the joint.

Starts and stops are made and craters are filled as practiced in previous jobs.

Weld both edges of the joint. Practice traveling in all directions.

Operations

1. Obtain plates; check the print for quantity and size.

2. Obtain electrodes of each quantity, type, and size specified in the job print.

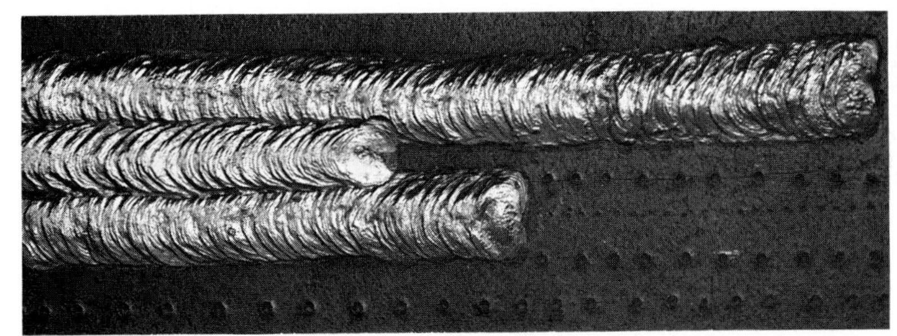

20-26. *Method of marking parallel lines and typical appearance of weaved beads welded in the flat position with DCRP shielded-arc electrodes.*

20-27. *Typical appearance of weaved beads welded in the flat position with the E6011 a.c. type of shielded-arc electrode.*

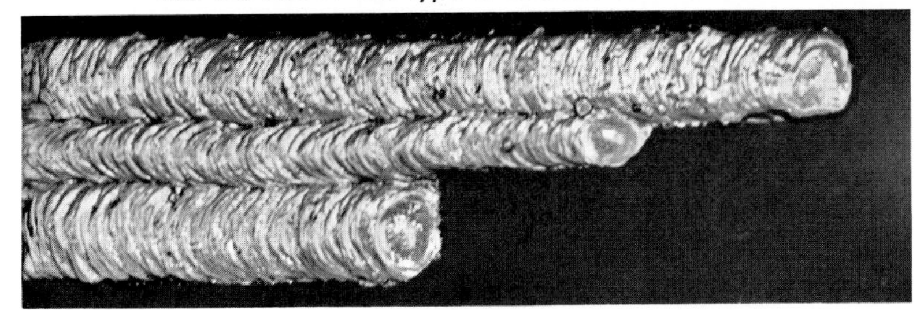

3. Set a d.c. power source for straight polarity at 110 to 190 amperes or an a.c. power source at 120 to 210 amperes.

4. Set up the plates and tack weld them at each corner.

5. Place the joint in the flat position on the welding table. Make sure it is well-grounded.

6. Make welds as shown on job print. Hold electrode as shown in Figs. 20-28 and 20-29.

7. Chip the slag from the welds, brush, and inspect. Refer to Inspection.

8. Practice these welds until you can produce good welds consistently with $\frac{3}{16}$-inch diameter electrodes.

Inspection

Compare the welds with Fig. 20-30 and check them for the following characteristics:

Width and height: Uniform

Appearance: Smooth with close ripples; free of voids and high spots. Restarts should be difficult to locate.

Size: Refer to the job print.

Face of welds: Slightly convex

Edges of welds: Good fusion, no overlap, no undercut, no weld metal hanging over the edges of the joint

Starts and stops: Free of depressions and high spots

Beginnings and endings: Full size, craters filled

Penetration and fusion: To plate surfaces

Surrounding plate surfaces: Free of spatter

Slag formation: Full coverage, easily removable

Disposal

Run stringer beads on the unused sides of the joints or put the completed joints in the scrap bin so that they are available for stringer bead practice.

JOB 20-J10 WELDING AN EDGE JOINT

Objective

To weld an edge joint in the flat position with DCRP and/or

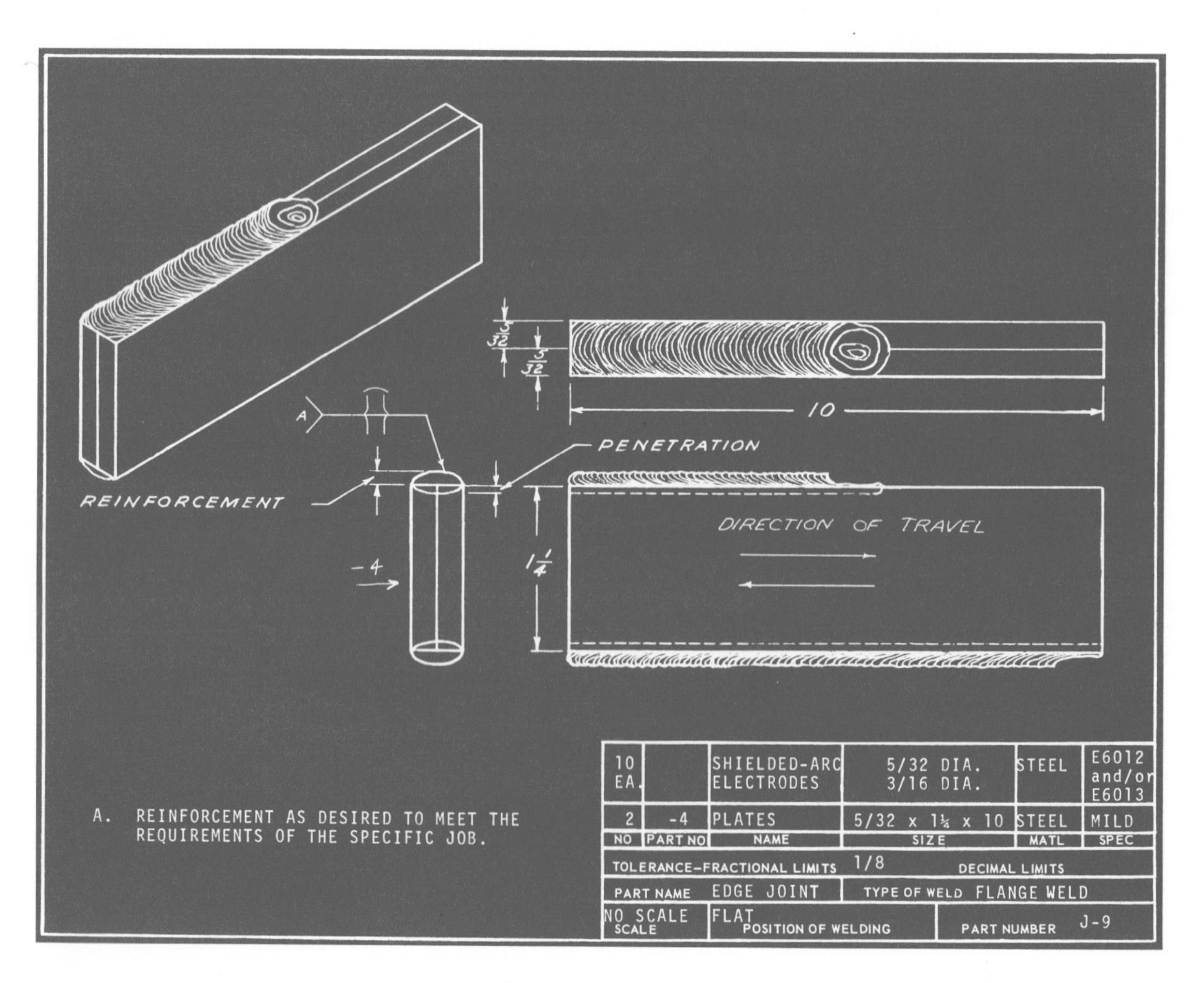

a.c. shielded-arc electrodes (AWS-ASTM E6010-E6011).

General Job Information

This joint is the same as that welded in the previous job except that a different type of electrode is used.

Welding Technique

The edge joint can be welded over a wide range of current settings. The electrode position should be approximately that shown in Fig. 20-28, page 598. Use the same basic welding technique practiced in Job 20-J9. The arc gap will be less and the current adjustment will be slightly lower than with straight polarity. Welds will not have very high buildup, and slag will be less.

Practice starts and stops carefully. Weld in all directions.

Operations

1. Obtain plates. Check the job print for quantity and size.
2. Obtain electrodes of each quantity, type, and size specified in the job print.
3. Set a d.c. power source for reverse polarity at 110 to 170 amperes or an a.c. power source at 110 to 190 amperes.
4. Set up the plates and tack weld them at each corner.
5. Place the joint in the flat position on the welding table. Make sure it is well-grounded.
6. Make welds as shown on the job print. Hold the electrode as shown in Fig. 20-28, page 598.
7. Chip the slag from the welds, brush, and inspect. Refer to Inspection, below.
8. Practice these welds until you can produce good welds consistently with both the E6010 and E6011 electrodes and with $^3/_{16}$-inch diameter electrodes.

Inspection

Compare the welds with Fig. 20-31 and check them for the following characteristics:

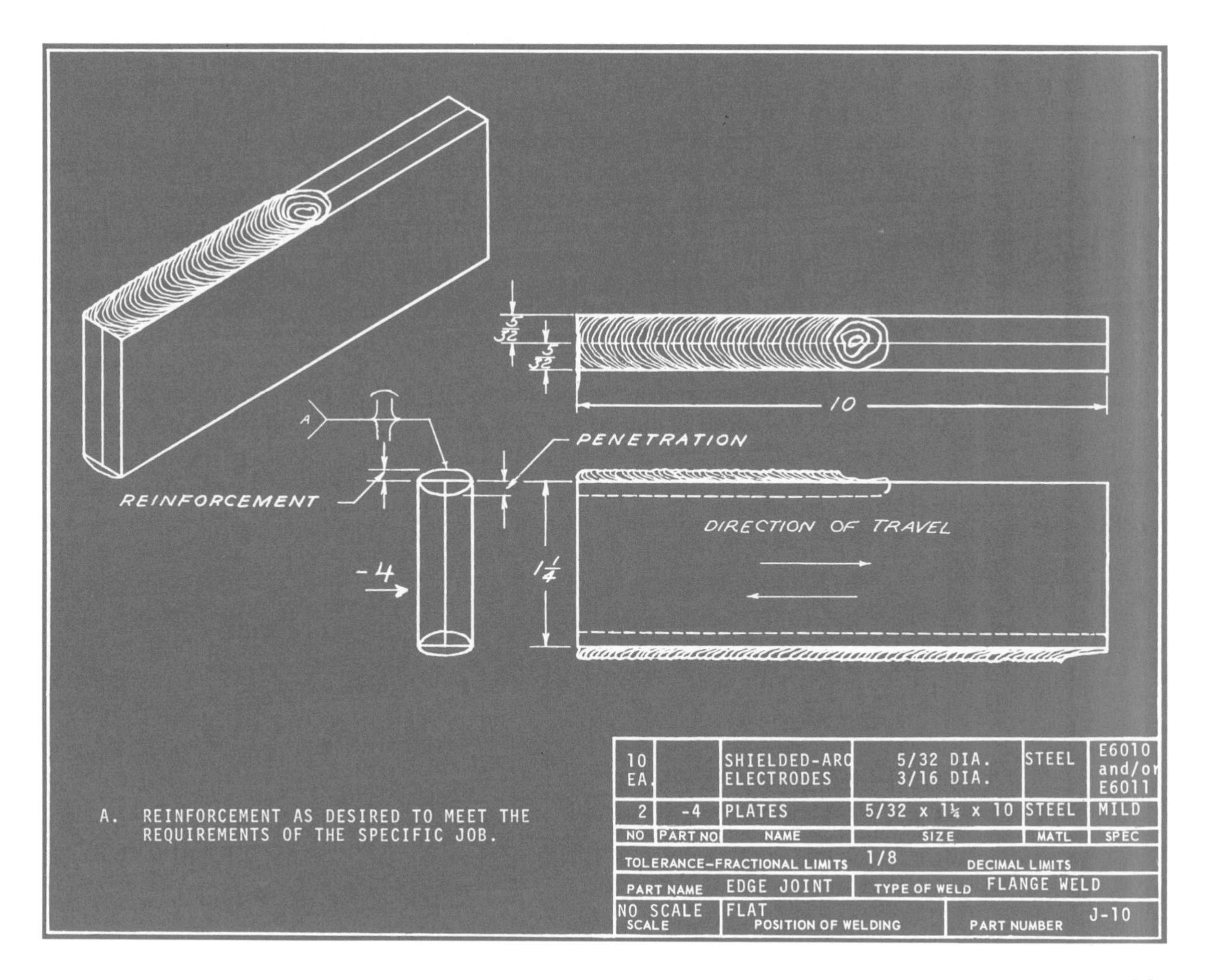

NO	PART NO	NAME	SIZE	MATL	SPEC
10 EA.		SHIELDED-ARC ELECTRODES	5/32 DIA. 3/16 DIA.	STEEL	E6010 and/or E6011
2	-4	PLATES	5/32 x 1¼ x 10	STEEL	MILD

TOLERANCE–FRACTIONAL LIMITS 1/8 DECIMAL LIMITS
PART NAME EDGE JOINT TYPE OF WELD FLANGE WELD
SCALE NO SCALE POSITION OF WELDING FLAT PART NUMBER J-10

Width and height: Uniform
Appearance: Smooth with close ripples; free of voids and high spots. Restarts should be difficult to locate.
Size: Refer to the job print.
Face of welds: Slightly convex
Edges of welds: Good fusion, no overlap, no undercut, no weld metal hanging over the edges of the joint
Starts and stops: Free of depressions and high spots
Beginnings and endings: Full size, craters filled
Penetration and fusion: To plate surfaces
Surrounding plate surfaces: Free of spatter
Slag formation: Full coverage, easily removable

Disposal

Run stringer beads on the unused sides of the joints or put the completed joints in the scrap bin so that they are available for stringer bead practice.

JOB 20-J11 WELDING A LAP JOINT

Objective

To weld a lap joint in the horizontal position by means of a single-pass fillet weld with DCSP and/or a.c. shielded-arc electrodes (AWS-ASTM E6012-E6013).

General Job Information

The lap joint is used extensively in industry for tank, structural, and shipyard construction. It is an economical joint because it needs very little joint preparation and fitup. The lap joint is strongest when it is double-lap welded on both sides. This job and the jobs that follow specify a single pass. For heavier plate, several passes must be made with both stringer and weaved beads.

Welding Technique

The adjustment of current should not be too high. The electrode position should be approximately that shown in Fig. 20-14 and 20-15, page 584. A fairly close arc gap is necessary, and it should be directed towards the root of the joint and towards the flat plate surface. Use a slight back-and-forth motion as needed along the line of weld. This preheats the joint ahead of the weld, results in full reinforcement, and keeps the slag washed back over the deposited metal.

Complete penetration must be secured at the root of the weld, and good fusion must be obtained with the two plate surfaces. The top edge of the top plate has a tendency to burn away. This can be prevented by making sure that the bead formation is full and that the arc is not played along the top surface. Undercutting is not a problem with the E6012 and E6013 electrodes.

There should be a straight line of fusion with the top and bottom plates and a smooth transition along the edge of the weld where it enters the plate. A rate of travel that is too slow causes an excess deposit of weld metal. When it rolls over on the bottom plate, it changes the contour of the weld abruptly and increases the possibility of poor fusion to the plate surface. Excess weld metal also weakens the joint by causing stress concentration at this point.

Practice stops and starts. Travel in all directions.

Operations

1. Obtain plates; check the job print for the correct quantity and size.

2. Obtain electrodes of each quantity, type, and size specified in the job print.

20-28. *Electrode position when welding an edge joint.*

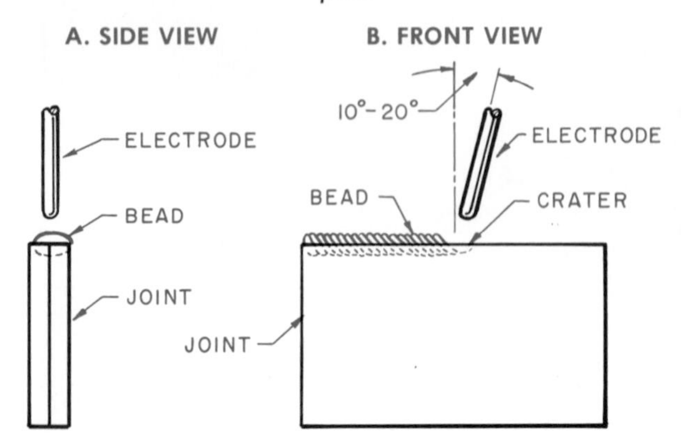

20-29. *Welding an edge joint. The electrode and electrode holder are positioned as instructed in Fig. 20-28.*

3. Set a d.c. power source for straight polarity at 110 to 190 amperes or an a.c. power source at 120 to 210 amperes.

4. Set up the plates and tack weld them at each corner.

5. Place the joint in the flat position on the welding table. Make sure that the plates are well-grounded.

6. Make welds as shown on the job print. Hold the electrode as shown in Figs. 20-14 and 20-15, page 584.

7. Chip the slag from the welds, brush, and inspect. Refer to Inspection.

8. Practice these welds until you can produce good welds

20-30. *Typical appearance of an edge joint welded in the flat position with DCSP or a.c. shielded-arc electrodes.*

20-31. *Typical appearance of an edge joint welded in the flat position with DCRP or a.c. shielded-arc electrodes.*

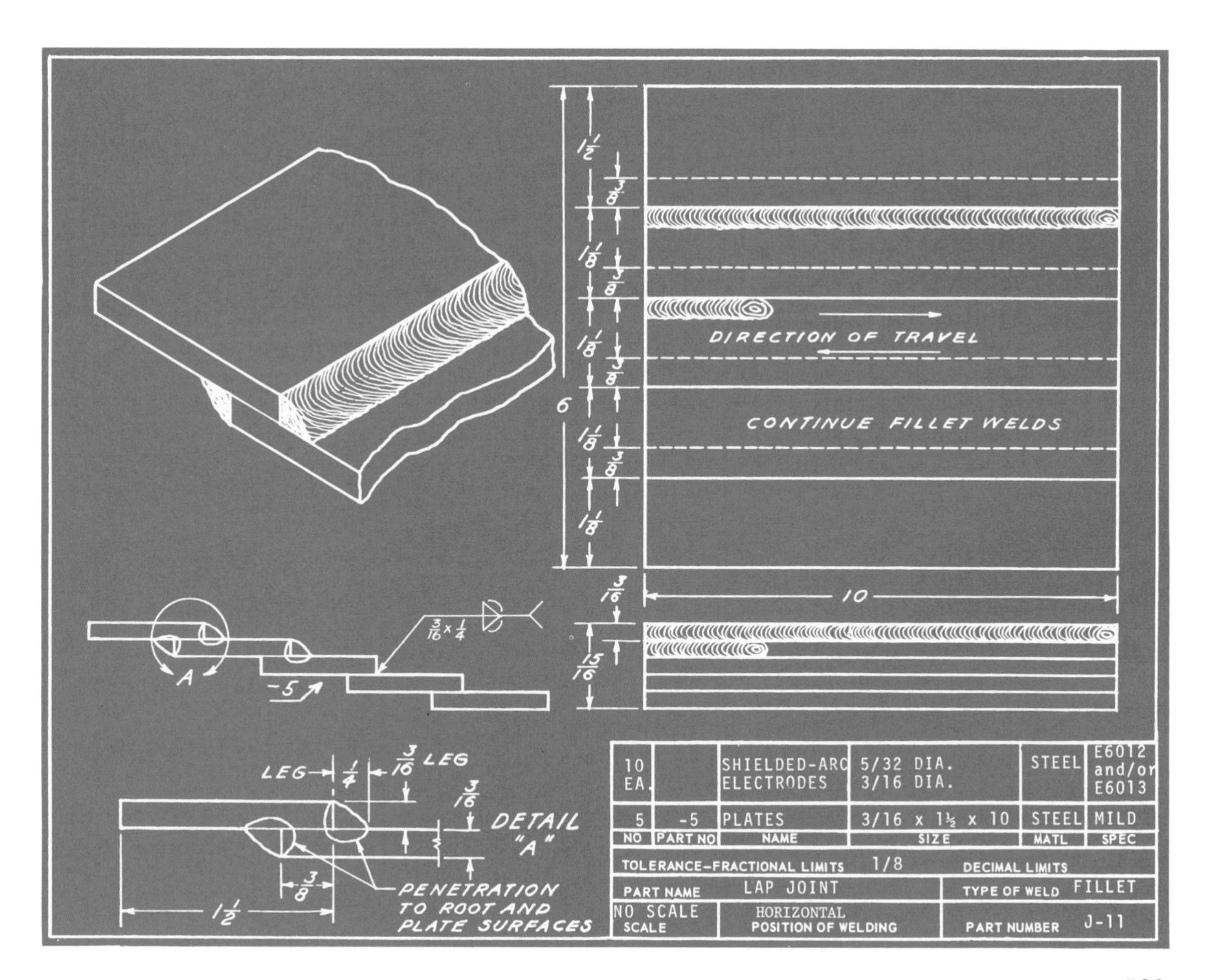

consistently with both types of electrodes and with $^3/_{16}$-inch diameter electrodes.

Inspection

Compare the welds with Fig. 20-32 and check them for the following characteristics:

Width and height: Uniform
Appearance: Smooth with close ripples; free of voids and high spots. Restarts should be difficult to locate.
Size: Refer to the job print. Check with a fillet weld gauge.
Face of welds: Slightly convex
Edges of welds: Good fusion, no overlap, no undercut
Starts and stops: Free of depressions and high spots
Beginnings and endings: Full size, craters filled
Penetration and fusion: To root of joint and plate surfaces
Surrounding plate surfaces: Free of spatter
Slag formation: Full coverage, easily removable

Check Test

After you are able to make welds that are satisfactory in appearance, make up a specimen similar to that shown in Fig. 20-17, page 585. Use the plate thickness, welding procedure, and electrode size specified for this job. Weld on one side only.

Break the joint as shown in Fig. 20-18, page 585. Examine the surface of the fracture for soundness. The weld must not be porous and must show good fusion and penetration at the root and to both plate surfaces. The joint should break through the throat of the weld, and the weld should not peel away from the plate.

Disposal

Put completed joints in the scrap bin so that they will be available for further use. Unwelded plate surfaces can be used for beading.

JOB 20-J12 WELDING A LAP JOINT

Objective

To weld a lap joint in the horizontal position by means of a single-pass fillet weld with DCRP and/or a.c. shielded-arc electrodes (AWS-ASTM E6010-E6011).

General Job Information

This joint is the same as that practiced in Job 20-J11 except that a different type of electrode is used.

Welding Technique

Current adjustment must not be too high. The electrode position should be approximately that shown in Figs. 20-14 and 20-15, page 584. The welding technique is basically the same as that used for the previous job. You should, however, hold a shorter arc. Be careful to avoid undercut both on the flat plate surface and along the top edge of the upper plate.

Remember that the arc movement must be within the limits of the desired weld size.

Practice stops and starts. Travel in all directions.

Operations

1. Obtain plates; check the job print for size.
2. Obtain electrodes of each quantity, type, and size specified in the job print.
3. Set a d.c. power source for reverse polarity at 110 to 170 amperes or an a.c. power source at 110 to 190 amperes.
4. Set up the plates and tack weld them at each corner.
5. Place the joint in the flat position on the welding table. Make sure it is well-grounded.
6. Make welds as shown on the job print. Hold the electrode as shown in Figs. 20-14 and 20-15, page 584.
7. Chip the slag from the welds, brush, and inspect. Refer to Inspection, below.
8. Practice these welds until you can produce good welds consistently with both types of electrodes and with $^3/_{16}$-inch diameter electrodes.

Inspection

Compare the welds with Fig. 20-33 and check them for the following characteristics:

Width and height: Uniform
Appearance: Smooth with close ripples; free of voids and high spots. Restarts should be difficult to locate.
Size: Refer to the job print. Check with a fillet weld gauge.
Face of weld: Flat
Edges of weld: Good fusion, no overlap, no undercut

20-32. *Typical appearance of a fillet weld in a lap joint welded in the horizontal position with DCSP or a.c. shielded-arc electrodes.*

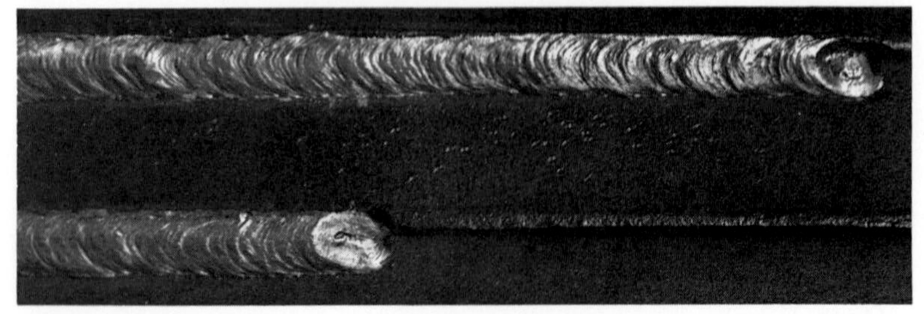

Starts and stops: Free of depressions and high spots
Beginnings and endings: Full size, craters filled
Penetration and fusion: To root of joint and to plate surfaces
Surrounding plate surfaces: Free of spatter
Slag formation: Full coverage, easily removed

Check Test

After you are able to make welds that are satisfactory in appearance, make up a specimen similar to that shown in Fig. 20-17, page 585. Use the plate thickness, welding technique, and electrode type and size used in this job. Weld on one side only. Break the joint as shown in Fig. 20-18, page 585. Examine the surfaces for soundness. The weld must not be porous, and it must show good fusion and penetration at the root and to the plate surfaces. The joint should break through the throat of the weld.

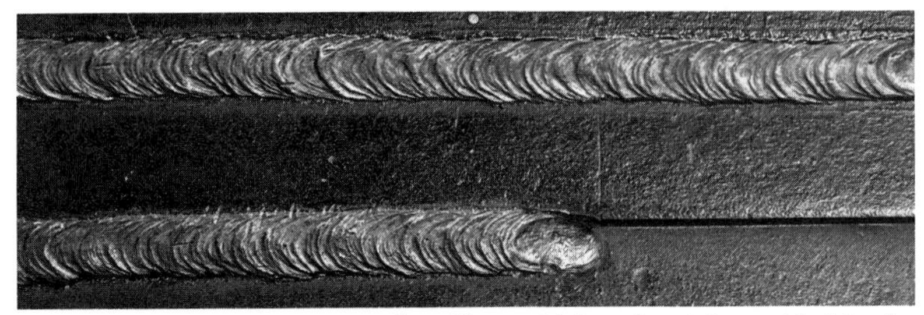

20-33. *Typical appearance of a fillet weld in a lap joint welded in the horizontal position with DCRP or a.c. shielded-arc electrodes.*

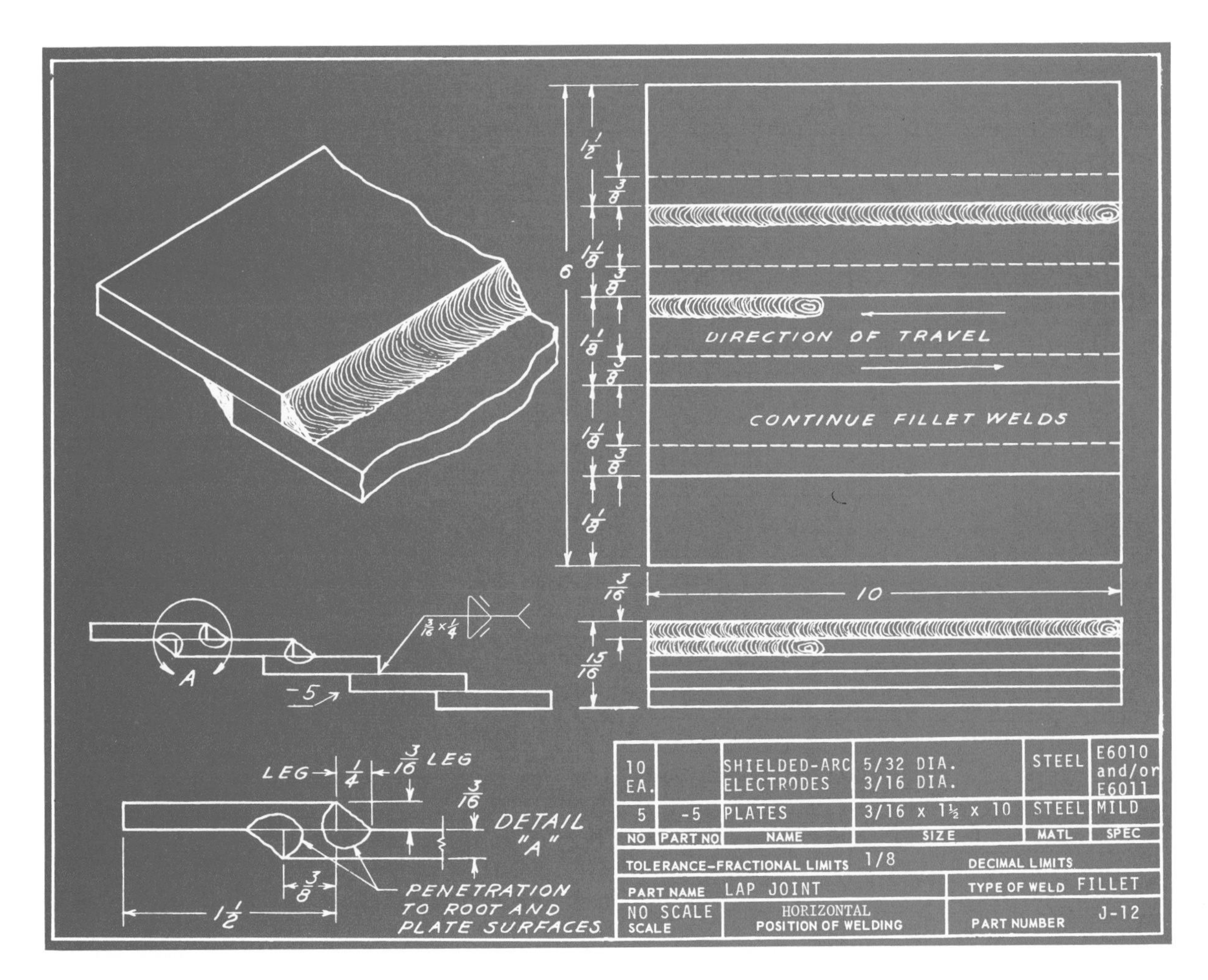

Disposal

Put completed joints in the scrap bin so that they will be available for further use. Unwelded plate surfaces can be used for beading.

JOB 20-J13 STRINGER BEADING

Objective

To deposit horizontal stringer beads on flat steel plate in the vertical position with DCRP and/or a.c. shielded-arc electrodes (AWS-ASTM E6010-E6011).

General Job Information

Horizontal groove welds and fillet welds are welded with stringer beads. The finishing layer is most frequently made in the same manner practiced on this job.

Welding Technique

Current adjustment must not be too high. The electrode position should be approximately that shown in Figs. 20-34 and 20-35. Hold the electrode at a 15-degree angle downward from the horizontal, Fig. 20-34A and 15 degrees forward in the direction of travel, Fig. 20-34B. The electrode is manipulated in a roughly circular motion known as *oscillation.* The actual positions of the electrode and holder are shown in Fig. 20-35.

This job provides practice in running a straight weld bead without the guidance of a line to follow. It also provides practice in fusing a weld bead to the plate and to another weld bead.

Practice laying beads across the plate in the horizontal position. Begin at the bottom of the plate with the first bead. Remove the slag thoroughly after each pass and deposit another bead along the side. It should be fused into the preceding bead for one-third to one-half of its width.

You will find it necessary to use a slightly lower welding current and to hold a shorter arc than when welding in the flat position. Watch the weld carefully so that it does not sag on the bottom edge. Sagging can be prevented by moving the elec-

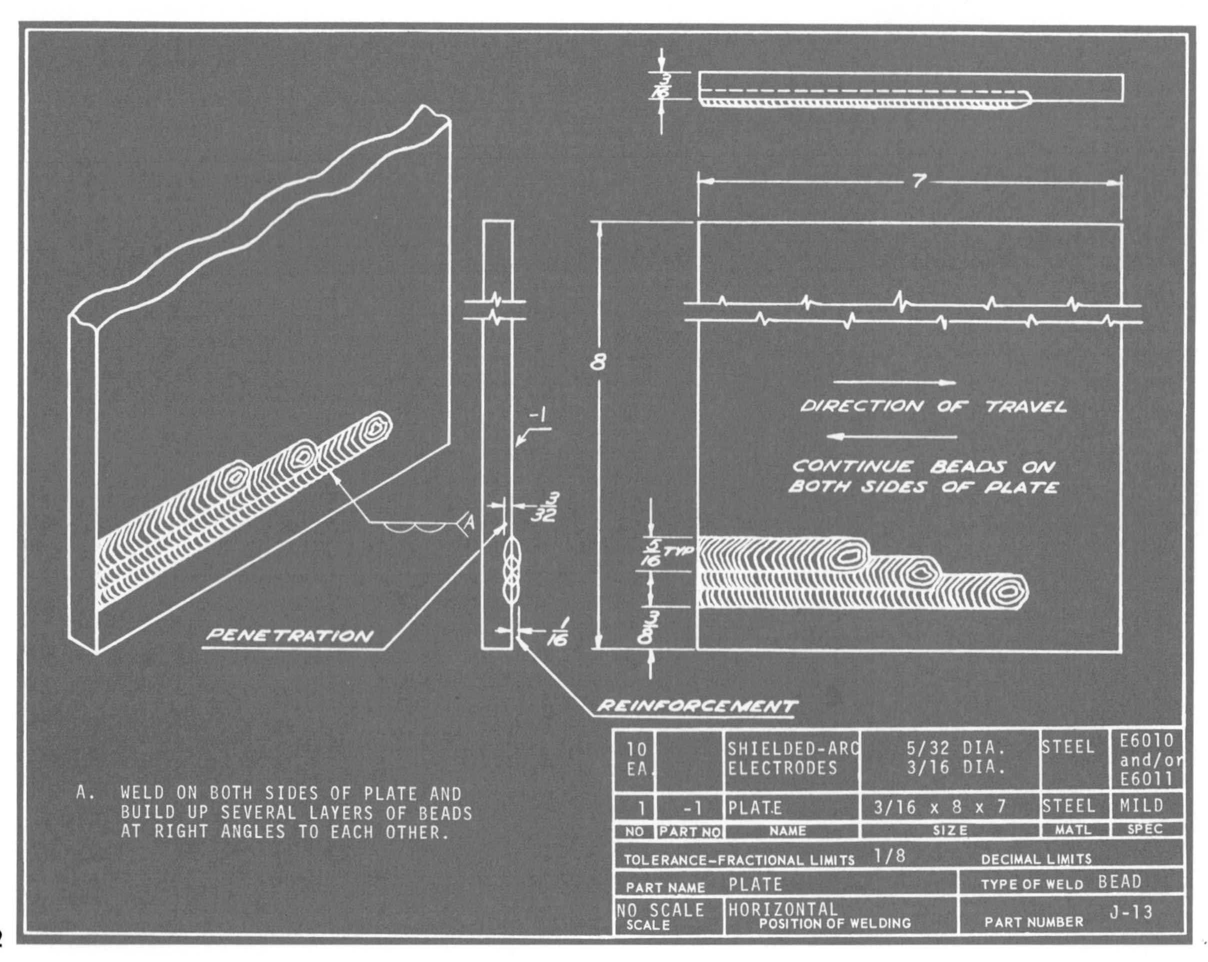

trode back and forth. Be sure to get good penetration into the plate as well as good fusion between beads. Clean the slag from each bead before applying the next bead. Each bead should lap over the preceding bead to the extent of 40 to 50 percent. To prevent undercut, keep electrode movement within the bead size.

Practice starts and stops. Practice travel from left to right and from right to left. Weld on both sides of the plate and lay several layers on each side.

Operations

1. Obtain plate; check the job print for size.
2. Obtain electrodes in each quantity, type, and size specified in the job print.
3. Stand the plate in the vertical position on the welding bench. Make sure that it is well-grounded.
4. Set a d.c. power source for reverse polarity at 110 to 160 amperes or an a.c. power source at 110 to 180 amperes.
5. Make stringer beads as shown on the job print. Hold the electrode as shown in Figs. 20-34 and 20-35.
6. Chip the slag from the beads, brush, and inspect. Refer to Inspection.
7. Practice these beads until you can produce uniform beads consistently with both types of electrodes and with 3/16-inch diameter electrodes.

Inspection

Compare the beads with Fig. 20-36 and check them for the following characteristics:

Width and height: Uniform
Appearance: Smooth with close ripples; free of voids and high spots. Restarts should be difficult to locate.
Size: Refer to the job print.
Face of beads: Slightly convex
Edges of beads: Good fusion, no overlap, no undercut
Starts and stops: Free of depressions and high spots
Beginnings and endings: Full size, craters filled
Penetration and fusion: To plate surfaces and adjacent beads
Surrounding plate surfaces: Free of spatter
Slag formation: Full coverage, easily removable

Disposal

Discard completed plates in the waste bin. Plates must be filled on both sides.

JOB 20-J14 STRINGER BEADING

Objective

To deposit stringer beads on flat steel plate; vertical position, travel down; DCRP and/or a.c. shielded-arc electrodes (AWS-ASTM E6010-E6011).

General Job Information

Welding downhill on a vertical surface is frequently done in industry, especially on noncritical work. This method of welding is often used on overland pipeline and on light gauge sheet metal. On flat surfaces penetration is not too good, and there is danger of slag inclusions.

Welding Technique

Pay particular attention to current adjustment. The electrode position should be approximately that shown in Fig. 20-37. Tilt the electrode 30 degrees downward in the direction of travel, Fig. 20-37A, and at a 90-degree angle with the plate surface Fig. 20-37B. A very short arc

20-34. *Electrode position when welding stringer beads in the horizontal position.*

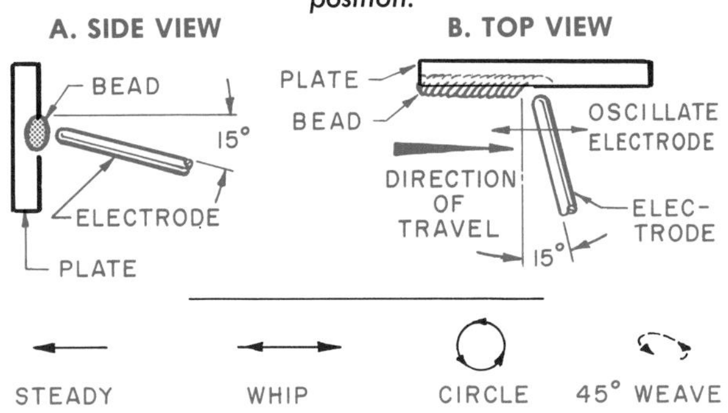

20-35. *Welding beads in the horizontal position. The electrode and electrode holder are positioned as instructed in Fig. 20-34.*

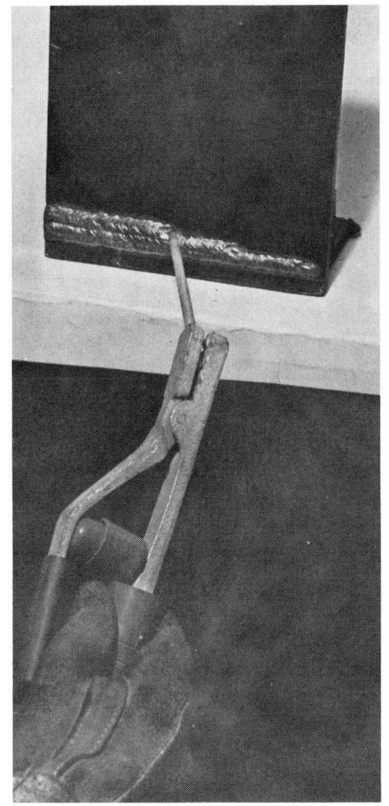

gap must be maintained so that the slag will not run ahead of the weld deposit. Weaving is not generally used, and the speed of travel is fast. This results in beads that are narrow and do not have very high reinforcement. Undercutting is not a problem.

Practice stops and starts. Weld on both sides of the plate.

Operations

1. Obtain plate; check the job print for size.
2. Obtain a square head and scale, dividers, scribe, and center punch from the toolroom.
3. Lay out parallel lines as shown on the job print.
4. Mark lines with the center punch.
5. Obtain electrodes of each quantity, type, and size specified in the job print.
6. Stand the plate in the vertical position on the welding table. Make sure that it is well-grounded.
7. Set a d.c. power source for reverse polarity at 110 to 170 amperes or an a.c. power source at 110 to 190 amperes.
8. Make stringer beads on

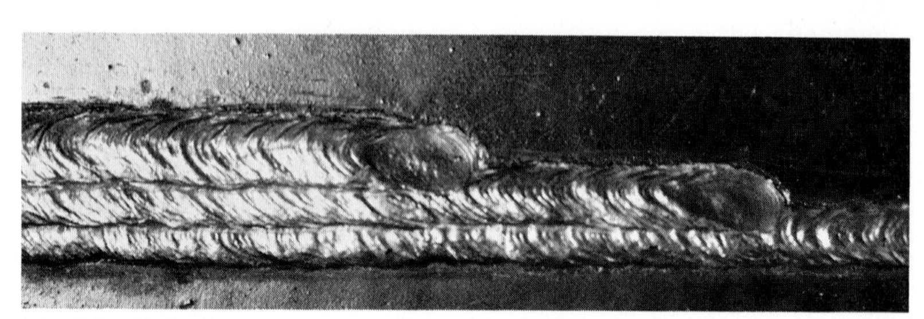

20-36. *Typical appearance of stringer beads welded in the horizontal position with DCRP or a.c. shielded-arc electrodes.*

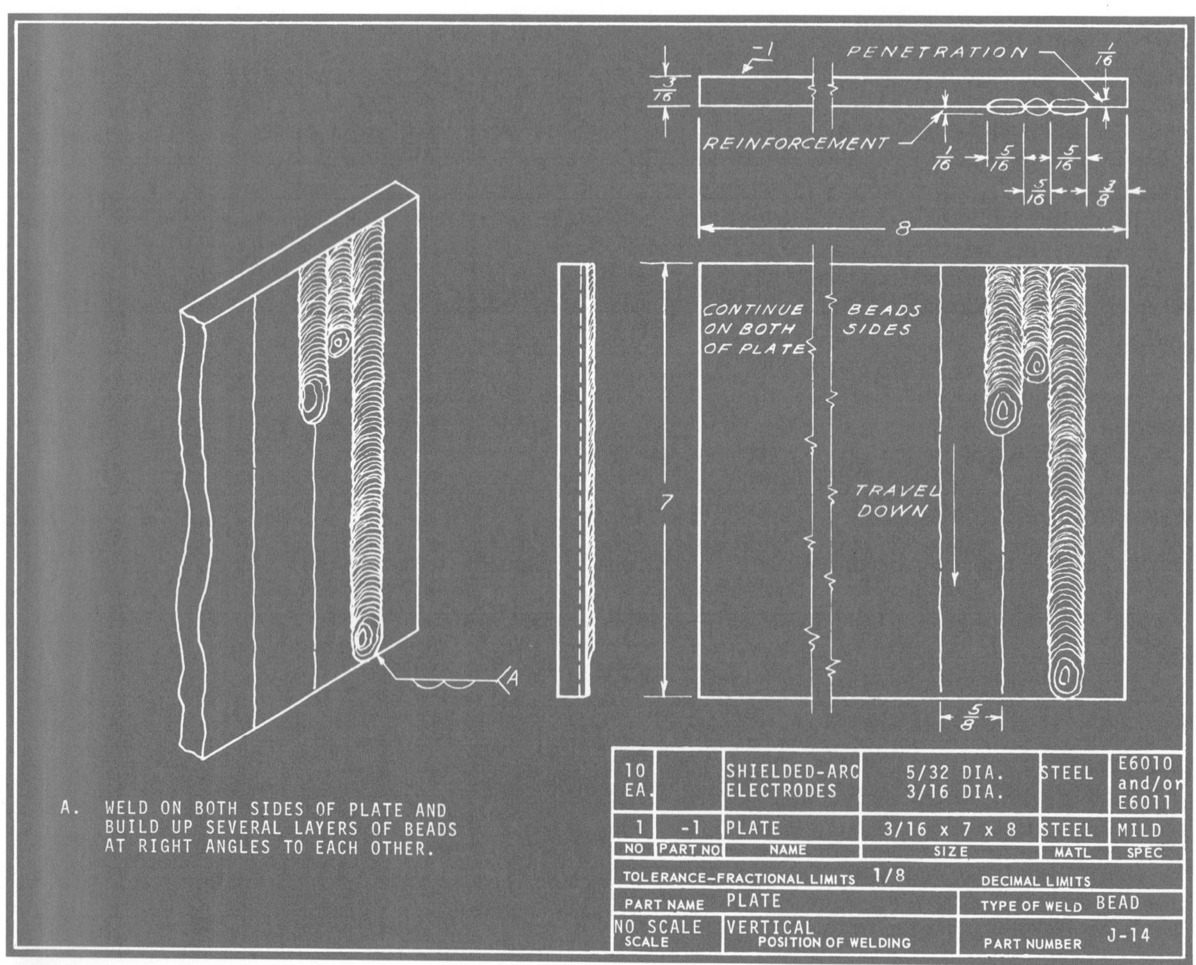

PENETRATION
REINFORCEMENT
CONTINUE BEADS ON BOTH SIDES OF PLATE
TRAVEL DOWN
A
A. WELD ON BOTH SIDES OF PLATE AND BUILD UP SEVERAL LAYERS OF BEADS AT RIGHT ANGLES TO EACH OTHER.
10 EA. SHIELDED-ARC ELECTRODES 5/32 DIA. 3/16 DIA. STEEL E6010 and/or E6011
1 -1 PLATE 3/16 x 7 x 8 STEEL MILD
NO PART NO NAME SIZE MATL SPEC
TOLERANCE-FRACTIONAL LIMITS 1/8 DECIMAL LIMITS
PART NAME PLATE TYPE OF WELD BEAD
NO SCALE SCALE VERTICAL POSITION OF WELDING PART NUMBER J-14

⅝-inch center lines as shown on the job print. Hold the electrode as shown in Fig. 20-37.

9. Chip the slag from the beads, brush, and inspect. Refer to Inspection, below.

10. Make additional stringer beads between already deposited beads as shown on the job print.

11. Chip the slag from the beads, brush, and inspect. Refer to Inspection, below.

12. Practice these beads until you can produce uniform beads consistently with both types of electrodes and with $\frac{3}{16}$-inch diameter electrodes.

Inspection

Compare the beads with Fig. 20-38 and check them for the following weld characteristics:

Width and height: Uniform

Appearance: Smooth with close ripples; free of voids and high spots. Restarts should be difficult to locate.

Size: Refer to the job print.

Face of weld: Flat with very little reinforcement

Edges of weld: Good fusion, no overlap, no undercut

Starts and stops: Free of depressions and high spots

20-37. *Electrode position when making stringer beads, vertical position, travel down.*

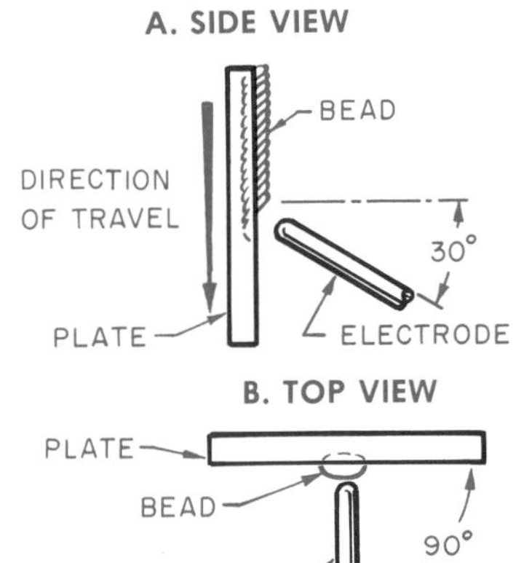

Beginnings and endings: Full size, craters filled

Penetration and fusion: To plate surfaces and adjacent beads

Surrounding plate surfaces: Free of spatter

Slag formation: Full coverage, easily removable

Disposal

Discard completed plates in the waste bin. Plates must be filled on both sides.

JOB 20-J15 WELDING A LAP JOINT

Objective

To weld a lap joint in the vertical position by means of a single-pass fillet weld, travel down, with DCSP and/or a.c. shielded-arc electrodes (AWS-ASTM E6012-E6013).

20-38. *Typical appearance of stringer beads on flat plate welded in the vertical position, travel down, with DCRP or a.c. shielded-arc electrodes.*

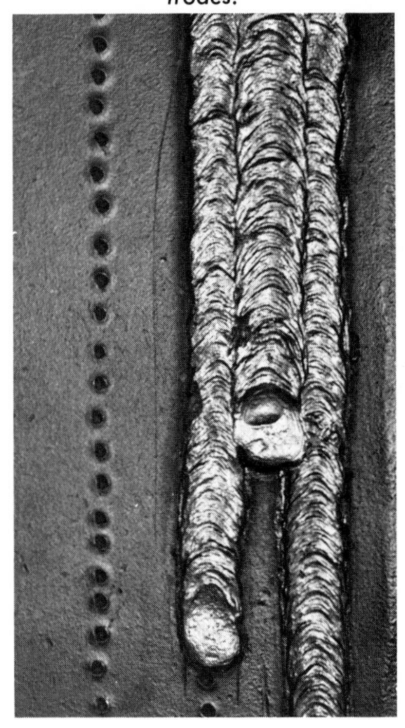

General Job Information

You will find that lap joints are often welded downhill on noncritical work. This is especially true on lighter plate construction such as auto body work and sheet metal weldments. This joint should not be used when maximum reinforcement and strength are required. It is usually made with straight polarity electrodes in order to take advantage of the greater reinforcement and absence of undercut that are characteristic of these electrodes.

Welding Technique

Current adjustment should be high. Tilt the electrode 30 degrees downward in the direction of travel, Fig. 20-39A. Hold it at a 45-degree angle to the plate surface as shown in Fig. 20-39B. The correct position for an actual weld is shown in Fig. 20-40. Hold a close arc gap and proceed at a uniform rate of travel. Do not weave the electrode. A slight oscillating (circular) motion may be used. Slag must be kept back on

20-39. *Electrode position when welding lap joints in the vertical position, travel down.*

A. SIDE VIEW

JOINT

DIRECTION OF TRAVEL

30°

ELECTRODE

B. TOP VIEW

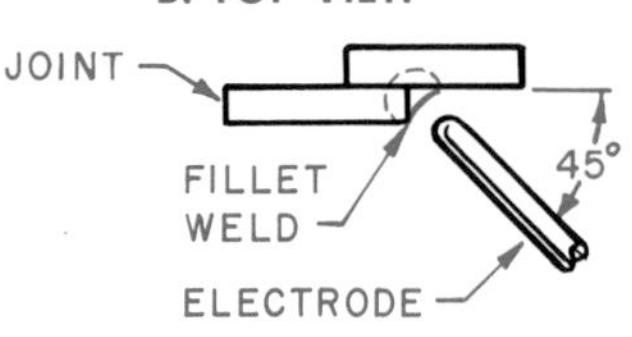

HOLD CLOSE ARC. DO NOT LET SLAG RUN AHEAD OF WELD.

weld deposit and not allowed to run ahead of weld. Good fusion and penetration are difficult and should be watched very closely. Fusion to the plate surfaces and penetration at the root of the joint are essential. Undercut will not be a problem.

Practice starts and stops. Remember to clean the slag from the weld before making new starts.

20-40. *Welding lap joints in the vertical position, travel down. The electrode and electrode holder are positioned as instructed in Fig. 20-39.*

Operations

1. Obtain plates; check the job print for quantity and size.
2. Obtain electrodes of each quantity, type, and size specified in the job print.
3. Set a d.c. power source for straight polarity at 110 to 190 amperes or an a.c. power source at 120 to 200 amperes.
4. Set up the plates and tack weld them at each corner.
5. Stand the plates in the vertical position on the welding table. Make sure that they are well-grounded.
6. Make welds as shown on the job print. Hold the electrode as shown in Fig. 20-40.

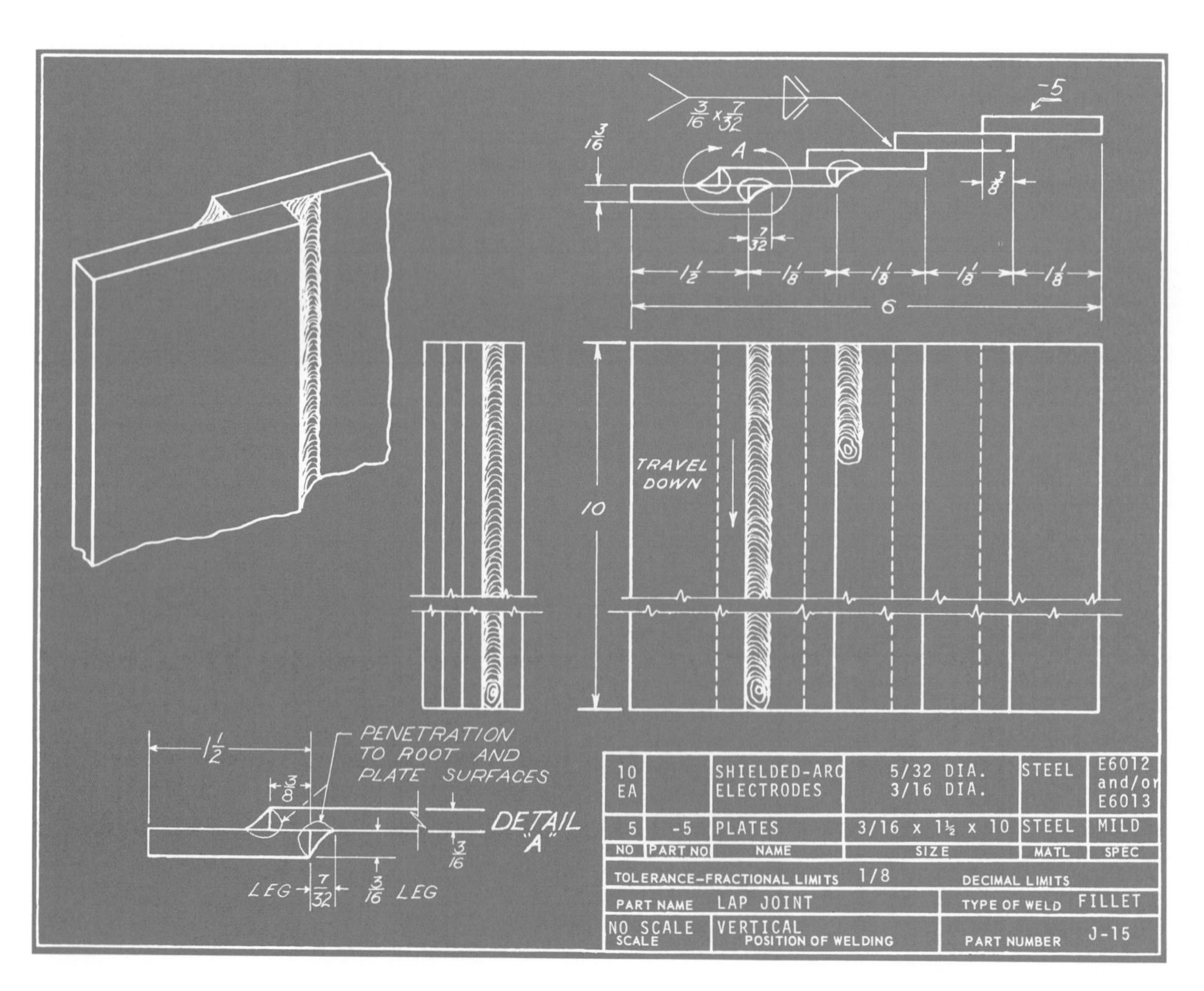

10 EA		SHIELDED-ARC ELECTRODES	5/32 DIA. 3/16 DIA.	STEEL	E6012 and/or E6013
5	-5	PLATES	3/16 x 1½ x 10	STEEL	MILD
NO	PART NO	NAME	SIZE	MATL	SPEC
TOLERANCE-FRACTIONAL LIMITS 1/8			DECIMAL LIMITS		
PART NAME LAP JOINT			TYPE OF WELD FILLET		
NO SCALE SCALE	VERTICAL POSITION OF WELDING		PART NUMBER J-15		

7. Chip the slag from the welds, brush, and inspect. Refer to Inspection, below.

8. Practice these welds until you can produce good welds consistently with both types of electrodes and with $^3/_{16}$-inch diameter electrodes.

Inspection

Compare the welds with Fig. 20-41 and check them for the following characteristics:

Width and height: Uniform
Appearance: Very smooth with fine ripples; free of voids and high spots. Restarts should be difficult to locate.
Size: Refer to the job print.
Face of weld: Slightly concave
Edges of weld: Good fusion, no overlap, no undercut
Starts and stops: Free of depressions and high spots
Beginnings and endings: Full size, craters filled
Penetration and fusion: To the root of the joint and both plate surfaces
Surrounding plate surfaces: Free of spatter
Slag formation: Full coverage, easily removable

Disposal

Put completed joints in the scrap bin so that they will be available for further use. Unwelded plate surfaces can be used for beading.

JOB 20-J16 WELDING A LAP JOINT

Objective

To weld a lap joint in the horizontal position by means of a single-pass fillet weld with DCSP and/or a.c. shielded-arc electrodes (AWS-ASTM E6012-E6013).

General Job Information

This position is similar to that used in Jobs 20-J14 and 15. The horizontal position is frequently necessary for tank and structural work.

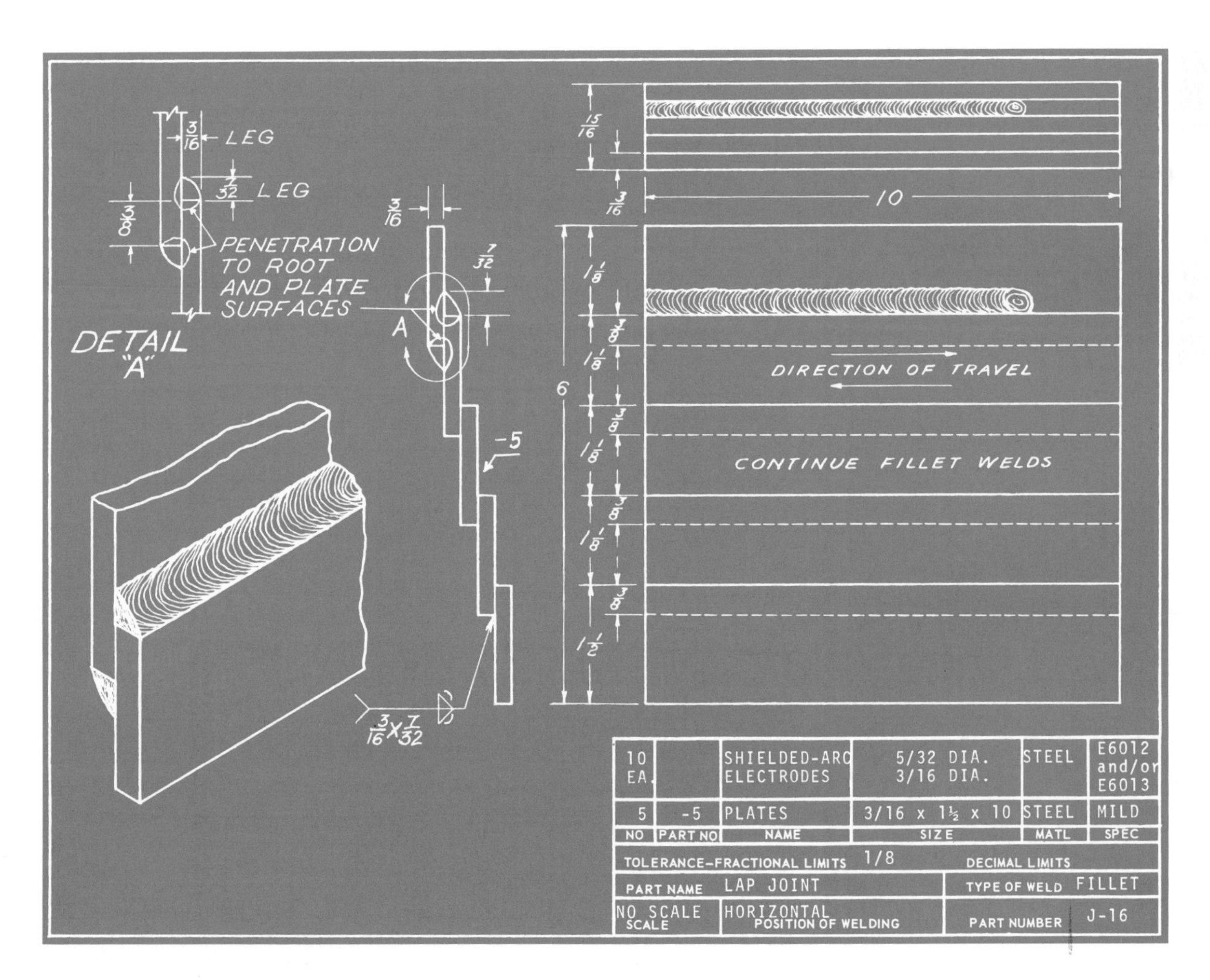

Welding Technique

Current adjustment must not be too high. Electrode position should be approximately that shown in Fig. 20-42. Direct the electrode at the root of the weld and at the vertical plate surface. Tilt the electrode 30 degrees up from the horizontal position, Fig. 20-42A, and 30 degrees toward the direction of travel, Fig. 20-42B. A slight back-and-forth motion may be used with very little side movement. The molten pool must not become too hot since it makes the weld spill off. Be sure to manipulate the electrode so that the plate edge is not burned away and so that the arc does not come in contact with the surface of the vertical plate beyond the weld lines, thus causing undercut.

Practice starts and stops. Travel in all directions.

Operations

1. Obtain plates; check the job print for quantity and size.
2. Obtain electrodes of each quantity, type, and size specified in the job print.
3. Set a d.c. power source for straight polarity at 110 to 190 amperes or an a.c. power source at 115 to 200 amperes.
4. Set up the plates and tack weld them at each corner.
5. Stand the plates in the vertical position on the welding table. Make sure that they are well-grounded.
6. Make welds as shown on the job print. Hold the electrode as shown in Fig. 20-42.
7. Chip the slag from the welds, brush, and inspect. Refer to Inspection, below.
8. Practice these welds until you can produce good welds consistently with both types of electrodes and with $^{3}/_{16}$-inch diameter electrodes.

Inspection

Compare the welds with Fig. 20-43 and check them for the following characteristics:

Width and height: Uniform
Appearance: Smooth with close ripples; free of voids. Restarts should be difficult to locate.
Size: Refer to the job. Check with a fillet weld gauge.

20-41. *Typical appearance of a fillet weld in a lap joint welded in the vertical position, travel down, with DCSP or a.c. shielded-arc electrodes.*

20-42. *Electrode position when welding lap joints in the horizontal position.*

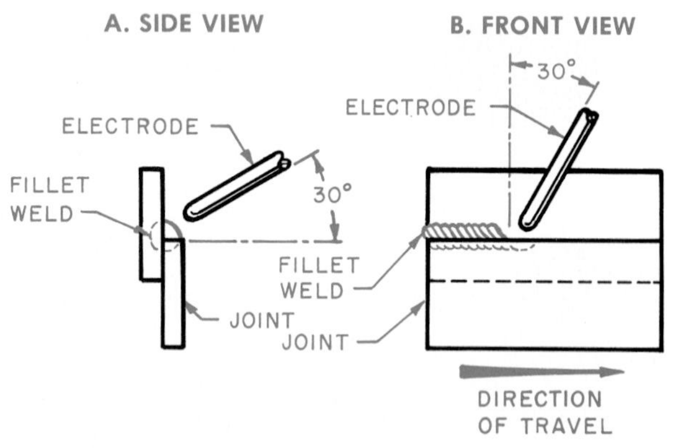

20-43. *Typical appearance of a fillet weld in a lap joint welded in the horizontal position with DCSP or a.c. shielded-arc electrodes.*

Face of weld: Slightly convex
Edges of weld: Good fusion, no overlap or hangdowns, no undercut
Starts and stops: Free of depressions and high spots
Beginnings and endings: Full size, craters filled
Penetration and fusion: To root of joint and plate surfaces
Surrounding plate surfaces: Free of spatter
Slag formation: Full coverage, easily removable

Check Test

Perform a check test, following the procedures outlined in Job 20-J12.

Disposal

Put completed joints in the scrap bin so that they will be available for further use. Unwelded plate surfaces can be used for beading.

JOB 20-J17 WELDING A T-JOINT

Objective

To weld a T-joint in the flat position by means of single-pass fillet weld with DCSP and/or a.c. shielded-arc electrodes (AWS-ASTM E6012-E6013).

General Job Information

The T-joint requires a fillet weld. It is widely used in all forms of welded fabrication, Fig. 20-44. It is a very effective joint when welded from both sides or when complete penetration is secured from one side. Welding is done with all types of electrodes in all positions and can be single or multiple pass.

Welding Technique

Current adjustment should be high. Place the joint on the welding table as shown in Fig. 20-45A and hold the electrode at a 90-degree angle to the table. Tilt it 5 to 20 degrees in the direction of travel, Fig. 20-45B. The actual position for welding is illustrated in Fig. 20-46.

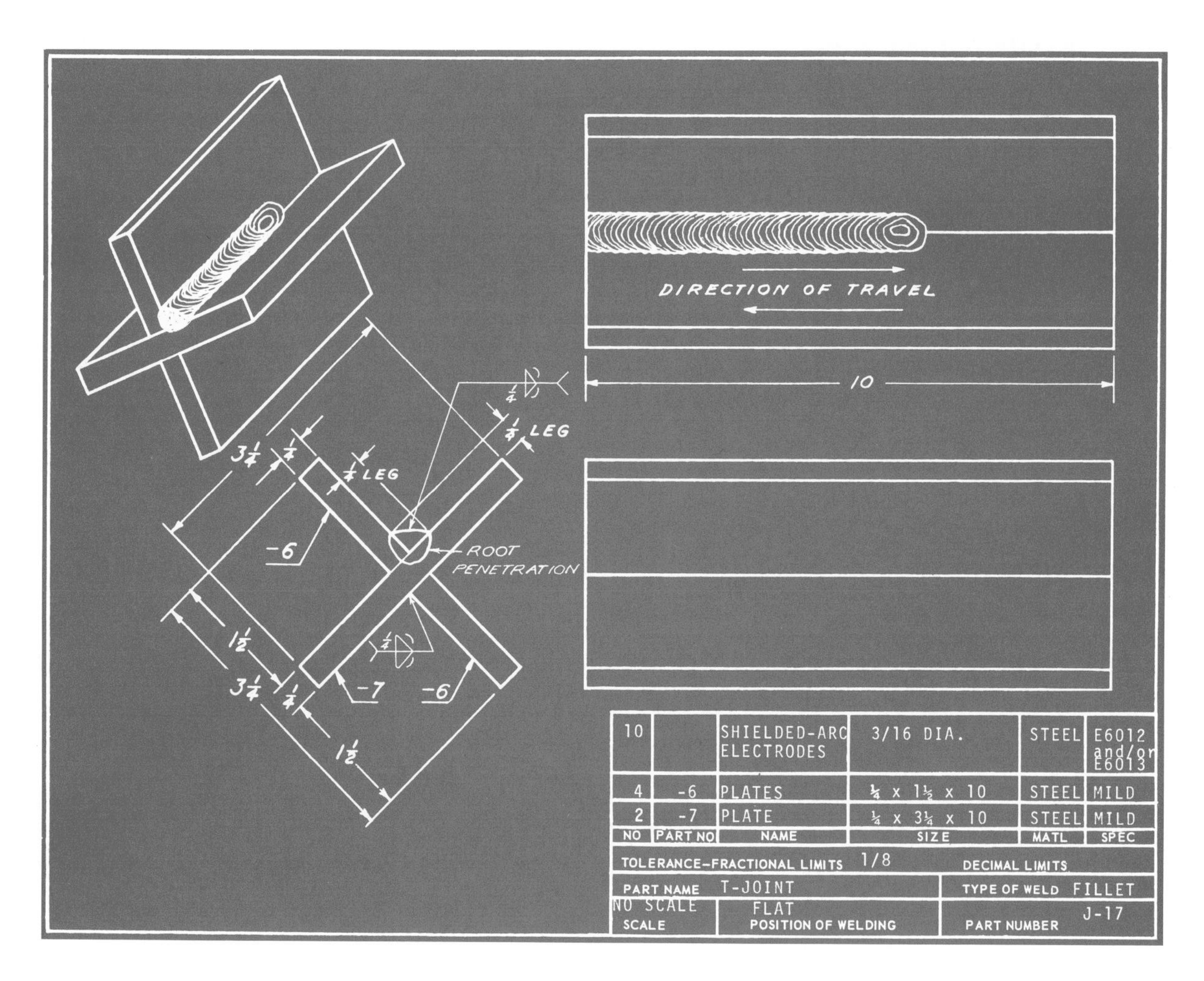

When you start to weld, direct the electrode over the edge of the plate. After the plate edge has been preheated by holding a long arc for an instant, deposit a bead of the desired width and height. Then advance the electrode along the line of weld with a slight back-and-forth motion. This helps to preheat the root of the weld and prevents the slag from running ahead of the weld. Direct the electrode precisely at the root of the weld. Never allow the arc to contact the plate surface outside the width of the weld bead formation. Keep a short arc to obtain good fusion at the root of the joint. Never deposit too much metal in one pass to avoid porosity.

20-44. *A woman welding truss braces in industry. This job is similar to the practice jobs on T-joints.*

Practice many starts and stops and be sure to fill the craters at the end of the weld. Travel in all directions.

Operations

1. Obtain plates; check the job print for quantity and size.
2. Obtain electrodes of each quantity, type, and size specified in the job print.
3. Set a d.c. power source for straight polarity at 160 to 220 amperes or an a.c. power source at 180 to 230 amperes.
4. Set up the plates and tack weld them at each corner.
5. Place the joint on the welding table in the position shown in Fig. 20-46.
6. Make welds in the flat position as shown in the job print. Hold the electrode as shown in Figs. 20-45 and 20-46.
7. Chip the slag from the welds, brush, and inspect. Refer to Inspection, below.
8. Practice these welds until you can produce good welds consistently with both types of electrodes.

Inspection

Compare the welds with Fig. 20-47 and check them for the following characteristics:

Width and height: Uniform
Appearance: Smooth with close ripples; free of voids and high spots. Restarts should be difficult to locate.
Size: Refer to the job print. Check with a fillet weld gauge.
Face of welds: Flat to very slightly convex
Edges of welds: Good fusion, no overlap, no undercut
Starts and stops: Free of depressions and high spots
Beginnings and endings: Full size, craters filled
Penetration and fusion: To root of joint and plate surfaces
Surrounding plate surfaces: Free of spatter
Slag formation: Full coverage, easily removable

Check Test

After you are able to make welds that are satisfactory in ap-

20-45. *Electrode position when welding a T-joint with a single-pass fillet weld in the flat position.*

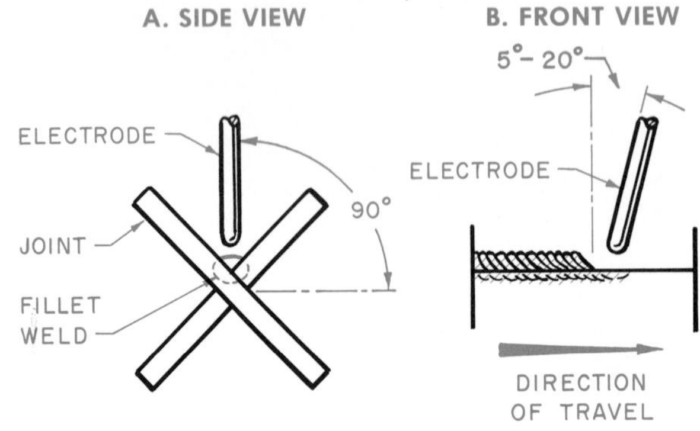

20-46. *Welding a T-joint in the flat position. The electrode and electrode holder are positioned as instructed in Fig. 20-45.*

pearance, make up a specimen like the one shown in Fig. 20-48. Use the plate thickness, welding technique, and electrode type and size specified for this job. Weld on one side only. Break the finished weld as shown in Fig. 20-49. Examine the surfaces for soundness. The weld must not be porous and must show good fusion and penetration at the root and to the plate surfaces. The joint should break evenly through the throat of the weld.

Disposal

Save the completed single-pass joints for your next job, which will be multipass fillet welding.

JOB 20-J18 WELDING A T-JOINT

Objective

To weld a T-joint in the flat position by means of multipass fillet welds, weaved beading technique, with DCSP and/or a.c. shielded-arc electrodes (AWS-ASTM E6012-E6013).

General Job Information

The welding of T-joints in the flat position often requires more than one pass. Single-pass fillet welds should not be more than $\frac{1}{16}$ to $\frac{1}{8}$ inch larger than the size of the electrode being used. On multipass fillet welds, not more than $\frac{1}{8}$ inch thickness of weld metal should be deposited on each layer. A T-joint is a convenient joint on which to practice weaved beading in grooves since the T-joint in the flat position forms a groove that is similar to the groove formed by beveled-butt welds in heavy plate.

You should understand that a T-joint made of $\frac{1}{4}$-inch plate does not require three passes to develop maximum strength. The additional passes are only for practice purposes and to make full and economical use of materials.

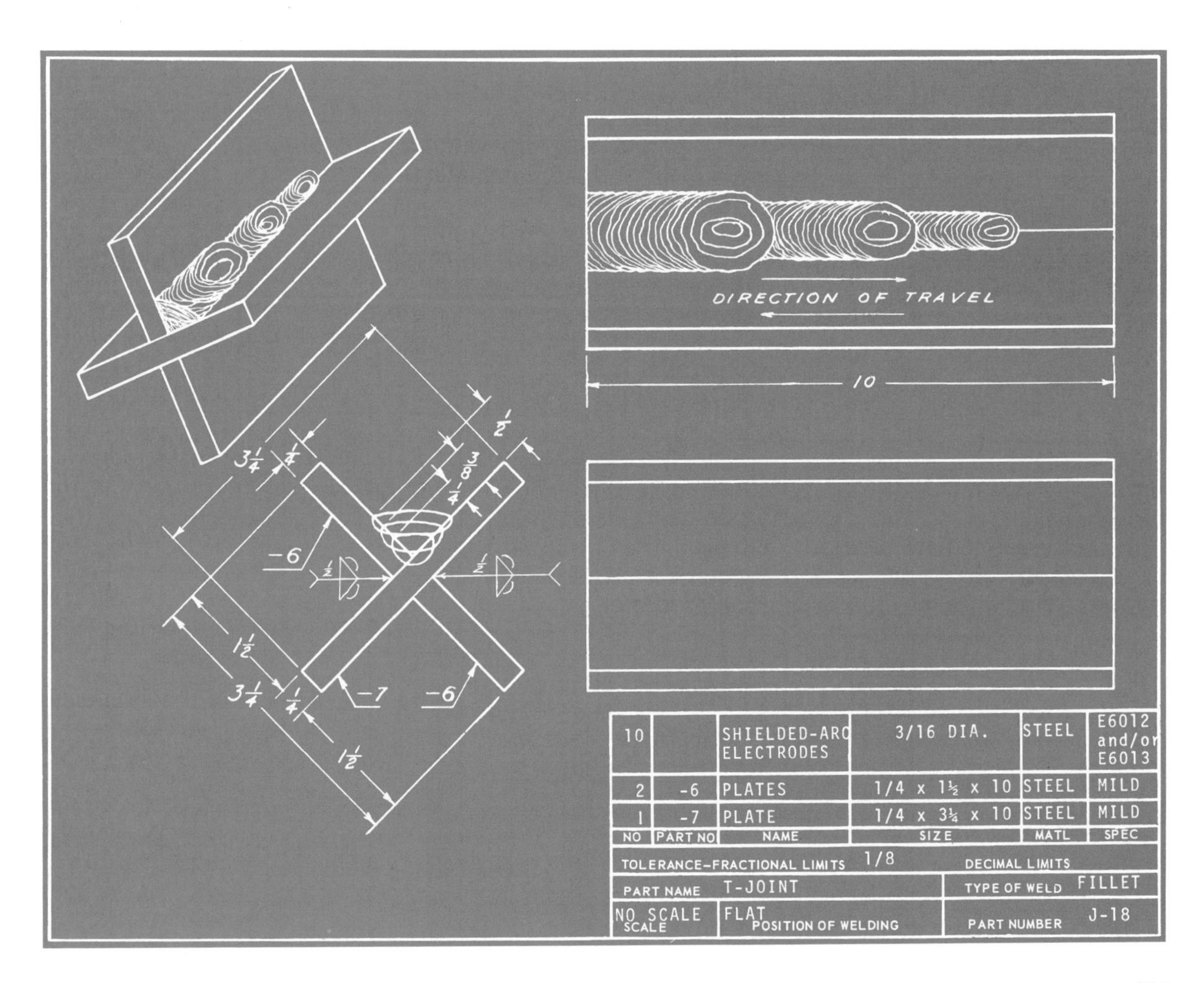

Welding Technique

Current adjustment should be selected with care. Electrode position should be approximately that shown in Fig. 20-50. Hold the electrode at a 90-degree angle to the table, Fig. 20-50A, and tilt it 5 to 20 degrees in the direction of travel, Fig. 20-50B. It is assumed that you are using the practice joints welded for Job 20-J17. The second and final pass must be weaved from side to side to insure complete fusion and size, Fig. 20-50A. Bear in mind that the weave should not be as wide as the width of the desired weld. It should be only as wide as the underneath pass. Hesitate slightly (maybe half a second) on each side, and move the electrode smoothly and quickly across the face of the underneath weld. Remember that you cross the middle of the weld twice as often as you touch either side. To produce an even deposit across the weld, you must have a fast crossover speed. Otherwise, you will pile up weld metal in the middle.

It is very important to remove slag thoroughly between each pass. Failure to do this will result in slag inclusions.

Practice many starts and stops and be sure to fill the craters at the end of the weld. Travel in all directions.

It is advisable to have several joints available so that one joint may be cooling while the other is being worked on. Joints must not become too overheated.

Operations

1. Select good quality joints welded in Job 20-J17.
2. Obtain electrodes of each quantity, type, and size specified in the job print.
3. Set a d.c. power source for straight polarity at 150 to 240 amperes or an a.c. power source at 165 to 260 amperes.
4. Place the joint in the flat position on the welding table. Make sure that the plates are well-grounded.
5. Make the second pass as shown on the job print. Hold the electrode as shown in Fig. 20-50.
6. Chip the slag from the weld, brush, and inspect. Refer to Inspection, below.
7. Make the third pass as shown on the job print.
8. Chip the slag from the weld, brush, and inspect. Refer to Inspection, below.
9. Practice these welds until you can produce good welds consistently with both types of electrodes.

Inspection

Compare each pass with Figs. 20-51 and 20-52 and check it for the following characteristics:

Width and height: Uniform
Appearance: Smooth with close ripples; free of voids and high spots. Restarts should be difficult to locate.
Size: Refer to the job print. Check with a fillet weld gauge.
Face of welds: Flat to slightly convex
Edges of welds: Good fusion, no overlap, no undercut
Starts and stops: Free of depressions and high spots
Beginnings and endings: Full size, craters filled
Penetration and fusion: To root of joint, preceding passes, and plate surfaces
Surrounding plate surfaces: Free of spatter
Slag formation: Full coverage, easily removable

Disposal

Put completed joints in the scrap bin so that they will be

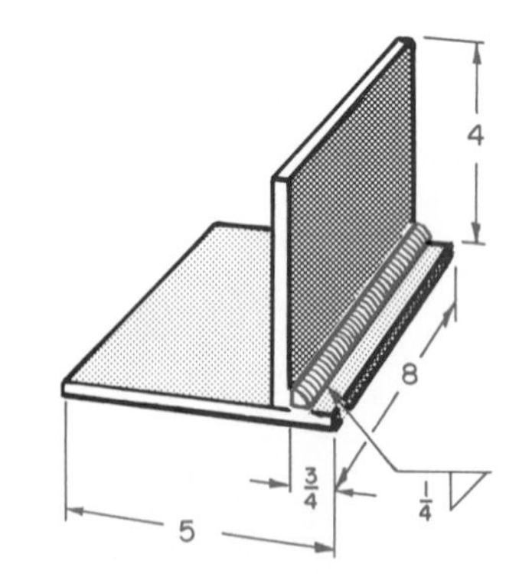

20-48. *Joint for fillet weld test specimen.*

20-49. *Method of breaking check-test specimen.*

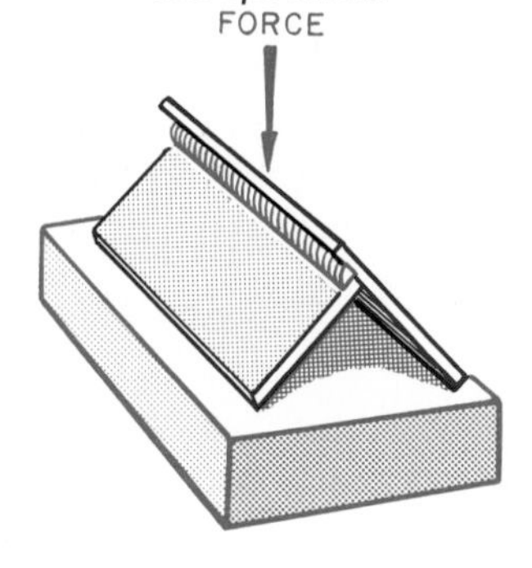

20-47. *Typical appearance of a single-pass fillet weld in a T-joint welded in the flat position with DCSP or a.c. shielded-arc electrodes.*

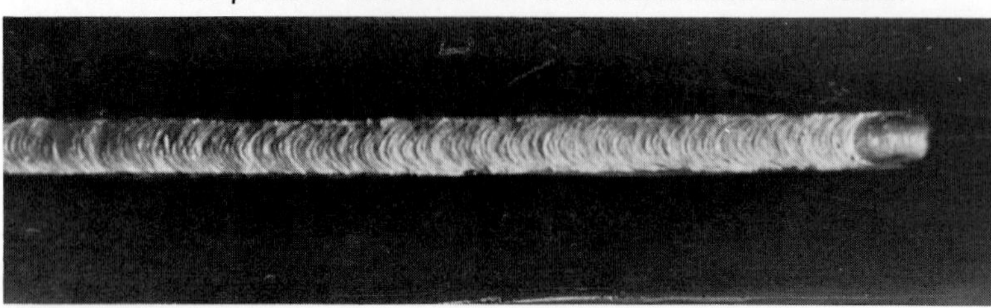

available for further use. Square edges can be butted together to make square butt joints, and unwelded plate surfaces can be used for beading.

JOB 20-J19 WELDING A T-JOINT

Objective

To weld a T-joint in the flat position by means of a single-pass fillet weld, with DCRP and/or a.c. shielded-arc electrodes (AWS-ASTM E6010-E6011).

General Job Information

This job is similar to Job 20-J17, except that welding is done with reverse polarity electrodes.

Welding Technique

Current adjustment should be high. Electrode position should be approximately that shown in Fig. 20-45, page 610. Use the welding technique practiced in Job 20-J17. The arc gap will be slightly less, and you will have to

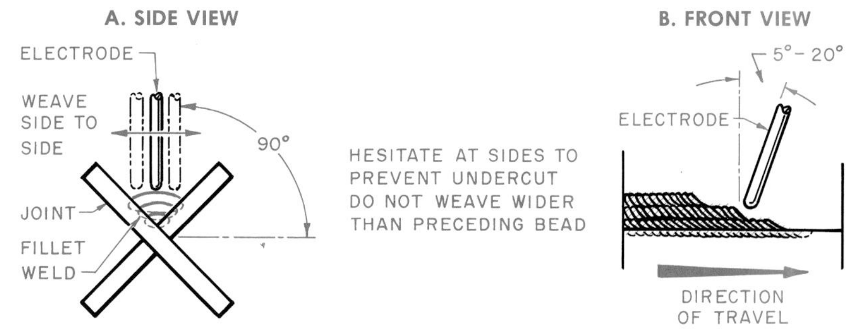

20-50. *Electrode position when welding a T-joint with multipass fillet welds in the flat position.*

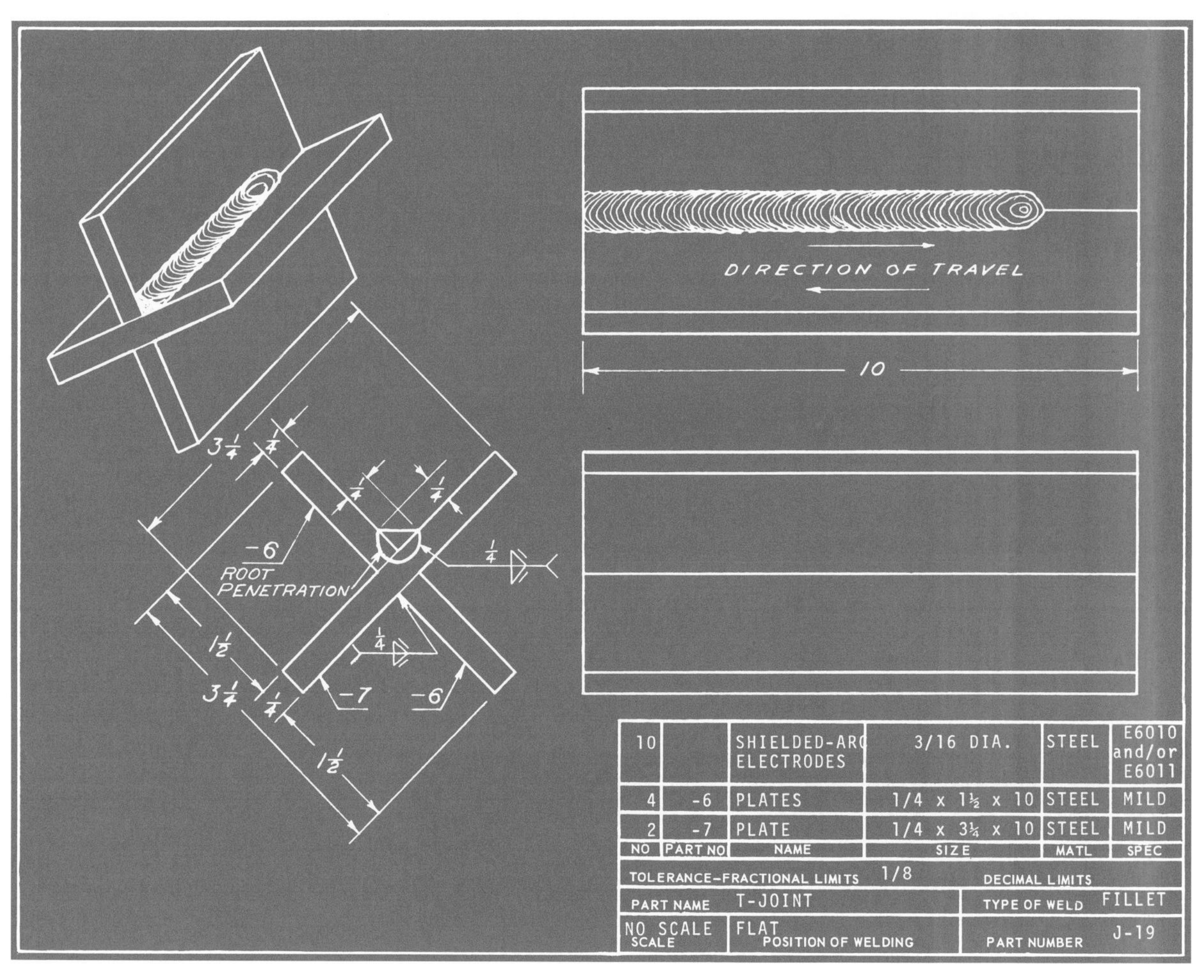

NO	PART NO	NAME	SIZE	MATL	SPEC
10		SHIELDED-ARC ELECTRODES	3/16 DIA.	STEEL	E6010 and/or E6011
4	-6	PLATES	1/4 x 1½ x 10	STEEL	MILD
2	-7	PLATE	1/4 x 3¼ x 10	STEEL	MILD

TOLERANCE–FRACTIONAL LIMITS 1/8		DECIMAL LIMITS	
PART NAME	T-JOINT	TYPE OF WELD	FILLET
SCALE: NO SCALE	POSITION OF WELDING: FLAT	PART NUMBER	J-19

be more careful to avoid undercut than when using straight polarity electrodes. Slag will not be as heavy, but it must be removed thoroughly.

Practice starts and stops and be sure to fill the craters at the end of the weld. Travel in all directions.

Operations

1. Obtain plates; check the job print for quantity and size.
2. Obtain electrodes of each quantity, type, and size specified in the job print.
3. Set the power source for 140 to 215 amperes. Set a d.c. power source for reverse polarity.
4. Set up the plates and tack weld them at each corner.
5. Place the joint in the flat position on the welding table. Make sure that the plates are well-grounded.
6. Make the weld as shown on the job print. Hold the electrode as shown in Fig. 20-45, page 610.
7. Chip the slag from the weld, brush, and inspect. Refer to Inspection.
8. Practice these welds until you can produce good welds consistently with both types of electrodes.

Inspection

Compare the welds with Fig. 20-53 and check them for the following characteristics:

Width and height: Uniform
Appearance: Smooth with close ripples; free of voids and high spots. Restarts should be difficult to locate.
Size: Refer to the job print. Check with a fillet weld gauge.
Face of welds: Flat
Edges of welds: Good fusion, no overlap, no undercut
Starts and stops: Free of depressions and high spots
Beginnings and endings: Full size, craters filled
Penetration and fusion: To root of joint and plate surfaces
Surrounding plate surfaces: Free of spatter
Slag formation: Full coverage, easily removable

Check Test

After you are able to make welds that are satisfactory in appearance, make up a test specimen like that made in Job 20-J17. Use the same plate thickness, welding technique, and electrode type and size as in this job. Weld on one side only.

Break the completed weld and examine the surfaces for soundness. The weld must not be porous. It must show good fusion and penetration at the root and to the plate surfaces. The joint should break evenly through the throat of the weld.

Disposal

Save the completed joints for your next job, which will be multipass fillet welding.

JOB 20-J20 WELDING A T-JOINT

Objective

To weld a T-joint in the flat position by means of multipass fillet welds, weaved beading technique, with DCRP and/or a.c. shielded-arc electrodes (AWS-ASTM E6010-E6011).

General Information

This job is similar to Job 20-J18, except that reverse polarity electrodes are used for welding. The greater amount of multipass welding is done with reverse polarity electrodes. This is especially true of code welding.

Welding Technique

Current adjustment should be selected with care. The electrode position should be approximately that shown in Fig. 20-50, page 613. Use the welding technique outlined in Job 20-J18. Arc gap should be slightly less. To prevent undercut, hesitate longer at the sides of the weld than when welding with straight polarity electrodes and control the width of your weave carefully.

Clean slag thoroughly between each pass. Practice starts and stops and be sure to fill the

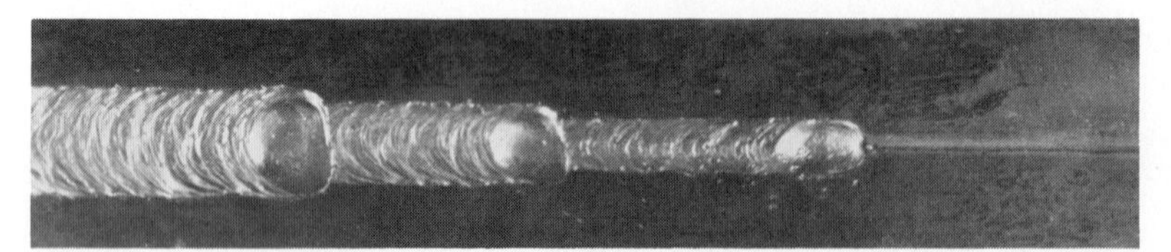

20-51. *Typical appearance of a multipass fillet weld in a T-joint welded in the flat position with DCSP shielded-arc electrodes.*

20-52. *Typical appearance of a multipass fillet weld in a T-joint welded in the flat position with a.c. type (E6013) shielded-arc electrodes.*

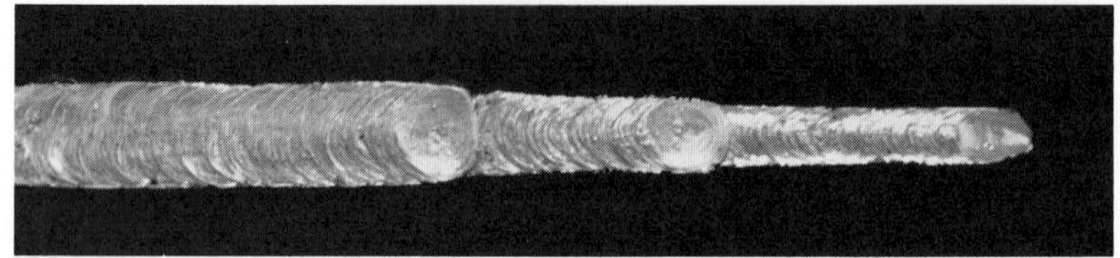

craters at the end of the weld. Travel in all directions.

It is advisable to have several joints available so that one joint may be cooling while the other is being worked on. Joints must not become overheated.

Operations

1. Select good quality joints welded in Job 20-J19.
2. Obtain electrodes of each quantity, type, and size specified in the job print.
3. Set the power source for 140 to 215 amperes. Set a d.c. power source for reverse polarity.
4. Place the joint in the flat position on the welding table. Make sure it is well-grounded.
5. Make a second pass as shown on the job print.
6. Chip the slag from the weld, brush, and inspect. Refer to Inspection.
7. Make a third pass as shown on the job print.
8. Chip slag from the weld, brush, and inspect. Refer to Inspection.
9. Practice these welds until you can produce good welds consistently with both types of electrodes.

Inspection

Compare each pass with Figs. 20-54 and 20-55 and check it for these weld characteristics:

Width and height: Uniform

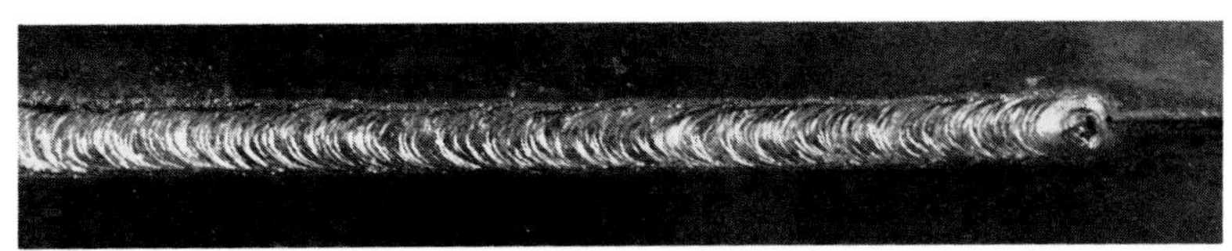

20-53. *Typical appearance of a single-pass fillet weld in a T-joint welded in the flat position with DCRP or a.c. shielded-arc electrodes.*

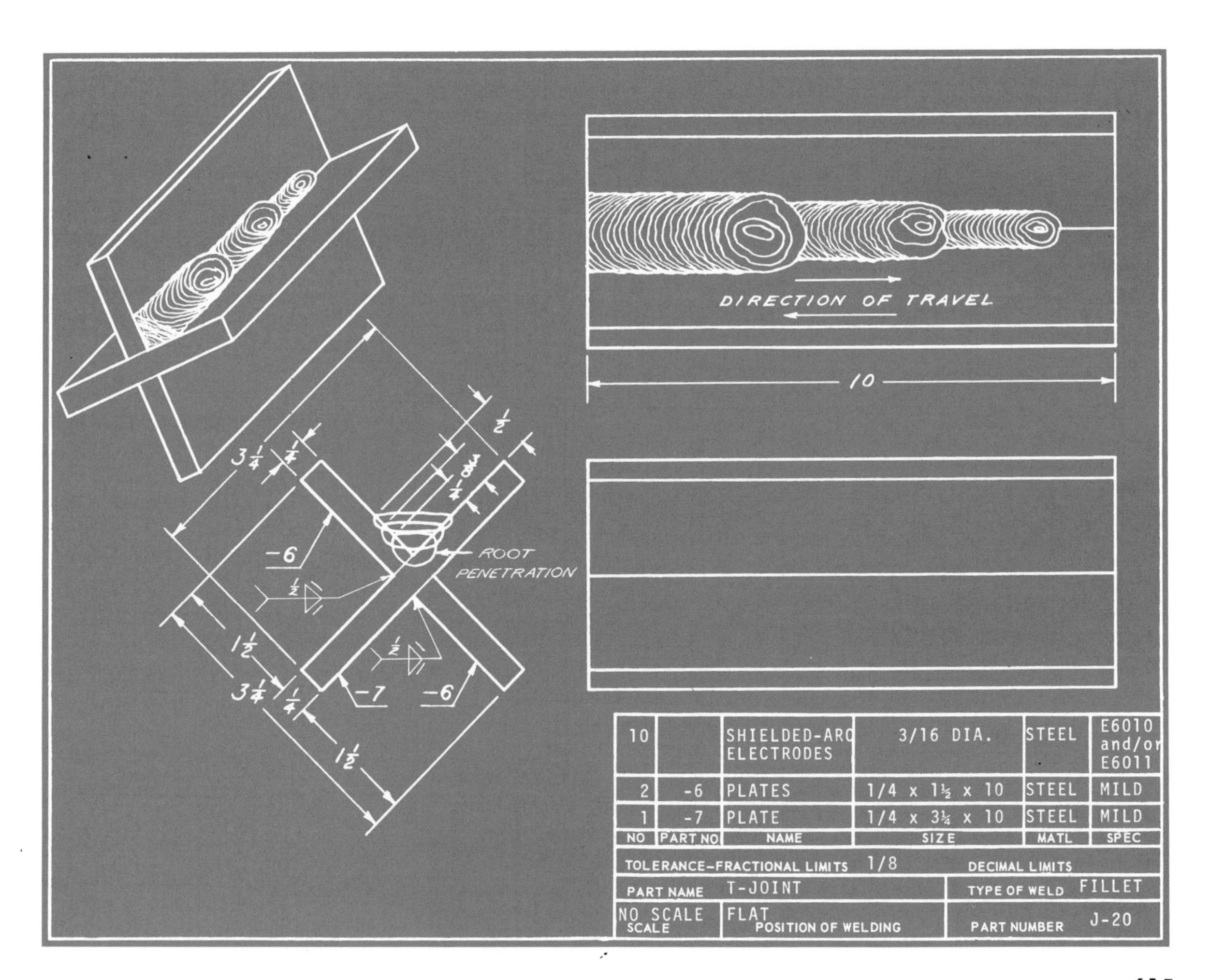

10		SHIELDED-ARC ELECTRODES	3/16 DIA.	STEEL	E6010 and/or E6011
2	-6	PLATES	1/4 x 1½ x 10	STEEL	MILD
1	-7	PLATE	1/4 x 3¼ x 10	STEEL	MILD
NO	PART NO	NAME	SIZE	MATL	SPEC

TOLERANCE-FRACTIONAL LIMITS 1/8	DECIMAL LIMITS
PART NAME T-JOINT	TYPE OF WELD FILLET
SCALE NO SCALE · POSITION OF WELDING FLAT	PART NUMBER J-20

Appearance: Smooth with close ripples; free of voids and high spots. Restarts should be difficult to locate.
Size: Refer to the job print. Check with a fillet weld gauge.
Face of welds: Flat
Edges of welds: Good fusion, no overlap, no undercut
Starts and stops: Free of depressions and high spots
Beginnings and endings: Full size, craters filled
Penetration and fusion: To the root of the joint, preceding passes, and plate surfaces
Surrounding plate surfaces: Free of spatter
Slag formation: Full coverage, easily removable

Disposal

Put completed joints in the scrap bin so that they will be available for further use. Square edges can be butted together to make square butt joints, and unwelded plate surfaces can be used for beading.

JOB 20-J21 STRINGER BEADING

Objective

To deposit stringer beads on flat steel plate in the vertical position, travel up, with DCRP and/or a.c. shielded-arc electrodes (AWS-ASTM E6010-E6011).

General Job Information

For the most part, vertical welding is usually done by traveling up, especially on critical work. Pressure piping, shipbuilding, pressure vessels, and structural steel are some of the fields of welding using this procedure. The technique presented here is like that required for a lap joint and the first pass on multipass fillet welds.

Welding Technique

Current adjustment must not be too high. Tilt the electrode downward about 5 to 15 degrees, Fig. 20-56A, and hold it at an angle of 90 degrees to the plate, Fig. 20-56B. Basic welding motions used in vertical welding (travel up) are shown in Fig. 20-57. Figure 20-58 illustrates the actual welding position. The arc gap should be short when depositing weld metal. Lengthen the gap on the upward stroke, and do not break the arc. This type of motion permits the deposited weld metal to solidify so that a shelf is formed on which additional metal is deposited. Some operators prefer to move the electrode back and forth slightly as it is advanced so that it stays in the molten pool. If you use this method, you will find it difficult to keep the weld from building up too high and spilling off. The up-and-down rocking movement shown in Fig. 20-56A produces flatter welds, and there is less danger from slag inclusions.

Practice starts and stops. Weld on both sides of the plate.

Operations

1. Obtain plate; check the job print for size.
2. Obtain a square head and scale, dividers, scribe and center punch from the toolroom.
3. Lay out parallel lines as shown on the job print.
4. Mark the lines with a center punch.
5. Obtain electrodes of each quantity, size, and type specified in the job print.
6. Stand the plate in the vertical position on the welding table. Make sure that it is well-grounded.

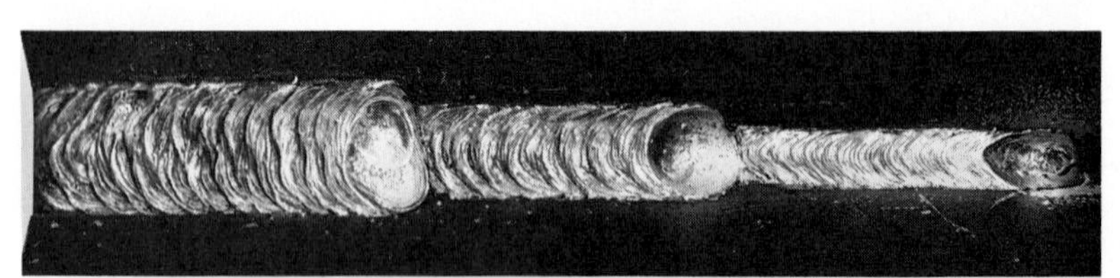

20-54. *Typical appearance of a multipass fillet weld in a T-joint welded in the flat position with DCRP shielded-arc electrodes.*

20-55. *Typical appearance of a multipass fillet weld in a T-joint welded in the flat position with a.c. type (E0011) shielded-arc electrodes.*

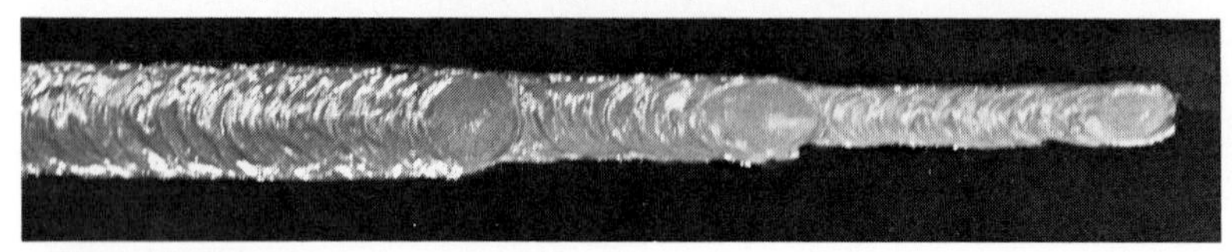

20-56. *Electrode position for making stringer beads in the vertical position, travel up.*

A. SIDE VIEW

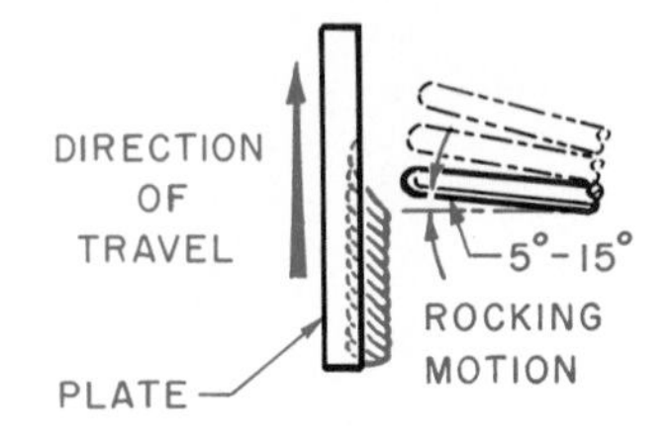

B. TOP VIEW

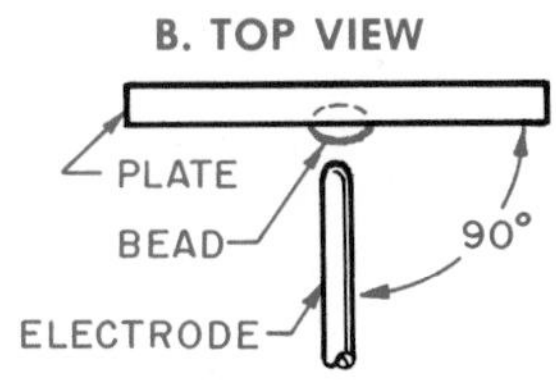

7. Set the power source for 75 to 125 amperes. Set a d.c. power source for reverse polarity.

8. Make stringer beads on ⅝-inch center lines. Manipulate the electrode as instructed in Figs. 20-56 through 20-58.

9. Chip slag from the beads, brush, and inspect. Refer to Inspection.

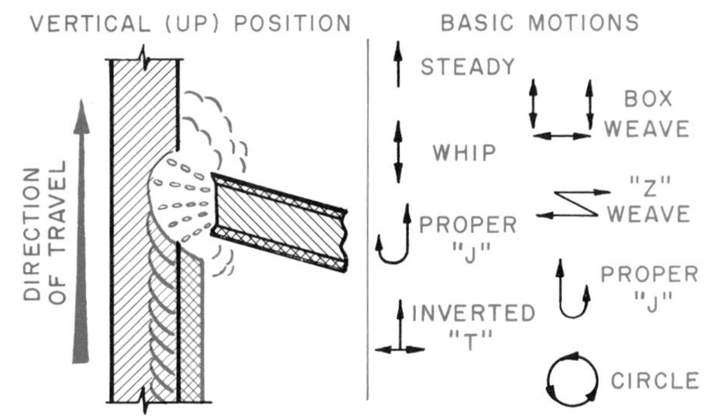

20-57. *Basic motions that can be employed in welding in the vertical position, travel up.*

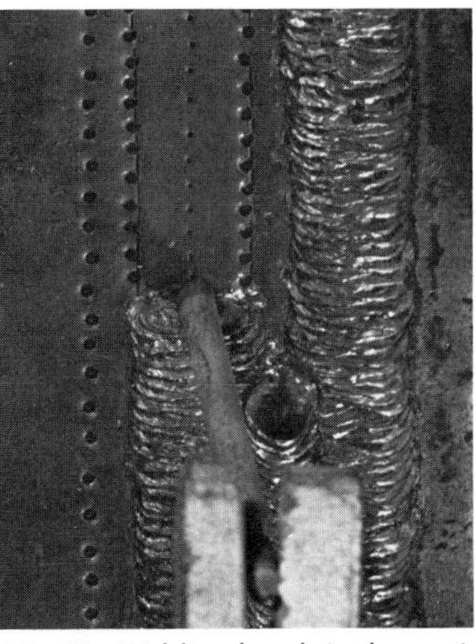

20-58. *Welding beads in the vertical position, travel up. The electrode and electrode holder are positioned as instructed in Fig. 20-56.*

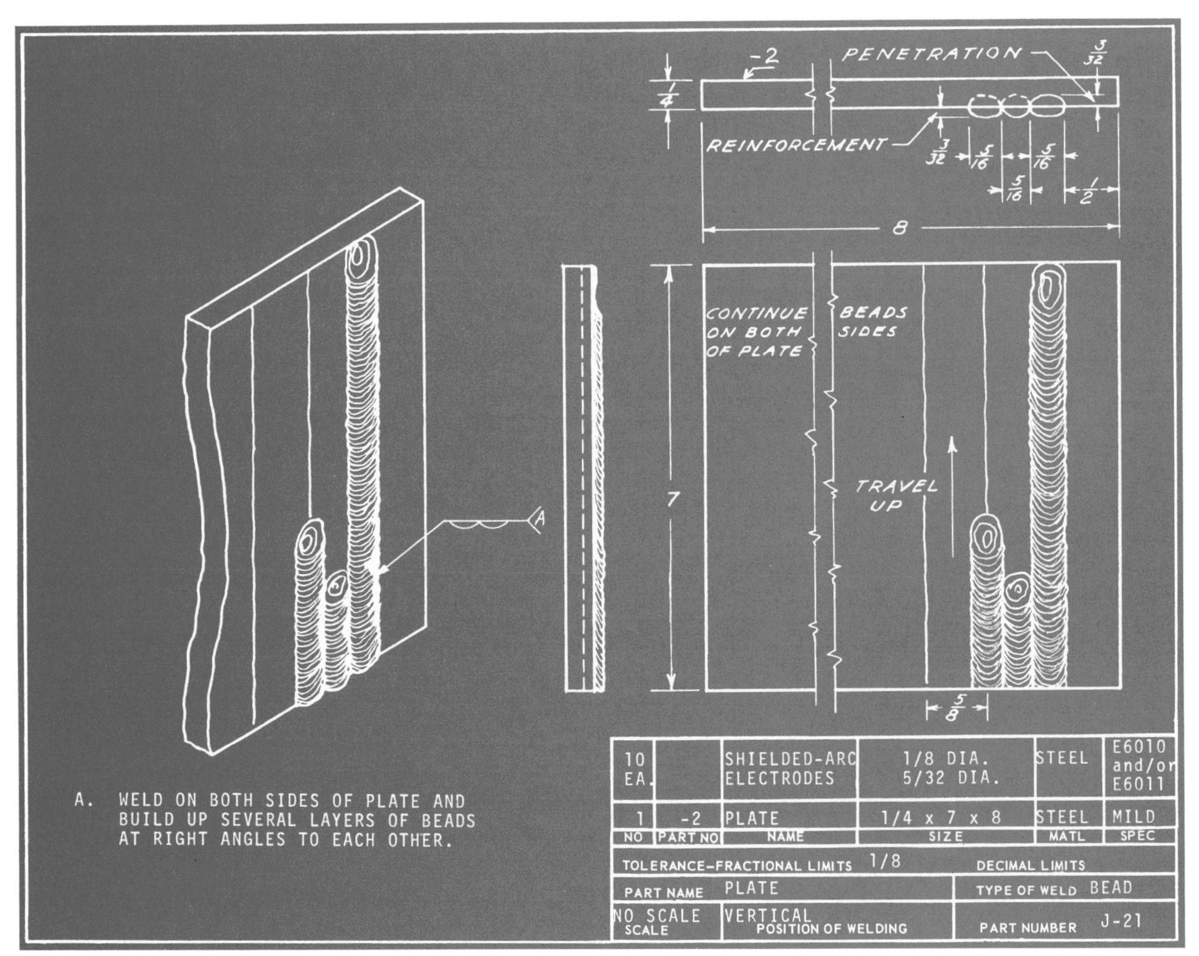

NO	PART NO	NAME	SIZE	MATL	SPEC
10 EA.		SHIELDED-ARC ELECTRODES	1/8 DIA. 5/32 DIA.	STEEL	E6010 and/or E6011
1	-2	PLATE	1/4 x 7 x 8	STEEL	MILD

TOLERANCE–FRACTIONAL LIMITS 1/8	DECIMAL LIMITS
PART NAME PLATE	TYPE OF WELD BEAD
SCALE NO SCALE — POSITION OF WELDING VERTICAL	PART NUMBER J-21

10. Make additional stringer beads between the beads already deposited as shown on the job print.

11. Chip the slag from the beads, brush, and inspect. Refer to Inspection, below.

12. Practice these beads until you can produce good beads consistently with both types of electrodes and with $^5/_{32}$-inch diameter electrodes.

Inspection

Compare the beads with Fig. 20-59 and check them for the following weld characteristics:

Width and height: Uniform
Appearance: Smooth with close ripples; free of voids and high spots. Restarts should be difficult to locate.
Size: Refer to the job print.
Face of weld: Convex with slight reinforcement
Edges of weld: Good fusion, no overlap, no undercut
Starts and stops: Free of depressions and high spots
Beginnings and endings: Full size, craters filled
Penetration and Fusion: To plate surfaces and adjacent beads
Surrounding plate surfaces: Free of spatter
Slag formation: Full coverage, easily removable

20-59. *Typical appearance of stringer beads on flat plate welded in the vertical position, travel up, with DCRP or a.c. shielded-arc electrodes.*

Disposal

Discard completed plates in the waste bin. Plates must be filled on both sides.

JOB 20-J22 WEAVED BEADING

Objective

To deposit weaved beads on flat plate in the vertical position, travel up, with DCRP and/or a.c. shielded-arc electrodes (AWS-ASTM E6010-E6011).

General Job Information

When welding in the vertical position and when it is necessary to weave the bead, the direction of travel is usually up. This technique is used on pressure piping and pressure vessels, and in shipbuilding and structural welding. Reverse polarity electrodes are generally selected. The travel-up method is employed in the welding of multipass groove and fillet welds in the vertical position.

Welding Technique

Current adjustment must not be too high, but the arc must be hot enough to insure good fusion. Tilt the electrode downward about 10 degrees from the horizontal position, Fig. 20-60A, and hold it at a 90-degree angle to the plate, Fig. 20-60B. A shelf is formed at the bottom of the plate that is the desired width and height. Weave the electrode across the face of the weld, hesitating at the sides of the weld to eliminate undercut, Fig. 20-60C. You may advance the weld by keeping the electrode in the pool, but be careful that the molten metal does not become so fluid that it spills out. When this occurs, run the electrode up on either side of the weld. Lengthen the arc, but do not break it. Do not keep the arc out of the crater any longer than necessary since this allows the crater to cool and causes excessive spatter ahead of the weld. *Remember that undercut is the gouging out of the base metal by the arc over a wider surface than is being covered by weld deposit. Do not weave the electrode beyond the*

20-60. *Electrode position when making weaved beads in the vertical position, travel up.*

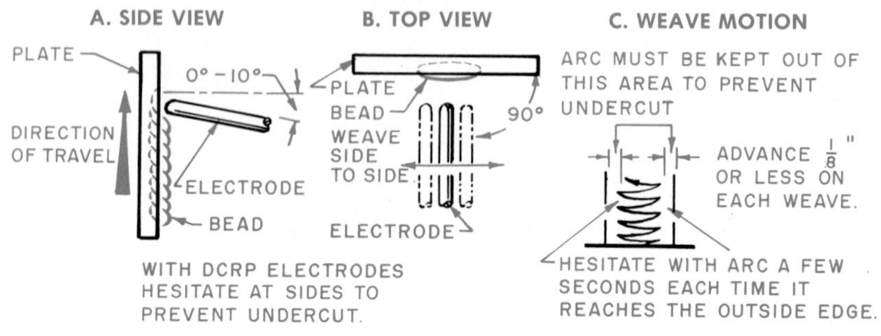
A. SIDE VIEW
PLATE
0° – 10°
DIRECTION OF TRAVEL
ELECTRODE
BEAD
B. TOP VIEW
PLATE
BEAD
90°
WEAVE SIDE TO SIDE
ELECTRODE
WITH DCRP ELECTRODES HESITATE AT SIDES TO PREVENT UNDERCUT.
C. WEAVE MOTION
ARC MUST BE KEPT OUT OF THIS AREA TO PREVENT UNDERCUT
ADVANCE 1/8" OR LESS ON EACH WEAVE.
HESITATE WITH ARC A FEW SECONDS EACH TIME IT REACHES THE OUTSIDE EDGE.

desired width of the bead, Fig. 20-60C.

Practice starts and stops. Weld on both sides of plate.

Operations

1. Obtain plate; check the job print for quantity and size.
2. Obtain a square head and scale, dividers, scribe, and center punch from the toolroom.
3. Lay out parallel lines as shown on the job print.
4. Mark lines with the center punch.
5. Obtain electrodes of each quantity, type, and size specified in the job print.
6. Stand the plate in the vertical position on the welding table. Make sure that it is well-grounded.
7. Set the welding power source for 75 to 125 amperes. Set a d.c. power source for reverse polarity.
8. Make weaved beads between the parallel lines which are ¾ inch apart as shown on the job print. Manipulate the electrode as instructed in Fig. 20-60.
9. Chip the slag from the beads, brush, and inspect. Refer to Inspection.
10. Make stringer beads between the weaved beads already deposited. Manipulate the electrode as instructed in Figs. 20-56 and 20-57, page 616.
11. Chip the slag from the beads, brush, and inspect. Refer to Inspection.
12. Practice these beads until you can produce good beads consistently with both types of electrodes and with 5/32-inch diameter electrodes.

Inspection

Compare the beads with Fig. 20-61 and check them for the following weld characteristics:

Width and height: Uniform

Appearance: Smooth with close ripples; free of voids and high spots. Restarts should be difficult to locate.

Size: Refer to the job print.

Face of weld: Convex with slight reinforcement

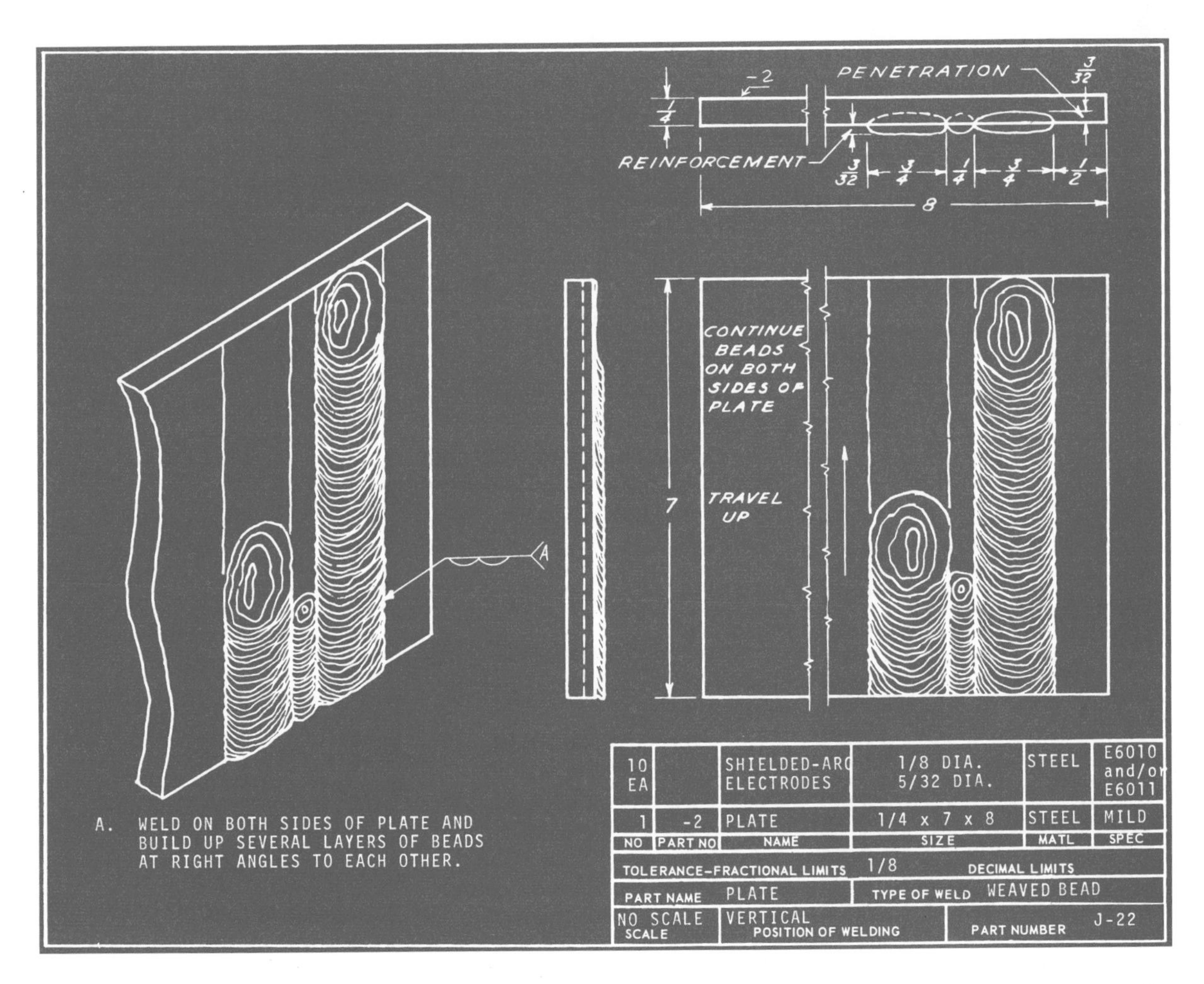

NO	PART NO	NAME	SIZE	MATL	SPEC
10 EA		SHIELDED-ARC ELECTRODES	1/8 DIA. 5/32 DIA.	STEEL	E6010 and/or E6011
1	-2	PLATE	1/4 x 7 x 8	STEEL	MILD

TOLERANCE-FRACTIONAL LIMITS 1/8	DECIMAL LIMITS
PART NAME PLATE	TYPE OF WELD WEAVED BEAD
SCALE NO SCALE	POSITION OF WELDING VERTICAL	PART NUMBER J-22

Edges of weld: Good fusion, no overlap, no undercut
Starts and stops: Free of depressions and high spots
Beginnings and endings: Full size, craters filled
Penetration and fusion: To plate surface and adjacent beads
Surrounding plate surfaces: Free of spatter
Slag formation: Full coverage, easily removable

Disposal

Discard completed plates in the waste bin. Plates must be filled on both sides.

JOB 20-J23 WEAVED BEADING

Objective

To deposit weaved beads on flat plate in the vertical position, travel up, with DCSP and/or a.c. shielded-arc electrodes (AWS-ASTM E6012-E6013).

General Job Information

This job is like Job 20-J22, except that straight polarity electrodes are specified for welding. Although the use of straight polarity electrodes for vertical work is not very extensive, there is enough of it to justify your spending some time on the application.

Welding Technique

Current adjustment will be slightly higher than for reverse polarity electrodes. The electrode position should be approximately that shown in Fig. 20-58, page 617. Because of the higher buildup and the greater amount of slag, the speed of travel should be faster than with reverse polarity electrodes. Undercutting is not a problem so that it is unnecessary to hesitate as long at the sides of the weld. Stringer beads should be deposited as instructed in Job 20-J21.

Practice starts and stops. Weld on both sides of the plate.

Operations

1. Obtain plate; check the job print for size.
2. Obtain a square head and scale, dividers, scribe, and center punch from the toolroom.
3. Lay out parallel lines as shown on the job print.
4. Mark the lines with a center punch.
5. Obtain electrodes of each quantity, type, and size specified in the job print.
6. Stand the plate in the vertical position on the welding table. Make sure that it is well-grounded.
7. Set power source for 80 to 140 amperes. Set a d.c. power source for straight polarity.
8. Make weaved beads between the parallel lines which are $\frac{3}{4}$ inch apart as shown on the job print. Manipulate the electrode as instructed in Fig. 20-57, page 617, and Fig. 20-60, page 618.
9. Chip the slag from the beads, brush, and inspect. Refer to Inspection.
10. Make stringer beads between the weaved beads already deposited. Manipulate the electrode as instructed in Figs. 20-56 and 20-57, pages 616, 617.
11. Chip the slag from the beads, brush, and inspect. Refer to Fig. 20-61 and Inspection.
12. Practice these beads until you can produce good beads consistently with both the E6012 and E6013 electrodes and with $\frac{5}{32}$-inch diameter electrodes.

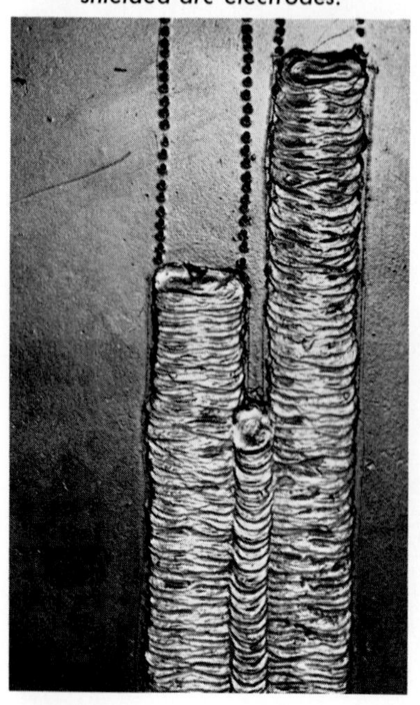

20-61. *Typical appearance of weaved and stringer beads on flat plate welded in the vertical position, travel up, with DCRP or a.c. shielded-arc electrodes.*

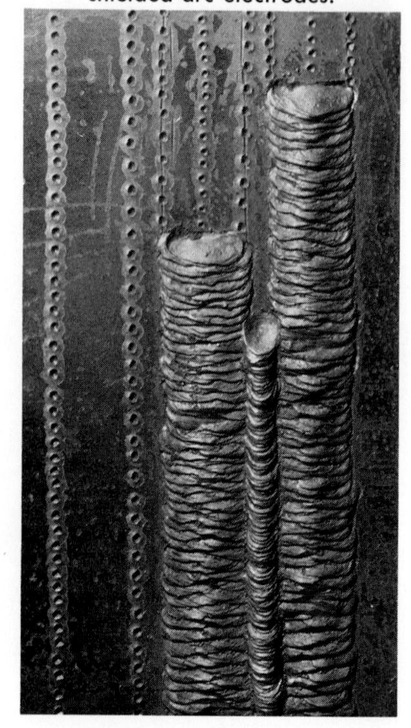

20-62. *Typical appearance of weaved and stringer beads on flat plate welded in the vertical position, travel up, with DCSP or a.c. shielded-arc electrodes.*

Inspection

Compare the beads with Fig. 20-62 and check them for the following weld characteristics:

Width and height: Uniform
Appearance: Smooth with close ripples; free of voids and high spots. Restarts should be difficult to locate.
Size: Refer to the job print.
Face of weld: Convex with slight reinforcement
Edges of weld: Good fusion, no overlap, no undercut
Starts and stops: Free of depressions and high spots
Beginnings and endings: Full size, craters filled
Penetration and fusion: To plate surfaces and adjacent beads
Surrounding plate surfaces: Free of spatter
Slag formation: Full coverage, easily removable

Disposal

Discard completed plates in the waste bin. Plates must be filled on both sides.

JOB 20-J24 WELDING A LAP JOINT

Objective

To weld a lap joint in the vertical position by means of a single-pass fillet weld, travel up, with DCRP and/or a.c. shielded-arc electrodes (AWS-ASTM E6010-E6011).

General Job Information

On critical work the direction of travel for lap joints in the vertical position is usually up. This is true for the shipbuilding, pressure vessel, and structural steel industries. The technique is used for single-pass lap welds and for the first pass on multipass lap welds. When welding heavier plate, the lap joint is often welded with the multipass stringer bead procedure. The first bead should be put into the

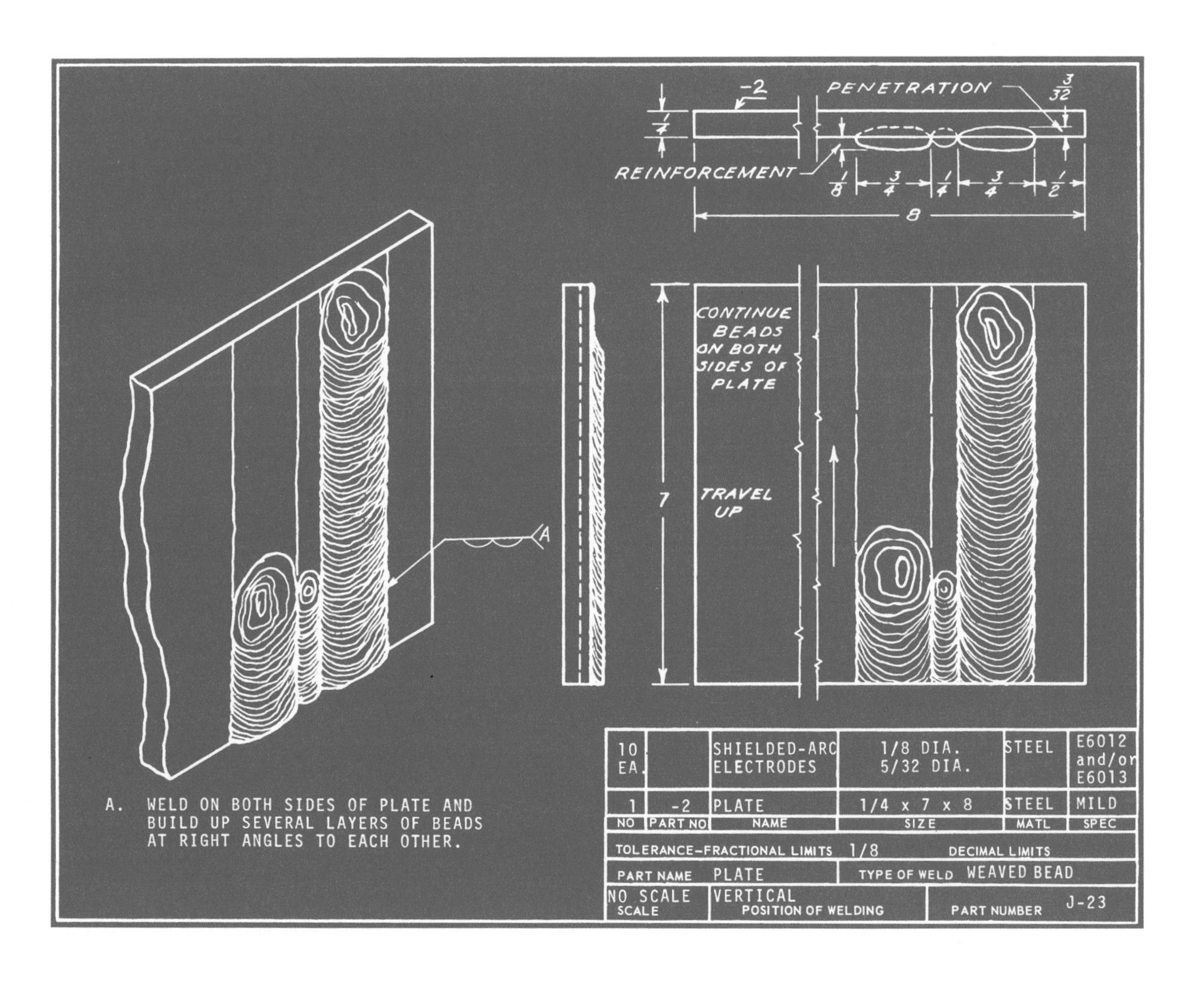

root with good penetration. The second bead should fuse thoroughly into $\frac{3}{4}$ of the first bead and extend $\frac{1}{4}$ inch out on the bottom plate. The third bead should fuse with the first two and extend to the edge of the top plate.

The lap joint may also be welded with the weaved fillet weld procedure for the last one or two passes, depending upon the thickness of the material.

Welding Technique

Current adjustment must not be too high. Hold the electrode 10 degrees downward from the horizontal position, Fig. 20-63A, and at a 45-degree angle to the plate surface shown in Fig. 20-63B. Form a shelf at the bottom of the joint which is the required size of the weld. Manipulate the electrode with a rocking back-and-forth motion. Deposit metal with a short arc gap and lengthen the arc gap on the upward stroke. Do not break the arc on the upward stroke. When the electrode is in the molten pool, you may use a slight circular motion to form a full bead. Keep the electrode motion within the confines of the weld width so that the edge of the upper plate is not burned away and so that undercut does not occur on the surface of the lower plate.

Operations

1. Obtain plates; check the job print for quantity and size.
2. Obtain electrodes of each quantity, type, and size specified in the job print.
3. Set a d.c. power source for reverse polarity at 120 to 170 amperes or an a.c power source at 130 to 190 amperes.
4. Set up the plates and tack weld them at each corner.
5. Stand the plates in the vertical position on the welding table. Make sure that they are well-grounded.
6. Make welds as shown on the job print. Manipulate the electrode as instructed in Fig. 20-63.
7. Chip the slag from the welds, brush, and inspect. Refer to Inspection.
8. Practice these welds until

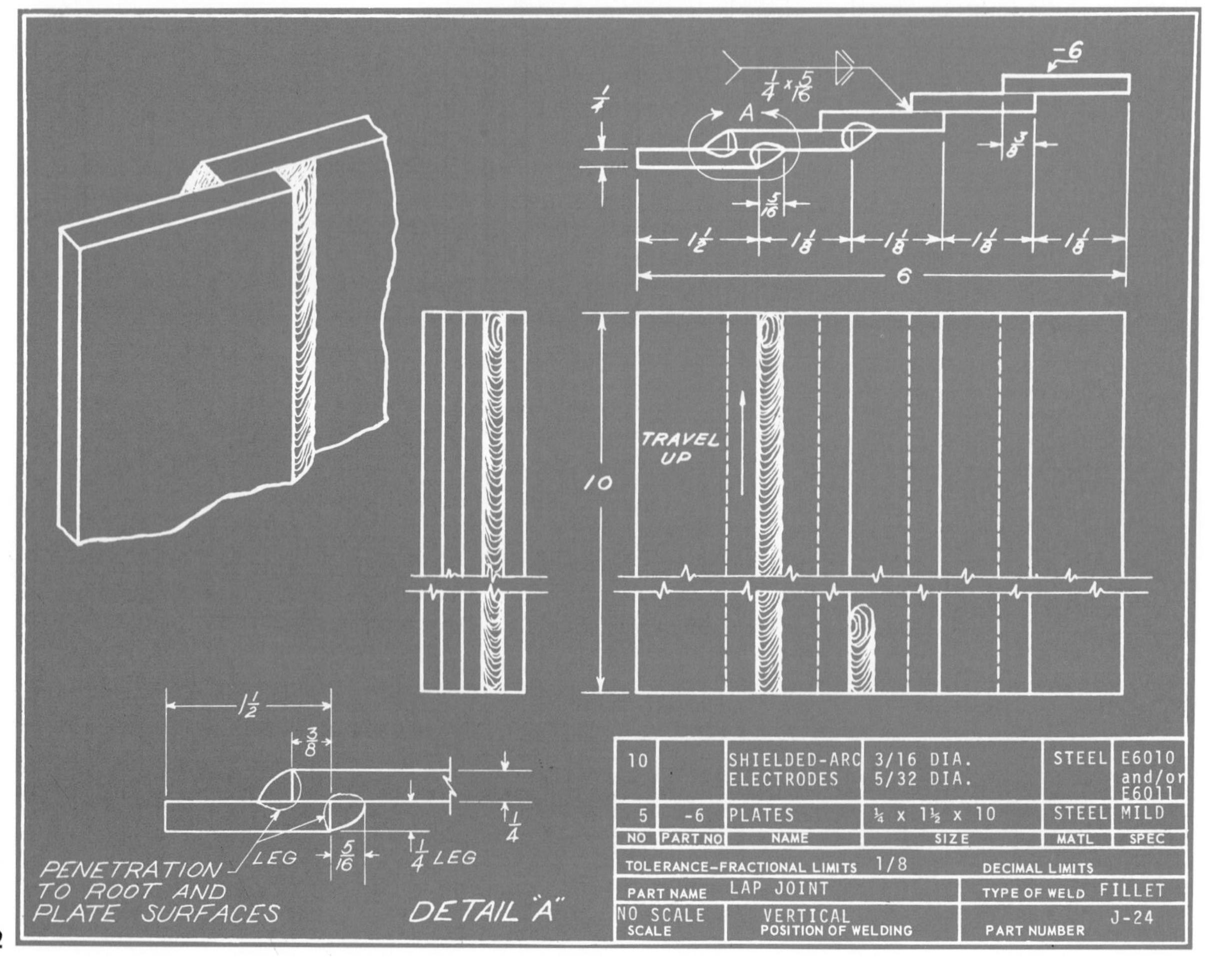

10		SHIELDED-ARC ELECTRODES	3/16 DIA. 5/32 DIA.	STEEL	E6010 and/or E6011
5	-6	PLATES	¼ x 1½ x 10	STEEL	MILD
NO	PART NO	NAME	SIZE	MATL	SPEC

TOLERANCE-FRACTIONAL LIMITS 1/8 — DECIMAL LIMITS
PART NAME: LAP JOINT — TYPE OF WELD: FILLET
SCALE: NO SCALE — POSITION OF WELDING: VERTICAL — PART NUMBER: J-24

you can produce good welds consistently with both types of electrodes.

Inspection

Compare the weld with Fig. 20-64 and check it for the following characteristics:

Width and height: Uniform

Appearance: Smooth with close ripples; free of voids and high spots. Restarts should be difficult to locate.

Size: Refer to the job print. Check with a fillet weld gauge.

Face of weld: Slightly convex and full

Edges of weld: Good fusion, no overlap, no undercut

Starts and stops: Free of depressions and high spots

Beginnings and endings: Full size, craters filled

Penetration and fusion: To the root of the joint and both plate surfaces

Surrounding plate surfaces: Free of spatter

Slag formation: Full coverage, easily removable

Check Test

After you are able to make welds that are satisfactory in appearance, make up a specimen like that made in Job 20-J4, Fig. 20-17, page 585. Use the thickness of plate, welding technique, and electrode types and size specified for this job. Weld on one side only. Break the joint as shown in Fig. 20-18, page 585. Inspect the weld for lack of penetration and fusion at the root of the joint and to the plate surfaces. The weld metal should be sound and show no evidence of porosity, slag inclusions, or gas pockets. The weld should break evenly through the throat.

Disposal

Put completed joints in the scrap bin so that they will be available for further use. Unwelded plate surfaces can be used for beading.

JOB 20-J25 WELDING A T-JOINT

Objective

To weld a T-joint in the horizontal position by means of a single-pass fillet weld, with DCSP and/or a.c. shielded-arc electrodes (AWS-ASTM E6012-E6013).

General Job Information

Horizontal fillet welding is a large part of the work that the operator will do in the field. Welding procedures may include single-pass and multipass techniques with all types of electrodes. The final size of a fillet weld determines whether it is to be single pass or multipass. Multipass procedures may require either the weaved or stringer bead welding techniques. It is particularly important for you to become proficient in the next few jobs.

Welding Technique

Current should be high enough to maintain a fluid puddle. Direct the electrode at the root of the weld and toward the vertical plate, Fig. 20-65A, and tilt it 30 degrees in the direction of travel, Fig. 20-65B. The position for an actual weld is shown in Fig. 20-66. Advance the electrode at a uniform rate of travel.

20-63. *Electrode position when welding lap joints in the vertical position, travel up.*

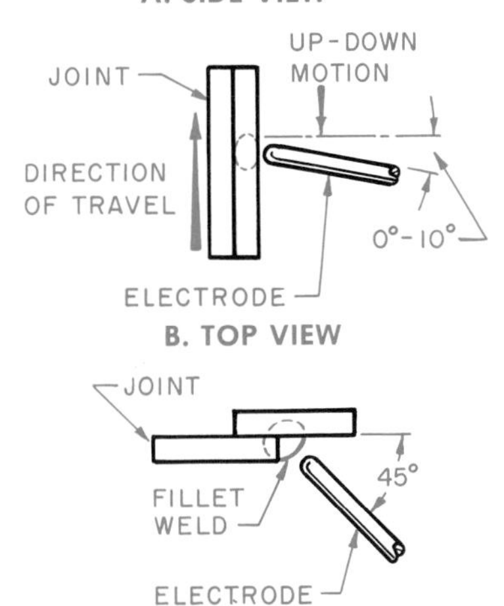

20-64. *Typical appearance of a fillet weld in a lap joint welded in the vertical position, travel up, with DCRP or a.c. shielded-arc electrodes.*

20-65. *Electrode position when welding a T-joint in the horizontal position.*

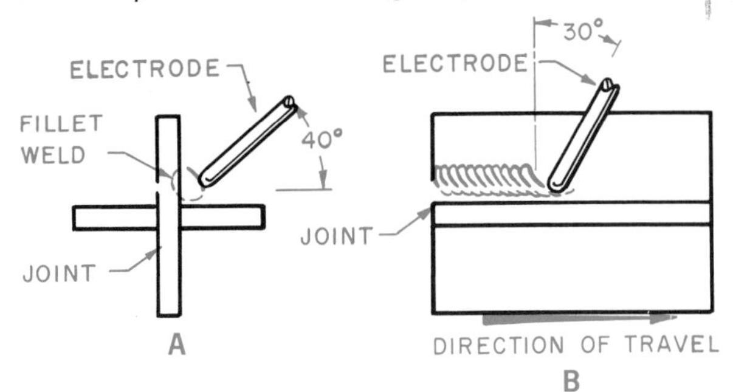

You may keep it in the weld pool for the entire length of the weld. Some operators, however, prefer to use a slight back-and-forth motion, which preheats the root ahead of the weld, aids the formation of the weld deposit against the vertical plate, and helps to keep the slag from running ahead of the weld.

Undercut is not a serious problem with a straight polarity electrode. If the arc is too long or if the angle of the electrode is incorrect, however, undercutting may result. Direct the arc about equally against both legs of the fillet weld with just enough direction against the vertical plate to permit the weld metal to build up on it. Control the speed to obtain good weld contour and a full throat section of the fillet weld. Hold the arc in the crater long enough to build up the molten weld metal to the desired height.

Practice starts and stops. Travel in all directions.

Operations

1. Obtain plates; check the job print for quantity and size.
2. Obtain electrodes of each quantity, type, and size specified in the job print.
3. Set a d.c. power source for straight polarity at 110 to 190 amperes or an a.c. power source at 120 to 210 amperes.
4. Set up the plates and tack weld them at each corner.
5. Place the joint in the horizontal position on the welding table. Make sure that it is well-grounded.
6. Make welds as shown on the job print. Manipulate the electrode as in Fig. 20-65.
7. Chip the slag from the welds, brush, and inspect. Refer to Inspection.
8. Practice these welds until you can produce good welds consistently with both types of electrodes and with $^3/_{16}$-inch diameter electrodes.

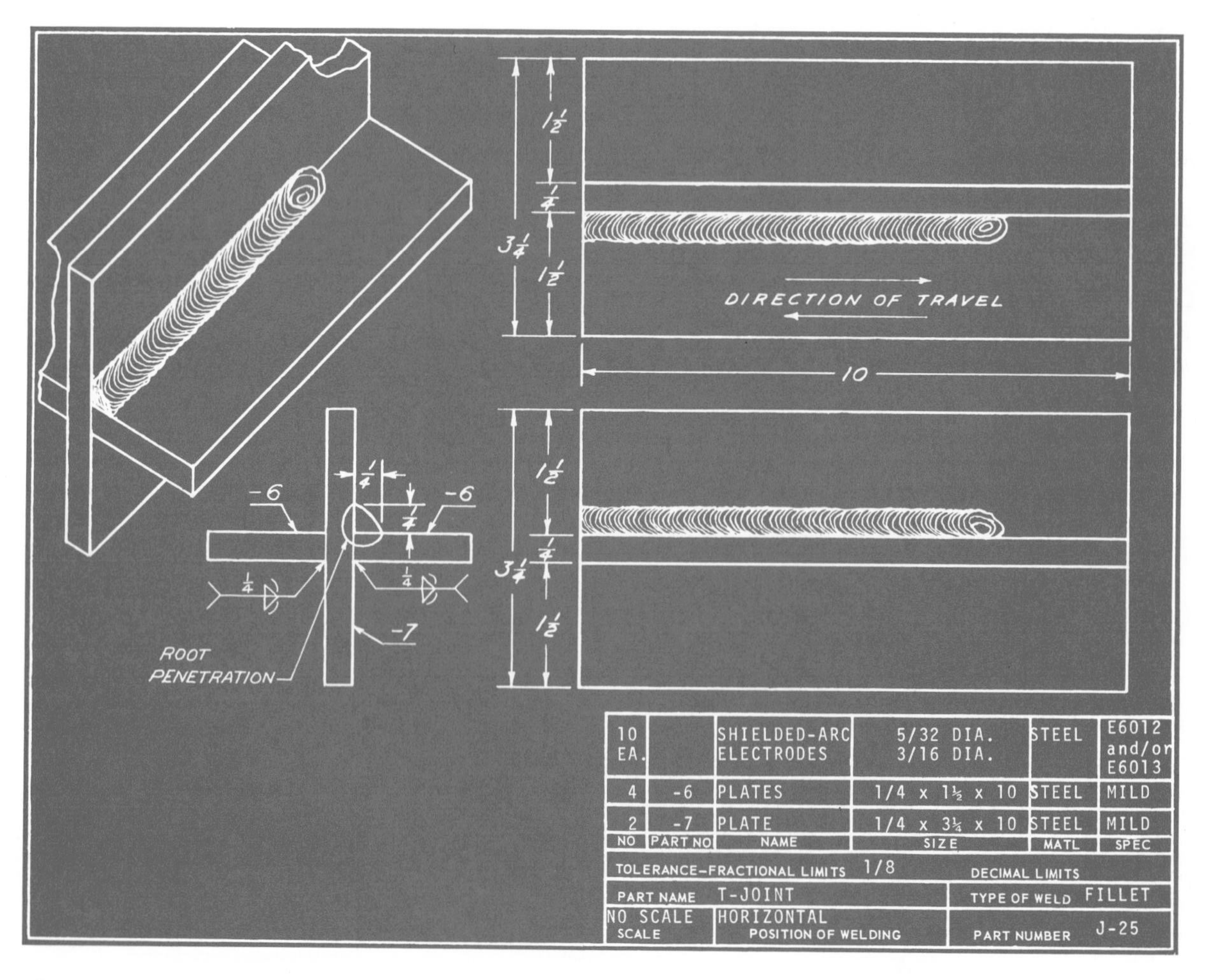

NO	PART NO	NAME	SIZE	MATL	SPEC
10 EA.		SHIELDED-ARC ELECTRODES	5/32 DIA. 3/16 DIA.	STEEL	E6012 and/or E6013
4	-6	PLATES	1/4 x 1½ x 10	STEEL	MILD
2	-7	PLATE	1/4 x 3¼ x 10	STEEL	MILD

TOLERANCE–FRACTIONAL LIMITS 1/8	DECIMAL LIMITS
PART NAME T-JOINT	TYPE OF WELD FILLET
SCALE NO SCALE · POSITION OF WELDING HORIZONTAL	PART NUMBER J-25

Inspection

Compare the welds with Fig. 20-67 and check them for the following characteristics:

Width and height: Uniform
Appearance: Smooth with close ripples; free of voids. Restarts should be difficult to locate.
Size: Refer to the job print. Check with a fillet weld gauge.
Face of weld: Slightly convex
Edges of weld: Good fusion, no overlap, no undercut
Starts and stops: Free of depressions and high spots
Beginnings and endings: Full size, craters filled
Penetration and fusion: To root of joint and plate surfaces
Surrounding plate surfaces: Free of spatter
Slag formation: Full coverage, easily removable

Check Test

After you are able to make welds that are satisfactory in appearance, make up a specimen similar to that made in Job 20-J17, Fig. 20-48, page 612. Use the plate thickness, welding technique, and electrode type and size specified for this job. Weld on one side only. Break the finished weld as shown in Fig. 20-49, page 612, and examine the surfaces for soundness. The weld must not be porous and must show good fusion and penetration at the root of the joint and to the plate surfaces. The joint should break evenly through the throat of the weld.

Disposal

Save the completed joints for your next job, which will be multipass fillet welding.

JOB 20-J26 WELDING A T-JOINT

Objectives

1. To weld a T-joint in the horizontal position by means of multipass fillet welds using the stringer bead technique with DCSP and/or a.c. shielded-arc electrodes (AWS-ASTM E6012-E6013).

2. To weld a cover pass in the horizontal position by means of a lacing bead with DCRP and/or a.c. shielded-arc electrodes (AWS-ASTM E6010-E6011).

General Job Information

Large fillets are often welded with multipass stringer beads. On many jobs the final layer is made up of stringer beads. On other jobs a lacing bead is applied as the final bead. This is especially true in piping and pressure vessel work.

For practical purposes $\frac{1}{4}$-inch plate would not require more than one pass. The additional passes specified in this job are for practice purposes and the full and economical use of material.

Welding Technique

Current adjustment should be similar to that used for the first pass. Electrode position varies with the sequence of passes. The electrode is held at a 30-degree angle for the second pass, a 40-degree angle for the third pass, a 50-degree for the fourth pass, and a 40-degree angle for the fifth and sixth passes, Fig. 20-68. It is also tilted at a 30-degree angle in the direction of travel. Electrode movement is the same as for the preceding job. Beads should be so proportioned that the size of the final weld is accurate and to specification.

Be very critical of the angle of the electrode and the arc length to make sure that undercut does not develop in the vertical plate. The second pass should lay about half on the flat plate and half on the first bead. For the third pass, lay down another

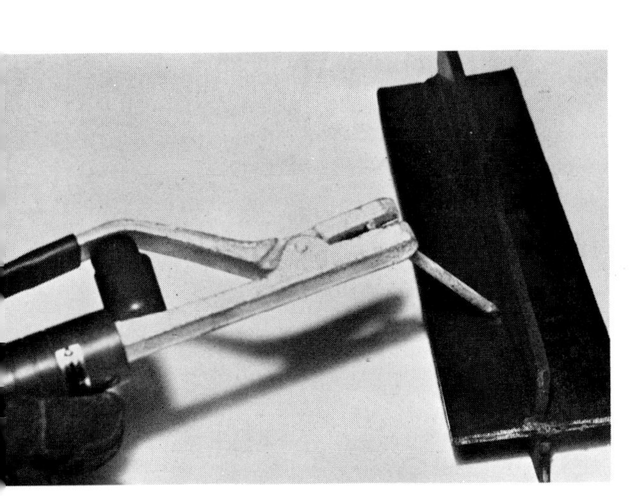

20-66. *Welding a T-joint in the horizontal position. The electrode and electrode holder are positioned as instructed in Fig. 20-65.*

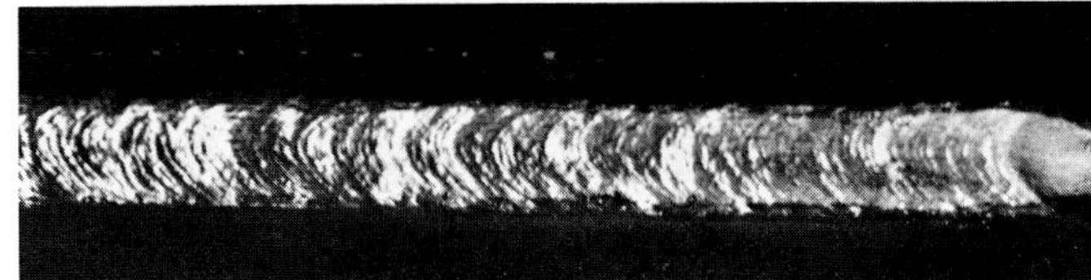

20-67. *Typical appearance of a single-pass fillet weld in a T-joint welded in the horizontal position with DCSP or a.c. shielded-arc electrodes.*

bead which covers the other half of the first pass and also fuses with the vertical plate and the second pass, Fig. 20-68. Follow this same procedure in making the next layer of three passes, one at the bottom, one in the middle, and one at the top as shown in Fig. 20-68.

The technique of lacing is difficult. Current adjustment must not be too low. A smaller electrode is used because it is hard to handle a large amount of weld metal with this technique. An out-of-position electrode is also necessary.

The electrode position should be approximately that shown in Fig. 20-69. The electrode is tilted at a 35-degree angle to the flat plate, Fig. 20-69A, and in a 30-degree angle toward the direction of travel, Fig. 20-69B. Weld metal is deposited only on the downward stroke. The upward stroke should be fast and the arc should be lengthened, but it should not go out. The top and bottom edges of the weld should be your guide for

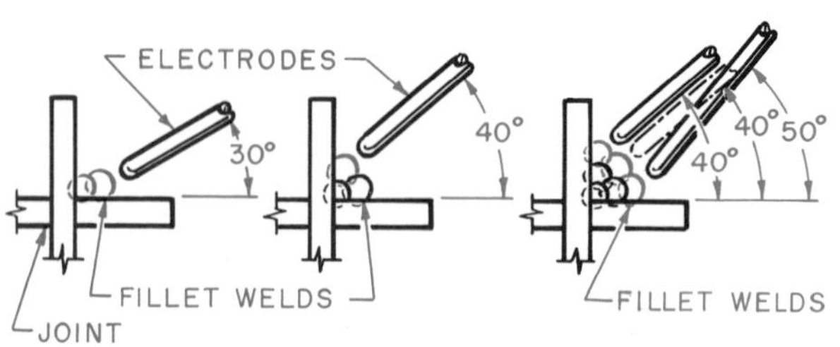

20-68. *Electrode position when welding a T-joint with multipass fillet welds using the stringer bead technique.*

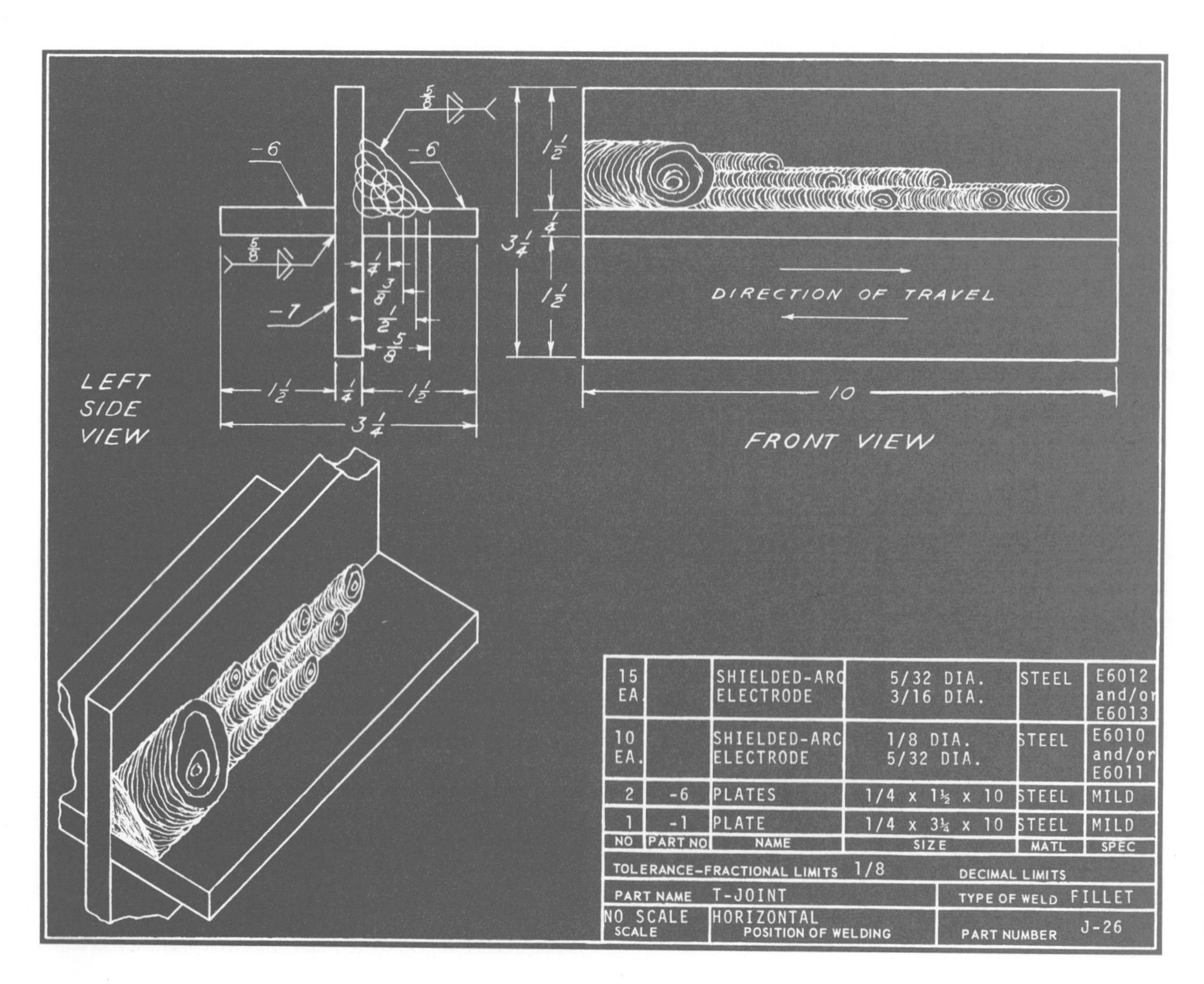

NO	PART NO	NAME	SIZE	MATL	SPEC
15 EA.		SHIELDED-ARC ELECTRODE	5/32 DIA. 3/16 DIA.	STEEL	E6012 and/or E6013
10 EA.		SHIELDED-ARC ELECTRODE	1/8 DIA. 5/32 DIA.	STEEL	E6010 and/or E6011
2	-6	PLATES	1/4 x 1½ x 10	STEEL	MILD
1	-1	PLATE	1/4 x 3¼ x 10	STEEL	MILD

TOLERANCE–FRACTIONAL LIMITS 1/8 DECIMAL LIMITS

PART NAME T-JOINT TYPE OF WELD FILLET

SCALE NO SCALE POSITION OF WELDING HORIZONTAL PART NUMBER J-26

the width of the electrode movement. Split the weld edge with your electrode. Hesitate an instant at the top and bottom edges of the weld to eliminate undercut.

Bear in mind that all multipass welding requires thorough cleaning between passes. Practice starts and stops. Travel in all directions.

Operations

1. Select good quality joints welded in Job 20-J25.
2. Obtain electrodes of each quantity, type, and size specified in the job print.
3. Set a d.c. power source for straight polarity at 100 to 190 amperes or an a.c. power source at 125 to 210 amperes.
4. Place the joint in the horizontal position on the welding table. Make sure that it is well-grounded.
5. Make passes two through six with 5/32-inch electrodes as shown on the job print. Hold the electrode as instructed in Fig. 20-68 for each pass.
6. Chip the slag from the weld, brush, and inspect between each pass. Refer to Inspection.
7. Set a d.c. power source for reverse polarity at 80 to 140 amperes or an a.c. power source at 80 to 160 amperes.
8. Make the lacing bead with 1/8-inch electrodes as shown on the job print. Manipulate the electrode as instructed in Fig. 20-69.
9. Chip the slag from the weld, brush, and inspect. Refer to Inspection, below.
10. Practice these welds until you can produce good welds consistently with both the E6012 and E6013 electrodes in 5/32- and 3/16-inch diameters for the stringer beads and the E6010 and E6011 electrode in 1/8- and 5/32-inch diameters for the lacing pass.

Inspection

Compare passes two through six with Fig. 20-70 and the lacing pass with Fig. 20-71. Check each pass for the following weld characteristics:

Width and height: Uniform
Appearance: Smooth with close ripples; free of voids. Restarts should be difficult to locate.
Size: Refer to the job print. Check with a fillet weld gauge.
Face of weld: Flat
Edges of weld: Good fusion, no overlap, no undercut
Starts and stops: Free of depressions and high spots
Beginnings and endings: Full size, craters filled
Penetration and fusion: To root of joint, adjacent beads, and plate surfaces
Surrounding plate surfaces: Free of spatter
Slag formation: Full coverage, easily removable

Disposal

Put completed jobs in the scrap bin so that they can be used again later. Plate edges can be used for square butt welding, and unwelded plate surfaces can be used for beading.

20-69. *Electrode position when welding the final pass on a T-joint using the lacing bead technique.*

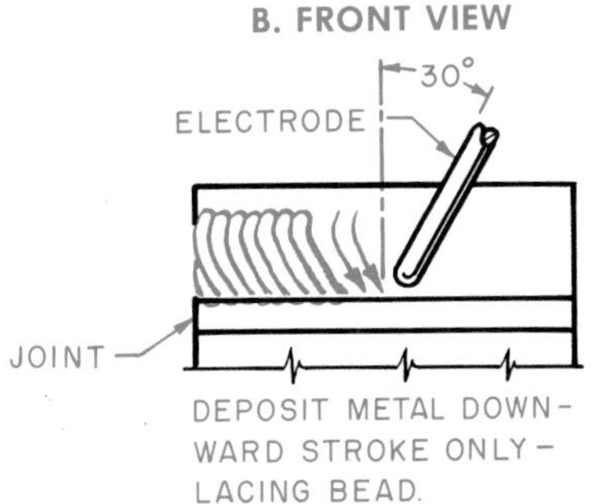

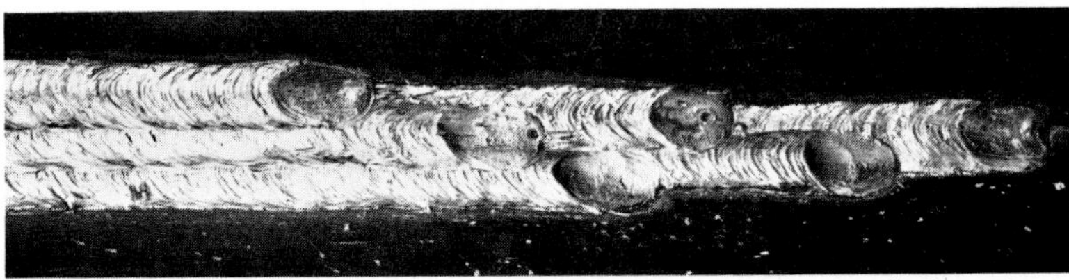

20-70. *Typical appearance of a multipass fillet weld done with the stringer bead technique in a T-joint welded in the horizontal position with DCSP or a.c. shielded-arc electrodes.*

20-71. *Typical appearance of a multipass fillet weld done with the lacing bead technique for the cover pass.*

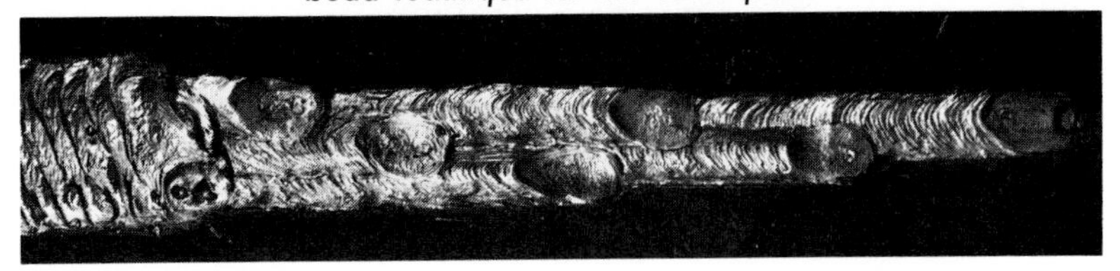

JOB 20-J27 WELDING A T-JOINT

Objective

To weld a T-joint in the horizontal position by means of a single-pass fillet weld with DCRP and/or a.c. shielded-arc electrodes (AWS-ASTM E6010-E6011).

General Job Information

This is the same type of joint practiced in Job 20-J25 except that reverse polarity electrodes are used. Reverse polarity electrodes are not the best choice for single-pass fillet welds, but since this is the practice on many jobs, you should master the technique. Beads have a flat face, and there is the tendency toward undercutting.

Welding Technique

Current adjustment should be somewhat lower than for straight polarity electrodes. More care should be given to electrode position. Hold the electrode at a 35-degree angle to the flat plate, Fig. 20-72A, and at an angle of 40 degrees in the direction of travel, Fig. 20-72B. The welding technique is similar to that used for straight polarity except that the arc gap should be closer and more care must be exercised to prevent undercut.

If the arc is too long or if the angle of the electrode is incorrect, undercutting will result. Direct the arc at the root of the weld and about equally against both legs of the fillet with just enough direction against the vertical plate to permit the weld metal to build up on it. Control the speed of welding to obtain good weld contour and the full throat section of the fillet. Hold the arc in the crater long enough to build up the molten metal to the desired height and travel at uniform speed.

Practice starts and stops. Travel in all directions.

Operations

1. Obtain plates; check the job print for quantity and size.

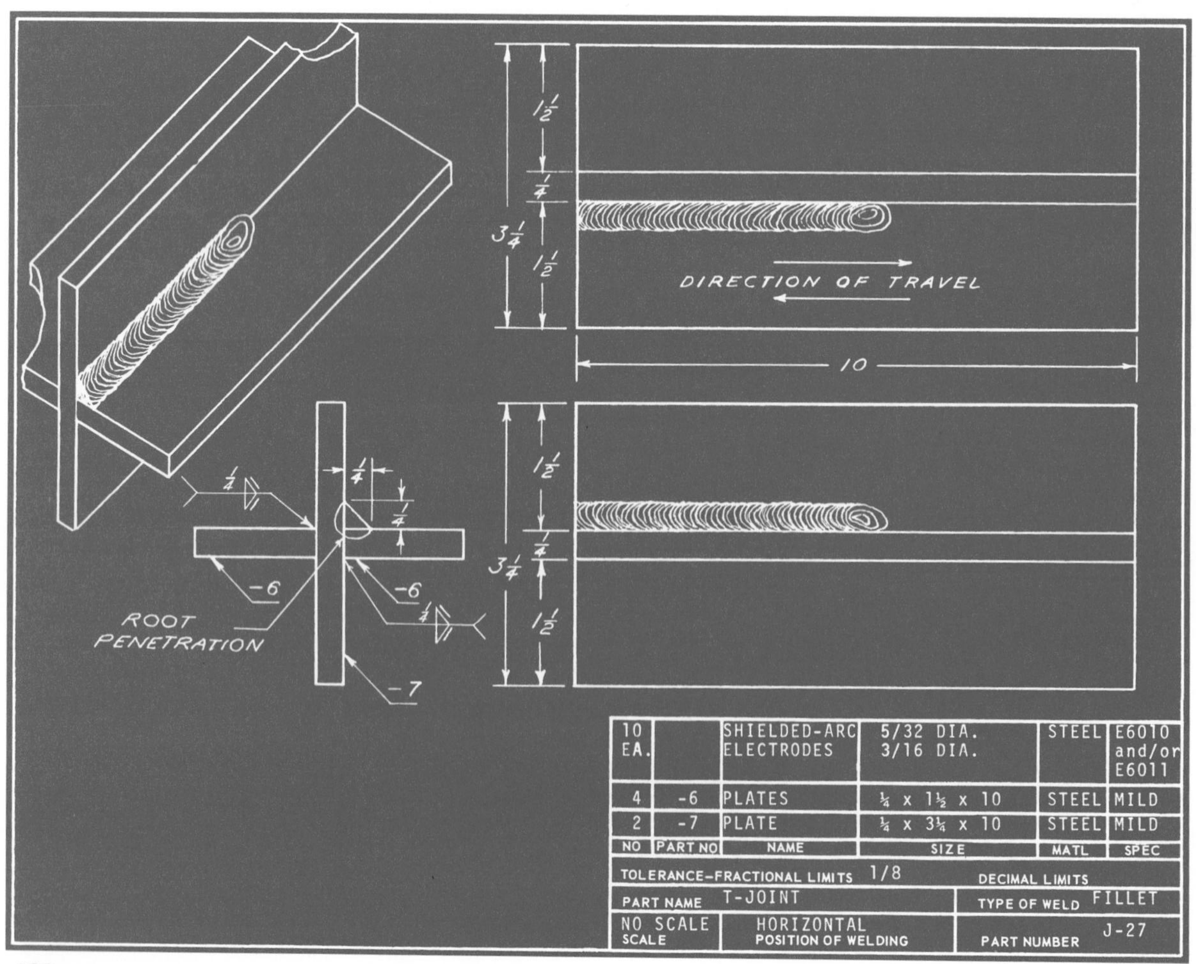

ROOT PENETRATION
-6
-6
-7
DIRECTION OF TRAVEL
10
1/2
1/4
3 1/4
10 EA. | SHIELDED-ARC ELECTRODES | 5/32 DIA. 3/16 DIA. | STEEL | E6010 and/or E6011
4 | -6 | PLATES | ¼ x 1½ x 10 | STEEL | MILD
2 | -7 | PLATE | ¼ x 3¼ x 10 | STEEL | MILD
NO | PART NO | NAME | SIZE | MATL | SPEC
TOLERANCE—FRACTIONAL LIMITS 1/8
DECIMAL LIMITS
PART NAME T-JOINT
TYPE OF WELD FILLET
NO SCALE
SCALE
HORIZONTAL
POSITION OF WELDING
PART NUMBER J-27

2. Obtain electrodes of the quantity, type, and size specified in the job print.

3. Set a d.c. power source for reverse polarity at 110 to 170 amperes or an a.c. power source at 115 to 190 amperes.

4. Set up the plates and tack weld them at each corner.

5. Place the joint in the horizontal position on the welding table. Make sure that it is well-grounded.

6. Make welds as shown on the job print. Manipulate the electrode as instructed in Fig. 20-72.

7. Chip the slag from the welds, brush, and inspect. Refer to Inspection, below.

8. Practice these welds until you can produce good welds consistently with both types of electrodes.

Inspection

Compare welds with Fig. 20-73 and check them for the following characteristics:

Width and height: Uniform
Appearance: Smooth with close ripples; free of voids. Restarts should be difficult to locate
Size: Refer to the job print. Check with a fillet weld gauge.
Face of weld: Flat
Edges of weld: Good fusion, no overlap, no undercut
Starts and stops: Free of depressions and high spots
Beginnings and endings: Full size, craters filled
Penetration and fusion: To the root of the joint and plate surfaces
Surrounding plate surfaces: Free of spatter
Slag formation: Full coverage, easily removable

Disposal

Save the completed joints for your next job, which will be multipass fillet welding.

Check Test

After you are able to make welds that are satisfactory in appearance and are approved by your instructor, make up a test specimen like that shown in Fig. 20-48, page 612. Use the plate thickness, welding technique, and electrode type and size specified for this job. Weld on one side only.

Break the completed weld as shown in Fig. 20-49 and examine the surfaces for soundness. They must not be porous and must show fusion and penetration at the root of the weld and to the plate surfaces. The weld should break through the throat. It should not peel off the plate surfaces.

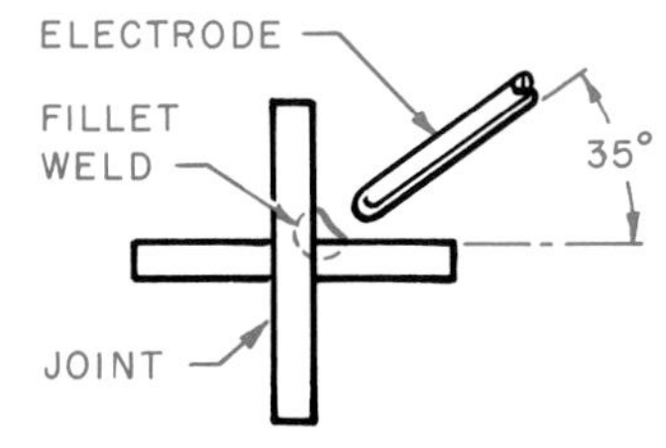

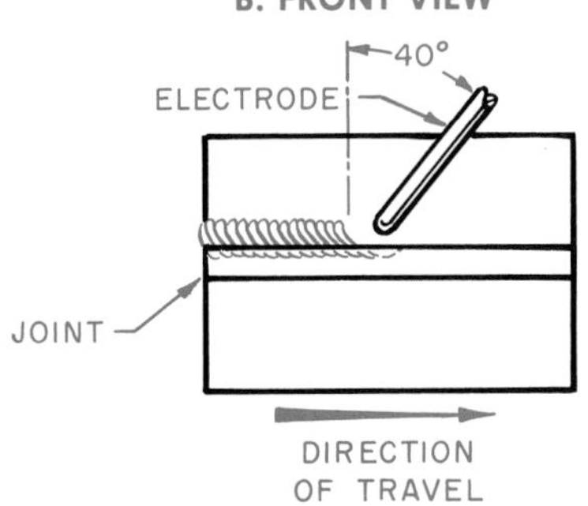

20-72. *Electrode position when welding a T-joint with a single-pass fillet weld in the horizontal position.*

JOB 20-J28 WELDING A T-JOINT

Objectives

1. To weld a T-joint in the horizontal position by means of multipass fillet welds and the stringer bead technique with DCRP and/or a.c. shielded-arc electrodes (AWS-ASTM E6010-E6011).

2. To weld an overlay in the horizontal position by means of a lacing bead with DCRP and/or a.c. shielded-arc electrodes (AWS-ASTM E6010-E6011).

General Job Information

This job is similar to Job 20-J26 except that a different type of electrode is used.

Welding Technique

Current adjustment should be somewhat lower than for straight polarity electrodes. More care should be given to electrode position. The angle of the electrodes is slightly different with reverse polarity. It is held at a 45-degree angle to the flat plate to deposit the second bead, Fig. 20-74A. The angle is changed to 35 degrees for the third bead,

20-73. *Typical appearance of a single-pass fillet weld in a T-joint welded in the horizontal position with DCRP or a.c. shielded-arc electrodes.*

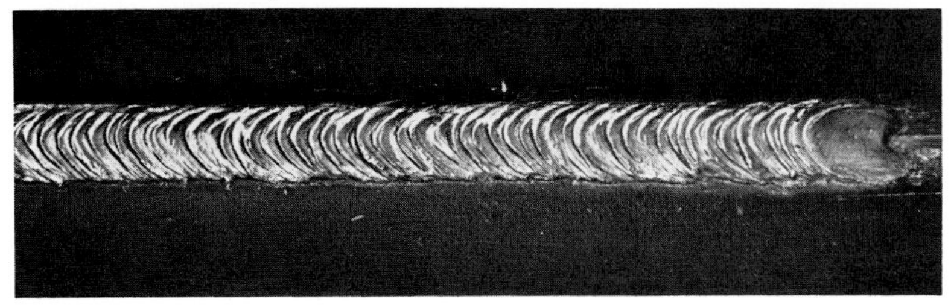

Fig. 20-74B. The fourth bead is deposited at a 45-degree angle, and the angle is changed to 35 degrees for the fifth and sixth beads, Fig. 20-74C. The electrode is also tilted at a 30-degree angle in the direction of travel. The welding technique for both stringer and lacing beads is similar to that practiced in Job 20-J26. Refer to Fig. 20-69, page 627, for deposition of the lacing bead. Undercut is a problem and can be remedied in the usual manner. Review the welding technique described in Job 20-J26.

Clean all passes thoroughly. Practice starts and stops. Travel in all directions.

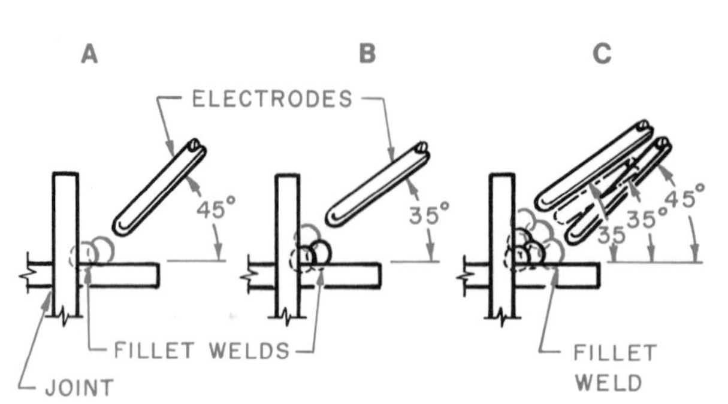

20-74. *Electrode position when welding a T-joint with the stringer bead technique.*

Operations

1. Use joints welded for Job 20-J27.
2. Obtain electrodes of each quantity, type, and size specified in the job print.

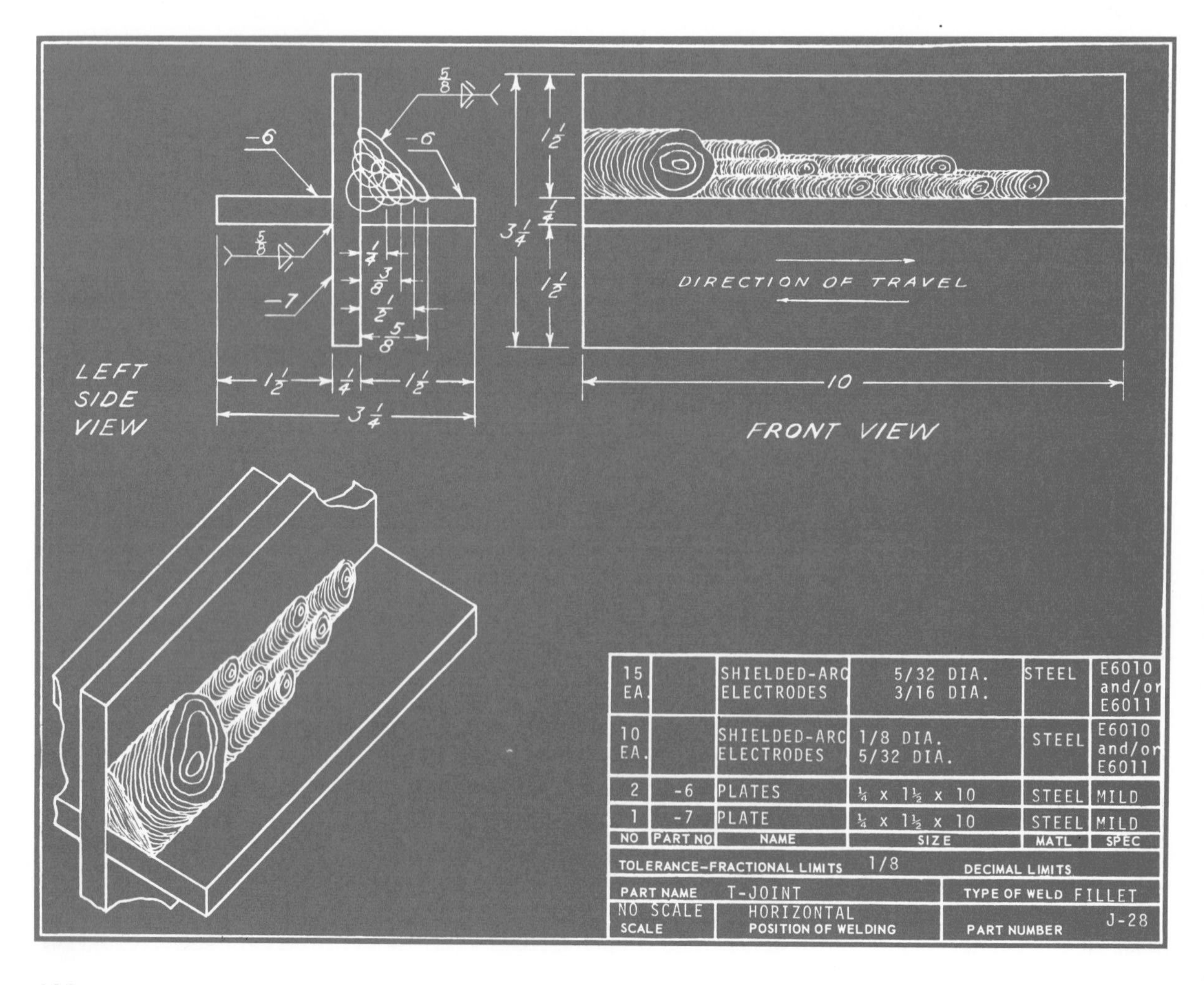

LEFT SIDE VIEW
FRONT VIEW
DIRECTION OF TRAVEL
-6
-6
-7
5/8
1/4
3/8
1/2
1 1/2
3 1/4
10
15 EA. SHIELDED-ARC ELECTRODES 5/32 DIA. 3/16 DIA. STEEL E6010 and/or E6011
10 EA. SHIELDED-ARC ELECTRODES 1/8 DIA. 5/32 DIA. STEEL E6010 and/or E6011
2 -6 PLATES ¼ x 1½ x 10 STEEL MILD
1 -7 PLATE ¼ x 1½ x 10 STEEL MILD
NO PART NO NAME SIZE MATL SPEC
TOLERANCE-FRACTIONAL LIMITS 1/8 DECIMAL LIMITS
PART NAME T-JOINT
TYPE OF WELD FILLET
NO SCALE
SCALE
HORIZONTAL
POSITION OF WELDING
PART NUMBER J-28

3. Set a d.c. power source for reverse polarity at 110 to 170 amperes or an a.c. power source at 115 to 190 amperes.

4. Place the joint in the horizontal position on the welding table. Make sure that it is well-grounded.

5. Make passes two through six with $\frac{5}{32}$-inch electrodes as shown on the job print. Hold the electrode as instructed in Fig. 20-74 for each pass.

6. Chip the slag from the weld, brush, and inspect between each pass. Refer to Inspection.

7. Set a d.c. power source for reverse polarity at 75 to 125 amperes or set an a.c. power source at about the same heat value.

8. Make the lacing bead with $\frac{1}{8}$-inch electrodes as shown on the job print. Manipulate the electrode as shown in Fig. 20-69, page 627.

9. Chip the slag from the weld, brush, and inspect. Refer to Inspection.

10. Practice these welds until you can produce good welds consistently. In addition, practice with $\frac{3}{16}$-inch diameter electrodes for the stringer beads and $\frac{5}{32}$-inch diameter electrodes for the lacing pass.

Inspection

Compare each pass with Fig. 20-75 and check it for the following weld characteristics:

Width and height: Uniform
Appearance: Smooth with close ripples; free of voids. Restarts should be difficult to locate.
Size: Refer to the job print. Check with a fillet weld gauge.
Face of weld: Flat
Edges of weld: Good fusion, no overlap, no undercut
Starts and stops: Free of depressions and high spots
Beginnings and endings: Full size, craters filled
Penetration and fusion: To the root of the joint, adjacent beads, and plate surfaces
Surrounding plate surfaces: Free of spatter
Slag formation: Full coverage, easily removable

Disposal

Put completed joints in the scrap bin so that they can be used again later. Plate edges can be used for square butt welding, and unwelded plate surfaces can be used for beading.

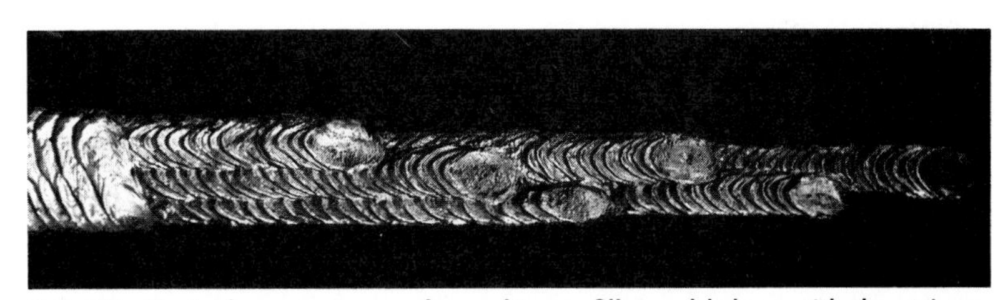

20-75. *Typical appearance of a multipass fillet weld done with the stringer bead technique and a lacing overlay in a T-joint welded in the horizontal position with DCRP or a.c. shielded-arc electrodes.*

REVIEW QUESTIONS

1. Are bare electrodes used extensively in industry? Why are they used in this course?
2. How is the welding arc started?
3. What causes the electrode to freeze to the plate? How may it be freed?
4. Explain the methods for striking an arc.
5. How can fusion be assured at the beginning of the weld?
6. How may the operator tell if he or she is holding the correct arc gap?
7. Traveling too fast will result in what type of beads?
8. Traveling too slowly will result in what type of beads?
9. Welding with too long an arc gap will result in what type of beads?
10. What are the results of heat adjustment that is too hot? Too cold?
11. Explain the technique for restarting a weld bead.
12. Explain the technique for eliminating the crater at the end of a bead.

(Continued on page 633)

Table 20-7
Course Outline: Shielded Metal-Arc Welding Practice: Jobs 20-J 1-28 (Plate)

Job no.	Joint	Type of weld	Type of electrode: Position	DCRP	DCSP	AC	AWS specif. no. d.c. (and/or)	AWS specif. no. a.c.	Text reference
20-J1	Flat plate	Striking the arc—short beading	Flat		X Bare		E4510		576
20-J2	Flat plate	Beading—continuous stringer	Flat		X Bare		E4510		579
20-J3	Flat plate	Beading—weaved	Flat		X Bare		E4510		581
20-J4	Lap joint	Fillet welding—single pass	Flat		X Bare		E4510		582
20-J5	Flat plate	Beading—stringer	Flat		X	X	E6012	E6013	586
20-J6	Flat plate	Beading—weaved	Flat		X	X	E6012	E6013	589
20-J7	Flat plate	Beading—stringer	Flat	X		X	E6010	E6011	592
20-J8	Flat plate	Beading—weaved	Flat	X		X	E6010	E6011	593
20-J9	Edge joint	Beading	Flat		X	X	E6012	E6013	595
20-J10	Edge joint	Beading	Flat	X		X	E6010	E6011	596
20-J11	Lap joint (plates flat)	Fillet welding—single pass	Hor.		X	X	E6012	E6013	598
20-J12	Lap joint (plates flat)	Fillet welding—single pass	Hor.	X		X	E6010	E6011	600
20-J13	Flat plate	Beading—stringer	Hor.	X		X	E6010	E6011	602
20-J14	Flat plate	Beading—stringer—travel down	Ver.	X		X	E6010	E6011	603
20-J15	Lap joint	Fillet welding—single pass—travel down	Ver.		X	X	E6012	E6013	605
20-J16	Lap joint (plates vert.)	Fillet welding—single pass	Hor.		X	X	E6012	E6013	607
20-J17	T-joint	Fillet welding—single pass	Flat		X	X	E6012	E6013	609
20-18J	T-joint	Fillet welding—weaved multipass	Flat		X	X	E6012	E6013	611
20-J19	T-joint	Fillet welding—single pass	Flat	X		X	E6010	E6011	613
20-J20	T-joint	Fillet welding—weaved multipass	Flat	X		X	E6010	E6011	614
20-J21	Flat plate	Beading-stringer—travel up	Ver.	X		X	E6010	E6011	616
20-J22	Flat plate	Beading-weaved—travel up	Ver.	X		X	E6010	E6011	618

Table 20-7 (Continued)

Job no.	Joint	Type of weld	Type of electrode						Text reference
			Position	DCRP	DCSP	AC	AWS specif. no. d.c. and/or	AWS specif. no. a.c.	
20-J23	Flat plate	Beading-weaved—travel up	Ver.		X	X	E6012	E6013	620
20-J24	Lap joint	Fillet welding—single pass-travel up	Ver.	X		X	E6010	E6011	621
20-J25	T-joint	Fillet welding—single pass	Hor.		X	X	E6012	E6013	623
20-J26	T-joint	Fillet welding—stringer—multipass and lacing pass	Hor.		X	X	E6012	E6013	625
20-J27	T-joint	Fillet welding—single pass	Hor.	X		X	E6010	E6011	628
20-J28	T-joint	Fillet welding—stringer—multipass and lacing pass	Hor.	X		X	E6010	E6011	629

(Continued from page 631)

13. What technique is employed to make a wide bead?
14. What are the characteristics of good weld beads?
15. In welding a lap joint it is important to obtain good penetration and fusion with which parts of the joint?
16. What is the function of the shielded arc?
17. For flat welding the tendency in industry is toward what size electrodes?
18. What are the effects of too long an arc gap on the weld?
19. What are the effects of too close an arc gap on the weld?
20. There are two basic techniques employed for making stringer beads. Explain each.
21. Explain the technique of restarting a weld bead.
22. Explain the technique for running weaved beads.
23. How may undercut be prevented?
24. What is the approximate setting for a $^{3}/_{16}$-inch reverse polarity electrode when welding in the flat position? A $^{3}/_{16}$-inch straight polarity electrode?
25. Is there any difference in the deposit of a straight polarity and a reverse polarity electrode? Explain.
26. On what type of work is the edge joint generally used?
27. List some of the advantages of the edge joint.
28. List some of the disadvantages of the edge joint.
29. Explain the technique of welding the edge joint.
30. In welding an edge joint it is important to obtain good penetration and fusion with which parts of the joint?
31. Explain the technique for running a fillet weld in a lap joint in the horizontal position.
32. Give the approximate electrode position in welding a lap joint in the horizontal position.
33. What welding technique is employed to keep from burning away the top edge of the plate?
34. What is the name of the weld used for a lap joint?
35. The finished lap joint should conform to what appearance standards?
36. In welding a lap joint, it is important to

obtain good penetration and fusion with which parts of the joint?

37. Large fillet welds in T-joints in the horizontal position are sometimes accomplished in what manner?

38. The last layer of a horizontal weld is sometimes welded in what manner? Explain the technique.

39. When welding horizontal fillet welds, is undercut a problem with straight polarity electrodes? With reverse polarity electrodes?

40. How can undercut be prevented?

Welding student preparing to do shielded metal arc welding in the overhead position. Note the protective clothing and equipment she is wearing.

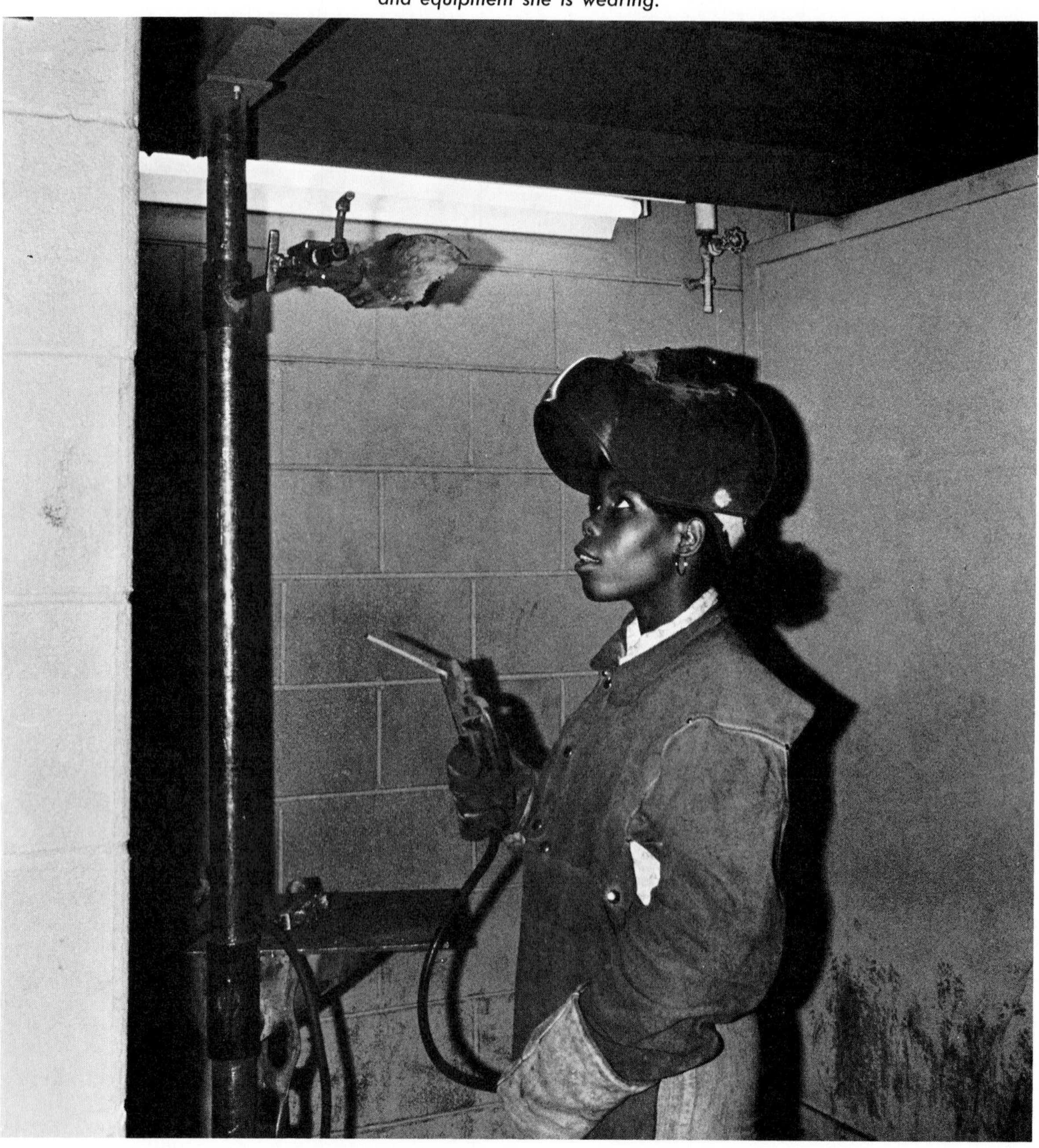

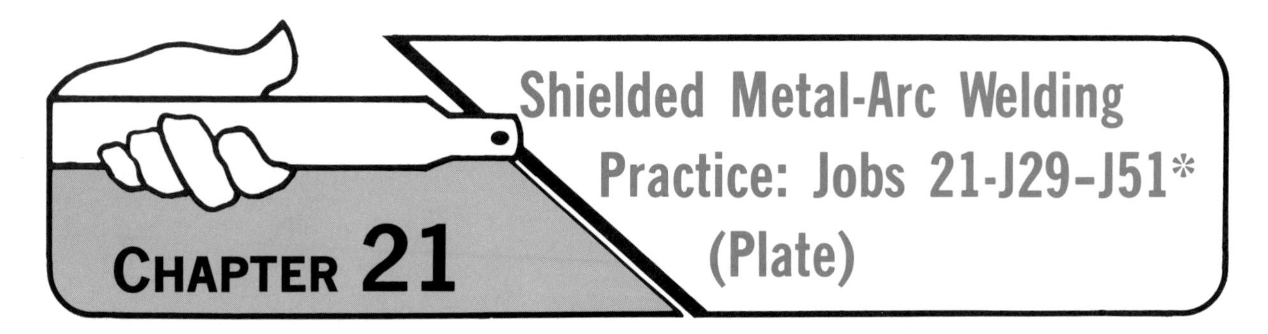

Chapter 21 Shielded Metal-Arc Welding Practice: Jobs 21-J29–J51* (Plate)

INTRODUCTION

This second section of the course in shielded metal-arc welding (SMAW) provides practice in a number of weld joints that are more difficult to weld and require greater skill than those experiences that you have had up to this point. You will also have the opportunity to practice with a series of new electrodes of the low hydrogen and iron powder types. These are often referred to in the trade as a *free-flowing class* of electrodes because they are highly liquid and form a very smooth bead with fine ripples. Very often the appearance is like that of glass.

This chapter will also provide considerable practice in welding various types of basic joints in the flat, horizontal, and vertical positions. You will have your first experience welding in the overhead position. In the shop every effort is made to position the work so that welding can be in the flat or horizontal positions, but it is not always possible to position all of the welding in this way. A portion of the work will always need to be welded in the vertical and overhead positions, Fig. 21-1. A greater amount of out-of-position welding is encountered in field work. It is not possible to position a building or all of the joints on a pipeline. When you have achieved a high degree of skill in welding in the vertical and overhead positions, you are well on your way to becoming a skilled welder able to perform the tasks required by industry.

*The jobs in Units 20–22 are numbered consecutively because these units constitute a course in shielded metal-arc welding.

JOB 21-J29 STRINGER BEADING

Objective

To deposit stringer beads on flat steel plate in the overhead position with DCRP and/or a.c. shielded-arc electrodes (AWS-ASTM E6010-E6011).

General Job Information

Although the tendency today in shops is to eliminate as much overhead welding as possible by positioning the work, you must be able to weld in this position. Much of the work in the piping, shipbuilding, and structural fields, is overhead. Overhead welding will seem quite difficult at first, mainly because of the position that you must assume. In order to minimize fatigue, you should be as comfortable as possible. If you are standing, drape the heavy welding cable over your shoulders to reduce the weight on the arm you are welding with.

Welding Technique

Current adjustment should be about the same as for vertical welding. Hold the electrode at a 90-degree angle to the plate, Fig. 21-2A and at an angle of 5 to 15 degrees in the direction of travel, Fig. 21-2B. Move the electrode back and forth in an oscillating manner. Figure 21-3 shows the actual position for welding. You should be in such position that you can see the metal being deposited behind the arc if the direction of travel is away from you. It is often necessary, however, that the direction of travel be towards the operator. This is usually true of pipe welding.

21-1. *Making a weld in the vertical position. Note the protective clothing.*

Manpower Magazine, U.S. Department of Labor

Hold a close arc gap, and do not let the pool become too molten. A slight oscillating motion back and forth along the line of weld preheats the base metal ahead of the weld and keeps the slag washed back on the weld. If the weld pool gets too hot, move forward faster. If it becomes too cold, slow down your rate of travel. The molten weld metal is held up by molecular attraction and surface tension until each drop solidifies. Study the basic motions shown in Fig. 21-4.

Practice starts and stops. Travel should be in all directions. Weld on both sides of the plate.

21-2. *Electrode position when making stringer beads in the overhead position.*

Operations

1. Obtain plate; check the job print for size.
2. Obtain a square head and scale, dividers, scribe, and center punch from the toolroom.

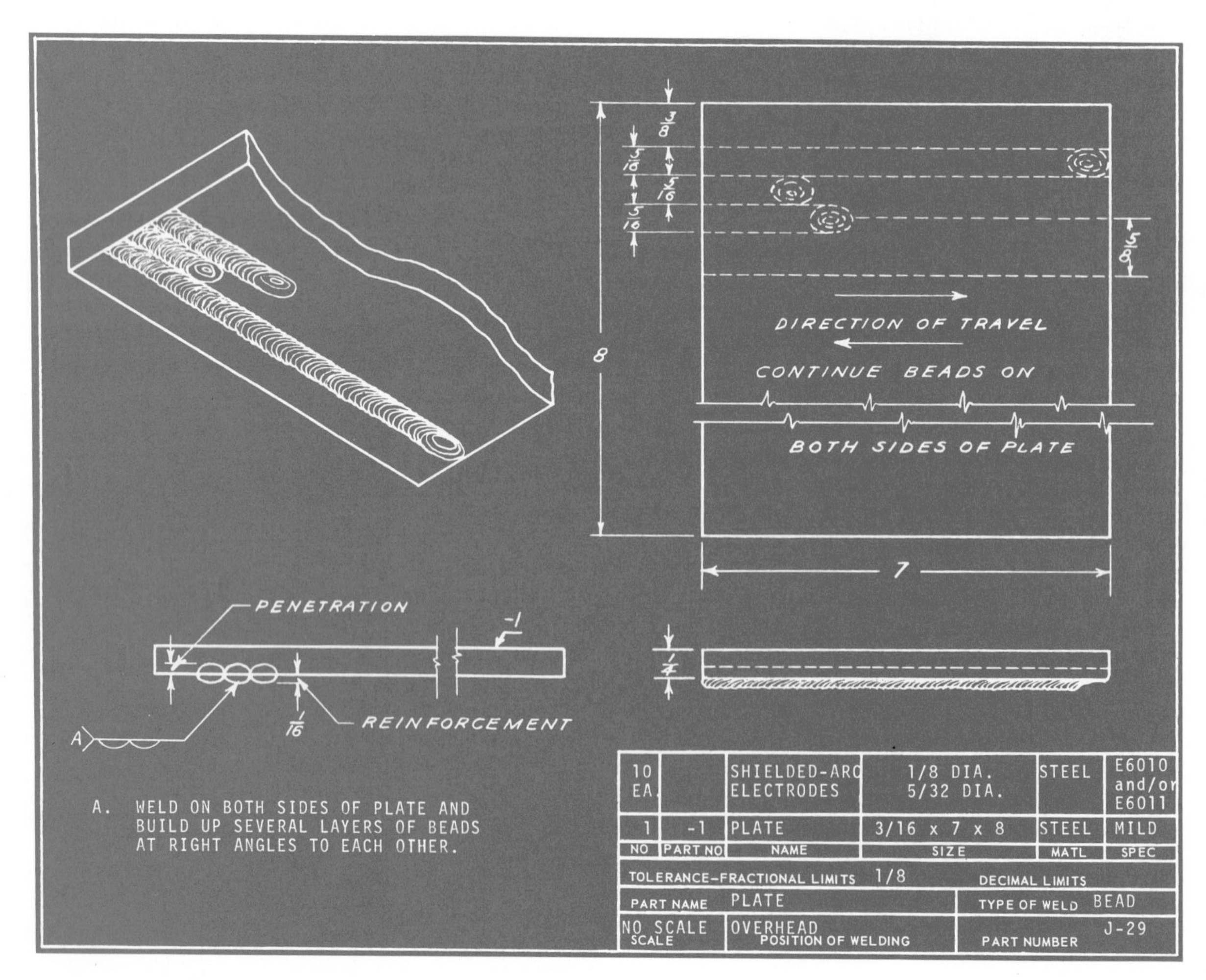

3. Lay out parallel lines as shown on the job print.
4. Mark the lines with a center punch.
5. Obtain electrodes of each quantity, type, and size specified in the job print.
6. Fasten the plate in the overhead position. Use an overhead welding jig, Fig. 21-5.
7. Set the power source at 110 to 170 amperes. Set a d.c. power source for reverse polarity.
8. Make stringer beads on $\frac{5}{8}$-inch center lines as shown on the job print. Manipulate the electrode as instructed in Figs. 21-2 through 21-4.
9. Chip the slag from the beads, brush, and inspect. Refer to Inspection, below.
10. Make additional stringer beads between the beads already deposited as shown on the job print.
11. Chip the slag from the beads, brush, and inspect. Refer to Inspection, below.
12. Practice these beads until you can produce good beads consistently with both types of electrodes and with $\frac{5}{32}$-inch diameter electrodes.

Inspection

Compare the beads with Fig. 21-6 and check them for the following weld characteristics:

Width and height: Uniform
Appearance: Smooth with close ripples; free of voids and high spots. Restarts should be difficult to locate.
Size: Refer to the job print.

21-3. *Position of the electrode and electrode holder when welding in the overhead position: travel toward the operator.*

21-5. *Overhead welding jig.*

21-4. *Basic motions when welding in the overhead position.*

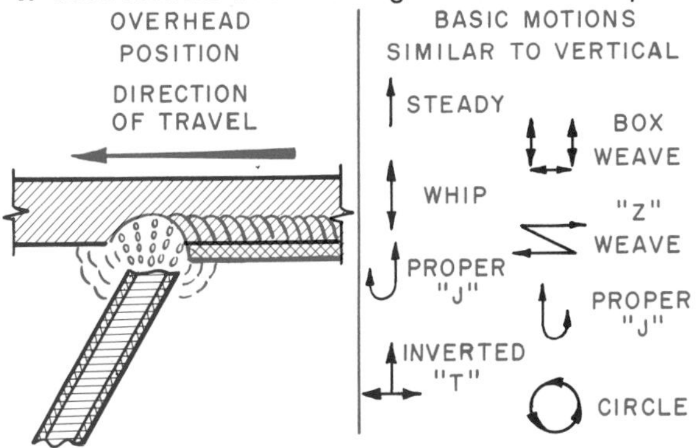

Face of beads: Slightly convex
Edges of beads: Good fusion, no overlap, no undercut
Starts and stops: Free of depressions and high spots
Beginnings and endings: Full size, craters filled
Penetration and fusion: To plate surface and adjacent beads
Surrounding plate surfaces: Free of spatter
Slag formation: Full coverage, easily removable

Disposal

Discard completed plates in the waste bin. Plates must be filled on both sides.

JOB 21-J30 STRINGER BEADING

Objective

To deposit stringer beads on flat steel plate in the overhead position with DCSP and/or a.c. shielded-arc electrodes (AWS-ASTM E6012-E6013).

General Job Information

While it is a fact that most overhead welding is done with

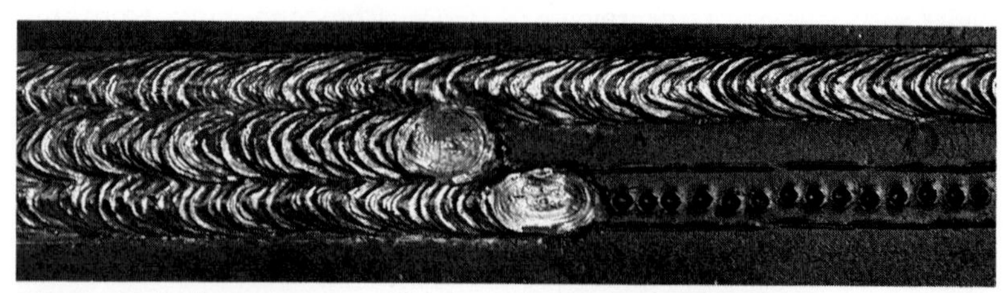

21-6. *Typical appearance of stringer beads welded in the overhead position with DCRP or a.c. shielded-arc electrodes.*

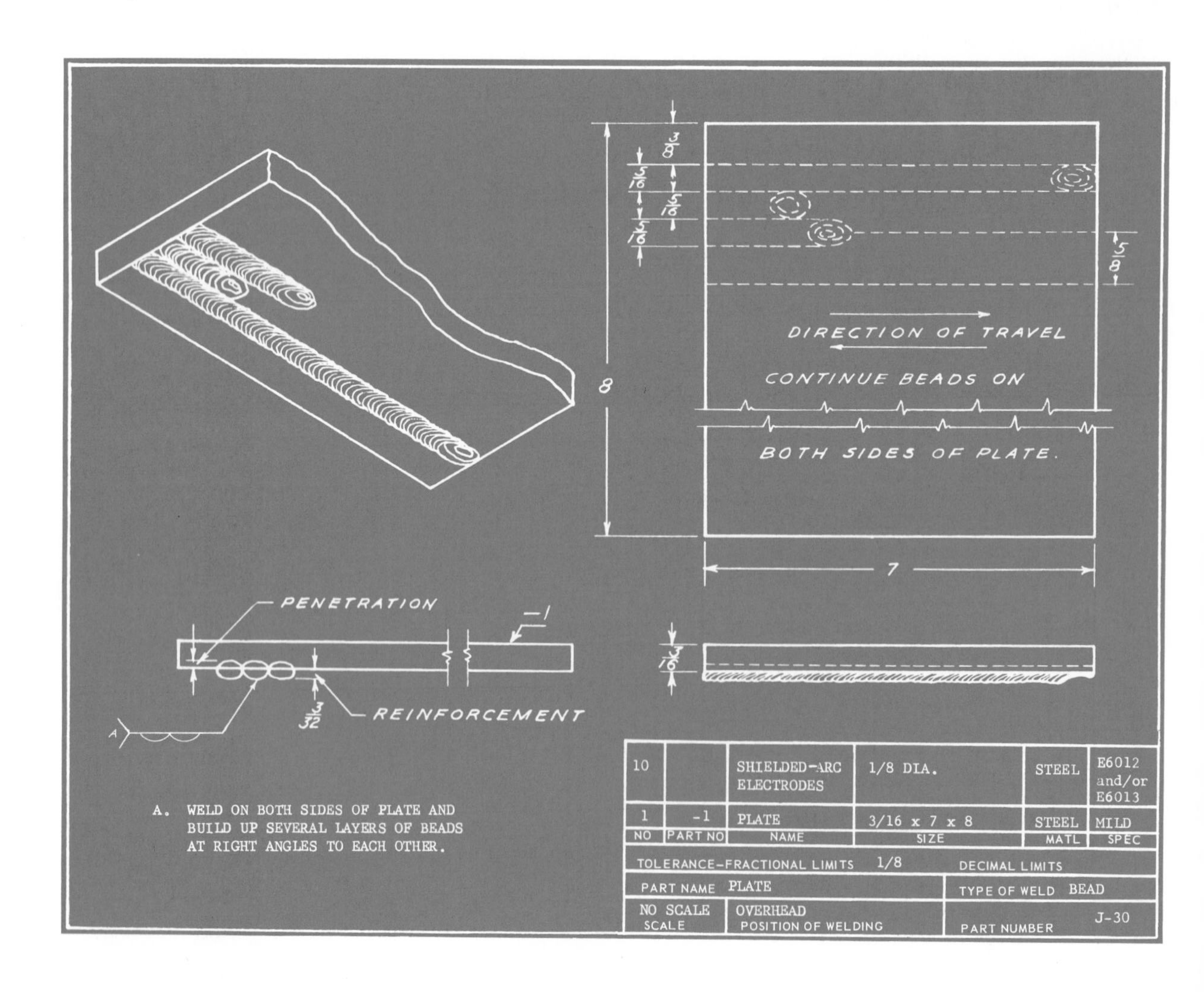

reverse polarity electrodes, there are enough straight polarity electrodes used to justify the training of a student on this electrode in this position.

Welding Technique

Current adjustment should be slightly higher than for reverse polarity electrodes. The electrode position should be approximately that shown in Fig. 21-2, page 636. The arc gap must be close, but it may be somewhat longer than with reverse polarity electrodes. Some operators prefer to advance the electrode with very little movement. Study the basic motions shown in Fig. 21-4, page 637. High reinforcement and heavy slag is a problem and should be watched carefully. There is no difficulty with undercut.

Practice starts and stops. Travel in all directions. Fill both sides of the plate.

Operations

1. Obtain plate; check the job print for size.
2. Obtain a square head and scale, dividers, scribe, and center punch from the toolroom.
3. Lay out parallel lines as shown on the job print.
4. Mark lines with a center punch.
5. Obtain electrodes of each quantity, type, and size specified in the job print.
6. Fasten the plate in the overhead position. Use an overhead welding jig.
7. Set a d.c. power source for straight polarity at 110 to 190 amperes or an a.c. power source at 115 to 210 amperes.
8. Make stringer beads on ⅝-inch center lines as shown on the job print. Manipulate the electrode as shown in Figs. 21-2 through 21-4, pages 636, 637.
9. Chip the slag from the beads, brush, and inspect. Refer to Inspection, below.
10. Make additional stringer beads between the beads already deposited as shown on the job print.
11. Chip the slag from the beads, brush, and inspect. Refer to Inspection, below.
12. Practice these beads until you can produce good beads consistently with both types of electrodes, if possible.

Inspection

Compare the beads with Fig. 21-7 and check them for the following weld characteristics:

Width and height: Uniform
Appearance: Smooth with close ripples; free of voids and high spots. Restarts should be difficult to locate.
Size: Refer to the job print.
Face of beads: Convex (more so than with reverse polarity electrodes)
Edges of beads: Good fusion, no overlap, no undercut
Starts and stops: Free of depressions and high spots
Beginnings and endings: Full size, craters filled
Penetration and fusion: To plate surfaces and adjacent beads
Surrounding plate surfaces: Free of spatter
Slag formation: Full coverage, easily removable

Disposal

Discard completed plates in the waste bin. Plates must be filled on both sides.

JOB 21-J31 WEAVED BEADING

Objective

To deposit weaved beads on flat steel plate in the overhead position with DCRP and/or a.c. shielded-arc electrodes (AWS-ASTM E6010-E6011).

General Job Information

On many jobs it is necessary to make weaved beads in the overhead position. This is more difficult than stringer beading and has to be practiced diligently. The technique is commonly used in pipe welding.

Welding Technique

Current adjustment must not be too high. Hold the electrode at a 90-degree angle to the plate, Fig. 21-8A, and at an angle no more than 10 degrees in the direction of travel, Fig. 21-8B. Weave the electrode from side to side and hesitate at the sides to prevent undercut.

It is essential that the arc gap be close and that the gap be uniform for the entire width of the bead. You may advance the weld by keeping the electrode in the pool, but you must prevent the molten metal from becoming so

21-7. *Typical appearance of stringer beads welded in the overhead position with DCSP or a.c. shielded-arc electrodes.*

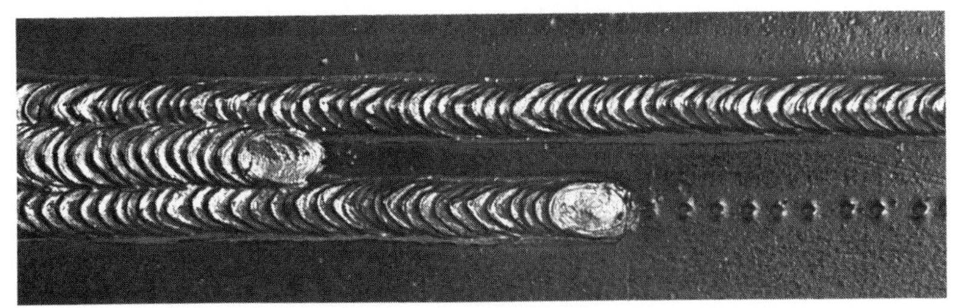

fluid that it sags. If sagging occurs, advance the electrode along either side of the weld in the box weave shown in Fig. 21-4, page 637. Lengthen the arc but do not break it when you correct the sag. Do not keep the arc out of the crater any longer than necessary. Keeping it out too long will cause excess spatter ahead of the weld. Move the electrode across the face of the weld rapidly so that you do not overheat the metal deposited in the middle of the weld and cause it to sag. Making beads wider than $\frac{3}{4}$ inch is not good welding practice.

Practice starts and stops. Travel in all directions. Fill both sides of the plate.

Operations

1. Obtain plate; check the job print for size.

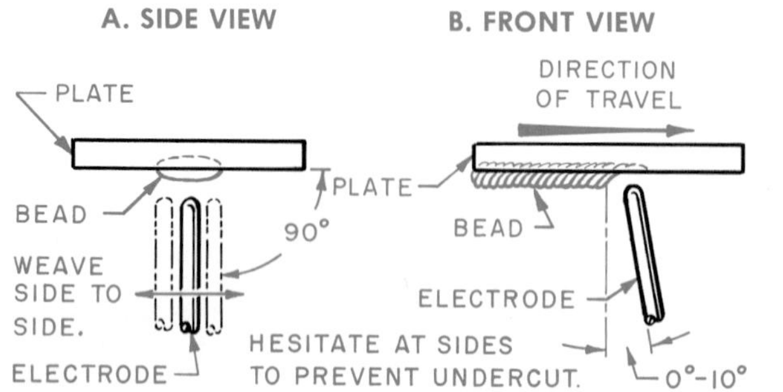

21-8. *Electrode position when making weaved beads in the overhead position.*

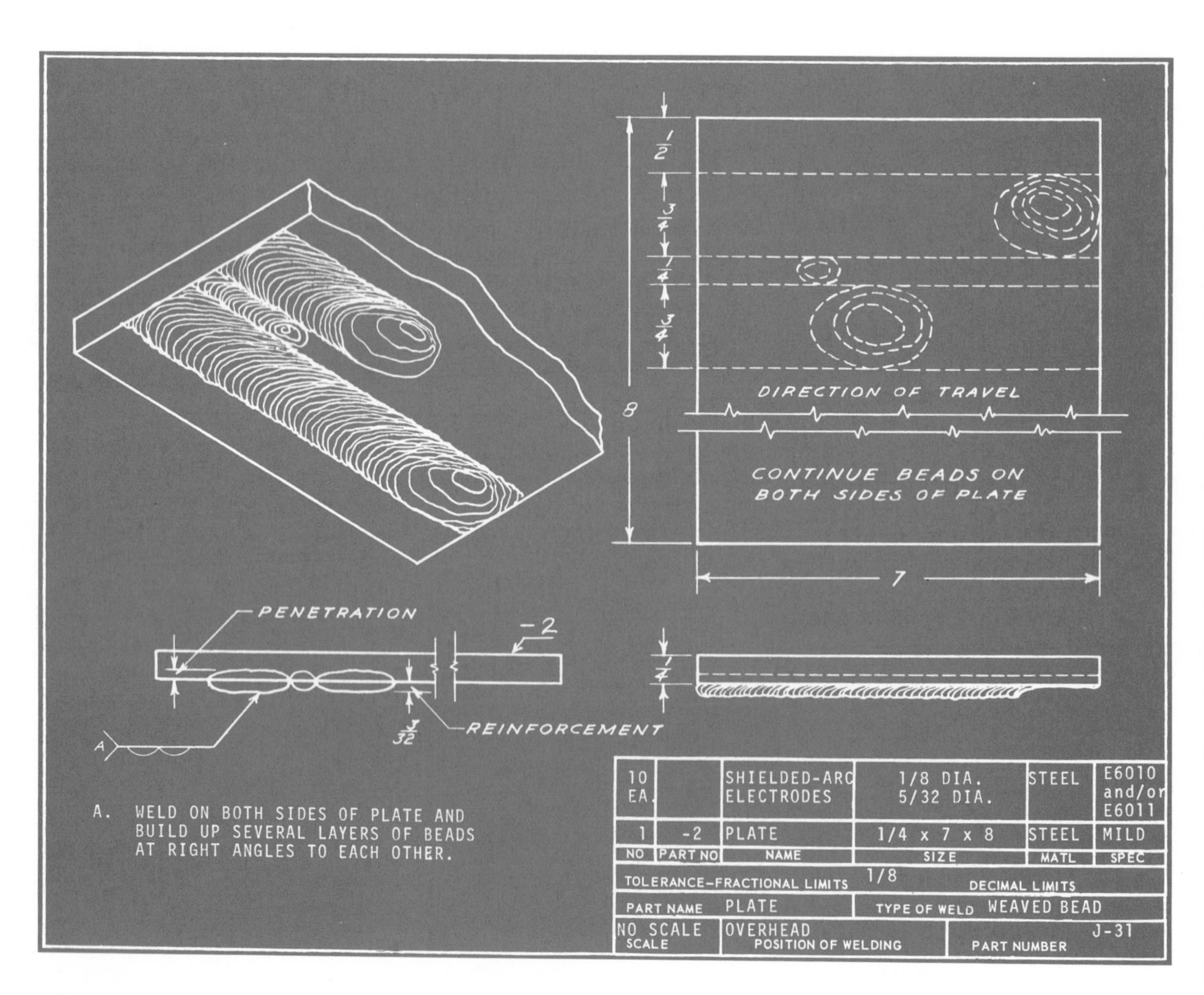

2. Obtain a square head and scale, dividers, scribe, and center punch from the toolroom.

3. Lay out parallel lines as shown on the job print.

4. Mark the lines with a center punch.

5. Obtain electrodes of each quantity, type, and size specified in the print.

6. Fasten the plate in the overhead position. Use an overhead welding jig.

7. Set the power source at 75 to 125 amperes. Set a d.c. power source for reverse polarity.

8. Make weaved beads between the $\frac{3}{4}$-inch parallel lines. Manipulate the electrode as instructed in Fig. 21-8.

9. Chip the slag from the beads, brush, and inspect. Refer to Inspection, below.

10. Make stringer beads between the weaved beads already deposited. Current should be set a little higher (110- to 170-ampere range). Manipulate the electrode as in Figs. 21-2 through 21-4, pages 636, 637.

11. Chip the slag from the beads, brush, and inspect. Refer to Inspection, below.

12. Practice these beads until you can produce good beads consistently with both types of electrodes and with $\frac{5}{32}$-inch diameter electrodes.

Inspection

Compare the beads with Fig. 21-9 and check them for the following weld characteristics:

Width and height: Uniform
Appearance: Smooth with close ripples: free of voids and high spots. Restarts should be difficult to locate.
Size: Refer to the job print.
Face of beads: Slightly convex
Edges of beads: Good fusion, no overlap, no undercut
Starts and stops: Free of depressions and high spots
Beginnings and endings: Full size, craters filled
Penetration and fusion: To plate surfaces and adjacent beads
Surrounding plate surfaces: Free of spatter
Slag formation: Full coverage, easily removable

Disposal

Discard completed plates in the waste bin. Plates must be filled on both sides.

JOB 21-J32 WELDING A SINGLE-V BUTT JOINT (BACKUP STRIP CONSTRUCTION)

Objective

To weld a single-V butt joint assembled with backup strips in the flat position by means of multipass groove welds and weaved beading technique with DCRP and/or a.c. shielded-arc electrodes (AWS-ASTM E6010-E6011).

General Job Information

This form of joint construction is used quite frequently in piping, pressure vessel, and shipyard construction. The backing material creates the effect of welding a joint with a chill ring. The single-V butt joint with backup strips is identical to the test specified by code authorities such as the Bureau of Shipping, Department of the Navy, and recommended by the American Welding Society. The root opening varies with different welding shops. If it is as large as $\frac{3}{8}$ inch, two or three root passes are needed to weld the joint.

Welding Technique

FIRST PASS. Current adjustment can be high. Hold the electrode at a 90-degree angle to the joint. Fig. 21-10A, and at a 10- to 20-degree angle in the direction of travel, Fig. 21-10B. Use the familiar technique for running stringer beads with a very slight weave. Secure good fusion with the backup strip and the root faces of the bevel plates. Keep the face of the weld as flat as possible.

SECOND AND THIRD PASSES. Current adjustment can be high. Electrode position should be approximately that shown in Fig. 21-10. Weave wider beads than for the first pass. Do not advance the electrode rapidly. This will cause coarse ripples and even voids. Keep the side-to-side motion within the limits of the finished bead width to avoid undercutting. Close arc gap control is important. Remove the slag thoroughly so that good fusion may be obtained with preceding passes and with the walls of the beveled plate. Use the edges of the plate as a guide for determining the width of the last pass.

Practice starts and stops. Travel in all directions.

21-9. *Typical appearance of weaved beads welded in the overhead position with DCRP or a.c. shielded-arc electrodes.*

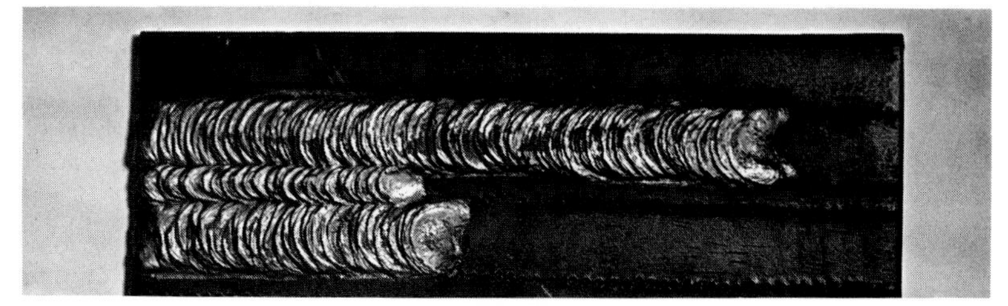

Operations

1. Obtain plates; check the job print for quantity and size.
2. Obtain electrodes of each quantity, type, and size specified in the job print.
3. Set a d.c. power source for reverse polarity at 110 to 170 amperes or an a.c. power source at 115 to 190 amperes.
4. Set up the plates and tack weld them as shown on the job print and Fig. 21-11.
5. Place the joint in the flat position on the welding table. Make sure it is well-grounded.
6. Make the first pass with a $^5/_{32}$-inch electrode. Manipulate the electrode as instructed in Fig. 21-10.
7. Chip the slag from the welds, brush, and inspect. Refer to Inspection.
8. Readjust the power source for 140 to 215 amperes.
9. Make the second and third passes with a $^3/_{16}$-inch electrode.
10. Chip the slag from the welds, brush, and inspect.
11. Practice these welds until you can produce good welds consistently with both types of electrodes and with $^3/_{16}$-inch electrodes for the first pass and $^1/_4$-inch electrodes for the others.

Inspection

Compare the weld with Figs. 21-11 and 21-12 and check each pass for the following weld characteristics:

Width and height: Uniform

Appearance: Smooth with close ripples; free of voids and high spots. Restarts should be difficult to locate.

Size: Refer to the job print: check with butt weld gauge.

Face of welds: First two passes flat; last pass slightly convex

Edges of welds: Good fusion, no overlap, no undercut

Starts and stops: Free of depressions and high spots

Beginnings and endings: Full size, craters filled

Penetration and fusion: To the backup strip, preceding passes, and plate surfaces

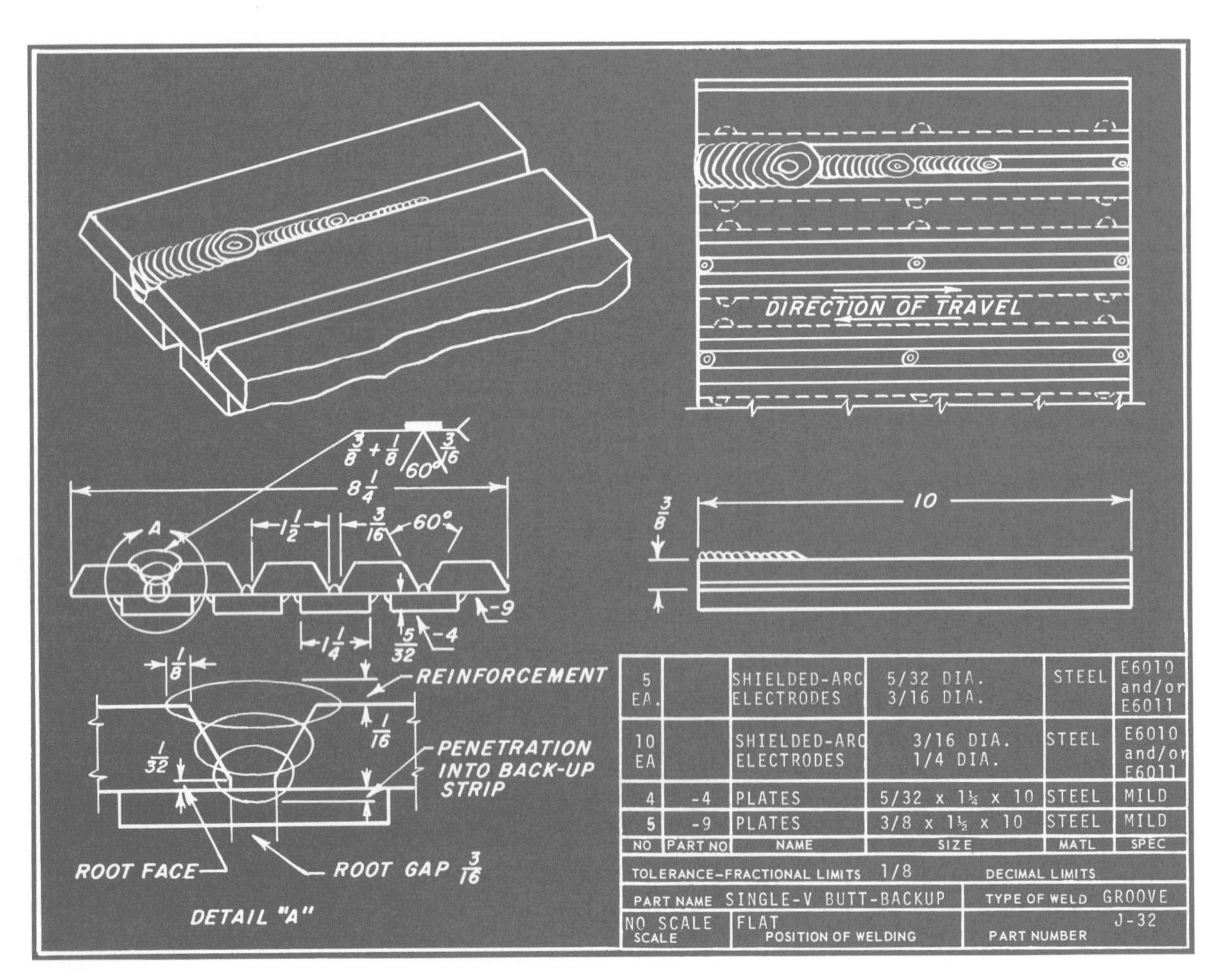

NO	PART NO	NAME	SIZE	MATL	SPEC
5 EA.		SHIELDED-ARC ELECTRODES	5/32 DIA. 3/16 DIA.	STEEL	E6010 and/or E6011
10 EA		SHIELDED-ARC ELECTRODES	3/16 DIA. 1/4 DIA.	STEEL	E6010 and/or E6011
4	-4	PLATES	5/32 x 1¼ x 10	STEEL	MILD
5	-9	PLATES	3/8 x 1½ x 10	STEEL	MILD

TOLERANCE–FRACTIONAL LIMITS 1/8 | DECIMAL LIMITS

PART NAME SINGLE-V BUTT-BACKUP | TYPE OF WELD GROOVE

SCALE NO SCALE | POSITION OF WELDING FLAT | PART NUMBER J-32

Surrounding plate surfaces: Free of spatter
Slag formation: Full coverage, easily removable

Disposal

Put completed joints in the scrap bin so that they will be available for further use. The plate can be cut and beveled between welds for further butt welding, and unwelded plate surfaces can be used for beading.

Qualifying Test

At this point you should be able to pass Test 1 in the flat position. (See Chapter 22, pages 714–717). This is identical to Test 1, Department of the Navy Requirements, and the test recommended by the American Welding Society.

It will also be desirable for you to pass Test 3 (fillet weld) in the flat position. (See Chapter 22, pages 717, 718).

JOB 21-J33 WELDING A T-JOINT

Objective

To weld a T-joint in the horizontal position by means of multipass fillet welds, weaved-bead technique, with DCRP and/or a.c. shielded-arc electrodes (AWS-ASTM E6010-E6011).

General Job Information

It is often the practice to make large fillet welds in the horizontal position with few passes and the weaved-bead technique. This is especially true in shipyard and construction work. This technique can deposit more weld metal in a single pass than other methods.

Welding Technique

Current adjustment should be high. Hold the electrode at a 40-degree angle to the flat plate, Fig. 21-13A, and at a 30-degree angle in the direction of travel Fig. 21-13B. The first pass should be run in the same manner as any other single-pass fillet weld. Keep the face of weld as flat as possible. Make sure that penetration to the root of the joint is secured and that there is no undercut.

The second pass is weaved over the first pass. Direct the electrode towards the vertical

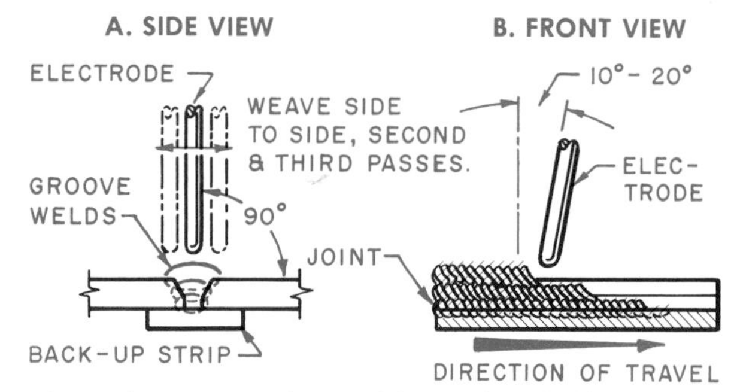

21-10. *Electrode position when welding a beveled-butt joint in the flat position.*

21-11. *Backing strip applied to the back side of a beveled-butt joint.*

21-12. *Joint construction and appearance of a typical multipass weld in a beveled-butt joint with a backup strip welded in the flat position with DCRP or a.c. shielded-arc electrodes.*

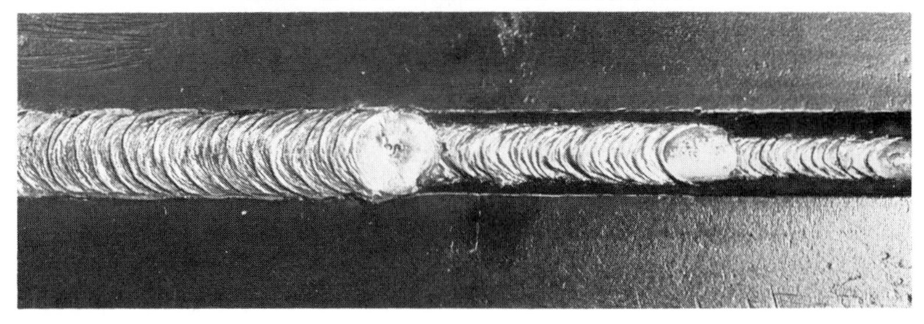

plate so that the weld metal is forced up on its surface. To prevent undercut, keep the electrode within the limits of the desired weld width. Use a very short arc and pause at the top of the bead against the vertical plate. Then bring the arc toward the bottom plate at a normal rate of travel and at a 30-degree angle. Return the arc to the top of the fillet weld with a semicircular, counterclockwise motion. Secure good fusion with the preceding pass and the plate surfaces.

Practice starts and stops. Travel in all directions.

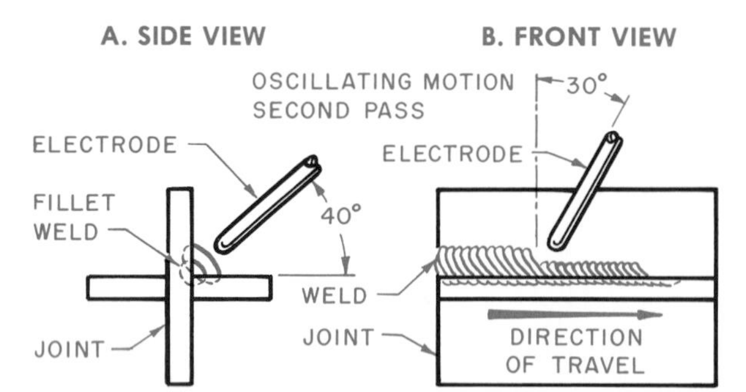

21-13. *Electrode position when welding a T-joint in the horizontal position.*

Operations

1. Obtain plates; check the job print for quantity and size.
2. Obtain electrodes of the type and size specified in the job print.
3. Set a d.c. power source for reverse polarity at 140 to 215

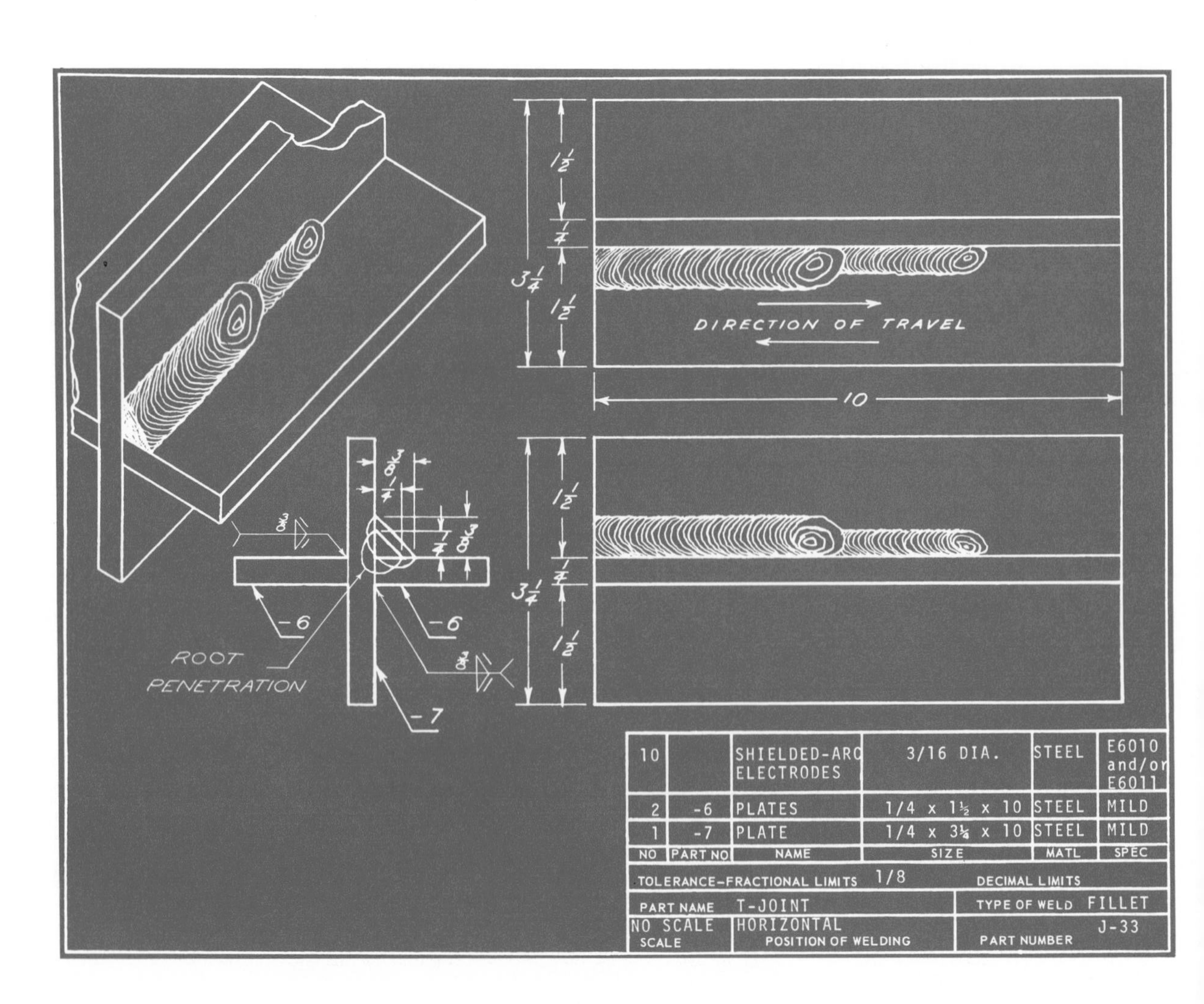

NO	PART NO	NAME	SIZE	MATL	SPEC
10		SHIELDED-ARC ELECTRODES	3/16 DIA.	STEEL	E6010 and/or E6011
2	-6	PLATES	1/4 x 1½ x 10	STEEL	MILD
1	-7	PLATE	1/4 x 3¼ x 10	STEEL	MILD

TOLERANCE-FRACTIONAL LIMITS 1/8 DECIMAL LIMITS
PART NAME T-JOINT TYPE OF WELD FILLET
SCALE NO SCALE POSITION OF WELDING HORIZONTAL PART NUMBER J-33

amperes or an a.c. power source at 140 to 225 amperes.

4. Set up the plates as shown on the job print and tack weld them at each corner.

5. Place the joint in the horizontal position on the welding table. Make sure that it is well-grounded.

6. Make the first pass as shown on the job print. Manipulate the electrode as instructed in Fig. 21-13.

7. Chip the slag from the weld, brush, and inspect. Refer to Inspection, below.

8. Make the second pass as shown on the job print. Be careful with your current setting and electrode movement.

9. Chip the slag from the weld, brush, and inspect. Refer to Inspection, below.

10. Practice these welds until you can produce good welds consistently with both types of electrodes.

Inspection

Compare each pass with Fig. 21-14 and check it for the following weld characteristics:

Width and height: Uniform
Appearance: Smooth with close ripples, free of voids. Restarts should be difficult to locate.
Size: Refer to the job print. Check with a fillet weld gauge.
Face of weld: Flat to slightly convex
Edges of weld: Good fusion, no overlap, no undercut
Starts and stops: Free of depressions and high spots
Beginnings and endings: Full size, craters filled
Penetration and fusion: To the root of the joint and plate surfaces
Surrounding plate surfaces: Free of spatter
Slag formation: Full coverage, easily removable

Disposal

Put completed joints in the scrap bin so that they will be available for further use. Square edges can be butted together for square butt joints, and unwelded surfaces can be used for beading.

JOB 21-J34 WELDING A SINGLE-V BUTT JOINT (BACKUP STRIP CONSTRUCTION)

Objective

To weld a single-V butt joint (backup strip construction) in the horizontal position by means of multipass groove welds, stringer bead technique, with DCRP and/or a.c. shielded-arc electrodes (AWS-ASTM E6010-E6011).

General Job Information

This joint design and position will be encountered in pipe, tank, and shipyard welding. The single-V butt joint is identical to the joint which is required by various code authorities such as the Bureau of Shipping, Department of the Navy, and recommended by the American Welding Society. The root opening varies with different welding shops. If it is as large as $^3/_8$ inch, two or three root passes are needed to weld the joint.

Welding Technique

FIRST PASS. Current adjustment must be high. Hold the electrode at a 90-degree angle to the joint, Fig. 21-15A. Use the stringer bead technique. Make sure that you are obtaining fusion with the backup strip and root faces of the plates.

ALL OTHER PASSES. Current adjustment can be higher than for the first pass. Hold the electrode at a 30-degree angle to the lower plate for the second pass (Fig. 21-15B) and a 45-degree angle for the third pass (Fig. 21-15C). Stringer bead technique may be used. It is important that all passes be fused to the preceding passes and plate surfaces. Avoid sagging and undercut.

Four cover passes are needed to fill up the groove completely. They must be at least flush with the plate and should have a slight reinforcement. Hold the

21-14. *Typical appearance of multipass fillet welds in a T-joint welded in the horizontal position, modified weaved technique, with DCRP or a.c. shielded-arc electrodes.*

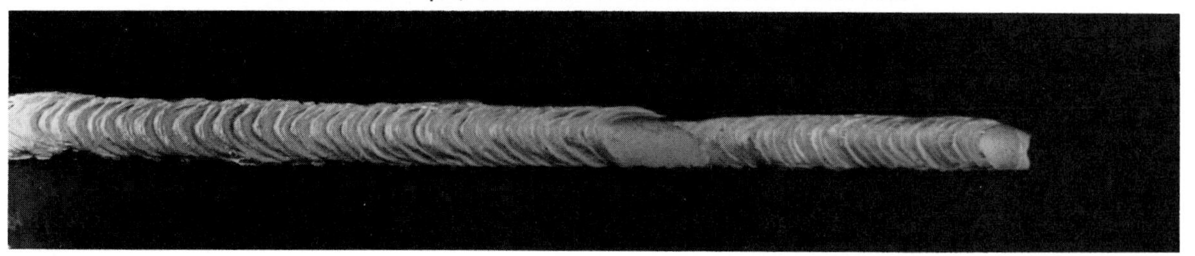

electrode at an angle of 5 to 15 degrees from the horizontal for the fourth, fifth, sixth, and seventh passes, Fig. 21-15D. Make sure that the fourth and seventh passes do not extend more than $\frac{1}{8}$ inch beyond the edge of the beveled plate. In this way the edge of the beveled plate is your guide for the final width of the cover passes.

21-15. *Electrode position when welding a beveled-butt joint in the horizontal position.*

For all passes tilt the electrode 5 to 15 degrees in the direction of travel.

Practice starts and stops. Travel from left to right and from right to left. Very thorough cleaning between passes is essential.

Operations

1. Obtain plates; check the job print for quantity and size.
2. Obtain electrodes of each quantity, type, and size specified in the job print.
3. Set a d.c. power source for reverse polarity at 110 to 170

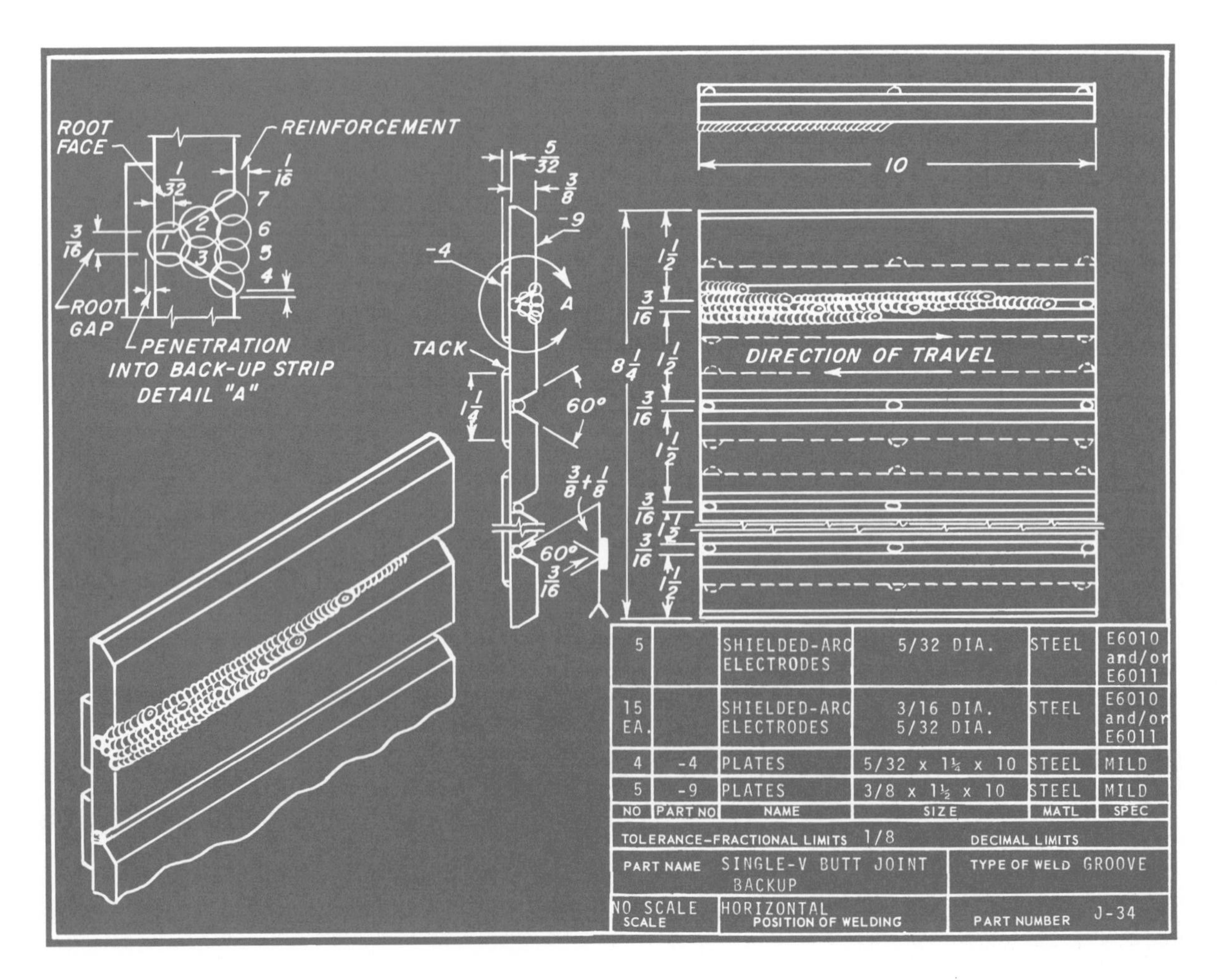

NO	PART NO	NAME	SIZE	MATL	SPEC
5		SHIELDED-ARC ELECTRODES	5/32 DIA.	STEEL	E6010 and/or E6011
15 EA.		SHIELDED-ARC ELECTRODES	3/16 DIA. 5/32 DIA.	STEEL	E6010 and/or E6011
4	-4	PLATES	5/32 x 1¼ x 10	STEEL	MILD
5	-9	PLATES	3/8 x 1½ x 10	STEEL	MILD

TOLERANCE-FRACTIONAL LIMITS 1/8 DECIMAL LIMITS
PART NAME SINGLE-V BUTT JOINT BACKUP
TYPE OF WELD GROOVE
SCALE NO SCALE
POSITION OF WELDING HORIZONTAL
PART NUMBER J-34

amperes or an a.c. power source at 110 to 190 amperes.

4. Set up the plates and tack weld them as shown on the job print.

5. Place the joint in the horizontal position on the welding table. Make sure that it is well-grounded.

6. Make the first, second, and third passes with $^{5}/_{32}$-inch electrodes as shown on the job print. Manipulate the electrode as instructed in Fig. 21-15A, B, and C.

7. Chip the slag from the welds, brush, and inspect between each pass. Refer to Inspection, below.

8. Increase the current setting to 140 to 215 amperes.

9. Make the fourth, fifth, sixth, and seventh passes with $^{3}/_{16}$ inch electrodes as shown on the job print. Manipulate the electrode as instructed in Fig. 21-15D and E.

10. Chip the slag from the welds, brush, and inspect between each pass. Refer to Inspection, below.

11. Practice these welds until you can produce good welds consistently with both types of electrodes. Practice with $^{3}/_{16}$-inch electrodes for all passes.

Inspection

Compare each pass with Fig. 21-16 and check it for the following weld characteristics:

Width and height: Uniform

Appearance: Smooth with close ripples; free of voids and high spots. The lapover of the beads should be well-proportioned. Restarts should be difficult to locate.

Size of weld: Refer to the job print. Check convexity with a butt weld gauge.

Face of weld: Some reinforcement

Edges of weld: Good fusion, no overlap, no undercut

Starts and stops: Free of depressions and high spots

Beginnings and endings: Full size, craters filled

Penetration and fusion: To the backup strip, preceding passes, and plate surfaces

Surrounding plate surfaces: Free of spatter

Slag formation: Full coverage, easily removable

Disposal

Put completed joints in the scrap bin so that they will be available for further use. The plates can be cut and beveled between welds for further butt welding, and unwelded plate surfaces can be used for beading.

Qualifying Test

At this point you should be able to pass Test 1 in the horizontal position. (See Chapter 22, pages 714–717). This is identical to Test 1, Department of the Navy Requirements, and the test recommended by the American Welding Society.

It will also be desirable for you to pass Test 3 (fillet weld) in the horizontal position. (See Chapter 22, pages 717, 718).

JOB 21-J35 WELDING A SQUARE BUTT JOINT

Objective

To weld a square butt joint in the flat position from both sides of the plate with DCRP and/or a.c. shielded-arc electrodes (AWS-ASTM E6010-E6011).

General Job Information

The square butt joint is widely used in industry on ordinary work. When welded from both sides on thicknesses of metal not exceeding $^{1}/_{4}$ inch, it is a highly efficient joint. The usual practice, however, is to weld the joint from one side only. If this procedure is followed, the strength of the joint varies with the depth of penetration which, in turn, depends on the size of electrode, the amount of current, the amount of gap when setting up the plates, and the thickness of plates. When welding from one side, complete penetration without gapping is doubtful on plates heavier than $^{3}/_{16}$ inch.

Welding Technique

The heat setting should be high. Hold the electrode at a 90-degree angle to the plates. Fig. 21-17A, and tilt it 10 to 20 de-

21-16. *Joint construction and typical appearance of a multipass groove weld in a beveled-butt joint welded in the horizontal position with DCRP or a.c. shielded-arc electrodes.*

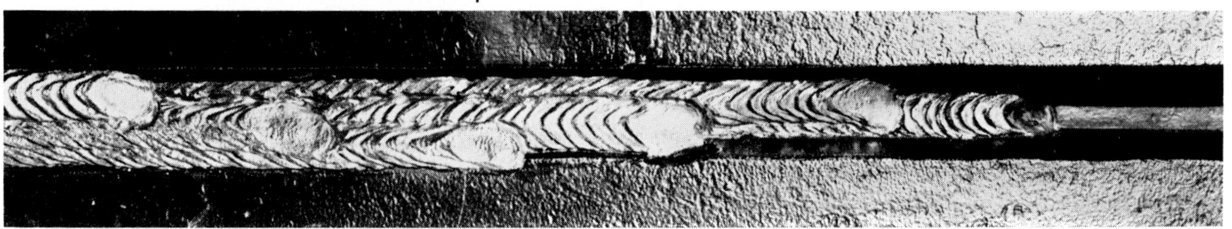

grees in the direction of travel, Fig. 21-17B. Move the electrode back and forth along the line of weld. This preheats the metal ahead of the weld and minimizes the tendency to burn through. It also forces the slag back over the top of the weld, thus lessening the danger of forming slag inclusions. It is important that the rate of travel and the arc gap be steady and uniform. If travel is too fast, undercut will result from the insufficient buildup of weld metals. If travel is too slow, the pool will become too hot and burn through. A long arc gap causes poor appearance, poor penetration, excess spatter, and weld metal with poor physical characteristics. The arc must not be held so short, however, that the slag touches the molten pool.

Practice starts and stops. Weld travel should be in all directions.

Operations

1. Obtain plates; check the job print for quantity and size.
2. Obtain electrodes of each quantity, type, and size specified in the job print.
3. Set a d.c. power source for reverse polarity at 160 to 190 amperes or an a.c. power source at 160 to 210 amperes.
4. Set up the plates and tack weld them as shown on the job print and Fig. 21-18.
5. Place the joint in the flat position on the welding table. Make sure the plates are well-grounded.
6. Make the first pass as shown on the job print. Manipulate the electrode as shown in Fig. 21-17.
7. Chip the slag from the welds, brush, and inspect. Refer to Inspection.
8. Clean the back side of the first pass before welding the second pass.
9. Make the second pass as shown on the job print. Current should be higher.
10. Chip the slag from the welds, brush, and inspect. Refer to Inspection.

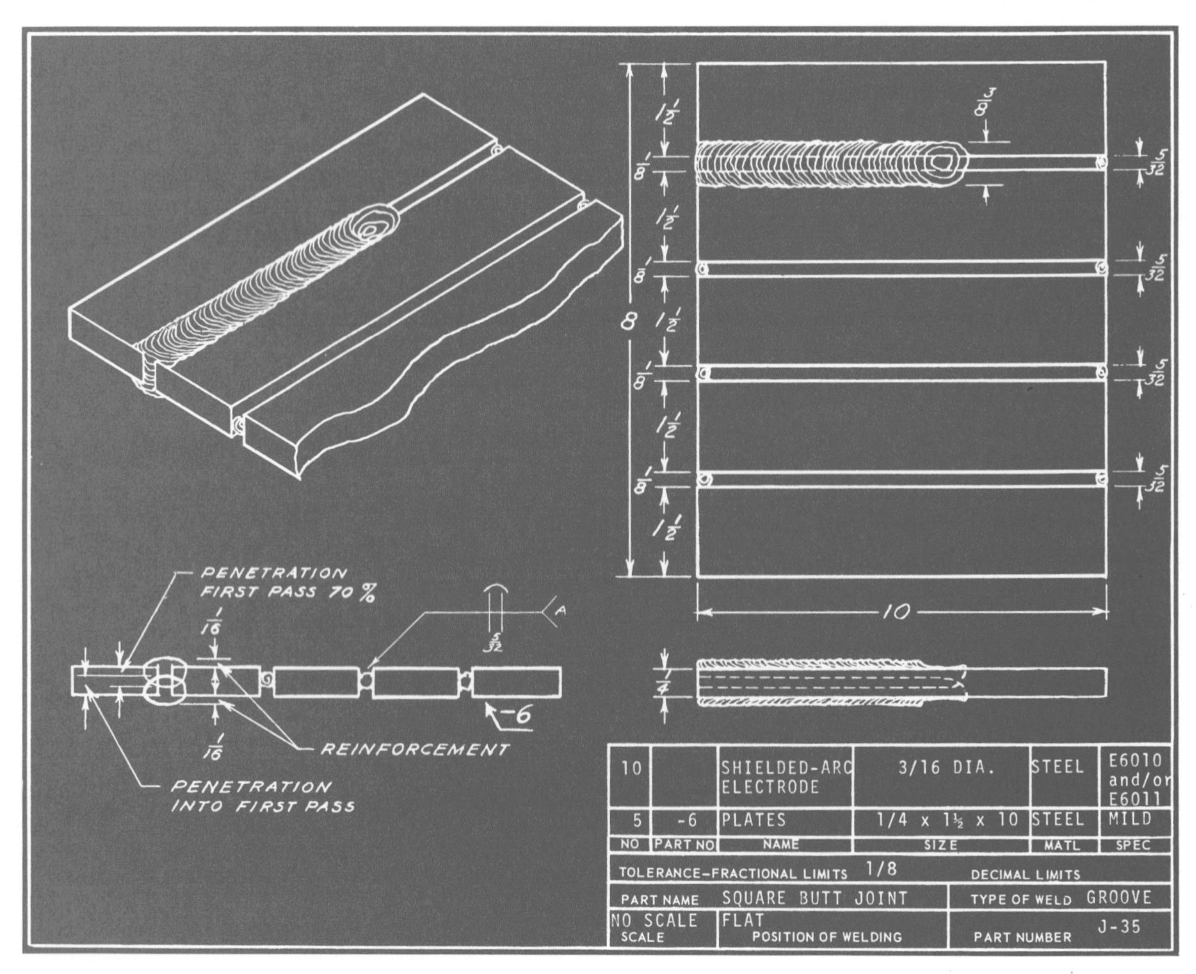

PENETRATION FIRST PASS 70 %
REINFORCEMENT
PENETRATION INTO FIRST PASS
10
8
10 | SHIELDED-ARC ELECTRODE | 3/16 DIA. | STEEL | E6010 and/or E6011
5 | -6 | PLATES | 1/4 x 1½ x 10 | STEEL | MILD
NO | PART NO | NAME | SIZE | MATL | SPEC
TOLERANCE–FRACTIONAL LIMITS 1/8
DECIMAL LIMITS
PART NAME SQUARE BUTT JOINT
TYPE OF WELD GROOVE
NO SCALE
SCALE
FLAT
POSITION OF WELDING
PART NUMBER J-35

11. Practice these welds until you can produce good welds consistently with both types of electrodes.

Inspection

Compare the welds with Fig. 21-18 and check them for the following characteristics:

Width and height: Uniform
Appearance: Smooth with close ripples; free of voids and high spots. Restarts should be difficult to locate.
Size: Refer to the job print. Check with a butt weld gauge.
Face of welds: Slightly convex
Edges of welds: Good fusion, no overlap, no undercut
Starts and stops: Free of depressions and high spots
Beginnings and endings: Full size, craters filled
Penetration and fusion: To the root of the weld and plate surfaces
Surrounding plate surfaces: Free of spatter
Slag formation: Full coverage, easily removable

Disposal

Put completed plates in the scrap bin. They can be used for beading practice.

JOB 21-J36 WELDING AN OUTSIDE CORNER JOINT

Objective

To weld an outside corner joint in the flat position by means of multipass groove welds, weaved beading technique, with DCRP and/or a.c. shielded-arc electrodes (AWS-ASTM E6010-E6011).

General Job Information

The outside corner joint is not welded as frequently as butt, lap, and T-joints. It is a good joint to practice on since it can be prepared quickly. Welding conditions are similar to those for V-groove butt joints. To develop maximum strength, full penetration must be secured through the back side. The addition of a fillet weld in the inside corner adds greatly to efficiency of the corner joint. Joint preparation is inexpensive, but electrode costs are high for heavy plate.

Welding Technique

FIRST PASS. Current adjustment should not be too high. Hold the electrode at a 90-degree angle to the line of weld, Fig. 21-19A, and tilt it 10 to 20 degrees in the direction of welding, Fig. 21-19B. Use the stringer bead welding technique to deposit the first pass in the bottom of the groove. This weld must penetrate through to the back side and have good fusion to both plates, Fig. 21-20. A close arc is essential. The correct combination of welding current, electrode position, and speed of travel will produce penetration through to the back side. To achieve complete penetration, form a little hole at the leading edge of the weld crater right under the tip of the electrode. Learn how to keep this hole throughout the entire welding operation. Do not let it get so large that you lose control and burn through the joint. The presence of this hole during welding is your assurance that you are melting through to the back side of the groove.

If you do burn through the joint, move the electrode in a whiplike motion to control the size of the weld pool and the rate at which weld metal is added. If the pool is too hot and burning through, whip ahead along one of the plate edges about ½ inch and whip back into the pool to

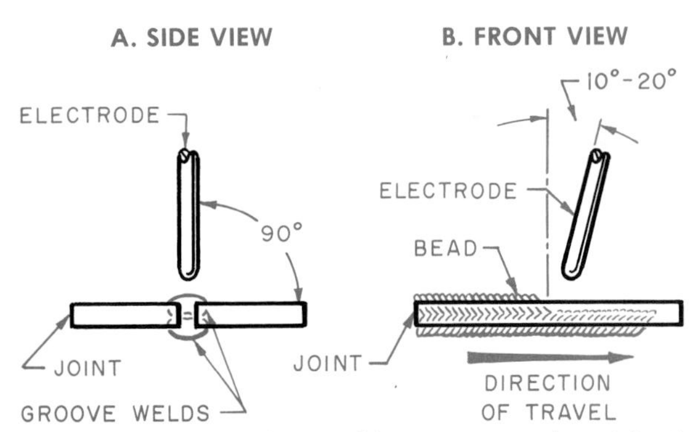

21-17. *Electrode position when welding a square butt joint in the flat position.*

21-18. *Joint construction and typical appearance of a groove weld in a square butt joint welded in the flat position with DCRP or a.c. shielded-arc electrodes.*

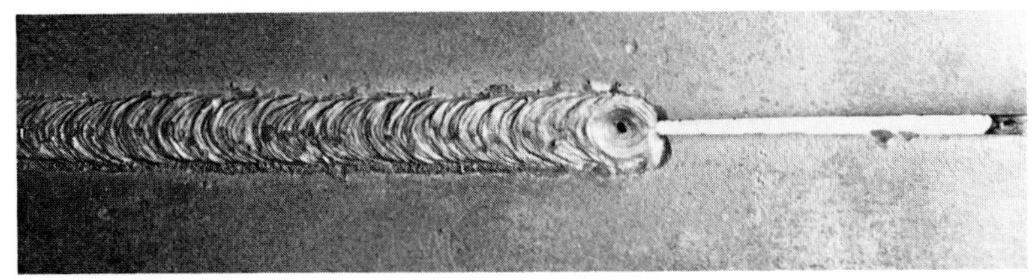

deposit more metal. Alternate this whiplike motion along opposite plate edges.

If the bottom edges of the groove do not seem to be heating and melting enough and if you are having difficulty maintaining the hole, keep the electrode directly in the bottom of the V-groove, slow the forward movement, and maintain a circular motion.

SECOND AND THIRD PASSES. Current adjustment can be hot. Electrode position should be approximately that shown in Fig. 21-19. The weaved beading technique may be used. The third pass requires a wider weave than the second. Do not weave past the edges of the plate and hesitate at the sides to eliminate burning away of the plate edges. Make certain that you are securing good fusion with the preceding passes and with both plate surfaces. Do not build reinforcement too high on the last pass. Remove the slag thoroughly between each pass. Practice starts and stops. Travel in all directions.

Operations

1. Obtain plates; check the job print for the correct quantity and size.
2. Obtain electrodes of each quantity, type, and size specified in the job print.
3. Set the power source for 75 to 125 amperes for the first pass. Set a d.c. power source for reverse polarity.
4. Set up the plates and tack weld them as shown on the job print and in Fig. 21-21.
5. Place the joint in the flat position on the welding table. Make sure it is well-grounded.
6. Make the first pass with a $\frac{1}{8}$-inch electrode as shown on the job print. Manipulate the electrode as instructed in Fig. 21-19.

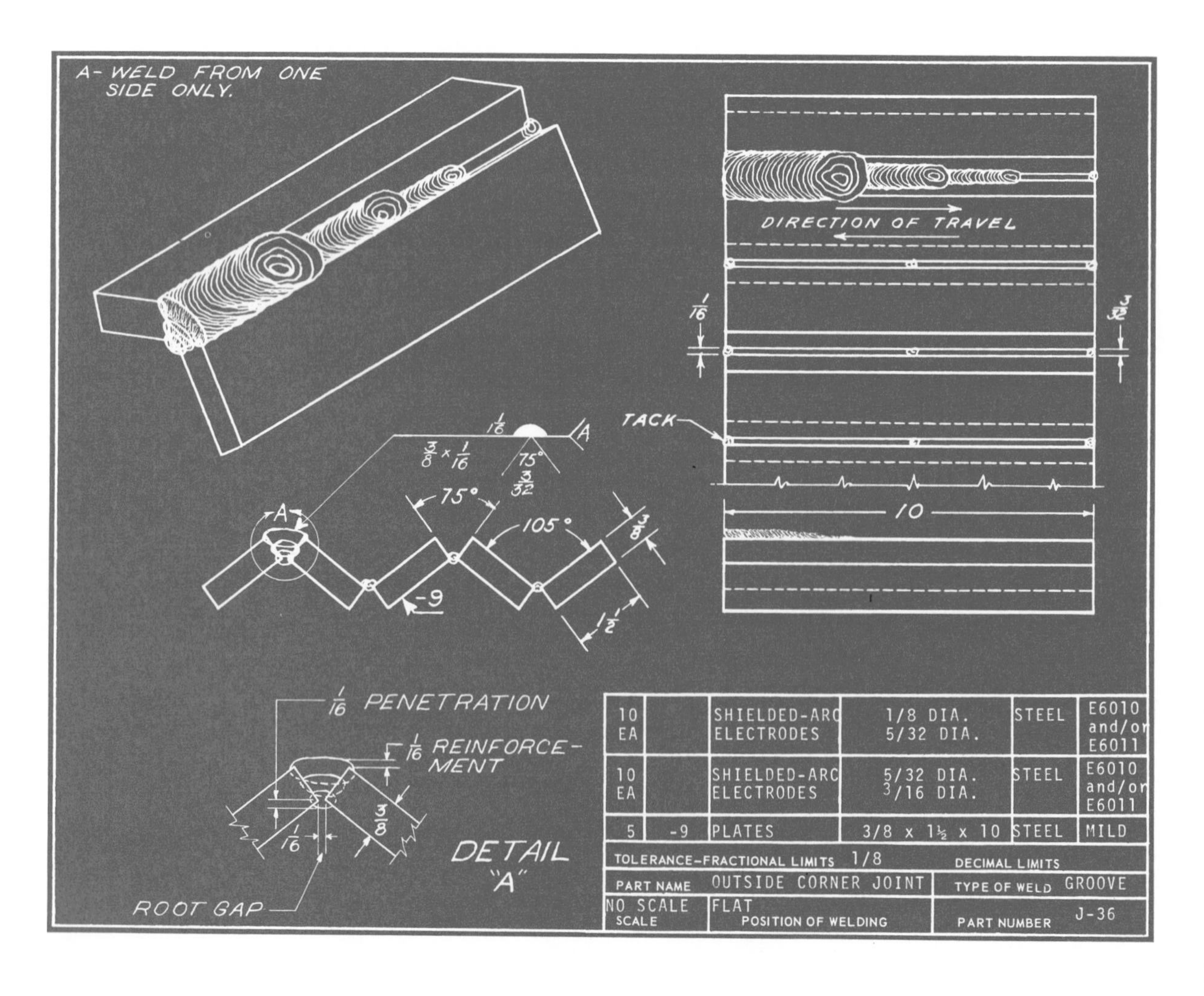

7. Chip the slag from the welds, brush, and inspect. Refer to Inspection.

8. Readjust the power source for 110 to 170 amperes.

9. Make the second and third passes with a $\frac{5}{32}$-inch electrode as shown on the job print. Manipulate the electrode as instructed in Fig. 21-19.

10. Chip the slag from the welds, brush, and inspect. Refer to Inspection.

11. Practice these welds until you can produce good welds consistently with both types of electrodes. Also practice with $\frac{5}{32}$-inch diameter electrodes for the first pass, and $\frac{3}{16}$-inch diameter electrodes for the other passes.

Inspection

Compare the back side of the first bead with Fig. 21-20 and all passes with Fig. 21-21. Check them for the following weld characteristics:

Width and height: Uniform
Appearance: Smooth with close ripples; free of voids and high spots. Restarts should be difficult to locate.
Size: Refer to the job print.
Face of welds: First two passes flat; last pass slightly convex
Edges of welds: Good fusion, no overlap, no undercut; no excess metal at edges of joint
Starts and stops: Free of depressions and high spots
Beginnings and endings: Full size, craters filled
Back side: Complete and uniform penetration (See Fig. 21-20.)
Penetration and fusion: Through the back side, preceding passes, and plate surfaces
Surrounding plate surfaces: Free of spatter
Slag formation: Full coverage, easily removable

Disposal

Put completed joints in the scrap bin so that they will be available for further use. Inside corners can be used for practicing fillet welds, and unwelded plate surfaces can be used for beading.

JOB 21-J37 WELDING A T-JOINT

Objective

To weld a T-joint in the flat position by means of multipass fillet welds, weaved-beading technique, with low hydrogen, iron powder, shielded-arc electrodes (AWS-ASTM E7028).

General Job Information

The E7028 classification is one of the series of electrodes for mild steel that has a low hydrogen coating containing 50 percent iron powder. The E7028 has many of the characteristics of the E7018 with a few exceptions. The E7018 is an all-position elec-

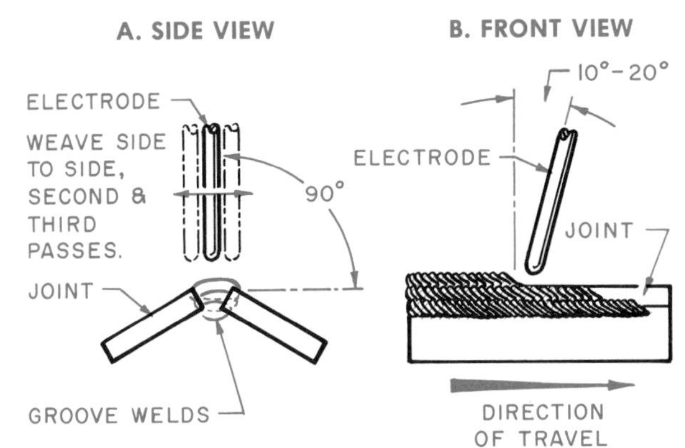

21-19. *Electrode position when welding an outside corner joint in the flat position.*

21-20. *Penetration required through the back side of a corner joint. Note the penetration hole at the leading edge of the weld bead.*

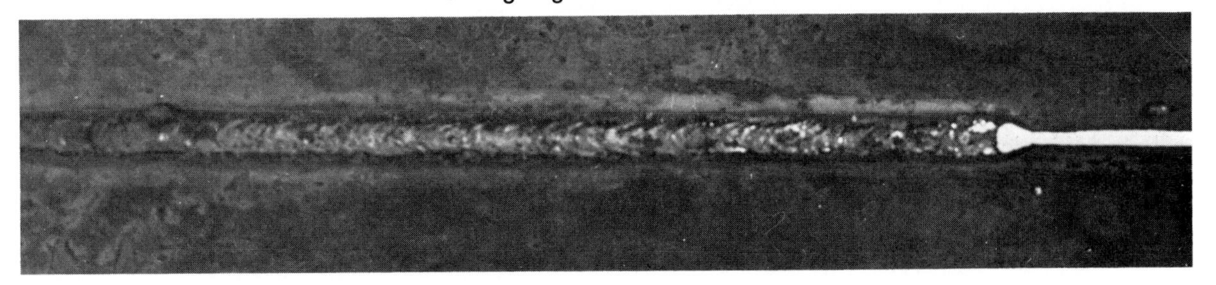

trode, whereas the E7028 electrode is suitable only for horizontal fillet and flat position welding. The covering on the E7028 electrode is approximately 50 percent of the weight of the electrode and is thicker than the covering on the E7018 electrodes. The deposition rate is considerably higher than that of the E7018 due to the iron powder content. The E7028 is a fast-fill type with a soft arc that gives slight penetration and very little spatter. The slag is heavy and highly fluid. It covers the weld completely, but not always uniformly. The face of the weld may be flat or slightly concave with a smooth, glossy appearance. The E7028 can be used for welding mild steels; high strength, low alloy steels; high carbon steels; and high sulfur steels.

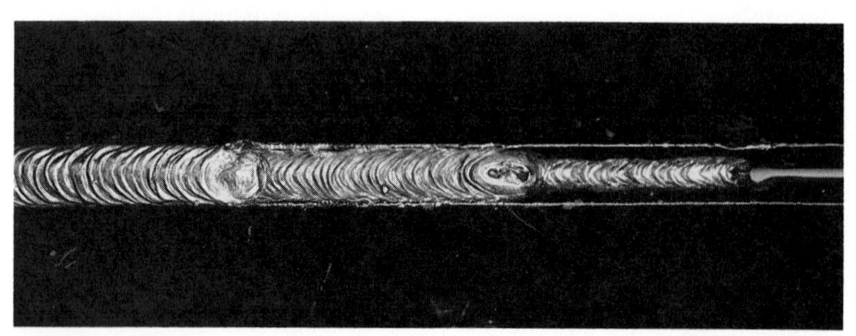

21-21. *Joint construction and typical appearance of a multipass groove weld in an outside corner joint welded in the flat position with DCRP or a.c. shielded-arc electrodes.*

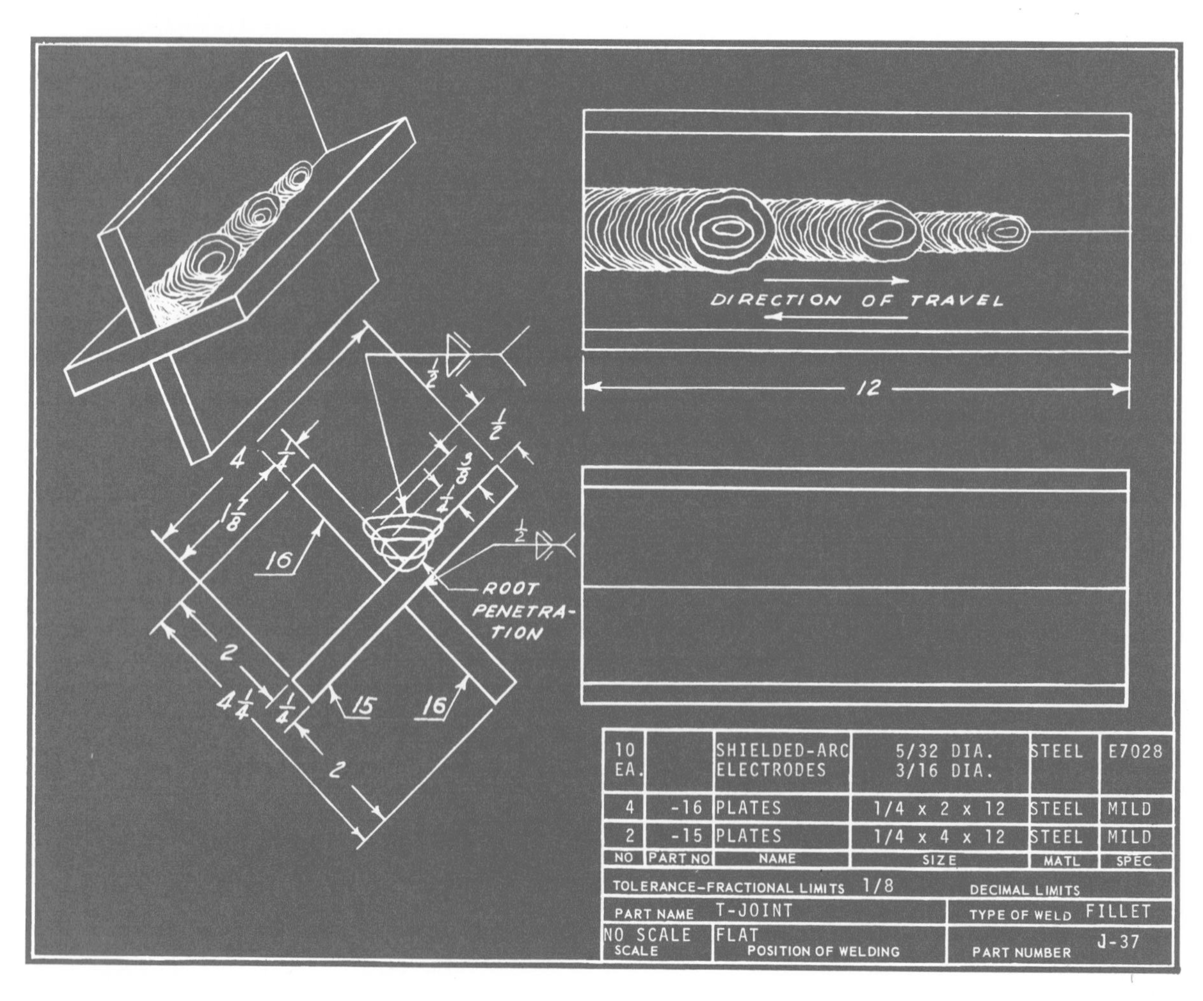

DIRECTION OF TRAVEL
12
ROOT PENETRATION
16
15
2
4
4 1/4
1/4
1/8
1/2
3/8
10 EA | | SHIELDED-ARC ELECTRODES | 5/32 DIA. 3/16 DIA. | STEEL | E7028
4 | -16 | PLATES | 1/4 x 2 x 12 | STEEL | MILD
2 | -15 | PLATES | 1/4 x 4 x 12 | STEEL | MILD
NO | PART NO | NAME | SIZE | MATL | SPEC
TOLERANCE-FRACTIONAL LIMITS 1/8 DECIMAL LIMITS
PART NAME T-JOINT TYPE OF WELD FILLET
NO SCALE SCALE FLAT POSITION OF WELDING PART NUMBER J-37

You are reminded that the coatings of low hydrogen electrodes have a tendency to absorb moisture during storage and prolonged exposure to high humidity on the job. Before using these electrodes, some shops bake them at temperatures up to 700 degrees F. for a period as long as 1 to 3 hours.

Welding Technique

Use alternating current or direct current, reverse polarity. With alternating current the arc characteristics and restrike ability of the E7028, either hot or cold, surpass every other electrode of this type. The E7028 has spray transfer while the E7018 has a globular-type transfer. Hold the electrode at a 90-degree angle to the intersection of the plates, Fig. 21-22A, and tilt it 30 degrees in the direction of travel, Fig. 21-22B.

The first pass can be welded with relatively high current settings and the drag technique. The arc gap should be short. Higher currents can be used since much of the electrical energy passing through the electrode is used to melt the iron powder. As the electrode melts, the iron powder and the core wire dissolve into the weld metal. Since the slag is very heavy and fluid, care must be taken to make sure the slag does not run ahead of the weld pool. Do not permit the arc to touch the sides of the plates since this will cause undercutting. Take care with the first pass. Consider it as the finish pass.

The second and third passes are made with larger size electrodes. A slight weave may be necessary in welding the third pass. Very little weave is necessary in welding the second pass since a larger electrode is used with higher currents and higher deposition rates. Hold a short arc and keep the electrode in the weld pool.

E7028 electrodes are designed primarily for large, heavy weldments. Since the practice joint is relatively small, the joint may become overheated so that the third pass has small pinholes on the surface of the weld. Working with two joints on an alternate basis may avoid overheating.

Slag must be removed thoroughly between each pass. Practice many starts and stops. Travel in all directions.

Operations

1. Obtain plates; check the job print for the correct number and size.
2. Obtain electrodes of each quantity, type, and size specified in the job print.
3. Set a d.c. power source for reverse polarity at 180 to 250 amperes. If an a.c. power source is used, set it somewhat higher than for direct current.
4. Set up two sets of plates as shown in the job print and tack weld them at each corner.
5. Place the joint in the flat position on the welding table. Make sure it is well-grounded.
6. Make the first pass with $^{5}/_{32}$-inch diameter electrodes as shown on the job print. Manipulate the electrode as instructed in Fig. 21-22. Weld on both joints. (This is to keep the joint from getting too hot.)
7. Chip the slag from the welds, brush, and inspect. Make sure that the instructor inspects this pass. Refer to Inspection.
8. Readjust a d.c. power source for 230 to 305 amperes. Set an a.c. power source somewhat higher.
9. Make the second and third passes with $^{3}/_{16}$-inch diameter electrodes as shown on the job print. Manipulate the electrode as instructed in Fig. 21-22. Alternate your welding between both joints to avoid overheating.
10. Chip slag from the welds, brush, and inspect. Refer to Inspection, below.
11. Practice these welds until you can produce good welds consistently. If possible, change your procedure by using $^{3}/_{16}$-inch diameter electrodes for the first pass, and E7018 or E7024 electrodes for the second and third passes.

Inspection

Compare each pass with Fig. 21-23 and check it for the following weld characteristics:

Width and height: Uniform

Appearance: Very smooth with fine ripples; free of voids and

21-22. *Electrode position when welding a T-joint in the flat position with low hydrogen, iron powder electrodes.*

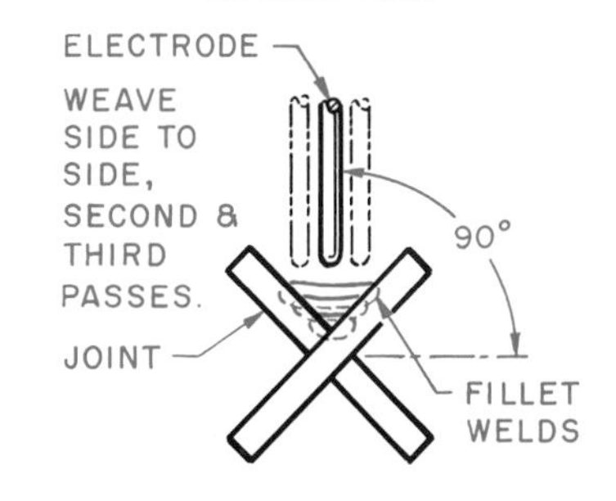

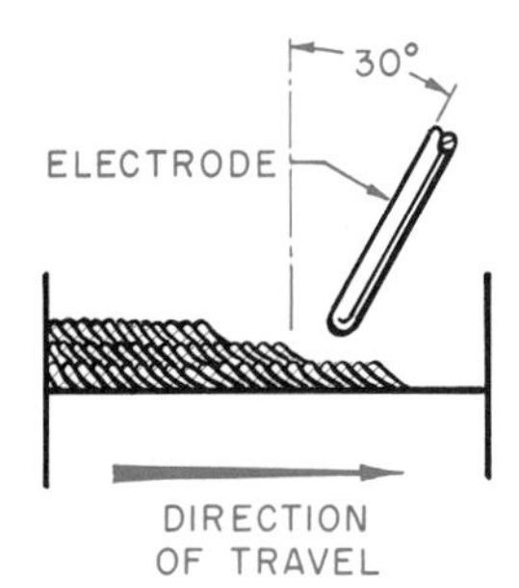

high spots. Restarts should be difficult to locate.

Size: Refer to the job print. Check with a fillet weld gauge.

Face of weld: Flat to slightly convex

Edges of weld: Good fusion, no overlap, no undercut

Starts and stops: Free of depressions and high spots

Beginnings and endings: Full size, craters filled

Penetration and fusion: To the root of the weld and to plate surfaces

Surrounding plate surfaces: Free of spatter

Slag formation: Full coverage, easily removable

Disposal

Put completed joints in the scrap bin so that they will be available for further use. The plate edges can be beveled for welding beveled-butt joints, and unwelded plate surfaces can be used for beading.

JOB 21-J38 WELDING A T-JOINT

Objective

To weld a T-joint in the flat position by means of multipass fillet welds, weaved-beading technique, with free-flowing, heavy slag, shielded-arc electrodes: DCSP, DCRP, or a.c. (AWS-ASTM E6020).

General Job Information

The E6020 electrode is used on all types of pressure and nonpressure vessels and heavy weldments.

It has a heavy mineral type of coating and is outstanding for high speed, high quality downhand welds. The heavy coating permits high welding currents without undue spatter and provides very rapid deposition and burnoff rates. Arc action is forceful and provides deep penetration, especially at high currents.

The E6020 electrode is designed for making deep groove welds, such as V-, U-, and J-grooves in the downhand welding position as well as fillet welds in the horizontal and flat positions. The bead contour in deep grooves is concave and flat in horizontal fillets. The thick slag covering is porous, brittle, and very easy to remove. The weld metal has good tensile strength and impact qualities, high ductility, and freedom from porosity.

This type of electrode may be used with either alternating current or direct current with straight or reverse polarity. When direct current is used, reverse polarity is recommended for groove welding; and straight polarity, for horizontal fillet welding.

Welding Technique

FIRST PASS. Current adjustment should be higher than for the same size electrode of the reverse or straight polarity type. The position of the electrode should be approximately like that shown in Fig. 21-22, page 653. Hold a close arc. Advance the electrode at a uniform rate of travel and keep it in the weld pool for the entire length of the weld. This type of electrode produces a heavy slag that is very fluid, and it should not be permitted to flow ahead of the arc at any time.

SECOND AND THIRD PASSES. Current adjustment can be about the same as that used for the first pass. Electrode position should be approximately that shown in Fig. 21-22. The weaving action is somewhat different than that employed with the other electrodes you have practiced with. Keep the side-to-side movement within the limits of the specified weld width. To pre-

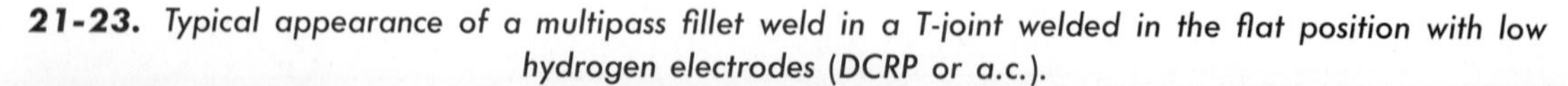

21-23. *Typical appearance of a multipass fillet weld in a T-joint welded in the flat position with low hydrogen electrodes (DCRP or a.c.).*

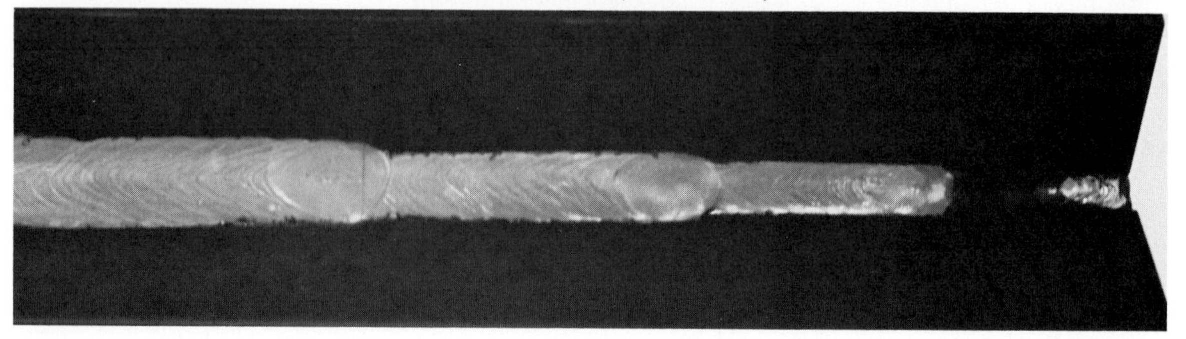

vent undercut, never let the arc touch the beveled surfaces of the plate. Let the heat radiate against the side walls until they begin to melt. Give careful attention to fusion to the side walls of the joint and to the preceding weld.

Practice starts and stops. Travel in all directions.

Operations

1. Obtain plates; check the job print for quantity and size.
2. Obtain electrodes of each quantity, type, and size specified in the job print.
3. Set a d.c. power source for reverse polarity at 175 to 250 amperes. Set an a.c. power source at a higher current value.
4. Set up the plates as shown on the job print and tack weld them at each corner.
5. Place the joint in the flat position on the welding table. Make sure the plates are well-grounded.
6. Make welds as shown on the job print. Manipulate the electrode as instructed in Fig. 21-22.
7. Chip the slag from the welds, brush, and inspect between each pass. Refer to Inspection, below.
8. Practice these welds until you can produce good welds consistently.

Inspection

Compare each pass with Fig. 21-24 and check it for the following weld characteristics:

Width and height: Uniform

Appearance: Very smooth with fine ripples; free of voids and high spots. Restarts should be difficult to locate.

Size: Refer to the job print. Check with a fillet weld gauge.

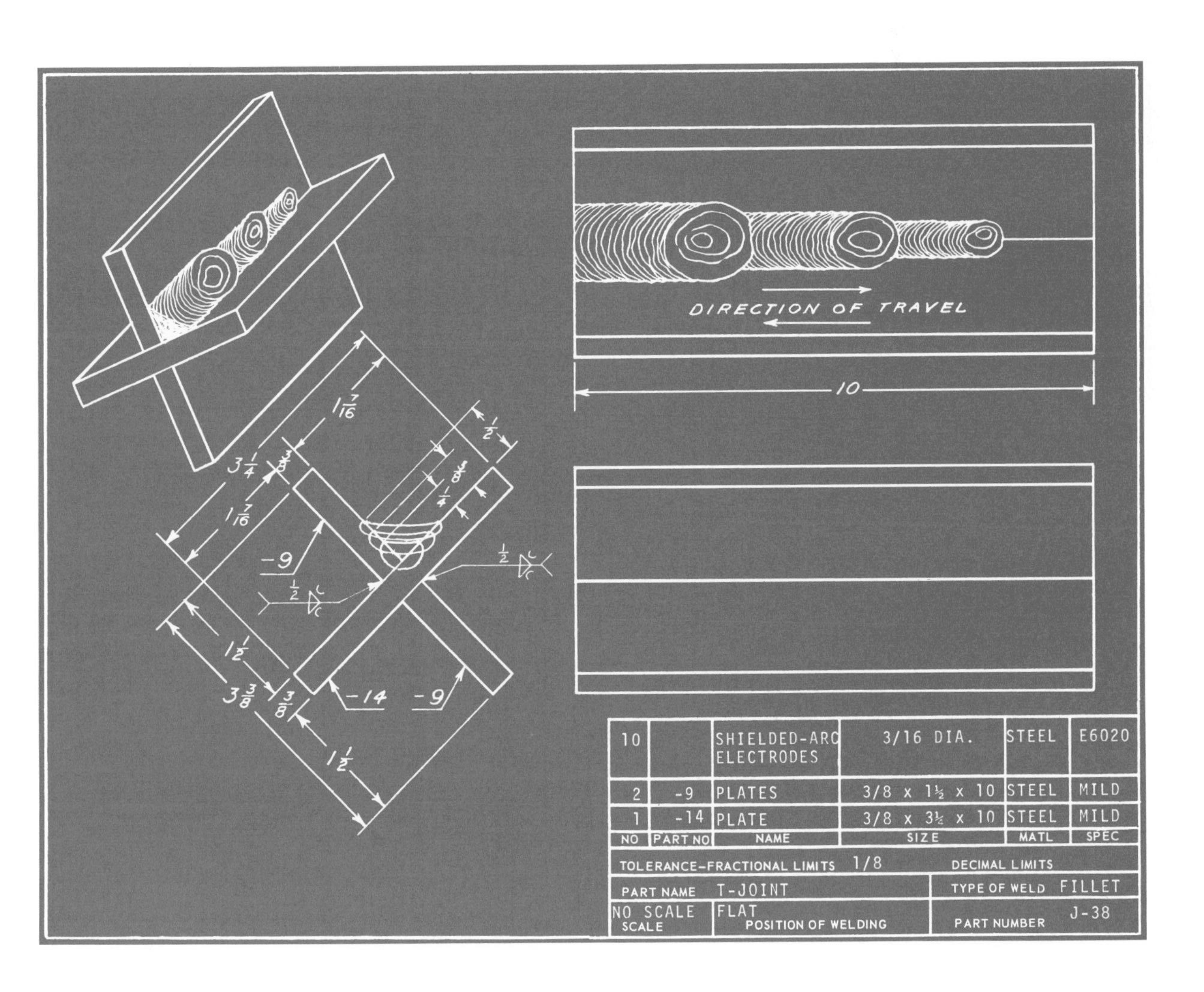

DIRECTION OF TRAVEL
10
-9
-14
-9
10 | SHIELDED-ARC ELECTRODES | 3/16 DIA. | STEEL | E6020
2 | -9 | PLATES | 3/8 x 1½ x 10 | STEEL | MILD
1 | -14 | PLATE | 3/8 x 3¼ x 10 | STEEL | MILD
NO | PART NO | NAME | SIZE | MATL | SPEC
TOLERANCE–FRACTIONAL LIMITS 1/8
DECIMAL LIMITS
PART NAME T-JOINT
TYPE OF WELD FILLET
NO SCALE
SCALE
FLAT
POSITION OF WELDING
PART NUMBER J-38

Face of weld: Concave
Edges of weld: Good fusion, no overlap, no undercut
Starts and stops: Free of depressions and high spots
Beginnings and endings: Full size, craters filled
Penetration and fusion: To the root of the weld and plate surfaces
Surrounding plate surfaces: Free of spatter
Slag formation: Full coverage, easily removable

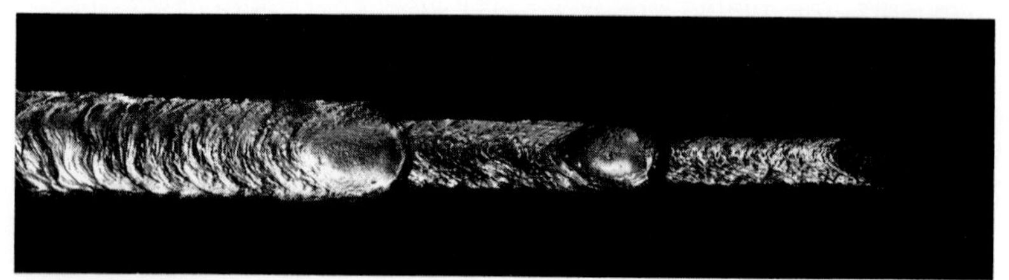

21-24. *Typical appearance of a multipass fillet weld in a T-joint welded in the flat position with free-flowing, shielded-arc electrodes (DCRP or a.c.).*

Disposal

Put completed joints in the scrap bin so that they will be available for further use. Plate edges can be beveled for welding beveled-butt joints, and unwelded surfaces can be used for beading.

JOB 21-J39 WELDING A T-JOINT

Objective

To weld a T-joint in the horizontal position by means of multipass fillet welds, straight beading technique, with free-flowing, heavy slag, shielded-arc electrodes: DCSP, DCRP, or a.c. (AWS-ASTM E6020).

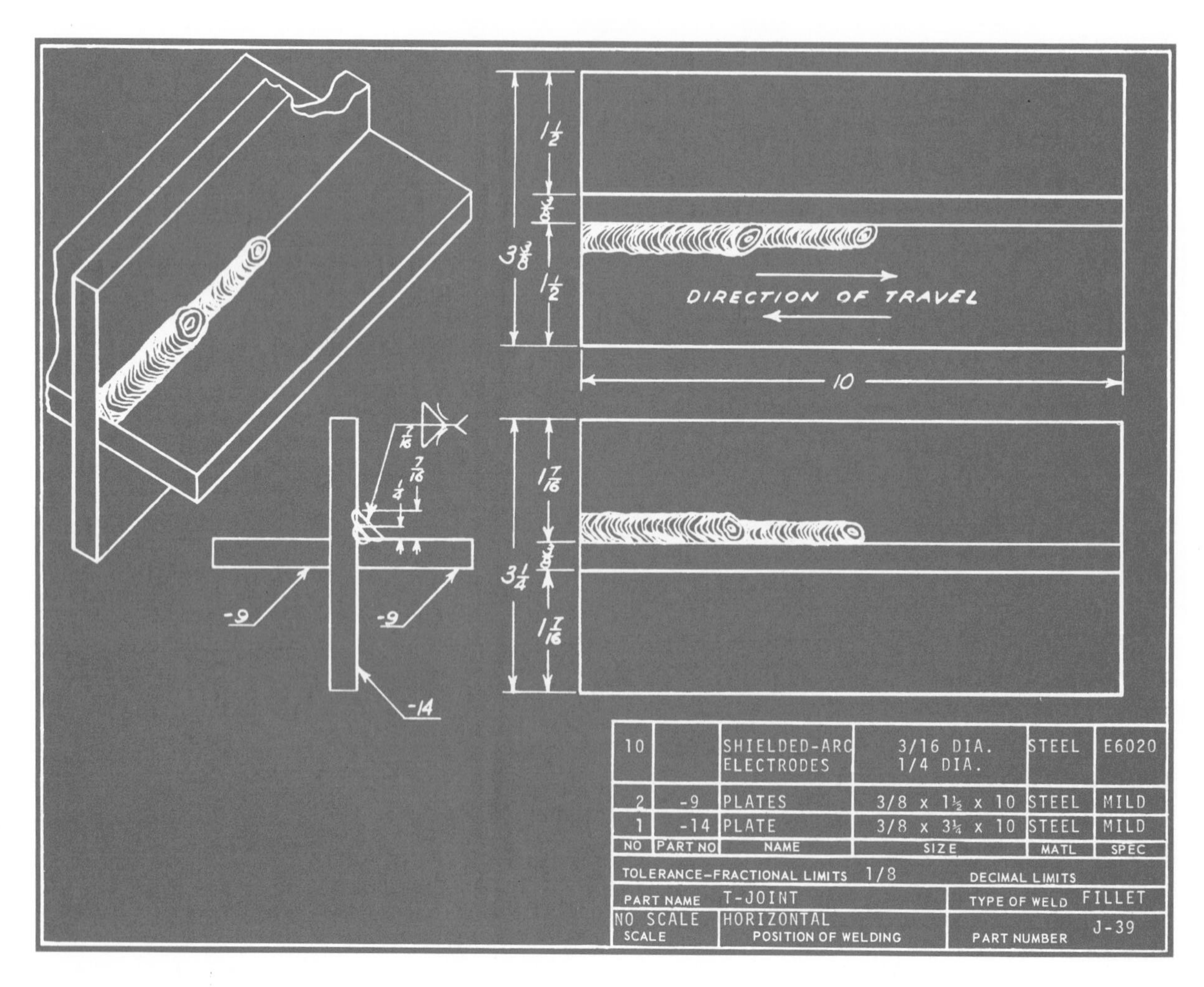

General Job Information

You will often be called upon to use the E6020 electrode in the horizontal position, usually for single-pass fillet welds. Succeeding passes are applied with a larger sized electrode. Weaving is rarely used.

Welding Technique

FIRST PASS. Current adjustment should be high. Hold the electrode at a 40- to 45-degree angle to the flat plate. Fig. 21-25A, and tilt it 30 degrees in the direction of travel, Fig. 21-25B. Hold a fairly close arc gap. Limit electrode motion and advance the weld at a uniform rate of travel. If travel is not too rapid, undercut will not be a problem. Make certain that the slag does not run ahead of the weld deposit.

SECOND PASS. Current adjustment should be high. The electrode angle should be about the same as that used for first pass. Again, very little motion will be necessary since a larger electrode is being used for this pass. Close attention must be given to fusion and the condition of the slag. The rate of travel will be determined by the extent of bead formation.

Practice starts and stops. Travel in all directions.

Operations

1. Obtain plates; check job print for quantity and size.
2. Obtain electrodes of each quantity, type, and size specified in the job print.
3. Set power source at 175 to 250 amperes. Set a d.c. power source for straight polarity.
4. Set up two sets of plates as shown on the job print and tack weld them at each corner.
5. Place the joint in the horizontal position on the welding table. Make sure that it is well-grounded.
6. Make the first pass with $^3/_{16}$-inch electrodes as shown on the job print. Manipulate the electrode as instructed in Fig. 21-25.
7. Chip the slag from the welds, brush, and inspect. Refer to Inspection, below.
8. Increase the setting on the d.c. power source to 250 to 350 amperes. Set an a.c. power source at a higher current value.
9. Make the second pass with $^1/_4$-inch electrodes as shown on the job print.
10. Chip the slag from the welds, brush, and inspect. Refer to Inspection, below.
11. Practice these welds until you can produce good welds consistently. If you want to make further use of the joint before discarding it, make a third pass with $^1/_4$-inch electrodes in the flat position. Use the weaved-beading technique practiced in the previous job.

Inspection

Compare each pass with Fig. 21-26 and check it for the weld characteristics which are designated here:

Width and height: Uniform
Appearance: Very smooth with fine ripples; free of voids and high spots. Restarts should be difficult to locate.
Size: Refer to the job print. Check with a fillet weld gauge.
Face of weld: Concave
Edges of weld: Good fusion, no overlap, no undercut
Starts and stops: Free of depressions and high spots
Beginnings and endings: Full size, craters filled
Penetration and fusion: To the root of the weld and plate surfaces

A. SIDE VIEW
FILLET WELDS
40°–45°
JOINT

B. FRONT VIEW
30°
ELECTRODE
FILLET WELDS
DO NOT ALLOW SLAG TO RUN AHEAD OF WELD.
JOINT
DIRECTION OF TRAVEL

21-25. *Electrode position when welding a T-joint in the horizontal position with free-flowing electrodes.*

21-26. *Typical appearance of a multipass fillet weld in a T-joint welded in the horizontal position with free-flowing, shielded-arc electrodes (DCSP or a.c.).*

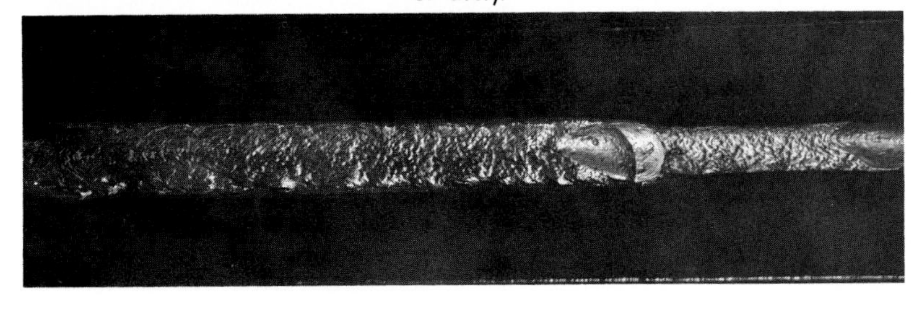

Surrounding plate surfaces: Free of spatter

Slag formation: Full coverage, easily removable

Disposal

Put completed joints in the scrap bin so that they will be available for further use. Plate edges can be beveled for beveled–butt welding, and unwelded surfaces can be used for beading.

JOB 21-J40 WELDING A SINGLE-V BUTT JOINT

Objective

To weld a single-V butt joint in the flat position by the means of multipass groove welds, weaved-beading technique, with DCRP and/or a.c. shielded-arc electrodes (AWS-ASTM E6010-E6011).

General Job Information

The single-V butt joint is used in pipe welding and critical plate welding. The joint structure and welding procedure are similar to those in the ASME piping and plate tests.

Welding Technique

FIRST PASS. Current adjustment must not be too high. Hold the electrode at a 90-degree angle to the plates, Fig. 21-27A, and tilt it 10 to 20 degrees in the direction of travel, Fig. 21-27B. Hesitate at the sides of the weld to prevent undercut. Hold a close arc. Use a back-and-forth motion and keep the electrode within the space provided by the root gap. The arc must not be broken on the forward motion. Do not let drops of metal be deposited ahead of the weld because they may interfere with the progress of the weld. You will recall the need to keep a small hole at the leading edge of the weld crater. A small bead must be formed on the reverse side of the groove, Fig. 21-28. The face of the weld must be kept as flat as possible.

SECOND PASS. Use the weaved-beading technique. A uniform weave and rate of travel

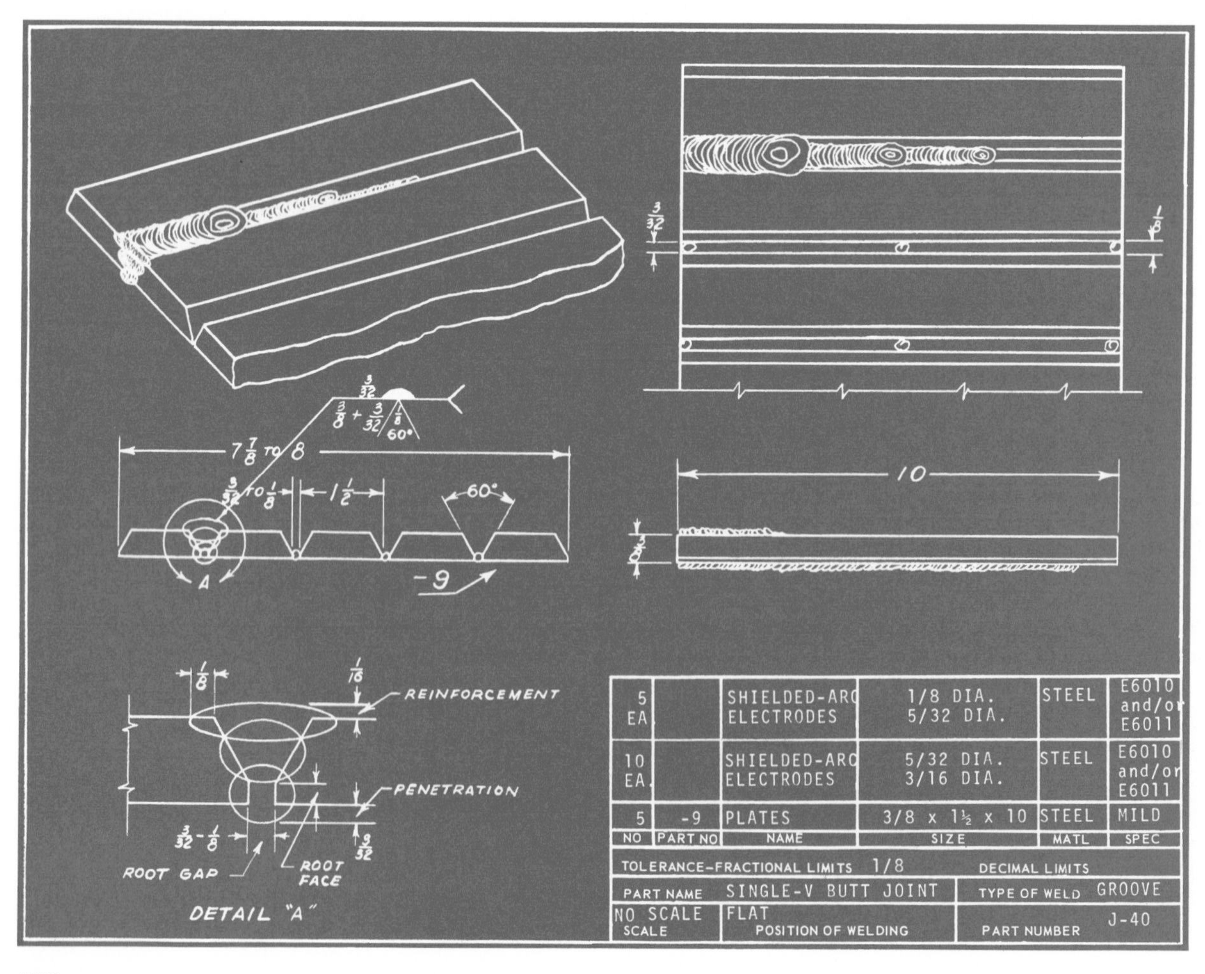

is important to obtain a weld of uniform quality and appearance. Avoid undercutting. Obtain fusion with the underneath pass and the side walls of the plate. The face of the weld must be flat. Remove the slag thoroughly.

THIRD PASS. Use the same type of weave as for the second pass, but widen the path of the electrode. Use the edges of the beveled plate as a guide for bead width. The finished weld should be slightly convex.

Practice starts and stops on all passes. Travel in all directions.

Operations

1. Obtain plates; check the job print for quantity and size.
2. Obtain electrodes of each quantity, type, and size specified in the job print.
3. Set the power source at 75 to 125 amperes. Set a d.c. power source for reverse polarity.
4. Set up the plates and tack weld them as shown on the job print and Fig. 21-29.
5. Place the joint in the flat position on the welding table. Make sure it is well-grounded.
6. Make the first pass with $\frac{1}{8}$-inch diameter electrodes as shown on the job print. Manipulate the electrode as shown in Fig. 21-27.
7. Chip the slag from the weld, brush, and inspect. Refer to Inspection.
8. Increase the power supply setting to 110 to 170 amperes.
9. Make the second and third passes with a $\frac{5}{32}$-inch electrode as shown on the job print.
10. Chip the slag from the welds, brush, and inspect. Refer to Inspection.
11. Practice these welds until you can produce good welds consistently with both types of electrodes. You are also asked to practice this job using $\frac{5}{32}$-inch diameter electrodes for the first pass and $\frac{3}{16}$-inch diameter electrodes for the second and third passes.

Inspection

Compare the back side of the first bead with Fig. 21-28 and all passes with Fig. 21-29. Check them for the following weld characteristics:

Width and height: Uniform
Appearance: Smooth with close ripples; free of voids and high spots. Restarts should be difficult to locate.
Size: Refer to the job print; check with a butt weld gauge.
Face of welds: First two passes flat; last pass slightly convex
Edges of welds: Good fusion, no overlap, no undercut
Starts and stops: Free of depressions and high spots

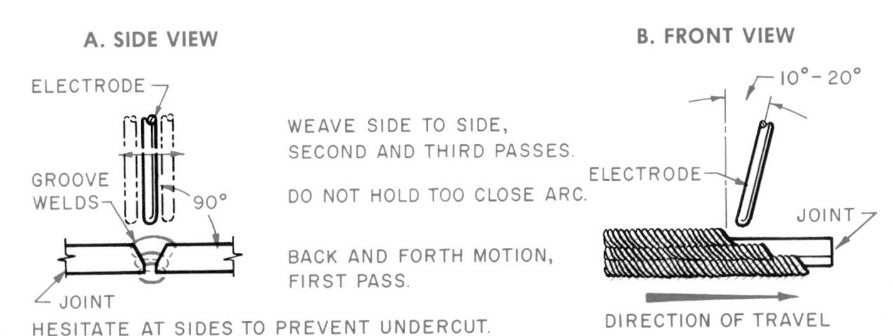

21-27. *Electrode position when welding a beveled-butt joint in the flat position.*

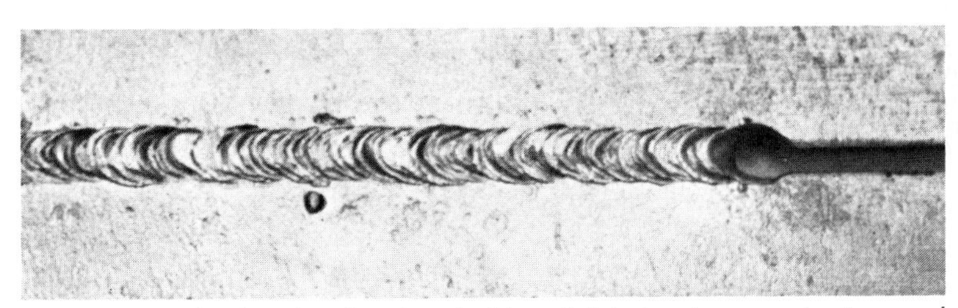

21-28. *The back side of a beveled-butt joint showing penetration and fusion. Note the penetration hole at the leading edge of the weld.*

21-29. *Joint construction and typical appearance of a multipass groove weld in a single-V butt joint welded in the flat position with DCRP or a.c. shielded-arc electrodes.*

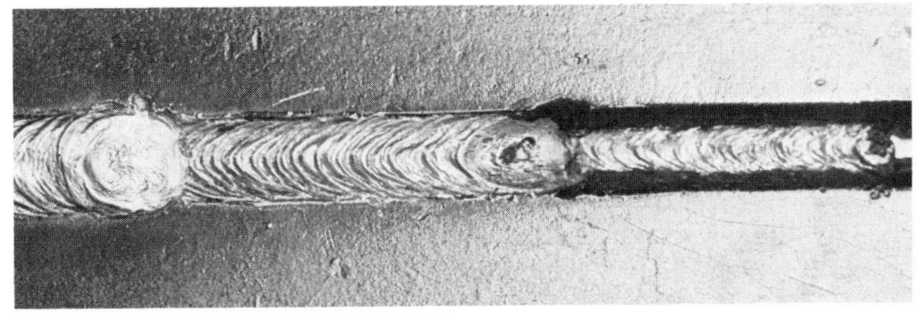

Beginnings and endings: Full size, craters filled
Penetration and fusion: Through the back side and all plate surfaces, Fig. 21-28
Surrounding plate surfaces: Free of spatter
Slag formation: Full coverage, easily removable

Disposal

Return completed joints to the scrap bin so that they will be available for further use. The plate can be beveled between welds for further butt welding, and unwelded plate surfaces can be used for beading.

Qualifying Test

At this point, you should be able to pass Test 2 in the flat position. (See Chapter 22, pages 716, 717). This test is similar to the ASME piping and plate tests.

JOB 21-J41 WELDING A T-JOINT

Objective

To weld a T-joint in the flat position by means of multipass fillet welds, weaved-beading technique, with free-flowing, iron powder, shielded-arc electrodes (AWS-ASTM E6024 and/or E7024).

General Job Information

The E6024 and E7024 are heavily coated, iron powder electrodes. They are used in the flat and horizontal positions. Alternating current produces the best speeds and operating characteristics, but direct current, reverse polarity may also be used. DCRP increases arc blow, however, and makes control of the molten puddle somewhat more difficult. These electrodes have high deposition rates. They produce exceptionally smooth beads and have a thick, dense slag which tends to peel off the weld. They penetrate lightly so that there is little pickup of alloy from the base metal. This makes them highly suitable for medium carbon, crack-sensitive steels.

Welding Technique

Much of the work done with the E6024 and E7024 electrodes is single-pass fillet welding. They may be also used for multipass fillet and groove welds with weaved-beading techniques. A drag technique may be used for making single beads. Most operators prefer an open, short arc, which permits weaving. The electrode position should be about 30 degrees in the direction of travel. (See Fig. 21-22, page 653). The electrode should be advanced at a uniform rate of travel, just fast enough to stay $\frac{1}{4}$ to $\frac{3}{8}$ inch ahead of the molten slag. The arc is forceful and assists the operator in controlling the slag. Current values are higher than those used with previous electrodes.

Make sure that the weaving action is within the width of the desired finished weld. The arc must not touch the side walls of the plates or undercutting will occur. Let the heat radiate against the side walls until they begin to melt.

The first pass is welded with $\frac{5}{32}$-inch diameter electrodes and the second and the third passes with $\frac{3}{16}$-inch diameter electrodes.

Practice starts and stops and travel in all directions.

Operations

1. Obtain plates; check the job print for quantity and size.
2. Obtain electrodes of each quantity, type, and size specified in the job print.
3. Set an a.c. power source at 180 to 225 amperes. Direct current, reverse polarity may be used if alternating current is not available.
4. Set up two sets of plates as shown on the job print and tack weld them at each corner.
5. Place the joint in the flat position on the welding table. Make sure it is well-grounded.
6. Make the first pass with $\frac{5}{32}$-inch diameter electrodes as shown on the job print. Use both types of electrodes. Weld on both joints. (This is to keep the joint from getting too hot.)
7. Chip the slag from the welds, brush, and inspect. Refer to Inspection, below.
8. Readjust the power source to 200 to 280 amperes.
9. Make the second and third passes with $\frac{3}{16}$-inch diameter electrodes as shown on the job print. Use both types of electrodes. Alternate your welding between both joints.
10. Chip the slag from the welds, brush, and inspect. Refer to Inspection, below.
11. Practice these welds until you can produce good welds consistently. Change your procedure by using $\frac{3}{16}$-inch diameter electrodes for the first pass.

Inspection

Compare the welds with Fig. 21-30 and check all passes for these weld characteristics:
Width and height: Uniform
Appearance: Very smooth with fine ripples; free of voids and high spots. Restarts should be difficult to locate.
Size: Refer to the job print. Check with a fillet weld gauge.
Face of weld: Flat to slightly convex
Edges of weld: Good fusion, no overlap, no undercut
Starts and stops: Free of depressions and high spots

Beginnings and endings: Full size, craters filled
Penetration and fusion: To the root of the weld and to plate surfaces
Surrounding plate surfaces: Free of spatter
Slag formation: Full coverage, easily removable

Disposal

Put completed joints in the scrap bin so that they will be available for further use. Plate edges can be beveled for beveled–butt welding, and unwelded plate surfaces can be used for beading.

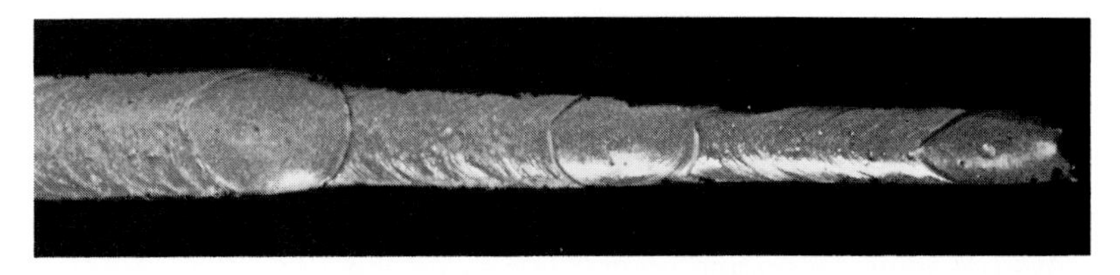

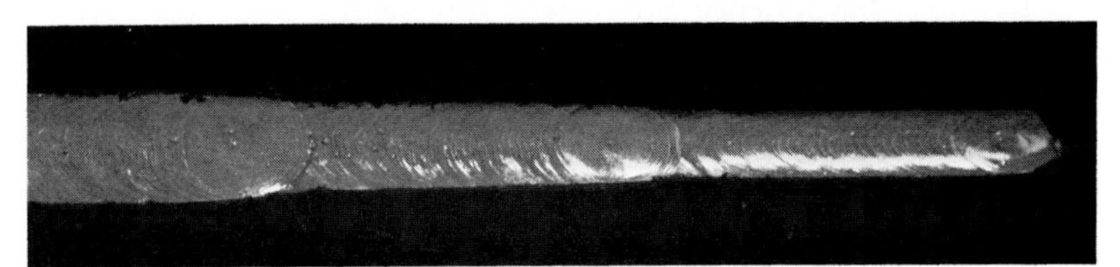

21-30. *Multipass fillet welds in T-joints welded in the flat position with iron powder, shielded-arc electrodes (DCRP or a.c.). The joint above was welded with E6024 electrodes and the one below with E7024 electrodes. Note the similarities in appearance.*

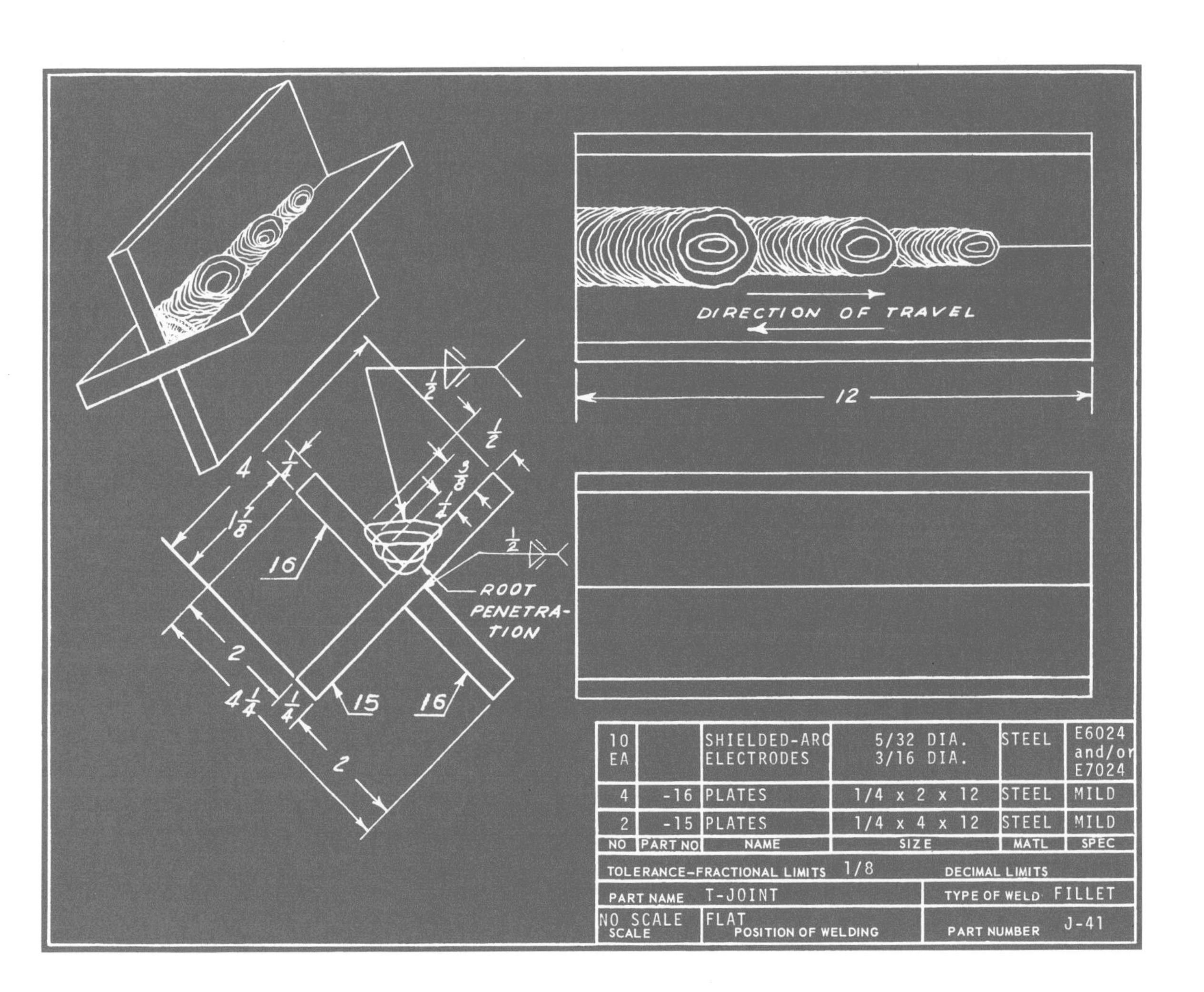

NO	PART NO	NAME	SIZE	MATL	SPEC
10 EA		SHIELDED-ARC ELECTRODES	5/32 DIA. 3/16 DIA.	STEEL	E6024 and/or E7024
4	-16	PLATES	1/4 x 2 x 12	STEEL	MILD
2	-15	PLATES	1/4 x 4 x 12	STEEL	MILD

TOLERANCE–FRACTIONAL LIMITS 1/8 DECIMAL LIMITS
PART NAME T-JOINT — TYPE OF WELD FILLET
SCALE NO SCALE — POSITION OF WELDING FLAT — PART NUMBER J-41

JOB 21-J42 WELDING A SINGLE-V BUTT JOINT

Objective

To weld a single-V butt joint in the flat position by means of multipass groove welds, weaved-beading technique, with iron powder, shielded-arc electrodes (AWS-ASTM E26010, E6011, E6027).

General Job Information

The E6027 electrode is similar to the E6024 and E7024 used in Job 21-J41 in respect to welding characteristics and appearance. This electrode may be used with a.c., DCRP, or DCSP. It is recommended only for deep groove joints. It finds its greatest use in pressure piping and vessel work. The welding procedure for this job is similar to that used in pressure work and in the Hartford Appearance Test. The coverings of the E6027 electrodes are very heavy. They contain a high percentage of iron powder and other ingredients that are in the E6020 electrodes. The coating is about 50 percent of the weight of the electrodes. E6027 electrodes are well-suited for the welding of heavy sections.

E6027 electrodes are designed to produce high quality fillet or groove welds in the flat position with a.c., DCRP, or DCSP and horizontal fillet welds with a flat or concave face with either a.c. or DCSP. Alternating current produces the highest speeds and best operating characteristics. It is possible to use higher heats with almost complete absence of arc blow. The E6027 is particularly suited for multipass, deep groove welding.

The electrode has a spray type of metal transfer and is capable of high lineal speeds with medium penetration and very low spatter loss. The slag is fluid and covers the weld completely. It is crumbly and is easily removed from the joint. Welds have a flat to slightly concave profile with a

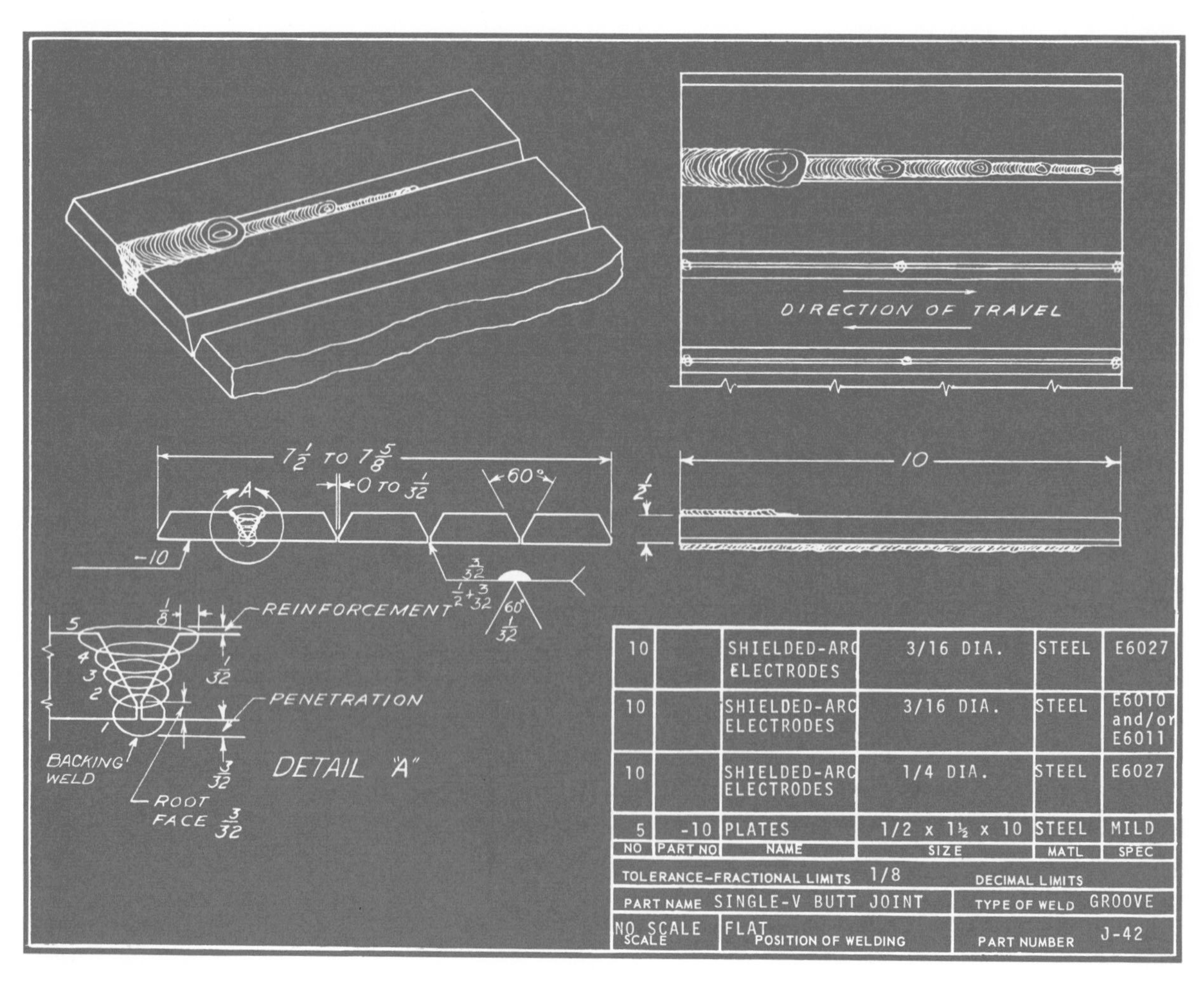

smooth, even ripple and a good metal wash up the joint sides. Because of the iron powder content of the coating, current values can be high.

These electrodes are considered the iron powder equivalent to the E6020 electrodes. They are used for heavy structural work, high speed production work, and when fast deposition and medium penetration is desired.

Welding Technique

FIRST PASS. Make the first pass on the back side of the joint. Current adjustment should be high. The electrode position is the same as for stringer beading. Limit electrode motion and use standard stringer bead technique. Keep the bead flat and secure as much penetration as possible. The back side of the bead must be chipped out before applying other passes.

SECOND PASS. Current adjustment should be high. Hold the electrode at a 90-degree angle to the plates, Fig. 21-31A, and tilt it 30 degrees in the direction of welding, Fig. 21-31B. This pass need not be weaved; it should be advanced at a uniform rate of travel. Never let the arc touch the beveled edges of the base metal. Let it radiate its heat against the side walls of the groove until they melt. Remove the slag before depositing the next pass.

THIRD TO FIFTH PASSES. It will be necessary to weave these passes. Use the edges of the preceding pass as your guide for width. Remember to keep the weave within the limits of the finished weld and do not touch the side walls with the arc. The face of the weld should be concave, and the edges should be washed up well on the side walls of the plate without undercut. *Note:* This joint is often welded by applying the first pass to the groove side of the plate with a free-flowing electrode. Then the back side is chipped out, and the back side pass is applied. Another type of joint construction requires a backup strip.

Travel is usually toward the operator. Practice starts and stops.

Operations

1. Obtain plates; check the job print for quantity and size.
2. Obtain electrodes of each quantity, type, and size specified in the job print.
3. Set a d.c. power source for reverse polarity at 140 to 220 amperes. Set an a.c. power source at a higher current value.
4. Set up the plates and tack weld them as shown on the job print.
5. Place the joint back side up in the flat position on the welding table. Make sure that it is well-grounded.
6. Make the first pass with a $^{3}/_{16}$-inch electrode (E6010 or E6011) as shown on the job print. Use the stringer bead technique.
7. Chip the slag from both sides of the weld, brush, and inspect. Refer to Inspection, below.
8. Set a d.c. power source for straight polarity at 180 to 250 amperes or an a.c. power source at 200 to 260 amperes.
9. Make the second, third, and fourth passes with $^{3}/_{16}$-inch iron powder electrodes as shown on the job print. Manipulate the electrode as instructed in Fig. 21-31.
10. Chip the slag from the welds, brush, and inspect between each pass. Refer to Inspection, below.
11. Set a d.c. power source at 275 to 325 amperes. Set an a.c. power source somewhat higher.
12. Make the last pass with $^{1}/_{4}$-inch iron powder electrodes as shown on the job print.
13. Chip the slag from the weld, brush, and inspect. Refer to Inspection, below.

Inspection

Compare the back side of the first bead with Fig. 21-32 and all passes with Fig. 21-33. Check them for the following weld characteristics:

Width and height: Uniform
Appearance: Very smooth with fine ripples; free of voids and high spots. Restarts should be difficult to locate.
Size: Refer to the job print.
Face of weld: Concave on all but the last pass, which should have a little reinforcement.
Edges of weld: Good fusion, no overlap, no undercut
Starts and stops: Free of depressions and high spots
Beginnings and endings: Full size, craters filled

21-31. *Electrode position when welding a beveled-butt joint with free-flowing electrodes.*

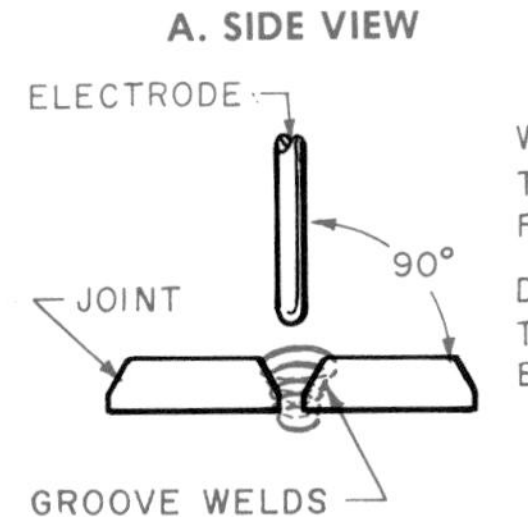

WEAVE SIDE TO SIDE, THIRD, FOURTH AND FIFTH PASSES.

DO NOT WEAVE WIDER THAN PRECEDING BEAD.

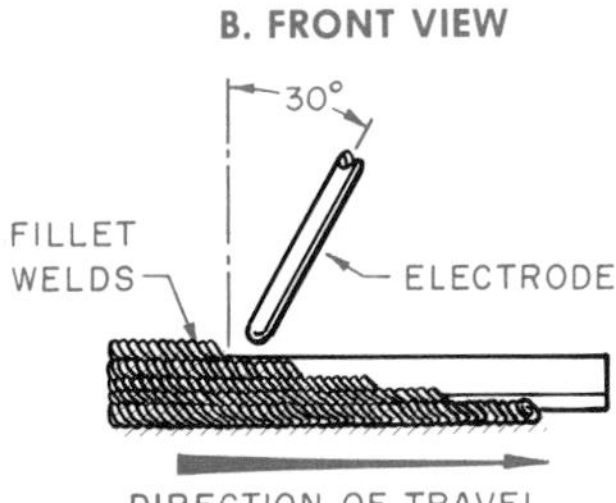

Penetration and fusion: To the root of the joint and all plate surfaces

Surrounding plate surfaces: Free of spatter

Slag formation: Full coverage, easily removable

Disposal

Put completed joints in the scrap bin so that they will be available for further use. Bevel the plate between welds for further butt welding, and use unwelded plate surfaces for beading.

JOB 21-J43 WELDING A T-JOINT

Objective

To weld a T-joint in the horizontal position by means of multipass fillet welds, stringer bead technique, with low hydrogen, shielded-arc electrodes (AWS-ASTM E7016).

General Job Information

All low hydrogen electrodes produce deposits that are practically free of hydrogen. This reduces underbead and microcracking on low alloy steels and thick weldments. Underbead cracks usually occur in base metal just below the weld metal and are caused by hydrogen absorption from the arc atmosphere. Low hydrogen electrodes produce sound welds on troublesome steels such as the high sulfur, high carbon, and low alloy grades. They are recommended for welds that are to be porcelain enameled.

The E7016 electrode produces welds of highest quality and may be used for practically all code work. The welds are high in tensile strength and ductility. Their excellent physical qualities reduce the tendency for cracking at the root of the weld, which is caused by shrinkage stresses. This in turn reduces the need for preheat and stress relief requirements so that the job is done quickly and at lower cost.

Welding Technique

Use alternating current or direct current, reverse polarity. Some manufacturers recommend DCRP whenever possible with electrodes having diameters of $^5/_{32}$ inch and smaller and a.c. with larger sizes. Although the E7016 is an all-position electrode, sizes larger than $^1/_8$ inch or $^5/_{32}$ inch cannot be used for vertical and overhead welding.

The weld metal freezes rapidly even though the slag stays relatively fluid. The arc is quiet with medium penetration and little spatter. The slag is moderately heavy: it produces good weld metal protection, excellent bead appearance, and easy cleaning. The beads are slightly convex and have distinct ripples.

The currents used with these electrodes generally are higher than those recommended for the more conventional electrodes of the same diameter. As short an arc as possible should be used in all positions for best results. Stringer beads or small weave passes are preferred to wide weave passes. A long arc and whipping generally causes porosity and trapped slag. Use lower

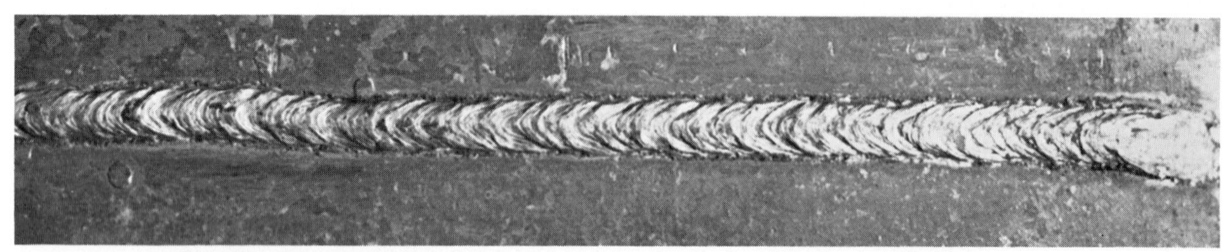

21-32. *Appearance of the weld bead on the back side of a beveled-butt joint.*

21-33. *Typical appearance of a multipass fillet weld in a beveled-butt joint welded in the flat position with E6027 iron powder, shielded-arc electrodes.*

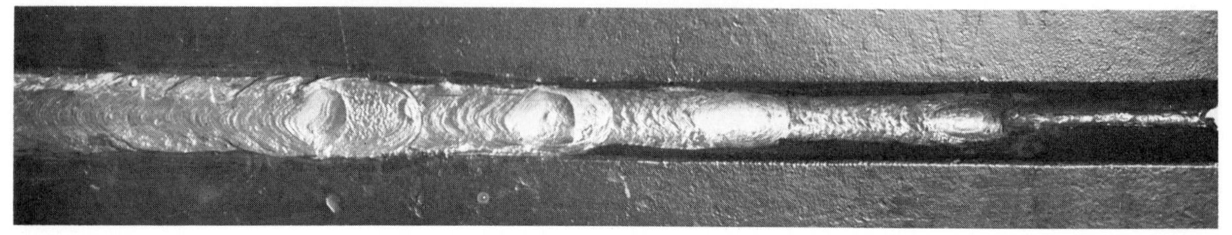

currents with direct current than with alternating current. Point the electrode directly into the joint and tip the holder forward slightly in the direction of travel. See Fig. 20-73, page 629. Govern travel speed by the desired bead size.

Current adjustment should be somewhat higher for the first pass than for the other stringer passes. The electrode position will vary with the sequence of passes, Fig. 20-68, page 626. The beads should be so proportioned that the size of the final welds is accurate. Recall the lacing technique, and refer to Job 20-J26 for details. Be sure to clean all the passes thoroughly. Practice starts and stops. Travel in all directions.

Operations

1. Obtain plates; check the job print for the correct quantity and size.
2. Obtain electrodes of each quantity, type, and size specified in the job print.
3. Set a d.c. power source for reverse polarity at 140 to 200 amperes. If an a.c. power source is used, set it somewhat higher than for d.c.
4. Set up the plates as shown on the job print and tack weld them at each corner.
5. Place the joint in the horizontal position on the welding table. Make sure that it is well-grounded.
6. Make all welds (first to sixth pass inclusive) with $\frac{5}{32}$-inch electrodes as shown on the job print. Manipulate the electrode as instructed in Fig. 20-68, page 626.
7. Chip the slag from the weld, brush, and inspect between each pass. Refer to Inspection.
8. Set a d.c. power source for reverse polarity at 80 to 140 amperes. If an a.c. power source is used, set it at a somewhat lower amperage.
9. Make the lacing pass with a $\frac{1}{8}$-inch electrode as shown on the job print. Manipulate the

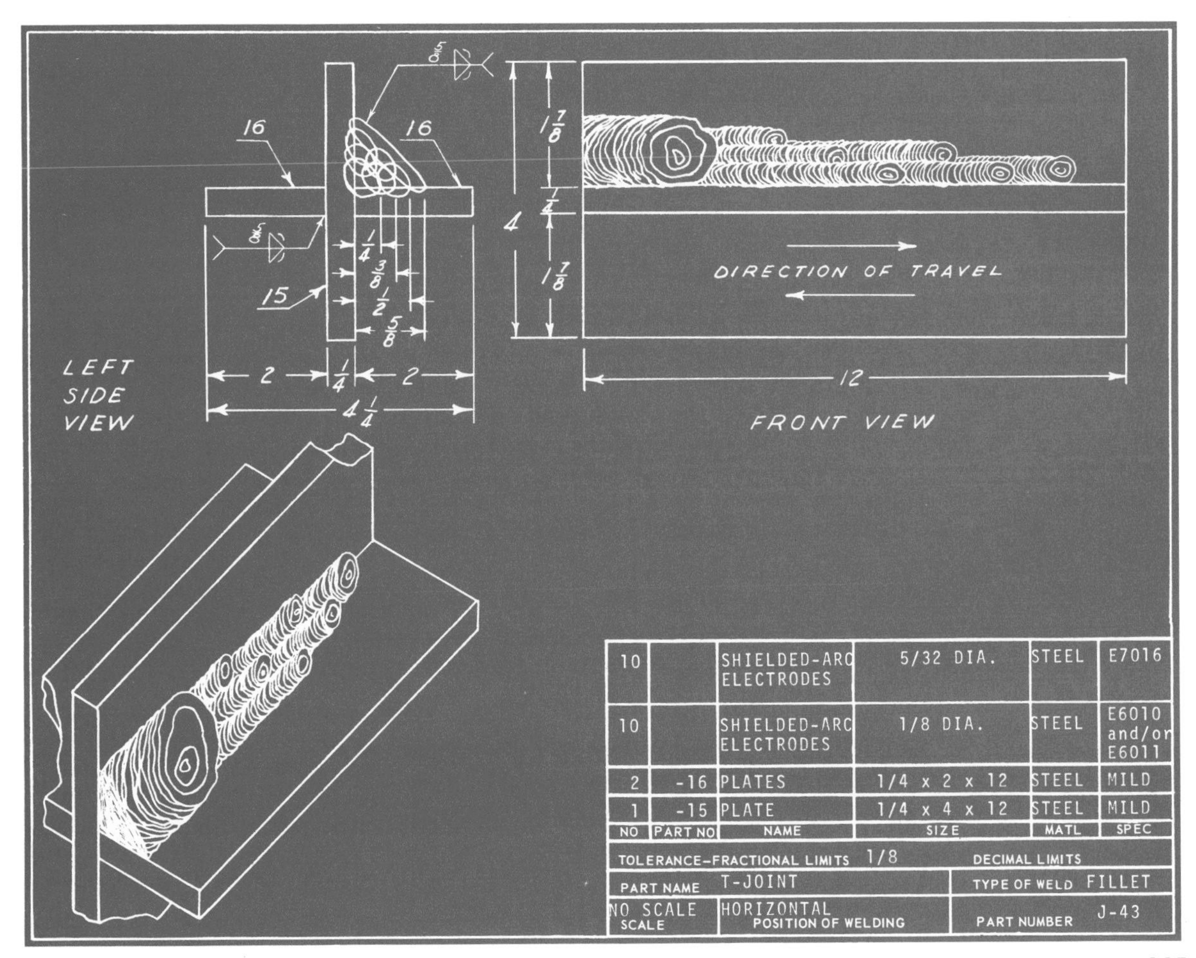

electrode as shown in Fig. 20-69, page 627.

10. Chip the slag from the welds, brush, and inspect. Refer to Inspection, below.

11. Practice these welds until you can produce good welds consistently.

Inspection

Compare each pass with Fig. 21-34 and check for the following weld characteristics:

Width and height: Uniform
Appearance: Very smooth with fine ripples; free of voids and high spots. Restarts should be difficult to locate.
Size: Refer to the job print. Check with a fillet weld gauge.
Face of weld: Flat to slightly convex
Edges of weld: Good fusion, no overlap, no undercut
Starts and stops: Free of depressions and high spots
Beginnings and endings: Full size, craters filled
Penetration and fusion: To the root of the weld and plate surfaces
Surrounding plate surfaces: Free of spatter
Slag formation: Full coverage, easily removable

Disposal

Put completed joints in the scrap bin so that they will be available for further use. Plate edges can be beveled for beveled-butt welding, and unwelded plate surfaces can be used for beading.

JOB 21-J44 WELDING A T-JOINT

Objective

To weld a T-joint in the vertical position by means of a single-pass fillet weld, travel up, with DCRP and/or a.c. shielded-arc electrodes (AWS-ASTM E6010-E6011).

General Job Information

A T-joint in the vertical position is frequently used in the fabricating and structural steel industries. Welding is usually up, but welding down is sometimes allowed. Whether single or multiple passes are specified depends upon the use of the joint and the thickness of the plate.

Welding Technique

Current adjustment should be high enough to insure good fusion and penetration to the root of the joint and to the plate surfaces. Hold the electrode at a 90-degree angle to the intersection of the plates, Fig. 21-35A, and tilt it no more than 10 degrees down from the horizontal position, Fig. 21-35B. Figure 21-36 shows the actual welding position. Use the familiar up-and-down motion recommended for the vertical position. It is necessary to leave the crater at short intervals to prevent the metal from becoming overheated. Overheating makes it spill off and run down. Hold a close arc gap when depositing metal. The arc gap can be lengthened on the upward motion, but it must not be broken. Keep the electrode in the weld pool long enough to insure fusion and penetration, to form the required weld contour, and to prevent undercut.

Practice starts and stops.

Operations

1. Obtain plates; check the job print for quantity and size.
2. Obtain electrodes of each quantity, type, and size specified in the job print.
3. Set a d.c. power source for reverse polarity at 110 to 170 amperes or an a.c. power source at 110 to 180 amperes.
4. Set up the plates as shown on the job print and tack weld them at each corner.
5. Stand the joint in the vertical position on the welding table. Make sure the joint is well-grounded.
6. Make welds as shown on the job print. Manipulate the electrode as instructed in Fig. 21-35.
7. Chip the slag from the welds, brush, and inspect. Refer to Inspection, below.
8. Practice these welds until you can produce good welds consistently with both types of electrodes, if possible, and with $\frac{3}{16}$-inch diameter electrodes.

Inspection

Compare the weld with Fig. 21-37 and check it for the following characteristics:

Width and height: Uniform
Appearance: Smooth with close ripples; free of voids and high spots. Restarts should be difficult to locate.

21-34. *Typical appearance of a multipass fillet weld in a T-joint welded in the horizontal position with E7016 iron powder electrodes for the stringer passes and E6010 DCRP or a.c. shielded-arc electrodes for the lacing cover pass.*

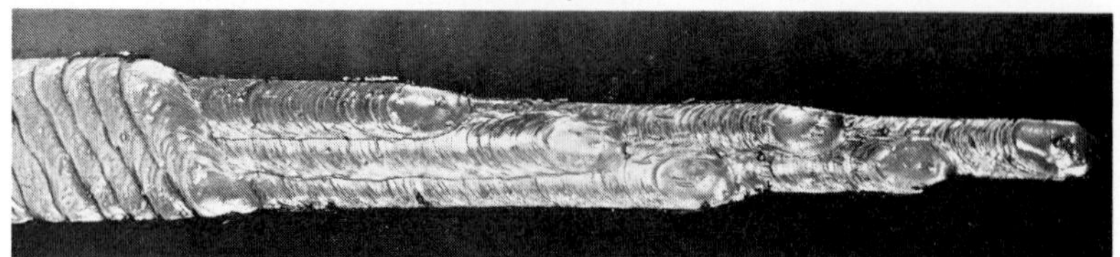

Size: Refer to job print. Check with a fillet weld gauge.
Face of weld: Flat
Edges of weld: Good fusion, no overlap, no undercut
Starts and stops: Free of depressions and high spots
Beginnings and endings: Full size, craters filled
Penetration: To the root of the joint and both plate surfaces
Surrounding plate surfaces: Free of spatter
Slag formation: Full coverage, easily removable

Check Test

After you are able to make welds that are satisfactory in appearance, make up a test specimen similar to that made as a check test in Job 20-J17, Figs. 20-48 and 20-49, page 612. Use the same plate thickness, welding technique, and electrode type and size used in this job. Weld on one side only.

Break the completed weld and examine the surfaces for sound-

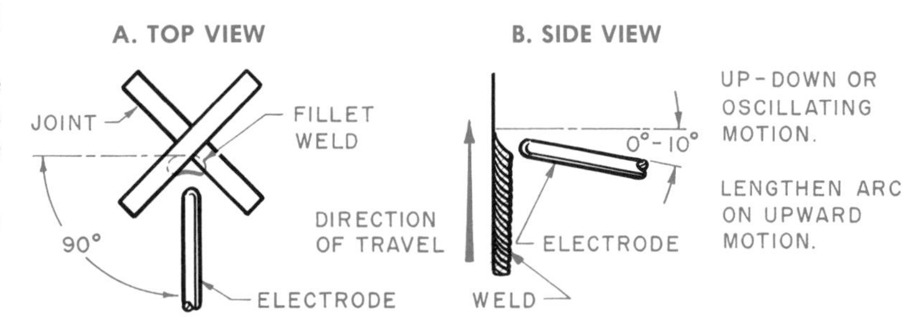

21-35. *Electrode position when welding a single pass in a T-joint in the vertical position, travel up.*

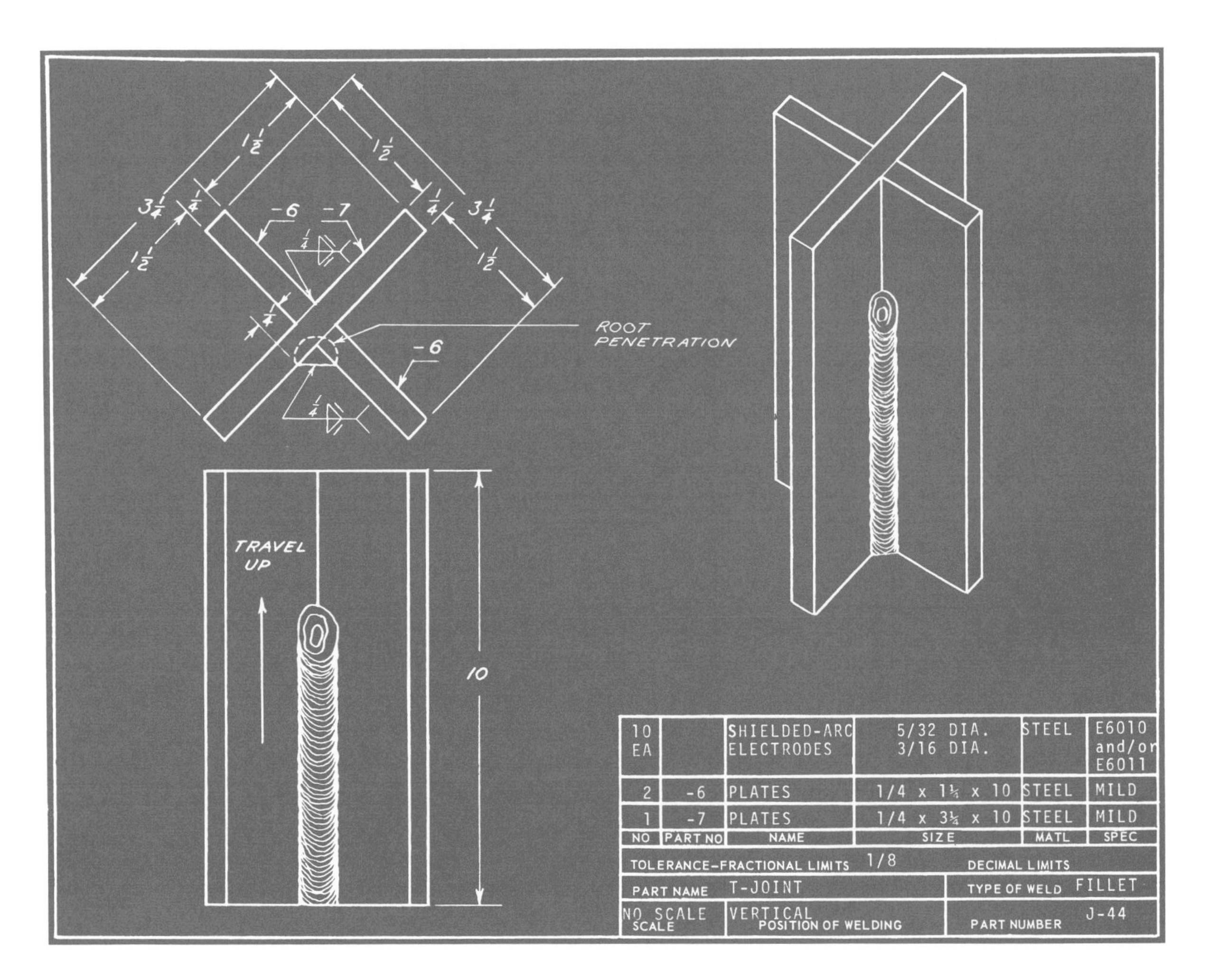

ness. The weld must not be porous and must show good fusion and penetration at the root of the joint and to the plate surfaces. The joint should break evenly through the throat of the weld.

Disposal

Save the completed joints for your next job which will be multipass fillet welding in the vertical position.

JOB 21-J45 WELDING A T-JOINT

Objective

To weld a T-joint in the vertical position by means of multipass fillet welds, weaved-beading technique, travel up, with DCRP and/or a.c. shielded-arc electrodes (AWS-ASTM E6010-E6011).

21-36. *Electrode and electrode holder positioned as instructed in Fig. 21-35.*

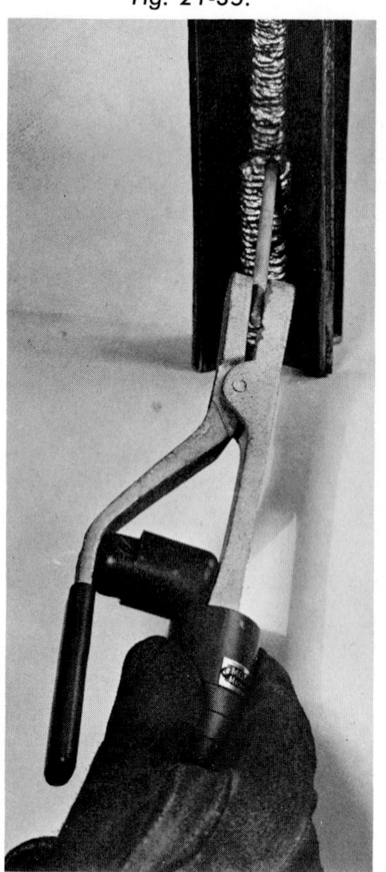

General Job Information

For practical purposes, a joint in $\frac{1}{4}$-inch plate would not require three passes. Three passes are required in this job to permit full use of the plate and to develop the ability to weld weaved beads in the vertical position. Multipass fillet welds in the vertical position are usually welded up with reverse polarity electrodes. Straight polarity electrodes are sometimes used.

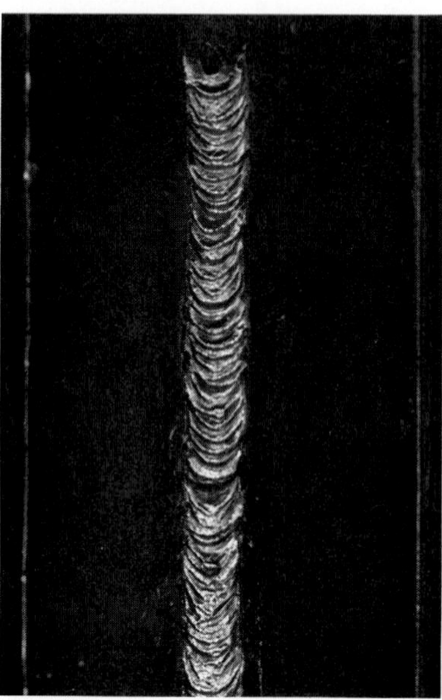

21-37. *Typical appearance of a single-pass fillet weld in a T-joint welded in the vertical position, travel up, with DCRP or a.c. shielded-arc electrodes.*

Welding Technique

SECOND PASS. Use the joint from the previous job. Current adjustment should be high enough to insure good fusion and penetration. Hold the electrode at a 90-degree angle to the intersection of the plate, Fig. 21-38A, and tilt it no more than 10 degrees down from the horizontal position, Fig. 21-38B. Fig. 21-36 shows the actual position. You will find that it is possible to keep the electrode in the weld pool and advance it by weaving from side to side. A smooth, even motion across the face of the weld will form a flat face, and hesitation at the sides of the weld will prevent undercut. Hold a short arc gap but do not touch the molten metal with the electrode.

In welding both this second pass and the third pass, the upward movement when you reach the bead edges controls the appearance and thickness of the bead being deposited. A large upward movement will result in wider spaced ripples and a thin-

21-38. *Electrode position when welding a T-joint in the vertical position with weaved beading, travel up.*

A. TOP VIEW

JOINT
WELDS
DIRECTION OF TRAVEL
90°
WEAVE SIDE TO SIDE.
ELECTRODE

B. SIDE VIEW

0°-10°
HESITATE AT SIDES TO PREVENT UNDERCUT.
ELECTRODE
HOLD CLOSE ARC.
WELDS

ner buildup. A small upward movement will produce more closely spaced ripples and a thicker buildup.

THIRD PASS. Current adjustment will have to be higher for the larger electrode. The electrode position is the same as for previous passes. When using larger electrodes and higher heats, run the electrode up quickly and away from the pool at each side. Do not break the arc on this movement and lengthen it on the upward stroke. When the weld pool has cooled enough, bring the electrode back to the crater and deposit additional metal. Keep the width of the weave within the limits of the final bead width.

Practice starts and stops.

Operations

1. Use the joints welded in Job 21-J44.
2. Obtain electrodes of each quantity, type, and size specified in the job print.
3. Set the power source at 110 to 170 amperes. Set a d.c. power source for reverse polarity.
4. Stand the joint in the vertical position on the welding table. Make sure that it is well-grounded.
5. Make the second pass with a $\frac{5}{32}$-inch electrode as shown on the job print. Manipulate the electrode as instructed in Fig. 21-38.
6. Chip the slag from the weld, brush, and inspect. Refer to Inspection.
7. Increase the power source setting to 140 to 215 amperes.
8. Make the third pass with $\frac{3}{16}$-inch electrodes as shown on the job print. If you find it difficult with $\frac{3}{16}$-inch electrodes, practice this pass with $\frac{5}{32}$-inch electrodes at 110 to 170 amperes.
9. Chip the slag from the weld, brush, and inspect. Refer to Inspection.

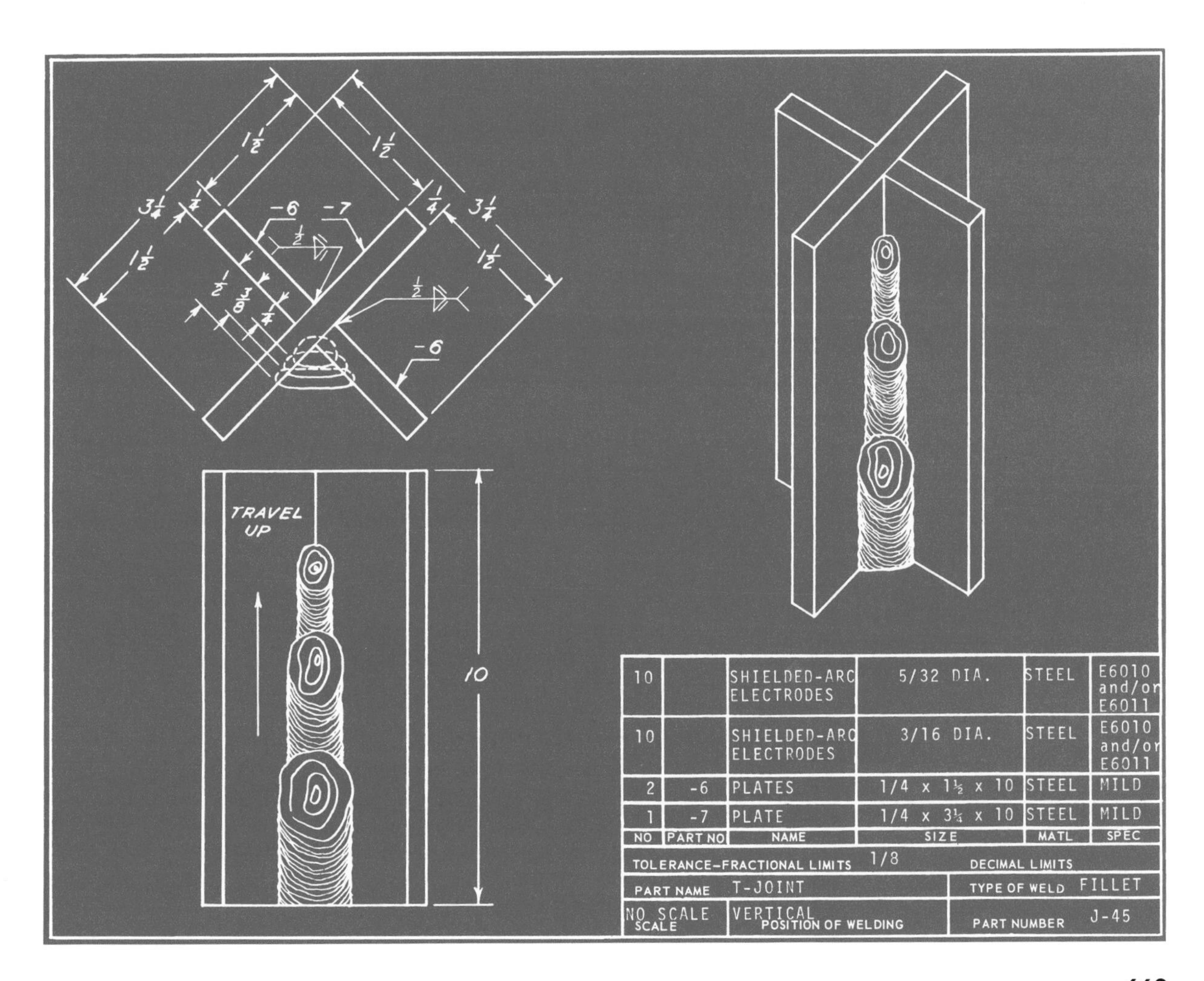

NO	PART NO	NAME	SIZE	MATL	SPEC
10		SHIELDED-ARC ELECTRODES	5/32 DIA.	STEEL	E6010 and/or E6011
10		SHIELDED-ARC ELECTRODES	3/16 DIA.	STEEL	E6010 and/or E6011
2	-6	PLATES	1/4 x 1½ x 10	STEEL	MILD
1	-7	PLATE	1/4 x 3¼ x 10	STEEL	MILD

TOLERANCE–FRACTIONAL LIMITS 1/8 DECIMAL LIMITS
PART NAME T-JOINT TYPE OF WELD FILLET
SCALE NO SCALE POSITION OF WELDING VERTICAL PART NUMBER J-45

10. Practice these welds until you can produce good welds consistently with both types of electrodes.

Inspection

Compare each pass with Fig. 21-39 and check it for the following weld characteristics:

Width and height: Uniform
Appearance: Smooth with close ripples; free of voids and high spots. Restarts should be difficult to locate.
Size: Refer to the job print. Check with a fillet weld gauge.
Face of weld: All passes flat
Edges of weld: Good fusion, no overlap, no undercut
Starts and stops: Free of depressions and high spots
Beginnings and endings: Full size, craters filled
Penetration and fusion: To preceding passes and plate surfaces
Surrounding plate surfaces: Free of spatter
Slag formation: Full coverage, easily removable

21-39. *Typical appearance of a multipass fillet weld in a T-joint welded in the vertical position, travel up, with DCRP or a.c. shielded-arc electrodes.*

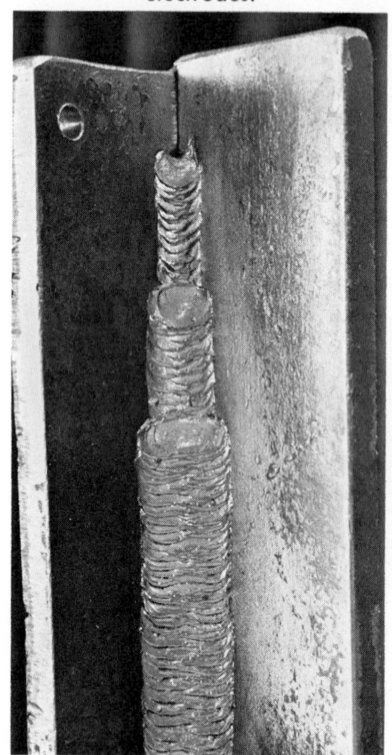

Disposal

Put completed joints in the scrap bin so that they will be available for further use. Plate edges can be used for square butt welding, and unwelded plate surfaces can be used for beading.

JOB 21-J46 WELDING A T-JOINT

Objective

To weld a T-joint in the vertical position by means of multipass fillet welds, weaved-beading technique, travel up, with low hydrogen, iron powder, shielded-arc electrodes (AWS-ASTM E7018).

General Job Information

The addition of 25 to 40 percent iron powder to a low hydrogen electrode covering has increased its usage for mild steel and the low alloy, high sulfur, and high carbon grades. Iron powder also increases the deposition rate and gives better restrike characteristics than the E7016 type. The arc action is smoother and more stable. The deposit wets more readily so that undercutting is prevented. The bead is smoother and slag removal is easier than with most E7016 electrodes. All of the desirable low hydrogen characteristics are retained in this electrode. Review page 206 for additional information.

The coatings containing iron powder are thicker than regular low hydrogen coatings.

Welding Technique

Use a.c. or DCRP. You will find that the arc is smooth and quiet with shallow penetration and little spatter. Since the coating of these electrodes is heavier than normal, vertical and overhead welding is usually limited to smaller diameters. Currents used are somewhat higher than for the E6010s of corresponding size. A short arc must be maintained at all times.

Vertical-up welding is done with either a $\frac{1}{8}$-inch or $\frac{5}{32}$-inch electrode. Do not use a whip technique or take the electrode out of the molten pool. Use a small triangular weave. Build up a small shelf of weld metal and deposit layer upon layer of metal with the weave as the weld progresses up the joint. Point the electrode directly into the joint and slightly upward to permit the arc force to assist in controlling the puddle. See Fig. 21-38, page 668. Travel slow enough to maintain the shelf without causing the molten metal to spill off. Use currents in the lower portion of the quoted range. Keep the width of the weave within the confines of the bead width. You have learned to make sure that you obtain good fusion at the root of the joint with the first pass and that all other passes are thoroughly fused with the underneath pass and the side walls of the joint.

Be sure to clean all passes thoroughly. Practice starts and stops.

Operations

1. Obtain plates; check the job print for quantity and size.
2. Obtain electrodes in each quantity, type, and size specified in the job print.
3. Set a d.c. power source for reverse polarity at 150 to 220 amperes. An a.c. power source

may be used at a somewhat higher heat.

4. Set up two sets of plates as shown in the job print and tack weld them at each corner.

5. Stand the joint in the vertical position on the welding table. Make sure it is well-grounded.

6. Make all passes with 5/32-inch diameter electrodes as shown in the job print. Manipulate the electrode as instructed in Fig. 21-38, page 668. Weld both joints. (This is to keep the joint from getting too hot.)

7. Chip the slag from the welds, brush, and inspect between each pass. Refer to Inspection.

8. Practice these welds until you can produce good welds consistently.

Inspection

Compare each pass with Fig. 21-40 and check it for the following weld characteristics:

Width and height: Uniform

Appearance: Very smooth with fine ripples; free of voids and high spots. Restarts should be difficult to locate.

Sizes: Refer to the job print. Check with a fillet weld gauge.

Face of weld: Flat to slightly convex

Edges of weld: Good fusion, no overlap, no undercut

Starts and stops: Free of depressions and high spots

Beginnings and endings: Full size, craters filled

Penetration and fusion: To the root of the weld and plate surfaces

Surrounding plate surfaces: Free of spatter

Slag formation: Full coverage, easily removable

Disposal

Put completed joints in the scrap bin so that they can be used again. Plate edges can be beveled for beveled-butt welding, and unwelded plate surfaces can be used for beading.

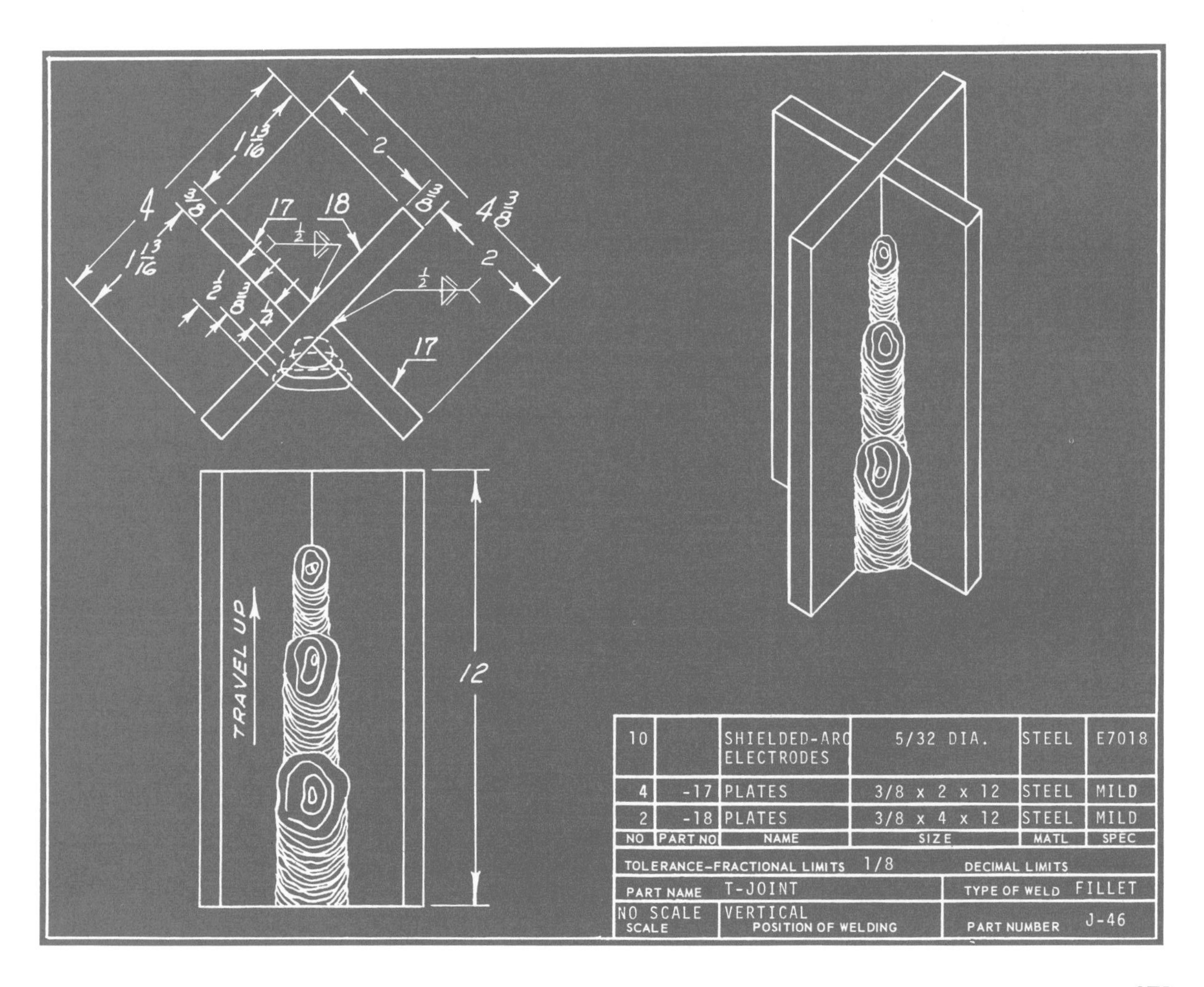

TRAVEL UP
12
17
18
10 SHIELDED-ARC ELECTRODES 5/32 DIA. STEEL E7018
4 -17 PLATES 3/8 x 2 x 12 STEEL MILD
2 -18 PLATES 3/8 x 4 x 12 STEEL MILD
NO PART NO NAME SIZE MATL SPEC
TOLERANCE-FRACTIONAL LIMITS 1/8 DECIMAL LIMITS
PART NAME T-JOINT
TYPE OF WELD FILLET
NO SCALE SCALE
VERTICAL POSITION OF WELDING
PART NUMBER J-46

JOB 21-J47 WELDING A SINGLE-V BUTT JOINT (BACKUP STRIP CONSTRUCTION)

Objective

To weld a single-V butt joint (backup strip construction) in the vertical position by means of multipass groove welds, weaved-beading technique, travel up, with DCRP and/or a.c. shielded-arc electrodes (AWS-ASTM E6010-E6011).

General Job Information

The single-V butt joint with a backup strip is found quite frequently in piping, pressure vessel, and ship construction. This type of joint is required by Test 1 of the Department of the Navy and other code authorities. It is also recommended by the American Welding Society. The backing material creates the effect of welding a joint with a chill ring. The root opening varies with different welding shops. If it is as large as $^3/_8$ inch, two or three root passes are necessary because of the greater separation between the plates. Another welding procedure is to use stringer beads.

21-40. *Typical appearance of multipass fillet weld in a T-joint welded in the vertical position, travel up, with low hydrogen, iron powder, shielded-arc electrodes (DCRP or a.c.).*

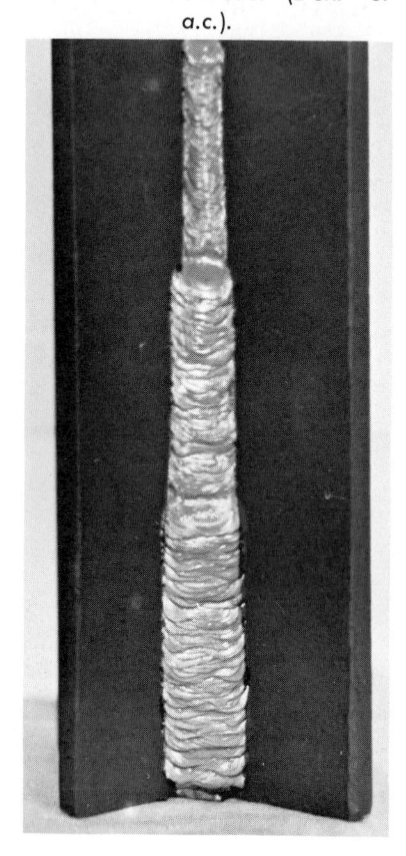

Welding Technique

FIRST PASS. Current adjustment can be high. Tilt the electrode no more than 10 degrees down from the horizontal position, Fig. 21-41A, and hold it at a 90-degree angle to the surface of the plates, Fig. 21-41B. Use an up-and-down motion for running stringer beads in the vertical position. The weld deposit must be fused into the backup strip and the root surfaces of the two plates. Keep the face of the weld flat. If the joint is to be set up with a wider root gap, two or three root passes should be made. Be sure that all passes are fused with each other.

SECOND PASS. The current setting must not be too high. Check Fig. 21-41 for the electrode position. Use the technique for running weaved beads in the vertical position. Hesitate at the sides to prevent undercut. Use the edges of the first bead as a guide for the width of the new pass. Keep the face of the weld flat. Reinforcement should not be so high that it covers the sharp edges of the beveled plate since these edges may be used as a guide for the last pass.

THIRD PASS. Current adjustment must be high enough to insure good fusion and penetration together with smooth appearance. The electrode position is the same as for the other passes. Use the technique for running weaved beads in the vertical position. Do not weave beyond the sharp edges of the plate bevel. Hesitate long enough at the sides to prevent undercut. Travel must be fast enough to prevent high convexity.

Practice starts and stops. Clean slag thoroughly between each pass.

Operations

1. Obtain plates; check the job print for quantity and size.
2. Obtain electrodes of each quantity, type, and size specified in the job print.
3. Set the power source at 110 to 170 amperes. Set a d.c. power source for reverse polarity.
4. Set up the plates as shown on the job print. Tack

21-41. *Electrode position when welding a beveled-butt joint in the vertical position, travel up.*

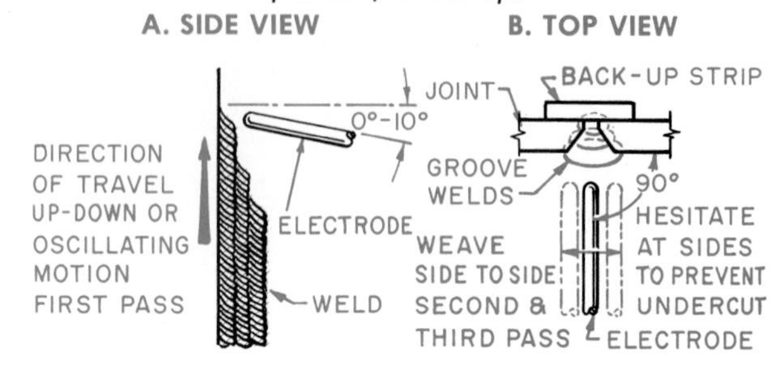

weld them at each corner and at the center of the back side.

5. Stand the joints in the vertical position on welding table. Make sure that they are well-grounded.

6. Make welds as shown on the job print. Manipulate the electrode as instructed in Fig. 21-41.

7. Chip the slag from the welds, brush, and inspect. Refer to Inspection.

8. Practice these welds until you can produce good welds consistently with both types of electrodes and with $\frac{3}{16}$-inch diameter electrodes.

Inspection

Compare each pass with Fig. 21-42 and check it for the weld characteristics which are designated here:

Width and height: Uniform

Appearance: Smooth with close ripples; free of voids and high spots. Restarts should be difficult to locate.

Size: Refer to the job print. Check the last pass with a butt weld gauge.

Face of weld: First two passes: flat; last pass: slight reinforcement

Edges of weld: Good fusion, no overlap, no undercut

Starts and stops: The weld should be free of depressions and high spots

Beginnings and endings: Full size, craters filled

Penetration and fusion: To the backup strip, preceding passes, and plate surface

Surrounding plate surfaces: Free of spatter

Slag formation: Full coverage, easily removable

Disposal

Put completed joints in the scrap bin so that they will be available for further use. The plate can be beveled between

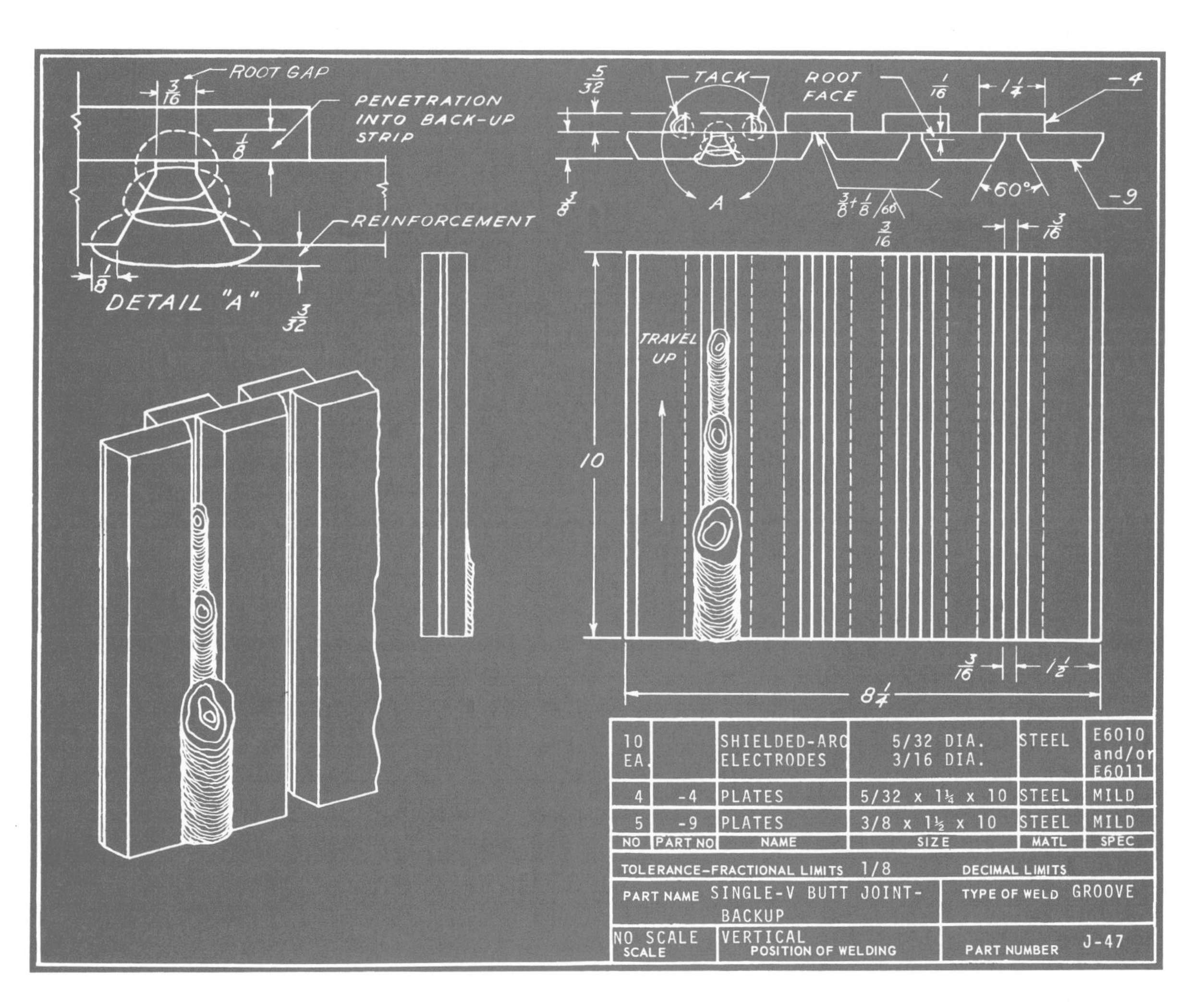

welds for further butt welding, and unwelded plate surfaces can be used for beading.

Qualifying Test

At this point you should be able to pass Test 1 in the vertical position. (See Chapter 22, pages 714–717.) This is identical to Test 1, Department of the Navy Requirements, and AWS recommendations. It will also be desirable for you to pass Test 3 (fillet weld) in the vertical position.

JOB 21-J48 WELDING A SQUARE BUTT JOINT

Objective

To weld a square butt joint in the vertical position by means of multipass groove welds, stringer and weaved beading techniques, travel up, with DCRP and/or a.c. shielded-arc electrodes (AWS-ASTM E6010-E6011).

21-42. *Typical appearance of a multipass groove weld in a beveled-butt joint with a backing strip welded in the vertical position, travel up, with DCRP or a.c. shielded-arc electrodes.*

General Job Information

The purpose of this job is to develop the ability of depositing the first pass in the root of an open butt joint. This job should be practiced with great care. It is the same procedure for making the first pass in a butt weld in pipe. This type of joint is assigned since it lends itself readily to practice and it can be prepared quickly and economically.

Welding Technique

FIRST PASS. The current setting must not be too high, and yet the arc must be hot enough to penetrate the back side of the plate thoroughly and to melt away the root faces of the plates. Tilt the electrode 10 degrees down from the horizontal position, Fig. 21-43A, and hold it at a 90-degree angle to the surface of the plates, Fig. 21-43B. Stringer beading technique may be used. Be sure to maintain the small hole at the leading edge of the weld pool. A small bead should be formed on the back side of the weld, Fig. 21-44.

SECOND PASS. Current adjustment and electrode position are the same as for first pass. Do not weave beads that are too wide. Adjust the speed of travel to avoid excessive convexity and undercut.

Practice starts and stops, especially on the first pass.

Operations

1. Obtain plates; check the job print for quantity and size.
2. Obtain electrodes of each quantity, type, and size specified in the job print.
3. Set the power source for 90 to 120 amperes. Set a d.c. power source for reverse polarity.
4. Set up the plates and tack weld as shown on the job print.
5. Stand the plates in the vertical position on the welding table. Make sure that they are well-grounded.
6. Make the first pass as shown on the job print. Manipulate the electrode as instructed in Fig. 21-43.
7. Chip the slag from the weld, brush, and inspect. Refer to Inspection.
8. Make the second pass as shown on the job print.
9. Chip the slag from the weld, brush, and inspect. Refer to Inspection.
10. Practice these welds until you can produce good welds consistently with both types of electrodes.

21-43. *Electrode position when welding a square butt joint in the vertical position, travel up.*

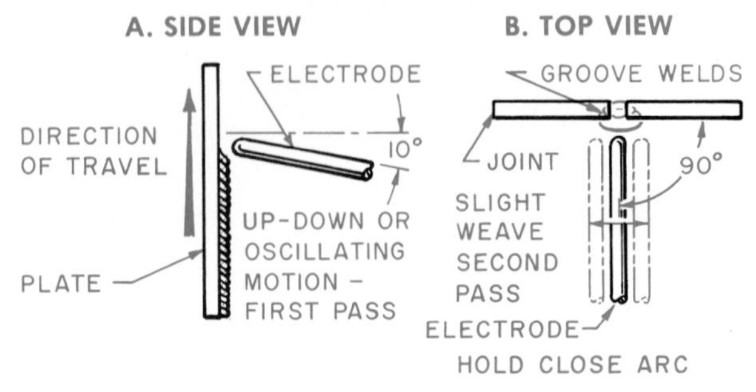

Inspection

Compare the back side of the first bead with Fig. 21-44 and both passes with Fig. 21-45. Check each pass for the weld characteristics which are designated here:

Width and height: Uniform
Appearance: Smooth with close ripples; free of voids and high spots. Restarts should be difficult to locate.
Size: Refer to the job print. Check the last pass with a butt weld gauge.
Face of weld: Very slight reinforcement
Edges of weld: Good fusion, no overlap, no undercut
Starts and stops: Free of depressions and high spots
Beginnings and endings: Full size, craters filled
Back side: Complete and uniform penetration
Penetration and fusion: Through the back side and plate surfaces. See Fig. 21-44.
Surrounding plate surfaces: Free of spatter
Slag formation: Full coverage, easily removable

Disposal

Put completed joints in the scrap bin so that they will be available for further use. Joints can be cut between the welds for further butt welding, and unwelded plate surfaces can be used for beading.

JOB 21-J49 WELDING AN OUTSIDE CORNER JOINT

Objective

To weld an outside corner joint in the vertical position by means of multipass groove welds, weaved-beading technique, travel up, with DCRP and/or a.c. shielded-arc electrodes (AWS-ASTM E6010-E6011).

General Job Information

Although the outside corner joint is often welded in industry,

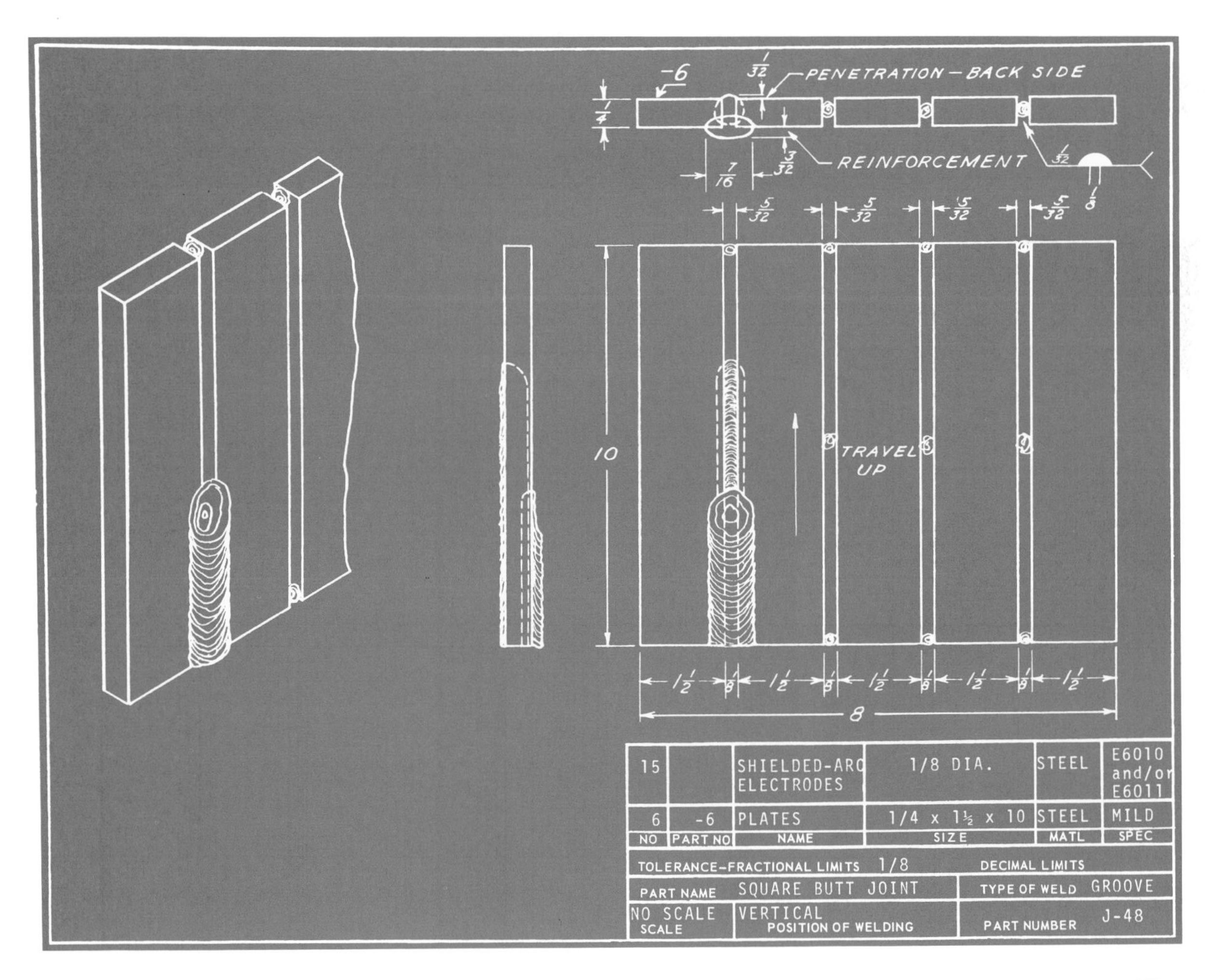

it is included in this text chiefly because of its all-around utility. It offers practice in depositing root passes and in multipass weaved technique, and it requires little edge preparation since it has a square plate edge.

Welding Technique

FIRST PASS. The current adjustment must not be too high. Tilt the electrode no more than 10 degrees down from the horizontal position, Fig. 21-46A, and hold it at a 90-degree angle to the intersection of the plates, Fig. 21-46B. Use the standard root pass technique.

SECOND AND THIRD PASSES. The current adjustment can be higher than for the first pass. Check the electrode position with Fig. 21-46. Use the weaved-beading technique. Secure good fusion with the preceding passes and the walls of the plate. Do not undercut. The face of welds should be kept flat.

FOURTH PASS. The current adjustment and electrode position are approximately those used for previous passes. Use the weaved-beading technique. Let the edges of the plate be your guide for bead width. The face of the weld should be slightly convex to provide full reinforcement.

Practice starts and stops.

Operations

1. Obtain plates; check the job print for quantity and size.
2. Obtain electrodes of each quantity, type, and size specified in the job print.
3. Set the power source for 75 to 125 amperes for the first pass. Set a d.c. power source for reverse polarity.
4. Set up the plates and tack weld them as shown on the job print.
5. Stand the joints in the vertical position on the welding table. Make sure that they are well-grounded.
6. Make the first pass with a $\frac{1}{8}$-inch electrode as shown on the job print. Manipulate the electrode as instructed in Fig. 21-46.
7. Chip the slag from the weld, brush and inspect. Refer to Inspection.
8. Readjust the power source for 110 to 170 amperes.
9. Make the second, third, and fourth passes with $\frac{5}{32}$-inch electrodes as shown on the job print.
10. Chip the slag from the welds, brush, and inspect between each pass. Refer to Inspection.
11. Practice these welds until you can produce good welds consistently with both types of electrodes.

21-44. *Penetration and fusion through the back side of a square butt joint welded in the vertical position, travel up.*

21-45. *Typical appearance of a multipass groove weld in a square butt joint welded in the vertical position, travel up, with DCRP or a.c. shielded-arc electrodes.*

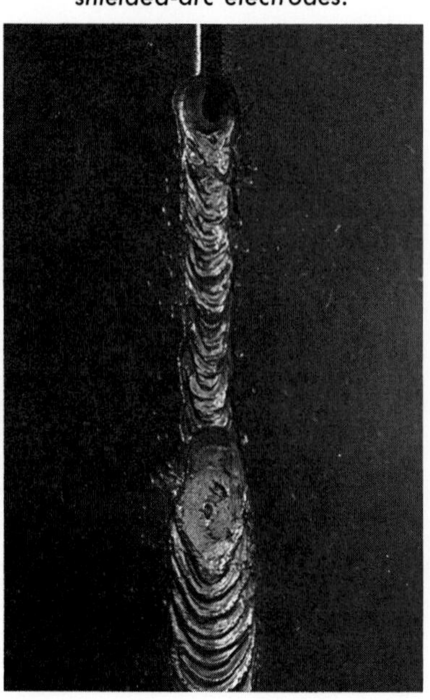

21-46. *Electrode position when welding a corner joint in the vertical position, travel up.*

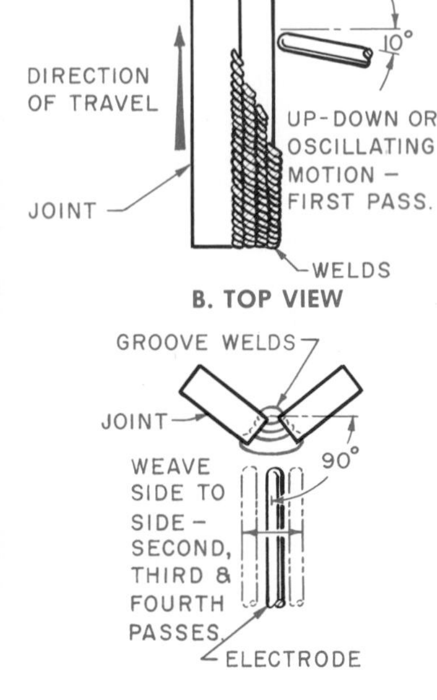

Inspection

Compare the back side of the first bead with Fig. 21-47 and all passes with Fig. 21-48. Check each pass for the weld characteristics which are designated here:

Width and height: Uniform
Appearance: Smooth with close ripples; free of voids and high spots. Restarts should be difficult to locate.
Size: Refer to the job print.
Face of weld: First three passes: flat; last pass: slightly convex
Edges of weld: Good fusion, no overlap, no undercut
Starts and stops: Free of depressions and high spots
Beginnings and endings: Full size, craters filled
Back side: Complete and uniform penetration
Penetration and fusion: Through the back side and to all plate surfaces. See Fig. 21-47.
Surrounding plate surfaces: Free of spatter
Slag formation: Full coverage, easily removable

Disposal

Put completed joints in the scrap bin so that they will be available for further use. Fillet welds can be run on the reverse side, or the flat unwelded surfaces can be used for beading.

JOB 21-J50 WELDING A T-JOINT

Objective

To weld a T-joint in the horizontal position by means of multipass fillet welds, weaved-beading technique, with low hydrogen, iron powder, shielded-arc electrodes (AWS-ASTM E7018).

General Job Information

The purpose of this job is to practice making large size fillet welds with low hydrogen, iron powder electrodes. Review all of the material concerning this type of electrode in Chapter 7 and Job 21-J46. Note that the weld sym-

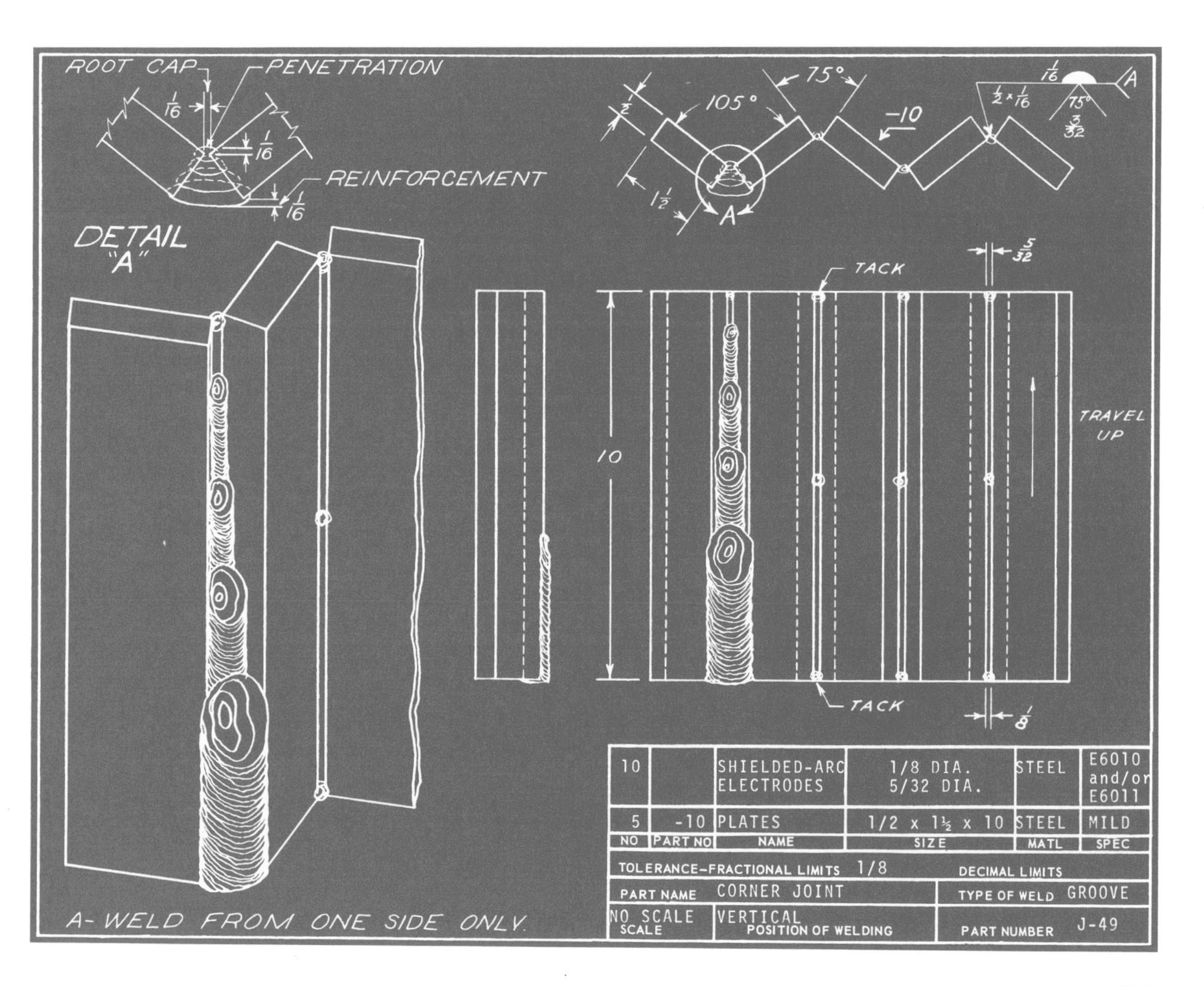

NO	PART NO	NAME	SIZE	MATL	SPEC
10		SHIELDED-ARC ELECTRODES	1/8 DIA. 5/32 DIA.	STEEL	E6010 and/or E6011
5	-10	PLATES	1/2 x 1½ x 10	STEEL	MILD

TOLERANCE–FRACTIONAL LIMITS 1/8 DECIMAL LIMITS
PART NAME CORNER JOINT TYPE OF WELD GROOVE
SCALE NO SCALE POSITION OF WELDING VERTICAL PART NUMBER J-49

bol in the job print carries a diagonal line (/) to indicate a flat face and *12* for the length of the weld.

Welding Technique

FIRST PASS. Current adjustment should be somewhat higher than for flat welding. Hold the electrode at a 40-degree angle to the flat plate, Fig. 21-49A, and tilt it 30 degrees in the direction of travel, Fig. 21-49B. As with other low hydrogen electrodes, a short arc should be maintained at all times. Keep the electrode in the pool so that the slag does not run ahead of the weld. The E7018 electrode can be used with the coating in contact with the work. This is known as the *drag technique.* Use very little weave movement for the first pass. The face of the weld should be kept as flat as possible. Make sure that penetration and fusion is secured at the root of the joint and that there is no undercutting along the side walls.

SECOND PASS. Deposit the second pass over the first pass with a slight oscillating motion, Fig. 21-49B. Direct the electrode towards the vertical plate in order to force the weld metal up on its surface. Keep a short arc and restrict electrode movement within the limits of the desired weld width to prevent undercut. The circular motion should be counterclockwise. Secure good fusion with the underneath pass and the plate surfaces.

Practice starts and stops. Travel in all directions. These electrodes pick up moisture; be sure to keep them dry.

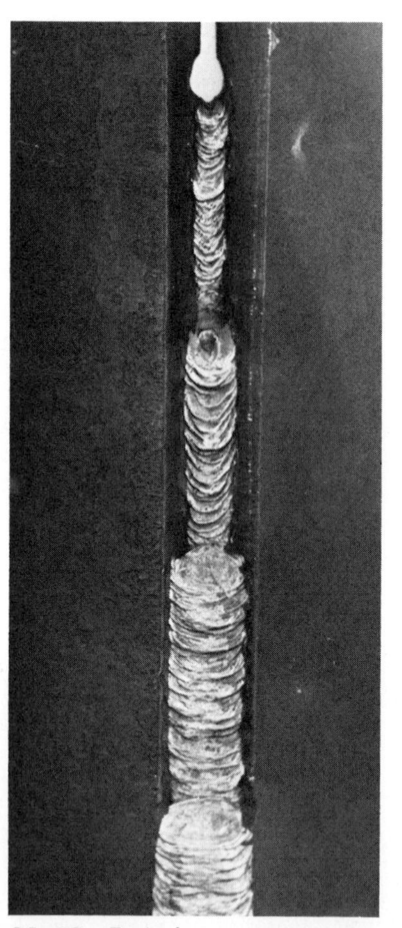

21-48. *Typical appearance of a multipass groove weld in an outside corner joint welded in the vertical position, travel up, with DCRP or a.c. shielded-arc electrodes. Note the penetration hole at the forward edge of the root pass.*

21-47. *Penetration and fusion through the back side of the outside corner joint shown in Fig. 21-48.*

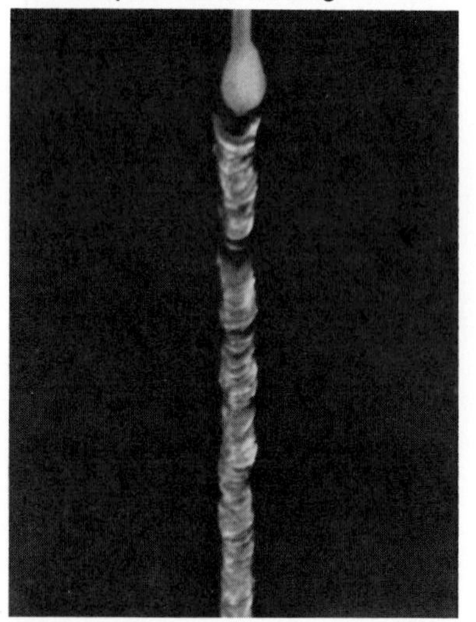

Operations

1. Obtain plates; check the job print for quantity and size.
2. Obtain electrodes of each quantity, type, and size specified in the job print.
3. Set a d.c. power source for reverse polarity at 150 to 220 amperes. An a.c. power source may be used at a somewhat higher heat.
4. Set up two sets of plates as shown on the job print and tack weld them at each corner.
5. Place the joints in the horizontal position on the welding table. Make sure that they are well-grounded.
6. Make the first pass with $\frac{5}{32}$-inch diameter electrodes as shown on the job print. Manipu-

21-49. *Electrode position when welding a T-joint in the horizontal position.*

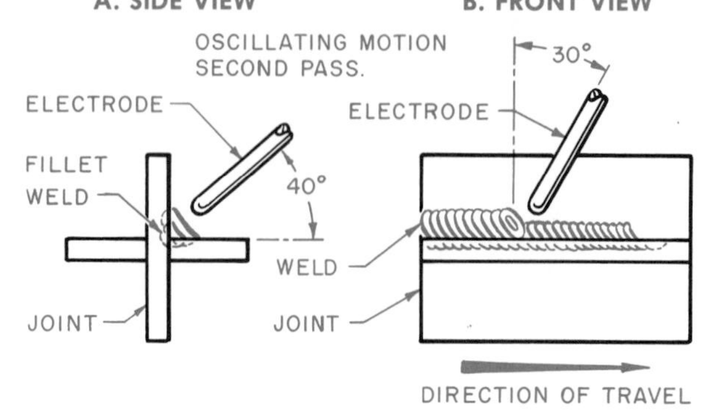

late the electrodes as instructed in Fig. 20-49. Weld on both joints. (This is to keep the joint from getting too hot.)

7. Chip the slag from the welds, brush, and inspect. Refer to Inspection.

8. Readjust the d.c. power source to 200 to 280 amperes. Set an a.c. power source somewhat higher.

9. Make the second pass with $^{3}/_{16}$-inch diameter electrodes as shown on the job print. Alternate your welding between both joints.

10. Chip the slag from the welds, brush, and inspect. Refer to Inspection.

11. Practice these welds until you can produce good welds consistently. Change your procedure by using $^{3}/_{16}$-inch diameter electrodes for the first pass. You may also wish to weld a third pass with the $^{3}/_{16}$-inch size.

Inspection

Compare both passes with Fig. 21-50 and check them for the following weld characteristics:

Width and height: Uniform

Appearance: Very smooth with fine ripples; free of voids and

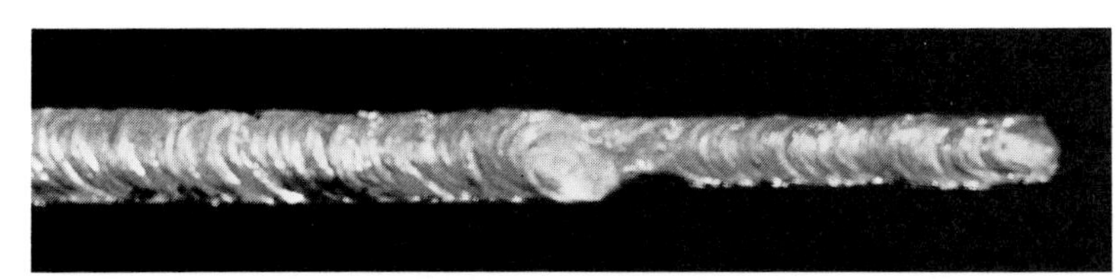

21-50. *Typical appearance of a multipass fillet weld in a T-joint welded in the horizontal position with low hydrogen, iron powder, shielded-arc electrodes (DCRP or a.c.).*

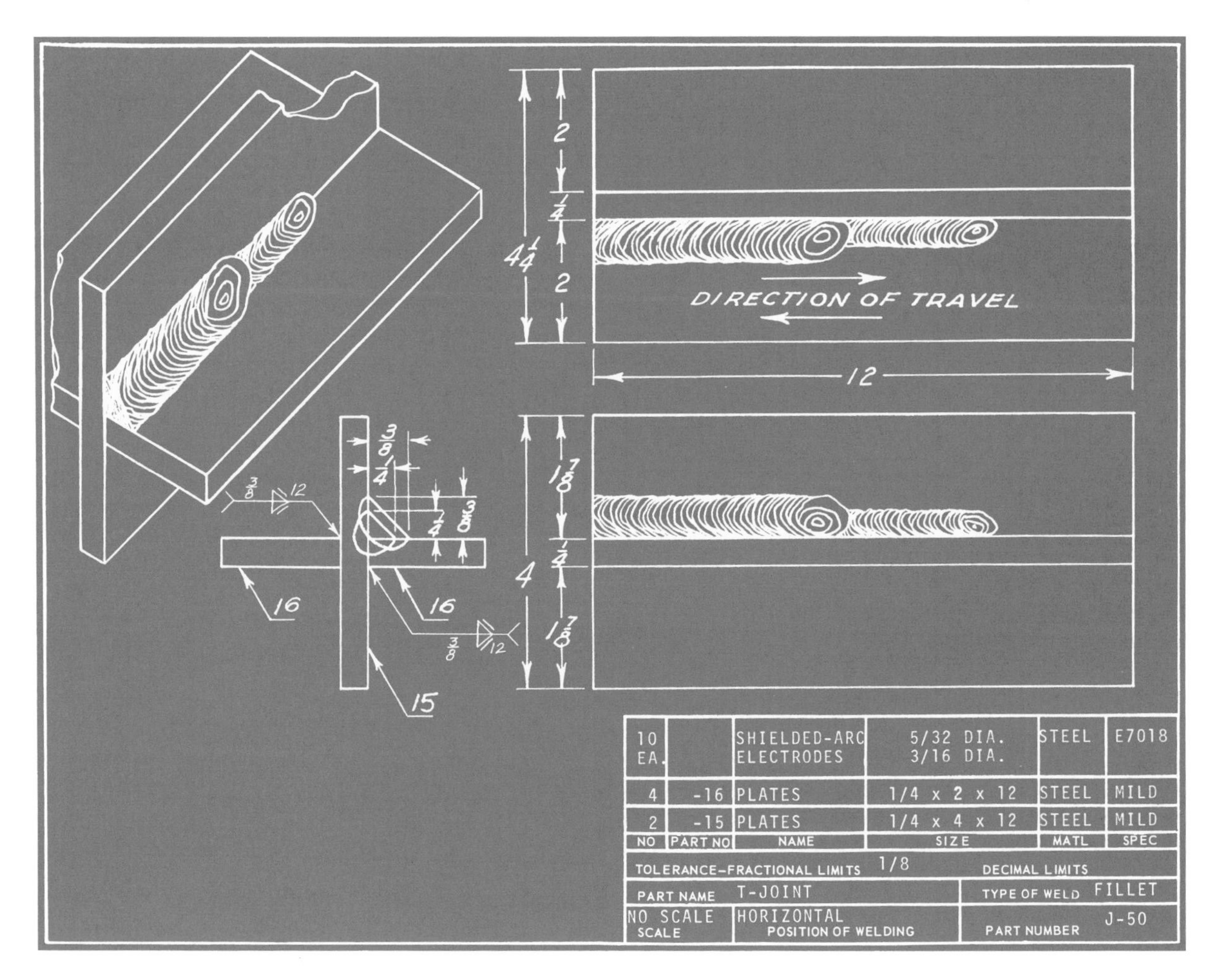

high spots. Restarts should be difficult to locate.

Size: Refer to the job print. Check with a fillet weld gauge.

Face of weld: Flat to slightly convex

Edges of weld: Good fusion, no overlap, no undercut

Starts and stops: Free of depressions and high spots

Beginnings and endings: Full size, craters filled

Penetration and fusion: To the root of the weld and plate surfaces

Surrounding plate surfaces: Free of spatter

Slag formation: Full coverage, easily removable

Disposal

Put completed joints in the scrap bin so that they will be available for further use. Plate edges can be beveled for beveled-butt welding, and unwelded plate surfaces can be used for beading.

JOB 21-J51 WELDING A SINGLE-V BUTT JOINT (BACK UP STRIP CONSTRUCTION)

Objective

To weld a single-V butt joint with backup strip construction in the vertical position by means of multipass groove welds, the weaved-beading technique, travel down, with DCRP and/or a.c. shielded-arc electrodes (AWS-ASTM E6010-E6011).

General Job Information

The single-V butt joint with backup strip, welded by the downhill method, is used extensively in pipeline, tank, and shipyard work. It is suitable for both plate and pipe. The backup strip in plate is comparable to the inner liner which acts as a backup for butt joints in pipe. Generally, the joint is used with pipe 12 inches or larger. The downhill welding technique is faster than the uphill method.

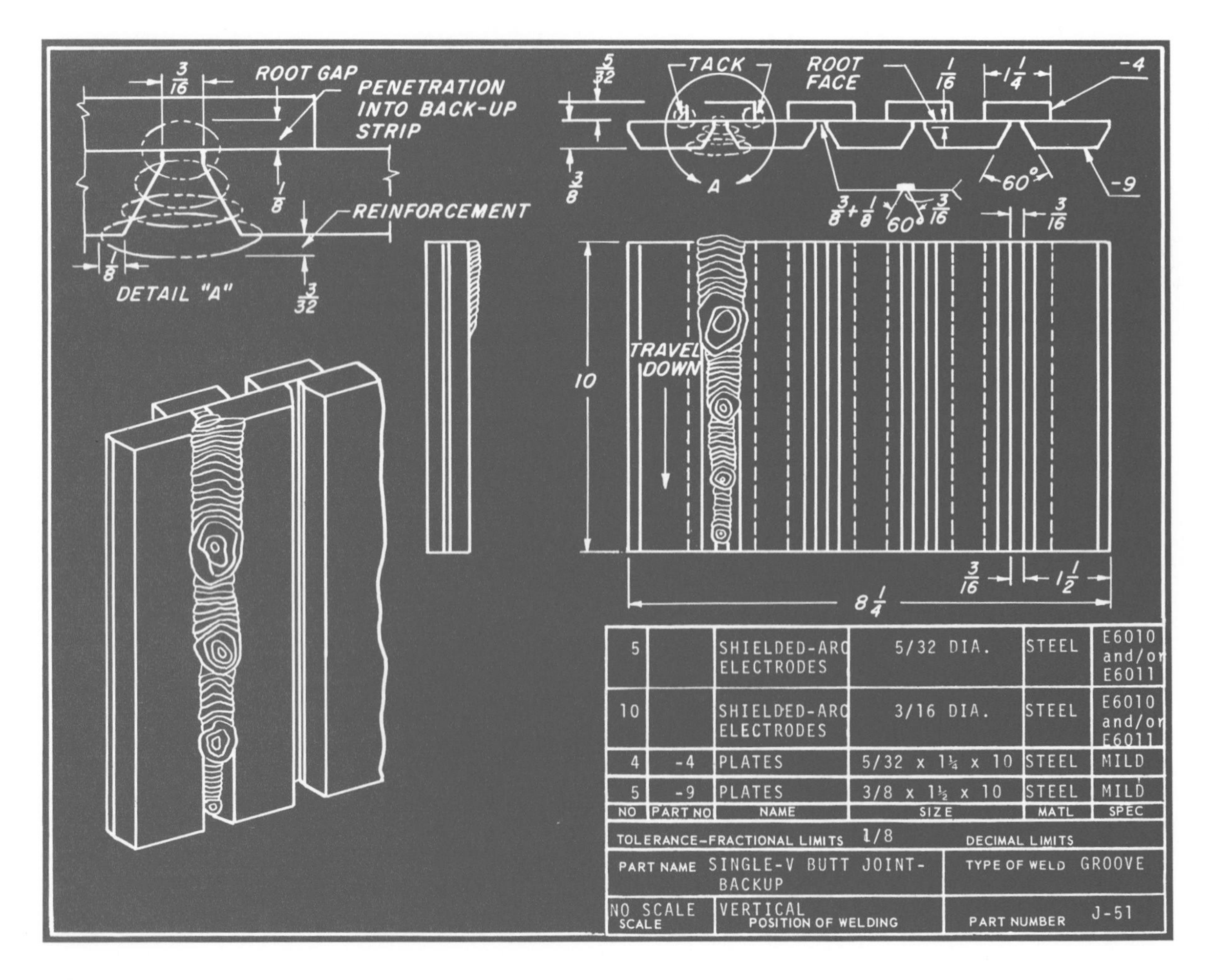

NO	PART NO	NAME	SIZE	MATL	SPEC
5		SHIELDED-ARC ELECTRODES	5/32 DIA.	STEEL	E6010 and/or E6011
10		SHIELDED-ARC ELECTRODES	3/16 DIA.	STEEL	E6010 and/or E6011
4	-4	PLATES	5/32 x 1¼ x 10	STEEL	MILD
5	-9	PLATES	3/8 x 1½ x 10	STEEL	MILD

TOLERANCE–FRACTIONAL LIMITS 1/8 DECIMAL LIMITS

PART NAME SINGLE-V BUTT JOINT-BACKUP TYPE OF WELD GROOVE

SCALE NO SCALE POSITION OF WELDING VERTICAL PART NUMBER J-51

The skill learned in the practice of this joint will assist you in learning to weld pipe, which is presented in Chapter 23.

Welding Technique

Downhill welding is used in the interest of high production. In the piping field two welders usually work on a joint; each welder takes one side of the joint. They are able to weld as many as five 12-inch pipe joints in an hour. Many welders in the field use $^{3}/_{16}$-inch electrodes for all passes. For our initial practice the first bead will be done with $^{5}/_{32}$-inch electrodes.

The current adjustment should be somewhat higher than for vertical upwelding. Tilt the electrode 15 to 20 degrees down from the horizontal position, Fig. 21-51B, and hold it at a 90-degree angle to the surface of the plates, Fig. 21-51C.

ROOT PASS. The first pass should be run with a fairly close arc, without weaving, and with a higher than normal rate of travel. The weld pool and the slag always have a tendency to run ahead of the arc. When this happens, it can be overcome by using a shorter arc, increasing the rod angle, and increasing the rate of travel. Usually one or a combination of all three must be employed. There is no problem with undercutting along the sides of the weld. Make sure that fusion and penetration are secured at the root of the weld with the backup strip and the root faces of the beveled plates. Keep the face of the weld as flat as possible. There may be some trouble with holes or voids in the face of the weld. They are usually caused by slag piling up ahead of the arc. Watch arc length, electrode angle, and speed of travel.

FILLER AND COVER PASSES. The second, third, and fourth passes should be run like the first pass. For each pass the side-to-side motion should be a little wider than for the preceding pass. Watch carefully for holes and voids in the face of the weld. There is also a tendency for the face of the weld to become concave. This can be avoided by pausing for a shorter time at the edges of the weld in making the weave motion. Because undercutting is not a problem, the pause at the sides of the weld should be short.

21-51. *Electrode position when welding a beveled-butt joint with backing in the vertical position, travel down.*

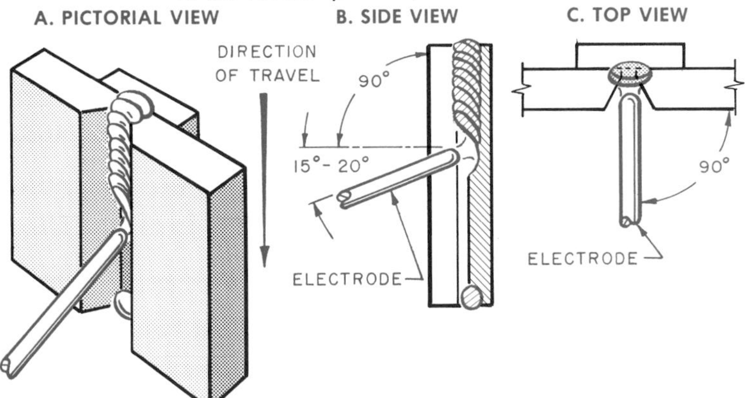

Make sure that fusion and penetration are secured between weld passes and into the beveled walls of the plate. The face of each completed pass should be as flat as possible. The final finish pass should be slightly convex in order to make sure that it is full and not below the surface of the plate. In making the final pass, do not weave beyond the sharp edges of the plate at each side.

Practice varying the heat setting, the angle of the electrode, the length of the arc, and the speed of travel. Starts and stops should also be practiced.

Operations

1. Obtain plates; check the job print for quantity and size.
2. Obtain electrodes of each quantity, type, and size specified in the job print.
3. Set the power source for 125 to 200 amperes. Set a d.c. power source for reverse polarity.
4. Set up the plates as shown on the job print and tack weld them at each corner.
5. Stand the joint in the vertical position on the welding table. Make sure that it is well-grounded.
6. Make the first pass with $^{5}/_{32}$-inch electrodes as shown on the job print. Manipulate the electrode as instructed in Fig. 21-51.
7. Chip the slag from the weld, brush, and inspect. Refer to Inspection.
8. Readjust the power source to 130 to 225 amperes.
9. Make the second, third, and fourth passes with $^{3}/_{16}$-inch electrodes as shown on the job print.
10. Chip the slag from the welds, brush, and inspect. Refer to Inspection.
11. Practice these welds until you can produce good welds

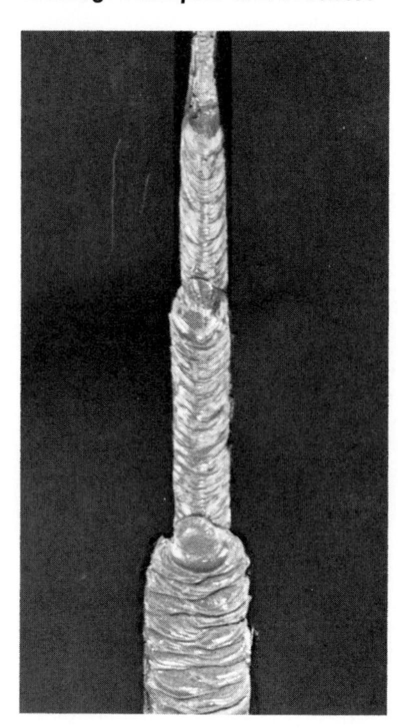

21-52. *Typical appearance of a multipass groove weld in a beveled-butt joint with a backup strip welded in the vertical position, travel down, with DCRP or a.c. shielded-arc electrodes.*

consistently. Change your procedure by using $^{3}/_{16}$-inch electrodes for the first pass.

Inspection

Compare all passes with Fig. 21-52 and check them for the following weld characteristics:

Width and height: Uniform

Appearance: Very smooth with fine ripples; free of voids and high spots. Restarts should be difficult to locate.

Size: Refer to the job print. Check with a fillet weld gauge.

Face of weld: Flat to slightly convex

Edges of weld: Good fusion, no overlap, no undercut

Starts and stops: Free of depressions and high spots

Beginnings and endings: Full size, craters filled

Penetration and fusion: To the root of the weld and plate surfaces

Surrounding plate surfaces: Free of spatter

Slag formation: Full coverage, easily removable

Disposal

Put completed joints in the scrap bin so that they will be available for further use. Plate edges can be beveled for beveled–butt welding, and unwelded plate surfaces can be used for beading.

REVIEW QUESTIONS

1. Is it possible to weld a square butt joint in $^{1}/_{2}$-inch plate and secure maximum strength? Explain.
2. Why is spacing necessary?
3. What is the maximum thickness of plate (used in a square butt joint) that can be welded effectively from one side?
4. What are free-flowing electrodes?
5. Name the weld used for a T-joint.
6. In overhead welding, the metal is deposited in front of the arc. True. False.
7. Explain the technique for running stringer beads in the overhead position.
8. Explain the technique for running weaved beads in the overhead position.
9. Which method of welding—stringer or weaved beading—is more difficult in the overhead position?
10. What is the greatest weld width that should be weaved in the overhead position?
11. In welding a T-joint, it is important to obtain good penetration and fusion with what parts of the joint?
12. Which types of fabrications frequently require the use of single-V butt joints, backup strip construction?

13. How is it possible to develop full strength in a corner joint?
14. In making the first pass in an open corner joint in flat position with $5/32$-inch reverse polarity electrodes, what is the approximate heat setting?
15. In welding a corner joint, it is important to obtain good penetration and fusion with which parts of the joint?
16. The single-V butt joint, backup strip construction, in $3/8$-inch plate is used in what important test?
17. The single-V butt joint, without backup strip, in $3/8$-inch plate is similar to what important test?
18. Is it necessary to gap the plates for single-V butt joints? What should the gap be if a backing strip is used? What should the gap be if no backing strip is used?
19. What is the included angle of the V in single-V butt joints?
20. In welding the single-V butt joint with backing strip, is it necessary to obtain fusion with the backing strip? Explain.
21. In welding the single-V butt joint without backing strip it is not necessary to obtain fusion and penetration through the back side. True. False.
22. Welding the first pass of the single-V butt joint, without backing strip, in the flat position is accomplished with what size electrode? What is the heat setting?
23. The single-V butt joint is used in pipe welding and critical plate welding. True. False.
24. What instrument is used to check the size of a butt weld?
25. In welding a single-V butt joint, it is important to obtain good penetration and fusion with what parts of the joint?
26. Explain the technique for welding a T-joint with free-flowing, iron powder electrodes.
27. What size electrode should be used for the first pass? The remaining passes?
28. The heat setting should be 180 to 225 amperes for each pass. True. False.
29. In welding a square butt joint, it is important to obtain good penetration and fusion with what parts of the joint?
30. Explain how to make a large sized fillet weld in a T-joint in the horizontal position.
31. How are single-V butt welds in the horizontal position usually made?
32. A single-V butt joint in $3/8$-inch plate, welding in horizontal position, stringer bead technique, would require how many passes?
33. Can free-flowing electrodes be used with reverse polarity?
34. A fillet weld is to be run in a T-joint of $3/8$-inch metal in the flat position with $3/16$-inch free-flowing electrodes. What should be the approximate heat setting?
35. The slag formation of free-flowing electrodes is the same as for reverse and straight polarity electrodes. True. False.
36. What will be the face appearance of a fillet weld made with free-flowing electrodes?
37. A large sized fillet weld in a T-joint in the horizontal position, made with free-flowing electrodes, is accomplished in what manner?
38. What major industry uses free-flowing electrodes extensively?
39. Explain the procedure for making groove welds in beveled plate with free-flowing electrodes. How is the width of the weave determined?
40. Is there any difference in the free-flowing electrodes used for fillet welds and those used for groove welds? Give AWS classification numbers.
41. How many passes of free-flowing electrodes are necessary in welding a single-V butt joint in $1/2$-inch plate?
42. What weld characteristics should be checked when inspecting a finished weld?
43. Are free-flowing electrodes used frequently for vertical welding? Explain your answer.
44. The E7016 electrode produces welds of highest quality and may be used for practically all code work. True. False.
45. Draw a simple sketch showing the position of the electrode for downhill travel. For uphill travel.
46. List the sizes of electrodes that are recommended for vertical welding.
47. Explain the technique used when making a single-pass fillet weld in the vertical position.
48. Explain the technique used when making a multipass fillet weld in the vertical position.
49. When welding in the vertical position, what factors must be kept in mind if undercut is to be prevented?
50. When welding a single-V butt joint with backing strip in the vertical position, is it necessary to secure complete penetration and fusion with the backing strip? Explain.
51. Is it important to learn overhead welding? Why or why not?
52. A single-V butt joint with a backup strip is similar to what joint in the piping industry?

Table 21-1
Course Outline: Shielded Metal-Arc Welding Practice: Jobs 21-J 29-51 (Plate)

Job no.	Joint	Type of weld	Type of electrode						Text reference
			Position	DCRP	DCSP	a.c.	AWS specif. no. d.c. and/or	AWS specif. no. a.c.	
21-J29	Flat plate	Beading—stringer	Over.	X		X	E6010	E6011	635
21-J30	Flat plate	Beading—stringer	Over.		X	X	E6012	E6013	638
21-J31	Flat plate	Beading—weaved	Over.	X		X	E6010	E6011	639
21-J32	Single-V butt joint backup strip	Groove welding—weaved—multipass	Flat	X		X	E6010	E6011	641
21-J33	T-joint	Fillet welding—weaved—multipass	Hor.	X		X	E6010	E6011	643
21-J34	Single-V butt joint backup strip	Groove welding—stringer—multipass	Hor.	X		X	E6010	E6011	645
21-J35	Square butt joint	Groove welding—multipass	Flat	X		X	E6010	E6011	647
20-J36	Outside corner joint	Groove welding—weaved—multipass	Flat	X		X	E6010	E6011	649
21-J37	T-joint	Fillet welding—weaved—multipass	Flat	X		X	E7028	E7028	651
21-J38	T-joint	Fillet welding—weaved—multipass	Flat		X	X	E6020	E6020	654
21-J39	T-joint	Fillet welding—weaved—multipass	Hor.		X	X	E6020	E6020	656
21-J40	Single-V butt joint	Groove welding—weaved—multipass	Flat	X		X	E6010	E6011	658
21-J41	T-joint	Fillet welding—weaved—multipass	Flat	X		X	E6024 E7024	E6024 E7024	660
21-J42	Single-V butt joint	Groove welding—weaved—multipass	Flat	X X	 X	X X	E6010 E6027	E6011 E6027	662
21-J43	T-joint	Fillet welding—stringer—multipass	Hor.	X		X	E7016	E7016	664
21-J44	T-joint	Fillet welding—single pass—travel up	Ver.	X		X	E6010	E6011	666
21-J45	T-joint	Fillet welding—weaved—multipass—travel up	Ver.	X		X	E6010	E6011	668
21-J46	T-joint	Fillet welding—weaved—multipass—travel up	Ver.	X		X	E7018	E7018	670
21-J47	Single-V butt joint backup strip	Groove welding—weaved—multipass—travel up	Ver.	X		X	E6010	E6011	672
21-J48	Square butt joint	Groove welding—multipass—travel up	Ver.	X		X	E6010	E6011	674
21-J49	Outside corner joint	Groove welding—weaved—multipass—travel up	Ver.	X		X	E6010	E6011	675
21-J50	T-joint	Fillet welding—weaved—multipass	Hor.	X		X	E7018	E7018	677
21-J51	Single-V butt joint backup strip	Groove welding—weaved—multipass—travel down	Ver.	X		X	E6010	E6011	680

CHAPTER 22 Shielded Metal-Arc Welding Practice: Jobs 22-J52–J65 (Plate)

INTRODUCTION

This third chapter in shielded metal-arc welding (SMAW) practice may be considered as advanced welding. It provides practice in joints that are basic to code welding required for pipe and building construction. (As stated previously, the practice jobs are numbered consecutively through the three units.) Successful completion of these practice jobs is the final step in development of skill as a welder. A welder who is able to satisfactorily weld an open butt joint with complete penetration through to the back side in the vertical and overhead positions has achieved the manipulative skill that provides the basis for all types of welding application.

The chapter also provides considerable experience in the use of both the hand and machine oxyacetylene cutting torches in the preparation of practice joints and test specimens. In addition, it adds to the student's experience in the use of various hand and machine tools.

The abilities to weld, to use the various cutting processes, to read job prints and sketches, to do simple layout, and to be able to use the basic metalworking tools will have been acquired through the completion of these practice jobs. If the student welder is able to perform these skills satisfactorily, he can feel confident that he has the necessary skills to satisfy the demands of industry for those jobs in which shielded metal-arc welding is performed, Fig. 22-1.

JOB 22-J52 WELDING A SINGLE-V BUTT JOINT

Objective

To weld a single-V butt joint in the vertical position by means of multipass groove welds, weaved beading technique, travel up, with DCRP and/or a.c. shielded-arc electrodes (AWS-ASTM E6010-E6011).

General Job Information

The single-V butt joint in the vertical position will be encountered in pipe and critical plate welding. The joint design and welding procedure are similar to those required in the ASME plate and pipe tests.

Welding Technique

FIRST PASS. The current setting should not be too low. Hold the electrode at a 90-degree angle to the plate surfaces and no more than 10 degrees down from the horizontal, Fig. 22-2A. Use the root pass technique in which an

22-1. *A former welding student on his first job in industry. He is tack welding parts held in place by another worker before he welds them.*

up-and-down motion is employed. On the upward motion do not allow the electrode to travel much more than 2 inches and do not extinguish the arc.

Make sure that the small hole is present at the point of welding during the welding operation. This is your guarantee that you are securing penetration through to the back side. Complete penetration is necessary for maximum strength. It is also necessary when making a test weld. Keep the face of the weld as flat as possible.

SECOND AND THIRD PASSES. Current adjustment must be increased. Hold the electrode at a 90-degree angle to the plate surfaces and weave it from side to side, Fig. 22-2B. A uniform weave and rate of travel is important to obtain a weld of uniform quality and appearance. Movement across the face of the weld must be fast to prevent excessive convexity in the center of the weld. Maintain a close arc gap at all times. Keep electrode

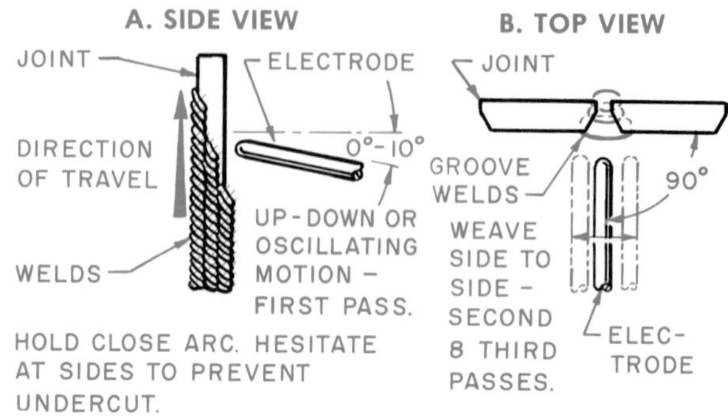

22-2. *Electrode position when welding a butt joint in the vertical position.*

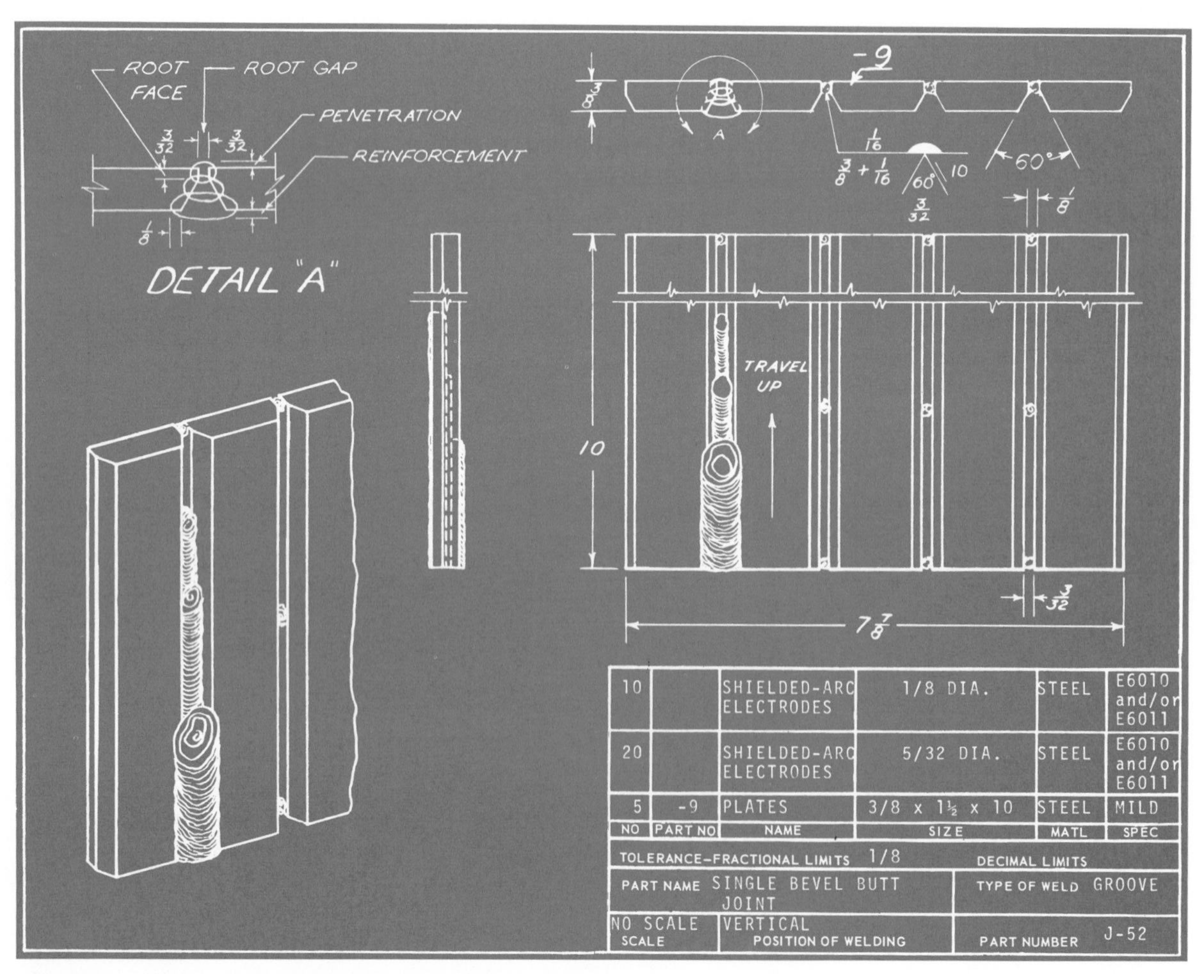

NO	PART NO	NAME	SIZE	MATL	SPEC
10		SHIELDED-ARC ELECTRODES	1/8 DIA.	STEEL	E6010 and/or E6011
20		SHIELDED-ARC ELECTRODES	5/32 DIA.	STEEL	E6010 and/or E6011
5	-9	PLATES	3/8 x 1½ x 10	STEEL	MILD

TOLERANCE-FRACTIONAL LIMITS 1/8	DECIMAL LIMITS
PART NAME SINGLE BEVEL BUTT JOINT	TYPE OF WELD GROOVE
SCALE NO SCALE / POSITION OF WELDING VERTICAL	PART NUMBER J-52

movement within the limits of the bead width and hesitate at the sides of the weld to avoid undercut.

Practice starts and stops on all passes.

Operations

1. Obtain plates; check the job print for quantity and size.
2. Obtain electrodes of each quantity, type, and size specified in the job print.
3. Set the power for 75 to 125 amperes. Set a d.c. power source for reverse polarity.
4. Set up the plates and tack weld them as shown on the job print.
5. Place the joint in the vertical position on the welding table. Make sure that it is well-grounded.
6. Make the first pass with ⅛-inch electrodes as shown on the job print.
7. Chip the slag from the weld, brush, and inspect. Refer to Inspection, below.
8. Increase the current to 130 to 160 amperes.
9. Make the second and third passes with $^5/_{32}$-inch electrodes as shown on the job print.
10. Chip the slag from the weld, brush, and inspect. Refer to Inspection, below.
11. Practice these welds until you can produce good welds consistently with both types of electrodes. You should also practice with $^5/_{32}$-inch electrodes for the first pass.

Inspection

Compare the back side of the first pass with Fig. 22-3 and all passes with Fig. 22-4. Check them for the following weld characteristics:

Width and height: Uniform
Appearance: Smooth with close ripples; free of voids and high spots. Restarts should be difficult to locate.
Size: Refer to the job print. Check with a butt weld gauge.
Face of weld: First two passes, flat; last pass, slight convexity
Edges of weld: Good fusion, no overlap, no undercut
Starts and stops: Free of depressions and high spots
Beginnings and endings: Full size, craters filled
Back side: Complete and uniform penetration
Penetration and fusion: Through the back side, preceding

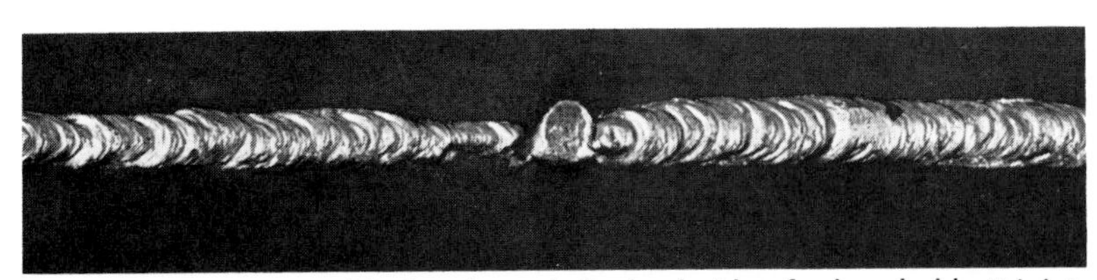

22-3. *Penetration and fusion through the back side of a beveled-butt joint. Note the spot in the center of the bead where there is a break in the penetration. This can be caused by failure to burn through a tack or a poor tie-in on a restart.*

22-4. *(A) Typical appearance of a multipass groove weld in a beveled-butt joint without a backup strip welded in the vertical position, travel up, with DCRP or a.c. shielded-arc electrodes. (B) Closeup of the weld bead.*

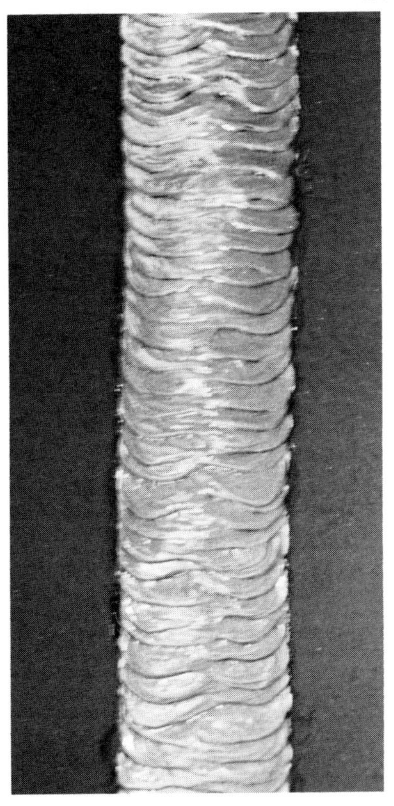

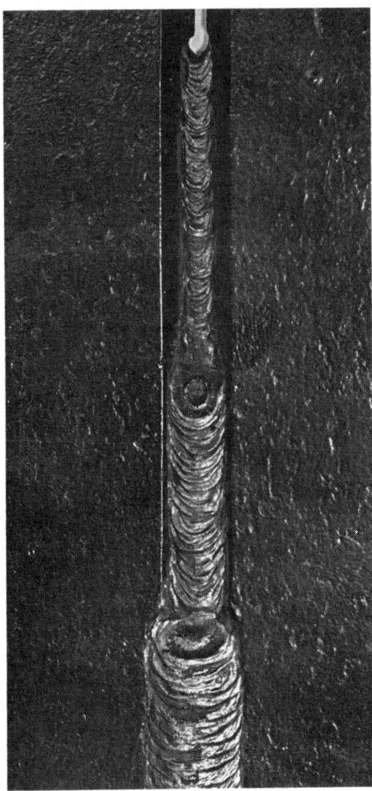

passes, and all plate surfaces. See Fig. 22-3.

Surrounding Plate Surfaces: Free of spatter

Slag formation: Full coverage, easily removable

Disposal

Put completed joints in the scrap bin so that they will be available for further use. Joints can be cut and beveled between welds for further butt welding, and unwelded plate surfaces can be used for beading.

Qualifying Test

At this point you should be able to pass Test 2 in the vertical position. (See pages 716, 717). This test is similar to the test for pipe or plate developed by the ASME.

JOB 22-J53 WELDING A T-JOINT

Objective

To weld a T-joint in the overhead position by means of a single-pass fillet weld, stringer bead technique, with DCRP and/or a.c. shielded-arc electrodes (AWS-ASTM E6010-E6011).

General Job Information

The T-joint welded in the overhead position is used extensively in shipyard and structural work.

Welding Technique

Current adjustment should be high. Hold the electrode at a 20- to 30-degree angle to the vertical plate, Fig. 22-5A, and tilt it no more than 15 degrees in the direction of travel, Fig. 22-5B. Figure 22-6 shows the actual welding position. Use a back-and-forth motion along the line of weld. Hold a close arc when depositing metal. The arc must not be extinguished on the forward motion. Pay close attention to fusion and penetration at the root of the joint and to the

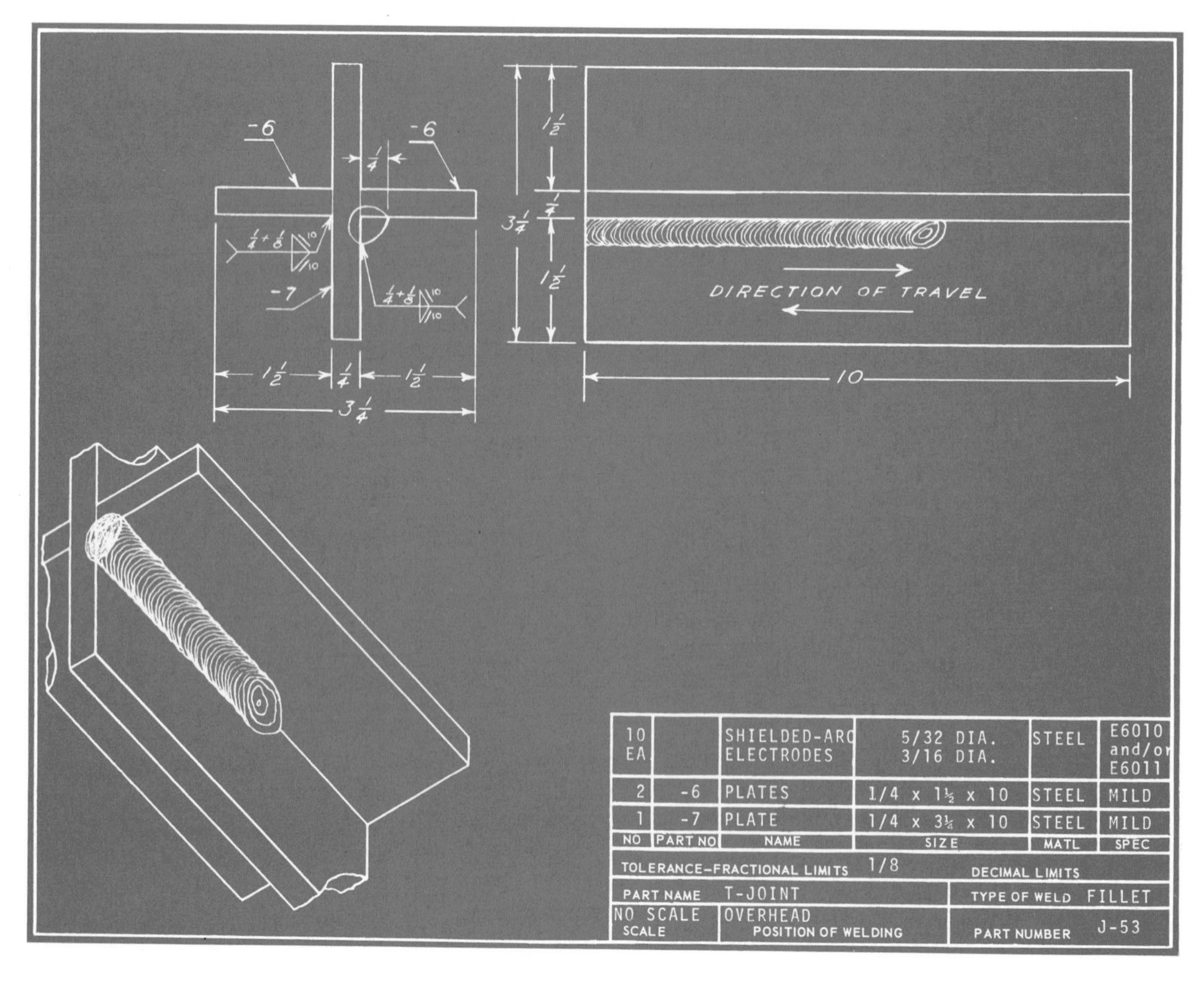

plate surfaces. Slag must not be permitted to run ahead of the metal being deposited. Avoid overheating the weld pool to prevent excessive convexity and overlap. While undercut is a problem, it can be overcome in the usual manner. Watch your speed of travel; the bead must be permitted to build up to its full size.

Practice starts and stops. Travel from left to right and from right to left.

Operations

1. Obtain plates; check the job print for quantity and size.
2. Obtain electrodes of the quantity, type, and size specified in the job print.
3. Set the power source for 120 to 170 amperes. Set a d.c. power source for reverse polarity.
4. Set up the plates as shown on the job print and tack weld them at each corner.
5. Fasten the joint in the overhead welding jig. See Fig. 22-6.
6. Make the weld as shown on the job print. Manipulate the electrode as instructed in Fig. 22-5.
7. Chip the slag from the weld, brush, and inspect. Refer to Inspection, below.
8. Practice these welds until you can produce good welds consistently with both types of electrodes and with $^3/_{16}$-inch electrodes.

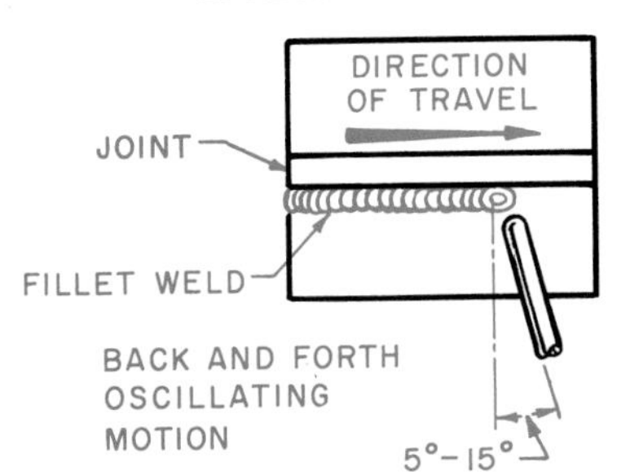

22-5. *Electrode position when welding a T-joint in the overhead position.*

22-6. *Electrode and electrode holder positioned as instructed in Fig. 22-5. Note how the joint is held by the jig.*

Inspection

Compare the weld with Fig. 22-7 and check it for the following characteristics:

Width and height: Uniform
Appearance: Smooth with close ripples; free of voids and high spots. Restarts should be difficult to locate.
Size: Refer to the job print. Check with a fillet weld gauge.
Face of weld: Flat
Edges of weld: Good fusion, no overlap, no undercut
Starts and stops: Free of depressions and high spots
Beginnings and endings: Full size, craters filled

22-7. *Typical appearance of a single-pass fillet weld in a T-joint welded in the overhead position with DCRP or a.c. shielded-arc electrodes.*

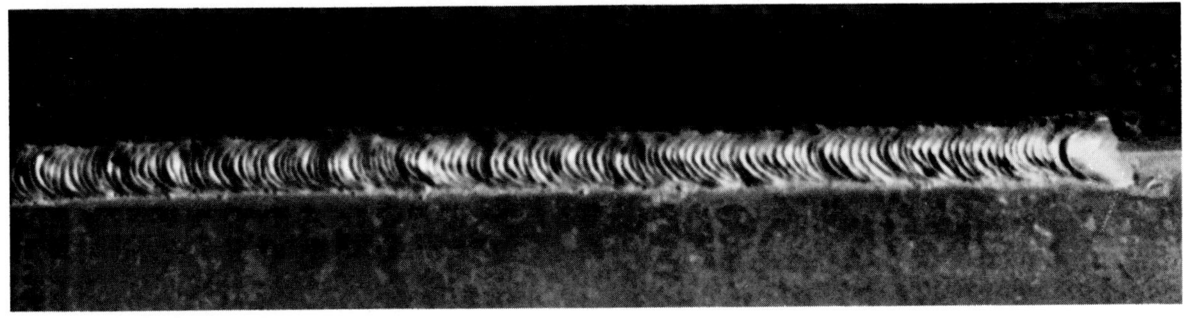

Penetration and fusion: To the root of the joint and both plate surfaces

Surrounding plate surfaces: Free of spatter

Slag formation: Full coverage, easily removable

Disposal

Save the completed joints for your next job, which will be multipass fillet welding in the overhead position.

Check Test

After you are able to make welds that are satisfactory in appearance, make up a test specimen similar to that made in Job 20-J17, (Refer to Figs. 20-48 and 20-49, page 612.) Use the plate thickness, welding technique, and electrode type and size specified for this job. Weld on one side only.

Break the completed weld and examine the surfaces for soundness. The weld must not be porous. It must show good fusion and penetration at the root of the joint and to the plate surfaces. The joint should break evenly through the throat of the weld.

JOB 22-J54 WELDING A T-JOINT

Objective

To weld a T-joint in the overhead position by means of multipass fillet welds, stringer bead technique, with DCRP and/or a.c. shielded-arc electrodes (AWS-ASTM E6010-E6011).

General Job Information

When more than one pass is necessary for fillet welds in the overhead position, it is usually the practice to apply stringer beads. It is to be understood that the thickness of plate used in this job would not require a $\frac{1}{2}$-inch fillet weld. The plate is merely being used as much as possible.

Welding Technique

Current adjustment can be high. The angle of the electrode to the vertical plate changes ac-

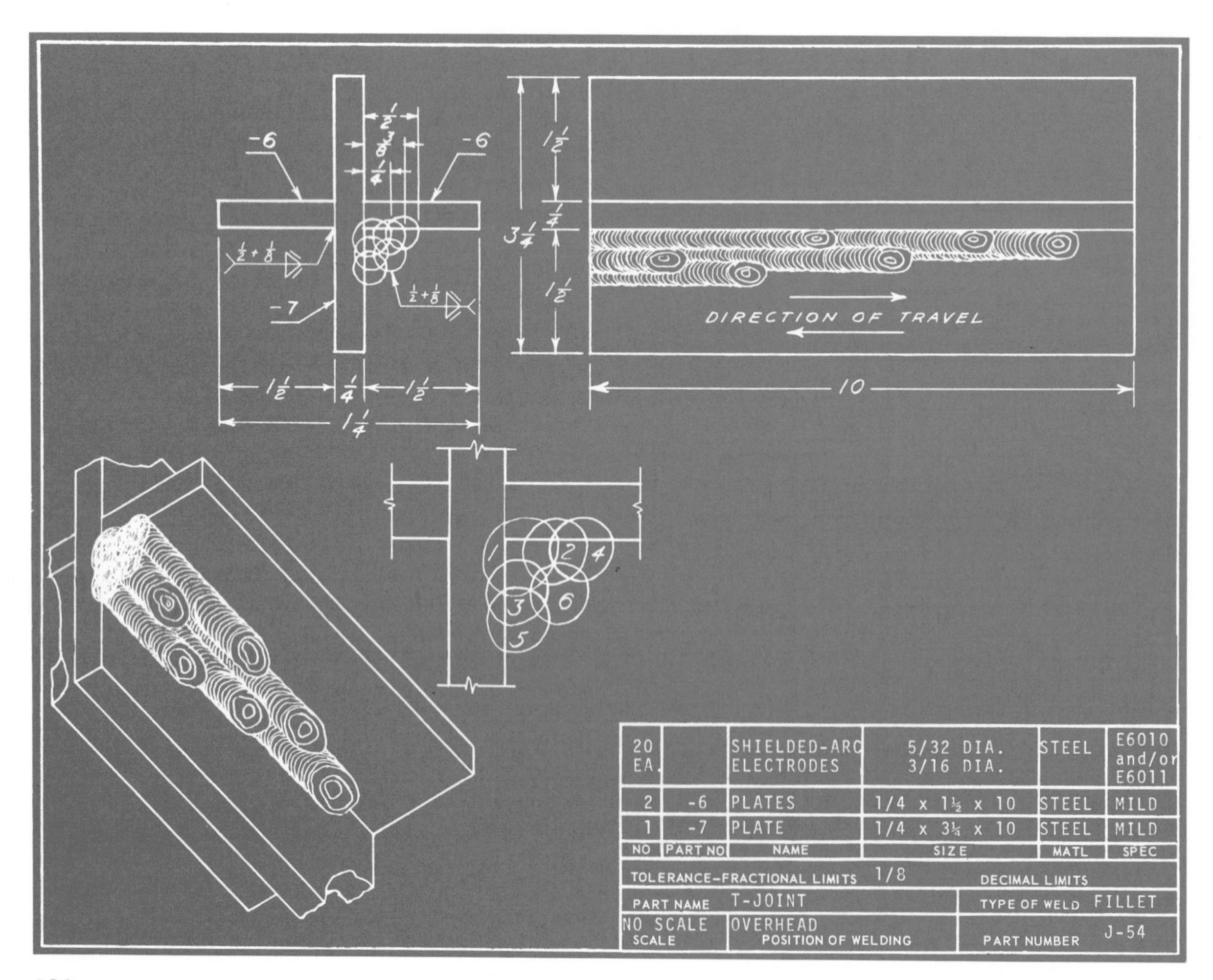

cording to the pass being deposited, Fig. 22-8. It is 20 to 30 degrees for the second pass, 35 to 45 degrees for the third and fourth passes, 45 degrees for the fifth pass, and 20 to 30 degrees for the sixth pass. The electrode is tilted no more than 15 degrees in the direction of travel for all passes. Use the stringer bead technique and deposit each pass in the sequence shown on the job print. You will have difficulty in obtaining the proper proportions, but practice will overcome it. Each bead must be fused with the preceding bead and the plate surface. The face of each layer should be as flat as possible.

Practice starts and stops. Travel from left to right and from right to left. Be sure to remove slag from each bead before starting the next bead.

Operations

1. Use the joints welded in Job 22-J53.
2. Obtain electrodes of the quantity, type, and size specified in the job print.
3. Set the power source for 120 to 160 amperes. Set a d.c. power source for reverse polarity.
4. Fasten the joint in the overhead position with an overhead welding jig.
5. Make welds as shown on the job print. Manipulate the electrode as shown in Fig. 22-8.
6. Chip the slag from the welds, brush, and inspect between each pass. Refer to Inspection, below.
7. Practice these welds until you can produce good welds consistently with both types of electrodes and with $^{3}/_{16}$-inch electrodes.

Inspection

Compare each pass with Fig. 22-9 and check it for the following weld characteristics:

Width and height: Uniform
Appearance: Smooth with close ripples; free of voids and high spots. Restarts should be difficult to locate.
Size: Refer to the job print. Check with a fillet weld gauge.
Face of weld: Flat
Edges of weld: Good fusion, no overlap, no undercut
Starts and stops: Free of depressions and high spots
Beginnings and endings: Full size, craters filled
Penetration and fusion: To preceding beads and both plate surfaces
Surrounding plate surfaces: Free of spatter
Slag formation: Full coverage, easily removable

Disposal

Put completed joints in the scrap bin so that they will be available for further use. Plate edges can be used for square butt welding, and unwelded surfaces can be used for beading.

JOB 22-J55 WELDING A LAP JOINT

Objective

To weld a lap joint in the overhead position by means of a single-pass fillet weld with DCRP and/or shielded-arc electrodes (AWS-ASTM E6010-E6011).

General Job Information

The overhead lap joint is found very frequently in tank and shipyard work. Because of the size and nature of these objects, it is impractical to position them. The lap joint can be welded with three stringer passes instead of a single weaved bead.

Welding Technique

Current adjustment must not be too high. Hold the electrode at a 30-degree angle to the lower

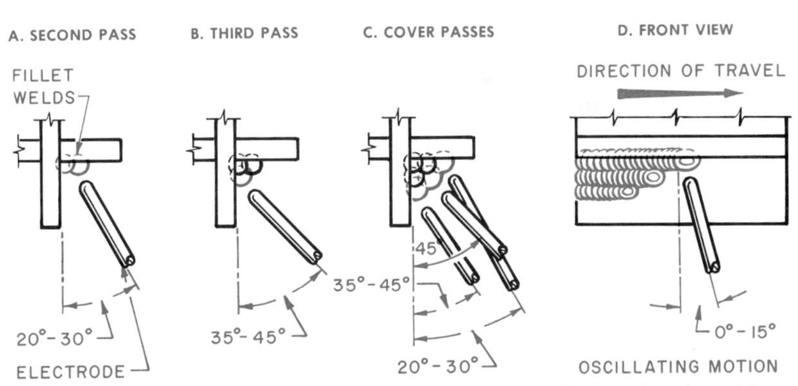

22-8. *Electrode position when welding a T-joint in the overhead position using the stringer bead technique.*

22-9. *Typical appearance of a multipass fillet weld in a T-joint welded in the overhead position (stringer bead technique) with DCRP or a.c. shielded-arc electrodes.*

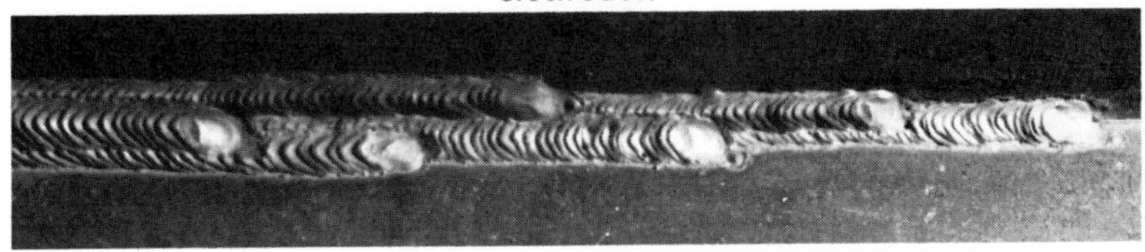

plate, Fig. 22-10A, and tilt it 20 degrees in the direction of travel, Fig. 22-10B. Figure 22-11 shows the actual position for welding. A back-and-forth whipping motion may be used. Do not extinguish the arc on the forward motion. The whiplike action helps to prepare the root of the joint for the weld deposit by preheating it, allows the weld pool to cool off so that excessive convexity and spilling are prevented, and washes the slag back over the weld deposit. Do not allow the arc to contact the light edge of the outer plate. At no time should the arc touch the sections of the plate surface that are outside the final bead width.

Practice starts and stops. Travel from left to right and from right to left.

Operations

1. Obtain plates; check the job print for quantity and size.
2. Obtain electrodes of the quantity, type, and size specified in the job print.
3. Set the power source for 140 to 215 amperes. Set a d.c. power source for reverse polarity.
4. Set up the plates and tack weld them as shown on the job print.
5. Fasten the joints in the overhead position with an overhead welding jig. See Fig. 22-11.
6. Make welds as shown on the job print. Manipulate the electrode as shown in Fig. 22-10.
7. Chip the slag from the welds, brush, and inspect. Refer to Inspection, below.
8. Practice these welds until you can produce good welds consistently with both types of electrodes and by using the stringer bead technique with $\frac{5}{32}$-inch electrodes.

Inspection

Compare the welds with Fig. 22-12 and check them for the following weld characteristics:

Width and height: Uniform

Appearance: Smooth with close ripples; free of voids and high spots

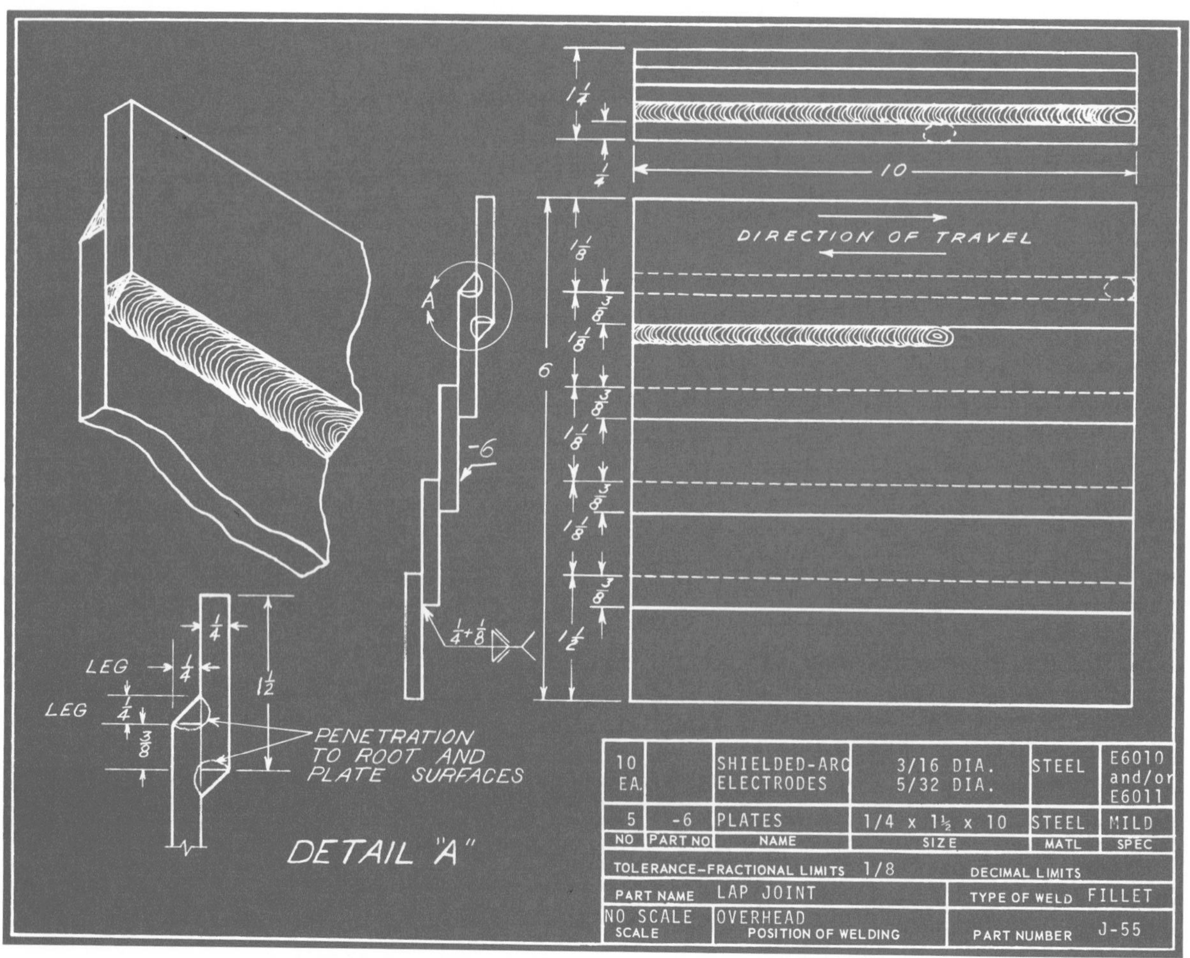

Size: Refer to the job print. Check with a fillet weld gauge.
Face of weld: Flat
Edges of weld: Good fusion, no overlap, no undercut. The edge of the upper plate must not be burned away.
Starts and stops: Free of depressions and high spots
Beginnings and endings: Full size, craters filled
Penetration and fusion: To the root of the joint and both plate surfaces
Surrounding plate surfaces: Free of spatter
Slag formation: Full coverage, easily removable

Disposal

Put completed joints in the scrap bin so that they will be available for further use. Unwelded plate surfaces may be used for beading.

Check Test

Make up test specimens and carry out the usual lap joint (fillet-weld) test.

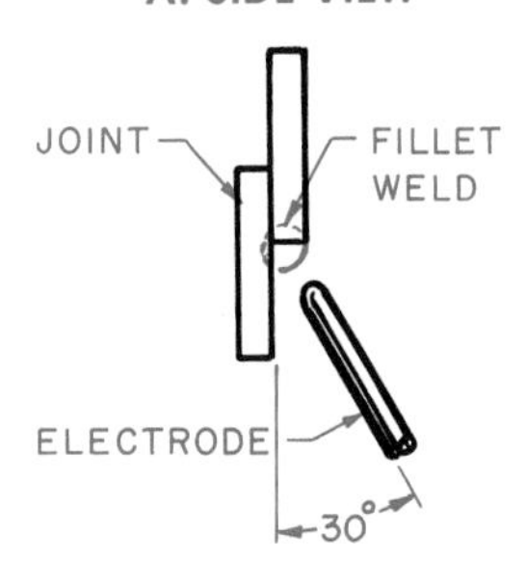

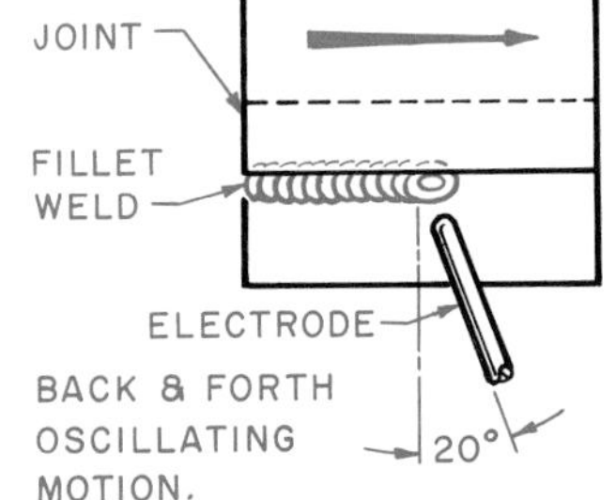

22-10. *Electrode position when welding a lap joint in the overhead position.*

22-11. *Electrode and electrode holder positioned as instructed in Fig. 22-10. Note how the joints are held by the jig.*

22-12. *Typical appearance of a single-pass fillet weld in a lap joint welded in the overhead position with DCRP or a.c. shielded-arc electrodes.*

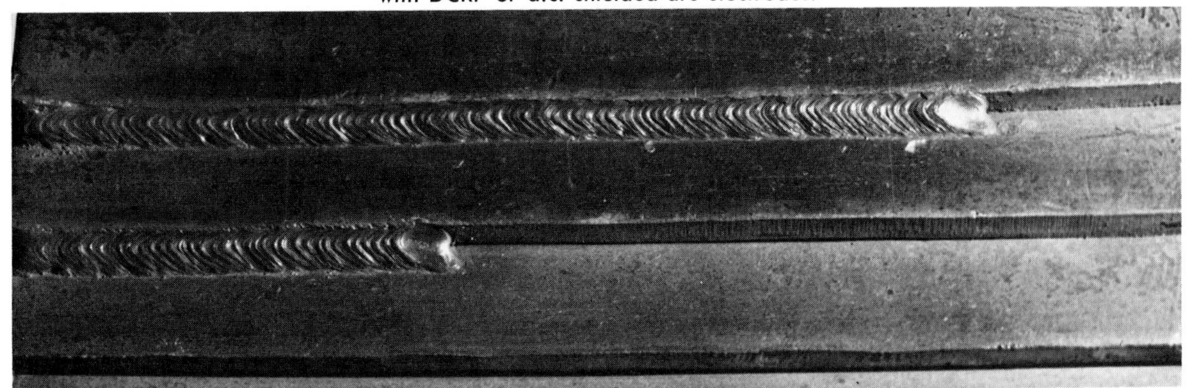

JOB 22-J56 WELDING A LAP JOINT

Objective

To weld a lap joint in the overhead position by means of a single-pass fillet weld with DCSP and/or shielded-arc electrodes (AWS-ASTM E6012-E6013).

General Job Information

This job is similar to Job 22-J55 except that straight polarity electrodes are used. Most of this type of work is done with reverse polarity electrodes, but there are enough straight polarity applications to justify its practice.

Welding Technique

Current adjustment may be somewhat higher than for reverse polarity. See Fig. 22-10, page 693, for the correct electrode position. Use the welding technique which you practiced in the previous job. The slag will be somewhat more difficult to control than with reverse polarity. The bead will have a tendency to become convex and to run down the vertical plate surface. Undercut is not a problem in this job.

Practice starts and stops. Travel from left to right and from right to left.

Operations

1. Obtain plates; check the job print for quantity and size.
2. Obtain electrodes of the quantity, type, and size specified in the job print.
3. Set the power source for 140 to 230 amperes. Set a d.c. power source for straight polarity.
4. Set up the plates and tack weld them as shown on the job print.
5. Fasten the joints in the overhead position with an overhead welding jig.
6. Make welds as shown on the job print. Manipulate the

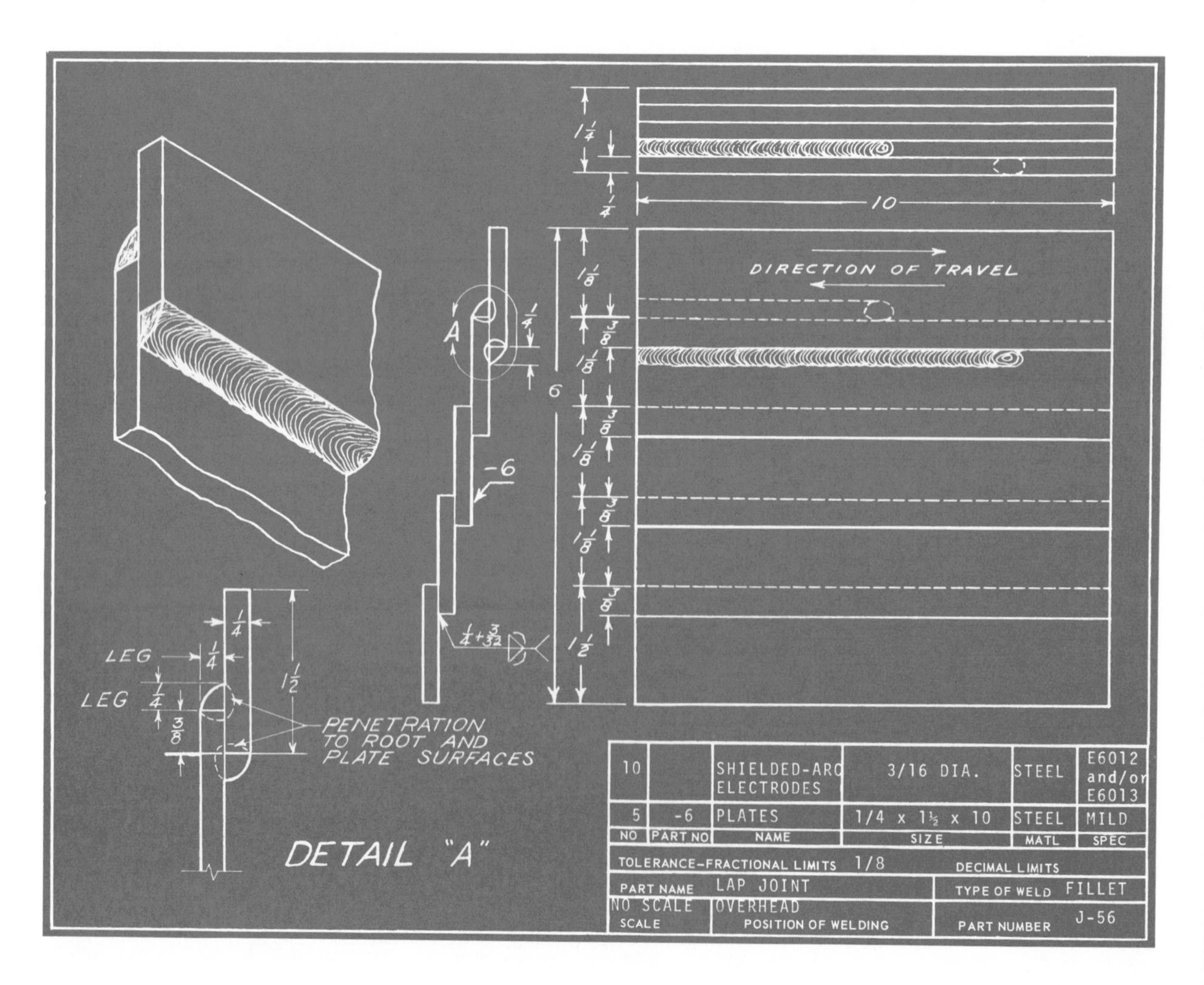

electrode as instructed in Fig. 22-10, page 693.

7. Chip the slag from the welds, brush, and inspect. Refer to Inspection, below.

8. Practice these welds until you can produce good welds with both types of electrodes.

Inspection

Compare the welds with Fig. 22-13 and check them for the following characteristics:

Width and height: Uniform

Appearance: Smooth with close ripples; free of voids and high spots. Restarts should be difficult to locate.

Size: Refer to the job print. Check with a fillet weld gauge.

Face of weld: Slightly convex

Edges of weld: Good fusion, no overlap, no undercut. The edge of the upper plate must not be burned away.

Starts and stops: Free of depressions and high spots

Beginnings and endings: Full size, craters filled

Penetration and fusion: To the root of the joint and both plate surfaces

Surrounding plate surfaces: Free of spatter

Slag formation: Full coverage, easily removable

Disposal

Put completed joints in the scrap bin so that they will be available for further use. Unwelded plate surfaces can be used for beading.

Check Test

Make up a test specimen and carry out the usual lap joint (fillet weld) test.

JOB 22-J57 WELDING A SINGLE-V BUTT JOINT

Objective

To weld a single-V butt joint in the vertical position by means of multipass groove welds, weaved-beading technique, travel down, with DCRP and/or a.c. shielded-arc electrodes (AWS-ASTM E6010-E6011).

General Job Information

The single-V butt joint is welded by the downhill method in pipeline work, tank, and shipyard welding. When used in pipe, it is usually limited to sizes under 14 inches in diameter. While the downhill technique is somewhat difficult to master, it has the advantage of speed. When butt joints in pipe are welded by the downhill method without an inner liner, the process is somewhat slower than when a liner is used. The cost of the liner, however, and the time spent installing it erases the advantage of speed when applied to pipe 10 inches and under.

The skill acquired in the practice of this joint will assist you in learning to weld pipe, which is presented in the following chapter.

Welding Technique

Current adjustment should be a little higher than that used for a joint with a backup strip and much higher than that used for uphill welding. Electrode angle, arc length, and speed of travel are very important. Tilt the electrode 15 to 20 degrees in the direction of travel, Fig. 22-14A, and hold it at a 90-degree angle to the surface of the plates, Fig. 22-14B.

FIRST PASS. The first pass must be run with a close arc. Some welders like to drag the coating in contact with the work and proceed downward at a rapid rate. Very little side-to-side motion is necessary unless there is a tendency to burn through and cause an enlarged hole. The back side of the joint should show a weld of uniform width and height without holes or excessive burnthrough. The face side of the weld should be flat with good fusion into the side walls of the plate. There will be no real problem with undercut. If you are having difficulty with this pass, practice with different current settings. Some welders can handle a wider gap than others can.

22-13. *Typical appearance of a single-pass fillet weld in a lap joint welded in the overhead position with DCSP or a.c. shielded-arc electrodes.*

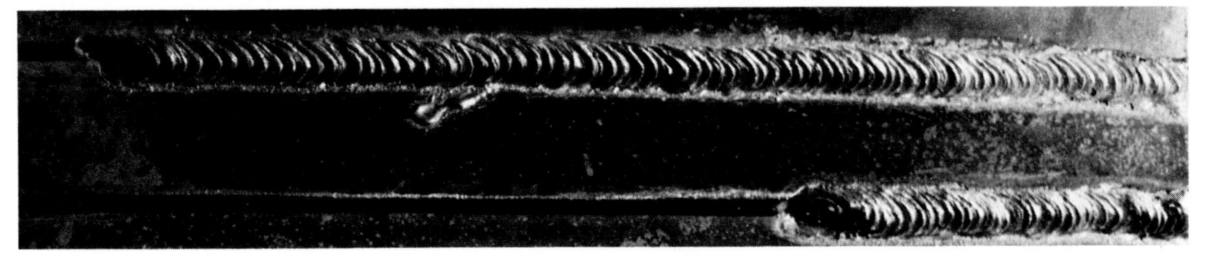

SECOND AND THIRD PASSES. This joint should be welded in three passes. The second and third passes should be run with a larger electrode. Weave the electrode slightly and increase the side-to-side motion with each pass. Watch carefully for holes and voids in the face of the weld. There is also a tendency for the face of the weld to become concave. This can be avoided by pausing for a shorter time at the edges of the weld when making the weave motion. Since there is no problem with undercutting, it is not necessary to pause at the side of the weld to any extent.

Make sure that fusion and penetration are secured between beads and into the beveled walls of the plate. The face of each completed bead should be as flat as possible. The final bead should have a slight contour in order to make sure that it is full and not below the surface of the plate. When making the final pass, do not weave beyond the sharp edges of the plate at each side.

Review Job 21-J51 for additional information on downhill welding. Practice varying the heat setting, electrode angle, arc length, and speed of travel. Starts and stops should also be practiced.

Operations

1. Obtain plates; check the job print for quantity and size.
2. Obtain electrodes of each quantity, type, and size specified in the job print.
3. Set the power source for 125 to 200 amperes. Set a d.c. power source for reverse polarity.
4. Set up the plates and tack weld them as shown on the job print.
5. Stand the joint in the vertical position on the welding table. Make sure that it is well-grounded.
6. Make the first pass with $\frac{5}{32}$-inch electrodes as shown on the job print. Manipulate the electrode as instructed in Fig. 22-14.
7. Chip the slag from the weld, brush, and inspect. Refer to Inspection.

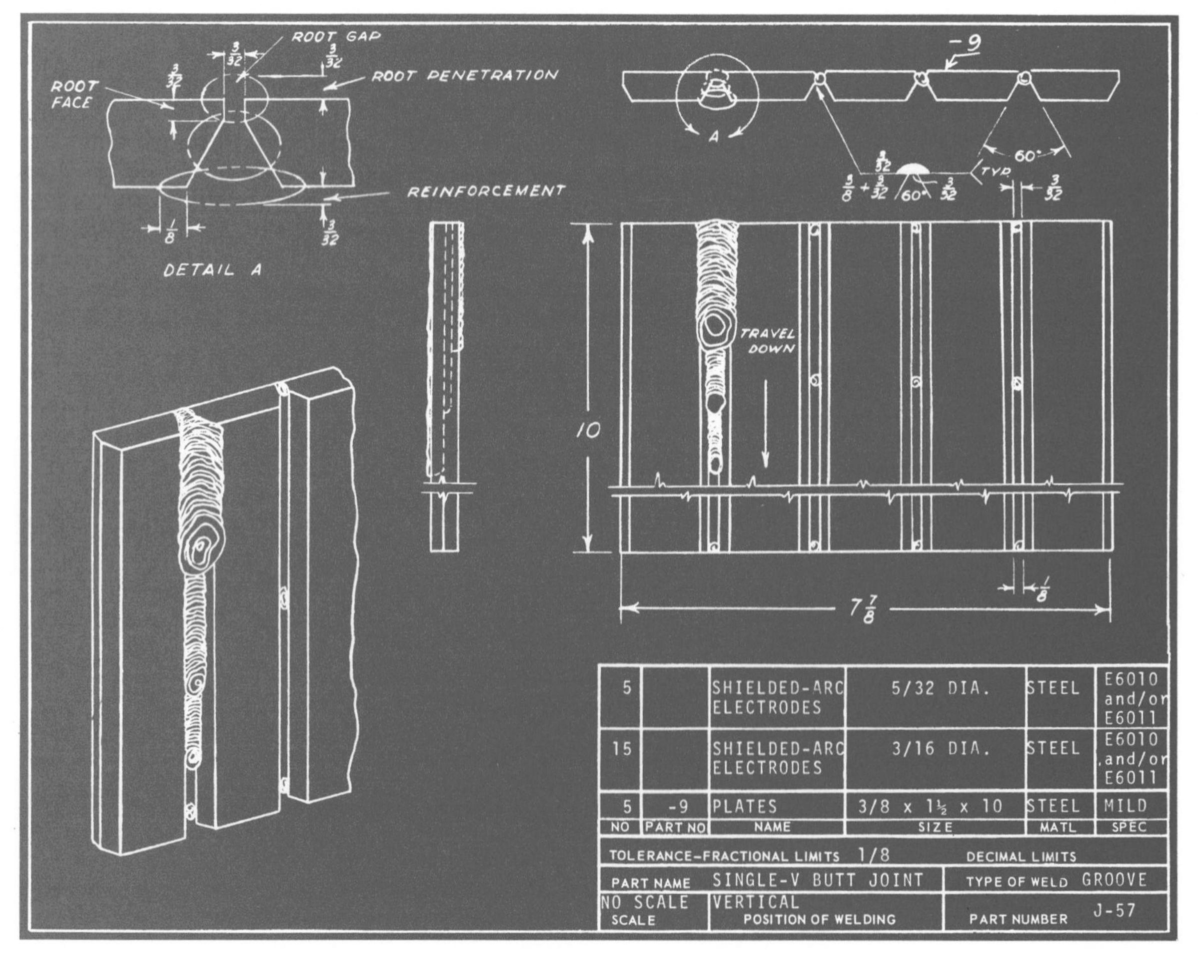

8. Readjust the power source to 130 to 225 amperes.

9. Make the second and third passes with 3/16-inch electrodes as shown on the job print.

10. Chip the slag from the welds, brush, and inspect. Refer to Inspection, below.

11. Practice these welds until you can produce good welds consistently. Change your procedure by using 3/16-inch electrodes for the first pass.

Inspection

Compare each pass with Fig. 22-15 and check it for the following weld characteristics:

Width and height: Uniform
Appearance: Very smooth with fine ripples; free of voids and high spots. Restarts should be difficult to locate.
Size: Refer to the job print. Check with a butt weld gauge.
Face of weld: Flat to slightly convex
Edges of weld: Good fusion, no overlap, no undercut
Starts and stops: Free of depressions and high spots
Beginnings and endings: Full size, craters filled
Penetration and fusion: To the root of the weld and plate surfaces
Surrounding plate surfaces: Free of spatter
Slag formation: Full coverage, easily removable

Disposal

Put completed joints in the scrap bin so that they will be available for further use. The plate edges can be beveled for bevel-butt welding, and unwelded plate surfaces can be used for beading.

Qualifying Test

At this point you should make up test plates. Cut face and root bevel test coupons from the plates and test them. (See Test 2, page 716.) This test is similar to the ASME test requirements and AWS recommendations.

JOB 22-J58 WELDING A T-JOINT

Objective

To weld a T-joint in the overhead position by means of multipass fillet welds, weaved-beading technique, with DCRP and/or a.c. shielded-arc electrodes. (AWS-ASTM E6010-E6011).

General Job Information

The welder is frequently asked to make large fillets in the overhead position with a minimum number of passes and large electrodes. This is especially true of shipyard and heavy fabrication welding.

Welding Technique

FIRST PASS. Current setting can be high. Hold the electrode at a 30-degree angle to the vertical plate, Fig. 22-16A, and tilt it 5 to 15 degrees in the direction of travel, Fig. 22-16B. Use the welding technique learned in Job 22-J53.

SECOND PASS. The current is increased to burn the larger electrode. The electrode position is the same as for the first pass. Weave the electrode slightly with

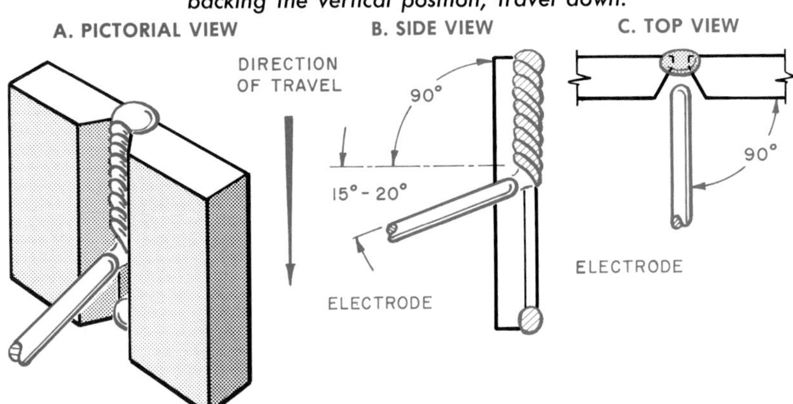

22-14. *Electrode position when welding a beveled-butt joint without backing the vertical position, travel down.*

22-15. *Typical appearance of a multipass groove weld in a beveled-butt joint without a backup strip welded in the vertical position, travel down, with DCRP or a.c. shielded-arc electrodes.*

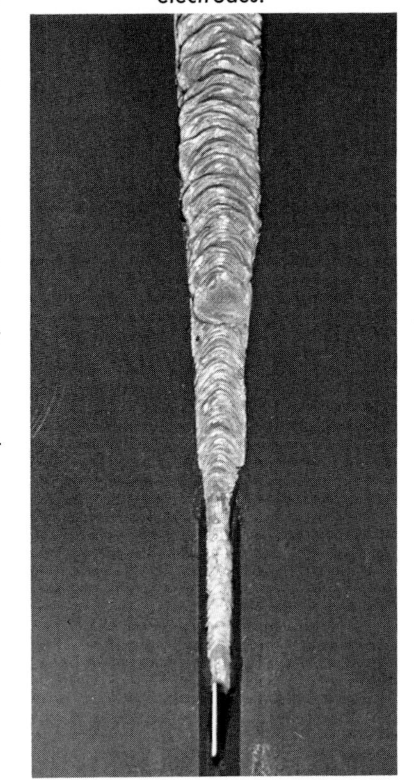

a whiplike oscillating motion. Remember that because the electrode is larger, more metal is being deposited in the crater and so there is more slag present. Keep the electrode motion within the limits of the weld size to prevent undercut.

Note: If it is desirable to get more use out of the plate, a third layer may be applied. This should consist of stringer beads deposited according to the procedures practiced in making beads 4, 5, and 6 in Job 22-J54.

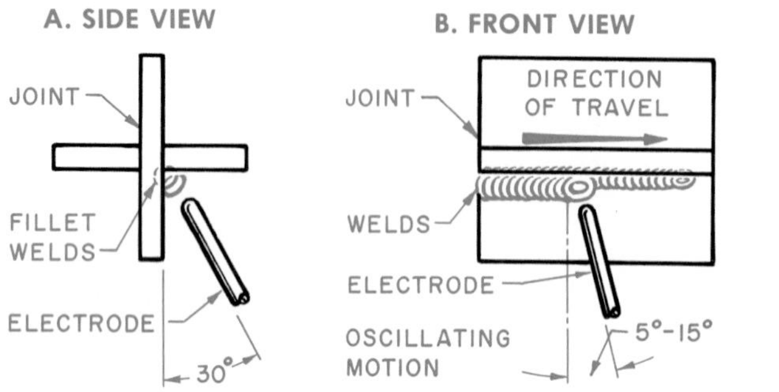

22-16. *Electrode position when welding a T-joint in the overhead position.*

Operations

1. Obtain plates; check the job print for the correct quantity and size.
2. Obtain electrodes of each quantity, type, and size specified in the job print.
3. Set the power source for 120 to 170 amperes. Set a d.c. power source for reverse polarity.
4. Set up the plates as shown on the job print and tack weld them at each corner.
5. Fasten the joint in the

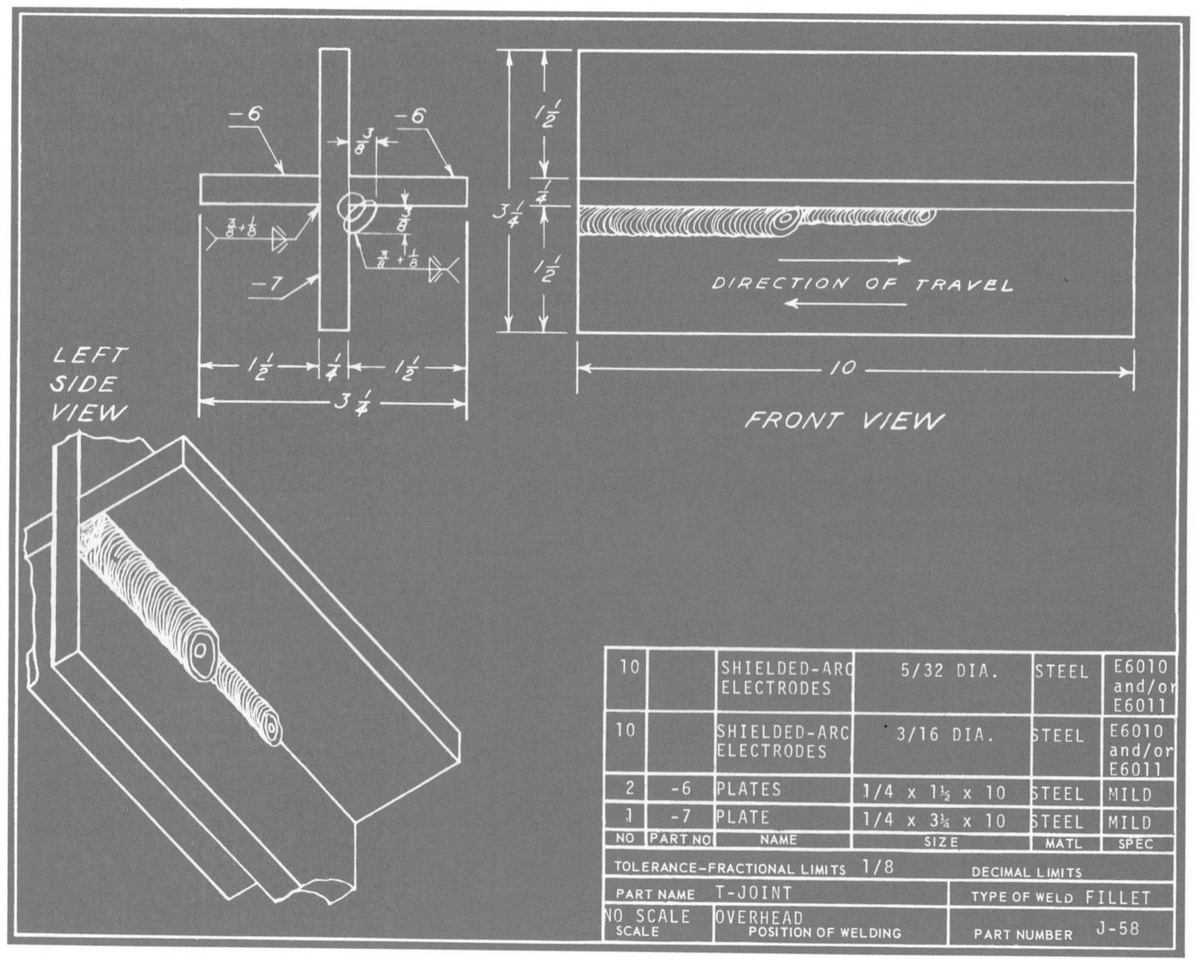

overhead position with an overhead welding jig.

6. Make the first pass with $^{5}/_{32}$-inch electrodes as shown on the job print. Manipulate the electrode as instructed in Fig. 22-16.

7. Chip the slag from the weld, brush, and inspect. Refer to Inspection, below.

8. Increase the power source setting to 140 to 225 amperes.

9. Make the second pass with $^{3}/_{16}$-inch electrodes as shown on the job print.

10. Chip the slag from welds, brush, and inspect. Refer to Inspection, below.

11. Practice these welds until you can produce good welds consistently with both types of electrodes.

Inspection

Compare each pass with Fig. 22-17 and check it for the following characteristics:

Width and height: Uniform

Appearance: Smooth with close ripples; free of voids and high spots. Restarts should be difficult to locate.

Size: Refer to the job print. Check with a fillet weld gauge.

Face of weld: Flat

Edges of weld: Good fusion, no overlap, no undercut

Starts and stops: Free of depressions and high spots

Beginnings and endings: Full size, craters filled

Penetration and fusion: To the root of the joint and both plate surfaces

Surrounding plate surfaces: Free of spatter

Slag formation: Full coverage, easily removable

Disposal

Put completed joints in the scrap bin so that they will be available for further use. The plate edges can be used for square butt welding, and unwelded surfaces can be used for beading.

JOB 22-J59 WELDING A T-JOINT

Objective

To weld a T-joint in the overhead position by means of multipass fillet welds, weaved-beading technique, with DCRP and/or a.c. shielded-arc electrodes (AWS-ASTM E6010-E6011).

General Job Information

This joint is used extensively in marine construction, especially barge construction, and in heavy fabrication. It differs from the previous job only in that the size of the electrode is larger, the size of the weld is larger, and more weld metal is carried in each pass.

Welding Technique

FIRST PASS. Current adjustment can be high. See Fig. 22-16, page 698, for the correct electrode position. Maintain a close arc gap. Weave the electrode by moving it forward with a rapid whipping motion and at the same time lengthen the arc slightly. Do not break the arc and return it to the crater when the crater has solidified. This prevents the molten pool from becoming too hot and spilling off, preheats the root of the joint ahead of the deposit, and forces the slag over the deposit. Because better control of weld metal and arc is possible, undercut, overlap, and convexity are reduced, and the appearance is uniform.

SECOND PASS. The same technique is used for the second pass as for the first pass, but more metal is deposited. You will need practice to increase the deposition.

Practice starts and stops. Travel in all directions.

Note: If it is desirable to get more use out of the plate, a third layer can be applied. This should consist of stringer beads deposited according to the procedures practiced in making beads 4, 5, and 6 in Job 22-J54.

Operations

1. Obtain plates; check the job print for quantity and size.

2. Obtain electrodes of the quantity, type, and size specified in the job print.

3. Set the power source for 140 to 215 amperes. Set a d.c. power source for reverse polarity.

22-17. *Typical appearance of a multipass fillet weld in a T-joint welded in the overhead position (Weaved-bead technique) with DCRP or a.c. shielded-arc electrodes.*

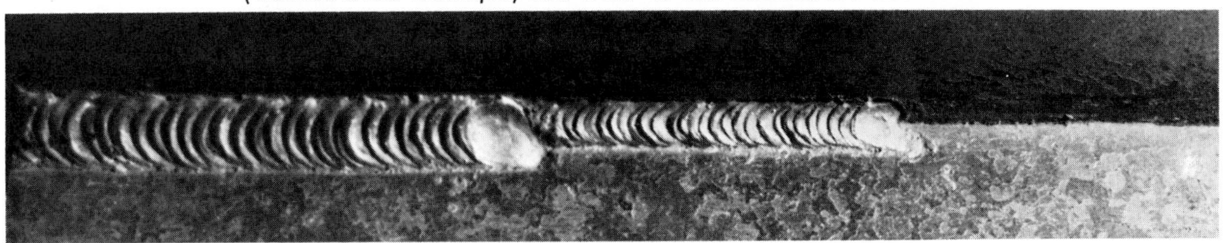

4. Set up the plates as shown on the job print and tack weld them.

5. Fasten the joint in the overhead position with an overhead welding jig.

6. Make the first pass as shown on the job print. Manipulate the electrode as instructed in Fig. 22-16, page 698.

7. Chip the slag from the weld, brush, and inspect. Refer to Inspection.

8. Make the second pass as shown on the job print.

9. Chip the slag from the weld, brush, and inspect. Refer to Inspection.

10. Practice these welds until you can produce good welds consistently with both types of electrodes.

Inspection

Compare both passes with Fig. 22-18 and check them for the following characteristics:

Width and height: Uniform
Appearance: Smooth with close ripples; free of voids and high spots. Restarts should be difficult to locate.
Size: Refer to the job print. Check with a fillet weld gauge.
Face of weld: Flat
Edges of weld: Good fusion, no overlap, no undercut
Starts and stops: Free of depressions and high spots
Beginnings and endings: Full size, craters filled
Penetration and fusion: To the root of the joint and both plate surfaces
Surrounding plate surfaces: Free of spatter
Slag formation: Full coverage, easily removable

Disposal

Put the completed joints in the scrap bin so that they will be available for further use. The plate edges can be used for square butt welding, and the unwelded surfaces can be used for beading.

Check Test

After you are able to make welds that are satisfactory in appearance, make up a specimen similar to that made in previous jobs. Use the plate thickness, welding technique, and electrode

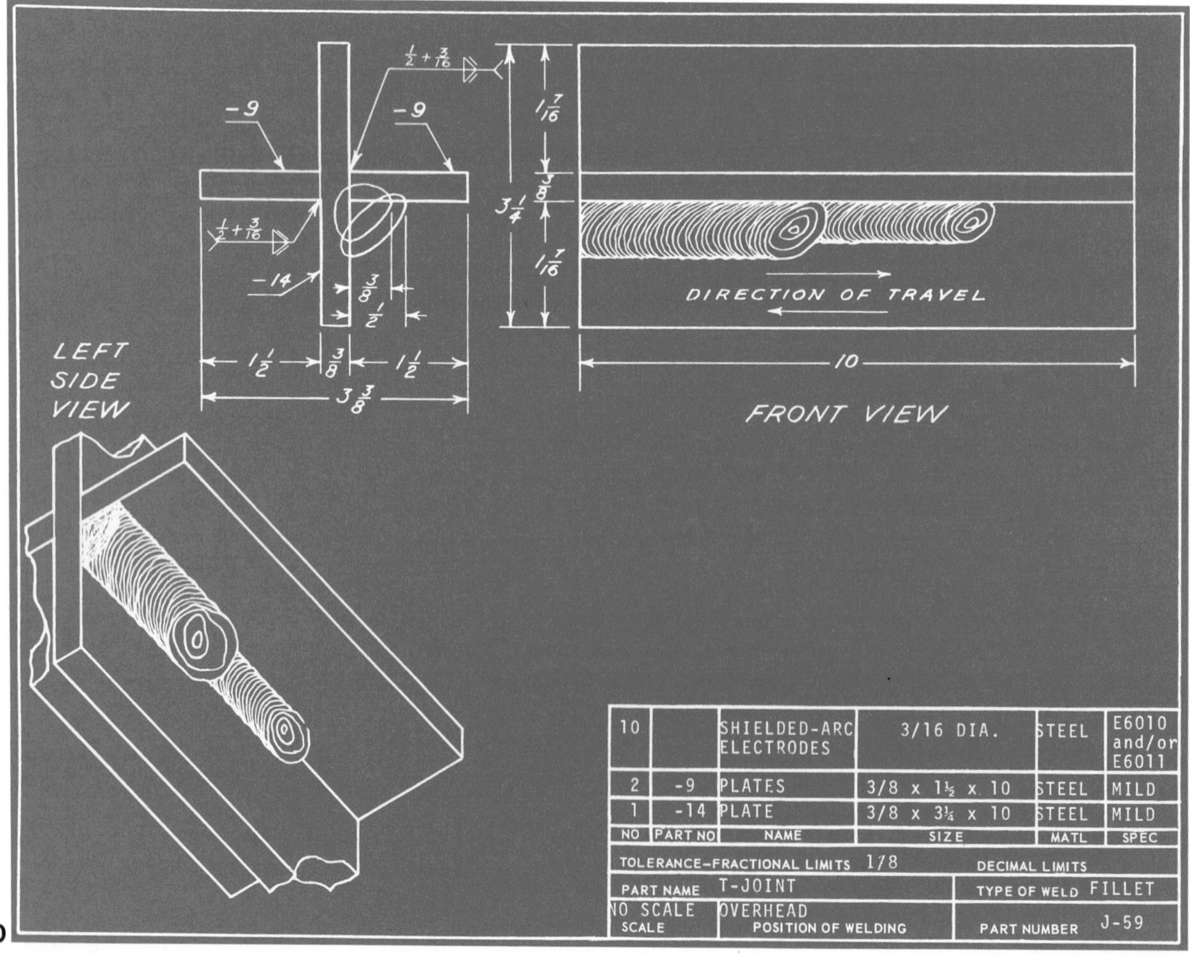

NO	PART NO	NAME	SIZE	MATL	SPEC
10		SHIELDED-ARC ELECTRODES	3/16 DIA.	STEEL	E6010 and/or E6011
2	-9	PLATES	3/8 x 1½ x 10	STEEL	MILD
1	-14	PLATE	3/8 x 3¼ x 10	STEEL	MILD

TOLERANCE-FRACTIONAL LIMITS 1/8 — DECIMAL LIMITS
PART NAME T-JOINT — TYPE OF WELD FILLET
SCALE NO SCALE — POSITION OF WELDING OVERHEAD — PART NUMBER J-59

type and size specified for this job. Weld the first pass only and on one side.

Break the finished weld and examine the surfaces for soundness. The weld must not be porous. It must show good fusion and penetration at the root and to the plate surfaces. The joint should break evenly through the throat of the weld.

JOB 22-J60 WELDING A SINGLE-V BUTT JOINT

Objective

To weld a single-V butt joint in the horizontal position by means of multipass groove welds, stringer bead technique, with DCRP and/or a.c. shielded-arc electrodes (AWS-ASTM E6010-E6011).

General Job Information

The beveled-butt joint in the horizontal position is used in pipe and critical plate welding. The joint and the welding procedure are similar to those required for ASME plate and pipe tests. Some companies demand that the finish bead be laced. However, most companies prefer stringer beading.

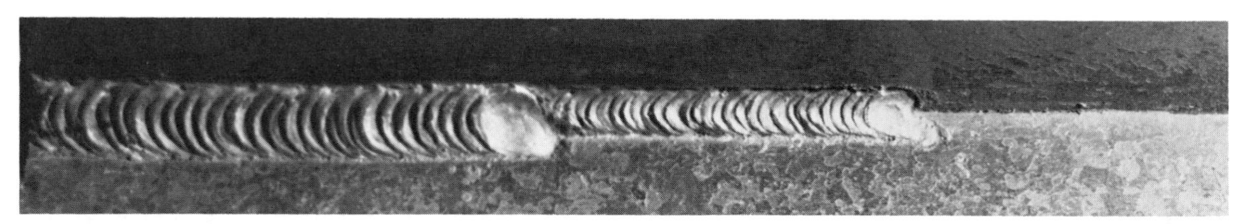

22-18. *Typical appearance of a multipass fillet weld in a T-joint welded with a slight whiplike motion in the overhead position with large DCRP or a.c. shielded-arc electrodes.*

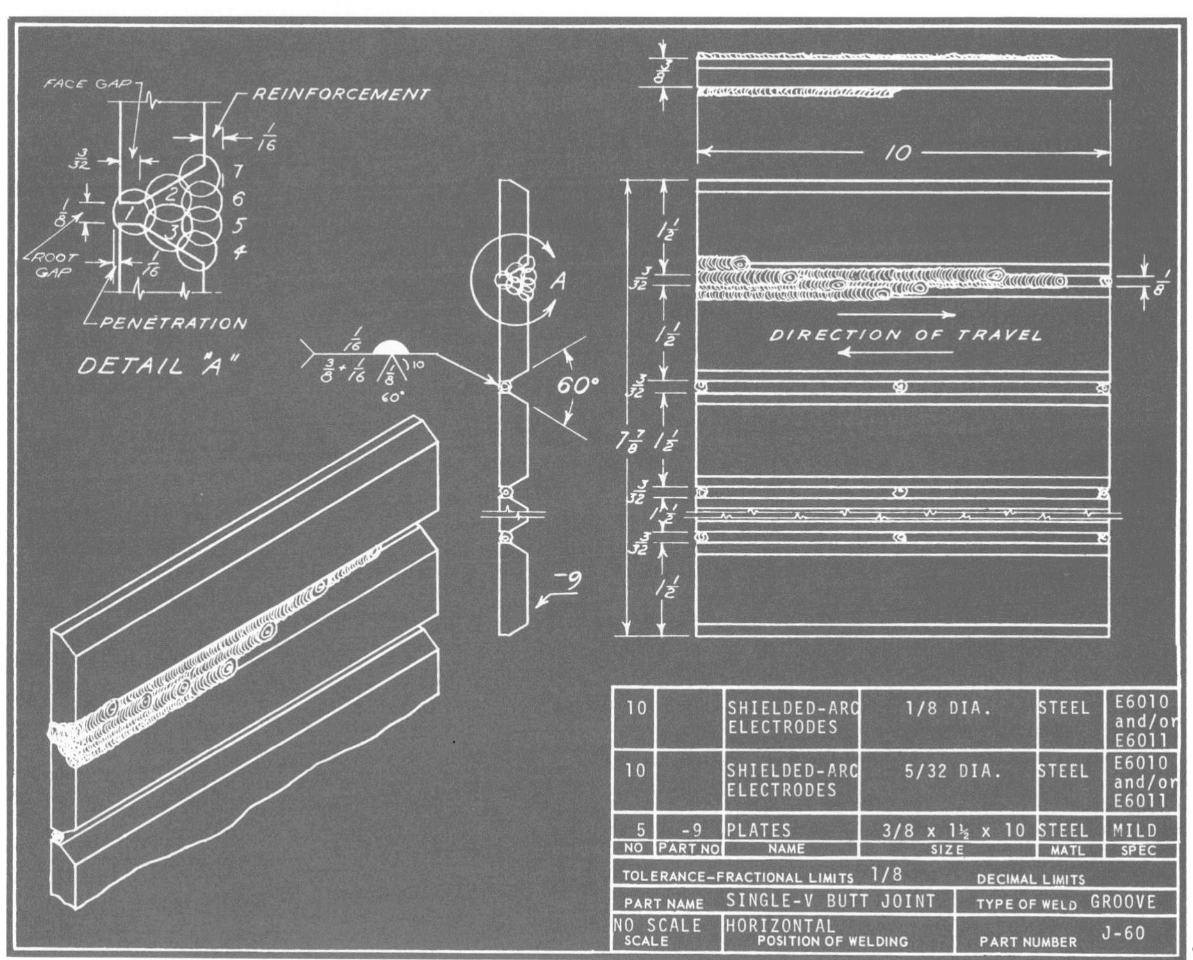

Welding Technique

The angle of the electrode changes according to the pass, Fig. 22-19. The angle is 90 degrees to the top plate for the first pass, 30 degrees to the bottom plate for the second pass, and 45 degrees to the top plate for the third pass. For the cover passes, the electrode should be held at an angle of 5 to 15 degrees to the bottom plate. The electrode should lean from 5 to 15 degrees in the direction of travel for all passes.

FIRST PASS. The current control must not be too high. Hold a short arc gap so that weld metal is forced into the gap at the root of the joint. A back-and-forth motion may be used. Do not break the arc on the forward motion. This movement preheats the metal ahead of the deposit and allows the deposited metal to cool to prevent sagging on the bottom plate.

Complete penetration must be secured through the back side, Fig. 22-20. Keep in mind that the correct combination of welding current, electrode position, and speed of travel produces penetration through to the back side. Penetration is complete if there is a little hole at the leading edge of the weld crater right under the tip of the electrode. Learn how to form and keep this hole throughout the welding operation without letting it get too large and causing you to lose control and burn through the plate. The presence of this hole during welding is your assurance that you are melting through to the back side of the groove.

If you are burning through the plate, you may use a whiplike or circular motion to control the size of the weld pool and the rate at which weld metal is added. If the pool is too hot and burning through, whip ahead along one of the plate edges about ½ inch and whip back into the pool to deposit more metal. Alternate this whipping motion along opposite plate edges for each forward movement.

If the root faces of the plates do not seem to be heating and melting enough and you are having difficulty maintaining the hole, keep the electrode directly in the bottom of the *V*, slow the forward movement, and maintain a circular motion.

ALL OTHER PASSES. Current control can be much higher. Electrode position should be approximately that shown in Fig. 22-19B through E below. The stringer bead technique may be used. It is important that each pass be fused to the preceding passes and to the plate surfaces. Avoid sagging and undercut. Thorough cleaning is essential. Travel from left to right and from right to left.

Operations

1. Obtain plates; check the job print for quantity and size.
2. Obtain electrodes of each quantity, type, and size specified in the job print.
3. Set the power source for 80 to 130 amperes. Set a d.c. power source for reverse polarity.
4. Set up the plates and tack weld them as shown on the job print.
5. Place the joints in the horizontal position on the welding table. Make sure that it is well-grounded.
6. Make the first, second, and third passes with ⅛-inch electrodes as shown on the job print. Manipulate the electrode as shown in Fig. 22-19.
7. Chip the slag from the welds, brush, and inspect between each pass. Refer to Inspection, below.
8. Increase the power source setting to 112 to 170 amperes.
9. Make the fourth, fifth, sixth and seventh passes with 5/32-inch electrodes as shown on the job print.
10. Chip the slag from the welds, brush, and inspect between each pass. Refer to Inspection, below.
11. Practice these welds until you can produce good welds consistently with both types of electrodes.

Inspection

Compare the back side of the first pass with Fig. 22-20 and all passes with Fig. 22-21. Check all passes for the following weld characteristics:

Width and height: All passes uniform

Appearance: Smooth with close ripples; free of voids and high spots. The lapover of the beads should be well-proportioned. Restarts should be difficult to locate.

Size: Refer to the job print. Check convexity with a butt weld gauge.

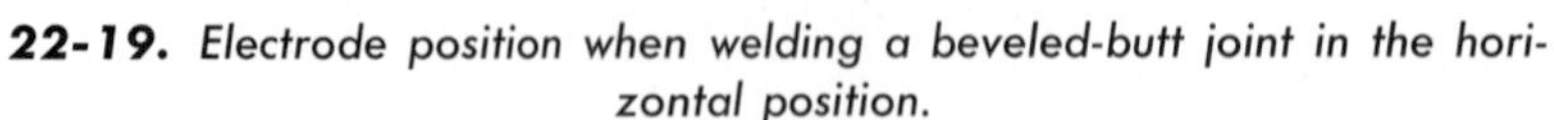

22-19. *Electrode position when welding a beveled-butt joint in the horizontal position.*

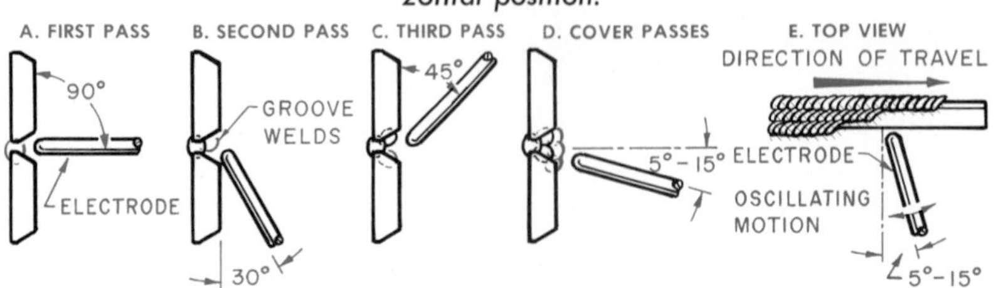

Face of weld: Some reinforcement
Edges of weld: Good fusion, no overlap, no undercut
Starts and stops: Free of depressions and high spots
Beginnings and endings: Full size, craters filled
Back side: Complete and uniform penetration
Penetration and fusion: Through the back side and to all plate surfaces.
Surrounding plate surfaces: Free of spatter
Slag formation: Full coverage, easily removable

Disposal

Put completed joints in the scrap bin so that they will be available for further use. The plate can be cut and beveled between welds for further butt welding, and unwelded plate surfaces can be used for beading.

Qualifying Test

At this point you should be able to pass Test 2 in the horizontal position. (See pages 716, 717.) This test is similar to the ASME test in pipe or plate.

JOB 22-J61 WELDING A COUPLING TO A FLAT PLATE

Objective

To weld pipe or coupling to a flat plate in the flat position by means of a single-pass horizontal fillet weld with DCSP and/or a.c. shielded-arc electrode (AWS-ASTM E6012-E6013).

General Job Information

This job is similar to those found in tank work in which couplings and fittings are welded into the shell, heads, and bottom. It is also included in this chapter to develop flexibility in positioning the electrode and to give experience in welding in close quarters.

Welding Technique

The current adjustment should be rather high. Hold the electrode at a 45-degree angle to the plate, Fig. 22-22A, and tilt it 10 to 20 degrees in the direction of travel, Fig. 22-22B. The actual position for welding is shown in Fig. 22-23. The welding technique is not different than that for running stringer beads and single-pass fillet welds in the horizontal position. The position of the electrode is constantly changing in respect to the operator as it travels around the outside of the coupling, however, and space is limited.

Practice starts and stops. Travel in all directions.

Operations

1. Obtain plate and pieces of pipe; check the job print for quantity and size.
2. Obtain electrode of each quantity, type, and size specified in the job print.
3. Set the power source for 110 to 190 amperes. Set a d.c. power source for straight polarity.
4. Place the plate in the flat position on the welding table. Make sure the plate is well-grounded.
5. Tack weld the small pipe to the plate.
6. Make inside and outside fillet welds with $^5/_{32}$-inch electrodes as shown on the job print. Manipulate the electrode as instructed in Fig. 22-22.
7. Chip the slag from the

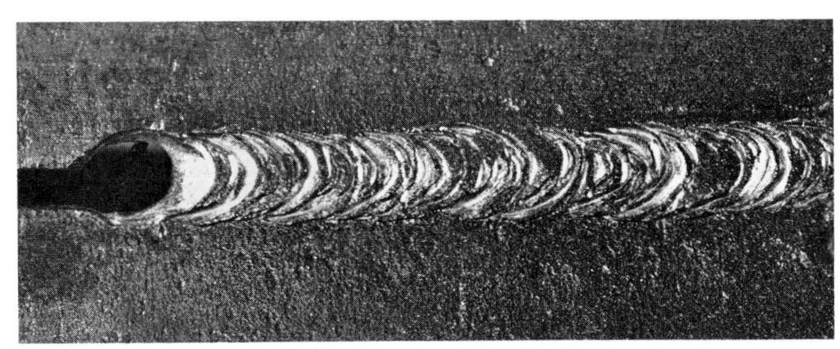

22-20. *Penetration and fusion through the back side of a beveled-butt joint. Note the hole at the leading edge of the weld.*

22-21. *Typical appearance of a multipass groove weld in a beveled-butt joint without a backup strip welded in the horizontal position with DCRP or a.c. shielded-arc electrodes.*

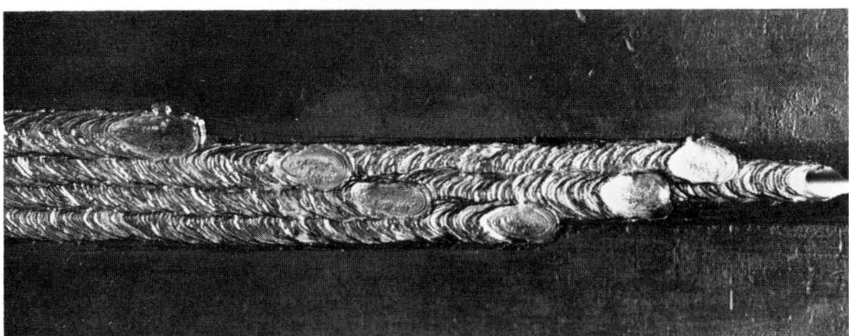

welds, brush, and inspect. Refer to Inspection.

8. Increase the power source setting to 140 to 240 amperes.

9. Tack weld the large pipe to the plate.

10. Make inside and outside fillet welds with $^3/_{16}$-inch electrodes as shown on the job print.

11. Chip the slag from the welds, brush, and inspect. Refer to Inspection, below.

12. Practice these welds until you can produce good welds consistently with both types of electrodes.

22-22. *Electrode position when welding couplings with straight polarity.*

Inspection

Compare the welds with Fig. 22-23 and check them for the following characteristics:

Width and height: Uniform

Appearance: Smooth with close ripples; free of voids. Restarts should be difficult to locate.

Size: Refer to the job print. Check with a fillet weld gauge.

Face of weld: Slightly convex

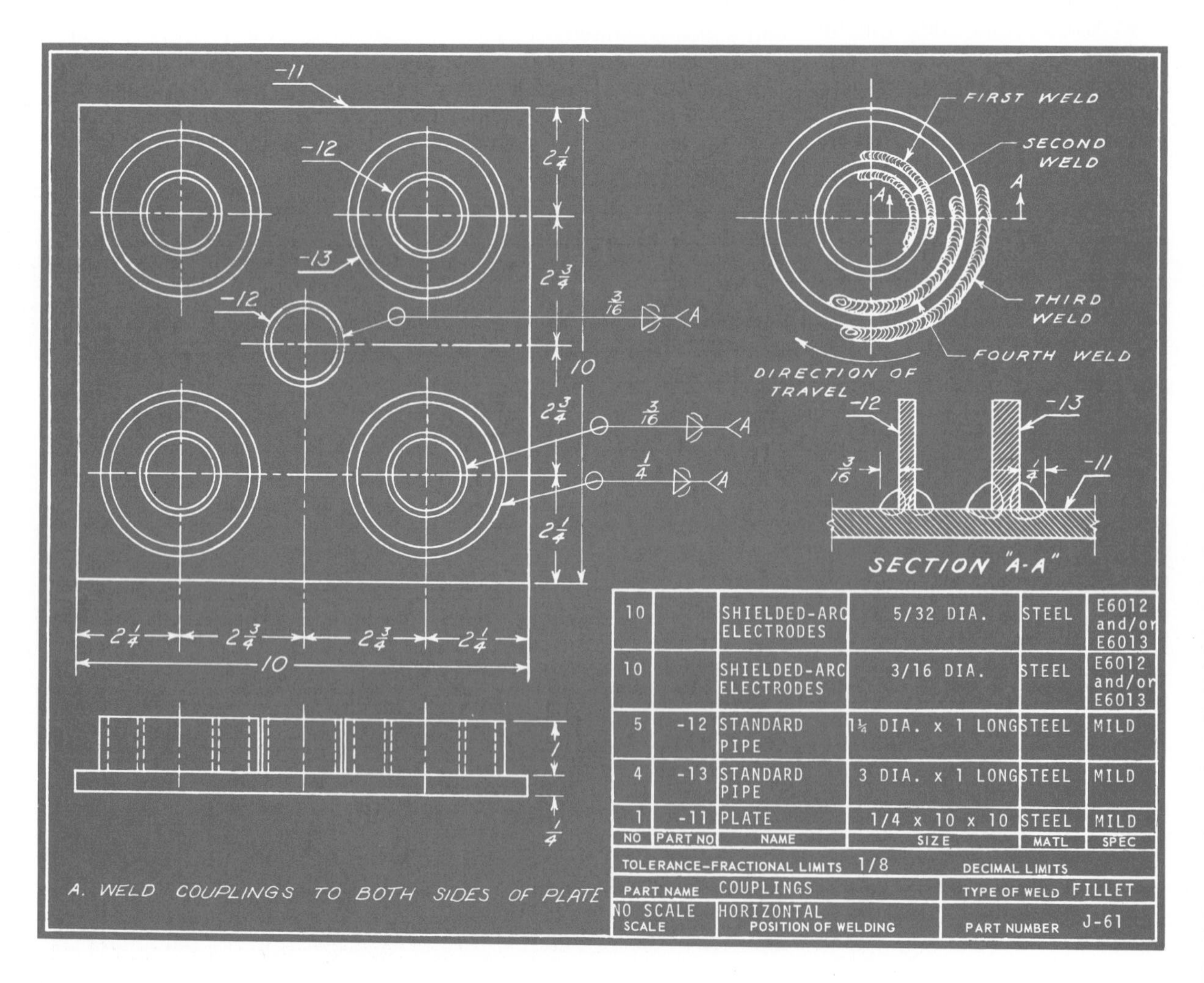

NO	PART NO	NAME	SIZE	MATL	SPEC
10		SHIELDED-ARC ELECTRODES	5/32 DIA.	STEEL	E6012 and/or E6013
10		SHIELDED-ARC ELECTRODES	3/16 DIA.	STEEL	E6012 and/or E6013
5	-12	STANDARD PIPE	1¼ DIA. x 1 LONG	STEEL	MILD
4	-13	STANDARD PIPE	3 DIA. x 1 LONG	STEEL	MILD
1	-11	PLATE	1/4 x 10 x 10	STEEL	MILD

TOLERANCE-FRACTIONAL LIMITS 1/8 DECIMAL LIMITS
PART NAME COUPLINGS TYPE OF WELD FILLET
SCALE NO SCALE POSITION OF WELDING HORIZONTAL PART NUMBER J-61

Edges of weld: Good fusion, no overlap, no undercut
Starts and stops: Free of depressions and high spots
Beginnings and endings: Full size, craters filled. Pay particular attention to points where the end of the weld overlaps the starting point.
Penetration and fusion: To the root of the joint and plate and pipe surfaces.
Surrounding plate surfaces: Free of spatter
Slag formation: Full coverage, easily removable

Disposal

Put completed plate in the scrap bin so that it will be available for further use. The unwelded plate surfaces between the couplings can be used for beading.

JOB 22-J62 WELDING A COUPLING TO A FLAT PLATE

Objective

To weld pipe or coupling to a flat plate in the flat position by means of a single-pass horizontal fillet weld with DCRP and/or a.c. shielded-arc electrode (AWS-ASTM E6010-E6011).

General Job Information

This job is identical to Job 22-J61 except that reverse polarity electrodes are used for welding.

Welding Technique

The current adjustment cannot be as high as with straight polarity electrodes. More care must be taken with electrode position. Undercut is a possibility, and arc blow may be a problem. Note in Fig. 22-24 that the electrode is held slightly closer to the plate. The angle is 40 degrees instead of 45 degrees to prevent undercut. Also note that the face of the weld is flat instead of convex. Review the material concerning arc blow, page 571. Use a welding technique like that used in the previous job, but hold a closer arc gap.

Practice starts and stops. Travel in all directions.

Operations

1. Obtain plate and pieces of pipe; check the job print for quantity and size.
2. Obtain electrodes in each quantity, type, and size specified in the job print.
3. Set the power source for 110 to 170 amperes. Set a d.c. power source for reverse polarity.
4. Place the plate in the flat position on the welding table. Make sure the plate is well-grounded.
5. Tack weld the small pipe to the plate.
6. Make inside and outside fillet welds with 5/32-inch electrodes as shown on the job print. Manipulate the electrode as instructed in Fig. 22-22.
7. Chip the slag from the welds, brush, and inspect. Refer to Inspection, below.
8. Increase the power source setting to 140 to 240 amperes.
9. Tack weld the large pipe to the plate.
10. Make inside and outside fillet welds with 3/16-inch electrodes as shown on the job print.
11. Chip the slag from the welds, brush, and inspect. Refer to Inspection, below.
12. Practice these welds until you can produce good welds consistently with both types of electrodes.

Inspection

Compare the welds with Fig. 22-25 and check them for the following characteristics:

Width and height: Uniform
Appearance: Smooth with close ripples; free of voids. Restarts should be difficult to locate.
Size: Refer to the job print. Check with a fillet weld gauge.
Face of weld: Flat

22-23. *Electrode and electrode holder positioned as instructed in Fig. 22-22. Note the typical appearance of the fillet welds in couplings welded in the flat position with DCSP or a.c. shielded-arc electrodes.*

Edges of weld: Good fusion, no overlap, no undercut

Starts and stops: Free of depressions and high spots

Beginnings and endings: Full size, craters filled. Pay particular attention to points where the end of the weld overlaps the starting point.

Penetration and fusion: To the root of the joints and plate and pipe surfaces

Surrounding plate surfaces: Free of spatter

Slag formation: Full coverage, easily removable

Disposal

Put completed plate in the scrap bin so that it will be available for further use. The unwelded plate surfaces between the couplings can be used for beading.

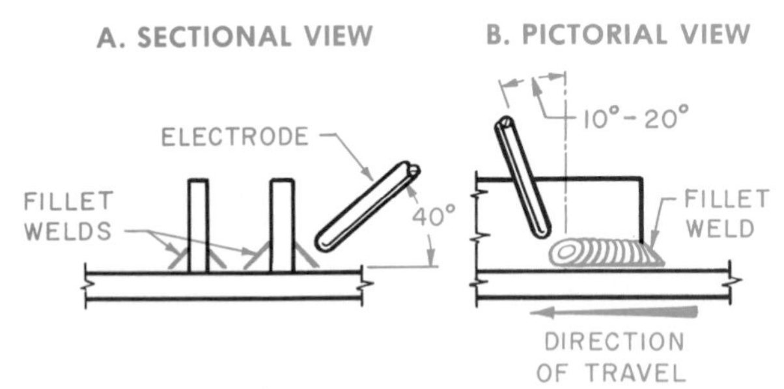

22-24. *Electrode position when welding couplings with DCRP.*

JOB 22-J63 WELDING A SINGLE-V BUTT JOINT (BACKUP STRIP CONSTRUCTION)

Objective

To weld a single-V butt joint assembled with a backup strip in the overhead position by means of multipass groove welds,

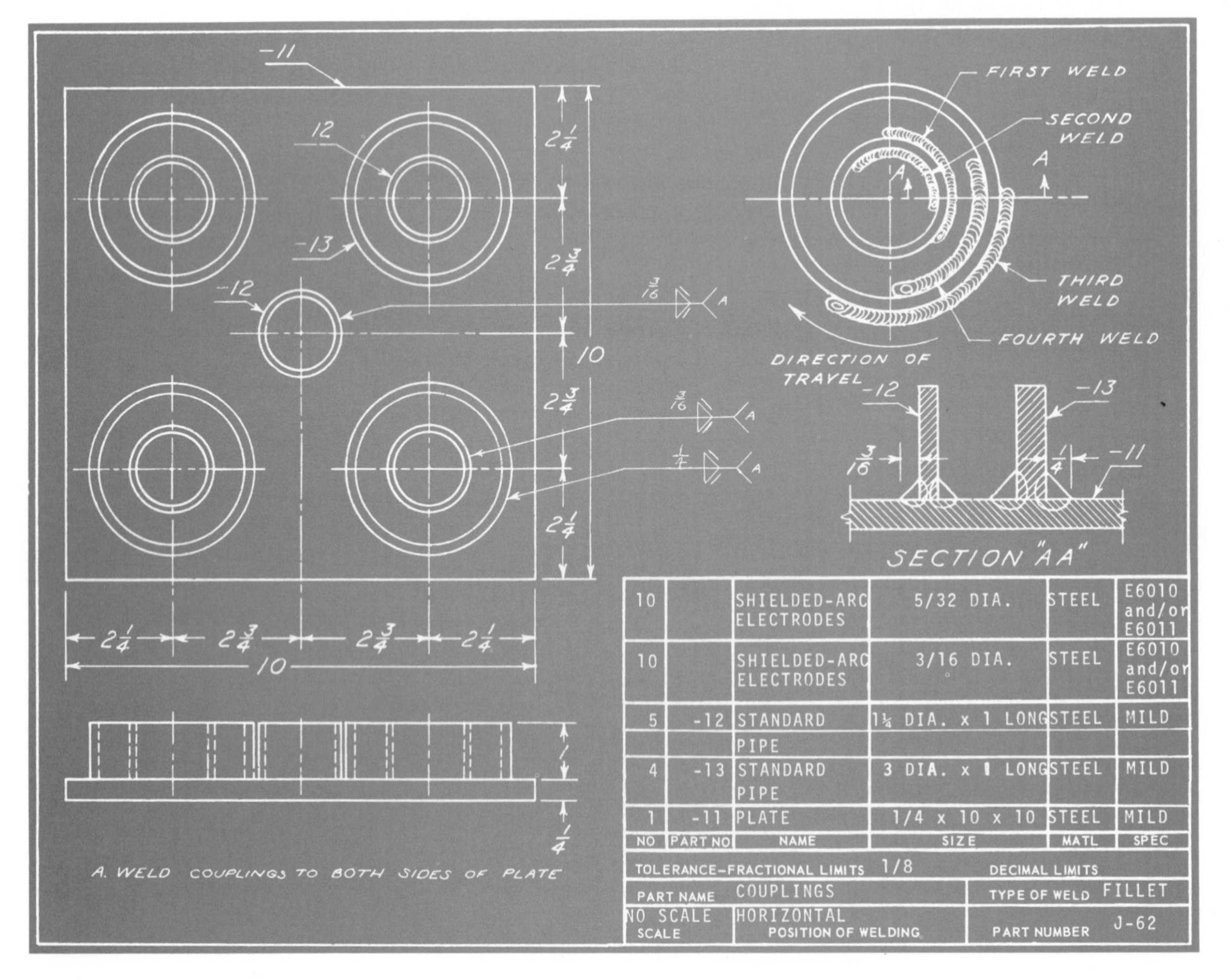

NO	PART NO	NAME	SIZE	MATL	SPEC
10		SHIELDED-ARC ELECTRODES	5/32 DIA.	STEEL	E6010 and/or E6011
10		SHIELDED-ARC ELECTRODES	3/16 DIA.	STEEL	E6010 and/or E6011
5	-12	STANDARD PIPE	1¼ DIA. x 1 LONG	STEEL	MILD
4	-13	STANDARD PIPE	3 DIA. x 1 LONG	STEEL	MILD
1	-11	PLATE	1/4 x 10 x 10	STEEL	MILD

stringer bead technique, with DCRP and/or a.c. shielded-arc electrodes (AWS-ASTM E6010-E6011).

General Job Information

The beveled-butt joint in plate with a backup strip is similar to joints in pipe and tanks welded with a backing ring or chill ring. This job is identical to the joint and position required for Test No. 1 of the Department of the Navy Requirements and that used by various code authorities and recommended by the American Welding Society. Often the test joint is set up with a $^3/_8$-inch root gap. Two or three root passes are necessary when welding a joint with this gap. More passes are necessary to complete the joint because of greater separation between the plates.

Welding Technique

Tilt the electrode 5 to 10 degrees in the direction of travel, Fig. 22-26A. Hold it at 90 degrees to the surface of the plates for the first pass (Fig. 22-26B), 20 to 35 degrees to the surface of each plate for the second and third passes (Fig. 22-26C and D), and 10 to 20 degrees to the surface of each plate for the fourth and fifth passes (Fig. 22-26E). Like the first pass, the last pass is made at a 90-degree angle to the surface of the plates, Fig. 22-26F.

FIRST PASS. Current adjustment should be high. A short arc is essential to insure proper metal transfer. A whipping motion should be employed. Make sure that you are obtaining fusion to the backup strip and the root faces of the two plates. Keep the weld as flat as possible.

SECOND THROUGH SIXTH PASSES. Current adjustment should be high. Use the stringer bead technique and decrease the whipping motion. If the metal becomes too hot, lengthen the arc and whip the electrode forward until the crater has cooled. Fusion must be secured with the preceding passes and the walls of the groove. Be sure to remove the slag between each pass. Avoid undercut for all passes.

Practice starts and stops. Travel in all directions.

Operations

1. Obtain plates; check the job print for quantity and size.
2. Obtain electrodes of each quantity, type, and size specified in the print.
3. Set the power source for 110 to 170 amperes. Set a d.c. power source for reverse polarity.
4. Set up the plates and tack weld them as shown on the job print.
5. Fasten the joint in the overhead position with an overhead welding jig.
6. Make all passes with $^5/_{32}$-inch electrodes as shown on the job print. Manipulate the electrode as instructed in Fig. 22-26.
7. Chip the slag from the welds, brush, and inspect between each pass. Refer to Inspection, below.
8. Practice these welds until you can produce good welds consistently with both types of electrodes.

Inspection

Compare each pass with Fig. 22-27 and check it for the following weld characteristics:

Width and height: Uniform

Appearance: Smooth with close ripples; free of voids and high

22-25. *Typical appearance of fillet welds on couplings welded in the flat position with DCRP or a.c. shielded-arc electrodes.*

spots. Restarts should be difficult to locate.

Size: Refer to the job print. Check the last pass with a butt weld gauge.

Face of weld: Slight reinforcement

Edges of weld: Good fusion, no overlap, no undercut

Starts and stops: Free of depressions and high spots

Beginnings and endings: Full size, craters filled

Penetration and fusion: To the backup strip, preceding passes, and plate surfaces

Surrounding plate surfaces: Free of spatter

Disposal

Put completed joints in the scrap bin so that they will be available for further use. The plate can be cut and beveled between the welds for further butt welding, and unwelded plate surfaces can be used for beading.

Qualifying Test

At this point you should be able to pass Test 1, pages 714–717, in the overhead position. This is identical to Test 1 of the Department of the Navy Requirements and AWS recommendations.

It will also be desirable for you to take Test 3 (fillet weld) in the overhead position (pages 717, 718).

JOB 22-J64 WELDING A SQUARE BUTT JOINT

Objective

To weld a square butt joint in the overhead position by means of multipass groove welds, with DCRP and/or a.c. shielded-arc electrodes (AWS-ASTM E6010-E6011).

General Job Information

It is doubtful whether a square butt joint will ever be welded in

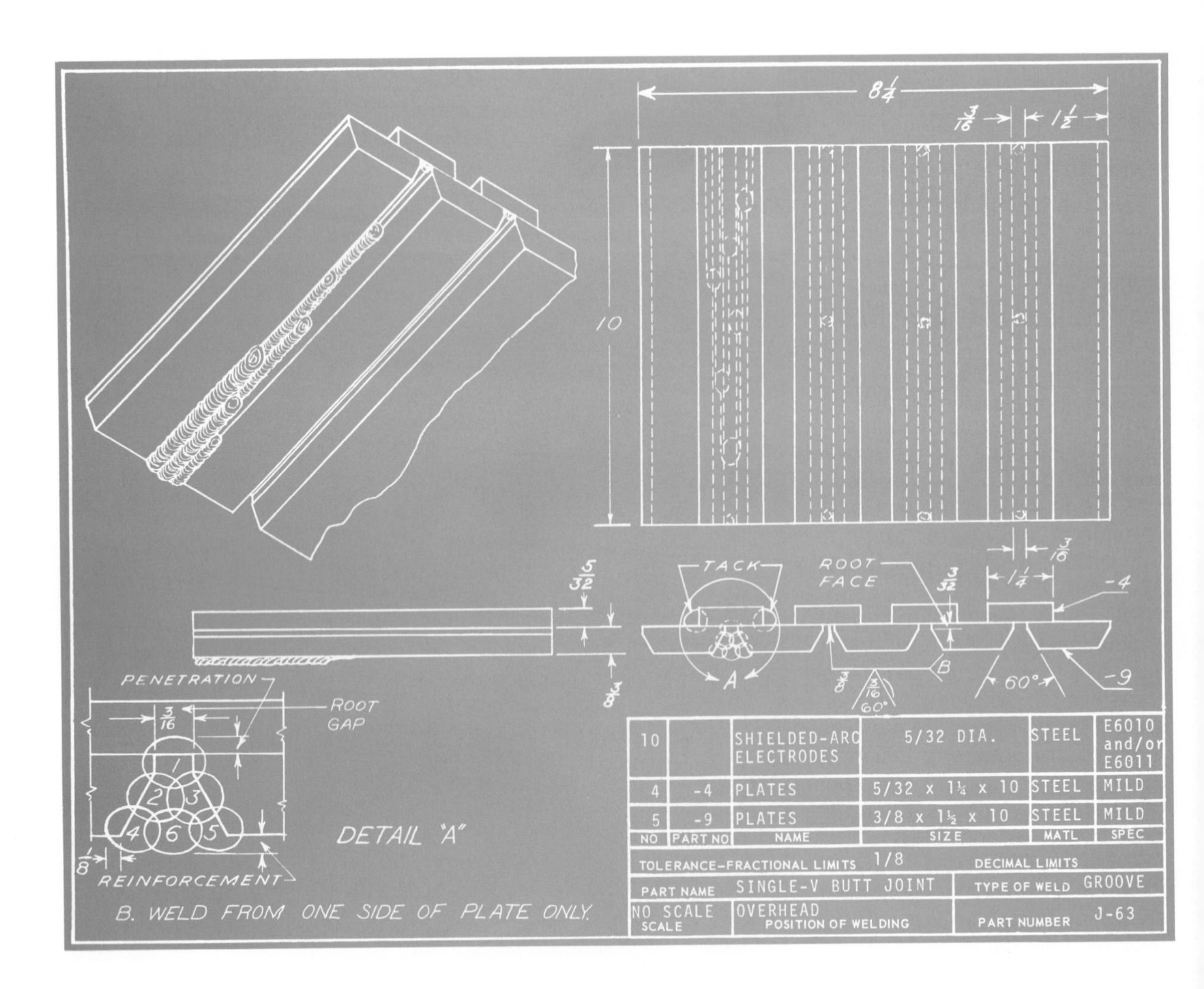

the overhead position on the job. It is assigned for the purpose of developing the ability to deposit the first pass in the root of an open butt joint in overhead position. This is a difficult operation and requires much practice. This job will also develop the ability to make the first pass in pipe. The square butt joint requires very little joint preparation and is economical.

Welding Technique

Hold the electrode at a 90-degree angle to the surface of the plates. Fig. 22-28A, and tilt it in the direction of travel no more than 10 degrees, Fig. 22-28B.

FIRST PASS. The current setting should not be too high. The arc gap must be close. The electrode should be kept within the groove in order to secure complete penetration through the back side (Fig. 22-29) and with the root faces of the plate. A back-and-forth motion is employed. In Fig. 22-29, note that a large tack could not be melted through. Tacks that are too large can cause a test failure.

SECOND PASS. The current setting must not be too high. A slight oscillating, whipping motion is used. The side weave must not be too wide. Hold a close arc gap.

Practice starts and stops. Travel in all directions.

Operations

1. Obtain plates; check the job print for quantity and size.
2. Obtain electrodes of each quantity, type, and size specified in the job print.
3. Set the power source for 75 to 125 amperes. Set a d.c. power source for reverse polarity.
4. Set up the plates and tack weld as shown on the job print.
5. Fasten the joints in the overhead position with an overhead welding jig.
6. Make the first pass with ⅛-inch electrodes as shown on the job print. Manipulate the electrode as shown in Fig. 22-28, page 711.
7. Chip the slag from the weld, brush, and inspect. Refer to Inspection.
8. Increase the power source setting to 110 to 170 amperes.
9. Make the second pass with $\frac{5}{32}$-inch electrodes as shown on the job print.
10. Chip the slag from the weld, brush, and inspect. Refer to Inspection.
11. Practice these welds until you can produce good welds consistently with both types of electrodes.

Inspection

Compare the back side of the first pass with Fig. 22-29 and both passes with Fig. 22-30. Check them for the following weld characteristics:

Width and height: Uniform
Appearance: Smooth with close ripples; free of voids and high spots. Restarts should be difficult to locate.
Size: Refer to the job print. Check with a butt weld gauge.
Face of weld: First pass: flat; last pass: slight reinforcement

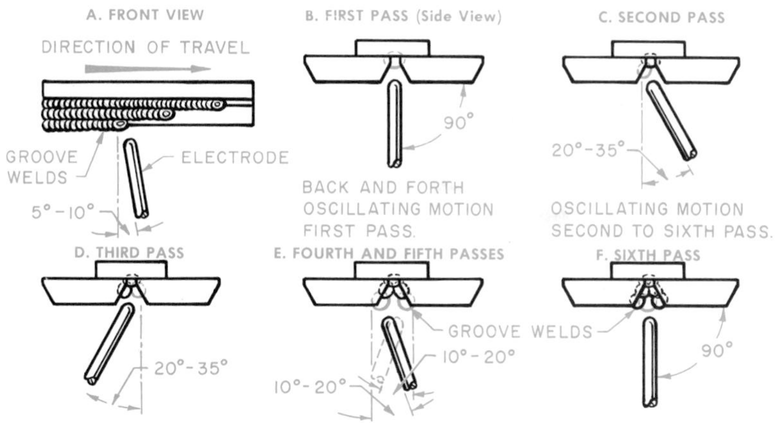

22-26. *Electrode position when welding a beveled-butt joint in the overhead position.*

22-27. *Typical appearance of a multipass groove weld in a beveled-butt joint with a backup strip welded in the overhead position with DCRP or a.c. shielded-arc electrodes.*

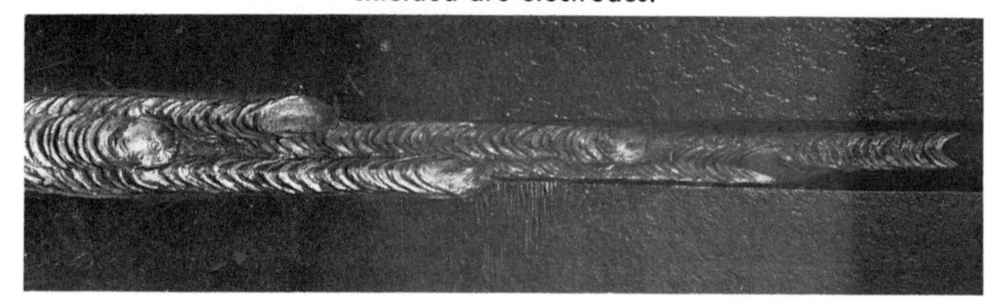

Edges of weld: Good fusion, no overlap, no undercut
Starts and stops: Free of depressions and high spots
Beginnings and endings: Full size, craters filled
Back side: Complete and uniform penetration
Penetration and fusion: Through the back side and plate surfaces, Fig. 22-29.
Surrounding plate surfaces: Free of spatter
Slag formation: Full coverage, easily removable

Disposal

Put completed joints in the scrap bin so that they will be available for further use. Joints can be cut between welds for further butt welding, and unwelded plate surfaces can be used for beading.

JOB 22-J65 WELDING A SINGLE-V BUTT JOINT

Objective

To weld a single-V butt joint in the overhead position by means of multipass groove welds, weaved-beading technique, with DCRP and/or a.c. shielded-arc electrodes (AWS-ASTM E6010-E6011).

General Job Information

The single-V butt joint in the overhead position will be encountered in pipe and critical plate welding. Both joint construction and welding procedure are similar to those in the ASME test.

Welding Technique

FIRST PASS. Current adjustment should be high enough to insure penetration through the back side. Hold the electrode as shown in Fig. 22-31A and use the root pass technique practiced in the preceding job. A small uniform bead should be formed on the back side (Fig. 22-32), and the face of the weld should be flat. Keep a small hole at the forward edge of the weld pool as you proceed with the weld pool. If the hole gets too large, move forward along the bevel until the

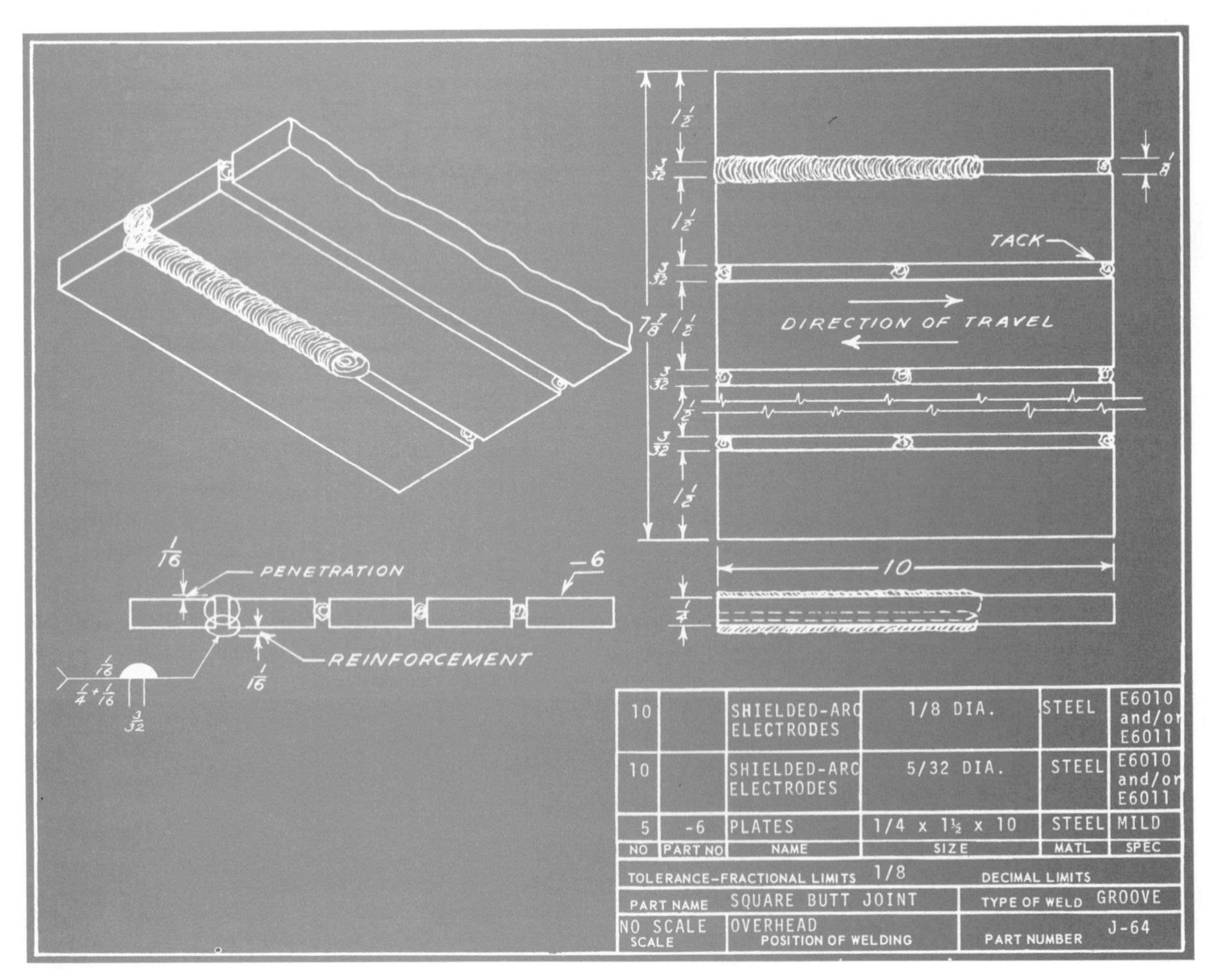

puddle freezes. Then go back to the root and proceed with the weld as before.

SECOND AND THIRD PASSES. Current adjustment can be high. Hold the electrode as shown in Fig. 22-31B and use weaved beading technique. Move the electrode across the face of the weld with a rapid crossover movement so that you do not deposit too much metal in the center of the bead, causing it to sag. The side-to-side movement

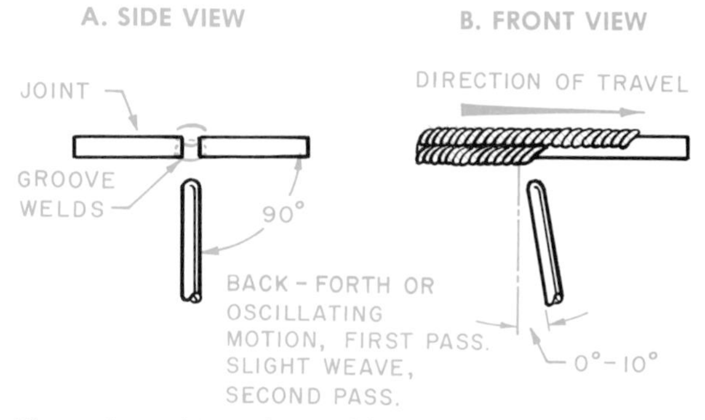

22-28. *Electrode position when welding a square butt joint in the overhead position.*

22-29. *Penetration and fusion through the back side of a square butt joint. Note that one tack weld was too large to be melted through. Poor tacking causes many operator qualification test failures.*

22-30. *Typical appearance of a multipass groove weld in a square butt joint welded in the overhead position with DCRP or a.c. shielded-arc electrodes.*

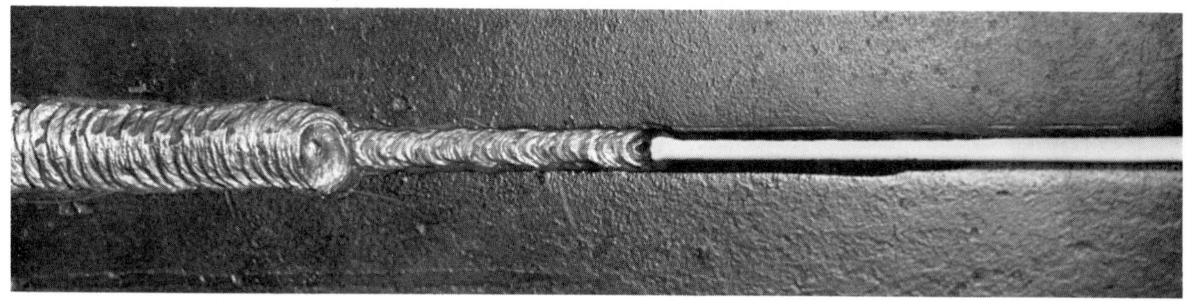

should be kept within the limits of the required weld size. Hesitate at the sides of the weld to prevent undercut. Remember that undercutting is caused by gouging out the plate with the arc and failing to fill it up with deposited weld metal.

Practice starts and stops on all passes.

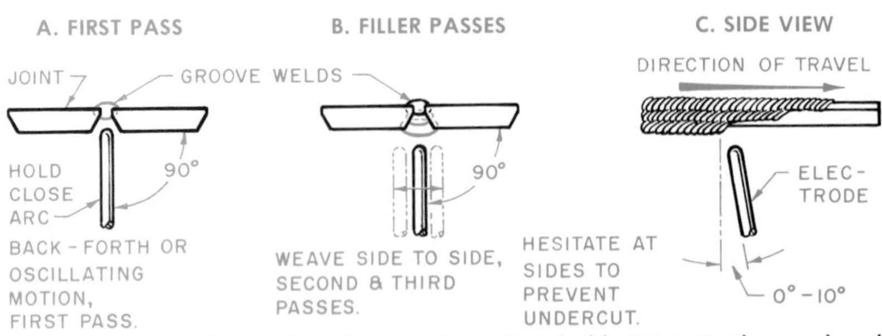

22-31. *Electrode position when welding a beveled-butt joint in the overhead position.*

Operations

1. Obtain plates; check the job print for quantity and size.
2. Obtain electrodes of each quantity, type, and size specified in the job print.
3. Set the power source for 75 to 125 amperes. Set a d.c. power source for reverse polarity.
4. Set up the plates and tack weld them as shown on the job print.
5. Fasten the joints in the overhead position with an overhead welding jig.
6. Make the first and second passes with $\frac{1}{8}$-inch electrodes as shown on the job print.
7. Chip the slag from the

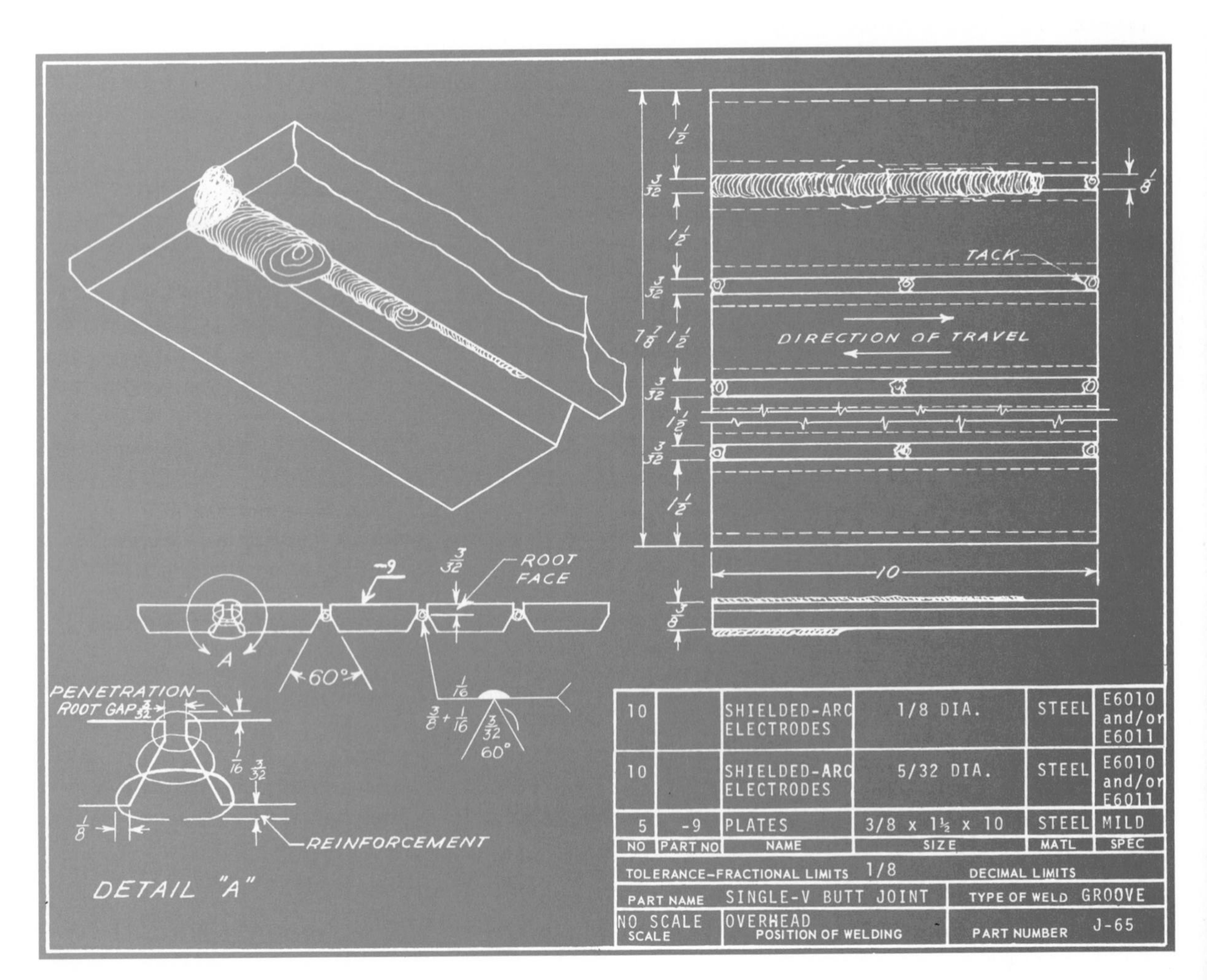

welds, brush, and inspect. Refer to Inspection, below.

8. Increase the setting of the power source to 110 to 170 amperes.

9. Make the third pass with $^{5}/_{32}$-inch electrodes as shown on the job print.

10. Chip the slag from the weld, brush, and inspect. Refer to Inspection, below.

11. Practice these welds until you can produce good welds consistently with both types of electrodes.

Inspection

Compare the back side of the first pass with Fig. 22-32 and all passes with Fig. 22-33. Check each pass for the following weld characteristics:

Width and height: Uniform
Appearance: Smooth with close ripples; free of voids and high spots. Restarts should be difficult to locate.
Size: Refer to the job print. Check with a butt weld gauge.
Face of weld: First two passes flat; last pass slightly convex
Edges of weld: Good fusion, no overlap, no undercut
Starts and stops: Free of depressions and high spots
Beginnings and endings: Full size, craters filled
Back side: Complete and uniform penetration
Penetration and fusion: Through the back side, preceding passes, and all plate surfaces. See Fig. 22-32.
Surrounding plate surfaces: Free of spatter
Slag formation: Full coverage, easily removable

Disposal

Put completed joints in the scrap bin so that they will be available for further use. Joints can be cut and beveled between welds for further butt welding, and unwelded surfaces can be used for beading.

Qualifying Test

At this point you should be able to pass Test 2 in the overhead position. (See page 716.) This test is similar to the ASME test for pipe and plate.

You should also take Test 4, page 720.

TESTS

General Information

Each test in this section will be conducted after the student has completed the job in which it is assigned and at the conclusion of the course. No student should be considered as having completed the course who has not fulfilled the requirements of Tests 1, 3, and 4. Only students of exceptional ability are expected to meet the requirements of Test 2.

Types of Tests

The tests which follow are especially devised to determine the student's ability to produce sound and pressure-tight welds. These tests are standard tests used by various code authorities to qualify welders for welding of high quality and performance. They are also similar to those recommended by the American Welding Society.

The fillet weld soundness test is prescribed for the testing of fillet welds. For the testing of groove welds, the tests require the welding of beveled-butt joints with and without a backing plate. Specimens are taken from the

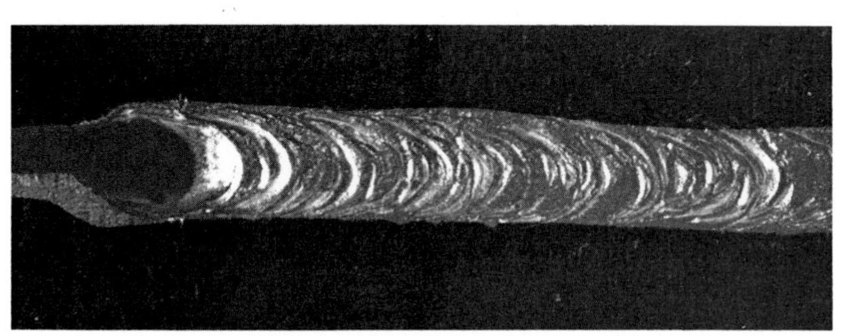

22-32. *Penetration and fusion through the back side of a beveled-butt joint without a backup strip.*

22-33. *Typical appearance of a multipass groove weld in a beveled-butt joint welded without a backup strip in the overhead position (weaved beading technique) with DCRP or a.c. shielded-arc electrodes.*

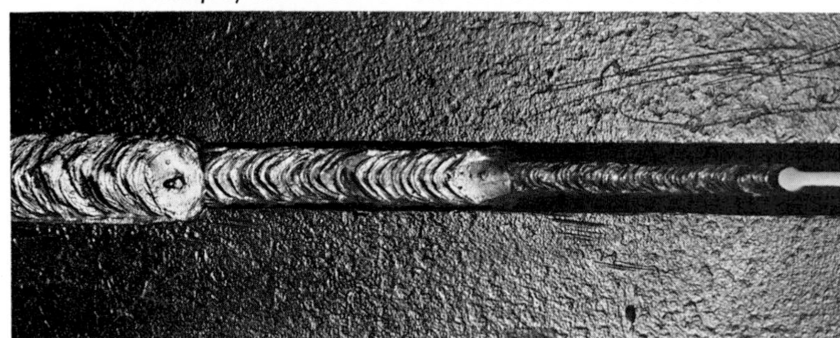

test joint and are subjected to the standard face- and root-bend test procedures.

In order to determine the student's ability to produce welds that are pressure tight and do not leak, he will fabricate a special small vessel completely by welding. It will be subjected to a pressure of 2,500 p.s.i. internal pressure. All of the types of electrodes used in the course are used in making the testing unit.

Base Metal and Its Preparation

The base metal is mild steel plate. For each type of joint, the length of the weld and the dimensions of the base metal are such as to provide enough material for the test specimens required.

- *Groove welds with backup strip:* The preparation of the base metal for welding is for a single-V groove butt joint meeting the requirements of Test 1, Fig. 2 (page 714).
- *Groove welds without a backup strip:* The preparation of the base metal for welding is for a single-V groove butt joint meeting the requirements of Test 2, Fig. 2 (page 716).
- *Fillet welds:* The preparation of the base metal for welding is shown in Test 3, Fig. 2 (page 718).
- *Pressure testing:* The preparation of the base metal for welding is shown in Test 4 (page 720).

Position of Test Welds

The positions for the tests are those illustrated in Test 1, Fig. 1; Test 2, Fig. 1 (page 716); Test 3, Fig. 1 (page 718); and Test 4 (page 720). If the student passes the tests in more difficult positions, he need not be tested in the less difficult positions. He should be considered as having passed such tests in accordance with the following provisions:

- *Groove welds:* If the student passes the tests in the horizontal, vertical, or overhead position, he need not be tested in the flat position.

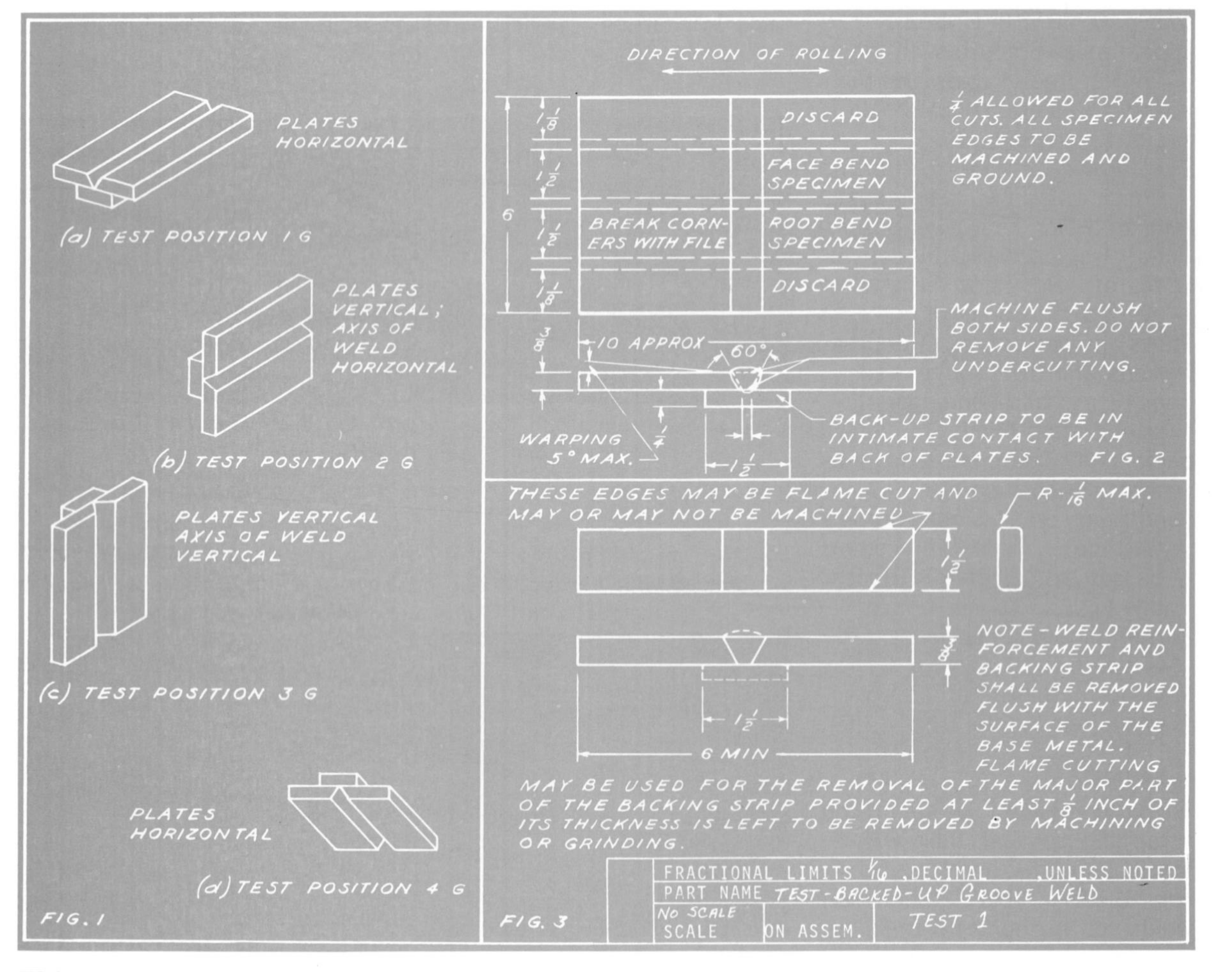
PLATES HORIZONTAL
(a) TEST POSITION 1 G
PLATES VERTICAL; AXIS OF WELD HORIZONTAL
(b) TEST POSITION 2 G
PLATES VERTICAL AXIS OF WELD VERTICAL
(c) TEST POSITION 3 G
PLATES HORIZONTAL
(d) TEST POSITION 4 G
FIG. 1
DIRECTION OF ROLLING
DISCARD
FACE BEND SPECIMEN
BREAK CORNERS WITH FILE
ROOT BEND SPECIMEN
DISCARD
1/4 ALLOWED FOR ALL CUTS. ALL SPECIMEN EDGES TO BE MACHINED AND GROUND.
MACHINE FLUSH BOTH SIDES. DO NOT REMOVE ANY UNDERCUTTING.
10 APPROX
60°
WARPING 5° MAX.
BACK-UP STRIP TO BE IN INTIMATE CONTACT WITH BACK OF PLATES.
FIG. 2
THESE EDGES MAY BE FLAME CUT AND MAY OR MAY NOT BE MACHINED
R-1/16 MAX.
NOTE-WELD REINFORCEMENT AND BACKING STRIP SHALL BE REMOVED FLUSH WITH THE SURFACE OF THE BASE METAL. FLAME CUTTING MAY BE USED FOR THE REMOVAL OF THE MAJOR PART OF THE BACKING STRIP PROVIDED AT LEAST 1/8 INCH OF ITS THICKNESS IS LEFT TO BE REMOVED BY MACHINING OR GRINDING.
6 MIN
FIG. 3
FRACTIONAL LIMITS 1/16 DECIMAL UNLESS NOTED
PART NAME TEST-BACKED-UP GROOVE WELD
NO SCALE
SCALE ON ASSEM.
TEST 1

- *Fillet welds:* If the student passes the tests in the horizontal, vertical, or overhead position, he need not be tested in the flat position. If he passes the tests in the vertical or overhead position, he need not be tested in either the flat position or the horizontal position.
- *Pressure welds:* The vessel must be welded as specified in Table 22-1, below.

Number of Test Welds Required

- *Groove welds:* One test weld as shown in Test 1, Fig. 2, and/or Test 2, Fig. 2 (page 716) is made for each position in which the student is to be tested.
- *Fillet welds:* One test weld as shown in Test 3, Fig. 2 (page 718), is made for each position in which the student is to be tested.
- *Pressure welds:* One vessel which conforms to Test 4 (page 720) is constructed.

Welding Procedure

The test welds are made in accordance with the following illustrations and jobs.

- *Groove welds* (Test 1: groove weld with backup strip—$3/16$-inch root gap)
  1. Flat position: See Fig. 22-34 below and review Job 21-J32 (page 641).
  2. Horizontal position: See Fig. 22-35 (page 716) and review Job 21-J34 (page 645).
  3. Vertical position: See Fig.

22-34. *Beveled-butt test joint with a backup strip and a $3/16$-inch root opening welded in the flat position. Note the sequence of weld beads and the appearance*

Table 22-1

Welding procedures for Test 4.

Joints	Operation	Welding technique	Position	DCRP	DCSP	a.c.	AWS specification number
1. T-joint	Fillet welding, single pass, travel up	Job 21-J44	Vertical	X		X	E6010–E6011
2. Lap joint	Fillet welding, single pass, travel up	Job 20-J24	Vertical	X		X	E6010–E6011
3. Outside corner joint	Groove welding, multipass, weaved, travel up	Job 21-J49	Vertical	X		X	E6010–E6011
4. Single-V butt joint	Groove welding, multipass, weaved, travel up	Job 22-J52	Vertical	X		X	E6010–E6011
5. Outside corner joint	Groove welding, multipass, weaved	Job 21-J36	Flat	X		X	E6010–E6011
6. Lap joint	Fillet welding, single pass	Job 20-J11	Horizontal		X	X	E6012–E6013
7. T-joint	Fillet welding, single pass	Job 22-J53	Overhead	X		X	E6010–E6011
8. Single-V butt joint	Groove welding, multipass, weaved	Job 22-J65	Overhead	X		X	E6010–E6011
9. T-joint	Fillet welding, single pass	Job 21-J39	Horizontal		X	X	E6020
10. T-joint	Fillet welding, multipass, travel down	Job 20-J15	Vertical	X		X	E6010–E6011
11. T-joint	Fillet welding, single pass	Job 21-J41	Flat	X		X	E6024–E7024
12. T-joint	Fillet welding, multipass, travel up	Job 21-J45	Vertical	X		X	E6010–E6011
13. T-joint	Fillet welding, single pass	Job 21-J37	Horizontal	X		X	E7028
14. T-joint	Fillet welding, multipass, weaved, travel up	Job 21-J46	Vertical	X		X	E7018
15. T-joint	Fillet welding, single pass	Job 22-J59	Overhead	X		X	E6010–E6011
16. T-joint	Fillet welding, single pass	Job 20-J26	Horizontal		X	X	E6012–E6013
17. Coupling	Fillet welding, single pass	Job 22-J61	Horizontal		X	X	E6012–E6013

22-36 (page 717) and review Job 21-J47 (page 672).
4. Overhead position: See Fig. 22-37 (page 717) and review Job 22-J63 (page 706).

- *Groove welds* (Test 1: groove weld with backup strip—3/8-inch root gap)
  1. Flat position: See Fig. 22-38 (page 718) and review Job 21-J32 (page 641).
  2. Horizontal position: See Fig. 22-39 (page 719) and review Job 21-J34 (page 645).
  3. Vertical position: See Fig. 22-40 (page 719) and review Job 21-J47 (page 672).
  4. Overhead position: See Fig. 22-41 (page 719) and review Job 22-J63 (page 706).
- *Groove welds* (Test 2: groove weld without backup strip—standard root gap). Figure 22-42 (page 719) shows the back side of a joint. It is absolutely necessary to obtain this type of pene-

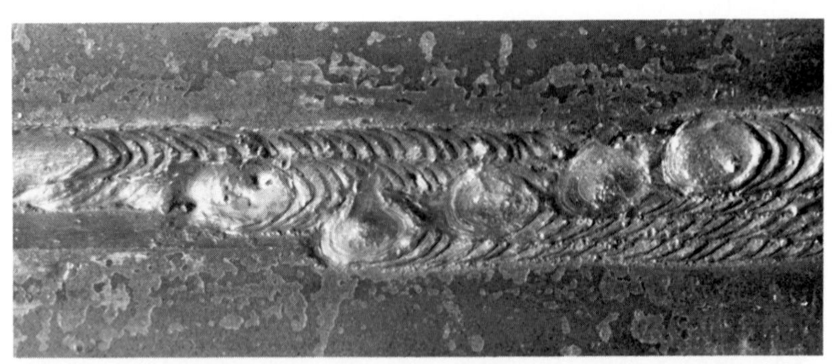

22-35. *Beveled-butt test joint with a backup strip and a 3/16-inch root opening welded in the horizontal position. Note the sequence of weld beads and the appearance.*

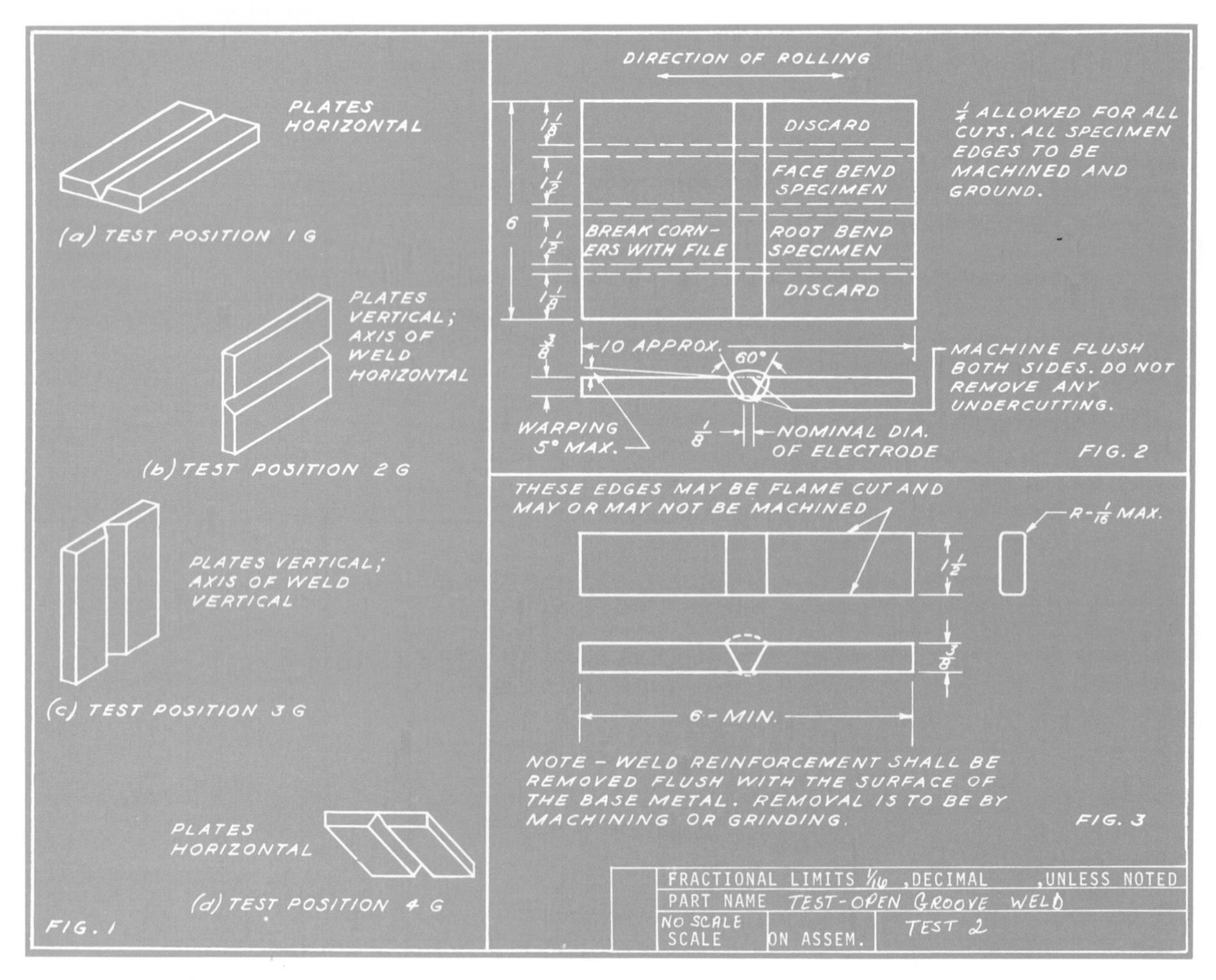

tration through the back side, or root-bend test specimens will fracture upon bending.

1. Flat position: See Fig. 22-43 (page 720) and review Job 21-J40 (page 658).

22-36. *Beveled-butt test joint with a backup strip and a 3/16-inch root opening welded in the vertical position, travel up. Note the sequence of weld beads and the appearance.*

2. Horizontal position: See Fig. 22-44, (page 721) and review Job 22-J60 (page 701).
3. Vertical position: See Fig. 22-45 (page 721) and review Job 22-J52 (page 685).
4. Overhead position: See Fig. 22-46 (page 721) and review Job 22-J65 (page 710).

- *Fillet welds* (Test 3)

1. Flat position: See Fig. 22-47 (page 721) and review Job 20-J20 (page 614).
2. Horizontal position: See Fig. 22-48 (page 722) and review Job 20-J12 (page 600) and Job 20-J27 (page 628).
3. Vertical position: See Fig. 22-49 (page 722) and review Jobs 20-J24 (page 621) and 21-J44 (page 666).
4. Overhead position: See Fig. 22-50 (page 722) and review Job 22-J53 (page 688) and Job 22-J55 (page 691).

- *Pressure welds* (Test 4. See Table 22-1 [page 715] and Fig. 22-51 [page 722]).

Test Specimens—Number, Type, and Preparation

- *Groove welds:* one root-bend specimen and one face-bend specimen are prepared from the finished test weld as shown in Test 1, Fig. 3 (page 714) and Test 2, Fig. 3 (page 716).
- *Fillet welds:* two fillet-weld soundness specimens are prepared from the finished test weld as shown in Test 3, Fig. 3 (page 718).
- *Pressure welds:* the finished vessel, Test 4 (page 720), is the test specimen.

Method of Testing Specimens

GROOVE AND FILLET WELD SPECIMENS. Each specimen is bent in a jig having the contour shown in Fig. 22-52, page 723. Any convenient means—mechanical, electrical, or hydraulic—may be used for moving the male member with relation to the female member.

The specimen is placed on the female member of the jig with the weld at midspan. Face-bend specimens are placed with the face of the weld directed toward the gap. Root-bend and fillet-weld–soundness specimens are placed with the root of the weld directed toward the gap. The two members of the jig are forced together until the curvature of the specimen is such that a wire 1/32 inch in diameter cannot be passed between the curved portion of the male member and the specimen. Then the specimen is removed from the jig and carefully examined.

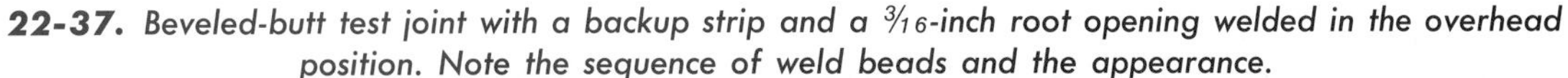

22-37. *Beveled-butt test joint with a backup strip and a 3/16-inch root opening welded in the overhead position. Note the sequence of weld beads and the appearance.*

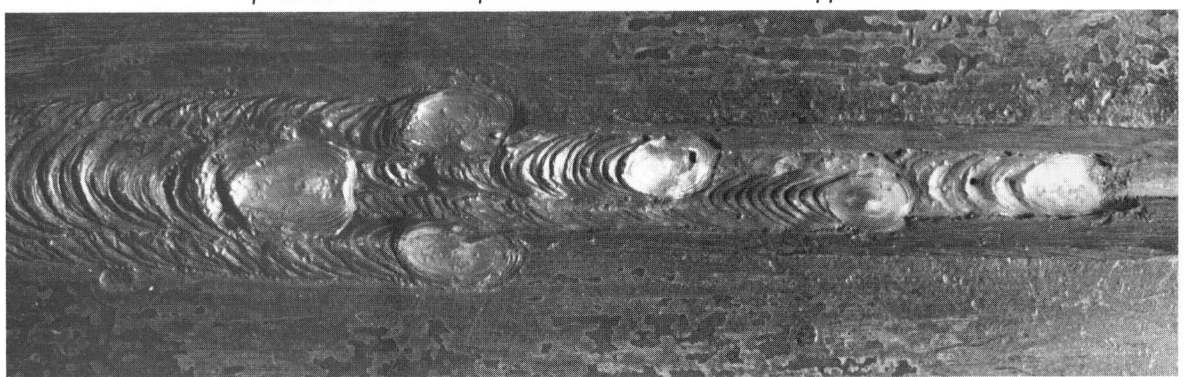

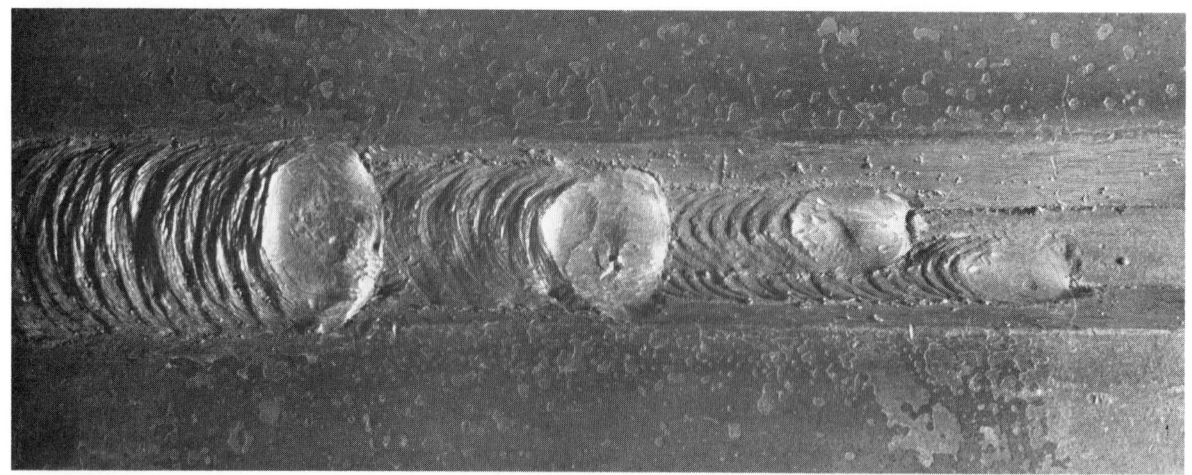

22-38. *Beveled-butt test joint with a backup strip and a 3/8-inch root opening welded in the flat position. Note the sequence of weld beads and the appearance.*

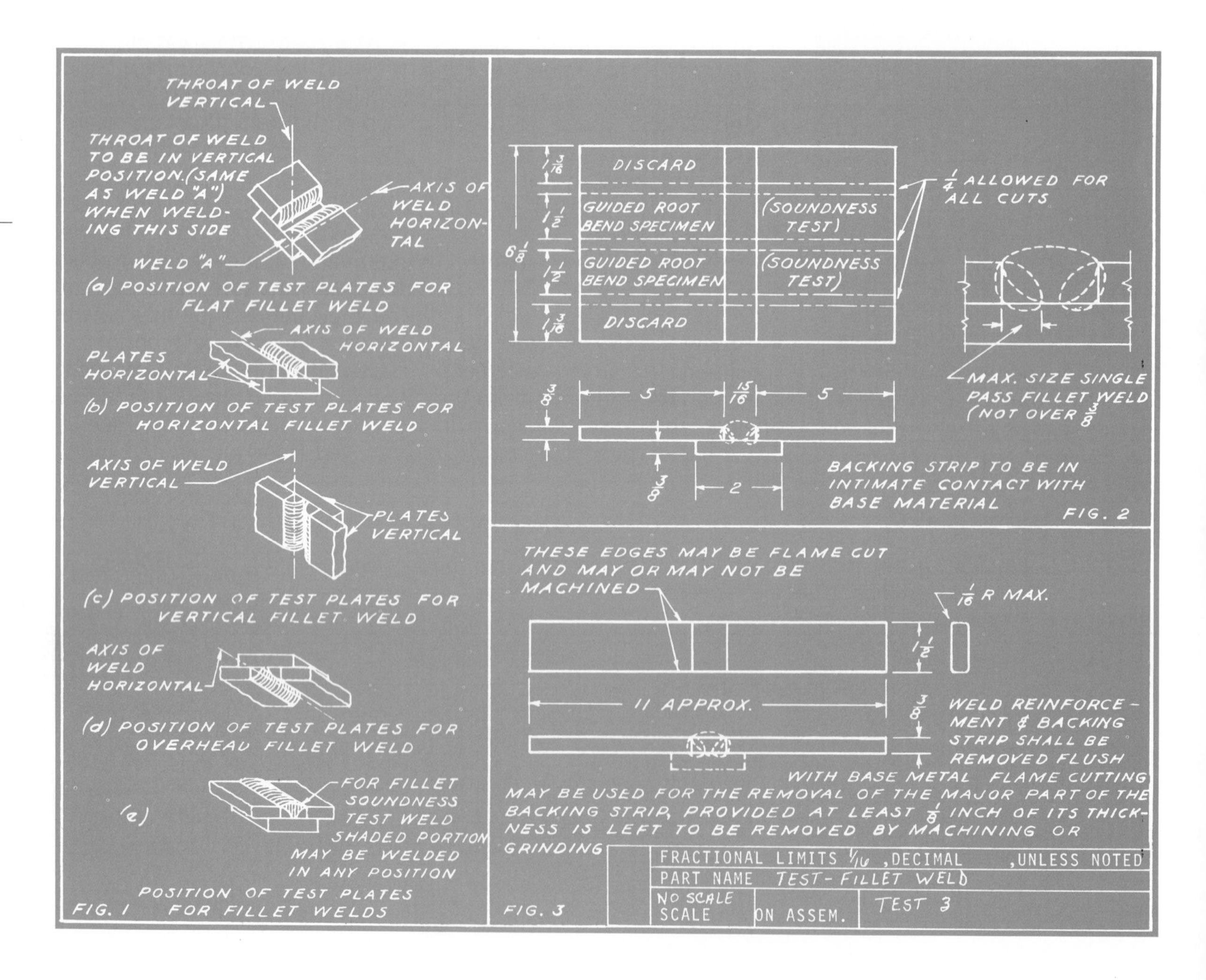

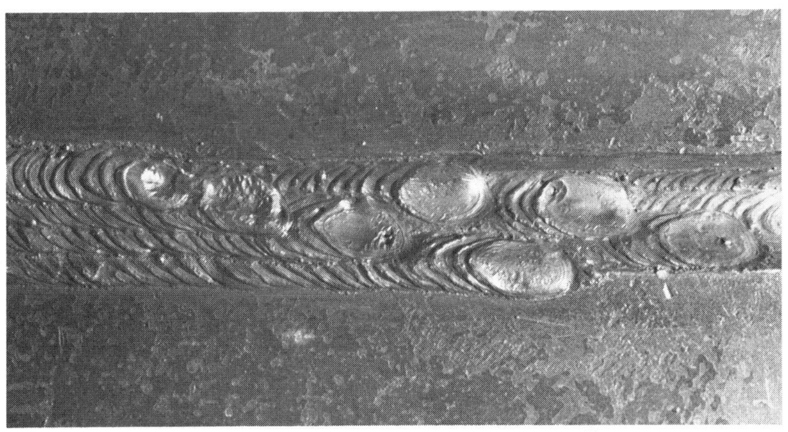

22-39. *Beveled-butt test joint with a backup strip and a 3/8-inch root opening welded in the horizontal position. Note the sequence of weld beads and the appearance.*

PRESSURE WELDS. The finished vessel is subjected to 2,500 p.s.i. internal pressure. The vessel begins to bulge at about 2,800 p.s.i. pressure. *Do not use air.* Use water or oil as the liquid. Water is preferred because it is cleaner to work with. Any hand- or motor-powered liquid pump may be used as the means of producing pressure.

Evaluation of Test Results

GROOVE AND FILLET WELD SPECIMENS. The outside surface of the specimens is examined for the appearance of

22-40. *Beveled-butt test joint with a backup strip and a 3/8-inch root opening welded in the vertical position, travel up. Note the sequence of weld beads and the appearance.*

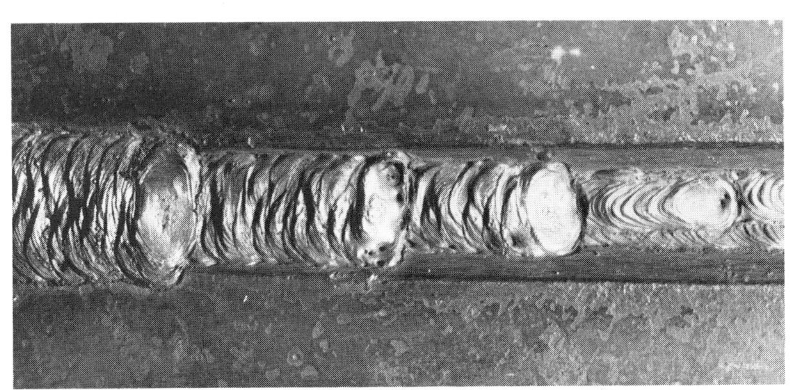

22-41. *Beveled-butt test joint with a backup strip and a 3/8-inch root opening welded in the overhead position. Note the sequence of weld beads and the appearance.*

22-42. *Penetration through the back side of a beveled-butt joint without a backup strip. The formation of a small bead without undercut is absolutely necessary for successful root-bend test specimens.*

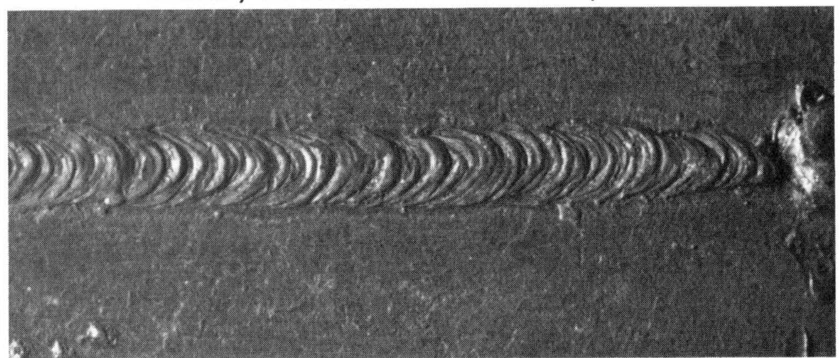

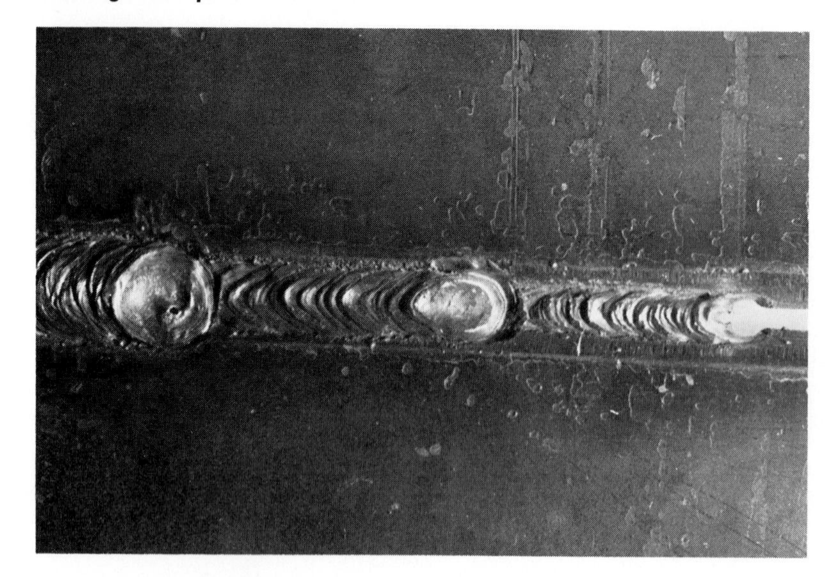

22-43. *Beveled-butt test joint without a backup strip welded in the flat position. Note the sequence of weld beads and the appearance.*

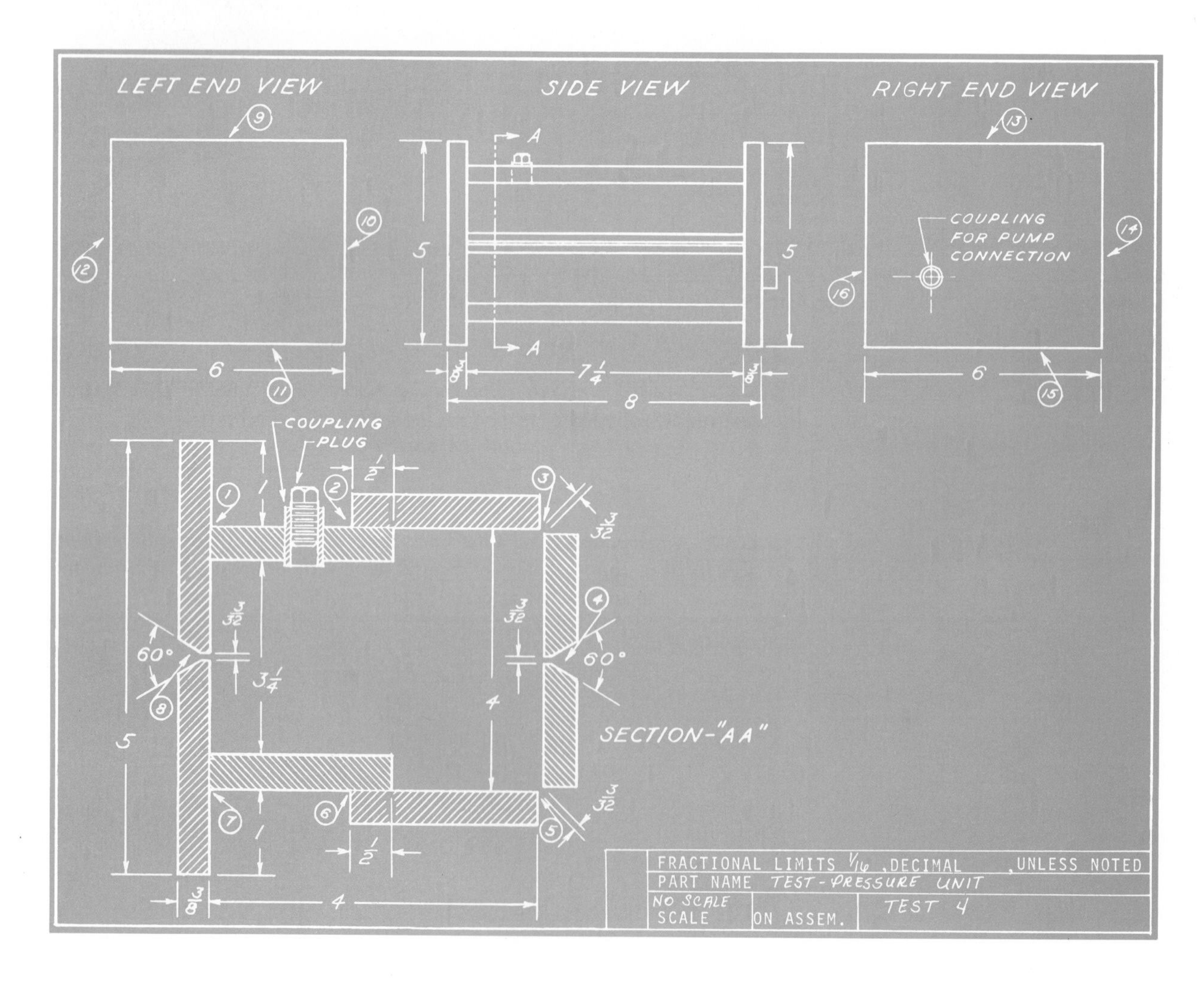

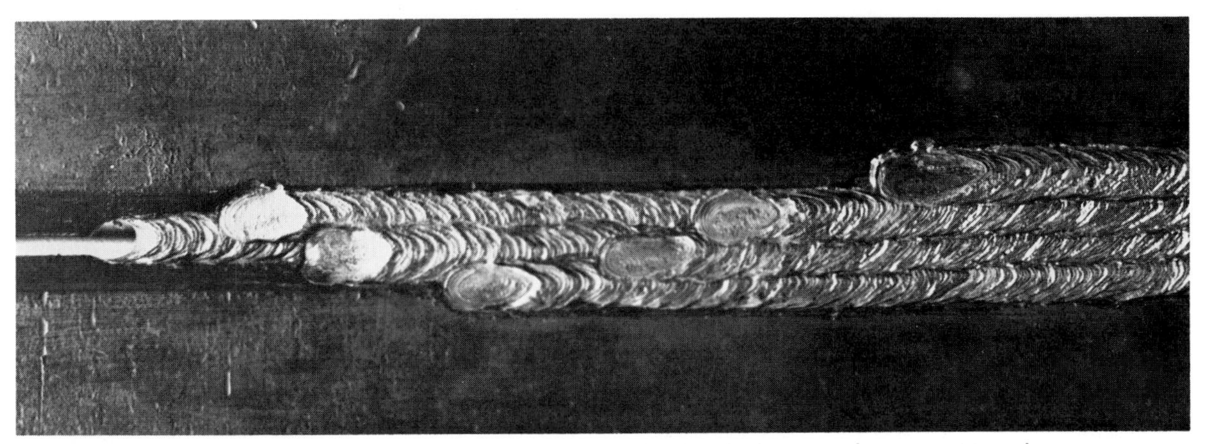

22-44. *Beveled-butt test joint without a backup strip welded in the horizontal position. Note the sequence of weld beads and the appearance.*

22-45. *Beveled-butt test joint without a backup strip welded in the vertical position, travel up. Note the sequence of weld beads and the appearance.*

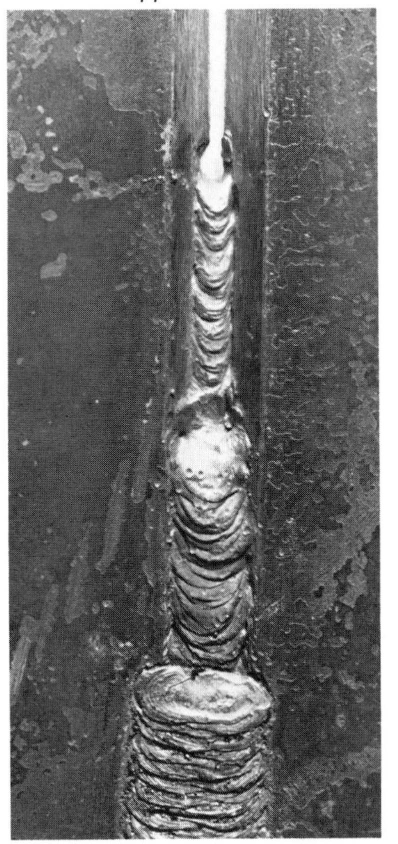

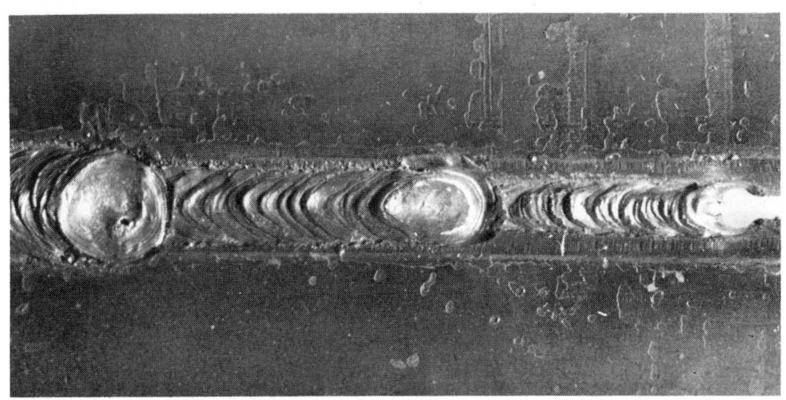

22-46. *Beveled-butt test joint without a backup strip welded in the overhead position. Note the sequence of weld beads and the appearance.*

22-47. *Joint for the fillet weld soundness test welded in the flat position. Note the sequence of weld beads and the appearance.*

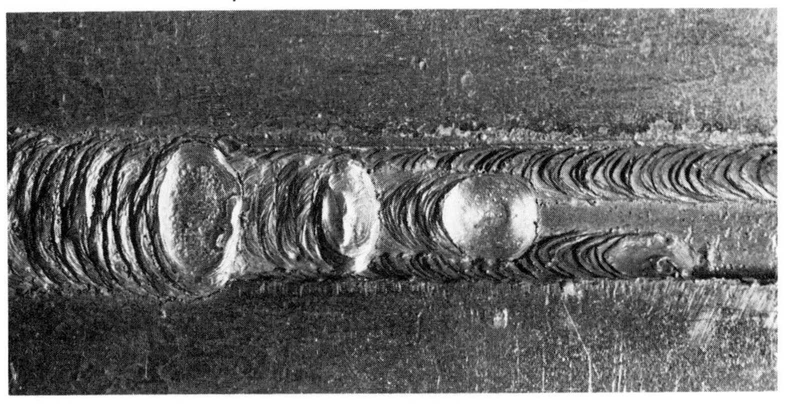

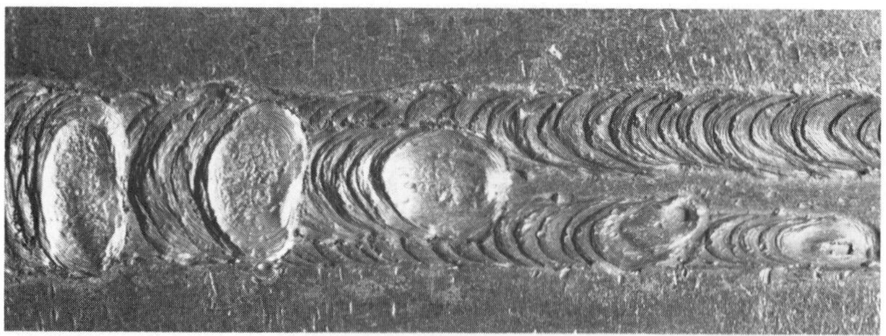

22-48. *Joint for the fillet weld soundness test welded in the horizontal position. Note the sequence of weld beads and the appearance.*

22-51A. *One side of the pressure vessel test unit: a corner joint, lap joint, and T-joints are visible. Welds include both fillet and groove welds.*

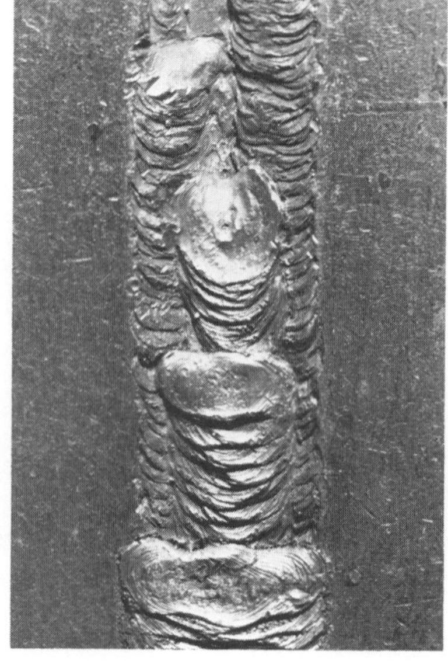

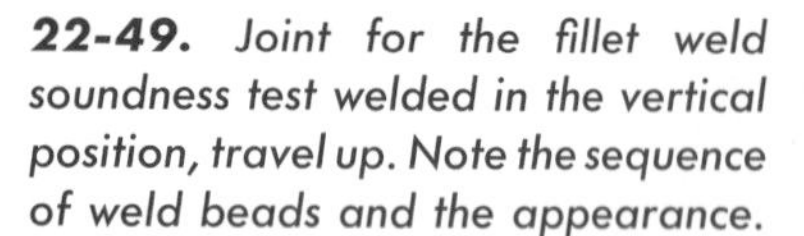

22-49. *Joint for the fillet weld soundness test welded in the vertical position, travel up. Note the sequence of weld beads and the appearance.*

22-51B. *Another side of the pressure vessel test unit: a corner joint, open beveled-butt joint, and a T-joint are visible. Welds include both fillet and groove welds.*

22-50. *Joint for the fillet weld soundness test welded in the overhead position. Note the sequence of weld beads and the appearance.*

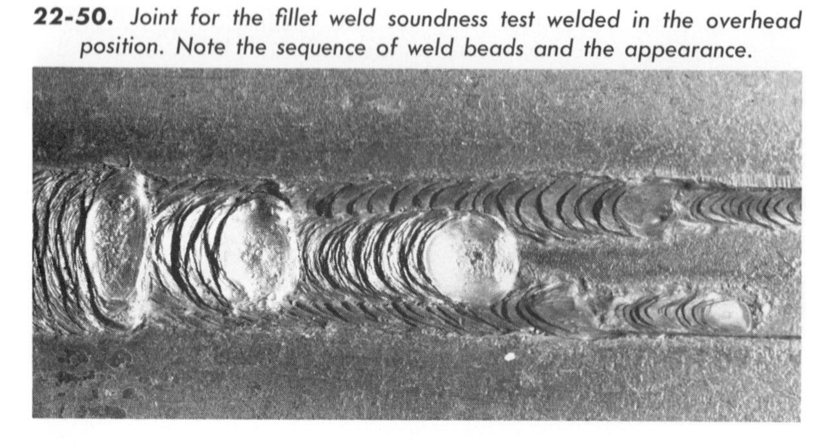

cracks or other open defects. Any specimen in which there is such a defect exceeding $\frac{1}{8}$ inch measured in any direction after bending, is considered as having failed. Cracks occurring on the corners of the specimen during testing are not considered.

PRESSURE WELDS. The internal pressure within the vessel is brought to 2,500 p.s.i. All welds are tapped with the ball peen end of a machinist's hammer. Any leaks or evidence of sweating is considered as a test failure.

Requirements for Passage of Tests

The student is required to pass all the tests in this section in at least one position. The positions in which these tests have been passed will be stated in the student's record.

(A) Front view of a shop-built testing unit.

(B) Side view of the testing unit showing the gauge mount.

22-52. *A shop-built testing machine. The gauge and jack may be purchased, and the rest of the unit can be built in the school shop. Tensile testing is done between the first two levels. Compression testing is carried on between the second and third crossheads. A standard root- and face-bend jig is mounted between these levels. The maximum tensile pull depends on the size of the jack. The unit illustrated has a capacity of 60,000 p.s.i.*

REVIEW QUESTIONS

1. What type of weld is used to weld a coupling to a tank or flat surface?
2. A single-V butt joint, in the vertical position, is used for what types of welding jobs? How many passes are required?
3. In welding a coupling to a plate surface it is important to obtain good penetration and fusion with which parts of the joint?
4. Find the mistake(s) in the following suggested procedure: It is desired to weld a single-V butt joint, without backing strip, in $\frac{3}{8}$-inch plate in the horizontal position without root gap and with a $\frac{1}{4}$-inch straight polarity electrode.
5. Lap joints cannot be welded in the overhead position. True. False.
6. What type of joint is used for Test No. 1, Navy Department Requirements?
7. Explain the technique for making the first pass in an open single-V butt joint.
8. Would a $\frac{1}{4}$-inch electrode be a desirable size to use for the first pass in an open single-V butt joint?
9. When welding a single-V butt joint without backing strip in the vertical position, is it necessary to secure complete penetration through the back side? Explain.
10. By now, you are familiar with the various types of welded joints. Name the welds used for the following joints: beading on flat plate, square butt joint, lap joint, corner joint, single-V butt joint, T-joint, edge joint, double-bevel butt joint.
11. It is desired to weld a single-V butt joint on $\frac{3}{8}$-inch plate, without backing strip, in the vertical position, travel down. Explain the entire welding procedure including root condition, number of passes, size of electrode, and welding heats.
12. How are single-V butt joints in the vertical position tested?
13. When vertical welding is to be done in a hurry, is it necessary to clean slag from each pass of a multipass weld?
14. Penetration and fusion are necessary to which parts of the following joints: beading on flat plate, edge joint, square butt joint, T-joint, lap joint, corner joint, single-V butt joints with backing strips, and single-V butt joints without backing strips?

15. Is it necessary to maintain a close arc gap when welding in the overhead position?

16. Name two industries where overhead welding is necessary.

17. Name two methods that may be used to weld a single-V butt joint in the overhead position.

18. What is the purpose of welding a coupling on a flat plate?

Table 22-2

Course Outline: Shielded Metal-Arc Welding Practice: Jobs 22-J 52-65 (*Plate*)

Job no.	Joint	Type of weld	Type of electrode						
			Position	DCRP	DCSP	a.c.	AWS specif. no. d.c.	and/or AWS specif. no. a.c.	Text reference
22-J52	Single-V butt joint	Groove welding—weaved—multipass—travel up	Ver.	X		X	E6010	E6011	685
22-J53	T-joint	Fillet welding—single pass	Over.	X		X	E6010	E6011	688
22-J54	T-joint	Fillet welding—stringer—multipass	Over.	X		X	E6010	E6011	690
22-J55	Lap joint	Fillet welding—single pass	Over.	X		X	E6010	E6011	691
22-J56	Lap joint	Fillet welding—single pass	Over.		X	X	E6012	E6013	694
22-J57	Single-V butt joint	Groove welding—weaved—multipass—travel down	Ver.	X		X	E6010	E6011	695
22-J58	T-joint	Fillet welding—multipass	Over.	X		X	E6010	E6011	697
22-J59	T-joint	Fillet welding—weaved—multipass	Over.	X		X	E6010	E6011	699
22-J60	Single-V butt joint	Groove welding—stringer—multipass	Hor.	X		X	E6010	E6011	701
22-J61	Fittings	Fillet welding—single pass	Hor.		X	X	E6012	E6013	703
22-J62	Fittings	Fillet welding—single pass	Hor.	X		X	E6010	E6011	705
22-J63	Single-V butt joint backing strip	Groove welding—stringer—multipass	Over.	X		X	E6010	E6011	706
22-J64	Square butt joint	Groove welding—weaved—multipass	Over.	X		X	E6010	E6011	708
22-J65	Single-V butt joint	Groove welding—weaved—multipass	Over.	X		X	E6010	E6011	710

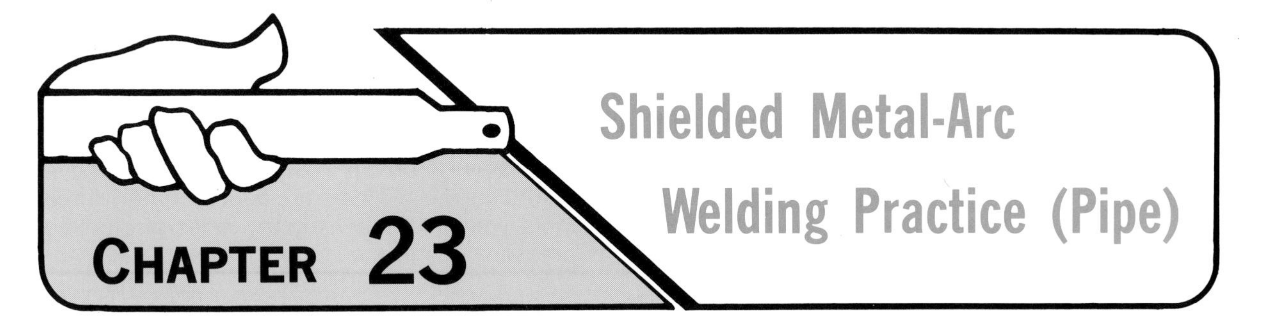

Chapter 23 Shielded Metal-Arc Welding Practice (Pipe)

INTRODUCTION

The next several chapters present information and practice jobs in shielded metal-arc welding carbon steel pipe and in gas tungsten-arc welding and gas metal-arc welding steel, stainless steel, and aluminum plate and pipe. You should undertake these jobs only after you have successfully mastered all of the skills practiced in the previous jobs and passed all of the welding tests given in the previous chapter. Many of the welding skills necessary to master these new welding processes are very similar to those you have already learned. You should have very little difficulty learning these new skills.

You are strongly urged to make every effort to master pipe welding skills. There is a need for people with pipe welding skills, and this need will increase in the years ahead.

In the Ninth AWS "Plummer Lecture," which was presented in San Francisco, California in the fall of 1973 at the Golden Gate Welding and Metals Conference, Porter E. Thompson, Senior Vice President and Director of Bechtel Corporation, said "There is no question that the shortage of welders and support personnel needed to staff our heavy construction projects over the next 15 years is a very serious situation." In discussing plant construction for our industry for the next 15 years, Mr. Thompson forecast the percent of increase in needed manpower that is shown in Table 23-1.

This increase is the number of pipe welders (other types of welders will also experience the same increase) needed if we are to build the number of new nuclear power generating plants required by 1985.

The History of Pipe Welding

The first piping system in recorded history was used to carry water. Early piping systems were made of stone or wood. Welded piping was of considerable importance in the growth of the modern pipeline industry.

The first attempt at welding pipeline involved the oxyacetylene process and took place in 1911 near Philadelphia. In 1922 the first attempt was made to arc weld a pipeline. The job involved 150 joints on a 12-inch line in Mexico. A major advancement came in 1926 when large diameter, seamless steel pipe was made available. The coated electrode was introduced to pipeline welding in 1930 when it was used with great success on a 32-mile, 20-inch line in Kansas.

The early pipelines had small diameters, but in 1942 the "Big Inch," a 30-inch diameter line, was completed from Houston, Texas to Linden, New Jersey. Since then diameters of 36 and 40 inches have been used in welded cross-country lines.

Table 23-1

Projected need for pipefitter welders and skilled construction welders.

	Demand for labor[1] (in thousands)			% increase
	1970	1980	1985	
Skilled pipefitter welders	42.5	71	85	100
Total skilled construction welders	55	91	109	98

Porter E. Thompson

[1] It is assumed that productivity will improve and that the skilled crafts required as a percent of the whole will increase.

Pipe Welding Today

Today piping is made out of many materials and in many sizes to meet the demands of modern industry. The ferrous metals include wrought iron, carbon steels, chromium-molybdenum alloy steels, low temperature steels, stainless steels, and various lined and clad steels. The nonferrous metals include aluminum and aluminum alloys (Fig. 23-1), nickel and nickel alloys, titanium and titanium alloys, copper and copper alloys and lead. Nonmetallic materials include polyvinyl chloride, fiberglass, and asbestos. In addition, metal piping may be lined with nonmetallic materials such as glass, plastics, cement, and wood.

The number of materials required for piping systems is expanding constantly due to the need for better materials to meet increasingly severe operating conditions. Industries requiring special piping include power plants, nuclear plants, refineries, chemical and petrochemical plants, paper mills, textile mills, aerospace plants, and shipyards.

The most common welded piping in buildings is low pressure steam and hot water systems for heating. A large building or manufacturing plant needs piping for plumbing, sprinkler systems, air conditioning systems, gas and air lines, and lines that carry the various materials used in the manufacturing process. These pipelines must service a combination of low and high pressure and temperature conditions.

Power plant piping systems and nuclear piping systems, Fig. 23-2, are usually under very high temperature and pressure and often contain radioactive fluids. Joint tightness is absolutely essential because of the dangers of radioactive contamination. The temperature in these installations may exceed 1,000 degrees F., and the pressure may exceed 2,500 p.s.i. Highly corrosive materials and conditions make it necessary to use piping of stainless steel, nickel steels, and various nonferrous materials when high temperature or pressure is not involved.

Welded pipeline, Fig. 23-3, forms an underground network for transporting natural gas, crude oil, and refined petroleum products to all sections of the United States. There are well over a million miles of pipeline. Eighty percent of these lines are welded, and all new pipelines are completely welded. On short pipelines, diameters of 80 and 96 inches have become common. Water pipelines are not as long as the lines for gas and other products, but many of them are much larger. The Hoover Dam penstocks, for example, are 30 feet in diameter and have a wall thickness up to $2^3/_4$ inches. A line named Capline is a giant hauler. It is an underground steel pipe 630 miles long and 40 inches in diameter which is capable of moving 400,000 barrels of oil a day into Illinois from the Louisiana oilfields. If it were gasoline, a day's supply would keep an average family car on the road for 30,000 years.

23-1. *Aluminum piping fabricated in the shop for a large chemical plant.*
Crane Co.

There are approximately three times as many miles of pipeline in the United States as there are miles of railroad track. The lines start in the gas and oil fields and feed natural gas into towns and cities and crude oil into big refineries. Petroleum products flow out of the refineries into still more miles of pipelines. Gasoline, for example, flows through pipe to terminals where it is picked up by tank trucks and delivered to service stations.

A pipeline system can be compared with the Mississippi River. It begins as a multitude of tiny tributaries and ends in a maze of channels at the delta. The entire interstate system moves more than 230 billion gallons of oil and 13 trillion cubic feet of gas around our country every year. This amount is about 20 percent of the nation's total overland cargo. The petroleum products moved by pipelines every day would fill a fleet of tank trucks lined up bumper to bumper over 400 miles long.

Several products may flow through a pipeline at the same time, each in a stream that might be several miles long.

A batch of premium gasoline may be followed by a batch of regular gasoline, and that batch may be followed by several miles of kerosene. Different products are kept apart by separators called pigs. These may be round rubber balls or other devices.

Building a Pipeline

Running a pipeline across the country has been reduced to a smooth step-by-step production process that gets the job done with a minimum of effort and time. Usually the men who build pipelines are specialists who wander from country to country, following the jobs wherever they are.

On the job site the bulldozers and scrapers clear and grade the right-of-way. Then come the trucks, bearing sections of pipe which may be as long as 80 feet. Caterpillar® tractors with side booms lay out the pipe along the trail. If there is a curve in the right-of-way, a tractor with huge jaws grabs a section of the pipe and bends it.

The tractor is followed by a ditching machine, which scoops out the trench at the rate of a mile a day. Any solid rock that is in the way of the digger is blasted out by the dynamite crew. The pipe is strung out along the trench where the crews of welders join the pipe joints together to make a continuous pipeline, Fig. 23-4. The welding crews may consist of men who line up the pipe, a crew of tackers, and a crew of welders who apply the root pass, the intermediate pass, and the finishing pass. X-ray men check the completed welds. If a weld shows any faults, it is cut out and rewelded.

The pipe is ready to be put into the trench after it has been coated with an enamel-like base, wrapped with fiberglass, and wrapped again with feltlike paper. The line receives its final check and is placed in the trench.

When put into service, the line is continually checked for leaks by metering devices at each end of the line to check input and output. Leaks can also be spotted from the air. The big pipeline companies have fleets of small planes. A pilot can patrol 100 miles of line in an hour and find a spot of oil no bigger than a man's hat.

The use of pipelines as cargo carriers is in its infancy. If a rubber ball or an electronic sensing device can be sent through a pipeline, someday a capsule loaded with mail, ore, oranges, or even people may be transported this way.

Welded-Pipe Application

In the past, pipe was joined together only by mechanical means such as threaded joints,

23-2. *Inside a pipe-fabricating shop. Several pipe fabrications for a power plant are shown: two pipe headers and a butt weld in a long length of pipe.*

Crane Co.

23-3. *Overland pipeline being welded in the field. Note that a welder works from each side of the pipe.*

Mobil Oil Corp.

flanges with bolts and gaskets, and lead and hemp joints. At the present time, practically all pipe 2 inches in diameter and larger are joined by welding. The materials to be handled, the operating temperature, and the internal pressure required make any other method of joining out of the question for many piping systems. Many of today's industrial processes would not have been possible with threaded joints in piping installations. The discovery, development, and application of nuclear energy certainly would not have taken place.

The important advantages of welded piping installations include:

- Permanent installation
- Improved overall strength
- Low maintenance costs because of leakproof, care-free joints
- Ease of erection
- Improved flow characteristics
- Reduction in weight due to compactness and lighter weight fittings
- More pleasing appearance
- More economical application of insulation

COST. Welded fittings are available that meet or are lower than the installed cost of threaded malleable iron fittings of comparable size. It is estimated that if all joints on a piping system are welded, a saving of 50 percent can be realized on the cost of material alone. Threading cuts the effective thickness of pipe wall practically in half. When welding fittings are used, pipe with much thinner walls can be used to reduce cost without lowering the efficiency of the piping system. Thus welded piping systems make sense not only from the standpoint of service durability and safety but also from the standpoint of cost.

STRENGTH. Welded fittings have the same thickness as the pipe with which they are used, Fig. 23-5. They are joined to the pipe by an equal thickness of weld. There is no loss of strength caused by thread cutting. The life expectancy of a welded system is double that of threaded pipe.

JOINT SECURITY. Welded fittings and pipe are permanently joined by a metal bond so that pipe, fittings, and welds all become integral parts of the system. The security of the joints does not depend upon the strength used in wrenching nor the amount of dope applied. There is no chance for leaks due to vibration or stress.

FLOW. The inside diameters of welded fittings and pipe meet exactly. There are no step-ups or step-downs as in a threaded system. Pockets that create turbulence, clogging, and erosion are

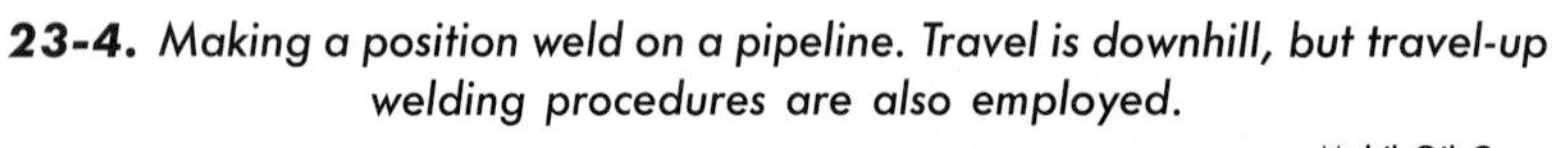

23-4. *Making a position weld on a pipeline. Travel is downhill, but travel-up welding procedures are also employed.*

Mobil Oil Corp.

23-5. *Welded piping improves utility and appearance. It is also leakproof.*

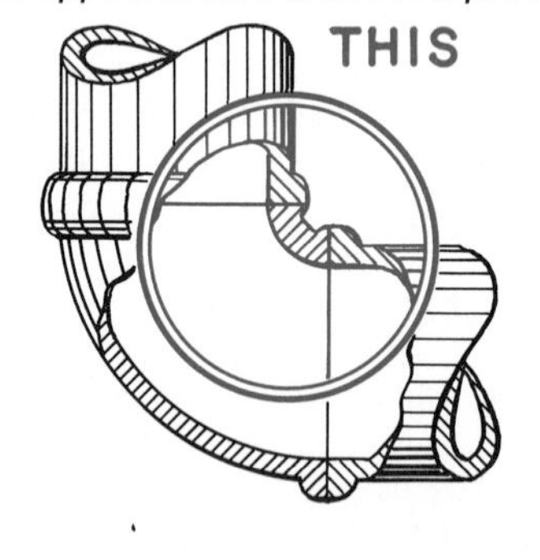

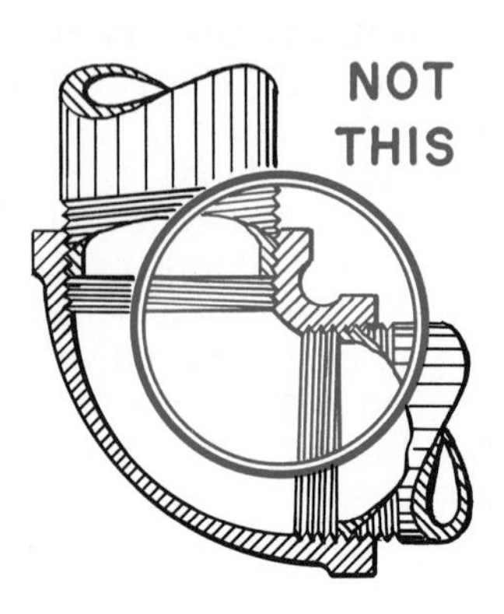

eliminated so that water or coolant flows freely, Fig. 23-6.

INSTALLATION. When assembling a welded piping system, there is no need to calculate overlaps for threading nor to allow for differences in the depth of joints. Roughing-in is quicker because the pipe is simply cut to length, placed in position, and welded. Assemblies can be fabricated conveniently in the shop or at the floor level of a construction job, and then they can be quickly tied into the system. Less manual labor is required than for a threaded system because wrenching is not required. Welding equipment is lighter and easier to handle than threading equipment.

TESTING AND REPAIR. When a contractor installs a threaded piping system, he knows that there will be some leaks which must be tightened before the system goes into operation. He also knows that in some cases he will have to repair leaks after the system has been completed. He may have to take out three or more fittings or repair a leaking joint. He even may have to recut some of the pipe. As a rule of thumb, contractors estimate that 5 to 10 percent of their cost for labor will be for this kind of work. Such repair is not necessary on welded piping because no hemp, gaskets, or other sealing materials are used.

EASY INSULATION. The outside diameters of the pipe and fittings are identical, so that the application of insulation is simple. There is just one smooth surface and no bulky fittings or difficult angles. Sleeve or tubular insulation can be slipped on over the pipe and fittings without difficulty. Tape or wraparound insulation can be applied over joints as smoothly as over pipe lengths, Fig. 23-7. Once applied, insulation never has to be removed to get at loosened and leaking joints.

WEIGHT. Since the ends of the fittings need not encompass welded pipe, considerable weight is saved through the reduced diameter and shorter length of each fitting. The fact that pipe with thinner walls can be used also cuts down on the weight of the system.

APPEARANCE. The smooth lines of a welded piping system present a much more professional appearance than a threaded system. There are no unsightly bulges and exposed threads.

SPACE. Welded-pipe systems are ideal for installations where space is a problem. Fittings are compact, and very little clearance is needed when the fabricator is welding instead of swinging a wrench or turning a fitting. A space capsule is an example of a limited area in which piping systems are required.

23-6. *Welded piping improves the flow of materials.*

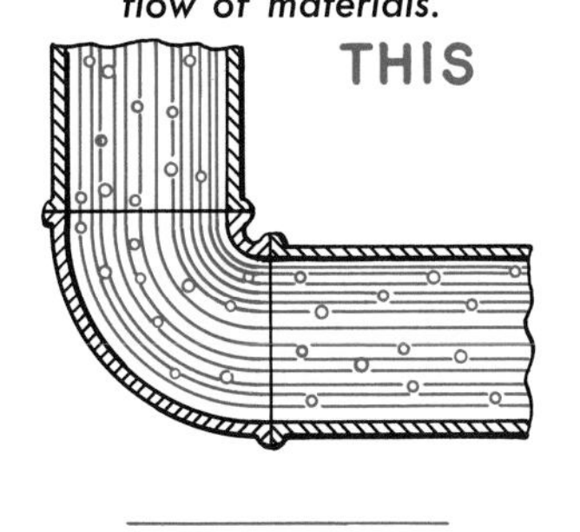

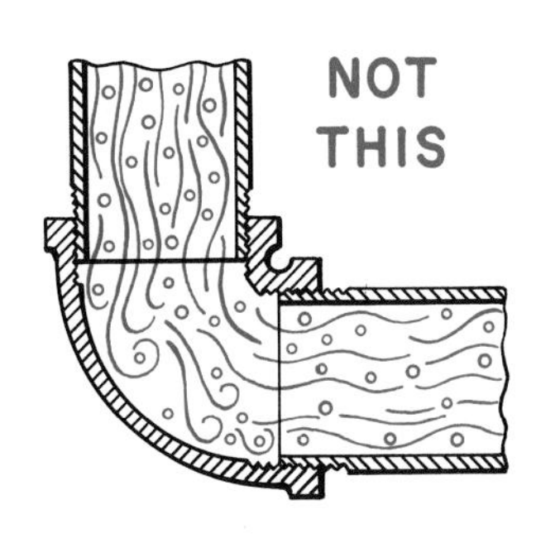

EQUIPMENT. If a threaded system with a full range of pipe sizes from 2 to 6 inches is being installed, the worker may need up to three different threading outfits, a dozen pipe wrenches, expensive dies, reamers, and pipe cutters. Welding, on the contrary, requires only one unit for pipes of all sizes. The exact equipment will vary, depending on the welding process.

Pipe Specifications and Materials

The American Society for Testing Materials (ASTM) and the American Petroleum Institute (API) issue specifications for ferrous and nonferrous pipe. Study Table 23-2 which lists standard pipe sizes.

CARBON STEEL PIPE. Carbon steel is by far the most widely used material in piping systems

23-7. *Welded piping insures the easy application of insulation.*

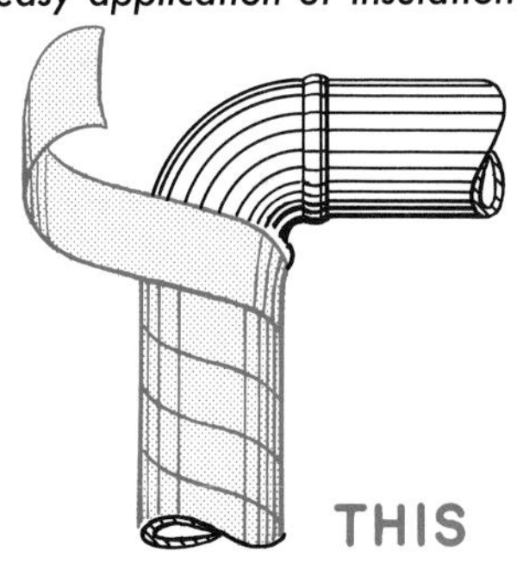

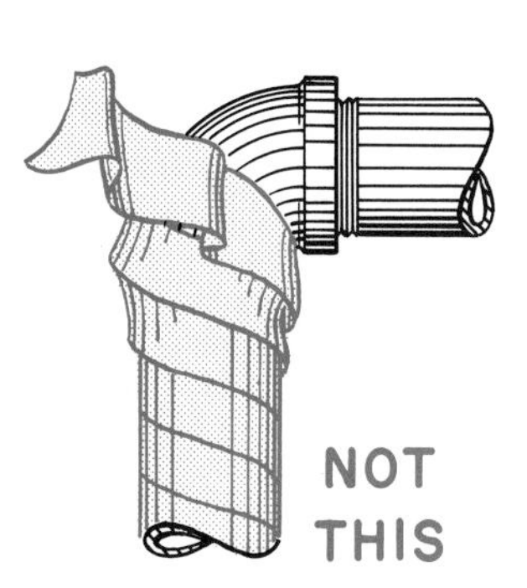

today. The relatively low cost of carbon steel pipe and its excellent forming and welding properties demand its use except for extreme conditions. Carbon steel pipe with a tensile strength up to 70,000 p.s.i. is available for high temperature, high pressure service.

HIGH YIELD-STRENGTH PIPE. High yield-strength pipe has been developed for high pressure transmission lines because of the cost advantage of using lighter weight pipe for long, cross-country lines. This pipe must be readily weldable in the field. Therefore carbon steels or high strength, low alloy steels are preferred.

WROUGHT IRON PIPE. Wrought iron pipe is widely used in medium and low pressure installations. It is preferred by many engineers for maritime, underground, condensate, and radiant heating piping. It has good resistance to some types of corrosion and fatigue, and it can take on and hold protective coatings such as galvanizing or paint.

Although the outside diameter of wrought iron pipe is identical with that of steel pipe, the wall thicknesses of wrought iron pipe are slightly thicker to compensate for the difference in weights per cubic foot of wrought steel and wrought iron.

CHROME-MOLY PIPE. The addition of chromium and molybdenum inhibits graphitization and oxidation and increases high temperature strength. These alloying elements have advanced the useful temperature limits of piping systems, but there are certain drawbacks. Chrome steels have a pronounced tendency to air-harden, particularly when the amount of chromium is 2 percent or more. In such cases, special care in fabrication, special welding procedures, and carefully controlled annealing are necessary.

CHROME-NICKEL (STAINLESS STEEL) PIPE. Special consideration must be given to 18-percent chrome, 8-percent nickel steels when they are to be welded, hot formed, or used for corrosion resistance. At temperatures from 800 to 1,500 degrees F., chromium carbides are formed along the grain boundaries and reduce corrosion resistance. This characteristic is minimized by the addition of small amounts of stabilizing elements such as columbium or titanium or by other methods.

LOW TEMPERATURE PIPE. There is an ever-increasing demand for pipe that can provide satisfactory service at low temperatures. At the present time there are six grades of steel pipe for low temperature service: one carbon steel and three nickel grades. Other grades of less commonly used steel may be furnished to the general requirements of this specification.

NONFERROUS PIPE. Nonferrous pipe is used for such extreme conditions as very low temperatures, very high temperatures, and corrosive conditions beyond the abilities of ferrous metals. The most common nonferrous pipe materials are nickel, copper, and aluminum and their many alloyed compositions: Monel®, Inconel,® Hastelloy,® red brass, yellow brass, and cupro-nickel.

SHIELDED METAL-ARC WELDING OF PIPE

Shielded metal-arc welding is the principal process for welding pipe both in the shop and in the field. Welds of X-ray quality are produced on a production basis. This process may be used for nearly all ferrous and nonferrous piping. Standard welding power sources which produce alternating or direct current such as a rectifier, transformer, motor-generator, or an engine-driven machine may be used. A wide range of electrode types and sizes are available.

Welding may be done in all positions, and the direction of welding may be up or down.

Pipe Welding Electrodes

E6010 CLASSIFICATION. This type of shielded-arc electrode is generally used in the welding of carbon steel pipe. It is also used to weld the stringer bead on higher strength pipe. The deposits of electrodes in this classification have a tensile strength of 50,000 to 65,000 p.s.i.

Some electrode manufacturers make two types of electrodes under the E6010 classification: one that meets the general classification and another which is especially designed for pipe. The electrode for pipe has a more accurately directed arc and improved wash-in action on the inside of the pipe. These features permit a slightly narrower gap but still produce a flat inside bead. The narrower gap reduces the volume of weld metal in the joint so that overall welding speed is increased. This type also reduces undercut on the back side of the joint and generally improves X-ray quality. Some welders feel they can carry a slightly higher current and larger puddle on the stringer pass.

E6011 CLASSIFICATION. The operating characteristics, mechanical properties, and welding applications of the E6011 resemble those of the E6010, but the E6011 is used with alternating current. Although it may also be used with direct current reverse polarity, it loses many of its beneficial characteristics with this current and polarity.

E7010 CLASSIFICATION. E7010 electrodes are used for higher strength pipe than E6010 electrodes. Welds made with these electrodes will equal or exceed the minimum yield strength of high strength pipe. They do not necessarily exceed the ultimate tensile strength of these materials since the pipe will frequently sustain 100,000 p.s.i. The deposits of these electrodes have a

Table 23-2

Pipe dimensions.

Nominal pipe size	Outside diameter	Nominal wall thickness for schedule numbers									
		Schedule 10	Schedule 20	Schedule 30	Schedule 40	Schedule 60	Schedule 80	Schedule 100	Schedule 120	Schedule 140	Schedule 160
1/8	0.405	...	...	...	**0.068**	...	**0.095**	...	...	...	...
1/4	0.540	...	...	...	**0.088**	...	**0.119**	...	...	...	...
3/8	0.675	...	...	...	**0.091**	...	**0.126**	...	...	...	...
1/2	0.840	...	...	...	**0.109**	...	**0.147**	...	...	...	0.187
3/4	1.050	...	...	...	**0.113**	...	**0.154**	...	...	...	0.218
1	1.315	...	...	...	**0.133**	...	**0.179**	...	...	...	0.250
1 1/4	1.660	...	...	...	**0.140**	...	**0.191**	...	...	...	0.250
1 1/2	1.900	...	...	...	**0.145**	...	**0.200**	...	...	...	0.281
2	2.375	...	...	...	**0.154**	...	**0.218**	...	...	...	0.343
2 1/2	2.875	...	...	...	**0.203**	...	**0.276**	...	...	...	0.375
3	3.5	...	...	...	**0.216**	...	**0.300**	...	...	...	0.437
3 1/2	4.0	...	...	...	**0.226**	...	**0.318**	...	...	...	...
4	4.5	...	...	...	**0.237**	...	**0.337**	...	0.437	...	0.531
5	5.563	...	...	...	**0.258**	...	**0.375**	...	0.500	...	0.625
6	6.625	...	...	...	**0.280**	...	**0.432**	...	0.562	...	0.718
8	8.625	...	0.250	**0.277**	**0.322**	0.406	**0.500**	0.593	0.718	0.812	0.906
10	10.75	...	0.250	**0.307**	**0.365**	**0.500**	0.593	0.718	0.843	1.000	1.125
12	12.75	...	0.250	**0.330**	0.406	0.562	0.687	0.843	1.000	1.125	1.312
14 OD	14.0	0.250	0.312	0.375	0.437	0.593	0.750	0.937	1.062	1.250	1.406
16 OD	16.0	0.250	0.312	0.375	0.500	0.656	0.843	1.031	1.218	1.437	1.562
18 OD	18.0	0.250	0.312	0.437	0.562	0.718	0.937	1.156	1.343	1.562	1.750
20 OD	20.0	0.250	0.375	0.500	0.593	0.812	1.031	1.250	1.500	1.750	1.937
24 OD	24.0	0.250	0.375	0.562	0.687	0.937	1.218	1.500	1.750	2.062	2.312
30 OD	30.0	0.312	0.500	0.625	...	...	...	...	...	...	...

tensile strength of 70,000 to 78,000 p.s.i.

Here again manufacturers make several types with special characteristics. One type is specifically designed for vertical-down pipe welding and is used on all passes. It produces less slag interference so that it minimizes the problem of pinholes caused by slag running under the arc in downhill welding.

E7018, E9018G, AND E11018G CLASSIFICATIONS. EXX18 electrodes are low hydrogen electrodes used to weld high tensile and low alloy pipe. In addition to the low hydrogen characteristics, they may also contain small amounts of chromium and molybdenum. They may be used in all positions.

Electrodes in the EXX18 classification produce hydrogen-free weld deposits. The lack of hydrogen reduces the tendency for underbead cracking and microcracking when welding low alloy steel. It also reduces the amount of preheat required. The preheat temperature may be as much as 300 degrees less than that required for other electrodes.

Low hydrogen electrodes are used to weld the low alloy pipe used in many inner plant high pressure and high temperature applications. Mild steel pipe with wall thickness over ½ inch is sometimes welded with these electrodes because of their superior cleaning properties and an overall speed advantage over E6010 electrodes. The E7018 electrodes produce weld deposits with a tensile strength of 65,000 to 80,000 p.s.i.; the E9018G electrodes, 90,000 to 105,000 p.s.i.; and the E11018G electrodes, 110,000 to 115,000 p.s.i.

23-8. *Typical 30-inch headers for a gas conditioning system. The headers are made of pipe with a 30-inch outside diameter and 5/16-inch wall thickness. Nozzles of the same material have outside diameters of 30 inches, 24 inches, and 12¾ inches. Welding saddles were used for all nozzles except the 30-inch size, which were reinforced with welded rings.*

Crane Co.

JOINT DESIGN

The most common types of joint in welded piping systems are the *circumferential butt joint* and the *socket* or *(lap) joint.* A groove weld is used for the butt joint, and a fillet weld is used for the socket joint. Fillet-welded joints often join flanges, valves, and fittings to pipe 2 inches and smaller in diameter. The fabrication shown in Fig. 23-8 is composed of several of the basic types of joints and welds used in the fabrication of piping systems.

Butt Joint

The butt joint, Fig. 23-9, is readily welded by all of the standard welding processes. Butt joints may not have inside support or they may be set up with an inner liner called a *backing ring* or *chill ring.* The butt joint is not difficult to prepare for welding, and it can be welded in all positions without great difficulty. This joint design provides good stress distribution and has maximum strength while permitting the unobstructed flow of materials through the pipe. It is also pleasing in appearance. Its general field of application is pipe to pipe, pipe to flanges, pipe to valves, and pipe to other types of fittings. The butt joint can be used for any size or thickness of pipe and for any type of service.

While there are no universally accepted welding grooves, the bevels specified in Table 23-3 page 734, are considered, for all practical purposes, as standard for industrial piping. Bevels may be made by mechanical means or flame cut.

Socket Joints

Socket joints are fillet welded. They are generally used for joining pipe to pipe, pipe to flanges, pipe to valves, and pipe to socket joints in pipe about 3 inches in diameter and under. This type of joint permits unrestricted fluid flow in small pipe. Figure 23-10 illustrates three typical fillet-welded socket joints. Adequate penetration of the pieces being joined is an absolute requirement.

Socket joints are not always acceptable. Butt welds with complete penetration to the inside of the piping are required for piping systems containing radioactive solutions or gases and those subjected to corrosive service with materials that are likely to cause crevice or stress corrosion.

23-9. *This welding student is making a circumferential butt joint in pipe in the horizontal fixed position.*

Job 5 in the Course Outline

Intersection Joints

Intersection joints such as medium and large *Ts*, laterals, *Ys*, and openings in vessels are usually the most difficult to weld. Factory-wrought fittings are preferred for the layout of the piping system since they are the equivalent in strength to the pipe being used. Moreover, their installations involves butt welds only. Cutting and beveling may be done manually or with a machine beveler designed for that purpose.

Figure 23-11 illustrates the three forms of preparation for 90-degree intersection joints. In type A, the header opening is equal to the inside diameter of the branch, and only the branch is beveled. This type of branch permits the use of a specially shaped backing ring if necessary. In type B, the header opening is large enough to permit insertion of the branch so that only the header opening must be beveled. Type C is a third form of preparation in which both the header and the branch are beveled. This form of unreinforced branch is adequate only when the pipe is to be used at pressures of 50 to 75 percent of the full pressure that the pipe can withstand. Reinforcement or manufactured welded fittings are required to make 90-degree intersections equivalent in strength to the pipe from which they are fabricated. Intersection joints at an angle of 45 degrees or less produce a condition at the heel of the intersection that makes it difficult, if not impossible, to secure the degree of penetration and soundness of weld deposit that is necessary for severe service conditions. Wherever possible, intersection joints should be made under shop conditions where the work can be positioned for welding in the flat position.

Welded Fittings

Pipe manufacturers have kept pace with design for welding by providing the industry with a line of seamless and welded fittings especially for welding. They are available in most grades of material suitable for welding and in many combinations of size and thickness. They combine the best characteristics for unimpaired flow conditions with wide availability, ease of welding, and maximum strength.

Insulation can be applied to ready-made fittings without difficulty. The system can be installed in less space than when fittings are fabricated for the job,

23-10. *Typical fillet weld joints.*

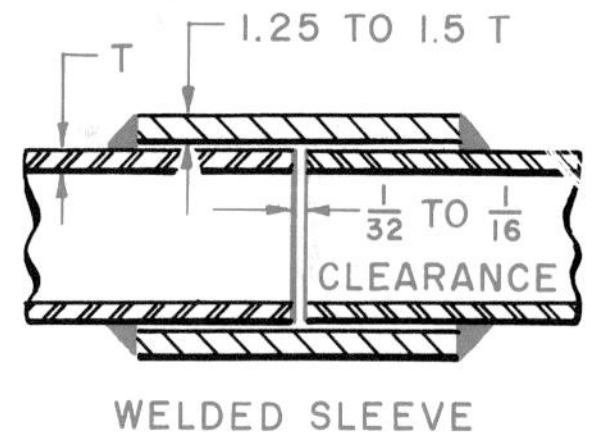

WELDED SLEEVE COUPLING

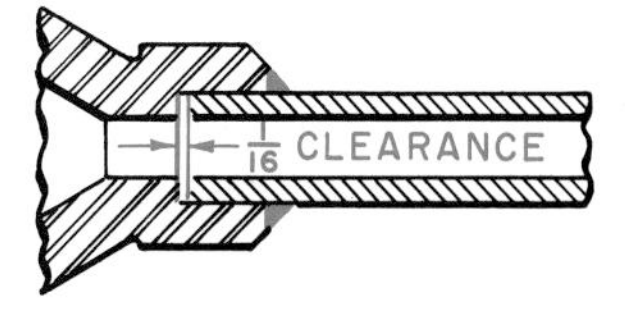

SOCKET DETAIL FOR SMALL WELDING END VALVE.

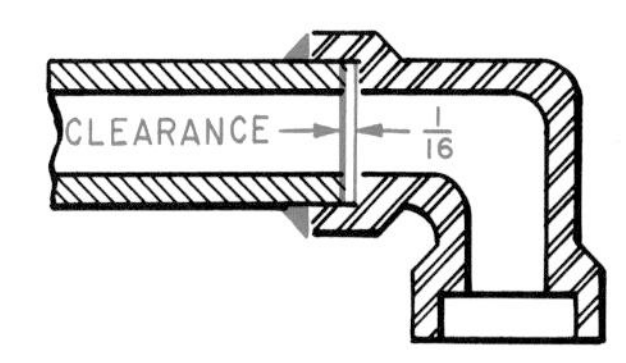

SOCKET END WELDING ELBOW.

and system changes can be made without difficulty. The piping requires less maintenance, and new systems are easy to design. The use of these fittings also saves a great deal of time and eliminates a considerable amount of the cutting, tacking, and fitting required in the hand fabrication of various types of joints. They also enhance the appearance of the job. They make the use of hand-mitered construction obsolete.

Manufactured fittings for welding that are available are shown in Fig. 23-12. The application of welding to flanges is illustrated in Fig. 23-13. The sequence of manufacturing steps from flat plate to the finished fitting is shown in Fig. 23-14.

BACKING RINGS. The term *backing ring* is applied to a ring-shaped structure, Fig. 23-15, which is fitted to the inside surface of the pipe before welding. Its functions are to assist the welder in securing complete penetration and fusion without burn-through (Fig. 23-16), to prevent spatter and slag from entering the pipe at the joint, and to prevent the formation on the inside of the joint of irregularly shaped masses of metal, sometimes called *icicles* or *grapes*. The prevention of icicles is especially important in pipe joints ahead of such units as valves and turbines. Backing rings also help to secure proper alignment of the pipe ends. There are two types of backing rings: split and solid. Split backing rings are designed to fit on the inside of the pipe without machining. They can be expanded or contracted to fit the inside diameter of the pipe. The solid type is machined and requires that the pipe also be machined to receive it. Its use is usually restricted to pipe that is to sustain severe service conditions. The installation cost of solid rings is higher than that of split rings due to the higher cost of the ring itself and the cost of preparing the pipe. A solid ring

Table 23-3

Recommended pipe bevels.

Type of material	Position of welding	Direction of welding	No backing ring	With backing ring	Bevel angle[1]	Root face[2]	Root opening[3]
Carbon steel	All	Up	x		30° to $37\frac{1}{2}$°	$\frac{1}{16}$	$\frac{1}{8}$
Carbon steel	All	Up		x	30° to $37\frac{1}{2}$°	$\frac{1}{8}$	$\frac{3}{16}$
Carbon steel	All	Down	x		$37\frac{1}{2}$°	$\frac{1}{16}$	$\frac{1}{16}$

[1]Bevel angle may be $\pm 2\frac{1}{2}$°.
[2]Root face may be $\pm \frac{1}{32}$.
[3]Root opening may be $\pm \frac{1}{16}$.

23-11. *Acceptable types of preparation for 90-degree full size and 90-degree reduced-size branch connections.*

"B" MAY BE INCREASED WHEN BACKING RING IS USED.

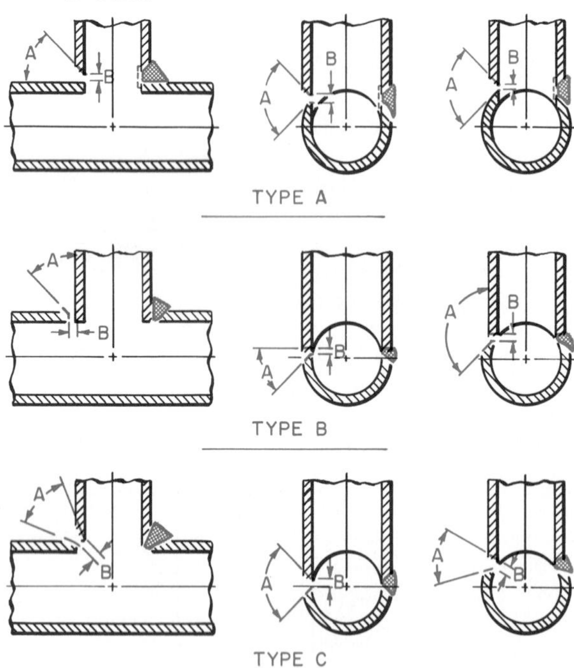

ANGLE "A" TO BE NOT LESS THAN 45° FOR ANY WALL THICKNESS OF PIPE.

"B" TO BE NOT LESS THAN 1/16" NOR MORE THAN 1/4".

ACCEPTABLE TYPES OF PREPARATION FOR 90 DEG FULL SIZE AND 90 DEG REDUCED SIZE BRANCH CONNECTIONS.

is shown in Fig. 23-15A; and a split ring, in Fig. 23-15B.

Correct spacing between the pipe ends is essential to securing sound welds. Most backing rings contain a series of small nubs, ranging from $\frac{1}{16}$ to $\frac{3}{4}$ inch in length, on the outer surface of the ring. Backing rings for pipe sizes 4 inches and under are $\frac{3}{32}$ inch thick by $\frac{5}{8}$ inch wide. For pipe sizes 5 inches and over, they are $\frac{1}{8}$ inch thick by 1 inch wide. Root-spacer nubs may be $\frac{1}{8}$ to $\frac{3}{16}$ inch in diameter. The material of the ring should be substantially the same chemical analysis as the pipe or tube that is to be welded.

Consumable insert rings, Fig. 23-17, improve the quality of the weld. They must be of the proper composition and dimensions. In piping for atomic reactors, stainless steel piping is best installed with consumable insert rings. These rings provide the most favorable weld contour to resist cracking from weld metal shrinkage and *hot shortness* (brittleness in hot metal), and they eliminate notches at the weld root. They also insure weld metal composition with the best possible properties of strength, ductility, and toughness. See Fig. 23-18 for end preparation of pipe for consumable insert rings.

PIPE CLAMPS. One difficulty encountered in assembling pipe to be fabricated is the positioning of the pipe before tacking. One method is to clamp the pipe in a fixture in the exact position desired for welding. The joint must be carefully mounted in the proper alignment in the pipe clamp or fixture. Pipe clamps of various types and sizes are available to align and hold pipe joints in preparation for tacking. Figures 23-19 through 23-25 (Pages 738, 739) show various pipe

23-12. *Machine-beveled manufactured welding fittings.*

Tube Turns

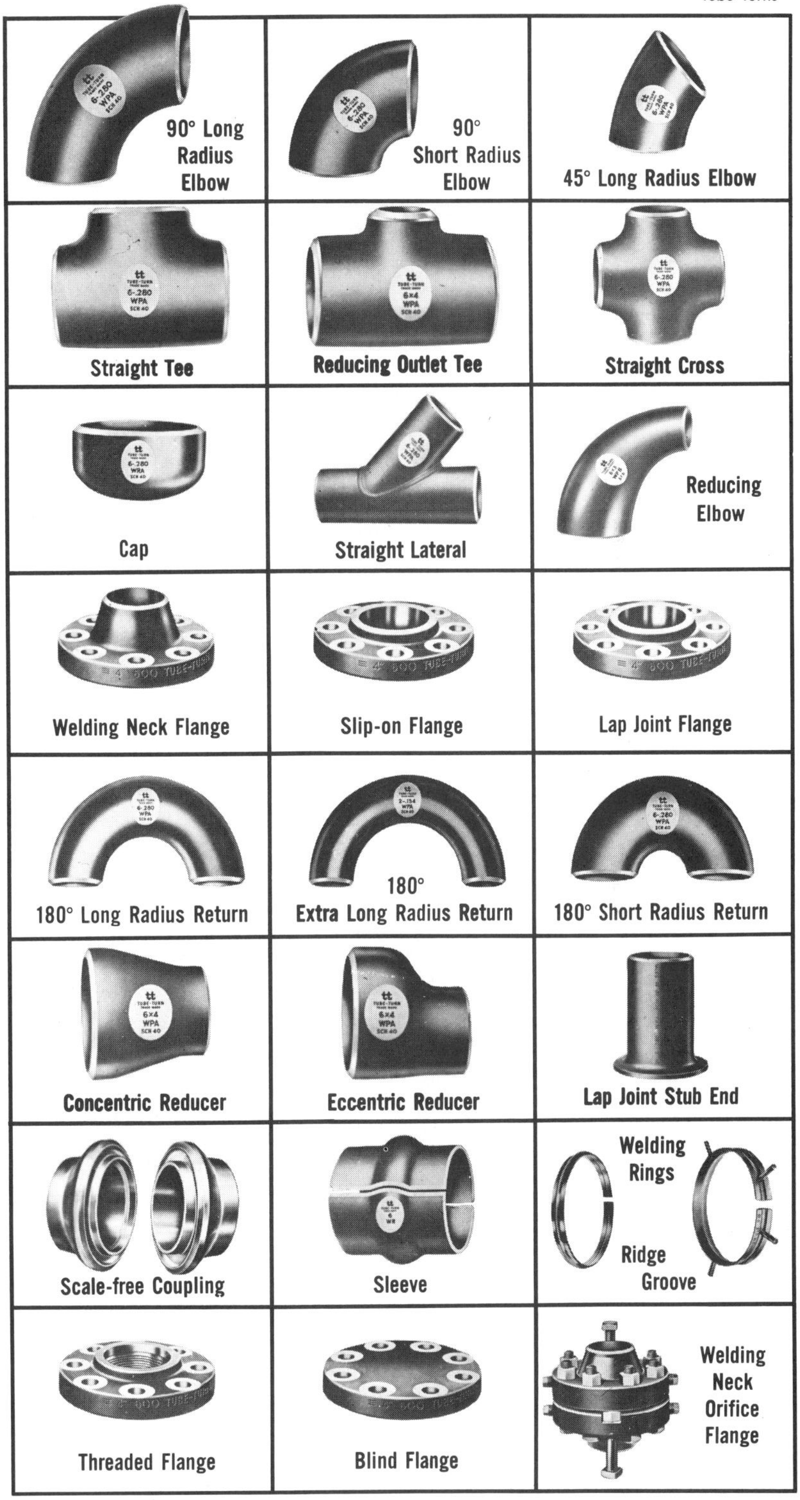

joints properly clamped and ready for tacking.

SPECIAL FABRICATIONS. Sometimes manufactured welded fittings are not available, and the welder must fabricate his own fittings or make special connections on the job. There are a variety of special designs in welded-pipe construction that may be fabricated when necessary. Among these are *Ls* and laterals, side outlet fittings, mitered joints, elbows, *Ys*, and expansion joints, Figs. 23-26 and 23-27 (Page 740). Joints should be carefully laid out with templates and standard layout curves. These fittings may be formed on the job from straight lengths of pipe by hand or machine flame cutting.

CODES AND STANDARDS

You will recall that piping systems are subject to wide varia-

23-13. *Typical welding flanges.*

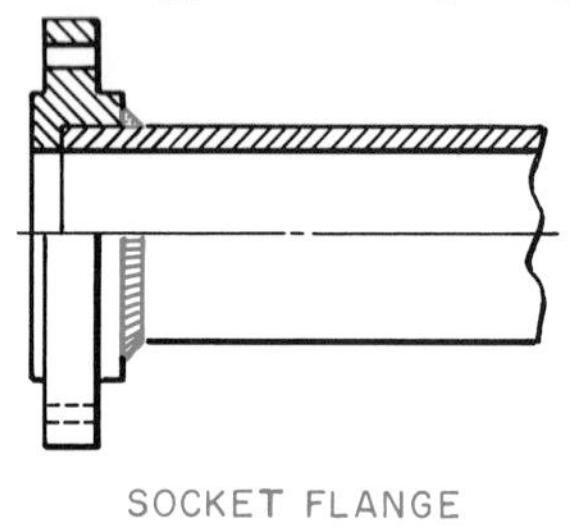

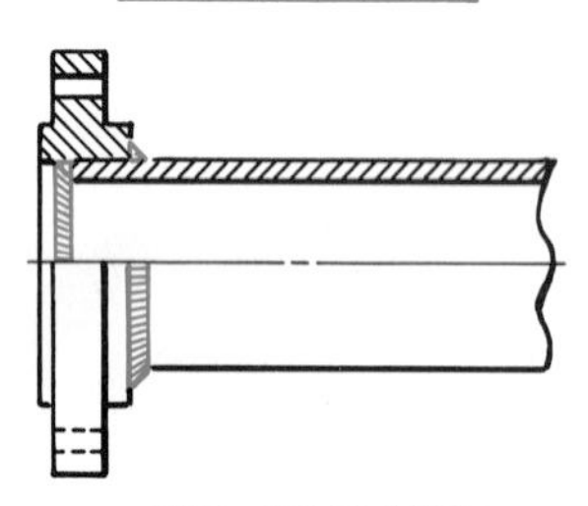

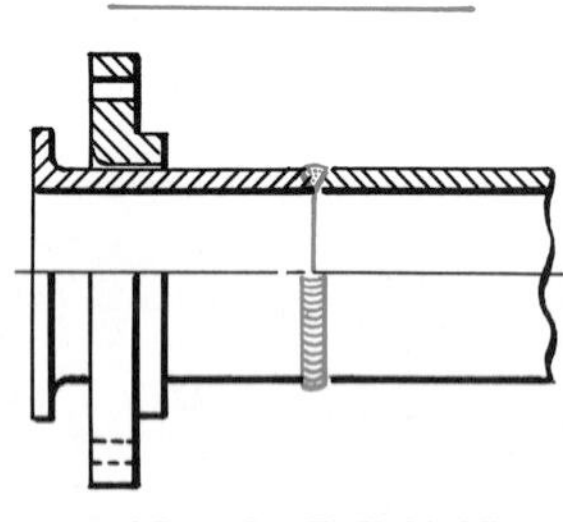

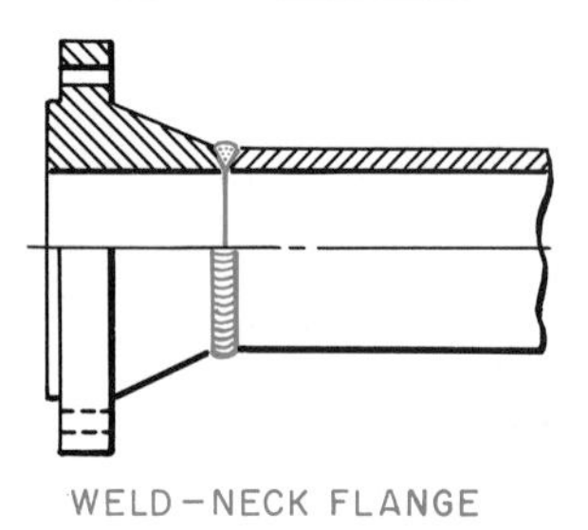

Crane Co.

23-14. *The manufacturing steps necessary to fabricating a welded L with machined ends.*

23-15. *(A) A machined ring. (B) A ring with chamfered nubs for quick "strike off" sets the pipe gap for the root pass. Nubs melt with the weld metal.*

Robvan Backing Ring Co.

A

B

tions in pressure and temperature, that corrosion is a problem in many systems, and that many of the materials carried are dangerous to life and property. Piping systems must be insurable against any loss resulting from the failure of the pipe or the welding.

Code-Making Organizations

In order to insure uniform practices in the design, installation, and testing of piping systems, various codes and standards applicable to welded piping systems have been prepared by committees of leading engineering societies, trade associations, and standardization groups. These are generally written to cover minimum requirements of quality and safety. A few insurance companies, manufacturers, and the military set up their own codes to cover the fabrication of welded piping systems. All of these codes have been set up for the purpose of establishing welding procedures and inspecting and testing methods to provide assurance that the work meets the purchaser's specifications and to give protection against accident. Our prime concern in this chapter will be with the qualification tests for welding operators rather than with the procedure qualification tests.

Following is a brief description of a few of the principal code-making organizations

THE AMERICAN SOCIETY OF MECHANICAL ENGINEERS (ASME). This society is responsible for the *ASME Boiler and Pressure Vessel Code,* which covers piping connected to boilers. It is recognized in almost all states and is a prerequisite to acceptance of the installation by the states and insurance companies. It covers the installation of piping systems in connection with power boilers, nuclear vessels, and unfired pressure vessels.

THE AMERICAN PETROLEUM INSTITUTE (API). This group is concerned with the standards for welding pipelines and related facilities. The standard includes regulation of the gas and arc welding of piping used in the compression, pumping, and transmission of crude petro-

23-16. *The root side of a pipe joint welded without a backing ring. Note the burn-through in several places.*

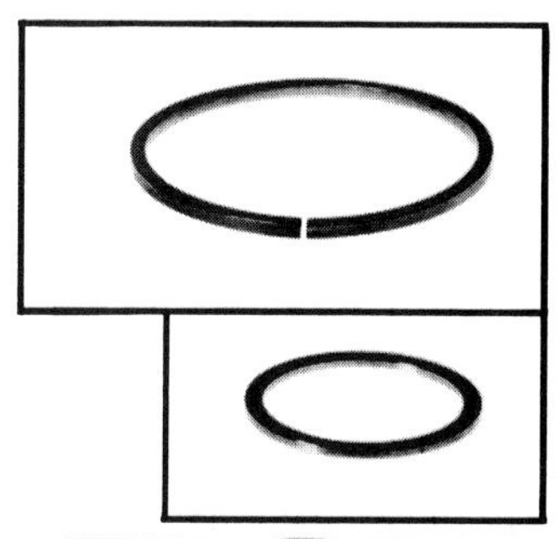

Robvan Backing Ring Co.

23-17. *The consumable insert ring becomes part of the root weld bead and insures complete penetration and smoother flow of materials inside the pipe. Fitup is easy.*

23-18. *Typical pipe end preparation for consumable insert rings for pipe wall thicknesses up to 1 inch.*

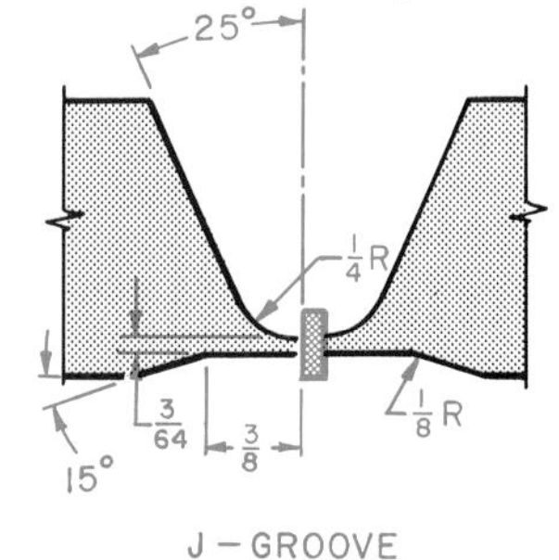

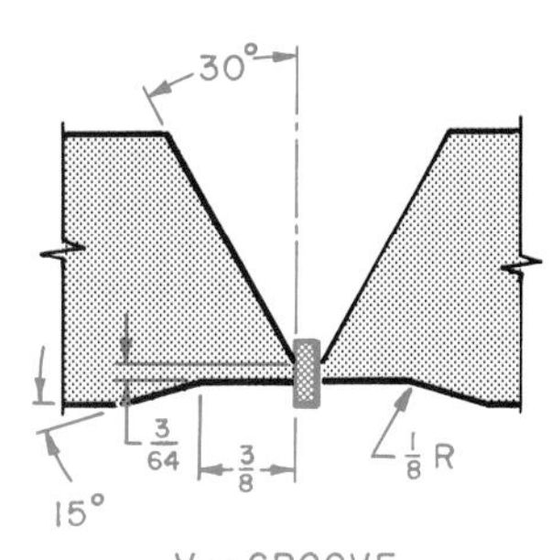

leum, petroleum products, and fuel gases, and certain distribution systems.

THE AMERICAN STANDARDS ASSOCIATION (ASA). This body has issued a code for pressure piping that is concerned with power piping, industrial gas and air piping, petroleum refinery piping, oil transportation piping, refrigeration piping, chemical industry process piping, nuclear power piping, and gas transmission and distribution piping systems.

THE AMERICAN WATER WORKS ASSOCIATION. This group has issued standards covering the fabrication of piping for water purification plants.

THE HEATING, PIPING, AND AIR CONDITIONING CONTRACTORS NATIONAL ASSOCIATION. This piping contractors association has set up standard welding procedures for the installation of piping systems by their member contractors. The National Certified Pipe Welding Bureau is one of several organizations which

Jewel Manufacturing Co.

23-19. *Standard butt joint clamped and ready for tacking.*

Jewel Manufacturing Co.

23-21. *A flange is being set up with a length of pipe for a butt joint.*

23-20. *A manufactured L is being set up with a length of pipe for a butt joint.*

Jewel Manufacturing Co.

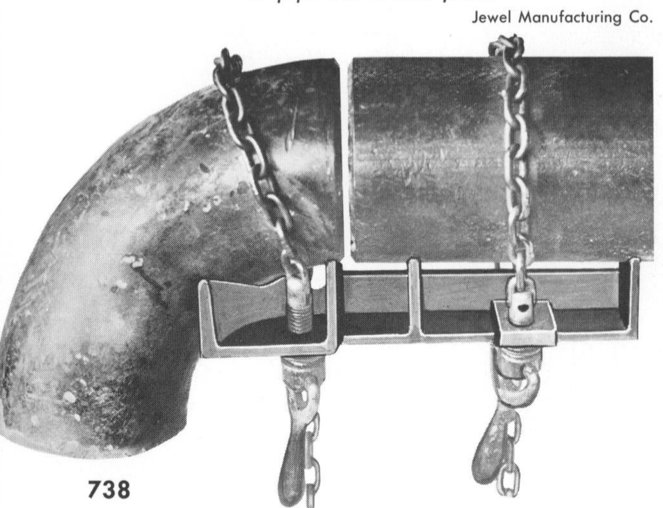

23-22. *Method of clamping a T-joint. This is usually referred to as putting a branch into a header.*

Jewel Manufacturing Co.

supervises and certifies welder qualification tests in a uniform manner in accordance with the Association's standard procedures.

THE PIPE FABRICATION INSTITUTE. Welding standards prepared by this group cover joint preparation, preheating, and welding.

THE AMERICAN WELDING SOCIETY (AWS). This organization is responsible for a great many of the uniform practices in the field of welding. It was one of the early pioneers in the development of uniform training and testing practices for welding operators. Today the society is one of the outstanding authorities in all matters pertaining to welding.

These are the principal code bodies concerned with the design, installation, and testing of piping systems. There are a number of other groups concerned with special purpose installations.

It is suggested that you obtain a copy of each of the following publications in order to become thoroughly familiar with all of the information in regard to the common codes and testing situations:

- The American Welding Society, *Standard Methods for Mechanical Testing of Welds.*
- American Petroleum Institute, *Standard for Welding Pipe Lines and Related Facilities* (API Std. 1104).
- The American Society of Mechanical Engineers, *Section IX—ASME Boiler and Pressure Vessel Code—Welding Qualifications.*

Procedure and Welding Qualification

Code welding requires the setting up and acceptance of the qualifications for welding procedures and for welders who are

23-23. *A pipe claw is a simple quick tool for perfect joint alignment.*
Ransome Co.

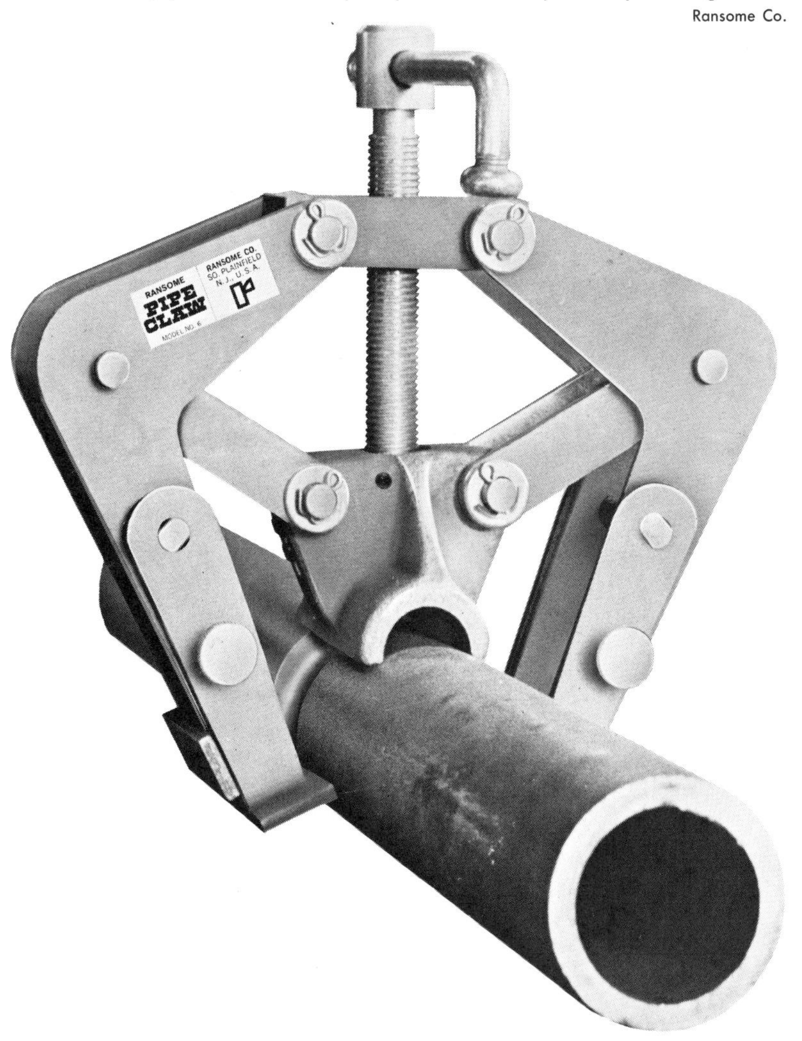

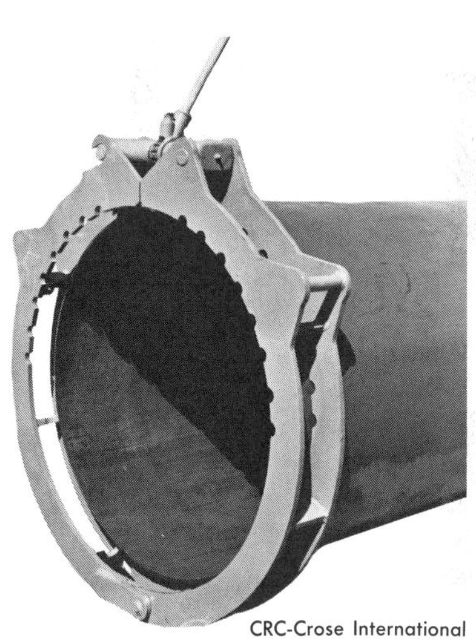

CRC-Crose International
23-24. *A 10-ratchet outside line-up clamp secures positive alignment.*

23-25. *Pipe welding clamps are available with a hand lever or a hydraulic jack.*
CRC-Crose International

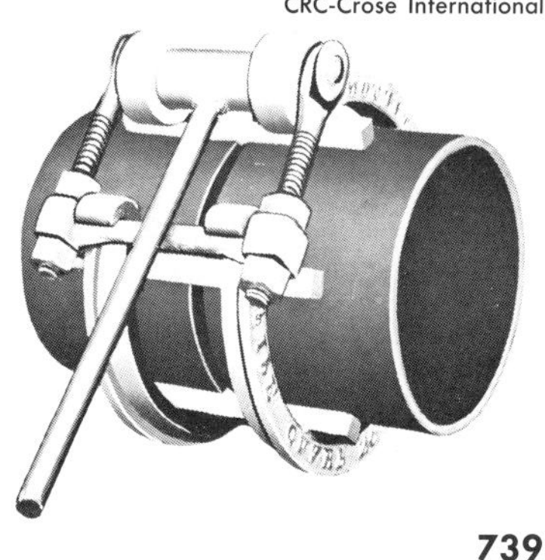

going to fabricate the piping installation following the welding procedure adopted.

PROCEDURE QUALIFICATION TESTS. The welding procedure qualification is set up and approved before the start of production welding. It establishes the fact that the procedure can produce welds having suitable mechanical properties and soundness.

The details of each qualified procedure are recorded and include a record of the following:

- Material for pipe and fitting
- Diameter and wall thickness of the pipe
- Joint design
- Joint preparation, including root gap and bevel angle if necessary
- Position of welding
- Welding process
- Type and size of electrode or filler rod
- Type of current
- Current setting
- Number of passes
- Welding technique
- Use of a backup strip if necessary
- Preheat, between-pass heat, and postheat temperatures
- Special information peculiar to each job.

A new welding procedure must be established if there is a change in any of the items listed above. Figure 23-28 shows the location of the test coupons that are to be removed and indicates the type of testing to which they must be subjected in accordance with the API code. Figures 23-29 and 23-30 give this same information in order to qualify the work under the ASA and ASME codes. The preparation for these specimens for the ASA and ASME codes are shown in Chapter 12 under Preparation of Test Specimens, pages 376–378. The special nature of the test coupons for the API code are shown in Figs. 23-31 through 23-35 (Page 742 through 744). Consult Table 23-4 for type and number of test specimens for procedure qualification test requirements for API code and Table 23-5 (Page 746) for the ASME code. A tested procedure can produce quality welding only if the welder has the ability to apply the procedure. Each welder must be properly qualified by demonstrating his ability to make acceptable welds with the tested procedure that has been ac-

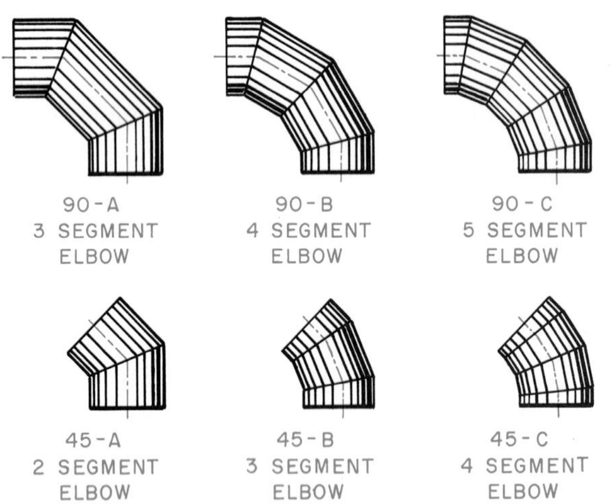

23-26. *Hand-fabricated segmental 45- and 90-degree Ls.*

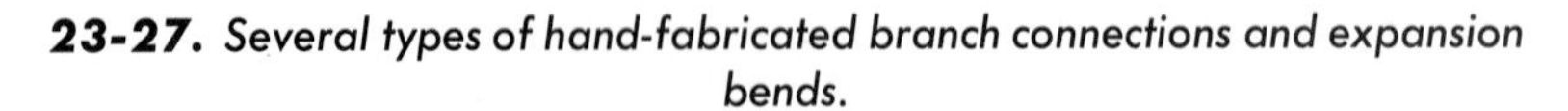

23-27. *Several types of hand-fabricated branch connections and expansion bends.*

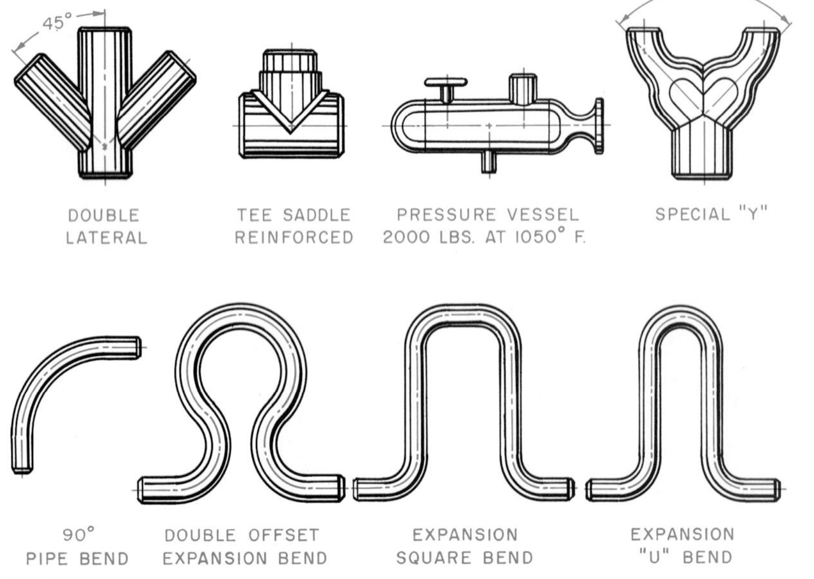

cepted for a particular job. Welder qualification is usually recognized only on a particular job or for the current employer and only in the procedure in which the welder has qualified. A few piping contractors require the welders in their employ to retest every six months.

THE OPERATOR QUALIFICATION TESTS. The operator qualification tests are the ones that the student welder is asked to perform and pass. The purpose of the welder qualification test is to determine the ability of the welder to make sound welds using a previously qualified welding procedure.

If the welder wishes to qualify under the API code, test coupons should be removed from the pipe as indicated in Fig. 23-36 (Page 744). The preparation of the test coupons for the ASA and ASME codes is shown in Figs. 23-29 and 23-30. The preparation for these specimens is described in Chapter 12 under Preparation of Test Specimens, pages 376–378. Consult Table 23-6 (Page 746) for type and number of test specimens for operator qualification test requirements for the API code and Table 23-7 for the ASME code.

The above information concerns butt welds only. The procedure and operator qualifications for fillet welds is as follows. For the API code, Table 23-8 lists the type and number of test specimens required. Figure 23-37 (Page 745) indicates the location of these test specimens in the pipe joint, and Fig. 23-38 indicates the nature of the preparation of the test specimens.

Figure 23-39 (Page 748) shows a sample form that can be used as a record of testing. This form comes from the API code and can be used for both procedure and operator qualification on butt welds. The record form in Fig. 23-40 (Page 749) is from the ASME code and is used for operator qualification for fillet welds.

For qualification, the exposed surfaces of the specimen must show complete penetration and no more than 6 gas pockets per square inch of surface area with the greatest dimension not to exceed $^1/_{16}$ inch. Slag inclusions must not be greater than $^1/_{32}$ inch in depth, or greater than $^1/_8$ inch or $^1/_2$ the nominal wall thickness of the thinner member in length, whichever is the smaller. Such inclusions must be separated by at least $^1/_2$ inch of sound metal.

Study Chapter 12 for fillet weld testing for the ASME code.

It should also be noted that a particular specimen qualifies within the limits of essential variables. If any of these variables are changed, the welder must requalify by being tested on the new procedure. The following modifications call for retesting:

- A change from one welding process to any other welding

23-28. *Location of butt weld test specimens for procedure qualification.*

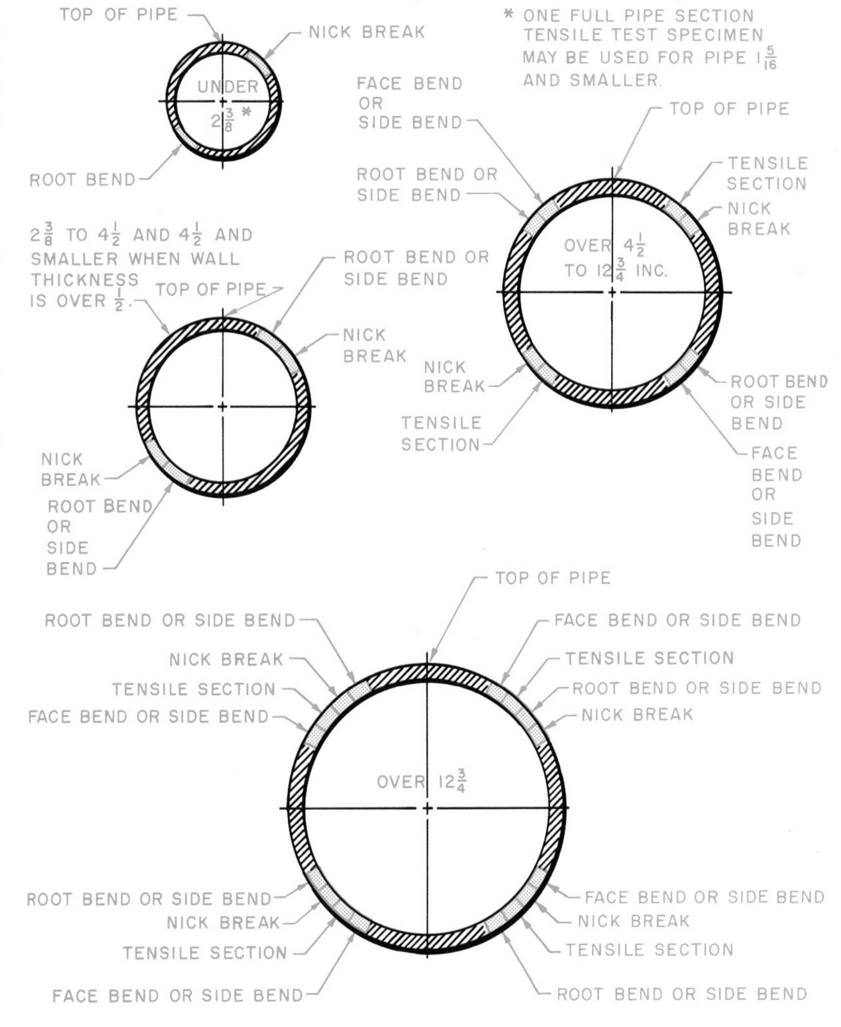

process or combination of welding processes

- A change in the direction of welding from vertical-up to vertical-down or vice-versa
- A change in filler metal from one classification group to another
- A change in pipe diameter from one group to another
- A change in pipe wall thickness from one group to another
- A change in position other than that already qualified
- A change in the joint design (from backing strip to no backing strip or from V-bevel to U-bevel)

After you have passed these tests, you will have earned the right to look upon yourself as a skilled welder with better than average knowledge and skill. Code pipe welding is one of the most demanding forms of the welding trade, and it is the highest paid of all of the fields of welding.

MAKING THE TEST WELD. A skilled welder may fail a test because of reasons that have noth-

Table 23-4

Test requirements for procedure qualification in pipeline according to the API code.

Type and Number of Test Specimens for Procedure Qualification Test

Pipe size, outside diameter—inches	Tensile	Nick-break	Root-bend	Face-bend	Side-bend	Total
	Wall thickness—½ Inch and under					
Under $2\frac{3}{8}$	0	2	2	0	0	4*
$2\frac{3}{8}$ to $4\frac{1}{2}$ inclusive	0	2	2	0	0	4
Over $4\frac{1}{2}$ to $12\frac{3}{4}$ inclusive	2	2	2	2	0	8
Over $12\frac{3}{4}$	4	4	4	4	0	16
	Wall thickness—over ½ inch					
$4\frac{1}{2}$ and smaller	0	2	0	0	2	4
Over $4\frac{1}{2}$ to $12\frac{3}{4}$ inclusive	2	2	0	0	4	8
Over $12\frac{3}{4}$	4	4	0	0	8	16

American Petroleum Institute

*One nick-break and one root-bend specimen from each of two test welds or for pipe $1\frac{5}{16}$ in. and smaller, one full pipe section tensile specimen.

23-29. *Order of removal of test specimens from welded pipe 1/16 to 3/4 inch thick.*

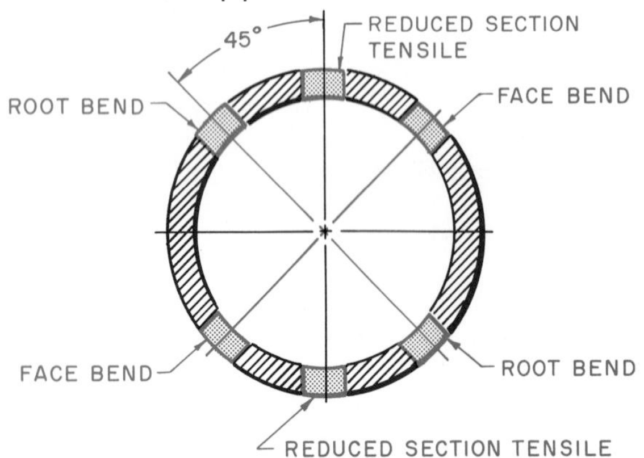

23-30. *Order of removal of test specimens from welded pipe over 3/4 inch thick. This order may also be used for pipe 3/8 to 3/4 inch thick.*

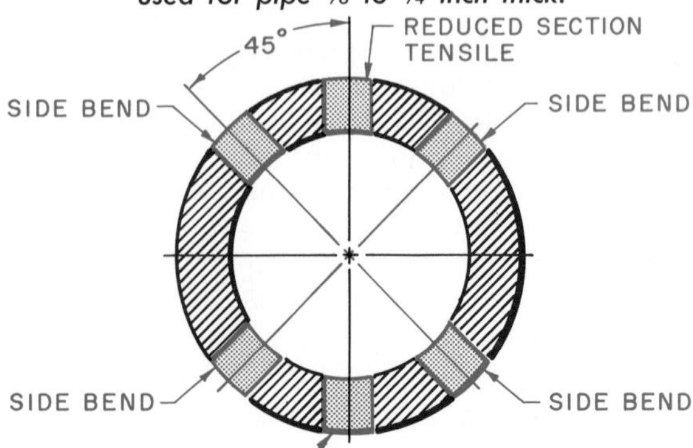

ing to do with his ability to deposit a sound weld bead. Following are test conditions that must be given careful attention to insure satisfactory test results.

- The test plate should be of a material similar to that used on the job.
- Proper plate preparation is a must. Correct angle of bevel, root face, root gap, and proper tacking are essential.
- Plate strength must be similar to that on the job. If it is greater than the weld metal, the weld will break when subjected to the face- and root-bend tests.
- Use an electrode similar to that used on the job.
- Sound root penetration is essential.
- Do not become overly concerned about weld appearance. Good penetration and fusion are more important.
- Welding procedure should be the same as that employed on the job.
- Keep the plates hot during welding and allow them to cool slowly after the completion of the test joint. Never quench the plates in water nor blow air on them. If test is taken outside, provide a shield from the wind.

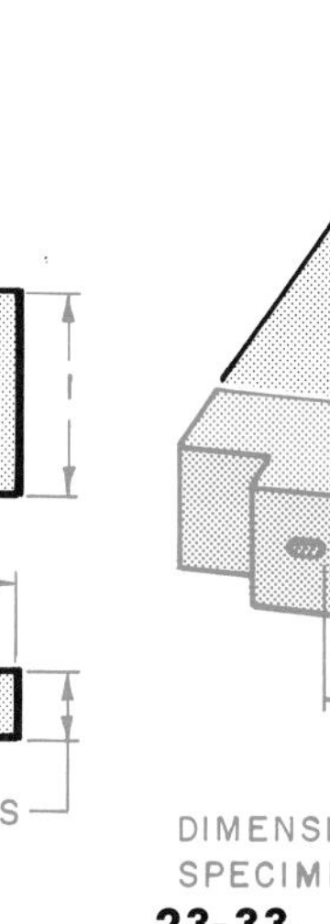

23-31. *Tensile test specimen.*

23-33. *Allowable defects in the specimen shown in Fig. 23-32.*

23-32. *Nick-break test specimen.*

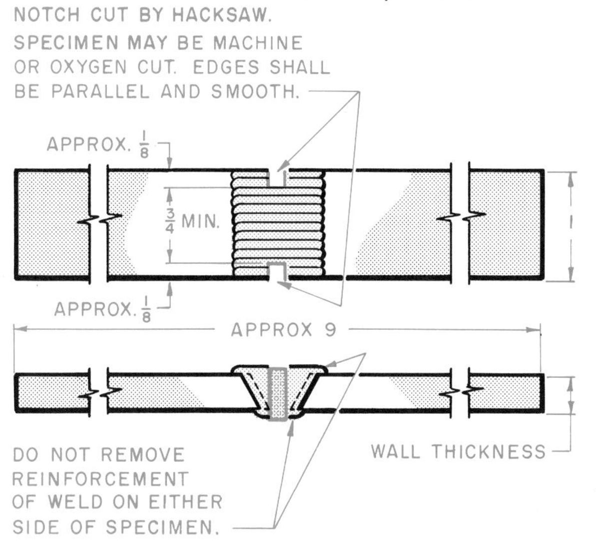

23-34. *Root- and face-bend test specimen.*

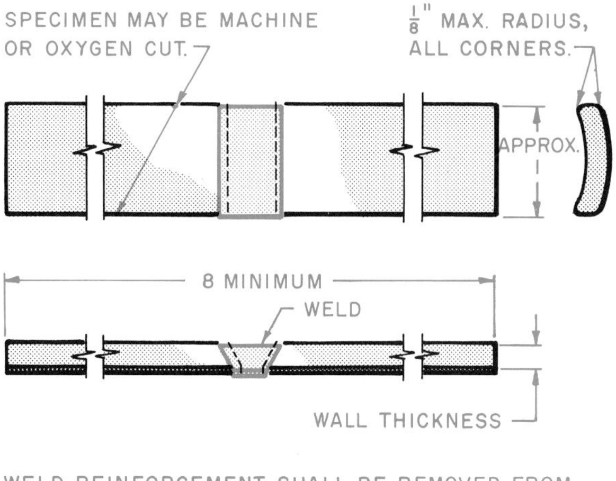

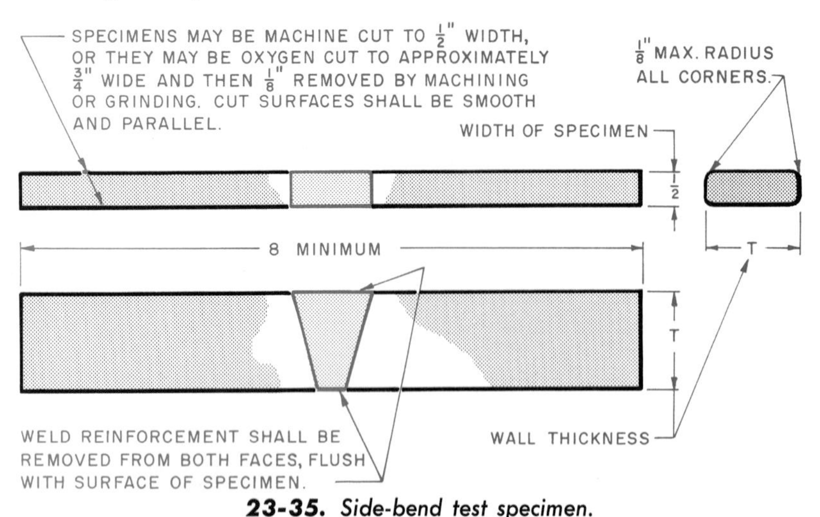

23-35. *Side-bend test specimen.*

23-36. *Location of butt weld test specimens for operator qualification.*

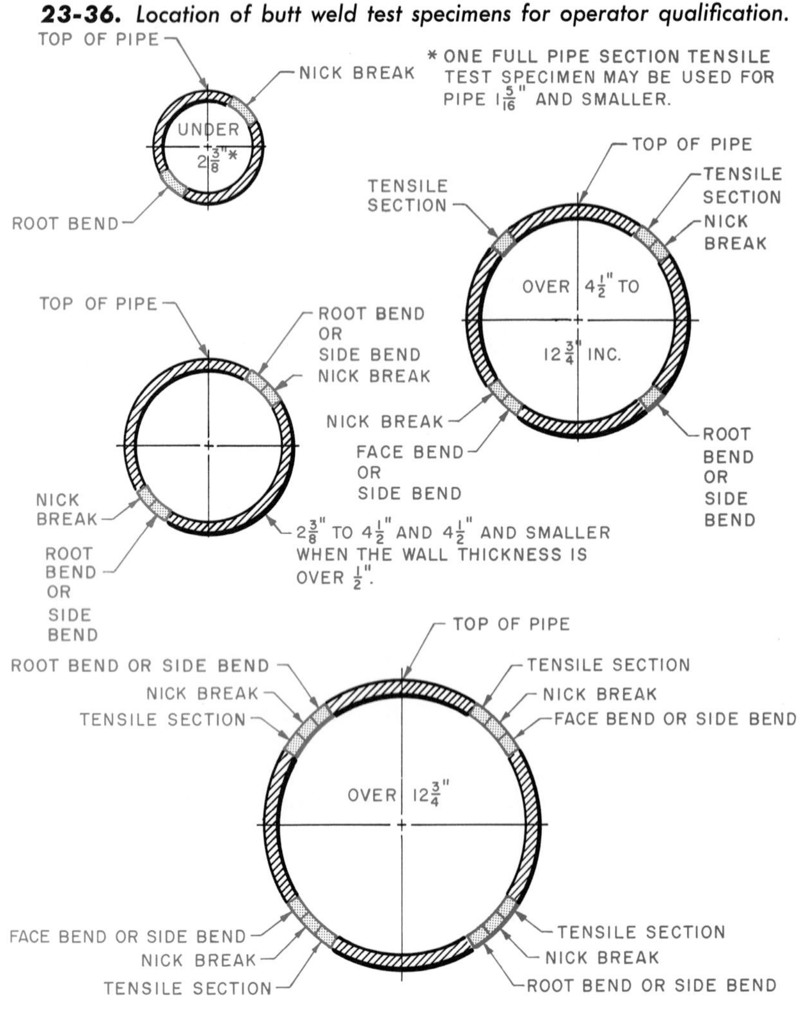

- The test specimen must be free of nicks or deep scratches. All grinding or machining must be lengthwise on the specimen.
- Remove all face and root reinforcement and round the edges of the specimen. Do not quench in water after grinding.

Methods of Testing and Inspection

A variety of testing methods are used in determining the quality of pipe joints and piping systems, and fall into two major categories: nondestructive and destructive testing. Following is a listing of the tests used. For additional information you are urged to review Chapter 12.

VISUAL INSPECTION. Although visual inspection is not actually a form of testing, it is a control of the work being performed. While the work is in progress, inspectors in the field inspect individual joints to make sure that each joint has been properly prepared, set up, and fitted. During welding, frequent observations are made to determine whether or not the prescribed welding procedure is being followed by each welder. Welding current, number of passes, interpass temperature control, and cleaning between passes are particularly important and frequently checked.

Completed welds are inspected visually for general appearance. The amount of reinforcement, the presence or absence of undercutting, the nature of the penetration, and any signs indicating a lack of fusion are noted. These are the same characteristics that you have learned to look for in your previous welding practice.

RADIOGRAPHIC INSPECTION. One of the most important and reliable nondestructive testing tools is radiographic inspection.

On a commercial basis this is done with X-ray or with radioactive isotopes such as cobalt 60. Generally, X-ray work is done in the shop where equipment can be under better control, but portable X-ray equipment such as the pipeline crawler is also available for field inspection. The X-ray pipeline crawler is an automated, self-propelled X-ray machine for inspecting circumferential welds in cross-country pipelines. The crawler is battery operated and requires no external cables or connectors.

The crawler is positioned at each weld by means of a radioactive isotope locator placed at a predetermined position on the outside of the pipeline. The machine then stops at the joint to by X-rayed, Fig. 23-41 (Page 750). The X-ray beam produces a latent image of the weld on a strip of X-ray film wrapped around the outside of the joint. Defects are revealed when the film is developed. When the exposure is completed, the crawler moves along on command from the isotope locator, Fig. 23-42, which is placed at the next weld, and the process is repeated.

The crawler can be used in pipelines with diameters ranging from 18 to 60 inches. It can crawl along at a top speed of 39 feet per minute and can take X-ray photos of up to a mile of pipeline without having its nickel-cadmium batteries recharged, Fig. 23-43 (Page 751). It can climb or descend a 30-degree incline.

ULTRASONIC INSPECTION. Ultrasonic inspection for defects in pipe welds is not used extensively. The problem is that few testing laboratories have developed the large number of tests and correlated standards necessary to interpret the many different conditions that can occur in a

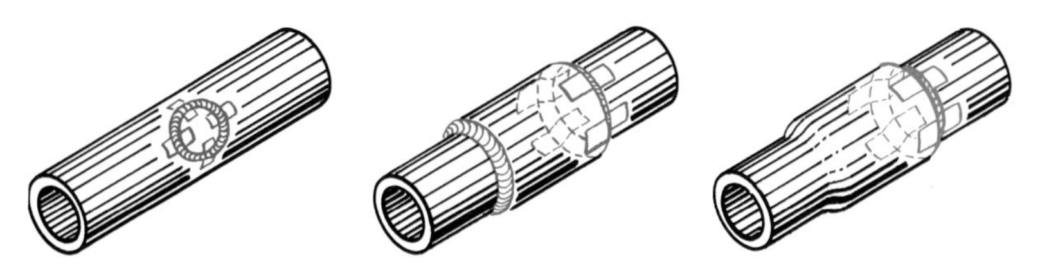

$2\frac{3}{8}$" TO $12\frac{3}{4}$" INCLUSIVE

FOR JOINTS UNDER $2\frac{3}{8}$" CUT NICK BREAK SPECIMENS FROM SAME GENERAL LOCATION BUT REMOVE TWO SPECIMENS FROM EACH OF TWO TEST WELDS.

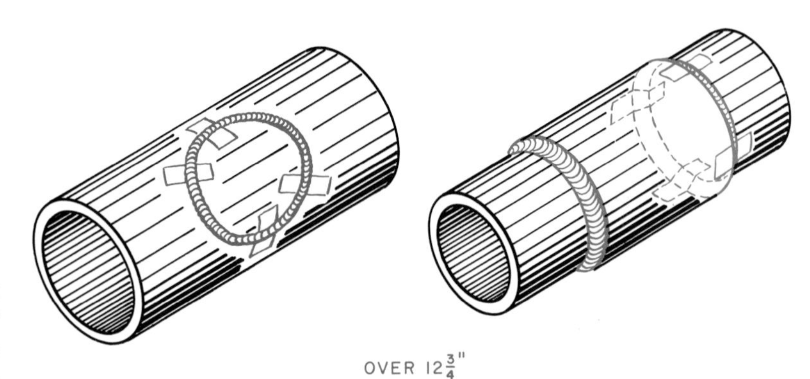

OVER $12\frac{3}{4}$"

23-37. *Location of nick-break test specimens in fillet welds for both procedure and operator qualification.*

23-38. *Preparation of fillet weld specimens.*

American Petroleum Institute

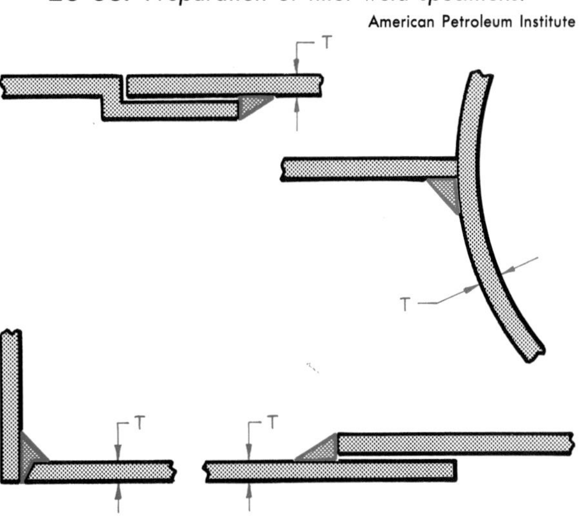

T = WALL THICKNESS

MAY BE MACHINE CUT OR OXYGEN CUT.
SPECIMENS SHOULD BE AT LEAST 2" LONG & 1" WIDE.
SPECIMENS MAY BE HACKSAW NOTCHED IN THE WELD.

weld. Only a highly experienced inspector may be able to interpret the difference between a serious defect and minor, noncritical evidence of slag, porosity, and changes in surface contour.

LIQUID PENETRANT INSPECTION. Dye penetrant inspection is used extensively where narrow cracks are suspected. This method is extensively used on nonmagnetic materials where magnetic particle inspection processes could not be used.

MAGNETIC PARTICLE INSPECTION. Often referred to as Magnaflux® testing, this method can be used only on magnetic materials. Fluorescent magnetic particles in a liquid medium (usually water) are applied to the work for the detection of extremely narrow cracks.

HYDROSTATIC TESTING. Hydrostatic testing is used extensively for welded-piping components and pipelines. The nondestructive testing of sections between valves is often done. This form of testing is also applied to shop-fabricated pipe assemblies. For most work clean water is used at about the atmospheric temperature to prevent sweating of the pipe and weld joints. Pressure is applied gradually by

Table 23-5

Test requirements for procedure qualification in pressure piping according to the ASME code.

Thickness, t, of test plate or pipe as welded, in.	Range of thickness of materials qualified by test plate or pipe, in.		Type and number of tests required[3]			
	Min	Max[1,2]	Tension	Side-bend	Face-bend	Root-bend
$\frac{1}{16}$ to $\frac{3}{8}$, incl.	0.5 t (Note 4)	2 t	2		2	2
Over $\frac{3}{8}$ but less than $\frac{3}{4}$ (Note 3)	$\frac{3}{16}$	2 t	2		2	2
$\frac{3}{4}$ and over	$\frac{3}{16}$	2 t	2	(4 (Note 3)		

American Society of Mechanical Engineers

[1] The maximum thickness qualified in oxyacetylene gas welding is the thickness of the test plate or pipe.
[2] The maximum thickness qualified with pipe smaller than 5 in. in diameter is two times the thickness of the pipe, but not more than $\frac{3}{4}$ in.
[3] Either the face- and root-bend tests or the side-bend tests may be used for thicknesses from $\frac{3}{8}$ to $\frac{3}{4}$ in.

Table 23-6

Test requirements for operator qualification in pipeline according to the API code.

Type and number of test specimens for welder qualification test and for destructive testing of production welds—butt welds

Pipe size, outside diameter—inches	Number of specimens: Tensile	Nick-break	Root-bend	Face-bend	Side-bend	Total
	Wall thickness—½ Inch and under					
Under $2\frac{3}{8}$	0	2	2	0	0	4*
$2\frac{3}{8}$ to $4\frac{1}{2}$ inclusive	0	2	2	0	0	4
Over $4\frac{1}{2}$ to $12\frac{3}{4}$ inclusive	2	2	2	0	0	6
Over $12\frac{3}{4}$	4	4	2	2	0	12
	Wall thickness—over ½ Inch					
$4\frac{1}{2}$ and smaller	0	2	0	0	2	4
Over $4\frac{1}{2}$ to $12\frac{3}{4}$ inclusive	2	2	0	0	2	6
Over $12\frac{3}{4}$	4	4	0	0	4	12

*Obtain from two welds or one full pipe section tensile specimen for pipe $1\frac{5}{16}$ in. and smaller.

American Petroleum Institute

means of a pump and registers on a gauge. The test pressure should be that called for by the governing code or contract. It should be at least $1\frac{1}{2}$ times the maximum working pressure to which the piping will be subjected. For power piping the test pressure should never be less than 50 p.s.i. nor more than twice the service pressure of any valve included in the test.

The use of compressed air or other gas is not recommended because of its explosive force. A failure could be highly destructive and a hazard to human life. Air testing is used to a limited extent by certain small tank fabricators and at low pressure. Freon tests and soap bubble tests are also falling into disuse.

Destructive hydrostatic testing, Fig. 23-44 (Page 751), is usually carried out under controlled conditions in a shop or laboratory on a particular component of the piping system. Oil or water may be used as the internal liquid and subjected to high pressure. The pressure is applied to the yield point in the material or even to the bursting point. This is for the purpose of determining the stress concentrations, strengths, and weaknesses.

SERVICE TESTS. After the piping installation has been completed, a service test, which is essentially a trial operation of a system under normal service conditions, is performed. Care must be taken to make sure that all valves and regulating devices are set for proper operation and that all hangers and anchor structures are secured.

The customer expects that the system is free from leaks and is able to perform as called for in the specifications. If the welder has done his work well, the system will perform as designed. This is a serious responsibility since the contractor is responsible for any leaks and any damage due to leaks.

COUPON TESTING. The most common type of destructive testing that you will be concerned with as a pipe welder is done by taking test coupons from pipe joints that you have welded in accordance with the requirements and specifications of whatever code you are working under. Coupons may be cut accurately in the field by means of a test coupon cutter, Fig. 23-45 (Page 751). The cutter uses an oxyacetylene cutting torch. It may be used for both plate and pipe.

MISCELLANEOUS TESTS. Several other destructive test methods are also employed to maintain control over weld quality.

Trepanning is generally done by removing a cylindrical plug from the weld by means of a power-driven hole saw. This per-

Table 23-8

Fillet weld test specimen requirements for procedure qualification according to the API code.

Pipe size, OD in.	Number of root-bend specimens
Under $2\frac{3}{8}$	4 (Obtain from 2 welds)
$2\frac{3}{8}$ to $12\frac{3}{4}$ inclusive	4
Over $12\frac{3}{4}$	6

American Petroleum Institute

Table 23-7

Test requirements for operator qualification in pressure piping according to the ASME code.

(Performance qualification)

Type of joint	Thickness, *t*, of test plate or pipe as welded, in.	Range of thickness of materials qualified by test, in.		Type and number of tests required[1] guided-bend tests			
		Min	Max[4]	Side-bend	Face-bend	Root-bend	Joint
Groove	$\frac{1}{16}$ to $\frac{3}{8}$, incl.	0.5 *t* (Note 3)	2 *t*		1	1	
Groove[2]	Over $\frac{3}{8}$ but less than $\frac{3}{4}$	$\frac{3}{16}$	2 *t*		1	1	
Groove[3]	Over $\frac{3}{8}$ but less than $\frac{3}{4}$	$\frac{3}{16}$	2 *t*	2			
Groove	$\frac{3}{4}$ and over	$\frac{3}{16}$	Max to be welded (Note 4)	2 (Note 2)			

American Society of Mechanical Engineers

[1] A total of four specimens are required to qualify for position 5G.
[2] Either the face- and- root-bend tests or the side-bend tests may be used for thicknesses from $\frac{3}{8}$ to $\frac{3}{4}$ in.
[3] Not less than $\frac{1}{16}$ in.
[4] The maximum thickness qualified in oxyactylene gas welding is the thickness of the test plate or pipe.

23-39. *Sample record form for a weld coupon test report.*

American Petroleum Institute

(Sample form)

Coupon test report

Test No. . . . .
Location . . . . Date . . . .
Contractor . . . . Sub-contractor . . . .
Schedule . . . . Gang . . . . Inspector . . . .
Date . . . . State . . . . Roll Weld . . . . Fixed position weld . . . . . .
Welder . . . . Mark . . . .
Welding time . . . . Time of day . . . . M. Temperature . . . . F.
Weather condition . . . .
Wind break used . . . . voltage . . . . amperage . . . .
Type of welding machine . . . . Size . . . .
Filler metal . . . .
Size of reinforcement . . . .
Pipe mfr. . . . . Kind . . . .
Wall thickness . . . . . . Dia. O.D. . . . . . . Wt./ft. . . . . . . Joint length . . . .

	1	2	3	4	5	6	7
Bead No.	. . . .	. . . .	. . . .	. . . .	. . . .	. . . .	. . . .
Size of Electrode	. . . .	. . . .	. . . .	. . . .	. . . .	. . . .	. . . .
No. of Electrode	. . . .	. . . .	. . . .	. . . .	. . . .	. . . .	. . . .

	1	2	3	4	5	6	7
Coupon stenciled	. . . .	. . . .	. . . .	. . . .	. . . .	. . . .	. . . .
Original							
Dimension: of plate . . .	. . . .	. . . .	. . . .	. . . .	. . . .	. . . .	. . . .
Orig. area of plate in^2 .	. . . .	. . . .	. . . .	. . . .	. . . .	. . . .	. . . .
Maximum load	. . . .	. . . .	. . . .	. . . .	. . . .	. . . .	. . . .
Tensile S/in. plate area	. . . .	. . . .	. . . .	. . . .	. . . .	. . . .	. . . .
Fracture location	. . . .	. . . .	. . . .	. . . .	. . . .	. . . .	. . . .

☐ Procedure ☐ Qualifying Test ☐ Qualified
☐ Welder ☐ Line Test ☐ Disqualified

Max. tensile . . . . Min. tensile . . . . Avg. tensile . . . .
Remarks on tensile . . . .
1 . . . .
2 . . . .
3 . . . .
4 . . . .
Remarks on Bend Tests . . . .
1 . . . .
2 . . . .
3 . . . .
4 . . . .
Remarks on Nick Tests . . . .
1 . . . .
2 . . . .
3 . . . .
4 . . . .
Test make at . . . . Date . . . .
Tested by . . . . Supervised by . . . .

(Use back for additional remarks.)

Note: Can be used to report both Procedure Qualification test and Welder Qualification tests.

mits checking the inside weld surface for porosity, cracking, penetration, and fusion.

Sometimes narrow strips are removed with a saw or high speed cutting wheels. After the specimen has been removed, it is ground and polished, and the prepared surface is etched. Examination of the etched surface discloses such defects as cracks, lack of fusion, and slag inclusions.

Common Defects in Pipe Welds

Following is a list of weld defects that occur frequently in pipe welds. The welding process is dependable and will consistently produce work that is sound, Fig. 23-46. The variable is the welder. Much poor welding is caused not by the welder's lack of skill, but by his carelessness or lack of interest in the work. A skilled mechanic is a concerned mechanic who is careful about everything he does.

Essentially, inadequate penetration is defined as the incomplete filling of the weld root with weld metal. Incomplete fusion is defined as the lack of bond between beads or between the weld metal and base metal. Inadequate penetration and incomplete fusion are separate and distinct conditions which occur in different forms as follows.

- *Inadequate penetration.* Inadequate penetration that is not caused by a high or low condition is defined as incomplete filling of the weld root, Fig. 23-47 (Page 752).
- *Inadequate penetration due to a high-low condition.* High-low is defined as a condition in which the pipe and/or fitting surfaces are misaligned, Fig. 23-48A. The weld in a high-low misalignment is permissible provided that the

23-40. *Recommended form for the manufacturer's record of operator performance qualification tests on fillet welds.*

American Society of Mechanical Engineers

Recommended form manufacturer's record of welder performance qualification tests on fillet welds

Welder Name ______________ Clock No. ________ Stamp No. ____________

Welding Process ______________

Position (If vertical state whether upward or downward) ______________
(Flat, horizontal, vertical, or overhead; see Pars. & Figs. Q-2 & Q-3, or QN-2 & QN-3)

In accordance with Procedure Specification No. ______________

Material—Specification ________ to ________ of P-No. ________ to P-No. ________

Diameter and Wall Thickness (if pipe) otherwise Joint Thickness ______________

Filler metal

Specification No. ________ Group No. F ______________

Describe Filler Metal if not included in Table Q-11.2 or QN-11.2 ______________

____

Is Backing Strip Used? ______________

For information only

Filler Metal Diameter and Trade Name ________ Flux for Submerged Arc or Gas for Inert Gas Shielded Arc ______________ Welding ______________

Test results (see Figs. Q-9 and QN-9)

FractureTest ______________
(Describe the location, nature and size of any crack or tearing of the specimen.)

Length and Percent of Defects ________ inches ________ %

Macro Test—Fusion ______________

Appearance—Fillet Size ________ in. × ________ in. Convexity or Concavity ________ in.

Test Conducted by ______________ Laboratory—Test No. ______________
per ______________

We certify that the statements in this record are correct and that the tests welds were prepared, welded and tested in accordance with the requirements of Section IX of the ASME Code.

Signed: ______________
(Manufacturer)

Date ______________ By: ______________

(Detail of record of tests are illustrative only and may be modified to conform to the type and number of tests required by the Code.
NOTE: Any essential variables in addition to those above shall be recorded.

root of the adjacent pipe and/or fitting joints is completely bonded by weld metal, Fig. 23-48B.

- *Internal concavity.* This applies to a bead which is properly fused to and completely penetrates the pipe wall thickness along both sides of the bevel, but the center of the bead lacks buildup and is somewhat below the inside surface of the pipe wall, Fig. 23-49.
- *Incomplete fusion.* Incomplete fusion is the lack of bond at the root of the joint or at the top of the joint between the weld metal and the base metal, Fig. 23-50.
- *Incomplete fusion due to cold lap.* Incomplete fusion due to cold lap, also called discontinuity, occurs between two adjacent weld beads or between a weld bead and the base metal, Fig. 23-51.

PRACTICE JOBS

Instructions for Completing Practice Jobs

In practicing the jobs presented in the Course Outline, pages 771, 772, follow the same general procedures presented in Chapters 20 through 22. For example, the welding procedures used in practice Job 23-J2 are essentially the same as those used in practice Job 21-J40. The Course Outline lists the jobs in the previous chapters which give the techniques to be followed in completing the jobs in this chapter. Review the job print, the General Job Information, and the Welding Technique in the job to which the Course Outline refers you. Then study the information on the pages in this chapter to which the Course Outline refers you.

The sequence followed in completing the basic operations for the jobs in the Course Outline should be as follows.

Joint Preparation

The preparation of the pipe to be welded should be done so as to conform as nearly as possible to the specifications as shown in Table 23-3, page 734, and Fig. 23-52. Poor beveling and poor preparation and setup of the parts to be welded are responsible for many weld failures. A joint properly prepared is insurance against failure.

1. Refer to Table 23-3, page 734, for the specifications that refer to the joint and position being practiced.
2. Bevel the pipe by hand, machine, or machine oxyacetylene cutting. Be careful to leave the inside edge of the pipe intact so that it serves as a point of contact in preserving proper alignment of the joint when it is set up. Make sure that the bevel angle conforms to the specifications. An insufficient angle of bevel will make it very difficult to secure the proper root penetration and fusion.
3. Clean the bevel face and the pipe surface for at least 1 inch from the edge of the welding groove to remove rust, scale, paint, oil, and grease. Grind flame-cut surfaces smooth enough to remove all traces of scale and any cutting irregularities. Surface finishing may be

23-41. *The operator holds a signal device that causes the pipeline crawler to move into position to X-ray a welded joint in pipelines.*

Picker Industrial

23-42. *The isotope locator for the pipeline crawler is adjusted before it X-rays the welds in a cross-country pipeline. Signals for this detector unit position the crawler at the weld to be X-rayed, activate the X-ray beam, and then move the crawler to the next weld where the procedure is repeated.*

Picker Industrial

done with a file, scraper, emery cloth, chipper, power grinder, or sandblasting equipment.

Setting Up and Tacking

Tack welding should be done with great care and precision. The tacks must hold the joint in proper alignment and keep distortion at a minimum. If improperly applied, tack welds contribute to weld failure. There must be enough tack welds of the proper size to hold the pipe in place.

1. Refer to Figs. 23-19 through 23-25, page 738, to see how the various joints are clamped and prepared for tacking.

2. Make sure that the proper root opening between the pipe ends has been set up. This is very important because an opening that is too small frequently causes weld failure on the root- and side-bend tests. The root opening should be $3/32$ to $1/8$ inch. The simplest way to secure and

23-43. *The battery-operated locomotive that propels the pipeline crawler X-ray machine through cross-country pipelines where it looks for defects in welded joints.*

Picker Industrial

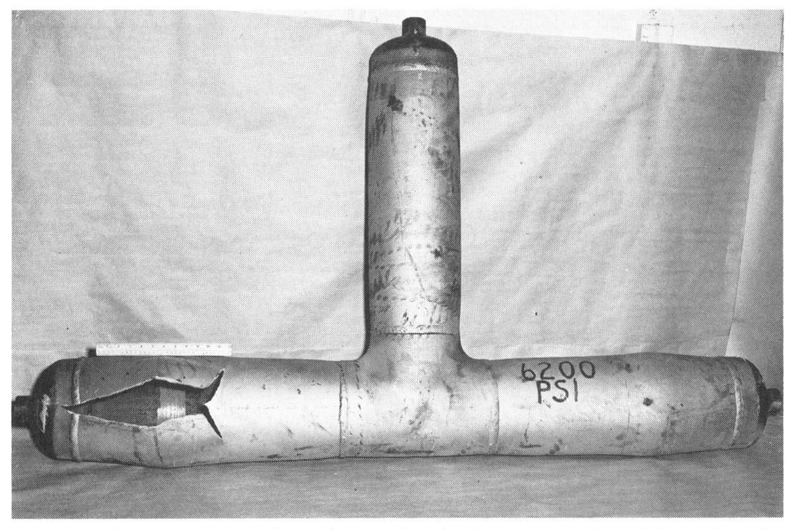

23-44. *A unit subjected to destructive hydrostatic testing. Note that the bursting point was in the pipe and that the welds are intact. The pipe burst at 6,200 p.s.i., a pressure which is well beyond service requirements.*

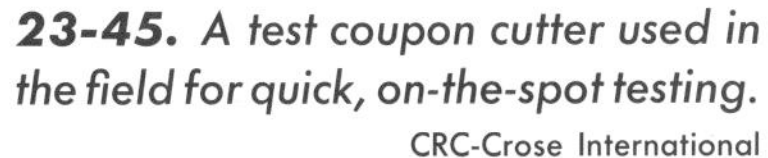

23-45. *A test coupon cutter used in the field for quick, on-the-spot testing.*

CRC-Crose International

23-46. *A pipe fabrication that fractured in service. Note that the welds are intact.*

Crane Co.

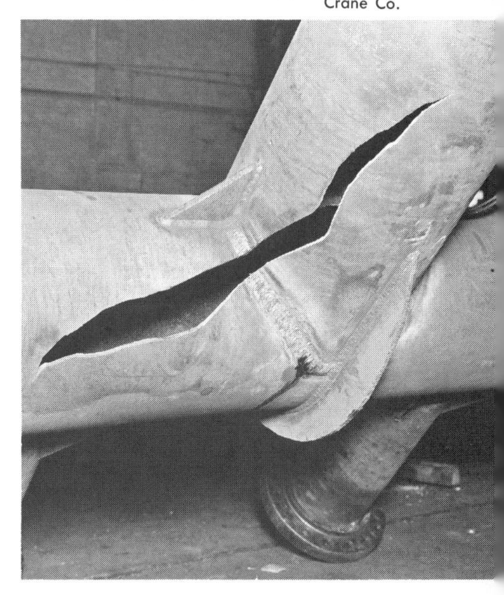

maintain an accurate root opening is to use a $\frac{1}{8}$-inch gas welding wire of the proper size as shown in Fig. 27-15, page 860. By crossing the wire, you have accurate spacing at four points along the circumference of the joint if you have maintained a square edge around the pipe.

3. Tack the joint. On piping up to and including 4 inches in diameter, make two $\frac{1}{2}$-inch long tacks on opposite sides of the joint. On larger pipe sizes, make tack welds at 4- to 6-inch intervals around the circumference of the pipe. In no case should tack welds be longer than 1 inch. They should penetrate the joint and fuse through the full thickness of the land. Make sure that the faces of the welds are flat.

4. Clean the tack welds thoroughly and remove any high spots or crater cracks by grinding.

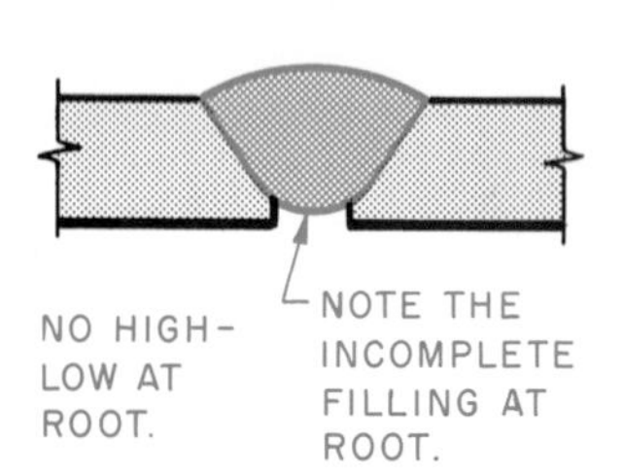
NO HIGH-
LOW AT
ROOT.
NOTE THE
INCOMPLETE
FILLING AT
ROOT.
ONE OR BOTH BEVELS MAY
BE INADEQUATELY FILLED
AT INSIDE SURFACE.

23-47. *Inadequate penetration of the weld groove.*

23-48. *Inadequate penetration due to high-low pipe postions.*

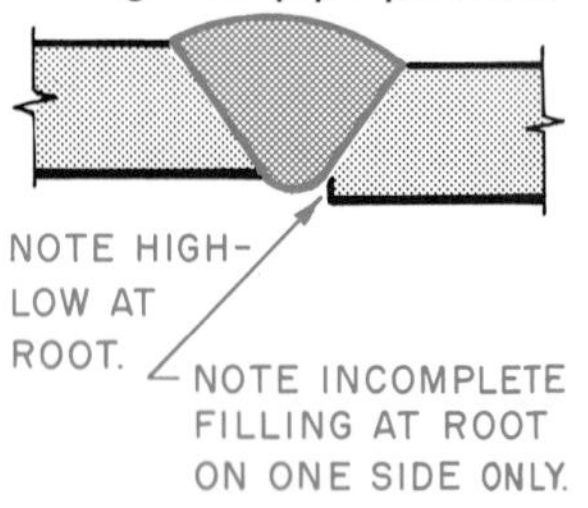
NOTE HIGH-
LOW AT
ROOT.
NOTE INCOMPLETE
FILLING AT ROOT
ON ONE SIDE ONLY.
A

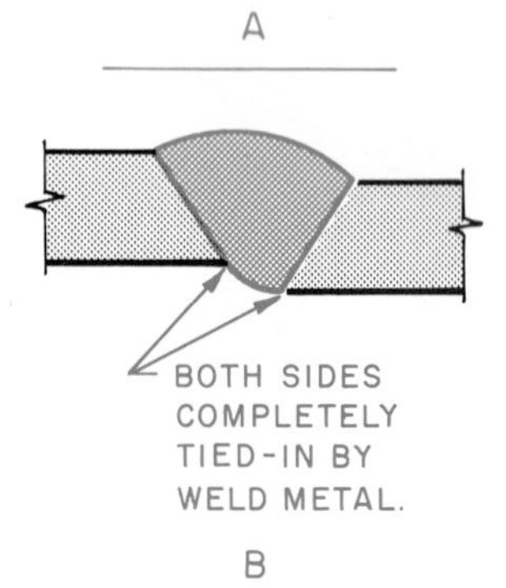
BOTH SIDES
COMPLETELY
TIED-IN BY
WELD METAL.
B

Welding Technique

It is well at this point to review a few of the important welding techniques that you will need to practice if you are going to master the welding of pipe in all positions.

DIRECTION OF TRAVEL. Two methods of welding used are vertical-down travel and vertical-up travel. *Vertical-down welding* requires higher welding current and faster travel speeds than vertical-up so that the joint is made with several smaller beads. Root openings are less than those required for vertical-up welding, or there may be no root openings at all. The bevel angle of the joint is also less than in vertical-up. The vertical-down method is faster and more economical on pipe under $\frac{1}{2}$ inch in wall thickness. Since most cross-country pipelines have walls less than $\frac{1}{2}$ inch thick, they are usually welded by this method. The welds are of excellent quality and able to meet ordinary test requirements.

Vertical-up welding uses lower current and slower travel speed to produce a joint with fewer but heavier beads. Since there are fewer beads to clean, cleaning time is greatly reduced, giving the vertical-up method a speed advantage on heavy wall pipe. The slower travel speed of vertical-up welding and the highly liquid puddle melts out gas holes more effectively than vertical-down welding. Welds made by this method are better able to meet the X-ray requirements for the high pressure, high tempera-

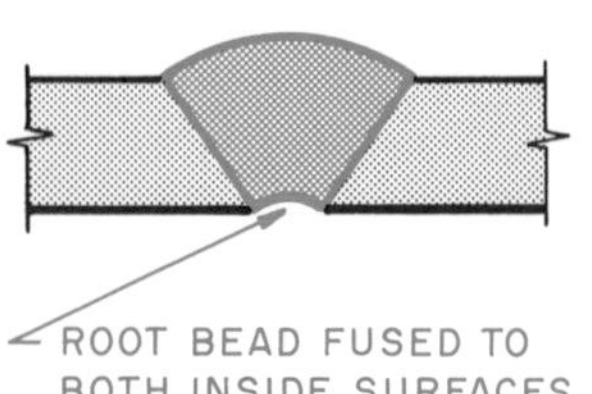
ROOT BEAD FUSED TO
BOTH INSIDE SURFACES
BUT CENTER OF ROOT
PASS IS SLIGHTLY BE-
LOW INSIDE SURFACE
OF PIPE.

23-49. *Inadequate penetration due to internal concavity.*

23-50. *Incomplete fusion at the root of the bead and the top of the joint.*

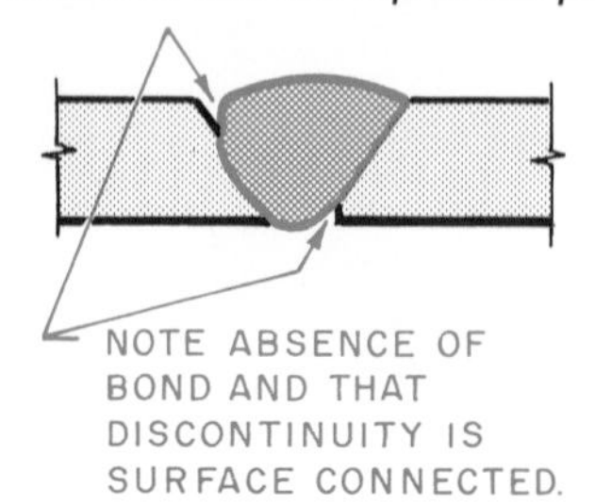
NOTE ABSENCE OF
BOND AND THAT
DISCONTINUITY IS
SURFACE CONNECTED.

23-51. *Incomplete fusion due to cold lap.*

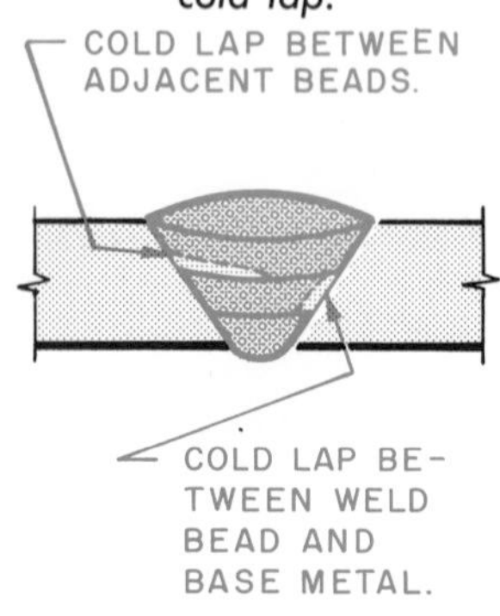
COLD LAP BETWEEN
ADJACENT BEADS.
COLD LAP BE-
TWEEN WELD
BEAD AND
BASE METAL.
NOTE: COLD LAP IS NOT
SURFACE CONNECTED.

ture piping found in refineries and power plants. Vertical-up welding requires a larger root spacing and bevel angle than vertical-down. Thus more electrode is required per joint.

Compare the specifications given in Figs. 23-53 and 23-54. The vertical-down method requires 55 to 80 more amperes than the vertical-up method. Although the number of passes is the same, larger electrode sizes are specified for vertical-down. The bevel angle is 15 degrees less for vertical-down, and the root opening is $\frac{1}{16}$ inch less. Thus the beads are smaller for welds of equal size, and less filler metal is needed. For the joint design shown, the travel speed for vertical-down is more than twice that for vertical-up.

NUMBER OF PASSES. The number of passes required for welding various joints in ferrous piping varies with the following factors:

- Diameter and wall thickness of the pipe
- Joint design
- Position of welding
- Size and type of electrode
- Direction of travel
- Adjustment of the welding current employed

When welding in the rolled and fixed horizontal positions, a generally accepted technique is to deposit one layer for each $\frac{1}{8}$ inch of pipe wall thickness. The size of electrodes may vary from $\frac{1}{8}$ to $\frac{5}{32}$ inch in diameter for the root pass. The electrode for the intermediate and final passes should have a diameter of $\frac{5}{32}$ inch. For pipeline welding using the downhill method and occasionally for other piping systems, electrodes with diameters of $\frac{3}{16}$ or even $\frac{1}{4}$ inch are used for the intermediate and final passes. In welding medium carbon and alloy steels, the number of layers is increased by the use of smaller electrodes and thinner passes. This reduces the heat concentration and ensures complete grain refinement of the weld metal.

When the pipe is in the fixed vertical position and welding is on a horizontal plane, the weld metal is deposited in series of overlapping stringer beads for the intermediate and final passes. Occasionally, a lacing bead is used for the final pass. Few welders are able to master this technique, however, and its use is discouraged. The root pass is made with $\frac{1}{8}$- or $\frac{5}{32}$-inch electrodes. The intermediate and cover passes may be made with $\frac{5}{32}$- and even $\frac{3}{16}$-inch electrodes.

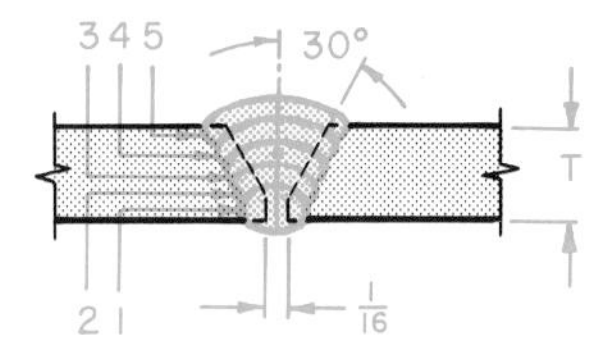

PASS	ELECTRODE	AMPS
1	5/32	150 - 175
2	5/32	170 - 200
3	3/16	170 - 205
4	3/16	170 - 205
5	3/16	170 - 205

23-53. *Vertical-down welding technique.*

23-52. *Recommended bevel sizes for pipe joints. (Left) Recommended standard U-bevel of welding ends for thickness (T) greater than ¾". (Right) Recommended standard straight bevel of welding ends for thickness (T) 3/16" to ¾" inclusive.*
The recommended practice for the detail of welding bevel is as follows:
For wall thicknesses 3/16" to ¾", inclusive, 37½° ± 2½°, straight bevel; land 1/16" ± 1/32".
For wall thicknesses greater than ¾", 20° ± 2½° U-bevel, 3/16" radius; land 1/16" ± 1/32".
Welding ends having thicknesses less than 3/16" are prepared with a slight chamfer or square in accordance with manufacturer's practice.

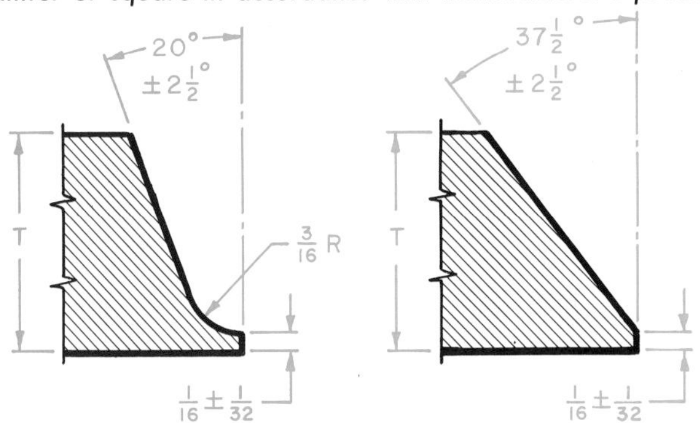

23-54. *Vertical-up welding technique.*

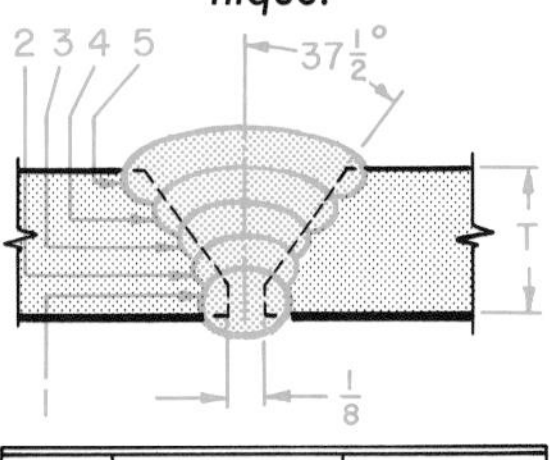

PASS	ELECTRODE	AMPS
1	1/8	85/95
2	5/32 E-6010	115/125
3	5/32 E-6010	115/125
4	5/32 E-6010	115/125
5	5/32 E-6010	115/125

CONTROL OF THE ARC. Before beginning to weld, it is essential to establish a good ground and make sure that both cable connections to the machine are tight. The ground should be attached close to the pipe joint being welded. Current control is extremely important in welding the root pass. A high current setting causes excessive penetration on the inside of the pipe and makes control of the molten weld pool impossible. If the current setting is too low, there is little penetration and fusion at the root.

You will recall that the length of the arc determines the voltage across the arc. The voltage and the amperage provide the heat to melt the electrode and the base metal to form the weld pool. High voltage caused by a long arc gap provides a wide weld bead with little directional control and excessive weld spatter. Low voltage caused by a short arc length causes the bead to pile up in the groove with little fusion and penetration. Low voltage also causes an uneven burnoff rate of the electrode coating so that the slag covers the weld bead inadequately. Poor slag coverage causes porosity in the weld metal.

The speed of travel is another important element in the formation of the weld bead that is deposited. A high speed tends to cause undercutting and a high, narrow bead. Undercutting produces a stress concentration at the point of undercut and can result in joint failure. A low speed causes too much metal to pile up. This causes poor fusion along each edge of the weld and seriously affects the soundness of the weld.

ELECTRODE ANGLE. The angle of the electrode while welding has a significant effect on the final weld. The angle is especially critical in pipe welding since it changes constantly as the weld progresses around the pipe, Fig. 23-55. The welder can maintain control of the weld pool and reduce the erratic effects of arc blow by varying the angle of the electrode to meet the conditions of welding.

Refer to Figs. 22-34 through 22-51, pages 715–722, which show properly made welds.

BEADING AND MAKING A BUTT WELD ON PIPE IN THE ROLLED HORIZONTAL POSITION: JOBS 23-J1 AND J2. Before proceeding with jobs 23-J1 and 2, review the welding procedures and photographs for Jobs 20-J7, 20-J8, 21-J40, and 21-J42. Also study Fig. 23-54.

Beading practice around the outside of the pipe in Job 23-J1 is for the purpose of getting used to the changing position of the torch while following the contour around the pipe. This is an essential technique in pipe welding. The practice should be both downhill and uphill, and both stringer and weaved beads should be deposited.

23-55. *Angle of the electrode as the welder follows the contour of the pipe.*

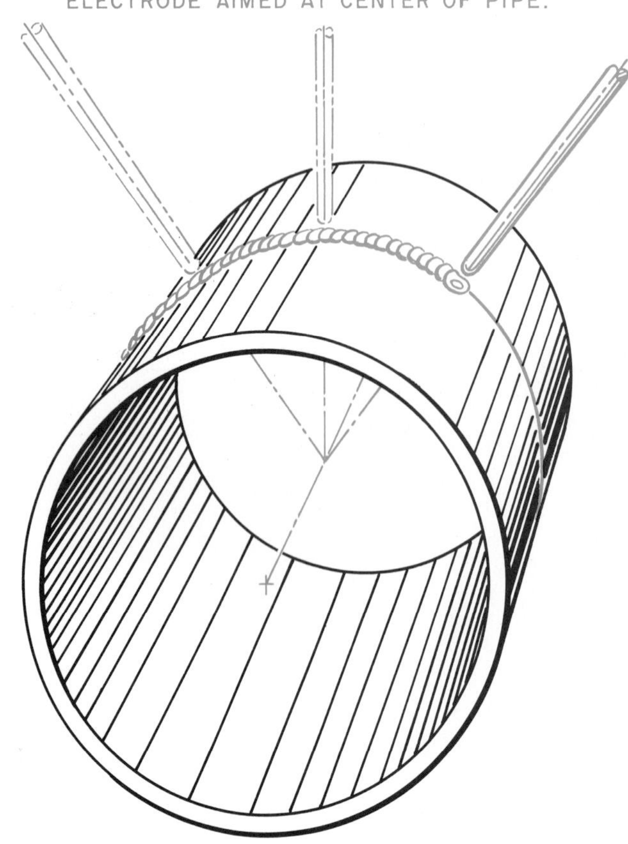

23-56. *Butt joint preparation.*

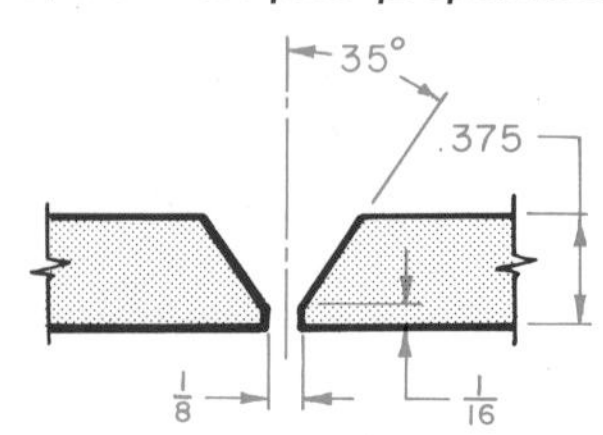

For the butt weld, prepare the pipe for welding as shown in Fig. 23-56. Clean, set up, and tack the pieces of pipe as previously directed. The root pass is welded with a $\frac{1}{8}$-inch electrode and a current setting of 90 to 125 amperes. The vertical-up and vertical-down directions of welding should be practiced. Restarting the weld bead after a stop may cause you some trouble. Practice the procedure outlined on pages 752, 753.

Root Pass: Vertical-Up Method. Start just below the 2 o'clock point and weld uphill to the 12 o'clock point. Make sure that you are getting penetration through the back side of the pipe. The back side and face side of the bead should be flat (without undercut or excessive buildup) and show uniform fusion without undercut.

Rotate the pipe clockwise so that the weld crater at the end of the completed weld bead is just below the 2 o'clock point. After making sure that the crater is clean, restart the weld bead and weld uphill to the 12 o'clock point. Figure 23-57 shows the pass sequence on butt-welded pipe. There is a hole formation at the tip of the root pass. This is referred to as the "keyhole." If the keyhole is allowed to get too big, it will cause internal undercut or burn through. The size of the keyhole is affected by the amount of current, electrode angle, and electrode pressure. Excessive current must be avoided. It is necessary to achieve this formation as you weld in order to secure complete penetration on the inside of the pipe, Fig. 23-58. Proper bead formation inside the pipe insures maximum strength and the inside of the pipe is smooth to permit the free flow of materials. Clean the weld thoroughly.

Root Pass: Vertical-Down Method. Start at about the 12 o'clock point and weld downhill to just below the 2 o'clock point. Rotate the pipe counterclockwise so that the crater of the previous weld is at the 12 o'clock point. After making sure that the crater is clean, restart the weld bead and weld downhill to the 2 o'clock point. Make sure that the slag does not run ahead of the weld and prevent fusion. Check the weld for the characteristics described under the Vertical-Up Method. Clean the weld thoroughly.

Filler and Cover Passes. The direction of welding and rotation of the pipe are the same as for the root pass. Filler and cover passes, however, may be made with $\frac{5}{32}$-inch electrodes and a current setting of 120 to 160 amperes. All passes must be cleaned thoroughly before proceeding with the next pass. Each pass must fuse and penetrate the underneath pass and the side walls of the pipe. Make sure that the pool fills up at each side of the weld and does not leave an undercut. Travel must be rapid enough to avoid an excessively convex contour. Every effort should be made to produce welds with a flat face.

Inspection and Testing. After the weld has been completed, clean it by brushing. Inspect the weld for the characteristics listed in Inspection, Job 21-J40, page 658. Compare the weld appearance with Figs. 23-57 and 23-58 and note any defects in your weld.

Continue to practice this joint until you can consistently produce sound welds of good appearance. When you feel that you have mastered the welding of

23-57. *Vertical weld in a butt joint in pipe welded in the horizontal fixed position (travel up). Note the sequence of passes.*

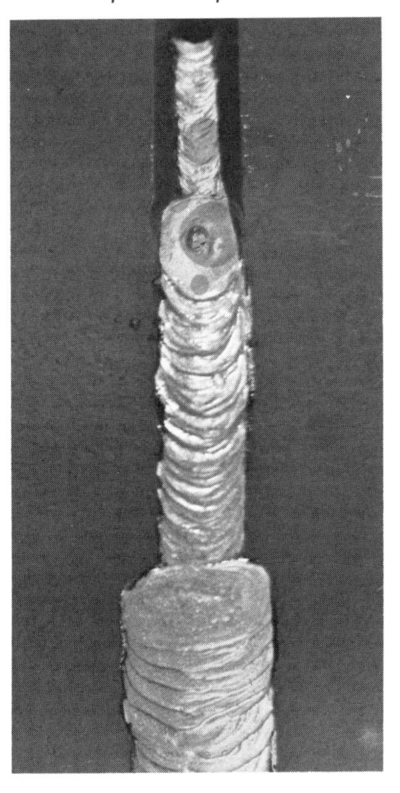

23-58. *Root penetration on the inside of the pipe. Note that there is complete fusion and penetration without excess burn-through.*

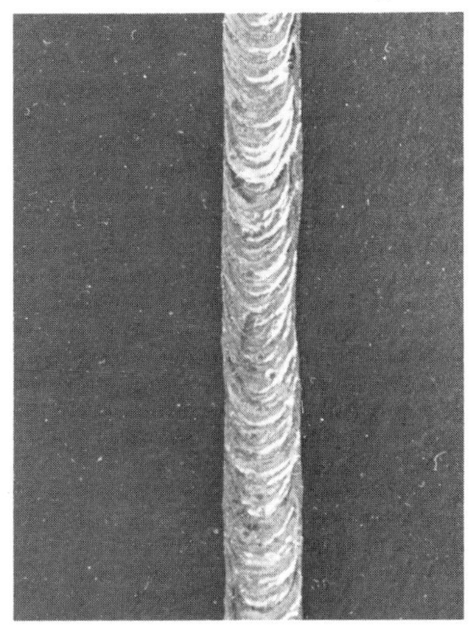

this joint in the semi-flat position, cut two test coupons. Remove the test coupons as instructed in Fig. 23-36, page 744, and subject them to face- and root-bend tests as described under Root-, Face-, and Side-Bend Soundness Tests, pages 380, 381. Examine the bended coupons carefully for signs of lack of fusion and slag inclusions. Any opening more than $\frac{1}{8}$ inch long in any direction is considered a failure.

HORIZONTAL BUTT WELD ON PIPE IN THE FIXED VERTICAL POSITION: JOBS 23-J3 AND J4. Before proceeding with Jobs 23-J3 and 4, review the procedures and photographs for Job 22-J60, page 701. Prepare the two pieces of pipe for welding as shown in Fig. 23-59. Set up the joint and tack it as previously instructed.

Root Pass. The root pass and the two intermediate passes are made with an $\frac{1}{8}$-inch electrode and a current setting of 80 to 110 amperes. You may use $\frac{5}{32}$-inch electrodes if you can handle this size. Insert the electrode well into the bottom of the groove. Direct the arc toward the top and bottom bevel of the joint at the same time in order to obtain fusion. Manipulate the electrode with a slight circular or back-and-forth motion and play most of the heat on the upper bevel. The angle of electrode should vary from 5 to 15 degrees from the horizontal and lean from 30 to 45 degrees in the direction of welding, Fig. 23-60. Control the puddle by making small changes in the electrode angle as you proceed around the pipe.

23-59. *Joint preparation in the vertical fixed position.*

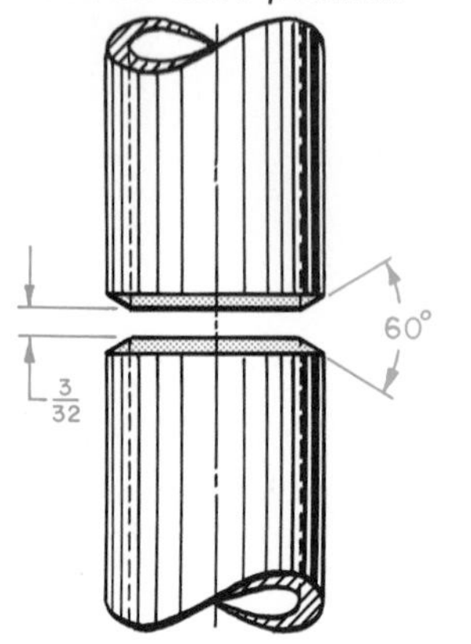

As the weld progresses, inspect the root pass for signs of not enough penetration, too much penetration, undercut, and poor fusion. If you observe any of these faults, check the speed of travel, the angle of the electrode, the heat setting, the arc gap, and the welding technique.

A procedure may call for a hot pass to be run as the second pass. This is needed if the root pass is small or if it has a number of minor defects in the face of the weld. The hot pass will cause more metal to be deposited in the groove and will eliminate the defects.

Filler Passes. Make intermediate passes like those in Job 22-J60. The second and third passes are usually made with a $\frac{1}{8}$-inch electrode although a $\frac{5}{32}$-inch size may be used. The remaining intermediate passes should be made with $\frac{5}{32}$-inch electrodes at 120 to 160 amperes. In making these passes, be sure that complete fusion is obtained with the underneath passes and the wall of the pipe. The second pass must be fused with the root pass and the lower pipe bevel. The third pass is placed immediately above the second pass and must be fused to the second pass, the root pass, and the upper pipe bevel. Each layer is started at the lower side of the groove and built upward.

If the travel is too slow or the heat setting too low, poor fusion and excessive buildup will occur. If the travel is too fast or the current is too high, poor bead formation and undercut are likely to occur. Care must be taken to eliminate deep valleys along the edges of each bead as it is applied. Valleys are caused by convexity in the center or poor fillup along the edges of the weld. They are hard to clean and increase the chances of trapping slag. Careful cleaning is essential.

Cover Pass. The final layer should be practiced with both the stringer bead technique and the weaved bead technique required for Job 20-J27, page 628. Procedure qualifications differ.

When practicing stringer beads, be sure to start each bead in a slightly different area around the pipe so that you do not start and end each bead in the same location. Weld with a $\frac{5}{32}$-inch electrode and a current setting of 120 to 160 amperes. Undercut is a problem along the weld edges, especially along the edges of the first and last stringer bead. This can be overcome by careful attention to the speed of travel and current adjustment.

In practicing the lacing bead technique, be sure to pause at the top of the weave in order to prevent undercut. Sag is also a problem and can be prevented by weaving at an angle and a rapid advance toward the bottom of the weave.

Inspection and Testing. After the weld has been completed, clean and inspect it as instructed for Jobs 23-J1 and 2. Compare it with Figs. 23-61 and 23-62. After you have made a number of these welds and they seem satisfactory on visual inspection, cut some test coupons and subject them to face- and root-bend tests.

BUTT WELD ON PIPE IN THE FIXED HORIZONTAL POSITION (BELL-HOLE) (TRAVEL-UP): JOB 23-J5. Two welding procedures are used to weld pipe in the fixed horizontal position. On building construction the usual method is to start the weld at the bottom and progress upward. For pipeline work and for pipe of thin or medium thickness, the weld is started at the top, and progress is downhill. Heavy pipe for pipelines may be welded travel-up. You will be required to learn both procedures. Before proceeding with Jobs 23-J5 and 6, review the procedures and photographs for Job 22-J57, (pages 694–697), and Job 22-J65 (pages 710–713). Prepare the two pieces of pipe as shown in Fig. 23-56, page 754. Set up and tack the joint as previously instructed.

Root Pass. Practice welding from the bottom to the top. The root pass is the most important weld that must be made in completing this joint. Fusion must be complete on both pipe sections, and penetration must be through the inside of the pipe, Fig. 23-63. If there is insufficient buildup inside the pipe, the weld will fail in service and also when subjected to the root-bend test. Too much buildup on the inside will restrict the flow of materials inside the pipe and may also cause failure of root-bend specimens under test.

The root pass should be made with a ⅛-inch electrode and a current setting of 80 to 110 amperes. Generally, best results are obtained by using currents on the low side of the range.

The root pass should never be started at the absolute bottom of the pipe since this is the location from which one of the test coupons is taken, and the possibility of failure is increased. Start the root pass at either the 5 o'clock or 7 o'clock position and proceed across the bottom to the top of the pipe joint, Fig. 23-64. Never stop at the top center of the pipe since this is another test location. Stop either at the 1 o'clock or the 11 o'clock position. Take care in welding over the tack welds in the joint. They must be completely fused and become a part of the weld bead. Chip out or grind any unsound or large tacks before welding the root pass. As you carry the pass around the pipe, a small hole (keyhole) of about $^{3}/_{16}$ inch in diameter should precede the weld pool, Figs. 23-65 and 23-66. Make sure the hole is not too large. The hole at the root of the joint is caused by the complete melting down of the bevel edge. It can be maintained by the use of a circular movement of the electrode tip or by moving back and forth in a straight line. The angle of electrode should be about 10 degrees in the direction of travel as shown in Fig. 23-67.

Changing Electrodes. You will probably have some trouble at first with restarting after a stop to change electrodes. The bead inside the pipe will have a crater or depression at the end which is lower than the inside wall of the pipe and lower than the bead already laid at the point where the new weld must join the weld just completed. This can cause a future failure in the pipeline when it is in service and certainly will cause failure of the root- and side-bend test specimens. The following precautions are necessary to make a smooth, uninterrupted bead on the inside wall of the pipe at a start-and-stop point.

1. When an electrode is used up, enlarge the hole (keyhole) preceding the weld pool to ¼ inch in diameter before breaking the arc, Fig. 23-68 (Page 760).

2. In restarting the weld bead with a new electrode, strike the arc ½ to 1 inch back on the previous weld and run slowly up to the end of the weld. This heats the weld metal and the base

23-60. *Angle of the electrode for a horizontal groove weld in pipe in the vertical fixed position.*

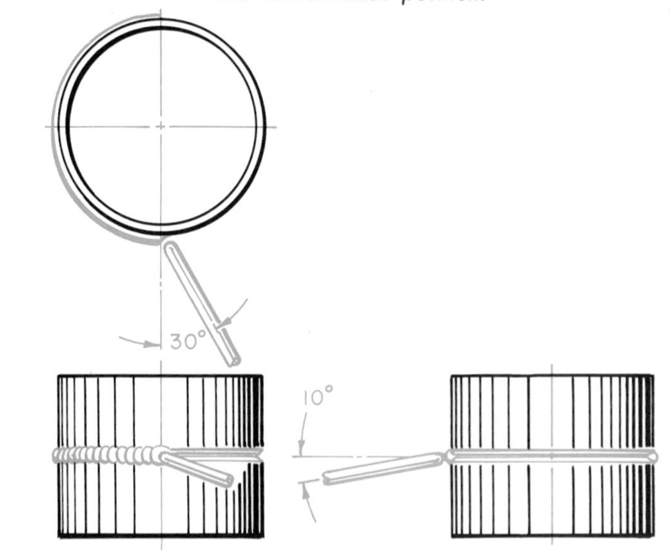

metal enough to insure a hot and fluid start, Fig. 23-69.

3. As you reach the hole at the end of the previously laid weld bead, insert the tip of the electrode clear through the hole into the pipe and incline the electrode at an angle of 45 degrees in the direction of travel. Maintain this position only long enough to deposit new weld metal on the end of the previous weld bead on the inside of the pipe, Fig. 23-70. Resume the normal electrode position and proceed as before.

4. Inspect the back side of the completed root pass carefully. It should be no more than $\frac{1}{16}$ inch higher than the inside wall of the pipe, and it should have good fusion along the edges of the weld. There should be no evidence of icicles, grapes (globular deposits of metal), or undercut inside the pipe. The face of the pass should be fairly flat with smooth, even ripples, and there should be no undercut along the weld edges.

Filler Pass. Before making the next pass, clean the face of the root pass carefully. Chip off any high spots or weld spatter, and chip out any holes or depressions along the edges of the weld. Start the second bead in the same general area as the first bead, but try to avoid the exact starting point. Start on the opposite side of the joint or just ahead of or in back of the previous starting point.

The single filler or intermediate pass is applied with a weaving technique using a $\frac{5}{32}$-inch electrode a current setting of 120 to 160 amperes.

The current should be hot enough to insure complete fusion with the face of the root pass and both beveled faces of the pipe. You should watch for any tendency to undercut or to trap slag in the weld. If either defect is not remedied, it can become the cause of failure of the weld joint when in service. Avoid piling up the weld metal. The face of the weld bead should be flat, and the bead should not be thicker than $\frac{1}{8}$ inch. In joining the bead from one side to that deposited from the other side of the pipe, be sure to fill the crater by slowly withdrawing the electrode and moving backward over the finished bead for a short distance.

Cover Pass. The third or cover pass is made with a $\frac{5}{32}$-inch electrode and a current setting of 120 to 160 amperes. Be sure to clean the preceding weld and chip off spatter and excess weld metal. The cover pass is also a weave pass, and the welding technique is similar to that used for the second pass. Avoid undercutting by hesitating at the

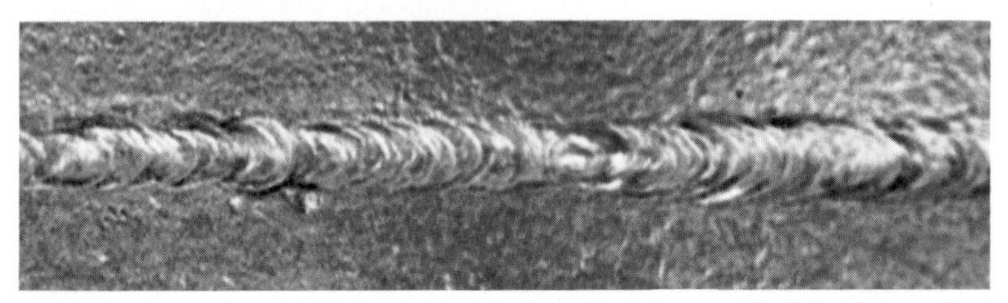

23-61. *Root penetration on the inside of a pipe welded in the vertical fixed position.*

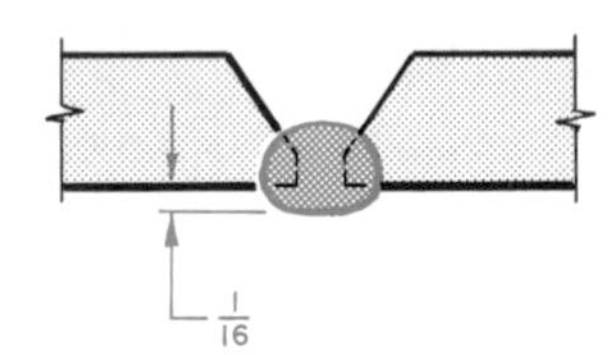

23-63. *Proper contour of the root pass.*

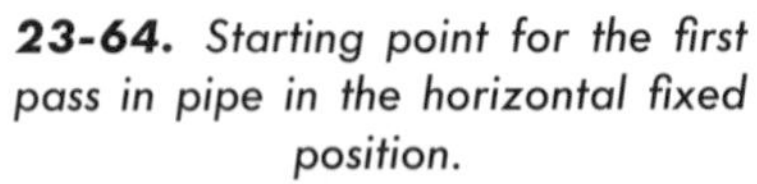

23-64. *Starting point for the first pass in pipe in the horizontal fixed position.*

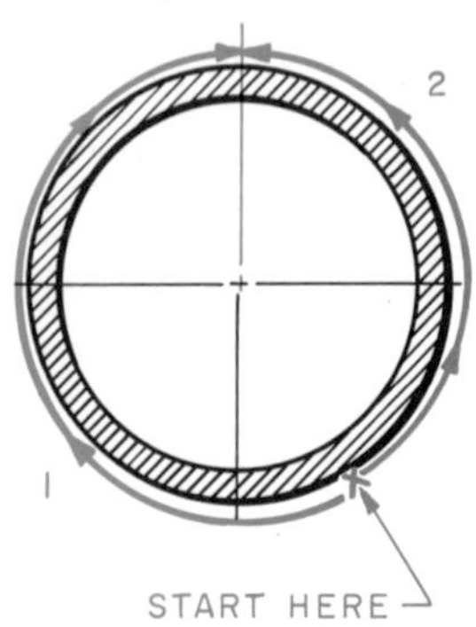

23-62. *Cover passes on pipe welded in the vertical fixed position with stringer bead technique.*

sides of the weld as you weave. There is also the tendency to widen the pass as you proceed with the weld. The weld bead should extend about $\frac{1}{16}$ to $\frac{1}{8}$ inch beyond the edge of the groove on each side. The face of the weld should have an oval shape with the bead no higher than $\frac{1}{16}$ inch above the surface of the pipe at the center of the weld. The edges of the weld should flow smoothly into the surface of the pipe. Note the even width and close ripples of the bead shown in Fig. 23-71.

Inspection and Testing. After you have completed this weld, use the visual inspection procedures described previously. The weld should look like that shown in Fig. 23-72. After you have made a number of these joints that are visually satisfactory, cut a few test coupons from the joint and subject them to the usual face- and root-bend tests.

BUTT WELD ON PIPE IN THE HORIZONTAL FIXED POSITION (TRAVEL-DOWN): JOB 23-J6. This job involves practice on the same pipe joint as in Job 23-J5, but welding is from the top to the bottom. Vertical-down welding is a cross-country pipeline technique. (Study the photographs of the rolling butt weld made in Job 23-J2, page 755.)

Root Pass. Here again you are cautioned to take extra precaution in making the root pass. Always keep in mind that the root pass is the basis for success or failure in making a test weld. It is also the basis of a sound weld. The root pass should be made with a $\frac{5}{32}$-inch electrode and a current setting of not more than 200 amperes nor less than 150 amperes.

Correct joint preparation is important. Internal undercut will occur if the root face is too small, the root opening is too large, or a severe high and low condition of the pipe ends exists. Refer to Table 23-3, page 734. Joint preparation determines the welding heat. A small root face and large gap requires a low heat setting. A large root face and minimum gapping requires a high heat setting.

Start the pass at the 11 o'clock or 1 o'clock position. Weld across the top of the pipe and downward past the 6 o'clock position to the 7 o'clock or 5 o'clock position, depending upon the side you are welding from. Make the stringer bead with a drag technique. Rest the elec-

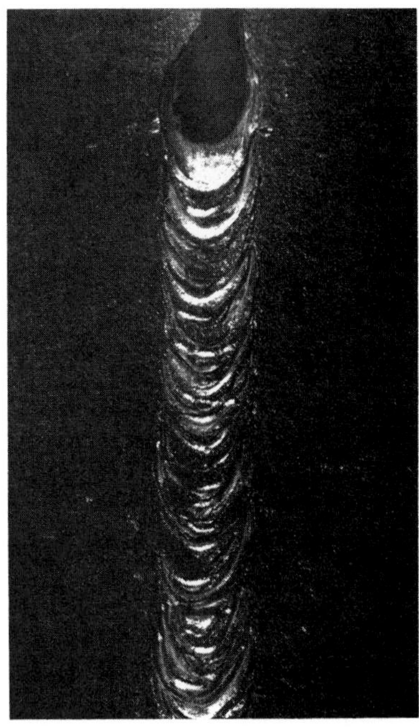

23-66. Back side of the root pass welded in the horizontal fixed position. Note the hole formation.

23-65. Method of applying the root pass.

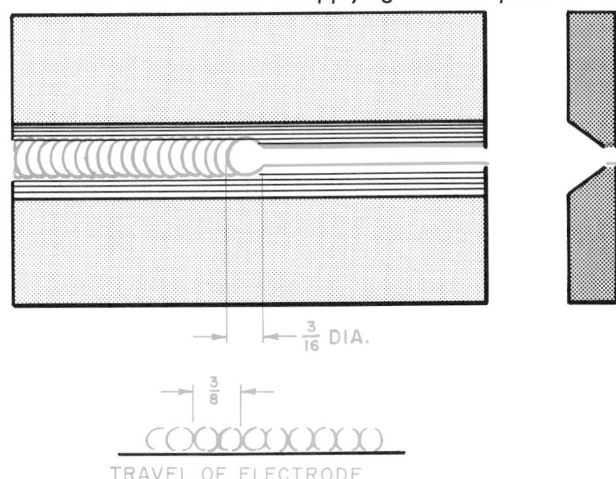

23-67. Angle of the electrode for pipe in the horizontal fixed position.

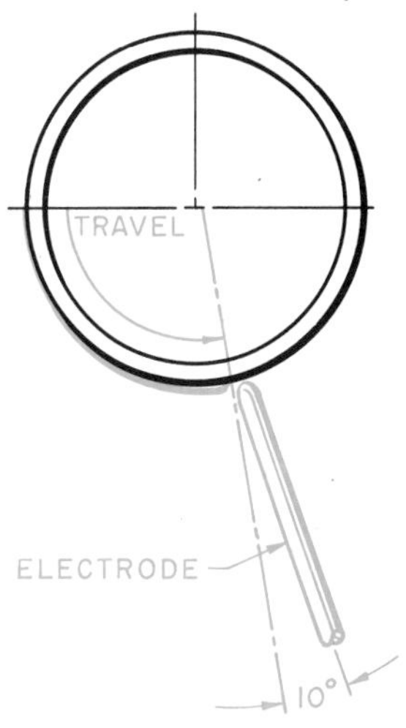

trode coating on the bevel as you drag the electrode downhill around the pipe.

The electrode angle should be about 30 to 45 degrees from the horizontal center line, and it must vary little as you travel around the pipe. Maintain a close arc but be careful that you do not push the electrode through the back side of the pipe.

To secure adequate penetration on the inside of the pipe, it is necessary to maintain a small visible keyhole at all times. If the hole becomes too large, it will cause internal burn-through, sag, or undercut. Make sure that enough heat is applied against the cold bevel of the pipe to insure good fusion.

The speed of travel is faster than that used in welding uphill, and it changes from time to time, depending upon the joint condition and weld formation. You can also maintain control of the weld pool by varying the arc gap, the speed of travel, electrode angle, and electrode movement. After the root pass is completed on one side of the pipe, weld the other side in the same manner. Be sure to overlap the weld on the other side about ½ inch at the top and bottom.

One method of determining whether or not there is complete penetration is to listen to the sound of the arc. A sound somewhat like compressed air being released inside the pipe may be heard when penetration is complete. Insufficient penetration or lack of fusion may be caused by an excessive root face, a root opening that is not wide enough, or too low a welding current. In this event current settings will have to be increased.

23-68. *The hole at the front edge of the weld pool is enlarged before breaking the arc at the finish of the electrode.*

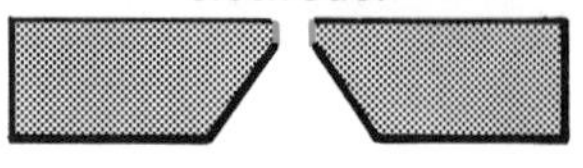

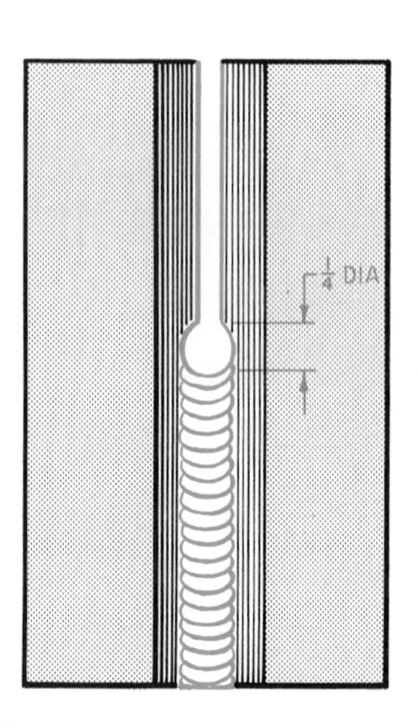

23-69. *Method for starting the new electrode at the end of the previous weld bead.*

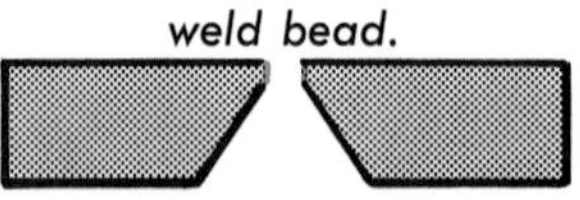

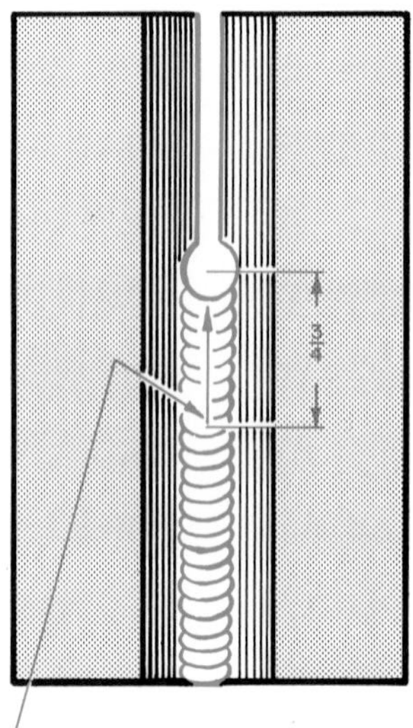

Insufficient gap or low welding current can result in lack of penetration through the root of the joint. Other defects that can occur are lack of fusion between weld and base metal, burn-through, and globular deposits (*grapes*) on the inside of the pipe.

Burning away the side walls of the groove results in undercut. External undercut along the edges of the weld bead is called *"wagon tracks."* Undercut can also occur on the inside of the pipe. Internal undercut can be repaired only from inside the pipe which is either impossible or a costly procedure.

If a tendency to undercut is observed while welding, it can be overcome by tilting the electrode just a few degrees toward the undercut side. Burn-through, metal deposits on the inside of the pipe, and external and internal undercut are usually caused by too high a current, too wide a root gap, or too little root face. Holes and slag pockets are common faults which can be avoided only by the correct heat setting and complete control of electrode manipulation. When inspecting the root pass, look for the same satisfactory characteristics discussed previously and refer to Fig. 23-72.

Hot Pass. Be sure to remove all slag before making the hot pass (second pass). Undercut along the edges of the root pass makes slag removal difficult. An old power hacksaw blade, taped on one end to serve as a handle, is a good cleaning tool. The bead should be thoroughly brushed, preferably with a power brush. Any surface defects or excessive spatter should be removed by chipping.

The second pass not only helps to fill up the groove, but it also reinforces the root pass. The hot pass is welded with a current setting similar to that used for the root pass: about 160 to 190 amperes. Make the second pass with a 5/32-inch electrode or a 3/16-inch electrode if you can handle the larger size. The hot pass should be started within 5 minutes of the completion of the root pass.

Welding should start at the top of the joint outside of the area of the previous starting point, proceed downhill, and stop at the bottom outside the area of the previous stopping point.

The angle of the electrode should be about 30 to 45 degrees down from the horizontal position. A slight electrode movement is necessary, especially in the 4 to 8 o'clock sections of the pipe. A close arc gap, uniform speed of travel, and a steady electrode angle are extremely important. Failure to do so causes holes and permits the slag to run ahead and extinguish the arc.

The hot pass must be applied with enough heat to insure good penetration into the root pass. This is necessary to burn out wagon tracks and to float any remaining slag to the surface. Too fast a rate of travel results in surface pinholes.

Make sure that the weld is started and finished away from previous starting and stopping points and outside the area from which the test coupons are going to be cut.

Stripper Pass. Welds made with the downhill welding technique may be thin at the 2 to 4 o'clock and the 8 to 10 o'clock positions. A weld pass known as a *stripper pass* is used to build these sections up to the same height as the rest of the weld bead. Use a welding technique similar to that used in making the hot pass and a slightly longer arc.

Filler Passes. Since the beads applied when welding downhill are thinner than beads welded in

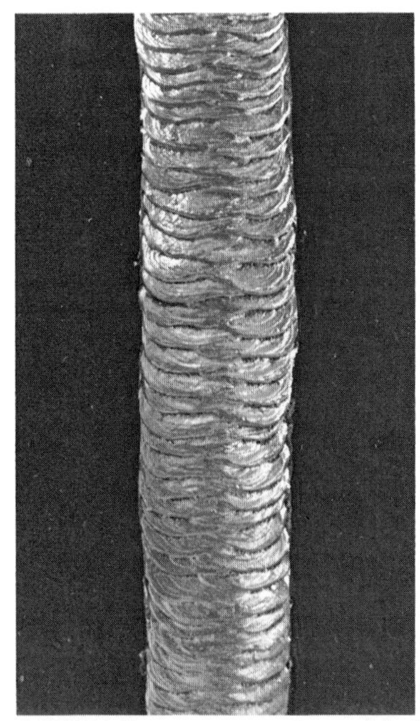

23-71. *Cover pass on pipe welded in the bell-hole (horizontal fixed) position.*

23-70. *The method used in restarting the weld. Weld metal is deposited at the end of the previous weld on the inside of the pipe.*

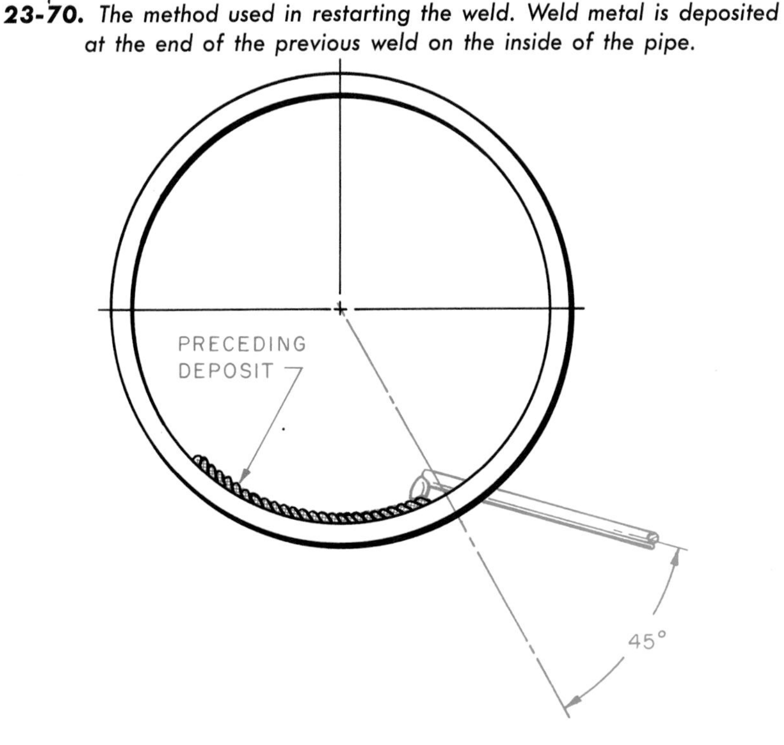

23-72. *Root penetration on the inside of a pipe welded with the vertical-down technique.*

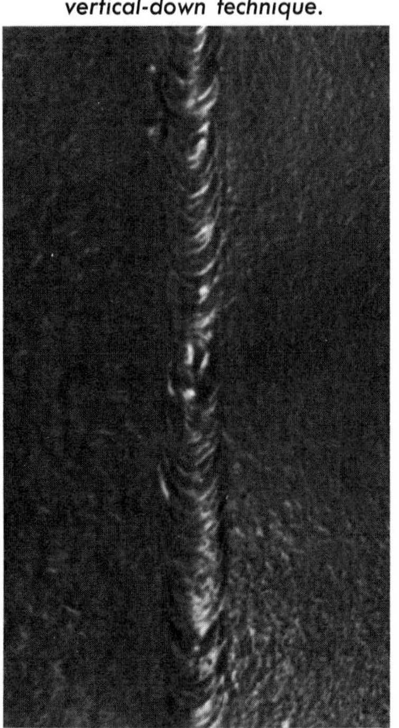

the vertical-up direction, additional passes are necessary to complete the pipe joint. The tendency for these beads to be concave in the center may cause problems for the other passes. This can be overcome by reducing the heat used, by slowing the rate of travel, or by a slower weave motion.

The filler passes should be made with a $^{3}/_{16}$-inch electrode and a current setting of 170 to 210 amperes. Pay close attention to the angle of electrode and the length of the arc gap. Use a slight side-to-side weave and make sure that the weld deposit fills the groove and fuses into the side walls. Improper electrode manipulation or too high a rate of travel can cause porosity, undercut, and a concave face. Travel that is too slow causes the weld deposit to pile up and run ahead of the slag.

Cover Pass. This pass is also referred to as a cap pass. Before applying the cover pass, it is sometimes necessary to make the concave portions of the weld flush with the stripper passes. Concavity usually occurs in the 2 to 4 o'clock and 8 to 10 o'clock positions on the pipe.

The cover pass is made with a $^{5}/_{32}$-inch electrode and a current setting of 130 to 160 amperes. A $^{3}/_{16}$-inch electrode may be used with a heat setting of about 160 to 180 amperes. Be sure to start and stop in an area outside of the starting and stopping points of the previous weld bead. A weave motion with some hesitation at each side is used to prevent undercut. Take great care and maintain the same electrode angle, arc gap, and speed of travel practiced previously. The finished weld should be from $^{1}/_{32}$ inch to $^{1}/_{16}$ inch higher than the pipe wall and should overlap the groove by $^{1}/_{16}$ inch to $^{1}/_{8}$ inch on each side. Better stress distribution is obtained by keeping the cover pass as flat and narrow as possible. A narrow bead is also less susceptible to surface porosity. If the bead is wider than $^{3}/_{4}$ inch, use two weave beads around the pipe.

Inspection and Testing. After you have completed this weld, use the visual inspection procedures described previously. The root pass should look like that shown in Fig. 23-72, page 761, and the cover pass should look like that in Fig. 23-71, page 761. After you have made a number of these joints and they are visually satisfactory, cut a few test coupons from the joint and subject them to the usual face- and root-bend tests.

PIPE FABRICATION PRACTICE: JOBS 23-J7 THROUGH J17. Because of the availability of manufactured fittings for practically every installation purpose, the amount of hand fabrication performed in the field is kept at a minimum. Custom fittings are more costly and do not meet the service requirements of pipe fabricated with manufactured fittings. Since it is still a practice in the field, it is necessary for the student welder to practice this

23-73. *A 45-degree lateral in 20-inch diameter pipe. Note the lacing technique for the cover pass.*

Crane Co.

23-74. *A 45-degree branch in extra heavy pipe with a diameter of 36 inches.*

Crane Co.

type of construction. It is not the purpose of this book to teach the layout necessary to hand fabricate fittings. It is assumed this will be taught in a related class using a textbook that presents all phases of pipe fabrication. The cost of manufactured welded fittings is very high, and this rules out their use for school practice of *Ts, Ys,* and laterals.

The 90-degree branch connections and 45-degree lateral connections (Jobs 23-J7 through J12) are merely a series of lap and fillet welds. The manufacturer's procedure determines whether or not beveling is required and whether the branch connection or the hole in the header is beveled to insure complete fusion into the inside of the pipe. Welding should be done with $\frac{5}{32}$-inch electrodes. The weld is started at a point about 2 inches on either side of the intersection at the heel of the joint. The arc should be in continuous operation when passing over the intersection and the stert of the bead. Branch and lateral connections may be made with either the stringer bead or the weaved bead technique and a lacing pass as the finish pass.

The welding of 45- and 90-degree *Ls* and other fittings (Jobs 23-J13 through J17) are groove welds, and the welding technique is similar to that used in making butt welds in straight pipe sections.

Use the visual inspection techniques described previously and strive for welds that meet these standards. You may also cut out test coupons from the fillet lap and butt sections of the joints for macro-etching. The specimens must be ground, polished, and etched with acid. Inspect them thoroughly for such flaws as lack of fusion at the root and to the surface of the pipe, slag inclusions, gas pockets, and undercutting.

Figures 23-73 through 23-78 illustrate some of the various pipe units that can be fabricated in a modern pipe-fabricating shop.

TOOLS FOR PIPE FABRICATION

Although every effort is made to fabricate as much pipe as possible in the shop, there is still a considerable amount done in the field. The following power and hand tools speed up this work and improve the accuracy of the work.

Power Tools

PIPE SAWS. Pipe saws consist of a motor driving either a circular saw blade or a milling cutter through a friction clutch with a worm and gear train. The motor and gears of the saw are totally enclosed so that water that usually collects in a trench does not damage the working of the machine.

Pipe saws will cut cast iron and steel pipes used for sewage, gas, water, and other systems. They can be used for cutting pipes to length before laying and also for cutting pipes already laid. The cutting of finished pipe frequently has to be done under very difficult conditions, and it must be done as quickly as possible when fractures occur.

Cuts are smooth and clean. The ends of the pipes are free from flaws and cracks. The cutting tool does not dig in, and the

23-75. A cap joined to the end of a pipe and a 90-degree branch welded into a header. A saddle is set over the weld and will be welded to the branch. Crane Co.

23-76. Tacking up a fabricated L. It is 18 inches in diameter and has a wall thickness of 2 inches. Crane Co.

saw requires little space in which to operate, Fig. 23-79.

PIPE-BEVELING MACHINES. Pipe ends that are to be welded require a smooth, metallically clean surface that is free of oxide. All types of bevels can be prepared economically and so that they are metallically clean with the new beveling machine, Fig. 23-80. These machines can be used to bevel both plate and pipe. The welding edge to be prepared may be straight, convex, or concave, and it does not matter whether it runs in a vertical, horizontal, or any other direction. Pipe-beveling machines are available as electric or pneumatic tools. An oxyacetylene pipe-beveling machine may be made in the school shop. (See Fig. 16-12, page 475.) The driving power is an old worm-gear arrangement, the rolls are cold-rolled shafting, and the bearing ends are old welder bearings or automobile axle bearings. The torch mechanism may be purchased from any welding equipment distributor.

Figures 23-81 and 23-82 are photographs of commercial oxyacetylene pipe cutting and beveling machines. These portable machines are built to cut and bevel any size pipe with speed, economy, and accuracy, either in the field or in the shop. A welding student is using a commercial pipe-cutting machine in Fig. 23-83.

PORTABLE POWER HACKSAW. Power hacksaws, Fig. 23-84, are highly portable tools which make it possible to cut pipe at any location. They are electric powered and pneumatic powered for areas with explosion hazards. Saw blades are made of high speed steel and super high-speed steel for stainless steel and tough ferrous metals. The machine can be used for bevel-cutting angles up to 45 degrees on pipe 5 inches and smaller. Straight cuts can be made on pipe 12 inches in diameter. These machines are of particular value when space is limited and when it is difficult to cut materials by hand.

CUTTER PIPE SAW. The cutter pipe saw, Fig. 23-85, travels around the pipe and is adjustable to all pipe sizes 6 inches through 72 inches. The machine is held to the pipe by a tensioned timing chain. The chain acts as a flexible ring gear and provides the guide and feed for the machine. It can cut and bevel all grades of steel. The fact that it is powered by air motors makes it possible to use underwater. It can both cut and bevel for weld preparation in only one trip around the pipe.

ABRASIVE POWER SAW. The abrasive power saw, Fig. 23-86, is a light gasoline-powered tool. It cuts 4-inch to 12-inch iron, cast iron, and cement-lined pipe. With a diamond abrasive wheel, Fig. 23-87 (Page 768), it is highly useful in cutting asbestos-cement pipe.

23-77. *Cross section of an elbow with an outside diameter of 20 inches and a wall thickness of 4 1/16 inches. It was formed from chrome-molybdenum plate, and 150 passes were required for welding the seam.*
Crane Co.

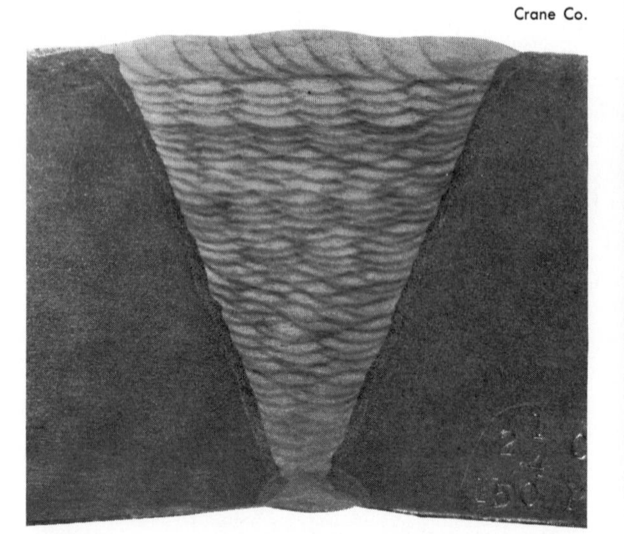

23-78. *A large diameter Y with walls 2 inches thick during fabrication. Note the series of multipass welds.*
Crane Co.

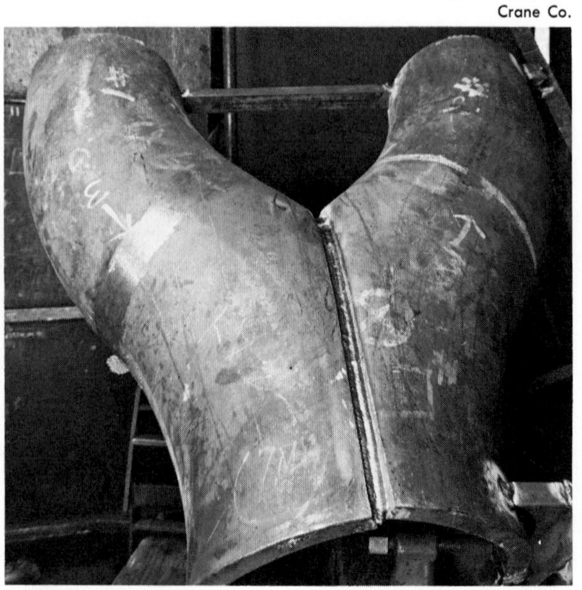

GUILLOTINE PIPE SAW. The guillotine pipe saw, Fig. 23-88, is a highly portable cutter that speed cuts in tight places. It is available with an electric or pneumatic motor. The saw can be used to cut steel, cast iron, stainless steel, and alloy steel pipe as well as bar stock, structurals, and rail.

A chain pipe vise clamps the saw to pipe so that it is ready for cutting in a matter of seconds. The machined cast-steel V-saddle base assures a square cut at right angle to pipe. The guillotine pipe saw can cut from 2 to 8 inches in diameter.

BEV-L GRINDER.® The Bev-L Grinder,® Fig. 23-89, is a fast, accurate grinding machine that produces clean-faced, uniform weld bevels without causing any change in the physical properties of the pipe. It is possible to get good line-up and fit in the field. The cutting action is fast, accurate, and easy to control. The operator simply rotates the grinder head around the spindle.

The grinder is available with an electric or pneumatic motor. It can be used on all steel, stainless steel, high alloy, and aluminum pipe from 3 to 18 inches in diameter.

23-79. *Cutting a large diameter pipe with the automatic pipe saw.*

Widder Corp.

Hand Tools

HAND CUTTING TORCH. A great deal of pipe welding is done in maintenance work and in the renewing of piping systems. The standard oxyacetylene hand cutting torch is an ideal tool to take out sections of pipe that must be replaced. You are again urged to practice hand cutting as much as possible in order to develop and perfect your skill, Fig. 23-90.

PIPE LAYOUT TOOLS. A number of tools have been developed to make the job of laying out and fabricating pipe faster and with a greater degree of accuracy. These tools prevent layout mistakes and pipe that is out of alignment.

Contour Marker. Figure 23-91 shows the contour marker being used to lay out a section of pipe. The marker prevents errors in layout and takes no longer than conventional layout tools, regardless of the type or size of the pipe joint to be welded. With an adaptor the four sides of the marker can be used on pipe up to 30 inches. With it the fabricator can work compound angles in one setting of the tool, and mark over nipples and old welds without having to break the wrap far back on wrapped pipe. The contour marker eliminates "cut and try" and the need for calculation.

Dial-Angle Flange Level. Figure 23-92 shows the dial-angle flange level being used to level a pipe flange. This tool insures that flange holes will be level and in alignment. It is a casting equipped with a protractor and a level and two $\frac{3}{4}$-inch pins held to the body of the level on a sliding scale with wingnuts. It can be used on any size flange.

Circle-Ellipse Projector. Figure 23-93 shows the circle-ellipse projector marking out a header to receive the branch. This tool is designed for marking perfect circles, ellipses, and oblongs. The tool can be used for work on pipe, tanks, boilers, cones, and any other irregular, flat, or round surface.

The projector has a strong magnetic base to hold it in position while the marking is being done. It has sliding rods with a protractor and locking lever to adjust it for the desired layout and a chalkholder for round or flat soapstone that will also receive a steel scribe or pencil.

The tool can be used on circles $1\frac{1}{2}$ to 18 inches in diameter, and it can be adjusted from 0 to 70 degrees.

Centering Head. Figure 23-94 shows the use of the centering head to locate the center of the pipe. The tool determines the center line at any degree and measures the degree of declivity. It has a dial-set level and center punch.

23-80. *A pipe-beveling machine preparing pipe with a heavy wall thickness.*

Widder Corp.

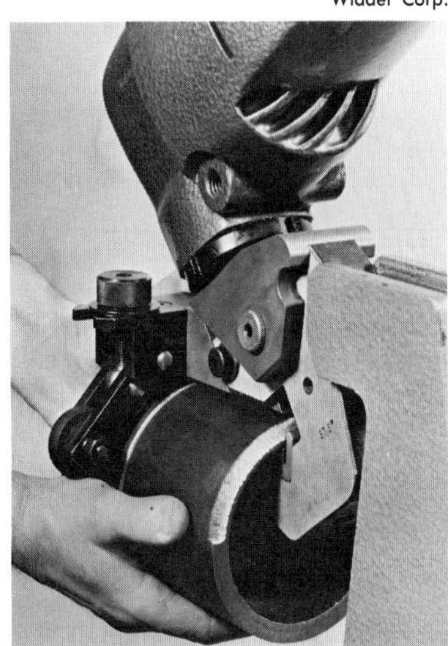

CRC-Crose International

23-81. *Cutting and beveling machine that can be used in both shop and field. An out-of-round attachment permits the torch carrier to follow the contour of the pipe. It may be operated manually or motor driven.*

23-83. *Welding student using a commercial pipe-beveling machine.*

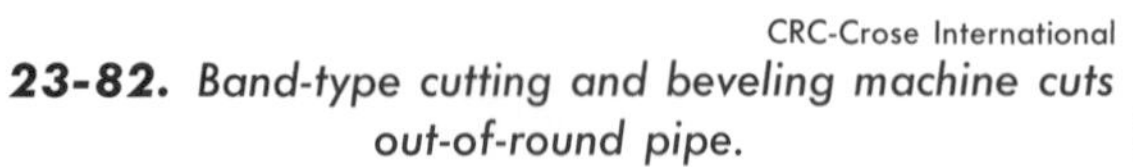

CRC-Crose International

23-82. *Band-type cutting and beveling machine cuts out-of-round pipe.*

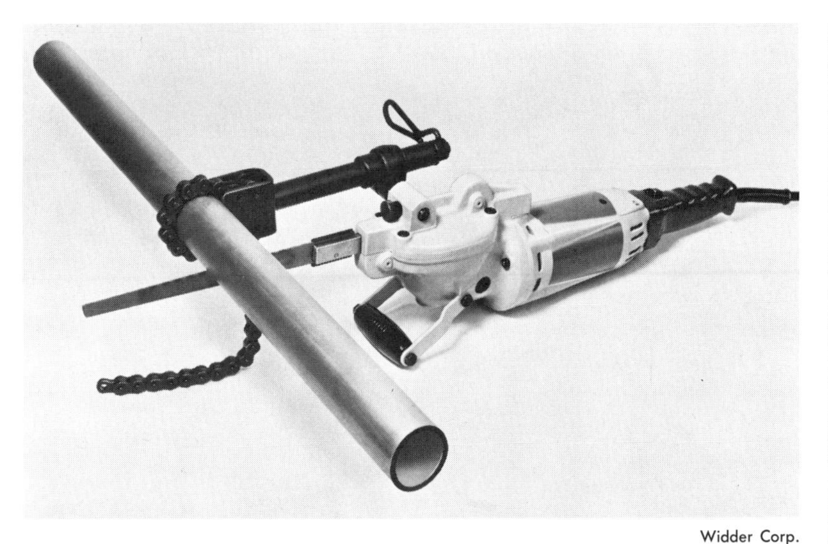

Widder Corp.

23-84. *A portable power hacksaw has the advantages of being light and highly portable. This one is shown with 3-inch stainless steel pipe.*

E. H. Wachs Co.

23-86. *A power saw with an abrasive cutting wheel.*

E. H. Wachs Co.

23-85. *Air-operated cutter and pipe saw. This tool cuts and bevels in one operation.*

E. H. Wachs Co.

23-87. *An abrasive power saw with a diamond cutting wheel.*

23-88. *This guillotine pipe saw cuts pipe 2 to 18 inches in diameter. It is highly useful for the field fabrication of stainless steel pipe.*

E. H. Wachs Co.

E. H. Wachs Co.

23-89. *The Bev-L-Grinder® for beveling pipe in the field.*

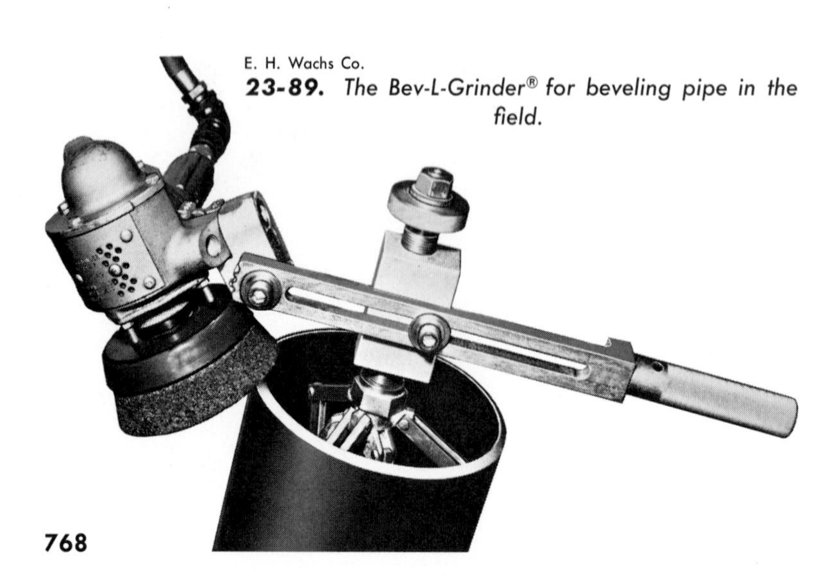

23-90. *Student using an oxyacetylene hand cutting torch to flame cut and bevel pipe. A great deal of field fabrication is cut with this method.*

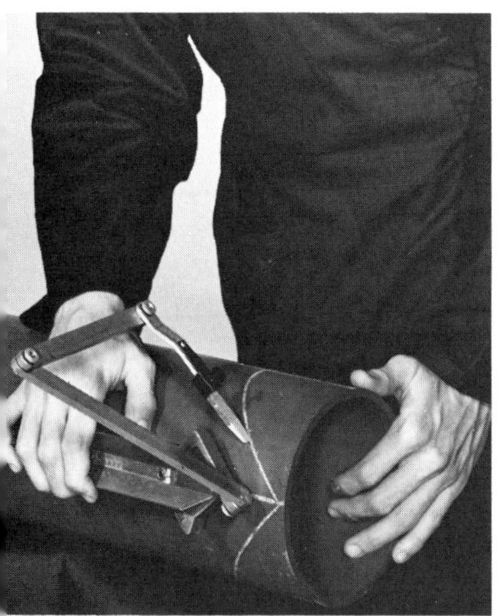

Contour Sales Corp.

23-91. *Contour marker for marking off pipe joints and structural angle cuts.*

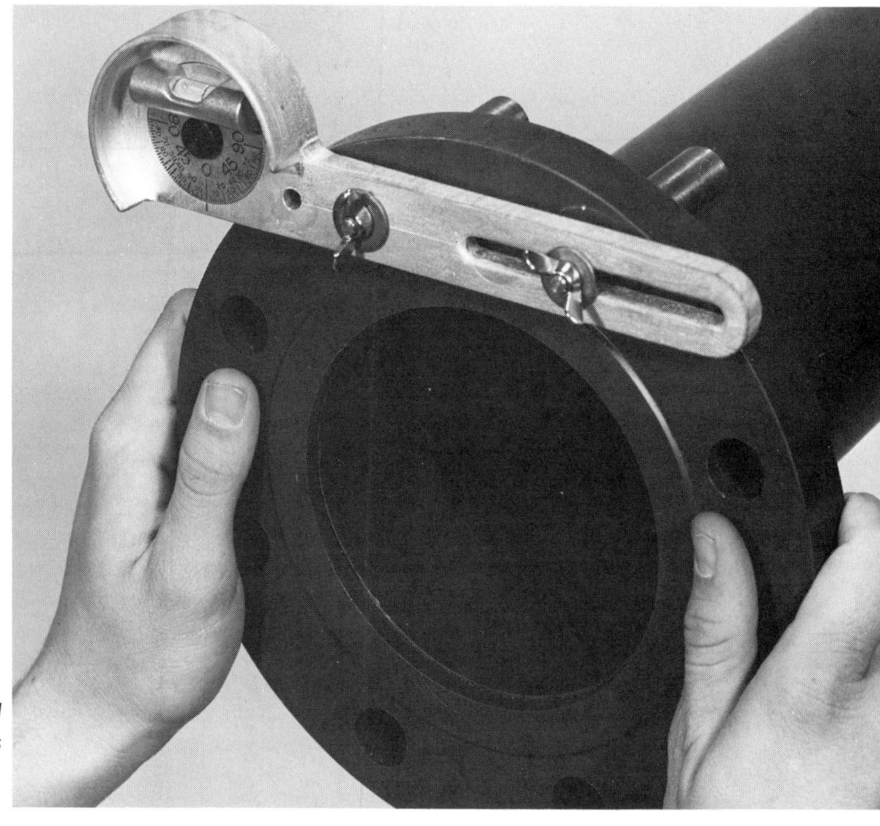

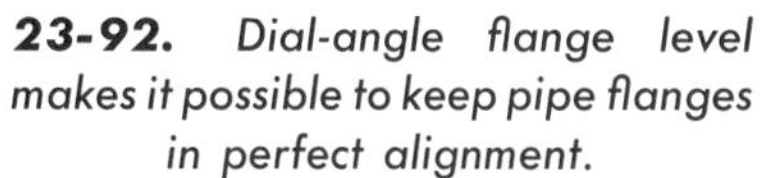

23-92. *Dial-angle flange level makes it possible to keep pipe flanges in perfect alignment.*

Contour Sales Corp.

23-93. *Circle-ellipse projector for marking circles, ellipses, and oblongs.*

Contour Sales Corp.

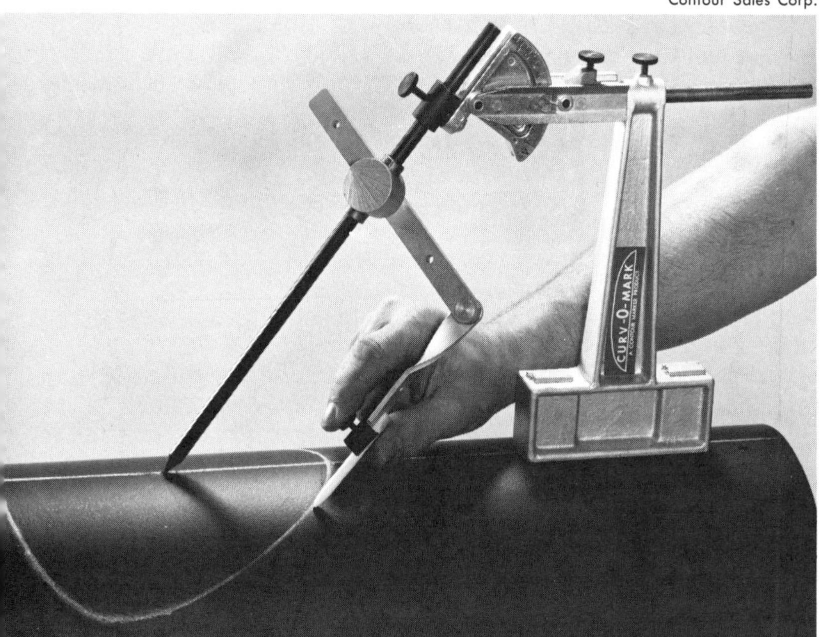

23-94. *Center head equipped with a dial-set level and center punch for locating the center line of pipe.*

Contour Sales Corp.

Contour Sales Corp.

23-95. *Wrap-A-Round® for marking off square pipe ends.*

Wrap-A-Round.® Figure 23-95 shows the Wrap-A-Round® being used for marking around a pipe. It is also used as a straightedge. This device assures square-cut pipe ends. It is made of flexible gasket material which is reasonably resistant to heat and cold. The Wrap-A-Round® is available in sizes for pipe 1 to 10 inches in diameter.

Miscellaneous Pipe Tools. The following tools are also very useful to the pipe welder in the fabrication of pipe in the field.

- *Fitter welder protractors* determine branch angles.
- Pipe-flange aligners automatically align with the pipe axis and align pipe flanges quickly.
- *Radius markers* mark circles.
- *Magnetic protractor levels* solve angle and setting problems.
- *Multi-trammel heads* make large circles.

REVIEW QUESTIONS

1. List five advantages of welded piping systems over threaded pipe.
2. The practice of welding pipe began about 1930. True. False.
3. At the present time oil is the only product that flows through pipe. True. False.
4. One of the disadvantages of pipelines is that only one type of product can be carried at one time because of cleaning problems. True. False.
5. List the stages in building a pipeline.
6. Name at least five materials used for pipe.
7. What types of joints are used in joining pipe? What angle of bevel is used for butt welds in pipe?
8. Name twelve types of manufactured pipe fittings.
9. Name the various types of fabricated pipe joints.
10. What tests are used for the physical testing of pipe joints? Indicate the number of coupons required and the section of pipe from which the test coupons are taken.
11. Name four of the various code authorities that are concerned with pipe welding.
12. Draw a simple sketch showing the welding procedure used for the following pipe thicknesses and positions of welding:
    a. Horizontal roll position, butt joint, $\frac{1}{4}$- and $\frac{3}{8}$-inch wall
    b. Horizontal fixed-position, butt joint, $\frac{3}{8}$- and $\frac{1}{2}$-inch wall
    c. Vertical fixed position, butt joint, $\frac{3}{8}$- and $\frac{1}{2}$-inch wall
13. Why is complete penetration necessary in welding a pipe joint?
14. What is the reason for the use of a backing ring in the welding of butt welds in pipe?
15. In welding a butt joint in the horizontal fixed position, indicate the starting point for each weld pass, assuming that three passes are necessary to complete the joint.
16. The best type of electrode to use for pipe welding is the E6012 classification. True. False.
17. The use of manufactured fittings is growing. True. False.

18. What is a stripper pass?
19. What is a hot pass?
20. Name at least six power tools used in the field in the fabrication of pipelines.
21. Name at least six hand tools used by the pipe welder.
22. Name at least seven different methods of testing of pipe welds.
23. Section I of the ASME code applies to the welding of boilers and pressure vessles. True. False.
24. Name at least five variables that require retesting of the welder if they are changed.
25. A welder who has been qualified on plate does not have to retest for pipe welding. True. False.

Table 23-9
Course Outline: Shielded Metal-Arc Welding Practice (Pipe)

Job no.	Type of joint	Type of weld	Welding position	Welding technique	Pipe specifications		Electrode specifications			Text references	
					Dia.	Weight	Type*	Size	Polarity	This unit	Other jobs
23-J1	Pipe	Bead	Horizontal roll	Beading stringer & weaved.	4″ to 10″	Standard	E6010	1/8 & 5/32	Reverse	754	20-J7 & 8
23-J2	Bevel butt	Groove	Horizontal roll	1st pass stringer; 3 passes weaved.	6″ to 10″	Standard	E6010	1P-1/8 Others 5/32	Reverse	754	21-J40 21-J42
23-J3	Bevel butt	Groove	Vertical fixed	7 passes stringer.	6″ to 10″	Standard	E6010	1 to 3P-1/8 4 to 7P-5/32	Reverse	756	22-J60
23-J4	Bevel butt	Groove	Vertical fixed	7 passes stringer. Cover pass laced.	6″ to 10″	Standard	E6010	1 to 3P-1/8 4 to 7P-5/32 Lacing 5/32	Reverse	756	20-J27
23-J5	Bevel butt	Groove	Horizontal fixed	1st pass stringer; 2 passes weaved; travel up.	6″ to 10″	Standard	E6010	1P-1/8 2P-5/32 3P-5/32	Reverse	757	22-J57 & 65
23-J6	Bevel butt	Groove	Horizontal fixed	1st pass stringer; 3 passes weaved; travel down.	6″ to 10″	Standard	E6010	1P-1/8 2P-5/32 3-4P-3/16	Reverse	759	22-J57 & 65

At this point the student should hand-cut, bevel, and weld several pipe joints in all three positions.
The student should be able to pass bend tests of pipe coupons cut from pipe welded in the horizontal fixed and vertical fixed positions.

Job no.	Type of joint	Type of weld	Welding position	Welding technique	Pipe specifications		Electrode specifications			Text references	
					Dia.	Weight	Type*	Size	Polarity	This unit	Other jobs
23-J7	90° branch	Groove	Top (branch)	Multipass stringer.	4″ to 10″ Small to large. Size on size.	Standard	E6010	1/8 & 5/32	Reverse	762	21-J40 20-J28
23-J8	90° branch	Groove	Horizontal (branch)	Multipass: 1st pass stringer; others weaved.	4″ to 10″ Small to large. Size on size.	Standard	E6010	1/8 & 5/32	Reverse	762	22-J52 21-J45
23-J9	90° branch	Groove	Bottom (branch)	Multipass: Last pass laced.	4″ to 10″ Small to large. Size on size.	Standard	E6010	1/8 & 5/32	Reverse	762	22-J65 20-J27
23-J10	45° branch	Groove	Top (branch)	Multipass stringer	4″ to 10″ Small to large. Size on size.	Standard	E6010	1/8 & 5/32	Reverse	762	21-J40 20-J28

(Continued on page 772)

Table 23-9 (Continued)

Job no.	Type of joint	Type of weld	Welding position	Welding technique	Pipe specifications		Electrode specifications			Text references	
					Dia.	Weight	Type*	Size	Polarity	This unit	Other jobs
23-J11	45° branch	Groove	Horizontal (branch)	Multipass: 1st pass stringer; others weaved.	4″ to 10″ Small to large. Size on size.	Standard or heavy	E6010	$\frac{1}{8}$ & $\frac{5}{32}$	Reverse	762	22-J52 21-J45
23-J12	45° branch	Groove	Bottom (branch)	Multipass: Last pass laced.	4″ to 10″ Small to large. Size on size.	Standard	E6010	$\frac{1}{8}$ & $\frac{5}{32}$	Reverse	762	22-J65
23-J13	45° *L*	Groove	Horizontal fixed	Multipass: 1st pass stringer; others weaved.	4″ to 10″	Standard	E6010	$\frac{1}{8}$ & $\frac{5}{32}$	Reverse	762	22-J52–65
23-J14	90° *L*	Groove	Horizontal fixed	Multipass. 1st pass stringer; others weaved.	4″ to 10″	Standard	E6010	$\frac{1}{8}$ & $\frac{5}{32}$	Reverse	762	22-J52–65
23-J15	Blunt pipe head	Groove	Horizontal fixed	Multipass.	4″ to 6″	Standard	E6010	$\frac{1}{8}$ & $\frac{5}{32}$	Reverse	762	22-J52–65
23-J16	Orange peel head	Groove	Horizontal fixed	Multipass.	4″ to 6″	Standard	E6010	$\frac{1}{8}$ & $\frac{5}{32}$	Reverse	762	21-J40 22-J52–65
23-J17	90° Y	Groove	Fixed	Multipass.	4″ to 6″	Standard	E6010	$\frac{1}{8}$ & $\frac{5}{32}$	Reverse	762	21-J40 22-J52–65

At this point the student should repeat those jobs on which additional practice is necessary, using pipe of different diameters and wall thicknesses. The student is now ready to take the official API or ASME Code Certification Test for Carbon Steel Pipe. Welding is to be in the horizontal fixed and the vertical fixed positions.

Students who qualify on standard carbon steel pipe for both the electric arc process and the oxy-acetylene process may take additional training in arc and gas welding of heavy, extra heavy, and alloy pipe and in TIG and MIG welding of alloy and aluminum pipe.

*E7010 electrodes should also be used for practice.
E6011 electrodes should be used with a.c. current.
E7018, E9018, and E10018 electrodes can also be used for low alloy steel pipe.

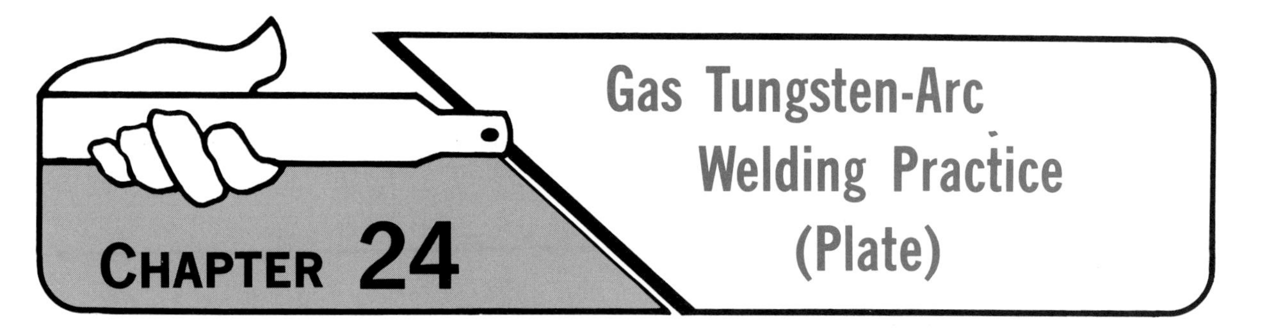

Gas Tungsten-Arc Welding Practice (Plate)

GAS TUNGSTEN-ARC WELDING OF VARIOUS METALS

The gas tungsten-arc welding (GTAW) process can be used for nearly all types of metals. The welds are of the highest physical and chemical quality obtainable. Study the following material carefully so that you will be familiar with the GTAW characteristics of each metal.

Aluminum

Aluminum is one of the most abundant metals. It has the characteristics of being light and strong, highly corrosion resistant, ductile, and malleable. It has good electrical conductivity and is relatively inexpensive.

Aluminum melts at 1,218 degrees F. and its oxide melts at about 3,700 degrees F. (Steel melts at about 2,800 degrees F., and copper melts at approximately 1,980 degrees F.) Aluminum oxidizes readily at room temperatures. The oxide is removed when welding with the oxyacetylene or metal arc process with a flux. Flux is not necessary with gas tungsten-arc welding.

Selection of the arc welding method for joining aluminum depends largely on the individual application. The thickness of the metal, the design of the parts and assemblies, production quantities, and equipment available must be considered.

The best welding methods for aluminum are the gas tungsten-arc (GTAW, or TIG) and the gas metal-arc (GMAW, or MIG) processes. Due to the shielding gas, each offers good protection for the weld pool. The gas shield is transparent so that the welder can see the fusion zone. This helps in making neater and sounder welds.

One disadvantage in the welding of aluminum with other processes is the fact that it does not change color when it approaches the melting point as most other metals do. The flux required when welding with other processes also causes a glare in the weld pool and emits a considerable amount of smoke. When TIG welding, there is no glare or smoke, and the weld pool is clearly visible to the welder.

The gas tungsten-arc process is preferred for welding aluminum sections less than 1/8 inch in thickness. This method can also be used on heavier sections, but the gas metal-arc process is usually chosen for its higher welding speed and economy. In this chapter we are concerned with the TIG welding of aluminum. The welding of aluminum with the MIG process is included in a later chapter.

All types of aluminum alloys can be welded with the TIG process, including those alloys in the 1000, 1100, 3000, 5000, 6000, and 7000 series. (Refer to Table 3-20, page 105 for the classification system of aluminum alloys.) Alternating current with high frequency stabilization is recommended. Welding is possible with direct current, straight polarity, but it is not as successful as with stabilized alternating current. The shielding gas is usually argon.

PLATE PREPARATION. Aluminum plates may be prepared for welding by mechanical or thermal cutting processes. Gas tungsten-arc cutting (GTAC), metal-arc cutting (MAC), and plasma-arc cutting (PAC) are three thermal processes that are used a great deal. Mechanical processes include machining, shearing, sawing, chipping, and filing. Any contamination such as oil and grease must be removed from the prepared surface.

In addition to the contamination that may be on the surface of the plate due to joint preparation, there is also an oxide film. It is necessary to remove this material from the plate before welding. Contamination may be removed from the surface with caustic soda, acids, and certain solutions. Mechanical methods include wire brushing, scraping, filing, and the use of steel wool. The interval between cleaning and welding should be as short as possible since aluminum oxidizes so rapidly.

When aluminum is welded from one side only, some type of

backup may be desirable to control penetration. Backup strips may be made of steel, copper, or stainless steel. For certain welding jobs, especially pipe, the backing may consist of an inert gas shield.

PREHEATING. Aluminum alloys are of two types, the work-hardenable alloys such as EC, 1100, 3003, 5052, 5083, and 5086; and the heat-treatable alloys such as 6061, 6062, 6063, 7005, and 7039. Alloys in the 2000 and 7000 series can also be welded with either the gas metal-arc process or resistance welding.

Heat-treatable alloys may be preheated to keep cracking at a minimum. These alloys are heated at temperatures above 900 degrees F. and then given a low temperature aging treatment above 300 degrees F. Heat treatment is not generally required in the welding of work-hardenable aluminum alloys.

Preheating is necessary if the mass of the base metal is such that heat is conducted away from the joint so fast that the welding arc cannot supply the heat required to produce fusion. Insufficient heat results in poor fusion of the weld bead and inadequate melting of the base metal. Preheating the parts being joined helps to produce a satisfactory weld, reduces distortion and cracking in the finished product, and increases welding speed. In gas tungsten-arc welding, preheating is necessary when welding plate over $\frac{1}{2}$ inch thick. One of the advantages of gas metal-arc welding is that preheat is seldom required, regardless of plate thickness.

WELDING TECHNIQUE. Because aluminum alloys have relatively high coefficients of thermal expansion as compared with most weldable metals, there may be a problem with distortion whenever unequal expansion or contraction occurs. You will recall that control of distortion is a matter of proper design, joint preparation and fitup, choice of welding process, and the use of a proper welding sequence.

In planning the sequence of welding, individual pieces and members must have freedom of movement. Joints that are likely to contract a great deal should be welded first. Welding should be done on both sides of the structure at the same time and alternated from one side to the other to equalize stresses. The weld should be completed as rapidly as possible without interruption.

Due to the high thermal conductivity of aluminum, a very high heat input must be maintained in the weld zone to balance the heat loss to the adjacent metal. Because of the intense temperature of the tungsten arc, little difficulty is encountered.

The molten pool of aluminum has a high degree of surface tension and solidifies rapidly so that it is possible to weld in all positions.

When welding aluminum, it is good practice to "ball" or round the end of the electrode before welding to keep the arc steady. If the end is uneven, the arc may move from side to side. The method for balling the end of the electrode is given in Chapter 9, page 275. Consult Table 24-1 for the proper operating conditions for the welding of aluminum with alternating current.

Table 24-1

Operating conditions with stabilized a.c. current and argon shielding gas: aluminum

Work thickness	Welding current—amperes			Dia. tungsten	Filler rod size	Cup size	Gas flow cu. ft./hr.
	Flat	Vertical	Overhead				
1/16"	60–90	60–90	60–90	1/16"	1/16"	1/4, 5/16, 3/8	15
1/8"	125–160	115–135	120–160	3/32"	3/32"	3/8, 7/16	17
3/16"	190–240	190–220	180–210	1/8"	1/8"	7/16, 1/2	21
1/4"	260–340	220–260	210–250	3/16"	1/8", 3/16"	1/2, 5/8, 3/4	25
3/8"	330–400	250–300	250–300	3/16", 1/4"	3/16", 1/4"		29
1/2"	400–470	290–350	250–375	3/16", 1/4"	3/16", 1/4"		31

Eutectic Corp.

Note: Optimum conditions for each application should be determined by trial.

Carbon and Low Alloy Steels

The Course Outline provides for the welding of carbon steels after you have mastered the welding of aluminum because they are a little easier to weld than the stainless steels. There is a great deal of similarity, however, and the practice gained in welding these steels will make it easier for you to learn to weld the stainless steels. There is also the matter of cost to consider. By practicing on carbon and low alloy steels, a great deal of experience can be gained with the gas tungsten-arc process at the lowest possible cost.

TIG welding of carbon steels is usually done on light gauge materials which require relatively less heat than heavy plate. A given size electrode also requires less heat in the welding of carbon steel than the same size used in the welding of aluminum. Lower current results in a smaller arc. The small arc is more difficult to see than the arc for aluminum welding. For this reason a lighter shade of welding lens is recommended when welding carbon steels.

Best results in the welding of carbon steels are obtained with direct current, straight polarity. Use argon as the shielding gas and a thoriated electrode with the end sharpened to a point. The sharpened electrode makes it easier to control the arc and the weld bead size. Since the current is DCSP, it takes a lot of heat to melt the sharpened electrode. Make sure that the electrode does not touch the weld pool. The molten metal will contaminate it and cause a wild arc. Use a filler rod that matches the base metal.

High frequency is used only to start the arc instead of being continuous as when welding aluminum. Welding could begin with the touch start, but this would destroy the sharp end of the electrode and contaminate it.

Less heat is needed to start the weld puddle in carbon steel than aluminum. Since the machine heat setting may be held fairly steady from start to finish, a foot-operated heat control is not absolutely necessary. The welding heat can be controlled by varying travel speed. In order to obtain the highest weld quality, however, a heat control should be used. Consult Table 24-2 for the proper operating conditions for the welding of low carbon steel with DCSP.

STAINLESS STEEL. Stainless steel is one of the most widely used of all of the alloys. It is suitable for all types of welded fabrications in which strength and resistance to high temperatures, pressures, and corrosion are desired. Root passes laid down by gas tungsten-arc welding are always specified for X-ray quality welds in large, heavy-walled pipe for nuclear power plants and other critical services. Stainless steel is made in all standard steel shapes and forms.

You will recall from Chapter 3 that there are three general types of stainless steel, and all can be welded with the gas tungsten-arc process. Steels in the 300 series, which are referred to as the *austenitic* type, are chromium steel alloys and chromium-nickel-manganese steel alloys. They are highly weldable. These steels are hardenable only by cold working. A second type is the *martensitic* steels. They are straight-chromium stainless steels in the 400 and 500 series. These steels are hardenable by rapid cooling from a high temperature. A third type in the 400 series is *ferritic*. It has a straight-chromium content of 14 percent or above. Ferritic steels are non-hardenable in certain high carbon-chromium combinations. Lower chromium grades may be air hardened.

Stainless steels have a high resistance to corrosion and high temperatures, excellent strength-to-weight ratios, and a high degree of ductility. They can be fabricated by the same methods as carbon steel. Their thermal conductivity, however, is about 50 percent less than that of carbon steel so that heat stays in the weld zone. Because their thermal expansion is about 50 percent greater than that of carbon steel, there is a problem with distortion. Distortion and heat retention can be minimized through the use of proper jigs and fixtures.

The gas tungsten-arc process makes it possible to weld stainless steels without the problems of flux and weld spatter. The dangers involved in multipass welding due to heavy slag are not present.

Stainless steel can be successfully welded with either direct current, straight polarity or alternating current with high frequency stabilization (ACHF). Much greater penetration and welding speed can be obtained with direct current, straight polarity. When welding with DCSP, high frequency is usually used

Table 24-2

Operating conditions with DCSP and argon shielding gas: carbon steel plate.

Plate thick.	Tungsten diam.	Ceramic gas cup size	Argon flow c.f.h.	Welding current DCSP—amperes	Filler rod size
1/16″	1/16″	3/8″	10–12	80–120	1/16″
3/32″	3/32″	1/2″	12–14	100–140	3/32″
1/8″	3/32″	1/2″	12–16	100–140	3/32″
3/16″	3/32″	1/2″	12–18	100–140	1/8″
1/4″	3/32″	1/2″	14–18	130–175	3/16″
3/8″	1/8″	1/2″	16–20	170–200	3/16″
1/2″	1/8″	1/2″	18–22	200–250	1/4″

Hobart Brothers Co.

only to start the weld. Consult Table 24-3 for the proper operating conditions for the welding of stainless steel with DCSP. High frequency alternating current is recommended for welding thin materials since the lower heat input reduces the tendency to burn through. Very thin materials may be welded with DCRP. Because of high electrode consumption with DCRP, the electrode size should be at least four times as large as that for DCSP and the same current value. A 1 percent or preferably a 2 percent thoriated-tungsten electrode is used with all three currents.

Either argon or helium may be used as a shielding gas. Argon provides smoother arc action and gives good results when welding the thinner gauges. Helium shielding produces a hotter arc than argon and permits higher welding speeds and deeper penetration, particularly on heavier materials. Welding research has determined that argon-hydrogen mixtures of about 15 to 20 percent hydrogen are the equivalent of helium and produce sound welds in austenitic stainless steels. Shielding the back side of the weld may be necessary to prevent oxidation and to promote maximum corrosion resistance. One method is to introduce argon or a back-up powder under the weld with some type of backing device to confine the gas to the weld area.

One of the principal differences between welding stainless steel and aluminum is that aluminum requires more heat and a faster speed of travel. Take care in the selection of filler rod size. If the rod diameter is too large, it will soak up a good deal of heat and make welding more difficult.

It is important that the filler rod match the type of stainless steel being welded. The rod must also deposit metal that has the physical and chemical properties that the job requires.

Travel speed is slow with relatively low heat, and it takes considerable skill to keep from burning through thin materials. Because stainless steel distorts much more than carbon and low alloy steels during welding, proper tacking and clamping are important.

Magnesium

Many people mistake magnesium for aluminum because they are similar in appearance and characteristics. Magnesium is a very light material. It is about two-thirds the weight of aluminum and one-quarter the weight of steel. The melting point of magnesium is 1,204 degrees F.; the melting point of aluminum, 1,218 degrees F. The strength-to-weight ratio of magnesium alloys is good. The thermal conductivity is relatively high—a little less than aluminum.

Welding magnesium is similar to welding aluminum. The same type of joints and method of joint preparation are used. Careful cleaning of the workpiece is always required since magnesium oxidizes readily. The parts should be degreased by chemically cleaning and/or mechanically cleaned with abrasives.

Rolled or extruded magnesium sections can be joined easily to each other or to castings. All welds are made without flux and have the approximate strength of the base metal.

Either alternating current with high frequency stabilization or direct current, reverse polarity may be used. A generator with voltage adjustment is preferred for direct current so that the open circuit voltage can be set to suit the work. The shielding gas may be argon or helium.

ACHF has certain advantages over DCRP. It is easier to start the arc, and a longer arc can be used. Since more heat is in the weld zone and less in the electrode, faster welding is possible with less current, and smaller electrodes can be used to obtain greater penetration. Less pickup of weld metal on the tungsten electrode provides for longer electrode life.

Consult Table 24-4 for the proper operating conditions for the welding of magnesium with alternating current.

Table 24-3

Operating conditions with DCSP and argon shielding gas: stainless steel

Work thickness	Welding current—amperes			Dia. tungsten	Filler rod size	Cup size	Gas flow cu. ft./hr.
	Flat	Vertical	Overhead				
1/16"	80–110	70–100	70–100	1/16"	1/16"	1/4", 5/16", 3/8"	11
3/32"	100–130	90–120	90–120	1/16"	3/32"	1/4", 5/16", 3/8"	11
1/8"	120–150	110–135	105–140	1/16", 3/32"	3/32"	1/4", 5/16", 3/8"	11
3/16"	200–275	150–225	150–225	3/32", 1/8"	1/8"	3/8", 7/16", 1/2"	13
1/4"	275–375	200–275	200–275	1/8"	3/16"		13
1/2"	350–375	225–280	225–280	1/8", 3/16"	1/4"		15

Eutectic Corp.

Note: Optimum conditions for each application should be determined by trial.

COPPER AND ITS ALLOYS. Copper is one of the oldest metals used by man. It is not a scarce metal and is readily mined. It has a melting point of 1,981 degrees F. Copper lends itself to all of the modern methods of fabrication. The weldability of each copper-alloy group depends largely upon the alloying elements. Copper is widely used in all industries and has a special application in the fabrication of electrical equipment.

Copper is one of the best conductors of electricity and heat. It is highly resistant to corrosion. When mixed with certain other elements, copper is both ductile and wear resistant, and it can be heat treated.

Copper has the following disadvantages for welding.

- It is brittle at high temperatures and presents problems in jigging.
- It is strongest when cold worked. The high welding temperatures cause it to lose strength.
- Stress reversals cause the metal to become hard and brittle so that it breaks from fatigue.

While pure copper presents no great difficulty in welding, many of its alloys require special treatment. The following copper alloys lend themselves to welding.

- Copper-silicon alloys such as Everdur® and Herculoy®
- The copper-aluminum alloys known as aluminum bronze
- The copper-phosphorus alloys known as phosphor bronze
- Copper-nickel alloys

Those copper alloys containing zinc, tin, or lead are either difficult or impossible to weld. These materials have a low melting point that causes them to volatize under the intense heat of the arc. They can be successfully brazed, however, with the oxyacetylene process.

DCSP is used for welding pure copper and most copper alloys. ACHF or DCRP is recommended, however, for beryllium copper, aluminum bronze, and copper alloys less than 20 gauge thick. Current settings should be higher for copper and its alloys than for most other metals because of their high thermal conductivity. Always provide good ventilation when welding beryllium copper and when using certain fluxes. Copper has a tendency to form oxides which must be removed just before welding.

Consult Table 24-5 for the proper operating conditions for welding deoxidized copper with DCSP.

NICKEL AND NICKEL-BASE ALLOYS. Nickel has wide application in those industries in which corrosion and low and high temperatures are encountered. It is very ductile and can be worked readily. Its tensile strength, elasticity, melting point, and magnetic properties are similar to those of steel. Fol-

Table 24-4

Operating conditions with stabilized a.c. current and argon shielding gas: magnesium

Work thickness	Welding current—amperes flat position		Dia. tungsten	Filler rod size	Cup size	Gas flow cu. ft./hr.
	Backup	No backup				
1/16"	60	35	1/16"	3/32"	1/4", 5/16", 3/8"	13
3/32"	90	60	1/16"	1/8"	1/4", 5/16", 3/8"	16
1/8"	115	85	1/16"	1/8"	1/4", 5/16", 3/8"	19
	One pass	Two passes				
3/16"	120	75	1/16"	5/32"	1/4", 5/16", 3/8"	19
1/4"	130	85	3/32"	5/32"		19
3/8"	180	100	3/32"	3/16"		24
1/2"	—	260	5/32"	3/16"		24
3/4"	—	370	3/16"	1/4"		36

Eutectic Corp.

Note: Optimum conditions for each application should be determined by trial.

Table 24-5

Operating conditions with DCSP and argon shielding gas: deoxidized copper

Work thickness	Welding current flat position	Preheat temp.	Dia. tungsten	Filler rod size	Cup size	Gas flow cu. ft./hr.
1/16"	110–150	—	1/16"	1/16"	3/8"	15
1/8"	175–250	—	3/32"	3/32"	3/8", 1/2"	15
3/16"	250–325	500°F	1/8"	1/8"	1/2"	15
1/4"	300–375	600°F	1/8"	1/8"	1/2"	15
3/8"	375–450	800°F	3/16"	3/16"	1/2"	17
1/2"	500–700	900°F	3/16"	1/4"	1/2", 5/8"	17

Eutectic Corp.

Note: Optimum conditions for each application should be determined by trial.

lowing is a list of the important high nickel alloy metals.

- The nearly pure nickels and *Duranickel®*, a high strength, low alloy nickel
- The *Monels®*, which are about two-thirds nickel and one-third copper
- The *Inconels®*, which are higher in nickel and iron content than the Monels. They are outstanding in their ability to resist damage from abrupt changes in temperature.
- The *Nimonics®*, which contain approximately 80 percent nickel and 20 percent chromium. These alloys are used in gas turbine engines.
- The *Hastelloys®* are alloys of nickel, molybdenum, and iron. They have a high resistance to acids.

While nickel can be welded with the shielded metal-arc welding process, gas tungsten-arc welding has the advantage of eliminating slag entrapment in the weld. Generally, DCSP is recommended. On thin material, however, ACHF has the advantage of lower heat input. The current values and electrode sizes are similar to those used to weld carbon steel. Helium is recommended for most applications, but argon is recommended for thin materials.

Consult Table 24-6 for the proper operating conditions for the welding of the Hastelloys® with DCSP.

Welding Dissimilar Metals

Many industrial fabrications make it necessary to join dissimilar metals. While both the oxyacetylene and the shielded metal-arc welding processes may be used, a greater variety of materials can be joined with gas tungsten-arc welding.

Consult Table 24-7 for the listing of dissimilar metals that can be joined and the type of current and filler rod to use. The filler rod should be selected with care. Table 24-7 lists one company's recommendations.

Hard Surfacing

Hard surfacing, or *hard facing* as it is sometimes called, is a process of applying a hard, wear-resistant layer of metal to the surfaces or edges of parts. Hard-facing rods are deposited for the purpose of improving resistance to impact or abrasion, or both. This may be done to build up worn areas to make them as good as new or to put hard wear-resistant cutting edges on soft, ductile materials. Hard facing is used for stone-crushing equipment, power shovel buckets, farm implements, and many other applications. While both the oxyacetylene and the shielded metal-arc welding processes may be used, somewhat better results can be obtained with the gas tungsten-arc process. Consult Table 24-8 for the listing of materials that may be hard surfaced and the proper operating conditions.

Filler Metals

The composition of filler metals for the various metals welded by the gas tungsten-arc process are selected on the basis of the following criteria:

- The requirements of the specific job
- The type of metal being welded
- Ease of welding
- Characteristics desired such as strength, ductility, hardness, machinability, and resistance to corrosion, impact, abrasion, and oxidation
- Preheat and postheat requirements
- Color matching when this is important
- Type of shielding gas being used

You are urged to become familiar with the AWS specifications for filler metals. On most welding jobs the type of filler metal will be specified by the engineering department. It is necessary, however, for the welder to know the types of filler metals and their uses.

Table 24-6
Operating conditions with DCSP and argon shielding gas: Hastelloy® alloys

Work thickness	Welding current—amperes		Dia. tungsten	Filler rod size	Gas flow cu. ft./hr.
	Flat	Vertical			
1/16″	60–90	55–75	1/16″	1/16″, 3/32″	20
1/8″	100–125	80–110	3/32″	3/32″	25
3/16″	130–175	100–140	3/32″	3/32″	30
1/4″	130–175	100–140	3/32″	1/8″	35
3/8″	140–200	110–160	3/32″	1/8″	40
1/2″	200–250	160–250	1/8″	5/32″	40

Eutectic Corp.

Note: Optimum conditions for each application should be determined by trial.

JOINT DESIGN AND PRACTICES

There is no limit to the types of joints that may be welded with the gas tungsten-arc process. The basic types of joints used for plate welding are butt, lap, edge, corner, and T-joints. They are similar to those used with other welding processes. Selection of the proper design for a particular application depends primarily on the following factors:

- Physical properties desired in the weld
- Cost of preparing the joint and making the weld
- Type of metal being welded
- Size, shape, and appearance of the assembly to be welded.

Other considerations include the following:

- Number and size of tacks
- Purging and shielding gas
- Root land, root opening, and bevel
- Number of passes required
- Size of filler rod
- Whether or not rod is required for the first pass and whether or not a wash pass is permitted
- Method of striking and breaking the arc
- Whether the direction of travel is up or down
- Allowable protrusion and reinforcement
- Type of electrode
- Type of welding power
- Tolerances of fitup and alignment

While gas tungsten-arc welding is particularly suited to the welding of materials up to 1/8 inch in thickness, it may also be used to weld heavier thicknesses of metals. The MIG process may be used on heavy stock with better results. The nature and application of the joint is a major consideration. Cost must also be considered in determining the welding process for a particular job.

Usually, filler rod need not be used for thinner materials. Careful consideration should be given to the welding of heavy carbon steel since it is quite possible that the metal arc or MIG process is more suitable.

No matter what type of joint is used, proper cleaning of the work before welding is essential if welds of good appearance and sound physical properties are to be obtained. This is of special importance in welding some alloys such as aluminum and magnesium. Welds in these metals will be defective if even minute quantities of foreign material contaminate the inert gas atmosphere.

On small assemblies, manual cleaning with a wire brush, steel wool, or a chemical solvent is usually sufficient. Do not grind aluminum or magnesium on an emery wheel. Be sure to remove completely all oxide, scale, oil, grease, dirt, and rust from the work surfaces.

Precautions should be taken when using certain chemical solvents for cleaning purposes. The fumes from some chlorinated solvents break down in the heat of the electric arc and form a toxic gas. Ventilating equipment should be provided to remove fumes and vapor from the work area.

Weld Backup

The joint should be backed up on many gas tungsten-arc applications. Backing protects the underside of the weld from atmospheric contamination. At-

Table 24-7
Welding dissimilar metals.

Materials to be joined	Type of current	Rod type
Stainless steel to cast iron	DCSP	OXWELD® 26
	DCSP	Nickel and stainless steel
Stainless to carbon or low alloy steel		310 stainless
Copper to stainless steel	DCSP	None required
Copper to Everdur®		OXWELD® 26
Cupro-nickel to Everdur®		OXWELD® 26
Nickel to steel	DCSP	Nickel
HASTELLOY® Alloy C to steel	DCSP	HASTELLOY® W; Inconel;® Nickel; or 310 stainless
Aluminum to steel	ACHF	No. 25M bronze rod; B.T. silver brazing alloy; OXWELD® 14 Al. Rod
Stainless steel to Inconel®	DCSP	310 stainless
Tungsten to molybdenum		Platinum
Copper and Everdur® to steel	DCSP	Copper or 26 Everdur®

Linde Division, Union Carbide Corp.

mospheric contamination causes weld porosity, poor surface appearance, cracking, and burn-through.

The weld may be backed up by (1) metal backup bars, (2) an inert gas atmosphere on the weld underside, (3) a combination of the first two methods, or (4) flux painted on the weld underside.

Metal backup bars should not actually touch the weld zone. (See Figs. 24-1 through 24-4 for typical backup for butt, lap, and T-joints.) The material used for making a backup bar is determined by the composition of the material being welded. A copper bar may be used to back up welds in stainless steel. For the welding of aluminum or magnesium, the bar should be made of stainless steel or steel. Carbon steel can be used for carbon steel welding. Very often, backup bars are water cooled to carry off

Table 24-8
Hard Facing and Surfacing.

Base metal	Surfacing material	Current		Rod	Welding technique	Argon flow c.f.h.	Deposit	Remarks
		Type	Amps.	Type			Rc hardness	
Mild & stainless steels	Haynes Stellite alloys	ACHF[1]		Stellite #1	Backhand	25	54	
		ACHF[1]		Stellite #6	Backhand	25	39	
		ACHF[1]		Stellite #12	Backhand	25	47	
		ACHF[1]		Stellite #93	Backhand	25	62	
		ACHF[1]		Hascrome	Backhand	25	23–43	Extruded rod has better weld characteristics than rolled rod.
Copper	Stellite #6 alloy	DCSP	180–230 for 3/16-in. material	Stellite #6	Forehand	15	42	Arc directed mainly at welding rod.
Steel, copper & silicon bronze	Aluminum bronze	DCSP		Aluminum-bronze rods	Forehand	10	150–300	
Mild steel & cast iron	Bronze & copper	ACHF or DCSP	150 for 1/2-in. material	Aluminum-bronze & copper rods	Forehand	10		
Stainless steel	Silver	ACHF	160 for 1/2-in. material		Either	10		Plates pickled prior to surfacing.
Mild steel	Stainless steel	ACHF or DCSP			Forehand	10		
Mild steel	Lead	DCSP	75		Forehand	10		Steel ground or pickled and then coated with liquid soldering flux before surfacing.
Carbon & alloy tool steels	Tungsten carbide	DCSP	300–375	Tube of 8/15 mesh tungsten particles		30		

[1]ACHF develops maximum hardness values; DCSP will permit higher welding speeds.

Linde Division, Union Carbide Corp.

the heat of the welding operation.

When the final weld composition must conform to extremely rigid specifications, extra care must be taken to exclude all atmospheric contamination from the weld. This is accomplished by introducing an atmosphere of inert gas on the back side of the weld. Nitrogen may be used for stainless steels. Argon should be used for aluminum, magnesium, and other metals that oxidize readily or react with nitrogen at high temperatures. Helium may also be used as a purging gas. Review Chapter 9 for a detailed presentation of shielding gases.

SETTING UP THE EQUIPMENT

Gas tungsten-arc welding is a precision technique. Care must be taken to make sure that the equipment is set up in the proper way and that all of the variables are correct for the particular welding job that is to be performed. The procedure for setting up the equipment before welding requires the checking of every detail. You should make the checks described in the following paragraphs before starting to weld. Consult Chapter 9 to review many of these points.

1. Make sure that you have a torch of the proper type and size to meet the requirements of the welding job.

2. Check the size, appearance, and position of the tungsten electrode in the torch. It should have the diameter recommended for the amount of current and the electrode holder used. See Table 9-5, page 276, and Table 24-9. The end of the electrode should be clean and smooth. A dirty electrode end indicates that during a previous use the inert gas was shut off

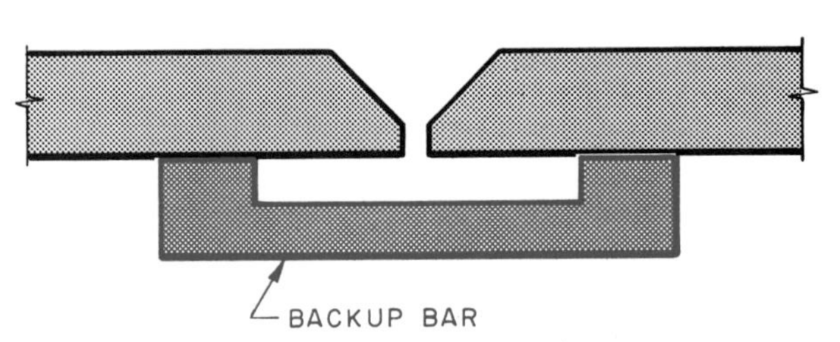
BACKUP BAR

24-1. *Grooved backup bar for square and beveled-butt joints.*

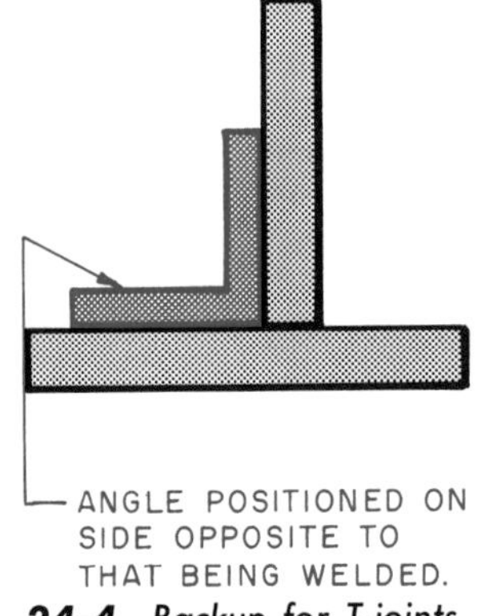
ANGLE POSITIONED ON SIDE OPPOSITE TO THAT BEING WELDED.

24-4. *Backup for T-joints.*

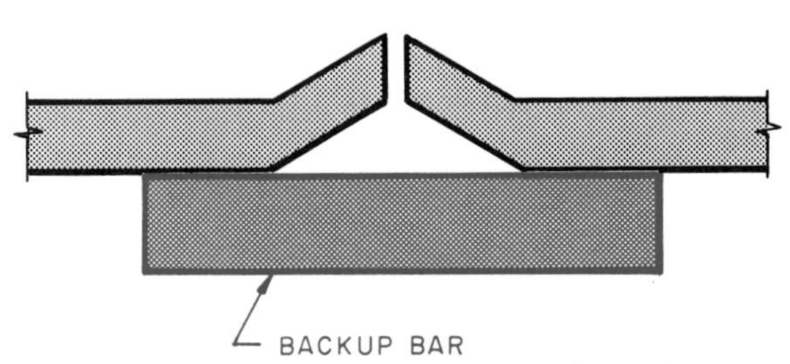
BACKUP BAR

24-2. *Flat backup bar for upset butt joints.*

24-3. *Backing for lap joints.*

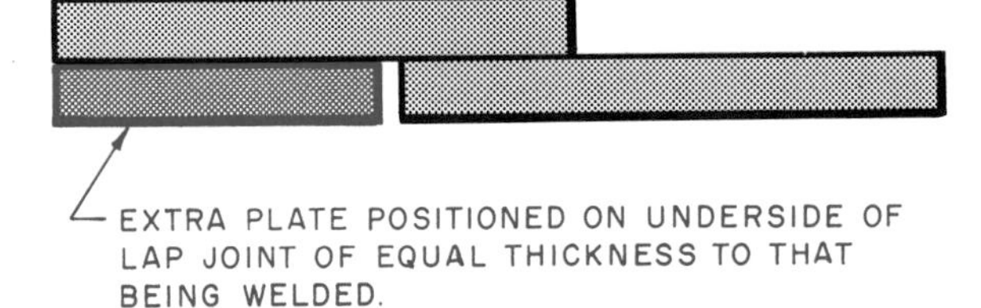
EXTRA PLATE POSITIONED ON UNDERSIDE OF LAP JOINT OF EQUAL THICKNESS TO THAT BEING WELDED.

Table 24-9

General sizes of TIG electrode holders and diameters of the electrodes which fit them.

General sizes of TIG electrode holders	Diameters of electrodes
Small holder	0.010″, 0.020″, 0.040″, $\frac{1}{16}$″ & $\frac{3}{32}$″
Medium holder	0.020″, 0.040″, $\frac{1}{16}$″, $\frac{3}{32}$″, & $\frac{1}{8}$″
Large or heavy duty holder	0.020″, 0.040″, $\frac{1}{16}$″, $\frac{3}{32}$″, $\frac{1}{8}$″ $\frac{5}{32}$″, $\frac{3}{16}$″, & $\frac{1}{4}$″ (Some special holders—$\frac{5}{16}$″, $\frac{3}{8}$″)

before the electrode cooled, that there was an air leakage in the gas supply system or the torch, or that the electrode tip was contaminated by touching metal. If the tip is not too rough, it may be cleaned with a fine emery cloth. When welding aluminum, the tip should be rounded and when welding stainless steel, it should be ground to a needle point.

3. Check the torch for the proper gas cup type and size. Make sure that it is clean and free of spatter. If it is important to see the weld clearly, you may wish to use a glass cup.

4. Check all of the connections on the gas supply for leaks. If there is some reason to doubt a connection, check it with soapy water.

5. Check all ground connections. Pay special attention to the location of the ground connection to the work and the location of the ground in relation to the joint design and the direction of welding.

6. Open the main shutoff valve of the cylinder of inert gas and adjust the gas flow to meet the needs of the particular job. See Chapter 9, pages 258, 259.

7. Before turning on the water supply, make sure that the water pressure is not higher than that recommended by the torch manufacturer. Make sure that there are no leaks in the water supply.

8. Adjust the current range for the joint being welded and the size of the tungsten electrode. The position of welding also makes a difference here. As with the shielded metal-arc process, vertical and overhead welding generally requires a lower current setting than other positions.

9. It is assumed that you have the proper eye and skin protection for welding and that you will take the necessary safety precautions.

SAFE PRACTICES

In welding with the gas tungsten-arc process, observe the same precautions and safe practices that would apply to any other electric welding operation. In any form of electric welding, there are potential shock hazards, burn hazards, and fire hazards. In addition, welding operations on certain metals and alloys may produce unpleasant or dangerous fumes. For a detailed treatment of this subject, you are urged to secure materials from such organizations as the American Welding Society and various welding equipment companies.

The operator should also be properly protected from the rays of the arc. This requires suitable clothing to cover all exposed skin surfaces and a welder's helmet with the proper shade of glass to protect the eyes and face. The shade of the glass lens depends on the intensity of the arc. Table 24-10 lists the recommended lens shade for different current ranges.

Gas Tungsten-Arc Welding Safety Rules

Observe the following rules when welding with the TIG process.

Table 24-10

Correct lenses for various current settings.

Glass no.	Welding current
6	Up to 30 amperes
8	30 to 75 amperes
10	75 to 200 amperes
12	200 to 400 amperes
14	Above 400 amperes

- Always use a welding headshield equipped with a colored lens of a shade number suitable to the welding current you are using. The headshield should be mounted on an adjustable headband or, if there are overhead hazards, on a safety cap.
- Cover all skin surfaces with leather, heavy clothing, or other adequate protection against burns from sparks, the arc ray, and spatter, Figs. 24-5 and 24-6. Recommended personal protective equipment includes the following:

  Headshield
  Safety glasses
  Leather cape
  Sleeves
  Apron
  Gauntlet gloves
  Heavy, flameproof shirt and pants (without cuffs)
  Safety shoes
- Do not weld in or near flammable gases, powders, or liquids or on untreated containers that have held such materials.
- Remove or protect all combustible material in the welding area. If uncertain of the adequacy of the protection, have a worker stand by with a fire extinguisher.
- The welding area should be dry and uncluttered to avoid electric shock and falls.
- Be sure you have adequate ventilation. Use special precautions when toxic fumes, whether from cleaning fluids, coatings, or the metal itself, are given off.
- Use nonreflective welding curtains to protect others in the area. Even short exposure to tungsten-arc rays can burn the eyes, and skin burns from arc rays or molten metal spatter must be prevented.
- The TIG torch, power supply unit, service leads, ground connections, inert gas equipment, and protective equipment should

be inspected and in good condition before welding.

- Never change polarity under load, alter current settings under load if prohibited by the instructions, nor in any way overload any of the electrical equipment.
- Know and obey all safety rules pertaining to the particular area in which the welding is to be done, shop practices, and personal protection.

ARC STARTING

Arc starting with the gas tungsten-arc welding process is somewhat more difficult than with shielded metal-arc welding. Poor arc starting can result from the following:

- Improper cup size
- Dirty or burned electrode tip
- Erratic high frequency current
- Improper shielding gas flow rates
- Dirty work surface
- Low welding current
- Low welding voltage
- Faulty connections in the equipment

Each condition should be checked carefully and the proper corrections made.

Three arc starting methods are generally used in TIG welding: high frequency starting, touch starting, and high voltage starting.

High Frequency, No-Touch Starting

In welding with alternating current, the electrode does not have to touch the work to start the arc. The superimposed high frequency current establishes a path for the welding current to follow by jumping the gap between the electrode and the work. To strike an arc, first turn on the welding current and hold the torch in a horizontal position about 2 inches above the workpiece as shown in Fig. 24-7. Then quickly swing the end of the torch down toward the work, so that the end of the electrode is about $\frac{1}{8}$ inch above the plate as shown in Fig. 24-8. The arc will then start. For most applications, an arc length of approximately one electrode diameter may be maintained. With argon the arc length is much less critical than with helium.

The high frequency, no-touch method may also be used with direct current, straight or reverse polarity, if the d.c. welding machine is equipped with a high frequency unit. The high frequency current is used only for starting the arc and not for welding. The high frequency is automatically turned off by means of a current relay when the arc is started. With a.c. welding the high frequency current is maintained for the welding operation.

Touch Starting

In welding with direct current without high frequency, the same motion is used for striking the arc. In this case, however, the electrode must touch the work in order for the arc to start. Never use a carbon starting block. The arc can be started on a separate plate of aluminum, copper, or steel and carried to the workpiece. As soon as the arc is struck, withdraw the electrode approximately $\frac{1}{8}$ inch above the work to avoid contaminating the electrode in the molten puddle.

High Voltage Starting

Another method of starting is through the use of a high surge of voltage as the electrode is brought close to the base metal. After the arc is established, the high voltage is cut off, and the power returns to its normal voltage value for welding.

It is a common practice to warm the electrode and the noz-

Miller Electric Manufacturing Corp.

24-5. *This operator is risking injury to his hands by welding with the gas tungsten-arc process without wearing gloves.*

24-6. *Student is properly protected and is in the proper position for bench welding.*

zle on a practice piece to give better starting results on the job.

Arc Blow

Arc blow, a condition in which the arc wavers as it leaves the electrode, may occur in gas tungsten-arc welding. It is caused by the following problems:

- A current setting that is too low
- Contamination of the electrode with carbon
- Magnetic effects
- Drafts in the work area

A low current setting and carbon contamination are distinguished by a very rapid movement of the arc from side to side during welding. When the current is correct, the entire end of the electrode is molten and completely covered by the arc.

In DCSP welding, a current setting that is too low generates only enough heat to make part of the electrode end molten. The arc emerges only from this small molten spot which continually shifts, pulling the arc with it.

When welding with DCRP and alternating current, a current setting that is too low has a different effect. The electrode end melts into a molten ball that is too large to be held by the surface tension of the tungsten and drops off. When these conditions prevail, raise the operating current level.

Carbon contamination is the result of striking the arc with a carbon pencil or on a carbon block. This should never be done.

Magnetic influences usually draw the arc to one side or the other along the entire length of the weld. Magnetic effects great enough to create serious disturbances are not encountered frequently. However, when such influences do occur, shifting the position of the ground will give some relief.

Air movement causes varying amounts of arc wandering. Great care must be taken to make sure that the welding operation is shielded from drafts.

WELDING TECHNIQUE

You will find that many of the techniques learned in mastering the gas and shielded metal-arc welding processes will assist you in learning how to weld with the gas tungsten-arc process. The objective is to complete the weld without disturbing the molecular structure of the surrounding base metal. In order to accomplish this, you should work with the lowest machine setting possible for a given condition. The right amount of heat must be applied in the right place at the right time without exceeding a safe time limit.

When welding light gauge carbon and stainless steel or when heat must be concentrated on a limited area, the electrode should be sharpened to a needle point with a long, smooth taper. This insures that the arc will jump off

24-7. *Torch position for the starting swing in high frequency starting.*

Linde Division, Union Carbide Corp.

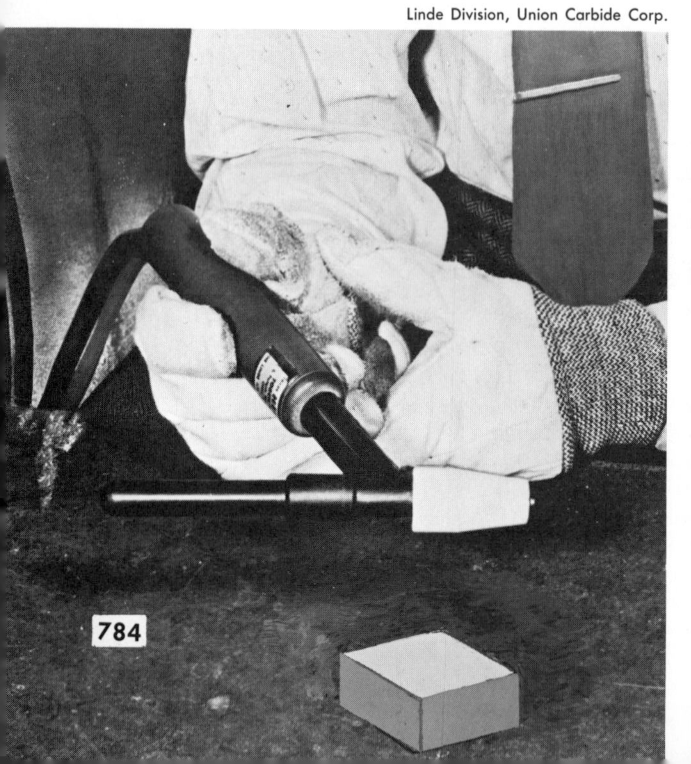

24-8. *End of swing to draw an arc in high frequency starting.*

Linde Division, Union Carbide Corp.

the designated place each time you strike an arc and will continue arcing in the same direction during the entire operation. If you are welding aluminum or magnesium with alternating current, ball the end of the tungsten. If the electrode becomes contaminated or deformed, do not increase the machine setting. Reshape the tungsten instead.

A short arc should be maintained. This helps to confine the heat to the immediate area. Be careful to direct the arc only in the exact spot where it is desired to liquefy the metal. To apply heat to any other spot only increases distortion and deterioration. Play the arc on the seam to be welded until a molten puddle appears. Then proceed as rapidly as possible and make sure that both pieces to be joined become molten under the arc. If the puddle is slow in forming on both pieces, it may be necessary to use a slight oscillating motion to distribute the metal evenly. Care must be taken not to leave the molten puddle, not to stay in one spot too long, and not to remelt metal that has just been melted.

If the addition of filler rod is necessary, choose a rod size close to the thickness of the material being welded. Do not play the rod into the arc. Instead, move the rod out with the arc and then carry the correct amount back to the weld zone in order to make a uniform distribution. Correct motions to use in different positions are very important for successful results. A different, distinct motion is used in each position; each of these must be learned from an experienced operator by demonstration. In the case of corner welds, apply the heat in the corner first and then work away from the corner toward the outside.

The extension of the tungsten electrode beyond the cup is governed by the shape of the object and the type of joint. The longer the extension, the less effective the shielding. Consequently, some adjustment in rate of gas flow is necessary where shapes require longer extension and loss of shielding occurs. The cup size is governed by the type of joint, the shape of the object, and the amount of current used.

All torch connections must be kept tight to prevent siphoning of air through connections. A small amount of contamination can be serious.

General Recommendations

Study Table 24-11, The Trouble Shooting Guide, thoroughly before you begin welding practice. Look up your defects in the table during practice and correct them as instructed before beginning the next job. You will find the following recommendations helpful.

- Keep the tungsten electrode clean and free of contamination and keep the point reasonably sharp.
- Remove all surface oxides by wire brushing or grinding before welding over any previous weld.
- Prevent shrinkage holes from forming in the weld craters. Do not finish a weld by breaking the arc abruptly. Taper off the weld bead by carrying it off to the side of the weld and by increasing the travel speed so that the weld puddle diminishes in size before breaking the arc.
- Electrodes are normally operated at the highest possible current density in order to obtain maximum arc stability. This often results in electrode contamination and excessive electrode consumption. The problem is reduced by using the next size larger electrode and grinding it to a long, needlelike point.
- Control the length of the arc carefully. A long arc length—$\frac{1}{8}$ to $\frac{3}{16}$ inch—gives poor penetration and wide weld beads. A long arc length is particularly harmful on the first pass of fusion welds. A short arc length—$\frac{1}{32}$ to $\frac{3}{32}$ inch—gives good penetration and a narrow weld bead. This arc length is excellent for the first pass in welding plate or pipe.
- Whether cover passes are made with stringer beads or weaved beads depends on the following considerations:
  - The kind of material: weave beads cause cracks in some metals.
  - The thickness of the material: excessive penetration occurs in thin metals.
  - The position of welding
  - The type of current: for example, since DCSP causes deep penetration, weaved beads increase the possibility of excessive penetration.

PRACTICE JOBS

The practice jobs in this chapter require the welding of the three metals that are most commonly welded by the gas tungsten-arc process. The major application of TIG welding is in those fields that make use of aluminum, carbon and low alloy steels, and stainless steels. Few welders will have an occasion to weld the other metals that can be welded by the process. If you are called upon to weld such materials as magnesium or copper, however, the techniques that you have learned in welding the three major metals can be applied with little difficulty.

Welding practice will begin with aluminum since this metal is one of the easiest to weld with the TIG process, it has the widest

Table 24-11

Trouble shooting guide.

Problem	Cause	Remedy
Wasteful electrode consumption	1. Improper inert shielding (resulting in oxidation of electrode)	1. Clean nozzle; bring nozzle closer to work; step up gas flow.
	2. Operating on reverse polarity	2. Employ larger electrode or change to straight polarity.
	3. Improper size electrode for current required	3. Use larger electrode.
	4. Excessive heating in holder	4. Ground-finish electrodes; change collet; check for improper collet contact.
	5. Contaminated electrode	5. Remove contaminated portion—erratic results will continue as long as contamination exists.
	6. Electrode oxidation during cooling	6. Keep gas flowing after stoppage of arc for at least 10–15 seconds. Rule: 1 second for each 10 amperes.
Erratic arc	1. Base material is dirty, greasy.	1. Use appropriate chemical cleaners, wire brush, or abrasives.
	2. Too narrow joint	2. Open joint groove; bring electrode closer to work; decrease voltage.
	3. Electrode is contaminated.	3. Remove contaminated portion of electrode.
	4. Too large diameter electrode	4. Use smaller electrode—use smallest diameter needed to properly handle current.
	5. Arc too long	5. Bring holder closer to work to shorten arc.
Porosity	1. Entrapped gas impurities (hydrogen, nitrogen, air, water vapor)	1. Blow-out air from all lines before striking arc; remove condensed moisture from lines; use welding grade (99.995%) inert gas.
	2. Possible use of old acetylene hoses	2. Use only new hoses. Acetylene impregnates hose.
	3. Gas and water hoses interchanged	3. Never interchange water and gas hoses. (Color coding helps).
	4. Oil film on base material	4. Clean with chemical cleaner not prone to break-up in arc; DO NOT WELD WHILE WET.
Tungsten contamination of workpiece	1. Contact-starting with electrode	1. Use hi-frequency starter; use copper striker plate.
	2. Electrode melting and alloying with base plate	2. Use less current or larger electrode; use thoriated or zirconated tungsten (these run cooler).
	3. Shattering of electrode by thermal shock	3. Make certain electrode ends are not slivered or cracked when using high current values. Use embrittled tungsten to facilitate easy and clean breakage.

Eutectic Corp.

application in industry, and it creates a high degree of interest for the student. The welding of the basic weld joints in aluminum is followed by practice in the welding of mild steel and stainless steel in that order.

Instructions for Completing Practice Jobs

Since the gas tungsten-arc process may be used to weld so many different materials and the type of welding current varies with the material, it is very difficult to work out a series of jobs that gives the exact current range for all situations. Also, since the students are accomplished gas and arc welders, it is unnecessary to go into detailed instructions in regard to welding technique. Only those characteristics that are peculiar to TIG welding are presented.

The jobs listed in the Course Outline, page 800, are presented for you to perform. They provide experiences in all positions of welding. As in your previous practice, the simplest and easiest jobs are presented first, followed by those of increasing difficulty. Each new weld contains techniques previously learned plus new techniques. It is very important to master each new technique before going on to the next job. Failure to do this will seriously hamper you in your efforts to master the gas tungsten-arc process.

Consult the Course Outline before you begin welding each joint. The outline presents the type of joint, the sequence of welding, and the proper settings for making each weld. Then turn to the pages listed in the Text Reference column for a discussion of the welding technique recommended for the job.

Weld Beading on Flat Plate: Job 24-J1

After the arc has been struck, hold the torch at about a 75-degree angle to the surface of the work. Preheat the starting point by moving the torch in small circles until a molten puddle is formed, Fig. 24-9. The size of the pool is determined by the electrode size and the welding current. (The metal thickness determines the electrode size and current value.) Pool size can be increased by widening the circular motion of the torch.

The end of the electrode should be held approximately $\frac{1}{8}$ inch above the work. When the puddle becomes bright and fluid, move the torch slowly and steadily along the plate at a speed that will produce a bead of uniform width. The welding technique is similar to that used with gas welding.

Filler rod must be held within the shielding envelope of inert gas while it is hot, or it must be withdrawn quickly beyond the heat of the arc to prevent excessive oxidation. Hold the filler rod at an angle of about 15 degrees to the work and about 1 inch away from the starting point, Fig. 24-10. When the puddle is fluid, move the arc to the rear of the puddle and add filler rod to the leading edges of the pool at one side of the center line, Fig. 24-11. Move both hands together with a slight backward and forward motion along the joint.

Make sure that fusion takes place between the weld metal and the base metal. By watching the edges of the weld pool, you

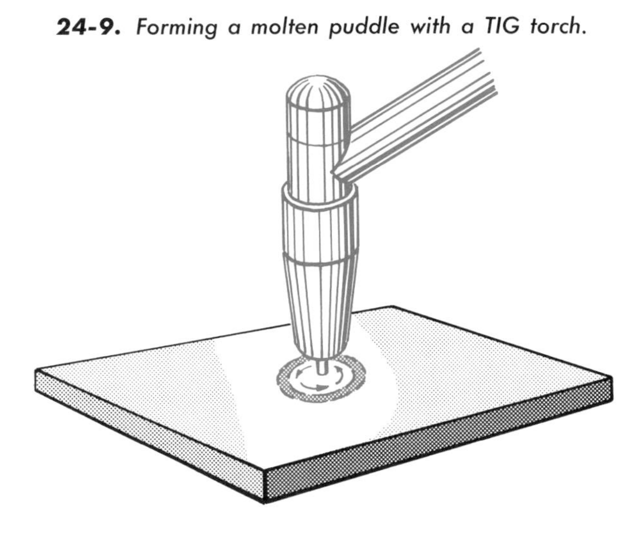

24-9. *Forming a molten puddle with a TIG torch.*

24-10. *Correct positions of the torch and filler rod at the start of bead welding.*
Kaiser Aluminum & Chemical Corp.

will learn to judge its fluidity and the extent of buildup and fusion into the base metal. Incorrect torch angle, improper manipulation, too high a welding current, or too slow a welding speed can cause undercutting in the base plate along one or both edges of the weld bead. Make sure that the tungsten electrode does not touch the filler rod and do not withdraw the hot end of the filler rod from the gas shield.

A short arc length must be maintained to obtain adequate penetration and to avoid undercutting and excessive width of the weld bead. It is important that you maintain control of penetration and weld contour. One rule is to use an arc length approximately equal to the diameter of the tungsten electrode. It is also important that you see the arc and weld puddle. Always use a gas cup or nozzle that is as small as possible for a given weld and still affords adequate gas shielding.

There are two different techniques for manipulating the torch and filler rod. In the two-step technique, addition of the filler rod to the weld pool is alternated with torch movement. In the uninterrupted technique, the torch is moved forward steadily, and the filler rod is fed intermittently as the weld pool needs it. This latter technique gives a somewhat better appearing weld bead which needs little or no finishing.

For beading on a vertical surface, the torch is held perpendicular to the work. The weld is usually made from top to bottom. When the filler rod is used, it is added from the bottom (leading edge) of the puddle in the same manner as described previously. Figure 24-12 shows the correct positioning of the rod and torch relative to the workpiece.

Various methods can be used to break the arc or stop welding. If the arc is broken abruptly and the torch removed from the weld area, shrinkage cracks will probably occur in the weld crater and cause a defective weld. Cracking can be prevented by one of the following methods.

- Move the torch faster until the metal is no longer molten.
- Backtrack the bead slightly before breaking the arc.
- Lengthen the arc gradually and break and restrike it while adding filler metal to the crater.
- Reduce the current by operating a foot control.

After the arc has been broken, it is necessary to let the gas continue to flow for about 30 seconds to permit the end of the electrode to cool. The tungsten electrode requires an inert gas shield until it is cool to prevent oxidation.

INSPECTION AND TESTING. Compare the appearance of the beads you have made with those in Fig. 24-13B. Beads should be somewhat convex in contour. The face should be smooth; and the ripples, close together. There should be good fusion at the weld edges and no surface porosity. Figure 24-13 shows both the surface appearance and etched cross-sections of three beads on a flat plate. The welding current selected for each weld determines its quality. The weld bead shown in *A* indicates that the current is too high. In *B* the current is correct, and in *C* the current is too low.

24-11. *Filler metal is fed by hand in a manner similar to that used when welding with the oxyacetylene process.*

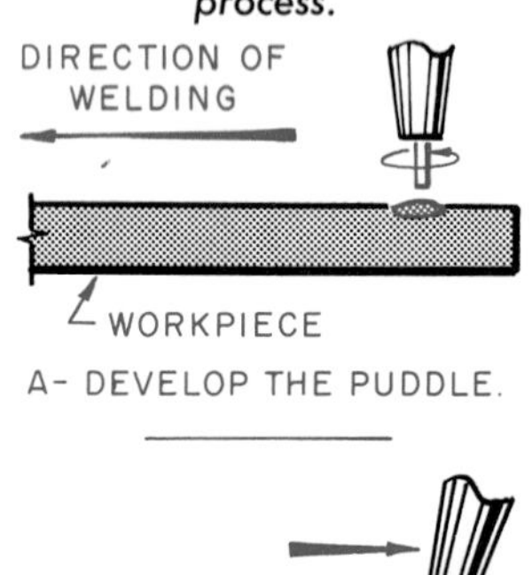

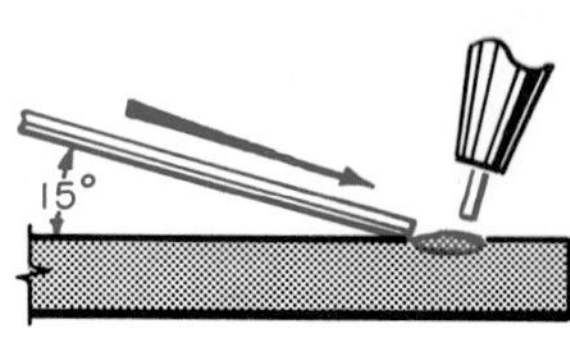

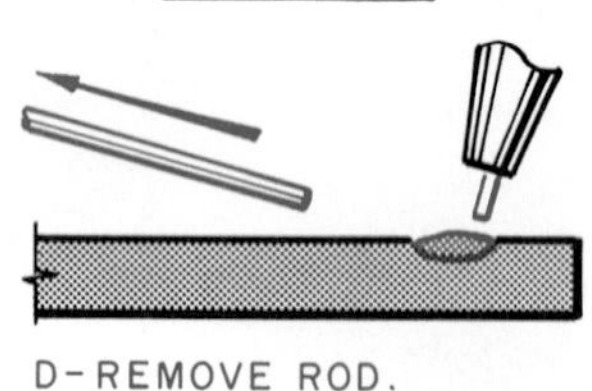

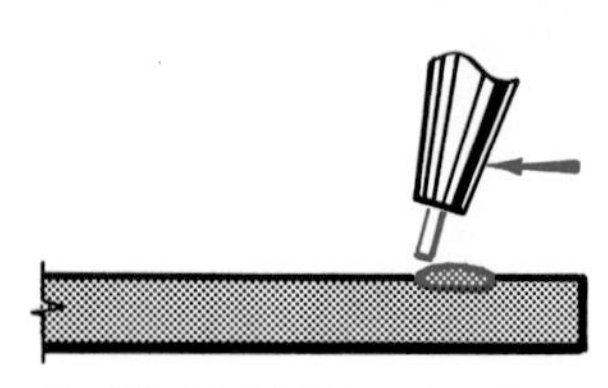

24-12. *Addition of filler metal when welding in the vertical position.*

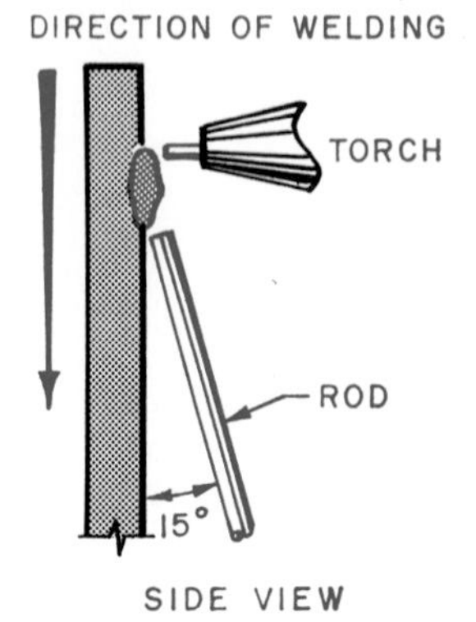

Weld beads made with sufficient and insufficient shielding gas are illustrated in Fig. 24-14. Insufficient shielding gas produces an unsound weld bead having a great deal of porosity and very poor appearance, Fig. 24-14B. Using too much shielding gas is wasteful.

Continue to practice until you can make beads that compare favorably with Fig. 24-13B and that are acceptable to your instructor.

Welding an Edge Joint: Job 24-J2

The edge joint is the easiest type of weld to make with the gas tungsten-arc process. The technique is similar to that used with the gas and shielded metal-arc processes. Edge joints should be used only on light gauge metal, and they usually require no filler rod. Preparation is simple, and welding is economical and fast. An edge joint should not be used where direct tension or bending stresses will be applied because it may fail at the root under low stress loads. The joint may be used for small tank bottoms.

Uniform speed of travel will produce beads that are smooth and even. Joints may be welded with or without filler rod. A welding speed that is too slow causes the bead to pile up and roll over the edges of the plates. Too high a speed causes skimming of the surface and poor penetration in the joint.

Make sure that the abutting edges of the two plates are fused along the center line and that the entire edge surface of the plates are fused. The position of the torch is shown in Fig. 24-15.

In welding an edge joint in aluminum, be sure to "ball," or round, the end of the electrode before welding to keep the arc steady. In welding carbon steels and stainless steel, sharpen the end of the electrode in order to secure pinpoint concentration of heat. The welding technique for stainless steel usually requires a slower travel speed than for aluminum or carbon steel.

INSPECTION AND TESTING. Complete a number of edge joints in all positions and compare them with the bead shown in Fig. 24-16. Pry the plates apart to see if you are getting penetration at the inside corners of the plate and fusion to the surface of the plate edges. Welds should pass the inspection of your instructor.

Welding a Corner Joint: Job 24-J3

Corner joints are used in the manufacture of boxes, pans, guards, and all types of containers. An open corner joint may be used for thicknesses up

24-13. *Surface appearance and etched cross section of three aluminum weld beads: A. Welding current too high. B. Correct welding current. C. Welding current too low.*

Kaiser Aluminum & Chemical Corp.

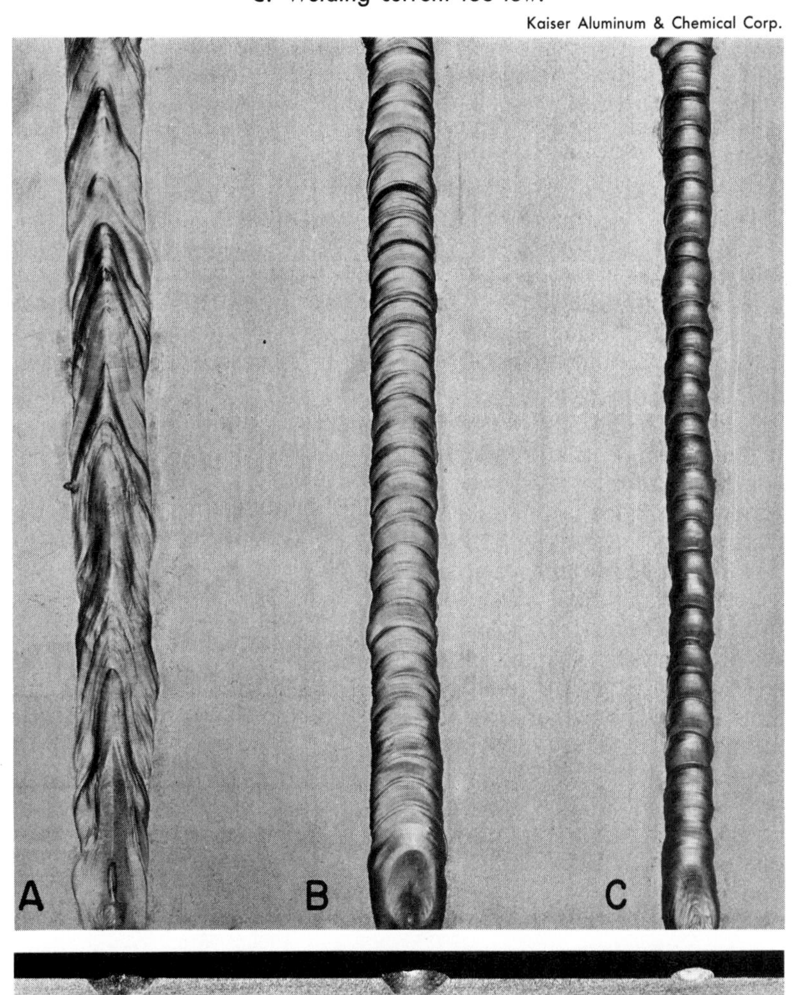

to $^5/_{32}$ inch. Filler rod is required. Heavier materials require the beveling of one of the plates in order to secure complete penetration. Both pieces must be in good contact all along the seam. The number of passes required depends on the thickness of the material and the bevel angle.

Welding this joint is somewhat similar to making an edge joint, but it is usually necessary to add filler rod during welding. The amount of filler rod required depends on the thickness of the material and the degree to which it is a full open joint design. It is necessary to get complete fusion through the root of the joint. The inside buildup should not be high, but it must be complete. The position of the welding torch is like that used in the welding of an edge joint.

INSPECTION. Practice making corner joints until you are satisfied with their general appearance (Fig. 24-17) and you obtain fusion through to the back side. Place a welded joint on an anvil and hammer it flat. You should be able to flatten the joint without breaking the weld through the throat.

24-14. *Surface appearance and etched cross-section of aluminum weld beads made with gas shielding: A. Sufficient shielding gas. B. Insufficient shielding gas.*

Kaiser Aluminum & Chemical Corp.

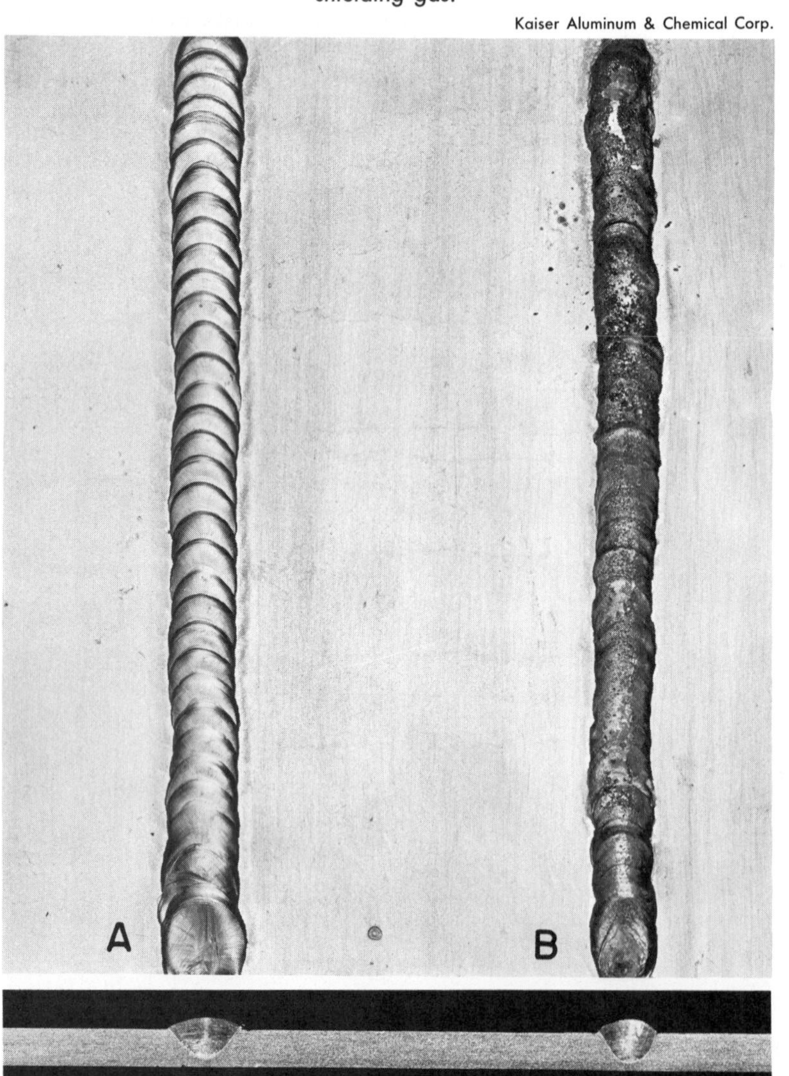

Welding a Lap Joint: Job 24-J4

A lap joint does not require edge preparation, but the plates must be in close contact along the entire length of the joint. On materials less than $^3/_{16}$ inch thick, filler rod may not be required. The general practice is to use a filler rod in making the weld. This type of joint is not recommended for material more than $^1/_4$ inch thick. The thickness of the material determines the number of passes.

The fillet weld in a lap joint is started by first forming a puddle on the bottom sheet. When the puddle becomes bright and fluid, shorten the arc. Play the torch over the upper sheet. The puddle will be V-shaped. The center of the puddle is called the *notch,* Fig. 24-18. The speed at which the notch travels determines how fast the torch can be moved ahead. In order to secure complete fusion and good penetration into the corner and on both plates, this notch will have to be filled in for the entire length of the seam.

Adding filler rod increases the speed of welding. Be sure to get complete fusion, and not merely lay in bits of filler rod on cold, unfused base metal. The filler rod should be alternately dipped into the weld pool and withdrawn $^1/_4$ inch or so. It is very important to control the melting rate of the edge of the top plate. If this edge is melted too rapidly and too far back, a uniform weld is impossible.

There are two conditions necessary to obtain a uniform bead which has the proper proportions and good penetration.

- The top plate must be kept molten along the edge to the corner, and the bottom plate must be molten on the flat surface to the corner.
- Just the right amount of filler metal must be added where it is needed, Fig. 24-19.

INSPECTION AND TESTING. Practice making lap joints until you are satisfied with their general appearance (Fig. 24-20) and feel that you are getting good fusion and penetration. Make up a test joint and pry it apart. The top plate should be capable of bending 90 degrees without breaking, and there should be evidence of complete penetration at the root of the joint.

Welding a T-Joint: Job 24-J5

All T-joints require the addition of filler rod to provide the necessary buildup and to give the strength required. The number of passes depends on the thickness of material and the size of the weld desired. The joint may be welded from one or both sides. Complete penetration may require a beveled edge on the web plate. When there is no edge preparation, be sure that current values are high enough for the thickness of the web plate.

The procedure for making a fillet weld in a T-joint is like that described above for a lap weld. It is important to proportion the

24-15. *Welding an edge joint in the flat position. No filler rod added.*
Linde Division, Union Carbide Corp.

Kaiser Aluminum & Chemical Corp.
24-16. *Aluminum edge joint. Note the smooth appearance of the bead.*

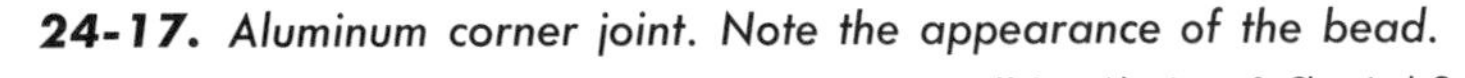

24-17. *Aluminum corner joint. Note the appearance of the bead.*
Kaiser Aluminum & Chemical Corp.

weld properly between the two plates so that each leg of the fillet weld is of equal length.

The design of the joint makes it somewhat difficult to avoid undercutting the vertical plate. If filler rod is used, it can be fed to the weld pool in such a manner that it provides protection for the upper plate. See Figs. 24-21 and 24-22 for the proper positions of the torch and filler rod. Undercutting can also be caused by using too high a welding current or by traveling too fast.

Make sure that you obtain complete penetration to the root of the joint and that complete fusion is taking place on both plate surfaces. Incomplete penetration is caused by not forming the weld pool in the base metal before adding filler rod to the leading edge of the pool. This defect can also be caused by welding with inadequate current or by welding too fast.

When welding a T-joint in aluminum, care should be taken to avoid penetration fall-through. In starting the weld, begin about 1 inch away from one edge and weld to the nearest edge. Then return to the start of this weld and begin welding in the opposite direction. Start the second weld puddle about $\frac{1}{4}$ inch into the short bead. This is necessary in order to prevent cracking the plate along the edge of the weld or through the center of the weld.

Do not travel too fast or add filler rod too often. Filler rod absorbs a good deal of heat. If a large amount is added quickly, the weld puddle chills and causes a loss of penetration and fusion.

A good fillet weld looks shiny, and every ripple is evenly spaced. (See Fig. 24-21.) A dull-colored weld indicates that the weld was made either too hot or too cold. If there is penetration fall-through, the weld was run too hot. If the weld was run too cold, the edges of the weld overlap the plate surface, and the ripples are coarse and rough.

In making a fillet weld in the vertical position, you must make sure that the current setting is not too high and that the weld pool does not become too large. A large weld pool is difficult to handle and spills over. A slight

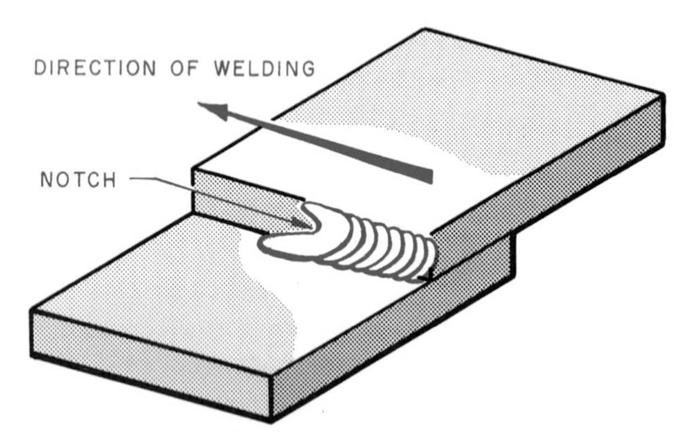

24-18. *Lap welding technique.*

24-19. *Method of adding filler rod to a lap weld.*

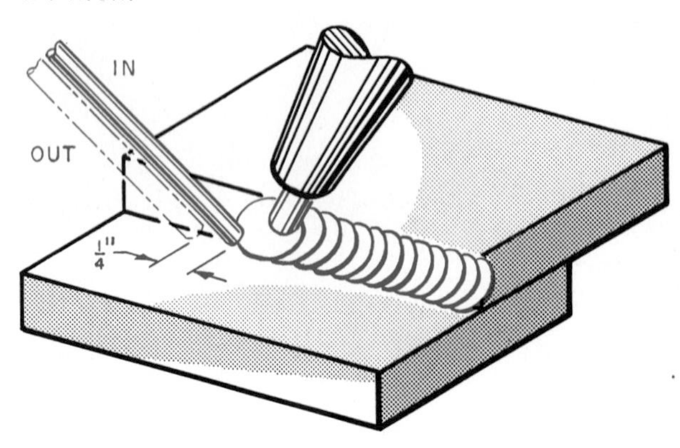

24-20. *A lap joint in aluminum plate. Note the complete fusion at the toe of the weld bead.*
Kaiser Aluminum & Chemical Corp.

weave assists in smoothing out the bead. Keep the face of the weld as flat as possible. When weaving, a slight hesitation at each side prevents undercut. See Fig. 24-23 for torch and filler rod position.

When making a fillet weld in the overhead position, a lower current setting and slower travel speed are used than in flat position welding, but the flow of shielding gas is higher. Care must be taken to avoid sagging and poor penetration. These defects result from adding too much filler rod and carrying too large a weld pool. Let the pool wet out enough before adding more filler rod. You may find the overhead position awkward. Therefore, try to get in as comfortable and relaxed a position as possible when welding. This helps in maintaining steady, even manipulation of the torch and filler rod. Figure 24-24 illustrates the correct torch and filler rod positions.

MULTIPASS WELDING. Multipass welding is generally necessary for material over $\frac{3}{16}$ inch thick. The number of passes required depends on the thickness of the material, the design of the joint, the position of welding, and the nature of the assembly being fabricated. The first pass is always to be considered as the "root weld." It must provide complete fusion and penetration at the bottom of the joint. Subsequent passes can be welded at higher current values since the root pass acts as a backup. Complete fusion into the root weld and the side walls of the joint are necessary to prevent areas of lack of fusion. It is also important to provide for clear flow of the weld pool to prevent inclusions. Both weaved and stringer bead techniques may be used.

Practice making both single- and multiple-pass fillet welds. By making multipass welds you get the fullest use out of the material. Figure 24-25 gives the sequence of passes for each of the positions of welding.

INSPECTION AND TESTING. Practice making T-joints until you are satisfied with their general appearance and feel that you are getting good fusion to the plate surfaces and good penetration at the root of the weld. Make a test joint in aluminum like that shown in Fig. 24-26. Bend the specimen back, and etch a cross section of the weld. Examine the cross section for root tie-in, porosity, skips and folds, and inclusions. Fusion must be to the root of the weld. The legs of the fillet must show fusion to the base plates for a distance that is at least equal to the thickness of the plates. For fillet welds in plates of different thicknesses, the fillet leg should at least equal the thickness of the thinner plate. The weld convexity or concavity must not exceed $\frac{1}{16}$ inch beyond a flat face.

Welding a Butt Joint: Jobs 24-J6 and J7

Butt joints are common to all welding. The square butt joint is the easiest to prepare. It can be

24-21. *The correct positioning of the torch and filler rod to prevent undercut and obtain fusion in making a horizontal fillet weld.*

Kaiser Aluminum & Chemical Corp.

24-22. *A closeup of the positions shown in Fig. 24-21. Note the relationship of torch, rod, and crater.*

Kaiser Aluminum & Chemical Corp.

welded with or without filler rod, depending on the thickness of the material. Complete penetration can be secured. You must try to avoid both lack of penetration and burn-through.

The single-V butt joint is used when complete penetration is required on material thicknesses ranging between 3/8 and 1 inch. Filler rod is necessary for multipass welding. The angle of bevel should be approximately 30 degrees. The root face should measure from 1/8 to 3/16 inch, depending on the type and thickness of the metal being welded.

A double-V butt joint is generally used on material thicker than 1/2 inch when welding can be done from both sides. The angle of bevel is also 30 degrees, and the root face varies from 1/8 to 3/16 inch. With proper welding, complete penetration and fusion are assured.

The procedure for making a butt joint is similar to that described previously for beading on flat plate. There is the added problem, however, of securing complete penetration through the root of the joint to the back side. It is, of course, necessary to make sure that the burn-through on the backside is not excessive. Refer to Figs. 24-27 through 24-30 for the correct position of the torch and rod when welding aluminum plate in the different positions.

In welding aluminum, complete penetration is desirable for plate thicknesses 1/8 inch or under. The back side should look smooth and bright and have close ripples. A good aluminum butt weld on thin plate has a thickness through the throat of the weld equal to twice the thickness of the material. Fall-through is a problem. Hold the torch forward, and add filler rod when you see that the puddle is about to fall through. Do not start the

24-23. *Positions of the torch and filler rod when making a vertical fillet weld in aluminum plate.*

Kaiser Aluminum & Chemical Corp.

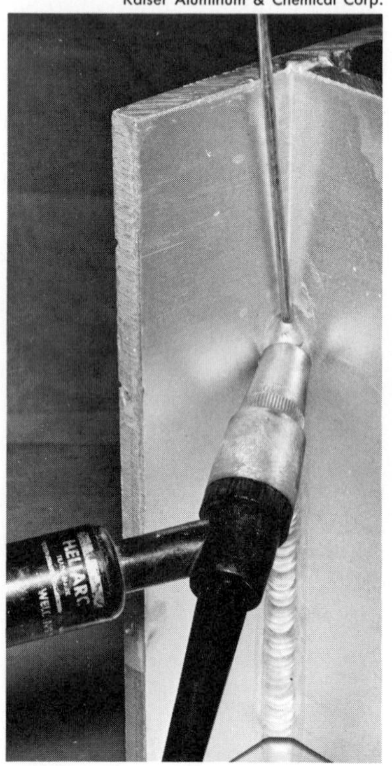

24-24. *Positions of the torch and filler rod when making an overhead fillet weld in aluminum plate.*

Kaiser Aluminum & Chemical Corp.

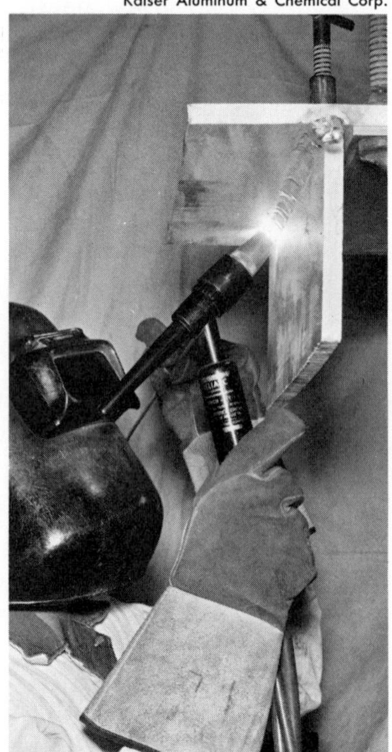

24-25. *Weld pass sequence for (A) horizontal, (B) vertical, and (C) overhead positions.*

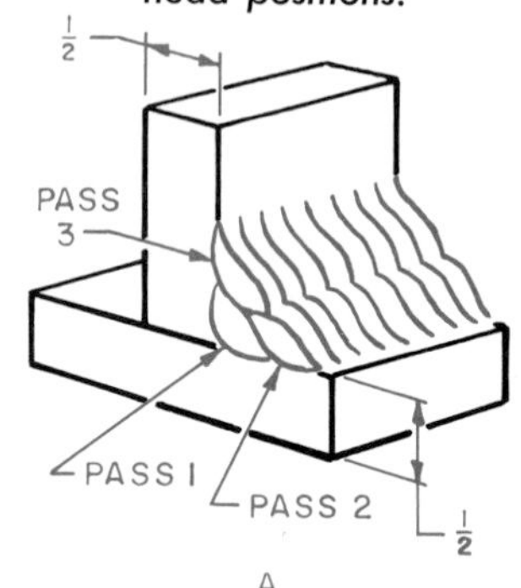

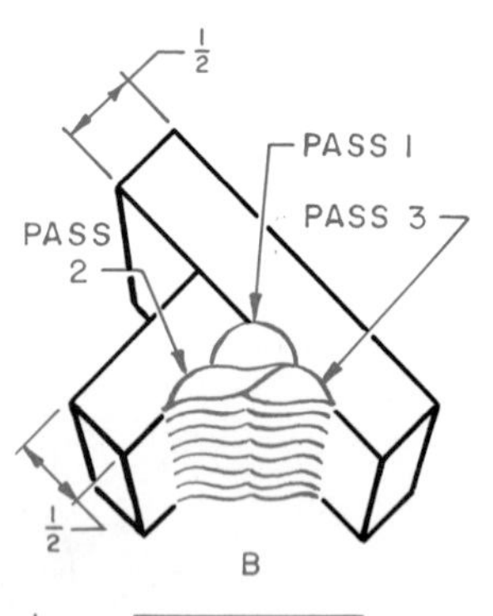

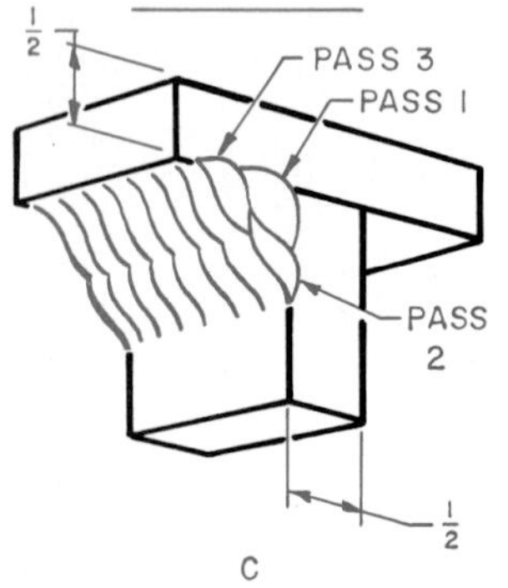

weld at the edge of the plate. Fill the crater hole at the finish of the weld with the technique described earlier. When high quality welds are desired, an inert gas backup keeps the air away from the underside of the weld to prevent contamination.

Practice making both single- and multiple-pass butt welds. Figure 24-31 shows the sequence of passes for multipass welds in

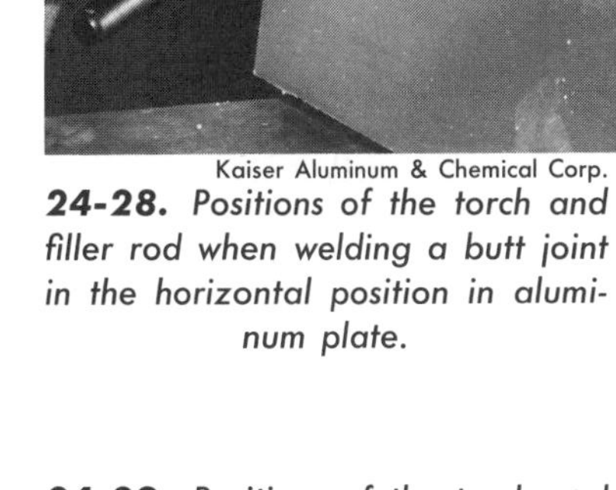

Kaiser Aluminum & Chemical Corp.

24-28. *Positions of the torch and filler rod when welding a butt joint in the horizontal position in aluminum plate.*

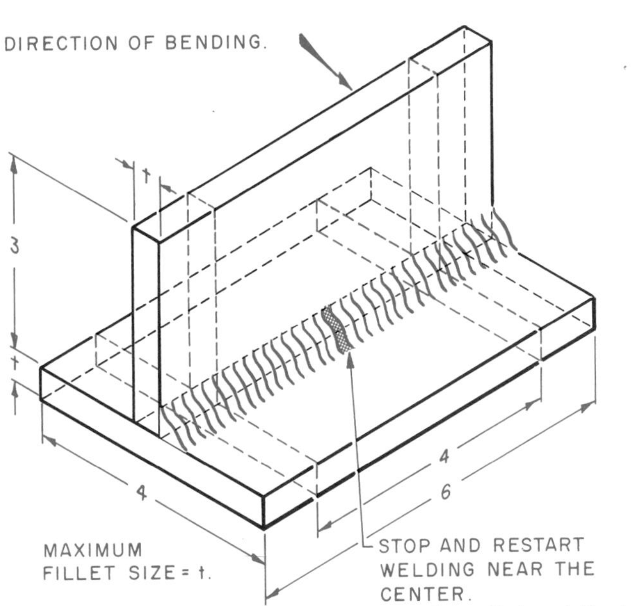

Kaiser Aluminum & Chemical Corp.

24-26. *Fillet weld test specimen.*

24-29. *Positions of the torch and filler rod when welding a butt joint in the vertical position in aluminum plate.*

Kaiser Aluminum & Chemical Corp.

24-27. *Positions of the torch and filler rod when welding a butt joint in the flat position in aluminum plate.*

Linde Division, Union Carbide Corp.

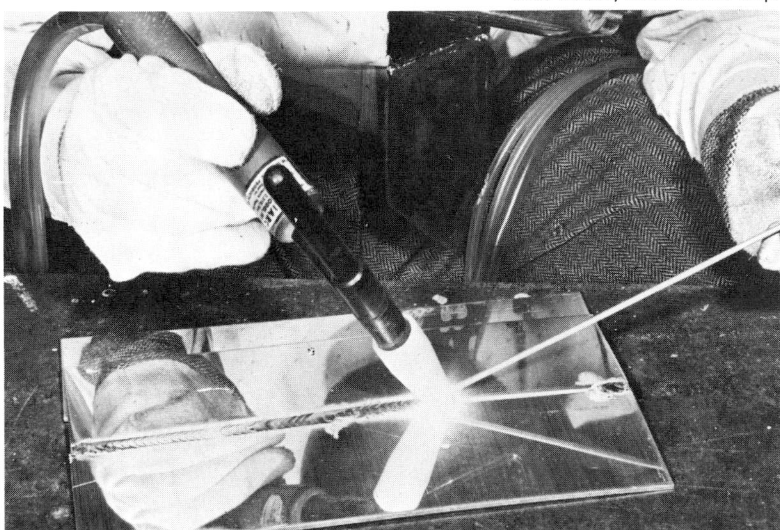

heavy plate for each position of welding.

INSPECTION AND TESTING. Practice making butt joints until you are satisfied with their general appearance and there is evidence that you are getting complete penetration without excessive buildup on the back side. Make up a test weld in $\frac{3}{16}$- to $\frac{3}{8}$-inch aluminum plate, Fig. 24-32. Prepare two specimens sectioned from the test plate for bend testing as instructed in Fig. 24-33. The specimens must bend 180 degrees without cracking or fracturing in a fixture as specified in Fig. 24-34.

The hydraulic guided-bend test fixture shown in Fig. 24-35 can be made in the school. The hydraulic jack and pressure gauge will need to be purchased. Bending the specimens in a fixture of this kind assures full control of the testing situation and provides a uniform test procedure for all specimens tested. Students have complete confidence in the procedure.

Carbon and Stainless Steel Practice: Jobs 24-J8 through 24-J19

In welding carbon and stainless steel, follow much the same

24-31. *Sequence of weld passes for butt welds in (A) flat, (B) horizontal, (C) vertical and (D) overhead positions in heavy aluminum plate.*

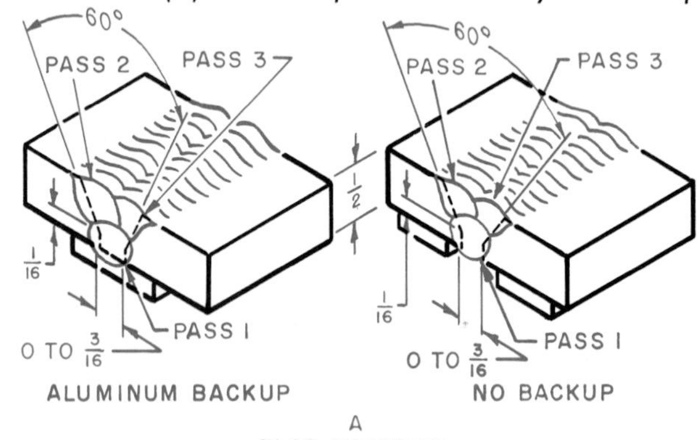

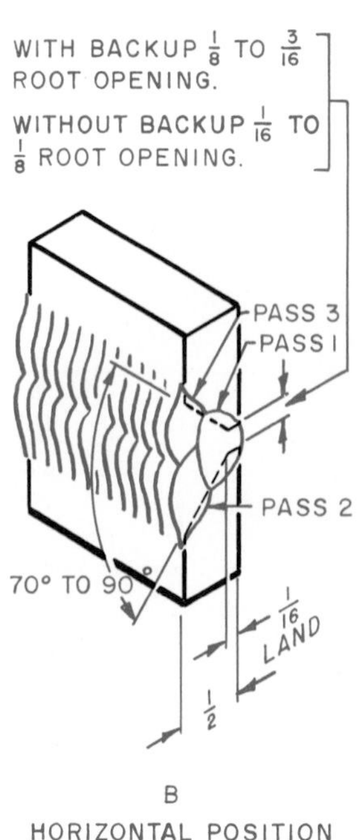

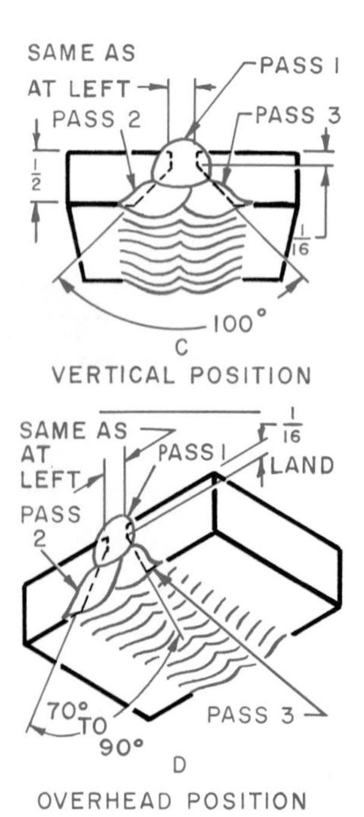

24-30. *Positions of the torch and filler rod when welding a butt joint in the overhead position in aluminum plate.*

Kaiser Aluminum & Chemical Corp.

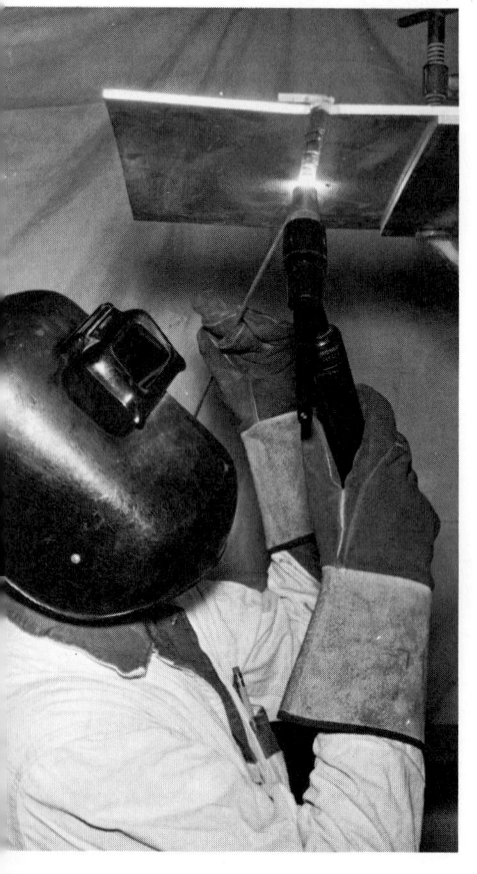

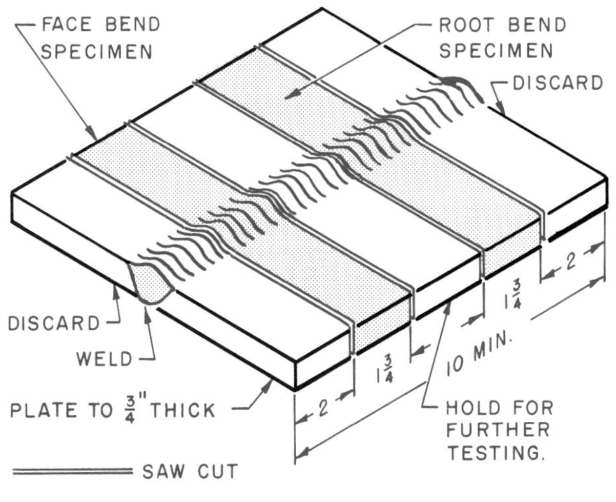

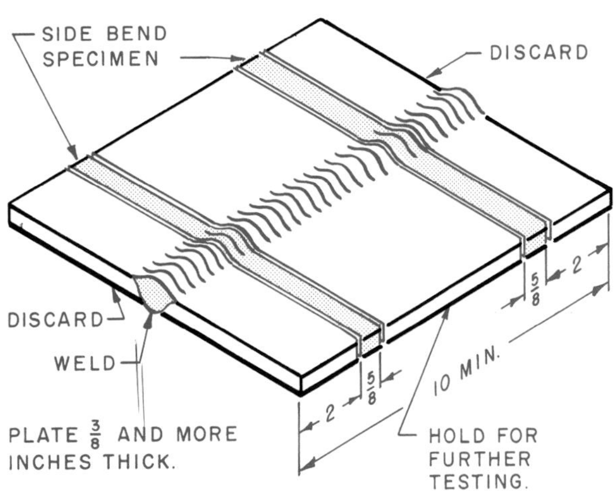

24-32. *Test plates for butt welds in aluminum.*

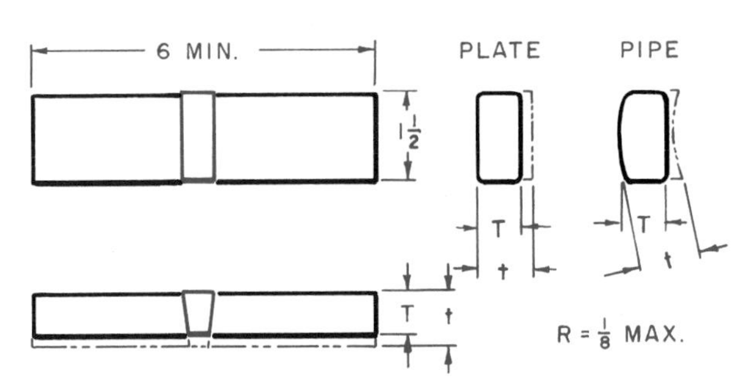

24-33. *Shapes of face- and root-bend specimens.*

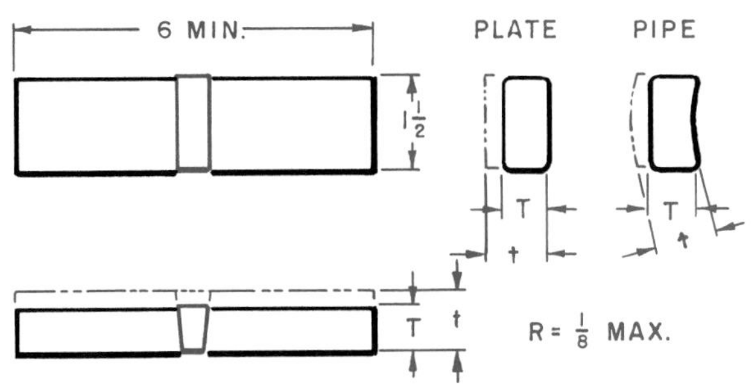

procedure that you followed in the welding of aluminum. Penetration fall-through is a problem because it is harder to control when welding steel than when welding aluminum. The larger aluminum puddle makes it easier to judge when the weld pool is about to fall through. The steel weld pool is about half as large as the aluminum pool. A copper backup strip prevents fall-through. It may be necessary to heat the copper backup before beginning the weld. This method is used for operator certification when penetration fall-through is a cause for failure.

When welding stainless steel, you will notice that the puddle seems to stay in the same place until you move the torch. You will have to push the weld pool with the torch. Other than this, the welding technique is not much different than that used for carbon steel or aluminum.

Stainless steel requires less heat than aluminum. The tungsten electrode should be sharpened to a point at the end. The choice of the filler rod size is crucial. If the rod diameter is too large, it will soak up a good deal of heat and make welding more difficult.

Distortion is greater in stainless steels than in aluminum and carbon steel. Proper clamping is very important. Although travel speed can be increased by increasing the heat setting, high heat requires great skill on the part of the welder. The adding of filler rod must be timed perfectly, or the weld puddle will overheat and burn through.

Butt welds in stainless steel are made with exactly the same technique as butt welds in carbon steel except for the slower travel speed. A dark purplish or a dark purplish-blue bead with hardly noticeable ripples indicates too much heat. Figure 24-36 shows stainless steel welds.

INSPECTION AND TESTING. Practice the jobs listed in the Course Outline for each type of metal. When you are satisfied with their appearance and are getting good penetration and fusion on all types of joints, make up the usual test plates and test the specimen in the usual manner.

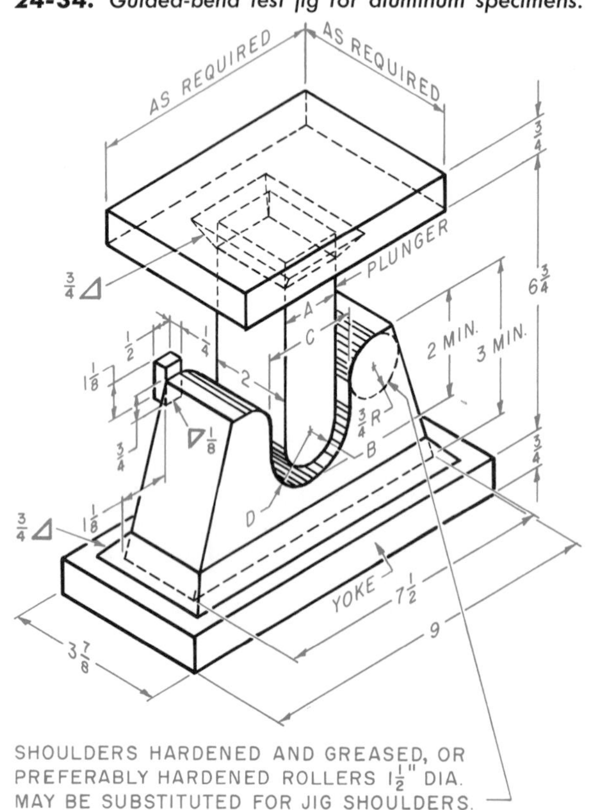

24-34. *Guided-bend test jig for aluminum specimens.*

24-35. *Hydraulic guided-bend test fixture. An aluminum test coupon is being bent in the jig shown in Fig. 24-34.*

McDonnell Douglas Corp.

24-36. *Gas tungsten-arc welds in stainless steel. These are parts for a spacecraft.*

REVIEW QUESTIONS

1. List at least eight safety precautions required for the gas tungsten-arc process.
2. Only certain types of joints can be welded with the gas tungsten-arc process. True. False.
3. Name and explain the various types of weld backup.
4. List five factors necessary to good TIG welding technique.
5. List four causes of arc blow.
6. In the gas tungsten-arc welding of aluminum, carbon dioxide is the best shielding gas. True. False.
7. When welding stainless steels with the gas tungsten-arc process, the weld is sound, but it loses much of its high resistance to corrosion. True. False.
8. Aluminum should never be welded with a.c. current. True. False.
9. What are the proper filler rod and torch angles for gas tungsten-arc welding?
10. Name and explain the various starting methods.
11. Explain the method that should be followed in breaking the arc to stop welding.
12. Fillet welds require a higher rate of gas flow than other joints. True. False.
13. Give two reasons why direct current, straight polarity is recommended for welding stainless steel.
14. A greater flow of shielding gas is necessary when welding stainless steel than other metals. True. False.
15. Why is it necessary to provide more jigs and fixtures when welding stainless steel?
16. Stainless steel requires less heat than aluminum for welding. True. False.

Table 24-12
Course Outline: Gas Tungsten-Arc Welding Practice (Plate)

Job no.	Type of joint	Type of weld	Position	Material			Current		Shielding gas		Cup size	Filler rod size	Text references
				Type	Thickness	Tungsten size	Type	Amperes	Type	Gas flow c.f.h.			
24-J1	Flat plate	Beading	Flat	Aluminum	1/8	3/32	ACHF	90–140	Argon	12–20	3/8	3/32	787
24-J2	Edge	Beading	Flat	Aluminum	1/8	3/32	ACHF	90–140	Argon	12–20	3/8	3/32	789
24-J3	Outside corner	Fillet	Flat	Aluminum	1/8	3/32	ACHF	100–140	Argon	12–20	3/8	3/32	789
24-J4	Lap	Fillet	Flat	Aluminum	1/8	3/32–1/8	ACHF	90–140	Argon	12–20	3/8	3/32	790
24-J5	T	Fillet	Flat	Aluminum	1/8	3/32–1/8	ACHF	90–150	Argon	12–20	5/16	3/32	791
24-J6	Square butt	Groove	Flat	Aluminum	1/8	3/32	ACHF	80–130	Argon	12–20	3/8	3/32	793
24-J7	Single-V butt	Groove	Flat	Aluminum	1/4	3/16	ACHF	190–280	Argon	22–28	1/2	3/32–1/8	793
24-J8	Flat plate	Beading	Flat	Mild steel	1/8	1/16–3/32	DCSP-ex* ACHF-g*	90–120	Argon	10–14	1/4–3/8	3/32	796
24-J9	Edge	Beading	Flat	Mild steel	1/8	1/16–3/32	DCSP-ex ACHF-g	90–120	Argon	10–14	1/4–3/8	3/32	796
24-J10	Outside corner	Fillet	Flat	Mild steel	1/8	1/16–3/32	DCSP-ex ACHF-g	90–120	Argon	10–14	1/4–3/8	3/32	796
24-J11	Lap	Fillet	Flat	Mild steel	1/8	1/16–3/32	DCSP-ex ACHF-g	100–130	Argon	10–14	1/4–3/8	3/32	796
24-J12	T	Fillet	Flat	Mild steel	1/8	1/16–3/32	DCSP-ex ACHF-g	100–130	Argon	10–14	1/4–5/16	3/32	796
24-J13	Flat plate	Beading	Flat	Stainless steel	1/8	1/16–3/32	DCSP-ex ACHF-g	80–130	Argon	10–12	1/4–3/8	3/32	796
24-J14	Outside corner	Fillet	Flat	Stainless steel	1/8	1/16–3/32	DCSP-ex ACHF-g	80–130	Argon	10–12	1/4–3/8	3/32	796
24-J15	Lap	Fillet	Flat	Stainless steel	1/8	1/16–3/32	DCSP-ex ACHF-g	80–140	Argon	10–12	1/4–3/8	3/32	796
24-J16	T	Fillet	Flat	Stainless steel	1/8	1/16–3/32	DCSP-ex ACHF-g	80–140	Argon	10–12	1/4–3/8	3/32	796
24-J17	Square butt	Groove	Flat	Stainless steel	1/8	1/16–3/32	DCSP-ex ACHF-g	70–120	Argon	10–12	1/4–3/8	3/32	796
24-J18	Single-V butt	Groove	Flat	Stainless steel	3/16	3/32	DCSP-ex ACHF-g	90–150	Argon	12–15	5/16–3/8	3/32–1/8	796
24-J19	Single-V butt	Groove	Flat	Stainless steel	1/4	1/8	DCSP-ex ACHF-g	110–180	Argon	12–15	1/2	1/8–5/32	796

NOTE:
1. Also practice in the vertical and overhead positions.
2. For vertical and overhead welding, reduce the amperage by 10 to 20 amperes.
3. When two rod and tungsten sizes are listed, the smaller is for vertical and overhead position welding.
4. Ceramic or glass cups should be used for currents to 250 amperes.
5. Water-cooled cups should be used for currents above 250 amperes.
6. The gas flow should be set at the maximum rate for vertical and overhead welding.
7. The type of filler rod must always match the type of base metal being welded.
8. This table is typical. Tungsten electrode size, filler rod size, and welding current will vary with the welding situation and the operator's skill.

* In each case ex designates excellent and g good.

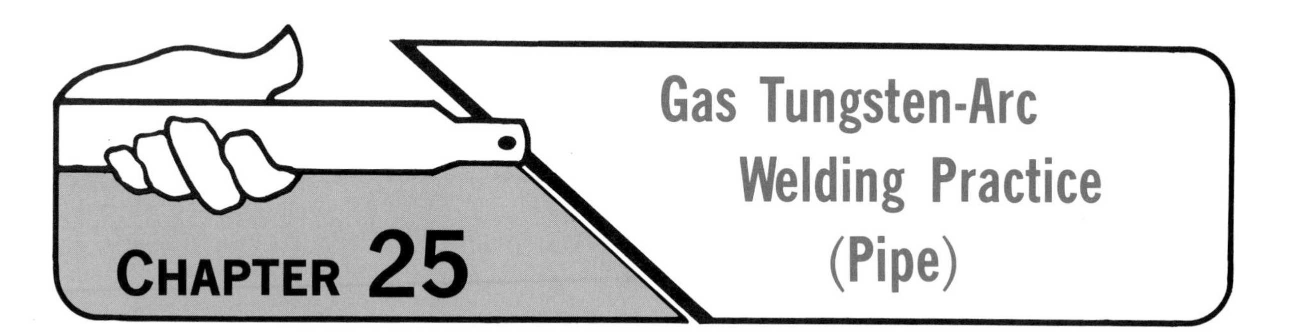

Chapter 25 Gas Tungsten-Arc Welding Practice (Pipe)

After you have mastered the gas tungsten-arc welding of commercial metals and their alloys in the form of plate, you may gain further skill in the process by practicing on pipe. Gas tungsten-arc welding produces welds in pipe that are unusually smooth, fully penetrated, and free from obstructions and crevices on the inside. These welds have maximum strength and are highly resistant to corrosion.

The process may be used to weld both ferrous and nonferrous piping materials. It is a highly desirable process for the welding of nonferrous materials.

Gas tungsten-arc welding is used in the fabrication of aluminum, stainless steel, nickel, nickel alloys, and alloy steel piping. See Figs. 25-1 and 25-2. For temperatures above 750 degrees F. and for power plant work, low alloy steel pipe containing small amounts of chromium and molybdenum is used. Stainless steel piping is used for high and low temperature work, for chemical and process piping, and in those situations in which the pipe must resist corrosive chemicals. Copper, nickel, and their alloys are selected for resistance to chemical attack, especially by saltwater. Aluminum pipe has the advantages of being light in weight, resisting corrosion, and not forming toxic chemicals in the processing of

25-1. *Making cover passes on heavy walled, carbon-moly pipe. Note the welding machine, shielding gas cylinders, regulator and flowmeter, and the positions of the torch and filler rod.*

Crane Co.

food and drugs. A comparatively recent application is in the welding of nuclear piping systems, reactor vessels, and their auxiliary vessels. In these applications it is necessary to contain radioactive fluids under high pressures at both low and high temperatures. Joint tightness and cleanliness are absolutely essential because of the dangers of radioactive contamination.

Substantial quantities of carbon steel pipe are welded with the gas tungsten-arc process for the root pass only.

The thinnest section that can be manually welded is approximately $\frac{1}{32}$ inch. For pipe wall thicknesses from $\frac{1}{4}$ to $\frac{3}{8}$ inch, it is generally more economical to complete the pipe weld, after the gas tungsten-arc root pass, with the gas metal-arc or submerged-arc process. The shielded metal-arc welding process may also be used. Some type of gas or metal backing improves the quality of the weld and makes the job of the welder somewhat easier.

Accurate joint preparation and good fitup are essential. Cleaning in preparation for welding must be carefully performed, and in some cases the base metal may require chemical treatment. Some type of gas or metal backing improves the conditions for welding. Welding may be done in all positions.

25-2. *Welding a butt joint in aluminum pipe, in the roll position. The welding machine is a 300-ampere a.c.-d.c. rectifier with high frequency.*
Westinghouse Electric Corp.

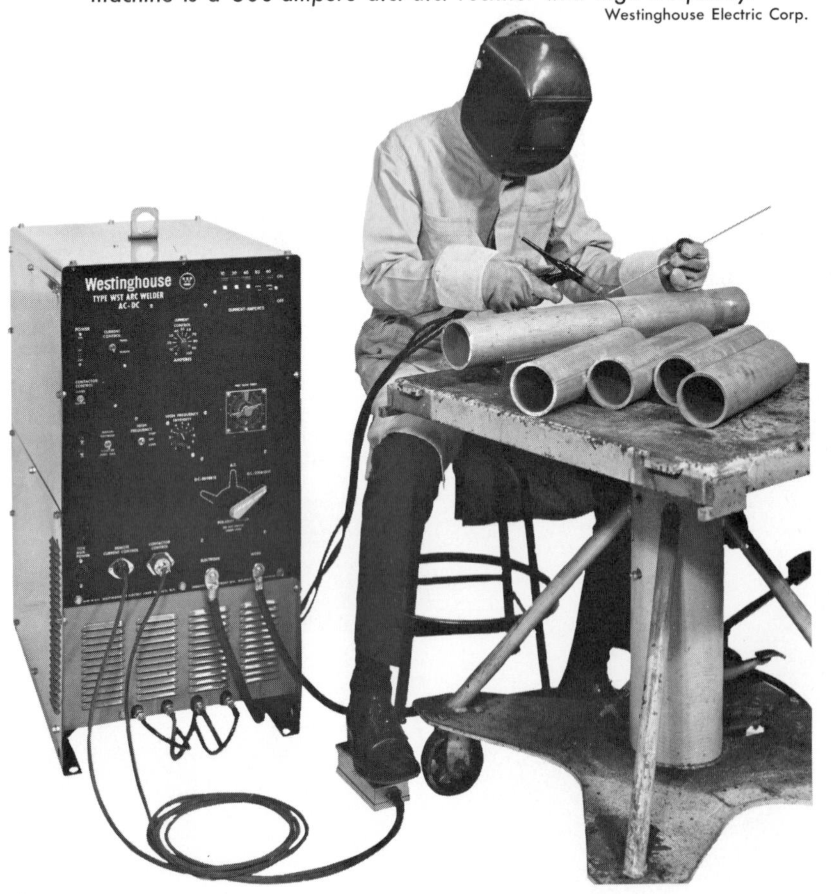

JOINT DESIGN

The following information may be applied to the welding of steel, aluminum, and stainless steel pipe. Five basic joints are in general use in the gas tungsten-arc welding of pipe. Selection of a particular design depends on the type of material, the thickness of the material, and the nature of the installation. It is recommended that all pipe with a wall thickness of more than $\frac{1}{8}$ inch be beveled to insure sound welds.

Standard V-Joint

The V-joint, Fig. 25-3, with the 75-degree bevel and a root face of $\frac{1}{16}$ inch is considered standard. On noncritical jobs the joint may be butted together and welded without a filler rod on the root pass. Code quality welding requires that the joint be spaced at the root from $\frac{1}{16}$ to $\frac{3}{32}$ inch as shown.

Sharp V-Joint

This joint, Fig. 25-4, does not have a root face like the standard V-joint. It can be prepared by cutting with an oxyacetylene cutting torch. The root pass must be welded with filler rod.

U-Groove Joint

The U-groove joint, Fig. 25-5, is used when uniform welds of high quality are required. This joint is particularly adaptable to position welding because it is possible to obtain complete and uniform penetration. Root passes should be made with filler rod, although good penetration can be secured without it.

Consumable Insert Joint

This type of joint, Fig. 25-6, produces welds of the highest quality to specification. The insert may be thought of as a ring of filler metal, which may have a special composition to prevent porosity and to enable the weld metal to meet specific requirements. A disadvantage of the consumable insert joint is the close control of dimensions and fitup needed to produce satisfactory results.

Rolled-Edge Joint

The rolled-edge joint, Fig. 25-7, produces welds of the highest quality. Preparation is somewhat less costly than that for the consumable insert joint. Fitup conditions are also less critical. The nose of the U-joint is rolled towards the inside of the pipe. When the root face is melted, the upset metal flows into the weld puddle to produce weld reinforcement inside the pipe. This type of joint is recommended for high quality work such as that required for refinery piping, pressure piping, nuclear piping, and other quality applications.

JOINT PREPARATION

When pipe is of such thickness that beveling is necessary, it should conform to the specifications given in Figs. 25-3 through 25-7 and Table 25-1. All beveling should be done by machine. Grinding may deposit impurities on the surface of the metal, which contaminate the completed weld and make the welding operation more difficult. Prevention of surface contamination is especially important when welding aluminum pipe.

Backing or Purging

In order to obtain the smooth interior surface specified in Fig. 25-8 and shown in Fig. 25-9, it is necessary to protect the back as well as the face of the weld bead from oxidation. This may be done by introducing argon or helium into the pipe. This operation is known as *purging.* The gas shields the molten pool on the inside of the pipe. It should flow at a rate of about 6 c.f.m. during the welding operation.

A fixture similar to that shown in Fig. 25-10 may be used for purging. This fixture can be made with materials usually found in the school welding shop, and it may be fabricated and welded by the students.

The conical pipe inserts *D* and *E* may be made with any desired diameter. They will fit all sizes of pipe up to the maximum diameter of the cone at its largest end. The cone *D* is welded to the tube and is fixed while cone *E* is adjustable by slipping the clamp *F* along the tube *H* in order to adjust to various lengths of pipe.

The nut adjustment *B* permits positioning of the weld at any point around the circle. The pipe support *C* may be inserted into a larger fixed pipe in a stand, or it may be attached to the welding table so that the height of the unit can be adjusted. The purging gas passes from a flow valve through the welding hose *A* and into the tube which is plugged at the far end. This forces the gas to flow through the gas outlet holes *G* in the tubing and inside the pipe joint for purging.

In order to be able to weld under the most favorable of con-

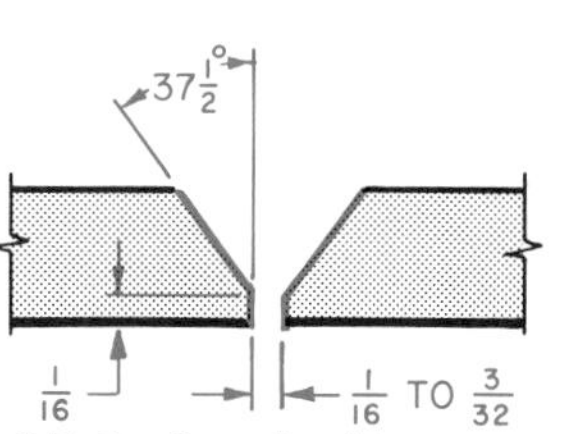

25-3. *Standard V-joint.*

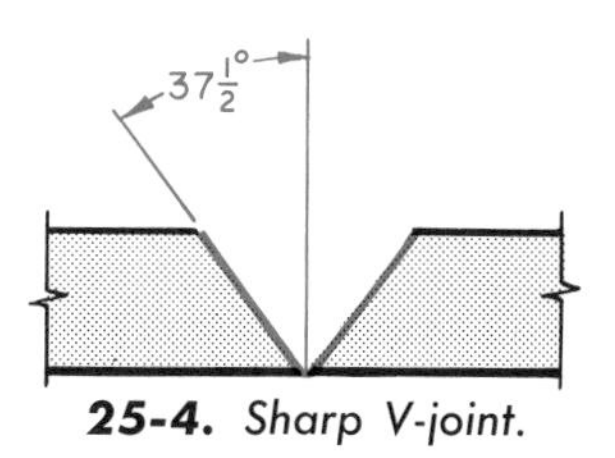

25-4. *Sharp V-joint.*

25-5. *U-groove joint.*

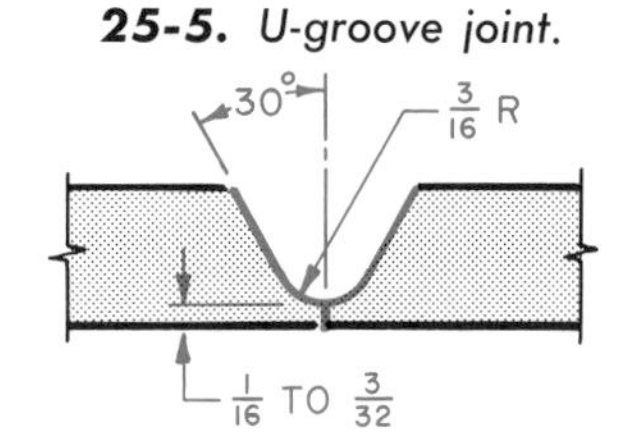

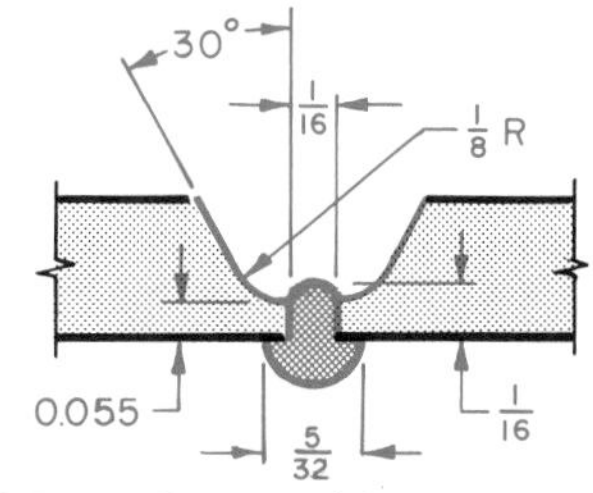

25-6. *Consumable insert joint.*

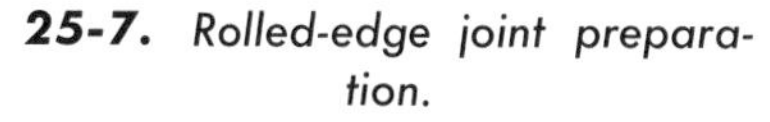
25-7. *Rolled-edge joint preparation.*

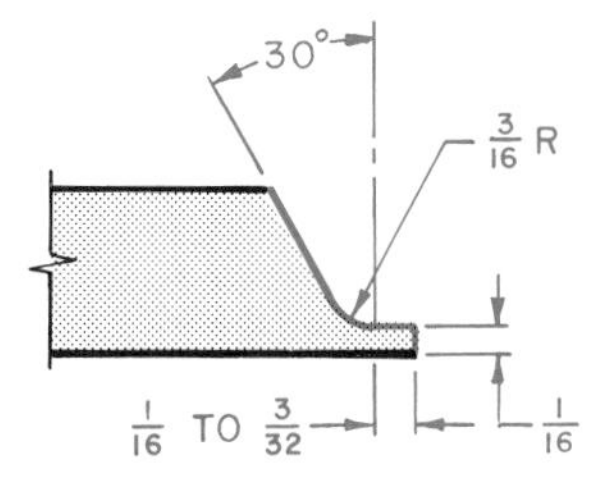

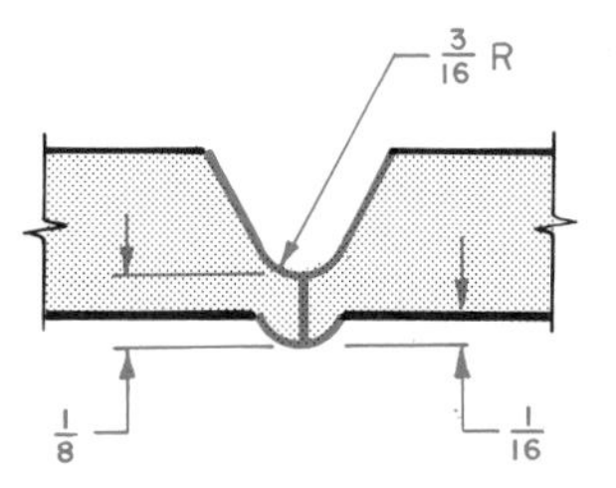

ditions, some kind of pipe-rotating device is necessary for the rolled-position welds. Figure 25-11 illustrates a manually operated unit that may be made quickly in the school shop. The gas-purging unit for the pipe interior, shown in Fig. 25-10, may also be attached to the rotating unit. This unit may be driven by an electric motor having a control unit with a speed control range up to 10 r.p.m.

The inside of the pipe may also be protected by painting with the proper flux. Welds backed up by flux are high quality, and they are almost equal in physical properties to those made with gas purging. Flux permits more penetration through the back side, whereas gas purging reduces the inside buildup and causes a very smooth inside bead.

Table 25-1

Joint preparation for the welding of pipe with the gas tungsten-arc process.

Type of material	Position of welding	No backing ring	With backing ring	Type of gas shielding	Bead angle[1]	Root face[2]	Root opening[3]
Aluminum	Vert. fixed & hor. fixed	x		Helium	37½°	1/16	None
Aluminum	All		x	Helium	37½°	1/16	3/16
Aluminum	Hor. fixed bottom only	x		Helium	60°	1/16	None
Aluminum	Hor. rolled	x		Argon	37½°	1/16	1/16
Aluminum	Hor. fixed & Vert. fixed	x		Argon	55°	1/16	1/16
Aluminum	Hor. rolled		x	Argon	37½°	1/16	3/16
Aluminum	Hor. fixed & Vert. fixed		x	Argon	55°	1/16	3/16
Steel & stainless steel							
1/16 to 5/32 thick	All	x		Argon	none	total	1/16
	All		x	Argon	none	total	1/8
5/32 to 1/4 thick	All	x		Argon	37½°	1/16	3/32
	All		x	Argon	37½°	1/16	3/16

[1] Bevel angle may be ±2½°.
[2] Root face may be ±1/32 inch.
[3] Root opening may be ±1/16 inch.

25-8. *Specifications for penetration bead contours in a standard V-joint.*

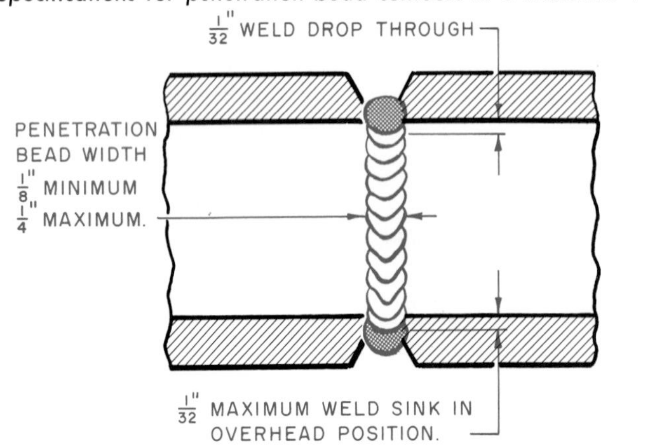

Tack Welds

Tack welds are used to hold the sections of pipe together in preparation for welding. In pipe welding by the gas tungsten-arc process, tack welding must be done with care and skill because poor tacks are often the cause of defects in the final weld.

Inert gas purging or flux is applied inside the pipe, and the tack is made as small as possible. The torch is drawn to the side of the bevel until the pool diminishes in size, and then it is broken.

If filler rod is to be used for the root pass, you should use filler rod for the tack weld. Tack welds should be about 3/8 to 1 inch long. Four to six tacks per joint, depending on the pipe size, should be spaced evenly around the pipe. Too large a tack weld interferes with the weld that is to follow. If the torch is withdrawn too rapidly, the tack weld will crack.

When the root pass is welded, the tack weld must be re-fused and become a part of the weld.

POROSITY IN GAS TUNGSTEN-ARC WELDS

It is important at this time to call your attention again to the matter of porosity in weld metal. There are two main causes: (1) that resulting from the composition of the base metal, and (2) contamination from the surface of the joint or from the surrounding air.

Base Metal Porosity

Some steels contain iron oxides that react with the carbon in the steel during welding to release carbon monoxide and carbon dioxide. These gases cause porosity in the weld. Gas formation can be controlled by the use of deoxidizing filler rod, powders, or slurries.

Porosity from Contamination

Contamination may be caused by iron oxides on the joint surface. Heavy rust or scale on the surface produces a condition like that produced by oxides within the material. Therefore, rust and scale inside and outside the pipe should be removed for a distance of at least 2 inches on each side.

Oil, grease, and moisture on the joint surface also contaminate the weld. These materials vaporize when heated and produce gases that may be trapped in the weld and cause porosity. Pipe surfaces must be cleaned with care.

Porosity may also result from atmospheric contamination. Inadequate gas shielding leaves the weld pool open to attack. Inadequate shielding has the following causes:

- Gas flow that is too slow
- An arc that is too long
- Too much electrode extending beyond the end of the gas cup
- Air currents in the work area

PRACTICE JOBS

The equipment needed is the same as that used for the gas tungsten-arc welding of plate, Fig. 25-12.

Instructions for Completing Practice Jobs

Complete the jobs listed in the Course Outline, page 818, as assigned by your instructor. Before you begin a job, study the specifications given in the Course Outline.

After you have demonstrated that you are able to make satisfactory welds in carbon steel pipe

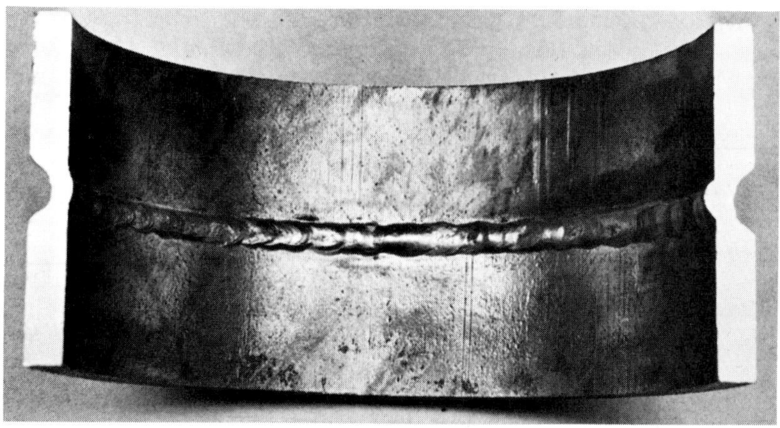

Worthington Corp.

25-9A. *The underside of a root pass in steel pipe welded with the gas tungsten-arc process. Filler metal has been added. The joint shows complete penetration free of notches, icicles, crevices, and other obstructions to flow. Reinforcement is between 1/32 and 3/32 inch.*

25-9B. *The back side of a root pass in stainless steel pipe. This is the inside of the pipe.*

25-10. *Conical fixture used for purging the inside of piping and tubing. This can be made in the school welding shop.*

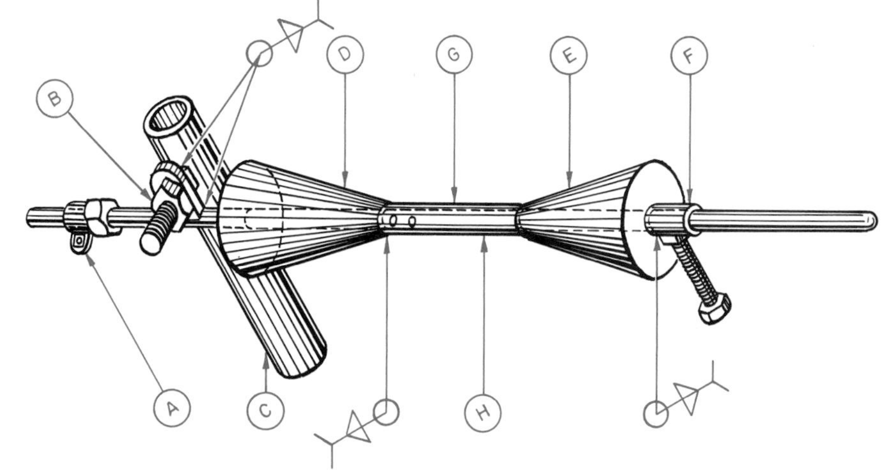

of various diameters, you are ready to proceed with the other pipe jobs listed in the Course Outline. These jobs are on aluminum and stainless steel pipe. The welding technique is similar to that used in your earlier practice on plate.

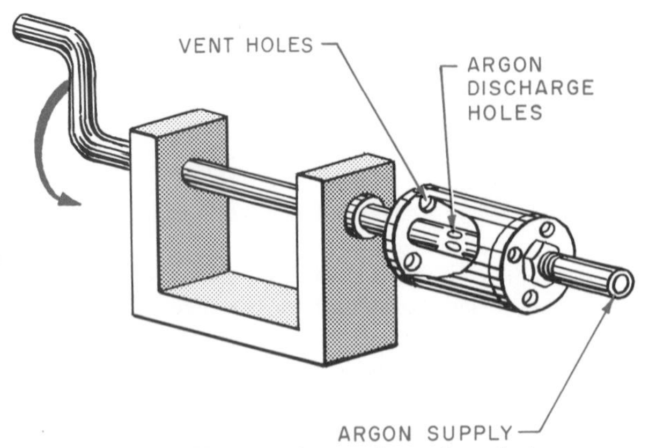

25-11. *Pipe-rotating and purging device.*

Practice on steel, aluminum, and stainless steel pipe in sizes ranging from 2 to 6 inches in diameter. You are expected to complete the jobs calling for mild steel pipe before practicing with aluminum and stainless steel pipe. Welding should be done in the roll, horizontal fixed, and vertical fixed positions. Use the same general welding techniques you learned in the previous chapter for metal of the same thickness.

The procedures used in welding these practice joints in pipe are those that are practiced on the job. No single point is overlooked in making sure that the procedure will produce welds that are sound. Piping installations are such that a weld failure

25-12. *A typical gas tungsten arc pipe-welding installation.*

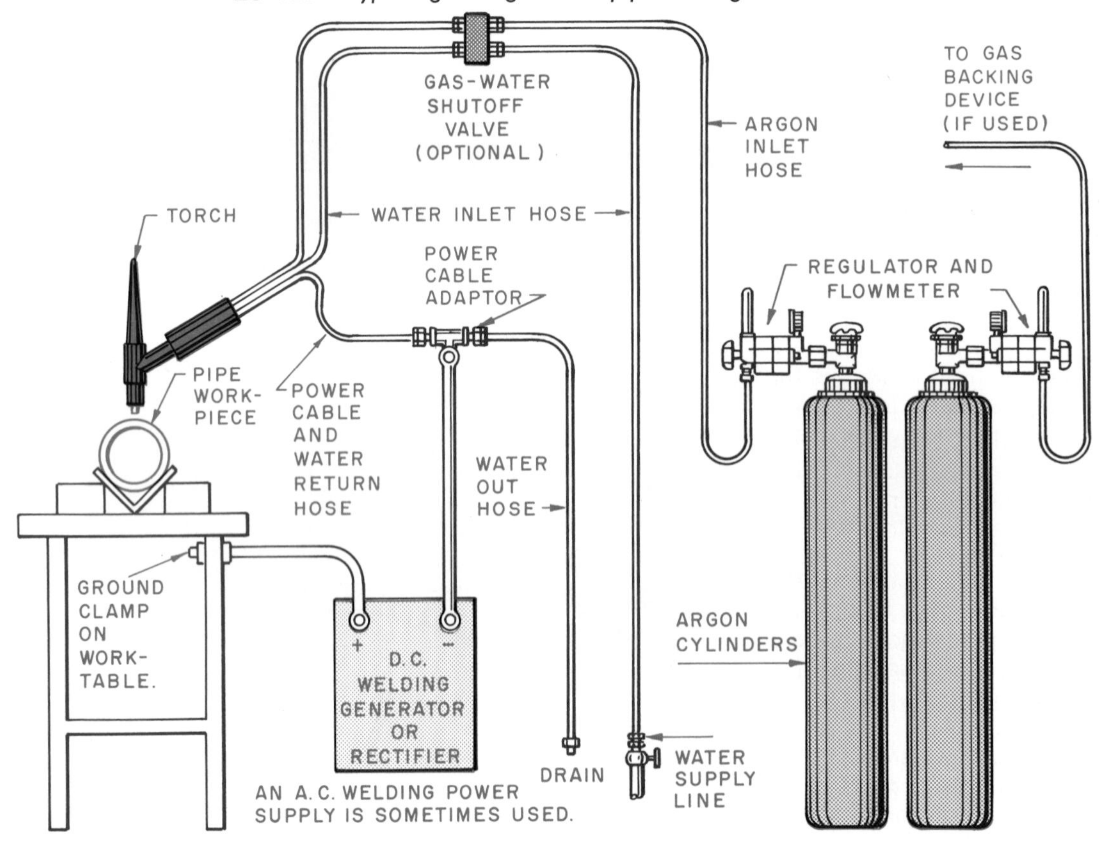

would not only damage property but also endanger the safety of human beings.

Study the various joint designs and weld procedures carefully. The nature of the material, the size of pipe, and the position of welding are the three important variables that determine the joint design, the pass sequence, and all that this implies. Quality welding requires close control. It is truly a scientific process that demands a great deal of skill in its application.

It is highly important that the student who feels that he wants to specialize in pipe welding become highly proficient in the gas tungsten-arc welding of aluminum and stainless pipe, since it is extensively used for piping made of these materials and alloys of steel.

Roll Butt Welds: Jobs 25-J1 through J7

ROOT PASS. Practice on pipe 4 to 6 inches in diameter. After you have practiced stringer and weaved beading around the outside of the pipe (Job 25-J1), begin practice on joints with the standard V-joint (Job 25-J2). Prepare the edges as instructed in Fig. 25-3, page 803. Tack the pipe in about three places with small tacks so that they do not interfere with welding. Brush and clean tacks before proceeding with the welding. Set up the pipe so that it can be purged with argon to protect the underside of the weld bead and to assist in making a smooth bead. Flux may be used instead of gas purging. A 2 percent thoriated-tungsten electrode with a diameter of $^3/_{32}$ or $^1/_8$ inch is preferred for welding. It should be ground to a point as shown in Fig. 25-13.

Point the electrode toward the center of the pipe, Fig. 25-14. The arc length should be $^1/_{32}$ to $^1/_{16}$ inch long. Be careful to manipulate the torch so that the tungsten electrode does not become contaminated.

Weld the root pass with the stringer bead technique. Start just below the 2 o'clock position and weld upward to the 12 o'clock position. Rotate the pipe clockwise so that the crater of the bead just deposited is just below the 2 o'clock position. After making sure that the weld crater is clean, continue welding as before.

The filler rod should be added almost tangent to the pipe surface, and the torch should be slanted about 15 to 20 degrees toward the rod with an arc length of about $^1/_{16}$ inch, Fig. 25-15. When the puddle increases to about $^1/_8$ inch in thickness, remove the rod and allow the puddle to flatten out. When penetration appears to be complete, as shown by the shape of the puddle, add more filler rod and advance the weld. Hold the tungsten electrode perpendicular to the surface of the work in all positions. Excessive angulation may result in inadequate shielding and reduce control over the reinforcement contour.

A fluid weld puddle changes shape when certain conditions are present. You should learn how to "read" the weld puddle by observing its changing contours as the associated conditions of penetration and bead formation change.

When welding begins, hold the electrode stationary and point it toward the center line of the pipe. Preheating is done by moving the torch in small circles over the two bevels. The puddle continues to grow until sharp points

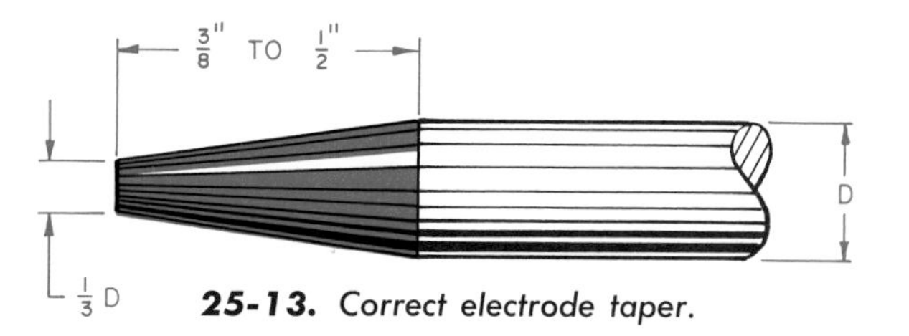

25-13. *Correct electrode taper.*

25-14. *Position of the torch for a roll weld.*

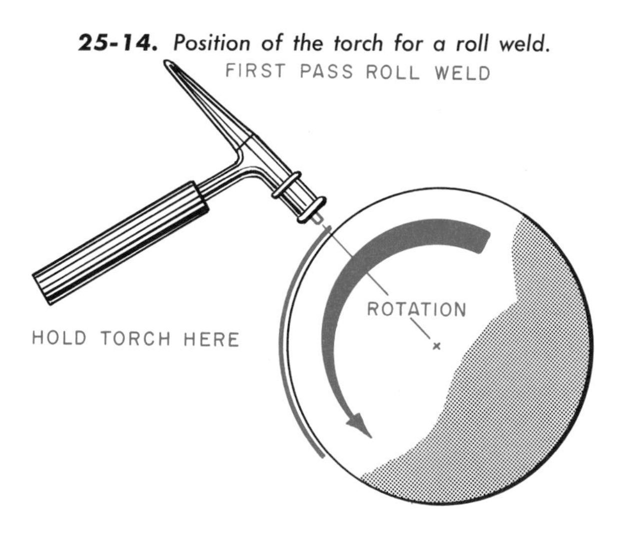

are formed on opposite sides of the weld puddle at the bottom of the *V*. When you see these points, advance the electrode in a straight line. The sharp point on the leading edge of the weld puddle indicates whether or not the electrode is proceeding in a straight line. Figure 25-16 shows the point formation after a portion of the pipe has been welded. As long as this point formation is present, penetration is being obtained on the inside of the pipe. If you advance the electrode too rapidly, the point on the weld puddle becomes rounded, or it may even form a notch (re-entrant angle), Fig. 25-17. The presence of the notch in the weld pool indicates the absence of penetration on the inside of the pipe.

There are other signs that indicate penetration and contour of the weld bead on the inside of the pipe. When the puddle is first formed, the surface across the center is raised (convex). When the point formation is obtained on the lead edge of the weld puddle, observe the convex surface carefully. Soon after the point formation is obtained, the convex surface of the puddle suddenly becomes flat. At this instant, advance the electrode along the line of weld until this surface again appears to be convex. If the puddle is allowed to pass through the flat stage to the concave stage, too much penetration will result.

Excessive penetration can be overcome by adding more filler rod to the puddle. Slant the torch toward the filler rod so that more rod is melted and increase the speed of welding. This directs more heat to the rod and less to the pipe surface to reduce the amount of heat in the weld pool.

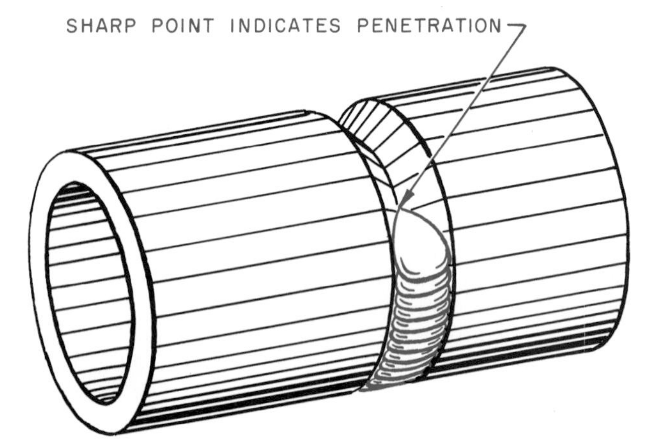

25-16. *Contour of the weld pool when penetration is obtained.*

25-15. *The relative position of the torch and filler rod when making a roll weld on pipe. The pipe is stainless steel.*

Tube Turns, Inc.

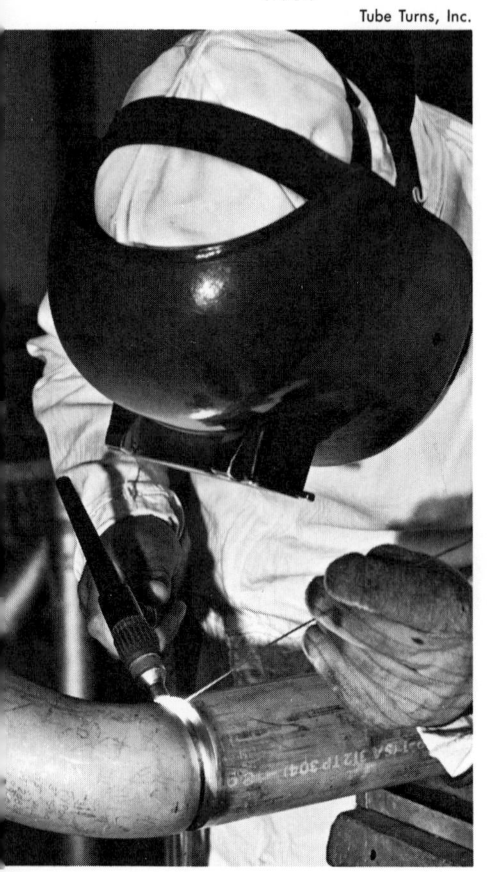

25-17. *Contour of the weld pool when penetration is not obtained.*

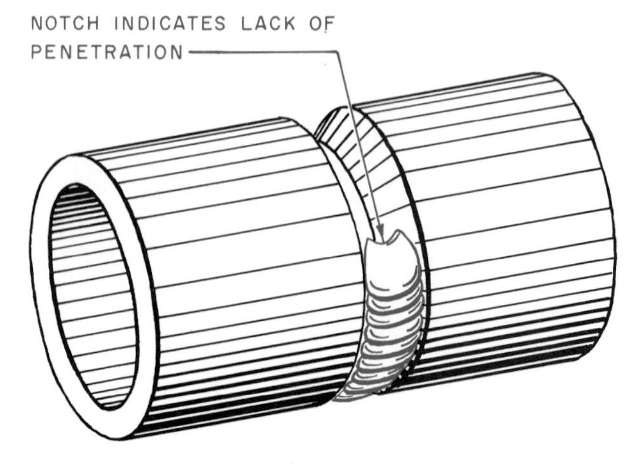

Additional control of penetration may be obtained by lengthening the arc to reduce the amount of penetration when required or by shortening the arc to drive penetration through the root of the joint.

This first pass is critical, and it must be mastered before the welding of the pipe joint may be considered satisfactory. Control of penetration is the most important factor in successful root-pass welding. Such control can be obtained only by repeated practice. Satisfactory service of the pipe joint can be obtained only if the root pass is solid. See Fig. 25-9B, page 805, for the proper appearance of the back side of the root weld and Fig. 25-18 for the appearance of the face of the weld.

25-18. *Face contour of a satisfactory root pass in stainless steel pipe.* Crane Co.

Improper termination and pickup of the weld beads increase the probability of defects in the weld. Figure 25-19 and 25-20 show the proper termination of the weld bead. The weld bead is overlapped approximately three times the width of the bead, and the size of the weld puddle is reduced by rapidly increasing the weld travel speed. The weld puddle is directed off the side of the joint so that it may be terminated in a section that is relatively heavy in wall thickness. This decreases the chill on the weld puddle and prevents cracks in the weld crater. In starting the weld after the interruption, begin the puddle $\frac{1}{4}$ inch or more back from the end of the previous bead to insure good fusion of the two beads, Fig. 25-21.

FILLER PASSES. Filler passes are used to fill the pipe joint to within approximately $\frac{1}{16}$ inch of the top surface of the pipe. Heavier passes can be used for the filler pass when making a rolled weld than when welding in a fixed position since the weld pool does not have a tendency to sag out of the joint due to gravity.

Follow the same procedure as for the root pass. Either stringer beads or weave beads can be used for the filler passes in carbon and low alloy steel pipe welded in the rolled or fixed horizontal position.

The stringer bead technique should be used for welding aluminum and stainless steel pipe in any position when the wall thickness is more than $\frac{1}{4}$ inch. Stringer beads should also be used for carbon and low alloy steel welded in the vertical fixed position.

Each stringer bead should be made completely around the pipe before starting the next bead. Stringer beads on stainless steel should be kept small since the weld bead solidifies quickly. Fast freezing prevents the precipitation of carbides in these steels and reduces the tendency for hot-short cracking.

Care must be taken so that succeeding welding operations do not interfere with, or destroy, the inside surface of the root pass weld. For the root pass, the torch is directed toward the center line of the pipe. When making

25-19. *Proper ending of the weld bead prevents crater cracks.*

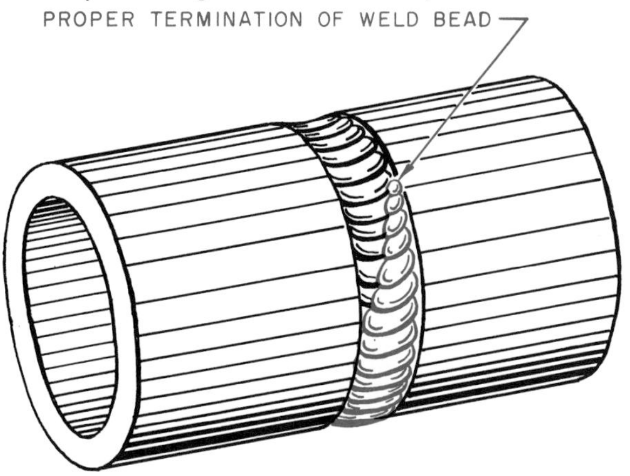

filler passes on pipe, however, you must incline the torch at an angle of approximately 45 degrees to a point tangent to the surface of the pipe as shown in Fig. 25-22.

25-20. *Face of a root pass which shows proper termination where the bevels meet. This is a V-joint in stainless steel pipe.*

Clean the root pass and each succeeding pass carefully with a stainless steel wire brush.

The filler layer may be made with a single weave bead or with two stringer beads. If a weave bead is used, weld with a $^{3}/_{32}$-inch filler rod. Clean the first pass thoroughly in order to remove surface oxides. The electrode should be held as instructed above. Move the arc from side to side, and add filler rod as rapidly as it melts off. Direct the arc on the side walls of the pipe to avoid burning through the first pass, Fig. 25-23A. If you have difficulty with weave beads technique, run stringer beads. Make the welds along the sides of the pipe with $^{1}/_{16}$-inch filler rod and overlap at the center, Fig. 25-23B. Advance the torch with a slight oscillating motion in the direction of welding, and add filler rod as rapidly as possible. Making stringer beads reduces the problem of burning through due to excessive heat in the center of the first pass. Figure 25-24 shows the proper appearance of stringer beads.

COVER PASSES. Smooth, uniform finish passes can be made only if the filler passes are smooth and free of undercut at the edges, and the pipe joint is filled to a uniform level. Care spent in making smooth filler passes will result in finished welds of excellent appearance as well as quality.

Follow the procedure described for the root and filler passes. Stringer beads or weave beads may be used for carbon and low alloy steel pipe in the rolled or fixed horizontal position. Stringer beads must be used for carbon and low alloy steel pipe in the vertical fixed position, and for stainless or aluminum pipe in any position.

Finished cover passes should be about $^{3}/_{32}$ inch wider than the beveled edges on each side of the joint, and the overlap should be spaced evenly on each side of the joint. The face of the weld should be slightly convex, and the crown should be about $^{1}/_{16}$ inch above the surface of the pipe. Weld edges should be straight and without any under-

25-22. *Proper relation of the electrode to the pipe for additions of filler rod on the filler and cover passes.*

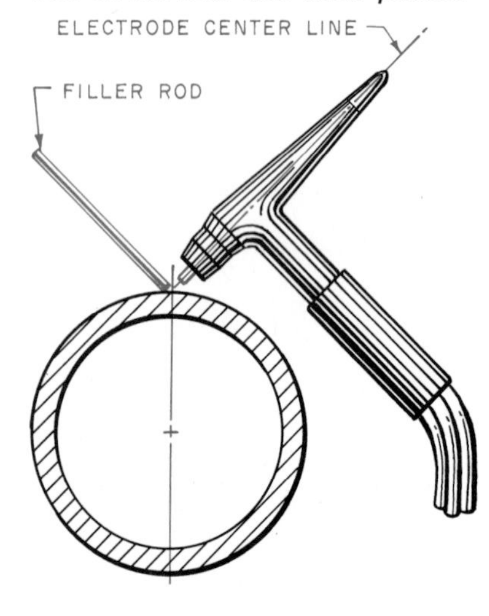

25-21. *Restarting a weld bead.*

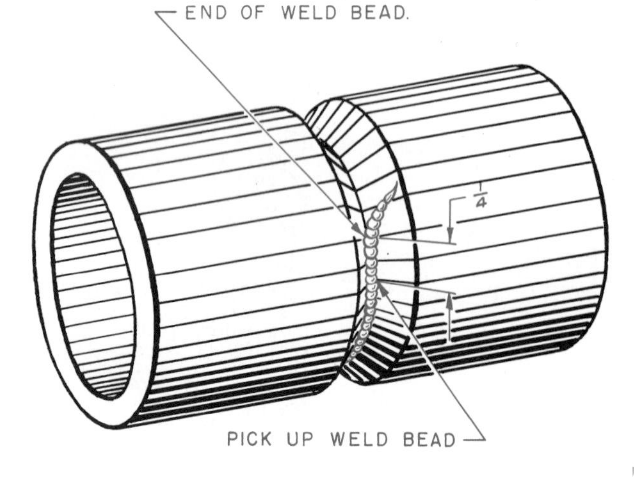

cutting. Figure 25-25 shows the proper appearance of the cover pass. Also see Fig. 25-43B

When you are able to make welds that have good penetration through to the back side, good appearance, and are satisfactory to your instructor, make up a test weld and remove the usual test coupons for root- and face-bend testing.

Horizontal Fixed Position with 180-Degree Rotation

This weld is another version of Job 25-J2, but welding is done on the top side only. The pipe is in the horizontal position, and the weld is vertical. The weld is completed in two halves. Start the weld slightly below the center line of the pipe as shown in Fig. 25-26, and weld toward the top and beyond top center. The other side of the pipe is started in the same way and proceeds to the top. The second weld overlaps the first by about three widths of the weld. The filler and cover passes are welded as previously described with either the weave or stringer bead technique.

When the cover passes are completed, rotate the pipe 180 degrees and weld the second half as described above. Be sure that the welds in the first pass overlap at both their start and finish to avoid incomplete fusion.

Vertical Fixed Position: Jobs 25-J8, J9, J10, and J11

ROOT PASS. The pipe is in the vertical position, and the joint is in the horizontal position. The first pass is made in four quarters. You should change position as each quarter is completed to make sure that you can observe the leading edge of the weld as it progresses. The first pass should overlap at each start

25-23. *Filler pass welding techniques.*

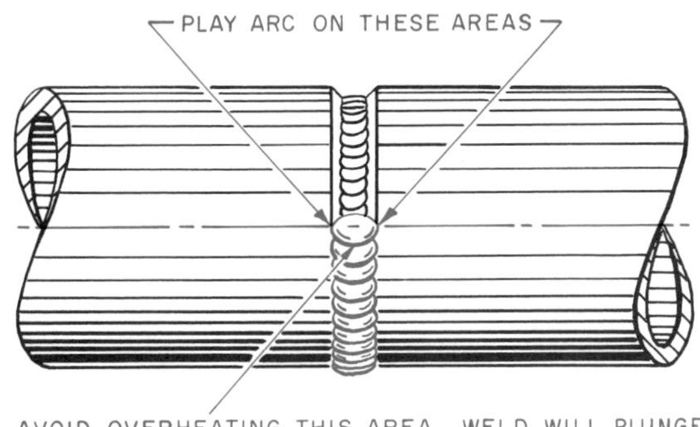
SECOND PASS ROLL WELD
PLAY ARC ON THESE AREAS
AVOID OVERHEATING THIS AREA. WELD WILL PLUNGE THROUGH IF IT GETS TOO HOT.
A

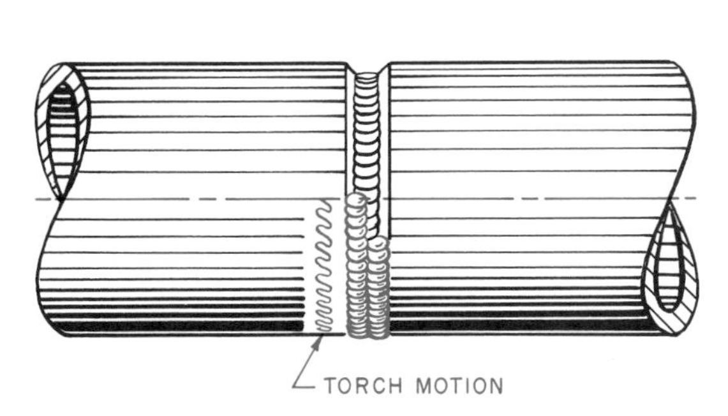
TORCH MOTION
TRY 2 STRINGER BEADS OVER FIRST PASS USING 1/16" ROD. DO NOT WEAVE.
B

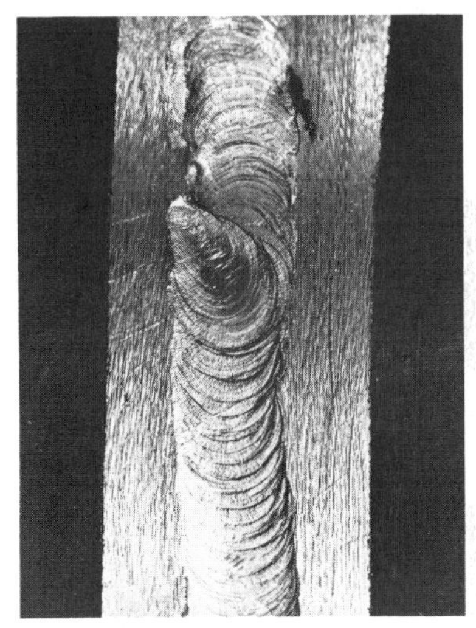

25-24. *Face appearance of a filler pass in a V-joint in stainless steel pipe.*

25-25. *Face appearance of a cover pass in stainless steel pipe.*

and at the finish, Fig. 25-27. The first pass and the cover passes should be made as shown in Fig. 25-28.

Although the shape of the puddle and the flow of the molten pool for the vertical fixed position are the same as for rolled pipe, special techniques are required to compensate for the sagging of the puddle due to gravity.

Position your torch as shown in Fig. 25-29. The weld puddle is formed on the upper pipe bevel and is kept slightly above the center line of the joint. Move the torch in small circles. Proceed from the top of the puddle, around the puddle to the bottom, and then up to the other side to the top. Do not let the arc dwell too long on the bottom, but let it favor the top of the weld. This circular motion ensures fusion of the bottom of the joint with the filler rod, and yet it does not undercut the upper side of the weld bead.

When the pipe is prepared with a U-groove, consumable insert, or rolled edge, the root pass may be welded without filler rod, Figs. 25-30 and 25-31.

FILLER AND COVER PASSES. The filler and cover passes should be made as shown in Fig. 25-32. The first filler pass should be a stringer bead made with $\frac{1}{16}$- or $\frac{3}{32}$-inch filler rod and a slight oscillating technique. This pass should cover all or two-thirds of the root pass, depending on the thickness of pipe wall. The tungsten electrode should be pointed straight in or slightly upward for this pass.

The rest of the passes, including the cover passes, should overlap the preceding passes. The torch should be pointed up or down, depending on the pass sequence. Keep the welds small and make sure that you are securing good fusion with the preceding welds and the pipe surfaces. The puddle must be kept fluid, and the forward edge should flow without being forced by the filler rod or torch. Add the filler rod sparingly so that the puddle does not get too large and so that cold rod is not deposited on the bead or pipe surfaces. Be sure to clean each pass thoroughly before applying the next pass.

When you are able to make welds that show good penetration through to the back side, have good appearance, and are satisfactory to your instructor, make up a test weld and remove the usual test coupons for root- and face-bend testing.

Horizontal Fixed Position (Bell Hole Weld): Jobs 25-J12 through J17

The pipe is in the horizontal position, and the weld is vertical, Figs. 25-33 and 25-34.

25-26. *Welding the top side of the pipe in the horizontal fixed position.*

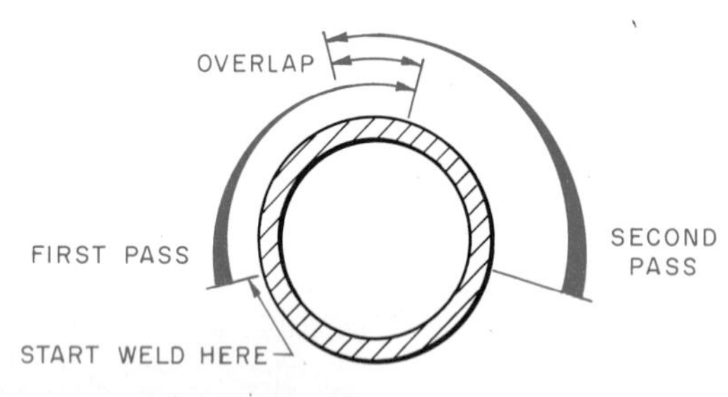
FIXED POSITION
WELD TOP HALF ONLY
OVERLAP
FIRST PASS
SECOND PASS
START WELD HERE
PLACE ONE WEAVE OR TWO STRINGERS OVER THE FIRST PASS BEFORE TURNING OVER THE PIPE.

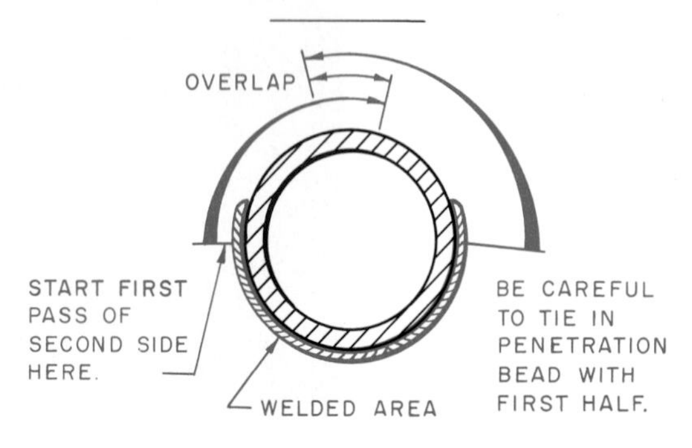
OVERLAP
START FIRST PASS OF SECOND SIDE HERE.
WELDED AREA
BE CAREFUL TO TIE IN PENETRATION BEAD WITH FIRST HALF.

25-27. *Welding procedure for the root pass in pipe in the vertical fixed position.*

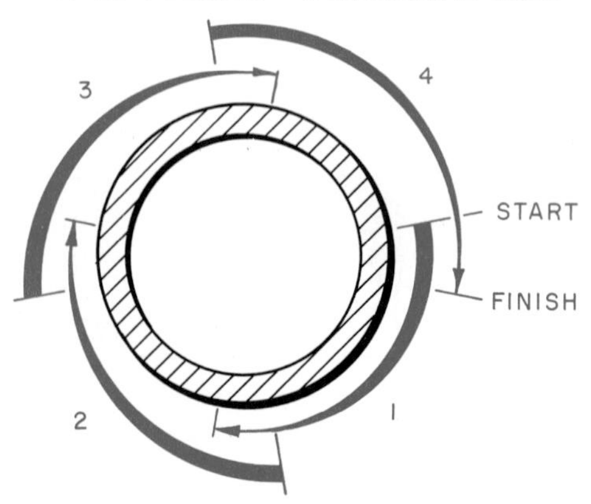
FIXED POSITION – HORIZONTAL WELD
3
4
START
FINISH
2
1
WELD ON QUARTERS WITH GOOD OVERLAP OF FIRST PASS TO INSURE GOOD PENETRATION BEAD.

When making a pipe weld in the horizontal fixed position, you will practice all of the welding positions: flat, vertical, and overhead. The weld will be made at a constantly changing angle. The ability to make welds of this nature that pass the regular tests marks the welder as a craftsman. Figures 25-35 through 25-37 show the change of position from bottom to top.

ROOT PASS. The same general welding technique followed previously for the welding of rolled pipe may also apply to welding in the horizontal fixed position. You must control the weld puddle more closely, however, to get the proper penetration without weld

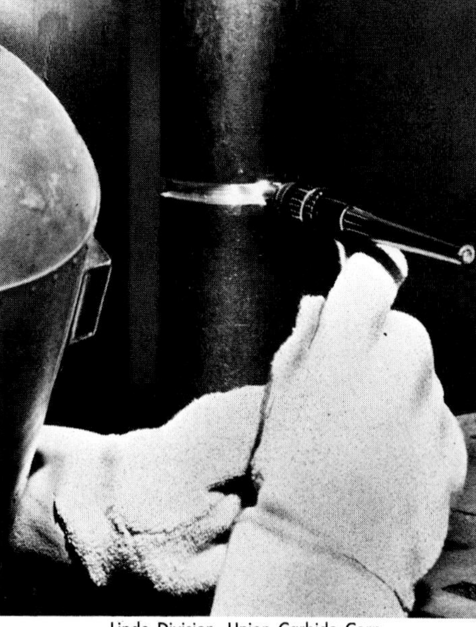

Linde Division, Union Carbide Corp.

25-30. A clean, fully penetrating root pass is being made on carbon steel pipe in the vertical fixed position. No filler rod is needed because of the U-groove joint preparation.

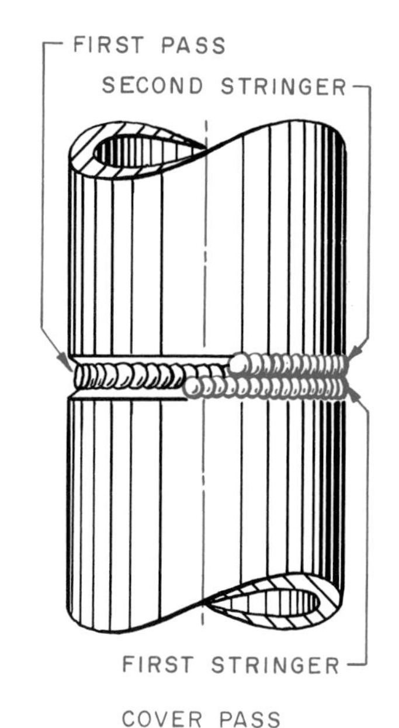

25-28. Welding procedure for the first pass and the cover pass on pipe in the vertical fixed position.

25-31. Welding the root pass without the use of a filler rod. Joint preparation is U-groove, and the pipe is being rolled in the horizontal position.

Crane Co.

25-29. Torch position for welding pipe in the vertical position.

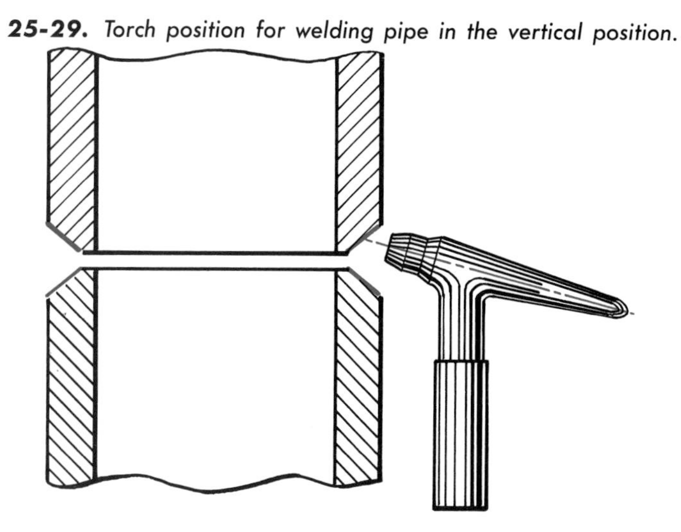

"sink" when you are welding overhead. Your response to the demands of the puddle must be immediate. You may find it helpful to weld with lower current values in order to secure a smaller weld puddle. Keep the face of the weld as flat as possible and free of undercut.

The weld bead should be started past bottom center. The welding should progress downhill across the bottom of the pipe and then uphill to the top of the pipe past the 12 o'clock point, Fig. 25-38. If the welding is interrupted in order to change position as you travel around the pipe, be careful to overlap the bead at each start after removing the surface oxides.

The second half of the first pass should overlap the starting point of the first half of the weld. The welding should then proceed to the top and overlap the previous weld at the top, Fig. 25-38. There will be a tendency for the middle of the first pass to be concave and wide. Avoid welding with too little current. Travel as fast as possible.

This joint should also be practiced by welding the root pass from top to bottom on each side and all other passes from bottom to top. This is a procedure that is very often used.

THE FILLER AND/OR COVER PASSES. The filler and/or cover passes should be made as shown in Fig. 25-39. These passes may be either stringer or weave beads. The first filler pass must

25-32. *Stringer beads on pipe in the vertical position and welding sequence (top).*

Linde Division, Union Carbide Corp.

WELDING SEQUENCE

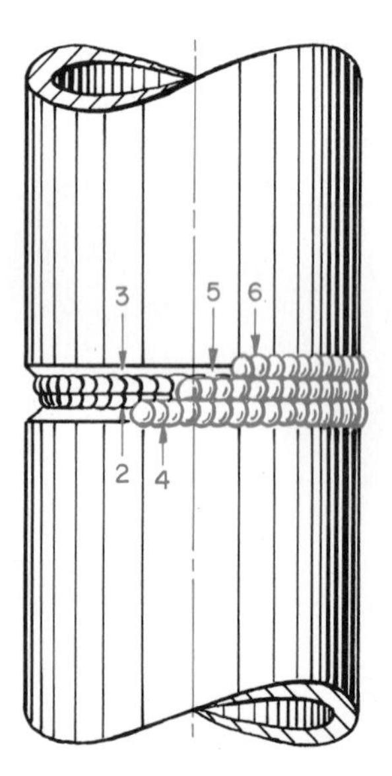

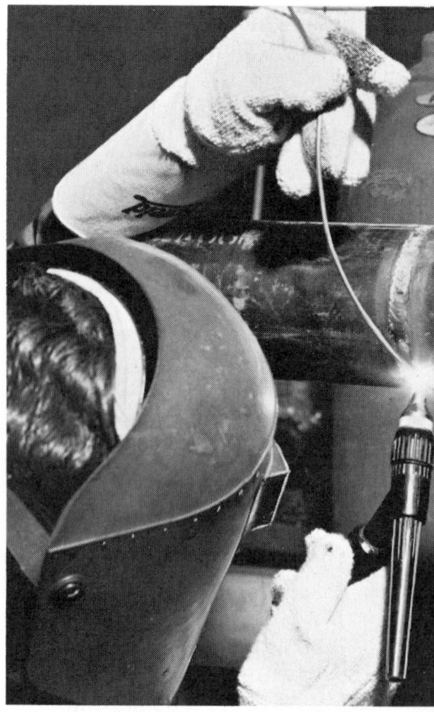

Linde Division, Union Carbide Corp.

25-33. *Welding of carbon steel pipe in the horizontal fixed position. The weld is started at the bottom, and travel is up.*

25-34. *Welding aluminum pipe in the horizontal fixed position in a tight situation. Travel is up.*

Tube Turns Corp.

be applied with great care because the root pass made with the gas tungsten-arc process is likely to be thin, and the danger of burning through is always present. Keep the filler rod in contact with the bottom of the joint, and do not remove it from the weld puddle. This limits the penetration into the root pass and helps to prevent burn-through. Do not let the arc dwell in the center of the bead. Keep it on one side when making stringer beads, or move it quickly from one side of the joint to the other for weave beads. Make sure that the puddle blends into the sides of the joint and that it is fused to the bead underneath. Be sure to clean each pass thoroughly before applying the next pass.

25-35. *Welding aluminum pipe in the bottom position of a bell-hole joint. This is overhead welding, travel up.*

Kaiser Aluminum & Chemical Corp.

When you are able to make welds that show good penetration through to the back side, have good appearance, and are satisfactory to your instructor, make up a test weld and remove the usual test coupons for root- and face-bend testing.

Inspection and Testing

When completing the remaining jobs assigned from the Course Outline, inspect each pass carefully for weld defects. Compare root passes with the specifications given in Fig. 25-8 and the actual appearance shown in Fig. 25-9, page 805. Also inspect the outside surface appearance of the root pass and compare it with Figs. 25-18, page 809, and 25-20, page 810.

Finished welds in stainless and aluminum pipe should have the appearance shown in Figs. 25-40 through 25-43. Look for essentially the same good weld characteristics that you have been striving for in shielded metal-arc welding practice and gas tungsten-arc welding practice on plate.

Testing to determine welder qualification for stainless steel pipe is usually done by welding together two 6-inch lengths of stainless steel pipe 1 inch in diameter. One joint is welded in the horizontal fixed position and a second is welded in the vertical fixed position. The specimens are subjected to a tensile pull test. The weld reinforcement is removed. Successful specimens will reach a load limit of more than 75,000 p.s.i. before breaking. Failure may take place in the weld or the pipe wall.

Testing may also be done on larger diameter pipe. To test your practice welds in both stainless steel and aluminum, remove

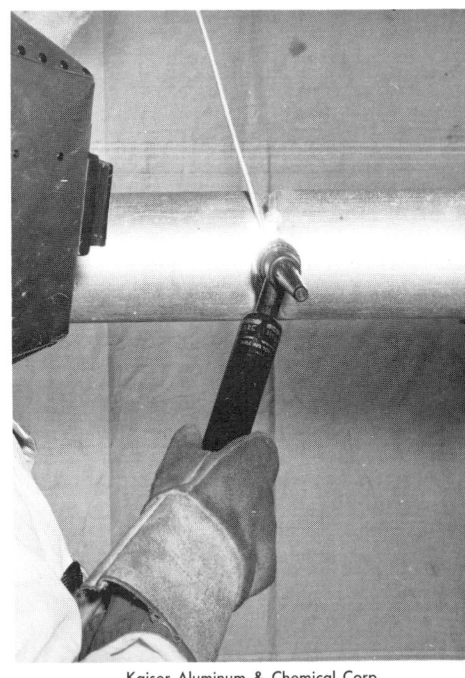

Kaiser Aluminum & Chemical Corp.

25-36. *Welding aluminum pipe in the side position of a bell-hole joint. The welding position changes from the overhead to the vertical position.*

25-37. *Welding aluminum pipe in the top position of a bell-hole joint. The welding position changes from the vertical to the flat position.*

Kaiser Aluminum & Chemical Corp.

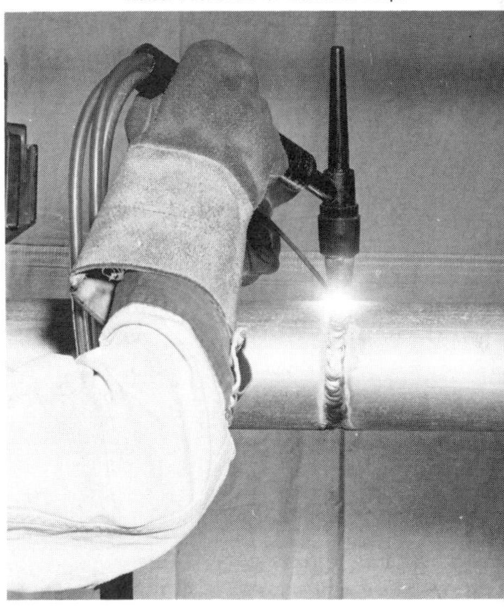

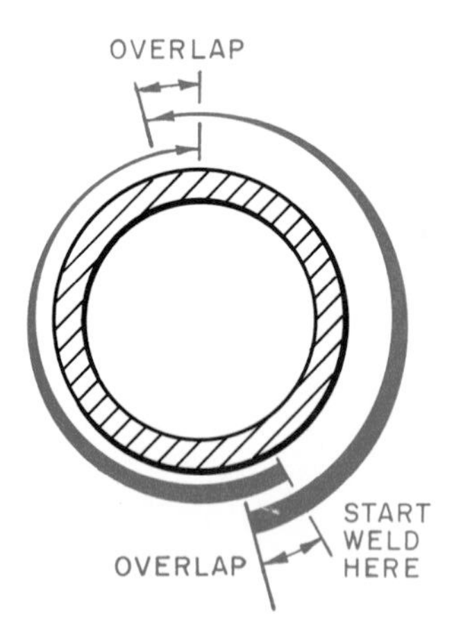

25-38. *Welding procedure for root pass on pipe in bell-hole position.*

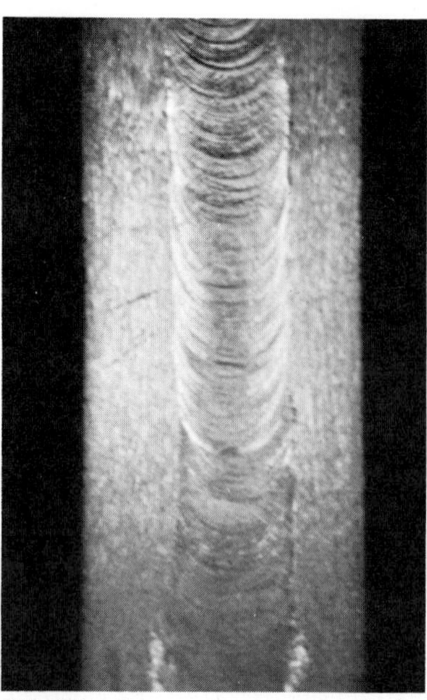

25-40. *Face appearance of the cover pass in a butt joint in stainless steel pipe welded with the TIG process. Slight reinforcement.*

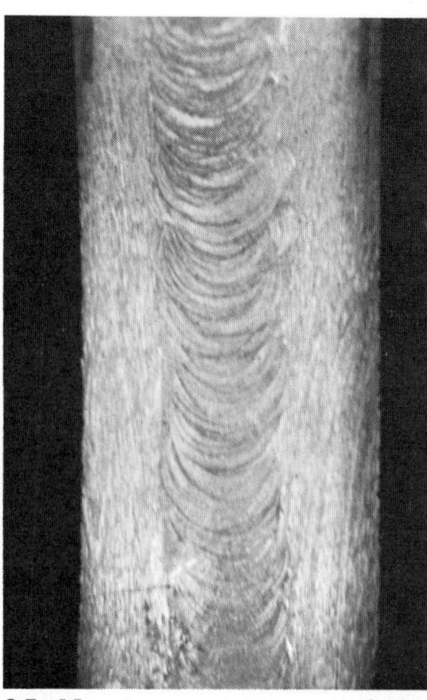

25-41. *Face appearance of the cover pass in a butt joint in stainless steel pipe welded with the gas tungsten-arc process. Relatively flat.*

25-39. *Welding procedure for filler and cover pass on pipe in the bell-hole position.*

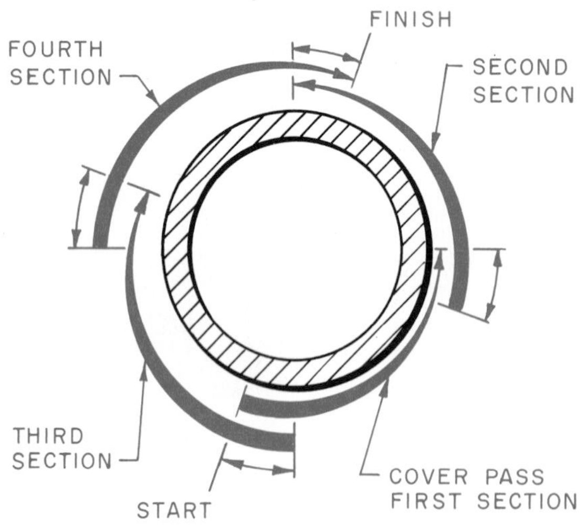

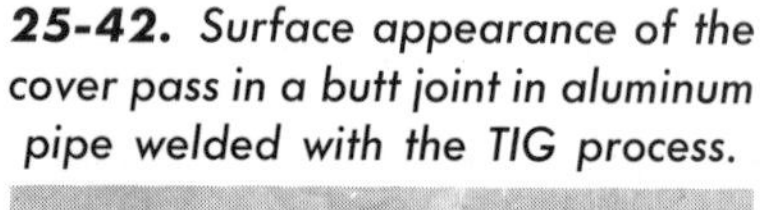

25-42. *Surface appearance of the cover pass in a butt joint in aluminum pipe welded with the TIG process.*

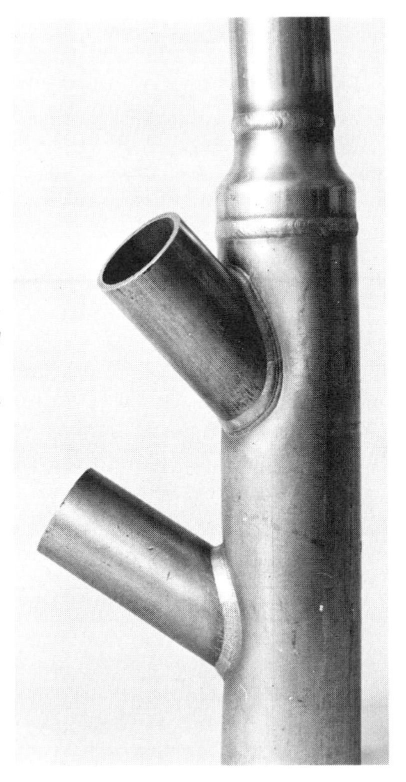

Pipefitters Union, St. Louis, Mo.

25-43A. *Butt and branch welds made with the gas tungsten-arc process. Pipe material is Inconel®. Diameters are 2½ and 3½ inches; wall thickness is ¼ inch. Joint gap for butt welds was ⅛ inch, and for the laterals, 3/16 inch. Helium was used both for purging and as a shielding gas. (Argon can also be used.) Tungsten electrode size was 3/32, filler rod size was 3/32, and welding heat was about 55 to 75 amperes. Stringer bead technique was used throughout.*

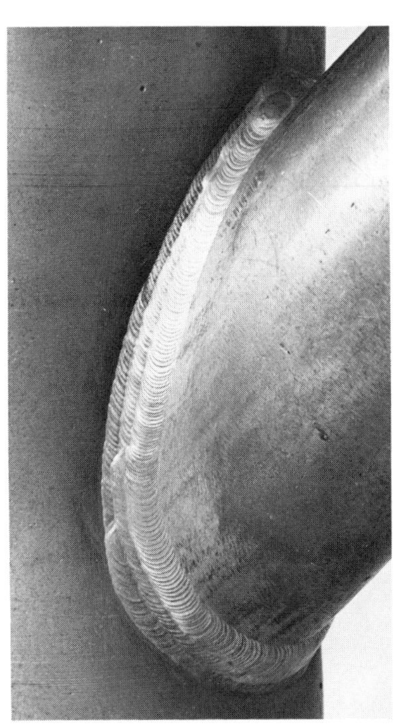

Pipefitters Union, St. Louis, Mo.

25-43C. *Close-up of branch weld shown in Fig. 25-43A.*

face- and root-bend specimens from the top, bottom, and sides of the pipe and subject them to bending stress. This test is similar to the testing procedures for the shielded metal-arc welding of plate and pipe and for the gas tungsten-arc welding of plate.

Branch Welds

After you have demonstrated that you are able to make satisfactory butt welds in pipe of various materials, you may wish to practice a number of branch (lateral) joints. Both 90 and 45 degree branches are suggested. These are similar to those done with the shielded metal-arc process, using the stringer bead technique. Completed welds should have the appearance shown in Figs. 25-43A and C.

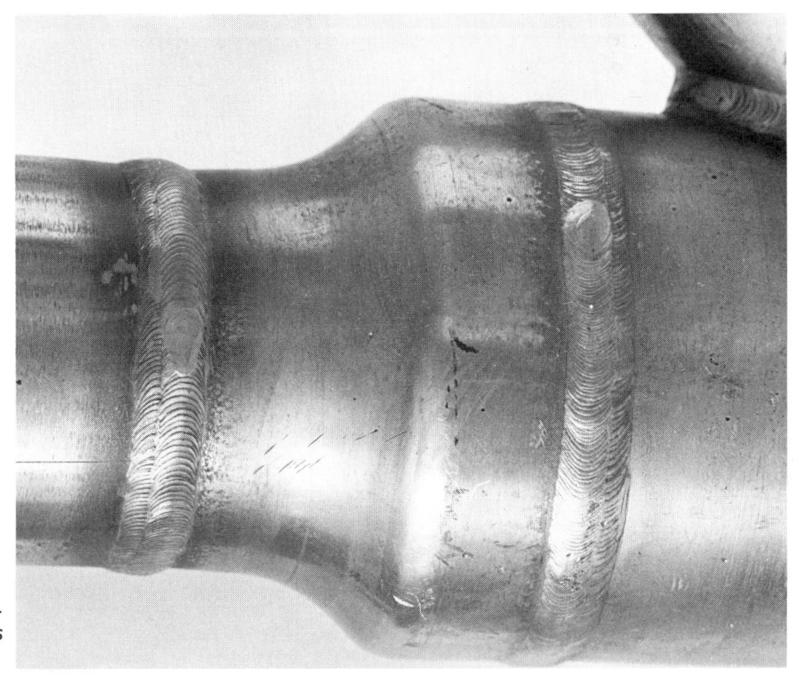

Pipefitters Union, St. Louis, Mo.

25-43B. *Close-up of butt welds shown in Fig. 25-43A.*

REVIEW QUESTIONS

1. List the five basic joints used in the gas tungsten-arc welding of pipe.
2. Contamination is one cause of porosity in the weld metal. Explain.
3. What is meant by purging? What is its purpose?
4. Explain the terms root pass, filler pass, and cover pass.
5. In making a pipe weld, how can you tell if you are getting penetration through to the root of the joint?
6. What is the most critical point in root pass welding?
7. List the important factors in root pass welding.
8. List the important factors in filler pass welding.
9. List the important factors in cover pass welding.
10. Test welds in stainless steel pipe must have a tensile strength that is greater than 75,000 p.s.i. True. False.
11. Explain the proper welding technique when ending a weld bead.
12. Explain the general rule for determining the width of a cover pass weld bead on pipe.
13. The height of the reinforcement of the cover pass on a pipe joint should be
    a. 3/8 inch c. flat with the surface
    b. 1/4 inch d. 1/16 inch
14. Gas tungsten-arc welds in a branch joint are similar to those done with the shielded metal-arc process, using the stringer bead technique. True. False.

Table 25-2
Course Outline: Gas Tungsten-Arc Welding Practice (Pipe)

Recommended job order[1]	Number in text	Type of joint	Type of weld	Welding position	Welding technique	Pipe specifications: Material	Dia.	Thickness	Tungsten electrode size	Current: Type	Amperes[2]	Shielding Gas: Gas type	Gas flow (cfh)[3]	Cup size	Filler rod size[4]	Text reference
1st	25-J1	—	Beading	Horizontal roll	Stringer—up and weaved—up	Steel	4″–6″	1/4″	3/32	DCSP	140–160	Argon	8–10	3/8	3/32–1/8	807
2nd	25-J2	V butt	Groove	Horizontal roll	1 stringer—up 1 weaved—up	Steel	4″–6″	1/4″	3/32	DCSP	120–140 140–160	Argon	8–10	3/8	3/32–1/8	807
3rd	25-J8	V butt	Groove	Vertical fixed	5 stringer	Steel	4″–6″	1/4″	3/32	DCSP	120–160	Argon	8–10	3/8	3/32–1/8	807
4th	25-J12	V butt	Groove	Horizontal fixed	1 stringer—up 1 weaved—up	Steel	4″–6″	1/4″	3/32	DCSP	120–140	Argon	8–10	3/8	3/32–1/8	807
5th	25-J3	—	Beading	Horizontal roll	Stringer—up Weaved—up	Aluminum	4″–6″	1/4″	1/8	ACHF	150–280	Argon	22–28	1/2	3/32–1/8	807
6th	25-J4	V butt	Groove	Horizontal roll	1 stringer—up 1 weaved—up	Aluminum	4″–6″	1/4″	1/8	ACHF	150–280 160–300	Argon	22–28	1/2	3/32–1/8	807
7th	25-J9	V butt	Groove	Vertical fixed	5 stringer	Aluminum	4″–6″	1/4″	1/8	ACHF	150–200 160–210	Argon	25–30	1/2	3/32–1/8	807
8th	25-J13	V butt	Groove	Horizontal fixed	1 straight—up 2 weaved—up	Aluminum	4″–6″	1/4″	1/8	ACHF	120–200 140–220	Argon	25–30	1/2	3/32–1/8	811
9th	25-J14	V butt	Groove	Horizontal fixed	1 semi-weaved	Aluminum	2″–3″	1/8″	3/32	ACHF	120–145	Argon	15–20	3/8	3/32–1/8	811
10th	25-J15	V butt	Groove	Horizontal fixed	1 semi-weaved	Aluminum	2″–3″	1/8″	3/32	ACHF	100–120	Argon	12–22	3/8	3/32–1/8	811
11th	25-J10	V butt	Groove	Vertical fixed	1 semi-weaved	Aluminum	2″–3″	1/8″	3/32	ACHF	100–120	Argon	15–20	3/8	3/32–1/8	811
12th	25-J5	—	Beading	Horizontal roll	Stringer—up weaved—up	Stainless	4″–6″	1/4″	3/32	DCSP	120–160	Argon	12–15	1/2	3/32–1/8	812
13th	25-J6	V butt	Groove	Horizontal roll	1 stringer—up 1 weaved—up	Stainless	4″–6″	1/4″	3/32	DCSP	110–150 110–160	Argon	12–15	1/2	3/32–1/8	812
14th	25-J11	V butt	Groove	Vertical fixed	5 stringer	Stainless	4″–6″	1/4″	3/32	DCSP	110–150 110–160	Argon	12–15	1/2	3/32–1/8	812
15th	25-J16	V butt	Groove	Horizontal fixed	1 stringer—up 2 weaved—up	Stainless	4″–6″	1/4″	3/32	DCSP	110–150 110–160	Argon	14–18	1/2	3/32–1/8	812
16th	25-J7	Square butt	Groove	Horizontal roll	1 semi-weaved	Stainless	2″–3″	1/8″	1/16–3/32	DCSP	60–120	Argon	12–15	1/4–3/8	3/32–1/8	812
17th	25-J17	Square butt	Groove	Horizontal fixed	1 semi-weaved	Stainless	2″–3″	1/8″	1/16–3/32	DCSP	80–140	Argon	12–15	1/4–3/8	3/32–1/8	812

[1] It is recommended that the student do the jobs in this order. In the text, the jobs are grouped according to the type of operation to avoid repetition.
[2] Ceramic or glass cups should be used for currents up to 250 amperes. Water-cooled cups should be used for currents above 250 amperes.
[3] The gas flow should be set for the maximum rate for the vertical and overhead positions. Gas backing is optional for steel. It is required for aluminum and stainless.
[4] The type of filter rod must always match the type of base metal being welded.
The conditions given here are basic, and they vary with the job situation, the results desired, and the skill of the welder.

CHAPTER 26 Gas Metal-Arc Welding Practice With Solid Core Wire, (Plate)

The practice jobs outlined on the following pages include gas metal-arc (GMAW, or MIG) welding practice on a variety of materials in all positions, Fig. 26-1. Since you are already an accomplished welder, having mastered the gas welding, shielded metal-arc and gas tungsten-arc processes, it will not be very difficult to master this new process. It is estimated that most students will become competent in 30 to 45 hours of practice. Each job requires some of the skills already learned. This new process and the jobs provided also require the learning of new skills and the application of the technical information given in this chapter and Chapter 10. As with all previous practice, each job should be mastered before going on to the next.

26-1. *This student is practicing gas metal-arc welding.*

OPERATING VARIABLES THAT AFFECT WELD FORMATION

Welding variables are those factors that affect the operation of the arc and the weld deposit. Sound welding of good appearance results when the variables are in balance. In order for you to develop a feel for the process and understand the arc characteristics and metal formation, it is necessary to become familiar with all the variables and study their effect on the weld deposit through experience. As an advanced student, you will be familiar with some of the variables since they are present in all welding processes to a certain extent. You will find, however, that each electric welding process has its own arc characteristics.

Type of Current

DIRECT CURRENT, REVERSE POLARITY (DCRP). Reverse polarity is generally used for gas metal-arc welding. Because it provides maximum heat input into the work, it allows relatively deep penetration to take place, Fig. 26-2. It also assists in the removal of oxides from the plate, which contributes to a clean weld deposit of high quality. Low current values produce the globular transfer of metal from the electrode. A gradual increase in current increases the electrode melting rate so that at relatively high current values the spray transfer of metal is produced.

The welding of ferrous metals with DCRP, bare filler wire, and pure argon causes the arc to be unstable and introduces porosity into the weld metal. This condition may be corrected by the addition of 2 to 5 percent oxygen to the gas mixture.

DIRECT CURRENT, STRAIGHT POLARITY (DCSP). Straight polarity has a limited use in the welding of thin gauge materials. The greatest amount of heat occurs at the electrode tip. The wire meltoff rate is a great deal faster than with DCRP. The arc is not stable at the end of the filler wire, and it becomes very difficult to direct the transfer of weld metal where it is desired. The erratic arc results in poor fusion and a considerable amount of spatter. Penetration is also less than with reverse polarity, Fig. 26-3.

The unstable arc can be corrected to a considerable extent by the use of a shielding gas mixture of approximately 5 percent oxygen added to argon. The normally high electrode meltoff rate, however, is reduced substantially when oxygen is added so that any advantage of DCSP is cancelled out.

ALTERNATING CURRENT. Alternating current is seldom used in gas metal-arc welding. The arc is unstable because current and voltage both pass through the zero point many times each second as the current reverses. This causes the arc to be extinguished each time. It is necessary to use a rather high open-circuit voltage. Since alternating current is a combination of straight and reverse polarity, the rate of metal transfer and the depth of penetration fall between those of both polarities.

Shielding Gas

Argon and helium were first used as the shielding gases for the gas metal-arc process, and they continue to be the basic gases. Argon is used more than helium on ferrous metals to keep spatter at a minimum. Argon is also heavier than air and therefore gives better weld coverage. Oxygen or carbon dioxide is added to the pure gases to improve arc stability, minimize undercut, reduce porosity, and improve the appearance of the weld.

Helium may be added to argon to increase penetration with little effect on metal transfer characteristics. Hydrogen and nitrogen are used for only a limited number of special applications in which their presence will not cause porosity or embrittlement of the weld metal.

Carbon dioxide is becoming increasingly popular as a shielding medium, because of the following advantages:

- Low cost
- High density, resulting in low flow rates
- Less burn-back problems because of its shorter arc characteristics.

Review Chapters 9 and 10 for additional information regarding the shielding gases.

The following recommendations concerning specific metals are helpful.

Aluminum alloys: argon. With direct current, reverse polarity, argon removes surface oxide.

Magnesium and aluminum alloys: 75 percent helium, 25 percent argon. Correct heat input reduces the tendency toward porosity and removes surface oxide.

Stainless steels: argon plus oxygen. When direct current, straight polarity is used, 5 percent oxygen improves arc stability.

Magnesium: argon. With direct current, reverse polarity, argon removes surface oxide.

Deoxidized copper: 75 percent helium, 25 percent argon preferred. Good wetting and increased heat input counteract high thermal conductivity.

Low alloy steel: argon, plus 2 percent oxygen. Oxygen eliminates undercutting tendencies and removes oxidation.

Mild steel: carbon dioxide (dip transfer). Carbon dioxide promotes high quality. It is suitable for low current, out-of-position welding. There is little spatter.

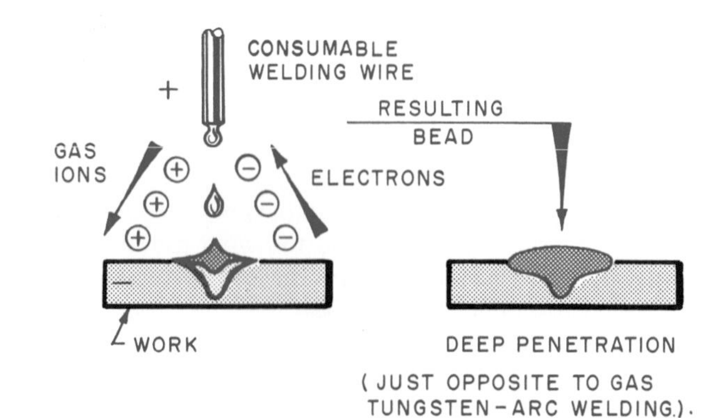

26-2. *Gas metal-arc DCRP welding: wire positive, work negative.*

26-3. *Gas metal-arc DCSP welding: wire negative, work positive.*

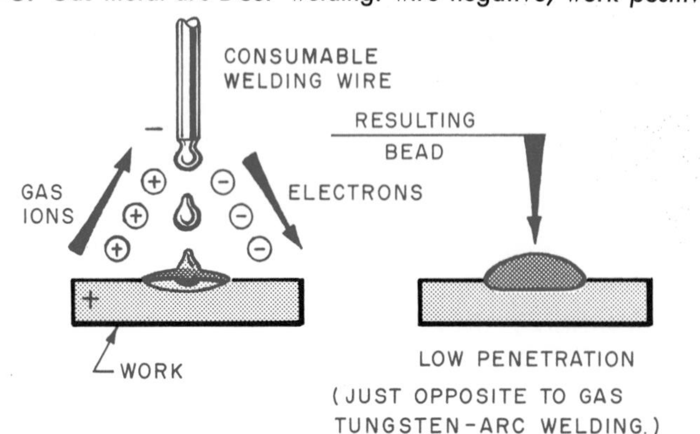

Nickel, Monel®, and Inconel®: argon. Good wetting decreases the fluidity of the weld metal.

Titanium: argon. Argon reduces the size of the heat-affected zone and improves metal transfer.

Silicon bronze: argon. Argon reduces crack sensitivity on this hot-short material.

Aluminum bronze: argon. Argon reduces penetration of the base metal. It is commonly used as a surfacing material.

Joint Preparation

Joint designs like those used with other arc welding processes may be used with the gas metal-arc process. Costs can be reduced, however, by employing somewhat different designs. Any joint design should provide for the most economical use of filler metal. The correct design for a particular job depends on the type of material being welded, the thickness of the material, the position of welding, the welding process, the final results desired, the type and size of filler wire, and welding technique.

The arc in gas metal-arc welding is somewhat more penetrating and narrower than the arc in shielded metal-arc welding. Therefore heavier root faces and smaller root openings may be used for groove welds. It is also possible to provide a narrower groove for MIG welding. The angle in a single-V or double-V butt weld is about 75 degrees for shielded metal-arc welding. For gas metal-arc welding, this angle may be reduced to 30 to 45 degrees, Fig. 26-4. These changes in joint design increase the speed of welding, cut the time necessary for joint preparation, and reduce the amount of weld metal that is required. Thus they lower the cost of materials and labor.

The penetration achieved with shielded-arc electrodes is about $\frac{1}{8}$ inch maximum in steel. With the MIG process 100 percent penetration may be secured in $\frac{1}{4}$-inch plate in a square butt joint welded from both sides.

For 60-degree single- or double-V butt joints, no root face is recommended, and the root opening should range from 0 to $\frac{3}{32}$ inch. Double-V joints may have wider root openings than single-V joints. Poor fitup and root overlap should be avoided. If the root opening is large, a backup strip should be used.

Plates thicker than 1 inch should have U-groove preparation. They require considerably less weld metal. The root face thickness should be less than $\frac{3}{32}$ inch, and the root spacing should be between $\frac{1}{32}$ and $\frac{3}{32}$ inch.

In multipass welding, the absence of slag insures easier cleaning and so reduces the problem of porosity in the weld metal.

In making fillet welds with the MIG process, advantage is taken of the deeper penetration by depositing smaller weld beads on the surface of the material. The throat area of the weld is not reduced since a greater part of the weld bead is beneath the surface of the base metal, Fig. 26-5.

Certain types of joints are backed up to prevent the weld from projecting through the back side. The usual materials include blocks, strips, and bars of copper or steel.

Electrode Diameter

The electrode diameter influences the size of the weld bead, the depth of penetration, and the speed of welding. As a general rule, for the same current (wire-feed speed setting), the arc becomes more penetrating as the electrode diameter decreases. At the same time the speed of welding is also affected because the deposition (burnoff) rate also increases.

When welding with wires below $\frac{1}{16}$ inch, the smaller wire operating at a given current density burns off faster than the larger wire. To get the maximum deposition rate at a given current density, use the smallest wire possible that is consistent with the underbead shape.

26-4. *V-butt joint comparison.*

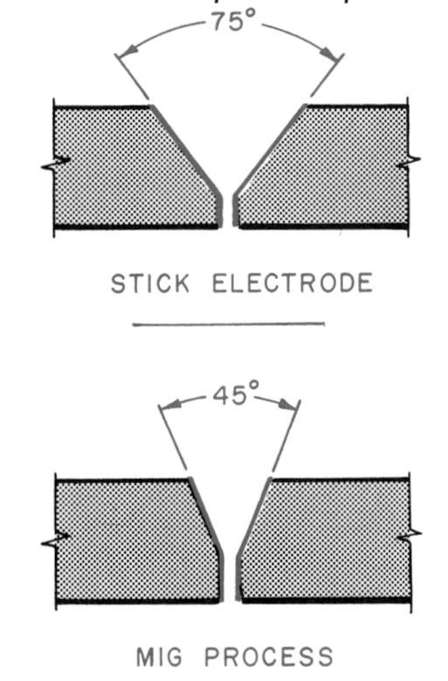

26-5. *Comparison of penetration in a fillet weld: carbon dioxide shielded MIG weld vs. coated electrode weld.*

f = FILLET SIZE
t = THROAT – COATED ELECTRODE
t' = THROAT – CO_2 PROCESS

With wires $1/16$ inch and larger for a given welding current, the next size larger wire provides a higher deposition rate. This is because the effects of preheating on the larger wires are less. Large wires deposit wider beads than small wires do under identical conditions.

As in all welding, the selection of the type and size of wire is important. Generally, filler wires should be of the same composition as the materials being welded. The position of welding or other special conditions may affect the size of the electrode. For most purposes, however, filler wires with diameters of 0.020, 0.030, and 0.035 inch are best for welding thin materials. Diameters of 0.045 inch or $1/16$ inch are used for medium thickness, and a diameter of $1/8$ inch is best for heavy materials. Filler wires with small diameters are recommended for welding in the vertical and overhead positions. Large diameter wires are desirable for those applications in which penetration is undesirable. Hard surfacing, overlays, and buildup work are examples.

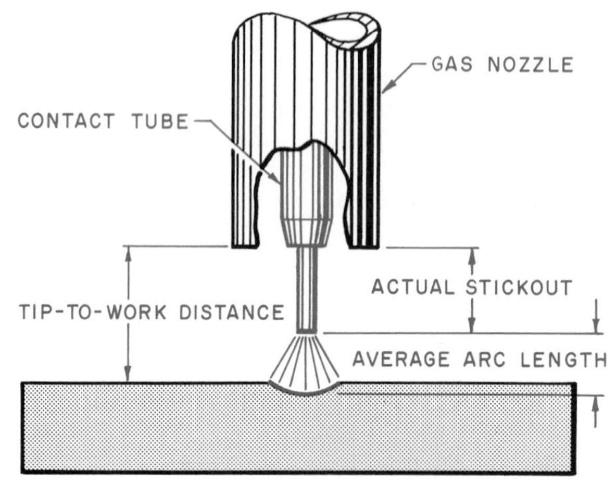

26-6. *Nomenclature of area between nozzle and workpiece.*

26-7. *The long stickout attachment for semi-automatic, self-shielded, open-arc welding is simply a longer tube that provides guidance and protection for the extra length of electrode.*

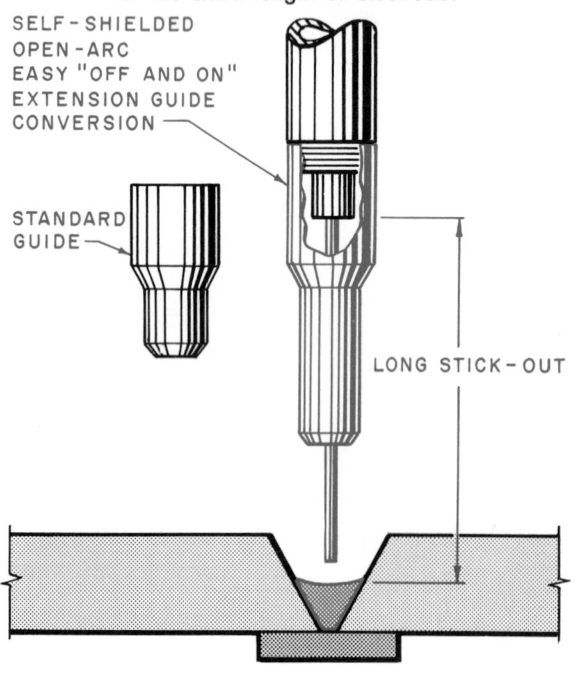

Stickout

Stickout is that length of filler wire which extends past the gas nozzle and contact tube, Fig. 26-6. This is the area where preheating of the filler wire occurs. Stickout is also called the *electrode extension.* It controls the dimensions of the weld bead since the length of the stickout affects the burnoff rate. Stickout exerts an influence on penetration through its effect on the welding current. As the stickout length is increased, the preheating of the wire increases and the current is reduced. The current reduction in turn decreases the amount of penetration into the work. Stickout distance may vary from $1/8$ to $1\frac{1}{4}$ inches.

Short stickout ($1/8$ to $3/8$ inch) is desirable for wire $1/32$ to $3/64$ inch in diameter and low current applications. Stickout of $1/4$ to $1/2$ inch is generally necessary when using slope and prereactance to get proper arc starting characteristics. Longer stickout, as much as 1 to $1\frac{1}{4}$ inches, is necessary on wire $5/64$ and $3/32$ inch in diameter and high current applications such as welding with flux-cored wire.

Long stickout, as much as 5 inches, is used with semiautomatic submerged-arc welding. A guide tip, Fig. 26-7, is attached to the nozzle end for long stickout. The increase in stickout from the point of electrical contact in the gun to the arc permits the entire length of electrode to be preheated by its resistance to the current passing through it. It reaches a temperature approaching the melting point so that only a small amount of arc heat is needed for it to become molten at the point of welding.

The relationship which exists between stickout and current and stickout and penetration should be understood, Table 26-1. Tests have indicated that when stickout is increased from $^3/_{16}$ to $^5/_8$ inch, the welding current then drops approximately 60 amperes. The current is reduced because of the change in the amount of preheating that takes place in the wire. As the stickout is increased, the preheating of the wire increases. Thus less welding current is required from the power source at a given feed rate. Due to the self-regulating characteristics of the constant voltage power source, the welding current is decreased. As the welding current is decreased, the depth of penetration also decreases. (Increased stickout also increases the weld deposition rate.) On the other hand, if the stickout decreases, the preheating of the filler wire is reduced, and the power source furnishes more current in order to melt the wire at the required rate. This increase in welding current causes an increase in penetration.

Position of the Gun

The position of the welding gun with respect to the joint is expressed by two angles: the longitudinal (nozzle) angle and the transverse (head) angle, Fig. 26-8.

The bead shape, as well as the penetration pattern, can be changed by changing the direction of the wire as it goes into the joint in the line of travel.

NOZZLE ANGLE. The nozzle angle can be compared to the angle of the electrode in shielded metal-arc welding. The *trail* and *lead* nozzle angles are shown in

Table 26-1

Effect of stickout length on weld characteristic.

Increase stickout	Increase deposition bead height
	Decrease welding current penetration bead width
Decrease stickout	Decrease deposition bead height
	Increase welding current penetration bead width

26-8. *Transverse and longitudinal nozzle angles.*

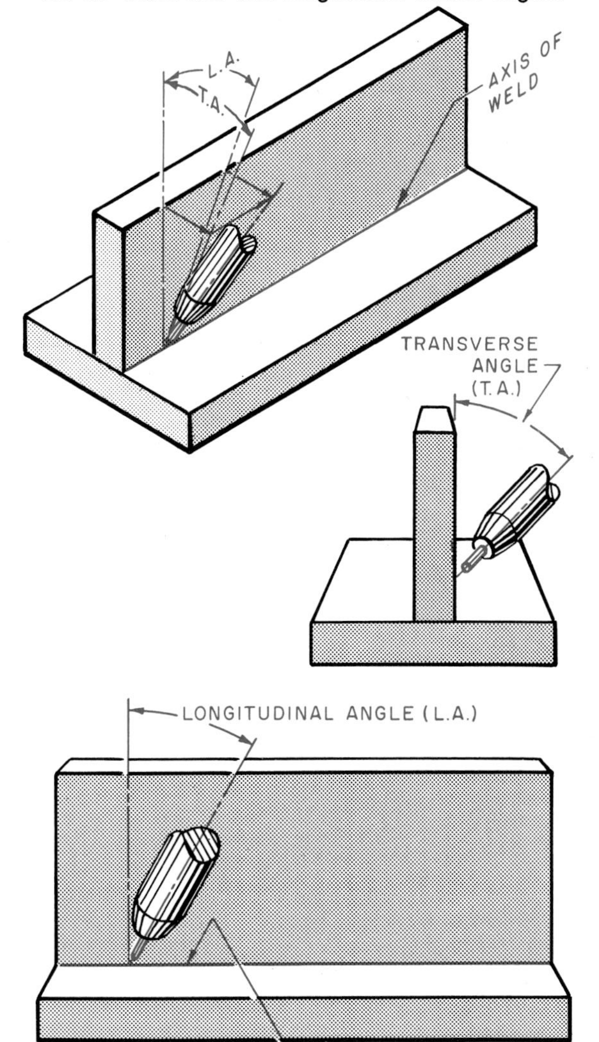

Fig. 26-9. The trail technique results in a high, narrow bead with relatively deeper penetration. The penetration is deeper because the arc tends to run into the puddle and create a greater concentration of heat. The force of the arc pushes the molten metal back for the rounded-bead contour. Maximum penetration is obtained when a trailing angle of 10 degrees is used. As the trailing angle is reduced, the bead height decreases, and the width increases.

The arc in the leading technique strikes cold base metal so that penetration is shallow. The force of the arc pushes the metal ahead of the bead and flattens the contour of the bead. Increased travel speeds are a characteristic of the lead technique.

HEAD ANGLE. The head angle (transverse angle) refers to the position of the wire to the joint in a plane perpendicular to the line of travel, Figs. 26-8 and 26-10. For fillet weld joints, the head angle is normally half of the included angle between the plates forming the joint. For butt welds the head angle is normally 90 degrees to the surface of the plate being joined.

The head angle utilizes the natural arc force to push (wash) the weld metal against a vertical surface to prevent undercut and provide good bead contour. This has particular significance in welding lap and T-joints. High travel speeds usually require greater head angles to insure the proper washing action.

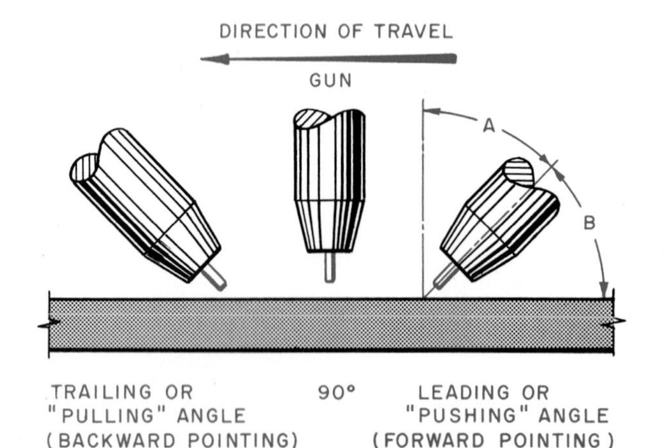

26-9. *Trail and lead nozzle angles.*

26-10. *Head and nozzle angles.*

Arc Length

The constant voltage welding machine used for gas metal-arc welding provides for the self-adjustment of the arc length. The power source supplies enough current to burn off the filler wire as fast as it is being fed to maintain the arc length appropriate to the voltage setting.

If for any reason the arc length is shortened, the arc voltage will be reduced. This increases the current so that the filler wire melts at an increased rate until the correct arc length is reestablished. But if the arc length is lengthened, the arc voltage will be increased. This reduces the current and slows down the melting rate of the filler wire until the correct arc length is reestablished. No change in the wire-feed speed occurs. The arc length is corrected by the automatic increase or decrease of the burnoff rate of the filler wire. The operator has complete control of the welding current and the arc length by setting the wire-feed speed on the wire feeder and the open-circuit voltage on the welding machine.

Arc Voltage

In the gas metal-arc process, the voltage remains constant

as set by the person doing the welding. Furthermore, the burn-off rate of the metal electrode is constant at a given voltage setting.

In the gas metal-arc process, the arc voltage has a decided effect upon penetration, bead height, and bead width. The chief function of voltage is to stabilize the welding arc and to provide a smooth, spatter-free weld bead. Thus, for any given welding current there is a particular voltage that will provide the smoothest possible arc and the deepest penetration.

A higher or lower arc voltage causes the arc to become unstable and affects the penetration. High arc voltage produces a wider, flatter bead. Excessive voltage also increases the possibility of porosity in the weld metal and also increases the spatter. In fillet welds it increases undercut and produces a concave fillet subject to cracking. Low voltage causes the bead to be high and narrow. Extremely low voltage causes the wire to stub on the plate. Therefore, changes in voltage have opposite effects on bead height and bead width. As the arc voltage increases, bead height decreases and bead width increases. (See Figs. 26-11 and 26-12 and Table 26-2.) This does not change the overall size of the bead, as in the case of welding current and travel speed, but it changes the contour and shape of the bead.

Generally speaking, high arc voltages result in *drop transfer* of metal from the wire to the weld puddle. Drop transfer is spatter-prone and reduces deposition efficiency. High arc voltage also reduces burnoff rates because of greater radiation losses.

For a fixed deposition rate, low arc voltages allow faster travel speeds and downhill operation. High arc voltages necessitate slower travel rates to allow the weld deposit to accumulate properly. A long arc may also cause contamination of the gas field since the shielding gas may not be completely contained, and the contaminating air is permitted to enter the gas shield. The proper voltage has a sharp, crackling sound. It produces good penetration and fusion, and the weld bead formation is excellent.

26-11. *The relationship of arc length to weld bead width.*

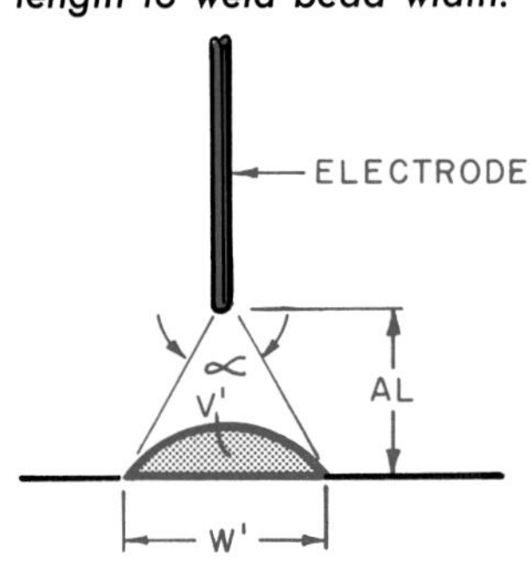

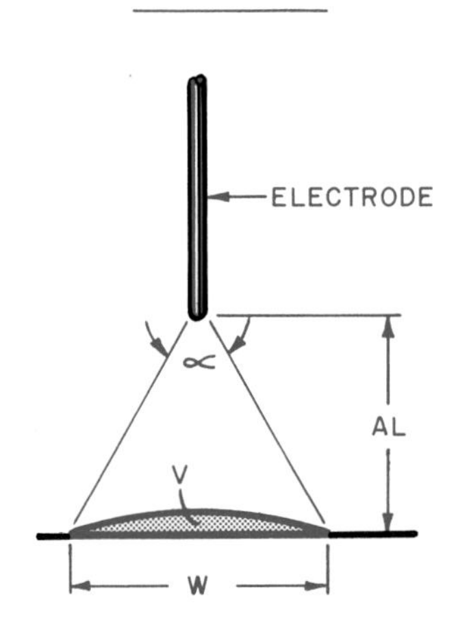

W = BEAD WIDTH AT HIGH VOLTAGE.
W' = BEAD WIDTH AT LOW VOLTAGE.
V' & V = VOLUME OF BEAD V = V'.
AL = ARC LENGTH.
∝ = INCLUDED ANGLE OF ARC STREAM.

Wire-Feed Speed

The special wire feeder and the constant voltage welding machine constitute the heart of the MIG welding process. There is a fixed relationship between the rate of filler wire burnoff and

26-12. *Shallow penetration results when arc voltage is too high for travel speed. Deep penetration occurs when arc voltage is too low for speed. With proper arc voltage for speed, smooth profile in center illustration is obtained.*

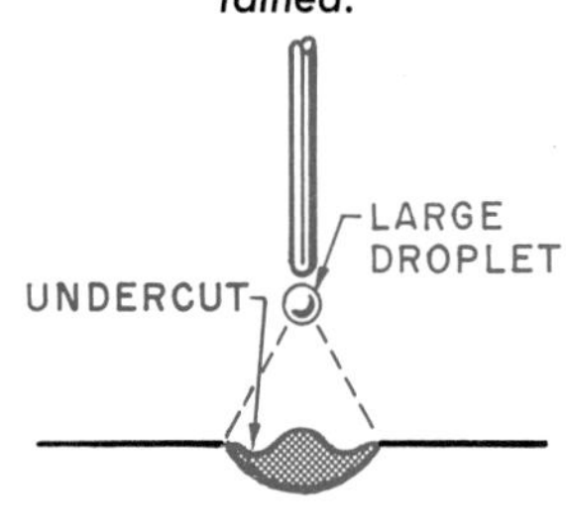

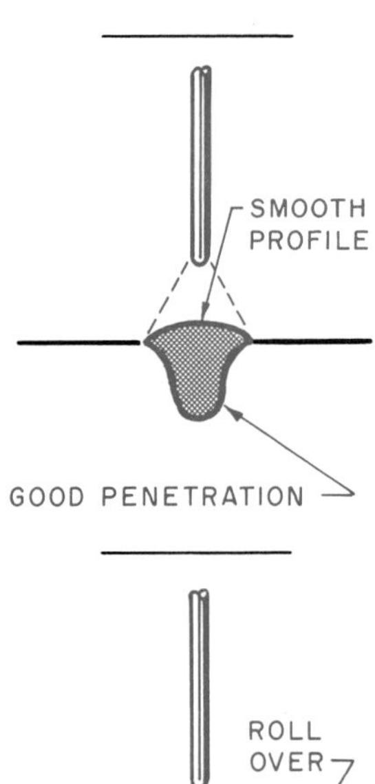

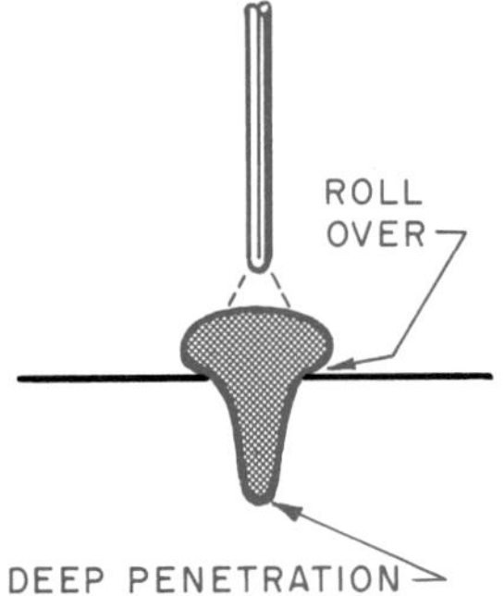

welding current. The electrode wire-feed speed determines the welding current. Thus current is set by the wire-feed speed control on the wire feeder. The welding machine supplies the amount of current (amperes) necessary to melt the electrode at the rate required to maintain the preset voltage and resultant arc length.

An increase in the electrode-feed speed requires more electrode to be melted to maintain the preset voltage and arc length. Increased wire-feed speed causes higher current to be supplied by the welding machine, and the melting rate and deposition rate increase. More weld metal and more heat applied to the weld joint produce deeper penetration and larger weld beads. If wire-feed speeds are excessive, the welding machine cannot put out enough current to melt the wire fast enough, and "stubbing" or "roping" of the wire occurs, Fig. 26-13. An excessive wire-feed speed causes convex weld beads and poor appearance. A decrease in electrode wire-feed speed results in less electrode being melted. The welding machine supplies less current and so reduces the deposition rate. Less weld metal and less heat are applied to the weld joint so that penetration is shallow and the weld bead is smaller.

Generally, for a given filler wire size, a high setting of the filler wire speed rate results in a short arc. A slow speed setting contributes to a long arc.

Welding Current

The setting at the wire-feed speed control determines the amount of current that will be delivered at the arc. The term current is often related to current density. *Current density* is the amperage per square inch of cross-sectional area of the electrode. Thus, at a given amperage the current density of an electrode 0.035 inch in diameter is higher than that of an electrode 0.045 inch in diameter.

Each type and size of electrode has a minimum and maximum current density. The best working range lies between them.

The depth of penetration, bead formation, filler wire burn-off, speed of travel, and the size and appearance of the weld reinforcement are all affected by the amount of welding current.

There is direct relationship between the welding current and penetration. In general, for any change in welding current, there is a corresponding change of penetration. As the welding current increases, the penetration increases, and as the welding current decreases, the penetration decreases. Increasing the current also increases the wire meltoff rate and the rate of deposition. If the current is too low for a given electrode size, metal transfer is sluggish, and poor penetration and poor fusion result. The bead is rough, and the reinforcement is high. If the current is too high, penetration may be too deep. Excessive penetration causes burn-through and undercut. The weld bead has poor reinforcement and buildup. Undercut is a particular problem in making fillet welds with current settings that are too high.

Each filler wire size and type has a minimum and maximum current range that give best results. Avoid excessive current on a wire of a given size. Switch to a larger wire if you think that the current is too high for the wire. Study Table 26-3 for comparative current ranges. If the current is too high, there is the possibility of electrode burn-back into the contact tube, the arc is unstable, and gas shielding is disturbed. Spatter results, and the deposited weld metal has poor physical characteristics. Increases in current increase bead height and width. As the current is increased, the voltage must also be increased.

If the current is too low, the filler wire may become short cir-

Table 26-2

Recommended variable adjustments for MIG welding.

Change of welding variables			Effect on weld deposit			
Variables	Increase	Decrease	Penetration	Deposition	Bead height	Bead width
Voltage	x		Increase[1]		Decrease	Increase
Voltage		x	Decrease		Increase	Decrease
Current[3]	x		Increase	Increase	Increase	Increase[2]
Current		x	Decrease	Decrease	Decrease	Decrease
Travel speed	x		Decrease		Decrease	Decrease
Travel speed		x	Increase		Increase	Increase
Stickout	x		Decrease	Increase	Increase	Decrease
Stickout		x	Increase	Decrease	Decrease	Increase
Nozzle angle (trailing-max. 25°)			Maximum		High	Narrow
Nozzle angle (leading)			Decrease		Decrease	Increase

[1]Up to an optimum arc voltage—raising or lowering beyond this point reduces penetration.
[2]Up to an optimum current setting—raising or lowering beyond this point reduces bead width.
[3]Adjusted by wire-feed speed control.

cuited to the work, the electrode may become red hot, and the arc will be extinguished. If the welding can be maintained, the arc is unstable, and poor fusion occurs. The bead formation is high and limited to the surface of the base metal with little or no penetration. This destroys the advantage of gas metal-arc welding, which owes its success to the sharp concentration of high current density at the electrode wire tip.

Too much or too little current affects the physical properties of the weld metal. Tensile strength and ductility are reduced. Porosity may occur. Excessive oxides and other impurities may be present. Bead formation is also affected.

Travel Speed

Travel speed has a decided effect on penetration, bead size, and appearance.

At a given current density, slower travel speeds provide proportionally larger weld beads and more heat input in the base metal per unit length of weld. The longer heating time of the base metal increases the depth of penetration, and the increased weld deposit per unit length results in a higher and wider bead contour. If the speed of travel is too slow, unusual weld buildup occurs. Excessive buildup causes poor fusion, increased penetration, porosity, slag inclusions and a rough, uneven bead.

Progressively increased travel speeds have opposite effects. Less weld metal is deposited with lower heat input per unit length of weld. This produces a narrower weld bead and less penetration. Excessively fast speeds cause high spatter and undercutting. The bead is irregularly shaped because there is too little weld metal deposited per unit length of weld.

Travel speed is a variable that is as important as wire-feed speed and arc voltage. Travel speed is influenced by the thickness of the metal being welded, joint design, cleanliness, joint fitup, and welding position. As travel speed is increased, it is necessary to increase the wire-feed speed, which in turn increases the current and the burnoff rate, to produce the same weld cross section. Excessive travel speed produces spatter, undercutting, and a low rate of weld metal deposit. A speed of travel that is too low produces excessive fusion of base metal, excessive penetration, and even burn-through.

Summary of Operating Variables

You will recall that the height and width of the bead obtained with the gas metal-arc process depend on the adjustment and control of these variables:

- Size and type of filler wire
- Characteristics of the shielding gas
- Joint preparation
- Gas flow rate
- Voltage
- Speed of travel
- Arc length
- Polarity
- Stickout
- Nozzle angle
- Wire-feed speed (current)

These variables are adjusted on the basis of the type of material being welded, the thickness of the material, the position of welding, the deposition rate required, and the final weld specifications. Proper adjustment and control determine formation of the weld bead by affecting such things as penetration, bead width, bead height, arc stability, deposition rate, weld soundness, and appearance. An understanding of these variables and their control is essential if you are to master the gas metal-arc process.

Welding current and *travel speed* have a similar effect on both bead height and width. Each variable increases or decreases both bead height and

Table 26-3
Typical MIG current ranges.

Electrode diameter	Welding current range
0.030″	50–140 amperes
0.035″	65–160 amperes
0.045″	100–220 amperes
1/16″	165–375 amperes
3/32″	210–550 amperes
1/8″	375–600 amperes

Miller Electric Manufacturing Co.

26-13. *Effect of wire-feed speeds.*

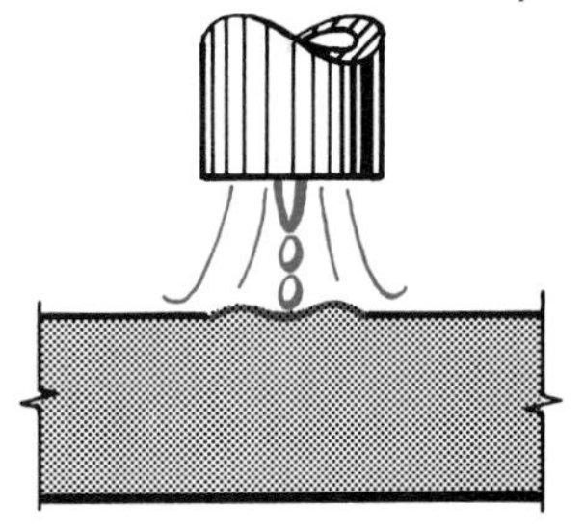

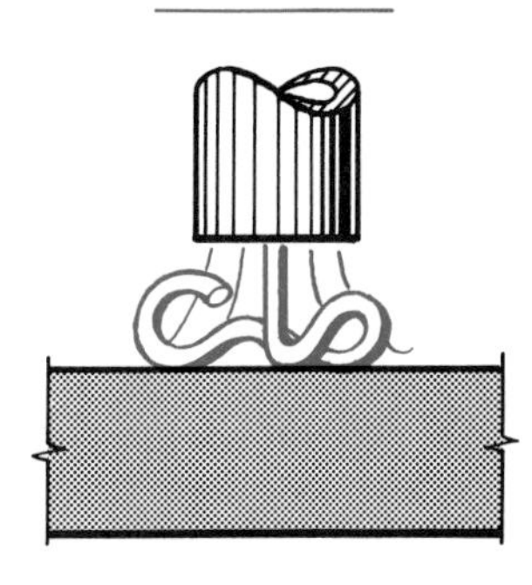

width at the same time. If travel speed is decreased, both bead height and width are increased. Any change in these variables affects the amount of filler metal being deposited per a given length of joint. Therefore, if a given travel speed and welding current (wire-feed speed) are not providing enough weld metal to fill a particular joint, either the travel speed must be decreased or the welding current (wire-feed speed) must be increased.

In multipass welding, excessive bead height interferes with fusion between weld beads, especially along the bead edges. This condition causes voids in the joint, which appear as wagon tracks in X-ray. If the bead width is too small, similar defects can occur due to the lack of fusion at the side of the weld. This is one of the reasons why this text has stressed multipass beads with a flat face.

Arc voltage has the opposite effect on bead height that it has on bead width. As the arc voltage increases, bead height decreases and bead width increases. While changes in travel speed and welding current (wire-feed speed) change the overall size of the bead, changes in arc voltage affect the shape of the bead. (Refer to Fig. 26-11, page 825.) For example, if a weld bead has excessive convexity and shows crowning, it can be corrected by increasing the arc voltage and thereby flattening the bead.

Nozzle angle also affects bead contour. A trailing nozzle angle tends to produce a high, narrow bead. The bead height decreases and the width increases as the trailing angle is reduced. This effect continues into the leading angle range. If the leading angle is increased too far, the bead starts to become narrow again. Refer to Table 26-2, page 826, for a review of the effect of variable adjustments on the physical characteristics of the weld bead.

WELD DEFECTS

Each completed job requires the close and careful inspection that you have followed with all the other welding processes presented in this text. High quality welds require the application of precise welding procedures. The defects found in welds made by the gas metal-arc process are similar to those found in other welding processes. However, the causes and the corrective action recommended are entirely different.

Lack of Penetration

Lack of penetration is the result of too little heat input in the weld area. This can be corrected by increasing the wire-feed speed and reducing the stickout distance to obtain maximum current for the particular wire-feed setting. If the wire-feed speed and stickout are correct, check the speed of travel. It may be too fast.

Lack of penetration may also be caused by improper welding techniques. It is important that the arc be maintained on the leading edge of the weld pool in order to secure maximum penetration into the base metal.

Excessive Penetration

Excessive penetration usually causes burn-through. It is the result of too much heat in the weld area. This can be corrected by reducing the wire-feed speed to obtain lower amperage or increasing the speed of travel.

Improper joint design is another cause of excessive penetration. If the root opening is too wide or if the root face is too small, burn-through is likely to occur. This difficulty can be prevented before welding by checking the bevel angle if there is one, the root opening, and the root face to make sure that they are correct for the position of welding. Excessive penetration can be remedied during welding by increasing the stickout distance as far as good working practice will allow and weaving the gun.

Whiskers

Whiskers are short lengths of electrode wire sticking through the weld on the root side of the joint. They are caused by pushing the electrode wire past the leading edge of the weld pool. Whiskers can be prevented by reducing the travel speed, reducing the wire-feed speed, increasing the stickout distance, and weaving the gun. It is important that the operator does not allow the electrode wire to get ahead of the weld pool.

Voids

Voids are sometimes referred to as *wagon tracks* because of their resemblance in X-rays to ruts in a dirt road. They may be continuous along both sides of the weld deposit. They are found in multipass welding. When the underneath pass has a bead with a large contour or a bead with too much crown or undercut, the next bead may not completely fill the void between the previous pass and the plate. Voids may be prevented by making sure that the edges of all passes are filled in so that undercut cannot take place. A large crown can be reduced by welding the next pass with a slightly higher arc voltage. Voids can also be corrected by increasing the travel speed on the next pass and making sure that the arc melts the previous bead and fuses into the sides of the joint.

Lack of Fusion

Lack of fusion, also referred to as *cold lap,* is largely the result of improper torch handling, low heat, and a high speed of travel. It is important that the arc be directed at the base metal and the leading edge of the puddle.

To prevent this defect, give careful consideration to the following.

- Direct the arc so that it covers all areas of the joint. The arc, not the puddle, should do the fusing.
- Keep the electrode at the leading edge of the puddle.
- Reduce the size of the puddle as necessary by reducing either the travel speed or wire-feed speed.
- Check current values carefully.

Porosity

The most common defect in welds produced by any welding process is porosity. Porosity that exists on the face of the weld is readily detected, but porosity in the weld metal below the surface must be determined by X-ray or other testing methods. The causes of most porosity are contamination by the atmosphere, a change in the physical qualities of the filler wire, and improper welding technique. Porosity is also caused by entrapment of the gas evolved during weld metal solidification. The intense heat of the arc separates water vapor and other hydrogen-bearing compounds from the metal. Because these compounds are lighter than molten metal, they tend to rise to the surface as gas bubbles before solidification. If they do not reach the surface, they will be entrapped as internal porosity. If they reach the surface, they appear as external porosity. Study the following causes of porosity carefully.

- Travel is so fast that part or all of the shielding gas is lost, and atmospheric contamination occurs.
- The shielding gas flow rate is too low so that the gas does not fully displace all of the air in the arc area.
- The shielding gas flow rate is too high. This draws the air into the arc area and causes turbulence, which reduces the effectiveness of the shield.
- The shielding gases must be of the right type for the metal being welded. The gases must be pure and dry.
- The gas shield may be blown away by wind or drafts. The operator should protect the weld area with a wind break or the position of his body. In some cases increasing the gas flow rate may be helpful. Gas flow may be increased as much as 50 to 100 percent for welding outdoors.
- There may be defects in the gas system. This could be the result of spatter clogging the nozzle, a broken gas line, defective fittings in the gas system, a defective gas valve, or a frozen regulator. It is important that the filler wire is in the center of the shielding gas flow. If the wire is off center, it can produce an erratic arc that can cause porosity.
- Excessive current for the electrode diameter can overheat the wire so that it loses its deoxidizers and alloying elements. This not only causes porosity, but it also seriously affects the physical characteristics of the weld.
- Foreign material such as oil, dirt, rust, grease, and paint is on the wire or material to be welded. Electrode wire should be stored in clean, dry areas. The material to be welded may be cleaned with chemicals, sandblasted, brushed, or scraped.
- Improper welding techniques are used. Excessive stickout coupled with an improper torch angle and movement is a serious cause of porosity. Removal of the gun and the shielding gas before the weld pool has solidified causes crater porosity and cracks. The gun should be held at its normal position at the end of the weld until the wire feed and the flow stop. Travel speed should not be so fast that the molten metal does not solidify under the gas shield.

Other Defects

WARPAGE. Warpage occurs when the forces of expansion and contraction are poorly controlled. A thorough understanding of these forces and their relationship to each other will enable you to reduce warpage in a weldment to a minimum.

WELD CRACKING. Weld cracking is brought about by compositional problems, poor joint design, and poor welding technique. The manganese content of the filler metal may be too low, or the sulfur content may be too high. The weld bead may be too small to withstand expansion and contraction movements in a joint, or the speed of travel may be too fast. Cracking can be prevented by making sure that the filler metal has a composition suitable for the base metal and by providing for expansion and contraction forces during welding.

SPATTER. Spatter is made up of very fine particles of metal on the plate surface adjoining the weld area. It is usually caused by high current, a long arc, an irregular and unstable arc, improper shielding gas, or a clogged nozzle.

IRREGULAR WELD SHAPE. Irregular welds include those that are too wide or too narrow, those that have an exessively convex or concave surface, and those that have coarse, irregular ripples. Such characteristics may be caused by poor torch manipulation, a speed of travel that is too fast or too slow, current that is too high or too low, improper arc voltage, improper stickout, or improper shielding gas.

Undercutting

Undercutting is a cutting away of the base material along the edge of the weld. It may be present in the cover pass weld bead or in multipass welding. This condition is usually the result of high current, high voltage, excessive travel speed, low wire-feed speed, poor torch technique, improper gas shielding, or the wrong filler wire. To correct undercutting move the welding gun from side to side in the joint, and hesitate at each side before returning to the opposite side.

Producing High Quality Welds

You are urged to study the material concerning weld defects carefully and to be ready to recognize poor welding when you see it. You must take great care and have pride of workmanship to produce quality welds. The gas metal-arc process is able to produce high quality welds in a wide variety of materials. Make sure that you understand the interrelationships of all of the variables and that you apply this knowledge with the manual skill of precise torch technique. Study the characteristics of these defects and their correction.

SAFE PRACTICES

In any industrial activity, safety is a most important consideration to both the worker and the employer. If the worker wishes to enjoy a long and profitable career, he must have a concern for his own health and well-being. Accidents and a disregard of the basic safeguards for good health cut short thousands of careers each year. Disability resulting from accidents and poor health costs the employer thousands of dollars in compensation and increased insurance costs. In these times of serious skill shortages, the loss to the workforce is also costly to the nation.

Welding is no more dangerous than other industrial operations. The safety precautions and protective equipment required for the MIG process are essentially the same as for any other electric welding process. There are a few factors, however, that need attention. You are urged to review the general precautions contained in Chapter 15 in addition to observing the specific precautions listed here.

Eye, Face, and Body Protection

The welding helmets and protective clothing worn when working with the other electric welding processes are necessary. The radiant energy, particularly in the ultraviolet range, that is produced by the gas-shielded process, is 5 to 30 times more intense than that produced by shielded metal-arc welding. The lowest intensities are produced by the gas tungsten-arc; the highest, by the gas metal-arc. Argon produces greater intensities than helium does. Radiant energy is particularly dangerous to the bare skin and unprotected eyes. The greater intensity of ultraviolet radiation also causes rapid disintegration of cotton clothing. There is, however, less spatter from hot metal and slag, and there is less danger of clothing fires than with shielded metal-arc welding. The following clothing regulations should be enforced during practice.

- Standard arc welding helmets with lenses ranging in shade from no. 6 for work using up to 30 amps to no. 14 for work using more than 400 amps should be worn. A glass that is too dark should be avoided to prevent eyestrain. When welding in a shop or a dark area, a heavier shade lens should be used than is normally used for welding outdoors. The arc should never be viewed with the naked eye when standing closer than 20 feet. The intensity at 2 feet away from the arc is high enough to burn the eyes in a few seconds. Good safety practice also requires that safety glasses be worn under the welding helmet. If these have a light tint, they protect the welder from getting a flash from the work of other welders working in the same area.
- The skin should be covered completely to prevent burns and other damage from ultraviolet light. The skin can be burned if exposed to the arc for a few minutes at a distance of 2 feet from the arc. The back of the head and neck should be protected from radiation reflected from bright surfaces such as aluminum. Gloves should always be worn, and when welding in all positions, special leather jackets are appropriate.
- Shirts should be dark in color to reduce reflections, thereby preventing ultraviolet burns to the face and neck underneath the helmet. The shirt should have a tight collar and long sleeves, and it should not be made of cotton. All outer clothing should be made of materials other than cotton. Leather, wool, and aluminum-coated cloth withstand the action of radiant energy reasonably well.

Handling of Gas Cylinders

The shielding gases are provided in steel cylinders. The same general precautions should be observed in the handling of these cylinders as those containing oxygen and acetylene. (Refer to Chapters 15 and 17.)

- Stored cylinders should be in a protected area away from fire, cold, and grease and away from the general shop activity.
- Cylinders must be secured to equipment to prevent their being knocked over.
- The proper regulators and flowmeters must be used with each special type of cylinder. High pressure cylinders must be treated with extreme care.
- Cylinders should not be dropped, used as rollers, lifted with magnets, connected into an electrical circuit, or handled in any other way that might damage the cylinder or regulator. When cylinders are empty, they should be stored in an upright position with the valve closed.

Ventilation

The ventilation requirements for shielded metal-arc welding are ample for gas metal-arc welding.

- Ozone is generated in small quantities, generally below the allowable limits of concentration. Adequate ventilation insures complete safety.
- Nitrogen dioxide is also present around the area of the arc in quantities below allowable limits. The only point where high concentrations may be found is in the fumes 6 inches from the arc. With only natural ventilation, this concentration is reduced quickly to safe levels. Special ventilation should be provided during the MIG cutting of stainless steel and when using nitrogen as a shielding gas.
- Carbon dioxide shielding may create a hazard from carbon monoxide and carbon dioxide if the welder's head is in the path of the fumes or if welding is done in a confined space. In these conditions, special ventilation to control carbon dioxide fumes should be provided.
- Eye, nose, and throat irritation can be produced when welding near such degreasers as carbon tetrachloride, trichlorethylene, and perchloroethylene. These degreasers break down into phosgene under the action of the powerful rays from the arc. It is necessary to locate degreasing operations far away from the welding activities to remove this hazard.
- During welding certain metals emit toxic fumes that may cause respiratory irritation and stomach upset. The most common toxic metal vapors are those given off by the welding of lead, cadmium, copper, zinc, and beryllium. These fumes can be controlled by general ventilation, local exhaust ventilation, or respiratory protective equipment.

Electrical Safety

The electrical hazard associated with gas metal-arc welding is less than that with the shielded metal-arc process because the open circuit voltage is considerably less. In some types of equipment, however, a 115-volt control circuit may be utilized in the welding gun. This is no different from the voltage present in many other hand tools, and ordinary precaution insures safety.

Electrical maintenance of the power source and wire feeder should be done only by a qualified person. The equipment should never be worked on in the electrical "hot" condition. Routine maintenance such as gun cleaning, blowing out the cable assembly, alignment of the cable assembly to the wire feeder, and alignment and adjustment of drive rolls does not involve any hazard and can be handled by the welding operator. Properly maintained welding equipment is safe welding equipment.

Fire Safety

Here again, some of the same hazards are present in gas metal-arc welding that are encountered in other welding processes. There is less spatter, and the spatter that does occur is smaller and does not travel as far away from the arc zone as in shielded metal-arc welding. Good practice, however, dictates certain precautions.

- Welding should not be done near areas where flammable materials or explosive fumes are present.
- Paint spray or dipping operations should not be located close to any welding operation.
- Combustible material should not be used for floors, walls, welding tables, or in the immediate vicinity of the welding operation.
- When welding on containers that have previously contained combustible materials, special precautions should be taken.

CARE AND USE OF EQUIPMENT

The efficient operation of welding equipment depends on its proper application and care. Although gas metal-arc equipment is somewhat similar to that used for manual shielded metal-arc welding, there are some major differences.

Do not push the gun into the arc like an electrode. If you do, you will melt the nozzle tip off. The wire driver pushes the wire into the arc—sit back and let the machine do the work. All you have to do is lead it down the joint. You may proceed straight down the seam, or you may weave the arc back and forth as you would do with a stick electrode. Fill craters at the end of the weld just as you always have.

Either a forehand or backhand technique can be used. The forehand, or leading arc technique, lends itself to tracking joints accurately at fast travel speeds. Under the same set of conditions, a forehand technique produces a flatter bead shape than a backhand technique, Fig. 26-14.

A backhand technique lends itself to producing high beads and fillets. The arc force holds the molten metal back from the crater. This technique is especially effective when using slow-cooling electrodes such as flux-cored wires.

Whenever possible, welding should be done in the downhand (flat) position to take advantage of the increased penetration and deposition rate that are characteristic of the MIG process. This is especially true of large fillets and large grooves.

Small fillets and butt welds should be positioned so that the arc can run slightly downhill. Travel speed is faster, and a flat-to-concave bead contour is produced. An angle of 5 to 15 degrees is sufficient. Sheet metal that is 14 gauge and under can be welded vertical-down.

The equipment has to be kept clean, in proper adjustment, and in good mechanical condition. The wire-feeding system requires special attention. Maladjustment leads to erratic wire feeding which in turn causes porosity.

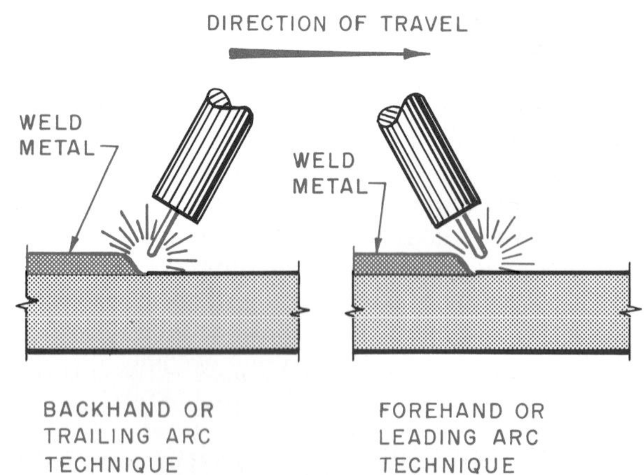

26-14. *Both the forehand and backhand methods are used in semiautomatic welding. The backhand, or trailing arc, produces large wide beads; the forehand, or leading arc, produces a flatter bead shape.*

Care of Nozzles

Keep the gun nozzle, contact tube, and wire-feeding system clean to eliminate wire-feeding stoppages. The nozzle itself is a natural spatter collector, Fig. 26-15. It surrounds the contact tube and provides a good target area as the spatter flies out of the arc stream.

If the spatter builds up thick enough, it can actually bridge the gap and electrically connect the insulated nozzle to the contact tube. If you accidentally touch the nozzle to a grounded surface when this happens, there will be a blinding flash. It is quite likely the nozzle will be ruined. If the face of the nozzle is burned back, the shielding gas pattern over the arc may be changed so much that the nozzle must be replaced.

To remove spatter, use a soft, blunt tool such as an ice cream stick for prying. If the nozzle is kept clean, shiny, and smooth, the spatter almost falls out by itself. An antispatter compound may be applied to the gun nozzle and contact tube end.

26-15. *This plugged gun nozzle is an example of improper care of welding equipment, which can damage parts.*
Harnischfeger Corp.

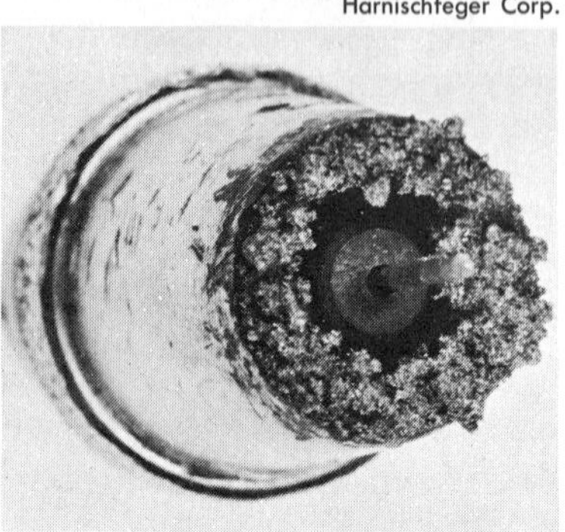

26-16. *Nozzles with a slight inside taper are the easiest to clean. They do not perform as well, however, in directing the shielding gas over the arc.*
Harnischfeger Corp.

Nozzles with a slight inside taper are the easiest to clean, Fig. 26-16. They do not perform as well as the straight bore nozzle, however, in directing the shielding gas over the arc. For this reason, inside taper nozzles are usually used with carbon dioxide; and straight bore nozzles, with argon and other mixtures.

Do not clean guns or torches by tapping or pounding them on a solid object. Bent gun nozzles distort the gas pattern. Threads are damaged so that it is difficult to replace nozzle tips and contact tubes. The high temperature insulation in the nozzle breaks, and damaged insulation causes arcing in the handles. Trigger switches are broken, and in the case of water-cooled guns, leaks are sure to develop.

Care of Contact Tubes

The contact tube transfers the welding current to the electrode wire. New contact tubes have smooth, round holes of the proper minimum diameter. The hole has to be big enough to allow a wire with a slight cast to pass through easily.

With use, the wire wears the hole to an oval shape, especially on the end closest to the arc. The wire slides more easily, but the transfer of current is not as good. Arcing in the tube results, and movement of the wire can actually be stopped. When this happens, the contact tube has to be replaced. It is not a good idea to salvage these tubes.

When the hole in the contact tube becomes too big through wear, spatter flies up into the bore and wedges against the wire. The wire slows down because of the friction. Spatter in the tube also produces a long arc that burns back up to the contact tube and fuses the wire to the tube. Worn contact tubes also allow the wire to flop around as it emerges from the gun so that it is difficult to track a seam.

Care of Wire-Feed Cables

The wire-feed conduit is not unlike an automobile speedometer cable. It is a flexible steel tube that does not stretch. Because wire-feed cables are the main source of friction in the wire-feed system, they should always be as clean and straight as possible. When welding, do not lay the cables in a loop or bend them around a sharp corner. Bending pinches the wire.

Under normal operating conditions, steel and copper chips loosened by the driving action of the feed rolls rub off the wire and collect in the wire-feed cables. If these chips are allowed to collect for any period of time, they fill up the clearance between the wire and the cable and jam the wire. It is not uncommon for this type of stoppage to completely ruin a feed cable. Jerky wire feeding is an indication of chip buildup.

Wire-feed cables should be cleaned with dry compressed air. The conduit should be removed at the feed roll end, and the air nozzle placed on the tip of the contact tube. This results in a reverse flush. Lubricating a blown-out wire-feed cable with dry powdered graphite reduces the friction in the cable and results in smooth wire feeding. Wear safety glasses when cleaning and lubricating the cable.

Do not put so much lubricant in the cable that there is not room for chips. Pour about $\frac{1}{4}$ teaspoon of graphite into the feed end of the cable. Blow the graphite through the cable and reassemble the cable and wire driver. Wire-feed cables should be cleaned every time a spool or coil is changed.

Liquid lubricating compounds should be avoided because a paste is formed with the chips that tends to pack very lightly. If a liquid lubricant has been applied, cleaning with compressed air does not work. The cable must be washed with a solvent. This is not only messy, but time-consuming.

Bird Nesting

Occasionally, the wire coils sideways between the wire-feed cable and the drive rolls. This happens with greater frequency when small wires are pushed because the column strength (stiffness) of the small wire is not great enough to withstand the push of the drive motor. Worn contact tubes and other sources of friction previously discussed increase the chances of a "bird nest."

Bird nesting can be prevented by accurate alignment of the wire-feed cable inlet guide. The inlet guide should be aligned exactly with the rollers so that the wire does not have to make a reverse bend, Fig. 26-17A, B, and C. It should be as close to the rollers as possible without touching them, Fig. 26-17D. The notch in the drive rolls must be in perfect alignment to provide smooth passage for the wire, Fig. 26-17E and F.

Cleanliness of the Base Metal

Although the gas metal-arc process produces quality welds free of porosity and cracks even when welding on steels with rust, scale, burned edges, and chemical coating, the quantity and type of the contamination that can be tolerated varies from application to application.

These contaminants by and large are gas producers. On high speed applications the weld metal solidifies before the gas has a chance to float to the surface and escape. Porosity is the result. On slow speed applications porosity is not as common. The gas has a chance to escape from the molten metal, and some of the contaminant actually has time to burn away in front of the crater because of the intense heat of the arc. This reduces the amount of gas formed quite effectively.

A good general rule to follow is to clean the area thoroughly before welding. Good ventilation is essential when welding galvanized, zinc-coated, aluminum-coated, cadmium-plated, lead-plated and red-lead-painted surfaces to prevent breathing toxic fumes.

Arc Blow

You will recall from your practice with shielded metal-arc welding that arc blow is always a problem when welding with direct current whether it is straight or reverse polarity.

To understand what arc blow is, think of the ionized gases carrying the arc from the end of the electrode wire to the work field as a flexible conductor. This conductor has a magnetic field around it. If it is placed in a location such as the corner of a joint or the end of a plate, the magnetic field is distorted and pulled in another direction. The magnetic field reacts to this force by trying to return to a state of equilibrium. The magnetic field actually moves the flexible conductor to another location where the magnetic forces are in balance. Thus we have a condition of pull and counter-pull: the arc is "blown" to one side or the other as if by a draft. Hence, the term *arc blow.*

Arc blow does not occur with a.c. welding arcs because the forces exerted by the magnetic field on the flexible conductor are reversed 120 times (60 cycles) per second. This tends to keep the magnetic field and the flexible conductor in a constant state of equilibrium.

GROUNDING TO MINIMIZE ARC BLOW. The return path for the welding power from the workpiece to the power source, called the *ground,* is just as important as the path from the power source to the electrode. The following suggestions will shorten the trial-and-error process of locating the best ground location.

- Attach the ground lead or leads directly on the workpiece if possible. Do not ground through ball or roller bearings, hinges, a hold-down clamp, or swivels. Remove rust or paint on the weldment at the point of the ground connection.
- Ground both ends of long, narrow weldments.
- Use electrical conductors of the proper length. Extra long cables that must be coiled up act as a reactor. Stray magnetic fields that affect the action of the arc are also set up.
- Weld away from the ground connection.
- On parts that rotate, use a rotating ground or allow the ground cable to wind up no more than one or two turns.
- In making longitudinal welds on cylinders, use two grounds—one on each side of the seam as close as possible to the point of starting.
- If multiple grounds are necessary, make sure that the cables are the same size and length and that they have identical terminals. Both cables should run from the workpiece to the power source.
- On multiple-head installations, all heads should weld in the same direction and away from the ground.
- Use individual ground circuits on multiple-head installations.
- Do not place two or more arcs close to one another on weldments that are prone to magnetic disturbance with one arc such as tubes or tanks requiring longitudinal seams.

Setting Up the Equipment

Examine the photographs of gas metal-arc equipment in Chapter 10. Figure 26-18 shows a MIG welding station in the plant. The following equipment is required:

- Constant voltage d.c. rectifier or generator. The welding machine may be engine or motor driven.
- Wire-feeding mechanism with controls and spooled or reeled filler wire mounted on a fixture
- Gas-shielding system consisting of one or more cylinders of compressed gas, depending on the gas mixture, pressure-reducing cylinder regulator, flowmeter assembly. (The regulator and flowmeter may be separate or a combination device.)
- Combination gas, water, wire, and cable control assembly and welding gun of the type and size for the particular job
- Connecting hoses and cables, ground lead, and ground clamp
- Face helmet, gloves, sleeves (if necessary), and an assortment of hand tools.

The following safety precautions are assumed.

- The welding equipment has been installed properly.
- The welding machine is in a dry location, and there is no water on the floor of the welding booth.
- The welding booth is lighted and ventilated properly.
- All connections are tight, and all hoses and leads are arranged so that they cannot be burned or damaged.

● Gas cylinders are securely fastened so that they cannot fall over.

Starting Procedure

1. Check the power cable connections. Connect the gun cable to the proper welding terminal on the welding machine. Make sure that the ground cable end is connected to the proper terminal on the welding machine. (For reverse polarity the cable should be connected to the negative terminal.) Connect the ground cable clamp to either the work or the work table.

2. Start the welding machine by pressing the *on* button or, in the case of an engine drive, start the engine.

3. Turn on the wire-feed unit. Make sure that the wire and feed rollers are clean.

4. Check the gas-shielding supply system. Make sure that it is properly connected and that all connections are tight. Open the gas cylinder valve. Open the flowmeter valve slowly, and at the same time squeeze and hold down the gun trigger. Adjust the gas flow rate for the particular welding operation you are about to perform. It is usually 15 to 25 c.f.h. Release the gun trigger.

5. Check the water flow if the gun is water cooled.

6. Set the wire-feed speed control for the type and size of filler wire and for the particular job. Remember that the wire-feed speed determines the welding current in a constant voltage system. If you find that more current is required during welding, increase the wire-feed speed. If less current is required, decrease the wire-feed speed. The current value can be checked on the ammeter only while welding.

7. The voltage rheostat should be set to conform to the type and thickness of material being welded the diameter of the filler wire, the type of shielding gas, and the type of arc. Voltage may range from 18 to as high as 32 volts. The lower voltages are normally used for thinner metals at low amperages. For heavier materials the higher voltage and amperage values apply. Adjust the voltage rheostat until you get a smooth arc.

8. Adjust for the proper stickout (length of electrode wire extending beyond the contact tube). This varies slightly to meet different welding conditions. Usually a minimum of $\frac{1}{4}$ inch and a maximum of $\frac{5}{8}$ inch are satisfactory. The tip of the tube is sometimes flush with the nozzle surface to maintain normal stickout. An extended tube is seldom used and only for very

26-17. *Wire-feed alignment.*

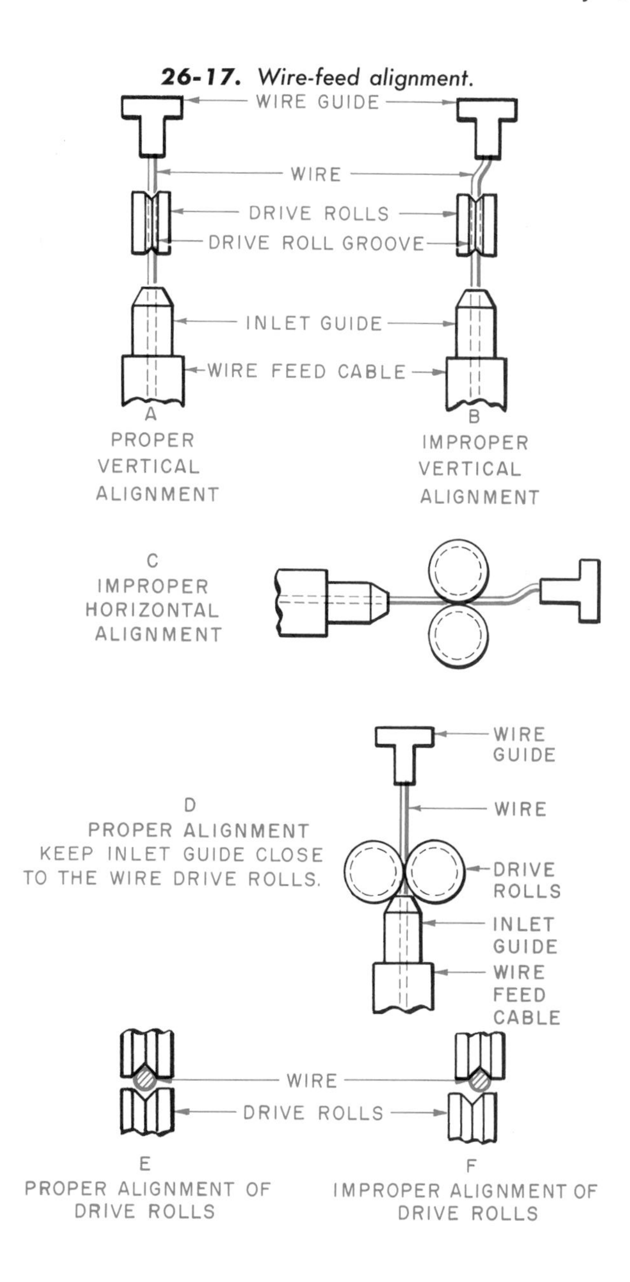

low amperages. The tube is normally recessed for a distance of no more than 1/8 inch from the nozzle opening. The distance from the nozzle to the work may vary from 1/2 inch to 1 inch maximum.

9. To start the arc, touch the end of the electrode wire to the proper place on the weld joint, usually just ahead of the weld bead, with the current shut off. Lower your helmet and then press the gun trigger on the torch. Depressing the gun trigger on the torch causes the wire to feed, the current to flow, and the shielding gas to flow. Do not press the trigger on the torch before you are ready to start welding. If this happens, cut off the excessive filler wire extending out of the nozzle with a pair of wire cutters or pliers. A steady crackling or hissing sound of the arc is a good indicator of correct arc length and wire feed.

Shutting Down the Equipment

1. Stop welding and release the gun trigger
2. After returning the feed speed to the zero position, turn off the wire feeder.
3. Close the gas outlet valve in the top of the gas cylinder.
4. Squeeze the welding gun trigger, hold it down, and bleed the gas lines.
5. Close the gas flowmeter valve until it is finger-tight. This will prevent the possibility of breaking the glass tube on the flowmeter when the gas cylinder is turned on again.
6. Shut off the welding machine.
7. Hang up the welding gun and cable assembly.

WELDING TECHNIQUE

Make sure that the filler wire is of the correct type and size for the type of metal, type of joint, and thickness of metal being welded. Shielding gas flow, power supply, and wire-feed speed should be adjusted for the welding procedure employed. Voltage and wire-feed speed (current) are interdependent with a constant voltage power supply. Adjustment of one may

26-18. *Typical gas metal-arc welding station in industry. Note the equipment—the welding machine, argon gas cylinder, gas regulator and flowmeter, and the wire feeder in the gun.*

Miller Electric Manufacturing Co.

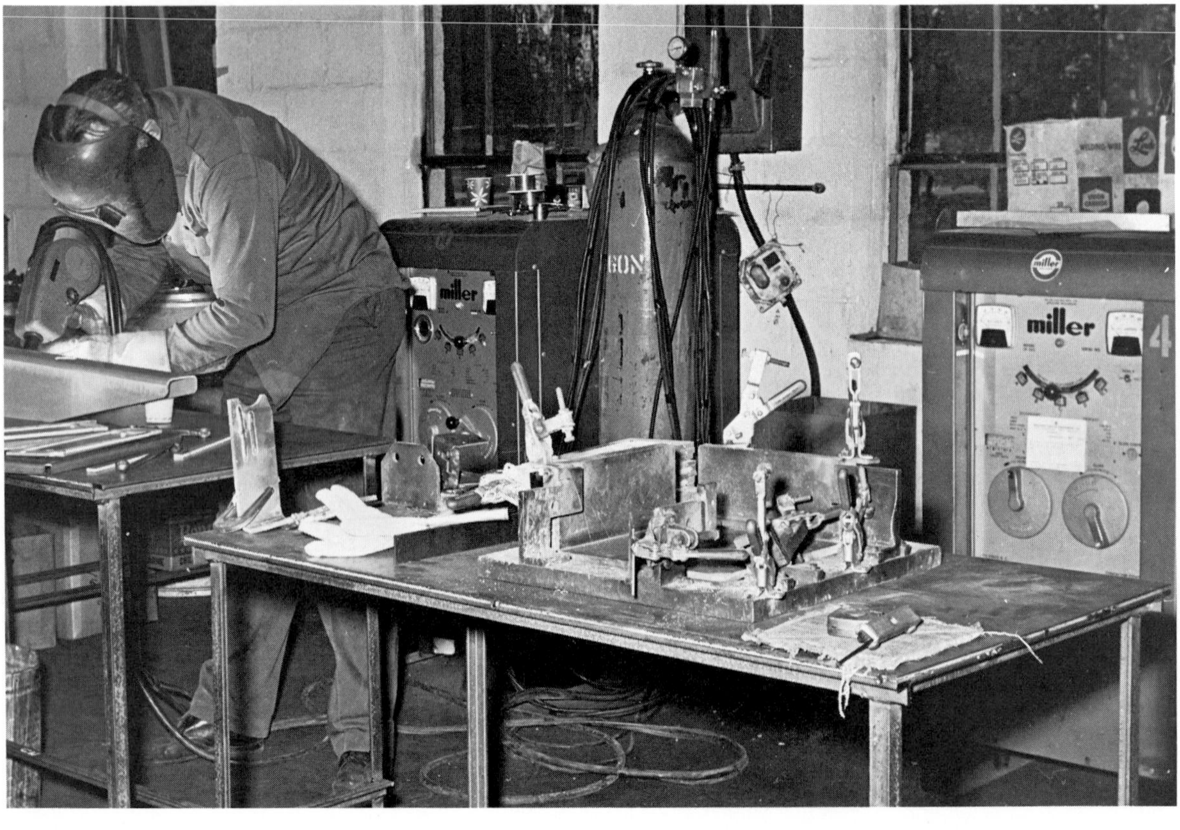

require readjustment of the other.

Starting the Weld

The arc is struck with either a *running start* or a *scratch start.* In the running start, the arc is started at the beginning of the weld. The electrode end is put in contact with the base metal, and the trigger on the torch is pressed. Because this type of start tends to be too cold at the beginning of the weld, it may cause poor bead contour, lack of penetration, and porosity at this point.

The scratch start usually gives somewhat better results. In this type of start, the arc is struck approximately 1 inch ahead of the beginning of the weld. Then the arc is quickly moved back to the starting point of the weld, the direction of travel is reversed, and the weld is started. The arc may also be struck outside of the weld area on a starting tab.

Finishing the Weld

When completing the weld bead, the arc should be manipulated to reduce the penetration depth and weld pool size. This decreases the final shrinkage area. The reduction is accomplished by rapidly increasing the speed of welding for approximately 1 to 2 inches of the weld length. Then the trigger is released. This stops the wire feed and interrupts the welding current. The gun should be kept over the weld until the gas stops flowing in order to protect the weld until the metal has solidified.

Torch Angle

A forehand torch angle (lead nozzle angle) of 5 to 15 degrees is generally employed when welding in the flat position. Care should be taken that the forehand angle is not changed or increased as the end of the weld is approached.

When welding uniform thicknesses, the electrode-to-work angle should be equal on all sides; that is, the nozzle angle should equal the head angle. When welding in the horizontal position, best results are obtained by pointing the gun upward slightly. When welding thick-to-thin joints, it is helpful to direct the arc toward the heavier section. A slight backhand angle (trail nozzle angle) may help when welding thin sections.

Control of the Arc

Keep in mind that arc travel speed has a major effect on bead size. Arc voltage is also very important to satisfactory welding. It controls penetration, bead contour, and to some degree, such defects as undercutting, porosity, and weld discontinuities. Because a long arc is smoother and quieter than a short arc, it is frequently used at the expense of quality and economy. The arc should be noisy for most applications.

PROCESS AND EQUIPMENT PROBLEMS

You will probably experience very little difficulty in welding with the MIG process. All of the newer equipment in use today is in an advanced stage of development and is dependable over a long service life. All things of a mechanical nature, however, are subject to wear and unpredictable breakdown. As a welder in the shop, it will be to your advantage to understand thoroughly the process and the equipment that you are working with. Study the two charts that are presented here very carefully. Table 26-4 lists problems with the MIG short arc process and their correction. Table 26-5 lists problems with the MIG process and equipment, their causes, and possible remedies.

PRACTICE JOBS

Instructions for Completing Practice Jobs

You will practice gas metal-arc welding on mild steel, aluminum, and stainless steel. Complete the practice jobs according to the specifications given in the Course Outline, pages 851, 852, in the order assigned by your instructor. Before beginning each job, study the pages listed under Text Reference in the outline.

These jobs should provide approximately 25 hours practice, depending on your skill. In addition, practice other forms of joints in all positions. Use various types and sizes of filler wire and different shielding gases. (Review pages 322–328 for information on filler wire classifications.) Compare the spray arc and short arc transfer. Experiment with a variety of current values. Make sure that the practice plates are properly prepared, are clean, and have been tack welded with great care.

MIG Welding of Carbon Steel

The bulk of all welding is done on carbon steel. In practically every industry in which welding is a part of the fabrication process, the use of MIG as the welding process is on the increase. The process produces welds that are of the highest quality, and the cost is somewhat lower than with other processes. Operators find it relatively easy to master the process, and they can consistently produce sound welds at a high rate of speed.

GROOVE WELDS: JOBS 26-J1 AND J2. Plate up to $\frac{1}{8}$ inch thick may be butt welded with square edges with a root opening of 0 to $\frac{1}{16}$ inch. Heavier plate thicknesses of $\frac{3}{16}$ and $\frac{1}{4}$ inch may be welded without beveling the edges if a $\frac{1}{16}$ to $\frac{3}{32}$-inch opening is provided. The bead should be wider than the root spacing for proper fusion. Two passes, one from each side, are usually necessary.

For code welding and when the requirments are unusually high, plate thicknesses from $\frac{3}{16}$ to 1 inch should be beveled. A 60-degree single- or double-V without a root face is recommended. A root opening of 0 to $\frac{1}{16}$ inch should be maintained. Wider root openings may be provided for double-V joints than for single-V joints. In single-V grooves a sealing pass from the reverse side is generally required unless the fitup is uniform. If possible, the double-V joint should be used. Less filler metal is needed to fill the joint, and less distortion results when welding from both sides of the joint. Uniform penetration can be ob-

Table 26-4

Short arc process trouble shooting.

Observed fault \ Possible action	Voltage	Current	Slope	Inductance	Polarity[1]	Wire composition	Gas type	Torch manip.	Puddle size	Travel speed	Fitup	Stickout	Wind[2]	Gas shielding[3]	Conduit[4]	Wire-feed system	Contact tip[5]	Oil, rust, paint	Type of steel[6]	Surface treatment	Slag, glass	Arc blow	Procedure
Porosity, internal						✓								✓				✓	✓	✓			
Porosity, visible													✓	✓				✓		✓			
Lack of penetration					✓			✓			+	−										✓	
Too much penetration							✓					+											
Lack of fusion	✓	✓			✓			✓															
Wagon tracks							✓	✓	✓												✓		✓
Cold lapping					✓	✓		✓	✓	+													✓
Cracks																			✓				✓
Whiskers	✓	✓						✓		−	−												
Spatter	✓	✓	✓	+			✓																
Stubbing	+		−																				
Blobbing	−											−											
Bad starts	+		−	−								−									✓		
Unstable arc	✓												✓	✓	✓	✓	✓				✓		
Effervescent puddle													✓	✓				✓					
Too fluid puddle				−		✓	✓																
Too viscous puddle				+		✓	✓																
Crowned bead	+	+		+	✓			✓		✓													
Undercutting	−			−				✓		✓												✓	

+ Raise or increase
− Lower or decrease
✓ Check
[1]Should be DCRP. DCSP looks very similar in short arc welding.
[2]Windshields are recommended for winds over 5 mph.
[3]Clogged nozzles, loose connections, and damaged hoses.
[4]Wrong size, clogged.
[5]Wrong size, damaged.
[6]Rimming grades require highly deoxidized wires.

Table 26-5

Troubleshooting: MIG process and equipment.

INTRODUCTION

When troubleshooting gas metal-arc welding process and equipment problems, it is well to isolate and classify them as soon as possible into one of the following categories:

1. Electrical
2. Mechanical
3. Process

This eliminates much needless lost time and effort. The data collected here for your benefit discusses some of the common problems of gas metal-arc welding processes. A little thought will probably enable you to solve your particular problem through the information provided.

The assumption of this data is that a proper welding condition has been achieved and has been used until trouble developed. In all cases of equipment malfunction, the manufacturer's recommendations should be strictly adhered to and followed.

Problem	Probable cause	Possible remedy
A. Feeder-control stops feeding electrode wire while welding.	1. Power source. a) Fuse blown in power source primary. b) Control circuit fuse blown.	1. Power source. a) Replace fuse. b) Replace fuse.
	2. Primary line fuse blown.	2. Replace fuse.
	3. Wire-feeder control. a) Control relay defective. b) Protective fuse blown. c) Wire drive rolls misaligned. d) Drive roll pressure too great. e) Spindle friction too great. f) Excess loading of drive motor. g) Drive rolls worn; slipping. h) Drive motor burned out.	3. Wire-feeder control. a) Replace contact relay. b) Replace fuse. Find overload cause. c) Realign drive rolls. d) Loosen and readjust drive rolls. e) Loosen and readjust nut pressure. f) Clear restriction in drive assembly. g) Replace drive rolls. h) Test motor; replace if necessary.
	4. Torch and casing assembly. a) Casing liner dirty, restricted. b) Broken or damaged casing or liner. c) Torch trigger switch defective or wire leads broken. d) Contact tube orifice restricted; burnback of electrode. e) Friction in torch.	4. Torch and casing assembly. a) Remove liner, blow out with compressed air. b) Replace faulty part. c) Replace switch; check connections. d) Replace. e) Check wire passage; clean, replace parts as required.
B. Electrode wire feeds but is not energized. No welding arc.	1. Primary line fuse blown.	1. Replace line fuse.
	2. Power source. a) Contactor plug not tight in socket. b) Contactor control leads broken. c) Remote-standard switch defective or in wrong position. d) Primary contactor coil defective. e) Primary contactor points defective. f) Welding cables loose on terminals. g) Ground connection loose.	2. Power source. a) Tighten plug in receptacle. b) Repair or replace. c) Repair or replace; position correctly. d) Replace. e) Replace points or contactor. f) Tighten connections. g) Connect to work; tighten connection.
	3. Wire-feeder control. a) Contactor plug not properly seated. b) Contact relay defective.	3. Wire-feeder control. a) Tighten plug in receptacle. b) Repair or replace.
C. Porosity in the weld deposit.	1. Dirty base metal; heavy oxides, mill scale.	1. Clean base metal before welding.

(Continued on page 840)

Table 26-5 (Continued)

Problem	Probable cause	Possible remedy
C. Porosity in the weld deposit.	2. Gas cylinder and distribution system. a) Gas cylinder valve off. b) Regulator diaphragm defective. c) Flowmeter cracked or broken. d) Gas supply hose connections loose. e) Gas supply hose leaks. f) Insufficient shielding gas flow. g) Moisture in shielding gas. h) Freezing of carbon dioxide regulator/flowmeter.	2. Gas cylinder and distribution system. a) Turn cylinder valve on. b) Replace diaphragm or regulator. c) Replace and repair. d) Tighten fittings. e) Repair or replace. f) Increase flow rate of gas. g) Replace gas cylinder or supply. h) Thaw unit; install gas line heater.
	3. Wire-feeder control. a) Gas solenoid defective. b) Gas hose connections loose.	3. Wire-feeder control. a) Replace solenoid. b) Tighten connections.
	4. Torch and casing assembly. a) Gun body and/or accessories aspirating atmosphere or air. b) Contact tube extended too far. c) Nozzle-to-work distance too great. d) Improper torch angle. e) Welding speed too fast. f) Electrode not centered in nozzle.	4. Torch and casing assembly. a) Test; replace or repair faulty units. b) Distance from nozzle end maximum $\frac{3}{32}$″. c) Should be as recommended by equipment manufacturer. d) Use correct torch angle; approximately 15-degree angle. e) Adjust welding condition for slower speed. f) Adjust contact tube, nozzle and wire.
	5. Improper electrode wire composition.	5. Obtain and use correct electrode wire.
D. Welding electrode wire stubs into workpiece.	1. Power source. a) Excessive slope numerical values set. b) Arc voltage too low.	1. Power source. a) Reduce slope settings as required. b) Increase open circuit voltage at power source.
	2. Wire-feeder control. a) Excess wire-feed speed.	2. Wire-feeder control. a) Reduce wire-feed speed rate.
E. Excessive spatter while welding.	1. Shielding gas system. a) Excessive gas flow rates. b) Insufficient gas flow.	1. Shielding gas system. a) Adjust gas flow rate as required. b) Adjust gas flow rate as required.
	2. Power source. a) Excessive arc voltage. b) Insufficient slope setting value.	2. Power source. a) Reduce voltage at power source. b) Increase slope setting as required.
	3. Torch contact tube recessed in nozzle too far.	3. Replace contact tube with longer one.
	4. Improper electrode.	4. Obtain and use correct electrode wire.
F. Weld bead appearance indicates need for more amperage and/or larger bead.	1. Power source. a) Volt-ampere condition too low.	1. Power source. a) Increase open circuit voltage slowly with applicable increase in wire feed.
G. Weld bead appearance indicates need for less amperage and/or smaller bead.	1. Power source. a) Volt-ampere condition too high.	1. Power source. a) Reduce open circuit voltage and wire feed as required.

Miller Electric Manufacturing Co.

tained in joints having no root face if the root opening is held between 0 and $^{3}/_{32}$ inch.

U-grooves should be employed on plate thicker than 1 inch. Root spacing between $^{1}/_{32}$ and $^{3}/_{32}$ inch should be maintained. A root face of $^{3}/_{32}$ inch or less should be provided to assure adequate penetration. The U-joint requires less filler metal than the V-joint.

For spray arc welding of carbon steel, an argon-oxygen mixture containing 1 to 5 percent oxygen is generally recommended. The addition of oxygen produces a more stable arc, improves the flow of weld metal, and reduces the tendency to undercut.

Straight carbon dioxide and argon-carbon dioxide mixtures are sometimes used. With carbon dioxide, the arc is not a true spray arc. For short arc welding of carbon steel, a mixture of 75 percent argon and 25 percent carbon dioxide may be used. Straight carbon dioxide is becoming increasingly popular for MIG small wire welding.

Code welding requires great care in its application. Good fusion and a minimum of porosity are necessary. The following precautions should be observed when doing the practice jobs.

- Avoid excessive current values. If the current seems to be too high and cannot be reduced without affecting the transfer of metal, switch to a larger size wire.
- Check your welding speed. Welding speeds that are too high cause porosity, and speeds that are too low may cause a lack of fusion.
- Make sure that the gas flow is adequate. The entire weld area must have the protection of the gaseous shield.
- Keep the wire centered in the gas pattern and in the center of the joint. Make sure that the correct electrode angle is maintained at all times.
- Select the proper filler wire for material being welded and for such special situations as rust, scale, and excessive oxygen.
- When welding from both sides of the plate, be sure that the root pass on the first side is deeply penetrated by the root pass on the second side.

FILLET WELDS: JOBS 26-J3 THROUGH J10. Fillet welds are used in T-joints, lap joints, and corner joints. Much of the welding done with the gas metal-arc process is fillet welding. The deposit rate and rate of travel are high, and there is deep penetration. Since the strength of a fillet weld depends on the throat area, the deep penetration that gas metal-arc welding provides permits smaller fillet welds than is possible with stick electrode welding.

The position of the nozzle and the speed of welding are important. You must make sure that penetration is secured at the root of the weld in order to take advantage of the deep penetration characteristics. The welding technique should provide protection for the vertical plate to avoid undercutting. Many of the skills learned in making fillet welds with stick electrodes and the TIG process can be applied to MIG welding.

Welding may be single pass or multipass, depending on the requirements of the job. Multipass welding may be done with weaved beads or stringer beads. The sequence of weld passes is the same as that used with stick electrodes. Each pass must be cleaned carefully. In making each pass, fusion must be secured with the underneath pass and the surface of the plate. Review the precautions listed above for butt welds.

INSPECTION AND TESTING. After each weld has been completed, use the same inspection and testing methods that you have used in previous welding practice. Look for surface defects. Keep in mind that it is important to have good appearance and uniform weld contour. These characteristics usually indicate that the weld was made properly and that the weld metal is sound throughout. Observe the appearance of the welds shown in Figs. 26-19 through 26-25.

- Corner joint in the flat position: Figs. 26-19 and 26-20
- Lap joint in the horizontal position: Fig. 26-21
- Lap joint in the vertical position; direction of travel, down: Fig. 26-22
- T-joint in the horizontal position: Fig. 26-23
- Beveled-butt joint in the flat position; no backup strip; Figs. 26-24 and 26-25. Note that penetration must be secured through the back side in open root joints like that required when welding with other processes.

To determine the soundness of your welds, test lap, T-, and butt joints in the usual manner.

MIG Welding of Aluminum

You will recall that aluminum is readily joined by welding, brazing, soldering, adhesive bonding, and mechanical fastening. Aluminum is light weight, and yet some of its alloys have strengths comparable to mild steel. Pure aluminum can be alloyed readily with many other metals to produce a wide range of physical and mechanical properties. It is highly ductile and retains that ductility at subzero temperatures. It has high resistance to corrosion, forms no colored salts, and is not toxic. Aluminum has good electrical and thermal conductivity and high reflectivity to both heat and light. It is nonsparking and nonmagnetic.

Aluminum is easy to fabricate. It can be cast, rolled, stamped, drawn, spun, stretched, and roll formed. The metal may also be hammered, forged, and extruded into a wide variety of shapes. Machining ease and speed are important factors in using aluminum parts. Aluminum may also be given a wide variety of mechanical, electrochemical, chemical, and paint finishes.

Pure aluminum melts at 1,220 degrees F., and aluminum alloys have an approximate melting range of 900 to 1,220 degrees F., depending upon the alloying elements. Aluminum does not change color when heated to the welding or brazing range. This makes it difficult to judge when the metal is near the melting point during welding. The high thermal conductivity necessitates high heat input for fusion welding.

Aluminum and its alloys rapidly develop a tenacious, refractory oxide film when exposed to air. The oxide film must be removed or broken up during welding to permit the base and filler material to flow together properly when fusion welding or to permit flow in brazing or soldering. The oxide may be removed by fluxes, by the action of the welding arc in an inert gas atmosphere, or by mechanical and chemical means.

Aluminum is used by all industries, particularly the automotive, aircraft, electrical, chemical, and food industries. Review Chapter 3 for a more detailed discussion of aluminum and its properties.

MIG and TIG welding have all but replaced stick electrode welding for aluminum and its alloys. A small percentage of aluminum welding is still being done with stick electrodes. The gas tungsten-arc process is generally used in welding the lighter gauges of aluminum. Heavier gauges are welded with the gas metal-arc process. The type of joint and the position of welding determine to a great extent the process to be used on thicknesses $\frac{1}{8}$ inch and under.

The following factors make gas metal-arc welding a desirable joining process for aluminum.

- Cleaning time is reduced because there is no flux on the weld.
- The absence of slag in the weld pool eliminates the possibility of entrapment.
- The weld pool is highly visible due to the absence of smoke and fumes.
- Welding can be done in all positions.

26-19. *Outside corner joint in steel plate welded with the gas metal-arc process in the flat position.*

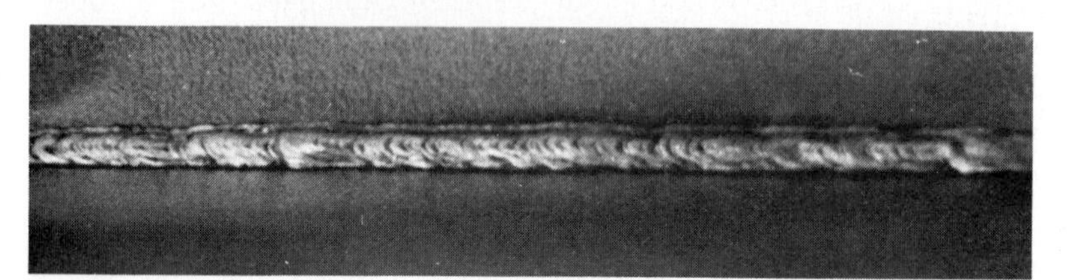

26-20. *Penetration through the back side of a corner joint welded in the flat position.*

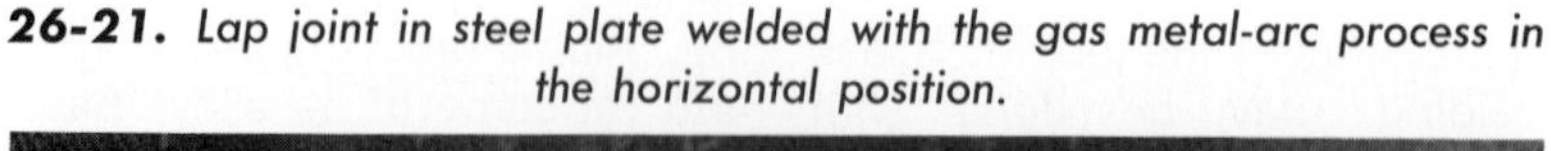

26-21. *Lap joint in steel plate welded with the gas metal-arc process in the horizontal position.*

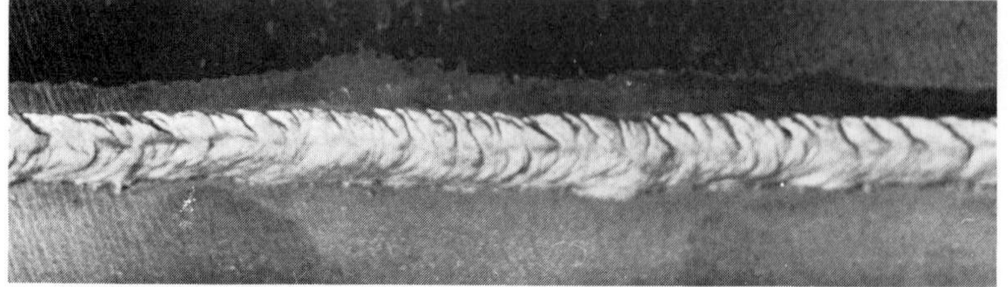

JOINT PREPARATION. Generally, welded joints for aluminum are designed like those for steel. Because of the higher fluidity of aluminum under the welding arc, narrower joint spacing and lower welding currents are generally used. The joint spacing should never be so narrow as to prevent complete penetration.

Foreign substances such as oil, grease, paints, and refractory oxide film must be removed if quality welds are to be produced. Foreign substances are wiped off or removed by vapor degreasing with a suitable commercial solvent. Whenever possible, plate surfaces should be degreased before shearing or machining. In using any solvent, proper safety procedures must be followed.

The oxide film may be removed by both chemical and mechanical cleaning methods. Cleaning by mechanical means such as filing and scraping may not be uniform, but it is usually satisfactory if properly done. When doing multipass welding, clean each pass with a stainless steel wire brush. Once the parts have been cleaned, they should be welded as soon as possible before the oxide film has a chance to form again. Oxides also form on the surface of aluminum welding wire, and it should be checked often.

Sheared edges can also cause poor quality welds. During the shearing operation, dirt and oxide are rolled over and trapped in the sheared edge. In many cases this results in weld inclusions and porosity. To eliminate this problem, sheared edges should be thoroughly degreased

26-22. *Lap joint in steel plate welded with the gas metal-arc process in the vertical position travel down. Note the holes caused by poor gas shielding.*

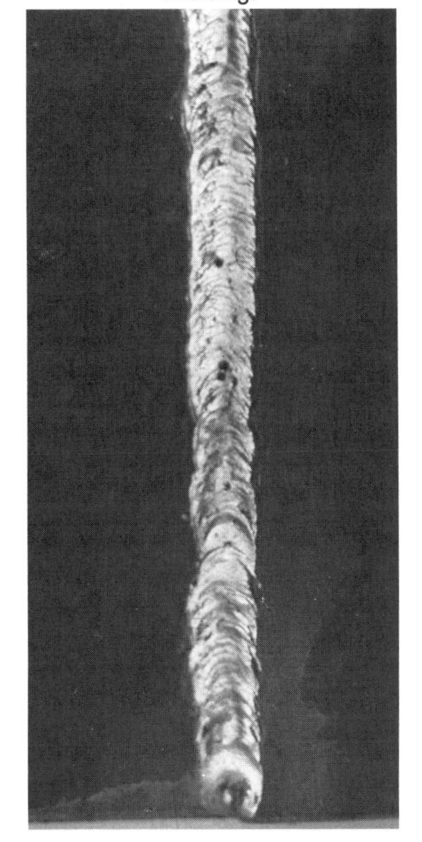

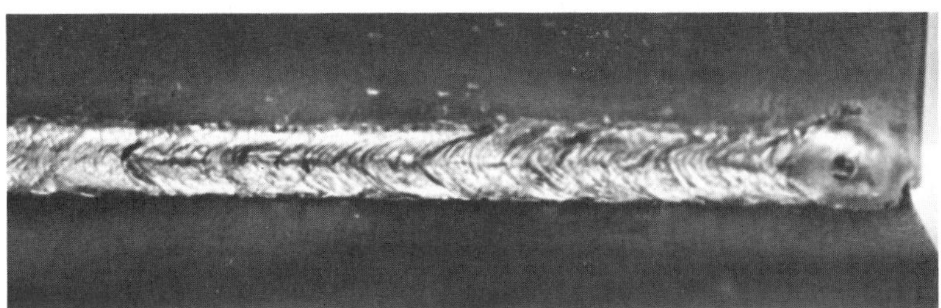

26-23. *Fillet weld in a T-joint welded in the horizontal position with the gas metal-arc process in steel plate.*

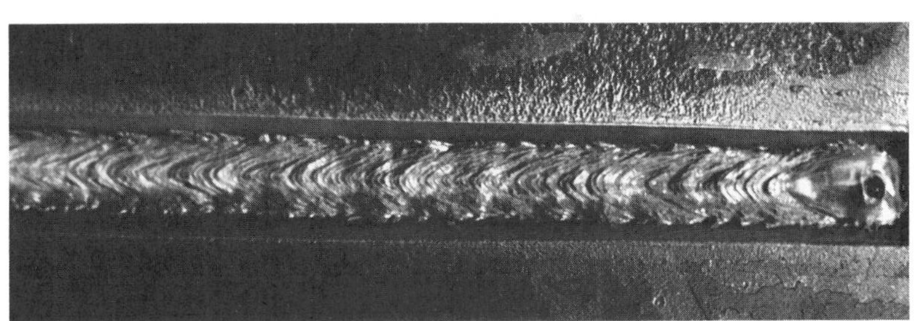

26-24. *The first pass of a groove weld in a beveled-butt joint welded in the flat position with the gas metal-arc process in steel plate.*

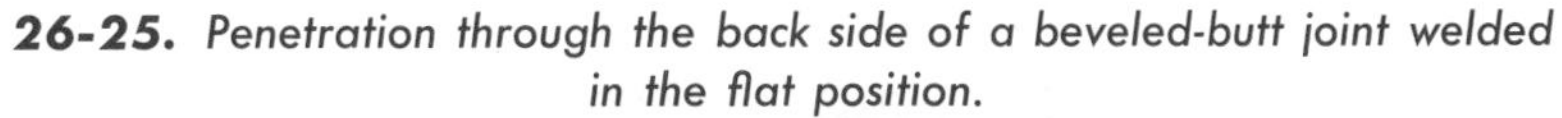

26-25. *Penetration through the back side of a beveled-butt joint welded in the flat position.*

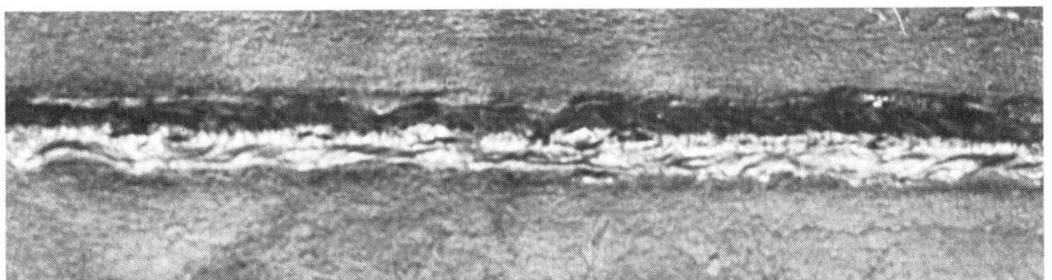

and, if possible, mechanically cleaned before welding.

SHIELDING GAS. Argon is preferred for welding aluminum plate thicknesses up to 1 inch. When compared to helium, argon provides better metal transfer and better arc stability, thus reducing spatter.

For the welding of plate thicknesses from 1 to 2 inches, the following shielding may be used: pure argon, a mixture of 50 percent argon and 50 percent helium, or a mixture of 75 percent argon and 25 percent helium. Helium provides a high heat input rate, and argon provides excellent cleaning action.

26-26. *Aluminum weld made with insufficient shielding gas.*
Kaiser Aluminum & Chemical Corp.

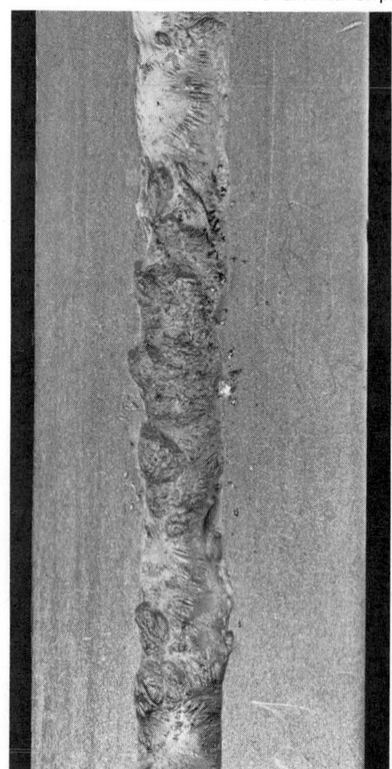

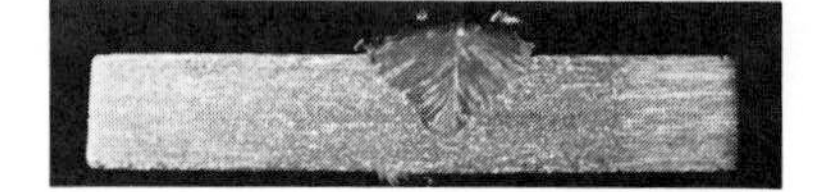

For the welding of plate thicknesses from 2 to 3 inches, a mixture of 50 percent argon and 50 percent helium or 25 percent argon and 75 percent helium may be used, depending upon the job conditions. For aluminum thicker than 3 inches, a mixture of 25 percent argon and 75 percent helium is preferred. This amount of helium provides a high heat input for welding thick sections. High heat input is necessary to minimize porosity.

You will recall the need to keep the welding area adequately shielded with the shielding gas. An inadequate shield causes a weld to be porous and appear dirty, Fig. 26-26. Using too much shielding gas is wasteful and may cause weld turbulence and porosity. Figure 26-27, shows a smooth, porosity-free aluminum weld made with the proper amount of shielding gas.

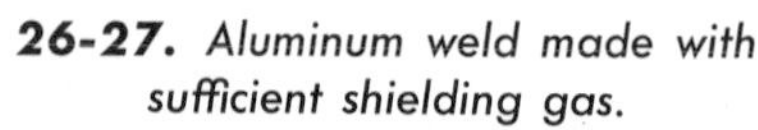

26-27. *Aluminum weld made with sufficient shielding gas.*
Kaiser Aluminum & Chemical Corp.

SHORT ARC WELDING. The short arc type of metal transfer, which deposits weld metal when the filler wire shorts to the base material, has a much colder arc than the spray type of arc, and so it permits the weld puddle to solidify rapidly. It is generally employed in the welding of plate thicknesses of 20 gauge to $\frac{3}{32}$ inch. The short arc permits the welding of aluminum in all positions. Argon shielding gas is recommended for its arc stability.

SPRAY ARC WELDING. Spray arc metal transfer differs from short arc transfer in that no short circuit occurs between the filler wire and the base material. Weld metal is deposited continuously. More arc energy and greater heat is provided for melting the filler wire and base material. Thus, thick sections are more easily welded. Helium, helium-argon mixtures, and argon may be used as shielding gases. The choice of gas is dependent upon the type of material, its thickness, and the welding position. Welding can be done in all positions with the spray-arc type of transfer, but out-of-position welding is somewhat more difficult than with the short arc technique. For out-of-position welding, a 75 percent helium and 25 percent argon mixture, straight helium, or straight argon is the shielding gas.

OUT-OF-POSITION WELDING. Out-of-position welding of aluminum with the gas metal-arc welding process is no more difficult than when welding out of position with any one of the other welding processes.

Horizontal Position. In welding butt joints and T-joints in the horizontal position, Figs. 26-28 and 26-29, care must be taken to penetrate to the root of the joint. Overheating in any one area causes sagging or undercutting. The weld metal should be directed against the upper plate. In multipass welding be sure that there is no lack of fusion between passes.

Kaiser Aluminum & Chemical Corp.

26-28. *Position of the MIG torch when welding a T-joint in aluminum plate in the horizontal position.*

26-29. *Position of the MIG torch when welding a beveled-butt joint in aluminum plate in the horizontal position.*

Kaiser Aluminum & Chemical Corp.

Vertical Position. Fillet and groove welds in the vertical position are usually welded with the travel-up technique, Figs. 26-30 and 26-31. Do not use too high a welding current nor deposit too large a weld bead. If the molten pool is too large, the effect of gravity makes it difficult to control. A slight side-to-side motion may be helpful. In multipass welding make sure that there is no lack of fusion between passes.

Overhead Position. Fillet and groove welds are made in the over-head position without difficulty. Welding current and travel speed are lower than for the flat position. Because the shielding gas has a tendency to leave the weld area, the gas flow rate is higher. Extreme care must be taken to avoid sagging and poor penetration. Trying to deposit too much metal and carrying too large a weld pool are direct causes of such conditions. You may find overhead welding with the MIG torch somewhat awkward. Assume as comfortable and relaxed a position as possible, Figs. 26-32 and 26-33. This will help to keep the gun steady, which is necessary for quality welding.

BUTT JOINTS: JOBS 26-J11 AND J12. Butt joints are easy to design, require a minimum of base material, present good appearance, and perform better under fatigue loading than other types of joints. They require accurate alignment and edge preparation, and it is usually necessary to bevel the edge on thicknessess of $\frac{1}{4}$ inch or more to permit satisfactory root pass penetration. On heavier plate chipping the back side and welding the back side with one pass are recommended to ensure complete fusion. Sections with different thicknesses should be beveled before welding.

LAP JOINTS: JOB 26-J13. Lap joints are more widely used on aluminum alloys than on most

Kaiser Aluminum & Chemical Corp.

26-30. *Position of the MIG torch when welding a T-joint in aluminum plate in the vertical position, travel up.*

26-31. *Position of the MIG torch when welding a beveled-butt joint in aluminum plate in the vertical position, travel up.*

Kaiser Aluminum & Chemical Corp.

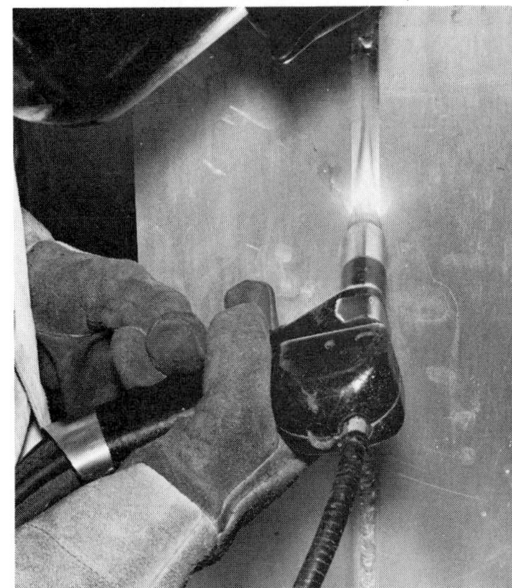

other materials. In thicknesses of aluminum up to $\frac{1}{2}$ inch, it is more economical to use double-welded, single-lap joints than double-welded butt joints. Lap joints require no edge preparation, are easy to fit, and require less jigging than butt joints do.

T-JOINTS: JOBS 26-J14, J15, AND J16. T-joints have several of the advantages of lap joints. They seldom require edge preparation on material $\frac{1}{4}$ inch or less in thickness. This is because the fillet welds on T-joints, as on lap joints, are fully penetrated if the weld is fused into the root of the joint. Edge preparation may be used on thick material to reduce welding costs and minimize distortion. T-joints are easily fitted and normally require no back chipping. Any necessary jigging is usually quite simple. Welding a T-joint on one side only is not ordinarily recommended. Although this type of joint may have adequate shear and tensile strength, a weld on one side acts as a hinge under load so that it is very weak. It is better to put a small, continuous fillet weld on each side of the joint, rather than a large weld on one side of intermittent welds on both sides. Continuous fillet welding is recommended over intermittent welding for longer fatigue life.

EDGE AND CORNER JOINTS. These joints are economical from the standpoint of preparation, base metal used, and welding requirements. However, they are harder to fit up and are prone to fatigue failure. The edges do not require preparation.

INSPECTION AND TESTING. After the weld has been completed, inspect it carefully for defects. Use the same inspection and testing procedures that you learned in previous practice. Look for surface defects. Keep in mind that it is important to have good appearance and uniform weld contour. These characteristics usually indicate that the weld was made properly and that it is sound throughout. Remember that high quality welds in aluminum can be produced only if proper welding conditions and good cleaning procedures have been established and maintained. Figures 26-34 through 26-36 show some acceptable and unacceptable aluminum weld beads.

The following weld defects are found most often in the welding of aluminum.

- Cracking in the weld metal or in the heat-affected zone. Weld metal cracks are generally in crater or longitudinal form. Crater cracks often occur when the arc is broken sharply and leaves a crater. Manipulating the torch properly eliminates this problem. Longitudinal cracks are caused by
  - Incorrect weld metal composition
  - Improper welding procedure
  - High stresses imposed during welding by poor joint design or poor jigging.
- Porosity is a major concern. A small amount of porosity scattered uniformly throughout the weld has little or no influence on the strength of joints in aluminum. Clusters or gross porosity can adversely affect the weld joint. The main causes of porosity in aluminum welds are
  - Hydrogen in the weld area
  - Moisture, oil, grease, or heavy oxides in the weld area

26-32. *Position of MIG torch when welding a T-joint in aluminum plate in the overhead position.*
Kaiser Aluminum & Chemical Corp.

26-33. *Position of MIG torch when welding a beveled-butt joint in aluminum plate in the overhead position.*
Kaiser Aluminum & Chemical Corp.

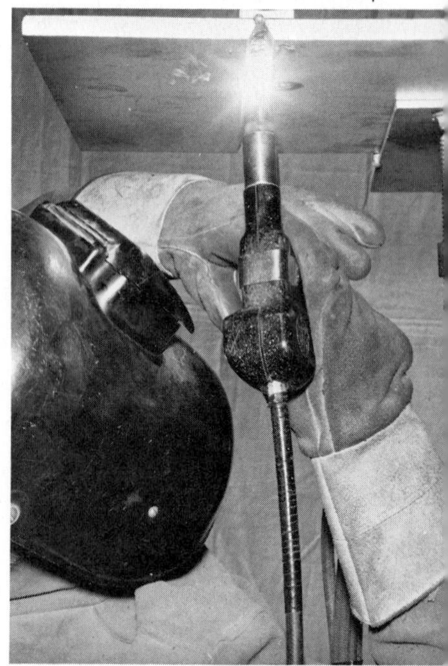

- Improper voltage or arc length
- Improper or erratic wire feed
- Contaminated filler wire
- Leaky torch
- Contaminated or insufficient shielding gas.

● Incomplete fusion of the weld metal with the base metal. The major causes are

- Incomplete removal of the oxide film before welding
- Unsatisfactory cleaning between passes
- Insufficient bevel or back chipping
- Improper amperage or voltage.

● Inadequate penetration at the root of the weld and into the side walls of the joint. This is generally caused by

- Low welding current
- Improper filler metal size
- Improper joint preparation
- Too fast travel speeds for the selected wire-feed speed.

● Aluminum welds may have metallic and nonmetallic inclusions. These may be caused by

- Copper inclusions caused by burn-back of the electrode to the contact tube
- Metallic inclusions from cleaning the weld with a wire brush which leaves bristles in the weld
- The nonmetallic inclusions from poor cleaning of the base metal.

MIG Welding of Stainless Steel

You will recall that stainless steel is a heat- and corrosion-resistant alloy which is made in a wide variety of compositions. It always contains a high percentage of chromium in addition to nickel and manganese. Stainless

26-34. *Aluminum weld bead made with current that is too high. Note the flat appearance and excessive penetration.*

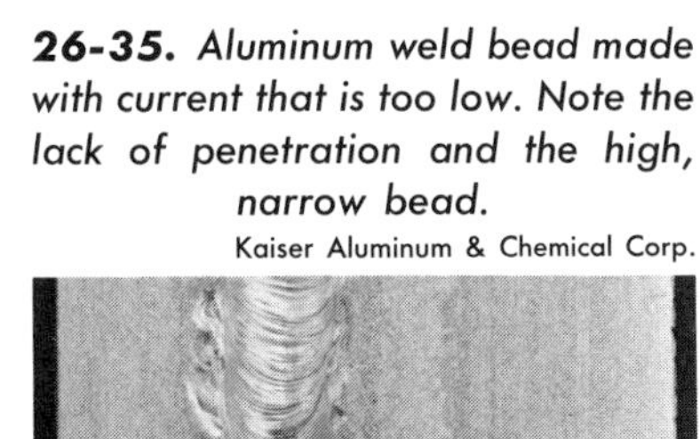

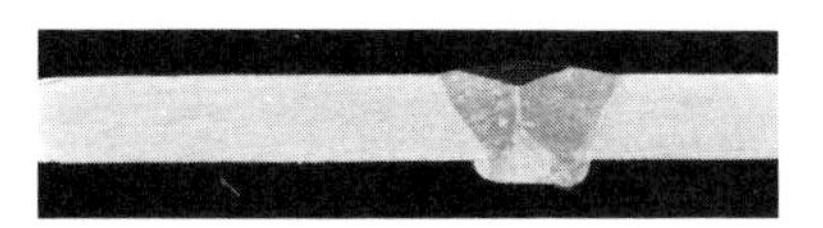

26-35. *Aluminum weld bead made with current that is too low. Note the lack of penetration and the high, narrow bead.*

Kaiser Aluminum & Chemical Corp.

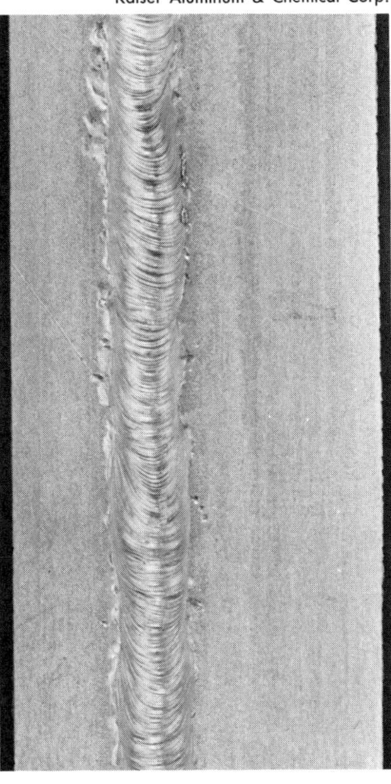

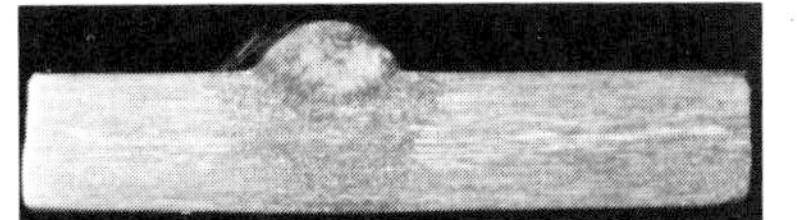

26-36. *Aluminum weld bead made with the correct current. Note the smooth, even ripple; smooth contour, and even penetration.*

Kaiser Aluminum & Chemical Corp.

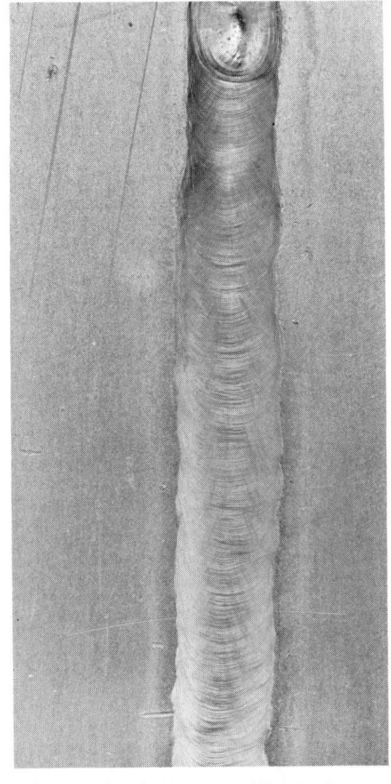

steel has excellent strength-to-weight ratios, and many of the alloys possess a high degree of ductility. It is widely used in products such as tubing and piping, kitchen equipment, heating elements, ball bearings, and processing equipment for a wide variety of industries. Stainless steel is supplied in sheets, strip, plate, structural shapes, tubing, pipe, and wire extrusions in a wide variety of alloys and finishes.

Because stainless steel has a lower rate of thermal conductivity than carbon steel, the heat is retained in the weld zone much longer. On the other hand, its thermal expansion is much greater than that of carbon steel, thus causing greater shrinkage stresses and the possibility of warpage. These difficulties can be overcome by the proper use of jigs and fixtures. Stainless steel also has a tendency to undercut which must be provided for in the welding procedure.

All standard forms of joints are used in stainless steel fabrication. Sheet up to $3/16$ inch can be square-edge butt-welded from one side. Plates $3/16$ inch and thicker are beveled to provide access to the root of the joint. Butt joint designs for stainless steel include the single-V groove with 60-and 90-degree included angles, the double-V groove, and the single-U groove. Standard lap, corner, and T-joint designs are also employed.

Copper backup strips are necessary for welding stainless steel sections up to $1/16$ inch thick. Backup is also needed when welding plate $1/4$ inch and thicker from only one side. No air must be permitted to reach the underside of the weld while the weld puddle is solidifying. The oxygen and nitrogen in the air weaken molten stainless steel during cooling. If it is difficult to use a backup strip, argon should be used as a backup gas shield.

Although the shielded metal-arc process is still very popular for welding stainless steel, the MIG and TIG welding of stainless is on the increase. Light gauge materials are welded with TIG, and heavier materials are welded with the MIG process. MIG welding stainless steel has the following advantages.

- The absence of slag-forming flux reduces cleaning time and makes it possible to observe the weld pool.
- Continuous wire feed permits uninterrupted welding.
- MIG lends itself to automation.
- Welding may be performed with the short-circuiting arc, spray arc, or the pulsed arc.

SPRAY ARC WELDING. Electrode diameters as large as $3/32$ inch can be used for stainless steel. Usually $1/16$-inch wire is used with high currents to create the spray arc transfer of metal. Approximately 300 to 350 amperes are required for an electrode $1/16$ inch in diameter, depending on the shielding gas and type of stainless wire. The amount of spatter is determined by the composition and flow rate of the shielding gas, wire-feed speed, and the characteristics of the welding power supply. DCRP is used for most stainless steel welding. For metal thicknesses up to and including $3/16$ inch, a mixture of argon and 1 to 2 percent oxygen is used. This argon-oxygen mixture improves the rate of transfer and arc stability. It produces a more fluid and controllable weld puddle and good coalescence and bead contour while minimizing undercutting. For metal thicknesses greater than $3/16$ inch, a mixture of argon and 2 percent oxygen is used. This mixture provides better arc stability, weld coalescence, and higher welding speed on heavier materials. It is recommended for single-pass welding.

A forehand welding technique should be employed on plate $1/4$ inch thick or more. The torch should be moved back and forth in the direction of travel and, at the same time, moved slightly from side to side. On thinner metal, only the back-and-forth motion along the joint is used.

SHORT ARC WELDING. Short arc welding requires a low current ranging from 20 to 175 amperes, a low voltage of 12 to 20 volts, and small diameter wires. Metal transfer occurs when the filler wire short circuits with the base metal. The stable arc with low energy and heat input that results is ideally suited for most stainless steel welding on thicknesses from 16 gauge to $3/16$ inch. The short arc transfer should also be employed for the first pass in those situations in which fitup is poor or copper backing is unsuitable. It is also very desirable in the vertical and overhead positions for the first pass.

For short arc welding of stainless steel in light gauges, the argon-oxygen shielding gas mixtures do not produce the best coalescence, A triple mixture consisting of 90 percent helium, $7\frac{1}{2}$ percent argon, and $2\frac{1}{2}$ percent carbon dioxide gives good arc stability and excellent coalescence. It does not lower corrosion resistance, and it produces a small heat-affected zone that eliminates undercutting and reduces distortion. The flow rates must be increased because of the lower density of the helium gas.

HOT CRACKING. Some stainless steels have a tendency toward hot shortness and hot cracking. Type 347 is an example. When these metals are welded, more welding passes are needed. Stringer beads are recommended instead of weave beads. Stringer beads reduce contraction stresses, and cooling is more rapid through the hot-short temperature range.

In welding sections 1 inch or thicker, bead contour and hot cracking can be reduced by preheating to about 500 degrees F. Hot cracking may also be reduced by short arc welding. The short arc prevents excessive dilution of the weld metal with the base metal, a condition which produces strong cracking characteristics.

INSPECTION AND TESTING: JOBS 26-17 THROUGH J23. After each weld has been completed, inspect it carefully for defects. Use the inspection and testing procedures that you have learned in previous welding practice. Look for surface defects. Keep in mind that it is important to have good appearance and uniform weld contour. These characteristics usually indicate that the weld was made properly and that it is sound throughout. Be conscious of the tendency of stainless steel to undercut along the edges of the weld.

Examine the appearance of fillet welds made on heavy stainless steel plate, Figs. 26-37 through 26-39.

GAS METAL-ARC WELDING OF OTHER METALS

Earlier in this chapter it was stated that the MIG process is capable of welding any metal or alloy that can be welded by the other arc and gas welding processes. Thus most aluminum, magnesium, iron, nickel, and copper alloys, as well as titanium and zirconium, can be MIG welded. Welds of the highest quality are produced at production welding speeds in these metals by the MIG process.

The Course Outline on pages 851, 852 includes practice jobs on three of the major metals that are likely to be encountered by the new welder in industry: carbon steel, aluminum, and stainless steel. A new welder in a plant is not likely to be called upon to weld the other metals listed here. It is the purpose of this text to make sure that the student welder has an opportunity to become familiar with the often used metals. You are urged, however, to secure pieces of metals not included in the course and practice with them. The following information will provide you with the necessary information about these metals, and the instructor can readily demonstrate the welding procedure.

Copper and Its Alloys

Copper can be alloyed with zinc, tin, nickel, aluminum, magnesium, iron, beryllium, lead and other metals. Copper and many of its alloys, including manga-

26-37. *Lap joint in 3/8-inch stainless steel plate welded in the flat position with the gas metal-arc process.*

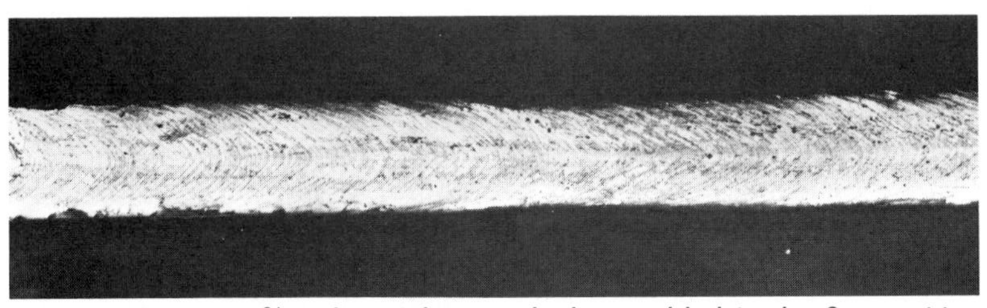

26-38. *T-joint in 3/8-inch stainless steel plate welded in the flat position with the gas metal-arc process.*

26-39. *T-joint in 3/8-inch stainless steel plate welded in the horizontal position with the gas metal-arc process.*

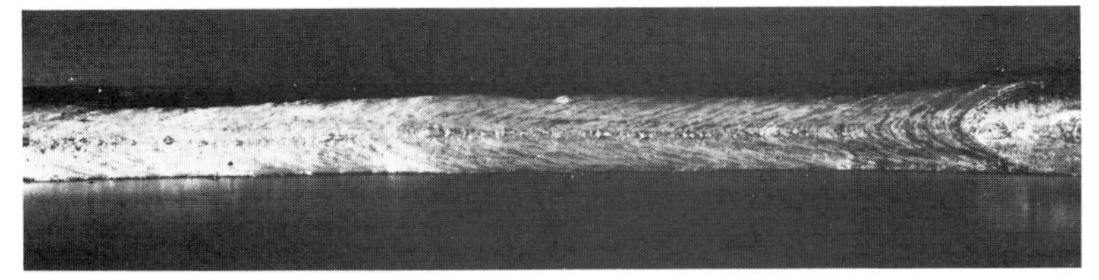

nese-bronze, aluminum-bronze, silicon-bronze, phosphor-bronze, cupro-nickel, and some of the tin bronzes may be welded successfully by the gas metal-arc process. Electrolytic copper can be joined by using special techniques, but its weldability is not good. The various grades of deoxidized copper are readily weldable with the MIG process. Deoxidized filler wires are necessary for welding deoxidized copper. For welding other copper-base alloys, with the exception of the zinc-bearing type, filler wires of approximately matching chemistry are generally used. Copper-zinc alloys are not suitable as filler wire because zinc boils at a low temperature (1,663 degrees F.) and vaporizes under the intense heat of the electric arc. These alloys can be welded, however, with aluminum-bronze filler wires.

Argon is the preferred shielding gas for welding material 1 inch and thinner. A flow of 50 c.f.h. is sufficient. For heavier materials, mixtures of 65 percent helium and 35 percent argon are used.

Joint design is like that for any other metal. Steel backup is usually necessary for sheets $1/8$ inch and thinner. Backup is not needed on plates more than $1/8$ inch thick. Because of the high heat conductivity of copper, welding currents on the high side are required. Preheat is not required when welding thicknesses of $1/4$ inch or less. Preheating to 400 degrees F. has proved to be helpful when welding copper $3/8$ inch or more in thickness.

Always provide good ventilation when welding copper and its alloys. This is of particular importance when welding beryllium-copper. The dust, fumes, and mist produced by beryllium compounds are highly toxic. Precautions should be taken to reduce the dust, fumes and mist to zero.

Nickel and Nickel-Copper Alloys

Nickel, nickel-copper alloy (Monel®), nickel-chromium-iron (Inconel®), and most other nickel alloys can be welded using the gas metal-arc process. Always remove all foreign material in the vicinity of the weld or heated area. Nickel alloys are susceptible to severe embrittlement and cracking when heated in contact with such foreign materials as lead, phosphorus, and sulfur.

Argon is generally preferable for welding nickel and most nickel alloys up to about $3/8$ inch in thickness. Above that thickness, argon-helium mixtures are usually more desirable. The higher heat input of 50 percent and 75 percent helium mixtures offsets the high heat conductivity of heavier gauges. Oxygen should not be added to the inert shielding gases because it produces oxide films and inclusions in the weld and rough, heavily oxidized weld surfaces.

Joint preparation is like that used with other metals.

Magnesium

Magnesium is a silvery white metal which is two-thirds the weight of aluminum and one-quarter the weight of steel. It has a melting point of 1,204 degrees F., which is near that of aluminum. Its strength-to-weight ratio is high when compared to that of steel.

Welding techniques for magnesium are like those for aluminum. The rate of expansion of magnesium is greater than that of aluminum. This must be taken into consideration when preparing the joint for welding and in choosing the type of restraint to put upon the assembly. Severe warpage will result if proper precautions are not taken. As with aluminum, care must be taken that the surface is clean before welding. The surface may be mechanically cleaned with abrasives or chemically cleaned.

The arc characteristics of helium and argon are somewhat different with magnesium than they are with other metals. The burnoff rates of the wires are equal for both gases at the same current. Penetration is greater with argon-helium mixtures. Argon is recommended in most cases because of the excellent cleaning action obtained. The argon-helium mixtures might be preferred in multipass welding in which the rounded type of penetration pattern is most desirable.

Titanium and Zirconium

Titanium is a bright white metal which burns in air, and it is the only element which burns in nitrogen. It has a melting point of about 3,500 degrees F. Its most important compound is titanium dioxide, which is used extensively in welding electrode coatings. Titanium is also used extensively as a stabilizer in stainless steel.

Zirconium is a bright gray metal with a melting point above 4,500 degrees F. It is very hard and brittle and readily scratches glass. Because of its hardness, it is sometimes used in hard-facing materials. Zirconium is often alloyed with iron and aluminum.

Both titanium and zirconium and many of their alloys may be welded by the gas metal-arc process. Special precautions must be taken, however, to protect the welding operation during the period when the metal is hot and susceptible to atmospheric contamination. Welding may be done in an enclosed chamber filled with inert gas, or other special gas shielding methods may be necessary to insure adequate inert gas coverage. Argon or helium-argon mixtures may be used.

REVIEW QUESTIONS

1. List five operating variables that affect weld formation when welding with the gas metal-arc process.
2. Name the three types of current that may be used with the gas metal-arc process. Explain the effect of each.
3. ________________ is that length of filler wire that extends past the gas nozzle and contact tube.
4. Flood angle refers to the position of the wire to the joint in a plane perpendicular to the line of travel. True. False.
5. Wire diameters below $\frac{1}{16}$ inch are seldom used when welding with the gas metal-arc process. True. False.
6. The amount of amperage set on the controls of the constant voltage welding machine determines the wire-feed speed. True. False.
7. Name and explain at least five defects that may be found in gas metal-arc welds.
8. List and explain at least six causes of porosity in welds made with the gas metal-arc process.
9. List the equipment needed for the gas metal-arc welding process.
10. The physical properties of gas metal-arc welds are somewhat inferior to welds made with either the shielded metal-arc or the gas tungsten-arc process. True. False.
11. List and explain at least six practices that keep arc blow at a minimum.
12. Briefly give the starting procedures for gas metal-arc welding.
13. The gas metal-arc process is seldom used in the welding of aluminum. True. False.
14. It is very difficult to weld in the overhead position with the gas metal-arc process. True. False.
15. Alternating current is best suited for the welding of stainless steel. True. False.
16. List the metals that may be welded with the gas metal-arc process.
17. Stainless steel plate $\frac{1}{8}$ inch thick should be welded with (a) 0.035, (b) 0.065, (c) $\frac{1}{16}$, or (d) $\frac{1}{8}$-inch electrodes.
18. Stainless steel does not require backing when welding with the gas metal-arc process. True. False.
19. List the four causes of poor penetration when welding aluminum with the gas metal-arc process.
20. The best shielding gas for the welding of aluminum with the gas metal-arc process is helium mixed with oxygen. True. False.

Table 26-6
Course Outline: Gas Metal-Arc Welding Practice with Solid Core Wire (Plate)

Recommended job order[1]	Number in text	Material		Type of weld	Type of joint	Weld position	No. of passes	Electrode		Shielding gas	Gas flow CFH	Welding current DCRP		Text reference
		Type	Thickness					Type AWS-ASTM	Size inch			Volts	Amperes	
1st	26-J1	Carbon steel	$\frac{1}{8}$	Groove	Square butt	Flat	2	[2]E70S-3	0.035	Carbon dioxide	20–25	19–21	110–160	837
2nd	26-J3	Carbon steel	$\frac{3}{16}$	Fillet	Lap	Flat	1	E70S-3	0.035	Carbon dioxide	20–25	19–21	110–170	841
3rd	26-J4	Carbon steel	$\frac{3}{16}$	Fillet	Lap	Vertical-down	1	E70S-3	0.035	Carbon dioxide	20–25	19–21	120–160	841
4th	26-J5	Carbon steel	$\frac{3}{16}$	Fillet	Lap	Overhead	1	E70S-3	0.035	Carbon dioxide	20–25	19–21	120–160	841
5th	26-J6	Carbon steel	$\frac{1}{8}$	Fillet	T	Horizontal	1	E70S-3	0.035	Argon 75% carbon dioxide 25%	20–25	19–21	110–170	841

(Continued on page 852)

Table 26-6 (Continued)

Recommended job order[1]	Number in text	Material		Type of weld	Type of joint	Weld position	No. of passes	Electrode		Shielding gas	Gas flow CFH	Welding current DCRP		Text reference
		Type	Thickness					Type AWS-ASTM	Size inch			Volts	Amperes	
6th	26-J7	Carbon steel	1/4	Fillet	T	Horizontal	1	E70S-3	1/16	Argon 98% Oxygen 2%	40–50	26–33	220–325	841
7th	26-J8	Carbon steel	3/8	Fillet	T	Vertical-up	3	E70S-3	0.035	Carbon dioxide	20–25	21–23	150–160	841
8th	26-J9	Carbon steel	3/8	Fillet	T	Vertical-up	3	E70S-3	3/32	Argon 95% Oxygen 5%	40–50	26–33	275–400	841
9th	26-J10	Carbon steel	3/8	Fillet	T	Overhead	6	E70S-3	0.035	Carbon dioxide	20–25	21–23	160–180	841
10th	26-J2	Carbon steel	1/2	Groove	V-butt 60°	Vertical-down	2	E70S-3	0.035	Carbon dioxide	20–25	21–23	160–190	837
						Vertical-up	2				20–25	21–23	160–180	
11th	26-J11	Aluminum	1/8	Groove	Square butt	Flat	2	ER1100	3/64	Argon	30–35	20–24	150–190	845
12th	26-J13	Aluminum	1/8	Fillet	Lap	Horizontal	1	ER1100	3/64	Argon	30–35	20–24	150–190	845
13th	26-J14	Aluminum	3/16	Fillet	T	Horizontal	3	ER1100	3/64	Argon	30–35	21–25	160–200	846
14th	26-J15	Aluminum	1/4	Fillet	T	Vertical-up	2	ER4043	3/64	Argon	30–35	21–25	160–190	846
15th	26-J16	Aluminum	1/4	Fillet	T	Overhead	3	ER4043	3/64	Argon	30–35	21–25	170–190	846
16th	26-J12	Aluminum	3/8	Groove	V-butt 60% back-up	Vertical	4	ER4043	1/16	Argon 75% Helium 25%	30–40	22–26	200–250	845
17th	26-J17	Stainless	1/8	Beading	Plate	Flat	Cover plate	ER308	0.035	Helium 90% Argon 7½% Carbon dioxide 2½%	22–24	24–26	130–160	849
18th	26-J18	Stainless	1/4	Beading	Plate	Flat	Cover plate	ER308	0.045	Argon 98% Oxygen 2%	25–35	24–27	130–190	849
19th	26-J19	Stainless	1/8	Fillet	Lap	Horizontal	1	ER308	0.035	Helium 90% Argon 7½% Carbon dioxide 2½%	22–24	24–26	140–160	849
20th	26-J20	Stainless	1/8	Fillet	T	Horizontal	1	ER308	0.035	Helium 90% Argon 7½% Carbon dioxide 2½%	22–24	24–26	150–180	849
21st	26-J21	Stainless	1/4	Fillet	T	Horizontal	6	ER308	0.045	Argon 95% Oxygen 5%	25–35	24–27	180–240	849
22nd	26-J22	Stainless	1/4	Fillet	T	Vertical-up	6	ER308	0.035	Argon 98% Oxygen 2%	25–35	20–26	140–200	849
23rd	26-J23	Stainless	3/8	Groove	V-butt 60° back-up	Vertical-up	1 down 5 up	ER308	0.035	Argon 98% Oxygen 2%	25–35	20–28	140–200	849

NOTE: The conditions indicated here are basic. They will vary with the job situation, the results desired, and the skill of the welder.

[1] It is recommended that the student do the jobs in this order. In the text, the jobs are grouped according to the type of operation to avoid repetition.

[2] Solid wire electrode classification E70S-2 may also be used for this job.

CHAPTER 27 Gas Metal-Arc Welding Practice With Flux-Cored Wire (Plate), Submerged-Arc Welding, and Related Processes

FLUX-CORED WIRE WELDING

Flux-cored arc welding (FCAW) has been growing very rapidly. This growth has kept pace with the growth of other gas metal-arc welding processes. The biggest use is in the fabrication of medium-to-heavy weldments of carbon and alloy steel. Flux-cored wire increases welding speeds and deposition rates considerably. Figure 27-1 shows an application in the field; and Figure 27-2, an application in a fabricating plant.

All of the major producers of welding equipment and suppliers are involved in the process. Hobart refers to flux-cored wire welding as FabCo®; Airco, as Fluxcore®; Lincoln, as Innershield®; National Cylinder Gas, as Dual Shield®; and Linde, as Flux Core®.

There are two basic types of flux-cored wire welding—the *gas-shielded*, flux-cored arc welding process and the *self-shielded*, flux-cored arc welding process. These processes can produce welds of the highest quality even on scaled and rusted plate. They combine the advantage of MIG solid wire welding with high tolerance to poor fitup, high deposition rates, good performance on fillets, and the ability to weld in other than the flat position. Welds have excellent appearance and meet the requirements of many welding codes.

Gas-shielded flux-cored wire welding is a gas metal-arc process employing a consumable electrode that has its core filled with flux and alloying agents, Fig. 27-3. The solid metal portion of the electrode comprises about 80 to 85 percent of its weight. The core material makes up the remainder. The core material performs the following functions, Fig. 27-4.

- Acts as a deoxidizer and cleans the weld metal

27-1. *Gas metal-arc welding of structural steel with flux-cored wire and a carbon dioxide shield. The power source is located elsewhere. Because of the long cable, careful consideration must be given to the correct cable size.*

Hobart Brothers Co.

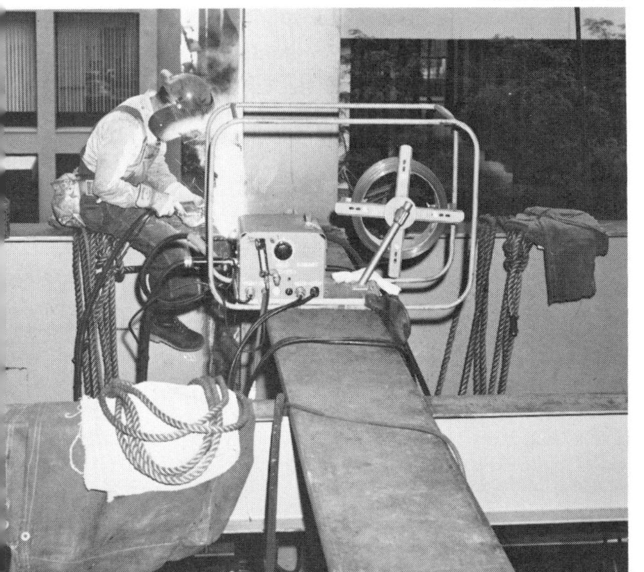

27-2. *Gas metal-arc welding of gear blanks for large presses with flux-cored wire and a carbon dioxide shield. The weldment is mounted on a welding positioner for downhand welding.*

Hobart Brothers Co.

- Forms a protective slag to cover the deposited weld metal until it has solidified
- Stabilizes the arc so that it is smooth and reduces spatter
- Adds alloying elements to the weld metal to increase strength and provide other desirable weld properties
- Provides shielding gas in addition to that supplied externally

Flux-cored wires are made from a continuous tube. Steel strip is formed into a U-shape to hold core flux powders. After flux loading, the *U* is closed, and the wire is drawn to the required size.

Flux-cored filler metal is, in effect, an inside-out covered electrode. This filler metal deposits about 85 to 90 percent of its mass into usable weld metal and achieves deposition rates as high as 20 lbs./hr. The metal of the cored electrode wire is transferred through the intense heat (approximately 12,000 degrees F.) of the arc column to the work. These electrodes may be used with or without shielding gas and with semi-automatic or automatic equipment.

The flux-cored electrode was first developed about 1954 and introduced in its present form in 1957. Flux-cored electrodes are available in diameters of $\frac{1}{16}$, $\frac{5}{64}$, $\frac{3}{32}$, $\frac{7}{64}$, and $\frac{1}{8}$ inch. Electrodes are available for the welding of mild steel. There are also low alloy grades for high strength structural steel, low nickel wire for low temperature applications, and electrodes for a few of the stainless steels. Hard-surfacing electrodes are also available.

Flux-cored filler metals are produced on the basis of three general groupings.

- The so-called *single-pass filler metals,* which operate in carbon dioxide shielding gas, are intended for welding rusted or mill-scaled plate.
- The so-called *multipass filler metals,* which also operate in carbon dioxide shielding gas, provide ductile weld metal with high impact strength at both low and high tensile strengths.
- *Self-shielding filler metals* are used without auxiliary shielding gas. They offer a convenient simplicity of equipment which is efficient for mass production operations like those in the automotive industry.

The flux-cored processes are particularly suited for those applications in which rust and poor fitup are problems, and when larger size fillet welds than those provided by solid core wire are desired. They are also replacing other processes in root pass work in which weld metal drop-through or a lack of visibility is a problem.

In addition, the gas-shielded, flux-cored arc welding process has the following desirable characteristics.

- It has high deposition rates with little electrode loss.
- It can be adapted to semi-automatic or full automatic operation.
- Welds of high quality that can pass radiographic tests are produced, which are suitable for code work.
- The shielding gas, carbon dioxide, is low in cost.
- Deep penetration reduces weld size.
- The highly stable arc reduces spatter loss.
- Slag can be removed with a minimum of labor.
- Weld appearance is highly desirable.
- Small wires can be used in all positions.

SHEATH
CORE
ALLOYING MATERIALS
FLUX
SLAG FORMERS

27-3A. *Flux-cored electrode cross section.*

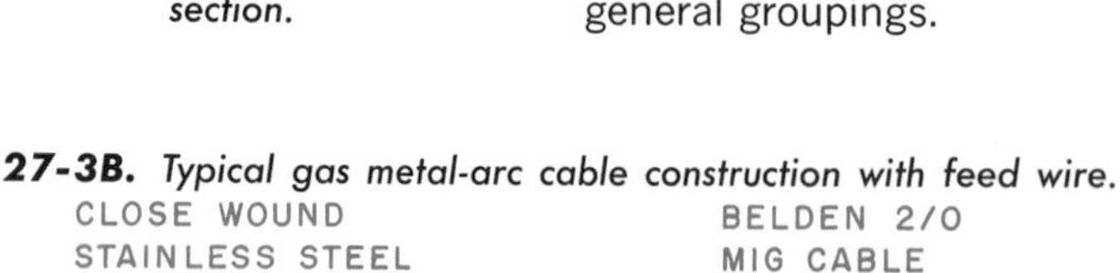

27-3B. *Typical gas metal-arc cable construction with feed wire.*

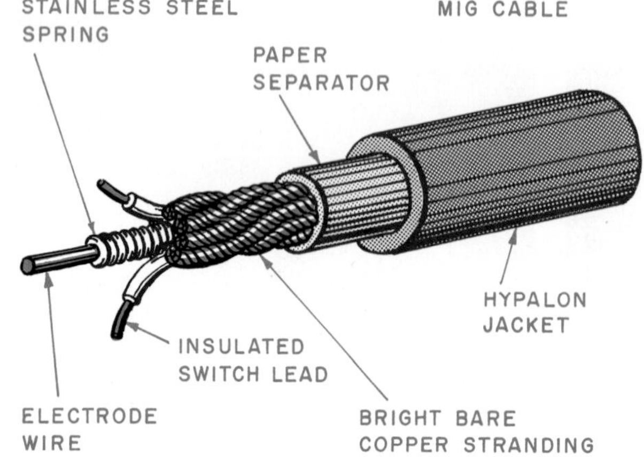

- All mild and low alloy steels can be welded.
- Cost is lower than that of other processes.

The major MIG equipment components are shown in Fig. 27-5:

- Shielding gas and control unit if carbon dioxide is used
- Power source
- Wire-feeding mechanism and controls
- Electrode wire
- Welding gun and cable assembly.

A constant voltage welder with DCRP is generally used with the continuous-feed electrode process. It can be a direct current rectifier, Figs. 27-6 and 27-7, or a generator, Fig. 27-8. The generator may be either motor or engine driven. Alternating current and direct current, straight polarity, are used for special situations to a limited extent.

You will recall that in MIG welding with a constant voltage power supply, current output is determined by the wire feeder adjustment. A specially designed wire-feed and control unit is necessary, Figs. 27-9 and 27-10 (Page 858). There are a variety of gun styles, Figs. 27-11 and 27-12. Guns may be air or water cooled. Review Chapter 10 as necessary concerning the fundamentals of the gas metal-arc welding (GMAW) process.

When flux-cored wires are used with a shielding gas such as carbon dioxide or argon plus carbon dioxide, ductility, penetration, and toughness of the weld metal are improved. Flux-cored wire is also superior on dirty or corroded base metal. See Table 27-1 for a comparison of the gas metal-arc processes.

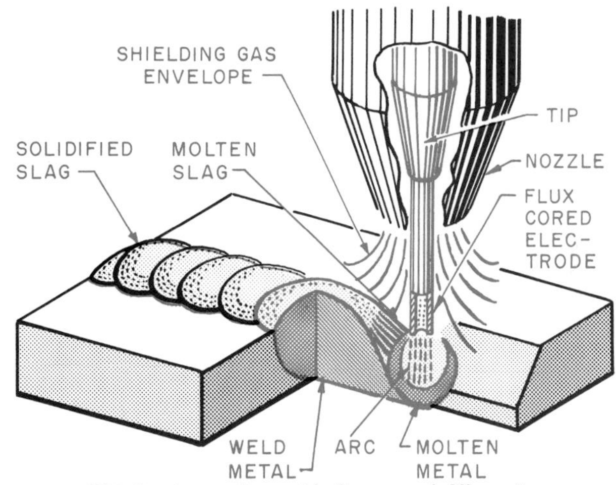

27-4. *Arc action with flux-cored filler wire.*

27-5. *Equipment needed for gas metal-arc welding with flux-cored wire.*

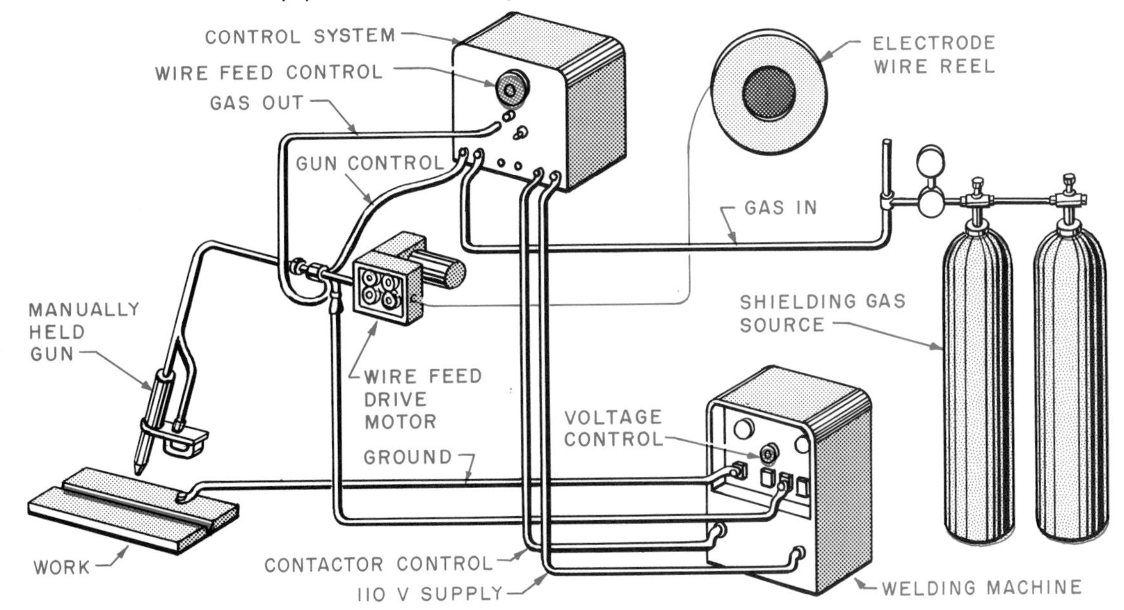

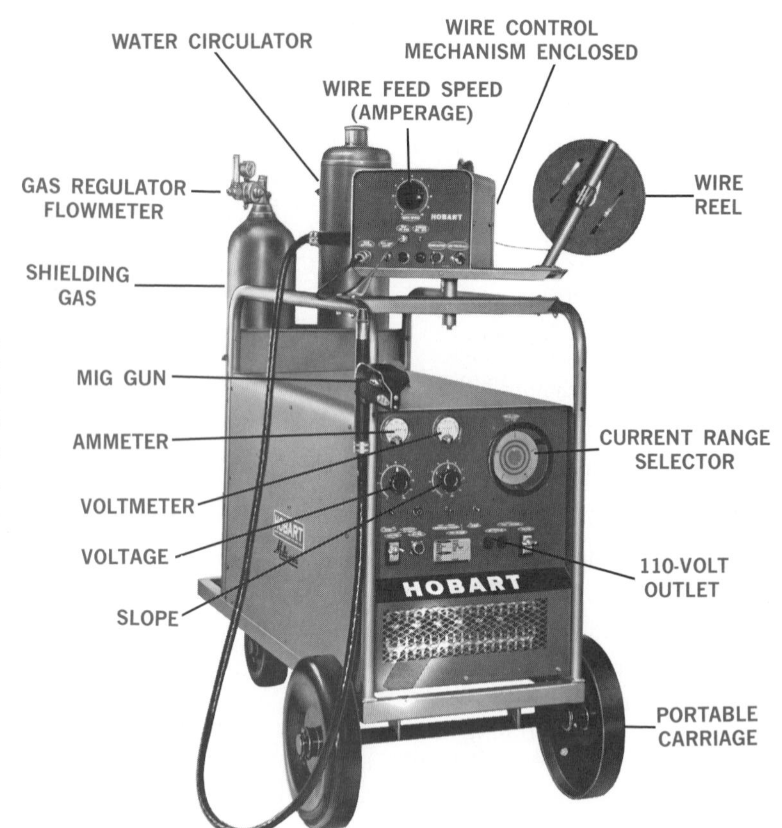

Hobart Brothers Co.

27-6. *Constant voltage, multi-purpose, 500-ampere, d.c. rectifier arc welding machine. It can be used for MIG solid and flux-cored wire welding and submerged-arc welding. Other components are also shown.*

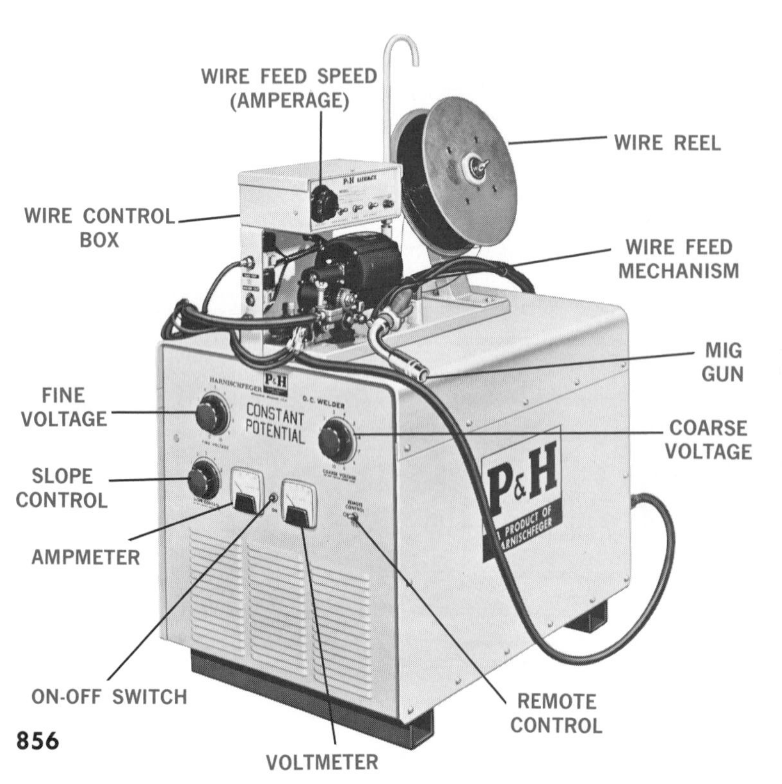

Harnischfeger Corp.

27-7. *Constant voltage, 600-ampere d.c. rectifier arc welding machine. It can be used for MIG welding with both solid and flux-cored wire. It is designed for remote control. Note the wire feeder and control box, the wire reel, and the gun.*

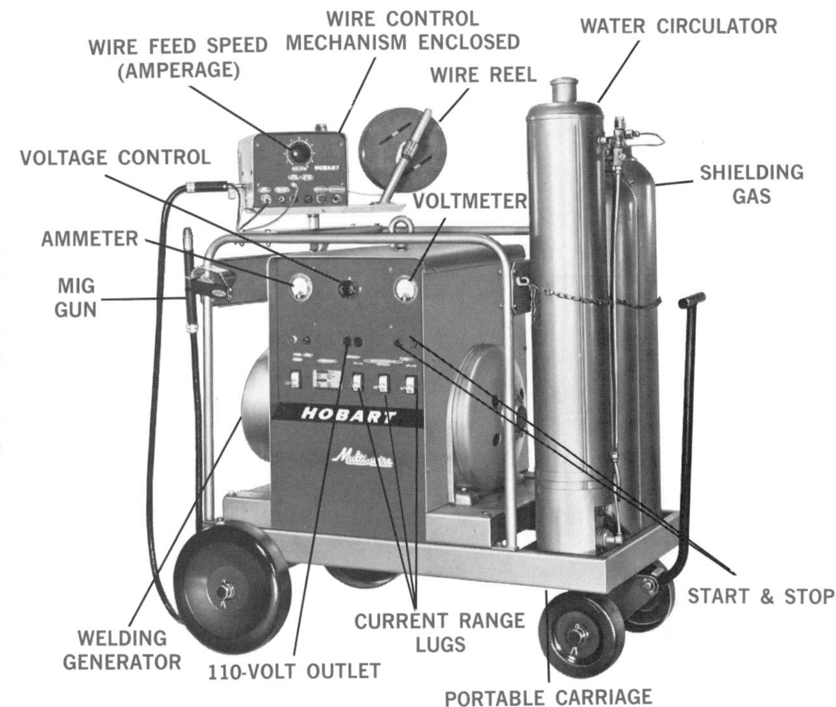

Hobart Brothers Co.

27-8. *Constant voltage, multi-purpose, 500 ampere d.c. motor-generator arc welding machine. It can be used for MIG welding with both solid and flux-cored wire and for submerged-arc welding. Other components are also shown.*

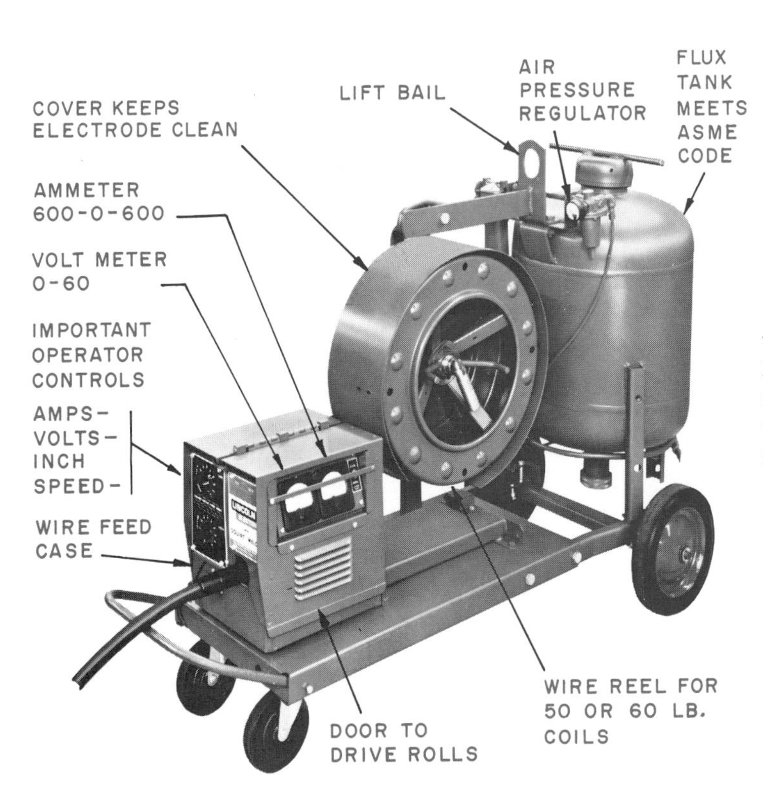

The Lincoln Electric Co.

27-9. *This wire feeder can be used for both solid and flux-cored wire with the gas metal-arc process and for submerged-arc welding. It feeds a wide range of wire sizes.*

The Lincoln Electric Co.

27-10. *The wire feeder shown in Fig. 27-9 being used on the job with flux-cored wire.*

Flux-cored electrodes require currents in the range of 350 to 750 amperes. Because of their deep penetrating qualities and the highly fluid weld pool, the best welds are obtained by welding in the flat and horizontal positions. Flux-cored electrodes are especially intended for large single- or multiple-pass fillet welds in either the flat or horizontal position with DCRP. The wires are also suitable for long groove welds in heavy plate. Flux-cored electrodes often replace the E7018, E6027, and E7028 shielded-metal-arc electrodes.

The flux-cored wire weld deposit is fully covered by a dense, easily removed slag. The combination of full slag coverage and the gas-shielded arc results in X-ray sound welds with excellent mechanical properties. The chemical composition of the weld is constant because alloying elements are built into the cored electrode.

Typical applications include farm equipment, truck bodies,

27-11. *Internal construction of a typical flux-cored wire MIG welding gun.*

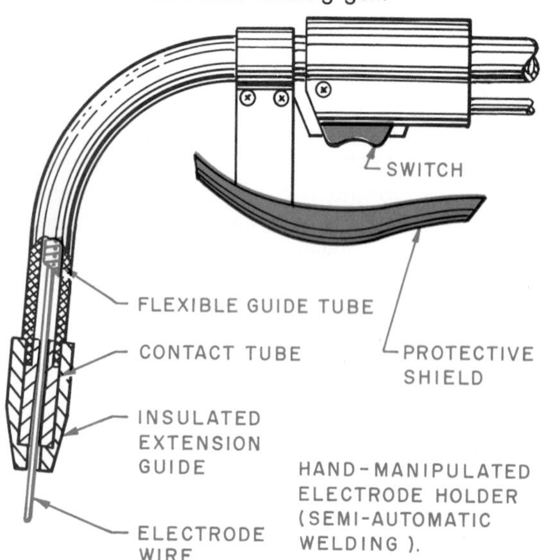

27-12. *Gun for heavy duty, cored-wire, carbon-dioxide-shielded welding up to 600 amps. It can be easily adapted for 0.035-, 0.045- and 1/16-inch solid wire welding of steel and aluminum.*

Bernard Welding Equipment Co.

earthmoving equipment, ships, pressure vessels, railroad cars, machine bases, structural steel, storage tanks, and repair of castings.

Wire Classifications

Wire classifications for flux-cored electrodes of both the self-shielded and gas-shielded types are covered in the *AWS A5.20-69* tentative specifications. They include E60T-7, E60T-8, E70T-1, E70T-2, E70T-3, E70T-4 E70T-5, E70T-6, and E70T-G. Gas shielding is used

Table 27-1

Comparison of MIG wire operating characteristics.

Characteristic	Solid wire	Gas-shielded flux-cored wire	Self-shielded flux-cored wire
Penetration	Shallow to deep	Medium to deep	Shallow to medium
Deposition rate	Low to high	Medium to high	Medium to high
Deposition efficiency	95 to 98%	85 to 92%	82 to 87%
Fluidity of puddle	Viscous to fluid	Fluid	Fluid
Bead width	Narrow	Wide	Wide
Bead shape	Convex	Flat to convex	Flat to convex
Wire size range	0.030″ to 0.125″	0.045″ to $\frac{5}{32}$″	$\frac{3}{32}$″ to $\frac{5}{32}$″
Shielding gas			None
Variety	Medium	Wide	Few
Stock thickness	0.025/0.030″	0.089/0.0187″	0.089/0.0187″
Travel speeds	Slow to fast	Slow to medium	Slow to fast
Fitup	Good to poor	Medium	Excellent

Table 27-2

Characteristics of continuous flux-cored filler wires.

Electrodes	Comments	Shielding gas	Current and polarity	Minimum tensile strength, p.s.i.	Minimum yield strength at 0.2 pct offset, p.s.i.
Flux-cored wires					
E60T-7	Used with internally applied gas shielding. Single-and multiple-pass welds.	None	DCSP	67,000	55,000
E60T-8	Same as for E60T-7. Less tensile strength.	None	DCRP	62,000	50,000
E70T-1	One of the most popular flux-cored wires. It is fast and produces little spatter. Requires clean surfaces. Single-and multiple-pass welds.	Carbon dioxide	DCRP	70,000	60,000
E70T-2	Another very popular flux-cored wire. For single-pass welds. Some overlap between this wire and E70T-1.	Carbon dioxide	DCRP	70,000	not req'd
E70T-3	Open arc welding; no gas shielding is used. For single-pass welds on light gage steel.	None	DCRP	70,000	not req'd
E70T-4	The most popular gasless cored wire. Single and multipass welds. Crack insensitive.	None	DCRP	70,000	60,000
E70T-5	Can be used with or without gas shielding.	None, CO_2	DCRP	70,000	60,000
E70T-6	Similar to E70T-5 without externally applied shielding gas.	None	DCRP	72,000	60,000
E70T-G	Composite electrodes not included in preceding classes. With or without gas shielding. Single or multiple-pass welds.	With or without	Not specified	72,000	60,000
Emissive-coated wire					
E70U-1	Premium wire. No spatter	Argon-oxygen, argon	DCSP	70,000	60,000

Elongation is 20 pct without gas shielding and 22 pct with it.

The Iron Age (March 31, 1966), page 71.

with E70T-1 and E70T-2. The E70T-5 and E70T-G can be used with or without external gas shielding. The E70T-6 electrode is used without gas shielding. Table 27-2.

In the designation E60T-7, the prefix *E* indicates an arc welding electrode. The number *60* indicates minimum arc-welded tensile strength of 60,000 to 69,000 p.s.i. The letter *T* indicates it is a tubular wire. The suffix *7* designates a particular grouping based on the chemical composition of the deposited weld metal, the type of current, polarity, and whether it can be used with or without gas.

Joint Design

Flux-cored wire can save time and weld metal. Part of the savings results from continuous welding with high deposition rates. Further savings are achieved by designing joints to take full advantage of the deep penetration and sound weld metal.

The volume of weld metal required to complete a butt joint can be effectively reduced by reducing the root opening, increasing the root face, and using smaller bevel angles. Deep penetration enables the operator to compensate for poor fitup, and strong fillets are possible with fewer passes and less metal. Smaller joint openings can be used in metals thicker than $\frac{1}{2}$ inch. Welding is usually on material thicknesses greater than $\frac{1}{8}$ inch. Thicknesses up to $\frac{1}{2}$ inch are weldable with no edge preparation, and with edge preparation maximum thickness is practically unlimited.

Fillet welds can be reduced in exterior size and retain comparable or greater strength because of the deep penetration of the arc, Fig. 27-13.

Figures 27-14 to 27-19 show the various joint designs and procedures for welding butt joints.

Shielding Gas

The variety of shielding gases that may be used for flux-cored wire welding is limited. In addition to carbon dioxide, a mixture consisting of 98 percent argon and 2 percent oxygen or 95 percent argon and 5 percent oxygen may be used. A mixture of 75 percent argon and 25 percent

DUAL SHIELD PROCESS THROAT DIMENSION
MANUAL ARC THROAT DIMENSION
FILLET WELD
LEG DIMENSIONS
FILLET THROAT DIMENSIONS

27-13. *A fillet weld made with shielded-arc electrodes has less penetration at the root than a weld made with flux-cored wire. A fillet weld made with flux-cored wire may be reduced in size to secure maximum strength.*

27-14. *Butt joints with backing in plate up to ½ inch thick do not require beveling. Thicknesses up to ¼ inch can be welded with one pass.*

SQUARE GROOVE
1/4
1/4" MAX.
31 VOLTS, 450 AMPERES
12 INCHES PER MINUTE – ONE PASS

27-15. *Butt joints with backing in plate more than ½ inch thick require beveling. Plates ¼ to ½ inch thick are welded with two passes.*

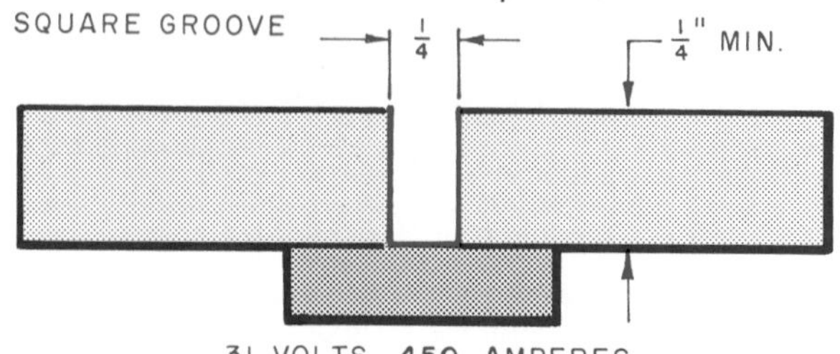

carbon dioxide is also satisfactory. In general, the argon-rich mixtures offer no advantage over carbon dioxide, and usually they are more expensive. A possible exception is when welding with an electrode with a very small diameter (0.045 inch).

When using carbon dioxide, a gas flow rate of 30 to 45 c.f.m. must be maintained. It may be necessary to increase flow rates 25 to 50 percent when welding outdoors or in drafty locations. To maintain this flow rate, two or more standard cylinders should be manifolded together. A pressure-reducing regulator and flowmeter are needed for controlling the shielding gas in the weld zone.

Weld porosity is a result of poor shielding of the arc and molten weld pool when the gas flow is inadequate. Excessive gas flow can cause turbulence of the arc. This causes porosity, weld spatter, and irregular beads.

Welding Variables

You will recall the effect that changes in the various variables had on solid wire welding. The effect on flux-cored wire is similar. Consistently good welds throughout a wide range of conditions are easily obtained when the variables are understood and controlled. All variables must be in balance for sound welds of good appearance.

Review the material in Chapter 26 concerning the effects changing the welding variables has on the weld bead.

GUN ANGLES. Gun angle affects flux-cored wire welding as it does solid wire welding, Fig. 27-20.

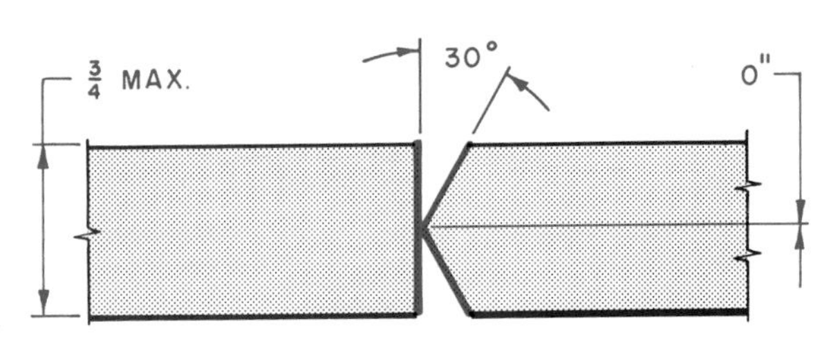

27-16. *Butt joints without backing in plate up to ¾ inch thick require 30-degree double bevels on one plate only.*

27-17. *Butt joints without backing in plate over ¾ inch thick require 45-degree double bevels on one plate only.*

27-18. *Butt joints with backing in plate up to 1 inch thick require a 30-degree single bevel and no root face.*

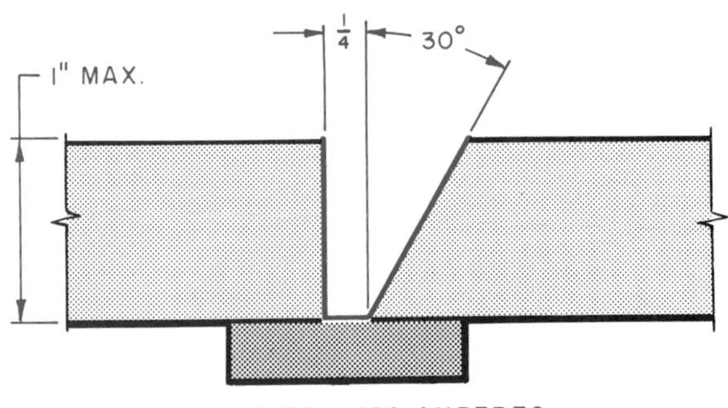

27-19. *Butt-welded plates, 1 inch and thicker, may have a heavy root face with no opening at the root and a 22½-degree double bevel on each plate.*

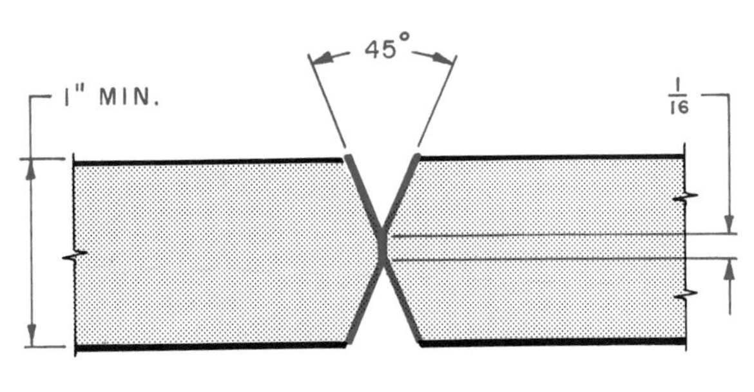

A leading nozzle angle causes the gas shield to be directed over the molten pool. A portion of the arc is insulated from the base metal by the molten pool. Less penetration results, and a higher and narrower weld bead is deposited. A leading angle of 2 to 15 degrees is usually recommended on heavy weldments. Operators usually prefer to use the leading angle because a better view of the arc action and the weld being deposited is obtained.

A *lagging nozzle angle* causes the gas shield to be directed ahead of the molten pool. The arc stream plays ahead on the cold base metal and results in a wide weld bead with shallow penetration. A lagging angle is usually applied on thin gauge metals to reduce the concentrated heating of the thin plate and to avoid burn-through.

ELECTRODE STICKOUT. You will recall from the previous chapter that electrode stickout is the length of the electrode wire extending from the tip of the contact tube. This extended length is subject to resistance heating called *electrode preheat.* Electrical stickout has an effect on weld quality, penetration, arc stability, and deposition rate.

When welding with gas-shielded, flux-cored wire, stickout has an effect on the weld characteristics similar to that produced by solid core wire. There is a difference, however, in

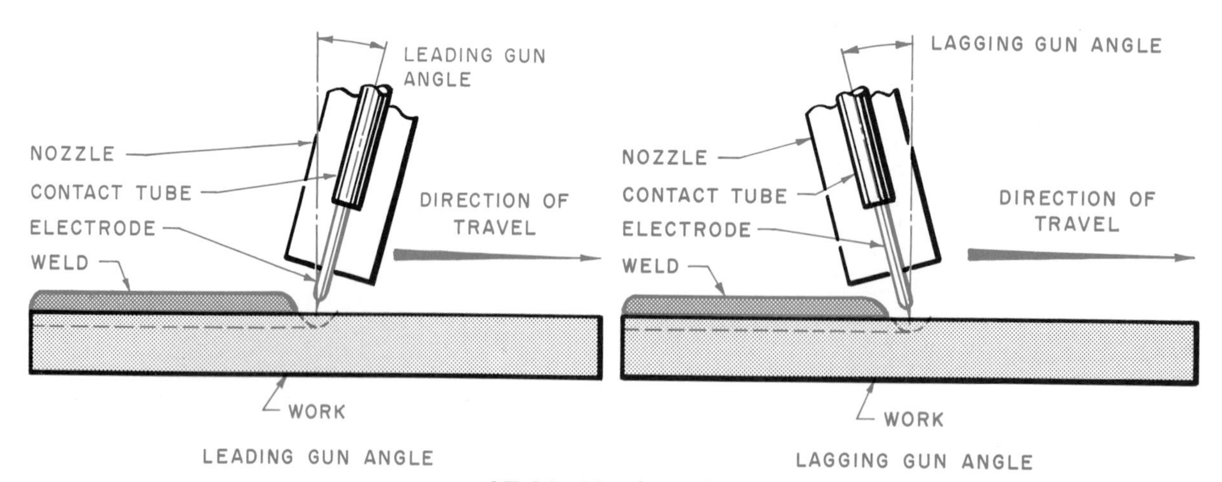

27-20. *Nozzle angles.*

27-21. (A) *Stickout for the auxiliary gas-shielded method and* (B) *the self-shielding method of flux-cored MIG welding.*

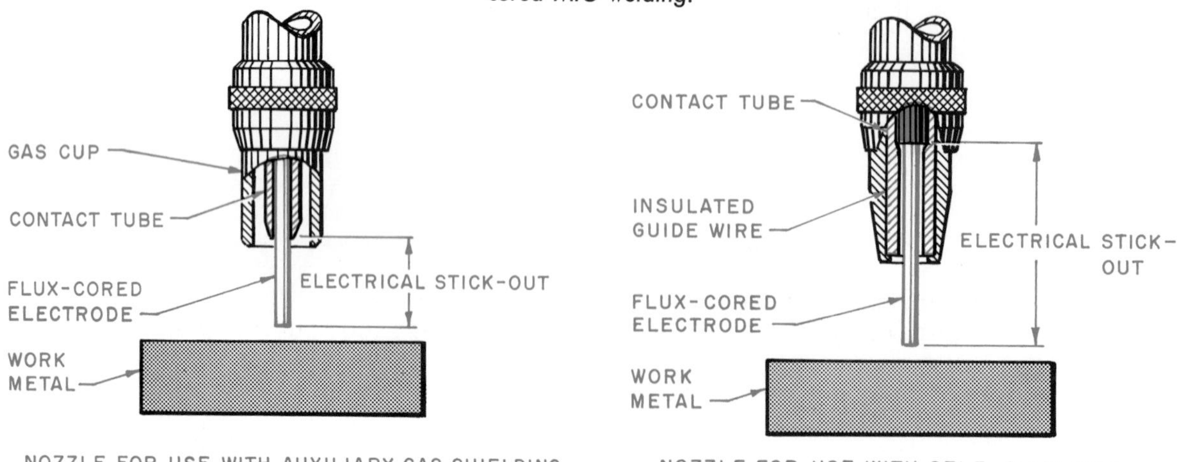

the length of stickout. Gas-shielded, flux-cored welding is usually done with a minimum stickout of 1/2 to 3/4 inch and a maximum stickout of 1 1/2 inches. Spatter, irregular arc action, and some loss of gas shielding occur when the electrode stickout is too long. Stickout that is too short causes fast spatter buildup on the nozzle and contact tube of the gun. At a given wire-feed speed, a short stickout results in deeper penetration than a long stickout.

It is also recommended when using a long stickout that the contact tube tip be recessed a distance of 1/2 to 3/4 inch from the end of the gas cup to reduce the spatter buildup and the possibility of overheating the contact tube. Spatter buildup can be reduced by smearing an antispatter silicon compound on the outside of the contact tube end and on the bottom edge of the welding nozzle before use each working day. The correct cup-to-work distance insures complete gas shielding and proper electrode preheating. A correct setting at the nozzle of the welding gun assures maximum weld quality, penetration, deposition, and appearance.

Stickout is also adjusted for auxiliary gas shielding. It is less if auxiliary gas shielding such as carbon dioxide is used, Fig. 27-21.

GAS-SHIELDED, FLUX-CORED WIRE WELDING PRACTICE JOBS

Instructions for Completing Practice Jobs

Complete the practice jobs according to the specifications given in the Course Outline, pages 882, 883, in the order assigned by your instructor. Before beginning welding practice, turn to Chapter 26 and review the steps for setting up the equipment and the welding recommendations.

These jobs should provide about 20 to 30 hours of practice, depending upon the skill of the individual student. After you have completed these jobs, you may wish to practice with other forms of joints, other sizes of filler wire, and a wide range of current values. You should also practice with flux-cored wire without auxiliary gas shielding. If you have any particular trouble in your practice, consult Tables 27-3 and 27-4.

Welding Technique

Welding current is DCRP. The stickout is about 1/2 to 1 inch. The nozzle angle of the gun is about 60 degrees from the lower plate when welding a lap joint and about 45 degrees from the lower plate when welding a T-joint. A copper backup strip is required when welding a butt joint. The bevel angle for butt joints is about 30 degrees.

Check the stickout distance carefully. Excessive stickout reduces the gas shield and overheats the wire. The arc and the welding area must be properly shielded from drafts which can blow the shielding gas away from the weld area. Do not let the weld metal overheat. If the weld gets too hot, the flux on the bead surface is hard to remove. Take particular care not to overheat the weld metal when making multiple-pass welds. Also be very careful to remove all of the flux from the underneath passes when making multiple-pass welds. Make use of all the welding skills that you have learned up to this point in order to insure sound welds and good appearance.

You will find that the arc is smooth, steady, and essentially spatter-free when operated at the proper wire-speed feed and voltage settings. It is also forceful and penetrating. The arc transfer has a semi-spray characteristic, and the arc appears to be buried in the puddle. The weld deposits are even lower in hydrogen than those made with low hydrogen stick electrodes. The chief source of any hydrogen and porosity in the weld deposit is moisture absorbed from the metal surfaces.

Spatter may be caused by an arc that is too long. In such a case the voltage setting is too high. Another possible cause of spatter is poor arc stability at low wire-feed speeds. This can be corrected by increasing the wire-feed speed or by shortening the stickout. Normal voltage is 29 to 35 volts, but higher voltage may be needed if the leads are long.

Penetration is directly related to the travel speed. At a given wire-feed speed, penetration decreases if travel speed is increased. Penetration is deeper with the forehand welding technique than with the backhand. You will also find that clean plate will be more likely to be undercut than oxidized plate.

GROOVE WELDING. You are reminded to make sure that smooth, even penetration is obtained at the root of the weld. The bead should be equally proportioned and fused to the root face and to the bevel face of each beveled workpiece. Pay particular attention to the width of the bead formation so that it is not more than 1/8 inch on each side beyond the width of the bevel. Regulate your speed of travel so that the face of the weld is not more than 1/8 inch higher than the plate surface.

FILLET WELDING. The angle of the electrode is most important in making a fillet weld. The electrode wire is pointed at the bottom plate, close to the corner of the joint, so that the weld metal will wash up on the vertical plate and form a weld that does not have undercut along its top edge. Here again, speed of travel is important in bead formation. Travel speed that is too fast for the current forms an undersized bead that has poor penetration at the root, undercut along the edges of the weld, and convexity at the center. Travel speed that is too slow for the current causes excessive pile up of weld metal, slag entrapment, and porosity.

Table 27-3

"Troubleshooting"

If a weld should be unsatisfactory, there is one or more welding conditions which must be changed or corrected in order to obtain a satisfactory weld. The following table lists some of the weld faults and the possible cause of these troubles.

Trouble	Possible Cause
Burn-through	Current too high Excessive gap between plates Travel speed too slow Bevel angle too large Nose too small Wire size too small Insufficient metal holddown or clamping
Crown (reinforcement) too high or too low	Current too high or low Voltage too high or low Travel speed too high Improper weld backing Improper spacing in welds with backing Workpiece not level
Penetration too deep or too shallow	Current too high or low Voltage too high or low Improper gap between plates Improper wire size Travel speed too slow or fast
Porosity and gas pockets	Flux too shallow Improper cleaning Contaminated weld backing Improper fitup in welds with manual backing Insufficient penetration in double welds
Reinforcement narrow and steep-sloped (pointed)	Insufficient width of flux Voltage too low
"Mountain range" reinforcement	Flux too deep
Undercutting	Travel speed too high Improper wire position (fillet welding) Improper weld backing
Voids and cracks	Improper cooling Failure to preheat Improper fitup Concave reinforcement (fillet weld)

Hobart Brothers Co.

Inspection and Testing

After the weld is completed, use the same inspection and testing procedures that you have learned in previous arc welding practice. Examine Figs. 27-22, 27-23 and 27-24 carefully for weld formation and appearance. Welds made with flux-cored wire are often so smooth that the ripples are not apparent. The weld has a molded appearance. To be satisfactory, practice joints should be free of the defects described in Chapter 26, pages 828–830. As usual, any surface defects should be noted and avoided in further practice. It is important to have good appearance and uniform weld contour. These characteristics are an indication that the weld was made under the proper conditions and that the weld metal is sound throughout.

SELF-SHIELDED, FLUX-CORED ARC WELDING

The AWS accepted term for this process is self-shielded, flux-cored arc welding. It is often referred to as open-arc "squirt" welding, vapor-shielded welding, or cored-electrode welding. It is a semi-automatic welding process. One form is known in the trade by the name of Innershield®, a process introduced by the Lincoln Electric Company. In this process a vapor produced by a special electrode shields the molten weld metal during the welding operation. The flux-cored electrode is a continuous wire which also serves as a filler wire, Fig. 27-25.

The equipment needed to carry on this process is essentially the same as that used for other gas metal-arc processes: constant voltage power source, control station, and a continuous wire-feed mechanism. The equipment can be fully automatic or semi-automatic. Auxiliary gas shielding is not used.

The tubular steel filler wire contains all of the necessary ingredients for shielding, deoxidizing, and fluxing. These materials melt at a lower temperature than the steel electrode metal and form a vapor shield around the arc and molten weld metal. Also included in the electrode are the alloying materials needed to provide high grade weld metal.

Formerly, the process was limited to welding in the flat and horizontal positions. Recent modifications in filler wire permit the use of the process in the vertical-down, the vertical-up, and the overhead positions.

Process Advantages

The following information on the Innershield® process has been provided by the Lincoln Electric Company. These characteristics may also be applied to all flux-cored wire welding without auxiliary gas shielding.

- It offers much of the simplicity, adaptability, and uniform weld quality that accounts for the continuing popularity of manual stick electrode welding.
- It is a visible arc process which allows the operator to place the weld metal accurately and to control the puddle visually for maximum weld quality.
- It operates in all positions including vertical-up, vertical-down, and overhead.
- Welding can be done outdoors and in drafty locations without wind screens because the shielding does not blow away.
- The job needs only simple fixturing or none at all. Often existing stick electrode fixtures can be used.
- Simple wire-feeding equipment is not encumbered by flux-feeding systems or gas bottles. Installation is quicker and

Table 27-4

Troubleshooting adjustments for welding flux-cored electrodes.

Problem	Solution*				
	Current	Voltage	Speed	Stickout	Drag angle
Porosity	5↑	1↓	4↓	2↑	3↑
Spatter	4↓†	1↑	5↓	3↓	2↓
Convexity	4↓	1↑	5↓	2↓	3↑
Back Arc blow	4↓	3↓	5↓	2↑	1↑
Insufficient penetration	2↑	3↓	4↑	1↓	5↑
Not enough follow	4↑	1↓	5↓	2↑	3↑
Stubbing	4↓	1↑		3↓	2↓

The Lincoln Electric Co.

*Arrows indicate the need to increase or decrease the setting to correct the problem. Numbers indicate order of importance.

†With E70T-G electrodes, increasing the current reduces droplet size and decreases spatter.

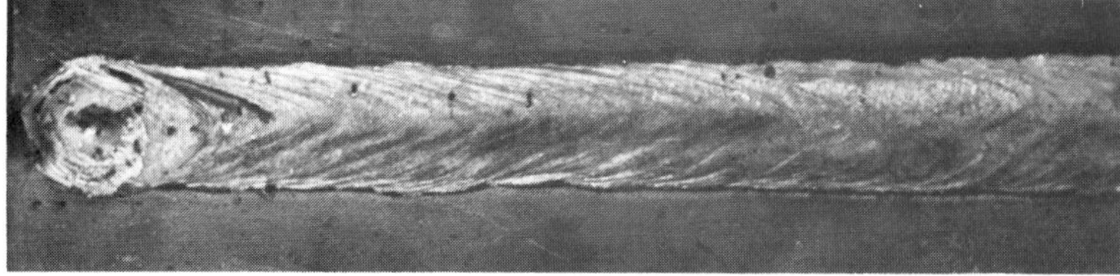

St. Louis Car Co.

27-22. *Square butt joint in ⅜-inch mild steel plate with a backup strip. The joint was welded in the flat position, and the weld was made with flux-cored filler wire 3/32 inch in diameter. Welding required 450 amperes and 32 volts.*

27-23. *Lap joint in ⅜-inch mild steel plate welded in the flat position. The horizontal fillet weld was made in a single pass with flux-cored filler wire 3/32 inch in diameter. Welding required 450 amperes and 32 volts.*

St. Louis Car Co.

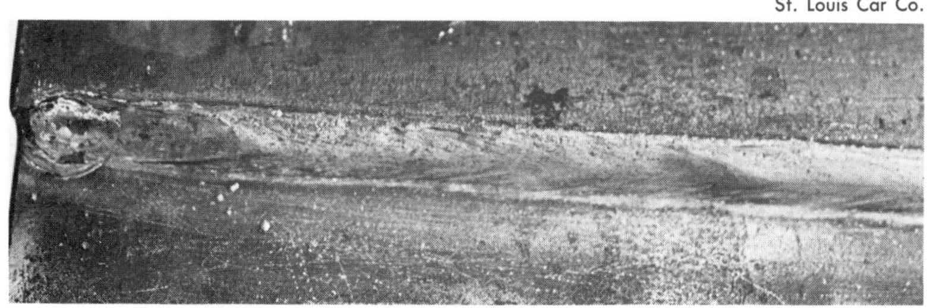

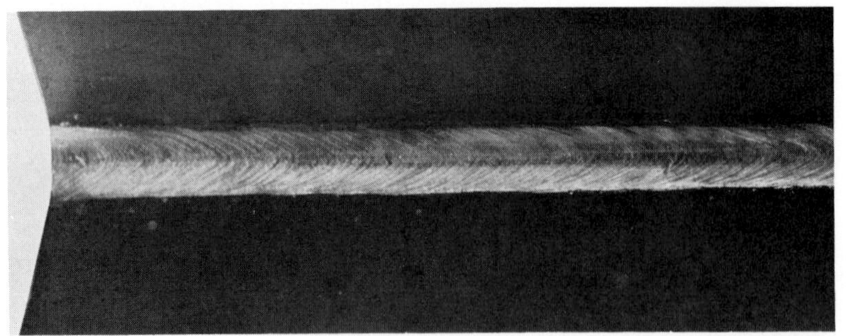

St. Louis Car Co.

27-24. *T-joint in 3/8-inch mild steel plate. The horizontal fillet weld was made in a single pass with flux-cored filler wire 3/32 inch in diameter. Welding required 450 amperes and 32 volts.*

27-25. *The self-shielded, flux-cored arc welding process.*

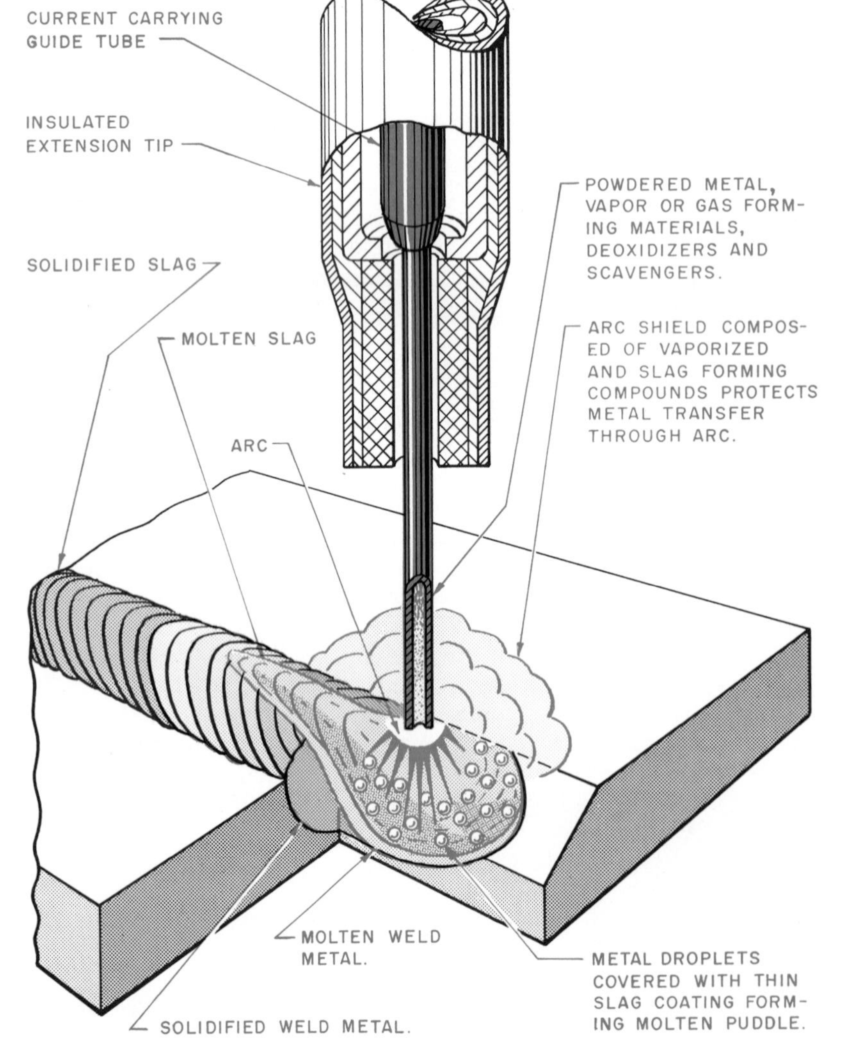

more flexible. Welding rigs are more portable.

- Operator fatigue is minimized during sustained welding operations because the guns are lightweight, flexible, and easy to handle.
- Compact guns fit into places where other semi-automatic guns or stick electrode holders with the long electrodes cannot go.
- Uninterrupted wire feeding results from special electrode lubrication, freedom from spatter-clogged guns, and the ability to resist crushing of the tubular electrode by high drive roll pressure.
- Arc starts are quick and positive without sticking, skipping, or excessive spatter. The process corrects poor fitup and resists cracking in many crack-sensitive applications.
- Welding costs are cut because higher practical welding currents provide increased deposition rates and travel speeds.
- The process has been job-proven on many applications including repair welding, machinery fabrication, assembly welding, ship and barge building, field welding of storage tanks, and erection of structural steel for small and large buildings.

SELF-SHIELDED, FLUX-CORED WIRE WELDING PRACTICE JOBS

Instructions for Completing Practice Jobs

Complete the practice jobs according to the specifications given in the Course Outline, pages 882, 883, in the order assigned by your instructor. Set up for welding as you did when practicing with gas-shielded, flux-cored arc electrodes.

These jobs should provide about 20 to 30 hours of practice, depending upon the skill of the individual student. After completing these jobs, you may wish to practice with other forms of joints, various thicknesses of material, and different filler wire sizes. A wide range of current values should be employed. Develop a good understanding of the results of changes in the basic variables. Study Table 27-4, page 865, carefully.

Power Sources

Constant voltage d.c. transformer-rectifiers of the flat-slope type or electric motor-driven or engine-driven generators are used for welding with self-shielded, flux-cored electrodes.

Wire Feeders

Wire feeders are very similar to continuous wire feeders for other gas metal-arc processes. They are the pull-push variety.

Guns

Guns for semi-automatic welding should be light and maneuverable to facilitate high speed work. They should be equipped with a small guide tip for reaching into deep grooves. Guns are made in light, medium, and heavy duty models to provide for different current ranges and electrode diameters. The electrode in the gun is cold until the trigger is pressed. Medium and heavy duty guns have a shield to protect the operator's hand from excessive heat.

STICKOUT. Stickout is set before welding starts. Electrical stickout may range from $\frac{1}{2}$ to $3\frac{3}{4}$ inches from the contact tip. (See Fig. 27-21B, page 862.) The *visible* stickout is the length of electrode extending from the end of the nozzle on the gun. A $2\frac{3}{4}$-inch electrical stickout may be only a $1\frac{3}{8}$-inch visible stickout. The specified electrical stickouts are obtained by using the proper guide tip and visible stickout on the welding gun. When a long guide tip provides a $3\frac{3}{4}$-inch electrical stickout, both the voltage and amperage must be increased. When the electrical stickout is $\frac{1}{2}$ to 1 inch, a guide tip is not used.

Welding Technique

Before starting, check all welding control settings. Drive rolls and wire guide tubes should be correct for the wire size being used. The gun, cable, and nozzle contact tip should also be correct for the wire size and stickout.

STARTING THE ARC. After the proper stickout has been set, the tip of the electrode is positioned just above the work or so that it is lightly touching the work. The trigger is pressed to start the arc. The mechanical feed will take care of advancing the electrode. Welding is stopped by releasing the trigger or quickly pulling the gun from the work.

GUN ANGLES. The gun leans in the direction of travel at about the same angle as for stick electrode welding. The electrode-to-joint angle varies with the type of joint and thickness of material. For horizontal fillets $\frac{5}{16}$ inch and larger, the electrode points at the bottom plate at about a 45-degree angle. This causes the weld material to be washed up on the vertical plate. For fillets $\frac{1}{4}$ inch and smaller, the electrode is pointed directly into the corner of the joint at an angle of about 40 degrees. When welding sheet or thin plate, visible stickout is $\frac{1}{2}$ to 1 inch. A stringer bead should be applied with steady travel to avoid burn-through. Burn-through makes the metal sag on the underside of the joint and causes weld porosity. Joint fitup should be tight.

For out-of-position welding with E70T-G wire, the work is positioned downhill (vertical-down). Stringer beads are applied with settings in the middle to high range. The gun leans in the direction of travel so that the arc force keeps the molten weld from spilling out of the joint.

In vertical-up and overhead welding, whipping, breaking the arc, moving out of the weld puddle, or moving too fast in any direction should be avoided. Currents are in the low range.

Operating Variables

Four major variables affect the welding performance with self-shielded, flux-cored electrodes: arc voltage, current, travel speed, and electrical stickout. These are all interdependent, and if one is changed, one or more of the others will require adjustment. Review the material in Chapter 26 concerning the effects changing these variables has on the weld bead. Also study Table 27-4, page 865, for an understanding of the corrections that you should make when problems arise.

Inspection and Testing

After the weld is completed use the same inspection and testing procedures that you have learned in previous welding practice. Examine the welds for bead formation and fine ripple appearance. It is impossible to determine the physical characteristics of a weld by its appearance. However, a weld that shows good fusion along the edges, has normal buildup, is free of undercut and surface defects, and has fine, smooth ripples usually meets the physical requirements.

MAGNETIC-FLUX COATED WIRE WELDING

Magnetic-flux coated wire welding is a gas metal-arc weld-

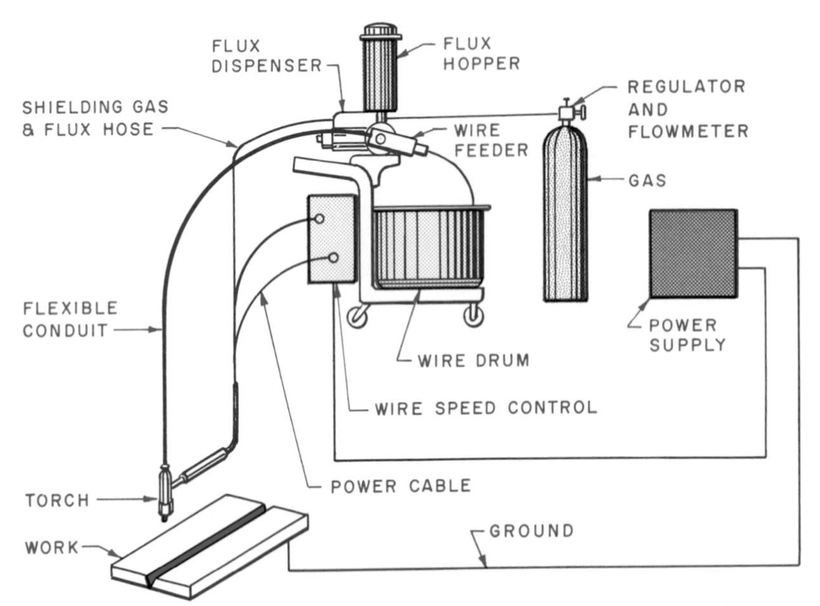

27-26. *The equipment for magnetic flux, carbon-dioxide-shielded arc welding.*

27-27. *Nozzle assembly and weld for magnetic flux, carbon-dioxide-shielded welding.*

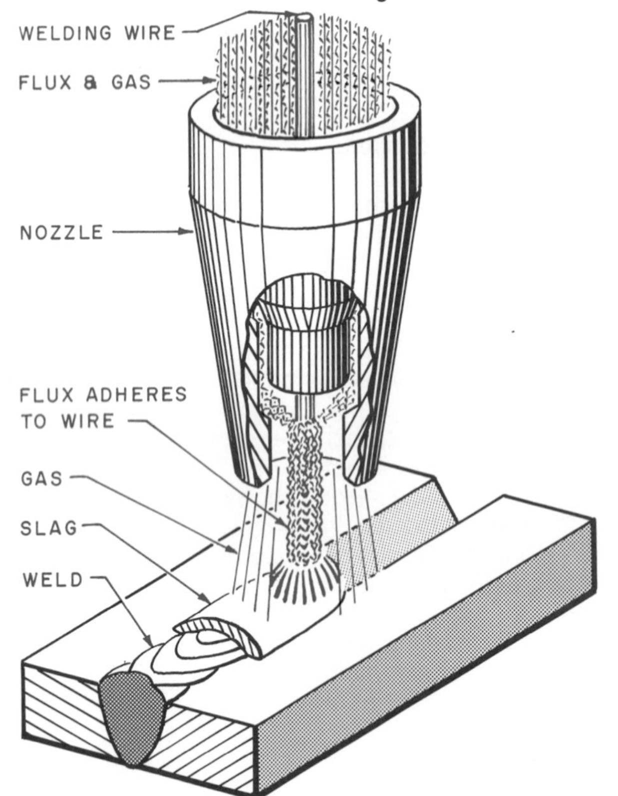

ing process. It makes use of a wire electrode that is magnetically coated with flux and shielded by carbon dioxide gas. The essential components are bare steel welding wire, magnetic flux, carbon dioxide, and a power supply, Fig. 27-26. During welding, the wire and the flux, which is suspended in gas, are fed simultaneously to the torch. The welding current flowing through the wire in the torch establishes a magnetic field that attracts the flux to the wire as shown in Fig. 27-27. As a result, the electrode is flux coated. The flux stabilizes the arc, refines and protects the weld pool, controls the formation of the weld, and improves coalescence. The shielding gas carries the flux to the torch and also provides the usual shielding qualities.

A constant voltage, direct current power supply is usually used with reverse polarity. The torch is similar to a gas metal-arc welding torch, but it is not water cooled. Wire $\frac{3}{32}$ inch in diameter is generally used for welding in the flat and horizontal positions. Wire $\frac{3}{64}$ inch in diameter can be used in all positions.

The arc characteristics of this process are similar to those produced by covered electrodes and carbon-dioxide-shielded, flux-cored wire. Metal transfer is the spray type, and very little metal spatter is given off.

AUTOMATIC WELDING PROCESSES

Industry is continually finding new applications for automatic welding processes, Fig. 27-28. These processes have economically replaced many other joining methods such as rivets, bolts, resistance welding, and castings. It is estimated that by 1980 at least 30 percent of all welding operations will be automatic.

Automatic arc welding has many characteristics that are highly desirable for manufacturing:

- Repeatability
- High quality
- High production
- Low welding costs
- Uniform welds
- Desirable weld appearance
- Continuous output
- Ability to conform to various welding conditions
- Almost 100 percent duty cycle.

Each company needs to study the individual job and the general plant conditions in making the decision to use automatic welding. A few considerations:

- Can the job be redesigned for automatic welding?
- How does the cost compare with the costs of other processes?
- Are new work processes such as handling the material introduced that will offset the saving in welding time?
- Which automatic system is best for the job?
- Are there sufficient floor space and power available for the new equipment?
- Is skilled labor available to operate the automatic system?
- Is the work repetitive enough to gain the full advantage of automatic welding?

27-28. *Automatic gas-shielded, metal-arc welding of a brewery vessel.*

Hobart Brothers Co.

Submerged-Arc Welding (SAW)

Submerged-arc welding may be automatic or semi-automatic. Welding takes place beneath a blanket of granular, fusible flux. Bare filler wire is fed automatically, and the arc is completely hidden by a mound of flux. Equipment includes the power source, an automatic wire-feeding device, a flux-feed system, a gun in semi-automatic welding, and flux pickup, Fig. 27-29.

The filler wire is not in actual contact with the workpiece. The current is carried across the gap through the flux. The weld pool is completely covered at all times, and the welding operation is without sparks, spatter, smoke, or flash, Fig. 27-30. Thus protective shields, helmets, smoke collectors, and ventilating systems are not needed. Goggles should be worn, however, for safety.

Welds made under the protective layer of flux have unusually

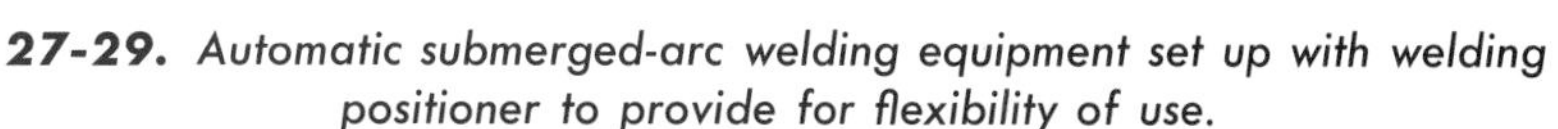

27-29. *Automatic submerged-arc welding equipment set up with welding positioner to provide for flexibility of use.*

Rexarc

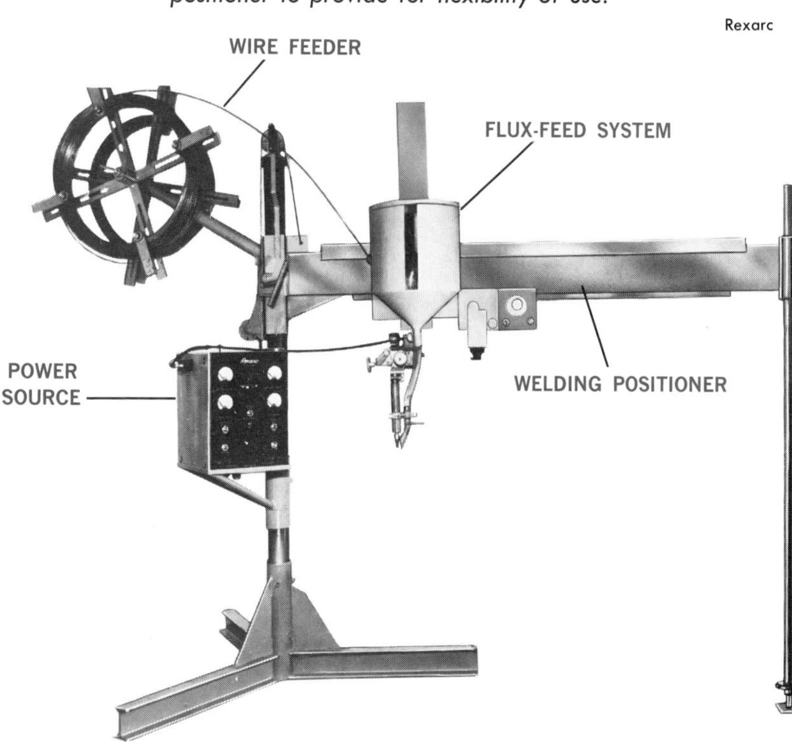

good ductility, impact strength, uniformity, density, corrosion resistance, and low nitrogen content. Physical properties are equal or better than the base metal when the base metal is free from rust, scale, moisture, and other surface impurities. In heavy sections, cracking can be minimized by the use of multipass welding.

Automatic submerged-arc welding is used extensively in the welding of carbon steel, chromium-molybdenum alloy steel, and stainless steel piping. Applications include the longitudinal seam welding of rolled-plate piping, circumferential welds on large piping, and straight fillet welds on fabricated plate attachments. The process can be used to make butt welds, fillet welds, and plug welds in both production and repair work, Figs. 27-31 and 27-32. It is also suitable for hard surfacing and the welding of castings.

Because the bed of flux must be supported, welding is performed in the flat position. Welds can be made in one or two passes in any thickness of steel from 16 gauge to 3 inches or more. High welding speeds are possible, and the current may be as high as 4,000 amperes at 55 volts. Backing rings are often used for groove welds. When fitup is poor, the first pass should be made by manual shielded metal-arc welding. When a backing ring is not used, the first pass may be made by the gas tungsten-arc welding process. In the welding of alloy pipe, the alloying elements should be contained in the filler wire and not in the flux.

POWER SOURCE. Welding power may be provided by a d.c. motor-generator, a rectifier, or an a.c. transformer. Welding with direct current provides versatile control over the bead shape, penetration, and speed, and arc starting is easier. DCRP provides the best control of bead shape and maximum penetration. DCSP provides the highest deposition rates and minimum penetration. Figure 27-33 shows a suitable d.c. power supply. Alter-

27-30. *Cutaway view of the welding zone in a single-V groove welded with the submerged-arc welding process.*

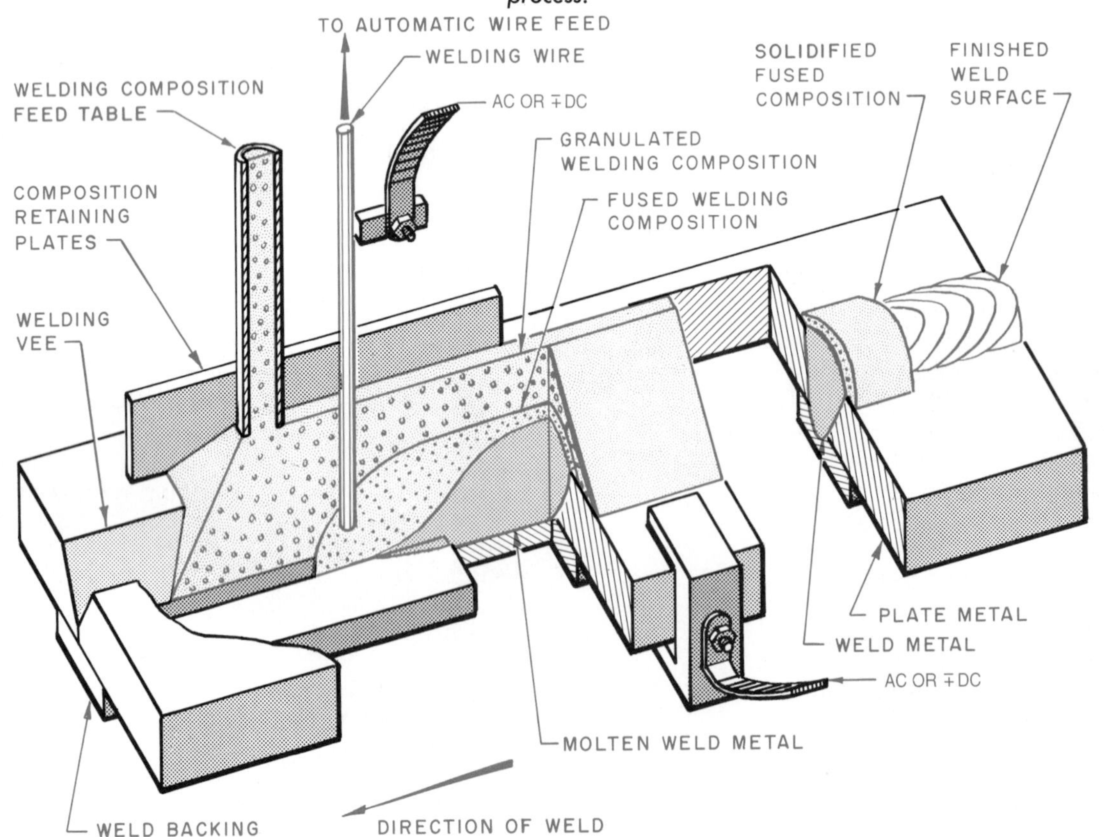

nating current provides penetration that is less than DCRP but greater than DCSP. It minimizes arc blow at high amperage, and it is preferred for single-electrode welding with high currents and for multiple-wire, multiple-power welding. The cost of producing current with transformers is less than with generators.

Both constant current (variable voltage) and constant voltage power supplies are used for automatic submerged-arc welding. In the *constant current* welding machine, a voltage-sensitive wire control system maintains a constant welding current. The wire feeder varies the wire-feed speed to produce a constant arc voltage and keep the desired arc length. This system is generally used for welding with high currents and large electrodes ($^5/_{32}$ inch and over). It is also a good choice for hard-surfacing and welding alloys.

The *constant voltage* welding machine is used with the constant-feed wire control system. This system maintains a constant-speed wire feeder. Voltage is selected at the machine and is held constant regardless of current demand. Constant voltage, constant speed systems are preferred for small diameter filler wires and for high speed welding of thin materials.

WIRE AND FLUX CLASSIFICATIONS. Wire and flux specifications for the submerged-arc welding process are covered in the *AWS A5.17-69* tentative specifications. These specifications list 9 welding wires and 10 fluxes.

The wires are divided into three groups according to manganese content. The low-manganese steel wires are EL8, EL8K, and EL12. In the medium-manganese steel category are EM5K, EM12, EM12K, EM13K,

27-31. *Welding heavy wall pipe with the submerged-arc process and d.c. motor-generator welding machines (in background). The pipe is mounted on a turntable to permit downhand welding.*

27-32. *The submerged-arc process being used to weld a pipe flange to a length of pipe. Note the manipulator, positioner, and gripper on a coordinated unit.*

Pandjiris Weldment Co.

and EM15K. The single high manganese wire is EH14.

The classification system follows the standard AWS pattern for filler metal specifications. For example, let us consider the classification EL8K. The prefix *E* designates an electrode as in other specifications. The letter *L* indicates that this is an electrode which has a comparatively low manganese content (0.60 percent maximum). The letters *M* and *H* indicate a medium (1.25 percent maximum) and high (2.25 percent maximum) manganese content respectively. The number 8 in the designation indicates the nominal carbon content of the electrode. The letter *K*, which appears in some designations, indicates that the electrode is made from a heat of steel that has been silicon killed.

The companion fluxes for these wires are F60 through F64 for welds with a tensile strength of 62,000 to 80,000 p.s.i. A second set of wires, F70 through F74, are for welds with a tensile strength of 72,000 to 95,000 p.s.i. These fluxes may also be designated with the companion electrode.

Fluxes are classified on the basis of the mechanical properties of a weld deposit made with the flux in combination with any of the electrodes classified in the specification. For example, let us consider the classification F60-EH14. The prefix designates a flux. The number 6, which immediately follows this prefix, designates the required minimum tensile strength. In this case it is 60,000 p.s.i. The second digit after the prefix desig-

27-33. *Multi-process 400-ampere, d.c. motor-generator welding machine with constant and variable voltage. It can be used for flux-cored and solid core wire MIG welding, stick electrode welding, and submerged-arc welding. (A) Variable voltage control sets the open circuit voltage. It is a fine adjustment of welding current for stick electrode or submerged-arc welding. (B) Constant voltage control sets open circuit voltage when welding with the flux-cored wire process or other open arc processes. (C) Toggle switch selects constant or variable voltage output. (D) Current control is a reactor for continuous current control when set for variable voltage welding. It provides variable inductance control for low voltage, open arc semi-automatic processes. (E) Polarity switch selects DCRP or DCSP. (F) There are two electrode studs—one for stick welding and one for automatic processes—plus a common ground stud. The stick electrode cable can remain connected for alternate tacking and semi-automatic submerged-arc welding.*

The Lincoln Electric Co.

Table 27-5

Impact requirements of the weld deposit according to classification for submerged-arc flux.

Classification	Required minimum impact strength
0	No requirement
1	20 ft-lbs. at 0°F.
2	20 ft-lbs. at −20°F.
3	20 ft-lbs. at −40°F.
4	20 ft-lbs. at −60°F.

nates the required minimum impact strength, Table 27-5. The zero in the example means that there are no impact requirements. The suffix which is placed after the hyphen gives the electrode classification with which the flux will meet the specified mechanical property requirements. The electrode in the example is in the high manganese classification.

TYPES OF FLUXES. The fluxes used with submerged-arc welding have the special characteristic of being able to carry the high welding currents required by the process. In all other respects they produce results like those provided by other fluxes.

- They protect the weld pool from the surrounding air in an envelope of molten flux.
- They act as a cleaning agent for the base metal.
- They may satisfy special metallurgical or chemical needs.
- They may provide minerals or alloys to the weld metal.
- They guard against porosity caused by rusty plate.
- They provide maximum resistance to weld cracking.
- Specific fluxes are designed to work best with certain electrodes, materials, and welding conditions.
- They improve weld appearance.

These fluxes are granular, fusible mineral compounds. They contain various amounts of silicon, manganese, aluminum, titanium, zirconium, and deoxidizers, which are bound together with a binder.

WIRE SIZES. Submerged-arc electrodes are furnished in continuous lengths wound into coils or drums or on liners. Standard sizes include 1/16, 5/64, 3/32, 1/8, 5/32, 3/16, 7/32, 1/4, 5/16, and 3/8 inch. Wire 1/16 inch in diameter is used for making speed welds on steel ranging from 14 gauge to 1/4 inch thick. Wire 5/64 inch in diameter is used for welding material 12 gauge and thicker. Wires 3/32 inch and thicker are used when the gun is carried mechanically. The stiff wire decreases the flexibility of the cable, and the large pool of molten metal is hard to handle when welding is semi-automatic. Large sizes require high current and provide high deposition rates. They can bridge the gaps when fitup is poor. Arc starting, however, becomes more difficult as wire size increases.

POLARITY. Reverse polarity is recommended for most applications. It produces smoother welds and better bead shape. It has greater penetration and better resistance to porosity. Fillet welds also have deep penetration.

Straight polarity has meltoff rates that are about one-third greater and provide less penetration than reverse polarity. It is suitable for the following applications:

- Conventional fillets in clean and rust-free plate
- Hard surfacing
- Hard-to-weld steels in which cracking and porosity, due to the admixture resulting from deep penetration, must be controlled
- Prevention of cracking due to deep penetration and heavy buildup in the root pass of deep groove welds.

When changing from DCRP to DCSP at the same current, increase the voltage about 4 volts to maintain a similar bead shape.

ALTERNATING CURRENT. A.c. is recommended for two specific applications:

- Tandem arc welding for increased welding speed.
- Single-arc applications in which arc blow cannot be overcome and travel speed is slowed.

To maintain good arc stability, a higher current density is needed for a.c. than for d.c.

JOINT DESIGN. Basic joints common to all welding are used for submerged-arc welding. The process is a deep, penetrating one. To avoid burn-through, the plates are generally either butted tightly together or a backup strip is used.

CLEANING. Rust, scale, and moisture cause porosity. A thoroughly cleaned joint gives the best welding results. All substances on the joint edges must be removed. If not, foreign matter will become entrapped in the weld zone and cause porosity on the surface and possibly gas pockets beneath the surface. Use clean, rust-free wire and flux that has been screened to remove large particles and foreign matter.

POSITION OF WELDING. Practically all welding should be done with the work level or in the flat position. Some sheet metal may be welded slightly downhill. Heavy, deep-groove welding is sometimes done with the plate at an uphill angle of 2 to 5 degrees. This helps to keep the molten metal from running ahead of the arc.

FITUP. Joint fitup should be uniform and accurate since it affects the appearance and quality of the weld. When establishing welding conditions, the seams should be butted tightly unless a root gap is specified. A gap may be required to secure penetration or to prevent weld cracking or distortion of the plates. An insufficient gap gives a weld an excessive crown or reinforcement. If the gap or bevel angle is excessive, burn-through or a concave weld will result.

Gaps greater than $^1/_{16}$ inch may be filled manually with stick electrodes. Although the base metal dictates the type of electrode, an E6010 type is usually used for sealing square-butt joints; and an E6020, for beveled-butt joints.

STARTING AND STOPPING TABS. It is general practice when welding long seams on a tank to tack-weld pieces of steel at each end of the seam. By starting and ending in these tabs, the possibility of cracking and weld craters may be eliminated. They enable the welding conditions to stabilize and maintain uniformity at the beginning and end of the seam.

In general, tabs should be similar in material and design to the weld joint. They should be large enough to support the flux and molten metal properly. Tabs are welded or supported on the base metal in a manner that will prevent molten metal from dropping through any gaps.

FLUX COVERAGE. Insufficient flux coverage permits the arc to flash-through and does not provide proper shielding. Excessive flux produces a narrow hump bead. For applications like roundabout, edge, and horizontal welds, a support may be needed to hold the flux around the arc while welding. It can be a piece of fire-resistant material clamped to the nozzle or flux dams tacked or clamped to the work, Fig. 27-34.

FLUX DEPTH. A proper amount of flux is required to establish the best welding conditions. A suitable depth of flux gives a fast, quiet welding action. If the layer

27-34. *Various types of flux supports for submerged-arc welding.*

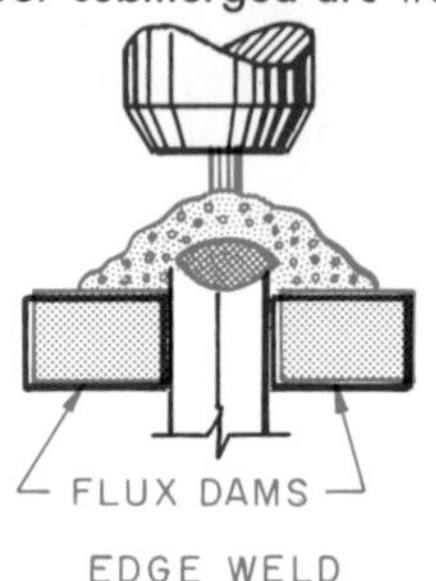

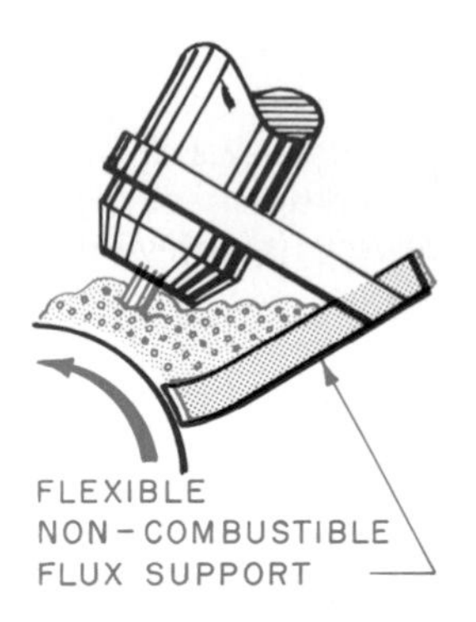

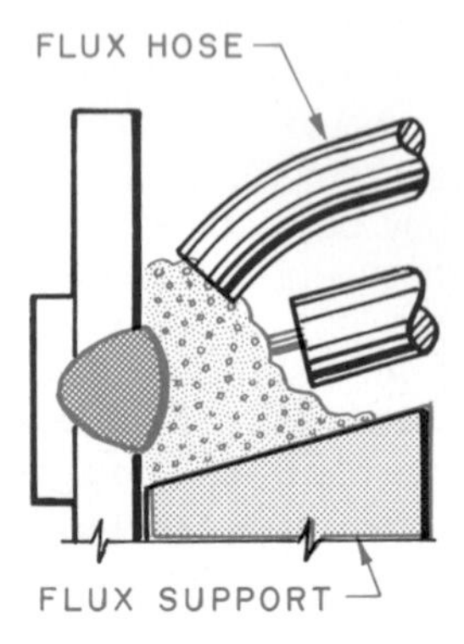

27-35. *A. Normal stickout. B. Use of nozzle extension for long stickout.*

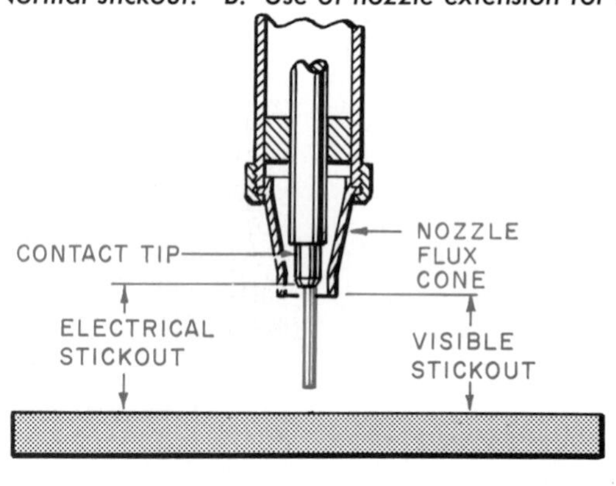

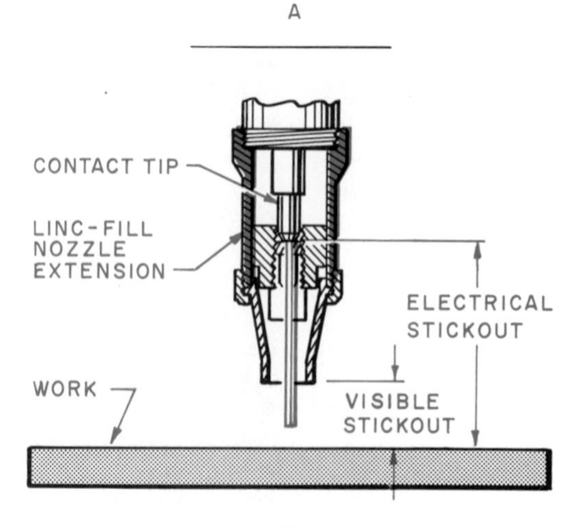

of flux is too deep, the gases generated during welding cannot escape. The weld may have a shape somewhat like a mountain range—rough and uneven—and it will be porous. A shallow layer of flux permits arc flashing and a porous weld.

It is very important that the flux be kept clean and dry. Dirty or damp flux produces an unsound, porous weld. Damp flux should be dried before welding.

WELD BACKING. The relatively high heats used in automatic submerged-arc welding cause a large pool of molten metal to be formed. Because the weld pool is highly fluid, it will fall through if the joint is not supported on the back side. The most common forms of weld backing are steel backup bars, weld metal backup, copper, or flux. When steel bars are used, they usually remain as part of the weldment. Weld metal backup in the form of a weld bead on the back side also remains as part of the weldment. Copper is one of the best materials to use. It is nonfusible and is a good conductor of heat. The copper backing bar is often liquid cooled. Flux may be placed at the bottom of the groove to act as a support for the molten weld pool.

OPERATING VARIABLES. The welds produced by the submerged-arc welding process and the gas metal-arc processes are affected by similar changes and adjustments in the operating variables (voltage, current, wire-feed speed, travel speed, stickout and nozzle angle). Review Chapter 26 as necessary.

There is some difference in regard to stickout. Normal stickout, Fig. 27-35A, may be used for most applications. Some welding engineers, however, recommend long stickout. Nozzle attachments are available that increase the electrical stickout to either $2\frac{1}{4}$ or $3\frac{1}{4}$ inches, Fig. 27-35B. The welding current passing through the longer lengths of exposed filler wire preheats the wire so that it melts more quickly in the arc. This reduces costs by increasing the deposition rate and the speed of welding.

ELECTRODE SIZE. At a fixed current setting, electrode size affects the depth of penetration. Penetration decreases as the size of electrode increases. Smaller filler wires are generally used in semi-automatic equipment.

For automatic submerged-arc welding, filler wires are larger to take advantage of higher currents and higher deposition rates. Sizes of $\frac{5}{64}$ to $\frac{7}{32}$ inch are generally used with currents from 200 to 1,200 amperes. For special situations, filler wire sizes of $\frac{1}{4}$ to $\frac{3}{8}$ inch with currents from 700 to 4,000 amperes are used. Large electrodes have hard starting characteristics.

With all other conditions held constant, an increase in wire size reduces the deposition rate and penetration. An increase in current corresponding to an increase in electrode size increases the deposition rate and penetration.

MULTIPLE-WIRE TECHNIQUES. Multiple arcs, Fig. 27-36, increase meltoff rates and direct the arc blow to provide an in-

27-36. *Butt welding with the automatic submerged-arc process and the multiple-arc, two-wire method.*

The Lincoln Electric Co.

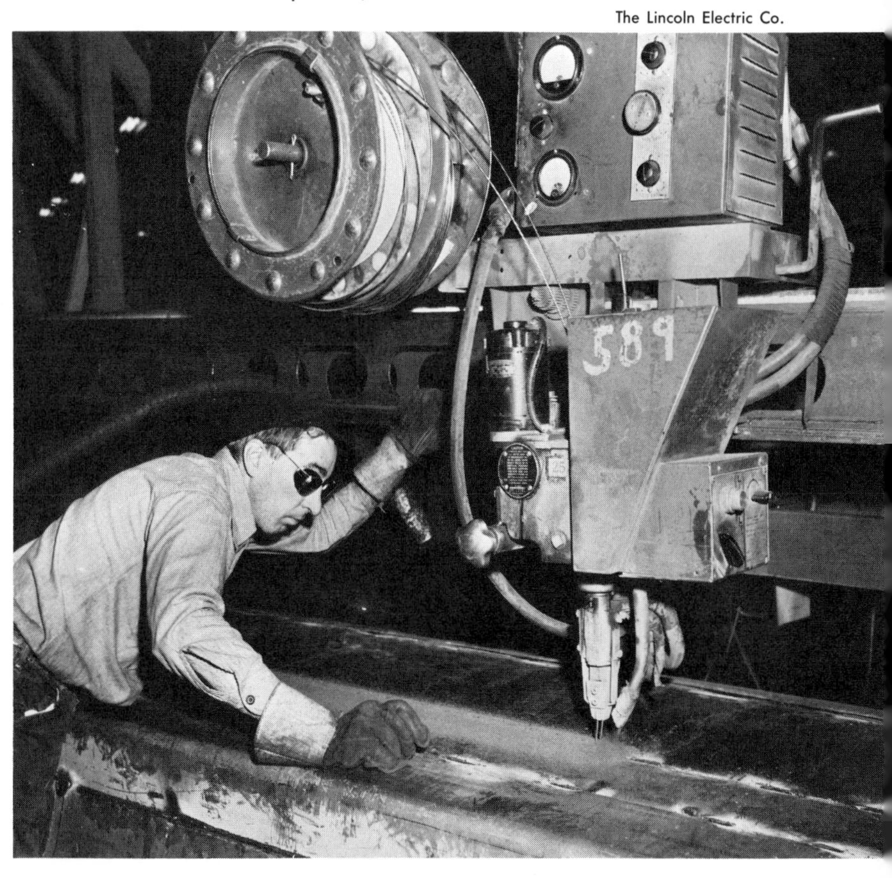

crease in welding speed. Two electrodes fed through the same jaws from one power source increase the deposition rate by 50 percent on work such as large, flat-position fillets and wide groove welds in which fill-in is a major consideration. The two arcs pull together, causing back blow at the front arc and forward blow at the trailing arc.

Multiple-wire, multiple-power arcs, in which two or more electrodes each have a separately controlled power source, provide high speed on both fill-in and square groove welds. The procedure is difficult to set up, Fig. 27-37, and can be justified only on long welding runs or on production work. At least one arc should be a.c.

The *two-wire series* power technique obtains high deposition rates with minimum penetration into the base metal. It is used extensively for hard surfacing materials. Each filler wire operates independently; it has its own feed motor and voltage control. The power supply cable is connected to one welding head, and the return power cable is connected to the second welding head instead of to the workpiece. The two filler wires are in series. The welding current travels from one electrode to the other through the weld puddle and surrounding material. The two electrodes are mounted at 45 degrees to each other. Either alternating or direct current can be used, depending upon the application. Alternating current is preferred for mild steel or stainless steel, whereas direct current should be used for nonferrous metals. DCRP gives deeper penetration than DCSP.

Thinner materials may be welded with a square edge. A backup strip is often necessary. Heavy materials may be beveled. The material may be only partially beveled so that a relatively heavy root face remains. Welding is often done from both sides. Fillet welds up to $\frac{3}{4}$ inch may be made without beveling. Fillet weld penetration is deep.

The flux is supplied from a hopper which is either mounted directly on the welding head or connected to the head by tubing. The bare filler rod is fed into the head in straight lengths or from a coil or rod mounted on a rod reel. Filler wire up to $\frac{1}{2}$ inch in diameter and welding currents up to 4,000 amperes are used.

27-37. *Welding a girth seam on the outside of a tank with the multiple-wire, submerged-arc process. Note the welding rolls that make it possible to rotate the tank under the arc. The welding unit is mounted on a manipulator, and the welder works from a weld elevator.*

The Lincoln Electric Co.

SEMI-AUTOMATIC SUBMERGED-ARC WELDING. Semi-automatic submerged-arc welding is rapidly being replaced by solid wire and flux-cored wire welding. It is still used to some extent for hard surfacing. Two examples of this process are Squirt Welding®, a Lincoln process, and Union Melt®, a Linde process. If the school welding shop has this type of equipment, it would be an added experience for the advanced student to practice a number of butt, lap, and T-joints with the process.

Semi-automatic submerged-arc welding can be used on those jobs for which submerged-arc welding would be desirable, but the fully automatic process would not be suited for economic reasons or physical limitations.

The semi-automatic process is similar to the automatic process in that the welding operation takes place beneath a blanket of flux, but the hand gun is guided manually. The electrode is a continuous wire which is fed through the center of a flexible welding cable and through a gun to the arc. The flux is deposited by gravity on the joint from a welding gun or dipper. Equipment includes the welding machine, the conical flux container and welding nozzle, wire reel, wire-feeding mechanism, and the control unit for the control of wire feed and arc voltage, Fig. 27-38.

Semi-automatic submerged-arc welding is essentially a small wire process. It is possible to weld $\frac{3}{4}$-inch plate having a square edge with filler wire $\frac{5}{64}$ inch in diameter and 600 amperes of welding current. Each side is welded with one pass. The resultant weld is smooth, penetrating, and spatter-free. Appearance is similar to that obtained with flux-cored electrode welding.

Process Characteristics. In comparision with automatic submerged-arc welding, the semi-automatic process has the following characteristics.

- It can follow irregular shapes
- Welding can be done without fixtures or with only simple fixtures.
- Equipment is easily portable, and the process is highly versatile.
- The cost of equipment is lower than for the automatic process.

The semi-automatic process can be used for any of the following broad applications:

- When the gun can be dragged along the joint, providing accurate guiding
- When the work can be rotated and the gun held in position by hand
- When the work can be rotated and the gun held in a simple locating fixture
- When both the gun and the work can be moved by special fixtures.

Electroslag Welding (ESW)

Electroslag welding, one of the newer processes, was developed for the welding of vertical plates, ranging in thickness from $1\frac{1}{4}$ to 14 inches, with a single pass. The plate edges require no preparation.

Electroslag welding is an automatic process. The equipment used for electroslag welding consists of a carriage assembly,

27-38. *Schematic diagram of component units needed for automatic or semi-automatic submerged-arc welding.*

Hobart Brothers Co.

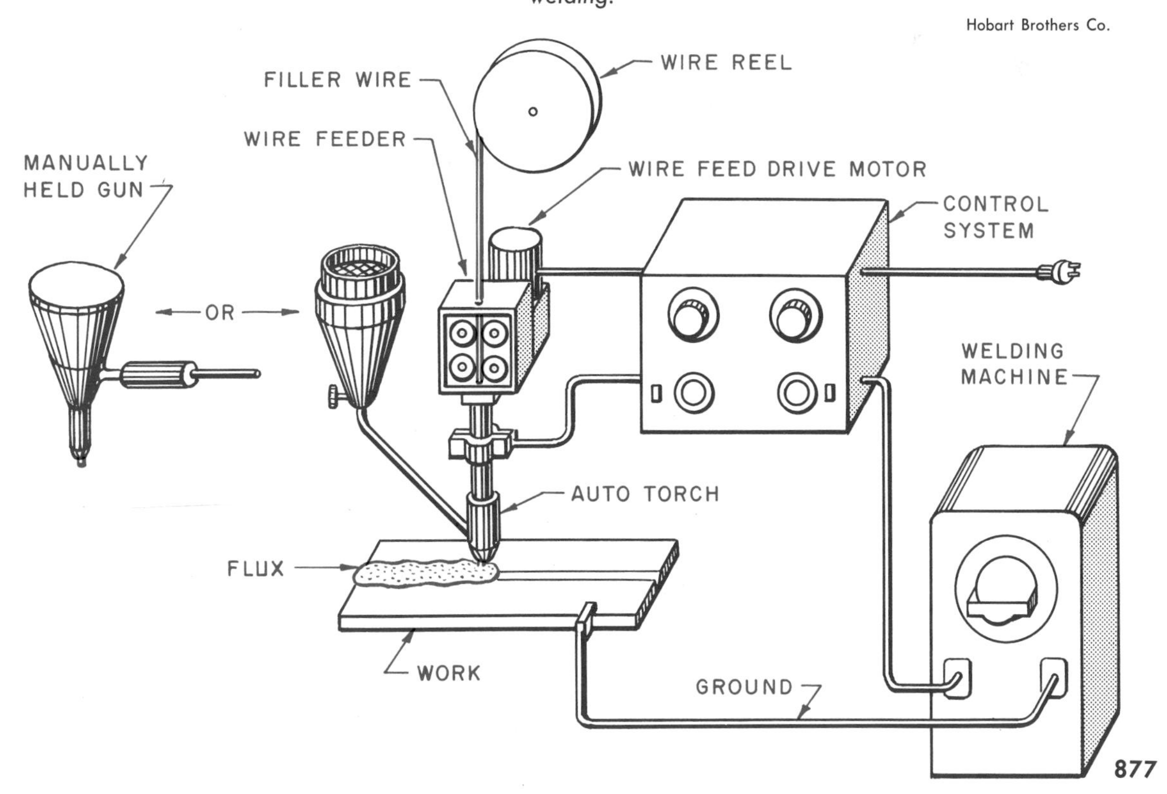

which moves upward along the joint, and a multiple set of feed-wire guide assemblies which can be made to oscillate horizontally, Fig. 27-39. Copper shoes are positioned against the joint to act as a dam. The filler wire is fed to the weld zone through a nozzle in a vertically down feed. The welding operations are controlled from a panel board and may be automatic or manually controlled.

The fusion of the base metal and the continuously fed filler wires takes place under a heavy layer of high temperature, electrically conductive molten flux. The filler wires may be either solid or flux cored. For welding plates up to 5 inches in thickness, only one electrode may be used. Two electrodes are generally used for plates 4 to 10 inches thick, and three electrodes for 10- to 14-inch thicknesses. When necessary, the electrodes may be oscillated to provide better distribution of weld metal. It is a high heat process. Voltage ranges from 42 to 52; and amperage, from 500 to 640, depending on the thickness of metal. The power is alternating current.

The plates are set up in a vertical plane with the square edges spaced from 1 to $1\frac{5}{16}$ inches, according to the thickness of the plate. Water-cooled copper shoes form a mold around the joint gap and give form to the weld. The shoes are mechanized so that they can move vertically upward as the weld proceeds. A prepared block is placed under the plate edges to close the joint cavity. Granular flux is poured into the cavity, and the arc is established

27-39. *Basic components of an electroslag welding operation. Note the travel rail which provides the vertical up-and-down movement.*

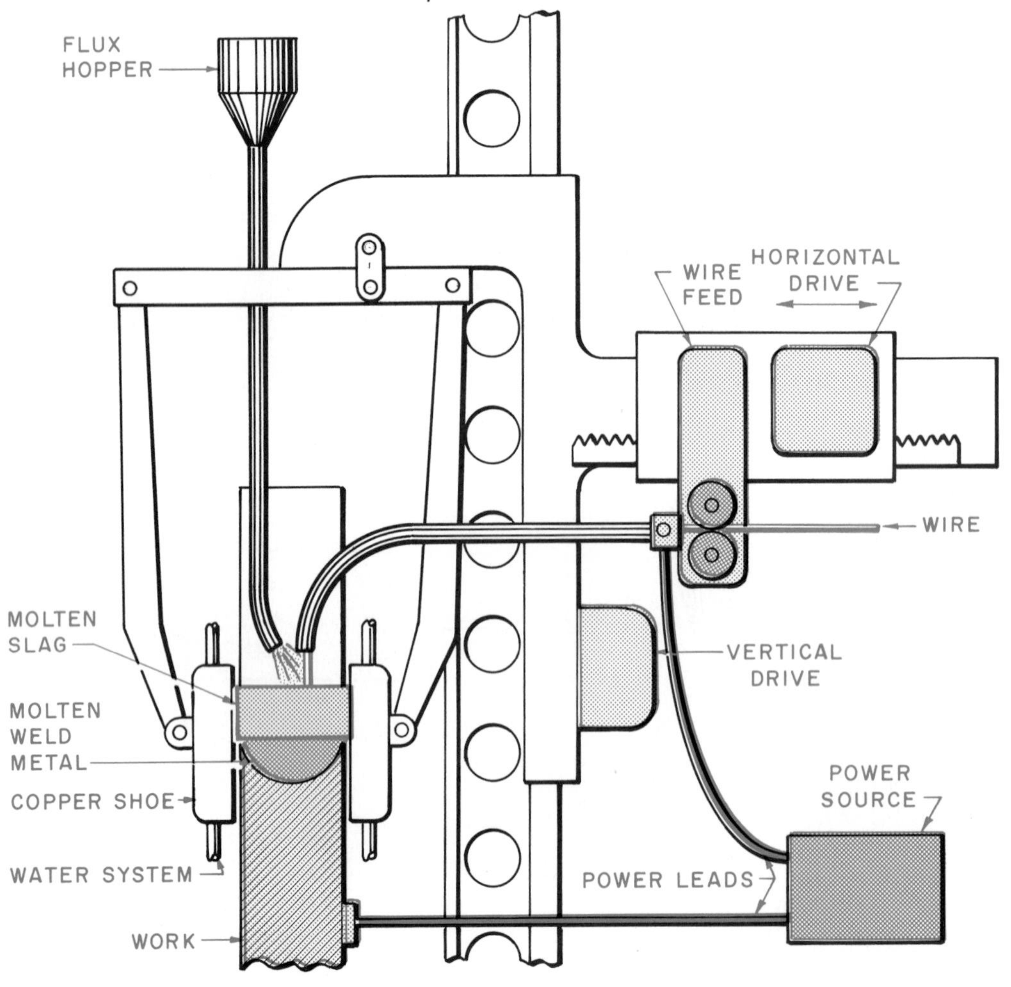

with the filler wires. At first only the flux is fused into molten slag. The extreme heat produced by the molten slag, the welding arc, and the molten filler wire cause the base metal to melt. The weld is formed by the water-cooled plates. It is homogeneous and has good penetration into the base metal and smooth, clean weld faces, Fig. 27-40.

ELECTROSLAG WELDING WITH A CONSUMABLE GUIDE. Consumable guide welding, or *CG* as it is referred to in the trade, is a relatively new process that is on the increase, especially in shipbuilding. It is used to weld vertical beam joints in ship construction.

A tube, which is coated with slag-forming and alloying elements, guides a filler wire from the wire-feeding unit into a bath formed by the two sides of the joint being welded and two water-cooled copper shoes, Fig. 27-41. This bath contains the molten weld metal and slag. The tube, or guide, is connected to the positive side of a rectifier power source. The heat necessary to melt the guide, the filler wire, and the joint edges being welded is generated by the passage of the welding current through the ionized slag bath. The guide tube and filler wire melt at a rate that determines the welding speed.

The consumable guide process operates equally well on d.c.

27-40. *A square butt joint made with the electroslag process in steel plate 3 inches thick. It had a 1-inch gap and was welded in 1 pass. The completed weld is 1¾ inches wide with ⅛-inch surface reinforcement. Welding was done in the vertical position, travel up.*

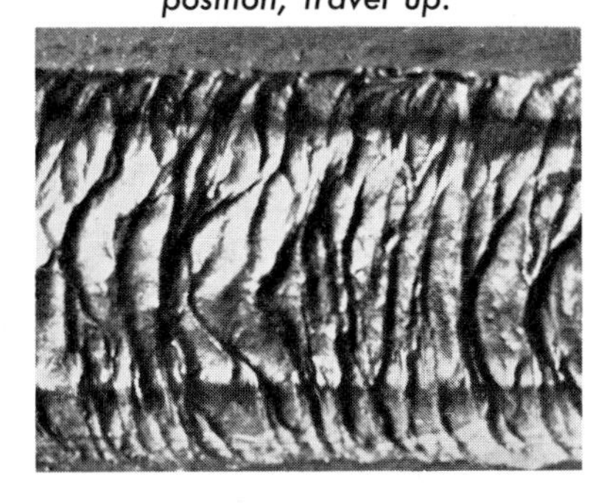

27-41. *Essentials of consumable guide welding.*

ESAB

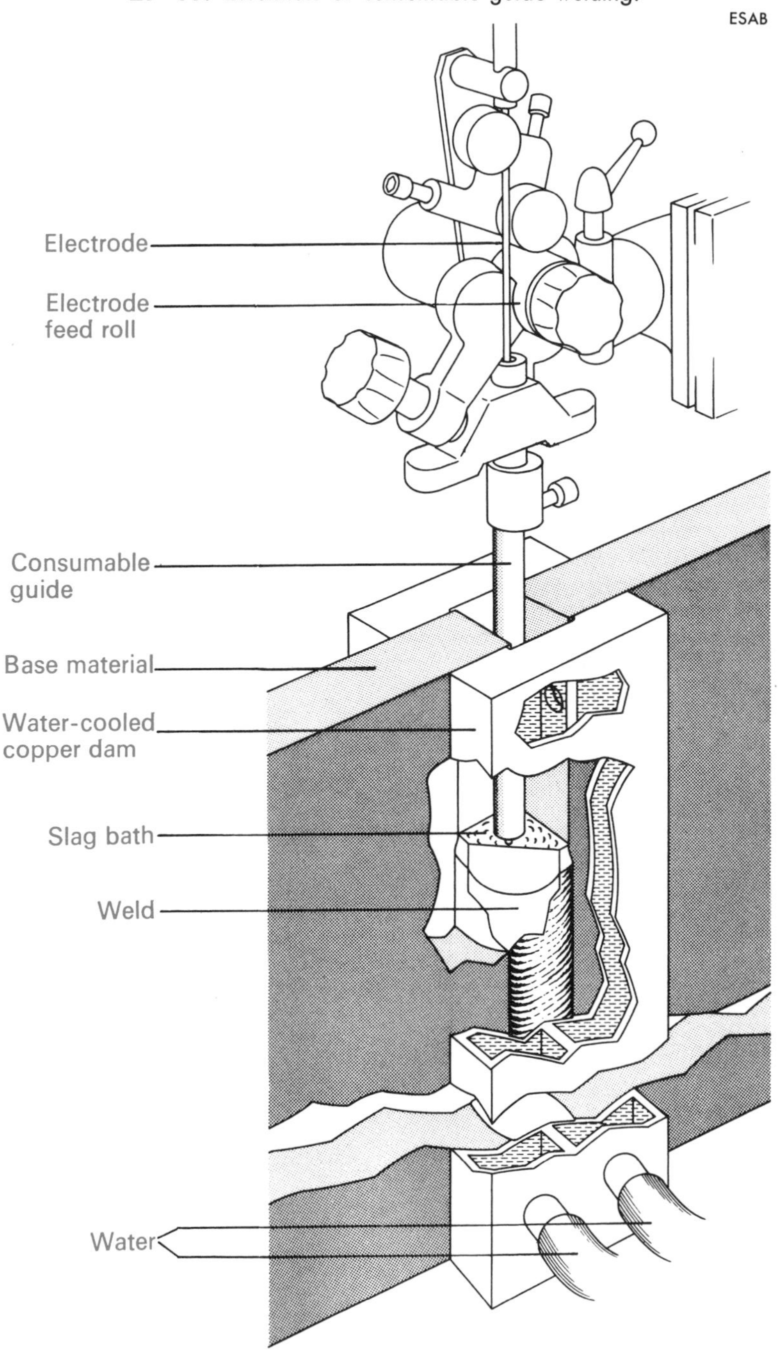

or a.c. For d.c. the constant voltage rectifier is recommended. For a.c. an auxiliary control box is connected to an a.c. power source with drooping-current characteristics.

Consumable guide welding is not a true submerged-arc welding process. It uses an arc only at the start of the process to generate heat for the melting of the slag. As soon as the bath of molten slag is established, the slag causes the arc to be extinguished automatically. Another characteristic of the slag is low conductivity. This increases resistance heating from the passage of the electrical current.

Due to resistance heating, the length of the tube is limited to approximately 40 inches. The tube can be extended, however, if a movable current contact is used.

The consumable guide process has the following advantages over conventional electroslag welding.

- The welding machine is portable, lightweight, and relatively easy to operate.
- Welding of thinner metals is possible.
- An increase in welding speed increases productivity and improves the mechanical properties of the weld metal and the heat-affected zones. The weld metal is free from porosity and slag inclusions. There are fewer problems with residual stresses and plate deformation.

Electrogas Welding

There are two electrogas welding processes: (1) gas metal-arc electrogas welding (GMAW-EG), in which solid core filler wire is used and (2) flux-cored arc electrogas welding (FCAW-EG), in which flux-cored filler wire is used.

The electrogas welding process is similar to electroslag welding. It is a fully automated method for the welding of butt, corner, and T-joints in the vertical position. It differs only in the welding current and the medium used for protecting the weld pool from atmospheric contamination. Direct current, reverse polarity is used instead of alternating current, and shielding gas instead of granular flux is fed into the weld pool. It is possible to weld metal sections of ½ inch to over 2 inches in thickness with a single pass and without any edge preparation.

Here again, water-cooled copper shoes span the joint cavity and form a dam to contain the molten weld metal. Flux-cored or solid filler wire is fed into the cavity by means of a curved guide, Fig. 27-42. An electric arc is established and continuously maintained between the filler wire and the weld pool. Helium, argon, carbon dioxide, or mixtures of these gases are fed continuously into the cavity to provide a suitable atmosphere for shielding the arc and the weld pool. The flux core of the filler wire provides deoxidizers and slagging materials for cleaning the weld metal. The base metal is melted and fused as a result of the high preheat temperature from the ionized shielding gas and the molten slag. The molten slag forms a protective coating between the shoes and the gases of the weld and a seal between the shoes and the surfaces of the work to prevent air from entering the weld pocket. As the welding progresses, the copper shoes automatically move upward. Welding is done with DCRP.

27-42. *Basic components for electrogas welding. Note the point of entrance of the shielding gas.*

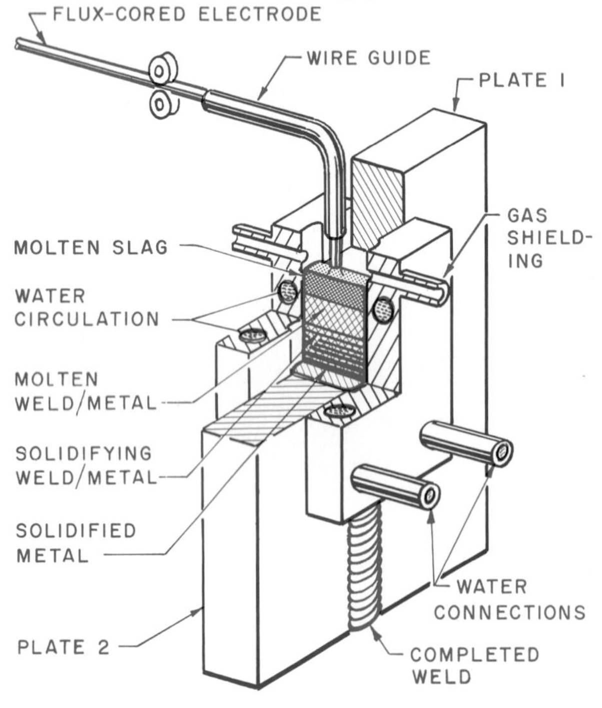

The edges are square cut and are spaced from $^{11}/_{16}$ to $^{7}/_{8}$ inch or more. The minimum spacing must be large enough to admit the wire guide and permit it to oscillate without arcing on the plate surfaces. Spacing that is too wide requires excessive filler wire and increases the welding time. Welding is done within a range of 500 to 700 amperes. Filler wire $^{1}/_{8}$ inch in diameter is used.

The electrogas process is used in the field erection of pressure vessels and liquid storage tanks and in the shop fabrication of large pressure vessels and heavy structures. Carbon, low-alloy, high-tensile, medium-alloy, air-hardening, and chrome-nickel stainless steels are successfully welded by this process. In field welding, the work is enclosed in an all-weather shelter to protect it from the elements and air currents which might disturb the gas shield during welding. Alloy steels do not require preheat and may be welded at 32 degrees F., providing there is no frost. Shrinkage, warpage, and distortion are avoided by this process.

CHOICE OF WELDING PROCESS

In general, the selection of the welding process (manual, semi-automatic, or automatic) depends on the proper evaluation of each job. The three processes have the following applications and advantages.

Hand welding applications

- Out-of-position welding in which a large, highly fluid weld pool would spill
- Relatively short welds
- Light and heavy gauge metals
- Nonrepetitive jobs
- Jobs which are costly or difficult to fixture
- Jobs in which fitup cannot be controlled
- Jobs in which it would be difficult to retain flux because of the shape of the work
- Jobs where obstructions of one kind or another make it impossible to make an uninterrupted weld

Hand welding advantages

- Can be done indoors or outdoors
- Can be done in any position and inaccessible locations
- Can weld a wide range of alloys and dissimilar metals
- Low cost, portable equipment

Semi-automatic applications

- Jobs in which you can take advantage of the additional meltoff provided by currents higher than those possible with manual equipment
- Jobs which are repetitive enough so that a high degree of skill can be acquired
- Medium and heavy metals
- When the continuous wire feed increases the welding time (duty cycle)
- When complicated shapes or extremely large weldments make fixturing for the automatic process too difficult
- When the penetration, which is deeper than that produced by the hand processes, is an advantage
- When the contour of the work is irregular and fitup is not accurate enough for fully automatic guiding

Semi-automatic advantages

- Produces welds of desirable appearance and high quality weld metal characteristics
- Higher welding speeds than hand welding
- Less slag and weld spatter
- Reduced costs

Automatic welding application

- Highly repetitive jobs that can be fixtured

Automatic welding advantages

- Pushbutton automatic arc striking
- High rate of weld metal deposition
- Welds have smooth, even appearance
- Heavier construction than manual and semi-automatic equipment
- Higher current capacity than manual or semi-automatic equipment
- Increased welding speeds
- Self-contained travel mechanism
- Reduced electrode loss
- Machine precision
- Minimum slag removal
- Reduced problem of heat distortion
- Accurate, continuous control and fewer weld rejects due to automation
- High mechanical properties of welds

REVIEW QUESTIONS

1. Welds made with flux-cored wire are inferior to those made with solid core wires. True. False.
2. One of the disadvantages of flux-cored wire is that joint fitups must be almost perfect. True. False.
3. The commercial use of flux-cored wires began about ten years ago. True. False.
4. What is the function of the flux core?
5. A constant current welding machine is used for flux-cored wire welding. True. False.
6. Because of the delicate nature of the flux-

ing materials in a flux-cored wire, high welding heats are never used. True. False.

7. A joint welded with flux-cored wire requires more filler wire than with other processes. True. False.

8. Since a flux-cored wire provides its own shielding, a shielding gas is never used. True. False.

9. Name and explain the effect of the various welding variables on flux-cored wire welding.

10. What are the effects of short stickout?

11. Give a brief explanation of magnetic-flux coated wire welding.

12. What is vapor shield welding?

13. Give a brief explanation of the difference among the electroslag, the electroslag with consumable guide, and the electrogas welding processes.

14. Give a brief explanation of the submerged-arc welding process.

15. Submerged-arc welding is always an automatic process. True. False.

16. Alternating current gives the best results for submerged-arc welding. True. False.

17. List at least five advantages of the automatic submerged-arc welding process.

18. Argon is the best shielding gas for submerged-arc welding. True. False.

19. Because the weld metal deposited by the submerged-arc welding process loses some of its physical properties, the process is never used for code work. True. False.

20. Manual submerged-arc welding is on the increase. True. False.

21. List at least 5 advantages of the self-shielded, flux-cored arc welding process.

22. What is the basic difference between gas-shielded, flux-cored arc welding and self-shielded, flux-cored arc welding?

23. List at least five desirable characteristics of gas-shielded, flux-cored arc welding.

24. A constant current welding machine with straight polarity gives best results when welding with flux-cored welding wires. True. False.

Table 27-6

Course Outline: Gas Metal-Arc Welding Practice with Gas-Shielded, Flux-Cored Wire (Plate)

Recommended job order	Material	Plate thickness	Joint	Weld	Weld position	No. of passes	Electrode		Shielding gas		Welding current DCRP	
							Type AWS-ASTM	Size inch	Type	Gas flow c.f.h.	Volts	Amperes
1st	Steel	1/4	Flat plate	Beading	Flat	Cover plate	*E70T-1	1/16	Carbon dioxide	30–40	25–30	350–400
2nd	Steel	1/4	T	Fillet	Flat	2	E70T-1	3/32	Carbon dioxide	30–40	25–30	350–400
3rd	Steel	1/4	Lap	Fillet	Horizontal	1	E70T-1	1/16	Carbon dioxide	30–40	25–30	280–350
4th	Steel	3/8	T	Fillet	Horizontal	6	E70T-1	1/16	Carbon dioxide	35–40	28–32	325–450
5th	Steel	3/16	Square butt backup	Groove	Flat	1	E70T-1	1/16	Carbon dioxide	30–40	24–26	240–320
6th	Steel	3/8	V-butt backup	Groove	Flat	2	E70T-1	3/32	Carbon dioxide	35–40	25–32	375–400
7th	Steel	3/8	T	Fillet	Flat	2	E70T-1	3/32	Carbon dioxide	35–40	28–32	450–525
8th	Steel	1/2	V-butt backup	Groove	Vertical	3	E70T-1	0.045	Carbon dioxide	35–40	22–24	160–200

NOTE: The conditions given here are basic and will vary with the job situation, the results desired, and the skill of the welder.
*Electrode type E70T-2 may also be used with excellent results.
The electrode stickout should be about 3/4 to 1 1/2 inches.

Table 27-7

Course Outline: Gas Metal-Arc Welding Practice with Self-Shielded, Flux-Cored Wire (Plate)

Recommended job order	Material	Plate thickness	Joint	Weld	Weld position	No. of passes	Electrode		Shielding Gas	Welding Current	
							Type AWS-ASTM	Size Inch		Volts	Amperes
1st	Steel	3/16	Flat plate	Beading	Flat	Cover plate	E70T-G	1/16	None	24-26	280–350
2nd	Steel	3/16	Square butt backup	Groove	Flat	1	E70T-G	1/16	None	24-26	280–350
3rd	Steel	3/16	T	Fillet	Horizontal	1	E70T-4	1/16	None	28-30	250–300
4th	Steel	10 Ga.	T backup	Fillet	Horizontal	1	E70T-3	1/16	None	25-26	250–310
5th	Steel	1/4	T	Fillet	Vertical	1	E70T-G	1/16	None	18-20	125–180
6th	Steel	3/16	Lap	Fillet	Flat	1	E70T-3	1/16	None	28-30	350–400
7th	Steel	10 Ga.	Lap backup	Fillet	Flat	1	E70T-3	1/16	None	25-26	250–310
8th	Steel	1/4	T	Fillet	Overhead	1	E60T-7	5/64	None	18-20	125–150
9th	Steel	3/8	V-butt backup	Groove	Flat	2	E70T-G	1/16	None	18-20	150–200
10th	Steel	1/2	V-butt	Groove	Horizontal	1 2–3 4–6	E60T-7 E60T-7 E60T-7	5/64 5/64 5/64	None None None	18–20 19–20 18–20	150–175 200–250 225–325
11th	Steel	1/2	V-butt	Groove	Vertical	1–3	E60T-7	5/64	None	18–20	150–175

The electrode stickout should be about 3/8 to 1 1/4 inches.

CHAPTER 28

Gas Metal-Arc Welding Practice With Solid Core Wire (Pipe)

This chapter deals only with the techniques for gas metal-arc welding standard (schedule 40) and heavy (schedule 80) wall carbon steel pipe.

INDUSTRIAL APPLICATIONS OF MIG PIPE WELDING

Spurred on by the rapid growth of nuclear power, space and rocket exploration, marine construction, and the chemical, oil, and gas industries, welded-pipe fabrication is increasing at a tremendous rate. Over 90 percent of all piping installations are welded. There are over one million miles of pipeline in this country. Over 10 percent of the steel produced in the United States is used in the production of pipe. The Hoover Dam piping installations required pipe 30 feet in diameter with a wall thickness up to $2\frac{3}{4}$ inches.

There is a considerable increase in the use of the gas metal-arc process in the field for pipeline and building construction (Figs. 28-1 and 28-2) and in shop fabrication (Figs. 28-3 and 28-4). Much of this is being done with the *fine-wire process.* The practice jobs in this chapter specify fine-wire techniques for carbon steel pipe. Pipe is available in such ferrous metals as low and high strength carbon steels, wrought iron, carbon-molybdenum steels, chromium-molybdenum steels, low temperature steels, stainless steels, and clad and lined steels. Pipe is also available in such nonferrous metals as aluminum, nickel, copper, titanium, lead, and their alloys.

28-1. *MIG welding large diameter pipe in the field; d.c. engine-powered generator welders are used. Shielding gas is carbon dioxide. Note the welding equipment on the truck. The wire feeder and wire reel are near the job.*

Tube Turns Co.

Welding may be done in all positions. The direction of welding for vertical welds may be up or down. The usual practice is to weld the first pass down followed by passes that may be welded up or down.

It is not difficult to obtain X-ray quality welds. The MIG pipe welding process is used extensively on both noncritical and critical piping in which the weld must meet requirements of the codes for such critical applications as nuclear piping, steam power plant piping, and chemical process piping.

It is assumed that you are already skilled in the welding of pipe with the shielded metal-arc process and the gas tungsten-arc process. You are urged to take the utmost care in the practice of these jobs. Gas metal-arc pipe welding is a growing field and offers many opportunities for those who develop a high degree of craftsmanship in this welding process.

Advantages

The piping industry is one of the most progressive of all industries. The installation of piping is costly, and for this reason the industry is alert to new developments in the fabrication of pipe. The gas metal-arc welding of pipe was first studied in the laboratory and then tested in field experiments. The results led to the use of MIG as a fabrication

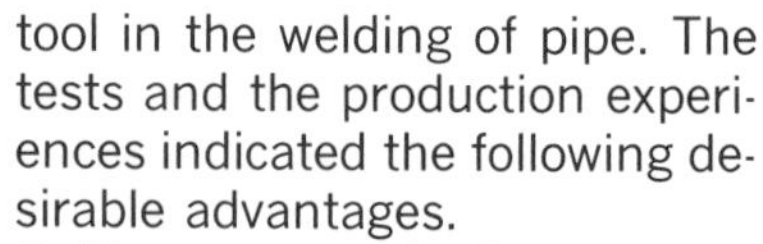

tool in the welding of pipe. The tests and the production experiences indicated the following desirable advantages.

- The process is fast and in many instances faster than other welding processes.
- The elimination of flux and slag reduces the cleaning time considerably.
- Heavier passes can be made, thus cutting down the number of passes per joint.
- Since the electrode is continuous, fewer starts and stops are necessary, and there is no electrode stub loss.
- The process can be used on all pipe sizes and pipe wall thicknesses.
- Weld metal is high quality and meets requirements of most codes.
- Weld appearance is good.
- Good penetration, fusion, and a smooth weld bead can be produced inside the pipe on the underside of the root pass.
- Backup rings may not be required.
- Due to the concentration of heat when welding, distortion and warpage are reduced.

The root pass is thicker and stronger than that produced by the TIG and stick electrode process so that the danger of first pass cracking is reduced.

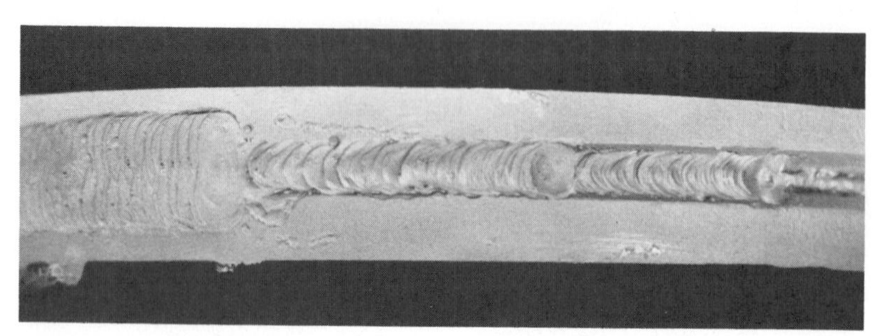

Plumbers and Pipefitters Union, Alton, Ill.

28-2. *Typical butt-welded pipe joint used in building construction. It was welded with the MIG process—travel up. Travel-down procedures are also used.*

28-3. *Welding carbon steel pipe in the shop with the MIG short-arc process. The power source is a constant voltage motor-generator, and the shielding gas is carbon dioxide. Note the portable outfit: the welder, shielding gas, gas regulator, wire feeder, and wire reel are in one compact unit.*

Crane Co.

USE OF EQUIPMENT AND SUPPLIES

Equipment

The equipment needed for the gas metal-arc welding of pipe does not differ in any way from that required for other forms of MIG welding. Review the procedures used in setting up and starting the equipment in Chapter 26, page 834. Your procedure will vary depending on whether the welding machine is a motor-generator, engine-driven generator, or a d.c. rectifier. Refer to

Fig. 28-5 for the equipment in the MIG system. It includes the following:

- D.c. power source, constant voltage type (engine- or motor-driven generator or rectifier)
- Wire-feeding mechanism with controls, spooled electrode wire, and spool support
- Gas shielding system: one or more cylinders of carbon dioxide (CO_2) or a mixture of 75 percent argon and 25 percent carbon dioxide, whichever is used, and a combination pressure-reducing regulator-flowmeter assembly
- Welding gun and cable assembly
- Connecting hoses and cables
- Face helmet, gloves, protective clothing, and hand tools

The engine-driven generator is suitable for outdoor welding applications such as pipelines and bridge and building construction, and in locations where easy access to electric power is not available.

The motor-driven generator is used mostly for inside fabrication. It is preferred over the d.c. rectifier in locations where input power fluctuates frequently. It is more stable than the rectifier.

The d.c. rectifier, Fig. 28-6, is found in fabricating shops where the input power supply fluctuates very little. It is quieter than the other power sources, and it draws current only during welding.

It is an advantage to use a machine with variable slope control and variable inductance control. It should also be equipped with a "hot start" feature, which enables the operator to select a higher voltage for a timed period at the beginning of a pass or at a new start. A hot start ensures good bead contour and the elimination of faults caused by a cold start. All other equipment is the same as that used for welding plate with the gas metal-arc process.

Shielding Gas

Usually carbon dioxide is the shielding gas for welding carbon steel pipe. Other gases and gas mixtures may also be used. One of these is a mixture of 75 percent argon and 25 percent carbon dioxide. When welding stainless steel pipe, this mixture provides a smooth, good-appearing weld. A mixture of carbon dioxide and helium is used in the welding of chrome-moly pipe. Argon is the shielding gas for welding galvanized piping.

Carbon dioxide produces a broad, deep penetration pattern which helps to prevent such defects as lack of penetration and lack of fusion. Bead contour is good, and there is no tendency to undercut. Carbon dioxide also costs less than an inert shielding gas, and less carbon dioxide is needed for full protection than inert gas. Carbon dioxide is contained in the weld area since it is the heaviest of the shielding gases and so has the best resistance to crossdrafts.

Be sure that you have a carbon dioxide regulator and flowmeter. The calibrations of such equipment for other gases are different.

Gas Flow Rate

The gas flow rate is the key to good gas shielding. A rate of 15 to 25 cubic feet per hour is adequate in most indoor welding sit-

28-4. *Welding carbon steel pipe in the shop with the MIG short-arc process. The pipe is being rolled with a pipe positioner. Note the tacked-up joints.*

Crane Co.

uations. When welding on the construction site or inside the shop with the doors open, drafts may disturb the gas shielding. Increasing the gas flow rate by 5 to 10 c.f.h. may have a slightly beneficial effect, but generally it is necessary to erect draft shields made from canvas or like material.

Gas shielding in a windy location may also be improved by using a short contact tip for the torch. Because of the small electrode extensions, it is usual in short-arc welding to select a contact tip that extends approximately $\frac{1}{8}$ inch beyond the end of the gas nozzle. This allows the welder to have a better view of the puddle in most situations. A contact tip which is $\frac{1}{4}$ inch shorter than the usual standard with the normal electrode extension, however, brings the gas nozzle approximately $\frac{1}{4}$ inch closer to the work. There is a definite improvement in wind resistance with the shortened shielding gas column, Fig. 28-7.

Filler Wire

Pipe is usually welded with E70S-3 filler wire. A filler wire diameter of 0.035 inch gives the best results. It enables the welder to control the puddle better than other sizes do and gives him more leeway in torch manipulation. Adjustment of the power supply is easy. An added advantage of the 0.035-inch diameter wire is that it costs somewhat less than 0.030-inch diameter wire. If the welder wishes to use 0.030-inch diameter wire, he may use the same welding procedures, but he must increase the wire-feed speed to obtain the same level of welding current. Filler wires of $\frac{3}{64}$-inch diameter are available, but they are not recommended for all-position short-arc pipe welding. The weld puddle tends to be very large and fluid, and it is hard to manage so that it is difficult to do code quality work.

WELDING OPERATIONS

Edge Preparation

The surface of the pipe end must be smooth, and the edge must be square with the pipe length. A poorly prepared pipe end can be the cause of an unsatisfactory weld.

A bevel of $37\frac{1}{2}$ degrees is widely used in industry for ASME Boiler Code welding, Fig. 28-8. The API (The American Petroleum Institute) code recom-

28-5. *Schematic diagram of the MIG system for welding pipe.*

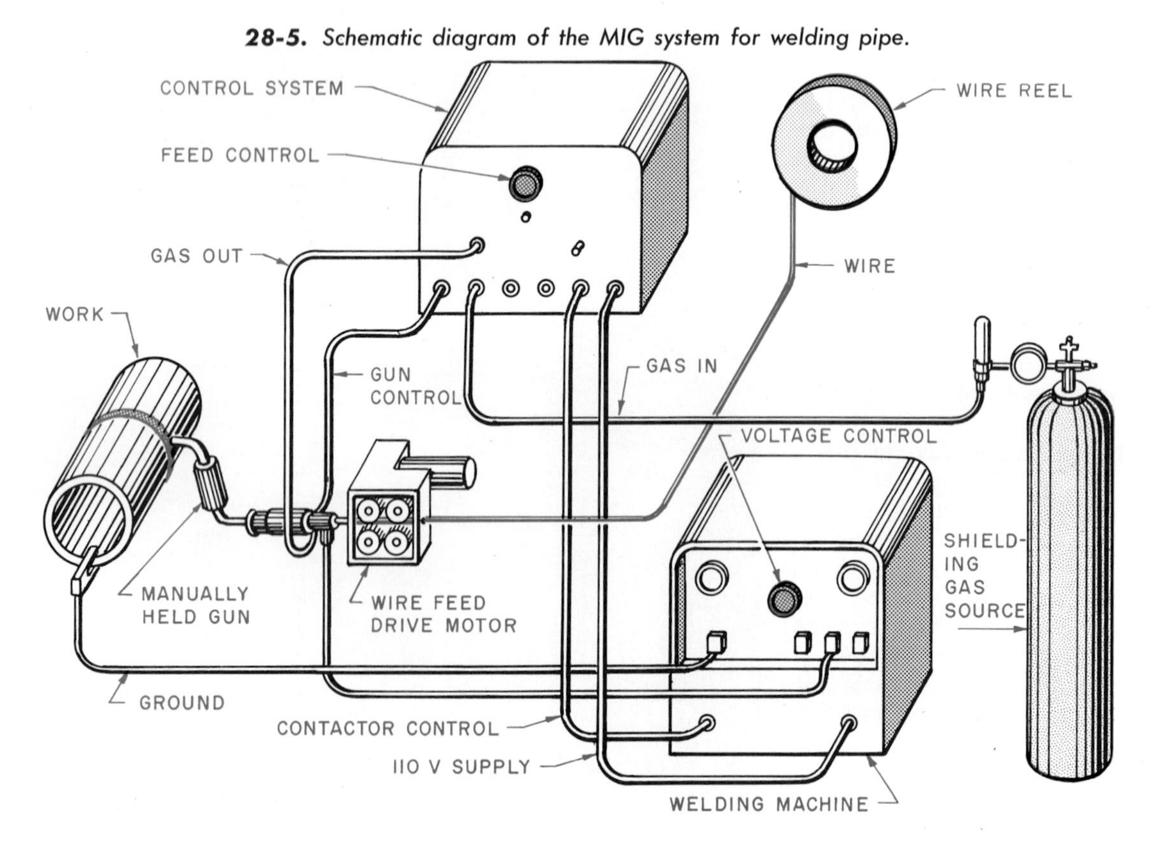

mends a 30-degree bevel, but a $37\frac{1}{2}$ degree bevel is sometimes used, Fig. 28-9.

Beveling may be done by any suitable means such as machining, oxyacetylene cutting, abrasive wheel cutting, or cutting on a lathe. Care must be taken with the root face dimension. If the root face is greater than $\frac{1}{16}$ inch, adequate penetration cannot be obtained. Some welders like to use a sharp V-joint and increase the root opening. You will recall from your past practice that the root bead is the most important bead and that proper edge preparation is necessary to insure good root beads.

It is important that the beveled edges and the pipe within 1 to 3 inches of the joint be free of oil, paint, scale, rust, and any other foreign material which could cause porosity in the weld.

Fitup and Tacking

In setting up the weld joint for work according to the ASME code, make sure that the root opening is no less than $\frac{3}{32}$ inch and no more than $\frac{1}{8}$ inch to insure adequate penetration, Fig. 28-8. On pipeline work according to the API code, it is common to use higher currents than in the shop. A root opening of $\frac{1}{16}$ to $\frac{3}{32}$ inch is used in order to limit burn-through on the inside of the pipe, Fig. 28-9.

Tack welds should be a minimum of $\frac{3}{4}$ inch long and preferably about 1 inch long. For highly critical work welded in the shop, gas tungsten-arc welding is often used for tacking and may be used for the first pass.

Both ends for all tacks should be "feathered" by grinding as shown in Fig. 28-10. There are two important reasons for feathering: (1) to remove any possible defects in the ends of the tack, and (2) to reduce the mass of

28-6. *A d.c. constant voltage rectifier welder equipped with 7 fixed-slope settings. Note the wire drive, wire reel, gun, cylinder of shielding gas, gas regulator and flowmeter.*

Miller Electric Co.

28-7. *Long and short contact tips. Note the effect on the shielding gas column and on the nozzle-to-work distance.*

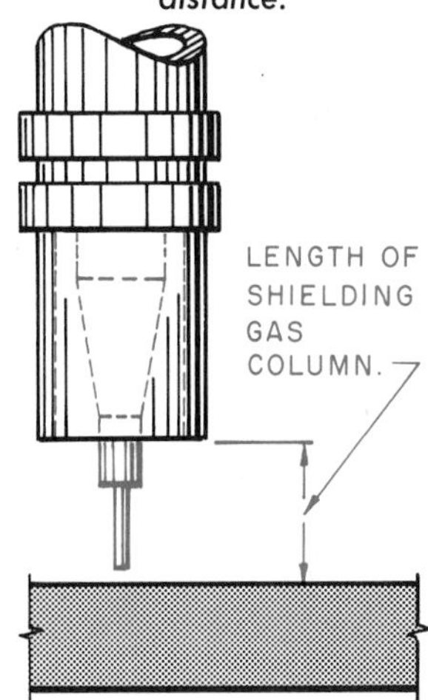

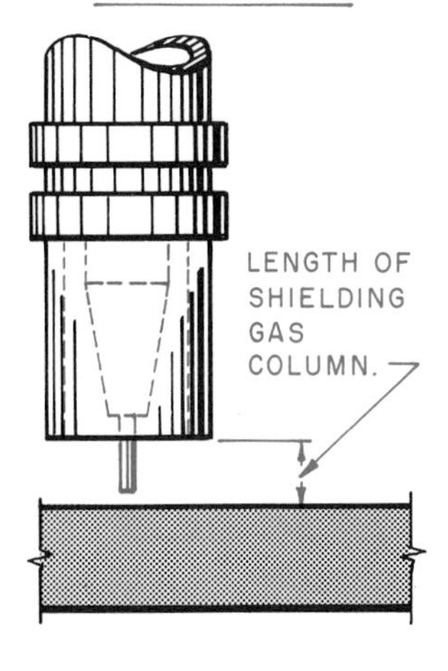

metal at the ends of the tack and, therefore, insure good fusion of the root bead to the tack. Four tacks are all that are necessary for diameters up to twelve inches.

Some welding procedures require a backing ring to be used. In this case the root opening should be not less than $\frac{3}{16}$ inch, and a sharp beveled edge should be prepared as shown in Fig. 28-11. This provides for excellent fusion of the root bead to the backing ring.

Interpass Cleaning

You will recall from your previous welding practice that it is very important to clean between each pass in multipass welding. Although there will not be a heavy flux deposit to remove, there will be a little dust and a black glassy material in spots on the surface of the bead. This material should be removed with a power wire brush. If the materials are not removed, they will become entrapped in the weld, causing porosity, and the arc will be unstable.

PRACTICE JOBS

Instructions for Completing Practice Jobs

Your instructor will assign appropriate practice in the gas metal-arc welding of pipe from the jobs listed in the Course Outline, page 906. Before you begin a job, study the specifications given in the Course Outline. Then turn to the pages indicated in the Text Reference column for that job and study the welding technique described.

Note that the specifications are basic. Specific materials and techniques vary with the job situation, the results desired, and the skill of the welder.

Study Table 28-1, Troubleshooting for the MIG Process, carefully before you begin welding practice. Look up your defects in the table during practice and correct them as instructed before beginning the next job.

Beading Practice: Jobs 28-J1 and J2

Beading practice around the outside of the pipe is for the purpose of getting used to changing the position of the torch while following the contour around the pipe. The direction of travel should be both downhill and uphill.

Practice with various stickout distances. With a wire stickout of about $\frac{1}{4}$ inch, you will notice that the penetration is deep and that you may even burn through the wall of the pipe. With a stickout distance of $\frac{1}{2}$ to $\frac{5}{8}$ inch, you will notice that the penetration is not as deep. A longer stickout will also bridge a gap with less burn-through.

Take a piece of pipe of the size specified in the Course Outline. Hold your gun at an angle of 20 to 25 degrees in the direction of travel, Fig. 28-12, and make a stringer bead. Start at the 12 o'clock position and weld downhill to the 3 o'clock position. Reposition the pipe by turning it counterclockwise and continue this until the bead is joined at the starting point. Move over about $\frac{1}{4}$ to $\frac{3}{8}$ inch and make a second bead. Then make a weave pass between the stringer beads, Fig. 28-13. Weave the gun and pause at the side of each bead to per-

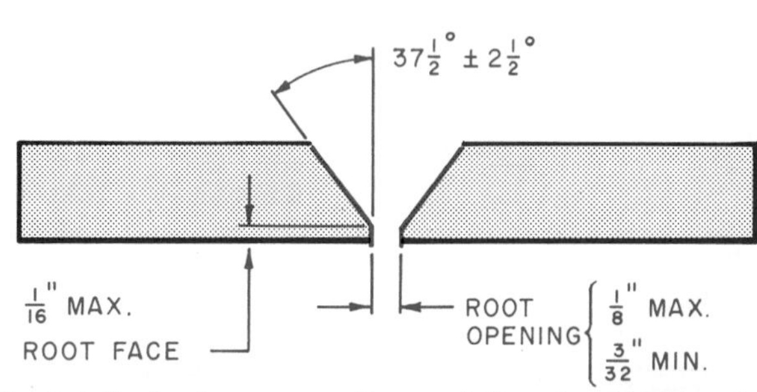

28-8. *Detail of edge preparation and fitup for ASME code work.*

28-9. *Detail of edge preparation and fitup for API code work.*

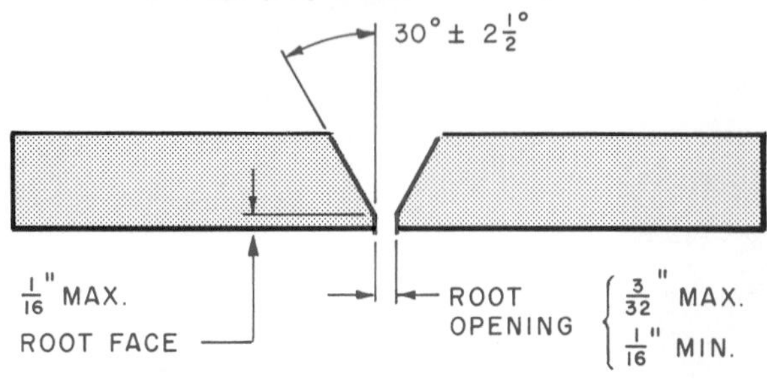

28-10. *Detail of pipe showing "feathered" tack.*

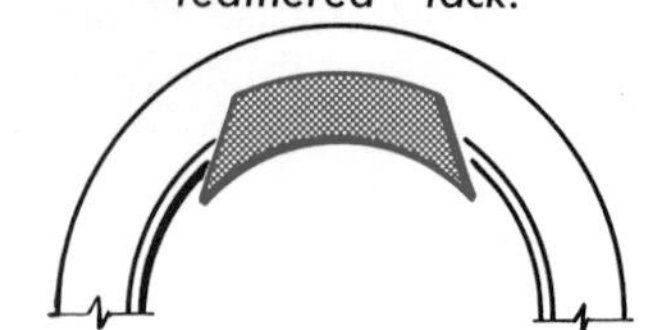

mit fusion to take place. This pass is like the second and third passes in a groove weld or a cap pass.

Repeat this procedure, but weld uphill for both the stringer and weaved passes. When welding uphill, hold the gun nozzle at a 90-degree angle to the pipe surface. Start at about the 2 o'clock position and weld to just beyond the 12 o'clock position. Here again, experiment with various stickout distances and with various voltage and amperage settings. Be very careful that the weld metal does not pile up and run off.

Compare the beads with those shown in Fig. 28-14 and inspect them carefully for the usual defects. Pay particular attention to weld contour and surface porosity.

Butt Joint

The butt joint is the most commonly used pipe joint in welded pipe systems. In the field the welder is called upon to weld this joint in all positions.

Pressure piping usually requires heavy wall pipe. The bevel angle is $37\frac{1}{2}$ degrees, and the direction of travel is uphill. Cross-country and distribution piping usually has thinner walls. The bevel angle is 30 degrees, and the direction of travel is downhill. As in previous pipe practice, the beveling may be done by flame-cutting or machine cutting.

Job 28-J3 will be the welding of a butt joint in pipe in the horizontal roll position, travel down, from the 12 o'clock to the 3 o'clock position.

HORIZONTAL ROLL POSITION: JOB 28-J3. Root Pass. Select two lengths of pipe of the size and weight specified in the Course Outline. Set up the two nipples in the vertical position as shown in Fig. 28-15 and tack weld at four equally spaced places (the 12, 6, 3, and 9 o'clock positions in that order). Make sure that the pipe remains equally spaced all the way around. Also make sure

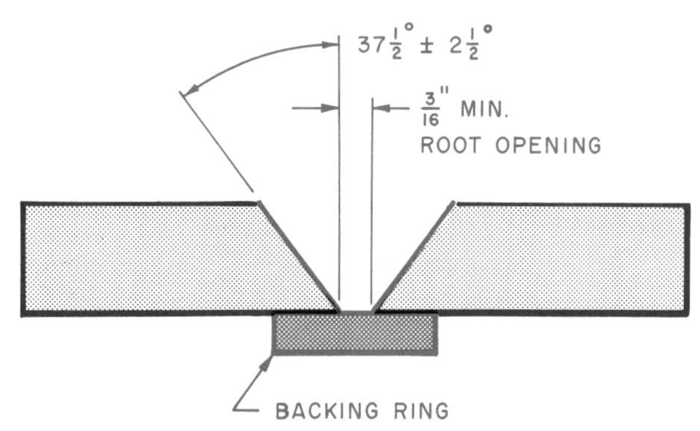

28-11. *Edge preparation and fitup if backing ring is used.*

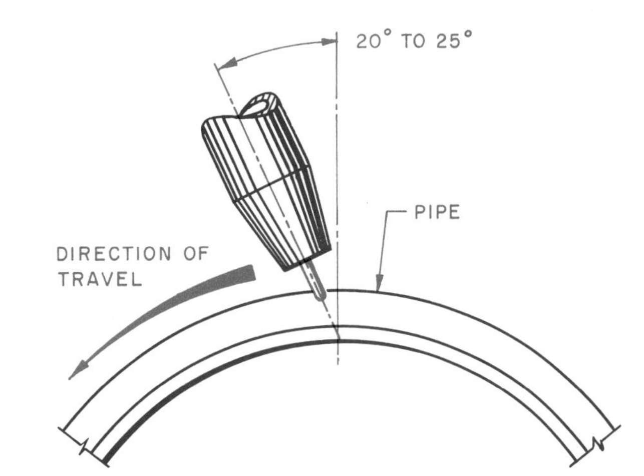

28-12. *Position of gun for beading or butt welding around pipe the direction of travel is down, and the pipe is rolled.*

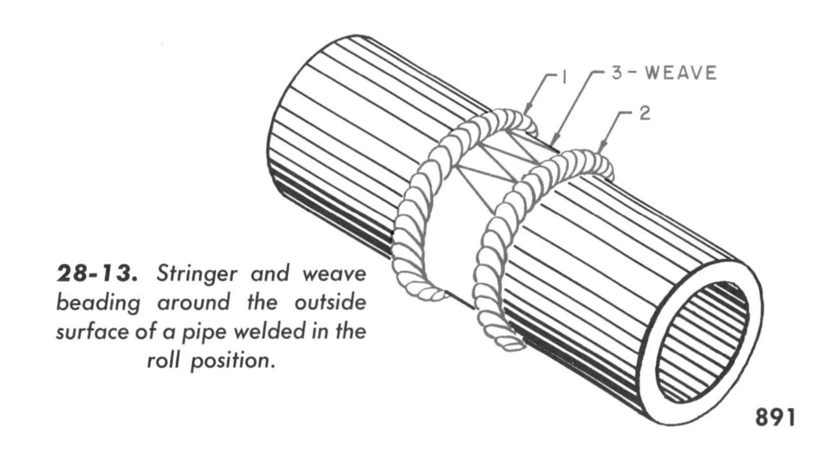

28-13. *Stringer and weave beading around the outside surface of a pipe welded in the roll position.*

Table 28-1

Troubleshooting with the MIG process.

Weld defects	Causes	Remedies
1. Convex bead	Arc voltage too low	Raise voltage
	Oscillation too narrow	Widen oscillation
2. Gross porosity	Breeze blowing shielding gas away	Erect wide shields
3. Scattered porosity	Failure to remove large islands of black "glass"	Use power wire brush to remove "glass"
	Oil or other foreign material on pipe	Clean prior to welding
4. Wagon tracks	Failure to remove "glass" from edges of previous pass	Remove "glass"
	Convex root or fill pass	Raise voltage or widen oscillation
5. Undercutting in overhead position	Voltage too high	Lower voltage
	Insufficient dwell at edge of bead	Increase dwell
	Welding current too high	Reduce wire feed speed
6. Cold lapping	Welding current and thus disposition rate too high	Lower wire feed speed
	Welding speed too low	Increase forward speed
	Arc not on leading edge of puddle	Keep arc ahead of puddle on base metal
7. Burn-through in root	Root opening too wide	Reduce root opening
		Oscillate torch
		Increase backhand angle
8. Lack of fusion in root	Root opening too narrow	Increase root opening
	Arc voltage too low	Increase voltage
9. Lack of penetration or "suckback" in overhead position	Travel speed too slow	Speed up travel
	Root opening too wide	Decrease root opening
10. Lack of fusion to ends of tacks	Root opening too narrow	Increase root opening
	Tacks not adequately feathered	Feather tacks properly
11. Unstable arc	"Glass" left on previous pass	Remove "glass"
	Voltage too high or too low	Use correct voltage
	Contact tip clogged	Renew contact tip
12. Erratic start	Hot start voltage too high	Use a hot start voltage of 2 volts over welding voltage for about 1 second

Linde Division, Union Carbide Corp.

that tack welds have good penetration and are not too heavy. Feather them by grinding, filing, or chiseling until they are just thick enough to hold the two pipe nipples together as the joint is being welded. See Fig. 28-10, page 890.

Place the pipe in the horizontal position and weld the root pass with the stringer bead technique. The torch should always be on the weld and the motion is U-shaped. Start at the 12 o'clock position and weld downhill to the 3 o'clock position. At this point stop and rotate the pipe in a counterclockwise direction until the weld crater is again in the 12 o'clock position. Then repeat the welding procedure. The electrode wire stickout should be between $\frac{1}{4}$ and $\frac{3}{8}$ inch. At certain times it may be as long as $\frac{5}{8}$ inch because of the 45-degree angle. The stringer bead should be no more than $\frac{1}{8}$ inch thick. The root opening often varies. If the opening is wider than $\frac{3}{32}$ inch, it may be necessary to weave the torch slightly, Fig. 28-16.

Be careful when welding over a tack. The tack must be completely fused and become a part of the weld bead. Many welders have failed qualification tests because of their failure to completely remelt the tack. As you approach the leading edge of the tack, position the gun about 20 to 25 degrees from the perpendicular, Fig. 28-17. As you leave the tack, position the gun about 45 to 55 degrees from the perpendicular, Fig. 28-18.

Each time the weld is restarted, it is necessary to "tie-in" with the previously deposited bead carefully. As you move up into the puddle, let the metal wash up against the sides of the groove, and then move downhill. Also be sure to reposition the pipe when stopping the weld. A stop in welding a root pass in an open joint can cause shrinkage, cracks, cavities, and cratering. The weld should be carried slowly $\frac{1}{4}$ to $\frac{3}{8}$ inch to the pipe bevel wall, Fig. 28-19.

Inspect the weld carefully for penetration through the back side (Fig. 28-20) and fusion to the side walls. Note the absence of undercut along the edges of the weld and along the side walls of the pipe. The surface of the weld should be flat, Fig. 28-21 (Page 896). Brush the weld before applying the next pass.

Filler Pass. Start the second pass about 2 inches past the original starting point for the root pass. Begin welding by moving your gun from side to side, Fig. 28-22. Pause briefly at each edge of the root pass in order to permit the weld metal to fill up the groove and obtain proper fusion. Do not depend on the weld metal to wash up on the sides of the groove. This may result in a cold lap. You must direct the filler wire where you want complete fusion without undercut. As with the root pass, weld from the 12 to the 3 o'clock position before rotating the pipe. Repeat your procedures for making good tie-ins and stops. Inspect the pass and clean it.

Cover Pass. Make the cover pass like the filler pass. Be sure to pause briefly at each edge of the filler pass to obtain proper fusion. The bevel edges on each

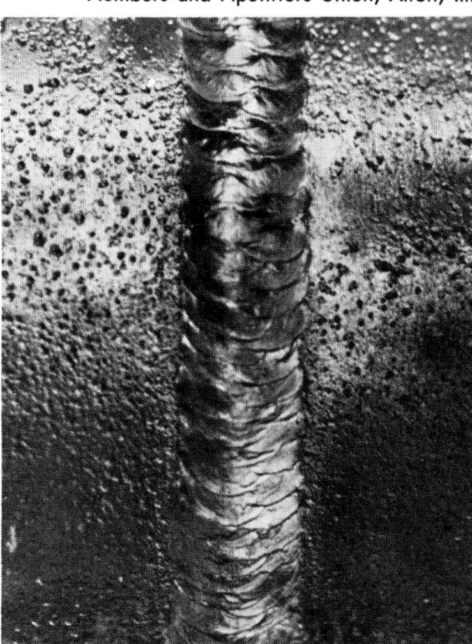

28-14. *Sample of a weaved finish pass on a butt joint in pipe welded in the roll position.*

Plumbers and Pipefitters Union, Alton, Ill.

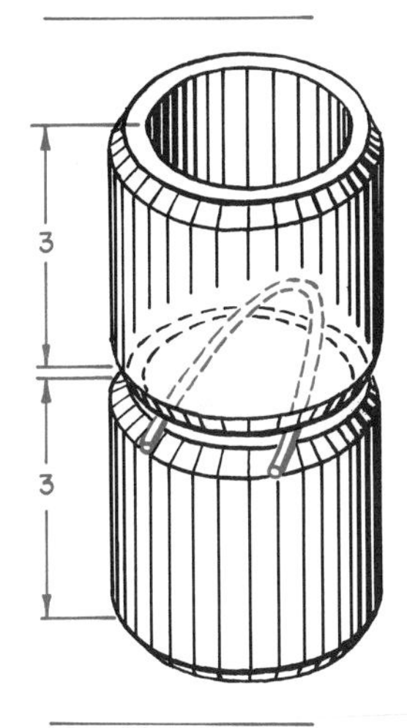

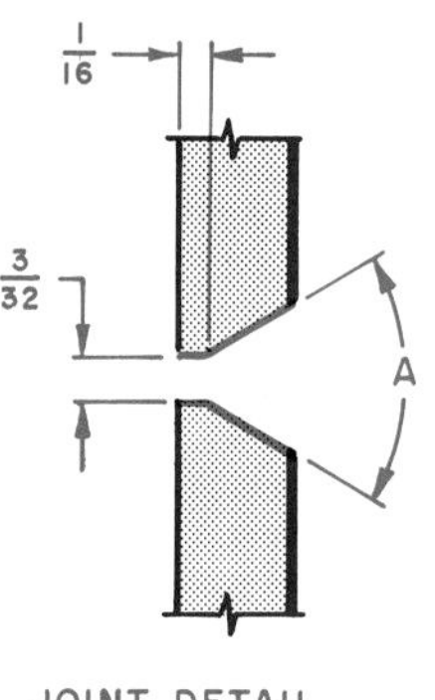

28-15. *Setting up and spacing the pipe before tack welding. The diameter of the spacing wire determines the pipe spacing.*

side should be your guide to determine the final width of the cover pass, Fig. 28-23. The bead should be about $\frac{1}{16}$ inch at the crown and should taper out to the edges of the bead.

Inspect the cover pass for smooth tie-in, lack of undercut, good fusion, good appearance, and contour, Fig. 28-21.

Inspection and Testing. Weld a butt joint, following the same procedures that you have used in completing your practice. Inspect the weld for visual defects and cut the usual face- and root-bend test coupons from four positions on the pipe: top, bottom, and each side. After preparing the test coupons, bend them on the standard test jig. The welds should show no separations of any kind.

VERTICAL FIXED POSITION: JOBS 28-J4 AND J5. The pipe is fastened in the vertical fixed position, and the weld is horizontal. This position is found mostly in power piping and construction installations. Very little cross-country transmission line piping is welded in this position.

Select two pieces of pipe 4 to 6 inches long of the size and weight specified in the Course Outline. Bevel the edges to 30 or $37\frac{1}{2}$ degrees as directed by your instructor and leave a $\frac{1}{16}$-inch root face on each bevel. Refer to Figs. 28-8 and 28-9, page 890. Tack weld the two nipples together as shown in Fig. 28-15, page 893. Place the tack-welded nipples in a welding fixture in the vertical position.

Root Pass. Weld the root pass with the stringer bead technique. Start welding at a tack. Hold the gun nozzle at a 10- to 15-degree angle from a point perpendicular to the center of the pipe and pull your gun in the direction of travel, Fig. 28-24A. It is also important to lower your gun 5 degrees from the 90-degree position, Fig. 28-24B. Carry your wire on the leading edge of the puddle to get complete penetration. Make sure that fusion is taking place along the root edges of each pipe bevel. Also make sure that the weld does not sag along

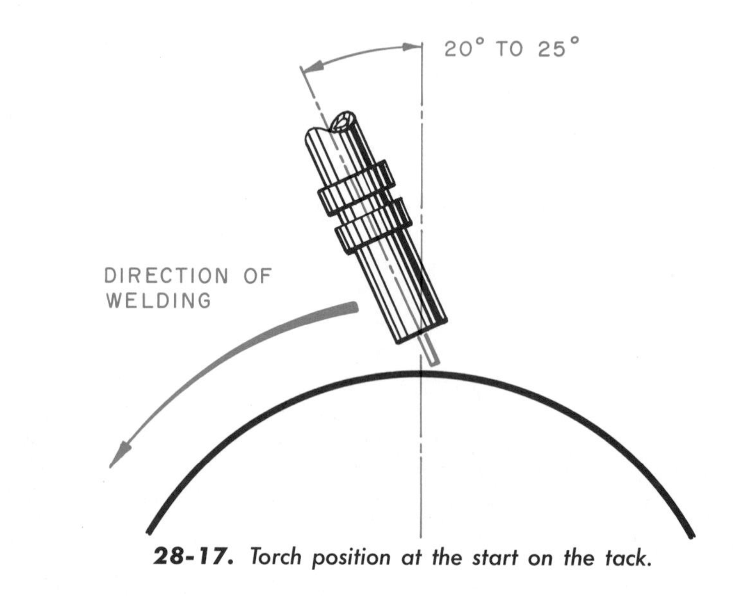

28-17. *Torch position at the start on the tack.*

28-18. *Torch position after leaving the tack at the beginning of the root pass.*

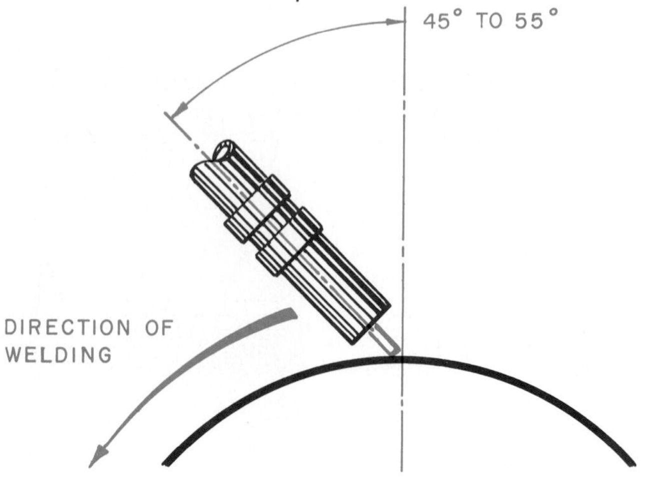

28-16. *Suggested pattern for torch oscillation for root pass welding.*

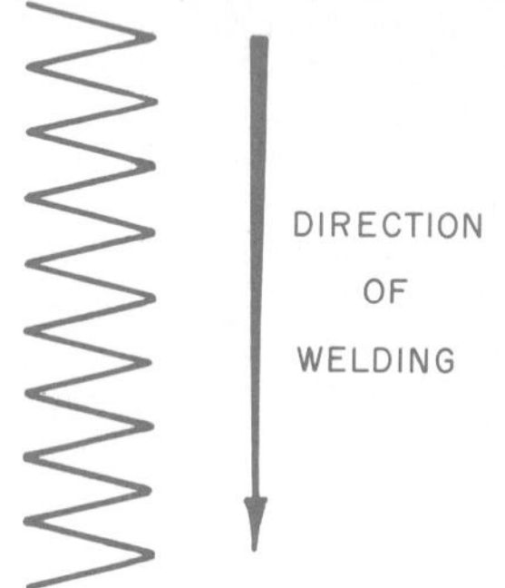

the bottom bevel so that it causes overlap and lack of fusion on the bottom edge. Also be careful not to undercut the top bevel. Penetration through the back side is necessary, Figs. 28-25 and 28-26 (Page 898).

Torch oscillation, if necessary, is best carried out according to the pattern illustrated in Fig. 28-27. When completing a section of the root pass and approaching a feathered tack, change the torch angle gradually over a distance of $\frac{1}{4}$ inch before the tack from the 20-degree trailing angle to a 5-degree leading angle. This gun position ensures a good tie-in at the tack, Fig. 28-28.

Filler Passes. The second and intermediate passes may be made with the stringer bead technique shown in Fig. 28-29 or the weaved-bead technique shown in Fig. 28-30 (Page 899). The choice of technique depends upon the requirements for the particular job. If there is no preference, then the welder should use a technique that he can handle with good results. Starts and stops should be outside the area of previous starts and stops. The bead sequence that may be used for Schedule 40 pipe is shown in Fig. 28-29B, and the sequence for Schedule 80 pipe is shown in Fig. 28-29C.

In making a circumferential weld around the pipe, you will have to make a number of starts and stops as required by the change of position. You are again cautioned to make your stops and new starts as instructed in Fig. 28-19, below. Tie-ins require that the electrode be started back in the heel of the puddle and that the weld metal be washed up against the sides of the groove until the tie-in is made.

Cover Passes. Use the stringer bead technique and about the same electrode position as described for the root pass. Make sure that the arc is directed toward the surface that is being welded. Proceed at a uniform rate of travel so that good fusion and penetration can take place with the bevel surface and the previous welds without undercut and with good bead formation. Each layer of weld metal should not be thicker than $\frac{1}{8}$ inch. Brush each pass so that it is clean and free from foreign materials.

Inspection and Testing. Inspect each pass as you complete it. The final layer of stringer beads should show good fusion with each other and be equally spaced, Figs. 28-31 and 28-32 (Page 899).

Use the same testing procedure as that used for the roll position.

HORIZONTAL FIXED POSITION, DOWNHILL TRAVEL: JOB 28-J6. In this position, commonly referred to as the *bell-hole position,* the pipe is in the horizontal position and is not turned as the welding progresses. The bell-hole position is encountered in all types of pipe installations. Travel may be uphill or downhill. The downhill welding technique is employed in cross-country transmission line piping, and the uphill welding technique is employed in pressure and power piping and in general construction. Either technique pro-

28-19. *Recommended technique for eliminating crater cracks when welding is stopped in an open joint.*

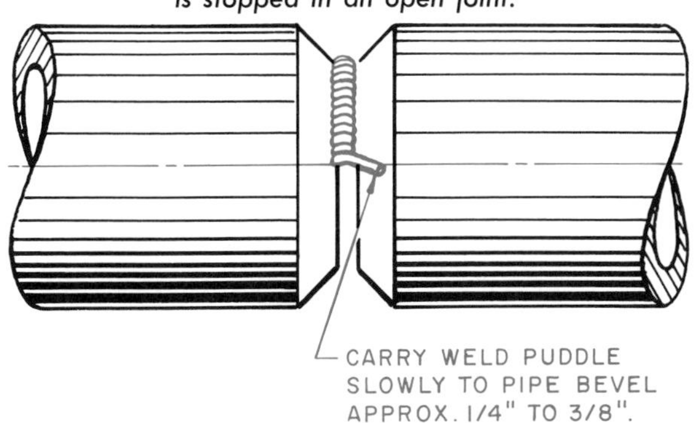
CARRY WELD PUDDLE
SLOWLY TO PIPE BEVEL
APPROX. 1/4" TO 3/8".

28-20. *Root penetration through the back side of an open butt joint welded with the downhill technique.*
Plumbers and Pipefitters Union, Alton, Ill.

duces sound welds, but the downhill technique is faster. When welding downhill, some welders find it difficult to keep the weld metal from running ahead of the weld pool. You should weld several joints in this position and practice various welding techniques. Do Job 28-J6 with the downhill technique.

Select two pieces of pipe 4 to 6 inches long of the size and weight specified in the Course Outline. Bevel the edges 30 or $37\frac{1}{2}$ degrees as directed by your instructor and leave a $\frac{1}{16}$-inch root face on each bevel. Refer to Figs. 28-8 and 28-9, page 890. Tack weld the two nipples together as shown in Fig. 28-15, page 893. Place the tack-welded pipe in a welding fixture in the horizontal position.

Special care must be taken when welding in the horizontal fixed position. Before making the tack welds and welding the pipe joint, check the stickout carefully ($\frac{1}{4}$ to $\frac{3}{8}$ inch) and regulate the heat until the arc is smooth and active. Make sure that the pipe has a root opening of $\frac{3}{32}$ inch all around. This root opening is most important and must be maintained. Feather the tack welds so that they do not become an obstruction when you pass over them with the root pass. Tack welds that are too thick cause a lack of penetration and fusion so that the finished weld is not X-ray quality.

Root Pass. The root pass should be started at the 11 or 1 o'clock position, continued across the top of the pipe, and carried down past the 6 o'clock position to the 5 or 7 o'clock position, depending on the side being welded.

A lagging gun angle of approximately 45 to 55 degrees to the vertical should be maintained as the weld progresses past the feathered end of the tack, Fig. 28-18, page 894. It may be necessary to oscillate the gun narrowly during a part or the whole of the weld, Fig. 28-16, page 894. Side movement is necessary to make sure that fusion is taking place on the root face of each pipe section.

When you reach that part of the pipe between the 4 and 6 o'clock positions, you may need to decrease the lagging angle of the gun to approximately 20 to 25 degrees. A uniform travel speed is important. If the travel speed is too slow, the base metal becomes highly fluid and sags so that there is a lack of penetration inside the pipe. You are again cautioned to keep the thickness of each pass to about $\frac{1}{8}$ inch.

Use care when stopping a weld and restarting a weld. If for any reason you find it necessary to stop the root pass, there is

28-22. *Detail of a joint with the root pass in place, showing the correct width of the torch movement for a filler pass.*

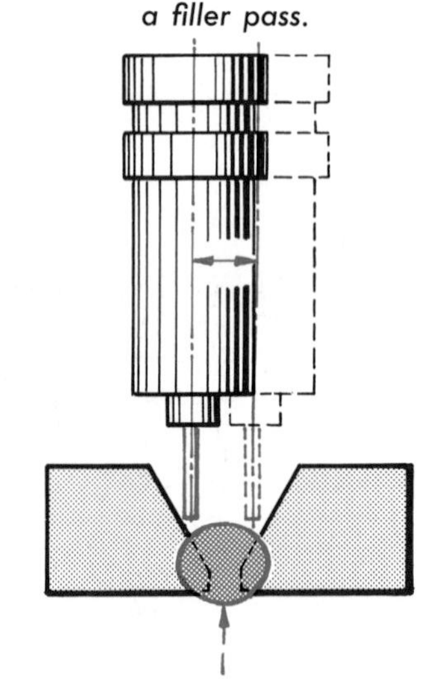

28-23. *Detail of a joint with the final fill pass in place, showing its correct relationship to the outside diameter of the pipe. The correct width of the torch movement for the cover pass is also shown.*

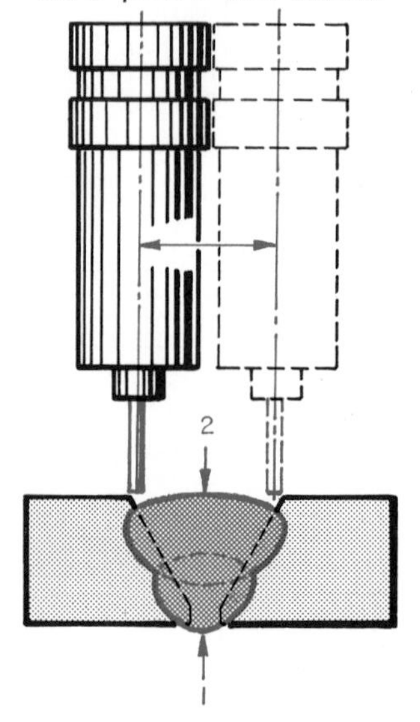

28-21. *Typical face appearance of root and cover passes in an open butt joint in pipe welded with the downhill technique. Note the flat face.*

Plumbers and Pipefitters Union, Alton, Ill.

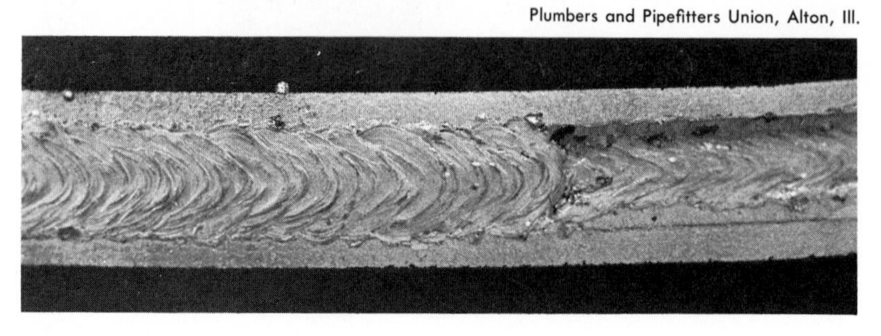

always the danger that cracks will form in the crater of the weld pool. To eliminate these cracks, start the weld just above the crater. This preheats the bead so that the crater is hot when you reach it, Fig. 28-33 (Page 900). This insures a good tie-in with the weld, and there is no danger of the cracks remaining in the deposit.

After the first half of the root pass is completed, inspect it thoroughly and clean it. Return to the top of the pipe and weld the other half of the pipe to the 6 o'clock position with the same procedure as for the first half. Make the tie-in at the top and at the bottom of the pipe very carefully. These are very critical points and often cause trouble. Compare Figs. 28-34 and 28-35.

The downhill technique of welding is preferred for the root pass, but if the fitup is close, the uphill technique may be employed. Follow the procedure previously described but hold the gun at a 90-degree angle to the groove, Fig. 28-36. The welding technique is like that for welding pipe with the shielded metal-arc process and plate with the gas metal-arc process.

Filler Passes. You are now ready to weld the filler passes. Recheck your electrode stickout and heat setting. The current may have to be increased slightly.

Start welding at the top, outside of the area of the previous starting point, and travel downhill. Stop at the bottom, outside of the area of the previous stopping point. The second pass is made with a slight weave by moving the gun from side to side and pausing briefly at each edge of the underneath pass, Fig. 28-37. Do not depend on the metal to wash up on the sides of the groove. In order to avoid cold lap and to secure complete fusion, the electrode wire must be directed over the surface to be joined. Clean and inspect the weld.

To weld the second pass on the other side of the pipe, start at the top at the end of the previous weld and tie into the end of the previous weld at the bottom.

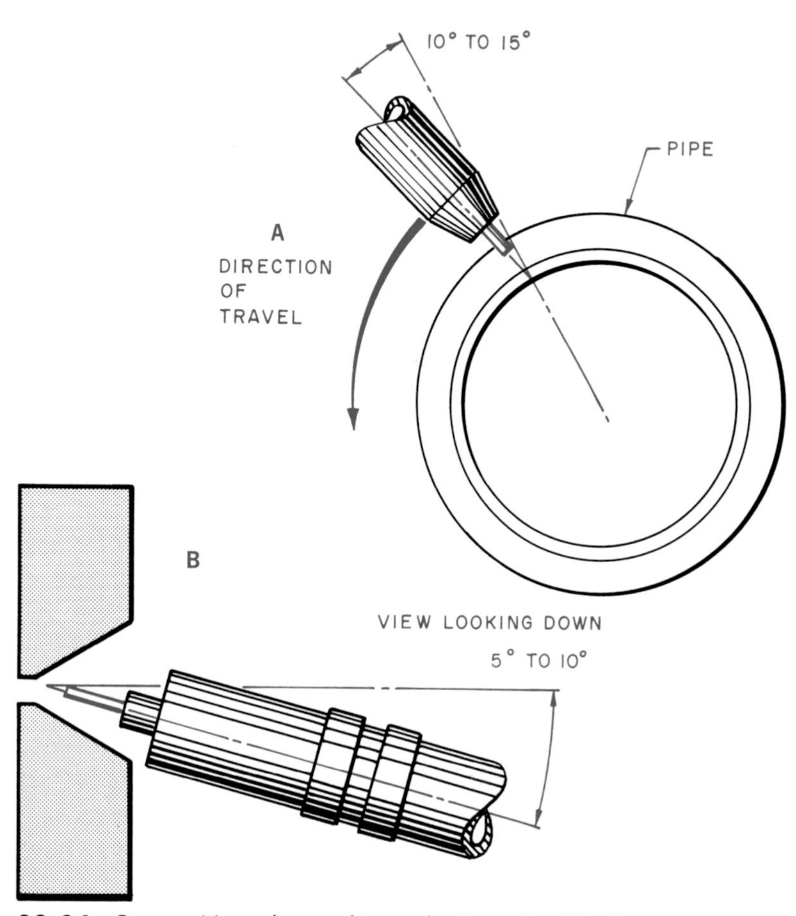

28-24. *Gun position when making a horizontal weld. The pipe is in the vertical fixed position.*

28-25. *Closeup view of root pass penetration on the inside of a pipe obtained when making a horizontal weld on pipe in the vertical position.*

Plumbers and Pipefitters Union, Alton, Ill.

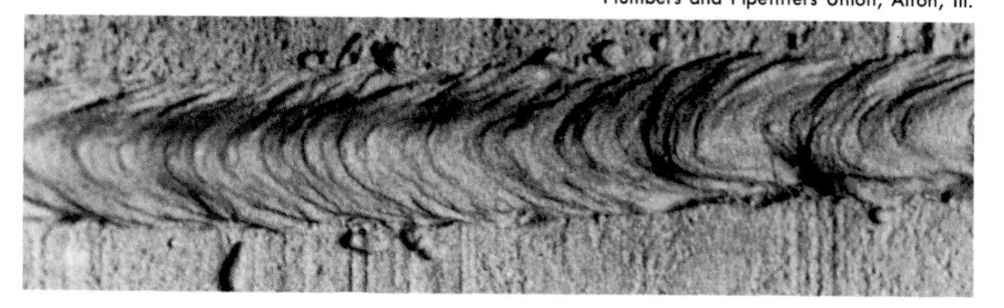

Manipulate the gun as instructed for the first half of the pass. All tie-ins should be staggered so that they are not all made at the same point, one over another. The proper technique is shown in Fig. 28-38. Clean and inspect the weld, Fig. 28-39.

Make a third filler pass like the second pass. Clean and inspect.

Cover Pass. The cover pass is welded like the second and third passes. Use the edge of each bevel as a guideline to determine the width of the weld. To prevent undercut, hesitate at the sides to permit the weld pool to fill up. The bead should be about $\frac{1}{16}$ inch high at the crown, and it should taper out to the edges of the bead.

It is important to keep the arc ahead of the puddle when doing filler and cover passes downhill. Any attempt to slow down the travel speed in order to deposit more weld metal allows the molten pool to run ahead of the arc and causes cold lapping or lack of fusion. Keep the metal thin. The maximum pass width that can be handled with any degree of success is $\frac{1}{2}$ inch. Split passes are necessary beyond this width.

Practice starts and stops so that you will develop the skill to perform the technique with a minimum of problems. You will find this somewhat difficult when making filler and cover passes. Avoid the tendency to add too much weld deposit, because it results in poor fusion and very poor appearance.

Inspection. Inspect your completed welds carefully. Keep in mind the weld characteristics that are indicative of sound welding. Study Figs. 28-40 and 28-41 (Page 902) which show the penetration inside the pipe. Fig-

Worthington Corp.

28-26. *Underside of a root pass in steel pipe welded with the MIG process and a short-circuiting arc in the vertical fixed position. Note the absence of obstruction to flow.*

DIRECTION OF WELDING

28-27. *Suggested pattern for torch manipulation in making the root pass for a horizontal weld.*

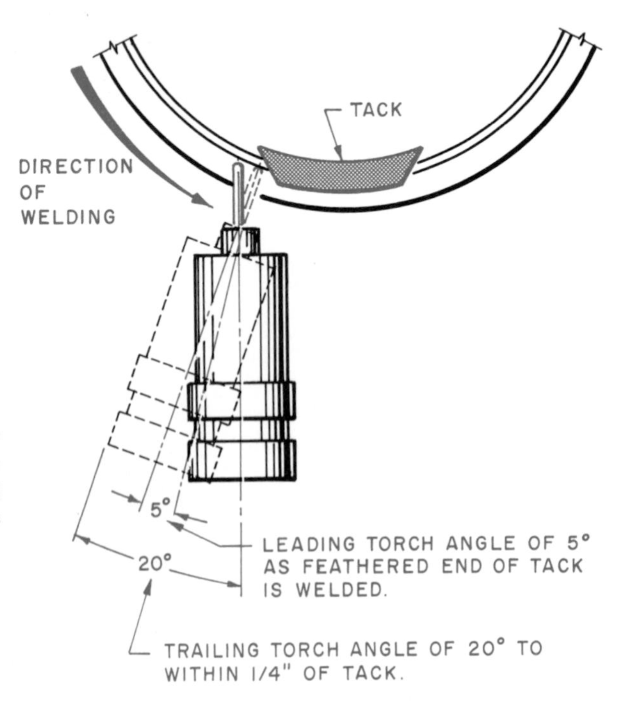
TACK
DIRECTION OF WELDING
5°
20°
LEADING TORCH ANGLE OF 5° AS FEATHERED END OF TACK IS WELDED.
TRAILING TORCH ANGLE OF 20° TO WITHIN 1/4" OF TACK.

28-28. *Top view of a pipe in the vertical fixed position, showing rotation of the torch from a trailing angle of 20 degrees to a leading angle of 5 degrees as a tack weld is approached when welding a root pass.*

ure 28-42 shows the desirable face appearance of a root pass in a weldment. Figures 28-43 and 28-44 illustrate the faces of the root, filler, and cover passes when welding in the vertical downhill direction.

HORIZONTAL FIXED POSITION, UPHILL TRAVEL: JOBS 28-J7 AND J8. After you have mastered the downhill welding technique, practice Jobs 28-J7 and J8 with the downhill welding technique for the root pass and the uphill welding technique for all other passes.

Select the pipe nipples specified in the Course Outline. Prepare them as previously instructed, tack them, and set them up in the welding fixture.

28-29. *Weld sequence for various pipe wall thicknesses welded with the MIG short arc process. The pipe is in the vertical position for horizontal groove welding.*

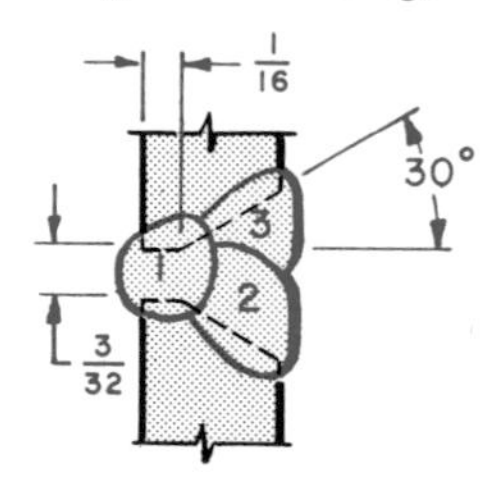

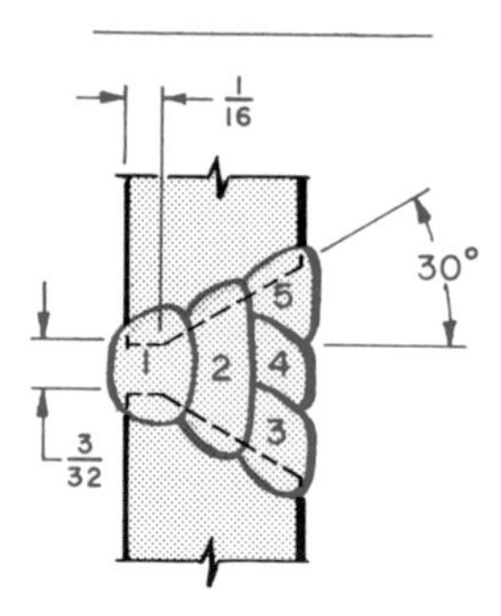

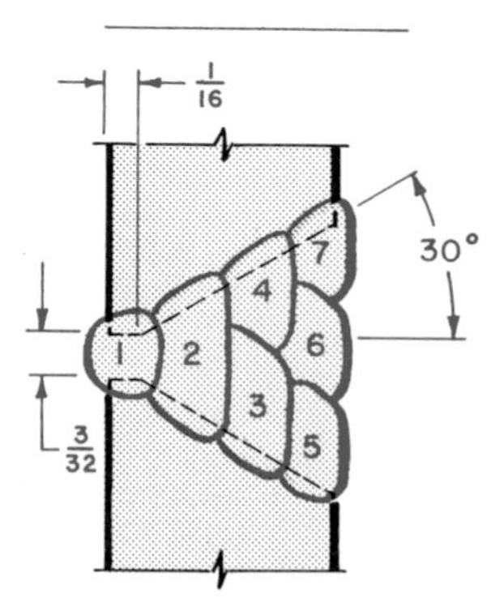

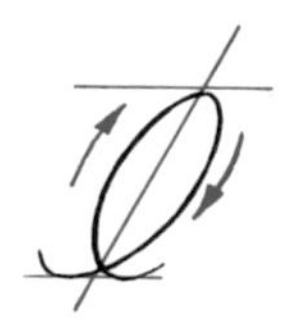

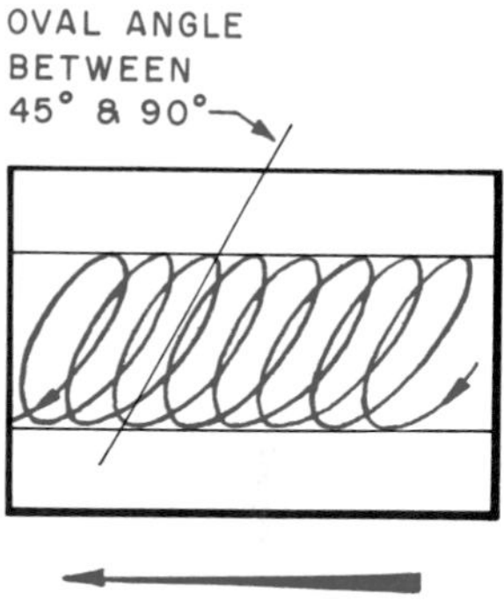

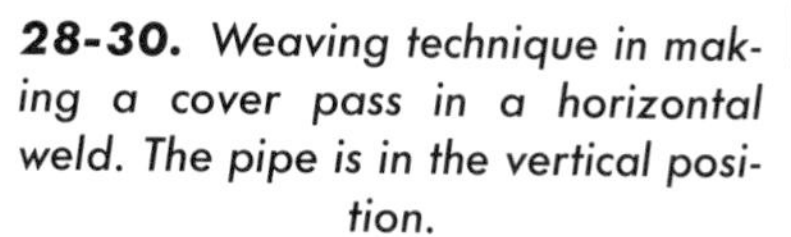

28-30. *Weaving technique in making a cover pass in a horizontal weld. The pipe is in the vertical position.*

Plumbers and Pipefitters Union, Alton, Ill.

28-31. *Typical appearance of stringer beads welded in the proper sequence in a horizontal groove weld with the pipe in the vertical position. The weld was made with the short-arc MIG process.*

28-32. *Closeup view of a completed sequence in a weld like that shown in Fig. 28-31.*

Plumbers and Pipefitters Union, Alton, Ill.

Root Pass. Weld the root pass downhill with the welding technique practiced in Job 28-J6. Clean and inspect the weld for defects.

Filler and Cover Passes. Weld the filler and cover passes with the uphill welding technique. Reduce the wire-feed speed to reduce the current. Adjust the voltage until you have a smooth arc. Start at the bottom at the 5 or 7 o'clock position and weld across the 6 o'clock position up toward the 11 or 1 o'clock posi-

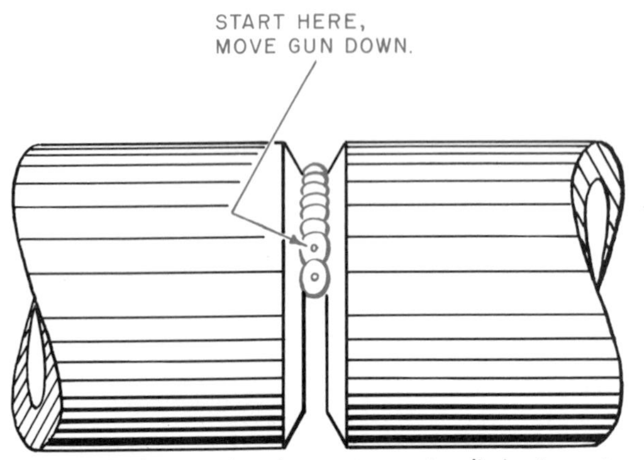
START HERE,
MOVE GUN DOWN.

28-33. *Method of restarting the root pass to eliminate crater cracks.*

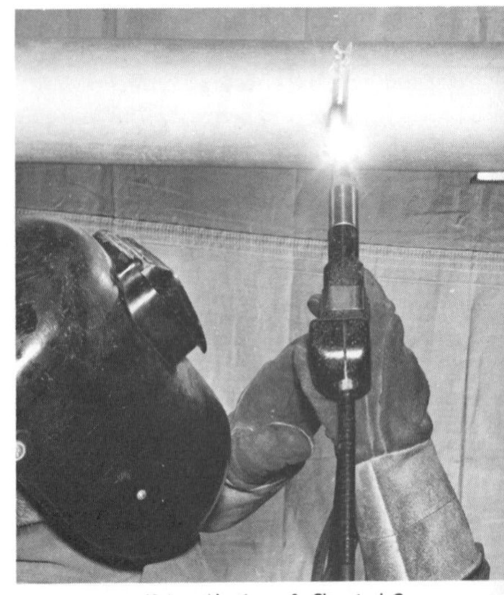

Kaiser Aluminum & Chemical Corp.

28-36A. *MIG welding pipe in the horizontal fixed position. Care must be taken to follow the contour of the pipe. The welder is just leaving the bottom.*

28-34. *The underside of the root pass on the inside of a pipe welded in the horizontal fixed position. Note that the welder failed to make the proper tie-in at the bottom of the pipe. This would cause a test or service failure.*

Plumbers and Pipefitters Union, Alton, Ill.

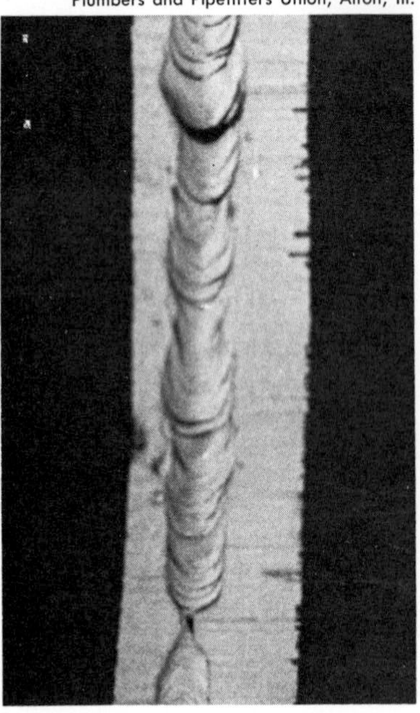

28-35. *Proper penetration on the inside of a pipe welded with downhill travel.*

Plumbers and Pipefitters Union, Alton, Ill.

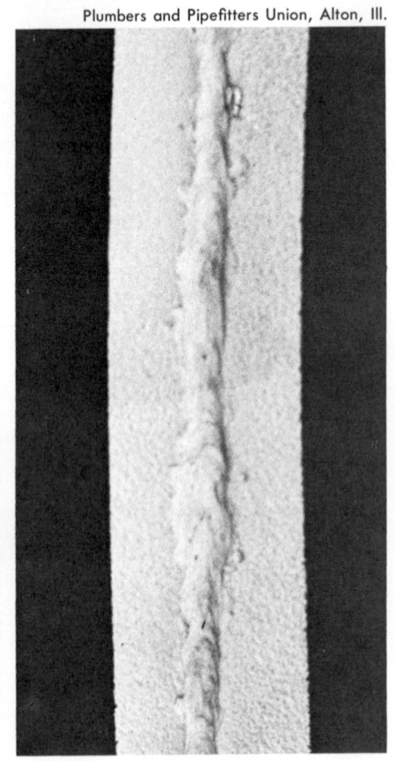

28-36B. *The torch is in the 3 o'clock position. Some welders find this a difficult position. It involves a mixing of the overhead and vertical positions.*

Kaiser Aluminum & Chemical Corp.

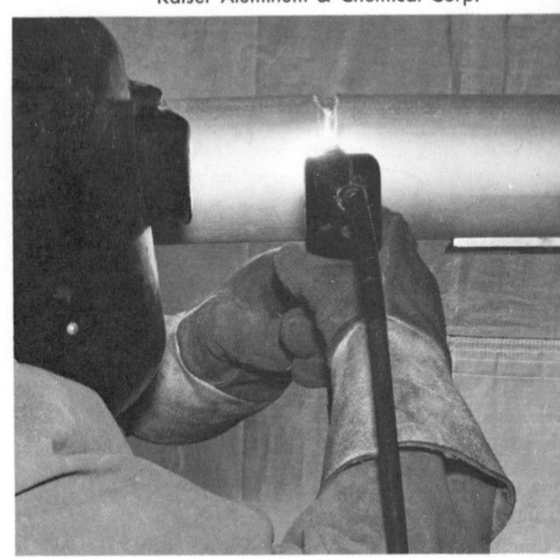

tion at the top of the pipe, depending on the side of the pipe you are welding from. Hold the gun as illustrated in Fig. 28-36. Weave the gun from side to side and pause at each edge of the previous weld. This gives the weld metal a chance to fuse to the metal surface and allows the edges to fill up so that undercut will not take place. Weld passes should not be more than $\frac{1}{8}$ inch thick. The filler passes should have a flat face. The cover pass may have a slight crown about $\frac{1}{16}$ inch high. You are again cautioned to be very careful when making stops and tie-ins. Use the technique shown in Fig. 28-45. Clean and inspect for welding defects. Compare your welds with those shown in Fig. 28-46 (Page 904).

Whether the direction of travel is up or down makes some difference in the face appearance of the welds. Generally, uphill travel produces a weld with close ripples. Compare the appearance of the welds shown in Fig. 28-47 (downhill travel) and Fig. 28-48 (Page 904) (uphill travel).

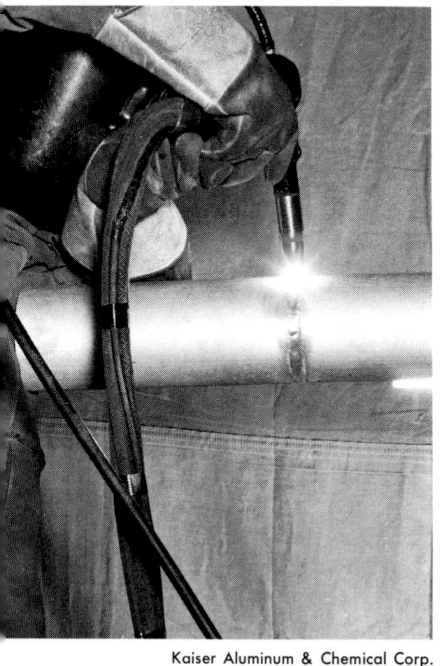

Kaiser Aluminum & Chemical Corp.

28-36C. *Welder has reached the top of the pipe.*

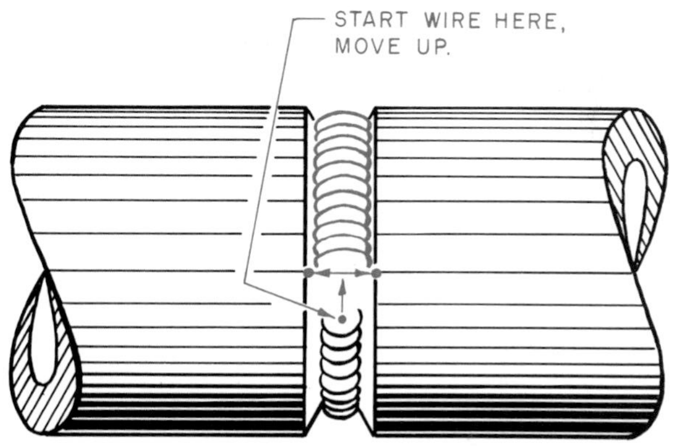
START WIRE HERE,
MOVE UP.

28-38. *Welding technique for making tie-ins when the direction of travel is downhill.*

28-37. *Gun movement when welding filler passes in pipe in the horizontal fixed position when the direction of travel is downhill.*

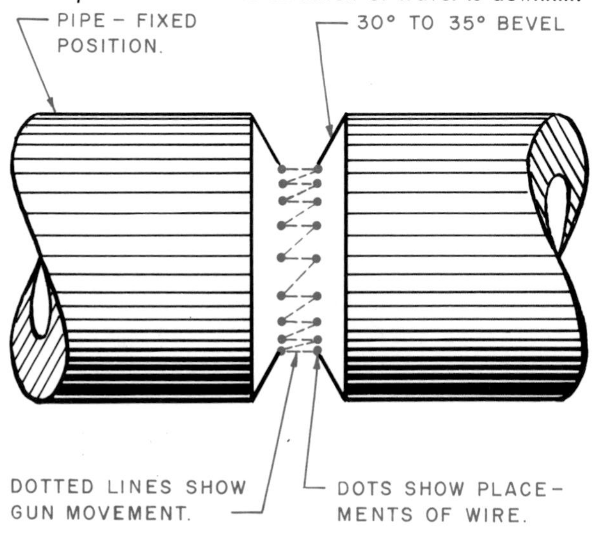
PIPE – FIXED
POSITION.
30° TO 35° BEVEL
DOTTED LINES SHOW
GUN MOVEMENT.
DOTS SHOW PLACE-
MENTS OF WIRE.

28-39. *Typical face appearance of a filler pass and a root pass when the direction of travel is downhill.*

Plumbers and Pipefitters Union, Alton, Ill.

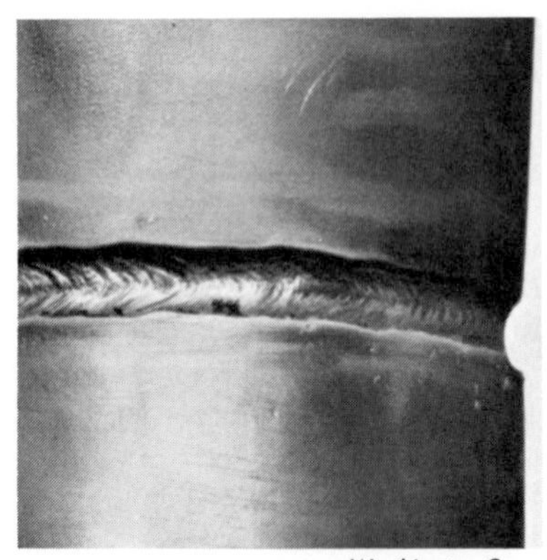

Worthington Corp.

28-40. *Underside of a root pass in steel pipe. The joint shows complete penetration. The underside is free of notches, icicles, crevices, and other obstructions to flow. Reinforcement is between 1/32 and 3/32 inch.*

28-41. *Underside of a root pass deposited in the pressure vessel shown in Fig. 28-42. The joint shows complete penetration free of all obstruction. Reinforcement is between 1/32 and 3/32 inch.*

Worthington Corp.

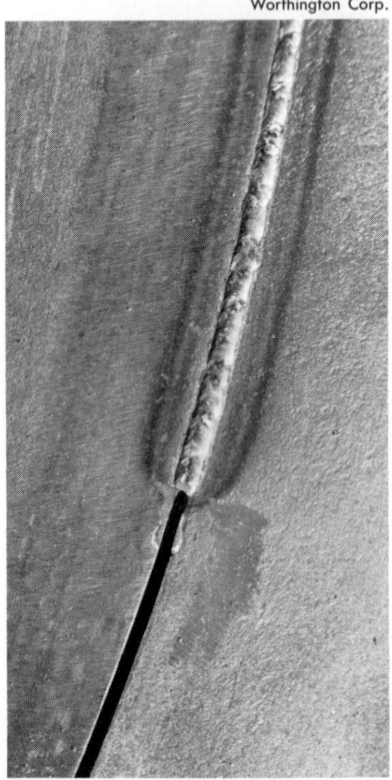

Inspection and Testing. Make up two test joints. One should be welded with the travel-down technique, and the other should be welded with the travel-up technique. Cut the usual test coupons from these joints, prepare them carefully, and subject them to the usual face- and root-bend tests. Inspect the surface of the bends carefully for any cracks or voids. None greater than $\frac{1}{16}$ inch in any direction should be present.

INTERSECTION JOINTS: JOBS 28-J9 THROUGH J14. Intersection joints such as *T*'s, *Y*'s, and laterals are found in pressure and powerhouse installations and in most general construction work. As you have already discovered when welding these joints with other processes, they are difficult to fabricate and even more difficult to weld. Great care must be taken in beveling and fitup.

You will recall that prefabricated fittings are available that require only the application of a butt joint for installation. These fittings are expensive, however, and many companies prefer to have the welder fabricate the

28-42. *The short-circuiting arc technique was used to weld the root pass in this 1 ½-inch thick, mild steel pressure vessel. The joint members were spaced 1/8 to 5/32 inch apart to give proper penetration and reinforcement buildup. Tack welds for holding parts in proper alignment were removed by grinding before the root pass was completed. Ends of the root pass increments were tapered with an 8-inch wide disc grinder wheel at the same time tack welds were removed to insure complete fusion at the tie-in.*

Worthington Corp.

joints on the job. It is always a problem to determine which method is less costly.

Practice Jobs 28-J9 through 14 as specified in the Course Outline. The intersection joint requires the welding of fillet and lap welds. The usual progression is from a fillet weld to a lap weld, back to fillet, then to a lap, and finally back to the original fillet weld that is the starting point.

Follow the general procedure previously recommended for gas metal-arc pipe welds. Prepare the pipe properly, tack weld it, and place it in the position called for in the Course Outline. Welding may be performed with both the downhill and uphill techniques. Both stringer and weaved beads are used, depending on the position of the joint and the recommended procedures on the job. The root pass is always a stringer bead, and travel may be downhill or uphill. The intermediate and cover passes may be stringer or weaved beads, depending on the position of the pipe and requirements of the job. You have already learned the techniques required because the jobs in the Course Outline have provided practice in all of them.

You are again instructed to take extreme care in making the root pass. Be sure that there is penetration to the root of the joint and fusion to the pipe surfaces. All other passes must be fused to the underneath pass and to the pipe wall. When multiple-pass welds are necessary, make sure that all starts and stops are in different areas. The starts and stops must not be at the same point.

The welds should be started on either side of the intersection at the heel of the joint. The heel is the coldest point of the joint. Because it requires the greatest amount of heat to weld, it should be welded after the joint has been heated from the welding of other areas. This procedure permits the welding of the heel intersection in a preheat condition when the arc is in continuous operation.

28-43. *Face appearance of root, filler, and cover passes in pipe welded in horizontal fixed position, travel down.*

Plumbers and Pipefitters Union, Alton, Ill.

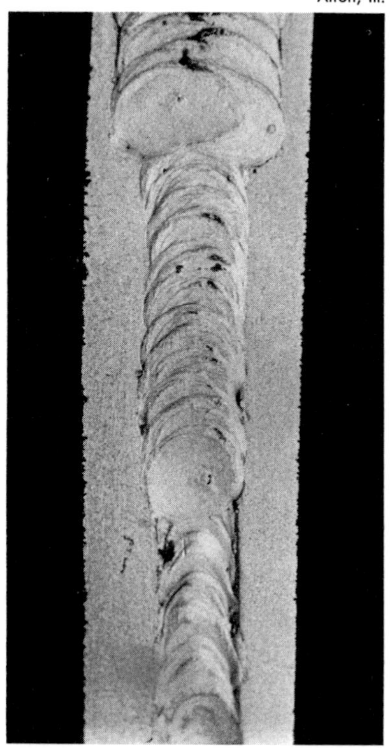

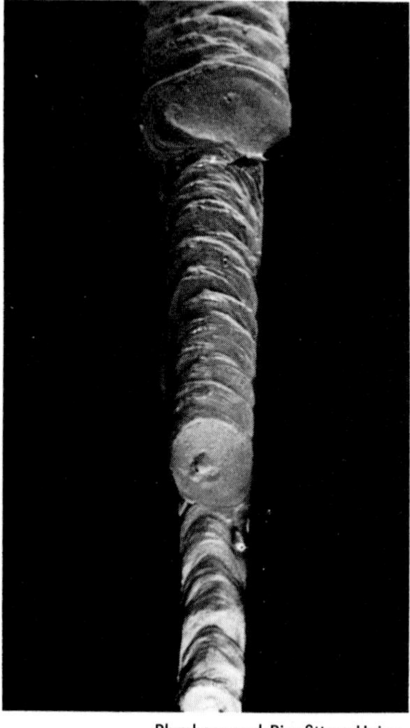

Plumbers and Pipefitters Union, Alton, Ill.

28-44. *Face appearance of root, filler and cover passes, welding with pipe in the horizontal fixed position, travel down. Travel was somewhat slower than welding shown in Fig. 28-43 so welds are heavier.*

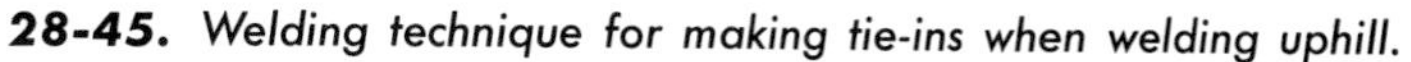

28-45. *Welding technique for making tie-ins when welding uphill.*

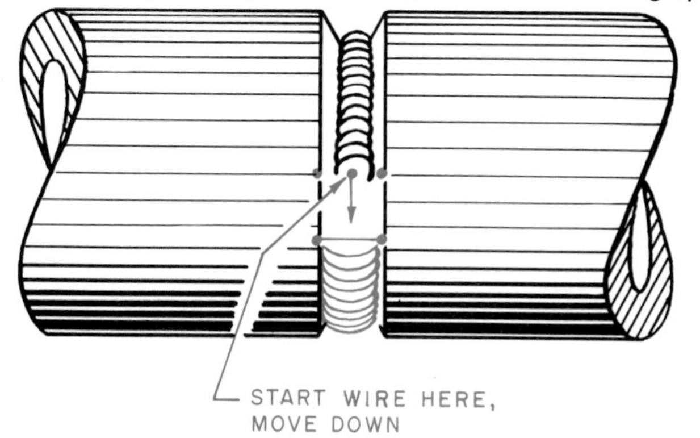
START WIRE HERE,
MOVE DOWN

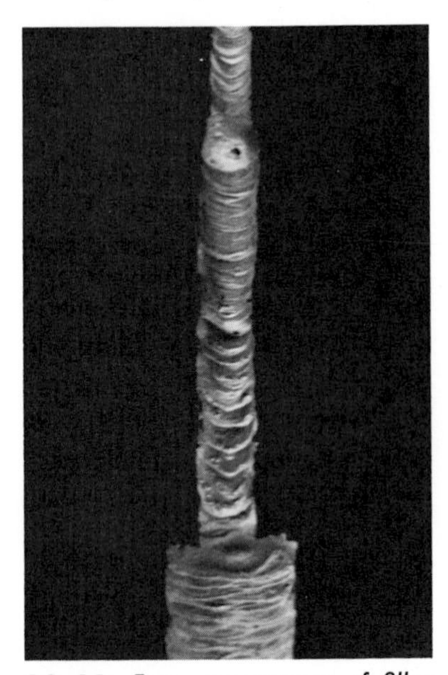

28-46. *Face appearance of filler passes and cover pass welded with pipe in the horizontal fixed position, travel uphill.*

In welding these joints, there is a tendency to undercut at the heel and the sides. Making sure that the weld pool fills up eliminates this possibility.

Clean and inspect each pass. Make sure that the welds are free of surface defects. After the joint is completed and you have carefully inspected its surface, cut a number of sections from the joint and macro-etch them. Inspect the sections for evidence of lack of penetration, lack of fusion, porosity, and undercut.

Practice on Aluminum Pipe

HORIZONTAL FIXED POSITION: JOB 28-J15. The opportunity to weld aluminum pipe with the gas metal-arc process is offered as a bonus for those welding students who have mastered the welding of pipe with the stick electrode and TIG processes with a high degree of skill.

Secure two lengths of 5- or 6-inch 6061 aluminum alloy pipe and 4043 or 5356 alloy filler wire $\frac{1}{16}$ inch in diameter. The pipe should be prepared, tacked, and set up as shown in Fig. 28-49. Brush the beveled surfaces thoroughly to remove the protective oxide coating and other foreign materials.

Shielding gas should be argon or helium or a mixture of argon and helium. Gas flow should be approximately 60 c.f.h. The high gas flow is necessary because of the tendency for the gas to leave the weld area when welding in the overhead position. The welding current should be set at about 150 to 190 amperes. Follow the welding pass sequence shown in Fig. 28-49.

In making the root pass, a slight motion may be necessary to properly control the metal. Special care must be taken to direct the arc so that overheating is not caused in any one area. This burns away the root face of the pipe. It also causes excessive penetration on the inside of the pipe and excessive buildup and sagging of the weld bead.

In welding the filler and cover passes, a slight weave may help you to control the metal. Be careful to keep the molten pool under control. If the pool is too large, the effect of gravity on the molten metal causes it to spill over and sag. Bead size, weld speed, and bead sequence must be such that there is no lack of fusion between passes.

Clean the welds and inspect the inside of the pipe for the root pass appearance and the outside of the pipe for the face appearance of the cover pass. Look for the same weld characteristics that you have learned are indicative of a sound weld, Fig. 28-50.

28-47. *Face appearance of a cover pass in steel pipe welded in the horizontal position, travel down. The weld has a coarse appearance compared to that shown in Fig. 28-46.*

Plumbers and Pipefitters Union, Alton, Ill.

28-48. *Face appearance of a cover pass in steel pipe welded in the horizontal position, travel up. Note that the ripples are close together. Compare with the weld shown in Fig. 28-47. Travel up will usually produce a smoother weld.*

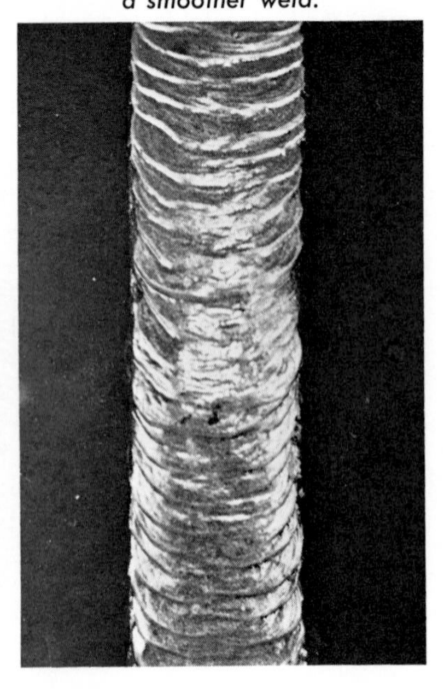

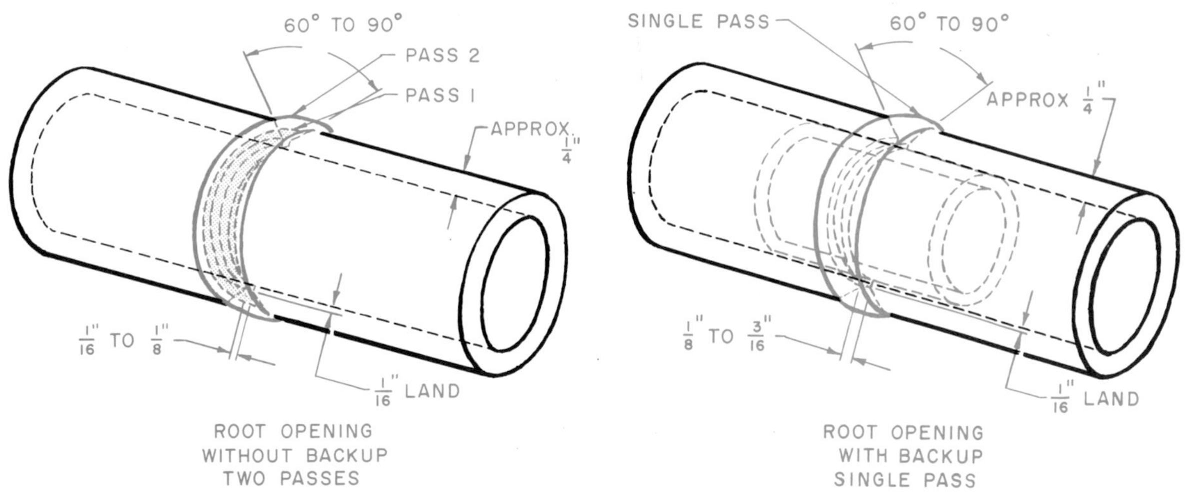

28-49. *Joint design and weld pass sequence for MIG welding of aluminum pipe in the horizontal fixed position.*

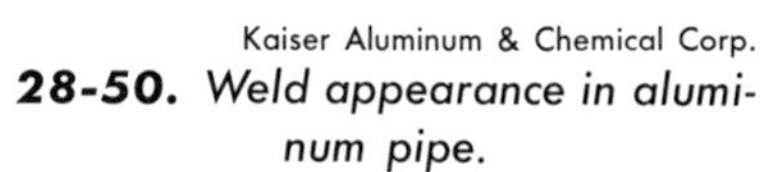

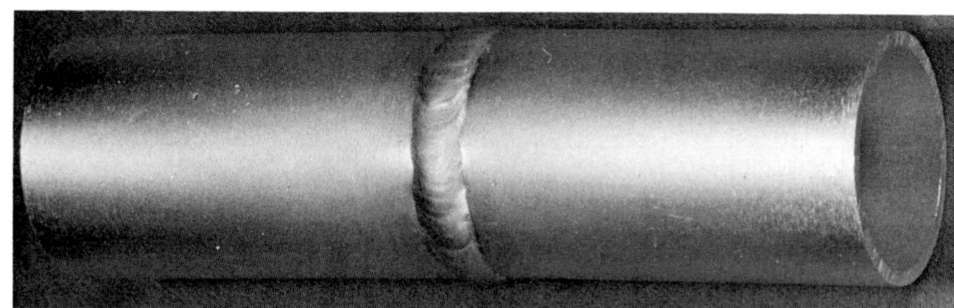

Kaiser Aluminum & Chemical Corp.

28-50. *Weld appearance in aluminum pipe.*

REVIEW QUESTIONS

1. The piping industry has grown at a great pace. Indicate the approximate number of miles of pipe in the U.S.: (a) 200,000, (b) 850,000, (c) 1,000,000, (d) 2,000,000.

2. List eight advantages of welding pipe with the gas metal-arc process.

3. List the special equipment needed for welding pipe in addition to that needed for welding plate.

4. One of the problems that keeps the welding of pipe with the gas metal-arc process from growing is the lack of electrical current on construction jobs. True. False.

5. Name the gas that is finding increased use as a shielding gas.

6. Shielding gas flow rates are usually increased when welding outdoors. Explain.

7. A filler wire of $\frac{1}{8}$-inch diameter is usually used for pipe welding. True. False.

8. Pipe should not be welded in the downhill direction of travel with the gas metal-arc process. True. False.

9. The maximum bead width which can be handled with any degree of success is $\frac{1}{2}$ inch. True. False.

10. List the welding conditions for making a beveled-butt weld in 8-inch carbon steel pipe in the horizontal fixed position.

11. Draw a simple sketch of the pass sequence for welding a beveled-butt joint in pipe with a $\frac{3}{8}$-inch wall thickness in the vertical fixed position.

12. Shortening the contact tip of the welding gun reduces the amount of shielding gas needed. True. False.

13. Gas metal-arc pipe welding is best performed with a constant voltage welding machine. True. False.

Table 28-2
Course Outline: Gas Metal-Arc Welding Practice with Solid Core Wire (Pipe)

Job no.	Type of joint	Type of weld	Welding position	Welding technique	Pipe specifications				Electrode specifications		Welding current DCRP		Shielding gas	Gas flow c.f.h.	Text reference
					Material	Dia.	Weight schedule	Wall thickness	Type	Size	Volts	Amperes			
28-J1	—	Beading	Horizontal roll	Stringer down Weaved down	Carbon steel	4″ to 8″	40	$\frac{1}{4}$ to $\frac{5}{16}$	*E70S-3	0.035	18–21	150–160 130–150	Carbon dioxide	15–25	890
28-J2	—	Beading	Horizontal roll	Stringer up Weaved up	Carbon steel	4″ to 8″	80	$\frac{5}{16}$ to $\frac{1}{2}$	E70S-3	0.035	19–23	130–160	Carbon dioxide	15–20	890
28-J3	V-butt	Groove	Horizontal roll	1 Stringer—down 2 Weaved—down	Carbon steel	6″	40	$\frac{5}{16}$	E70S-3	0.035	18–21	130–140 140–160	Carbon dioxide	15–20	891
28-J4	V-butt	Groove	Vertical fixed	5 Stringer	Carbon steel	6″	40	$\frac{5}{16}$	E70S-3	0.035	18–21	1–130 to 140 4–140 to 160	Carbon dioxide	15–25	894
28-J5	V-butt	Groove	Vertical fixed	7 Stringer	Carbon steel	8″	80	$\frac{1}{2}$	E70S-3	0.035	19–23	1–140 to 150 6–150 to 170	Carbon dioxide	15–25	894
28-J6	V-butt	Groove	Horizontal fixed	1 Stringer—down 3 Weaved—down	Carbon steel	8″	40	$\frac{5}{16}$	E70S-3	0.035	19–21	120–130 130–150	Carbon dioxide	15–20	895
28-J7	V-butt	Groove	Horizontal fixed	1 Stringer—down 2 Weaved—up	Carbon steel	8″	40	$\frac{5}{16}$	E70S-3	0.035	19–21 19–23	120–130 110–120	Carbon dioxide	15–20	899
28-J8	V-butt	Groove	Horizontal fixed	1 Stringer—down 2 Weaved—up	Carbon steel	8″	80	$\frac{1}{2}$	E70S-G	0.035	19–21 19–24	120–130 110–125	Carbon dioxide	15–20	899
28-J9	90° Branch	Groove	Header horizontal fixed; Branch vertical fixed top	3 Stringer	Carbon steel	6″ to 8″	40	$\frac{5}{16}$	E70S-3	0.035	18–21	130–140	Carbon dioxide	15–20	902
28-J10	90° branch	Groove	Header vertical fixed; Branch horizontal fixed	1 Stringer—down 2 Weaved up	Carbon steel	6″ to 8″	40	$\frac{5}{16}$	E70S-3	0.035	18–21	120–130 110–120	Carbon dioxide	15–20	902
28-J11	90° branch	Groove	Header horizontal fixed; Branch vertical fixed bottom	3 Stringer	Carbon steel	6″ to 8″	40	$\frac{5}{16}$	E70S-3	0.035	18–21	120–130	Carbon dioxide	15–20	902
28-J12	45° branch	Groove	Header horizontal fixed; Branch top	3 Stringer	Carbon steel	6″ to 8″	40	$\frac{5}{16}$	E70S-3	0.035	18–21	130–140	Carbon dioxide	15–20	902
28-J13	45° branch	Groove	Header vertical fixed; branch side	1 Stringer—down 2 Weaved—up	Carbon steel	6″ to 8″	40	$\frac{5}{16}$	E70S-3	0.035	18–21 18–23	120–130 110–120	Carbon dioxide	15–20	902
28-J14	45° branch	Groove	Headers horizontal fixed; branch bottom	5 Stringer	Carbon steel	6″ to 8″	40 80	$\frac{5}{16}$ to $\frac{1}{2}$	E70S-G	0.035	18–23	130–140	Carbon dioxide	15–20	902
28-J15	V-butt	Groove	Horizontal roll	1 Stringer 1 Weaved	Aluminum	5″ to 6″		$\frac{1}{4}$	ER4043	$\frac{3}{64}$	20–24	180–190 190–210	Argon	30–40	904

*Electrode types E70S-2 and E70S-1B may also be used with excellent results.

NOTE: The conditions indicated here are basic and will vary with the job situation, the results desired and skill of the welder.

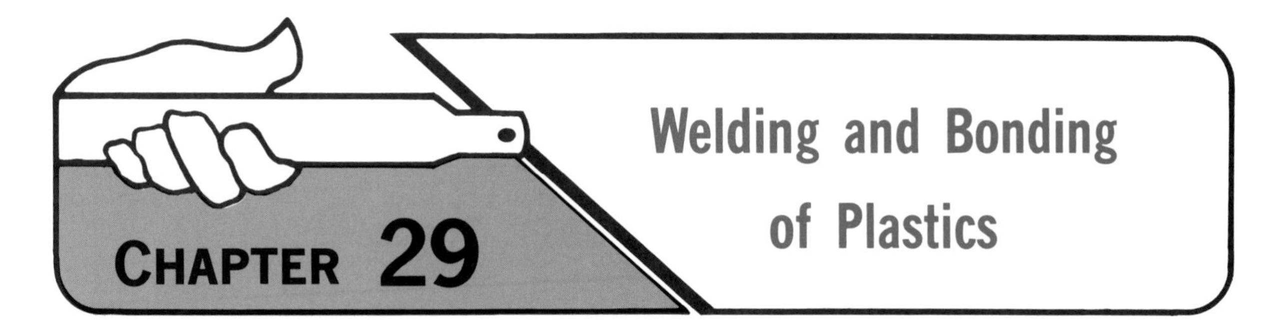

Welding and Bonding of Plastics

The cooperation of the Kamweld Products Company, the Seelye Plastics Company, the Laramy Products Company, and the ABS Institute in providing information and photographs for this chapter is gratefully acknowledged.

Thermoplastic was developed over 100 years ago, but no serious research was done until the Germans began searching for material substitutes in the early 1930s. They discovered that when thermoplastics were heated, they could be pressed together and a permanent bond obtained. In 1938 this principle became the basis for the hot gas welding technique. World War II spurred the Germans' use of these materials and processes. It was not until after the war, however, that thermoplastics and hot gas welding came into use in America.

CHARACTERISTICS OF PLASTICS

There are two basic types of plastics: (1) thermosetting plastics and (2) thermoplastics. Thermosetting plastics harden under heat. Through chemical reaction they are formed into permanent shapes that cannot be changed either by reapplying heat or chemical reaction. These plastics are used for such products as wall panels, countertops, and fiberglass boats. They cannot be welded. Thermoplastics, however, soften when heated and solidify when cooled with no chemical change. These plastics can be machined, formed, and welded. Heating and cooling can take place many times without changing their appearance or chemical makeup.

Types of Thermoplastics

POLYVINYL CHLORIDE (PVC). Polyvinyl chloride, usually known by the abbreviation PVC, is one of the most popular materials of construction because of excellent physical properties, ease of fabrication, relatively low cost, and the ability to be formed into a wide range of products such as sheeting and roll goods.

PVC has wide forming-temperature ranges and self-extinguishing properties. This is important when flammability is a major consideration in the selection of materials. The primary limitation of PVC is its recommended working temperature range of 140 to 150 degrees F. This range limits its structural strength and its corrosion resistance, which decrease rapidly as the temperature increases.

Rigid Polyvinyl Chloride (PVC). There are two broad classifications for rigid PVC products. Type I has normal resistance to impact and high resistance to corrosion within its working temperature range. Type II PVC is modified with rubber to increase the impact resistance.

Both type I and type II PVC plastics are the best materials for general corrosion protection because of their physical properties, chemical resistance, and low cost. They can be hot-air welded, cemented, or assembled by machining processes. Type I is slightly higher in corrosion resistance and is considered the best to use for this purpose.

Both types may be found in such shapes as sheets, pipe, valves, fittings, structural shapes, round and hex bars, and slab stock. Flexible lining materials are available for tanks, large ducts, and troughs. They are applied with adhesives. Flexible joining strips are used for weld-seams and corners on the inside of line tanks.

MODIFIED HIGH IMPACT RIGID POLYVINYL CHLORIDE (PVC). This class of rigid polyvinyl chloride (PVC) was developed for intermediate corrosion service, primarily in fume exhaust systems. It is readily formed in press and vacuum operations and can be worked and welded at the same temperature as regular polyvinyl chloride. Its availability is limited to 4×8 foot sheets up to $\frac{1}{2}$ inch thick. Corrosion resistance is somewhat less than for type II PVC, but it is superior to ABS formulations. It oxidizes but does not burn. It cannot be exposed to very strong acids and the usual solvents that attack rigid polyvinyl chloride. It has slightly better impact strength than unmodified PVC.

Neither pipe nor structural shapes are made from high impact PVC since the availability of unmodified rigid PVC structural shapes is sufficient to meet the limited need. It can be welded to type I or type II PVC, and the same welding rod as that used to weld type I PVC can be used. It is joined in this manner for the lower corrosion areas in an overall exhaust system. Modified high impact PVC is not recommended generally for tanks, either lined or self-supporting. The weld strength is slightly less than for unmodified rigid polyvinyl chloride, and welding speeds are somewhat slower.

POLYETHYLENE (PE). Polyethylene, abbreviated PE, is available in three classes of material. All three classes are the same chemically. Density and physical properties differ, however, according to the packing of the resins. The three types are referred to as low density, medium density, and high density. In general, the main differences in going from low to high density materials are in corrosion resistance, working temperature, and tensile strength. These characteristics increase from low density to high density. Various forms such as sheet, structural shapes, and pipe are available.

Low Density (Branched) Polyethylene. Polyethylene is lighter than metal and floats in water. It also burns. This plastic is produced as film, sheet, rod, tubing, and block stock. It is furnished in natural and black.

Low density polyethylene offers reasonably good corrosion resistance. Its main drawback is that its working temperatures and tensile strengths are lower than those of the other classes of polyethylene. It is generally utilized when stiffness, structural strength, and high working temperatures are not required. It can be readily formed by vacuum- or press-forming operations. Although it cannot be joined by cement, it can be welded using the same class of polyethylene rod. Dry nitrogen is usually recommended as a source for hot gas welders.

Welding rod is available in diameters of $1/8$, $5/32$, and $3/16$ inch. It is supplied in spools of 1–15 pounds. Straight rod, approximately 4–5 feet long, is also supplied.

Medium Density Polyethylene. This plastic is produced as film, sheet, rod, tubing, and block. Corrosion resistance is somewhat better than that of low density (conventional) polyethylene. It is not cementable, and it will burn. The tensile strength and maximum working temperature are not as great as for high density (linear) polyethylene. It is used for both pressure and conduit tubing and pipe. Applications for structural fabrication from sheet are limited. Impact strength is good, and it can be both vacuum and press formed.

Hot gas welding is done with dry nitrogen and medium density welding rod.

High Density (Linear) Polyethylene. This class of materials may also be referred to as *low pressure polyethylene*. It is much lighter than metal. It is combustible. Although it cannot be cemented, it can be welded. Because of its strength and resistance to high temperature, this material has replaced some of the other polyethylene classes. It has the highest working stress factor and best corrosion resistance of all three classes of polyethylene. The material has a reasonably high working temperature under low load conditions. At high heats, low load conditions must be maintained for satisfactory usage.

This material is used more than the other two classes of polyethylene for structural fabrication. Applications include tanks, fume exhaust hoods, ductwork, valve bodies, and pipe. It can be vacuum formed. Film, sheet, rod, tubing, pipe, and block are available. Welding is done with dry nitrogen and welding rod that matches the material.

POLYPROPYLENE (PP). This plastic has lower impact strength than polyethylene, but its tensile strength is much higher and its working temperatures are superior. It also offers more resistance to organic solvents and degreasing agents than polyethylene. While the material can be joined by welding, it cannot be cemented. Such forms as film, rod, sheet, tubing, and blocks are available. This material is used in the manufacture of hoods, plating barrels, ductwork, and other industrial corrosion-resistant applications.

Welding rod is available in $1/8$-, $5/32$-, and $3/16$-inch coil and flat stock. Dry nitrogen is recommended for welding. Some welding is being done with dry air.

ACRYLONITRILE BUTADIENE STYRENE (ABS). There are two classifications of rigid ABS plastics. Type I is designed for normal temperatures, and there are various mixtures for desired physical properties. Type II is for use in higher temperatures. Different compounds are available for specific manufacturing operations such as calendaring, press laminating, extrusion, and injection molding. There is also a military specification for both type I and type II. Pipe materials are classified as type I with grades 1, 2, and 3, which express increasing working stress values. A type-II plastic is intermediate in working stress between grades 2 and 3 of type I.

ABS is much lighter than rigid polyvinyl chloride and only slightly heavier than water. While cementing is the main joining method, the materials can be hot gas welded with nitrogen. This plastic offers good corrosion resistance except with oxidizing and very strong acids and organic solvents that normally attack thermoplastic materials such as PVC. The material supports combustion, and the flame is characterized by a smoky soot. The type II material is used mostly for higher temperature applications, and pipe is available in both types. The strength of the weld is limited to 60 to 80 percent of the original strength of the plastic.

A major use of ABS is in heat-formed structural parts in which reasonable corrosion resistance is desirable, along with high impact strength. Semiflexible and rigid sheet, extruded pipe, and many injection-molded items are available in a wide range of colors.

ACRYLICS. Because acrylic plastics are transparent, they are widely used as a substitute for glass. The sheet comes both plain and masked for protection in handling. Premium grades are preshrunk before shipment. Certain grades have built-in properties such as ultraviolet absorption, resistance to scratching and/or shattering, and other specialty requirements. Due to its clarity, acrylic plastic is often used in connection with regular thermoplastic structural fabrications. When ordering, it is necessary to specify corrosion resistance, crazing characteristics, and other specifications desired. The material can be cemented and welded.

WELDING AS A METHOD OF JOINING PLASTICS

Thermoplastics, particularly polyvinyl chloride, polyethylene, and polypropylene, are enjoying increased use today because of their many advantages. Such advantages include light weight, resistance to chemicals and moisture, strength, pleasant appearance, ease of forming, and low cost. The development of hot gas welding has also helped to spur their use. In manufacturing, fabricating, construction, and plant maintenance, the possibilities of utilizing thermoplastics are as unlimited as the imagination.

Welding is the ideal fabrication process for making such plastic industrial products as chemical-resistant tanks, hoods, vents, ductwork, special containers, and piping systems, Figs, 29-1 and 29-2. The welding of plastic pipe is increasing in oil refineries and chemical plants. Because there is no need to take the line apart, downtime and replacement costs are saved. Welding seals leaks instantly in new or old installations. Plastic tank linings are installed and repaired by welding. The welding of thermoplastics is somewhat similar to the gas welding of metals. All of the basic joint designs are used, and welding is possible in all positions, Fig. 29-3. As with all other welding applications, the joint must be prepared properly, and the welding procedure must be carefully applied. The weld is not difficult to make. It can be done quickly, and it is permanent.

The Preparation of Plastics

The preparation of thermoplastics for assembly by welding is similar to the procedures for metal fabrications. The pieces are laid out, cut, machined, and joined with the same tools, equipment, and skills employed in the metalworking trades. There are, however, some special working requirements.

LAYOUT. All layout work should be done directly on the plastic sheet in pencil, soapstone, or china marker. Do not use wax pencil since the wax cannot be removed after the plastic is heated. Because of the high notch-sensitivity of plastics, a scribe or other sharp pointed tool causes fracture and so should not be used for marking.

SHRINKAGE. Where accurate dimensional control is required, the sheet should be preshrunk for approximately 20 minutes at 250 degrees F., depending on the gauge of material. The cooling of the sheet must be controlled so that it does not buckle and deform.

FORMING. Plastic sheet can be heated with a heat gun and formed around metal forms to make round duct and other curved shapes.

CUTTING. Thermoplastics can be cut with the same hand or power tools used to cut wood or metal. Most cutting jobs are shop operations requiring power tools.

SAWING. In sawing plastic sheet, there is likely to be concentrated heat buildup in the saw blade because of the poor heat conductivity of the plastic. To allow for this, the blade must be selected in accordance with the gauge of material. Thin materials require the use of fine pitch blades, and blades for thicker materials should be heavier and hollow ground. The number of teeth per inch may vary from 8 to 22 teeth. In order to make a slicing cut in the material, the teeth should have negative rake with little or no set. The saw blade should just show through the material to keep friction at a minimum, and the feed should be slow.

Circular and band saws may be used. Band saw blades should have a slight set and finer pitch than circular saw blades. There is less heating with band saw blades because the length of the blade allows the teeth to cool. Special blades can be obtained for the cutting of plastics.

SHEARING. Shears can be used for the cutting of light gauge sheets. A $\frac{1}{8}$-inch sheet of type I PVC can be sheared easily. Heavier gauge type I PVC shows stress marks. The other plastic types shear better and to a higher gauge. All shearing should be done at room temperature. A cold sheet will crack or shatter.

ROUTING. Beveling is quickly done with a power router, a file, or a rotary sander. The router can also be used for rimming the edges of sheets or for shaping and recessing. The feed must be slow and continuous, and the swarf must be removed by compressed air.

A woodworking jointer may also be used to square or bevel the edge of a sheet. The table and fence must be clean and smooth so that the plastic will not be scratched. The jointer knives should be sharp and finely honed.

OTHER WORKING PROCESSES. The work processes previously described involve those processes that the welder is likely to use in preparing the material for welding. A number of other working processes are used in the fabrication of various plastic units. These processes, which include drilling, punching, machining, milling, threading, knurling, riveting, and bolting, are also used in fabricating steel and other metals.

SAFETY. In machining thermoplastics with power tools, the same safety rules that apply in metal working must be followed.

The use of guards on all equipment is vital. Hold-down clamps are used to secure the work to the worktable. Hands must be kept as far away as possible from the cutting tool. The safety switch for shutting off power tools must be in a conspicuous place, and its location must not be changed.

Proper clothing should be worn. No loose shirt sleeves or

29-1. Laboratory with plastic installations. In this case, resistance to corrosion is the principal advantage of plastic fabrication.

Industrial Plastics Fabricators

29-2. Large welded plastic duct.

Industrial Plastics Fabricators

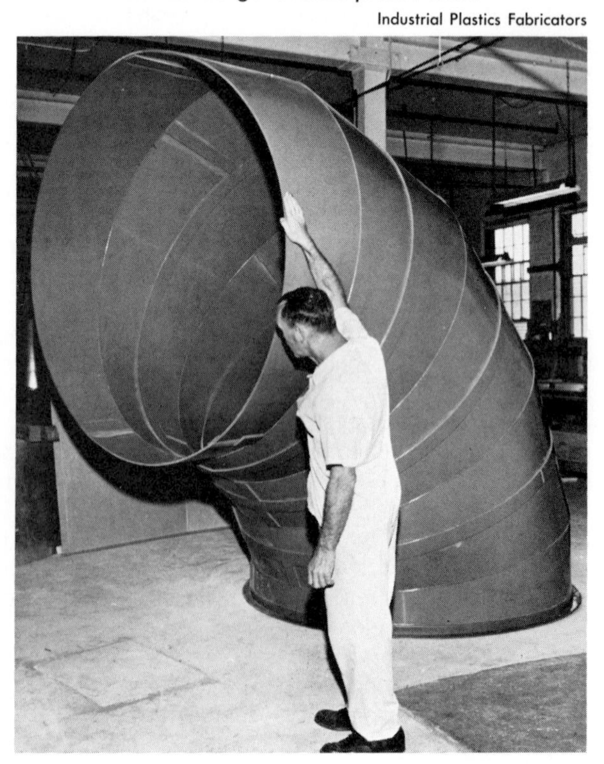

neckties can be allowed. Shop coats, properly fastened, are recommended.

Hot Gas Welding

As with metals, plastics are welded by applying enough localized heat to produce fusion of the areas that are to be joined. Several different methods of welding have been developed. They include hot gas welding, heated tool welding, induction welding, and friction welding (also known as spin welding). The major difference among these various methods of welding is in the means of applying heat.

Hot gas welding is one of the principal methods. The two basic requirements for hot gas welding are a heat source and a welding rod that aids in fusion of the weld to the base material. The joints used in thermoplastic welding are identical to those in metal welding, namely, butt, lap, edge, corner, and T-joints. The same material preparation, fitup, root gap, and beveling are required in plastic welding as in metal welding. Beveling of plastic edges is essential to obtaining a satisfactory weld. Flux is not required in the welding of plastics. Some plastics, however, are welded more satisfactorily in an atmosphere of dry nitrogen. Procedures for determining the quality of the completed weld are like the inspection and testing methods in metal welding.

Because of the differences in the physical characteristics of thermoplastics and metals, there are corresponding differences between welding techniques for metals and plastics. In the welding of metal, the welding rod and the base metal become molten and fuse into the required bond to form the welded joint. There is a sharply defined melting point in metal welding. This is not the case in thermoplastic welding.

Because they are poor heat conductors, plastic materials are difficult to heat uniformly. When heat is incorrectly applied, the surface of the plastic welding rod and the base material can char, melt, or decompose before the material immediately below the surface becomes fully softened. You will, therefore, have to develop skill in working within temperature ranges that are narrower than those in metal welding. You will also have to become accustomed to the welding rod not becoming molten throughout. In fact, the exposed surface of the rod will seem unchanged except for flow lines on either side. The familiar molten crater in metal welding is not found in plastic welding, Fig. 29-4. Only the lower surface of the welding rod becomes fusible while the inner core of the rod merely becomes flexible. Hence, the materials do not flow to-

29-3. *Basic joints that may be formed and welded with round rod and triangular welding rods.*

Kamweld Products Co.

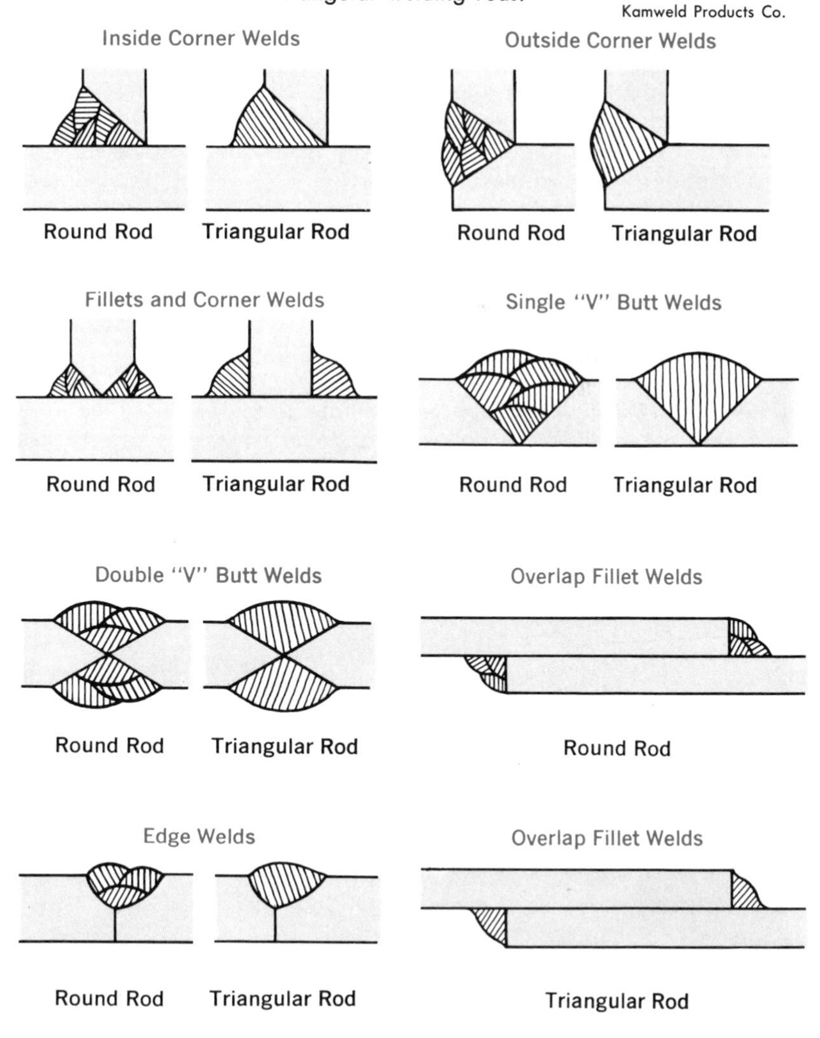

gether as a liquid. You must apply pressure on the welding rod to force the fusible portion into the joint and make the permanent bond.

STRETCHING AND DISTORTION. Regardless of the skill of the welder, some stretching of the welding rod will always occur. Stretching should not exceed 15 percent. The welding rod stretches too much when it leans away from the direction of welding. In speed welding, stretching is caused by too much pressure on the rod or by plastic residue on the shoe and in the preheating tube. When a thermoplastic rod is heated enough to form a weld, it becomes soft and tends to stretch. Contraction during cooling causes stresses that produce cracks and checks across the face of the weld. The amount of stretch in a completed weld can be determined by measuring the length of the rod before and after welding.

In multilayer welds, deposited beads are reheated in the process of laying new beads one on another. Stretching in multilayer welds must be held to a minimum since checks and cracks caused by stretching show up as voids in the finished weld and cannot be detected by visual inspection. When making multilayer welds, allow ample time for each weld pass to cool before proceeding with additional welds. To save time and give added strength to multilayer welds, alternate welds from one side of the groove to the other. Thermoplastics are notably poor conductors of heat, and stresses in plastic are confined to a much smaller area than those in metal. Shrinkage of a weld upon cooling is greater near the crown than at the root. As with steel, when welding single-V and fillet welds, the work should be offset before welding to compensate for distortion. Preheating the general area to be welded also helps. Distortion can also be reduced by using speed welding and triangular welding rod instead of making multilayer welds with round welding rod.

29-4. *Basic welding procedure for welding plastics. The use of the torch and filler rod is similar to gas welding. Note, however, the distinctive appearance of the weld deposit. The joint is an outside corner joint, and the weld is a fillet weld.*

Seelye Plastics

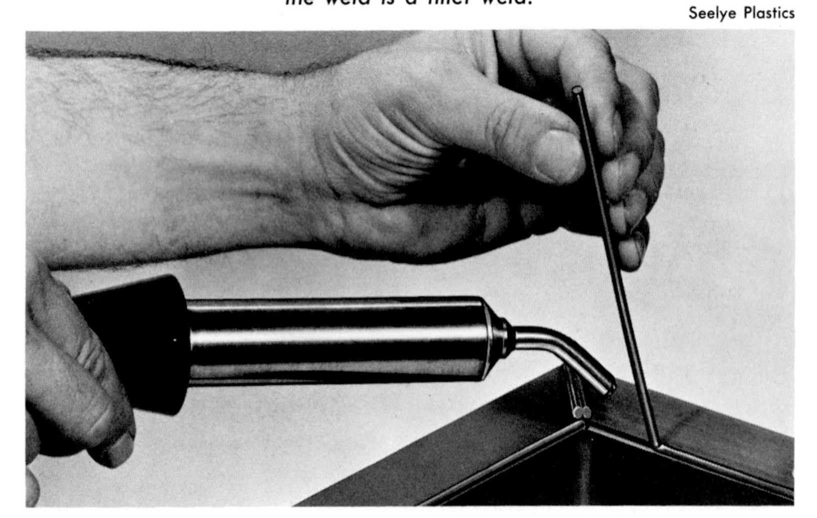

WELDING PVC. The material must be kept clean at all times. It can be cleaned by wiping with MEK or a similar solvent. Freshly beveled edges do not need to be cleaned. Welding edges are beveled or offset to provide areas for the welding rod and to permit better adhesion by removing polished outer surfaces. Cut bevels with a jointer, sander, router, or plane. Clamp the sections firmly in place or tack weld. Avoid backup materials that are good heat conductors. Allow a root gap in most assembly procedures except when tack welding. Thickness, shape, size, good construction techniques, and the strength required dictates the type of weld to use.

WELDING POLYETHYLENE AND POLYPROPYLENE. Procedures are the same for these materials except for the welding gas and welding temperatures. These should be varied according to Table 29-1.

In welding polyethylene and polypropylene, a number of precautions should be followed:

- The base material should be freshly cut or scraped and clean.
- The welding rod and the material must be of the same density.
- Polyethylene and polypropylene are subject to stress cracking. The base material and welding rod should not be overheated nor should the welding rod be stretched. Use only 1 foot of rod for 1 foot of weld.
- If the welded joint will be under stress in service, the weld will be subject to chemical attack that would not occur under normal circumstances. This is called "environmental stress cracking." It causes welds to fail and cracks to radiate into the sheet from the weld.

- Polyethylene and polypropylene welding rods tend to loop in the direction of the weld. Do not let this cause you to force the rod and add undue strain on the weld.

Plastic Welding Equipment

Modern plastic welding torches are divided into two basic types: electrically heated and gas heated. In both types, welding gas (compressed air or nitrogen) passes over a heat source which raises its temperature to 450–800 degrees.

Electric torches, Fig. 29-5, are used in manufacturing plants and wherever current is available. Many welders prefer this type of torch because it is more compact and easier to handle. Gasheated torches are used primarily in field operations where electricity is not available.

Figure 29-6 shows the internal construction of the typical electrically heated torch. The electricity heats a stainless steel barrel. A gentle flow of compressed air or nitrogen passes over the heating element. This heats the air rapidly to the desired temperature. The hot air leaves the nozzle from the lower end of the barrel. From there it goes through a tip or high speed tool to heat both the base material and the welding rod simultaneously. The heat softens the rod and base material so that they become one.

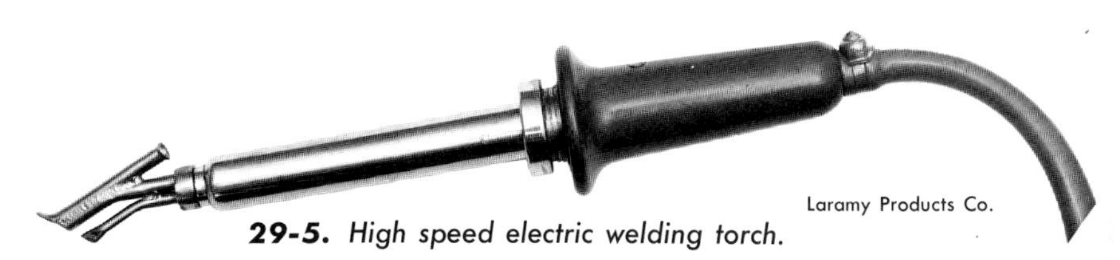

Laramy Products Co.

29-5. *High speed electric welding torch.*

Table 29-1
Thermoplastic welding chart.

	PVC type I	PVC type II	PVC plasticized	Polyethylene regular	Polyethylene linear	Polypropylene	Chlorinated polyether	FEP fluorocarbon	Acrylic
Welding temp. °F[1]	500 to 550	475 to 525	500 to 800	500 to 550	550 to 600	550 to 600	600 to 650	550 to 650	600 to 650
Welding gas	Air	Air	Air	Inert	Inert	Inert	Air	Air	Air
Butt-weld strength-%	75–90	75–90	75–90	80–95	50–80	65–90	65–90	80–95	75–85
Maximum continuous service temp. °F	160	145	150	140	210	230	250	250	140
Bending & forming temperature	250	250	100	245	270	300	350	550	280
Cementable	yes	yes	yes	no	no	no	no	no	yes
Specific gravity	1.35	1.35	1.35	.91	.95	.90	1.4	2.15	1.19
Support combustion	no	no	no	yes	yes	yes	no	no	yes
Odor under flame	HCL	HCL	HCL	Wax	Wax	Wax	Sweet chlorine	Pungent	Sweet
Color[2]	Grey	Light grey	Black	Translucent or black	White or black	Cream to amber	Olive drab	Bluish translucent	Transparent

Laramy Products Co.

[1]Measured 1/4″ from welding tip.
[2]These are most commonly used colors, and are subject to change. For complete details refer to specifications of individual plastic manufacturers.

Figures 29-7, 29-8, and 29-9 show the simple outfit that is needed for plastic welding. The equipment in Fig. 29-7 includes

A. A double-jacketed and insulated stainless steel heating tube which encloses a 110-volt, a.c.-d.c. heating element. These elements are available in three sizes: (1) 350 watts, with a temperature range of 420 to 580 degrees F., (2) 500 watts, with a temperature range of 550 to 800 degrees F., and (3) 650 watts, with a temperature range of 700 to 1,000 degrees F. Torches are also available that use a fuel-gas flame to heat the gas.

B. Lightweight nylon handle designed for coolness and heat resistance; proportioned for handling ease and comfort.

C. Twenty feet of neoprene-insulated, three-wire grounded electric cord (110 volts, a.c.-d.c., 7 amps) inside a neoprene air hose. This prevents kinking or collapsing, which restricts gas flow and interferes with proper welding temperatures.

D. Self-relieving air regulator and easy-to-read gauge. Primary pressure is 200 p.s.i., and can be regulated to 0.15 p.s.i.

E. Welding tip. These come in a variety of sizes and types.

Figure 29-9 shows a complete portable plastic welding outfit with its own built-in air compressor. It requires only electrical power for operation. The gas and power control unit shown in Fig. 29-10 makes it possible to shift from gas to compressed air when the torch is temporarily not in use. It is used with electric torches.

The welder of plastics should have a number of miscellaneous tools such as a wire brush, sharp knife, rotary sander, rasp, bending spring, files, saws, and C-clamps. He also uses a heat gun, which is an electrically heated blower, for softening plastics for bending and forming before welding. Unlike metal welding, plastic welding requires no gloves, aprons, or goggles since welding temperatures are low, and visibility is not obstructed by a flame or arc.

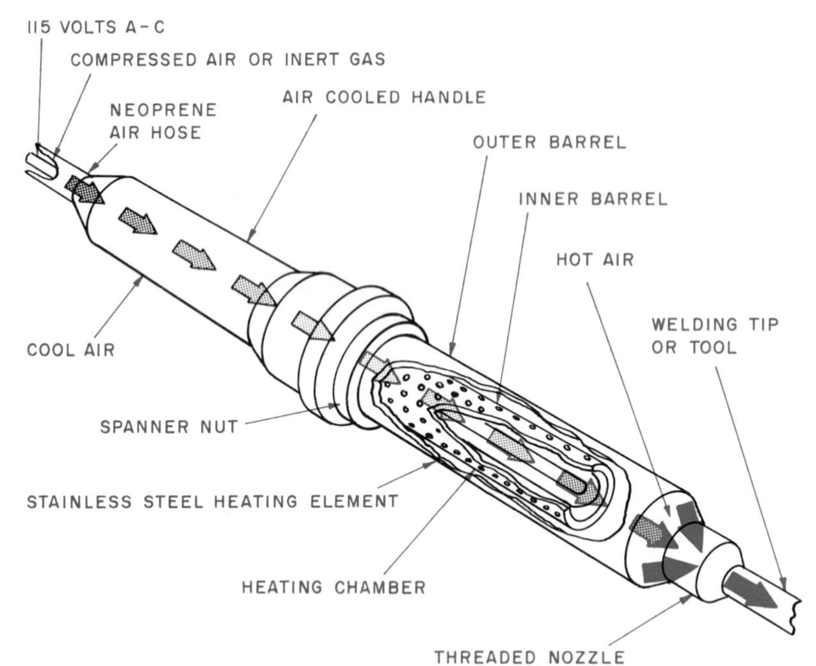

29-6. *Internal construction of a typical electric welding torch.*

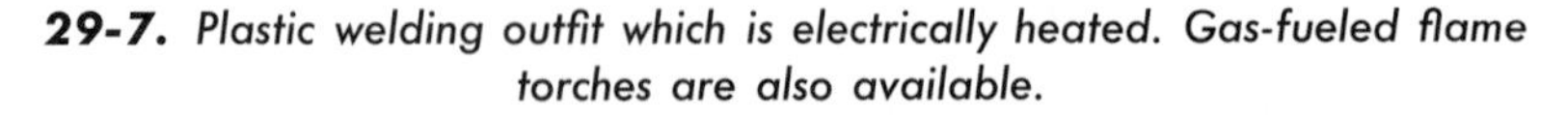

29-7. *Plastic welding outfit which is electrically heated. Gas-fueled flame torches are also available.*

Seelye Plastics Co.

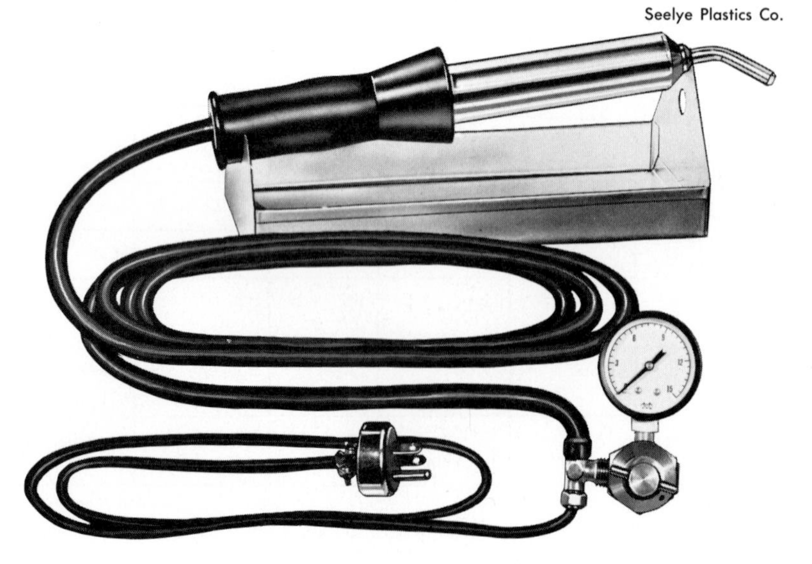

Setting up the Equipment

1. Make sure that you have the proper type of torch for the work at hand.
2. Select the proper heating element.
3. Relieve the regulator-adjusting screw to prevent damage to the regulator due to sudden excessive air pressure.
4. Connect the welder to the air or nitrogen supply and adjust the regulator for about 3 pounds pressure.
5. Connect the torch with a 115-volt electrical outlet.
6. Let the torch warm up for 3 or 4 minutes. Make sure that the compressed air or nitrogen is flowing continuously through the barrel of the torch to prevent burning out the heating element.
7. Select the proper tip or high-speed welding tool for the type of work to be welded, Fig. 29-11. As with other types of gas welding, tips can be changed while the torch is hot. Avoid touching the hot barrel of the torch.
8. Select the proper air pressure for the size of the heating element (in watts) and for the temperature desired at the tip end.

Rules for Welding with Electric Welding Torches

- Be sure the welding gas supply is clean. Moisture or oil in the welding gas may prevent a satisfactory bond in welding and cause a short circuit in the heating element of the torch.
- Never leave electricity on when the welding gas is turned off. Always turn gas on first and turn it off last.
- Always ground the torch to prevent a short circuit, possible electric shock, and damage to the heating element.
- The volume of welding gas passing over the heating element determines the welding temperature. To increase the temperature, reduce the gas volume. To decrease the temperature, increase the gas volume. To determine the temperature of the heated air, hold a thermometer $\frac{1}{4}$ inch from the end of the welding tip.
- Never touch the end of the torch barrel or welding tip when the torch is turned on.
- To obtain maximum life from the heating element, always use the recommended welding temperature.
- Read the manufacturer's operating instructions before using a torch for the first time.

Rules for Welding with Gas Welding Torches

- Be sure the torch is equipped with the proper jet for the heating gas being used.
- Be sure the welding gas supply is clean. Moisture or oil in the welding gas may prevent a satisfactory bond in the welding.
- When regulating welding temperatures, reduce the volume of welding gas or increase the pressure of the heating gas to raise the temperature. To lower the temperature, increase the volume of welding gas or reduce the pressure of heating gas.
- Never touch the end of the torch barrel or welding tip when torch is turned on.
- Always turn the welding gas on before lighting torch.

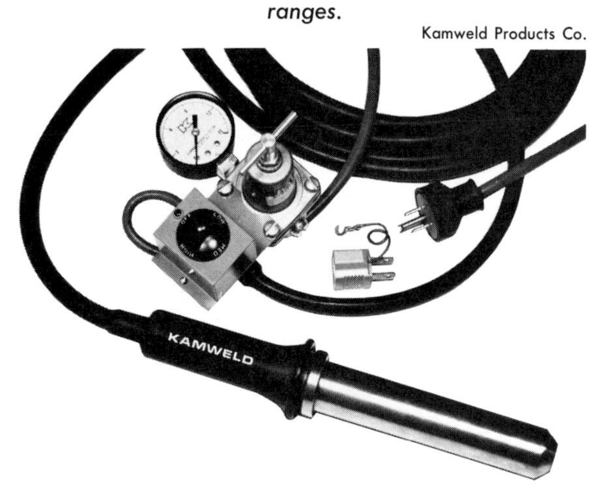

29-8. *Electrically heated torch, air regulator/pressure gauge, and neoprene air hose with electrical cable inside. The torch has low, medium, and high heat ranges.*

Kamweld Products Co.

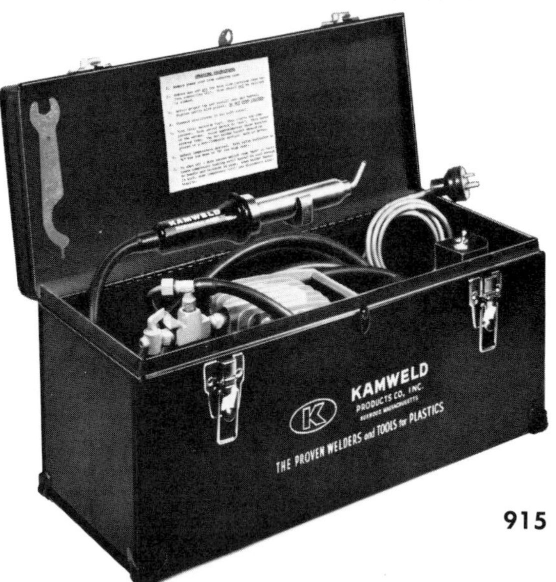

29-9. *Plastic welding outfit includes an air compressor, and it can be used in both shop and field.*

Kamweld Products Co.

- Never leave the torch lighted when the welding gas is off. Always turn off flame before shutting off the welding gas.
- Always read the manufacturer's instructions before using a torch for the first time.

INSPECTION AND TESTING

Good plastic welding can be achieved only if there is an adequate program of inspection and testing. You will recall the importance of testing in metal welding procedures in order to determine fusion, penetration, porosity, and appearance. These same weld characteristics are important to plastic welds, Fig. 29-12.

In general, the strength of a plastic weld is dependent on a combination of six interrelated factors:

- Strength of the base material
- Temperature and type of welding gas
- Pressure on the welding rod during welding
- Proper weld and joint selection
- Proper material preparation before welding
- Skill of the welder

Unlike metal welds, dressing plastic welds decreases the strength of the completed welds by approximately 25 percent. Welds that are equivalent to less than 75 percent of the material strength are considered to be unsatisfactory.

Faulty Welds

Faulty welds are the result of the following errors that sometimes occur in plastic welding:

- Overheating the base material or the plastic filler rod, Fig. 29-13
- Underheating the base material or the plastic filler rod, Fig. 29-14
- Improper penetration through entire root of weld, Fig. 29-15
- Porosity caused by air inclusions or dirt, Fig. 29-15
- Stretching the filler rod
- Incorrect handling of the welding torch
    - Wrong torch or tip angle
    - Too slow or too fast travel
    - Lack of or faulty fanning motion of the torch
    - Heat at the torch tip too close or too far away from work
    - Heat at the torch tip not centered on the weld bead

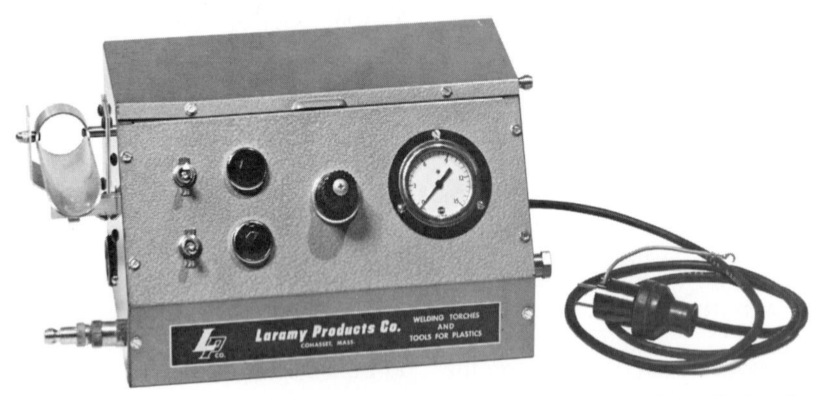

Laramy Products Co.

29-10. *Gas and power control unit.*

29-11. *Welding tips and high speed tools. KT—tacker tip; KR—round tip; KF—flat strip tip; KV—flat corner tip; KS-1—high speed tip for round rod; KST-1—high speed tip for triangular rod; KS-3—high speed tip for flat and corner strips*

Kamweld Products Co.

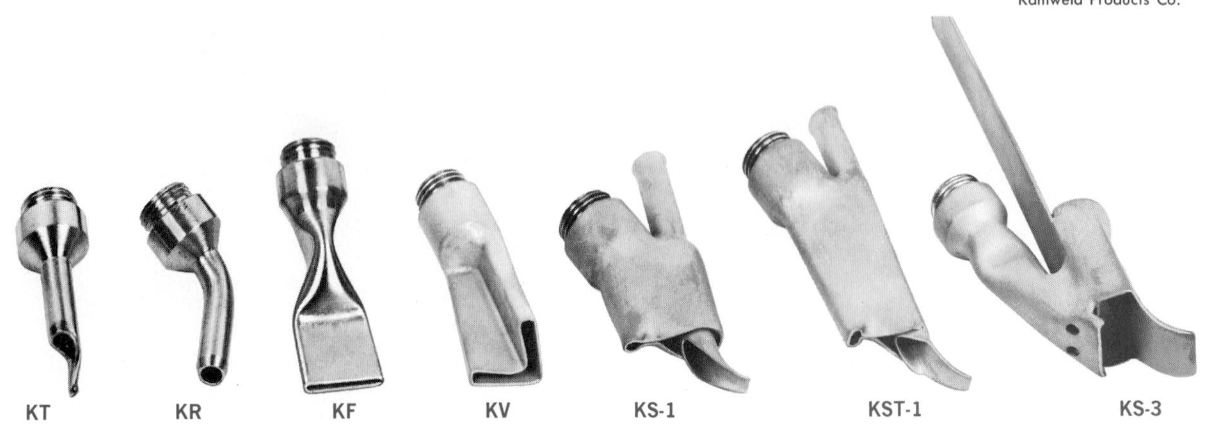

A good plastic weld requires the following:

- Thorough root penetration
- Proper balance between the heat used on the weld and the pressure exerted on the welding rod
- Correct handling of the welding torch
- Correct preparation of the joint to be welded

Study Table 29-2, which presents in greater detail the causes of common plastic welding troubles and how to correct them.

Visual Inspection

All welding should be inspected before the welded part is put into service. Visual inspection permits only partial evaluation of the weld bead. Such internal defects as incomplete fusion and penetration, air inclusions, and cracks cannot be determined by visual inspection.

A great deal, however, can be learned through visual inspection. A good weld has flowlines (little wavelike lines) along each side of the deposited weld rod and shows no signs of decomposition. If the flowlines are present, continuous, and uniform, it is visual evidence of a good weld. The continuity of these flowlines indicates that there was enough heat on the filler rod to create flow. The continuity also shows whether or not the welder applied the correct pressure to the rod, so that the hot, viscous material was forced out of the weld bed and bonded the plastic parts together.

Visual examination of multilayer welds can be accomplished by cutting across the axis of the weld and polishing the cross-section. Close inspection will reveal faults such as voids, scorching, and notching.

Testing of Welds

The procedures for testing plastic welds have been established by the American Society of Testing Materials and the Socie-

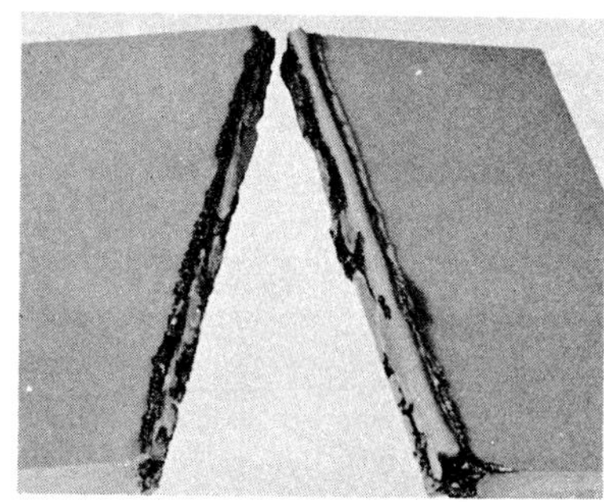

Kamweld Products Co.

29-13. *Overheated weld.*

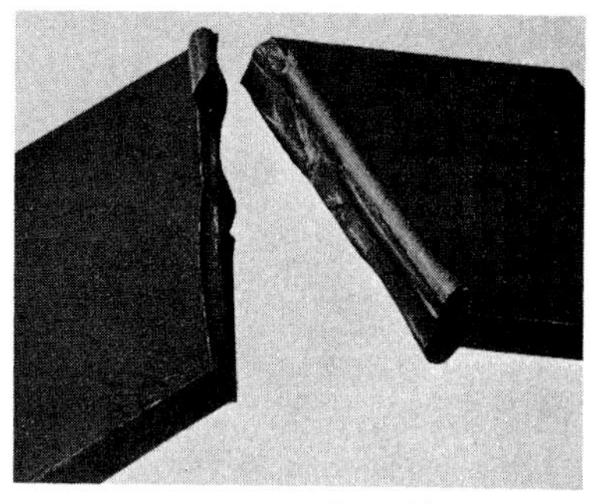

Kamweld Products Co.

29-14. *Cold weld.*

29-12. *Good and faulty welds.*

Kamweld Products Co.

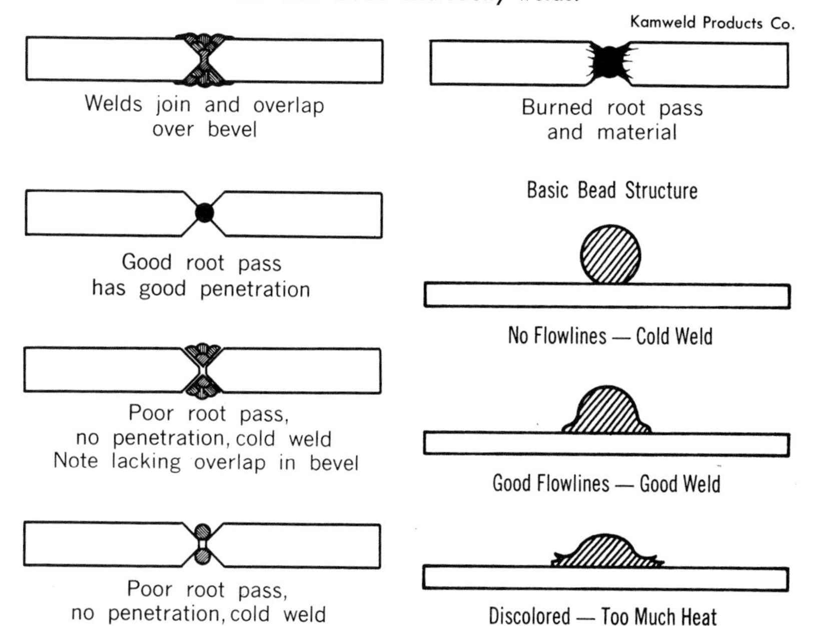

29-15. *Voids (1) and scorching (2) can be seen in this broken weld. There is poor penetration throughout.*

Laramy Products Co.

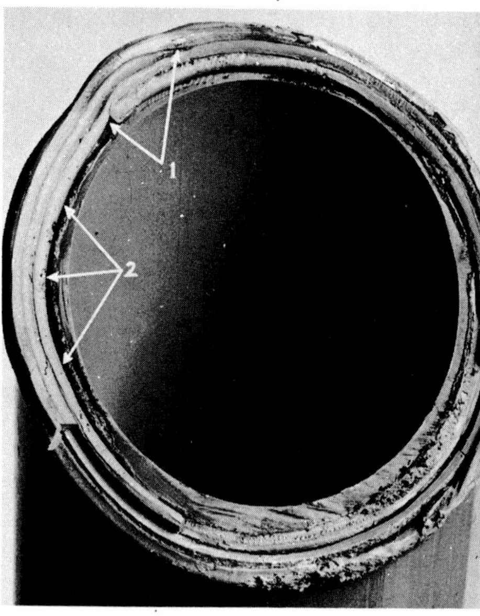

Table 29-2
Causes of faulty welds.

POROUS WELD

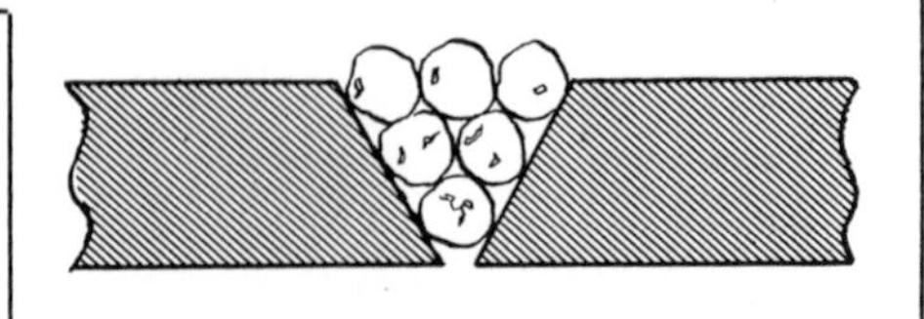

WHY:
1. Porous weld rod
2. Balance of heat on rod
3. Welding too fast
4. Rod too large
5. Improper starts or stops
6. Improper crossing of beads
7. Stretching rod

CORRECTION:
1. Inspect rod
2. Use proper fanning motion
3. Check welding temperature
4. Weld beads in proper sequence
5. Cut rod at angle, but cool before releasing
6. Stagger starts and overlap splices ½″

POOR PENETRATION

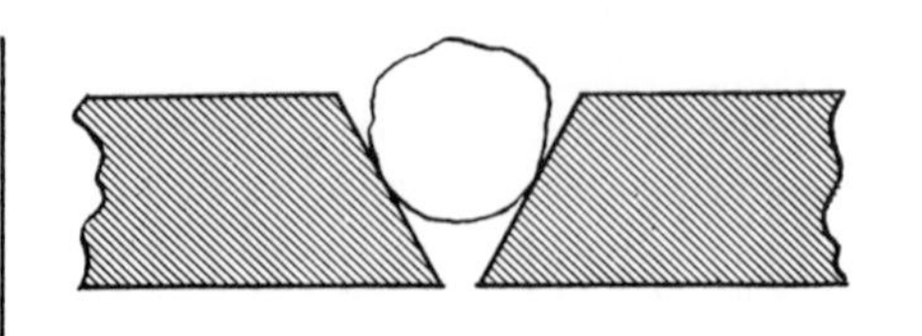

WHY:
1. Faulty preparation
2. Rod too large
3. Welding too fast
4. Not enough root gap

CORRECTION:
1. Use 60° bevel
2. Use small rod at root
3. Check for flow lines while welding
4. Use tacking tip or leave $\frac{1}{32}$″ root gap and clamp pieces

SCORCHING

WHY:
1. Temperature too high
2. Welding too slow
3. Uneven heating
4. Material too cold

CORRECTION:
1. Increase air flow
2. Hold constant speed
3. Use correct fanning motion
4. Preheat material in cold weather

DISTORTION

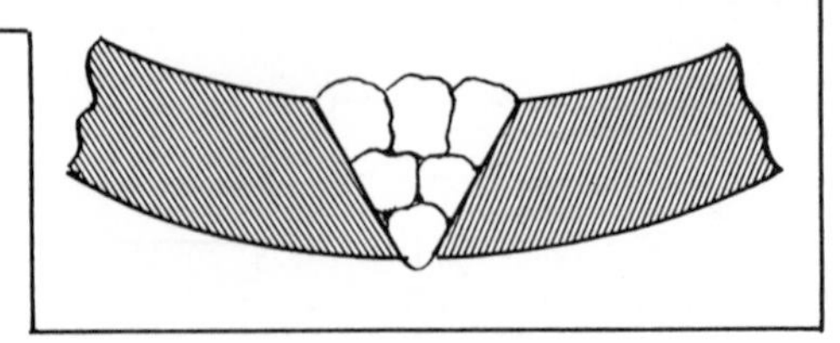

WHY:
1. Over-heating at joint
2. Welding too slow
3. Rod too small
4. Improper sequence

CORRECTION:
1. Allow each bead to cool
2. Weld at constant speed—use speed tip
3. Use larger sized or triangular shaped rod
4. Offset pieces before welding
5. Use double V or back-up weld
6. Back-up weld with metal

WARPING

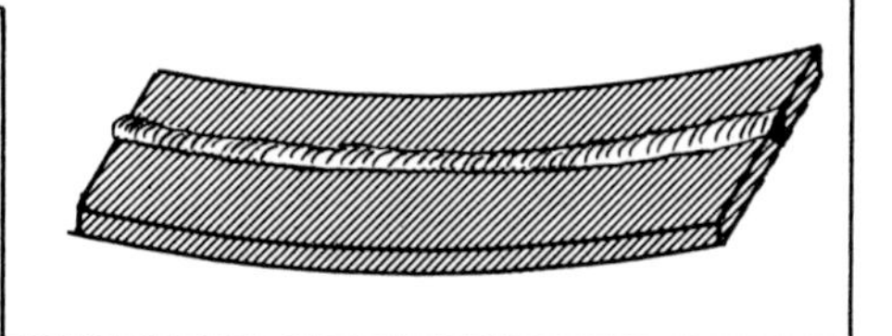

WHY:
1. Shrinkage of material
2. Overheating
3. Faulty preparation
4. Faulty clamping of parts

CORRECTION:
1. Preheat material to relieve stress
2. Weld rapidly—use back-up weld
3. Too much root gap
4. Clamp parts properly—back-up to cool
5. For multilayer welds—allow time for each bead to cool

POOR APPEARANCE

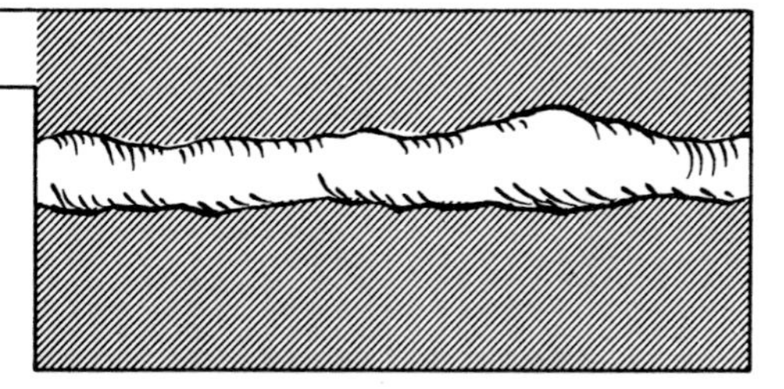

WHY:
1. Uneven pressure
2. Excessive stretching
3. Uneven heating

CORRECTION:
1. Practice starting, stopping and finger manipulation on rod
2. Hold rod at proper angle
3. Use slow uniform fanning motion, heat both rod and material

(For Speedwelding: use only moderate pressure, constant speed, keep shoe free of residue)

STRESS CRACKING

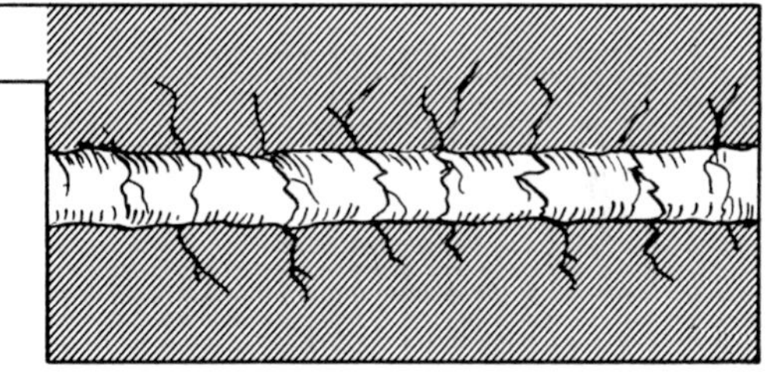

WHY:
1. Improper welding temperature
2. Undue stress on weld
3. Chemical attack
4. Rod and base material not same composition
5. Oxidation or degradation of weld

CORRECTION:
1. Use recommended welding temperature
2. Allow for expansion and contraction
3. Stay within known chemical resistance and working temperatures of material
4. Use similar materials and inert gas for welding
5. Refer to recommended application

POOR FUSION

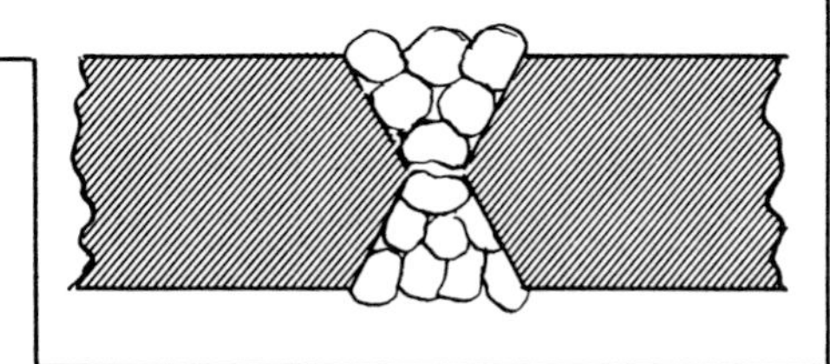

WHY:
1. Faulty preparation
2. Improper welding techniques
3. Wrong speed
4. Improper choice of rod size
5. Wrong temperature

CORRECTION:
1. Clean materials before welding
2. Keep pressure and fanning motion constant
3. Take more time by welding at lower temperatures
4. Use small rod at root and large rods at top—practice proper sequence
5. Preheat materials when necessary
6. Clamp parts securely

Laramy Products Co.

ties of Plastic Industries and Plastic Fabricators. Testing of plastic welds is divided into three general categories: destructive, nondestructive, and chemical.

DESTRUCTIVE TESTING. Tensile Test. This test is used primarily to evaluate butt welds on rigid sheet. Cut two 4 × 6-inch pieces from $\frac{3}{16}$-inch sheet and make a 30-degree double bevel with a $\frac{1}{64}$-inch root face. Clamp each piece to the bench, leaving a root gap of $\frac{1}{64}$ inch. Apply two or three beads on each side of the specimen. Use small diameter rod for root welding and larger rod to complete the weld. Cut the test section into $\frac{1}{2}$-inch wide specimens and pull each in a tensile tester at the rate of 0.025 inch per minute. The welding value may be calculated as follows:

$$\frac{\text{Breaking strength of weld} \times 100}{\text{Original tensile strength of material}} = \text{Welding value}$$

Bending Test. Weld a test specimen like that used for tensile testing. While it is still hot, bend it double along the axis of the weld. The weld should be fused to the beveled surface of the plastic sheet. Another bend test may be conducted 24 hours after welding. In this test, bend the specimen 90 degrees by hand. It should resist breaking, and the base material should break outside of the weld.

Burst Test. This is the most effective way of testing pipe butt welds and fillet welds on fabricated fittings and couplings. The open ends of the pipe are capped, and the weld is subjected to hydrostatic presure with water as the liquid.

Impact Test. The weld is subjected to a sudden impact by hitting it with a hammer. The broken joint is examined for faulty bonding, voids, and scorching.

NONDESTRUCTIVE TESTING. It should be noted here that visual inspection is considered by many to be a nondestructive form of testing.

Spark Coil Test. A high frequency, high voltage spark-coil tester, Fig. 29-16, detects pores and cracks in a plastic weld that are not visible by any other inspection method. In using a spark-coil tester, the plastic weld is grounded with a metal backing, and the pencil-like tip of the tester is run along the weld area. If there is porosity or cracking, a straight line of sparks passes through the weld to the metal ground.

Radiography. This is perhaps the most efficient method of plastic weld inspection because it shows defects in a plastic weld that are not readily discernible by any other inspection method. It gives a complete, detailed picture of the internal characteristics of the weld joint and a permanent record of the weld. The one disadvantage of this type of testing is the high cost.

Chemical Tests. A PVC weld test specimen is immersed in acetone for two to four hours. Faulty welds will separate from the base material, and strains and stresses in the weld will be indicated by the swelling of the material. The dye penetrant method is another chemical inspection method suitable for plastic welds. A dye penetrant is painted or sprayed on the weld. A poor weld will be disclosed by the penetrant seeping through and discoloring the weld.

29-16. *High frequency spark tester.* Laramy Products Co.

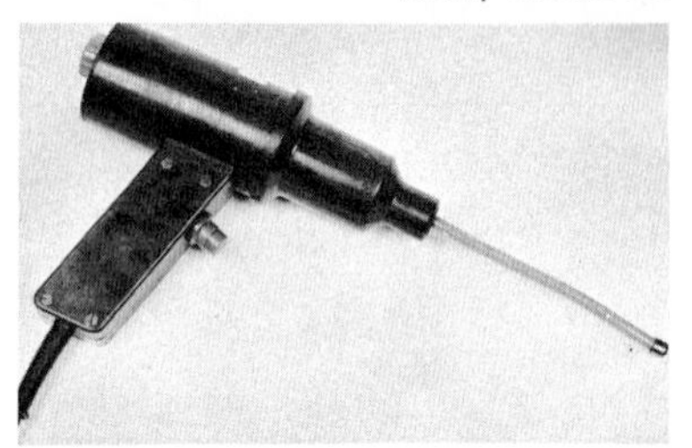

PRACTICE JOBS

Instructions for Completing Practice Jobs

Welding plastics is very similar to welding other materials. You must learn, however, a definite knack, or feel, that is somewhat different from that experienced when welding metal. Most mechanics get the feel for plastic welding in a few hours, but it can take as much as 40 hours of diligent practice before you make consistently good thermoplastic welds.

Learn the types of joints and what to look for in good and poor welds. Practice starts, stops, and tie-ins. Be careful that the base material and the rod are not exposed to overheating and underheating.

There are five important "musts" concerned with plastic welding:

- Small beads should form along each side of the weld where the rod meets the base material.
- The rod should hold its basic round shape.
- Neither the rod nor the base material should char or discolor.
- It is very important that the rod does not stretch over the weld. The length of rod used should be no more nor less than the length of the weld. If the rod is forced ahead in the direction of welding, instead of straight down into the weld point, fewer feet of the rod will be used than the length of the weld. This stretches the hot filler rod so that it breaks open at the top of the

weld bead. The broken bead is a special problem when the weld is reheated to add additional beads to a multipass weld.

- Do not use oxygen or other flammable gases.

The plastic filler rod must be of the same composition as the type of plastic being welded. Plastic rods for the different types of plastic are available in such shapes as round, oval, triangular, and flat strips. The sizes available are $^3/_{32}$ inch, $^1/_8$ inch, $^5/_{32}$ inch and $^3/_{16}$ inch. The type of plastic, the joint design, the thickness of the material, the position of welding, and the type of equipment are considerations in choosing a filler rod.

Heat for plastic welding is supplied by a heated gas instead of a flame. Compressed air, nitrogen, or an inert gas is used. (Never use oxygen or other flammable gases.) The gas passes through the torch where it is heated by the heating element and is then directed through the torch tip to the surface of the joint.

The torch is held in one hand, and the filler rod is held and fed with the other hand, Fig. 29-4, page 912. This is similar to the gas welding position. The rod can also be fed automatically with the use of a high speed welding tip, Fig. 29-17. This tip increases the speed of welding. It feeds the welding rod automatically in the right position and produces a uniform weld bead. One hand is left free to steady the work and insert new rods. Flexible PVC plastics are welded with a special strip feeder and both flat and corner tips. Welding filler rod for flexible plastics is in strip form.

The temperature of the welding gas is regulated by increasing or decreasing the volume of gas supplied to the torch. A slow movement of the gas over the heating element produces more heat. Therefore, a decrease in the welding gas volume increases the temperature of the gas. Conversely, a rapid movement of the gas over the heating element produces less heat, and so an increase in the welding gas volume decreases the temperature of the gas. The tip and the heating element may be changed to alter the heating capacity of the torch further.

Tack Welding

As with metal welding, tacking is used to assemble the parts to be welded quickly. Tacking ensures the proper alignment of all pieces. It holds the sections together during welding so that clamps, jigs, or additional manpower are not needed for this purpose. Tack-welded sections can be broken apart easily for reassembly and retacking.

PROCEDURE FOR TACK WELDING.

1. Attach the tack welding tip to the torch.
2. Wait for one or two minutes so that the tip can reach its proper temperature.
3. Hold the tip at an angle of approximately 90 degrees and place it directly on the joint to be tacked.
4. Draw the tacker tip along the joint for the desired length, Fig. 29-18. Tacks should be about $^1/_2$ to 1 inch long. Pieces that are large or unwieldy should be tacked at regular intervals as when welding metal.
5. The unit is now ready for continuous welding. The tacks should not show any brown or burned spots. This is an indication that too much heat was applied, that the rate of travel was too slow, or a combination of both. Tacks that do not hold the workpieces together are caused by lack of fusion due to insufficient heat, incorrect travel speed, or a combination of both.
6. Practice welds until you are able to make satisfactory tacks. The material must not be burned and should show good fusion.

Hand Welding (Beading)

The purpose of hand welding is to join two or more pieces permanently together with a rod or strip as a filler. This is similar to other forms of metal welding.

29-17. *Using the high speed welding tip. A plastic weld is being run on a girth seam of a tank.*

Seelye Plastics Co.

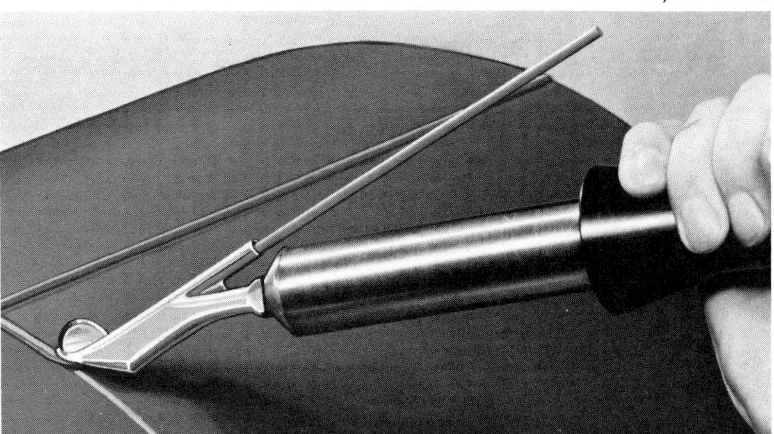

The fusion that takes place is the result of the proper combination of heat and pressure. The welder applies pressure on the filler rod with one hand while he applies heat to the rod and base material with hot gas from the welding torch, Fig. 29-19. Both pressure and heat must be kept constant and in proper balance. Too much pressure on the rod tends to stretch the bead and cause poor fusion and irregular bead buildup. Too much heat chars, melts, and distorts the material.

PROCEDURE FOR HAND WELDING PVC PLASTICS WITH A ROUND TIP.

1. Install a heating element that produces from 450 to 500 degrees F.
2. Attach a round tip to the torch.
3. Set the air pressure according to the recommendations by the manufacturer of the equipment.
4. Obtain a flat piece of PVC about 6 inches long, 4 inches wide, and at least $\frac{3}{32}$ inch thick. Make sure that the surface is clean. Clamp the piece to the workbench.
5. Secure a PVC filler rod $\frac{1}{8}$ inch in diameter and cut the end at a 60-degree angle with cutting pliers.
6. Check for correct temperature by holding the tip $\frac{1}{4}$ inch from the material and counting off 4 seconds. At the count of four, the material should show a faint yellowish tinge.
7. Hold the torch $\frac{1}{4}$ to $\frac{3}{4}$ inch from the material to be welded and preheat the starting area on the base material and rod until it appears shiny and becomes tacky. The rod is held at an angle of 90 degrees to each side of the base material, Fig. 29-20.
8. Move the torch up and down with a fanning or weaving motion in order to heat both the filler rod and base material equally, Fig. 29-21. About 60 percent of the heat should be directed toward the base material and 40 percent to the filler rod.
9. A good start is essential. It should take only a few seconds for the material and rod to be hot enough. Once the weld has been started, continue to fan the torch from the rod to the base material at the rate of two full oscillations per second. Exert only as much pressure on the rod as is necessary to cause fusion to take place.

- Too much forward pressure causes stretching which will lead to cracking as you weld.
- You should notice a small bead forming along both edges of the welding bead and a small roll forming under the welding rod, Fig. 29-22.
- Too much heat in the rod softens it so that pressure bends the rod rather than forcing it into the base material. Too little heat causes it to lay on the surface of the material without being fused to it.
- A slight yellowing of the rod and base material is caused by slight overheat. Run a number of welds that are alternately too hot and too cold in order to be able to recognize this condition.
- When a weld is to be ended, stop all forward motion and direct a quick heat directly at the intersection of the rod and base material. Remove the heat and maintain downward pressure for several seconds until rod is cool. Then release

29-18. *Tack welding.*
Kamweld Products Co.

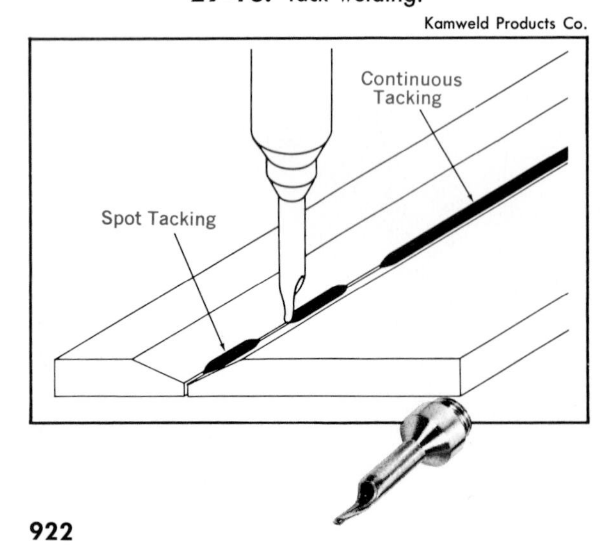

29-19. *Hand welding with a round tip.*
Kamweld Products Co.

downward pressure. Twist the rod with your fingers until it breaks. If a continuation weld is to be made, the deposited bead should be cut at an angle of 30 degrees with a sharp knife or cutting pliers.

BEAD TEST. Test the strength of your bead by trying to pick the rod off the base material with a pair of pliers while the bead is still hot. If fusion has taken place, the rod will stretch and may break, but it will not part from the base material. If the rod lifts readily from the base material, there may have been not enough heat or too much heat.

After you have run a series of satisfactory beads on a plastic pad with $^{1}/_{8}$-inch rod, repeat the procedure with different size welding rods. If the beads are laid so that they overlap the preceding bead, change the angle of the welding rod to 45 degrees. Always allow cooling time between beads. Welds may be cooled with a wet sponge or a jet of cool air.

Hand Welding of Joints

When you have mastered making good bead welds, you are ready to practice welding plastic sheet together.

There are three basic types of butt joints generally used in plastic construction: (1) the square butt joint, (2) the single-V butt joint, and (3) the double-V butt joint.

SQUARE BUTT JOINTS. This type of joint is generally made in light gauge sheets as thick as $^{3}/_{32}$ inch. No preparation of the edge is required. A root gap of approximately $^{1}/_{64}$ inch is necessary to permit full penetration through the back side. Welding is from both sides when possible, Fig. 29-23.

The square butt joint is acceptable when the work is not of a critical nature and when cost is a consideration.

1. Obtain two pieces of plastic sheet 6 inches long, 3 inches wide, and $^{3}/_{32}$ inch thick.
2. Set up the pieces with a root gap of $^{1}/_{64}$ inch to allow the semimolten plastic to flow through to the back side of the joint. Make sure the back side is suspended above the table.
3. Use the same welding technique described for beading. Weld one pass on each side of the plate.
4. Inspect the weld carefully for faults. Test the weld according to the procedure given under Butt Joint Test, page 924.

SINGLE-V BUTT JOINTS. Generally, single-V butt joints are used when only one side of the plastic sheet is accessible. If the back side is accessible, a single bead on the back side provides additional strength.

1. Obtain two pieces of plastic sheet 6 inches long, 3 inches wide, and $^{1}/_{8}$ inch thick.
2. Prepare the pieces with a 30-degree bevel and a $^{1}/_{32}$-inch flat face at the root. This is like the preparation necessary for steel welds.
3. Set up the pieces with a root gap of about $^{1}/_{64}$ inch to allow the semimolten plastic to flow through to the back side of the work. Make sure that the back side is suspended above the table. Use the same welding technique described for beading.
4. Weld the first pass along the root of the weld. Make sure that you obtain penetration through the back side, Fig. 29-24.
5. Weld two additional passes along the edge of each sheet. Make sure that the weld is built up evenly. The joint must be completely filled with overlaps on the top beveled edges, Fig. 29-24.

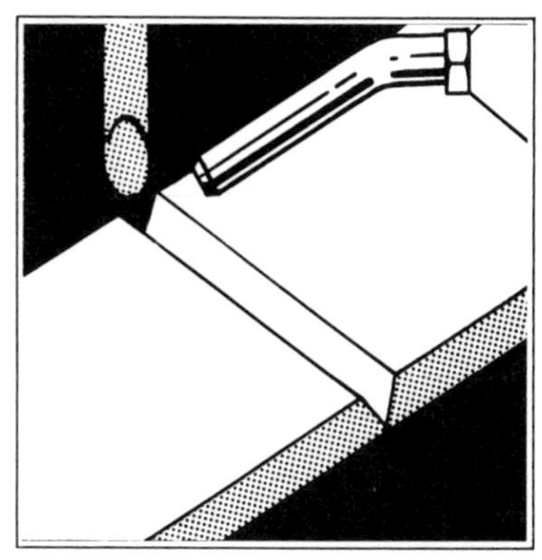

Seelye Plastics Co.

29-20. *Starting the welding operation.*

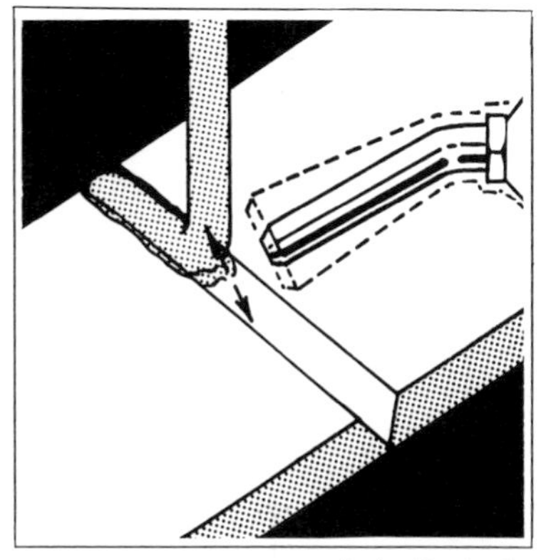

Seelye Plastics Co.

29-21. *The torch motion during welding.*

29-22. *Note the bead being formed along both edges of the weld.*

Seelye Plastics Co.

6. For the sake of additional practice and after you have checked the back side for faults, weld a bead on the back side. A single-V weld, reinforced in this manner, is somewhat stronger than a double-V butt weld.

7. Inspect the weld carefully for faults. Test the weld according to the procedure given under Butt Joint Test.

DOUBLE-V BUTT WELDS. The double-V butt weld has a high strength factor only slightly less than the single-V joint with reinforcement on the back side. Both sides of the joint must be accessible for welding.

1. Obtain two pieces of plastic sheet 6 inches long, 3 inches wide, and 1/8 inch thick.

2. Prepare the edges of the sheets to be welded with a 30-degree bevel. Allow a root gap of 1/32 to 1/64 inch.

3. Weld the first pass along the root of the weld on one side. Make sure that you obtain penetration through to the back side, Fig. 29-25.

4. Weld two additional passes along the edge of each sheet. Make sure that the weld is built up evenly, Fig. 29-25.

5. Inspect your welds carefully for good fusion and appearance.

6. Turn the plate over and repeat the above procedure on the other side.

7. Inspect the welds carefully for faults. Test the weld according to the procedure given under Butt Joint Test.

BUTT JOINT TEST. A square butt joint may be tested by fracturing it. Place the joint in the jaws of a vise with the weld bead facing away from you and about 3/16 inch above and parallel to the top of the vise jaws. Cover it with a cloth to prevent injury due to flying chips and pieces. A blow with a hammer on the weld side usually breaks off the top pieces. If a break occurs through the weld bead with some portion of the weld on each piece, the weld may be considered good. A 100-percent weld breaks through the base material. If there is overheating or underheating, the material usually separates at that point. For most plastic fabrication, a weld strength of 75 percent is acceptable, but every effort should be made to achieve 100 percent strength.

FILLET WELDS. Fillets and corner welds are used to attach two sheets of plastic at 90 degrees to each other in T-joints and corner joints. The vertical plate should be beveled. If welding is to be from one side only, the vertical plate must be beveled from one side only as shown in Fig. 29-26A. If, on the other hand, welding can take place from both sides, the vertical plate must be beveled from both sides as shown in Fig. 29-26B. Welding from both sides produces the stronger of the two welds.

LAP WELDS. Very often two sheets are joined by forming a lap joint, Fig. 29-27. Welding a plastic angle to a sheet also forms a lap joint. The welds may be made by fusing the overlapping areas with a flat tip, or they may be made with the round tip and filler rod.

EDGE WELDS. Edge welds are used to weld heads and bottoms into tanks or boxes and to weld an angle to the edge of a sheet. Two flat surfaces are set up side by side, and the two edges are welded together. If the material is heavy, the edge of each plate is chamfered and the weld is a groove, Fig. 29-28. The procedure for welding is similar to that used in making V-groove butt welds. *Note:* If you are having trouble making satisfactory welds, review Table 29-2, page 918, that lists contributing factors to faulty welding.

High Speed Welding

The high speed welding tip increases the average welding speed to over 4 feet per minute

Kamweld Products Co.

29-23. *Square butt joint welded from both sides.*

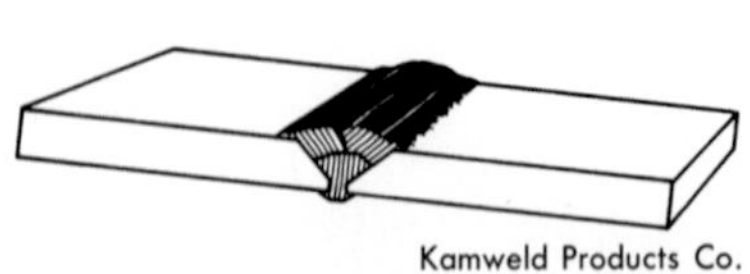

Kamweld Products Co.

29-24. *Single-V butt joint welded from one side only.*

29-25. *Double-V butt joint welded from both sides.*

Kamweld Products Co.

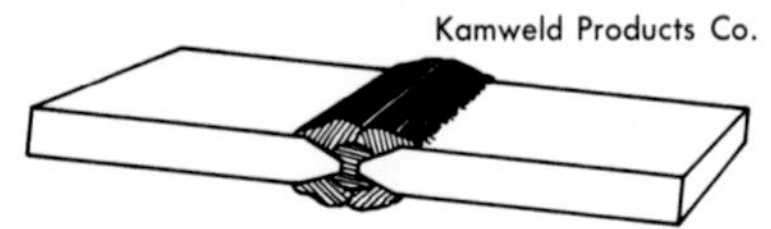

29-26. *Fillet welds for T-joints.*

Kamweld Products Co.

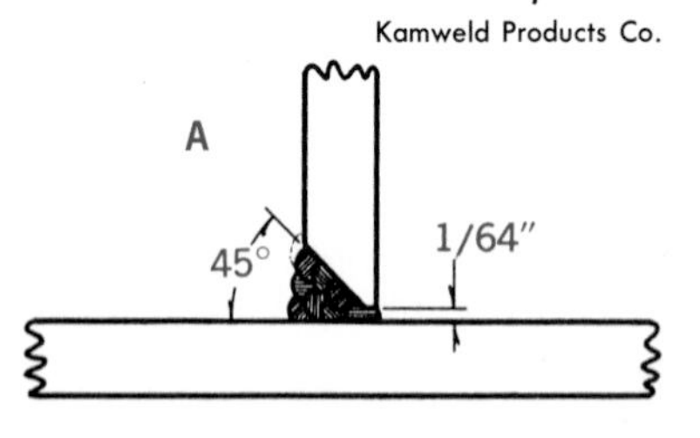

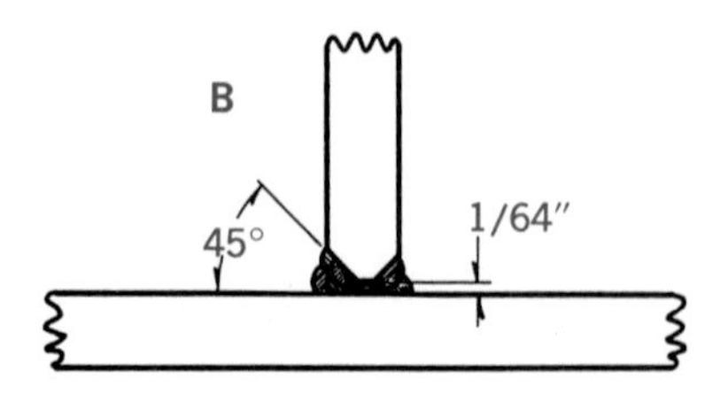

on flat or curved surfaces. It feeds the welding rod automatically in the right position and produces a uniform weld head, Fig. 29-29. One hand is left free to steady or turn the work and insert new rods. A handy cutting blade is attached to the tip. The 500-watt heating element is recommended for high speed welding.

PROCEDURE FOR HIGH SPEED WELDING WITH ROUND ROD.

1. Secure two pieces of PVC about 18 inches long, 3 inches wide, and $\frac{3}{32}$ inch thick, and clamp them to the workbench.
2. Select the high speed tool designed for the diameter of filler rod to be used. Prepare the rod by cutting one end at a 60-degree angle.
3. Set up the equipment and allow the unit to warm up.
4. Hold the welding unit straight down at a 90-degree angle to the work.
5. Hold the shoe of the high speed tool about $\frac{1}{2}$ to $\frac{3}{4}$ inch above the surface of the workpiece, and hold it at the starting point, Fig. 29-30.
6. Insert the beveled filler rod into the preheated tube and push it into the softened base material until the rod bends slightly backwards, Fig. 29-30.
7. Change the angle of the tip to about 60 degrees in the direction of welding. Apply pressure on the top surface of the rod sticking out until it starts to fuse to the surface of the material, Fig. 29-31.
8. Continue to exert pressure with the shoe and start pulling the torch in the direction of welding. At the same time exert light pressure on the rod in the preheating tube.
9. Continue to press on the top surface of the rod with the shoe as you proceed with the weld. Do not move forward faster than the fusion of the rod to the base material will permit.
10. Once the weld is started, there can be no hesitation. As you continue with the weld, the rod will feed through, being pulled by its adhesion to the base material. The following conditions must be given careful attention:
    - The speed of the weld can be increased by lowering the angle of the welder to about 45 degrees.
    - Note the flowlines which are similar to those visible in hand welding. The absence of flowlines indicates a "cold" and unsatisfactory weld. The crown is higher than in hand welding.
    - If the tip is not moved forward fast enough, the rod softens and bunches up in the preheating tube. Sometimes it chars or burns. The emerging end of the rod softens, flattens out, stretches, and breaks. When this happens, withdraw the tip and cut the rod. Any residue in the preheat tube can be removed by pushing a cold rod back and forth through the tube until it is cleared.
    - Observe the emerging rod constantly so that any corrective action that may be necessary can be taken immediately.
11. If the rod is stretching, the weld is going too slowly and the rod is overheating. When stretching occurs, withdraw the tip, cut off the rod, and make a new start before the point where the rod started to stretch.
12. If flowlines do not appear, the weld is going too fast, and adequate bonding is not taking place. Bring back the tip to the 90-degree angle temporarily in order to slow down the welding rate. Then move it to the desired angle for proper welding speed. The rate at which the weld proceeds is governed by the temperature, the consistency of the rod, and the angle of the welder.

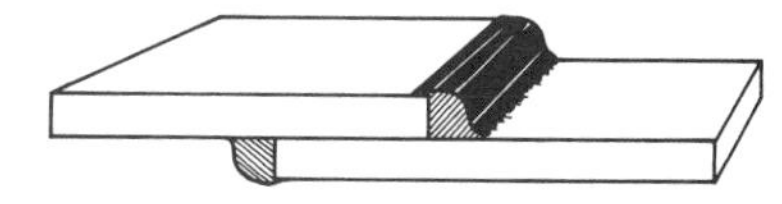

Kamweld Products Co.

29-27. *Method of welding lap joints.*

29-28. *Method of welding edge joints.*

Kamweld Products Co.

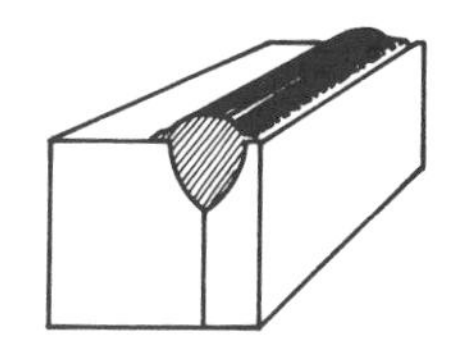

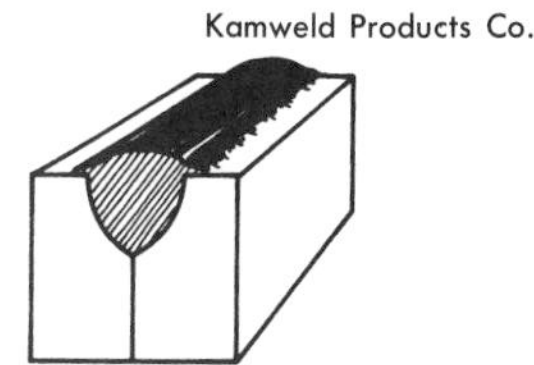

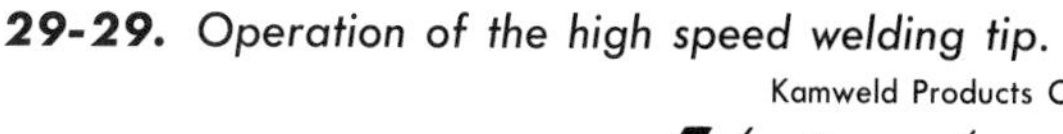

29-29. *Operation of the high speed welding tip.*

Kamweld Products Co.

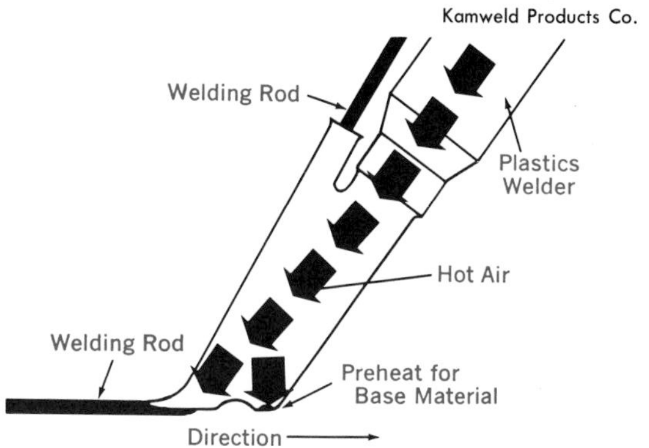

13. Make sure that the preheater hole and the shoe are always in line with the direction of the weld so that only the material in front of the shoe is preheated.

14. To stop the welding process, (a) withdraw the tip quickly until the rod is out of the tube, and (b) bring the tip quickly to the 90-degree angle and cut off the rod with the end of the shoe. Remove the rod from the preheating tube immediately so that it does not clog up the tube. The preheating tube must be kept clean, and the shoe should be cleaned with a soft wire brush to remove residue.

15. A good speed weld in a V-joint has a slightly higher crown than the normal hand weld, and it is more uniform. It should appear smooth and shiny, with a slight bead on each side. Study the weld characteristics shown in Table 29-2, page 918, and compare with your weld.

HIGH SPEED WELD TESTS. When you and your instructor feel that you have acquired the skill of making this type of plastic weld, make up test specimens and subject them to the following test procedures:

- Cut through the joint and inspect for complete bonding. Strips should also be cut from the work and subjected to tensile and bending stress tests.
- If a pressure test is desired, make up a small box and subject it to a water-pressure test. The box can be similar to that used for metal arc welding.

HIGH SPEED WELDING WITH A PLASTIC STRIP. Flexible PVC plastic is welded with a flexible plastic strip instead of a round rod. Plastic strips come in different shapes such as the flat strip and the corner V-strip and are generally supplied in roll form. Strip welding is generally used for tank linings and similar applications.

Only one pass is necessary with a strip. The technique of welding is like that of welding with round rod with the following exceptions:

- The strip must be precut in length, and 1 to 2 inches must be allowed for trimming.
- Start the weld by tamping with the broad shoe of the high speed tool on the top of the first inch of strip in the direction of the weld. Do not drag the tip until the first inch of the strip adheres firmly to the base material. The tip should be held at an angle of about 80 degrees, and some pressure should be applied by hand to hold the top of the strip down.
- Guide the strip by hand and continue the weld at sufficient speed. The usual flowlines should appear on both sides of the strip, Fig. 29-31. If travel speed is too slow, excessive heat causes the strip to soften and stretch. This can be corrected by a quick tamping motion with the shoe like that used to start the weld. Then proceed in the direction of the weld.
- To stop the weld, simply remove the tip and allow the remaining strip to pull through the high speed tool.

If a strip weld is to be made from corner to corner, such as on the bottom of a tank, the weld should be started in one corner and proceed only halfway across. Then another weld is started in the other corner and overlaps the scarfed end of the first weld. Where each strip butts to the other or in corners, additional flat strip overlays are necessary. A flat tip is used to preheat the area while the flat strip, held in a gloved hand, is pressed firmly into the corner. All high corners can be heated and pushed down with the blunt end of a knife.

Welding Plastic Pipe

After you have mastered welding plastic sheet, you should practice making butt welds in plastic pipe. The preparation and welding of pipe is similar to that used for flat material. There is one major difference. In welding sheet, the welding proceeds in a forward direction only. In welding pipe, the torch and filler rod must follow the direction of the round shape.

Solvent Weld Process (Bonding)

Plastic pipe may be joined by a method referred to as a *solvent weld* or *bonding* process. Pipe and fittings are solvent welded with a solvent known as "MEK" (methylethyl-ketone). The solvent chemically etches the surface of both the pipe and fittings so that when joined, the two surfaces are fused into each other like brazed copper or welded steel. Cutting through a solvent-welded joint shows that the original line of division no longer exists. This joint is now stronger than either the pipe or fitting.

PROCEDURE FOR BONDING PIPE.

1. Use Schedule 40 ABS-DWV pipe and MEK solvent. Cut the pipe and remove all burrs. Be sure both pipe and fitting are clean. Wipe clean with a cloth if necessary. Grease or oil may be removed with paint thinner.

2. In applying solvent, use a brush large enough to pass around the pipe end or fitting socket quickly. Apply solvent to the fitting socket first and then to the pipe end. Use solvent moderately, but spread it thoroughly and evenly. Excess solvent serves no purpose.

3. Insert the pipe into the fitting and position it with a quick rotating motion of a quarter turn or so. This insures a full, even spread of the solvent. Do not attempt to realign the pipe and fitting after the set has begun. If you notice a mistake, remove the fitting quickly and re-apply the solvent as before. Do not attempt to turn the fitting after the set has taken place. This will destroy the joint. In dry weather, joints will dry in $1\frac{1}{2}$ to 2 minutes. Cold, wet weather slows the drying period, but in no event will it exceed 5 minutes. Three minutes is the typical setting period.

4. After full set, water tests may be applied immediately. Pressure systems require a longer drying period because of higher pressure tests.

5. Not all plastics nor solvents react the same. Some plastic piping requires solvent which has a considerably longer setting and drying period. A two-stage process for cleaning and cementing, which increases the time required, is necessary with some solvents. Solvent can be removed from the hands with thinner. Wash hands afterwards with soap and water.

6. Test bonded joints by cutting through the joint and inspect for thorough bonding. Strips should also be cut from the work and subjected to tensile and bending stress tests.

Care of Plastic Pipe

As with all welding, there are certain material characteristics that must be provided for.

- Be careful in storing pipe and fittings. They should never be stored in the sun. These materials heat very slowly. Therefore the side exposed to the sun's rays absorbs the heat faster than the unexposed side, and uneven expansion causes the pipe to "bow." Both pipe and fittings should be kept as free of dirt as possible.
- Care must be taken in cutting pipe. Plastic pipe may be cut with pipe cutters or any crosscut saw. Special pipe cutter wheels for plastic are available to fit all standard pipe cutters. Lightweight cutters designed exclusively for plastic piping are also available. Do not use dull cutters; they may cause too many burrs. All burrs must be removed from the pipe. Care must be taken to insure a square cut. All socket fittings require a good square pipe to insure a good fit. Keep in mind that the better the fit, the better the joint.
- Pipe must be laid out and cut with a high degree of accuracy because errors cannot be rectified with stress, heat, or a hammer. A fitting socket must be measured to the full depth of the socket.
- Plastic pipe must be supported properly. Avoid tight supports, which tend to compress or cut the pipe or fitting. Horizontal piping should be supported every 4 or 5 feet and on both sides at vital turns and large fittings to relieve strain. Vertical stacks should be supported at the base and between floor and ceilings. Provide for expansion on runs over 50 feet in length.

29-30. *Welding positions for the high speed tip.*
Kamweld Products Co.

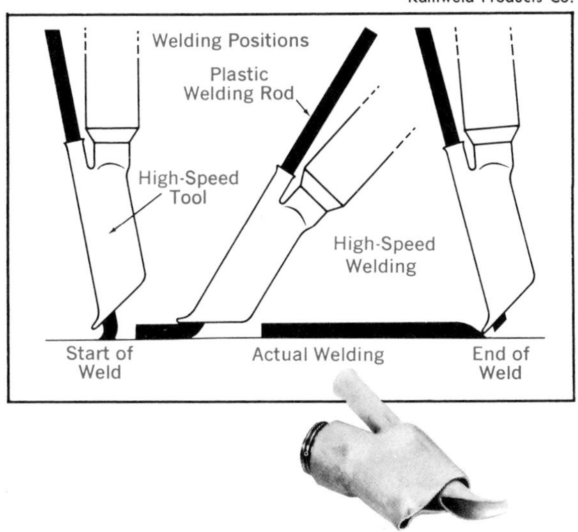

29-31. *Adhesion of flexible strip to base material during high speed welding.*
Kamweld Products Co.

Plasticized Strip
Position of Welder
High-Speed Tool
Flowlines
Base Material
High-Speed Welding

- Never attempt to heat and bend plastic pipe.
- Plastic pipe may be placed underground with safety since it is impervious to the soil acids in this country. It is the only material that can be safely installed in a cinder fill. It requires no painting or coating. Make sure that pipe is fully supported all around before completing the fill in order to protect pipe alignment. The pipe is impervious to corrosion by cement so that no coating or wrapping is required.
- Plastic piping lends itself to prefabrication. Its light weight coupled with its strong welded joints permits preassembly of large assemblies and easy movement from shop to field. Assemblies can be moved five minutes after the last joint is made. Always test the assembly before moving it to the field.
- Plastic pipe can be joined to pipe made of other materials only with the appropriate adapter fittings. A complete line of adapter fittings is available. Lead can be poured and caulked directly into a cast iron hub without damage to the plastic pipe.

REVIEW QUESTIONS

1. How long ago were thermoplastics developed?
2. Where were thermoplastics put into general use and why?
3. Name five industrial applications of plastics.
4. Name the various types of plastics.
5. Name the three grades of polyethylene and the characteristics of each.
6. What form of plastic is transparent?
7. Explain the various work processes that can be applied to plastics.
8. Name the four methods of welding plastics and describe the principal method.
9. Name the two basic types of plastic welding torches.
10. List the equipment needed for plastic welding.
11. List the six interrelated factors that offset the strength of a plastic weld.
12. List the causes of faulty welds.
13. Name the three types of testing that are used to test plastic welds, and briefly explain each.
14. The temperature of the welding gas is increased by increasing the electrical current. True. False.
15. A rapid movement of the gas over the heating element produces more heat. True. False.
16. Plastic joints should never be held together with tack welds. True. False.
17. Plastic welds must have a strength of 100 percent. True. False.
18. What precautions must be taken in storing plastic pipe?
19. Plastic pipe can be heated and bent. True. False.
20. Plastic pipe can never be used in connection with pipe made of other materials. True. False.

Table of Contents for Appendix

Table A-1

U.S. Standard Gauges and their fractional, decimal, and metric equivalents.

U.S. Standard Gauge	Fractional equivalent (inches)			Decimal equivalent (inches)	Metric equivalent (mm)
	Major fraction	32nds	64ths		
30	...	...	...	0.0125	0.3175
29	...	...	...	0.0140625	0.3572
28	...	...	1	0.015625	0.3969
27	...	...	...	0.0171875	0.4366
26	...	...	...	0.01875	0.4762
25	...	...	...	0.021875	0.5556
24	...	...	...	0.025	0.6350
23	...	...	...	0.028125	0.7144
22	...	1	2	0.03125	0.7938
21	...	...	...	0.034375	0.8731
20	...	...	...	0.0375	0.9525
19	...	...	...	0.04375	1.1112
...	...	...	3	0.046875	1.1906
18	...	...	...	0.05	1.2700
17	...	...	...	0.05625	1.4288
16	1/16	2	4	0.0625	1.5875
15	...	...	...	0.0703125	1.7859
14	...	...	5	0.078125	1.9844
13	...	3	6	0.09375	2.3812
12	...	...	7	0.109375	2.7781
11	1/8	4	8	0.125	3.1750
10	...	...	9	0.140625	3.5719
9	...	5	10	0.15625	3.9688
8	...	...	11	0.171875	4.3656
7	3/16	6	12	0.1875	4.7625
6	...	...	13	0.203125	5.1594
5	...	7	14	0.21875	5.5562
4	...	...	15	0.234375	5.9531
3	1/4	8	16	0.25	6.3500
2	...	...	17	0.265625	6.7469
1	...	9	18	0.28125	7.1438
...	...	...	19	0.296875	7.5406
0	5/16	10	20	0.3125	7.9375
...	...	...	21	0.328125	8.3344
2/0	...	11	22	0.34375	8.7312
...	...	...	23	0.359375	9.1281
3/0	3/8	12	24	0.375	9.5250
...	...	...	25	0.390625	9.9219
4/0	...	13	26	0.40625	10.3188
...	...	...	27	0.421875	10.7156
5/0	7/16	14	28	0.4375	11.1125
...	...	...	29	0.453125	11.5094
6/0	...	15	30	0.46875	11.9062
...	...	...	31	0.484375	12.3031
7/0	1/2	16	32	0.5	12.7000

Table A-2

S.I. metric and English (customary) equivalent linear measurements.

English (customary) to metric	Metric to English (customary)
1 inch = 25.40 millimetres	1 millimetre = 0.03937 inch
1 inch = 2.540 centimetres	1 centimetre = 0.3937 inch
1 inch = 0.0254000 metres	1 metre = 39.37 inches
1 foot = 30.480 centimetres	1 metre = 3.2808 feet
1 foot = 0.3048 metre	1 metre = 1.0936 yards
1 yard = 91.440 centimetres	1 kilometre = 0.62137 mile
1 yard = 0.9144 metre	
1 mile = 1609.35 metres	
1 mile = 1.609 kilometres	

Table A-3

S.I. metric and English (customary) equivalent measurements of area.

English (customary) to metric	Metric to English (customary)
1 sq. inch = 645.16 sq. millimetres	1 sq. millimetre = 0.00155 sq. inch
1 sq. inch = 6.4516 sq. centimetres	1 sq. centimetre = 0.1550 sq. inch
1 cu. inch = 16.387 cu. centimetres	1 cu. centimetre = 0.0610 cu. inches
1 sq. foot = 929.03 sq. centimetres	1 sq. metre = 10.7640 sq. feet
1 sq. foot = 9.290 sq. decimetres	1 sq. metre = 1.196 sq. yards
1 sq. foot = 0.0929 sq. metre	1 cu. metre = 35.314 cu. feet
1 cu. foot = 0.02832 cu. metre	1 cu. metre = 1.308 cu. yards
1 sq. yard = 0.836 sq. metre	1 sq. hectometre = 2.471 acres
1 cu. yard = 0.765 cu. metre	1 hectare = 2.471 acres
1 acre = 0.4047 sq. hectometre	1 sq. kilometre = 0.386 sq. mile
1 acre = 0.4047 hectare	
1 sq. mile = 2.59 sq. kilometres	

Table A-4

S.I. metric and English (customary) equivalent measurements of mass (weight).

English (customary) to metric	Metric to English (customary)
1 ounce (dry) = 28.35 grams	1 gram = 0.03527 ounce
1 pound = 0.4536 kilogram	1 kilogram = 2.2046 pounds
1 short ton (2000 lb.) = 907.2 kilograms	1 metric ton = 2204.6 pounds
1 short ton (2000 lb.) = 0.9072 metric ton	1 metric ton = 1.102 tons (short)

Table A-5

S.I. metric and English (customary) equivalent measurements of volume (capacity).

English (customary) to metric	Metric to English (customary)
1 fluid ounce = 2.957 centilitres	1 centilitre = 10 cubic centimetres
1 fluid ounce = 29.57 cubic centimetres	1 centilitre = 0.338 fluid ounces
1 pint (liq.) = 4.732 decilitres	1 decilitre = 100 cubic centimetres
1 pint (liq.) = 473.2 cubic centimetres	1 decilitre = 0.0528 pint (liq.)
1 quart (liq.) = 0.9463 litre	1 litre = 1 cubic decimetre
1 quart (liq.) = 0.9463 cubic decimetre	1 litre = 1.0567 quarts (liq.)
1 gallon (liq.) = 3.7853 litres	1 litre = 0.26417 gallon (liq.)
1 gallon (liq.) = 3.7853 cubic decimetres	1 hectolitre = 26.417 gallons (liq.)

Table A-6

Conversion of decimal inches to millimetres and fractional inches to decimal inches and millimetres.

Inches dec.	mm.	Inches dec.	mm.
0.01	0.2540	0.51	12.9540
0.02	0.5080	0.52	13.2080
0.03	0.7620	0.53	13.4620
0.04	1.0160	0.54	13.7160
0.05	1.2700	0.55	13.9700
0.06	1.5240	0.56	14.2240
0.07	1.7780	0.57	14.4780
0.08	2.0320	0.58	14.7320
0.09	2.2860	0.59	14.9860
0.10	2.5400	0.60	15.2400
0.11	2.7940	0.61	15.4940
0.12	3.0480	0.62	15.7480
0.13	3.3020	0.63	16.0020
0.14	3.5560	0.64	16.2560
0.15	3.8100	0.65	16.5100
0.16	4.0640	0.66	16.7640
0.17	4.3180	0.67	17.0180
0.18	4.5720	0.68	17.2720
0.19	4.8260	0.69	17.5260
0.20	5.0800	0.70	17.7800
0.21	5.3340	0.71	18.0340
0.22	5.5880	0.72	18.2880
0.23	5.8420	0.73	18.5420
0.24	6.0960	0.74	18.7960
0.25	6.3500	0.75	19.0500
0.26	6.6040	0.76	19.3040
0.27	6.8580	0.77	19.5580
0.28	7.1120	0.78	19.8120
0.29	7.3660	0.79	20.0660
0.30	7.6200	0.80	20.3200
0.31	7.8740	0.81	20.5740
0.32	8.1280	0.82	20.8280
0.33	8.3820	0.83	21.0820
0.34	8.6360	0.84	21.3360
0.35	8.8900	0.85	21.5900
0.36	9.1440	0.86	21.8440
0.37	9.3980	0.87	22.0980
0.38	9.6520	0.88	22.3520
0.39	9.9060	0.89	22.6060
0.40	10.1600	0.90	22.8600
0.41	10.4140	0.91	23.1140
0.42	10.6680	0.92	23.3680
0.43	10.9220	0.93	23.6220
0.44	11.1760	0.94	23.8760
0.45	11.4300	0.95	24.1300
0.46	11.6840	0.96	24.3840
0.47	11.9380	0.97	24.6380
0.48	12.1920	0.98	24.8920
0.49	12.4460	0.99	25.1460
0.50	12.7000	1.00	25.4000

Inches frac.	Inches dec.	mm.	Inches frac.	Inches dec.	mm.
1/64	0.015625	0.3969	33/64	0.515625	13.0969
1/32	0.031250	0.7938	17/32	0.531250	13.4938
3/64	0.046875	1.1906	35/64	0.546875	13.8906
1/16	0.062500	1.5875	9/16	0.562500	14.2875
5/64	0.078125	1.9844	37/64	0.578125	14.6844
3/32	0.093750	2.3812	19/32	0.593750	15.0812
7/64	0.109375	2.7781	39/64	0.609375	15.4781
1/8	0.125000	3.1750	5/8	0.625000	15.8750
9/64	0.140625	3.5719	41/64	0.640625	16.2719
5/32	0.156250	3.9688	21/32	0.656250	16.6688
11/64	0.171875	4.3656	43/64	0.671875	17.0656
3/16	0.187500	4.7625	11/16	0.687500	17.4625
13/64	0.203125	5.1594	45/64	0.703125	17.8594
7/32	0.218750	5.5562	23/32	0.718750	18.2562
15/64	0.234375	5.9531	47/64	0.734375	18.6531
1/4	0.250000	6.3500	3/4	0.750000	19.0500
17/64	0.265625	6.7469	49/64	0.765625	19.4469
9/32	0.281250	7.1438	25/32	0.781250	19.8437
19/64	0.296875	7.5406	51/64	0.796875	20.2406
5/16	0.312500	7.9375	13/16	0.812500	20.6375
21/64	0.328125	8.3344	53/64	0.828125	21.0344
11/32	0.343750	8.7312	27/32	0.843750	21.4312
23/64	0.359375	9.1281	55/64	0.859375	21.8281
3/8	0.375000	9.5250	7/8	0.875000	22.2250
25/64	0.390625	9.9219	57/64	0.890625	22.6219
13/32	0.406250	10.3188	29/32	0.906250	23.0188
27/64	0.421875	10.7156	59/64	0.921875	23.4156
7/16	0.437500	11.1125	15/16	0.937500	23.8125
29/64	0.453125	11.5094	61/64	0.953125	24.2094
15/32	0.468750	11.9062	31/32	0.968750	24.6062
31/64	0.484375	12.3031	62/64	0.984375	25.0031
1/2	0.500000	12.7000	1	1.000000	25.4000

For converting decimal-inches in "thousandths," move decimal point in both columns to left.

L. S. Starrett Co.

Table A-7

Conversion of millimetres to decimal inches.

mm	Inches	mm	Inches	mm	Inches	mm	Inches	mm	Inches
0.01	0.00039	0.41	0.01614	0.81	0.03189	21	0.82677	61	2.40157
0.02	0.00079	0.42	0.01654	0.82	0.03228	22	0.86614	62	2.44094
0.03	0.00118	0.43	0.01693	0.83	0.03268	23	0.90551	63	2.48031
0.04	0.00157	0.44	0.01732	0.84	0.03307	24	0.94488	64	2.51968
0.05	0.00197	0.45	0.01772	0.85	0.03346	25	0.98425	65	2.55905
0.06	0.00236	0.46	0.01811	0.86	0.03386	26	1.02362	66	2.59842
0.07	0.00276	0.47	0.01850	0.87	0.03425	27	1.06299	67	2.63779
0.08	0.00315	0.48	0.01890	0.88	0.03465	28	1.10236	68	2.67716
0.09	0.00354	0.49	0.01929	0.89	0.03504	29	1.14173	69	2.71653
0.10	0.00394	0.50	0.01969	0.90	0.03543	30	1.18110	70	2.75590
0.11	0.00433	0.51	0.02008	0.91	0.03583	31	1.22047	71	2.79527
0.12	0.00472	0.52	0.02047	0.92	0.03622	32	1.25984	72	2.83464
0.13	0.00512	0.53	0.02087	0.93	0.03661	33	1.29921	73	2.87401
0.14	0.00551	0.54	0.02126	0.94	0.03701	34	1.33858	74	2.91338
0.15	0.00591	0.55	0.02165	0.95	0.03740	35	1.37795	75	2.95275
0.16	0.00630	0.56	0.02205	0.96	0.03780	36	1.41732	76	2.99212
0.17	0.00669	0.57	0.02244	0.97	0.03819	37	1.45669	77	3.03149
0.18	0.00709	0.58	0.02283	0.98	0.03858	38	1.49606	78	3.07086
0.19	0.00748	0.59	0.02323	0.99	0.03898	39	1.53543	79	3.11023
0.20	0.00787	0.60	0.02362	1.00	0.03937	40	1.57480	80	3.14960
0.21	0.00827	0.61	0.02402	1	0.03937	41	1.61417	81	3.18897
0.22	0.00866	0.62	0.02441	2	0.07874	42	1.65354	82	3.22834
0.23	0.00906	0.63	0.02480	3	0.11811	43	1.69291	83	3.26771
0.24	0.00945	0.64	0.02520	4	0.15748	44	1.73228	84	3.30708
0.25	0.00984	0.65	0.02559	5	0.19685	45	1.77165	85	3.34645
0.26	0.01024	0.66	0.02598	6	0.23622	46	1.81102	86	3.38582
0.27	0.01063	0.67	0.02638	7	0.27559	47	1.85039	87	3.42519
0.28	0.01102	0.68	0.02677	8	0.31496	48	1.88976	88	3.46456
0.29	0.01142	0.69	0.02717	9	0.35433	49	1.92913	89	3.50393
0.30	0.01181	0.70	0.02756	10	0.39370	50	1.96850	90	3.54330
0.31	0.01220	0.71	0.02795	11	0.43307	51	2.00787	91	3.58267
0.32	0.01260	0.72	0.02835	12	0.47244	52	2.04724	92	3.62204
0.33	0.01299	0.73	0.02874	13	0.51181	53	2.08661	93	3.66141
0.34	0.01339	0.74	0.02913	14	0.55118	54	2.12598	94	3.70078
0.35	0.01378	0.75	0.02953	15	0.59055	55	2.16535	95	3.74015
0.36	0.01417	0.76	0.02992	16	0.62992	56	2.20472	96	3.77952
0.37	0.01457	0.77	0.03032	17	0.66929	57	2.24409	97	3.81889
0.38	0.01496	0.78	0.03071	18	0.70866	58	2.28346	98	3.85826
0.39	0.01535	0.79	0.03110	19	0.74803	59	2.32283	99	3.89763
0.40	0.01575	0.80	0.03150	20	0.78740	60	2.36220	100	3.93700

For converting millimetres in "thousandths," move decimal point in both columns to left.

L. S. Starrett Co.

Table A-8

Fractional and number drill sizes, decimal equivalents, and tap sizes.

Fraction or drill size	Decimal equivalent	Tap size	Fraction or drill size	Decimal equivalent	Tap size	Fraction or drill size	Decimal equivalent	Tap size
Number size drills 80	0.0135		26	0.1470		25/64	0.3906	7/16-20
79	0.0145		25	0.1495	10-24	X	0.3970	
1/64	0.0156		24	0.1520		Y	0.4040	
78	0.0160		23	0.1540		13/32	0.4062	
77	0.0180		5/32	0.1562		Z	0.4130	
76	0.0200		22	0.1570	10-30	27/64	0.4219	1/2-13
75	0.0210		21	0.1590	10-32	7/16	0.4375	
74	0.0225		20	0.1610		29/64	0.4531	1/2-20
73	0.0240		19	0.1660		15/32	0.4687	
72	0.0250		18	0.1695		31/64	0.4844	9/16-12
71	0.0260		11/64	0.1719		1/2	0.5000	
70	0.0280		17	0.1730		33/64	0.5156	9/16-18
69	0.0292		16	0.1770	12-24	17/32	0.5312	5/8-11
68	0.0310		15	0.1800		35/64	0.5469	
1/32	0.0312		14	0.1820	12-28	9/16	0.5625	
67	0.0320		13	0.1850	12-32	37/64	0.5781	5/8-18
66	0.0330		3/16	0.1875		19/32	0.5937	11/16-11
65	0.0350		12	0.1890		39/64	0.6094	
64	0.0360		11	0.1910		5/8	0.6250	11/16-16
63	0.0370		10	0.1935		41/64	0.6406	
62	0.0380		9	0.1960		21/32	0.6562	3/4-10
61	0.0390		8	0.1990		43/64	0.6719	
60	0.0400		7	0.2010	1/4-20	11/16	0.6875	3/4-16
59	0.0410		13/64	0.2031		45/64	0.7031	
58	0.0420		6	0.2040		23/32	0.7187	
57	0.0430		5	0.2055		47/64	0.7344	
56	0.0465		4	0.2090		3/4	0.7500	
3/64	0.0469	0-80	3	0.2130	1/4-28	49/64	0.7656	7/8-9
55	0.0520		7/32	0.2187		25/32	0.7812	
54	0.0550	1-56	2	0.2210		51/64	0.7969	
53	0.0595	1-64, 72	1	0.2280		13/16	0.8125	7/8-14
1/16	0.0625		Letter size drills A	0.2340		53/64	0.8281	
52	0.0635		15/64	0.2344		27/32	0.8437	
51	0.0670		B	0.2380		55/64	0.8594	
50	0.0700	2-56, 64	C	0.2420		7/8	0.8750	1-8
49	0.0730		D	0.2460		57/64	0.8906	
48	0.0760		1/4 E	0.2500		29/32	0.9062	
5/64	0.0781		F	0.2570	5/16-18	59/64	0.9219	
47	0.0785	3-48	G	0.2610		15/16	0.9375	1-12, 14
46	0.0810		17/64	0.2656		61/64	0.9531	
45	0.0820	3-56, 4-32	H	0.2660		31/32	0.9687	
44	0.0860	4-36	I	0.2720	5/16-24	63/64	0.9844	1 1/8-7
43	0.0890	4-40	J	0.2770		1	1.0000	
42	0.0935	4-48	K	0.2810		1 3/64	1.0469	1 1/8-12
3/32	0.0937		9/32	0.2812		1 7/64	1.1093	1 1/4-7
41	0.0960		L	0.2900		1 1/8	1.1250	
40	0.0980		M	0.2950		1 11/64	1.1719	1 1/4-12
39	0.0995		19/64	0.2968		1 7/32	1.2187	1 3/8-6
38	0.1015	5-40	N	0.3020		1 1/4	1.2500	
37	0.1040	5-44	5/16	0.3125	3/8-16	1 19/64	1.2968	1 3/8-12
36	0.1065	6-32	O	0.3160		1 11/32	1.3437	1 1/2-6
7/64	0.1093		P	0.3230		1 3/8	1.3750	
35	0.1100		21/64	0.3281		1 27/64	1.4219	1 1/2-12
34	0.1110	6-36	Q	0.3320	3/8-24	1 1/2	1.5000	
33	0.1130	6-40	R	0.3390				
32	0.1160		11/32	0.3437				
31	0.1200		S	0.3480				
1/8	0.1250		T	0.3580				
30	0.1285		23/64	0.3594				
29	0.1360	8-32, 36	U	0.3680	7/16-14			
28	0.1405	8-40	3/8	0.3750				
9/64	0.1406		V	0.3770				
27	0.1440		W	0.3860				

Pipe thread sizes

Thread	Drill	Thread	Drill
1/8-27	R	1 1/2-11 1/2	1 47/64
1/4-18	7/16	2-11 1/2	2 7/32
3/8-18	37/64	2 1/2-8	2 5/8
1/2-14	23/32	3-8	3 1/4
3/4-14	59/64	3 1/2-8	3 3/4
1-11 1/2	1 5/32	4-8	4 1/4
1 1/4-11 1/2	1 1/2		

L. S. Starrett Co.

Table A-9

Twist drill number sizes in inches, millimetres, and centimetres.

Number	Inches	Millimetres	Centimetres	Number	Inches	Millimetres	Centimetres
1	0.2280	5.7912	0.5791	41	0.0960	2.4384	0.2438
2	0.2210	5.6134	0.5613	42	0.0935	2.3622	0.2362
3	0.2130	5.4102	0.5410	43	0.0890	2.2606	0.2261
4	0.2090	5.3086	0.5309	44	0.0860	2.1844	0.2184
5	0.2055	5.2070	0.5207	45	0.0820	2.0828	0.2083
6	0.2040	5.1816	0.5182	46	0.0810	2.0574	0.2057
7	0.2010	5.1054	0.5105	47	0.0785	1.9812	0.1981
8	0.1990	5.0800	0.5080	48	0.0760	1.9304	0.1930
9	0.1960	4.9784	0.4978	49	0.0730	1.8542	0.1854
10	0.1935	4.9022	0.4902	50	0.0700	1.7780	0.1778
11	0.1910	4.8514	0.4851	51	0.0670	1.7018	0.1702
12	0.1890	4.8006	0.4801	52	0.0635	1.6256	0.1626
13	0.1850	4.6990	0.4699	53	0.0595	1.4986	0.1499
14	0.1820	4.6228	0.4623	54	0.0550	1.3970	0.1397
15	0.1800	4.5720	0.4572	55	0.0520	1.3208	0.1321
16	0.1770	4.4958	0.4496	56	0.0465	1.1684	0.1168
17	0.1730	4.3942	0.4394	57	0.0430	1.0922	0.1092
18	0.1695	4.2926	0.4292	58	0.0420	1.0668	0.1067
19	0.1660	4.2164	0.4216	59	0.0410	1.0414	0.1041
20	0.1610	4.0894	0.4089	60	0.0400	1.0160	0.1016
21	0.1590	4.0386	0.4039	61	0.0390	0.9906	0.0991
22	0.1570	3.9878	0.3988	62	0.0380	0.9652	0.0965
23	0.1540	3.9116	0.3912	63	0.0370	0.9398	0.0940
24	0.1520	3.8608	0.3861	64	0.0360	0.9144	0.0914
25	0.1495	3.7846	0.3785	65	0.0350	0.8890	0.0889
26	0.1470	3.7338	0.3734	66	0.0330	0.8382	0.0838
27	0.1440	3.6576	0.3658	67	0.0320	0.8128	0.0813
28	0.1405	3.5560	0.3556	68	0.0310	0.7874	0.0787
29	0.1360	3.4544	0.3454	69	0.0292	0.7366	0.0737
30	0.1285	3.2512	0.3251	70	0.0280	0.7112	0.0711
31	0.1200	3.0480	0.3048	71	0.0260	0.6604	0.0660
32	0.1160	2.9464	0.2946	72	0.0250	0.6350	0.0635
33	0.1130	2.8702	0.2870	73	0.0240	0.6096	0.0610
34	0.1110	2.8194	0.2819	74	0.0225	0.5588	0.0559
35	0.1100	2.7940	0.2794	75	0.0210	0.5334	0.0533
36	0.1065	2.6924	0.2692	76	0.0200	0.5080	0.0508
37	0.1040	2.6416	0.2642	77	0.0180	0.4572	0.0457
38	0.1015	2.5654	0.2565	78	0.0160	0.4064	0.0406
39	0.0995	2.5146	0.2515	79	0.0145	0.3556	0.0356
40	0.0980	2.4892	0.2489	80	0.0135	0.3302	0.0330

Table A-10

Standard pipe data.

Nominal pipe diameter in inches	Actual inside diameter in inches	Actual outside diameter in inches	Weight per foot-pounds	Nominal pipe diameter in inches	Actual inside diameter in inches	Actual outside diameter in inches	Weight per foot-pounds
1/8	0.269	0.405	0.244	2 1/2	2.469	2.875	5.793
1/4	0.364	0.540	0.424	3	3.068	3.500	7.575
3/8	0.493	0.675	0.567	3 1/2	3.548	4.000	9.109
1/2	0.622	0.840	0.850	4	4.026	4.500	10.790
3/4	0.824	1.050	1.130	4 1/2	4.560	5.000	12.538
1	1.049	1.315	1.678	5	5.047	5.563	14.617
1 1/4	1.380	1.660	2.272	6	6.065	6.625	18.974
1 1/2	1.610	1.900	2.717	8	7.981	8.625	28.554
2	2.067	2.375	3.652	10	10.020	10.750	40.483

Table A-11

Commercial pipe sizes and wall thicknesses.

The following table lists the pipe sizes and wall thicknesses currently established as standard, or specifically:

1. The traditional standard weight, extra strong, and double extra strong pipe.
2. The pipe wall thickness schedules listed in American Standard B36.10, which are applicable to carbon steel and alloys *other than* stainless steels.
3. The pipe wall thickness schedules listed in American Standard B36.19, and ASTM Specification A409, which are applicable *only* to corrosion resistant materials. (NOTE: Schedule 10S is also available in carbon steel in sizes 12″ and smaller.)

ASA-B36.10 and B36.19

Nominal pipe size	Outside diameter	Nominal wall thickness for														
		Schedule 5S*	Schedule 10S*	Schedule 10	Schedule 20	Schedule 30	Standard†	Schedule 40	Schedule 60	Extra strong‡	Schedule 80	Schedule 100	Schedule 120	Schedule 140	Schedule 160	XX Strong
1/8	0.405	—	0.049	—	—	—	*0.068*	0.068	—	*0.095*	0.095	—	—	—	—	—
1/4	0.540	—	0.065	—	—	—	*0.088*	0.088	—	*0.119*	0.119	—	—	—	—	—
3/8	0.675	—	0.065	—	—	—	*0.091*	0.091	—	*0.126*	0.126	—	—	—	—	—
1/2	0.840	0.065	0.083	—	—	—	*0.109*	0.109	—	*0.147*	0.147	—	—	—	0.188	0.294
3/4	1.050	0.065	0.083	—	—	—	*0.113*	0.113	—	*0.154*	0.154	—	—	—	0.219	0.308
1	1.315	0.065	0.109	—	—	—	*0.133*	0.133	—	*0.179*	0.179	—	—	—	0.250	0.358
1 1/4	1.660	0.065	0.109	—	—	—	*0.140*	0.140	—	*0.191*	0.191	—	—	—	0.250	0.382
1 1/2	1.900	0.065	0.109	—	—	—	*0.145*	0.145	—	*0.200*	0.200	—	—	—	0.281	0.400
2	2.375	0.065	0.109	—	—	—	*0.154*	0.154	—	*0.218*	0.218	—	—	—	0.344	0.436
2 1/2	2.875	0.083	0.120	—	—	—	*0.203*	0.203	—	*0.276*	0.276	—	—	—	0.375	0.552
3	3.5	0.083	0.120	—	—	—	*0.216*	0.216	—	*0.300*	0.300	—	—	—	0.438	0.600
3 1/2	4.0	0.083	0.120	—	—	—	*0.226*	0.226	—	*0.318*	0.318	—	—	—	—	—
4	4.5	0.083	0.120	—	—	—	*0.237*	0.237	—	*0.337*	0.337	—	0.438	—	0.531	0.674
5	5.563	0.109	0.134	—	—	—	*0.258*	0.258	—	*0.375*	0.375	—	0.500	—	0.625	0.750
6	6.625	0.109	0.134	—	—	—	*0.280*	0.280	—	*0.432*	0.432	—	0.562	—	0.719	0.864
8	8.625	0.109	0.148	—	0.250	0.277	*0.322*	0.322	0.406	*0.500*	0.500	0.594	0.719	0.812	0.906	0.875
10	10.75	0.134	0.165	—	0.250	0.307	*0.365*	0.365	0.500	*0.500*	0.594	0.719	0.844	1.000	1.125	1.000
12	12.75	0.156	0.180	—	0.250	0.330	*0.375*	0.406	0.562	*0.500*	0.688	0.844	1.000	1.125	1.312	1.000
14 O.D.	14.0	0.156	0.188	0.250	0.312	0.375	0.375	0.438	0.594	0.500	0.750	0.938	1.094	1.250	1.406	—
16 O.D.	16.0	0.165	0.188	0.250	0.312	0.375	0.375	0.500	0.656	0.500	0.844	1.031	1.219	1.438	1.594	—
18 O.D.	18.0	0.165	0.188	0.250	0.312	0.438	0.375	0.562	0.750	0.500	0.938	1.156	1.375	1,562	1.781	—
20 O.D.	20.0	0.188	0.218	0.250	0.375	0.500	0.375	0.594	0.812	0.500	1.031	1.281	1.500	1.750	1.969	—
22 O.D.	22.0	0.188	0.218	0.250	0.375	0.500	0.375	—	0.875	0.500	1.125	1.375	1.625	1.875	2.125	—
24 O.D.	24.0	0.218	0.250	0.250	0.375	0.562	0.375	0.688	0.969	0.500	1.218	1.531	1.812	2.062	2.344	—
26 O.D.	26.0	—	—	0.312	0.500	—	0.375	—	—	0.500	—	—	—	—	—	—
28 O.D.	28.0	—	—	0.312	0.500	0.625	0.375	—	—	0.500	—	—	—	—	—	—
30 O.D.	30.0	0.250	0.312	0.312	0.500	0.625	0.375	—	—	0.500	—	—	—	—	—	—
32 O.D.	32.0	—	—	0.312	0.500	0.625	0.375	0.688	—	0.500	—	—	—	—	—	—
34 O.D.	34.0	—	—	0.312	0.500	0.625	0.375	0.688	—	0.500	—	—	—	—	—	—
36 O.D.	36.0	—	—	0.312	0.500	0.625	0.375	0.750	—	0.500	—	—	—	—	—	—
42 O.D.	42.0	—	—	—	—	—	0.375	—	—	0.500	—	—	—	—	—	—

American Standards Association

All dimensions are given in inches.

The decimal thicknesses listed for the respective pipe sizes represent their nominal or average wall dimensions. The actual thicknesses may be as much as 12.5% under the nominal thickness because of mill tolerance. Thicknesses shown in light face for Schedule 60 and heavier pipe are not currently supplied by the mills, unless a certain minimum tonnage is ordered.

* Schedules 5S and 10S are available in corrosion resistant materials and Schedule 10S is also available in carbon steel.

† Thicknesses shown in italics are available also in stainless steel, under the designation Schedule 40S.

‡ Thicknesses shown in italics are available also in stainless steel, under the designation Schedule 80S.

Table A-12

Melting points of metals and alloys of practical importance.

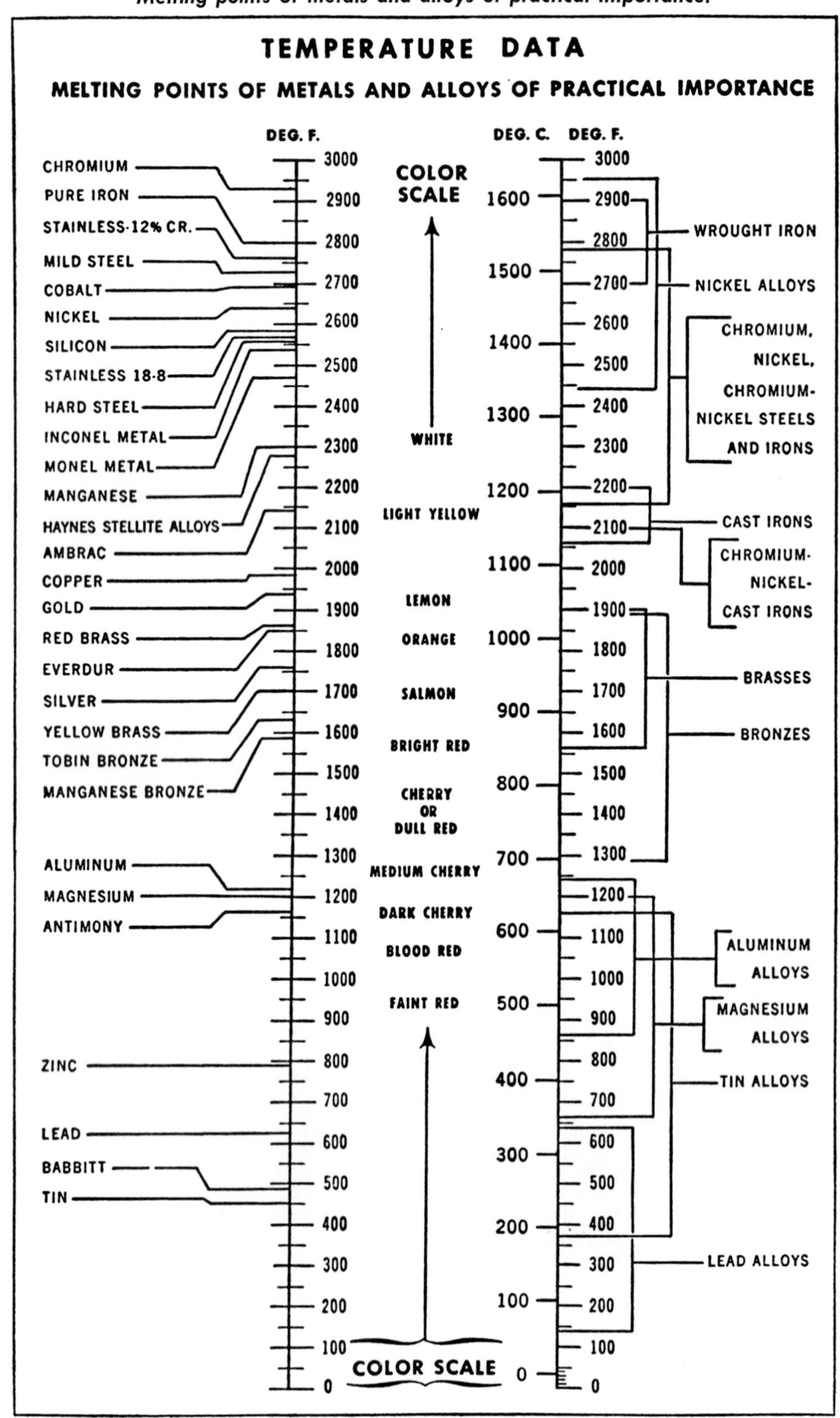

Linde Division, Union Carbide Corp.

Table A-13

Properties of elements and metal compositions.

Elements	Symbol	Density (specific gravity)	Weight per cu. ft.	Specific heat	Melting point	
					°C.	°F.
Aluminum	Al	2.7	166.7	0.212	658.7	1217.7
Antimony	Sb	6.69	418.3	0.049	630	1166
Armco iron	...	7.9	490.0	0.115	1535	2795
Carbon	C	2.34	219.1	0.113	3600	6512
Chromium	Cr	6.92	431.9	0.104	1615	3034
Columbium	Cb	7.06	452.54	...	1700	3124
Copper	Cu	8.89	555.6	0.092	1083	1981.4
Gold	Au	19.33	1205.0	0.032	1063	1946
Hydrogen	H	0.070*	0.00533	...	−259	−434.2
Iridium	Ir	22.42	1400.0	0.032	2300	4172
Iron	Fe	7.865	490.9	0.115	1530	2786
Lead	Pb	11.37	708.5	0.030	327	621
Manganese	Mn	7.4	463.2	0.111	1260	2300
Mercury	Hg	13.55	848.84	0.033	−38.7	−37.6
Nickel	Ni	8.80	555.6	0.109	1452	2645.6
Nitrogen	N	0.97*	0.063	...	−210	−346
Oxygen	O	1.10*	0.0866	...	−218	−360
Phosphorus	P	1.83	146.1	0.19	44	111.2
Platinum	Pt	21.45	1336.0	0.032	1755	3191
Potassium	K	0.87	54.3	0.170	62.3	144.1
Silicon	Si	2.49	131.1	0.175	1420	2588
Silver	Ag	10.5	655.5	0.055	960.5	1761
Sodium	Na	0.971	60.6	0.253	97.5	207.5
Sulfur	S	1.95	128.0	0.173	119.2	246
Tin	Sn	7.30	455.7	0.054	231.9	449.5
Titanium	Ti	5.3	218.5	0.110	1795	3263
Tungsten	W	17.5	1186.0	0.034	3000	5432
Uranium	U	18.7	1167.0	0.028		
Vanadium	V	6.0	343.3	0.115	1720	3128
Zinc	Zn	7.19	443.2	0.093	419	786.2
Bronze (90 percent Cu 10 percent Sn)	...	8.78	548.0	...	850–1000	1562–1832
Brass (90 percent Cu 10 percent Zn)	...	8.60	540.0	...	1020–1030	1868–1886
Brass (70 percent Cu 30 percent Zn)	...	8.44	527.0	...	900–940	1652–1724
Cast pig iron	...	7.1	443.2	...	1100–1250	2012–2282
Open-hearth steel	...	7.8	486.9	...	1350–1530	2462–2786
Wrought-iron bars	...	7.8	486.9	...	1530	2786

* Density compared with air.

Linde Division, Union Carbide Corp.

Table A-14
Identifying metals by spark testing.

Spark tests should be made on a high speed power grinder, and the specimen should be held so that the sparks will be given off horizontally. For most accurate results, the sparks should be examined against a dark background, preferably in a dark corner of the shop.

The color, shape, average length, and activity of the sparks are details which are characteristics of the material tested. Spark testing can be a very accurate method of identifying metals but it requires considerable practice and experience to become an expert. Several common sparks are given in the table. If the operator learns the technique for identifying these metals readily, he will soon be able to expand his experience to include others by observation and comparison with the sparks from known samples.

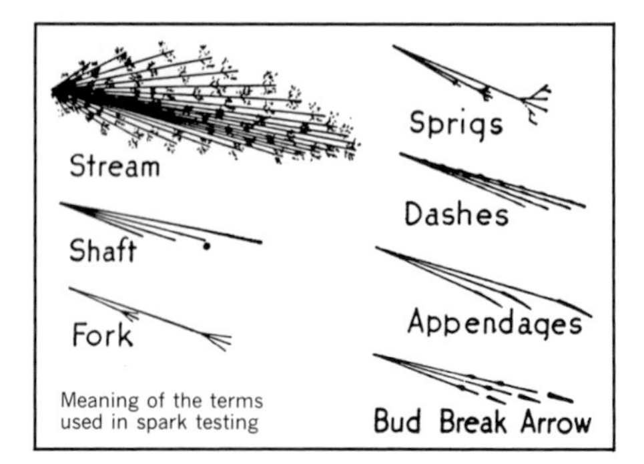

Meaning of the terms used in spark testing

Wrought Iron	Low-Carbon Steel*	High-Carbon Steel	Alloy Steel**
Color - straw yellow. Average stream length with power grinder-65 in. Volume-large. Long shafts ending in forks and arrowlike appendages. Color-white	Color-white. Average length of stream with power grinder - 70 in. Volume - moderately large. Shafts shorter than wrought iron and in forks and appendages. Forks become more numerous and sprigs appear as carbon content increases	Color-white. Average stream length with power grinder - 55 in. Volume - large. Numerous small and repeating sprigs	Color - straw yellow. Stream length varies with type and amount of alloy content. Shafts may end in forks, buds or arrows, frequently with break between shaft and arrow. Few, if any, sprigs. Color-white
White Cast Iron	**Gray Cast Iron**	**Malleable Iron**	**Nickel*****
Color - red. Color - straw yellow. Average stream length with power grinder-20 in. Volume - very small. Sprigs - finer than gray iron, small and repeating	Color - red. Color - straw yellow. Average stream length with power grinder - 25 in. Volume - small. Many sprigs, small and repeating	Color - straw yellow. Average stream length with power grinder - 30 in. Volume-moderate. Longer shafts than gray iron ending in numerous small, repeating sprigs	Color - orange. Average stream length with power grinder - 10 in. Short shafts with no forks or sprigs

Linde Division, Union Carbide Corp.

* These data apply also to cast steel
** Spark shown is for stainless steel
*** Monel metal spark is very similar to nickel

Glossary

If a student expects to make progress in a new field of training, it is vitally important that he understand the terms which describe the processes and equipment. Very often different terms may have the same meaning. Many terms developed out of expressions used in the field by welders and others working in the occupation.

The following list of terms and their definitions are those standardized and approved by the American Welding Society. Also included are a number of terms from the fields of electricity and metallurgy which are applicable to welding and weld testing. Only those terms that are concerned with the manual and semi-automatic processes are included because they are generally used in the field by welders and must be understood if the student expects to become competent in the craft.

Although these terms are presented in the Glossary for easy reference, the student is urged to begin the study of the definitions at the beginning of the training program and to continue the effort for the duration of the training period. Constant reference after entrance into the occupation is also recommended.

Those students who wish a more complete listing of terms, including those for automatic and resistance welding processes, should secure a copy of the AWS publication *Terms and Definitions* (AWS A3.0-69). The author is indebted to the American Welding Society for permission to use these definitions. In addition, the James F. Lincoln Welding Foundation supplied those definitions indicated by a single asterisk (*), and the Stanley G. Flagg Company supplied those definitions indicated by a double asterisk (**).

The figure numbers in the Glossary refer to the Illustrated Guide to Welding Terminology, pages 956–972.

A

Abradant*: a substance which causes abrasive wear, i.e., displacement or removal of surface material by a scratching action.

Actual throat (throat of fillet weld): the distance from the root to the face of the weld. See Fig. A-40.

Admixture*: the interchange of filler metal and base metal during welding resulting in weld metal of composition borrowed from both.

Air-hardening*: characteristic of a steel that becomes partially or fully hardened (martensitic) when cooled in air from above its critical point. May not apply when the object to be hardened has considerable thickness.

Alloy*: a material having metallic characteristics and made up of two or more elements, one of which is a metal.

All-weld metal test specimen: a test specimen wherein the portion being tested is composed wholly of weld metal.

Alternating current (a.c.): an electrical current that has positive and negative values alternately. The current flows in one direction during any half cycle, and reverses direction during the next half cycle, causing the alternation of direction of current flow.

Alternating current arc welding: an arc welding process wherein the power supply at the arc is alternating current.

Ampere: ampere is another name for current and is the unit of electrical rate measurement.

Angle of bevel: see bevel angle.

Annealing*: the process of softening a metal by heating it, usually above the upper critical temperature and then cooling it at a slow rate.

Anode: the positive pole in an arc. When welding with direct current, the work is the anode when using straight polarity, and the electrode is the anode when using reverse polarity.

Arc blow: the swerving of an electric arc from its normal path because of magnetic forces.

Arc cutting: a group of cutting processes wherein the severing of metals is effected by melting with the heat of an arc between an electrode and the base metal.

Arc gouging: an application of arc cutting wherein a chamfer or groove is formed.

Arc seam welding: an arc welding process wherein coalescence at the faying surfaces is produced continuously by heating with an electric arc between an electrode and the work. The weld is made without preparing a hole in either member. Filler metal or a shielding gas or flux may or may not be used. See Fig. A-57.

Arc spot welding: an arc welding process wherein coalescence at the faying surfaces is produced in one spot by heating with an electric arc between an electrode and the work. The weld is made without preparing a hole in either member. Filler metal or a shielding gas or flux may or may not be used. See Fig. A-58.

Arc voltage: the voltage across the welding arc.
Arc welder: an arc welding machine. (The term "welder" is often used to denote an operator of welding equipment.)
Arc welding: a group of welding processes wherein coalescence is produced by heating with an arc or arcs, with or without the application of pressure and with or without the use of filler metal.
Atom: the smallest unit of an element which can take part in a chemical reaction.
Austenite*: a solid solution of carbon or iron-carbide in face-centered cubic iron.
Automatic gas cutting: see machine oxygen cutting.
Automatic welding: welding with equipment that automatically controls the entire welding operation (including feed, speed, etc.).
Axis of a weld: a line through the length of a weld, perpendicular to the cross section at its center of gravity, parallel to the root. See Fig. A-1.

B

Backfire: the momentary recession of the flame into the torch tip following a pop of the torch, causing extinguishment of the flame or continued burning of the gases within the torch handle.
Backhand welding: a gas welding technique wherein the torch or gun is directed opposite to the progress of the welding. See Fig. A-68.
Backing: material (metal, asbestos, carbon, granular flux, etc.) backing up the joint during welding to facilitate obtaining a sound weld at the root.
Backing bead: see backing weld.
Backing ring: backing in the form of a ring generally used in the welding of pipe.
Backing weld: backing in the form of a weld. See Fig. A-70.
Backstep sequence: a longitudinal sequence wherein the weld bead increments are deposited in the direction opposite to the progress of welding the joint. See Fig. A-51.
Back weld: a weld deposited at the back of a single-groove weld. See Fig. A-71.
Bare electrode: a filler metal electrode used in arc welding, consisting of a metal wire with no coating other than that incidental to the drawing of the wire.
Base metal (parent metal): the metal to be welded or cut. Often called "parent metal."
Bead weld: a type of weld made by one passage of electrode or rod. See Figs. A-4 and A-17.
Beading (parallel beads): a technique of depositing weld metal without oscillation of the electrode, which generally results in a relatively narrow pass. See Fig. 46.
Bend test: see free-bend test and guided-bend test.
Bevel: an angular type of edge preparation.
Bevel angle: the angle formed between the prepared edge of a member and a plane perpendicular to the surface of the member. See Fig. A-32.
Beveling: a type of chamfering.
Block sequence: a combined longitudinal and buildup sequence for a continuous multiple-pass weld wherein separated lengths are completely or partially built up in cross section before intervening lengths are deposited. See Fig. 52.
Blowhole*: a defect in metal caused by hot metal cooling too rapidly when excessive gaseous content is present. Specifically, in welding, a gas pocket in the weld metal resulting from the hot metal solidifying without all of the gases having escaped to the surface.
Blowpipe: see welding torch or cutting torch.
Bond line: the junction of the weld metal and the base metal or the junction of the base metal parts when weld metal is not present. See Fig. A-73.
Bore:** (brazing) the inside diameter of the cup on a fitting. See Fig. A-78.
Bottle: see cylinder.
Braze welding: a method of welding whereby a groove, fillet, plug, or slot weld is made using a nonferrous filler metal having a melting point below that of the base metals but above 800 degrees F. The filler metal is not distributed in the joint by capillary attraction. (*Bronze welding*, formerly used, is a misnomer for this term.)
Brazed joint: a union of two or more members produced by the application of a brazing process.
Brazing: a group of welding processes wherein coalescence is produced by heating to suitable temperatures above 800 degrees F. and by using a nonferrous filler metal having a melting point below that of the base metals. The filler metal is distributed between the closely fitted surfaces of the joint by capillary attraction.
Brazing filler metal:** a nonferrous metal or alloy to be added in making a braze. Its melting temperature is above 800 degrees F, but below the melting temperatures of the base metals being joined.
Brittleness*: the tendency of a material to fail suddenly by breaking without any permanent deformation of the material before failure.
Bronze welding: a term erroneously used to de-

note braze welding. See preferred term, braze welding.
Buildup sequence: the order in which the weld beads of a multiple-pass weld are deposited with respect to the cross-section of the joint. See Fig. A-53.
Burner: see oxygen cutter.
Butt joint: a welded joint between two abutting parts lying in approximately the same plane. See Fig. A-2.

C

Capillary attraction: the phenomenon of surface tension setting up adhesion and cohesion forces which cause molten metals to flow between closely fitted solid surfaces. The term is usually used in connection with brazing processes.
Carbide*: the chemical combination of carbon with some other element. A metallic carbide takes the form of very hard crystals.
Carbon arc cutting (CAC): the process of severing metals by melting with the heat of an arc between a carbon electrode and the base metal.
Carbon arc welding (CAW): an arc welding process wherein coalescence is produced by heating with an arc between a carbon electrode and the work, and no shielding is used. Pressure may or may not be used, and filler metal may or may not be used.
Carbon electrode: a non-filler-metal electrode used in arc welding consisting of a carbon or graphite rod.
Carbon-electrode arc welding: a group of arc welding processes wherein carbon electrodes are used. See shielded carbon-arc welding, inert-gas carbon-arc welding, carbon-arc welding, and twin-carbon arc welding.
Carburizing (carbonizing or reducing flame): a gas flame having the property of introducing carbon into the metal heated. See Fig. A-50.
Cascade sequence: a combined longitudinal and build-up sequence wherein weld beads are deposited in overlapping layers. See Fig. A-54.
Cathode: the negative pole in an arc. When welding with direct current, the electrode is the cathode when using straight polarity, and the work is the cathode when using reverse polarity.
Caulk weld: See preferred term, seal weld.
Cementite*: iron carbide (Fe_3C), constituent of steel and cast iron. It is hard and brittle.
Chain intermittent fillet welds: two lines of intermittent fillet welding on a joint wherein the fillet increments welded in one line are approximately opposite to those in the other line. See Fig. A-20.
Chamfer: see edge preparation.
Chamfering: the preparation of a contour, other than for a square groove weld, on the edge of a member for welding.
Chemical dip brazing: a dip brazing process wherein the filler metal is added to the joint before immersion in a bath of molten chemicals.
Chemical flux cutting (FOC): an oxygen cutting process wherein a chemical flux is used to facilitate cutting.
Chill ring: see backing ring.
Coalescence: the growing together, or growth into one body, of the base metal parts.
Coated electrode: see covered shielded-arc electrode and lightly coated electrode.
Cold shortness*: the characteristic tendency of a metal toward brittleness at room temperature or lower.
Complete fusion: fusion which has occurred over the entire base metal surfaces exposed for welding and between all layers and passes.
Complete joint penetration: see joint penetration.
Composite electrode: a filler metal electrode used in arc welding consisting of more than one metal component combined mechanically. It may or may not include materials which protect the molten metal from the atmosphere, improve the properties of the weld metal, or stabilize the arc.
Composite joint: a joint wherein welding is used in conjunction with a mechanical joining process. (Welding and riveting.)
Composition*: the contents of an alloy, in terms of what elements are present and in what amount (by percentage of weight).
Compressive strength*: the resistance of a material to a force which is tending to deform or fail it by crushing.
Concave fillet weld: a fillet weld having a concave face. See Fig. A-41.
Concavity: the maximum distance from the face of a concave fillet weld perpendicular to a line joining the toes.
Concurrent heating: the application of supplemental heat to a structure during a welding or cutting operation.
Conductor: a conductor is an electrical path. Metals like copper and aluminum that offer little resistance to current flow are considered good conductors.
Cone: the conical part of a gas flame next to the orifice of the tip. See Fig. A-48.
Constant potential: constant potential and constant voltage are terms applied to welding machines. They indicate the ability of the machine

to produce welding current of stable voltage regardless of the amperage output.
Consumable electrode*: a metal electrode (filler wire) that establishes the arc and gradually melts away, being carried across the arc (deposited) to provide filler metal into the joint.
Continuous weld: a weld which extends without interruption for its entire length.
Convexity: the maximum distance from the face of a convex fillet weld perpendicular to a line joining the toes. See Figs. A-41 and A-47.
Convex fillet weld: a fillet weld having a convex face. See Fig. A-41.
Convexity ratio: See Fig. A-47.
Corner-flange weld: a flange weld with only one member flanged at the location of welding. See Fig. A-59.
Corner joint: a welded joint between two parts located approximately at right angles to each other in the form of an *L*. See Fig. A-3.
Cover glass (lens): a clear, transparent material used in goggles, hand shields, and helmets to protect the filter glass from spattering material.
Covered shielded-arc electrode: a filler metal electrode used in arc welding consisting of a metal core wire with a relatively thick covering which provides protection for the molten metal from the atmosphere, improves the properties of the weld metal, and stabilizes the arc.
CO_2 welding: see preferred term, gas metal-arc welding.
Crack propagation*: the development, growth, or progress of a crack through a solid.
Crater: in arc welding a depression at the termination of a weld bead, or in the weld pool beneath the electrode.
Creep*: the slow deformation (for example, elongation) of a metal under prolonged stress. Not to be confused with that deformation which results immediately upon application of a stress.
Cutting attachment: a device which is attached to a gas welding torch to convert it into an oxygen cutting torch.
Cutting process: a process wherein the severing or removing of metals is effected.
Cutting tip nozzle: that part of an oxygen cutting torch from which the gases issue.
Cutting torch (blowpipe): a device used in gas cutting for controlling the gases used for preheating and the oxygen for severing the metal.
Cycle: applied to a.c. current, it refers to the alternating flow of current during a unit of time, $\frac{1}{60}$ second. The cycle is repeated as long as current is flowing in the circuit.
Cylinder (bottle): a portable steel container for storage of a compressed gas.
Cylinder manifold: see manifold.

D

Decarburization*: the loss of carbon from a ferrous alloy in a reactive atmosphere at high temperatures.
Deoxidizing*: the removal of oxygen from the molten weld metal, usually by chemical combination with other elements forming inorganic compounds that float to the surface of the molten metal to form slag when cooled.
Deoxidizer: an element or compound, added to the electrode flux, or core wire, to remove oxygen and its derivatives from the weld.
Deposit sequence: see deposition sequence.
Deposited metal: filler metal that has been added during a welding operation.
Deposition rate*: the speed with which filler metal is added to a weld joint, usually stated in terms of volume of metal deposited per minute.
Deposition sequence: the order in which the increments of weld metal are deposited.
Depth of fusion: the distance that fusion extends into the base metal from the surface melted during welding. See Fig. A-42.
Differential hardening*: the characteristic of certain steels to develop high surface or case hardness and to retain a tough, soft core. Any hardening procedure designed to produce only localized or superficial hardness.
Direct current (d.c.): an electrical current that flows in one direction only and has either a positive or a negative polarity. It differs from a.c. current in that there is no change in the flow of current.
Direct current arc welding: an arc welding process wherein the power supply at the arc is direct current.
Double-bevel groove weld: a type of groove weld. See Fig. A-13.
Double-groove weld: see groove welds
Double-J groove weld: a type of groove weld. See Fig. A-15.
Double-U groove weld: a type of groove weld. See Fig. A-14.
Double-V groove weld: a type of groove weld. See Fig. A-12.
Double-welded butt joint: a butt joint welded from both sides.
Double-welded joint: in arc and gas welding, any joint welded from both sides. See Fig. A-72.
Double-welded lap joint: a lap joint in which the

overlapped edges of the members to be joined are welded along the edges of both members.
Downhand: see flat position.
Drag: the distance between the point of exit of the cutting oxygen stream and the projection, on the exit surface, of the point of entrance. See Fig. A-55.
Ductility*: the ability of a material to become permanently deformed without failure.

E

Edge-flange weld: a flange weld with two members flanged at the location of welding. See Fig. A-60.
Edge joint: a welded joint connecting the edges of two or more parallel or nearly parallel parts. See Fig. A-4.
Edge preparation: a prepared contour on the edge of a part to be joined by a groove weld. See Fig. A-26.
Elastic limit*: the maximum stress to which a material can be subjected without permanent deformation or failure by breaking.
Elasticity*: the ability of a material to return to original shape and dimensions after a deforming load has been removed.
Electrode: (arc welding): see bare electrode, carbon electrode, composite electrode, covered shielded-arc electrode, lightly coated electrode, metal electrode, flux-cored electrode and tungsten electrode.
Electrode holder: a device used for mechanically holding the electrode and conducting current to it.
Electrode lead: the electrical conductor between the source of arc welding current and the electrode holder.
Element*: a substance which can't be broken down into two other substances. Everything on earth is a combination of such elements, of which there are only 103.
Elongation*: the stretching of a material by which any straight line dimension increases.
Endurance limit*: the maximum stress that a material will support indefinitely under variable and repetitive load conditions.

F

Face of weld: the exposed surface of a weld on the side from which welding was done. See Fig. A-35.
Face reinforcement: reinforcement of the weld at the side of the joint from which welding was done.
Face shield: see hand shield.
Fatigue failure*: the cracking, breaking, or other failure of a material as the result of repeated or alternating stressing below the material's ultimate tensile strength.
Fatigue limit*: the maximum stress that a material will support indefinitely under variable and repetitive load conditions.
Fatigue strength*: the resistance of a material to repeated or alternating stressing without failure.
Ferrite*: pure iron of body-centered cubic crystal structure. It is soft and ductile.
Ferrous*: descriptive of a metallic material that is dominated by iron in its chemical composition.
Filler metal: material to be added in making a welded, brazed, or soldered joint.
Fillet:** (*brazing*) the small bead of alloy in the chamfer at the edge of the cup when a brazed joint is complete.
Fillet weld: a weld of approximately triangular cross section, as used in a lap joint, T-joint, or corner joint joining two surfaces approximately at right angles to each other. See Fig. A-16; also Figs. A-3, A-5, and A-6.
Fillet weld size: see size of weld.
Filter glass (lens): a glass, usually colored, used in goggles, helmets, and hand shields to exclude harmful light rays.
Fixture: a device designed to hold parts to be joined in proper relation to each other.
Flame gouging: see oxygen gouging.
Flare-bevel groove weld: a type of groove weld. See Fig. A-61.
Flare-V-groove weld: a type of groove weld. See Fig. A-61.
Flashback: a recession of the flame into or back of the mixing chamber of the torch.
Flash welding: a resistance welding process wherein coalescence is produced simultaneously over the entire area of abutting surfaces by the heat obtained from resistance to the flow of electric current between the two surfaces and by the application of pressure after heating is substantially completed. Flashing and upsetting are accompanied by expulsion of metal from the joint. See Fig. A-25.
Flat position (downhand): the position of welding wherein welding is performed from the upper side of the joint, and the face of the weld is approximately horizontal. See Fig. A-1.
Flush weld: a weld made with a minimum reinforcement.
Flux*: a fusible material or gas used to dissolve and/or prevent the formation of oxides, nitrides,

or other undesirable inclusions formed in welding.
Flux-cored arc welding (FCAW): an arc welding process wherein coalescence is produced by heating with an arc between a continuous filler metal (consumable) electrode and the work. Shielding is obtained from a flux contained within the electrode. Additional shielding may or may not be obtained from an externally supplied gas or gas mixture.
Flux-cored electrode: a continuous filler metal electrode consisting of a metal tube containing flux and various alloying ingredients.
Flux-oxygen cutting: an oxygen cutting process wherein severing of metals is effected by using a flux to facilitate the cutting.
Forehand welding: a welding technique wherein the welding torch or gun is directed toward the progress of welding. See Fig. A-69.
Forge welding (FOW), (blacksmith, roll, hammer)*: a group of pressure welding processes wherein the parts to be welded are brought to suitable temperature by means of external heating, and the weld is consummated by pressure or blow. Temperature is below that required for fusion.
Free-bend test specimen: a specimen which is tested by bending without constraint of a jig.
Free carbon*: this is the percentage of carbon that is not combined and usually exists as free atoms at grain boundaries or in the form of pure carbon flakes of graphite.
Freezing*: solidification of a hot, liquid metal. In the case of pure iron, this starts and ends at the same temperature. With the addition of carbon, freezing starts at one temperature and ends at a lower temperature.
Full fillet weld: a fillet weld whose size is equal to the thickness of the thinner member being joined.
Fusion: the melting together of filler metal and base metal or of base metal only, which results in coalescence. See depth of fusion.
Fusion line*: the junction between the metal that has been melted and the unmelted base metal.
Fusion welding: a group of welding processes in which metals are welded together without the application of mechanical pressure or blows. The metals are welded by bringing them to the molten state at the surfaces to be joined. Fusion welding processes can be carried out with or without the addition of filler metal.
Fusion welds*: welds in which the two metal members to be joined are fused directly to each other without the addition of filler metal from a consumable electrode or welding rod.
Fusion zone: the area of base metal melted as determined on the cross section of a weld. See Fig. A-73.

G

Gap:** (*brazing*) the space between the outside of the pipe and the inside of the cup. (If both are perfectly round, the gap will be one-half the difference in diameters.) For lap seams and other types of joints, the gap is the clearance space to be filled with alloy.
Gas brazing: a brazing process wherein the heat is obtained from a gas flame.
Gas cutter: see oxygen cutter.
Gas cutting: the process of severing ferrous metals by means of the chemical action of oxygen on elements in the base metal.
Gas gouging: see oxygen gouging.
Gas metal-arc cutting: an arc cutting process wherein the severing of metals is effected by melting with an electric arc between a metal (consumable) electrode and the work. Shielding is obtained from a gas, a gas mixture (which may contain an inert gas), or a mixture of a gas and a flux.
Gas metal-arc welding (GMAW): an arc welding process wherein coalescence is produced by heating with an electric arc between a filler metal (consumable) electrode and the work. Shielding is obtained from an externally supplied gas, or gas mixture (which may contain an inert gas), or a mixture of a gas and a flux. This process has sometimes been called MIG or CO_2 welding.
Gas pocket (blowhole): a weld cavity caused by entrapped gas.
Gas regulator: see regulator.
Gas-shielded arc welding: a general term used to describe gas metal-arc welding and gas tungsten-arc welding.
Gas torch: see welding torch and cutting torch.
Gas tungsten-arc cutting (GTAC): an arc cutting process wherein the severing of metals is effected by melting with an electric arc between a single tungsten (nonconsumable) electrode and the work. Shielding is obtained from a gas or gas mixture (which may contain an inert gas).
Gas tungsten-arc welding (GTAW): an arc welding process wherein coalescence is produced by heating with an electric arc between a single tungsten (nonconsumable) electrode and the

work. Shielding is obtained from a gas or gas mixture (which may contain an inert gas). Pressure may or may not be used and filler metal may or may not be used. This process is frequently called TIG welding.

Gas welding: a group of welding processes wherein coalescence is produced by heating with a gas flame or flames, with or without the application of pressure.

Globular transfer: a mode of metal transfer in gas metal-arc welding in which the comsumable electrode is transferred across the arc in large droplets.

Goggles: see welding goggles.

Gouging: the forming of a bevel or groove by material removal.

Grain*: the crystalline body which may be viewed under a microscope as having definable limits.

Graphitic carbon*: free, uncombined carbon existing in a metallic material in the form of flakes.

Groove (scarf): the opening provided between two members to be joined by a groove weld. See Fig. A-27.

Groove angle: the total included angle of the groove between parts to be joined by a groove weld. See Fig. A-33.

Groove face: that surface of a member included in the groove. See Fig. A-27.

Groove radius: the radius of a J- or U-groove. See Figs. A-14, A-15, and A-30.

Groove weld: a weld made in the groove between two members to be joined. The standard types of groove welds are as follows:

- Square groove weld, Fig. A-7
- Single-V groove weld, Fig. A-8
- Single-bevel groove weld, Fig. A-9
- Single-U groove weld, Fig. A-10
- Single-J groove weld, Fig. A-11
- Double-V groove weld, Fig. A-12
- Double-bevel groove weld, Fig. A-13
- Double-U groove weld, Fig. A-14
- Double-J groove weld, Fig. A-15

Ground connection: the connection of the work lead to the work.

Ground lead: see work lead.

Guided-bend test: a bending test wherein the specimen is bent to a definite shape by means of a jig.

Gun: (*arc welding*) in semi-automatic, machine, and automatic welding, a manipulating device to transfer current and guide the electrode into the arc. It may also include provisions for shielding and arc initiation.

H

Hand (face) shield: a protective device used in arc welding for shielding the face and neck; equipped with suitable filter glass lens and designed to be held by handle.

Hard crack*: see underbead crack.

Hard surfacing: see surfacing.

Heat-affected zone: that portion of the base metal which has not been melted, but whose mechanical properties or microstructures have been altered by the heat of welding or cutting. See Fig. A-77.

Helmet: a protective device used in arc welding for shielding the face and neck; equipped with suitable filter glass lens and designed to be worn on the head.

Horizontal fixed position: the position of a pipe joint wherein the axis of the pipe is approximately horizontal and the pipe is not rotated during welding.

Horizontal position: (*fillet weld*) the position of welding wherein welding is performed on the upper side of an approximately horizontal surface and against an approximately vertical surface. See Fig. A-1B. (*Groove weld*) the position of welding wherein the axis of the weld lies in an approximately horizontal plane and the face of the weld lies in an approximately vertical plane. See Fig. A-1A.

Horizontal rolled positions: (*pipe welding*) the position of a pipe joint wherein welding is performed in the flat position by rotating the pipe. See Fig. A-67.

Hot shortness*: the characteristic tendency of a material toward brittleness at elevated temperatures.

Hydrogen embrittlement: this condition is caused by the loss of ductility in weld metal as a result of the metal absorbing hydrogen. Loss of ductility causes cracking in some metals.

I

Impedance: the combination of resistance and reactance, which opposes the flow of current, in an alternating current circuit.

Inadequate joint penetration: joint penetration which is less than that specified.

Included angle: see groove angle.

Inclusions: these are usually nonmetallic particles such as slag that are trapped in the weld as a result of fast freezing or poor manipulation.

Incomplete fusion: fusion which is less than complete. See Fig. A-74.

Inert gas*: a gas such as helium or argon which does not chemically combine with other elements. Such a gas serves as an effective shield of the welding arc and protects the molten weld metal against contamination from the atmosphere until it freezes.
Inert-gas carbon-arc welding: see gas carbon-arc welding.
Inert-gas metal-arc welding: see preferred term, gas metal-arc welding.
Inert-gas tungsten-arc welding: see preferred term, gas tungsten-arc welding.
Insert**: (*brazing*) a ring of silver-brazing alloy set into a machined groove in the cup of a fitting.
Intermittent weld: a weld whose continuity is broken by unwelded spaces. See Figs. A-20 and A-21.
Intermittent welding: welding wherein the continuity is broken by recurring unwelded spaces. See Figs. A-20 and A-21.
Interpass temperature*: in a multiple-pass weld, the lowest temperature of the deposited weld metal before the next pass is started.

J

J-groove welds: see groove welds.
Joint: the location where two or more members are to be joined.
Joint design: the joint geometry together with the required dimensions of the welded joint.
Joint penetration: the minimum depth a groove weld extends from its face into a joint, exclusive of reinforcement. See Fig. A-42.
Joint (unwelded): the location where two or more members are to be joined by welding.
Joint (welded): see welded joint.
Joint welding procedure: the materials, detailed methods, and practices employed in the welding of a particular joint.
Joint welding sequence: see buildup sequence.

K

Kerf: the space from which metal has been removed by a cutting process. See Fig. A-55.

L

Lack of fusion: see incomplete fusion.
Lack of joint penetration: see inadequate joint penetration.
Land: see preferred term, root face.
Lap joint: a welded joint in which two overlapping parts are connected by means of fillet, plug, slot, spot projection, or seam welds. See Fig. A-5.
Layer: a stratum of weld metal consisting of one or more weld beads. See Figs. A-44.
Lead burning: a misnomer for the welding of lead.
Leg of a fillet weld: the distance from the root of the joint to the toe of the fillet weld. See Fig. A-39.
Lens: see filter glass.
Lightly coated electrode: a filler metal electrode used in arc welding consisting of a metal wire with a light coating applied subsequent to the drawing operation primarily for stabilizing the arc.
Liquidus**: (*brazing*) the lowest temperature at which the metal is completely liquid.
Load*: the amount of a force applied to a material or structure.
Local preheating: preheating a specific portion of a structure.
Local stress-relief heat treatment: stress-relief heat treatment of a specific portion of a structure.
Locked-up stress: see residual stress.
Longitudinal sequence: the order in which the increments of a continuous weld are deposited with respect to its length.
Lower critical temperature*: the temperature at which an alloy completes its transformation from one type of solid structure to another as it is cooling.

M

Machine oxygen cutting: oxygen cutting with equipment that performs the cutting operation under the constant observation and control of an operator. The equipment may or may not perform the loading and unloading of the work.
Malleability: the pliability of a material in forming operations. It is the characteristic of a metal to be deformed without rupturing.
Manifold: a multiple header for interconnecting several cylinders to one or more torch supply lines.
Manual oxygen cutting: oxygen cutting wherein the entire cutting operation is performed and controlled by hand.
Manual welding: welding wherein the entire welding operation is performed and controlled by hand.
Martensite*: a structure resulting from transformation of austenite at temperature considerably below the usual range; achieved by rapid cooling. It is made up of ultra-hard, needlelike crystals that are a supersaturated solid solution of carbon in iron.
Matrix*: the principal, physically continuous metallic constituent in which crystals or free atoms

of other constituents are embedded. It serves as a binder, holding the entire mass together.
Melt point:** (*brazing*) the temperature below which an alloy is substantially solid.
Melting rate: the weight or length of electrode melted in a unit of time.
Melting ratio: the ratio of the volume of weld metal below the original surface of the base metal to the total volume of the weld metal.
Melt-thru: complete joint penetration of weld metal in a joint welded from one side with visible root reinforcement. See Fig. A-62.
Metal arc cutting (MAC): an arc cutting process wherein the severing of metals is effected by melting with the heat of an arc between a metal electrode and the base metal. See shielded metal-arc cutting and gas metal-arc cutting.
Metal arc welding: an arc welding process wherein the electrode supplies the filler metal in the weld.
Metal electrode: a filler- or nonfiller-metal electrode used in arc welding consisting of a metal wire with or without a covering or coating.
Metallurgy*: the science and technology of extracting metals from their ores, refining them, and preparing them for use.
Metal powder cutting (POC): an oxygen cutting process wherein a metal powder such as iron is used to facilitate cutting.
Micro-cracks*: cracks or fissures in a metallic structure which cannot be seen except with the aid of a microscope.
Microstructure*: the detailed structure of a metal or alloy, as revealed by microscopic examination, showing the various continuous phases as well as any nonmetallic inclusions.
MIG welding: see preferred term, gas metal-arc welding.
Mixing chamber: that part of a gas welding or cutting torch wherein the gases are mixed for combustion.
Modulus of elasticity: the ratio of tensile stress to the strain it causes within that range of elasticity where there is a straight-line relationship between stress and strain. The higher the modulus, the lower the degree of elasticity.

N

Neutral flame: a gas welding flame wherein the portion used is neither oxidizing nor reducing. See Fig. A-48.
Nonferrous*: lacking iron in sufficient percentage to have any dominating influence on properties of the material.
Nonpressure welding: a group of welding processes wherein the weld is made without pressure (an obsolete term).
Nozzle: a device at the end of a gun which directs gas-shielding media.
Nucleus: the central core of the atom, which carries all of the positive charge.
Nugget: the weld metal joining the parts in resistance spot, resistance seam, or projection welds.
Nugget size: the diameter of width of the nugget measured in the plane of the interface between the pieces joined. See Fig. A-63.

O

Ohm: a unit of electrical resistance. There is one ohm resistance when a pressure of one volt causes a current of one ampere to flow in a conductor.
Open circuit voltage: the voltage reading at the terminals of a welding machine when it is turned on but is not engaged in a welding operation. The welding circuit is not complete, and no current is flowing.
Open joints: see root opening. See Fig. A-31.
Ore*: the rock or earth in which we find metals in their natural form.
Overhead position: the position of welding wherein welding is performed from the underside of the joint. See Fig. A-1.
Overheating*: sufficient exposure of a metal to an extremely high temperature for an undesirable coarse grain structure to develop. The structure often can be corrected by suitable heat treatment, cold working, or a combination of these.
Overlap: protrusion of weld metal beyond the bond at the toe or root of the weld. See Fig. A-56.
Oxide:** the scale that forms on metal surfaces when they are exposed to air and especially when they are heated.
Oxidize:** (of a metal) to combine chemically with oxygen, forming another composition which is called an oxide.
Oxidizing flame: a gas welding flame wherein the portion used has an oxidizing effect. See Fig. A-49.
Oxyacetylene cutting (OFC-A): an oxygen cutting process wherein the severing of metals is effected by means of the chemical reaction of oxygen with the base metal at elevated temperatures, the necessary temperature being maintained by means of gas flames obtained from the combustion of acetylene with oxygen.
Oxyacetylene welding (OAW): a gas welding process wherein coalescence is produced by heating

with a gas flame or flames obtained from the combustion of acetylene with oxygen with or without the application of pressure and with or without the use of filler metal.

Oxy-arc cutting (AOC): an oxygen cutting process wherein the severing of metals is effected by means of the chemical reaction of oxygen with the base metal at elevated temperatures, the necessary temperature being maintained by means of an arc between an electrode and the base metal.

Oxy-city gas cutting: an oxygen cutting process wherein the severing of metals is effected by means of the chemical reaction of oxygen with the base metal at elevated temperatures, the necessary temperature being maintained by means of gas flames obtained from the combustion of city gas with oxygen.

Oxy-fuel gas welding: see gas welding.

Oxygen cutter: one who is capable of performing a manual oxygen cutting operation.

Oxygen cutting operator: one who operates machine or automatic oxygen cutting equipment.

Oxygen gouging: an application of oxygen cutting wherein a chamfer or groove is formed.

Oxygen lance: a length of pipe used to convey oxygen to the point of cutting in oxygen lance cutting.

Oxygen lance cutting (LOC): an oxygen cutting process wherein only oxygen is supplied by the lance, and the preheat is obtained by other means.

Oxygen machining: a process of shaping ferrous metals by oxygen cutting or oxygen grooving.

Oxyhydrogen welding (OHW): a gas welding process wherein coalescence is produced by heating with a gas flame or flames obtained from the combustion of hydrogen with oxygen without the application of pressure and with or without the use of filler metal.

Oxy-natural gas cutting (OFC-N): an oxygen cutting process wherein the severing of metals is effected by means of the chemical reaction of oxygen with the base metal at elevated temperatures, the necessary temperature being maintained by means of gas flames obtained from the combustion of natural gas with oxygen.

Oxy-other fuel gas welding: a gas welding process wherein the welding heat is obtained from the combustion of oxygen and any fuel gas other than acetylene.

Oxy-propane cutting (OFC-P): an oxygen cutting process wherein the severing of metals is effected by means of the chemical reaction of oxygen with the base metal at elevated temperatures, the necessary temperature being maintained by means of gas flames obtained from the combustion of propane with oxygen.

P

Parallel beads: see beading.

Parent metal*: the metal to be welded or otherwise worked upon; also called the base metal.

Partial joint penetration: joint penetration which is less than complete.

Pass: the weld metal deposited by one general progression along the axis of a weld. See Fig. A-43.

Pass sequence: the method of depositing a weld with respect to its length.

Pearlite*: a continuous granular mixture composed of alternate plates or layers of cementite (pure iron-ferrite and iron-carbide).

Peel test: a destructive method of inspection wherein a lap joint is mechanically separated by peeling.

Peening: mechanical working of metal by means of impact blows.

Penetration: the penetration, or depth of fusion, of a weld is the distance from the original surface of the base metal to that point at which fusion ceases. See Fig. A-42.

Physical property*: an inherent physical characteristic of a material which is not directly an ability to withstand a physical force of any kind.

Pickup*: the absorption of base metal by the weld metal as the result of admixture. Depending upon the materials involved, this can be an asset and not a liability.

Plasma-arc cutting (PAC): an arc cutting process wherein severing of the metal is obtained by melting a localized area with a constricted arc and removing the molten material with a high velocity jet of hot, ionized gas issuing from the orifice.

Plasma-arc welding (PAW): an arc welding process wherein coalescence is produced by heating with a constricted arc between an electrode and the workpiece (transferred arc) or the electrode and the constricting nozzle (non-transferred arc). Shielding is obtained from the hot, ionized gas issuing from the orifice which may be supplemented by an auxiliary source of shielding gas. Shielding gas may be an inert gas or a mixture of gases. Pressure may or may not be used, and filler metal may or may not be supplied.

Plug weld: a weld made in a hole in one member of a lap joint, joining that member to that portion

of the surface of the other member which is exposed through the hole. See Fig. A-18.
Porosity*: the presence of gas pockets, inclusions, or voids in metal.
Positioned weld: a weld made in a joint which has been so placed as to facilitate making the weld.
Position of welding: see flat, horizontal, vertical, and overhead positions and horizontal rolled, horizontal fixed, and vertical pipe positions.
Postheating: heat applied to the base metal after welding or cutting for the purpose of tempering, stress-relieving, or annealing.
Preheating: the application of heat to the base metal prior to a welding or cutting operation.
Preinserted-ring type:** (*brazing*) describes the kind of fittings that have inserts to supply silver-brazing alloy to their joints.
Pressure welding: any welding process or method wherein pressure is used to complete the weld.
Procedure qualification: the demonstration that welds made by a specific procedure can meet prescribed standards.
Projection welding (RPW): a resistance welding process wherein localization of heat between two surfaces or between the end of one member and surface of another is effected by projections. See Fig. A-23.
Properties*: those features or characteristics of a metal that make it useful and distinctive from all others.
Proportional elastic limit*: the stress point beyond which an increase in stress results in permanent deformation.
Protective atmosphere: a gas envelope surrounding the part to be brazed or welded in which the gas composition is controlled with respect to chemical composition, dew point, pressure, flow rate, etc. Examples are inert gases, combusted fuel gases, hydrogen, and a vacuum

Q

Qualification: see welder qualification and procedure qualification.

R

Radial crack*: a crack originating in the fusion zone and extending into the base metal, usually at right angles to the line of fusion. This type of crack is due to the high stresses involved in the cooling of a rigid structure.
Radiographic quality*: soundness of a weld that shows no internal or underbead cracks, voids, or inclusions when inspected by X-ray or gamma ray techniques.
Rectifier: an electrical assembly used in rectifier welders to change alternating current to direct current.
Red hardness*: the property of metallic material to retain its hardness at high temperatures.
Reducing flame:** an oxyacetylene flame that has an excess of acetyelene gas in the mixture (contrasted with an oxidizing flame, which has an excess of oxygen in the mixture).
Refractory*: the quality of resistance to the effects of high temperatures, especially the ability of a material to maintain its hardness and shape. A "refractory" material has an exceptionally high melting point in contrast to a "fusible" material which has an exceptionally low melting point.
Regulator: a device for controlling the delivery of gas at some substantially constant pressure regardless of variation in the higher pressure at the source.
Reinforcement of weld: weld metal in excess of the metal necessary for the specified weld size. See Fig. A-37.
Residual stresses*: internal stresses that exist in a metal at room temperature as the result of (1) previous non-uniform heating and expansion, or (2) mechanical treatment. Both may be involved.
Resistance: an electrical unit that opposes the passage of current. The ohm is the unit of measure for resistance.
Resistance seam welding (RSEW): a resistance welding process wherein coalescence at the faying surfaces is produced by the heat obtained from resistance to the flow of electric current through the work parts held together under pressure by electrodes. The resulting weld is a series of overlapping resistance spot welds made progressively along a joint by rotating the electrodes. See Fig. A-64.
Resistance spot-welding (RSW): a resistance welding process wherein coalescence at the faying surfaces is produced in one spot by the heat obtained from the resistance to the flow of electric current through the work parts held together under pressure by electrodes. The size and shape of the individually formed welds are limited primarily by the size and contour of the electrodes. See Fig. A-64.
Resistance welding: a group of welding processes wherein coalescence is produced by the heat obtained from resistance of the work to the flow of electric current in a circuit of which the work is a part and by the application of pressure.
Reverse polarity (electrode positive): the arrangement of direct arc welding leads wherein the work

is the negative pole and the electrode is the positive pole of the welding arc. See Fig. A-75.

Root: see root of joint and root of weld.

Root crack: a crack in the weld or base metal occurring at the root of a weld.

Root edge: the edge of the part to be welded which is adjacent to the root. See Fig. A-28.

Root face (shoulder): the portion of the prepared edge of a part to be joined by a groove weld which has not been beveled or grooved. See Fig. A-29.

Root of joint: that portion of a joint to be welded where the members approach closest to each other. In cross section the root of the joint may be a point, a line, or an area. See Fig. A-34.

Root opening: the separation at the root between parts to be joined by a groove weld. See Fig. A-31.

Root penetration: the depth a groove weld extends into the root of a joint measured on the centerline of the root cross section. See Fig. A-34.

Root of weld: the point at the bottom of the weld. See Fig. A-34.

Root radius: the radius near the root portion of the prepared edge of a part to be joined by a U- or J-groove weld.

Root reinforcement: reinforcement of weld at the side other than that from which welding was done. See Fig. A-37.

Root surface: the exposed surface of a weld on the side other than that from which welding was done. See Fig. A-28.

S

Scarf: see groove or edge preparation.

Scarfing: see chamfering.

Seal weld: any weld used primarily to obtain tightness against leakage.

Seam welding: a resistance welding process wherein coalescence is produced by the heat obtained from resistance to the flow of electric current through the work parts held together under pressure by circular electrodes. The resulting weld is a series of overlapping spot welds made progressively along a joint by rotating the electrodes. See Fig. A-24.

Semi-automatic arc weld: a weld made with equipment which automatically controls only the feed of the electrode, the manipulation of the electrode being controlled by hand.

Shielded metal-arc cutting (SMAC): a method of metal-arc cutting wherein the severing of metals is effected by melting with the heat of an arc between a covered metal electrode and the base metal.

Shielded metal-arc welding (SMAW): an arc welding process wherein coalescence is produced by heating with an electric arc between a covered metal electrode and the work. Shielding is obtained from decomposition of the electrode covering. Pressure is not used, and filler metal is obtained from the electrode.

Shielding*: primarily the protection of molten metal by the arc from oxidizing or otherwise reacting with elements in the surrounding air. Usually, the shielding also stabilizes the arc.

Short-circuiting arc welding: a method of gas metal-arc welding using short-circuiting transfer.

Short-circuiting transfer: a mode of metal transfer in gas metal-arc welding in which the consumable electrode is deposited during repeated short circuits.

Shoulder:** (*brazing*) the machined ridge at the bottom or inner end of the cup where the end of the pipe seats. See Fig. A-78.

Shoulder: (*welding*) see root face.

Shrinkage stress: see residual stress.

Silver alloy brazing: (*silver soldering*) a brazing process wherein a silver alloy is used as a filler metal.

Silver-brazing alloy:** metal that contains silver in its composition and which is designed especially for relatively low-temperature joining of metal parts such as pipe, tubing, and fittings.

Single-groove welds: see groove welds.

Single-welded butt joint: a butt joint welded from one side only.

Single-welded joint: in arc and gas welding, any joint welded from one side only. See Fig. A-65.

Single-welded lap joint: a lap joint in which the overlapped edges of the members to be joined are welded along the edge of one member.

Size of weld:

Groove weld—the joint penetration (depth of chamfering plus the root penetration when specified).

Fillet weld—for equal-leg fillet welds, the leg length of the largest isosceles right triangle which can be inscribed within the fillet weld cross-section. See Fig. A-41.

Flange weld—the weld metal thickness measured at the root of the weld. For unequal leg fillet welds, the leg length of the largest right triangle which can be described within the fillet weld cross-section.

Skip sequence: see wandering sequence.

Slag*: the nonmetallic layer that forms on top of molten metal. It is usually a complex of chemi-

cals (oxides, silicates, etc.) that float to the top of the hot molten metal. When a bead of weld metal cools, the slag "cap" on the bead can be readily chipped or ground away.
Slag inclusion: nonmetallic solid material entrapped in weld metal or between weld metal and base metal.
Slope: slope is of importance when setting the current for the gas metal arc welding process. It indicates the shape of the volt-ampere output curve.
Slot weld: a weld made in an elongated hole in one member of a lap joint joining that member to that portion of the surface of the other member which is exposed through the hole. The hole may be open at one end, i.e., may extend to the edge of the member, and may or may not be filled completely with weld metal. See Fig. A-19.
Soaking*: prolonged heating of a metal at a selected temperature.
Solidus:** (*brazing*) the highest temperature at which the metal is completely solid.
Spacer strip: a metal strip or bar inserted in the root of a joint prepared for a groove weld to serve as a backing and to maintain root opening during welding.
Spatter: in arc and gas welding, the metal particles expelled during welding and which do not form a part of the weld.
Spatter loss: the difference in weight between the weight of the electrode deposited and the weight of the electrode consumed (melted); metal lost due to spatter.
Spool: a filler metal package type consisting of a continuous length of electrode wound on a cylinder (called the barrel) which is flanged at both ends. The flange extends below the inside diameter of the barrel and contains a spindle hole. Used for MIG welding.
Spot welding: a resistance welding process wherein coalescence is produced by the heat obtained from resistance to the flow of electric current through the work parts held together under pressure by electrodes. The size and shape of the individually formed welds are limited primarily by the size and contour of the electrodes. See Fig. A-22.
Spray transfer: a mode of metal transfer in gas metal-arc welding in which the consumable electrode is propelled axially across the arc in small droplets.
Square groove weld: see groove weld.
Stack cutting: oxygen cutting of stacked metal plates arranged so that all the plates are severed by a single cut.
Staggered intermittent fillet welds: two lines of intermittent fillet welding in a T- or lap joint in which the increments of welding in one line are staggered with respect to those in the other line. See Fig. A-21.
Static electricity: a term applied to stationary electricity which is nonactive; may be produced by friction between two or more bodies.
Stepback sequence: see backstep sequence.
Stick electrode: see preferred term, covered electrode.
Stitch welding: the use of intermittent welds to join two or more parts.
Straight polarity (electrode negative): the arrangement of direct current arc welding leads wherein the work is the positive pole and the electrode is the negative pole of the welding arc. See Fig. A-76.
Strain*: the physical effect of stress, usually evidenced by stretching or other deformation of the material.
Stress*: the load, or amount of a force, applied to a material tending to deform or break it.
Stress cracking*: cracking of a weld or base metal containing residual stresses.
Stress-relief heat treatment*: uniform heating of a structure or portion thereof to a sufficient temperature, below the critical range, to relieve the major portion of the residual stresses, followed by uniform cooling. (*Note:* Terms such as normalizing, annealing, etc., are misnomers for this application.)
String bead: a type of weld bead made without transverse oscillation. See Beading. See Fig. A-17.
String beading: the deposition of string beads.
Subcritical annealing*: heating of a metal to a point below the critical or transformation range, in order to relieve internal stresses.
Submerged-arc welding (SAW): an arc welding process wherein coalescence is produced by heating with an electric arc or arcs between a bare metal electrode or electrodes and the work. The arc is shielded by a blanket of granular, fusible material on the work. Pressure is not used, and filler metal is obtained from the electrode and sometimes from a supplementary welding rod.
Surfacing: the deposition of filler metal on a metal surface to obtain desired properties or dimensions. Also referred to as *buttering* and *clodding*.

Surfacing weld: a type of weld composed of one or more string or weave beads deposited on an unbroken surface to obtain desired properties or dimensions. See Fig. A-66.

T

Tack weld: a weld made to hold parts of a weldment in proper alignment until the final welds are made.
T-joint: a welded joint at the junction of two parts located approximately at right angles to each other in the form of a *T*. See Fig. A-6.
Temper*:

1. The amount of carbon present in the steel: 10 temper is 1.00% carbon.
2. The degree of hardness that an alloy has after heat treatment or coldworking, e.g. the aluminum alloys.

Temporary weld: a weld made to attach a piece or pieces to a weldment for temporary use in handling, shipping, or working on the weldment.
Tensile strength*: the resistance of a material to a force which is acting to pull it apart.
Theoretical throat of fillet weld: the distance from the root to the hypotenuse of the largest isosceles right triangle that can be inscribed within the weld cross section. See Fig. A-40.
Thermal stress: the stresses produced in a structure or member caused by differences in temperature or coefficients of expansion.
Thermit welding (TW): a group of welding processes wherein coalescence is produced by heating with superheated liquid metal and slag resulting from a chemical reaction between a metal oxide and aluminum with or without the application of pressure. Filler metal, when used, is obtained from the liquid metal.
Throat (of a groove weld): see preferred term, size of weld.
TIG welding: see preferred term, gas tungsten-arc welding.
Toe crack*: a crack originating at the junction between the face of the weld and the base metal. It may be any one of three types: (1) radial or stress crack; (2) underbead crack extending through the hardened zone below the fusion line; or (3) the result of poor fusion between the deposited filler metal and the base metal.
Toe of weld: the junction between the face of the weld and the base metal. See Fig. A-36.
Torch: see welding torch or cutting torch.
Torch brazing (TB): a brazing process wherein coalescence is produced by heating with a gas flame and by using a nonferrous filler metal having a melting point above 800 degrees F. but below that of the base metal. The filler metal is distributed in the joint capillary attraction.
Torch tip: see welding tip or cutting tip.
Transformation range*: unless otherwise specified, the temperature range during which a metal, when cooled, is changing from one type of crystal structure to the structure it will have permanently at room temperature.
Tungsten electrode: a non-filler-metal electrode used in arc welding consisting of a tungsten wire.

U

Ultimate tensile strength*: the maximum pulling force to which the material can be subjected without failure.
Underbead crack*: a crack in the heat-affected zone generally not extending to the surface of the base metal.
Undercut: a groove melted into the base metal adjacent to the toe of the weld and left unfilled. See Figs. A-38 and A-56.
Upper critical temperature*: the temperature at which an alloy begins to transform from one solid structure to another as it cools.
Unshielded metal-arc welding: a metal arc welding process wherein no shielding medium is used.

V

Vertical position: the position of welding wherein the axis of the weld is approximately vertical. See Fig. A-1.
Vertical position (pipe welding): the position of a pipe joint wherein the axis of the pipe is vertical, welding is performed in the horizontal position, and the pipe may or may not be rotated. See Fig. A-67.
Volt: the force that causes current to flow.
Voltage regulator: an automatic electrical control device for maintaining a constant voltage supply to the primary of a welding transformer.
Volt-ampere curve: curve which indicates the current output of a welding machine and is formed through the use of voltage and amperage values.

W

Wandering block sequence: a block sequence wherein successive blocks are completed at random after several starting blocks have been completed.
Wandering sequence: a longitudinal sequence wherein the weld bead increments are deposited at random.
Watt: a unit of electrical power measurement.

Weaving: a technique of depositing weld metal in which the electrode is oscillated. See Fig. A-45.
Weld: a localized coalescence of metal wherein coalescence is produced by heating to suitable temperatures with or without the application of pressure and with or without the use of filler metal. The filler metal either has a melting point approximately the same as the base metals or has a melting point below that of the base metals but above 800 degrees F.
Weldability: the capacity of a metal to be welded under the fabrication conditions imposed into a specific, suitably designed structure and to perform satisfactorily in the intended service.
Weld bead: a weld deposit resulting from a pass.
Weld crack: a crack in weld metal.
Weld gauge: a device designed for checking the shape and size of welds.
Weld line: see bond line.
Weld metal: that portion of a weld which has been melted during welding.
Weld metal area: the area of the weld metal as measured on the cross section of a weld.
Weld penetration: see joint penetration and root penetration.
Welded joint: a union of two or more members produced by the application of a welding process.
Welder: one who is capable of performing a manual or semi-automatic welding operation; also used to denote a welding machine.
Welder certification: certification in writing that a welder has produced welds meeting prescribed standards.
Welder qualification: the demonstration of a welder's ability to produce welds meeting prescribed standards.
Welding current: the current flowing through the welding circuit during the making of a weld.
Welding electrode: see electrode.
Welding generator: a generator used for supplying current for welding.
Welding goggles: goggles with tinted lenses used during welding, brazing, or oxygen cutting, which protect the eyes from harmful radiation and flying particles.
Welding ground: see work lead.
Welding leads: the work lead and electrode lead of an arc-welding circuit.
Welding machine: equipment used to perform the welding operation. For example, spot welding machine, arc welding machine, seam welding machine, etc.
Welding operator: one who operates machine or automatic welding equipment.
Welding procedure: the detailed methods and practices including joint welding procedures involved in the production of a weldment.
Welding process: a metal joining process wherein coalescence is produced by heating to suitable temperature with or without the application of pressure and with or without the use of filler metal. (See forge welding, thermit welding, flow welding, gas welding, arc welding, resistance welding, induction welding, and brazing.)
Welding rod: a form of filler metal, used for welding or brazing wherein the filler metal does not conduct the electrical current.
Welding sequence: the order of making the welds in a weldment.
Welding technique: the details of a manual, machine, or semi-automatic welding operation which, within the limitations of the prescribed joint welding procedure, are controlled by the welder or welding operator.
Welding tip: a welding torch tip designed for welding.
Welding torch (blowpipe): a device used in gas welding or torch brazing for mixing and controlling the flow of gases.
Welding transformer: a transformer used for supplying current for welding.
Weldment: an assembly whose component parts are joined by welding.
Weldor: see welder.
Wetting: the bonding or spreading of a liquid filler metal or flux on a solid base metal.
Work connection: the connection of the work lead to the work.
Work hardening*: the capacity of a material to harden as the result of cold rolling or other cold working involving deformation of the metal.
Work lead: the electric conductor between the source of arc welding current and the work. See Figs. A-75 and A-76.

Y

Yield strength*: the stress point at which permanent deformation results.

Illustrated Guide to Welding Terminology

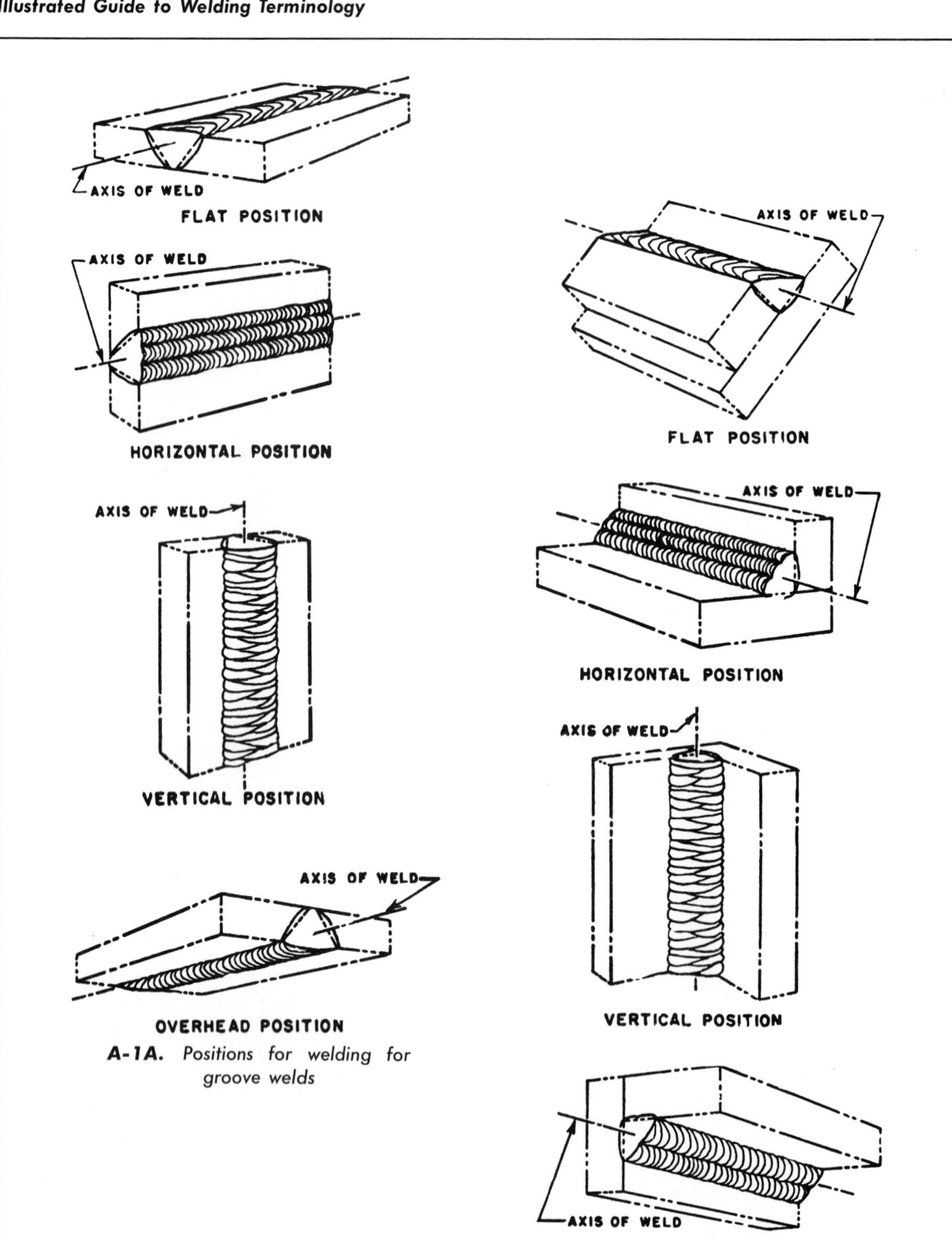

A-1A. *Positions for welding for groove welds*

A-1B. *Positions of welding for fillet welds*

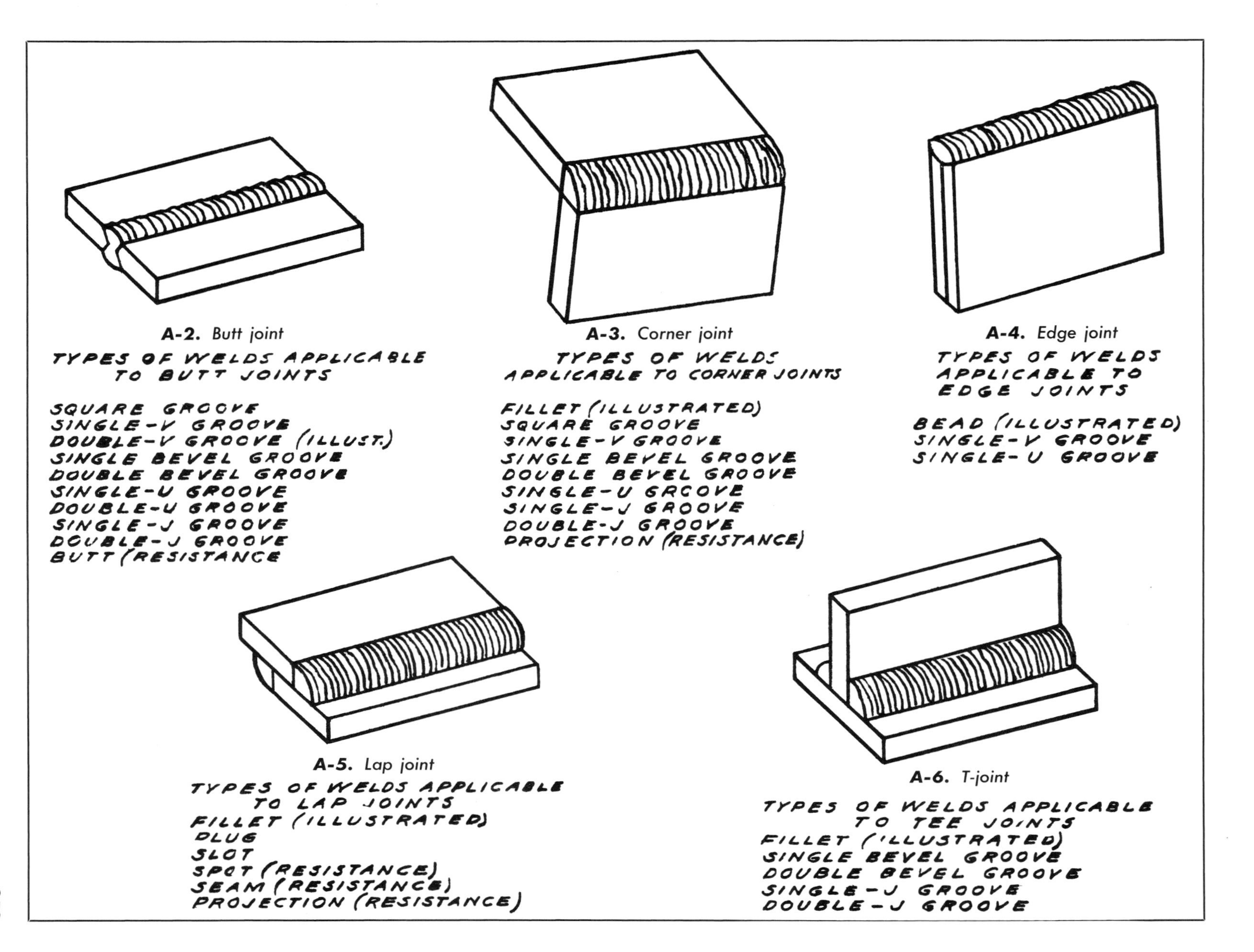

A-2. Butt joint

A-3. Corner joint

A-4. Edge joint

A-5. Lap joint

A-6. T-joint

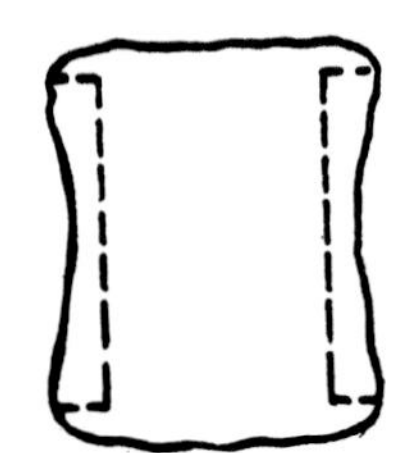

A-7. Square groove weld

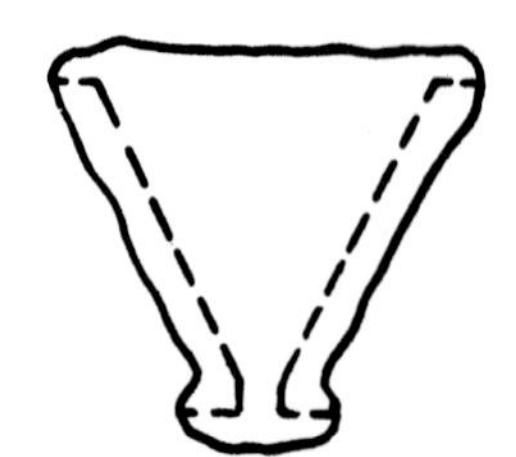

A-8. Single-V groove weld

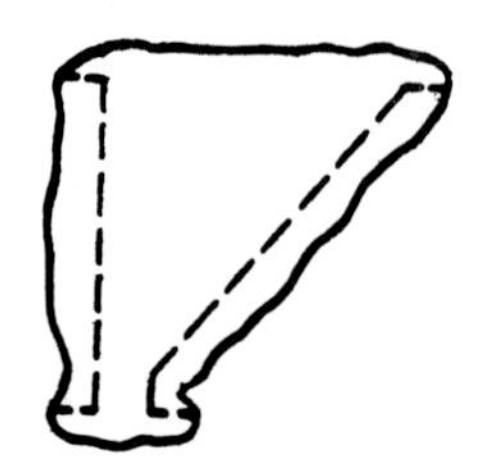

A-9. Single-bevel groove weld

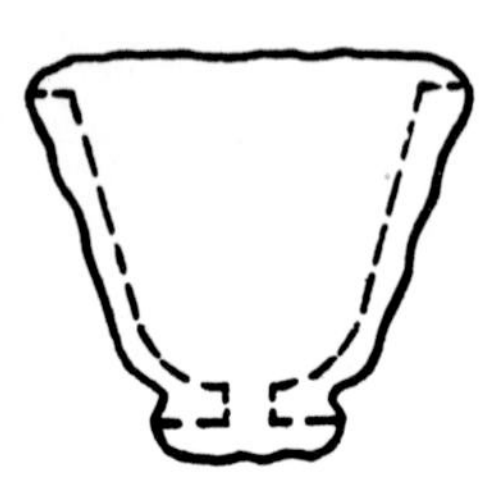

A-10. Single-U groove weld

A-11. Single-J groove weld

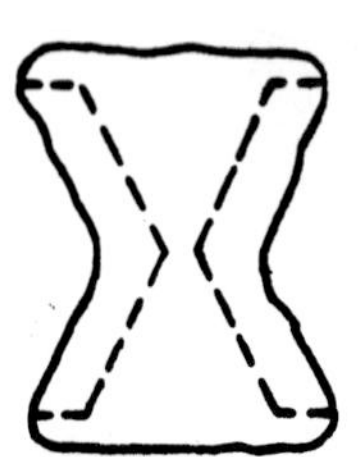

A-12. Double-V groove weld

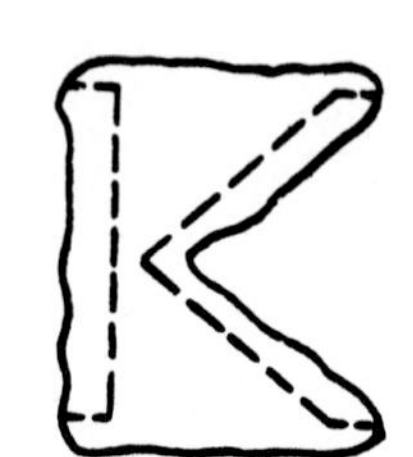

A-13. Double-bevel groove weld

A-14. Double-U groove weld

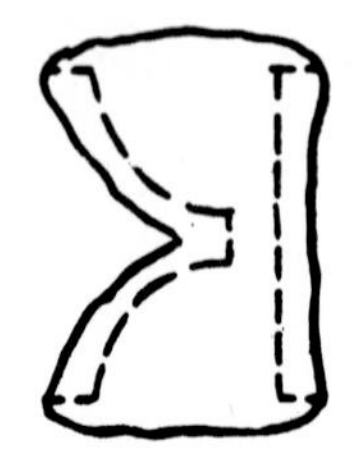

A-15. Double-J groove weld

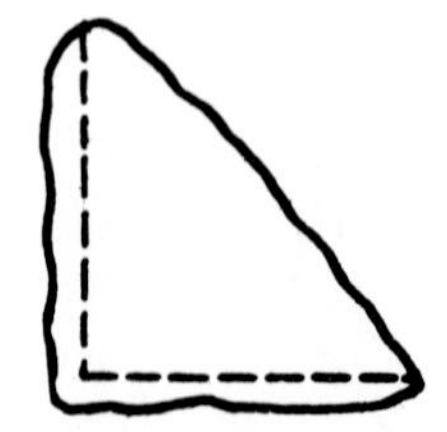

A-16. Fillet weld

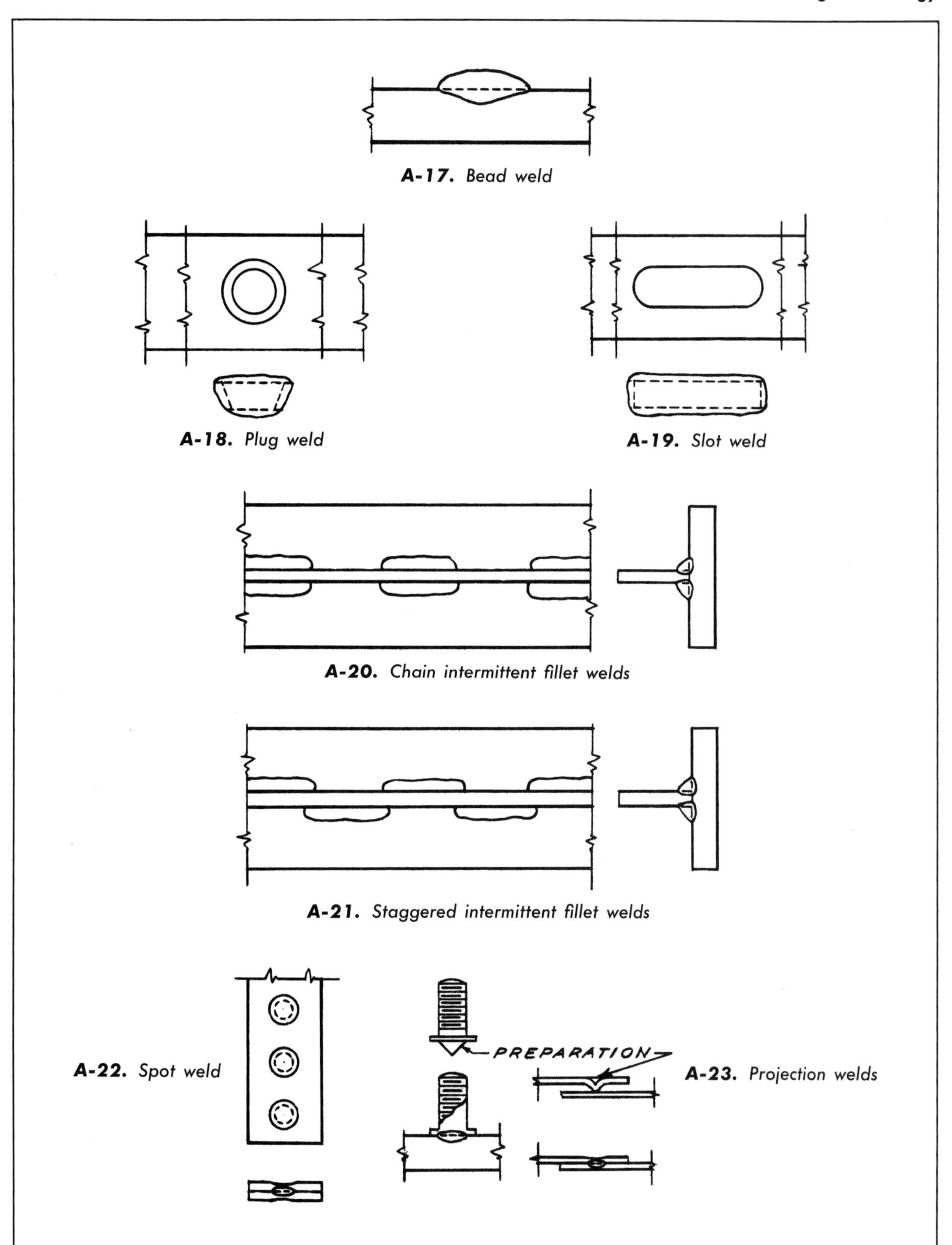

A-17. Bead weld

A-18. Plug weld

A-19. Slot weld

A-20. Chain intermittent fillet welds

A-21. Staggered intermittent fillet welds

A-22. Spot weld

A-23. Projection welds

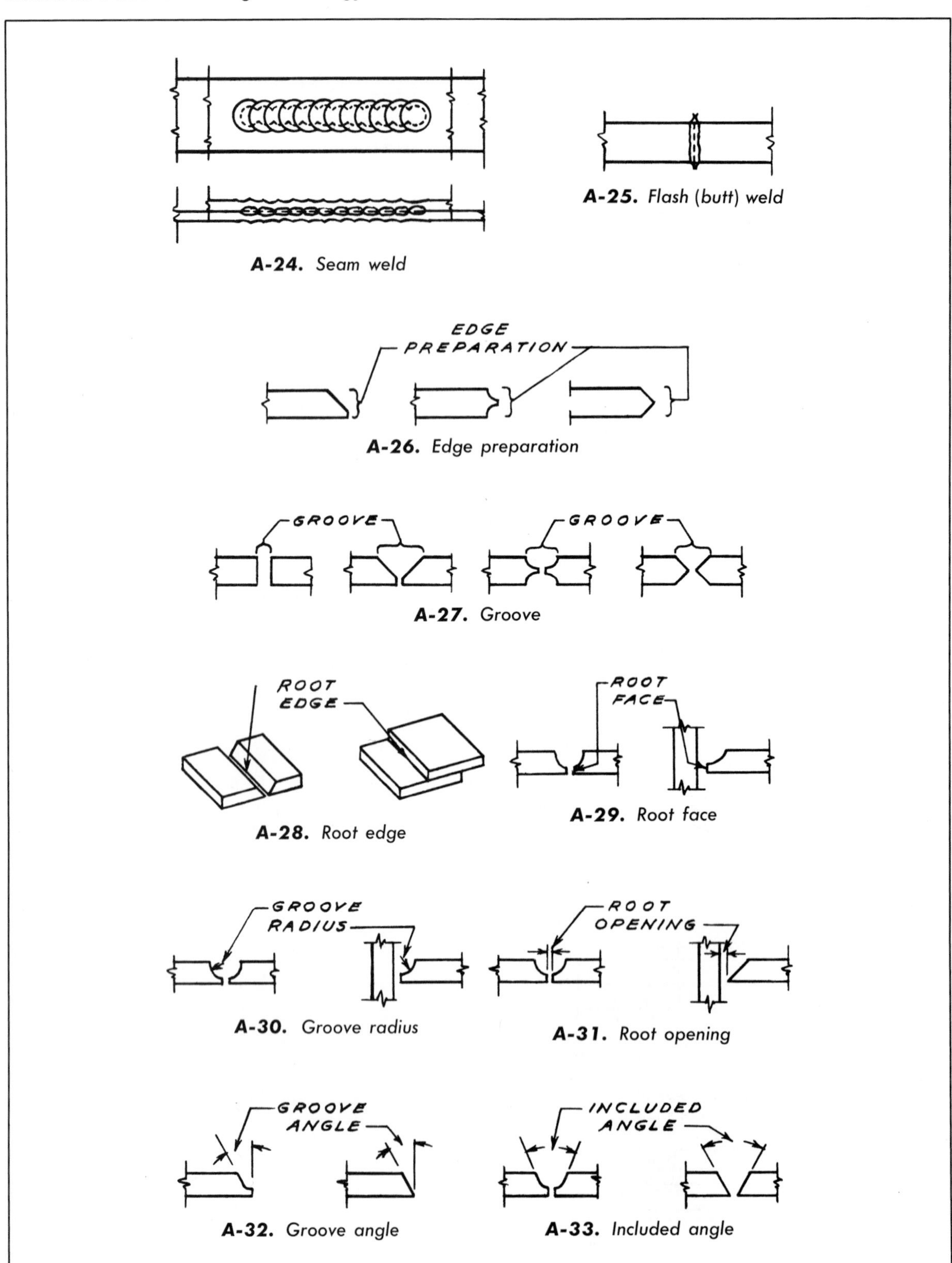

A-24. Seam weld

A-25. Flash (butt) weld

A-26. Edge preparation

A-27. Groove

A-28. Root edge

A-29. Root face

A-30. Groove radius

A-31. Root opening

A-32. Groove angle

A-33. Included angle

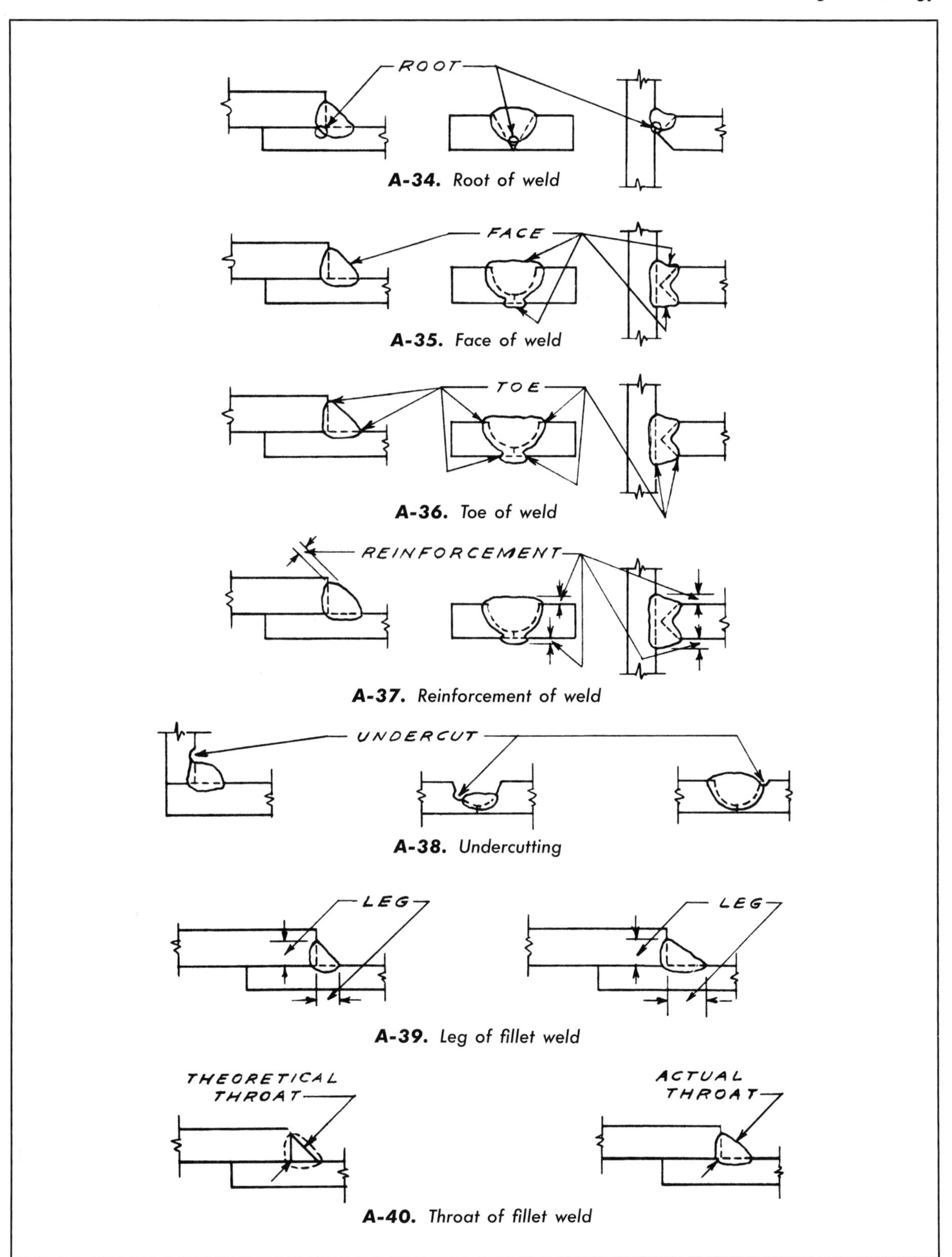

A-34. Root of weld

A-35. Face of weld

A-36. Toe of weld

A-37. Reinforcement of weld

A-38. Undercutting

A-39. Leg of fillet weld

A-40. Throat of fillet weld

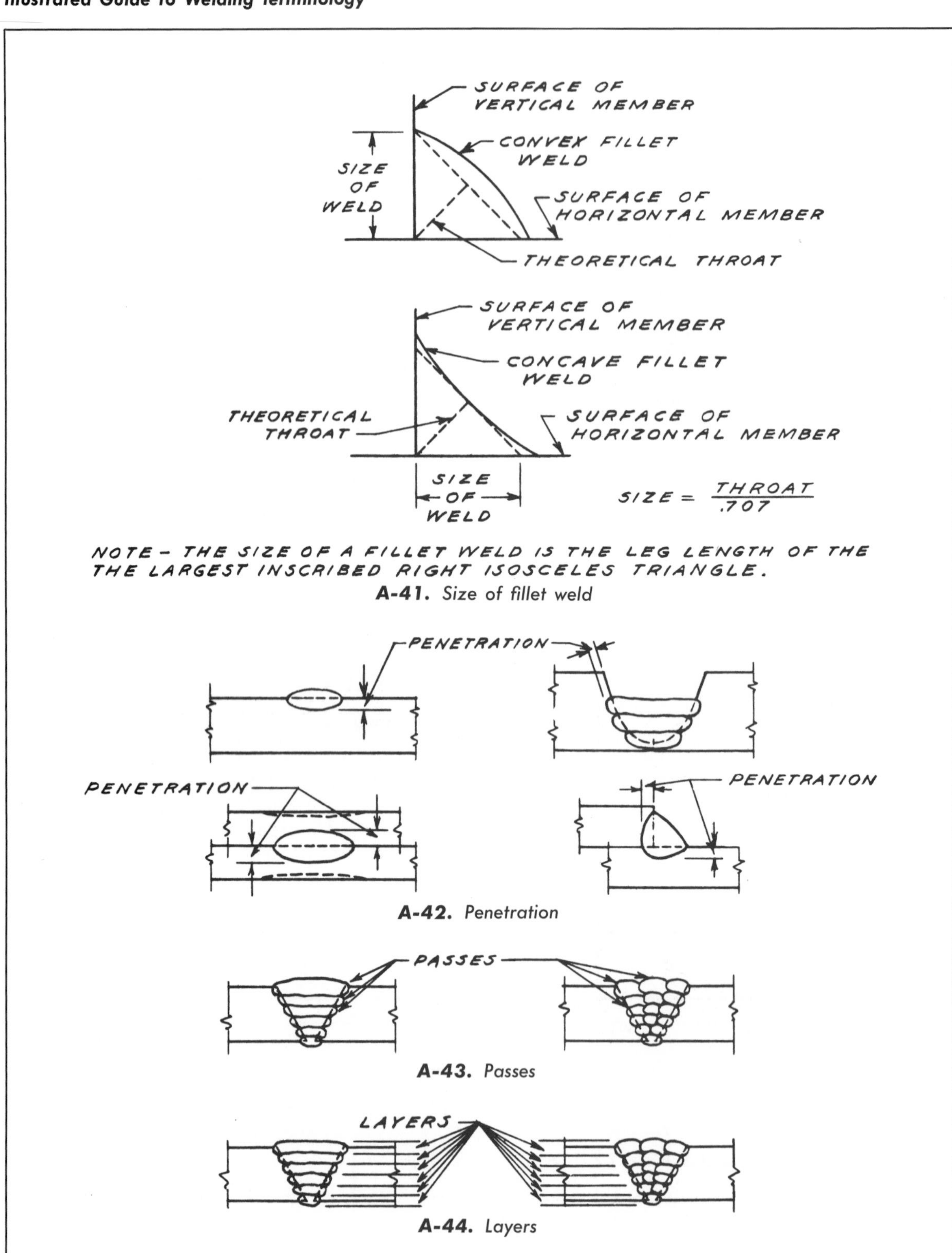

A-41. Size of fillet weld

A-42. Penetration

A-43. Passes

A-44. Layers

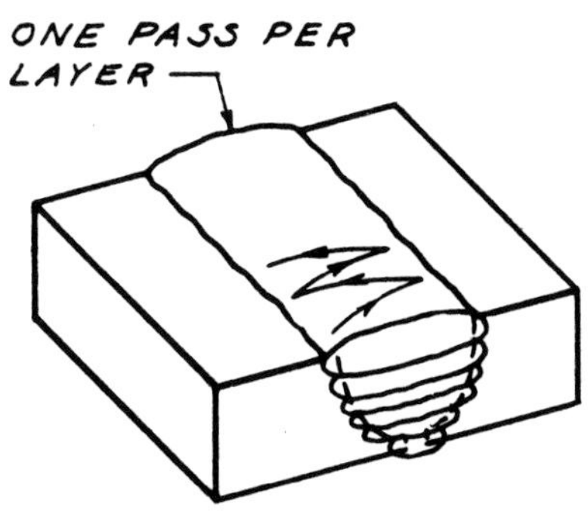

A-45. Weaving

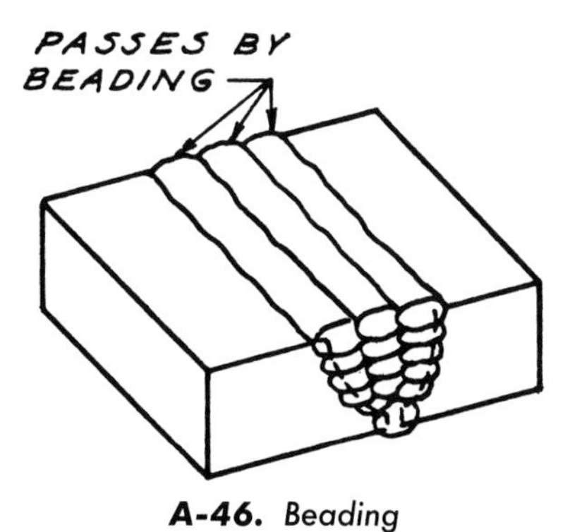

A-46. Beading

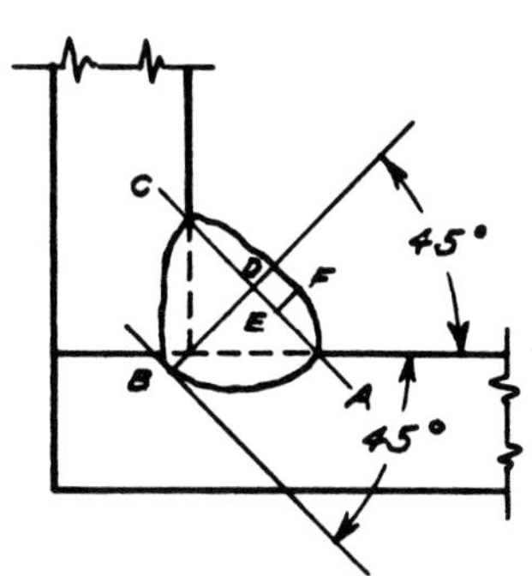

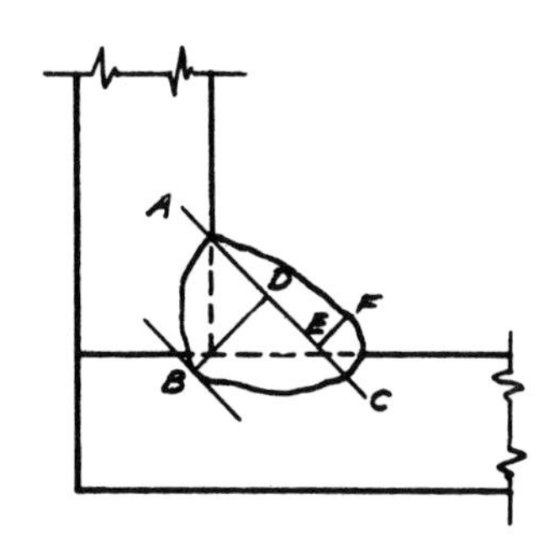

NOTE– LINE AC IS DRAWN INTERNALLY TANGENT TO THE INMOST POINT IN THE FACE OF THE FILLET.

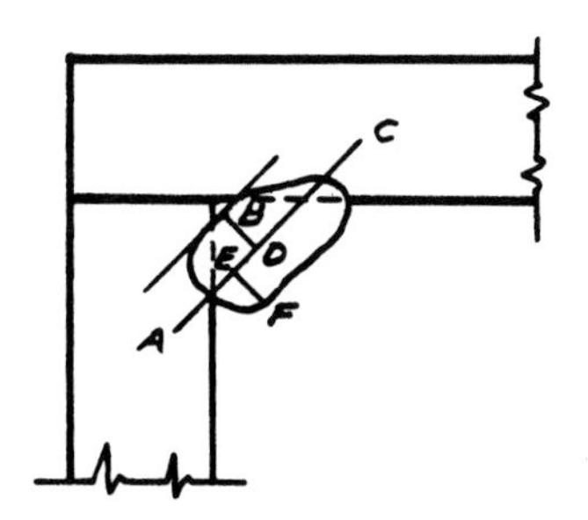

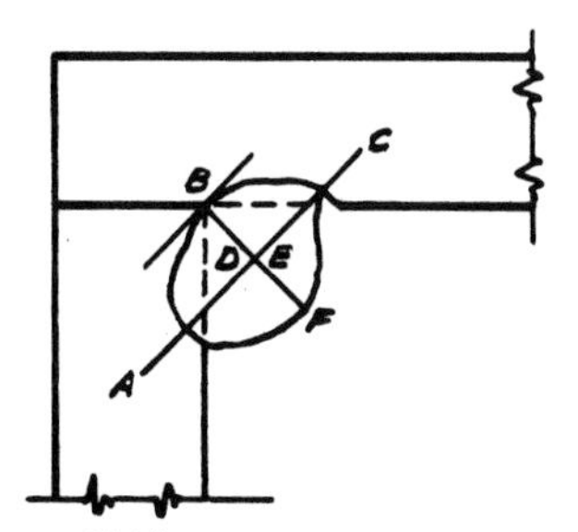

CONVEXITY RATIO = $\frac{EF}{BD}$

A-47. Convexity ratio

FLAME AS SEEN WITHOUT WELDERS GOGGLES

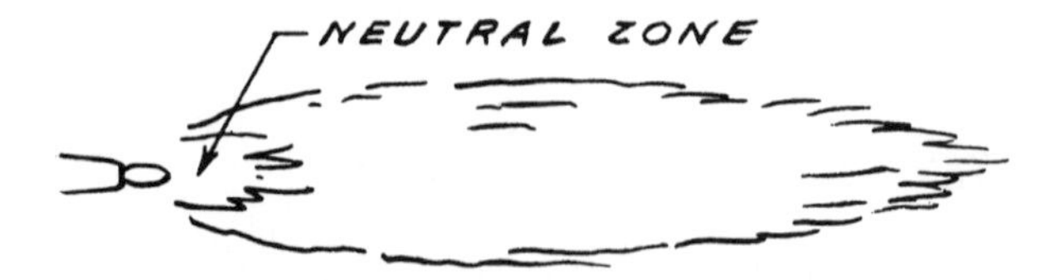

FLAME AS SEEN WITH WELDERS GOGGLES

A-48. Neutral flame

FLAME AS SEEN WITHOUT WELDERS GOGGLES

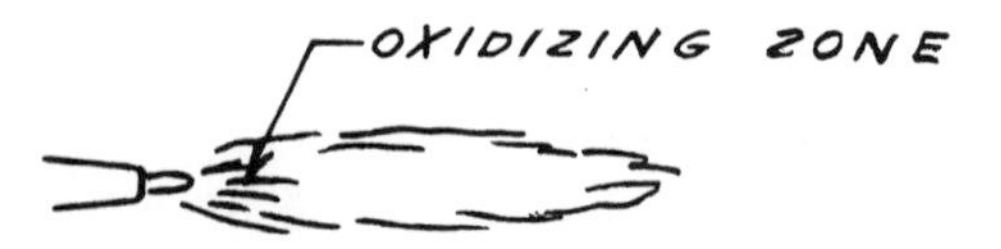

FLAME AS SEEN WITH WELDERS GOGGLES

A-49. Oxidizing flame

FLAME AS SEEN WITHOUT WELDERS GOGGLES

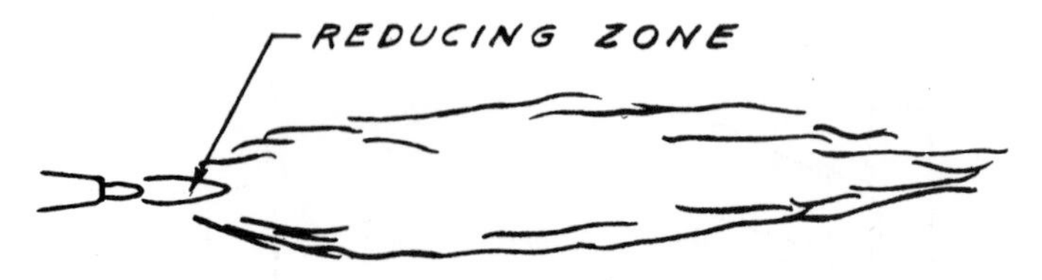

FLAME AS SEEN WITH WELDERS GOGGLES

A-50. Reducing flame

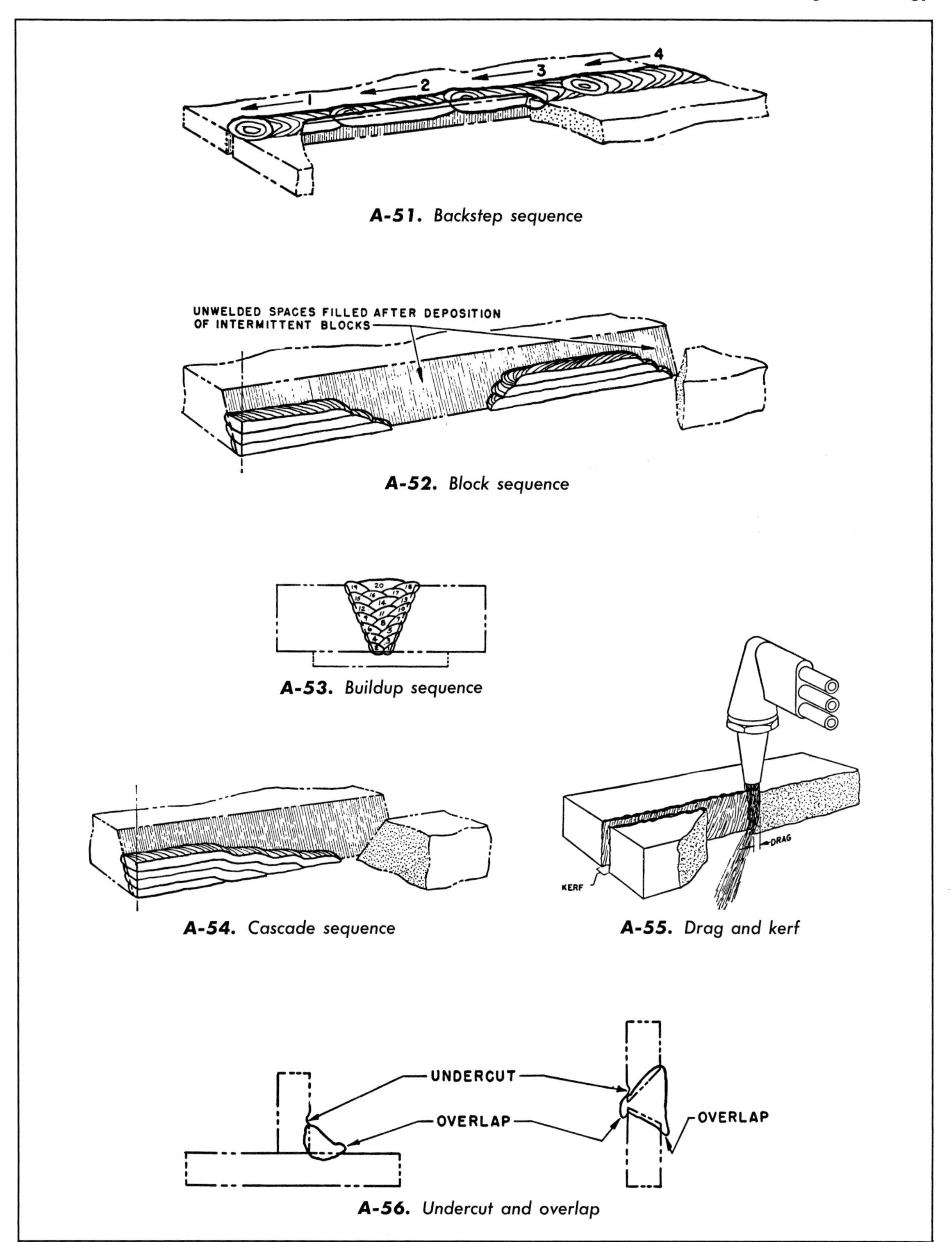

A-51. *Backstep sequence*

A-52. *Block sequence*

A-53. *Buildup sequence*

A-54. *Cascade sequence*

A-55. *Drag and kerf*

A-56. *Undercut and overlap*

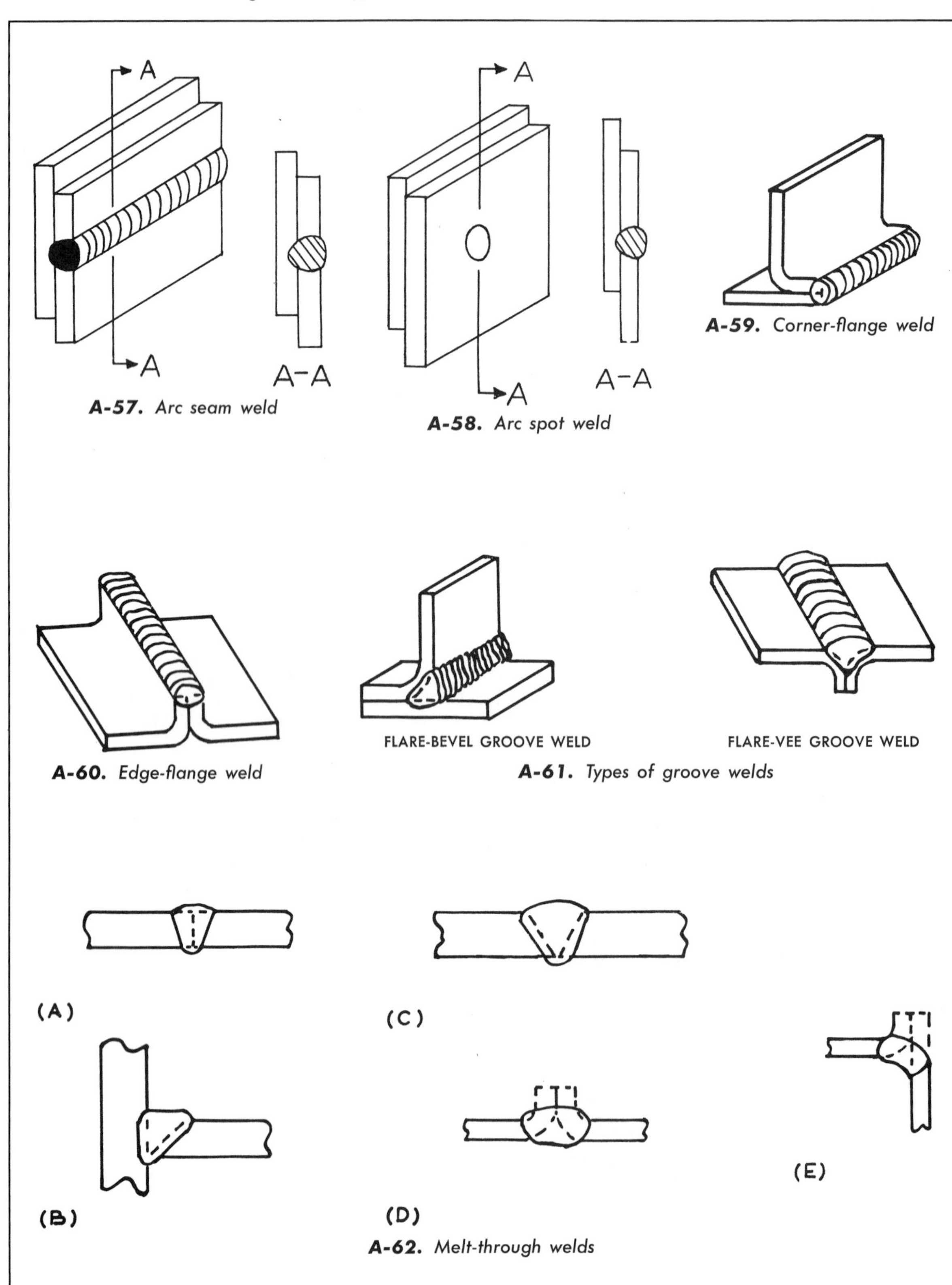

A-57. *Arc seam weld*

A-58. *Arc spot weld*

A-59. *Corner-flange weld*

A-60. *Edge-flange weld*

A-61. *Types of groove welds*

A-62. *Melt-through welds*

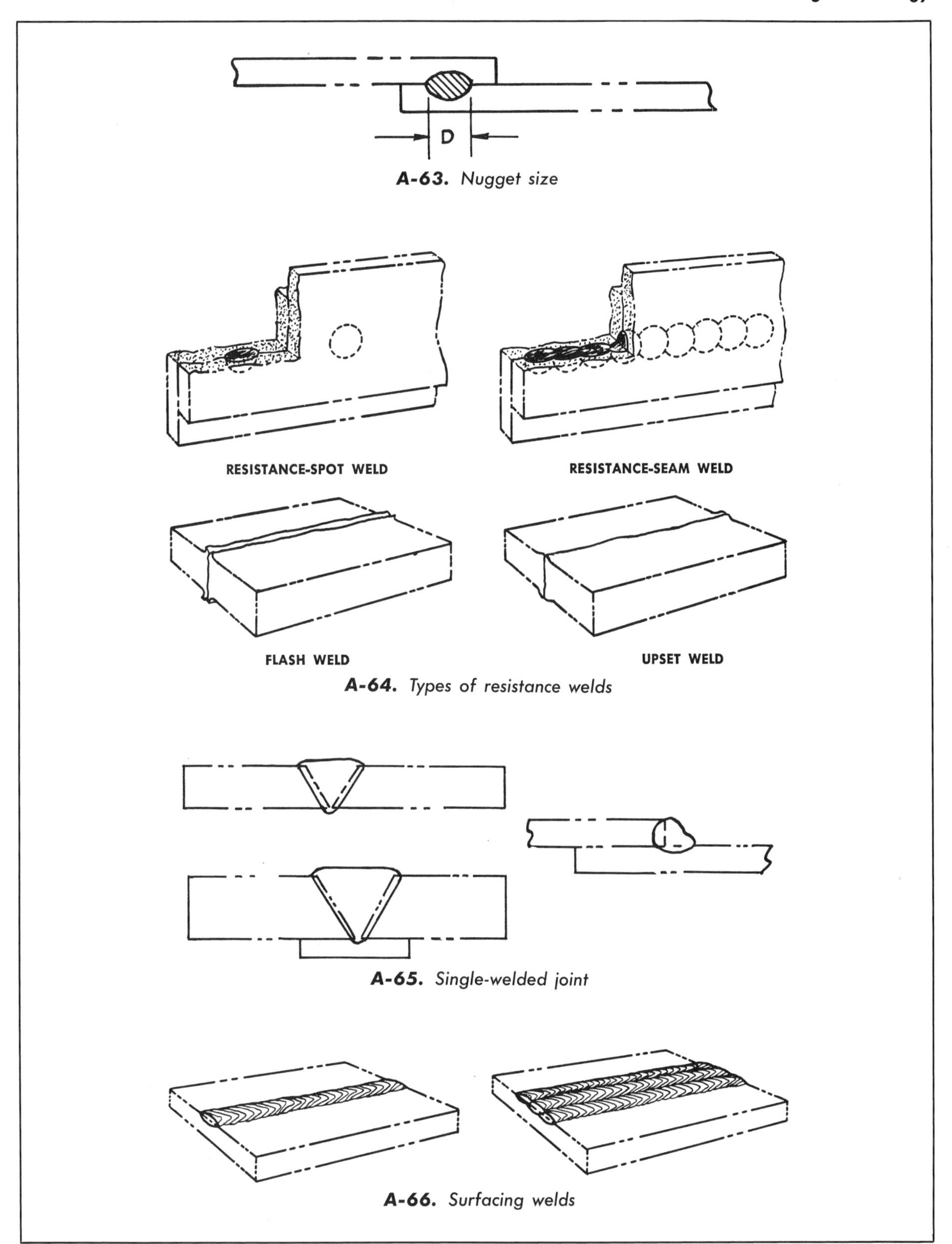

A-63. *Nugget size*

A-64. *Types of resistance welds*

A-65. *Single-welded joint*

A-66. *Surfacing welds*

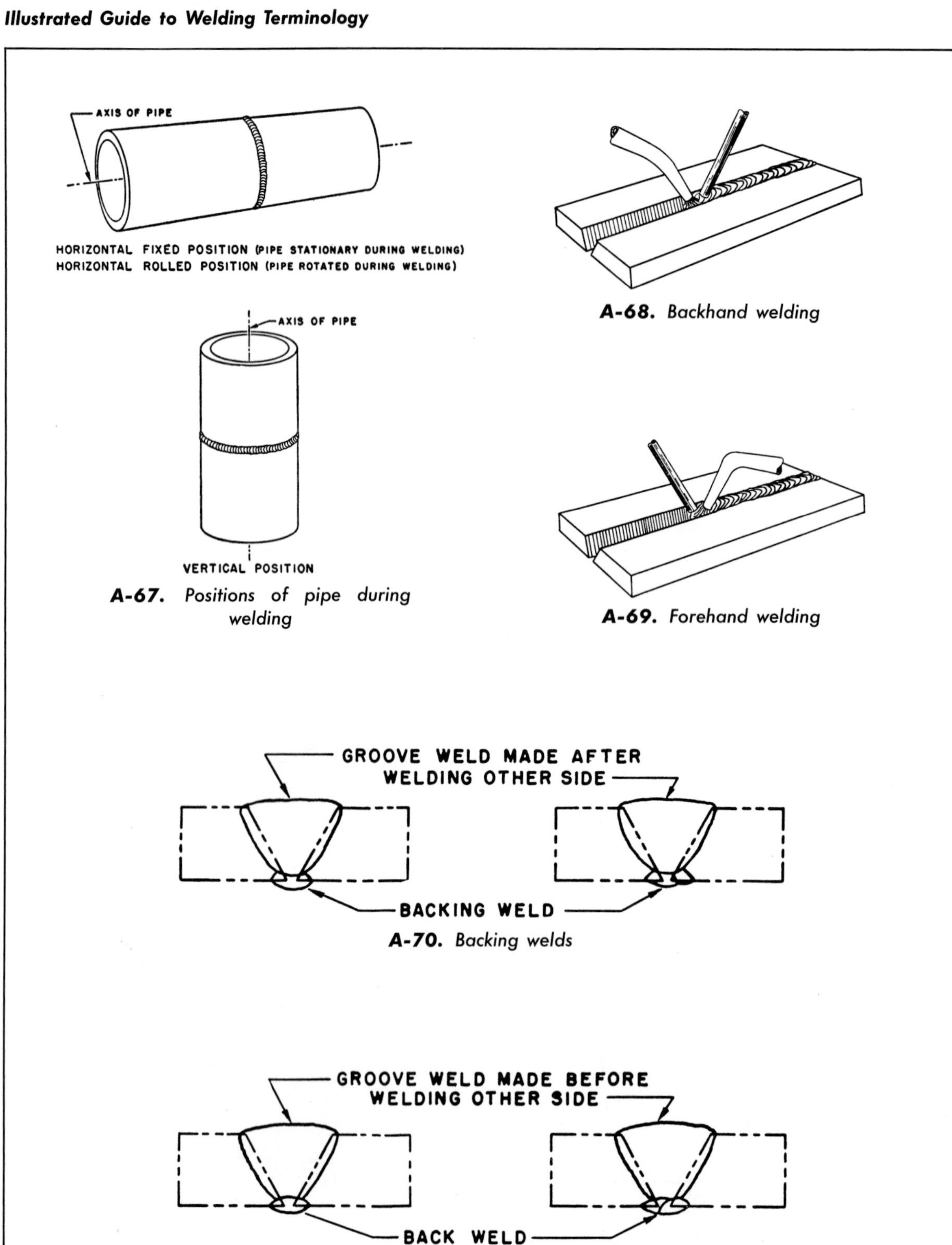

A-67. *Positions of pipe during welding*

A-68. *Backhand welding*

A-69. *Forehand welding*

A-70. *Backing welds*

A-71. *Back welds*

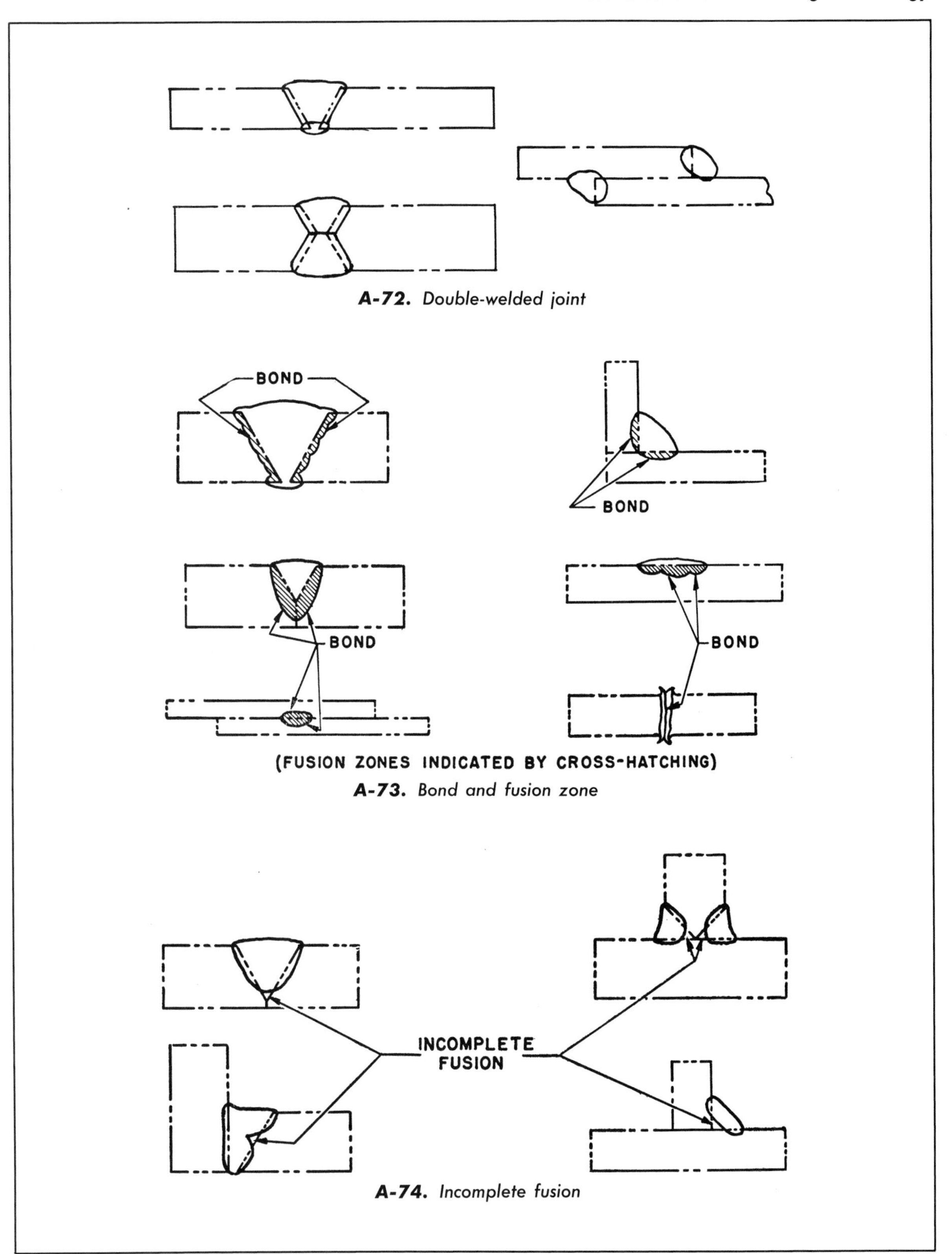

A-72. *Double-welded joint*

A-73. *Bond and fusion zone*

A-74. *Incomplete fusion*

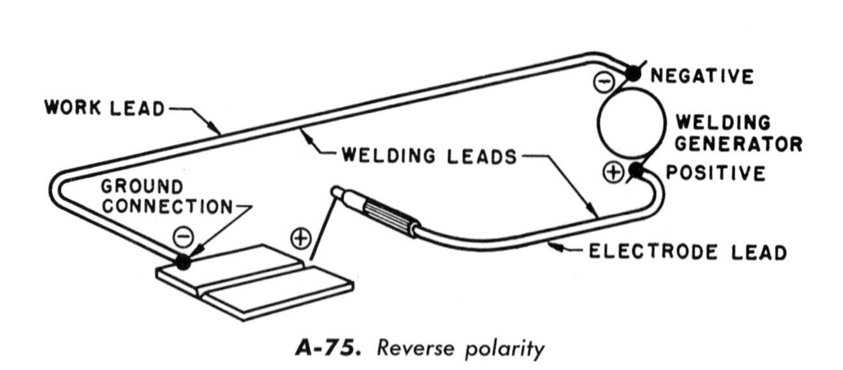

A-75. *Reverse polarity*

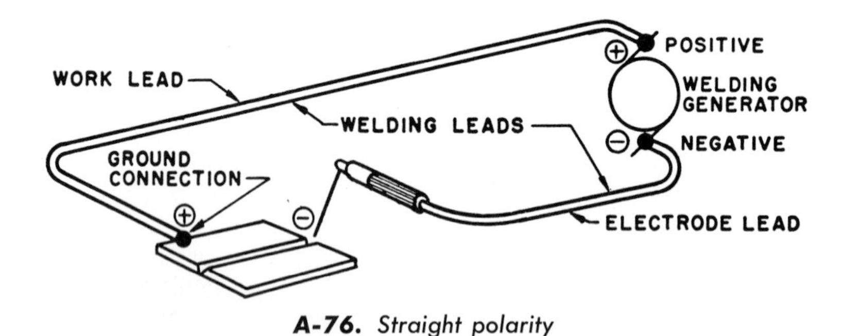

A-76. *Straight polarity*

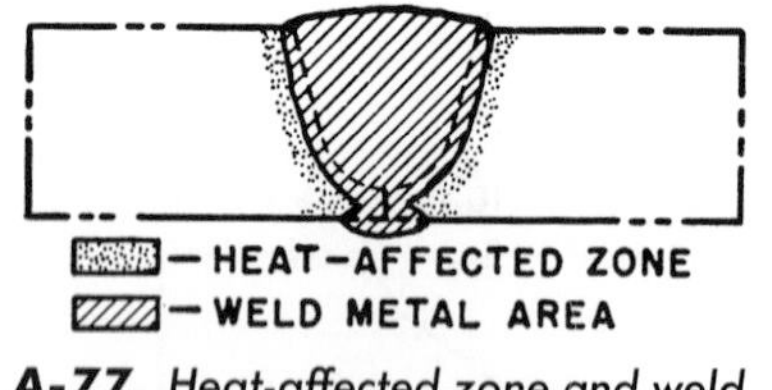

A-77. *Heat-affected zone and weld metal area*

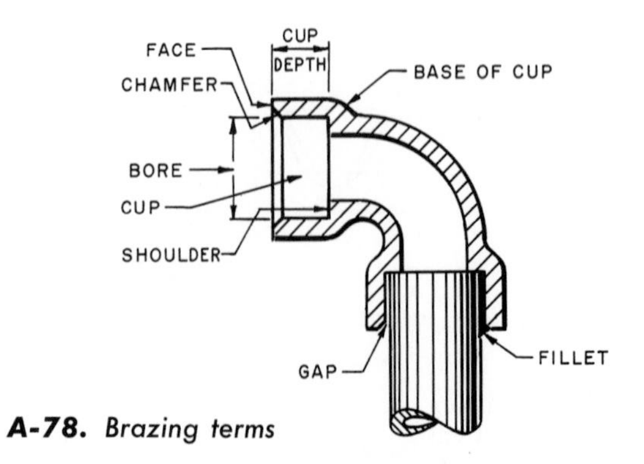

A-78. *Brazing terms*

Welding and Allied Processes

(These terms are American Welding Society standards)

WELDING PROCESSES

Oxyfuel Gas Welding (OFW)
- air acetylene welding . . . AAW
- oxyacetylene welding . . . OAW
- oxyhydrogen welding . . . OHW
- pressure gas welding . . . PGW

Arc Welding (AW)
- atomic hydrogen welding . . . AHW
- bare metal arc welding . . . BMAW
- carbon arc welding . . . CAW
  - —gas . . . CAW-G
  - —shielded . . . CAW-S
  - —twin . . . CAW-T
- gas metal arc welding . . . GMAW
  - —electrogas . . . GMAW-EG
  - —pulsed arc . . . GMAW-P
  - —short circuiting arc . . . GMAW-S
- gas tungsten arc welding . . . GTAW
  - —pulsed arc . . . GTAW-P
- plasma arc welding . . . PAW
- shielded metal arc welding . . . SMAW
- stud arc welding . . . SW
- submerged arc welding . . . SAW
  - —series . . . SAW-S
- flux cored arc welding . . . FCAW
  - —electrogas . . . FCAW-EG

Brazing (B)
- arc brazing . . . AB
- block brazing . . . BB
- diffusion brazing . . . DFB
- dip brazing . . . DB
- flow brazing . . . FLB
- furnace brazing . . . FB
- induction brazing . . . IB
- infrared brazing . . . IRB
- resistance brazing . . . RB
- torch brazing . . . TB
- twin carbon arc brazing . . . TCAB

Soldering (S)
- dip soldering . . . DS
- furnace soldering . . . FS
- induction soldering . . . IS
- infrared soldering . . . IRS
- iron soldering . . . INS
- resistance soldering . . . RS
- torch soldering . . . TS
- wave soldering . . . WS

Solid State Welding (SSW)
- cold welding . . . CW
- diffusion welding . . . DFW
- explosion welding . . . EXW
- forge welding . . . FOW
- friction welding . . . FRW
- hot pressure welding . . . HPW
- roll welding . . . ROW
- ultrasonic welding . . . USW

Resistance Welding (RW)
- flash welding . . . FW
- high frequency resistance welding . . . HFRW
- percussion welding . . . PEW
- projection welding . . . RPW
- resistance seam welding . . . RSEW
- resistance spot welding . . . RSW
- upset welding . . . UW

Other Welding
- electron beam welding . . . EBW
- electroslag welding . . . ESW
- flow welding . . . FLOW
- induction welding . . . IW
- laser beam welding . . . LBW
- thermit welding . . . TW

ALLIED PROCESSES

Adhesive Bonding (ABD)

Thermal Cutting (TC)

Oxygen Cutting (OC)
- chemical flux cutting . . . FOC
- metal powder cutting . . . POC
- oxyfuel gas cutting . . . OFC
  - —oxyacetylene cutting . . . OFC-A
  - —oxyhydrogen cutting . . . OFC-H
  - —oxynatural gas cutting . . . OFC-N
  - —oxypropane cutting . . . OFC-P
- oxygen arc cutting . . . AOC
- oxygen lance cutting . . . LOC

Arc Cutting (AC)
- air carbon arc cutting . . . AAC
- carbon arc cutting . . . CAC
- gas metal arc cutting . . . GMAC

gas tungsten arc cutting . . . . GTAC
metal arc cutting . . . . MAC
plasma arc cutting . . . . PAC
shielded metal arc cutting . . . . SMAC

Other Cutting

electron beam cutting . . . . EBC
laser beam cutting . . . . LBC

Thermal Spraying (THSP)*

electric arc spraying . . . . EASP
flame spraying . . . . FLSP
plasma spraying . . . . PSP

*Sometimes a welding process.

PROCESS METHODS

Automatic Welding . . . . AU
Machine Welding . . . . ME
Manual Welding . . . . MA
Semi-Automatic Welding . . . . SA

Major Agencies Issuing Codes and Specifications

Department of the Air Force
WPAFB (EWBFSA)
Wright-Patterson Air Force Base, OH 45433

Amer. Assn. of State Highway & Transportation Officials
444 N. Capitol
Washington, D.C. 20001

American Bureau of Shipping
45 Eisenhower Dr.
P.O. Box 910
Paramus, NJ 07653

Amer. Institute of Steel Construction
400 N. Michigan Ave., 8th Fl.
Chicago, IL 60611

(AISI) American Iron & Steel Inst.
1133 15th St., N.W.
Washington, D.C. 20005

American Petroleum Institute
1220 L. St. N.W.
Washington, D.C. 20005

American Society of Civil Engineers
345 East 47th Street
New York, NY 10017

American Society of Mechanical Engineers
345 E. 47th St.
New York, NY 10017

American Society for Testing & Materials
1916 Race Street
Philadelphia, PA 19103

American Water Works Assn.
6666 W. Quincy Ave.
Denver, CO 80235

American Welding Society
P.O. Box 351040
550 LeJeune Rd., N.W.
Miami, FL. 33135

Department of Navy
Naval Supply Depot
5801 Taber Avenue
Philadelphia, PA 19120

Society of Automotive Engineers
400 Commonwealth Dr.
Warrendale, PA 15096

Sources of Welding Information

Airco Plating Inc.
3650 N.W. 46th St.
Miami, FL 33142

Alloy Castings Institute
Cast Metals Fed. Bldg.
455 State St.
Des Plaines, IL 60016

American Foundrymen's Society
Golf and Wolfe Roads
Des Plaines, IL 60016

American National Standards Institute
1430 Broadway
New York, NY 10018

American Society for Metals
Metals Park, OH 44073

American Society for Nondestructive Testing
4153 Arlingate Plaza
Caller #28518
Columbus, OH 43228

American Welding Society
P.O. Box 351040
550 LeJeune Rd., N.W.
Miami, FL 33135

Copper Development Association
Greenwich Office Park 2
Box 1840
Greenwich, CT 06836

Ductile Iron Society
615 Sherwood Parkway
Mountainside, NJ 07092

Harris Calorific
Division of Emerson Electric Co.
2345 Murphy Blvd.
Gainesville, GA 30501

Hobart Brothers Co.
Hobart Square
Troy, OH 45373

Industrial Print Co.
1635 Coining Dr.
Toledo, OH 43612

Lincoln Electric Co.
22801 St. Clair Ave.
Cleveland, OH 44117

Metals Handbook
American Society for Metals
Metals Park, OH 44073

Miller Electric Manufacturing Co.
718 S. Bounds St.
Appleton, WI 54912

Modern Eng. Co.
P.O. Box 31725
St. Louis, MO 63131

National Association of Corrosion Engineers
P.O. Box 218340
Houston, TX 77218

National Certified Pipe Welding Bureau
5410 Grosvenor Ln., Suite 120
Bethesda, MD 20814

Steel Founders Society of America
Cast Metals Fed. Bldg.
455 State St.
Des Plaines, IL 60016

Tescom Corporation
12616 Industrial Blvd.
Elk River, MN 55330

Union Carbide Corporation
39 Old Ridgebury Rd.
Danbury, CN 06817

Directory of Reliable Sources of Metric Conversion Information

General

American National Metric Council
1010 Vermont Ave., N.W., #320
Washington, D.C. 20005

American National Standards Institute, Inc.
1430 Broadway
New York, NY 10018

U.S. Metric Assn. (Standards)
10245 Andasol Ave.
Northridge, CA 91325

Technical & Trade Associations

Air Conditioning & Refrigeration Inst.
1501 Wilson Blvd.
Arlington, VA 22209

American Paper Institute
260 Madison Avenue
New York, NY 10016

American Society for Abrasive Methods
118 Main St.
P.O. Box 441
Connoquenessing, PA 16027

The American Society of Mechanical Engineers
345 East 47th St.
New York, NY 10017

American Society for Metals
Metals Park, OH 44073

American Society for Testing & Materials
1916 Race St.
Philadelphia, PA 19103

American Welding Society
P.O. Box 351040
LeJeune Rd., N.W.
Miami, FL 33135

Association for Information & Image Management
1100 Wayne Ave., Suite 1100
Silver Springs, MD 20910

Illum. Engineering Soc. of North America
345 E. 47th St.
New York, NY 10017

Industrial Fasteners Inst.
1505 E. Ohio Bldg.
Cleveland, OH 44114

Institute of Electrical & Electronic Engineers
345 East 47th St.
New York, NY 10017

National Association of Manufacturers
1331 Pennsylvania Ave., N.W., Suite 1500 N.
Washington, D.C. 20004

National Fluid Power Association
3333 N. Mayfair Rd.
Milwaukee, WI 53222

National Forest Products Association
1250 Connecticut Ave., N.W.
Washington, D.C. 20036

National Tooling & Machining Association
9300 Livingston Rd.
Ft. Washington, MD 20744

Society of Automotive Engineers
400 Commonwealth Dr.
Warrendale, PA 15096

Steel Plate Fabricators Assn.
2400 S. Downing Ave.
Westchester, IL 60153

Index

F

G

N

O

P

R

S